CW00799068

International Handbook of
Earthquake and Engineering Seismology

To:

Susan Francis

Best wishes,

Willie Lee

This is Volume 81A in the
INTERNATIONAL GEOPHYSICS SERIES
A series of monographs and textbooks
Edited by RENATA DMOWSKA, JAMES R. HOLTON AND H. THOMAS ROSSBY

A complete list of books in this series appears at the end of this volume

International Handbook of Earthquake and Engineering Seismology

IASPEI

PART A

Edited by

William H. K. Lee, Hiroo Kanamori, Paul C. Jennings, and Carl Kisslinger

A project of the Committee on Education
International Association of Seismology and Physics of the Earth's Interior

in collaboration with
International Association for Earthquake Engineering

Published by Academic Press for
International Association of Seismology
and Physics of the Earth's Interior

Amsterdam Boston London New York Oxford Paris
San Diego San Francisco Singapore Sydney Tokyo

This book is printed on acid-free paper.

Academic Press
An Imprint of Elsevier Science
84 Theobald's Road, London WC1X 8RR, UK
http://www.academicpress.com

Academic Press
An Imprint of Elsevier Science
525 B Street, Suite 1900, San Diego, California 92101-4495, USA
http://www.academicpress.com

ISBN 0-12-440652-1
CD ISBN: 0-12-440653-X

Library of Congress Control Number: 2002103787

A catalogue record for this book is available from the British Library

Typeset by Newgen Imaging Systems (P) Ltd, Chennai, India
Printed and bound in China by RDC
02 03 04 05 06 07 RD 9 8 7 6 5 4 3 2 1

Contents

III. Observational Seismology

IV. Earthquake Geology and Mechanics

V. Seismicity of the Earth

VI. Earth's Structure

Contributors

Adams, Robin D.
International Seismological Centre
Thatcham, Berkshire RG19 4NS, UK

Agnew, Duncan
Institute of Geophysics and Planetary Physics
University of California at San Diego
La Jolla, CA 92093, USA

Aki, Keiiti
Observatoire Volcanologique du Piton de la Fournaise
4 Route Nationale 3 - 27eme km
97418 La Plaine des Cafres
La Reunion, 97418, France

Ambraseys, N.N.
Department of Civil Engineering
Imperial College of Science, Technology,
and Medicine
London, SW7 2BU, UK

Beeler, Nicholas M.
MS 977, U.S. Geological Survey
345 Middlefield Road
Menlo Park, CA 94025, USA

Benz, Harley M.
MS 967, U.S. Geological Survey
Box 25046, Denver Federal Center
Denver, CO 80225, USA

Bolton, Harold F.
Albuquerque Seismological Laboratory
U.S. Geological Survey
801 University SE, Suite 300
Albuquerque, NM 87106-4345, USA

Branum, David
California Division of Mines and Geology
801 K street
Sacramento, CA 95814, USA

Brune, James N.
Seismology Lab, MS 174
Mackay School of Mines
University of Nevada at Reno
Reno, NV 89557, USA

Cara, Michel
Institute de Physique du Globe de Strasbourg
5, rue René Descartes
F-67084 Strasbourg Cedex, France

Cecić, Ina
Ministrstvo za okolje in prostor
Agencija Republike Slovenije za okolje
Ljubljana, SI-1000, Slovenia

Chapman, Chris H.
Schlumberger Cambridge Research
High Cross, Madingley Road
Cambridge CB3 OEL, UK

Clévédé, Eric
Departement de Sismologie
Institut de Physique du Globe de Paris
Tour 24, 4eme Etage
4, Place Jussieu, case 89
75252 Paris cedex 05, France

Curtis, Andrew
Schlumberger Cambridge Research
High Cross, Madingley Road
Cambridge CB3 OEL, UK

Douglas, Alan
AWE Blacknest, Brimpton
Nr Reading, Berks, RG7 4RS, UK

Engdahl, E.R.
Center for Imaging the Earth's Interior
Department of Physics
University of Colorado
Boulder, CO 80309, USA

Fehler, Michael
Los Alamos Seismic Research Center
MS D443, Los Alamos National Laboratory
Los Alamos, NM 87545, USA

Feigl, Kurt L.
Dept. of Terrestrial and Planetary Dynamics
Centre National de la Recherche Scientifique
14, Ave. Edouard Belin
31400 Toulouse, France

Grant, Lisa B.
Department of Environmental Analysis and Design
University of California, Irvine
Irvine, CA 92697, USA

Guidoboni, Emanuela
S.G.A. Storia Geofisica Ambiente s.r.l.
Via Bellombra, 24/2
Bologna, 40136, Italy

Holcomb, L. Gary
Albuquerque Seismological Laboratory
U.S. Geological Survey
801 University SE, Suite 300
Albuquerque, NM 87106-4345, USA

Housner, George W.
211 Thomas Laboratory
California Institute of Technology
Pasadena, CA 91125, USA

Hutt, Charles R.
Albuquerque Seismological Laboratory
U.S. Geological Survey
801 University SE, Suite 300
Albuquerque, NM 87106-4345, USA

Igarashi, George
Laboratory for Earthquake Chemistry
The University of Tokyo
Tokyo, 113-0033, Japan

Jackson, James A.
Bullard Laboratories
University of Cambridge
Madingley Rise, Madingley Road
Cambridge, CB3 OEZ, UK

Johnston, Malcolm J.S.
MS 977, U.S. Geological Survey
345 Middlefield Road
Menlo Park, CA 94025, USA

King, Chi-Yu
Earthquake Prediction Research, Inc.
381 Hawthorne Ave.
Los Altos, CA 94022, USA

Klosko, Eryn
Department of Geological Sciences
Northwestern University
1847 Sheridan Road
Evanston, IL 60208, USA

Kulhánek, Ota
Seismology, Department of Earth Sciences
Uppsala University
Villavagen 16
SE-752 36, Uppsala, Sweden

Lay, Thorne
Earth Sciences Department
University of California at Santa Cruz
Santa Cruz, CA 95064, USA

Lee, William H.K.
MS 977, U.S. Geological Survey
345 Middlefield Road
Menlo Park, CA 94025, USA

Linde, Alan T.
Carnegie Institute of Washington
5241 Broad Branch Road, NW
Washington, D.C. 20015, USA

Lockner, David A.
MS 977, U.S. Geological Survey
345 Middlefield Road
Menlo Park, CA 94025, USA

Lognonné, Philippe
Departement de Géophysique Spatiale
et Planétaire
Institut de Physique du Globe de Paris
4 avenue de Neptune
94107 Saint Maur des Fosses Cedex, France

Madariaga, Raúl
Laboratoire de Géologie
Ecole Normale Supérieure
24 Rue Lhomond
75231 Paris Cedex 05, France

Majewski, Eugeniusz
Institute of Geophysics
Polish Academy of Sciences
Ul. Ksiecia Janusza St. 64
01-452 Warszawa, Poland

Malamud, Bruce D.
Department of Geography
King's College London, Strand
London, WC2R 2LS, UK

McGarr, Arthur
MS 977, U.S. Geological Survey
345 Middlefield Road
Menlo Park, CA 94025, USA

McNutt, Stephen R.
Alaska Volcano Observatory
Geophysical Institute
University of Alaska
Fairbanks, AK 99775, USA

Melville, Charles P.
Faculty of Oriental Studies
University of Cambridge
Cambridge, CB3 9DA, UK

Minshull, Timothy A.
Southampton Oceanography Centre
School of Ocean and Earth Science
European Way
Southampton SO14 3ZH, UK

Mochizuki, Kimihiro
Earthquake Research Institute
The University of Tokyo
Tokyo, 113-0032, Japan

Mooney, Walter D.
MS 977, U.S. Geological Survey
345 Middlefield Road
Menlo Park, CA 94025, USA

Mosegaard, Klaus
Niels Bohr Institute
Juliane Maries Vej 30
2100 Copenhagen, OE, Denmark

Musson, Roger M.W.
British Geological Survey
West Mains Road
Edinburgh, EH9 3LA, UK

Nur, Amos
Department of Geophysics
Stanford University
Stanford, CA 94305, USA

Okubo, Paul
Hawaiian Volcano Observatory
U.S. Geological Survey
Hawaii National Park, HI 96718, USA

Olsen, Kim Bak
Institute of Crustal Studies
University of California at Santa Barbara
Santa Barbara, CA 93106, USA

Pavlenkova, Nina I.
Institute of Physics of the Earth
Russian Academy of Science
Bol. Gruzinskaya str., 10
Moscow, 123810, Russia

Prodehl, Claus
Geophysical Institute
University of Karlsruhe
Karlsruhe, D-76187, Germany

Richards, Paul G.
Lamont-Doherty Earth Observatory
Columbia University
Palisades, NY 10964, USA

Romanowicz, Barbara
Berkeley Seismological Laboratory
University of California
Berkeley, CA 94720, USA

Ruff, Larry J.
Department of Geological Sciences
University of Michigan
Ann Arbor, MI 48109, USA

Satake, Kenji
Active Fault Research Center
National Institute of Advanced Industrial Science
& Technology
Tsukuba, 305-8567, Japan

Sato, Haruo
Department of Geophysics
Tohoku University, Aoba-ku
Sendai, 980-8578, Japan

Satyabala, S.P.
National Geophysical Research Institute
Uppal Road
Hyderabad, 500 007, India

Scherbaum, Frank
Institut fuer Geowissenschaften
Universitaet Potsdam
Postfach 601553
D-14415 Potsdam, Germany

Seeber, Leonardo
Lamont-Doherty Earth Observatory
Columbia University
Palisades, NY 10964, USA

Sibson, Richard H.
Department of Geology
University of Otago
P.O. Box 56, Dunedin, New Zealand

Simpson, David
The IRIS Consortium
1200 New York Avenue NW, Suite 800
Washington, D.C. 20005, USA

Sipkin, Stuart A.
MS 967, U.S. Geological Survey
Box 25046, Denver Federal Center
Denver, CO 80225, USA

Snieder, Roel
Department of Geophysics
Colorado School Mines
Golden, CO 80401, USA

Song, Xiaodong
Department of Geology
University of Illinois
Urbana, IL 61801, USA

Stauder, William U.
St. Louis University
Jesuit Hall, 3601 Lindell
St. Louis, MO 63108, USA

Stein, Seth
Department of Geological Sciences
Northwestern University
1847 Sheridan Road
Evanston, IL 60208, USA

Suyehiro, Kiyoshi
Deep Sea Research Department
Japan Marine Science & Technology Center
2-15 Natsushima-cho
Yokosuka, 237-0061, Japan

Tarantola, Albert
Institut de Physique du Globe de Paris
4 place Jussieu
75005 Paris, France

Teisseyre, Roman
Institute of Geophysics
Polish Academy of Sciences
Ul. Ksiecia Janusza St. 64
01-452 Warszawa, Poland

Thatcher, Wayne
MS 977, U.S. Geological Survey
345 Middlefield Road
Menlo Park, CA 94025, USA

Toppozada, Tousson
California Division of Mines and Geology
801 K street
Sacramento, CA 95814, USA

Turcotte, Donald, L.
Department of Geological Science
Cornell University
Ithaca, NY 14853, USA

Udías, Agustín
Catedra de Geofisica
Fac de Ciencias Fisicas
Universita Complutense de Madrid
28040 Madrid, Spain

Usami, Tatsuo
Takatsukadai 1-8-1-6411,
Kawai-cho, Nara-ken 636-0071
Japan

Utsu, Tokuji
9-5-18, Kitami
Setagaya-ku
Tokyo 157-0067, Japan

Uyeda, Seiya
Earthquake Prediction Research Center
Tokai University
3-20-1, Orido
Shimizu, 424-8610, Japan

Villaseñor, Antonio
Department of Tectonophysics
Faculty of Earth Science
University of Utrecht
PO Box 80021
3501 TA, Utrecht, Netherlands

Webb, Spahr C.
Lamont-Doherty Earth Observatory
Columbia University
Palisades, NY 10964, USA

Wielandt, Erhard
Institute of Geophysics
University of Stuttgart
Richard-Wagner-Str. 44
D-70184 Stuttgart, Germany

Wu, Ru-Shan
Center for Study of Imaging and
Dynamics of the Earth
Institute of Geophysics and Planetary Physics
University of California at Santa Cruz
Santa Cruz, CA 95064, USA

Zoback, Mark D.
Department of Geophysics
Stanford University
Stanford, CA 94305, USA

Zoback, Mary Lou
MS 977, U.S. Geological Survey
345 Middlefield Road
Menlo Park, CA 94025, USA

Foreword

Seismology is, in the truest sense of the word, a global science, not limited to political boundaries. Thus, it can only be effectively practiced through international cooperation.

Seismology is primarily an observational science. Quantitative measurements of seismic ground motion as a function of time require physically well defined, continuously recording instruments. Suitable seismographs of different design principles became available at the end of the 19th century. In the following years, a considerable number of seismograph stations were installed at sites around the globe. At the dawn of this century, some of these stations celebrate their 100th anniversary, as does the International Association of Seismology and Physics of the Earth's Interior (IASPEI).

When it became evident that seismic signals propagate through the Earth over large distances, the need for an international organization to coordinate, collect and distribute the data, and promote earthquake research arose. During two "International Conferences on Seismology", in 1901 and 1903 in Strasbourg, the establishment of an "International Association of Seismology" (IAS) was proposed. The association was inaugurated in 1904 and signed by 18 nations. This can be considered the root of what finally became the "International Association of Seismology and Physics of the Earth's Interior".

So it is appropriate to state that the present Handbook reflects and encompasses one hundred years of a very dynamical and successful history of scientific seismology. Of course it has been a long and rocky path with changes and interruptions, mainly caused by the international political scene during and after the two World Wars. In 1919, the International Union of Geodesy and Geophysics (IUGG) was created with the "Seismological Section" as one of the founding components. In 1933, the name was changed to "International Association of Seismology" of the IUGG. With the expansion of its scope to include material properties of the Earth's interior, it was renamed in 1951 to "International Association of Seismology and Physics of the Earth's Interior", as it is known today.

Along with the progress in instrumentation, the theoretical basis of wave propagation in the Earth was being developed for the evaluation and interpretation of the observed travel-time data. By the end of the 1930s, a rough and still valid first model of the structure of a spherically symmetric Earth existed. This is equivalent to a static picture of the Earth.

While in the first half of the 20th century scientific seismology developed mainly in Europe and Japan, after World War II, the center of research activity shifted to the USA. A mighty impetus came with the installation of the first true worldwide network of homogeneous seismograph stations (WWSSN) by the US government. Originally justified for fundamental research in support of the search for means to detect and identify underground nuclear explosions, it turned out to be a powerful tool for a very fruitful period of basic scientific research on a wide range of problems in the science of the Earth.

In a third phase, modern seismology took full advantage of the great technical progress in digital technology from recent decades. Broadband digital seismographs with very high signal resolution provide a far superior data quality and make possible powerful application of digital data handling and analysis. Several hundred broadband digital seismographs now cover the globe and data from many of them are directly accessible in almost real time by fast telecommunication links.

A multitude of precise data has enabled us to resolve lateral heterogeneities in the Earth's structure by tomographic methods, and to study parameters of the seismic source. This empowered seismologists to deduce dynamic properties of the Earth such as fracture process, flow, temperature, and the stress field in the Earth. This has greatly improved our understanding of the driving mechanisms of plate tectonics and convection in mantle and core. In close interaction with experimental and theoretical high pressure/high temperature rock physics and chemistry, we have now also a realistic understanding of the material which constitutes the Earth's interior. Thus, the picture of the static Earth evolved to a picture of a dynamic Earth.

In addition to academic interest in the Earth's interior, earthquakes are also capable of paramount economical and sociological consequences. An important issue of seismology from the early times on has been the question of earthquake prediction. In the 1970s, some seismologists were quite optimistic that the problem could be solved fairly soon. A joint American–Soviet publication left only the choice between two alternative theoretical approaches. But with increasing database and experience, it became obvious that the problem is much more complex and fundamental, and general solutions are even now not in sight.

However, important branches of science and technology, such as Engineering Seismology and Earthquake Engineering, have developed in modern form during the past 50 years. These innovations provide powerful tools to estimate and to reduce building destruction, and consequently losses of lives. The seismic hazard or the potential ground shaking of an area by earthquakes can be estimated from the local seismicity and seismotectonics, and from records of strong-motion accelerographs. The dynamic response of technical construction elements is investigated experimentally and by computer simulations. Together with social and economical aspects, it allows assessment of seismic risk for a certain region. Recommendations for earthquake resistant constructions form the content of "seismic building codes".

This Handbook is a document of the great achievements that have been made during the past century by the very dynamically developing sciences of seismology and earthquake engineering. It presents the contributions of earthquake research to the scientific community and to our comprehension of the world. Great credit goes to the editors for attracting over 1200 of the most experienced and highly regarded experts from more than 50 countries to participate in this comprehensive and truly international work.

It is expected that this Handbook will become a standard work of seismology, which should be present in many libraries and research institutes. It is also a justification of the great amount of public funds that have been spent on earthquake research to date. It covers not only the history of our science but leads to the present day frontiers of our knowledge.

However, earthquakes and earthquake research will continue into the future as they have in the past. Earthquake recording and treatment of seismograms have to be carried out in a professional, scientific way. Therefore, the IASPEI Commission on Practice began some five years ago to write a "New Manual of Seismological Observatory Practice" (NMOSP) which will replace the 1979 edition. The old publication and the new manual in its latest available prepublication version are attached as computer readable files on the Handbook CD-ROM. It is a very impressive work by itself and the NMSOP editor and contributors deserve our sincere thanks and high appreciation. The NMSOP will also certainly play an important role in the "International Training Courses on Seismology and Seismic Risk Assessment for Developing Countries" regularly held in different countries and co-sponsored by IASPEI.

Hans Berckhemer, Past President of IASPEI, 1975–1979.
Hardtbergweg 13
61462 Koenigstein/Taunus
Germany

Preface

Modern scientific investigations of earthquakes began in the 1880s, and the *International Association of Seismology* was organized in 1901–4 to promote collaboration of scientists and engineers in studying earthquakes. It became a Section (later Association) of the *International Union of Geodesy and Geophysics* in 1919, and the present *International Association of Seismology and Physics of the Earth's Interior* (IASPEI) in 1951. With the rapid advances in science during the 20th century, many branches of seismology developed and some excellent text books and monographs have been published on various seismological subjects and disciplines. However, there is no reference book that summarizes our present knowledge about earthquake and engineering seismology as a whole.

We think it is appropriate that such a reference work, international in scope, be published on the occasion of the Association's centennial. It is our hope that in addition to documenting the advances in most aspects of seismology, with the exception of exploration seismology, over the past century, this Handbook will help to foster more communication between seismologists and earthquake engineers.

Our aims for the Handbook are: (1) to summarize the well-established facts of earthquake and engineering seismology, (2) to review relevant theories, (3) to survey useful methods and techniques, (4) to summarize the historical development and current status of seismology in the member countries of IASPEI, and (5) to document and archive some basic seismic data. In collaboration with the International Association of Earthquake Engineering (IAEE), we also include national reports from member countries of IAEE. Since participation of IASPEI and IAEE member countries was voluntary, we have extensive but not complete coverage.

The Handbook consists of two printed volumes (Part A and Part B) and three CDs containing supplementary materials. Part A contains Chapters 1 through 56 plus one attached CD, and Part B contains Chapters 57 through 90 plus three appendices and two attached CDs. The Handbook is intended as a general reference on earthquake and engineering seismology, and also as a comprehensive "resource library" for anyone interested in earthquakes and related subjects.

We attempted to achieve these goals through three types of articles: (1) review articles of topics in theoretical seismology, observational seismology, earthquake geology and mechanics, seismicity of the Earth, structure of the Earth's interior, strong-motion seismology, earthquake engineering, earthquake prediction, and hazards mitigation, (2) national and institutional reports on seismology, earthquake engineering, and physics of the Earth's interior, and (3) supplementary articles on history, general information, and seismological data.

The three CDs attached to the Handbook constitute a "digital" library of a few hundred volumes on seismology and earthquake engineering. Many authors of printed articles used the attached CDs to provide supplementary materials. For example, a few authors provided an expanded version of their printed-volume chapters, and several authors archived lengthy data sets, out-of-print books, or hard-to-find reports. In addition, there are chapters (summaries in print and details on CD) devoted to a global inventory of seismographic networks, a collection of seismological software, a global earthquake catalog of the 20th century, a collection of early seismic bulletins and some classic monographs and papers in seismology and earthquake engineering. We included many "historical" materials: histories and publication lists from many institutions in seismology and earthquake engineering, biographies of over 250 notable earthquake scientists and engineers now deceased, and biographical sketches from over 1,000 living active researchers. Finally, we included a technical glossary of terms that are frequently used in earthquake and engineering seismology in the printed Part B volume.

The one-page biographical sketches allow each individual to state concisely his or her major contributions or research interests, and a brief list of his or her most important publications. The more than 1,000 biographical sketches enhances and supplements the review chapters and national/institutional reports. We have also invited some prominent non-seismological researchers to contribute their biographical sketches in order to provide information that seismologists should be aware of.

The widespread use of the World Wide Web and the Internet beginning in the 1990s profoundly changes the way we communicate and disseminate information. In this Handbook, we have attempted to provide many useful web site addresses, and several data sets on the attached Handbook CDs can be explored using a Web browser. Although information on the CDs can not be changed, CDs provide an inexpensive way of archiving data and much faster access than via the Internet.

This Handbook is the result of a six-year project undertaken by the *IASPEI Committee on Education* in collaboration with the *International Association for Earthquake Engineering*. Planning, writing, and editing this Handbook involved four editors, 15 associate editors, 11 coordinators, over 250 authors, and an advisory board of over 40 members. More than 1,200 scientists and engineers from over 50 countries participated in this project. Manuscripts for the printed Handbook were reviewed similarly to those submitted to a scientific journal. We are grateful to over 200 reviewers worldwide for improving the manuscripts. Thanks to the efforts of all these people we have largely achieved our goal of compiling a comprehensive reference work in earthquake and engineering seismology. However, given such a broad undertaking it is perhaps inevitable that articles on some topics (e.g., intraplate seismicity, deep earthquakes, general seismotectonics, recent advances in earthquake location, and earthquake resistant design) intended for inclusion in the Handbook were not, in the end, available for publication.

We thank Frank Cynar, the publishing editor at Academic Press, Jennifer Helé, his assistant, and Roopa Baliga, the production editor, for their patience in transforming several thousand manuscript pages into the final printed volume. We also thank the following publishers: Academic Press, the American Geophysical Union, the Annual Reviews Inc., the Earthquake Engineering Research Institute, and the Seismological Society of America for permission to scan their copyrighted materials into computer readable files to form a significant part of a "digital library" on the attached CDs.

This Handbook project is a labor of love by volunteers, as no financial support was available from IASPEI or any funding agencies. We declined royalties and financial support from the publisher in order to retain the copyright by the IASPEI's Committee on Education and to keep the price of this Handbook as low as possible.

William H.K. Lee, Hiroo Kanamori, Paul C. Jennings,
and Carl Kisslinger, Editors
March 15, 2002

I

History and Prefatory Essays

1

History of Seismology

Duncan Carr Agnew
University of California at San Diego, La Jolla, California, USA

1. Introduction

At present seismology is the study of seismic sources (mostly earthquakes), the waves they produce, and the properties of the media through which these waves travel. In its modern form the subject is just over 100 years old, but attempts to understand earthquakes go back to the beginnings of science. The course of seismology, more than that of many other sciences, has been affected by its object of study: From Lisbon in 1755 through Kobe in 1995, destructive earthquakes have provoked scientific interest, and, quite often, social support for seismic studies. Table 1 lists some earthquakes (and one explosion) that have had an impact on seismology.

This article describes the history of seismology up to about 1960, with a brief sketch of major themes since then. To cover this history in the space available requires a fair amount of selection. Any reading of the older literature shows that a great many ideas were suggested long before they became generally accepted: For example, the ideas that shaking is a wave propagated from a source, that some earthquakes (at least) are

TABLE 1 Some Events of Significance to the History of Seismology

Name	Date	Location	Magnitude	Importance
	143	Gansu, China	7	Possibly first instrumental record of unfelt shock
Lisbon	1755 Nov. 1	Azores-Cape St Vincent Ridge	8+	Widespread seiching and tsunami: basis for theories of wave propagation
Calabria	1783 Feb. 5	Southern Italy	7+	First event studied by a scientific commission
Basilicata (Neapolitan)	1857 Dec. 16	Southern Italy	7	Detailed investigation by Mallet
Yokohama	1880 Feb. 22	Near south coast of Honshu, Japan	5.8	Led to Seismological Society of Japan
Casamicciola (Ischia)	1883 Jul. 28	Tyrrhenian Sea	6	Led to foundation of Italian seismological service
Vyernyi (Alma-Alta)	1887 Jun. 9	Lake Issyk-Kul, Krygyzstan	7.3	Stimulated Russian study of earthquakes
North Canterbury	1888 Aug. 31	South Island, New Zealand	7+	First scientific observation of strike-slip faulting
	1889 Apr. 18	Near south coast of Honshu, Japan?	7+?	First teleseismic recording
Nobi (Mino-Owari)	1891 Oct. 28	Western Honshu, Japan	8.4	Large surface rupture; led to Imperial Earthquake Investigation Committee
Assam	1897 Jun. 12	Northeastern India	8.7	Recognition of teleseismic primary and secondary phases
San Francisco	1906 Apr. 18	Central California	7.7	Large surface rupture; geodetic detection of off-fault motions; led to Seismological Society of America
Kulpa Valley	1909 Oct. 8	Northwestern Balkan region	6	Discovery of crustal phases
Kanto (Kwanto)	1923 Sep. 1	Near south coast of Honshu, Japan	7.9	Led to foundation of Earthquake Research Institute (Tokyo)
Buller (Murchison)	1929 Jun. 16	South Island, New Zealand	7.8	Stimulated local earthquake recording in New Zealand; inner-core phases in Europe
	1946 Apr. 1	Unimak Island region	7.2	Led to tsunami warning system
Ashkhabad	1948 Oct. 5	Turkmenistan–Iran border region	7.2	Stimulated Soviet earthquake program
	1952 Nov. 4	Off east coast of Kamchatka	9.0	First suggested observation of free oscillations
RAINIER	1957 Sep. 19	Southern Nevada	4.0	First underground nuclear explosion, helped stimulate VELA-UNIFORM program
Chilean	1960 May. 22	Off coast of Central Chile	9.5	First detection of free oscillations
Alaskan	1964 Mar. 28	Southern Alaska	9.2	Stimulated US earthquake-hazards program
Niigata	1964 Jun. 16	Near west coast of Honshu, Japan	7.5	Stimulus for modern Japanese earthquake programs
Xintiang	1966 Mar. 22	Northeastern China	7.0	Stimulated Chinese earthquake program

INTERNATIONAL HANDBOOK OF EARTHQUAKE AND ENGINEERING SEISMOLOGY, VOLUME 81A ISBN: 0-12-440652-1

caused by faulting, and that the Earth contains a liquid core. I have therefore focused less on the earliest occurrence of an idea than the time when it became something seriously considered within the seismological community. To make the narrative more readable, the sources for particular statements are given in a separate set of notes, referenced through footnote numbers; these notes, the reference list, and a bibliography of the history of seismology, are all included on the attached Handbook CD.[1]

2. Early Ideas about Earthquakes

The most common explanation given for earthquakes in early cultures was the same as for any other natural disaster: they were a manifestation of divine wrath. In European thought this idea did not disappear from scientific discussion until well into the 18th century. But two premodern cultures, the Chinese and the Greek, also developed naturalistic explanations for seismic shaking. Greek natural philosophers suggested a variety of causes for earthquakes; the most influential (and extensive) treatment extant was by Aristotle (ca. 330 BCE), who attributed earthquakes to winds (driven by an "exhalation," the pneuma) blowing in underground caverns. The classical authors also attempted to classify earthquakes by different types of shaking: something that remained a mainstay of seismology for a long time, and still survives in popular terminology.[2]

Chinese ideas about earthquakes were put forth by various thinkers at roughly the same time as the Greek ones, the dominant idea also being that shaking was caused by the blocking of a subtle essence (the *qi*). After the Han dynasty (200 BCE), Imperial state orthodoxy associated natural disasters with dynastic decline; this led to systematic preservation of accounts of earthquakes in the official annals. China's technology also produced the first seismoscope, invented in AD 132 by Zhang Heng. This well-known device, which reportedly signaled the direction of an earthquake as well as the occurrence of shaking, is supposed on at least one occasion to have responded to unfelt shaking; the mechanism of its operation remains obscure.[3]

The Aristotelian view of earthquakes (as of many other aspects of the world) remained the primary theory during the medieval periods of both Islam and Europe. With the decline of Aristotelian thought in early modern Europe, other ideas were put forward, though many of the writers were from northern Europe and so (unlike the Greeks) had little direct experience of earthquakes. They did, however, know about gunpowder: This new technology of chemical explosives suggested that earthquakes might be explosions in the Earth (or in the air); of the various chemical theories put forward, the most popular involved the combustion of pyrites or a reaction of iron with sulfur. Such theories also explained volcanic action; that the most seismic part of Europe, namely Italy, was also volcanic helped to support this association. In the 18th century the development of theories of electricity, and especially their application to lightning, provoked several theories that related earthquakes to electrical discharges. In England, much of this theorizing was stimulated by the occurrence of several damaging earthquakes in the year 1750.

However, the greatest stimulus to seismological thinking was undoubtedly the Lisbon earthquake of 1755, partly for its destructiveness, but even more for providing evidence of motion at large distances: It caused seiches over much of Europe. At least two writers, J. Michell (1761) and (much more obscurely) J. Drijhout (1765), proposed that this distant motion was caused by a wave propagating from a specific location, thus more clearly than before separating the earthquake source from the effects it produced. The type of wave envisaged was like a traveling wrinkle in a carpet; Michell also suggested that the vibrations close to the source were related to waves propagated through the elasticity of the rocks, as sound waves were known to propagate through the elasticity of the air. The elasticity and pressure of gases, more specifically of high-temperature steam, also provided Michell's driving force for the earthquake itself, which he took to be caused by water vaporized by sudden contact with underground fires. (This steam also supported the propagation of waves to great distances.) Michell attempted to locate this "place of origin" by comparing the times of the seiche-inducing wave and the observed sea wave, and also hazarded a guess at the depth. While these ideas were not forgotten, they did not lead to any additional research, and certainly did not replace older theories of earthquakes.[4]

3. The Nineteenth Century to 1880

The expansion and professionalization of science in the 19th century meant that earthquake studies, like many other parts of science, could become a specialization, at least for a few scientists for part of their careers. One type of research that began in this period was the accumulation of large volumes of data, with the aim of finding underlying patterns—a style that has been called "Humboldtean" when applied to the Earth, but was in fact much more general. For earthquake studies, this meant the first systematic catalogs of shocks (as opposed to lists of catastrophes); leaders in this were K.A. Von Hoff and A. Perrey, the latter being a disciple of A. Quetelet, one of the founders of statistics. Many of these compilations were used to look for possible correlations between earthquake occurrence and astronomical cycles or meteorological events; this was one of the main topics of 19th-century seismology. Along with these catalogs came studies of individual shocks: Most European earthquakes after about 1820 stimulated some sort of special study, by individuals (often local professors) or by commissions set up by governments or local scientific societies. The first such commission was established after the Calabrian earthquake of 1783; a century later, one earthquake (in Andalusia, 25 December 1884) would bring forth three commissions, one each from Spain, France, and Italy.

These special studies developed many of the tools and vocabulary still used to describe the felt effects of a large earthquake. One such tool, quite in keeping with the overall trend in science toward quantification, was scales of intensity of shaking: the first by P. Egen in 1828, followed by many others, notably those of M. de Rossi, F. Forel, these two working together, and G. Mercalli. The first cartographic application of an intensity scale, creating the isoseismal map, was by J. Nöggerath in 1847; in turn, the accumulation of maps stimulated questions about why the distribution of shaking was as observed, and to what extent it could be explained by waves radiating from a central source.

This period also saw the first efforts to relate earthquakes to other geological processes. Von Hoff was explicitly interested in this relationship and in the English-speaking world it was promoted most assiduously by C. Lyell, whose program of reducing all past geological change to current causes was aided by showing that earthquakes could cause vertical motions over large areas. Prominent examples of such motion in Lyell's treatment were the 1819 Rann of Cutch (India) and 1822 Chilean earthquakes, and later the 1835 Chilean and 1855 Wairarapa (New Zealand) earthquakes. The 1819 and 1855 earthquakes produced some of the earliest known examples of a break at the surface, though the first scientific observations of fault rupture did not take place until much later, by A. McKay in 1888 (North Canterbury) and by B. Koto in 1891 (Nobi).[5]

In retrospect, the basis for a different approach to earthquake study can be seen to have begun in 1829 to 1831, with investigations by S.D. Poisson into the behavior of elastic materials. He found that in such materials wave motions could occur, and were propagated at two speeds; the slower wave had particle motion perpendicular to the direction of propagation, and the faster one included dilatation of the material. A wave motion with transverse vibrations was of great interest because it offered a model for the recently discovered polarization of light, and many of the 19th-century investigations into elastic wave propagation were in fact made with optical observations in mind, attempting to explain light as transverse waves in an elastic "luminiferous ether." Notable examples were the study of G. Green (1838) into wave transmission across a boundary, and of G.G. Stokes (1850) on radiation from a limited source.[6]

These results were applied to earthquake studies by W. Hopkins (1847), and by R. Mallet (1848 onwards). Hopkins showed how timed observations of wave arrivals could be used to locate an earthquake, but went no further than this purely theoretical exercise. Mallet, a polymathic engineer, not only coined the term seismology but tried to develop it systematically as a science of earthquakes, observed through the waves they generate. Mallet constructed one of the most complete earthquake catalogs to date, which he summarized in a map (Color Plate 1) that clearly delineates the seismic and aseismic regions of the world (1858). But Mallet aimed to do more than describe: Whenever possible he argued for the application of quantitative mechanical principles to determine how much, and in what direction, the ground moved in an earthquake. This quantitative emphasis is perhaps most notable in his 1862 study of the 1857 Basilicata (Neapolitan) earthquake, in which he attempted to estimate the direction of arrival of the shaking at many points, and so infer the location (and depth) of the source. It is also apparent in his earlier attempts (1851) to measure actual wave velocities from explosions and compare these with known elastic constants; he obtained much lower values than expected, which he attributed to inhomogeneity but which were more likely caused by insensitive instruments. Like Michell, Mallet believed that earthquakes were caused by the sudden expansion of steam as water met hot rock; because of the explosive nature of such a source, he believed that the earthquake waves would be almost entirely compressional.[7]

What was lacking in Mallet's otherwise comprehensive program was an adequate method of recording earthquake motion; though he and others proposed possible ways to do this, few of these schemes were actually built and even fewer were used by more than the inventor—though it was recognized early (for example, by Hopkins) that a network of instruments was really what was needed. The first instrument to automatically record the time and some aspects of the shaking was the seismoscope of L. Palmieri (1856), used in Italy and Japan. However, the first network of instruments, set up in Italy starting in 1873, was not intended to record earthquake shaking. Rather, these "tromometers," developed by T. Bertelli and M. de Rossi, were used to measure ongoing unfelt small motions, looking for changes in the amplitude or period of these related either to weather or to earthquakes, a study called "endogenous meteorology."[8]

4. The Birth of the "New Seismology": 1880–1920

With its emphasis on background vibrations, the substantial Italian effort turned out to be less fruitful than what happened in Japan as a consequence of the Meiji restoration of 1868: the establishment of modern science in a very seismic region. This was begun by the Japanese government bringing foreign experts (*yatoi*) to Japan, including professors well trained in the latest methods of physics and engineering. Of these, the most important for seismology was John Milne, who arrived in Japan in 1876 to be a professor (aged 26) at the Imperial College of Engineering in Tokyo. He made seismology his main interest after the earthquake of 22 February 1880, which also led to the foundation of the Seismological Society of Japan, with Milne as its effective leader. Such organization among the foreign experts was soon paralleled by similar initiatives from the Japanese: the Meteorological Agency established (or rather, took over from Milne) a regular reporting system in 1883, and S. Sekiya became the world's first professor of seismology in 1886. The routine reporting of earthquakes allowed Sekiya's successor, F. Omori, to develop his law for the decay of aftershocks from data for the 1891 earthquake.

Even before the 1880 earthquake, attempts had been made by foreign scientists in Japan to record the time history of felt shaking. This developed into a rivalry between J.A. Ewing at the University of Tokyo, and Milne (with his colleague T. Gray). Ewing applied the horizontal pendulum to get the first good records of ground shaking (at what would now be regarded as the lower limit of strong motion) in 1880–1881. These records showed the motion to be much smaller than had been assumed, and also much more complicated: nothing like a few simple pulses and not the purely longitudinal motion envisaged by Mallet. Thus, as soon as seismologists had records of ground motion, they faced the problem that has been central to the science ever since: Explaining observed ground motion and deciding how much of the observed complication comes from the earthquake and how much from the complexities of wave propagation in the Earth.

While Milne may not have been the first to record earthquake shaking, he soon became a leading seismologist, not so much from any new ideas he brought to the subject as from his energy and flair for organization. Like Mallet, Milne aimed to study all aspects of earthquakes and elastic waves, but he added to Mallet's quantitative emphasis a regular use of instrumental measurements, often designing the instruments for the occasion. This regular use of quantitative instrumental records (not just by Milne) led contemporaries to term these activities "the new seismology."[9]

These instrumental improvements were largely focused on local shaking, but they formed the basis of quite different studies under the stimulus of an unexpected result. In Germany E. von Rebeur-Paschwitz had built sensitive horizontal pendulums for measuring tidal tilts, his interest being primarily astronomical. These showed transient disturbances, and he was able to correlate one of these, on 18 April 1889 (Fig. 1) with a Japanese earthquake reported in *Nature*. This demonstration that distant earthquakes could be recorded led to new developments in seismic instrumentation in Italy, by G. Agamemnone and A. Cancani, and in Japan, by Omori and Milne. Milne returned to England in 1895, bringing an enthusiasm for global seismology, a design for an inexpensive seismometer, and a long association with the British Association for the Advancement of Science (BAAS). Both he and von Rebeur-Paschwitz proposed in that year a global network of seismic instruments, but it fell to Milne, capitalizing in large part on the geographical reach of the British Empire and his BAAS connection, to install the first such network, with himself and an assistant as (in modern terms) data and analysis center combined.

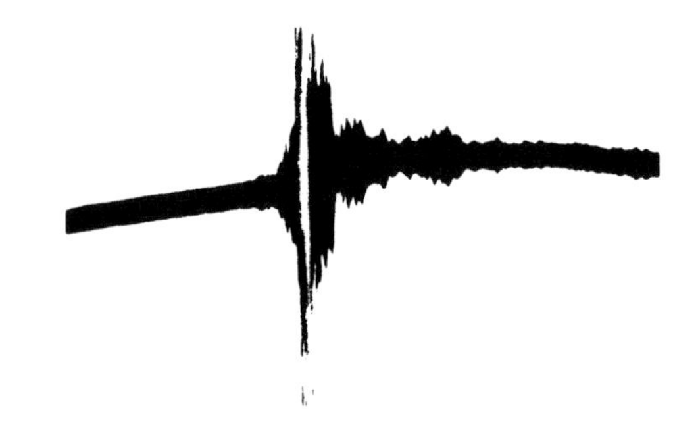

FIGURE 1 The first recording of a distant earthquake. This photo reproduction is from von Rebeur-Paschwitz (1895) and is clearer than the engraving used in his 1889 article. This earthquake is often said to have been "in Tokyo," since the one report available was from there, but the description (see also Knott 1889) makes it clear that the shaking was just barely felt, so the shock must have been offshore. The magnitude in Table 1 is estimated from this record, applying the instrumental constants in Abe (1994).

Unfortunately, the global scope of Milne's network was not matched by the quality of his instruments: These low-gain undamped sensors, registering on a low-speed record, were adequate to detect large shocks but not to show the details of the waves from them. Major advances in instrumentation came from two physicists who had turned to seismology. The first was E. Wiechert, who (following a study of Italian instruments) introduced in 1904 his inverted-pendulum sensor, the first seismometer to be properly damped—something Wiechert's own theoretical developments, themselves new, showed to be important. The second was B.B. Golicyn (Galitzin), who, beginning in 1906, applied electrodynamic sensors and photographically recording galvanometers to create instruments of unprecedentedly high sensitivity and accuracy.[10]

As these and other instruments were installed at observatories around the world, seismologists faced a new problem: sorting out the different "phases" observed, and relating them to different kinds of waves propagating inside the Earth. Theorizing about the Earth's interior was an active subject in the 19th century, but (rather like cosmology today) one in which a maximum of ingenuity was applied to minimal amounts of data. It was agreed that the depths of the Earth were hot and dense, but what parts were solid or liquid (or even gaseous) was the subject of much debate, though by the 1890s there was general agreement (from Kelvin and G.H. Darwin's tidal studies) that a large part of the Earth must be solid, and thus capable of transmitting the two known types of elastic waves, along with the elastic surface wave proposed by Rayleigh in 1885. But it was not known which of these wave types would actually occur, and how much they might become indistinguishably confused during propagation. Given that the Earth was an inhomogeneous body, and that rocks were anisotropic, seismologists had a wide range of options available to explain the observations.[11]

The most obvious distinction was between the large "main phase" and the preceding "preliminary tremors," which were early suggested to be surface waves and body waves, respectively. In 1900 R.D. Oldham, using measurements of the 1897 Assam earthquake, classified the preliminary tremors into two phases, which he identified with longitudinal and transverse body waves, the main phase being a surface wave. That this is the same identification that would be now made should not obscure its controversial nature at the time. For one thing, measurements of particle motion in the main phase showed motions quite different from Rayleigh's theory; for another, there was disagreement about the extent to which the records reflected tilting as opposed to horizontal motion. Perhaps because of these disagreements, in the proposal by Wiechert and van dem Borne in 1904 for designators, the letters *P*, *S*, and *L* referred only to the timing or form of the waves, not to their type.

This was by no means Wiechert's only contribution to global seismology; it would be fair to say that the work he and his students at Göttingen did between 1900 and 1920 made them the leaders of the subject. Already expert in optics and electrodynamics, Wiechert was able to apply his knowledge to waves in the Earth. With G. Herglotz, he developed the first solution to the geophysical inverse problem of deducing wave velocities from travel times (the forward version of this had been discussed by H. Benndorf). Another student, K. Zöppritz, determined a set of travel times that were used for some years, as well as working out the equations for transmission of elastic waves at an interface (earlier solved for wave energy by C.G. Knott). B. Gutenberg used the Herglotz–Wiechert method to determine a velocity profile—and in particular, to find the radius of the core (a region of decreased wave speed first proposed by Oldham in 1906) at a value very close to the modern one. And L. Geiger showed how to determine earthquake locations from distant observations using least-squares. Many of these names are still familiar, attesting to the extent to which Wiechert's group laid foundations for the field that have in many ways endured.

Of course, there were others contributing to wave-propagation studies. One was A. Mohorovičić, who used data from the 1909 Kulpa Valley earthquake in Croatia to study travel times at relatively short distances, and found additional phases which he explained by assuming a velocity discontinuity at about 50 km depth. A related development was the demonstration in 1911 by the mathematician A.E.H. Love that surface waves with particle motion transverse to the direction of propagation were possible in a layered Earth, thus satisfying the characteristics of the main phase in seismograms.[12]

5. The "Classical" Period: 1920–1960

In the forty years from 1880 to 1920, seismology had thus gone from being primarily a descriptive natural history of earthquakes to having a large (and mathematical) component concerned with the wave propagation within the Earth. The next forty years (roughly) might well be called the "classical" period: A time during which many of the ideas first developed before 1920 were refined and improved, but without any substantial changes in aims or techniques. This period spans what is often taken to be a huge change in science (especially in the United States), from low-tech poverty before World War II to high-tech affluence after it; but while this change did occur for physics, it does not seem to have applied to much of seismology.

5.1 General Developments

Despite setbacks from both World Wars, the number of seismic stations increased substantially over these four decades; and the wider use of radio time signals after 1920 made timing more accurate (even if still poor by later standards). However, this growth was not at all standardized: individual organizations would set up one, or in a few cases, several new stations. The most "global" organization engaged in seismology at this time was probably the Jesuits, some of whose schools included seismological observatories. In the United States these observatories formed their own organization (the Jesuit Seismological Association), which provided one of the first rapid determinations of global epicenters. There were also international organizations: The International Seismological Association was founded to promote international cooperation in 1904, though it ceased to exist with the start of World War I. When revived in 1922, it created what was perhaps the most important entity for global seismic research, namely the International Seismological Summary (ISS), in many ways the continuation of Milne's efforts in earthquake location, which had been continued after his death by H.H. Turner and the BAAS. Through the occasional exchange of records, the regular exchange of station bulletins, and the ISS, information collected worldwide was available to individual researchers. Seismologists thus had what would now perhaps be called a "virtual" global network, something which greatly stimulated studies of seismic waves at great distances.[13]

While such wave-propagation studies were a great research opportunity for many scientists, they were also a departure from seismology's previous focus on earthquakes and their effects—and were viewed by some as an abandonment of the most useful part of the subject. One response was to develop instruments appropriate for studying nearby earthquakes, a strategy pursued by H.O. Wood in southern California in the 1920s. This resulted in a new and more sensitive instrument developed by himself and J.A. Anderson, as well as more sensitive electromagnetic seismometers, and the strain seismometer (both invented by H. Benioff). The Wood–Anderson, in particular, made southern California a model local seismic network, in a style soon copied from northern California to New Zealand. The quality of data available in California helped to attract the most productive of Wiechert's pupils, B. Gutenberg, to a position in Pasadena in 1930, a move which (as in other parts of science) transplanted the outstanding quality of German science to an American setting.[14]

Another shift from Germany to America happened in this period, namely the application of seismology to subsurface exploration, usually for oil. The pioneer here was yet another student of Wiechert's, L. Mintrop, who had for his thesis developed a portable seismograph to measure waves generated by a falling weight. After wartime experiences in artillery location (also influential for many Americans), he formed an exploration company, which began work along the Gulf Coast (Mexico and United States) at about the same time as several US groups. From this beginning, exploration seismology grew rapidly and became a field dominated by US companies.[15] Another "applied" area in which much early work was done in the United States was the recording of strong ground motion, begun in the 1930s (and treated in detail in an accompanying article).

5.2 First-Arrival Seismology: Global and Local Earth Structure

The main problem of instrumental seismology in this period remained that of relating observed bursts of energy on the seismic record ("arrivals") to elastic waves in the Earth. This was, for global seismology, a period of rapid progress: By 1940 the seismic-wave velocity structure of the mantle and core had been worked out in terms that remain little altered. Many travel-time investigations were "special studies" of global recordings from particular earthquakes: usually those that could be well located from other data. H. Jeffreys and K.E. Bullen took a quite different approach, using global data from the International Seismological Summary to iteratively construct improved epicenters and travel times, with due attention to statistical problems throughout. This analysis (and improved timekeeping) showed that many of the complications invoked by earlier seismologists were superfluous: The Earth is in fact a nearly spherically symmetric body, with only a few major internal discontinuities. One of these, the core–mantle boundary, was shown by Jeffreys in 1926 to separate a solid from a fluid region, based on the difference between mantle rigidities found seismically and whole-earth rigidities from earth tides. That the core itself contained a discontinuity was shown by I. Lehmann in 1936, using records of New Zealand earthquakes antipodal to the densely spaced stations of Europe. This finding of the inner core may be said to have completed the discovery phase of exploration of the inside of the Earth, though not its detailed mapping.

On more local scales there were a number of studies of crustal velocity structure using earthquakes and the occasional large explosion (usually detonated for some other purpose, or an accident). The results were interpreted in the framework given by Mohorovičić, finding the depth of crust–mantle discontinuity and sometimes of other velocity discontinuities (notably one first suggested by V. Conrad) as well. This is one branch of seismology that did change significantly after World War II: There were then several programs that used large-scale seismic refraction to determine crustal structure. In the United States the pioneering postwar land-based program was by M. Tuve and H. Tatel. A program of land-based "deep-seismic sounding" was begun in the USSR before World War II by V.A. Gamburtsev, and greatly expanded after it. A similar pattern, of modest prewar beginnings and huge postwar expansion, also applied to seismic measurements at sea, led by M.N. Hill in Great Britain, and M. Ewing and R.W. Raitt in the United States.[16]

5.3 Seismic Geography

While the compilation of earthquake catalogs in the 19th century reached its culmination in the massive (and mostly unpublished) lists of F. Montessus de Ballore, the global picture of seismicity stood to be greatly improved by instrumental recording—though for many decades errors in the data and travel times combined to give a picture that, while more complete, was rather blurred. The first major advance from instrumental measurements came from the dense local network in Japan, which enabled K. Wadati to distinguish (in 1927) between deep and shallow earthquakes. Wadati's results, combined with global travel-time data (which had suggested great depths for some earthquakes) cleared up the longstanding problem of earthquake depth, and showed that deep earthquakes were relatively restricted in their distribution. The other advance in seismic geography grew out of the sensitivity of the local network in southern California, which detected many small earthquakes that had to be clearly distinguished from larger, damaging ones. This led C.F. Richter, transforming another idea of Wadati's, to develop a "magnitude" scale for the size of the earthquakes: As any seismologist who has had to explain it to the public knows, it is not such an obvious concept as it might seem. Gutenberg and Richter soon extended it from southern California shocks to earthquakes throughout the world. The magnitude scale, combined with the distinction between earthquakes of various depths, and the more reliable locations possible with improved travel times, came together in Gutenberg and Richter's study of the seismicity of the Earth (1941, and later revisions). This delineated, better than before, the belts of major seismicity and large aseismic regions, giving a synoptic view of current global activity that influenced tectonic theories. The magnitude scale also made possible estimates of the distribution of earthquakes by size: This magnitude–frequency relation and its parametric description (the *b*-value) remain a basic relationship of seismology.[17]

5.4 Earthquake Mechanism

If earthquake distribution was greatly clarified between 1920 and 1960, the same could not be said of seismologists' understanding of what actually happened at the source of seismic waves. By the 1920s there was little doubt that earthquakes and faults were closely related, though a respectable minority of seismologists maintained that the faulting associated with earthquakes was not the cause but an effect. In this matter seismologists tended to be influenced by the earthquakes and faults they knew best, so for Americans the paradigmatic event was the 1906 California earthquake. This exhibited large motions not just on the fault but also at a distance, leading H.F. Reid to state (1910) the theory of elastic rebound, according to which the earthquake is caused by the release of stress built up along a fault.

The problem for seismologists was, as it still is, to relate what happened at the earthquake source to the seismograms observed. Given that Golicyn (Galitzin) had shown that the first motion of the *P* wave pointed directly toward (or away from) the epicenter, seismologists looked for patterns in the first motion as recorded at an observatory, or seen at several observatories around an earthquake. The latter approach required a large number of stations, which were available only (for local records) in Japan; Japanese seismologists for some time were the leaders in studying this topic. The first observational results were from T. Shida, who in 1917 showed a pattern of first motion divided into quadrants separated by nodal lines—though he found other patterns as well (Fig. 2). The first theoretical treatment was by H. Nakano in 1923, in a paper which examined the first motion from isolated forces and couples. Of little influence in Japan (nearly all copies were destroyed in the fire after the 1923 Kanto earthquake), a copy of this paper reached P. Byerly in California. By 1938 Byerly had developed a method (extended distances) for plotting the distribution of first motions recorded globally in such a way as to identify nodal planes at the source. Apparently because of its intuitive match with the elastic rebound concept, he chose from Nakano's source models the single couple as representative of faulting.

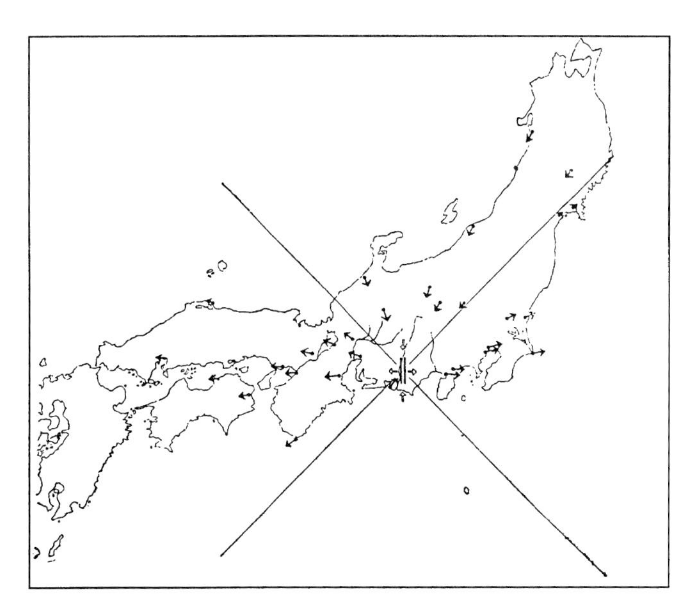

FIGURE 2 An early plot of first motions, from Shida (1929), showing one example used in his unpublished 1917 work. Note that the source pictured, while equivalent to a double-couple, is not shown as one.

Starting from Shida's initial investigation, there was a rapid development of first-motion studies in Japan, both theoretical and observational. It was soon observed, by S.T. Nakamura and S.I. Kunitomi, that the nodal lines coincided with geotectonic trends, strengthening the idea that seismic waves were produced by faulting. H. Honda showed that one implication of the simplest interpretation was false, namely, that the amplitudes of first motions would be largest close to the plane corresponding to the fault. Honda also argued that a double-couple system of forces was required to represent stress relaxation at the times of earthquakes; such a mechanism would produce two indistinguishable nodal planes. This view was itself soon challenged by M. Ishimoto, who argued that the distributions of first motions for many earthquakes were in better accord with conical nodal surfaces rather than planes, implying a distribution of forces that would now be called a compensated linear vector dipole. Ishimoto explained this as being consistent with magma intrusion causing both the earthquake and the faulting.

Ishimoto's interpretation of nodal lines created a controversy restricted to Japanese seismologists, perhaps because the associated magma theory was unappealing elsewhere. But other questions of earthquake mechanism were debated more widely—though with a tendency for the positions taken to coincide with nationality. After World War II earthquake-mechanism studies in Japan, led by Honda, favored a double-couple of forces as the source representation. An independent Soviet program on this subject began after 1948 under the leadership of V.I. Keilis-Borok, and developed methods for mechanism determination that made use, in principle, of first motions and amplitudes of both *P* and *S* waves; their results were interpreted as showing, most often, a single-couple source (for which the *S*-wave distribution, unlike that of the *P*-wave, differs from that for a double couple). Both Soviet and Japanese seismologists worked with local earthquakes. The systematic application of Byerly's techniques to large global earthquakes was undertaken by J. Hodgson and co-workers in Canada, with the results largely being interpreted both in terms of fault planes (taken to imply a single couple) or stresses (a double couple). Attempts to use *S* waves to discriminate between these models were not especially successful, nor were efforts to compile maps of inferred stress directions and relate these to regional tectonics.[18]

5.5 Surface Waves

Though the period from 1920 through 1960 can be viewed as the heyday of "travel-time seismology," it was not without attempts to interpret other parts of the seismogram. Once it was clear that what had been called the "principal part" was a combination of Love and Raleigh waves, it became possible to use measurements of their velocities to determine shallow structure—a possibility made difficult by the computational burden of finding wave velocities for even very simple structures. Early results by G.H. Angenheister (yet another of Wiechert's students) showed Rayleigh-wave velocities that were higher along oceanic than along continental paths. Subsequent studies through 1940 confirmed that oceanic paths showed faster velocities, implying a thinner crust—though the estimates of ocean crustal thickness varied, from 40 km to less than 10 km. Much about the observations, notably the long coda often associated with Rayleigh waves, remained unexplained.

The 1950s saw further developments in surface-wave studies, most notably in the program led by M. Ewing and F. Press. Ewing had had considerable experience with dispersive wave propagation through his ocean-acoustics work during World War II, and also had seismic refraction evidence against a thick crust under the ocean. He and Press were able to show that the structure seen from refraction indeed fit observations quite well, even explaining the coda. They and their coworkers went on to study surface waves over a variety of paths, to develop a new design of seismometer for better recording of the longest periods, to install several such instruments in a global network intended for surface-wave studies, and to publish a treatise covering the entire subject of elastic waves propagating along boundaries. This treatise included, though it did not emphasize, the improved computational method that N.A. Haskell had developed for computing dispersion in multilayered media—a method that was to become the basis for many later developments in seismic-wave computation.[19]

6. From 1960 on: The Modern Era

In many ways the 1960s brought major changes, and much growth, to seismology. Rather than trying to cover every subject, I sketch new trends that can be seen, retrospectively, to be important; I have not tried to associate ideas with individuals. The most important new development is that seismology became what physics had been since the 1940s, a science viewed as relevant to national security—in this case not for the weapons it could build, but for those it could detect. The first nuclear explosion, in 1945, was detected seismically, and US investigations of detection included seismology from their beginnings in 1947; but in an era of atmospheric testing seismic methods took third place to infrasound and radionuclide collection (though even third place was enough to produce significant funding in the United States). With the pressure to move testing underground, seismic methods became more important—and the results from the RAINIER test, and the US–Soviet debates over how to interpret them and data from other underground tests, showed the inadequacy of existing knowledge. A US government panel (led by L.V. Berkner) recommended a large-scale program of "fundamental research in seismology." This resulted in the creation in 1960 of the VELA-UNIFORM program—which, though a project funded by the US government, provided support to a large number of seismologists outside the United States. The interest in this project, and in seismic means of detection, only increased with the Limited Test Ban Treaty of 1963, which moved most tests underground. A large fraction of the VELA-UNIFORM funds went for improved instrumentation, including a considerable amount for seismic array development (also pursued by test-detection groups in the UK and the USSR). VELA-UNIFORM's most important instrumental contribution to seismology was certainly the World Wide Standard Seismograph Network (WWSSN). This provided seismologists, for the first time, with easy access to records from standardized and well-calibrated sensors spread around the world. These data improved old results (e.g., on earthquake distribution) and made possible new ones; every subsequent global network has been modeled on the WWSSN style of operation.[20]

Much of what seismologists did with WWSSN data would not have been possible without the other tool that became common at the time: rapid computation. As computing costs fell, and available funds rose, seismologists were able to speed up calculations that previously had taken up much time (epicenter location) and begin to do things that the labor of computation had ruled out before, such as compute surface wave dispersion in realistic structures. But the effect of the computer was not just to allow seismologists to model complex structures; it also gave them new ways to look at data. The ideas of signal processing and Fourier analysis, developed largely by statisticians and electrical engineers, began to make their way into seismology, to show what could be done with waveforms beyond timing them.

The year 1960 brought an impressive demonstration of this new style of seismology, with the first detection of the Earth's free oscillations. The oscillations of an elastic sphere had been investigated by S.D. Poisson in 1829 (before he studied elastic waves), and some approximate periods of oscillation for a somewhat realistic (but homogeneous) Earth were worked out by H. Lamb (1883) and A.E.H. Love (1911). Following a large earthquake in 1952 H. Benioff believed that he had observed the gravest mode of oscillation, with a period of 52 minutes. But to see how this result compared with that expected it was necessary to compute the periods for a realistic Earth. This was done in 1959 by Z. Alterman, H. Jarosch, and C.L. Pekeris, using an early electronic computer. The occurrence of the 1960 Chilean earthquake spurred several groups to apply the novel techniques of Fourier analysis to their records. The demonstration, at the 1960 IUGG meeting, that the peaks in these transformed records matched the computed periods of oscillation, is a marker of the advent of new seismological techniques. Free-oscillation investigations flourished from then on, contributing to more precise Earth models, to the advent of digital recording of seismic data, and (by being the first substantial demonstration of geophysical inverse theory) to the methodology of geophysics: All important aspects of the last few decades of seismology.

But perhaps the most important change for seismology since 1960 was, as for most of the earth sciences, the change in our picture of the Earth created by sea-floor spreading and plate tectonics. Though most of the data that went into this theory came from nonseismic measurements, earthquakes did play a role in two ways. The first was through oceanic seismicity: The more accurate epicenter locations got, and the smaller the magnitude threshold became, the greater the extent to which oceanic earthquakes appeared to occupy only a very narrow and continuous zone along the ocean ridges. The match between this narrow belt of earthquakes and (in some places) a deep median rift helped to focus attention on these narrow zones. The second contribution, especially important for the development of plate tectonics, came from focal-mechanism studies. The debate between single-couple and double-couple mechanisms was settled in 1964 by the demonstration that only the latter was compatible with a dislocation source (first applied to earthquakes by A.V. Vvedenskaya in 1956), and the WWSSN data made possible much more reliable solutions, for smaller earthquakes, with much less effort. Awareness of sea-floor spreading stimulated a study of the mechanism of oceanic earthquakes, which showed that oceanic fracture zones behaved as transform faults, connecting segments of spreading ridge. On a much larger scale, the first published paper suggesting plate tectonics showed that slip vectors of earthquakes around the North Pacific were consistent with the rigidity of this large area. This emphasis on earthquakes as indicators of motion rather than stress rapidly became the norm: only ten years from the time when focal mechanisms were difficult to estimate and confusing to interpret, they had become a routine tool for elucidating tectonics. Indeed, the largest effect of plate tectonics on seismology was to integrate earthquake occurrence (in many areas at least) with other evidence of deformation in ways not done before: Seismicity was finally felt to have been explained, not just described. This was perhaps most true for deep earthquakes: Though the physics of such deep sources was not resolved, the concept of subduction zones changed deep shocks from something that just happened to occur in some places to a consequence of the geometry and history of plate motion.

These conceptual breakthroughs gave a sense that what was going on at the earthquake source was far better understood than before. This, and the increasing level of damage earthquakes produced in a more urbanized world (in which growth has outpaced applications of seismically resistant construction), has created a steadily stronger desire to apply seismology to problems of seismic hazard. One consequence of this was the construction of many local seismic networks, to detect smaller earthquakes and so improve the description of seismicity. Most such networks use telemetered data, a technique first applied by P.G. Gane and others in 1949 to a network for studying rockbursts in South Africa. The increased interest in hazard, and the increased understanding of seismicity, created new interest in the possibility of earthquake prediction. Prediction studies (and the problem of discriminating earthquakes from explosions) have made studies of the earthquake source, and the mechanics of rock deformation associated with it, into a major part of the seismology of the last forty years, though earthquake prediction belongs to the future, not the history, of seismology.[21]

Acknowledgments

I should like to thank Ben Howell, Carl Kisslinger, S. Miyamura, Carl-Henry Geschwind, David Oldroyd, Frank Evison, Jim Dewey, Bruce Bolt, Teruyuki Kato, Niko Shimazaki, Naomi Oreskes, Robin Adams, Marta Hanson, and R.E.W. Musson for comments, corrections, and additional material—and I apologize for any errors that remain.

Editor's Note

The numbered footnotes and bibliography are given on the attached Handbook CD under the directory \01Agnew. Readers are encouraged to read: Chapter 2, Historical View of Earthquake Engineering, by Housner; Chapter 3, The Jesuit Contribution to Seismology, by Udias and Stauder; Chapter 4, International Seismology by Adams; and Chapter 79 on centennial national reports to IASPEI, especially from Germany, Japan, Russia, the United Kingdom, and the United States, edited by Kisslinger. Biography of many persons mentioned in this chapter is given in Chapter 89, Biography of Notable Earthquake Scientists and Engineers, edited by Howell.

Plate Section

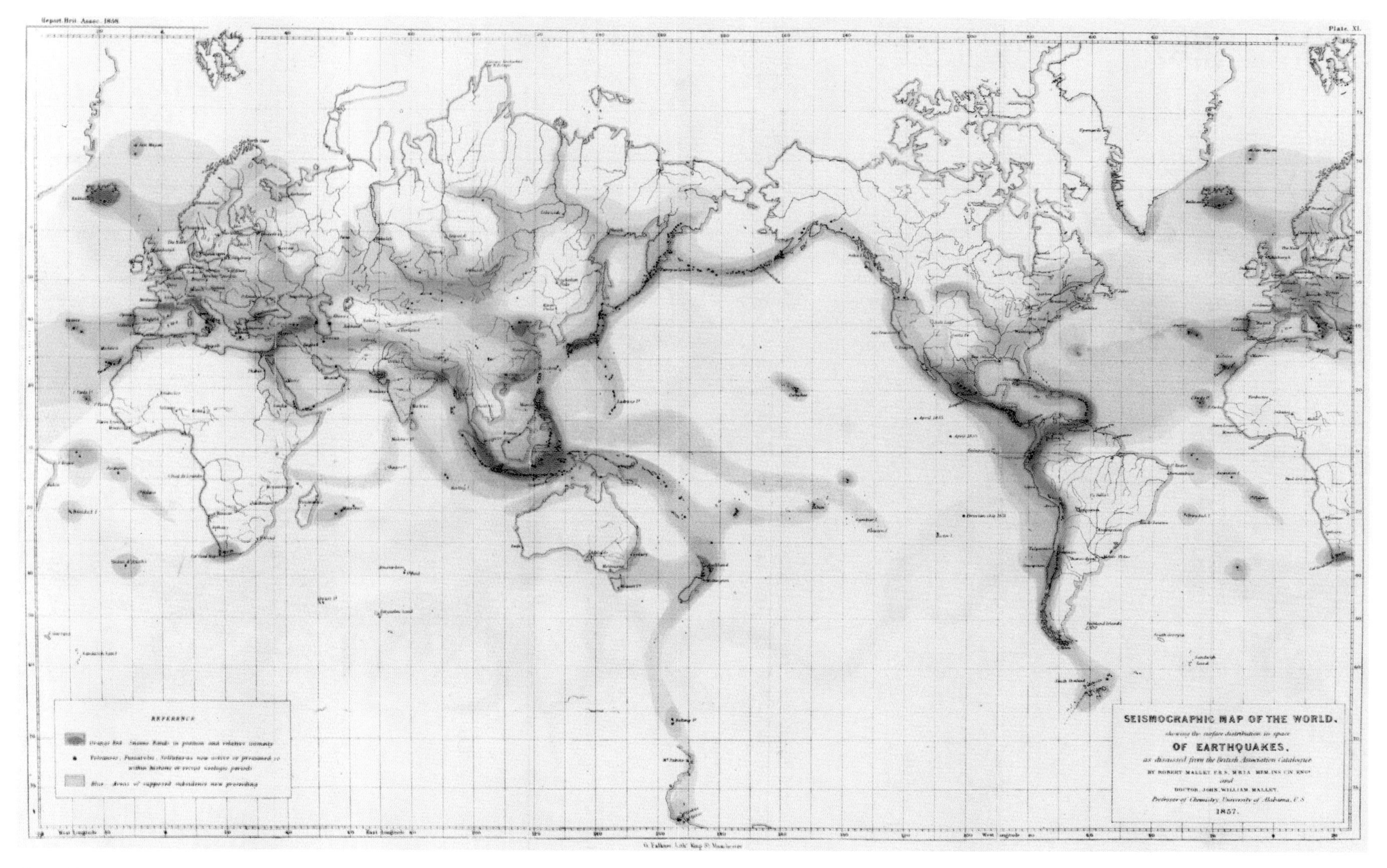

Color Plate 1 Map of the global incidence of earthquakes, published in Mallet (1858), from the catalog compiled by himself and his son J.W. Mallet. Based entirely on felt reports, this clearly shows the major non-marine belts of seismic activity; Mallet noted the association of these with large mountain ranges.

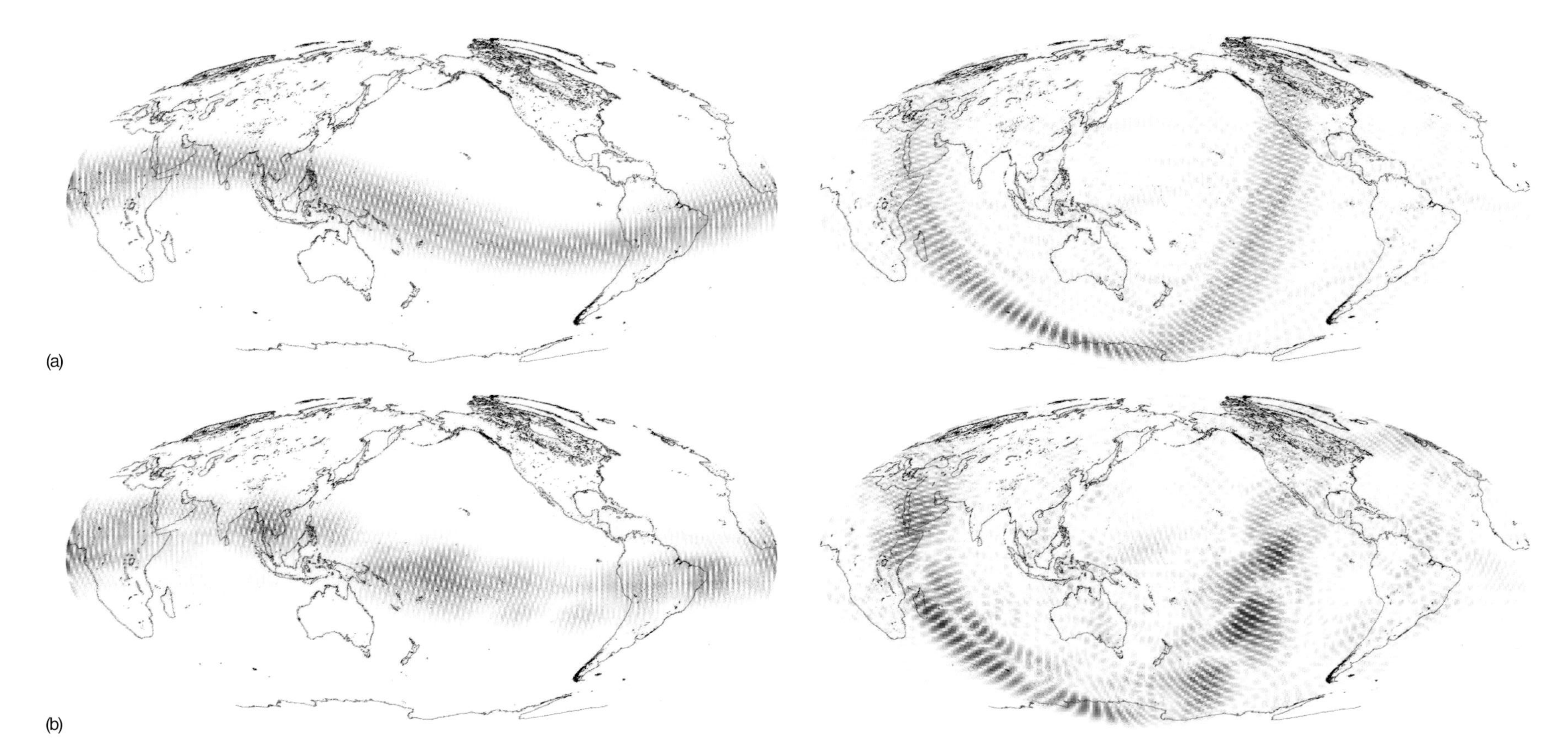

Color Plate 2 (a) Zeroth order perturbation contribution in the vertical amplitude, for two singlets of the fundamental spheroidal multiplet ${}_0S_{60}$, computed for the model SAW12D of Li and Romanowicz (1996). The left case corresponds to the singlet with positive deviation of 9.8‰ with respect to the reference frequency, the right case corresponds to the singlet with positive deviation of 3.7‰ with respect to the reference frequency. Note that for the two singlets, the amplitude is localized around a great circle, with $l = 60$ oscillations. (b) Higher order perturbations contribution in the vertical amplitude, for the two modes depicted in (a). This amplitude is the result of coupling terms, when coupling along the dispersion branch is taken into account for the 16 closest multiplets. Note that the amplitude variations are large where lateral variations are strong, i.e. in the transition zone between oceanic and continental structure for example. Note also that the amplitude is no longer neither symmetric nor antisymmetric. The contribution of these terms is of same order as the amplitude of the zeroth-order contribution.

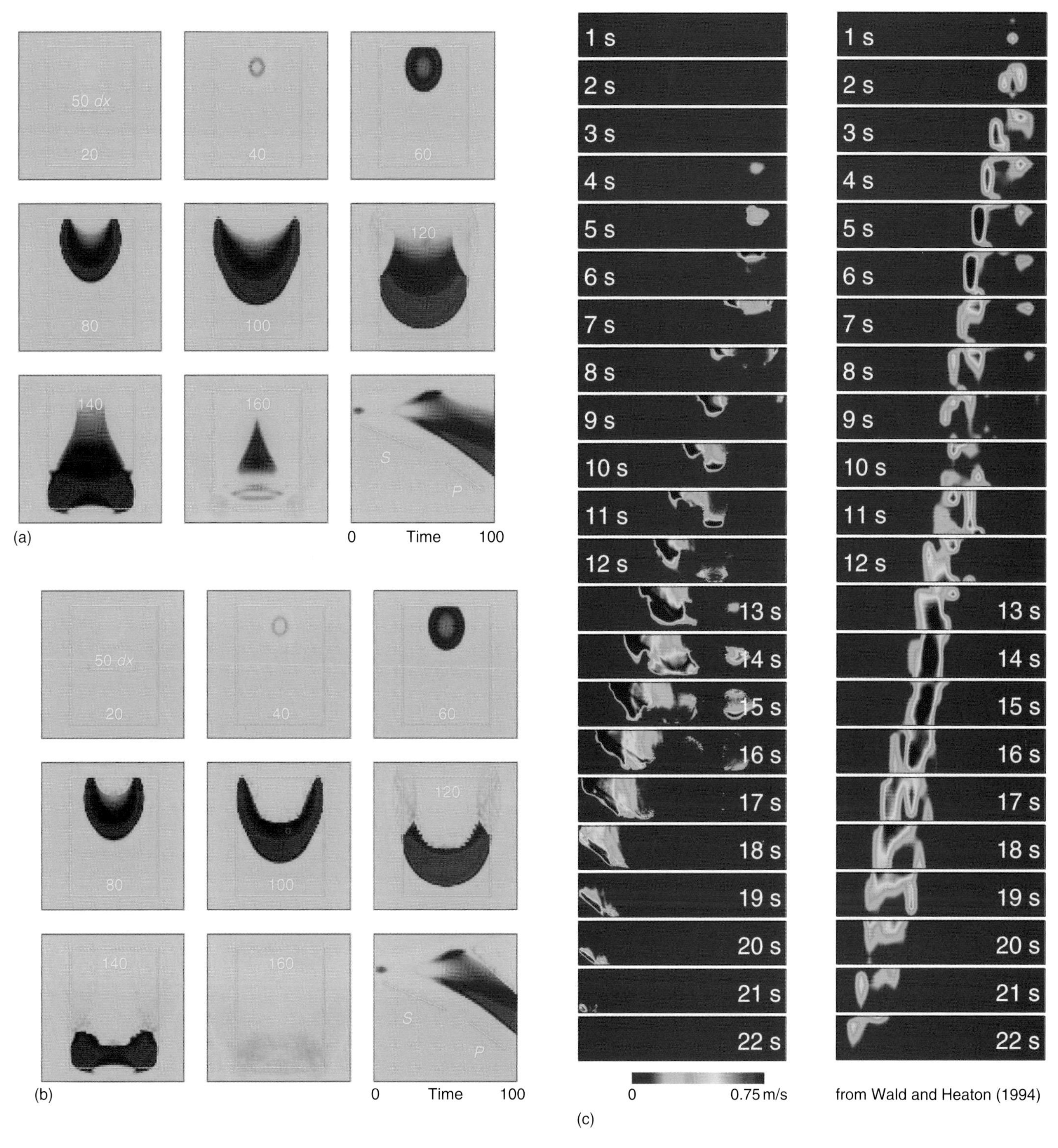

Color Plate 3 (a) Snapshots of slip rate at successive instants of time for the spontaneous rupture of an overloaded asperity inside a rectangular fault (solid line). Hotter colors depict larger slip rates. No velocity weakening friction is included. The nondimensional time is listed for each snapshot. The lower right panel shows the slip rate versus time along a line in the in-plane direction intersecting the asperity. The slope of the slanted lines depicts the *P*- and *S*-wave velocity. (Reproduced with permission from Madariaga *et al.* 1998, fig. 10, p. 1193.) (b) Same as part (a) but with velocity weakening friction included in the simulation. (Reproduced with permission from Madariaga *et al.*, 1998; fig. 11, p. 1194.) (c) Comparison of snapshots of the dynamic rupture simulation (left) of the 1992 Landers earthquake on the fault plane and the kinematic model (right) recomputed from Wald and Heaton (1994) along a single fault plane. The snapshots depict the horizontal slip rate in 1-second time slices.

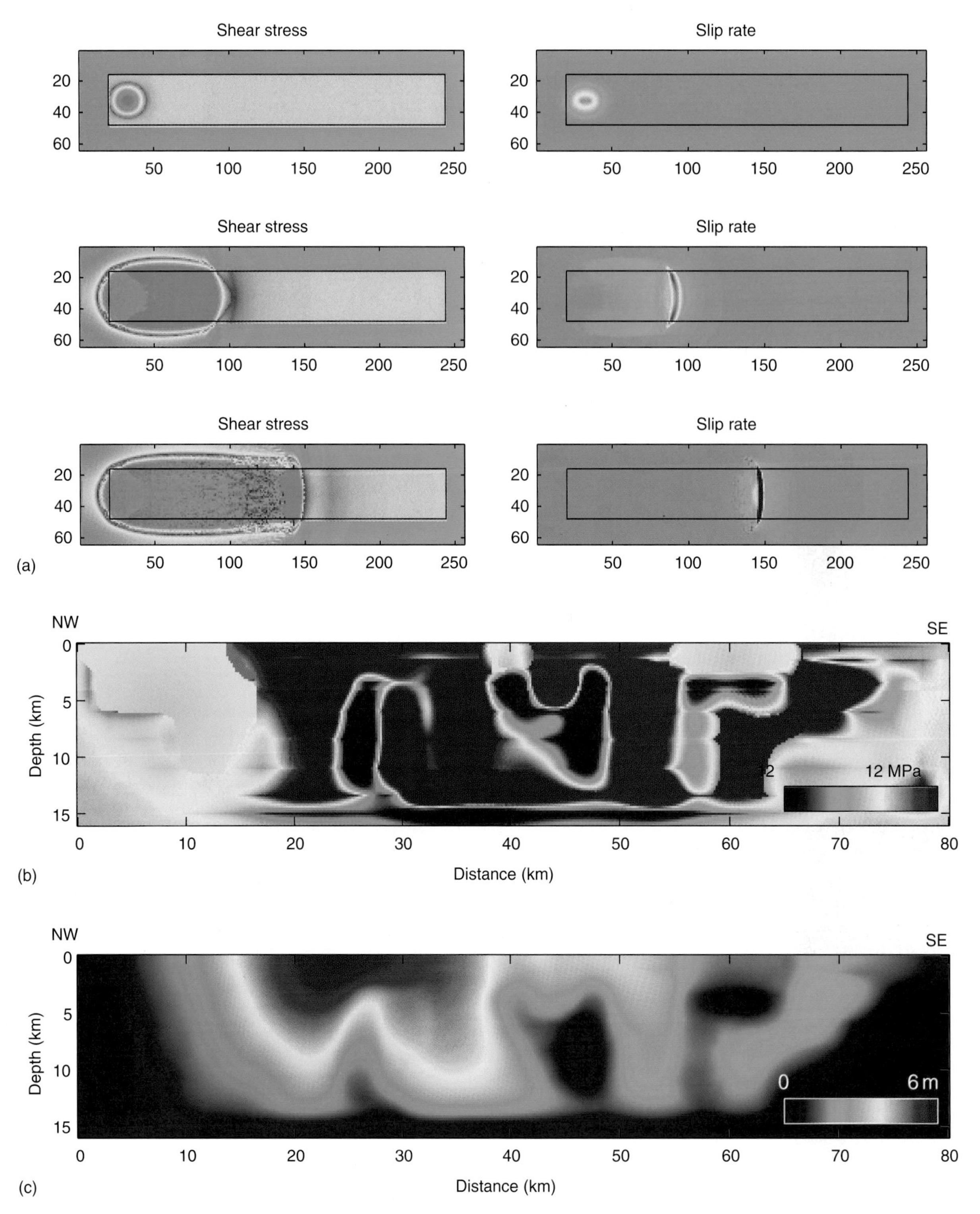

Color Plate 4 (a) Snapshots of the shear stress and slip rate fields on a fault containing a very long and narrow asperity (yellow in the stress snapshots). The panels show (from top to bottom) ruptures at sub-critical, slightly super critical and super critical values of the parameter κ. (b) Initial stress field on the Landers fault obtained by trial and error. (c) Slip distribution on the fault for the preferred dynamic model of the Landers 1992 earthquake.

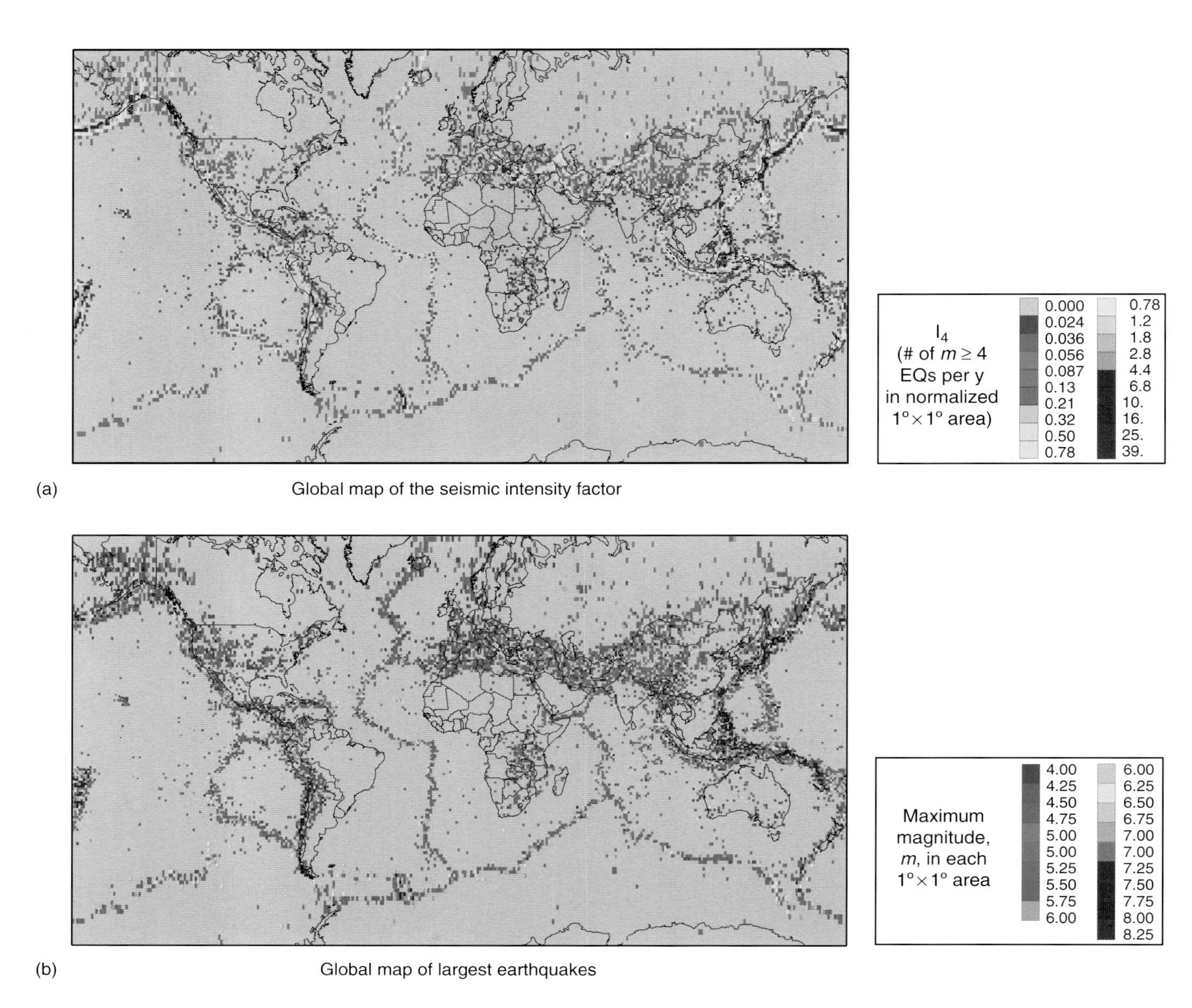

Color Plate 5 (a) Global map of the seismic intensity factor, I_4, the average annual number of earthquakes per year during 1964–1995 with magnitudes $m \geq 4$ in each normalized $1^\circ \times 1^\circ$ cell (Kossobokov *et al.*, 2000). (b) The maximum magnitude earthquake that occurred in each normalized $1^\circ \times 1^\circ$ cell during the period 1900–1997. Data for both (a) and (b) are from the NEIC Global Hypocenter Database (GHDB, 1989).

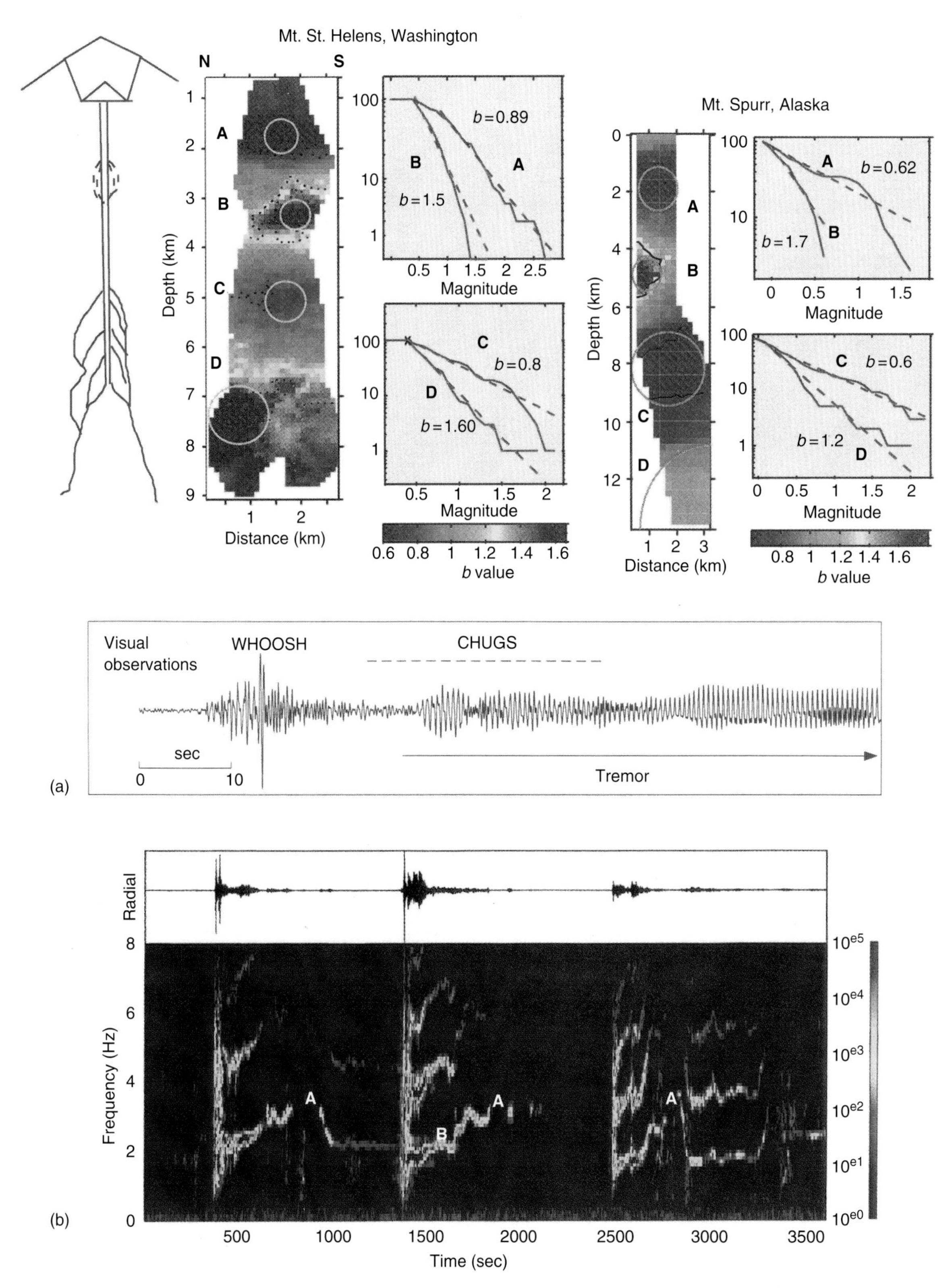

Color Plate 6 Top (Chapter 25, Section 4.3). Images showing the distribution of *b* values beneath Mt. St. Helens and Mt. Spurr. Each grid element shows the *b* value for the nearest 100 earthquakes, coloring low *b* values as blue, normal values as green and yellow, and high values as red. Four samples labeled A–D are plotted to the right of each cross section. These show the actual data used to compute each *b* value. Note the anomalies (red) at 3–4 km for each volcano and 7–9 km for Mt. St. Helens and 10–13 km for Mt. Spurr. The deeper anomalies are near magma chambers and the shallower ones are at depths of vesiculation of gases. The black line drawing represents a schematic view of the Mt. St. Helens plumbing system (modified from Wiemer and McNutt, 1997). **Bottom** (Chapter 25, Section 5.3). (a) An example seismogram of a small eruption (whoosh) at Arenal volcano, followed by rhythmic degassing (chugs), followed by harmonic tremor. The eruption ejected ash to 500 m above the crater and the chugs were audible at 2.8 km for 25 sec. (b) Seismogram and spectrogram from April 1994 showing three small eruptions. The blue colors show low signal levels, the yellow intermediate, and the red strongest. The spectra are dominated by narrow peaks spaced regularly in frequency, each of which changes frequency systematically with time. Spectrogram vertical scale is 0 to 8 Hz, and horizontal scale is 3600 sec (1 h) (modified from Benoit and McNutt, 1997).

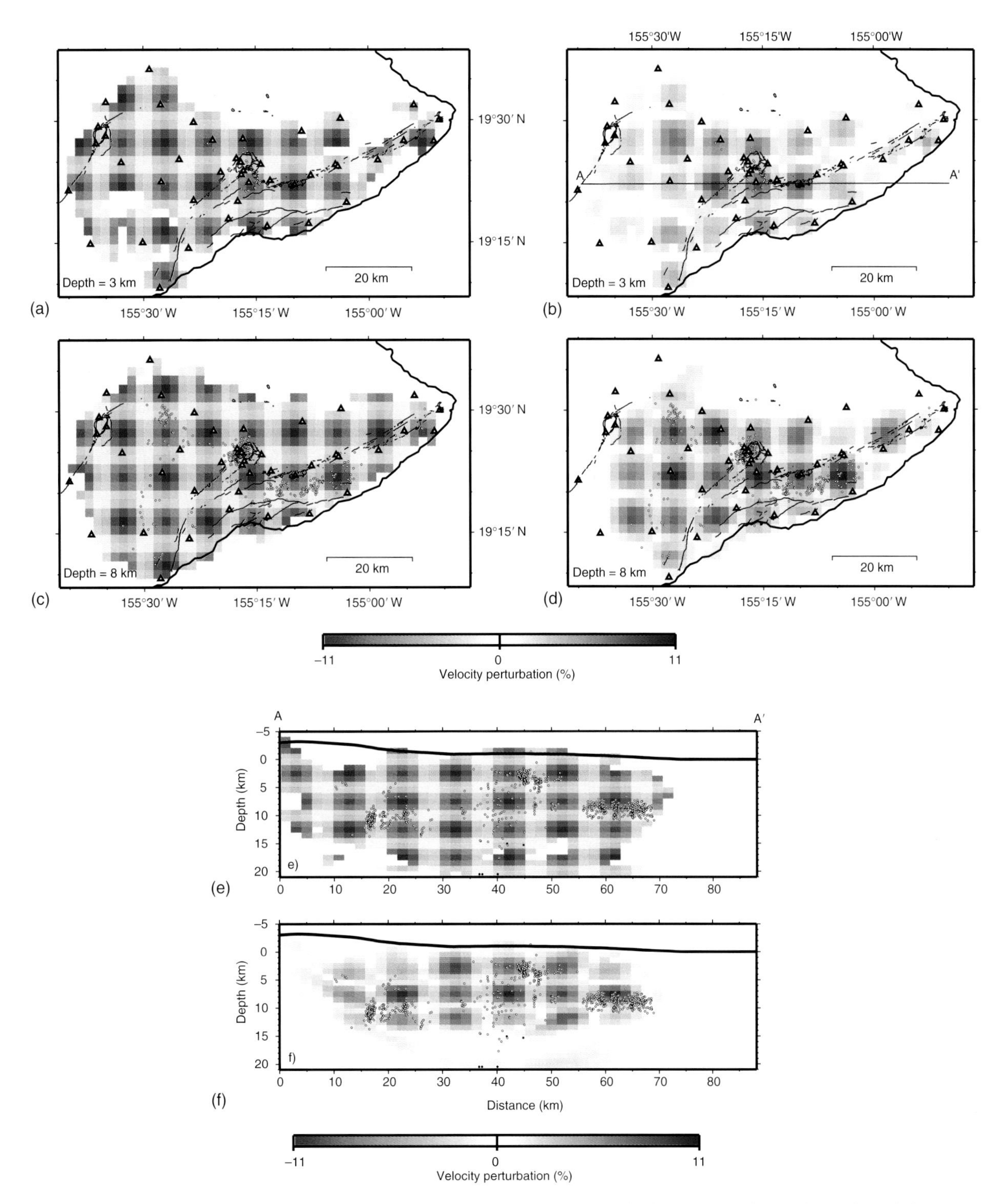

Color Plate 7 Map view of synthetic *P*-wave checkerboard model (a,c) and the synthetic reconstruction (b,d) at selected depths. East-west cross-section of a synthetic *P*-wave checkerboard model (e) and the synthetic reconstruction (f). The location of the cross-section is shown in (b). To aid in the comparison of the "true" and reconstructed velocity models, cells that are not sampled in calculating the synthetic arrival-times and in the synthetic reconstruction are blanked out (white). Earthquakes that are located within the layer are plotted (open circles). Earthquakes that lie within ± 2 km of the cross-section are plotted (filled circles). See the text for a description on the generation of the checkerboard velocity model and calculation of the synthetic arrival-times.

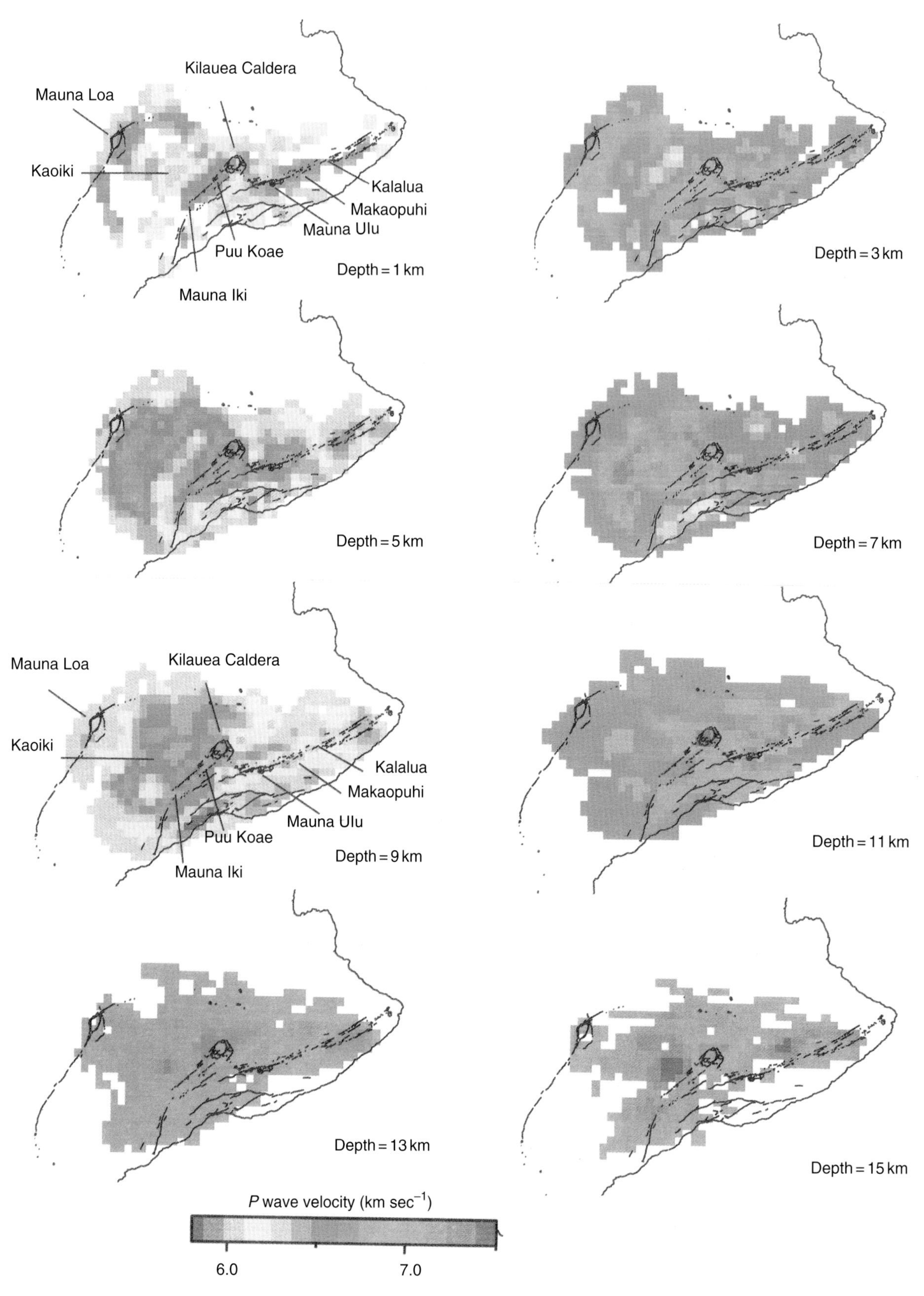

Color Plate 8 Map view of the *P*-wave velocities at selected depths between 1 to 15 km. The seismic stations used to monitor seismicity and major faults, fissures, vents and calderas are shown for reference. Cells not sampled by any ray are not plotted.

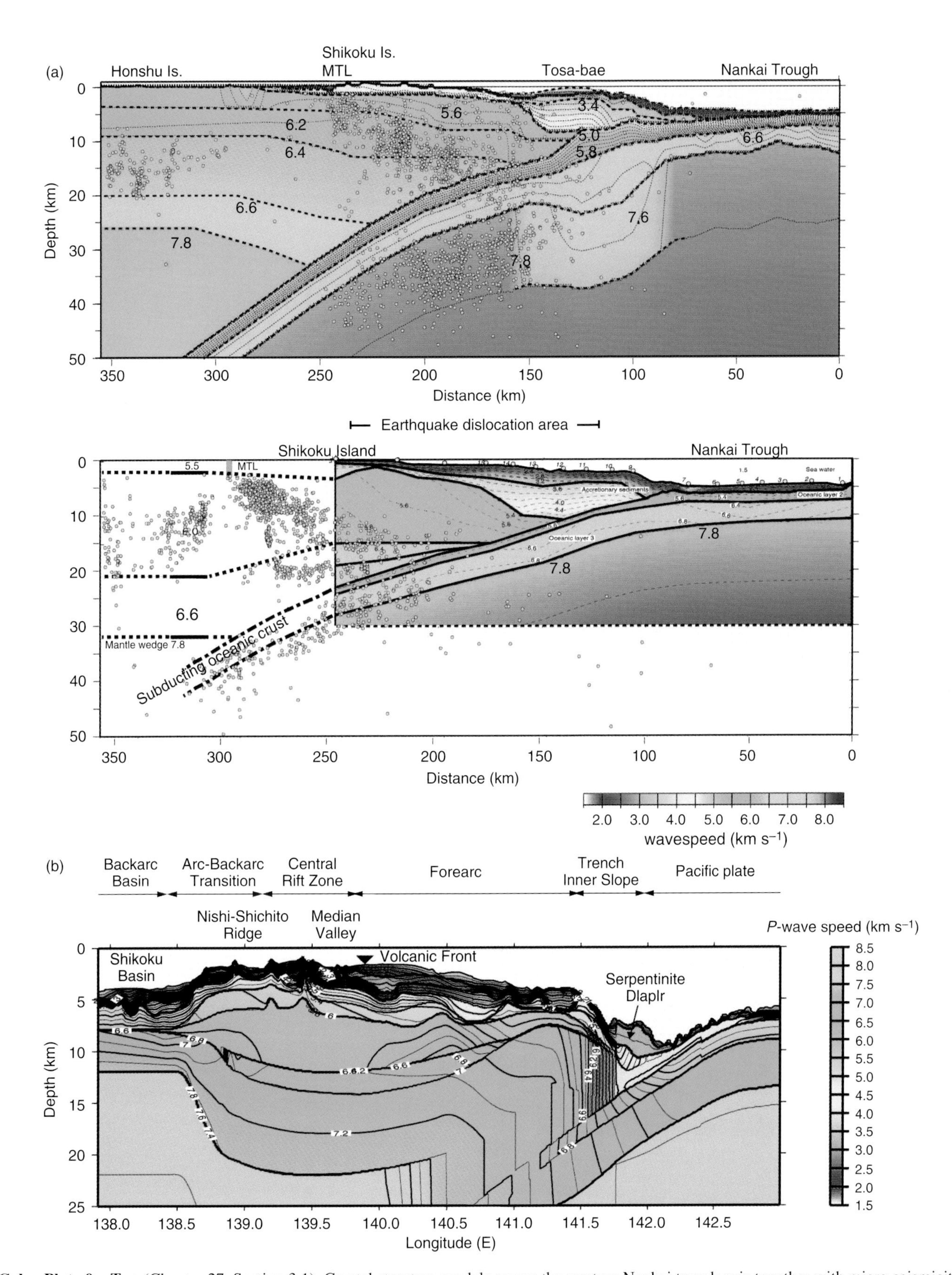

Color Plate 9 **Top** (Chapter 27, Section 3.1). Crustal structure models across the western Nankai trough axis together with micro-seismicity determined by land network (see Fig. 1B) (from Kodaira *et al.*, 2000a,b). Also, estimated rupture portion along plate boundary of the 1946 Nankaido Earthquake (M_W 8.2) is shown. Anomalous structure beneath Tosa-bae (top panel) is interpreted to be due to seamount subduction. Bottom panel (about 50 km westward) has no such feature. **Bottom** (Chapter 27, Section 3.2). Across-arc crustal model of northern Izu–Ogasawara arc (from Suyehiro *et al.*, 1996). Combined OBS and MCS investigations revealed the 2D image of the Pacific plate subduction and oceanic island arc crust. Copyright American Association for the Advancement of Science.

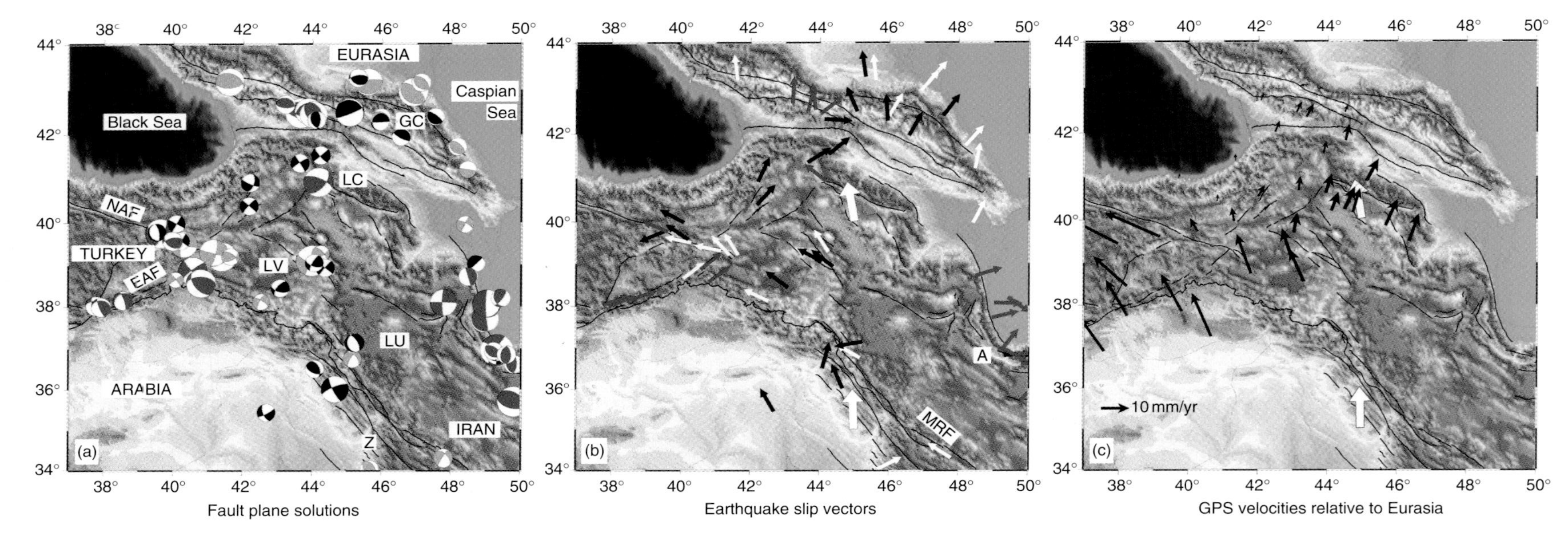

Color Plate 10A (a) Fault plane solutions in eastern Turkey, NW Iran and the Caucasus. Red focal spheres are those constrained by body-wave modeling, black ones are Harvard CMT solutions for additional earthquakes with $M_W \geq 5.3$ and with more than 70% double-couple component (see Appendix), and gray spheres are other earthquakes whose mechanisms are based on long-period first-motion polarities. Important patterns here include: (1) low angle thrusting on both sides of the Greater Caucasus (GC) and in the SW corner of the Caspian Sea; (2) NW–SE right-lateral strike-slip faulting in a belt from the northern Zagros mountains (Z), through the regions between Lakes Urumiyeh (LU) and Van (LV) to join with the North Anatolian Fault (NAF). The East Anatolian Fault is marked EAF. (b) Earthquake slip vectors taken from the mechanisms in (a), with the same colour code for red and black arrows and with white arrows representing the slip vectors from the first-motion (gray) solutions. Some slip vectors have been omitted, where the choice of nodal plane is uncertain. Examples of the separation of strike-slip and convergent components of motion ('partitioning') occur on different scales. At the largest scale, the overall oblique convergence between Arabia and Eurasia (large white arrows) is separated into strike-slip in the south and shortening in the eastern Great Caucasus, with approximately orthogonal slip vectors. At smaller scales, nearly perpendicular slip vectors on adjacent strike-slip and thrust earthquakes are seen in the Alborz (A) and near the Main Recent Fault (MRF) in the NW Zagros. (c) Velocities of various points relative to Eurasia, determined by GPS (from McClusky *et al.*, 2000).

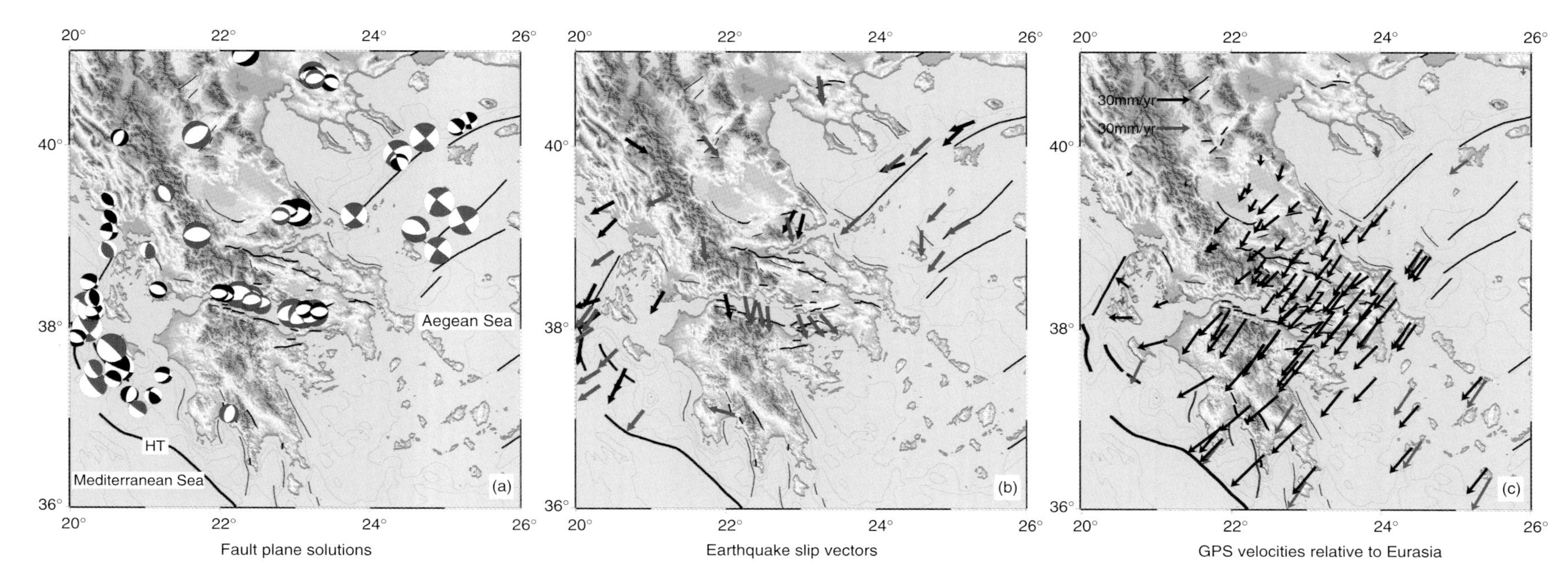

Color Plate 10B (a) Fault plane solutions in central Greece and the western Aegean. Red focal spheres are those constrained by body-wave modeling. Black focal spheres are Harvard CMT solutions for additional earthquakes with $M_w = 5.3$ and with more than 70% double-couple component (see Appendix). Plio-Quaternary and active faults trends are marked, with those thought to be currently active shown in thicker lines. HT is the Hellenic Trench. Bathymetric contours are shown at 500, 1000, 2000 and 3000 m. Note how the NE–SW right-lateral strike-slip faults in the northern Aegean change abruptly to E–W normal faults in central Greece. (b) Slip vectors for earthquakes shown in (a), with red arrows from body-wave solutions and black ones from CMT solutions. Note the abrupt change in slip vector direction where the strike-slip faulting changes to normal faulting in central Greece. (c) Velocities relative to Eurasia, determined by GPS. Black arrows are from Clarke *et al*. (1998), red arrows are from McClusky *et al*. (2000). Note that the change in slip vector direction in central Greece is not seen in the velocity azimuths.

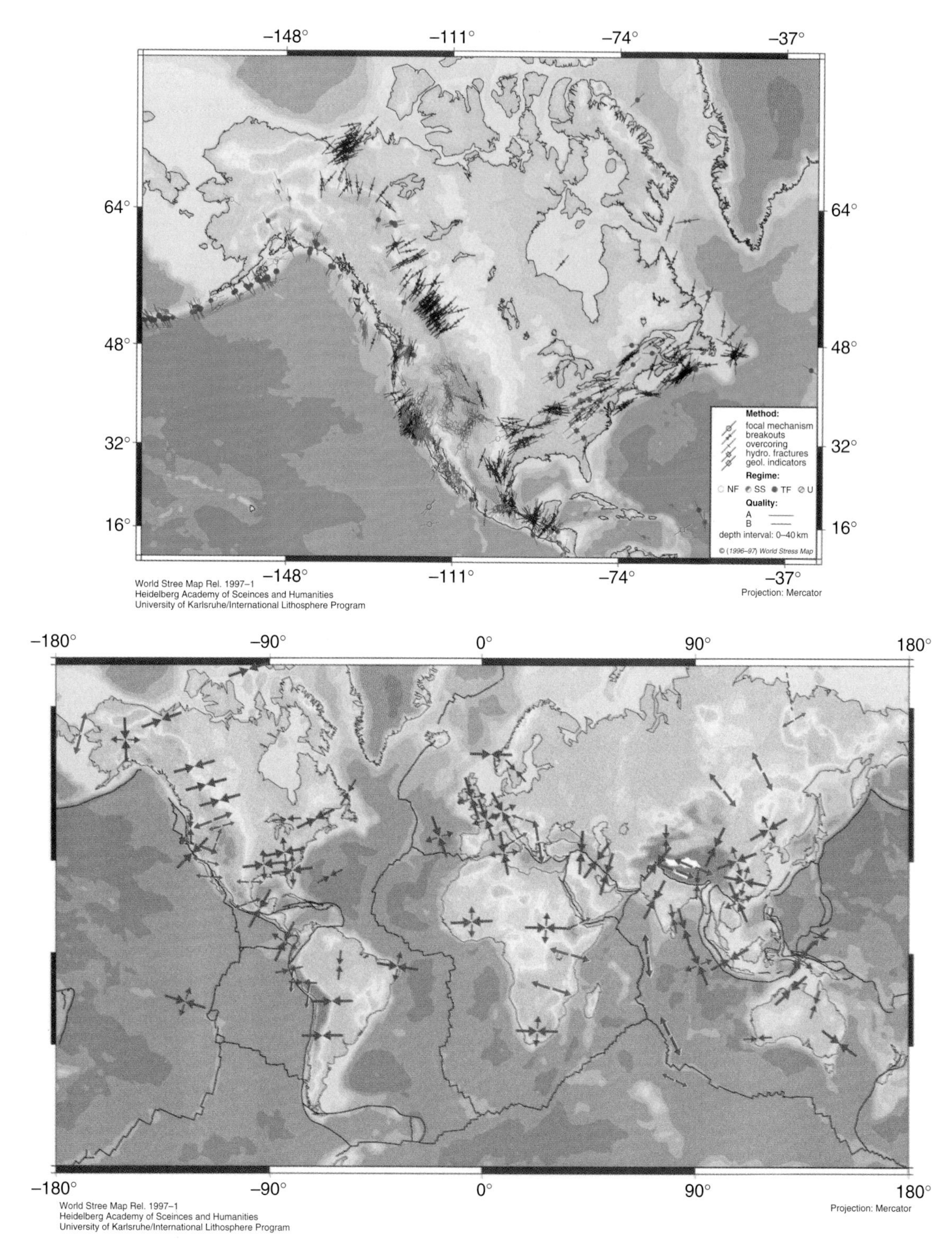

Color Plate 11 **Top** (Chapter 34, Section 3): Directions of maximum horizontal stress from the World Stress Map database superimposed on global topography and bathymetry. Only A and B quality data are shown. Data points characteristic of normal faulting are shown in red, strike-slip areas are shown in green, reverse faulting areas are shown in blue and unknown stress indicators are shown in black. **Bottom** (Chapter 34, Section 4): Generalized world stress map based on the data in Color Plate 12 and similar to that of Zoback (1992). Inward pointed arrows indicate high compression as in reverse faulting regions. Paired inward and outward arrows indicate strike-slip faulting. Outward directed red arrows indicate areas of extension. Note that the plates are generally in compression and that areas of extension are limited to thermally uplifted areas.

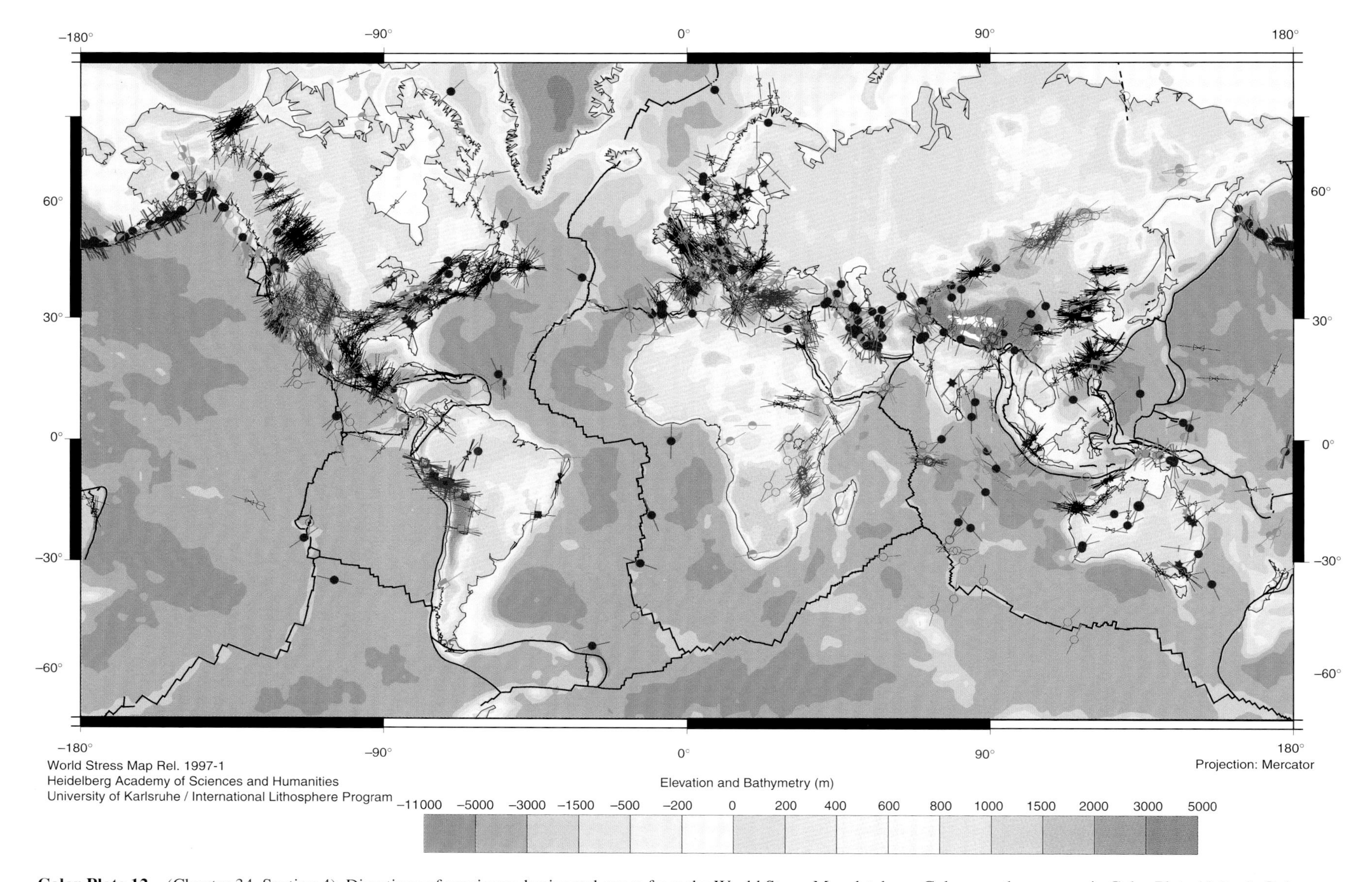

Color Plate 12 (Chapter 34, Section 4). Directions of maximum horizontal stress from the World Stress Map database. Colors are the same as in Color Plate 11 (top). Only A and B quality data are shown.

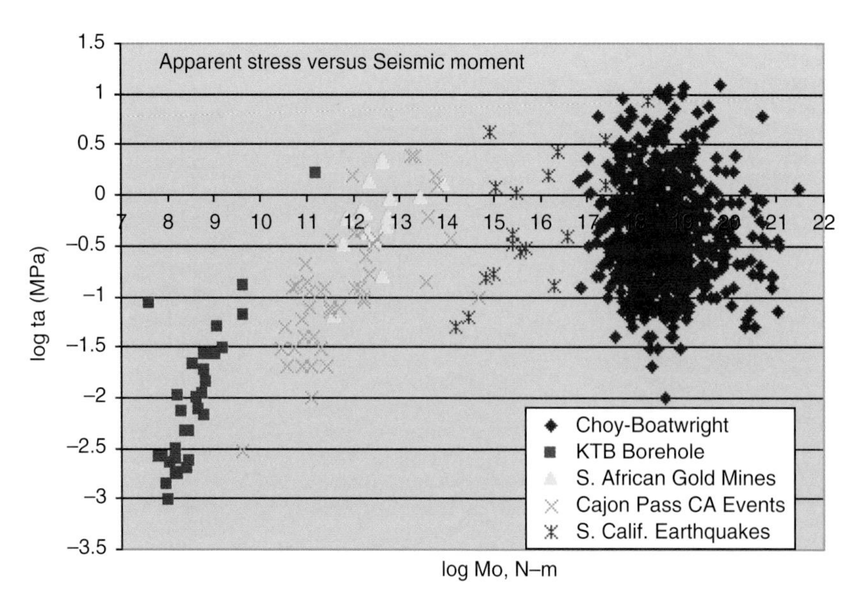

Color Plate 13 Apparent stress versus seismic moment (M_0). Data sources: KTB borehole events, Zoback and Hartjes (1997); South African mine tremors, McGarr, 1999; Cajon Pass, Abercrombie (1995); Southern California, Kanamori *et al.* (1993). Modified from McGarr (1999).

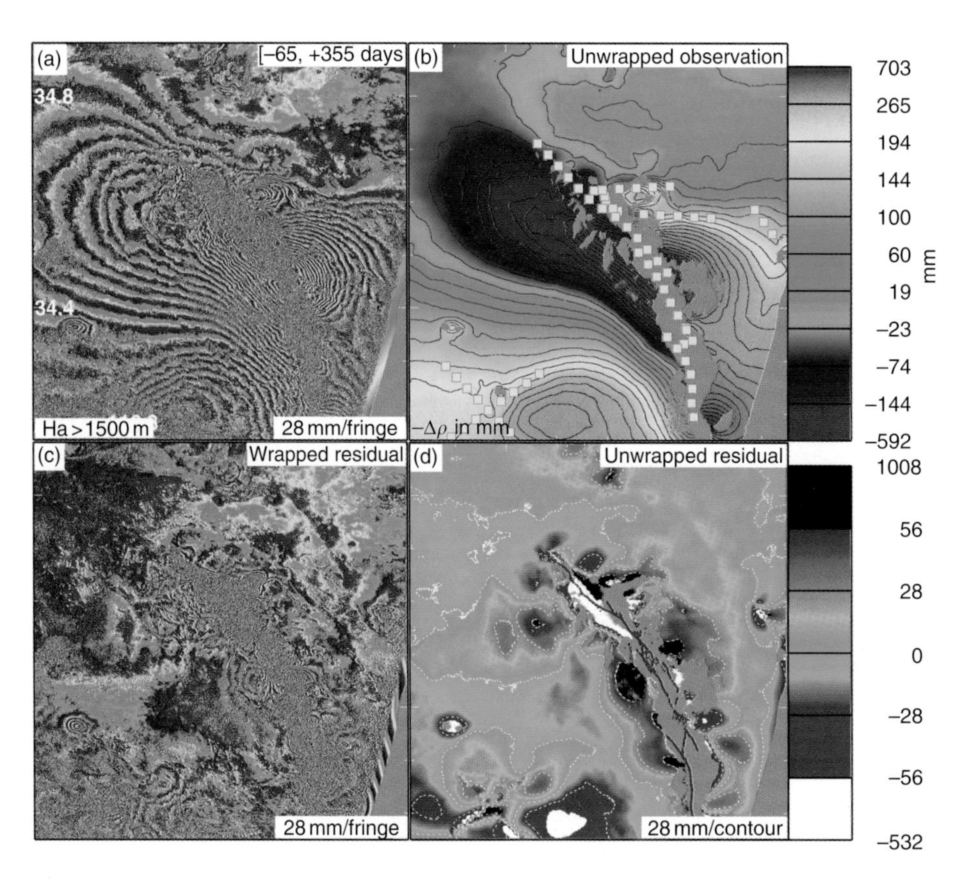

Color Plate 14 Coseismic deformation field for the Landers earthquake measured by INSAR. (a) Observed "wrapped" interferogram, shown as 28-mm fringes. (b) Observed "unwrapped" interferogram in mm (Trouvé, 1996; Trouvé *et al.*, 1998). (c) Residual (observed minus calculated) wrapped interferogram, shown as 28-mm fringes. (d) Residual unwrapped interferogram in mm. The interferogram is calculated from ERS-1 SAR images taken before (April 24, 1992) and after (June 18, 1993) the earthquake (Massonnet *et al.*, 1994). Each fringe in parts a and c denotes 28 mm of change in range. Here, the altitude of ambiguity h_a exceeds 1500 m.

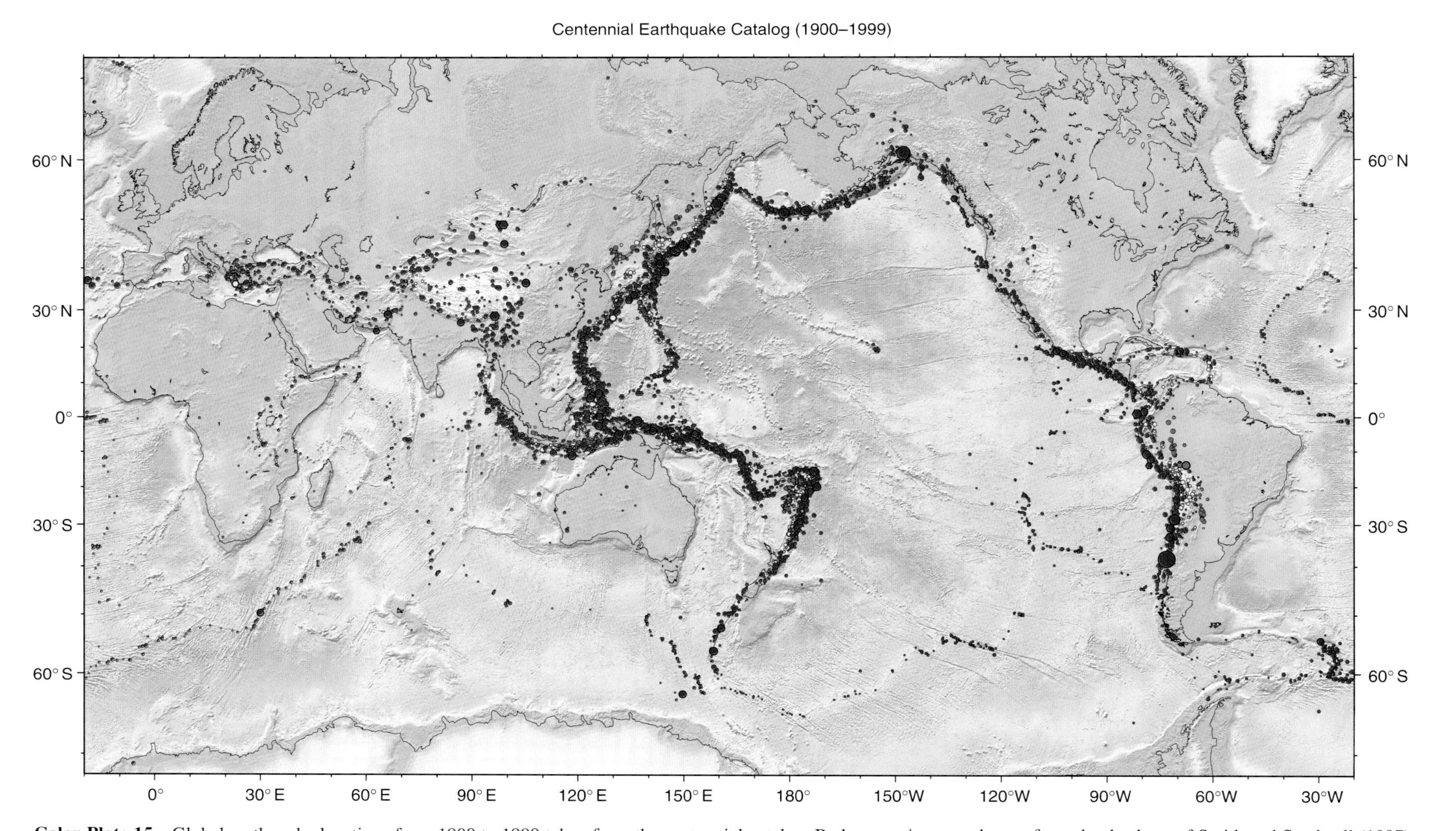

Color Plate 15 Global earthquake locations from 1900 to 1999 taken from the centennial catalog. Bathymetry/topography are from the database of Smith and Sandwell (1997). Earthquakes relocated in this study are shown by filled circles and unrelocated earthquakes by filled hexagons. Symbol fill is color-coded according to focal depth h: red = shallow events ($h < 70$ km); yellow = intermediate ($70 \leq h < 350$ km); and blue = deep ($h \geq 350$ km). A thick symbol outline is used for events with magnitudes greater or equal than 8.0.

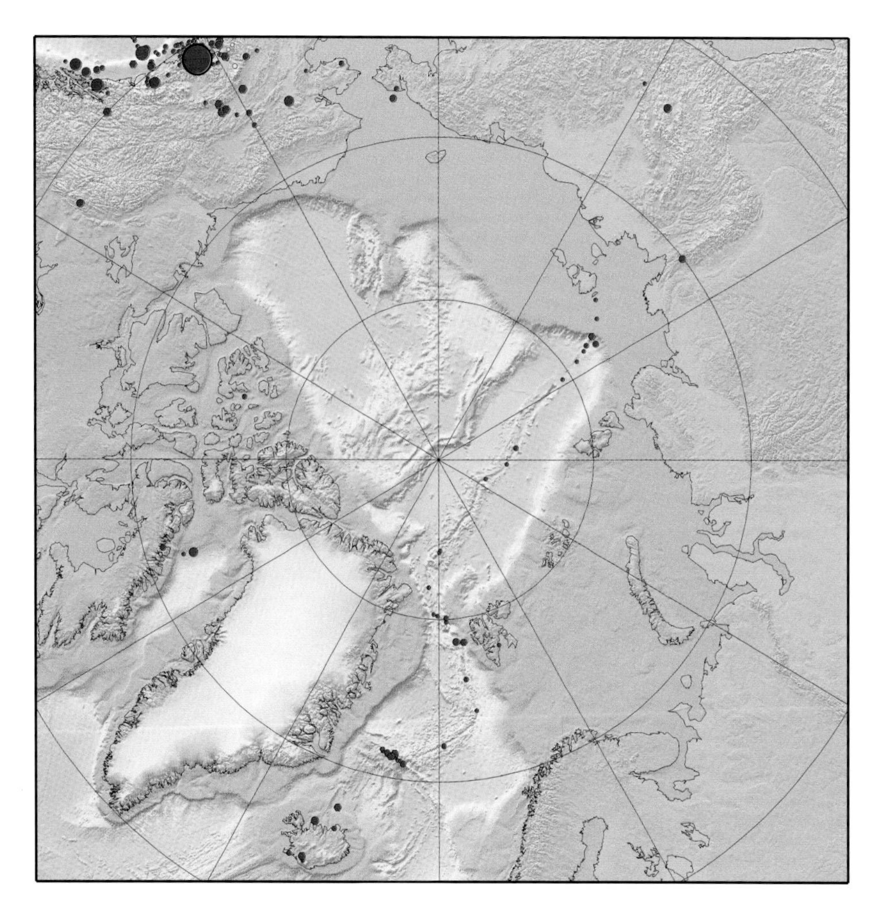

Color Plate 16 Earthquake locations from 1900 to 1999 for the Arctic region, taken from the centennial catalog. Bathymetry/topography are from the database of Jakobsson *et al*. (2000). Same symbols as in Color Plate 15.

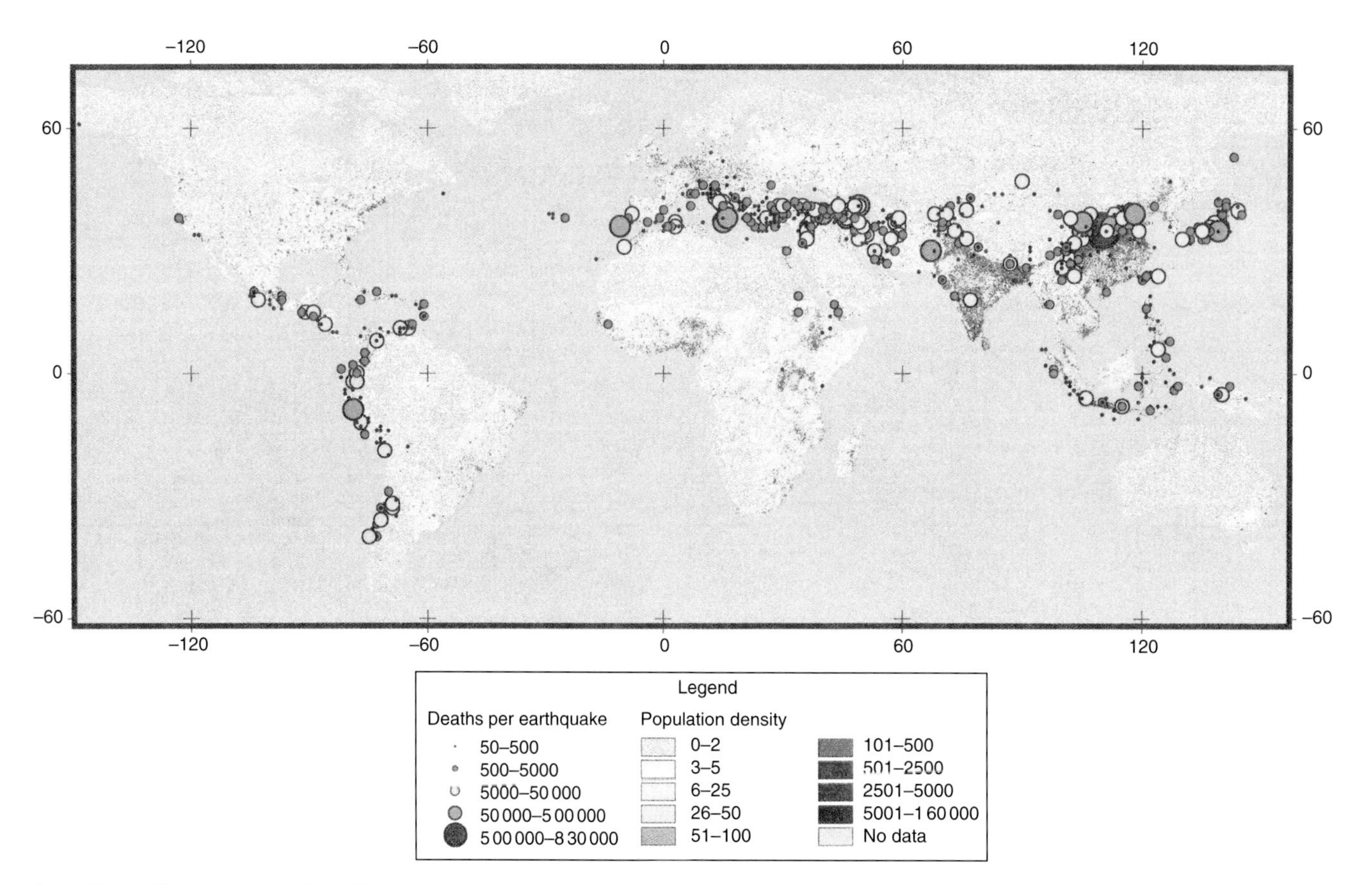

Color Plate 17 Locations of deadly earthquakes of the world: 1500–2000. Population density data was from the US Oak Ridge National Laboratory, 1999.

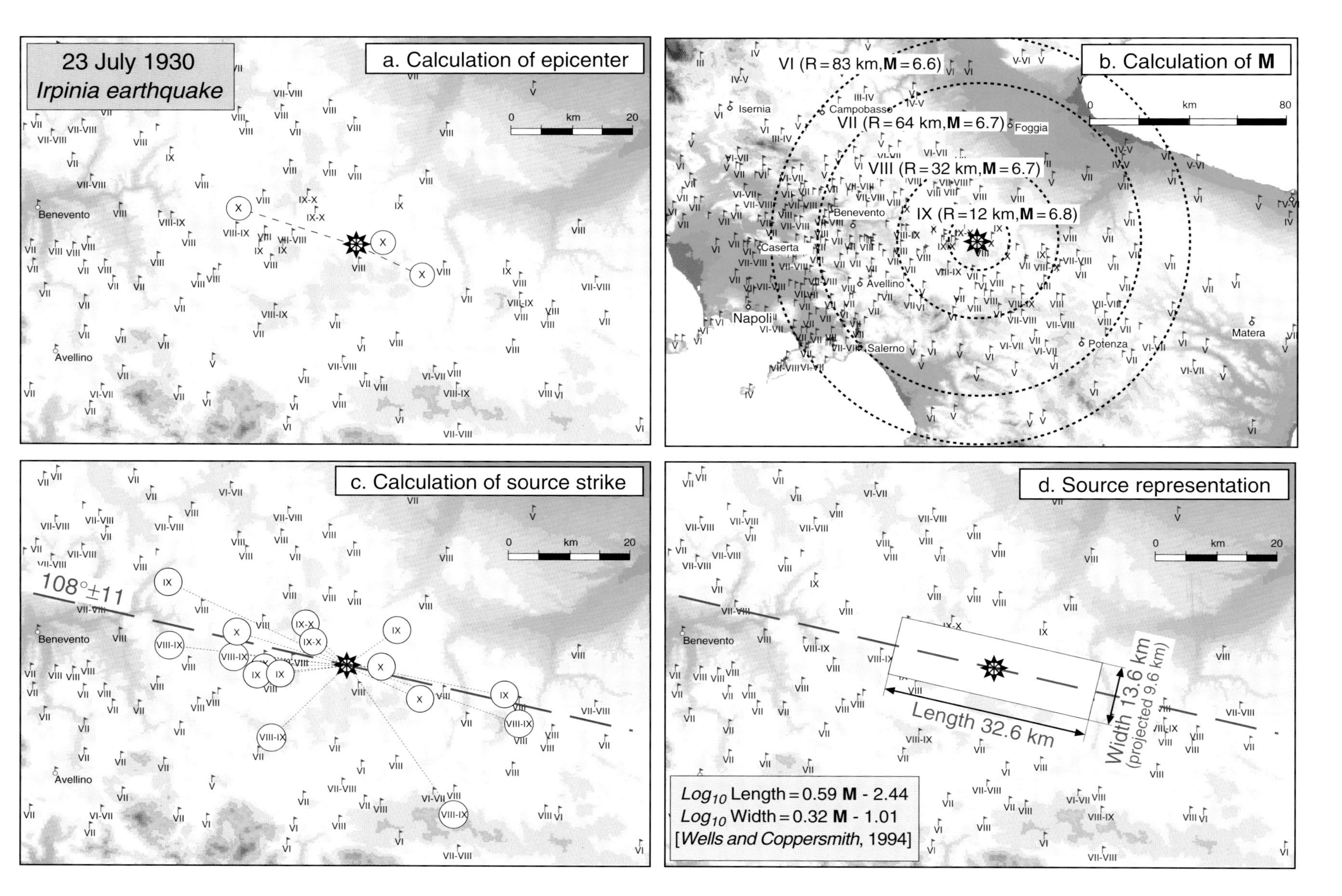

Color Plate 18 A method to assess parameters (Gasperini *et al.*, 1998).

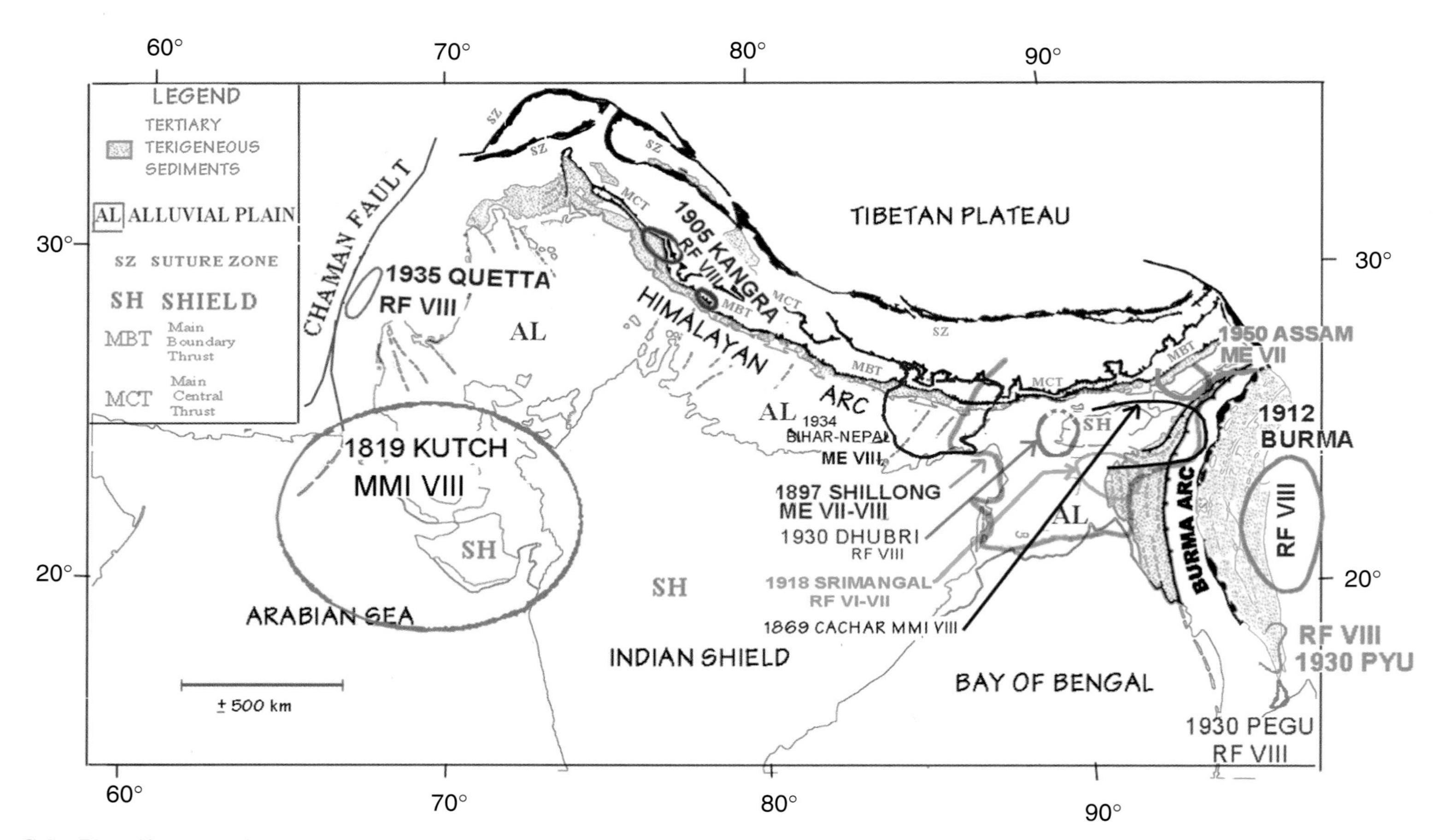

Color Plate 19 Approximate RF VIII or equivalent isoseismal areas of some major historical earthquakes of India. Generalized tectonic features are adapted from Gansser (1964, 1983). (*On account of the large scale of the map, the isoseismal areas are only approximate. For exact details, the reader must refer the figures prepared for the individual earthquakes.*)

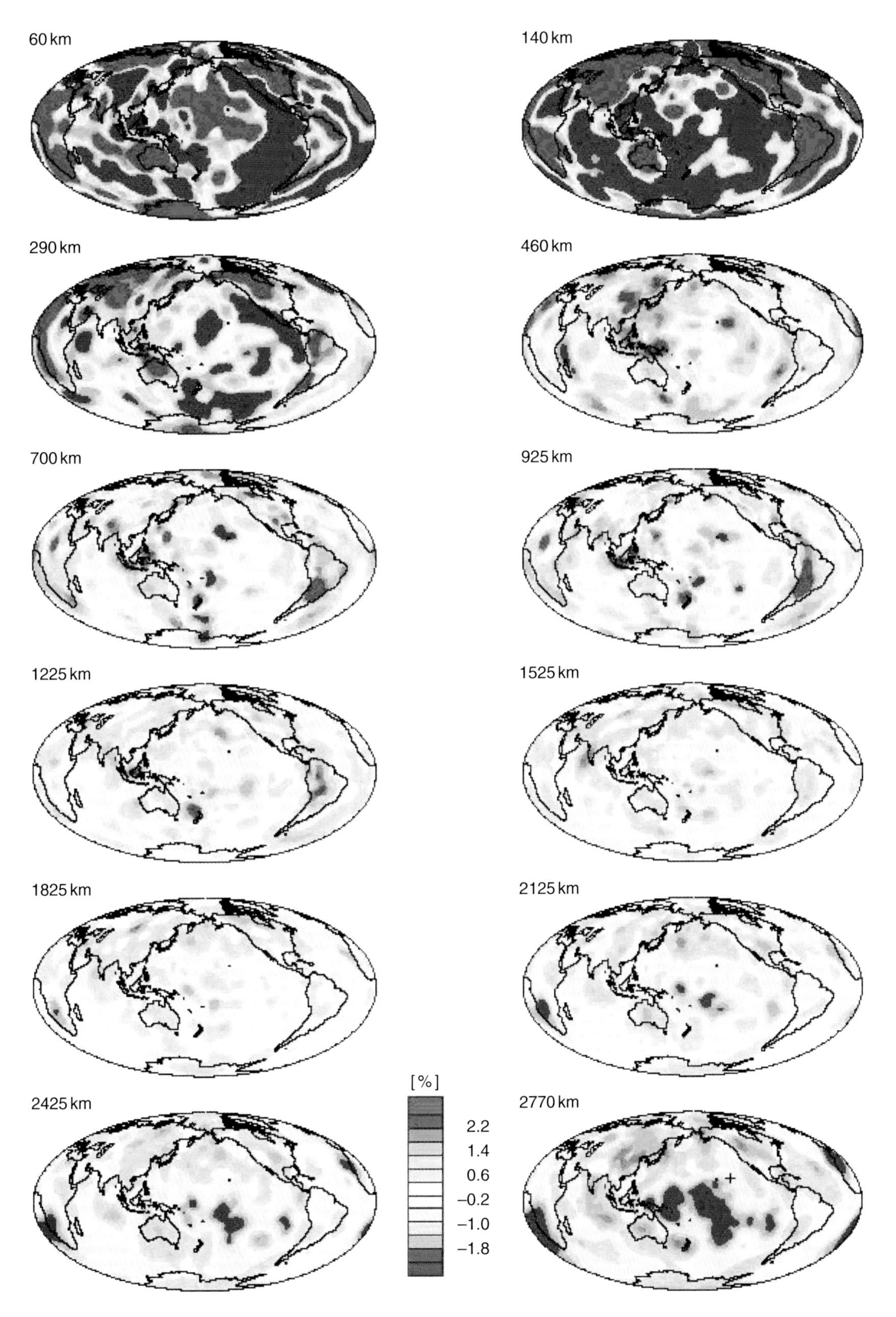

Color Plate 20 The global shear velocity tomography model SB4l18, obtained using a large data set of body wave differential times, surface wave dispersion measurements, and free oscillation splitting functions. (Adapted from G. Masters *et al*., 1999.)

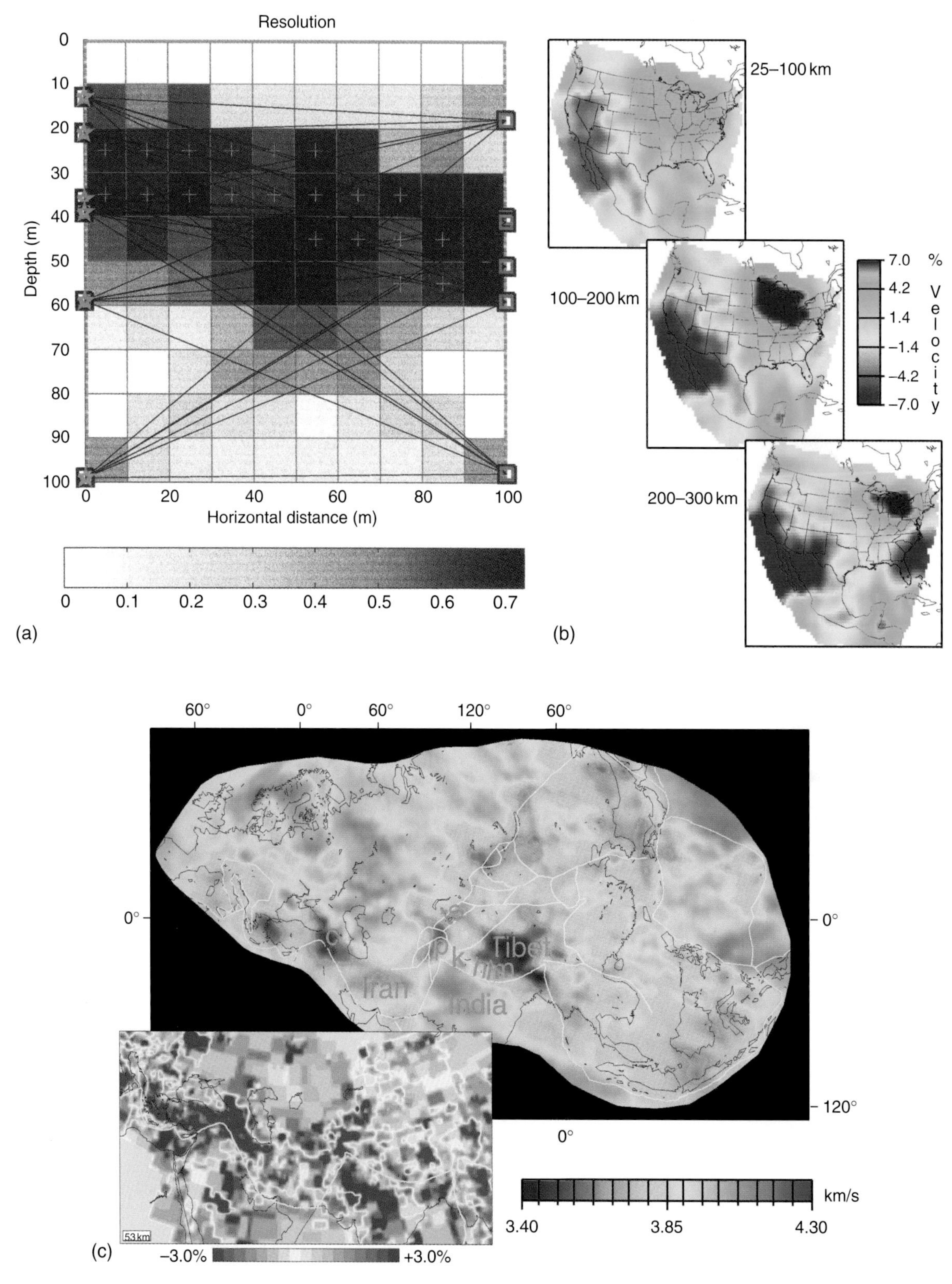

Color Plate 21 (a) Best focused crosswell survey geometry found. Figure key similar to Figure 4 of Chapter 52, but here shading also illustrates the resolution of each model cell (best resolution in darker shades). In this example only six sources (boxes with stars) and receivers (empty boxes) were used. Crosses show model cells on which the survey was focused. (b) The relative shear velocity perturbation under North America in three layers at depths between 25 and 300 km as determined by Alsina *et al.* (1996). (c) 40 s period fundamental mode Rayleigh phase velocity map from Curtis *et al.* (1998). Inset: slice of the *P*-velocity anomaly through the model of Bijwaard and Spakman (2000) at 53 km depth. Major tectonic boundaries shown in yellow, c – Caucassus, p – Pamir, k – Karakoram, ts – Tien Shan, him – Himalayan arc.

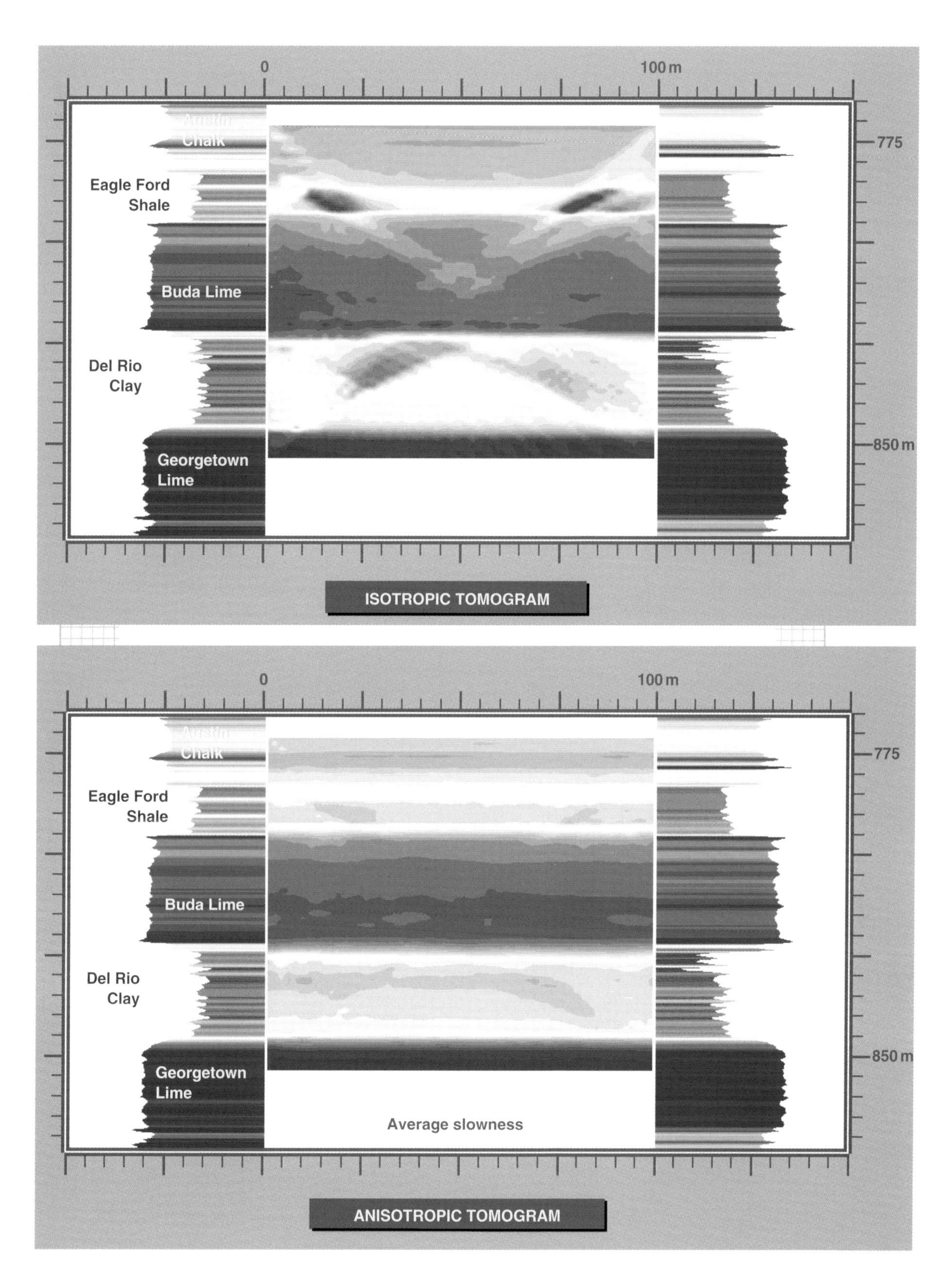

Color Plate 22 Isotropic components of an isotropic (top) and a transversely anisotropic (bottom) tomographic cross-well inversion. Wells are approximately vertical at 0 and 100 m. Left and right hand sides show slowness logs from left-hand well (structures are roughly flat-lying). Colour coding represents slowness (blue ≈ 0.2 ms/m, red ≈ 0.34 ms/m).

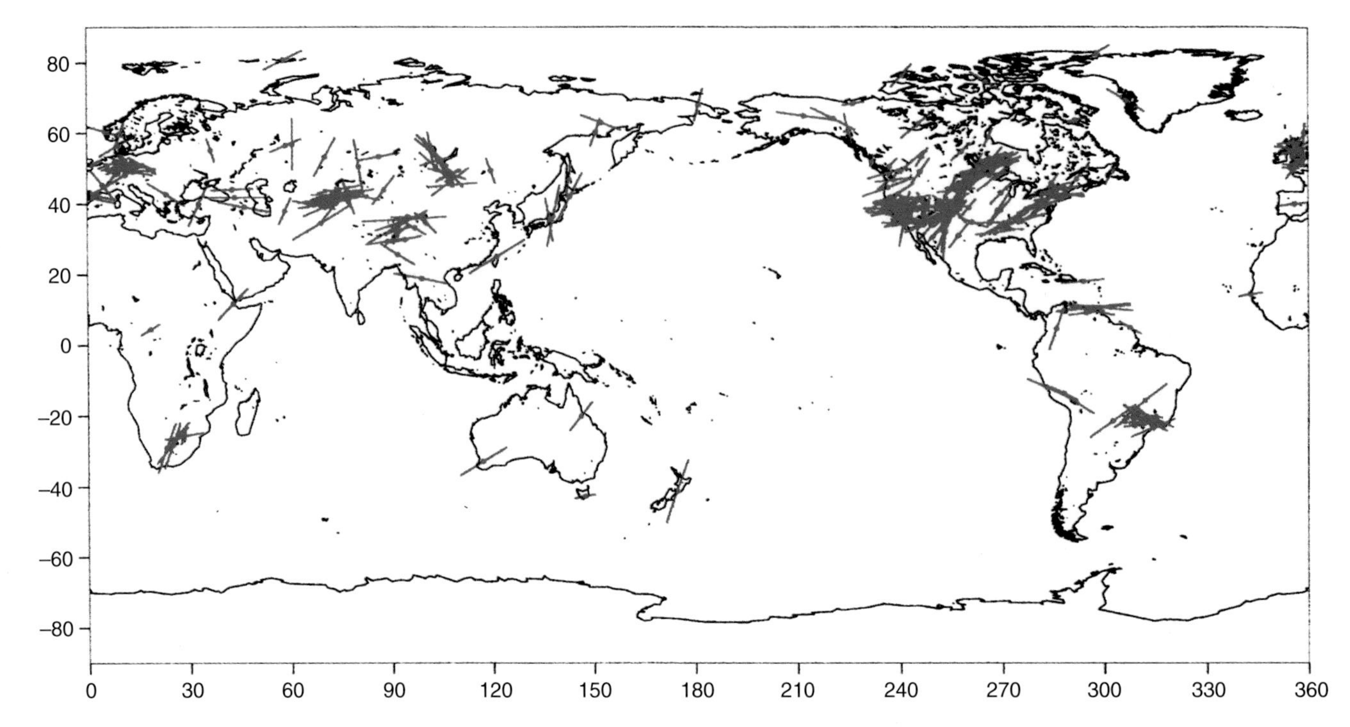

Color Plate 23 Examples of observed directions of polarization of the fast split SKS waves. The delay time between the two split SKS waves is given by the length of the segments (from Silver, 1996).

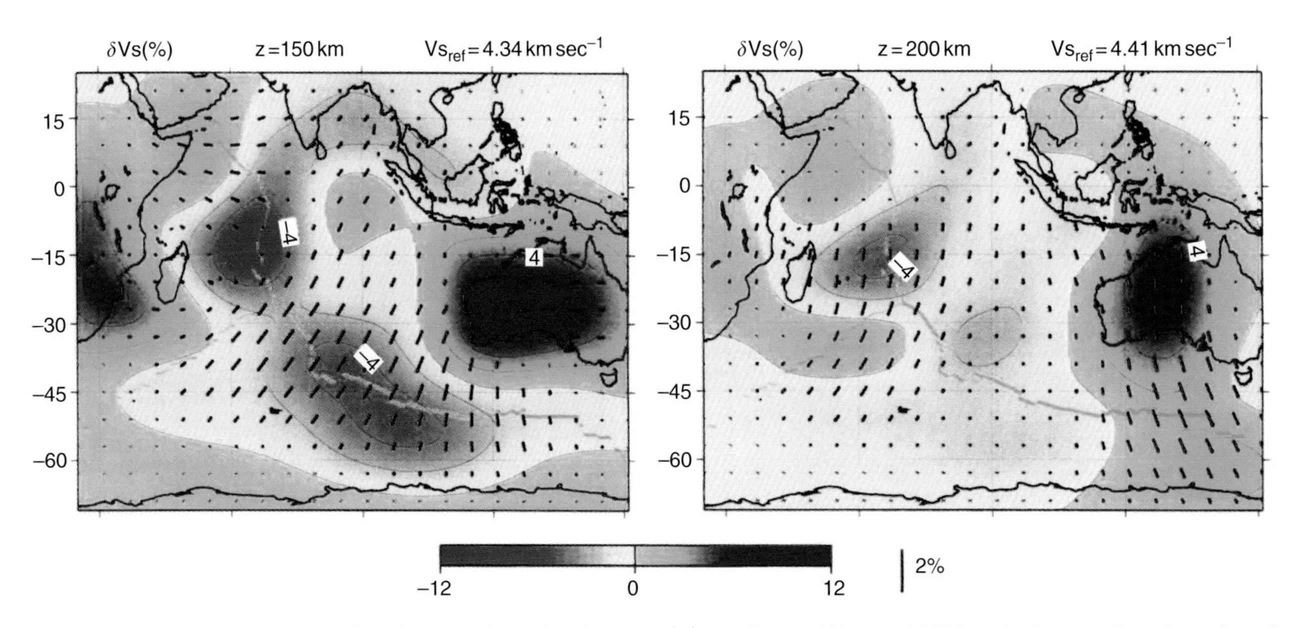

Color Plate 24 Fast SV-wave directions beneath the Indian Ocean and Australia at 150 km and 200 km depth (waveform inversion of the surface-waves by Lévêque *et al.*, 1998). The length of each segment is proportional to the amount of azimuthal anisotropy in velocity. Contours indicate the isotropic part of the lateral variations of SV velocity (from −12% to +12%).

2

Historical View of Earthquake Engineering[1]

G.W. Housner
California Institute of Technology, Pasadena, California, USA

1. Introduction

Earthquake engineering is a 20th-century development, so recent that it is yet premature to attempt to write its history. Many persons in many countries have been involved in the development of earthquake engineering and it is difficult, if not impossible to identify the contributions of each. Many advances in the subject are not well-documented in the literature, and some of the documentation is misleading, or even incorrect. For example, in some instances, earthquake requirements were adopted in building codes but were not used by architects and engineers. And in other instances earthquake design was done by some engineers before seismic requirements were put in the code. A history of earthquake engineering written now could not present a satisfactory account because of poorly documented facts and, in addition, there are still many people that remember relevant information and would be severe critics of a history. To write an acceptable history, it is necessary to wait till most of the poorly known facts have disappeared from memory and literature, then, with a coherent blend of fact and fiction, history can be written.

Although 1984 is too soon to write a definitive history, it is an appropriate time for a historical view of earthquake engineering development to see where we were, where we are now, and where we are going. In this regard, it is interesting to compare the Eighth World Conference with the First World Conference on Earthquake Engineering. In 1956, the 50th anniversary of the San Francisco earthquake, the First World Conference was held in the city of Berkeley, California. It is indicative of the very recent development of earthquake engineering that many of those pioneers who attended the first conference are also present, 28 years later, at the eighth conference. It is gratifying to see that the attendance at 8WCEE is more than 10 times as large as the number attending 1WCEE. In the Preface to the Proceedings of the First Conference, the President of EERI said "The World Conference on Earthquake Engineering was originated and planned by EERI for the purpose of (1) observing by an appropriate technical meeting the 50th anniversary year of the destructive San Francisco earthquake of 1906 and, (2) bringing together the scientists and engineers from major seismic areas of the world in order that their knowledge of earthquakes and developments in the science and art of earthquake-resistant design and construction might be pooled for the benefit of all mankind." And this still represents the purpose of the eighth conference. However, a big change has occurred in the number of papers presented. The Proceedings of the First Conference had 40 papers and the Proceedings of the Eighth Conference contain 844 papers. This 20 times increase in the number of persons seriously studying earthquake engineering is indicative of the increased importance of the subject in the seismic countries of the world. The authors in the First Conference came from 11 countries and the authors in the Eighth Conference came from 42 countries. Very few seismic countries are not represented at the 1984 Conference and this indicates that there are few seismic countries that are not actively trying to protect against destructive earthquakes, and this is a great change from the situation in 1956.

2. Early Days of Earthquake Engineering

In viewing the early days of earthquake engineering, it is not appropriate to consider developments in pure seismology but, rather, restrict consideration to developments which were made by engineers or which have a special relevance to earthquake engineering. It is surprising to learn that in the early days the most prominent men in earthquake engineering were almost all

[1] See Editor's Note.

natives of England, a country of low seismicity. This can be attributed to the Industrial Revolution, in which England was a pioneering country. The intellectual excitement associated with rapid developments in all of engineering between the years 1700–1900 attracted the attention of many able men and some developed an interest in earthquakes. Robert Hooke (1635–1703), the discoveror of Hooke's Law, which is well-known in engineering, gave a series of lectures at the Royal Society in 1667–68 which were published in book form in 1705 with the title "Lectures and Discourses of Earthquakes and Subterranean Eruptions." This was before the days of earthquake engineering and Hooke was actually considering geological matters when he argued that the raising of sub-aqueous land into mountains was caused by earthquakes, so it might be said that Hooke took the first step along the path that led to the theory of plate tectonics. Also, Thomas Young (1773–1829), of Young's modulus, in his book "Lectures on Natural Philosophy," Vol. 2, 1807, gave what appears to be the first European bibliography of earthquake publications. It is interesting that both Hooke and Young, who are so well known by engineers, should have studied earthquakes, though in those early days it was premature to think about earthquake engineering.

In the 19th century, a number of English engineers developed a keen interest in earthquakes, including Robert Mallet (1810–1881), a civil engineer, John Milne (1850–1913), a mining engineer, James Ewing (1855–1935) and Thomas Gray (1850–1908), both mechanical engineers. In the last century no distinction was made between seismology and earthquake engineering. In fact, the word "seismology" derived from the Greek work "*seismos*" (=shaking), was invented by the engineer Robert Mallet and covered all the various interests in earthquakes: earthquake occurrence, ground shaking, earthquake damage, etc. It seems that the name "seismology" (=shake-knowledge) originally designated what we would call earthquake engineering, and it was only during later developments that the name came to designate the studies of the nonengineering aspects of the subject. Robert Mallet also coined the terms "epicenter," "seismic focus," "isoseismal line," and "meizoseismal area."

Robert Mallet published his first paper, "On the Dynamics of Earthquakes" in the Transactions of the Irish Academy, Vol. 2, 1848. In this paper he discusses earthquake effects and considers seismic waves and tsunamis, and he also describes his invention of the electromagnetic seismograph (see Fig. 1). This instrument sat at rest till it sensed the arrival of the seismic waves which activated it and the response of the instrument was recorded on a rotating drum. Mallet did not build such an instrument but a modified seismograph along these lines was built in 1855 by Luigi Palmieri (1867–1896) in Naples, Italy which actually made some earthquake records. Mallet also invented the "rocking blocks" (or falling pins) intensity meter, a form of which for many years was used by the construction industry to measure the intensity of ground motion generated by blasting; and he also compiled a seismic map of the world which was in use for many years. Mallet also studied the destructive Naples earthquake of 16 December 1857 and wrote a detailed report which included carefully drawn isoseismal lines. He also compiled a 600-page catalog of earthquakes which he said was the "first attempt to complete a catalog that shall embrace all recorded earthquakes." Robert Mallet, I believe, can be called the Primeval Earthquake Engineer.

Milne, Ewing and Gray (Thomas Gray later emigrated to the United States and from 1888 to 1908 was professor of engineering mechanics at Rose Polytechnic Institute in Indiana) developed an interest in earthquakes while teaching at Imperial College of Engineering in Tokyo (later merged into Tokyo University). In addition to studying earthquake damage and other phenomena, they were pioneers in the design and construction of sensitive seismographs and the study of the seismograms. The work of these men, together with Japanese seismologists such as Seikei Sekiya (1855–1896), the world's first officially appointed professor of seismology, and Fusakichi Omori (1868–1923) led to modern seismology. The Seismological Society of Japan was organized by these men in 1880 and this first earthquake society was the forerunner of the many National Societies of Earthquake Engineering that make up the International Association for Earthquake Engineering.

In the latter part of the 19th century and the early part of the 20th century, some large and important earthquakes occurred that aroused the interests of engineers and seismologists and marked an important phase of earthquake engineering. These were the 1891 Mino-Owari, Japan; the 1906 San Francisco, California; and the 1908 Messina, Italy earthquakes. The Mino-Owari earthquake left a prominent fault scarp which is still shown in books on seismology and the San Francisco earthquake focused attention on the San Andreas fault and its displacement. Although these two large earthquakes received worldwide attention, the time was not yet ready for earthquake engineering. In his 1907 ASCE paper "The Effects of the San Francisco Earthquake of 18 April 1906 on Engineering Construction," Professor Charles Derleth said "An attempt to calculate earthquake stress is futile. Such calculations could lead to no practical conclusions of value." Engineering thinking was still based in a static world and dynamics seemed yet to be unthinkable. Despite the fact that in 1906 California had a small population with no great cities, the damage caused by the earthquake was $10–20 billion (1980) though the number of deaths was only about 1000. This did not shock engineers into developing earthquake engineering. In 1908, however, a large earthquake devastated the city of Messina, Italy and surrounding area, with a loss of life of 83 000, and this disaster was responsible for the birth of practical earthquake design of structures.

It appears that prior to 28 December 1908 engineering thinking was not ready for grappling with the engineering design of structures to resist earthquakes. In most seismic regions, the common type of construction was masonry

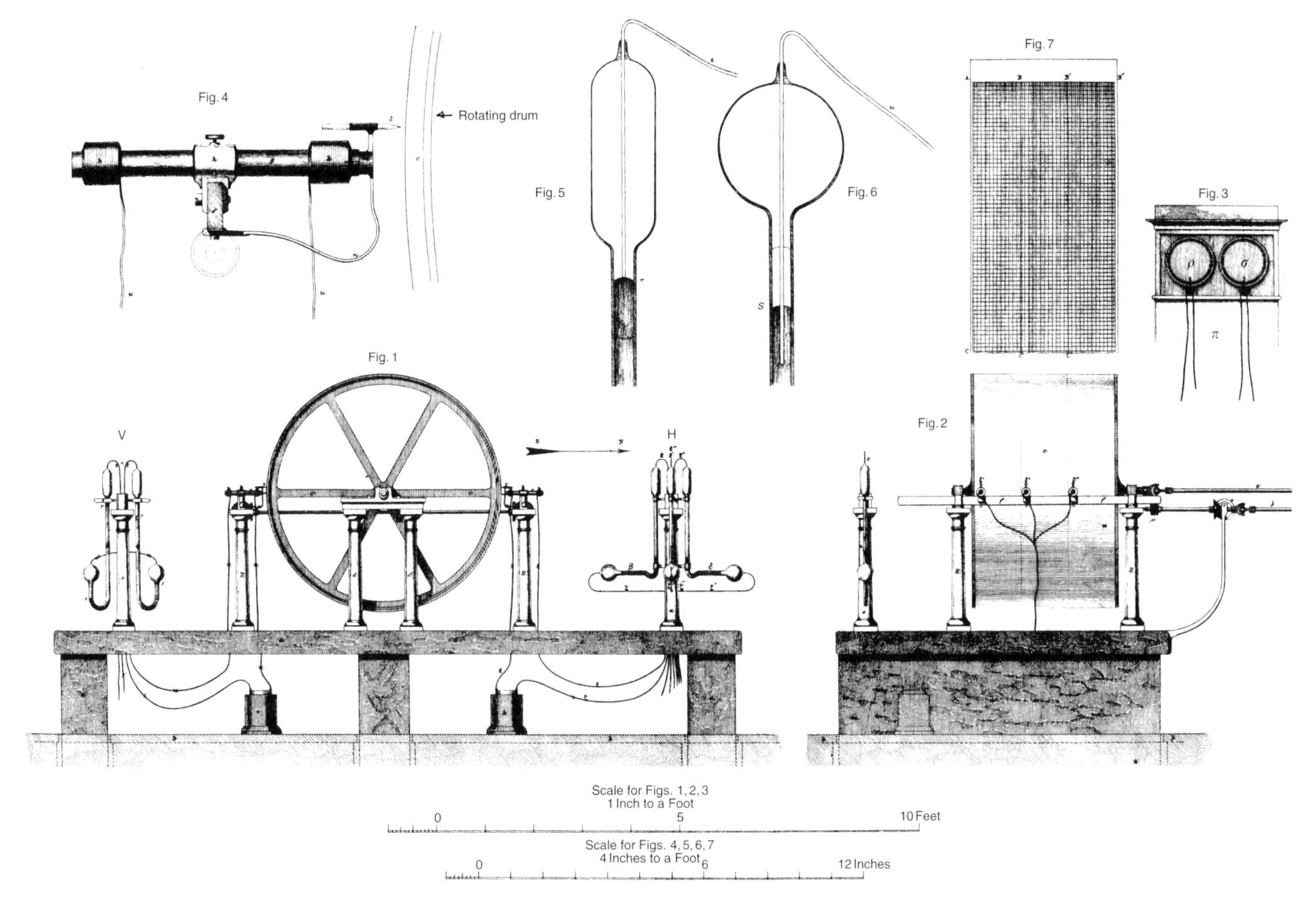

FIGURE 1 Drawing of Robert Mallet's proposed seismograph (1848). The seismic elements, two horizontal and one vertical, are tubes of mercury which are excited into oscillation by the earthquake. The oscillation of the mercury makes and breaks an electric circuit and this activates a spring-loaded solenoid to press a pencil against a rotating drum, thus recording the time of making or breaking the circuit. This would give some information about the oscillations of the mercury.

buildings, low in height. At that date, the use of reinforced concrete, and the use of structural steel, was still in its infancy, and the education of engineers was not of a type to encourage thinking about earthquake forces and stresses.

3. Messina, Italy Earthquake of 28 December 1908

As the population of the world increases, the number of structures at risk also increases, and the number of people exposed to earthquake hazards increases. This leads to the possibility of great disasters. The 83 000 death toll of the Messina earthquake was the greatest number ever from a European earthquake. Even the famous 1775 Lisbon, Portugal earthquake had fewer deaths (60 000). The government of Italy responded to the Messina earthquake by appointing a special committee composed of nine practicing engineers and five professors of engineering to study the earthquake and to make recommendations. The report of this committee appears to be the first engineering recommendation that earthquake-resistant structures be designed by means of the equivalent static method (%g method). This portion of the report appears to have been the contribution of M. Panetti, Professor of Applied Mechanics in Turin, and he recommended that the first story be designed for a horizontal force equal to 1/12 the weight above and the second and third stories to be designed for 1/8 of the building weight above. He stated that the problem is really one of dynamics which, however, is so complicated that it is necessary to have recourse to a static method. Also, in 1909, A. Danusso, Professor of Structural Engineering at Milan, won a prize with his paper, "Statics of Anti-Seismic Construction." The method recommended by Panetti and explained by Danusso, gradually spread to seismic countries around the world. First it was used by progressive engineers and later was adopted by building codes. Until the 1940s it was the standard method of design required by building codes. In Japan, the method was applied successfully to reinforced concrete buildings by Professor Tachu Naito prior to the 1923 Tokyo earthquake and, in the late 1920s, it was applied by

Professor R.R. Martel in the design of a 12-story steel-frame building in Los Angeles. Following the Tokyo earthquake the static method of design with seismic coefficient of 10% was adopted by the Japanese building ordinance and, following the 1933 Long Beach earthquake, the city of Los Angeles adopted the method with a coefficient of 8%. On 1 January 1943, the City of Los Angeles changed its earthquake requirements so that the seismic coefficient varied over the height of the building and was also a function of the total height (i.e., the period of vibration). This was the first time that the seismic requirements of a building code took into account the flexibility of a building as well as its mass; and these requirements were based on dynamic analyses of structures, carried out by R.R. Martel and his students, under research grants made by the Los Angeles County Department of Building and Safety.

The 1923 Tokyo earthquake was also responsible for the establishment of the Earthquake Research Institute at Tokyo University with an eminent engineer, Professor Kyoji Suyehiro, as the first director. This was the first research group formed to study earthquake engineering and seismology.

4. Recording and Analyzing Strong Earthquake Shaking

The recording of earthquake ground accelerations was often recommended by engineers who studied the problem of designing for earthquakes, including John Milne, Kyoji Suyehiro, R.R. Martel and others. However, it seems that prior to 1925 the technology of seismic instruments was not adequate to building strong-motion accelerographs and even in the later 1920s the mental and financial inertia was too great for an accelerograph to be developed. We owe the first accelerographs to the eminent engineer John R. Freeman (1855–1932) who became interested in earthquakes in 1925 at the age of 70, and who attended the 1929 World Engineering Conference in Tokyo where he met and became friends with Martel and Suyehiro. He immediately understood the need for a strong-motion accelerograph and strongly recommended to the Secretary of Commerce that such an instrument should be designed and constructed. On 11 March 1930, Secretary Lamont, an engineer, stated in a letter to Freeman that it would be done. The first accelerographs were installed by the Seismological Field Survey of the US Coast and Geodetic Survey in late 1932, just in time to record the strong ground shaking of the destructive 10 March 1933 Long Beach earthquake. This was a most important step in the development of earthquake engineering. For the first time engineers could see the nature of strong ground shaking: the amplitude of motion, the frequency characteristics, and the duration of shaking. These were items of great interest and they cleared up much confusion, as the literature prior to 1933 contained many erroneous estimates of these quantities. We, who live in 1984, can hardly conceive of the difficulty of estimating these quantities 100 years ago when the world had not yet moved into the modern age. For example, in the Transactions of the Seismological Society of Japan a paper by J. MacGowan on "Earthquakes in China" describes the earthquake of 12 June 1878 which occurred near Suchow, and he said that it was reported that in Suchow the "shaking was felt for the space of time taken in swallowing 1/2 bowl of rice." At the time of the First World Conference in 1956, fewer than 70 strong-motion accelerographs were installed in the world, and at that time, few destructive earthquakes occurred close to an accelerograph. It is significant that now most seismic countries have installed networks of accelerographs and that important records are being obtained in many countries. We now have a much better understanding of earthquake ground motions and their effects. In addition, many accelerographs are now installed on structures to record their motions during earthquakes, and these have demonstrated the dynamic responses of multistory buildings, bridges, dams, etc.

The development of computers, first the analog type and then the digital type, was very important for earthquake engineering. These made possible the practical analysis of accelerograms, for without computers the analysis was exceedingly slow and laborious. Computers made possible the development of the response spectrum of earthquake motions and the design spectrum which have played important roles in earthquake engineering and have been adopted in other branches of engineering also. Computers have also made possible the calculation of the dynamic response of structures to earthquake ground shaking and this has greatly clarified our understanding of structural dynamics. We are now able to make dynamic analyses of complex structures. The finite element method of analysis has also been an important development for earthquake engineering. The widespread installation of strong-motion accelerographs, together with the development of powerful computers, has provided large amounts of data and this poses problems of data acquisition, data analysis, data storage, data retrieval, and also data understanding.

In retrospect, it can be seen that in the last twenty years earthquake engineering has been strongly influenced by the implementation of major projects, such as: nuclear power plants, high-rise buildings, major dams, offshore oil drilling platforms, longspan bridges, LNG storage tanks, etc. The large cost of these projects and the need for a high degree of safety required a level of earthquake analysis and design much higher than for ordinary structures. As a consequence, the development of earthquake engineering was accelerated by the needs of these special projects.

5. Development of Building Codes

Special structures such as high-rise buildings, major dams, LNG storage tanks, etc., because of their critical importance are usually analyzed and designed by making use of the latest research developments. On the other hand, ordinary structures which compose the bulk of modern construction are usually designed according to the requirements of a building code. Because the code is a legal document that specifies a minimum level of design that must be attained by structures, and because it has a large socio-economic impact, substantial changes in code requirements are made slowly and cautiously. In addition, because the building code affects so many agencies, groups, individuals, etc. there is a great inertia against change and developments in the building code tend to lag behind developments in research. In many instances, needed changes in the code are deferred until the occurrence of a destructive earthquake; for example, 1908 Messina, Italy; 1923 Tokyo, Japan; 1933 Long Beach, California; 1971 San Fernando, California; 1976 Tangshan, China. The development of the building code thus illustrates the development of applied earthquake engineering for ordinary construction. The development of earthquake requirements in the Los Angeles city building code is a good example. Until 1933, earthquake design was not required; then the deaths and damage caused by the *M* 6.3 10 March 1933 Long Beach earthquake produced the requirement that every building must be designed for a minimum horizontal force of 8% of the weight, without consideration of height, shape, rigidity, material of construction, use, foundation condition, or degree of seismic hazard. The code was changed at intervals and the requirements of the present code do take into account the foregoing items. In addition, following the 1971 San Fernando earthquake a requirement was added that every structure over 160 ft in height shall have strength sufficient to resist the effects of earthquakes as determined by a dynamic analysis, and that this analysis shall be based on the ground shaking prescribed by a soil-geology-seismology report. The 1984 Los Angeles code is a great improvement over the 1934 code. The past fifty years, which is a short time in the life of a building code, have seen great improvements in building codes worldwide; and these improvements will certainly reduce loss of life and property damage in future earthquakes. During the past 100 years, over one million persons have been killed by earthquakes and it is the responsibility of earthquake engineers to ensure that this does not happen again during the coming 100 years. Earthquake engineers should examine the building codes in their countries to make sure that the requirements are indeed appropriate to the seismic risk.

6. Nontechnical Earthquake Engineering

A history of earthquake engineering should also examine the development of thinking in governmental agencies, and the thinking of the public. Relatively recent developments in some cities, such as Tokyo and Los Angeles, include improved thinking in city government agencies. National and state government agencies are also giving earthquake hazards increasing consideration. In some regions, earthquake preparedness measures are being taken to reduce the risk of life and property, to prepare to handle the emergency when the coming earthquake occurs, and to mitigate the effects of future earthquake disasters upon the functioning of the city and the impact on the public. These are noteworthy developments in nontechnical earthquake engineering, if we define earthquake engineering broadly to encompass all nontechnical, as well as technical, efforts directed toward minimizing those harmful effects of earthquakes.

Public understanding of earthquake risk is also important. A great advance in earthquake knowledge possessed by the average citizen has occurred over the past thirty years, and this has resulted in greater support for earthquake preparedness measures. However, even today, the average citizen has a relatively poor understanding of earthquake risks and earthquake engineering benefits. Because of the way the news media handle the topic of earthquakes, the public tends to oscillate between excessive alarm and excessive complacency. Improvements need to be made in public education about earthquake risks and earthquake preparedness, beginning with children in the public schools.

7. Summary

Over the past fifty years, there has been remarkable progress in earthquake engineering research. Knowledge of earthquake ground shaking and earthquake vibrations of structures has undergone a great expansion. Advances in methods of dynamic analysis enable the earthquake response of planned structures to be calculated. Experimental research is providing valuable data on the physical properties of structures and structural elements. The increase in number of research papers published each year is indicative of the progress being made. We have now attained a good understanding of the elastic earthquake response of typical buildings. However, there are many special structures, industrial equipment, etc. whose earthquake survival is important and which need special earthquake engineering research. Also, the question of "maximum credible" or "maximum probable" earthquake needs to be better defined by research, as well as the question of appropriate level of design of structures for such events of low likelihood. And more research needs to be done on designing for controlled damage in the event of large, infrequent earthquakes.

Building codes have also undergone a big development over the past fifty years. The 1984 building codes now handle the earthquake design of structures in a much improved way over the 1934 building codes. However, the real test of a building code comes when a city experiences strong ground shaking. Actual structures, as distinguished from ideal structures, are so complex that their behavior must be tested by strong shaking to establish their adequacy or to reveal inadequacies. In 1984,

TABLE 1

Earthquake Location	Date	Magnitude	Approx. No. of Deaths
Hokkaido, Japan	30 Dec. 1730	?	137 000
Calcutta, India	1737	?	300 000
Lisbon, Portugal	1 Nov. 1755	8+	60 000
Syria	30 Oct. 1759	?	30 000
Calabria, Italy	5 Feb. 1783	?	30 000
Peru–Ecuador	4 Feb. 1797	?	40 000
Kangr, India	4 Apr. 1905	?	19 000
San Francisco	18 Apr. 1906	8.3	1 000
Santiago, Chile	17 Aug. 1906	8.6	20 000
Messina, Italy	28 Dec. 1908	7.5	83 000
Avezzano, Italy	13 Jan. 1915	7	30 000
Kansu, China	16 Dec. 1920	8.6	100 000
Tokyo, Japan	1 Sep. 1923	8.3	100 000
Bihar, India	15 Jan. 1934	8.4	11 000
Taiwan	20 Apr. 1935	7.1	3 000
Quetta, Pakistan	30 May 1935	7.5	30 000
Chile	25 Jan. 1939	8.3	28 000
Erzincan, Turkey	26 Dec. 1939	7.9	30 000
Ambato, Ecuador	5 Aug. 1949	6.8	6 000
Agadir, Morocco	29 Feb. 1960	5.8	10 000
Quazin, Iran	1 Sep. 1962	7.0	12 000
Tangshan, China	28 Jul. 1976	7.8	250 000+

conclusive tests of modern building codes have not yet been made. It is important that, in the future, destructive earthquakes be studied with a view to assessing the adequacy of the building code. Many earthquakes have been inspected in the past, and many reports have been written, but too few valuable conclusions have been deduced. Earthquake engineers should prepare ahead of time to learn from coming destructive earthquakes. Thought must be given ahead of time to what can be learned, and preparations should be made ahead of time for the learning process.

When the earthquake comes, everything connected to the Earth either directly, or indirectly, will be shaken. All those items whose survival is very important must be given special attention in design and construction to insure against unacceptable damage, and structures must also be designed to protect against injury and loss of life. An earthquake disaster requires three things: (1) the occurrence of an earthquake sufficiently large to produce strong ground shaking; (2) the earthquake must be sufficiently close so that a city experiences strong shaking; and (3) the city must be unprepared for an earthquake, with numerous weak buildings. When these three coincide there is a disaster. Such coincidences were not uncommon in the past (see Table 1, for example).

The list in Table 1 is incomplete; many other destructive earthquakes occurred where death tolls numbered in the thousands. Disastrous earthquakes can be expected also in future years, for many existing cities are poorly prepared to resist earthquakes. In addition, inasmuch as the world's population is increasing by 80 million persons per year and cities are correspondingly expanding, the earthquake risk is increasing. In coming years large earthquakes can be expected to occur close to large cities more frequently than in the past. It is the responsibility of earthquake engineers to insure that the new construction in these cities is earthquake resistant, and that the greatest hazards from old weak buildings are eliminated. The discipline of earthquake engineering is now entering its golden age with greatly expanded knowledge and capabilities, but it is also facing important new problems, for in coming years there will be construction of large, complex, and costly structures, industrial facilities, and socioeconomic projects which the earthquake engineer must make safe against destructive ground shaking.

References

Berg, G.V. (1982). "Seismic Codes and Procedures." EERI.

Davison, Charles (1927). "The Founders of Seismology." Cambridge University Press, Cambridge.

Dewey, J. and P. Byerley (1969). The early history of seismometry (to 1900). *Bull. Seism. Soc. Am.* **59,** 183–227.

Freeman, John R. (1932). "Earthquake Damage and Earthquake Insurance." McGraw-Hill, New York.

Richter, C.F. (1958). "Elementary Seismology." Freeman, San Francisco.

Runcorn, K. (Ed.) (1967). "International Dictionary of Geophysics," Pergamon Press, Oxford.

Transactions of the Seismological Society of Japan, (1880–96) Vols. 1–16, Yokohama.

Editor's Note

This article was a Conference Lecture published in the Post-Conference Volume of the Proceedings of the Eighth World Conference on Earthquake Engineering, San Francisco, 1984, copyright 1986 by Earthquake Engineering Research Institute (EERI). We thank EERI for permission to reprint it. This article has been reformatted and a few typographic errors have been corrected. For further readings, please see chapters in VII. Strong-Motion Seismology, VIII. Selected Topics in Earthquake Engineering, and Chapter 80, Centennial National and Institutional Reports: Earthquake Engineering. For a more complete list of destructive earthquakes from 1500–2000, please see Chapter 42 by Utsu.

3

The Jesuit Contribution to Seismology

Agustín Udías
Universidad Complutense, Madrid, Spain
William Stauder
Saint Louis University, St. Louis, Missouri, USA

1. Introduction

The contribution to seismology of the Society of Jesus as an institution through its colleges and universities, and through its members as individual scientists, forms an important chapter in the history of this science. This is especially so in the early years of its development. Several papers have described in part the work of Jesuits in seismology: Sánchez Navarro-Neumann (1928, 1937), Heck (1944), who limits himself to American Jesuits, and the interesting but short summaries of Linehan (1970, 1984). No recent or comprehensive work, however, exists on the topic. Recently, moreover, many Jesuit seismographic stations have been closed and the number of Jesuits actually working in seismology has been greatly reduced. To a certain extent, apart from a very few academic departments and research institutes associated with Jesuit universities, it can be said that this is a chapter which is coming to a close. The interest of Jesuits has moved in other directions and it is not likely that seismology will become again an important aspect of the work of individual Jesuits as it was in the past. For this reason we feel that it will be of interest to present an overall picture of the extent of the Jesuit involvement in seismology.

It may be intriguing to some that a religious order dedicated so much effort to a science like seismology. From the very early years of its foundation in the 16th century by Ignacio de Loyola, the Society of Jesus dedicated itself primarily to educational work through its many colleges and universities. From the beginning of these institutions, science was an important subject in the curriculum. A key figure in this development was Christopher Clavius (1537–1612), Professor of Mathematics in the Collegio Romano. Clavius was instrumental in incorporating a serious program of mathematics, astronomy, and natural sciences not only in his own college but also in all Jesuit colleges and universities (MacDonnell, 1989). Secondly, in the 17th and 18th centuries a number of astronomical observatories were established in these institutions. In a number of these, meteorological observations also were made. Finally, in a particularly notable page of this history, Jesuits were appointed Directors of the Astronomical Observatory in Beijing, China (Udías, 1994). This tradition forms the background of modern Jesuit scientific work. Since the middle of the 19th century, as many as 40 geophysical observatories were created by Jesuits around the world and in many of these seismological stations were installed (Udías and Stauder, 1991).

As a science, seismology is relatively young. Davison (1927) dates its beginning to 1750 with the Lisbon earthquake and the work of John Michell (1724–1793). Although pendulums were used to detect the ground motion due to earthquakes at the beginning of the 19th century, the earliest seismographs were those developed in Italy by Luigi Palmieri (1807–1896) in 1855 and by Timoteo Bertelli (1826–1905) in 1872. Seismographs with continuous recording were developed in Japan from 1881 to 1882 by John Milne (1850–1913) and Thomas Gray (1850–1908). One of the first studies of earthquakes is that of Robert Mallet (1810–1881), who investigated the Naples earthquake of 1857 (Davison, 1927). The first record of a long-distance earthquake is a fortuitous recording of an earthquake in Japan by pendulums set up in Potsdam in 1889 by E. A. von Reuber-Paschwitz (1861–1895). From these dates and events, seismology began to develop not only as a study of the nature of earthquakes but also, through the analysis of the seismic waves, as a method to investigate the structure of the interior of the Earth.

A series of circumstances and interests involved Jesuits in the development of this new science from its inception. This interest, certainly, was consonant with the tradition of Jesuits in science dating from the 16th century, which developed, as

INTERNATIONAL HANDBOOK OF EARTHQUAKE AND ENGINEERING SEISMOLOGY, VOLUME 81A ISBN: 0-12-440652-1

has been mentioned, out of their work in colleges and universities. The character of seismology as a public service to mitigate the destructive effects of earthquakes was another influential factor. Especially in undeveloped countries, Jesuits were in many instances the first to install seismographic stations and to carry out seismicity and seismic risk studies.

Two trends may be distinguished in this involvement of Jesuits in seismology. In the United States, emphasis was on the cooperation of Jesuit institutions in the establishment of seismographic stations organized first as the Jesuit Seismological Service and subsequently as the Jesuit Seismological Association. In other countries, especially in mission lands, the movement developed out of the activity of individual institutions in establishing seismic observatories, usually as a complement to the recording of other geophysical data.

2. Jesuit Seismographic Stations

Table 1 presents a list of 38 seismographic stations installed and maintained by Jesuits. In each case dates of the installation of different types of seismographs are given (OM indicates "own manufacture"). The distribution by continents is as follows: Europe 6, Asia 4, Africa 2, Australia 1, North America 18, and South America 7 (Fig. 1). Most of these stations were founded before 1920, and many ceased operation in the 1960s and 1970s. At present, there are only 8 functioning regularly. Initially, the preferred instruments were Wiechert and Mainka mechanical seismographs, and later, Galitzin–Wilip and Sprengnether electromagnetic instruments. In 1962, 10 Jesuit stations became part of the 125-station global World Wide Standard Seismographic Network (WWSSN) supported

TABLE 1 Jesuit Seismographic Stations

Station	Dates	Instruments
Manila, Philippines	1868–	1868: OM Seismoscopes Z, H 1882: Cecchi, Bertelli, Rossi 1889: Gray–Milne, Ewing 1911: Wiechert, NE, 1000 kg
Baguio		1911: Vicentini, Omori 1951: Sprengnether SP ZNE 1962: WWSSN
Guam y Butuam		1912: Wiechert NE 200 kg
Ambulong		1912: Vicentini, Agamennone 1930: Galitzin–Wilip, ZNE
Tagaytay		1930: Wiechert NE
Davao		1962: WWSSN
Manila		1962: Sprengnether SP ZNE
Puebla, Mexico	1877–1914	1877: OM 3 H Seismoscopes
Tusculano Frascati, Italy	1888–1920	1888: OM Seismoscope (Egidi) 1908: Cancani, Agammenone
Tanarive, Madascar	1899–1992	1899: Cecchi, Bertelli 1927: Mainka ZNE 460 kg
John Carroll, Cleveland, OH, JSA	1900–1992	1900: OM Seismoscope (Odenbach) 1908: Wiechert NE 80 kg 1946: Sprengnether LP-ZNE, SP Z
Cartuja, Granada, Spain	1902–1971	1902: Vincentini NE 305 kg, Z 245 kg; Stiattesi NE 1908: OM Z 208 kg, NE (Cartuja) 1920: OM NE 3000 kg (Cartuja) 1924: OM ZNE Electromag (Neumann)
Ebro, Tarragona, Spain	1904–	1904: Vicentini ZNE, Grablowitz 1914: Mainka Z 1500 kg, NE 300 kg 1965: Benioff SP ZNE 1969: Sprengnether LP ZNE
Zikawei, Shanghai, China	1904–1949	1904: Omori 1909: Weichert NE 1200 kg 1913: Galitzin Z 1932: Galitzin–Wilip ZNE
Belen, Havana, Cuba	1907–1920	1907: Bosch–Omori NE
Santa Clara, CA	1907–1958	1907: OM seismograph 1909: Wiechert ZNE 80 kg 1930: Galitzin–Wilip ZNE 1931: Wood Anderson NE
Stonyhurst, Lancs., England	1908–1947	1908: Milne NE 1924: Milne–Shaw NE
Mungret, Limerick, Ireland	1908–1915	1908: OM (O'Leary, inv. pend.)
Riverview, Australia	1909–1985	1909: Wiechert ZNE 1910: Mainka NE 1941: Galitzin–Wilip ZNE 1953: Sprengnether SP-Z 1962: WWSSN
Regis, Denver, CO, JSA	1909–1988	1909: Wiechert NE 80 kg 1946: Sprengnether LP-NE, SP-Z 1962: Sprengnether SP-ZNE
Gonzaga, Spokane, WA, JSA	1909–1970	1909: Wiechert NE 80 kg 1946: Wood Anderson NE
Holy Cross, Worcester, MA	1909–1934	1909: Wiechert NE 80 kg
Marquette, Milwaukee, WI, JSA	1909–1951	1909: Wiechert NE 80 kg
Georgetown, Washington, DC, JSA	1910–1972	1910: Wiechert NE 80 kg 1912: Bosch–Omori NE 25 kg 1912: Bosch photog. NE 200 kg 1912: Wiechert ZNE 200 kg 1912: Mainka NE 135 kg 1923: Galitzin ZNE 1962: WWSSN
Canisius, Buffalo, NY, JSA	1910–	1910: Wiechert ZNE 80 kg 1932: Galitzin–Wilip Z 1946: Sprengnether LP-NE, SP-Z
Saint Louis, St. Louis, MO, JSA	1910–	1910: Wiechert NE 80 kg 1928: Wood Anderson NE

(*continued*)

TABLE 1 *(continued)*

Station	Dates	Instruments
		1946: Sprengnether LP-NE, SP-Z 1960: Benioff SP-Z 1963: Sprengnether LP ZNE, Benioff SP N
Florissant	1928–1974	1928: Galitzin ZNE 1928: Wood Anderson NE 1962: WWSSN
French Village	1974–	1972: WWSSN
Cape Girardeau, MO	1938–	1938: Wood Anderson NE 1980: Benioff SP-ZNE
Little Rock, AR	1930–1958	1930: Wood Anderson NE
Cathedral Cave, Onandago State Pk	1991–	1991: IRIS Station
Saint Boniface, Manitoba, Canada	1910–1922	1910: Wiechert NE 80 kg
Fordham, New York, NY, JSA	1910–1977	1910: Wiechert NE 80 kg 1924: Milne–Shaw NE 1927: Galitzin–Wilip ZNE 1932: Wood Anderson NE 1936: Benioff SP-Z 1950: Sprengnether SP-ZNE
Loyola, New Orleans, LA, JSA	1910–1960	1910: Wiechert NE 80 kg 1946: Sprengnether LP-NE, SP-Z
Spring Hill, Mobile, AL, JSA	1910–1989	1910: Wiechert NE 80 kg 1941: McComb–Romberg NE 1962: WWSSN
Ksara, Bekka, Lebanon	1910–1979	1910: Mainka NE 135 kg 1921: Mainka NE 460 kg 1933: Galitzin–Wilip ZNE 1957: APX SP-ZNE (Grenet Coulomb)
Loyola, Chicago, IL, JSA	1913–1960	1912: Wiechert NE 80 kg 1957: Sprengnether ZNE 1983: Kinemetrics SP-Z
San Calixto, La Paz, Bolivia	1913–	1913: OM Z 1500 kg, NE 2000 kg 1930: Galitzin–Wilip ZNE 1962: WWSSN 1972: HGLP-ASRO
Sucre, Bolivia	1915–1948	1915: OM Z 1500 kg, NE 3000 kg
Rathfarnham Castle, Ireland	1916–1961	1916: OM Z (O'Leary) 1500 kg 1932: Milne–Shaw NE
San Bartolome, Bogotá, Colombia	1923–1940	1923: OM NE 2000 kg (Cartuja) 1926: Wiechert NE 2000 kg 1930: OM NE 1000 kg (Cartuja)
Xavier, Cincinnati, OH, JSA	1927–1986	1927: Galitzin–Wilip ZNE 1949: OM ZNE (electromag)
Weston, Boston, MA, JSA	1928–	1930: Bosch–Omori NE 1934: Wiechert NE 2000 kg 1936: Benioff SP-ZNE 1962: WWSSN
Saint Louis, Island of Jersey	1936–1979	1936: Mainka NE 1000 kg
Saint George's, Kingston, Jamaica	1940–1975	1940:
Instituto Geofísico, Bogotá, Colombia	1941–	1941: Benioff SP-Z 1943: Wiechert NE 1946: Sprengnether NE 1962: WWSSN 1973: ASRO
Galerazamba		1949: Sprengnether ZNE
Chinchina		1949: Sprengnether ZNE
San Luis, Antofagasto, Chile	1949–1965	1949: Bosch Omori 1960: Wilson–Lamison
San Francisco, CA	1950–1964	1960: Sprengnether LP-NE, SP-Z
S. Jean de Brébeuf, Montreal, Canada	1952–	1952: Wilmore 1952: Benioff SP-ZNE 1961: Press–Ewing LP-ZNE 1961: Geotech SP-Z
Addis Ababa, Ethiopia	1957–1978	1957: 1962: WWSSN

by the US Government. Of these two, one in Colombia and the other in Bolivia, were later upgraded to become Seismological Research Observatory (SRO) and High Gain Long Period-Adapted Seismological Research Observatory (HGLP-ASRO) stations, respectively.

The first seismographic station installed by Jesuits was about 1868 in the Observatory of Manila. This consisted of two pendulums, one vertical the other horizontal, and were what we would today call seismoscopes. They were designed by Juan Ricart, Professor of Sciences at the Manila Jesuit College. They functioned intermittently until 1877 when regular, continuous-recording instruments were installed. After the Manila earthquakes of 1880, which were recorded on the old seismographs and studied by Federico Faura (1847–1897), the station was better equipped with Cecchi, Bertelli, and Rossi seismographs made in Italy. Additional stations were installed at other points of the Philippines, namely at Baguio, Ambulong, Butuam, Tagaytay, and the island of Guam (Saderra-Masó, 1895, 1915; Su, 1988). Unfortunately, all seismographic records were lost in the destruction of the observatory during World War II. The continuous catalog of Philippine earthquakes, however, was salvaged and was published shortly after the war by Repetti (1946). After the war, new seismographs were installed in Manila, Baguio, and Davao. The last two sites became WWSSN stations in 1962.

The first seismograph installed by Jesuits in Europe was a seismoscope made by Giovanni Egidi (1835–1897) and installed in the meteorological observatory in Tuscolano, Frascati, Italy, founded in 1868 (Egidi, 1888). An important station was installed in 1902 in Granada, the most seismically active region of Spain. Due to a lack of funds most seismographs were made patiently under the direction of Manuel

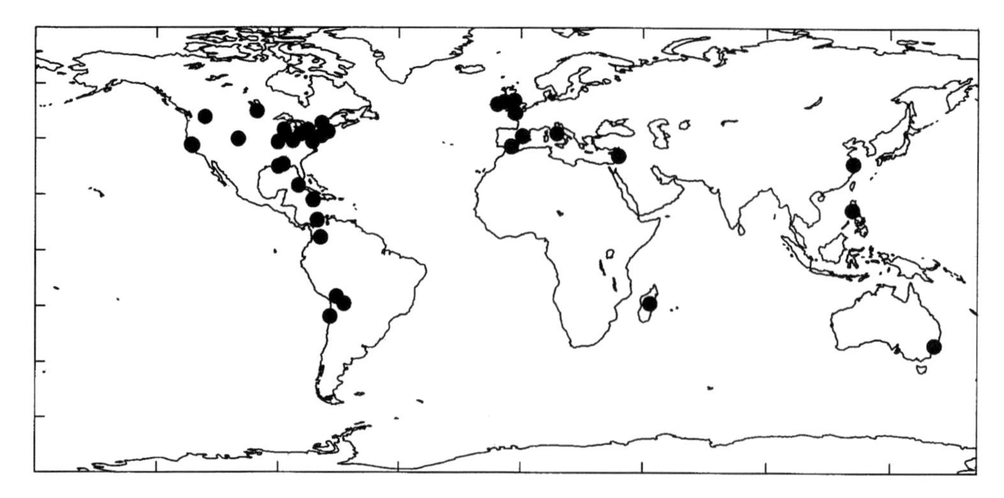

FIGURE 1 Map showing the location of Jesuit seismographic stations.

Sánchez Navarro-Neumann (1867–1941), who reproduced with some improvements the Omori, Wiechert, and Galitzin seismographs (Sánchez Navarro-Neumann, 1928). In Ireland, William J. O'Leary installed the first seismograph, also of local design, at Mungret College in 1908. In 1916 it was moved to Rathfarnham Castle, where it was in operation until 1961. This was an inverted pendulum suspended by steel wires with a mass of 600 kg recording on smoked paper on two drums (Murphy, 1995). In the Ebro Observatory, Spain, a seismological station has functioned continuously from 1904 to the present. At Stonyhurst, England, the station operated from 1908 to 1947. A seismographic station was also established by Jesuits on the Island of Jersey, operating from 1936 to 1979.

Very early, in 1899, Jesuits installed a seismographic station in Madagascar using instruments of Italian design. In 1927 a three-component Mainka seismograph was installed. This station operated under Jesuit supervision until 1967 and may have been one of the earliest seismographic stations in Africa. In 1904 Jesuits installed an Omori seismograph donated by the Japanese government in the observatory of Zikawei. This may have been the first such station in China. Updated with Wiechert and Galitzin–Wilip instruments in 1909 and 1932, respectively, Zikawei continued as a first-class station up to the time the Jesuits were expelled from China in 1949 (Gherzi, 1950). In Ksara, Lebanon, seismographs were installed in 1910. This was an important station due to lack of stations in the Middle East. It operated uninterruptedly until 1979 with Wiechert instruments. In Australia, the Jesuit seismographic station of Riverview was initiated in 1909 with Wiechert instruments. In 1962 it became a WWSSN station and operated until 1985. For many years this was the best known and best equipped station in Australia (Drake, 1980; Doyle and Underwood, 1965).

At the 2nd General Assembly of the International Seismological Association, Manchester, 1911, a resolution was passed recommending that the Jesuits install a seismic station in the central part of South America. In response to this recommendation, a seismological station was installed in 1913 at La Paz, Bolivia, with the name of Observatory de S. Calixto, by Pierre M. Descotes (1877–1964), with instruments of local design. In 1930, the station was upgraded with Galitzin–Wilip seismographs. From 1964 to 1993, the station was directed by Ramón Cabré. The Observatory of S. Calixto has been one of the most reliable stations in South America (Cabré, 1988; Coenraads, 1993). The first seismograph in Colombia was installed by Jesuits in 1923 in Bogotá. In 1941 the Instituto Geofísico de los Andes Colombianos (today the Instituto Geofísico, Universidad Javeriana) was founded by Jesus E. Ramirez (1904–1983). This soon became one of the best seismological research institutions in South America (Ramirez, 1977; Goberna, 1988). In 1962, the stations of La Paz and Bogotá became WWSSN stations. Later La Paz became an HGLP-ASRO station (1972) and Bogotá became an SRO station (1973). This upgrade is a clear recognition of the work done by Jesuits in these two stations.

Other Jesuit seismographic stations in Cuba and Chile functioned only a few years. In 1940 a seismographic station was installed in St. George's College, Kingston, Jamaica, dependent on Weston Observatory. In Montreal, Canada, a seismographic station was installed in 1952 (Buist, 1983). The Montreal station was a modern station with WWSSN-type instruments. It was the last new station installed under direct Jesuit auspices. Maurice Buist was its director for 31 y until his retirement in 1983. In Ethiopia, Haile Salassie invited the Jesuits to undertake the administration of the National University in Addis Ababa. Although not a Jesuit station in the strict sense, the associated Geophysical Observatory of Addis Ababa was directed by the Canadian Jesuit Pierre Gouin from 1957 to 1978. In 1962, a WWSSN seismographic station was installed.

3. The Jesuit Seismological Association

The history of the work of Jesuits in seismology in the United States is linked to the Jesuit Seismological Association (Macelwane, 1926, 1950). The first Jesuit to install a seismograph in the United States was Frederik L. Odenbach (1857–1933) in 1900 at John Carroll University, Cleveland, Ohio, with two seismoscopes of his own design. In 1908 Odenbach conceived the notion that the system of Jesuit colleges and universities distributed throughout the United States offered an opportunity to establish a network of similar seismographic stations. In 1906 the International Seismological Center had been established in Strasbourg, France, where data were centrally reported and epicenters determined. Odenbach envisioned a network of Jesuit stations that could contribute significant data to this international enterprise.

Odenbach sold the presidents of the colleges and universities and the American Jesuit Provincial Superiors on the idea. In 1909 sixteen identical horizontal Wiechert seismographs of 80 kg mass were purchased in Germany and were distributed to 15 colleges in the United States and one in Canada. A typical station was that of Georgetown (Fig. 2). These stations formed the Jesuit Seismological Service. Individual stations were to process their own seismograms and send the readings to the Central Station in Cleveland. The data would then be collated and forwarded to the International Seismological Center in Strasbourg. These stations, in effect, constituted the first seismological network of continental scale with uniform instrumentation.

Many of the first stations, for a variety of reasons, foundered early, and the cooperative effort was never fully established. In 1925 James B. Macelwane (1883–1956) returned to Saint Louis University after completing his doctoral studies at the University of California. One of Macelwane's early efforts was to revitalize the Jesuit seismographic network. The impetus to this came not only from his own interest but also from the urging of scientists of the National Research Council and the Carnegie Institution in Washington, and with the further encouragement of Sánchez Navarro-Neumann of the Spanish Observatory at Cartuja.

Thus, in the summer of 1925 the stations were reorganized into the Jesuit Seismological Association. The 14 member stations are indicated as JSA in Table 1. The Presidents of the Association have been J.B. Macelwane of Saint Louis University, 1925–1956; J.J. Lynch of Fordham University, 1957–1970; and D. Linehan of Weston Observatory, 1970–1986. W. Stauder of Saint Louis University has been President from 1986 to the present.

Saint Louis University became the Central Station in 1925. Through its recently established Department of Geophysics it became a resource for graduate education in seismology for a number of Jesuits who then returned as directors to their own

FIGURE 2 Instruments of the Georgetown seismographic station, about 1920. The Wiechert is on the far right.

institutions. The Central Station assumed as well the responsibility on behalf of the JSA of collecting data from member stations and from other stations around the world and of locating earthquake epicenters and publishing these to the worldwide seismological community. The Central Station continued this service until the early 1960s when, with the advent of computer determination of epicenters, it was discontinued as an unnecessary duplication of the determinations by the US Coast and Geodetic Survey (later by the US Geological Survey) in the United States and by other international agencies.

Most of the JSA seismographic stations continued in regular operation until relatively recent time. Florissant (Saint Louis), Weston, Georgetown, and Spring Hill became WWSSN stations in 1962. At present only Saint Louis and Weston continue as seismological research institutes. Both conduct active research programs and operate regional networks, and are members of IRIS, the Incorporated Research Institutes for Seismology. Saint Louis has installed a broadband station of the IRIS system at Cathedral Cave, Missouri and is deploying other similar broadband stations in the Midwest.

4. Jesuit Seismologists

From 1868, the approximate date of the installation of the first seismograph by the Jesuits in Manila, to the present, many members of this religious order have dedicated their time and efforts to seismology. In this short article it is hardly possible to do more than mention even the most important of them. However, a few words must be said about those that occupy an important place in the history of seismology. We will mention only the work of past Jesuits, although at present there are still Jesuits actively working in seismology (e.g., L. Drake, presently at La Paz, P. Gouin in Montreal, W. Stauder in St. Louis, S. Su in Manila, R. Van Hissenhoven in Bogotá, and A. Udías in Madrid).

The first Jesuit to be mentioned is Federico Faura, who published a study about the destructive Manila earthquakes of 1880 (Faura, 1880). In the paper he reproduced the records obtained by the instruments (seismoscopes) in the Observatory of Manila. Faura continued his interest in seismology, improving the seismological instrumentation of the Observatory and publishing a seismological bulletin. Beginning in 1877, Giovanni Egidi, director of the Observatory Tuscolano in Italy, collaborated with M.S. De Rossi in the seismological observations that were published in *Bullettino del Vulcanismo Italiano* (Davison, 1927, p. 100).

Two Jesuits contributed very early to the study of the seismicity and seismotectonics of the Philippines and Spain. Manuel Saderra-Masó dedicated himself to the study of the seismicity of the Philippines, interpreting it in terms of seismotectonic lines and relating it to the geological structure of the archipelago in a very early work of this type (Saderra-Masó, 1895; Saderra-Masó and Smith, 1913). M. Sánchez Navarro-Neumann, Director of the Observatory of Cartuja, Spain, compiled the first modern earthquake catalogue of Spain and published numerous studies on the seismicity of that region (Sánchez Navarro-Neumann, 1919). He also published an early paper on the energy in earthquakes (Sánchez Navarro-Neumann, 1915).

The most renowned Jesuit seismologist was without doubt James B. Macelwane (Fig. 3) (Byerly and Stauder, 1958). Macelwane obtained his doctoral degree at the University of California, Berkeley, in 1923, with the first dissertation on a seismological topic in the United States. In 1925 he became the first Director of the Department of Geophysics at Saint Louis University and reorganized the Jesuit Seismological Association. Travel times of seismic waves, the constitution of the interior of the earth, and the nature of microseisms and their relation to atmospheric storms were a few of the topics of his research papers (e.g., Macelwane and Dahm, 1937; Macelwane, 1939, 1946). In 1936 he published the first textbook on seismology in America (Macelwane, 1936). In 1928–29 he was President of the Seismological Society of America and in 1953–56 of the American Geophysical Union. In 1944 he was elected to the National Academy of Sciences. Throughout his career Macelwane took an active part in the committees and commissions of these societies, as well as in projects of the National Research Council and the International Union of Geodesy and Geophysics. He was also always interested in promoting educational programs and in encouraging young geophysicists. For the latter reason, in 1962 the American Geophysical Union created a medal in his honor for recognition of significant contribution to the geophysical sciences by a young scientist of outstanding ability.

FIGURE 3 James B. Macelwane, S.J.

Two Jesuit students of Macelwane deserve to be mentioned. William C. Repetti (1884–1966) studied the interior of the Earth from the travel times of body waves, inferring the existence of several discontinuities in its interior. In 1928 he joined the staff of the Manila Observatory, where he assumed charge of the seismological section. Among other contributions he compiled a catalog of earthquakes of the Philippines (Repetti, 1946). Jesus E. Ramirez (1904–1983) worked on the problem of microseisms and storms, designing a tripartite station system to track the center of tropical hurricanes. In 1941 he founded in Bogotá the Instituto Geofísico de los Andes Colombianos, where he conducted and published a large number of studies of Colombian seismicity (Ramirez, 1967). He was a leading figure in the seismicity of South America.

J. Joseph Lynch (1894–1987) became Director of the Seismographic Station of Fordham University in 1920. This was the beginning of a long career as a seismologist that led to a variety of seismological studies, among them a field study of the Dominican Republic earthquake of 1946. He has left us a lively account of his involvement in seismology over a period of 50 y (Lynch, 1970). Daniel Linehan (1904–1987) was Professor of Geophysics and Director of Weston Observatory for 32 y. A prolific writer in many aspects of seismology, he dedicated himself especially to seismic exploration. In 1950 he carried out, accompanied by Lynch, a shallow seismic exploration survey under St. Peter's Basilica in Rome for archeological purposes. He participated in three expeditions to the Antarctic, one to the Arctic, and several UNESCO seismological missions in Africa, Asia, and South America. Two other Jesuits participated in separate expeditions to the Antarctic during the International Geophysical Year: Edward Bradley of Xavier University, Cincinnati, and Henry Birkenhauer of John Carroll University, Cleveland.

Of European Jesuit seismologists, Richard E. Ingram (1916–1967), Director of Rathfarnham Castle, Ireland merits mention for his theoretical papers (Ingram, 1963).

Among the different topics in seismology, the study of microseisms attracted the special interest of Jesuit seismologists. We have already mentioned the work of Macelwane and Ramirez, with a first paper in 1935. Probably the first suggestion of the relation between microseisms and storms was made by José Algue (1859–1930), Director of Manila Observatory, as early as 1894 in his study of Philippine Islands typhoons (Deppermann, 1951). Ernesto Gherzi (1886–1976), Director of Zikawei Observatory, published several papers on the relation between microseisms and atmospheric conditions (Gherzi, 1924). In 1952 a seminar was organized by the Pontifical Academy of Sciences on the problem of microseisms in which Gherzi, Macelwane, and Due-Rojo (Cartuja) participated, along with a select group of other specialists.

Jesuits participated in the early stages of the organization of seismological associations. R. Cirera (1864–1932), first Director of Ebro Observatory, represented Spain as one of two delegates in the second meeting of the International Seismological Conference in 1903 in Strasbourg (Rothé, 1981). Three Jesuits, Berloty (Ksara), Sánchez Navarro-Neumann (Cartuja) and Stein (Vatican) were present at the first General Assembly of the International Association of Seismology in the Hague in 1907. Jesuits have participated actively in the International Association of Seismology and the Physics of the Earth's Interior since its establishment in 1922 as part of the International Union of Geodesy and Geophysics (IUGG).

Jesuits have had a special relation with the Seismological Society of America. One of the 13 participants at the first meeting for its founding in 1906 was Jerome S. Ricard (1850–1930), Director of Santa Clara Observatory, who was elected

FIGURE 4 Participants in the joint meeting of the Eastern Section of the Seismological Society of America and the Jesuit Seismological Association held in St. Louis, 1937. *First row (l to r)*: Anthony J, Westland, S.J., Archie Blake, Ernest A. Hodgson, Capt. Nicholas H. Heck, James B. Macelwane, S.J., J. Joseph Lynch, S.J. *Second row (l to r)*: J. Emilio Ramirez, S.J., Alphonse R. Schmitt, S.J., Louis B. Schlichter, Alton C. Chick. *Third row (l to r)*: H.M. Rutherford, Florence Robertson, Paul Weaver, —?, Victor G. Stechschulte, S.J., Theodore Zegers, S.J. *Fourth row (l to r)*: Daniel Linehan, S.J., Albert J. Frank, Arthur C. Ruge, Msgr. Joseph A. Murray, John P. Delaney, S.J., Ross Heinrich.

a member of the first Board of Directors (Byerly, 1964). J.B. Macelwane also served on the Board from 1925 to 1956 and was elected President of the Society in 1928. W. Stauder served on the Board from 1962 to 1967 and was President in 1966. Several Jesuits have been Chairman of the Eastern Section of the Seismological Society of America: J.B. Macelwane, 1926 (first Chairman); J.J. Lynch, 1930; V.C. Stechschulte, 1933; D. Linehan, 1954; V. Blum, 1955; H. Birkenhauer, 1956; and W. Stauder, 1963. For many years the Jesuit Seismological Association met jointly with the Eastern Section. In 1937, for example, at the joint meeting in St. Louis, of 22 participants 9 were Jesuits (Fig. 4); in 1948 at the meeting in Cleveland, of 29 participants 8 were Jesuits; and in 1961 at the joint meeting in Cincinnati, of more than 80 participants 9 were Jesuits.

5. Conclusion

As we have seen, Jesuits contributed to organizational, experimental, and theoretical aspects of seismology. Their principal contribution has been providing seismological data for research into the constitution of the Earth and the processes of generation of earthquakes. To accomplish this more effectively, Jesuit stations regularly endeavored to update the quality of their instrumentation (Table 1). This was particularly of significance in the early times, between 1910 and 1960, when the number and quality of seismological stations worldwide was rather limited. Jesuit stations in South America, Africa, and Asia were particularly important in those early times. In some instances, they were the only reliable stations in a region for many years. Establishment of modern national seismological networks and research institutions has made their work no longer necessary and explains the closing of many Jesuit stations.

The reporting of data has been a service of Jesuit observatories. Although many of the stations have been closed, those still active keep up this tradition. More and more, though, the contributions of Jesuits to this now very modern and developed science are through the research centers, principally Saint Louis University, Weston Observatory, Instituto Geofísico (Bogotá), Observatorio de S. Calixto (La Paz), and Manila Observatory. There are also contributions by individual Jesuits associated with Jesuit stations or working in or with other institutions or associations. They have continued to play an important part in the development of the theory of plate tectonics, in the study of the deep constitution of the Earth and of the mechanism of earthquakes, and in the earthquake hazard reduction programs of their various national efforts and cooperative international initiatives. They do indeed stand even today in the tradition of the early Jesuit pioneers.

In order to preserve a recognition of the contribution of Jesuits to this science, in 1989 the Jesuit Seismological Association approached the Eastern Section of the Seismological Society of America with the offer to fund the establishment of an award to honor an individual who has contributed notably to observational seismology. The Eastern Section accepted the proposal in 1991. The award is now conferred annually and bears with it a plaque and a small monetary prize.

References

Buist, M. (1983). L'Observatoire de Géophysique. *Bulletin de Géophysique*, Collège Jean-de-Brébeuf, Montreal.

Byerly, P. (1964). History of the Seismological Society of America. *Bull. Seismol. Soc. Am.* **54**, 1723–1741.

Byerly, P. and W. Stauder (1958). James B. Macelwane, S.J. *Biograph. Memo. Natl. Acad. Sci. USA* **31**, 254–281.

Cabré, R. (1988). 75 años en la vanguardia de la sismología. *Rev. Geofísi.* **29**, 1–3.

Coenraads, R.R. (1993). The San Calixto Observatory in La Paz, Bolivia, eighty years of operation, Director Dr. L. Drake, S.J. *J. Proc. R. Soc. New South Wales* **126**, 191–198.

Davison, C. (1927). "The Founders of Seismology." Cambridge University Press.

Deppermann, C.E. (1951). Father José Algué and microseisms. *Bull. Seismol. Soc. Am.* **41**, 301–302.

Doyle, H., and R. Underwood (1965). Seismological stations in Australia. *Aust. J. Sci.* **28**, 40–43.

Drake, L.A. (1980). Riverview Observatory. In: "St. Ignatius' Centennial 1880–1980" (C. Fracer and E. Lea Scarlett, Eds.), pp. 210–214. St. Ignatius College, Lane-Cove NSW.

Egidi, G. (1888). "Applicazioni delle asta vibranti e oscillanti alle osservazioni dei moti sismici." Rome.

Faura, F. (1880). "Observaciones seismométricas de los terremotos del mes de Julio de 1880." La Oceanía, Manila.

Gherzi, E. (1924). Étude sur les microséismes. *Observatory of Zikawei, Notes Seismologiques* **5**, 1–16.

Gherzi, E. (1950). The scientific work of the Catholic Church at the Zikawei Observatory in Shanghai. *Int. Portugues de Hongkong* **3**, 435–457.

Goberna, R. (1988). The historical seismograms of Colombia. In: "Historical Seismograms and Earthquakes of the World" (W.K. Lee, H. Meyers, and R. Shimazaki, Eds.), pp. 467–473. Academic Press, New York.

Heck, N.H. (1944). The Jesuit contribution to seismology in the U.S.A. *Thought* **19**, 221–228.

Ingram, R.E. (1963). Focal mechanism of couples without moment. *Bull. Seismol. Soc. Am.* **53**, 817–820.

Linehan, D. (1970). Jesuits in seismology. *Earthq. Inform. Bull.* **2**, 14–17.

Linehan, D. (1984). Jesuits in seismology. *Earthq. Inform. Bull.* **16**, 156–165.

Lynch, J.J. (1970). "Watching our Trembling Earth for 50 Years." Dodd, Mead and Co., New York.

MacDonell, J. (1989). "Jesuit Geometers." Vatican Observatory Publications, Vatican City, 56 pp.

Macelwane, J.B. (1926). The Jesuit seismographic stations in the United States and Canada: A retrospect. *Bull. Seismol. Soc. Am.* **16**, 187–193.

Macelwane, J.B. (1936). "Introduction to Theoretical Seismology, Part I, Geodynamics." Wiley, New York.

Macelwane, J.B. (1939). Evidence of the interior of the Earth derived from seismic sources. In: "Internal Constitution of the Earth," pp. 219–290. McGraw-Hill, New York.

Macelwane, J.B. (1946). Storms and the origin of microseisms. *Ann. Géophys. France* **2**, 281–289.

Macelwane, J.B. (Ed.) (1950). "Jesuit Seismological Association, 1925–1950." *Twenty-fifth Anniversary Commemorative Volume*. Saint Louis University, St. Louis, Missouri.

Macelwane, J.B. and C.G. Dahm (1937). Revised travel-time tables. *Publ. Bureau Central Seismol. Int. Series A, Trav. Scientifique* **15**, 38.

Murphy, T. (1995). "A Treatise on the Seismology Observatories of Mungret and Rathfarnham" (MS), Ireland.

Ramirez, J.E. (1967). Historia de los terremotos en Colombia. *Bol. Bibl. de Geof. y Ocean. Am.* **4**, 1–146.

Ramirez, J.E. (1977). "Historia del Instituto Geofísico al Conmemorar sus 35 años." Universidad Javeriana, Bogotá.

Repetti, W.C. (1946). Catalogue of Philippine earthquakes. *Bull. Seismol. Soc. Am.* **36**, 133–322.

Rothé, J.P. (1981). Fifty years of history of the International Association of Seismology (1901–1951). *Bull. Seismol. Soc. Am.* **71**, 905–923.

Saderra-Masó, M. (1895). "La Sismología en Filipinas." Imp. Ramirez, Manila.

Saderra-Masó, M. (1915). "Historia del Observatorio de Manila, 1865–1915." McCullough, Manila.

Saderra-Masó, M. and W.D. Smith (1913). The relation of seismic disturbances in the Philippines to the geological structure. *Bull. Seismol. Soc. Am.* **3**, 151–186.

Sánchez Navarro-Neumann, M.N. (1915). Travail produit par un tremblement de terre. *Boll. Soc. Sismol. Italiana* **34**, 87–89.

Sánchez Navarro-Neumann, M.N. (1919). Ensayo sobre la sismicidad del suelo español, *Bol. R. Soc. Hist. Nat.* **13**, 83–108.

Sánchez Navarro-Neumann, M.N. (1928). La estación sismológica de Cartuja y su labor científica, 1903–1928. *Razón y Fe* **27**, 115.

Sánchez Navarro-Neumann, M.N. (1928). Actual cooperación de la Compañia de Jesus a los estudios sismológicos. *Iberica* **15**, 58–62.

Sánchez Navarro-Neumann, M.N. (1937). Os Jesuitas e a sismologia. *Broteria* **24**, 145–151.

Su, S. (1988). Historical seismograms of the Manila Observatory. In: "Historical Seismograms and Earthquakes of the World" (W.K. Lee, H. Meyers, and K. Shimazaki, Eds.), pp. 490–496. Academic Press, New York.

Udías, A. (1994). Jesuit astronomers in Beijing, 1601–1805. *Q. J. R. Astron. Soc.* **35**, 463–478.

Udías, A. and W. Stauder (1991). Jesuit geophysical observatories. *EOS Trans. Am. Geophys. Union* **72**, 185–187.

Editor's Note

This article was published in the *Seismological Research Letters*, Vol. 67, No. 3, pp. 10–19; May/June 1996. Reprinted (with reformatting) by permission of the Seismological Society of America. We thank Bob Herrmann and Melanie Whittington for supplying us with scanned image files of Figs. 2–4. Please see also Chapter 1, History of Seismology, by Agnew, and Chapter 88, Old Seismic Bulletins, by Schweitzer and Lee. Biography for many Jesuits mentioned in this chapter may be found in Chapter 89, Biography of Notable Earthquake Scientists and Engineers.

4

International Seismology

R.D. Adams
International Seismological Centre, UK

1. Introduction

Among the sciences, seismology is probably that which benefits most by people working together, and from the very beginning of our science this need was recognized. The advent of instrumental recording late in the 19th century made the need for such cooperation even more evident, and it was necessary for seismologists to work together in two main ways. On the practical side it is necessary to exchange details of earthquake recordings made in different countries to establish the earthquakes' positions with reliability, and it is also important for seismologists to meet together to plan activities and discuss results.

This account gives particular emphasis to the early history of the international bodies involved in seismology and global earthquake location and traces their development. Much information is taken from Rothé's (1981) history of the early days of seismology, and there is also much useful and complementary information in Chapter 1 by Agnew and Chapter 88 by Schweitzer and Lee. It is impossible to refer to all international bodies and activities, but it is hoped that those chosen are representative, and there are no major omissions. Fuller details of the organization and activities of many present-day bodies, including their contact addresses, can be found in other chapters of this publication.

2. Historical Developments

2.1 Early Organizations

Early efforts were made by Frenchmen Alexis Perrey and Count Ferdinand de Montessus de Ballore in the second half of the 19th century to make international collections of natural phenomena, including earthquakes, and by this time many national seismological groups had grown up, including those in Germany, Switzerland, Italy as well as in Japan and the United States. In many of these countries instruments were being developed to record earth movement, and the first reported observation of a teleseism was that made by Rebeur-Paschwitz at Potsdam from a Japanese earthquake in 1889. The time was ripe for international cooperation to develop.

The first serious discussions on international needs of seismology were held under the auspices of the geographers. The Sixth International Congress of Geography, held in London in 1895, acknowledged "the scientific necessity of an international system of stations for the observation of earthquakes," while it was at the Seventh International Congress, held in Berlin in 1899, that Dr G.C.K. Gerland, Professor of Geography at Strasbourg, suggested the establishment of an "International Society of Seismology."

As a consequence, the First International Conference on Seismology was held in Strasbourg, 11–13 April 1901. It was at this meeting that the formal setting up of an International Association of Seismology was discussed in detail. This may be considered as the first meeting of our Association. Professor Gerland presided and the Secretary of the meeting was Dr E. Rudolph, also from Germany. In addition there were members from Austria, Belgium, Germany, Hungary, Italy, Japan, Russia and Switzerland. Gerland was himself in favour of a society of individuals, but the majority wanted an association of states and the German government was asked to make the necessary diplomatic approaches to interested countries.

By the time of the Second International Conference on Seismology, also held in Strasbourg, 24–28 July 1903, twenty countries were officially represented. In addition to those in Europe there were representatives from Argentina, Chile, Congo, Japan, Mexico and the United States.

The International Association of Seismology (IAS) was formally established on 1 April 1904, with 18 member countries, but some others such as Argentina, Austria, Denmark, France, Great Britain, Serbia and Sweden declined or would only join on condition that some changes be made to the constitution.

The first meeting under the new constitution was the Third International Conference on Seismology, at Berlin on 15 August 1905. At this meeting Professor Palazzo from Rome

INTERNATIONAL HANDBOOK OF EARTHQUAKE AND ENGINEERING SEISMOLOGY, VOLUME 81A ISBN: 0-12-440652-1

became President, and the Hungarian Baron von Kövesligethy became Secretary-General. The Association was now firmly established.

2.2 Instrumental Cooperation—Milne's Network

While administrative discussions were progressing for establishing a formal International Association of Seismology, the Englishman John Milne was setting up the first worldwide recording network and analysis centre. Milne had been Professor of Geology and Mining at the Imperial College of Engineering in Tokyo until 1895, when, at the relatively young age of 45, he returned to England and set up a seismological observatory at Shide on the Isle of Wight. While in Japan, he had developed with the help of others a reliable horizontal pendulum seismograph, which was to become the standard instrument of his network. Money was made available from the British Association for the Advancement of Science (BAAS) and from 1899 onwards instruments were installed.

The network grew to a total of about 30 instruments, an international network unrivalled until the World Wide Standard Seismograph Network sponsored by the United States in the 1960s. Milne also received registers of readings from these stations, which from 1899 were published by the BAAS as the *Shide Circulars*. The first circular contains readings from stations at Shide, Kew, San Fernando (Spain), Cape of Good Hope, Mauritius, Cairo, Calcutta, Madras, Bombay, Tokyo and Batavia. The intermediate-period Milne seismographs were not good at recording short-period body waves, and early circulars reported times of the commencement (of "preliminary tremors") and first and second maxima, maximum amplitude and duration. Despite only a very primitive knowledge of travel times, attempts were made at location, which improved as knowledge expanded; from 1907 tables prepared by Zöppritz were available for the main phases. Between 1911 and 1913 the BAAS, of which Milne was the Secretary, published bulletins containing global epicentres derived from readings at the Milne stations for the period 1899–1910. Milne died in 1913, and his work was carried on at Shide under the British Association's Seismological Committee until it moved to Oxford University under Professor H.H. Turner in 1916.

Figure 1 shows a contemporary map of Milne's network in 1899, with the stations shown by their names, and catalogue numbers giving approximate positions of earthquakes.

2.3 International Association of Seismology 1904–1916

Independently of Milne, the newly formed IAS was setting up its own analysis organization in Europe. A Central Bureau was established in Strasbourg, advised by a Permanent Commission. The Bureau was responsible for collecting and analysing station readings, and published macroseismic and instrumental catalogues for the years 1903 to 1908. Work was also carried out on instrumental development. Naturally, because of their similar aims, rivalry grew up between the Central Bureau at Strasbourg and Milne's centre in Britain; this continued for many decades.

The Association was administered by a Permanent Commission, consisting of the Director of the Central Bureau and one member from each adhering state. Committees were set up to study seismic instrumentation and intensity scales. Further meetings of the Association were held in 1907 (The Hague) and in 1911 (Manchester), at which numerous papers were read, and seismological problems were discussed. The next Assembly was planned to be held at St Petersburg from 30 August to 6 September 1914, but the outbreak of hostilities in Europe at the beginning of August forced its postponement. The Secretary-General, Baron von Kövesligethy, who had travelled to St Petersburg to prepare for the meeting with the President, Prince Galitzin, had to undertake a perilous return journey to Hungary through Sweden.

The international convention setting up the IAS was for only ten years, so independently of the war, the Association lapsed on 1 April 1916.

2.4 Seismology Section of the International Union of Geodesy and Geophysics 1922–1930

Soon after the First World War meetings were held in London and Paris to discuss the organization of international scientific unions. As a result of these, the International Union of Geodesy and Geophysics was created in Brussels in 1919, and was to hold its first General Assembly in Rome in May 1922, with a new Seismology Section meeting for the first time. Immediately before this, in April 1922, a "Dissolution Assembly" of the old International Association of Seismology was held at the Central Bureau in Strasbourg. The Assembly decided to refund all remaining funds to the contributing countries and donate all its records and other scientific documents to the new organization.

The first meeting of the Seismology Section was attended by delegates from 33 countries, who elected Professor H.H. Turner (Great Britain) as President, and Professor E. Rothé (France) as Secretary. There were also decisions to be made about the international responsibility for earthquake location. After Milne's death in 1913, his work had been carried on, first at Shide and later at Oxford, with the support of BAAS. The 1922 meeting in Rome set up the International Seismological Summary (ISS) to continue at Oxford under Professor Turner, to produce the definitive global catalogues from the data year 1918 onwards, under the auspices of IUGG. At Strasbourg, Professor Rothé continued to produce the Bulletins of the Bureau Central International de Séismologie (BCIS) forming a valuable additional source of global earthquake information.

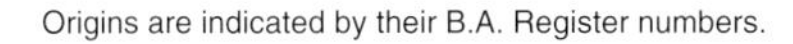

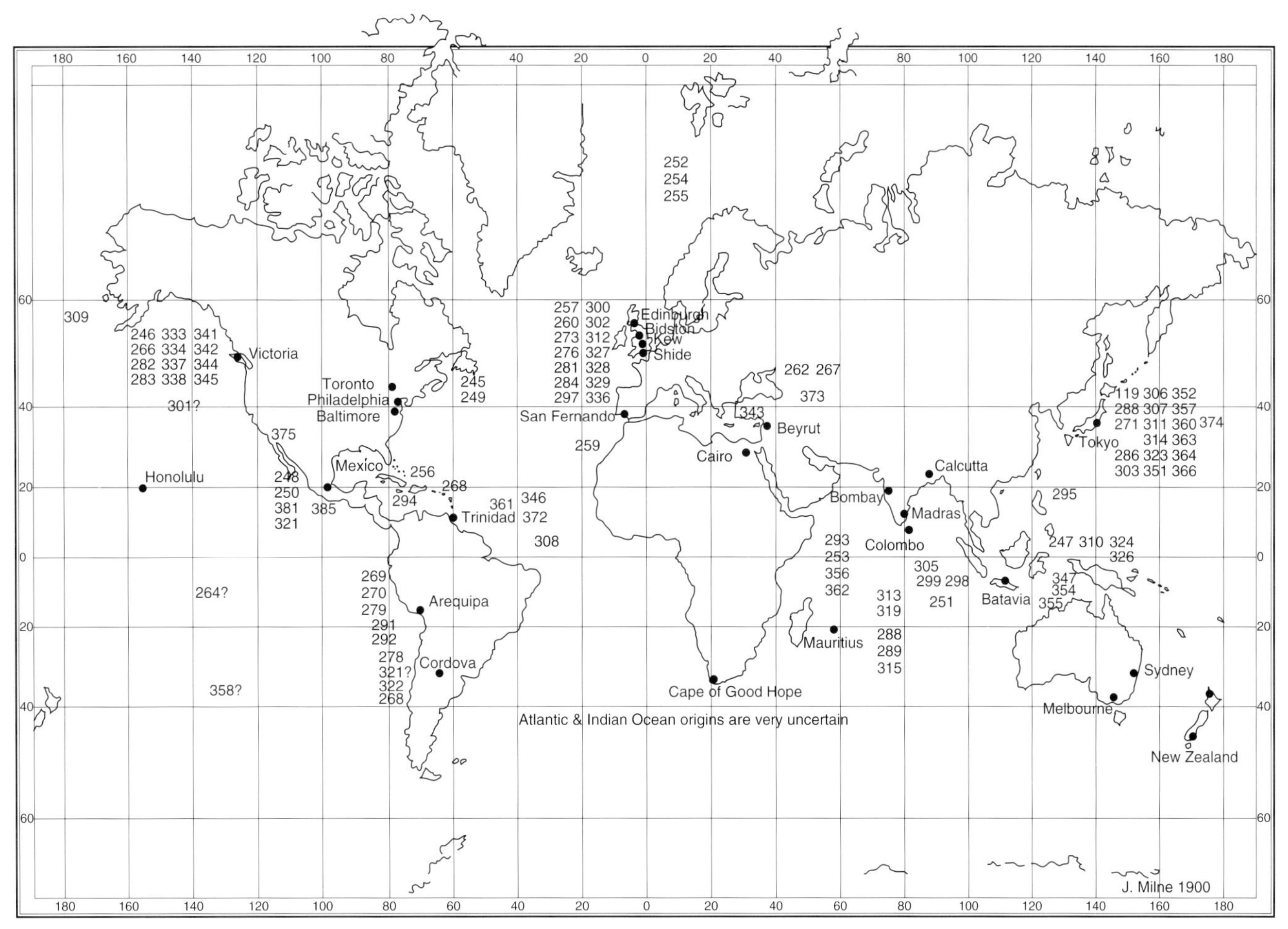

FIGURE 1 Global earthquakes and seismograph stations in 1899, as published by Milne. Numbers refer to earthquakes listed in Milne's catalogue, and show approximate positions. (Originally published in J. Milne (1900) Fifth Report of the Committee on Seismological Investigations, Plate II. British Association for the Advancement of Science, London.)

The Seismology Section held further meetings at Madrid (1924), Prague (1927) and Stockholm (1930), at General Assemblies of IUGG. Professor Turner remained President for these three meetings, but tragically was to die during the last, shortly after giving his Presidential address, in which he had praised the burgeoning seismological activity in the United States, with the US Coast and Geodetic Survey (USCGS) and Jesuit Seismological Association both being extremely active. He also praised the research being carried out in Japan, particularly by Wadati on deep earthquakes.

2.5 International Association of Seismology 1930–1951

During the Stockholm Assembly, sections of IUGG were renamed as Associations, and the name International Association of Seismology was once more adopted. The first meeting under the new statutes was held in Lisbon in 1933, with Professor Oddone (Italy) being elected President. There were many scientific presentations, but much discussion was also given to the perilous financial state of ISS—a situation that was to continue for many years. The next Assembly took place in Edinburgh in 1936, at which Captain Heck of the USCGS was elected President, the first from outside Europe.

Throughout this time the ISS had continued to produce its bulletin of global earthquake locations, with ever-increasing scientific standards. In 1934 the first travel-time tables of Harold Jeffreys were used in the analysis of earthquakes of 1930. In 1946 the ISS came under the control of the UK Meteorological Office and moved to Kew, and from 1948 the full Jeffreys–Bullen tables were used in the analysis.

At the beginning of the Second World War history was to repeat itself. An Assembly had been arranged for Washington DC, to start on 4 September 1939, and the Secretary-General,

Professor Rothé, had arrived a few days beforehand. As in 1914, hostilities began in Europe in the few days before the planned opening of the Assembly. Professor Rothé was called back to France, but the meeting continued, although with greatly reduced attendance.

Throughout the war, Strasbourg, on the border between France and Germany, suffered greatly from the hostilities, and the University moved to Clermont-Ferrand in Central France, although the seismological station remained in operation. Monthly Bulletins of BCIS continued to be produced until June 1944. The war years also saw changes in the officers of the Association. The President, Captain Heck, became ill and resigned, and Professor Whipple, the Vice-President, died, leaving Dr R. Stoneley (UK) as Acting President. Professor E. Rothé died in 1942, with his work being continued by his son, J.-P. Rothé. After the war it fell to Dr Stoneley and the younger Professor Rothé to reorganize the Association.

Dr Stoneley and Professor Rothé were confirmed as President and Secretary-General respectively at the first post-war Assembly in Oslo in 1948. At the next Assembly, in Brussels in 1951, the Association acknowledged the expanding area of its interest and was renamed the International Association of Seismology and Physics of the Earth's Interior. Thus IASPEI was established in its present form.

2.6 International Association of Seismology and Physics of the Earth's Interior 1951–

Since 1951 IASPEI has continued to expand its work and interests. A pattern has emerged for IASPEI to hold Assemblies every four years, alternately with its parent body, the International Union of Geodesy and Geophysics, and either on its own or jointly with some other IUGG Association. Under a succession of Presidents, Professor J.-P. Rothé remained Secretary-General until 1971, completing a remarkable period during which he and his father had held the post since 1922, almost 50 years. The new Secretary-General was Professor J.-C. De Bremaecker (USA), who although originally from Europe became the first Secretary-General to be based outside Europe. The post returned to Europe in 1979, with the appointment of Dr R.D. Adams, originally from New Zealand but then based at the International Seismological Centre in England. At this time Professor B.A. Bolt (USA) became President and undertook a major restructuring of the IASPEI Commissions to deal with the changing interests of the Association. From this time there also began a fruitful collaboration with UNESCO, involving cosponsoring of many sessions and meetings of mutual interest, particularly in the fields of hazard reduction and training. In 2002 the President is Professor B.L.N. Kennett (Australia) and the Secretary-General Dr E.R. Engdahl (USA). Full lists of Presidents and Secretaries-General are given in an Appendix.

3. International Earthquake Location

3.1 Early International Location

While changes were taking place within the Association, there were also changes in the organization of international earthquake location. In addition to the internationally controlled ISS and the European-based BCIS, many national agencies were undertaking this work. Foremost among them were agencies of the US Government. The first was the US Coast and Geodetic Survey, which from 1928 undertook collection of global earthquake readings and their analysis. This "Preliminary Determination of Epicenters" (PDE) service became one of the accepted standards of excellence and completeness for many decades. Global earthquake location was also undertaken by other national agencies, particularly the Institute of Earth Physics at Moscow.

New interest developed in reliable earthquake location in the aftermath of the Second World War, especially in the need to detect and identify nuclear explosions. Seismologists were early to realize the scientific benefits of being able to use such explosions as known sources of energy to improve their knowledge of travel times of seismic waves, but the nuclear arms race had an even greater effect in that it prompted governments to invest more in seismological recording networks. Foremost among these was the World Wide Standard Seismograph Network (WWSSN) set up by the US Government in the early 1960s. At its peak it comprised more than 100 identical stations placed widely around the globe, each of three-component short-period and long-period instruments. Readings from these stations were used by the PDE service, and made freely available to those interested, as were copies of seismograms.

Interest still remained, however, in an internationally controlled service. The ISS, now at Kew, was continuing to experience financial difficulties and was falling farther behind in its analysis. There was a new development in 1961, when Professor Bolt successfully demonstrated his new computer location program to find a location for the deep Spanish earthquake of 1954. This program was used by ISS for its analysis of earthquakes from 1954.

In its later stages the ISS had a succession of eminent Directors, such as Sir Harold Jeffreys and Dr R. Stoneley, who acted mainly in an advisory and supervisory role, while the day-to-day analysis was undertaken by a small dedicated permanent staff. Particular contributions were made by Miss E.F. Bellamy, and Mr J.S. Hughes, who worked as computer and analyst at ISS from 1923 until his death in 1965.

3.2 International Seismological Centre

Early in the 1960s UNESCO became interested in helping with the problem of international earthquake location, and called an international meeting to consider the setting up of a body to collect and store worldwide seismological information and produce definitive bulletins of relocated earthquakes. Thus the International Seismological Centre (ISC) was established to come into operation in 1964, taking over the tasks of the ISS. At this stage the funding was still uncertain, with money coming from IASPEI, some national agencies and UNESCO itself.

The founding Director of ISC, Dr P.L. Willmore, insisted that its operation be entirely computer oriented, and requested that all data be submitted in computer-readable form, and that the analysis be undertaken entirely by computer. It was agreed that ISC wait two years before carrying out its analysis in order to make its data files as complete as possible. After some initial difficulties, the Bulletins of ISC appeared monthly, and have remained the definitive global earthquake listings ever since. As with any new procedure, the early program produced some errors in analysis, mainly by associating readings from earthquakes in different parts of the world into a single fictitious event, and by interpreting later phases as first arrivals. With increasing experience, staff at ISC have learned to recognize these, and to override the program's association where necessary.

The new funding arrangements were not working well, so UNESCO called a further meeting of national representatives in 1969. Delegates at this meeting were not keen to set up an official intergovernmental body, but rather opted for an independent Non-Governmental Organization to be established under Scottish law at Edinburgh. These new ISC statutes allowed for a Governing Council of representatives from national seismological agencies that subscribed a minimum subvention. The first meeting of the Governing Council was held at the Royal Society, London, in November 1970, with representatives present from subscribing organizations in six countries, viz. Canada, New Zealand, Soviet Union, Sweden, the United Kingdom and the United States. In addition there was a representative from UNESCO and observers from several other countries that later became members. The Governing Council has met regularly at least every two years, and has now grown to include representatives from more than 50 countries. In 1970 Dr Willmore was replaced as Director by Dr E.P. Arnold from the US Geological Survey. Dr Arnold carried out substantial revision of the Centre's procedures and analysis programs, which with only minor changes remained in use until the late 1990s. In 1975 the Centre moved from Edinburgh to Newbury in southern England to be closer to the large computer at the Rutherford Laboratory, which it was using at that time; in 1986 a further move of a few miles took it to its present location in its own premises at Thatcham.

Dr Arnold returned to the USGS in 1975, as the Centre moved to Newbury, and his place was taken by Mr A.A. Hughes, son of Mr J.S. Hughes. Mr Hughes retired in 1998. It is a tribute to the work of Dr Arnold and others that the analysis systems and programs that they set up in the early 1970s have remained in operation for more than 25 years. Throughout this time, however, there has been a continual honing of techniques, particularly in analyst assistance to the program in poorly-recorded events, and in deciding the reality of poor associations of stations into events. As well as revising earthquakes already located by other agencies, ISC seeks new, previously unlocated, events, and includes these in its listings after careful examination. During this period the Centre had the benefit of the practical experience of Dr R.D. Adams, who came to the Centre in 1978 from the New Zealand Seismological Observatory, and remained until his retirement in 1995.

Although ISC has continued in its present type of analysis for nearly 30 years, the time has now come for a major restructuring of its procedures, making use of more modern computer systems. This task will fall on Mr Hughes's successor, Dr R.J. Willemann, who brings with him experience from the prototype International Data Centre near Washington, DC, which determines earthquake locations for the intergovernmental organizations involved in the monitoring of nuclear test ban treaties.

4. Present Situation

4.1 Non-Governmental Organizations—ICSU

Figure 2 shows diagrammatically the relationship among various international organizations involved in seismology. On one side we have Non-Governmental Organizations (NGOs), of which the principal is the International Council of Scientific Unions (ICSU). The "ICSU Family" comprises more than twenty Unions covering a wide range of sciences, including IUGG, and hence its constituent Associations, such as IASPEI. Closely related unions include the International Union of Geological Sciences (IUGS), with which IUGG has a joint venture in the Inter-Union Commission on the Lithosphere (ICL). ICSU has other interunion committees, such as the Scientific Committee for Antarctic Research (SCAR) and the Scientific Committee for Oceanographic Research (SCOR). Note, however that the early International Association of Seismology, before the First World War, was in fact an intergovernmental body.

ICSU bodies are generally controlled by national representatives who are elected by their peers in their respective countries, and are meant to be free to act independently, without government influence. Among other NGOs outside the ICSU family are the International Seismological Centre and some engineering bodies such as the International Association for Earthquake Engineering (IAEE), which cooperate closely with ICSU bodies in their relevant disciplines.

4.2 Major International Projects

A major boost to international cooperation came in 1957–1958 with the International Geophysical Year (IGY). This project, run by a special committee of ICSU with the main initiative

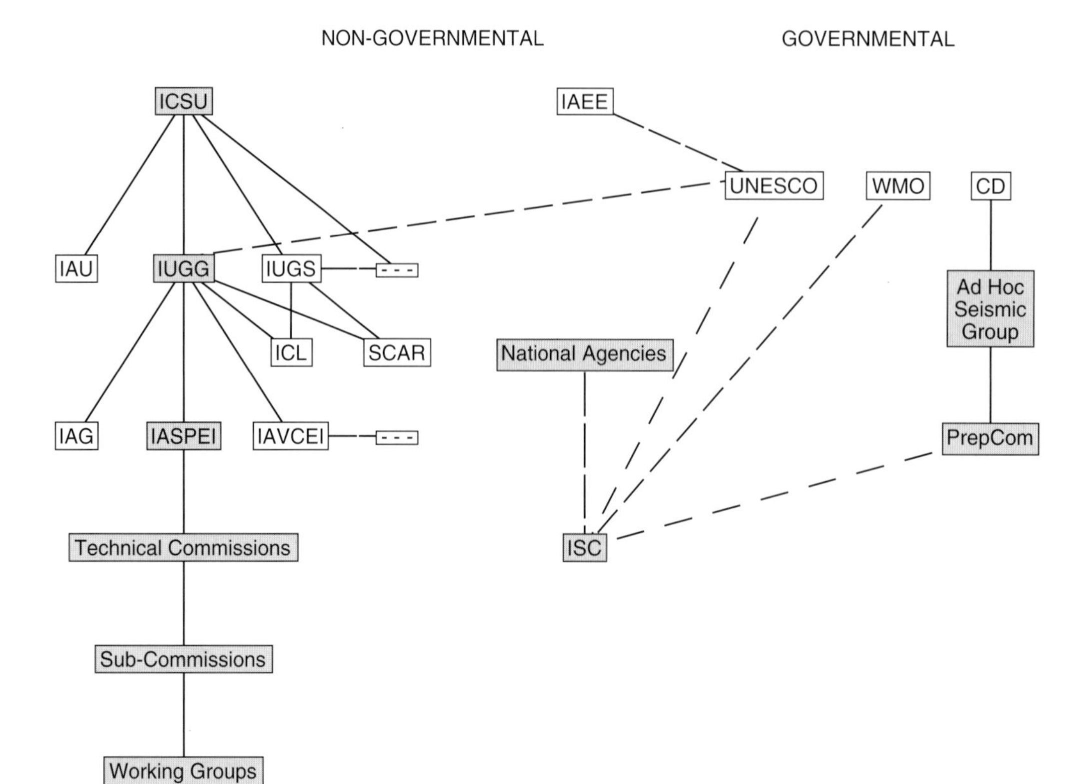

FIGURE 2 The organization of international scientific bodies, with those directly involved in seismology shown shaded. Firm administrative links are shown by full lines; broken lines show advisory and scientific links.

from IUGG, embraced not only seismology but many other aspects of geophysics, such as geomagnetism, ionosphere and upper atmosphere studies. International Data Centres were also set up to ensure the free availability and exchange of information. Particular emphasis was placed on improving global coverage of monitoring, and it is from this time that extensive geophysical recording was established in the Antarctic, with many nations setting up bases there, some of which still operate.

The work started in IGY continued afterwards, and the then President of IUGG, Professor V.V. Beloussov of the Soviet Union, suggested an International Upper Mantle Project, which ran from 1962 to 1970. It was organized at the initiative of IUGG, but with the active support and participation of the International Union of Geological Sciences (IUGS) and other appropriate Unions of ICSU. The Upper Mantle Project was followed from 1970 to 1980 by the International Geodynamics Programme. It is these activities which have led to the setting up of the present Inter-Union Commission on the Lithosphere (ICL), now jointly run by IUGG and IUGS.

4.3 Intergovernmental Organizations

Foremost among the intergovernmental organizations with an interest in seismology is the United Nations Educational, Scientific and Cultural Organization (UNESCO) whose Division of Earth Sciences has always played a major part in fostering seismological work, particularly under its early Director, Dr E.M. Fournier d'Albe. Since the early 1980s UNESCO has had a particularly close association with ISC and with IASPEI. Among many joint ventures have been cosponsored sessions at Assemblies, special workshops and training courses. One particularly successful venture was a joint IASPEI/UNESCO/ICL training course in seismology held in Zambia in 1988. UNESCO has also given special support for IASPEI Regional Assemblies, held in India in 1985, Kenya in 1990, Brazil in 1994 and China in 1996. UNESCO's role is to advise, provide expertise and encourage activities. The provision of full-scale programs for equipment and instrumentation is the responsibility of other UN agencies such as United Nations Development Programme (UNDP). The UN Department of Humanitarian Affairs (formerly UN Disaster Relief Organization (UNDRO)) is also active in earthquake hazard mitigation and relief work.

The World Meteorological Organization (WMO) is a UN agency with an operational branch, which has also been involved in seismological matters, particularly data transmission, hazard reduction and technology exchange. This involvement is not surprising, as over a large part of the world, particularly in Africa and Asia, seismology and seismological recording are undertaken by national meteorological agencies.

There is much mutual benefit from the cooperation between UN agencies and ICSU bodies. Organizations such as UNESCO rely on scientific advice from the scientific unions and associations to help plan their programmes, and at the same time can often provide funding to assist the scientists in their work.

4.4 Committee on Disarmament

The United Nations, through its Committee on Disarmament (CD), has devoted much attention to the setting up of treaties to limit and control the use of nuclear explosions for military and civil engineering purposes. Seismology plays a crucial role in the detection and identification of such explosions and the CD therefore set up in Geneva in the late 1970s an Ad Hoc Group of Scientific Experts to devise means of monitoring any treaty. This Group carried out a series of projects and experiments leading to the establishment of a global network of digitally-recording seismographs with their recordings continuously available for analysis. An experimental International Data Center was set up near Washington DC by the US Government to produce bulletins of global earthquakes since the beginning of 1995. Preliminary solutions are available within a few hours, and the final Revised Event Bulletins (REB) within a day or so, with the readings on which they are based. The results of the IDC, although primarily intended to help in treaty monitoring, are freely available to all participating governments.

Figure 3 shows the distribution of seismic recording stations of this network, which has developed from the ideas first implemented by Milne nearly one hundred years earlier.

The success of the work of IDC has led to the establishment in Vienna of a full-scale organization to monitor by seismological and other means the Comprehensive Test Ban Treaty when it comes into force. The first phase, known as the Preparatory Commission (PrepCom), is now in place.

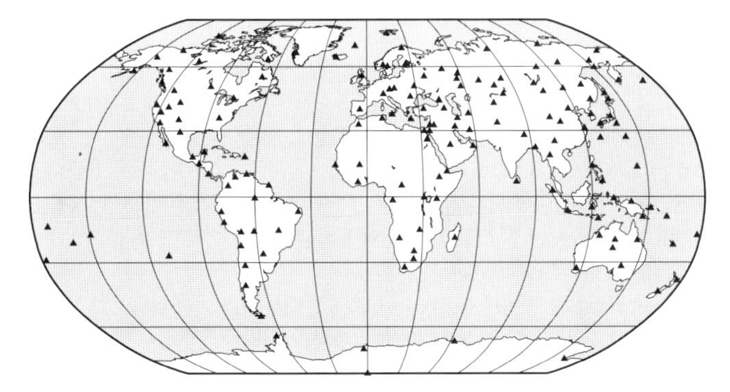

FIGURE 3 Seismological recording network used in 1999 by the International Data Center to monitor the Comprehensive Test Ban Treaty, including both primary and auxiliary stations.

4.5 Responsibilities of International Location Agencies

There are thus now three levels of international earthquake location, carried out by three different agencies with differing aims. The IDC produces a speedy location, based on a few widely distributed sensitive stations, so that any event of special interest may be quickly identified and investigated further. For this purpose its locations need not be particularly accurate. Next the NEIC of USGS produces its series of publications ranging from the Preliminary Determination of Epicenters files, available in a few days, to their Monthly Bulletins, which appear after about six months. These listings include many more small, regional, events and a fuller listing of station readings than is available to IDC. Finally the ISC produces its publications after about two years, employing and listing all available information for the definitive location of events for tectonic and hazard studies, and station phase readings for studies of earth structure. The three agencies involved fulfil complementary functions, each meeting the needs of a different section of the international seismological community.

4.6 Broadband Seismological Recording

The international location agencies have tended to rely on readings from short-period instruments to give accurate arrival times for the main seismological phases. In recent years, however, seismological recording has moved toward broadband digital instruments, which provide much additional information that can be used for research purposes, as well as for location. To coordinate the setting up and operation of such instruments, in 1986 there was established a Federation of Digital Broadband Seismograph Networks, under the joint auspices of IASPEI and ICL.

Most new stations are now broadband, but apart from the network of the Committee of Disarmament, there is no global network under international control. Many countries, however, operate networks of regional or global extent, usually for research purposes rather than for providing readings mainly for earthquake location. Prominent among these are the French GEOSCOPE network, and the Italian MedNet. A consortium of US institutions, Incorporated Research Institutions for Seismology (IRIS) operates an extensive global network and data exchange facility. Data from broadband recording in Europe are coordinated by ORFEUS (Observatories and Research Facilities for European Seismology) based in The Netherlands.

4.7 International Decade for Natural Disaster Reduction

Following a suggestion by Professor F. Press, an eminent seismologist and then President of the US National Academy of Sciences, the United Nations designated the 1990s an International Decade for Natural Disaster Reduction (IDNDR), encouraging national, regional and international programmes to alleviate the effects of natural disasters in fields such as earthquake, volcanic eruption, wind storm, flood, landslide, drought, wild fire and insect infestation. This programme produced much focused work relating to seismic hazard assessment, earthquake engineering measures, disaster preparation and management.

5. Regional Cooperation

In addition to the global international activities discussed above, seismology is widely carried out at a regional level. The European Seismological Commission (ESC) is one of the oldest and most active of the IASPEI commissions, with its own statutes and holding regular assemblies. Since 1996 there has also been an Asian Seismological Commission (ASC), with interest in a wide area from the Middle East to the Southwest Pacific. Following the Regional Seismological Assembly in Kenya there was set up an International Commission for Earth Sciences in Africa (ICESA), which also has seismological interests.

Seismological recording in Europe continues to be coordinated by the European-Mediterranean Seismological Centre (EMSC), or Centre Sismologique Euro-Méditerranéen (CSEM), that continued in Strasbourg after the main responsibility for global earthquake location was taken over by ISC. It is now located at the Laboratoire de Détection et Géophysique (LDG) in Paris.

A particularly successful example of regional seismological cooperation is the East African Regional Seismological Working Group. Rather than just discuss policy, the members of this group actively work together. Originally funded by Scandinavian aid agencies, it meets regularly to undertake joint analysis of seismograms from the nine participating countries, which extend from Eritrea and Ethiopia in the north to South Africa. The Group thus produces a combined bulletin of events for the entire region, of a quality much higher than any of the participating countries could produce on their own.

There are many other examples of regional cooperation. The Centro Regional de Sismología para América del Sur (CERESIS) has coordinated South American seismological activities from its offices in Lima for many decades. Similarly, there have been seismological and earthquake engineering organizations for South East Asia with headquarters in Indonesia and the Philippines, and seismological cooperation in the Arab countries has benefited through the Arab Geological Union. These are but a few examples of the regional cooperation that exists around the world.

6. Conclusions

It may be seen that seismology has always been a science that can only prosper if there is full cooperation among agencies at national, regional and global levels. The above discussion traces the development of the main international agencies for seismological organization and earthquake location, and gives some examples of the many projects that have taken place in the past and some of those that operate today. Seismology is also a science that is continually changing to meet new needs, so in a few years the picture may be somewhat different. The spirit of cooperation that has endured in the past is sure to continue.

Reference

Rothé, J.-P. (1981) Fifty years of history of the International Association of Seismology (1901–1951). *Bull. Seismol. Soc. Am.* **71**, 905–923.

Editor's Note

Please see also Chapter 81, International Organization Reports to IASPEI, edited by Kisslinger. Biography of many persons mentioned in this chapter is given in Chapter 89, Biography of Notable Earthquake Scientists and Engineers. Rothé (1981) is reproduced in full on the attached Handbook CD under the directory \04 Adams, and we thank the Seismological Society of America for permission.

Appendix: Officers of IASPEI and Earlier Bodies

Presidents

1901–1905	G.C.K. Gerland (Germany)
1905–1907	L. Palazzo (Italy)
1907–1911	A. Schuster (UK)
1911–1916	B.B. Galitzin (Russia)
1916–1922	G. Lecointe (Belgium)
1922–1930	H.H. Turner (UK)
1930–1936	E. Oddone (Italy)
1936–1940	N.H. Heck (USA)
1940–1951	R. Stoneley (UK)
1951–1954	B. Gutenberg (USA)
1954–1957	K.E. Bullen (Australia)
1957–1960	H. Jeffreys (UK)
1960–1963	P. Byerly (USA)
1963–1967	J.H. Hodgson (Canada)
1967–1971	K. Wadati (Japan)
1971–1975	V.A. Magnitsky (USSR)
1975–1979	H. Berckhemer (Germany)
1979–1983	B.A. Bolt (USA)
1983–1987	Z. Suzuki (Japan)
1987–1991	S. Müller (Switzerland)
1991–1995	A.V. Nikolaev (Russia)

1995–1999 C. Froidevaux (France)
1999– B.L.N. Kennett (Australia)

Secretaries-General

1901–1905 E. Rudolph (Germany)
1905–1922 R. von Kövesligethy (Hungary)
1922–1942 E. Rothé (France)
1942–1971 J.-P. Rothé (France)
1971–1979 J.-C. De Bremaecker (USA)
1979–1991 R.D. Adams (UK)
1991– E.R. Engdahl (USA)

5

Synthesis of Earthquake Science Information and Its Public Transfer: A History of the Southern California Earthquake Center

Keiiti Aki
University of Southern California, and Observatoire Volcanologique du Piton de la Fournaise, La Réunion, France

Where do we come from?
What are we?
Where are we going?

Paul Gauguin

1. Introduction

"A natural disaster will hit us at around the time when we forget about the last one" are household words in Japan coined by Torahiko Terada, a physicist, essayist, and leading intellectual of the early 20th-century Japan, who was also one of the founders of the Earthquake Research Institute, Tokyo University. This statement was an effective warning to the public as well as to officials in charge of public safety through the early part of the 20th century, although it did not contain a bit of scientific information about the phenomena causing the disaster. It was effective because everyone agreed with the statement.

On the other hand, a public warning made on a scientific basis that is still controversial may be ineffective and can bring personal disaster to individual scientists involved. An example is the well known story of Omori and Imamura (e.g., Aki, 1980). Imamura believed in a kind of gap theory (Fedotov, 1965) and anticipated a major earthquake in Sagami Bay near Tokyo, which belongs to a belt of major earthquakes, but lacked historical records. He wrote an article in 1906 warning of the lack of fire-preventing facilities in Tokyo and estimating the death of 100 000 by fire when the Sagami Bay earthquake hit the city. This article was criticized severely by Omori as causing social unrest without solid scientific ground. Imamura had a miserable life for 17 y after the publication of the article until the disaster he predicted actually happened in 1923, causing the loss of 140 000 lives. At the time of this great Tokyo earthquake, Omori was attending the second Pan-Pacific Science Congress in Australia. During his return trip by sea, his health declined sharply and he died shortly after arrival in Japan.

The lesson we learn from wise Terada, courageous Imamura, and cautious Omori is that we must build a consensus among scientists for an effective transfer of the earthquake science information to the public, and the consensus must represent the true current status of the scientific community that will not suppress an Imamura's view. Terada's warning had no scientific information, which may have correctly represented the status of the scientific community of his time. Imamura had a specific warning based on his concrete idea about earthquakes, which was rejected by a more prominent scientist of his time. A more sensible message from the scientific community at that time might have been to present Imamura's warning with a likelihood of, say, 50% to the public, because it is supported by one scientist, and opposed by another. This may sound meaningless from the scientific viewpoint, but if the public could have accepted such a warning, with a certain degree of response, the disaster of 1923 might have been reduced significantly. It is not easy to implement such a scheme of information transfer when the science is at a rudimentary stage of development and a small number of scientists are involved in earthquake problems of a given region.

Earthquake science has made a great stride in the last three decades with the plate tectonic revolution, wide acceptance of the fault model, and advances in computer technology. The increasing societal need for mitigating earthquake hazard has also been an important factor for its development. Time has come to find an effective framework through which we integrate all scientific information relevant to earthquakes in a region as a consensus of the scientific community, and present

INTERNATIONAL HANDBOOK OF EARTHQUAKE AND ENGINEERING SEISMOLOGY, VOLUME 81A ISBN: 0-12-440652-1

it to the public in the region. The Southern California Earthquake Center (SCEC) was founded for this purpose in 1991, as one of the centers of the US National Science Foundation (NSF) Science and Technology Center program.

The first word "synthesis" in the title of this chapter has a more intense meaning than used normally. A traditional way of synthesizing scientific results was, for example, a joint publication of contributions by various authors, like the present Handbook. There is very little interaction among the authors in this form of synthesis. A traditional way of achieving strong interaction among experts on a complex issue has been to form a working group with a task of producing a consensus report. An example is the Working Group on California Earthquake Probabilities organized by the US Geological Survey (USGS). These working groups have limited lifetimes and disappear after completion of the report. SCEC may be considered as a permanent working group addressing the problem of earthquake hazard in Southern California, participated in by scientists of various disciplines who will constantly interact among themselves for consensus building through workshops, seminars, and report writing.

In the present chapter, I shall describe the concept of a "master model," which is proposed as a framework for unifying multidisciplinary observations pertinent to earthquakes in Southern California. A first-generation master model was born in the process of creating a public policy document in response to the public need for knowing the probability of major earthquakes in Southern California after the Landers earthquake of 1992. For a healthy growth of a science, the joy and excitement of doing creative science are as essential as effective public transfer of its product. I shall describe how the concept of a master model served the dual purpose at the SCEC.

2. Master Model Concept as a Framework for Unifying Multidisciplinary Observations on Earthquakes

In May 1989, a workshop was held at Lake Arrowhead, California attended by those who were concerned with seismic hazard in Southern California, and they decided to create SCEC and seek its funding from the Science and Technology Center program of NSF. The University of Southern California (USC) was selected as the host institution for the NSF proposal, and I agreed to serve as the Principal Investigator for the proposal on the condition that T. Henyey would help me in administration and management. During the workshop, a subgroup addressing scientific issues came up with the idea of a "master model," which appears to allow an orderly inclusion of all the research projects proposed by individual participants as its elements. A consensus was quickly developed to put the master model concept as a central idea of the proposal to be submitted to NSF. As such, the master model was a very vague concept when it was born.

One of my long-term dreams has been to develop a model of the Earth's lithosphere in which we predict the occurrence of earthquakes on the basis of the space–time distribution of tectonic stress calculated using various geophysical observables, as the atmospheric scientists forecast weather by computer on the basis of observed pressure, temperature, wind speed, etc. Such a model would have an image of physical concreteness and determinism, and could be described in a straightforward manner in a proposal to NSF. Unfortunately, a proposal along this line was obviously too ambitious and premature, because we do not really know enough about the parameters of material properties and constitutive laws as well as boundary and initial conditions needed to define the problem and it was hopeless to produce anything useful for the public during the lifetime of the center (at most 11 y with three cycles of the evaluation–renewal process) along this line. We called the master model along this line a "physical master model," and I discuss its development at the SCEC later in this chapter.

One thing clear to me at that time was that the master model must be able to absorb all observations pertinent to earthquakes in Southern California. In other words, it must be characterized by parameters that may be effectively constrained by multidisciplinary data including geologic, geodetic, seismological, and historical data on earthquakes in Southern California. After several workshops for preparing the proposal, we recognized the importance of clearly defining the end product expected from the master model. We decided that the end product would be the space–time-dependent probabilities of earthquakes in Southern California and associated ground motions. The master model was defined as a framework through which we should produce this end product by integrating all available information from earth sciences. This scheme is similar to the engineering practice called Probabilistic Seismic Hazard Analysis (PSHA) originated by Cornell (1968) in the western world, but an important difference is that whereas PSHA receives earth science information as given by experts, we generate earth science information by ourselves, and the outcome of PSHA is used to define critical scientific issues to be investigated at the center. Thus, center activities follow a usual scientific cycle of observation, model development, testing the model by comparing predicted with observed seismic hazard, and designing new observations to improve the model. We refer to this model as "hazard master model."

It was easy for me to accept PSHA as a key element of the master model, because I served as the chairman of a panel that evaluated the usefulness of PSHA upon the request of the US National Research Council. Panel members were divided between earth scientists and hazard analysts, and we had to overcome a major disagreement before reaching a consensus report.

The disagreement was concerned with assigning likelihood or probability to the validity of a scientific hypothesis, by procedures that are used by hazard analysts to arrive at a decision needed today, when the data are not sufficient for a unique solution. It was not so difficult for me to accept this because I accepted earlier, in my own seismological studies, the stochastic inverse solution in which a model is considered as a stochastic process. In the end all earth scientist members accepted it as a necessary evil.

In the introduction of this chapter, I suggested that giving 50% probability to the validity of Imamura's prediction might have been more sensible than rejecting it. Here is the essence of the hazard master model. If we know the Earth and earthquakes well enough to construct a physical master model, we do not need the hazard master model. At present we do not have enough facts, though we have competing ideas and hypotheses that will be tested against observations in the long run; but the public needs advice today. The master model approach intends to advance science by model testing with observations and to transfer deliverable information to the public at the same time. We all want to indulge in the joy and excitement of creative science, but it is necessary to deliver our product to the public for continued support of science.

Facing societal problems that require multidisciplinary approaches to their solution, we tend to fault the narrowness of a single scientific discipline. But this very narrowness enabled us to dig deeply into a simplified and isolated problem, leading often to a breakthrough. For example, a PhD thesis that tries to cover a broad area is usually less satisfactory than one that gives a definitive solution to a small problem. We recognized the importance of disciplinary independence in the operation of SCEC in order to keep science alive, and created permanent working groups according to disciplines.

The following eight Working Groups were set up as a balanced representation of the research areas of those interested in participating in the activities of the center:

Group A: Master model and seismic hazard analysis
Group B: Strong ground motion
Group C: Earthquake geology
Group D: Subsurface imaging
Group E: Crustal deformation
Group F: Seismicity and source parameters
Group G: Earthquake physics
Group H: Engineering applications

Although participation in the center activities was open to anyone interested in earthquakes in Southern California, most of the active participants were from universities, government agencies, and private consulting companies in Southern California, for the reason of obvious ease in attending workshops and seminars. This regional limitation of participants was not a problem for Southern California, because we found some of the best scientists in the country in each of the above working groups. This is a direct or indirect inheritance of the long history of excellent earthquake science at the California Institute of Technology (Caltech) Seismological Laboratory, which has been a world center of seismology not only in its disciplinary sense (defined, for example, by Aki and Richards (1980) as a science based on records of seismographs), but also in a broader sense of the physics and geology of earthquakes.

The tasks to be carried out by these working groups for constructing the hazard master model were clear for most groups. Groups C, E, and F would contribute to the characterization of earthquake sources in Southern California, and Groups B, D, and H would contribute in calculating ground motion expected from these sources. Group A would integrate these results into products useful for the public. On the other hand, the role of Group G, earthquake physics, was intended to explore, under the protective umbrella of SCEC, paths toward the physical master model, our ultimate goal.

The proposal to NSF was successful, and SCEC began its activities at the beginning of 1991. It quickly established three data centers, the Seismic Data Center (Caltech, USGS) for waveform data and earthquake parameters; the GPS Data Center [University of California at Los Angeles (UCLA), University of California at San Diego (UCSD), and Massachusetts Institute of Technology (MIT)] for GPS survey data; and the Strong Motion Data Center [University of California at Santa Barbara (UCSB)] for strong motion data.

The funding was nearly equally divided between the infrastructure support and research funding. The infrastructure included the operation of the above three centers, visitors program, expenses for meetings, education and outreach programs required for the NSF centers, and the management of the center.

Research funds were, and still are, distributed in the following manner. The process starts with an Annual Meeting of SCEC, where the tasks to be carried out in the coming year are discussed and decided. Each Working Group then discusses its contribution to the decided tasks, and gives priorities to research areas to be investigated in the coming year. Conclusions from the annual meeting are summarized in the form of RFP (request for proposal) and sent to all SCEC members. Short proposals submitted in response to the RFP are reviewed by leaders of the Working Group for the quality of science, and by the Science Director for the contribution to the SCEC tasks. The funding decision is made by a committee composed of the leaders of the Working Group and the board members who represent the core institutions of SCEC [USC, Caltech, UCLA, UCSB, UCSD, University of California at Santa Cruz (UCSC), Columbia University, and USGS until 1998, and now including California State University at San Diego and University of Nevada at Reno].

The above funding decision process implies a matrix approach made of horizontal lines of tasks and vertical columns of disciplines. The task at the highest level may be the response to a request from the broadest public, and those at lower levels may be in response to more restricted professional and technical user groups. Interactions among different disciplines occur at all levels of tasks, but the more creative, and cutting-edge interactions occur at lower levels, while the consensus-building for public documents is done at higher levels. One may compare the former interaction to the fermentation process and the latter to the distillation, as both processes are needed to produce a fine liquor. The uniqueness of SCEC is this very combination of producing cutting-edge research and passing it along in a timely manner to users.

The Landers earthquake of 1992, the largest earthquake in Southern California since the Kern County earthquake of 1952, occurred in the second year of SCEC, when it had established its working style. Perhaps for the first time in the history of earthquake science, we were ready for a major earthquake, and it occurred just in time. In the next section, I describe how the earthquake helped to pull scientists together to create the first-generation master model for earthquakes in Southern California.

3. The First-Generation Master Model of Earthquake Sources in Southern California

On the morning of 28 June 1992, Southern California was awakened by a very large ($M = 7.3$) earthquake which occurred near the town of Landers, located 30 km north of the San Andreas fault. It was followed by another large one ($M = 6.5$) a few hours later occurring in the Big Bear area 40 km to the west of Landers. People were concerned what this series of earthquakes meant with regard to future earthquakes in Southern California, particularly to the one anticipated on the San Andreas fault, nicknamed the Big One. I felt the need for a quick response to the concerns of the public, and organized a one-day workshop at USC, two weeks after the earthquake, to discuss the implications of these events and SCEC's response. It was attended by SCEC members and key people from the US Geological Survey and the California Department of Conservation's Division of Mines and Geology. It concluded with a decision to produce two reports, Phase 1 and Phase 2. The Phase 1 report would address the short term effects of the Landers–Big Bear sequence on nearby active faults—principally the San Andreas, and the level of ground shaking from potential earthquake scenarios involving the San Andreas fault. The Phase 2 report would extend the study area to the whole of Southern California, and consider the seismic hazard for a longer term. In both reports, the starting point would be the conclusions given earlier in a report entitled "Probabilities of Large Earthquakes Occurring in California on the San Andreas Fault" prepared by the Working Group on California Earthquake Probabilities (WGCEP, 1988).

The Phase 1 report was completed quickly in the fall of 1992, but the Phase 2 report took a long time partly because its innovative nature required a lot of explanation to the broader community of earthquake scientists and to the user community. It was finally published in the April 1995 issue of the *Bulletin of the Seismological Society of America* (WGCEP, 1995). The time spent for the completion was, however, not in vain. The ensemble of the model parameters given in the report is now widely accepted as the first-generation master model of earthquakes in Southern California.

The most radical innovation in the report was the use of crustal strain estimated from the GPS data for seismic hazard estimation as proposed by Ward (1994). Southern California was divided into 65 source zones, and the strain accumulating in each zone was translated into seismic moment according to the Kostrov (1974) formula, thus making seismic moment a central parameter of the master model reducible from geologic, geodetic, and seismological data. Since the GPS network in principle can cover the whole of Southern California uniformly, it can give information on earthquake probabilities in the area where geologic information on faults is lacking. The crustal strain estimated from the GPS data showed more diffused pattern than that inferred from geologic data. This comparison stimulated the group of geologists led by D. Schwartz to come up with the consensus parameters of faults in all of the 65 source zones, making the Phase 2 report go beyond the 1988 Working Group report, which was concerned only with the San Andreas and San Jacinto faults. The importance of this extended areal coverage in the report is clearly demonstrated by the Northridge earthquake of 17 January 1994, the most costly (loss of $20 billion) earthquake in the history of United States, which occurred off the San Andreas fault.

The Northridge earthquake was no surprise from the perspective of the Earthquake source model described in the Phase 2 report. The earthquake occurred in a source zone characterized by relatively high earthquake potential, falling in the top 13% of Southern California in terms of moment rate per unit area. While the specific fault had not been known prior to the event, the location, magnitude, and style of earthquake were consistent with our expressed understanding of the regional geology and tectonics.

In order to integrate geologic, geodetic, and seismological data, it was important to classify the 65 source zones into Types A, B, and C according to the availability of these data. Type A zones contained faults for which paleoseismic data sufficed to estimate conditional probabilities of earthquake occurrence in the manner of the 1988 Working Group. We allowed, however, for failure over multiple segments in the form of a "cascade" model, in which the amount of slip is characteristic of a segment, but all earthquakes that can occur by combining contiguous segments are allowed. Type B zones contained faults for which geologists had documented possible characteristic earthquakes but information was insufficient for the conditional probability calculation. For Type C zones, we lacked direct evidence of a preferred characteristic earthquake and we set the seismicity rate to the average of the smoothed catalogue seismicity rate (Kagan and Jackson, 1994) and that corresponding to the geodetic moment rate. By the use of the above classification, we were able to combine existing data from seismology, geology, and geodesy in order to arrive at the preferred model for the whole of Southern California.

Once the preferred model was determined, we could predict the frequency distribution of magnitude for the whole of Southern California. We found that the predicted frequency was about a factor of 2 greater than that observed in the past 150 y. This discrepancy can be explained in the following three ways. First, the seismicity in the past 150 y was anomalously low as compared to the long-term average rate expected from geologic and geodetic data. Secondly, some part of geodetic strain accumulating in the brittle part (assumed to be 11 km thick in this report) may be released aseismically. Thirdly, the maximum magnitude assigned to a source zone using geologic data may be underestimated. All these problems are important from the point of a fundamental understanding of earthquake processes, but we now recognize their importance in practical evaluation of seismic hazard in Southern California and focus our future research on these issues.

Thus, our experience with preparing the Phase 2 report demonstrated all the attributes of the master model, which we described as a framework for integrating multidisciplinary data in the preceding section. Furthermore, the Phase 2 report included the estimation of ground motion and its exceedance probabilities, but was incomplete in this regard because we had neglected the effect of local site conditions on ground motion, which will be the subject of discussion in the next section.

4. Delineation of Local Site Effects on Strong Ground Motion

Once a distinguished earthquake engineer complained to me that engineers read what seismologists write, but seismologists never read what engineers write. So, when I was asked by the geotechnical engineering community to write a review paper on local site effects on strong ground motion, I tried to read papers written on the subject by engineers and engineering seismologists in the United States (e.g., Mohraz, 1976; Seed *et al.*, 1976; Trifunac, 1976a,b; Boore *et al.*, 1980; Joyner and Boore, 1980) and Japan (e.g., Hayashi *et al.*, 1971; Kuribayashi *et al.*, 1972; Katayama *et al.*, 1978; Kawashima *et al.*, 1986). I found remarkably consistent results common to both countries. Soil sites show higher amplification than rock sites for periods longer than about 0.2 sec, while the relation is reversed for shorter periods. This frequency dependence of site amplification is reflected in the site dependence of peak ground motions. Peak ground velocity and displacement show higher amplifications at soil sites than rock sites, while peak ground acceleration is roughly independent of the site classification.

The conclusion of my review (Aki, 1988) was that the above apparent disappearance of site effect at high frequencies is due to the failure of the broad classification (soil vs. rock) of site condition in capturing the essential factor controlling the site effect at high frequencies. A meaningful microzonation is possible for the frequency range, at least, from 1 to 10 Hz, because the geographic variation of site-specific amplification factor varies by a factor of 10, while the standard error of its variation at a given site for different directions of incident waves is less than a factor of 2.

At the time I wrote the above review, there were no systematic studies on the frequency dependence of site effect on weak motions, and it required another review (Aki, 1993) to conclude the difference between weak and strong motion in this respect. The weak motion amplification is higher for soil sites than for rock sites at least up to 12 Hz in California (Su and Aki, 1995), distinctly different from the frequency dependence of strong motion: a clear indication of the pervasiveness of nonlinearity of ground response in earthquake strong motion.

At the time of my review in 1988, I was aware of only one strong motion record demonstrating the unequivocal effect of nonlinearity of soil, namely, the record of the Niigata earthquake of 1964 obtained near a building that failed completely owing to the liquefaction of soil composed of water-saturated sand. There were other cases suggesting nonlinear soil response. For example, strong ground motion observed in Pasadena during the San Fernando earthquake of 1971 by Hudson (1972) did not show any site dependence, while the amplification factor mapped by Gutenberg (1957) using weak motions of small near earthquakes or large distant earthquakes was distinctly higher at soil sites than at rock sites. These examples, however, can be explained by combination of various linear effects, such as frequency-dependent radiation pattern, absorption, impedance, and scattering effects. I agreed, then, with Esteva (1977), who stated that the influence of nonlinearities is often overshadowed by the overall patterns of shock generation and propagation. In other words, the seismological detection of the nonlinear site effect requires a simultaneous understanding of the effects of earthquake source, propagation path and local geologic site conditions.

The Loma Prieta earthquake of 1989 was the first earthquake for which the simultaneous determination of the above three effects was attempted by Chin and Aki (1991), who found that the amplification factor estimated from weak motion overestimates the peak ground acceleration for sites within 50 km from the epicenter. The amount of overestimation appeared to be in the range expected from geotechnical engineering studies. This discovery changed my position with regard to the soil nonlinearity by 180 degrees, and I began to see pervasive nonlinear effects in various seismic observations, as summarized in Aki (1993).

This change in my position was reflected in SCEC's relation with the geotechnical engineering community. The seismological community has been skeptical about the nonlinear effect (which reduces the estimate of ground acceleration) claimed by the geotechnical engineers primarily on the basis of experiments on small soil samples in the laboratory. There has been very little interaction between the two communities because of the diametrically opposed difference on this most fundamental issue on the site effect. Because of my acceptance of nonlinearity, it was easy for SCEC to have a smooth cooperation between the two disciplines from its inception. We also found, especially through many observations made during the Northridge earthquake of 1994, that the site effect is a much more complicated problem than seismologists used to think, and certainly deserves such a cooperative research. For example, a timely and ingenious deployment of dense seismograph networks by a UCLA group (Gao *et al.*, 1996) after the earthquake revealed that the concentration of damage in Santa Monica was not due to site conditions at shallow depths normally concerned by geotechnical engineers, but to a wave focusing effect by a geologic structure at greater depth. The Phase 3 report, currently under preparation under the leadership of N. Abrahmson, is intended to include the consensus on site effects, in addition to the updates of earthquake source characterization given in the Phase 2 report.

5. Toward a Physical Master Model of Earthquakes like that Used for Weather Forecasting by Computer

As mentioned in an earlier section, we wanted to develop at SCEC a physical master model, a model of the Earth's lithosphere in which we could predict the occurrence of earthquakes based on the space–time distribution of tectonic stress calculated using various geophysical observables as the atmospheric scientists forecast weather by computer, based on observed pressure, temperature, wind speed, etc. For practical reasons, however, we took the approach of a hazard master model that unified observations in the end product, namely, PSHA.

Our practical approach based on PSHA tried to include all accomplishments of earthquake science made in the past three decades on the basis of plate tectonics and the fault model of earthquakes. The concept of seismic moment unified the geologic, geodetic, and seismological data under the constraint from plate tectonics (Aki, 1966; Brune, 1968; and Kostrov, 1974). The concept of characteristic earthquake introduced by Schwartz and Coppersmith (1984) from paleoseismological observations played a major role in translating observed fault parameters, such as the slip rate, amount of slip, and style of fault into earthquake probabilities. The concept of "cascade" was introduced by D. Jackson into the SCEC Phase 2 report to allow a less restrictive characteristic earthquake model.

The closest thing to what I had in mind as a physical master model for Southern California came from the work by Bird and Kong (1994), done outside of SCEC activities. Their model consisted of a brittle upper crust and a ductile lower crust. Their upper crust model includes all faults with annual slip greater than 1 mm. The deformation of the upper crust is governed by frictional rheology containing a pore pressure effect and specified by different coefficients of friction on the fault and off the fault. The deformation of the lower crust follows the dislocation creep law with the temperature dependence parametrized by the activation energy. The subsurface temperature was estimated from the heat flow at the surface. The boundary condition is given by a standard global model of plate motion. The data to fit with the model are long-term fault slip rates measured geologically, trilateration and very-long-baseline radio interferometry data, and stress orientations from focal mechanisms and *in situ* measurements. The coefficient of friction of the upper crust and the activation energy of the lower crust were free parameters, and they concluded that the coefficients of friction for faults is as low as 0.17, and that a lower activation energy improves the model fit. However, they concluded that the disagreement between the observed fault slip rate and the predicted rate by the best fitting model is too large to allow use of the computer model for practical purposes of seismic hazard estimation. Furthermore, the model of Bird and Kong (1994) was intended for studying a long-term (order of 1000 y) behavior of the Earth's crust, while SCEC is interested in the behavior on shorter time scales.

Rock friction behavior in the time scale of interest to SCEC has been systematically studied in laboratory experiments by Dieterich (1972, 1979, 1981), with results generalized in the form of a rate- and state-dependent friction law by Ruina (1983). This law includes a characteristic slip distance L for evolution of surface state and slip weakening. In other words, there is a unique length scale associated with a fault governed by this law specified by a given set of parameters. This length scale unique to a given fault must play a fundamental role in creating the characteristic slip that should repeatedly occur according to the characteristic earthquake model. The SCEC Working Group G (Earthquake Physics) addressed this fundamental problem from the start. For example, Rice (1993) analyzed a 3D static problem of a vertical fault governed by the rate- and state-dependent friction law by discretizing equations involved with the cell size h. He found that the simulated slip shows spatiotemporal complexity when h is greater than a critical size h^*, called "nucleation size," that scales with the characteristic slip distance L. When h is less than h^*, the complexity disappears in favor of simple periodically repeated large earthquakes, namely, the characteristic earthquake. This conclusion highlights the importance of length scale unique to a fault for our goal of constructing a computer model of earthquakes.

Another fundamental contribution from SCEC to the fault-specific length scale is an unequivocal confirmation of the existence of seismic guided waves trapped in the fault zone of the Landers earthquake of 1992 by Li *et al.* (1994), who estimated a fault zone width of about 200 m, with shear velocity of 2.0–2.2 km sec^{-1} and Q of about 50 by comparing the observed waveform with the synthetic one. This size of the fault width is consistent with the size of breakdown zone (comparable to the nucleation size mentioned above) estimated by Papageorgiou and Aki (1983) from the upper cutoff frequency (source-controlled f_{max}) of strong motion acceleration spectra for several major earthquakes in California.

Interestingly, a similar estimate of the width of the Landers earthquake fault zone was made by an entirely different method. From a detailed study of tension cracks on the surface, Johnson *et al.* (1994) concluded that the Landers fault rupture was not a distinct slip across a fault plane but a belt of localized shearing spread over a width of 50–200 m. They suggested that this might be a common structure of an earthquake fault, which might have been unrecognized previously because the shearing is small and surficial material is usually not as brittle as in the Landers area. We identify this shear zone with the low-velocity, low-Q zone found from the trapped modes because they occur exactly at the same place. Since the trapped modes were observed from aftershocks with focal depths greater than 10 km, we conclude that the shear zone found by Johnson *et al.* (1994) extends to the same depth.

Repeated measurements of fault-zone trapped waves using artificial sources by Li *et al.* (1998) led to the first discovery of fault healing by a controlled field experiment. The travel times for identical shot–receiver pairs decreased by 0.5–1.5% from 1994 to 1996, with the larger changes at stations closer to the fault zone. These observations indicated that the fault strengthened after the main shock, most likely owing to the closure of cracks that were opened by the 1992 earthquake. In the past, indirect evidence such as time-dependent variations in source properties of repeating earthquakes (Marone *et al.*, 1995) showed some consistency to the healing predicted by the rate- and state-dependent friction law. The new discovery from the trapped mode offers direct evidence for a rather broad fault zone associated with the Landers earthquake. The observation of the fault-zone trapped mode will enable direct estimation of the friction law parameters, serving as a link between the laboratory and the field.

Suppose we have a model of Southern California with fault zones delineated in 3D and characterized by a friction law with known parameters; we need to know how this model is loaded. Some researchers assign the loading at the periphery of the model as the displacement given by the global plate motion, and others give it as a drag exerted on the brittle part of the lithoshere by the deformation in its ductile part. Castle and Gilmore (1992), for example, found that recent major earthquakes in Southern California are located in the zone of steep lateral gradient in uplift, and hypothesized a localized creep below the zone of steep gradient along the detachment surface between the brittle and ductile part of the lithosphere. The GPS network, currently expanding in Southern California, will give improved information on this matter. The abundantly and easily available seismic information on coda waves of local earthquakes may also be useful for estimating the localized creep, because the observed correlation between coda Q and seismicity has been interpreted by a model of creep along the brittle–ductile boundary (Jin and Aki, 1989, 1993), and there is an indication of low coda Q (Ouyang, 1997) along the zones of steep gradient in uplift found by Castle and Gilmore (1992). In this regard, Aki and Ferrazzini (2000) have found a more powerful method of coda analysis than the traditional coda Q measurement for mapping localized heterogeneities from observations at the Piton de la Fournaise volcano on the Reunion island in the Indian Ocean. The new method appears to give better correlation with the creeping zone geodetically assessed by Castle and Gilmore (1992) when applied to Southern California (A. Jin, personal communication), offering a promising way of studying deformation in the brittle–ductile transition zones in an earthquake region.

When we combine Rice's (1993) analysis based on the rate and state-dependent friction law with the fault zone width estimated from field observations for the Landers earthquake by Li *et al.* (1994, 1998), we will be able to generate a large characteristic earthquake, but unable to generate smaller earthquakes, which were abundantly observed as aftershocks and background seismicity of the region. Among what are missing in Rice's analysis, the following three elements stand out: (1) dynamics of rupture that may produce nonuniform slip and stress roughening on the fault plane; (2) preexisting irregular geometry of the fault, such as segmentation and branching; and (3) preexisting inhomogeneity in the frictional properties including pore fluid distribution.

The contribution from earthquakes other than characteristic earthquakes to seismic hazard was important in Southern California as described in the Phase 2 report, and our physical master model must somehow account for them if it ever is to replace the hazard master model. The only way to include the above three elements into our physical model would be via a stochastic approach, because their detailed specification in any deterministic manner would be impossible now. Thus, our strategy for constructing the physical master model will be separated into a deterministic part dealing with characteristic earthquakes, and a stochastic part dealing with non-characteristic earthquakes. For the deterministic part, the tomographic monitoring of fault zones by trapped mode observation and the monitoring of the loading process at the brittle–ductile transition zone by the geodetic and seismic observations will supply basic parameters of the physical master model to be used in computer simulation. For the stochastic part, we may learn from the approaches taken by atmospheric scientists in dealing with effects of land surface heterogeneity that spans a wide range of scales. Giorgi and Avissar (1997) reviewed recent studies in this subject area, and suggest that these approaches may be applicable to earthquake faults with inhomogeneous frictional strength, surface roughness, porosity, and permeability that vary across a variety of length scales.

6. Transportability of the Master Model Concept to Other Regions of the World

It is true that the SCEC was possible because of the funding from the federal and local governments and other sources available in the United States, and also because of the scientific manpower available in Southern California. The concept of a master model, however, can be useful for any region with any available funding and manpower. An example is early 20th-century Japan as discussed in the introduction. If we had had a master model including Imamura's hypothesis, and had assigned a certain likelihood to his assessment of damage, and if the government official in charge of the disaster mitigation had accepted it as a consensus from the scientific community, the loss of lives in 1923 might have been reduced significantly. At present, there are many earthquake countries where only a few Imamuras exist. They can apply the master model concept by searching for all the theories and models applicable to their region and taking advantage of the knowledge about them accumulated worldwide.

The advantage of the master model concept comes from the fact that model construction may not be as expensive and time-consuming as data gathering. As mentioned earlier, the construction of the first-generation master model for Southern California was possible only after we divided the region into zones A, B, and C according to the availability of data, adopting different ways of using available data for different zones. As the data gathering proceeds, some of the C zones will be upgraded to B zones, and some of the B zones to A zones. Since the public needs advice today, we must accept the existing data and make the best use of it. This, however, should not be used as an excuse for quitting the data gathering, which is essential for the advancement of science. Keeping a proper balance between the data gathering and model construction is an important part of the master model maintenance, as explained earlier using the SCEC example.

The concept of the master model, however, will be most productive in countries with past extensive developments in earthquake studies, such as Japan, where the emphasis has been placed on multidisciplinary data gathering without much effort in synthesizing the data as a consensus of the research community. The master model concept will help to develop a quantitative model integrating the data from historical and instrumental seismology, geodesy, and geology, in the manner of the SCEC example. The model will not only give consensus and up-to-date policy documents needed by the public, but also help the scientific community to identify the focus of future research, as the SCEC identified the three most important issues for earthquakes in Southern California, namely, (1) the maximum magnitude expected in a region, (2) the fraction of aseismic strain in geodetically determined deformation, and (3) significance of the long-term change in seismic activity of a region. These issues have been discussed in Japan by individual scientists, but never brought out as a consensus focus of future study by the scientific community. Consensus building among scientists using the master model approach will quench the futile controversies that have been going on among scientists in Japan regarding earthquake prediction research. Earthquake prediction research in Japan has traditionally not included the component of strong ground motion prediction and consequently Japan has not developed a well-established bridge between the community of earthquake engineers and that of scientists, despite the effort of those working in the interface. The situation in the United States has been similar to that in Japan, but we saw some improvement at the SCEC as explained earlier.

The ability to reach a true consensus is a strength of a community, and the inability is a weakness. If the scientific community is divided, nonscientists will take the lead in research and may kill science eventually. As explained in the SCEC example, the master model concept will avoid such a disaster, giving the leadership to scientists in doing the best in creative science and, at the same time, giving the best to the public.

7. Conclusion

The SCEC successfully developed a new concept called "master model" for earthquakes in Southern California, which functions as a framework for integrating all multidisciplinary data on earthquakes in Southern California. The first-generation master model was constructed in response to the public's need for information on future earthquakes in Southern California after the Landers earthquake of 1992. It not only gave the public a consensus view of the scientific community on future earthquakes in Southern California, but also stimulated creative science through vigorous interactions among various disciplines involved in earthquake studies. The first-generation master model unifies the multidisciplinary data through its end product, namely, the PSHA (Probabilistic Seismic Hazard Analysis). Our ultimate goal, however, is to unify the data through a physical model of the Earth's lithosphere containing seismogenic faults. The SCEC made some progress toward this goal of constructing a "physical master model" both theoretically and observationally. The discovery of seismic guided waves trapped in the fault zone of the Landers earthquake of 1992 and the temporal change in their properties associated with possible healing of the fault measured in repeated controlled experiments will offer a field estimation of the parameters of the rate- and state-dependent friction law, which can be used to develop a computer model of earthquake occurrence relevant to the time scale of interest to SCEC. This deterministic approach may be used for simulating the characteristic earthquake, while noncharacteristic earthquakes may require a stochastic approach for simulation. Thus, the master model approach stimulates the cutting edge research for earthquake science. Furthermore, it is highly transportable to developing countries, where the funding and manpower for research are lacking, because a master model can make use of plausible hypotheses that may compensate for the lack of data. It is, however, most effective for countries where extensive data gathering has been done but little has been done in their synthesis for public use and for the sake of science itself.

Acknowledgment

I thank Paul Jennings, Lucy Jones, Carl Kisslinger, Leon Knopoff, and an anonymous reviewer for their advice and comments on the original manuscript, which were valuable in revising the manuscript.

The idea of forming a center for earthquake research in Southern California was first discussed at a dinner table among Jim Dieterich, Lynn Sykes, Rob Wesson, and myself during the US–Japan earthquake prediction conference at Morro Bay, California in the fall of 1988. After my initial call for a meeting to create the center failed, Ralph Archuleta, Bill Stuart, and I composed an appeal for the center, and circulated it at the AGU fall 1988 meeting in San Francisco. Rob Wesson's persuasion as the chief of the USGS Office of Earthquakes finally succeeded in organizing the Lake Arrowhead workshop in May, 1989 as mentioned in the text. During the Lake Arrowhead workshop, the name "Southern California Earthquake Research Center" was initially proposed, but Brian Tucker insisted that we drop the word "Research" from it. This turned out to be an extremely important decision as it expressed our commitment to everything about earthquakes in Southern California, including outreach and education. For two years before the official inception of SCEC, Leon Knopoff and I organized monthly seminars on earthquake prediction in which future SCEC members participated. This helped a quick development of cooperation among the members after the official start of SCEC in the beginning of 1991.

During the first site visit by NSF/USGS in 1990 a diagram of the master model drawn by Bernard Minster was used in every presentation. He served as the vice chairman of the SCEC Board and, among other things, made smooth and complete the transition to the new leadership of Tom Henyey (Center Director) and Dave Jackson (Science Director) after my resignation as Science Director in 1996.

In the workshop held two weeks after the Landers earthquake, strong opposition to the Phase 2 report was expressed by some, who felt that the Phase 1 report, a more traditional public document by scientists in response to an earthquake, was enough. Dave Jackson supported the Phase 2 vigorously, and took leadership in consensus building and writing.

Geoff Martin organized several workshops in which we explained the Phase 2 report to potential users. The subsequent wide acceptance of the report is due largely to his effort. The Advisory Council to SCEC went beyond mere advising, and participated in the growth of SCEC. Its insistence on the need for strategic planning led to the task-discipline matrix approach mentioned earlier. The Council was chaired by Barbara Romanowicz in the formative years of SCEC, who was then succeeded by John Rundle. Its members were C.B. Crouse, Jim Davis, Jim Dieterich, Paul Flores, I.M. Idriss, Tom Jordan, Shirley Mattingly, Dennis Mileti, Bill Petak, Kaye Shedlock, Bob Smith, Cheryl Tateishi, and Susan Tubessing.

Bill Petak used to tell me that "SCEC is a miracle," watching many people gathered at various SCEC meetings, who competed with each other in the past but now cooperate. The miracle occurred thanks to three incredibly timely earthquakes. The Loma Prieta earthquake of 1989 occurred at the time when the original SCEC proposal was being reviewed at NSF. The Landers earthquake of 1992, in the second year of SCEC, pulled the researchers together to develop the first-generation master model, just in time for the first renewal proposal to NSF. Finally, shortly after the Northridge earthquake of 1994, NSF notified us that the funding increase requested in our renewal proposal was approved. The last earthquake also justified our research focus given to the Los Angeles basin rather than the San Andreas fault.

Important decisions at SCEC were made by the Steering Committee, composed of the Board members representing core institutions (Kei Aki, Ralph Archuleta, Rob Clayton, Tom Heaton, Dave Jackson, Karen McNally, Bernard Minster, Jim Mori, Nano Seeber), the Working Group leaders (Duncan Agnew, Kei Aki, Ralph Archuleta, Rob Clayton, Steven Day, Egill Hauksson, Dave Jackson, Leon Knopoff, Geoff Martin, Kerry Sieh), Executive Director (Tom Henyey), Director for Knowledge Transfer (Jill Andrews), Director for Education (Curt Abdouch), and Director for Administration (John McRaney). The last four mentioned above have formed the core of the SCEC operation, implementing the decisions made by the Steering Committee to run a highly complicated organization.

Last but not least, Jim Whitcomb served as the NSF program director for SCEC.

References

Aki, K. (1980). Possibilities of seismology in the 1980s. *Bull. Seismol. Soc. Am.* **70**, 1969–1976.

Aki, K. (1988). Local site effect on ground motion. In: "Earthquake Engineering and Soil Dynamics. II: Recent Advances in Ground Motion Evaluation" (J.L. Von Thun, Ed.), American Society of Civil Engineering Geotechnical Spec. Publ., **20**, 103–155.

Aki, K. (1993). Local site effects on weak and strong ground motion. *Tectonophysics* **218**, 93–111.

Aki, K. (1996). Generation and propagation of G waves from the Niigata earthquake of June 16, 1964. Part 2. Estimation of earthquake moment, released energy, and stress-strain drop from G wave spectrum. *Bull. Earthq. Res. Inst., Tokyo Univ.* **44**, 73–88.

Aki, K. and V. Ferrazzini (2000). Seismic monitoring and modeling of an active volcano for prediction. *J. Geophys. Res.* **105**, 16617.

Aki, K. and P. Richards (1980). "Quantitative Seismology: Theory and Methods", W.H. Freeman, New York.

Bird, P. and X. Kong (1994). Computer simulations of California tectonics confirm very low strength of major faults. *Geol. Soc. Am. Bull.* **106**, 159–174.

Boore, D.M., W.B. Joyner, A.A. Oliver III, and R.A. Page (1980). Peak acceleration, velocity, and displacement from strong motion records. *Bull. Seismol. Soc. Am.* **70**, 305–321.

Brune, J.N. (1968). Seismic moment, seismicity, and rate of slip along major fault zones. *J. Geophys. Res.* **73**, 777–784.

Castle, R.O. and T.D. Gilmore (1992). A revised configuration of the southern California uplift. *Geol. Soc. Am. Bull.* **104**, 1577–1591.

Chin, B.-H. and K. Aki (1991). Simultaneous study of the source, path and site effects on strong ground motion during the Loma Prieta earthquake: a preliminary result on pervasive nonlinear site effects. *Bull. Seismol. Soc. Am.* **81**, 1859–1884.

Cornell, C.A. (1968). Engineering seismic risk analysis. *Bull. Seismol. Soc. Am.* **58**, 1583–1606.

Dieterich, J.H. (1972). Time-dependent friction in rocks. *J. Geophys. Res.* **77**, 3690–3697.

Dieterich, J.H. (1979). Modeling of rock friction, 1. Experimental results and constitutive equations. *J. Geophys. Res.* **84**, 2161–2168.

Dieterich, J.H. (1981). Constitutive properties of faults with simulated gauge. Mechanical behavior of crustal rocks. *Monograph 24* (Carter, N.L., Friedman, M., Logan, J.M., and Stearn, D.W., Ed.), *Am. Geophys. Union,* pp. 103–120.

Esteva, L. (1977). Microzoning: models and reality. *Proc. 6th World Congr. Earth. Eng.*, New Delhi.

Fedotov, S.A. (1965). Regularities of the distribution of strong earthquakes in Kamchatka, the Kurile Islands, and north-east Japan. *Tr. Inst. Fiz. Zemli Akad. Nauk SSSR* **36**, 66–93.

Gao, S., H. Liu, P.M. Davis, and L. Knopoff (1996). Localized amplification of seismic waves and correlation with damage due to the Northridge earthquake: evidence for focusing in Santa Monica. *Bull. Seismol. Soc. Am.* **86**, 209–230.

Giorgi, F. and R. Avissar (1997). Representation of heterogeneous effects in earth system modeling: experience from land surface modeling. *Rev. Geophys.* **35**, 413–438

Gutenberg, B. (1957). Effects of ground on earthquake motion. *Bull. Seismol. Soc. Am.* **47**, 221–250.

Hayashi, S., H. Tsuchida, and E. Kurata (1971). Average response spectra for various subsoil conditions. *3rd Joint Meeting, US-Japan Panel on Wind and Seismic effects*, UJNR, Tokyo, May 10–12.

Hudson, D.E. (1972). Local distribution of strong earthquake ground motions. *Bull. Seismol. Soc. Am.* **62**, 1765–1786.

Jin, A. and K. Aki (1989). Spatial and temporal correlation between coda Q and seismicity and its physical mechanism. *J. Geophys. Res.* **94**, 14041–14059.

Jin, A. and K. Aki (1993). Temporal correlation between coda Q and seismicity—Evidence for a structural unit in the brittle–ductile transition zone. *J. Geodyn.* **17**, 5–119.

Johnson, A.M., R.W. Fleming, and K.M. Cruikshank (1994). Shear zones formed along long, straight traces of fault zones during the 28 June 1992 Landers, California, earthquake. *Bull. Seismol. Soc. Am.* **84**, 499–510.

Joyner, W.B. and D.M. Boore (1981). Peak horizontal acceleration and velocity from strong motion recording including records from the 1979 Imperial Valley, California, earthquake. *Bull. Seismol. Soc. Am.* **71**, 2011–2038.

Kagan, Y.Y. and D.D. Jackson (1994). Long term probabilistic forecasting of earthquakes. *J. Geophys. Res.* **99**, 13685–13700.

Katayama, T., T. Iwasaki, and M. Saeki (1978). Statistical analysis of earthquake acceleration response spectra. *Collected Papers* **275**, Japanese Society of Civil Engineering, pp. 29–40.

Kawashima, K., K. Aizawa, and K. Takahashi (1986). Attenuation of peak ground acceleration, velocity, and displacement based on multiple regression analysis of Japanese strong motion records. *Earthq. Eng. Struct. Dyn.* **14**, 199–215.

Kostrov, B.V. (1974). Seismic moment and energy of earthquakes, and seismic flow of rock. *Izv. Acad. Sci. USSR Phys. Solid Earth* **1**, 23–40.

Kuribayashi, E., T. Iwasaki, Y. Iida, and K. Tuji (1972). Effects of seismic and subsoil conditions on earthquake response spectra. *Proc. Int. Conf. on Microzonation*, pp. 499–512.

Li, Y.G., J.E. Vidale, K. Aki, C.J. Marone, and W.H.K. Lee (1994). Fine structure of the Landers fault zone; segmentation and the rupture process. *Science* **256**, 367–370.

Li, Y.G., J.E. Vidale, K. Aki, F. Xu, and T. Burdette (1998). Evidence of shallow fault zone strengthening after the 1992 M7.5 Landers, California, earthquake. *Science* **279**, 217–219.

Marone, C., J.E. Vidale, and W.L. Ellsworth (1995). Fault healing inferred from time dependent variations in source properties of repeating earthquakes. *Geophys. Res. Lett.* **22**, 3095–3098.

Mohraz, B. (1976). A study of earthquake response spectra for different geologic conditions. *Bull. Seismol. Soc. Am.* **66**, 915–935.

Ouyang, H. (1997). Spatial and temporal characteristics of coda Q as a tectonic parameter in Southern California. Ph.D. Thesis, University of Southern California, Los Angeles, CA.

Papageorgiou, A.S. and K. Aki (1983). A specific barrier model for the quantitative description of inhomogeneous faulting and the prediction of strong motion. Part 2, Applications of the model. *Bull. Seismol. Soc. Am.* **73**, 953–978.

Rice, J.R. (1993). Spatio-temporal complexity of slip on a fault. *J. Geophys. Res.* **98**, 9885–9908.

Ruina, A.L. (1983). Slip instability and state variable friction laws. *J. Geophys. Res.* **88**, 10359–10370.

Schwartz, D.P. and K.J. Coppersmith (1984). Fault behavior and characteristic earthquakes: examples from the Wasatch and San Andreas fault zones. *J. Geophys. Res.* **89**, 5681–5698.

Seed, H.B., C. Ugas, and J. Lysmer (1976). Site-dependent spectra for earthquake-resistant design. *Bull. Seismol. Soc. Am.* **66**, 221–243.

Su, F. and K. Aki (1995). Site amplification factors in Central and Southern California determined from coda waves. *Bull. Seismol. Soc. Am.* **85**, 452–466.

Trifunac, M.D. (1976). Preliminary empirical model for scaling Fourier amplitude spectra of strong ground acceleration in terms of earthquake magnitude, source-to-station distance and recording site condition. *Bull. Seismol. Soc. Am.* **66**, 1343–1373.

Trifunac, M.D. (1976). Preliminary analysis of the peaks of strong ground motion dependence on earthquake magnitude, epicentral distance and recording site conditions. *Bull. Seismol. Soc. Am.* **66**, 18–219.

Ward, S.N. (1994). A multidisciplinary approach to seismic hazard in Southern California. *Bull. Seismol. Soc. Am.* **84**, 1293–1309.

WGCEP (Working Group on California Earthquake Probabilities) (1988). Probabilities of large earthquakes occurring in California on the San Andreas fault. *U.S. Geol. Surv. Open-File Rept.* **88**, 398.

WGCEP (Working Group on California Earthquake Probabilities). (1995). Seismic hazards in southern California: probable earthquakes, 1994–2014. *Bull. Seismol. Soc. Am.* **85**, 379–439.

6

Continental Drift, Sea-Floor Spreading, and Plate/Plume Tectonics

Seiya Uyeda
International Frontier Research Group on Earthquakes, Riken, Japan

1. Introduction

The study of the solid Earth began with land geology at the time of the industrial revolution about 200 years ago. With a wealth of knowledge on orogenic zones like the Alps and Appalachians available early, grand concepts were developed like those of geosynclines. From around 1900 onward, with the aid of physics and new technologies, geophysical research such as geodesy and seismology began to flourish, and the principle of isostasy and views on the Earth's major internal layered structure were established. Despite all this and its long history, however, geodynamics remained a highly speculative subject (Scheidegger, 1963).

The theory of continental drift proposed by Alfred Wegener in 1912 was largely discredited until paleomagnetic research (e.g., Runcorn, 1962) accomplished its dramatic revival after World War II. At almost the same time in the late 1950s, marine geology and geophysics (marine G&G) were established as disciplines and led rapidly to important discoveries on the sea floor. These discoveries in turn led to new ideas, such as the world-encircling oceanic rift system (Ewing and Heezen, 1956) and the sea-floor spreading hypothesis (Dietz, 1961; Hess, 1962; Vine and Matthews, 1963). The late 1960s brought their glorious verification.

A further turning point was the proposal of transform faults by Wilson (1965a) and of plate tectonics. Morgan (1968), McKenzie and Parker (1967), and Le Pichon (1968) worked out the fundamental part of the kinetic theory of plate tectonics. In a narrow sense, plate tectonics postulates that the Earth's surface is divided into ten or so blocks moving in large-scale horizontal motions with little deformation (see figure 8 of Chapter 7 by Stein and Klosko). If plates are moving without deformation, their motion can be described by rotation around a pole, called the Euler pole (see also Chapter 7). There are three main types of plate boundaries: divergent, convergent, and sliding boundaries, representing spreading ocean ridges, subducting and colliding zones, and transform faults, respectively. There are two kinds of ocean–continent boundaries: the active and the passive margins. An active margin is a (convergent) plate boundary, whereas a passive margin is not a plate boundary. Building on these postulates, plate tectonics in the broader sense provided, for the first time, a unified explanation of major geologic phenomena in terms of plate formation, evolution, and interaction (e.g., Dewey and Bird, 1970). Wilson (1968c) stressed that what was going on in the 1960s was a scientific revolution or paradigm change (Kuhn, 1962).

However, plate tectonics explains only the orogenic phenomena that have occurred in the superficial parts of the Earth during its later history. To understand the long-term changes in orogenic characteristics, an understanding of the mantle dynamics underneath the lithospheric plates was needed. Based mainly on the deep seismic tomography that has indicated the existence of possible upwelling superplumes beneath the South Pacific and Africa and a downwelling superplume beneath central Asia, the so-called plume tectonics has been postulated (e.g., Maruyama, 1994). Now, at the threshold to the 21st century, a changing view of the Earth from plate tectonics towards plume tectonics seems to be well underway. In the following, a brief attempt will be made to reflect how these views of the Earth have been developed. References are not exhaustive but often limited to reviews and textbooks, through which more in-depth information and original references can be traced. Examples are Wyllie (1971), Bird and Isacks (1972), Le Pichon *et al.* (1973), Cox (1973), Uyeda (1978), Bird (1980), Turcotte and Schubert (1982), Cox and Hart (1986), Allegre (1988), Kearey and Vine (1990), Sullivan (1991), and Sleep and Fujita (1997).

INTERNATIONAL HANDBOOK OF EARTHQUAKE AND ENGINEERING SEISMOLOGY, VOLUME 81A ISBN: 0-12-440652-1

2. Continental Drift Theory—Its Birth, Death, and Revival

2.1 Wegener's Theory

Alfred Wegener (1880–1930) was not the first to notice that continental outlines fit together like the pieces of a jigsaw puzzle (Fig. 1), but it was he who made this idea part of modern earth science. His major work, *Die Entstehung der Kontinente und Ozeane* in 1915, went through three revisions before his death (Wegener, 1929). Probably because Wegener was basically a meteorologist, many of his arguments related to the solid Earth are erroneous from the perspective of present-day knowledge, yet it was his book that broke down the wall between geophysics and geology. It makes fascinating reading even today.

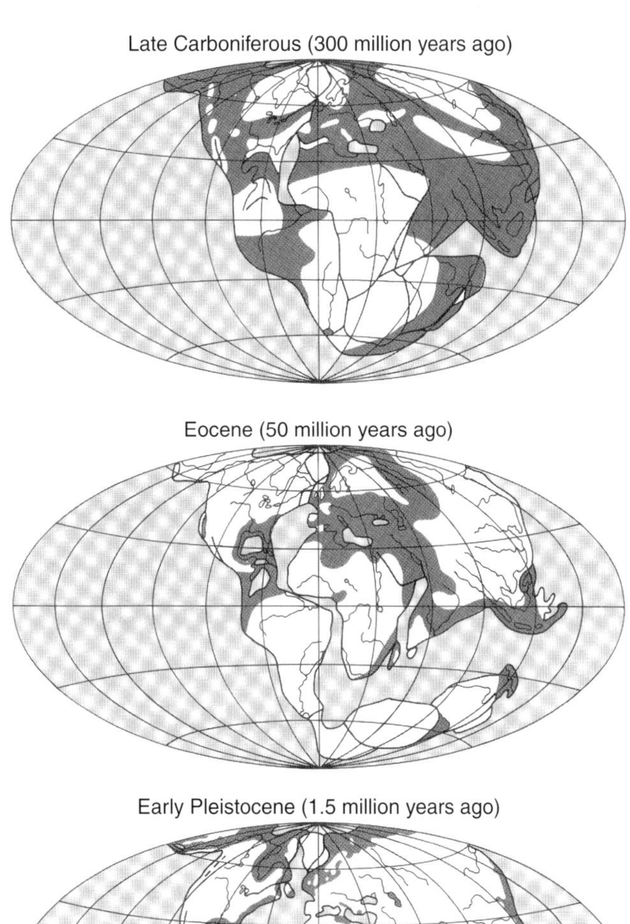

FIGURE 1 Reconstruction of the map of the world for three periods according to Wegener's theory of continental drift. The dark-shaded areas represent shallow seas.

In addition to fitting the "pieces of the puzzle," Wegener used a variety of arguments to demonstrate a past linkage of continents that are now separated. Indeed, geologists had found supporting evidence of past connection between far-apart continents. This included fossils of animals that could not have crossed vast oceans and clear signs of the Permo-Carboniferous glaciation in different continents. The Austrian geologist, Eduard Suess (1831–1914) had already given the name "Gondwanaland" to the imaginary protocontinent in the Southern Hemisphere. However, geologists simply assumed that continents now separated by oceans had once been connected by land bridges. Wegener refuted this idea, arguing that the principle of isostasy would not allow for the submergence of such vast land bridges.

Furthermore, Wegener held great insight into orogenesis, or the mountain building process, which is one of the most important problems in solid earth science. In his time, the popular view about the folded structure seen in great mountain ranges was to assume that the Earth had contracted as it cooled. Wegener rejected this view because it could not be quantitatively justified. Instead, largely referring to the then pioneering works of Swiss geologist Emile Argand (1879–1940) on large mountain ranges such as the Alps and the Himalayas, he contended that they were created by collisions of continents.

Wegener's theory of continental drift unleashed a heated controversy. It could not, however, overturn the belief in the "fixist" (static) Earth that was prevalent. After Wegener's tragic death during his 1930 Greenland expedition, his theory went into decline and was almost completely forgotten by the 1940s. The main objection had been that the forces that allegedly moved the continents could not be explained. The forces proposed during Wegener's days included tidal forces exerted by the Moon, and the so-called "*Polfluchtkraft* (centrifugal force from the poles)." However, the British theoretical geophysicist Sir Harold Jeffreys (1891–1989) demonstrated that these forces are many orders of magnitude smaller than those required to move continents. Wegener himself lamented that "the Newton of the Continental Drift Theory is yet to arrive." The mantle convection theory proposed by another British scientist, Arthur Holmes (1890–1965) as early as in 1929 provided the only hope (see, e.g., Holmes and Holmes, 1978), but Holmes was a minority of one.

2.2 Paleomagnetism and the Revival of Continental Drift Theory

The main geomagnetic field can be approximated by a geocentric magnetic dipole and several higher spherical harmonics, representing regional nondipole magnetic fields. They are caused by electric currents within the Earth's fluid outer core and exhibit secular variations, while local geomagnetic anomalies are caused by the magnetization of crustal rocks. Paleomagnetism attempts to reconstruct the geomagnetic fields of the past by studying the natural remanent magnetism (NRM) of rocks. The NRM can preserve a record of the geomagnetic vector present at the time when the rocks formed (e.g., Nagata, 1961). Paleomagnetic research was pioneered in the 1950s by British and American scientists (e.g., Irving, 1964). Their main accomplishment was the establishment of (1) geomagnetic reversals, (2) geomagnetic polar wandering, and (3) continental drift.

It had been known since the early 20th century that the NRM of some Cenozoic igneous rocks is oriented opposite to the present geomagnetic field. On the basis of these observations, scientists began considering the possibility that the geomagnetic field has occasionally reversed its polarity in the geologic past. To test this hypothesis, Cox and others launched vigorous study of geologically young igneous rocks all over the world. Their results indicated that there were as many as four reversals during the last 5 My (Cox, 1973). Each polarity epoch, named after the pioneers of geomagnetic research, lasted for about 10^6 years. Within a polarity epoch are a few shorter (less than 10^5 years) polarity events each identified by the region where first found. Ever since the Vine–Matthew–Morley hypothesis in 1963 used the geomagnetic reversals to explain the stripe-patterned geomagnetic anomalies of the ocean floors, the concept of geomagnetic reversal has played a major, albeit unexpected, role in the plate tectonic revolution. Today, the history of geomagnetic reversals stretches back almost 190 My (see Section 3).

In a paleomagnetic study, the past position of the geomagnetic pole can be calculated from the direction of NRM with the assumption that the Earth's magnetic field has always been dipolar. The pole position thus obtained is called a "virtual" geomagnetic pole (VGP). The geomagnetic pole after averaging the fluctuation due to secular variations is called a paleomagnetic pole. It is the latter pole that is important in our discussion. The paleomagnetic poles of the Cenozoic Period obtained from all continents are concentrated around the present geographic pole. For older ages, paleomagnetic poles systematically deviate from this position. The locus of the paleomagnetic pole is called the apparent polar wander (APW) path. It is "apparent" because the wander is relative to the continent or plate from where the rocks were sampled. Figure 2a shows the APW paths for the European and North American continents with the assumption that continents did not move. If the geomagnetic field had always been dipolar and if the continents had not drifted, the two APW paths should coincide. But it is obvious in the figure that they are two distinct curves. On turning the North American continent counterclockwise around the North Pole so that the Atlantic Ocean closes, as in Figure 2b, the APW paths overlap for a long period from the Silurian (Paleozoic) through the Triassic (Mesozoic) Period. This agrees with the assumption that Pangea was one continent during that period, and that the drifting apart of the North American and European continents began in the late Triassic–Jurassic Period, or thereabouts, when the two paths began to deviate.

It was mainly Runcorn, Irving, and other British scientists of the 1950s who provided, with these results, the most

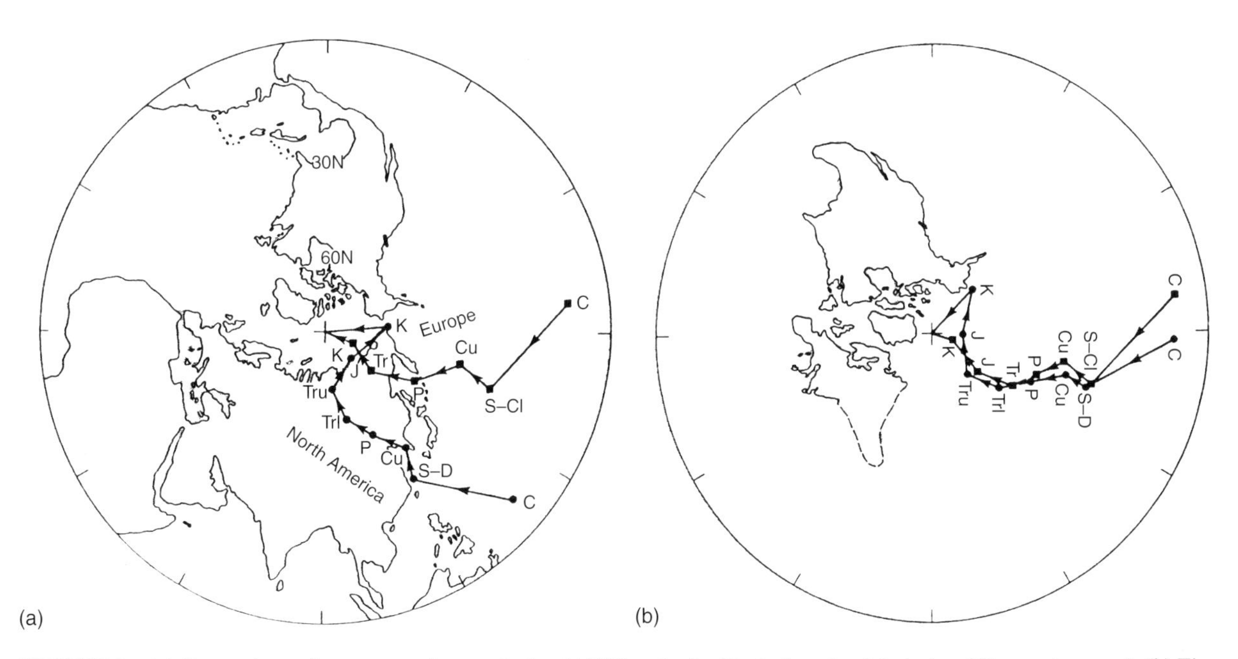

FIGURE 2 (a) Comparison of apparent polar wandering (APW) paths for North America (circles) and Europe (squares). (b) The two APW paths after closing the Atlantic Ocean. K, Cretaceous; J, Jurassic; Tr, Triassic; Tru, Upper Triassic; Trl, Lower Triassic; P, Permian; Cu, Upper Carboniferous; S–D, Silurian–Devonian; S–Cl, Silurian–Lower Carboniferous; C, Cambrian.

significant evidence for the continental drift since Wegener's time. Thus by the end of the 1950s the continental drift theory, which had seemed all but dead for decades, was suddenly revived in the geoscience community, especially in the British paleomagnetic community. It was, however, a far from unanimous acceptance even in the United Kingdom. When the present author visited there in 1958, he was constantly asked, "Do you believe in continental drift?" Many of those who posed this question at that time could have been half-doubters. It took some years for the revival of the continental drift to be globally accepted and become the principal topic in earth science. For instance, Hurley (1968) at MIT demonstrated that the age distribution of ancient rocks in South America and Africa coincides with the Pangea reconstruction obtained by the use of the then newly introduced electronic computer at Cambridge University (Bullard *et al.*, 1965).

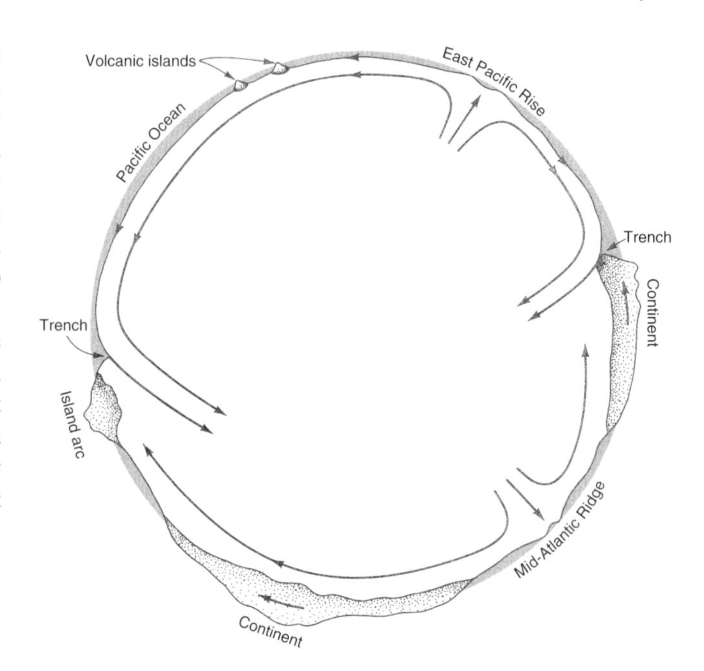

FIGURE 3 Schematic cross-section of the Earth based on the sea-floor spreading hypothesis.

3. Sea-Floor Spreading Hypothesis and its Verification

3.1 Hess–Dietz Proposal

With unexpected help from paleomagnetism in the late 1950s, the continental drift theory, little by little, gained wide acceptance. As for the driving force for the motion of continents, mantle convection arose as a promising possibility. At about the same time, as mentioned in the Introduction, marine geology and geophysics (marine G&G) became very active and many new discoveries began to emerge. Based on these marine surveys, Ewing and Heezen (1956) at Lamont Geological Observatory introduced the idea of world-encircling rift system (e.g., Heezen, 1960). A breakthrough came when Hess (1962) and Dietz (1961) introduced the insightful hypothesis of sea-floor spreading. Hess focused his attention on the facts about the mid-oceanic ridges then coming to light, such as the high heat flow, the median rift valley, the shallow earthquake activity, and the presence of anomalously low seismic wave velocity in the upper mantle. Hess proposed that when mantle convection rises under a mid-oceanic ridge, it produces new oceanic crust, which then divides and migrates to either side of the ridge (Fig. 3). When mantle convection rises under a continent, the continent will split and drift "passively." New sea floor is produced in the wakes of the continental fragments. Where the leading edge of the moving continent collides with a mantle current, the mantle subducts and an oceanic trench is formed. Since the continents are too buoyant to sink, they are deformed and produce a mountain range when they collide.

For the first time, Hess' theory explained why the thickness of the ocean sediment was one order of magnitude thinner than the estimate obtained by multiplying the age of the sea by the sedimentation rate. It also explained why rock older than Cretaceous had never been discovered on the sea floor. These facts were no longer mysterious; the entire sea floor sinks into the mantle in a matter of 200–300 My with new sea floor constantly replacing it. Dietz's paper (1961), in which he coined the term "sea-floor spreading" was not as geopoetic, but was even more convincing. He was already proposing, for example, that a hard lithosphere approximately 70 km thick lay over a softer asthenosphere.

As can be guessed from the remark by Hess in the foreword of his paper, "I should consider this paper an essay in geopoetry," the whole idea was quite remote from the common sense in the geoscience community. Those geoscientists with mobilist views welcomed these papers, but the wider majority, for whom even to accept the continental drift theory was sufficiently risky, were skeptical. Researchers of the day would greet one another with, "Do you believe in sea-floor spreading?" Actually, Holmes had suggested a similar idea back in 1929 as mentioned earlier, but little attention was paid to it. As in the case of Wegener's continental drift theory, scientists usually distrust innovative ideas.

3.2 Supporting Evidence

With the appearance of the sea-floor spreading hypothesis, solid earth science of the late 1960s was abuzz with discussion of it. Findings from marine geology and geophysics were generally consistent with the hypothesis, but some reported that their findings were not.

Oceanic ridges and trenches are in contrast with regard to many features. The former sit in the middle of the ocean, while the latter are found on the margins. The seismic activity (see Chapter 7 by Stein and Klosko) directly below oceanic ridge axis is limited to small magnitude ($M < 6$) shallow ($d < 10$ km) earthquakes, apparently caused by the tensile stress associated with sea-floor spreading. In contrast, earthquakes at trench areas include shallow low-angle thrust types (including the $M > 8$ class great earthquakes) and deep-focus earthquakes, associated with the subduction of the sea floor. The contrast in heat flow (see Chapter 81.4 by Cermak), which is anomalously high over the mid-oceanic ridges, and low in the trench areas, is also supportive of the hypothesis.

By far the most decisive support of sea-floor spreading was provided by the geomagnetic anomalies of the sea floor. The development of electronic devices during the 1950s made it remarkably easy to make continuous and accurate measurements of the total geomagnetic force at sea. Armed with new tools, scientists, mainly from the Scripps Institution of Oceanography of the University of California, led by V. Vacquier, began extensive surface ship geomagnetic surveys over the eastern Pacific, and made a startling discovery. As demonstrated by Raff and Mason (1961), for instance, there are striking zebra-striped patterns of geomagnetic anomalies, with widths of 20–30 km and lengths of several hundred kilometers in the northeastern Pacific off the coast of North America. And that was not all. Here and there across topographic discontinuities called fracture zones, the stripes showed displacements as great as many tens of kilometers. Such strange geomagnetic anomalies have never been seen on land. Vine and Matthews (1963) provided an intriguing answer to the question of the origin of these striped patterns. If the sea floor spreads, the newly produced sea floor would be magnetized parallel to the geomagnetic field of the time. If at the same time the geomagnetic field keeps reversing its polarity, the sea floor would inevitably be magnetized with positive and negative stripes. This model, which can be likened to a magnetic tape recorder, has proven to provide the most powerful time frame for plate tectonics and today leaves little room for doubt. But it was regarded as extreme and too speculative when it was first proposed. To be exact, it would better be called the Vine–Matthews–Morley Hypothesis, including L.W. Morley from Canada, who submitted a paper with almost identical content to scientific journals independently, but was unable to get it published. As this episode tells, the tape recorder model was initially not taken seriously by the general geophysical community.

However, there was a real breakthrough at about 1965 (Vine, 1968). Assuming that the crust is magnetized alternately in stripes, the magnetic field on the ocean surface can be calculated. In early 1965, Pitman obtained a high-quality magnetic profile, the famous Eltanin-19 profile, over the East Pacific Rise south of Easter Island. It was then demonstrated that the pattern of the geomagnetic intensity anomalies on either side of the Rise agreed perfectly with the model based on the geomagnetic reversal history that had been obtained from the paleomagnetic studies of rocks on land (Pitman and Heirtzler, 1966). Vine and Wilson (1965) demonstrated the same for the Juan de Fuca Ridge off Vancouver Island. For instance, Figure 4 shows the remarkable agreement between calculated and observed geomagnetic field anomaly profiles of the Pacific–Antarctic Ridge. In the meantime, the geomagnetic reversal history was pursued in a different way (e.g., Opdyke, 1972). The history of the geomagnetic field for several million years could be continuously analyzed by looking at the magnetization of some 10 m of core samples from sea floor—since the sedimentation rate on the deep sea floor is extremely slow. Their results also quantitatively agreed with the reversal history based on volcanic rocks on land. Thus, identical results were obtained from three independent sources—volcanic rocks on land, deep-sea sediments, and the marine magnetic anomalies. This "trinity" was irrefutable evidence not only of geomagnetic reversals but also of sea-floor spreading.

By the time the Vine–Matthews–Morley hypothesis was established for the past four or so million years, attempts to extend this hypothesis to older sea floors were already underway by Heirtzler and others of the Lamont-Doherty

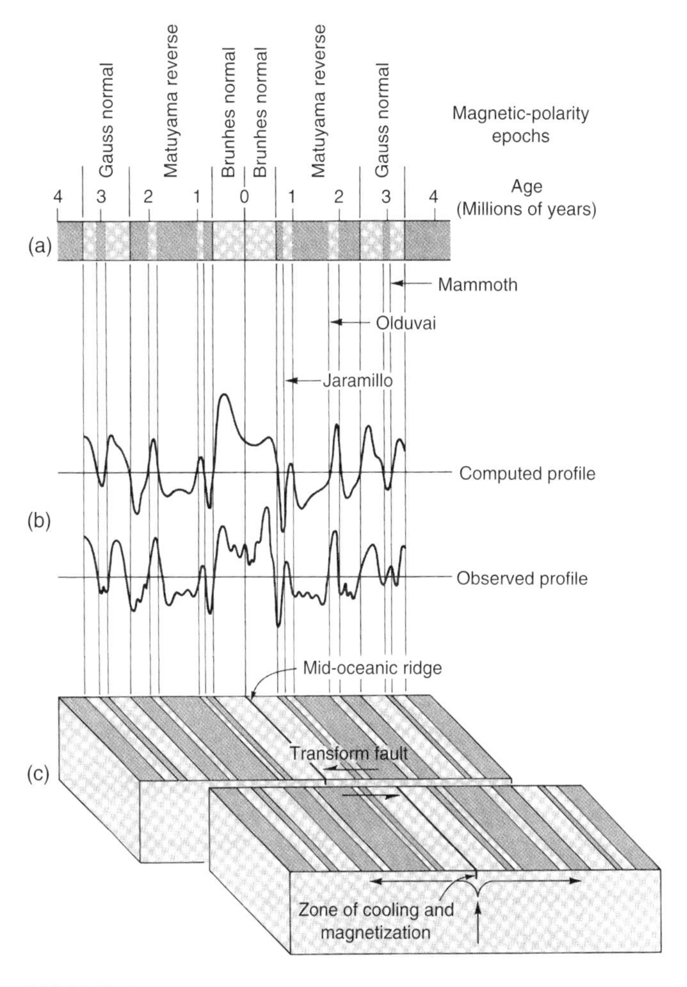

FIGURE 4 The magnetic tape recorder model. (a) Geomagnetic reversal history according to Cox (1973). (b) Comparison of computed and observed anomaly profiles for Eltanin-19 according to Vine (1966). (c) The pattern of normal and reverse magnetizations of oceanic crust with a transform fault.

Geological Observatory of Columbia University. Such research was possible only at that institution, thanks to the enormous amount of surface ship geomagnetic data that had been amassed under the leadership of Maurice Ewing. They analyzed the long and numerous geomagnetic profiles for each of the world oceans and, by 1968, had demonstrated that magnetic stripe patterns in different oceans are correlated as far back to some 80 My old sea floor (Heirtzler *et al.*, 1968).

As each stripe represents sea floor produced at the same time, it is called an isochron or chron. Some prominent isochrons were given numbers. The problem they had was that the ages of isochrons were not known for the period older than 4.5 My. Heirtzler and his colleagues, however, extrapolated the age by assuming that the sea-floor spreading rate of the South Atlantic was constant. Their assumption was proved to be correct by the DSDP (Deep Sea Drilling Project) in 1968 (Maxwell and von Herzen, 1969). In this way, a magnetic lineation map consisting of isochrons was translated into a map showing the age of the ocean floor. Thus, by about 1974, the age of almost the entire world's ocean floor was determined (Fig. 5; Pitman *et al.*, 1974).

There are several features to be noted in relation to Figure 5. (1) The oldest sea floor of Jurassic age lies in an area east of the Mariana Arc in the western Pacific. The present Pacific Plate was born there about 190 Ma (Hilde *et al.*, 1977). Of course, this does not mean that the Pacific Ocean at that time was zero in area. On the contrary, the Ocean was much larger but was occupied by other, now subducted, plates. As the present Pacific Plate grew, the spreading ridge (East Pacific Ridge, EPR) migrated eastward and the sea floor produced in the eastern side of the ridge was continuously subducted under North America. Finally, the ridge itself collided with North America and was subducted. The collision of ridges with trenches, however, was rather puzzling from the view that regards oceanic ridges as upwelling zones and oceanic trenches as downwelling zones of mantle convection. How can an upwelling flow go down? Beloussov (1979) used this apparent contradiction to attack the "new view of the Earth." But the observation does require the subduction of spreading ridges, with some important implications. (2) Although not shown in Figure 5, the vast area shown as Cretaceous age is characterized by absence of striped magnetic anomaly lineations. It is therefore called the Magnetic Quiet Zone (MQZ). This vast sea floor is interpreted to have been produced during an abnormally long absence of geomagnetic reversals. Taking into consideration the width of the MQZ and its age span, the rate of sea-floor spreading seems to have been 50–75% faster during the Cretaceous Period (124–83 Ma) than during other periods. Larson and Pitman (1972) called this phenomenon a pulse. When spreading is faster, the area with young age increases. Given the fact that the sea depth increases with age, they proposed that the sea floor must have risen during the pulse period, causing the Cretaceous "transgression of the sea," a long-known fact in geology. This pulse of spreading, with a long period of no geomagnetic reversal, was speculated as related to mantle superplume activity (Larson, 1991). This was the dawn of plume tectonics.

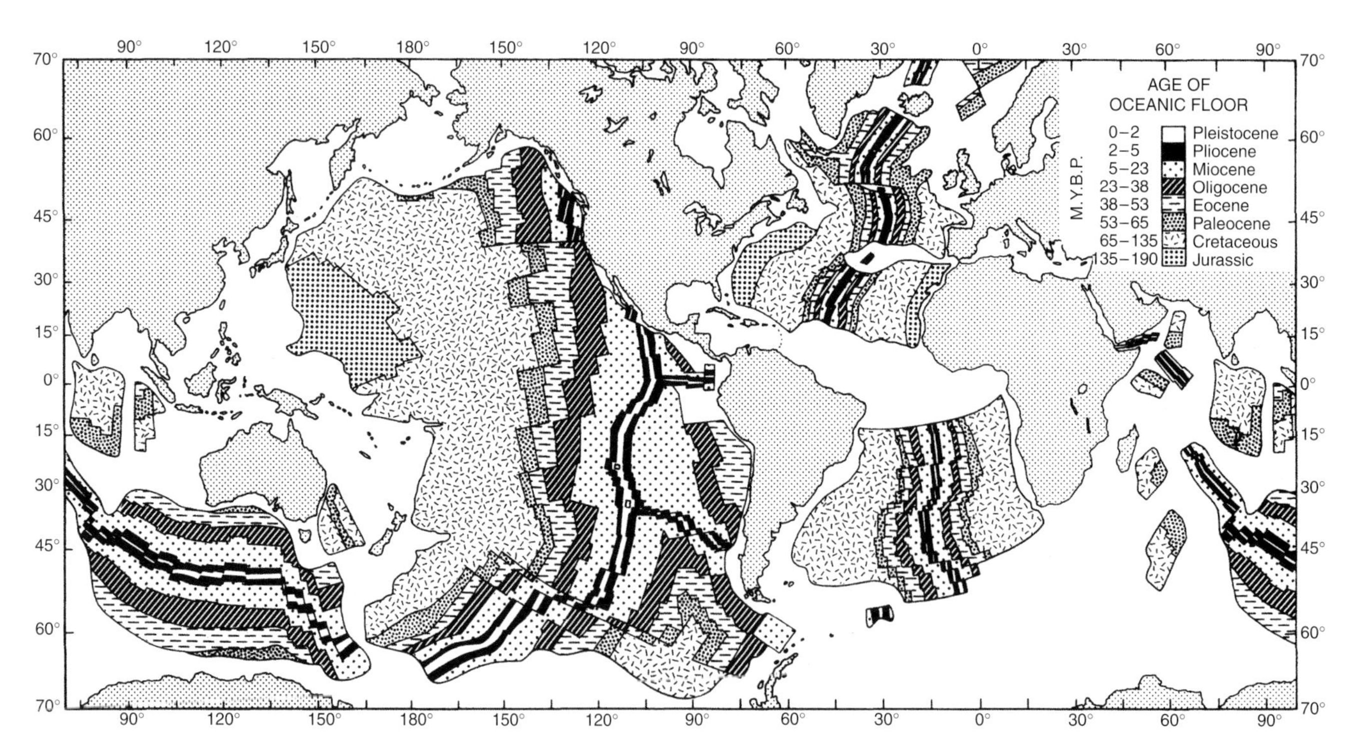

FIGURE 5 Age of sea floor as established by W.C. Pittman and others at Lamont in 1974.

4. Advent of Plate Tectonics

4.1 Transform Faults

As mentioned in Section 3, geomagnetic stripes are displaced many tens of kilometers (sometimes as much as 1400 km) at many places across fracture zones. According to the idea of transform faults (Wilson, 1965a), the offsets do not represent relative motions between crustal blocks. Wilson, visiting Cambridge University at the time, is said to have demonstrated his elegant new idea by using a simple paper cutout. As a typical example, let us consider the case shown in Figure 6b. This kind of fault is often seen on mid-oceanic ridges. If the spreading rate is the same on both sides of ridge, we can see that (1) the relative motion occurs only between the two oceanic ridges (b–b′) and not along the fracture zones (b–f, b′–f′); (2) the direction of the slip ("left-lateral" in this case) between b and b′ will be opposite to the direction of the displacement of the ridge, i.e., of the magnetic stripe pattern. In an ordinary transcurrent fault situation, as in Figure 6a, neither of the above would hold. Sykes (1967) convincingly demonstrated, through seismological means, that both (1) and (2) actually occur (see Chapter 7 by Stein and Klosko; Isacks *et al.*, 1968). This new type of fault—the transform fault—cannot be understood without the concepts of sea-floor spreading and rigid tectonic plates. But if one accepts these concepts, transform faults are an inevitable consequence. In active transcurrent faults (Fig. 6a), the displacement B–B′ increases with time. In the transform faults (Fig. 6b), on the other hand, the ridge displacement b–b′ does not change with time, implying that the offset has been there from the beginning. This may sound somewhat strange, but Wilson argued that the crust may have cracked at the time of the continental split along some "weak zones" that had already existed, and that the form of the cracks was later reflected in the displacement of the oceanic ridge.

Wilson further proposed that the mobile zones, such as oceanic ridges, and orogenic belts, were not isolated features, but that they form a network that divides the Earth's outermost layer into blocks—the plates. Transform faults play a key role in this network, by transforming, for example, a segment of a ridge to another kind of plate boundary. This realization was truly the beginning of plate tectonics. Other types of transform faults, such as those connecting an oceanic ridge and a trench, or two oceanic trenches, are also possible. Depending on the type of transform faults, the displacements, i.e., the length of faults, can remain constant as in the case of Figure 6b, or increase or decrease with time.

A remarkable example of transform faults on land is the San Andreas Fault in California. It was long known that among the active zones of the circum-Pacific Belt, the western coast of North America alone lacked oceanic trenches and deep-focus earthquakes. Instead, it had what looked like an enormous transcurrent fault and accompanying "right-lateral" earthquakes. The cause for these anomalous characters of the western coast of North America was totally unknown. The mystery can be neatly explained, however, when we consider a scenario in which East Pacific Rise becomes a transform fault as it enters the Gulf of California from the south (Fig. 7). If this is true, there should be a spreading ridge at the northwestern end of the San Andreas Fault. Sure enough, Juan de Fuca and Gorda Ridges were found exactly where expected. With the introduction of the transform fault, the stage for plate tectonics was ready.

4.2 Rigid Body Rotation

Using Euler's Fixed Point Theorem, one can conveniently express the displacement of a plate by a rotation around its Euler pole (see Chapter 7 by Stein and Klosko). To describe the motion of, say, the North American Plate, one does not have to specify how many kilometers San Francisco moved in which direction, or what happened to New York, etc. Let us write the rotation of Plate A relative to Plate B as ARB, its Euler pole as APB, and relative velocity and relative angular velocity as AVB, and $A\omega B$. When plate tectonics was introduced, its goal was to substantiate the supposed rigid motion of plates. For this purpose, the proof that actual plate motions satisfied Euler's theorem was sufficient. Consider the case of two plates A and B

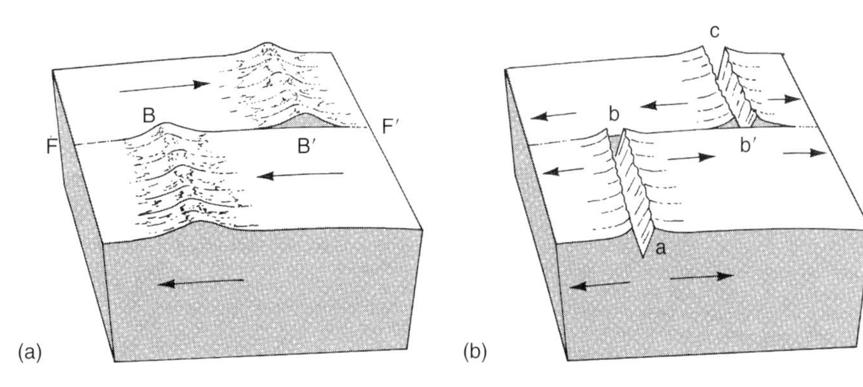

FIGURE 6 (a) Ordinary transcurrent fault. (b) Transform fault.

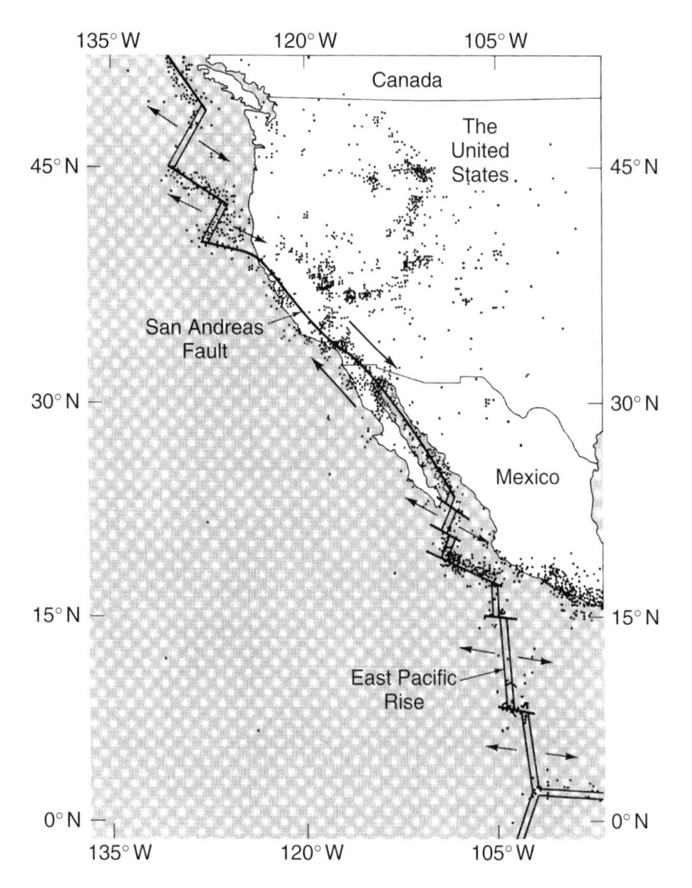

FIGURE 7 The San Andreas Fault as a transform fault. The double lines are the spreading ridges; dots are earthquake epicenters.

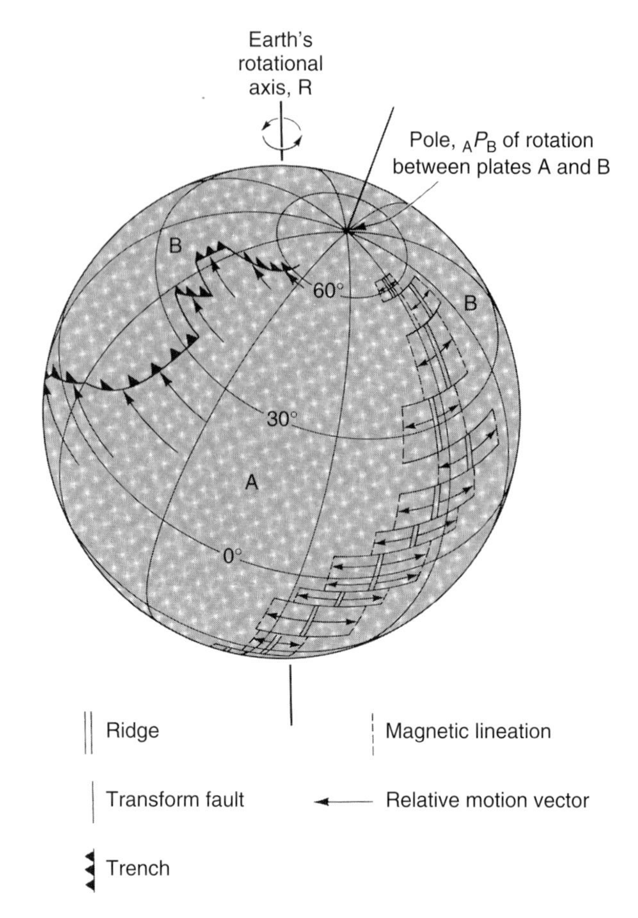

FIGURE 8 Relative motion of plates A and B, represented by a rotation around the Euler pole.

(Fig. 8). The plate tectonic theory predicts the following: (1) The direction of the relative motion between A and B, i.e., the direction of the transform fault, *should* be parallel to the small circle on the globe with A*P*B as its pole; (2) the magnitude of the relative velocity *should* be in proportion to sin(q), where q *is* the angular distance from A*P*B. A more intuitive picture would be a world map using a Mercator projection, but with the Euler pole A*P*B as its pole. We see that the direction of the relative motion (i.e., the transform fault) is parallel to the lines of "latitude," and the length of the vector that represents the relative velocity is constant regardless of "latitude." Confirmation of these predictions was the starting point of plate tectonics.

McKenzie and Parker (1967) investigated the relative motion of the Pacific Plate and the North American Plate, and saw that the projections onto the horizontal plane of slip vectors of several earthquakes at the boundaries of these two plates were nearly parallel to the lines of "latitude" in a Mercator projection with Euler pole at 50°N, 85°W. The Euler pole position was determined from the directions of San Andreas Fault and the average slip vectors of aftershocks of the 1964 great Alaskan earthquake in the Kodiak Island region of the Aleutians.

Morgan (1968) began independently with the same concept, but expanded it to encompass the global view. He took into consideration not only slip vectors of earthquakes and the directions of transform faults, but also the distribution of the spreading rate estimated from the width of geomagnetic stripes. For example, he determined the Euler pole of the relative motion between the American and African Plates (1) from the points of intersection of great circles perpendicular to both transform faults and slip vectors of earthquakes associated with the Mid-Atlantic Ridge and (2) by matching the distribution of spreading rates with the relations of sin(q). These independent estimates agreed well.

Le Pichon (1968) was inspired by Morgan's work and enthusiastically proceeded to undertake a systematic verification of the plate theory. He divided the Earth's surface into six large plates, and tried to find, by the method of least squares, the set of plate motions that would best fit with both the width of geomagnetic stripes and the strikes of the transform faults. He then drew a Mercator projection with the Euler pole thus obtained as its pole, and demonstrated that the two "*shoulds*" stated above were satisfied.

A natural extension of the plate tectonic methodology is its use for quantitative reconstruction of continents and ocean basins in the geologic past. Where dated sea-floor spreading magnetic lineation patterns are available on both sides of a ridge, the part of the sea floor between the ridge and an isochron can be resorbed by appropriately rotating the plates and this gives past positions of the plates (continents and ocean basins). Le Pichon (1968) pioneered this type of reconstruction, which has been followed by many, including McKenzie and Sclater (1971) for the Indian Ocean and Pitman and Talwani (1972) for the Atlantic Ocean. Where subduction plays major role, namely, the Pacific Ocean, it is not possible to estimate the past position of continents from magnetic lineations alone, but the evolution of the Pacific Ocean itself and its margin can be discussed (Hilde *et al.*, 1977).

4.3 Triple Junctions

A closed surface cut into more than two plates will inevitably produce a point where three plates meet, called a triple junction. Triple junctions are endowed with many interesting characteristics (McKenzie and Morgan, 1969). For example, some triple junctions maintain their configurations indefinitely, while others can exist for only an instant in time. They are called, respectively, stable and unstable triple junctions. Let us first look at the example seen in the bottom right corner of Figure 7, where three spreading ridges meet. Such a triple junction, called an "R-R-R" type, is stable because it can persist without causing any change in plate configuration. Consider next a "T-T-T" triple junction, consisting of three trenches (Fig. 9a). Take plate A as the reference and consider the motions of plates B and C relative to A. The arrows in this figure represent the relative motion vectors across the trenches. Assume that at a trench, the plates without tooth marks subduct beneath the plates which have tooth marks. After a certain length of time, the leading edges of Plates B and C will reach the positions marked by the broken lines in Figure 9b, causing changes in the configuration of plates around the triple junction: namely J_1 becomes J_2. This change means that J_1 was an unstable triple junction. But as J_2 continues to move north along the trench AB, the plate configuration no longer changes. The triple junction has stabilized.

Another interesting point is that an observer at point P, for instance, watching Plate B subducting beneath Plate A (on which he stands), will see, after a few moments in geologic time, that he is now watching Plate C subducting in front of him. Which plate is subducting from which direction is a matter of importance for the tectonics of the region, yet this drastic change—from Plate B to C—can occur while there is no change in the motion of the three plates. As plate boundaries also include transform faults (F), various combinations of R, T, and F can be considered as types of triple junctions. McKenzie and Morgan (1969) made an elegant general treatment of the stability of triple junctions using the velocity space presentation. Their paper also demonstrated that the concept of the triple junction was more than a theoretical game by applying it to explain how the San Andreas Fault came to exist.

4.4 Relative and Absolute Plate Motions

In plate tectonics, clearly definable motion is the relative motion (rotation) between two adjacent plates. The major data sources are the width of magnetic lineations and directions of transform faults and slip vectors of interplate earthquakes. Except for the last item, the data represent the features concerning geologic time scale of millions of years. Therefore, it was not certain how plates are actually moving today in a human time scale. However, recent dramatic developments in the so-called space geodesy, notably the Global Positioning System (GPS), has proven that plates are moving almost exactly as plate tectonics postulates. GPS is actually bringing about a revolutionary progress in earth science. The beauty of GPS is its extreme accuracy for its simplicity and inexpensiveness, as laymen can easily realize through automobile navigation systems.

The plate motion models have been improved over the last decades, the most recent being NUVEL-1A (Fig. 10a; DeMets *et al.*, 1994). If there were points fixed to the main body of the Earth, motion relative to them could be termed absolute motion. As the candidate of such points, hot spots like Hawaii

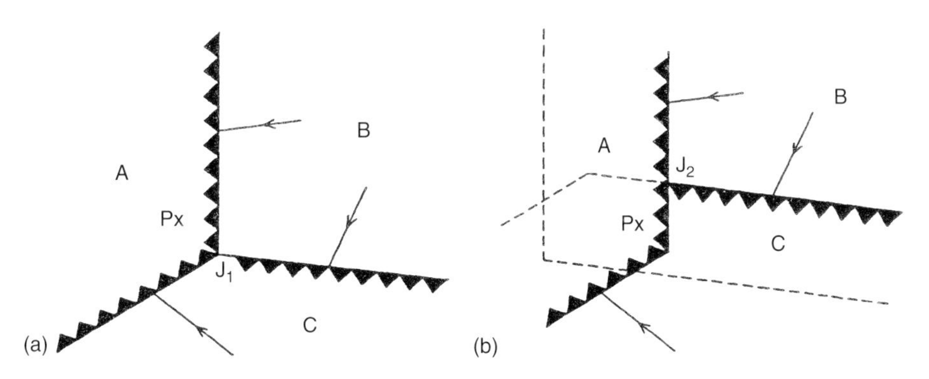

FIGURE 9 T-T-T type triple junction (after McKenzie and Morgan, 1969). Arrows are velocity vectors of subducting plates relative to the upper plates. (a) At time = 0; (b) after some time.

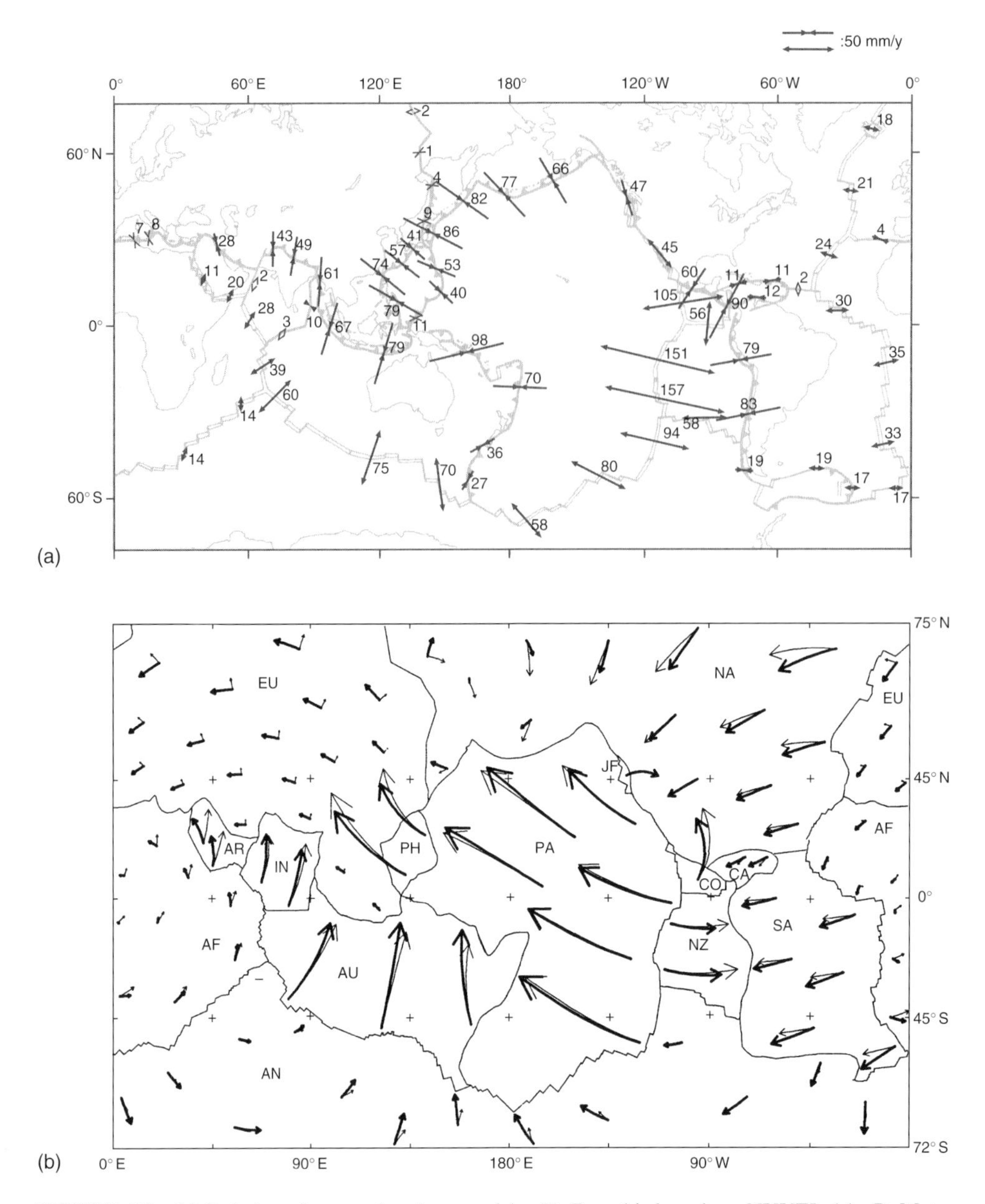

FIGURE 10 (a) Relative plate motion (prepared by K. Tamaki, based on NUVEL-1A; DeMets *et al.*, 1994). (b) Plate velocities relative to the hotspots by models HS2NUVEL1 (thick arrows) and AM1-2 (Minster and Jordan, 1978) (thin arrows). Arrow length and shape show the trace a hotspot would make over 50 My. (After Gripp and Gordon, 1990.)

and Iceland have been suggested (Wilson, 1965b; Morgan, 1971). Although the definition is not rigorous, a hot spot is a place where volcanic activity has been taking place for a long period of time (throughout the Cenozoic Era, for instance). It is also a topographic high with a radius of more than, say, 100 km. Long-lived volcanoes in these hot spots may be the outlets of deep-rooted magma associated with mantle plumes that may be regarded immobile compared to surface plates. If this conjecture holds, the age progression of the Hawaii–Emperor Seamount chain could be the result of the absolute motion of the Pacific Plate, because they may be the trace of the Hawaiian hot spot imprinted on the surface of the Pacific Plate. The most recent plate motion model relative to hot spots is HS2-NUVEL1 (Fig. 10b; Gripp and Gordon, 1990).

Some highlights common to all the plate motion models are: (1) The East Pacific Ridge, among numerous oceanic ridges, shows an especially large spreading rate, reaching 17 cm y^{-1}, while the spreading rate of the Mid-Atlantic Ridge is small at approximately 4 cm y^{-1}; (2) the absolute velocity of oceanic plates that possess a subduction-type boundary (the Pacific,

Nazca, Cocos Plates) is large, and that of the African and Antarctic Plates, which are surrounded by oceanic ridges, is small. Even though they are not surrounded by oceanic ridges, the absolute velocities of the South and North American Plates and the Eurasian Plate, which have no subduction zones, are small. The Australian–Indian Plate, with Sunda subduction boundary, is moving with considerable speed, although it carries large continents.

Absolute motions can provide us with key information on their driving force. Based on the observations outlined above, Forsyth and Uyeda (1975) demonstrated that, among various possible plate-driving forces, the gravitational pull of the subducted slab is most important. This view now seems to be generally supported.

It is difficult to prove the immobility of hot spots (e.g., Molnar and Stock, 1987). There is another possible model of "absolute" plate motions, which is obtained by imposing no-net-rotation constraint on relative motion model. The most recent of such models is NNR-NUVEL1 (Argus and Gordon, 1991). The difference between the hotspot model and the no-net-rotation model is significant, but not large enough to alter the importance of slab pull in driving plates. Which of the two models is better as motion model relative to the deep mantle is still an unsolved problem.

4.5 What is Lithosphere (Plate)?

A plate is the part of the Earth's surficial layer that moves the same way as the continents and sea floor. The motion within the mantle, however, cannot be measured at present, so the thickness of the plate defined this way cannot be determined. If a plate moves as a rigid body, perhaps it glides on the softer layer (asthenosphere). The plate may be likened a piece of ice sheet floating on water. It has long been known that below approximately 75 km depth there exists the so-called upper mantle low-velocity zone or LVZ (see Chapter 51 by Lay). The LVZ may also be low in viscosity. Thus, the layer above the LVZ has come to be called the lithosphere (plate) and the LVZ the asthenosphere.

Oceanic plate is also considered as the cooled (hardened) surface boundary layer of the mantle convection system. Various thermal models have been proposed to explain its main features, such as variation of water depth and heat flow with sea floor age, t (e.g., Turcotte and Schubert, 1982). Essentially, early models led to the result that the water depth changes as $\sqrt{t}$ and heat flow as $1/\sqrt{t}$. There were two problems that modelers worried about. One was that $\sqrt{t}$ increase of water depth did not agree with observation for sea floor older than 70 My. It becomes flatter (Parsons and Sclater, 1977). A constant plate thickness model (McKenzie, 1967) and its later variants tried to explain this disagreement, though not so convincingly. The second problem was that the heat flow from young sea floor was much lower than $1/\sqrt{t}$. This problem was solved by the discovery that observed heat flow from young, poorly sedimented sea floors is biased toward low value because of hydrothermal advective heat transfer that escapes ordinary measurement (see Chapter 81.4 by Cermak). Considering the uncertainty of the real significance of deviations from $\sqrt{t}$ relations, a simple model of a cooling semi-infinite body seems to be adequate at this time (Davis and Lister, 1974; Davies, 1980). Then, the problem becomes the one Lord Kelvin (1863) solved when he calculated the age of the Earth. In the present case, the temperature of the bottom surface of a plate can be that of the solidus or the critical temperature (probably lower than the solidus in general) above which the non-rigid body deformation can occur on the geological time scale. From the Kelvin-type classical solution, the $\sqrt{t}$ laws concerning the plate thickness, water depth, and heat flow, can all be deduced.

5. Plate Tectonic Revolution of Our View of the Earth

5.1 Plate Boundary Processes

There are basically three types of plate boundary: divergent, convergent, and transform fault types. The transform faults displacing mid-oceanic ridges are under water, but the San Andreas Fault in North America, Alpine Fault in New Zealand and North Anatolia Fault in Turkey are the most prominent transform plate boundaries on land characterized by strike-slip seismicity. However, their role in global tectonics is not quite as important as that of divergent and convergent boundaries.

The representatives of divergent plate boundaries are spreading mid-oceanic ridges. The peaks of spreading ridges are approximately 2500 m below sea level. This height roughly represents an equipotential surface, the mantle geoid, as the fluidal asthenosphere is practically exposed there. The width of the mid-oceanic ridges, however, varies drastically, depending on the spreading rate, agreeing with the $\sqrt{t}$ relation. The production of plate is nothing but submarine volcanic activity. This means that most (~80%) of Earth's volcanic output is occurring at the bottom of the sea. As already mentioned, the results of heat flow measurements in oceanic regions did not agree with the expected $1/\sqrt{t}$ relation. The heat flow was high around the ridge axis, but measured values were scattered widely, and even extremely low values were not rare. In natural science, when theory and observation do not agree, it is customary to doubt the theory. Because plate theory had been so successful in other aspects, however, researchers chose to doubt the observations, and this led them to the right answer—that the ordinary method cannot measure much of the heat released by active hydrothermal circulation within young, heavily fissured oceanic crust (Anderson and Skillbeck, 1981). As the age of the sea floor advances and the impermeable sedimentary cover thickens, the hydrothermal circulation within the crust ceases, and the observed values converge to the theoretical values. Later research, including observation from deep-sea submersibles, revealed the existence of hydrothermal activity with emission of high-temperature fluids and colonies of organisms such as giant tubeworms on the ridge axis area.

A convergent plate boundary consists of subduction zones and collision zones. In the development of the sea-floor spreading hypothesis, mid-oceanic ridges played a major role. From the global tectonics point of view, however, the role of convergent boundaries is equally important. The idea that continental crust was thrusting over oceanic crust at trenches was introduced in the 1930s (e.g., Griggs, 1939). The seismicity of subduction zones may be summarized as follows (e.g., Yoshii, 1979). Approaching from the ocean side, earthquakes on the oceanic plate are mostly shallow normal-fault type, probably caused by bending. There is occasionally another kind of normal-fault earthquake deeper in the slab, caused probably by tensile forces due to the negative buoyancy of the subducted slab. Sometimes their magnitude becomes 8 class and they seem to tear the plate almost completely (Kanamori, 1971). Thrust earthquakes are very common farther landward, where the depth of the upper surface of the slab reaches 20 km or so. Great earthquakes that occur here are mostly low-angle reverse-fault types, which are attributed to the elastic rebound of the upper plate. These interplate earthquakes do not occur in regions deeper than about 60 km. Mantle earthquakes in the upper plate also cease to occur at about the same depth. These facts prompted Yoshii (1979) to name the boundary between the aseismic and seismic mantle the "aseismic front." Deeper than about 60 km along the subducting slab, there are Wadati–Benioff zone earthquakes. Their source mechanisms indicate that they are intraplate earthquakes, occurring within the slab itself. The physical mechanism of the deep earthquakes is still not completely clear (Frohlich, 1994; Green and Houston, 1995). The Wadati–Benioff zone under some convergent boundaries has a double structure (Hasegawa *et al.*, 1978).

Compared with seismicity, other thermomechanical aspects, such as igneous activity, heat flow, and backarc basin spreading, are more difficult to explain by subduction models (see, e.g., Bebout *et al.*, 1996). Active volcanoes are distributed in a zone of which the oceanward limit, the "volcanic front," is well defined. Compared with the igneous rocks on the ocean floor, the arc igneous rocks are much more diverse, reflecting their complicated generation process (e.g., Wyllie, 1988). From simple geophysical point of view, the basic question concerning the arc magmatism is how subduction of a cold oceanic plate could produce temperatures high enough to generate magma in the mantle beneath arcs (e.g., Peacock, 1996). Distribution of heat flow (see Chapter 81.4 by Cermak) raises similar questions. Heat flow is low in the oceanic side of the arc, but high in the landward side, extending throughout the entire backarc basin region. The transition from low to high heat flow zones takes place between the aseismic and the volcanic fronts. How is such a distribution derived by subduction of a cold plate? The origin of backarc basins such as the Japan Sea and the Philippine Sea is another problem. At least some of them are believed to have been opened by extensional tectonics relatively recently (Cenozoic backarc spreading). Why such extensional tectonics operates in the backarc region is not self-evident. Various models, invoking the effect of friction between converging plates, mantle wedge flow, and dehydration of subducted slab and so forth, have been proposed to try to solve these enigmas.

Under such circumstances, one method of countering the difficulty would be to compare and analyze the characteristics of subduction zones in the world, and examine which factors are effective in which way. This is "comparative subductology" (e.g., Uyeda, 1982). Subduction zones can be classified into two types, depending on whether or not their backarc basins are presently actively spreading: Mariana-type and Chilean-type. Crustal stress, inferred from earthquake source mechanisms, is tensional in the backarc regions of the Mariana-type, while it is compressional in the Chilean-type. Although interplate thrust earthquakes seem to occur similarly at all subduction zones, most of the seismic energy is released at the Chilean-type subduction zones and not at Mariana-type. This fact indicates that mechanical coupling of oceanic and landward plates in the Chilean-type zones is much stronger than that in the Mariana-type. Interestingly, the interplate convergence rate does not seem to greatly affect the strength of coupling. The difference in the mechanical coupling is reflected in many other phenomena and helps in solving the enigmas mentioned above. The two endmember subduction modes may be schematically described as in Figure 11, but what causes such a difference is not entirely clear. A suggestion is that the older the subducting plate, the more advanced its cooling and the higher its density. As a consequence, the subduction would become more "voluntary" and the coupling would weaken, thus creating a Mariana-type subduction zone. This proposal explains the distribution of subduction types—Mariana-types are more common in the Western Pacific and Chilean-types are more common in the Eastern Pacific. An alternative suggestion is that the contrast in the subduction modes between the East and West Pacific margins may stem from the direction of flow in the asthenosphere with respect to the lithosphere; namely, the eastward asthenospheric flow would steepen the dip of the Mariana slab and flatten the Chilean slab (Uyeda and Kanamori, 1979). Since the plate motion models relative to hot spots have a net westward component relative to no-net-rotation models, this mechanism may be viable (Argus and Gordon, 1991; Ricard *et al.*, 1991).

5.2 Orogenesis

During the 18th century, the orthodox school of geological thought was Abraham G. Werner's (1750–1817) "neptunism" which was challenged by James Hutton's (1726–1797) "plutonism," which in turn was challenged by the idea that the great

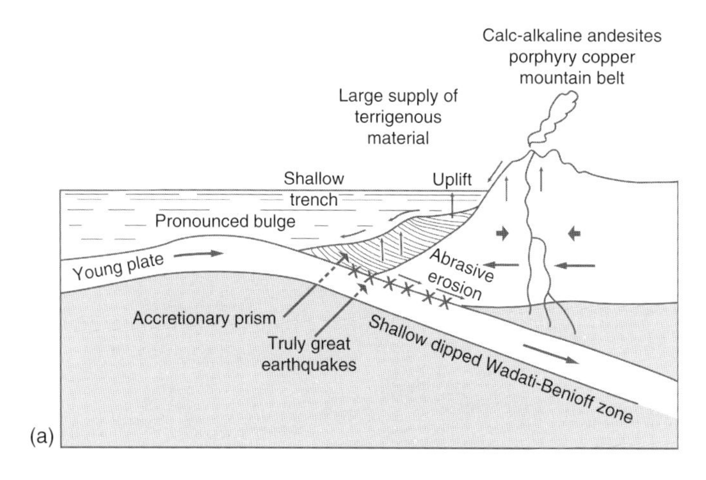

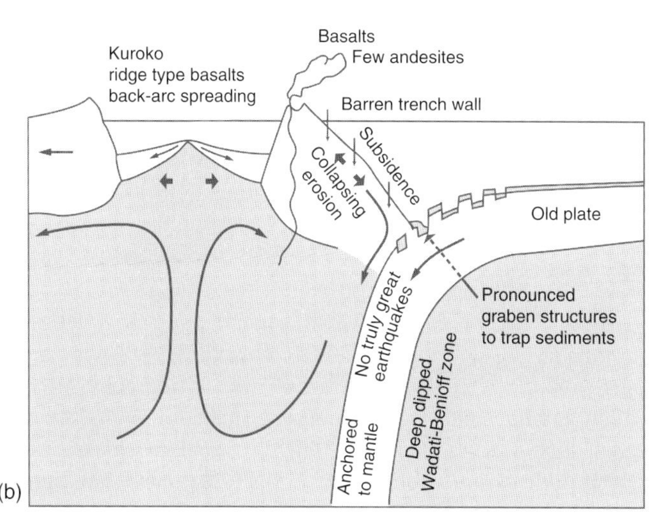

FIGURE 11 Schematic diagrams showing the two typical modes of subduction. (a) High-stress (Chilean) type; (b) low-stress (Mariana) type.

mountain ranges were created by lateral pressure (Eduard Suess, 1831–1914). Whether the origin of the great mountain ranges is igneous activity or lateral pressure is not completely solved even at present. A concept that played an important role in classical orogenesis was that of "geosyncline" (James Hall (1811–1898), 1859; James Dana (1813–1894), 1873). It was believed that sedimentary layers were formed along with subsidence of long narrow zones, where orogenic activities including igneous and metamorphic processes would eventually occur. The strata would then be deformed, elevated and finally create great mountain ranges. Yet there was no physical mechanism to explain this grand scenario.

Orogenesis based on the idea of a mobilist view of the Earth goes back to Wegener, although he had little support. With the advent of plate tectonics, however, many scientists have energetically attempted to apply this theory to orogenesis. Their theories are similar in principal aspects, but probably the most influential paper was that of Dewey and Bird (1970). In a simplified manner, tectonic conditions seen in today's Earth are exhibited in the schematic cross-sectional diagrams in Figure 12. The most basic situation is shown in Figure 12a: The Atlantic Ocean spreads and at the western margin of the South American continent the East Pacific (Nazca) Plate is subducting, and the Andes range is being formed (Cordilleran or Pacific type orogeny). Figure 12b is the cross-section of the Pacific Ocean—also a basic situation: close to its western margin, the trench-island arc-backarc system of Japan develops. Figure 12c is a cross-section of the region from North America (San Andreas fault) to the Mariana Arc, Philippine Sea, and China Sea. If the condition shown in this cross-section persists, the Philippine Sea will eventually disappear and the Mariana Arc and the Philippine Arc will collide unless backarc spreading in the Philippine Sea overcomes the loss due to subduction. If the condition is like Figure 12d, in which the South China Sea subducts eastward, it will be the South China Sea that will disappear. Thus, if the sea floor between continents and/or island arcs subducts, the eventual collision will result (collision type orogeny; Fig. 12g). There are, thus, two basic types of orogenesis: Cordilleran or Pacific-type and collision-type. The causal factors may be mainly thermal (igneous) for the former and dynamic (lateral pressure) for the latter.

As one may recognize from Figure 12, there is a life cycle of oceans. An ocean is born when a continent begins to split and grows by sea-floor spreading (Atlantic stage). When the Atlantic stage ocean spreads, the Pacific Ocean shrinks although sea-floor spreading is going on there, too. The latter is called the Pacific stage, because the Atlantic Ocean will go into this stage after spreading to the extreme. Then the spreading ridge of the Pacific stage ocean will subduct and the sea at last goes into the Mediterranean stage until it is closed by continental collision. When the next continental splitting takes place, the next cycle will start. This concept was presented early in the game (Wilson, 1968b) and was later named the Wilson cycle.

When, during subduction, features like islands, seamounts, plateaus, and oceanic ridges reach trenches, they collide with island arcs or continents. Subduction of a spreading ridge may have significant tectonic effects. If features are too light (buoyant) to subduct, they will accrete to the landward plate, to form the so-called allochtonous terranes, or simply terranes (e.g., Nur and Ben-Avraham, 1982). Even without such salient topographic features, oceanic crust and sediments (terrigenous or pelagic) at the bottom of trenches often accrete, forming an accretionary complex. An accretionary complex is the major agent for the oceanward growth of arcs. Extending these views still farther, geologists have shown that the seemingly extensive one-piece land of Asia is in fact an accumulation of a number of blocks of all sizes called microplates, that have collided and accreted one after another.

One of the prominent facts of today's collision process is relative lack of truly great earthquakes in collision zones compared with Chilean type subduction zones (Kanamori, 1977). In collision zones where the plates engage tightly, the

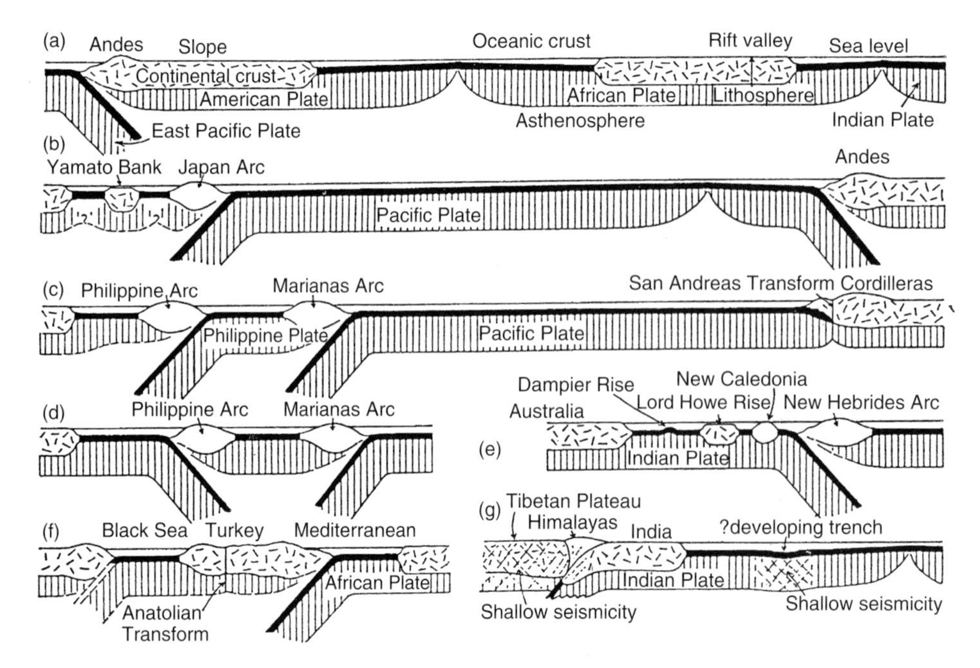

FIGURE 12 Schematic sections, showing the major tectonic regimes of the present Earth according to Dewey and Bird (1970).

stress may be high, but a large-scale seismogenic displacement may become more difficult when the coupling gets too tight, probably making truly great earthquakes less frequent. In a sense, a collision zone is no longer an ordinary plate boundary and the stress tends to be relaxed by scattered local fault motions or nonelastic deformation. At the collision zone, the upper plate, that had been an active margin, seems to yield; one typical example is the extrusion tectonics in the region north of the Himalayas and Tibet (Tapponnier *et al.*, 1982). It now seems more appropriate to state the basics of plate tectonics as "plates move without deformation until they meet strong stress," rather than "plates are rigid."

6. Beyond Plate Tectonics—Plume Tectonics

Almost all major aspects of orogenic phenomena appear to be explained by plate tectonics. However, it is also evident that plate tectonics can explain the phenomena occurring in the superficial layer of the Earth for only a fraction of its whole history. As well as the long-term secular changes in the global tectonic regime with time scales of billions of years, such events as formation/dispersion of supercontinents (the Wilson cycle) and the Cretaceous pulse of spreading and enhanced magmatism (Large Igneous Province) appear to be difficult to explain by normal steady-state plate tectonics. To understand the whole story, spatial and temporal insights beyond plate tectonics are required. With the aid of new advances in various fields of earth science, such as ultrahigh-pressure mineral physics, the theory of the origin of the Earth, simulation of mantle convection, and global seismic tomography, what is called "plume tectonics" is emerging. In the following, we will touch upon it, although it is still highly speculative and requires rigorous checking before taken seriously as a new paradigm.

About a decade ago, there was much discussion of two-layered mantle convection vs. whole-mantle convection (see, e.g., Silver *et al.*, 1988). The former was based on mainly geochemical data that appeared to require separate material reservoirs in the mantle, whereas the latter was based on seismological information related to deep penetration of subducted slabs. Upwelling plumes were inferred to rise from the deep mantle to supply magmas to the immobile hot spots. Plumes then were envisaged as thin upwelling jets of columnar shape but were regarded as playing a rather minor role in the whole framework of plate tectonics. Global-scale seismic tomography, pioneered by Adam Dziewonski and colleagues at Harvard University (see, e.g., Chapter 52 by Curtis and Snieder), however, began to clarify the three-dimensional mantle structures at about the same time. Here we refer to global tomography results and their interpretation (Fig. 13) by Fukao (1992) and Fukao *et al.* (1994). There are two low-velocity regions with enormous size ($\sim$3000 km across) in the lower mantle under the south Pacific and Africa. Moreover, a large region of the lower mantle characterized by high-velocity anomaly exists beneath central Asia where the large amount of subducted oceanic lithosphere has been accumulated. These regions of anomalously low and high velocities are high-temperature upwelling superplumes and low-temperature

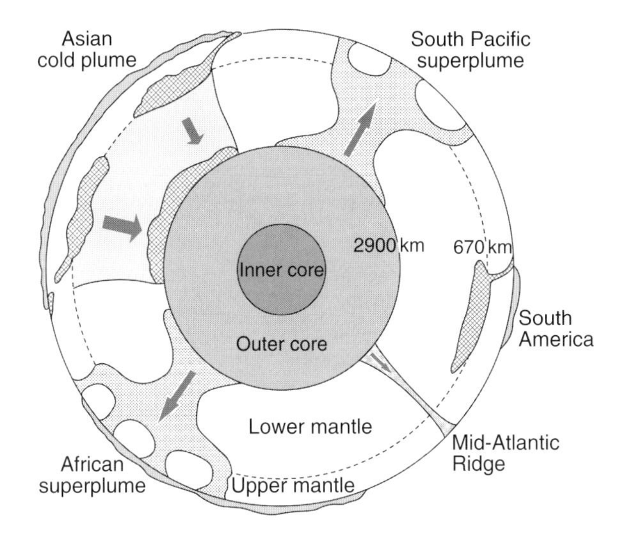

FIGURE 13 Schematic cross-section of the Earth showing the global material circulation according to Fukao *et al.* (1994).

downwelling superplumes. Surface expressions of the upwelling superplumes are the well-known superswell in the south Pacific (300 m higher and 2000 km across) and the east African Rift region, both with several hot spots. The superplumes seem to branch into upper mantle plumes for hot spots at the 670 km discontinuity. Whether or not the slabs penetrate through the 670 km transition has long been discussed. Tomographic images show complex behaviors. Slabs behave as if they meet a strong resistance at the 670 km discontinuity, bend horizontally, thicken, and accumulate (the megaliths of Ringwood and Irifune, 1988). The negative Clapeyron *P–T* slope of the phase transition at the 670 km discontinuity can explain this stagnation behavior. When the volume of the cold stagnant slab material becomes large after continued subduction, it will fall down in big blobs intermittently at some 100–400 My intervals. Numerical modeling supports this view (e.g., Honda *et al.*, 1993).

Maruyama (1994) and Kumazawa and Maruyama (1994), based on these and other data and conjectures encompassing a wide range of earth and planetary science, are rather far-reaching accounts of the new plume tectonics at this stage. The origin of the Pacific superplume may go back to 750–700 Ma, when a supercontinent called Rodinia was broken up. A new ocean, the Pacific was born in the wakes of the dispersed continental fragments. The African superplume was formed probably when the supercontinent Pangea was rifted ~200 Ma. The present Earth is composed of three domains with different tectonics; they are growth tectonics in the core driven by the energy released from the growing solid inner core, plume tectonics in the lower mantle characterized by large-scale vertical plume movements, and plate tectonics in the upper mantle where horizontal movement of rigid plates dominates. These domains are usually decoupled but are episodically coupled when a massive injection of upper mantle material into the lower mantle takes place. The steady-state plate movement basically controls the surface dynamics at normal times. The Earth after the Pangean breakup corresponds to this state, except for the Cretaceous pulse period. The intermittent catastrophic avalanche of stagnant slabs at subduction zones, resulting in unusually active sea-floor spreading and magmatism, activates strong mantle convection. This corresponds to the Cretaceous pulse. If catastrophic collapse of the whole downwelling superplume took place, induced mantle overturn would be of global scale. It may occur once in a 400 My or longer time interval, resulting in the formation and breakup of supercontinents. Plume theory says that major events like the Cretaceous pulse and the formation and dispersal of supercontinents were driven by the activation of superplumes, which was triggered by the avalanche of cold slab material into the lower mantle. It is interesting to note that all these geodynamic processes are understood as driven by downgoing cold material, which is inherent to the convective cooling process of the earth. These interesting ideas of "plume tectonics" are, however, still highly speculative at this stage as noted earlier.

References

Allegre, C. (1988). "The Behavior of the Earth—Continental and Sea-floor Mobility." Harvard University Press, Cambridge, MA.

Anderson, R.N. and J.N. Skillbeck (1981). Oceanic heat flow. In: "The Sea," Vol. 7 (C. Emiliani, Ed.) pp. 489–523. Wiley-Interscience, New York.

Argus, D.F. and R.G. Gordon (1991). No-net-rotation model of current plate velocities incorporating plate motion model NUVEL-1. *Geophys. Res. Lett.* **18**, 2039–2042.

Bebout, G.E., D.W. Scholl, S.H. Kirby, and A.B. Platt (Eds.) (1996). "Subduction Top to Bottom," *Geophys. Monogr. 96*. American Geophysical Union, Washington, DC.

Beloussov, V.V. (1979). Why do I not accept plate tectonics. *EOS* **60**, 207–211.

Bird, J. (Ed.) (1980). "Plate Tectonics." American Geophysical Union, Washington, DC.

Bird, J. and B. Isacks (Eds.) (1972). "Plate Tectonics." American Geophysical Union, Washington, DC.

Bullard, E.C., J.E. Everett, and A.G. Smith (1965). The fit of the continents around the Atlantic. In: "A symposium on Continental Drift" (P.Ms.S. Blackett, E. Bullard, and S.K. Runcorn, Eds.). *Philos. Trans. R. Soc. Lond.* **A258**, 41–51.

Cox, A. (Ed.) (1973). "Plate Tectonics and Geomagnetic Reversals," W.H. Freeman, San Francisco.

Cox, A. and B. Hart (1986). "Plate Tectonics—How It Works." Blackwell, Oxford.

Davies, G.F. (1980). Review of oceanic and global heat flow estimates. *Rev. Geophys. Space Phys.* **18**, 718–722.

Davis, E.E. and C.R.B. Lister (1974). Fundamentals of ridge crest topography. *Earth Planet. Sci. Lett.* **21**, 405–413.

DeMets, C., R. Gordon, D.F. Argus, and S. Stein (1994). Effect of recent revisions to the geomagnetic reversal time scale on estimates of current plate motions. *Geophys. Res. Lett.* **21**, 2191–2194.

Dewey, J.F. and J.M. Bird (1970). Mountain belts and the new global tectonics. *J. Geophys. Res.* **75**, 2625–2647.
Dietz, R. (1961). Continent and ocean evolution by spreading of the sea floor. *Nature* **190**, 854–857.
Ewing, M. and B. Heezen (1956). Some problems of Antarctic submarine geology. In: "Antarctica in the International Geophysical Year" (A. Crary, I.M. Gould, E.O. Hurlburt, H. Odishaw, and W.E. Smith, Eds.) pp. 75–81. *Geophys. Monogr.* American Geophysical Union, Washington, DC.
Forsyth, D.W. and S. Uyeda (1975). On the relative importance of driving forces of plate motion. *Geophys. J. R. Astron. Soc.* **43**, 163–200.
Frohlich, C. (1994). A break in the deep. *Nature* **368**, 190–191.
Fukao, Y. (1992). Seismic tomogram of the earth's mantle: geodynamic implications. *Science* **258**, 625–630.
Fukao, Y., S. Maruyama, M. Obayashi, and K. Inoue (1994). Geologic implication of the whole mantle P-wave tomography. *Geol. Soc. Japan* **100**, 4–23.
Green, H.W. and H. Houston (1995). The mechanics of deep earthquakes. *Ann. Rev. Earth Planet. Sci.* **23**, 169–213.
Griggs, D.T. (1939). A theory of mountain building. *Am. J. Sci.* **237**, 611–650.
Gripp, A.E. and R.G. Gordon (1990). Current plate velocities relative to the hotspots incorporating the NUVEL-1 global plate motion model. *Geophys. Res. Lett.* **17**, 1109–1112.
Hasegawa, A., N. Umino, and A. Takagi (1978). Double planed deep seismic zone and upper mantle structure in the Northeastern Japan arc. *Geophys. J. R. Astron. Soc.* **54**, 281–296.
Heezen, B.C. (1960). The rift in the ocean floor. *Sci. Am.* **203**, 98–110.
Heirtzler, J.R., G.O. Dickson, E.M. Herron, W. Pitman, and X. Le Pichon (1968). Marine magnetic anomalies, geomagnetic field reversals, and motions of the ocean floor and continents. *J. Geophys. Res.* **73**, 2119–2136.
Hess, H.H. (1962). History of ocean basins. In: "Petrologic Studies," Buddington Volume, pp. 599–620. Geological Society of America, New York.
Hilde, T.W.C., S. Uyeda, and L. Kroenke (1977). Evolution of the western Pacific and its margin. *Tectonophysics* **38**, 145–168.
Holmes, A. and D.L. Holmes (1978). "Principles of Physical Geology," 3rd ed. Nelson & Sons, Sunbury-on-Thames.
Honda, S., D.A. Yuen, S. Balachandar, and D. Reuteler (1993). Three-dimensional instabilities of mantle convection with multiple phase transitions. *Science* **259**, 1308–1311.
Hurley, P.M. (1968). The confirmation of continental drift. *Sci. Am.*, April, 53–64.
Irving, E. (1964). "Paleomagnetism." Wiley, New York.
Isacks, B., J. Oliver, and L.R. Sykes (1968). Seismology and the new global tectonics. *J. Geophys. Res.* **73**, 5855–5899.
Kanamori, H. (1971). Seismological evidence for a lithospheric normal faulting—the Sanriku earthquake of 1933. *Phys. Earth Planet. Inter.* **4**, 289–300.
Kanamori, H. (1977). The energy release in great earthquakes. *J. Geophys. Res.* **82**, 2981–2987.
Kearey, P. and F. Vine (1990). "Global Tectonics." Blackwell, Oxford.
Kuhn, T. (1962). "The Structure of Scientific Revolution." University of Chicago Press, Chicago.
Kumazawa, M. and S. Maruyama (1994). Whole Earth tectonics. *J. Geol. Soc. Japan* **100**, 81–102.
Larson, R. (1991). Geological consequences of superplumes. *Geology* **19**, 963–966.
Larson, R.L. and W.C. Pitman (1972). World-wide correlation of Mesozoic magnetic anomalies and its implications. *Geol. Soc. Am. Bull.* **83**, 3627–3644.
Le Pichon, X. (1968). Sea-floor spreading and continental drift. *J. Geophys. Res.* **73**, 3661–3697.
Le Pichon, X., J. Francheteau, and J. Bonnin (1973). "Plate Tectonics." Elsevier, Amsterdam.
Maruyama, S. (1994). Plume tectonics. *J. Geol. Soc. Japan* **100**, 24–29.
Maxwell, A.E. and R.P. von Herzen (1969). The Glomar Challenger completes Atlantic Track—Highlights of Leg III. *Ocean Ind.* **4**, 5.
McKenzie, D.P. (1967). Some remarks on heat flow and gravity anomalies. *J. Geophys. Res.* **72**, 6262–6271.
McKenzie, D.P. and W.J. Morgan (1969). Evolution of triple junction. *Nature* **224**, 125–133.
McKenzie, D.P. and R.L. Parker (1967). The north Pacific: an example of tectonics on a sphere. *Nature* **216**, 1276–1280.
McKenzie, D.P. and J.G. Sclater (1971). The evolution of the Indian ocean since the late Cretaceous. *Geophys. J. R. Astron. Soc.* **25**, 437–528.
Minster, J.B. and T.H. Jordan (1978). Present-day plate motions. *J. Geophys. Res.* **83**, 5331–5354.
Molnar, P. and J. Stock (1987). Relative motions of hot spots in the Pacific, Atlantic, and Indian Oceans since Late Cretaceous time. *Nature* **327**, 587–591.
Morgan, W.J. (1968). Rises, trenches, great faults and crustal blocks. *J. Geophys. Res.* **73**, 1959–1982.
Morgan, W.J. (1971). Convection plumes in the lower mantle. *Nature* **230**, 42–43.
Nagata, T. (1961). "Rockmagnetism." Maruzen, Tokyo.
Nur, A. and Z. Ben-Avraham (1982). Oceanic plateaus, the fragmentation of continents, and mountain building. *J. Geophys. Res.* **87**, 3644–3661.
Opdyke, N.D. (1972). Paleomagnetism of deep-sea cores. *Rev. Geophys. Space Phys.* **19**, 213–249.
Parsons, B. and J.G. Sclater (1977). An analysis of the variation of ocean floor with age. *J. Geophys. Res.* **82**, 803–827.
Peacock, S.M. (1996). Thermal and petrologic structure of subduction zones. In: "Subduction; Top to Bottom" (G.E. Bebout *et al.*, Eds.), pp. 119–133. *Geophys. Monogr. 96*. American Geophysical Union, Washington, DC.
Pitman, W.C. and J.R. Heirtzler (1966). Magnetic anomalies over the Pacific-Antarctic ridge. *Science* **154**, 1164–1171.
Pitman, W.C. and M. Talwani (1972). Sea-floor spreading in the North Atlantic. *Geol. Soc. Am. Bull.* **83**, 619–646.
Pitman, W.C., R.L. Larson, and E.M. Herron (1974). Age of the ocean basins determined from magnetic anomaly lineations. *Geol. Soc. Am. Map and Chart Series*, MC-6.
Raff, A.D. and R.G. Mason (1961). Magnetic survey off the coast of North America 40° N to 50° N. *Geol. Soc. Am. Bull.*, **72**, 1267–1270.
Ricard, Y., C. Doglioni, and R. Sabadini (1991). Differential rotation between lithosphere and mantle: a consequence of lateral mantle viscosity variations. *J. Geophys. Res.* **95**, 8407–8415.

Ringwood, A.E. and T. Irifune (1988). Nature of the 650-km seismic discontinuity: implications for mantle dynamics and differentiation. *Nature* **331**, 131–136.

Runcorn, S.K. (Ed.) (1962). "Continental Drift." Academic Press, New York.

Scheidegger, A.E. (1963). "Principles of Geodynamics." Academic Press, New York.

Silver, P.G., R.W. Carlson, and P. Olson (1988). Deep slabs, geochemical heterogeneity, and the large scale structure of mantle convection. *Annu. Rev. Earth Planet. Sci.* **16**, 477–541.

Sleep, N.H. and K. Fujita (1997). "Principles of Geophysics." Blackwell Science, Oxford.

Sullivan, W. (1991). "Continents in Motion." American Institute of Physics, New York.

Sykes, L.R. (1967). Mechanism of earthquakes and nature of faulting on the mid-oceanic ridges. *J. Geophys. Res.* **75**, 5041–5055.

Tapponnier, P. *et al.* (1982). Propagating extrusion tectonics in Asia: new insights from simple experiments with plasticine. *Geology* **19**, 679–688.

Turcotte, D.L. and G. Schubert (1982). "Geodynamics." Wiley, New York.

Uyeda, S. (1978). "The New View of the Earth." W.H. Freeman, San Francisco.

Uyeda, S. (1982). Subduction zones: an introduction to comparative subductology. *Tectonophysics* **81**, 133–159.

Uyeda, S. and H. Kanamori (1979). Backarc opening and the mode of subduction. *J. Geophys. Res.* **84**, 1049–1061.

Vine, F.J. (1966). Spreading of the ocean floor: New evidence. *Science* **154**, 1405–1415.

Vine, F.J. (1968). Magnetic anomalies associated with mid-oceanic ridges. In: "History of the Earth's Crust," (R.A. Phinney, Ed.), pp. 73–89. Princeton University Press, Princeton.

Vine, F.J. and D.H. Matthews (1963). Magnetic anomalies over oceanic ridges. *Nature* **199**, 947–949.

Vine, F.J. and J.T. Wilson (1965). Magnetic anomalies over young oceanic ridge off Vancouver Island. *Science* **150**, 485–489.

Wegener, A. (1929). "The Origin of Continents and Oceans," (4th edn.), Dover paperback, New York.

Wilson, J.T. (1965a). A new class of faults and their bearing on continental drift. *Nature* **207**, 343–347.

Wilson, J.T. (1965b). Submarine fracture zones, aseismic ridges and the International Council of Scientific Unions Line: proposed western margin of the East Pacific Ridges. *Nature* **297**, 907–911.

Wilson, J.T. (1968a). A revolution in earth science. *Geotimes* **13**(10), 10–16.

Wilson, J.T. (1968b). Did the Atlantic close and then re-open again? *Nature* **211**, 676–681.

Wilson, J.T. (1968c). Static or mobile earth: The current scientific revolution, Gondwanaland revisited. *Am. Phil. Soc. Proc.* **112**, 309–320.

Wyllie, P.J. (1971). "The Dynamic Earth." Wiley, New York.

Wyllie, P.J. (1988). Magma genesis, plate tectonics and chemical differentiation of the earth. *Rev. Geophys.* **26**, 370–404.

Yoshii, T. (1979). A detailed cross-section of the deep seismic zone beneath northeastern Honshu, Japan. *Tectonophysics.* **565**, 349–360.

Editor's Note

Please see also the personal historical account by Jack Oliver and by Lynn Sykes in the institution report of Lamont-Doherty Earth Observatory in Chapter 79, Centennial National and Institutional Reports: Seismology, edited by Kisslinger. Biography of many persons mentioned in this chapter is given in Chapter 89, Biography of Notable Earthquake Scientists and Engineers.

7

Earthquake Mechanisms and Plate Tectonics

Seth Stein and Eryn Klosko
Northwestern University, Evanston, Illinois, USA

1. Introduction

Earthquake seismology has played a major role in the development of our current understanding of global plate tectonics and in making plate tectonics the conceptual framework used to think about most large-scale processes in the solid Earth. During the dramatic development of plate tectonics, discussed from the view of participants by Uyeda (1978, and this volume), Cox (1973), and Menard (1986), the distribution of earthquakes provided some of the strongest evidence for the geometry of plate boundaries and the motion on them (e.g., Isacks *et al.*, 1968). More than thirty years later, earthquake studies retain a central role, as summarized here.

Because earthquakes occur primarily at the boundaries between lithospheric plates, their distribution is used to map plate boundaries and their focal mechanisms provide information about the motion at individual boundaries.

Plate boundaries are divided into three types (Fig. 1). Oceanic lithosphere is formed at *spreading centers*, or mid-ocean ridges, and is destroyed at *subduction zones*, or trenches.

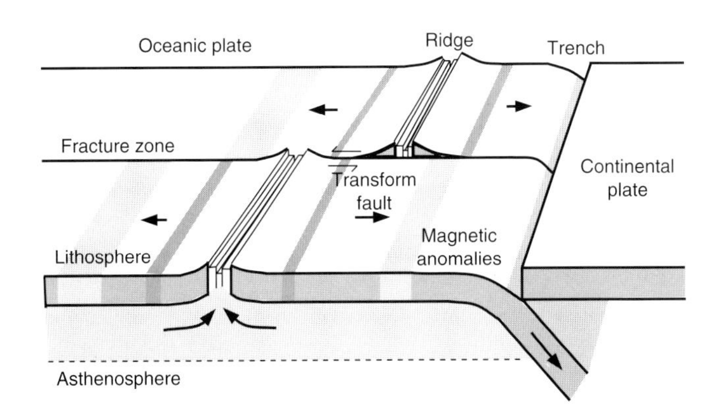

FIGURE 1 Plate tectonics at its simplest. Plates are formed at ridges and subducted at trenches. At transform faults, plate motion is parallel to the boundaries. Each boundary type has typical earthquakes.

Thus, at spreading centers plates move away from the boundary, whereas at subduction zones the subducting plate moves toward the boundary. At the third boundary type, *transform faults*, plate motion is parallel to the boundary. The *slip vectors* of the earthquakes on plate boundaries, which show the motion on the fault plane, reflect the direction of relative motion between the two plates.

The basic principle of plate kinematics is that the relative motion between any two plates can be described as a rotation on a sphere about an *Euler pole* (Fig. 2). Specifically, at any point along the boundary between plates i and j, with latitude λ and longitude μ, the linear velocity of plate j with respect to plate i is

$$\mathbf{v}_{ji} = \boldsymbol{\omega}_{ji} \times \mathbf{r} \qquad (1)$$

the usual formulation for rigid body rotations in mechanics. The vector $\mathbf{r}$ is the position vector to the point on the boundary,

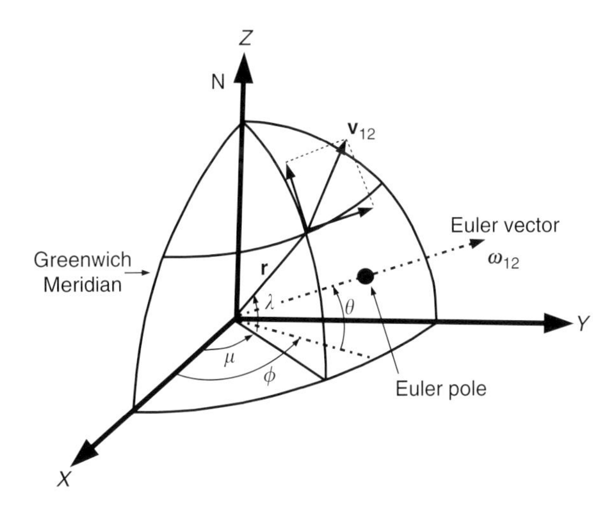

FIGURE 2 Geometry of plate motions. At any point $\mathbf{r}$ along the boundary between plate i and plate j, with geopraphic latitude λ and longitude μ, the linear velocity of plate j with respect to plate i is $\mathbf{v}_{ji} = \boldsymbol{\omega}_{ji} \times \mathbf{r}$. The Euler pole at latitude θ and longitude ϕ is the intersection of the Euler vector $\boldsymbol{\omega}_{ji}$ with the Earth's surface.

INTERNATIONAL HANDBOOK OF EARTHQUAKE AND ENGINEERING SEISMOLOGY, VOLUME 81A ISBN: 0-12-440652-1

and $\boldsymbol{\omega}_{ji}$ is the rotation vector or *Euler vector*. Both are defined from an origin at the center of the Earth.

The direction of relative motion at any point on a plate boundary is a small circle, a parallel of latitude *about the Euler pole* (not a geographic parallel about the North Pole!). For example, in Figure 3a the pole shown is for the motion of plate 2 with respect to plate 1. The first-named plate ($j=2$) moves counterclockwise about the pole with respect to the second ($i=1$). The segments of the boundary where relative motion is parallel to the boundary are transform faults. Thus, transforms are small circles about the pole and earthquakes occurring on them should have pure strike-slip mechanisms. Other segments have relative motion away from the boundary,

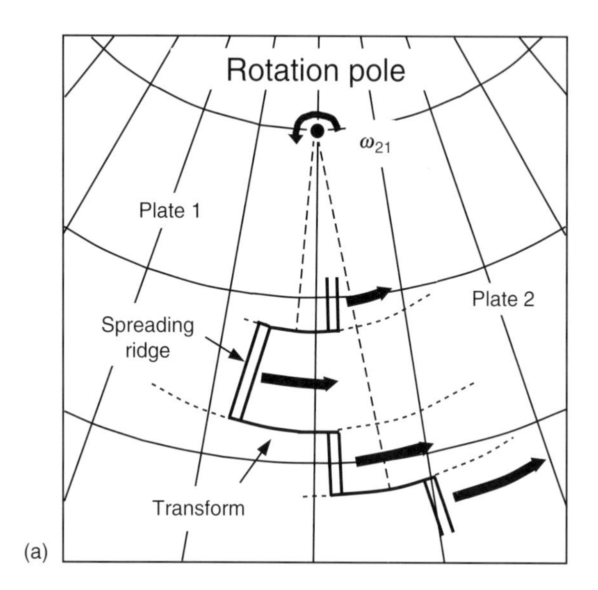

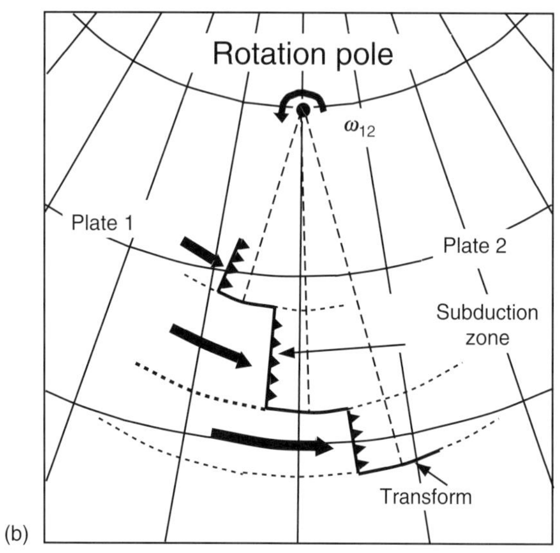

FIGURE 3 Relationship of motion on plate boundaries to the Euler pole. Relative motion occurs along small circles about the pole; the rate increases with distance from the pole. Note the difference the sense of rotation makes: $\boldsymbol{\omega}_{ji}$ is the Euler vector corresponding to the rotation of plate *j* counterclockwise with respect to *i*.

and are thus spreading centers. Figure 3b shows an alternative case. The pole here is for plate 1 ($j=1$) with respect to plate 2 ($i=2$), so plate 1 moves toward some segments of the boundary, which are subduction zones. Note that the ridge and subduction zone boundary segments are not small circles.

The magnitude, or rate, of relative motion increases with distance from the pole, since

$$|\mathbf{v}_{ji}| = |\boldsymbol{\omega}_{ji}||\mathbf{r}| \sin \gamma \tag{2}$$

where γ is the angle between the Euler pole and the site (corresponding to a colatitude about the pole.) Thus, although all points on a plate boundary have the same angular velocity, the linear velocity varies.

If we know the Euler vector for any plate pair, we can write the linear velocity at any point on the boundary between the plates in terms of the local E–W and N–S components by a coordinate transformation. With this, the rate and azimuth of plate motion become

$$rate = |\mathbf{v}_{ji}| = \sqrt{(\mathbf{v}_{ji}^{\mathrm{NS}})^2 + (\mathbf{v}_{ji}^{\mathrm{EW}})^2} \tag{3}$$

$$azimuth = 90 - \tan^{-1}\left(\frac{\mathbf{v}_{ji}^{\mathrm{NS}}}{\mathbf{v}_{ji}^{\mathrm{EW}}}\right) \tag{4}$$

such that azimuth is measured in degrees clockwise from North.

Given a set of Euler vectors with respect to one plate, those with respect to others are found by vector arithmetic. For example, the Euler vector for the reverse plate pair is the negative of the Euler vector

$$\boldsymbol{\omega}_{ij} = -\boldsymbol{\omega}_{ji} \tag{5}$$

Euler vectors for other plate pairs are found by addition

$$\boldsymbol{\omega}_{jk} = \boldsymbol{\omega}_{ji} + \boldsymbol{\omega}_{ik} \tag{6}$$

so, given a set of vectors all with respect to plate *i*, any Euler vector needed is found from

$$\boldsymbol{\omega}_{jk} = \boldsymbol{\omega}_{ji} - \boldsymbol{\omega}_{ki} \tag{7}$$

For further information on plate kinematics see an introductory text such as Cox and Hart (1986). As discussed there, motions between plates can be determined by combining three different types of data from different boundaries. The rate of spreading at ridges is given by sea-floor magnetic anomalies, and the directions of motion are found from the orientations of transform faults and the slip vectors of earthquakes on transforms and at subduction zones. As is evident, earthquake slip vectors are only one of three types of plate motion data available. Euler vectors are determined from the relative motion data, using geometrical conditions. Since slip vectors and transform faults lie on small circles about the pole, the pole must lie on a line at right angles to them (Fig. 3). Similarly, the rates of plate motion increase with the sine of the distance from the pole. These constraints make it possible to locate the poles. Determination of Euler vectors for all the plates can thus be treated as an overdetermined least-squares problem, and the best solution found using the generalized inverse to derive *global plate motion models* (Chase, 1972; Minster and Jordan, 1978; DeMets *et al.*, 1990, 1994). Because these models use magnetic anomaly data, they describe plate motion averaged over the past few million years.

New data have become available in recent years due to the rapidly evolving techniques of space-based geodesy. These techniques (Gordon and Stein, 1992) (very long baseline radio interferometry (VLBI), satellite laser ranging (SLR), the global positioning system (GPS), and DORIS (similar to GPS, but using ground transmitters)) use space-based technologies to measure the positions of geodetic monuments to accuracies of better than a centimeter, even for sites thousands of kilometers apart. Hence measurements of positions over time yield relative velocities to precisions almost unimaginable during the early days of plate tectonic studies. A series of striking results, first with VLBI and SLR (e.g., Robbins *et al.*, 1993), and now with GPS (Argus and Heflin, 1995; Larson *et al.*, 1997), show that plate motion over the past few years is generally quite similar to that predicted by global plate motion model NUVEL-1A. This agreement is consistent with the prediction that episodic motion at plate boundaries, as reflected in occasional large earthquakes, will give rise to steady motion in plate interiors due to damping by the underlying viscous asthenosphere (Elsasser, 1969). As a result, the earthquake mechanisms can be compared to the plate motions predicted by both global plate motion models and space-based geodesy.

2. Oceanic Spreading Center Focal Mechanisms

Earthquake mechanisms from the mid-ocean ridge system reflect the spreading process. Figure 4 schematically shows a portion of a spreading ridge offset by transform faults. Because new lithosphere forms at the ridges and then moves away, the relative motion of lithosphere on either side of a transform is in opposing directions. The direction of transform offset, not the spreading direction, determines whether there is right or left lateral motion on the fault. This relative motion, defined as transform faulting, is not what produced the offset of the ridge crest. In fact, if the spreading at the ridge is symmetric (equal rates on either side), the length of the transform will not change with time. This is a very different geometry from a transcurrent fault, where the offset is produced by motion on the fault and the length of the offset between ridge segments would increase with time.

The model is illustrated by focal mechanisms. Figure 5a shows a portion of the Mid-Atlantic Ridge composed of

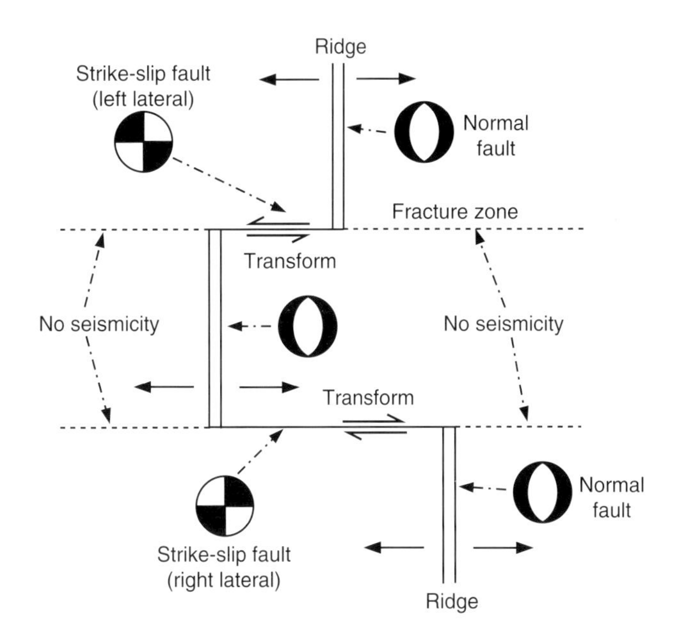

FIGURE 4 Possible tectonic settings of earthquakes at an oceanic spreading center. Most events occur on the active segment of the transform and have strike-slip mechanisms consistent with transform faulting. On a slow spreading ridge, like the Mid-Atlantic, normal fault earthquakes occur. Very few events occur on the inactive fracture zone.

north–south trending ridge segments, offset by transform faults, such as the Vema Transform, which trend approximately east–west. Both the ridge crest and the transforms are seismically active. The mechanisms show that the relative motion along the transform is right–lateral. Sea-floor spreading on the ridge segments produces the observed relative motion. For this reason, earthquakes occur almost exclusively on the active segment of the transform fault between the two ridge segments, rather than on the inactive extension, known as a *fracture zone*. Although no relative plate motion occurs on the fracture zone it is often marked by a distinct topographic feature, due to the contrast in lithospheric ages across it. Unfortunately, some transform faults named before this distinction became clear, such as the Vema, are known as "fracture zones" along their entire length. Earthquakes also occur on the spreading segments. Their focal mechanisms show normal faulting, with nodal planes trending along the ridge axis.

The seismicity is different on fast spreading ridges. Figure 5b shows a portion of the Pacific–Antarctic boundary along the East Pacific Rise. Here, strike-slip earthquakes occur on the transforms, but we do not observe the ridge crest normal faulting events. These observations can be explained by the thermal structure of the lithosphere, because fast spreading produces younger and thinner lithosphere than slow spreading. The axis of a fast ridge has a larger magma chamber than the slow ridge, and the lithosphere moving away from a fast spreading ridge is more easily replaced than for a slow ridge. Thus, in contrast to the axial valley and normal faulting earthquakes on a slow ridge, a fast ridge has an axial high and absence of earthquakes.

The mechanisms are consistent with the predictions of plate kinematics. The area in Figure 5a is a portion of the boundary between the South American and Nubian (West African) plates. An Euler vector for Nubia with respect to South America with a pole at 62° N, 37.8° W and a magnitude of 0.328 degrees My^{-1} predicts that at 0° N, 20° W Africa is moving N81°E, or almost due East, at 33 mm y^{-1} with respect to South America. The Vema is a boundary segment parallel to this direction, and so is a transform fault characterized by strike-slip earthquakes with directions of motion along the trace of the transform. The short segments essentially at right angles to the direction of relative motion are then spreading ridge segments. The spreading rate determined from magnetic anomalies, and thus the slip rate across the transform, is described by the Euler vector.

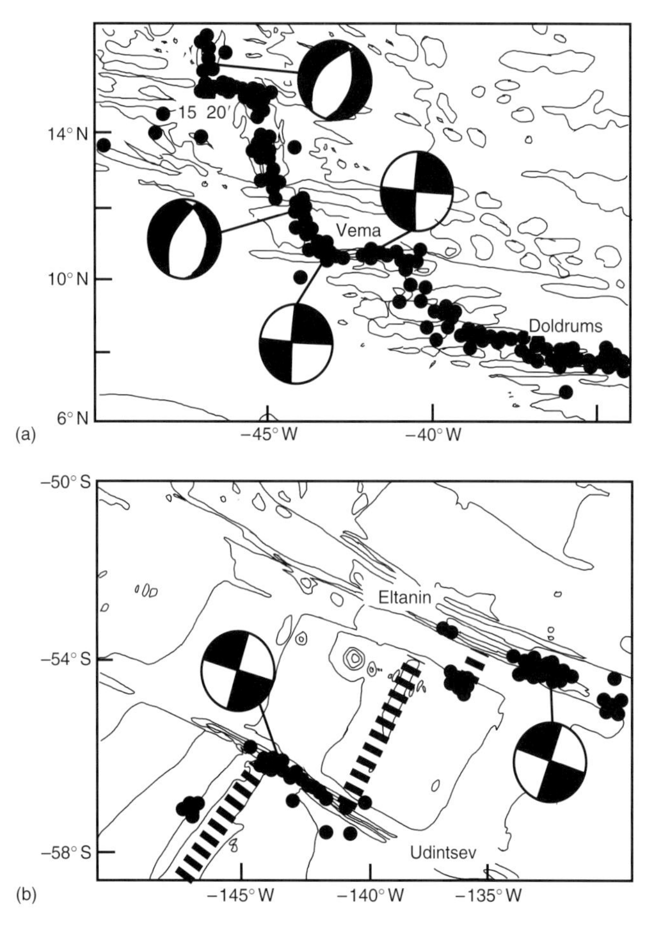

FIGURE 5 Maps contrasting faulting on slow and fast spreading centers. (a) The slow Mid-Atlantic ridge has earthquakes both on the active transform and ridge segment. Strike-slip faulting on a plane parallel to the transform azimuth is characteristic. On the ridge segments, normal faulting with nodal planes parallel to the ridge trend is seen. (b) The fast East Pacific Rise has only strike-slip earthquakes on the transform segments. Mechanisms from Engeln *et al.* (1986), Huang *et al.* (1986), and Stewart and Okal (1983).

3. Subduction Zone Focal Mechanisms

Both the largest earthquakes and the majority of large earthquakes occur at subduction zones. Their focal mechanisms reflect various aspects of the subduction process. Figure 6 is a composite cartoon showing some of the features observed in different subduction zones.

Most of the large, shallow, subduction zone earthquakes indicate thrusting of the overriding plate over the subducting lithosphere. The best such examples are the two largest ever recorded: the 1960 Chilean (M_0 2.7 × 10^{30}, M_s 8.3) and 1964 Alaskan (M_0 7.5 × 10^{29}, M_s 8.4) earthquakes. These were impressive events; in the Chilean earthquake 24 m of slip occurred on a fault 800 km long along-strike and 200 km long down-dip. Smaller, but large, thrust events are characteristic. For example, Figure 7a shows the focal mechanisms of large shallow earthquakes along a portion of the Peru–Chile Trench, where the Nazca Plate is subducting beneath the South American Plate. The mechanisms along the trench show thrust faulting on fault planes with a consistent geometry; parallel to the coast, which corresponds to the trench axis, with shallow dips to the northeast.

These thrust events directly reflect the plate motion. At a point on the trench (17° S, 75° W), global plate motion model NUVEL-1A (DeMets *et al.*, 1994) predicts motion of the Nazca plate with respect to South America at a rate of 68 mm y^{-1} and an azimuth of N76°E. The direction of motion is toward the trench, as expected at a subduction zone. The major thrust earthquakes at the interface between subducting and overriding plates thus directly reflect the subduction, and slip vectors from their focal mechanisms can be used to determine the direction of plate motion. The rate of subduction is harder to assess. Although the rate can be computed from global plate motion models or space geodesy, not all of the plate motion is always

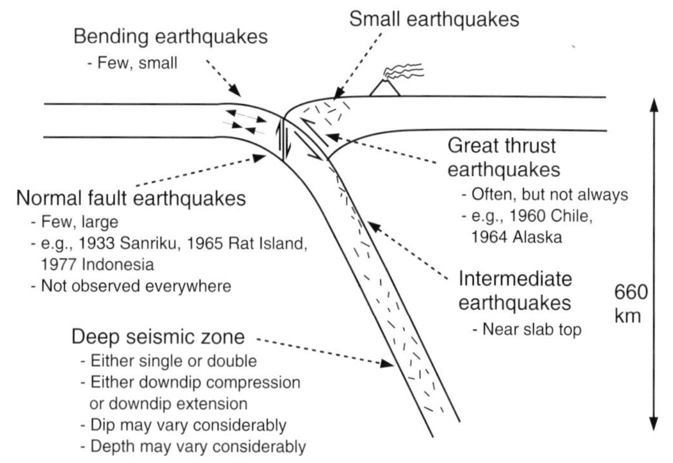

FIGURE 6 Schematic of some of the features observed at subduction zones. Not all features are seen at all subduction zones.

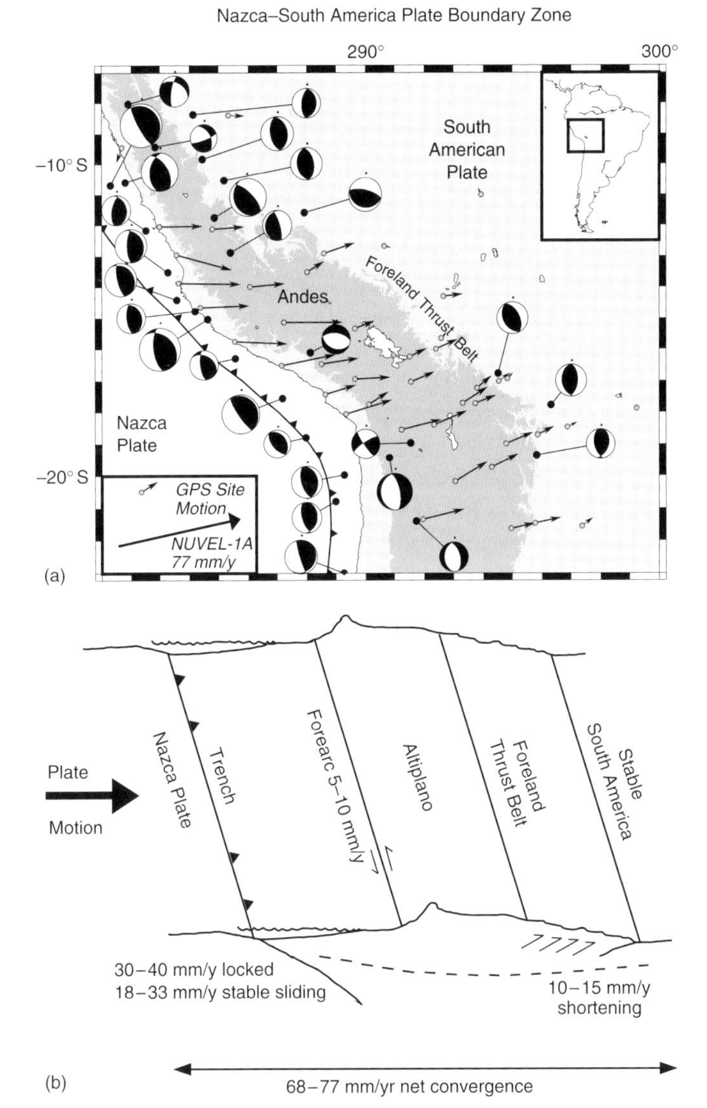

FIGURE 7 (a) GPS site velocities relative to stable South America (Norabuena *et al.*, 1998), and selected earthquake mechanisms in the boundary zone. Rate scale is given by the NUVEL-1A vector. (b) Cross-section across Andean orogenic system showing velocity distribution inferred from GPS data.

released seismically in earthquakes (Kanamori, 1977). In this case, the seismic slip rate estimated from seismic moments can be only a fraction of the real plate motion. Nonetheless, it is useful to determine the seismic slip rate to assess the fraction of seismic slip, as it reflects the mechanics of the subduction process. It is also interesting to know how this seismic slip varies as a function of time and position along a subduction zone.

Figure 6 also shows other types of shallow subduction zone earthquakes. An interesting class of subduction zone earthquakes result from the flexural bending of the downgoing plate as it enters the trench. Precise focal depth studies show a pattern of normal faulting in the upper part of the plate to a depth of 25 km and thrusting in its lower part, between 40 and 50 km. These observations constrain the position of the neutral surface separating the upper extensional zone from the lower flexural zone, and thus provide information on the mechanical state of the lithosphere. Occasionally, trenches are the sites of large normal fault earthquakes (e.g., Sanriku 1933 and Indonesia 1977). There has been some controversy whether to interpret these earthquakes as bending events in the upper flexural sheet or as "decoupling" events showing rupture of the entire downgoing plate due to "slab pull."

The deeper earthquakes, which form the Wadati–Benioff zone, go down to depths of 700 km within the downgoing slab. Their mechanisms provide important information about the physics of the subduction process. The essence of the process is the penetration and slow heating of a cold slab of lithosphere in the warmer mantle. This temperature contrast has important consequences. The subducting plate is identified by the locations of earthquakes in the Wadati–Benioff zone below the zone of thrust faulting at the interface between the two plates. Earthquakes occur to greater depths than elsewhere because the slab is colder than the surrounding mantle. The mechanisms of earthquakes within the slab similarly reflect this phenomenon. The thermal evolution of the downgoing plate and its surroundings is controlled by the relation between the rate at which cold slab material is subducted and that at which it heats up, primarily by conduction as it equilibrates with the surrounding mantle. In addition, adiabatic heating due to the increasing pressure with depth and phase changes contribute.

Numerical temperature calculations show that the downgoing plate remains much colder than the surrounding mantle until considerable depths, where the downgoing slab heats up to the ambient temperature. Comparison of calculated temperatures, the observed locations of seismicity, and images from seismic tomography shows that the earthquakes occur in the cold regions of the slab. The thermal structure also helps explain their focal mechanisms. The force driving the subduction is the integral over the slab of the force due to the density contrast between the denser subducting material and the density of "normal" mantle material outside. This force, known as "slab pull," is the plate driving force due to subduction. Its significance for stresses in the downgoing plate and for driving plate motions depends on its size relative to the resisting forces at the subduction zone. There are several such forces. As the slab sinks into the viscous mantle, material must be displaced. The resulting force depends on the viscosity of the mantle and the subduction rate. The slab is also subject to drag forces on its sides and resistance at the interface between the overriding and downgoing plates. The latter, of course, is often manifest as the shallow thrust earthquakes.

One way to study the relative size of the negative buoyancy and resistive forces is to use focal mechanisms to examine the state of stress in the downgoing slab. Earthquakes above 300 km generally show stress axes corresponding to extension directed down the slab dip, whereas those below

300 km generally show downdip compression. A proposed explanation is that there are two basic processes operating: near the surface the slab is being extended by its own weight; at depth the slab begins to "run into" stronger material and downdip compression occurs. Another crucial effect may be buoyancy due to mineral phase changes that occur at different depths in the cold slab and in the surrounding mantle. Numerical models of stress in downgoing slabs, using these assumptions, can reproduce the shallow down-dip tension and deep downdip compression (Vassiliou, 1984; Bina, 1996).

Finally, it is worth noting that not all features shown in the schematic (Fig. 6) have been observed at all places. For example, the dips and shapes of subduction zones vary substantially. Some show double planes of deep seismicity; some do not. Even the very large thrust earthquakes, considered characteristic of subduction zone events, are not observed in all subduction zones. In recent years, considerable effort has been made to understand such variations.

4. Diffuse Plate Boundary Earthquake Focal Mechanisms

Although the basic relationships between plate boundaries and earthquakes apply to continental as well as oceanic lithosphere, the continents are more complicated. The continental crust is much thicker, less dense, and has very different mechanical properties from the oceanic crust. Because continental crust and lithosphere are not subducted, the continental lithosphere records a long, involved tectonic history. In contrast, the oceans record only the past 200 million years. One major result of these factors is that plate boundaries in continents are often diffuse, rather than the idealized narrow boundaries assumed in the rigid plate model, which are a good approximation to what we see in the oceans. The initial evidence for this notion comes from the distribution of seismicity and the topography, which often imply a broad zone of deformation between the plate interiors.

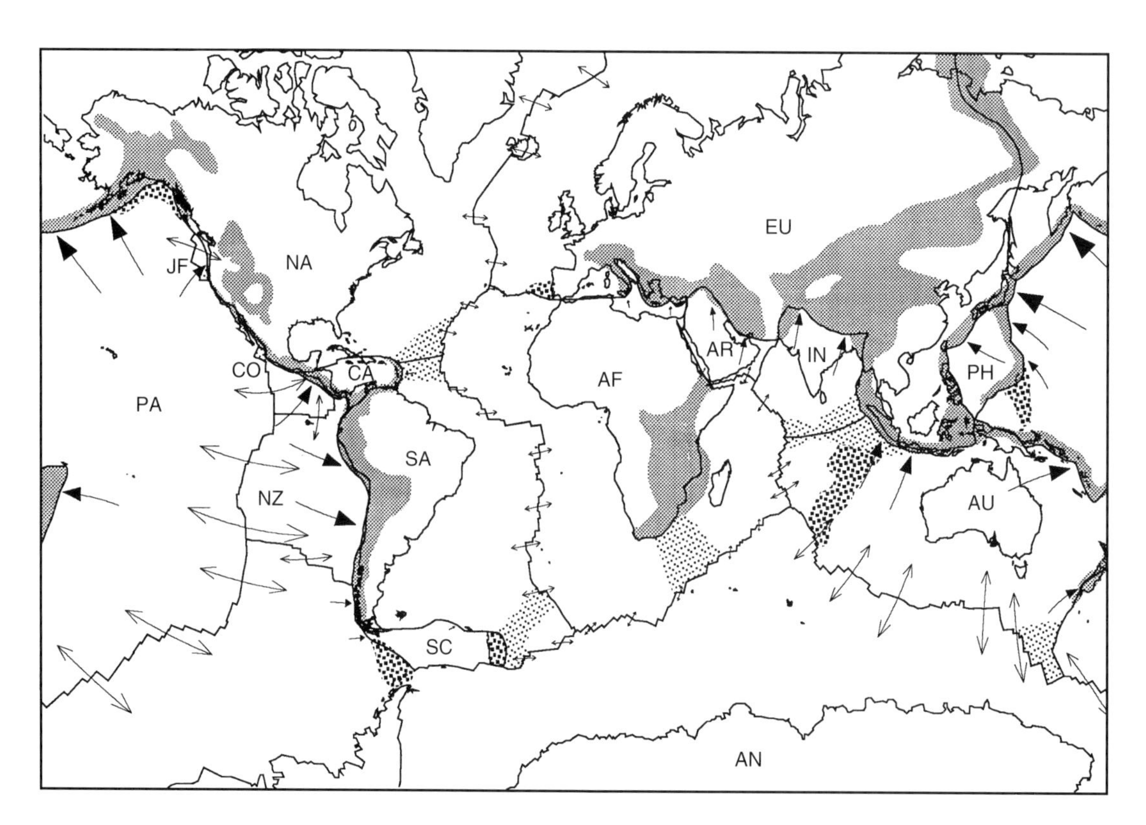

FIGURE 8 Comparison of the idealized rigid plate geometry to the broad boundary zones implied by seismicity, topography, or other evidence of faulting. Fine stipple shows mainly subaerial regions where the deformation has been inferred from seismicity, topography, other evidence of faulting, or some combination of these. Medium stipple shows mainly submarine regions where the nonclosure of plate circuits indicates measurable deformation; in most cases these zones are also marked by earthquakes. Coarse stipple shows mainly submarine regions where the deformation is inferred mainly from the presence of earthquakes. These deforming regions form wide plate boundary zones, which cover about 15% of the Earth's surface. The precise geometry of these zones, and in some cases their existence, is under investigation. Plate motions shown are for the NUVEL-1 global relative plate motion model. Arrow lengths are proportional to the displacement if plates maintain their present relative velocity for 25 My. Divergence across mid-ocean ridges is shown by diverging arrows. Convergence is shown by single arrows on the underthrust plate. (After Gordon and Stein, 1992.)

This effect is especially evident in continental interiors, such as the India–Eurasia collision zone in the Himalayas or the Pacific–North America boundary zone in the Western US. Plate boundary zones (Fig. 8), indicated by earthquakes, volcanism, and other deformation, appear to cover about 15% of the Earth's surface (Gordon and Stein, 1992; Stein, 1993).

Insight into plate boundary zones is being obtained by combining focal mechanisms with geodetic, topographic, and geological data. Although plate motion models predict only the integrated motion across the boundary, GPS, geological, and earthquake data can show how this deformation varies in space and time. Both variations are of interest. Possible spatial variations include a single fault system taking up most of the motion (e.g., Prescott *et al.*, 1981), a smooth distribution of motion (e.g., England and Jackson, 1989), or motion taken up by a few relatively large microplates or blocks (e.g., Acton *et al.*, 1991; Thatcher, 1995). Each of these possibilities appears to occur, sometimes within the same boundary zone. The distribution of the motion in time is of special interest because steady motion between plate interiors gives rise to episodic motion at plate boundaries, as reflected in occasional large earthquakes, and in some cases steady creep (Fig. 9). The detailed relation between plate motions and earthquakes is complicated and poorly understood and hence forms a prime target of present studies.

For example, Figure 7a shows focal mechanisms and vectors derived from GPS illustrating the distribution of motion within the boundary zone extending from the stable interior of the oceanic Nazca plate, across the Peru–Chile trench to the coastal forearc, across the high Altiplano and foreland thrust belt, and into the stable interior of the South American continent. The GPS site velocities are relative to stable South America, so if the South American plate were rigid and all motion occurred at the boundary, they would be zero. Instead, they are highest near the coast and decrease relatively smoothly from the interior of the Nazca plate to the interior of South America. Figure 7b shows an interpretation of these data. In this, about half of the plate convergence (30–40 mm y^{-1}) is locked at the plate boundary thrust interface, causing elastic strain that is released in large interplate trench thrust earthquakes. Another 18–30 mm y^{-1} of the plate motion occurs aseismically by smooth stable sliding at the trench. The rest occurs across the sub-Andean fold-and-thrust belt, causing permanent shortening and mountain building, as shown by the inland thrust fault mechanisms. Comparison of strain tensors derived from GPS and earthquake data shows that the shortening rate inferred from earthquakes is significantly less than indicated by the GPS, implying that much of the shortening occurs aseismically. The focal mechanisms also indicate some deformation within the high Andes themselves. There may be some (at most 5–10 mm y^{-1}) motion of a forearc sliver distinct from the overriding plate, a phenomenon observed in some areas where plate convergence is oblique to the trench, making earthquake slip vectors at the trench trend between the trench-normal direction and the predicted convergence direction (McCaffrey, 1992).

Another broad plate boundary zone is the Pacific–North America boundary in western North America. Figure 10 shows the boundary zone, in a projection about the Euler pole. The relative motion is parallel to the small circle shown. Thus the

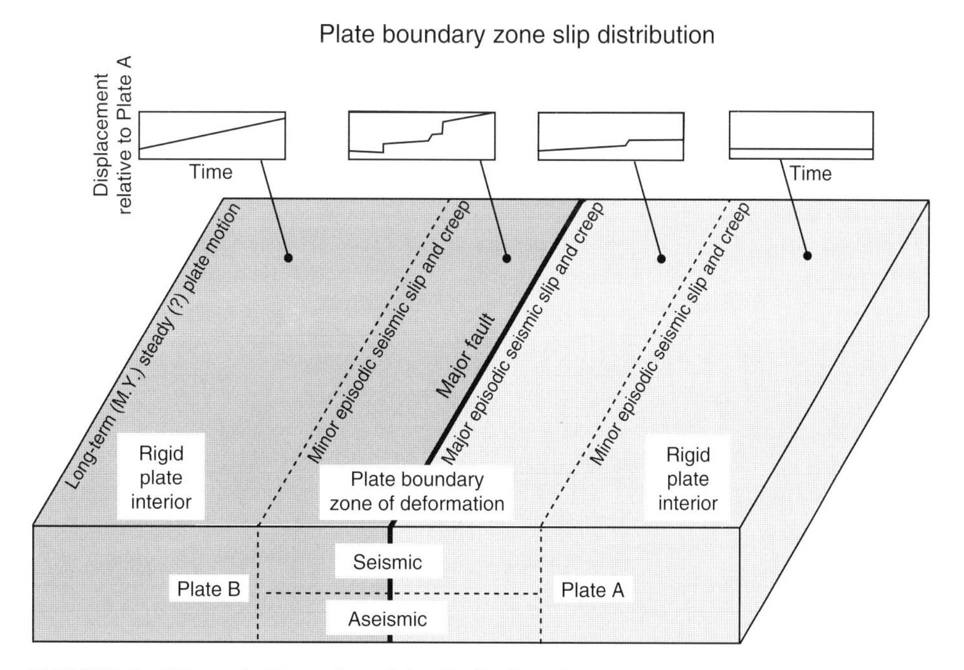

FIGURE 9 Schematic illustration of the distribution of motion in space and time for a strike-slip boundary zone between two major plates (Stein, 1993).

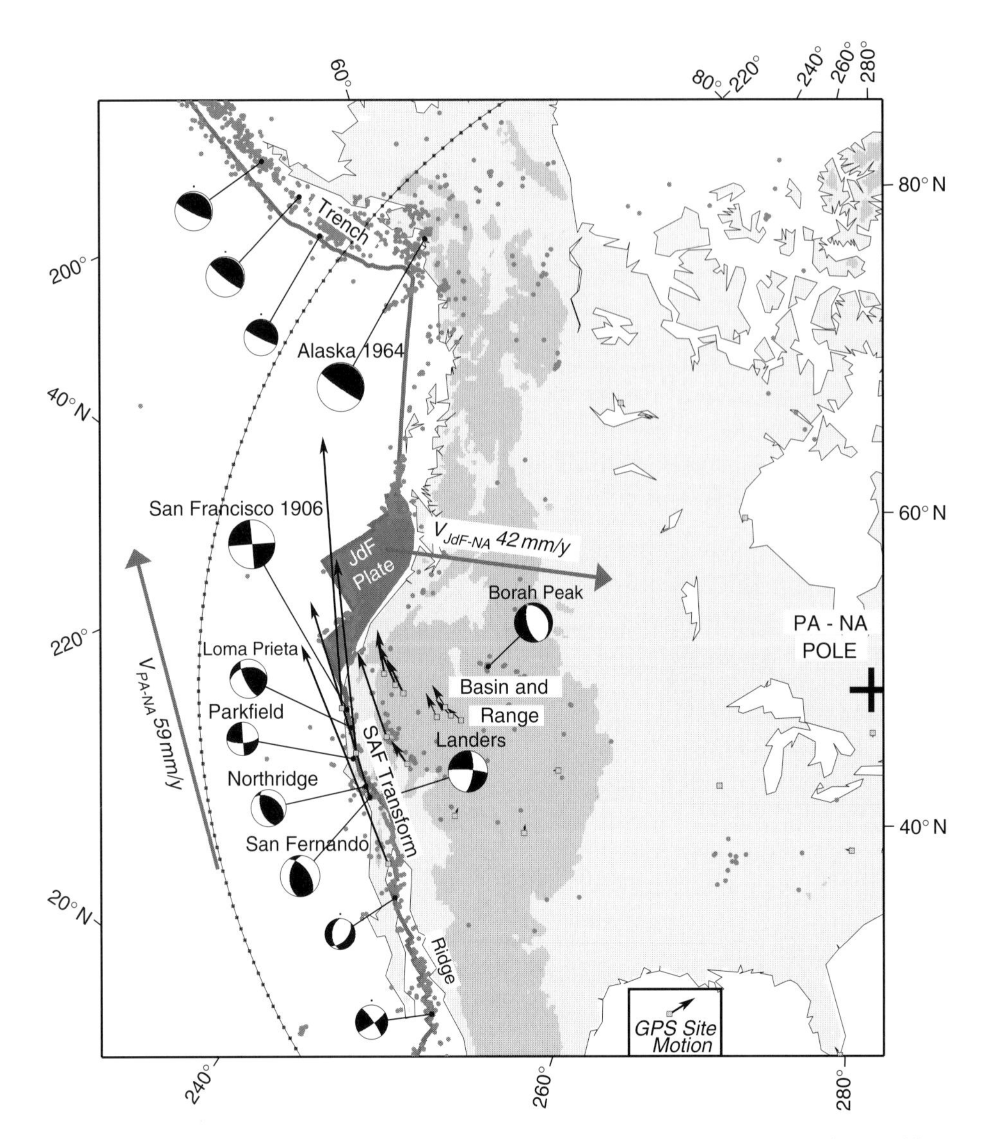

FIGURE 10 Geometry and focal mechanisms for a portion of the North America–Pacific boundary zone. Dot-dash line shows small circle, and thus direction of plate motion, about the Pacific–North America Euler pole. The variation in the boundary type along its length from extension, to transform, to convergence, is shown by the focal mechanisms. The diffuse nature of the boundary zone is shown by seismicity (small dots), focal mechanisms, topography (1000 m contour shown shaded), and vectors showing the motion of GPS and VLBI sites with respect to stable North America (Bennett *et al.*, 1999; Newman *et al.*, 1999).

boundary is extensional in the Gulf of California, essentially a transform along the San Andreas fault system, and convergent in the eastern Aleutians. The focal mechanisms reflect these changes. For example, in the Gulf of California we see strike-slip along oceanic transforms and normal faulting on a ridge segment. The San Andreas has both pure strike-slip earthquakes (Parkfield) and earthquakes with some dip-slip motion (Northridge, San Fernando, and Loma Prieta) when it deviates from pure transform behavior. The plate boundary zone is also broad, as shown by the distribution of seismicity. Although the San Andreas fault system is the locus of most of the plate motion and hence large earthquakes, seismicity extends as far eastward as the Rocky Mountains. For example, the Landers earthquake shows strike-slip east of the San Andreas, and the Borah Peak earthquake illustrates Basin and Range faulting. The diffuse nature of the boundary is also illustrated by vectors showing the motion of GPS and VLBI sites with respect to stable North America. Net motion across the zone is essentially that predicted by global plate motion model NUVEL-1A. The site motions show that most of the strike-slip occurs along the San Andreas fault system, but significant motions occur for some distance eastward.

5. Intraplate Deformation and Intraplate Earthquakes

A final important use of earthquake mechanisms is to study the internal deformation of major plates. Although idealized plates would be purely rigid, the existence of intraplate earthquakes reflect the important and poorly understood tectonic processes of intraplate deformation. One such example is the New Madrid area in the central United States, which had very large earthquakes in 1811–1812. The seismicity of such regions is generally thought to be due to the reactivation of preexisting faults or weak zones in response to intraplate stresses. Because motion in these zones are at most a few mm y^{-1}, compared to the generally much more rapid plate boundary motions, seismicity is much lower (Fig. 10). Similarly, major intracontinental earthquakes occur substantially less frequently than plate boundary events; recurrence estimates for 1811–1812 type earthquakes average 500–1000 years. Efforts are being made to combine geodetic data, which indicate deviations from rigidity, to the earthquake data. For example, comparison of the velocities for permanent GPS sites in North America east of the Rocky Mountains to velocities predicted by modeling these sites as being on a single rigid plate shows that the interior of the North American plate is rigid at least to the level of the average velocity residual, less than 2 mm y^{-1} (Dixon *et al*., 1996; Newman *et al*., 1999). Similar results emerge from geodetic studies of other major plates, showing that plates thought to have been rigid on geological time scales are quite rigid on decadal scales. Moreover, geological data suggest that such intraplate seismic zones may be active for only a few thousands of years, even though plate motions have been steady for millions of years. As a result, understanding how these intraplate seismic zones operate is a major challenge. A special case of this phenomenon occurs at passive margins, where continental and oceanic lithosphere join. Although these areas are in general tectonically inactive, magnitude 7 earthquakes can occur, as on the eastern coast of North America. Such earthquakes are thought to be associated with stresses at the continental margin, including those due to the removal of glacial loads, which reactivate the faults remaining along the continental margin from the original rifting.

References

Acton, G.D., S. Stein, and J.F. Engeln (1991). Block rotation and continental extension in Afar: a comparison to oceanic microplate systems. *Tectonics* **10**, 501–526.

Argus, D.F. and M.B. Heflin (1995). Plate motion and crustal deformation estimated with geodetic data from the Global Positioning System. *Geophys. Res. Lett.* **22**, 1973–1976.

Bennett, R.A., J.L. Davis, and B.P. Wernicke (1999). Present-day pattern of Cordilleran deformation in the Western United States. *Geology* **27**, 371–374.

Bina, C.R. (1996). Phase transition buoyancy contributions to stresses in subducting lithosphere. *Geophys. Res. Lett.* **23**, 3563–3566.

Chase, C.G. (1972). The *n*-plate problem of plate tectonics. *Geophys. J. R. Astron. Soc.* **29**, 117–122.

Cox, A. (1973). "Plate Tectonics and Geomagnetic Reversals." W.H. Freeman, San Francisco.

Cox, A. and R.B. Hart (1986). "Plate Tectonics: How it Works." Blackwell Scientific, Palo Alto.

DeMets, C., R.G. Gordon, D.F. Argus, and S. Stein (1990). Current plate motions. *Geophys. J. Int.* **101**, 425–478.

DeMets, C., R.G. Gordon, D.F. Argus, and Stein, S. (1994). Effect of recent revisions to the geomagnetic reversal time scale on estimates of current plate motion. *Geophys. Res. Lett.* **21**, 2191–2194.

Dixon, T.H., A. Mao, and S. Stein (1996). How rigid is the stable interior of the North American plate? *Geophys. Res. Lett.* **23**, 3035–3038.

Elsasser, W.M. (1969). Convection and stress propagation in the upper mantle. In: "The Application of Modern Physics to the Earth and Planetary Interiors" (S.K. Runcorn, Ed.), pp. 223–246. Wiley, New York.

Engeln, J.F., D.A. Wiens, and S. Stein (1986). Mechanisms and depths of Atlantic transform earthquakes. *J. Geophys. Res.* **91**, 548–577.

England, P. and J. Jackson (1989). Active deformation of the continents. *Annu. Rev. Earth Planet. Sci.* **17**, 197–226.

Gordon, R.G. and S. Stein (1992). Global tectonics and space geodesy. *Science* **256**, 333–342.

Huang, P.Y., S.C. Solomon, E.A. Bergman, and J.L. Nabelek (1986). Focal depths and mechanisms of Mid-Atlantic Ridge earthquakes from body waveform inversion. *J. Geophys. Res.* **91**, 579–598.

Isacks, B., J. Oliver, and L.R. Sykes (1968). Seismology and the new global tectonics. *J. Geophys. Res.* **73**, 5855–5899.

Kanamori, H. (1977). Seismic and aseismic slip along subduction zones and their tectonic implications. In: "Island Arcs, Deep-sea Trenches and Back-arc Basins, Maurice Ewing Series" (M. Talwani and W.C. Pitman, Eds.), Vol. III, pp. 163–174. American Geophysical Union, Washington, DC.

Larson, K.M., J.T. Freymueller, and S. Philipsen (1997). Global plate velocities from the Global Positioning System. *J. Geophys. Res.* **102**, 9961–9981.

McCaffrey, R. (1992). Oblique plate convergence, slip vectors, and forearc deformation. *J. Geophys. Res.* **97**, 8905–8915.

Menard, H.W. (1986). "The Ocean of Truth: A Personal History of Global Tectonics," Princeton Series in Geology and Paleontology (A.G. Fischer, Ed.), Princeton University Press, Princeton.

Minster, J.B. and T.H. Jordan (1978). Present-day plate motions. *J. Geophys. Res.* **83**, 5331–5354.

Newman, A., S. Stein, J. Weber, J. Engeln, A. Mao, and T. Dixon (1999). Slow deformation and lower seismic hazard at the New Madrid Seismic Zone. *Science* **284**, 619–621.

Norabuena, E., L. Leffler-Griffin, A. Mao, T. Dixon, S. Stein, I.S. Sacks, L. Ocala, and M. Ellis (1998). Space geodetic observations of Nazca–South America convergence along the Central Andes. *Science* **279**, 358–362.

Prescott, W.H., M. Lisowski, and J.C. Savage (1981). Geodetic measurements of crustal deformation on the San Andreas,

Hayward, and Calaveras faults, near San Francisco, California. *J. Geophys. Res.* **86**, 10853–10869.

Robbins, J.W., D.E. Smith, and C. Ma (1993). Horizontal crustal deformation and large scale plate motions inferred from space geodetic techniques. In: "Contributions of Space Geodesy to Geodynamics: Crustal Dynamics, Geodynamics Series" (D.E. Smith and D.L. Turcotte, Eds.), Vol. 23, pp. 21–36. American Geophysical Union, Washington, DC.

Stein, S. (1993). Space geodesy and plate motions. In: "Space Geodesy and Geodynamics, Geodynamics Series" (D.E. Smith and D.L. Turcotte, Eds.), Vol. 23, pp. 5–20. American Geophysical Union, Washington, DC.

Stewart, L.M. and E.A. Okal (1983). Seismicity and aseismic slip along the Eltanin Fracture Zone. *J. Geophys. Res.* **88**, 10495–10507.

Thatcher, W. (1995). Microplate versus continuum descriptions of active tectonic deformation. *J. Geophys. Res.* **100**, 3885–3894.

Vassiliou, M.S. (1984). The state of stress in subducting slabs as revealed by earthquakes analyzed by moment tensor inversion. *Earth Planet. Sci. Lett.* **69**, 195–202.

Uyeda, S. (1978). "The New View of the Earth." W.H. Freeman, San Francisco.

Editor's Note

Please see also Chapter 6, Continental Drift, Sea-Floor Spreading, and Plate/Plume Tectonics, by Uyeda.

II

Theoretical Seismology

8

Theoretical Seismology: An Introduction

Agustín Udías
Universidad Complutense de Madrid, Madrid, Spain

1. Introduction

Theoretical seismology is based on the principles and equations of the mechanics of continuous media. This is an idealization in which the granular structure of the materials of the Earth and their molecular and atomic nature are not considered. From this point of view, the Earth is assumed to be an imperfectly elastic, heterogeneous, anisotropic body of approximate ellipsoidal shape in which vibrations are produced by earthquakes. The mathematical analysis of these vibrations is performed by finding adequate solutions for the wave equation. Complete descriptions include the finite dimensions of the Earth and are given by the sum of the normal modes of its eigenvibrations. If wavelengths are small compared to the Earth's dimensions, solutions can be established as propagating waves in an unbounded medium. Flat Earth approximations are sufficient for some problems, while others require the consideration of its nearly spherical shape. In the simplest models, the Earth is considered as a homogeneous, isotropic, perfectly elastic medium. Heterogeneity in the Earth's materials can be treated using layered media with different elastic coefficients or letting these properties vary with the spatial coordinates. Ray theory provides a useful high-frequency approximation to wave propagation in heterogeneous media. In many cases, the assumption of a spherical Earth with radial symmetry provides a close approximation to the real situation, but more realistic models must include three-dimensional heterogeneities. Lack of perfect elasticity is accounted for by introducing dissipation of energy and attenuation factors. As a first approximation, isotropic models are adopted, but further analysis needs to consider anisotropic conditions. By proceeding through these successive modifications in Earth models, its imperfect elasticity, heterogeneity, and anisotropy can be adequately considered.

The generation of earthquakes is considered by the theory of source mechanisms. The processes of fracture on near-surface geological faults or of other types at greater depths are represented by mechanical models of increasing complexity. When observations at long distance and low frequencies are used, the source can be represented by simple models consisting of systems of forces or dislocations acting at a point. In this case, the first-order seismic moment tensor provides a useful description of a general type of point source. Dimensions of the source and rupture propagation can be treated from kinematic and dynamic approaches. In a more complete description, complex dynamic models with variable distributions of stress and friction on the rupture surface provide more realistic descriptions of earthquake source processes. Theoretical seismology starts with the recognition by J. Michell in 1761 and T. Young in 1807 that motion produced by earthquakes propagates in the Earth in the form of waves with finite velocity. Developments in the theory of elasticity by, among others, Bernoulli, Euler, Navier, Poisson, Cauchy, Stokes, Lord Rayleigh, Lamb, Volterra, and Love, began to be applied to the problems of seismic wave generation and propagation by the first generation of seismologists such as Oldham, Wiechert, Omori, Mohorovičić, Zöppritz, and Herglotz and later, before 1939, by Gutenberg, Jeffreys, Bullen, Nakano, Stoneley, and others (Ben Menahem, 1995). Theoretical developments of seismology after 1939 are summarized in chronological order by Ben Menahem (1995). The theory of seismology can be found in the general seismology textbooks such as Macelwane and Sohon (1936), Bullen (1947), Sawarensky and Kirnos (1960), Pilant (1979), Aki and Richards (1980), Ben Menahem and Singh (1981), Bullen and Bolt (1985), Lay and Wallace (1995), Dahlen and Tromp (1998), and Udías (1999). References not given in this chapter can be found in these texts.

In this chapter the most basic concepts necessary for the understanding of the excitation and propagation of elastic waves in the Earth are presented. More advanced developments and special topics can be found in other chapters of this Handbook; for example, seismic ray theory in Chapter 9 by Chapman, scattering and attenuation of seismic waves in

INTERNATIONAL HANDBOOK OF EARTHQUAKE AND ENGINEERING SEISMOLOGY, VOLUME 81A *ISBN: 0-12-440652-1*

Chapter 13 by Sato, Fehler, and Wu, and earthquake dynamics in Chapter 12 by Madariaga and Olsen.

2. The Earth as an Imperfectly Elastic Medium

2.1 Constitutive Relations

The first approximation to the mechanical behavior of the Earth's material is that corresponding to an elastic body. According to Hooke's law, for a perfect elastic body and infinitesimal deformations, stress and strain are linearly related. Cauchy's formulation is given by the tensor relation

$$\tau_{ij} = C_{ijkl} e_{kl} \quad (1)$$

where τ_{ij} is the stress tensor, e_{kl} is the infinitesimal strain tensor, and C_{ijkl} is the fourth-order tensor of elastic coefficients. Repeated subindices are summed over their three values. The strain energy is given by

$$W = \tfrac{1}{2} C_{ijkl} e_{ij} e_{kl} \quad (2)$$

In terms of the derivatives of the elastic displacements, since due to the symmetry of τ_{ij} and $e_{kl}, C_{ijkl} = C_{klij}$, Eq. (1) gives

$$\tau_{ij} = C_{ijkl} u_{k,l} \quad (3)$$

where the comma in the subscript represents partial deriatives with respect to the coordinates indicated by the following index. For a heterogeneous medium the components of C_{ijkl} are functions of the space coordinates. For a perfect elastic body in the general case, the tensor has 21 independent elements. If in the medium there is some kind of symmetry, the number of independent components is reduced. For example, for orthorhombic symmetry it is reduced to 9, for hexagonal symmetry to 5, and for cubic symmetry to 3. A simplifying assumption is that of isotropy, where the coefficients are reduced to two:

$$C_{ijkl} = \lambda \delta_{ij} \delta_{kl} + \mu(\delta_{ik}\delta_{jl} + \delta_{il}\delta_{jk}) \quad (4)$$

where λ and μ are Lamé coefficients, μ is the shear modulus, and λ is related to the bulk modulus K, which relates changes in volume to applied normal stresses, $K = (\tau_{11} + \tau_{22} + \tau_{33})/3(e_{11} + e_{22} + e_{33})$. According to Eq. (2), the strain energy is given by

$$W = \tfrac{1}{2}\lambda(e_{11} + e_{22} + e_{33})^2 + \mu e_{ij} e_{ij} \quad (5)$$

Most problems in seismology can be solved considering the Earth as an isotropic medium. Furthermore, equations are simplified for the special case of $\lambda = \mu$ (Poisson solid), approximately satisfied in the materials of the Earth's interior. Although deviations from isotropy are small, they are observed in wave propagation in the Earth. Many problems can be approximated by assuming hexagonal symmetry. In this case, there are five elastic coefficients, A, C, F, L, and N (Love, 1945). This symmetry has a principal axis and is called transverse symmetry since any direction normal to this axis has the same properties. Strain energy for a transverse isotropic medium is given by

$$W = \tfrac{1}{2}A(e_{11}^2 + e_{22}^2) + \tfrac{1}{2}Ce_{33}^2 + F(e_{11} + e_{22})e_{33} + (A - 2N)e_{22}e_{11} + \tfrac{1}{2}L(e_{13}^2 + e_{31}^2) + Ne_{12}^2 \quad (6)$$

Imperfect elasticity of Earth's materials is a consequence of energy dissipation in its deformation due to internal friction. One way to introduce the nonelastic behavior is to add a viscous component of deformation where stress depends on the time rate of strain. Mechanical viscoelastic models have the viscous element in series (Kelvin–Voight body), in parallel (Maxwell body), or both (standard linear solid). The effect of the viscous element is to introduce a relaxation time (T) into the stress–strain relation. In the case of the standard linear solid, strain is relaxed with relaxation time T_e under constant stress and stress is relaxed with T_t under constant strain.

Another way to introduce anelasticity is to make the elastic coefficients complex numbers. For an isotropic medium the shear and bulk moduli are

$$\mu' = \mu + i\mu*, \qquad K' = K + iK* \quad (7)$$

The real and imaginary parts of μ' are related to the quality factor Q that is a measure of the energy lost in wave propagation:

$$\frac{1}{Q_\mu} = 2\left(\frac{\mu*}{\mu}\right)^{1/2} \quad (8)$$

In a similar manner we can define Q_K. In the viscoelastic models Q can be related to the relaxation times and the frequency ω of the harmonic excitation; for a standard linear solid this is given by

$$\frac{1}{Q} = \frac{\omega(T_e - T_t)}{1 + \omega^2 T_e T_t} \quad (9)$$

The elastic and anelastic properties of Earth's materials are usually represented by the elastic coefficients and the quality factor.

2.2 The Equation of Motion

The dynamic behavior of an elastic body is given by the equation of motion (Newton's second law). For a volume V surrounded by a surface S (Fig. 1), according to Euler's formulation, this equation is given by

$$\int_V F_i \, dV + \int_S T_i \, dS = \frac{d}{dt}\int_V \rho v_i \, dV \quad (10)$$

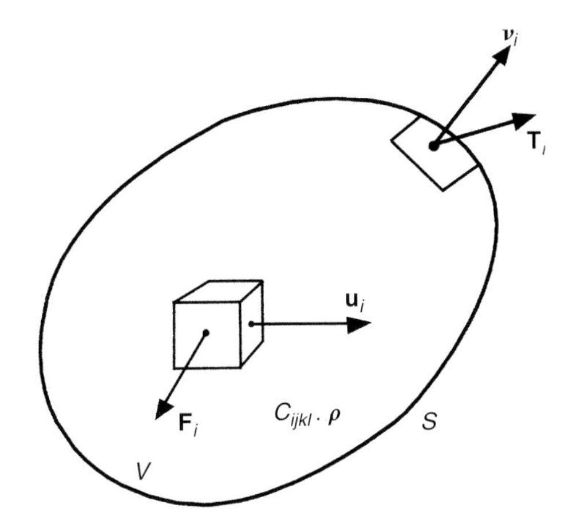

FIGURE 1 Elastic medium of volume V surrounded by surface S, forces $\mathbf{F}$, and displacements $\mathbf{u}$ at each element of volume and stress $\mathbf{T}$ at each element of surface of normal $\boldsymbol{\nu}$. (From Udías, 1999.)

where $\mathbf{F}$ are forces acting at volume elements dV or body forces, $\mathbf{T}$ are stresses (tractions) acting across surface elements dS, and v is the velocity at each point of the volume (Fig. 1). Expressing the vector $\mathbf{T}$ by the tensor τ_{ij}, using $T_i = \tau_{ij}\nu_j$, where ν_j is the normal to the surface element dS, and applying Gauss's theorem to the surface integral to convert it into a volume integral, we obtain, for an infinite medium with density constant with time, the differential equation

$$\frac{\partial \tau_{ij}}{\partial x_j} + F_i = \rho \frac{dv_i}{dt} \tag{11}$$

This equation can be written in terms of elastic displacements by the substitution of Eq. (3), approximating the total time derivatives by partial ones in the Lagrangian formulation,

$$\frac{\partial}{\partial x_j}\left(C_{ijkl}u_{k,l}\right) + F_i = \rho \frac{\partial^2 u_i}{\partial t^2} \tag{12}$$

For an isotropic material, substituting Eq. (4) in Eq. (12), we obtain

$$\left[\lambda \delta_{ij} u_{k,k} + \mu\left(u_{i,j} + u_{j,i}\right)\right]_{,j} + F_i = \rho \ddot{u}_i \tag{13}$$

where the dots represent the time derivatives. Equations (12) and (13) are the fundamental second-order differential equations for the elastic displacements in an infinite medium for the general case [Eq. (12)] and isotropic conditions [Eq. (13)]. In heterogeneous media, these equations include the spatial derivatives of the elastic coefficients and are very difficult to solve. For a homogeneous medium, we can write Eq. (13), in vector form:

$$(\lambda + \mu)\boldsymbol{\nabla}(\boldsymbol{\nabla} \cdot \mathbf{u}) + \mu \nabla^2 \mathbf{u} + \mathbf{F} = \rho \ddot{\mathbf{u}} \tag{14}$$

This expression, known as the Navier equation, represents the equation of motion in terms of displacements for a continuous, homogeneous, isotropic, infinite, elastic medium. This equation can be easily transformed into

$$\alpha^2 \boldsymbol{\nabla}(\boldsymbol{\nabla} \cdot \mathbf{u}) - \beta^2 \boldsymbol{\nabla} \times (\boldsymbol{\nabla} \times \mathbf{u}) + \frac{\mathbf{F}}{\rho} = \ddot{\mathbf{u}} \tag{15}$$

In this equation we have introduced the parameters α and β whose values in terms of the elastic coefficients are

$$\alpha^2 = \frac{\lambda + 2\mu}{\rho} = \frac{K + \frac{4}{3}\mu}{\rho} \tag{16}$$

$$\beta^2 = \frac{\mu}{\rho} \tag{17}$$

The parameter α is related to the divergence of $\mathbf{u}$ and in consequence to changes in volume and β to the curl of $\mathbf{u}$, that is, to changes in form without changes in volume. Equations (14) and (15), if there are no body forces ($\mathbf{F} = 0$), become homogeneous Navier differential equations.

Equation (15) can be separated into two equations each depending only on α or β using the potential functions for the displacements and the forces, according to Helmholtz's theorem,

$$\mathbf{u} = \boldsymbol{\nabla}\phi + \boldsymbol{\nabla} \times \boldsymbol{\psi} \tag{18}$$

$$\mathbf{F} = \nabla\Phi + \boldsymbol{\nabla} \times \boldsymbol{\Psi} \tag{19}$$

where ϕ and Φ are scalar potentials and $\boldsymbol{\psi}$ and $\boldsymbol{\Psi}$ vector potentials with zero divergence. Applying Eqs. (18) and (19) to Eq. (15) we obtain

$$\alpha^2 \nabla^2 \phi + \frac{\Phi}{\rho} = \frac{\partial^2 \phi}{\partial t^2} \tag{20}$$

$$\beta^2 \nabla^2 \boldsymbol{\psi} + \frac{\boldsymbol{\Psi}}{\rho} = \frac{\partial^2 \boldsymbol{\psi}}{\partial t^2} \tag{21}$$

The vector differential equation of motion [Eq. (14) or (15)] is now expressed in terms of the potentials by a scalar and a vector equation of a much more simple form. Equation (20) is related to the elastic perturbations that imply changes in volume, and Eq. (21) to perturbations with only changes in form. This separation of the equation of motion is very important, because it greatly facilitates its solution. Unfortunately, this is only possible in homogeneous media.

Equations (14), (15), (20), and (21) are second-order differential equations for partial derivatives of the displacements or their potentials with respect to spatial coordinates and time with an independent term formed by the body forces or their potentials. If we specify these forces, the solution of these equations gives us the elastic displacement field for an infinite homogeneous isotropic medium. These equations are very important in seismology, since many problems can be solved using this approximation.

2.3 The Green Function and the Representation Theorem

The solution of many problems in seismology can be formulated using the Green function of elastodynamics, which represents the response of the medium to a unit impulsive force in space and time with arbitrary direction. This force may be represented mathematically by means of the Dirac delta function:

$$\mathbf{F}_n(x_j, t) = \delta(x_j - \xi_j)\,\delta(t - \tau)\mathbf{e}_n \tag{22}$$

The force $\mathbf{F}_n$ is applied at the point of coordinates ξ_j and time τ, is null outside this point and time, and its orientation is given by the unit vector $\mathbf{e}_n$. If we substitute this force in the equation of motion (10), its solutions are the elastic displacements for every point of coordinates x_j for every time t, given by a second-order tensor $G_{ni}(x_j, \xi_j, t, \tau)$, where subindex n refers to the direction of the exciting force. Replacing stress in terms of the derivatives of the displacements (3) for a homogeneous medium in the equation of motion (10), we obtain

$$\int_V \rho \ddot{G}_{ni}\, dV - \int_S C_{ijkl} G_{nk,l} \nu_j\, dS = \int_V \delta(x_s - \xi_s)\delta(t - \tau)\delta_{ni}\, dV \tag{23}$$

The solutions of Eq. (23) are the Green functions of the medium and depend on its characteristics, elastic coefficients and density, shape of volume V, and boundary conditions on S. For each medium there is a different Green function that describes how this medium reacts mechanically to an impulsive excitation and is, therefore, a proper characteristic of each medium.

The Green function can be used to find the elastic displacements in a medium corresponding to a specified system of body forces in an alternative way to solving the equation of motion (11). For this purpose we use the Green–Volterra formulation of Betti's reciprocity theorem. Assuming causality conditions, that is, that the medium is at rest until a time when motion starts, the elastic displacements at each point of the volume V are given by

$$u_n(x_s, t) = \int_{-\infty}^{\infty} d\tau \int_V F_i G_{ni}\, dV + \int_{-\infty}^{\infty} d\tau \int_S (G_{ni} T_i - C_{ijkl} u_i G_{nk,l} \nu_j)\, dS \tag{24}$$

This equation is generally referred to as the representation theorem. It gives us the elastic displacements inside a volume V as the sum of two double integrals over time and space. The first is the volume integral of the body forces multiplied by the Green function and the second the surface integral of the stress (tractions) across each element dS multiplied by the Green function plus the displacements times the derivatives of the Green function (Fig. 2). Equation (24) allows us to determine elastic displacements produced by a system of body forces defined in a volume or by a system of stresses and displacements defined over a surface by means of the appropriate Green functions. In this equation, which is very useful in source studies and the generation of synthetic seismograms, the Green function acts as a propagator which propagates the effects of forces, stresses, or displacements defined on coordinates (ξ_i, t) to determine the elastic displacements in points of coordinates (x_i, t). In general, the determination of Green's function is not an easy problem. The most simple case is for an infinite, homogeneous, isotropic, elastic medium, but solutions for other media such as layered half-space or a radially symmetric sphere have been determined and can be computed numerically.

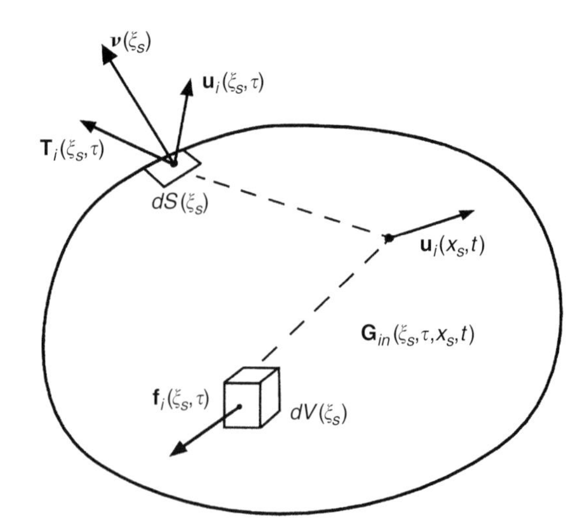

FIGURE 2 Forces acting inside a volume and stresses and displacements at its surface and displacements **u** at any point in terms of Green's function. (From Udías, 1999.)

3. Seismic Waves

3.1 Wave Equations and Their Solutions

Equations (20) and (21) in the absence of body forces become wave equations, where the parameters α and β are the wave velocities. The wave equation can be written in a general form, for a velocity c as,

$$\nabla^2 f(x_i, t) = \frac{1}{c^2} \frac{\partial^2}{\partial t^2} f(x_i, t) \tag{25}$$

The dependence on time of Eq. (25) can be eliminated if f has a harmonic dependence on time:

$$f(x_i, t) = \hat{f}(x_i) e^{-i\omega t} \tag{26}$$

Substituting in Eq. (25), we obtain Helmholtz's equation,

$$(\nabla^2 + k^2)\hat{f}(x_i) = 0 \tag{27}$$

where $k=\omega/c$ is the wavenumber. For harmonic waves, the form of the solutions of Eq. (27) depends on the geometry of the wave front and can be found expressing the Laplacian in Cartesian, cylindrical, or spherical coordinates.

In Cartesian coordinates, solutions for monochromatic waves of frequency ω can be written as

$$f(x_1,x_2,x_3,t) = A\ \exp\{i[k(x_1\nu_1+x_2\nu_2+x_3\nu_3)-\omega t+\varepsilon]\} \tag{28}$$

Waves propagate in the direction of the unit vector ν_i (normal to the wave front) with velocity $c=\omega/k$. The wavenumber can be separated into its three components, $k_i=k\nu_i$. This solution represent waves with plane wave fronts of infinite extension, which implies that their source is at an infinite distance. This condition does not correspond to any real problem; however, if we are interested only in wave propagation and can assume that the source is at very great distance, plane waves are a good approximation in the far field.

Many problems of wave propagation in seismology—for example, those concerning surface waves—have axial symmetry and are solved using cylindrical coordinates. For the most general cases with dependence on the three coordinates, $f(r,\phi,z,t)$, using the method of separation of variables, solutions are given in terms of $J_n(k_r r)$ (Bessel functions of order n) and harmonic functions of z, ϕ, and t. The wavenumber has components in the z and r directions and $k^2=k_r^2+k_z^2$. The general solution is given by the sum over all values of n. For symmetry with respect to ϕ and large values of r, we can express Bessel functions in asymptotic form in terms of harmonic functions, obtaining

$$f(r,z,t) = \left(\frac{2}{\pi k_r r}\right)^{1/2} A\ \exp\left[i\left(k_r r + k_z z - \omega t - \frac{\pi}{4}\right)\right] \tag{29}$$

In this equation, we find the dependence of wave amplitudes with the inverse of the square root of the distance r. Wave fronts are cylindrical surfaces whose areas increase with increasing r, and energy generated at the source is distributed over an increasing area. As a consequence, amplitudes per unit area decrease with distance. This decrease of amplitude with distance is called "geometric spreading." This term has not appeared in plane waves, since their wave fronts are always infinite planes and their source is at an infinite distance.

Propagation of seismic waves generated at a point leads to the consideration of spherical waves $f(r,\theta,\phi,t)$. In spherical coordinates for propagating waves, complex solutions are given by

$$f(r,\theta,\phi,t) = A_{nm}\left(\frac{2k}{\pi}\right)^{1/2} h_n(kr)P_n^m(\cos\,\theta)\ \exp[i(\pm m\phi-\omega t)] \tag{30}$$

where $h_n(kr)=j_n(kr)+in_n(kr)$, are Hankel spherical functions, ($j_n(kr)$ and $n_n(kr)$ are Bessel spherical functions of first and second kind) and $P_n^m(\cos\,\theta)$ the Legendre associate functions. As in the previous case, the general solution is the sum for all values of n and m. For each value of n, there are values of m from $m=-n$ to $m=n$. For symmetry with respect to ϕ and large values of r, using the asymptotic expression for $h(kr)$ in terms of harmonic functions, the solution may be written approximately as,

$$f(r,\theta,t) = A_n\frac{1}{r}\left(\frac{2}{k\pi}\right)^{1/2} P_n(\cos\,\theta)\exp\left[i\left(kr-\omega t-\frac{n+1}{2}\pi\right)\right] \tag{31}$$

Wave amplitudes decay with distance as $1/r$, which is the geometrical spreading factor for spherical waves.

Seismic waves are not monochromatic harmonic waves, but are usually pulses of different shapes or trains of waves with different frequencies. Waves with an arbitrary dependence of time can be represented by the sum or integral of harmonic waves of different frequencies using Fourier's theorem. For plane waves,

$$f(x_i,t) = \frac{1}{2\pi}\int_{-\infty}^{\infty} F(\omega)\ \exp\{i[k(\omega)\nu_k x_k - \omega t]\}\,d\omega \tag{32}$$

where $F(\omega)$ is a complex function termed the complex spectrum of $f(x_i,t)$ and can be represented as

$$F(\omega) = R(\omega) + iI(\omega) = A(\omega)e^{i\Phi(\omega)} \tag{33}$$

where $A(\omega)$ is the amplitude spectrum and $\Phi(\omega)$ is the phase spectrum. These two functions of frequency represent the contributions of each harmonic component to the amplitude and initial phase of the resulting function of time. Phase velocity $c(\omega)=\omega/k(\omega)$ is now a function of frequency, that is, each harmonic component may have different phase velocity. Motion due to a propagating wave at a point is given by a function of time, $f(t)$. The same information can be represented as a function of frequency $F(\omega)$, that is, obtained by means of the Fourier integral

$$F(\omega) = \int_{-\infty}^{\infty} f(t)e^{-i\omega t}\,dt \tag{34}$$

where $f(t)$ is a real function that represents amplitudes of waves at each time for a particular point of space. Its Fourier transform $F(\omega)$ is a complex function that can be represented by its amplitudes $A(\omega)$ and initial phases $\Phi(\omega)$ for each frequency. In this way, problems of wave propagation can be studied in the domain of time or frequency.

3.2 Body Waves

According to Eqs. (20) and (21) we obtain for an infinite homogeneous isotropic elastic medium two types of waves with velocities α and β. Since the potential ϕ is related to the divergence of the displacement $\mathbf{u}$ and $\boldsymbol{\psi}$ to its curl, α is the velocity of longitudinal waves (P waves) and β that of transverse or shear waves (S waves). Since these waves travel inside the medium, they are called body waves. The total elastic displacement of body waves $\mathbf{u}$ is the sum of the displacements of P and S waves, $\mathbf{u} = \mathbf{u}^P + \mathbf{u}^S$. In terms of the potentials (18), they can be written as

$$\mathbf{u}^P = \nabla\phi \tag{35}$$

$$\mathbf{u}^S = \nabla \times \boldsymbol{\psi} \tag{36}$$

Depending on the geometry of the problem, we will have solutions for the potentials ϕ and $\boldsymbol{\psi}$, like those discussed above [Eqs. (28) to (31)], with wavenumbers $k_\alpha = \omega/\alpha$ and $k_\beta = \omega/\beta$. For line sources and cylindrical waves, amplitudes decrease with distance as $1/\sqrt{r}$ and for point sources and spherical waves as $1/r$.

Displacement components of P and S waves in the Earth are usually referred to geographical axes (North, East, Down) (u_1^P, u_2^P, u_3^P) and (u_1^S, u_2^S, u_3^S). The S-wave displacement is on a plane normal to the ray and can be separated into two components SH (on the horizontal plane and normal to the ray) and the SV (on the incidence or vertical plane). The unit vectors $\mathbf{r}$ (direction of the ray), $\mathbf{SH}$ and $\mathbf{SV}$ form a system of axes in respect of which the displacement vector is (u_P, u_{SH}, u_{SV}).

It is convenient for many problems to separate the P–SV motion from the SH motion, selecting the plane (x_1, x_3) as the plane containing the ray. The components of displacements of P–SV motion (u_1 and u_3), can be expressed in terms of scalar potentials ϕ and ψ, the SH displacement (u_2) being kept apart:

$$u_1 = \phi_{,1} - \psi_{,3} = u_1^P + u_1^{SV} \tag{37a}$$

$$u_3 = \phi_{,3} - \psi_{,1} = u_3^P + u_3^{SV} \tag{37b}$$

Since the ray is contained in plane (x_1, x_3), we can define c as the apparent velocity of propagation in the direction of x_1, and write

$$\phi = (Ae^{ikrx_3} + De^{-ikrx_3})\exp[ik(x_1 - ct)] \tag{38a}$$

$$\psi = (Be^{iksx_3} + Ee^{-iksx_3})\exp[ik(x_1 - ct)] \tag{38b}$$

$$u_2 = (Ce^{iksx_3} + Fe^{-iksx_3})\exp[ik(x_1 - ct)] \tag{38c}$$

where, r and s are given by

$$r = \sqrt{\frac{c^2}{\alpha^2} - 1} = \tan e = \cot i \tag{39}$$

$$s = \sqrt{\frac{c^2}{\beta^2} - 1} = \tan f = \cot j \tag{40}$$

where e and f are the angles of emergence and i and j are the angles of incidence for P and S waves, respectively. For real values of r and s ($c > \alpha > \beta$), waves propagate in the x_3 direction, up and down, with apparent velocities c/r and c/s, respectively.

3.3 Reflection and Refraction

The presence in an elastic medium of any kind of surface across which there are changes in the properties, and especially, of a free surface, introduces a series of phenomena that must be considered in the study of wave propagation. First, there are body-wave reflections and refractions; that is, in the surface separating two media, the energy of incident waves is partly reflected into the same medium and partly transmitted or refracted into the second medium.

If e and f are the angles of emergence of incident and reflected P and S waves in medium M of velocities, α, β and e', f' are those of refracted or transmitted waves in medium M′ with velocities α' and β', then Snell's law is given by (Fig. 3)

$$\frac{\cos e}{\alpha} = \frac{\cos f}{\beta} = \frac{\cos e'}{\alpha'} = \frac{\cos f'}{\beta'} = \frac{1}{c} = p \tag{41}$$

where p is the ray parameter. As a function of the velocities, tangents of e and f are given by Eqs. (39) and (40). Reflection R and transmission T coefficients are given by the ratios of incident to reflected or refracted amplitudes and are found applying the boundary conditions at the surface of separation of the two media, namely, continuity of stress and displacement. Equations relating incident amplitudes to those transmitted and reflected are known as Zöppritz's equations. For SH waves and incidence from M to M′: $R_{SH} = C/C_0$ and $T_{SH} = C'/C_0$, where C_0 is the incident amplitude, C reflected and C' transmitted or refracted [in the notation of Aki and Richards (1980), $R_{SH} = \acute{S}\grave{S}$ and $T_{SH} = \acute{S}\acute{S}$]. The matrix expressing the total SH motion transmitted and reflected between two elastic media

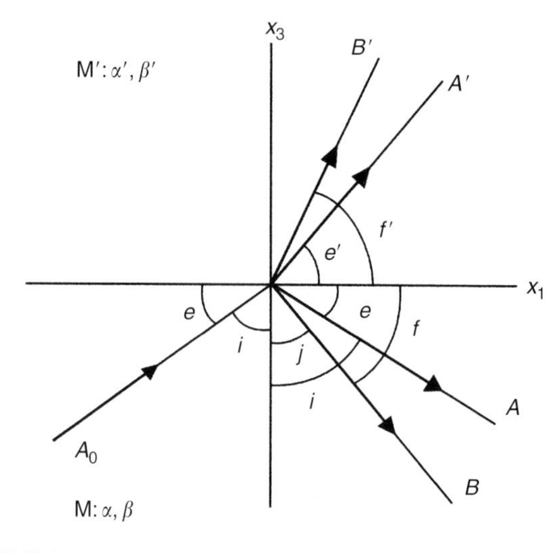

FIGURE 3 Incident P wave and reflected and refracted P and S waves. (From Udías, 1999.)

with incidence from both media (R'_{SH} and T'_{SH} for waves traveling from M′ to M, equivalent to $\acute{S}\grave{S}$ and $\acute{S}\acute{S}$), or the scattering matrix, is given by

$$\mathbf{SH} = \begin{pmatrix} R_{SH} & R'_{SH} \\ T_{SH} & T'_{SH} \end{pmatrix} \tag{42}$$

In a similar way for *P* and *SV* incident waves, we have the reflection and refraction coefficients,

$$R_{PP} = \frac{A}{A_0}; \quad R_{PS} = \frac{B}{A_0}; \quad T_{PP} = \frac{A'}{A_0}; \quad T_{PS} = \frac{B'}{A_0} \tag{43a}$$

$$R_{SS} = \frac{B}{B_0}; \quad R_{SP} = \frac{A}{B_0}; \quad T_{SS} = \frac{B'}{B_0}; \quad T_{SP} = \frac{A'}{B_0} \tag{43b}$$

The scattering matrix for the general *P–SV* motion generated in two elastic media in contact with incident waves traveling in both directions has 16 elements: 8 for incident *P* and *SV* waves traveling from medium M and another 8 for those traveling from medium M′.

$$\mathbf{P\text{–}SV} = \begin{pmatrix} R_{PP} & R_{SP} & R'_{PP} & R'_{SP} \\ R_{PS} & R_{SS} & R'_{PS} & R'_{SS} \\ T_{PP} & T_{SP} & T'_{PP} & T'_{SP} \\ T_{PS} & T_{SS} & T'_{PS} & T'_{SS} \end{pmatrix} \tag{44}$$

SH and **P–SV** scattering matrices can be combined into a total 6×6 matrix. Reflection and transmission coefficients for incident *P* and *S* waves in isotropic media are functions of the elastic coefficients λ and μ, and density ρ of the two media and their angles of incidence. Equations are simplified for the case when $\lambda = \mu$.

Under certain conditions, for angles of incidence greater than the critical, r or s become imaginary ($c < \alpha$, or $c < \beta$). Then waves represented by Eqs. (39) to (41) do not propagate in the x_3 direction but only along a direction parallel to the surface (x_1) and their amplitudes decrease with depth from the surface. These waves are called inhomogeneous or evanescent waves. For example, for *P* waves traveling from M to M′, if $\alpha' > \alpha$, $\sin(i_c) = \alpha/\alpha'$, where i_c is the critical angle. Incident waves with $i > i_c$ are not transmitted into medium M′, but are totally reflected into medium M (total reflection). However, inhomogeneous waves do appear in medium M′. The reflection and transmission coefficients represent amplitude quotients (displacements for *SH* and potentials for *P–SV*) and must be modified to represent the energy partition at the boundary. The analysis of these phenomena is very important for the interpretation of seismic reflection–refraction data in studies of the Earth's structure.

3.4 Surface Waves

A phenomenon related to the presence of surfaces separating media of different characteristics, and especially of a free surface, is the generation of what, in general, are called surface waves. These waves are produced by constructive interference of body waves in connection with such surfaces. To generate surface waves we need at least two types of waves that can interfere in a constructive way. Their main characteristics are that they propagate parallel to the surface and their amplitudes decrease with depth from the surface. In seismology, the most important are those related to the presence of the Earth's free surface and those that are also affected by other surfaces of contact between layers of different elastic properties in its interior.

The most basic problem is the study of the existence of surface waves in an elastic, homogeneous half-space. Since we are considering surface waves, their amplitudes must decrease with depth ($-x_3$) [in Eqs. (39) to (41), $A = B = C = 0$, and r and s must be both imaginary ($c < \beta < \alpha$)]. The boundary conditions are that the three stress components across the free surface (tractions) must be null. For the τ_{32} component the condition leads to $F = 0$, and in this case surface waves do not have transverse component (u_2). The other two conditions ($\tau_{31} = \tau_{33} = 0$) lead to two homogeneous equations whose solution require that the determinant of the system must be null. This condition results in the Rayleigh equation for the phase velocity c, namely

$$\left(2 - \frac{c^2}{\beta^2}\right)^2 = 4\sqrt{1 - \frac{c^2}{\alpha^2}}\sqrt{1 - \frac{c^2}{\beta^2}} \tag{45}$$

For $\lambda = \mu$ ($\alpha = \sqrt{3}\beta$), the solution compatible with the existence of surface waves ($c/\beta < 1$) is $c = 0.9194\beta$, the phase velocity of Rayleigh waves. These waves have displacement components u_1 and u_3, shifted in phase by $\pi/2$, with elliptical particle motion with vertical major axis and amplitude decreasing with depth (Fig. 4).

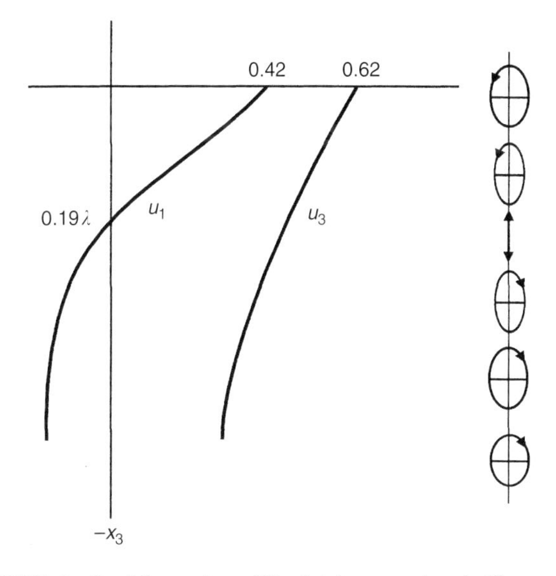

FIGURE 4 Particle motion of Rayleigh waves in a half-space and dependence of displacement components with depth. (From Udías, 1999.)

The propagation of surface waves in layered media of isotropic material with a free surface introduces new elements with respect to the half-space problem. Inside each layer there may exist waves traveling upward and downward, so that in Eqs. (39) to (41) s is real and r is real or imaginary and in the half-space both r and s must be imaginary. Boundary conditions are null stress components across the free surface, and continuity of displacement and stress across the surfaces of contact between layers and the last layer and the half-space. The solutions result, first, in two types of waves with different velocity, Rayleigh (P–SV) and Love (SH). Second, phase velocities depends on frequency $c(\omega)$, that is, waves are dispersed. Third, there are different modes of propagation due to the relation between layer thickness and wavelengths.

The above-mentioned characteristics appear in the basic problem of a single layer over a half-space. For a layer of S velocity β', rigidity μ', and thickness H over a half-space of velocity β and rigidity μ with $\beta > \beta'$, the solution of the problem for the SH displacement results in the existence of Love waves with phase velocity c ($\beta > c > \beta'$) that satisfies the dispersion equation

$$\frac{\mu\sqrt{1 - c^2/\beta^2}}{\mu'\sqrt{c^2/\beta'^2 - 1}} = \tan\left(kH\sqrt{c^2/\beta'^2 - 1}\right) \tag{46}$$

The phase velocity of Love waves is a function of frequency, $c(\omega)$, or wavenumber, $c(k)$. Since the tangent function has positive values between zero and infinity for different intervals of its argument (the first between 0 and $\pi/2$, the second between π and $3\pi/2$, and so on), each of them corresponds to a mode of propagation, the fundamental mode to the first interval ($0 < kHs' < \pi/2$) and higher modes to the rest. Higher modes have a cut-off frequency and their displacements inside the layer have a number of nodes equal to the mode number (Fig. 5).

The problem for P–SV motion in a layer over a half-space is more complicated and its outcome is the existence of Rayleigh waves. The dispersion equation follows from the expansion of a 6×6 determinant resulting from the six boundary conditions (null stress components at the free surface and continuity of stress and displacement between the layer and the half-space). Phase velocity is a function of frequency and modes are now separated into pairs of symmetric and antisymmetric modes. Limits of Rayleigh wave phase velocity for the fundamental mode are $\beta > c > \beta'$ and for higher modes $c_R > c > c'_R$ (c_R and c'_R are the Rayleigh wave velocities corresponding to the half-space and the layer).

The multilayer problem involves solving the equations resulting from the boundary conditions, free stresses across the free surface, and continuity of stress and displacement at each boundary, with r and s imaginary in the half-space. The problem is formulated in matrix form according to the Thomson–Haskell method or using the propagator matrix

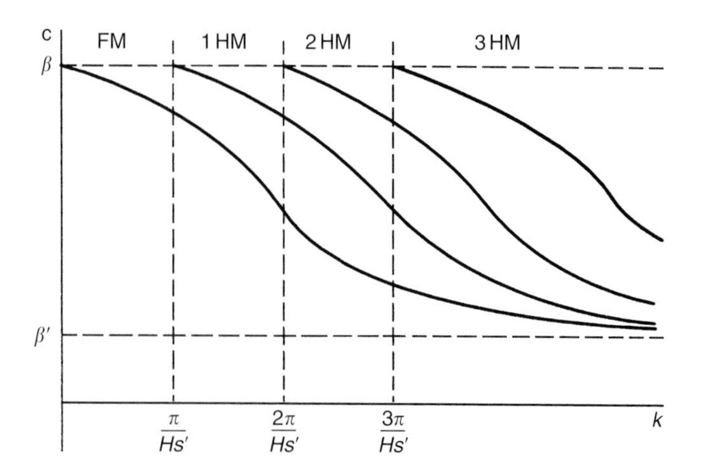

FIGURE 5 Dispersion curves for Love waves in a layer over a half-space for fundamental (FM) and higher modes (HM). (From Udías, 1999.)

(Section 3.6). Surface waves in spherically stratified media introduce certain modifications into the phase velocities with respect to the flat case.

3.5 Wave Dispersion. Phase and Group Velocity

We have seen that surface waves in layered media are dispersed, that is, their phase velocity is a function of frequency (or period). Then, they are formed by trains of waves, with different frequencies arriving at different times. Arrival times, amplitudes, and phases for each frequency depend on the dispersion equation [for example, Eq. (46)]. If phase velocity is a function of frequency $c(\omega)$ and we can also write $k(\omega)$ and $\omega(k)$, then the group velocity corresponding to the velocity of energy propagation is given by $U = d\omega/dk$ and in terms of the phase velocity,

$$U = c + k\frac{dc}{dk} \tag{47}$$

Amplitudes of dispersed waves are a function of frequency and can be written as in Eq. (32) as an integral over all frequencies. In the integrand, the amplitude $F(\omega)$ varies very slowly compared with the variation of the phase ($\Phi(\omega) = kx - \omega t$). Then the integral has only significant values for frequencies corresponding to stationary values of the phase, that is, for $d\Phi/d\omega = 0$. For a wave propagating in the x direction, taking derivatives of the phase with respect to the frequency we obtain

$$\frac{d\Phi}{d\omega} = \frac{d}{d\omega}(kx - \omega t) = \frac{x}{U} - t \tag{48}$$

For the frequency ω_0 that makes the phase stationary ($d\Phi/d\omega = 0$), the group velocity is $U = x/t$. Group velocity is, then,

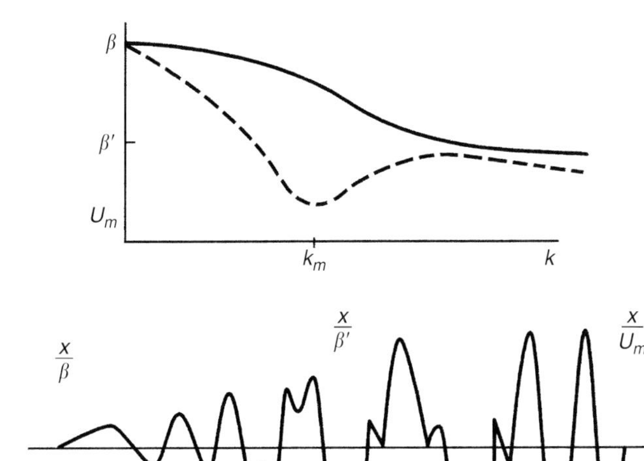

FIGURE 6 Phase and group velocity curves and a dispersed wave train. (From Udías, 1999.)

the velocity corresponding to the frequency that makes the phase stationary for a given distance and time. Using the principle of stationary phase, the integral of Eq. (32) for waves propagating in the x direction with variable phase velocity $c(k)$ results in

$$u(x,t) = A(k_0)\left[\frac{2\pi}{\dfrac{u}{U}\dfrac{dU}{dk}}\right]^{1/2} \cos\left(k_0 x - \omega_0 t \pm \frac{\pi}{4}\right) \tag{49}$$

Then, for a given time and distance, energy is contained in a harmonic wave of wavenumber k_0 and frequency ω_0, values corresponding to the stationary phase. Since dU/dk is in the denominator, largest amplitudes correspond to $dU/dk = 0$ (minimum of the group velocity curve), or the Airy phase (Fig. 6).

3.6 Wave Propagation in Layered Media

Many problems of wave propagation in seismology, not only propagation of surface waves, can be solved by representing the Earth as a stratified or layered medium, that is, formed by layers of a certain thickness and with different mechanical properties. For certain problems we can use a flat approximation of parallel horizontal layers and the problem is reduced to two dimensions. Layers of constant properties may be considered as an approximation for media where elastic coefficients vary in continuous form with depth. In layered or stratified media, problems are presented in discrete form and may be treated using matrix formulation (Kennett, 1983). In this formulation, we make use of the displacement–stress vector and, as in Eqs. (39) to (41), we separate the vertical and the horizontal dependence of propagation. Considering only the dependence with depth, putting $z = x_3$, for *SH* and *P–SV* we can write,

$$\mathbf{SH} = [u_2(z), \tau_{32}(z)] \tag{50}$$

$$\mathbf{P–SV} = [u_1(z), u_3(z), \tau_{31}(z), \tau_{33}(z)] \tag{51}$$

It can be shown that the derivatives with respect to z of the components of the displacement–stress vector are linear functions of the same components. The fundamental equations for the development of matrix methods of resolution of wave propagation in layered media are based on this property and can be written in a general matrix form as,

$$\frac{d\mathbf{b}(z)}{dz} = \mathbf{A}(c,z)\mathbf{b}(z) \tag{52}$$

where $\mathbf{b}$ is the displacement–stress vector defined in Eqs. (50) and (51) and $\mathbf{A}$ is a square matrix whose elements depend on the elastic parameters and density of the medium functions of depth ($\alpha(z)$, $\beta(z)$, $\rho(z)$), frequency ω and horizontal velocity c. In the case of layers of constant properties within each layer, the only variables are c and ω. The displacement–stress vector at a level z_n is related to that of a reference level z_0 by means of a so-called propagator matrix $\mathbf{P}(z, z_0)$ by

$$\mathbf{b}(z_n) = \prod_{i=1}^{n} \mathbf{P}(z_i, z_{i-1})\mathbf{b}(z_0) \tag{53}$$

The propagator matrix $\mathbf{P}$ is related to the solution of Eq. (52) and, in general, it can be computed numerically. The matrix formulation is a powerful method for solving wave propagation problems in layered media such as surface waves and *SH* and *P–SV* reflected and refracted waves.

4. Ray Theory

In a heterogeneous medium, elastic coefficients and density, and consequently wave velocities, are functions of the spatial coordinates. In Eqs. (12) and (13) we will have derivatives of the elastic parameters, resulting in complicated expressions even for isotropic media. If we assume that elastic properties change very slowly, so that over a distance of one wavelength the wave velocity changes much less than the wave velocity itself, then, we can use the approximation of ray theory. Ray theory is a powerful method used extensively in seismology for the study of the Earth's interior and calculation of synthetic seismograms. In this section a very short introduction is given to its basic principles. An extended treatment of the subject can be found in Chapter 9 by Chapman. Ray theory is a high-frequency approximation for the solution of the wave equation that holds for media with properties that vary very slowly. Under these conditions we can write an approximate wave equation such as Eq. (25), where $c(x_i)$ is a function of coordinates. For harmonic time dependence, according to Eq. (27), we have

$$\left(\nabla^2 + \frac{\omega^2}{c^2(x_i)}\right)u(x_i) = 0 \tag{54}$$

Solutions of this equation can be written as

$$u(x_i) = A(x_i)\ \exp[i\omega T(x_i)] \tag{55}$$

where the amplitudes $A(x_i)$ and the travel times $T(x_i)$ are functions of the coordinates. Ray theory approximation implies that spatial gradients of A and T are small in comparison with the functions themselves. Solutions to this problem in terms of ray theory are given by the eikonal equation and the WKBJ (Wentzel, Kramers, Brillouin, and Jeffreys) method. Substituting Eq. (55) into Eq. (54), with the approximation $|A''/A| \ll \omega/c$, we obtain

$$\left(\frac{\partial T}{\partial x_1}\right)^2 + \left(\frac{\partial T}{\partial x_2}\right)^2 + \left(\frac{\partial T}{\partial x_3}\right)^2 = \frac{1}{c^2} \tag{56}$$

This equation is known as the eikonal equation or the equation of Hamilton's characteristic functions. The equation relates the value of the square of the gradient of the travel time at a point of a medium of variable velocity to the square of the inverse of the velocity c at the same point. The velocity c is in the direction normal to the wave fronts or along the ray. Since the gradient of the wave front $S(x_i) = \omega T(x_i)$ represents the direction of propagation or ray trajectory, Eq. (56) shows how this trajectory changes as wave fronts propagate through a medium of variable velocity. If we know how c varies in the medium, the eikonal equation can be used to trace the rays.

4.1 Ray Paths and Travel Times

In a spherical medium where velocity varies continuously with the radius according to Snell's law and the Benndorf relation, we can write for each point on a ray,

$$\frac{r\ \sin i}{v} = p = \frac{dt}{d\Delta} \tag{57}$$

where i is the incident angle at a point of radial distance r where velocity is v and $t(\Delta)$ is the travel time at angular distance Δ. In the point of the ray path where it turns upward, $i = \pi/2$ and $r_p/v_p = p$. In terms of the slowness η (defined as $\eta = r/v$), $p = \eta_p$. Equation (57) relates the velocity distribution $v(r)$ with the ray path $i(r)$ and the travel times $t(\Delta)$. By integration we can obtain expressions for angular distance Δ and travel time t in terms of the velocity distribution. If the focus is at the surface

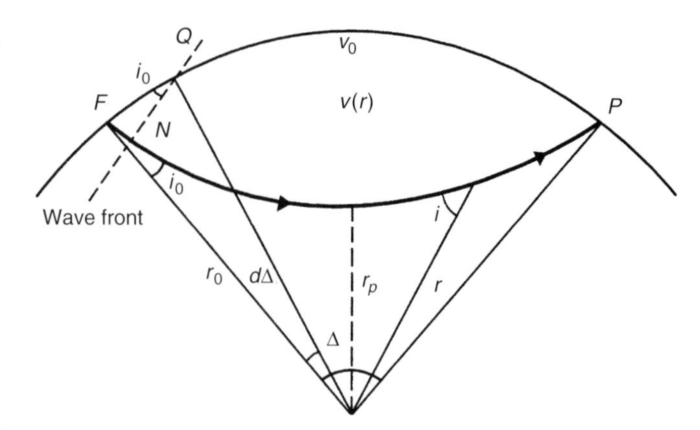

FIGURE 7 Ray path in a spherical medium with velocity increasing with depth. (From Udías, 1999.)

we obtain (Fig. 7),

$$\Delta = 2\int_{r_p}^{r_0} \frac{p\ dr}{r\sqrt{\eta^2 - p^2}} \tag{58}$$

$$t = 2\int_{r_p}^{r_0} \frac{\eta^2\ dr}{r\sqrt{\eta^2 - p^2}} \tag{59}$$

These expressions are the fundamental equations of kinematic ray theory. They allow the calculation of theoretical travel times $t(\Delta)$ from velocity models of the Earth with radial symmetry, $v(r)$, and the inversion of observed data of travel times to obtain the corresponding velocity distributions.

5. Normal Mode Theory

We have seen that perturbations produced in an unbounded elastic body have the form of waves that propagate in its interior. If the medium is perfectly elastic, homogeneous and isotropic, there are only two types of waves (P and S), which propagate with constant velocities (α and β) that depend on the elastic coefficients and density with no conditions imposed on their frequency. The Earth has finite dimensions and is bounded by a free surface; therefore, we have to consider the elastic behavior of an elastic body of finite dimensions. This consideration leads us into normal mode theory or free oscillations of the Earth (see also Chapter 10 Lognonné and Clévédé).

From the point of view of free oscillations, the Earth reacts to an earthquake by vibrating as a whole, in the same way as a bell when struck. Vibrations of a finite elastic body are the sum of an infinite number of modes (harmonics) that correspond to frequencies with values proportional to the inverse of the body characteristic dimension. For an elastic body of finite dimensions, wave propagation and free vibrations are two different approaches to the study of the same phenomenon. If wavelengths are very small compared with the dimensions of the body, we can approximate the problem by wave propagation in an unbounded medium, but we cannot if they are not so. Free eigenvibrations or normal mode theory includes the complete phenomenon, but for short wavelengths we need a sum of many modes, which is not very practical. Thus, in seismology, wave propagation is used for high frequencies and normal modes for lower ones. The study of the Earth's free oscillations is based on the theory of vibrations of an elastic sphere. The problem increases in complexity as we proceed from an isotropic homogeneous sphere to models with radial distribution of elastic parameters and density, three-dimensional heterogeneities, and lack of perfect elasticity, and those that include the effects of gravity, rotation, and ellipticity.

A first approximation to the problem of free oscillations of the Earth is that of a homogeneous elastic sphere. The problem is solved in spherical coordinates where displacements are derived from a scalar Φ and a vector $\boldsymbol{\psi}$ potential as we saw in Eq. (18). Vector potential $\boldsymbol{\psi}$ can be separated into two potentials that are products of the unit vector in the radial direction and a scalar potential,

$$\boldsymbol{\psi} = \nabla \times S\mathbf{r} + T\mathbf{r} \tag{60}$$

$$\mathbf{u} = \nabla\Phi + \nabla \times \nabla \times S\mathbf{r} + \nabla \times T\mathbf{r} \tag{61}$$

In Eq. (61), displacements are now separated into three parts, the first term (Φ) representing P-wave motion, the second (S) SV and the third (T) SH. In relation to surface waves, Φ and S represent Rayleigh-wave motion and T Love wave. Thus, S is called spheroidal potential and T toroidal potential.

In the absence of body forces and without considering rotation and gravity, Navier's equation (15) is given by

$$\alpha^2\nabla(\nabla \cdot \mathbf{u}) - \beta^2\nabla \times \nabla \times \mathbf{u} = \ddot{\mathbf{u}} \tag{62}$$

The problem is solved in spherical coordinates by expressing $\mathbf{u}$ in terms of potentials Φ, S, and T according to Eq. (61). This leads for the harmonic time dependence (26) to Helmholtz equations (27) for Φ, S, and T. As in Section 3.1, real solutions for standing waves convergent at the origin (the center of the Earth) are given in terms of Bessel spherical functions $j_n(kr)$, Legendre associate functions $P_n^m(\cos\theta)$, and harmonic functions $e^{\pm im\phi}$. As a consequence, we obtain for potential Φ, a solution as a sum,

$$\Phi(r,\theta,\phi) = \sum_{l=0}^{\infty} j_l(k_\alpha r) \sum_{m=-l}^{l} Y_l^m(\theta,\phi) \tag{63}$$

where

$$Y_l^m(\theta,\phi) = (-1)^m \left[\frac{2l+1}{4\pi}\frac{(l-m)!}{(l+m)!}\right] P_l^m(\cos\theta) \times \left[C_l^m \cos m\phi + S_l^m \sin m\phi\right] \tag{64}$$

are the associate spherical harmonic functions of index (l, m). Similar expressions are found for S and T depending on k_β. Subindex l is the angular and m is the azimuthal number. For each value of l there are $2l+1$ values of m. The solution for each value of l and m is called a mode of oscillation and the general solution is the sum over all their values. The problem is separated into toroidal modes involving potential T and spheroidal modes involving Φ and S. Thus we have toroidal modes T_l^m and spheroidal modes S_l^m of oscillation.

5.1 Toroidal Modes

The part of the problem corresponding to potential $T(r, \theta, \phi, t)$ leads to toroidal modes T_l^m or modes of SH motion. Boundary condition at the free surface of the sphere ($r = a$) is $\tau_{r\phi} = 0$. This condition leads to a relation for the part of the solution depending on Bessel spherical functions that may be written for a homogeneous medium as

$$x j_{l+1}(x) = (l-1) j_l(x) \qquad \text{for } x = k_\beta a \tag{65}$$

This equation has an infinite number of solutions, for which we use subindex n. Thus, boundary conditions introduce a third index in the problem. For each value of l, the first value of n ($n = 0$) corresponds to the fundamental mode and the rest to higher modes, harmonics, or overtones, and modes are represented by ${}_nT_l^m$.

For toroidal modes index l refers to the order of Legendre associate functions and consequently to the distribution of displacements on spherical surfaces depending on angle θ. For $l = 1$, the fundamental mode implies whole-body oscillations around the axis origin of θ, which are not possible as free vibrations. They are possible for higher modes since internal parts oscillate in different senses and there are no changes in total angular momentum. For $l = 2$ each hemisphere oscillates in an opposite sense, and for higher values ($l > 3$) there are as many zones oscillating in opposite senses as the order number of the mode (Fig. 8). Index m refers to the dependence of displacements on azimuthal angle ϕ and takes values from $-l$ to l. Index n refers to the number of roots in the r dependence of displacements for each configuration of l and m. For a homogeneous sphere, eigenfrequencies do not depend on index m. Thus, different displacements, according to values of m, correspond to the same frequencies; that is, the problem presents degeneracy. The same happens if elastic properties depend only on the radius, but not for the completely heterogeneous problem.

Theoretical values for eigenperiods of toroidal modes have been calculated for different models of the Earth, generally with radial symmetry. The longest period (43.82 min for a particular Earth model) corresponds to the fundamental mode of the second-order mode ${}_0T_2$. With increasing order number, mode periods decrease in value.

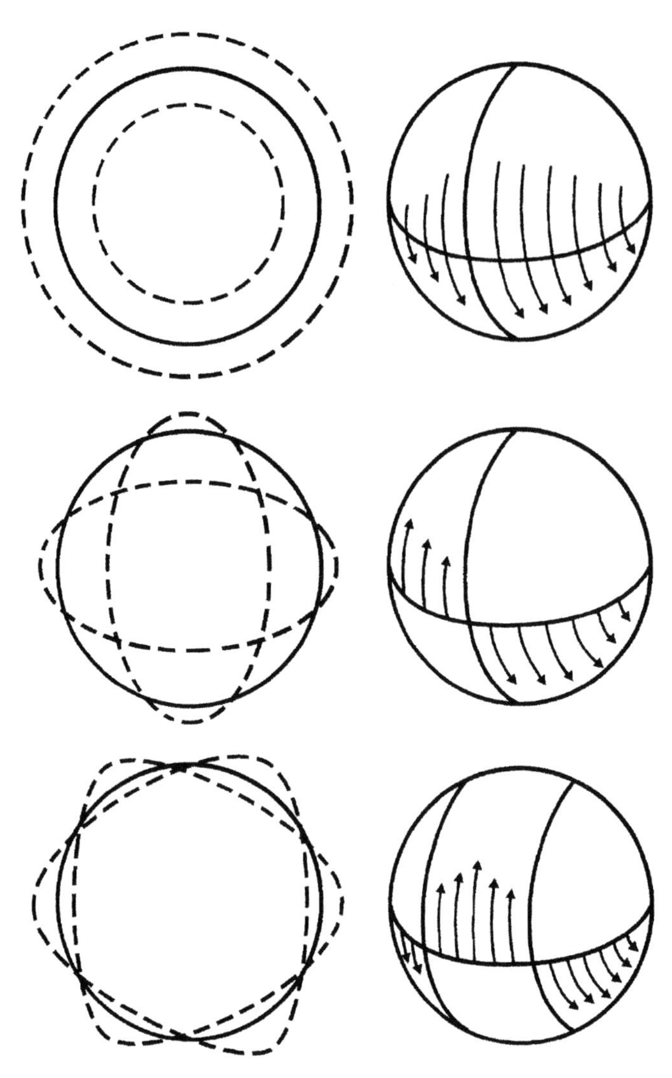

FIGURE 8 Spheroidal S_0, S_2, and S_3 and toroidal modes T_1, T_2, T_3. (From Udías, 1999.)

5.2 Spheroidal Modes

The problem of spheroidal modes implies solutions for potentials Φ and S and displacements correspond to *P–SV* motion. Boundary conditions at the free surface are that the stresses are null ($\tau_{rr} = \tau_{r\theta} = 0$). Solutions for potentials Φ and S are of the type of Eqs. (63) and (64). As in the case of toroidal modes, spheroidal modes are represented by ${}_nS_l^m$. For the lowest order $l = 0$, according to Eq. (63), potentials Φ and S are functions only of r. Then there are only displacements u_r, and $u_\theta = u_\phi = 0$. These modes are called radial modes. The fundamental mode ${}_0S_0$, corresponds to an expansion and contraction of the sphere without changing its form. For higher values of n, there are inside the sphere as many nodal surfaces at which the motion is null and changes sign, as the order number.

For $l = 1$, the fundamental mode ${}_0S_1$ does not exist, since a displacement of the center of mass is not possible in free vibrations. However, there are values for the higher modes ${}_1S_1$, ${}_2S_1$, etc. For $l = 2$, displacements are symmetric with respect to the plane normal to the axis from which the angle θ is measured. The Earth takes an ellipsoidal form, alternately flattened and elongated at the poles (Fig. 8). Fundamental mode ${}_0S_2$ has the lowest frequency of all modes (53.80 min for a particular Earth model) and no nodes of displacement in the interior of the sphere. For higher modes ($n > 1$), as in toroidal modes, there are in the interior of the sphere a number of nodal surfaces equal to the order of the mode. The longest period corresponds to mode ${}_0S_2$, which is greater than the period of corresponding toroidal mode ${}_0T_2$. Thus, this is the longest period for free oscillations of the Earth. For a mode of the same order, periods of spheroidal modes are longer than those of toroidal modes.

5.3 Nonspherical Symmetry

The Earth has properties that we have not treated in the above discussion. First, we have the effect of gravity, which affects mainly the spheroidal modes. Second, the presence of a liquid core results in spheroidal modes with periods significantly different from those in models without such a core and the existence of toroidal modes only in the mantle. This influence is greater for low-order modes. These two effects do not change the spherical symmetry of the problem.

Three additional effects that break the spherical symmetry derive from the Earth's rotation, the ellipticity of its shape, and its lateral heterogeneity (dependence of the elastic properties and density on θ and ϕ). The problem is no longer degenerate and the values of eigenfrequencies are modified. For each value of l and n, there are now several values of frequencies depending on the index m with a maximum of $2l + 1$, which are called multiplets. The most important of these effects is a consequence of the Earth's rotation, which introduces centrifugal and Coriolis forces, which have axial symmetry. This effect produces splitting of eigenfrequencies. If angle ϕ is measured on a plane normal to the Earth's rotation axis, a perturbation, due principally to Coriolis force, displaces eigenfrequencies associated to modes of angular order l by a quantity related to the index m and to the angular frequency of the Earth's rotation. This effect is small for high eigenfrequencies and can only be observed in low frequencies.

Two other effects are due to the Earth's ellipticity and to deviations of elastic properties and density from radial symmetry. These effects are small and also produce splitting in eigenfrequencies. If the lack of radial symmetry in the Earth's composition were very large, besides splitting of eigenfrequencies we could have a coupling of toroidal and spheroidal modes. Although their effects are small, heterogeneities in the Earth's composition can be detected in the analysis of free oscillations.

6. Effects of Anisotropy and Anelasticity in Wave Propagation

6.1 Anisotropy

Consideration of the influence of the lack of isotropy or anisotropy in wave propagation in the Earth has recently become an important subject of seismology and can be related to geodynamic processes. Although deviations from the conditions of isotropy are small, their effect on the propagation of seismic waves can be observed and provide important information (Babuska and Cara, 1991; Chapter 53 by Cara). Hexagonal or cylindrical symmetry is often assumed in seismological problems. As discussed in Section 2.1, this symmetry, called transverse isotropy, has a principal axis and properties in any direction normal to this axis are the same. The elastic tensor has five independent components, namely, A, C, F, L, and N.

Wave propagation in an isotropic medium with hexagonal symmetry has the following properties. For waves propagating along the principal axis of symmetry (x_3), there are two values of velocity, which correspond to P and S waves, namely, $\alpha=\sqrt{C/\rho}$ and $\beta=\sqrt{L/\rho}$. Wave propagation in this case is similar to that of an isotropic medium. For waves propagating in a direction (x_1) perpendicular to the principal axis (x_3), there are three different waves, namely, P waves with velocity $\alpha=\sqrt{A/\rho}$, different from that of waves propagating along the principal axis, and two types of S waves with different velocities, $\beta_1=\sqrt{N/\rho}$ (displacement in x_2 direction) and $\beta_2=\sqrt{L/\rho}$ (displacement in x_3 direction) (Fig. 9). If x_3 is the vertical direction, for waves traveling in x_1 direction, SH and SV components propagate with different velocities and are two different waves. This phenomenon is known as S-wave splitting, since SV and SH components arrive with a time delay between them. Due to this type of symmetry, this phenomenon takes place for any orientation of propagation in the plane normal to the principal axis.

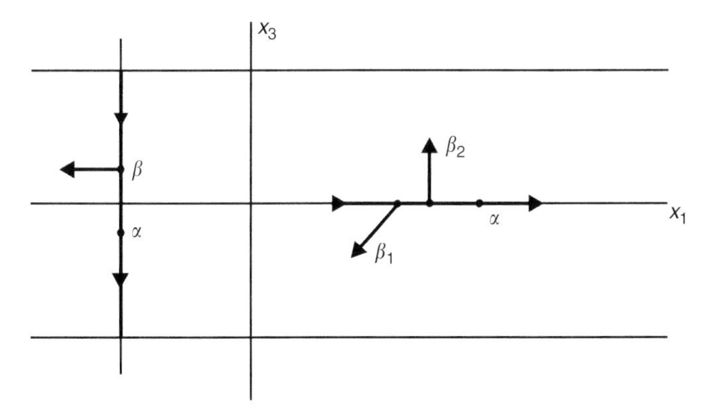

FIGURE 9 Propagation of P and S waves in an anisotropic medium with hexagonal symmetry (x_3 principal axis) for rays traveling along the principal axis and a perpendicular direction. (From Udías, 1999.)

In general, for any type of anisotropy, there are always three types of body waves propagating with three different velocities. Choosing the three components of displacement appropriately, they are called quasi-P, quasi-SH and quasi-SV waves. Velocities for these three types of waves change according to the type of symmetry present in the medium. According to these properties, anisotropy can be detected by observations of changes in P-wave velocity along two perpendicular directions and by observations of S-wave splitting.

The main effect of anisotropy on surface waves is that they cannot always be separated into Rayleigh and Love waves as in isotropic media. Radial and transverse components of motion are coupled, forming generalized dispersed surface waves that are a combination of Rayleigh and Love waves. We can distinguish three main effects of anisotropy in the propagation of surface waves. First, there is a discrepancy in the relation between the phase velocities of Rayleigh and Love waves with respect to those in isotropic media. Second, there are discrepancies in the phase velocities found for trajectories along different azimuths in the same region. Third, there is a departure of the polarization plane of Rayleigh waves from the vertical orientation.

6.2 Anelasticity

Wave propagation in imperfectly elastic media results in the attenuation of the amplitudes of wave motion (Fig. 10). Attenuation can be expressed in terms of the quality factor $Q(\omega)$, in general, a function of frequency, that is defined as

$$\frac{1}{Q(\omega)}=\frac{1}{2\pi}\frac{\Delta E}{E} \tag{66}$$

Thus, $1/Q$ represents the ratio between the elastic energy ΔE dissipated during one cycle of harmonic motion of frequency ω and the maximum or the mean energy E accumulated in the same cycle.

Wave attenuation can also be considered by assigning complex values to frequency and wavenumber, $k'=k+ik*$ and $\omega'=\omega-i\omega*$. In this form we can now define the quality factors of propagation in time and space, Q_t and Q_s, in terms of the imaginary parts of frequency or wavenumber:

$$\frac{1}{Q_t}=\frac{2\omega*}{\omega} \tag{67}$$

$$\frac{1}{Q_s}=\frac{2k*}{k} \tag{68}$$

From this it follows that phase velocity ($c'=\omega'/k'$) also has complex values, $c'=c+ic*$. Attenuation of body waves can then also be expressed by taking complex values for velocities of P and S waves, namely, $\alpha'=\alpha+i\alpha*$ and $\beta'=\beta+i\beta*$. Since observations of the attenuation of body waves are measured from amplitudes at different distances, the imaginary

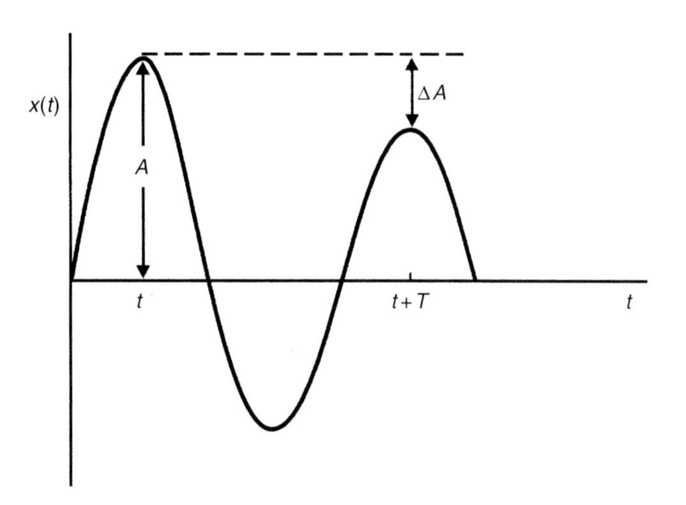

FIGURE 10 Amplitude attenuation due to anelasticty during a period. (From Udías, 1999.)

part of velocity is related to the spatial quality factor Q_s. In a similar way as in Eqs. (67) and (68), we can define quality factors for P and S waves as, $1/Q_\alpha = 2\alpha*/\alpha$ and $1/Q_\beta = 2\beta*/\beta$. These factors are related to Q_μ and Q_K [Eq. (8)] that are defined in a similar way, giving complex values to coefficients μ and K. For values of Q_α and Q_β larger than unity this relation is

$$\frac{1}{Q_\beta} = \frac{1}{Q_\mu} \tag{69}$$

$$\frac{1}{Q_\alpha} = \frac{4}{3}\left(\frac{\beta}{\alpha}\right)^2 \frac{1}{Q_\mu} + \left[1 - \frac{4}{3}\left(\frac{\beta}{\alpha}\right)^2\right] \frac{1}{Q_K} \tag{70}$$

In most seismological problems, it is assumed that there is no dissipation of energy in purely compressive or dilatational processes and therefore $Q_K = \infty$. Then, for $\lambda = \mu$, $Q_\alpha = \frac{9}{4} Q_\beta$.

In the ray theory approximation, we are interested in attenuation with distance along the ray path. Amplitude attenuation of a monochromatic P wave of frequency ω is given by

$$A = A_0 \exp\left(-\frac{\omega s}{2\alpha Q_\alpha}\right) = A_0 \exp(-\omega t*) \tag{71}$$

where A and A_0 are amplitudes at two points separated by a distance s along the ray path. For a homogeneous medium, $t* = t/2Q_\alpha$, where $t = s/\alpha$ is the travel time of P waves. A similar relation can be written for S waves using β and Q_β. In a spherical medium of radial symmetry with velocity $v(r)$, and quality factor $Q(r)$, both depending on the radius, for a ray with surface focus and ray parameter p along the complete path, we obtain

$$t* = \int_{r_p}^{r_0} \frac{\eta(r)^2 \, dr}{rQ(r)\sqrt{\eta(r)^2 - p^2}} \tag{72}$$

where $\eta(r)$ is the slowness ($\eta = r/v$). If $\bar{Q}$ is the mean value of $Q(r)$ along the ray and t is the travel time, an approximation is $t* = t/(2\bar{Q})$. In the Earth, for surface foci and epicentral distances between 30° and 90°, $t*$ is practically constant with a value of 1 sec for P waves and 4 sec for S waves; that is, S waves attenuate faster than P waves.

Anelastic attenuation of surface waves with distance and time can be expressed by coefficients γ_s and γ_t, related to spatial and temporal quality factors by $\gamma_s = \omega/2cQ_s$ and $\gamma_t = \omega/2Q_t$. Since, in a dispersed wave, energy propagates with group velocity U, for amplitudes corresponding to instantaneous frequencies, phase velocity must be substituted by group velocity. Then, the relation between Q_t and Q_s is

$$\frac{1}{Q_t} = \frac{U}{c}\frac{1}{Q_s} \tag{73}$$

For dispersed waves, temporal and spatial quality factors are different.

The influence of anelasticity on the free oscillations of the Earth affects the peaks of the spectral amplitudes. In a perfectly elastic medium, spectral peaks for each mode are delta functions at corresponding frequencies. In the presence of anelasticity, the spectrum has maximum amplitudes for each mode centered at the corresponding frequency ${}_n\omega_l$ with a certain width $\Delta\omega$ (width of the spectrum when its amplitude is half its maximum value). The quality factor is related to the width of the spectral peak by $1/Q_t = \Delta\omega/{}_n\omega_l$. Since attenuation increases with $1/Q$, the larger the attenuation, the wider the peaks of the spectrum. This is, in fact, a very general effect and can also be applied to observations in the time domain of propagating pulses in imperfectly elastic media.

7. Representation of the Source Mechanism

In seismology the problem of the source mechanism consists in relating observed seismic waves to the parameters that describe the source. In the direct problem theoretical seismic wave displacements are determined from source models, and in the inverse problem parameters of source models are derived from observed wave displacements. The first step in both problems is to define the seismic source by a mechanical model that represents the physical phenomena that corresponds to a fracture in the Earth's crust. These models or representations of the source are defined by parameters whose number depends on its complexity (Kasahara, 1981; Kostrov and Das, 1988; Koyama, 1997).

7.1 Green and Sommigliana Tensors

The representation theorem [Eq. (24)] gives the elastic displacement in terms of the Green tensor. As we saw, this tensor corresponds to the displacements of an impulsive force acting at a point. Then, finding solutions for the Green tensor is the first step in solving the problem of source mechanism. The simplest case of the Green tensor is that corresponding to an infinite, homogeneous isotropic elastic medium,

$$G_{ij} = \frac{1}{4\pi\rho}\left[\frac{1}{r^3}(3\gamma_i\gamma_j - \delta_{ij})\int_{r/\alpha}^{r/\beta}\tau\delta(t-\tau)\,d\tau + \frac{1}{r\alpha^2}\gamma_i\gamma_j\delta\left(t-\frac{r}{\alpha}\right) - \frac{1}{r\beta^2}(\gamma_i\gamma_j - \delta_{ij})\delta\left(t-\frac{r}{\beta}\right)\right] \quad (74)$$

where the force is in an arbitrary direction indicated by the index j, γ_i are the direction cosines of the line from the source to the observation point, and r the distance. The right side of Eq. (74) is composed of three terms: The first depends on $1/r^3$ and on both α and β; this is called the near field, since it decays rapidly with distance. The second and third terms depend on $1/r$; they constitute the far field and each depends only on α or β. Then, P and S motion is mixed in the near field and separated in the far field. Equation (74) is a very important result in elastodynamics that gives the elastic displacement field for the most fundamental type of source and medium. It constitutes the basic building block in seismic source studies.

A similar fundamental problem is the static solution for a constant force acting at a point in an arbitrary direction. For an infinite, homogeneous, isotropic, elastic medium and a unit force, static displacements are given by,

$$S_{ij} = \frac{1}{8\pi\rho r}\left[\left(\frac{1}{\beta^2} - \frac{1}{\alpha^2}\right)\gamma_i\gamma_j + \left(\frac{1}{\beta^2} + \frac{1}{\alpha^2}\right)\delta_{ij}\right] \quad (75)$$

Subindex j indicates the direction of the force and, as in Eq. (74), displacements are given by a tensor. This expression, known as the Somigliana tensor, is the fundamental equation in elastostatics. The Somigliana tensor can be considered as the limit of the Green tensor for infinite time and the static terms come only from the near-field displacement.

7.2 Equivalent Body Forces

The first mathematical formulation of the mechanism of earthquakes was presented by Nakano in 1923, using ideas already developed by Lamb and Love. Nakano used the point source approximation, valid if observation points are at a sufficiently long distance compared with source dimensions and long wavelengths. Thus, he represented the source by a system of body forces acting at a point. Since these forces must represent the phenomenon of fracture, they are called equivalent forces.

The problem may be stated as follows. Let us consider an elastic medium of volume V surrounded by a surface S. In its interior there is a small region of volume V_0, surrounded by a surface Σ, that we will call the focal region, where fracture takes place (Fig. 11). This process can be represented by a distribution of body forces $\mathbf{F}(\xi_i, t)$ acting per unit volume inside V_0. If it is assumed that no other body forces are present (gravity, etc.), and that on its surface S stresses and displacements are null, using the representation theorem in terms of the Green tensor [Eq. (24)], the elastic displacements in an infinite medium are given by

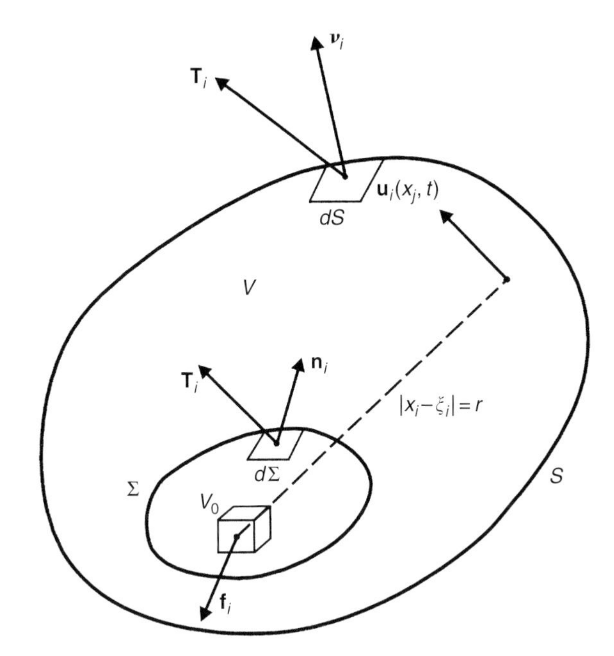

FIGURE 11 Focal region V_0, forces **f**, stresses **T**; and displacements **u** outside the focal region. (From Udías, 1999.)

$$u_i(x_s, t) = \int_{-\infty}^{\infty} d\tau \int_{V_0} F_k(\xi_s, \tau)G_{ki}(x_s, t; \xi_s, \tau)\,dV \quad (76)$$

The Green tensor **G** acts as a "propagator" of the effects of forces **F**, from the points where they are acting (ξ_i inside V_0) to points x_i outside V_0 where elastic displacements u_i are produced. For a point source at the origin of coordinates we have

$$u_i(x_s, t) = \int_{-\infty}^{\infty} F_k(\tau)G_{ki}(x_s, t-\tau)\,d\tau \quad (77)$$

Elastic displacements are given now by the time convolution of the forces acting at the focus with the Green function for the medium. Internal sources must be in equilibrium so that they should satisfy the condition that their resulting total force and moment are null. If we want to represent the shear motion at a fault, the equivalent system of forces is that of two couples with no resulting moment or double-couple (DC) model (Fig. 12). If the couples are oriented in the direction of unit vector **l** and **n**, where $\mathbf{n}\cdot\mathbf{l} = 0$, and their moment is $M = sF$, where s is the arm of the couple, displacements are given by

$$u_i^{DC} = \int_{-\infty}^{\infty} M(l_k n_l + n_k l_l)G_{ik,l}\,d\tau \quad (78)$$

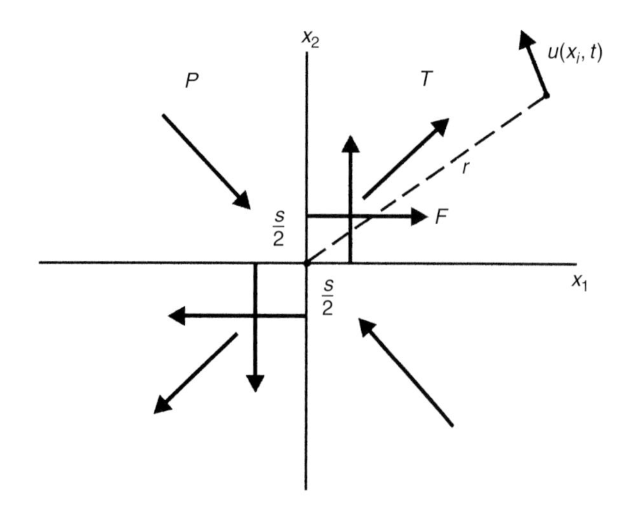

FIGURE 12 Double-couple model and equivalent system of pressure and tension forces. (From Udías, 1999.)

An equivalent system is that formed by two perpendicular linear dipoles with the opposite sign rotated 45° (linear dipole with forces in positive direction corresponds to tension **T** and in negative direction to pressure **P**) for which the elastic displacements are,

$$u_i^{TP} = \int_{-\infty}^{\infty} M(T_k T_k - P_l P_l) G_{ik,l}\, d\tau \tag{79}$$

For the two equivalent systems, relations between the unit vectors **P**, **T** and **n**, **l** are $\mathbf{P} = 1/\sqrt{2}(\mathbf{n} - \mathbf{l})$ and $\mathbf{T} = 1/\sqrt{2}(\mathbf{n} + \mathbf{l})$.

7.3 Shear Dislocations

If an earthquake is produced by a fracture of the Earth's crust, a mechanical representation of its source can be given in terms of fractures or dislocations in an elastic medium. The theory of elastic dislocations was developed by Volterra in 1907 and was first applied to the problem of the seismic source in the 1950s by Vvedenskaya, Keylis-Borok, Steketee, Maruyama, and others. A displacement dislocation consists of an internal surface inside an elastic medium across which there exists a discontinuity of displacement but stress is continuous. The focal region consists in an internal surface Σ with two sides (positive and negative). This surface can be considered as derived from a certain focal volume V_0 that is flattened to form a surface with both sides stuck together without any volume. Coordinates on this surface are ξ_k and the normal at each point $n_i(\xi_k)$. From one side to the other of this surface there is a discontinuity in displacement or slip,

$$u_i^+(\xi_k, t) - u_i^-(\xi_k, t) = \Delta u_i(\xi_k, t) \tag{80}$$

where the superscript plus and minus signs refer to the displacement at each side of the surface Σ. If there are no body forces ($\mathbf{F} = 0$), and the stresses are continuous through Σ (their integral is null) for an infinite medium, Eq. (24) results in

$$u_n(x_s, t) = \int_{-\infty}^{\infty} d\tau \int_{\Sigma} \Delta u_i(\xi_s, \tau) C_{ijkl} n_j(\xi_s) G_{nk,l}(x_s, t; \xi_s, \tau)\, dS \tag{81}$$

Consequently, the seismic source is represented by a dislocation or discontinuity in displacement given by the slip vector $\Delta\mathbf{u}$ on the surface Σ, which corresponds to the relative displacement between the two sides of a fault. This is, then, a nonelastic displacement that, once produced, does not go back to the initial position. In the most general case, $\Delta\mathbf{u}(\xi_k, \tau)$ can have a different direction for each point ξ_k of the surface Σ and at each of these points varies with time, starting with a zero value at $t = 0$, to a maximum value after a certain time. The normal to the surface Σ, given by unit vector $\mathbf{n}(\xi_k)$, can have different direction at points of the surface, but usually is considered to be constant, that is, Σ is a plane. Green's function **G** includes the propagation effects of the medium from points (ξ_k) of surface Σ to points (x_k) where elastic displacements (u_i) are evaluated. To solve the problem, according to Eq. (81), we must first know the derivatives of Green's function for the medium, which are also known as excitation functions.

Equation (81) corresponds to a kinematic model of the source, that is, a model in which elastic displacements **u** are derived from slip vector $\Delta\mathbf{u}$, which represents nonelastic displacement of the two sides of a fault of surface Σ. Slip is assumed to be known and is not derived from stress conditions in the focal region as it is done in dynamic models. In Eq. (81) derivatives of the Green function include derivatives of delta functions. If we change the order of integration, integral on time of the product of $\Delta\mathbf{u}$ with the derivatives of the delta function results in the time derivative of $\Delta\mathbf{u}$, that is, the slip velocity. Thus, elastic displacements depend not on slip but on slip velocity. This means that the source radiates elastic energy only while it is moving; when motion at the source stops, it ceases to radiate energy.

Let us consider the seismic source represented by a shear dislocation fracture, with fault plane Σ of area S and normal **n**, slip $\Delta\mathbf{u}(\xi_i, t)$ in the direction of unit vector **l**, contained on the plane so that **l** and **n** are perpendicular ($\mathbf{n} \cdot \mathbf{l} = 0$). For an infinite, homogeneous isotropic medium, displacement according to Eq. (81) is (Fig. 13),

$$u_k = \int_{-\infty}^{\infty} d\tau \int_{\Sigma} \mu\, \Delta u (l_i n_j + l_j n_i) G_{ki,j}\, dS \tag{82}$$

If the distance from observation point to the source is long in comparison with the source dimensions ($r \gg \Sigma$) and the wavelengths are also long ($\lambda \gg \Sigma$), the problem can be approximated by a point source and Eq. (82) takes the simpler form

$$u_k(t) = \mu S (l_i n_j + l_j n_i) \int_{-\infty}^{\infty} \Delta u(\tau) G_{ki,j}(t - \tau)\, d\tau \tag{83}$$

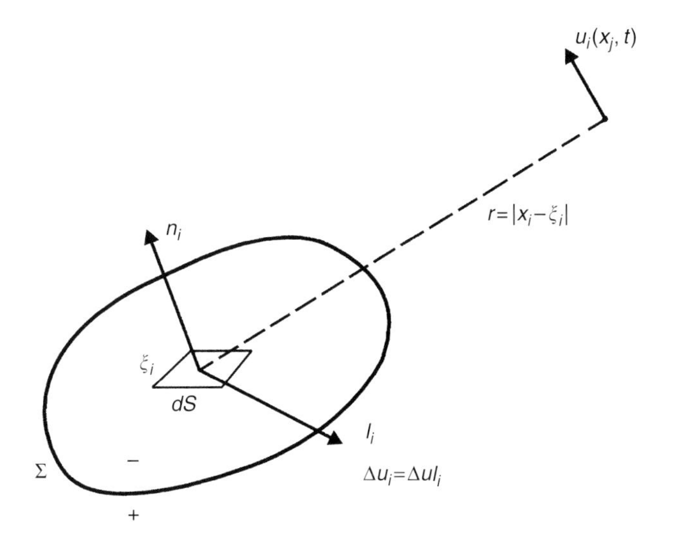

FIGURE 13 Representation of a shear dislocation. (From Udías, 1999.)

Displacements are given by temporal convolution of slip with the derivatives of the Green function. The geometry of the source is now defined by the orientation of the two unit vectors **n** and **l**. These two vectors, referred to the geographical system of axes (North, East, Nadir), define the orientation of the source, namely, **n** orientation of the fault plane and **l** that of slip.

If we substitute the derivatives of the Green function [Eq. (74)], taking only the terms for the far field, we obtain for the displacements of the P and S waves:

$$u_j^P = \frac{M_0}{4\pi\rho\alpha^3 r}(n_k l_i + n_i l_k)\gamma_i\gamma_k\gamma_j f\left(t - \frac{r}{\alpha}\right) \quad (84)$$

$$u_j^S = \frac{M_0}{4\pi\rho\beta^3 r}(n_k l_i + n_i l_k)\left(\delta_{ij} - \gamma_i\gamma_j\right)\gamma_k f\left(t - \frac{r}{\beta}\right) \quad (85)$$

where $M_0 = \mu\Delta\dot{u}S$, is the scalar seismic moment and $f(t)$, the source time function (STF) that represents the time dependence of the slip rate $\Delta\dot{u}$. Therefore, in the far field, P and S waves are pulses that arrive at a distance r at times, $t=r/\alpha$ and $t=r/\beta$, with a form depending on the STF and they are distributed in the space according to the radiation pattern that is a function of n_i, l_i, and γ_i.

8. Seismic Moment Tensor

In the formulation of the theory of source mechanism, an important concept is the seismic moment tensor M_{ij}, and the moment tensor density per unit volume or unit surface m_{ij}:

$$M_{ij} = \int_V m_{ij}\, dV = \int_S m_{ij}\, dS \quad (86)$$

The seismic moment tensor was first proposed by Gilbert in 1970, who related it to the total stress drop in earthquakes. Backus and Mulcahy in 1976 showed that the moment tensor represents only that part of the internal stress drop dissipated in nonelastic deformations at the source. The moment tensor is related to the equivalent body forces by

$$F_i = -\frac{\partial m_{ij}}{\partial x_j} \quad (87)$$

The elastic displacements in an infinite medium can now be expressed according to Eq. (76) by a distribution of moment tensor density over the source volume V_0. In the absence of external forces and torques, the sums of all internal forces and moments are null; then by an appropriate choice of the origin of coordinates, $m_{ij}G_{ij}=0$, and we obtain

$$u_i = \int_{-\infty}^{\infty} d\tau \int_{V_0} m_{kj} G_{ik,j}\, dV \quad (88)$$

If the moment tensor is defined only on a surface Σ, we obtain a similar expression in terms of the moment tensor density per unit surface. Equation (88) shows that elastic displacements outside the focal region can be derived from the seismic moment tensor and the derivatives of Green's function integrated over the focal region (V_0 or Σ). Since we have not specified its form, m_{ij} can represent a very general type of source. Each component of m_{ij} is a dipole and, provided the net effect of their sum and the sum of their moments are null, it can correspond to any system of internal body forces. The moment tensor is, thus, a very convenient form to represent the source of an earthquake in a general way. For a point source, Eq. (88) can be written in a compact form using an asterisk to express temporal convolution,

$$u_i = M_{kj} * G_{ik,j} \quad (89)$$

If we compare Eqs. (88) and (81), we can define the moment tensor density corresponding to a dislocation with slip Δu on a surface Σ of normal **n**:

$$m_{ij} = C_{ijkl}\,\Delta u_k\, n_l \quad (90)$$

For an isotropic medium, if the slip direction is given by unit vector **l**, we obtain

$$m_{ij} = \Delta u\left[\lambda l_k n_k \delta_{ij} + \mu\left(l_i n_j + l_j n_i\right)\right] \quad (91)$$

In this expression the first term represents changes in volume; if l_k and n_k are perpendicular, this term is zero and the source represents a shear fracture. With the moment tensor we can represent various types of sources. Equation (89) shows that the elastic displacements are linear functions of the components of the moment tensor. Thus, we can obtain these components from observations of displacements by linear inversion.

8.1 Separation of the Moment Tensor

Since the moment tensor is a symmetric tensor, we can obtain its real eigenvalues σ_1, σ_2, σ_3 and its eigenvectors ν_i^1, ν_i^2, ν_i^3, which are mutually orthogonal and define the principal axes. With respect to these axes, the moment tensor has the form

$$M_{ij} = \begin{pmatrix} \sigma_1 & 0 & 0 \\ 0 & \sigma_2 & 0 \\ 0 & 0 & \sigma_3 \end{pmatrix} \tag{92}$$

In this system, the moment tensor is formed by three linear dipoles in the direction of the principal axes and thus represent the principal stresses. If we order the eigenvalues $\sigma_1 > \sigma_2 > \sigma_3$, then, σ_1 corresponds to the greatest stress, σ_3 to the least stress, and σ_2 to intermediate stress. The sum of the elements of the principal diagonal is the first invariant of the tensor and has the same value for any reference system:

$$M_{11} + M_{22} + M_{33} = \sigma_1 + \sigma_2 + \sigma_3 = 3\sigma_0 \tag{93}$$

This sum represents the change in volume and σ_0 is the isotropic part of the moment tensor. If we subtract this from tensor M_{ij}, we obtain the deviatoric tensor M'_{ij} whose sum of the diagonal elements is always zero and does not include changes in volume,

$$M'_{ij} = M_{ij} - \delta_{ij}\sigma_0 \tag{94}$$

Thus, the moment tensor can be separated into two tensors, one isotropic, $M^0_{ij} = \delta_{ij}\,\sigma_0$, representing changes in volume and the other deviatoric M'_{ij}. The moment tensor that represents an explosive source is purely isotropic and that for a shear fracture is purely deviatoric.

Earthquake sources correspond to shear fractures, or nearly so. However, this may not always be the case, so that the presence of changes in volume cannot be completely ruled out. Methods of inversion of the moment tensor from observations that do not impose any condition sometimes result in the presence of certain amounts of isotropic components (changes in volume). More often, the deviatoric part does not correspond exactly to a double-couple source. Then it is convenient to separate the moment tensor into three parts: an isotropic part corresponding to changes in volume, a part of pure shear fracture or double couple (DC), and a third that may represent different mechanisms. This analysis is called partition or separation of the moment tensor and can be expressed by

$$M = M^0 + M^{DC} + M^R \tag{95}$$

In many cases the isotropic part is assumed to be null and is introduced as a condition in the inversion. The deviatoric part is separated into a shear fracture or double-couple source M^{DC}, considered as the standard model for the source of earthquakes, and other effects that are represented by M^R. The partition of the deviatoric part that isolates the double-couple part is not unique. Usually, a partition is sought for which M^{DC} represents the greatest part of the DC component possible (this is called the best double-couple). A separation that satisfies this condition, introduced by Knopoff and Randall in 1970, is

$$\begin{pmatrix} \sigma_1 & 0 & 0 \\ 0 & \sigma_2 & 0 \\ 0 & 0 & \sigma_3 \end{pmatrix} = \begin{pmatrix} \frac{1}{2}(\sigma_1 - \sigma_3) & 0 & 0 \\ 0 & 0 & 0 \\ 0 & 0 & -\frac{1}{2}(\sigma_1 - \sigma_3) \end{pmatrix} + \begin{pmatrix} -\frac{1}{2}\sigma_2 & 0 & 0 \\ 0 & \sigma_2 & 0 \\ 0 & 0 & -\frac{1}{2}\sigma_2 \end{pmatrix} \tag{96}$$

As before, $\sigma_2 = -\sigma_1 - \sigma_3$. The first term is a DC source. The second is called a compensated linear vector dipole (CLVD). Its physical meaning is a sudden change in the shear modulus in a direction normal to the fault plane, without changes in volume. Deviation from a pure DC is sometimes represented by $\delta = |\sigma_3/\sigma_1|$, the ratio of least to greatest eigenvalues. For pure DC, $\delta = 1$ and the non-DC part is given by $1 - \delta$.

9. Models of Fracture

Fracture processes can be approached in two different ways—kinematic and dynamic. Kinematic models of the source consider the slip of the fault without relating it to the stresses that cause it. The fracture process is described purely by the slip vector as a function of coordinates on the fault plane and time. From this type of model, it is a relatively simple problem to determine the corresponding elastic displacement field. Dynamic models consider the complete fracture process relating fault slip to stress acting on the focal region. A complete dynamic description must be able to describe fracture from material properties of the focal region and stress conditions. Dynamic models present greater difficulties and their solutions, in many cases, can only be found by numerical methods.

9.1 Kinematic Models

Let us consider the general characteristics of the kinematic models of extended sources represented by a surface Σ over which a shear dislocation $\Delta u(\xi_i)$ propagates with velocity $v_i(\xi_i)$, from a point ($\xi_i = 0$) to the edge of Σ (Aki and Richards, 1980). In some cases the velocity of fracture propagation is assumed to be constant and less than the velocity of wave propagation ($v < \beta < \alpha$), that is, subsonic fractures (a common value is $v = 0.7\beta$). Displacements of P waves in the far field for an infinite, homogeneous, isotropic medium can be written as

$$u_i^P(x_j, t) = \frac{\mu}{4\pi\rho\alpha^3} \int_\Sigma \frac{R(n_k, l_k, \gamma_k)}{r} \Delta\dot{u}\left(\xi_k, t - \frac{r}{\alpha}\right) dS \tag{97}$$

where $r = |x_i - \xi_i|$ is the distance from the point of observation x_i to a point of the source ξ_i, where slip Δu is located at each moment, and $R(n_k, l_k, \gamma_k)$ is the radiation pattern that depends on the orientation of the source (l_k, n_k) and position of the observation point (γ_k).

A simple kinematic model of finite dimensions, known as Haskell's model, is a rectangular fault of length L and width W, where slip Δu propagates only along the L direction with constant velocity v (slip moves instantaneously along W). The coordinate along L is ξ, with origin in one end of the fault, and Δu has only one component (Fig. 14). Fractures that propagate only in one sense (from 0 to L) are called unilateral fractures and those that propagate in both senses (from 0 to $L/2$ and from 0 to $-L/2$) are called bilateral. For unilateral fracture, not considering radiation pattern, dependence on distance, and the other factors of Eq. (97), the form of P waves in the far field, if slip moves in the positive direction of ξ with constant fracture velocity v, $[\Delta u(\xi, t) = \Delta u(t - \xi/v)]$, is given by

$$u(x_i, t) = W \int_0^L \Delta \dot{u} \left[t - \frac{r_0}{\alpha} - \frac{\xi}{\alpha} \left(\frac{\alpha}{\nu} - \cos \theta \right) \right] d\xi \tag{98}$$

Passing from the domain of time to that of frequency and integrating, the Fourier transform of elastic displacements of P waves $U(x_i, \omega)$, is

$$U(x_i, \omega) = WL\omega\, \Delta U(\omega) \frac{\sin \omega X}{\omega X} \exp\left[-i \left(\frac{\omega r_0}{\alpha} - \omega X - \frac{\pi}{2} \right) \right] \tag{99}$$

where

$$X = \frac{L}{2\alpha} \left(\frac{\alpha}{\nu} - \cos \theta \right) \tag{100}$$

The form of the amplitude spectrum depends on the factor $\sin(\omega X)/\omega X$ and the frequency dependence of $\Delta U(\omega)$, the Fourier transform of the STF. The function $\sin(\omega X)/\omega X$ has a value of unity for $\omega X = 0$, roots for ωX equal to integer multiples of π, and an envelope that decreases as $1/\omega X$. If the STF has a rise time τ, which $\Delta u(t)$ takes to reach its maximum value, then the form of the spectrum has a constant value for low frequencies and for high frequencies starting from a frequency ω_c, called the corner frequency, the envelope of $U(\omega)$ decreases with frequency as $1/\omega^2$ (Fig. 15). Dimensions of the fault are inversely proportional to ω_c.

Another effect of Eq. (99) on the form of the radiation pattern, if wavelengths are of the same order as source dimensions ($\lambda' \sim L$), is that amplitudes are affected by the factor $\sin(\omega X)/\omega X$, which depends on θ. Since this factor is maximum for $\theta = 0$ and minimum for $\theta = \pi$, amplitudes are larger in the same direction as the fracture propagation ($\theta = 0$) and smaller in the opposite direction ($\theta = \pi$). This directivity effect is due to the focusing of energy in the direction of fracture propagation and is a phenomenon present in all propagating sources.

Another fundamental model of an extended seismic source is that of a circular fault, known as Brune's model. This model consists of a circular fault plane with a finite radius (a) on which a shear stress pulse is applied instantaneously. Since this model specifies stress on the fault, this is not exactly a kinematic model. Because the stress pulse is applied instantaneously on the whole fault area, there is no fracture propagation. The form of the spectrum is similar to that of Haskell's model, with corner frequency depending on the radius, $\omega_c = 2.33\beta/a$.

Haskell's model does not include the effect of either the beginning or nucleation of rupture or its stopping or arresting at the fault edge. The first kinematic model that includes both effects was proposed by Savage in 1966. This model consists of an elliptical fault in which slip begins in one of

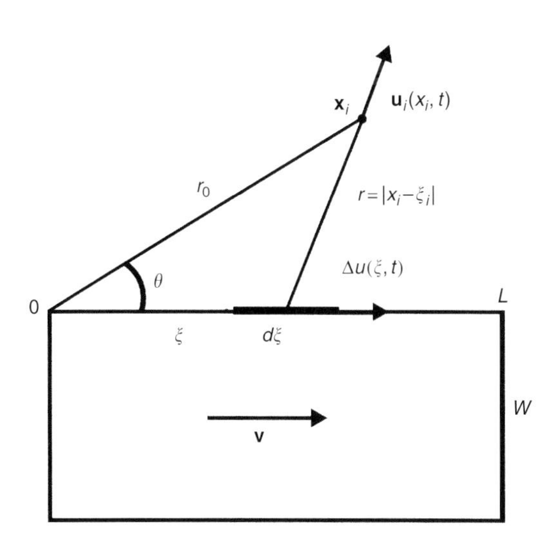

FIGURE 14 Rectangular fault in Haskell's model. (From Udías, 1999.)

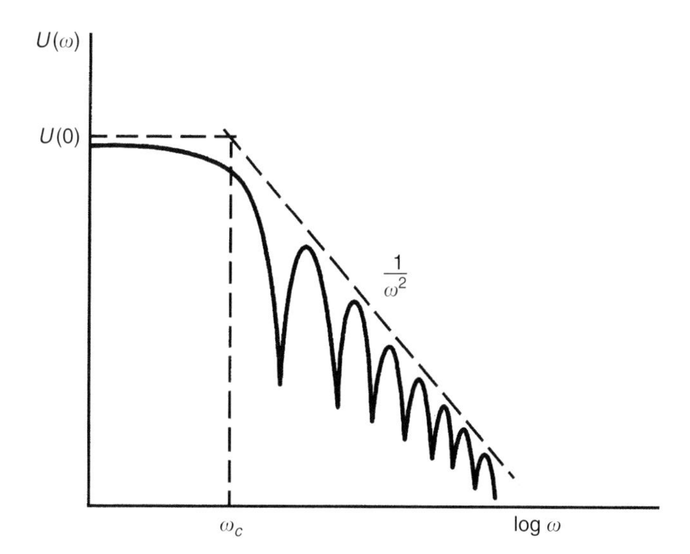

FIGURE 15 Form of the amplitude spectrum of seismic waves from an extended source. (From Udías, 1999.)

the foci and stops when it reaches the border of the ellipse. The result for the elastic displacements is the appearance of a stopping phase when fracture stops at its edge. As rupture propagates outward, motion on the fault stops when it reaches the border of the fault. This way of stopping motion on the fault is not physically possible, since it implies that the information of fault arrest at the edges instantaneously reaches (with infinite velocity) all points of the fault. This can be solved by introducing a finite velocity, for example, velocity of P waves, to bring the information of the cessation of the rupture at the border to all points of the interior of the fault. This wave is called the healing front, since it heals the fault by stopping its motion. In kinematic models, healing of the rupture inside the fault must be introduced in a somewhat artificial way.

9.2 Dynamic Models

Kinematic models are, naturally, a simplification of real fractures and include certain arbitrary properties in the definition of slip and conditions at the fault edge. The physical problem of fracture is a dynamic problem in which slip has to be considered as a result of changes in stress conditions and the strength of material in the focal region. Dynamic models of the seismic source take these conditions into consideration and are based on the theory of fracture generation and propagation in stressed media. Basic notions are presented here; a more detailed discussion of dynamic models is given in Chapter 12 by Madariaga and Olsen, and the physics of the process in Chapter 15 by Teisseyre and Majewski. From the dynamic point of view, the mechanism of an earthquake is represented by a shear fracture produced by the stress drop in the focal region. Fracture begins at a point of the fault when stress acting on the fault plane exceeds a critical value, propagates with a certain velocity, and finally stops when conditions impede its further propagation. A complete dynamic model must, then, include the whole fracture process, its initiation or nucleation, and propagation and arrest, derived from stress conditions and material properties at the focal region. Two determinant factors are tectonic stresses that are the consequence of lithospheric plate motion and mechanical properties of rocks in the fault region.

The dynamic problem of fracture propagation is centered in the energy balance near the rupture front (Kostrov and Das, 1988). For the fracture front to advance, new fracture surface must be created and a certain amount of energy must be consumed. For this, a certain elastic energy flux from the part that has not yet been fractured must reach the fracture front. When the energy flux to the rupture front equals the energy necessary to create new fracture surface or Griffith's energy, then fracture advances. This is called Griffith's fracture condition. For a perfect brittle fracture in elastic material, the material ahead of the fracture front (not yet fractured) is continuous $(u_i^+ = u_i^-)$ and that behind (already fractured) is discontinuous $(u_i^+ - u_i^- = \Delta u_i)$. Stress drops by $\Delta\sigma = \sigma_0 - \sigma_f$, where σ_0 is the initial stress and σ_f is the final stress due to friction. There is, then, a discontinuity at the rupture front and there is no transition between unfractured and fractured material.

The relation between stress drop and slip can be stated in a simplified form as follows. We start with the solution of the equation of motion (11) for the relative displacement (slip) of the two sides of the fault plane $\Delta u_i(x_k, t) = u_i^+ - u_i^-$, for $x < vt$, where x is the direction of rupture propagation with velocity v. Boundary conditions require that, as fracture front advances on the fault plane, stress drops by $\Delta\sigma$. If the fracture front is a plane and propagates in the direction of x, with constant velocity v, slip inside the fault can be written as

$$\Delta u(x,t) = \frac{\Delta u}{a}\sqrt{v^2t^2 - x^2} \qquad x < vt \qquad (101)$$

and slip velocity as

$$\Delta \dot{u}(x,t) = \frac{V}{\sqrt{v^2t^2 - x^2}} \qquad x < vt \qquad (102)$$

If stress drop is total ($\sigma_f = 0$), stress inside the fault is zero and outside,

$$\sigma(x,t) = \frac{K}{\sqrt{x^2 - v^2t^2}} \qquad x > vt \qquad (103)$$

where V is the velocity intensity dynamic factor and K is the stress intensity dynamic factor. As the rupture front advances (vt increases in the x direction), slip (101) increases from zero ahead of the rupture front to a constant value inside, slip velocity (102) becomes infinite at the rupture front, and stress (103) becomes infinite when it approaches the rupture front from outside and has either a constant value or zero inside (Fig. 16). Dynamic factors K and V are related to energy flux from unfractured material to the rupture front. For a circular fracture that grows from its center, the relation between Δu and $\Delta\sigma$, known as Kostrov's formula, is

$$\Delta u(r,t) = \frac{\Delta\sigma}{\mu} C(v) v \sqrt{t^2 - \frac{r^2}{v^2}} \qquad t > r/v \qquad (104)$$

where $v < \beta$, and $C(v)$ is a factor with value near unity. Growth of fracture is assured by constant energy flux from the as yet unfractured material. If the medium is homogeneous, rupture once started cannot stop and grows indefinitely. This is due to the constant conditions of the material ahead of the rupture front.

Since in a homogeneous medium rupture once started cannot stop by itself, stopping must be introduced as a condition; for example, in a circular fault by imposing a limit at a predetermined value of the radius. The beginning of fracture must also be introduced as an added condition; for example,

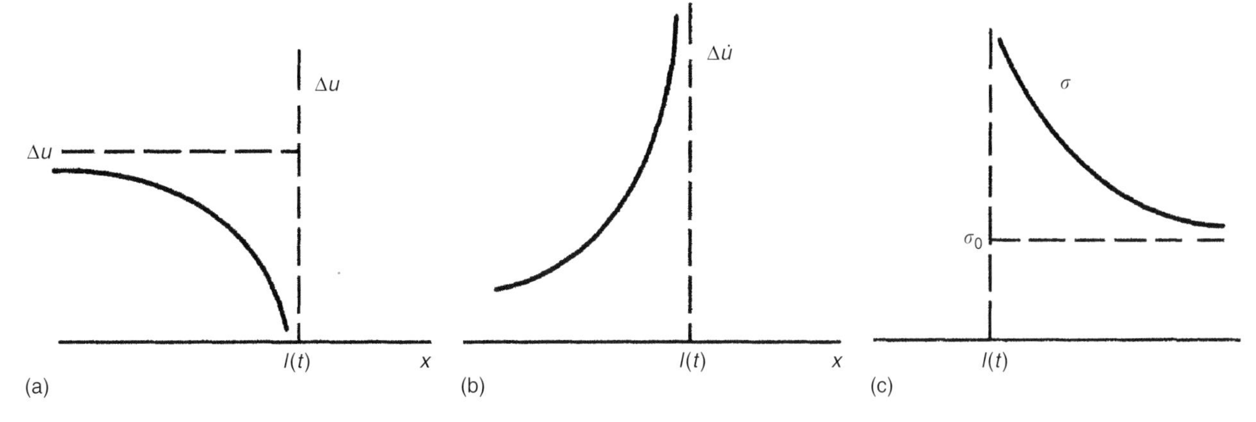

FIGURE 16 Dependence of slip and slip rate inside the fault and stress outside for total stress drop; $l(t)$ is the position of the rupture front. (From Udías, 1999.)

fracture starts when applied stress exceeds a certain critical value, namely, the maximum stress which the material can support without breaking. Dynamic models of fracture allow determination of slip and its propagation over the fault plane from specified stress drop. Thus, we can find a solution for the elastic displacement field from the dynamic conditions in the source region. The solution of dynamic problems, even for simple cases, is difficult and they must always be solved numerically.

9.3 Complexity of Fracture

Under homogeneous conditions, the dynamic problem of rupture propagation implies certain nonrealistic border conditions concerning stress and slip velocity and nucleation and arrest. To solve some of these problems, we must introduce heterogeneities in the medium and complexities in the fracture process. According to Eqs. (102) and (103) in the homogeneous model, there is a discontinuity of slip velocity and stress at the rupture front that follows from the fact that material is either purely elastic (unfractured) ahead of the front or fractured behind it. To avoid this situation, we must consider the existence of a transition zone immediately ahead of the fracture front where material behaves in a nonelastic form. As introduced by Barenblatt in 1959, this transition zone is called the cohesive zone. In this zone cohesive forces act to oppose the advance of fracture and hold stress finite immediately ahead of the rupture front, eliminating the stress singularity. Slip does not become zero in the cohesive zone and there is no singularity in the slip velocity at the rupture front. In the slip weakening model, cohesive stress is taken to be dependent on slip.

The simple fact that rupture must stop at the edge of the fault indicates that conditions cannot be homogeneous. In the Earth, faults cross rocks of different strength, often change direction, and have jumps, joints, and bends. The analysis of observed wave forms from earthquakes also shows greater complexity than expected from homogeneous fractures. This is especially present in the complex form of high-frequency waves in the near field and observed values of stress drops, which are really average values over the entire fault zone. Two models have been proposed to explain this complexity, namely, barriers and asperities models.

The barrier model, introduced by Das and Aki in 1977, assumes that fracture takes place under uniform conditions of stress on the fault fracture, but with different strength in the material on the fault surface. Zones with high strength form barriers that make fracture propagation difficult or impede it. Once the fracture process has finished on the whole fault plane, stress is released in fractured zones and accumulated in the unbroken barriers. These can be broken later by aftershocks. The model of asperities, presented by Kanamori in 1978, consists of a fault with a heterogeneous distribution of stress on its surface, with zones of high values (asperities) and low values. This model takes into account the previous history of stress accumulation on certain zones of the fault (asperities) and of stress release on other weaker zones by small earthquakes (foreshocks). Breaking of strong zones or asperities, where high stress has accumulated, constitutes the occurrence of the main large earthquake. In both models, earthquakes are produced by complex fracture processes consisting of the breaking of several asperities or zones between barriers. Since each model cannot explain both foreshocks and aftershocks, mixed models must be considered where both asperities and barriers are present (Kostrov and Das, 1988). Barriers are zones that remain unbroken after the main earthquake and asperities are those that break with high stress drops. Stress distribution on the fault plane is heterogeneous before and after an earthquake.

An important factor in the heterogeneity on the fault surface is the distribution of friction. High and low values of friction along the fault plane contribute to accumulation of stress. Recent dynamic models of fracture give great importance to the problem of friction. The search for more realistic models of earthquake sources leads to a consideration of complexities in the fracture process, heterogeneities in the distribution of stress, strength, and friction over the fault surface, and the existence of a transition zone at the rupture front. These considerations increase the number of parameters necessary to define source models, introducing dimensions of asperities, distance between barriers, cohesive zone, critical slip, friction distribution, and so on. We must also consider geometric irregularities of the fault surface such as branching, stepping, bending, and junctions that depart from the simple planar model.

10. Conclusions

Theoretical seismology is based on the application of the equations of the mechanics of continuous bodies to the propagation and generation of vibrations produced by earthquakes in the Earth. Basic equations are obtained by assuming perfect elasticity and homogeneous conditions in isotropic material. Heterogeneous conditions in wave propagation and generation require more complex formulations. Anisotropy introduces more than two elastic coefficients and lack of perfect elasticity produces the attenuation of wave amplitudes and normal modes and can be taken into account by means of the quality factor Q. More realistic models of Earth structure and fracture processes require highly heterogeneous three-dimensional models of both source and medium. Solutions for these cases are usually found by means of complex numerical methods.

Acknowledgments

The content of this chapter is taken from my textbook "Principles of Seismology" published by Cambridge University Press. All figures are reproduced from this textbook with permission of the Cambridge University Press. I wish to thank Professors R. Madariaga, Ecole Normal Superieur, Paris, S. Das, Oxford University, and E. Buforn, Universidad Complutense de Madrid, for helpful suggestions.

References

Aki, K. and P.G. Richards (1980). "Quantitative Seismology. Theory and Methods," 2 vols. Freeman, San Francisco.

Babuska, V. and M. Cara (1991). "Seismic Anisotropy in the Earth." Kluwer Academic, Dordrecht.

Ben Menahem, A. (1995). A concise history of mainstream seismology: origins legacy and perspectives. *Bull. Seismol. Soc. Am.* **85**, 1202–1225.

Ben Menahem, A. and S.J. Singh (1981). "Seismic Waves and Sources." Springer-Verlag, Berlin.

Bullen, K. (1947). "An Introduction to the Theory of Seismology." Cambridge University Press, Cambridge.

Bullen, K. and B.A. Bolt (1985). "An Introduction to the Theory of Seismology." Cambridge University Press, Cambridge.

Dahlen, F.A. and J. Tromp (1998). "Theoretical Global Seismology." Princeton University Press, Princeton, NJ.

Kasahara, K. (1981). "Earthquake Mechanics." Cambridge University Press, Cambridge.

Kennett, B.L.N. (1993). "Seismic Wave Propagation in Stratified Media." Cambridge University Press, Cambridge.

Kostrov, B. and S. Das (1988). "Principles of Earthquake Source Mechanics." Cambridge University Press, Cambridge.

Koyama, J. (1997). "The Complex Faulting Process of Earthquakes." Kluwer Academic, Dordrecht.

Lay, T. and T.C. Wallace (1995). "Modern Global Seismology." Academic Press, San Diego.

Love, A.E.H. (1945). "The Mathematical Theory of Elasticity," 4th edn. Cambridge University Press, Cambridge.

Macelvane, J.B. and F.W. Sohon (1936). "Introduction to Theoretical Seismology: Part I *Geodynamics*; Part II, *Seismometry*." Wiley, New York.

Sawarenski, E.F. and D.P. Kirnos (1960). "Elemente der Seismologie und Seismometrie." Akademie-Verlag, Berlin.

Udías, A. (1999). "Principles of Seismology." Cambridge University Press, Cambridge.

9

Seismic Ray Theory and Finite Frequency Extensions

Chris H. Chapman
Schlumberger Cambridge Research, Cambridge, UK

DEDICATION

Franta Hron (University of Alberta) and Carl Spencer (Schlumberger Cambridge Research), both friends and colleagues, died suddenly and prematurely while I was writing this chapter. Franta taught me much about ray theory and Carl about the Kirchhoff integral method. I should like to dedicate this chapter to their memory—I miss you both.

1. Introduction

The application of ray theory to the propagation of elastodynamic waves in the Earth has a long history and an extensive literature. Although ray theory is only a high-frequency approximation, most seismic interpretation relies on it. The classic work of Jeffreys and Bullen, and Gutenberg, in the first half of the 20th century, used ray theory to determine whole Earth structure and locate earthquakes. Ray theory is used to determine crustal structure and dominates interpretation in controlled source seismology, both refraction and reflection. Despite the impressive developments in normal mode and surface wave studies in the last 30 years, ray theory remains the dominant tool for interpretation.

The main advantages of ray theory are that the results, if not the theory, are relatively straightforward and intuitive, and correspond to our everyday experience. Essentially all communications rely on ray theory and the concepts of rays, wavefronts, ray tubes, reflection, and transmission are easily understood. If ray theory were not valid for most seismic signals, our knowledge of the internal structure of the Earth would be much poorer. Ray theory has the added advantage that it is also relatively easy and inexpensive to compute theoretical results.

Despite the fact that historically most seismic studies have treated the Earth as isotropic, in this chapter we have chosen first to develop ray theory in an anisotropic medium. The reason for this is twofold: first, in all branches of seismology there is now considerable interest in and evidence for anisotropy; and second, anisotropic ray theory is in fact simpler when it is nondegenerate, and it can then be specialized to isotropy. Therefore, in Section 2 we develop anisotropic ray theory. There is a considerable literature on this (for example, Lewis, 1965; Červený, 1972; Hanyga, 1982a; Gajewski and Pšenčík, 1987). The objective of the section is not to introduce anything new, nor to provide a comprehensive review, but to cover all aspects of the theory in a consistent notation and straightforward manner and to obtain the ray theory Green function. The section covers the eikonal and transport equations, kinematic and dynamic raytracing, paraxial rays, interaction with a curved interface, Snell's law, and reflection/transmission coefficients, and finally the ray Green function and its reciprocity. Many alternative but equivalent results have appeared in the literature, but we have made no attempt to review or include all these. Nevertheless, the section should contain sufficient information to compute all parts of the Green function.

Section 3 specializes the results of Section 2, initially to isotropic media and then to 1D or 2D models. In isotropic media, ray theory is degenerate and determining the shear-wave polarization is more complicated. However, other expressions are obviously much simpler in isotropic media. Similarly, simplifying to a 1D medium reduces the ray equations significantly and the standard ray integrals are obtained.

Although ray theory is valid in many circumstances, it breaks down in common situations. Section 4 outlines three useful extensions that remain valid at the singularities of basic ray theory. First, Maslov asymptotic ray theory remains valid at

INTERNATIONAL HANDBOOK OF EARTHQUAKE AND ENGINEERING SEISMOLOGY, VOLUME 81A ISBN: 0-12-440652-1

caustics where rays are focused and the amplitudes of normal ray theory diverge. It also models headwaves on plane interfaces. Only the results of normal raytracing are neeeded, and the final result is efficiently evaluated in the time-domain as an integral over different ray directions. The result describes the frequency-dependent signals near caustics and diffracted into geometrical shadows. Second, we outline the use of ray theory in the Kirchhoff integral method. This technique models, at least approximately, diffractions from roughness or discontinuities on interfaces. Again, only the results of normal raytracing are needed, to and from the interface, and the final expression is efficiently evaluated in the time-domain as an integral over the interface. The result describes the frequency-dependent, diffracted signals caused by the interface geometry. Finally, we outline quasi-isotropic ray theory. Isotropic and anisotropic ray theories have coexisted and been used for many years in seismology, despite the fact that anisotropic ray theory does not reduce to isotropic theory in the isotropic limit. Since, to a first approximation, the Earth has been treated as isotropic, this fact is surprising and troubling. Normal anisotropic ray theory is only correct when the medium is homogeneous or the anisotropy is strong and we are far from degeneracies. Recently, quasi-isotropic ray theory has been developed, which describes the strong coupling that must exist between shear rays in anisotropic media near degeneracies. Again, using the results of normal raytracing, a simple result is obtained for the frequency-dependent, quasi-shear rays.

2. Asymptotic Ray Theory for 3D, Anisotropic Media

In this section we first develop the basic equations of asymptotic ray theory, the eikonal and transport equations. From these we derive the kinematic and dynamic ray equations, the latter providing solutions for paraxial rays and polarizations. We also solve for the reflection/transmission (*r/t*) coefficients needed at interfaces in the ray expansion and, combining all these, obtain the ray theory Green function and prove its reciprocity.

The equations for vector/tensor waves can either be written using subscript notation or using vectors, etc. The notation in this chapter is a compromise between these which, at least subjectively, is compact but readable (Woodhouse, 1974; Chapman and Coates, 1994). The operators L, M, and N are close to those in Červený (1972), etc.

The equation of motion is

$$\frac{\partial \mathbf{v}}{\partial t} = \frac{1}{\rho} \frac{\partial \mathbf{t}_j}{\partial x_j} \tag{1}$$

and the constitutive equation for elastic media is

$$\frac{\partial \mathbf{t}_j}{\partial t} = \mathbf{c}_{jk} \frac{\partial \mathbf{v}}{\partial x_k} \tag{2}$$

where $\mathbf{v}$ is the particle velocity, $\mathbf{t}_j$ are the traction vectors (the stress components are $(\mathbf{t}_j)_i = \sigma_{ij}$), $\mathbf{c}_{jk}$ are stiffness matrices (the elastic parameters are $(\mathbf{c}_{jk})_{il} = c_{ijkl}$ and $\mathbf{c}_{jk}^{\mathrm{T}} = \mathbf{c}_{kj}$), and ρ is the density. These equations are the standard equations of motion ($\rho u_{,tt} = \sigma_{ij,j}$) and anisotropic constitutive equation ($\sigma_{ij} = c_{ijkl} u_{k,l}$) (Aki and Richards, 1980, Eqs. (2.17) and (2.18), respectively), rewritten using this vector/matrix notation and to emphasize symmetry. No source term is included as this is not an intrinsic part of ray theory. Later (Section 2.9), we will introduce a general point force to obtain the Green function by matching the ray solution to the known Green function in a homogeneous medium, e.g., Aki and Richards (1980, Ch. 4) for an isotropic medium. More general sources can then be generated by differentiation and superposition (see, for instance, Aki and Richards, 1980, Ch. 4) exploiting the linearity of the wave equations. Eqs. (1) and (2) are easily rewritten in the frequency domain and in this chapter the Fourier transform is defined such that $\partial/\partial t \rightarrow -i\omega$.

2.1 The Ray Expansion

The ray ansätze are

$$\begin{pmatrix} \mathbf{v} \\ \mathbf{t}_j \end{pmatrix}(\omega, \mathbf{x}) = f(\omega) \sum_n e^{i\omega T(\mathbf{x}, \mathscr{L}_n)} \sum_{m=0}^{\infty} \frac{1}{(-i\omega)^m} \begin{pmatrix} \mathbf{V}^{(m)} \\ \mathbf{T}_j^{(m)} \end{pmatrix}(\mathbf{x}, \mathscr{L}_n) \tag{3}$$

where $f(\omega)$ is an arbitrary spectrum depending on the source. The notation $\mathscr{L}_n$ is used to indicate the "ray path"—various paths may exist to the point $\mathbf{x}$ and these are enumerated by the index n. The path in general is a function of the source, receiver, and other parameters along the ray, e.g., the *r/t* history.

From studies in homogeneous media, it is well known that elastic waves propagate without dispersion with a velocity independent of frequency. At interfaces between homogeneous media, plane waves satisfy Snell's law. In inhomogeneous media, we expect similar behavior at high frequencies when the wavelength is short compared with the scale of heterogeneities in the medium. The solution is therefore written as a series in *amplitude coefficients*, $\mathbf{V}^{(m)}$ and $\mathbf{T}_j^{(m)}$, which are independent of frequency, and a phase factor linearly dependent on the frequency ω and the *travel time*, T, which again is independent of frequency. At high frequencies, the solution is therefore nondispersive, but the asymptotic series in inverse powers of $-i\omega$ allows for pulse distortion at low frequencies.

Thus the frequency, or time dependence, only enters through the linear phase term and inverse powers in the (asymptotic) expansion. We allow the amplitude coefficients to be complex but require the travel time to be real. Strictly, if the amplitude coefficients are complex, they must depend on the sign of frequency. To obtain a real time series for the solution, we must have $\mathbf{V}^{(m)*}(\omega) = \mathbf{V}^{(m)}(-\omega)$ where * indicates the complex conjugate. Since this is the only frequency dependence, we omit it from the amplitude coefficients and, where necessary, understand the value for positive frequencies. For generality we have included an arbitrary function of frequency $f(\omega)=f^*(-\omega)$, which in practice will be closely related to the source spectrum.

It is straightforward to write the ansätze of asymptotic ray theory (ART) in the time domain. Inverting the Fourier transform, we obtain

$$\mathbf{V}(\mathbf{x},t)=\mathrm{Re}\left[F(t)*\sum_n\sum_{m=0}^{\infty}\mathbf{V}^{(m)}(\mathbf{x},\mathscr{L}_n)\delta^{(m)}(t-T(\mathbf{x},\mathscr{L}_n))\right] \tag{4}$$

where $F(t)=f(t)+i\bar{f}(t)$ is the analytic time series corresponding to $f(t)$, and $\delta^{(m)}(t)$ is the mth integral of the Dirac delta function, i.e., $\delta^{(0)}(t)=\delta(t)$ the Dirac delta function, $\delta^{(1)}(t)=H(t)$ the Heaviside function, and $\delta^{(m+1)}(t)=t^mH(t)/m!$ for $m>1$. In the time domain, it is obvious that the series (4) is not convergent (unless all terms for $m>1$ are zero, or terms cancel when summed over rays, n). If the series (4) is terminated at $m=M$, then at large times the Mth term will dominate and the series will diverge as $t^{M-1}H(t)/(M-1)!$. This is physically impossible (to conserve energy the velocity must decay as $t\to\infty$). Therefore the series is only asymptotic. Only in special cases, e.g., an isotropic, homogeneous medium, when the series terminates and/or terms cancel in the ray summation, is the series convergent.

Taking the Fourier transformation of Eqs. (1) and (2), substituting Eq. (3) and setting the coefficient of each power of ω to zero, we obtain for $m\geq 0$ (defining $\mathbf{V}^{(-1)}=\mathbf{T}_j^{(-1)}=\mathbf{0}$),

$$\frac{1}{\rho}\frac{\partial\mathbf{T}_j^{(m-1)}}{\partial x_j}=\mathbf{V}^{(m)}+\frac{1}{\rho}p_j\mathbf{T}_j^{(m)} \tag{5}$$

$$\mathbf{c}_{jk}\frac{\partial\mathbf{V}^{(m-1)}}{\partial x_k}=\mathbf{T}_j^{(m)}+p_k\mathbf{c}_{jk}\mathbf{V}^{(m)} \tag{6}$$

where

$$\mathbf{p}=\boldsymbol{\nabla}T \tag{7}$$

is the *slowness vector*, and we have omitted the argument $(\mathbf{x},\mathscr{L}_n)$. Eliminating $\mathbf{T}_j^{(m)}$ between (5) and (6), we obtain

$$(p_jp_k\mathbf{c}_{jk}-\rho\mathbf{I})\mathbf{V}^{(m)}=p_j\mathbf{c}_{jk}\frac{\partial\mathbf{V}^{(m-1)}}{\partial x_k}-\frac{\partial\mathbf{T}_j^{(m-1)}}{\partial x_j} \tag{8}$$

which can be solved for the travel time and amplitude coefficients. This equation can be rewritten in shorthand as

$$\mathsf{N}(\mathbf{V}^{(m)})-\overline{\mathsf{M}}(\mathbf{V}^{(m-1)},\mathbf{T}_j^{(m-1)})=\mathbf{0} \tag{9}$$

where

$$\mathsf{N}(\mathbf{V}^{(m)})=(p_jp_k\mathbf{c}_{jk}-\rho\mathbf{I})\mathbf{V}^{(m)} \tag{10}$$

$$\overline{\mathsf{M}}(\mathbf{V}^{(m)},\mathbf{T}_j^{(m)})=p_j\mathbf{c}_{jk}\frac{\partial\mathbf{V}^{(m)}}{\partial x_k}-\frac{\partial\mathbf{T}_j^{(m)}}{\partial x_j} \tag{11}$$

Substituting for $\mathbf{T}_j^{(m)}$ using Eq. (6), we write

$$\overline{\mathsf{M}}(\mathbf{V}^{(m)},\mathbf{T}_j^{(m)})=\mathsf{M}(\mathbf{V}^{(m)})-\mathsf{L}(\mathbf{V}^{(m-1)}) \tag{12}$$

where

$$\mathsf{M}(\mathbf{V}^{(m)})=p_j\mathbf{c}_{jk}\frac{\partial\mathbf{V}^{(m)}}{\partial x_k}+\frac{\partial}{\partial x_j}\left(p_j\mathbf{c}_{jk}\mathbf{V}^{(m)}\right) \tag{13}$$

$$\mathsf{L}(\mathbf{V}^{(m)})=\frac{\partial}{\partial x_j}\left(\mathbf{c}_{jk}\frac{\partial\mathbf{V}^{(m)}}{\partial x_k}\right) \tag{14}$$

Note $\overline{\mathsf{M}}(\mathbf{V}^{(0)},\mathbf{T}_j^{(0)})=\mathsf{M}(\mathbf{V}^{(0)})$.

2.2 The Eikonal Equation ($m=0$)

For $m=0$, Eq. (9) reduces to the eigenvector (Christoffel) equation

$$\frac{1}{\rho}\mathsf{N}(\mathbf{V}^{(0)})=(p_jp_k\mathbf{c}_{jk}/\rho-\mathbf{I})\mathbf{V}^{(0)}=\mathbf{0} \tag{15}$$

The 3×3 matrix, $p_jp_k\mathbf{c}_{jk}/\rho$, is real and symmetric, and so has orthogonal eigenvectors (we discuss later how to define these in degenerate cases). We denote the orthonormal eigenvectors as $\hat{\mathbf{g}}_I(\mathbf{x},\mathbf{p})$, $I=1$, 2, or 3, which are defined at all $(\mathbf{x},\mathbf{p})$. We can define a Hamiltonian

$$H_I(\mathbf{x},\mathbf{p})=\tfrac{1}{2}(p_jp_k\hat{\mathbf{g}}_I^{\mathrm{T}}\mathbf{c}_{jk}\hat{\mathbf{g}}_I/\rho-1) \tag{16}$$

(with no summation over I). The eigenvalue of $p_jp_k\mathbf{c}_{jk}/\rho$ is unity and eigensolutions of Eq. (15) exist provided

$$|p_jp_k\mathbf{c}_{jk}/\rho-\mathbf{I}|=0 \tag{17}$$

which is the *eikonal* equation, defining the *slowness surface*. This is equivalent to the constraint

$$H_I(\mathbf{x},\mathbf{p})=0 \tag{18}$$

Substituting Eq. (7) in Eq. (18) shows that the latter is equivalent to the Hamilton–Jacobi equation. Thus the position and slowness must satisfy the Hamilton equations (the *kinematic ray equations*)

$$\frac{dx_i}{dT}=\frac{\partial H}{\partial p_i}=c_{ijkl}p_k\hat{g}_j\hat{g}_l/\rho=p_k\hat{\mathbf{g}}_I^{\mathrm{T}}\mathbf{c}_{ik}\hat{\mathbf{g}}_I/\rho=U_i \tag{19}$$

$$\frac{dp_i}{dT}=-\frac{\partial H}{\partial x_i}=-\frac{1}{2}\frac{\partial(c_{jklm}/\rho)}{\partial x_i}p_jp_m\hat{g}_k\hat{g}_l \tag{20}$$

(note that derivatives of $\hat{\mathbf{g}}_I(\mathbf{x},\mathbf{p})$ do not occur in these expressions as for a normalized vector, the derivative is necessarily

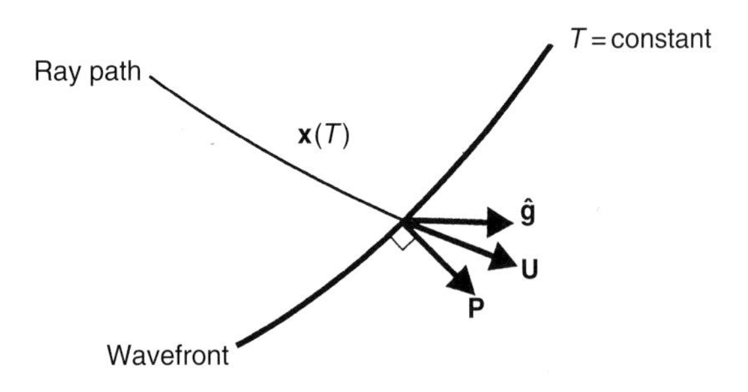

FIGURE 1 A ray path and wavefront, and slowness, group, and polarization vectors.

orthogonal). These equations are 6th-order ordinary differential equations that, given initial conditions $\mathbf{x}_0$ and $\mathbf{p}_0$ at $T = T_0$, and a ray type I, can be solved numerically, or occasionally analytically, for the *ray path*, $\mathbf{x}(T)$. This procedure is called *ray shooting*. With the constraint (18), $\mathbf{p}_0$ has only two degrees of freedom, i.e., to define its direction. Later we parametrize this direction with ray parameters q_1 and q_2. Equation (19) defines the *ray* or *group velocity*

$$\frac{d\mathbf{x}}{dT} = \mathbf{U} \tag{21}$$

The useful relationship

$$\mathbf{p} \cdot \mathbf{U} = 1 \tag{22}$$

can be obtained from the geometry of the wavefront (the slowness vector $\mathbf{p} = \nabla T$ is normal to the wavefront, while Eq. (22) holds whatever the ray direction, $\mathbf{U}$). It can also be obtained algebraically from Eq. (19) using the constraint (18). Figure 1 illustrates a ray path found by solving the kinematic ray equations, a wavefront defined by constant travel time, T, and the slowness, group, and polarization vectors, $\mathbf{p}$, $\mathbf{U}$, and $\hat{\mathbf{g}}$, respectively.

2.3 The Transport Equation ($m \geq 1$)

First we consider the general transport equation and then specialize to find the zeroth-order amplitude coefficients. The equations are solved iteratively so we assume we have solved for the amplitude coefficients to order $m-1$, and are solving for order m.

Let us write $\mathbf{V}^{(m)}$ in terms of the orthogonal eigenvectors

$$\mathbf{V}^{(m)} = V_i^{(m)} \hat{\mathbf{g}}_i \tag{23}$$

Premultiplying Eq. (9) by $\hat{\mathbf{g}}_I^{\mathrm{T}}$ and substituting Eq. (23), we obtain

$$\begin{aligned} \hat{\mathbf{g}}_I^{\mathrm{T}} \overline{\mathsf{M}}(\mathbf{V}^{(m-1)}, \mathbf{T}_j^{(m-1)}) &= \hat{\mathbf{g}}_I^{\mathrm{T}} \mathsf{N}(\mathbf{V}^{(m)}) \\ &= \hat{\mathbf{g}}_I^{\mathrm{T}} (p_j p_k \mathbf{c}_{jk} - \rho \mathbf{I}) V_i^{(m)} \hat{\mathbf{g}}_i \\ &= 2\rho V_I^{(m)} H_I \end{aligned} \tag{24}$$

(no summation over I) using the orthogonality of the eigenvectors, $\hat{\mathbf{g}}_i$, of the matrix $p_j p_k \mathbf{c}_{jk}$.

In general H_I will be zero when I corresponds to the ray type, and nonzero for the other indices. In degenerate cases, $H_I = 0$ for two indices. We denote indices for which $H_I = 0$ by $I = E$, and $H_I \neq 0$ by $I = N$ (summation over an upper-case index is restricted to one or two terms. For instance, E might be the index 2 and N the set of indices 1 and 3. In a degenerate case, E might be the set 1 and 2, and N just 3).

We can solve Eq. (24) for the $V_N^{(m)}$, i.e.,

$$V_N^{(m)} = \frac{1}{2\rho H_N} \hat{\mathbf{g}}_N^{\mathrm{T}} \overline{\mathsf{M}}(\mathbf{V}^{(m-1)}, \mathbf{T}_j^{(m-1)}) \tag{25}$$

(no summation over N). These components are called the *additional terms* and we write them as

$$\mathbf{W}^{(m)} = V_N^{(m)} \hat{\mathbf{g}}_N \tag{26}$$

(summation over N). They can be found from the known, lower-order terms (25). Thus Eq. (23) is

$$\mathbf{V}^{(m)} = V_E^{(m)} \hat{\mathbf{g}}_E + \mathbf{W}^{(m)} \tag{27}$$

$V_E^{(m)}$ are called the *principal component(s)*.

First let us consider the nondegenerate case, where there is only one E index. Later we show how, with the correct choice of eigenvectors, the same method can be used to find the principal components in the degenerate case (where there are two E indices).

Setting $I = E$ and $m - 1 \rightarrow m$ in Eq. (24), we have (since $H_E = 0$)

$$\hat{\mathbf{g}}_E^{\mathrm{T}} \overline{\mathsf{M}}(\mathbf{V}^{(m)}, \mathbf{T}_j^{(m)}) = 0$$

Separating the part that is known using Eqs. (12) and (27), we obtain

$$\hat{\mathbf{g}}_E^{\mathrm{T}} \mathsf{M}(V_E^{(m)} \hat{\mathbf{g}}_E) = \hat{\mathbf{g}}_E^{\mathrm{T}} \left(\mathsf{L}(\mathbf{V}^{(m-1)}) - \mathsf{M}(\mathbf{W}^{(m)}) \right) = \xi^{(m)} \tag{28}$$

say. The RHS, $\xi^{(m)}$, can, at least in principle, be calculated from the known terms, $\mathbf{V}^{(m-1)}$ and $\mathbf{W}^{(m)}$. Overall, this equation is a scalar and so we can transpose the second term in M, and multiply by $V_E^{(m)}$ to reduce it, using Eq. (19), to a simple differential:

$$\nabla \cdot \left(\rho V_E^{(m)2} \mathbf{U} \right) = V_E^{(m)} \xi^{(m)} \tag{29}$$

This can be rewritten

$$\frac{d}{dT} \ln \left(\rho V_E^{(m)2} \right) = \frac{\xi^{(m)}}{\rho V_E^{(m)}} - \nabla \cdot \mathbf{U} \tag{30}$$

using $\mathbf{U} \cdot \nabla = d/dT$. This [Eq. (30)], or variants thereof, is known as the *transport equation*.

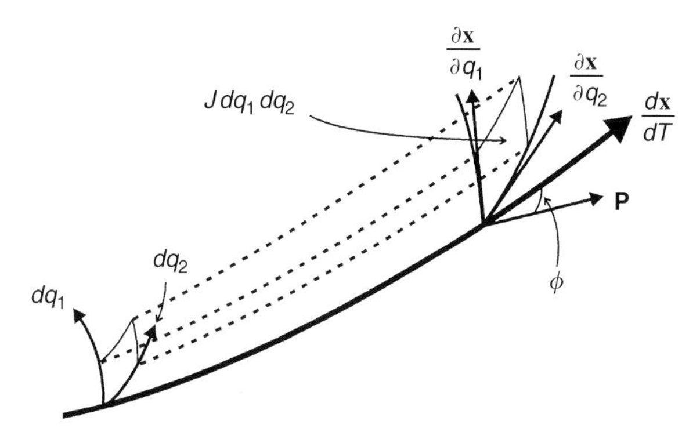

FIGURE 2 A ray tube formed by perturbations dq_1 and dq_2.

Suppose the ray paths are $\mathbf{x}=\mathbf{x}(T, q_1, q_2)$, where, q_1 and q_2 are two parameters describing the initial conditions. Smirnov's lemma (Thomson and Chapman, 1985) gives that if Eq. (21) holds, then

$$\frac{d}{dT}\left(\ln\frac{\partial(x_1, x_2, x_3)}{\partial(T, q_1, q_2)}\right)=\nabla\cdot\mathbf{U}$$

The Jacobian is derived considering differentials of the ray with respect to the initial conditions. A ray tube is defined by perturbations dq_1 and dq_2 to the ray parameters. Figure 2 illustrates a ray tube formed by perturbations dq_1 and dq_2. The cross-section of the ray tube in the wavefront is defined as $J\,dq_1\,dq_2$. The Jacobian is therefore

$$\frac{\partial(x_1, x_2, x_3)}{\partial(T, q_1, q_2)}=\frac{d\mathbf{x}}{dT}\cdot\left(\frac{\partial\mathbf{x}}{\partial q_1}\times\frac{\partial\mathbf{x}}{\partial q_2}\right)=UJ\cos\phi=uJ \quad (31)$$

where u is the phase velocity ($u|\mathbf{p}|=1$), U is the group velocity ($U=|\mathbf{U}|$), and ϕ is the angle between the slowness vector, $\mathbf{p}$, and the ray velocity, $\mathbf{U}$ [Eq. (22) gives $U\cos\phi=u$]. Thus

$$\frac{d}{dT}\ln(uJ)=\nabla\cdot\mathbf{U} \quad (32)$$

Combining Eqs. (30) and (32), we obtain

$$\frac{d}{dT}\ln\left(\rho u V_E^{(m)2}J\right)=\frac{\xi^{(m)}}{\rho V_E^{(m)}}$$

or

$$\frac{d}{dT}\left(V_E^{(m)}\sqrt{\rho u|J|}\right)=\frac{1}{2}\sqrt{\frac{u|J|}{\rho}}\,\xi^{(m)} \quad (33)$$

This is simply solved as

$$V_E^{(m)}=\frac{1}{\sqrt{\rho u|J|}}\left(\text{constant}+\frac{1}{2}\int_{\mathscr{L}_n}\sqrt{\frac{u|J|}{\rho}}\,\xi^{(m)}\,dT\right) \quad (34)$$

It is now trivial to specialize these results to the zeroth-order coefficients. With $m=0$, the additional terms (25) are zero and $\mathbf{W}^{(0)}=\mathbf{0}$. Thus the polarization of the leading term in the ray expansion (3) is parallel to the corresponding eigenvector, $\hat{\mathbf{g}}_E$. For the principal component, $\xi^{(0)}=0$ and Eq. (34) reduces to

$$V_E^{(0)}=\frac{\text{constant}}{\sqrt{\rho u|J|}} \quad (35)$$

The zeroth-order transport equation (29) can be written

$$\nabla\cdot\mathbf{N}=0 \qquad \text{where} \qquad N_j=-\mathbf{V}^{(0)\,\mathrm{T}}\mathbf{T}_j^{(0)}=\rho V^{(0)\,2}U_j \quad (36)$$

and $\mathbf{N}$ is the *energy flux vector* (cf. the Poynting vector in electromagnetism). The amplitude coefficients may be complex due to caustics or total reflections, etc. Eq. (36) does not hold at caustics or r/t points.

2.4 Paraxial Ray Equations

In this section we derive the equations needed to obtain the amplitude coefficient, in particular the function J. The *dynamic ray equations* are obtained by perturbing the ray position and slowness in the kinematic ray equations (19) and (20). This can be written

$$\frac{d\mathbf{y}}{dT}=\begin{pmatrix}\mathbf{T}^{\mathrm{T}} & \mathbf{R}\\ \mathbf{S} & -\mathbf{T}\end{pmatrix}\mathbf{y}=\mathbf{D}\mathbf{y} \quad (37)$$

say, where $\mathbf{y}=(\delta\mathbf{x} \quad \delta\mathbf{p})^{\mathrm{T}}$ is the 6-vector perturbation of the ray position and slowness, and $T_{ij}=\partial^2H/\partial x_i\partial p_j$, $R_{ij}=\partial^2H/\partial p_i\partial p_j$ and $S_{ij}=\partial^2H/\partial x_i\partial x_j$. In principle, this 6th-order system has six solutions, but only four of them are of interest: perturbations in the ray direction at the source; and perturbations of the source position perpendicular to the slowness direction at the source. Another solution, the perturbation along the ray, can be obtained directly from the kinematic ray results, and the slowness constraint (18) eliminates the other. However, first let us consider the complete 6th-order system.

The dynamic equations (37) have symplectic symmetry, which can be used to prove the reciprocity of the ray results and provides a useful check on numerical solutions. It is convenient to define

$$\mathbf{I}_1=\begin{pmatrix}\mathbf{0} & \mathbf{I}\\ -\mathbf{I} & \mathbf{0}\end{pmatrix},\quad \mathbf{I}_2=\begin{pmatrix}\mathbf{0} & \mathbf{I}\\ \mathbf{I} & \mathbf{0}\end{pmatrix}\quad\text{and}\quad \mathbf{I}_3=\begin{pmatrix}-\mathbf{I} & \mathbf{0}\\ \mathbf{0} & \mathbf{I}\end{pmatrix} \quad (38)$$

Because $\mathbf{R}$ and $\mathbf{S}$ are symmetric, it is easily shown that

$$\mathbf{D}=\mathbf{I}_1\,\mathbf{D}^{\mathrm{T}}\mathbf{I}_1=-\mathbf{D}^{\dagger} \quad (39)$$

say. We call this a *symplectic symmetry* (and $\mathbf{D}^{\dagger}$ a *symplectic transform*).

Consider a propagator solution of Eq. (37) from T_0 to T, i.e., $\mathbf{P}(T, T_0)$, which satisfies Eq. (37) and $\mathbf{P}(T_0, T_0)=\mathbf{I}$. Expanding the derivative of $\mathbf{P}^{\mathrm{T}}\mathbf{I}_1\mathbf{P}$ and using Eq. (37), its transpose,

$\mathbf{I}_1^2 = -\mathbf{I}$, and Eq. (39), we obtain

$$\frac{d}{dT}(\mathbf{P}^{\mathrm{T}}\mathbf{I}_1\mathbf{P}) = 0, \qquad \text{i.e.,} \qquad \mathbf{P}^{\mathrm{T}}\mathbf{I}_1\mathbf{P} = \mathbf{I}_1 \tag{40}$$

from the value at T_0. Multiplying by $-\mathbf{I}_1$, this establishes that

$$\mathbf{P}^{-1}(T, T_0) = -\mathbf{I}_1\mathbf{P}(T, T_0)\mathbf{I}_1 = \mathbf{P}^{\dagger}(T, T_0) = \begin{pmatrix} \mathbf{P}_{pp}^{\mathrm{T}} & -\mathbf{P}_{xp}^{\mathrm{T}} \\ -\mathbf{P}_{px}^{\mathrm{T}} & \mathbf{P}_{xx}^{\mathrm{T}} \end{pmatrix} \tag{41}$$

where $\mathbf{P}$ has been written as 3×3 submatrices

$$\mathbf{P} = \begin{pmatrix} \mathbf{P}_{xx} & \mathbf{P}_{xp} \\ \mathbf{P}_{px} & \mathbf{P}_{pp} \end{pmatrix} \tag{42}$$

Thus the inverse propagator, or the propagator in the reverse direction, $\mathbf{P}(T_0, T) = \mathbf{P}^{-1}(T, T_0)$, can be obtained by symplectic transformation, which is significantly easier than inverting a 6×6 matrix. The particular result we will need is

$$\mathbf{P}_{xp}(T_0, T) = -\mathbf{P}_{xp}^{\mathrm{T}}(T, T_0) \tag{43}$$

2.5 Rays at an Interface

The above results describe rays in media without discontinuities. If the medium contains interfaces, i.e., discontinuities in the material properties ρ or $\mathbf{c}_{jk}$, then the ray theory solution breaks down due to discontinuities in the solution or its derivatives. It is necessary to impose the discontinuities on the solution at the interface before continuing the ray solution, and this is investigated in this section.

Let us define the interface separating two different media by $S(\mathbf{x}) = 0$. For simplicity, let us label these media 1 and 2, where the wave is incident from medium 1. A normal to the interface is defined by $\hat{\mathbf{n}} = \epsilon \boldsymbol{\nabla} S$ where $\epsilon = \pm|\boldsymbol{\nabla} S|^{-1}$ and by convention the sign is taken so the vector $\hat{\mathbf{n}}$ points into medium 1. In order to solve the continuity conditions at the interface, it is convenient to transform the elastic parameters into an interface coordinate system, where the direction $\hat{\mathbf{n}}$ is an axis. We complete an orthogonal basis at the point where the ray intersects the interface by defining two vectors, $\hat{\mathbf{l}}$ and $\hat{\mathbf{m}}$, in the tangent plane of the interface. The orientation of these is arbitrary but in an isotropic medium it is convenient to choose one in the plane of the ray and the normal, and one normal to this ray plane, i.e.,

$$\hat{\mathbf{m}} = \frac{\hat{\mathbf{n}} \times \mathbf{U}}{|\hat{\mathbf{n}} \times \mathbf{U}|} \qquad \text{and} \qquad \hat{\mathbf{l}} = \hat{\mathbf{m}} \times \hat{\mathbf{n}} \tag{44}$$

in order to separate the SV and SH components of the shear wave. In anisotropic media, we might as well use the same basis. When Eq. (44) degenerates (when the ray is normal to the interface), any choice will do. Defining an orthonormal transformation matrix $\mathbf{L} = (\hat{\mathbf{l}}\ \hat{\mathbf{m}}\ \hat{\mathbf{n}})$, we can transform any vector/tensor components to the interface basis, e.g.,

$$c'_{i'j'k'l'} = L_{i'i}L_{j'j}L_{k'k}L_{l'l}c_{ijkl} \tag{45}$$

where the prime indicates values in this interface basis.

Assuming the interface is welded, the boundary condition is that the particle velocity, $\mathbf{v}$, and the interface traction, $\mathbf{t}_n = \mathbf{t}'_3 = \hat{n}_j\mathbf{t}_j$, should be continuous. For later use, we define a 6-vector of the amplitude coefficients of these continuous field variables,

$$\mathbf{w} = \begin{pmatrix} \mathbf{V}^{(0)} \\ \mathbf{T}_n^{(0)} \end{pmatrix} = \begin{pmatrix} \mathbf{V}^{(0)} \\ \mathbf{T}_3^{(0)\prime} \end{pmatrix} \tag{46}$$

In general, continuity of $\mathbf{w}$ is impossible for just one ray incident and one ray leaving the interface. The continuity of the six components of $\mathbf{w}$ require that six new rays be generated by one incident ray: Three rays radiated away from the interface in each medium, *reflected* in medium 1, and *transmitted* in medium 2. The six continuity conditions determine the amplitudes of the six generated rays. We denote any property of these rays by two subscripts, e.g., $\mathbf{p}_{kl}$ are the slowness vectors with $k = 1$ or 2 indicating the medium, and $l = 1, 2$ or 3, the ray type. The ordering of the ray types is not crucial but it is convenient to use $l = 1$ to indicate the slower qS wave, 2 the faster qS wave, and 3 the qP wave (i.e., in order of increasing velocity). We indicate the incident wave with a single, ray-type subscript j, i.e., $\mathbf{p}_j$. All rays will intersect the interface at the same point, $\mathbf{x} = \mathbf{x}_j = \mathbf{x}_{kl}$. In general, the existence of all six generated rays is necessary to satisfy the continuity of $\mathbf{w}$, but, having solved for the starting conditions of the generated rays, in practice we follow each generated ray separately. A complete ray consists of a sequence of *ray segments* between interfaces, and the *ray history* or *signature* describes the sequence of choices defining the generated ray type at each interface. The incident and generated ray choice at an interface is indicated by the *coefficient triplet* $\{j, k, l\}$. The complete response consists of a sum of rays with all possible histories—this is enumerated by the index n in Eq. (3).

Throughout the rest of this section, we assume variables are in the interface basis, e.g., (45), and for brevity we omit the prime.

2.6 Kinematic Continuity—Snell's Law

Continuity of $\mathbf{w}$ at an interface requires that T and its gradient in the plane of the interface are the same for all rays. Continuity of T is trivial, just imposing an initial condition on the new ray segment. Continuity of the gradient of T [Eq. (7)] in the plane of the interface leads to the slowness vectors for the generated rays and the initial conditions for the kinematic ray equations of the generated ray.

The slowness in the plane of the interface is given by

$$\mathbf{p}_{\perp} = \mathbf{p} - (\mathbf{p} \cdot \hat{\mathbf{n}})\hat{\mathbf{n}} = p_1\hat{\mathbf{l}} + p_2\hat{\mathbf{m}} \tag{47}$$

This must be the same for all the rays at the interface and can be calculated for the incident ray. Thus the generated rays will have slownesses

$$\mathbf{p}_{kl} = p_n\hat{\mathbf{n}} + \mathbf{p}_{\perp} = p_3\hat{\mathbf{n}} + \mathbf{p}_j - (\mathbf{p}_j \cdot \hat{\mathbf{n}})\hat{\mathbf{n}} \tag{48}$$

where p_3 must be found from the constraint (18), i.e.,

$$H_{kl}(\mathbf{x}, \mathbf{p}_{kl}) = 0 \tag{49}$$

is solved for p_3 (below in Eq. (67), we give a 6th-order eigen-equation for $\mathbf{w}$ that can be solved for the eigenvalue p_3). In general, there will be six solutions for p_3 and we must choose the solutions corresponding to the waves propagating away from the interface. From the $\mathbf{p}_{kl}$, we derive the ray velocities $\mathbf{U}_{kl}$ using Eq. (19) and choose the correct solutions using the constraints

$$\hat{\mathbf{n}} \cdot \mathbf{U}_{1l} > 0 \qquad \text{and} \qquad \hat{\mathbf{n}} \cdot \mathbf{U}_{2l} < 0 \tag{50}$$

which follow from the convention used to define the sign of $\hat{\mathbf{n}}$ ($\hat{\mathbf{n}} \cdot \mathbf{U}_j < 0$, necessarily). They can be ordered in l using the velocity criteria suggested above. Note that the ray velocity, $\mathbf{U}_{kl}$, rather than the slowness, $\mathbf{p}_{kl}$, should be used in these conditions (50)—in some situations in anisotropic media, the two conditions differ (Garmany, 1988).

In some circumstances, some of the solutions for p_3 may be complex, i.e., the waves may be evanescent. While these waves should be included to solve for the amplitude continuity condition, the corresponding rays are not normally included in the ray solution.

The relationship between the slowness vectors, $\mathbf{p}_j$ and $\mathbf{p}_{kl}$, is known as *Snell's law*. This and the continuity of $\mathbf{x}$ and T give the required conditions for kinematic raytracing through an interface.

2.7 Paraxial Ray Discontinuity

In order to continue the solution of the dynamic ray equations (37) through an interface, we must connect the derivatives of $\mathbf{x}$, $\mathbf{p}_j$, and $\mathbf{p}_{kl}$ (see Figure 3). Results have been given by Červený *et al.* (1974) (for isotropic media) and Gajewski and Pšenčík (1990) (for anisotropic media). These have been modified by Farra and Le Bégat (1995) to maintain the symplectic symmetries (41) and in this section we summarize their results.

In order to study the derivatives, it is crucial to distinguish the values in the wavefront (as given by the dynamic ray equations (37) with constant T), and those on the interface ($S = 0$). These are related by

$$\frac{\partial}{\partial q_\nu}\begin{pmatrix}\mathbf{x}\\ \mathbf{p}\end{pmatrix}\Bigg|_S = \frac{\partial}{\partial q_\nu}\begin{pmatrix}\mathbf{x}\\ \mathbf{p}\end{pmatrix}\Bigg|_T + \frac{\partial T}{\partial q_\nu}\Bigg|_S\begin{pmatrix}\dot{\mathbf{x}}\\ \dot{\mathbf{p}}\end{pmatrix} \tag{51}$$

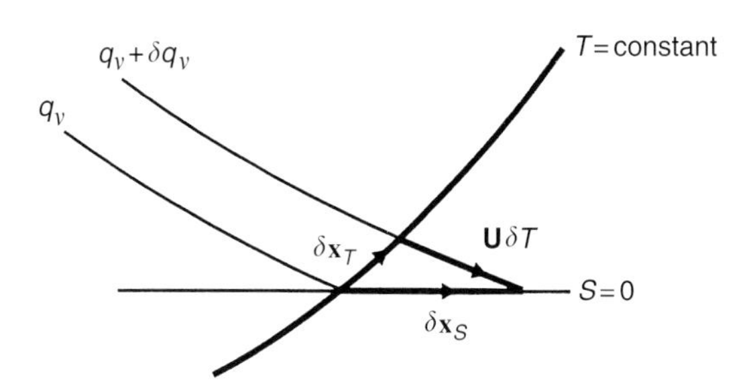

FIGURE 3 A ray and a paraxial ray at an interface, showing the connection between wavefront and interface perturbations, i.e., Eq. (51).

where for brevity we have written $d\mathbf{x}/dT = \dot{\mathbf{x}}$, etc., given by Eqs. (19) and (20) (Greek subscripts, e.g., ν, are restricted to 1 and 2). The position derivative $\partial\mathbf{x}/\partial q_\nu|_S$ must be in the interface, so premultiplying by $\hat{\mathbf{n}}^{\mathrm{T}}$, we can solve for the travel time interface derivative

$$\frac{\partial T}{\partial q_\nu}\Bigg|_S = -\frac{\hat{\mathbf{n}}^{\mathrm{T}}}{\dot{\mathbf{x}}_j^{\mathrm{T}}\hat{\mathbf{n}}}\frac{\partial \mathbf{x}}{\partial q_\nu}\Bigg|_T \tag{52}$$

Substituting in Eq. (51) we can convert the wavefront perturbation to the interface

$$\mathbf{y}_j|_S = \begin{pmatrix}\mathbf{F}_1 & \mathbf{0}\\ \mathbf{F}_2 & \mathbf{I}\end{pmatrix}\mathbf{y}_j|_T \tag{53}$$

where

$$\mathbf{F}_1 = \mathbf{I} - \frac{\dot{\mathbf{x}}_j\hat{\mathbf{n}}^{\mathrm{T}}}{\dot{\mathbf{x}}_j^{\mathrm{T}}\hat{\mathbf{n}}} \qquad \text{and} \qquad \mathbf{F}_2 = -\frac{\dot{\mathbf{p}}_j\hat{\mathbf{n}}^{\mathrm{T}}}{\dot{\mathbf{x}}_j^{\mathrm{T}}\hat{\mathbf{n}}} \tag{54}$$

The position of the incident ray on the interface and the generated ray are identical, i.e.,

$$\mathbf{x}_j|_S = \mathbf{x}_{kl}|_S \tag{55}$$

so perturbations are also equal. A useful expression can be obtained using Eq. (51) in the expression for the ray-tube cross-section [Eq. (31) with $\dot{\mathbf{x}} = \mathbf{U}$ (19)]:

$$\frac{u_j J_j}{\mathbf{U}_j^{\mathrm{T}}\hat{\mathbf{n}}} = \frac{u_{kl}J_{kl}}{\mathbf{U}_{kl}^{\mathrm{T}}\hat{\mathbf{n}}} \tag{56}$$

However, the connection between the slowness vectors is more complicated. The difference between the generated and incident slownesses is normal to the interface [Eq. (48)], i.e.,

$$\mathbf{p}_{kl} - \mathbf{p}_j = \hat{\mathbf{n}}^{\mathrm{T}}(\mathbf{p}_{kl} - \mathbf{p}_j)\hat{\mathbf{n}} \tag{57}$$

This can be differentiated with respect to q_ν. To evaluate the differentials of the RHS, we use the differentials of the Hamiltonians

$$\frac{\partial H}{\partial q_\nu} = -\dot{\mathbf{p}}^{\mathrm{T}}\frac{\partial \mathbf{x}}{\partial q_\nu}\Bigg|_S + \dot{\mathbf{x}}^{\mathrm{T}}\frac{\partial \mathbf{p}}{\partial q_\nu}\Bigg|_S = 0 \tag{58}$$

for either the incident ray, H_j, or generated ray, H_{kl}, with Eqs. (19) and (20). Premultiplying the differential of Eq. (57) by $\dot{\mathbf{x}}_{kl}^{\mathrm{T}}$ and using Eq. (58) for H_{kl}, we can solve for $\partial[\hat{\mathbf{n}}^{\mathrm{T}}(\mathbf{p}_{kl} - \mathbf{p}_j)]/\partial \mathbf{q}_\nu|_S$. With

$$\left.\frac{\partial \hat{\mathbf{n}}}{\partial q_\nu}\right|_S = \boldsymbol{\nabla}^{\mathrm{T}}\hat{\mathbf{n}} \left.\frac{\partial \mathbf{x}}{\partial q_\nu}\right|_S \tag{59}$$

the differentials of Eqs. (57) and (55) can be written

$$\mathbf{y}_{kl}|_S = \begin{pmatrix} \mathbf{I} & \mathbf{0} \\ \mathbf{F}_3 & \mathbf{F}_4 \end{pmatrix} \mathbf{y}_j|_S \tag{60}$$

where

$$\mathbf{F}_3 = \frac{\hat{\mathbf{n}}}{\dot{\mathbf{x}}_{kl}^{\mathrm{T}}\hat{\mathbf{n}}}(\dot{\mathbf{p}}_{kl} - \dot{\mathbf{p}}_j)^{\mathrm{T}} + \hat{\mathbf{n}}^{\mathrm{T}}(\mathbf{p}_{kl} - \mathbf{p}_j)\left(\mathbf{I} - \frac{\hat{\mathbf{n}}\dot{\mathbf{x}}_{kl}^{\mathrm{T}}}{\dot{\mathbf{x}}_{kl}^{\mathrm{T}}\hat{\mathbf{n}}}\right)\boldsymbol{\nabla}^{\mathrm{T}}\hat{\mathbf{n}} \tag{61}$$

$$\mathbf{F}_4 = \mathbf{I} - \frac{\hat{\mathbf{n}}(\dot{\mathbf{x}}_{kl} - \dot{\mathbf{x}}_j)^{\mathrm{T}}}{\dot{\mathbf{x}}_{kl}^{\mathrm{T}}\hat{\mathbf{n}}} \tag{62}$$

Note that these matrices contain factors that combine to give Eq. (58) for H_j, i.e., zero in the perturbation, but are included in the propagator (60) so it satisfies the symplectic symmetry (41) (Farra and Le Bégat, 1995). Finally, we use Eqs. (51) and (52) to convert the perturbation (60) onto the wavefront, i.e.,

$$\mathbf{y}_{kl}|_T = \mathbf{y}_{kl}|_S + \begin{pmatrix} \mathbf{F}_5 & \mathbf{0} \\ \mathbf{F}_6 & \mathbf{0} \end{pmatrix} \mathbf{y}_j|_T \tag{63}$$

where

$$\mathbf{F}_5 = \frac{\dot{\mathbf{x}}_{kl}\hat{\mathbf{n}}^{\mathrm{T}}}{\dot{\mathbf{x}}_j^{\mathrm{T}}\hat{\mathbf{n}}} \quad \text{and} \quad \mathbf{F}_6 = \frac{\dot{\mathbf{p}}_{kl}\hat{\mathbf{n}}^{\mathrm{T}}}{\dot{\mathbf{x}}_j^{\mathrm{T}}\hat{\mathbf{n}}} \tag{64}$$

Overall

$$\mathbf{y}_{kl}|_T = \mathbf{F}\ \mathbf{y}_j|_T \tag{65}$$

where

$$\mathbf{F} = \begin{pmatrix} \mathbf{F}_1 + \mathbf{F}_5 & \mathbf{0} \\ \mathbf{F}_3\mathbf{F}_1 + \mathbf{F}_4\mathbf{F}_2 + \mathbf{F}_6 & \mathbf{F}_4 \end{pmatrix} \tag{66}$$

propagates the perturbation from the incident to the generated ray. It satisfies the symplectic symmetry (41) (Farra and Le Bégat, 1995).

2.8 Reflection/Transmission Coefficients

At an interface, the 6-vector $\mathbf{w}$ [Eq. (46)] is continuous. We can use this condition to calculate the magnitudes of the generated rays relative to the incident ray, i.e., the *reflection/transmission coefficients*. Eliminating the discontinuous $\mathbf{T}_\nu^{(0)}$'s from Eqs. (5) and (6), we obtain

$$\mathbf{A}\mathbf{w} = p_3\mathbf{w} \tag{67}$$

where

$$\mathbf{A}_{22} = \mathbf{A}_{11}^{\mathrm{T}} = -p_\eta \mathbf{c}_{\eta 3}\mathbf{c}_{33}^{-1} \tag{68}$$

$$\mathbf{A}_{12} = -\mathbf{c}_{33}^{-1} \tag{69}$$

$$\mathbf{A}_{21} = p_\eta p_\nu \mathbf{c}_{\eta\nu} - \rho\mathbf{I} - p_\eta p_\nu \mathbf{c}_{\eta 3}\mathbf{c}_{33}^{-1}\mathbf{c}_{3\nu} \tag{70}$$

This eigenequation (67) can be solved for six eigenvalues p_3. Let us write the eigensolutions as

$$\mathbf{A}\mathbf{W} = \mathbf{W}\mathbf{p}_3 \tag{71}$$

where the columns of $\mathbf{W}$ are the eigenvectors $\mathbf{w}$, and $\mathbf{p}_3$ is the diagonal matrix of eigenvalues. We order the eigenvectors so the first three correspond to waves propagating in the positive $\hat{\mathbf{n}}$ direction, i.e., $\hat{\mathbf{n}} \cdot \mathbf{U} > 0$, and the last three are traveling in the opposite direction, i.e., $\hat{\mathbf{n}} \cdot \mathbf{U} < 0$. Within the triplets, we order them with increasing velocity as suggested above. In general, some of the eigenvalues for the generated waves may be complex, corresponding to evanescent waves. The eigenvectors should be ordered according to their evanescent decay.

The various *r/t* experiments can be described by a matrix equation,

$$\mathbf{W}_1\begin{pmatrix} \mathscr{T}_{11} & \mathscr{T}_{12} \\ \mathbf{I} & \mathbf{0} \end{pmatrix} = \mathbf{W}_2\begin{pmatrix} \mathbf{0} & \mathbf{I} \\ \mathscr{T}_{21} & \mathscr{T}_{22} \end{pmatrix} \tag{72}$$

where $\mathbf{W}_1$ and $\mathbf{W}_2$ are the eigenvectors in the two media, $\mathscr{T}$ is a 6×6 matrix of *r/t* coefficients, and the 3×3 submatrices in Eq. (72), $\mathscr{T}_{ki}$, are the coefficients where the incident waves' medium is indicated by i, and the generated waves' medium is indicated by k (the unit matrices $\mathbf{I}$ represent the incident waves). The elements of the full matrices and submatrices are related by

$$(\mathscr{T}_{ki})_{lj} = \mathscr{T}_{l+3(k-1)\,j+3(i-1)} \tag{73}$$

where j (column index) and l (row index) indicate the incident and generated wave types, respectively (the symbols correspond to the coefficient triplet $\{j, k, l\}$ used in Sections 2.5, 2.6, and 2.7, with $i = 1$ by design). Figure 4 illustrates the *r/t* experiments represented by Eq. (72). Solving this equation we have

$$\begin{aligned} &\begin{pmatrix} \mathscr{T}_{11} & \mathscr{T}_{12} \\ \mathbf{I} & \mathbf{0} \end{pmatrix}\begin{pmatrix} \mathbf{0} & \mathbf{I} \\ \mathscr{T}_{21} & \mathscr{T}_{22} \end{pmatrix}^{-1} \\ &= \begin{pmatrix} \mathscr{T}_{12} - \mathscr{T}_{11}\mathscr{T}_{21}^{-1}\mathscr{T}_{22} & \mathscr{T}_{11}\mathscr{T}_{21}^{-1} \\ -\mathscr{T}_{21}^{-1}\mathscr{T}_{22} & \mathscr{T}_{21}^{-1} \end{pmatrix} = \mathbf{W}_1^{-1}\mathbf{W}_2 \\ &= \mathbf{Q} \end{aligned} \tag{74}$$

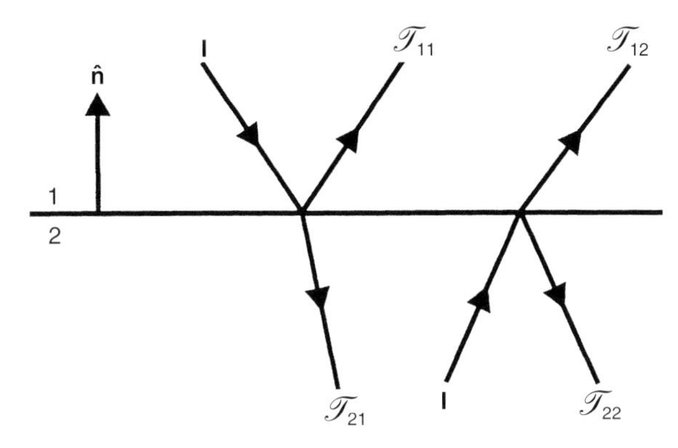

FIGURE 4 Reflection/transmission experiments—each represents three incident ray types and six generated rays (36 *r*/*t* coefficients in total).

say. This can be solved for the coefficient matrix

$$\mathscr{T} = \begin{pmatrix} \mathscr{T}_{11} & \mathscr{T}_{12} \\ \mathscr{T}_{21} & \mathscr{T}_{22} \end{pmatrix} = \begin{pmatrix} \mathbf{Q}_{12}\mathbf{Q}_{22}^{-1} & \mathbf{Q}_{11} - \mathbf{Q}_{12}\mathbf{Q}_{22}^{-1}\mathbf{Q}_{21} \\ \mathbf{Q}_{22}^{-1} & -\mathbf{Q}_{22}^{-1}\mathbf{Q}_{21} \end{pmatrix} \tag{75}$$

where $\mathbf{Q}_{ij}$ are the 3×3 submatrices of $\mathbf{Q}$. The matrix $\mathbf{Q}$ is defined by Eq. (74) in terms of the known eigenvectors in the two media, so this allows us to solve for all the *r*/*t* coefficients. Computing $\mathbf{Q}$ seems to involve inverting a 6×6 matrix $\mathbf{W}_1$, but fortunately this can be avoided.

We note that $\mathbf{I}_2\mathbf{A}$ is symmetric and, taking the transpose of Eq. (71) times $\mathbf{I}_2$ [Eq. (38)], we can reduce it to

$$(\mathbf{W}^{\mathrm{T}}\mathbf{I}_2)\mathbf{A} = \mathbf{p}_3(\mathbf{W}^{\mathrm{T}}\mathbf{I}_2)$$

Postmultiplying this by $\mathbf{W}$, premultiplying Eq. (71) by $\mathbf{W}^{\mathrm{T}}\mathbf{I}_2$, and subtracting, we find that

$$\mathbf{W}^{\mathrm{T}}\mathbf{I}_2\mathbf{W} = \mathbf{K} \tag{76}$$

say, must be diagonal. The eigenvector columns of $\mathbf{W}$ are

$$\mathbf{w} = w_E \begin{pmatrix} \hat{\mathbf{g}}_E \\ -p_k \mathbf{c}_{3k} \hat{\mathbf{g}}_E \end{pmatrix} \tag{77}$$

where w_E is an arbitrary normalization. From Eq. (76), we find that the diagonal elements of $\mathbf{K}$ are

$$K_E = -(2\rho U_3 w^2)_E \tag{78}$$

where Eq. (19) defines U_3. Thus the required inverse matrix in Eq. (74) can be computed simply as

$$\mathbf{W}^{-1} = \mathbf{K}^{-1}\mathbf{W}^{\mathrm{T}}\mathbf{I}_2 \tag{79}$$

If $w_E = 1/\sqrt{\pm 2\rho U_3}$, with the positive sign for $E = 1$ to 3, and the negative sign for $E = 4$ to 6, then $\mathbf{K} = \mathbf{K}^{-1} = \mathbf{I}_3$. In general, especially for evanescent waves, it is simpler to take $w_E = 1$, when $\mathbf{K}$ is defined by Eq. (78). It is important to remember that the normalization w_E (77) effects the numerical value of the *r*/*t* coefficient. An incident plane wave of unit amplitude $\hat{\mathbf{g}}_j$ converts as

$$\hat{\mathbf{g}}_j \rightarrow \frac{w_\ell}{w_j} \mathscr{T}_{\ell j}\, \hat{\mathbf{g}}_{kl} \tag{80}$$

where $\ell = l + 3(k-1)$, and $\{j, k, l\}$ is the coefficient triplet defined in Section 2.5.

Finally we need to prove the reciprocity of the *r*/*t* coefficients $\mathscr{T}$ [Eq. (75)] (Chapman, 1994). In the reciprocal rays, the slownesses in the plane of the interface are reversed and the matrix $\mathbf{A}$ becomes

$$\mathbf{A}' = \mathbf{A}(-p_\nu) \tag{81}$$

The eigensolution becomes

$$\mathbf{A}'\mathbf{W}' = -\mathbf{W}'\mathbf{p}_3 \tag{82}$$

where the change of sign occurs as the slowness surface has point symmetry. The revised eigenvectors, $\mathbf{W}'$, are related by

$$\mathbf{W}' = -\mathbf{I}_3\mathbf{W} \tag{83}$$

i.e., the traction components change sign. Importantly, the propagation directions of the columns of $\mathbf{W}'$ are reversed, so Eq. (72) becomes

$$\mathbf{W}'_1 \begin{pmatrix} \mathbf{I} & \mathbf{0} \\ \mathscr{T}'_{11} & \mathscr{T}'_{12} \end{pmatrix} = \mathbf{W}'_2 \begin{pmatrix} \mathscr{T}'_{21} & \mathscr{T}'_{22} \\ \mathbf{0} & \mathbf{I} \end{pmatrix} \tag{84}$$

Taking the transpose of Eq. (84) and multiplying by $\mathbf{I}_1$ times (72) we obtain

$$\begin{pmatrix} \mathbf{I} & \mathscr{T}'^{\mathrm{T}}_{11} \\ \mathbf{0} & \mathscr{T}'^{\mathrm{T}}_{12} \end{pmatrix} \mathbf{W}'^{\mathrm{T}}_1 \mathbf{I}_1 \mathbf{W}_1 \begin{pmatrix} \mathscr{T}_{11} & \mathscr{T}_{12} \\ \mathbf{I} & \mathbf{0} \end{pmatrix} = \begin{pmatrix} \mathscr{T}'^{\mathrm{T}}_{21} & \mathbf{0} \\ \mathscr{T}'^{\mathrm{T}}_{22} & \mathbf{I} \end{pmatrix} \mathbf{W}'^{\mathrm{T}}_2 \mathbf{I}_1 \mathbf{W}_2 \begin{pmatrix} \mathbf{0} & \mathbf{I} \\ \mathscr{T}_{21} & \mathscr{T}_{22} \end{pmatrix} \tag{85}$$

Using Eq. (83), it is straightforward to simplify this as

$$\mathbf{W}'^{\mathrm{T}}\mathbf{I}_1\mathbf{W} = -\mathbf{W}^{\mathrm{T}}\mathbf{I}_3^{\mathrm{T}}\mathbf{I}_1\mathbf{W} = \mathbf{W}^{\mathrm{T}}\mathbf{I}_2\mathbf{W} = \mathbf{K} = \begin{pmatrix} \mathbf{K}_+ & \mathbf{0} \\ \mathbf{0} & \mathbf{K}_- \end{pmatrix} \tag{86}$$

where we have expanded the matrix (76) into diagonal 3×3 submatrices for the positive and negative traveling waves. Then expanding Eq. (85), it is seen to be equivalent to

$$\mathscr{T}' = \begin{pmatrix} \mathbf{K}_{1-}^{-1} & \mathbf{0} \\ \mathbf{0} & -\mathbf{K}_{2+}^{-1} \end{pmatrix} \mathscr{T}^{\mathrm{T}} \begin{pmatrix} -\mathbf{K}_{1+} & \mathbf{0} \\ \mathbf{0} & \mathbf{K}_{2-} \end{pmatrix} \tag{87}$$

which is the reciprocity result for the *r*/*t* coefficients. If the eigenvectors are normalized so that $\mathbf{K}=\mathbf{I}_3$, then this simplies to $\mathcal{T}'=\mathcal{T}^{\mathrm{T}}$.

2.9 The Ray Green Function

In practice, we do not need to solve for the complete dynamic propagator [Eq. (42)]. For a point source, all we need is solutions with initial conditions

$$\mathbf{y}(T_0)=\begin{pmatrix}0\\ \partial\mathbf{p}_0/\partial q_\nu\end{pmatrix} \tag{88}$$

i.e., perturbations of the direction of the initial slowness $\mathbf{p}_0$. The ray parameters, q_ν, parametrize the source slowness surface, e.g., the polar angles. The solutions obtained directly are then

$$\frac{\partial\mathbf{x}}{\partial q_\nu}=\mathbf{P}_{xp}\frac{\partial\mathbf{p}_0}{\partial q_\nu} \tag{89}$$

Alternatively, we can use a basis vector in the wavefront, when the dynamic equations (37) can be rewritten as a 4×4 system, but numerically it is probably just as simple to use cartesian coordinates and find two or four solutions of the 6×6 system of Eq. (37). This equation can also be reduced to a nonlinear equation for J, but again it is probably easier to work with system (37), which is relatively simple algebraically and computationally. Furthermore, quantities other than J alone are needed for Maslov asymptotic ray theory.

Near a point source, the approximate solution of Eq. (37) gives $\mathbf{P}_{xp}\simeq\mathbf{R}_0(T-T_0)$. Using this in Eqs. (89) and (31), Kendall *et al.* (1992) have shown, by comparison with the exact solution for a point force $\mathbf{f}_0$ in homogeneous media (Buchwald, 1959; Lighthill, 1960; Duff, 1960; Burridge, 1967), that Eq. (35) becomes

$$V_E^{(0)}=\frac{1}{4\pi}\left(\left|\frac{\partial\mathbf{p}_0}{\partial q_1}\times\frac{\partial\mathbf{p}_0}{\partial q_2}\right|\Big/\rho_0\rho u U_0 J\right)^{1/2}\hat{\mathbf{g}}_0^{\mathrm{T}}\mathbf{f}_0 \tag{90}$$

Numerically, this is all that is needed as Eq. (89) gives the factor uJ [Eq. (31)]. Although sufficient for the numerical solution, theoretically it is useful to analyze the system further to eliminate the arbitrary parametrization q_ν, and prove reciprocity. At the receiver, Kendall *et al.* (1992) have shown that with Eq. (89), the required expression Eq. (31) can be written

$$uJ=\mathbf{U}\cdot\left(\frac{\partial\mathbf{x}}{\partial q_1}\times\frac{\partial\mathbf{x}}{\partial q_2}\right)=\frac{1}{U_0}\left|\frac{\partial\mathbf{p}_0}{\partial q_1}\times\frac{\partial\mathbf{p}_0}{\partial q_2}\right|\mathcal{R} \tag{91}$$

where

$$\mathcal{R}(\mathbf{x},\mathbf{x}_0)=\mathbf{U}_0^{\mathrm{T}}\,\mathrm{adj}(\mathbf{P}_{xp})\mathbf{U}=|\mathbf{P}_{xp}|\mathbf{U}_0^{\mathrm{T}}\mathbf{P}_{xp}^{-1}\mathbf{U} \tag{92}$$

Combining these results (90) and (91), and introducing the product of appropriate *r*/*t* coefficients for the ray history, the amplitude coefficient for the dyadic Green function is

$$\underline{\mathbf{V}}^{(0)}=\frac{e^{-i(\pi/2)\mathrm{sgn}(\omega)\sigma(\mathbf{x},\mathcal{L}_n)}\left(\prod\mathcal{T}_{\ell j}\right)\hat{\mathbf{g}}\,\hat{\mathbf{g}}_0^{\mathrm{T}}}{4\pi\left(\rho(\mathbf{x})\rho(\mathbf{x}_0)|\mathcal{R}|\right)^{1/2}} \tag{93}$$

(we use an underline to indicate a Green function, which for the velocity field for a point force can be treated as a 3×3 matrix). The KMAH index, $\sigma(\mathbf{x},\mathcal{L}_n)$, is introduced as J in Eq. (90) and $\mathcal{R}$ in Eq. (91) may be negative. Symbolically we have written the product of *r*/*t* coefficients as simply $\prod\mathcal{T}_{\ell j}$ where $\mathcal{T}_{\ell j}$ is an element of the matrix $\mathcal{T}$ with the normalization $w_E=1/\sqrt{\pm2\rho U_3}$—the values of ℓ and j [as defined in Eqn. (80), etc.] and the elastic parameters used for the $\mathcal{T}_{\ell j}$ depend on the ray history and the ray path $\mathcal{L}_n$.

It is interesting to verify the behavior of the Green function (93) just before and just after an interface. The relevant factors in Eq. (93) convert as

$$\frac{\hat{\mathbf{g}}_j}{\left(\rho_j|\mathcal{R}_j|\right)^{1/2}}\longrightarrow\mathcal{T}_{\ell j}\frac{\hat{\mathbf{g}}_{kl}}{\left(\rho_k|\mathcal{R}_{kl}|\right)^{1/2}} \tag{94}$$

where $\{j,k,l\}$ is the coefficient triplet. Using Eqs. (56) and (91), it is straightforward to see that this is exactly equivalent to Eq. (80).

The dynamic ray equations, e.g., Eq. (32), break down at caustics where $J=0$. Only by using a more complete wave theory or an extension of ray theory, e.g., Maslov asymptotic ray theory (Section 4.1), can we determine how the ray solutions connect through the caustic. In fact Eq. (90) remains valid provided the correct root is taken for $J^{-1/2}$ and this is introduced by the KMAH index. The KMAH index counts the caustics along the ray (incrementing by 1 at each first-order zero—line caustic—of $\mathbf{P}_{xp}$, and by 2 at a second-order zero—point caustic). In anisotropic media, it may also decrease depending on the properties of the matrix $\mathbf{R}$ in the dynamic system (37) (Lewis, 1965; Garmany, 1988; Klimeš, 1997; Bakker, 1998). Although the KMAH index counts zeros of the spreading matrix $\mathbf{P}_{xp}$, i.e., changes in sign of eigenvalues, result (43) is not sufficient to establish reciprocity of the KMAH index as the signature of the matrix only counts zeros in modulo arithmetic (although, as only $\exp(-i\pi\sigma/2)$ is needed, it is sufficient for reciprocity of the Green function). Nevertheless, we can easily argue that the KMAH index is reciprocal. Consider a short ray with no caustics, e.g., with the receiver near the source so we can approximate by straight rays. The KMAH index is obviously reciprocal (normally zero). If we gradually move the receiver away from the source along the ray path $\mathcal{L}_n$, result (43) guarantees that the KMAH index for the ray and its reciprocal will increment at the same time (note that this is not a trivial argument as the caustic will always occur at the receiver point, different for the ray and its reciprocal). Thus although Eq. (43) only guarantees that the KMAH indices are equal in modulo arithmetic, growing the ray and its reciprocal from nothing proves that the KMAH index is exactly reciprocal.

Substituting Eq. (93) in Eq. (4), we obtain the time-domain, zeroth-order ray theory Green function,

$$\underline{\mathbf{v}}(t, \mathbf{x}; \mathbf{x}_0) \simeq \frac{1}{4\pi}\mathrm{Re}\left[\Delta(t - T(\mathbf{x}, \mathscr{L}_n))e^{-i(\pi/2)\sigma(\mathbf{x},\mathscr{L}_n)} \times \left(\prod \mathscr{T}_{\ell j}\right) (\rho(\mathbf{x})\rho(\mathbf{x}_0)|\mathscr{R}|)^{-1/2}\, \hat{\mathbf{g}}\, \hat{\mathbf{g}}_0^{\mathrm{T}}\right] \tag{95}$$

where $\Delta(t) = \delta(t) - i/\pi t$ is the analytic delta function. In a homogeneous, isotropic medium this reduces to the standard result (Aki and Richards, 1980, Ch. 4)

$$\underline{\mathbf{v}}(t, \mathbf{x}; \mathbf{x}_0) = \frac{\delta(t - T)\hat{\mathbf{g}}\, \hat{\mathbf{g}}_0^{\mathrm{T}}}{4\pi\rho U^2 R} \tag{96}$$

where $R = |\mathbf{x} - \mathbf{x}_0|$ and $T = R/U$.

The reciprocity of Eq. (95) is established as $T(\mathbf{x}, \mathscr{L}_n)$ is obviously reciprocal, the KMAH index is reciprocal from the argument above, the r/t coefficients are reciprocal from Eq. (87), and $\mathscr{R}$ is reciprocal from Eq. (43). Thus,

$$\underline{\mathbf{v}}(t, \mathbf{x}_0; \mathbf{x}) = \underline{\mathbf{v}}^{\mathrm{T}}(t, \mathbf{x}; \mathbf{x}_0) \tag{97}$$

In using Eq. (95), it is important that a consistent rule is used to define the sign of the polarization $\hat{\mathbf{g}}$ such that $\hat{\mathbf{g}} \to \hat{\mathbf{g}}_0$ as $\mathbf{x} \to \mathbf{x}_0$—the sign is ambiguous in the eigenvector equation (15) or Hamiltonian (16). Thus, typically we choose $\hat{\mathbf{g}} \simeq u\mathbf{p}$ for the qP ray, but this means that in the reciprocal ray (97), both polarizations change sign.

3. Specializations to Simpler Earth Models

3.1 Isotropic Ray Theory

We now specialize the general, anisotropic results to those for an isotropic model. Although it simplifies the kinematic ray equations, the shear waves are now degenerate.

In an isotropic medium, the elastic tensor is given by

$$c_{ijkl} = \lambda\delta_{ij}\delta_{kl} + \mu\,(\delta_{ik}\delta_{jl} + \delta_{il}\delta_{jk}) \tag{98}$$

where λ and μ are the Lamé parameters. The matrices $\mathbf{c}_{jk}$ reduce to

$$\mathbf{c}_{11} = \begin{pmatrix} \lambda + 2\mu & 0 & 0 \\ 0 & \mu & 0 \\ 0 & 0 & \mu \end{pmatrix}, \qquad \mathbf{c}_{23} = \begin{pmatrix} 0 & 0 & 0 \\ 0 & 0 & \lambda \\ 0 & \mu & 0 \end{pmatrix}, \quad \text{etc.} \tag{99}$$

3.1.1 The Isotropic Eikonal

As the medium is isotropic, we can choose any propagation direction to determine the Hamiltonian. Let us choose

$$\mathbf{p} = (0 \quad 0 \quad p)^{\mathrm{T}} \tag{100}$$

Then

$$p_j p_k \mathbf{c}_{jk} = p^2\mathbf{c}_{33} = \begin{pmatrix} \mu p^2 & 0 & 0 \\ 0 & \mu p^2 & 0 \\ 0 & 0 & (\lambda + 2\mu)p^2 \end{pmatrix} \tag{101}$$

Thus the Hamiltonians are

$$H_1 = H_2 = \tfrac{1}{2}(\beta^2 p_i p_i - 1), \qquad H_3 = \tfrac{1}{2}(\alpha^2 p_i p_i - 1) \tag{102}$$

where the P and S wave velocities are

$$\alpha = \sqrt{\frac{\lambda + 2\mu}{\rho}} \quad \text{and} \quad \beta = \sqrt{\frac{\mu}{\rho}} \tag{103}$$

(for definiteness, we always choose $I = 3$ as the P wave).

Using the isotropic Hamiltonian (102), the kinematic ray equations (19) and (20) simplify to

$$\frac{d\mathbf{x}}{dT} = U^2\mathbf{p} \tag{104}$$

$$\frac{d\mathbf{p}}{dT} = -\frac{\nabla U}{U} \tag{105}$$

The P wave solution requires $u = U = \alpha$ and $\hat{\mathbf{g}}_3 = \alpha\mathbf{p}$, i.e., the wave is longitudinal. The S wave solution requires $u = U = \beta$ and the polarizations are perpendicular to the ray direction, i.e., $\hat{\mathbf{g}}_\nu \cdot \mathbf{p} = 0$ and the wave is transverse. However, the eigenequation (15) is degenerate and does not determine the S polarization uniquely. To find the shear wave polarizations in isotropic media, we must consider the equation (9) with $m = 1$.

For future use, let us give some useful relationships involving the polarization vectors and elastic matrices:

$$p_j\mathbf{c}_{jk}\hat{\mathbf{g}}_3 = (\lambda + \mu)p_k\hat{\mathbf{g}}_3 + \frac{\mu}{\alpha}\,\hat{\mathbf{e}}_k \tag{106}$$

$$p_j\mathbf{c}_{kj}\hat{\mathbf{g}}_3 = 2\mu p_k\hat{\mathbf{g}}_3 + \frac{\lambda}{\alpha}\,\hat{\mathbf{e}}_k \tag{107}$$

$$p_j\mathbf{c}_{jk}\hat{\mathbf{g}}_\nu = \mu p_k\hat{\mathbf{g}}_\nu + \lambda(\hat{\mathbf{g}}_\nu \cdot \hat{\mathbf{e}}_k)\mathbf{p} \tag{108}$$

$$p_j\mathbf{c}_{kj}\hat{\mathbf{g}}_\nu = \mu p_k\hat{\mathbf{g}}_\nu + \mu(\hat{\mathbf{g}}_\nu \cdot \hat{\mathbf{e}}_k)\mathbf{p} \tag{109}$$

where $\hat{\mathbf{e}}_k$ are the unit cartesian coordinate vectors. These results can easily be derived using Eq. (99).

3.1.2 The Isotropic Transport Equation and Shear Ray Polarization

In the equation (9) with $m=1$, we expand $\mathbf{V}^{(1)}$ according to Eq. (23). For a shear ray, the equation reduces to

$$\overline{\mathsf{M}}(\mathbf{V}^{(0)}, \mathbf{T}_j^{(0)}) = \rho\left(\frac{\beta^2}{\alpha^2} - 1\right) V_3^{(1)} \hat{\mathbf{g}}_3 \quad (110)$$

Premultiplying by $\hat{\mathbf{g}}_\nu^{\mathrm{T}}$, and considering one shear wave, $\mathbf{V}^{(0)} = V_1{}^{(0)}\hat{\mathbf{g}}_1$, say, without loss in generality we obtain

$$\hat{\mathbf{g}}_\nu^{\mathrm{T}} \mathsf{M}(V_1^{(0)}\hat{\mathbf{g}}_1) = 0 \quad (111)$$

Using Eqs. (108) and (109) in (111) and expanding, we obtain

$$\delta_{1\nu}\,(2\mu\mathbf{p}\cdot\boldsymbol{\nabla}V_1^{(0)} + V_1^{(0)}\mathbf{p}\cdot\boldsymbol{\nabla}\mu + \mu V_1^{(0)}\boldsymbol{\nabla}\cdot\mathbf{p}) + 2\rho V_1^{(0)}\,\hat{\mathbf{g}}_\nu^{\mathrm{T}}\,\frac{d\hat{\mathbf{g}}_1}{dT} = 0 \quad (112)$$

where we have used the orthonormality of the eigenvectors, and $\beta^2\mathbf{p}\cdot\boldsymbol{\nabla} = d/dT$ to simplify the final term.

When $\nu = 1$, this gives

$$\begin{aligned} &2\mu\mathbf{p}\cdot\boldsymbol{\nabla}V_1^{(0)} + V_1^{(0)}\mathbf{p}\cdot\boldsymbol{\nabla}\mu + \mu V_1^{(0)}\boldsymbol{\nabla}\cdot\mathbf{p} \\ &= \frac{1}{V_1^{(0)}}\,\boldsymbol{\nabla}\cdot(\rho\beta^2 V_1^{(0)2}\mathbf{p}) = 0 \end{aligned} \quad (113)$$

which is the isotropic transport equation equivalent to Eq. (29) with $\mathbf{U} = \beta^2\mathbf{p}$ and $\xi^{(0)} = 0$.

When $\nu = 2$, Eq. (112) is simply

$$2\rho V_1^{(0)}\,\hat{\mathbf{g}}_2^{\mathrm{T}}\,\frac{d\hat{\mathbf{g}}_1}{dT} = 0 \quad (114)$$

For $\hat{\mathbf{g}}_1$ and $\hat{\mathbf{g}}_2$ to be the polarizations of the independent shear waves, they must satisfy this differential equation (114). Thus the change in $\hat{\mathbf{g}}_1$, which is necessarily orthogonal to itself (as the vector is normalized), is also orthogonal to $\hat{\mathbf{g}}_2$. It can only be in the $\hat{\mathbf{g}}_3$ direction, i.e.,

$$\frac{d\hat{\mathbf{g}}_\nu}{dT} = a\hat{\mathbf{g}}_3 \quad (115)$$

(generalized for both $\nu = 1$ and 2). Thus the shear polarization only changes in the ray direction and does not twist about it. It remains to find a in Eq. (115), which is determined from the geometrical condition that $\hat{\mathbf{g}}_\nu$ remain transverse:

$$a = \beta\mathbf{p}^{\mathrm{T}}\,\frac{d\hat{\mathbf{g}}_\nu}{dT} = -\beta\,\hat{\mathbf{g}}_\nu^{\mathrm{T}}\,\frac{d\mathbf{p}}{dT} = \hat{\mathbf{g}}_\nu^{\mathrm{T}}\boldsymbol{\nabla}\beta \quad (116)$$

using Eq. (105). Thus,

$$\frac{d\hat{\mathbf{g}}_\nu}{dT} = (\hat{\mathbf{g}}_\nu^{\mathrm{T}}\boldsymbol{\nabla}\beta)\,\hat{\mathbf{g}}_3 \quad (117)$$

In the literature this result has been proved indirectly using the equations for the normal and binormal of the ray path, and relating the shear polarization to the torsion of the ray (Popov and Pšenčík, 1978; Červený and Hron, 1980; Červený, 1985). The above direct proof is due to Pšenčík (personal communication, 1998) and avoids ever having to discuss these other matters.

Having achieved the decoupling of the shear wave amplitude coefficients by using the shear wave polarization vectors as the basis, the analysis for higher-order terms in anisotropic media remains valid for isotropic media (Červený and Hron, 1980): for the P ray, we have two additional terms and one principal component; for S rays, we have one additional term, and two principal components, both determined using Eq. (34).

In isotropic media, the first-order additional terms [Eq. (25) with $m=1$] can be written in relatively simple forms (Eisner and Pšenčík, 1996). For P we obtain

$$\begin{aligned} V_\nu^{(1)} &= \frac{\alpha^2\hat{\mathbf{g}}_\nu^{\mathrm{T}}}{\rho(\beta^2-\alpha^2)}\,\mathsf{M}(V_3^{(0)}\hat{\mathbf{g}}_3) \\ &= -\hat{\mathbf{g}}_\nu^{\mathrm{T}}\left(\alpha\,\frac{\partial V_3^{(0)}}{\partial g_\nu} + \frac{V_3^{(0)}}{\alpha^2-\beta^2}\left[(\alpha^2-\beta^2)\boldsymbol{\nabla}\alpha - 4\alpha\beta\,\boldsymbol{\nabla}\beta \right.\right. \\ &\qquad \left.\left. + \,\alpha(\alpha^2-2\beta^2)\,\boldsymbol{\nabla}(\ln\rho)\right]\right) \end{aligned} \quad (118)$$

(g_ν is a coordinate in the direction $\hat{\mathbf{g}}_\nu$). Similarly for S waves, the additional term is

$$\begin{aligned} V_\nu^{(1)} &= \frac{\beta^2\hat{\mathbf{g}}_3^{\mathrm{T}}}{\rho(\alpha^2-\beta^2)}\,\mathsf{M}(V_\nu^{(0)}\hat{\mathbf{g}}_\nu) \\ &= \hat{\mathbf{g}}_3^{\mathrm{T}}\left(\beta\,\frac{\partial V_\nu^{(0)}}{\partial g_\nu} + \frac{V_\nu^{(0)}}{\alpha^2-\beta^2}\left[(\alpha^2+3\beta^2)\boldsymbol{\nabla}\beta + \beta^3\boldsymbol{\nabla}(\ln\rho)\right]\right) \end{aligned} \quad (119)$$

Apart from using Eqs. (106) to (109), (117), the orthonormality of the polarization vectors, and kinematic ray results, it is useful to note

$$\frac{d\hat{\mathbf{g}}_3}{dT} = -\,(\hat{\mathbf{g}}_\nu\cdot\boldsymbol{\nabla}U)\,\hat{\mathbf{g}}_\nu \quad (120)$$

$$\boldsymbol{\nabla}\cdot\hat{\mathbf{g}}_\nu = \frac{1}{U}\,(\hat{\mathbf{g}}_\nu\cdot\boldsymbol{\nabla}U) \quad (121)$$

3.2 1D and 2D Media

Naturally the ray equations simplify if the heterogeneity in the model is restricted to one or two dimensions. The appropriate ray equation, Eq. (20) or Eq. (105), immediately reduces to the conservation of the corresponding slowness component:

$$p_\nu = \text{constant} \quad (122)$$

Although there is some simplification of the other equations in 2D, only in 1D media are these really significant, circumventing the solution of ordinary differential equations. Thus if the medium depends only on x_3, then Eq. (19) is

$$\frac{dx_i}{dT} = U_i(p_1, p_2, x_3) \tag{123}$$

as p_3 can be found from the constraint (18). Then for $\nu = 1$ or 2 we have

$$x_\nu(p_1, p_2, x_3) = \int \frac{U_\nu}{U_3}\, dx_3 \tag{124}$$

$$T(p_1, p_2, x_3) = \int \frac{1}{U_3}\, dx_3 \tag{125}$$

where the depth integral is over all segments of the ray arranged so that dx_3/U_3 is positive (in anisotropic media it is possible for U_ν to be negative—Shearer and Chapman, 1988).

In isotropic media, we have axial symmetry and, without loss in generality, we can take $p_2 = x_2 = 0$. Then Eqs. (124) and (125) reduce to the simple *ray integrals*,

$$x_1(p_1, 0, x_3) = \int \frac{p_1}{p_3}\, dx_3 = \int \tan i(x_3)\, dx_3 \tag{126}$$

$$T(p_1, 0, x_3) = \int \frac{1}{U^2 p_3}\, dx_3 = \int \frac{\sec i(x_3)}{U(x_3)}\, dx_3 \tag{127}$$

used in layered media. The slowness vector is

$$\mathbf{p} = \frac{1}{U(x_3)} \begin{pmatrix} \sin i(x_3) \\ 0 \\ \cos i(x_3) \end{pmatrix} \tag{128}$$

where p_1 is conserved and $i(x_3)$ is the angle the ray makes with the vertical. Using $q_1 = p_1$ and $q_2 = \phi$ (the azimuth), it is straightforward to reduce the amplitude coefficient (90) to

$$V_E^{(0)} = \frac{1}{4\pi} \left(\rho_0 \rho U_0 U \cos i_0 \cos i \frac{x_1}{p_1} \left| \frac{dx_1}{dp_1} \right| \right)^{-1/2} \hat{\mathbf{g}}_0^{\mathrm{T}}\, \mathbf{f}_0 \tag{129}$$

where dx_1/dp_1 can be obtained from Eq. (126) [but care must be taken differentiating (126) if the lower limit is a turning point where $p_3 = 0$, i.e., when $p_1 U(x_3) = 1$ and $i(x_3) = \pi/2$]. This expression (129) remains valid at $x_1 = 0$ using l'Hopital's rule to replace $x_1/p_1 \rightarrow dx_1/dp_1$. It is interesting to note that on a symmetry plane in a 2D model, i.e., 2.5D wave propagation, the expression (129) is still valid provided the factor x_1/p_1 is replaced by $dx_2/dp_2 = \int U^2\, dT$ (Brokešova, 1992—the kinematic and dynamic differential equations need only be solved in 2D), which is identical in 1D.

The equivalent results in a spherically symmetric Earth can be obtained by the conformal, Earth flattening transformation $x_1 = r_0 \theta$, $r/r_0 = \exp(x_3/r_0)$, and $U(r) = rU(z)/r_0$ (r_0 is a reference radius). The conserved "horizontal" slowness is

$$p_\theta = \frac{r \sin i(r)}{r_0 U(r)} \tag{130}$$

corresponding to the angular slowness, and the angular range and travel time are

$$\theta(p_\theta, r) = \int \tan i(r)\, \frac{dr}{r} \tag{131}$$

$$T(p_\theta, r) = \int \frac{\sec i(r)}{U(r)}\, dr \tag{132}$$

These correspond to the ray integrals of Bullen (1963, §7.2.2) where $r_0 p_\theta$ is Bullen's ray parameter.

4. Generalizations to Ray Singularities

Although ray theory has wide application, there are many circumstances in which it breaks down. Nevertheless, ray theory extensions can often be used to fill these gaps with minimal extra computation or expense. In this section we briefly discuss three generalizations of ray theory: Maslov asymptotic ray theory, which remains valid at caustics, i.e., when $\mathscr{R} = 0$ in Eq. (93); Kirchhoff surface integral method, which models reflections from rough surfaces, i.e., $\left(\prod \mathscr{T}_{\ell j}\right)/|\mathscr{R}|^{-1/2}$ varies rapidly in $\underline{\mathbf{V}}^{(0)}$ [Eq. (93)]; and quasi-isotropic ray theory, which remains valid when the velocities of the qS rays are similar, i.e., $H_N \rightarrow 0$ in Eq. (25). Necessarily, the discussion is brief and further details can be found in the specialist publications. Another useful generalization, the Born scattering volume integral method and its generalizations, is not included but is a natural extension of the Kirchhoff method from surface to volume integrals (Chapman and Coates, 1994; de Hoop and Bleistein, 1997; Ursin and Tygel, 1997; Spencer *et al.*, 1997).

4.1 Maslov Asymptotic Ray Theory

As already noted, the ray theory Green function (95) breaks down at caustics where $\mathscr{R} = 0$ and rays are focused to a point or line. Maslov asymptotic ray theory avoids this problem by generalizing the ansatz of ray theory with an integration over neighboring rays, by analogy with transform methods. In this section, we follow the development in Chapman and Keers (2002) and write the ansatz for Maslov asymptotic ray theory as

$$\underline{\mathbf{v}}(\omega, \mathbf{x}; \mathbf{x}_0) \simeq |\omega| f(\omega) \sum_n \iint \widetilde{\underline{\mathbf{V}}}^{(0)}(\mathbf{q}, \mathscr{L}_n)\, \mathbf{e}^{i\omega\theta(\mathbf{q}, \mathscr{L}_n)}\, d\mathbf{q} \tag{133}$$

where $\mathbf{q} = (q_1\ q_2)^{\mathrm{T}}$ is a ray parameter vector. Note that the summation is now only over ray types not rays, i.e., as defined by the sequence of ray segments and r/t coefficients. If multipathing of rays of the same type exists, i.e., at different take-off angles, then it is contained within the integral. For brevity, in what follows we drop the ray type summation. The factor $|\omega|$ is introduced as with hindsight it makes $\widetilde{\underline{\mathbf{V}}}^{(0)}$ independent of frequency (we assume frequency is real and the modulus is required so $\underline{\mathbf{v}}(t, \mathbf{x}; \mathbf{x}_0)$ is real).

The Maslov integral is normally written as an integral over (components of) slowness at the receiver. However, difficulties arise later as multiple rays may exist with the same slowness. It is simpler to start with an integral that is single valued, i.e., there is one-to-one mapping between $\mathbf{q}$ and the rays. Locally, the integrands will only differ by a Jacobian, but using $\mathbf{q}$ we avoid later complications. The integral (133) is only an ansatz, so our choice of variable is not critical.

How do we define $\underline{\widetilde{\mathbf{V}}}^{(0)}$ and θ in Eq. (133)? There are many choices (see Kendall and Thomson, 1993, etc.). All that is certainly required is that for high enough frequency, Eq. (133) agrees (asymptotically) with ray theory [Eq. (3)], when the latter is valid. And for high enough frequency, ray theory is valid almost everywhere as it only breaks down on caustic surfaces and points. So first we choose $\underline{\widetilde{\mathbf{V}}}^{(0)}$ and θ so that asymptotically Eq. (133) agrees with ray theory almost everywhere as $\omega \to \infty$. Then we argue that because with this choice of $\underline{\widetilde{\mathbf{V}}}^{(0)}$ and θ, the integral is valid everywhere and behaves smoothly at finite frequencies, as must the exact solution, so Eq. (133) will be a useful solution at caustics and at finite frequencies.

A more sophisticated argument using pseudo-differential and Fourier integral operators can be developed by substituting Eq. (133) in the wave equations and finding the asymptotic behaviour for $\underline{\widetilde{\mathbf{V}}}$ (see Kendall and Thomson, 1993, and references therein). The result for the leading term, $\underline{\widetilde{\mathbf{V}}}^{(0)}$, must be the same using either argument. Higher-order terms are rarely useful or even evaluated. Therefore we follow the simpler approach.

Shooting a ray with parameters $\mathbf{q}$ to a target surface we obtain a traveltime, $\widetilde{T}(\mathbf{q}, \mathscr{L}_n)$ and an end-point, $\widetilde{\mathbf{X}}(\mathbf{q}, \mathscr{L}_n)$. Normally the target surface is a plane through the receiver(s), but some other end-condition is possible. Note that $\widetilde{T}$ and $\widetilde{\mathbf{X}}$ are functions of the source and ray parameters, as in ray shooting. In general we use a tilde to indicate variables on the target surface, e.g., $\tilde{\mathbf{p}}(\mathbf{q}, \mathscr{L}_n)$, the ray slowness on the target surface.

One choice for θ (by analogy with transform methods) is

$$\theta(\mathbf{q}, \mathscr{L}_n) = \widetilde{T}(\mathbf{q}, \mathscr{L}_n) + \tilde{\mathbf{p}}^{\mathrm{T}}(\mathbf{q}, \mathscr{L}_n)\left(\mathbf{x} - \widetilde{\mathbf{X}}(\mathbf{q}, \mathscr{L}_n)\right) \quad (134)$$

The integral (133) has a stationary-phase point when $\mathbf{x} = \widetilde{\mathbf{X}}(\mathbf{q}, \mathscr{L}_n)$, at which θ is the travel time, i.e., $T(\widetilde{\mathbf{X}}, \mathscr{L}_n) = \theta(\mathbf{q}, \mathscr{L}_n) = \widetilde{T}(\mathbf{q}, \mathscr{L}_n)$, i.e., $\mathbf{q}$ are the ray parameters that solve the two-point ray problem. Evaluating (133) by the second-order, stationary-phase method, and comparing with the ray result (3), we find

$$\underline{\widetilde{\mathbf{V}}}^{(0)}(\mathbf{q}, \mathscr{L}_n) = 2\pi \underline{\mathbf{V}}^{(0)}(\widetilde{\mathbf{X}}, \mathscr{L}_n) \| \boldsymbol{\nabla}_{\mathbf{q}}(\boldsymbol{\nabla}_{\mathbf{q}}\theta)^{\mathrm{T}} \|^{1/2} \times e^{-i(\pi/4)\mathrm{sgn}(\omega \boldsymbol{\nabla}_{\mathbf{q}}(\boldsymbol{\nabla}_{\mathbf{q}}\theta)^{\mathrm{T}})} \quad (135)$$

where the signature (sgn) of the matrix is the number of positive eigenvalues minus the number of negative eigenvalues (2, 0, or -2 as the matrix is 2×2—we treat $\boldsymbol{\nabla}_{\mathbf{q}}$ as 2×1). Thus with Eqs. (134) and (135) we have defined the integrand of (133) in such a way that it reduces, asymptotically, to ray theory [Eq. (3)] when the latter is valid. We now discuss some of the practical aspects of computing these terms.

Using the results in Section 2, it is straightforward to compute θ and $\underline{\mathbf{V}}^{(0)}$ for any $\mathbf{q}$. The more complicated term is the 2×2 matrix

$$\boldsymbol{\nabla}_{\mathbf{q}}(\boldsymbol{\nabla}_{\mathbf{q}}\theta)^{\mathrm{T}} = -(\boldsymbol{\nabla}_{\mathbf{q}}\tilde{\mathbf{p}}^{\mathrm{T}})\left(\boldsymbol{\nabla}_{\mathbf{q}}\widetilde{\mathbf{X}}^{\mathrm{T}}\right)^{\mathrm{T}} \quad (136)$$

as $\tilde{\mathbf{p}}$ and $\widetilde{\mathbf{X}}$ are on the target surface not the wavefront. Using the results of dynamic raytracing, we can compute the wavefront differentials

$$\begin{pmatrix} \mathbf{J}_{xp} \\ \mathbf{J}_{pp} \end{pmatrix} = \begin{pmatrix} \mathbf{P}_{xp} \\ \mathbf{P}_{pp} \end{pmatrix} \begin{pmatrix} \dfrac{\partial \mathbf{p}_0}{\partial q_1} & \dfrac{\partial \mathbf{p}_0}{\partial q_2} \end{pmatrix} \quad (137)$$

[cf. Eq. (89)—remember we normally compute the LHS directly from the paraxial ray equations (37) with appropriate initial conditions (88), rather than with the propagator]. The matrices $\mathbf{J}_{xp}$ and $\mathbf{J}_{pp}$ are 3×2. In order to compute the target differentials in Eq. (136), the wavefront differentials must be modified for the extra ray path. If $\hat{\mathbf{n}}$ is the normal to the target surface, then [cf. matrix $\mathbf{F}_1$ (54)]

$$\left(\boldsymbol{\nabla}_{\mathbf{q}}\widetilde{\mathbf{X}}^{\mathrm{T}}\right)^{\mathrm{T}} = \left(\mathbf{I} - \frac{\dot{\mathbf{x}}\hat{\mathbf{n}}^{\mathrm{T}}}{\dot{\mathbf{x}}^{\mathrm{T}}\hat{\mathbf{n}}}\right)\mathbf{J}_{xp} \quad (138)$$

where the second term is designed to be in the ray direction, $\mathbf{U}$, and to make the differential in the target surface, i.e., orthogonal to $\hat{\mathbf{n}}$. The slowness differential is modified by the extra ray path, and using Eq. (20) we obtain the 2×3 matrix [cf. matrix $\mathbf{F}_2$ (54) and (53)]

$$\boldsymbol{\nabla}_{\mathbf{q}}\tilde{\mathbf{p}}^{\mathrm{T}} = \left(\mathbf{J}_{pp} - \frac{\dot{\mathbf{p}}\hat{\mathbf{n}}^{\mathrm{T}}}{\dot{\mathbf{x}}^{\mathrm{T}}\hat{\mathbf{n}}}\mathbf{J}_{xp}\right)^{\mathrm{T}} \quad (139)$$

Combining Eqs. (138) and (139) for (136), we can compute Eq. (135).

Having obtained θ and $\underline{\widetilde{\mathbf{V}}}^{(0)}$ using the results of normal ray shooting, they can be used to evaluate (133). Rather than evaluate the spectrum, it is easier to compute the time-domain response directly. Evaluating the inverse Fourier transform of Eq. (133), we can change the order of integration and reduce the triple integral without approximation, assuming θ is real. The result is

$$\underline{\mathbf{v}}(t, \mathbf{x}; \mathbf{x}_0) = -\frac{d}{dt}\mathrm{Im}\left[F(t) * \int_{t=\theta} \frac{\underline{\widetilde{\mathbf{V}}}^{(0)}}{|\boldsymbol{\nabla}_{\mathbf{q}}\theta|}\, dq\right] \quad (140)$$

$$\simeq -\frac{1}{2\Delta t}\frac{d}{dt}\mathrm{Im}\left[F(t) * \iint_{t=\theta \pm \Delta t} \underline{\widetilde{\mathbf{V}}}^{(0)}\, d\mathbf{q}\right] \quad (141)$$

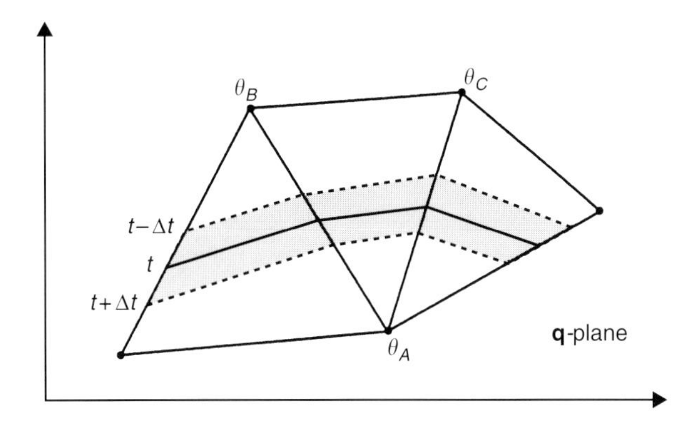

FIGURE 5 The bandlimited Maslov integral. The amplitude, $\widetilde{\mathbf{V}}^{(0)}$, is integrated over the strips defined by $t \pm \Delta t$ of the θ function. Within the triangular ray tube (the vertices of the triangles correspond to rays), the θ and $\widetilde{\mathbf{V}}^{(0)}$ functions are linearly interpolated. The same diagram applies to the Kirchhoff integral, where the triangular elements form the interface.

where Eq. (140) is the impulse result and Eq. (141) is the bandlimited result. The integral in Eq. (140) is evaluated along the line in the **q** domain where $t = \theta$. The bandlimited result has been convolved with a boxcar $2\Delta t$ long. The integral in Eq. (141) is evaluated over the strip defined by $t = \theta \pm \Delta t$ in the **q** domain (Fig. 5). Because $\underline{\widetilde{\mathbf{V}}}^{(0)}$ normally varies smoothly, this area integral is easily evaluated. An efficient algorithm, described in Spencer *et al.* (1997), is to linearly interpolate within triangular ray tubes in the **q** domain (Fig. 5). Details of the algorithm are given in the Appendix.

4.2 Kirchhoff Surface Integral Method

The ray method describes the high-frequency behavior of reflections from interfaces. It takes into account the amplitude changes due to the r/t coefficients, and the spreading changes that may occur due to the change in ray type and the curvature of the interface. However, diffracted signals that occur due to roughness and discontinuities in the interface are not modeled. The Kirchhoff surface integral method is a useful generalization of ray theory that models, at least approximately, such signals by integrating over non-specular reflections on the interface. Rays are traced from the source to the interface, and from the interface to the receiver (or by reciprocity, from the receiver to the interface) without satisfying Snell's law at the interface. The basic method has been described by Bleistein (1984, p. 282), and Frazer and Sen (1985) and references therein to earlier papers have described its application to seismology for acoustic or isotropic media. Recent papers, e.g., de Hoop and Bleistein (1997) and Ursin and Tygel (1997), have extended the method to anisotropic media.

The representation (Betti's) theorem gives the reflected signal in terms of a surface integral (Aki and Richards, 1980, p. 28). Thus, following the standard method, we have for the scattered signal

$$\underline{\mathbf{v}}(\omega, \mathbf{x}; \mathbf{x}_0) = -i\omega \int_S \Big[\underline{\mathbf{t}}_m^{\mathrm{T}}(\omega, \boldsymbol{\zeta}; \mathbf{x})\underline{\mathbf{v}}(\omega, \boldsymbol{\zeta}; \mathbf{x}_0) - \underline{\mathbf{v}}^{\mathrm{T}}(\omega, \boldsymbol{\zeta}; \mathbf{x})\underline{\mathbf{t}}_m(\omega, \boldsymbol{\zeta}; \mathbf{x}_0)\Big] \hat{n}_m \, d\mathbf{s} \quad (142)$$

where the point $\boldsymbol{\zeta}$ lies on the surface S of the integral and $d\mathbf{s}$ is the surface area element. The Green functions from the receiver—argument $(\omega, \boldsymbol{\zeta}; \mathbf{x})$—are the free-space Green functions (Bleistein, 1984, p. 282), which we approximate by ray theory (Section 2.9),

$$\begin{pmatrix} \underline{\mathbf{v}} \\ \underline{\mathbf{t}}_m \end{pmatrix}(\omega, \boldsymbol{\zeta}; \mathbf{x}) \simeq \frac{e^{i\omega T(\boldsymbol{\zeta}, \mathscr{L}_r) - i(\pi/2)\mathrm{sgn}(\omega)\sigma(\boldsymbol{\zeta}, \mathscr{L}_r)}}{4\pi(\rho(\mathbf{x})\rho_r(\boldsymbol{\zeta})|\mathscr{R}(\boldsymbol{\zeta}, \mathscr{L}_r)|)^{1/2}} \left(\prod \mathscr{T}_{ij}(\mathscr{L}_r)\right) \times \begin{pmatrix} \hat{\mathbf{g}} \\ -\mathbf{Z}_m\hat{\mathbf{g}} \end{pmatrix}(\boldsymbol{\zeta}, \mathscr{L}_r)\hat{\mathbf{g}}^{\mathrm{T}}(\mathbf{x}, \mathscr{L}_r) \quad (143)$$

where

$$\mathbf{Z}_m(\boldsymbol{\zeta}, \mathscr{L}_r) = p_j(\boldsymbol{\zeta}, \mathscr{L}_r)\mathbf{c}_{mj}(\boldsymbol{\zeta}) \quad (144)$$

[from Eq. (6) with $m = 0$)] is a *matrix impedance* (Chapman and Coates, 1994). We use the shorthand $\mathscr{L}_r$ to indicate the ray path from $\mathbf{x}$ to $\boldsymbol{\zeta}$, subscript r to indicate the generated ray type, and for simplicity, omit a summation over multipathing and ray type. The product of coefficients is for interfaces between $\mathbf{x}$ and $\boldsymbol{\zeta}$ but not the interface at $\boldsymbol{\zeta}$.

We approximate $\mathbf{v}(\omega, \boldsymbol{\zeta}; \mathbf{x}_0)$ and $\mathbf{t}_m(\omega, \boldsymbol{\zeta}; \mathbf{x}_0)$ within the surface integral (142) by the ray theory values. Normally this is restricted to the reflected ray that matches the ray type to the receiver, but more generally we can include the incident wave and all three reflections (to model the complete, continuous displacement and traction on the interface). Thus using the free-space Green function from the source $\mathbf{x}_0$, we have

$$\begin{pmatrix} \underline{\mathbf{v}} \\ \underline{\mathbf{t}}_m \end{pmatrix}(\omega, \boldsymbol{\zeta}; \mathbf{x}_0) \simeq \frac{e^{i\omega T(\boldsymbol{\zeta}, \mathscr{L}_s) - i(\pi/2)\mathrm{sgn}(\omega)\sigma(\boldsymbol{\zeta}, \mathscr{L}_s)}}{4\pi(\rho(\mathbf{x}_0)\rho_0(\boldsymbol{\zeta})|\mathscr{R}_s(\boldsymbol{\zeta}, \mathscr{L}_s)|)^{1/2}} \times \left(\prod \mathscr{T}_{ij}(\mathscr{L}_s)\right) \begin{pmatrix} \widetilde{\mathbf{V}}_s \\ \widetilde{\mathbf{T}}_{sm} \end{pmatrix}(\boldsymbol{\zeta}, \mathscr{L}_s)\hat{\mathbf{g}}^{\mathrm{T}}(\mathbf{x}_0, \mathscr{L}_s) \quad (145)$$

with

$$\widetilde{\mathbf{V}}_s(\boldsymbol{\zeta}, \mathscr{L}_s) = \hat{\mathbf{g}}_s + \sum_{l=1}^{3} \mathscr{T}_{ls} \left|\frac{\rho_s \mathscr{R}_s}{\rho_r \mathscr{R}_{1l}}\right|^{1/2} \hat{\mathbf{g}}_{1l} \quad (146)$$

$$\widetilde{\mathbf{T}}_{sm}(\boldsymbol{\zeta}, \mathscr{L}_s) = -\mathbf{Z}_{sm}\hat{\mathbf{g}}_s - \sum_{l=1}^{3} \mathscr{T}_{ls} \left|\frac{\rho_s \mathscr{R}_s}{\rho_r \mathscr{R}_{1l}}\right|^{1/2} \mathbf{Z}_{1lm}\hat{\mathbf{g}}_{1l} \quad (147)$$

where in these expressions we have omitted the argument $(\boldsymbol{\zeta}, \mathscr{L}_s)$, have used the subscript notation for incident and reflected waves from Section 2.8, e.g., Eq. (94), and have used the subscript s to indicate the source ray type. Note that the matrix impedances for each term in the tractions differ as the slowness vectors, $\mathbf{p}_s$ and $\mathbf{p}_{1l}$, differ. The ratio of the spreading functions for the incident and generated rays can be replaced by the inverse ratio of the normalization factors used in the eigenvectors (77) for the r/t coefficients [Eq. (94) is equivalent to Eq. (80)]. Again the product of coefficients is for interfaces between $\mathbf{x}_0$ and $\boldsymbol{\zeta}$ but not the interface at $\boldsymbol{\zeta}$. Expressions (146) and (147) contain the coefficients of the interface at $\boldsymbol{\zeta}$.

Combining the ray theory approximation for the wave from the source to the interface (145) with the free-space Green function (143), in the surface integral (142), we have

$$\underline{\mathbf{v}}(\omega, \mathbf{x}; \mathbf{x}_0) = -i\omega \int e^{i\omega\theta(\mathbf{x}, \mathscr{L}_r, \mathscr{L}_s)} \underline{\mathbf{A}}(\boldsymbol{\zeta}, \mathscr{L}_r, \mathscr{L}_s)\, d\mathbf{s} \tag{148}$$

where

$$\theta(\boldsymbol{\zeta}, \mathscr{L}_r, \mathscr{L}_s) = T(\boldsymbol{\zeta}, \mathscr{L}_r) + T(\boldsymbol{\zeta}, \mathscr{L}_s) \tag{149}$$

$$\begin{aligned} \underline{\mathbf{A}}(\boldsymbol{\zeta}, \mathscr{L}_r, \mathscr{L}_s) = &- \frac{e^{-i(\pi/2)\mathrm{sgn}(\omega)(\sigma(\boldsymbol{\zeta}, \mathscr{L}_r) + \sigma(\boldsymbol{\zeta}, \mathscr{L}_s))}}{16\pi^2(\rho(\mathbf{x})\rho(\mathbf{x}_0)\rho_s(\boldsymbol{\zeta})\rho_r(\boldsymbol{\zeta}))^{1/2}} \\ &\times \left(\prod \mathscr{T}_{\ell j}(\mathscr{L}_r)\right)\left(\prod \mathscr{T}_{\ell j}(\mathscr{L}_s)\right) \\ &\times \hat{\mathbf{g}}^{\mathrm{T}}(\boldsymbol{\zeta}, \mathscr{L}_r)\Big[\mathbf{Z}_{\mathrm{m}}^{\mathrm{T}}(\boldsymbol{\zeta}, \mathscr{L}_r)\tilde{\mathbf{V}}_s(\boldsymbol{\zeta}, \mathscr{L}_s) \\ &+ \tilde{\mathbf{T}}_{sm}(\boldsymbol{\zeta}, \mathscr{L}_s)\Big]\hat{n}_m \hat{\mathbf{g}}(\mathbf{x}, \mathscr{L}_r)\hat{\mathbf{g}}^{\mathrm{T}}(\mathbf{x}, \mathscr{L}s) \end{aligned} \tag{150}$$

For numerical purposes, it is straightforward to transform Eq. (148) into the time-domain:

$$\mathbf{v}(t, \mathbf{x}; \mathbf{x}_0) = \frac{d}{dt}\mathrm{Re}\left[\Delta(t) * \int_{t=\theta} \underline{\mathbf{A}}(\boldsymbol{\zeta}, \mathscr{L}_r, \mathscr{L}_s)\frac{ds}{|\boldsymbol{\nabla}_{\mathbf{s}}\,\theta(\boldsymbol{\zeta}, \mathscr{L}_r, \mathscr{L}_s)|}\right] \tag{151}$$

$$\simeq \frac{1}{2\Delta t}\frac{d}{dt}\mathrm{Re}\left[\Delta(t) * \int_{t=\theta\pm\Delta t} \underline{\mathbf{A}}(\boldsymbol{\zeta}, \mathscr{L}_r, \mathscr{L}_s)d\mathbf{s}\right] \tag{152}$$

where ds is the line element. The integral in the impulse response (151), is evaluated along lines on the surface where $t = \theta(\boldsymbol{\zeta}, \mathscr{L}_r, \mathscr{L}_s)$. The bandlimited result (152), which is in a form suitable for numerical evaluation, has surface integrals over strips defined by $t = \theta(\boldsymbol{\zeta}, \mathscr{L}_r, \mathscr{L}_s) \pm \Delta t$. If the surface is divided into triangular elements, where the ray results are known at the apexes, then exactly the same algorithm used for Maslov seismograms can be used to evaluate these surface integrals efficiently (Spencer *et al.*, 1997; Figure 5 and Appendix).

The surface integral (148) has stationary points [and (151) has corresponding singularities] when

$$\begin{aligned} \boldsymbol{\nabla}_{\mathbf{s}}\theta(\boldsymbol{\zeta}, \mathscr{L}_r, \mathscr{L}_s) &= \boldsymbol{\nabla}_{\perp}T(\boldsymbol{\zeta}, \mathscr{L}_r) + \boldsymbol{\nabla}_{\perp}T(\boldsymbol{\zeta}, \mathscr{L}_s) \\ &= \mathbf{p}_{\perp}(\boldsymbol{\zeta}, \mathscr{L}_r) + \mathbf{p}_{\perp}(\boldsymbol{\zeta}, \mathscr{L}_s) = \mathbf{0} \end{aligned} \tag{153}$$

which corresponds to the Snell's law condition for a specular reflection [cf. Eq. (57)—remember that $\mathbf{p}(\boldsymbol{\zeta}, \mathscr{L}_r)$ is reversed compared with the combined ray]. The appropriate reflected ray in Eqs. (146) and (147) will match the receiver ray at this point. Usually only this term is retained in the Kirchhoff integral and the other parts of the wavefield on the interface (146) and (147) are dropped. However, if they are retained, to first order they make no contribution to the integral. Let us consider the scattering term in $\underline{\mathbf{A}}$ (150), i.e.,

$$\begin{aligned} &-\hat{\mathbf{g}}^{\mathrm{T}}(\boldsymbol{\zeta}, \mathscr{L}_r)\Big[\mathbf{Z}_m^{\mathrm{T}}(\boldsymbol{\zeta}, \mathscr{L}_r)\tilde{\mathbf{V}}_s(\boldsymbol{\zeta}, \mathscr{L}_s) + \tilde{\mathbf{T}}_{sm}(\boldsymbol{\zeta}, \mathscr{L}_s)\Big]\hat{n}_m \\ &= \sum_{l=0}^{3} \mathscr{T}_{ls}(\mathscr{L}_s)\hat{n}_m \left|\frac{\rho_s\mathscr{R}_s}{\rho_r\mathscr{R}_{1l}}\right|^{1/2} \hat{\mathbf{g}}^{\mathrm{T}}(\boldsymbol{\zeta}, \mathscr{L}_r)[\mathbf{Z}_{1lm}(\boldsymbol{\zeta}, \mathscr{L}_s) \\ &\quad - \mathbf{Z}_{rm}^{T}(\boldsymbol{\zeta}, \mathscr{L}_r)]\hat{\mathbf{g}}_{il}(\boldsymbol{\zeta}, \mathscr{L}_s) \end{aligned} \tag{154}$$

$$\begin{aligned} &= \sum_{l=0}^{3} \mathscr{T}_{ls}(\mathscr{L}_s)\left|\frac{\rho_s\mathscr{R}_s}{\rho_r\mathscr{R}_{1l}}\right|^{1/2} [\hat{n}_j p_m(\boldsymbol{\zeta}, \mathscr{L}_s)_{1l} \\ &\quad - \hat{n}_m p_j(\boldsymbol{\zeta}, \mathscr{L}_r)_r]\hat{\mathbf{g}}^{\mathrm{T}}(\boldsymbol{\zeta}, \mathscr{L}_r)\mathbf{c}_{jm}(\boldsymbol{\zeta})\hat{\mathbf{g}}_{il}(\boldsymbol{\zeta}, \mathscr{L}_s) \end{aligned} \tag{155}$$

using Eqs. (144), (146), and (147), where for brevity we have indicated the incident wave by $l = 0$, i.e., subscript $1l \rightarrow s$ and $\mathscr{T}_{0s} = 1$. This expression (155) contains the directionality of the *obliquity factor*. At the saddle point defined by (153), when the appropriate reflection matches the receiver ray, the contribution to (155) from the other three terms (incident ray and two reflections) is zero using the orthogonality relationship in Eq. (79), and it reduces to

$$\begin{aligned} &-\hat{\mathbf{g}}^{\mathrm{T}}(\boldsymbol{\zeta}, \mathscr{L}_r)\Big[\mathbf{Z}_m^{\mathrm{T}}(\boldsymbol{\zeta}, \mathscr{L}_r)\tilde{\mathbf{V}}_s(\boldsymbol{\zeta}, \mathscr{L}_s) + \tilde{\mathbf{T}}_{sm}(\boldsymbol{\zeta}, \mathscr{L}_s)\Big]\hat{n}_m \\ &= 2\mathscr{T}_{rs}(\mathscr{L}_s)\mathbf{U}_{1r}^{\mathrm{T}}\hat{\mathbf{n}}\left(\rho_s\rho_r\left|\frac{\mathscr{R}_s}{\mathscr{R}_{1r}}\right|\right)^{1/2} \end{aligned} \tag{156}$$

where we have let $l = r$ for the matching ray. $\mathscr{T}_{rs}(\mathscr{L}_s)$ is the reflection coefficient of the interface at $\boldsymbol{\zeta}$. The saddle point contribution can be evaluated using the stationary phase method (Bleistein, 1984, p. 88) and reduces to the ray theory result (93), where the saddle point values give

$$T(\mathbf{x}, \mathbf{x}_0) = \theta(\boldsymbol{\zeta}, \mathscr{L}_r, \mathscr{L}_s) = T(\boldsymbol{\zeta}, \mathscr{L}_r) + T(\boldsymbol{\zeta}, \mathscr{L}_s) \tag{157}$$

$$\sigma(\mathbf{x}, \mathbf{x}_0) = \sigma(\boldsymbol{\zeta}, \mathscr{L}_r) + \sigma(\boldsymbol{\zeta}, \mathscr{L}_s) + 1 - \tfrac{1}{2}\mathrm{sgn}(\boldsymbol{\nabla}_{\mathbf{s}}(\boldsymbol{\nabla}_{\mathbf{s}}\theta)^{\mathrm{T}}) \tag{158}$$

(Ursin and Tygel, 1997). Using the symmetries of the dynamic propagator (41), the factor $|\mathscr{R}_{\mathrm{r}}\mathscr{R}_{1r}|$ can be related to $|\mathscr{R}|$ for the complete ray using $\|\boldsymbol{\nabla}_{\mathbf{s}}(\boldsymbol{\nabla}_{\mathbf{s}}\theta)^{\mathrm{T}}\|$ (Coates and Chapman, 1990a; Ursin and Tygel, 1997).

4.3 Quasi-Isotropic Ray Theory

In addition to the degeneracy that always occurs in isotropic media, degeneracy may occur in anisotropic media. In certain directions, the quasi-shear (qS) waves degenerate and their velocities are equal. In these directions, degenerate theory must be used.

Near these degenerate directions, and in the case of weak, heterogeneous anisotropy, in all directions, standard anisotropic ray theory breaks down as the qS velocities are similar (H_N small) and the additional terms [Eq. (25)] blow up (errors in asymptotic ray theory are bounded by the next term). In homogeneous media, the two qS waves propagate independently as they do in anisotropic media when the velocities are significantly different (making the additional term small). They interact when these conditions fail as the time separation of the two waves is small compared with the pulse period. This occurs when $L\,\Delta U/U^2 < 2\pi/\omega$, where ΔU is the difference in qS velocities and L is a characteristic length of the heterogeneities. If we use anisotropic ray theory in weakly anisotropic media or near degeneracies, and assume the qS waves propagate independently, then absurd results can be obtained. Rapid changes in polarization are predicted for each independent ray that are clearly not allowed physically.

To use ray methods at or near degeneracies and in weakly anisotropic media, we use a quasi-isotropic (QI) ray theory introduced by Pšenčík (1998) using a method by Kravtsov and Orlov (1990, p. 233). The effects of anisotropy and differences in the S velocities are modeled as perturbations to the isotropic solution by treating the anisotropic part of the model as another small parameter, in addition to $1/\omega$, in the asymptotic expansion.

We factor the elastic parameters into isotropic and anisotropic parts, e.g.,

$$c_{ijkl} = c^0_{ijkl} + \Delta c_{ijkl} \tag{159}$$

where c^0_{ijkl} is isotropic, i.e., like Eq. (98),

$$c^0_{ijkl} = \lambda\delta_{ij}\delta_{kl} + \mu(\delta_{ik}\delta_{jl} + \delta_{il}\delta_{jk}) \tag{160}$$

We assume that Δc_{ijkl} is small and of order $1/\omega$. Substituting Eq. (159) in the constitutive equation (2), and using the same ray ansätze [Eq. (3)], we obtain a revised equation (6):

$$\begin{aligned}\mathbf{c}^0_{jk}\frac{\partial \mathbf{V}^{(m-1)}}{\partial x_k} - i\omega\,\Delta\mathbf{c}_{jk}\frac{\partial \mathbf{V}^{(m-2)}}{\partial x_k} \\ = \mathbf{T}^{(m)}_j + p_k\mathbf{c}^0_{jk}\mathbf{V}^{(m)} - i\omega p_k\,\Delta\mathbf{c}_{jk}\mathbf{V}^{(m-1)}\end{aligned} \tag{161}$$

Note that the new terms in this equation are frequency dependent and have been included as $O(\omega\,\Delta c_{ijkl}) \sim 1$. Equation (5) remains valid, and as before $\mathbf{T}^{(m)}_j$ can be eliminated between Eqs. (5) and (161) to obtain a revised version of Eq. (9),

$$\begin{aligned}\mathsf{N}^0(\mathbf{V}^{(m)}) - \overline{\mathsf{M}}^0(\mathbf{V}^{(m-1)}, \mathbf{T}^{(m-1)}_j) \\ = i\omega p_j\Delta\mathbf{c}_{jk}\left(p_k\mathbf{V}^{(m-1)} - \frac{\partial \mathbf{V}^{(m-2)}}{\partial x_k}\right)\end{aligned} \tag{162}$$

the operators N^0 and $\overline{\mathsf{M}}^0$ being in the isotropic model.

4.3.1 The Quasi-Isotropic Eikonal

The $m=0$ terms are as before for the isotropic part of the model, c^0_{ijkl}, i.e., Eq. (162) reduces to Eq. (15) when $m=0$. Thus the rays are traced in the isotropic part of the model (see Section 3.1). The P ray is longitudinal, i.e., $\hat{\mathbf{g}}_3 = \alpha\mathbf{p}$. The S rays are degenerate and the polarizations are in the plane perpendicular to the ray.

4.3.2 The Quasi-Isotropic Transport Equation

Letting $m=1$ in expression (162) and proceeding as before, the transport equation (36) is modified to

$$\nabla\cdot\mathbf{N} = -i\omega\mathbf{V}^{(0)\mathrm{T}}p_jp_k\,\Delta\mathbf{c}_{jk}\mathbf{V}^{(0)} \tag{163}$$

For future use, we define

$$B_{\mu\nu} = \frac{\hat{p}_j\hat{p}_k\hat{\mathbf{g}}^{\mathrm{T}}_\mu\,\Delta\mathbf{c}_{jk}\hat{\mathbf{g}}_\nu}{\rho} \tag{164}$$

***P* Quasi-Isotropic Rays**. As the isotropic eikonal applies, the polarization is still longitudinal, i.e., $\mathbf{V}^{(0)} = V^{(0)}_3\hat{\mathbf{g}}_3$ where $\hat{\mathbf{g}}_3 = \alpha\mathbf{p}$.

For P waves, (163) reduces to [cf. Eq. (30)]

$$\frac{d}{dT}\ln\left(\rho V^{(0)2}_3\right) = -\nabla\cdot\mathbf{U} - i\omega\frac{B_{33}}{\alpha^2} \tag{165}$$

where B_{33} is defined in Eq. (164), Combining with Eq. (32),

$$\frac{d}{dT}\ln\left(\rho u V^{(0)2}_3 J\right) = -i\omega\frac{B_{33}}{\alpha^2} \tag{166}$$

and the solution is

$$\begin{aligned}V^{(0)}_3 &= \frac{\text{constant}}{\sqrt{\rho u|J|}}\exp(i\omega\Delta T_3), \qquad \text{where} \\ \Delta T_3 &= -\int_{\mathscr{L}_n}\frac{B_{33}}{2\alpha^3}\,ds\end{aligned} \tag{167}$$

is the travel time shift due to the anisotropic perturbation (Červený, 1982; Hanyga, 1982b; Červený and Jech, 1982).

The additional terms are evaluated as before, Eq. (25), including the extra QI term

$$V^{(1)}_N = \frac{1}{2\rho H_N}\hat{\mathbf{g}}^{\mathrm{T}}_N\left(\overline{\mathsf{M}}(\mathbf{V}^{(0)}, \mathbf{T}^{(0)}_j) + i\omega p_jp_k\,\Delta\mathbf{c}_{jk}\mathbf{V}^{(0)}\right) \tag{168}$$

Specializing to the quasi-isotropic P wave ($\mathbf{V}^{(0)} = V^{(0)}_3\hat{\mathbf{g}}_3$), we obtain the isotropic terms (118) plus additional QI terms:

$$\begin{aligned}V^{(1)}_\nu = -\,\hat{\mathbf{g}}^{\mathrm{T}}_\nu\Bigg(\alpha\frac{\partial V^{(0)}_3}{\partial g_\nu} + \frac{V^{(0)}_3}{\alpha^2-\beta^2}\big[(\alpha^2-\beta^2)\nabla\alpha - 4\alpha\beta\,\nabla\beta \\ + \alpha(\alpha^2-2\beta^2)\,\nabla(\ln\rho)\big]\Bigg) + \frac{i\omega\alpha^2\hat{\mathbf{g}}^{\mathrm{T}}_\nu}{\rho(\beta^2-\alpha^2)} \\ \times\left(p_j\left(\mathbf{c}^0_{jk} + \mathbf{c}^0_{kj}\right)\mathbf{V}^{(0)}\frac{\partial(\Delta T_3)}{\partial x_k} + p_jp_k\,\Delta\mathbf{c}_{jk}\mathbf{V}^{(0)}\right)\end{aligned} \tag{169}$$

Using Eqs. (106) and (107), and definition (164), we obtain

$$V_\nu^{(1)} = -\hat{\mathbf{g}}_\nu^{\mathrm{T}}\left(\alpha\frac{\partial V_3^{(0)}}{\partial g_\nu} + \frac{V_3^{(0)}}{\alpha^2-\beta^2}\left[(\alpha^2-\beta^2)\nabla\alpha - 4\alpha\beta\nabla\beta + \alpha(\alpha^2-2\beta^2)\nabla(\ln\rho)\right]\right) - i\omega V_3^{(0)}\left(\alpha\frac{\partial(\Delta T_3)}{\partial g_\nu} + \frac{B_{\nu 3}}{\alpha^2-\beta^2}\right) \tag{170}$$

The two extra QI terms were described by Pšenčík (1998). They are proportional to frequency, so overall their contribution is independent of frequency (and the perturbed polarization due to anisotropy remains linear in contrast to additional terms due to heterogeneity). The first QI term corrects the polarization due to the perturbation in **p** from the anisotropy. The second term corrects the polarization for deviations from the slowness vector in the anisotropic medium. It has also been obtained using the perturbation method (Jech and Pšenčík, 1989; Pšenčík and Gajewski, 1998).

***S* Quasi-Isotropic Rays**. Now the polarization is degenerate. The shear polarization can be written as a combination of any two orthogonal vectors in the transverse plane. Various choices suggest themselves, e.g., the isotropic shear polarizations, $\hat{\mathbf{g}}_\nu^0$, or the polarizations in the anisotropic medium, themselves perhaps estimated by perturbation theory. Pšenčík (1998) has considered both cases and argues that the former case is easier. In this, the polarizations are easily calculated using Eq. (117) and in the QI limit, never generate strong coupling between the solutions. The latter choice requires that the anisotropic polarizations be found and, as they may vary rapidly, will generate strong coupling between the two solutions. The second case is equivalent to that of Coates and Chapman (1990b), who investigated this coupling between qS waves. A final possibility is to use the normal and binormal vectors as a basis, but given that the isotropic polarizations, $\hat{\mathbf{g}}_\nu^0$, are known, this offers no advantages. We therefore follow Pšenčík (1998).

Thus, we write

$$\mathbf{V}^{(0)} = V_\nu^{(0)}\hat{\mathbf{g}}_\nu^0 \tag{171}$$

where $\hat{\mathbf{g}}_\nu^0$ are found by solving Eq. (117) in the isotropic part of the model. Premultiplying Eq. (162) by $\hat{\mathbf{g}}_\eta^{\mathrm{T}}$, we modify Eq. (111) to

$$\hat{\mathbf{g}}_\eta^{\mathrm{T}}\mathsf{M}(V_\nu^{(0)}\hat{\mathbf{g}}_\nu) = -\frac{i\omega\rho}{\beta^2}B_{\eta\nu}V_\nu^{(0)} \tag{172}$$

Then Eq. (112) is modified to

$$\delta_{\nu\eta}(2\mu\mathbf{p}\cdot\nabla V_\nu^{(0)} + \mathbf{V}_\nu^{(0)}\mathbf{p}\cdot\nabla\mu + \mu V_\nu^{(0)}\nabla\cdot\mathbf{p}) + 2\rho V_\nu^{(0)}\hat{\mathbf{g}}_\eta^{0\mathrm{T}}\frac{d\hat{\mathbf{g}}_\nu^0}{dT} = -\frac{i\omega\rho}{\beta^2}B_{\eta\nu}V_\nu^{(0)} \tag{173}$$

As we are using the isotropic shear polarizations, the last term on the LHS is zero [Eq. (117)], and this equation simplifies to

$$\frac{1}{V_\eta^{(0)}}\nabla\cdot(\rho\beta^2V_\eta^{(0)}\mathbf{p}) = -\frac{i\omega\rho}{\beta^2}B_{\eta\nu}V_\nu^{(0)} \tag{174}$$

(no summation over η). We define new amplitude coefficients with the behavior (35) removed,

$$\widetilde{V}_\nu^{(0)} = V_\nu^{(0)}\sqrt{\rho\beta|J|} \tag{175}$$

and Eq. (174) reduces to

$$\frac{d\widetilde{\mathbf{V}}^{(0)}}{dT} = -\frac{i\omega}{2\beta^2}\mathbf{B}\widetilde{\mathbf{V}}^{(0)} \tag{176}$$

where we have used Eq. (32). $\widetilde{\mathbf{V}}^{(0)}$ is the 2-vector with components $\widetilde{V}_\nu^{(0)}$ and **B** is the 2×2 matrix with components $B_{\eta\nu}$. This equation (176) models the frequency-dependent coupling between the isotropic shear polarizations. In the isotropic limit, the RHS is zero and $\widetilde{\mathbf{V}}^{(0)}$ is constant, i.e., the components are independent in an isotropic model.

In order to analyze the solution further, we need the eigenvalues of the matrix **B**. These are

$$b_\nu = \tfrac{1}{2}(B_{11} + B_{22} \mp B), \qquad \text{where}$$

$$B = \left[(B_{11} - B_{22})^2 + 4B_{12}^2\right]^{1/2} \tag{177}$$

(as before, let us choose $\nu = 1$ for the slower qS ray which corresponds to the negative sign, and $\nu = 2$ with the positive sign—this differs from Pšenčík (1998) as we have ordered the solutions with increasing velocity). Solutions of Eq. (176) can be considered in two regimes, corresponding to low and high frequencies. When the elements of **B** are small (in the sense that $\omega Tb/\beta^2 \ll 1$, which is equivalent to the QI condition given above), $\widetilde{\mathbf{V}}^{(0)}$ varies slowly and Eq. (176) is relatively easy to solve numerically. Any of the standard techniques for finding a propagator can be used. Assuming **B** can be approximated as piecewise constant, the propagator is easily found from the eigenvector matrix

$$\mathbf{\Phi} = \begin{pmatrix}\cos\phi & -\sin\phi \\ \sin\phi & \cos\phi\end{pmatrix} \tag{178}$$

(the first column corresponds to b_1 and the second to b_2) where the rotation angle, ϕ, is found from

$$\tan 2\phi = \frac{2B_{12}}{B_{22} - B_{11}} \tag{179}$$

The propagator for a constant **B** interval T to $T + \Delta T$ is then

$$\mathscr{V}(T + \Delta T, T) = \mathbf{\Phi}\mathbf{\Lambda}(T + \Delta T, T)\mathbf{\Phi}^{-1} \tag{180}$$

where

$$\mathbf{\Lambda}(T+\Delta T, T) = \begin{pmatrix} e^{i\omega\Delta T_1} & 0 \\ 0 & e^{i\omega\Delta T_2} \end{pmatrix} \quad \text{with}$$

$$\Delta T_\nu = -\int_T^{T+\Delta T} \frac{b_\nu}{2\beta^2}\, dT \tag{181}$$

The angle ϕ is the rotation of the anisotropic qS polarizations, and ΔT_ν are the travel time perturbations, relative to the isotropic results (by degenerate perturbation theory—Jech and Pšenčík, 1989). Expanding Eq. (180) and inverting the Fourier transform, we can increment $\widetilde{\mathbf{V}}^{(0)}(t, T)$ using

$$\begin{aligned} &\widetilde{\mathbf{V}}^{(0)}(t, T+\Delta T) \\ &= \begin{pmatrix} \cos^2\phi & \cos\phi\,\sin\phi \\ \cos\phi\,\sin\phi & \sin^2\phi \end{pmatrix} \widetilde{\mathbf{V}}^{(0)}(t+\Delta T_1, T) \\ &+ \begin{pmatrix} \sin^2\phi & -\cos\phi\,\sin\phi \\ -\cos\phi\,\sin\phi & \cos^2\phi \end{pmatrix} \widetilde{\mathbf{V}}^{(0)}(t+\Delta T_2, T) \end{aligned} \tag{182}$$

The time-domain solution is attractive as usually the response is compact in the time domain.

When the QI condition is not satisfied, i.e., in the high-frequency limit except at degeneracies, we would expect the quasi-shear rays to propagate independently (Section 2). Expanding the solution in terms of anisotropic rays,

$$\widetilde{\mathbf{V}}^{(0)} = \mathbf{\Phi}\mathbf{\Lambda}(T, T_0)\overline{\mathbf{V}}^{(0)} \tag{183}$$

the differential equation (176) becomes

$$\frac{d\overline{\mathbf{V}}^{(0)}}{dT} = -\frac{d\phi}{dT}\mathbf{\Lambda}^{-1}\mathbf{I}_1\mathbf{\Lambda}\overline{\mathbf{V}}^{(0)} \tag{184}$$

or

$$\frac{d\overline{\mathbf{V}}^{(0)}}{d\phi} = \begin{pmatrix} 0 & -e^{-i\omega\Delta T_{12}} \\ e^{-i\omega\Delta T_{21}} & 0 \end{pmatrix} \overline{\mathbf{V}}^{(0)} \tag{185}$$

where

$$\begin{aligned} \Delta T_{12} = -\Delta T_{21} = \Delta T_1 - \Delta T_2 &= -\int_{T_0}^{T} \frac{b_1 - b_2}{2\beta^2}\, dT \\ &= \int_{T_0}^{T} \frac{B}{2\beta^2}\, dT \end{aligned} \tag{186}$$

Assuming ΔT_{12} can be approximated as constant, the propagator for Eq. (185) is

$$\begin{aligned} &\overline{\mathbf{V}}^{(0)}(\omega, T+\Delta T) \\ &= \begin{pmatrix} \cos\Delta\phi & -e^{-i\omega\Delta T_{12}}\sin\Delta\phi \\ e^{-i\omega\Delta T_{21}}\sin\Delta\phi & \cos\Delta\phi \end{pmatrix} \overline{\mathbf{V}}^{(0)}(\omega, T) \end{aligned} \tag{187}$$

whatever the size of $\Delta\phi$, i.e., even if $d\phi/dT$ is large, such as near a singularity. Again the time-domain solution

$$\begin{aligned} \overline{V}_1^{(0)}(t, T+\Delta T) &= \cos\Delta\phi\overline{V}_1^{(0)}(t, T) - \sin\Delta\phi\overline{V}_2^{(0)}(t-\Delta T_{12}, T) \\ \overline{V}_2^{(0)}(t, T+\Delta T) &= \sin\Delta\phi\overline{V}_1^{(0)}(t-\Delta T_{21}, T) + \cos\Delta\phi\overline{V}_2^{(0)}(t, T) \end{aligned} \tag{188}$$

is attractive. This is exactly equivalent to the solution obtained by Coates and Chapman (1990b) by a different technique.

Finally, we need the first-order additional term for shear waves. Equation (119) is modified to include the QI term

$$\begin{aligned} V_3^{(1)} = \hat{\mathbf{g}}_3^{\mathrm{T}}&\left[\beta\frac{\partial V_\nu^{(0)}}{\partial g_\nu} + \frac{V_\nu^{(0)}}{\alpha^2-\beta^2}\left((\alpha^2+3\beta^2)\boldsymbol{\nabla}\beta + \beta^3\boldsymbol{\nabla}(\ln\rho)\right)\right] \\ &- i\omega V_\nu^{(0)}\frac{B_{3\nu}}{\alpha^2-\beta^2} \end{aligned} \tag{189}$$

As before, overall the QI term is independent of frequency and describes the correction to the shear polarization, in agreement with degenerate perturbation theory (Jech and Pšenčík, 1989).

Acknowledgments

Most of this chapter was written while the author was visiting Queen's University, Kingston, Canada in 1998. He thanks Professor Colin Thomson for arranging the visit and for many useful discussions, and Schlumberger Cambridge Research for permitting the visit and publication of this chapter. Reviewers Professor Vlastislav Červený and Leo Eisner provided many useful suggestions.

Appendix: Triangular Integration

The bandlimited Maslov and Kirchhoff seismograms (Sections 4.1 and 4.2) both reduce to area integrals [Eqs. (141) and (152)]. If the corresponding domain, ray tubes or interface, is triangulated (Figs. 5 and 6), these integrals can be efficiently evaluated using an algorithm described in Spencer *et al.* (1997).

Consider a ray tube (triangle) ABC, with phases θ_A, etc. and amplitudes $\mathbf{A}_A$, etc. Let us assume that the phases are ordered $\theta_A < t < \theta_B < \theta_C$. The line DD$'$ is where $t = \theta$ and B' is where $\theta = \theta_B$ on the side AC. The area of the triangle ADD' is

$$\Delta_{ADD'} = \frac{(t-\theta_A)^2}{(\theta_B-\theta_A)^2}\Delta_{ABB'} \tag{190}$$

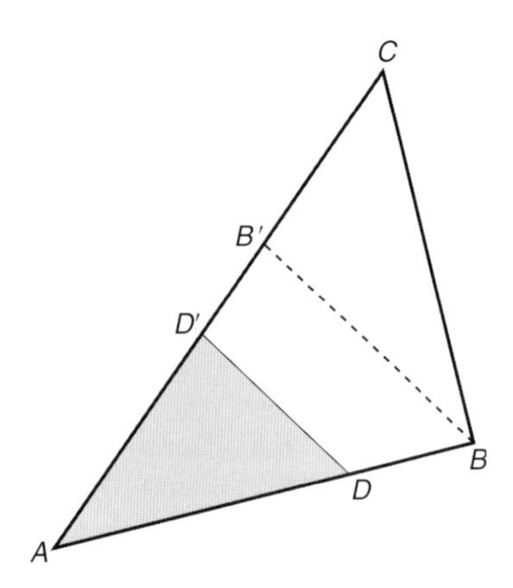

FIGURE 6 A triangular element. The phase at B' is the same as B. For a phase between A and B, e.g., D, the triangle ADD' is similar to ABB' and its area is easily calculated.

To obtain the integral of $\mathbf{A}$ over this triangle with linear interpolation, we just need the mean of the amplitude at A, $\mathbf{A}_A$, and at D and D', i.e.,

$$\mathbf{A}_D = \mathbf{A}_A + \frac{\mathbf{A}_B - \mathbf{A}_A}{\theta_B - \theta_A}(t - \theta_A) \tag{191}$$

$$\mathbf{A}_{D'} = \mathbf{A}_A + \frac{\mathbf{A}_C - \mathbf{A}_A}{\theta_C - \theta_A}(t - \theta_A) \tag{192}$$

Thus,

$$\begin{aligned} &\iint_{\mathbf{q}_A}^{t=0} \mathbf{A}\, d\mathbf{q} \\ &= \left[\mathbf{A}_A + \frac{1}{3}\left(\frac{\mathbf{A}_B - \mathbf{A}_A}{\theta_B - \theta_A} + \frac{\mathbf{A}_C - \mathbf{A}_A}{\theta_C - \theta_A}\right)(t - \theta_A)\right] \Delta_{ADD'} \\ &= \left[\mathbf{A}_A + \frac{1}{3}\left(\frac{\mathbf{A}_B - \mathbf{A}_A}{\theta_B - \theta_A} + \frac{\mathbf{A}_C - \mathbf{A}_A}{\theta_C - \theta_A}\right)(t - \theta_A)\right] \\ &\quad \times \frac{(t - \theta_A)^2}{(\theta_B - \theta_A)^2} \Delta_{ABB'} \end{aligned} \tag{193}$$

where $\mathbf{q}$ is either the ray parameter or interface variable. Similar results hold for $t > \theta_B$. Repeating at intervals of Δt, results for strips are obtained by subtraction.

References

Aki, K. and P.G. Richards (1980). "Quantitative Seismology—Theory and Methods," Vol. 1. W.H. Freeman, San Francisco.

Bakker, P.M. (1998). Phase shift at caustics along rays in anisotropic media. *Geophys. J. Int.* **134**, 515–518.

Bleistein, N. (1984). "Mathematical Methods for Wave Phenomena." Academic Press, Orlando.

Buchwald, V.T. (1959). Elastic waves in anisotropic media, *Proc. R. Soc. Lond. A*, **253**, 563–580.

Bullen, K.E. (1963). "An Introduction to the Theory of Seismology," 3rd Ed. Cambridge University Press, Cambridge.

Brokešova, J. (1992). High frequency 2.5D modelling of seismic wavefield applied to a finite extent source simulation. In: "Proceedings of XXIII General Assembly of the European Seismological Commission."

Burridge, R. (1967). The singularity on the plane lids of the wave surface of elastic media with cubic symmetry. *Q. J. Mech. Appl. Math.* **20**, 41–56.

Červený, V. (1972). Seismic rays and ray intensities in inhomogeneous anisotropic media. *Geophys. J. R. Astron. Soc.* **29**, 1–13.

Červený, V. (1982). Direct and inverse kinematic problems for inhomogeneous anisotropic media—linearization approach. *Contr. Geophys. Inst. Slov. Acad. Sci.* **13**, 127–133.

Červený, V. (1985). The application of ray tracing to the numerical modeling of seismic wavefields in complex structures. In: "Handbook of Geophysical Exploration, Section I: Seismic Exploration" (K. Helbig and S. Treitel, Eds.), Vol. 15, "Seismic Shear Waves" (G. Dohr, Ed.), pp. 1–124. Geophysical Press, London.

Červený, V. and F. Hron (1980). The ray series and dynamic ray tracing system for three-dimensional inhomogeneous media. *Bull. Seismol. Soc. Am.* **70**, 47–77.

Červený, V. and J. Jech (1982). Linearized solutions of kinematic problems of seismic body waves in inhomogeneous slightly anisotropic media. *J. Geophys.* **51**, 96–104.

Červený, V., J. Langer, and I. Pšenčík (1974). Computations of geometrical spreading of seismic body waves in laterally inhomogeneous media with curved interfaces. *Geophys. J. R. Astron. Soc.* **38**, 9–19.

Chapman, C.H. (1994). Reflection/transmission coefficient reciprocities in anisotropic media. *Geophys. J. Int.* **116**, 498–501.

Chapman, C.H. and R.T. Coates (1994). Generalized Born scattering in anisotropic media. *Wave Motion* **19**, 309–341.

Chapman, C.H. and H. Keers (2002). Application of the Maslov seismogram method in three dimensions. *Studia Geophys. Geodet.* (submitted).

Coates, R.T. and C.H. Chapman (1990a). Ray perturbation theory and the Born approximation. *Geophys. J. Int.* **100**, 379–392.

Coates, R.T. and C.H. Chapman (1990b). Quasi-shear wave coupling in weakly anisotropic 3-D media. *Geophys. J. Int.* **103**, 301–320.

de Hoop, M.V. and N. Bleistein (1997). Generalized Radon transform inversions for reflectivity in anisotropic elastic media. *Inverse Problems* **13**, 669–690.

Duff, G.F.D. (1960). The Cauchy problem for elastic waves in an anisotropic media. *Phil. Trans. R. Soc. Lond. A* **252**, 249–273.

Eisner, L. and I. Pšenčík (1996). Computation of additional components of the first-order ray approximation in isotropic media. *Pageoph* **148**, 227–253.

Farra, V. and S. Le Bégat (1995). Sensitivity of qP-wave traveltimes and polarization vectors to heterogeneity, anisotropy and interfaces, *Geophys. J. Int.* **121**, 371–384.

Frazer, L.N. and M.K. Sen (1985). Kirchhoff–Helmholtz reflection seismograms in a laterally inhomogeneous multi-layered elastic medium—I. Theory. *Geophys. J. R. Astron. Soc.* **80**, 121–147.

Gajewski, D. and I. Pšenčík (1987). Computation of high-frequency seismic wavefields in 3-D laterally inhomogeneous anisotropic media. *Geophys. J. R. Astron. Soc.* **91**, 383–411.

Gajewski, D. and I. Pšenčík (1990). Vertical seismic profile synthetics by dynamic ray tracing in laterally varying layered anisotropic structures. *J. Geophys. Res.* **95**, 11301–11315.

Garmany, J. (1988). Seismograms in stratified anisotropic media: 1. WKBJ theory. *Geophys. J.* **92**, 365–377.

Hanyga, A. (1982a). Dynamic ray tracing in an anisotropic medium. *Tectonophysics* **90**, 243–251.

Hanyga, A. (1982b). The kinematic inverse problem for weakly laterally inhomogeneous anisotropic media. *Tectonophysics* **90**, 253–262.

Jech, J. and I. Pšenčík (1989). First-order perturbation method for anisotropic media. *Geophys. J. Int.* **99**, 369–376.

Kendall, J.-M. and C.J. Thomson (1993). Maslov ray summation, pseudo-caustics, Lagrangian equivalence and transient seismic waveforms. *Geophys. J. Int.* **113**, 186–214.

Kendall, J.-M., W.S. Guest, and C.J. Thomson (1992). Ray-theory Green's function reciprocity and ray-centred coordinates in anisotropic media. *Geophys. J. Int.* **108**, 364–371.

Klimeš, L. (1997). Phase shift of the Green function due to caustics in anisotropic media. *SEG Expanded Abstract* **ST 14.2**, 1834–1837.

Kravtsov, Yu. A. and Yu. I. Orlov (1990). "Geometrical Optics of Inhomogeneous Media." Springer-Verlag, Berlin.

Lewis, R.M. (1965). Asymptotic theory of wave-propagation. *Arch. Ration. Mech. Anal.* **20**, 191–250.

Lighthill, M.J. (1960). Studies on magneto-hydrodynamic waves and other anisotropic wave motions. *Phil. Trans. R. Soc. Lond. A* **252**, 397–430.

Popov, M.M. and I. Pšenčík (1978). Computation of ray amplitudes in inhomogeneous media with curved interfaces. *Stud. Geophys. Geod.* **22**, 248–258.

Pšenčík, I. (1998). Green's functions for inhomogeneous weakly anisotropic media. *Geophys. J. Int.* **135**, 279–288.

Pšenčík, I. and D. Gajewski (1998). Polarization, phase velocity and NMO velocity of *qP* waves in arbitrary weakly anisotropic media. *Geophysics* **63**, 1754–1766.

Shearer, P.M. and C.H. Chapman (1988). Ray tracing in anisotropic media with a linear gradient. *Geophys. J.* **94**, 575–580.

Spencer, C.P., C.H. Chapman, and J.E. Kragh (1997). A fast, accurate, integration method for Kirchhoff, Born and Maslov synthetic seismogram generation. *SEG Expanded Abstract* **ST 14.3**, 1838–1841.

Thomson, C.J. and C.H. Chapman (1985). An introduction to Maslov's asymptotic method. *Geophys. J. R. Astron. Soc.* **83**, 143–168.

Ursin, B. and M. Tygel (1997). Reciprocal volume and surface scattering integrals for anisotropic elastic media. *Wave Motion* **26**, 31–42.

Woodhouse, J.H. (1974). Surface waves in laterally varying media. *Geophys. J. R. Astron. Soc.* **37**, 29–51.

10

Normal Modes of the Earth and Planets

Philippe Lognonné and Eric Clévédé
Institut de Physique du Globe de Paris, France

1. Introduction and Historical Key Dates

During almost a century, and up to 1960, normal modes were only theoretical concepts. Normal modes of the Earth were first described by Lord Kelvin (Thomson, 1863a,b) with a computation of the lowest fundamental spheroidal mode ${}_0S_2$ frequency for a homogeneous Earth model, either fluid or rigid. This early attempt was followed by a complete description of the normal modes of a spherical nonrotating elastic isotropic (SNREI) Earth in the early works of Rayleigh (1894, 1906) and Love (1911). The first computations of normal mode frequencies for realistic models of the Earth (at that time obtained from body wave arrival time data) were performed independently by Jobert (1956, 1957), Pekeris and Jarosh (1958), and Takeuchi (1959), as well as by Alterman *et al.* (1959) with a more complete calculation of eigenfrequencies and eigenmodes, and the search for normal mode signatures started in the very late 1950s (Benioff *et al.*, 1959).

The theory and the deployment in the late 1950s of the first real long-period sensors, such as strainmeters, tiltmeters, or gravimeters, were therefore just in time for the Chile 1960 earthquake. The analyses of records performed by Benioff *et al.* (1961), Ness *et al.* (1961), Alsop *et al.* (1961), Bolt and Marussi (1962), and Connes *et al.* (1962) showed an excellent fit with the theoretical frequencies, as well as splitting related to the rotation of the Earth (Backus and Gilbert, 1961; Pekeris *et al.*, 1961). These first early observations were just the starting point, followed later by a continuous increase of the number of observations, from about 250 modes (Derr, 1969a; Dziewonski and Gilbert, 1972, 1973) to more than 1000 modes (Gilbert and Dziewonski, 1975; Dziewonski and Anderson, 1981). The inversion of the two last sets of data produced, respectively, the 1066 A-B and PREM models, still today used often as the reference model for normal mode and internal structure studies. These two models were also the first seismic models to measure directly the density profile of the Earth, through the sensitivity of normal modes to gravitation. Only a few new modes were later discovered by the analysis of subsequent earthquakes, such as the lowest toroidal mode, ${}_0T_2$, for which the first observation was made only after the 1989 Macquarie Rise earthquake (Widmer *et al.*, 1992b), almost 30 years later than ${}_0S_2$.

From a theoretical point of view and after the pioneering work of Backus and Gilbert (1961), most of the attention after 1970 was given to the effect of elastic lateral variations and ellipticity. Perturbation theory appeared very rapidly as a powerful tool for the treatment first of the Earth rotation and ellipticity (Dahlen, 1968) and then of the lateral variations in the isolated multiplet case by Zharkov and Lyubimov (1970a,b) and Madariaga (1972). Later, refined perturbation theories were used: quasi-degenerate theory for the rotation and ellipticity (Dahlen, 1969) or for lateral variations (Woodhouse, 1980), and higher-order perturbation theory (Lognonné and Romanowicz, 1990a; Park, 1990b). The development of computers also allowed the brute-force resolution of the problem with variational methods (Park and Gilbert, 1986; Hara *et al.*, 1991, 1993). In parallel with these theoretical and numerical studies, observations and inversions were performed to constrain the lateral variations of the Earth with early works of Jordan (1978) and Silver and Jordan (1981), the latter leading to the discovery of the elastic degree-two pattern of the Earth's mantle (Masters *et al.*, 1982) and of the anomalous splitting of some modes depicting a large sensitivity to the core (Masters and Gilbert, 1981). However, after almost two decades of discussions, a complete explanation of all anomalous modes and of body wave travel times in the core is still lacking. Normal modes, as intermediate functions in the computation of seismograms, were also used for the first 3D tomographic model of Woodhouse and Dziewonski (1984).

INTERNATIONAL HANDBOOK OF EARTHQUAKE AND ENGINEERING SEISMOLOGY, VOLUME 81A ISBN: 0-12-440652-1

The late 1980s and 1990s saw the first observation of attenuation lateral variation (e.g., Romanowicz, 1990, 1994; Roult *et al.*, 1990) and the final refinements in the theory of normal modes. The theoretical framework for a nonhydrostatic prestressed Earth was completed by Woodhouse and Dahlen (1978) and Valette (1986), but the signature of prestress in seismic data has not yet been reported. Theory taking into account the attenuation was also developed, first in a non-rotating Earth (Lognonné, 1989), and then in a rotating anelastic Earth (Lognonné, 1991), in the latter case correcting and completing earlier work of Dahlen (1981) and Park and Gilbert (1986).

The last and most recent step was performed in 1998, with the discovery of the permanent excitation of the normal modes of the Earth. Despite an early suggestion of Benioff *et al.* (1959), normal modes were always sought only after strong earthquakes. Nawa *et al.* (1998), Suda *et al.* (1998), and Kobayashi and Nishida (1998) showed that normal modes were in fact permanently excited, at the level of the nanogal (10^{-11} m sec^{-2}), too small for their observation on a single record but sufficiently enhanced by stacking of at least one year's signals.

The aim of this chapter is to provide an overview of the theory of normal modes and of observations and inversions of normal modes performed since 1961. A much more detailed normal mode description can be found in Dahlen and Tromp (1998). The final results of these normal mode observations, i.e., the tomographic models of the Earth, are reviewed by Romanowicz (1991) and Ritzwoller and Lavely (1995). Despite the fact that the Earth is the only telluric planet where normal modes have been observed, Lognonné *et al.* (1996) and Kobayashi and Nishida (1998) have shown that they may be observed on Mars and even possibly on Venus. See Lognonné and Mosser (1993) for a review of the planetary seismology and normal modes of the telluric and giant planets.

After description of the gravito-elastic equation, the normal modes of a SNREI model are briefly described. Numerical codes for the computation of these SNREI modes are provided on the attached Handbook CD, by courtesy of G. Masters. The perturbation theory for the modeling of the asphericity of the Earth is then presented, as well as the method of normal mode summation, which allows use of the aspherical modes for the computation of seismograms in a complex Earth. This chapter therefore describes the theory necessary for the numerical computation of the normal modes of a rotating, aspherical, anelastic, and anisotropic Earth. The software for the computation of the aspherical, anelastic, and anisotropic normal modes and seismograms is provided on the attached Handbook CD. See Chapter 85.16 by Clévédé and Lognonné for a summary, as well as a more detailed software manual on the attached Handbook CD. This software will provide the user with a tool for the computation of normal modes and seismograms in such a complex Earth.

2. The Gravito-Elastic Equation of Normal Modes in a Linear Aspherical, Anisotropic, and Anelastic and Rotating Earth

2.1 General Properties

Let us consider a general model of the Earth or another planet, including if necessary an ocean and an atmosphere, and let us suppose that all equations are expressed in the inertial frame, rotating with the angular velocity $\boldsymbol{\Omega}$. In the state of equilibrium, the prestress (or static pressure for fluid parts) is compensated by the gravity, the centrifugal force, as well as the Reynolds stress and thermal stress associated with convection and temperature gradient. We will focus on a hydrostatic prestressed Earth, where only gravity and centrifugal force are significant, which is used in most if not all data analysis. The general case can, however, be found in Dahlen and Tromp (1998), following works of Dahlen (1972, 1973), Woodhouse and Dahlen (1978), Valette (1986), and Vermeersen and Vlaar (1991), and must be used for attempts related to the search for the prestress deviatoric in the Earth.

In the state of equilibrium, the static stress of the Earth $\mathbf{T}_0$ at radius point $\mathbf{r}$ follows

$$\nabla \cdot \mathbf{T}_0 + \rho_0(\mathbf{g}_0 - \boldsymbol{\Omega} \times (\boldsymbol{\Omega} \times \mathbf{r})) = 0 \tag{1}$$

where ∇ is the derivative operator, $\mathbf{g}_0$ is the gravity of the Earth, and ρ_0 the density, both at equilibrium . We have added here the inertial force related to the rotation of the Earth frame, with an instantaneous rotation vector $\boldsymbol{\Omega}$. Let us now assume that the Earth is vibrating, as the result of some excitation. Denote by ρ the density in the nonequilibrium state and assume that, at time t, a particle initially at position $\mathbf{r}$ will then be at

$$\mathbf{r}'(t) = \mathbf{r} + \mathbf{u}(\mathbf{r}, t) \tag{2}$$

The principle of mechanics, expressed at the point $\mathbf{r}'$ can be written as

$$\rho \frac{d\mathbf{v}}{dt} = \nabla \cdot \mathbf{T} + \rho(\mathbf{g} - \boldsymbol{\Omega} \times (\boldsymbol{\Omega} \times \mathbf{r}')) - 2\rho\boldsymbol{\Omega} \times \mathbf{v} \tag{3}$$

where $\mathbf{T}$, $\mathbf{g}$, and $\mathbf{v}$ are the stress tensor, gravity field, and particle velocity, all expressed at the position $\mathbf{r}'$ changing with time, and therefore in an Eulerian point of reference. Note that discontinuities of the Earth model will also be deformed with time. The Eulerian density and gravity can be expressed easily by considering a virtual time during which all deformations take place and has to follow the principle of mass conservation. Therefore, to first order,

$$\rho - \rho_0 + \mathrm{div}(\rho_0 \mathbf{u}) = 0 \tag{4}$$

These density variations are affecting the gravity field of the planet with potential Φ, such that $\mathbf{g} = -\nabla\Phi$ has to be solution of

$$\nabla^2\Phi = 4\pi\mathcal{G}\rho \quad (5)$$

where $\mathcal{G}$ is the gravitational constant. From Eq. (4), this gravity field can therefore be decomposed into the equilibrium gravity field and a perturbation, related to a potential Φ_1, called the redistribution potential, and such that $\mathbf{g} = \mathbf{g}_0 - \nabla\Phi_1$ with

$$\nabla^2\Phi_1 = -4\pi\mathcal{G}\,\mathrm{div}(\rho_0\mathbf{u}) \quad (6)$$

We now have to obtain the expression of the Eulerian stress at $\mathbf{r}'$. This stress can be defined as the superposition of a transported stress and of a second term, related to stresses due to compression and shear acting on the media:

$$\begin{aligned}\mathbf{T}(\mathbf{r}') &= \mathbf{T}_0(\mathbf{r}' - \mathbf{u}(\mathbf{r}, t)) + \delta\mathbf{T}_{\text{elastic}}(\mathbf{r}, t)\\ &= \mathbf{T}_0(\mathbf{r}') - \mathbf{u}\cdot\nabla\mathbf{T}_0 + \delta\mathbf{T}_{\text{elastic}}\end{aligned} \quad (7)$$

which can be simplified in the hydrostatic case by replacing $\mathbf{T}_0$ by $-p_0\mathbf{I}$, where p_0 is the static pressure and $\mathbf{I}$ is the identity tensor. The second part is related to elastic or anelastic stresses. In the linear assumption, valid for the typical amplitudes of normal modes, we have for an elastic and linear model,

$$\delta T^{ij}_{\text{elastic}}(\mathbf{r}, t) = C^{ijkl} D_k u_l(\mathbf{r}, t) \quad (8)$$

where C^{ijkl} is the elastic tensor, and D_k the derivative operator along direction k, while, for anelastic and linear models, we have instead a generalization in terms of convolution:

$$\delta T^{ij}_{\text{anelastic}}(\mathbf{r}, t) = \int_0^{+\infty} C^{ijkl}(\tau)\, D_k u_l(\mathbf{r}, t-\tau)\, d\tau \quad (9)$$

In general, the tensor has 21 independent coefficients in the triclinic case and is symmetric, obeying $C^{ijkl} = C^{jilk} = C^{klij} = C^{ijlk}$. In the isotropic case, we have two coefficients, and the tensor is such that

$$C^{ijkl} = (\kappa - \tfrac{2}{3}\mu) g^{ij} g^{kl} + \mu(g^{ik} g^{jl} + g^{il} g^{jk}) \quad (10)$$

where κ and μ are the bulk and shear modulus respectively, and where g^{ij} is the metric tensor of the projection basis. For a cartesian basis, it is equal to the Kronecker tensor, i.e., to 1 for $i = j$ and to 0 for other terms. For fluids, μ cancels.

2.2 Boundary Conditions

In the Eulerian description, boundaries are moving with time, as well as the normal to these surfaces. At solid/solid interfaces, the displacement is continuous as soon as the boundaries are welded. For liquid/solid or liquid/liquid boundaries, only the displacement perpendicular to the surface is continuous if no cavity at the surface is expected to occur (note here that the viscosity of ocean, atmosphere, and outer core is sufficiently small to be neglected in all the boundary conditions, for any frequency used in seismology).

Let us now focus on the stress continuity relations. A surface element, initially described with a surface element $d\Sigma_0$ and normal vector $\mathbf{n}_0$ is therefore deformed and associated with a new surface $d\Sigma$ and a new normal vector $\mathbf{n}$. Assuming that the surface has no mass, we get the general continuity relation:

$$\mathbf{n}^+\cdot\mathbf{T}^+\, d\Sigma^+ = \mathbf{n}^-\cdot\mathbf{T}^-\, d\Sigma^- \quad (11)$$

where $\mathbf{T}$ is the stress expressed at position $\mathbf{r}'$, with expressions estimated above the surface for index $+$ and below it for index $-$. Both the normal vector and the surface element can be expressed by considering a local basis moving with the deformation and therefore associated with position $\mathbf{r}'$. By differentiating Eq. (2), we get

$$dX'^i = (g^i_j + D_j u^i)\, dX^j = F^i_j\, dX^j \quad (12)$$

where X'^i are the components of $\mathbf{r}'$, D_j are the j component of the derivative operator and F^i_j is the matrix associated with the coordinate change. The three rows of $\mathbf{F}^{-1}$ give the basis vectors, and for the normal vector we then get

$$\mathbf{n} = \mathbf{n}_0 - \nabla_\Sigma \mathbf{u}(\mathbf{n}_0) \quad (13)$$

where Σ denotes the tangential derivation. In the same way, it is possible to define the surface change, and we have

$$d\Sigma = d\Sigma_0(1 + \nabla_\Sigma\cdot\mathbf{u}) \quad (14)$$

From these two equations, and the expression (7) for the stress, we finally get the continuity relation of stresses, as shown in Table 1 for the hydrostatic case. The surface of the Earth model can be considered in different ways. The most common is a free surface at the base of the atmosphere, which leads to neglect of the seismic energy lost by acoustic conversion in the atmosphere of the Earth. This is clearly not valid when the seismic source is in the atmosphere, as for an atmospheric eruption or the permanent excitation of normal modes. In that case, a radiating surface at the top of the atmosphere is more adequate, as shown by Lognonné *et al.* (1998).

2.3 The Elastodynamic Equation of Normal Modes

The elastodynamic equation can now be detailed. Starting from Eq. (3) expressed at the position $\mathbf{r}'$, neglecting the advection term in the time derivation, and subtracting Eq. (1), we finally get

$$\begin{aligned}\rho_0(\partial_{t^2}\mathbf{u} + 2\boldsymbol{\Omega}\times\partial_t\mathbf{u}) &= \nabla\cdot(\delta\mathbf{T}_{\text{elastic}} - \mathbf{u}\cdot\nabla\mathbf{T}_0)\\ &\quad - \mathrm{div}(\rho_0\mathbf{u})\mathbf{g}'_0 - \rho_0\nabla\Phi_1\end{aligned} \quad (15)$$

where $\mathbf{g}'_0 = \mathbf{g}_0 - \boldsymbol{\Omega}\times(\boldsymbol{\Omega}\times\mathbf{r}')$. Let us take the Laplace transform of the equation, defined here as $u(\sigma) = \int_0^{+\infty} e^{-i\sigma t} u(t)\, dt$, where $i^2 = -1$ and where $\sigma = \omega + i\alpha$ is the complex frequency (here in the upper part of the complex plane). We then obtain

TABLE 1 Summary of the boundary conditions for different types of surfaces and discontinuities. The function $F(\omega)$ is given by Lognonné *et al.* (1998) and is associated with the radiating surface

Type of Surface	Stress	Displacement
Solid–solid	$[-p_0\mathbf{n}_0\nabla_\Sigma\cdot\mathbf{u}+p_0\nabla_\Sigma\mathbf{u}(\mathbf{n}_0)+\delta\mathbf{T}_{\text{elastic}}\cdot\mathbf{n}_0]_-^+$	$[\mathbf{u}]_-^+$
Solid–liquid	$[-p_0\mathbf{n}_0\nabla_\Sigma\cdot\mathbf{u}+p_0\nabla_\Sigma\mathbf{u}(\mathbf{n}_0)+\delta\mathbf{T}_{\text{elastic}}\cdot\mathbf{n}_0]_-^+$	$[\mathbf{u}\cdot\mathbf{n}_0\,\mathbf{n}_0]_-^+$
Liquid–liquid	$[-p_0\mathbf{n}_0\nabla_\Sigma\cdot\mathbf{u}+p_0\nabla_\Sigma\mathbf{u}(\mathbf{n}_0)+\delta\mathbf{T}_{\text{elastic}}\cdot\mathbf{n}_0]_-^+$	$[\mathbf{u}\cdot\mathbf{n}_0\,\mathbf{n}_0]_-^+$
Free solid surface	$\delta\mathbf{T}_{\text{elastic}}\cdot\mathbf{n}_0=0$ and $p_0=0$	
Free atmospheric surface	$\delta p=\kappa\nabla\cdot\mathbf{u}=0$	
Radiating atmospheric surface	$\delta p=F(\omega)\,\rho(\mathbf{g}_0\cdot\mathbf{n}_0)(\mathbf{u}\cdot\mathbf{n}_0)$	

the expression for the elastodynamic normal modes in the complex frequency domain in operator form:

$$-\sigma^2\mathbf{u}(\sigma)+\sigma\mathbf{B}\mathbf{u}(\sigma)+\mathbf{A}(\sigma)\mathbf{u}(\sigma)=0 \qquad (16)$$

where the Coriolis operator **B** and the elastodynamic-operator **A** are expressed by

$$\begin{aligned}\mathbf{B}\mathbf{u}(\sigma)&=2i\mathbf{\Omega}\times\mathbf{u}\\ \mathbf{A}(\sigma)\mathbf{u}(\sigma)&=\frac{1}{\rho_0}[-\nabla\cdot(\mathbf{C}:\nabla\mathbf{u}-\mathbf{u}\cdot\nabla\mathbf{T}_0)+\operatorname{div}(\rho_0\mathbf{u})\,\mathbf{g}_0'\\ &\quad+\rho_0\nabla\Phi_1(\mathbf{u})]\end{aligned} \qquad (17)$$

Now the only difference between the elastic and non-elastic case is the possible frequency dependence of the stiffness tensor **C**. Following Valette (1986), the operator **A**, together with the boundary conditions, can be rewritten in the hydrostatic case and for a free surface as

$$\begin{aligned}\langle\mathbf{v}|\mathbf{A}\mathbf{u}\rangle&=\int dV\,\mathbf{v}\cdot\mathbf{A}\mathbf{u}=\int dV\left[d^{ijkl}D_ku_lD_iv_j-u^kD_iv_jD_kT_0^{ij}\right.\\ &\quad\left.+\operatorname{div}(\rho_0\mathbf{u})\mathbf{g}_0'\cdot\mathbf{v}+\rho_0\mathbf{v}\cdot\nabla\Phi_1\right]\end{aligned} \qquad (18)$$

where

$$d^{ijkl}=C^{ijkl}+p_0g^{ij}g^{kl}-p_0g^{ik}g^{jl}-p_0g^{jk}g^{il} \qquad (19)$$

with p_0 being the initial pressure and where we have defined the bracket bilinear form

$$\langle\mathbf{v}|\mathbf{u}\rangle=\int_V\rho\,dV\,\mathbf{u}(\mathbf{r})\cdot\mathbf{v}(\mathbf{r}) \qquad (20)$$

Other expressions for the operator can be found, such as

$$\begin{aligned}\langle\mathbf{v}|\mathbf{A}\mathbf{u}\rangle&=\int dV\,[d^{ijkl}D_ku_lD_iv_j+\rho S\,(\mathbf{u}\cdot\mathbf{g}\operatorname{div}(\mathbf{v})\\ &\quad-\mathbf{v}\cdot\nabla(\mathbf{u}\cdot\mathbf{g}))]-\int\frac{\nabla\Phi(\mathbf{u})\cdot\nabla\Phi(\mathbf{v})}{4\pi\mathcal{G}}\\ &\quad-\int_\Sigma S[\mathbf{u}\cdot\mathbf{n}(\mathbf{v})\cdot\nabla_\Sigma p_0]\,d\Sigma\end{aligned} \qquad (21)$$

where $S(f(\mathbf{u},\mathbf{v}))$ is the symmetric form $S=\frac{1}{2}(f(\mathbf{u},\mathbf{v})+f(\mathbf{v},\mathbf{u}))$. Expression (21) shows that the gravito-anelastic operator is always symmetric for the bilinear product (20), the latter being a scalar product for real functions only. The Coriolis operator is antisymmetric for this form. For a hermitian scalar product, where the function **v** in Eq. (20) is replaced by its complex conjugate, the operator **A** is self-adjoint only in the elastic case, while the Coriolis operator is self-adjoint. The elastic, rotating case is therefore a hermitian eigenproblem, while the anelastic case is symmetric. The anelastic, rotating case is more complex and is described in Section 5.

3. Normal Modes of a Spherical NonRotating Elastic Isotropic Earth Model

Normal modes of a spherical nonrotating elastic isotropic (SNREI) Earth constitute for many purposes the basis of normal mode theory, even in an aspherical Earth. They indeed constitute a complete basis of all functions verifying the same boundary and continuity conditions on the spherical surface of the Earth and internal discontinuities, respectively, as shown by Valette (1987) in the elastic case. Normal modes of a spherical Earth can therefore be used to describe the normal modes of all Earth models with any lateral variations but the same spherical shape and discontinuities.

The modern theory for a SNREI model was described by Takeuchi and Saito (1972) and Phinney and Burridge (1973). Let us recall from these works the solution of the SNREI equation

$$\sigma^2\mathbf{u}=\mathbf{A}_0(\mathbf{u}) \qquad (22)$$

where the operator $\mathbf{A}_0$ is defined for a spherically symmetric model such as PREM (Dziewonski and Anderson, 1981). As a consequence of the symmetry, by using group theory arguments, or by a separation of variables, it can be shown that the eigenfrequencies depend on two integer indices and can be noted $\omega_{l,n}$ and that the normal modes can be expressed in the two following forms:

$$\mathbf{u}_{n,l,m}(\mathbf{r}) = U_{n,l}(r)\, Y_l^m(\theta,\phi)\, \mathbf{e}_r + V_{n,l}(r)\, \nabla Y_l^m(\theta,\phi) \quad \text{or}$$

$$\mathbf{u}_{n,l,m}(\mathbf{r}) = W_{n,l}(r)\, \mathbf{e}_r \times \nabla Y_l^m(\theta,\phi) \tag{23}$$

Here $\mathbf{e}_r$ is the radial basis vector, n is the radial order, l is a positive integer called the angular order, m is the azimuthal order, such that $-l \le m \le l$ and $Y_l^m(\theta,\phi)$ are the spherical harmonic functions. The functions $U_{n,l}$, $V_{n,l}$, or $W_{n,l}$ as well as the real eigenfrequency associated to a given mode depend only on l, n. SNREI normal modes are invariant by a 2π rotation and by a symmetry around the center of the Earth, the latter defining the parity of the mode. Therefore, for the same angular order, we have two types of modes. The first are the spheroidal normal modes, have a null curl and a parity equal to that of l. The second are the toroidal modes, have a null divergence and a parity opposite to that of l.

Theoretically, the basis of SNREI normal modes is made up of all spheroidal and toroidal eigensolutions, including the secular modes with zero eigenfrequency associated with rotation of the fluid parts of the Earth, the gravity modes associated with positive Brunt–Väisälä frequencies in the liquid part and the toroidal core modes. Generally, the seismic modes have a positive index n while the gravity modes have a negative index. However the spectrum is discrete in the seismic band, a continuous spectrum exists in the case of a rotating Earth or if the liquid part of the Earth has a nonzero Brunt–Väisälä frequency (Valette 1989a,b). Only that complete set of eigenfunctions constitutes the complete basis. It is, however, generally accepted that the aspherical seismic normal modes (and seismograms) can be described with the seismic part of this SNREI basis only, i.e., all modes with frequency greater than 0.1 mHz.

The relation (23) involves the gradient of the spherical harmonics functions. To overcome this, Phinney and Burridge (1973) have developed a canonical basis that is more convenient than the spherical basis and more closely related to group theories. In this basis, the vectors are described with complex coordinates and complex basis vectors, and the gradients are incorporated in new functions, called generalized spherical harmonics. We then have

$$\mathbf{u} = u^+\mathbf{e}_+ + u^0\mathbf{e}_0 + u^-\mathbf{e}_- \tag{24}$$

where

$$\mathbf{e}_- = \frac{1}{\sqrt{2}}(\mathbf{e}_\theta - i\mathbf{e}_\phi) \qquad \mathbf{e}_0 = \mathbf{e}_r \qquad \mathbf{e}_+ = \frac{1}{\sqrt{2}}(-\mathbf{e}_\theta - i\mathbf{e}_\phi)$$
$$u^- = \frac{1}{\sqrt{2}}(u_\theta + iu_\phi) \qquad u^0 = u^r \qquad u^+ = \frac{1}{\sqrt{2}}(-u_\theta + iu_\phi) \tag{25}$$

with $\mathbf{e}_\phi$ and $\mathbf{e}_\theta$ being the longitude and co-latitude basis vectors. The eigenfunctions are now expressed with the generalized spherical harmonics of Phinney and Burridge (1973). Both the spherical harmonic or the generalized spherical harmonics are complex functions.

For numerical modeling, a useful further simplification is achieved if the SNREI normal modes are chosen as pure real or imaginary eigenfunctions as well as of a symmetry around the Greenwich plane defining the zero longitude (Lognonné and Romanowicz, 1990a). Such decomposition is generally much more convenient for numerical applications. Table 2 summarizes the different cases of SNREI modes.

The computation of SNREI modes is detailed by Woodhouse (1988) and is briefly summarized here. In the nongravitating case, each spheroidal normal mode is characterized by a vector of four functions, $\mathbf{v} = (\sqrt{2}V, U, \sqrt{2}T_s^{+0}, T_s^{00})$. The first two are the displacements, and the other two are related to the projection of stress on the radial direction. For the

TABLE 2 Summary of the different SNREI modes in the Phinney and Burridge (1973) basis, for the complex and hermitian case and for the real and symmetric ones

Complex Spheroidal Modes	$u^0 = U_{l,n}(r)\ Y_l^{0m}$	$u^\pm = V_{l,n}(r)\ Y_l^{\pm m}$
Complex Toroidal Modes	$u^0 = 0$	$u^\pm = \pm i\ W_{l,n}(r)\ Y_l^{\pm m}$
Real spheroidal modes		
$m > 0$	$u^0 = U_{l,n}(r)\frac{1}{\sqrt{2}}(Y_l^{0m} + (-1)^m\ Y_l^{0-m})$	$u^\pm = V_{l,n}(r)\frac{1}{\sqrt{2}}(Y_l^{\pm m} + (-1)^m\ Y_l^{\pm -m})$
$m = 0$	$u^0 = U_{l,n}(r)Y_l^{00}$	$u^\pm = V_{l,n}(r)Y_l^{\pm 0}$
$m < 0$	$u^0 = U_{l,n}(r)\frac{-i}{\sqrt{2}}(Y_l^{0m} - (-1)^m\ Y_l^{0-m})$	$u^\pm = V_{l,n}(r)\frac{-i}{\sqrt{2}}(Y_l^{\pm m} - (-1)^m\ Y_l^{\pm -m})$
Real toroidal modes		
$m > 0$	$u^0 = 0$	$u^\pm = W_{l,n}(r)\frac{\pm i}{\sqrt{2}}(Y_l^{\pm m} + (-1)^m\ Y_l^{\pm -m})$
$m = 0$	$u^0 = 0$	$u^\pm = \pm iW_{l,n}(r)Y_l^{\pm 0}$
$m < 0$	$u^0 = 0$	$u^\pm = W_{l,n}(r)\frac{\pm 1}{\sqrt{2}}(Y_l^{\pm m} - (-1)^m\ Y_l^{\pm -m})$

toroidal modes, the set reduces to two functions, $\mathbf{v} = (\sqrt{2}W, \sqrt{2}T_l^{+0})$, being the horizontal displacement and the radial projection of stress, respectively. In the gravitating case, the set is expanded into functions for the spheroidal modes only, in order to account for the mass redistribution. In both cases, the normal mode equation (3) and the constitutive equation (8) yield a propagator equation defining the normal modes:

$$\frac{d\mathbf{v}}{dr} = \mathbf{M}\mathbf{v} \tag{26}$$

where $\mathbf{v}$ is continuous across boundaries. The propagator matrix $\mathbf{M}$ is given by Phinney and Burridge (1973) for both the spheroidal and toroidal cases. The resolution of normal modes is generally performed by starting from the center of the Earth (or from a shallower depth for modes confined more closely to the surface) for two initial conditions corresponding to $\mathbf{v} = (0, 0, 1, 0)$ and $\mathbf{v} = (0, 0, 0, 1)$, respectively. The propagation of these functions toward the surface produces two vectors $\mathbf{v}_1$ and $\mathbf{v}_2$, both depending on the frequency. Eigenmodes and eigenfrequencies are then obtained by finding the frequency ω and real number β such that

$$v_{1,3}(a,\omega) + \beta v_{2,3}(a,\omega) = 0, \qquad v_{1,4}(a,\omega) + \beta v_{2,4}(a,\omega) = 0 \tag{27}$$

where $v_{1,i}$ and $v_{2,i}$ are the ith components of vectors $\mathbf{v}_1$ and $\mathbf{v}_2$, respectively. The MINEOS software (G. Masters, personal communication, 1999), necessary for the computation of normal modes for a SNREI model can be found on the attached Handbook CD, and was developed by Gilbert, Masters, and Woodhouse. See Woodhouse (1988) for more details. It allows the computation of modes for a SNREI model with a free surface at the top of the model (either solid or fluid). An estimate of the quality factor Q is also computed following first-order perturbation theory.

Figure 1 shows the different eigenfrequencies found for an Earth model with a solid part described by the PREM model and an atmospheric part described by the US Standard atmospheric model (US Standard Atmosphere, 1976). Four different types of normal modes are found: The first ones are related to the spheroidal modes of the solid Earth, with the fundamental branch $_0S_l$. The second are related to the Tsunami modes, which have almost all their energy localized in the oceanic layer of PREM. Note that the group velocity of these modes depends weakly on the frequency, in contrast to the fundamental spheroidal branch. The third and fourth are related to the acoustic and gravity modes of the atmosphere, the latter generally described with negative radial index. With the exception of the fundamental branch of acoustic modes, these modes have their eigenfrequency varying with the height of the top of the model atmosphere, as well as with the type of boundary condition—free, rigid, or radiative (Watada, 1995).

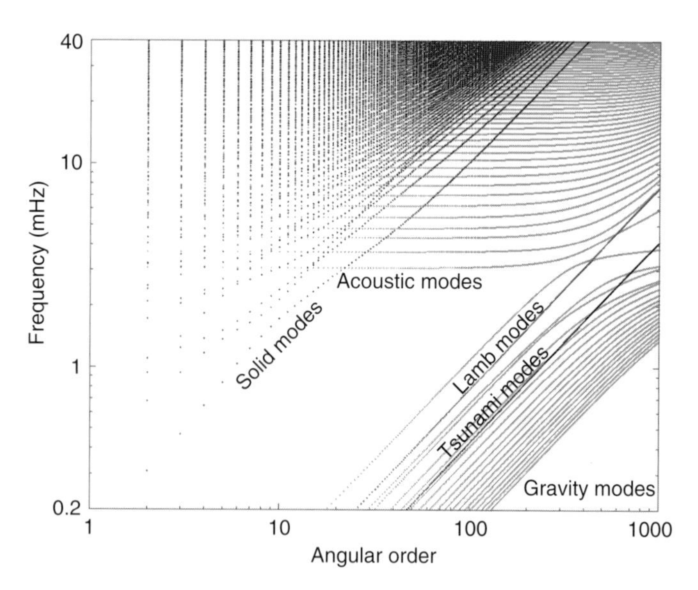

FIGURE 1 The different eigenfrequencies found for an Earth model with a solid part described by the PREM model and an atmospheric part described by the US Standard Atmospheric Model (1976) in logarithmic scale. Note that the group velocity of tsunamis, and the group velocity of the Lamb modes are constant. The intersection between the fundamental acoustic mode and the Lamb mode is an artifact of the boundary condition, here a free surface at the top of atmosphere artificially limited to 150 km of altitude. The spacing between the acoustic modes depends also on the height of the atmospheric model.

The solid Earth mode frequencies, either spheroidal (Fig. 2a), or toroidal (Fig. 2b) have frequencies and amplitudes that are practically unaffected by the energy transferred in the atmosphere. Their properties in terms of propagation in the solid Earth are therefore adequately described by an Earth with a free surface at the base of the atmosphere. Figure 3 shows the vertical amplitude with depth of the normal modes $_nS_{20}$, for increasing radial order n.

4. Normal Modes of a Spherical Nonrotating Anelastic Isotropic Earth Model

Attenuation or a radiating surface affect the normal modes due to the nonconservation of energy. For intrinsic attenuation and at a given frequency, this implies an imaginary part of the stiffness tensor, as shown by Liu *et al.* (1976) and Anderson and Given (1982). For causality reasons originally demonstrated for similar problems by Kramers and Kronig, this also implies a frequency dependence of the stiffness tensor (Anderson, 1989; Anderson and Minster, 1979), which in seismology is necessary to reconcile normal mode and body

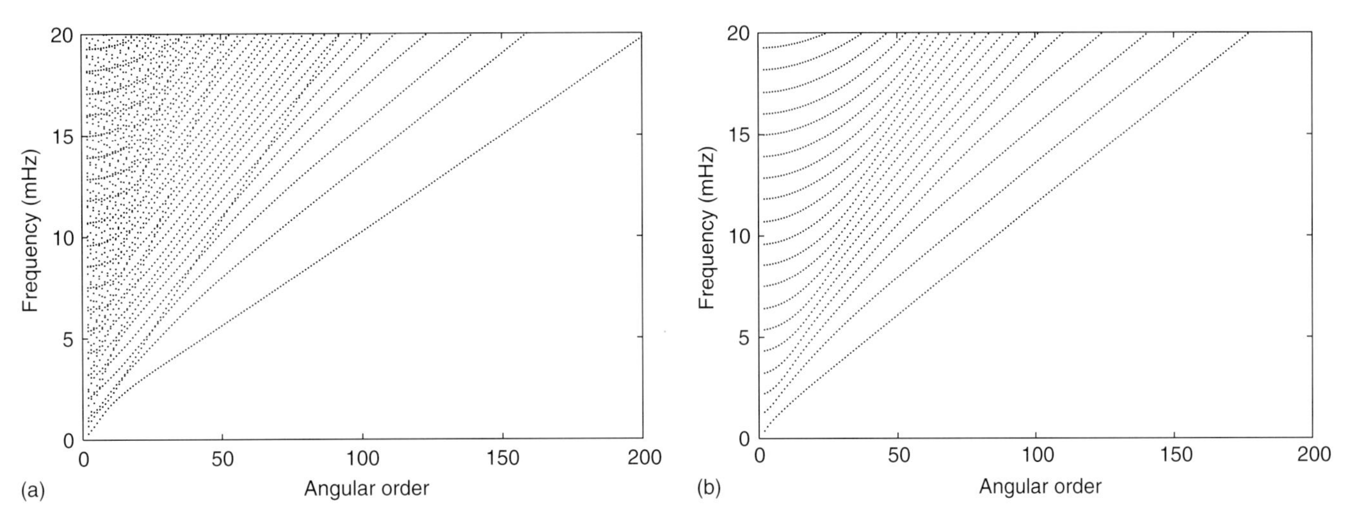

FIGURE 2 The solid Earth eigenfrequencies for (a) spheroidal modes and (b) toroidal modes in linear scale. Note for the spheroidal mode the bifurcation between the surface wave type modes and the two Stoneley modes branch, both with a constant group velocities, and respectively associated to modes trapped at the CMB and ICB boundary. The dispersion branches above the ICB Stoneley mode branch show also several bifurcations, related to the superposition of three types of mode family, respectively associated to *ScS*, PKIKP, and core-associated modes (Dahlen and Tromp, 1998).

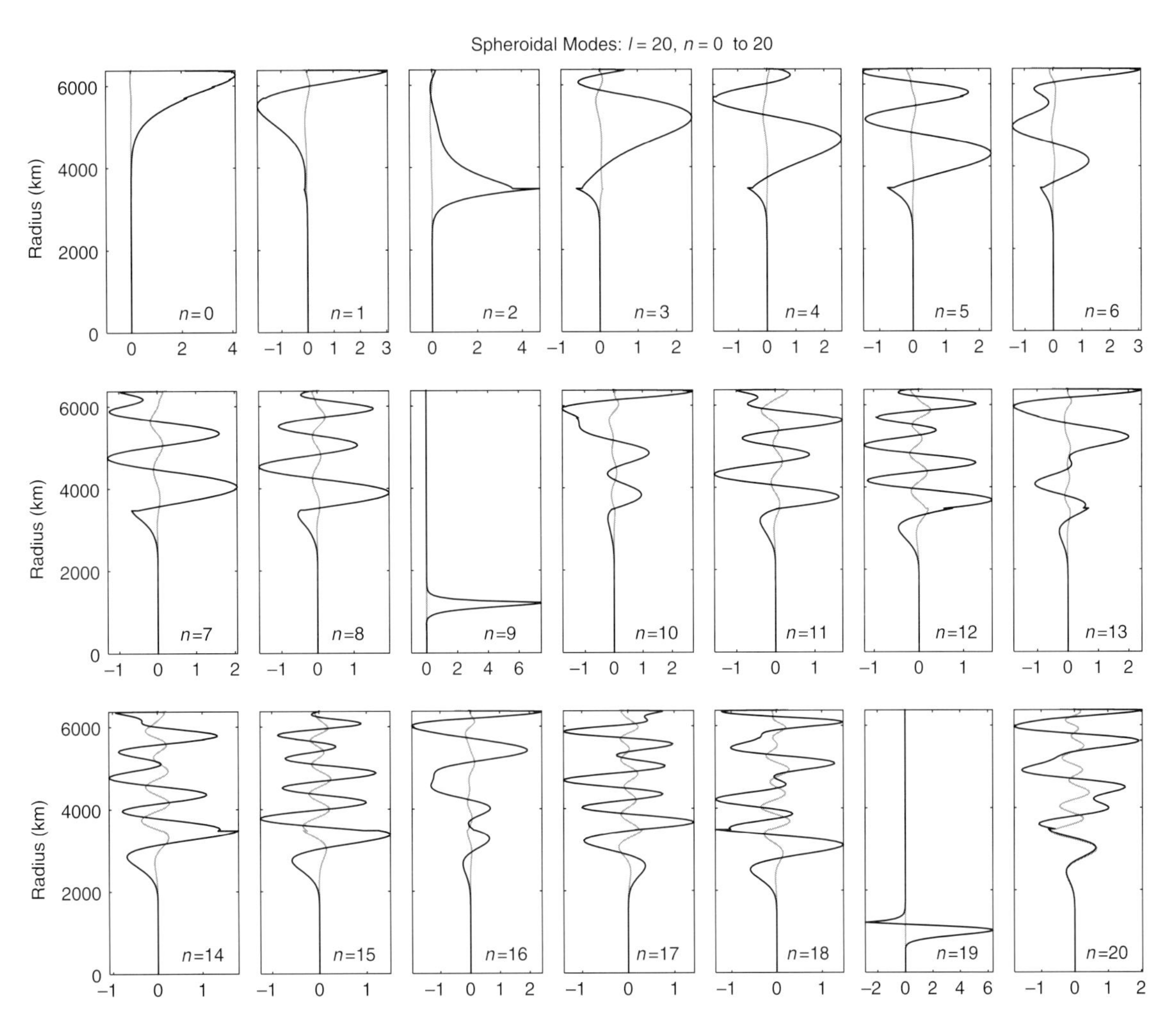

FIGURE 3 Vertical amplitude (heavy line) and imaginary amplitude (light line) for modes of angular order $l = 20$ and increasing radial order n. Both amplitudes are multiplied by the square root of the density, and are therefore proportional to the square root of the kinetic energy distribution. Amplitude for the imaginary part is multiplied by 10.

wave mean velocity models (Kanamori and Anderson, 1977; Montagner and Kennett, 1996).

The computation of the normal modes of a SNRAI Earth can be performed using the complete basis of the SNREI normal modes (Lognonné, 1989; Tromp and Dahlen, 1990; Clévédé, 1991). The solution, for each n and l value, is then expressed as a summation of these functions:

$$\mathbf{v} = \sum_{n'} c_{n'\ n'}\mathbf{u}_l \tag{28}$$

where the indexes n, l are omitted in $\mathbf{v}$ and $c_{n'}$ for simplicity. The spherical anelastic perturbation implies coupling between modes with different values of n, but the same value of l and m. This formulation can then be used to find the normal modes, e.g., the coefficients c'_n and the eigenfrequencies. Starting from

$$\omega^2\mathbf{v} = \mathbf{A}(\omega)\mathbf{v} \tag{29}$$

by a second-order Taylor expansion around the frequency ω_0 we first get

$$\omega^2\mathbf{v} = [\mathbf{A}(\omega_0) + (\omega - \omega_0)\partial_\omega\mathbf{A}(\omega_0) + \tfrac{1}{2}(\omega - \omega_0)^2\partial_{\omega^2}\mathbf{A}(\omega_0)]\mathbf{v} \tag{30}$$

and we can then express the second-order problem as a first-order matrix problem. See Park and Gilbert (1986) for another example of this technique. We finally get

$$\omega\left(1 - \frac{1}{2}\partial^2_\omega A\right)\begin{pmatrix}\omega\mathbf{v}\\ \mathbf{v}\end{pmatrix} = \begin{pmatrix}\partial_\omega A - \omega_0\partial^2_\omega A & A(\omega_0) - \omega_0\partial_\omega A + \dfrac{\omega_0^2}{2}\partial^2_\omega A\\ 1 - \dfrac{1}{2}\partial^2_\omega A & 0\end{pmatrix}\begin{pmatrix}\omega\mathbf{v}\\ \mathbf{v}\end{pmatrix} \tag{31}$$

Classical EISPACK routines (Smith *et al.*, 1976) can be used to solve this problem. This improved theory affects the value of the attenuation factor Q of the modes slightly, by 2–3%, as well as the amplitudes of modes by comparable relative effects. It also produces an initial delay of the modes, typically of 0.5–2 sec, which may have to be taken into account for very precise source analysis.

The effect of the atmosphere is comparable and also leads to the escape of seismic energy. It is described in detail by Lognonné *et al.* (1998), and is related to a boundary term having a frequency dependence (Table 1). Although not affecting frequency, attenuation, or modal amplitudes in the solid Earth in an observable way, it must obviously be taken into account when modes are excited by atmospheric sources or when seismic signals are observed in the atmosphere (Artru *et al.*, 2001). This is the case for explosions and atmospheric stochastic turbulences. Figure 4 shows both the effects of the atmosphere and of the PREM attenuation model, when the dispersion law of Liu *et al.* (1976) is taken into account in the computation of normal mode fundamental branches. The atmospheric displacement of normal modes is at the origin of the atypical excitation of normal modes by strong atmospheric explosion, such as the Pinatubo explosion (Kanamori and Mori, 1992; Kanamori *et al.*, 1994; Zurn and Widmer, 1996).

5. Normal Modes and Seismograms in the General Aspherical Case

5.1 Normal Mode Orthonormality Relations

In the general case, normal modes cannot be described by a single spherical harmonic and have a more complicated shape on the sphere. Many general properties remain, however. In the general case, following Lognonné (1991), let us first consider the right eigenmodes $|\mathbf{u}_k\rangle$ and associated eigenfrequencies σ_k satisfying

$$-\sigma_k^2|\mathbf{u}_k\rangle + \sigma_k\mathbf{B}|\mathbf{u}_k\rangle + \mathbf{A}(\sigma_k)|\mathbf{u}_k\rangle = 0 \tag{32}$$

and the left eigenmodes, with left eigenfunctions $\langle\mathbf{v}_k|$ and their associated eigenfrequencies σ_k verifying

$$-\sigma_k^2\langle\mathbf{v}_k| + \sigma_k\langle\mathbf{v}_k|\mathbf{B}^\dagger + \langle\mathbf{v}_k|\mathbf{A}^\dagger(\sigma_k) = 0 \tag{33}$$

Here † denotes the left dual operator, defined here as $\langle\mathbf{v}\,|\,\mathbf{A}\mathbf{u}\rangle = \langle\mathbf{A}^\dagger\mathbf{v}\,|\,\mathbf{u}\rangle$. As shown by Lognonné (1991), and due to the antisymmetry of operator $\mathbf{B}$ and symmetry of operator $\mathbf{A}$, the left or dual normal modes are solutions of the right eigenproblem with the inverted rotation

$$-\sigma_k^2|\mathbf{v}_k\rangle - \sigma_k\mathbf{B}|\mathbf{v}_k\rangle + \mathbf{A}(\sigma_k)|\mathbf{v}_k\rangle = 0 \tag{34}$$

and have the same eigenvalues as the primal normal modes, despite different eigenfunctions. The relation between the primal and dual mode is trivial only in a few cases (Table 3), such as the elastic case, where the dual eigenmode is simply the complex conjugate of the primal mode, or in the axisymmetric rotating and anelastic case, where $|\mathbf{v}_k\rangle = |\mathbf{S}\mathbf{u}_k\rangle$, S being an orthogonal symmetry around a plane containing the rotation axis. In the general case, eigenfrequencies exist also with either a positive or negative real part of the frequency. However, the gravito-elastic operator is such that $\mathbf{A}^*(\sigma) = \mathbf{A}(-\sigma^*)$, so that the complex conjugate of (32) shows that each eigenfrequency σ_k and associated eigenmode $|\mathbf{u}_k\rangle$

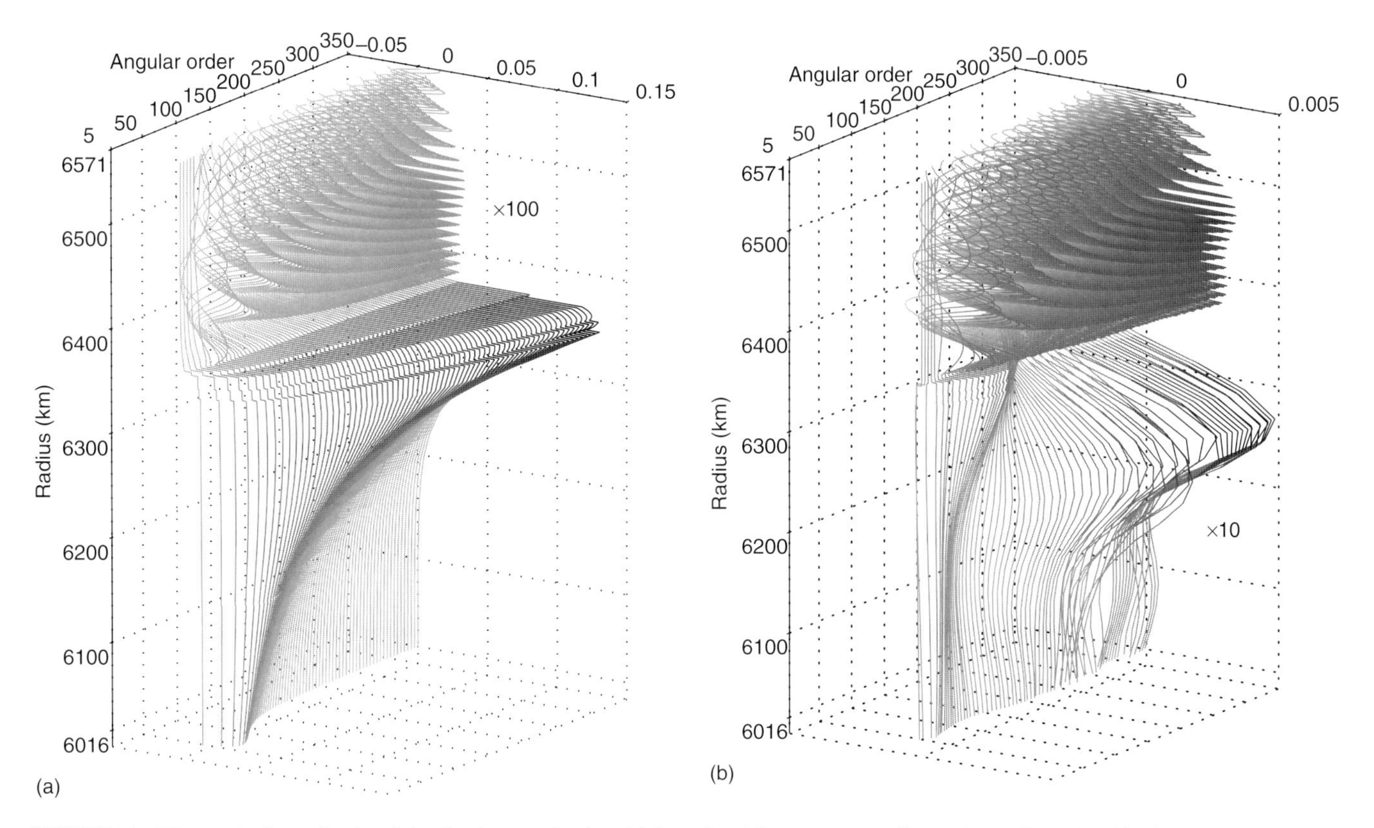

FIGURE 4 The vertical amplitude of the fundamental spheroidal mode with respect to radius and angular order l in the upper mantle and atmosphere, for a radiative boundary condition at an altitude of 150 km. (a) Real part multiplied by 100 in the atmosphere; (b) imaginary part multiplied by 10 in the solid Earth. The amplitude in both examples is multiplied by the square root of density, and is therefore proportional to the square root of the kinetic energy distribution. Note the transition between trapped atmospheric extensions and untrapped ones. In the solid Earth, note the shallow increasing amplitudes with the increasing angular order as well as the increasing amplitude in the low-velocity–high-attenuation zone of the upper mantle.

TABLE 3 Summary of the different cases for the dual mode and orthogonality relations depending on the inclusion of attenuation and rotation in the theory

Case	Normal Mode Equation	Dual Mode	Orthogonality
Nonrotating elastic	$[-\sigma^2 + \mathbf{A}_0] \mid \mathbf{u}\rangle = 0$		$\langle \mathbf{v}_{k'} \mid \mathbf{u}_k\rangle = \delta_{kk'}$
Complex case		$\mid \mathbf{u}^*\rangle$	
Real case		$\mid \mathbf{u}\rangle$	
Rotating elastic	$[-\sigma^2 + \sigma\mathbf{B} + \mathbf{A}_0] \mid \mathbf{u}\rangle = 0$	$\mid \mathbf{u}^*\rangle$	$\langle \mathbf{v}_{k'} \vert \mathbf{u}_k\rangle - \frac{1}{\sigma_k + \sigma_{k'}} \langle \mathbf{v}_{k'} \vert \mathbf{B}\mathbf{u}_k\rangle = \delta_{kk'}$
Nonrotating anelastic	$[-\sigma^2 + \mathbf{A}_0(\sigma)] \mid \mathbf{u}\rangle = 0$	$\mid \mathbf{u}\rangle$	$\langle \mathbf{v}_{k'} \vert \mathbf{u}_k\rangle - \frac{\langle \mathbf{v}_{k'} \vert \mathbf{A}(\sigma_{k'})\mathbf{u}_k\rangle - \mathbf{v}_{k'} \vert \mathbf{A}(\sigma_k)\mathbf{u}_k\rangle}{\sigma_{k'}^2 - \sigma_k^2} = \delta_{kk'}$
Axisymmetric rotating anelastic	$[-\sigma^2 + \sigma\mathbf{B} + \mathbf{A}(\sigma)] \mid \mathbf{u}\rangle = 0$	$\mid \mathbf{S}\mathbf{u}\rangle$	Eqs. (35)–(36)
Aspherical rotating anelastic	$[-\sigma^2 + \sigma\mathbf{B} + \mathbf{A}(\sigma)] \mid \mathbf{u}\rangle = 0$	Solution of Eq. (34)	Eqs. (35)–(36)

imply that the pair $-\sigma_k^*$ and $|\mathbf{u}_k^*\rangle$ are also eigenfrequency, and eigenmode, respectively.

Let us consider two singlets k and k' with different eigenfrequencies, multiply the relations (32) and (33) by $\langle \mathbf{v}_{k'} \mid$ and $\mid \mathbf{u}_k\rangle$ respectively, make the difference of the two obtained relations and divide the result by $\sigma_{k'}^2 - \sigma_k^2$. We finally get the following bi-orthogonality relation, with bracket product related to the aspherical density:

$$\langle \mathbf{v}_{k'} \vert \mathbf{u}_k\rangle - \frac{1}{\sigma_k + \sigma_{k'}} \times \left[\langle \mathbf{v}_{k'} \vert \mathbf{B}\mathbf{u}_k\rangle + \frac{\langle \mathbf{v}_{k'} \vert \mathbf{A}(\sigma_{k'})\mathbf{u}_k\rangle - \langle \mathbf{v}_{k'} \vert \mathbf{A}(\sigma_k)\mathbf{u}_k\rangle}{\sigma_{k'} - \sigma_k} \right] = 0 \qquad (35)$$

The relation (35) defines the general bi-orthogonality relation between the modes associated with two different eigenfrequencies. The limit for two identical modes defines the normalization, so that

$$\begin{aligned}\langle \mathbf{v}_k|\mathbf{u}_k\rangle - \frac{1}{2\sigma_k}[\langle \mathbf{v}_k|\mathbf{B}\mathbf{u}_k\rangle + \langle \mathbf{v}_k|\partial_\sigma \mathbf{A}(\sigma_k)\mathbf{u}_k\rangle] \\ = -\frac{1}{2\sigma_k}\langle \mathbf{v}_k|\partial_\sigma \mathbf{H}(\sigma_k)\mathbf{u}_k\rangle = 1 \qquad (36)\end{aligned}$$

where $\mathbf{H} = -\sigma^2 + \sigma\mathbf{B} + \mathbf{A}(\sigma)$. The elastic limit gives the orthogonality relation obtained by Dahlen and Smith (1975) for a rotating elastic case:

$$\langle \mathbf{v}^*_{k'}|\mathbf{u}_k\rangle - \frac{1}{\omega_k + \omega_{k'}}\langle \mathbf{v}^*_{k'}|\mathbf{B}\mathbf{u}_k\rangle = \delta_{kk'} \qquad (37)$$

6. Mode Summation and Seismograms

In the general case, as well as in the different simplified cases (e.g., Gilbert, 1970, in the elastic case), normal modes can be used for computing the response of the Earth produced by an equivalent body force $|\mathbf{f}(t)\rangle$. In the frequency domain, the solution has to satisfy

$$[-\sigma^2 + \sigma\mathbf{B} + \mathbf{A}]|\mathbf{u}(\sigma)\rangle = |\mathbf{f}(\sigma)\rangle \qquad (38)$$

and can moreover be expressed as a projection in the complete basis of eigenfunctions such as

$$|\mathbf{u}(\sigma)\rangle = \sum_{k'} c_{k'}(\sigma)|\mathbf{u}_{k'}\rangle \qquad (39)$$

As noted above, we will restrict the summation to the seismic modes (e.g., with a period less than 1 h), but for both the positive and negative frequencies. The time expression of the seismogram, given by

$$|\mathbf{u}(t)\rangle = \frac{1}{2\pi}\int_{i\epsilon-\infty}^{i\epsilon+\infty} |\mathbf{u}(\sigma)\rangle\, e^{i\sigma t}\, d\sigma \qquad (40)$$

can then be obtained by closing the integration path, using relation (35) and the Cauchy theorem. Following Lognonné (1991) we then get

$$|\mathbf{u}(t)\rangle = H(t)\sum_k e^{i\sigma_k t} \lim_{\sigma\to\sigma_k} i(\sigma - \sigma_k)c_k(\sigma)|\mathbf{u}_k\rangle \qquad (41)$$

For an impulsive source, the limit is found to be

$$\lim_{\sigma\to\sigma_k} i(\sigma-\sigma_k)c_k(\sigma) = \frac{i\langle \mathbf{v}_k|\mathbf{f}\rangle}{\langle \mathbf{v}_k|\partial_\sigma \mathbf{H}(\sigma_k)\mathbf{u}_k\rangle} = \frac{\langle \mathbf{v}_k|\mathbf{f}\rangle}{2i\sigma_k} \qquad (42)$$

The response of the Earth to an impulsive equivalent body force is obtained from the summation over all positive and negative indices, i.e., positive frequencies as well as negative frequencies. This finally allows recovery of a real amplitude for the Green tensor, equal to

$$|\mathbf{u}(t)\rangle = H(t)\sum_{k>0} \Re e\left(\frac{1}{i\sigma_k}\langle \mathbf{v}_k|\mathbf{f}\rangle e^{i\sigma_k t}|\mathbf{u}_k\rangle\right) \qquad (43)$$

where $\Re e$ denotes the real part. Table 4 describes several cases that can be derived from this Green's function by a convolution. Note also that Eqs. (38) and (39) can also be used to compute the amplitude produced by nonlocal sources, such as tidal forces or atmospheric forces, by a double convolution in space and time.

The last step corresponds to instrumental transfer function between the ground displacement, as expressed by relations in Table 4, and the digital data. This transfer function is defined as the receiver function. It therefore clearly depends on the type of instruments. For seismometers, the signal is the sum of all inertial forces (ground acceleration and gravity forces), projected along the measurement direction and, when appropriate, filtered by the frequency transfer function of the instrument. The measurement direction $\nu(t)$ as well as the position of the instrument varies with time due to the tilt, so that $\nu(t) = \nu_0 - \nabla_\Sigma \mathbf{u}(\nu_0)$. Without the filtering term, we therefore have to first order:

$$\begin{aligned}\langle \mathbf{R}|\mathbf{u}\rangle &= \nu(t)\cdot(\mathbf{g}(t,\mathbf{r}') - [\partial_t^2\mathbf{u} + 2\boldsymbol{\Omega}\times\partial_t\mathbf{u}]) - \nu_0\cdot\mathbf{g}_0(\mathbf{r}) \\ &= -\nu_0\cdot[\partial_t^2\mathbf{u} + 2\boldsymbol{\Omega}\times\partial_t\mathbf{u}] - \nu_0\cdot\nabla\Phi_1 + \nu_0\cdot(\mathbf{u}\cdot\nabla\mathbf{g}_0) \\ &\quad - \mathbf{g}_0\cdot\nabla_\Sigma\mathbf{u}(\nu_0) \qquad (44)\end{aligned}$$

TABLE 4 Amplitude of the ground displacement as described by a summation of normal modes for a local double couple. The index s denotes the position of the source

Source Function	Displacement Expression
Seismic tensor with general source function $\|\mathbf{f}(t)\rangle = -\mathbf{M}(t)\nabla\delta(\mathbf{r}-\mathbf{r}_s)$	$\|\mathbf{u}(t)\rangle = \sum_{k>0}\Re e\left(\frac{1}{i\sigma_k}\int_0^t dt'\,\mathbf{M}(t'):\nabla\mathbf{v}_k(\mathbf{r}_s)e^{i\sigma_k(t-t')}\|\mathbf{u}_k\rangle\right)$
Seismic tensor with Heaviside source function $\|\mathbf{f}(t)\rangle = -\mathbf{M}_0 H(t)\nabla\delta(\mathbf{r}-\mathbf{r}_s)$	$\|\mathbf{u}(t)\rangle = H(t)\sum_{k>0}\Re e\left(\frac{1}{\sigma_k^2}\mathbf{M}_0:\nabla\mathbf{v}_k(\mathbf{r}_s)(1-e^{i\sigma_k t})\|\mathbf{u}_k\rangle\right)$

where $\mathbf{g}_0$ includes both the gravitational and centrifugal terms. The contributions of the nonground acceleration terms are significant for normal modes with angular orders less than 10 generally, and are therefore related to the Coriolis acceleration, the perturbation of gravity, free air correction, and surface tilt, respectively. The last term related to free air correction is in phase with the acceleration, while the term related to the mass redistribution is generally in opposition phase. Table 5 gives the projected terms along the horizontal and vertical components.

The source function in Table 4 can be expressed in the same way. For a seismic tensor moment, we have

$$\begin{aligned}\langle \mathbf{v}_k|\mathbf{S}\rangle &= -\epsilon_{\alpha\beta}M^{\alpha\beta}\\ &= \epsilon_{00}M^{00} + 2\epsilon_{+-}M^{+-} + 4\Re e(\epsilon_{+0}M^{+0}) + 4\Re e(\epsilon_{++}M^{++})\end{aligned} \quad (45)$$

where $M^{\alpha\beta}$ are the canonical covariant components of the seismic moment tensor in the Phinney and Burridge (1973) basis, and where $\epsilon_{\alpha\beta}$ are the contravariant components of the dual stress tensor, defined as $e_{\alpha\alpha'}e_{\beta\beta'}\epsilon^{\alpha'\beta'}$. We used the symmetry relations $M^{\alpha\beta} = (-1)^{\alpha+\beta}M^{-\alpha-\beta*}$. For an elastic Earth, $|\mathbf{v}\rangle = |\mathbf{u}^*\rangle$, and then $\epsilon_{\alpha\beta} = (\epsilon^{\alpha\beta})^*$. Table 6 gives the expressions of the seismic tensor for a centroid moment tensor source.

For atmospheric sources, a seismic moment can also be used for the representation of the source density in the fluid parts of the Earth, ocean, or atmosphere, by using the generalization of Lognonné *et al.* (1994). In the latter case, we can decompose the excess pressure $\Delta p = p_{\text{true}} - p_{\text{Hooke}}$ into nonlaminar and laminar contributions: $\Delta p = \delta p - \frac{1}{2}\rho v^2$, and then get

$$M^{\alpha\beta} = -\delta p e^{\alpha\beta} + \rho\left(\frac{1}{2}|v|^2 e^{\alpha\beta} - v^\alpha v^\beta\right) \quad (46)$$

where v^α is the α component of the velocity of the fluid in the atmosphere and $|v|$ is the amplitude of the velocity. This generalizes the incomplete source proposed by Tanimoto (1999) for the explanation of the permanent excitation of the atmosphere.

TABLE 5 Expression of the receiver functions: u_r, u_θ, u_ϕ are the components of $\mathbf{u}$ in a spherical basis, and θ is the colatitude. Ω_0 is the rotation vector of the Earth. ϕ_r, ϕ_θ, ϕ_ϕ are the components of $\nabla\phi$

Z-axis seismometer	$\langle \mathbf{R}_Z \mid \mathbf{u}_k\rangle$	$\sigma_k^2 u_r - 2i\Omega_0\sigma_k u_\theta \sin\theta - \phi_r + \frac{2u_r g_0}{r}$
S-axis seismometer	$\langle \mathbf{R}_S \mid \mathbf{u}_k\rangle$	$\sigma_k^2 u_\theta - 2i\Omega_0\sigma_k(-\sin\theta\, u_r - \cos\theta\, u_\phi) - \phi_\theta + g_0 D_\theta u_r$
W-axis seismometer	$\langle \mathbf{R}_W \mid \mathbf{u}_k\rangle$	$\sigma_k^2 u_\phi - 2i\Omega_0\sigma_k \cos\theta\, u_\theta - \phi_\phi + g_0 D_\phi u_r$

TABLE 6 Relation between the canonical component of the seismic tensor and those in the spherical basis

M^{00}	M_{rr}
M^{+-}	$-\frac{1}{2}(M_{\theta\theta} + M_{\phi\phi})$
M^{0+}	$1/\sqrt{2}(-M_{r\theta} + iM_{r\phi})$
M^{++}	$1/2(M_{\theta\theta} - M_{\phi\phi}) - iM_{\theta\phi})$

7. Normal Modes of a Rotating, Aspherical, Anelastic, and Anisotropic (RA_3) Earth Model: Perturbation Theory

As noted in the introduction, the early observations almost immediately pointed out the limits of the spherical Earth theory and showed that the departure of sphericity and rotation resulted in an observable degeneracy of the singlets. For each normal mode multiplet, the frequencies of the $2l+1$ singlets are therefore slightly different, and each has excitation and observation amplitude depending on the position of both the source and station and on the mechanism of the earthquake. Therefore, the superposition of all singlet peaks is done differently for all receiver–source pairs. As a consequence, the shape and maximum of the spectral peaks associated with a given normal mode varies with the source–receiver pair, which provides a unique source of information for the determination of lateral variations of the Earth (Fig. 5). Let us now therefore describe the forward problem necessary for the computation of seismograms, spectra or secondary data for a general RA_3 Earth model.

This 3D model can be seen as the superposition of a mean model and a 3D de-meaned structure. For all realistic Earth models, the integrated power spectrum of the 3D lateral variations is small compared to the 1D structure, even if large 3D variations may exist locally. In that respect, 3D lateral variations act as small perturbations superimposed on the reference spherical elastic Earth, the latter being associated with a given elastodynamic operator $\mathbf{A}_0$. This reference model is not necessarily the mean Earth model of the RA_3, and the perturbation has therefore generally a mean component, as for example for the couple PREM (Dziewonski and Anderson, 1981) and M84A/C of Woodhouse and Dziewonski (1984). Note that the eigenfunctions of the reference operator are defined generally for the spherical volume corresponding to the spherical Earth, e.g., for a sphere with 6371 km radius in the case of PREM. Because the perturbations related to lateral variations and asphericity are relatively small, perturbation theory is therefore well suited to model these effects. Note that such perturbation theory is also extensively used in quantum mechanics (e.g., Cohen-Tannoudji *et al.*, 1980).

7.1 The Perturbation of Asphericity, Rotation, and Lateral Heterogeneities

The first and theoretically most constraining perturbation is related to the asphericity of the boundary conditions, especially

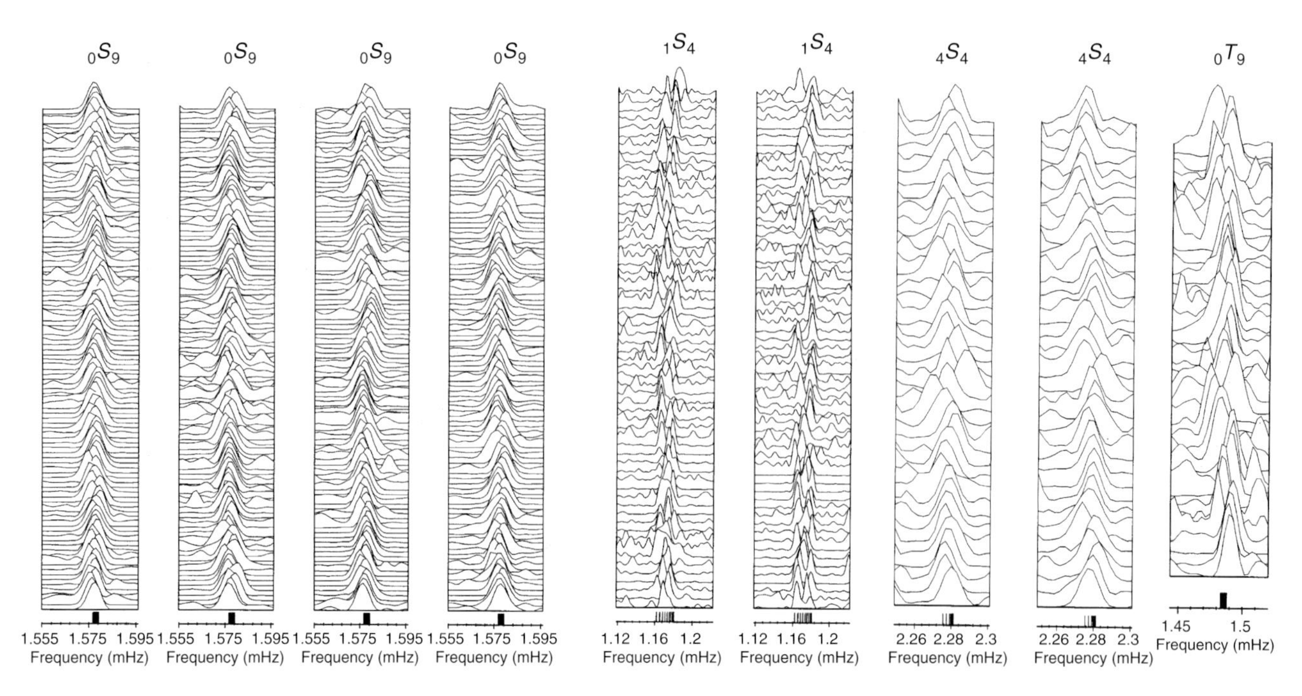

FIGURE 5 Selection of normal modes peaks for $_0S_9$, $_1S_4$, $_4S_4$, and $_0T_9$. The power spectra are obtained by a FFT performed on seismograms corrected from the drift and tides, and then tapered by an Hanning window. The selection was performed from a series of records of the IDA (Agnew *et al.*, 1976), GEOSCOPE (Romanowicz *et al.*, 1991), and IRIS networks, for earthquakes with M_s magnitudes greater than 7 (Nawab, 1993). Bars on the frequency axis indicate the frequencies of the singlets for a rotating elliptical spherical earth.

the free surface and all liquid–solid boundary conditions. Indeed, all perturbation theories or variational theories use as complete basis the eigenfunctions of the reference operator. With the latter, the eigenfunctions of the perturbed operator are computed, but only for the same volume as the spherical Earth and for the same set of discontinuities.

The first step is therefore to map the aspherical shape of the 3D Earth model into a spherical shape. Today, this is well described only for an ellipsoidal Earth. For such an ellipsoidal Earth, with a shape obtained with modern refinement of the Clairaut equation (Clairaut, 1743; Jeffreys, 1971; Bullen, 1973), each iso-density surface ($\rho(r')$ = constant value) in the real aspherical volume (where r' is the radius) is mapped to an iso-radius surface (r = constant value) in the mapped spherical volume. To first order, the mapped elastodynamic operator can be written as (Woodhouse and Dahlen, 1978; Woodhouse, 1980):

$$\mathbf{A}(r', \theta', \phi') = \mathbf{A}_0(r, \theta, \phi) + \delta\mathbf{A}_e(r, \theta, \phi) \tag{47}$$

where $\mathbf{A}_0$ is the spherical operator, and where the perturbation $\delta\mathbf{A}_e$, given by Woodhouse (1980), expresses the effect of shape and discontinuities asphericities. Note here that on the left side, r' is the real radius, and θ' and ϕ' are the geographical coordinates. On the right side, however, θ and ϕ are the geocentric coordinates, and r is the mapped radius. For an elliptical Earth, the geocentric and geographic longitudes are equal, while $0 < r < a$ and $\theta = \arctan\{[1 + 2\epsilon(r)]\tan\theta'\}$, where ϵ is the flattening coefficient. In all the following expressions, including the computation of seismograms, only geocentric coordinates and mapped radius will be used. The lateral variations in the shape of the free surface or internal discontinuities produce a perturbation, for which the first-order expression is given by Woodhouse (1980) and which will be noted hereafter $\delta\mathbf{A}_h(r, \theta, \phi)$. The generalization to higher orders of the geographic/geocentric mapping for such an aspherical Earth is nontrivial and has not yet been achieved.

With this mapping, the SNREI or SNRAI normal modes can be used as a complete basis of functions defined in the unperturbed volume $\mathcal{V}_0$. Any function verifying the boundary conditions in $\mathcal{V}_0$ can then be written as

$$|\mathbf{u}\rangle = \sum_{l,m,n} c_{l,m,n} |l, m, n\rangle \tag{48}$$

where $|l, m, n\rangle$ are the spherical eigenfunctions. This decomposition will be the basis of perturbation theories or variational methods for the resolution of the aspherical problem.

The last part of the perturbation is related to volumetric perturbations, such as lateral variations of elastic or anelastic structure. Finally, the operator can be written as

$$\begin{aligned}\mathbf{A}(r', \theta', \phi') &= \mathbf{A}_0(r, \theta, \phi) + \delta\mathbf{A}_e(r, \theta, \phi) + \delta\mathbf{A}_s(r, \theta, \phi) \\ &\quad + \delta\mathbf{A}_h(r, \theta, \phi) \\ &= \mathbf{A}_0(r, \theta, \phi) + \delta\mathbf{A}(r, \theta, \phi)\end{aligned} \tag{49}$$

where $\delta\mathbf{A}_h$ and $\delta\mathbf{A}_s$ are the perturbations of boundaries and volumetric lateral heterogeneities, respectively. Note here that the lateral heterogeneities are those after mapping, i.e., at their geocentric coordinates. The summation of the three terms is the perturbation of the operator $\mathbf{A}$. It is linear with respect to the structure and linearized for the interface and ellipticity perturbations.

The last perturbation is now related to the introduction of rotation, which was neglected in the SNRAI model. We end up with the expression of the eigenproblem of an aspherical, laterally heterogeneous, and anelastic Earth, which can be written in Hamiltonian form as

$$\sigma^2|\mathbf{u}\rangle = [\mathbf{H}_0 + \delta\mathbf{H}(\sigma)]|\mathbf{u}\rangle \quad (50)$$

Around a frequency σ_0, this can be rewritten with the following operators:

$$\begin{aligned} \mathbf{H}_0 &= \mathbf{A}_0(\sigma_0), \\ \delta\mathbf{H}(\sigma) &= -\sigma^2\delta\mathbf{K} + \sigma\mathbf{B} + \mathbf{A}_0(\sigma) - \mathbf{A}_0(\sigma_0) + \delta\mathbf{A}(\sigma) \end{aligned} \quad (51)$$

where we recognize in the perturbations the terms related to the density perturbation $\delta\mathbf{K} = \delta\rho$, to the Coriolis operator with perturbed density $\mathbf{B}$, to the physical dispersion of the spherical operator, and lastly to the perturbation in structure, shape, and boundary $\delta\mathbf{A}$.

8. Higher-Order Perturbation Theory

Let us now consider a singlet k and note its eigenfrequency σ_k. The associated primal eigenfunction $|\mathbf{u}_k\rangle$ and dual eigenfunction $|\mathbf{v}_k\rangle$ verify

$$\begin{aligned} \lambda_k|\mathbf{u}_k\rangle &= \mathbf{H}(\lambda_k)|\mathbf{u}_k\rangle \\ \lambda_k|\mathbf{v}_k\rangle &= \hat{\mathbf{H}}(\lambda_k)|\mathbf{v}_k\rangle \end{aligned} \quad (52)$$

where $\lambda_k = \sigma_k^2$ is the squared frequency and $\hat{\mathbf{H}}$ is the dual operator with reversed rotation. They have to follow the bi-orthogonality relation (35) and to be normalized with (36). All the singlets of the multiplet K characterized by l and n in the reference model will define a subspace of the spectrum noted S_K and with a $2l + 1$ dimension. In the nonrotating, anelastic case, $\mathbf{H}$ is a self-dual operator and the dual and primal eigenmodes are identical for the real scalar product. Let us now concentrate on the primal eigenfunction (all dual expressions might be obtained by substituting $\mathbf{H}$ by $\hat{\mathbf{H}}$ and the primal modes by their duals, the eigenfrequencies of the dual and primal modes, as well as all perturbations of eigenfrequencies being identical). Assuming that all eigenmodes and eigenfrequencies can be expanded in terms of a power series of a small parameter ϵ, related to the perturbation $\delta\mathbf{H}$, we can thus write

$$\lambda_k = \sigma_k^2 = \lambda_K + \delta_1\lambda_k + \delta_2\lambda_k + \cdots \quad (53)$$

$$|\mathbf{u}_k\rangle = |\mathbf{u}_k^{(0)}\rangle + |\mathbf{u}_k^{(1)}\rangle + |\mathbf{u}_k^{(2)}\rangle + \cdots \quad (54)$$

where $\lambda_k = \sigma_k^2$ is the squared frequency of the spherical normal mode in the reference operator, $\delta_n\lambda_k$ and $|\mathbf{u}_k^{(n)}\rangle$ are the nth-order perturbations in the squared eigenfrequency and eigenmode of the singlet k, respectively. All the eigenmode perturbations will be described in the SNRAI/SNREI basis by

$$|\mathbf{u}_k^{(p)}\rangle = \sum_j c_{n,l,m}^p |n, l, m\rangle \quad (55)$$

Following Lognonné (1991), we leave the perturbation path open, by decomposing it into several incremental perturbations, and introduce a renormalization operator. This gives

$$\begin{aligned} \delta\mathbf{H}(\lambda) &= \delta\mathbf{H}(\lambda_K) + (\lambda - \lambda_K)\frac{\partial\delta\mathbf{H}}{\partial\lambda}(\lambda_K) + \cdots \\ &= \delta_1\mathbf{H} + \delta_2\mathbf{H} + \cdots \end{aligned} \quad (56)$$

where

$$\begin{aligned} \delta\mathbf{H}(\lambda_K) &= -\sigma_K^2\,\delta\mathbf{K} + \sigma_K\mathbf{B} + \delta\mathbf{A}(\sigma_K), \\ \frac{\partial\delta\mathbf{H}}{\partial\lambda}(\lambda_K) &= -2\sigma_K\delta K + [\mathbf{B} + \partial_\sigma\mathbf{A}(\sigma_K)] \end{aligned}$$

and where $\delta_1\mathbf{H}$, $\delta_2\mathbf{H}$ are the succession of perturbation paths of increasing orders. Note the Coriolis operator in the coupling terms, which is producing strong coupling between toroidal and spheroidal modes at long period, as shown by Masters *et al.* (1983).

For the sake of simplicity, we assume here that the perturbation of the bi-orthonormality relation is weaker than the Hamiltonian, and can be written as

$$1 = \mathbf{N}_0 + \delta_2\mathbf{N} + \delta_3\mathbf{N} + \cdots \quad (57)$$

where the renormalization operator $\mathbf{N}_0$ incorporates potential first-order perturbations in the subspace $\mathscr{S}_K$, but with all coupling perturbations in $\delta_2\mathbf{N}$. The general case is shown by Lognonné (1991), and here we can assume that the second-order perturbation operators $\delta_2\mathbf{H}$ and third-order one $\delta_3\mathbf{N}$ and higher orders can be assumed nonzero only in the subspace $\mathscr{S}_K$. Note that a slightly different method, valid to the second order, was developed using subspace projection (e.g., Dahlen 1987; Park, 1990b).

The only constraint imposed by Eqs. (56)–(57) at this stage is to have a sum equal to the real perturbation. We will see later that this freedom is crucial for the efficiency of the perturbation procedure. Let us indeed solve the equation of normal modes by developing the perturbation terms by increasing order and focus on the first- and second-order perturbations. Starting from

$$\lambda_k|\mathbf{u}_k\rangle = [\mathbf{H}_0 + \delta\mathbf{H}]|\mathbf{u}_k\rangle \tag{58}$$

we get, after development, the following expression

$$\begin{aligned} 0 &= [\lambda_k - \mathbf{H}_0 - \delta\mathbf{H}]|\mathbf{u}_k\rangle \\ &= [\lambda_K - \mathbf{H}_0]|\mathbf{u}_k^{(0)}\rangle \\ &\quad + [\delta_1\lambda_k\mathbf{N}_0 - \delta_1\mathbf{H}]|\mathbf{u}_k^{(0)}\rangle + [\lambda_K - \mathbf{H}_0]|\mathbf{u}_k^{(1)}\rangle \\ &\quad + [\delta_2\lambda_k\mathbf{N}_0 - \delta_2\mathbf{H}]|\mathbf{u}_k^{(0)}\rangle + [\delta_1\lambda_k\mathbf{N}_0 - \delta_1\mathbf{H}]|\mathbf{u}_k^{(1)}\rangle \\ &\quad + [\lambda_K - \mathbf{H}_0]|\mathbf{u}_k^{(2)}\rangle + \cdots \end{aligned} \tag{59}$$

where we have used the renormalization freedom corresponding to operator $\mathbf{N}_0$. The first line must be equal to zero, as well as all other lines. For the first line, we therefore have

$$\lambda_K|\mathbf{u}_k^{(0)}\rangle = \mathbf{H}_0|\mathbf{u}_k^{(0)}\rangle \tag{60}$$

which shows that the zeroth-order perturbation $|\mathbf{u}_k^{(0)}\rangle$ belongs to $\mathscr{S}_K$, the subspace mapped by the spherical eigenmodes. Let us now first project Eq. (59) into the space $\mathscr{S}_K$, which is the kernel of the operator $[\lambda_K - \mathbf{H}_0]$. We then get from the second line,

$$\delta_1\lambda_k\mathbf{N}_0|\mathbf{u}_k^{(0)}\rangle = \mathscr{P}_K\delta_1\mathbf{H}|\mathbf{u}_k^{(0)}\rangle \tag{61}$$

where $\mathscr{P}_K$ is the projection into $\mathscr{S}_K$. Relation (61) shows that the first-order perturbation in frequency and the zeroth-order in eigenmodes are the eigensolutions of a $2l+1$ dimension eigenproblem. The same applies for the dual mode, therefore:

$$\delta_1\lambda_k\mathbf{N}_0|\mathbf{v}_k^{(0)}\rangle = \mathscr{P}_K\delta_1\widehat{\mathbf{H}}|\mathbf{v}_k^{(0)}\rangle \tag{62}$$

By multiplying the third line of Eq. (59) by $\langle\mathbf{v}_k^{(0)}|$ and using (61), we cancel the last term and therefore obtain the second-order frequency perturbation of the singlet k:

$$\begin{aligned} \delta_2\lambda_k &= \frac{\langle\mathbf{v}_k^{(0)}|\delta_2\mathbf{H}\mathbf{u}_k^{(0)}\rangle + \langle\mathbf{v}_k^{(0)}|\delta_1\mathbf{H}[1-\mathscr{P}_K]\mathbf{u}_k^{(1)}\rangle}{\langle\mathbf{v}_k^{(0)}|\mathbf{N}_0\mathbf{u}_k^{(0)}\rangle} \\ &= \langle\mathbf{v}_k^{(0)}|\delta_2\mathbf{H}\mathbf{u}_k^{(0)}\rangle + \langle\mathbf{v}_k^{(0)}|\delta_1\mathbf{H}[1-\mathscr{P}_K]\mathbf{u}_k^{(1)}\rangle \end{aligned} \tag{63}$$

where we have used the normalization of the zeroth-order singlets, $\langle\mathbf{v}_{k\prime}^{(0)}||\mathbf{N}_0\mathbf{u}_k\rangle$. Let us now multiply Eq. (59) by the projector into a subspace $\mathscr{S}_{K'}$, associated with a different multiplet. We then get the expression of the perturbations in eigenmodes:

$$\mathscr{P}_{K'}|\mathbf{u}_k^{(1)}\rangle = \frac{1}{\lambda_K - \lambda_{K'}}\mathscr{P}_{K'}\delta_1\mathbf{H}|\mathbf{u}_k^{(0)}\rangle \tag{64}$$

$$\begin{aligned} \mathscr{P}_{K'}|\mathbf{u}_k^{(2)}\rangle &= \frac{1}{\lambda_K - \lambda_{K'}}[-\mathscr{P}_{K'}[\delta_1\lambda_k\mathbf{N}_0 - \delta_1\mathbf{H}]|\mathbf{u}_k^{(1)}\rangle \\ &\quad + \mathscr{P}_{K'}\delta_2\mathbf{H}|\mathbf{u}_k^{(0)}\rangle + \mathscr{P}_{K'}\delta_1\mathbf{H}|\mathbf{u}_k^{(1)}\rangle] \end{aligned} \tag{65}$$

The last missing terms are the projection of the eigenmode perturbations inside the perturbed space $\mathscr{S}_K$. As shown by Lognonné (1991), this term appears in the cancellation of the bracket product of (59) with a singlet k' inside the same multiplet as singlet k, which gives

$$\begin{aligned} 0 &= \langle\mathbf{v}_{k'}|[\lambda_k - \mathbf{H}]|\mathbf{u}_k\rangle \\ &= \langle\mathbf{v}_{k'}^{(0)}|[\lambda_K - \mathbf{H}_0]|\mathbf{u}_k^{(0)}\rangle + \mathbf{v}\langle_{k'}^{(1)}|[\lambda_K - \mathbf{H}_0]|\mathbf{u}_k^{(0)}\rangle \\ &\quad + \langle\mathbf{v}_{k'}^{(0)}|[\lambda_K - \mathbf{H}_0]|\mathbf{u}_k^{(1)}\rangle + \langle\mathbf{v}_{k'}^{(0)}|[\delta_1\lambda_k\mathbf{N}_0 - \delta_1\mathbf{H}]|\mathbf{u}_k^{(0)}\rangle \\ &\quad + \mathbf{v}\langle_{k'}^{(1)}|[\lambda_K - \mathbf{H}_0]|\mathbf{u}_k^{(1)}\rangle + \langle\mathbf{v}_{k'}^{(1)}|[\delta_1\lambda_k\mathbf{N}_0 - \delta_1\mathbf{H}]|\mathbf{u}_k^{(0)}\rangle \\ &\quad + \mathbf{v}\langle_{k'}^{(0)}|[\delta_1\lambda_k\mathbf{N}_0 - \delta_1\mathbf{H}]|\mathbf{u}_k^{(1)}\rangle - \langle\mathbf{v}_{k'}^{(0)}|\delta_2\mathbf{H}|\mathbf{u}_k^{(0)}\rangle \end{aligned} \tag{66}$$

Using the orthogonality between the zeroth-order singlets $|\mathbf{u}_k^{(0)}\rangle$ and decomposing each perturbation into its parts inside and perpendicular to the subspace $\mathscr{S}_K$,

$$|\mathbf{u}_k^{(1)}\rangle = \mathscr{P}_K|\mathbf{u}_k^{(1)}\rangle + \sum_{K'\neq K}\mathscr{P}_{K'}|\mathbf{u}_k^{(1)}\rangle \tag{67}$$

we then rewrite Eq. (61), and with the help of relations (59), (61), (62), and (63), respectively, we obtain

$$\begin{aligned} 0 &= \langle\mathbf{v}_{k'}|[\lambda_k - \mathbf{H}]|\mathbf{u}_k\rangle \\ &= 0+0+0+0+\langle\mathbf{v}_{k'}^{(0)}|\delta_1\mathbf{H}\,\boldsymbol{\Delta}\delta_1\mathbf{H}|\mathbf{u}_k^{(0)}\rangle - \langle\mathbf{v}_{k'}^{(0)}|\delta_1\mathbf{H}\,\boldsymbol{\Delta}\delta_1\mathbf{H}\mathbf{u}_k^{(0)}\rangle \\ &\quad + \left[(\delta_1\lambda_k - \delta_1\lambda_{k'})\langle\mathbf{v}_{k'}^{(0)}|\mathbf{u}_k^{(1)}\rangle - \langle\mathbf{v}_{k'}^{(0)}|\delta_1\mathbf{H}\,\boldsymbol{\Delta}\delta_1\mathbf{H}\mathbf{u}_k^{(0)}\rangle\right] \\ &\quad - \langle\mathbf{v}_{k'}^{(0)}|\delta_2\mathbf{H}\mathbf{u}_k^{(0)}\rangle \end{aligned} \tag{68}$$

where

$$\boldsymbol{\Delta} = \sum_{K'\neq K}\frac{1}{\lambda_K - \lambda_{K'}}\mathscr{P}_{K'}$$

This finally gives a first expression for the first-order perturbation inside the multiplet K:

$$\begin{aligned} &\langle|\mathbf{v}_{k'}^{(0)}|\mathbf{u}_k^{(1)}\rangle \\ &= \frac{1}{\delta_1\lambda_k - \delta_1\lambda_{k'}}\left[\langle\mathbf{v}_{k'}^{(0)}|\delta_1\mathbf{H}\,\boldsymbol{\Delta}\delta_1\mathbf{H}\mathbf{u}_k^{(0)}\rangle + \langle\mathbf{v}_{k'}^{(0)}|\delta_2\mathbf{H}\mathbf{u}_k^{(0)}\rangle\right] \end{aligned} \tag{69}$$

The difference in frequency between two singlets can be as small as a few micro-Hertz and expression (69) therefore leads to very large amplitudes for the first-order perturbation of the eigenmode inside the multiplet K. A similar problem can be found in quantum mechanics (Landau and Lifschitz, 1965).

This is a source of numerical errors and leads practically to the divergence of the perturbation sequence. A perturbation path such that the first-order perturbation of the eigenmode is orthogonal to the subspace K must therefore be chosen, which implies cancellation of the right part of (64). As shown by Lognonné and Romanowicz (1990a), the perturbation theory can then be practically iterated if the iterations are orthogonal to the subspace K. The zeroth-order eigenmode $|\mathbf{u}_k^{(0)}\rangle$ therefore contains all the perturbations in the subspace $\mathscr{S}_K$. This implies, from relations (56) and (69) choosing the perturbation path such that

$$\begin{aligned} \langle \mathbf{u}_{k'}^{(0)}|[\delta_2\mathbf{H} + \delta_1\mathbf{H}\,\boldsymbol{\Delta}\delta_1\mathbf{H}]|\mathbf{u}_k^{(0)}\rangle = 0 \\ \delta\mathbf{H}(\lambda_K) + \delta_1\lambda_k\frac{\partial\delta\mathbf{H}}{\partial\lambda}(\lambda_K) - \delta_1\mathbf{H} - \delta_2\mathbf{H} = 0 \end{aligned} \tag{70}$$

Let us now take the orthogonality relation (35). We can develop this relation with the perturbation sequence, and get, to the first order,

$$-2\sigma_K\langle \mathbf{v}_{k'}^{(0)}|\frac{\partial\mathbf{H}}{\partial\lambda}(\lambda_K)|\mathbf{u}_k^{(0)}\rangle + \langle \mathbf{v}_{k'}^{(1)}|\mathbf{u}_k^{(0)}\rangle + \langle \mathbf{v}_{k'}^{(0)}|\mathbf{u}_k^{(1)}\rangle + \cdots \tag{71}$$

This allows us to get the first-order path $\mathbf{N}_0$ and $\delta_1\mathbf{H}$ of the sequence of perturbation operators, namely:

$$\begin{aligned} \mathbf{N}_0 &= -\mathscr{P}_K\frac{\partial\mathbf{H}}{\partial\lambda}\mathscr{P}_K \\ \delta_1\mathbf{H} &= \delta\mathbf{H}(\lambda_K) + \mathscr{P}_K\,\delta\mathbf{H}\,\boldsymbol{\Delta}\delta\mathbf{H}\,\mathscr{P}_K \\ \delta_2\mathbf{H} &= -\mathscr{P}_K\,\delta\mathbf{H}\,\boldsymbol{\Delta}\delta\mathbf{H}\,\mathscr{P}_K \end{aligned}$$

where we have omitted the dispersion correction. Lognonné (1991) gives the expression of the second- and third-order perturbation paths. The final expressions for the perturbations are then summarized in Table 7, with detailed expressions from Lognonné (1991). It is noteworthy that the second-order perturbation in the squared frequency cancels, so that all the frequency perturbation is included in the isolated multiplet eigenvalue up to second order.

9. Kernels and Computation of Splitting Matrices

All previous expressions need the computation of the coupling matrices expressing the coupling between two singlets of the reference model. As an example, this is the case of the first-order perturbation, which can be rewritten as

$$\begin{aligned} |\mathbf{u}_k^{(1)}\rangle &= \boldsymbol{\Delta}\delta\mathbf{H}|\mathbf{u}_k^{(0)}\rangle \\ &= \sum_{K'\neq K}\frac{1}{\lambda_K-\lambda_{K'}}\sum_{m',m}|l',m',n'\rangle\langle l',m',n'|\delta\mathbf{H}|l,m,n\rangle\langle l,m,n|\mathbf{u}_k^{(0)}\rangle \end{aligned} \tag{72}$$

where the multiplets $K=(n,\,l)$ and $K'=(n',\,l')$ have dimensions $2l'+1$ and $2l+1$ and where $|\,l,\,m,\,n\rangle$ are the SNREI or SNRAI singlets, associated with a symmetric eigenproduct and with self-dual modes. The coupling matrix between the multiplets K and K' has dimension $(2l'+1)\times(2l+1)$ and can be divided into three parts, namely, for the coupling terms between two singlets $k=(n,\,l,\,m)$ and $k'=(n',\,l',\,m')$:

$$\begin{aligned} \langle l',m',n'|\delta\mathbf{H}|l,m,n\rangle &= H^{kk'} \\ &= H^{kk'(\mathrm{ell})} + H^{kk'(\mathrm{rot})} + H^{kk'(\mathrm{3D})} \end{aligned} \tag{73}$$

where $H^{kk'(\mathrm{ell})}$, $H^{kk'(\mathrm{rot})}$ and $H^{kk'(\mathrm{3D})}$ are respectively the ellipticity, rotation, and lateral variation contributions. Assuming a rotating Earth in hydrostatic equilibrium, the coupling terms $[H^{kk'(\mathrm{ell})}+H^{kk'(\mathrm{rot})}]$ are nonzero only for $m'=m$ and can be expressed as a second-order polynomial of m (Dahlen, 1968).

TABLE 7 Zero-, first-, and second-order expressions of the perturbations in eigenmode and eigenfrequency. The equivalent frequency perturbation, to the second order is given by $\delta\sigma_k = \dfrac{\delta_1\lambda_k}{2\sigma_K} - \dfrac{\delta_1\lambda_k^2}{8\sigma_K^3}$

	Eigenfrequency and Direction of Perturbation in K
Eigenproblem	$\delta_1\lambda_k\mathbf{N}_0\lvert\mathbf{u}_k^{(0)}\rangle = \mathscr{P}\delta_1\mathbf{H}\lvert\mathbf{u}_k^{(0)}\rangle$
Eigenoperators	$\mathscr{P}\delta_1\mathbf{H}\mathscr{P} = \mathscr{P}[\delta\mathbf{H} + \delta\mathbf{H}\,\boldsymbol{\Delta}\delta\mathbf{H} + \delta\mathbf{H}\,\boldsymbol{\Delta}\delta\mathbf{H}\,\boldsymbol{\Delta}\delta\mathbf{H}]\mathscr{P}$
	$\mathbf{N}_0 = \mathscr{P}\left[\mathbf{I} - \dfrac{\mathbf{B}+\partial_\sigma\mathbf{A}(\sigma_K)}{2\sigma_K} + \delta\mathbf{K} + \delta\mathbf{H}\boldsymbol{\Delta}^2\delta\mathbf{H} + \delta\mathbf{H}\boldsymbol{\Delta}\left[\delta\mathbf{K} - \dfrac{\mathbf{B}+\partial_\sigma\mathbf{A}(\sigma_K)}{2\sigma_K}\right] + \left[\delta\mathbf{K} - \dfrac{\mathbf{B}+\partial_\sigma\mathbf{A}(\sigma_K)}{2\sigma_K}\right]\boldsymbol{\Delta}\delta\mathbf{H}\right]\mathscr{P}$
	$\delta_2\lambda_k = 0$
	Coupling Terms
First-order	$\lvert\mathbf{u}_k^{(1)}\rangle = \boldsymbol{\Delta}\delta\mathbf{H}\lvert\mathbf{u}_k^{(0)}\rangle$
Second-order	$\lvert\mathbf{u}_k^{(2)}\rangle = \dfrac{\delta_1\lambda_k}{2\sigma_K}\boldsymbol{\Delta}[\mathbf{B} + \partial_\sigma\mathbf{A}(\sigma_K) - 2\sigma_K\delta\mathbf{K} - 2\sigma_K\mathbf{K}_0\boldsymbol{\Delta}\delta\mathbf{H}]\lvert\mathbf{u}_k^{(0)}\rangle + \boldsymbol{\Delta}\delta\mathbf{H}\,\boldsymbol{\Delta}\delta\mathbf{H}\lvert\mathbf{u}_k^{(0)}\rangle$

The self-coupling part can be expressed as

$$H^{kk\,(\mathrm{ell})} + H^{kk\,(\mathrm{rot})} = a + bm + cm^2 \tag{74}$$

where a, b, c are given by (Dahlen and Smith, 1975). The last term includes the effect of the departure from sphericity in the structure,

$$H^{kk'\,(3\mathrm{D})} = \langle l', m', n' | \delta \mathbf{A} | l, m, n \rangle \tag{75}$$

This matrix is a linear expression of the structure and may be linearized with respect to the topography of boundaries. If volumetric perturbations are described in term of complex spherical harmonics,

$$\delta p(r,\theta,\phi) = \sum_{s,t} \delta p_s^t(r) Y_s^t(\theta,\phi) \tag{76}$$

where p can be the density ρ, the bulk modulus κ, or the shear modulus μ, and if the same is applied for the topography of boundaries,

$$\delta h_i(\theta,\phi) = \sum_{s,t} \delta h_{i\,s}^{t} Y_s^t(\theta,\phi) \tag{77}$$

where $\delta h_i(\theta,\phi)$ is the topography of the ith boundary at radius r_i, the matrix can therefore be developed from the separation of variables as

$$H^{kk'} = \sum_s {}_s^t\gamma_{ll'}^{mm'}\; {}_{KK'}c_s^t \tag{78}$$

where

$${}_s^t\gamma_{ll'}^{mm'} = \int dS\, \bar{Y}_{l'}^{m'} Y_s^t Y_l^m \tag{79}$$

and

$$\begin{aligned} {}_{KK'}c_s^t = \int_0 & \left[{}_{KK'}R_s(r)\delta\rho_s^t(r) + {}_{KK'}K_s(r)\delta\kappa_s^t(r) \right. \\ & \left. + {}_{KK'}M_s(r)\delta\mu_s^t(r) \right] r^2\, dr - \sum_i r_i^2 h_{i\,s\,KK'}^t B_s(r_i) \end{aligned} \tag{80}$$

where the kernels ${}_{KK'}R_s$, ${}_{KK'}K_s$, ${}_{KK'}M_s$ and ${}_{KK'}B_s$ are functions of the radially dependent parts of the eigenfunctions and are given by Woodhouse and Dahlen (1978). ${}_{KK'}c_s^t$ are the so-called structure coefficients and are associated with the vertical integration. $\gamma_{ll'}^{mm'}$ are related to the integration on the sphere of the product between spherical harmonics and are subject to selection rules. In the isolated multiplet case, the expression for ${}_s^t\gamma_{ll}^{mm'}$ is

$${}_s^t\gamma_{ll}^{mm'} = (-1)^m (2l+1) \left[\frac{2s+1}{4\pi} \right]^{\frac{1}{2}} \begin{vmatrix} l & s & l \\ -m & t & m' \end{vmatrix} \begin{vmatrix} l & s & l \\ 0 & 0 & 0 \end{vmatrix} \tag{81}$$

where the last two terms are the Wigner $3-j$ symbols (Edmonds, 1960).

The splitting matrix in Eq. (75) can also be computed by a direct integration, as shown by Lognonné and Romanowicz (1990a). We recall here this method can be used for either smooth or rough (Park, 1989, 1990a; Lognonné and Romanowicz, 1990b) models of lateral variations and does not require a development of the structure in spherical harmonics. The lateral variations related to variations in the stiffness tensor correspond to the following splitting matrix:

$$H^{kk'} = \int dS \int r^2\, dr\, D^\alpha \bar{u}_{k'}^\beta\, e_{\alpha,\alpha'} e_{\beta,\beta'}\, \delta C^{\alpha'\beta'\gamma'\delta'}\, e_{\gamma,\gamma'} e_{\delta,\delta'}\, D^\gamma u_k^\delta \tag{82}$$

where $\delta C^{\alpha'\beta'\gamma'\delta'}$ is the anisotropic perturbation in stiffness tensor. When SNREI modes are used for the description of the splitting matrices, expression (82) can be rewritten as

$$H^{kk'} = \sum_{N,N'} \int dS\, \bar{Y}_{l'}^{N'm'} \mathcal{N}_{NN'}^{KK'} Y_l^{Nm} \tag{83}$$

where here the kernels are functions of latitude and longitude and are given by Table 8. Note that the software on the attached Handbook CD uses the spectral method described above.

TABLE 8 Expression of the Kernels for expression (83). $E^{\alpha,\beta}$ are the strain tensor component and $C^{\alpha\beta\gamma\delta}$ those of the stiffness tensor. See Phinney and Burridge (1973) and Mochizuki (1986) for the detailed relations. The software included in the attached Handbook CD was developed in this framework, and allows computation of the splitting matrix for any anisotropic and anelastic Earth model

$\mathcal{N}_{0,0}$	$\int r^2\, dr\, (\bar{E}_K^{00} E_{K'}^{00} C^{0000} + 2\bar{E}_K^{00} E_{K'}^{+-} C^{00+-} + 2\bar{E}_K^{+-} E_{K'}^{00} C^{+-00} + 4\bar{E}_K^{+-} E_{K'}^{+-} C^{+-+-})$
$\mathcal{N}_{0,\pm}$	$-\int r^2\, dr (2\bar{E}_K^{00} E_{K'}^{\pm 0} C^{00\mp 0} + 4\bar{E}_K^{+-} E_{K'}^{\pm 0} C^{+-\mp 0})$
$\mathcal{N}_{0,\pm 2}$	$\int r^2\, dr (\bar{E}_K^{00} E_{K'}^{\pm\pm} C^{00\mp\mp} + 2\bar{E}_K^{+-} E_{K'}^{\pm\pm} C^{+-\mp\mp})$
$\mathcal{N}_{\pm,0}$	$\int r^2\, dr (2\bar{E}_K^{\pm 0} E_{K'}^{00} C^{0\pm 00} + 4\bar{E}_K^{-0} E_{K'}^{+-} C^{0\pm +-})$
$\mathcal{N}_{-1,\pm}$	$-\int r^2\, dr (4\bar{E}_K^{-0} E_{K'}^{\pm 0} C^{0-0\mp})$
$\mathcal{N}_{-1,\pm 2}$	$\int r^2\, dr (2\bar{E}_K^{-0} E_{K'}^{\pm\pm} C^{0-\mp\mp})$
$\mathcal{N}_{1,\pm}$	$-\int r^2\, dr (4\bar{E}_K^{+0} E_{K'}^{\pm 0} C^{0+0\mp})$
$\mathcal{N}_{1,\pm 2}$	$\int r^2\, dr (2\bar{E}_K^{+0} E_{K'}^{\pm\pm} C^{0+\mp\mp})$
$\mathcal{N}_{\pm 2,0}$	$\int r^2\, dr (\bar{E}_K^{\pm\pm} E_{K'}^{00} C^{\pm\pm 00} + 2\bar{E}_K^{\pm\pm} E_K^{+-} C^{\pm\pm +-})$
$\mathcal{N}_{-2,\pm 1}$	$-\int r^2\, dr (2\bar{E}_K^{--} E_{K'}^{\pm 0} C^{--0\mp})$
$\mathcal{N}_{-2,\pm 2}$	$\int r^2\, dr (\bar{E}_K^{--} E_{K'}^{\pm\pm} C^{--\mp\mp})$
$\mathcal{N}_{2,\pm}$	$-\int r^2\, dr (2\bar{E}_l^{++} E_{K'}^{\pm 0} C^{++0\mp})$
$\mathcal{N}_{2,\pm 2}$	$\int r^2\, dr (\bar{E}_K^{++} E_{K'}^{\pm\pm} C^{++\mp\mp})$

10. Lateral Variation: Normal Mode Synthetics, Observables and Seismograms

The first effect of lateral variations is to split the degenerated eigenfrequencies. The splitting for the model SAW12D of Li and Romanowicz (1996) is shown in Figure 6 for the spheroidal and toroidal fundamental branches. The splitting of normal modes increases with the angular order, as the consequence of both the increased sensitivity of modes to the shallow structure and the stronger lateral variations in the shallow structures. A shift of the mean frequencies of the singlets is also observed, jointly related to a departure of the reference model (here PREM) from the mean Earth structure and to the nonlinear effect related to lateral variations.

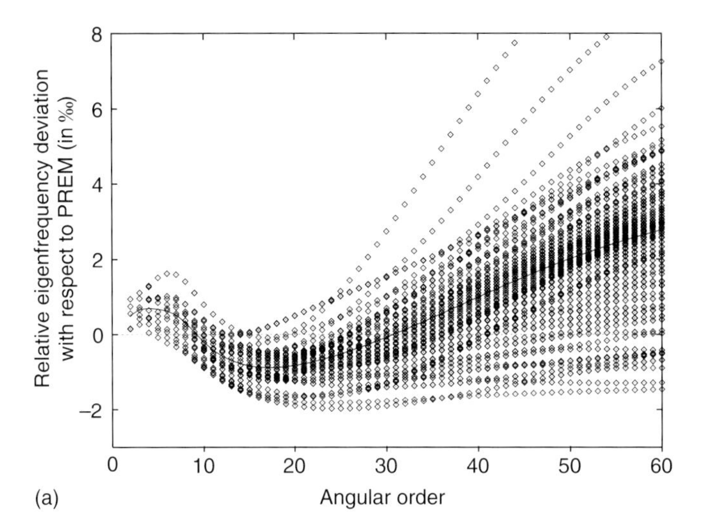

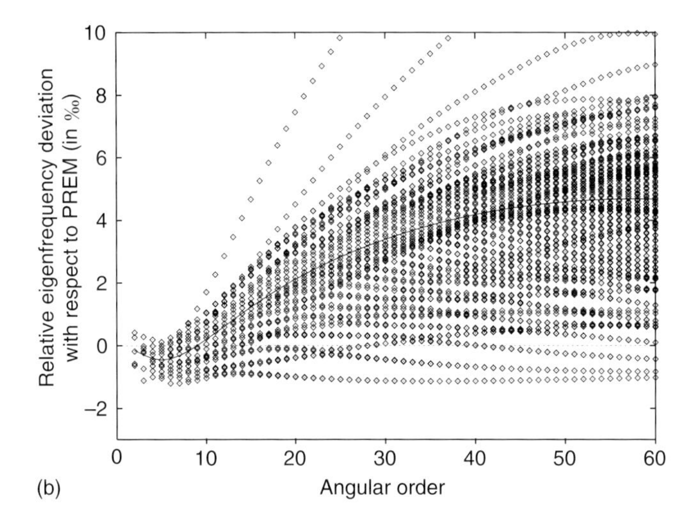

FIGURE 6 Splitting of the fundamental spheroidal modes (a), and toroidal modes (b), for the model SAW12D of Li and Romanowicz (1996). The continuous line corresponds to the mean frequency, and its deviation from the reference PREM frequency increases with increasing angular order.

The amplitude effects are depicted in Color Plate 2 (upper part) for the isolated case and for two singlets of the fundamental mode ${}_0S_{60}$. The effect of lateral variations indeed changes the pattern of the singlets significantly, as compared to the delocalized spherical harmonics. In contrast, the amplitude of the singlets in a 3D Earth is localized along a great circle for which two singlets are associated. They have almost identical frequencies, and the geographical patterns of the first are mainly shifted from the other along the great circle. This pattern corresponds to a closed and oscillating band along a great circle. Dahlen and Henson (1985) and Henson and Dahlen (1986) have shown that all these trajectories correspond to $2l+1$ local extrema of the phase perturbation along a closed path on the sphere. The higher-order perturbations of the amplitude of eigenmodes produce focusing and defocusing effects along the great circle associated to the singlet, as well as slight departure from the closed shape of the isolated case band (Color plate 2, lower part). In the asymptotic case, this corresponds to a nonclosed trajectory. For normal modes, these effects are the counterpart of the phase delay of the surface wave packets, the focusing/defocusing effects of the surface wave packets, and the off-path departure of polarization.

Let us now focus on how to observe these effects of lateral variations on normal modes, and how to proceed with data inversion for the retrieval of lateral variations of the Earth. The inversion methods of normal modes are separated into two main families.

The first inversion method consists in the extraction from the data of a set of secondary data corresponding to each multiplet, and then of the inversion of these secondary data for the retrieval of the structure. Let us first illustrate this method, by considering only one multiplet in a perfectly bandpassed seismogram around the frequency σ_K:

$$s(t) = \sum_{m=-l,+l} \langle \mathbf{R}|\mathbf{u}_m\rangle \, e^{i\sigma_m t} \, \langle \mathbf{v}_m|\mathbf{S}\rangle \tag{84}$$

where $|\mathbf{u}_m\rangle$, $\langle \mathbf{v}_m|$, and σ_m are the primal, dual amplitudes and eigenfrequency for the singlet m, respectively. Note here that m no longer corresponds to the number of nodes in longitude but can be used as a ranking of the frequency.

In the frequency domain, each of the singlets corresponds to a spectral peak with a minimum width at half power amplitude of ω_K/Q where Q is the quality coefficient of the modes. For the high-Q, very low-order modes, the first and most powerful method is the direct measurement of the frequency of all singlets. For these modes, indeed, the frequency spacing between the singlets is large enough for its resolution, and several peaks can be observed on the data. Several powerful methods were developed for this purpose (Lindberg and Park, 1987; Park *et al.*, 1987). For modes with a higher angular order, Q is lower and the relative frequency spacing between two successive singlets is also reduced: the singlets become unresolved. The contribution of all the singlets from a multiplet n, l can, however, be expressed by using the mean frequency of the multiplet as a carrier:

$$\sum_{m=-l,+l} \langle \mathbf{R}|\mathbf{u}_m\rangle \, e^{i\sigma_m t} \, \langle \mathbf{v}_m|\mathbf{S}\rangle = A(t) e^{i\sigma_K t} \tag{85}$$

where $A(t)$ is a slow amplitude modulation function crystallizing all the 3D perturbations and defined as

$$A(t) = \sum_{m=-l}^{m=l} \langle \mathbf{R}|\mathbf{u}_m\rangle \, e^{i\delta\sigma_m t} \, \langle \mathbf{v}_m|\mathbf{S}\rangle \tag{86}$$

For a spherical Earth, this function is a constant, and for the short-time approximation it can be expressed as

$$\begin{aligned} A(t) &= \left[\sum_{m=-l}^{m=l} u_m + it \sum_{m=-l}^{m=l} u_m \, \delta\sigma_m - \frac{t^2}{2} \sum_{m=-l}^{m=l} u_m \, \delta\sigma^2 \right] \\ &= u_0 \, e^{i\delta\sigma t} \, e^{-\beta t^2/2} \end{aligned} \tag{87}$$

where $u_m = \langle \mathbf{R}|\mathbf{u}_m\rangle \langle \mathbf{v}_m|\mathbf{S}\rangle$ and

$$u_0 = \sum_{m=-l}^{m=l} u_m, \qquad \delta\sigma = \frac{1}{u_0} \sum_{m=-l}^{m=l} u_m \, \delta\sigma_m, \qquad \beta = \frac{1}{u_0} \sum_{m=-l}^{m=l} u_m (\delta\sigma_k^2 - \delta\sigma^2) \tag{88}$$

u_0 is the amplitude of the normal mode at the initial time and $\delta\sigma$ is called the local frequency of the mode (Jordan, 1978; Dahlen, 1979). β corresponds to a scattering constant. This shows that the effect of lateral heterogeneities is complex, and consists of

(i) a perturbation of the initial amplitude and a slow change of the latter with time, and
(ii) a perturbation of the mean frequency of the peak, which will depend on the excitation of the singlets and therefore the source and receiver.

The two effects above are associated with the main secondary data easily extracted from the normal mode peaks, namely, their initial amplitude and mean frequency. These secondary data can also be computed more directly by asymptotic theory (Jordan, 1978; Romanowicz and Roult, 1986; Park, 1987; Pollitz, 1990). For the fundamental mode, the statistical distribution of the frequency of resonant peaks typically shows a splitting of several per mile (0.1–0.5%). The local frequency depends on the position of the source and receiver, as well as on the source mechanism. In the isolated multiplet hypothesis and asymptotic limit, however, it is equal to the integrated frequency shift along the great circle, and therefore gives information on the shape of lateral variations (Fig. 7). The second-order effects are the exponential decay of the mode and the time variation of the amplitude associated with β, which immediately shows the difficulty inherent in any attenuation study.

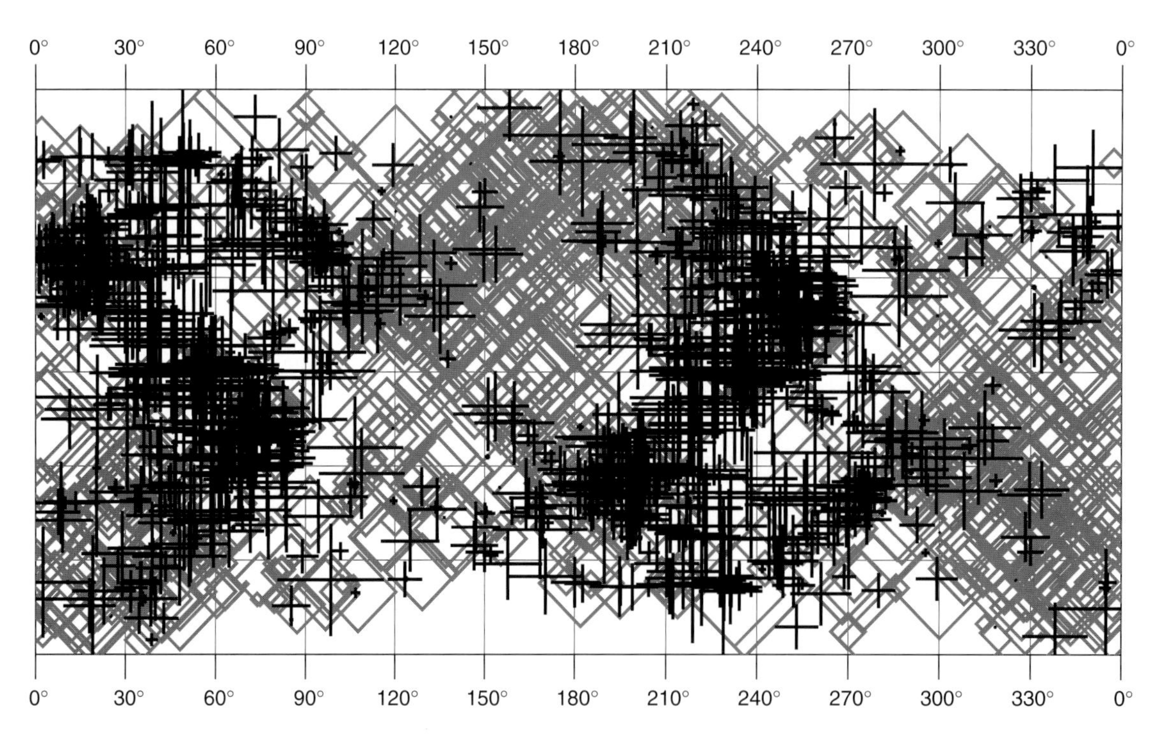

FIGURE 7 Map of the local frequency for the mode ${}_0S_{25}$ (Roult and Clévédé, 2000). Relative deviations of the frequencies (%): ◇, $d\omega/\omega = -0.5\%$; +, $d\omega/\omega = +0.5\%$.

For longer times, the short-time approximation is no longer valid. Attempts to develop alternative secondary data, such as a time-varying local frequency (Smith and Masters, 1989) or a spline expression for the modulation function (Nawab and Lognonné, 1994), have been proposed. Note here that a spline interpolation is also used in the software proposed for the computation of seismograms in the aspherical Earth. More sophisticated methods were also developed, allowing a direct inversion from the data of the structure coefficients described in relation (78). Numerous studies of normal mode observations were therefore performed for the retrieval of these structure coefficients, on the lower part of the seismic spectrum, up to around 5 mHz, where the spectral peaks are fairly well isolated. Two techniques to estimate the c_s^t can be distinguished: "singlet stripping" and "spectral fitting."

In the singlet stripping procedure (Gilbert, 1971; Buland *et al.*, 1979; Ritzwoller *et al.*, 1986), observed data (narrow-band) spectra are summed with appropriate weights in order to enhance a target singlet within a given multiplet. This technique works under the assumption that the singlets are sensitive only to axisymmetric structure (i.e., $t=0$). This assumption appears to be robust for low harmonic degree, high-Q multiplets (Ritzwoller *et al.*, 1986; Widmer *et al.*, 1992a).

In the spectral fitting technique (Ritzwoller *et al.*, 1986, 1988; Giardini *et al.*, 1987, 1988; Li *et al.*, 1991a,b; Widmer *et al.*, 1992a,b; Tromp and Zanzerkia, 1995; He and Tromp, 1996) the dependence of displacement on the structure coefficients and the frequency perturbation is linearized, and the ${}_{kk'}c_s^t$ are estimated by iteratively fitting the spectra in the frequency band surrounding the multiplet. In this case, no a priori knowledge on the aspherical structure is required and nonlinearity is taken into account by iterative procedure. Smith and Masters (1989) developed the technique to incorporate uncoupled hybrid multiplets, and more recently a generalized spectral fitting technique (Resovsky and Ritzwoller, 1995, 1998) has been developed incorporating inter-multiplet coupling. This extension of the spectral fitting technique allows inversion for odd-degree structure. This method has been used to produce a degree-8 shear velocity mantle model using normal mode observations below 3 mHz (Resovsky and Ritzwoller, 1999).

The second type of inversion is the direct inversion, which directly extracts the structure from the waveform or waveform packet synthesized by normal mode summations. The early works were based on a summation of SNREI modes with a phase correction computed by ray theory (e.g., Woodhouse, 1983; Woodhouse and Dziewonski, 1984), following theory described by Woodhouse and Wong (1986).

Using the SNREI modes for the computation of synthetic seismograms, asymptotic methods have provided better modeling of effects such as focusing or concentration of body-wave sensitivity (e.g., Romanowicz, 1987; Li and Tanimoto, 1993). Inversions (Li and Romanowicz, 1995, 1996) with these theories provided new and improved tomographic models. More sophisticated methods were then developed, with the computation of seismograms from aspherical normal modes, as described above. Such methods can indeed be used for the computation of seismograms in rather sophisticated models: Rough models (Lognonné and Clévédé, 1997), anisotropic models, and anelastic models. Compared to perturbation theories, alternative methods can also be used in the computation of the normal modes necessary for the summation, such as variational methods (Morris *et al.*, 1987; Hara *et al.*, 1991). The theory for the computation of the Fréchet derivatives was also developed, either in the framework of perturbation theory (Clévédé and Lognonné, 1996) or for variational theory (Geller and Ohminato, 1994).

Only one attempt has been done to perform a direct structure inversion from such complete waveforms, by Hara *et al.* (1993). Clévédé *et al.* (2000), however, confirmed that the impact of the theory errors in the tomographic models already developed can be quite significant, and will justify the stronger computational effort in the future.

11. Conclusion and Perspectives

Almost 40 years after their first observations, the results published on normal modes have clearly demonstrated that observing and inverting the normal modes of a planet is one of the most powerful methods for recovering its internal structure. It is also the only way to get from seismic data a direct sensitivity to density, including lateral variations (Tanimoto, 1991), as all other waves are sensitive mostly to the seismic velocities. The theory is now very mature, and almost completely described, with only a few last refinements missing, such as the nonlinear effect of the shape of discontinuities and boundary conditions. Very likely, the present models and theories are reaching a challenging maturity and give a satisfactory explanation of the frequency-related features of the normal modes. Amplitude-related features, including attenuation, are much less well explained.

In parallel, the already developed theories and the numerical software, such as those included on the attached Handbook CD (see, Chapter 85.16, by Clévédé and Lognonné), give the opportunity to use normal modes as one of the most powerful tools for the computation of seismograms in a complex and realistic Earth. The complete sequence, with first the mode computation of the SNREI, then the mode computation for an aspherical, anisotropic, and attenuating Earth, and finally the normal mode summation for the retrieval of long-period seismometers, is therefore provided. We can therefore expect in the near future the elaboration of a new generation of tomographic models with waveform modeling from fully coupled normal mode summations.

Future space missions to telluric planets of the solar system can also give us the opportunity to rediscover the excitement of the 1960s, but now with the first observations of normal modes of other planets. For the time being, only computations for speculative a priori structure have been performed. For the Moon, modes have been computed in different a priori models for the structure by Carr and Kovach (1962), Takeuchi *et al.* (1961), Saito *et al.* (1966), Bolt and Derr (1969), and Derr (1969b). Similarly, normal modes have been computed for Mars by Bolt and Derr (1969) and for Venus by Okal and Anderson (1978), Lognonné and Mosser (1993), Lognonné *et al.* (1996), and Sohl and Spohn (1997). All these studies demonstrate the high potential of observing normal modes. However, with the absence of strong tectonics on the Moon, the instrument noise will very likely be too high for their detection. Mars, as shown by Lognonné *et al.* (1996), may have a seismic activity compatible with normal mode detection, and moreover has an atmosphere allowing a permanent excitation of normal modes (Kobayashi and Nishida, 1998). Their detection will be one of the scientific objectives of the seismic experiment (Lognonné *et al.*, 2000) to be deployed by the NetLander mission, expected for launch in 2007 (Harri *et al.*, 1999; Lognonné *et al.*, 2000).

Acknowledgments

This work was sponsored by several CNRS/INSU grants, including ASP Tomographie, Programme Intérieur de la Terre, and Programme National des Risques Naturels. We thank J. Artru for her contribution, as well as B. Romanowicz, F. Pollitz, and an anonymous reviever for their comments and corrections. This is IPGP contribution 1708.

References

Agnew, D.C., J. Berger, R. Buland, W. Farrel, and F. Gilbert (1976). International deployment of accelerometers: a network for very long period seismology. *EOS Trans. Am. Geophys. Union* **57**, 180–188.

Alsop, L.E., G.H. Sutton, and M. Ewing (1961). Free oscillations of the Earth observed on strain and pendulum seismographs. *J. Geophys. Res.* **66**, 631–641.

Alterman, Z., H. Jarosh, and C.L. Pekeris (1959). Oscillations of the Earth. *Proc. R. Soc. Astron.* **252**, 80–95.

Anderson, D.L. (1989). "Theory of the Earth," Blackwell Scientific, Boston.

Anderson, D.L. and J.W. Given (1982). Absorption band *Q* model for the Earth. *J. Geophys. Res.* **87**, 3893–3904.

Anderson, D.L. and J.B. Minster (1979). The frequency dependence of *Q* in the Earth and implications for mantle rheology and Chandler wobble. *Geophys. J. R. Astron. Soc.* **58**, 431–440.

Artru J., P. Lognonné, and E. Blanc (2001). Normal modes modeling of post-seismic ionospheric oscillations. *Geophys. Res. Lett.* **28**, 697–700.

Backus, G.E. and J.F. Gilbert (1961). The rotational splitting of the free oscillations of the Earth. *Proc. Natl. Acad. Sci. USA* **47**, 362–371.

Benioff, H., J.C. Harrison, L. Lacoste, W.H. Munk, and L.B. Slichter (1959). Searching for the Earth's free oscillations. *J. Geophys. Res.* **64**, 1334–1337.

Benioff, H., F. Press, and S.W. Smith (1961). Excitation of the free oscillations of the Earth by earthquake. *J. Geophys. Res.* **66**, 605–620.

Bolt, B.A. and J.S. Derr (1969). Free bodily vibrations of the terrestrial planets. *Vistas Astron.* **11**, 69–102.

Bolt B.A. and A. Marussi (1962). Eigenvibrations of the Earth observed at Trieste. *Geophys. J. R. Astron. Soc.* **6**, 299–311.

Buland, R., J. Berger, and F. Gilbert (1979). Observation from the IDA network of attenuation and splitting during a recent earthquake. *Nature* **277**, 358–362.

Bullen, K.E. (1973). The ellipticities of surfaces of equal density inside the Earth. *Phys. Earth Planet. Int.* **7**, 199–202.

Carr, R.E. and R.L. Kovach (1962). Toroidal oscillation of the Moon. *Icarus* **1**, 75.

Clairaut, A.C. (1743). "Théorie de la Figure de la Terre, Tirée des Principes de l'Hydrostatique." David Fils, Paris.

Clévédé, E. (1991). Modes Propres d'une Terre Sphérique Anélastique, D.E.A. Analyse Numérique, Université Paris VI and Paris XIII.

Clévédé, E. and P. Lognonné (1996). Fréchet derivatives of coupled seismograms with respect to an anelastic rotating Earth. *Geophys. J. Int.* **124**, 456–482.

Clévédé, E., C. Mégnin, B. Romanowicz, and P. Lognonné (2000). Seismic waveform modeling and surface wave tomography in a three-dimensional Earth: asymptotic and non-asymptotic approaches. *Phys. Earth Planet. Int.* **119**, 37–56.

Cohen-Tannoudji, C., B. Diu, and F. LaLoë (1980). "Mécanique quantique," Vols. 1 and 2. Herman, Paris.

Connes, J., P.A. Blum, and N. Jobert (1962). Observation des oscillations propres de la Terre. *Annales de Géophysique* **18**, 160–268.

Dahlen, F.A. (1968). The normal modes of a rotating, elliptical Earth. *Geophys. J. R. Astron. Soc.* **16**, 329–367.

Dahlen, F.A. (1969). The normal modes of a rotating, elliptical Earth. *Geophys. J. R. Astron. Soc.* **18**, 397–436.

Dahlen, F.A. (1972). Elastic dislocation theory for a self-gravitating elastic configuration with an initial static stress field. *Geophys. J. R. Astron. Soc.* **28**, 357–383.

Dahlen, F.A. (1973). Elastic dislocation theory for a self-gravitating elastic configuration with an initial static stress field; II, Energy release. *Geophys. J. R. Astron. Soc.* **31**, 469–483.

Dahlen, F.A. (1979). The spectra of unresolved split normal mode multiplets. *Geophys. J. R. Astron. Soc.* **58**, 1–33.

Dahlen, F.A. (1981). The free oscillations of an anelastic aspherical Earth. *Geophys. J. R. Astron. Soc.* **66**, 1–22.

Dahlen, F.A. (1987). Multiplet coupling and the calculation of synthetic long-period seismograms. *Geophys. J. R. Astron. Soc.* **91**, 241–254.

Dahlen, F.A. and I.H. Henson (1985). Asymptotic normal modes of laterally heterogeneous Earth. *J. Geophys. Res.* **90**, 12653–12681.

Dahlen, F.A. and M.L. Smith (1975). The influence of rotation on the free oscillations of the Earth. *Phil. Trans. R. Soc. Lond. A* **279**, 583–627.

Dahlen, F.A. and J. Tromp (1998). "Theoretical Global Seismology." Princeton University Press, Princeton, NJ.

Derr, J.S. (1969a). Internal structure of the Earth inferred from free oscillations. *J. Geophys. Res.* **74**, 5202–5220.

Derr, J S. (1969b). Free oscillations of new lunar models. *Phys. Earth Planet. Int.* **2**, 61–98.

Dziewonski A. and D.L. Anderson (1981). Preliminary reference Earth model. *Phys. Earth Planet. Inter.* **25**, 297–356.

Dziewonski, A.M. and F. Gilbert (1972). Observations of normal mode from 84 recordings of the Alaskan earthquake of 28 March 1964. *Geophys. J. R. Astron. Soc.* **27**, 293–446.

Dziewonski, A.M. and F. Gilbert (1973). Observations of normal mode from 84 recordings of the Alaskan earthquake of 28 March 1964: Part II—Spheroidal overtones. *Geophys. J. R. Astron. Soc.* **35**, 401–437.

Edmonds, A.R. (1960). "Angular Momentum and Quantum Mechanics." Princeton University Press, Princeton, NJ.

Geller, R.J. and T. Ohminato (1994). Computation of synthetic seismograms and their partial derivatives from heterogeneous media with arbitrary natural boundary conditions using the direct solution method. *Geophys. J. Int.* **116**, 421–446.

Giardini, D., X.-D. Li, and J.H. Woodhouse (1987). Three-dimensional structure of the Earth from splitting in free-oscillation spectra. *Nature* **325**, 405–411.

Giardini, D., X.-D. Li, and J.H. Woodhouse (1988). Splitting functions of long-period normal modes of the Earth. *J. Geophys. Res.* **93**, 13716–13742.

Gilbert, F. (1970). Excitation of the normal modes of the Earth by earthquake sources. *Geophys. J. R. Astron. Soc.* **22**, 223–226.

Gilbert, F. (1971). The diagonal sum rule and averaged eigenfrequencies. *Geophys. J. R. Astron. Soc.* **23**, 119–123.

Gilbert, F. and A.M. Dziewonski (1975). An application of normal mode theory to the retrieval of structure parameters and source mechanisms for seismic spectra. *Phil. Trans. R. Soc. Astron.* **278**, 187–269.

Hara, T., S. Tsuboi, and R.J. Geller (1991). Inversion for laterally heterogeneous Earth structure using a laterally heterogeneous starting model; preliminary results. *Geophys. J. Int.* **104**, 523–540.

Hara, T., S. Tsuboi, and R.J. Geller (1993). Inversion for laterally heterogeneous upper mantle S-wave velocity structure using iterative waveform inversion. *Geophys. J. Int.* **115**, 667–698.

Harri, A.-M., O. Marsal, P. Lognonné *et al.*, and the NetLander Science Team (1999). Network Science Landers for Mars. *Adv. Space. Res.* **23**, 1915–1924.

Henson, I.H. and F.A. Dahlen (1986). Asymptotic normal modes of laterally heterogeneous Earth 2. Further results. *J. Geophys. Res.* **91**, 12467–12481.

He, X. and J. Tromp (1996). Normal-mode constraints on the structure of the Earth. *J. Geophys. Res.* **101**, 20053–20082.

Jeffreys, H. (1971). On the ellipticity correction in seismology. *Collected Papers of Sir Harold Jeffreys on Geophysics and Other Sciences* **1**, 187–190.

Jobert, N. (1956). Evaluation de la période d'oscillation d'une sphère hétérogène, par application du principe de Rayleigh. *C. R. Acad. Sci. Paris* **243**, 1230–1232.

Jobert, N. (1957). Sur la période propre des oscillations sphéroïdales de la Terre. *C. R. Acad. Sci. Paris* **244**, 921–922.

Jordan, T.H. (1978). A procedure for estimating lateral variations from low-frequency eigenspectra data. *Geophys. J. R. Astron. Soc.* **52**, 441–455.

Kanamori, H. and D.L. Anderson (1977). Importance of physical dispersion in surface wave and free oscillation problems: Review. *Rev. Geophys. Space Phys.* **15**, 105–112.

Kanamori, H. and J. Mori (1992). Harmonic excitation of mantle Rayleigh waves by the 1991 eruption of mount Pinatubo, Philippines. *Geophys. Res. Lett.* **19**, 721–724.

Kanamori, H., J. Mori, and D.G. Harkrider (1994). Excitation of atmospheric oscillations by volcanic eruptions. *J. Geophys. Res.* **22**, 21947–21961.

Kobayashi, N. and K. Nishida (1998). Continuous excitation of planetary free oscillations by atmospheric disturbances. *Nature* **395**, 357–360.

Landau, L.D. and E. Lifschitz (1965). "Quantum Mechanics, Non-relativistic Theory, Relativistic Theory." Pergamon Press, Oxford.

Li, X.-D. and B. Romanowicz (1995). Comparison of global waveform inversions with and without considering cross-branch modal coupling. *Geophys. J. Int.* **121**, 695–709.

Li, X.-D. and B. Romanowicz (1996). Global mantle shear velocity model developed using nonlinear asymptotic coupling theory. *J. Geophys. Res.* **101**, 22245–22272.

Li, X.-D. and T. Tanimoto (1993). Waveforms of long-period body waves in a slightly aspherical Earth model. *Geophys. J. Int.* **112**, 92–102.

Li, X.-D., D. Giardini, and J.H. Woodhouse (1991a). The relative amplitudes of mantle heterogeneity in *P*-velocity, *S*-velocity and density from free-oscillation data. *Geophys. J. Int.* **105**, 649–657.

Li, X.-D., D. Giardini, and J.H. Woodhouse (1991b). Large-scale three-dimensional even-degree structure of the Earth from splitting of long-period normal modes. *J. Geophys. Res.* **96**, 551–557.

Lindberg, C.R. and J. Park (1987). Multiple-taper spectral analysis of terrestrial free oscillations: Part II. *Geophys. J. R. Astron. Soc.* **91**, 795–836.

Liu, H.P, D.L. Anderson, and H. Kanamori (1976). Velocity dispersion due to anelasticity; implications for seismology and mantle composition. *Geophys. J. R. Astron. Soc.* **47**, 41–58.

Lognonné, P. (1989). Modélisation des modes propres de vibration dans une Terre anélastique et hétérogène: théorie et application. Thèse de Doctorat, Université de Paris VII, France.

Lognonné, P. (1991). Normal modes and seismograms in an anelastic rotating Earth. *J. Geophys. Res.* **96**, 20309–20319.

Lognonné, P. and E. Clévédé (1997). Diffraction of long-period Rayleigh waves by a slab: effects of mode coupling. *Geophys. Res. Lett.* **24**, 1035–1038.

Lognonné, P. and B. Mosser (1993). Planetary seismology. *Surv. Geophys.* **14**, 239–302.

Lognonné, P. and B. Romanowicz (1990a). Fully coupled Earth's vibrations: the spectral method. *Geophys. J. Int.* **102**, 365–395.

Lognonné, P. and B. Romanowicz (1990b). Effect of a global plume distribution on Earth normal modes. *Geophys. Res. Lett.* **17**, 1493–1496.

Lognonné, P., B. Mosser, and F.A. Dahlen (1994). Excitation of the Jovian seismic waves by the Shoemaker-Levy 9 cometary impact. *Icarus* **110**, 186–195.

Lognonné, P., J. Beyneix, W.B. Banerdt, S. Cacho, J.-F. Karczewski, and M. Morand (1996). Ultra broad-band seismology on InterMarsnet. *Planet. Space Sci.* **44**, 1237–1249.

Lognonné, P., E. Clévédé, and H. Kanamori (1998). Computation of seismograms and atmospheric oscillations by normal-mode summation for a spherical Earth model with realistic atmosphere, *Geophys. J. Int.* **135**, 388–406.

Lognonné, P., D. Giardini, B. Banerdt, *et al.* (2000). The NetLander Very Broad band seismometer. *Planet. Space Sci.* **48**, 1289–1302.

Love, A.E H. (1911). "Some Problems of Geodynamics." Cambridge University Press, Cambridge.

Madariaga, R.I. (1972). Spectral splitting of toroidal free oscillations due to lateral heterogeneities of the Earth's structure. *J. Geophys. Res.* **11**, 4421–4431.

Montagner, J.-P. and B.L.N. Kennett (1996). How to reconcile body-wave and normal-mode reference Earth models. *Geophys. J. Int.* **125**, 229–248.

Masters, G. and F. Gilbert (1981). Structure of the inner core inferred from observations of its spheroidal shear modes. *Geophys. Res. Lett.* **8**, 569–571.

Masters, G., T.H. Jordan, P.G. Silver, and F. Gilbert (1982). Aspherical Earth structure from fundamental spheroidal-mode data. *Nature* **298**, 609–613.

Masters, G., J. Park, and F. Gilbert (1983). Observations of coupled spheroidal and toroidal modes. *J. Geophys. Res.* **88**, 10285–10298.

Mochizuki, E. (1986). The free oscillations of an anisotropic and heterogeneous Earth. *Geophys. J. R. Astron. Soc.* **86**, 167–176.

Morris, S.P., R.J. Geller, H. Kawakatsu, and S. Tsuboi (1987). Variational normal modes computations for three laterally heterogeneous earth models. *Phys. Earth Planet. Inter.* **47**, 275–318.

Nawa, K., N. Suda, Y. Fukao, T. Sato, Y. Aoyama, and K. Shibuya (1998). Incessant excitation of the Earth's free oscillations. *Earth, Planets Space* **50**, 3–8.

Nawab, R. (1993). Observation et modulation des modes propres de la terre: contraintes sur la terre profonde. Thèse de Doctorat, Université de Paris 7, France.

Nawab, R. and P. Lognonné (1994). A new technique in normal modes demodulation. *Phys. Earth Planet. Inter.* **84**, 139–160.

Ness, N.F., C.J. Harrison, and L.F. Schlicher (1961). Observation of the free oscillation of the Earth. *J. Geophys. Res.* **66**, 621–629.

Okal, E. and D. Anderson (1978). Theoretical models for Mars and their seismic properties. *Icarus* **33**, 514–528.

Park, J. (1987). Asymptotic coupled-mode expressions for multiplet amplitude anomalies and frequency shifts on a laterally heterogeneous Earth. *Geophys. J. R. Astron. Soc.* **90**, 129–170.

Park, J. (1989). Roughness constraints in surface wave tomography. *Geophys. Res. Lett.* **16**, 1329–1332.

Park, J. (1990a). Radial mode observation from the 5/23/89 Macquarie ridge earthquake. *Geophys. Res. Lett.* **17**, 1005–1008.

Park, J. (1990b). The subspace projection method for constructing coupled mode synthetic seismograms. *Geophys. J. Int.* **101**, 111–123.

Park, J. and F. Gilbert (1986). Coupled free oscillations of an aspherical dissipative rotating Earth: Galerkin theory. *J. Geophys. Res.* **91**, 7241–7260.

Park, J., C.R. Lindberg, and D. Thomson (1987). Multiple-taper spectral analysis of terrestrial free oscillations: Part I. *Geophys. J. R. Astron. Soc.* **91**, 755–794.

Pekeris, C.L. and H. Jarosh (1958). The free oscillations of the Earth. "Contributions in Geophysics in Honor of Beno Gutenberg," pp. 171–192. Pergamon Press, NW.

Pekeris, C.L., Z. Alterman, and H. Jarosh (1961). Rotational multiplets in the spectrum of the Earth. *Phys. Rev.* **122**, 1692–1700.

Phinney, R.A. and R. Burridge (1973). Representation of the elastic-gravitational excitation of a spherical Earth model by generalized spherical harmonics. *Geophys. J. R. Astron. Soc.* **34**, 451–487.

Pollitz, F.F. (1990). Recovery of aspherical earth structure from observations of normal mode amplitudes. *Geophys. J. Int.* **102**, 313–339.

Rayleigh, J.W. (1894). "The Theory of Sound." Cambridge.

Rayleigh, J.W. (1906). On the dilational stability of the Earth. *Proc. R. Soc. Lond., A* **77**, 486–499.

Resovsky, J.S. and M.H. Ritzwoller (1995). Constraining odd-degree mantle structure with normal modes, *Geophys. Res. Lett.* **22**, 2301–2304.

Resovsky, J.S. and M.H. Ritzwoller (1998). New and refined constraints on three-dimensional Earth structure from normal modes below 3 mHz. *J. Geophys. Res.* **103**, 783–810.

Resovsky, J.S. and M.H. Ritzwoller (1999). A degree 8 mantle shear velocity model from normal mode observations below 3 mHz. *J. Geophys. Res.* **104**, 993–1014.

Ritzwoller, M.H. and E.M. Lavely (1995). Three-dimensional seismic models of the Earth's mantle. *Rev. Geophys.* **33**, 1–66.

Ritzwoller, M.H., G. Masters, and F. Gilbert (1986). Observations of anomalous splitting and their interpretation in terms of aspherical structure. *J. Geophys. Res.* **91**, 10203–10228.

Ritzwoller, M.H., G. Masters, and F. Gilbert (1988). Constraining aspherical structure with low harmonic degree interaction coefficients: application to uncoupled multiplets. *J. Geophys. Res.* **93**, 6269–6396.

Romanowicz, B. (1987). Multiplet-multiplet coupling due to lateral heterogeneity: asymptotic effects on the amplitude and frequency of the Earth's normal modes. *Geophys. J. R. Astron. Soc.* **90**, 75–100.

Romanowicz, B. (1990). The upper mantle degree two: constraints and inferences on attenuation tomography from global mantle wave measurements. *J. Geophys. Res.* **95**, 11051–11071.

Romanowicz, B. (1991). Seismic tomography of the Earth's mantle. *Annu. Rev. Earth Planet. Sci.* **19**, 77–99.

Romanowicz, B. (1994). Anelastic tomography: a new perspective in upper mantle thermal structure. *Earth. Planet. Sci. Lett.* **128**, 113–121.

Romanowicz, B. and G. Roult (1986). First order asymptotic for the eigen-frequencies of the Earth and application to the retrieval of large-scale lateral variations of structure. *Geophys. J. R. Astron. Soc.* **87**, 209–239.

Romanowicz, B., J.F. Karczewski, M. Cara, *et al.* (1991). The GEOSCOPE program: present status and perspectives. *Bull. Seismol. Soc. Am.* **81**, 2221–2243.

Roult, G. and E. Clévédé (2000). New refinements in attenuation measurements from free oscillation and surface wave observations. *Phys. Earth Planet. Inter.* **121**, 1–37.

Roult, G., B. Romanowicz, and J.P. Montagner (1990). 3D upper mantle shear velocity and attenuation from fundamental mode free oscillation data. *Geophys. J. Int.* **101**, 61–80.

Saito, M., H. Takeuchi, and N. Kobayashi (1966). Free oscillation of the Moon—effect of the liquid core. *Jishin, Seismol. Soc. Jpn.* **19**, 235–236.

Silver, P.G. and T.H. Jordan (1981). Fundamental spheroidal mode observations of aspherical heterogeneity. *Geophys. J. R. Astron. Soc.* **64**, 605–634.

Smith B.T., J.M. Boyle, J.J. Dongarra, B.S. Garbow, Y. Ikebe, V.C. Klema, and C.B. Moler (1976). "Matrix Eigensystem Routines-EISPACK Guide," 2nd edn., Vol. 6 of Lecture Notes in Computer Science. Springer-Verlag, Berlin.

Smith, M.F. and G. Masters (1989). Aspherical structure constraints from free oscillation frequency and attenuation measurements. *J. Geophys. Res.* **94**, 1953–1976.

Sohl, F. and D. Spohn (1997). The interior structure of Mars: implications from the SNC meteorites. *J. Geophys. Res.* **102**, 1613–1635.

Suda, N., K. Nawa, and Y. Fukao (1998). Incessant excitation of the Earth's free oscillations. *Science* **279**, 2089–2091.

Takeuchi, H. (1959). Torsional oscillations of the Earth and some related problems. *Geophys. J. R. Astron. Soc.* **2**, 89–100.

Takeuchi, H. and M. Saïto (1972). Seismic surface waves. In: "Methods in Computational Physics," Vol. 11 (B.A. Bolt, Ed.), pp. 217–295. Academic Press, New York.

Takeuchi, H., M. Saito, and N. Kobayashi (1961). Free oscillations of the Moon. *J. Geophys. Res.* **66**, 3895.

Tanimoto, T. (1991). Waveform inversion for three-dimensional density and *S* wave structure. *J. Geophys. Res.* **96**, 8167–8189.

Tanimoto, T. (1999). Excitation of normal modes by atmospheric turbulence: source of long-period seismic noise. *Geophys. J. Int.* **136**, 395–402.

Thomson, W. (1863a). On the rigidity of the Earth. *Phil. Trans. R. Soc. Lond.* **153**, 573–582.

Thomson, W. (1863b). Dynamical problems regarding elastic spheroidal shells and spheroids of incompressible liquid. *Phil. Trans. R. Soc. Lond.* **153**, 583–616.

Tromp J. and F.A. Dahlen (1990). Free oscillations of a spherical anelastic Earth. *Geophys. J. Int.* **103**, 707–723.

Tromp, J. and E. Zanzerkia (1995). Toroidal splitting observations from the great 1994 Bolivia and Kuril Islands earthquakes. *Geophys. Res. Lett.* **22**, 2297–3000.

US Standard Atmosphere (1976). "Committee on the Extension of the Standard Atmosphere." US Government Printing Office, Washington DC.

Valette, B. (1986). About the influence of pre-stress upon adiabatic perturbation of the Earth. *Geophys. J. R. Astron. Soc.* **85**, 179–208.

Valette, B. (1987). Spectre des oscillations libre de la Terre: aspects mathématiques et géophysiques. Thèse de Doctorat d'État, Université Paris VI.

Valette, B. (1989a). Spectre des vibrations propres d'un corps élastique, auto-gravitant, en rotation uniforme et contenant une partie fluide. *C. R. Acad. Sci. I* **309**, 419–422.

Valette, B. (1989b). Etude d'une classe de problèmes spectraux. *C. R. Acad. Sci. I* **309**, 785–788.

Vermeersen, L.L.A. and N.J. Vlaar (1991). The gravito-elasto-dynamics of a pre-stressed elastic Earth. *Geophys. J. Int.* **104**, 555–563.

Watada, S. (1995). Part I: Near-source acoustic coupling between the atmosphere and the solid Earth during volcanic eruptions. PhD thesis, California Institute of Technology.

Widmer, R., G. Masters, and F. Gilbert (1992a). Observably split multiplets-data analysis and interpretation in terms of large-scale aspherical structure. *Geophys. J. Int.* **111**, 559–576.

Widmer, R., W. Zurn, and G. Masters (1992b). Observations of low-order toroidal modes from the 1989 Macquarie Rise event. *Geophys. J. Int.* **104**, 226–236.

Woodhouse, J.H. (1980). The coupling and attenuation of nearly resonant multiplets in the Earth's free oscillation spectrum. *Geophys. J. R. Astron. Soc.* **61**, 261–283.

Woodhouse, J.H. (1983). The joint inversion of seismic wave forms for lateral variations in Earth structure and earthquake source parameters. *Proc. Enrico Fermi Int. Sch. Phys.* **85**, 366–397.

Woodhouse, J.H. (1988). The calculation of the eigenfrequencies and eigenfunctions of the free oscillations of the Earth and the Sun. In: "Seismological Algorithms" (D.J. Doornbos, Ed.), pp. 321–370. Academic Press, New York.

Woodhouse, J.H. and Y.K. Wong (1986). Amplitude, phase and path anomalies of mantle waves. *Geophys. J. R. Astron. Soc.* **87**, 753–773.

Woodhouse, J.H. and F.A. Dahlen (1978). The effect of a general aspherical perturbation on the free oscillations of the Earth. *Geophys. J. R. Astron. Soc.* **53**, 335–354.

Woodhouse, J.H. and A.M. Dziewonski (1984). Mapping of the upper-mantle: three dimensional modeling of Earth structure by inversion of seismic waveforms. *J. Geophys. Res.* **89**, 5953–5986.

Zurn, W. and R. Widmer (1996). World wide observation of bichromatic long-period Rayleigh-waves excited during the June 15, 1991 Eruption of Mt. Pinatubo. In: "Fire and Mud, Eruptions of Mount Pinatubo, Philippines" (C. Newhall, and R. Punongbayan, Eds.), pp. 615–624. Philippine Institute of Volcanology and Seismology, Quezo City and University of Washington Press.

Zharkov, B.N. and V.M. Lyubimov (1970a). Theory of toroidal vibrations for a spherically asymmetrical model of the Earth. *Izv. Acad. Sci. USSR Phys. Solid Earth* **10**(2), 71–76.

Zharkov B.N. and V.M. Lyubimov (1970b). Theory of spheroidal vibrations for a spherically asymmetrical model of the Earth. *Izv. Acad. Sci. USSR Phys. Solid Earth* **10**, 613–618.

11

Inversion of Surface Waves: A Review

Barbara Romanowicz
University of California, Berkeley, California, USA

1. Introduction

In what follows, we attempt to review progress made in the last few decades in the analysis of teleseismic and regional surface wave data for the retrieval of earthquake source parameters and global and regional Earth structure. This review is by no means exhaustive. We will rapidly skip over the early developments of the 1950s and 1960s that led the foundations of normal mode and surface wave theory as it is used today. We will not attempt to provide an exhaustive review of the vast literature on surface wave measurements and the resulting models, but rather focus on describing key theoretical developments that are relevant and have been applied to inversion. Since surface wave theory is closely related to that of the Earth's normal modes, we will discuss the latter when appropriate. However, we make no attempt to extensively review normal mode theory, as this subject is addressed in a separate contribution (see Chapter 10 by Lognonné and Clévédé).

2. Background

Most of the long-period energy (periods greater than 20 s) generated by earthquakes and recorded at teleseismic distances propagates as surface waves. Most clearly visible on long-period seismograms are the successive, Earth-circling, dispersed wave trains of the fundamental mode. For moderate size earthquakes recorded at teleseismic distances ($M \sim 5.5$), only the surface waves propagating along the direct great circle path between the epicenter and the station have significant signal-to-noise ratio, mostly between 20 and 100 s period, and the dispersive and attenuative properties of these wave trains have been used extensively, since the 1950s to infer crust and upper mantle structure in different regions of the Earth. For earthquakes of magnitude 7 or larger, successive Earth-circling surface wave trains can be followed for many hours (Fig. 1), and are then either analyzed individually or combined to produce a spectrum of the Earth's free oscillations by Fourier analysis of long time-series.

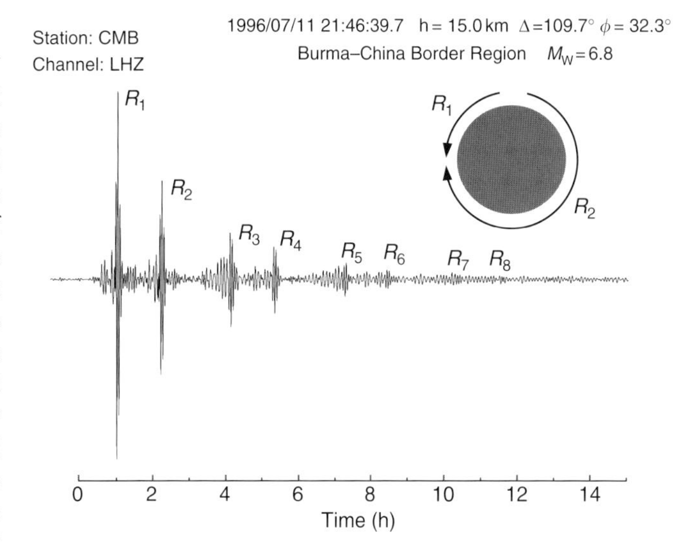

FIGURE 1 Example of vertical component record showing many Earth-circling mantle Rayleigh wave trains over a time window of 14 h. This record was recorded at station CMB of the Berkeley Digital Seismic Network (BDSN) and corresponds to a channel with a sampling rate of 1 sample/sec. The earthquake is shallow and the epicentral distance $\Delta = 109.7°$. Because the distance is close to 90° the wave packets corresponding to even and odd order trains are well separated from each other. (Courtesy of Joseph Durek and Lind Gee.)

Most studied are fundamental mode Rayleigh waves, which correspond to P–SV energy and have elliptical particle motions in the vertical plane containing the direction of propagation. These waves are well recorded on the quieter vertical component seismographs (Fig. 2). On the other hand, Love waves, which carry SH energy, and are polarized horizontally in a direction perpendicular to the direction of propagation, require rotating the two horizontal records to extract the transverse component of motion. Love wave studies have suffered,

INTERNATIONAL HANDBOOK OF EARTHQUAKE AND ENGINEERING SEISMOLOGY, VOLUME 81A ISBN: 0-12-440652-1

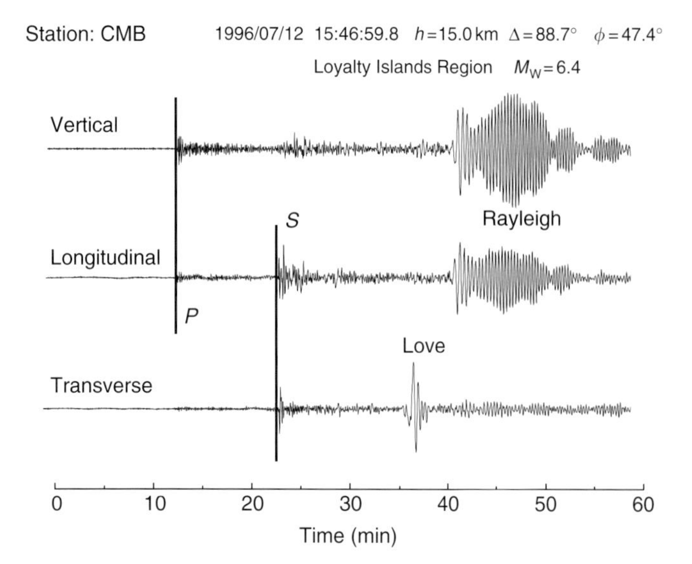

FIGURE 2 Three-component seismograms observed at station CMB for a shallow earthquake at a distance of $\Delta = 88.7°$. The horizontal components have been rotated to the longitudinal and transverse directions, clearly exhibiting fundamental mode Love and Rayleigh waves on the transverse and vertical/longitudinal components, respectively. (Courtesy of Joseph Durek and Lind Gee.)

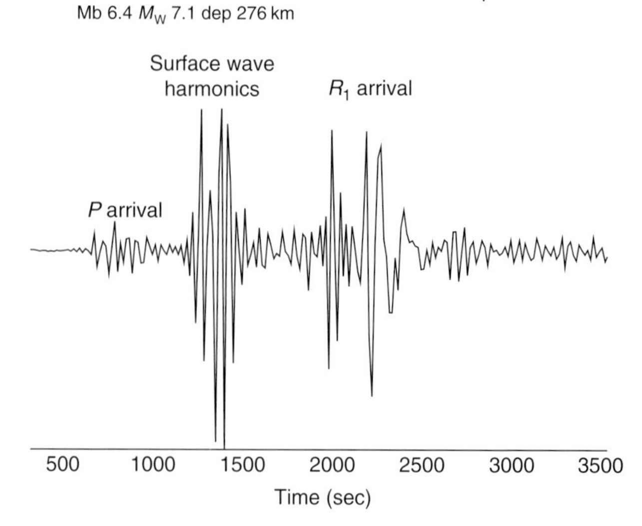

FIGURE 3 Example of longitudinal component seismogram recorded at IRIS/GSN station SUR showing the arrivals of multiply reflected body wave phases forming a higher-mode Rayleigh wave train in front of the fundamental mode (R_1). The Airy phase, corresponding to the group velocity minimum around 230 sec, is well visible in the R_1 train. The event occurred on 14 Jan. 1997 in southern Bolivia, at a depth of 276 km. The epicentral distance is 8419 km. The seismogram has been bandpass filtered with cut-off frequencies at 35 and 400 sec. (Courtesy of Yuancheng Gung.)

especially in the early days of analog recordings, from the more complex data processing required, and from the higher levels of background noise on horizontal components at long periods, due primarily to the influence of atmospheric pressure variations, inducing ground tilts. Fundamental mode Love and Rayleigh waves are generally well separated from other phases on the seismograms, and well excited by shallow, crustal earthquakes, while overtones travel at higher group velocities, appear as packets of mixed overtones (e.g., *X* phases, Jobert *et al.*, 1977), and are better excited by deeper earthquakes (Fig. 3).

Surface waves recorded at teleseismic distances contain information about both the characteristics of the earthquake source and the structure of the Earth along the source station path. Separating these two effects has been one of the long-standing challenges faced by seismologists.

Studies of the structure of the crust and upper mantle progressed rapidly in the 1950s and early 1960s, as the tools to measure group and phase velocities, and interpret them in terms of layered mantle and crust models, became readily available (e.g., Ewing *et al.*, 1957; Brune *et al.*, 1961a,b; Alsop, 1963). In these studies, source effects were generally eliminated by considering propagation between two or more stations aligned along the same great circle path, or, at longer periods, observation of consecutive Earth-circling wave trains at the same station. On the other hand, in early studies of earthquake sources, propagation effects were assumed to be known, and amplitudes were "equalized" to obtain the source radiation pattern and infer information about the fault orientation (e.g., Aki, 1960) and its directivity (Alterman *et al.*, 1959; Ben-Menahem, 1961; Kanamori, 1970). At that time, the theoretical formulation for the excitation of surface waves and normal modes of the Earth was developed (Sato *et al.*, 1962; Harkrider, 1964; Haskell, 1964), much stimulated by the occurrence of the great Chilean earthquake of 22 May 1960, and more quantitative studies of the effects of the earthquake source on spectra of surface waves followed. The association of a normal mode formalism (e.g., Gilbert, 1971) to compute dispersion and excitation of surface waves (and complete seismograms) with a moment tensor formalism to describe the earthquake source (e.g., Backus and Mulcahy, 1976; Mendiguren, 1977) has led to the present-day commonly used expressions and to a rapid development of source studies based on surface waves in the 1980s. A computational method (Takeuchi and Saito, 1972), following the theoretical approach of Saito (1967) based on Runge–Kutta matrix integration, has long been the main reference for the practical calculation of excitation for surface waves and normal modes in laterally homogeneous, elastic, flat or spherical Earth models. Later, different schemes, using different mathematical approaches (variational method) were developed (Wiggins, 1976; Buland and Gilbert, 1984). Today, another widely used code for spherical geometry and efficient to relatively short periods (10 s) is based on a propagator matrix approach, in which minors of sets of solutions are used (Gilbert and Backus, 1966; Woodhouse, 1980a, 1988).

These theoretical advances were first applied to the analog data of the World Wide Standard Seismic Network (WWSSN) accumulated in the 1960s and 1970s. The IDA (International Deployment of Accelerometers) network, established in the mid-1970s (Agnew *et al.*, 1976), provided the first long-period digital data, along with several stations installed and operated by the French (Jobert and Roult, 1976). The digital recording greatly facilitated the simultaneous analysis of many records, paving the way for large-scale tomographic studies of global structure and systematic teleseismic source studies. A major drawback, however, was the limited dynamic range of the IDA instruments, so that first-arriving low-frequency R_1 and G_1 wave trains would saturate for large earthquakes. This problem disappeared in the 1980s with the deployment by France of the high dynamic range, digital broadband GEOSCOPE network (Romanowicz *et al.*, 1984, 1991) and by the United States of the IRIS Global Seismic Network (e.g., Smith, 1986), gradually complemented by many broadband stations contributed by other countries through the Federation of Digital Seismic Networks (FDSN), which, starting in 1986, established high-level standards for broadband seismic sensors, recording systems, and data formats (e.g., Romanowicz and Dziewonski, 1986). The accumulation of high-quality data from numerous broadband stations has greatly contributed to the successes of global tomography and source moment tensor studies of the last 15 years.

3. Source Studies Using Surface Waves

To obtain the frequency spectrum of a single mode surface wave train from the expression of a seismogram obtained by summation of normal modes on a spherically symmetric Earth, one uses Poisson's formula (e.g., Gilbert, 1976; Aki and Richards, 1980), which decomposes the modes into infinite trains of propagating surface waves traveling in opposite directions around the Earth. In this process, a high-frequency approximation is used, in which the phase velocity of a surface wave is related to the corresponding normal mode frequency by Jeans's formula:

$$C(\omega) = \frac{a\omega_l}{l + 1/2} \tag{1}$$

where l is the angular order of the mode and ω_l its eigenfrequency, and a is the radius of the Earth. In this high-frequency approximation, the surface waves propagate along the great circle path between the epicenter and the station and are sensitive to structure only along this great circle.

The most widely used theoretical framework for the interpretation of surface wave data was thus established in the 1970s. It is derived from a normal mode formalism using a high-frequency zeroth-order approximation, and leads to a simple expression for the spectrum of a single mode propagating surface wave at distance Δ, azimuth θ, and angular frequency ω. Following Kanamori and Stewart (1976), and Nakanishi and Kanamori (1982):

$$U(\Delta, \theta, \omega) = U_s(\theta, \omega)S(\Delta)U_p(\Delta, \theta, \omega)F(\omega, \theta_0)D(\omega)I(\omega) \tag{2}$$

where U_s is the source spectrum, U_p contains propagation effects, I is the instrument response, $S(\Delta)$ the geometrical spreading term, and F and D express the source process as clarified below.

The propagation term U_p can be expressed as (e.g., Romanowicz and Monfret, 1986)

$$\begin{aligned} U_p(\Delta, \theta, \omega) = {} & \frac{1}{(\sin\Delta)^{1/2}} \exp(i\pi/4)\exp(im\pi/2) \\ & \times \exp[-i\omega\Delta/C(\omega, \theta)]\exp[-\eta(\omega, \theta)\Delta] \end{aligned} \tag{3}$$

where m denotes the number of polar passages and $C(\omega, \theta)$, $\eta(\omega, \theta)$ are, respectively, the average phase velocity and attenuation coefficient along the source–station path.

The source spectrum U_s is a linear combination of the moment tensor elements M_{ij} of the source. In the notation of Kanamori and Stewart (1976), we have for Rayleigh waves:

$$\begin{aligned} U_s^R(\theta, \omega) = {} & \tfrac{1}{3}(S_R + N_R)M_{zz} + \tfrac{1}{6}(2N_R - S_R)(M_{xx} + M_{yy}) \\ & + \tfrac{1}{2}P_R(M_{yy} - M_{xx})\cos 2\theta - P_R M_{xy}\sin 2\theta \\ & + iQ_R(M_{xz}\cos\theta + M_{yz}\sin\theta) \end{aligned} \tag{4}$$

where N_R, S_R, P_R, Q_R are the excitation functions, which are nonlinear functions of ω and of the depth h of the source.

Likewise, for Love waves:

$$\begin{aligned} U_s^L(\theta, \omega) = {} & P_L\left[\tfrac{1}{2}(M_{xx} - M_{yy})\sin 2\theta - M_{xy}\cos 2\theta\right] \\ & + iQ_L(-M_{xz}\sin\theta + M_{yz}\cos\theta) \end{aligned} \tag{5}$$

with depth-dependent excitation functions P_L, Q_L. $D(\omega)$ expresses the delay τ_D of the main faulting from the initial break (e.g., Nakanishi and Kanamori, 1982):

$$D(\omega) = \exp(-i\omega\tau_D) \tag{6}$$

and $F(\omega, \theta_0)$ expresses the directivity term arising for the propagation of rupture. Ben Menahem (1961) derived an approximate expression for F in the case of unilateral faulting, which has been shown to be exact if an equatorial coordinate system is used (Dziewonski and Romanowicz, 1977):

$$F(\omega, \theta_0) = \frac{\sin X}{X}\exp(-iX) \tag{7}$$

where

$$X = \frac{\omega L}{2V} - \frac{\omega L}{2C}\cos(\theta - \theta_0) \tag{8}$$

and L is the length of fault, C is the average phase velocity at frequency ω, V is the rupture velocity, and θ_0 is the azimuth of the rupture direction with respect to the fault strike.

As seen from Eqs. (2) to (8), fundamental mode surface wave spectra contain information about the source moment tensor, source depth (centroid), source process time, and, under favorable circumstances for very large earthquakes, source directivity. However, propagation effects U_p have to be known or effectively eliminated.

In order to correct for propagation, different approaches need to be taken depending on the size of the earthquake and period range considered. At very long periods ($T > 180$ sec) and for large earthquakes ($M > 6.5$), propagation effects can be accounted for approximately in the framework of a spherically symmetric reference Earth model. At shorter periods and for smaller earthquakes, corrections on individual source–station paths need to be known much more accurately. We thus discuss these two domains separately.

3.1 Teleseismic Studies of Moderate Size Events ($M \sim 5$–6)

Today significant efforts are still expended toward surface wave "path calibration," in particular in the framework of CTBT research (e.g., Stevens and McLaughlin, 2000). However, several methods have been developed to deal with poorly known path effects, with particular applications to the study of moderate earthquakes, for which good signal-to-noise records of 20–100 sec surface waves are obtained at teleseismic distances. In this case, source duration and directivity can generally be neglected, since their effect is significant only at shorter periods. A standard procedure is to use a "reference event," for which a reliable mechanism has been obtained independently, to obtain path corrections to specific stations. These path corrections are then used to infer source parameters of other neighboring events of interest. Following the approach of Weidner and Aki (1973), Patton (1977) developed an iterative method to infer simultaneously depth, source mechanism, and propagation effects for a group of closely located events observed teleseismically. On the path to a given station, the propagation effects are assumed to be the same for all events. In adjacent steps of the procedure, path effects and source characteristics are alternately improved. In the source improvement step, the linear system in M_{ij} [Eqs. (4) and (5)] is solved successively for different depth values, and a solution is declared corresponding to the depth that provides the best fit to the data in a least-squares sense. The imaginary parts of Eqs. (4) and (5) generally do not contribute to the depth determination, as, for shallow earthquakes, the eigenfunctions corresponding to the moment tensor elements M_{xz} and M_{yz} are very small (and go to zero at the free surface). This also results in poor constraints on these two moment tensor elements for shallow ($h < 20$ km) earthquakes if only fundamental mode surface waves are used.

The method of Patton (1977) is limited by the fact that it requires several earthquakes located in the same area to be well recorded teleseismically at the same set of stations. Moreover, Aki and Patton (1979) have shown that, in order to obtain reliable moment tensor solutions from data in the period range 20–100 sec, it is necessary to know average phase velocities on source station paths with an accuracy of 0.5%. When path corrections in the phase are inaccurate, the method breaks down because a clear minimum in the residuals/depth curves cannot be defined. On the other hand, because path effects on amplitudes are less coherent, an average-Q model is sufficient if only amplitude data are used (Mendiguren, 1977). Tsai and Aki (1971) first showed how the amplitude spectrum of Rayleigh waves (and to a lesser extent Love waves) contained the signature of depth of source in the period range 10–100 sec, in the form of a "hole" in the spectrum, which appears at a given period for a specific source mechanism. However, when amplitudes only are used, the inverse problem becomes nonlinear in the moment-tensor, requiring the knowledge of a reasonable starting solution.

A procedure that greatly relaxes the constraint of accurate path corrections in the phase was proposed by Romanowicz (1982a), based on the following observation: For a given source and at a given frequency ω_k, the real (α) and imaginary (β) parts in Eqs. (4) and (5) are functions of the azimuth θ only, and can be written in the form

$$\begin{aligned}\alpha(\omega_k, \theta) &= A_k + B_k \cos 2\theta + C_k \sin 2\theta \\ \beta(\omega_k, \theta) &= D_k \cos\theta + E_k \sin\theta\end{aligned} \tag{9}$$

where A_k, B_k, C_k, D_k, and E_k depend on frequency. By virtue of the uniqueness of the Fourier decomposition of continuous functions, at each frequency ω_k, given a set of values of α and β for different azimuths θ, these coefficients are uniquely determined. The Fourier expansions of α and β in azimuth contain other coefficients, of degree $n > 2$, that arise from imperfect knowledge of path corrections. In solving Eqs. (9), these other terms are eliminated. Thus only the very long-wavelength ($n \leq 2$) terms of the phase velocity "map" in the region containing sources and stations need to be known accurately, an increasingly reachable goal today, thanks to improvements in global surface wave tomography.

The inversion procedure thus proceeds in "two steps": First, Eqs. (9) are solved at a set of frequencies $\omega_1, \ldots, \omega_k$, and second, the following system of equations is solved for depth h, searching, as previously, for a minimum in the residuals/depth curve. Thus, for example, for Rayleigh waves:

$$\begin{aligned}A_k &= \tfrac{1}{2} S_r(\omega, h) M_{zz} \\ B_k &= \tfrac{1}{2} P_r(\omega, h)(M_{yy} - M_{zz}) \\ C_k &= -P_r(\omega, h) M_{xy} \\ D_k &= Q_r(\omega, h) M_{xz} \\ E_k &= Q_r(\omega, h) M_{yz}\end{aligned} \tag{10}$$

Here we have assumed that there is no volume change at the source so that $\sum_i M_{ii} = 0$ and $M_{xx} + M_{yy}$ is replaced by

$-M_{zz}$ in Eq. (2). While justified for most earthquakes, this assumption needs to be relaxed when a volumetric component of the source is sought. To resolve the latter requires multimode observations, and Eqs. (2) to (8) show that resolving M_{xx}, M_{yy}, and M_{zz} separately involves many trade-offs with structure effects (e.g., Dufumier and Rivera, 1997). Romanowicz (1982a) showed that this procedure allows the determination of accurate source parameters for small events ($M < 5.5$), for which only surface wave observations are available teleseismically, provided one event of larger magnitude is available, with a well-constrained mechanism, in a source region ~1500 km in aperture. This method was later extended to the nonlinear inversion of surface wave amplitude data (Romanowicz and Suarez, 1983).

3.2 Global Studies of Large Earthquakes

At very long periods ($150 < T < 320$ sec) and for large earthquakes ($M > 6.5$), spherically symmetric Earth models can be used to correct for propagation effects in the phase of "mantle" waves (as surface waves are called in this period range), and the biases introduced by neglecting lateral heterogeneity are relatively small, except for very shallow events, for which the source phase needs to be known with greater accuracy. In the last 15 years, the availability of global 3D tomographic models of the upper mantle of increasing accuracy has not only facilitated very long-period source studies but has also made it possible to extend the period range to shorter periods (down to ~120 sec) and to efficiently make use of alternative methodologies, based on time-domain waveform inversion, that are no longer restricted to the fundamental mode, thus providing more accurate estimation of the source depth. Such a waveform approach was introduced by Dziewonski *et al.* (1981), who combined waveforms of mantle waves at periods greater than 120 sec with waveforms of body waves at periods greater than 80 sec. This forms the basis of a now routine procedure that serves to construct the widely used Harvard centroid moment tensor (CMT) catalog.

The development of methodologies to invert fundamental mode mantle wave data has nevertheless continued, following Kanamori and Given (1981), who showed how the spectra of mantle waves, sampled at only a few frequencies, could be used to rapidly determine moment tensors of large earthquakes, when the depth of the event is known. Romanowicz and Guillemant (1984) extended the approach of Romanowicz (1982a) to show how centroid depth could be accurately determined using mantle waves even in a spherically symmetric reference Earth model. For the size of earthquakes considered here, the source process time (Nakanishi and Kanamori, 1982) cannot be neglected and there can be significant trade-offs with source depth. However, the source process time can be estimated, as a function of frequency, prior to the inversion by computing phase differences between three wave trains at each individual station (e.g., Furumoto, 1979; Furumoto and Nakanishi, 1982). For example, if Rayleigh wave trains R_{2n}, R_{2m+1} and R_{2n+2} are used, with R_{2m+1} traveling in the opposite direction to the two other trains, we have, for the corresponding phases Φ_i:

$$\begin{aligned}&\Phi_{2n} + \Phi_{2m+1} + (n+m)(\Phi_{2n} - \Phi_{2n+2}) + 2\Phi_{\text{instr}}\\&\quad = -\omega\tau + k2\pi\end{aligned} \tag{11}$$

where k is an arbitrary integer and the source time τ is defined as [Eqs. (6), (7), and (8)]

$$\tau = \frac{L}{V} + 2\tau_D \tag{12}$$

Expression (11) is independent of structure within the framework of the high-frequency approximation in which it is derived. This procedure requires that at least three mantle wave trains traveling in opposite directions have adequate signal-to-noise ratio, which can be rather restrictive. Romanowicz and Monfret (1986) proposed an approach that requires the availability of only one mantle wave train.

Noting from Eqs. (3) to (9) that, if U_c is the observed spectrum corrected for instrument response and propagation, then,

$$U_c(\theta,\omega)e^{i\Phi_\tau} = U_s(\theta,\omega) = \alpha + i\beta \tag{13}$$

where

$$\Phi_\tau = \frac{\omega\tau}{2 + \delta\phi} \tag{14}$$

$\delta\phi$ is the residual phase shift due to inaccurate propagation corrections in the phase, and α and β are the real and imaginary parts of Eqs. (4) and (5), respectively. For specified, incremental values of $\tau_j, j = 1, \ldots, p$ of τ, the system of Eqs. (4) and (5) is solved for different frequencies ω_k, the squared residuals of this inversion for each frequency are summed to obtain a squared residual as a function of τ, and the solution is found for the value of τ corresponding to the minimum value of this residual. The second step of the inversion proceeds as described above [Eqs. (9) and (10)]. Unlike the three-train method of Furumoto and Nakanishi (1982), because of the trade-off with Earth structure as seen in Eq. (14), the determination of the source time by this method depends on the accuracy of the Earth model used, and specifically, on errors in the Earth model that contribute a constant phase shift as a function of azimuth, that is, long-wavelength features. A regionalized global model, such as that of Okal (1977), was shown to be sufficient to obtain stable source time estimates. With the increased precision of currently available 3D tomographic models, this is no longer an issue. This method can be extended to the case of large earthquakes with significant directivity, by including a parameter search over the azimuth θ_0 of the fault and the rupture length L. For very large earthquakes, Kuge *et al.* (1996) showed that source complexity (spatiotemporal changes in the source mechanism) sometimes needs to be invoked to reconcile source mechanisms obtained separately using Love and Rayleigh waves. The spectral domain inversion of fundamental mode mantle waves was further tested against Earth models and applied to large earthquakes by Zhang and Kanamori (1988a,b), showing generally consistent results with the Harvard CMT solutions. An alternative approach to study the spatiotemporal characteristics of large earthquakes from surface wave data has been proposed by Bukchin (1995), who used a representation of the source in terms of higher-order moments.

3.3 Regional Distance Source Studies

While most of the modern formalism and methodology for intermediate-period surface wave inversion for source parameters was in place by the mid-1980s, they have only been recently adapted to the case of earthquakes observed at regional distances. This has been made possible by the rapid expansion, in the last ten years, of regional broadband networks in seismically active regions, such as, for example, TERRAscope (Thio and Kanamori, 1995) or the Berkeley Digital Seismic Network (BDSN; Romanowicz *et al.*, 1993) in southern and northern California, respectively, MEDNET in the Mediterranean region (Giardini *et al.*, 1993), and also in Japan. In the regional case, target frequent earthquakes have smaller magnitude (generally $M < 4.5$) and the period range of interest is ~10–60 sec (Fig. 4). Here, an important aspect is to obtain information about the earthquake source rapidly to guide local emergency response, and this has led to the development of automated procedures (e.g., Pasyanos *et al.*, 1993). The theoretical framework is essentially the same as was developed for teleseismic observations at intermediate (Romanowicz, 1982a) or long periods (Kanamori and Given, 1981; Romanowicz and Monfret, 1986). Some automatic procedures involve the comparison of a spectral domain inversion of fundamental mode surface waves with a time domain complete waveform inversion, at periods longer than 10 sec (e.g., Pasyanos *et al.*, 1993).

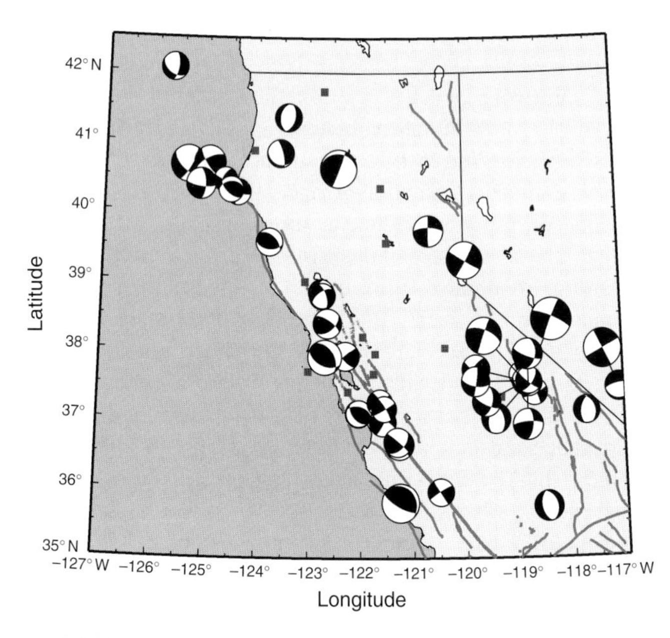

FIGURE 4 Moment tensor solutions obtained for all earthquakes of magnitude $M_W > 3.5$ in northern California for the period 08/98 to 09/99, using data from the Berkeley Digital Seismic Network. The size of the beachball is proportional to magnitude. (Courtesy of Hrvoje Tkalčić.)

With the improvement of global upper mantle models, it is now also possible to extend to smaller magnitudes ($M_W > 4.5$) the time-domain CMT inversion methodology developed by Dziewonski *et al.* (1981), making use of intermediate-period surface waves (Ritsema and Lay, 1993). Arvidsson and Ekström (1998) use a low-pass cutoff at ~45 sec and consider the fundamental mode surface waves as traveling waves. The phase is corrected for propagation effects using recent global phase velocity maps.

4. Structure Studies Using Surface Waves

4.1 Fundamental Mode Studies

Fundamental mode surface waves are well suited to study the elastic structure of the crust and upper mantle, which can be deduced from their group and/or phase dispersion properties. They allow the sampling of vast areas of the globe that are otherwise devoid of seismic stations and sufficiently strong earthquake sources, such as the oceans.

It is beyond the scope of this chapter to review the numerous studies that have used surface waves to infer lateral variations of seismic velocities in the crust and upper mantle, since the 1950s and up to this day. Many early studies documented the correlation of seismic velocity variations with surface tectonic features, using regional measurements of phase and group velocities of fundamental mode Love and Rayleigh waves in the period range 20–100 sec (e.g., Knopoff, 1972; Kovach, 1978) or at longer periods (e.g., Toksöz and Anderson, 1966; Dziewonski, 1971; Wu, 1972). Orginally, group velocity dispersion was obtained by measuring the times of arrival $t(\omega)$ of peaks and troughs of waves in a dispersed wave train. The time between two successive peaks would give the half period ($T/2 = \pi/\omega$), and the group velocity $U(\omega)$ would be computed as

$$U(\omega) = \frac{X}{t(\omega)} \tag{15}$$

where X is epicentral distance in kilometers and $t(\omega)$ is measured with respect to the earthquake's origin time. Since the early 1970s, the computation of "energy diagrams," as

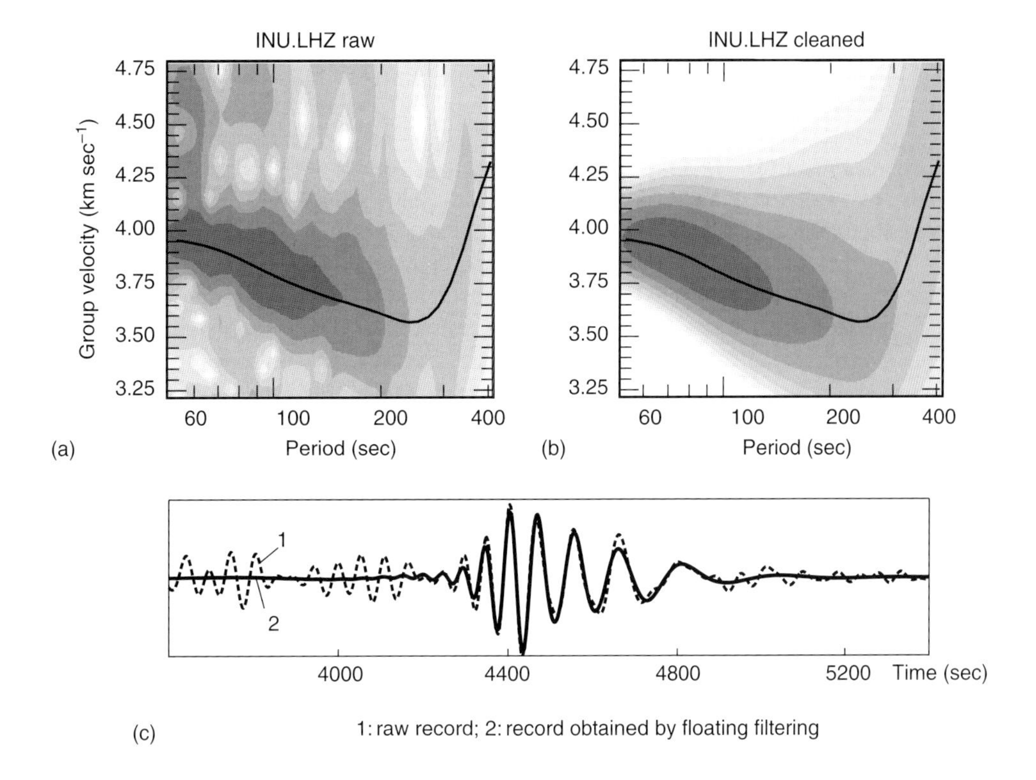

FIGURE 5 Example of group velocity dispersion diagrams obtained for the fundamental Rayleigh wave mode on the vertical record of the Chile M_S 6.8 earthquake of 15 Oct. 1997 at Geoscope station INU, using a multiple filtering approach as perfected in the FTAN method (e.g., Lander, 1989; Levshin *et al.*, 1994). (a) Dispersion diagram before filtering; (b) same after filtering. The group velocity dispersion curve obtained follows the maxima of energy as delineated by the gray scale contours. (c) Corresponding time-domain seismogram before and after variable filtering. The "raw" seismogram is bandpass filtered in the period 50–400 sec. (Courtesy of Boris Bukchin.)

described below, has greatly facilitated group velocity dispersion measurements (e.g., Fig. 5).

On the other hand, phase velocity $C(\omega)$ is obtained from the phase Φ of the Fourier spectrum of a dispersed wave train that has been corrected for the contribution of the source and the instrument [Eq. (2)]:

$$C(\omega) = \frac{X}{t_0 - [\Phi - N - (m/4) - 1/8]/\omega} \tag{16}$$

where t_0 is the start time of the Fourier window with respect to the event's origin time, m is the number of polar passages, and N is an integer arising from the 2π indeterminacy of the phase. This integer is determined first at long periods to obtain reasonable values of phase velocity compatible with well-constrained global models (e.g., PREM; Dziewonski and Anderson, 1981). It is then successively obtained at decreasing periods in such a way as to obtain a smooth phase velocity curve. This can become a problem at periods shorter than 30 sec, where small variations in phase velocity correspond to rapid cycling of the phase. The source phase also needs to be accurately corrected for.

Most early studies circumvented the issue of separating source and propagating effects by making measurements using the "two-station method," in which dispersion was measured between two stations approximately aligned with the epicenter on a great circle path, thus eliminating the common source phase. If, in addition, the two stations were located within a relatively homogeneous geological province, such measurements were called "pure path" and led directly to the determination of elastic velocity structure beneath that province. An extension of this method to "many stations" to infer structure beneath an array of stations spanning a geologically homogeneous region was also devised and used extensively in the 1960s (e.g., Knopoff *et al.*, 1967). To access remote regions devoid of stations but with relatively frequent earthquakes of magnitude 5.5 or larger, such as the Tibet Plateau, a "two-event" method was also devised. The success of this approach, which relied on an accurate independent knowledge of the source phase, was only possible under very restrictive conditions. In particular, the alignment of the three points (two epicenters, one receiver) had to be within 3° of the great circle path (e.g., Romanowicz, 1982b).

At longer periods, especially beyond the group velocity minimum at 200–250 sec, fundamental mode dispersion measurements were hampered by interference with higher modes (a problem also in the vicinity of the group velocity maximum around 60 sec), and the fact that the longest-period waves are not sufficiently well dispersed on first-arriving trains. In the early 1970s sophisticated filtering techniques were developed (Dziewonski *et al.*, 1969; Cara, 1973a,b) to isolate a surface wave mode along its group velocity curve (Fig. 5). This approach, later perfected by Levshin and collaborators in the FTAN method (e.g., Lander, 1989; Levshin *et al.*, 1994), involves two steps. In the first step, an "energy diagram" is formed, in which the energy contained in the seismogram is plotted as a function of period and time (i.e., group velocity). The maxima of the 2D diagram thus obtained delineate the group velocity curve of the dispersed surface wave modes present in the seismogram in the time window and frequency window considered. In a second step, the time-domain seismogram is then filtered using multiple filters centered, at each point in time, on the frequency corresponding to the maximum of energy at that time. The possibility of simultaneously inverting group and phase velocity data has been discussed by Yanovskaya and Ditmar (1990).

By allowing the extraction of a mode branch over many consecutive wave trains, these techniques also resulted in better measurements of the spheroidal and toroidal eigenfrequencies for the fundamental and several higher modes (e.g., Jobert and Roult, 1976). With the advent of digital recording in the mid-1970s and the expansion of global digital long-period and later broadband networks, the processing of the relatively long time-series needed to measure surface waves became much easier and opened the way in the 1980s to large-scale and global studies of upper mantle structure.

Large-scale studies first proceeded according to a regionalization scheme, in which it was assumed that the depth variation of seismic velocities is the same throughout each tectonic province. These studies confirmed and extended to longer periods (and hence larger depths) early results on the age dependence of structure in the oceans (e.g., Forsyth, 1975; Montagner and Jobert, 1983; Nishimura and Forsyth, 1989). The constraint of regionalization was soon relaxed, after Dziewonski *et al.* (1977) introduced the expansion of lateral heterogeneity into a basis of spherical harmonics (applied first, however, to the lower mantle using body wave travel time data).

On the global scale, as in source studies, two different approaches emerged, one based on dispersion measurements in the frequency domain, the other based on time-domain waveform inversions. Dispersion measurements in the frequency domain have been focused primarily on the fundamental mode, and thus limited in resolution beyond depths of 350–400 km. This approach has led to several generations of collections of global phase velocity maps at discrete frequencies. These maps can then be inverted jointly, in a second step, to infer 3D S-velocity structure of the upper mantle. This approach was pioneered by Nakanishi and Anderson (1983) and Nataf *et al.* (1986), and is now well established, with different research groups introducing various improvements in the measurement technique leading to increasingly high-resolution maps (e.g., Trampert and Woodhouse, 1995; Laske and Masters, 1996; Zhang and Lay, 1996; Ekström *et al.*, 1997). Trampert and Woodhouse (1995) designed an automatic method to measure phase and amplitude of surface waves in the period range 40–150 sec and invert the resulting dataset to obtain global phase velocity maps expanded in spherical harmonics up to degree 40. On the other hand, Ekström *et al.* (1997) used phase-matched filters to isolate the fundamental mode and make global dispersion measurements in the period range 35–150 sec and obtain models of phase velocity also up to degree 40 in spherical harmonics. In these studies, lateral variations of structure seem to be well resolved down to wavelengths of ~1000–2000 km. At the long-period end the period range is limited by the difficulty of separating consecutive wave trains, and at the short period end by the increased complexity of surface wave propagation in the strongly heterogeneous crust and uppermost mantle, resulting in lateral refractions and multipathing. The latter are not taken into account by the simple high-frequency, great circle propagation assumptions underlying the interpretation procedures.

Still in the frequency domain, but aiming at exploiting both the real and imaginary parts of the spectrum, Dziewonski and Steim (1982) developed a method to retrieve both phase velocity and attenuation averaged over complete great circle paths. These authors considered the frequency-dependent transfer function $T(\omega)$ between two consecutive fundamental mode Rayleigh wave trains traveling in the same direction along the great circle linking the epicenter and the station (say, R_n and R_{n+2}) and related it to the average dispersion and attenuation over the great circle path:

$$T(\omega) = \frac{R_{n+2}}{R_n}$$
$$T(\omega) = \exp\left\{-\pi a\left[i\left(2k(\omega) - \frac{1}{a}\right) + \frac{\omega q(\omega)}{2U(\omega)}\right]\right\} \tag{17}$$

where $k(\omega) = \omega/C$ is the real wavenumber, $U(\omega)$ is group velocity, $q(\omega)$ is the inverse of the quality factor, and the extra term $i\pi$ accounts for polar passages. The transfer function $T(\omega)$ is linearized and compared to that predicted by a reference model, to retrieve perturbations in phase velocity C and attenuation q along the great circle.

In order to proceed to a time-domain waveform inversion, synthetic seismograms needed to be computed in a reference model and perturbed. First-order perturbation theory had been developed in the 1970s, culminating in the work of Woodhouse and Dahlen (1978) and Woodhouse (1980b). Without going into detail beyond the scope of this chapter, we will briefly point out the key steps that led to the formalism routinely used today. An asymptotic expression relating the observed frequency shift $\delta\hat{\omega}_k$ of a normal mode k observed on a given source station path to the underlying average elastic structure over the corresponding great circle path was derived by Jordan (1978) and Dahlen (1979):

$$\delta\hat{\omega} = \frac{1}{2\pi}\int_0^{2\pi} \delta\omega_k(s)\, ds \tag{18}$$

where the local frequency shift $\delta\omega_k(s)$ represents an integral over depth of the difference between the local elastic structure and the reference Earth model at point s along the source–station great circle, weighted by the depth sensitivity kernels of mode k (e.g., Woodhouse and Girnius, 1982). The local frequency shift can be related to the phase velocity perturbation using Jeans's formula. A perturbed synthetic seismogram could then be formed as a sum over normal modes:

$$u(\Delta, t) = \mathrm{Re}\left[\sum_k A_0^k(\Delta)\, \exp(i(\omega_k + \delta\hat{\omega}_k)\, \exp(-(\alpha_k + \delta\hat{\alpha}_k)t)\right] \tag{19}$$

where Δ is epicentral distance in radians, and the amplitude term A_0^k, frequency ω_k, and attenuation α_k are computed in the reference spherically symmetric Earth model. $\delta\hat{\omega}_k$ and $\delta\hat{\alpha}_k$ are the real and imaginary parts, respectively, of the average frequency shift along the complete great circle path. However, this expression only reflects the effect of heterogeneity that is symmetric with respect to the center of the Earth, in agreement with the fact that, to zeroth order in the asymptotic approximation, normal modes are only sensitive to that part of lateral heterogeneity (e.g., Woodhouse, 1983).

Relating observed waveforms of a single (say, the first arriving) surface wave to a synthetic normal mode seismogram presented a challenge. From the practical point of view, this stumbling block was removed by Woodhouse and Dziewonski (1984), who introduced a "distance shift" that depended on the structure of the minor arc in the computation of the seismogram in a slightly heterogeneous Earth, as follows:

$$u(\Delta, t) = \mathrm{Re}\left[\sum_k A_0^k(\Delta + \delta\Delta)\, \exp(i(\hat{\omega}_k + \delta\hat{\omega}_k)\, \exp(-\alpha_k t)\right] \tag{20}$$

where the summation is over all modes k, and

$$\delta\Delta = \frac{\omega_k a}{(l + 1/2)U}(\delta\tilde{\omega} - \delta\hat{\omega}) \tag{21}$$

Here U is the group velocity of the mode k, a is the radius of the Earth, and $\delta\tilde{\omega}$ is the minor arc average of the local frequency shift $\delta\omega$, defined as

$$\delta\tilde{\omega} = \frac{1}{\Delta}\int_0^{\Delta} \delta\omega(s)\, ds \tag{22}$$

Theoretical proofs of Eqs. (20) and (21) were later given independently by Romanowicz (1987), Park (1987), and Mochizuki (1986a,b), in the framework of zeroth-order asymptotic coupling theory (Woodhouse, 1983; Tanimoto, 1984). The minor arc average $\delta\tilde{\omega}$ arises as a consequence of coupling of neighboring modes along a single dispersion branch, due to lateral heterogeneity, in contrast to the great circle average, which is obtained through coupling within isolated mode multiplets (Jordan, 1978; Woodhouse, 1983; Romanowicz and Roult, 1986).

The frequency-domain approach involves a two-step procedure: First, determine phase velocity maps of a single surface wave mode branch at individual frequencies; second, invert the dispersion curves obtained at every point on the globe to obtain the laterally varying depth distribution of shear velocities (since surface waves are primarily sensitive to shear velocity). On the other hand, the waveform approach is a single-step approach that obtains the 3D structure directly from the seismograms. Moreover, it can also be used for multimode waveforms, the approach favored by Woodhouse and Dziewonski (1984) and later extended, in combination with body wave travel time measurements, to whole-mantle tomography (e.g., Su *et al.*, 1994).

Whatever the approach, a significant issue regarding the inversion of fundamental mode surface waves is that of crustal corrections. Indeed, surface waves are sensitive to shallow crustal structure even at long periods (e.g., Dziewonski, 1971). Most studies perform crustal corrections in the framework of linear perturbation theory. However, Montagner and Jobert (1988) showed that the effect of strongly varying crustal structure is nonlinear and proposed a more accurate correction procedure based on a tectonic regionalization. The nonlinear part comes primarily from very large lateral variations in depth to Moho and other crustal discontinuities (for example, depth to Moho can vary by a factor of 4 between oceans and continents). The crustal contribution is computed in two steps. First, $i = 1 - n$ regional reference models are considered, and phase velocities as well as partial derivatives with respect to elastic parameters and discontinuity depths are computed for each of these models. In a second step, linear corrections are applied for each point along the source–station path, taking into account perturbations of the actual crustal elastic parameters and discontinuity depths with respect to the local tectonic model. The contribution $\Delta\Phi$ to the observed phase, due to crustal structure, is thus of the form:

$$\frac{\Delta\Phi}{\omega} = \sum_i \left[\int_S^R ds\,(M)\,\frac{\delta_i(M)}{C_i} - \int_S^R ds\,(M)\,\frac{\delta_i(M)}{C_i^2} \times \left(\sum_r \frac{\partial C}{\partial p_{ir}}\,\delta p_{ir} + \sum_k \frac{\partial C}{\partial h_{ik}}\,\delta h_{ik}\right)\right] \tag{23}$$

where $\delta_i(M)$ is 1 if point M belongs to a tectonic region i, and 0 otherwise; C_i is the phase velocity in crustal model i; h_{ik} is the depth of the kth discontinuity of the tectonic model i; and δh_{ik}

is the perturbation in that depth at point M. Also, p_{ir} is the rth elastic parameter in tectonic model i, and δp_{ik} is the perturbation in that parameter at point M. Montagner and Jobert (1988) give a slightly different expression for the crustal contribution, arguing that only the lateral variations in discontinuity depths need to be treated in a nonlinear fashion.

Developing accurate crustal models (e.g., Mooney *et al.*, 1998) remains a challenge for large-scale surface wave inversions for structure, as does the accurate treatment of crustal effects.

4.2 Higher-Mode Surface Wave Inversion

While well separated in the time domain from other mode branches and therefore well suited for single mode analysis techniques, fundamental mode surface waves have several shortcomings: at intermediate periods (say, 20–150 sec) their sensitivity to structure below about 200 km is poor, whereas longer-period mantle waves, which reach down to the top of the transition zone, have poor spatial resolution. In any case, resolving structure in the upper mantle transition zone, which is also poorly sampled by body waves, requires the analysis of higher-mode surface waves, whose sensitivity is larger at these depths (Fig. 6). They are also a powerful tool for investigating structures where low velocity zones may be present (e.g., Kovach and Anderson, 1964).

In some specific frequency windows, and for some specific source excitations, it has been possible to isolate and measure the dispersion of the first higher Rayleigh wave modes, either at very short period, where they are well separated on the seismogram (e.g., Crampin, 1964), or with the help of time-variable filtering at periods between 100 and 200 sec (e.g., Roult and Romanowicz, 1984). In general, however, higher-mode surface waves overlap in the time–frequency domain, and single mode dispersion methods therefore cannot be applied. For example, in the period range 80–150 sec, Rayleigh modes 3, 4, and 5 are well excited by intermediate-depth earthquakes, and are observed on seismograms as single energetic wavepackets, labeled "X-phase" by Jobert *et al.* (1977).

In the 1970s, in order to isolate higher modes of surface waves, similar array methods were developed independently by Nolet (1975) and Cara (1973b, 1978) and applied in the period range 20 to 100 sec to paths across Eurasia and the Pacific Ocean, respectively.

These methods require a linear regional array of stations approximately aligned with the epicenter (and not in a nodal direction) to separate modes in the $(\omega\ k)$ domain, where k is wavenumber. After correction for the instrument response, an array stack is formed (e.g., Nolet, 1975):

$$S(\hat{k},\omega) = \left| \sum_n F_n(\omega)\ \exp(i\Phi_n) H_n(\hat{k},\omega) \right|^2 \tag{24}$$

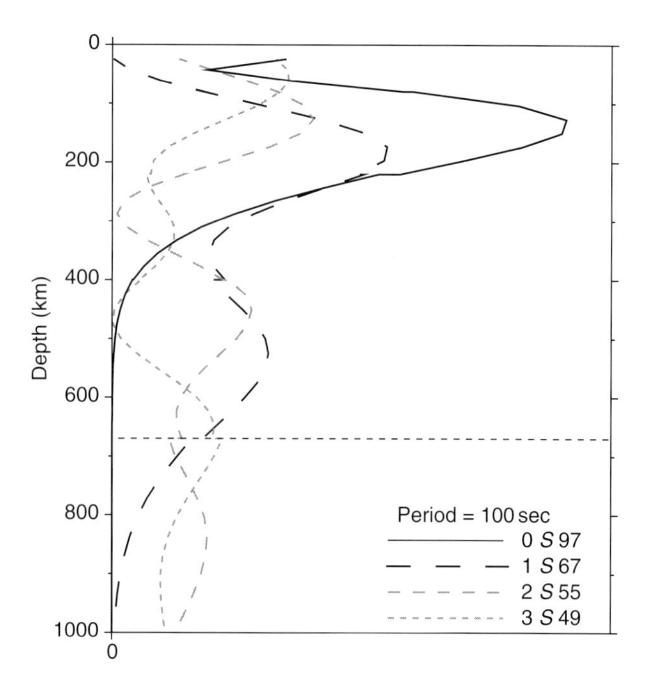

FIGURE 6 Depth profiles of partial derivatives of Rayleigh wave fundamental and first three higher modes with respect to S velocity at a period of ~100 sec, computed in the PREM model (Dziewonski and Anderson, 1981). At this period, the fundamental mode sensitivity peaks around 150 km depth and is negligible below 300 km depth. On the other hand, higher modes have significant sensitivity throughout the transition zone, and even, for modes 2 and 3, at the top of the lower mantle. (Courtesy of Yuancheng Gung.)

where the "array response" is

$$|H_n(\hat{k},\omega)|^2 = \left| \frac{1}{N} \sum_j \exp\left[i(k_n(\omega) - \hat{k})\Delta_j\right] \right|^2 \tag{25}$$

and $F_n(\omega)$, $\Phi_n(\omega)$, and $k_n(\omega)$ are, respectively, the amplitude spectrum, initial phase, and wavenumber of mode n, Δ_j is the epicentral distance to station j, and the spectrum of the multimode signal at station j is

$$W_j(\omega) = \sum_n F_n(\omega)\ \exp\left[ik_n(\omega)\Delta_j + i\Phi_n(\omega)\right] \tag{26}$$

The array-response has a peak at $\hat{k} = k_n$. These peaks are measured from plots of contours of the function $S(\hat{k},\omega)$ in the $(\omega,\hat{k})$ plane, from which dispersion curves $k_n(\omega)$ are then derived (Fig. 7).

This method is nevertheless limited in its application to a few regions of the world with relatively dense, linear arrays. The condition of alignment of the array in a narrow azimuthal range was later relaxed by Okal and Jo (1985, 1987), by correcting for the azimuthal variations of the initial phase $\Phi_n(\omega)$. This approach still suffered from lack of accuracy in the reading of the maxima in the $S(\hat{k},\omega)$ plots, in particular due to the presence of large side-lobes.

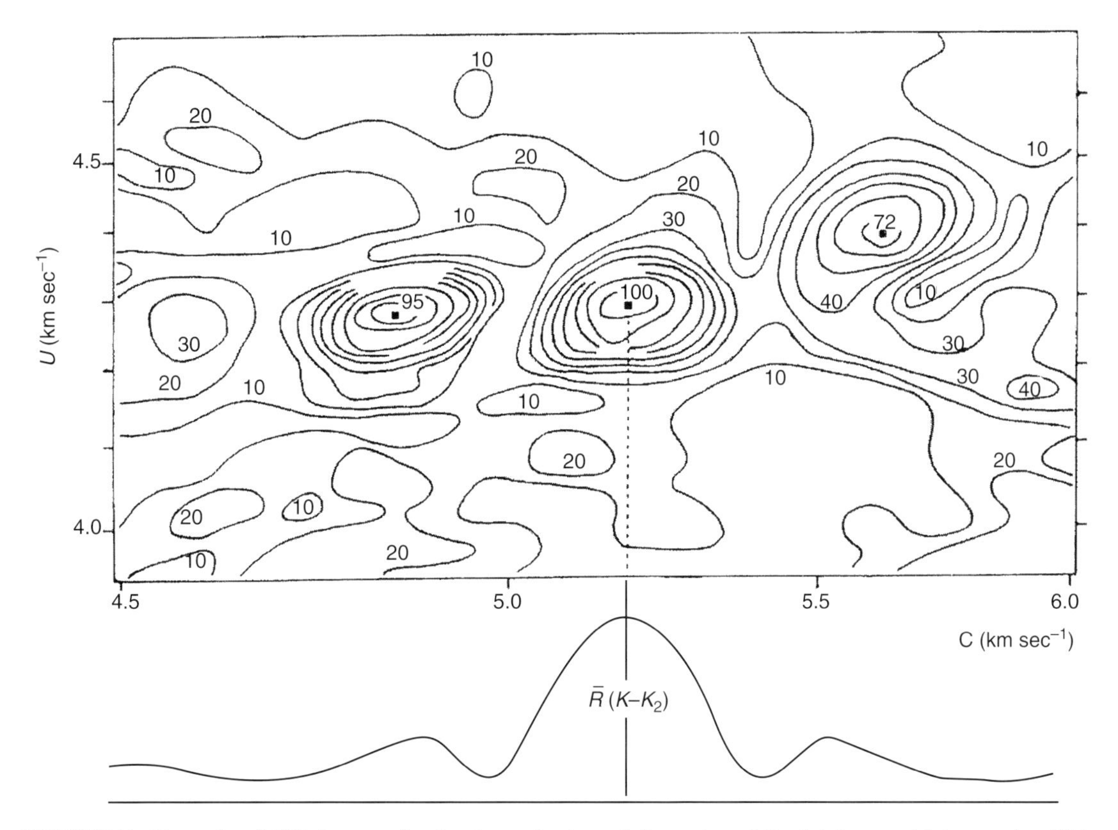

FIGURE 7 Example of UC diagram for the determination of dispersion of Rayleigh wave higher modes. This diagram was obtained from records of WWSSN stations in North America, at a period of 35 sec, for an event in Fiji-Tonga, which occurred on 13 Oct. 1969 at a depth of 246 km. The numbers indicate amplitude levels in percent of maximum. The maxima correspond to modes 1, 2, and 3, traveling with similar group velocities, but well separated in phase velocity. (Reproduced with permission from Cara, 1978.)

In the meantime, following a cross-correlation technique first proposed by Dziewonski *et al.* (1972), a waveform-based method involving the comparison of single record observed and synthetic seismograms was developed by Lerner-Lam and Jordan (1983) and later improved by Cara and Lévêque (1987) and Lévêque *et al.* (1991). In this approach, branch cross-correlation functions (bccf's) are formed between a particular single-mode synthetic and the observed seismogram, as follows. The observed seismogram $s(t)$ is written as the sum over overtone branches $u_n(t)$,

$$s(t) = \sum_n u_n(t) \tag{27}$$

and a synthetic seismogram $\tilde{s}(t)$ is computed for a reference spherically symmetric model suitable as a starting model for the average structure along the source–station path considered:

$$\tilde{s}(t) = \sum_n \tilde{u}_n(t) \tag{28}$$

The matched filter is the synthetic seismogram $\tilde{u}_m(t)$ for a particular mode branch m, so that the objective function to be minimized is formed as

$$\emptyset_n(\tau) = (s(t) - \tilde{s}(t)) * \tilde{u}_m(t) \tag{29}$$

The objective function should be peaked near $\tau = 0$, and the displacement of the peak from $\tau = 0$ is a function of the difference between the observed and the computed dispersion.

Partial derivatives with respect to model parameters can be formed to then invert for elastic model perturbations and obtain an average structure along the specific source–station path. Cara and Lévêque (1987) added a bandpass filter to further improve the resolution, and Lévêque *et al.* (1991) added a procedure to invert the envelope and phase of the filtered cross-correlograms. Indeed, this formalism lends itself also to the derivation of secondary observables such as group or phase velocities, which have a more linear dependence on the elastic structure, an approach discussed by Cara and Lévêque (1987) and also further developed by Gee and Jordan (1992). The main drawback of this cross-correlation methodology is the contamination of the single mode objective function by interference from other modes.

Other waveform inversion approaches have been developed that do not try to separate individual higher modes. Nolet (1990) introduced the "partitioned waveform inversion" approach in which inversion for elastic structure proceeds in two steps. Path integral parameters are defined and retrieved by nonlinear waveform fitting and, in a second step, inverted linearly for elastic structure. The main goal of this approach is to reduce the number of parameters to be fit in the nonlinear part of the inversion, and thereby to simplify the computations and increase their robustness. This method has subsequently been applied by Zielhuis and Nolet (1994) to retrieve upper mantle structure under central Europe, by van der Lee and Nolet (1997) in North America, and by Zielhuis and van der Hilst (1996) in Australia.

Alternatively, following the waveform modeling approach of Woodhouse and Dziewonski (1984), Li and Romanowicz (1995, 1996) introduce a global waveform modeling approach, based on a theoretical normal mode formalism that includes coupling across different mode branches (Li and Tanimoto, 1993). This formalism is particularly appropriate for body waves, in that it involves the computation of broadband sensitivity functions centered around the ray path, a more correct approach than the standard surface wave path average approximation, which averages kernels laterally between the source and the receiver. However, for higher-mode surface wave trains with sensitivity to the top of the lower mantle, across-branch mode coupling also starts to matter if increased resolution of structure is to be attained (e.g., Mégnin and Romanowicz, 1998). A similar mode-coupling formalism, using a propagating wave approach in the frequency domain, has been derived by Marquering and Snieder (1995) and Zhao and Jordan (1998).

Efforts to extract single mode observables using a waveform approach continued in the 1990s. Stutzmann and Montagner (1993, 1994) directly compare observed and synthetic waveforms of multimode wavepackets for paths between several earthquakes approximately in the same location, but at different depths, and for a given station. They set up an iterative, nonlinear inversion, in which they assume that the only unknowns are perturbations of the phase velocities of the fundamental mode and first few higher modes (in general three) along each path. They thus retrieve path-dependent dispersion properties, which, in a second step, they combine in a global inversion for structure. The drawback of this method is the constraint on path selection imposed by the requirement of finding several well-recorded neighboring earthquakes of different depths, which limits the number of paths that can be processed around the Earth, and therefore the resolution of the model.

Recently, van Heijst and Woodhouse (1997, 1999) proposed a "mode-branch stripping" method based on a bccf approach combined with a frequency-stepping procedure. These authors proceed iteratively to extract phase velocity and amplitude information, as a function of frequency, for the well-excited modes contained in a higher-mode wavepacket. They start at low frequency and proceed to higher frequencies, taking advantage of the smoothness of dispersion as a function of frequency, and thus are able to avoid the indeterminate 2π phase shifts that can be a problem at higher frequencies. By computing synthetics, they also determine the respective strengths of the different mode branches, and start the procedure by fitting the strongest mode branch. The improved synthetics for this mode branch are then subtracted from the data and the procedure is repeated for the next strongest mode with data and synthetics "stripped" of the already fitted preceding mode. At each step, the objective function to be minimized for mode q is then

$$\varnothing_q(\tau) = (s^{-p_1 p_2 \cdots}(t) - \tilde{s}^{-p_1 p_2 \cdots}) * \tilde{u}_q(m) \tag{30}$$

where $s^{-p_1 p_2 \cdots}(t)$ and $\tilde{s}^{-p_1 p_2 \cdots}$ are the observed and synthetic seismograms, respectively, stripped of the modes $p_1, p_2, \ldots$ that have already been fit.

van Heijst and Woodhouse (1997) argue that this approach not only provides better resolution for a given path than the original bccf approach, but, by allowing the determination of single mode dispersion and amplitude with many fewer restrictions on the paths than the method of Stutzmann and Montagner (1993), it also leads to better sampling of the globe and, not being limited to theoretical assumptions on great circle path propagation, leaves open the possibility of extracting information on off-path propagation.

4.3 Upper Mantle Anisotropy from Fundamental Mode Surface Waves

The incompatibility of dispersion curves measured for Love and Rayleigh waves has provided some of the earliest evidence for anisotropy in the crust and upper mantle (e.g., Anderson, 1961; Aki and Kaminuma, 1963; McEvilly, 1964). Its widespread character was confirmed in the 1970s and 1980s, mainly for fundamental modes, and in the oceans (e.g., Forsyth, 1975; Schlue and Knopoff, 1977; Mitchell and Yu, 1980; Montagner, 1985), but also for higher modes (Levêque and Cara, 1983). This discrepancy can be explained by introducing a transversely isotropic medium with vertical symmetry axis, and it is in this framework that the widely used Preliminary Reference Earth Model (PREM) has been constructed (Dziewonski and Anderson, 1981). Studies of the Love/Rayleigh discrepancy have been extended to the global scale and at long periods (100–250 sec) by Nataf *et al.* (1984), who first mapped the global lateral variations of transverse isotropy as a function of depth, expanded in spherical harmonics up to degree 6.

Recently, Ekström and Dziewonski (1998) used global Love and Rayleigh wave dispersion measurements in the period range 35–300 sec, complemented by long period waveform and travel time data, to invert separately for V_{SH} and V_{SV} in the upper mantle, and thus obtain a measure of the global distribution of transverse isotropy. Their model ("S20"), which is expanded laterally in spherical harmonics up to degree 20, confirms the widespread presence of this type of anisotropy in the upper mantle, and particularly singles out a strong anomaly in the central Pacific Ocean.

In addition, surface wave dispersion has also been shown to vary with azimuth, another indication of anisotropy (e.g., Forsyth, 1975; Suetsugu and Nakanishi, 1987). The global azimuthal variations of Rayleigh and Love wave dispersion at long periods (100–250 sec) were first mapped by Tanimoto and Anderson (1984, 1985), who showed that the fast direction appears to correlate with flow directions in the mantle.

Surface waves thus show two manifestations of anisotropy: (1) variations of phase and group velocities with azimuth (azimuthal anisotropy), and (2) inconsistent dispersion curves for azimuthally averaged Love and Rayleigh waves (transverse isotropy).

The theoretical expressions for the propagation of surface waves in an anisotropic plane layered medium was studied by Crampin (1970) and Smith and Dahlen (1973). The latter provided expressions for the azimuthal dependence of Love and Rayleigh waves in a slightly anisotropic medium. This formalism was extended to the case of a spherical Earth in a normal mode and spherical harmonics framework by Tanimoto (1986) and Mochizuki (1986b). Romanowicz and Snieder (1988) developed an equivalent formalism that does not require a spherical harmonics expansion and thus is applicable to regional studies, and Park (1997) generalized this to include source terms and compute complete synthetic seismograms in the Born approximation.

Here, we only present the basic asymptotic expressions that relate dispersion to anisotropic elastic structure. To first order in anisotropy, and at frequency ω, the azimuthal variation of phase velocity (Love or Rayleigh wave) is of the form:

$$C(\omega,\theta) = A_1(\omega) + A_2(\omega)\cos 2\theta + A_3(\omega)\sin 2\theta + A_4(\omega)\cos 4\theta + A_5(\omega)\sin 4\theta \qquad (31)$$

where θ is the azimuth of the wavenumber vector defined clockwise from north.

Montagner and Nataf (1986) provided expressions for the coefficients $A_i(\omega)$, which are depth integral functions, in terms of the following combinations of standard cartesian elastic coefficients C_{ij}, for both Love and Rayleigh waves:

Constant term (A_1):

$$\begin{aligned} A &= \rho V_{PH}^2 = \tfrac{3}{8}(C_{11} + C_{22}) + \tfrac{1}{4}C_{12} + \tfrac{1}{2}C_{66} \\ C &= \rho V_{PV}^2 = C_{33} \\ F &= \tfrac{1}{2}(C_{13} + C_{23}) \\ L &= \rho V_{SV}^2 = \tfrac{1}{2}(C_{44} + C_{55}) \\ N &= \rho V_{SH}^2 = \tfrac{1}{8}(C_{11} + C_{22} - \tfrac{1}{4}C_{12} + \tfrac{1}{2}C_{66}) \end{aligned} \qquad (32)$$

2θ azimuthal term:

$$\begin{array}{ll} (A_2) & (A_3) \\ \cos 2\theta & \sin 2\theta \\ B_c = \tfrac{1}{2}(C_{11} - C_{22}) & B_s = C_{16} + C_{26} \\ G_c = \tfrac{1}{2}(C_{55} - C_{44}) & G_s = C_{54} \\ H_c = \tfrac{1}{2}(C_{13} - C_{23}) & H_s = C_{36} \end{array} \qquad (33)$$

4θ azimuthal term:

$$\begin{array}{ll} (A_4) & (A_5) \\ \cos 4\theta & \sin 4\theta \\ E_c = \tfrac{1}{8}(C_{11} + C_{22}) - \tfrac{1}{4}C_{12} - \tfrac{1}{2}C_{66} & E_s = \tfrac{1}{2}(C_{16} - C_{26}) \end{array} \qquad (34)$$

The term A_1, independent of azimuth, involves only the five independent combinations of elastic coefficients needed to describe a transversely isotropic medium with vertical symmetry axis, labeled A, C, F, L, N (Love, 1927; Takeuchi and Saito, 1972). Montagner and Nataf (1986) showed that the partial derivatives of all the other terms only require the computation of partial derivatives with respect to the five parameters of a reference transversely isotropic Earth, thus providing the means to invert the azimuthal variations of surface wave phase velocities. They also estimated the different terms for realistic upper mantle models and showed that for Love waves, the N kernel is much larger than L and that the 4θ term is dominant, making it difficult to use Love waves to constrain azimuthal anisotropy on the global scale, whereas for Rayleigh waves the 2θ term prevails. Finally, because Rayleigh waves are sensitive both to shallow crustal and deeper mantle anisotropy, it is important to use a wide frequency range to resolve the depth dependence of anisotropy using surface waves.

Montagner and Nataf (1988) showed how this formalism could be used for the inversion of surface wave dispersion data including the azimuthal terms, under the assumption that the material possesses a symmetry axis (orthotropic medium). In this case the 3D model can be described using seven parameters (plus density): the five parameters A, C, F, L, N describing transverse isotropy, and two angles describing the orientation in space of the axis of symmetry. These authors used the inversion algorithm of Tarantola and Valette (1982) in its continuous form, as adapted to surface waves by Montagner (1986) with the introduction of an appropriate spatial correlation function, providing the description of the model and its error distribution on a grid of points (rather than a global basis function expansion), suitable for regional studies. This methodology, called "vectorial tomography," was first applied by Montagner and Jobert (1988) to retrieve lateral heterogeneity and anisotropy, described at each point by an amplitude scalar, and a direction vector (under the hypothesis of orthotropy), to the study of the Indian Ocean, and the results were interpreted in terms of flow in the mantle.

This procedure involves two steps: (1) determining maps of azimuthal variations of anisotropy at individual frequencies, and (2) inverting the retrieved coefficients locally for heterogeneity and anisotropy variations with depth. A similar approach was subsequently applied to many other regions (e.g., Mocquet *et al.*, 1989; Roult *et al.*, 1994; Griot *et al.*, 1998; Pillet *et al.*, 1999), and extended to the global study of heterogeneity and anisotropy in the upper mantle by Montagner and Tanimoto (1990, 1991). In this latter case, an approximation was introduced for the calculation of the data covariance matrix in the Tarantola and Valette (1982) framework, to make the computations manageable for a global dataset. While some questions linger about trade-offs between lateral heterogeneity and anisotropy in this type of inversion, Montagner and collaborators have shown that they can explain their datasets with fewer parameters when azimuthal anisotropy is considered than when it is ignored.

In order to circumvent the trade-off between lateral heterogeneity and azimuthal anisotropy, Park and Yu (1992, 1993) have looked for other diagnostic effects of anisotropy in long-period surface waves, such as waveform anomalies caused by Rayleigh–Love coupling, which generates "quasi-Love" waves on vertical components and "quasi-Rayleigh" waves on transverse components. Yu and Park (1994) have documented such observations, best seen in nodal directions of strike-slip sources, in the Pacific Ocean and inferred small scale variations in anisotropy related to tectonic features.

4.4 Effects of Scattering and Non-Great Circle Path Propagation

Until now, most regional and global models using surface waves have been derived using the standard "path-average" approximation, which is a high-frequency asymptotic approximation in which the propagation is assumed to be confined to the great circle path between the source and the receiver. This is valid only if the wavelength of lateral variations of structure is long with respect to that of the surface waves considered.

In fact, over the years, there have been many observations indicating that lateral heterogeneity is strong enough to cause departures from this simple hypothesis, as evidenced, for example, at short periods, where 20 sec surface waves sensitive to shallow crustal structure consistently show multipathing (e.g., Capon, 1970; Bungum and Capon, 1974; Fig. 8), or at long periods ($T > 100$ sec), where amplitude anomalies have been widely documented, with, for example, later-arriving trains showing larger amplitudes than the ones preceding them (Fig. 9), which cannot be explained by lateral variations of anelasticity (e.g., Lay and Kanamori, 1985; Roult *et al.*, 1986; Park, 1987).

Several approaches have been developed to try to explain these effects and exploit them to obtain better constraints on lateral variations of structure. The principal ones are cast in the

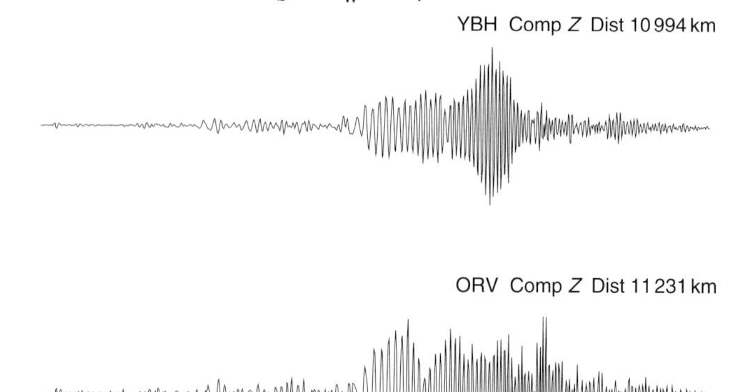

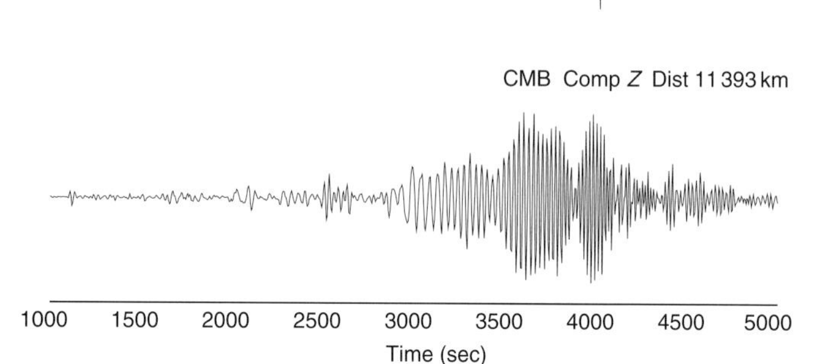

FIGURE 8 Vertical component records at BDSN stations YBH, ORV, and CMB for the 12/06/2000 earthquake in Turkmenia (M_W 7.0; depth 33 km). The records have been bandpass filtered between 0.001 and 0.03 Hz. These records show clear evidence of multipathing of 20–30 sec surface waves. The corresponding great circle path has an azimuth coming from the north which runs quasi parallel to the major structural boundaries in the crust in northern California (including the coast line). Multipathing is more severe at more southerly stations ORV and CMB, indicating that it does indeed originate on the station side. (Courtesy of Yuancheng Gung.)

framework of three different formalisms, depending on the application: ray theory, scattering theory, and a coupled-mode formalism. Each of these methods lends itself to various degrees and, under specific circumstances, to inversion.

Working in a 2D cartesian framework for the description of surface wave propagation in a smoothly laterally varying medium, Woodhouse (1974) introduced the concept of local modes. These are the surface wave modes corresponding to a laterally homogeneous model, which locally has the depth distribution of the laterally varying model. If the medium is smooth, the local mode branches propagate as independent wave trains, the dispersion and displacement of which are modified according to the evolution of the local modes. However, when the lateral variations are sharp (for example, in the presence of a structural discontinuity such as an ocean–continent boundary), the coupling of the local modes cannot be neglected, and its strength depends on the width of the structural transition zone (Kennett, 1972). Kennett (1984)

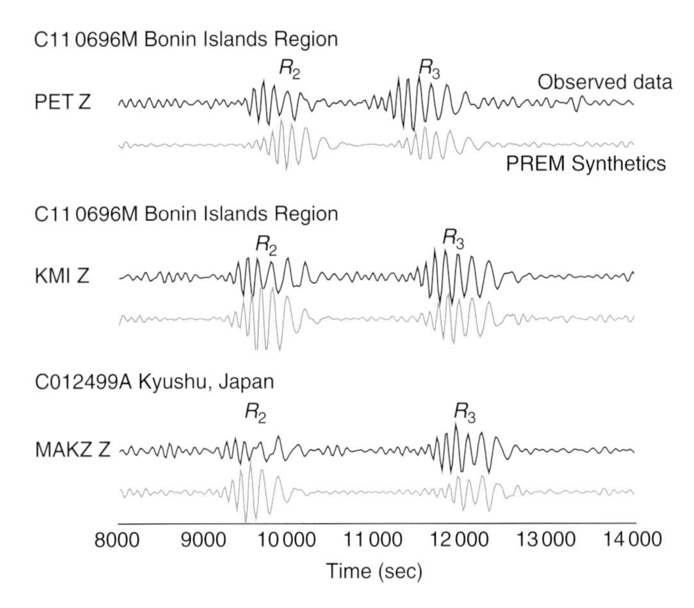

FIGURE 9 Examples of amplitude anomalies due to focusing for Earth-circling Rayleigh waves. Top traces are observed waveforms and bottom traces are synthetics computed in the spherically symmetric PREM model (Dziewonski and Anderson, 1981). In the synthetics, the later-arriving trains always have smaller amplitudes than the observed ones. In these examples, the data exhibit larger amplitudes in R_3 than in R_2. The data have been bandpassed filtered between 90 and 400 sec (event C110696M) and 90 and 300 sec (event C012499A). (Courtesy of Yuancheng Gung.)

derived a formalism for mode-coupling in 2D laterally varying structures using a representation of the displacement field on the modes of a reference, laterally homogeneous structure, which still requires continuity in the lateral variations. The coupled, reference mode approach has been extended to the 3D scattering case and for certain types of structures, in cartesian coordinates, by Bostock (1991, 1992) and Bostock and Kennett (1992).

To overcome the restriction of smooth lateral variations, coupled local modes need to be considered. In this case, the wavefield is expanded in terms of a laterally varying local mode basis. A specific case, that of Love waves in a one-layered laterally varied structure, was studied by Odom (1986), whereas Maupin (1988) generalized the coupled local mode approach to any 2D structure. In this case, no reference laterally homogeneous structure is needed.

While 2D mode coupling can provide exact computations of coupling and conversion, and is appropriate for certain applications such as the study of continental margins, a scattering formalism has been, so far, more easily implemented in the context of inversion for the study of 3D heterogeneity in relatively smooth media.

Several applications of ray theory have been proposed to study the departure of surface wave paths from the conventional great circle approximation. Theoretical calculations were proposed by Woodhouse (1974), Babich *et al.* (1976), and Hudson (1981). Using a simple transformation from cartesian to spherical coordinates, Jobert and Jobert (1983) traced low-frequency Earth-circling mantle waves in a smoothly varying heterogeneous model using a combination of spherical Earth normal mode theory and a Gaussian beam computation scheme (Červeny *et al.*, 1982), and showed significant amplitude and travel time anomalies for a model of heterogeneity with strength of $\sim 5\%$. At the regional, shorter period scale, Tanimoto (1990) showed how to compute the full wavefield of surface waves in a smoothly varying heterogeneous medium using an approximate scalar wave equation, and applied this to illustrate the distortion of surface waves propagating across California. Numerical computations showing the strong effects of lateral heterogeneity on the surface wavefield, in the framework of ray theory, were also performed by Yanovskaya and Roslov (1989). Tromp and Dahlen (1992a,b) developed a JWKB theory for the propagation of monochromatic surface waves in an Earth model with smooth lateral variations. These studies, however, did not result in any inversion procedures.

On the other hand, Yomogida and Aki (1985) used the Gaussian beam approach to compute intermediate period (20–80 sec) fundamental mode surface waveforms. They showed that, in this Born approximation framework, the amplitudes of surface waves depend on the second spatial derivatives of the phase velocity distribution, whereas the ray path depends on the first spatial derivatives and the phase on the phase velocity distribution itself. Based on this approach, they developed an inversion method for amplitude and phase data and applied it to retrieve phase velocity anomalies in the Pacific Ocean (Yomogida and Aki, 1987). The same dependence of phase and amplitude anomaly on the underlying phase velocity structure was derived independently by Woodhouse and Wong (1986) using ray theory based on the formalism of Woodhouse (1974). These authors showed that amplitude anomalies, as observed for long-period mantle waves on multiple Earth-circling paths, can be caused by focusing or defocusing due to lateral heterogeneity. Wong (1989) applied this theory to the retrieval of very long-period global mantle wave phase velocity maps from measured phase and amplitude anomalies. Since they depend on transverse gradients of structure along the propagation path, the amplitude anomalies can help constrain smaller-scale lateral variations of structure. The expression for the phase ($\delta\Phi$), amplitude ($\delta A/A_0$), and polarization (ν) anomalies obtained in this framework, at epicentral distance Δ, are of the forms, respectively:

$$\delta\Phi = \frac{1}{\Delta}\int_0^{\Delta}\frac{\delta c(\pi/2,\phi)}{c_0}\,d\phi \tag{35}$$

$$\frac{\delta A}{A_0} = \left(\frac{1}{2\sin\Delta}\right)\int_0^{\Delta}\left[\sin(\Delta-\phi)\partial_\theta^2\left(\frac{\delta c}{c_0}\right)\sin\phi - \partial_\phi\left(\frac{\delta c}{c_0}\right)\cos\phi\right]d\phi \tag{36}$$

$$\nu(\Delta) = -\text{cosec}(\Delta)\int_0^\delta \sin\phi\,\partial_\theta\left[\frac{\delta c(\pi/2,\phi)}{c_0}\right]d\phi \qquad (37)$$

In these equations, the great circle path under consideration has been rotated to lie along the Equator, along which the angle ϕ is measured ($\phi=0$ at the epicenter and $\phi=\Delta$ at the receiver). ν is related to the azimuth Θ along the great circle path by $\nu=\tan(\Theta)$. The local perturbation in phase velocity at a point along the source receiver path is $\delta c/c_0$. We note that the phase $\delta\Phi$ depends only on the phase velocity along the source station path [Eq. (35)]. On the other hand, the amplitude and polarization depend on first and second derivatives of the phase velocity, that is, on gradients of the phase velocity distribution, which in turn depend on short-wavelength features of that distribution. The expression under the integration sign in Eq. (36) actually corresponds to the transverse gradient in the direction perpendicular to the source station great circle (e.g., Romanowicz, 1987). Thus, amplitude and polarization anomalies potentially provide constraints on shorter-wavelength structure. Romanowicz (1987) showed that these expressions are equivalent, for relatively short propagation times, to those obtained for the perturbation to a low-frequency seismogram using a normal mode formalism, under an asymptotic approximation up to order $(1/l)$ in the development of spherical harmonics Y_l^m. In particular, the normal mode derived expression for the amplitude perturbation is

$$\frac{\delta A}{A} = -\frac{a\Delta}{U}\left(\frac{\tilde{D}}{2(l+1/2)}\right) \qquad (38)$$

where U is group velocity and $\tilde{D}$ is the minor-arc average of the second transverse derivative of the local frequency perturbation along the source–station great circle path. These expressions for amplitude anomalies were considered by Romanowicz (1990) and Durek *et al.* (1993) in attempts to separate the effects of focusing and intrinsic attenuation on the global scale.

Laske (1995) and Laske and Masters (1996) used expression (37) for the azimuth anomaly to interpret observed polarization anomalies, measured using a multitaper technique (Park *et al.*, 1987), in terms of lateral variations of phase velocities on the global scale and thus retrieve shorter-scale variations than can be obtained with the same global distribution of observations using phase data alone. Polarization analysis using the same multitaper technique had been proposed previously for the study of lateral heterogeneity by Lerner-Lam and Park (1989) and Paulssen *et al.* (1990).

Expression (35) for phase anomaly is correct to first order in lateral heterogeneity. Pollitz (1994) calculated the second-order contribution to the phase and concluded that it is unimportant for global phase velocity models expanded up to degree 12 but is potentially important for rougher models (degree of expansion ≥ 16). That study suggested systematic bias in phase velocity maps that do not account for the second-order effect, though the potential magnitude of that bias requires further exploration. This effect arises from structure gradients perpendicular to the great circle path, suggesting that it should be considered jointly with polarization measurements in global phase velocity inversions.

In the framework of scattering theory, the single-scattering Born approximation, developed for surface waves by Snieder (1986) in a flat Earth geometry and Snieder and Nolet (1987) in a spherical Earth geometry, has been the subject of many studies. Born scattering is well suited for inversion since the scattered wavefield depends linearly on structural perturbations. Indeed, following Snieder (1988a), the surface wave displacement field can be decomposed into a "path average" part u_0^{pava}, computed in a classical fashion, and a Born perturbation δu^{born}:

$$u = u_0^{\text{pava}} + \delta u^{\text{born}} \qquad (39)$$

where

$$\delta u^{\text{born}} = \iint \sum_{\nu,\sigma} \mathbf{p}^\nu \frac{e^{i(k_\nu a\Delta_2+\pi/4)}}{(\sin\Delta_2)^{1/2}} S^{\nu\sigma} \frac{e^{i(k_\sigma a\Delta_1+\pi/4)}}{(\sin\Delta_2)^{1/2}} \times (\mathbf{E}^\sigma : \mathbf{M})\sin\theta\,d\theta\,d\phi \qquad (40)$$

where σ, ν are the excited and scattered modes, respectively, $\mathbf{p}^\nu$ is the polarization vector at the receiver, Δ_1 and Δ_2 are the angular distances to the scatterer from the source and the receiver, respectively, and the scattering matrix $\Sigma_{\nu,\sigma}$ depends on elastic structural parameters via couping kernels K:

$$S_{\nu\sigma} = \int_0^a (K_\mu^{\nu\sigma}\delta\mu + K_\rho^{\nu\sigma}\delta\rho + K_\lambda^{\nu\sigma}\delta\lambda)\,dr \qquad (41)$$

where $\delta\mu$, $\delta\rho$, and $\delta\lambda$ are 3D perturbations to the reference laterally homogeneous model described by $\mu(r)$, $\rho(r)$, $\lambda(r)$. For intermediate- and long-period surface waves, perturbations in S velocity β need only be considered, and $S_{\nu\sigma}$ reduces to

$$S_{\nu\sigma} = \int_0^a K_\beta^{\nu\sigma}\,d\beta \qquad (42)$$

If $d\beta$ is expanded using 3D spatial basis functions, this leads to the linear expression

$$\delta u^{\text{born}} = \sum_i \gamma_i a_i \qquad (43)$$

where a_i can be evaluated in the reference medium, and γ_i are the expansion coefficients of $\delta\beta$ in the 3D spatial basis.

Snieder (1988a) showed how to set up a regional tomographic inversion using many fundamental surface waveforms. This method has been applied to the area of Europe and the Mediterranean by Snieder (1988b) and to North America by Alsina *et al.* (1996), and has been extended to the case of multimode wave trains by Meier *et al.* (1997). In the latter two cases, the partitioned waveform method of Nolet (1990) has been used in conjunction with Born scattering, to minimize the computational effort involved.

Snieder and Romanowicz (1988) generalized this scattering formalism in the case of spherical earth geometry, and in a normal mode context, by applying an operator formalism as introduced by Romanowicz and Roult (1986) and Romanowicz (1987), avoiding expansion in spherical harmonics of the 3D Earth model, as previously employed (Woodhouse and Girnius, 1982), and thus making it applicable to the case of any single scatterer (and not only to smooth heterogeneity). In this formalism, the addition theorem of spherical harmonics is used:

$$\sum_m Y_l^{m*}(\theta_s,\phi_s)Y_l^m(\theta,\phi) = \gamma_l Y_l^0(\lambda)\sum_m Y_l^{m*}(\theta,\phi)Y_l^m(\theta_r,\phi_r) = \gamma_l Y_l^0(\beta) \quad (44)$$

where l is the angular order of a normal mode k, (θ_s, ϕ_s), (θ_r, ϕ_r) and (θ, ϕ) are the coordinates of the epicenter, the receiver, and the scattering point, respectively (Fig. 10), and $\gamma_l = (2l + 1/4\pi)^{1/2}$. Angular distances λ and β are defined in Figure 10. The interaction coefficient $Z_{K,K'}^{m,m'}$ between modes K, K' and their singlets m, m' can then be written simply as

$$Z_{K,K'}^{m,m'} = \sum_{i=0}^{i=2}\iint Op_i(X_l^0(\lambda))\delta\omega_{KK'}^i(\theta,\phi)Op_i(X_l'^0(\beta)\,d\Omega \quad (45)$$

where the operators Op_i are linear combinations of differential operators acting on the coordinates (θ, ϕ) of the running point,

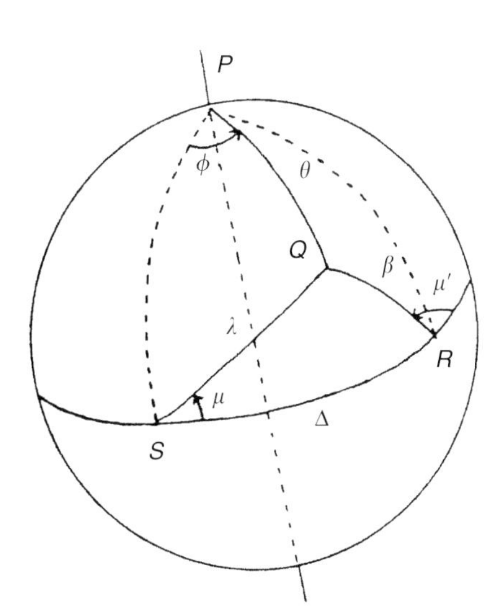

FIGURE 10 Geometry for single scattering corresponding to Eq. (45). S and R are the locations of the epicenter and the receiver, respectively. Q is a generic point on the surface of the sphere, and P is the pole of the source receiver great circle γ. (Reproduced with permission from Romanowicz, 1987.)

and incorporate source and receiver effects, computed in the reference spherically symmetric model. The integral is over the unit sphere, and $\delta\omega_{KK'}^i$ are local frequency perturbations, describing depth-integrated effects of perturbations in the elastic model. To zeroth order in an asymptotic expansion in orders of $1/l$, the summation over i reduces to only one term $(i = 0)$. Expression (45) has the same form as Eq. (40) and they are equivalent in the surface wave, high-frequency limit (Snieder and Romanowicz, 1988). Romanowicz and Snieder (1988) extended the same formalism to the case of a general anisotropic perturbation.

The Born approach is flexible as it is applicable to general 3D structures. However, it is a poor approximation in the case of strong heterogeneity, or when the region of scattering is large, in which case multiple scattering can play an important role.

On the other hand, Friederich *et al.* (1993) proposed an enhanced scattering theory for surface waves in a cartesian geometry, in which multiple forward scattering and single backward scattering are included. They use a potential formalism that can be summarized through the consideration of a scattering kernel K_p^{nm} that includes all the details of the regional scattering problem relevant to two modes n and m, depending only on the structural parameters. The surface wave displacement u for a mode m is decomposed as

$$u_m(x, y, z) = U_m(z)\Phi^m(x, y) \quad (46)$$

where Φ^m is the displacement potential and U is a vertical eigenfunction

$$\Phi_S^{nm}(x, y) = \iint dx'\,dy'\,K_p^{nm}(x, y|x', y')\Phi^m(x', y') \quad (47)$$

where (x, y), (x', y') are the coordinates of the observation point and the scattering point respectively, Φ^m is the displacement potential of mode m and Φ^{nm} is the scattered potential of mode n generated by mode m. The computation is simplified by discretizing the model into cells and assuming a piecewise plane wave approximation in each cell. When Φ^m is the potential corresponding to a reference 1D model of mode m, this is equivalent to single scattering theory. Friederich *et al.* (1993) use the reference potential for the row of cells closest to the source, compute the scattered potential $\sum\Phi_S^{nm}$ at all other rows, add the scattered wavefield and use this as starting potential to compute the scattered wavefield in the next row, and so on. They show that the computation thus includes all multiply forward scattered waves, but is incomplete for the backscattered field. This is not a problem in most cases, however, except when sharp, reflecting discontinuities are present. Subsequently, Friederich (1999) extended this approach to the case of spherical geometry and normal mode summation in a global Earth framework, using the operator formalism for the mode coupling terms of Romanowicz (1987). To limit the amount of computation, they compute only the seismic displacement wavefield between the S wave and the end of the first-arriving surface wave, at epicentral distances shorter than 50°.

Wielandt (1993) showed that the measured (or dynamic) phase velocity of teleseismic surface waves in a regional context is different from the structural phase velocity in the region considered. The former depends strongly on the geometry of the incoming wavefield, which can be distorted by structure along the propagation path outside of the area of study. Based on the work of Friederich *et al.* (1993), Friederich and Wielandt (1995) proposed a method to invert jointly for incoming wavefields and the heterogeneous phase velocity structure within a study region. In this approach, they include the multiple forward and backward scattering terms. They use a plane wave decomposition of the incoming wavefield and they reduce the nonuniqueness of the joint inversion by applying an "energy criterion," constraining the energy of the total wavefield averaged over all the events considered to be equal to the mean squared amplitude at the stations, also averaged over all events. This method has been applied to fundamental mode surface waves in southern Germany by Friederich (1998) and in northern California by Pollitz (1999). Notably, Pollitz (1999) showed how to relate the inverted 2D phase velocity maps to 3D structure, taking into account both isotropic [$i = 1$ in Eq. (45)] and nonisotropic scattering interactions. The application of these promising methodologies is still in an early development stage, as they present many computational challenges.

4.5 Surface Wave Attenuation Measurements and Inversion for Upper Mantle Anelastic Structure

Surface wave amplitude measurements can be used to try to retrieve information about the anelastic structure of the crust and upper mantle. In the absence of perturbing effects due to scattering and focusing, the amplitude spectrum of a monomode wave train i, usually the fundamental mode, can be written as

$$A_i(\omega) = A_0(\omega)\,\exp(-\eta_i(\omega))X_i \tag{48}$$

where X_i is the epicentral distance in kilometers and $A_0(\omega)$ represents the amplitude at the source. The attenuation coefficient can be related to Q through (e.g., Aki and Richards, 1980)

$$\eta(\omega) = \frac{\omega}{2C(\omega)Q(\omega)} \tag{49}$$

where C is phase velocity. Surface wave measurements provide the primary constraints on attenuation structure (both variations with depth and lateral variations) in the crust and uppermost mantle. To investigate the radial structure in Q over a wider depth range, long-period surface wave or normal mode attenuation measurements can be used. Both the radial and the lateral structures in Q are, however, to this day, less well constrained than elastic structure, due to a large extent to the contamination of amplitude data by scattering and focusing effects due to wave propagation in a heterogeneous elastic Earth, as described in the previous section. There still exists an apparent discrepancy (on the order of 15%, Fig. 11) between the measurements of fundamental mode Q obtained using a standing wave approach (e.g., Masters and Gilbert, 1983; Smith and Masters, 1989; Widmer *et al.*, 1991) and those using a propagating wave approach (Dziewonski and Steim, 1982; Romanowicz, 1990, 1995; Durek *et al.*, 1993; Durek and Ekström, 1996). This was investigated by Durek and Ekström (1997), who proposed that the discrepancy could be due to measurement techniques, with the presence of noise leading to an overestimation of Q from normal modes, which require the computation of spectra over long time-series (typically 24 hours). On the other hand, Masters and Laske (1997) have questioned the accuracy of surface wave measurements of Q at very long periods, in particular due to the difficulty of selecting appropriate time windows to isolate the fundamental

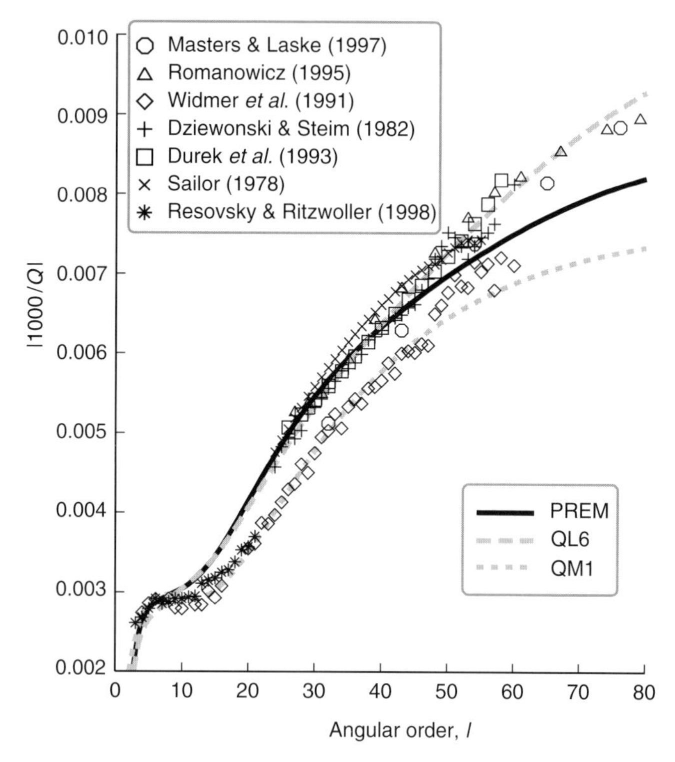

FIGURE 11 Fits of various mantle Q_μ model predictions to spheroidal mode data (Sailor and Dziewonski, 1978; Widmer *et al.*, 1991; Resovsky and Ritzwoller, 1998) and Rayleigh wave data (Dziewonski and Stein, 1982; Durek *et al.*, 1993; Romanowicz, 1995; Masters and Laske, 1997), illustrating the discrepancy between measurements obtained using standing wave and propagating wave approaches, respectively. Models QL6 and QM1 were constructed by Durek and Ekstrom (1996) and Widmer *et al.* (1991) respectively. (Reproduced with permission from Romanowicz and Durek, 2000.)

mode, in the presence of overlapping wave trains. This issue remains to be resolved.

Regional studies of amplitudes of fundamental mode surface waves in the period range 5–100 sec (Anderson *et al.*, 1965; Mitchell, 1975; Canas and Mitchell, 1978, 1981) and of *Lg* waves (Xie and Mitchell, 1990a,b) have long established the existence of large variations of Q in the crust and uppermost mantle, correlated with tectonic provinces and in particular with the age of the oceans, and with time elapsed since the latest tectonic activity on continents. A recent review can be found in Mitchell (1995). At longer periods, large lateral variations also exist (e.g., Nakanishi, 1978; Dziewonski and Steim, 1982). These lateral variations can be an order of magnitude larger than those in elastic velocity.

Progress in using surface wave data to constrain global 3D anelastic structure of the upper mantle has been slow because, as mentioned previously, of the inherent difficulty of measuring attenuation in the presence of focusing and scattering effects that can be as large as or larger than anelastic effects and depend strongly on the short-wavelength details of the elastic structure. Indeed, as discussed previously [Eq. (36)] to first order these effects depend on the transverse gradients of structure along the great circle path linking epicenter and station, which in turn depend on terms of the form s^2C_{st} where C_{st} are the coefficients of an expansion in spherical harmonics of the elastic model (s and t are the degree and order of the corresponding spherical harmonic Y_s^t). If the elastic structure of the Earth were accurately known to short wavelengths, one could first correct for its effects on the amplitude.

In the meantime, indirect methods have been used to minimize contamination of amplitudes by unwanted elastic effects. Romanowicz (1990) and Durek *et al.* (1993) took advantage of the fact that, in the linear approximation, focusing and anelastic effects could be separated by combining measurements over several consecutive wave trains, since attenuation effects are always additive whereas focusing depends on the direction of propagation. However, the longer the wave path the more the waves are affected by 3D elastic structure, and the harder it is to account for that in a simple approximate fashion. Another source of bias comes from uncertainties in the amplitude at the source. Romanowicz (1994) developed a method that involves computing attenuation coefficients in two ways: (1) using first-arriving trains only; (2) using the first three wave trains. The first measurement involves a source bias, but is less contaminated by elastic effects. On the other hand, in the second measurement, the source effect has been canceled out; however, the attenuation measurement is less accurate, and generally shows large variations with frequency, due to increased elastic effects over the longer paths. Comparison of the two $\eta(\omega)$ curves allows determination of a source correction factor and thus allows one to obtain a relatively accurate attenuation measurement using first- and second-arriving trains only.

Progress in the retrieval of accurate Q information from long-period surface wave data for global modeling is coupled to the development of efficient techniques for the modeling of elastic scattering effects. Recently, Billien *et al.* (2000) have proposed an approach in which phase and amplitude measurements of Rayleigh waves at intermediate periods (40–150 sec) are jointly inverted for elastic and anelastic structure, taking into account the first-order asymptotic focusing term [Eq. (36)]. It is still not clear, however, whether such a first-order, "smooth" approximation is sufficient to rule out biases due to scattering by small-scale heterogeneity.

Another issue is to account for dispersion in velocities due to attenuation, using an absorption band model for the Earth. This has been found to explain the discrepancy between average velocity models obtained using high-frequency body waves and surface waves (Kanamori and Anderson, 1977). However, most of the frequency dependence in surface waves is related to the effects of depth-dependent elastic structure. Surface waves alone cannot resolve the frequency dependence of Q. More detailed recent reviews on global Q structure can be found in Romanowicz (1998) and Romanowicz and Durek (2000).

5. Conclusions

We have reviewed the evolution of the main methodologies based on the inversion of surface wave data over the last several decades. We find that much progress has been made in the last 20 y in the development of surface wave methods to retrieve both source parameters and Earth structure using surface wave data. This progress has built upon fundamental theoretical developments of the 1960s and 1970s, during which the basis for the computation of earthquake source excitation as well as modeling of surface wave dispersion properties in terms of the Earth's structure was established. In the 1980s, surface wave studies have benefited from progress in computer speed and capacity, which, together with the deployment of a new generation of digital, broadband, high dynamic range global and regional seismic networks, has opened the way to large-scale source and structure inversions, with recent efforts to include anisotropic and anelastic contributions, as well as to systematically exploit information contained in overtones.

Most inversions, so far, have been performed under the assumption that the effects of propagation in the laterally heterogeneous Earth can be accounted for using the simple great circle path average approximation, which assumes that surface waves are sensitive only to the structure in the vertical plane containing the source and the receiver, and to the laterally averaged structure along this path. Beyond this approximation, there are two particularly important issues. One is the computation of more accurate kernels for propagation in the vertical plane containing the source and the receiver. This is important for waveform inversion that includes body waveforms, and is relevant, in the context of this review, to the inversion of overtone waveforms sensitive to the mantle transition zone and greater depths. The first step in this direction has been the incorporation of across-branch coupling effects computed asymptotically to zeroth order, both in a normal mode and in a propagating wave framework. This has started to be put into practice in the context of global waveform inversion (Li and Romanowicz, 1996; Mégnin and Romanowicz, 2000), and should help improve the resolution of shorter-wavelength features of tomographic models. The other issue is that of how to account for scattering and focusing effects, which can be viewed as departures of the wave path from the source–receiver great circle. Theoretical progress on how to describe these effects accurately has been steady, evolving from the consideration of single scattering to the more complex question of multiple scattering. Transition into practice (i.e., into large-scale inversions of real data) still presents many computational challenges. It is, however, a necessary step if the ever-increasing broadband data collection assembled in the last quarter of the 20th century is to be fully exploited to gain improved resolution at shorter wavelengths.

Acknowledgments

This paper was written while the author was on sabbatical leave from UC Berkeley, visiting at the Departement de Seismologie, Institut de Physique du Globe in Paris, France. The author thanks three reviewers who contributed to improving the quality of this review.

References

Agnew, D., J. Berger, R. Buland, W. Farrell, and F. Gilbert (1976). International Deployment of Accelerometers: a network for very long period seismology. *EOS* **57**, 180–188.

Aki, K. (1960). The use of Love waves for the study of earthquake mechanism. *J. Geophys. Res.* **65**, 323–331.

Aki, K. and K. Kaminuma (1963). Phase velocity of Love waves in Japan, Part I: Love waves from the Aleutian shock of March 1957. *Bull. Earthq. Res. Inst. Univ. Tokyo* **41**, 243–259.

Aki, K. and H. Patton (1979). Determination of seismic moment tensor using surface waves. *Tectonophysics* **49**, 213–222.

Aki, K. and P.G. Richards (1980). "Quantitative Seismology, Theory and Methods." W.H. Freeman, San Francisco.

Alterman, Z., H. Jarosch, and C.L. Pekeris (1959). Oscillations of the Earth. *Proc. R. Soc. Lond. A* **252**, 80–95.

Alsina, D., R.L. Woodward, and R. Snieder (1996). Shear wave velocity structure in North America from large-scale waveform inversions of surface waves. *J. Geophys. Res.* **101**, 15969–15986.

Alsop, L.E. (1963). Free spheroidal vibrations of the Earth. *Bull. Seismol. Soc. Am.* **53**, 483–502.

Anderson, D.L. (1961). Elastic wave propagation in layered anisotropic media. *J. Geophys. Res.* **66**, 2953–2963.

Anderson, D.L., A. Ben-Menahem, and C.B. Archambeau (1965). Attenuation of seismic energy in the upper mantle. *J. Geophys. Res.* **70**, 1441–1448.

Arvidsson, R. and G. Ekström (1998). Global CMT analysis of moderate earthquakes, $M_w \geq 4.5$, using intermediate period surface waves. *Bull. Seismol. Soc. Am.* **88**, 1003–1013.

Babich, V.M., B.A. Chikhalev, and T.B. Yanovskaya (1976). Surface waves in a vertically inhomogeneous elastic half-space with weak horizontal inhomogeneity. *Izv. Acad. Sci. USSR Phys. Solid Earth* **4**, 24–31.

Backus, G. and M. Mulcahy (1976). Moment tensors and other phenomenological descriptions of seismic sources. I—Continuous displacements. *Geophys. J. R. Astron. Soc.* **46**, 341–371.

Ben-Menahem, A. (1961). Radiation of seismic surface waves from finite moving sources. *Bull. Seismol. Soc. Am.* **51**, 401–435.

Billien, M. and J.J. Lévêque (2000). Global maps of Rayleigh wave attenuation for periods between 40 and 150 seconds. *Geophys. Res. Lett.* **27**, 3619–3622.

Bostock, M.G. (1991). Surface wave scattering from 3-D obstacles. *Geophys. J. Int.* **104**, 351–370.

Bostock, M.G. (1992). Reflection and transmission of surface waves in laterally varying media. *Geophys. J. Int.* **109**, 411–436.

Bostock, M.G. and B.L.N. Kennett (1992). Multiple scattering of surface waves from discrete obstacles. *Geophys. J. Int.* **108**, 52–70.

Brune, J.W., H. Benioff, and M. Ewing (1961a). Long period surface waves from the Chilean earthquake of May 22, 1960, recorded on linear strain seismographs. *J. Geophys. Res.* **66**, 2895–2910.

Brune, J.W., J.E. Nafe, and L.E. Alsop (1961b). The polar phase shift of surface waves on a sphere. *Bull. Seismol. Soc. Am.* **51**, 247–257.

Bukchin, B.G. (1995). Determination of stress glut moments of total degree 2 from teleseismic surface wave amplitude spectra. *Tectonophysics* **248**, 185–191.

Buland, R.P. and F. Gilbert (1994). Computation of free oscillations of the Earth. *J. Comput. Phys.* **54**, 95–114.

Bungum, H. and J. Capon (1974). Coda pattern and multipath propagation of Rayleigh waves at NORSAR. *Phys. Earth Planet. Inter.* **9**, 111–127.

Canas, J.A. and B.J. Mitchell (1978). Lateral variation of surface wave anelastic attenuation across the Pacific. *Bull. Seismol. Soc. Am.* **68**, 1637–1650.

Canas, J.A. and B.J. Mitchell (1981). Rayleigh-wave attenuation and its variation across the Atlantic Ocean. *Geophys. J. R. Astron. Soc.* **67**, 259–276.

Capon, J. (1970). Analysis of Rayleigh-wave multipath propagation at LASA. *Bull. Seismol. Soc. Am.* **60**, 1701–1731.

Cara, M. (1973a). Filtering of dispersed wavetrains. *Geophys. J. R. Astron. Soc.* **114**, 141–157.

Cara, M. (1973b). Méthodes d'analyse de trains d'ondes multimodes et observations d'ondes S_a de type SH. Thèse de 3e cycle, Université Pierre et Marie Curie, Paris.

Cara, M. (1978). Regional variations of higher Rayleigh-mode phase velocities: a spatial filtering method. *Geophys. J. R. Astron. Soc.* **54** 439–460.

Cara, M. and J.J. Lévêque (1987). Waveform inversion using secondary observables. *Geophys. Res. Lett.* **14**, 1046–1049.

Červeny, V., M.M. Popov, and I. Pšenčik (1982). Computation of wavefields in inhomogeneous media–Gaussian beam approach. *Geophys. J. R. Astron. Soc.* **70**, 109–128.

Crampin, S. (1964). Higher modes of seismic surface waves: phase velocities across Scandinavia. *J. Geophys. Res.* **69**, 4801–4811.

Crampin, S. (1970). The dispersion of surface waves in multilayered anisotropic media. *Geophys. J. R. Astron. Soc.* **21**, 387–402.

Dahlen, F.A. (1979). The spectra of unresolved split normal mode multiplets. *Geophys. J. R. Astron. Soc.* **58**, 1.

Dufumier, H. and L. Rivera (1997). On the resolution of the isotropic component in moment tensor inversion. *Geophys. J. Int.* **131**, 595–606.

Durek, J.J. and G. Ekstrom (1996). A radial model of anelasticity consistent with long-period surface-wave attenuation. *Bull. Seismol. Soc. Am.* **86**, 144–158.

Durek, J.J. and G. Ekstrom (1997). Investigating discrepancies among measurements of traveling and standing wave attenuation. *J. Geophys. Res.* **102**, 24529–24544.

Durek, J.J., M.H. Ritzwoller, and J.H. Woodhouse (1993). Constraining upper mantle anelasticity using surface wave amplitudes. *Geophys. J. Int.* **114**, 249–272.

Dziewonski, A.M. (1971). Upper mantle models from "pure path" dispersion data. *J. Geophys. Res.* **76**, 2587–2601.

Dziewonski, A.M. and D.L. Anderson (1981). Preliminary Reference Earth Model. *Phys. Earth Planet. Int.* **25**, 297–356.

Dziewonski, A.M. and B. Romanowicz (1977). An exact solution to the problem of excitation of normal modes by a propagating fault. *Lincoln Lab., MIT, Semi-annual Report.*

Dziewonski, A.M. and J. Steim (1982). Dispersion and attenuation of mantle waves through waveform inversion. *Geophys. J. R. Astron. Soc.* **70**, 503–527.

Dziewonski, A., M. Bloch, and M. Landisman (1969). A new technique for the analysis of transient seismic signals. *Bull. Seismol Soc. Am.* **59**, 427–444.

Dziewonski, A., J. Mills, and S. Bloch (1972). Residual dispersion measurement—a new method of surface-wave analysis. *Bull. Seismol. Soc. Am.* **62**, 129–139.

Dziewonski, A.M., B.H. Hager, and R.J. O'Connell (1977). Large scale heterogeneities in the lower mantle. *J. Geophys. Res.* **82**, 239–255.

Dziewonski, A.M., A.T. Chou, and J.H. Woodhouse (1981). Determination of earthquake source parameters from waveform data for studies of global and regional seismicity. *J. Geophys. Res.* **86**, 2825–2852.

Ekström, G. and A.M. Dziewonski (1998). The unique anisotropy of the Pacific upper mantle. *Nature* **394**, 168–172.

Ekström, G., J. Tromp, and E.W.F. Larson (1997). Measurements and global models of surface wave propagation. *J. Geophys. Res.* **102**, 8137–8157.

Ewing, W.M., W.S. Jardetsky, and F. Press (1957). "Elastic Waves in Layered Media." McGraw-Hill, New York.

Forsyth, D.W. (1975). The early structural evolution and anisotropy of the oceanic upper mantle. *Geophys. J. R. Astron. Soc.* **43**, 103–162.

Friederich, W. (1998). Wave-theoretical inversion of teleseismic surface waves in a regional network: phase velocity maps and a three-dimensional upper mantle shear-wave velocity model for southern Germany. *Geophys. J. Int.* **132**, 203–225.

Friederich, W. (1999). Propagation of seismic shear and surface waves in a laterally heterogeneous mantle by multiple forward scattering. *Geophys. J. Int.* **136**, 180–204.

Friederich, W. and E. Wielandt (1995). Interpretation of seismic surface waves in regional networks: joint estimation of wavefield geometry and local phase velocity—Method and tests. *Geophys. J. Int.* **120**, 731–744.

Friederich, W., E. Wielandt, and S. Stange (1993). Multiple forward scattering of surface waves: comparison with an exact solution and Born single-scattering methods. *Geophys. J. Int.* **112**, 264–275.

Furumoto, M. (1979). Initial phase analysis of R waves from great earthquakes. *J. Geophys. Res.* **84**, 6867–6874.

Furumoto, M. and I. Nakanishi (1982). Source times and scaling relations of large earthquakes. *J. Geophys. Res.* **88**, 2191–2198.

Gee, L. and T.H. Jordan (1992). Generalized seismological data functionals. *Geophys. J. Int.* **111**, 363–390.

Giardini, D., E. Boschi, and B. Palombo (1993). Moment tensor inversion from Mednet data (2). Regional earthquakes of the Mediterranean. *Geophys. Res. Lett.* **20**, 273–276.

Gilbert, F. (1971). Excitation of normal modes of the Earth by earthquake sources. *Geophys. J. R. Astron. Soc.* **22**, 326–333.

Gilbert, F. (1976). The representation of seismic displacements in terms of travelling waves. *Geophys. J. R. Astron. Soc.* **44**, 275–280.

Gilbert, F. and G.E. Backus (1966). Propagator matrices in elastic wave and vibration problem. *Geophysics* **31**, 326–333.

Griot, D.A., J.P. Montagner, and P. Tapponnier (1998). Confrontation of mantle seismic anisotrpy with two extreme models of strain, in central Asia. *Geophys. Res. Lett.* **25**, 1447–1480.

Harkrider, D.G. (1964). Surface waves in multi-layered elastic media. 1, Rayleigh and Love waves from buried sources in a multilayered elastic half-space. *Bull. Seismol. Soc. Am.* **54**, 627.

Haskell, B. (1964). Radiation pattern of surface waves from point sources in a multi-layered medium. *Bull. Seismol. Soc. Am.* **54**, 377.

Hudson, J.A. (1981). A parabolic approximation for surface waves. *Geophys. J. R. Astron. Soc.* **67**, 755–770.

Jobert, N. and G. Jobert (1983). An application of ray theory in the propagation of waves along a laterally heterogeneous spherical surface. *Geophys. Res Lett.* **10**, 1148–1151.

Jobert, N. and G. Roult (1976). Periods and damping of free oscillations observed in France after 16 earthquakes. *Geophys. J. R. Astron. Soc.* **45**, 155–176.

Jobert, N., R. Gaulon, A. Dieulin, and G. Roult (1977). Sur les ondes à très longue période, caractéristiques du manteau supérieur. *C. R. Acad. Sci. Paris B* **285**, 49–52.

Jordan, T.H. (1978). A procedure for estimating lateral variations from low frequency eigenspectra data. *Geophys. J. R. Astron. Soc.* **52**, 441–455.

Kanamori, H. (1970). Velocity and Q of mantle waves. *Phys. Earth Planet. Inter.* **27**, 8–31.

Kanamori, H. and D.L. Anderson (1977). Importance of physical dispersion in surface-wave and free-oscillation problems: review. *Rev. Geophys. Space Phys.* **15**, 105–112.

Kanamori, H. and J. Given (1981). Use of long-period surface waves for rapid determination of earthquake source parameters. *Phys. Earth Planet. Inter.* **27**, 8–31.

Kanamori, H. and G.S. Stewart (1976). Mode of the strain release along the Gibbs fracture zone, Mid-Atlantic ridge. *Phys. Earth Planet. Inter.* **11**, 312–332.

Kennett, B.L.N. (1972). Seismic waves in laterally heterogeneous media. *Geophys. J. R. Astron. Soc.* **27**, 301–325.
Kennett, B.L.N. (1984). Guided wave propagation in varying media, I. Theoretical development. *Geophys. J. R. Astron. Soc.* **79**, 235–255.
Knopoff, L. (1972). Observation and inversion of surface-wave dispersion. *Tectonophysics* **13**, 497–519.
Knopoff, L., M.J. Berry, and F.A. Schwab (1967). Tripartite phase velocity observations in laterally heterogenous regions. *J. Geophys. Res.* **72**, 2595–2601.
Kovach, R.L. (1978). Seismic surface waves and crustal and upper mantle structure. *Rev. Geophys. Space Phys.* **16**, 1–13.
Kovach, R.L. and D.L. Anderson (1964). Higher mode surface waves and their bearing on the structure of the Earth's mantle. *Bull. Seismol. Soc. Am.* **54**, 161–182.
Kuge, K., J. Zhang, and M. Kikuchi (1996). The 12 July 1993 Hokkaido-Nansei-Oki, Japan, Earthquake: effects of source complexity on surface wave radiation. *Bull. Seismol. Soc. Am.* **86**, 505–518.
Lander, A.V. (1989). Frequency-time analysis. In: "Seismic Surface Waves in a Laterally Inhomogeneous Earth" (V.I. Keilis-Borok, Ed.), pp. 153–163. Kluwer Academic, Dordrecht.
Laske, H. (1995). Global observations of off-great-circle propagation of long-period surface waves. *Geophys. J. Int.* **123**, 245–259.
Laske, G. and G. Masters (1996). Constraints on global phase velocity maps from long period polarization data. *J. Geophys. Res.* **11**, 16059–16075.
Lay, T. and H. Kanamori (1985). Geometric effects of global lateral heterogeneity on long-period surface wave propagation. *J. Geophys. Res.* **90**, 605–621.
Lerner-Lam, A. and T.H. Jordan (1983). Earth structure from fundamental and higher-mode waveform analysis. *Geophys. J. R. Astron. Soc.* **75**, 759–797.
Lerner-Lam, A. and J.J. Park (1989). Frequency-dependent refraction and multipathing of 10–100 second surface waves in the western Pacific. *Geophys. R. Lett.* **16**, 527–530.
Levshin, A., M. Ritzwoller, and L. Ratnikova (1994). The nature and cause of polarization anomalies of surface waves crossing northern and central Eurasia. *Geophys. J. Int.* **117**, 577–591.
Lévêque, J.J. and M. Cara (1983). Long-period Love wave overtone data in North America and the Pacific Ocean: new evidence for upper mantle anisotropy. *Phys. Earth Planet. Inter.* **33**, 164–179.
Lévêque, J.J., M. Cara, and D. Rouland (1991). Waveform inversion of surface wave data: test of a new tool for systematic investigation of upper mantle structures. *Geophys. J. Int.* **104**, 565–581.
Li, X.D. and T. Tanimoto (1993). Waveforms of long period body waves in a slightly aspherical earth. *Geophys. J. Int.* **112**, 92–112.
Li, X.-D. and B. Romanowicz (1995). Comparison of global waveform inversions with and without considering cross branch coupling. *Geophys. J. Int.* **121**, 695–709.
Li, X.D. and B. Romanowicz (1996). Global mantle shear velocity model developed using nonlinear asymptotic coupling theory. *J. Geophys. Res.* **101**, 22245–22273.
Love, A.E.H. (1927). "A Treatise on the Mathematical Theory of Elasticity," 4th ed. Cambridge University Press, Cambridge.
Marquering, H. and R. Snieder (1995). Surface-wave mode coupling for efficient forward modelling and inversion of body wave phases. *Geophys. J. Int.* **120**, 186–208.
Masters, G. and F. Gilbert (1983) Attenuation in the Earth at low frequencies. *Phil. Trans. R. Soc. Lond. A* **308**, 479–522.
Masters, G. and G. Laske (1997). On bias in surface wave and free oscillation attenuation measurements. *EOS Trans. Am. Geophys. Union* **78**, F485.
Maupin, V. (1988). Surface waves across 2-D structures: a method based on coupled local modes. *Geophys. J.* **92**, 173–185.
McEvilly, T.V. (1964). Central U.S. crust-upper mantle structure from Love and Rayleigh wave phase velocity inversion. *Bull. Seismol. Soc. Am.* **54**, 1997–2015.
Mégnin, C. and B. Romanowicz (1998). The effect of theoretical formalism and data selection scheme on mantle models derived from waveform tomography. *Geophys. J. Int.* **138**, 366–380.
Mégnin, C. and B. Romanowicz (2000). The 3D shear velocity structure of the mantle from the inversion of body, surface and higher mode waveforms. *Geophys. J. Int.* **143**, 709–728.
Meier, T., S. Lebedev, G. Nolet, and F.A. Dahlen (1997). Diffraction tomogaphy using multimode surface waves. *J. Geophys. Res.* **102**, 8255–8267.
Mendiguren, J.A. (1977). Inversion of surface wave data in source mechanism studies. *J. Geophys. Res.* **82**, 889–894.
Mitchell, B.J. (1975). Regional Rayleigh wave attenuation in north America. *J. Geophys. Res.* **80**, 4904–4916.
Mitchell, B.J. (1995). Anelastic structure and evolution of the continental crust and upper mantle from seismic surface wave attenuation. *Rev. Geophys.* **33**, 441–462.
Mitchell, B.J. and G.K. Yu (1980). Surface wave regionalized models and anisotropy of the Pacific crust and upper mantle. *Geophys. J. R. Astron. Soc.* **64**, 497–514.
Mocquet, A., B. Romanowicz, and J.P. Montagner (1989). Three dimensional structure of the upper mantle beneath the Atlantic Ocean inferred from long period Rayleigh waves, 1. Group and phase velocity distributions. *J. Geophys. Res.* **94**, 7449–7468.
Mochizuki, E. (1986a). Free oscillations and surface waves in an aspherical earth. *Geophys. Res. Lett.* **13**, 1478–1481.
Mochizuki, E. (1986b). The free oscillations of an anisotropic and heterogeneous earth. *Geophys. J. R. Astron. Soc.* **86**, 167–176.
Montagner, H.P. (1985). Seismic anisotropy of the Pacific Ocean inferred from long-period surface wave dispersion. *Phys. Earth Planet. Inter.* **38**, 28–50.
Montagner, J.P. (1986). Regional three-dimensional structures using long-period surface waves. *Ann. Geophys.* **4**, 283–294.
Montagner, J.P. and N. Jobert (1983). Variation with age of the deep structure of the Pacific Ocean inferred from very long-period Rayleigh wave dispersion. *Geophys. Res. Lett.* **10**, 273–276.
Montagner, J.P. and N. Jobert (1988). Vectorial tomography II. Application to the Indian Ocean. *Geophys. J.* **94**, 309–344.
Montagner, J.P. and H.C. Nataf (1986). A simple method for inverting the azimuthal anisotropy of surface waves. *J. Geophys. Res.* **91**, 511–520.
Montagner, J.P. and H.C. Nataf (1988). Vectorial Tomography—I. Theory. *Geophys. J. Int.* **94**, 295–307.
Montagner, J.P. and T. Tanimoto (1990). Global anisotropy in the upper mantle inferred from the regionalization of phase velocities. *J. Geophys. Res.* **95**, 4794–4819.
Montagner, J.P. and T. Tanimoto (1991). Global upper mantle tomography of seismic velocities and anisotropy. *J. Geophys. Res.* **96**, 20337–20351.

Mooney, W., G. Laske, and G. Masters (1998). CRUST-5.1: a global crustal model at $5^\circ \times 5^\circ$. *J. Geophys. Res.* **103**, 727–747.

Nakanishi, I. (1978). Regional differences in the phase velocity and the quality factor Q of mantle Rayleigh waves. *Science* **200**, 1379–1381.

Nakanishi, I. and D.L. Anderson (1983). Measurement of mantle wave velocities and inversion for lateral heterogeneity and anisotropy, 1. Analysis of great circle phase velocities. *J. Geophys. Res.* **88**, 10267–10283.

Nakanishi, I. and H. Kanamori (1982). Effects of lateral heterogeneity and source process time on the linear moment tensor inversion of long period Rayleigh waves. *Bull. Seismol. Soc. Am.* **72**, 2063–2080.

Nataf, H.C., I. Nakanishi, and D.L. Anderson (1984). Anisotropy and shear-velocity heterogeneities in the upper mantle. *Geophys. Res. Lett.* **11**, 109–112.

Nataf, H.C., I. Nakanishi, and D.L. Anderson (1986). Measurements of mantle wave velocities and inversion for lateral heterogeneities and anisotropy, Part III: Inversion. *J. Geophys, Res.* **91**, 7261–7307.

Nishimura, C.E. and D. Forsyth (1989). The anisotropic structure of the upper mantle in the Pacific. *Geophys. J. R. Astron. Soc.* **96**, 203–229.

Nolet, G. (1975). Higher-Rayleigh modes in western Europe. *Geophys. Res. Lett.* **2**, 60–62.

Nolet, G. (1990). Partitioned waveform inversion and two-dimensional structure under the network of autonomously recording seismographs. *J. Geophys. Res.* **95**, 8499–8512.

Okal, E. (1977). The effect of intrinsic oceanic upper-mantle heterogeneity on regionalization of long-period Rayleigh-wave phase velocities. *Geophys. J. R. Astron. Soc.* **49**, 357–370.

Okal, E. and B.-G. Jo (1985). Stacking investigations of higher-order mantle Rayleigh waves. *Geophys. Res. Lett.* **12**, 421–424.

Okal, E. and B.-G. Jo (1987). Stacking investigations of the dispersion of higher order mantle Rayleigh waves and normal modes. *Phys. Earth Planet. Inter.* **47**, 188–204.

Odom, R.I. (1986). A coupled mode examination of irregular waveguides including the continuum spectrum. *Geophys. J. R. Astron. Soc.* **86**, 425–453.

Park, J. (1987). Asymptotic coupled-mode expressions for multiplet amplitude anomalies and frequency shift on a laterally heterogeneous Earth. *Geophys. J. R. Astron. Soc.* **90**, 129–170.

Park, J. (1997). Free oscillations in an anisotropic earth: path-integral asymptotics. *Geophys. J. Int.* **129**, 399–411.

Park, J. and Y. Yu (1992). Anisotropy and coupled free oscillations: simplified models and surface wave observations. *Geophys. J. Int.* **110**, 401–420.

Park, J. and Y. Yu (1993). Seismic determination of elastic anisotropy and mantle flow. *Science* **261**, 1159–1162.

Park, J., C.R. Lindberg, and F.L. Vernon (1987). Multitaper spectral analysis of high-frequency seismograms. *J. Geophys. Res.* **92**, 12675–12684.

Pasyanos, M., D. Dreger, and B. Romanowicz (1993). Towards real time estimation of regional moment tensors. *Bull. Seismol. Soc. Am.* **86**, 1255–1269.

Patton, H.J. (1977). Reference point method for determining the source and path effects of surface waves. *J. Geophys. Res.* **82**, 889–894.

Paulssen, H., A.L. Levshin, A.V. Lander, and R. Snieder (1990). Time- and frequency-dependent polarization analysis: anomalous surface wave observations in Iberia. *Geophys. J. Int.* **103**, 483–496.

Pillet, R., D. Rouland, G. Roult, and D.A. Wiens (1999). Crust and upper mantle heterogeneities in the southwest Pacific from surface wave phase velocity analysis. *Phys. Earth Planet. Inter.* **110**, 211–234.

Pollitz, F. (1994). Global tomogaphy from Rayleigh and Love wave dispersion: effect of raypath bending. *Geophys. J. Int.* **102**, 8255–8267.

Pollitz, F. (1999). Regional velocity structure in northern California from inversion of scattered seismic waves. *J. Geophys. Res.* **104**, 15043–15072.

Ritsema, J. and T. Lay (1993). Rapid source mechanism determination of large ($M_w \geq 4.5$) earthquakes in the western United States. *Geophys. Res. Lett.* **20**, 1611–1614.

Resovsky, J.S. and M.H. Ritzwoller (1998). New and refined constraints on three-dimensional Earth structure from normal modes below 3 mHz. *J. Geophys. Res.* **103**, 783–810.

Romanowicz, B. (1982a). Moment tensor inversion of long-period Rayleigh waves: a new approach. *J. Geophys. Res.* **87**, 5395–5407.

Romanowicz, B. (1982b). Constraints on the structure of the Tibet Plateau from pure path phase velocities of Love and Rayleigh waves. *J. Geophys. Res.* **87**, 6865–6883.

Romanowicz, B. (1987). Multiplet–multiplet coupling due to lateral heterogeneity: asymptotic effects on the amplitude and frequency of the Earth's normal modes. *Geophys. J. R. Astron. Soc.* **90**, 75–100.

Romanowicz, B. (1990). The upper mantle degree 2: constraints and inferences from global mantle wave attenuation measurements. *J. Geophys. Res.* **95**, 11051–11071.

Romanowicz, B. (1994). On the measurement of anelastic attenuation using amplitudes of low-frequency surface waves. *Phys. Earth Planet. Inter.* **84**, 179–191.

Romanowicz, B. (1995). A global tomographic model of shear attenuation in the upper mantle. *J. Geophys. Res.* **100**, 12375–12394.

Romanowicz, B. (1998). Attenuation tomography of the Earths mantle: a review of current status. *Pageoph* **153**, 257–272.

Romanowicz, B. and J. Durek (2000). Seismological constraints on attenuation in the Earth: a review. *Geophysical Monograph Ser.* **117**, 161–180.

Romanowicz, B. and A.M. Dziewonski (1986). Towards a federation of broadband seismic networks. *EOS Trans. Am. Geophys. Union* **67**, 541–542.

Romanowicz, B. and P. Guillemant (1984). An experiment in the retrieval of depth and source parameters of large earthquakes using very long period Rayleigh wave data. *Bull. Seismol. Soc. Am.* **74**, 417–437.

Romanowicz, B. and T. Monfret (1986). Source process times and depths of large earthquakes by moment tensor inversion of mantle wave data and the effect of lateral heterogeneity. *Ann. Geophys.* **4**(B3), 271–283.

Romanowicz, B. and G. Roult (1986). First order asymptotics for the eigenfrequencies of the Earth and application to the retrieval of large scale variations of structure. *Geophys. J. R. Astron. Soc.* **87**, 209–239.

Romanowicz, B. and R. Snieder (1988). A new formalism for the effect of lateral heterogeneity on normal modes and surface waves, II. General anisotropic perturbation. *Geophys. J. Int.* **93**, 91–99.

Romanowicz, B. and G. Suarez (1983). An improved method to obtain the moment tensor depth of earthquakes from the amplitude spectrum of Rayleigh waves. *Bull. Seismol. Soc. Am.* **73**, 1513–1526.

Romanowicz, B., M. Cara, J.F. Fels, and D. Rouland (1984). Geoscope: a french initiative in long period three component seismic networks. *EOS Trans. Am. Geophys. Union* **65**, 753–754.

Romanowicz, B., J.F. Karczewski, M. Cara, *et al.* (1991). The Geoscope program: present status and perspectives. *Bull. Seismol. Soc. Am.* **81**, 243–264.

Romanowicz, B., D. Dreger, M. Pasyanos, and R. Uhrhammer (1993). Monitoring of strain release in central and northern California using broadband data. *Geophys. Res. Lett.* **20**, 1643–1646.

Roult, G. and B. Romanowicz (1984). Very long period data from the GEOSCOPE network: preliminary results on great circle averages of fundamental and higher Rayleigh and Love modes. *Bull. Seismol. Soc. Am.* **74**, 2221–2243.

Roult, G., B. Romanowicz, and N. Jobert (1986). Observations of departures from classical approximations on very long period GEOSCOPE records. *Ann. Geophys.* **4**(B3), 241–250.

Roult, G., D. Rouland, and J.P. Montagner (1994). Antarctica II: upper-mantle structure from velocities and anisotropy. *Phys. Earth Planet. Inter.* **84**, 33–57.

Sailor, R.V. and A.M. Dziewonski (1978). Measurements and interpretation of normal mode attenuation. *Geophys. J. R. Astron. Soc.* **53**, 559–581.

Saito, M. (1967). Excitation of free oscillations and surface waves by a point source in a vertically heterogeneous Earth. *J. Geophys. Res.* **72**, 3689.

Sato, Y., T. Usami, and M. Ewing (1962). Basic study on the oscillation of a homogeneous elastic sphere. *Geophys. Mag.* **31**, 237.

Schlue, J.W. and L. Knopoff (1977). Shear wave polarization anisotropy in the Pacific Basin. *Geophys. J. R. Astron. Soc.* **49**, 145–165.

Smith, M.F. and G. Masters (1989). Aspherical structure constraints from free oscillation frequency and attenuation measurements. *J. Geophys. Res.* **94**, 1953–1976.

Smith, M.L. and F.A. Dahlen (1973). The azimuthal dependence of Love and Rayleigh wave propagation in a slightly anisotropic medium. *J. Geophys. Res.* **78**, 3321–3333.

Smith, S.W. (1986). IRIS: a program for the next decade. *EOS Trans. Am. Geophys. Union* **67**, 213.

Snieder, R. (1986). 3-D linearized scattering of surface waves and a formalism for surface wave holography. *Geophys. J. R. Astron. Soc.* **84**, 581–605.

Snieder, R. (1988a). Large-scale waveform inversions of surface waves for lateral heterogeneity, 1. Theory and numerical examples. *J. Geophys. Res.* **93**, 12055–12065.

Snieder, R. (1988b). Large-scale waveform inversions of surface waves for lateral heterogeneity, 2. Application to surface waves in Europe and the Mediterranean. *J. Geophys. Res.* **93**, 12067–12080.

Snieder, R. and G. Nolet (1987). Linearized scattering of surface waves on a spherical Earth. *J. Geophys.* **61**, 55–63.

Snieder, R. and B. Romanowicz (1988). A new formalism for the effect of lateral heterogeneity on normal modes and surface waves—I. Isotropic perturbations, perturbations of interfaces and gravitational perturbations. *Geophys. J. R. Astron. Soc.* **92**, 207–222.

Stevens, J. and K.L. McLaughlin (2001). Optimization of surface wave identification and measurement. *Pageoph.* **158**(8), 1547–1582.

Stutzmann, E. and J.P. Montagner (1993). An inverse technique for retrieving higher mode phase velocity and mantle structure. *Geophys. J. Int.* **113**, 669–683.

Stutzmann, E. and J.P. Montagner (1994). Tomography of the transition zone from the inversion of higher-mode surface waves. *Phys. Earth Planet. Inter.* **86**, 99–115.

Su, W.-J., R.L. Woodward, and A.M. Dziewonski (1994). Degree 12 model of shear velocity heterogeneity in the mantle. *J. Geophys. Res.* **99**, 6945–6980.

Suetsugu, D. and I. Nakanishi (1987). Regional and azimuthal dependence of phase velocities of mantle Rayleigh waves in the Pacific Ocean. *Phys. Earth Planet. Inter.* **47**, 230–245.

Takeuchi, M. and M. Saito (1972). Seismic surface waves. In: "Seismology: Surface Waves and Free Oscillations" (B.A. Bolt, Ed.), Methods in Computational Physics, Vol. 11, pp. 217–295. Academic Press, New York.

Tanimoto, T. (1984). A simple derivation of the formula to calculate synthetic long-period seismograms in a heterogeneous Earth by normal mode summation. *Geophys. J. R. Astron. Soc.* **77**, 275–278.

Tanimoto, T. (1986). Free oscillations of an anisotropic and heterogeneous Earth. *Geophys. J. R. Astron. Soc.* **86**, 493–517.

Tanimoto, T. (1987). The three-dimensional shear wave structure in the mantle by overtone waveform inversion—I. Radial seismogram inversion. *Geophys. J. R. Astron. Soc.* **89**, 713–740.

Tanimoto, T. (1990). Modelling curved surface wave paths: membrane surface wave synthetics. *Geophys. J. Int.* **102**, 89–100.

Tanimoto, T. and D.L. Anderson (1984). Mapping convection in the mantle. *Geophys. Res. Lett.* **11**, 287–290.

Tanimoto, T. and D.L. Anderson (1985). Lateral heterogeneity and azimuthal anisotropy of the upper mantle: Love and Rayleigh waves 100–250 s. *J. Geophys. Res.* **90**, 1842–1858.

Tarantola, A. and B. Valette (1982). Generalized nonlinear inverse problems solved using least squares criterion. *Rev. Geophys. Space Phys.* **20**, 219–232.

Thio, H.K. and H. Kanamori (1995). Moment-tensor inversions for local earthquakes using surface waves recorded at TERRAscope. *Bull. Seismol. Soc. Am.* **85**, 1021–1038.

Toksöz, M.N. and D.L. Anderson (1966). Phase velocities of long period surface waves and structure of the upper mantle, 1. Great circle Love and Rayleigh wave data. *J. Geophys. Res.* **71**, 1649–1658.

Trampert, J. and J.H. Woodhouse (1995). Global phase velocity maps of Love and Rayleigh waves between 40 and 150 sec. *Geophys. J. Int.* **122**, 675–690.

Tromp, J. and F.A. Dahlen (1992a). Variational principles for surface wave propagation on a laterally heterogeneous Earth—I. Time domain JKWB theory. *Geophys. J. Int.* **109**, 581–598.

Tromp, J. and F.A. Dahlen (1992b). Variational principles for surface wave propagation on a laterally heterogeneous Earth—II. Frequency domain JWKB theory. *Geophys. J. Int.* **109**, 599–619.

Tsai, Y.B. and K. Aki (1969). Simultaneous determination of the seismic moment and attenuation of seismic surface waves. *Bull. Seismol. Soc. Am.* **59**, 275–287.

Tsai, Y.B. and K. Aki (1971). Amplitude spectra of surface waves from small earthquakes and underground nuclear explosions. *J. Geophys. Res.* **75**, 5729.

van der Lee, S. and G. Nolet (1997). Upper mantle *S* velocity structure of North America. *J. Geophys. Res.* **102**, 22815–22838.

van Heist, H. and J. Woodhouse (1997). Measuring surface-wave overtone phase velocities using a mode-branch stripping technique. *Geophys. J. Int.* **131**, 209–230.

van Heist, H. and J. Woodhouse (1999). Global high-resolution phase velocity distributions of overtone and fundamental mode surface waves determined by mode branch stripping. *Geophys. J. Int.* **137**, 601–620.

Weidner, D. and K. Aki (1973). Focal depth and mechanism of mid-ocean ridge earthquakes. *J. Geophys. Res.* **78**, 1818–1831.

Wielandt, E. (1993). Propagation and structural interpretation of non plane waves. *Geophys. J. Int.* **113**, 45–53.

Widmer, R., G. Masters, and F. Gilbert (1991). Spherically symmetric attenuation within the Earth from normal mode data. *Geophys. J. Int.* **104**, 541–553.

Wiggins, R. (1976). A fast new computational algorithm for free oscillations and surface waves. *Geophys. J. R. Astron. Soc.* **47**, 135–150.

Wong, Y.K. (1989). Upper mantle heterogeneity from phase and amplitude data of mantle waves. PhD thesis, Harvard University, Cambridge, MA.

Woodhouse, J.H. (1974). Surface waves in laterally varying structure. *Geophys. J. R. Astron. Soc.* **37**, 461–490.

Woodhouse, J.H. (1980a). Efficient and stable methods for performing seismic calculations in stratified media. In: "Physics of the Earth's Interior," Proc. Int. School of Physics "Enrico Fermi," Course LXXVIII, pp. 127–151.

Woodhouse, J.H. (1980b). The coupling and attenuation of nearly resonant multiplets in the Earth's free oscillation spectrum. *Geophys. J. R. Astron. Soc.* **61**, 261–283.

Woodhouse, J.H. (1983). The joint inversion of seismic waveforms for lateral variations in Earth structure and earthquake source parameters. In: "Earthquakes: Theory, Observation and Interprctation" (H. Kanamori and E. Boschi, Eds.), Proc. Int. School of Physics "Enrico Fermi," pp. 306–397.

Woodhouse, J.H. (1988). The calculation of eigenfrequencies and eigenfunctions of the free oscillations of the Earth and the Sun. In: "Seismological Algorithms" (D.J. Doornbos, Ed.), pp. 321–370. Academic Press, New York.

Woodhouse, J.H. and F.A. Dahlen (1978). The effect of a general aspherical perturbation on the free oscillations of the Earth. *Geophys. J. R. Astron. Soc.* **53**, 335–354.

Woodhouse, J.H. and A.M. Dziewonski (1984). Mapping the upper mantle: three dimensional modelling of the Earth structure by inversion of seismic waveforms. *J. Geophys. Res.* **89**, 5953–5986.

Woodhouse, J.H. and T.P. Girnius (1982). Surface waves and free oscillations in a regionalized earth model. *Geophys. J. R. Astron. Soc.* **68**, 653–673.

Woodhouse, J.H. and Y.K. Wong (1986). Amplitude, phase and path anomalies of mantle waves. *Geophys. J. R. Astron. Soc.* **87**, 753–773.

Wu, F.T. (1972). Mantle Rayleigh wave dispersion and tectonic provinces. *J. Geophys. Res.* **77**, 6445–6453.

Xie, J. and B.J. Mitchell (1990a). A back projection method for imaging large-scale lateral variations of *Lg* coda *Q* with application to continental Africa. *Geophys. J. Int.* **100**, 161–181.

Xie, J. and B.J. Mitchell (1990b). Attenuation of multiphase surface waves in the Basin and Range province, *I, Lg* and *Lg* coda. *Geophys. J. Int.* **102**, 121–137.

Yanovskaya, T.B. and Y.V. Roslov (1989). Peculiarities of surface wave fields in laterally inhomogeneous media in the framework of ray theory. *Geophys. J. Int.* **99**, 297–303.

Yanovskaya, T.B. and P.G. Ditmar (1990). Smoothness criteria in surface wave tomography. *Geophys. J. Int.* **102**, 63–72.

Yomogida, K. and K. Aki (1985). Waveform analysis of surface waves in a laterally heterogeneous earth by the Gaussian beam method. *J. Geophys. Res.* **90**, 7665–7688.

Yomogida, K. and K. Aki (1987). Amplitude and phase data inversions for phase velocity anomalies in the Pacific Ocean basin. *Geophys. J. R. Astron. Soc.* **88**, 161–204.

Yu, Y. and J. Park (1994). Hunting for azimuthal anisotropy beneath the Pacific Ocean region. *J. Geophys. Res.* **99**, 15399–15421.

Zhang, J. and H. Kanamori (1988a). Depths of large earthquakes determined from long-period Rayleigh waves. *J. Geophys. Res.* **93**, 4850–4868.

Zhang, J. and H. Kanamori (1988b). Source finiteness of large earthquakes measured from long-period Rayleigh waves. *Phys. Earth. Planet. Inter.* **52**, 56–84.

Zhang, Y.S. and T.H. Lay (1996). Global surface wave phase velocity variations. *J. Geophys. Res.* **101**, 8415–8436.

Zhao, L. and T.H. Jordan (1998). Sensitivity of frequency-dependent travel times to laterally heterogeneous, anisotropic Earth structre. *Geophys. J. Int.* **133**, 683–704.

Zielhuis, A. and G. Nolet (1994). Shear-wave velocity variations in the upper mantle beneath central Europe. *Geophys. J. Int.* **117**, 695–715.

Zielhuis, A. and R.D. van der Hilst (1996). Upper-mantle shear velocity beneath eastern Australia from inversion of waveforms from SKIPPY portable arrays. *Geophys. J. Int.* **127**, 1–16.

12

Earthquake Dynamics

Raúl Madariaga
Ecole Normale Supérieure, Paris, France
Kim Bak Olsen
University of California, Santa Barbara, California, USA

1. Introduction

Earthquake source dynamics provides key elements for the prediction of strong ground motion and for understanding the physics of earthquake initiation, propagation, and healing. Early studies pioneered our understanding of friction and introduced simple models of dynamic earthquake rupture, typically using homogeneous distributions of stress and friction parameters. Classical examples of such models are the mechanical spring-and-box models proposed by Burridge and Knopoff (1967), the rectangular dislocation model proposed by Haskell (1964), and the self-similar circular rupture model introduced by Kostrov (1966). Extensive research then followed to advance our understanding of seismic rupture propagation and stress relaxation. It became clear that the correct mathematical formulation of the problem of propagation and radiation by a seismic rupture was that of a propagating shear crack as proposed by Kostrov (1964, 1966). Very soon it became clear that friction also played a fundamental role in the initiation, the development of rupture, and the healing of faults. The classical Coulombian model of a sudden drop in friction from a static to a dynamic coefficient led to an impasse, with infinite stress singularities and many other physical problems. The reason is that this model lacks an essential length scale needed to define a finite energy release rate near the rupture front.

Better models of friction at low slip rates were studied in the laboratory by Dieterich (1978, 1979) and Ruina (1983), who proposed the model of rate- and state-dependent friction. Slip weakening friction laws were introduced in dynamic rupture modeling by Ida (1972) and Andrews (1976a,b) for plane (2D) ruptures and by Day (1982b) for 3D fault models. These authors showed that slip weakening regularizes the numerical model of the rupture front, distributing stress and slip concentrations over a distance controlled by the length scale in the friction law. Ohnaka and Kuwahara (1990), Ohnaka (1996), and Ohnaka and Shen (1999) concluded that their experiments could be explained with a simple slip weakening friction law. In fact, for many practical purposes, the rate-and-state and slip weakening friction laws can be reconciled by noting that both models contain a finite length scale that controls the behavior of the rupture front (see Okubo, 1989; Dieterich and Kilgore, 1996). Extensive reviews on rupture dynamics up to 1990 have been published by Kostrov and Das (1989) and Scholz (1989).

Recent studies of rupture processes for selected earthquakes have shed new light on our understanding of earthquake ruptures. These models suggest a complexity of the rupture process that the early models of rupture in a uniformly loaded medium were unable to explain. Although, in the late 1970s, Das and Aki (1977b), Mikumo and Miyatake (1978, 1979), Madariaga (1979), and Andrews (1980, 1981) pointed out the deficiencies of the classical dislocation and crack models, it was not until the late 1980s that good-quality near-field accelerometry became available for some large earthquakes. Simultaneously, new sophisticated and efficient numerical methods, such as boundary integral equations (BIE) and finite differences (FD), provided the tools to study realistic dynamic rupture propagation in a fault subject to a heterogeneous stress field and spatially varying friction.

Heaton (1990) noticed that rupture of large earthquakes was typically characterized by pulse-like behavior, where only a small part of the fault would rupture at a given instant. This result has been confirmed by a number of inversions of the slip-rate field for large earthquakes, such as the 1992 Landers earthquake in California (Cohee and Beroza, 1994; Wald and Heaton, 1994; Cotton and Campillo, 1995a). Cochard and Madariaga (1996) found that, at least for a simple velocity weakening friction law, heterogeneity could arise spontaneously in a two-dimensional homogeneous fault model as found earlier by Carlson and Langer (1989) for the Burridge and Knopoff model of sliding blocks connected by springs. Other authors studied complex fault models from a theoretical point of view (Harris and Day, 1993, 1997). Beroza and

INTERNATIONAL HANDBOOK OF EARTHQUAKE AND ENGINEERING SEISMOLOGY, VOLUME 81A ISBN: 0-12-440652-1

Mikumo (1996) found that dynamic models with heterogeneous fault parameters tend to generate short slip duration.

In a direct modeling approach, Olsen *et al.* (1997) and Peyrat *et al.* (2001) showed that rupture propagation in a dynamic model of the 1992 Landers earthquake, would follow a complex path, completely controlled by the spatial variation of the initial stress field. Ide and Takeo (1997) estimated the constitutive friction law parameters for the 1995 Kobe earthquake from their kinematic inversion results. Computations of dynamic stress changes for the 1992 Landers, 1994 Northridge, and 1995 Kobe earthquakes (Bouchon, 1997; Day *et al.*, 1998) showed highly variable distributions of stress drops. Spudich *et al.* (1998) detected coseismic changes in the slip direction for the 1995 Kobe earthquake. Nielsen *et al.* (2000) indicated that such complexity inherently arises as a result of many recurrent earthquakes on a single fault over a long time span.

In this chapter we review what we believe are the important results obtained to date in the field of earthquake rupture. In Section 2 we review the early models of earthquake rupture and discuss the elastic shear fault model and fundamental friction laws. We also briefly describe the BIE and FD methods for numerical modeling of dynamic rupture. In Section 3 we illustrate the most important phenomenology of simple rupture models with a single length scale for circular and rectangular fault models, including anisotropy and scaling of growth, generation of sub-shear and super-shear rupture speeds, and the numerical resolution of these models. Scaling laws for earthquake rupture are described in Section 4, including the complementary roles of friction, strength, and geometry. Section 5 shows the results of modeling the 1992 M 7.3 Landers event, including computation of a heterogeneous initial stress field and estimation of the frictional parameters. We compare the dynamic modeling results to those from kinematic models and strong motion data. Finally we discuss the importance of heterogeneity in the rupture process, including the necessity of multiple length scales, generation of self-healing pulses, and the origin of rupture complexity, and discuss the possibility of estimating friction from observations.

2. Fault Models and Friction

In this section we review some of the simpler models that have been used to model seismic ruptures: the Burridge–Knopoff (BK) model; one of its modern versions, the cellular automata (CA) model; and what is still the most useful kinematic description of an earthquake—the dislocation model. Then we introduce the theory behind the elastic shear fault model and basic friction laws. Finally, we briefly describe the concepts of the two numerical methods that have dominated the field of modeling of dynamic rupture: the boundary integral element (BIE) and finite-difference (FD) methods.

2.1 Classical Dynamic Model Assumptions: Burridge–Knopoff and its Successors

Burridge and Knopoff (1967) pioneered dynamic rupture behavior by studying a mechanical model composed of a chain of N blocks coupled by horizontal springs of stiffness κ sliding on a frictional surface that delays the motion of the blocks. The one-dimensional array of springs is connected by individual leaf springs of stiffness k to a rigid driving bar that moves horizontally with a constant velocity. For all reasonable friction laws in which friction decreases once slip starts, the blocks in the BK models move by stick–slip with long periods of stress accumulation and sudden jerky displacement. This model is an analogue or "toy" model of an earthquake fault that is loaded by slow plate motion and locked by friction except in brief intervals when the loading stress overcomes friction at the interface. When this model is loaded at sufficiently high stresses, rupture starts by slip on one of the blocks of the chain and spreads rapidly to neighboring blocks. Until the 1980s this model was a simple and curious analogue to an earthquake rupture, and most seismologists believed that slip episodes would always spread to all the blocks of the system. However, Cao and Aki (1984) and Carlson and Langer (1989) found numerically that this was not the case. Instead, they discovered that very complex rupture histories would develop in this model starting from a nominally homogeneous system. Actually, the BK model has two types of ruptures: Local events that tend to smooth the system; and long events that propagate along the whole chain and wrap around it when cyclic boundary conditions are used. These large, soliton-like events roughen the system, as shown by Schmittbuhl *et al.* (1996). Small events in this system obey a Gutenberg–Richter type power law for the number of events with respect to the number of sliding blocks that participate in any individual event. Large, macroscopic events have a completely different distribution centered around the total number of events in the chain. This model has become a paradigm for the dynamic origin of complexity on a fault, although several authors pointed out a number of reasons why this was not a very realistic model of earthquakes. The most serious problem is that it does not radiate and dissipate seismic energy. Rice and Ben Zion (1996) suggested that small event complexity in this system was probably due to the lack of continuum limit of this model.

An interesting class of models inspired from the BK model, but much simpler to compute, consists of the cellular automata models (see, e.g., Wolfram, 1986). Using different versions of cellular automata, it is possible to reproduce the power law distribution of earthquake statistics (Gutenberg and Richter law). More interestingly, certain interacting dynamical systems may spontaneously evolve into a critical state, producing earthquakes of all sizes. This is the so-called self-organized critical state (see, e.g., Bak *et al.*, 1987). A major problem with cellular automata models is that they lack scales for time and/or length. Time evolution occurs by discrete quasi-static steps so that—at least in the version known to us—they do not include dynamic effects. As will become apparent in a later section, length and time scales are essential for physically correct modeling of the forces underlying the dynamic behavior. Although these models are for the moment too simple for simulating the observed radiation from a real earthquake, there is no doubt that these concepts will play a major role in future earthquake studies.

In order to understand the physics of the dynamic behavior, we turn to the theory of an elastic shear fault, starting from the simplest approximation.

2.2 The Dislocation Model

In spite of much recent progress in understanding the dynamics of earthquake ruptures, the most widely used models for interpreting seismic radiation are the so-called dislocation models. In these models the earthquake is simulated as the kinematic spreading of a displacement discontinuity (slip or dislocation in seismological usage) along a fault plane. In its most general version, slip in a dislocation model is completely arbitrary and rupture propagation may be as general as desired. In this version the dislocation model is a perfectly legitimate description of an earthquake as the propagation of a slip episode on a fault plane. It must be remarked, however, that not all slip distributions are physically acceptable: as shown by Madariaga (1978), most dislocation models present unacceptable features such as interpenetration of matter, release of unbounded amounts of energy, and so on. For these reasons, dislocation models must be considered as a useful intermediate step in the formulation of a physically acceptable dynamic fault model.

Dislocation models are very useful in the inversion of near-field accelerograms (see, e.g., Cohee and Beroza, 1994; Wald and Heaton, 1994; Cotton and Campillo, 1995a; Mikumo and Miyatake, 1995; and many others). Radiation from a dislocation model can be written as a functional of the distribution of slip on the fault. In a simplified form a seismogram $u(x, t)$ at an arbitrary position x can be written as

$$u(r,t) = \int_0^t \int_{S_\xi} \Delta u(\xi,\tau) G(x-\xi, t-\tau)\, d\xi\, d\tau, \qquad (1)$$

where $\Delta u(\xi, \tau)$, the slip on the fault, is a function of space and time, and $G(x, t)$ is the Green tensor that may be computed using simple layered models of the crustal structure, or more complex numerical (for example, FD) simulations. Functional (1) is linear in slip amplitude but very nonlinear with respect to rupture propagation, which is implicit in the time dependence of Δu. For this reason, in most inversions the kinematics of the rupture process (position of rupture front as a function of time) is assumed to propagate at constant rupture velocity from the hypocenter. Different approaches have been proposed in the literature in order to invert approximately for variations in rupture speed about the assumed constant rupture velocity (see, e.g., Wald and Heaton, 1984; Cotton and Campillo, 1995a). The slip history on the fault can then be used to compute the history of stress on the fault by a procedure originally proposed by Mikumo and Miyatake (1979). This method has been used extensively in recent years to estimate the state of stress on a fault (see, e.g., Bouchon, 1997; Ide and Takeo, 1997; Olsen *et al.*, 1997; Day *et al.*, 1998; Guatteri and Spudich, 1998a,b; among many others).

The most important dislocation model was introduced by Haskell (1964). In this model, shown in Figure 1, a uniform displacement discontinuity spreads at constant rupture velocity inside a rectangular-shaped fault. At low frequencies, or wavelengths much longer than the size of the fault, this model is a reasonable approximation to a seismic rupture. In Haskell's model, at time $t = 0$ a line of dislocation of width W appears suddenly and propagates along the fault length at a constant rupture velocity until a region of length L of the fault has been broken. As the dislocation moves, it leaves behind a zone of constant slip D. The fault area $L \times W$ times the slip D and the rigidity μ of the medium defines the seismic moment $M_0 = \mu DLW$ of this model. Haskell's model captures some of the most important features of an earthquake and has been used extensively to invert for seismic source parameters from near-field and far-field seismic and geodetic data. The complete seismic radiation for Haskell's model was computed by Madariaga (1978) who showed that, because of the stress singularities around the edges, the Haskell model fails at high frequencies, as was noted by Haskell (1964) himself. All dislocation models with constant slip will have the same problems at high frequencies, although they can be improved by tapering the slip discontinuity near the edges of the fault, as proposed by Sato and Hirasawa (1973). Even better dislocation models can be obtained by taking into account the kinematics of rupture front propagation as proposed by Spudich and Hartzell (1984) and Bernard and Madariaga (1984). Kinematic models have played a mayor role in the quantification of earthquakes and in the inversion of seismic data, a subject that we cannot develop in depth in this chapter.

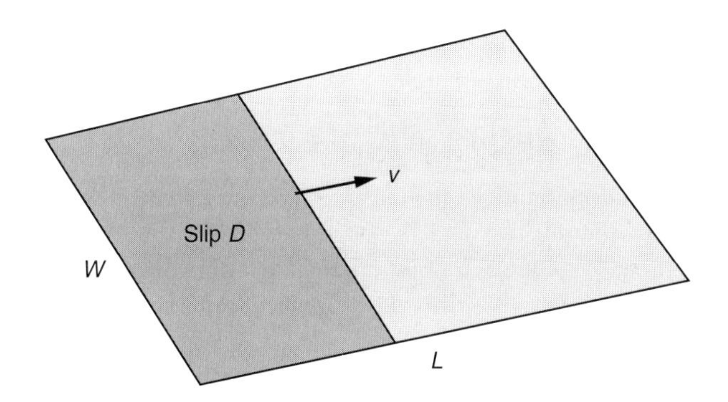

FIGURE 1 The Haskell kinematic earthquake model, probably the simplest possible earthquake model. The fault has a rectangular shape and a linear rupture front propagates from one end of the fault to the other at constant rupture speed v. Slip in the broken part of the fault is uniform and equal to D.

2.3 Elastic Shear Fault Model

Let us now study the main features of a properly posed source model in a simple homogeneous elastic model of the Earth. Expansion to more complex elastic media, including realistic wave propagation media, poses no major technical difficulties except, of course, that computation time may become very long.

Consider the 3D elastic wave equation:

$$\rho \frac{\partial^2}{\partial t^2} \mathbf{u} = \nabla \cdot \sigma, \tag{2}$$

where $\mathbf{u}(\mathbf{x}, t)$ is the displacement vector field, a function of both position $\mathbf{x}$ and time t, and $\rho(\mathbf{x})$ is the density of the elastic medium. Associated with the displacement field $\mathbf{u}$ the stress tensor $\sigma(\mathbf{x}, t)$ is defined by

$$\sigma = \lambda \nabla \cdot \mathbf{u}\, \mathbf{I} + \mu\left[(\nabla \mathbf{u}) + (\nabla \mathbf{u})^{\mathrm{T}}\right], \tag{3}$$

where $\lambda(\mathbf{x})$ and $\mu(\mathbf{x})$ are Lamé's elastic constants, $\mathbf{I}$ is the identity matrix, and superscript T indicates matrix transpose. We can transform this system into a more symmetric velocity–stress formulation (Madariaga, 1976; Virieux and Madariaga, 1982; Virieux, 1986):

$$\begin{aligned} \rho \frac{\partial}{\partial t} \mathbf{v} &= \nabla \cdot \sigma + \mathbf{f} \\ \frac{\partial}{\partial t} \sigma &= \lambda \nabla \cdot \mathbf{v}\, \mathbf{I} + \mu\left[(\nabla \mathbf{v}) + (\nabla \mathbf{v})^{\mathrm{T}}\right] + \dot{\mathbf{m}}, \end{aligned} \tag{4}$$

where $\mathbf{v}(\mathbf{x}, t)$ is the particle velocity vector, and $\mathbf{f}(\mathbf{x})$ and $\mathbf{m}(\mathbf{x})$ are the force and moment source distributions, respectively.

2.3.1 Boundary Conditions on the Fault

Assume that the earthquake occurs on a fault surface of normal n in the previous elastic medium. Due to frictional instability, a rupture zone can spread along the fault; let $\Gamma(t)$ be this rupture zone at time t. In general, $\Gamma(t)$ is a collection of one or more rupture zones propagating along the fault.

The main feature of a seismic rupture is that, at any point $\mathbf{x}$ inside the rupture zone $\Gamma(t)$, displacement and particle velocities are discontinuous. Let

$$\mathbf{D}(\mathbf{x}, t) = \mathbf{u}^{+}(\mathbf{x}^{+}, t) - \mathbf{u}^{-}(\mathbf{x}^{-}, t) \tag{5}$$

be the slip vector across the fault, i.e., the jump in displacement between the positive and the negative side of the fault. The notation $\mathbf{x}^{\pm}$ indicates a point immediately above or below the fault, and $\mathbf{u}^{\pm}$ are the corresponding displacements.

Slip $\mathbf{D}$ is associated with a *change in the traction* $\mathbf{T} = \sigma \cdot \mathbf{e}_z = [\sigma_{zx}, \sigma_{yz}, \sigma_{zz}]$ across the fault through the solution of the wave equation (4):

$$\Delta \mathbf{T}(\mathbf{x}, t) = \Delta \mathbf{\Sigma}[\mathbf{D}] \quad \text{for} \quad \mathbf{x} \in \Gamma(t), \tag{6}$$

where $\Delta \mathbf{\Sigma}[\mathbf{D}]$ is a shorthand notation for a functional of $\mathbf{D}$ and its temporal and spatial derivatives.

2.4 Friction

The main assumption in seismic source dynamics is that traction across the fault is related to slip at the same point through a *friction law* that can be expressed in the general form

$$\mathbf{T}(\mathbf{D}, \dot{\mathbf{D}}, \theta_i) = \mathbf{T}_{\text{total}} \quad \text{for} \quad \mathbf{x} \in \Gamma(t) \tag{7}$$

such that friction $\mathbf{T}$ is a function of at least slip, but an increasing amount of experimental evidence shows that it is also a function of slip rate $\dot{\mathbf{D}}$ and several state variables denoted by θ_i, $i = 1, \ldots, N$. The traction that appears in friction laws is the total traction $\mathbf{T}_{\text{total}}$ on the fault, which can be expressed as the sum of preexisting stress $\mathbf{T}^0(\mathbf{x})$ and the stress change $\Delta\mathbf{T}$ due to slip on the fault obtained from Eq. (6). The prestress is caused by tectonic load of the fault and will usually be a combination of purely tectonic loads due to internal plate deformation, plate motion, etc., and the residual stress field remaining from previous seismic events on the fault and its vicinity.

Using Eq. (6), we can now explicitly formulate the friction law on the fault [Eq. (7)]:

$$\mathbf{T}(\mathbf{D}, \dot{\mathbf{D}}, \theta_i) = \mathbf{T}^0(\mathbf{x}) + \Delta \mathbf{T}(\mathbf{x}, t) \quad \text{for} \quad \mathbf{x} \in \Gamma(t) \tag{8}$$

Friction as defined by Eq. (8) is clearly a vector. For the appropriate study of a shear fault we need to write Eq. (8) as a system of two equations. Archuleta and Day (1980), Day (1982a,b), and Spudich (1992) used a very simple approach that will certainly have to be refined in the future, assuming that slip rate and traction are antiparallel, i.e.,

$$\mathbf{T}(\mathbf{D}, \dot{\mathbf{D}}, \theta_i) = -T(D, \dot{D}, \theta_i)\mathbf{e}_V, \tag{9}$$

where $\mathbf{e}_V = \dot{\mathbf{D}}/\|\dot{\mathbf{D}}\|$ is a unit vector in the direction of instantaneous slip rate. With this assumption, the boundary condition reduces Eq. (8) to the special form

$$-T(D, \dot{D}, \theta_i)\mathbf{e}_V = \mathbf{T}^0(\mathbf{x}) + \Delta \mathbf{T}(\mathbf{x}, t) \quad \text{for} \quad \mathbf{x} \in \Gamma(t). \tag{10}$$

Figure 2a shows the vector diagram implied by this equation. The only fixed vector in this diagram is the prestress, which is assumed to be known. Friction and slip rate are collinear but antiparallel. Stress change $\Delta\mathbf{T}$ is in general collinear neither with prestress nor with friction. Recently, Spudich (1992), Cotton and Campillo (1995b), and Guatteri and Spudich (1998b) analyzed expression (10) and studied several recent earthquakes, showing that slip directions were not always parallel to stress drop. These rake rotations may also have information about the absolute stress levels (Spudich, 1992; Guatteri and Spudich, 1998b).

In most rupture models the above "vector" friction is simplified to a "scalar boundary condition," in which slip is only allowed in the direction of the initial stress, which is everywhere parallel to the x axis, i.e., $\mathbf{T}^0(\mathbf{x}) = [T_x^0(\mathbf{x}), 0]$ and $\mathbf{D}(\mathbf{x}, t) = [D_x(\mathbf{x}, t), 0]$; then the scalar components are simply related by

$$\Delta T_x(\mathbf{x}, t) = T(D, \dot{D}, \theta_i) - T_x^0(\mathbf{x}) \quad \text{for} \quad \mathbf{x} \in \Gamma(t). \tag{11}$$

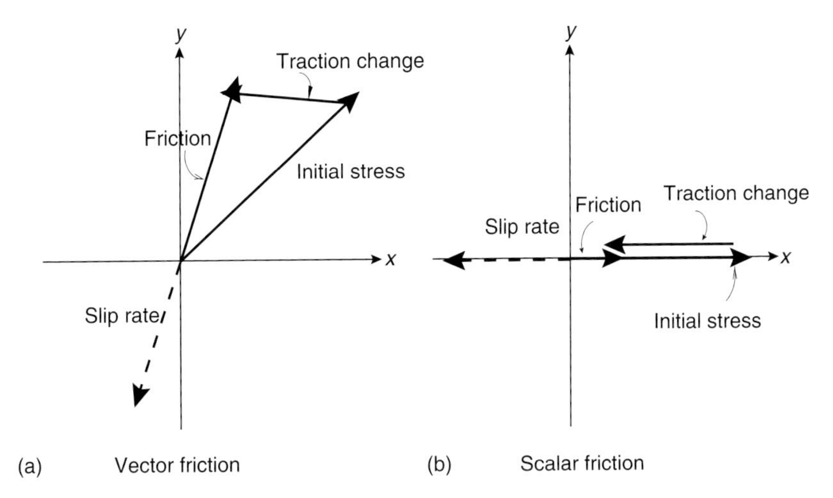

FIGURE 2 Diagram showing the relation between initial stress, slip rate, friction, and traction change for the vector (a) and scalar (b) approximations to friction on the fault plane. In the scalar case, traction change corresponds to the usual definition of stress drop. (Reproduced with permission from Madariaga *et al.*, 1998; fig. 1, p. 1185.)

This boundary condition, which may be graphically described as if the fault had "rails" aligned in the x direction, has been applied in most 3D source models, starting with Madariaga (1976).

2.5 Friction Laws

Both boundary conditions (10) and (11) require a friction law that relates scalar traction T to slip, its derivatives, and possible state variables. For more details, see Dieterich (1978, 1979) and Rice and Ruina (1983), but see also Ohnaka (1996) for an alternative point of view.

Let us first discuss the simple slip weakening friction law introduced by Ida (1972). It is an adaptation to shear faulting of the Barenblatt–Dugdale friction laws used in hydrofracturing and tensional (mode I) cracks. In this friction law, slip is zero until the total stress reaches a peak value (yield stress) that we denote by T_u. Once this stress has been reached, slip D starts to increase from zero and $T(D)$ decreases linearly to zero as slip increases:

$$T(D) = T_u\left(1 - \frac{D}{D_c}\right) + T_f \qquad \text{for} \qquad D < D_c$$
$$T(D) = T_f \qquad \text{for} \qquad D > D_c, \qquad (12)$$

where D_c is a characteristic slip distance and T_f is the residual friction at high slip rate, sometimes called the "kinematic" friction. There is a lot of discussion in the literature about how large this residual friction is. Many authors, following the observation that there is a very broad heat flow anomaly across the San Andreas fault in California, have proposed that faults are "weak," meaning that T_f is close to zero. Other authors propose that kinematic friction is high and faults are strong. We cannot go into any detail about this discussion here: interested readers may consult the papers by Scholz (2000) and Townend and Zoback (2000). For most applications of earthquake dynamics, only stress change is important, so that without loss of generality we can assume that $T_f = 0$ in much of the following. Let us remark, however, that Spudich (1992), Guatteri and Spudich (1998a), and Spudich *et al.* (1998) have found some evidence for absolute stress levels from nonparallel stress drop and slip of the 1995 Kobe earthquake in Japan.

The slip weakening friction law (12) has been used in numerical simulations of rupture by Andrews (1976a,b), Day (1982b), Harris and Day (1993), Fukuyama and Madariaga (1998), Madariaga *et al.* (1998), and many others. Slip weakening at small slip is *absolutely necessary for the friction law to be realizable*, otherwise stress would become unbounded near the rupture front, violating energy conservation so that seismic ruptures could only spread at either S, Rayleigh, or P wave velocities until they stop. Of course, in numerical implementations stress is never infinite, so that rupture velocity is numerically limited. In many earlier studies of earthquake dynamics, a simpler version of Eqs. (12) was used in which D_c was effectively zero. This numerical version of slip-weakening has been called the Irwin criterion by Das and Aki (1977a) and has been widely used by many authors although it is obviously grid-dependent (see, e.g., Virieux and Madariaga, 1982).

Once slip is larger than the slip weakening distance D_c, friction becomes a function of slip rate $\dot{D}$ and one or more state variables that represent the memory of the interface to previous slip. A very simple *rate-dependent friction law* was proposed by Burridge and Knopoff (1967) and has been used extensively in simulations by Carlson and Langer (1989) and their colleagues and by Cochard and Madariaga (1986):

$$T(\dot{D}) = T_s \frac{V_0}{V_0 + \dot{D}} + T_f, \qquad (13)$$

where V_0 is a characteristic slip velocity and $T_s \leq T_u$ is the limit of friction when slip rate decreases to zero. The applicability of rate weakening to seismic ruptures is much more controversial than slip weakening, although there is plenty of indirect evidence for its presence in seismic faulting. Heaton (1990) proposed that it was the cause of short rise times; rate-dependence at steady slip velocities is also an intrinsic part of the friction laws proposed by Dieterich (1978) and Ruina (1983), which will be reviewed in the following.

The slip weakening and slip rate weakening behavior described above can be combined if for any value of D and $\dot{D}$ the larger of expressions (12) and (13) is selected. Instead of writing a complex expression, it is simpler to show the friction law graphically in Figure 3 in the form of a law where friction depends on the two state variables slip and slip rate. The friction law described above allows rupture propagation that is completely controlled by the complex nonlinear interaction of

1. The initial stress field $\mathbf{T}^0$ in (8).
2. The distribution of yield frictional resistance T_u in Eqs. (12) and (13).
3. The parameters D_c, T_s, and V_0 of the friction laws in Eqs. (12) and (13).

As mentioned earlier, most recent work on friction has been concentrated in a class of friction laws that depend both on slip rate and state variables. These laws were developed from laboratory experiments at low slip rates by Dieterich (1978, 1979) and Ruina (1983).

Typically, a rate-and-state dependent friction law can be expressed by

$$T(\dot{D}, \theta) = \theta + \tau(\dot{D}), \tag{14}$$

where $\tau(\dot{D})$ is the instantaneous response of friction to slip rate changes ("the direct effect"). The state variable θ represents the weakening of the interface with time and in general it is considered to satisfy an evolution equation of the form

$$\dot{\theta} = \frac{V_0}{L} G(\dot{D}, \theta). \tag{15}$$

Here L is a *weakening distance* that measures how much slip will occur before the friction weakens to the steady state value; V_0 is a reference slip rate. There are many versions of these friction laws, but the main features are not very different from slip weakening, at least at the high slip rates that occur near the rupture front. Computations by Dieterich and Kilgore (1996) show that the slip weakening distance for rate-and-state dependent friction is roughly $D_c \simeq 4L$, an approximation also found by Gu and Wong (1991). Although rate-and-state dependent friction is very important for the study of rupture initiation and repeated ruptures on a fault surface, its features are indistinguishable from simpler slip-dependent and rate-dependent friction laws during the dynamic part of seismic ruptures.

2.6 The Boundary Integral Element Method

An essential requirement in studying dynamic faulting is an accurate and robust method for the numerical modeling of wave

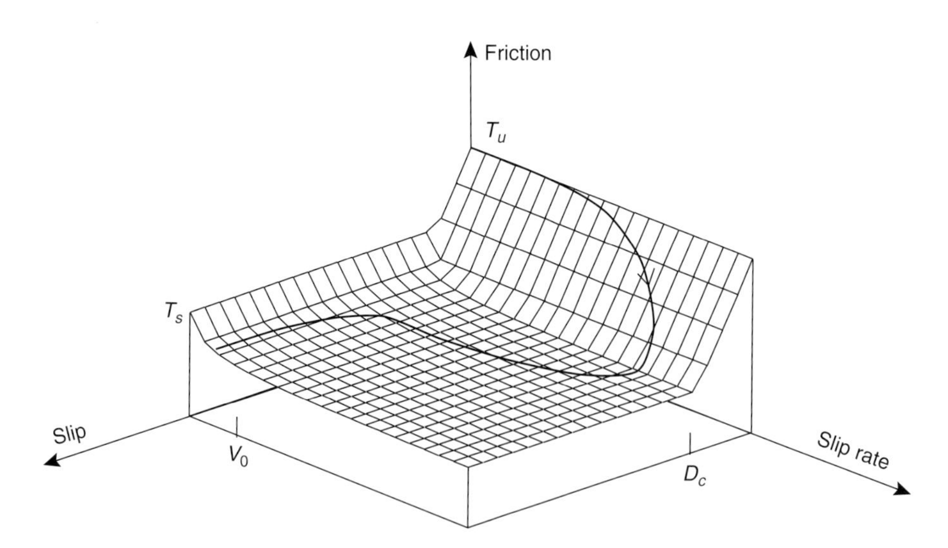

FIGURE 3 Slip- and slip rate-dependent friction law. For values of stress less than the peak static friction (T_u), slip and slip rate are zero. Once slip begins, stress is a function of both slip and slip rate described by the friction surface $T(D, \dot{D})$. Slip weakening is measured by D_c, rate weakening by V_0. The continuous curve shows the typical stress trajectory of a point on the fault. (Reproduced with permission from Madariaga *et al.*, 1998; fig. 2, p. 1185.)

propagation that can also accurately handle the nonlinear boundary conditions on the fault surface. One of the important methods is the boundary integral equation (BIE) method. The BIE method was pioneered in 2D by Das and Aki (1977a,b) in the so-called direct version. It was later extended to 3D by Das (1980, 1981) and improved by Andrews (1985). A new version of the BIE method, the so-called indirect, or displacement discontinuity, method, was proposed by Koller *et al.* (1992). The indirect method was improved by the removal of strong singularities by a number of authors (e.g., Fukuyama and Madariaga, 1995, 1998; Geubelle and Rice, 1995; Cochard and Madariaga, 1996; Bouchon and Streiff, 1997). Following Fukuyama and Madariaga (1998), the integral equations for a shear fault can be expressed as

$$
\begin{aligned}
T_\alpha(x_1, x_2, t) &= \frac{3\mu}{\pi} \iint_{S^+} \frac{r_{,\alpha}}{r^2} \int_1^{\kappa} \tau D_{\beta,\beta}(\xi_1, \xi_2, \|t - vr/c_T\|)\, d\tau\, d\xi_1\, d\xi_2 \\
&+ \frac{5\mu}{4\pi} \iint_{S^+} \frac{r_{,\alpha}}{r^2} D_{\beta,\beta}(\xi_1, \xi_2, \|t - r/c_T\|)\, d\xi_1\, d\xi_2 \\
&- \frac{\kappa^2 \mu}{\pi} \iint_{S^+} \frac{r_{,\alpha}}{r^2} D_{\beta,\beta}(\xi_1, \xi_2, \|t - r/c_L\|)\, d\xi_1\, d\xi_2 \\
&- \frac{\mu}{4\pi} \iint_{S^+} \frac{r_{,\beta}}{r^2} D_{\alpha,\beta}(\xi_1, \xi_2, \|t - r/c_T\|)\, d\xi_1\, d\xi_2 \\
&+ \frac{\mu}{4 c_T \pi} \iint_{S^+} \frac{r_{,\alpha}}{r} \dot{D}_{\beta,\beta}(\xi_1, \xi_2, \|t - r/c_T\|)\, d\xi_1\, d\xi_2 \\
&- \frac{\mu}{2 c_T} \dot{D}_\alpha(x_1, x_2, t),
\end{aligned}
\tag{16}
$$

where $D_{i,j}$ is the derivative with respect to T of the ith component of slip and the "dot" denotes the time derivative, μ is rigidity, c_L and c_T are P- and S-wave velocities, respectively, $\kappa = c_T/c_L$, ρ is density, $r = |\mathbf{x} - \mathbf{xi}|$, $r_{,i} = (x_i - \xi_i)/r$, and $\| a \|$ is a shorthand notation meaning that the slip functions D are evaluated only for positive values of a. S^+ represents the upper surface of the crack. The Greek indices α and β can be either 1 or 2. Each term in this system of boundary integral equations has a simple interpretation. The first four terms are near-field effects due to the horizontal gradient of slip. The fifth term represents "far-field" or high-frequency S-wave radiation. The last term is the radiation damping due to the emission of S-waves in the direction normal to the fault. In order to solve the integral equation (16) numerically for shear faults, a discretization using interpolation functions has to be introduced. Traditionally in BIE high-order spatial interpolation and very low-order interpolation in time are adopted (see, e.g., Hirose and Achenbach, 1989; Koller *et al.*, 1992). This was the method adopted for antiplane cracks by Cochard and Madariaga (1996) or by Fukuyama and Madariaga (1995, 1998) for 3D faulting. Since high-order spatial interpolations lead to implicit equations in time that are almost impossible to solve in the presence of nonlinear friction on the fault, several implementations use simpler piecewise constant interpolation of slip velocity. This leads to a formulation of crack problems that Cruse (1988) calls the displacement discontinuity method. With this interpolation, the slip gradient $D_{\beta,\alpha}$ is sharply localized at the boundaries of the grid elements.

BIE methods are excellent for the study of earthquake initiation and the transition from accelerated fault creep to fully dynamic rupture propagation. A problem with the BIE method for the simulation of dynamic earthquake ruptures is the relatively large requirement for computer memory (though not of computer time) and a need for explicit computation of the operator $\Delta\boldsymbol{\Sigma}$. Also, at least in their current implementations, BIE methods cannot be used in heterogeneous media, but can be used for complex fault geometries and homogeneous half-spaces. Aochi *et al.* (2000) have recently extended the BIEM to handle complex faults with noncoplanar segments. They studied rupture propagation for the 1992 Landers earthquake and showed that rupture can jump between segments under some restrictive conditions.

2.7 The Finite Difference (FD) Method

The other numerical method widely used for numerical modeling of dynamic wave propagation is the finite difference (FD) method. The FD method was introduced by Madariaga (1976) and Andrews (1976a,b) for the study of seismic ruptures and was developed by numerous authors (e.g., Miyatake, 1980, 1992; Day, 1982a,b; Mikumo *et al.*, 1987; Harris and Day, 1993; Olsen *et al.*, 1997; Madariaga *et al.*, 1998). The method can be used to study rupture propagation in heterogeneous elastic media and is very efficient. An advantage of finite differences is that the operator $\Delta\boldsymbol{\Sigma}$ in Eq. (6) does not to have to be computed explicitly. In the FD method, all that is needed is a numerical procedure that computes the stress change $\Delta\mathbf{T}$ given the slip distribution $\mathbf{D}$ at earlier times.

Numerous different implementations of the FD method have been presented in the literature, and it is beyond the scope of this chapter to cover them all. In general, FD methods used to model earthquakes can be divided into two types. The first type derives from the direct discretization of the second-order PDE obtained by substituting Eq. (3) into Eq. (2). Methods of this kind were derived and greatly developed by Mikumo and Miyatake (1979), Miyatake (1980), Mikumo *et al.* (1987), and Mikumo and Miyatake (1993). The other approach, the staggered grid velocity–stress formulation, was developed by Madariaga (1976) to study dynamic rupture problems and is based on the discretization of the system of Eq. (4). The staggered grid method is characterized by low numerical dispersion and the fact that no derivatives of media parameters are needed. The latter is a strong advantage over FD implementations of the second-order displacement formulation of the wave equation where accuracy is lost in the computation of derivatives of media parameters near significant gradients in the model. Olsen *et al.* (1995a) and Olsen and Archuleta (1996) demonstrated the efficiency of the fourth-order formulation of the velocity–stress

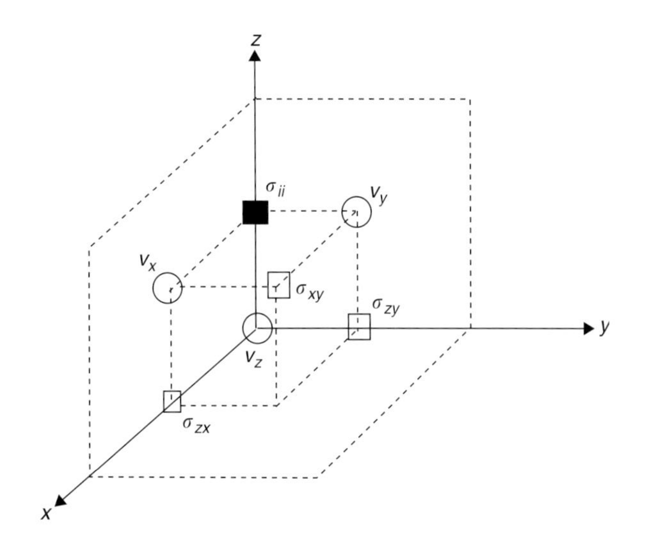

FIGURE 4 A cubic element of the 3D finite difference grid used in the dynamic modeling of a planar shear fault. σ and $\mathbf{v}$ depict the components of the stress tensor and particle velocity, respectively. (Reproduced with permission from Madariaga *et al.*, 1998; fig. 9, p. 1192.)

method by computing wave propagation around a kinematically defined rupture in a large-scale 3D model.

The velocity–stress FD method is illustrated in Figure 4; stress and velocities are defined at alternating half-integer time steps. At time $t_N = N\,\Delta t$, particle velocity v is computed from previously calculated stress components. At the next half-time step $t_{N+1/2} = \left(N + \frac{1}{2}\right)\Delta t$, stress σ is updated using the velocity field computed at time t_N. Thus as time increases, velocities and stresses are computed at alternate times. Because stress and velocities are computed from Eq. (4) using centered fourth-order finite differences, the grid is also staggered spatially as shown in Figure 4. Madariaga *et al.* (1998) extended the FD method presented by Olsen (1994), Olsen *et al.* (1995a), and Olsen and Archuleta (1996) in order to study dynamic rupture propagation on a planar shear fault embedded in a heterogeneous half-space. We will use their "thick fault" boundary condition to illustrate important dynamic rupture phenomenology in this section, although alternate implementations exist. Current developments include the coupling of a BIE solution for slip and stress on the fault to a finite difference computation of radiated waves (Olsen *et al.*, 2000).

3. Phenomenology of Rupture Models with a Single Length Scale

We start the study of seismic ruptures from a very simple earthquake model that is a sort of classic test model inspired by Kostrov's (1964) study of a self-similar circular shear crack. We study the spontaneous propagation of a seismic rupture starting from a circular asperity that is ready to break. The asperity is surrounded by a fault surface uniformly loaded at a stress level that is less than the peak stress in the slip weakening friction law [Eq. (12)]. These are very similar conditions to those used by Day (1982b) and Das (1981) to start rupture. There are two main reasons to proceed this way: First, if the asperity is too small, rupture will start and stop immediately. For rupture to expand, stress must be high over a finite zone, sometimes called the minimum rupture patch. The other reason why rupture has to start from a finite-size asperity is that, if the stress field were uniform, rupture would occur instantaneously or grow at the maximum possible velocity from an arbitrary point on the fault. This is unrealistic and not supported by observations. Finally, it is assumed that fault rupture must occur at stress levels that are below the yield stress except for a small number of isolated asperities. In the following examples there is only one such asperity, even though rupture starting from several locations is possible (e.g., Day, 1982b; Olsen *et al.*, 1997).

3.1 Dimensional Analysis

The following discussion uses nondimensional variables. This has the advantage of clearly showing how different variables scale with stresses and distances. We choose the following dimensional variables:

- Distances along the fault are measured in units of Δx, the grid interval.
- Wave velocities are measured in units of β, the shear wave velocity.
- Stress is measured in units of T_u, the peak frictional resistance (yield stress) in the friction laws described in Eqs. (12) and (13).

All other dimensions are determined by the previous three. In particular:

- Time is measured in units of $\Delta t = H\,\Delta x/\beta$, where β is the shear wave velocity. H is the so-called CFL (Courant–Friedrich–Lewy) parameter that controls stability of the numerical method. In our simulations it was usually taken as 0.25 in order to insure stability and sufficient accuracy.
- Displacement is measured in units of $T_u/\mu \times \Delta x$.
- Particle velocities are measured in units of $T_u/\mu \times \beta$.

Slip and slip rate are normalized by $2T_u/\mu \times \Delta x$ and $2T_u/\mu \times \beta$. The factor of 2 is not really necessary but follows a tradition in seismological publications. It is also assumed that the P-wave velocity α is equal to $\sqrt{3}\beta$. Finally, D_c, the slip weakening distance in Eq. (12) is measured in units of slip (i.e., $2T_u/\mu \times \Delta x$), and V_0, the rate weakening parameter, is measured in units of slip rate (i.e., $2T_u\mu \times \beta$). We have followed seismological tradition and used the grid length, Δx, to scale fault length, instead of a physical length such as the slip weakening distance D_c. The reason is that until recently most simulations of 3D seismic ruptures used a grid-dependent fracture criterion introduced by Das and Aki (1977a).

3.2 A Circular Fault Model

In our first model we study rupture propagation where the initial stress distribution is symmetric about the origin. We force rupture to stop once it reaches a circular distance R. Rupture resistance, represented by T_u and D_c, is perfectly uniform. We initiate the rupture from a finite asperity of radius $R_{\rm asp}$. Rupture velocity is not constant but is determined from the friction law. Our solutions are not self-similar and, as already illustrated by Das (1981), Day (1982b), Virieux and Madariaga (1982), and others, spontaneous ruptures do not maintain simple elliptical shapes as they grow.

The circular fault has a radius $R = 50 \times \Delta x$, starting from a concentric asperity of radius $R_{\rm asp} = 6 \times \Delta x$, $D_c = 4$, and stress inside the asperity was $T_{\rm asp} = 1.2 \times T_u$ and $T_{\rm ext} = 0.8 \times T_u$ outside. Snapshots of the slip rate are shown in Figure 5 at several successive instants of time. Time is measured in units of $\Delta t = H\,\Delta x/\beta$, where $H = 0.25$ as discussed earlier. From time steps $t = 17.5$ to 35, rupture is taking place inside the asperity, and is propagating away from the asperity for $t > 52.5$. We observe that rupture becomes spontaneously elongated in the vertical direction, which is also the direction of the initial stress. Thus, as already remarked by Das (1981) and Day (1982b), rupture tends to grow faster in the in-plane direction, which is dominated by mode II.

At time $t = 87.5$, rupture in Figure 5 has reached the unbreakable border of the fault in the in-plane direction, and at time $t = 105$ the stopping phases generated by the upper and bottom edges of the fault are moving toward the center of the fault. The snapshots after $t = 122.5$ show stopping phases propagating inward from all directions. The slipping patch in darker shading is now elongated in the antiplane direction, which is due to slower healing. At time $t = 140$ the in-plane stopping phases (moving in the vertical direction) have already reached the center of the fault and crossed each other. In the

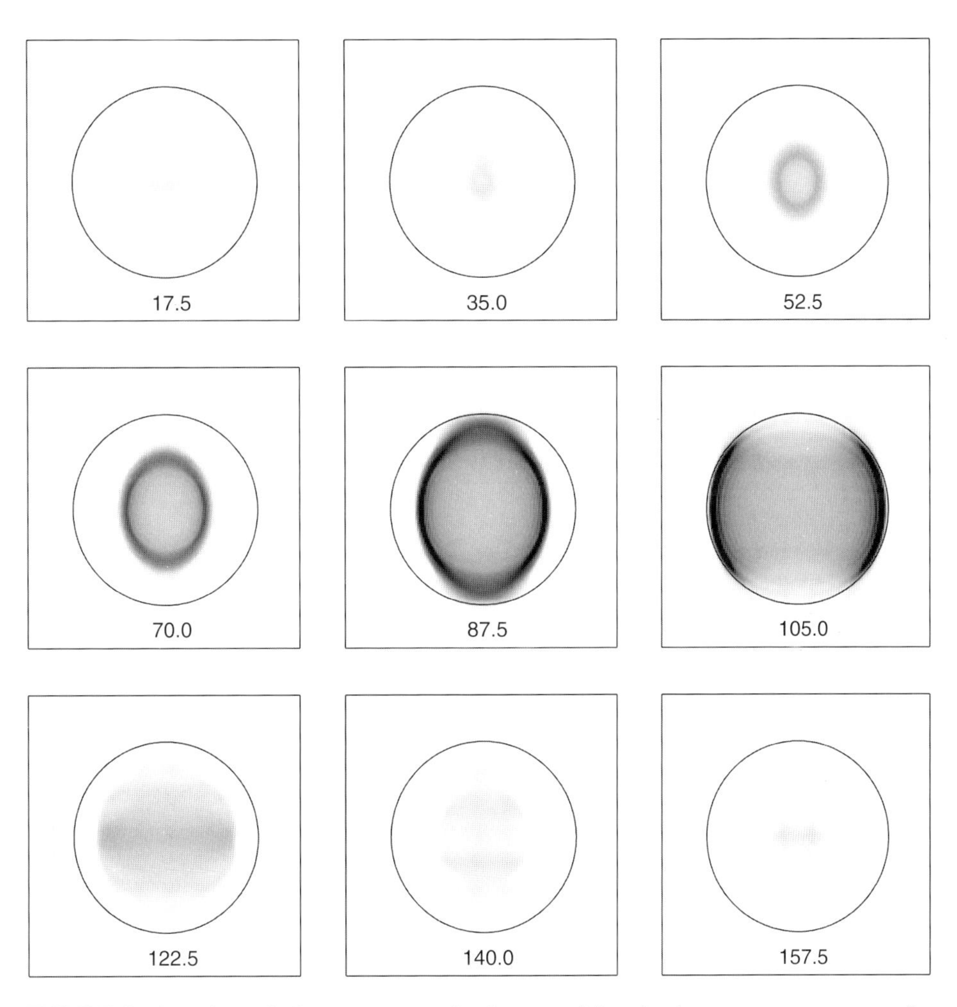

FIGURE 5 Snapshots of slip rate at successive instants of time for the spontaneous rupture of an overloaded asperity inside a circular fault (solid line). The nondimensional time for each snapshot is shown below each picture. (Reproduced with permission from Madariaga *et al.*, 1998; fig. 9, p. 1192.)

last two snapshots, rupture continues in a small patch near the center of the fault that coincides with the initial asperity. However, slip rate has decreased to such small values that it is very likely contaminated with numerical noise.

3.3 Rectangular Fault

Now we turn to a model that starts in the same way as the circular fault from an overloaded asperity. However, here the unbreakable barrier forces rupture to expand in essentially one direction along a rectangular fault. We build this model as a prototype of rupture along a shallow strike-slip fault and use similar values for the friction laws as those for the circular crack simulation ($H = 0.2$; slip weakening distance $D_c = 4$; initial stress inside and outside the asperity $T_{asp} = 1.2 \times T_u$ and $T_{ext} = 0.8 \times T_u$, respectively; radius of the initial asperity $R_{asp} = 6 \times \Delta x$; for the rate-weakening simulations we used $V_0 = 0.03$ and $T_s = T_u$). Unfortunately, as pointed out by Heaton (1990) and discussed by Cochard and Madariaga (1996), there is no information about velocity weakening at high slip rates. The value chosen for V_0 is arbitrary and corresponds to rapid healing when slip rate becomes about 3% of the peak slip rate.

Color Plate 3 shows snapshots of the slip rate on the fault plane for simulations using slip weakening and rate weakening friction, respectively. The prestress on the fault is directed along the vertical (long) axis of the fault. In the simulation of part (a) with slip weakening but no rate weakening, we see the rupture emerging from the asperity with relatively slow healing (long "tail" trailing from the front). Rupture starts out slowly, accelerates toward the S-wave speed and, at a mature stage near time $t = 80$, suddenly "jumps" to the P-wave speed. The transition to supershear rupture speeds is an instability that develops from the in-plane direction and spreads laterally along the rupture front, producing a "bulge" observed in the snapshots after $t = 80$. Stopping phases emitted from the sides of the fault clearly control the duration of slip, as shown in snapshots at $t = 120$ through 160. In the snapshot at $t = 160$, the stopping phases have reached the center of the fault just below the time label 160.

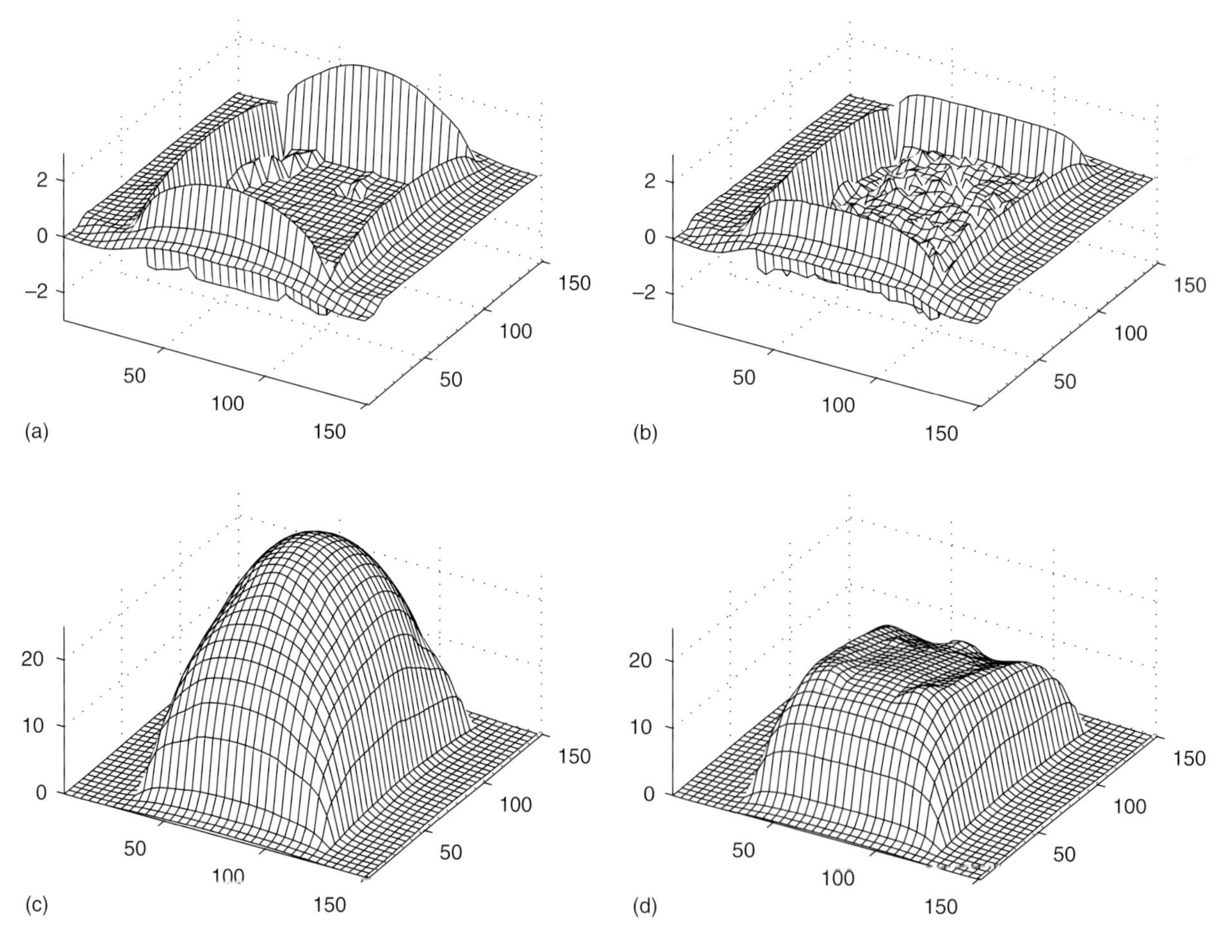

FIGURE 6 Final stress without (a) and with (b) rate weakening, and final slip without (c) and with (d) rate weakening, for the simulations shown in Color Plate 3(a) and (b). The plots clearly show a decrease in slip as well as the development of stress heterogeneity inside the fault due to rate weakening. (Reproduced with permission from Madariaga *et al.*, 1998; fig. 12, p. 1195.)

The situation is quite different with the use of a rate-dependent friction law, where the slip rate tends to concentrate in narrow patches (Part b). Compared to Part (a), the rupture front is narrower and clearly delineated. Well before the arrival of stopping phases from the sides of the fault, slip rate has become very small near the center of the asperity. Rupture takes the shape of a band. As time increases and rupture is controlled by the edges of the fault, the rupture front becomes narrow and localized. This is similar to the behavior predicted by Heaton (1990).

The other major difference introduced by rate-dependent friction is that, behind the rupture front, stress becomes heterogeneous. It appears that the rupture leaves a wake of complexity after its passage. This complexity is apparent in the distributions of both slip (Fig. 6c,d) and stress drop (Fig. 6a,b). The development of stress heterogeneity in the wake of the rupture front is an essential feature of rate weakening as proposed by Carlson and Langer (1989) for the BK model. It was shown by Cochard and Madariaga (1996) that stresses become complex because rate weakening promotes early healing of the fault. When the fault heals rapidly, all heterogeneities become frozen on the fault and cannot be eliminated until the fault slips again. This process of generating heterogeneity was similar to what Bak *et al.* (1987) had in mind when they proposed that earthquakes were an example of self-organized criticality. Finally, note that the faster healing caused by rate weakening friction decreases the slip significantly.

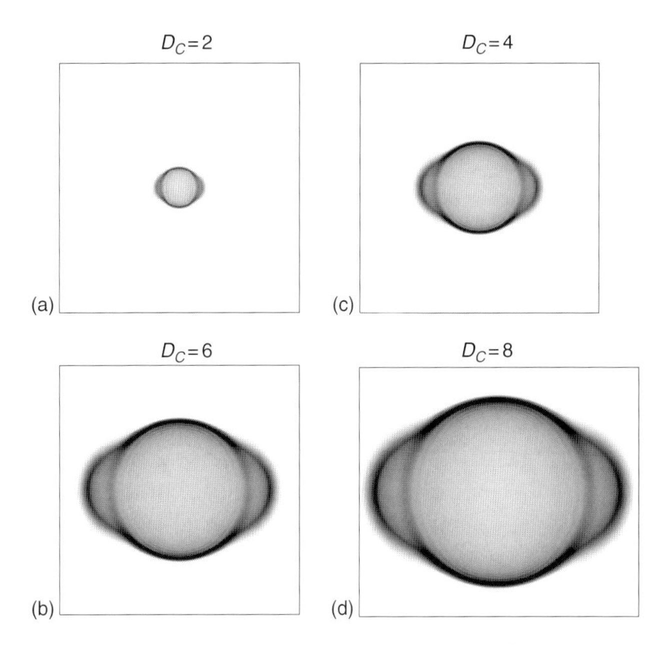

FIGURE 7 Scaling of rupture at constant load for spontaneous rupture starting from an overloaded asperity. The four snapshots show the distribution of slip rate on the fault at equivalent times for four different values of D_c. The initial asperity radius R as well as the instant of time of the snapshot all scale with D_c. (a) $D_c = 2$, $R = 3$, and $T = 140$; (b) $D_c = 4$, $R = 6$, and $T = 280$; (c) $D_c = 6$, $R = 9$, and $T = 420$; (d) $D_c = 8$, $R = 12$, and $T = 560$. (Reproduced with permission from Madariaga *et al.*, 1998; fig. 6, p. 1186.)

3.4 Numerical Resolution and Scaling with the Slip Weakening Distance

An essential requirement for an accurate numerical method is that the numerical solution becomes independent of grid size beyond the use of a certain number of grid points per wavelength. The shortest physical distances are the radius of the asperity R and the width of the rupture front. The latter depends on the slip weakening distance D_c as shown by Ida (1972), Andrews (1976b), and Day (1982b). For 2D faults and for the slip weakening law (12) this width, L_c, is

$$L_c = \frac{4\mu}{3\pi} \frac{T_u}{T_{\text{ext}}^2} D_c \tag{17}$$

We have assumed that the residual friction at high slip rates in Eq. (12) ($T_f = 0$) is zero. This expression is valid for a constant stress level T_{ext} outside the asperity.

In order to describe the convergence of the numerical method as the grid size is refined, consider a simple circular asperity, keeping all the parameters constant except the grid size and D_c. Stress inside the asperity is $T_{\text{asp}} = 1.8 \times T_u$, $T_{\text{ext}} = 0.8 \times T_u$ and $H = 0.20$. Replacing T_{ext} in Eq. (17) gives $L_c = 1.36 \times D_c$. Figure 7 shows snapshots of the slip rate as a function of position on the fault at equivalent instants of time. Since Δx is used as the scaling distance, both the radius of the asperity R as well as the time of the snapshot have to be increased for increasing values of D_c. The four parts of the figure show snapshots for (a) $t = 140$ ($D_c = 2$, $R = 3$), (b) $t = 280$ ($D_c = 4$, $R = 6$), (c) $t = 420$ ($D_c = 6$, $R = 9$), and (d) $t = 560$ ($D_c = 8$, $R = 12$). The external rectangles define the size of the grid, 256×256 for D_c from 2 to 6 and 300×256 for $D_c = 8$. Note the scaling of the figures—the snapshot for $D_c = 8$ is precisely twice as large as that for $D_c = 4$. Clearly, the degree of resolution improves as D_c increases.

From a close examination of these snapshots and several others, we concluded that fourth-order finite difference simulations are contaminated by numerical noise when $D_c < 4\Delta x$, and that numerical simulations are stable and reproducible for $D_c > 6\Delta x$ ($L_c > 7\Delta x$). For 3D simulations, this is a rather large number that requires the use of very dense grids for accurate simulation of spontaneous rupture.

4. Earthquake Scaling Laws

In the previous section we illustrated spontaneous rupture starting from a circular asperity of radius R_{asp} that is ready to break (with stress T_u) and is surrounded by a fault surface. For convenience in numerical modeling, we scaled all physical quantities by the grid interval Δx. Obviously this is not satisfactory in actual applications to earthquakes. The first earthquake scaling law was proposed by Aki (1967), who pointed out a relation between magnitudes at 1 Hz and at 20 sec, assuming that all earthquakes have a similar spectrum that depends on a single scaling variable, the size of the fault. This scaling law was later reformulated by Brune (1970) in the form of a universal shape for the spectrum of S-wave radiation.

The scaling laws can be derived by a simple dimensional analysis of the boundary value problem for an earthquake source in a uniform, infinite elastic medium. In such a simple medium there are only three independent physical dimensions: a length or geometric scale, a stress or dynamic scale, and the time scale. In elastodynamics the time scale is not independent of the length scale, since the two are connected by the speed of elastic waves. Adopting the shear wave speed β as the scale for velocities, and a measure of stress drop $\Delta\sigma$ as the stress scale, the only other free dimension is a length scale for the fault. The appropriate choice for the length scale is the overall size of the fault, its radius for a circular fault, or a characteristic dimension of stress distribution for complex sources. Once these variables are fixed, all other parameters can be scaled with respect to them. Thus, for instance, slip on the fault must scale like the characteristic length L times the ratio of the characteristic stress drop $\Delta\sigma$ to the elastic constant μ. Similarly, slip rate on the fault scales as the same ratio $\Delta\sigma/\mu$ times the wave speed β. All other physical quantities can be derived from these as shown in Table 1.

Studies of many earthquakes under completely different tectonic conditions, for shallow and deep sources, show that stress drop varies over at most two orders of magnitudes, while L varies over a wide range. As a consequence, seismic moment M_0 scales roughly with the cube of the fault size L for moments 10^{10}–10^{21} N m (moment magnitude $M_W = 8.0$). Beyond this magnitude there are serious uncertainties in the scaling law, but it is likely that very large earthquakes scale differently, in particular where ruptures are limited to a narrow zone near the surface of the Earth.

TABLE 1 Scaling of Different Physical Quantities from a Simple Fault Model as a Function of the Three Fundamental Parameters Length, Stress Drop, and Shear-wave Speed[a]

Variable	Symbol	Expression
Length	L	
Stress drop	$\Delta\sigma$	
Shear wave speed	β	
Slip	D	$\Delta\sigma/\mu \times L$
Slip rate	$\dot{D}$	$\Delta\sigma/\mu \times \beta$
Fault surface	S_0	L^2
Duration of radiation	T	L/β
S-wave corner frequency	f_c^S	β/L
P-wave corner frequency	f_c^P	α/L
Seismic energy	E_s	$\Delta\sigma^2/\mu \times L^3$
Seismic moment	M_0	$\Delta\sigma L^3$
Fracture energy	G	$\Delta\sigma D_c$

[a] Other groups of three fundamental units can be chosen, but this is the standard choice in seismology.

4.1 Other Length Scales

We have already seen in the discussion of numerical modeling that overall fault size is not the only length scale that is important in understanding earthquakes. The minimum slip patch—the minimum coherent zone of the fault that may rupture dynamically—is an important independent characteristic size of earthquake sources. Like all other length scales, this minimum patch size scales directly with the slip weakening distance D_c, the minimum slip necessary for friction to reduce to the kinematic friction on the fault. The exact nature of D_c is subject to debate, but the existence of such a length is absolutely necessary for the rupture problem to be physically well posed. Estimates of D_c and its associated minimum patch length scale $l = (\mu/\Delta\sigma)D_c$ vary widely, but it must be a small fraction of the overall length L. According to some authors, D_c is a property of the friction law to be determined from experiments on rock friction (see, e.g., Dieterich, 1978, 1979; Ruina, 1983); for others, D_c scales with the roughness of the fault surfaces, which in turn scales with earthquake size (Ohnaka and Kuwahara, 1990; Ohnaka and Shen, 1999). These two very opposite views emphasize different properties of earthquakes. For those authors who believe that D_c is a property of the fault zone, a universal friction law that describes the tribology of the fault must exist independently of the final size of the earthquake. This is consistent with most friction experiments carried out by the school of Dieterich, Ruina, and others.

For other authors, D_c is the result of rupture scaling over a broad range of magnitudes. Large earthquakes can occur only if D_c becomes large; otherwise they simply stop as first proposed by Aki (1979). Unfortunately it is not possible to settle this argument from seismic data alone because of the lack of high-quality near-field strong motion records. The current resolution of near-field observations is about 0.5 Hz, which translates into a shear wavelength of the order of 6–7 km and a slip weakening distance of $D_c \simeq 10$ cm (Ide and Takeo, 1997; Olsen *et al.*, 1997; Day *et al.*, 1998; Guatteri and Spudich, 1998b). This is too coarse to detect any scaling of D_c with earthquake size at the present time.

4.2 Scaling of Energy and Rupture Resistance

The previous discussion indicates that earthquake phenomena cannot be characterized by a single length scale unless, of course, the slip weakening distance scales with the size of the earthquake. Let us explore some of the consequences of small-size scaling for rupture dynamics through a simple fracture model.

Let us define rupture resistance, or energy release rate, which for the simple slip weakening friction model (12) is given by

$$G = \frac{1}{2} T_u D_c. \quad (18)$$

This is the amount of energy that is needed to produce a unit area of slipping fault. In most fracture studies G is assumed to be a material property. However, Ohnaka and Shen (1999) present strong evidence that this may not necessarily be the case.

4.2.1 Rupture Initiation

A sensitivity analysis of the effect of changing rupture resistance on rupture propagation made by Madariaga and Olsen (2000) shows that there are two regimes that are controlled by a single nondimensional number:

$$\kappa = \frac{T_e^2}{\mu T_u} \frac{R_{asp}}{D_c}, \quad (19)$$

where R_{asp} is the radius of the minimum asperity size (Madariaga and Olsen, 2000). This parameter can be derived from Andrews's (1976a) relation (17). As shown by Madariaga and Olsen (2000), using the calculations of Andrews (1976a), κ is a measure of the ratio of available strain energy ΔW to energy release rate G defined in Eq. (18). The strain energy change in a zone of radius R, uniformly loaded by an initial stress T_e is

$$\Delta W = \frac{1}{2} \langle D \rangle T_e \simeq \frac{T_e^2}{\mu R}, \quad (20)$$

where $\langle D \rangle$ is the average slip on a fault of radius R and stress drop T_e. Thus $\kappa \simeq \Delta W/G$. An essential requirement for rupture to grow beyond the asperity is that $\kappa > \kappa_c$, where the critical value of κ for the circular asperity can be derived from the study by Madariaga and Olsen (2000), who computed numerically the critical radius R_c for fixed T_u, T_e, and D_c. They found that

$$\kappa_c = 0.60. \quad (21)$$

κ_c defines a bifurcation of the problem as a function of parameter κ. Another estimate of κ_c was obtained earlier by Day (1982a,b), who found $\kappa_c = 0.91$. The reason Day's estimate is larger than ours is that Day assumed that rupture could start only when energy balance around the whole perimeter of the asperity allowed rupture to start. As shown by Das and Kostrov (1983), however, rupture may start from the edge of the asperity and then surround the fault before breaking away from the asperity. For $\kappa < \kappa_c$, rupture does not grow beyond the initial asperity. For $\kappa > \kappa_c$, rupture grows indefinitely at increasing speed. This is a simple example of a pitchfork bifurcation.

There is a complication, however: as shown by Andrews (1976b), the rupture front makes a sudden jump to speeds larger than the shear wave speed if T_{ext} is larger than a certain fraction of the peak stress drop T_u. Transition to super-shear speeds is a complex bifurcation that needs much further study. Andrews (1976b) found that the jump to super-shear speed was due to the formation of a stress peak that runs at the shear wave speed ahead of the rupture front. This mechanism does not seem to operate for the transitions observed in 3D by Madariaga and Olsen (2000). In any case, super-shear ruptures in mode II (in-plane shear) are well documented by a number of experiments made by Rosakis *et al.* (1999). Madariaga and Olsen showed that the jump to super-shear speed occurs for a value of $\kappa > 1.3\kappa_c$; further discussion of this issue may be found in their paper. A final important remark is that the problem of rupture propagation from an initial asperity in a homogeneous medium under uniform stress has no other nondimensional control number than κ because this problem has exactly the five independent parameters that appear in the expression for κ. Andrews (1976b) used a different way of plotting the condition for rupture propagation and for super-shear speeds. It can easily be shown that his diagram can be reduced to a single nondimensional number κ.

4.2.2 Sustained Rupture Propagation

To characterize the conditions for sustained rupture propagation in a heterogeneous initial stress field, we consider a very simple rectangular asperity model. It consists of a homogeneous initial stress field that contains a long asperity of width W loaded with a longitudinal shear stress T_e. The asperity is surrounded by an infinite fault plane where stress is low, only $0.1\,T_u$, where as before T_u is the peak frictional stress. At time $t = 0$, rupture is initiated by forcing rupture over a circular patch of radius $R > R_c$, where R_c is computed from Eq. (19) using $\kappa = 0.6$. Depending on the values of T_e and the width W, rupture either grows along the asperity or stops very rapidly. We are again in the presence of bifurcation with a critical value. Similarly to Eq. (19), we define a nondimensional number

$$\kappa = \frac{T_e^2}{\mu T_u} \frac{W}{D_c}, \quad (22)$$

where we have replaced the relevant length scale W in the numerator. We then verified numerically that ruptures stop for low values of $\kappa < \kappa_c$ and grow indefinitely for $\kappa > \kappa_c$, where κ_c is a numerically determined bifurcation point. As shown by Madariaga and Olsen (2000), $\kappa_c = 0.7$ for sustained rupture along a rectangular asperity. This value is slightly larger than $\kappa_c = 0.6$ for rupture initiation, a logical result explained by the fact that it is easier to propagate rupture on a uniformly loaded fault than on a fault loaded along a narrow asperity. Again, for a certain value of $\kappa > 1.3\kappa_c$, rupture grows initially at very high speeds and then jumps to a speed higher than the shear-wave velocity.

Color Plate 4, part (a) shows rupture propagation along the long rectangular asperity for three values of κ. Left and right panels show simultaneous snapshots of shear stress and slip, respectively. The top row shows snapshots for $\kappa < \kappa_c$. Rupture in this case starts near the asperity and then stops immediately. D_c is large relative to the value of W. The second row shows stress and slip rate when κ is slightly supercritical. The rupture propagates along the asperity at sub-shear speeds. As the rupture propagates, the rupture zone extends slightly outside the asperity, leaving an elongated final fault shape. Finally, the bottom row shows snapshots when κ is about 1.2 times critical. Now the rupture front is running faster than the shear wave producing a wake that spreads somewhat into the lower prestress zone. Thus we are again in the presence of a bifurcation: It is not enough to initiate a rupture, the stress field has to be high enough to maintain the continued rupture propagation.

5. Dynamic Model of a Major Earthquake: The 1992 *M* 7.3 Landers, California, Event

The 28 June 1992, magnitude 7.3 Landers earthquake occurred in a remotely located area of the Mojave Desert in Southern California (Figure 8), but the rupture process has been extensively studied due to its large size, proximity to the southern California metropolitan areas, and wide coverage by seismic instruments. Several studies inverted the rupture history of this event from a combination of seismograms, geodetic and geologic data, and the overall kinematics of the seismic rupture are thought to be understood (Campillo and Archuleta, 1993; Abercrombie and Mori, 1994; Cohee and Beroza, 1994; Dreger, 1994; Wald and Heaton, 1994; Cotton and Campillo, 1995a), making the Landers earthquake an appropriate test case for dynamic modeling. The work described in this section is a summary of work by Olsen *et al.* (1997) and Peyrat *et al.* (2001).

5.1 Estimation of Initial Stress and Frictional Parameters

The fault that ruptured during the Landers earthquake can be divided into three segments: The Landers/Johnson Valley (LJV) segment to the southeast where the hypocenter was located; the Homestead Valley (HV) segment in the central part of the fault; and the Camp Rock/Emerson (CRE) segment to the northwest. For the numerical simulations, the three segments of the fault were replaced by a single 78 km long vertical fault plane extending from the surface down to 15 km depth. A free-surface boundary condition was imposed at the top of the grid.

The most important parameter required for dynamic modeling is the initial stress on the fault before rupture starts; all other observables of the seismic rupture, including the motion

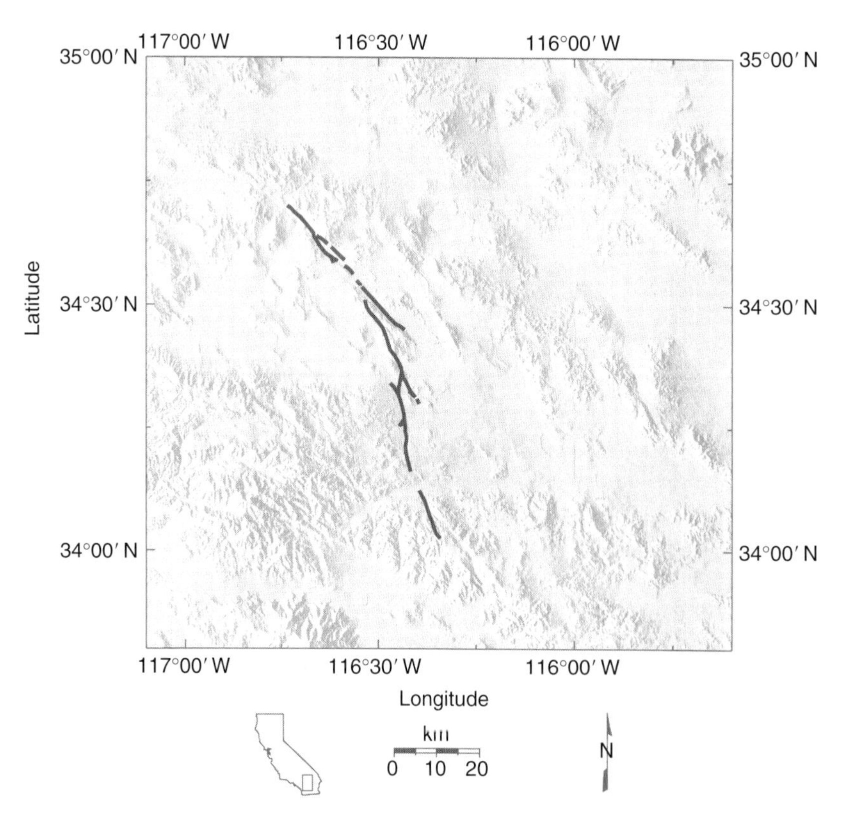

FIGURE 8 Topographic map showing the surface rupture from the 1992 Landers earthquake. The fault trace is depicted by the black lines.

of the rupture front, are determined by the friction law. An initial stress field was estimated by Olsen *et al.* (1997) from the slip distribution inverted by Wald and Heaton (1994). They simply computed the stress drop from the slip distribution; the initial stress was the sum of a stress baseline of 5 MPa plus the stress drop reversed in sign. The method of computing the initial stress field is similar to that introduced by Mikumo and Miyatake (1995) and used by Beroza and Mikumo (1996) and Bouchon (1997) to study the stress field of several earthquakes in California.

The simple slip weakening friction law discussed in an earlier section was used in the dynamic simulations. Slip was assumed to occur only along the long dimension of the fault, and it was found that a constant yield stress level of 12 MPa and $D_c = 0.80$ m produced a total rupture time and final slip distribution in agreement with kinematic inversion results. Before the simulation the initial stress $\mathbf{T}^0$ on the fault was constrained to values just below the specified yield level (12 MPa) in order to prevent rupture starting from several locations. The same regional 1D model of velocities and densities as in Wald and Heaton (1994) was used in the numerical simulation. Rupture was forced to initiate by lowering the yield stress in a small patch of radius 1 km inside a high-stress region near the hypocenter toward the southern end of the LJV fault strand, as inferred from the kinematic results.

5.2 Rupture Propagation in a Heterogeneous Stress Field

Olsen *et al.* (1997) presented the first study on spontaneous rupture propagation in a realistic heterogeneous stress field on the Landers fault. They found that the rupture propagated along a complex path, predominantly breaking patches of high stress and almost completely avoiding areas of low or negative stress. The general rupture pulse resembles the fast, almost instantaneous self-healing phase with a finite slip duration proposed by Heaton (1990) for large earthquakes. However, the healing of the pulse was controlled by the local length scale of stress distribution, and not by slip rate weakening or fault width. Olsen *et al.* (1997) also succeeded in reproducing the main features of the low-frequency ground motion for amplitude and waveform at four strong-motion stations. These results suggested that considerable new information could be obtained about rupture dynamics through studies of spontaneous rupture propagation in estimates of the heterogeneous stress field for well-recorded large earthquakes.

5.3 Inversion of Strong Motion Data

In a recent dynamic inversion, Peyrat *et al.* (2001) used the stress field constructed by Olsen *et al.* (1997) to start the inversion of accelerograms for the initial stress field. Peyrat *et al.* used the strong motion data recorded in the vicinity of the Landers fault to invert for the details of the rupture process by trial and error. For every initial stress distribution they computed the spontaneous rupture process, starting from the same initial asperity at the southwestern end of the fault. During the iterations, the geometry of the fault and the slip weakening friction law did not change, so that $T_u = 12$ MPa and $D_c = 0.8$ m for all the models they tested.

The preferred initial stress from the Peyrat *et al.* (2001) study is shown in Color Plate 4(b), and Color Plate 3(c) compares the rupture propagation for the dynamic model to that for the kinematic model computed by Wald and Heaton. In the case of the dynamic model, soon after initiation rupture propagates slowly downward. After 7 sec it appears that the rupture almost dies but, soon after, it suddenly accelerates upward. It again slows down at 11 sec before jumping to a northern part of the fault and continuing onward. Finally, the rupture finished on the shallow northwest part of the fault after about 21 sec, in agreement with the kinematic inversion. The rupture shows a confined band of slip propagating unilaterally toward the northwest along the fault, as pointed out by Heaton (1990). The finite width of the fault promotes the formation of a pulse by confining the rupture laterally, preventing the development of a cracklike rupture. The main differences between the kinematic and dynamic models occur within the first 10 sec of propagation. The slip rate peak at 5–6 sec for the kinematic model appears later (at 9 sec) for the dynamic model. Nevertheless, the main part of the rupture history (13–17 sec) is very similar for both models. Part (c) of Color Plate 4 shows the final slip distribution on the fault. The dynamic rupture model reproduces a smooth version of the slip pattern used to compute the initial stress distribution.

Accelerograms at the recording stations were computed using a frequency–wavenumber summation method that is more economical than using finite differences to propagate waves from the source to the stations. Figure 9 shows a comparison of the synthetic seismograms generated by the best model of Peyrat *et al.* (2001). Both synthetic and observed seismograms are low-pass filtered to frequencies below 0.5 Hz. The main features of the low-frequency ground motion for amplitude and waveform are reproduced by the synthetic seismograms for the relatively stronger ground motion recorded in the forward rupture direction. The fits for the back-azimuth stations were not as good because the effects of propagation and fault geometry are enhanced at these stations.

The dynamic rupture model of the Landers earthquake is controlled by several friction parameters that are not measured but that may eventually be determined by inversion of seismic and geodetic data. For instance, the rupture speed and healing of the fault are critically determined by the level of the yield stress and the slip weakening distance. If the slip weakening distance is chosen less than about 0.6–0.8 m, the rupture duration and therefore rise times are much shorter than those obtained from kinematic inversion (Campillo and Archuleta, 1993; Abercrombie and Mori, 1994; Cohee and Beroza, 1994; Dreger, 1994; Wald and Heaton, 1994; Cotton and Campillo, 1995),

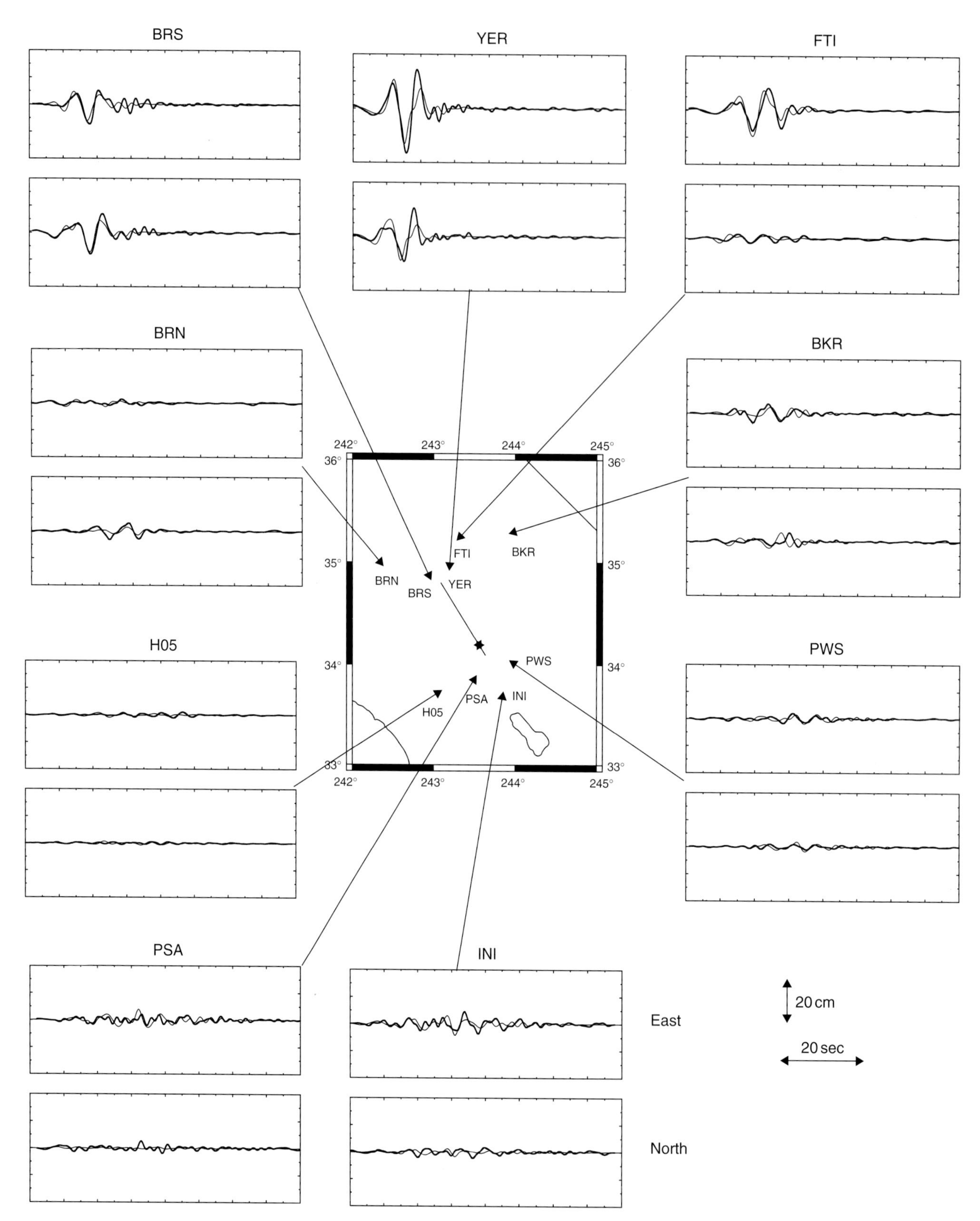

FIGURE 9 Comparison between observed ground displacements (thick traces) and those obtained for the preferred dynamic rupture model of the Landers 1992 earthquake (thin traces). For each station the upper trace is the east–west component of displacement and the bottom trace is the north–south component. The time window is 80 sec, and the amplitude scale is the same for each station.

while larger values produce a rupture resistance that prevents rupture from propagating at all. While the rupture duration and rise times are strongly related to the slip weakening distance, the final slip distribution remains practically unchanged for slip weakening distances that allow rupture propagation.

In addition to the asperity model, Peyrat *et al.* also inverted a barrier model with constant initial stress and variable yield stress. Although the dynamic parameters for the two models are inherently different, they both generate radiation in agreement with strong-motion data. Thus we have showed that the solution of the dynamic problem is non-unique.

6. Earthquake Heterogeneity and Dynamic Radiation

Initial models of dynamic rupture propagation (e.g., Burridge and Knopoff, 1967; Andrews, 1976a,b) studied the frictional instability of a uniformly loaded fault. Very rapidly it was realized that heterogeneity was an essential ingredient of seismic ruptures and that the simple uniformly loaded faults could not explain many significant features of seismic radiation. Two models of heterogeneity were proposed in the late 1970s, the "asperity" model of Kanamori and Stewart (1978), based on a study of the Guatemalan earthquake of 1976, and the barrier model of Das and Aki (1977b) and Aki (1979). The differences between the two models were discussed in some detail by Madariaga (1979), who pointed out that it would be very difficult to distinguish between these two models from purely seismic observations. This remains true today. In the asperity model, it is assumed that the initial stress field is very heterogeneous because previous events have left the fault in a very complex state of stress. In the barrier model, heterogeneity is produced by rapid changes in rupture resistance so that an earthquake would leave certain patches of the fault (barriers) unbroken. It was quickly realized that barriers and asperities were necessary in order to maintain a certain degree of heterogeneity on the fault plane that could explain the properties of high-frequency seismic wave radiation, and to leave highly stressed patches that would be the sites of aftershocks and future earthquakes. Andrews (1980, 1981) went much further and suggested that this heterogeneity was absolutely necessary, otherwise earthquakes would become dominated by very low frequencies and could not produce observed accelerograms. Heterogeneity was studied in many ways by a number of authors during the early 1980s (e.g., Day, 1982b; Das, 1980; Kostrov and Das, 1989).

6.1 Generation of Cracks Versus Self-Healing Pulses

Heaton (1990) noticed that the instantaneous rupture area for large earthquakes is seldom larger than about 10% of the total fault area, so that seismic ruptures look more like a patch propagating across the fault compared to Kostrov's (1964) model of a self-similar shear crack. Heaton explained the pulselike behavior with a self-healing mechanism due to velocity weakening friction. Similar pulselike behavior has been reported from modeling a rectangular fault with a large aspect ratio due to stopping phases from the edges of the fault (Day, 1982a; Cotton and Campillo, 1995a). More recently, Beroza and Mikumo (1996), Ide and Takeo (1997), and Day *et al.* (1998) have shown that short rupture pulses may also be generated by the so-called geometrical constraint in which these short durations may simply reflect the stress heterogeneity on the fault. If stress is concentrated in small areas of the fault, the rupture process will reflect this heterogeneity and produce rupture pulses controlled by the size of these stress asperities.

Zheng and Rice (1998) characterized the conditions for generation of self-healing pulses in terms of the steady-state strength relative to the background stress in a velocity-weakening regime. They found that pulselike rupture occurs only for rather restrictive conditions of rate dependence. They furthermore suggested that the reason why most earthquakes fail to grow to a large size may be that the crustal stresses are too low on average to allow cracklike modes and continued rupture. In other words, most earthquakes fail to maintain rupture propagation because the driving stresses are too low and stop before producing a significant moment release.

6.2 Memory of Earthquake Rupture: Recurrent Events on a Fault

In this chapter we have concentrated our attention on the study of a single event without concern about earthquake recurrence or earthquake distributions. One of the main results obtained by Carlson and Langer (1989) is that for certain friction laws stress heterogeneity will be self-sustained, i.e., every event initiates in a complex state and leaves the fault loaded with a complex stress distribution. Cochard and Madariaga (1996) showed that this mechanism could also occur on two-dimensional shear faults but for a limited set of rate dependent friction laws. Only if healing were fast enough could heterogeneity develop spontaneously. The mechanism studied by Cochard and Madariaga (1996) produced heterogeneity but could not explain the Gutenberg–Richter distribution of events as a function of moment (or moment magnitude).

Nielsen *et al.* (2000) found that recurrent ruptures for a single planar fault with aspect ratio close to 1 and sufficiently low rupture resistance tended to produce a periodic cycle, always breaking the entire fault. However, if the dimension of the fault was increased with the same aspect ratio and friction, the regime became more complex and periodicity was lost. In the case of a long and narrow fault, i.e., $L/W > 10$, the pulse width $l_p \approx W$ was smaller than the maximum fault length L, and a degree of complexity was observed. Indeed, after a transient regime affected by the initial conditions, the fault settled into a recurrence pattern in which no rupture would reach the entire length of the fault, and a wide spectrum of event sizes was produced as opposed to the periodic cycle of fault-wide events observed for faults with aspect ratio close to 1. In other words, the recurrent earthquakes on the fault would generate an inherent complexity of the stress field that completely controlled the rupture conditions for the following events.

6.3 Signature of Friction in Radiated Waves

Is it possible directly to retrieve information about friction from strong motion data? To answer this important question we need to know how (or whether) the friction manifests a signature in the radiated wave field from earthquake rupture. Ide and Takeo (1997) estimated a depth-varying slip weakening distance of 1 m near the surface and 0.5 m at depth for the 1995 Kobe earthquake using inverse techniques. Olsen *et al.* (1997) found that a slip weakening distance of 0.8 m and a yield stress of 12 MPa generated rupture durations and seismograms in good agreement with kinematic inversion results and data, respectively, for the 1992 Landers earthquake. On the other hand, Guatteri and Spudich (1998b) concluded that there is a strong trade-off between peak stress and slip weakening distances. Actually, Peyrat *et al.* (2001) found that the energy release rate $G \simeq T_u D_c$, a product of the peak stress and the slip weakening distance, is all that can be retrieved from seismic data.

If indeed the signature of friction is detectable in strong-motion data, the retrieval of such information will be possible only when data are available in sufficient quality and quantity. In this case, an estimate of friction using inverse methods may be feasible. We expect that this will be an important topic of future research.

7. Conclusions

Thanks to improvements in speed and memory capacity of parallel computers it is no longer a problem to model the propagation of seismic ruptures along a fault, or a fault system, embedded in an elastic 3D medium. The enhanced computational power can be used to improve classical models in order to determine the grid size necessary to do reproducible and stable earthquake simulations. We show that the conditions are that the slip-weakening zone near the rupture front must be sampled by at least six grid points. This is clearly a limiting condition, but it is already possible to model earthquakes of magnitudes from 6 to 7.5 without unsurmountable problems.

Recent inversions of earthquake slip distributions using kinematic source models have found that very complex source distributions require an extensive reappraisal of classical source models that were mostly based on Kostrov's model of a self-similar circular crack. Ruptures on a fault with a very heterogeneous load follow very tortuous rupture paths. While on the average the rupture propagates at a sub-shear speed from one end of the fault to the other, the rupture front may wander in any direction, following the areas of strong stress concentration and avoiding those areas with low stress or high rupture resistance. If this view of earthquake rupture is confirmed by future observations (we believe it will be) then many current arguments about earthquake complexity, narrow rupture pulses, earthquake distributions, etc., will be solved and we may concentrate on the interesting problem of determining which features of friction control stress complexity on the fault under all circumstances.

Acknowledgments

We thank Takeshi Mikumo and Joe Andrews for very detailed and helpful reviews. R. Madariaga's work was supported by CNRS (Centre National de la Recherche Scientifique) under contract 99PNRN13AS of the PNRN program. K.B. Olsen's work was supported by the Southern California Earthquake Center (SCEC). SCEC is funded by NSF Cooperative Agreement EAR-8920136 and USGS Cooperative Agreements 14-08-0001-A0899 and 1434-HQ-97AG01718. This is ICS contribution 378-116EQ and SCEC contribution number 539.

References

Abercrombie, R. and J. Mori (1994). Local observations of the onset of a large earthquake: 28 June 1992 Landers, California. *Bull. Seismol. Soc. Am.* **84**, 725–734.

Aki, K. (1967). Scaling law of seismic spectrum. *J. Geophys. Res.* **72**, 1217–1231.

Aki, K. (1979). Characterization of barriers of an earthquake fault. *J. Geophys. Res.* **84**, 6140–6148.

Andrews, J. (1976a). Rupture propagation with finite stress in anti-plane strain. *J. Geophys. Res.* **81**, 3575–3582.

Andrews, J. (1976b). Rupture velocity of plane strain shear cracks. *J. Geophys. Res.* **81**, 5679–5687.

Andrews, J. (1980). A stochastic fault model. I. Static case. *J. Geophys. Res.* **85**, 3867–3877.

Andrews, J. (1981). A stochastic fault model. II. Time-dependent case. *J. Geophys. Res.* **87**, 10821–10834.

Andrews, J. (1985). Dynamic plane-strain shear rupture with a slip-weakening friction law calculated by a boundary integral method. *Bull. Seismol. Soc. Am.* **75**, 1–21.

Aochi, H., E. Fukuyama, and M. Matsu'ura (2000). Spontaneous rupture propagation on a non-planar fault in 3D elastic medium. *Pageoph* **157**, 2003–2027.

Archuleta, R. and S.M. Day (1980). Dynamic rupture in a layered medium: the 1966 Parkfield earthquake. *Bull. Seismol. Soc. Am.* **80**, 671–689.

Bak, P., C. Tang, and K. Wiesenfeld (1987). Self-organized criticality. *Phys. Rev. A.* **38**, 364.

Bernard, P. and R. Madariaga (1984). A new asymptotic method for the modeling of near-field accelerograms. *Bull. Seismol. Soc. Am.* **74**, 539–557.

Beroza, G. and T. Mikumo (1996). Short slip duration in dynamic rupture in the presence of heterogeneous fault properties. *J. Geophys. Res.* **101**, 22449–22460.

Bouchon, M. (1997). The state of stress on some faults of the San Andreas system as inferred from near-field strong motion data. *J. Geophys. Res.* **102**, 11731–11744.

Bouchon, M. and D. Streiff (1997). Propagation of a shear crack on a nonplanar fault: a method of calculation. *Bull. Seismol. Soc. Am.* **87**, 61–66.

Burridge, R. and L. Knopoff (1967). Model and theoretical seismicity. *Bull. Seismol. Soc. Am.* **67**, 341–371.

Brune, J. (1970). Tectonic stress and the spectra of seismic shear waves from earthquakes. *J. Geophys. Res.* **75**, 4997–5009.

Campillo, M. and R. Archuleta (1993). A rupture model for the 28 June 1992 Landers, California, earthquake. *Geophys. Res. Lett.* **20**, 647–650.

Cao, T. and K. Aki (1984). Seismicity simulation with a mass-spring model and a displacement hardening-softening friction law. *Pageoph* **122**, 10–24.

Carlson, J. and J. Langer (1989). Mechanical model of an earthquake fault. *Phys. Rev. A* **40**, 6470–6484.

Cochard, A. and R. Madariaga (1996). Dynamic faulting under rate-dependent friction. *J. Geophys. Res.* **142**, 419–445.

Cohee, B. and G. Beroza (1994). Slip distribution of the 1992 Landers earthquake and its implications for earthquake source mechanics. *Bull. Seismol. Soc. Am.* **84**, 692–712.

Cotton, F. and M. Campillo (1995a). Frequency domain inversion of strong motions: application to the 1992 Landers earthquake. *J. Geophys. Res.* **100**, 3961–3975.

Cotton, F. and M. Campillo (1995b). Stability of rake: an evidence to partial stress-drop during the 1992 Landers earthquake? *Geophys. Res. Lett.* **22**, 1921–1924.

Cruse, M. (1988). "Boundary Element Method in Fracture Mechanics." Kluwer, Utrecht.

Das, S. (1980). A numerical method for the estimation of source time functions for general three-dimensional rupture propagation. *Geophys. J. Roy. Astron. Soc.* **62**, 591–604.

Das, S. (1981). Three-dimensional rupture propagation and implications for the earthquake source mechanism. *Geophys. J. Roy. Astron. Soc.* **67**, 375–393.

Das, S. and K. Aki (1977a). A numerical study of two-dimensional spontaneous rupture propagation. *Geophys. J. Roy. Astron. Soc.* **50**, 643–668.

Das, S. and K. Aki (1977b). Fault plane with barriers: a versatile earthquake model. *J. Geophys. Res.* **82**, 5658–5670.

Das, S. and D. Kostrov (1983). Breaking of a single asperity: rupture process and seismic radiation. *J. Geophys. Res.* **88**, 4277–4288.

Day, S.M. (1982a). Three-dimensional finite difference simulation of fault dynamics: rectangular faults with fixed rupture velocity. *Bull. Seismol. Soc. Am.* **72**, 795–727.

Day, S.M. (1982b). Three-dimensional simulation of spontaneous rupture: the effect of non-uniform prestress. *Bull. Seismol. Soc. Am.* **72**, 1881–1902.

Day, S.M., G. Yu, and D.J. Wald (1998). Dynamic stress changes during earthquake rupture. *Bull. Seismol. Soc. Am.* **88**, 512–522.

Dieterich, J. (1978). Time-dependent friction and the mechanics of stick-slip. *Pageoph* **116**, 790–806.

Dieterich, J. (1979). Modeling of rock friction. 1. Experimental results and constitutive equations. *J. Geophys. Res.* **84**, 2161–2168.

Dieterich, J. and B. Kilgore (1996). Implications of fault constitutive properties for earthquake prediction. *Proc. Natl. Acad. Sci. USA.* **93**, 3787–3794.

Dreger, D. (1994). Investigation of the rupture process of the 28 June 1992 Landers earthquake utilizing Terrascope. *Bull. Seismol. Soc. Am.* **84**, 713–724.

Fukuyama, E. and R. Madariaga (1995). Integral equation method for plane crack with arbitrary shape in 3D elastic medium. *Bull. Seismol. Soc. Am.* **85**, 614–628.

Fukuyama, E. and R. Madariaga (1998). Rupture dynamics of a planar fault in a 3D elastic medium: rate- and slip-weakening friction. *Bull. Seismol. Soc. Am.* **88**, 1–17.

Guatteri, M. and P. Spudich (1998a). Coseismic temporal changes of slip direction: the effect of absolute stress on dynamic rupture. *Bull. Seismol. Soc. Am.* **88**, 777–789.

Guatteri, M. and P. Spudich (1998b). What can strong motion data tell us about slip-weakening fault-friction laws? *Bull. Seismol. Soc. Am.* **90**, 98–116.

Geubelle, P. and J. Rice (1995). A spectral method for 3D elastodynamic fracture problems. *J. Mech. Phys. Solids* **43**, 1791–803.

Gu, Y. and T.-F. Wong (1991). *J. Geophys. Res.* **96**, 21677–21691.

Harris, R. and S. Day (1993). Dynamics of fault interaction: parallel strike-slip faults. *J. Geophys. Res.* **98**, 4461–4472.

Harris, R. and S. Day (1997). Effect of a low velocity zone on a dynamic rupture. *Bull. Seismol. Soc. Am.* **87**, 1267–1280.

Haskell, N.A. (1964). Total energy spectral density of elastic wave radiation from propagating faults. *Bull. Seismol. Soc. Am.* **54**, 1811–1841.

Heaton, T. (1990). Evidence for and implications of self-healing pulses of slip in earthquake rupture. *Phys. Earth. Planet. Inter.* **64**, 1–20.

Hirose, S. and J.D. Achenbach (1989). Time-domain boundary element analysis of elastic wave interaction with a crack. *Int. J. Num. Mech. Eng.* **28**, 629–644.

Ida, Y. (1972). Cohesive force across the tip of a longitudinal-shear crack and Griffith's specific surface energy. *J. Geophys. Res.* **77**, 3796–3805.

Ide, S. and M. Takeo (1997). Determination of the constitutive relation of fault slip based on wave analysis. *J. Geophys. Res.* **102**, 27379–27391.

Kanamori, H. and G.S. Stewart (1978). Seismological aspects of the Guatemala earthquake of February 4, 1976. *J. Geophys. Res.* **83**, 3427–3434.

Koller, M.G., M. Bonnet, and R. Madariaga (1992). Modeling of dynamical crack propagation using time-domain boundary integral equations. *Wave Motion* **16**, 339–366.

Kostrov, B. (1964). Self-similar problems of propagation of shear cracks, *J. Appl. Math. Mech.* **28**, 1077–1087.

Kostrov, B. (1966). Unsteady propagation of longitudinal shear cracks. *J. Appl. Math. Mech.* **30**, 1241–1248.

Kostrov, B. and S. Das (1989). "Principles of Earthquake Source Mechanics," Cambridge University Press.

Madariaga, R. (1976). Dynamics of an expanding circular fault. *Bull. Seismol. Soc. Am.* **66**, 639–667.

Madariaga, R. (1978). The dynamic field of Haskell's rectangular dislocation fault model, *Bull. Seismol. Soc. Am.* **68**, 869–887.

Madariaga, R. (1979). On the relation between seismic moment and stress drop in the presence of stress and strength heterogeneity. *J. Geophys. Res.* **84**, 2243–2250.

Madariaga, R. and K.B. Olsen (2000). Criticality of rupture dynamics in three dimensions. *Pageoph* **157**, 1981–2001.

Madariaga, R., K.B. Olsen, and R.J. Archuleta (1998). Modeling dynamic rupture in a 3D earthquake fault model. *Bull. Seismol. Soc. Am.* **88**, 1182–1197.

Mikumo, T. and T. Miyatake (1978). Dynamical rupture process on a three-dimensional fault with non-uniform friction and near-field seismic waves. *Geophys. J. Roy. Astron. Soc.* **54**, 417–438.

Mikumo, T. and T. Miyatake (1979). Earthquake sequences on a frictional fault model with non-uniform strength and relaxation times. *Geophys. J. Roy. Astron. Soc.* **59**, 497–522.

Mikumo, T. and T. Miyatake (1993). Dynamic rupture processes on a dipping fault, and estimates of stress drop and strength excess from the results of waveform inversion. *Geophys. J. Roy. Astron. Soc.* **112**, 481–496.

Mikumo, T. and T. Miyatake (1995). Heterogeneous distribution of dynamic stress drop and relative fault strength recovered from the results of waveform inversion. *Bull. Seismol. Soc. Am.* **85**, 178–193.

Mikumo, T., K. Hirahara, and T. Miyatake (1987). Dynamical fault rupture processes in heterogeneous media. *Tectonophysics* **144**, 19–36.

Miyatake, T. (1992). Reconstruction of dynamic rupture process of an earthquake with constraints of kinematic parameters. *Geophys. Res. Lett.* **19**, 349–352.

Miyatake, T. (1980). Numerical simulations of earthquake source process by a three-dimensional crack model. I. Rupture process. *J. Phys. Earth* **28**, 565–598.

Nielsen, S.B., J. Carlson, and K.B. Olsen (2000). The influence of friction and fault geometry on earthquake rupture. *J. Geophys. Res.* **105**, 6069–6088.

Okubo, P. (1989). Dynamic rupture modeling with laboratory-derived constitutive relations. *J. Geophys. Res.* **94**, 12321–12335.

Ohnaka, M. (1996). Non-uniformity of the constitutive law parameters for shear rupture and quasi-static nucleation to dynamic rupture: a physical model of earthquake generation processes. *Proc. Natl. Acad. Sci. USA* **93**, 3795–3802.

Ohnaka, M. and Y. Kuwahara (1990). Characteristic features of local breakdown near crack-tip in the transition zone from nucleation to dynamic rupture during stick-slip shear failure. *Tectonophysics* **175**, 197–220.

Ohnaka, M. and L.F. Shen (1999). Scaling of rupture process from nucleation to dynamic propagation: implications of geometric irregularity of the rupturing surfaces. *J. Geophys. Res.* **104**, 817–844.

Olsen, K.B. (1994). Simulation of three-dimensional wave propagation in the Salt Lake Basin. Ph.D. Thesis, University of Utah, Salt Lake City, UT.

Olsen, K., R. Archuleta, and J. Matarese (1995a). Three-dimensional simulation of a magnitude 7.75 earthquake on the San Andreas fault. *Science* **270**, 1628–1632.

Olsen, K.B., J.C. Pechmann, and G.T. Schuster (1995b). Simulation of 3-D elastic wave propagation in the Salt Lake Basin. *Bull. Seismol. Soc. Am.* **85**, 1688–1710.

Olsen, K.B. and R. Archuleta (1996). Three-dimensional simulation of earthquakes on the Los Angeles fault system. *Bull. Seismol. Soc. Am.* **86**, 575–596.

Olsen, K.B., R. Madariaga, and R. Archuleta (1997). Three dimensional dynamic simulation of the 1992 Landers earthquake. *Science* **278**, 834–838.

Olsen, K.B., E. Fukuyama, H. Aochi, and R. Madariaga (2000). Hybrid modeling of curved fault radiation in a 3D heterogeneous medium, Proc. 2nd ACES Meeting, Tokyo and Hakone.

Peyrat, S., K.B. Olsen, and R. Madariaga (2001). Dynamic modeling of the 1992 Landers earthquake. *J. Geophys. Res.* **106**, 26467–26482.

Rice, J. and Y. Ben-Zion (1996). Slip complexity in earthquake fault models. *Proc. Natl. Acad. Sci. USA* **93**, 3811–3818.

Rice, J.R. and A.L. Ruina (1983). Stability of steady frictional slipping, *Trans. ASME, J. Appl. Mech.* **50**, 343–349.

Rosakis, A.J., O. Samudrala, and D. Coker (1999). Cracks faster than the shear wave speed. *Science* **284**, 1337–1340.

Ruina, A. (1983). Slip instability and state variable friction laws. *J. Geophys. Res.* **88**, 10 359–10 370.

Sato, H. and T. Hirasawa (1973). Body wave spectra from propagating shear cracks. *J. Physics Earth* **84**, 829–841.

Scholz, C. (1989) "The Mechanics of Earthquake and Faulting." Cambridge University Press.

Scholz, C. (2000). Evidence for a strong San Andreas fault. *Geology* **28**, 163–166.

Schmittbuhl, J., J.-P. Vilotte, and S. Roux (1996). A dissipation-based analysis of an earthquake fault model. *J. Geophys. Res.* **101**, 27741–27764.

Spudich, P. and L.N. Hartzell (1984). Use of ray theory to calculate high-frequency radiation from earthquake sources having spatially variable rupture velocity and stress drop. *Bull. Seismol. Soc. Am.* **74**, 2061–2082.

Spudich, P. (1992). On the inference of absolute stress levels from seismic radiation. *Tectonophysics* **211**, 99–106.

Spudich, P., M. Guatteri, K. Otsuki, and J. Minagawa (1998). Use of fault striations and dislocation models to infer tectonic shear stress during the 1995 Hogyo-ken Nambu (Kobe) earthquake. *Bull. Seismol. Soc. Am.* **88**, 413–427.

Townend, J. and M.D. Zoback (2000). How faulting keeps the crust strong. *Geology* **28**, 399–402.

Virieux, J. (1986). P-SV wave propagation in heterogeneous media: velocity-stress finite-difference method. *Geophysics* **51**, 889–901.

Virieux, J. and R. Madariaga (1982). Dynamic faulting studied by a finite difference method. *Bull. Seismol. Soc. Am.* **72**, 345–369.

Wald, D. and T. Heaton (1994). Spatial and temporal distribution of slip for the 1992 Landers, California earthquake. *Bull. Seismol. Soc. Am.* **84**, 668–691.

Wolfram, S. (Ed.) (1986). "Theory and Applications of Cellular Automata: Including Selected Papers, 1983–1986." World Scientific, Singapore.

Zheng, G. and J.R. Rice (1998). Conditions under which velocity-weakening friction allows a self-healing versus a crack-like mode of rupture. *Bull. Seismol. Soc. Am.* **88**, 1466–1483.

13

Scattering and Attenuation of Seismic Waves in the Lithosphere

Haruo Sato
Tohoku University, Sendai, Japan
Mike Fehler
Los Alamos National Laboratory, New Mexico, USA
Ru-Shan Wu
University of California, Santa Cruz, California, USA

1. Attenuation of Seismic Waves

The scattering and attenuation of high-frequency seismic waves are important characteristics to be quantified and with which to physically characterize the Earth medium. We first compile measurements of attenuation of seismic waves, and discuss physical mechanisms of attenuation in the lithosphere: Intrinsic absorption and scattering loss due to distributed heterogeneities. As a model of attenuation, we introduce an approach for calculating the amount of scattering loss in a manner consistent with conventional seismological attenuation measurements.

1.1 Observed Attenuation of *P*- and *S*-Waves

The observed seismic wave amplitudes usually decay exponentially with increasing travel distance after correction for geometrical spreading, and decay rates are proportional to Q^{-1}, which characterizes the attenuation. For outgoing spherical *S* waves of frequency *f* in a medium with uniform velocity, the spectral amplitude at a travel distance *r* varies as

$$u^{S\,\mathrm{direct}}(r; f) \propto \frac{e^{-\pi r f Q_S^{-1}/\beta_0}}{r}$$

where β_0 is the *S*-wave velocity. We compile reported values of Q_S^{-1} and Q_P^{-1} for the lithosphere in Figures 1a and 1b,

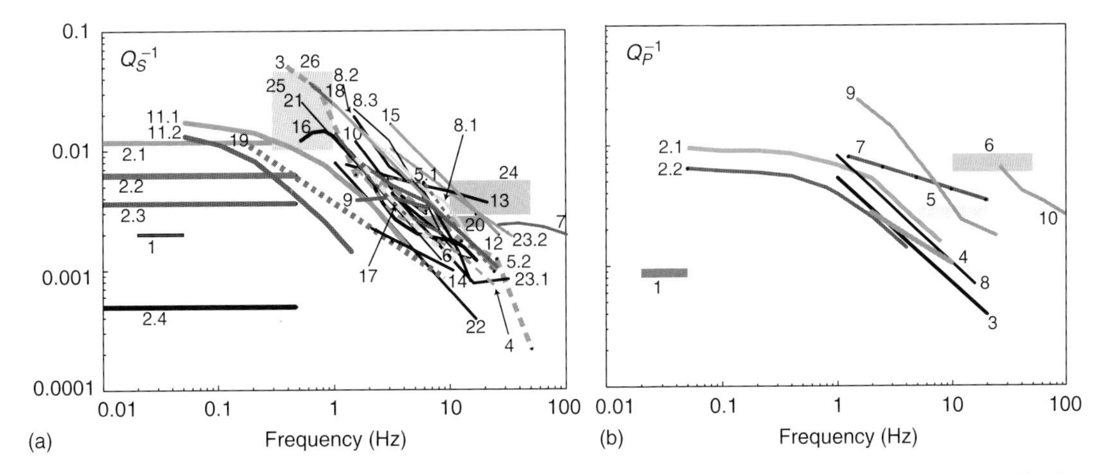

FIGURE 1 (a) Reported values of Q_S^{-1} for the lithosphere: surface wave analysis, 1–2; coda normalization method and multiple lapse time window analysis, 3–12; spectral decay analysis, 11–26. (b) Reported values of Q_P^{-1} for the lithosphere: surface wave analysis, 1; spectral decay analysis, 2–8; extended coda normalization method, 9–10. Detailed references are given by Sato and Fehler (1998). (Reprinted from Sato and Fehler (1998) with permission from Springer-Verlag.)

INTERNATIONAL HANDBOOK OF EARTHQUAKE AND ENGINEERING SEISMOLOGY, VOLUME 81A ISBN: 0-12-440652-1

respectively. Anderson and Hart (1978) proposed $Q_S^{-1} \approx 0.002$, $Q_P^{-1} \approx 0.0009$, and ratio $Q_P^{-1}/Q_S^{-1} \approx 0.5$ for frequencies <0.05 Hz for depths <45 km. Analyzing teleseismic *P* and *S* waves, Taylor *et al.* (1986) found that Q_S^{-1} was larger in the Basin and Range Province than in the North-American Shield. Analyzing records of microearthquakes in Kanto, Japan using the coda normalization method (see Section 2.4), Aki (1980) found that Q_S^{-1} decreases with increasing frequency as a power law $Q_S^{-1} \propto f^{-(0.6-0.8)}$ for 1–25 Hz. Kinoshita (1994) reported a decrease in Q_S^{-1} with decreasing frequency for frequencies less than about 0.8 Hz (curve 16 in Fig. 1a) in southern Kanto, Japan. Yoshimoto *et al.* (1993) estimated $Q_S^{-1} \approx 0.012 f^{-0.73}$ and $Q_P^{-1} \approx 0.031 f^{-0.5}$ for 1–32 Hz in Kanto, Japan for depth <100 km, and the resultant ratio Q_P^{-1}/Q_S^{-1} is larger than 1. Yoshimoto *et al.* (1998) found rather strong attenuation in the shallow crust, $Q_S^{-1} \approx 0.0034 f^{-0.12}$ and $Q_P^{-1} \approx 0.052 f^{-0.66}$ for 25–102 Hz. Carpenter and Sanford (1985) reported $Q_P^{-1}/Q_S^{-1} \approx 1.5$ for 3–30 Hz in the upper crust of the central Rio Grande Rift.

We may summarize the observed characteristics as follows: Q_S^{-1} is of the order of 10^{-2} at 1 Hz and decreases to the order of 10^{-3} at 20 Hz. It seems reasonable to write the frequency dependence of attenuation in the form of a power law as $Q_S^{-1} \propto f^{-n}$ for frequencies higher than 1 Hz, where the power n ranges from 0.5 to 1. The frequency dependence at $0.1 \sim 1$ Hz remains poorly understood because seismic measurements in this band are difficult to make. Results in Figure 1b show that Q_P^{-1} also decreases with increasing frequency for frequencies higher than 1 Hz. For frequencies lower than 0.05 Hz, the ratio Q_P^{-1}/Q_S^{-1} has been taken to be a constant less than 1. Many have assumed that the ratio for higher frequencies is the same as for low frequencies. However, recent observations have clearly shown that the ratio Q_P^{-1}/Q_S^{-1} ranges between 1 and 2 for frequencies higher than 1 Hz (Yoshimoto *et al.*, 1993).

There have been many attempts to derive attenuation tomogram images from the spectral decay analysis of recorded body waves (Clawson *et al.*, 1989; Al-Shukri and Mitchell, 1990; Scherbaum and Wyss, 1990; Ponko and Sanders, 1994). It is important to overcome the trade-off between the frequency dependences of source spectra and Q^{-1}, and to remove site amplification factors near the recording station for imaging the attenuation structure: Some supposed the ω^2 model for source spectra and frequency-independent Q^{-1} in the analyses.

Seismic attenuation is usually considered to be a combination of two mechanisms—scattering loss and intrinsic absorption. Measurements of attenuation of direct seismic waves give values for total attenuation. Scattering redistributes wave energy within the medium. Conversely, intrinsic absorption refers to the conversion of vibration energy into heat. Wu (1985) introduced the concept of seismic albedo B_0 as the ratio of scattering loss to total attenuation.

1.2 Intrinsic Absorption

Several review papers discuss proposed mechanisms for intrinsic absorption that lead to frequency-independent Q_P^{-1} and Q_S^{-1} (Knopoff, 1964; Jackson and Anderson, 1970; Mavko and Nur, 1979; Dziewonski, 1979). For seismic waves to remain causal in the presence of attenuation, the relationship between frequency-dependent Q^{-1} and velocity dispersion was discussed by Liu *et al.* (1976).

Many proposed mechanisms are based on the observation that crustal rocks have microscopic cracks and pores, which may contain fluids. These features have dimensions much smaller than the wavelengths of regional seismic phases. Walsh (1966) proposed frictional sliding on dry surfaces of thin cracks as an attenuation mechanism. Nur (1971) proposed viscous dissipation in a zone of partially molten rock to explain the low-velocity/high-attenuation zone beneath the lithosphere. Even though the addition of water reduces the melting temperature of rocks, it is unlikely that melted rock exists in most regions of the lithosphere. Mavko and Nur (1979) examined the effect of partial saturation of cracks on absorption: fluid movement within cracks is enhanced by the presence of gas bubbles. O'Connell and Budiansky (1977) proposed a model in which fluid moves between closely spaced adjacent cracks. Tittmann *et al.* (1980) measured an increase of Q_S^{-1} with increasing content of volatiles in dry rocks. They found that the rapid increase was due to an interaction between adsorbed water films on the solid surface by thermally activated motion. Thermally activated processes at grain boundaries have been proposed as an absorption mechanism for the upper mantle (Anderson and Hart, 1978; Lundquist and Cormier, 1980). Spatial temperature differences induced by adiabatic compression during wave propagation will be reduced by thermal diffusion (Zener, 1948; Savage, 1966a), which removes vibrational energy from the wave field. Grain-sized heterogeneities in a rock increase the amount of predicted absorption by this mechanism, which is termed the thermoelastic effect. Savage (1966b) investigated thermoelasticity caused by stress concentrations induced by the presence of cracks.

Most of the mechanisms discussed above can predict Q_S^{-1} having values in the range of 10^{-3}; however, the importance of various mechanisms varies with depth, temperature, fracture content, fracture aspect ratios, pressure, and the presence of fluids. Aki (1980) discussed the relation between physical dimensions and the observed and partially conjectured frequency-dependence of Q_S^{-1} with a peak on the order of 0.01 around 0.5 Hz. He preferred thermoelasticity as the most viable model at lithospheric temperatures since the required scales for rock grains and cracks along with the amount of attenuation are in closest agreement with observations.

1.3 Scattering Loss

Scattering due to heterogeneities distributed in the Earth also causes a decrease in amplitude with travel distance (Aki, 1980), where the characteristic frequency is determined by a characteristic spatial scale, such as the correlation length of random media or the crack length. We begin to study scattering using the scalar wave equation in inhomogeneous media. We suppose the wave velocity is written as $V(\mathbf{x}) = V_0[1 + \xi(\mathbf{x})]$, where V_0 is the background velocity and the small fractional fluctuation $\xi(\mathbf{x})$ is a homogeneous and isotropic random function of coordinate $\mathbf{x}$. We imagine an ensemble of random media $\{\xi(\mathbf{x})\}$. First, we define the autocorrelation function (ACF) $R(\mathbf{x}) \equiv \langle \xi(\mathbf{y})\xi(\mathbf{y}+\mathbf{x})\rangle$ as a statistical measure. The spatial scale and the strength of inhomogeneity are characterized by correlation length a and the mean-square fractional fluctuation ε^2, respectively. We divide the random medium into blocks of dimension larger than a. We imagine a scalar plane wave of angular frequency ω of unit amplitude incident upon a localized inhomogeneity. We calculate the generation of outgoing scattered waves using the Born approximation. Taking the ensemble average of the squared scattering amplitude at scattering angle ψ, we get the directional scattering coefficient, which is the scattering power per unit volume, at angular frequency ω as

$$g(\psi;\omega) = \frac{k^4}{\pi} P\left(2k \sin\frac{\psi}{2}\right) \tag{1}$$

where wavenumber $k = \omega/V_0$ and P is the power spectral density function (PSDF) of random inhomogeneity.

The fractional scattering loss of incident-wave energy per unit travel distance is the average of $g(\psi;\omega)$ over solid angle, which is the total scattering coefficient g_0 (Aki and Richards, 1980). Dividing g_0 by k, we get scattering loss

$$\begin{aligned} {}^{\mathrm{BSc}}Q^{-1}(\omega) &\equiv \frac{g_0}{k} = \frac{1}{4\pi k}\oint g(\psi;\omega) 2\pi \sin\psi \, d\psi \\ &= \frac{k^3}{2\pi}\int_0^{\pi} P\left(2k\sin\frac{\psi}{2}\right)\sin\psi \, d\psi \end{aligned} \tag{2}$$

In the case of exponential ACF $R(\mathbf{x}) = \varepsilon^2 e^{-r/a}$, with PSDF $P(\mathbf{m}) = 8\pi\varepsilon^2 a^3/(1 + a^2m^2)^2$, the asymptotic behavior is ${}^{\mathrm{Bsc}}Q^{-1} \approx 2\varepsilon^2 ak$, for $ak \gg 1$. As plotted by a dashed curve in Figure 2, Eq. (2) predicts attenuation larger than ε^2 for large ak even if ε^2 is small. The large scattering loss here predicted for high frequencies is caused by strong forward scattering in a cone around the forward direction, $\psi < 1/ak$. There is a disagreement between ${}^{\mathrm{Bsc}}Q^{-1}$ obtained from direct application of the Born approximation and observed Q_S^{-1}, which decreases with increasing frequency for high frequencies (see Fig. 1a).

Wu (1982) proposed a method for calculating the scattering loss by specifying a lower bound of scattering angle $\psi_C = 90°$ in Eq. (2) by arguing that this accounts only for the back-scattered energy, which is lost:

$$^{\mathrm{CSc}}Q^{-1}(\omega) = \frac{k^3}{2\pi}\int_{\psi_C}^{\pi} P\left(2k\sin\frac{\psi}{2}\right)\sin\psi \, d\psi \tag{3}$$

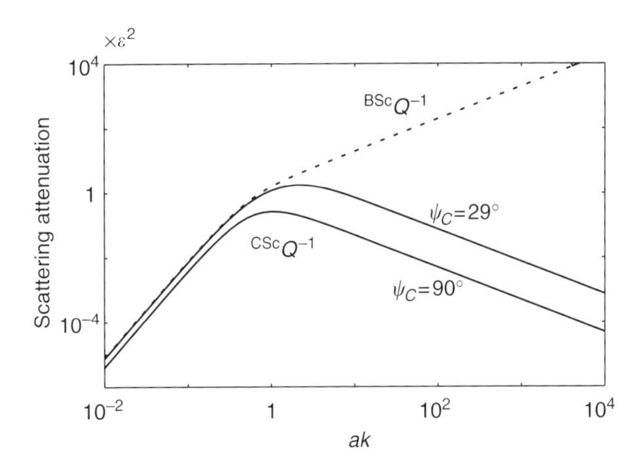

FIGURE 2 Scattering attenuation versus ak for scalar waves: dashed curve, the ordinary Born approximation; solid curves, the corrected Born approximation, where $k = \omega/V_0$. $\psi_C = 90°$ by Wu (1982) and $\psi_C = 29°$ by Sato (1982).

Sato (1982) suggested that the strong increase in attenuation predicted by the Born approximation for high frequencies is due to the travel time (phase) fluctuation caused by velocity fluctuation. He proposed a method for calculating the scattering loss after subtracting the travel time fluctuation caused by velocity fluctuation of which the wavelengths are larger than twice the wavelength of incident waves. This procedure ignores scattering loss for angles smaller than $\psi_C \approx 29°$ in Eq. (3). For an exponential ACF, corrected scattering loss ${}^{\mathrm{CSc}}Q^{-1}$ has a peak of the order of ε^2 at $ak \approx 1$, and becomes ${}^{\mathrm{CSc}}Q^{-1}(\omega) \propto \varepsilon^2/ak$, for $ak \gg 1$. Solid curves in Figure 2 show the scattering loss given by Eq. (3) against ak. Such a frequency dependence well explains the observed frequency dependence of attenuation as shown in Figure 1. Extending the above idea to elastic vector waves and fitting the predicted scattering loss curve to observed Q_S^{-1}, Sato (1984) estimated $\varepsilon^2 = 0.01$ and $a = 2$ km for the lithosphere. The choice of ψ_C was numerically examined by several investigators (Frankel and Clayton, 1986; Roth and Korn, 1993; Fang and Müller, 1996).

There have been many studies of scattering by distributed cracks and cavities (Varadan *et al.*, 1978; Kikuchi, 1981; Matsunami, 1990; Benites *et al.*, 1992, 1997; Kawahara and Yamashita, 1992). Scattering by cracks gives a peak in Q^{-1} when the wavelength is of the same order as the dimension of the crack; however, it is difficult to imagine open cracks having dimensions large enough to be comparable to regional seismic wavelengths deep in the Earth.

2. Seismogram Envelopes of Local Earthquakes

The appearance of coda waves in seismograms is one of the most prominent observations supporting the existence of small-scale random heterogeneities in the Earth (Aki, 1969).

We will now discuss characteristics of observed coda. Then we will introduce the radiative transfer theory, which can account for the effects of multiple scattering and model observed coda characteristics. We compile measurements of total scattering coefficient and coda attenuation, which characterize coda excitation and coda amplitude decay, respectively. We will then discuss the coda normalization method, which is widely used for the estimation of attenuation per travel distance, site amplification factors, and source spectra on the basis of the spatially uniform distribution of coda energy at a long lapse time. Finally, we will present the multiple lapse time window analysis method that has been developed for the measurement of scattering loss and intrinsic absorption based on the solution of the radiative transfer theory.

2.1 Coda Characteristics

We show seismograms of a typical local earthquake in Figure 3: the direct *S* wave is followed by complex wave trains, which are called "*S*-coda" or simply "coda." Clear *S*-coda waves have been identified on seismograms recorded at the bottom of deep boreholes (Sato, 1978; Leary and Abercrombie, 1994). The *f–k* analysis of array seismic data shows that *S*-coda is composed of many wavelets arriving from various directions. Recorded waveforms show a large variation in amplitude near the direct *S* arrival: however, the variation decreases as lapse time increases. *S*-coda waves have a common envelope shape at most stations near the epicenter after about twice the *S* wave travel time (Rautian and Khalturin, 1978). The logarithm of the trace duration of a local seismogram is generally proportional to earthquake magnitude. Trace duration has been used for the quick determination of earthquake magnitudes in many regions of the world (Tsumura, 1967).

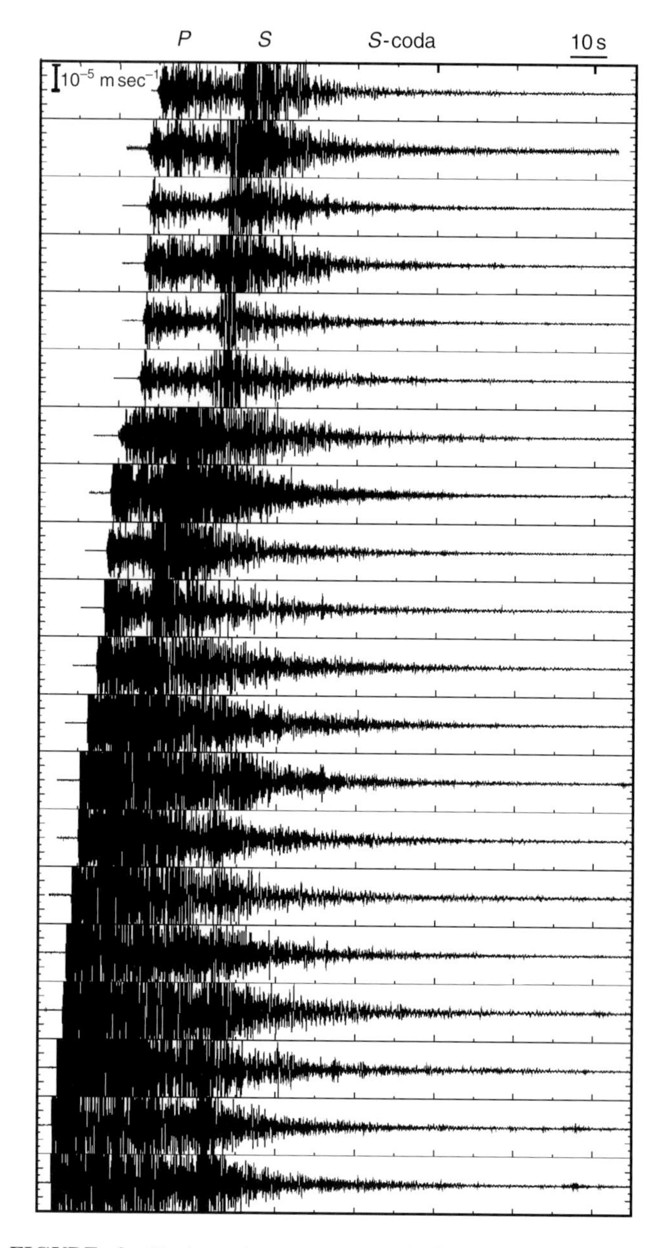

FIGURE 3 Horizontal component velocity seismograms of a crustal earthquake of $M_L = 4.6$ in Japan. Seismograms are arranged from bottom to top by increasing distance from the earthquake epicenter. The direct *S* wave is followed by *S*-coda waves. (Courtesy of K. Obara.)

2.2 Radiative Transfer Theory

Wu (1985) and Shang and Gao (1988) first introduced the radiative transfer theory into seismology; later Zeng *et al.* (1991) formulated the time-dependent multiple scattering process in 3D. Here, we introduce a theory for multiple isotropic scattering processes for a point shear-dislocation source (Sato *et al.*, 1997). We consider a nonabsorbing 3D scattering medium with background propagation velocity V_0, in which pointlike isotropic scatterers of cross section σ_0 are randomly and homogeneously distributed with density n, where the total scattering coefficient $g_0 \equiv n\sigma_0$. We assume an impulsive source located at the origin of energy W with radiation pattern $\Xi(\theta, \phi)$ (see Fig. 4a) at time zero. The scattering process is written as

$$E(\mathbf{x}, t) = W\Xi(\theta, \phi)G_E(\mathbf{x}, t) + g_0 V_0 \int_{-\infty}^{\infty}\int_{-\infty}^{\infty}\int_{-\infty}^{\infty}\int_{-\infty}^{\infty} G_E(\mathbf{x}-\mathbf{x}', t-t')E(\mathbf{x}', t')\, d\mathbf{x}'\, dt' \tag{4}$$

where the convolution integral means the propagation of energy from the last scattering point to the receiver. The first term is for the direct propagation, where

$$G_E(\mathbf{x}, t) = \frac{1}{4\pi V_0 r^2} H(t)\delta\left(t - \frac{r}{V_0}\right) e^{-g_0 V_0 t} \tag{5}$$

We can analytically solve Eqs. (4)–(5) by using the Fourier–Laplace transform in space–time, and a spherical harmonics

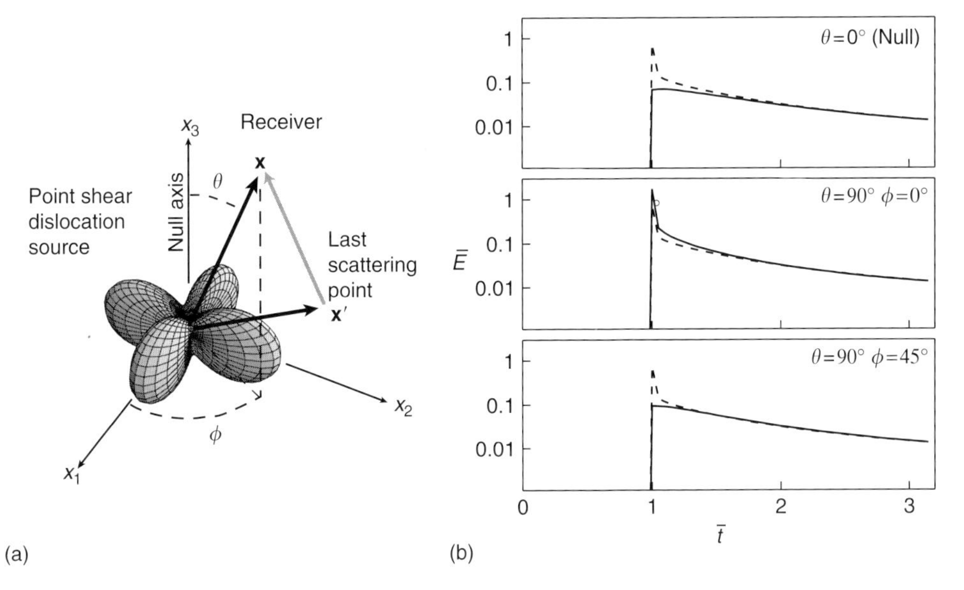

FIGURE 4 (a) Geometry of the multiple scattering process for a point-shear dislocation source, where the lobes show the radiation pattern of the S-wave energy. (b) Temporal change in the normalized energy density at $r = 1/g_0$ in different directions from a point shear-dislocation source. The broken curve corresponds to the spherical source radiation. (Reprinted from Sato *et al.* (1997) with permission from Elsevier Science.)

expansion in angle. For the case of spherical source radiation, $\Xi = 1$, Zeng *et al.* (1991) obtained

$$
\begin{aligned}
E(\mathbf{x}, t) = & \frac{W}{4\pi V_0 r^2}\delta\left(t - \frac{r}{V_0}\right)e^{-g_0V_0t}H(t) \\
& + \frac{Wg_0 e^{-g_0V_0t}}{4\pi r^2}\frac{r}{V_0 t}\ln\left[\frac{r + V_0 t}{r - V_0 t}\right]H\left(t - \frac{r}{V_0}\right) \\
& + Wg_0^2V_0^2\frac{1}{(2\pi)^2}\int_{-\infty}^{\infty}\int_{-\infty}^{\infty} d\omega\, dk\, e^{-i\omega t - ikr} \\
& \times \frac{ik}{2\pi r}\frac{\overline{\overline{G}}_{E0}(-k, -i\omega)^3}{1 - g_0V_0\overline{\overline{G}}_{E0}(-k, -i\omega)}
\end{aligned}
\tag{6}
$$

where $\overline{\overline{G}}_{E0}(k, s) = (1/kV_0)\tan^{-1}[kV_0/(s + g_0V_0)]$. The second term on the right hand side is the single scattering term (Kopnichev, 1975; Sato, 1977), which decreases according to the inverse square of lapse time at large lapse time as $Wg_0/2\pi V_0^2 t^2$ (Aki and Chouet, 1975).

We show temporal variations of normalized energy density ($\overline{E} = E/Wg_0^3$) against normalized lapse time ($\bar{t} = g_0V_0t$), at distance of one mean free path $r = g_0^{-1}$ for a point shear-dislocation source by solid curves (Sato *et al.*, 1997) in Figure 4b. The energy density faithfully reflects the source radiation pattern near the direct arrival; however, the azimuthal dependence diminishes with increasing lapse time. Each solid curve asymptotically converges to a broken curve for spherical source radiation. This simulation agrees qualitatively with the observed radiation pattern independence of coda amplitudes at long lapse time.

It should be mentioned that the energy flux model (Frankel and Wennerberg, 1987) and the use of Monte Carlo methods (Hoshiba, 1991; Gusev and Abubakirov, 1996) for envelope synthesis played important roles in the coda study.

2.3 Coda Attenuation and Scattering Coefficient

The coda energy density in a frequency band having central frequency f is the sum of the medium mass density times average square of particle velocity of S-coda around lapse time t, $\dot{u}_i^{S\text{-coda}}(t;f)$ as

$$
E^{S\text{-coda}}(t;f) \approx \sum_{i=1}^{3} \rho_0 \dot{u}_i^{S\text{-coda}}(t;f)^2
$$

where ρ_0 is the mass density. For practical analysis, we need to introduce an empirical parameter, known as coda attenuation Q_C^{-1}. In the case of single scattering model, the coda decay curve is given by

$$
E^{S\text{-coda}}(t;f) \approx \frac{Wg_0}{2\pi V_0^2 t^2} e^{-Q_C^{-1}(f)2\pi f t} \quad \text{for} \quad t \gg \frac{2r}{V_0} \tag{7}
$$

Figure 5a summarizes reported coda attenuation. In general, Q_C^{-1} is about 10^{-2} at 1 Hz and decreases to about 10^{-3} at 20 Hz. The frequency dependence within a region can be written as $Q_C^{-1} \propto f^{-n}$ for $f > 1$ Hz, where $n \sim 0.5$–1. Regional differences of Q_C^{-1} were extensively studied in relation to seismotectonic settings (Singh and Herrmann, 1983; Jin and Aki, 1988;

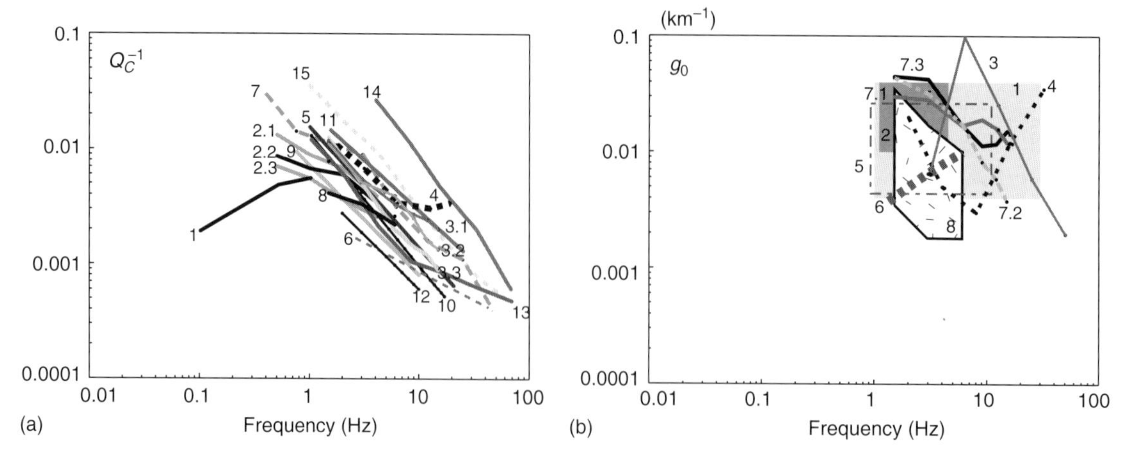

FIGURE 5 (a) Coda attenuation Q_C^{-1} against frequency for various regions. (b) Total scattering coefficient g_0 for *SS* scattering versus frequency, from regional measurements made throughout the world: results obtained using the single-scattering model are labeled 1–5 (plots include backscattering coefficient g_π); results based on the multiple lapse time window analysis (isotropic scattering is assumed) are labeled 6–8. Detailed references are given by Sato and Fehler (1998). (Reprinted from Sato and Fehler (1998) with permission from Springer-Verlag.)

Matsumoto and Hasegawa, 1989). We may say that Q_C^{-1} is smaller in tectonically stable areas and larger in active areas. Comparing the observed coda energy density with that predicted from theoretical coda models, we can estimate total scattering coefficient g_0 for *S*-to-*S* scattering. Figure 5b shows reported g_0, which is of the order of $10^{-2}\,\mathrm{km}^{-1}$ for frequencies 1–30 Hz, and each measurement has an error of about factor 2. Nishigami (1991) and later Revenaough (1995) inverted for the spatial distribution of scattering coefficient from the analysis of *S*- and *P*-coda amplitude residuals from the average coda decay. There have been various attempts to map coda attenuation in various scales (O'Doherty *et al.*, 1997; Mitchell *et al.*, 1997).

Seismic coda monitoring could provide information about the temporal change in fractures and attenuation caused by changes in tectonic stress during the earthquake cycle (Chouet, 1979). Jin and Aki (1986) reported a change in Q_C^{-1} associated with the occurrence of the Tangshan earthquake. Fehler *et al.* (1988) found differences in Q_C^{-1} before and after an eruption of Mt. St. Helens volcano. There have been many studies that indicate a correlation between temporal change in coda characteristics and the occurrence of large earthquakes. However, there have been criticisms, for example, concerning the possible influences of using different lapse times to establish the temporal change in coda characteristics and the effects of differing focal mechanisms and earthquake locations (Sato, 1988; Ellsworth, 1991; Frankel, 1991). Got *et al.* (1990) proposed the use of earthquake doublets. Recent measurements try to reflect those criticisms (Aster *et al.*, 1996). Analyzing short-period seismograms recorded at Riverside, California, between 1933 and 1987, Jin and Aki (1989) reported a temporal variability in Q_C^{-1} and a positive correlation with the seismic *b*-value. Gusev (1995) summarized coda observations made between 1967 and 1992 in Kamchatka by plotting coda magnitude residuals, and reported two prominent anomalies.

2.4 Coda Normalization Method

Coda waves provide a reliable way to isolate and quantify seismic propagation effects. Based on the single scattering model, we may write the average coda amplitude at a long lapse time t_c from the origin time at central frequency f as a product of the source, propagation, and site amplification as

$$\dot{u}^{S\text{-coda}}(t_c;f) \propto \sqrt{g_0(f)W(f)}N^S(f)\frac{e^{-Q_C^{-1}\pi f t_c}}{t_c}$$

where $N^S(f)$ is the site amplification factor, and $W(f)$ is the *S*-wave source energy. We suppose that $g_0(f)$ is constant in a region. The direct *S*-wave amplitude is written as

$$\dot{u}^{S\,\text{direct}}(r;f) \propto \sqrt{W(f)}N^S(f)\frac{e^{-Q_s^{-1}\pi f r/\beta_0}}{r}$$

where r is the hypocentral distance. Aki (1980) proposed a correction for source size and site amplification by normalizing direct *S*-wave amplitude by *S*-coda amplitude. Taking the logarithm of the ratio of the product of r and the direct *S*-wave amplitude to the averaged coda amplitude, where the common site amplification and source terms cancel, we get

$$\ln\left[\frac{r\,\dot{u}^{S\,\text{direct}}(r;f)}{\dot{u}^{S\text{-coda}}(t_c;f)}\right] = -\left(Q_S^{-1}(f)\frac{\pi f}{\beta_0}\right)r + \text{Const.} \tag{8}$$

We may smooth out the radiation pattern differences when the measurements are made over a large enough number of earthquakes. Plotting the LHS of Eq. (8) against r, the gradient gives the attenuation per travel distance. Aki (1980) first used this method for the estimation of Q_S^{-1} in Japan. Later, Yoshimoto *et al.* (1993) extended this method to measure Q_P^{-1}.

The ratio of coda amplitudes at different sites gives the relative site amplification factor since the source factor is common. Tsujiura (1978) found that estimates of site effects made using coda are similar to those obtained from direct wave measurements. He also found less variability in site effect measurements made for a single site using coda waves from various sources than from direct waves for the same sources. The ratio of coda amplitudes of different earthquakes at a single station gives the relative source radiation, since the site factor is common. These methods have been widely used around the world (Biswas and Aki, 1984; Dewberry and Crosson, 1995; Hartse *et al.*, 1995; Phillips and Aki, 1986).

2.5 Multiple Lapse Time Window Analysis

For the determination of seismic albedo, B_0, Fehler *et al.* (1992) and Hoshiba *et al.* (1991) developed a method based on two observations. (1) The early portion of an *S*-wave seismogram is dominated by the direct *S* wave whose amplitude change with distance is controlled by the total attenuation of the media; (2) *S*-coda level is composed entirely of scattered *S* waves whose amplitudes are controlled by the total scattering coefficient. Their method is based on the radiative transfer theory, in which scattering is assumed to be isotropic and radiation is spherically symmetric. They evaluate the integrated energy density as a function of source–receiver distance. Fehler *et al.* (1992) analyzed seismograms of local earthquakes in Kanto–Tokai, Japan in three frequency bands. Figure 6 shows the energy density integrated over three time windows for the 4–8 Hz band. The running means of the data over 15 km windows are plotted against distance using thin lines, and the broad lines show the best fit to the observed data from the theory. Later, Hoshiba (1993) proposed a single-station method for developing three curves, where he used average coda power at a fixed lapse time for the source energy. Variations of the methods have been used to make simultaneous measurements of attenuation and scattering coefficient. Seismic albedo B_0 has been found to vary from 0.2 to 0.8 and Q_S^{-1} decreases with increasing frequency over the range 1–20 Hz.

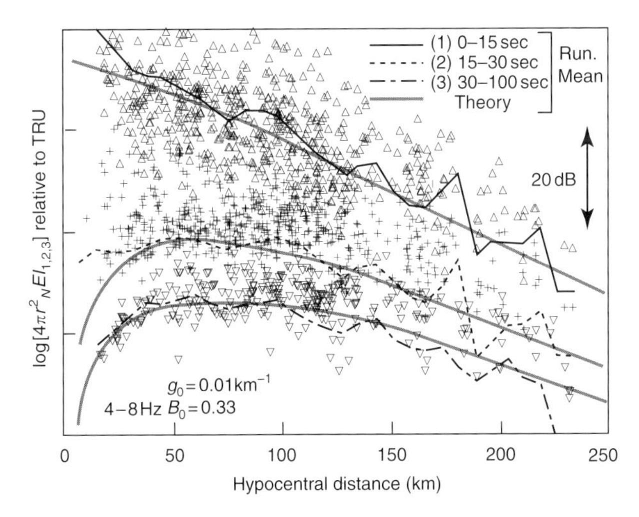

FIGURE 6 Normalized integrated energy density with geometrical spreading correction versus hypocentral distance in the Kanto–Tokai region, Japan, relative to the value at a hard rock borehole site for vertical component data. Average of data and best-fit theoretical curves are shown by thin lines and broad curves, respectively for the 4–8 Hz band. (Reprinted from Fehler *et al.* (1992) with permission from Blackwell Science.)

2.6 Other Studies of Seismogram Envelopes

The whole seismogram reflects not only the source information but also the scattering characteristics of the Earth medium. Sato (1984) proposed to synthesize three-component seismogram envelopes by summing single scattered wave energy, where frequency-dependent nonisotropic scattering amplitudes are calculated on the basis of the Born approximation. *S*-coda amplitudes are large on all three components, where *SS*-scattering contributes significantly for a wide range of lapse times, and pseudo *P*- and *S*-phases appear even in the null direction. Later, Yoshimoto *et al.* (1997a,b) developed an envelope synthesis including the reflection at the free surface.

The conventional waveform inversion for the source process was not successful when applied to high-frequency seismograms. The appropriate approaches would be to disregard the phase information and focus instead on seismogram envelopes. Zeng *et al.* (1993) mapped the high-frequency radiation from the fault plane using seismogram envelope analysis based on geometrical ray theory. Gusev and Pavlov (1991), and Kakehi and Irikura (1996) proposed using seismogram envelopes of small aftershocks as an empirical Green function. By using a solution of the radiative transfer equation, Nakahara *et al.* (1998) proposed an inversion method to estimate the spatial distribution of high-frequency energy radiation from the fault plane. Analyzing records of the 1994 off-Sanriku earthquake, Japan, they reported that the spatial distribution of high-frequency radiation does not always coincide with the slip distribution determined from longer-period waves.

The duration of the observed *S* wave packet at distances longer than 100 km is much longer than the source duration and the peak amplitude is delayed after the first arrival (Sato, 1989). It was initially proposed that this envelope broadening is a propagation effect due to diffraction and forward scattering caused by slowly varying velocity structure, which was modeled by employing a stochastic treatment of the parabolic wave equation (Lee and Jokipii, 1975a,b). Applying the theoretical prediction for the Gaussian ACF to envelop data observed at Kanto, Japan, Sato (1989), and Scherbaum and Sato (1991) estimated $\varepsilon^2/a \approx 10^{-(2.98\sim3.27)}\,\mathrm{km}^{-1}$. Analyzing the characteristics of seismogram envelopes from a larger region in Japan, Obara and Sato (1995) found that the random inhomogeneity in the back-arc side of the volcanic front is rich in short-wavelength components compared with the Gaussian spectra. Gusev and Abubakirov (1997) reported a decrease of scattering coefficient with increasing depth revealed from the envelope broadening of both *P* and *S* waves in Kamchatka.

3. Spatial Coherence of Seismic Array Data

The common practice in dealing with multiscale, broadband heterogeneities in seismology is to smooth both the observed wave field and the heterogeneity model. In this way the information about small-scale heterogeneities is ignored and the obtained image can only recover the slowly varying, large-scale heterogeneities.

Stochastic methods, on the other hand, can obtain some statistical characteristics of the small-scale heterogeneities from the statistics of the wave field fluctuations. Statistical parameters of the medium include the RMS perturbation of velocity distribution, characteristic scale length, power spectrum, or correlation function of velocity perturbations. Thus, deterministic and stochastic methods are complementary to each other when exploring multiscale complex media. In the overlapping spectral band, the deterministic and stochastic methods observe the same object from different aspects and using different simplifications during the analysis process.

The study of stochastic characteristics of random media using forward-scattered waves started a few decades ago. In the earlier studies (e.g., Nikolaev, 1975; Aki, 1973; Capon, 1974; Berteussen, 1975; Berteussen *et al.*, 1975, 1977; McLaughlin and Anderson, 1987), only variance and transverse coherence functions (TCF) of phase and amplitude fluctuations were used. Limited by the amount of information contained in these coherence functions, the medium model description is restricted to a single layer of uniform, isotropic random medium. Through the use of the scattering theory of Chernov (1960), several statistical parameters, such as the RMS velocity perturbation, the average scale length, and the total thickness of the layer, were inferred from the observed data. At the end of the 1980s, Flatté and Wu (1988) introduced a new statistical observable, the angular coherence function (AnCF), which significantly increased the statistical information in the reduced data sets, and allowed the authors to derive a simple model of a layered, multiscale random medium. More recently, Wu (1989), Wu and Flatté (1990), and Chen and Aki (1991) introduced the new joint coherence function (JCF) or joint transverse-angular coherence function (JTACF) and derived the theoretical relation between the joint coherence functions of array data and the heterogeneity spectrum (heterospectrum) of the random media. Wu and Xie (1991) conducted numerical experiments to test the performance of inverting JTACF to obtain the depth-dependent heterogeneity spectrum. The recent development in theory and methods greatly increases the amount of information in the statistical data set used and thereby significantly improves the model resolution, especially the depth resolution of the heterogeneity spectrum.

3.1 Observations of Amplitude and Phase Fluctuations and Their Coherences

As a result of wave diffraction and of focusing and defocusing effects caused by heterogeneities in a medium, wave front distortion and fluctuations of various parameters of the wave field such as amplitude, arrival time, and arrival angle may occur. Arrival time and amplitude fluctuations of waves crossing a seismic array such as NORSAR, LASA, and other local or regional arrays have been observed widely. The pattern of fluctuations may change drastically even between nearly colocated events.

3.1.1 Definitions of Various Coherence Functions

Coherence analysis is an effective method of describing statistical properties of the wave field. TCF defines the coherency (or similarity) of two transmitted wave fields as a function of horizontal separations between receivers; AnCF defines the coherency as a function of angles between incident waves. For the more general case, the JTACF gives the measure of coherency between two transmitted wave fields with different incident angles and observed at different stations (see Fig. 7).

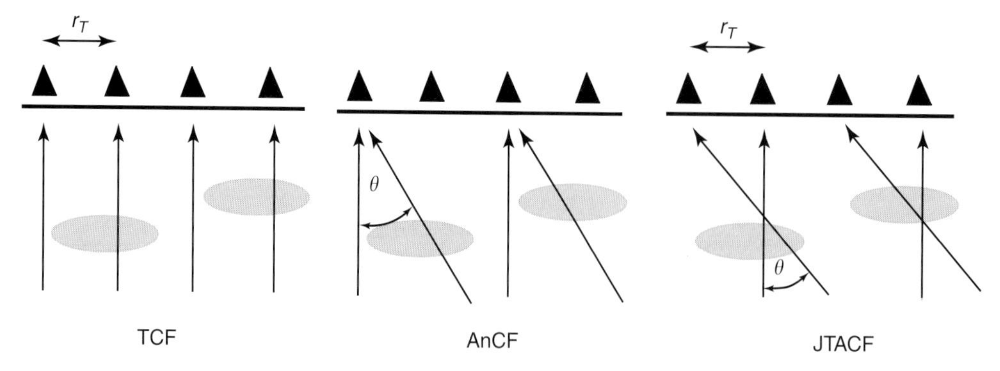

FIGURE 7 Comparison of the data reduction geometry for TCF (transverse coherence function), AnCF (angular coherence function), and JTACF (joint transverse angular coherence function).

Compared with TCF and AnCF, JTACF changes the coherence data from a 1D to a 3D matrix (a function of station separation, dip, and azimuth angles) and therefore increases tremendously the information content of the coherence data set, providing more constraints for the determination of medium properties under the array. However, the advantage of increasing the resolving power is offset by decreasing the statistical stability to some degree. For real array data sets, the compromise between resolution and stability depends on the amount of data and angular coverage of the events. In practical array measurements the influence of array aperture on the calculation of coherence functions has to be taken into consideration (Flatté and Xie, 1992).

3.2 Theoretical Basis of Coherence Analysis and Inversion

The theory of transverse coherence of the wave field after passing through a uniform random medium has long been available in the literature (Chernov, 1960; Tatarskii, 1971; Munk and Zacharasen, 1976; Flatté *et al.*, 1979). However, the formulations for angular coherence and for joint coherence functions have only recently been derived by Wu and Flatté (1990) using the Rytov approximation and Markov approximation. Chen and Aki (1991) independently derived similar formulas using the Born approximation. The theory of the joint coherence functions includes the TCF and AnCF as special cases. In the derivations of Wu and Flatté (1990), spectral representation of random media is used and the depth dependency of the spectrum is introduced. Thus, the new theory is more general and more suitable for the multiscale, depth-dependent Earth heterogeneities.

In transmission fluctuation analysis, only the initial P arrival (direct P, PKP or $PKIKP$) of a seismogram is used. By doing this, all the backscattered and large-angle scattered waves are neglected. Therefore, the problem becomes a forward scattering or small-angle scattering problem, for which the scalar wave approximation can be used (Wu and Aki, 1985; Wu, 1989). In such scattering problems, only the wave speed perturbations affect the scattered field. Let the wave speed be $V(\mathbf{x}) = V_0[1 + \xi(\mathbf{x})]$, where V_0 is the deterministic background velocity and $\xi(\mathbf{x})$ is the random perturbation. Let U_0 be the wave field in the absence of fluctuations, and define the field perturbation Ψ by

$$U = U_0 e^{\Psi} \qquad \text{and}$$
$$\Psi = \ln U - \ln U_0 = \ln \frac{A}{A_0} + i(\phi - \phi_0) = \nu + i\varphi \tag{9}$$

Substituting the above equation into the wave equation and taking the Rytov approximation and parabolic approximation, we derive the formulas for the *joint coherence functions* (for the detailed derivation, see Wu and Flatté, 1990)

$$\begin{aligned}
\langle \nu_1 \nu_2 \rangle (\mathbf{x}_T, \vec{\theta}) &= \frac{k^2}{4\pi^2} \int_0^{\mathrm{H}} dz \iint_{-\infty}^{\infty} d\mathbf{k}_T \, e^{i\mathbf{k}_T \cdot (\mathbf{x}_T + z\vec{\theta})} \\
&\quad \times \sin^2 \frac{k_T^2 z}{2k} P(\mathbf{k}_T, 0, z) \\
\langle \varphi_1 \varphi_2 \rangle (\mathbf{x}_T, \vec{\theta}) &= \frac{k^2}{4\pi^2} \int_0^{\mathrm{H}} dz \iint_{-\infty}^{\infty} d\mathbf{k}_T \, e^{i\mathbf{k}_T \cdot (\mathbf{x}_T + z\vec{\theta})} \\
&\quad \times \cos^2 \frac{k_T^2 z}{2k} P(\mathbf{k}_T, 0, z) \\
\langle \nu_1 \varphi_2 \rangle (\mathbf{x}_T, \vec{\theta}) &= \frac{k^2}{4\pi^2} \int_0^{\mathrm{H}} dz \iint_{-\infty}^{\infty} d\mathbf{k}_T \, e^{i\mathbf{k}_T \cdot (\mathbf{x}_T + z\vec{\theta})} \\
&\quad \times \sin \frac{k_T^2 z}{2k} \cos \frac{k_T^2 z}{2k} P(\mathbf{k}_T, 0, z)
\end{aligned} \tag{10}$$

where $\mathbf{x}_T = \mathbf{x}_{T1} - \mathbf{x}_{T2}$ is the receiver separation vector (transverse separation), $\vec{\theta} = (\vec{\theta}_1 - \vec{\theta}_2)$ is the incident-angle separation vector, H is the propagation range in the random medium, here equal to the thickness of the random layer, $k_T = |\mathbf{k}_T|$ is transverse wavenumber and $P(\mathbf{k}_T, k_z, z)$ is the 3D power spectrum of the random heterogeneities $\xi(\mathbf{x})$ at depth z. Equation (10) gives the general formulas for coherence analysis. Putting $\mathbf{x}_T = 0$ and $\vec{\theta} = 0$ in the above formulas, we obtain the variances and covariance of the fluctuations (magnitudes of fluctuations). For the transverse coherence of the fluctuations, we put $\vec{\theta} = 0$ into Eq. (10). For the angular coherence between fluctuations for two incident plane waves with angular separation of $\vec{\theta}$, we put $\mathbf{x}_T = 0$. In the past, the space domain formulation of Chernov (1960) was used for fluctuation analysis. Explicit expressions were derived for the case of a Gaussian ACF by Chernov (see also Sato and Fehler, 1998). The corresponding formulas can be obtained by substituting a Gaussian PSDF into Eq. (10).

3.3 Inversion for Statistical Characteristics of Earth Heterogeneities

3.3.1 Estimations of Turbidity in the Crust and Upper Mantle

In the 1960s, Russian scientists conducted extensive investigations using the log-amplitude fluctuation of P wave first motion from explosions and earthquakes to infer the crustal and upper mantle "turbidity coefficients" at different depths (see Nikolaev, 1975). The turbidity coefficient is defined as the variance of log-amplitude fluctuations produced by a unit travel distance. The depth of heterogeneities contributing to the measured turbidity was estimated by determining the seismic ray travel paths. In these measurements, the turbidity coefficients were rather phenomenological or apparent parameters, which might have included spatial variations of site factors and the variations of intrinsic attenuation. Nikolaev (1975) concludes that the turbidity for 5 Hz P waves in the crust and upper mantle is 0.0001–0.0025 km^{-1}, with an error factor of about 2.

3.3.2 Uniform Random Medium Model for the Lithosphere

For the single-layer Gaussian medium model, the model parameters are the RMS velocity perturbation ε, correlation length a, and the layer thickness H. Correlation length a can be estimated from a measurement of the transverse correlation of log-amplitude and phase. The layer thickness H and the wave speed perturbation ε can be obtained from the measured variance and covariance of ν and ϕ. This single layer isotropic Gaussian medium model has been used to analyze the data at LASA (Aki, 1973; Capon, 1974; Berteussen *et al.*, 1975), NORSAR (Capon and Berteussen, 1974; Berteussen *et al.*, 1977), and the Gauribindanur seismic array (GBA) in Southern India (Berteussen *et al.*, 1977). It is found that the estimate of correlation length a is much better constrained than the layer thickness H and perturbation ε. For LASA, $a \approx 10$–12 km, but H ranged from 60 to 120 km, with ε varying from 4 to 1.9% from different investigations.

3.3.3 Non-Gaussian Nature of the Heterogeneities

The Gaussian correlation function characterizes single-scale smoothly heterogeneous media, while real heterogeneities in the Earth are often multiscaled. Flatté and Wu (1988) showed that the exponential or Kolmogorov correlation functions fit the data much better than the Gaussian correlation function. The non-Gaussian nature of the lithospheric heterogeneities has been established also from velocity well-logging data (Sato, 1979; Wu, 1982).

3.3.4 Depth-Dependent Random Medium Model for the Crust and Upper Mantle

As many investigators have pointed out (Berteussen *et al.*, 1975; Flatté and Wu, 1988; Wu and Flatté, 1990), the use of only TCF resulted in poor determination of the random medium thickness and ambiguity in resolving the medium perturbation strength (variances) and the thickness. After introducing the AnCF, Flatté and Wu (1988) were able to invert both the TCF and AnCF for a more complex random medium model. They showed that a single-layer uniform random medium failed to explain the observed fluctuation coherences represented by both TCF and AnCF and proposed a two-layer (overlapping) random medium model for the crust and upper mantle beneath the NORSAR (see Fig. 8). Each layer has a different perturbation strength and a different power-law heterogeneity spectrum. The best model has the top layer extending to a depth of 200 km with a flat spectrum, representing the small-scale heterogeneities in the lithosphere, and the second layer located between 15 and 250 km with a k^{-4} power-law spectrum, where k is wavenumber. The latter spectrum characterizes the large-scale heterogeneities in the mantle. From the RMS travel time fluctuation (0.135 sec) and the RMS log-amplitude fluctuation (0.41 neper = 3.6 dB), the RMS P wave speed perturbation for the first layer is 0.9–2.2% and for the second layer is 0.5–1.3%.

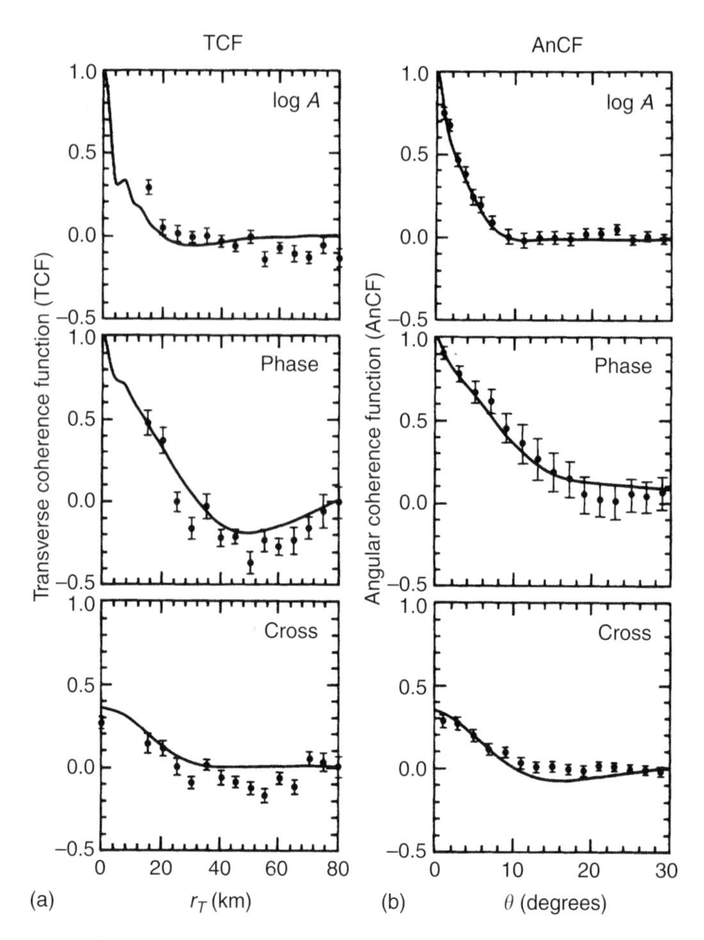

FIGURE 8 Comparison between the data and the prediction of the two-layered power-law random medium model at NORSAR: (a) TCF; (b) AnCF. (Reprinted from Flatté and Wu (1988) with permission from the American Geophysical Union.)

Based on the theory of general coherence analysis in random media, it is clear that TCF has no depth resolution and AnCF has only limited depth resolution that degenerates quickly with increasing depths. On the other hand, the joint coherency functions (JCF) increase tremendously the information content and provide high depth resolution. JTACFs have been calculated for the NORSAR data (Wu *et al.*, 1994) and the Southern California Seismic Network data (Liu *et al.*, 1994; Wu *et al.*, 1995), and some interesting findings were reported.

4. Numerical Modeling of Wave Propagation in Heterogeneous Media

We have presented some analytical methods for investigating wave propagation in heterogeneous media where only the random component of heterogeneity is modeled. Generally, it is difficult to model random heterogeneities with deterministic structures using the methods discussed. In addition, for studies of wave propagation in random media, we usually model the mean response of a suite of random media whose statistical characterizations are the same. Results from such modeling are often difficult to relate to real Earth observations. We now briefly discuss numerical modeling of wave propagation in heterogeneous media.

Numerous modeling studies have been undertaken to investigate wave propagation in heterogeneous media using numerical modeling. Numerical methods can model a wide range of Earth structures. Studies using finite differences (Frankel and Clayton, 1986; Shapiro and Kneib, 1993), the boundary integral approach (Benites *et al.*, 1992, 1997), homogeneous layer solutions (Richards and Menke, 1983), and Fourier domain methods (Spivack and Uscinski, 1989; Hoshiba, 1999; Wu *et al.*, 2000) have been made. In each case, the choice of numerical method was based on the type of wave phenomenon being investigated. Boundary integral and finite difference solutions provide reliable solutions including all wave field phenomena in strongly heterogeneous media; however, both methods are computationally expensive and it is difficult to investigate a range of models. For example, finite difference solutions require that grid spacing be chosen to minimize grid dispersion. As the propagation times and frequency being modeled increase, the spatial grid size and time increment must decrease to minimize grid dispersion (Holberg, 1987). As an example, Wu *et al.* (2000) required a grid spacing of 0.02 times the dominant wavelength to reliably model propagation to distances of 35 wavelengths. Boundary integral approaches have restrictions on the number of fictitious sources required along each boundary to adequately match boundary conditions. As velocity heterogeneity or frequency increases, the number of fictitious sources increases and the resulting computational cost increases.

Since several complete descriptions of finite difference and boundary integral techniques are available (Holberg, 1987; Benites *et al.*, 1992), we will discuss a class of numerical modeling techniques that has received little attention among seismologists. The methods are powerful in that they can model wave propagation faster than can be done using finite difference or boundary integral approaches. The methods have been used extensively in seismic exploration since the introduction of an intuitive approach for seismic wave modeling known as phase shift plus interpolation (PSPI) by Gazdag and Sguazzero (1984), and the work of Stoffa *et al.* (1990), who extended a wave equation-based method that was well known in acoustics (see, e.g., Jensen *et al.*, 1994). The method of Stoffa *et al.* (1990) is called the split-step Fourier (SSF) approach; its reliability has been quantitatively investigated by Huang and Fehler (1998). Extensions of the SSF method have been presented by Wu (1994), Huang *et al.* (1999a,b), Huang and Fehler (1999), and Wu *et al.* (2000). The methods are reliable for modeling wave propagation in the forward direction, but the effects of reverberations between scatterers are not included. The methods operate in the frequency domain and calculations are performed in the wavenumber and space domains. We will briefly introduce the SSF approach.

4.1 SSF Approach

The constant-density scalar wave equation in the frequency domain is

$$\left[\frac{\partial^2}{\partial x^2}+\frac{\partial^2}{\partial y^2}+\frac{\partial^2}{\partial z^2}+\frac{\omega^2}{V(\mathbf{x}_T,z)^2}\right]U(\mathbf{x}_T,z;\omega)=0 \quad (11)$$

where $\mathbf{x}_T\equiv(x,y)$, $U(\mathbf{x}_T,z;\omega)$ is the wave field in the frequency domain, $V(\mathbf{x}_T,z)$ is the velocity of the medium, and ω is the angular frequency. In some sense we can break the scalar wave equation into two parts, one representing downgoing waves and one representing upgoing waves. The equation for downgoing waves is

$$\frac{\partial}{\partial z}U(\mathbf{x}_T,z;\omega)=i\Lambda(\mathbf{x}_T,z;\omega)U(\mathbf{x}_T,z;\omega) \quad (12)$$

where the positive z direction is the propagation direction and the square-root operator Λ is defined by

$$\Lambda(\mathbf{x}_T,z;\omega)=\sqrt{\frac{\omega^2}{V(\mathbf{x}_T,z)^2}+\frac{\partial^2}{\partial x^2}+\frac{\partial^2}{\partial y^2}} \quad (13)$$

The formal split-step marching solution of the one-way wave equation is given by

$$U(\mathbf{x}_T,z_i+\Delta z;\omega)=\exp\left[i\int_{z_i}^{z_i+\Delta z}\Lambda(\mathbf{x}_T,z;\omega)\,dz\right]U(\mathbf{x}_T,z_i;\omega) \quad (14)$$

where we assume we know the wave field at z_i and we wish to compute the wave field at $z_i+\Delta z$. Huang and Fehler (1998) discuss how to use a small-angle approximation to evaluate the exponential of the integral in Eq. (14) to obtain

$$U(\mathbf{x}_T,z_{i+1};\omega)=\exp\left[i\omega\left(\int_{z_i}^{z_{i+1}}\Delta s(\mathbf{x}_T,z)dz\right)\right]U_0(\mathbf{x}_T,z_{i+1};\omega) \quad (15)$$

where the slowness heterogeneity $\Delta s\equiv 1/V(\mathbf{x}_\perp,z)-1/V_0(z)$ is assumed to be small, and $U_0(\mathbf{x}_T,z_{i+1};\omega)$ is the wave field at z_{i+1} obtained by propagation of the wave field across the interval Δz where the interval is assumed to have homogeneous velocity V_0 and

$$U_0(\mathbf{x}_T,z_{i+1};\omega)=F^{-1}_{\mathbf{k}_T}\left\{e^{ik_z(z_i)\Delta z}F_{\mathbf{x}_T}\{U(\mathbf{x}_T,z_i;\omega)\}\right\} \quad (16)$$

where $k_z(z_i) = \sqrt{k_0(z_i)^2 - \mathbf{k}_T^2}$ and $k_0 = \omega/V_0(z_i)$. $F_{\mathbf{x}_T}$ and $F_{\mathbf{k}_T}^{-1}$ are 2D Fourier and inverse Fourier transforms over $\mathbf{x}_T$ and $\mathbf{k}_T$, respectively. Wave propagation is done in two steps: a free space propagation in the wavenumber domain across each depth interval using the background slowness s_0 for the interval, followed by a correction for the heterogeneity described by Δs within the propagation interval, which is done in the space domain. The wave field is transferred between the space and wavenumber domains using a fast Fourier transform. Since the propagator depends only on the local medium properties, access to the entire velocity structure of a model is not required to propagate through a portion of the model. This gives the method a computational advantage over some other wave equation-based methods. The method is valid for extrapolating the wave field so long as the primary direction of propagation is nearly parallel to the positive z direction and the variation in velocity is small in the direction perpendicular to z. Later extensions mentioned above have systematically improved the accuracy for large-angle waves in strongly varying media.

The SSF method has been used by Hoshiba (1999) in an investigation of the influences of random structure on amplitudes of seismic waves. Wu *et al.* (2000) developed a method similar to the SSF method for modeling *Lg* propagation in heterogeneous media. Spivack and Uscinski (1989) present an analytic and numerical investigation of the accuracy of the SSF method and conclude that it is reliable for calculating wave fields in random media but that it is even more reliable when computing transverse correlations of the wave field. There is a stochastic treatment of the phase screen method, which is called the Markov approximation (Lee and Jokipii, 1975a,b; Sreenivasiah *et al.*, 1976). This method directly gives the wave envelope in random media, and is applied to the study of seismogram envelope characteristics (Sato, 1989; Scherbaum and Sato, 1991; Obara and Sato, 1995). Fehler *et al.* (2000) have compared waveforms calculated for random media using finite difference, an extension of the SSF method, and the Markov approximation.

TABLE 1 Fundamental observations that have led to advances in using stochastic seismology to model scattered waves, theory and methods used to explain the observations, and the parameters that may be inferred from the methods

Observation	Theory and Interpretation Method	Parameters Estimated
Existence of coda	• Phenomenological • Coda normalization method • Single scattering approximation • Energy-flux model	• Coda attenuation • Scattering coefficient • Relative site amplification • Relative source factors
Envelope shape of local earthquakes	• Radiative transfer theory • Multiple lapse time window analysis	• Scattering coefficient • Seismic albedo (scattering loss, intrinsic absorption) • High-frequency radiation from fault
Attenuation	• Spectral decay with distance • Coda normalization method • Born approximation	• Fractional velocity fluctuation • Scale length (correlation length) of heterogeneity • Spectra of heterogeneity • Spatial variation of stochastic characterization of medium
Array phase/amplitude characteristics	• Diffraction/forward scattering • Parabolic wave equation and Rytov approximation • Theory of coherence analysis	
Envelope broadening of regional seismograms	• Diffraction/forward scattering • Parabolic wave equation • Markov approximation	

5. Summary and Discussion

Beginning from visual observations of the character of recorded seismograms that led to new ways of thinking about seismic wave propagation, the study of seismic wave scattering and attenuation has led to new and improved observational tools for characterizing the Earth. One reason that the coda normalization method has been so useful is because station calibration and/or site amplification could be eliminated so that other medium parameters could be measured. Recent theories about seismic wave propagation in heterogeneous media can also be applied to investigate envelope shape and additional information about lithospheric structure can be obtained from broadband calibrated data.

Table 1 summarizes fundamental observations that have led to advances in our understanding of stochastic wave phenomena in the Earth's lithosphere. The model or interpretation method used to understand each fundamental observation is listed along with the parameters that can be estimated by application of the models to real data. Many of the parameters can be estimated using deterministic models. In most cases, observations made using deterministic and random-wave approaches are complementary. In many cases, the stochastic approach provides parameters characterizing the Earth's lithosphere that cannot be obtained from deterministic measurements.

Our understanding of the structure of heterogeneity of the Earth's lithosphere as a function of scale is limited. While some data can be explained using well-defined correlation functions that have a narrow band of heterogeneity, this does not mean that heterogeneity is limited to a narrow band. One fundamental issue facing those who investigate scattered waves is to find a unifying theory for a broad spectrum of seismic data that allows us to estimate the scale of heterogeneity over a broad range of scales.

Use of stochastic seismology has led to significant advances in our understanding of the character of seismic waveforms and enabled us to model portions of the waveforms that cannot be explained deterministically. The success of the models has improved our understanding of wave propagation in the Earth and led to the prediction of parameters that improve our understanding the structure and composition of the Earth's lithosphere.

References

[see Editor's Note]

Aki, K. (1969). Analysis of seismic coda of local earthquakes as scattered waves. *J. Geophys. Res.* **74**, 615–631.

Aki, K. (1980). Attenuation of shear-waves in the lithosphere for frequencies from 0.05 to 25 Hz. *Phys. Earth Planet. Inter.* **21**, 50–60.

Aki, K. (1973). Scattering of *P* waves under the Montana Lasa. *J. Geophys. Res.* **78**, 1334–1346.

Aki, K. and B. Chouet (1975). Origin of coda waves: Source, attenuation and scattering effects. *J. Geophys. Res.* **80**, 3322–3342.

Aki, K. and P. Richards (1980). "Quantitative Seismology—Theory and Methods," Vols. 1 and 2. W.H. Freeman, San Francisco.

Anderson, D.L. and R.S. Hart (1978). *Q* of the Earth. *J. Geophys. Res.* **83**, 5869–5882.

Benites, R.A., P.M. Roberts, K. Yomogida, and M. Fehler (1997). Scattering of elastic waves in 2-D composite media I. Theory and test. *Phys. Earth Planet Inter.* **104**, 161–173.

Berteussen, K.A., A. Christoffersson, E.S. Husebye, and A. Dahle (1975). Wave scattering theory in analysis of *P* wave anomalies at NORSAR and LASA. *Geophys. J. R. Astron. Soc.* **42**, 402–417.

Chernov, L.A. (1960). "Wave Propagation in a Random Medium" (Engl. trans. by R.A. Silverman). McGraw-Hill, New York.

Fehler, M., P. Roberts, and T. Fairbanks (1988). A temporal change in coda wave attenuation observed during an eruption of Mount St. Helens. *J. Geophys. Res.* **93**, 4367–4373.

Fehler, M., M. Hoshiba, H. Sato, and K. Obara (1992). Separation of scattering and intrinsic attenuation for the Kanto-Tokai region, Japan, using measurements of *S*-wave energy versus hypocentral distance. *Geophys. J. Int.* **108**, 787–800.

Fehler, M., H. Sato, and L.-J. Huang (2000). Envelope broadening of outgoing waves in 2D random media: a comparison between the Markov approximation and numerical simulations. *Bull. Seismol. Soc. Am.* **90**, 914–928.

Flatté, S.M. and R.-S. Wu (1988). Small-scale structure in the lithosphere and asthenosphere deduced from arrival-time and amplitude fluctuations at NORSAR. *J. Geophys. Res.* **93**, 6601–6614.

Frankel, A. and R.W. Clayton (1986). Finite difference simulations of seismic scattering: implications for the propagation of short-period seismic waves in the crust and models of crustal heterogeneity. *J. Geophys. Res.* **91**, 6465–6489.

Frankel, A. and L. Wennerberg (1987). Energy-flux model of seismic coda: separation of scattering and intrinsic attenuation. *Bull. Seismol. Soc. Am.* **77**, 1223–1251.

Gazdag, J. and P. Sguazzero (1984). Migration of seismic data by phase-shift plus interpolation. *Geophysics* **49**, 124–131.

Gusev, A.A. (1995). Baylike and continuous variations of the relative level of the late coda during 24 years of observation on Kamchatka. *J. Geophys. Res.* **100**, 20311–20319.

Gusev, A.A. and I.R. Abubakirov (1996). Simulated envelopes of non-isotropically scattered body waves as compared to observed ones: another manifestation of fractal heterogeneity. *Geophys. J. Int.* **127**, 49–60.

Hoshiba, M. (1991). Simulation of multiple-scattered coda wave excitation based on the energy conservation law. *Phys. Earth Planet. Inter.* **67**, 123–136.

Hoshiba, M. (1993). Separation of scattering attenuation and intrinsic absorption in Japan using the multiple lapse time window analysis of full seismogram envelope. *J. Geophys. Res.* **98**, 15809–15824.

Huang, L.-J. and M. Fehler (1998). Accuracy analysis of the split-step Fourier propagator: implications for seismic modeling and migration. *Bull. Seismol. Soc. Am.* **88**, 8–28.

Huang, L.-J., M. Fehler, and R.-S. Wu (1999a). Extended local Born Fourier migration. *Geophysics* **64**, 1524–1534.

Huang, L.-J., M. Fehler, P. Roberts, and C.C. Burch (1999b). Extended local Rytov Fourier migration method. *Geophysics* **64**, 1535–1545.

Jackson, D.D. and D.L. Anderson (1970). Physical mechanisms of seismic-wave attenuation. *Rev. Geophys. Space Phys.* **8**, 1–63.

Jin, A. and K. Aki (1986). Temporal change in coda *Q* before the Tangshan earthquake of 1976 and the Haicheng earthquake of 1975. *J. Geophys. Res.* **91**, 665–673.

Jin, A. and K. Aki (1989). Spatial and temporal correlation between coda Q^{-1} and seismicity and its physical mechanism. *J. Geophys. Res.* **94**, 14041–14059.

Kakehi, Y. and K. Irikura (1996). Estimation of high-frequency wave radiation areas on the fault plane by the envelope inversion of acceleration seismograms. *Geophys. J. Int.* **125**, 892–900.

Kinoshita, S. (1994). Frequency-dependent attenuation of shear waves in the crust of the southern Kanto, Japan. *Bull. Seismol. Soc. Am.* **84**, 1387–1396.

Kopnichev, Y.F. (1975). A model of generation of the tail of the seismogram. *Dok. Akad. Nauk SSSR* (Engl. trans.) **222**, 333–335.

Liu, H.P., D.L. Anderson, and H. Kanamori (1976). Velocity dispersion due to anelasticity; implications for seismology and mantle composition. *Geophys. J. R. Astron. Soc.* **47**, 41–58.

Mavko, G.M. and A. Nur (1979). Wave attenuation in partially saturated rocks. *Geophysics* **44**, 161–178.

Nakahara, H., T. Nishimura, H. Sato, and M. Ohtake (1998). Seismogram envelope inversion for the spatial distribution of high-frequency energy radiation on the earthquake fault: application to the 1994 far east off Sanriku earthquake ($M_W = 7.7$). *J. Geophys. Res.* **103**, 855–867.

Nikolaev, A.V. (1975). "The Seismics of Heterogeneous and Turbid Media" (Engl. trans. by R. Hardin). Israel Program for Science Translations, Jerusalem.

Nishigami, K. (1991). A new inversion method of coda waveforms to determine spatial distribution of coda scatterers in the crust and uppermost mantle. *Geophys. Res. Lett.* **18**, 2225–2228.

Obara, K. and H. Sato (1995). Regional differences of random inhomogeneities around the volcanic front in the Kanto-Tokai area, Japan, revealed from the broadening of *S* wave seismogram envelopes. *J. Geophys. Res.* **100**, 2103–2121.

O'Connell, R.J. and B. Budiansky (1977). Viscoelastic properties of fluid-saturated cracked solids. *J. Geophys. Res.* **82**, 5719–5735.

Phillips, W.S. and K. Aki (1986). Site amplification of coda waves from local earthquakes in central California. *Bull. Seismol. Soc. Am.* **76**, 627–648.

Rautian, T.G. and V.I. Khalturin (1978). The use of the coda for determination of the earthquake source spectrum. *Bull. Seismol. Soc. Am.* **68**, 923–948.

Roth, M. and M. Korn (1993). Single scattering theory versus numerical modeling in 2-D random media. *Geophys. J. Int.* **112**, 124–140.

Sato, H. (1977). Energy propagation including scattering effect: single isotropic scattering approximation. *J. Phys. Earth* **25**, 27–41.

Sato, H. (1982). Amplitude attenuation of impulsive waves in random media based on travel time corrected mean wave formalism. *J. Acoust. Soc. Am.* **71**, 559–564.

Sato, H. (1984). Attenuation and envelope formation of three-component seismograms of small local earthquakes in randomly inhomogeneous lithosphere. *J. Geophys. Res.* **89**, 1221–1241.

Sato, H. (1989). Broadening of seismogram envelopes in the randomly inhomogeneous lithosphere based on the parabolic approximation: Southeastern Honshu, Japan. *J. Geophys. Res.* **94**, 17735–17747.

Sato, H. and M. Fehler (1998). "Seismic Wave Propagation and Scattering in the Heterogeneous Earth." AIP Press/Springer Verlag, New York.

Sato, H., H. Nakahara, and M. Ohtake (1997). Synthesis of scattered energy density for nonspherical radiation from a point shear dislocation source based on the radiative transfer theory. *Phys. Earth Planet. Inter.* **104**, 1–14.

Scherbaum, F. and H. Sato (1991). Inversion of full seismogram envelopes based on the parabolic approximation: estimation of randomness and attenuation in southeast Honshu, Japan. *J. Geophys. Res.* **96**, 2223–2232.

Shapiro, S.A. and G. Kneib (1993). Seismic attenuation by scattering: theory and numerical results. *Geophys. J. Int.* **114**, 373–391.

Stoffa, P., J. Fokkema, R. de Luna Friere, and W. Kessinger (1990). Split-step Fourier migration. *Geophysics* **55**, 410–421.

Tsujiura, M. (1978). Spectral analysis of the coda waves from local earthquakes. *Bull. Earthq. Inst. Univ. Tokyo* **53**, 1–48.

Wu, R.-S. (1982). Attenuation of short period seismic waves due to scattering. *Geophys. Res. Lett.* **9**, 9–12.

Wu, R.-S. (1985). Multiple scattering and energy transfer of seismic waves—separation of scattering effect from intrinsic attenuation, I. Theoretical modeling. *Geophys. J. R. Astron. Soc.* **82**, 57–80.

Wu, R.-S. (1989). The perturbation method for elastic waves scattering. *Pure Appl. Geophys.* **131**, 605–637.

Wu, R.S. (1994). Wide-angle elastic wave one-way propagation in heterogeneous media and elastic wave complex-screen method. *J. Geophys. Res.* **99**, 751–766.

Wu, R.-S. and S.M. Flatté (1990). Transmission fluctuations across an array and heterogeneities in the crust and upper mantle. *Pure Appl. Geophys.* **132**, 175–196.

Wu, R.-S. and X.B. Xie (1991). Numerical tests of stochastic tomography. *Phys. Earth Planet. Inter.* **67**, 180–193.

Wu, R.-S., S. Jin, and X.-B. Xie (2000). Seismic wave propagation and scattering in heterogeneous crustal waveguides using screen propagators: I. SH waves. *Bull. Seismol. Soc. Am.* **90**, 401–413.

Yoshimoto, K., H. Sato, and M. Ohtake (1993). Frequency-dependent attenuation of *P* and *S* waves in the Kanto area, Japan, based on the coda-normalization method. *Geophys. J. Int.* **114**, 165–174.

Yoshimoto, K., H. Sato, and M. Ohtake (1997b). Short-wavelength crustal inhomogeneities in the Nikko area, central Japan, revealed from the three-component seismogram envelope analysis. *Phys. Earth Planet. Inter.* **104**, 63–74.

Zener, C. (1948). "Elasticity and Anelasticity of Metals." University of Chicago Press, Chicago.

Zeng, Y., F. Su, and K. Aki (1991). Scattering wave energy propagation in a random isotropic scattering medium 1. Theory. *J. Geophys. Res.* **96**, 607–619.

Zeng, Y., K. Aki, and T.L. Teng (1993). Mapping of the high-frequency source radiation for the Loma Prieta earthquake, California. *J. Geophys. Res.* **98**, 11981–11993.

Editor's Note

A complete set of references is given on the Handbook CD, under the directory of \13Sata.

14

Earthquakes as a Complex System

Donald L. Turcotte
Cornell University, Ithaca, New York, USA
Bruce D. Malamud
King's College London, UK

1. Introduction

Earthquakes are a direct consequence of the deformation of the Earth's crust. Parts of the crust are brittle, and significant deformation requires fractures to develop. Deformation across these fractures generates faults. Earthquakes are primarily associated with stick-slip behavior on these preexisting faults. Clearly these deformation processes are complex and occur on a wide range of scales. One approach to earthquake mechanics is to assume that displacements on a single (master) fault in a region are dominant and to neglect all other faults. Under this assumption, great earthquakes on the master fault occur on a near-periodic basis. This approach leads to the concept of seismic gaps—that regions where earthquakes have not occurred are regions where the next earthquake is expected (as was the case, for example, for the 1999 Izmit, Turkey earthquake).

An alternative approach to earthquake mechanics is to assume that the crust is a complex self-organizing system that can be treated by techniques developed in statistical physics; this is the main focus of this chapter. The basic hypothesis is that deformation processes interact on a range of scales from thousands of kilometers to millimeters or less. Evidence in support of this hypothesis comes from the universal validity of scaling relations. The most famous of these is the Gutenberg–Richter frequency–magnitude relation

$$\log \dot{N}_{CE} = -bm + \log \dot{a} \tag{1}$$

where $\dot{N}_{CE}$ is the cumulative number of earthquakes with a magnitude greater than m occurring in a specified area and time, and b and $\dot{a}$ are constants. In Section 2.1 we explore the validity of Eq. (1) for both regional and global earthquakes. The constant b, or "b-value," varies from region to region, but is generally in the range $0.8 < b < 1.2$ (Frohlich and Davis, 1993). The constant a is a measure of the regional level of seismicity. There are a variety of measures for the magnitude, m, including local, body-wave, surface-wave, and moment magnitude (Lay and Wallace, 1995). In general, for small-intensity earthquakes ($m < 5.5$), the different magnitude measures give approximately equivalent results.

Sections 2.1–2.4 explore the idea that complex phenomena often exhibit fractal (power-law) scaling (Mandelbrot, 1982; Turcotte, 1997) in magnitude, space, and time. For earthquakes, fractal scaling of their magnitudes implies the validity of the relation

$$\dot{N}_{CE} = C A_E^{-D/2} \tag{2}$$

where $\dot{N}_{CE}$ is the cumulative number of earthquakes with rupture area greater than A_E occurring in a specified area and time; C and D are constants, with D the fractal dimension. Aki (1981) showed that Eqs. (1) and (2) are equivalent with

$$D = 2b \tag{3}$$

The universal applicability of the Gutenberg–Richter relation implies that the number of earthquakes scale with their rupture area according to power-law (fractal) scaling.

Complex phenomena often also exhibit chaotic behavior (Section 3.1). Lorenz (1963) considered a set of nonlinear differential equations that approximate thermal convection in a fluid. He showed that the deterministic solutions to these equations exhibit an exponentially diverging sensitivity to initial conditions. This is the definition of deterministic chaos. Based on this work, it is now generally accepted that the behavior of the oceans and atmosphere is chaotic.

Burridge and Knopoff (1967) introduced the slider-block model as an analogue for the stick-slip behavior of faults. Slider-block models will be discussed in detail in Sections 3.1 and 3.2. Huang and Turcotte (1990a) showed that slider-block models also exhibit deterministic chaos. This is taken as evidence that the behavior of the Earth's crust is also chaotic. Carlson and Langer (1989) considered long linear arrays of

INTERNATIONAL HANDBOOK OF EARTHQUAKE AND ENGINEERING SEISMOLOGY, VOLUME 81A ISBN: 0-12-440652-1

slider blocks and concluded that they exhibit self-organized criticality.

The concept of self-organized criticality (Section 3.2) was introduced by Bak *et al.* (1988) as an explanation for the behavior of the cellular-automata "sandpile" model. Other models exhibiting self-organized critical behavior have been studied (Turcotte, 1999a). These models are characterized by a steady-state input and an output of "avalanches" that follow power-law (fractal) frequency–area distributions. Slider-block models clearly illustrate the transition from chaotic behavior (two slider blocks) to self-organized critical behavior (many slider blocks). The association of slider-block models with earthquakes provides strong evidence that earthquakes are an example of self-organized critical behavior in nature. One way to understand self-organized critical behavior is in terms of an inverse cascade of metastable clusters (Section 3.3), which grow and coalesce as strain occurs, giving a fractal frequency–area distribution.

Concepts of complexity have had direct applicability to probabilistic earthquake hazard studies (Section 4) and possibly to intermediate-term earthquake prediction (Section 5). The universal applicability of the Gutenberg–Richter relation, Eq. (1), provides one of the principal means of estimating the earthquake hazard. The rate of occurrence of small earthquakes provides a quantitative measure of the rate of occurrence of larger earthquakes. The concepts of complexity discussed in this chapter provide a rational basis for this extrapolation and possibly for intermediate-range earthquake prediction.

2. Fractal Distributions in Magnitude, Space, and Time

2.1 Gutenberg–Richter Relation

The best-known fractal (power-law) distribution in seismology is the Gutenberg–Richter frequency–magnitude relation, Eq. (1). This relation has been found to be valid both regionally and globally. As shown in Eq. (2), the b-value is equivalent to a fractal dimension D. The near universality of the b-value taking on values near unity is equivalent to a near-constant fractal dimension with $D \approx 2$. The fact that fractal dimensions are universal and not diagnostic is commonly found in applications (Turcotte, 1997). However, the constant a is a direct measure of the level of seismic activity in a region. This is one basis for estimating the seismic hazard, as will be shown in Section 4.

We first consider the worldwide number of earthquakes per year from the Harvard Centroid–Moment Tensor Database (1997) for the period 1977–1994. The seismic moments have been used to calculate moment magnitudes m and rupture areas A_E. In Figure 1, the cumulative number of earthquakes per year, $\dot{N}_{CE}$, with rupture area greater than A_E is given as

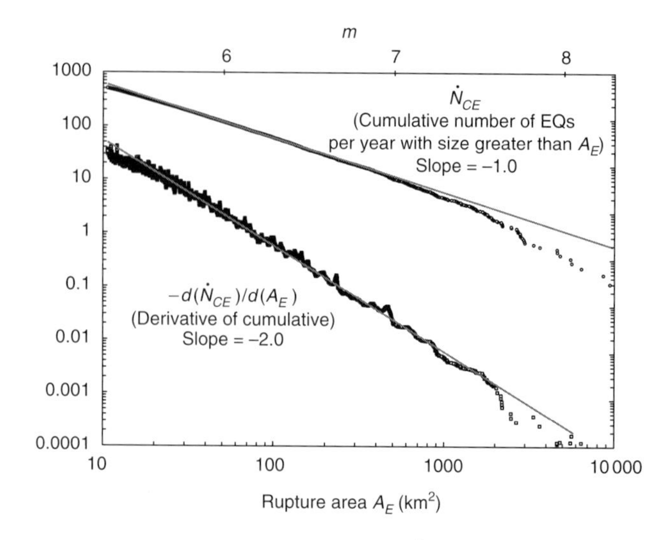

FIGURE 1 Worldwide cumulative, $\dot{N}_{CE}$, and noncumulative, $-d\dot{N}_{CE}/dA_E$, number of earthquakes per year with rupture area greater than A_E as a function of A_E. The data are from the Harvard Centroid–Moment Tensor Database (1997) for the years 1977–1994 with earthquake seismic moments converted to rupture areas A_E. The equivalent moment magnitudes m are also given.

a function of A_E, along with the equivalent moment magnitudes, m. The cumulative worldwide data per year correlate with Eq. (1) taking $b = 1.0$ ($D = 2.0$) and $\dot{a} = 10^8\ \text{y}^{-1}$.

The data in Figure 1 begin to deviate from the Gutenberg–Richter law, Eq. (1), for $m < 5.2$. This can be attributed to the resolution limits of the global seismic network. Regional studies indicate that good power-law correlations are obtained down to at least $m = 1$ (Aki, 1987). The deviation of data in Figure 1 from the Gutenberg–Richter relation for $m \geq 7.5$ is more controversial (Scholz, 1997). Clearly, there must be an upper limit to the size of an earthquake, but the deviations in Figure 1 can be attributed either to a real deviation from the correlation line or to the small number of very large earthquakes in the relatively short time span considered. There is a physical basis for a change in scaling for large earthquakes (Pacheco *et al.*, 1992). The rupture zone of smaller earthquakes is confined by the thickness of the seismogenic zone, say 20 km, whereas the length of the zone, l, can increase virtually without limit. Thus, for large earthquakes, $l \sim A_E$. The transition would be expected to occur for $A_E^{1/2} \approx 25$ km or $m \approx 7.0$.

Standard practice is to use cumulative distributions for earthquakes. However, this presents problems when making comparisons with models. Therefore, we wish to obtain a noncumulative distribution for global seismicity. Noncumulative distributions can be obtained by "binning" actual frequency–magnitude data and creating a histogram; however, care must be taken in how binning is carried out. If bins are linear, i.e., $r = 1$–2, 2–3, 3–4, . . ., and data have a power-law dependence of number of data in each bin on data size, $N_b \sim r^{-\alpha}$, then the resulting cumulative distribution will be $N_{Cb} \sim r^{-a+1}$; the exponents differ by 1.0. However, if bins are logarithmic, i.e., $r = 1$–2, 2–4, 4–8, ..., and data have a power-law dependence, $N_b \sim r^{-\beta}$, then the resulting cumulative distribution will be $N_{Cb} \sim r^{-\beta}$; the exponents are the same.

To avoid the difficulties of binning data, we use an alternative but equivalent approach to obtain a noncumulative distribution. We start with cumulative data, where N_{CE} is the number of earthquakes with area greater than A_E. The noncumulative distribution is the negative of the derivative of the cumulative distribution with respect to the area, $-dN_{CE}/dA_E$. The negative is because the cumulative distribution is summed from largest to smallest values. The derivative, $-dN_{CE}/dA_E$, is the slope from a specified number of adjacent cumulative data points. The noncumulative distribution given in Figure 1 uses five adjacent points to calculate the slopes of the cumulative distribution and a least-squares fit in log–log space. The noncumulative results given in Figure 1 correlate well with slope $b + 1 = 2$, as expected. We will return to noncumulative distributions when we consider models that exhibit self-organized critical behavior.

The remainder of our discussion will be restricted to cumulative distributions, as this is the standard practice in seismology when considering actual earthquakes. In our first example, we consider regional seismicity in Southern California. The cumulative frequency–magnitude distribution of seismicity in Southern California (SCSN Catalog, 1995) is given in Figure 2 for the period 1932–1994. Over the entire

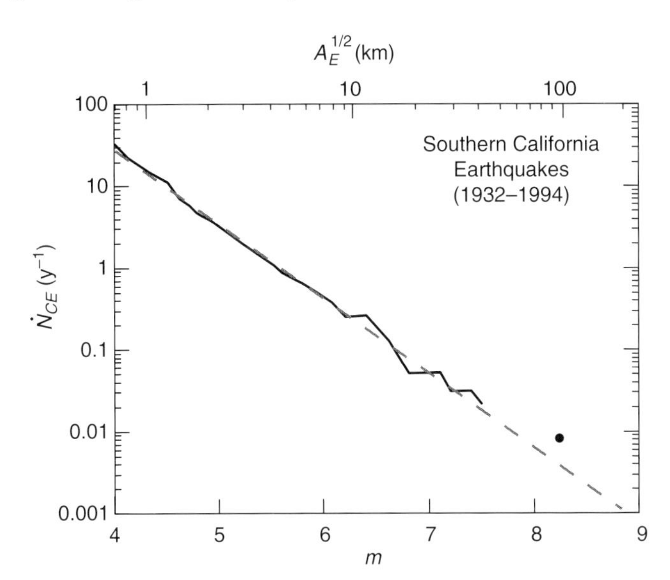

FIGURE 2 Number of earthquakes per year, $\dot{N}_{CE}$, occurring in Southern California with body-wave magnitudes greater than m as a function of m. Also given is the square root of the equivalent rupture area. The solid line is data for the period 1932–1994 (SCSN Catalog, 1995). The straight dashed line is the correlation with Eq. (1) taking $b = 0.92$ ($D = 1.84$) and $\dot{a} = 1.40 \times 10^5\ y^{-1}$. The solid circle is the observed rate of occurrence of great earthquakes in Southern California (Sieh *et al.*, 1989).

range $4 < m < 7.5$ the data are in excellent agreement with Eq. (1) taking $b = 0.92$ and $\dot{a} = 1.4 \times 10^5\ y^{-1}$.

Dates for ten large earthquakes on the southern section of the San Andreas Fault have been obtained from radiocarbon dating of faults, folds, and liquefaction features (Sieh *et al.*, 1989). In addition to historical great earthquakes in 1857 and 1812, additional great earthquakes were estimated to have occurred in 1480, 1346, 1100, 1048, 997, 797, 734, and 671. The mean repeat time is 132 y ($\dot{N}_{CE} = 0.0076\,y^{-1}$). Sieh (1978) estimates $m = 8.25$ for the 1857 earthquake. Taking these values for $\dot{N}_{CE}$ and m, we obtain the solid circle in Figure 2. An extrapolation of the Gutenberg–Richter statistics (the dashed line in Figure 2) appears to make a reasonable prediction of great earthquakes on this section of the San Andreas Fault. Acceptance of this hypothesis allows the regional background seismicity to be used in assessing seismic hazards (Turcotte, 1989). The regional frequency–magnitude statistics can be extrapolated to estimate recurrence times for larger-magnitude earthquakes. Unfortunately, no information is provided on the largest earthquake to be expected in a region.

An important question in seismology is whether the occurrence of large plate-boundary earthquakes, such as those at subduction zones, can be estimated by extrapolating the regional seismicity as was done above for Southern California. This is a subject of considerable controversy. Some authors argue that the large earthquakes occur more often than would be predicted by an extrapolation (Davison and Scholz, 1985; Pacheco *et al.*, 1992; Camelbeeck and Meghraoui, 1996; Scholz, 1997).

To further consider the time dependence of regional seismicity, i.e., the time dependence of $\dot{a}$ in Eq. (1), we consider yearly seismicity in Southern California. In Figure 3, the cumulative number of earthquakes $\dot{N}_{CE}$ in each year between 1980 and 1994 (SCSN Catalog, 1995) with magnitudes greater than m is given as a function of m. In general, there is good agreement with Eq. (1) taking $b = 1.05$ and $\dot{a} = 2.06 \times 10^5\ y^{-1}$. The exceptions can be attributed to the aftershock sequences of the 1987 Whittier, 1992 Landers, and 1994 Northridge earthquakes. The linear correlation line in Figure 2 has a slope (b-value) that is slightly greater than the linear correlation line in Figure 3. This is because the data in Figure 2 include aftershocks, resulting in a higher percentage of smaller-magnitude events. With aftershocks removed, there is near uniformity from year to year in the background seismicity in Southern California. This result strongly suggests a thermodynamic behavior for seismicity in a region. It also suggests that the Earth's crust is continuously on the brink of failure (Scholz, 1991).

Since the eastern United States is a plate interior, the concept of rigid plates would preclude seismicity in the region. However, plates act as stress guides and forces that drive plate tectonics are applied at plate boundaries. Because plates are

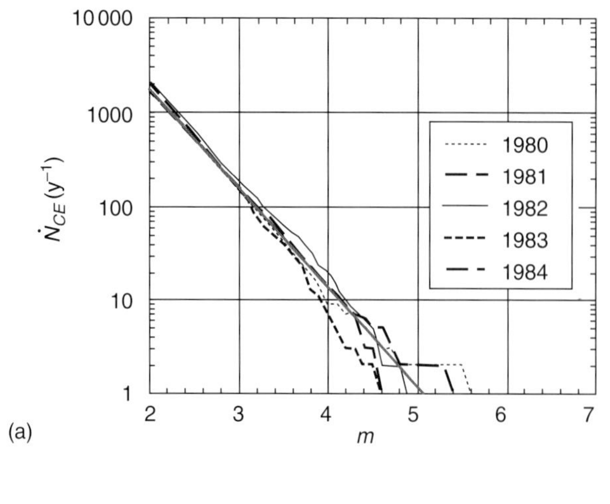

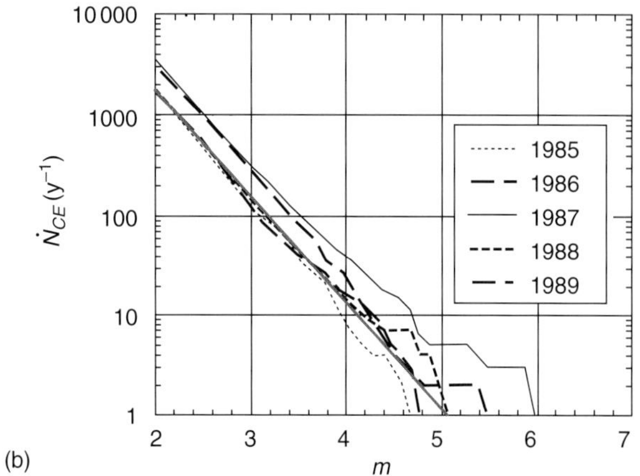

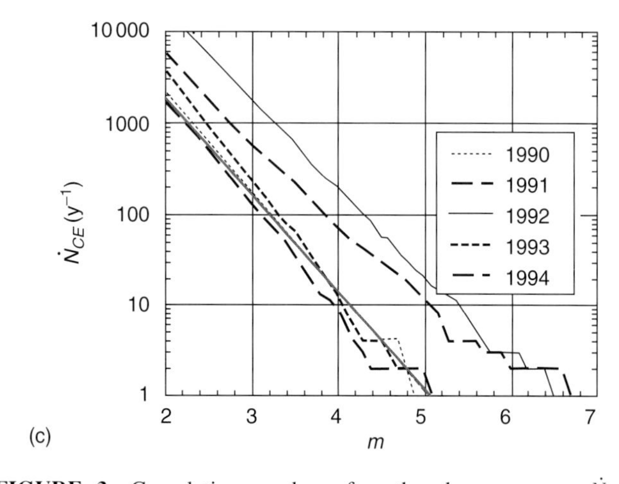

FIGURE 3 Cumulative number of earthquakes per year, $\dot{N}_{CE}$, occurring in Southern California with magnitudes greater than m as a function of m. Fifteen individual years are considered (SCSN Catalog, 1995): (a) 1980–1984; (b) 1985–1989; (c) 1990–1994. The larger numbers of earthquakes in 1987, 1992, and 1994 can be attributed to the aftershocks of the Whittier, Landers, and Northridge earthquakes, respectively. The solid straight line in (a) to (c) is the Gutenberg–Richter relation Eq. (1) with $b = 1.05$ and $\dot{a} = 2.06 \times 10^5\,\text{y}^{-1}$, the best fit to the data, excluding 1987, 1992, and 1994.

essentially rigid, these forces are transmitted through their interiors. The plate interiors have zones of weakness that will deform under these forces, resulting in earthquakes. Thus earthquakes occur within the interior of the tectonic plates, although the occurrence frequencies are much lower than at plate boundaries.

As an example of intraplate seismicity we consider three great earthquakes that occurred in the Memphis–St. Louis (New Madrid, Missouri) seismic zone during the winter of 1811–1812. Johnston and Schweig (1996) suggest that two of these earthquakes had moment magnitudes $m \approx 8$. Based on instrumental and historical records, Johnston and Nava (1985) have given the frequency–magnitude statistics for earthquakes in this area for the period 1816–1983 (Fig. 4). The data correlate well with Eq. (1) taking $b = 0.90$ ($D = 1.80$) and $\dot{a} = 2.24 \times 10^3\,\text{y}^{-1}$. Comparing Figures 4 and 2, the probability of having a moderate-sized earthquake in the Memphis–St. Louis seismic zone is about 1/60 the probability of having a similar magnitude earthquake in Southern California. Assuming that it is valid to extrapolate the data in Figure 4 to larger earthquakes, a magnitude $m \approx 8$ would have a recurrence time of about 7000 y.

We have considered two seismic regions, Southern California and the Memphis–St. Louis seismic zone, to illustrate the regional validity of the Gutenberg–Richter relation. Many other regions could have been considered. The universal validity of this relation is remarkable and any comprehensive theory for seismicity should explain it. We will now explore other fractal (power-law) relations in seismicity.

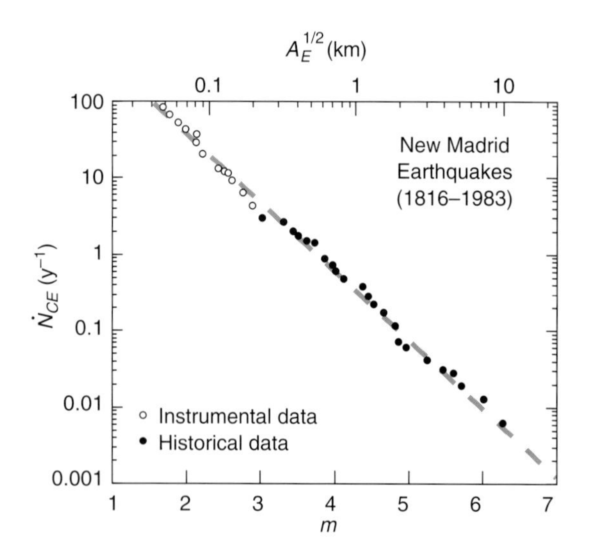

FIGURE 4 The cumulative number of earthquakes per year, $\dot{N}_{CE}$, occurring in the New Madrid, Missouri seismic zone (1816–1983) with magnitudes greater than m as a function of m. The open circles represent instrumental data and the solid circles historical data (Johnston and Nava, 1985). The straight dashed line is the Gutenberg–Richter relation Eq. (1) with $b = 0.90$ and $\dot{a} = 2.24 \times 10^3\,\text{y}^{-1}$.

2.2 Omori's Law

Earthquakes are universally followed by a sequence of aftershocks. A power-law decay in the number of aftershocks, $R(t)$, as a function of time t after the main shock takes the form

$$R(t) \sim \frac{1}{(t_0 + t)^{\rho}} \tag{4}$$

with t_0 and ρ being fitting constants. This relation is also found to be universally valid, with the constant ρ taking on values near unity, and is known as Omori's law (Utsu, 1961; Utsu *et al.*, 1995). Again this power-law relation requires a universal explanation.

2.3 Spatial Distribution of Faults

There are two end-member models that give fractal spatial distributions of earthquakes. The first is that the spatial distribution of faults is fractal and each fault has earthquakes of a single size (rupture area equal to fault area). The second is that each fault has a spatial distribution of earthquakes on it that is fractal. A variety of observations favor the first hypothesis. For example, on the northern and southern locked sections of the San Andreas Fault, there is no evidence for a fractal distribution of earthquakes. Great earthquakes occurred on the northern and southern sections in 1906 and 1857, respectively. Great earthquakes and their associated aftershock sequences occur, but between these great earthquakes, seismicity is essentially confined to secondary faults. A similar statement can be made about the Parkfield section of the San Andreas Fault, where moderate-sized earthquakes occurred in 1881, 1901, 1924, 1934, and 1966. There is no evidence for a fractal spatial distribution of events on this section of the San Andreas Fault. We therefore conclude that a reasonable working hypothesis is that each fault has earthquakes of a single size and a fractal spatial distribution of earthquakes implies a fractal spatial distribution of faults.

Although we can conclude that the frequency–size distribution of faults is fractal, the fractal dimension is not necessarily the same as that for earthquakes. Equal fractal dimensions would imply that the interval of time between earthquakes is independent of scale. This need not be the case. It is generally difficult to quantify the frequency–size distribution of faults, as many faults are not recognized until earthquakes occur on them. Coal mining areas provide access to faults and fractures at depth. The cumulative distributions of the number of faults N_{CF} with lengths greater than r are given in Figure 5 for two coal mining areas (Villemin *et al.*, 1995). Correlations with the fractal scaling relation, Eq. (2), are given with $D = 1.6$. Other compilations of number–length statistics of faults, and comparisons with power-law correlations, have been given by Gudmundson (1987), Hirata (1989), and Main *et al.* (1990).

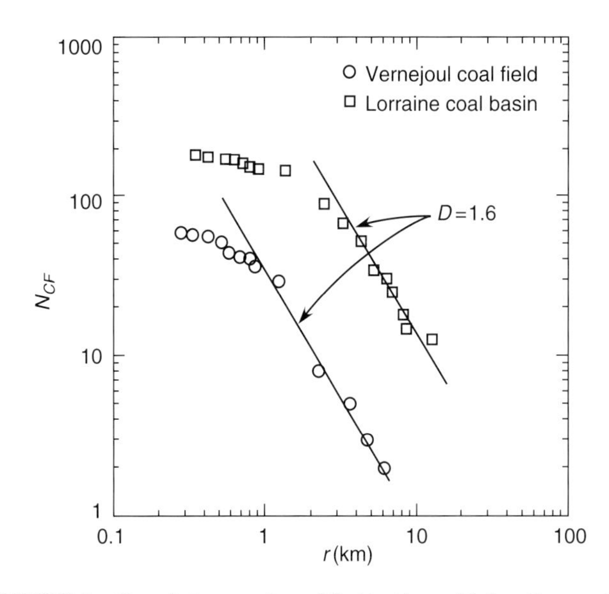

FIGURE 5 Cumulative number of faults, N_{CF}, with lengths greater than r as a function of r. The boxes are measurements in the Lorraine coal basin and the circles from the Vernejoul coal field (Villemin *et al.*, 1995). Correlations with the fractal scaling relation Eq. (2) are given with $D = 1.6$.

2.4 Spatial Distribution of Earthquakes

We next turn to the spatial distribution of earthquakes. A basic question is whether the spatial distribution is random (Poissonian) or fractal (scale-invariant). Studies of this problem require accurate locations of earthquakes with magnitudes above a specified minimum. Because of the deployment of high-density seismic arrays, the best spatial resolution data are for aftershock sequences. Robertson *et al.* (1995) have studied 3D (three-dimensional) spatial distributions of aftershocks using the box-counting method. These authors considered the aftershock sequences of the $m = 6.1$ Joshua Tree earthquake of 23 April 1992 (2600 events in a $20 \times 20 \times 19\,\text{km}^3$ volume in 160 days) and the $m = 6.2$ Big Bear earthquake of 28 June 1992 (818 events in a $20 \times 20 \times 17\,\text{km}^3$ volume in 375 days).

Figure 6 shows the number of cubes occupied by at least one earthquake as a function of cube linear dimension (0.5–20 km) for the two aftershock sequences. The data are in good agreement with Eq. (2) taking $D = 2.0$. A fractal dimension of 2 would be expected if the earthquakes lie on a plane; however, there is considerable 3D structure to the aftershock sequences. This led Robertson *et al.* (1995) to suggest that the earthquakes form the "backbone" of a percolation cluster, which we now discuss.

One of the standard models in statistical physics is the site-percolation model (Stauffer and Aharony, 1992). In the 2D (two-dimensional) version of this model a square array is made up of a matrix of boxes; each box may be either permeable or impermeable. The probability that a box is permeable is p_0, $0 \le p_0 \le 1$; the probability it is impermeable is $(1 - p_0)$.

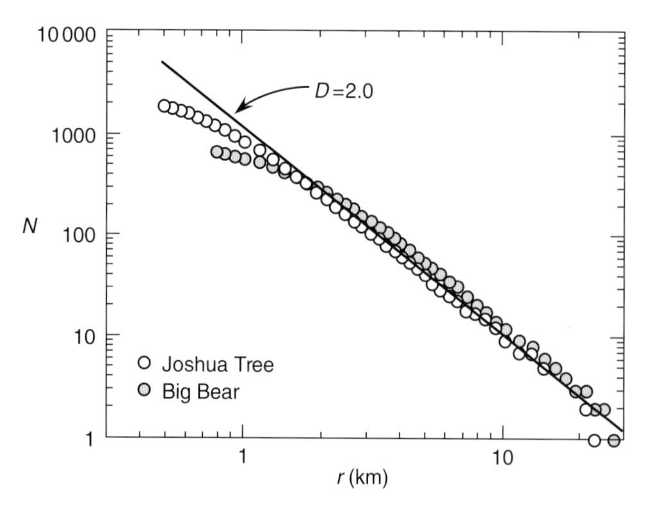

FIGURE 6 Three-dimensional spatial distributions of two aftershock sequences have been analyzed using the box-counting method (Robertson *et al.*, 1995). The aftershock sequence of the $m = 6.1$ Joshua Tree earthquake of 23 April 1992 (2600 events in a $20 \times 20 \times 19\,\mathrm{km}^3$ volume in 160 days), and the aftershock sequence of the $m = 6.2$ Big Bear earthquake of 28 June 1992 (818 events in a $20 \times 20 \times 17\,\mathrm{km}^3$ volume in 375 days) are considered. The number of cubes, N, occupied by one or more earthquakes is given as a function of the cube linear dimension, r, for the two aftershock sequences; cubes with $0.5 < r < 20\,\mathrm{km}$ were used. The correlation with the fractal scaling relation Eq. (2) is given with $D = 2.0$.

The question is whether the square array is permeable or impermeable. The array is defined to be permeable if there is a continuous path of permeable boxes from the top to the bottom of the array. This is clearly a statistical question since the actual distribution of permeable and impermeable boxes is random. For a specified value of p_0 there is a probability P that the square array of boxes is permeable. The critical probability for the percolation threshold is given by p^*. For large arrays, it is found that P is very small ($P \ll 1$) if $0 \leq p_0 \leq p^*$ and that P is near unity ($P \approx 1$) if $p^* < p_0 \leq 1$ (Stinchcombe and Watson, 1976). The value of p_0 that corresponds to the onset of flow through the array is the critical value and denoted by p^*. If $p_0 < p^*$, a large square grid will almost certainly be impermeable to flow; if $p_0 \geq p^*$, it will be permeable. Using the Monte Carlo approach, a large number of random realizations can be carried out and the probability $P(p_0)$ that the array is impermeable can be determined. For a square array with large n, numerical simulations find that the critical probability for the onset of flow in the array is $p^* = 0.59275$ (Stauffer and Aharony, 1992). The cluster of sites belonging to this permeable path is known as the percolation backbone.

The site-percolation threshold is a critical point and a variety of critical-point scalings are found (Stauffer and Aharony, 1992). This simple model can also be discussed in terms of fractures. As the stress on a thin sheet is increased, random ruptures will appear. Eventually these ruptures will coalesce to form a through-going rupture that crosses the sheet. The critical stress leading to the failure can be considered a critical point of the problem.

The site-percolation problem can also be considered in 3D where each box in a cubic array is assigned a probability of being permeable. The critical probability for the percolation backbone that crosses the cube is found to be $p^* = 0.3117$ (Stauffer and Aharony, 1992). This backbone has a fractal dimension close to 2. Sahimi *et al.* (1992, 1993) explained the fractal spatial distribution of aftershocks illustrated in Figure 6 as being the result of a 3D percolation backbone.

The box-counting technique has been applied to both the temporal and 2D spatial distributions of earthquakes in Japan by Bodri (1993). The temporal clustering of seismicity has been studied by Sadovsky *et al.* (1984) and by Smalley *et al.* (1987). The latter authors considered the temporal variations of seismicity in several regions in the New Hebrides island arc for the period 1978–1984. In each case, fractal temporal and spatial distributions were found.

3. Slider-Block Models: Chaos and Self-Organized Criticality

3.1 Deterministic Chaos

Lorenz (1963) discovered deterministic chaos. He derived three coupled total differential equations as an approximation for thermal convection in a fluid layer heated from below. He showed that solutions in a parameter range had an exponentially diverging sensitivity to initial conditions. This divergent behavior is the definition of chaotic behavior. The chaotic behavior of these equations has been taken as evidence that the atmosphere and oceans are chaotic.

Burridge and Knopoff (1967) introduced the slider-block model as a simple model for the behavior of a fault. One can show that a pair of connected slider blocks can exhibit deterministic chaos (Huang and Turcotte, 1990a; McCloskey and Bean, 1992). Consider the behavior of a pair of slider blocks as illustrated in Figure 7. A constant-velocity driver drags the blocks over the surface at a mean velocity, v. The two blocks are coupled to each other by a spring with spring constant k_c and to the driver plate by springs with spring constant k_p. Other model parameters are the block masses m and the frictional forces F_{s1} and F_{s2}. Solutions are governed by two parameters, the stiffness of the system k_c/k_p and the ratio of static to dynamic friction F_s/F_d. For some values of these parameters, deterministic chaos is found. Otsuka (1972) considered a 2D array of slider blocks and obtained a power law frequency–area distribution for slip events.

A modification to the analysis of a pair of slider blocks is to allow only one slider block to slip at a time. When a slider block becomes unstable, it is allowed to complete its harmonic motion before the stability of the second slider block is

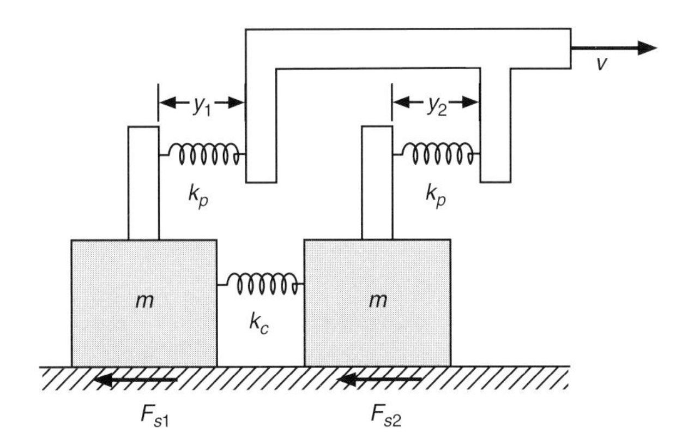

FIGURE 7 Behavior of a pair of slider blocks. A constant-velocity driver drags the blocks with mass m over the surface at a mean velocity v. The two blocks are coupled to each other by a spring with spring constant k_c and to the driver plate by springs with spring constant k_p. Frictional forces are F_{s1} and F_{s2}.

determined; if it is unstable, it is allowed to carry out its harmonic motion. The primary advantage of this modification is that slider-block displacements are given by analytical expressions and numerical solutions of differential equations are not required. Studies of this modified slider-block problem (Narkounskaia and Turcotte, 1992; Narkounskaia *et al.*, 1992) yield a variety of behaviors including periodic and chaotic solutions.

Huang and Turcotte (1990b) have applied the chaotic behavior of the asymmetric two-block system to two examples of interacting fault segments. In the first, the Philippine Sea plate descends beneath the Asian plate, resulting in the formation of the Nankai trough along the coast of southwestern Japan. The relative motion between the plates has resulted in a sequence of great earthquakes that have been documented through historical records, AD 684–1946. The sequence is marked by an irregular but somewhat repetitive pattern in which whole-section failures occur following several alternative failures of single segments. In the two-block model, the simultaneous slip of both blocks corresponds to an earthquake that ruptures the entire section, and single-block failures correspond to an earthquake on a single segment. Huang and Turcotte (1990b) found chaotic model behavior that strongly resembled the observed sequence of earthquakes in the Nankai trough.

The second example is the interaction between the Parkfield segment and the rest of the south-central locked segment of the San Andreas Fault in California. A sequence of $m \approx 6$ earthquakes occurred in 1857 and were also associated with a rupture on the Parkfield segment. Huang and Turcotte (1990b) found chaotic model behavior similar to that described above. A sequence of slip events on the weaker block often preceded the simultaneous slip of the weaker and stronger blocks. The model simulation suggested two alternative scenarios for a great Southern California earthquake following a sequence of Parkfield earthquakes. In the first scenario, a Parkfield earthquake will transfer sufficient stress to trigger the great Southern California earthquake; the Parkfield earthquake is thus essentially a foreshock for the great earthquake. In the second scenario, a small additional strain after a Parkfield earthquake will trigger an earthquake on the southern section, resulting in additional displacement on the Parkfield section. The evolution of the system is chaotic; its evolution is not predictable except in a statistical sense.

Slider-block models are a simple analogy to the behavior of faults in the Earth's crust. However, the chaotic behavior of low-dimensional analog systems often indicates that natural systems will also behave chaotically. Thus, it is reasonable to conclude that the interactions between faults that lead to fractal (power-law) frequency–magnitude distributions are an example of deterministic chaos. The prediction of earthquakes is not possible in a deterministic sense. Only a probabilistic approach to the occurrence of earthquakes will be possible.

3.2 Self-Organized Criticality

One explanation for fractal statistics is scale invariance. The power-law distribution is the only statistical distribution that does not require a characteristic length scale. Thus, natural phenomena that do not inherently have a natural length scale would be expected to obey power-law (fractal) statistics. However, there may be a more fundamental basis for the applicability of fractal statistics. In the past ten years, a variety of numerical models have been found to exhibit a universal behavior that has been called self-organized criticality. In self-organized criticality the "input" to a complex system is slow and steady; whereas, the output is a series of events or "avalanches" that follow power-law frequency–size statistics. Regional seismicity is often taken as a naturally occurring example of self-organized criticality. The input is the motion of the tectonic plates and the output is the earthquakes.

The concept of self-organized criticality was introduced by Bak *et al.* (1988) as an explanation for the behavior of the "sandpile" model. In this model, a square array of boxes is considered and at each time step a particle is dropped into a randomly selected box. When a box accumulates four particles, they are redistributed to the four adjacent boxes, or in the case of edge boxes they are lost from the grid. Because the redistributions involve only nearest-neighbor boxes, it is known as a cellular-automata model. Redistributions of particles can lead to further instabilities with, at each step, the possibility of avalanches of particles being lost from the grid. Each of the multiple redistributions during a time step contribute to the size of the model "avalanche".

This model was called a "sandpile" model because of the resemblance to an actual sandpile on a table. The randomly dropped particles in the model are analogous to the addition of particles to an actual sandpile. The model avalanches are analogous to sand avalanches down the sides of the sandpile.

In some cases the sand avalanches lead to the loss of particles off the table. Extensive numerical studies of the "sandpile" model were carried out by Kadanoff *et al.* (1989). They found the noncumulative frequency–area distribution of avalanches to have a power-law distribution with a slope near unity.

A second example of "self-organized criticality" is the behavior of large arrays of slider blocks. The two-block model was considered in the last section. The standard multiple slider-block model consists of a square array of slider blocks as illustrated in Figure 8. Each block with mass m is attached to the driver plate with a driver spring, spring constant k_p. Adjacent blocks are attached to each other with a connector spring, spring constant k_c. The constant-velocity driver plate is pulled forward with velocity v. A block remains stationary as long as the net force on the block is less than the static resisting force, F_s. Again, two parameters determine the behavior of the system, the ratio of static to dynamic friction, F_s/F_d, and the stiffness of the system, k_c/k_p.

Carlson and Langer (1989) considered long linear arrays of slider blocks with each block connected by springs to the two neighboring blocks and to a constant-velocity driver. They used a velocity weakening friction law and considered up to 400 blocks. Slip events involving large numbers of blocks were observed, the motions of all blocks involved in a slip event were coupled, and the applicable equations of motion had to be solved simultaneously. Although the system was completely deterministic, the behavior was apparently chaotic. Frequency–area statistics were obtained for slip events. The events fell into two groups: Smaller events obeyed a power-law (fractal) relationship with a slope near unity, but there were an anomalously large number of large events that included all the slider blocks. The observed behavior was characteristic of self-organized criticality. The motion of the driver plate is the steady input. The slip events are the "avalanches" with a power-law (fractal) frequency-size distribution.

Nakanishi (1990, 1991) studied multiple slider-block models using the cellular-automata approach. A linear array of slider blocks was considered but only one block was allowed to move in a slip event. The slip of one block could lead to the instability of either or both of the adjacent blocks, which would then be allowed to slip in a subsequent step or steps, until all blocks were again stable. Brown *et al.* (1991) proposed a modification of this model involving a 2D array of blocks. The use of cellular automata greatly reduces the complexity of the calculations and the results using the two approaches are generally very similar. A wide variety of slider-block models have been proposed and studied; these have been reviewed by Carlson *et al.* (1994) and Turcotte (1999a).

Huang *et al.* (1992) carried out a large number of simulations on a square array of blocks using stick-slip friction and the cellular-automata approach. Their noncumulative frequency–area statistics for model slip events are given in Figure 9. The number of slip events per time step N_e/N_T with area A_e is given as a function of A_e. The simulation results are well fit by the power-law scaling relation Eq. (2) with $D = 2.72$. For stiff systems (k_c/k_p large), the entire grid of slider blocks is strongly correlated and large slip events including all the blocks occur regularly. These are the peaks for $A_e = 400$, 900, and 1600 blocks, illustrated in Figure 9. For soft systems (k_c/k_p relatively small), no large events occur.

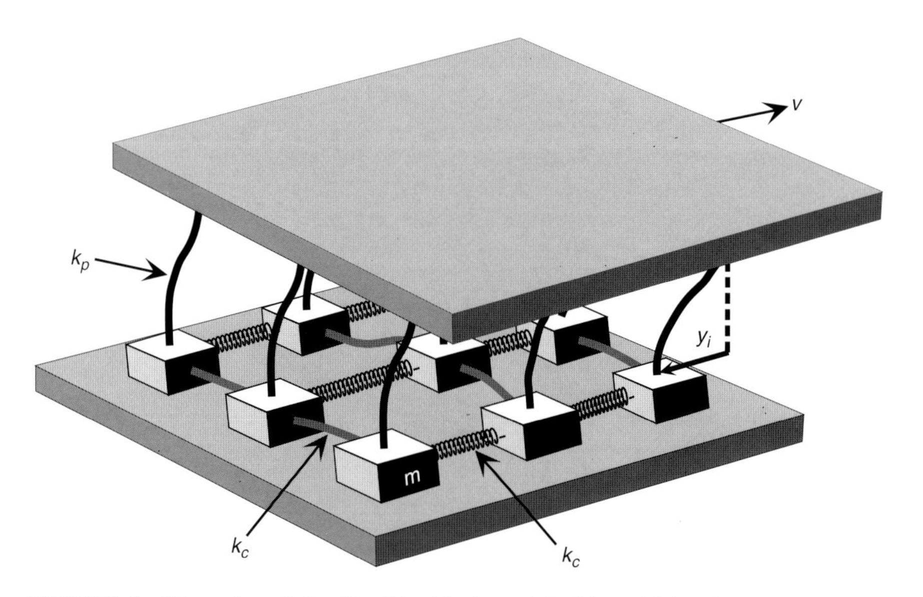

FIGURE 8 Illustration of the 2D slider-block model with multiple blocks. An array of blocks, each with mass m, is pulled across a surface by a driver plate at a constant velocity, v. Each block is coupled to adjacent blocks with either leaf or coil springs (spring constant k_c), and to the driver plate with leaf springs (spring constant k_p).

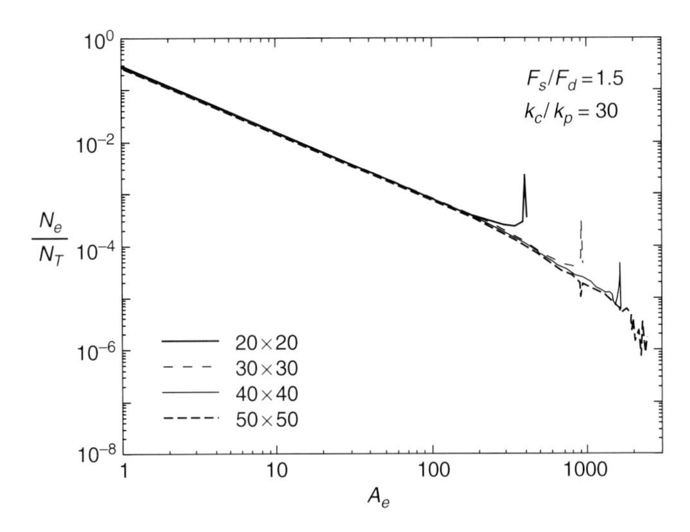

FIGURE 9 Results for a 2D slider-block model with multiple blocks (Huang *et al.*, 1992). The ratio of the number of slip events N_e with area A_e to the total number of time steps N_T is plotted against A_e, the number of blocks involved in an event. Results are for systems with stiffness $k_c/k_p = 30$, friction $F_s/F_d = 1.5$, and grid sizes 20 × 20, 30 × 30, 40 × 40, and 50 × 50 blocks. The peaks at $A_e = 400$, 900, and 1600 blocks correspond to catastrophic events involving the entire system.

3.3 Cascade Model

A simple inverse-cascade model can explain the self-organized critical behavior of slider-block arrays (Turcotte *et al.*, 1999; Turcotte, 1999b). A metastable cluster is the region over which a slip event propagates once it has been initiated. At any one time, there can be many metastable clusters on the grid. From Figure 9 the number of slip events N_e of size A_e can be related to the number of slip events that involve a single block N_{e0} by

$$N_e = \frac{N_{e0}}{A_e^{1.36}} \tag{5}$$

where A_e, the area, is the number of slider blocks participating in the slip event.

We assume that the probability that a slip event is triggered in a metastable cluster is proportional to the area of the cluster itself. Thus the number N_m of metastable clusters with area A_e on the grid at any one time is related to the number N_e of slip events over time with the same area A_e by

$$N_e \sim N_m A_e \tag{6}$$

It is also appropriate to assume that the number of metastable clusters satisfies the relation

$$N_m = \frac{N_{m0}}{A_e^{\gamma}} \tag{7}$$

where γ is a constant and N_{m0} is the number of metastable clusters with only one slider block. In order for Eqs. (5) and (7) to satisfy Eq. (6), the constant $\gamma = 2.36$. The cumulative number N_{Cm} of metastable clusters larger than A_e is given approximately by

$$N_{Cm} = N_{m0} \int_{A_e}^{\infty} \frac{dA}{A^{2.36}} = \frac{N_{m0}}{1.36 A_e^{1.36}} \tag{8}$$

and is inversely proportional to area to the 1.36 power.

The inverse-cascade model assumes that metastable clusters grow by combining. The process is referred to as an inverse cascade because it proceeds from the smallest to the largest scales. In the model it is assumed that the metastable clusters contain $A_n = 2^n$ blocks with $n = 0, 1, 2, 3, \ldots, N$ (i.e., $A_n = 1, 2, 4, 8, \ldots, 2^N$ blocks). This is equivalent to logarithmic binning. The number of clusters with A_n particles is denoted by N_n with $n = 0, 1, 2, 3, \ldots, N$. Differential equations are written for the set of N_n and take the form

$$\frac{dN_0}{dt} = C_0 - C_{0,1} N_0^2 - D_0 N_0 \tag{9}$$

$$\frac{dN_n}{dt} = \frac{1}{2} C_{n-1,n} N_{n-1}^2 - C_{n,n+1} N_n^2 - D_n N_n \tag{10}$$

The constant C_0 is the rate at which single blocks become metastable due to the driver plate motion. The net force on the block is greater than the dynamic friction. The transition probability $C_{0,1}$ is the rate at which single metastable blocks become pairs of metastable blocks at level $n = 1$. The constant D_0 is the rate at which clusters of one block are lost due to slip events. In the cascade model, it is assumed that metastable clusters of size 2^n are obtained by the merging of two clusters of size 2^{n-1}. The metastable clusters grow in size due to the motion of the driver plate. The rate at which metastable clusters of size 2^{n-1} merge to give metastable clusters of size 2^n is $\frac{1}{2}(C_{n-1,n} N_{n-1}^2)$. This is equivalent to a collision probability for metastable clusters; since each metastable cluster can merge with the N_{n-1} other metastable clusters, the probability is proportional to N_{n-1}^2. Similarly, metastable clusters of size 2^n are lost as they merge to form metastable clusters of size 2^{n+1}; the rate at which this occurs is $C_{n,n+1} N_n^2$. Metastable clusters of size 2^n are also lost in slip events at a rate $D_n N_n$; the loss rate is proportional to the number of metastable clusters of that size.

The cascade model is a renormalization group approximation in the sense that metastable clusters are binned into sizes 2^n. It is also necessary to specify how the transition probabilities $C_{n,n+1}$ and loss rate D_n depend on the metastable cluster area 2^n. For the transition probabilities we assume a power-law dependence on cluster area, but do not specify the exponent:

$$C_{n,n+1} = \frac{(2^n)^{\varepsilon} C_0}{N_R^2} \tag{11}$$

where ε and N_R are constants. We assume the loss rates are proportional to metastable cluster areas, so that

$$D_n = 2^n \beta \frac{C_0}{N_R} \quad (12)$$

where β is a constant.

We can absorb the constant C_0 into the definition of time with no loss of generality by introducing

$$\tau = C_0 t \quad (13)$$

Substitution of Eqs. (11) to (13) into (9) and (10) gives

$$\frac{dN_0}{d\tau} = 1 - \left(\frac{N_0}{N_R}\right)^2 - \beta\left(\frac{N_0}{N_R}\right) \quad (14)$$

$$\frac{dN_n}{d\tau} = \frac{1}{2}\left(2^{n-1}\right)^{\varepsilon}\left(\frac{N_{n-1}}{N_R}\right)^2 - \left(2^n\right)^{\varepsilon}\left(\frac{N_n}{N_R}\right)^2 - 2^n\beta\left(\frac{N_n}{N_R}\right) \quad (15)$$

The behavior of this inverse-cascade model is illustrated in Figure 10. The steady-state solution to this set of equations is obtained by setting $dN_n/d\tau = 0$, with the result

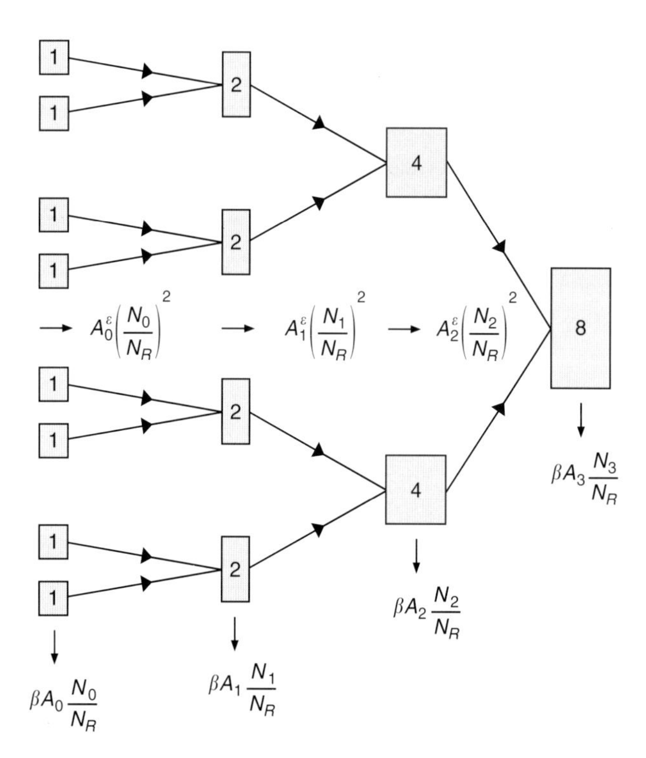

FIGURE 10 Schematic flow diagram for the inverse-cascade model. Clusters of size 2^0 combine to form clusters of size 2^1 with probability $\alpha A_0 N_0^2$ (where $A_0 = 2^0$) and clusters of size 2^0 are lost in slip events with a probability $\beta A_0 N_0$. Similarly, clusters of size 2^1 combine to form clusters of size 2^2 with probability $\alpha A_1 N_0^2$ (where $A_1 = 2^1$) and clusters of size 2^1 are lost in slip events with a probability $\beta A_1 N_0$.

$$\frac{N_0}{N_R} = \frac{1}{2}\left[-\beta + \left(\beta^2 + 4\right)^{1/2}\right] \quad (16)$$

$$\frac{N_n}{N_R} = \frac{1}{2^{n\varepsilon+2}}\left\{-2^{n+1}\beta + \left[2^{2n+2}\beta^2 + 2^{(2n-1)\varepsilon+3}\left(\frac{N_{n-1}}{N_R}\right)^2\right]^{1/2}\right\} \quad (17)$$

This system generates an inverse cascade if $\beta \ll 1$, and Eqs. (16) and (17) become

$$\frac{N_0}{N_R} = 1 \quad (18)$$

$$\frac{N_n}{N_R} = \frac{1}{2^{(\varepsilon+1)/2}}\left(\frac{N_{n-1}}{N_R}\right) \quad (19)$$

Using Eqs. (18) and (19) we can write

$$\frac{N_n}{N_R} = \frac{N_n}{N_0} = \left(\frac{N_1}{N_0}\right)\left(\frac{N_2}{N_1}\right)\left(\frac{N_3}{N_2}\right)\cdots\left(\frac{N_n}{N_{n-1}}\right) = \left(\frac{N_n}{N_{n-1}}\right)^n = \left(\frac{1}{2^{(\varepsilon+1)/2}}\right)^n = \left(\frac{1}{2^n}\right)^{(\varepsilon+1)/2} \quad (20)$$

and with $A_n = 2^n$ this result gives

$$\frac{N_n}{N_R} = \frac{1}{A_n^{(\varepsilon+1)/2}} \quad (21)$$

The logarithmic binning used in this model is equivalent to a cumulative distribution. Taking $\varepsilon = 1.72$, Eq. (21) is identical

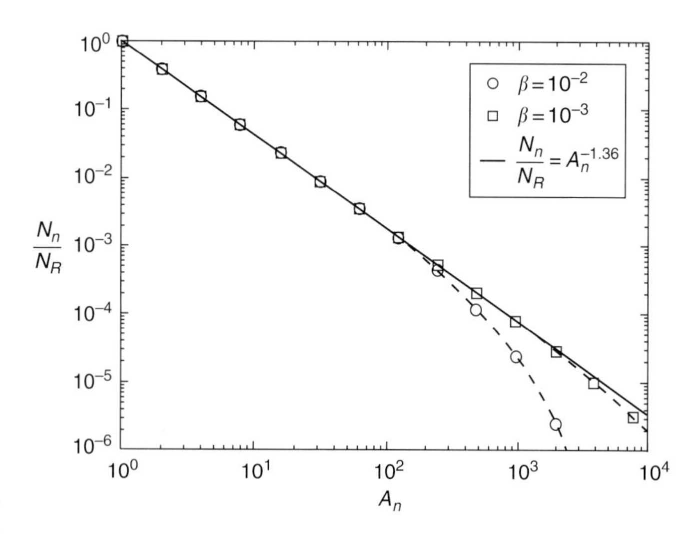

FIGURE 11 Frequency–area distribution for clusters obtained from the inverse-cascade model Eq. (17). The number of clusters N_n of size $A_n = 2^n$, divided by the reference number N_R, is given as a function of A_n for $\beta = 10^{-2}$ and 10^{-3}.

to those results obtained from the slider-block model, Eq. (8). Our simple inverse-cascade model reproduces the frequency–size distribution that is a characteristic of self-organized critical behavior.

As a specific example of the inverse-cascade model we take $\varepsilon = 1.72$ and in Figure 11 give the distribution of cluster sizes from Eq. (17) with $A_n = 1.72$ substituted. Distributions are given for $\beta = 10^{-2}$ and 10^{-3}. The parameter β plays the same role as the stiffness α in the results for the slider-block model given in Figure 9.

4. Earthquake Hazard Assessment

Presently, we cannot predict earthquakes in the way we predict volcanic eruptions. Seismic activation, surface uplift, and chemical emissions systematically precede major volcanic eruptions. No precursory phenomena have been recognized to occur systematically prior to earthquakes (Turcotte, 1991). However, we can provide rational assessments of the seismic hazard. To a first approximation, major earthquakes are restricted to the plate boundaries of plate tectonics. However, some of these boundaries are quite diffuse in continental areas (e.g., the western United States and much of China). Major earthquakes also occur within plate interiors (e.g., New Madrid, Missouri).

Seismic hazard maps have been prepared for many parts of the world. For the United States, the US Geological Survey's June 1996 maps (Frankel *et al.*, 1996) are available on the Internet. Seismic hazard maps generally give the probability of exceeding a specified peak ground acceleration during a specified time interval. The data used to formulate the maps include the historic seismicity discussed in this paper, maps of active faults, and levels of seismic attenuation. Techniques for seismic hazard assessment can be deterministic and probabilistic. Deterministic assessments are based on the previous occurrence of earthquakes of the magnitude to be considered. This occurrence can be based on recent instrumental records, older historic records, studies of the paleoseismic record using techniques such as trenching, and the presence of recently active "capable" faults. This approach follows from classical mechanics models for seismicity that leads to assumptions of quasi periodicity in earthquake occurrence.

Probabilistic hazard assessments are based on extrapolations of available data. The validity of the Gutenberg–Richter frequency–magnitude relation, Eq. (1), for regional seismicity has been recognized for some fifty years. The use of this relation to extrapolate the rate of occurrence of small earthquakes to larger earthquakes (Cornell, 1968) has been routinely incorporated into many regional seismic hazard assessments in a number of developed countries. Frankel (1995) proposed a systematic extrapolation technique for the Eastern United States. First, the region was divided into $11 \times 11\,\text{km}^2$ cells. Then, the number of earthquakes N_i in each cell i, with magnitudes greater than a given m was determined. The grid of N_i values was spatially smoothed using a Gaussian function with a correlation distance c. A map of smoothed 10^a values was prepared (Frankel, 1995, figure 4) taking $m = 3$ (1924–1991) and $c = 50\,\text{km}$. The optimal c-value of about 50 km was found through trial and error.

Turcotte (1999b) and Kossobokov *et al.* (2000) have proposed a similar technique for assessing the seismic hazard globally. Their extrapolation utilized the epicenters of earthquakes with magnitudes $m \geq 4$, where the data were obtained from the NEIC Global Hypocenter Database (GHDB, 1989) for the period 1964–1995. A magnitude 5 cut-off was also considered. The advantage of a magnitude 5 cut-off is that the database is globally complete for earthquakes $m \geq 5$ during this period (Engdahl *et al.*, 1998). The advantage of the magnitude 4 cut-off is that there are many more earthquakes on which to base statistics. Since the primary goal was to provide the best seismic hazard assessment in populous regions, and since these regions generally have seismic networks that provide a complete catalog for $m \geq 4$ during this period, this value was chosen for the cut-off.

The surface of the Earth was divided into $1^\circ \times 1^\circ$ regions and the number of earthquakes per year with magnitudes $m \geq 4$ in each region was determined. Each $1^\circ \times 1^\circ$ region varied in area with respect to changing latitude and they were therefore scaled by using the cosine of the latitude as a normalizing factor. The seismic intensity factor, I_4, is defined to be the (normalized) number of magnitude $m \geq 4$ earthquakes that have occurred in a given $1^\circ \times 1^\circ$ region per year. The minimum value of the seismic intensity factor considered is $I_4 = 1/32\,\text{y}^{-1}$, one magnitude $m \geq 4$ earthquake in the 32-year period considered. The maximum value of the seismic intensity factor is about $I_4 = 40\,\text{y}^{-1}$ (40 magnitude $m \geq 4$ earthquakes per year).

A global map of the seismic intensity factor is given in Color plate 5(a). The boundaries of plate tectonics are clearly defined. Seismicity is particularly intense in subduction zones (i.e., the ring of fire around the Pacific) as expected. A broadband of seismicity extends from southern Europe to southeast Asia; this is associated with the continent–continent collision zone between the Eurasian plate and the African, Arabian, and Indian plates.

In Figure 12, we give two regional maps of the seismic intensity factor, I_4. For the United States (Fig. 12a) the intense seismicity along the Pacific–North American plate boundary, including the San Andreas Fault, is clearly illustrated. The distributed seismicity associated with the intraplate deformation in the western United States (the Basin and Range province) is also shown. The most intense seismicity in the eastern United States is the Memphis–St. Louis seismic zone (36° N, 90° W). For Europe (Fig. 12b), the Aegean region has particularly intense seismicity, but high levels of seismicity extend throughout the Mediterranean area.

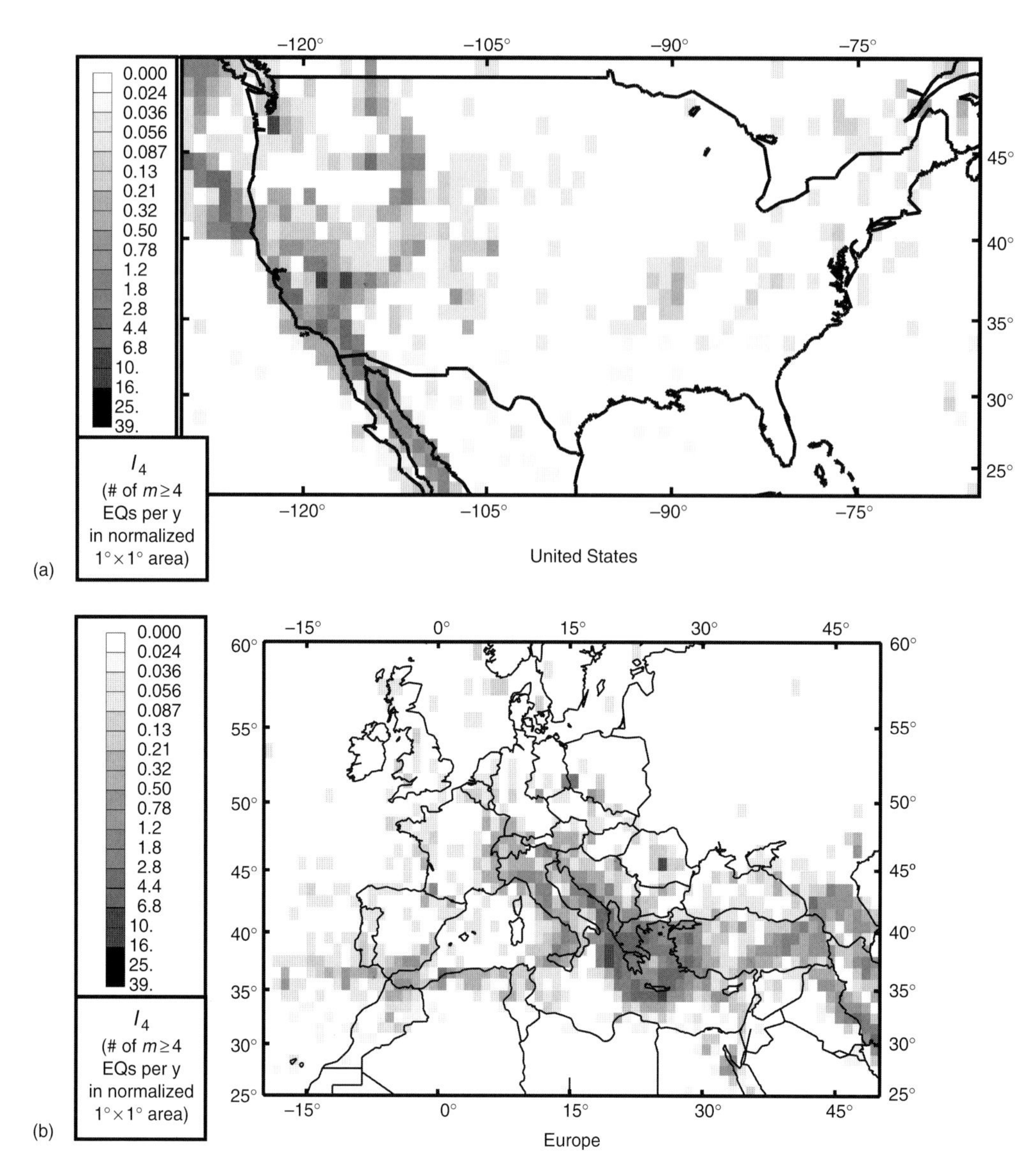

FIGURE 12 Regional maps of the seismic intensity factor, I_4, the average annual number of earthquakes per year during 1964–1995 with magnitudes $m \geq 4$ in each normalized $1° \times 1°$ cell (Kossobokov *et al.*, 2000) for (a) United States and (b) Europe. Data from the NEIC Global Hypocenter Database (GHDB, 1989).

The basis for using the seismic intensity factor I_4 is illustrated in Figure 13. For five values of I_4, the cumulative number of earthquakes per year, $\dot{N}_{CE}$, in a $1° \times 1°$ area, with magnitudes $m \geq 4$ are plotted as a function of m. The different lines are derived using the Gutenberg–Richter relation Eq. (1) with $b = 0.9$, and calculating $\dot{a}$ from $\log \dot{a} = 3.6 + \log I_4$. Given a region with $I_4 = 1\,\mathrm{y}^{-1}$, an earthquake with $m \geq 6$ has a return period of 63 y ($\dot{N}_{CE} = 0.016$) and one with $m \geq 8$ has a return period of 4000 y ($\dot{N}_{CE} = 0.00025$).

It can be argued that the use of earthquakes with $m \geq 4$ for such a short period (37 y) would not represent the long-term seismic hazard. To address this concern, we have used the NEIC Global Hypocenter Database (GHDB, 1989) to determine the largest earthquake occurring in each $1° \times 1°$ area for the period 1900–1997. These data are given in Color plate 5(b). Comparing parts (a) and (b) in Color plate 5, it is seen that there is a strong correlation between the value of the earthquake intensity factor, I_4, and the largest earthquake to have occurred in each $1° \times 1°$ area.

The primary advantage of this approach is that it is totally based on a generally accepted data set. There are no ambiguities with regard to the technique. This is not the case for

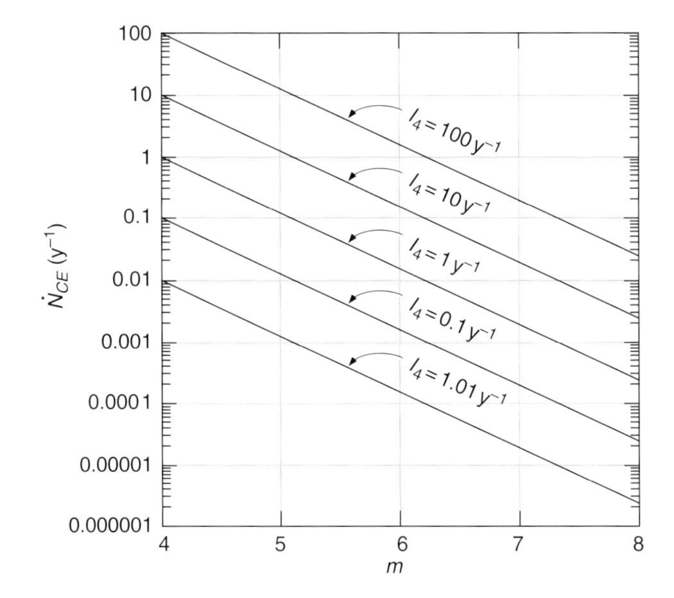

FIGURE 13 Similarity basis for extending the seismic intensity factor I_4 to higher earthquake magnitudes. For $I_4 = 0.01$, 0.1, 1, 10, and 100 y^{-1}, the cumulative number of earthquakes per year in a $1° \times 1°$ area, $\dot{N}_{CE}$, with magnitudes greater than m are plotted as a function of m. Each line is derived by using the Gutenberg–Richter relation, Eq. (1), with $b = 0.9$, and calculating the constant $\dot{a}$ from $\log \dot{a} = 3.6 + \log I_4$.

other approaches to assessing the seismic hazard, which combine geophysical and geological observations in arbitrary ways. Different studies give different weights to historical and paleoseismic data, and to the presence of "active" faults. Considering the many uncertainties regarding fault depth, seismic attenuation, available database, and the occurrence of an earthquake on a particular fault, this approach, which is based on the universal applicability of Gutenberg–Richter scaling, provides a reasonable basis for assessing the seismic hazard.

5. Intermediate-Term Earthquake Prediction

5.1 Pattern Recognition

Based on pattern recognition algorithms, a number of intermediate-range earthquake prediction algorithms were developed at the International Institute of Earthquake Prediction Theory and Mathematical Geophysics in Moscow (Keilis-Borok, 1990; Keilis-Borok and Kossobokov, 1990; Keilis-Borok and Rotwain, 1990; Kossobokov *et al.*, 1997). The pattern recognition includes quiescence (Schreider, 1990), increases in the clustering of events, and changes in aftershock statistics (Molchan *et al.*, 1990). The first algorithm, *M8*, was developed to make intermediate predictions of the largest earthquakes ($m > 8$). This method utilizes overlapping circles of seismicity with diameters of 384, 560, 854, and 1333 km for earthquakes, with magnitudes 6.5, 7.0, 7.5, and 8.0, respectively. Within each circle four quantities are determined. The first three are measures of intermediate levels of seismicity and the fourth is a measure of aftershock activity.

The first quantity that must be specified is the lower magnitude cut-off $m_{\min}$ for earthquakes to be considered in the circle. Two magnitude cut-offs are considered for each circular region. The long-term number of earthquakes per year N in the circle with magnitudes greater than $m_{\min}$ is determined; $m_{\min 10}$ corresponds to $N = 10$ and $m_{\min 20}$ corresponds to $N = 20$. The first quantity $N_1(t)$ is the number of earthquakes per year with magnitudes greater than $m_{\min 20}$. The second quantity is the 5-year trend in activity $L_1 = dN_1/dt$ and $L_2 = dN_2/dt$ for running 5-year windows. Clearly $N(t)$ and $L(t)$ are strongly correlated. The third quantity, $Z_1(t)$ or $Z_2(t)$, is the ratio of the average linear dimension of rupture to average separation between earthquakes for a year in a circle. The final measure is the number of aftershocks in a specified magnitude range and time window following a main shock. This quantity $B(t)$ is a measure of aftershock activation.

Since N, L, and Z are determined for both $N = 10$ and 20 y^{-1}, there are seven time-series to be considered. An earthquake alarm or time of increased probability (TIP) is issued if six of the seven quantities, including B, exceed their average values by a specified value of 75% for B and 90% for the others. In order to trigger an alarm these conditions must be satisfied for two successive time periods and the alarm lasts for 5 y. Details of this algorithm have been given by Keilis-Borok (1996). Two examples of the application of this algorithm are given in Figure 14 for the 1989 Loma Prieta and the 1992 Landers earthquakes.

The earthquake catalog used in the *M8* studies is taken from the NEIC Global Hypocenter Database (GHDB, 1989), complete for $m \geq 5$ since 1963. Since the application of the *M8* algorithm requires at least 12 y of data to establish stable mean values, it can be applied continuously since 1975 and is continuously updated. The second algorithm is the *MSc* or Mendocino Scenario. This algorithm was designed to localize predictions within the relatively large circles used in the *M8* algorithm. It uses similar approaches to the *M8* but with smaller magnitude cut-offs. Details of this algorithm have been given by Keilis-Borok (1996).

The algorithms described above were developed using pattern recognition approaches. Although they have had demonstrated predictive successes (Kossobokov *et al.*, 1999; Rotwain and Novikova, 1999), their use remains quite controversial. The main difficulty is that, although success-to-failure ratios of predictions is quite high, the time and spatial windows of alarms are also quite high.

5.2 Seismic Activation

A second approach to intermediate-range earthquake forecasting is to consider whether a power-law increase in seismicity

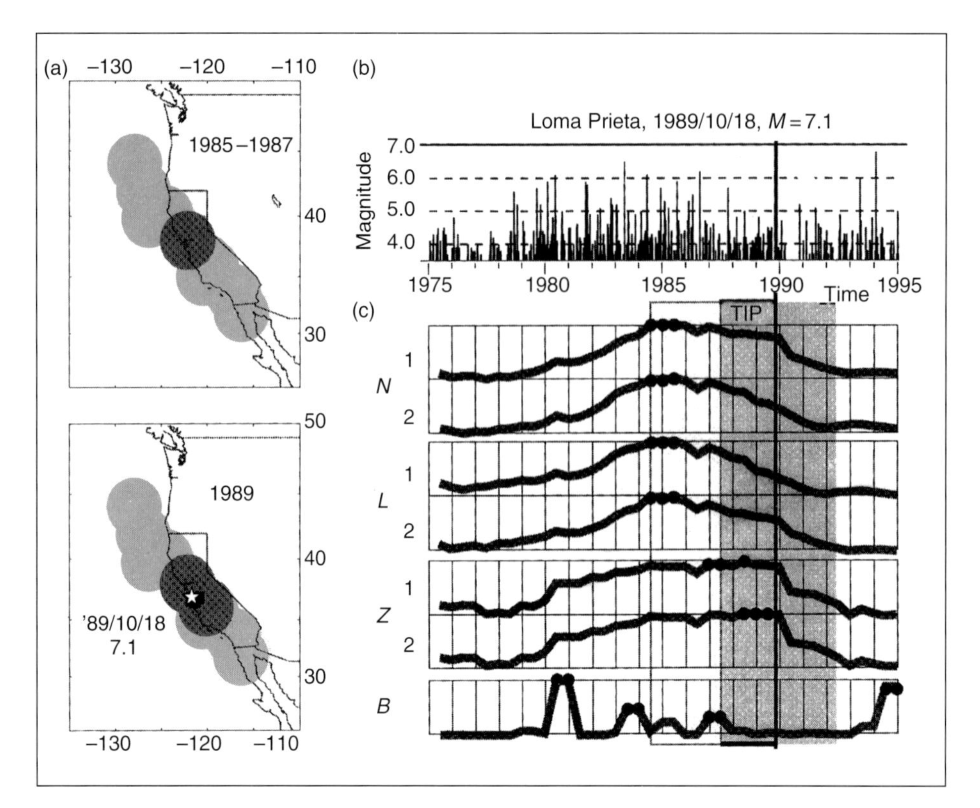

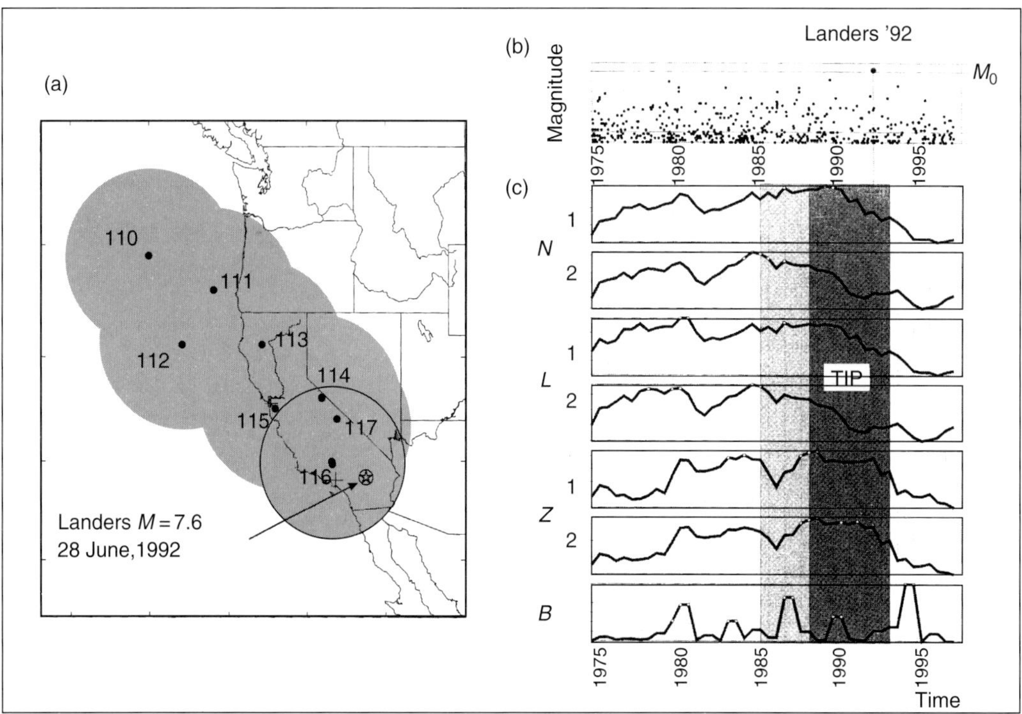

FIGURE 14 Applications of the *M8* algorithm to the $m = 7.1$ Loma Prieta earthquake (upper box) and the $m = 7.6$ Landers earthquake (lower box). In (a) the circular regions where TIPs have been declared are shown; in (b) the seismicity in the relevant region is shown and in (c) the values of the seven parameters N_1, N_2, L_1, L_2, Z_1, Z_2 and B are shown along with the time interval of the TIP.

occurs prior to a major earthquake. We will refer to this as the seismic activation algorithm. This was first proposed by Bufe and Varnes (1993). They considered the cumulative amount of Benioff strain (square root of seismic energy) in a specified region. They showed that an accurate retrospective prediction of the Loma Prieta earthquake could be made assuming a power-law increase in Benioff strain prior to the earthquake.

Systematic increases in intermediate-level seismicity prior to a large earthquake have been proposed by several authors (Sykes and Jaumé, 1990; Knopoff *et al.*, 1996; Brehm and Braile, 1998, 1999; Jaumé and Sykes, 1999). A systematic study of the optimal spatial region and magnitude range to obtain the power-law seismic activation has been carried out by Bowman *et al.* (1998). Four examples of their results are given in Figure 15. Clear increases in seismic activity prior to the 1952 Kern County, 1983 Coalinga, 1989 Loma Prieta, and 1992 Landers earthquakes are illustrated. In each case data for the cumulative Benioff strain ε are compared with the empirical relation (solid line),

$$\varepsilon[t] = \varepsilon_0 - B(t_0 - t)^s \tag{22}$$

where t is the time measured forward from the previous characteristic earthquake, t_0 is the time interval between characteristic earthquakes, ε_0 is the cumulative Benioff strain when the characteristic earthquake occurs, and B and s are positive constants used to fit the data. The comparison with the Kern County data is made with $s = 0.30$, Coalinga $s = 0.18$, Loma Prieta $s = 0.28$, and Landers $s = 0.18$. For the 12 earthquakes studied by Bowman *et al.* (1998), it was found that $s = 0.26 \pm 0.15$.

There are clearly very strong similarities between the *M8* and the seismic activation algorithms. Consider the Loma Prieta earthquake: the intermediate-sized events that led to the *M8* TIP (Fig. 14) were the same events that give the increase in Benioff strain (Fig. 15). Both these approaches are based on the concept that correlation lengths increase before major earthquakes (Harris, 1998).

Bowman *et al.* (1998) also found ξ, the optimal radius (the correlation length) for the precursory activation. This optimal radius is given as a function of earthquake magnitude in Figure 16. The dependence on the square root of the rupture area is also shown. The radius over which activation occurs is about

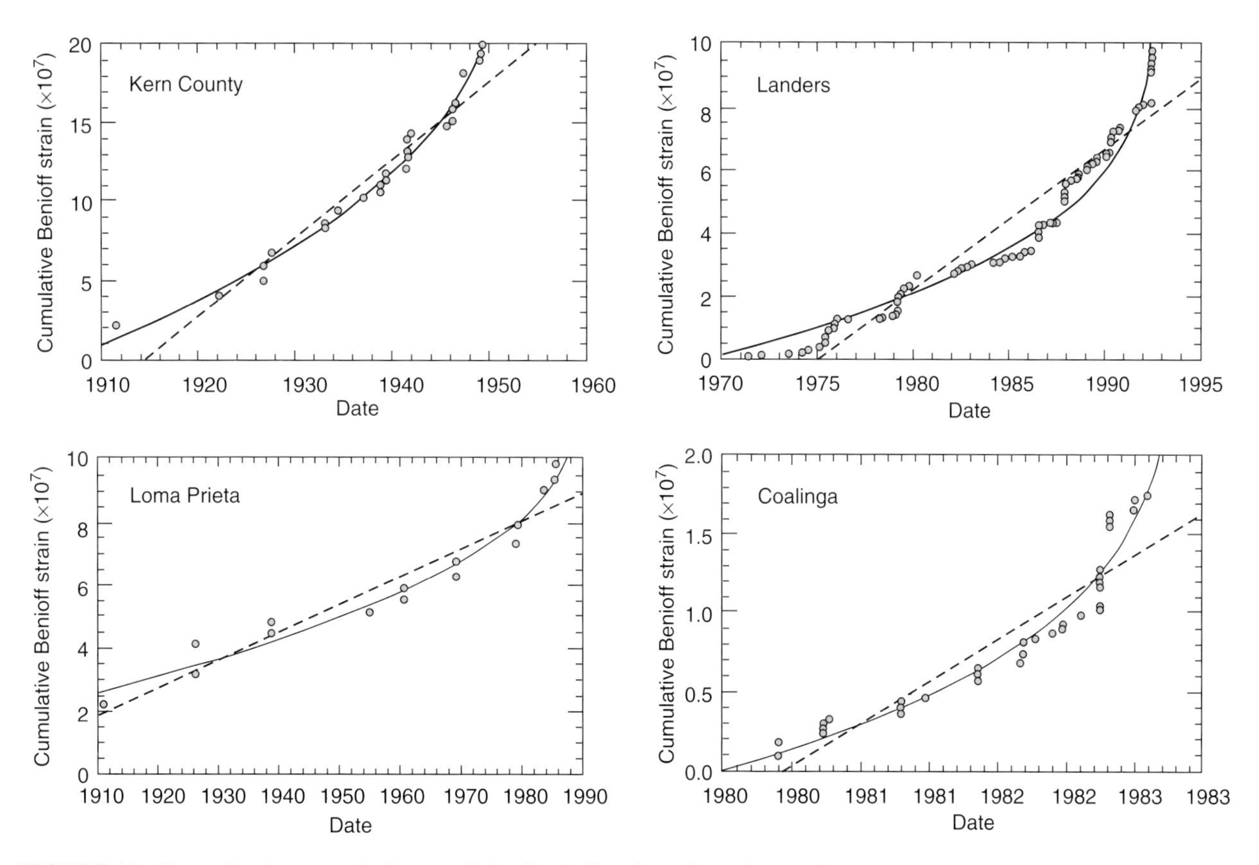

FIGURE 15 Power-law increases in the cumulative Benioff strains prior to four major earthquakes in California (Bowman *et al.*, 1998). Data points are cumulative Benioff strains $\varepsilon[t]$ prior to each earthquake. Clear increases in seismic activity prior to the Kern County (21 July 1952), Coalinga (2 May 1983), Loma Prieta (18 October 1989), and Landers (28 June 1992) earthquakes are illustrated. In each of the four examples, the data have been correlated (solid lines) with the power-law relation given in Eq. (22). Dashed straight lines represent a best-fit constant rate of seismicity.

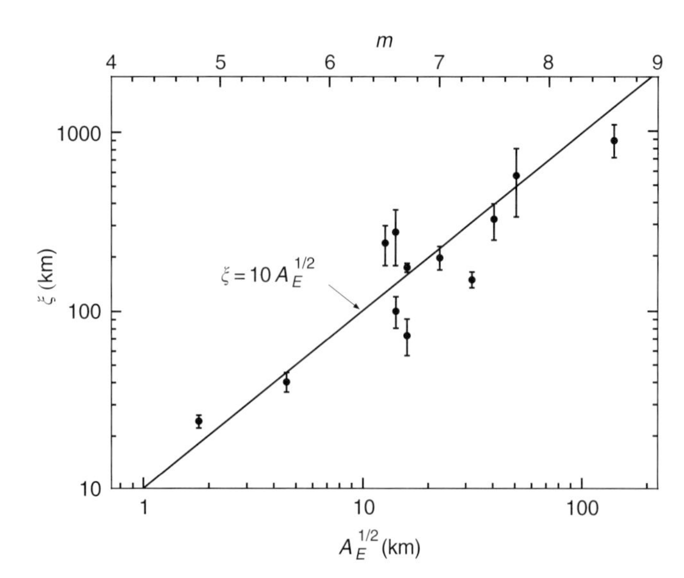

FIGURE 16 Optimum radius (correlation length) ξ for precursory seismic activation is given as a function of the square root of the rupture area $A_E^{1/2}$ and the magnitude m for 12 major earthquakes (Bowman *et al.*, 1998).

ten times the length of rupture $\xi \approx 10A_E^{1/2}$. Dobrovolsky *et al.* (1979) and Keilis-Borok and Kossobokov (1990) reported a similar scaling for the maximum distance between an earthquake and its precursors using pattern recognition techniques.

The observations of seismic activation given above are consistent with results obtained from studies of cellular-automata slider-block models using a mean-field approach (Rundle *et al.*, 1997, 1999, 2000). This approach involved concepts applied to equilibrium thermodynamics and the approach to a second-order phase transition through a spinoidal.

In the scaling associated with the approach to the critical point, the correlation length scales with the "rupture" length (square root of rupture area),

$$\xi \sim \langle A \rangle^{1/2} \tag{23}$$

as shown in Figure 16. In the vicinity of a critical point, it is appropriate to take (Stauffer and Aharony, 1992)

$$\xi \sim (t_0 - t)^{-1/2} \tag{24}$$

Combining Eqs. (23) and (24) we have

$$\langle A \rangle \sim (t_0 - t)^{-1} \tag{25}$$

We now use the results given above to obtain the power-law increase in Benioff strain prior to a characteristic earthquake. The mean seismic moment of the precursory earthquakes $\langle M \rangle$ is related to the mean of the slips $\langle \delta_s \rangle$ and the mean of the rupture areas $\langle A \rangle$ by

$$\langle M \rangle \sim \langle \delta_s \rangle \langle A \rangle \tag{26}$$

Under a wide variety of conditions it is found that the earthquake slip scales with the square root of the rupture area (Kanamori and Anderson, 1975),

$$\langle \delta_s \rangle \sim \langle A \rangle^{1/2} \tag{27}$$

so that we have

$$\langle M \rangle \sim \langle A \rangle^{3/2} \tag{28}$$

The rate at which precursory earthquakes generate moment, dM/dt, is given by

$$\frac{dM}{dt} = \langle M \rangle \frac{dN}{dt} \tag{29}$$

where dN/dt is the rate of occurrence of earthquakes. We assume that

$$\frac{dN}{dt} = \text{constant} \tag{30}$$

The rate at which precursory earthquakes occur is independent of time. This is consistent with observations of earthquakes. If there were a systematic increase in the rate of occurrence of precursory earthquakes prior to a main event, they could be used for earthquake prediction.

Substitution of Eqs. (25), (28), and (30) into Eq. (29) gives

$$\frac{dM}{dt} \sim A^{3/2} \tag{31}$$

Since the rate of change of Benioff strain is related to the rate of change of earthquake moment by

$$\frac{d\varepsilon}{dt} \sim \frac{dM^{1/2}}{dt} \sim \frac{1}{\langle M \rangle^{1/2}} \frac{dM}{dt} \tag{32}$$

substitution of Eqs. (25), (28), and (31) into Eq. (32) gives

$$\frac{d\varepsilon}{dt} \sim \langle A \rangle^{3/4} \sim \frac{1}{(t_0 - t)^{3/4}} \tag{33}$$

The cumulative Benioff strain as a function of time is given by

$$\begin{aligned} \varepsilon(t) &= \varepsilon_0 - \int_{\varepsilon}^{\varepsilon_0} d\varepsilon = \varepsilon_0 - \int_t^{t_0} \frac{d\varepsilon}{dt}\, dt \\ &= \varepsilon_0 - C \int_t^{t_0} \frac{dt}{(t_0 - t)^{3/4}} = \varepsilon_0 - B(t_0 - t)^{1/4} \end{aligned} \tag{34}$$

where C and B are constants. Comparing Eqs. (34) and (22), we see that the two equations are identical if we take $s = 0.25$. For the 12 earthquakes studied by Bowman *et al.* (1998), $\bar{s} = 0.26 \pm 0.15$. Thus excellent agreement is found between the critical point theory and actual earthquake data on seismic activation.

5.3 Log-Periodic Behavior

It has also been suggested that precursory seismic activity exhibits log-periodic behavior. Sornette and Sammis (1995) and Saleur *et al.* (1996) have used a generalized log-periodic relation to fit the strain accumulation data of Bufe and Varnes (1993) for the Loma Prieta earthquake. The data appear to exhibit a series of log-periodic fluctuations. An important question is whether this type of strain accumulation data can be used to predict earthquakes. Sornette and Sammis (1995) used the log-periodic fit to the data available prior to a cut-off date to predict when an earthquake would be expected. The predictions became increasingly accurate as the cut-off times approached the date of the Loma Prieta earthquake, 17 October 1989. Although the ability to make this retrospective prediction is encouraging, it remains to be demonstrated that this technique can be used successfully to predict earthquakes. A general discussion of log-periodicity has been given by Sornette (1998).

Another intriguing example of log-periodic behavior prior to an earthquake has been given by Johansen *et al.* (1996). These authors considered chlorine ion concentrations in bottled water taken from two 100 m deep wells located 20 km from the epicenter of the $m = 6.9$ Kobe (Japan) earthquake of 17 January 1995, prior to the earthquake. Again, a log-periodic fit to the data was found.

6. Implications

It is instructive to make comparisons between the behavior of the Earth's atmosphere and the behavior of the Earth's crust. Both are extremely complex on a very wide range of scales. From a practical point of view it is impossible, in either case, to make a sufficient number of measurements to fully specify the problem. Both have significant stochastic components. Both are undoubtedly chaotic. But what are the relative roles of stochastic versus chaotic behavior?

Massive numerical simulations are routinely used to forecast the weather. In many cases, they are quite accurate on time scales of 24–48 h, but on the scale of weeks they are of little value. The motions of the storm systems are relatively stable considering the complexity involved. In many cases the paths of major storms such as hurricanes can be predicted with considerable accuracy, but in other cases there are major uncertainties. One approach to establishing whether a particular predicted storm path is stable is to apply alternative numerical models. If the models all predict essentially the same path, then the path is taken to be stable. If the models predict radically different paths, then any forecast is considered to be suspect. This is also the way classical chaotic systems such as the Lorenz attractor behave. At many points on the evolving trajectory there is relatively little sensitivity to small perturbations, and at other points there is extreme sensitivity.

But what about the Earth's crust? Forecasting or predicting an earthquake is quite different from forecasting the path and intensity of a hurricane. The hurricane exists, but the earthquake does not exist until it happens. Geller *et al.* (1997) have argued, on the basis of the chaotic behavior of the Earth's crust, that "earthquakes cannot be predicted." This is certainly true in the sense that the exact time of occurrence of an earthquake cannot be predicted. But probabilistic forecasts of hurricane paths with a most probable path are routinely made and their use is of great value in terms of requiring evacuations and other precautions.

The essential question concerning earthquakes is whether similar, useful, probabilistic forecasts can be made. In fact, this is already being done in terms of hazard assessments. Certainly earthquakes do not occur randomly on the surface of the Earth. Also, the occurrence of large earthquakes can be associated with the occurrence of small earthquakes. In Section 4, we discussed a systematic approach to hazard assessment based on concepts of complexity.

In Section 5 we addressed the more intriguing question, whether useful forecasts of the temporal occurrence of earthquakes can be made, and in particular whether useful forecasts can be made based on concepts of complexity. In our opinion no definitive answer can be made at this time. We discussed two approaches to intermediate-range (month to several years) forecasting. There were a number of positive aspects to these approaches, but there is certainly no general consensus that useful intermediate-range forecasts can be made.

References

Aki, K. (1981). A probabilistic synthesis of precursory phenomena. In: "Earthquake Prediction" (D.W. Simpson and P.G. Richards, Eds.), pp. 566–574. American Geophysical Union, Washington DC.

Aki, K. (1987). Magnitude-frequency relation for small earthquakes: A clue to the origin of f_{max} of large earthquakes. *J. Geophys. Res.* **92**, 1349–1355.

Bak, P., C.J. Tang, and K. Wiesenfeld (1988). Self-organized criticality. *Phys. Rev.* **A38**, 364–374.

Bodri, B. (1993). A fractal model for seismicity at Izu-Tokai region, central Japan. *Fractals* **1**, 539–546.

Bowman, D.D., G. Ouillon, C.G. Sammis, A. Sornette, and D. Sornette (1998). An observational test of the critical earthquake concept. *J. Geophys. Res.* **103**, 24359–24372.

Brehm, D.J. and L.W. Braile (1998). Intermediate-term earthquake prediction using precursory events in the New Madrid seismic zone. *Bull. Seismol. Soc. Am.* **88**, 564–580.

Brehm, D.J. and L.W. Braile (1999). Intermediate-term earthquake prediction using the modified time-to-failure method in Southern California. *Bull. Seismol. Soc. Am.* **89**, 275–293.

Brown, S.R, C.H. Scholz, and J.B. Rundle (1991). A simplified spring-block model of earthquakes. *Geophys. Res. Let.* **18**, 215–218.

Bufe, C.G. and D.J. Varnes (1993). Predictive modeling of the seismic cycle of the greater San Francisco Bay region. *J. Geophys. Res.* **98**, 9871–9883.

Burridge, R. and L. Knopoff (1967). Model and theoretical seismicity. *Bull. Seismol. Soc. Am.* **57**, 341–371.

Camelbeeck, T. and K. Meghraoui (1996). Large earthquakes in northern Europe more likely than once thought. *EOS* **77**, 405.

Carlson, J.M. and J.S. Langer (1989). Mechanical model of an earthquake fault. *Phys. Rev.* **A40**, 6470–6484.

Carlson, J.M., J.S. Langer, and B.E. Shaw (1994). Dynamics of earthquake faults. *Rev. Mod. Phys.* **66**, 657–670.

Cornell, A.C. (1968). Engineering seismic risk analysis. *Bull. Seismol. Soc. Am.* **58**, 1583–1606.

Davison, F. and C.H. Scholz (1985). Frequency-moment distribution of earthquakes in the Aleutian Arc: A test of the characteristic earthquake model. *Bull. Seismol. Soc. Am.* **75**, 1349–1362.

Dobrovolsky, I.R., S.I. Zubkov, and V.I. Miachkin (1979). Estimation of the size of earthquake preparation zones. *Pure Appl. Geophys.* **117**, 1025–1044.

Engdahl, E.R., R. van der Hilst, and R. Buland (1998). Global teleseismic earthquake relocation with improved travel times and procedures for depth determination. *Bull. Seismol. Soc. Am.* **88**, 722–743.

Frankel, A.F. (1995). Mapping seismic hazard in the central and eastern United States. *Seismol. Res. Let.* **60**(4), 8–21.

Frankel, A.F., C. Mueller, T. Barnhard, *et al.* (1996). National Seismic Hazard Maps. *USGS Open-File Report 96–532.*

Frohlich, C. and S.D. Davis (1993). Teleseismic *b* values; or, much ado about 1.0. *J. Geophys. Res.* **98**, 631–644.

Geller, R.J., D.D. Jackson, Y.Y. Kagan, and F. Mulargia (1997). Earthquakes cannot be predicted. *Science* **275**, 1616–1617.

GHDB (1989). Global Hypocenter Database CD ROM and its updates through 1997. NEIC/USGS, Denver, CO.

Gudmundson, A. (1987). Geometry, formation and development of tectonic fractures on the Reykjanes Peninsula, southwest Iceland. *Tectonophysics.* **139**, 295–308.

Gutenberg, B. and C.F. Richter (1954). "Seismicity of the Earth and Associated Phenomenon," 2nd ed. Princeton University Press, Princeton, NJ.

Harris, R.A. (1998). Forecasts of the 1989 Loma Prieta, California, earthquake. *Bull. Seismol. Soc. Am.* **88**, 898–916.

Harvard Centroid-Moment Tensor Database (1997). Electronic Data. Department of Earth and Planetary Sciences, Harvard University, Cambridge, MA.

Hirata, T. (1989). Fractal dimension of fault systems in Japan: Fractal structure in rock fracture geometry at various scales. *Pure Appl. Geophys.* **131**, 157–170.

Huang, J. and D.L. Turcotte (1990a). Are earthquakes an example of deterministic chaos? *Geophys. Res. Lett.* **17**, 223–226.

Huang, J. and D.L. Turcotte (1990b). Evidence for chaotic fault interaction in the seismicity of the San Andreas Fault and Nankai trough. *Nature* **238**, 234–236.

Huang, J., G. Narkounskaia, and D.L. Turcotte (1992). A cellular-automata, slider-block model for earthquakes II. Demonstration of self-organized criticality for a 2-D system. *Geophys. J. Int.* **111**, 259–269.

Jaumé, S.H. and L.R. Sykes (1999). Evolving toward a critical point: a review of accelerating seismic moment/energy release prior to large and great earthquakes. *Pure Appl. Geophys.* **155**, 279–306.

Johansen, A., D. Sornette, H. Wakita, U. Tsunogai, W.I. Newman, and H. Saleur (1996). Discrete scaling in earthquake precursory phenomena: evidence in the Kobe earthquake, Japan. *J. Phys. I. France* **6**, 1391–1402.

Johnston, A.C. and S.J. Nava (1985). Recurrence rates and probability estimates for the New Madrid seismic zone. *J. Geophys. Res.* **90**, 6737–6753.

Johnston, A.C. and E.S. Schweig (1996). The enigma of the New Madrid earthquakes of 1811–1812. *Annu. Rev. Earth Planet. Sci.* **24**, 339–384.

Kadanoff, L.P., S.R. Nagel, L. Wu, and S.M. Zhou (1989). Scaling and universality in avalanches. *Phys. Rev.* **A39**, 6524–6533.

Kanamori, H. and D.L. Anderson (1975). Theoretical basis of some empirical relations in seismology. *Bull. Seismol. Soc. Am.* **65**, 1073–1096.

Keilis-Borok, V.I. (1990). The lithosphere of the Earth as a nonlinear system with implications for earthquake prediction. *Rev. Geophys.* **28**, 19–34.

Keilis-Borok, V.I. (1996). Intermediate-term earthquake prediction. *Proc. Natl. Acad. Sci. USA* **93**, 3748–3755.

Keilis-Borok, V.I. and V.G. Kossobokov (1990). Premonitory activation of earthquake flow. Algorithm *M8. Phys. Earth Planet. Inter.* **61**, 73–83.

Keilis-Borok, V.I. and I.M. Rotwain (1990). Diagnosis of time of increased probability of strong earthquakes in different regions of the world: algorithm CN. *Phys. Earth Planet. Inter.* **61**, 57–72.

Knopoff, L., T. Levshina, V.I. Keilis-Borok, and C. Mattoni (1996). Increased long-range intermediate-magnitude earthquake activity prior to strong earthquakes in California. *J. Geophys. Res.* **101**, 5779–5796.

Kossobokov, V.G., J.H. Healy, and J.W. Dewey (1997). Testing an earthquake prediction algorithm. *Pure Appl. Geophys.* **149**, 219–232.

Kossobokov, V.G., L.L. Romashkova, V.I. Keilis-Borok, and J.H. Healy (1999). Testing earthquake prediction algorithms: statistically significant advance prediction of the largest earthquakes in the Circum-Pacific, 1992–1997. *Phys. Earth Planet Inter.* **111**, 187–196.

Kossobokov, V.G., V.I. Keilis-Borok, D.L. Turcotte, and B.D. Malamud (2000). Implications of a statistical physics approach for earthquake hazard assessment and forecasting. *Pure Appl. Geophys.* **157**, 2323–2349.

Lay, T. and T.C. Wallace (1995). "Modern Global Seismology." Academic Press, San Diego.

Lorenz, E.N. (1963). Deterministic nonperiodic flow. *J. Atmos. Sci.* **20**, 130–141.

Main, I., S. Peacock and P.G. Meredith (1990). Scattering attenuation and the fractal geometry of fracture system. *Pure Appl. Geophys.* **133**, 283–304.

Mandelbrot, B.B. (1982). "The Fractal Geometry of Nature." W. H. Freeman, New York.

McCloskey, J. and C.J. Bean (1992). Time and magnitude predictions in shocks due to chaotic fault interactions. *Geophys. Res. Lett.* **19**, 119–122.

Molchan, G.M., O.E. Dmitrieva, I.M. Rotwain, and J. Dewey (1990). Statistical analysis of the results of earthquake prediction, based on bursts of aftershocks. *Phys. Earth Planet. Inter.* **61**, 128–139.

Nakanishi, H. (1990). Cellular-automaton model of earthquakes with deterministic dynamics. *Phys. Rev.* **A41**, 7086–7089.

Nakanishi, H. (1991). Statistical properties of the cellular automata model for earthquakes. *Phys. Rev.* **A43**, 6613–6631.

Narkounskaia, G. and D.L. Turcotte (1992). A cellular-automata, slider block model for earthquakes, I. Demonstration of chaotic behavior for a low order system. *Geophys. J. Int.* **111**, 250–258.

Narkounskaia, G., J. Huang, and D.L. Turcotte (1992). Chaotic and self-organized critical behavior of a generalized slider-block model. *J. Stat. Phys.* **67**, 1151–1183.

Otsuka, M. (1972). A simulation of earthquake occurrence. *Phys. Earth Planet. Inter.* **6**, 311–315.

Pacheco, J., C.H. Scholz, and L.R. Sykes (1992). Changes in frequency-size relationship from small to large earthquakes. *Nature* **355**, 71–73.

Robertson, M.C., C.G. Sammis, M. Sahimi, and A.J. Martin (1995). Fractal analysis of 3D spatial distributions of earthquakes with a percolation interpretation. *J. Geophys. Res.* **100**, 609–620.

Rotwain, I. and O. Novikova (1999). Performance of the Earthquake prediction algorithm CN in 22 regions of the world. *Phys. Earth Planet. Inter.* **111**, 207–213.

Rundle, J.B., S. Gross, W. Klein, C. Ferguson, and D.L. Turcotte (1997). The statistical mechanics of earthquakes. *Tectonophysics* **277**, 147–164.

Rundle, J.B., W. Klein, and S. Gross (1999). Physical basis for statistical patterns in complex earthquake populations: models, predictions and tests. *Pure Appl. Geophys.* **155**, 575–607.

Rundle, J.B., W. Klein, D.L. Turcotte, and B.D. Malamud (2000). Precursory seismic activation and critical-point phenomena. *Pure Appl. Geophys.* **157**, 2165–2182.

Sadovsky, A.M., T.V. Golubeva, V.F. Pisarenko, and M.G. Shnirman (1984). Characteristic dimensions of rock and hierarchical properties of seismicity. *Phys. Solid Earth* **20**, 87–95.

Sahimi, M., M.C. Robertson, and C.G. Sammis (1992). Relation between the Earthquake statistics and fault pattern, and fractals and percolation. *Physica* **A191**, 57–68.

Sahimi, M., M.C. Robertson, and C.G. Sammis (1993). Fractal distribution of earthquake hypocenters and its relation to fault patterns and percolation. *Phys. Rev. Lett.* **70**, 2186–2189.

Saleur, H., C.G. Sammis, and D. Sornette (1996). Discrete scale invariance, complex fractal dimensions, and log-periodic fluctuations in seismicity. *J. Geophys. Res.* **101**, 17661–17677.

Scholz, C.H. (1991). Earthquakes and faulting: self-organized critical phenomena with a character dimension. In: "Spontaneous Formation of Space Time Structure and Criticality" (T. Riste and D. Sherrington, Eds.), pp. 41–56. Kluwer, Amsterdam.

Scholz, C.H. (1997). Size distributions for large and small earthquakes. *Bull. Seismol. Soc. Am.* **87**, 1074–1077.

Schreider, S.Y. (1990). Formal definition of premonitory seismic quiescence. *Phys. Earth Planet. Inter.* **61**, 113–127.

SCSN Catalog (1995). Southern California Seismographic Network catalog in electronic format at the Southern California Earthquake Center, California Institute of Technology, Pasadena, CA.

Sieh, K.E. (1978). Slip along the San Andreas fault associated with the great 1857 earthquake. *Bull. Seismol. Soc. Am.* **68**, 1421–1448.

Sieh, K.E., M. Stuiver, and D. Brillinger (1989). A more precise chronology of earthquakes produced by the San Andreas fault in Southern California. *J. Geophys. Res.* **94**, 603–623.

Smalley, R.F., J.L. Chatelain, D.L. Turcotte, and R. Prevot (1987). A fractal approach to the clustering of earthquakes: applications to the seismicity of the New Hebrides. *Bull. Seismol. Soc. Am.* **77**, 1368–1381.

Sornette, D. (1998). Discrete-scale invariance and complex dimensions. *Phys. Rep.* **297**, 239–270.

Sornette, D. and C.G. Sammis (1995). Complex critical exponents from renormalization group theory of earthquakes: implications for earthquake predictions. *J. Phys. I France* **5**, 607–619.

Stauffer, D. and A. Aharony (1992). "Introduction to Percolation Theory," 2nd ed. Taylor and Francis, London.

Stinchcombe, R.B. and B.P. Watson (1976). Renormalization group approach for percolation conductivity. *J. Phys.* **C9**, 3221–3247.

Sykes, L.R. and S.C. Jaumé (1990). Seismic activity on neighboring faults as a long-term precursor to large earthquakes in the San Francisco Bay area. *Nature* **348**, 595–599.

Turcotte, D.L. (1989). A fractal approach to probabilistic seismic hazard assessment. *Tectonophysics*. **167**, 171–177.

Turcotte, D.L. (1991). Earthquake prediction. *Annu. Rev. Earth Planet. Sci.* **19**, 263–281.

Turcotte, D.L. (1997). "Fractals and Chaos in Geology and Geophysics," 2nd ed. Cambridge University Press, Cambridge.

Turcotte, D.L. (1999a). Self-organized criticality. *Rep. Prog. Phys.* **62**, 1377–1429.

Turcotte, D.L. (1999b). Seismicity and self-organized criticality. *Phys. Earth Planet. Inter.* **111**, 275–293.

Turcotte, D.L., B.D. Malamud, and G. Morein (1999). An inverse-cascade model for self-organized critical behavior. *Physica* **A268**, 629–643.

Utsu, T. (1961). A statistical study on the occurrence of aftershocks. *Geophys. Mag.* **30**, 521–605.

Utsu, T., Y. Ogata, and R.S. Matsu'ura (1995). The centenary of the Omori formula for a decay law of aftershock activity. *J. Phys. Earth* **43**, 1–33.

Villemin, T., J. Angelier, and S. Choon (1995). Fractal distribution of fault length and offsets: Implications of brittle deformation evaluation—the Lorraine coal basin. In: "Fractals in the Earth Sciences" (C.C. Barton and P.R. La Pointe, Eds.), pp. 205–226. Plenum Press, New York.

Editor's Note

For a general discussion of seismicity distribution, see Chapter 43, Statistical Features of Seismicity, by Utsu. Chapter 65, Seismic Hazards and Risk Assessment in Engineering Practice, by Sommerville and Moriwaki, and Chapter 74, The GSHAP Global Seismic Hazard Map by Giardini *et al.*, discuss earthquake hazard assessment. Chapter 72, Earthquake Prediction, by Kanamori, gives an overview on earthquake prediction.

15

Physics of Earthquakes

Roman Teisseyre and Eugeniusz Majewski
Institute of Geophysics, Polish Academy of Sciences, Warsaw, Poland

1. Introduction

The problem of earthquake processes is considered here at two levels. We start with an overview of the classical macroscopic approach to faulting processes, which is commonly used by seismologists. Next, we proceed with the microphysics of fracture in minerals and rocks. In the microscopic approach we emphasize the irreversible deformations resulting from dislocations, microcracks, and other microscopic defects. The microscopic formulation is finally applied in a qualitative description of earthquake processes. Some recent results of earthquake thermodynamics are also briefly reviewed. Theoretical results for the strain, stress, earthquake magnitude, seismic moment, and entropy change during an earthquake process are presented.

This chapter is organized as follows. In Section 2 there is a brief discussion of current knowledge about the classical macroscopic approach to fault zone dynamics, pointing out the advantages and disadvantages of this formulation. This is not intended to be a comprehensive review, as several of these have appeared in recent years. In Section 3 a microscopic approach to fracturing is reviewed, based mainly on microscopic fracture statistics. This is followed in Section 4 by a discussion of an elastic continuum medium with a defect content. Here the relevant theoretical mechanisms of microscopic deformation are described. Section 5 deals with earthquake thermodynamics. Finally, a qualitative discussion of various aspects of the earthquake models described is presented, and it is shown how physical effects such as self-organized criticality, chaos, and complexity arise. Some detailed considerations and further explanations are to be found in the Appendices on the attached Handbook CD.

2. The Macroscopic Approach to Fault Zone Dynamics

2.1 General Considerations

The seismological observations and the laboratory experimental data on fracturing processes are well described by the instantaneous development of slip on a fault (instantaneous crack propagation). The appropriate solutions of the elastodynamic equation $\operatorname{div} \mathbf{S}(\mathbf{u}) = \rho\ddot{\mathbf{u}}$ (where the stress tensor $\mathbf{S}$ is related to the displacement vector $\mathbf{u}$ by the constitutive law, body forces are neglected, and ρ is the density, constant within a fault zone) can be calculated with the boundary integral formulation on a fault surface. At each fault element, stress is continuous, while displacements can be opposite on the different sides of surface element during slip. It is assumed, that the applied stresses on the fault follow the values of dynamic friction when a slip nucleates; and further that friction obeys some weakening law (a source of instabilities) as determined from observations and experimental data. The singularities appearing in a crack field can be eliminated by defining a *cohesive force*, which acts only inside the crack near its tip (see Appendix I).

For a review of problems related to a crack theory, see Rice (1968a). A central problem of fracture mechanics is the knowledge of the stress and deformation fields near the advancing crack as its active plastic zone precedes the propagating crack (fracture toughness; Ponte Castañeda, 1987). In this context it is worth mentioning the phenomenon of stress corrosion which is related to the action of some chemical agents at a crack tip; the crack fracturing process becomes accelerated; and the stress intensity/crack velocity data can be described in the form of a power law (Atkinson and Meredith, 1987).

2.2 Instabilities and Friction Weakening Constitutive Laws

For a review of the instability problem, see Giovanola and Finnie (1984). The J-integral (Rice, 1968b) describes a potential energy variation by crack extension; its behavior relates to instabilities in crack motion and serves to estimate work done in the plastic process zone. Another source of instability may be due to pore fluid pressure and fluid percolation (formation of connectivity between voids). The instabilities can arise due to a dissipation of frictional heat (Lachenbruch, 1980; Shaw, 1995). Frictional sliding can be represented as a thermally activated creep process (Stesky, 1975). A general instability

INTERNATIONAL HANDBOOK OF EARTHQUAKE AND ENGINEERING SEISMOLOGY, VOLUME 81A ISBN: 0-12-440652-1

criterion states that instability in a nucleation volume occurs at a negative product of the stress rate and the sum of elastic and plastic strain rates $\dot{\mathbf{S}}(\dot{\mathbf{E}} + \dot{\boldsymbol{\varepsilon}})$ (see Hill, 1958).

Considering slip propagation along a fault, the frictional instabilities can be described by the rate and state formalism (Linker and Dieterich, 1992):

$$\mu = \mu_0 + a \ln\left(\frac{V}{V_0}\right) + b \ln\left(\frac{\psi}{\psi_0}\right), \qquad \frac{\partial \psi}{\partial t} = \frac{1}{t_0} - \frac{V\psi}{D_c} \tag{1}$$

where μ is the coefficient of friction; V is the macroscopic frictional slip velocity, ψ is the state variable; μ_0, a, b, and ψ_0 are constants; V_0 is a reference velocity; t_0 is a reference time; and D_c is a critical displacement on a fault (see Appendix I).

Ohnaka in a series of high-resolution laboratory experiments (Ohnaka, 1998; Ohnaka and Shen, 1999) demonstrated that the constitutive law for shear rupture should be primarily slip-dependent.

2.3 Elastodynamic Solutions

Good reviews of works on this topic are given by Aki and Richards (1980), Rice (1980), Ben-Menachem and Singh (1981), and Scholz (1990). Among more recent papers, we mention the analytic solutions for antiplane slip given by Campillo and Ionescu (1997). Their eigenvalue analysis associated the positive eigenvalues with the dominant part of the solution, which on the fault exhibits an exponential growth in time. This dominant part also decreases exponentially away from the fault in the same way as the eigenfunctions do. The other part of the solution associated with negative eigenvalues, the "wave part," rapidly becomes negligible when instability develops. This solution seems important in showing that the elastodynamic equation can be split into a slip and a wave part (see Section 4).

2.4 Earthquake Cycles on Tectonic Planes

Large earthquakes occur repeatedly along large-scale faults. An earthquake cycle includes tectonic stress accumulation, fracture nucleation, propagation, arrest, and fault-healing. Problems of such cycles can be approached by studying the constitutive rate- and state-dependent laws determining the fracturing, fracture nucleation, and healing conditions (Ben-Zion and Rice, 1995; Sleep, 1995). Matsu'ura (1995) considered a transition process from nucleation to dynamic rupture. Rundle (1988) introduced the idea that an earthquake represents a fluctuation about the long-term motion of the plates: the fluctuation hypothesis. Fault interactions in recurrent earthquake cycles have been studied by Senatorski (1995). The suggestion that strong heterogeneities in a fault zone act as barriers affecting seismicity and rupture arrest has been considered by Das and Aki (1977), Kanamori and Stewart (1978), Aki (1984), Rybicki and Yamashita (1998), and Yamashita (1999).

3. The Microscopic Approach to Fracturing

3.1 Microscopic Mechanisms

Depending on stress level, confining pressure, temperature, fluid content, and rock properties, the following different modes can accommodate the deformation: elastic distortion, plastic flow, phase changes, nucleation and growth of ductile microvoids, nucleation and growth of brittle microcracks, and nucleation and growth of shear instabilities (see Curran *et al.*, 1987). Fracture proceeds through the nucleation, growth, coalescence, and fragmentation phases. In the microfracturing processes the microcracks and voids (in which material bonds are broken) play the role of the main objects. However, considering more elementary objects of preparatory processes leading to fracturing, we will refer to dislocations and vacancies (in which bonds are unbroken and material consistency is preserved).

3.2 Nucleation and Growth

The noninteracting voids and cracks can be described according to experimental evidence (Curran *et al.*, 1987) by the following distribution:

$$N(R) = N_0 \exp\left(\frac{R}{R_0}\right) \tag{2}$$

where R is the size of the flaw, N is the number of flaws greater than or equal to R per unit volume, and R_0 is the characteristic size, specific for material and physical conditions. At high crack density, cracks are no longer independent objects. The evidence related to seismic events (Turcotte, 1986) points to a distribution relation of the fractal type:

$$N(R) \propto R^{-D} \tag{3}$$

This is a scale-invariant distribution law (Turcotte, 1992). Czechowski (1991) obtained this type of distribution using the kinetic theory of crack fusion (coagulation theory), finding a transition from the exponential relation toward a power-law relation [Eq. (3)] with $D = 3/2$.

For the equilibrium between stable and activated bonds, Boltzmann statistics can be applied (Curran *et al.*, 1987); for the breaking and healing of bonds, we can put (see Appendix II)

$$\dot{\eta} = \frac{\eta kT}{h} \exp\left(-\frac{\Delta G \mp W}{kT}\right) \tag{4}$$

where η is the number of bonds per unit area, ΔG is the energy difference between the activated and broken bonds, h is the Planck constant, and $W = S\,\delta\lambda/\eta$ is the external work required to break a bond (S is the stress field; $\delta\lambda$ is the stretching of a bond before it breaks; λ is the bond length); the upper sign is for bond breaking and the lower one for healing. The net rate of bond fracture is given by the difference; putting the number of bonds per unit volume $N = \eta/\lambda$, and introducing the bond breaking strain $\varepsilon = \delta\lambda/\lambda$, we get the rate of bond breaking per unit of microscopic flaw containing n bonds:

$$\dot{N} = \frac{S_s \varepsilon}{nh} \exp\left(\frac{S - S_f}{S_s}\right) \tag{5}$$

Two critical stress thresholds appear here: $S_s = NnkT/\varepsilon$ for the slow fracturing process, and $S_f = Nn\,\Delta/\varepsilon$ for the fast fracturing process (Curran *et al.*, 1987).

3.3 Plastic Failure

Shear instabilities lead to nucleation of shear bands; along shear planes the dislocation arrays are formed. The Orowan equation connects the plastic strain rate $\dot{\varepsilon}$ with the number of shear bands $\bar{N}$ intersecting a unit of surface and the average band velocity $\bar{V}$:

$$\dot{\varepsilon} = \bar{N}\,\bar{V}\,B \tag{6}$$

where B relates here to the jog accommodated by the band (it corresponds to a Burgers vector that had intersected the band).

Creep phenomena are regarded as thermally activated phenomena—Arrhenius's law; the Andrade law of creep gives a good representation of the experimental data in relation to stresses S. Both laws can be combined as (see Appendix II)

$$\dot{\varepsilon} = \dot{\varepsilon}_0\, t^m S^m \exp(-\Delta H/RT) \tag{7}$$

where for transient creep $m < 0$ and for steady-state creep $m = 0$, while for tertiary creep growth $m > 0$; ΔH is the activation enthalpy.

3.4 Coalescence Processes and Microcrack Formation

A further phase of shear deformation is based on the formation of slip bands, grain boundary sliding, dislocation pile-ups, dislocation-to-crack transition, and microcrack formation. The formation of dislocation arrays and the interaction between the groups of dislocations of opposite signs lead to the coalescence processes (see Section 4): microcracks nucleate and shear slip-bands merge. Stroh (1957) considered the energy balance related to the process of crack formation at the zone of stress concentration at the tip of the dislocation array; the "dislocation-to-crack" process has been considered by Teisseyre (1969) and Dmowska *et al.* (1972) as one of possible energy release processes. The premonitory and rebound theory (Teisseyre, 1995) describes the evolution of internal stress (stress concentration) due to dislocation pile-ups and further to their coalescence with the opposite dislocation arrays.

To continuum theories of fracturing we also add the crack fusion theories; the concept of crack fusion as an earthquake mechanism was investigated by Newman and Knopoff (1982, 1990). Percolation theories (e.g., Chelidze, 1982; Turcotte, 1992), following the original considerations on electric resistivity, use a network of sites with bonds between them. A critical process, like fracture, takes place when a cluster of the sites with broken bonds emerges.

4. Continuum with Defect Content

4.1 Equation of Motion for Elastodynamics with Defect Impact

In the elastodynamic solutions describing slip propagation along a fault, the condition for stresses at infinitely remote boundaries specifies the external stress load. A boundary condition on a fault incorporates the possibility of displacement discontinuity—a slip. This slip, related to the sought elastodynamic solutions, obeys the friction constitutive law, determined from the experimental data. This allows a successful use of the elastodynamic equations, in spite of the fact that these equations do not theoretically apply to an elastic continuum that includes any distribution of defects as objects with their own stress field. In an elastic continuum with defects (more exactly, a continuum with a continuous distribution of the self-strain nuclei: thermal nuclei, dislocations, and disclinations), we have to consider the elastic $\mathbf{E}$, $\mathbf{S}$ and self parts $\mathbf{E}^S$, $\mathbf{S}^S$ of the total strain and stress fields $\mathbf{E}^T$, $\mathbf{S}^T$, subjected to the same type of stress–strain constitutive relation:

$$\mathbf{E}^T = \mathbf{E} + \mathbf{E}^S, \qquad \mathbf{S}^T = \mathbf{S} + \mathbf{S}^S, \qquad \mathbf{S} = \lambda\,\mathrm{tr}(\mathbf{E})\mathbf{1} + 2\mu\mathbf{E} \tag{8}$$

where tr (. . .) means tensor contraction, and λ and μ are Lamé material constants.

We obtain the motion equation in an elastic continuum with defects (see Appendix III):

$$\mathrm{div}\,\mathbf{S} = \rho\frac{d}{dt}\mathbf{v}, \qquad \mathrm{div}\,\mathbf{S} = \rho\frac{d^2}{dt^2}\mathbf{u} - \rho\frac{d}{dt}\mathbf{v}^s \tag{9}$$

where ρ is a constant, $\mathbf{u}$ is the displacement, $\mathbf{v}$ is the velocity, and $\mathbf{v}^s$ is the self (plastic) velocity.

We could add that the commonly used stick-slip model, together with the elastodynamic equation, neglects the self-stress fields from objects on a fault plane. The self (plastic) parts are crucial as the self-stresses in an interaction with elastic stresses govern a slip evolution process. There exist direct relations between dislocations and stresses in two dimensions (Teissyre, 1997; Appendix III); for the antiplane (screw dislocations) and in-plane (edge dislocations) we obtain

$$\alpha - \frac{1}{2}\,\mathrm{tr}(\boldsymbol{\alpha})\mathbf{1} = \frac{1}{2\mu}\,\mathrm{curl}\left(\mathbf{S} - \frac{\nu}{1+\nu}\,\mathrm{tr}(\mathbf{S})\mathbf{1}\right), \qquad \boldsymbol{\alpha} = -\mathrm{curl}\,\boldsymbol{\beta}^s \tag{10}$$

where $\boldsymbol{\alpha}$ is the dislocation density defined in relation to the plastic distortion $\boldsymbol{\beta}^s$ (Kröner, 1958).

Assuming after Kossecka and DeWitt (1977) that the self (plastic) velocities $\mathbf{v}^s$ relate to motion of dislocations, we can consider the dislocation flow current $\mathbf{J}$ as the difference between the time and space derivatives of plastic distortion and plastic velocity, or equivalently, as the vector product of the dislocation density and its velocity $\mathbf{V}$:

$$\mathbf{J} = \frac{\partial}{\partial t}\boldsymbol{\beta}^s - \frac{\partial}{\partial \mathbf{x}}\mathbf{v}^s, \quad \mathbf{J} = c\boldsymbol{\alpha} \times \mathbf{V}, \quad \frac{\partial \boldsymbol{\alpha}}{\partial t} + \nabla \times \mathbf{J} = \boldsymbol{\Pi} \tag{11}$$

where $\boldsymbol{\Pi}$ is the source/sink function describing the interaction between the dislocations of the opposite signs (Teisseyre and Nagahama, 1998), and c is the shear wave velocity.

From Eqs. (10) and (11) we can derive the expression for plastic velocity $\mathbf{v}^s$; after differentiation and symmetrization of the motion equation [Eq. (9)], we come to the following form (for details, see Appendix III):

$$(\nabla \operatorname{div} \mathbf{S})_{\text{sym}} = \frac{\rho}{\mu}\left[\ddot{\mathbf{S}} - \frac{\lambda}{3\lambda + 2\mu}\operatorname{tr}(\ddot{\mathbf{S}})\mathbf{1}\right] + \rho c \frac{d}{dt}(\boldsymbol{\alpha} \times \mathbf{V})_{\text{sym}} \tag{12}$$

where "sym" means symmetrization procedure.

4.2 Splitting the Stress Motion Equations into Wave and Fault-Related Parts

Taking the antiplane problem (function independent of z) and assuming that the dislocations move only in the x direction, we obtain the motion equations (Teisseyre and Yamashita, 1999):

$$\frac{\partial^2 S_{zx}}{\partial x \partial y} + \frac{\partial^2 S_{zy}}{\partial y^2} - \frac{1}{c^2}\frac{\partial^2 S_{zy}}{\partial t^2} = \frac{\mu}{c}\frac{\partial}{\partial t}(\alpha V) - F,$$
$$\frac{\partial^2 S_{zx}}{\partial x^2} + \frac{\partial^2 S_{zy}}{\partial x \partial y} - \frac{1}{c^2}\frac{\partial^2 S_{zx}}{\partial t^2} = 0 \tag{13}$$

An elastic stress field $\mathbf{S}$ can now be presented as the difference $\mathbf{S}^T - \mathbf{S}^S$ between the total $\mathbf{S}^T$ field and the self $\mathbf{S}^S$ stress part, which is assumed to rapidly decrease away from the fault plane. We can identify these parts with a fault-related field $S^S_{zi} = \bar{S}_{zi}$ $(i = x, y)$ and a radiation field $S^T_{zi} = \tilde{S}_{zi}$ (total stresses are related to a displacement field).

On the fault plane (x, t) the fracturing processes cause the stress to change, while a seismic wave radiation will be observed in the considered (x, y, t) domain. The equation of motion [Eq. (13)] now splits into two parts. The first relates to a fault (x, t domain):

$$\frac{1}{c^2}\frac{\partial^2}{\partial t^2}\bar{S}_{zy} - \frac{1}{c}\frac{\partial}{\partial t}\left[V\frac{\partial}{\partial x}\bar{S}_{zy}\right] = F, \qquad \frac{1}{c^2}\frac{\partial^2}{\partial t^2}\bar{S}_{zx} - \frac{\partial^2}{\partial x^2}\bar{S}_{zx} = 0 \tag{14}$$

Here $\bar{\alpha} = -(1/\mu)\,\partial\bar{S}_{zy}/\partial x$ relates to a slip δ on a fault $\bar{\alpha} = (1/\Lambda)\,\nabla\delta$ (with some characteristic length Λ). For the $\bar{S}_{zx}$ component we can put $\bar{S}_{zx} = 0$, as it is subject to the homogeneous equation. For $\bar{S}_{zy}$ we may solve the equation (for details, see Appendix III)

$$\frac{1}{c^2}\frac{\partial^2}{\partial t^2}\bar{S}_{zy} + \frac{\mu}{cR}\operatorname{sgn}\bar{\alpha}\,\frac{\partial}{\partial t}\frac{\partial}{\partial x}\left[\frac{\bar{S}_{zy} - R}{|R^2 + (\bar{S}_{zy} - R)^2|^{1/2}}\frac{\partial\bar{S}_{zy}}{\partial x}\right] = \frac{\partial}{\partial t}\boldsymbol{\Pi} \tag{15}$$

Returning to the motion equation [Eq. (13)], we obtain for $\tilde{\mathbf{S}}(x, y, t)$ the homogeneous wave equations; assuming that the elastic stress field $\mathbf{S} = \tilde{\mathbf{S}} - \bar{\mathbf{S}}$ on a fault is small $(\mathbf{S} \ll \bar{\mathbf{S}})$ we have the boundary condition $\tilde{S}_{zy}(x, y = 0, t) = \bar{S}_{zy}(x, t)$. We put $\tilde{S}^0_{zx} = 0$ as $\bar{S}_{zx} = 0$.

For the dislocation stress resistance and for the friction on a fault, we may take the weakening constitutive laws introducing the instabilities in a preseismic phase (motion of dislocations), in a slip nucleation phase, and in a slip propagation phase. We note now that instead of the instability sources introduced by the weakening laws for all these phases, we may take the adequate source/sink functions $\boldsymbol{\Pi}$ (discussed by Teisseyre and Nagahama, 1998). The numerical simulations, performed according to Eq. (14), show some instabilities that might be identified with seismic events (Teisseyre, 1997).

5. Earthquake Thermodynamics

5.1 Thermodynamic Relationship

The equilibrium states are characterized by the extreme values of some thermodynamic functions (Majewski, 1993):

$$H = U - W, \qquad F = U - T\tilde{S}, \qquad G = H - T,$$
$$\text{where } dU + dW = dQ + S\,dE \tag{16}$$

where H is the enthalpy, F is the Helmholtz free energy, G is the Gibbs free energy, T is the absolute temperature, and $\tilde{S}$ is the entropy, and where internal energy U enters into the first law of thermodynamics; work done W relates here to shear load S and strain E.

Rice (1978) has estimated entropy production due to a stationary crack growth. Rundle (1989) reconsidered this question using a statistical approach (transitions to slip states approach a minimum of Helmholtz free energy). In a damage rheology model, Lyakhovsky *et al.* (1997a,b) applied a general thermodynamic formulation to account for viscous relaxation and evolution of damage with material degradation and healing. In a rate-dependent formulation, Majewski and Teisseyre (1998) considered anticrack-associated faulting and superplastic flow in deep subduction zones.

5.2 Dislocation Superlattice

Basic thermodynamic relations for line defects (dislocations and vacant dislocations) are derived under the assumption of a dense network of defects forming a kind of superlattice (Majewski and Teisseyre, 1997; Teisseyre and Majewski, 2001). Considering a system that contains a superlattice of dislocations and vacant dislocations with a superlattice parameter Λ ($\Lambda \gg \lambda$, where λ relates to a crystal lattice), we associate the thermodynamic functions with the vacant dislocations; the Gibbs free energy can have a minimum corresponding to the equilibrium concentration of the vacant dislocations (for details, see Appendix IV).

In a real body with n dislocations we may add to it the other $\hat{n}$ vacant dislocations; a whole set $N = n + \hat{n}$ corresponds to a superlattice; for the equilibrium state of density of dislocations and vacant dislocations we obtain ($\Delta s = \Lambda^2$):

$$\alpha = \frac{b}{\Lambda^2}\left[1 - \exp\left(-\frac{\hat{g}^f}{kT}\right)\right], \qquad \hat{\alpha} = \frac{b}{\Lambda^2}\exp\left(-\frac{\hat{g}^f}{kT}\right) \quad (17)$$

where $\hat{g}^f$ is the formation energy of a vacant dislocation.

For the vacant dislocations, a change in the Gibbs energy depends on the stress level and resistance stress ($\partial\hat{G}/\partial\hat{n} = \hat{g}^f = (S - R)b\lambda^2$; Kocks *et al.*, 1975). Similarly to the $CB\Omega$ model introduced by Varotsos and Alexopoulos (1986) for point defects, we postulate the approximate value for the formation energy, independent of stresses (C is a constant):

$$\hat{g}^f = C\mu\lambda\Lambda^2, \qquad \hat{n}^{\mathrm{eq}} = N\exp\left(-\frac{C\mu\lambda\Lambda^2}{kT}\right),$$
$$\hat{\alpha} = \frac{b}{\Lambda^2}\exp\left(-\frac{C\mu\lambda\Lambda^2}{kT}\right) \quad (18)$$

For the equilibrium state under a constant local shear S and temperature T, the Gibbs energy can reach its minimum. In a similar way, we can estimate the entropy changes.

5.3 Earthquake Thermodynamics

We approach the problem of earthquake thermodynamics with the principle of the local equilibrium condition; we assume that the state reached after a seismic event is near to an equilibrium distribution of vacant dislocations. However, the state before an earthquake cannot be described by a similar assumption; we may assume that in a zone of earthquake formation we have a very dense distribution of dislocations, forming arrays along the slip shear planes, while the number of vacant dislocations is near zero. However, we might be, formally, close to equilibrium when assuming that the Λ undergoes the changes; at least on the shear plane, an excess of dislocations destroys the Λ structure and the superlattice approaches the λ structure on slip planes (only).

Accordingly, we assume that before an earthquake a superlattice is almost completely filled with dislocations ($n \approx N$ and $\hat{n} \approx 0$), and we will also assume that only along the glide planes is the Λ structure destroyed by formation of the dislocation arrays. The maximum number of dislocations in arrays could reach the value (Λ/λ) per distance Λ; the maximum value of the Burgers vector becomes $b = \lambda(\Lambda/\lambda) = \Lambda$. After an earthquake, $\hat{n}$ will increase by $\Delta\hat{n}(\Lambda/\lambda)$ to nearly its equilibrium value [Eq. (18)]. The total number of line defects $N = n + \hat{n}$ can be estimated as the ratio of an earthquake source volume $\pi R^2 H$ (R is the seismic radius; H is the source thickness) and the parallelepiped volume $\lambda^2\Lambda$. Using the expression for a change in the Burgers vector and in the free energy values, we arrive at the formulae for the released energies, seismic moment, and entropy (for details, see Appendix IV).

6. Conclusions

The state of the art of the physics of earthquakes as well as of microscopic fracturing is reviewed. Our approach is to trace the development of the deformation from its roots at the atomic level to its macroscopic manifestation in earthquake faulting. Many examples of current works on the interplay between macroscopic rock fracture and microscopic line defect interactions in some earthquake processes are presented here. Models of crack nucleation are based on the hierarchical structure of the formation of slip-bands, grain boundary sliding, dislocation pile-ups, dislocation-to-crack transition, and microcrack formation. The formation of dislocation arrays and interactions between the groups of dislocations of opposite signs lead to the coalescence processes: microcracks nucleate and shear slip-bands merge. Many of these defects have been identified at the atomic level with the aid of measurements and microscopic observations.

Even with such a detailed understanding of the defects, it is still not possible to treat the full complexity of the problems encountered with the help of classical theories. As described by Anderson (1991), complementary views can be provided by theories of complexity and emergent phenomena, which include chaos and self-organized criticality. Self-organization and complexity theory were outlined in famous books by Nicolis and Prigogine (1977, 1989). Several aspects of complexity theory are beginning to have a major impact on the understanding of earthquake faulting and rock fracture processes, as discussed in Chapter 14 by Turcotte and Malamud.

In the last ten years it has become increasingly evident that there exist numerous examples of geophysical systems where spatial, temporal, or spatiotemporal structures arise out of chaotic states. These phenomena occur in systems in which there is an exchange of energy and matter with the environment, and these structures develop spontaneously and are self-organizing. In addition to thermodynamic instabilities, in such systems one can expect three stages of evolution: organization, self-organization, and chaos (Majewski, 1993). The transitions between these stages are the subject of study in the new interdisciplinary science named *synergetics*, first introduced by Haken (1978). It was shown in synergetics that numerous systems display striking similarities in their behavior when passing from the disordered to the ordered state and vice versa. This indicates that the functioning of these systems obeys the same general principles.

Recent years have brought a revival of work on self-organized space–time structures and criticality in earthquakes (Bak and Tang, 1989; Sornette and Sornette, 1989; Ito and Matsuzaki, 1990), which have been a source of fascination since their discovery. Preliminary investigation of the phenomenology of nonlinear phase transitions is also quite encouraging. However, it is clear that we have only begun to explore the structure of these new nonlinear theories (e.g., Lomnitz-Adler, 1993; Fisher *et al.*, 1997; Dahmen *et al.*, 1998). If history is a reliable guide to the future, then, as our understanding of these theories improves, new domains of nonlinear physics and new questions will appear.

Acknowledgments

This chapter was greatly improved by the remarks proposed by the reviewers: D.L. Turcotte, Y. Ben-Zion, and an anonymous reviewer; the authors express their thanks.

References

Aki, K. (1984). Asperities, barriers, characteristic earthquakes and strong motion prediction. *J. Geophys. Res.* **89**(B7), 5867–5872.

Aki, K. and P.G. Richards (1980). "Quantitative Seismology. Theory and Methods." W.H. Freeman, San Francisco.

Anderson, P.W. (1991). The science of complexity. *Phys. Today* **44**(7), 9–11.

Atkinson, B.K. and P.G. Meredith (1987). The theory of subcritical crack growth with application to minerals and rocks. In: "Fracture Mechanics of Rock" (B.K. Atkinson, Ed.), pp. 111–162. Academic Press, New York.

Bak, P. and C. Tang (1989). Earthquakes as a self-organized critical phenomenon. *J. Geophys. Res.* **94**, 15635–15637.

Ben-Menachem, A. and J. Singh (1981). "Seismic Waves and Sources." Springer-Verlag, New York.

Ben-Zion Y. and J.R. Rice (1995). Slip patterns and earthquake population along different classes of faults in elastic solids. *J. Geophys. Res.* **100**(B7), 12959–12983.

Campillo, M. and I.R. Ionescu (1997). Initiation of antiplane shear instability under slip dependent friction. *J. Geophys. Res.* **102**(B9), 20363–20371.

Chelidze, T.L. (1982). Percolation and fracture. *Phys. Earth Planet. Inter.* **28**, 93.

Curran, D.R., L. Seaman, and D.A. Shockey (1987). Dynamic failure of solids. *Phys. Rep.* **147**, 253–388.

Czechowski, Z. (1991). A kinetic model of crack fusion. *Geophys. J. Int.* **104**, 419–422.

Dahmen, K., D. Ertas, and Y. Ben-Zion (1998). Gutenberg–Richter and characteristic earthquake behavior in simple mean-field models of heterogeneous faults. *Phys. Rev.* **58**, 1494–1501.

Das, S. and K. Aki (1977). Fault planes with barriers: a versatile earthquake model. *J. Geophys. Res.* **82**, 5684–5670.

Dmowska, R., K. Rybicki, and R. Teisseyre (1972). Focal mechanism in connection with energy storage before crack formation. *Tectonophysics* **14**(3/4), 309–318.

Fisher, D.S., K. Dahmen, S. Ramanathan, and Y. Ben-Zion (1997). Statistics of earthquakes in simple models of heterogeneous faults. *Phys. Rev. Lett.* **78**, 4885–4888.

Giovanola, J.H. and I. Finnie (1984). A review of the use of the *J*-integral as a fracture parameter. *Solid Mech. Arch.* **9**(2), 197–225.

Haken, H. (1978). "Synergetics," 2nd edn. Springer-Verlag, Berlin.

Hill, R. (1958). A general theory of uniqueness and stability in elastic-plastic models. *J. Mech. Phys. Solids* **6**, 236–249.

Ito, K. and M. Matsuzaki (1990). Earthquakes as self-organized critical phenomena. *J. Geophys. Res.* **95**, 6853–6860.

Kanamori, H. and G.S. Stewart (1978). Seismological aspects of the Guatemala earthquake of February 4, 1976. *J. Geophys. Res.* **83**, 3427–3434.

Kocks, U.F., A.S. Argon, and M.F. Ashby (1975). "Thermodynamics and Kinetics of Slip." Pergamon Press, Oxford.

Kossecka, E. and R. DeWitt (1977). Disclination kinematics. *Arch. Mech.* **29**, 633–651.

Kröner, E. (1958). Kontinuums theories der Versetzungen und Eigenspannungen. *Ergeb. Angew. Math.* **5**, Berlin.

Lachenbruch, A.H. (1980). Frictional heating, fluid pressure, and the resistance to fault motion. *J. Geophys. Res.* **85**, 6097–6112.

Linker, M.F. and J.H. Dieterich (1992). Effects of variable normal stress on rock friction: observations and constitutive equations. *J. Geophys. Res.* **97**(B4), 4923–4940.

Lomnitz-Adler, J. (1993). Automaton models of seismic fracture: constraints imposed by the magnitude-frequency relation. *J. Geophys. Res.* **98**, 17745–17756.

Lyakhovsky, V., Y. Ben-Zion, and A. Agnon (1997a). Distributed damage, faulting, and friction. *J. Geophys. Res.* **102**(B12), 27635–27649.

Lyakhovsky, V., Z. Reches, R. Weinberger, and T.E. Scott (1997b). Non-linear elastic behaviour of damaged rocks. *Geophys. J. Int.* **130**, 157–166.

Majewski, E. (1993). Thermodynamic approach to evolution. In: "Dynamics of the Earth's Evolution" (R. Teisseyre, L. Czechowski, and J. Leliwa-Kopystynski, Eds.), pp. 390–445. Elsevier, Amsterdam.

Majewski, E. and R. Teisseyre (1997). Earthquake thermodynamics. *Tectonophysics* **277**, 219–233.

Majewski, E. and R. Teisseyre (1998). Anticrack-associated faulting in deep subduction zones. *Phys. Chem. Earth* **23**, 1115–1122.

Matsu'ura, M. (1995). Transition process from nucleation to dynamic rupture. In: "Theory of Earthquake Premonitory and Fracture Processes" (R. Teisseyre, Ed.), pp. 77–102. PWN, Warszawa.

Newman, W.I. and L. Knopoff (1982). Crack fusion dynamics: a model for large earthquakes. *Geophys. Res. Lett.* **9**, 735–738.

Newman, W.I. and L. Knopoff (1990). Scale invariance in brittle fracture and the dynamics of crack fusion. *Int. J. Fract.* **43**, 19–24.

Nicolis, G. and I. Prigogine (1977). "Self-Organization in Non-equilibrium Systems." Wiley, London.

Nicolis, G. and I. Prigogine (1989). "Exploring Complexity: An Introduction." W.H. Freeman, New York.

Ohnaka, M. (1998). Earthquake generation processes and earthquake prediction: implications of the underlying physical law and seismogenic environments. *J. Seismol. Soc. Japan, Ser. 2*, **50** (Special Issue), 129–155, 1998.

Ohnaka, M. and L. Shen (1999). Scaling of the shear rupture process from nucleation to dynamic propagation: implications of geometric irregularity of the rupturing surfaces. *J. Geophys. Res.* **104**, 817–844.

Ponte Castañeda, P. (1987). Plastic stress intensity factors in steady crack growth. *J. Appl. Mech.* **109**, 379–387.

Rice, J.R. (1968a). Mathematical analysis in the mechanics of fracture. In: "Fracture" (H. Liebowitz, Ed.), Vol. II, pp. 192–308. Academic Press, New York.

Rice, J.R. (1968b). A path-independent integral and the approximate analysis of strain concentration by notches and cracks. *J. Appl. Mech.* **35**, 379–386.

Rice, J.R. (1978). Thermodynamics of the quasi-static growth of Griffith cracks. *J. Mech. Phys. Solids* **26**, 61–78.

Rice, J.R. (1980). The mechanics of earthquake rupture. In: "Physics of the Earth's Interior" (A.M. Dziewonski and E. Boschi, Eds.), pp. 555–649. North-Holland, Amsterdam.

Rundle, J.B. (1988). A physical model for earthquakes. 1. Fluctuations and interactions. *J. Geophys. Res.* **93**(B6), 6237–6254.

Rundle, J.B. (1989). A physical model for earthquakes. 3. Thermodynamical approach and its relation to nonclassical theories of nucleation. *J. Geophys. Res.* **94**(B3), 2839–2855.

Rybicki, K.R. and T. Yamashita (1998). Faulting in vertically inhomogeneous media and its geophysical implications. *Geophys. Res. Lett.* **25**, 2893–2896.

Scholz, C.H. (1990). "The Mechanics of Earthquakes and Faulting." Cambridge University Press, Cambridge.

Senatorski, P. (1995). Spatio-temporal evolution of faults: deterministic model. *Physica D* **76**, 420–435.

Shaw, B.E. (1995). Frictional weakening and slip complexity in earthquake faults. *J. Geophys. Res.* **100**(B9), 18239–18251.

Sleep, N.H. (1995). Ductile creep, compaction, and rate and state dependent friction within major fault zones. *J. Geophys. Res.* **100**(B7), 13065–13080.

Sornette, A. and D. Sornette (1989). Self-organized criticality and earthquakes. *Europhys. Lett.* **9**, 197–202.

Stesky, R.M. (1975). The mechanical behavior of faulted rock at high temperature and pressure, PhD thesis, MIT, Cambridge, MA.

Stroh, A.N. (1957). A theory of the fracture of metals. *Adv. Phys. Philos. Mag. Suppl.* **6**, 418–465.

Teisseyre, R. (1969). Dislocation field dynamics as an approach to the physics of earthquake processes. *Publ. Dominion Obs.* **37**(7), 199–235.

Teisseyre, R. (Ed.) (1995). "Theory of Earthquake Premonitory and Fracture Processes." PWN, Warszawa.

Teisseyre, R. (1997). Dislocation–stress relations and evolution of dislocation fields. *Acta Geophys. Polon.* **45**(3), 205–214.

Teisseyre, R. and E. Majewski (Eds.) (2001). "Earthquake Thermodynamics and Phase Transformations in the Earth's Interior," International Geophysics Series. Academic Press, San Diego.

Teisseyre, R. and H. Nagahama (1998). Dislocation field evolution and dislocation source/sink function. *Acta Geophys. Polon.* **46**, 14–33.

Teisseyre, R. and T. Yamashita (1999). Splitting stress motion equations into seismic wave and fault related fields. *Acta Geophys. Polon.* **47**, 135–147.

Turcotte, D.L. (1986). Fractals and fragmentation. *J. Geophys. Res.* **91**, 1921–1926.

Turcotte, D.L. (1992). "Fractals and Chaos in Geology and Geophysics." Cambridge University Press, New York.

Varotsos, P.A. and K.D. Alexopoulos (1986). "Thermodynamics of Point Defects and their Relation with Bulk Properties." North-Holland, Amsterdam.

Yamashita, T. (1999). Pore creation due to fault slip in a fluid-permeated fault zone and its effect on seismicity. *Pure Appl. Geophys.* **155**, 625–647.

Editor's Note

The four appendices in this chapter are given in one computer readable file on the attached Handbook CD, under the directory \15Teisseyre.

16

Probabilistic Approach to Inverse Problems

Klaus Mosegaard
Niels Bohr Institute, Copenhagen, Denmark
Albert Tarantola
Institut de Physique du Globe, Paris, France

1. Introduction

In 'inverse problems' data from indirect measurements are used to estimate unknown parameters of physical systems. Uncertain data (possibly vague) prior information on model parameters, and a physical theory relating the model parameters to the observations are the fundamental elements of any inverse problem. Using concepts from probability theory, a consistent formulation of inverse problems can be made, and, while the most general solution of the inverse problem requires extensive use of Monte Carlo methods, special hypotheses (e.g., Gaussian uncertainties) allow, in some cases, an analytical solution to part of the problem (e.g., using the method of least squares).

1.1 General Comments

Given a physical system, the 'forward' or 'direct' problem consists, by definition, in using a physical theory to predict the outcome of possible experiments. In classical physics this problem has a unique solution. For instance, given a seismic model of the whole Earth (elastic constants, attenuation, etc. at every point inside the Earth) and given a model of a seismic source, we can use current seismological theories to predict which seismograms should be observed at given locations at the Earth's surface.

The 'inverse problem' arises when we do not have a good model of the Earth, or a good model of the seismic source, but we have a set of seismograms, and we wish to use these observations to infer the internal Earth structure or a model of the source (typically we try to infer both).

There are many reasons that make the inverse problem underdetermined (nonunique). In the seismic example, two different Earth models may predict the same seismograms,[1] the finite bandwidth of our data will never allow us to resolve very small features of the Earth model, and there are always experimental uncertainties that allow different models to be 'acceptable.'

The name 'inverse problem' is widely used. The authors of this chapter only like this name moderately, as we see the problem more as a problem of 'conjunction of states of information' (theoretical, experimental, and prior information). In fact, the equations used below have a range of applicability well beyond 'inverse problems': they can be used, for instance, to predict the values of observations in a realistic situation where the parameters describing the Earth model are not 'given' but only known approximately.

We take here a probabilistic point of view. The axioms of probability theory apply to different situations. One is the traditional statistical analysis of random phenomena, another one is the description of (more or less) subjective states of information on a system. For instance, estimation of the uncertainties attached to any measurement usually involves both uses of probability theory: Some uncertainties contributing to the total uncertainty are estimated using statistics, while some other uncertainties are estimated using informed scientific judgment about the quality of an instrument, about effects not explicitly taken into account, etc. The International Organization for Standardization (ISO) in *Guide to the Expression of Uncertainty in Measurement* (1993), recommends that the uncertainties evaluated by statistical methods are named 'type A' uncertainties, and those evaluated by other means (for instance, using Bayesian arguments) be named 'type B' uncertainties. It also recommends that former classifications, for instance into 'random' and 'systematic uncertainties,' should be avoided. In the present text, we accept ISO's basic point of view, and extend it by downplaying the role assigned by ISO to the particular Gaussian model for uncertainties (see Section 4.3) and by not assuming that the uncertainties are 'small.'

INTERNATIONAL HANDBOOK OF EARTHQUAKE AND ENGINEERING SEISMOLOGY, VOLUME 81A ISBN: 0-12-440652-1

In fact, we like to think of an 'inverse' problem as merely a 'measurement.' A measurement that can be quite complex, but the basic principles and the basic equations to be used are the same for a relatively complex 'inverse problem' as for a relatively simple 'measurement.'

We do not normally use, in this text, the term 'random variable,' as we assume that we have probability distributions over 'physical quantities.' This is a small shift in terminology that we hope will not disorient the reader.

An important theme of this paper is *invariant formulation* of inverse problems, in the sense that solutions obtained using different, equivalent, sets of parameters should be consistent, i.e., probability densities obtained as the solution of an inverse problem, using two different set of parameters, should be related through the well-known rule of multiplication by the Jacobian of the transformation.

This chapter is organized as follows. After a brief historical review of inverse problem theory, with special emphasis on seismology, we give a short introduction to probability theory. In addition to being a tutorial, this introduction also aims at fixing a serious problem of classical probability, namely the noninvariant definition of conditional probability. This problem, which materializes in the so-called Borel paradox, has profound consequences for inverse problem theory.

A probabilistic formulation of inverse theory for general inverse problems (usually called 'nonlinear inverse problems') is not complete without the use of Monte Carlo methods. Section 3 is an introduction to the most versatile of these methods, the Metropolis sampler. Apart from being versatile, it also turns out to be the most natural method for implementing our probabilistic approach.

In Sections 4, 5, and 6 time has come for applying probability theory and Monte Carlo methods to inverse problems. All the steps of a careful probabilistic formulations are described, including parametrization, prior information over the parameters, and experimental uncertainties. The hitherto overlooked problem of uncertain physical laws ('forward relations') is given special attention in this text, and it is shown how this problem is profoundly linked to the resolution of the Borel paradox.

Section 7 treats the special case of the mildly nonlinear inverse problems, where deterministic (non-Monte Carlo) methods can be employed. In this section, invariant forms of classical inversion formulas are given.

1.2 Brief Historical Review

For a long time scientists have estimated parameters using optimization techniques. Laplace explicitly stated the least absolute values criterion. This, and the least-squares criterion were later popularized by Gauss (1809). While Laplace and Gauss were mainly interested in overdetermined problems, Hadamard (1902, 1932) introduced the notion of an 'ill-posed problem,' which can be viewed in many cases as an underdetermined problem.

The late 1960s and early 1970s were a golden age for the theory of inverse problems. In this period the first uses of Monte Carlo theory to obtain Earth models were made by Keilis-Borok and Yanovskaya (1967) and by Press (1968). At about the same time, Backus and Gilbert, and Backus alone, in the years 1967–1970, made original contributions to the theory of inverse problems, focusing on the problem of obtaining an unknown *function* from discrete data. Although the resulting mathematical theory is elegant, its initial predominance over the more 'brute force' (but more powerful) Monte Carlo theory was only possible due to the quite limited capacities of the computers at that time. It is our feeling that Monte Carlo methods will play a more important role in the future (and this is the reason why we put emphasis on these methods in this chapter). An investigation of the connection between analog models, discrete models, and Monte Carlo models can be found in a paper by Kennett and Nolet (1978).

Important developments of inverse theory in the fertile period around 1970 were also made by Wiggins (1969), with his method of suppressing 'small eigenvalues,' and by Franklin (1970) by introducing the right mathematical setting for the Gaussian, functional (i.e., infinite dimensional) inverse problem (see also Lehtinen *et al.*, 1989). Other important papers from the period are those of Gilbert (1971) and Wiggins (1972).

A reference that may interest some readers is Parzen *et al.* (1998), where the probabilistic approach of Akaike is described.

To the 'regularizing techniques' of Tikhonov (1963), Levenberg (1944), and Marquardt (1970), we prefer, in this chapter, the approach where the a priori information is used explicitly.

For seismologists, the first bona fide solution of an inverse problem was the estimation of the hypocenter coordinates of an earthquake using the 'Geiger method' (Geiger, 1910), which present-day computers have made practical. In fact, seismologists have been the originators of the theory of inverse problems (for data interpretation), and this is because the problem of understanding the structure of the Earth's interior using only surface data is a difficult one.

3-D tomography of the Earth, using travel times of seismic waves, was developed by Keiiti Aki and his coworkers in a couple of well known papers (Aki and Lee, 1976; Aki, Christofferson and Husebye 1977). Minster and Jordan (1978) applied the theory of inverse problems to the reconstruction of the tectonic plate motions, introducing the concept of 'data importance.' Later, tomographic studies have provided spectacular images of the Earth's interior. Interesting papers on these inversions are by van der Hilst *et al.* (1997) and Su *et al.* (1992).

One of the major current challenges in seismic inversion is the nonlinearity of wave field inversions. This is accentuated by the fact that major experiments in the future most likely will allow us to sample the whole seismic wave field. For low frequencies, wave field inversion is linear. Dahlen (1976) investigated the influence of lateral heterogeneity on the free oscillations. He showed that the inverse problem of estimating lateral heterogeneity of even degree from multiplet variance and skewance is linear. At the time this was published, data accuracy and unknown ellipticity splitting parameters hindered its application to real data, but later developments, including the works of Woodhouse and Dahlen (1978) on discontinuous Earth models, led to present-day successful inversions of low-frequency seismograms. In this connection the works of Woodhouse, Dziewonski, and others spring to mind.[2] Later, the first attempts to go to higher frequencies and nonlinear inversion were made by Nolet *et al.* (1986), and Nolet (1990).

Purely probabilistic formulations of inverse theory saw the light around 1970 (see, for instance, Kimeldorf and Wahba, 1970). In an interesting paper, Rietsch (1977) made non-trivial use of the notion of a 'noninformative' prior distribution for positive parameters. Jackson (1979) explicitly introduced prior information in the context of linear inverse problems, an approach that was generalized by Tarantola and Valette (1982a,b) to nonlinear problems.

There are three monographs in the area of inverse problems (from the viewpoint of data interpretation). In Tarantola (1987), the general, probabilistic formulation for nonlinear inverse problems is proposed. The small book by Menke (1984) covers several viewpoints on discrete, linear, and nonlinear inverse problems, and is easy to read. Finally, Parker (1994) exposes his view of the general theory of linear problems.

Recently, the interest in Monte Carlo methods, for the solution of inverse problems, has been increasing. Mosegaard and Tarantola (1995) proposed a generalization of the Metropolis algorithm (Metropolis *et al.*, 1953) for analysis of general inverse problems, introducing explicitly prior probability distributions, and they applied the theory to a synthetic numerical example. Monte Carlo analysis was recently applied to real data inverse problems by Mosegaard *et al.* (1997), Dahl-Jensen *et al.* (1998), Mosegaard and Rygaard-Hjalsted (1999), and Khan *et al.* (2000).

2. Elements of Probability

Probability theory is essential to our formulation of inverse theory. This chapter therefore contains a review of important elements of probability theory, with special emphasis on results that are important for the analysis of inverse problems. Of particular importance is our explicit introduction of *distance* and *volume* in data and model spaces. This has profound consequences for the notion of *conditional probability density*, which plays an important role in probabilistic inverse theory.

Also, we replace the concept of conditional probability by the more general notion of 'conjunction' of probabilities, this allowing us to address the more general problem where not only the data, but also the physical laws, are uncertain.

2.1 Volume

Let us consider an abstract space $\mathcal{S}$, where a point $\mathbf{x}$ is represented by some coordinates $\{x^1, x^2, \dots\}$, and let $\mathcal{A}$ be some region (subspace) of $\mathcal{S}$. The measure associating a volume $V(\mathcal{A})$ to any region $\mathcal{A}$ of $\mathcal{S}$ will be denoted the *volume measure*

$$V(\mathcal{A}) = \int_{\mathcal{A}} d\mathbf{x}\, v(\mathbf{x}) \,, \qquad (1)$$

where the function $v(\mathbf{x})$ is the *volume density*, and where we write $d\mathbf{x} = dx^1\, dx^2 \dots$ The *volume element* is then[3]

$$dV(\mathbf{x}) = v(\mathbf{x})\, d\mathbf{x} \,, \qquad (2)$$

and we may write $V(\mathcal{A}) = \int_{\mathcal{A}} dV(\mathbf{x})$. A manifold is called a *metric manifold* if there is a definition of distance between points, such that the distance ds between the point of coordinates $\{x^i\}$ and the point of coordinates $\{x^i + dx^i\}$ can be expressed as[4]

$$ds^2 = g_{ij}(\mathbf{x})\, dx^i\, dx^j \,, \qquad (3)$$

i.e., if the notion of distance is 'of the L_2 type.'[5] The matrix whose entries are g_{ij} is the *metric matrix*, and an important result of differential geometry and integration theory is that the volume density of the space, $v(\mathbf{x})$, equals the square root of the determinant of the metric:

$$v(\mathbf{x}) = \sqrt{\det \mathbf{g}(\mathbf{x})} \,. \qquad (4)$$

Example 1. *In the Euclidean 3D space, using spherical coordinates, the distance element is* $ds^2 = dr^2 + r^2\, d\theta^2 + r^2 \sin^2\theta\, d\varphi^2$, *from which it follows that the metric matrix is*

$$\begin{pmatrix} g_{rr} & g_{r\theta} & g_{r\varphi} \\ g_{\theta r} & g_{\theta\theta} & g_{\theta\varphi} \\ g_{\varphi r} & g_{\varphi\theta} & g_{\varphi\varphi} \end{pmatrix} = \begin{pmatrix} 1 & 0 & 0 \\ 0 & r^2 & 0 \\ 0 & 0 & r^2 \sin^2\theta \end{pmatrix} . \qquad (5)$$

The volume density equals the metric determinant $v(r, \theta, \varphi) = \sqrt{\det \mathbf{g}(r, \theta, \varphi)} = r^2 \sin\theta$ *and therefore the volume element is* $dV(r, \vartheta, \varphi) = v(r, \vartheta, \varphi)\, dr\, d\vartheta\, d\varphi = r^2 \sin\theta\, dr\, d\vartheta\, d\varphi$.

2.2 Probability

Assume that we have defined over the space, not only the volume $V(\mathcal{A})$ of a region $\mathcal{A}$ of the space, but also its *probability* $P(\mathcal{A})$, which is assumed to satisfy the Kolmogorov axioms (Kolmogorov, 1933). This probability is assumed to be descriptible in terms of a probability density $f(x)$ through the expression

$$P(\mathcal{A}) = \int_{\mathcal{A}} d\mathbf{x}\; f(\mathbf{x}) \,. \qquad (6)$$

It is well known that, in a change of coordinates over the space, a probability density changes its value: it is multiplied by the Jacobian of the transformation (this is the *Jacobian rule*). Normally, the probability of the whole space is normalized to one. If it is not normalizable, we do not say that we have a probability, but a 'measure.' We can state here the following postulate.

Postulate 1. *Given a space $\mathcal{X}$ over which a volume measure $V(\cdot)$ is defined. Any other measure (normalizable or not) $M(\cdot)$ considered over $\mathcal{X}$ is* absolutely continuous *with respect to $V(\cdot)$, i.e., the measure $M(\mathcal{A})$ of any region $\mathcal{A} \subset \mathcal{X}$ with vanishing volume must be zero: $V(\mathcal{A}) = 0 \Rightarrow M(\mathcal{A}) = 0$.*

2.3 Homogeneous Probability Distributions

In some parameter spaces, there is an obvious definition of distance between points, and therefore of volume. For instance, in the 3D Euclidean space the distance between two points is just the Euclidean distance (which is invariant under translations and rotations). Should we choose to parametrize the position of a point by its Cartesian coordinates $\{x, y, z\}$, the volume element in the space would be $dV(x, y, z) = dx\, dy\, dz$, while if we choose to use geographical coordinates, the volume element would be $dV(r, \theta, \varphi) = r^2 \sin\theta\, dr\, d\vartheta\, d\varphi$.

Definition. *The* homogeneous probability distribution *is the probability distribution that assigns to each region of the space a probability proportional to the volume of the region.*

Then, which probability density represents such a homogeneous probability distribution? Let us give the answer in three steps.

- If we use Cartesian coordinates $\{x, y, z\}$, as we have $dV(x, y, z) = dx\, dy\, dz$, the probability density representing the homogeneous probability distribution is constant: $f(x, y, z) = k$.
- If we use geographical coordinates $\{r, \theta, \varphi\}$, as we have $dV(r, \theta, \varphi) = r^2 \sin\theta\, dr\, d\theta\, d\varphi$, the probability density representing the homogeneous probability distribution is $g(r, \theta, \varphi) = kr^2 \sin\theta$.
- Finally, if we use an arbitrary system of coordinates $\{u, v, w\}$, in which the volume element of the space is $dV(u, v, w) = v(u, v, w)\, du\, dv\, dw$, the homogeneous probability distribution is represented by the probability density $h(u, v, w) = kv(u, v, w)$.

This is obviously true, since if we calculate the probability of a region $\mathcal{A}$ of the space, with volume $V(\mathcal{A})$, we get a number proportional to $V(\mathcal{A})$.

From these observations we can arrive at conclusions that are of general validity. First, the homogeneous probability distribution over some space is represented by a constant probability density **only** if the space is flat (in which case rectilinear systems of coordinates exist) and if we use Cartesian (or rectilinear) coordinates. The other conclusions can be stated as rules:

Rule 1. *The probability density representing the homogeneous probability distribution is easily obtained if the expression of the volume element $dV(u_1, u_2, \ldots) = v(u_1, u_2, \ldots)\, du_1\, du_2 \ldots$ of the space is known, as it is then given by $h(u_1, u_2, \ldots) = kv(u_1, u_2, \ldots)$, where k is a proportionality constant (that may have physical dimensions).*

Rule 2. *If there is a metric $g_{ij}(u_1, u_2, \ldots)$ in the space, then the volume element is given by $dV(u_1, u_2, \ldots) = \sqrt{\det \mathbf{g}(u_1, u_2, \ldots)}\, du_1\, du_2 \cdots$, i.e., we have $v(u_1, u_2, \ldots) = \sqrt{\det \mathbf{g}(u_1, u_2, \ldots)}$. The probability density representing the homogeneous probability distribution is, then, $h(u_1, u_2, \ldots) = k\sqrt{\det \mathbf{g}(u_1, u_2, \ldots)}$.*

Rule 3. *If the expression of the probability density representing the homogeneous probability distribution is known in one system of coordinates, then it is known in any other system of coordinates, through the Jacobian rule.*

Indeed, in the expression above, $g(r, \theta, \varphi) = kr^2 \sin\theta$, we recognize the Jacobian between the geographical and the Cartesian coordinates (where the probability density is constant).

For short, when we say *the homogeneous probability density* we mean *the probability density representing the homogeneous probability distribution*. **One should remember that, in general, the homogeneous probability density is *not* constant.**

Let us now examine 'positive parameters,' like a temperature, a period, or a seismic wave propagation velocity. One of the properties of the parameters we have in mind is that they occur in pairs of mutually reciprocal parameters:

Period	$T = 1/\nu$	;	*Frequency*	$\nu = 1/T$
Resistivity	$\rho = 1/\sigma$	;	Conductivity	$\sigma = 1/\rho$
Temperature	$T = 1/(k\beta)$	;	Thermodynamic parameter	$\beta = 1/(kT)$
Mass density	$\rho = 1/\ell$	;	Lightness	$\ell = 1/\rho$
Compressibility	$\gamma = 1/\kappa$	;	Bulk modulus (uncompressibility)	$\kappa = 1/\gamma$
Wave velocity	$c = 1/n$	;	Wave slowness	$n = 1/c$.

When working with physical theories, one may freely choose one of these parameters or its reciprocal.

Sometimes these pairs of equivalent parameters come from a definition, like when we define frequency ν as a function of the period T, by $\nu = 1/T$. Sometimes these parameters arise when analyzing an idealized physical system. For instance, Hooke's law, relating stress $\sigma_{\ell j}$ to strain $\varepsilon_{\ell j}$ can be expressed as $\sigma_{ij} = c_{ij}{}^{k\ell} \varepsilon_{k\ell}$, thus introducing the stiffness tensor $c_{ijk\ell}$, or as $\varepsilon_{ij} = d_{ij}{}^{k\ell} \sigma_{k\ell}$, thus introducing the compliance tensor

$d_{ijk\ell}$, the inverse of the stiffness tensor. Then the respective eigenvalues of these two tensors belong to the class of scalars analyzed here.

Let us take, as an example, the pair conductivity–resistivity (which may be thermal, electric, etc.). Assume we have two samples in the laboratory S_1 and S_2 whose resistivities are respectively ρ_1 and ρ_2. Correspondingly, their conductivities are $\sigma_1 = 1/\rho_1$ and $\sigma_2 = 1/\rho_2$. How should we define the 'distance' between the 'electrical properties' of the two samples? As we have $|\rho_2 - \rho_1| \neq |\sigma_2 - \sigma_1|$, choosing one of the two expressions as the 'distance' would be arbitrary. Consider the following definition of 'distance' between the two samples:

$$D(S_1, S_2) = \left| \log \frac{\rho_2}{\rho_1} \right| = \left| \log \frac{\sigma_2}{\sigma_1} \right| . \quad (7)$$

This definition (i) treats symmetrically the two equivalent parameters ρ and σ and, more importantly, (ii) has an *invariance of scale* (what matters is how many 'octaves' we have between the two values, not the plain difference between the values). In fact, it is the only definition of distance between the two samples S_1 and S_2 that has an invariance of scale and is additive (i.e., $D(S_1, S_2) + D(S_2, S_3) = D(S_1, S_3)$).

Associated to the distance $D(x_1, x_2) = |\log(x_2/x_1)|$ is the distance element (differential form of the distance)

$$dL(x) = \frac{dx}{x} . \quad (8)$$

This being a 'one-dimensional volume,' we can now apply Rule 1 above to get the expression of the homogeneous probability density for such a positive parameter:

$$f(x) = \frac{k}{x} . \quad (9)$$

Defining the reciprocal parameter $y = 1/x$ and using the Jacobian rule, we arrive at the homogeneous probability density for y:

$$g(y) = \frac{k}{y} . \quad (10)$$

These two probability densities have the same form: the two reciprocal parameters are treated symmetrically. Introducing the logarithmic parameters

$$x^* = \log \frac{x}{x_0} ; \qquad y^* = \log \frac{y}{y_0} , \quad (11)$$

where x_0 and y_0 are arbitrary positive constants, and using the Jacobian rule, we arrive at the homogeneous probability densities:

$$f'(x^*) = k ; \qquad g'(y^*) = k . \quad (12)$$

This shows that the logarithm of a positive parameter (of the type considered above) is a 'Cartesian' parameter. In fact, it is the consideration of Eqs. (12), together with the Jacobian rule, that allows full understanding of the (homogeneous) probability densities (9) and (10).

The association of the probability density $f(u) = k/u$ with positive parameters was first made by Jeffreys (1939). To honor him, we propose to use the term *Jeffreys parameters* for all the parameters of the type considered above. The $1/u$ probability density was advocated by Jaynes (1968), and a nontrivial use of it was made by Rietsch (1977) in the context of inverse problems.

Rule 4. *The homogeneous probability density for a Jeffreys quantity u is $f(u) = k/u$.*

Rule 5. *The homogeneous probability density for a 'Cartesian parameter' u (like the logarithm of a Jeffreys parameter, an actual Cartesian coordinate in an Euclidean space, or the Newtonian time coordinate) is $f(u) = k$. The homogeneous probability density for an angle describing the position of a point in a circle is also constant.*

If a parameter u is a Jeffreys parameter with the homogeneous probability density $f(u) = k/u$, then its inverse, its square, and, in general, any power of the parameter is also a Jeffreys parameter, as it can easily be seen using the Jacobian rule.

Rule 6. *Any power of a Jeffreys quantity (including its inverse) is a Jeffreys quantity.*

It is important to recognize when we do *not* face a Jeffreys parameter. Among the many parameters used in the literature to describe an isotropic linear elastic medium we find parameters like the Lamé's coefficients λ and μ, the bulk modulus κ, the Poisson ratio σ, etc. A simple inspection of the theoretical range of variation of these parameters shows that the first Lamé parameter λ and the Poisson ratio σ may take negative values, so they are certainly not Jeffreys parameters. In contrast, Hooke's law $\sigma_{ij} = c_{ijk\ell}\,\varepsilon^{k\ell}$, defining a linearity between stress σ_{ij} and strain ε_{ij}, defines the positive definite stiffness tensor $c_{ijk\ell}$ or, if we write $\varepsilon_{ij} = d_{ijk\ell}\,\sigma^{k\ell}$, defines its inverse, the compliance tensor $d_{ijk\ell}$. The two reciprocal tensors $c_{ijk\ell}$ and $d_{ijk\ell}$ are 'Jeffreys tensors.' This is a notion whose development is beyond the scope of this paper, but we can give the following rule.

Rule 7. *The eigenvalues of a Jeffreys tensor are Jeffreys quantities.*[6]

As the two (different) eigenvalues of the stiffness tensor $c_{ijk\ell}$ are $\lambda_\kappa = 3\kappa$ (with multiplicity 1) and $\lambda_\mu = 2\mu$ (with multiplicity 5), we see that the incompressibility modulus κ and the shear modulus μ are Jeffreys parameters[7] (as are any parameters proportional to them, or any powers of them, including the inverses). If, for some reason, instead of working with κ and μ, we wish to work with other elastic parameters, for instance, the Young modulus Y and the

Poisson ratio σ, or the two elastic wave velocities, then the homogeneous probability distribution must be found using the Jacobian of the transformation (see Appendix H).

Some probability densities have conspicuous 'dispersion parameters,' like the σ's in the normal probability density $f(x) = k \exp\left(-\frac{(x-x_0)^2}{2\sigma^2}\right)$, in the log-normal probability $g(X) = \frac{k}{X} \exp\left(-\frac{(\log X/X_0)^2}{2\sigma^2}\right)$ or in the Fisher probability density (Fisher, 1953) $h(\vartheta, \varphi) = k \sin\theta \exp(\cos\theta/\sigma^2)$. A consistent probability model requires that when the dispersion parameter σ tends to infinity, the probability density tends to the homogeneous probability distribution. For instance, in the three examples just given, $f(x) \to k$, $g(X) \to k/X$, and $h(\theta, \varphi) \to k \sin\theta$, which are the respective homogeneous probability densities for a Cartesian quantity, a Jeffreys quantity, and the geographical coordinates on the surface of the sphere. We can state the following rule.

Rule 8. *If a probability density has some 'dispersion parameters,' then, in the limit where the dispersion parameters tend to infinity, the probability density must tend to the homogeneous one.*

As an example, using the normal probability density $f(x) = k \exp\left(-\frac{(x-x_0)^2}{2\sigma^2}\right)$, for a Jeffreys parameter is not consistent. Note that it would assign a finite probability to negative values of a positive parameter that, by definition, is positive. More technically, this would violate our Postulate 1. Using the log-normal probability density for a Jeffreys parameter is consistent.

There is a problem of terminology in the Bayesian literature. The homogeneous probability distribution is a very special distribution. When the problem of selecting a 'prior' probability distribution arises in the absence of any information, except the fundamental symmetries of the problem, one may select as prior probability distribution the homogeneous distribution. But enthusiastic Bayesians do not call it 'homogeneous,' but 'noninformative.' We cannot recommend using this terminology. The homogeneous probability distribution is as informative as any other distribution, it is just the homogeneous one (see Appendix D).

In general, each time we consider an abstract parameter space, each point being represented by some parameters $\mathbf{x} = \{x^1, x^2 \ldots x^n\}$, we will start by solving the (sometimes nontrivial) problem of defining a distance between points that respects the necessary symmetries of the problem. Only exceptionally this distance will be a quadratic expression of the parameters (coordinates) being used (i.e., only exceptionally our parameters will correspond to 'Cartesian coordinates' in the space). From this distance, a volume element $dV(\mathbf{x}) = v(\mathbf{x})\, d\mathbf{x}$ will be deduced, from where the expression $f(\mathbf{x}) = k\, v(\mathbf{x})$ of the homogeneous probability density will follow. Sometimes, we can directly define the volume element, without the need of a distance. We emphasize the need of defining a distance—or a volume element—in the parameter space, from which the notion of homogeneity will follow. With this point of view, we slightly depart from the original work by Jeffreys and Jaynes.

2.4 Conjunction of Probabilities

We shall here consider two probability distributions P and Q . We say that a probability R is a product of the two given probabilities, and is denoted $(P \wedge Q)$ if

- $P \wedge Q = Q \wedge P$;
- for any subset $\mathcal{A}$, $(P \wedge Q)(\mathcal{A}) \neq 0 \Rightarrow P(\mathcal{A}) \neq 0$ and $Q(\mathcal{A}) \neq 0$;
- if M denotes the homogeneous probability distribution, then $P \wedge M = P$.

The realization of these conditions leading to the simplest results can easily be expressed using probability densities (see Appendix G for details). If the two probabilities P and Q are represented by the two probability densities $p(\mathbf{x})$ and $q(\mathbf{x})$, respectively, and if the homogeneous probability density is represented by $\mu(\mathbf{x})$, then the probability $P \wedge Q$ is represented by a probability density, denoted $(p \wedge q)(\mathbf{x})$, that is given by

$$(p \wedge q)(\mathbf{x}) = k \frac{p(\mathbf{x})\, q(\mathbf{x})}{\mu(\mathbf{x})} , \tag{13}$$

where k is a normalization constant.[8]

The two left columns of Figure 1 represent these probability densities.

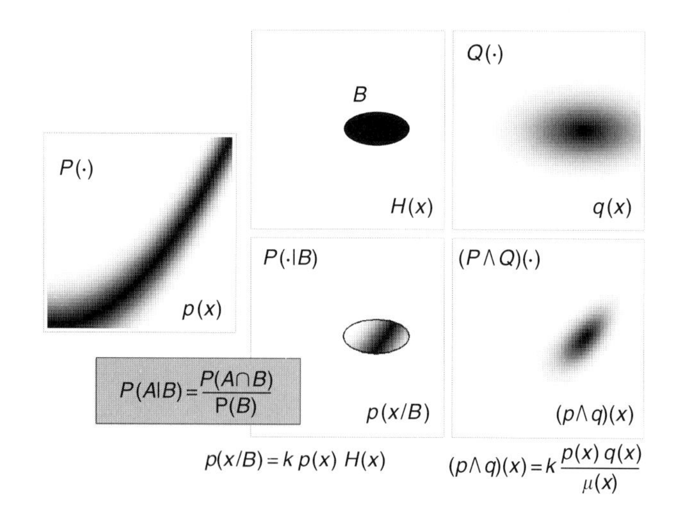

FIGURE 1 The two left columns of the figure illustrate the definition of conditional probability (see text for details). The right of the figure explains that the definition of the AND operation is a generalization of the notion of conditional probability. While a conditional probability combines a probability distribution $P(\ )$ with an 'event' B, the AND operation combines two probability distributions $P(\cdot)$ and $Q(\cdot)$ defined over the same space. See text for a detailed explanation.

Example 2. *On the surface of the Earth, using geographical coordinates (latitude* ϑ *and longitude* φ*), the homogeneous probability distribution is represented by the probability density* $\mu(\vartheta, \varphi) = \frac{1}{4\pi}\cos\vartheta$. *An estimation of the position of a floating object at the surface of the sea by an airplane navigator gives a probability distribution for the position of the object corresponding to the probability density* $p(\vartheta, \varphi)$, *and an independent, simultaneous estimation of the position by another airplane navigator gives a probability distribution corresponding to the probability density* $q(\vartheta, \varphi)$. *How do we 'combine' the two probability densities* $p(\vartheta, \varphi)$ *and* $q(\vartheta, \varphi)$ *to obtain a 'resulting' probability density? The answer is given by the conjunction of the two probability densities:*

$$(p \wedge q)(\vartheta, \varphi) = k\, \frac{p(\vartheta, \varphi)\, q(\vartheta, \varphi)}{\mu(\vartheta, \varphi)} \quad . \tag{14}$$

We emphasize here the following:

Example 2 is at the basis of the paradigm that we use below to solve inverse problems.

More generally, the conjunction of the probability densities $f_1(\mathbf{x}), f_2(\mathbf{x}) \ldots$ is

$$\begin{aligned} h(\mathbf{x}) &= (f_1 \wedge f_2 \wedge f_3 \cdots)(\mathbf{x}) \cdots \\ &= k\, \mu(\mathbf{x})\, \frac{f_1(\mathbf{x})}{\mu(\mathbf{x})}\, \frac{f_2(\mathbf{x})}{\mu(\mathbf{x})}\, \frac{f_3(\mathbf{x})}{\mu(\mathbf{x})} \cdots \quad . \end{aligned} \tag{15}$$

For a formalization of the notion of conjunction of probabilities, the reader is invited to read Appendix G.

2.5 Conditional Probability Density

Given a probability distribution over a space $\mathcal{X}$, represented by the probability density $f(\mathbf{x})$, and given a subspace $\mathcal{B}$ of $\mathcal{X}$ of lower dimension, can we, in a consistent way, infer a probability distribution over $\mathcal{B}$, represented by a probability density $f(\mathbf{x}|\mathcal{B})$ (to be named the conditional probability density 'given $\mathcal{B}$')? The answer is: Using only the elements given, NO, THIS IS NOT POSSIBLE.

The usual way to induce a probability distribution on a subspace of lower dimension is to assign a 'thickness' to the subspace $\mathcal{B}$, to apply the general definition of conditional probability (this time to a region of $\mathcal{X}$, not to a subspace of it) and to take the limit when the 'thickness' tends to zero. But, as suggested in Figure 2, there are infinitely many ways to take this limit, each defining a different 'conditional probability density' on $\mathcal{B}$. Among the infinitely many ways to define a conditional probability density, there is one that is based on the notion of distance between points in the space, and therefore corresponds to an intrinsic definition (see Fig. 2).

Assume that the space $\mathcal{U}$ has p dimensions, the space $\mathcal{V}$ has q dimensions, and define in the $(p+q)$-dimensional space $\mathcal{X} = (\mathcal{U}, \mathcal{V})$ a p-dimensional subspace by the p relations

$$\begin{aligned} v_1 &= v_1(u_1, u_2, \ldots, u_p) \\ v_2 &= v_2(u_1, u_2, \cdots, u_p) \\ \cdots &= \cdots \\ v_q &= v_q(u_1, u_2, \ldots, u_p) \ . \end{aligned} \tag{16}$$

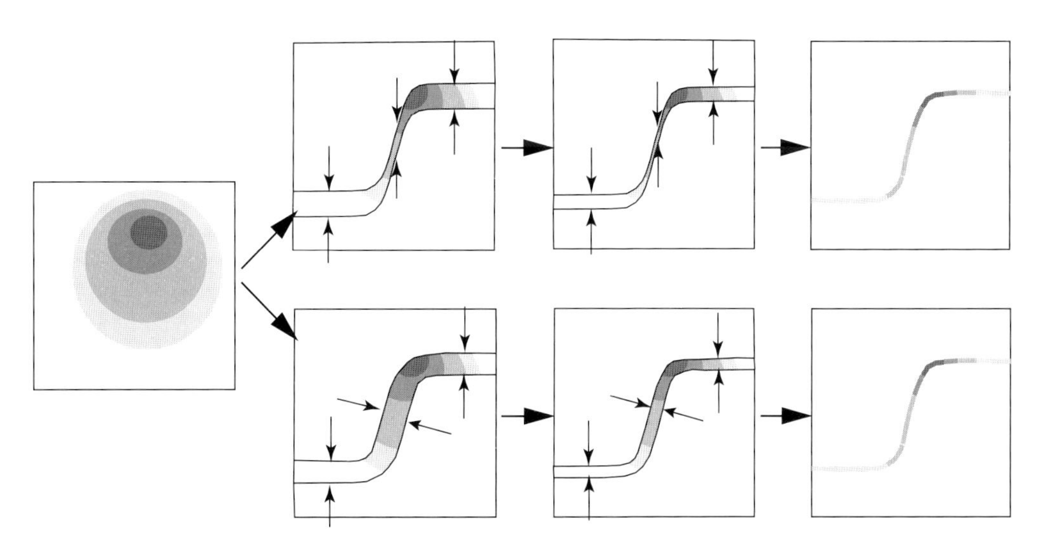

FIGURE 2 An original 2D probability density, and two possible ways (among many) of defining a region of the space whose limit is a given curve. At the top is the 'vertical' limit, while at the bottom is the normal (or orthogonal) limit. Each possible limit defines a different 'induced' or 'conditional' probability density. Only the orthogonal limit gives an intrinsic definition (i.e., a definition invariant under any change of variables). It is, therefore, the only one examined in this work.

The restriction of a probability distribution represented by the probability density $f(\mathbf{x}) = f(\mathbf{u}, \mathbf{v})$ into the subspace defined by the constraint $\mathbf{v} = \mathbf{v}(\mathbf{u})$, can be defined with all generality when it is assumed that we have a metric defined over the $(p+q)$-dimensional space $\mathcal{X} = (\mathcal{U}, \mathcal{V})$. Let us limit here to the special circumstance (useful for a vast majority of inverse problems[9]) where there the $(p+q)$-dimensional space $\mathcal{X}$ is built as the Cartesian product of $\mathcal{U}$ and $\mathcal{V}$ (then we write, as usual, $\mathcal{X} = \mathcal{U} \times \mathcal{V}$). In this case, there is a metric $\mathbf{g}_u$ over $\mathcal{U}$, with associated volume element $dV_u(\mathbf{u}) = \sqrt{\det \mathbf{g}_u}\, d\mathbf{u}$, there is a metric $\mathbf{g}_v$ over $\mathcal{V}$, with associated volume element $dV_v(\mathbf{v}) = \sqrt{\det \mathbf{g}_v}\, d\mathbf{v}$, and the global volume element is simply $dV(\mathbf{u}, \mathbf{v}) = dV_u(\mathbf{u})\, dV_v(\mathbf{v})$.

The restriction of the probability distribution represented by the probability density $f(\mathbf{u}, \mathbf{v})$ on the subspace $\mathbf{v} = \mathbf{v}(\mathbf{u})$ (i.e., the conditional probability density given $\mathbf{v} = \mathbf{v}(\mathbf{u})$) is a probability distribution *on* the submanifold $\mathbf{v} = \mathbf{v}(\mathbf{u})$. We could choose ad-hoc coordinates over this manifold, but as there is a one-to-one correspondence between the coordinates $\mathbf{u}$ and the points on the manifold, the conditional probability density can be expressed using the coordinates $\mathbf{u}$. The restriction of $f(\mathbf{u}, \mathbf{v})$ over the submanifold $\mathbf{v} = \mathbf{v}(\mathbf{u})$ defines the probability density (see Appendix B for the more general case)

$$f_{u|v(u)}(\mathbf{u}|\mathbf{v} = \mathbf{v}(\mathbf{u})) = k\, f(\mathbf{u}, \mathbf{v}(\mathbf{u})) \left. \frac{\sqrt{\det(\mathbf{g}_u + \mathbf{V}^T \mathbf{g}_v \mathbf{V})}}{\sqrt{\det \mathbf{g}_u}\, \sqrt{\det \mathbf{g}_v}} \right|_{\mathbf{v} = \mathbf{v}(\mathbf{u})}, \tag{17}$$

where k is a normalizing constant, and where $\mathbf{V} = \mathbf{V}(\mathbf{u})$ is the matrix of partial derivatives (see Appendix M for a simple explicit calculation of such partial derivatives)

$$\begin{pmatrix} V_{11} & V_{12} & \cdots & V_{1p} \\ V_{21} & V_{22} & \cdots & V_{2p} \\ \vdots & \vdots & \ddots & \vdots \\ V_{q1} & V_{q2} & \cdots & V_{qp} \end{pmatrix} = \begin{pmatrix} \frac{\partial v_1}{\partial u_1} & \frac{\partial v_1}{\partial u_2} & \cdots & \frac{\partial v_1}{\partial u_p} \\ \frac{\partial v_2}{\partial u_1} & \frac{\partial v_2}{\partial u_2} & \cdots & \frac{\partial v_2}{\partial u_p} \\ \vdots & \vdots & \ddots & \vdots \\ \frac{\partial v_q}{\partial u_1} & \frac{\partial v_q}{\partial u_2} & \cdots & \frac{\partial v_q}{\partial u_p} \end{pmatrix}. \tag{18}$$

Example 3. *If the hypersurface $\mathbf{v} = \mathbf{v}(\mathbf{u})$ is defined by a constant value of $\mathbf{v}$, say $\mathbf{v} = \mathbf{v}_0$, then Eq. (17) reduces to*

$$f_{u|v}(\mathbf{u}|\mathbf{v} = \mathbf{v}_0) = k\, f(\mathbf{u}, \mathbf{v}_0) = \frac{f(\mathbf{u}, \mathbf{v}_0)}{\int_{\mathcal{U}} d\mathbf{u}\; f(\mathbf{u}, \mathbf{v}_0)}. \tag{19}$$

Elementary definitions of conditional probability density are not based on this notion of distance-based uniform convergence, but use other, ill-defined limits. This is a mistake that, unfortunately, pollutes many scientific works. See Appendix P, in particular, for a discussion on the 'Borel paradox.'

Equation (17) defines the conditional $f_{u|v(u)}(\mathbf{u} \,|\, \mathbf{v} = \mathbf{v}(\mathbf{u}))$. Should the relation $\mathbf{v} = \mathbf{v}(\mathbf{u})$ be invertible, it would correspond to a change of variables. It is then possible to show that the alternative conditional $f_{v|u(v)}(\mathbf{v} \,|\, \mathbf{u} = \mathbf{u}(\mathbf{v}))$ is related to $f_{u|v(u)}(\mathbf{u} \,|\, \mathbf{v} = \mathbf{v}(\mathbf{u}))$ through the Jacobian rule. This is a property that elementary definitions of conditional probability do not share.

2.6 Marginal Probability Density

In the special circumstance described above, where we have a Cartesian product of two spaces, $\mathcal{X} = \mathcal{U} \times \mathcal{V}$, given a 'joint' probability density $f(\mathbf{u}, \mathbf{v})$, it is possible to give an intrinsic sense to the definitions

$$f_u(\mathbf{u}) = \int_{\mathcal{V}} d\mathbf{v}\; f(\mathbf{u}, \mathbf{v}) \,; \qquad f_v(\mathbf{v}) = \int_{\mathcal{U}} d\mathbf{u}\; f(\mathbf{u}, \mathbf{v}). \tag{20}$$

These two densities are called *marginal probability densities*. Their intuitive interpretation is clear, as the 'projection' of the joint probability density respectively over $\mathcal{U}$ and over $\mathcal{V}$.

2.7 Independence and Bayes Theorem

Dropping the index 0 in Eq. (19) and using the second of Eqs. (20) gives

$$f_{u|v}(\mathbf{u}|\mathbf{v}) = \frac{f(\mathbf{u}, \mathbf{v})}{f_v(\mathbf{v})}, \tag{21}$$

or, equivalently, $f(\mathbf{u}, \mathbf{v}) = f_{u|v}(\mathbf{u}|\mathbf{v})\, f_v(\mathbf{v})$. As we can also define $f_v | u(\mathbf{v}|\mathbf{u})$, we have the two equations

$$\begin{aligned} f(\mathbf{u}, \mathbf{v}) &= f_{u|v}(\mathbf{u}|\mathbf{v})\; f_v(\mathbf{v}) \\ f(\mathbf{u}, \mathbf{v}) &= f_{v|u}(\mathbf{v}|\mathbf{u})\; f_u(\mathbf{u}), \end{aligned} \tag{22}$$

that can be read as follows: 'When we work in a space that is the Cartesian product $\mathcal{U} \times \mathcal{V}$ of two subspaces, a joint probability density can always be expressed as the product of a conditional times a marginal.'

From these last equations there follows the expression

$$f_{u|v}(\mathbf{u}|\mathbf{v}) = \frac{f_{v|u}(\mathbf{v}|\mathbf{u})\; f_u(\mathbf{u})}{f_v(\mathbf{v})}, \tag{23}$$

known as the *Bayes theorem*, and generally used as the starting point for solving inverse problems. We do not think this is a useful setting, and we prefer here *not* to use the Bayes theorem (or, more precisely, not to use the intuitive paradigm usually associated with it).

It also follows from Eqs. (22) that the two conditions

$$f_{u|v}(\mathbf{u}|\mathbf{v}) = f_u(\mathbf{u}) \,; \qquad f_{v|u}(\mathbf{v}|\mathbf{u}) = f_v(\mathbf{v}) \tag{24}$$

are equivalent. It is then said that **u** and **v** are *independent parameters* (with respect to the probability density $f(\mathbf{u}, \mathbf{v})$). The term 'independent' is easy to understand, as the conditional of any of the two (vector) variables, given the other variable equals the (unconditional) marginal of the variable. Then, one clearly has

$$f(\mathbf{u}, \mathbf{v}) = f_u(\mathbf{u})\, f_v(\mathbf{v}) \qquad (25)$$

i.e., for independent variables, the joint probability density can be simply expressed as the product of the two marginals.

3. Monte Carlo Methods

When a probability distribution has been defined, we face the problem of how to 'use' it. The definition of 'central estimators' (such as the mean or the median) and 'estimators of dispersion' (such as the covariance matrix) lacks generality as it is quite easy to find examples (such as multimodal distributions in highly-dimensional spaces) where these estimators fail to have any interesting meaning.

When a probability distribution has been defined over a space of low dimension (say, from one to four dimensions) we can directly represent the associated probability density. This is trivial in one or two dimensions. It is easy in three dimensions, and some tricks may allow us to represent a four-dimensional probability distribution, but clearly this approach cannot be generalized to the high dimensional case.

Let us explain the only approach that seems practical, with help of Figure 3. At the left of the figure, there is an explicit representation of a 2D probability distribution (by means of the associated probability density or the associated (2D) volumetric probability). In the middle, some random points have been generated (using the Monte Carlo method about to be described). It is clear that, if we make a histogram with these points, in the limit of a sufficiently large number of points we recover the representation at the left. Disregarding the histogram possibility, we can concentrate on the individual points. In the 2D example of the figure we have actual points in a plane. If the problem is multidimensional, each 'point' may correspond to some abstract notion. For instance, for a geophysicist a 'point' may be a given model of the Earth. This model may be represented in some way, for instance, by a color plot. Then a collection of 'points' is a collection of such pictures. Our experience shows that, given a collection of randomly generated 'models,' the human eye–brain system is extremely good at apprehending the basic characteristics of the underlying probability distribution, including possible multimodalities, correlations, etc.

When such a (hopefully large) collection of random models is available, we can also answer quite interesting questions. For instance, a geologist might ask: *at which depth is that subsurface structure?* To answer this, we can make a histogram of the depth of the given geological structure over the collection of random models, and the histogram *is* the answer to the question. *What is the probability of having a low-velocity zone around a given depth?* The ratio of the number of models presenting such a low-velocity zone over the total number of models in the collection gives the answer (if the collection of models is large enough).

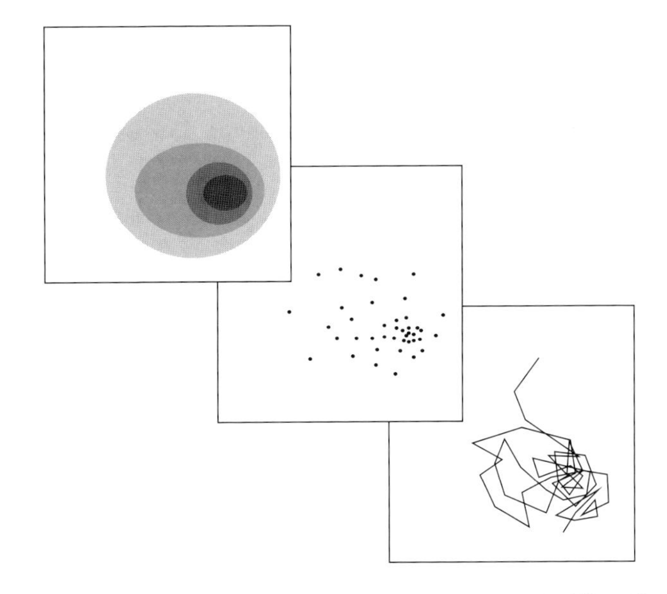

FIGURE 3 An explicit representation of a 2D probability distribution and the sampling of it, using Monte Carlo methods. While the representation at the top left cannot be generalized to high dimensions, the examination of a collection of points can be done in arbitrary dimensions. Practically, Monte Carlo generation of points is done through a 'random walk' where a 'new point' is generated in the vicinity of the previous point.

This is essentially what we propose: looking at a large number of randomly generated models in order to intuitively apprehend the basic properties of the probability distribution, followed by calculation of the probabilities of all interesting 'events.'

Practically, as we will see, the random sampling is not made by generating points independently of each other. Rather, as suggested in the last image of Figure 3, it is done through a 'random walk' where a 'new point' is generated in the vicinity of the previous point.

Monte Carlo methods have a random generator at their core. At present, Monte Carlo methods are typically implemented on digital computers, and are based on pseudo random generation of numbers.[10] As we shall see, any conceivable operation on probability densities (e.g., computing marginals and conditionals, integration, conjunction (the AND operation), etc.) has its counterpart in an operation on/by their corresponding Monte Carlo algorithms.

Inverse problems are often formulated in high-dimensional spaces. In this case a certain class of Monte Carlo algorithms, the so-called *importance sampling algorithms*, come to the rescue, allowing us to sample the space with a sampling density proportional to the given probability density. In this case excessive (and useless) sampling of low-probability areas of the space is avoided. This is not only important, but in fact vital in high-dimensional spaces.

Another advantage of the importance sampling Monte Carlo algorithms is that we need not have a closed-form mathematical expression for the probability density that we want to sample. Only an algorithm that allows us to evaluate it at a given point in the space is needed. This has considerable practical advantage in analysis of inverse problems where computer-intensive evaluation of, for example, misfit functions plays an important role in calculation of certain probability densities.

Given a probability density that we wish to sample, and a class of Monte Carlo algorithms that samples this density, which one of the algorithms should we choose? Practically, the problem here is to find the most efficient of these algorithms. This is an interesting and difficult problem for which we will not go into detail here. We will, later in this chapter, limit ourselves to only two general methods that are recommendable in many practical situations.

3.1 Random Walks

To escape the dimensionality problem, *any* sampling of a probability density for which point values are available only upon request has to be based on a *random walk*, i.e., in a generation of successive points with the constraint that point $\mathbf{x}_{i+1}$ sampled in iteration $(i+1)$ is in the vicinity of the point $\mathbf{x}_i$ sampled in iteration i. The simplest of the random walks are generated by the so-called Markov Chain Monte Carlo (MCMC) algorithms, where the point $\mathbf{x}_{i+1}$ depends on the point $\mathbf{x}_i$, but not on previous points. We will concentrate on these algorithms here.

If random rules have been defined to select points such that the probability of selecting a point in the infinitesimal 'box' $dx_1 \cdots dx_N$ is $p(\mathbf{x})\, dx_1 \cdots dx_N$, then the points selected in this way are called *samples* of the probability density $p(\mathbf{x})$. Depending on the rules defined, successive samples $i, j, k, \ldots$ may be dependent or independent.

3.2 The Metropolis Rule

The most common Monte Carlo sampling methods are the Metropolis sampler (described below) and the Gibbs sampler (Geman and Geman, 1984). As we believe that the Gibbs sampler is only superior to the Metropolis sampler in low-dimensional problems, we restrict ourselves here to the presentation of the latter.

Consider the following situation. Some random rules define a random walk that samples the probability density $f(\mathbf{x})$. At a given step, the random walker is at point $\mathbf{x}_j$, and the application of the rules would lead to a transition to point $\mathbf{x}_i$. By construction, when all such 'proposed transitions' $\mathbf{x}_i \leftarrow \mathbf{x}_j$ are always accepted, the random walker will sample the probability density $f(\mathbf{x})$. Instead of always accepting the proposed transition $\mathbf{x}_i \leftarrow \mathbf{x}_j$, we reject it sometimes by using the following rule to decide if it is allowed to move to $\mathbf{x}_i$ or if it must stay at $\mathbf{x}_j$:

- If $g(\mathbf{x}_i)/\mu(\mathbf{x}_i) \geq g(\mathbf{x}_j)/\mu(\mathbf{x}_j)$, then accept the proposed transition to $\mathbf{x}_i$.
- If $g(\mathbf{x}_i)/\mu(\mathbf{x}_i) < g(\mathbf{x}_j)/\mu(\mathbf{x}_j)$, then decide randomly to move to $\mathbf{x}_i$, or to stay at $\mathbf{x}_j$, with the following probability of accepting the move to $\mathbf{x}_i$:

$$P = \frac{g(\mathbf{x}_i)/\mu(\mathbf{x}_i)}{g(\mathbf{x}_j)/\mu(\mathbf{x}_j)} \, . \tag{26}$$

Then we have the following theorem.

Theorem 1. *The random walker samples the conjunction $h(\mathbf{x})$ of the probability densities $f(\mathbf{x})$ and $g(\mathbf{x})$*

$$h(\mathbf{x}) = k\, f(\mathbf{x})\, \frac{g(\mathbf{x})}{\mu(\mathbf{x})} = k\, \frac{f(\mathbf{x})\, g(\mathbf{x})}{\mu(\mathbf{x})} \tag{27}$$

(*see Appendix O for a demonstration*).

It should be noted here that this algorithm nowhere requires the probability densities to be normalized. This is of vital importance in practice, since it allows sampling of probability densities whose values are known only in points already sampled by the algorithm. Obviously, such probability densities cannot be normalized. Also , the fact that our theory also allows unnormalizable probability densities will not cause any trouble in the application of the above algorithm.

The algorithm above is reminiscent (see Appendix O) of the Metropolis algorithm (Metropolis *et al.*, 1953), originally designed to sample the Gibbs–Boltzmann distribution.[11] Accordingly, we will refer to the above acceptance rule as the *Metropolis rule*.

3.3 The Cascaded Metropolis Rule

As above, assume that some random rules define a random walk that samples the probability density $f_1(\mathbf{x})$. At a given step, the random walker is at point $\mathbf{x}_j$.

(1) Apply the rules that unthwarted would generate samples distributed according to $f_1(\mathbf{x})$, to propose a new point $\mathbf{x}_i$.
(2) If $f_2(\mathbf{x}_i)/\mu(\mathbf{x}_i) \geq f_2(\mathbf{x}_j)/\mu(\mathbf{x}_j)$, go to point 3; if $f_2(\mathbf{x}_i)/\mu(\mathbf{x}_i) < f_2(\mathbf{x}_j)/\mu(\mathbf{x}_j)$, then decide randomly to go to point 3 or to go back to point 1, with the following probability of going to point 3: $P = (f_2(\mathbf{x}_i)/\mu(\mathbf{x}_i))/(f_2(\mathbf{x}_j)/\mu(\mathbf{x}_j))$.

(3) If $f_3(\mathbf{x}_i)/\mu(\mathbf{x}_i) \geq f_3(\mathbf{x}_j)/\mu(\mathbf{x}_j)$, go to point 4; if $f_3(\mathbf{x}_i)/\mu(\mathbf{x}_i) < f_3(\mathbf{x}_j)/\mu(\mathbf{x}_j)$, then decide randomly to go to point 4 or to go back to point 1, with the following probability of going to point 4: $P = (f_3(\mathbf{x}_i)/\mu(\mathbf{x}_i))/(f_3(\mathbf{x}_j)/\mu(\mathbf{x}_j))$.

... ...

(n) If $f_n(\mathbf{x}_i)/\mu(\mathbf{x}_i) \geq f_n(\mathbf{x}_j)/\mu(\mathbf{x}_j)$, then accept the proposed transition to $\mathbf{x}_i$; if $f_n(\mathbf{x}_i)/\mu(\mathbf{x}_i) < f_n(\mathbf{x}_j)/\mu(\mathbf{x}_j)$, then decide randomly to move to $\mathbf{x}_i$, or to stay at $\mathbf{x}_j$, with the following probability of accepting the move to $\mathbf{x}_i$: $P = (f_n(\mathbf{x}_i)/\mu(\mathbf{x}_i))/(f_n(\mathbf{x}_j)/\mu(\mathbf{x}_j))$.

Then we have the following theorem.

Theorem 2. *The random walker samples the conjunction $h(\mathbf{x})$ of the probability densities $f_1(\mathbf{x}), f_2(\mathbf{x}), \ldots, f_n(\mathbf{x})$:*

$$h(\mathbf{x}) = k\, f_1(\mathbf{x})\, \frac{f_2(\mathbf{x})}{\mu(\mathbf{x})} \cdots \frac{f_n(\mathbf{x})}{\mu(\mathbf{x})} \,. \tag{28}$$

(see the supplementary materials to this chapter on the attached Handbook CD for a demonstration).

3.4 Initiating a Random Walk

Consider the problem of obtaining samples of a probability density $h(\mathbf{x})$ defined as the conjunction of some probability densitites $f_1(\mathbf{x}), f_2(\mathbf{x}), f_3(\mathbf{x}) \ldots,$

$$h(\mathbf{x}) = k\, f_1(\mathbf{x})\, \frac{f_2(\mathbf{x})}{\mu(\mathbf{x})}\, \frac{f_3(\mathbf{x})}{\mu(\mathbf{x})} \cdots \,, \tag{29}$$

and let us examine three common situations.

We start with a random walk that samples $f_1(\mathbf{x})$ (optimal situation): This corresponds to the basic algorithm where we know how to produce a random walk that samples $f_1(\mathbf{x})$, and we only need to modify it, taking into account the values $f_2(\mathbf{x})/\mu(\mathbf{x}), f_3(\mathbf{x})/\mu(\mathbf{x}) \ldots,$ using the cascaded Metropolis rule, to obtain a random walk that samples $h(\mathbf{x})$.

We start with a random walk that samples the homogeneous probability density $\mu(\mathbf{x})$: We can write Eq. (29) as

$$h(\mathbf{x}) = k \left(\left(\left(\mu(\mathbf{x})\, \frac{f_1(\mathbf{x})}{\mu(\mathbf{x})} \right) \frac{f_2(\mathbf{x})}{\mu(\mathbf{x})} \right) \cdots \right) . \tag{30}$$

The expression corresponds to the case where we are not able to start with a random walk that samples $f_1(\mathbf{x})$, but we have a random walk that samples the homogeneous probability density $\mu(\mathbf{x})$. Then, with respect to the example just mentioned, there is one extra step to be added, taking into account the values of $f_1(\mathbf{x})/\mu(\mathbf{x})$.

We start with an arbitrary random walk (worst situation): In the situation where we are not able to directly define a random walk that samples the homogeneous probability distribution, but only one that samples some arbitrary (but known) probability distribution $\psi(\mathbf{x})$, we can write Eq. (29) in the form

$$h(\mathbf{x}) = k \left(\left(\left(\left(\psi(\mathbf{x})\, \frac{\mu(\mathbf{x})}{\psi(\mathbf{x})} \right) \frac{f_1(\mathbf{x})}{\mu(\mathbf{x})} \right) \frac{f_2(\mathbf{x})}{\mu(\mathbf{x})} \right) \cdots \right) . \tag{31}$$

Then, with respect to the example just mentioned, there is one more extra step to be added, taking into account the values of $\mu(\mathbf{x})/\psi(\mathbf{x})$. Note that the closer $\psi(\mathbf{x})$ is be to $\mu(\mathbf{x})$, the more efficient will be the first modification of the random walk.

3.5 Convergence Issues

When has a random walk visited enough points in the space so that a probability density has been sufficiently sampled? This is a complex issue, and it is easy to overlook its importance. There is no general rule: Each problem has its own 'physics,' and the experience of the 'implementer' is crucial here.

Many methods that work for low dimension completely fail when the number of dimensions is high. Typically, a random walk select a random direction and, then, a random step along that direction. The notion of 'direction' in a high-dimensional space is far from the intuitive one we get in the familar three-dimensional space. Any serious discussion on this issue must be problem-dependent, so we do not even attempt one here.

Obviously, a necessary condition for adequate sampling is that any 'output' from the algorithm must 'look stationary.'

4. Probabilistic Formulation of Inverse Problems

A so-called 'inverse problem' arises when a usually complex measurement is made, and information on unknown parameters of the physical system is sought. Any measurement is indirect (we may weigh a mass by observing the displacement of the cursor of a balance), and therefore a possibly nontrivial analysis of uncertainties must be done. Any guide describing good experimental practice (see, for instance the ISO's *Guide to the Expression of Uncertainty in Measurement* (ISO, 1993) or the shorter description by Taylor and Kuyatt, 1994) acknowledges that a measurement involves, at least, two different sources of uncertainties: those estimated using statistical methods, and those estimated using subjective, common-sense estimations. Both are described using the axioms of probability theory, and this chapter clearly takes the probabilistic point of view for developing inverse theory.

4.1 Model Parameters and Observable Parameters

Although the separation of all the variables of a problem into two groups, 'directly observable parameters' (or 'data') and 'model parameters', may sometimes be artificial, we take this point of view here, since it allows us to propose a simple setting for a wide class of problems.

We may have in mind a given physical system, like the whole Earth or a small crystal under our microscope. The system (or a given state of the system) may be described by assigning values to a given set of parameters $\mathbf{m} = \{ m^1, m^2, \ldots, m^{NM} \}$, which we will name the *model parameters*.

Let us assume that we make observations on this system. Although we are interested in the parameters $\mathbf{m}$, they may not be directly observable, so we make indirect measurements such as obtaining seismograms at the Earth's surface for analyzing the Earth's interior, or making spectroscopic measurements for analyzing the chemical properties of a crystal. The set of (*directly*) *observable parameters* (or, by abuse of language, the set of *data parameters*) will be represented by $\mathbf{d} = \{ d^1, d^2, \ldots, d^{ND} \}$.

We assume that we have a physical theory that can be used to solve the *forward problem*, i.e., that given an arbitrary model $\mathbf{m}$, it allows us to predict the theoretical data values $\mathbf{d}$ that an ideal measurement should produce (if $\mathbf{m}$ were the actual system). The generally nonlinear function that associates with any model $\mathbf{m}$ the theoretical data values $\mathbf{d}$ may be represented by a notation such as

$$d^i = f^i(m^1, m^2, \ldots, m^{NM}) \; ; \qquad i = 1, 2, \ldots, ND \; , \qquad (32)$$

or, for short,

$$\mathbf{d} = \mathbf{f}(\mathbf{m}) \; . \qquad (33)$$

It is in fact this expression that separates the whole set of our parameters into the subsets $\mathbf{d}$ and $\mathbf{m}$, although sometimes there is no difference in nature between the parameters in $\mathbf{d}$ and the parameters in $\mathbf{m}$. For instance, in the classical inverse problem of estimating the hypocenter coordinates of an earthquake, we may put in $\mathbf{d}$ the arrival times of the seismic waves at seismic observatories, and we need to put in $\mathbf{m}$, besides the hypocentral coordinates, the coordinates defining the location of the seismometers—as these are parameters that are needed to compute the travel times—although we estimate arrival times of waves and coordinates of the seismic observatories using similar types of measurements.

4.2 Prior Information on Model Parameters

In a typical geophysical problem, the model parameters contain geometrical parameters (positions and sizes of geological bodies) and physical parameters (values of the mass density, of the elastic parameters, the temperature, the porosity, etc.).

The *prior information* on these parameters is all the information we possess independently of the particular measurements that will be considered as 'data' (to be described below). This prior probability distribution is generally quite complex, as the model space may be high-dimensional, and the parameters may have nonstandard probability densities.

To this generally complex probability distribution over the model space corresponds a probability density that we denote $\rho_m(\mathbf{m})$.

If an explicit expression for the probability density $\rho_m(\mathbf{m})$ is known, it can be used in analytical developments. But such an explicit expression is, by no means, necessary. Using Monte Carlo methods, all that is needed is a set of probabilistic rules that allows us to generate samples distributed according to $\rho_m(\mathbf{m})$ in the model space (Mosegaard and Tarantola, 1995).

Example 4. *Appendix E presents an example of prior information for the case of an Earth model consisting of a stack of horizontal layers with variable thickness and uniform mass density.*

4.3 Measurements and Experimental Uncertainties

Observation of geophysical phenomena is represented by a set of parameters $\mathbf{d}$ that we usually call data. These parameters result from prior measurement operations, and they are typically seismic vibrations on the instrument site, arrival times of seismic phases, gravity or electromagnetic fields. As in any measurement, the data are determined with an associated uncertainty, described by a probability density over the data parameter space, that we denote here $\rho_d(\mathbf{d})$. This density describes not only marginals on individual datum values, but also possible cross-relations in data uncertainties.

Although the instrumental errors are an important source of data uncertainties, in geophysical measurements there are other sources of uncertainty. The errors associated with the positioning of the instruments, the environmental noise, and the human factor (like for picking arrival times) are also relevant sources of uncertainty.

Example 5. Nonanalytic Probability Density. *Assume that we wish to measure the time t of occurrence of some physical event. It is often assumed that the result of a measurement corresponds to something like*

$$t = t_0 \pm \sigma \; . \qquad (34)$$

An obvious question is the exact meaning of the $\pm\sigma$. Has the experimenter in mind that she or he is absolutely certain that the actual arrival time satisfies the strict conditions

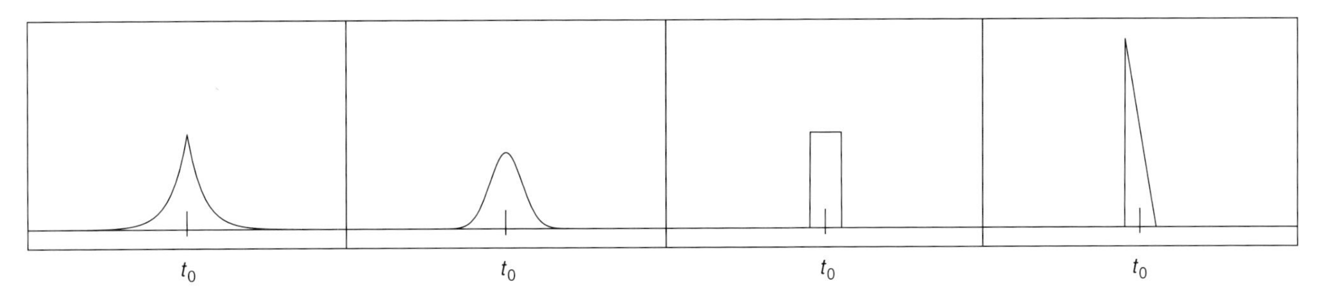

FIGURE 4 What has an experimenter in mind when she or he describes the result of a measurement by something like $t = t_0 \pm \sigma$?

$t_0 - \sigma \leq t \leq t_0 + \sigma$, or has she or he in mind something like a Gaussian probability, or some other probability distribution (see Fig. 4)? We accept, following ISO's recommendations (1993) that the result of any measurement has a probabilistic interpretation, with some sources of uncertainty being analyzed using statistical methods ('type A' uncertainties), and other sources of uncertainty being evaluated by other means (for instance, using Bayesian arguments) ('type B' uncertainties). But, contrary to ISO suggestions, we do not assume that the Gaussian model of uncertainties should play any central role. In an extreme example, we may well have measurements whose probabilistic description may correspond to a multimodal probability density. Figure 5 shows a typical example for a seismologist: the measurement on a seismogram of the arrival time of a certain seismic wave, in the case one hesitates in the phase identification or in the identification of noise and signal. In this case the probability density for the arrival of the seismic phase does not have an explicit expression like $f(t) = k \exp(-(t-t_0)^2/(2\sigma^2))$, but is a numerically defined function.

Example 6. *The Gaussian model for uncertainties. The simplest probabilistic model that can be used to describe experimental uncertainties is the Gaussian model*

$$\rho_d(\mathbf{d}) = k \exp\left(-\frac{1}{2}(\mathbf{d} - \mathbf{d}_{\text{obs}})^T \mathbf{C}_D^{-1} (\mathbf{d} - \mathbf{d}_{\text{obs}})\right). \tag{35}$$

It is here assumed that we have some 'observed data values' $\mathbf{d}_{\text{obs}}$ with uncertainties described by the covariance matrix $\mathbf{C}_D$. If the uncertainties are uncorrelated,

$$\rho_d(\mathbf{d}) = k \exp\left(-\frac{1}{2}\sum_i \left(\frac{d^i - d^i_{\text{obs}}}{\sigma^i}\right)^2\right), \tag{36}$$

where the σ^i are the 'standard deviations.'

Example 7. *The generalized Gaussian model for uncertainties. An alternative to the Gaussian model is to use the*

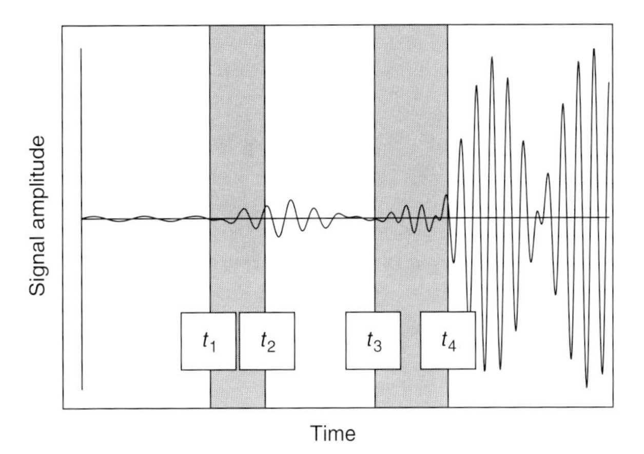

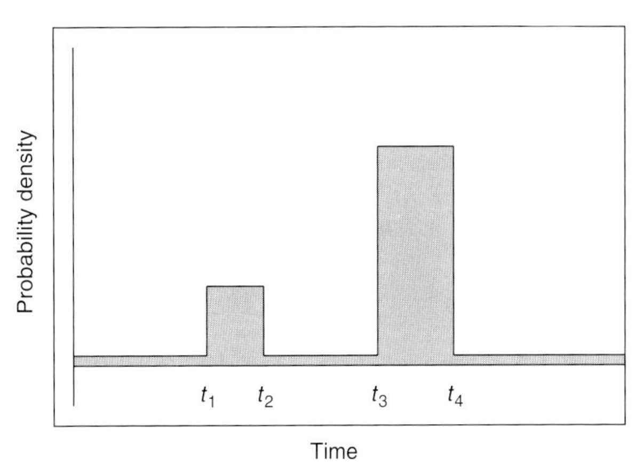

FIGURE 5 A seismologist tries to measure the arrival time of a seismic wave at a seismic station, by 'reading' the seismogram at the top of the figure. The seismologist may find quite likely that the arrival time of the wave is between times t_3 and t_4, and believe that what is before t_3 is just noise. But if there is a significant probability that the signal between t_1 and t_2 is not noise but the actual arrival of the wave, then the seismologist should define a bimodal probability density, as the one suggested at the bottom of the figure. Typically, the actual form of each peak of the probability density is not crucial (here, box-car functions are chosen), but the position of the peaks is important. Rather than assigning a zero probability density to the zones outside the two intervals, it is safer (more 'robust') to attribute some small 'background' value, as we may never exclude some unexpected source of error.

Laplacian (double exponential) model for uncertainties,

$$\rho_d(\mathbf{d}) = k \exp\left(- \sum_i \frac{|d^i - d^i_{\text{obs}}|}{\sigma^i} \right) . \tag{37}$$

While the Gaussian model leads to least-squares-related methods, this Laplacian model leads to absolute-values methods (see Section 4.5.2), well known for producing robust[12] *results. More generally, there is the* L_p *model of uncertainties*

$$\rho_p(\mathbf{d}) = k \exp\left(-\frac{1}{p} \sum_i \frac{|d^i - d^i_{\text{obs}}|^p}{(\sigma^i)^p} \right) \tag{38}$$

(see Fig. 6).

4.4 Joint "Prior" Probability Distribution in the $(\mathcal{M}, \mathcal{D})$ Space

We have just seen that the prior information on model parameters can be described by a probability density in the model space, $\rho_m(\mathbf{m})$, and that the result of measurements can be described by a probability density in the data space $\rho_d(\mathbf{d})$. As by 'prior' information on model parameters we mean information obtained *independently* from the measurements (it often represents information we had before the measurements were made), we can use the notion of independency of variables of Section 2.6 to define a joint probability density in the $\mathcal{X} = (\mathcal{M}, \mathcal{D})$ space as the product of the two 'marginals'

$$\rho(\mathbf{x}) = \rho(\mathbf{m}, \mathbf{d}) = \rho_m(\mathbf{m})\, \rho_d(\mathbf{d}) . \tag{39}$$

Although we have introduced $\rho_m(\mathbf{m})$ and $\rho_d(\mathbf{d})$ separately, and we have suggested building a probability distribution in the $(\mathcal{M}, \mathcal{D})$ space by the multiplication (39), we may have a more general situation where the information we have on $\mathbf{m}$ and on $\mathbf{d}$ is not independent. So, in what follows, let us assume that we have some information in the $\mathcal{X} = (\mathcal{M}, \mathcal{D})$ space, represented by the 'joint' probability density

$$\rho(\mathbf{x}) = \rho(\mathbf{m}, \mathbf{d}) , \tag{40}$$

and let us consider Eq. (39) as just a special case.

Let us in the rest of this chapter denote by $\mu(\mathbf{x})$ the probability density representing the homogeneous probability distribution, as introduced in Section 2.2. We may remember here the Rule 8, stating that the limit of a consistent probability density must be the homogeneous one, so we may formally write

$$\mu(\mathbf{x}) = \lim_{\text{infinite dispersions}} \rho(\mathbf{x}) . \tag{41}$$

When the partition (39) holds, then, typically (see Rule 8),

$$\mu(\mathbf{x}) = \mu(\mathbf{m}, \mathbf{d}) = \mu_m(\mathbf{m})\, \mu_d(\mathbf{d}) . \tag{42}$$

4.5 Physical Laws as Mathematical Functions

4.5.1 Physical Laws

Physics analyzes the correlations existing between physical parameters. In standard mathematical physics, these correlations are represented by 'equalities' between physical parameters (as when we write $\mathbf{F} = m\,\mathbf{a}$ to relate the force

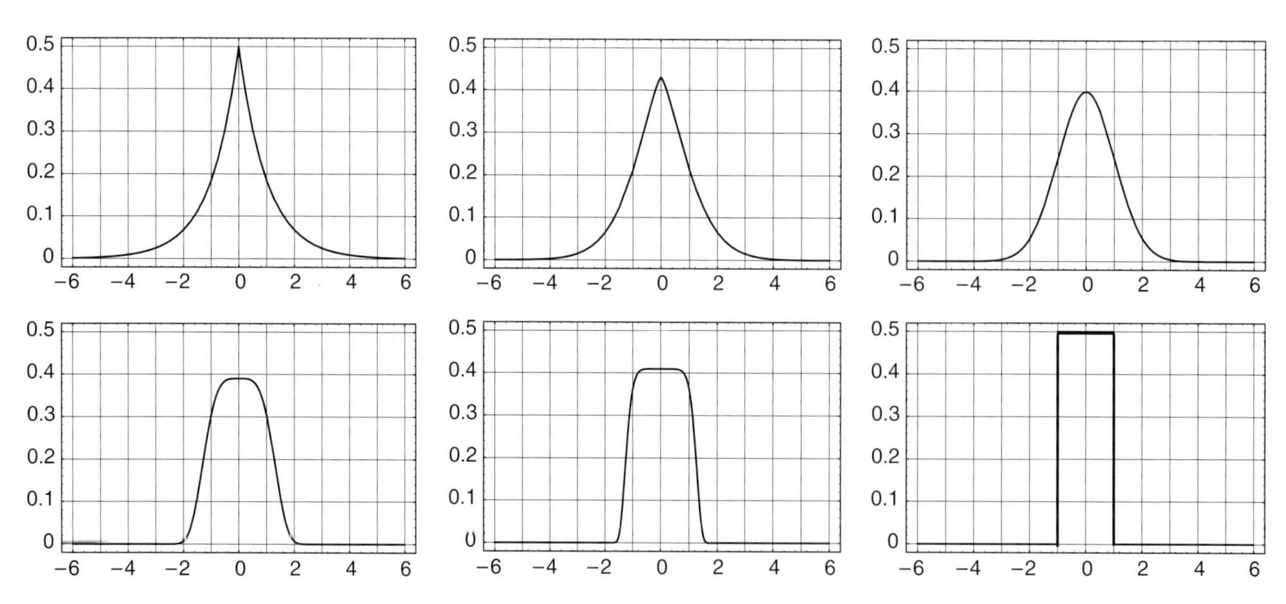

FIGURE 6 Generalized Gaussian for values of the parameter $p = 1, \sqrt{2}, 2, 4, 8$, and ∞ .

F applied to a particle, the mass m of the particle and the acceleration **a**). In the context of inverse problems, this corresponds to assuming that we have a function from the 'parameter space' to the 'data space' that we may represent as

$$\mathbf{d} = \mathbf{f}(\mathbf{m}) \ . \tag{43}$$

We do not mean that the relation is necessarily explicit. Given **m** we may need to solve a complex system of equations in order to get **d**, but this nevertheless defines a function $\mathbf{m} \to \mathbf{d} = \mathbf{f}(\mathbf{m})$.

At this point, given the probability density $\rho(\mathbf{m}, \mathbf{d})$ and given the relation $\mathbf{d} = \mathbf{f}(\mathbf{m})$, we can define the associated conditional probability density $\rho_{m|d(m)}(\mathbf{m} \mid \mathbf{d} = \mathbf{f}(\mathbf{m}))$. We could here use the more general definition of conditional probability density of Appendix B, but let us simplify the text by using a simplifying assumption: that the total parameter space $(\mathcal{M}, \mathcal{D})$ is just the Cartesian product $\mathcal{M} \times \mathcal{D}$ of the model parameter space $\mathcal{M}$ times the space of directly observable parameters (or 'data space') $\mathcal{D}$. Then, rather than a general metric in the total space, we have a metric $\mathbf{g}_m$ over the model parameter space $\mathcal{M}$ and a metric $\mathbf{g}_d$ over the data space, and the total metric is just the Cartesian product of the two metrics. In particular, then, the total volume element in the space, $dV(\mathbf{m}, \mathbf{d})$ is just the product of the two volume elements in the model parameter space and the data space: $dV(\mathbf{m}, \mathbf{d}) = dV_m(\mathbf{m})\, dV_d(\mathbf{D})$. Most inverse problems satisfy this assumption.[13] In this setting, the formulas of Section 2.4 are valid.

4.5.2 Inverse Problems

In the $(\mathcal{M}, \mathcal{D}) = \mathcal{M} \times \mathcal{D}$ space, we have the probability density $\rho(\mathbf{m}, \mathbf{d})$ and we have the hypersurface defined by the relation $\mathbf{d} = \mathbf{f}(\mathbf{m})$. The natural way to 'compose' these two kinds of information is by defining the conditional probability density induced by $\rho(\mathbf{m}, \mathbf{d})$ on the hypersurface $\mathbf{d} = \mathbf{f}(\mathbf{m})$,

$$\sigma_m(\mathbf{m}) \equiv \rho_{m|d(m)}(\mathbf{m}|\mathbf{d} = \mathbf{f}(\mathbf{m})) \ , \tag{44}$$

this gives (see Eq. (17))

$$\sigma_m(\mathbf{m}) = k\,\rho(\mathbf{m}, \mathbf{f}(\mathbf{m}))\, \frac{\sqrt{\det\left(\mathbf{g}_m + \mathbf{F}^T\, \mathbf{g}_d\, \mathbf{F}\right)}}{\sqrt{\det \mathbf{g}_m}\sqrt{\det \mathbf{g}_d}}\Bigg|_{\mathbf{d}=\mathbf{f}(\mathbf{m})} \ , \tag{45}$$

where $\mathbf{F} = \mathbf{F}(\mathbf{m})$ is the matrix of partial derivatives, with components $\mathbf{F}_{i\alpha} = \partial f_i / \partial m_\alpha$, where $\mathbf{g}_m$ is the metric in the model parameter space $\mathcal{M}$ and where $\mathbf{g}_d$ is the metric in the data space $\mathcal{D}$.

Example 8. *Quite often,* $\rho(\mathbf{m}, \mathbf{d}) = \rho_m(\mathbf{m})\, \rho_d(\mathbf{d})$. *Then, Eq. (45) can be written*

$$\sigma_m(\mathbf{m}) = k\,\rho_m(\mathbf{m}) \left(\frac{\rho_d(\mathbf{d})}{\sqrt{\det \mathbf{g}_m}}\, \frac{\sqrt{\det(\mathbf{g}_m + \mathbf{F}^T\, \mathbf{g}_d\, \mathbf{F})}}{\sqrt{\det \mathbf{g}_m}} \right)\Bigg|_{\mathbf{d}=\mathbf{f}(\mathbf{m})} . \tag{46}$$

Example 9. *If* $g_{\mathbf{m}}(\mathbf{m}) = constant$ *and* $g_{\mathbf{d}}(\mathbf{d}) = constant$, *and the nonlinearities are weak* $(\mathbf{F}(\mathbf{m}) = constant)$, *then Eq. (46) reduces to*

$$\sigma_m(\mathbf{m}) = k\ \rho_m(\mathbf{m})\, \frac{\rho_d(\mathbf{d})}{\mu_d(\mathbf{d})}\Bigg|_{\mathbf{d}=\mathbf{f}(\mathbf{m})} \ , \tag{47}$$

where we have used $\mu_d(\mathbf{d}) = k\, \sqrt{\det \mathbf{g}_d(\mathbf{d})}$ *(see Rule 2).*

Example 10. *We examine here the simplification that we arrive at when assuming that the 'input' probability densities are Gaussian:*

$$\rho_m(\mathbf{m}) = k\ \exp\left(-\frac{1}{2}(\mathbf{m} - \mathbf{m}_{\text{prior}})^t\ \mathbf{C}_M^{-1}\,(\mathbf{m} - \mathbf{m}_{\text{prior}}) \right) \tag{48}$$

$$\rho_d(\mathbf{d}) = k\ \exp\left(-\frac{1}{2}(\mathbf{d} - \mathbf{d}_{\text{obs}})^t\ \mathbf{C}_D^{-1}\,(\mathbf{d} - \mathbf{d}_{\text{obs}}) \right) . \tag{49}$$

In this circumstance, quite often, it is the covariance operators $\mathbf{C}_M$ and $\mathbf{C}_D$ *that are used to define the metrics over the spaces* $\mathcal{M}$ *and* $\mathcal{D}$ *. Then,* $\mathbf{g}_m = \mathbf{C}_M^{-1}$ *and* $\mathbf{g}_d = \mathbf{C}_D^{-1}$ *. Grouping some of the constant factors in the factor* k *, Eq. (45) becomes here*

$$\sigma_m(\mathbf{m}) = k\ \exp\left[-\frac{1}{2}\left((\mathbf{m} - \mathbf{m}_{\text{prior}})^t\ \mathbf{C}_M^{-1}\,(\mathbf{m} - \mathbf{m}_{\text{prior}}) + (\mathbf{f}(\mathbf{m}) - \mathbf{d}_{\text{obs}})^t\ \mathbf{C}_D^{-1}\,(\mathbf{f}(\mathbf{m}) - \mathbf{d}_{\text{obs}}) \right)\right] \times \frac{\sqrt{\det\left(\mathbf{C}_M^{-1} + \mathbf{F}^T(\mathbf{m})\ \mathbf{C}_D^{-1}\,\mathbf{F}(\mathbf{m})\right)}}{\sqrt{\det \mathbf{C}_M^{-1}}} \tag{50}$$

(the constant factor $\sqrt{\det \mathbf{C}_M^{-1}}$ *has been left for subsequent simplifications). Defining the misfit*

$$S(\mathbf{m}) = -2\,\log \frac{\sigma_m(\mathbf{m})}{\sigma_0} \ , \tag{51}$$

where σ_0 *is an arbitrary value of* $\sigma_m(\mathbf{m})$ *, gives, up to an additive constant,*

$$S(\mathbf{m}) = S_1(\mathbf{m}) - S_2(\mathbf{m}) \ , \tag{52}$$

where $S_1(\mathbf{m})$ is the usual least-squares misfit function

$$S_1(\mathbf{m}) = (\mathbf{m} - \mathbf{m}_{\text{prior}})^t\, \mathbf{C}_M^{-1}\, (\mathbf{m} - \mathbf{m}_{\text{prior}}) + (\mathbf{f}(\mathbf{m}) - \mathbf{d}_{\text{obs}})^t\, \mathbf{C}_D^{-1}\, (\mathbf{f}(\mathbf{m}) - \mathbf{d}_{\text{obs}}) \qquad (53)$$

and where[14]

$$S_2(\mathbf{m}) = \log\det\left(\mathbf{I} + \mathbf{C}_M\, \mathbf{F}^t(\mathbf{m})\, \mathbf{C}_D^{-1}\, \mathbf{F}(\mathbf{m})\right) . \qquad (54)$$

Example 11. *If, in the context of Example 10, we have*[15] $\mathbf{C}_M \mathbf{F}^t \mathbf{C}_D^{-1} \mathbf{F} \ll \mathbf{I}$, *we can use the low order approximation for $S_2(\mathbf{m})$, that is*[16]

$$S_2(\mathbf{m}) \approx \text{trace}\, \mathbf{C}_M\, \mathbf{F}^t(\mathbf{m})\, \mathbf{C}_D^{-1}\, \mathbf{F}(\mathbf{m}) . \qquad (55)$$

Example 12. *If in the context of Example 10 we assume that the nonlinearities are weak, then the matrix of partial derivatives* $\mathbf{F}$ *is approximately constant, and Eq. (50) simplifies to*

$$\sigma_m(\mathbf{m}) = k \exp\left[-\frac{1}{2}\left((\mathbf{m} - \mathbf{m}_{\text{prior}})^t\, \mathbf{C}_M^{-1}\, (\mathbf{m} - \mathbf{m}_{\text{prior}}) + (\mathbf{f}(\mathbf{m}) - \mathbf{d}_{\text{obs}})^t\, \mathbf{C}_D^{-1}\, (\mathbf{f}(\mathbf{m}) - \mathbf{d}_{\text{obs}})\right)\right] , \qquad (56)$$

and the function $S_2(\mathbf{m})$ is just a constant.

Example 13. *If the 'relation solving the forward problem'* $\mathbf{d} = \mathbf{f}(\mathbf{m})$ *happens to be a linear relation,* $\mathbf{d} = \mathbf{F}\mathbf{m}$, *then one gets the standard equations for linear problems (see Appendix F).*

Example 14. *We examine here the simplifications that we arrive at when assuming that the 'input' probability densities are Laplacian:*

$$\rho_m(\mathbf{m}) = k \exp\left(-\sum_\alpha \frac{|m^\alpha - m^\alpha_{\text{prior}}|}{\sigma_\alpha}\right) \qquad (57)$$

$$\rho_d(\mathbf{d}) = k \exp\left(-\sum_i \frac{|d^i - d^i_{\text{obs}}|}{\sigma_i}\right) . \qquad (58)$$

Equation (45) becomes here

$$\sigma_m(\mathbf{m}) = k \exp\left[-\left(\sum_\alpha \frac{|m^\alpha - m^\alpha_{\text{prior}}|}{\sigma_\alpha} + \sum_i \frac{|f^i(\mathbf{m}) - d^i_{\text{obs}}|}{\sigma_i}\right)\right] \Psi(\mathbf{m}) , \qquad (59)$$

where $\Psi(\mathbf{m})$ is a complex term containing, in particular, the matrix of partial derivatives $\mathbf{F}$. *If this term is approximately constant (weak nonlinearities, constant metrics), then*

$$\sigma_m(\mathbf{m}) = k \exp\left[-\left(\sum_\alpha \frac{|m^\alpha - m^\alpha_{\text{prior}}|}{\sigma_\alpha} + \sum_i \frac{|f^i(\mathbf{m}) - d^i_{\text{obs}}|}{\sigma_i}\right)\right] . \qquad (60)$$

The formulas in the examples above give expressions that contain analytic parts (like the square roots containing the matrix of partial derivatives $\mathbf{F}$) . What we write as $\mathbf{d} = \mathbf{f}(\mathbf{m})$ may sometimes correspond to an explicit expression; sometimes it may correspond to the solution of an implicit equation.[17] Should $\mathbf{d} = \mathbf{f}(\mathbf{m})$ be an explicit expression, and should the 'prior probability densities' $\rho_m(\mathbf{m})$ and $\rho_d(\mathbf{d})$ (or the joint $\rho(\mathbf{m}, \mathbf{d})$) also be given by explicit expressions (as when we have Gaussian probability densities), then the formulas of this section would give explicit expressions for the posterior probability density $\sigma_m(\mathbf{m})$.

If the relation $\mathbf{d} = \mathbf{f}(\mathbf{m})$ is a linear relation, then the expression giving $\sigma_m(\mathbf{m})$ can sometimes be simplified easily (as with the linear Gaussian case to be examined below). More often than not the relation $\mathbf{d} = \mathbf{f}(\mathbf{m})$ is a complex nonlinear relation, and the expression we are left with for $\sigma_m(\mathbf{m})$ is explicit, but complex.

Once the probability density $\sigma_m(\mathbf{m})$ has been defined, there are different ways of 'using' it. If the 'model space' $\mathcal{M}$ has a small number of dimensions (say between one and four) the values of $\sigma_m(\mathbf{m})$ can be computed at every point of a grid and a graphical representation of $\sigma_m(\mathbf{m})$ can be attempted. A visual inspection of such a representation is usually worth a thousand 'estimators' (central estimators or estimators of dispersion). But, of course, if the values of $\sigma_m(\mathbf{m})$ are known at all points where $\sigma_m(\mathbf{m})$ has a significant value, these estimators can also be computed.

If the 'model space' $\mathcal{M}$ has a large number of dimensions (say from five to many millions or billions), then an exhaustive exploration of the space is not possible, and we must turn to Monte Carlo sampling methods to extract information from $\sigma_m(\mathbf{m})$. We discuss the application of Monte Carlo methods to inverse problems, and optimization techniques, in Section 6 and 7, respectively.

4.6 Physical Laws as Probabilistic Correlations

4.6.1 Physical Laws

We return here to the general case where it is not assumed that the total space $(\mathcal{M}, \mathcal{D})$ is the Cartesian product of two spaces.

In Section 4.5 we have examined the situation where the physical correlation between the parameters of the problem are expressed using an exact, analytic expression $\mathbf{d} = \mathbf{g}(\mathbf{m})$. In this case, the notion of conditional probability density has been used to combine the 'physical theory' with the 'data' and the 'a priori information' on model parameters.

But, we have seen that in order to properly define the notion of conditional probability density, it has been necessary to introduce a metric over the space, and to take a limit using the metric of the space. This is equivalent to put some 'thickness' around the theoretical relation $\mathbf{d} = \mathbf{g}(\mathbf{m})$, and to take the limit when the thickness tends to zero.

But actual theories have uncertainties, and, for more generality, it is better to explicitly introduce these uncertainties. Assume, then, that the physical correlations between the model parameters $\mathbf{m}$ and the data parameters $\mathbf{d}$ are not represented by an analytical expression like $\mathbf{d} = \mathbf{f}(\mathbf{m})$ but by a probability density

$$\vartheta(\mathbf{m}, \mathbf{d}) \,. \tag{61}$$

Example: Realistic 'Uncertainty Bars' around a Functional Relation. In the approximation of a constant gravity field, with acceleration $\mathbf{g}$, the position at time t of an apple in free fall is $\mathbf{r}(t) = \mathbf{r}_0 + \mathbf{v}_0\, t + \frac{1}{2}\mathbf{g}\, t^2$, where $\mathbf{r}_0$ and $\mathbf{v}_0$ are, respectively, the position and velocity of the object at time $t = 0$. More simply, if the movement is 1D,

$$x(t) = x_0 + v_0\, t + \frac{1}{2} g t^2 \,. \tag{62}$$

Of course, for many reasons this equation can never be exact: air friction, wind effects, inhomogeneity of the gravity field, effects of the Earth rotation, forces from the Sun and the Moon (not to mention Pluto), relativity (special and general), and so on.

It is not a trivial task, given very careful experimental conditions, to estimate the size of the leading uncertainty. Although one might think of an equation $x = x(t)$ as a line, infinitely thin, there will always be sources of uncertainty (at least due to the unknown limits of validity of general relativity): looking at the line with a magnifying glass should reveal a fuzzy object of finite thickness. As a simple example, let us examine here the mathematical object we arrive at when assuming that the leading sources of uncertainty in the relation $x = x(t)$ are the uncertainties in the initial position and velocity of the falling apple. Let us assume that:

- the initial position of the apple is random, with a Gaussian distribution centered at x_0, and with standard deviation σ_x;
- the initial velocity of the apple is random, with a Gaussian distribution centered at v_0, and with standard deviation σ_v.

Then it can be shown that at a given time t, the possible positions of the apple are random, with probability density

$$\vartheta(x|t) = \frac{1}{\sqrt{2\pi}\sqrt{\sigma_x^2 + \sigma_v^2 t^2}} \times \exp\left(-\frac{1}{2} \frac{\left(x - (x_0 + v_0\, t + \frac{1}{2} g t^2)\right)^2}{\sigma_x^2 + \sigma_v^2 t^2} \right) \,. \tag{63}$$

This is obviously a conditional probability density for x, given t. Should we have any reason to choose some marginal probability density $\vartheta_t(t)$, then, the 'law' for the fall of the apple would be

$$\vartheta(x, t) = \vartheta(x|t)\, \vartheta_t(t) \,. \tag{64}$$

See Appendix C for more details.

4.6.2 Inverse Problems

We have seen that the result of measurements can be represented by a probability density $\rho_d(\mathbf{d})$ in the data space. We have also seen that the a priori information on the model parameters can be represented by another probability density $\rho_m(\mathbf{m})$ in the model space. When we talk about 'measurements' and about 'a priori information on model parameters,' we usually mean that we have a joint probability density in the $(\mathcal{M}, \mathcal{D})$ space, that is $\rho(\mathbf{m}, \mathbf{d}) = \rho_m(\mathbf{m})\rho_d(\mathbf{d})$. Let us consider the more general situation where for the whole set of parameters $(\mathcal{M}, \mathcal{D})$ we have some information that can be represented by a joint probability density $\rho(\mathbf{m}, \mathbf{d})$. Having well in mind the interpretation of this information, let us use the simple term 'experimental information' for it:

$$\rho(\mathbf{m}, \mathbf{d}) \qquad \text{(experimental information)} \,. \tag{65}$$

We have also seen that we have information coming from physical theories, that predict correlations between the parameters, and it has been argued that a probabilistic description of these correlations is well adapted to the resolution of inverse problems.[18] Let $\vartheta(\mathbf{m}, \mathbf{d})$ be the probability density representing this 'theoretical information':

$$\vartheta(\mathbf{m}, \mathbf{d}) \qquad \text{(theoretical information)} \,. \tag{66}$$

A quite fundamental assumption is that in all the spaces we consider, there is a notion of volume that allows us to give meaning to the notion of a 'homogeneous probability distribution' over the space. The corresponding probability density is not constant, but is proportional to the volume element of the space (see Section 2.2):

$$\mu(\mathbf{m}, \mathbf{d}) \qquad \text{(homogeneous probability distribution)} \,. \tag{67}$$

Finally, we have seen examples suggesting that the conjunction of the experimental information with the theoretical information corresponds exactly to the AND operation defined over the probability densities, to obtain the 'conjunction of information,' as represented by the probability density

$$\sigma(\mathbf{m},\mathbf{d}) = k\,\frac{\rho(\mathbf{m},\mathbf{d})\,\vartheta(\mathbf{m},\mathbf{d})}{\mu(\mathbf{m},\mathbf{d})} \qquad \text{(conjunction of information)}\,, \tag{68}$$

with marginal probability densities

$$\sigma_m(\mathbf{m}) = \int_{\mathcal{D}} d\mathbf{d}\,\sigma(\mathbf{m},\mathbf{d})\,; \qquad \sigma_d(\mathbf{d}) = \int_{\mathcal{M}} d\mathbf{m}\,\sigma(\mathbf{m},\mathbf{d})\,. \tag{69}$$

Example 15. *We may assume that the physical correlations between the parameters* **m** *and* **d** *are of the form*

$$\vartheta(\mathbf{m},\mathbf{d}) = \vartheta_{D|M}(\mathbf{d}|\mathbf{m})\,\vartheta_M(\mathbf{m})\,, \tag{70}$$

this expressing that a 'physical theory' gives, on the one hand, the conditional probability for **d** , *given* **m** , *and, on the other hand, the marginal probability density for* **m** . *See Appendix C for more details.*

Example 16. *Many applications concern the special situation where we have*

$$\mu(\mathbf{m},\mathbf{d}) = \mu_m(\mathbf{m})\,\mu_d(\mathbf{d})\,; \qquad \rho(\mathbf{m},\mathbf{d}) = \rho_m(\mathbf{m})\,\rho_d(\mathbf{d})\,. \tag{71}$$

In this case, Eqs. (68) and (69) give

$$\sigma_m(\mathbf{m}) = k\,\frac{\rho_m(\mathbf{m})}{\mu_m(\mathbf{m})}\int_{\mathcal{D}} d\mathbf{d}\,\frac{\rho_d(\mathbf{d})\,\vartheta(\mathbf{m},\mathbf{d})}{\mu_d(\mathbf{d})}\,. \tag{72}$$

If Eq. (70) holds, then

$$\sigma_m(\mathbf{m}) = k\,\rho_m(\mathbf{m})\,\frac{\vartheta_m(\mathbf{m})}{\mu_m(\mathbf{m})}\int_{\mathcal{D}} d\mathbf{d}\,\frac{\rho_d(\mathbf{d})\,\vartheta_{D|M}(\mathbf{d}|\mathbf{m})}{\mu_d(\mathbf{d})}\,. \tag{73}$$

Finally, if the simplification $\vartheta_M(\mathbf{m}) = \mu_m(\mathbf{m})$ *arises (see Appendix C for an illustration), then*

$$\sigma_m(\mathbf{m}) = k\,\rho_m(\mathbf{m})\int_{\mathcal{D}} d\mathbf{d}\,\frac{\rho_d(\mathbf{d})\,\vartheta(\mathbf{d}|\mathbf{m})}{\mu_d(\mathbf{d})}\,. \tag{74}$$

Example 17. *In the context of the previous example, assume that observational uncertainties are Gaussian:*

$$\rho_d(\mathbf{d}) = k\exp\left(-\frac{1}{2}(\mathbf{d}-\mathbf{d}_{\text{obs}})^t\,\mathbf{C}_D^{-1}(\mathbf{d}-\mathbf{d}_{\text{obs}})\right). \tag{75}$$

Note that the limit for infinite variances gives the homogeneous probability density $\mu_d(\mathbf{d}) = k$. *Furthermore, assume that uncertainties in the physical law are also Gaussian:*

$$\vartheta(\mathbf{d}|\mathbf{m}) = k\exp\left(-\frac{1}{2}(\mathbf{d}-\mathbf{f}(\mathbf{m}))^t\,\mathbf{C}_T^{-1}(\mathbf{d}-\mathbf{f}(\mathbf{m}))\right). \tag{76}$$

Here 'the physical theory says' that the data values must be 'close' to the 'computed values' $\mathbf{f}(\mathbf{m})$, *with a notion of closeness defined by the 'theoretical covariance matrix'* $\mathbf{C}_T$. *As demonstrated in Tarantola (1987, p. 158), the integral in Eq. (74) can be analytically evaluated, and gives*

$$\int_{\mathcal{D}} d\mathbf{d}\,\frac{\rho_d(\mathbf{d})\,\vartheta(\mathbf{d}|\mathbf{m})}{\mu_d(\mathbf{d})} = k\exp\left(-\frac{1}{2}(\mathbf{f}(\mathbf{m})-\mathbf{d}_{\text{obs}})^t(\mathbf{C}_D+\mathbf{C}_T)^{-1}(\mathbf{f}(\mathbf{m})-\mathbf{d}_{\text{obs}})\right). \tag{77}$$

This shows that when using the Gaussian probabilistic model, observational and theoretical uncertainties combine through addition of the respective covariance operators (a nontrivial result).

Example 18. *In the 'Galilean law' example developed in Section 4.61, we described the correlation between the position* x *and the time* t *of a free falling object through a probability density* $\vartheta(x,t)$. *This law says that falling objects describe, approximately, a space–time parabola. Assume that in a particular experiment the falling object explodes at some point of its space–time trajectory. A plain measurement of the coordinates* (x,t) *of the event gives the probability density* $\rho(x,t)$. *By 'plain measurement' we mean here that we have used a measurement technique that is not taking into account the particular parabolic character of the fall (i.e., the measurement is designed to work identically for any sort of trajectory). The conjunction of the physical law* $\vartheta(x,t)$ *and the experimental result* $\rho(x,t)$, *using expression (68), gives*

$$\sigma(x,t) = k\,\frac{\rho(x,t)\,\vartheta(x,t)}{\mu(x,t)}\,, \tag{78}$$

where, as the coordinates (x,t) *are 'Cartesian,'* $\mu(x,t) = k$. *Taking the explicit expression given for* $\vartheta(x,t)$ *in Eqs. (63) and (64), with* $\vartheta_t(t) = k$,

$$\vartheta(x,t) = \frac{1}{\sqrt{2\pi}\sqrt{\sigma_x^2+\sigma_v^2 t^2}} \times \exp\left(-\frac{1}{2}\frac{\left(x-(x_0+v_0\,t+\frac{1}{2}\,g\,t^2)\right)^2}{\sigma_x^2+\sigma_v^2 t^2}\right), \tag{79}$$

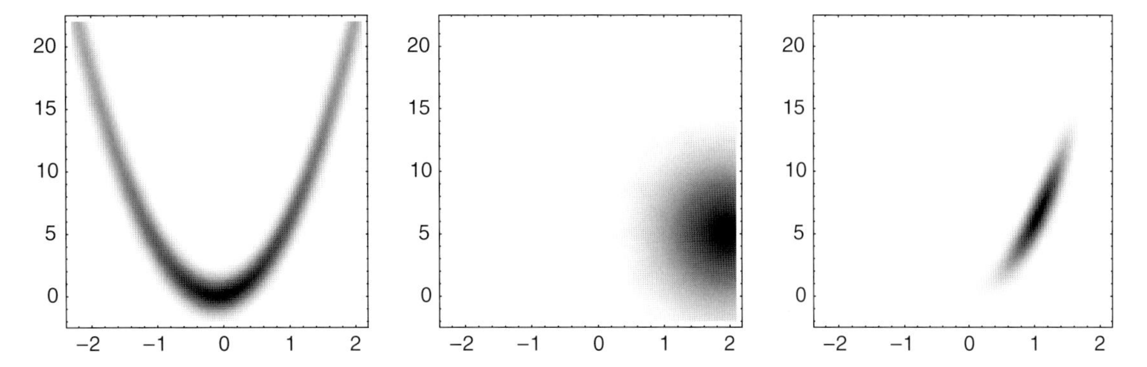

FIGURE 7 This figure has been made with the numerical values mentioned in Figure 17 (see Appendix C) with, in addition, $x_{obs} = 5.0\,\mathrm{m}$, $\Sigma_x = 4.0\,\mathrm{m}$, $t_{obs} = 2.0\,\mathrm{sec}$ and $\Sigma_t = 0.75\,\mathrm{sec}$.

and assuming the Gaussian form[19] *for* $\rho(x,t)$,

$$\rho(x,t) = \rho_x(x)\,\rho_t(t) = k \exp\left(-\frac{1}{2}\frac{(x-x_{\mathrm{obs}})^2}{\Sigma_x^2}\right)\exp\left(-\frac{1}{2}\frac{(t-t_{\mathrm{obs}})^2}{\Sigma_t^2}\right) \tag{80}$$

we obtain the combined probability density

$$\sigma(x,t) = \frac{k}{\sqrt{\sigma_x^2+\sigma_v^2 t^2}} \exp\left(-\frac{1}{2}\left(\frac{(x-x_{\mathrm{obs}})^2}{\Sigma_x^2} + \frac{(t-t_{\mathrm{obs}})^2}{\Sigma_t^2} + \frac{(x-(x_0+v_0 t+\frac{1}{2}gt^2))^2}{\sigma_x^2+\sigma_v^2 t^2}\right)\right) . \tag{81}$$

Figure 7 illustrates the three probability densities $\vartheta(x,t)$, $\rho(x,t)$, *and* $\sigma(x,t)$. *See Appendix C for a more detailed examination of this problem.*

5. Solving Inverse Problems (I): Examination of the Probability Density

The next two sections deal with Monte Carlo and optimization methods. The implementation of these methods takes some programming effort that is not required when we face problems with fewer degrees of freedom (say, between one and five).

When we have a small number of parameters we should directly 'plot' the probability density.

In Appendix K the problem of estimation of a seismic hypocenter is treated, and it is shown there that the examination of the probability density for the location of the hypocenter offers a much better possibility for analysis than any other method.

6. Solving Inverse Problems (II): Monte Carlo Methods

6.1 Basic Equations

The starting point could be the explicit expression (Eq. (46)) for $\sigma_m(\mathbf{m})$ given in Section 4.5.2:

$$\sigma_m(\mathbf{m}) = k\,\rho_m(\mathbf{m})\,L(\mathbf{m}) . \tag{82}$$

where

$$L(\mathbf{m}) = \left(\frac{\rho_d(\mathbf{d})}{\sqrt{\det \mathbf{g}_d(\mathbf{d})}}\frac{\sqrt{\det\left(\mathbf{g}_m(\mathbf{m})+\mathbf{F}^T(\mathbf{m})\,\mathbf{g}_d(\mathbf{d})\,\mathbf{F}(\mathbf{m})\right)}}{\sqrt{\det \mathbf{g}_m(\mathbf{m})}}\right)\Bigg|_{\mathbf{d}=\mathbf{f}(\mathbf{m})} . \tag{83}$$

In this expression the matrix of partial derivatives $\mathbf{F} = \mathbf{F}(\mathbf{m})$, with components $D_{i\alpha} = \partial f_i / \partial m_\alpha$, appears. The 'slope' $\mathbf{F}$ enters here because the steeper the slope for a given $\mathbf{m}$, the greater the accumulation of points we will have with this particular $\mathbf{m}$. This is because we use explicitly the analytic expression $\mathbf{d} = \mathbf{f}(\mathbf{m})$. One should realize that using the more general approach based on Eq. (68) of Section 4.6.2, the effect is automatically accounted for, and there is no need to explicitly consider the partial derivatives.

Equation (82) has the standard form of a conjunction of two probability densities, and is therefore ready to be integrated in a Metropolis algorithm. But one should note that, contrary to many 'nonlinear' formulations of inverse problems, the partial derivatives $\mathbf{F}$ are needed even if we use a Monte Carlo method.

In some weakly nonlinear problems, we have $\mathbf{F}^T(\mathbf{m})\,\mathbf{g}_d(\mathbf{d})\,\mathbf{F}(\mathbf{m}) \ll \mathbf{g}_m(\mathbf{m})$, and then Eq. (83) becomes

$$L(\mathbf{m}) = \frac{\rho_d(\mathbf{d})}{\mu_d(\mathbf{d})}\Bigg|_{\mathbf{d}=\mathbf{f}(\mathbf{m})} , \tag{84}$$

where we have used $\mu_d(\mathbf{d}) = k\sqrt{\det \mathbf{g}_d(\mathbf{d})}$ (see Rule 2).

This expression is also ready for use in the Metropolis algorithm. In this way, sampling of the prior $\rho_m(\mathbf{m})$ is modified into a sampling of the posterior $\sigma_m(\mathbf{m})$, and the Metropolis Rule uses the 'likelihood function' $L(\mathbf{m})$ to calculate acceptance probabilities.

6.2 Sampling the Homogeneous Probability Distribution

If we do not have an algorithm that samples the prior probability density directly, the first step in a Monte Carlo analysis of an inverse problem is to design a random walk that samples the model space according to the homogeneous probability distribution $\mu_m(\mathbf{m})$. In some cases this is easy, but in other cases only an algorithm (a *primeval random walk*) that samples an arbitrary (possibly constant) probability density $\psi(\mathbf{m}) \neq \mu_m(\mathbf{m})$ is available. Then the Metropolis rule can be used to modify $\psi(\mathbf{m})$ into $\mu_m(\mathbf{m})$ (see Section 3.4). This way of generating samples from $\mu_m(\mathbf{m})$ is efficient if $\psi(\mathbf{m})$ is close to $\mu_m(\mathbf{m})$, otherwise it may be very inefficient.

Once $\mu(\mathbf{m})$ can be sampled, the Metropolis Rule allows us to modify this sampling into an algorithm that samples the prior.

6.3 Sampling the Prior Probability Distribution

The first step in the Monte Carlo analysis is to temporarily 'switch off' the comparison between computed and observed data, thereby generating samples of the prior probability density. This allows us to verify statistically that the algorithm is working correctly, and it allows us to understand the prior information we are using. We will refer to a large collection of models representing the prior probability distribution as the 'prior movie' (in a computer screen, when the models are displayed one after the other, we have a 'movie'). The more models present in this movie, the more accurate the representation of the prior probability density.

6.4 Sampling the Posterior Probability Distribution

If we now switch on the comparison between computed and observed data using, e.g., the Metropolis rule for the actual Eq. (82), the random walk sampling the prior distribution is modified into a walk sampling the posterior distribution.

Since data rarely put strong constraints on the Earth, the 'posterior movie' typically shows that many different models are possible. But even though the models in the posterior movie may be quite different, all of them predict data that, within experimental uncertainties, are models with high likelihood. In other words, we must accept that data alone cannot have a preferred model.

The posterior movie allows us to perform a proper resolution analysis that helps us to choose between different interpretations of a given data set. Using the movie we can answer complicated questions about the correlations between several model parameters. To answer such questions, we can view the posterior movie and try to discover structure that is well resolved by data. Such structure will appear as 'persistent' in the posterior movie.

The 'movie' can be used to answer quite complicated questions. For instance, to answer the question '*Which is the probability that the Earth has this special characteristic, but not having this other special characteristic?*' we can just count the number n of models (samples) satisfying the criterion, and the probability is $P=n/m$, where m is the total number of samples.

Once this 'movie' is generated, it is, of course, possible to represent the 1D or 2D marginal probability densities for all or for some selected parameters: it is enough to concentrate one's attention on those selected parameters in each of the samples generated. Those marginal probability densities may have some pathologies (like being multimodal, or having infinite dispersions), but those are the general characteristics of the joint probability density. Our numerical experience shows that these marginals are, quite often, 'stable' objects, in the sense that they can be accurately determined with only a small number of samples.

If the marginals are, essentially, beautiful bell-shaped distributions, then, one may proceed to merely computing mean values and standard deviations (or median values and mean deviations), using each of the samples and the elementary statistical formulas.

Another, more traditional, way of investigating resolution is to calculate covariances and higher-order moments. For this we need to evaluate integrals of the form

$$R_f = \int_{\mathcal{A}} d\mathbf{m}\; f(\mathbf{m})\; \sigma_m(\mathbf{m}) \tag{85}$$

where $f(\mathbf{m})$ is a given function of the model parameters and $\mathcal{A}$ is an event in the model space $\mathcal{M}$ containing the models we are interested in. For instance,

$$\mathcal{A} = \{\mathbf{m}\,|\text{a given range of parameters in } \mathbf{m} \text{ is } \textit{cyclic}\} \; . \tag{86}$$

In the special case when $\mathcal{A}=\mathcal{M}$ is the entire model space, and $f(\mathbf{m})=m_i$, the R_f in Eq. (85) equals the mean $\langle m_i \rangle$ of the ith model parameter m_i . If $f(\mathbf{m})=(m_i-\langle m_i \rangle)(m_j-\langle m_j \rangle)$, R_f becomes the covariance between the ith and jth model parameters. Typically, in the general inverse problem we cannot evaluate the integral in Eq. (85) analytically because we have no analytical expression for $\sigma(\mathbf{m})$. However, from the samples of the posterior movie $\mathbf{m}_1, \ldots, \mathbf{m}_n$ we can approximate R_f by the simple average

$$R_f \approx \frac{1}{\text{total number of models}} \sum_{\{i \mid \mathbf{m}_i \in \mathcal{A}\}} f(\mathbf{m}_i) \ . \tag{87}$$

7. Solving Inverse Problems (III): Deterministic Methods

As we have seen, the solution of an inverse problem essentially consists of a probability distribution over the space of all possible models of the physical system under study. In general, this 'model space' is high-dimensional, and the only general way to explore it is by using the Monte Carlo methods developed in Section 3.

If the probability distributions are 'bell-shaped' (i.e., if they look like a Gaussian or like a generalized Gaussian), then one may simplify the problem by calculating only the point around which the probability is maximum, with an approximate estimation of the variances and covariances. This is the problem addressed in this section. Among the many methods available to obtain the point at which a scalar function reaches its maximum value (relaxation methods, linear programming techniques, etc.) we limit our scope here to the methods using the gradient of the function, which we assume can be computed analytically, or at least, numerically. For more general methods, the reader may have a look at Fletcher (1980, 1981), Powell (1981), Scales (1985), Tarantola (1987) or Scales *et al.* (1992).

7.1 Maximum Likelihood Point

Let us consider a space $\mathcal{X}$, with a volume element dV defined. If the coordinates $\mathbf{x} \equiv \{x^1, x^2, \ldots, x^n\}$ are chosen over the space, the volume element has an expression $dV(\mathbf{x}) = v(\mathbf{x})\, d\mathbf{x}$, and each probability distribution over $\mathcal{X}$ can be represented by a probability density $f(\mathbf{x})$. For any fixed small volume ΔV we can search for the point $\mathbf{x}_{\mathrm{ML}}$ such that the probability dP of the small volume, when centered around $\mathbf{x}_{\mathrm{ML}}$, attains a maximum. In the limit $\Delta V \to 0$ this defines the *maximum likelihood point*. The maximum likelihood point may be unique (if the probability distribution is unimodal), may be degenerate (if the probability distribution is 'chevron-shaped'), or may be multiple (as when we have the sum of a few bell-shaped functions).

The maximum likelihood point is *not* the point at which the probability density is maximum. Our definition implies that a maximum must be attained by the ratio between the probability density and the function $v(\mathbf{x})$ defining the volume element:[20]

$$\mathbf{x} = \mathbf{x}_{\mathrm{ML}} \iff F(\mathbf{x}) = \frac{f(\mathbf{x})}{v(\mathbf{x})} \quad \text{(maximum)} \ . \tag{88}$$

As the homogeneous probability density is $\mu(\mathbf{x}) = k\, v(\mathbf{x})$ (see Rule 2), we can equivalently define the maximum likelihood point by the condition

$$\mathbf{x} = \mathbf{x}_{\mathrm{ML}} \iff \frac{f(\mathbf{x})}{\mu(\mathbf{x})} \quad \text{(maximum)} \ . \tag{89}$$

The point at which a probability density has its maximum is, in general, not $\mathbf{x}_{\mathrm{ML}}$. In fact, the maximum of a probability density does not correspond to an intrinsic definition of a point: a change of coordinates $\mathbf{x} \mapsto \mathbf{y} = \psi(\mathbf{x})$ would change the probability density $f(\mathbf{x})$ into the probability density $g(\mathbf{y})$ (obtained using the Jacobian rule), but the point of the space at which $f(\mathbf{x})$ is maximum is not the same as the point of the space where $g(\mathbf{y})$ is maximum (unless the change of variables is linear). This contrasts with the maximum likelihood point, as defined by Eq. (89), which is an intrinsically defined point: no matter which coordinates we use in the computation we always obtain the same point of the space.

7.2 Misfit

One of the goals here is to develop gradient-based methods for obtaining the maximum of $F(\mathbf{x}) = f(\mathbf{x})/\mu(\mathbf{x})$. As a quite general rule, gradient-based methods perform quite poorly for (bell-shaped) probability distributions, as when one is far from the maximum the probability densities tend to be quite flat, and it is difficult to get, reliably, the direction of steepest ascent. Taking a logarithm transforms a bell-shaped distribution into a paraboloid-shaped distribution on which gradient methods work well.

The logarithmic volumetric probability, or *misfit*, is defined as $S(\mathbf{x}) = -\log(F(\mathbf{x})/F_0)$, where p' and F_0 are two constants, and is given by

$$S(\mathbf{x}) = -\log \frac{f(\mathbf{x})}{\mu(\mathbf{x})} \ . \tag{90}$$

The problem of maximization of the (typically) bell-shaped function $f(\mathbf{x})/\mu(\mathbf{x})$ has been transformed into the problem of minimization of the (typically) paraboloid-shaped function $S(\mathbf{x})$:

$$\mathbf{x} = \mathbf{x}_{\mathrm{ML}} \iff S(\mathbf{x}) \quad \text{(minimum)} \ . \tag{91}$$

Example 19. *The conjunction $\sigma(\mathbf{x})$ of two probability densities $\rho(\mathbf{x})$ and $\vartheta(\mathbf{x})$ was defined (Eq. (13)) as*

$$\sigma(\mathbf{x}) = p\, \frac{\rho(\mathbf{x})\, \vartheta(\mathbf{x})}{\mu(\mathbf{x})} \ . \tag{92}$$

Then,

$$S(\mathbf{x}) = S_\rho(\mathbf{x}) + S_\vartheta(\mathbf{x}) \ , \tag{93}$$

where

$$S_\rho(\mathbf{x}) = -\log\frac{\rho(\mathbf{x})}{\mu(\mathbf{x})}\,; \qquad S_\vartheta(\mathbf{x}) = -\log\frac{\vartheta(\mathbf{x})}{\mu(\mathbf{x})}\,. \tag{94}$$

Example 20. *In the context of Gaussian distributions we have found the probability density (see Example 12)*

$$\sigma_m(\mathbf{m}) = k\exp\left[-\frac{1}{2}\left((\mathbf{m}-\mathbf{m}_{\text{prior}})^t\,\mathbf{C}_M^{-1}(\mathbf{m}-\mathbf{m}_{\text{prior}}) + (\mathbf{f}(\mathbf{m})-\mathbf{d}_{\text{obs}})^t\,\mathbf{C}_D^{-1}(\mathbf{f}(\mathbf{m})-\mathbf{d}_{\text{obs}})\right)\right]. \tag{95}$$

The limit of this distribution for infinite variances is a constant, so in this case $\mu_m(\mathbf{m}) = k$. *The misfit function* $S(\mathbf{m}) = -\log(\sigma_m(\mathbf{m})/\mu_m(\mathbf{m}))$ *is then given by*

$$2\,S(\mathbf{m}) = (\mathbf{m}-\mathbf{m}_{\text{prior}})^t\,\mathbf{C}_M^{-1}(\mathbf{m}-\mathbf{m}_{\text{prior}}) + (\mathbf{f}(\mathbf{m})-\mathbf{d}_{\text{obs}})^t\,\mathbf{C}_D^{-1}(\mathbf{f}(\mathbf{m})-\mathbf{d}_{\text{obs}})\,. \tag{96}$$

The reader should remember that this misfit function is valid only for weakly nonlinear problems (see Examples 10 and 12). The maximum likelihood model here is the one that minimizes the sum of squares (96). This corresponds to the least-squares criterion.

7.3 Gradient and Direction of Steepest Ascent

One must not consider as synonymous the notions of 'gradient' and 'direction of steepest ascent.' Consider, for instance, an *adimensional* misfit function[21] $S(P,T)$ over a pressure P and a temperature T. Any sensible definition of the gradient of S will lead to an expression like

$$\text{grad}\ S = \begin{pmatrix} \frac{\partial S}{\partial P} \\ \\ \frac{\partial S}{\partial T} \end{pmatrix} \tag{97}$$

and this by no means can be regarded as a 'direction' in the (P,T) space (for instance, the components of this 'vector' does not have the dimensions of pressure and temperature, but of inverse pressure and inverse temperature).

Mathematically speaking, *the gradient of a function* $S(\mathbf{x})$ *at a point* $\mathbf{x}_0$ *is the linear function that is tangent to* $S(\mathbf{x})$ *at* $\mathbf{x}_0$. This definition of gradient is consistent with the more elementary one, based on the use of the first-order expansion

$$S(\mathbf{x}_0+\delta\mathbf{x}) = S(\mathbf{x}_0) + \widehat{\boldsymbol{\gamma}}_0^T\,\delta\mathbf{x} + \cdots \tag{98}$$

Here $\widehat{\boldsymbol{\gamma}}_0$ is called the gradient of $S(\mathbf{x})$ at point $\mathbf{x}_0$. It is clear that $S(\mathbf{x}_0) + \widehat{\boldsymbol{\gamma}}_0^T\,\delta\mathbf{x}$ is a linear function, and that it is tangent to $S(\mathbf{x})$ at $\mathbf{x}_0$, so the two definitions are in fact equivalent. Explicitly, the components of the gradient at point $\mathbf{x}_0$ are

$$(\widehat{\boldsymbol{\gamma}}_0)_p = \frac{\partial S}{\partial x^p}(\mathbf{x}_0)\,. \tag{99}$$

Everybody is well trained in computing the gradient of a function (event if the interpretation of the result as a direction in the original space is wrong). How can we pass from the gradient to the direction of steepest ascent (a bona fide direction in the original space)? In fact, the gradient (at a given point) of a function defined over a given space $\mathcal{E}$) is an element of the dual of the space. To obtain a direction in $\mathcal{E}$ we must pass from the dual to the primal space. As usual, it is the metric of the space that maps the dual of the space into the space itself. So if $\mathbf{g}$ is the metric of the space where $S(\mathbf{x})$ is defined, and if $\widehat{\boldsymbol{\gamma}}$ is the gradient of S at a given point, the *direction of steepest ascent* is

$$\boldsymbol{\gamma} = \mathbf{g}^{-1}\widehat{\boldsymbol{\gamma}}\,. \tag{100}$$

The direction of steepest ascent must be interpreted as follows: if we are at a point $\mathbf{x}$ of the space, we can consider a very small hypersphere around $\mathbf{x}_0$. The direction of steepest ascent points toward the point of the sphere at which $S(\mathbf{x})$ attains its maximum value.

Example 21. *In the context of least squares, we consider a misfit function* $S(\mathbf{m})$ *and a covariance matrix* $\mathbf{C}_M$. If $\widehat{\boldsymbol{\gamma}}_0$ *is the gradient of* S , *at a point* $\mathbf{x}_0$, *and if we use* $\mathbf{C}_M$ *to define distances in the space, the direction of steepest ascent is*

$$\boldsymbol{\gamma}_0 = \mathbf{C}_M\,\widehat{\boldsymbol{\gamma}}_0\,. \tag{101}$$

7.4 The Steepest Descent Method

Consider that we have a probability distribution defined over an n-dimensional space $\mathcal{X}$. Having chosen the coordinates $\mathbf{x} \equiv \{x^1, x^2, \ldots, x^n\}$ over the space, the probability distribution is represented by the probability density $f(\mathbf{x})$ whose homogeneous limit (in the sense developed in Section 2.2) is $\mu(\mathbf{x})$. We wish to calculate the coordinates $\mathbf{x}_{\text{ML}}$ of the maximum likelihood point. By definition (Eq. (89)),

$$\mathbf{x} = \mathbf{x}_{\text{ML}} \quad \Longleftrightarrow \quad \frac{f(\mathbf{x})}{\mu(\mathbf{x})} \quad (\text{maximum})\,, \tag{102}$$

that is,

$$\mathbf{x} = \mathbf{x}_{\text{ML}} \quad \Longleftrightarrow \quad S(\mathbf{x}) \quad (\text{minimum})\,, \tag{103}$$

where $S(\mathbf{x})$ is the misfit [Eq. (90)]

$$S(\mathbf{x}) = -k\log\frac{f(\mathbf{x})}{\mu(\mathbf{x})}\,. \tag{104}$$

Let us denote by $\widehat{\boldsymbol{\gamma}}(\mathbf{x}_k)$ the gradient of $S(\mathbf{x})$ at point $\mathbf{x}_k$, i.e. (Eq. (99)),

$$(\widehat{\gamma}_0)_p = \frac{\partial S}{\partial x^p}(\mathbf{x}_0) \ . \tag{105}$$

We have seen above that $\widehat{\boldsymbol{\gamma}}(\mathbf{x})$ should not be interpreted as a direction in the space $\mathcal{X}$ but as a direction in the dual space. The gradient can be converted into a direction using a metric $\mathbf{g}(\mathbf{x})$ over $\mathcal{X}$. In simple situations the metric $\mathbf{g}$ will be the one used to define the volume element of the space, i.e., we will have $\mu(\mathbf{x}) = k\,v(\mathbf{x}) = k\sqrt{\det \mathbf{g}(\mathbf{x})}$, but this is not a necessity, and iterative algorithms may be accelerated by astute introduction of ad-hoc metrics.

Given, then, the gradient $\widehat{\boldsymbol{\gamma}}(\mathbf{x}_k)$ (at some particular point $\mathbf{x}_k$) to any possible choice of metric $\mathbf{g}(\mathbf{x})$ we can define the direction of steepest ascent associated with the metric $\mathbf{g}$, by (Eq. (101))

$$\boldsymbol{\gamma}(\mathbf{x}_k) = \mathbf{g}^{-1}(\mathbf{x}_k)\,\widehat{\boldsymbol{\gamma}}(\mathbf{x}_k) \ . \tag{106}$$

The algorithm of steepest descent is an iterative algorithm passing from point $\mathbf{x}_k$ to point $\mathbf{x}_{k+1}$ by making a 'small jump' along the local direction of steepest descent,

$$\mathbf{x}_{k+1} = \mathbf{x}_k - \varepsilon_k\,\mathbf{g}_k^{-1}\,\widehat{\boldsymbol{\gamma}}_k \ , \tag{107}$$

where ε_k is an ad-hoc (real, positive) value adjusted to force the algorithm to converge rapidly (if ε_k is chosen too small, the convergence may be too slow; it is chosen too large, the algorithm may even diverge).

Many elementary presentations of the steepest descent algorithm just forget to include the metric $\mathbf{g}_k$ in expression (107). These algorithms are not consistent. Even the physical dimensionality of the equation is not assured. 'Numerical' problems in computer implementations of steepest descent algorithms can often be traced to the fact that the metric has been neglected.

Example 22. *In the context of Example 20, where the misfit function $S(\mathbf{m})$ is given by*

$$2\,S(\mathbf{m}) = (\mathbf{f}(\mathbf{m}) - \mathbf{d}_{\text{obs}})^t\,\mathbf{C}_D^{-1}\,(\mathbf{f}(\mathbf{m}) - \mathbf{d}_{\text{obs}}) + (\mathbf{m} - \mathbf{m}_{\text{prior}})^t\,\mathbf{C}_M^{-1}\,(\mathbf{m} - \mathbf{m}_{\text{prior}}) \ , \tag{108}$$

the gradient $\widehat{\boldsymbol{\gamma}}$, whose components are $\widehat{\gamma}_\alpha = \partial S/\partial m^\alpha$, is given by the expression

$$\widehat{\boldsymbol{\gamma}}(\mathbf{m}) = \mathbf{F}^t(\mathbf{m})\,\mathbf{C}_D^{-1}\,(\mathbf{f}(\mathbf{m}) - \mathbf{d}_{\text{obs}}) + \mathbf{C}_M^{-1}\,(\mathbf{m} - \mathbf{m}_{\text{prior}}) \ , \tag{109}$$

where $\mathbf{F}$ is the matrix of partial derivatives

$$F^{i\alpha} = \frac{\partial f^i}{\partial m^\alpha} \ . \tag{110}$$

An example of computation of partial derivatives is given in Appendix M.

Example 23. *In the context of Example 22 the model space $\mathcal{M}$ has an obvious metric, namely, that defined by the inverse of the 'a priori' covariance operator $\mathbf{g} = \mathbf{C}_M^{-1}$. Using this metric and the gradient given by Eq. (109), the steepest descent algorithm (107) becomes*

$$\mathbf{m}_{k+1} = \mathbf{m}_k - \varepsilon_k\left(\mathbf{C}_M\,\mathbf{F}_k^t\,\mathbf{C}_D^{-1}\,(\mathbf{f}_k - \mathbf{d}_{\text{obs}}) + (\mathbf{m}_k - \mathbf{m}_{\text{prior}})\right) \ , \tag{111}$$

where $\mathbf{F}_k \equiv \mathbf{F}(\mathbf{m}_k)$ and $\mathbf{f}_k \equiv \mathbf{f}(\mathbf{m}_k)$. The real positive quantities ε_k can be fixed after some trial and error by accurate linear search, or by using a linearized approximation.

Example 24. *In the context of Example 22 the model space $\mathcal{M}$ has a less obvious metric, namely, that defined by the inverse of the 'posterior' covariance operator, $\mathbf{g} = \widetilde{\mathbf{C}}_M^{-1}$.*[23] *Using this metric and the gradient given by Eq. (109), the steepest descent algorithm (107) becomes*

$$\mathbf{m}_{k+1} = \mathbf{m}_k - \varepsilon_k\left(\mathbf{F}_k^t\,\mathbf{C}_D^{-1}\,\mathbf{F}_k + \mathbf{C}_M^{-1}\right)^{-1}\left(\mathbf{F}_k^t\,\mathbf{C}_D^{-1}\,(\mathbf{f}_k - \mathbf{d}_{\text{obs}}) + \mathbf{C}_M^{-1}\,(\mathbf{m}_k - \mathbf{m}_{\text{prior}})\right) \ , \tag{112}$$

where $\mathbf{F}_k \equiv \mathbf{F}(\mathbf{m}_k)$ and $\mathbf{f}_k \equiv \mathbf{f}(\mathbf{m}_k)$. The real positive quantities ε_k can be fixed, after some trial and error, by accurate linear search, or by using a linearized approximation that simply gives[24] $\varepsilon_k \approx 1$.

The algorithm (112) is usually called a 'quasi-Newton algorithm.' This name is not well chosen: a Newton method applied to minimization of a misfit function $S(\mathbf{m})$ would be a method using the second derivatives of $S(\mathbf{m})$, and thus the derivatives $H^i_{\alpha\beta} = \frac{\partial^2 f^i}{\partial m^\alpha \partial m^\beta}$, that are not computed (or not estimated) when using this algorithm. It is just a steepest descent algorithm with a nontrivial definition of the metric in the working space. In this sense it belongs to the wider class of 'variable metric methods,' not discussed in this article.

7.5 Estimating Posterior Uncertainties

In the Gaussian context, the Gaussian probability density that is tangent to $\sigma_m(\mathbf{m})$ has its center at the point given by the iterative algorithm

$$\mathbf{m}_{k+1} = \mathbf{m}_k - \varepsilon_k\left(\mathbf{C}_M\,\mathbf{F}_k^t\,\mathbf{C}_D^{-1}\,(\mathbf{f}_k - \mathbf{d}_{\text{obs}}) + (\mathbf{m}_k - \mathbf{m}_{\text{prior}})\right) \ , \tag{113}$$

(Eq. (111)) or, equivalently, by the iterative algorithm

$$\mathbf{m}_{k+1} = \mathbf{m}_k - \varepsilon_k\left(\mathbf{F}_k^t\,\mathbf{C}_D^{-1}\,\mathbf{F}_k + \mathbf{C}_M^{-1}\right)^{-1}\left(\mathbf{F}_k^t\,\mathbf{C}_D^{-1}\,(\mathbf{f}_k - \mathbf{d}_{\text{obs}}) + \mathbf{C}_M^{-1}\,(\mathbf{m}_k - \mathbf{m}_{\text{prior}})\right) \tag{114}$$

(Eq. (112)). The covariance of the tangent Gaussian is

$$\widetilde{\mathbf{C}}_M \approx \left(\mathbf{F}_\infty^t\, \mathbf{C}_D^{-1}\, \mathbf{F}_\infty + \mathbf{C}_M^{-1}\right)^{-1}, \qquad (115)$$

where $\mathbf{F}_\infty$ refers to the value of the matrix of partial derivatives at the convergence point.

7.6 Some Comments on the Use of Deterministic Methods

7.6.1 Linear, Weakly Nonlinear and Nonlinear Problems

There are different degrees of nonlinearity. Figure 8 illustrates four domains of nonlinearity, calling for different

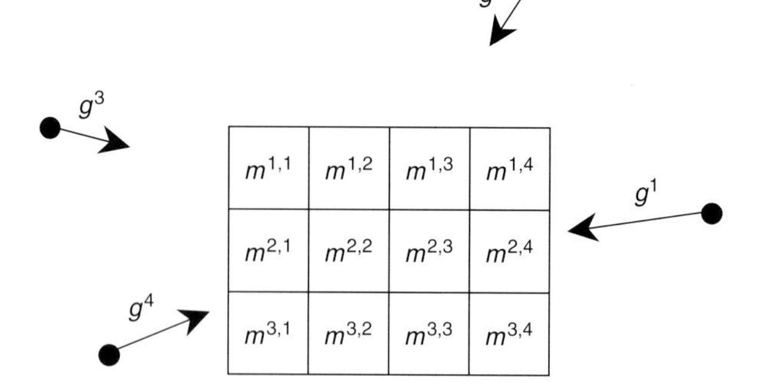

FIGURE 8 A simple example where we are interested in predicting the gravitational field **g** generated by a 2D distribution of mass.

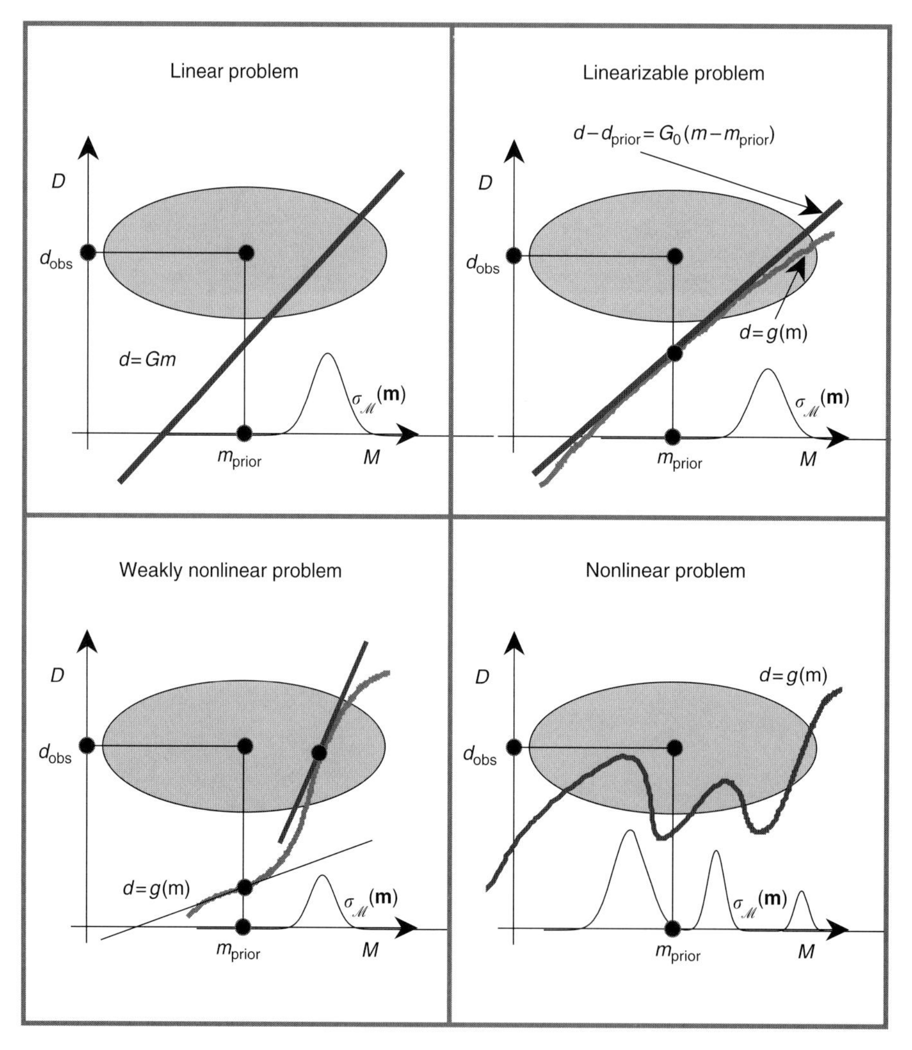

FIGURE 9 Illustration of the four domains of nonlinearity, calling for different optimization algorithms. The model space is symbolically represented by the abscissa, and the data space is represented by the ordinate. The gray oval represents the combination of prior information on the model parameters and information from the observed data. What is important is not an intrinsic nonlinearity of the function relating model parameters to data, but how linear the function is *inside the domain of significant probability*.

optimization algorithms. In this figure the abscissa symbolically represents the model space, and the ordinate represents the data space. The gray oval represents the combination of prior information on the model parameters, and information from the observed data.[25] It is the probability density $\rho(\mathbf{d}, \mathbf{m}) = \rho_d(\mathbf{d})\rho_m(\mathbf{m})$ seen elsewhere.

To fix ideas, the oval suggests here a Gaussian probability, but our distinction between problems according to their nonlinearity will not depend fundamentally on this.

First, there are strictly linear problems. For instance, in the example illustrated by Figure 8 the gravitational field $\mathbf{g}$ depends linearly on the masses inside the blocks.[26]

Strictly linear problems are illustrated at the top left of Figure 9. The linear relationship between data and model parameters, $\mathbf{d} = \mathbf{G\,m}$, is represented by a straight line. The prior probability density $\rho(\mathbf{d}, \mathbf{m})$ 'induces' on this straight line the posterior probability density[27] $\sigma(\mathbf{d}, \mathbf{m})$ whose 'projection' over the model space gives the posterior probability density over the model parameter space, $\sigma_m(\mathbf{m})$. Should the prior probability densities be Gaussian, then the posterior probability distribution would also be Gaussian: this is the simplest situation.

Quasi-linear problems are illustrated at the bottom left of Figure 9. If the relationship linking the observable data $\mathbf{d}$ to the model parameters $\mathbf{m}$,

$$\mathbf{d} = \mathbf{g}(\mathbf{m}) \,, \qquad (116)$$

is approximately linear *inside the domain of significant prior probability* (i.e., inside the gray oval of the figure), then the posterior distribution is just as simple as the prior distribution. For instance, if the prior is Gaussian the posterior is also Gaussian.

In this case also, the problem can be reduced to the computation of the mean and the covariance of the Gaussian. Typically, one begins at some 'starting model' $\mathbf{m}_0$ (typically, one takes for $\mathbf{m}_0$ the 'a priori model' $\mathbf{m}_{\text{prior}}$),[28] linearizing the function $\mathbf{d} = \mathbf{g}(\mathbf{m})$ around $\mathbf{m}_0$, and one looks for a model $\mathbf{m}_1$ 'better than $\mathbf{m}_0$.'

Iterating such an algorithm, one tends to the model $\mathbf{m}_\infty$ at which the 'quasi-Gaussian' $\sigma_m(\mathbf{m})$ is maximum. The linearizations made in order to arrive to $\mathbf{m}_\infty$ are so far not an approximation: the point $\mathbf{m}_\infty$ is perfectly defined, independently of any linearization and any method used to find it. But once the convergence to this point has been obtained, a linearization of the function $\mathbf{d} = \mathbf{g}(\mathbf{m})$ around this point,

$$\mathbf{d} - \mathbf{g}(\mathbf{m}_\infty) = \mathbf{G}_\infty(\mathbf{m} - \mathbf{m}_\infty) \,, \qquad (117)$$

allows to obtain a good approximation to the posterior uncertainties. For instance, if the prior distribution is Gaussian this will give the covariance of the 'tangent Gaussian.'

Between linear and quasi-linear problems there are the 'linearizable problems.' The scheme at the top right of Figure 9 shows the case where the linearization of the function $\mathbf{d} = \mathbf{g}(\mathbf{m})$ around the prior model,

$$\mathbf{d} - \mathbf{g}(\mathbf{m}_{\text{prior}}) = \mathbf{G}_{\text{prior}}(\mathbf{m} - \mathbf{m}_{\text{prior}}) \,, \qquad (118)$$

gives a function that, inside the domain of significant probability, is very similar to the true (nonlinear) function.

In this case, there is no practical difference between this problem and the strictly linear problem, and the iterative procedure necessary for quasi-linear problems is here superfluous.

It remains to analyze the true nonlinear problems that, using a pleonasm, are sometimes called *strongly nonlinear problems*. They are illustrated at the bottom right of Figure 9.

In this case, even if the prior distribution is simple, the posterior distribution can be quite complicated. For instance, it can be multimodal. These problems are in general quite complex to solve, and only a Monte Carlo analysis, as described in the previous chapter, is feasible.

If full Monte Carlo methods cannot be used, because they are too expensive, then one can mix a random part (for instance, to choose the starting point) and a deterministic part. The optimization methods applicable to quasi-linear problems can, for instance, allow us to go from the randomly chosen starting point to the 'nearest' optimal point. Repeating these computations for different starting points, one can arrive at a good idea of the posterior distribution in the model space.

7.6.2 The Maximum Likelihood Model

The *most likely model* is, by definition, that at which the volumetric probability (see Appendix A) $\sigma_\beta(\mathbf{m})$ attains its maximum. As $\sigma_\beta(\mathbf{m})$ is maximum when $S(\mathbf{m})$ is minimum, we see that the most likely model is also the 'best model' obtained when using a 'least-squares criterion.' Should we have used the double exponential model for all the

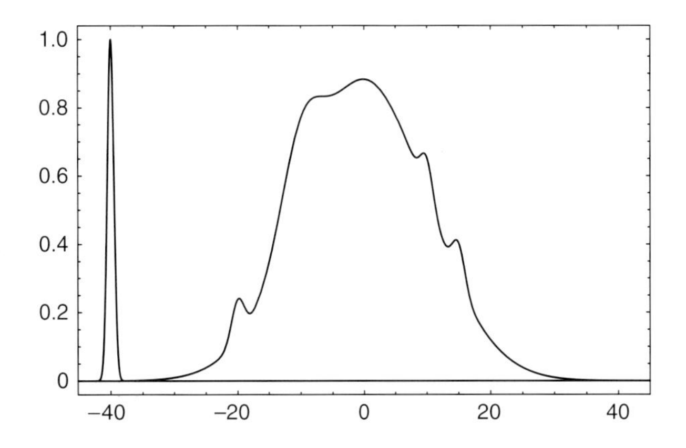

FIGURE 10 One of the circumstances where the 'maximum likelihood model' may not be very interesting is when it corresponds to a narrow maximum with small total probability, as the peak in the left part of this probability distribution.

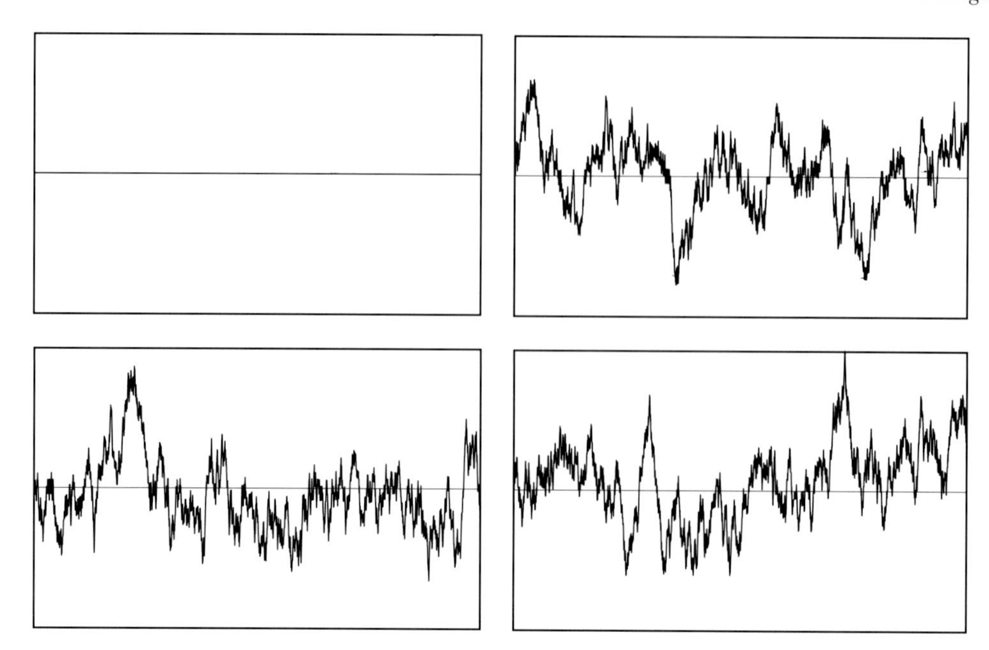

FIGURE 11 At the top right and bottom left and right, three random realizations of a Gaussian random function with zero mean and (approximately) exponential correlation function. The most likely function, i.e., the center of the Gaussian, is shown at the top left. We see that the most likely function is not a representative of the probability distribution.

uncertainties, then the most likely model would be defined by a 'least absolute values' criterion.

There are many circumstances where the most likely model is not an interesting model. One trivial example is when the volumetric probability has a 'narrow maximum,' with small total probability (see Fig. 10). A much less trivial situation arises when the number of parameters is very large, as for instance when we deal with a random function (that, strictly speaking, corresponds to an infinite number of random variables). Figure 11, for instance, shows a few realizations of a Gaussian function with zero mean and an (approximately) exponential correlation. The most likely function is the center of the Gaussian, i.e., the null function shown at the top left. But this is not a representative sample of the probability distribution, as any realization of the probability distribution will have, with a probability very close to one, the 'oscillating' characteristics of the other three samples.

8. Conclusions

Probability theory is well adapted to the formulation of inverse problems, although its formulation must be rendered intrinsic (introducing explicitly the definition of distances in the working spaces, by redefining the notion of conditional probability density, and by introducing the notion of conjunction of states of information). The Metropolis algorithm is well adapted to the solution of inverse problems, as its inherent structure allows us to sequentially combine prior information, theoretical information, etc., and allows us to take advantage of the 'movie philosophy.' When a general Monte Carlo approach cannot be afforded, one can use simplified optimization techniques (like least squares). However, this usually requires strong simplifications that can only be made at the cost of realism.

Acknowledgements

We are very indebted to our colleagues (Bartolomé Coll, Miguel Bosch, Guillaume Évrard, John Scales, Christophe Barnes, Frédéric Parrenin, and Bernard Valette) for illuminating discussions. We are also grateful to the students of the Geophysical Tomography Group, and the students at our respective institutes (in Paris and Copenhagen).

References

Aki, K. and W.H.K. Lee (1976). Determination of three-dimensional velocity anomalies under a seismic array using first *P* arrival times from local earthquakes. *J. Geophys. Res.* **81**, 4381–4399.

Aki, K., A. Christofferson, and E.S. Husebye (1977). Determination of the three-dimensional seismic structure of the lithosphere. *J. Geophys. Res.* **82**, 277–296.

Backus, G. (1970a). Inference from inadequate and inaccurate data: I. *Proc. Natl. Acad. Sci. USA* **65**(1), 1–105.

Backus, G. (1970b). Inference from inadequate and inaccurate data: II. *Proc. Natl. Acad. Sci. USA* **65**(2), 281–287.

Backus, G. (1970c). Inference from inadequate and inaccurate data: III. *Proc. Natl. Acad. Sci. USA* **67**(1), 282–289.

Backus, G. (1971). Inference from inadequate and inaccurate data. 'Mathematical Problems in the Geophysical Sciences,' Lectures in Applied Mathematics **14**. American Mathematical Society, Providence, Rhode Island.

Backus, G. and F. Gilbert (1967). Numerical applications of a formalism for geophysical inverse problems. *Geophys. J. R. Astron. Soc.* **13**, 247–276.

Backus, G. and F. Gilbert (1968). The resolving power of gross Earth data. *Geophys. J. R. Astron. Soc.* **16**, 169–205.

Backus, G. and F. Gilbert (1970). Uniqueness in the inversion of inaccurate gross Earth data. *Philos. Trans. R. Soc. London* **266**, 123–192.

Dahlen, F.A. (1976). Models of the lateral heterogeneity of the Earth consistent with eigenfrequency splitting data. *Geophys. J. R. Astron. Soc.* **44**, 77–105.

Dahl-Jensen, D., K. Mosegaard, N. Gundestrup, G.D. Clow, S.J. Johnsen, A.W. Hansen, and N. Balling (1998). Past temperatures directly from the Greenland Ice Sheet. *Science* (Oct. 9), 268–271.

Fisher, R.A. (1953). Dispersion on a sphere. *Proc. R. Soc. London, A* **217**, 295–305.

Fletcher, R. (1980). 'Practical Methods of Optimization,' Vol. 1: Unconstrained Optimization. Wiley, New York.

Fletcher, R. (1981). 'Practical Methods of Optimization,' Vol. 2: Constrained Optimization. Wiley, New York.

Franklin, J.N. (1970). Well posed stochastic extensions of ill posed linear problems. *J. Math. Anal. Appl.* **31**, 682–716.

Gauss, C.F. (1809). 'Theoria Motus Corporum Cœlestium in Sectionis Conicis Solem Ambientum,' Hamburg. (Also in 'Werke,' Vol. 7. Olmc-Verlag, 1981.)

Geiger, L. (1910). Herdbestimmung bei Erdbeben aus den Ankunftszeiten. *Nachrichten von der Königlichen Gesellschaft der Wissenschaften zu Göttingen* **4**, 331–349.

Geman, S. and D. Geman (1984). *IEEE Trans. Pattern Anal. Mach. Int.* **PAMI-6**(6), 721.

Gilbert, F. (1971). Ranking and winnowing gross Earth data for inversion and resolution. *Geophys. J. R. Astron. Soc.* **23**, 125–128.

Hadamard, J. (1902). Sur les problémes aux dérivées partielles et leur signification physique. *Bull. Univ. Princeton* **13**.

Hadamard, J. (1932). 'Le problème de Cauchy et les équations aux dérivées partielles linéaires hyperboliques.' Hermann, Paris.

ISO (1993). 'Guide to the Expression of Uncertainty in Measurement.' International Organization for Standardization, Switzerland.

Jackson, D.D. (1979). The use of a priori data to resolve non-uniqueness in linear inversion. *Geophys. J. R. Astron. Soc.* **57**, 137–157.

Jaynes, E.T. (1968). Prior probabilities. *IEEE Trans. Syst. Sci. Cybern.* **SSC-4**(3), 227–241.

Jeffreys, H. (1939). 'Theory of Probability.' Clarendon Press, Oxford. (Reprinted in 1961 by Oxford University Press.)

Kandel, A. (1986). 'Fuzzy Mathematical Techniques with Applications.' Addison-Wesley, Reading, MA.

Keilis-Borok, V.J. and T.B. Yanovskaya (1967). Inverse problems in seismology (structural review). *Geophys. J. R. Astron. Soc.* **13**, 223–234.

Kennett, B.L.N. and G. Nolet (1978). Resolution analysis for discrete systems. *Geophys. J. R. Astron. Soc.* **53**, 413–425.

Khan, A., K. Mosegaard, and K.L. Rasmussen (2000). A new seismic velocity model for the Moon from a Monte Carlo inversion of the Apollo lunar seismic data. *Geophys. Res. Lett.* **37**(11), 1591–1594.

Kimeldorf, G. and G. Wahba (1970). A correspondence between Bayesian estimation of stochastic processes and smooting by splines. *Ann. Math. Stat.* **41**, 495–502.

Kullback, S. (1967). The two concepts of information. *J. Am. Stat. Assoc.* **62**, 685–686.

Lehtinen, M.S., L. Päivärinta, and E. Somersalo (1989). Linear inverse problems for generalized random variables. *Inverse Prob.* **5**, 599–612.

Levenberg, K. (1944). A method for the solution of certain nonlinear problems in least-squares. *Q. Appl. Math.* **2**, 164–168.

Marquardt, D.W. (1963). An algorithm for least squares estimation of nonlinear parameters, SIAM J., **11**, 431–441.

Marquardt, D.W. (1970). Generalized inverses, ridge regression, biased linear estimation and non-linear estimation. *Technometrics* **12**, 591–612.

Menke, W. (1984). 'Geophysical Data Analysis: Discrete Inverse Theory.' Academic Press, New York.

Metropolis, N., A.W. Rosenbluth, M.N. Rosenbluth, A.H. Teller, and E. Teller (1953). Equation of state calculations by fast computing machines. *J. Chem. Phys.* **1**, 1087–1092.

Minster, J.B. and T.M. Jordan (1978). Present-day plate motions. *J. Geophys. Res.* **83**, 5331–5354.

Mosegaard, K. and C. Rygaard-Hjalsted (1999). Bayesian analysis of implicit inverse problems. *Inverse Probl.* **15**, 573–583.

Mosegaard, K. and A. Tarantola (1995). Monte Carlo sampling of solutions to inverse problems. *J. Geophys. Res.* **100**, 12,431–12,447.

Mosegaard, K., S.C. Singh, D. Snyder, and H. Wagner (1997). Monte Carlo analysis of seismic reflections from Moho and the W-reflector. *J. Geophys. Res. B* **102**, 2969–2981.

Nolet, G. (1990). Partitioned wave-form inversion and 2D structure under the NARS array. *J. Geophys. Res.* **95**, 8499–8512.

Nolet, G., J. van Trier, and R. Huisman (1986). A formalism for nonlinear inversion of seismic surface waves. *Geophys. Res. Lett.* **13**, 26–29.

Parker, R.L. (1994). 'Geophysical Inverse Theory.' Princeton University Press, Princeton, NJ.

Parzen, E., K. Tanabe, and G. Kitagawa (Eds.) (1998). 'Selected Papers of Hirotugu Akaike,' Springer Series in Statistics. Springer-Verlag, New York.

Powell, M.J.D. (1981). 'Approximation Theory and Methods.' Cambridge University Press, Cambridge.

Press, F. (1968). Earth models obtained by Monte Carlo inversion. *J. Geophys. Res.* **73**, 5223–5234.

Rietsch, E. (1977). The maximum entropy approach to inverse problems. *J. Geophys.* **42**, 489–506.

Scales, L.E. (1985). 'Introduction to Non-Linear Optimization.' Macmillan, London.

Scales, J.A., M.L. Smith, and T.L. Fischer (1992). Global optimization methods for multimodal inverse problems. *J. Comput. Phys.* **102**, 258–268.

Shannon, C.E. (1948). A mathematical theory of communication. *Bell Syst. Tech. J.* **27**, 379–423.

Su, W.-J., R.L. Woodward, and A.M. Dziewonski (1992). Deep origin of mid-oceanic ridge velocity anomalies. *Nature* **360**, 149–152.
Tarantola, A. and B. Valette (1982a). Inverse problems = quest for information. *J. Geophys.* **50**, 159–170.
Tarantola, A. and B. Valette (1982b). Generalized nonlinear inverse problems solved using the least-squares criterion. *Rev. Geophys. Space Phys.* **20**, 219–232.
Tarantola, A. (1984). Inversion of seismic reflection data in the acoustic approximation. *Geophysics* **49**, 1259–1266.
Tarantola, A. (1986). A strategy for nonlinear elastic inversion of seismic reflection data. *Geophysics* **51**, 1893–1903.
Tarantola, A. (1987). 'Inverse Problem Theory: Methods for Data Fitting and Model Parameter Estimation.' Elsevier, Amsterdam.
Taylor, A.E. and Lay, D.C. (1980). 'Introduction to Functional Analysis', John Wiley and Sons, New York.
Taylor, B.N. and C.E. Kuyatt (1994). Guidelines for evaluating and expressing the uncertainty of NIST measurement results. NIST Technical note 1297.
Tikhonov, A.N. (1963). Resolution of ill-posed problems and the regularization method (in Russian). *Dokl. Akad. Nauk SSSR* **151**, 501–504.
van der Hilst, R.D., S. Widiyantoro, and E.R. Engdahl (1997). Evidence for deep mantle circulation from global tomography. *Nature* **386**, 578–584.
Wiggins, R.A. (1969). Monte Carlo inversion of body-wave observations. *J. Geophys. Res.* **74**, 3171–3181.
Wiggins, R.A. (1972). The general linear inverse problem: implication of surface waves and free oscillations for Earth structure. *Rev. Geophys. Space Phys.* **10**, 251–285.
Woodhouse, J.H. and F.A. Dahlen (1978). The effect of general aspheric perturbation on the free oscillations of the Earth. *Geophys. J. R. Astron. Soc.* **53**, 335–354.

Notes

1. For instance, we could fit our observations with a heterogeneous but isotropic Earth model or, alternatively, with a homogeneous but anisotropic Earth.
2. Preliminary Earth Reference Model (PREM), Dziewonski and Anderson, PEPI, 1981. Inversion for Centroid Moment Tensor (CMT), Dziewonski, Chou and Woodhouse, JGR, 1982. First global tomographic model, Dziewonski, JGR, 1984.
3. The capacity element associated to the vector elements $d\mathbf{r}_1$, $d\mathbf{r}_2, \ldots d\mathbf{r}_n$ is defined as $d\tau = \varepsilon_{ij\ldots k}\, dr_1^i\, dr_2^j \ldots dr_n^k$, where $\varepsilon_{ij\ldots k}$ is the Levi-Civita capacity (whose components take the values $\{0, \pm 1\}$). If the metric tensor of the space is $\mathbf{g}(\mathbf{x})$, then $\eta_{ij\ldots k} = \sqrt{\det \mathbf{g}}\, \varepsilon_{ij\ldots k}$ is a true tensor, as it is the product of a density $\sqrt{\det \mathbf{g}}$ by a capacity $\varepsilon_{ij\ldots k}$. Then, the volume element, defined as $dV = \eta_{ij\ldots k}\, dr_1^i dr_2^j \ldots dr_n^k = \sqrt{\det \mathbf{g}}\, d\tau$, is a (true) scalar.
4. This is a property that is valid for any coordinate system that can be chosen over the space.
5. As a counterexample, the distance defined as $ds = |dx| + |dy|$ is not of the L_2 type (it is L_1).
6. This solves the complete problem for isotropic tensors only. It is beyond the scope of this text to propose rules valid for general anisotropic tensors: the necessary mathematics have not yet been developed.
7. The definition of the elastic constants was made before the tensorial structure of the theory was understood. Seismologists today should not use, at a theoretical level, parameters like the first Lamé coefficient λ or the Poisson ratio. Instead they should use κ and μ (and their inverses). In fact, our suggestion in this IASPEI volume is to use the true eigenvalues of the stiffness tensor, $\lambda_\kappa = 3\kappa$, and $\lambda_\mu = 2\mu$, which we propose to call the *eigen-bulk-modulus* and the *eigen-shear-modulus*, respectively.
8. Assume that $p(\mathbf{x})$ and $q(\mathbf{x})$ are normalized by $\int_\mathcal{X} d\mathbf{x}\, p(\mathbf{x}) = \mathbf{1}$ and $\int_\mathcal{X} d\mathbf{x}\, \mathbf{q}(\mathbf{x}) = \mathbf{1}$. Then, irrespective of the normalizability of $\mu(\mathbf{x})$ (as explained above, $p(\mathbf{x})$ and $q(\mathbf{x})$ are assumed to be absolutely continuous with respect to the homogeneous distribution), $(p \wedge q)(\mathbf{x})$ is normalizable, and its normalized expression is $(p \wedge q)(\mathbf{x}) = \dfrac{p(\mathbf{x})\, q(\mathbf{x})/\mu(\mathbf{x})}{\int_\mathcal{X} d\mathbf{x}\, p(\mathbf{x})\, q(\mathbf{x})/\mu(\mathbf{x})}$.
9. As a counter example, working at the surface of the sphere with geographical coordinates $(\mathbf{u}, \mathbf{v}) = (u, v) = (\vartheta, \varphi)$ this condition is **not** fulfilled, as $g_\varphi = \sin\theta$ is a function of ϑ: the surface of the sphere is not the Cartesian product of two 1D spaces.
10. That is, series of numbers that appear random if tested with any reasonable statistical test.
11. To see this, put $f(\mathbf{x}) = \mathbf{1}$, $\mu(\mathbf{x}) = \mathbf{1}$, and $g(\mathbf{x}) = \dfrac{\exp(-E(\mathbf{x})/T)}{\int \exp(-E(\mathbf{x})/T) d\mathbf{x}}$, where $E(\mathbf{x})$ is an 'energy' associated to the point $\mathbf{x}$, and T is a 'temperature'. The summation in the denominator is over the entire space. In this way, our acceptance rule becomes the classical Metropolis rule: point $\mathbf{x}_i$ is always accepted if $E(\mathbf{x}_i) \le E(\mathbf{x}_j)$, but if $E(\mathbf{x}_i) > E(\mathbf{x}_j)$, it is only accepted with probability $p_{ij}^{\text{acc}} = \exp\big(-\big(E(\mathbf{x}_i) - E(\mathbf{x}_j)\big)/T\big)$.
12. A numerical method is called robust if it is not sensitive to a small number of large errors.
13. It would be violated, for instance, if we use the pair of elastic parameters longitudinal wave velocity – shear wave velocity, as the volume element in the space of elastic wave velocities does not factorize (see Appendix H).
14. We use here the properties $\log \sqrt{\mathbf{A}} = \frac{1}{2} \log \mathbf{A}$, and $\det \mathbf{AB} = \det \mathbf{BA}$
15. Typically, this may happen because the derivatives $\mathbf{F}$ are small or because the variances in $\mathbf{C}_M$ are large.
16. We first use $\log\det \mathbf{A} = \operatorname{trace} \log \mathbf{A}$, and then the series expansion of the logarithm of an operator, $\log(\mathbf{I} + \mathbf{A}) = \mathbf{A} - \frac{1}{2}\mathbf{A}^2 + \cdots$
17. Practically, it may correspond to the output of some 'black box' solving the 'forward problem'.
18. Remember that, even if we wish to use a simple method based on the notion of conditional probability density, an analytic expression like $\mathbf{d} = \mathbf{f}(\mathbf{m})$ needs some 'thickness' before going to the limit defining the conditional probability density. This limit crucially depends on the 'thickness', i.e., on the type of uncertainties the theory contains.
19. Note that taking the limit of $\vartheta(x, t)$ or of $\rho(x, t)$ for infinite variances we obtain $\mu(x, t)$, as we should.
20. The ratio $F(\mathbf{x}) = f(\mathbf{x})\, v(\mathbf{x})$ is what we refer to as *the volumetric probability* associated to the probability density $f(\mathbf{x})$. See Appendix A.
21. We take this example because typical misfit functions are adimensional (have no physical dimensions) but the argument has general validity.

22. As shown in Tarantola (1987), if $\boldsymbol{\gamma}_k$ is the direction of steepest ascent at point $\mathbf{m}_k$, i.e., $\gamma_k = \mathbf{C}_M\, \mathbf{F}_k^t\, \mathbf{C}_D^{-1}\, (\mathbf{f}_k - \mathbf{d}_{\text{obs}}) + (\mathbf{m}_k - \mathbf{m}_{\text{prior}})$, then, a local linearized approximation for the optimal ε_k gives

$$\varepsilon_k = \frac{\boldsymbol{\gamma}_k^t \mathbf{C}_M^{-1} \boldsymbol{\gamma}_k}{\boldsymbol{\gamma}_k^t (\mathbf{F}_k^t\, \mathbf{C}_D^{-1}\, \mathbf{F}_k + \mathbf{C}_M^{-1})\, \boldsymbol{\gamma}_k} .$$

23. The 'best estimator' of $\tilde{\mathbf{C}}_M$ is

$$\tilde{\mathbf{C}}_M \approx \left(\mathbf{F}_k^t \mathbf{C}_D^{-1} \mathbf{F}_k + \mathbf{C}_M^{-1} \right)^{-1} . \tag{119}$$

See, e.g., Tarantola (1987).

24. While a sensible estimation of the optimal values of the real positive quantities ε_k is crucial for the algorithm 111, they can in many usual circumstances be dropped from the algorithm 113.
25. The gray oval is the product of the probability density over the model space, representing the prior information, and the probability density over the data space representing the experimental results.
26. The gravitational field at point $\mathbf{x}_0$ generated by a distribution of volumetric mass $\rho(\mathbf{x})$ is given by

$$\mathbf{g}(\mathbf{x}_0) = \int dV(\mathbf{y}) \frac{\mathbf{x}_0 - \mathbf{y}}{\|\mathbf{x}_0 - \mathbf{x}\|^3}\, \rho(\mathbf{x}) .$$

When the volumetric mass is constant inside some predefined (2D) volumes, as suggested in Figure 8, this gives

$$\mathbf{g}(\mathbf{x}_0) = \sum_A \sum_B \mathbf{G}^{\text{A,B}}(\mathbf{x}_0)\, m^{\text{A,B}} .$$

This is a strictly linear equation between data (the gravitational field at a given observation point) and the model parameters (the masses inside the volumes). Note that if instead of choosing as model parameters the total masses inside some predefined volumes one chooses the geometrical parameters defining the sizes of the volumes, then the gravity field is not a linear function of the parameters. More details can be found in Tarantola and Valette (1982b, page 229).

27. Using the 'orthogonal-limit' method described in Section 2.4.
28. The term 'a priori model' is an abuse of language. The correct term is 'mean a priori model'.

Editor's Note

Appendixes A–P are placed on the attached Handbook CD, under the directory \16Mosegaard. An introduction to probability concepts is given in Chapter 82, Statistical Principles for Seismologists, by Vere-Jones and Ogata. See also Chapter 52, Probing the Earth's Interior with Seismic Tomography, by Curtis and Snieder.

Observational Seismology

17

Challenges in Observational Seismology

W.H.K. Lee

US Geological Survey, Menlo Park, California, USA (retired)

1. Introduction

Earthquake seismology became a quantitative scientific discipline after instruments were developed to record seismic waves in the late 19th century (Dewey and Byerly, 1969; Chapter 1 by Agnew). Earthquake seismology is essentially based on *field* observations. The great progress made in the past several decades was primarily due to increasingly plentiful and high-quality data that are readily distributed. Our ability to collect, process, and analyze earthquake data has been accelerated by advances in electronics, communications, computers, and software (see Chapter 85 edited by Snoke and Garcia-Fernandez).

Instrumental observation of earthquakes has been carried out for a little over 100 years by seismic stations and networks of various sizes, from local to global scales (see Chapter 87 edited by Lahr and van Eck). The observed data have been used, for example, (1) to compute the source parameters of earthquakes, (2) to determine the physical properties of the Earth's interior, (3) to test the theory of plate tectonics, (4) to map active faults, (5) to infer the nature of damaging ground shaking, and (6) to carry out seismic hazard analysis. Construction of a satisfactory theory of the earthquake process has not yet been achieved within the context of physical laws. Good progress, however, has been made in building a physical foundation of the earthquake source process, partly as a result of research directed toward earthquake prediction.

This chapter is intended for a general audience. Technical details are not given, but relevant references and chapters in this Handbook are referred to. The first part of this chapter presents a brief overview of the observational aspects of earthquake seismology, concentrating on instrumental observations of seismic waves generated by earthquakes (i.e., seismic monitoring), and readers are referred to Chapter 49 by Musson and Cecic for noninstrumental observations. A few key developments and practices are summarized by taking a general view, since many national and regional developments have been chronicled in national and institutional reports (see Chapter 79 edited by Kisslinger). In the latter part of this chapter, the nature of seismic monitoring and some challenges in observational seismology are discussed from a *personal* perspective. Comments of a technical or *philosophical* nature are given in the Notes at the end of the chapter, and they are referenced by superscript numbers in the text.

2. Some Basic Information about Seismographs and Earthquakes

Besides geodetic data (see, e.g., Chapter 37 by Feigl), the primary instrumental data for the quantitative study of earthquakes are *seismograms*, records of ground motion caused by the passage of seismic waves generated by earthquakes. Seismograms are written by *seismographs*, instruments that detect and record ground motion with timing information. A seismograph usually consists of three components: (1) a seismometer that responds to ground motion and produces a signal proportional to acceleration, velocity, or displacement over a range of input motions in amplitude and in frequency; (2) a timing device; and (3) a recording device that writes seismograms (ground motion plus time marks) on papers or on electronic storage media. An *accelerograph* is a seismograph designed to record the time history of acceleration of strong ground motion on scale. Most modern seismographs are *velocigraphs* recording the time history of ground velocity. See Chapter 18 by Wielandt for a discussion of seismometry. A seismic network (or array) is a group of seismographs that are "linked" to a central headquarters. The link is by various methods of telemetry nowadays, and was by mail or telegraphy in the early days, or simply by manual collecting of the records. When we speak of a seismic *station*, it may be an observatory with multiple instruments in special vaults, or a small instrument package buried in a remote unmanned site.

INTERNATIONAL HANDBOOK OF EARTHQUAKE AND ENGINEERING SEISMOLOGY, VOLUME 81A ISBN: 0-12-440652-1

In 1935 C.F. Richter introduced the concept of *magnitude* to classify local earthquakes by their "size." See Chapter 44 by Utsu for a discussion of the various magnitude scales in use. Existing instruments and environments are such that the smallest natural earthquakes we routinely observe are about magnitude 0. The largest earthquake so far for which we have instrumental records is the magnitude 9.5 Chilean earthquake in 1960. Some commonly accepted adjectives used to describe the approximate size or magnitude (M) of an earthquake are "major" for $M \geq 7$ ("great" if $M \geq 8$), "moderate" for $5 \leq M < 7$, "small" for $3 \leq M < 5$, and "micro" for $M < 3$.

In 1941, B. Gutenberg and C.F. Richter discovered that over large geographic regions the frequency of earthquake occurrence is empirically related to magnitude by $\log N = a - bM$, where N is the number of earthquakes of magnitude M or greater, and a and b are numerical constants. Usually $b \approx 1$, implying, for example, that $M = 6$ earthquakes are about 10 times more frequent than $M = 7$ earthquakes. See Chapter 43 by Utsu for further detail. Engdahl and Villasenor in Chapter 41 show that there has been an *average* of about 15 major (i.e., $M \geq 7$) earthquakes per year over the past 100 y. A list of deadly earthquakes of the world for the past five centuries has been compiled by Utsu in Chapter 42. It shows that $M \geq 6$ earthquakes (about 150 in the world per year) can be damaging and deadly if they occur in populated areas and if their focal depths are shallow (e.g., <50 km). Strong ground motions above $0.1g$ in acceleration are mainly generated by $M \geq 6$ earthquakes.

3. Seismic Networks for Observing Earthquakes

Seismic waves from earthquakes have a vast range in amplitude and frequency (see Fig. 1 of Hutt *et al.* in Chapter 20—about 10 orders in amplitude of ground acceleration, from 10^{-7} to 10^{3} cm sec^{-2}, and about 7 orders in frequency, from 10^{-5} to 10^{2} Hz). Since no single instrument type can cover such vast ranges, seismographs and seismic networks have evolved from three different *optimizing* choices, as discussed in the next three subsections. National reports, including those from Germany, Japan, Russia, the United Kingdom, and the United States (collected in Chapter 79, edited by Kisslinger), contain detailed early history on a national basis. For example, Kisslinger and Howell give a detailed account of historical developments in the United States to about 1960 in the USA National Report.

3.1 Seismic Networks Optimized for Teleseisms

Teleseisms are *distant* earthquakes that are big enough to produce measurable seismic waves at great distances. A major earthquake occurs about once a month (or a potentially damaging earthquake occurs about *every week*) somewhere in the world (but rarely in one's own backyard), and is recorded on the seismograms of seismographs optimized for teleseisms. These seismograms also draw interest from research seismologists worldwide, as teleseisms are excellent sources of seismic waves for probing the Earth's interior (see, e.g., Chapter 11 by Romanowicz; Chapter 52 by Curtis and Snieder).

3.1.1 Early Years

In the beginning of instrumental seismology, observatories with various types of seismographs operated independently. The observatories were linked by mail, which could take months. Many seismological studies require seismograms or their readings from multiple stations. For example, arrival times of seismic waves from at least four well-distributed stations are needed to locate an earthquake satisfactorily. Even after one managed to get a few seismograms, it was difficult to work with records from different instruments with poorly synchronized time.

In the late 19th century the need for *standardization* and for *data exchange* was recognized by G. Gerland, J. Milne, and E. Rebeur-Paschwitz. With the support of the British Association for the Advancement of Science, over 30 Milne seismographs were placed at locations throughout the British Empire beginning in the late 1890s, and seismogram readings were reported to Milne's observatory at Shide on the Isle of Wight (see the UK National Report in Chapter 79). A global earthquake summary with seismogram readings was issued by John Milne beginning in 1899. These summaries are now known as the "Shide Circulars" (see Chapter 88 by Schweitzer and Lee). Milne seismographs were soon superseded by more advanced instruments,[1] and the headquarters of the International Association of Seismology was established in Strasbourg (see the German National Report in Chapter 79).

Seismographs for recording teleseisms were established at many observatories, especially meteorological and astronomical observatories. The early enthusiasts were academic professors, Jesuits, and gentleman scientists. See Chapter 89 edited by Howell for biographies of some notable pioneers. Revolutions and wars, however, frequently disrupted progress, especially in collecting and distributing earthquake information, during the first half of the 20th century.[2]

3.1.2 WWSSN and ESSN

In the late 1950s, attempts to negotiate a comprehensive test ban treaty failed, in part because of perceptions that seismic methods were inadequate for monitoring the underground environment for nuclear testing (see Chapter 24 by Richards). The influential Berkner report of 1959 advocated major support for seismology (see article by Kisslinger and Howell in the USA National Report in Chapter 79). As a result, the World Wide Standardized Seismograph Network (WWSSN) was created with about 120 continuously recording stations, located over much of the world (except China and USSR) in the early

1960s (Oliver and Murphy, 1971). Each WWSSN station was equipped with *identical* sets of short-period and long-period three-component seismographs and *accurate* chronometers. Seismograms were sent to the United States to be photographed onto 70-mm film chips for distribution (about $1 per chip). This network is credited with making possible rapid progress in global seismology, and with aiding the plate tectonic revolution in the Earth sciences in the late 1960s (see, e.g., Chapter 6 by Uyeda; article by Sykes in the USA National Report in Chapter 79).

At about the same time, the Unified System of Seismic Observations (ESSN) of the former USSR and its allied countries was established, consisting of almost 100 stations equipped with Kirnos short-period, broadband (1–20 sec displacement sensing) and long-period seismographs. See the Russian National Report in Chapter 79 for details.

Despite its great success, the WWSSN declined starting in the mid-1970s. By then it had produced 3 million analog seismograms, far more than seismologists could process and analyze. After about ten years of operation, funding for the WWSSN began to disappear. Although the initial costs were funded by the US Defense Advanced Research Projects Agency (DARPA), their emphasis is in research and not in long-term operation. Funding for continuing the WWSSN was then left to the National Oceanic and Atmospheric Administration (NOAA) and then to the US Geological Survey (USGS). Because of statutory restriction, USGS could not support global stations outside the United States. Although the US National Science Foundation (NSF) did pick up the funding for supporting foreign stations for some time, NSF also wanted to avoid funding any ongoing seismic networks. In addition, the emphasis in seismology was shifting to earthquake prediction at the USGS, then considered a new and promising venture.[3] Earthquake prediction, however, turned out to be far more difficult than anticipated, as reviewed for example, by Kanamori in Chapter 72.

3.1.3 The Digital Revolution and the GDSN

Since *analog* seismograms have a low dynamic range (about 3 orders or less in amplitude) and must be digitized before computer processing, some seismologists recognized that "digital" instrumentation should be developed to achieve a much higher dynamic range and for ease of computer processing. Many scientists and engineers in other disciplines had already been making great advances in that direction because of the emerging digital technology in the 1970s. Seismologists also recognized that the tandem use of short-period and long-period instruments was needed to avoid the natural seismic noise (see Chapter 19 by Webb). They realized that a new global seismic network should be rebuilt with (1) broadband, high dynamic range seismographs, (2) digital electronics, (3) communication by telemetry or a mass storage medium, and (4) processing by computers in mind.

According to Duncan Agnew (personal communication, 2001), the idea of digital recording goes back to 1960, but it was not practical until the late 1970s, and Block and Moore (1966) pioneered the use of the feedback gravimeter. The introduction of electronic force feedback to sealed inertial seismometers (Melton, 1976; Wielandt and Streckeisen, 1982) together with the application of high-resolution analog-to-digital converters made it possible to construct very broadband, large dynamic range seismograph systems. The first major digital broadband installation was the German Gräfenberg (GRF) array, the first station of which started recording in 1975 (Harjes and Seidl, 1978).

The International Deployment of Accelerometers (IDA) Project was created in the 1970s as a global *digital* seismic network to collect data for low-frequency seismology (Agnew *et al.*, 1976). Among the many contributions made by IDA are much improved studies of the Earth's free oscillation, long-period source mechanics for major earthquakes, and the aspherical structure of the Earth's interior (see Chapter 51 by Lay). Another notable effort in global digital seismic networks was the French GEOSCOPE Program (Romanowicz *et al.*, 1991), and the German GEOFON project.

With the availability of broadband, large dynamic range, force feedback seismometers and 24-bit digitizers, many of the WWSSN stations were replaced by broadband digital systems starting in the 1980s (see Chapter 20 by Hutt *et al.*). A global digital seismic network has emerged since the 1980s under the guidance of two effective organizations (the International Federation of Digital Broadband Seismographic Networks, FDSN, and the Incorporated Research Institutions for Seismology, IRIS).[4] Digital seismograms recorded by stations worldwide are now readily available via the Internet from the IRIS Data Management Center within tens of minutes of an $M \approx 6$ or larger earthquake occurring anywhere in the world (see Chapter 86 by Ahern), as well as through the Europen ORFEUS center at de Bilt, The Netherlands, the GEOFON center at the GeoForschungsZentrum, Postdam, Germany, and several other centers.

3.2 Seismic Networks Optimized for Regional Earthquakes

Another major development in seismic monitoring was the establishment of seismic networks optimized to record the *more frequent* but *smaller* regional and local earthquakes. In order to observe as many nearby earthquakes as possible, seismographs with high magnifications are adopted to record as small an earthquake as the technology and background noise allow. A consequence of this requirement, when applied to inexpensive sensors and telemetry with low dynamic range, is that the recorded amplitudes are saturated for earthquakes with $M > \sim 3$ within about 50 km. This is not a serious defect, because the emphasis is to obtain as many first-arrival times as possible, so that more earthquakes can be detected and located. Since seismic waves from small earthquakes are quickly attenuated with distance, it is also necessary to deploy many instruments with small station spacing (several to a few tens of kilometers), and to cover as large a territory as possible in order to record at least a few earthquakes every week. Since funding is finite, the regional seismic networks are usually optimized for a large number of stations rather than for high data quality.

3.2.1 A Brief History

In the 1910s, the Carnegie Institution of Washington (CIW)[5] was spending large sums of money building the world's largest telescope in southern California. Since astronomers were concerned about earthquakes that might disturb their telescopes, H.O. Wood was able to persuade CIW to support earthquake investigations. As a result, a regional network of about a dozen Wood–Anderson seismographs was established in southern California in the 1920s. See Goodstein (1991) for the early history leading to the establishment of the California Institute of Technology (Caltech) and its Seismological Laboratory. Many astronomers played important roles in getting seismic monitoring established in various regions of the world.

Regional networks using different seismographs were also established in many countries, such as Japan, New Zealand, and the USSR with its allies. In the 1960s, high-gain, short-period, telemetered networks were developed to study microearthquakes (see, e.g., Eaton, 1989). Over 100 microearthquake networks were implemented by the 1970s in various parts of the world for detailed studies of local earthquakes and especially for the purpose of earthquake prediction (Lee and Stewart, 1981). These microearthquake networks consisted of tens to hundreds of short-period seismometers with their signals telemetered into central recording sites for processing and analysis. High magnification was achieved by electronic amplification, which permitted recording of very small earthquakes (down to magnitude 0) at the expense that the recorded seismic wave amplitudes were saturated for earthquakes of $M > \sim 3$. Some microearthquake networks soon expanded into regional seismic networks.

As with the WWSSN, it was difficult to improve and sustain regional seismic networks for long in many countries. For example, by the 1980s the regional seismic networks in the United States were in decline. The 1989 Loma Prieta earthquake and the 1994 Northridge earthquake demonstrated that the existing regional seismic networks in the United States were not satisfactory, especially during large damaging earthquakes.[6] With new funding in response to these disastrous earthquakes, a new life began in the form of real-time seismology.

3.2.2 Some Recent Advances

Because of recent advances in electronics, communications, and microcomputers, it is now possible to deploy sophisticated digital seismograph stations at global, national, and local scales for *real-time* seismology (Kanamori *et al.*, 1997). Many such networks, including portable networks, have been implemented in many countries. In particular, various real-time and near real-time seismic systems began operation in the 1990s. For example, Mexico (Chapter 76 by Espinosa-Aranda), California (Chapter 77 by Gee *et al.*; Chapter 78 by Hauksson *et al.*), and Taiwan (Teng *et al.*, 1997).

For example, the Real-time Data (RTD) system operated by the Seismological Observation Center of the Central Weather Bureau in Taiwan is based on a network of telemetered digital accelerographs (see Chapter 64 by Shin *et al.*). This system, using pagers, e-mail and other techniques, has automatically and rapidly issued information on the hypocenter, magnitude, and shaking amplitude of felt earthquakes ($M > \sim 4$) in the Taiwan region since 1995. The disastrous Chi-Chi earthquake ($M_W = 7.6$) of 20 September 1999 caused 2471 deaths and an estimated economic losses of US \$14 billion. For this earthquake sequence, the RTD system delivered accurate information (102 sec after the main shock and about 50 sec for most aftershocks) to officials and proved to be useful for emergency response by the Taiwan government (Wu *et al.*, 2000; Goltz *et al.*, 2001).

3.3 Seismic Networks Optimized to Record Damaging Ground Shaking

Observing teleseisms at a spacing of several hundreds of kilometers does not yield the information about near-source strong ground shaking required for earthquake engineering purposes. A few hundred global stations, therefore, cannot provide the detailed data necessary to help in reducing seismic hazards. Broadband seismometers optimized to record earthquakes at great distances are not designed to perform well in the near field of a major earthquake. For example, during the 1999 Chi-Chi earthquake the *nearest* broadband station in Taiwan (epicentral distance of about 20 km) recorded mostly saturated amplitude data and stopped about one minute into the shock (see Kao and Angelier, 2001, for the recorded seismogram).

A regional seismic network with spacing of a few tens of kilometers cannot do the job either. The station spacing is still too large and, worse yet, the records are often saturated for earthquakes with magnitude $> \sim 3$, including a *big one* if it should occur. In his account of the early history of earthquake engineering, Housner in Chapter 2 credited John R. Freeman, an eminent engineer, for persuading the then US Secretary of Commerce to authorize a strong-motion program and in particular the design of an accelerograph for engineering purposes in 1930. In a letter to R.R. Martel (Housner's professor) at Caltech, Freeman wrote:

> I stated that the data which had been given to structural engineers on acceleration and limits of motion in earthquakes as a basis for their designs were all based on guesswork, that there had never yet been a precise

measurement of acceleration made. That of the five seismographs around San Francisco Bay which tried to record the Earthquake of 1906 not one was able to tell the truth.

Strong-motion recordings useful for engineering purposes are on-scale recordings of damaging earthquakes; in particular, recordings on or near structures in densely urbanized environments, within 20 km of the earthquake-rupture zone for sites on rock and within about 100 km for sites on soft soils. Recordings of motions at levels sufficient to cause damage at sites at greater distances also are of interest for earthquake engineering in areas likely to be affected by major subduction zone earthquakes or in areas with exceptionally low attenuation rates (Borcherdt, 1997).

Although several interesting accelerograms were recorded in southern California in the 1930s and 1940s, to the delight of earthquake engineers, most seismologists did not pursue strong-motion monitoring until much later. The 1971 San Fernando earthquake demonstrated the need for strong-motion data for engineering purposes (see Chapter 57 by Anderson). Two important programs merged in the United States—the National Strong-Motion Program (*http://nsmp.wr.usgs.gov/*), and the California Strong Motion Instrumentation Program (*http://docinet3.consrv.ca.gov/csmip/*). However, their budgets were and continue to be small in comparison with those of other earthquake programs. High levels of funding for strong-motion monitoring (comparable to that of the GDSN and the regional seismic networks) occurred in Taiwan and Japan in the early and mid-1990s, respectively (see Section 5.1). The Consortium of Organizations for Strong-Motion Observation Systems (*http://www.cosmos-eq.org/*) has recently been established to promote the acquisition and application of strong-motion data.

4. Record Keeping and Data Processing on a Global Scale

Many scientific advances are based on accurate and long-term observations. Because disastrous earthquakes in a given region recur on a time scale of tens to hundreds of years, special efforts are needed to ensure that seismic monitoring is carried out in a consistent manner in keeping records, detecting and timing all locatable events, and processing the observed and derived data. The amount of data in earthquake seismology is large and growing[7] and may be classified into six types (with their approximate annual output rate) as follows:

1. Information for instrument location, characteristics, and operational details: 2×10^7 bytes y^{-1}.
2. Raw observational data (continuous signals from seismographs): 10^{14} bytes y^{-1}.
3. Earthquake waveform data (containing seismic events): 5×10^{12} bytes y^{-1}.
4. Earthquake phase data, such as P, S, and secondary arrival times, maximum amplitude and period, first motion direction, signal duration, etc.: 5×10^8 bytes y^{-1}.
5. Event lists of origin time, epicenter coordinates, focal depth, magnitude, etc.: 2×10^7 bytes y^{-1}.
6. Scientific reports describing seismicity, focal mechanisms, etc.: 5×10^7 bytes y^{-1}.

Seismic monitoring for the entire world depends on continuous international cooperation as discussed in Chapter 4 by Adams, because seismic waves propagate throughout the Earth without regard to national boundaries. Seismologists need not only to exchange scientific results (via reports and papers published in journals or books), but also to rely on the exchange of primary data (i.e., seismograms) and their derived products (e.g., phase data). Each seismic observatory can interpret its recorded seismograms, but single-station data are insufficient for the study of earthquakes, especially those occurring some distance away.

The International Seismological Centre (ISC) is charged with the final collection, analysis, and publication of standard earthquake information from all over the world (Willemann and Storchak, 2001). The ISC bulletins (issued since 1964) are the definitive summary of reported earthquake phase data, and from these data the earthquake parameters are determined by a standard procedure. These bulletins are published with a time lag of about two years in order to incorporate as much information as possible from cooperating seismic stations. Engdahl and Villasenor in Chapter 41 prepared a comprehensive catalog of earthquakes of the world from 1900 to 1999 by relocating systematically many thousands of earthquakes using the ISS and ISC phase data. Their catalog and the phase data they used for location are archived on the attached Handbook CD.

For more rapid dissemination of earthquake information on a global scale, the US National Earthquake Information Service (NEIS) of the US Geological Survey (in Golden, Colorado) issues results based on its own network of stations as well as phase readings sent by other stations. Visit their Web site at *http:/neic.usgs.gov/*. Earthquake parameters for significant earthquakes are usually announced within about one hour, and summaries (now in computer files) are distributed on a weekly basis. The data collected by the NEIS are sent to the ISC for further analysis, as described in the preceding paragraph.

Moment tensor solutions for earthquakes of $M \geq \sim 5.5$ worldwide have been determined by the Harvard group since 1976 (see *http://www.seismology.harvard.edu/projects/CMT/*), and by the USGS since 1981 (see Chapter 50 by Sipkin). Preliminary solutions are produced within minutes after the NEIC's QED results. In addition to Harvard and USGS, there are several centers that determine moment tensor solutions (mostly for regional earthquakes) in near real time.

Most seismic observatories issue their own bulletins, and many national agencies publish national earthquake catalogs. However, their quality and contents vary greatly. The efforts in seismic monitoring in many countries can be found in their national reports (summarized in Chapter 79 edited by Kisslinger and presented in full on the attached Handbook CD). See also Chapter 87 edited by Lahr and van Eck for a global inventory of seismographic networks.

5. The Nature of Seismic Monitoring of Damaging Earthquakes

Seismic monitoring of earthquakes has been most successful on the global scale because there are (1) organizations (e.g., FDSN, IRIS, ISC, IASPEI, and CTBTO) that promote it, (2) hundreds of academic users, and (3) several major and/or damaging earthquakes every year in the world that draw public and academic attention. Field work is mostly cooperative, with a central organization supplying (or augmenting) the latest equipment, and the labor cost per station is a relatively small amount that can easily be absorbed by a cooperating local institution. On the other hand, seismic monitoring on a regional or local scale is far more difficult because there are only a few academic users, and usually no significant or destructive earthquakes occur for decades in most geographical areas.

5.1 Requirements for Station Site and Spacing

The station site and spacing requirements for seismological research are very different from those for earthquake engineering purposes. Seismologists (especially those who study Earth structure) want their stations to be located at quiet sites, as far away from any human activities as possible. On the other hand, earthquake engineers want instruments in the built environment of urban areas. Since the occurrence of major earthquakes in a given region is rare, we can understand why most seismologists would be reluctant to wait for decades for a few strong-motion records.

Studies indicate that we may need a station spacing of about 1 km or less in order to reduce the observed variances in strong ground motion to a factor less than 2 (Field and Hough, 1997; Evans, 2001). During 1991–1996, the Taiwan government deployed about 1200 accelerographs (at 640 free-field sites with a station spacing of 3–5 km, and in 56 buildings and bridges) nonuniformly in the urban areas of Taiwan (see Chapter 64 by Shin *et al.*). After the 1995 Kobe earthquake, the Japanese government deployed the 1000-station K-Net (see Chapter 64 by Kinoshita) at a uniform spacing of 25 km over Japan. In both cases, the total cost to deploy one station was about $30 000. Therefore, to achieve 1 km spacing in *one* major urban area would require a few thousand accelerograph stations, and would cost about one hundred million dollars to deploy. However, a more cost-effective alternative would be to selectively deploy strong-motion instruments at fewer sites that are representative of the local site conditions, rather than uniformly throughout an area.

In regard to strong-motion instrumentation needs of the United States, Borcherdt *et al.* (1997) derived estimates based on the National Seismic Hazard Maps, population exposure, and knowledge of the distribution of local geologic deposits. These estimates as reviewed and expanded to include the built environment provided the basis for consensus of a national workshop (Stepp, 1997). The estimates indicated that at least 7000 sites were needed to record strong ground shaking with station spacing for rock sites less than 7 km and preferably about 1.7 km in densely urbanized areas, such as San Francisco, California. Workshop consensus indicated that an additional 13 000 stations are needed to ensure that the next major earthquake is thoroughly documented on the built environment, with 7000 for buildings, 3000 for lifelines such as bridges and pipelines, and 3000 for critical facilities necessary for emergency response and near real time disaster assessment.

5.2 Total Cost

There are a number of costs associated with establishing and operating a seismic network: (1) deployment, (2) maintenance and operation, and (3) staffing. We should be aware that the deployment cost must include the capital cost for instruments (including equipment for telemetry, if any), and expenses for siting, site preparation, quality assurance, and administrative overhead in procurement and management. In both the recent Taiwan and Japan cases, the instrumentation cost was about 1/3 of the total deployment cost. Maintenance costs vary, but experience has shown that at least 20% of the instrument cost should be budgeted for parts, supplies, and repair services *every year* for satisfactory performance. Operating costs depend mostly on the telemetry method used for getting the data from the field to the headquarters, and vary greatly in different locations. Staffing costs for maintenance, operation, and research are most difficult to estimate and control over a long period of time. Therefore, the instrument cost is relatively minor in a project that must continue for *decades*.

Because governments and academics usually fund capital equipment out of the same budget, there is a tendency for some seismologists to regard inexpensive instruments as a panacea.[8] Experience has shown that the instrument cost constitutes only about 10% of the total project cost in 30 years, the length of time that is usually needed to accumulate some significant earthquake data in a high-seismicity area.

What one may do with four different levels of capital funding will be discussed next. Capital funding means a sufficient funding for the total deployment cost in the field and the cost for setting up a headquarters with a staff.

5.3 Very Low Capital Funding at $50 000

At this very low capital funding level, one may purchase a single low-end digital broadband system to monitor teleseisms, and operate it as a global seismic station. See Chapter 20 by Hutt *et al.* for more technical details. Deployment cost will be low if one simply installs it in the basement of one's building. Maintenance and operating cost will be low also, but a few thousand dollars per year are necessary for supplies and unexpected repairs. A part-time staff member is needed, but an existing staff member can usually serve in that capacity.

5.4 Low Capital Funding at $500 000

At this low capital funding level, one may consider setting up a ten-station telemetered network using digital accelerographs to cover an area of about 3000 km^2 at a station spacing of about 25 km. If the accelerograph allows four channels, then one would just add a vertical-component, short-period seismometer. However, if only three-component instruments are used, then to increase the number of detectable earthquakes one should replace the vertical-component accelerometer by a vertical-component, short-period seismometer (since engineers are interested mostly in the horizontal ground motions). Real-time telemetry may be achieved by leased telephone lines, radios, or satellites. If real-time performance is not needed, then it is cheaper to use the Internet or dial-up access, if they are available in the field. A few companies are selling this type of seismic network with a central processing system to monitor regional and local earthquakes and to provide some coverage for strong ground motion. Deployment cost will be highly dependent on where the network is to be located. The same is true for maintenance, operation, and staffing. One will need at least $100 000 per year to maintain and operate this kind of network, with at least one full-time staff member.

5.5 Medium Capital Funding at $5 000 000

At this level of capital funding one may consider setting up, for example: (1) a 100-station telemetered network using digital accelerographs to cover an area of about 10 000 km^2 at a station spacing of about 10 km, with the modification described in Section 5.4, for monitoring a modest area of high earthquake hazards; and (2) a few global broadband stations for academic research. There is always a trade-off between the area covered and the station spacing. Real-time telemetry may be by leased telephone lines, radios, or satellites. If real-time performance is not needed for all the stations, then one can make use of the Internet or dial-up connections, if available. Alternatively, one could set up a 100-station telemetered broadband network at a larger station spacing to cover more area, and supplement it with telemetered accelerometer signals and/or dial-up accelerographs. Deployment cost will be highly dependent on where the network is located. The same is true for maintenance and operation. One will need at least $1 000 000 per year and several full-time staff members to maintain and operate such a network.

5.6 High Capital Funding at $50 000 000

At this high capital cost, one may consider an instrumentation program similar to the TriNet in southern California (see Chapter 78 by Hauksson, *et al.*), the 1000-station K-Net in Japan (Chapter 63 by Kinoshita), or the Taiwan program (Chapter 64 by Shin, *et al.*). The annual budget for staff (10 or more), maintenance, and operation is typically a few million dollars or more, depending on many factors.

5.7 Remarks

It is important to do planning in advance and to visit some well-established seismic networks before one embarks on an instrumentation project. One must be clear as to what objectives can be accomplished within the expected long-term funding situation. This will also help to decide which option of the trade-offs discussed above is preferable. Some guidance in this respect is given in Bormann (2002). Experience indicates that, technically, telemetry is the weakest link in any telemetered seismic network, especially during and after a major earthquake. Therefore, the field units should have some on-board recording capabilities so that important data can be retrieved later if necessary. Some redundancy in the field units, alternative methods of telemetry, and backup data acquisition systems are absolutely necessary.

6. Some Difficulties in Seismic Monitoring for Hazard Mitigation

The major problem in seismic monitoring, especially for hazard mitigation on a long-term basis, is not technical but political and financial. Adequate funding over decades is necessary for the success of a seismic monitoring project. Major earthquakes may not be damaging if they occur in uninhibited or lightly populated areas. However, an earthquake need not be very large in magnitude to cause serious damage if it occurs in a heavily populated area, especially if the earthquake's focal depth is shallow. According to Munich Reinsurance Company (2000), economic losses due to earthquakes in the 20th century are very large. For example, the five largest losses are $100 billion for the 1995 Kobe (Japan) $M_W = 6.9$ earthquake; $44 billion for the 1994 Northridge (USA) $M_W = 6.7$ earthquake; $14 billion for the 1999 Chi-Chi (Taiwan) $M_W = 7.6$ earthquake; $14 billion for the 1988 Armenia $M_W = 6.7$ earthquake; and $12 billion for the 1999 Turkey $M_W = 7.6$ earthquake. Note that three of these five earthquakes are not called "major" earthquakes, because their magnitudes are below 7.

6.1 What Seismograms are Most Important to Engineering Designs?

According to Norm Abrahamson (private communication, 2000), designs of large engineering structures benefit most from strong-motion records of $M \geq 7$ earthquakes obtained within 20 km of fault ruptures. As of mid-1999, there were only eight such strong-motion records in the world after nearly 70 years of effort in strong-motion monitoring. As discussed in the next subsection, the odds for having an $M \geq 7$ earthquake occurring near a station are rare, and the number of strong-motion instruments at the free-field sites throughout the world was not large until about 1990. The Kocaeli, Turkey, earthquake of 17 August 1999 added five more such records. And the Chi-Chi, Taiwan, earthquake of 20 September 1999 added over 60 such records, thanks to an extensive strong-motion instrumentation program that was completed three years earlier (see Chapter 64 by Shin *et al.*).

6.2 Major Earthquakes are Rare in a Given Region

As noted in Section 2, the average annual number of major ($M \geq 7$) earthquakes worldwide is about 15. Most of the shallow major earthquakes occur in subduction zones off the coast, and large areas of the Earth have few or no earthquakes (see Color Plate 15 of this Handbook). Using the earthquake catalog for major earthquakes from 1900 to 1999 by Engdahl and Villasenor in Chapter 41, about 32% of the $M \geq 7$ earthquakes of the world occurred on land. If a seismic station is deployed on land (as deployment at sea is very expensive), what is the probability that an $M \geq 7$ earthquake will occur within a radius of 20 km of this station in one year?

A very rough estimate can be derived as follows. Since the total land area of the Earth is about $1.5 \times 10^8\,\mathrm{km}^2$, the probability is then about 4×10^{-5}, assuming that each $M \geq 7$ earthquake is a point source. We can improve this probability by, say, a factor of 10 by placing a seismic station in a known seismically active area based on geological and seismological information. A probability of 4×10^{-4} implies that we must wait about 2500 years for an $M \geq 7$ earthquake to occur within 20 km of the station. Alternatively, we need to deploy about 2500 stations in order for an $M \geq 7$ earthquake to occur within 20 km of one of the stations in one year. Observing potentially large damaging earthquakes at near-field distances will, therefore, require the deployment of many stations, or waiting for a long time, or being very lucky in selecting station locations.

6.3 Extending Observation Offshore is Expensive

As pointed out by T. Utsu (personal communication, 2001), it is important for seismologists to extend observation offshore, because more than 2/3 of all earthquakes occur at sea and some of them can be very damaging to lives and properties on land. An effective means to observe earthquakes at sea is to deploy a cable ocean-bottom seismograph (OBS) system. However, this is very expensive. For example, a cable OBS system including four seismic stations, 150 km of cable, and land facilities costs about $25 million according to N. Hamada (personal communication). As of 2001, there were seven cable OBS systems in operation in Japan and their signals are monitored in real time by the regional centers of the Japanese Meteorological Agency (JMA). These data greatly improve the determination of location, especially the focal depth, of offshore earthquakes in Japan.

6.4 Planning, Management, and Bureaucracy

Good seismic monitoring requires long-term planning and efficient management, but few seismologists have the experience or training necessary for the planning and management tasks. Several related technical and organizational topics are elaborated in Bormann (2002). Seismic monitoring involves deploying and operating seismic instruments in the field. If we wish to be good at it, it pays to study how field operations are conducted, such as military operations in a war (e.g., Sun-Tzu, 1993).[9] Seismic monitoring also needs good management, and it may be instructive to read two recent books by Andy Grove explaining how Intel became the most successful company in the world in the production of microprocessors (Grove, 1987; 1996).

Since funding for seismic monitoring is almost entirely from governments, we cannot avoid dealing with government bureaucracy. It helps to have some understanding of bureaucracy through reading books on this subject (e.g., Parkinson, 1957).[10] Seismic monitoring requires teamwork, and therefore a good leader. It is of utmost importance to have someone who knows how the bureaucracy operates in a particular organization and country.

6.5 Integration, Reorganization, and Data Loss

Seismic monitoring of earthquakes evolved over the years into three major branches as discussed in Section 3. One recent favored approach is to integrate all seismic monitoring into a single entity called real-time seismology. There are, of course, many advantages in this approach, especially from a management point of view. However, getting a large group of people to work together is by no means easy. Aki in Chapter 5 described an approach of getting people from many disciplines to work together under the Southern California Earthquake Center, and it may serve as a blueprint for integrated earthquake programs elsewhere.

As mentioned before, seismic monitoring is almost entirely funded by governments (directly or indirectly), except for some notable early efforts by nongovernment groups (e.g., the Jesuits as reported by Udias and Stauder in Chapter 3). There is no *bottom line* to speak of for government agencies, since there are no "profits" or "losses." There is a tendency for staff members to increase according to Parkinson's law to consume all available funding, and for staff quality to decrease with time as deadwood accumulates, for it is difficult to dismiss civil servants. Consequently, few government institutions can maintain a high-quality level of seismic monitoring for decades, before going out of existence or being reorganized.[11]

Each reorganization creates disruption, as new chiefs often have their own agendas and directions. Facing uncertainties in management and funding, the long-term preservation of seismic data has been at the bottom of any priority list in seismology. Millions of seismograms have been poorly kept or were inaccessible, and will soon disappear. This sad status was recognized in the 1970s, but little has been done (Lee *et al.*, 1988).

Major capital investments for seismic instrumentation occur infrequently, usually in response to disastrous earthquakes (e.g., K-Net), or military needs (e.g., WWSSN), or broader programs (e.g., microearthquake networks for earthquake prediction). There is a tendency to spend the new funds on equipment and more staff members, and to ignore the costs of operation and maintenance on a long-term basis. A few modern seismic networks have been idle because of this. It is tempting to buy the latest and most advanced equipment, without realizing the risk involved. It takes time and effort to master the latest pieces of equipment, and some of their new features often do not work well in the field.

7. Discussion

The history of earthquake seismology suggests that major advances take place shortly after accumulation of sufficient amounts of seismic data of a quality that surpasses that of previous data.[12] Major advances in earthquake seismology are expected in the near future because digital seismic data have become widely available on both local and global scales. In addition, advances in computers with increasing computing power at decreasing cost have been important to seismologists for more sophisticated data processing and analysis in order to gain insight from the increasing volume of seismic data collected (see Chapter 22 by Scherbaum; Chapter 85 edited by Snoke and Garcia-Fernandez).

In the 1990s, especially after the 1995 Hyogo-ken Nanbu (Kobe) earthquake, large amounts of funding became available in Japan for digital seismic and other instrumentation. Under the Headquarters for Earthquake Research Promotion, a "Fundamental Seismic Survey and Observation Plan" was carried out in Japan for:

1. Earthquake observations: (a) inland earthquake observation by high sensitivity seismographs (observation of microearthquakes), and (b) inland earthquake observation by broadband seismographs
2. Strong-motion observations
3. Observations of crustal deformation (GPS continuous observation)
4. Survey of inland and coastal active faults

These extensive instrumentation programs have now been completed and the data from the high-sensitivity seismograph (Hi-Net), nationwide broadband seismograph network (Freesia), and digital strong-motion seismograph (KiK-Net) are effectively distributed in near real time by the National Institute for Earth Science and Disaster Prevention (NIED) via the Internet (*http://www.bosai.go.jp/index.html*).

Extensive strong-motion instrumentation programs (typically with 1000 digital instruments) were implemented in Japan and Taiwan, for example, at costs of several tens of millions of US dollars each (see Chapter 63 by Kinoshita; Chapter 64 by Shin *et al.*). The extensive strong-motion data set recorded during the Chi-Chi (Taiwan) earthquake sequence of 20 September 1999 (Lee *et al.*, 2001) showed that these near-fault data will not only contribute to information needed for earthquake engineering but will also lead to a better understanding of the earthquake process (Lee and Shin, 2001; Teng *et al.*, 2001).[13]

Knowledge about the nature of strong ground motions expected from damaging earthquakes will become increasingly important as urbanization rapidly increases, because the lives and properties of ever greater numbers of people are exposed to seismic hazards (see Chapter 74 by Giardini *et al.*). Utsu in Chapter 42 compiled a list of deadly earthquakes in the world for the past five centuries and it is instructive to view his maps. A great challenge for seismologists will be the continued improvement of seismic monitoring to reduce seismic hazards. In particular, it is now technically possible to implement earthquake early-warning systems as discussed in Lee and Espinosa-Aranda (1998). These systems, in principle, provide lifeline operators and citizens with crucial seconds or minutes of lead time for taking some protective action before the strong shaking of a damaging earthquake begins. However, an earthquake early-warning system is expensive to implement and its benefit is difficult to evaluate.

Last, but not least, seismic arrays optimized to detect nuclear explosions (e.g., LASA and NORSAR) and their impact in the development of earthquake seismology have not been discussed, as readers can consult Chapter 23 by Douglas and Chapter 24 by Richards. Seismologists owe a great deal to the military for its generous support in advancing seismology,[14] although some may question its motives and policies regarding data analysis and exchange.

8. Concluding Remarks

Seismic monitoring of earthquakes is becoming a *big* science (Price, 1963; Weinberg, 1967). For example, the Japanese Fundamental Seismic Survey and Observation Plan had been executed successfully as discussed in Section 7. Two large projects are now underway in the United States: (1) the Advanced National Seismic System (ANSS) to integrate seismic monitoring of earthquakes (see *http://www.anss.org/*), and (2) the USArray, "a continental-scale seismic array to provide a coherent 3D image of the lithosphere and the deeper Earth" (see *http://www.earthscope.org/*). Since 1996, the Preparatory Commission for the Comprehensive Nuclear Test-Ban Treaty Organization (CTBTO) is establishing global monitoring systems (including seismological) worldwide (visit its website at *http://www.ctbto.org/*) (see also Chapter 24 by Richards).

The most direct argument for governments to support long-term seismic monitoring is to collect some relevant data for hazard mitigation. As noted in Section 6, economic losses from damaging earthquakes in the past decade are about $200 billion, and future losses will be even greater as rapid urbanization is taking place worldwide. For example, the recent Japanese Fundamental Seismic Survey and Observation Plan (costing several hundred million US dollars) is a direct response to the economic losses of about $100 billion due to the 1995 Kobe earthquake.

In addition to scientific and technological challenges in observational seismology, seismologists must pay attention to achieving (1) stable long-term funding,[15] (2) effective management and execution, and (3) delivery of useful products to the users. Observational seismologists must have perseverance in order to succeed.[16]

Acknowledgments

I owe my seismic training to colleagues at the USGS and at many other institutions. Roger Borcherdt and Yi-Ben Tsai convinced me of the importance of monitoring strong ground motion, and I wish to thank them for their patience. I was fortunate to have participated in the large-scale seismic instrumentation program of Taiwan from 1991 to 1996. I thank Tony Shin and his staff at the Central Weather Bureau, Taipei, for being gracious hosts.

I am grateful to Robin Adams, Duncan Agnew, Nick Ambraseys, Doc Bonilla, Dave Boore, Roger Borcherdt, Peter Bormann, Ken Campbell, Jim Cousins, John Evans, Jennifer Hele', Porter Irwin, Paul Jennings, Hiroo Kanamori, Carl Kisslinger, Fred Klein, John Lahr, Axel Plesinger, Paul Richards, Johannes Schweitzer, Tony Shakal, Shri Singh, Arthur Snoke, Chris Stephens, Ta-Liang Teng, Yi-Ben Tsai, and Erhard Wielandt for their comments and suggestions on the manuscript. I thank Tokuji Utsu for pointing out the importance of extending seismic networks to offshore zones in seismically active regions, and I am grateful to Nobuo Hamada for supplying information on such efforts in Japan. I also thank Lucy Jones and David Oppenheimer for answering my queries about the Northridge earthquake and the Loma Prieta earthquake, respectively.

References

Agnew, D., J. Berger, R. Buland, W. Farrell, and F. Gilbert (1976). International deployment of accelerometers: a network for very long period seismology. *EOS* **57**, 180–188.

Block, B. and R.D. Moore (1966). Measurements in the Earth mode frequency range by an electrostatic sensing and feedback gravimeter. *J. Geophys. Res.* **71**, 4361–4375.

Borcherdt, R.D. (Ed.) (1997). "Vision for the future of the US National Strong-Motion Program," The committee for the future of the US National Strong Motion Program. *US Geol. Surv. Open File Rept.* 97–530 B.

Borcherdt, R.D., A. Frankel, W.B. Joyner, and J. Bouabid (1997). Vision 2005 for earthquake strong ground-motion measurement in the United States. In: "Proceedings, Workshop, Vision 2005: An Action Plan for Strong Motion Programs to Mitigate Earthquake Losses in Urbanized Areas" (J.C. Stepp, Ed.), Monterey, CA, April, 1997, pp. 112–130.

Bormann, P. (Ed.) (2002). "New Manual of Seismological Observatory Practice," in preparation. [A pre-publication version is placed on the attached Handbook CD under the directory of \81\IASPEI\Training.]

Dewey, J. and P. Byerly (1969). The early history of seismometry (to 1900). *Bull. Seismol. Soc. Am.* **59**, 183–227.

Eaton, J.P. (1989). Dense microearthquake network study of northern California earthquakes. In: "Observatory Seismology" (J.J. Litehiser, Ed.), pp. 199–224. University of California Press, Berkeley.

Evans, J.R. (2001). Wireless monitoring and low-cost accelerometers for structures and urban sites. In: "Strong Motion Instrumentation for Civil Engineering Structures" (M. Erdik, M. Celebi, and V. Mihailov, Eds.), pp. 229–242. Kluwer Academic, Dordrecht.

Field, E.H. and S.E. Hough (1997). The variability of PSV response spectra across a dense array deployed during the Northridge aftershock sequence. *Earthq. Spectra* **13**, 243–258.

Geschwind, C.H. (2001). "California Earthquakes: Science, Risk and the Politics of Hazard Mitigation." Johns Hopkins University Press, Baltimore, MD.

Goltz, J.D., P.J. Flores, S.E. Chang, and T. Atsumi (2001). Emergency response and early recovery. In: "1999 Chi-Chi, Taiwan, Earthquake Reconnaissance Report." *Earthq. Spectra* **17** (Supplement A), 173–183.

Goodstein, J.R. (1991). "Millikan's School: A History of the California Institute of Technology." Norton, New York.

Grove, A.S. (1987). "One-on-One With Andy Grove: How to Manage Your Boss, Yourself and Your Co-Workers." Putnam, New York.

Grove, A.S. (1996). "Only the Paranoid Survive." Currency Doubleday, New York.

Harjes, H.-P. and D. Seidl (1978). Digital recording and analysis of broadband seismic data of the Gräfenberg (GRF) Array. *J. Geophys.* **44**, 511–523.

Kanamori H., E. Hauksson, and T. Heaton (1997). Real-time seismology and earthquake hazard mitigation. *Nature* **390**, 461–464.

Kao, H. and J. Angelier (2001). Data files from "Stress tensor inversion for the Chi-Chi earthquake sequence and its implication on regional collision." *Bull. Seismol. Soc. Am.* **91**, 1380 [and on the attached CD Supplement].

Kerr, R.A. (1991). A job well done at Pinatubo volcano. *Science* **253**, 514.

Lawson, A.C. (chair) (1908). "The California Earthquake of April 18, 1906—Report of the State Earthquake Investigation Commission." Carnegie Institution of Washington, Washington, DC.

Lee, W.H.K. (Ed.) (1994). "Realtime Seismic Data Acquisition and Processing." IASPEI Software Library, Vol. 1, 2nd edn. Seismological Society of America, El Cerrito.

Lee, W.H.K. and J.M. Espinosa-Aranda (1998). Earthquake early-warning systems: current status and perspectives. In: "Proceedings of International IDNDR-Conference on Early Warning Systems for the Reduction of Natural Disasters," Potsdam, Germany.

Lee, W.H.K. and T.C. Shin (2001). Strong-motion instrumentation and data. In: "1999 Chi-Chi, Taiwan, Earthquake Reconnaissance Report." *Earthq. Spectra* **17** (Supplement A), 5–18.

Lee, W.H.K. and S.W. Stewart (1981). "Principles and Applications of Microearthquake Networks." Academic Press, New York.

Lee, W.H.K., H. Meyers and K. Shimazaki (Eds.) (1988). "Historical Seismograms and Earthquakes of the World." Academic Press, San Diego.

Lee, W.H.K., T.C. Shin, K.W. Kuo, K.C. Chen, and C.F. Wu (2001). CWB free-field strong-motion data from the 921 Chi-Chi (Taiwan) earthquake. *Bull. Seismol. Soc. Am.* **91**, 1370–1376 [with data on the attached CD Supplement].

Melton, B.S. (1976). The sensitivity and dynamic range of inertial seismographs. *Rev. Geophys. Space Phys.* **14**, 93–116.

Munich Reinsurance Company (2000). "World of Natural Hazard," CD-ROM version (see *http://www.munichre.com/*).

Murray, T.L., J.A. Power, G. Davidson, and J.N. Marso (1996). A PC-based real-time volcano-monitoring data-acquisition and analysis system. In: "Fire and Mud" (C.G. Newhall and R.S. Punongbayan, Eds.), pp. 225–232. University of Washington Press, Seattle.

Oliver, J. and L. Murphy (1971). WWNSS: Seismology's global network of observing stations. *Science* **174**, 254–261.

Parkinson, C.N. (1957). "Parkinson's Law and Other Studies in Administration." Houghton Mifflin, Boston.

Price, D.J. de Solla (1963). "Little Science, Big Science." Columbia University Press, New York.

Richter, C.F. (1958). "Elementary Seismology." Freeman, San Francisco.

Romanowicz, B., J.F. Karczewski, M. Cara, *et al.* (1991). The GEOSCOPE Program: present status and perspectives. *Bull. Seismol. Soc. Am.* **81**, 243–264.

Salam, A. (1979). "Gauge unification of fundamental forces." Nobel Prize in Physics Award Address, Nobel foundation.

Stepp, J.C. (Ed.) (1997). "Vision 2005: An action plan for strong motion programs to mitigate earthquake losses in urbanized areas." In: "Proceedings of a Workshop, Monterey, CA," April, 1997 (posted at *http://www.cosmos-eq.org/vision2005.pdf*).

Sun-Tzu (R. Ames, Transl.) (1993). "The Art of Warfare." Ballantine Books, New York.

Teng, T.L., L. Wu, T.C. Shin, Y.B. Tsai, and W.H.K. Lee (1997). One minute after: strong motion map, effective epicenter, and effective magnitude. *Bull. Seismol. Soc. Am.* **87**, 1209–1219.

Teng, T.L., Y.B. Tsai, and W.H.K. Lee (Eds.) (2001). The 1999 Chi-Chi, Taiwan, Earthquake. *Bull. Seismol. Soc. Am.* **91**, 893–1395.

Weinberg, A.M. (1967). "Reflections on Big Science." MIT Press, Cambridge, MA.

Wielandt, E. and G. Streckeisen (1982). The leaf-spring seismometer: design and performance. *Bull. Seismol. Soc. Am.* **72**, 2349–2367.

Willemann, R.J. and D.A. Storchak (2001). Data collection at the International Seismological Centre. *Seismol. Res. Lett.* **72**, 440–453.

Wu, Y.M., W.H.K. Lee, C.C. Chen, T.C. Shin, T.L. Teng, and Y.B. Tsai (2000). Performance of the Taiwan Rapid Earthquake Information Release System (RTD) during the 1999 Chi-Chi (Taiwan) earthquake. *Seismol. Res. Lett.* **71**, 338–343.

Notes

1. Because of low magnification, slow recording speed, and no damping, Milne seismograms are useful only for recording $M \approx 8$ or larger earthquakes. Numerous seismographs were developed beginning in the late 1890s. Mechanical seismographs (e.g., Wiechert) are considered to be the first generation. Electromagnetic seismographs (developed by B. Galitzin in the first decade of the 20th century) are considered the second generation, and dominated the design for the next 60+ years. The third-generation seismographs developed in the 1970s are based on electronic force feedback.

2. Many dedicated individuals kept John Milne's vision alive for 50 difficult years. After Milne's death in 1913, the Shide Circulars were continued as bulletins by the British Association's Seismological Committee (i.e., J. H. Burgess and H.H. Turner). This publication became "The International Seismological Summary" (ISS) in 1922 under the direction of H.H. Turner and was supported by the International Union of Geodesy and Geophysics. After Turner's death in 1930, ISS volumes for 1927 to 1935 were issued by the University Observatory, Oxford. They were then issued by the Kew Observatory, Richmond, for the years 1936 to 1961. The ISS volumes for 1962 and 1963 were issued by the International Seismological Centre (ISC), which was organized in 1964 and continues to this day (Willemann and Storchak, 2001).

3. Earthquake prediction has always been a controversial subject. Proponents and skeptics debate the subject with an intensity comparable to that in religion or politics. Richter (1958, p. 8) wrote, "Prediction of earthquakes in any precise sense is not now possible. Any hope of such prediction looks toward a rather distant future. Cranks and amateurs frequently claim to predict earthquakes. They deceive themselves, and to some extent the public, ..." In the 1960s and 1970s, some seismologists claimed that they had good leads for predicting earthquakes, almost all in hindsight. Excitement soon faded because their claims could not be replicated. A few lonely voices even questioned the *usefulness* of a successful earthquake prediction. Nevertheless, earthquake prediction programs did advance earthquake seismology in many areas.

4. The concept of IRIS began in the late 1970s as a series of discussions to advance global seismology. One concern was about the failure of the WWSSN to modernize and

expand using digital technology (Paul Richards, personal communication, 2001).

5. After the 1906 San Francisco earthquake, government officials and civic leaders did not want to study earthquakes—bad for attracting business to the then-developing California (Geschwind, 2001). It was maintained that the *fire* destroyed San Francisco. The "Report of the State Earthquake Investigation Commission" on the 1906 earthquake (Lawson, 1908) in two volumes with an atlas was published by the private Carnegie Institution of Washington in Washington, DC.

6. Information about the location of the Loma Prieta and the Northridge earthquakes took ~1 hour and ~30 min, respectively, to reach the media and the public, although the locations were determined by the real-time systems quickly. According to David Oppenheimer (personal communication, 2001), "We lost A/C power and could not report out the solution [of the Loma Prieta earthquake location by the real-time system] to terminals or pagers." According to Lucy Jones (personal communication, 2001), the real-time system did not page the results because it flagged the location as a probable telemetry glitch in the Northridge case.

7. The amount of seismic data is *large* from the seismologists' point of view, but is *small* in comparison with that in many other disciplines. Unfortunately, most seismologists do not pay much attention to data archiving and, as a result, most seismograms and seismic data more than 20 years old are difficult to access. Current technology in computer storage can easily handle terabytes of data, but all storage media have finite lives and must be renewed in order to save the data permanently.

8. Some seismologists believe that it is cheaper to design and build your own instruments than to buy commercial ones. There is no doubt that designing and building an instrument oneself is instructive and useful for learning. It is also necessary in some cases to develop one's own instruments for certain applications when there are none available commercially. The argument that one needs only about 1/5 of the cost of a commercial instrument to buy the parts and that therefore one would save a lot of money, is false. By the time a prototype instrument is built and works in the field, it will usually cost more than the comparable commercial product when the salaries and overheads are taken into account.

9. Sun-Tzu was a general in the Kingdom of Wu (514–496 BC) in China. He gained an audience with the King by presenting his "Art of War." He had a distaste of war and urged using military action only as a last resort. He argued that the *cost* of military actions would always be expensive and questioned whether there would be any *financial* gains even if successful. He emphasized strategy, logistics, and discipline in the battle field. The "Art of War" has been widely circulated since his time, and bamboo strips containing it (dated about 200 BC) were discovered in 1974 in an ancient tomb.

10. C.N. Parkinson made a remarkable discovery that the number of subordinates in a government agency multiplies at a rate of 5–7% of the total staff per year, regardless of need. He used the staffing data from the British Admiralty and other British bureaus.

11. For example, the responsibility for earthquake monitoring in the United States government has already changed hands four times in less than a century—it began at the Weather Bureau in 1914, moved to the US Coast and Geodetic Survey in 1925, merged into the Environmental Science Services Administration (ESSA) and then the National Oceanic and Atmospheric Administration (NOAA) in the 1960s, and merged into the US Geological Survey (USGS) in the early 1970s. USGS, itself, narrowly escaped being eliminated by the US Congress in the mid-1990s.

12. For example, shortly after a few hundred seismographs were established around the world in the early 1900s, the gross structure of the Earth's interior was quickly established by the 1930s. The establishment of the World Wide Standardized Seismograph Network in the early 1960s enabled the study of global seismicity and focal mechanisms on a scale that was previously impossible. As a result, earthquake seismology made significant contributions to the development of the theory of plate tectonics in the late 1960s.

13. After the 1989 Loma Prieta earthquake, Yi-Ben Tsai realized the values of strong-motion records for correlating ground shaking with structural damage. In a visit to Taiwan, he met Dr. Ching-Yen Tsay, the then new Director-general of the Central Weather Bureau and proposed an extensive strong-motion instrumentation program in the urban areas of Taiwan. Dr. Tsay enthusiastically accepted the proposal and persuaded Dr. Chan-Hsuan Liu, the then Minister of Transportation and Communication to incorporate it as a part of the Six-year National Construction Programs, being planned at that time. An advisory board (chaired by Ta-Liang Teng) was established for the detailed planning and execution of the strong-motion instrumentation program, and subsequently a budget of about US $72 million was authorized by the Taiwan Legislature in 1991. I spent full time working in this program as an invited advisor, on loan from the USGS. This instrumentation program was completed slightly ahead of schedule in 1996 and for a little over US $40 million. Its successful completion was due largely to the efficient execution by Tony Shin and his staff at CWB, and

to a small extent to my desire to retire from the USGS in 1995. None of us expected that a major earthquake of magnitude 7.6 would occur so soon in Taiwan.

14. The following is just a personal case. I joined the Earthquake studies group of the US Geological Survey in 1967. This group had its origin in the Vela-Uniform program of the Defense Advanced Research Projects Agency (DARPA). The simple PC-based real-time system I developed in the 1980s (Lee, 1994) was a by-product of a project funded by DARPA to conduct experiments in a quarry, and I thank Bob Blandford for his kind support. Its early success was in monitoring volcanic eruptions (e.g., Kerr, 1991; Murray *et al.*, 1996). I was able to modify it quickly for the two key elements of the CWB strong-motion instrumentation program in Taiwan: (1) the RTD system (later refined by Y.M. Wu and his associates at CWB), and (2) the strong-motion array systems deployed in buildings and bridges. This simple PC-based real-time system with reference accelerometers and displacement gauges permitted me to evaluate commercial accelerographs on a shake table quickly, and made it possible to accelerate the procurement of accelerographs and accelerometers for deployment in Taiwan under open-bidding.
15. As pointed out by Duncan Agnew (personal communication, 2001), the funding system has a legitimate bias toward novelty and against doing the same thing as in long-term seismic monitoring. Seismology has to get by with occasional operational needs: whenever there is a damaging earthquake in a populated area, everyone gets very interested, but not for the rest of the time. One solution is for a government to impose a "tax" for funding seismic monitoring in the building permit fees; for example, the California Strong Motion Instrumentation Program has been funded this way.
16. In his Nobel lecture, Abdus Salam (1979) said he gave up experimental physics and started on quantum field theory because he recognized that the craft of experimental physics was beyond him, for "it was the sublime quality of patience—patience in accumulating data, patience with recalcitrant equipment" that he sadly lacked.

Editor's Note

For readers' convenience, two out-of-print books (Lee and Stewart, 1981; Lee, Meyers, and Shimazaki, 1988) are given as PDF files on the attached Handbook CD under the directory \17Lee1.

18

Seismometry

Erhard Wielandt
Institute of Geophysics, University of Stuttgart, Stuttgart, Germany

1. Scope of this Chapter and Historical Notes

1.1 Scope

Seismometry is the technical discipline concerned with the detection and measurement of seismic ground motion. It comprises the design of seismographs, their calibration, their installation, and the quantitative interpretation of seismograms in terms of ground motion. A seismograph basically consists of one or more seismometers, a clock or time-signal receiver, and a recorder. The present chapter concentrates on the theory and usage of seismometers. It does not describe their technical design in detail, but mentions some general design principles. Emphasis is put on precise methods for calibrating and testing. Although most commercial seismometers are well calibrated and normally need not be recalibrated by the user after installation, such methods are essential to ensure that an instrument is in good working condition.

1.2 Inertial and Strain Seismometers

There are two basic types of seismic sensors: *inertial* seismometers, which measure ground motion relative to an inertial reference (a suspended mass), and *strainmeters* or *extensometers*, which measure the motion of one point of the ground relative to another. The wavelength of seismic waves is so large that the differential motion of the ground within a vault is normally much smaller than the motion relative to an inertial reference; strainmeters are therefore generally less sensitive to earthquake signals. However, at very low frequencies it becomes increasingly difficult to maintain an inertial reference, and for the observation of low-order free oscillations of the Earth and tidal signals, strainmeters may outperform inertial seismometers. In the presence of gravity, inertial seismometers with a horizontal sensitive axis also respond to tilt, and the better ones are more sensitive to short-term tilt than the majority of dedicated tiltmeters (although inferior in their long-term stability). The principles of operation of horizontal seismometers and tiltmeters are identical. Instruments measuring the angular acceleration, although theoretically required for a complete description of ground motion, have not attained any significance in seismology because the rotational component of seismic signals is in general too small to be directly observed. (Even if the signals were strong enough, existing mechanical sensors would not be able to separate them from the associated large displacements.) We will treat only inertial seismometers in this article. Their theory will be presented as far as it is required for an understanding of specifications, calibration procedures, and operational requirements.

In contrast to most other sensors, inertial seismometers have an inherently frequency-dependent response that must be taken into account when the ground motion is restored from the recorded signal. This is because a suspended mass does not represent a perfect inertial reference. When the ground motion is slow, the mass will begin to follow it, and the output signal for a given ground displacement will therefore be diminished. The mechanical system forms a high-pass filter for the ground displacement. Recorders, on the other hand, normally have a constant gain up to some upper cut-off frequency, and contribute only a scale factor to the overall response. We will therefore not discuss their frequency response in detail.

1.3 Historical Seismographs

The term *seismograph* is today reserved for instruments recording the waveform of the ground motion versus time. In that sense, the first seismograph was built in Italy by Cecchi in 1875; it was, however, so insensitive that its first known seismogram dates from 1887. In the meantime, Ewing and colleagues in Japan had built several seismographs (Ewing, 1884) and recorded the first earthquake in 1880. In the same year one of the instruments was tested on a shake table! Von Rebeur-Paschwitz (1889) recognized seismic waves from an earthquake in Japan in the records of his tiltmeters at Potsdam and Wilhelmshaven, giving seismology a global dimension. The early history of seismometry to 1900 is described by Dewey

INTERNATIONAL HANDBOOK OF EARTHQUAKE AND ENGINEERING SEISMOLOGY, VOLUME 81A ISBN: 0-12-440652-1

and Byerly (1969) in an excellent article with many figures and references.

Going back in time, the "electromagnetic seismograph" built by Palmieri in 1856 was little short of being the first seismograph in a modern sense. It had motion-sensitive electric contacts whose closures were recorded on a strip of paper like Morse code. Earlier constructions, which were only designed to indicate the occurrence and direction of a seismic shock, would today be termed *seismoscopes*. The Chinese Chang Heng is reported to have built one in AD 132; models of his jar-shaped instrument are exhibited in many seismological institutes, but its inner mechanism is unknown.

At the beginning of the 20th century, the technical development concentrated on mechanical seismographs with smoke-paper recording. Viscous damping was introduced by Wiechert around 1900. To overcome the remaining solid friction, the mass had to be increased with the square of the magnification. The largest seismographs had masses from 10 to 20 tons, magnifications around 1000, and stable free periods up to 12 sec. Mainka and Wiechert seismographs served in many obervatories until after the Second World War, and a few of them are still (or again) operational. De Quervain and Piccard in Zürich (1924, 1927) built a mechanical three-component seismograph with a single mass of 21 tons whose position was stabilized with a water ballast—probably the first feedback-stabilized seismograph. Many of the seismographs of the early 20th century are described in Galitzin's lectures on seismometry (1914) and in a comprehensive article by Berlage (1932).

Photographic recording was occasionally used from the beginning, but the higher cost and lower quality of the record put the method at a disadvantage, at least until electric light was available. Later it became a practical alternative, for example, with the Wood–Anderson torsion seismograph on which the Richter magnitude scale is based (Anderson and Wood, 1925). The electromagnetic seismograph with galvanometer-photopaper recording, invented by Galitzin as early as 1904, remained for more than half a century the most sensitive long-period seismograph, but it had to wait for gradual improvements by LaCoste, Benioff, Press, Ewing, and Lehner (LaCoste, 1934; Benioff and Press, 1958; Press *et al.*, 1958; Lehner, 1959) before it was stable enough for wide deployment in the WWSSN (World Wide Standardized Seismograph Network; Oliver and Murphy, 1971).

The next generation of electromagnetic seismographs in the HGLP (High-Gain Long-Period) project (Savino *et al.*, 1972) was partially electronic, using galvanometer-phototube amplifiers. The SRO system (Seismic Research Observatory; Peterson *et al.*, 1976) had a fully electronic, broadband, force-balance sensor but did not record the broadband signal. The sensor of the original IDA network (International Deployment of Accelerometers; Agnew *et al.*, 1976, 1986) was a LaCoste–Romberg gravimeter with a slow electrostatic force-balance feedback; although this instrument was not useful as a general-purpose seismometer, its sensitivity in the free-mode band is unsurpassed. An eyewitness account of the emerging electronic era of seismometry from 1947 on is given by Melton (1981a,b).

In the time of transition from electromagnetic to electronic seismographs between 1960 and 1975, two opposite trends can be observed. As long as visible recording was the standard and magnetic tape recording was not much better, the gain could only be increased when the marine microseisms, at periods around 6 sec, were suppressed. This resulted in the development of high-gain, narrowband seismographs, which were excellent for studying ground noise and monitoring nuclear explosions but were easily saturated by earthquakes. On the other hand, several broadband seismographs with analog or digital magnetic tape recording were developed. They remained experimental because continuous broadband recording and digital or analog post-processing were too inconvenient for routine work. The first digital broadband seismograph was operated at the California Institute of Technology as early as 1962 (Miller, 1963) with the intention "to preserve the greatest spectrum, dynamic range, and sensitivity." The installation was discontinued because the digital technology was too inefficient at the time. Block and Moore (1970) built a small broadband quartz accelerometer that was the most sensitive broadband sensor of its time but not a very practical instrument; it required vacuum and a thermostat. An analog very-broadband seismograph was operated in Czechoslovakia from 1972 on (Plešinger and Horalek, 1976); its data archive was later converted to a digital standard format. The first practically successful digital broadband installation is the German GRF array (Harjes and Seidl, 1978; Buttkus, 1986), which has been operational since 1976. The present generation of digital very-broadband seismographs covering the full teleseismic bandwidth including the free-mode band was developed from 1984 on (Wielandt and Steim, 1986).

2. Mechanical Receivers, and Transducers

2.1 The Linear Pendulum

The simplest physical model for an inertial seismometer is a mass-and-spring system with viscous damping (Fig. 1). The mechanical elements are a mass of M kilograms, a spring with a stiffness S (measured in newtons per meter), and a damping element with a constant of viscous friction R (in newtons per meter per second). Let the time-dependent ground motion be $x(t)$, the absolute motion of the mass $y(t)$, and its motion relative to the ground $z(t) = y(t) - x(t)$. An acceleration $\ddot{y}$ of the mass results from any external force f acting on the mass, and from the forces transmitted by the spring and the damper:

$$M\ddot{y} = f - Sz - R\dot{z} \tag{1}$$

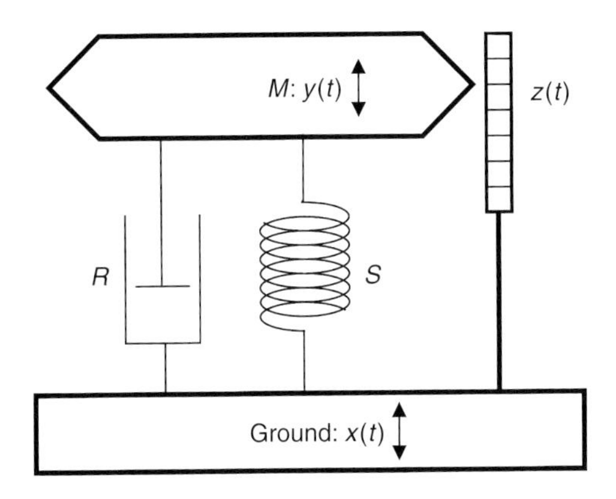

FIGURE 1 Damped harmonic oscillator.

Since we are interested in the relationship between z and x, we rearrange this into

$$M\ddot{z} + R\dot{z} + Sz = f - M\ddot{x} \qquad (2)$$

Before we solve this equation in the frequency domain, we observe that an acceleration $\ddot{x}$ of the ground has the same effect as an external force of magnitude $f = -M\ddot{x}$ acting on the mass in the absence of ground acceleration. We may thus simulate a ground motion x by applying a force $-M\ddot{x}$ to the mass while the ground is at rest. The force is normally generated by sending a current through an electromagnetic transducer, but it may also be applied mechanically.

2.2 Constraining the Motion

Although the mass-and-spring system of Figure 1 is a useful mathematical model for a seismometer, it is incomplete as a practical design. The suspension must suppress five out of the six degrees of freedom of the seismic mass (translational and rotational) but the mass must still move as freely as possible in the remaining direction. Some instruments achieve this with five tensioned threads or five tangential spokes (Geotech S13, Sensonics Mk III). In geophones, the seismic mass is normally suspended between labyrinth springs (which are stamped out of a circular elastic membrane). Most long-period suspensions are of the pendulum type, where the seismic mass rotates around a virtual axis defined by flexural hinges (Figs. 2–4). The point bearings shown in Figures 2 and 3 are for illustration only; crossed flexural hinges are normally used.

In principle it is also possible to let the mass move in all directions and observe its motion with three orthogonally arranged transducers, thus creating a three-component sensor with only one suspended mass. Indeed, some historical instruments have made use of this concept. However, it is difficult to reduce the restoring force and to suppress parasitic rotations of the mass when its translational motion is mechanically unconstrained. Modern three-component seismometers therefore have separate mechanical sensors for the three axes of motion.

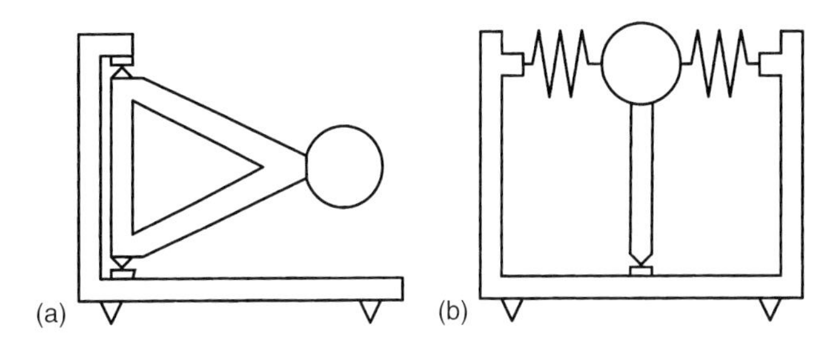

FIGURE 2 (a) Garden-gate suspension. (b) Inverted pendulum.

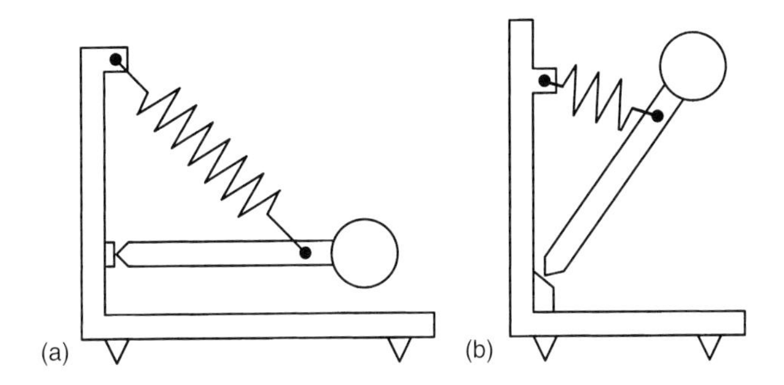

FIGURE 3 Lacoste suspensions.

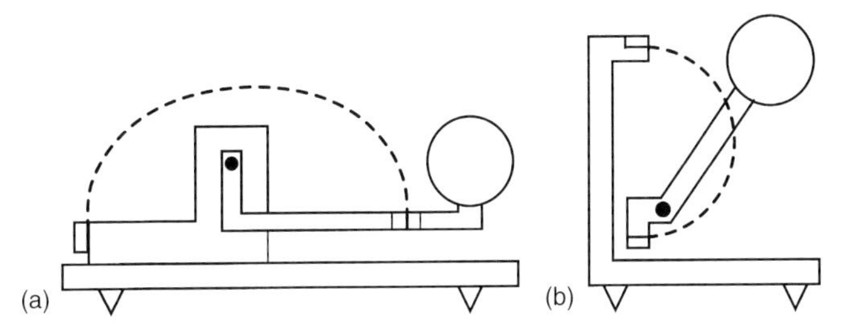

FIGURE 4 Leaf-spring astatic suspensions.

2.3 Pendulum Type Seismometers

These are sensitive not only to translational but also to angular acceleration. The rotational component of seismic shear waves is, however, too small to have a noticeable effect; its contribution to the output signal is of the order kl, where k is the horizontal wavenumber and l the length of the pendulum (Rodgers, 1969). In technical applications or on a shake table, effects of rotation may be noticeable.

For small translational ground motions, the equation of motion of a rotational pendulum is formally identical to Eq. (2) but z must then be interpreted as the angle of rotation. Since the rotational equivalents to the constants M, R, and S in Eq. (2) are of little interest in modern force-balance seismometers, we will not discuss them further and refer the reader instead to the older literature (Berlage, 1932).

2.4 Sensitivity of Horizontal Seismometers to Tilt

We have already seen [Eq. (2)] that a seismic acceleration of the ground has the same effect on the seismic mass as an external force. The largest such force is gravity. It is normally canceled by the suspension, but when the seismometer is tilted, the projection of the vector of gravity onto the axis of sensitivity changes, producing a force that is in most cases indistinguishable from a seismic signal. The effect is of second order (proportional to the square of the tilt) and therefore small in well-adjusted vertical seismometers but otherwise of first order. It not only modifies the amplitude with which the horizontal components of long-period *SV* and Rayleigh waves are recorded (Rodgers, 1968) but also introduces noise when the ground is tilted by moving or variable surface loads (traffic, people, wind, barometric pressure). Sensitivity to tilt is the reason why horizontal long-period seismic traces are generally noisier than vertical ones.

A short, impulsive tilt excursion is equivalent to a steplike change of the ground velocity and to a ramplike displacement. It will therefore cause a long-lasting transient in horizontal broadband seismograms. In the near field of a seismic source, the tilt has the same waveform as the displacement. The tilt signal can then be predicted from the vertical trace and removed from the horizontal traces (Wielandt and Forbriger, 1999).

2.5 Decreasing the Restoring Force

At low frequencies and in the absence of an external force, Eq. (2) can be simplified to $Sz = -M\ddot{x}$ and read as follows: A relative displacement of the seismic mass by $-\Delta z$ indicates a ground acceleration of magnitude

$$\ddot{x} = \frac{S}{M}\Delta z = \omega_0^2 \Delta z = \left(\frac{2\pi}{T_0}\right)^2 \Delta z \quad (3)$$

where ω_0 is the angular eigenfrequency of the pendulum and T_0 is its eigenperiod. If Δz is the smallest displacement that can be measured electronically, then the formula determines the smallest low-frequency ground acceleration that can be observed. For a given transducer, it is inversely proportional to the square of the free period of the suspension. A sensitive long-period seismometer therefore requires either a pendulum with a low eigenfrequency or a very sensitive transducer (for quantitative examples, see Section 5.2). Since the eigenfrequency of an ordinary pendulum is essentially determined by its size, and seismometers must be reasonably small, astatic suspensions have been invented that combine small overall size with a long free period.

The simplest astatic suspension is the "garden-gate" pendulum used for horizontal seismometers (Fig. 2a). The mass moves in a nearly horizontal plane around a nearly vertical axis. Its free period is the same as that of a mass suspended from the point where the plumb line through the mass intersects the axis of rotation. The period is infinite when the axis is vertical, and is usually adjusted by tilting the whole instrument. This is one of the earliest designs for long-period horizontal seismometers.

Another early design is the inverted pendulum held in stable equilibrium by springs or by a stiff hinge (Fig. 2b); a famous example is Wiechert's horizontal pendulum built around 1905 (Berlage, 1932).

An astatic spring geometry for vertical seismometers was invented by LaCoste (1934) (Fig. 3a). The mass is in neutral equilibrium when three conditions are met: the spring is prestressed to zero length (i.e., the spring force is proportional to the total length of the spring), its end points make a right angle from the hinge, and the mass is balanced in the horizontal position of the boom. A finite free period is obtained by making the angle slightly smaller, or by tilting the frame accordingly. By simply rotating the pendulum, astatic suspensions with a horizontal or oblique axis of sensitivity can also be constructed (Fig. 3b).

The astatic leaf-spring suspension (Fig. 4a) used in the STS1 seismometer (Wielandt, 1975; Wielandt and Streckeisen, 1982) has a limited range around its equilibrium position compared to a LaCoste suspension, but is much simpler to manufacture. A similar spring geometry is also used in the triaxial seismometer STS2 (Fig. 4b). The delicate equilibrium of forces in astatic suspensions makes them susceptible to external disturbances such as changes in temperature; they are difficult to operate without a stabilizing feedback system.

Apart from genuinely astatic designs, almost any seismic suspension can be made astatic with an auxiliary spring acting normal to the line of motion of the mass and pushing the mass away from its equilibrium; the Sensonics Mk III seismometer is an example. The long-period performance of such suspensions is, however, quite limited. Neither the restoring force of the original suspension nor the destabilizing force of the auxiliary spring can be made perfectly linear (i.e., proportional to the displacement). While the linear components of the force may cancel, the nonlinear terms remain and cause the oscillation to become anharmonic and even unstable at large amplitudes. Viscous and hysteretic behavior of the springs may also cause problems. The additional spring (which has to be soft) may introduce parasitic resonances. Modern seismometers do not use this concept and rely for their sensitivity either on a genuinely astatic spring geometry or on the sensitivity of electronic transducers.

2.6 Effects of Temperature and Pressure

The equilibrium between gravity and the spring force in a vertical seismometer is disturbed when the temperature changes. Although thermally compensated alloys such as Elinvar are available for springs, a self-compensated spring does not make a compensated seismometer. The geometry of the whole suspension changes with temperature; the seismometer must therefore be compensated as a whole. However, the different time constants involved prevent an efficient compensation at seismic frequencies. Short-term changes of temperature must therefore be suppressed by thermal insulation. Special caution is required with active seismometers: they heat themselves up when insulated and are then very sensitive to air drafts, so the insulation must suppress any possible convection. Long-term (seasonal) changes of temperature do not interfere with the seismic signal, but may drive the seismometer out of its operating range. Equation (3) can be used to calculate the thermal drift of a passive vertical seismometer when the temperature coefficient of the spring force is formally assigned to gravity.

Fluctuations of the barometric pressure (resulting from turbulent convection in the atmosphere) have multiple effects on seismic sensors. When the sensor is not sealed, its mass experiences a variable buoyancy. This is a large effect that is intolerable in a vertical broadband seismometer. Changes of pressure also disturb the temperature of the sensor even if it is thermally isolated. On the other hand, a sealed enclosure may be deformed by the barometric pressure and transmit stresses or tilt. The enclosure must therefore either be very rigid or mechanically decoupled from the sensor (see Section 6.2). Even when the sensor is perfectly shielded, there remain some effects of the barometric pressure that are independent of the sensor: ground tilt due to atmospheric loading, often enhanced by cavity effects (Beauduin *et al.*, 1996); the gravitational attraction of the atmosphere (Zürn and Widmer, 1995); and the continuous excitation of free modes by global atmospheric turbulence (Tanimoto, 1999). Only the first two effects are of local origin and can partially be removed by a correction for the local barometric pressure.

2.7 The Homogeneous Triaxial Arrangement

In order to observe ground motion in all directions, a triple set of seismometers oriented toward East, North, and upward (Z) has been the standard for a century. However, horizontal and vertical seismometers differ in their construction, and it costs some effort to make their responses equal. An alternative way of manufacturing a three-component set is to use three sensors of identical construction whose sensitive axes are inclined against the vertical like the edges of a cube standing on its corner (Fig. 5), by an angle of arctan $\sqrt{2}$, or 54.7°.

Presently only one commercial seismometer, the STS2, makes use of this concept, although it was not the first one to do so (Melton and Kirkpatrick, 1970). Since most seismologists want finally to see the conventional E, N, and Z components, the oblique components U, V, W of the STS2 are electrically recombined according to

$$\begin{pmatrix} X \\ Y \\ Z \end{pmatrix} = \frac{1}{\sqrt{6}} \begin{pmatrix} -2 & 1 & 1 \\ 0 & \sqrt{3} & -\sqrt{3} \\ \sqrt{2} & \sqrt{2} & \sqrt{2} \end{pmatrix} \begin{pmatrix} U \\ V \\ W \end{pmatrix} \quad (4)$$

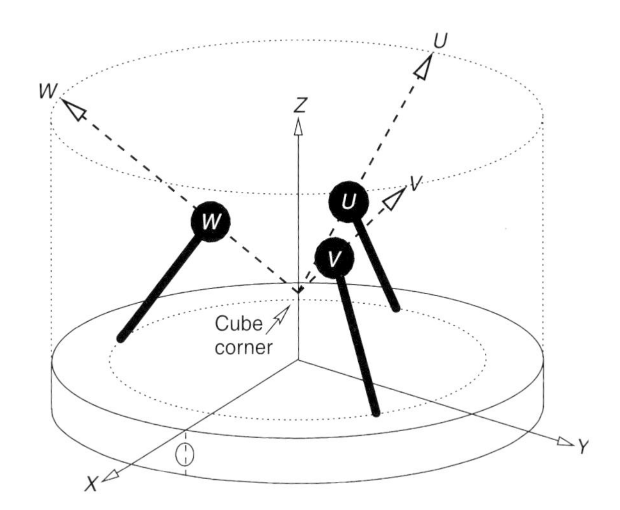

FIGURE 5 Geometry of the homogeneous triaxial seismometer STS2.

The X and Y axes are normally oriented toward E and N. Noise originating in one of the sensors of a triaxial seismometer will appear on all three outputs (except for Y being independent of U). Its origin can be traced by transforming the X, Y, and Z signals back to U, V, and W with the inverse (transposed) matrix. Disturbances affecting only the horizontal outputs are unlikely to originate in the seismometer, and are in general due to tilt.

2.8 Electromagnetic Velocity Sensing and Damping

The simplest transducer both for sensing motions and for exerting forces is an electromagnetic (electrodynamic) device where a coil moves in the field of a permanent magnet, as in a loudspeaker. Motion induces a voltage in the coil; a current flowing in the coil produces a force. From the conservation of energy it follows that the responsivity of the coil–magnet system as a force transducer, in newtons per ampere, and its responsivity as a velocity transducer, in volts per meter per second, are identical. The units are in fact the same (remember that 1 Nm = 1 J = 1 V A sec). When such a transducer is loaded with a resistor and thus a current is permitted to flow, it generates a force opposing the motion according to Lenz's law. This effect is used to damp the mechanical free oscillation of passive seismic sensors (geophones).

In comparison to technical vibrations, the seismic ground motion is slow and small most of the time. The signals delivered by electromagnetic velocity transducers are therefore normally quite small. For maximum sensitivity, the input stage of the electronic amplifier must be matched to the impedance of the coil (or vice versa). The matter is treated in detail by Riedesel *et al.* (1990), Rodgers (1992), and Rodgers (1993).

2.9 Electronic Displacement Sensing

At very low frequencies, the output signal of electromagnetic transducers becomes too small to be useful for seismic sensing. One then uses active electronic transducers where a carrier signal, usually in the audio frequency range, is modulated by the motion of the seismic mass. The basic modulating device is an inductive or capacitive half-bridge. Inductive half-bridges are detuned by a movable magnetic core. They require no electrical connections to the moving part and are environmentally robust; however, their sensitivity appears to be limited by the granular nature of magnetism, and they may push back on the seismometer mass. Capacitive half-bridges are realized as three-plate capacitors where either the central plate or the outer plates move with the seismic mass. Their sensitivity is limited by the ratio between the electric field strength and the electronic noise of the demodulator; it is typically a hundred times better than that of the inductive type. The comprehensive paper of Jones and Richards (1973) on the design of capacitive transducers still represents the state of the art in all essential aspects.

3. Mathematical Representation of the Response

3.1 The Transfer Function

We can give here only a very brief outline of the theory of linear systems. The reader who is not familiar with it should consult a textbook such as Oppenheim and Willsky (1983); for digital signal processing, Oppenheim and Schafer (1975) is a standard text. An instructive book on the Fourier transformation and its applications is that of Bracewell (1978). Seismological applications of the theory are treated in Plešinger *et al.* (1996) and in Scherbaum (1996).

Seismometers, amplifiers, and filters are designed as linear and time-invariant systems, i.e., the mathematical relationship between the time-dependent input and output signals is supposed to be a linear differential equation with constant coefficients. This has the mathematical consequence that sinusoidal input signals produce sinusoidal output signals. The response of such a system can be described by a complex gain factor T at each frequency: When the input signal is $x(t) = Xe^{j\omega t}$ and the output signal is $z(t) = Ze^{j\omega t}$, then the gain is $T = Z/X$. It may have a physical dimension (when the system is a transducer) and will in general depend on frequency. The function $T(\omega)$ is called the *complex frequency response*. Its absolute value $|T(\omega)|$ is the *amplitude response* and its phase is the *phase response* of the system. Signals of arbitrary time dependence can be represented as an integral or sum over sinusoidal signals of different frequency; knowledge of the response function $T(\omega)$ at all relevant frequencies is therefore sufficient to calculate the response of the system to any input signal. The decomposition of arbitrary signals into sinusoidal (time-harmonic) components is a *Fourier transformation*, and their synthesis from such components is an *inverse Fourier transformation*. For transient signals such as analog seismograms, the Fourier transformation is formulated as a pair of integral transformations:

$$f(t) = \frac{1}{2\pi} \int_{-\infty}^{\infty} a(\omega) e^{j\omega t} \, d\omega, \qquad a(\omega) = \int_{-\infty}^{\infty} f(t) e^{-j\omega t} \, dt \tag{5}$$

The most general class of signals for which the response of a linear and time-invariant system can be described by a gain factor are sinusoidal oscillations with exponentially growing or decaying amplitudes: $a(t) = Ae^{st}$ with complex $s = \sigma + j\omega$. (The complex exponential function is defined as $e^{(\sigma + j\omega)t} = e^{\sigma t}(\cos \omega t + j \sin \omega t)$). Again the system can be characterized by a complex gain factor $H(s)$: The input signal $x(t) = Xe^{st}$ produces the output signal $z(t) = Ze^{st} = H(s)Xe^{st}$, thus $Z = H(s)X$. Other signals are decomposed and synthesized with the *Laplace transformation*:

$$f(t) = \frac{1}{2\pi j} \int_{\sigma-j\infty}^{\sigma+j\infty} b(s) e^{st} \, ds \quad \text{for} \quad t > 0, \qquad b(s) = \int_{0}^{\infty} f(t) e^{-st} \, dt \tag{6}$$

The Fourier transformation may be considered as a special case of the Laplace transformation with a purely imaginary $s = j\omega$, although the mathematical concepts behind the two transformations are somewhat different. $H(s)$ is the *transfer function* of the system. It follows that $T(\omega) = H(j\omega)$; the functions $T(\omega)$ and $H(s)$ are related by a simple substitution of the frequency variable but are not identical. When their poles and zeros or other mathematical properties are discussed, it is important to state which one of the two functions is meant. The terminology is, however, not uniform in the literature; sometimes $T(\omega)$ is also referred to as a transfer function.

Transfer functions of seismometers and also of a wide class of other mechanical and analog electronic systems are rational functions of frequency with real coefficients: $H(s) = P(s)/Q(s)$, where $P(s)$ and $Q(s)$ are polynomials in s. This is a consequence of the fact that the differential equation of the system is transformed into an algebraic equation for the complex signal amplitudes when only signals with a time dependence of the form e^{st} or $e^{j\omega t}$ are admitted; Eq. (8) in the next paragraph is an example. A polynomial $P(s)$ is up to a factor determined by its zeros, i.e., by those complex values of s for which $P(s) = 0$. The zeros of $P(s)$ are at the same time those of $H(s)$; the zeros of $Q(s)$ are poles of $H(s)$. Rational transfer functions can therefore be specified with a limited number of numerical coefficients, by listing either the complex poles and zeros or the real polynomial coefficients. Different factorizations of the polynomials are possible in the latter case.

The essential options are:

1. The poles and zeros in the complex s plane are listed, together with a constant gain factor.
2. The polynomial coefficients of the nominator and denominator are listed (they are real when $s = j\omega$ is chosen as the frequency variable).
3. The polynomials are decomposed into normalized first- and second-order factors, each of which is defined by its corner frequency and, in case of second order, numerical damping. The individual factors can normally be attributed to physical subunits of the system. An overall gain factor is also required.

3.2 The Frequency Response of Geophones

Using time-harmonic signals, the solution of the differential equation (2) becomes very simple. Let the input signal be $x(t) = Xe^{j\omega t}$, the output signal $z(t) = Ze^{j\omega t}$, and the external force $f(t) = Fe^{j\omega t}$. Equation (2) then reduces to

$$(-\omega^2 M + j\omega R + S)Z = F + \omega^2 MX \tag{7}$$

$$Z = \frac{F/M + \omega^2 X}{-\omega^2 + j\omega R/M + S/M} \tag{8}$$

The mechanical pendulum is thus a second-order high-pass filter for displacements, and a second-order low-pass filter for accelerations and external forces. Its angular corner frequency is $\omega_0 = \sqrt{S/M}$. At this frequency, the ground motion X is amplified by a factor $\omega_0 M/R$ and phase shifted by $\pi/2$. The imaginary term in the denominator is usually written as $2j\omega\omega_0 h$, where $h = R/(2\omega_0 M)$ is the numerical damping, i.e., the ratio of the actual to the critical damping.

In order to convert the motion of the mass into an electrical signal, the mechanical pendulum is in the simplest case coupled with an electromagnetic velocity transducer (see Section 2.8) whose output voltage we denote by E. We then have an electromagnetic seismometer, also called a geophone when designed for seismic exploration. When the responsivity of the transducer is σ (V m^{-1} sec^{-1}; $E = \sigma j\omega Z$) we get

$$E = \frac{j\omega\sigma(F/M + \omega^2 X)}{-\omega^2 + 2j\omega\omega_0 h + \omega_0^2} \tag{9}$$

from which, in the absence of an external force, we obtain the complex response functions

$$T_d = \frac{E}{X} = \frac{j\omega^3\sigma}{-\omega^2 + 2j\omega\omega_0 h + \omega_0^2} \tag{10}$$

for the displacement;

$$T_v = \frac{E}{j\omega X} = \frac{\omega^2\sigma}{-\omega^2 + 2j\omega\omega_0 h + \omega_0^2} \tag{11}$$

for the velocity; and

$$T_a = \frac{E}{-\omega^2 X} = \frac{-j\omega\sigma}{-\omega^2 + 2j\omega\omega_0 h + \omega_0^2} \tag{12}$$

for the acceleration. The geophone is a second-order high-pass filter for the velocity, and a bandpass filter for the acceleration.

We have so far treated the damping as if it were a viscous effect in the mechanical receiver. Actually, only a small part h_m of the damping is due to mechanical causes. The main contribution normally comes from the electromagnetic transducer, which is suitably shunted for this purpose (see Section 2.8). Its contribution is

$$h_{el} = \frac{\sigma^2}{2M\omega_0 R_d} \tag{13}$$

where R_d is the total damping resistance (the sum of the resistances of the coil and of the external shunt). The total damping $h_m + h_{el}$ is preferably chosen as $1/\sqrt{2}$, a value that defines a second-order Butterworth filter characteristic, and gives a maximally flat response in the passband.

3.3 The Impulse Response

Alternatively, the transfer properties of a seismometer can be described in the time domain by its impulse response, which is the response of the system to an impulsive input signal. (An impulse in this sense is any signal whose time integral is undistinguishable from a unit step.) The impulse response and the transfer function are Laplace transforms of each other, so they offer mathematically equivalent descriptions of the system. In the same way, the complex frequency response is the Fourier transform of the impulse response. The impulse response can directly be calculated from the poles and zeros of the transfer function. For a practical specification, the impulse response is less suitable because it is a transcendental function of infinite duration that is inconvenient to formulate.

The response of a seismograph to an arbitrary input signal can in principle be computed as the convolution of that signal with the impulse response. However, due to the infinite length of the latter, this may not be an efficient procedure. Also, a sampled version of the impulse response may not represent the analytical form correctly when the system is not strictly band-limited. So computing the response of a system by convolution requires some precautions, and one would in most cases prefer either to do the computation in the frequency domain with the Fourier transformation, or to filter the input signal with a recursive filter that represents the seismograph, as explained in the next paragraph.

3.4 Representing a Seismograph by a Recursive Filter

For a general theory of recursive (or IIR) filters, we must refer the reader to the pertinent literature (Oppenheim and Schafer,

1975; Robinson and Treitel, 1980; Plešinger *et al.*, 1996). For mathematical reasons, recursive filters can only approximate, but not have, the rational transfer functions of seismographs. In a restricted sense, however, an exact equivalent exists. When we filter all signals with an antialias filter that has a rational transfer function and is sharp enough to practically prevent aliasing, then we can precisely model the overall transfer function with an "impulse-invariant" recursive algorithm (Schuessler, 1981). Remarkably, the rational antialias filter need not exist in hardware but only as part of the computer algorithm. So we can, in effect, precisely model any seismograph in a bandwidth that is by a factor of 2 or so smaller than theoretically permitted by the sampling theorem. Even this restriction can be overcome by sampling the signals twice as fast, or by resampling them numerically. The method is especially useful for the purpose of calibrating seismographs with arbitrary signals (see Section 9).

4. Force-Balance Accelerometers and Seismometers

4.1 The Force-Balance Principle

The precision of a conventional, passive seismometer depends on its two functional subunits: the mechanical suspension and the displacement or velocity transducer. An inertial seismometer basically measures the inertial force acting on the seismic mass in an accelerated local frame of reference [Eq. (2)]. The suspension converts the inertial force into a displacement of the mass, and the transducer converts this into an electrical signal. Neither one of these conversions is inherently precise. As discussed under Section 2.5, a sensitive seismometer must have a suspension with a small restoring force so that small accelerations produce noticeable displacements of the seismic mass. Then, of course, larger seismic signals or environmental disturbances produce large displacements that change the geometry of the spring and destroy the linear relationship between displacement and force. When the restoring force is diminished, undesired effects such as hysteresis and viscous behavior retain their absolute magnitudes and thus become relatively larger. Finally, it is difficult to build linear transducers with a large range. A passive seismic sensor therefore cannot be optimized for sensitivity and precision at the same time.

These problems are well known from the design of precision instruments, especially of laboratory balances. They are solved by compensating the unknown force with a known force, rather than determining it indirectly from the elongation of a spring. The compensating force is generated in an electromagnetic transducer and is controlled by a servo circuit (Fig. 6) that senses the position of the seismic mass and adjusts the force so that the mass returns to its center position. Such a system is most effective when it contains an integrator, in which case the offset of the mass is exactly nulled in the time average. Due to unavoidable delays in the feedback loop, servo systems have a limited bandwidth; however, at frequencies where they are effective, they force the mass to move with the ground by generating a feedback force strictly proportional to ground acceleration. When the force is proportional to the current in the transducer, then the current, the voltage across the feedback resistor R_1, and the output voltage are all proportional to ground acceleration. We have thus converted the acceleration into an electrical signal without relying on the mechanical precision of the spring. The suspension still serves as a detector but not as a converter, and may now be optimized for sensitivity without giving up precision.

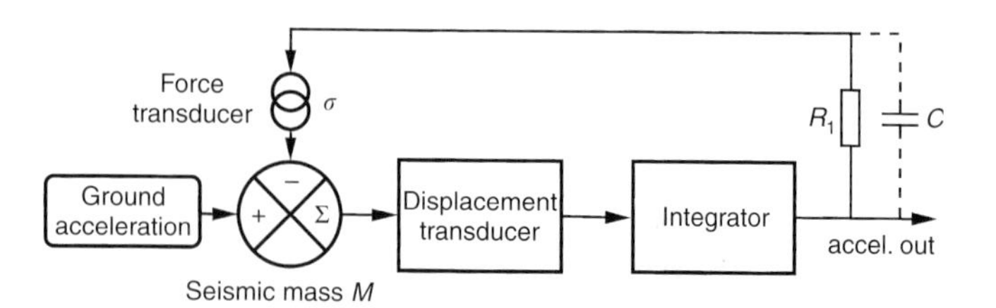

FIGURE 6 FBA feedback circuit.

The effectiveness of a servo system is measured by its loop gain L, which is the amplitude ratio of the feedback force to the uncompensated residual of the inertial force. If f is the inertial force and r the residual, then $r = f/(1 + L)$. A large loop gain implies a small residual and thus an output signal that represents the acceleration with a small error. The integrator provides a large loop gain and thus a high precision at low frequencies. At high frequencies, the loop gain is limited by stability conditions and cannot be increased arbitrarily; this topic is treated in textbooks on network analysis or system theory (Oppenheim and Willsky, 1983). Some minor modifications of the basic circuit, such as the addition of a small capacitor C parallel to the feedback resistor R_1, may be required to ensure stability.

The response of a servo system is approximately inverse to the gain of the feedback path. It can easily be modified by giving the feedback path a frequency-dependent gain. For example, if we make the capacitor C large so that it determines the feedback current, then the gain of the feedback path increases linearly with frequency, and we have a system whose responsivity to acceleration is inverse to frequency and thus flat to velocity over a certain passband. We will look more closely at this option in Section 4.3.

4.2 Force-Balance Accelerometers

By equating the inertial and the electromagnetic forces, it is easily seen that in the circuit of Figure 6 the factor of proportionality between the output voltage and the acceleration is

$$A = \frac{MR_1}{\sigma} \tag{14}$$

where M is the seismic mass, R_1 is the resistance of the feedback path (including the coil), and σ is the responsivity of the forcer (in $\mathrm{N\,A^{-1}}$). The conversion is determined by only three passive components, of which the mass is error-free by definition (it defines the inertial reference), the resistor is a nearly ideal component, and the force transducer can be very precise because the motion is small.

Figure 6 is the circuit of a force-balance accelerometer (FBA), a device that is widely used for earthquake strong-motion recording, measuring tilt, and inertial navigation. Since the dynamic range (see Section 5.1) of a feedback system is mainly determined by its feedback path, which is in this case composed of passive components whose range is not limited by semiconductor noise or clipping, FBAs can have a very large dynamic range (up to 160 dB). The operating range can conveniently be adjusted by changing the feedback resistor, which is external in some types. FBAs work down to zero frequency, but the servo loop becomes ineffective at some high frequency (typically between 100 and 1000 Hz), above which the arrangement acts like an ordinary inertial displacement sensor. FBAs are therefore low-pass filters for ground acceleration and high-pass filters for ground displacement.

4.3 Velocity Broadband Seismometers

For broadband seismic recording with high sensitivity, an output signal proportional to ground acceleration is unfavorable. At high frequencies, sensitive accelerometers are easily saturated by traffic noise or impulsive disturbances. At low frequencies, a system with a response that is flat to acceleration generates a voltage at the output as soon as the suspension is not completely balanced. Such a system would easily be saturated by the offset voltage resulting from thermal drift or tilt. What we need is a bandpass response in terms of acceleration, or equivalently a high-pass response in terms of ground velocity, like that of a normal electromagnetic seismometer but with a lower corner frequency. Essentially the same considerations are expressed in Section 5.5 as the rule that the response of a broadband seismometer should be approximately inverse to the spectral distribution of the noise.

The desired velocity broadband (VBB) response is obtained from the FBA circuit by adding paths for differential feedback and integral feedback (Fig. 7). The capacitor C is chosen so large that the differential feedback dominates throughout the desired passband. While the feedback current is still proportional to ground acceleration as before, the voltage across the capacitor C is a time integral of the current, and thus proportional to ground velocity. This voltage serves as the output signal. The factor of proportionality—the apparent generator constant of the feedback seismometer—is

$$V = \frac{M}{\sigma C} \tag{15}$$

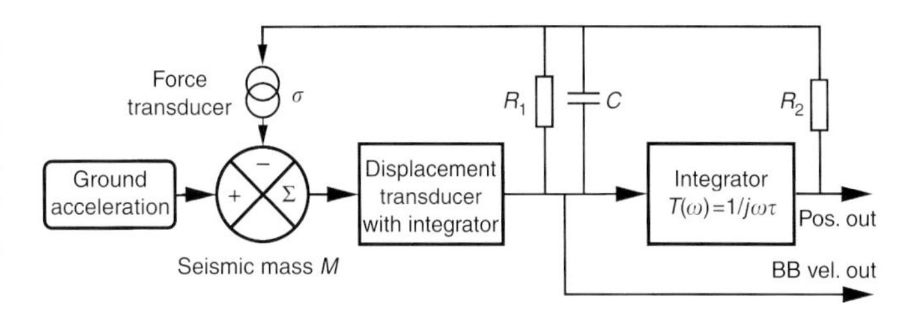

FIGURE 7 VBB feedback circuit.

Again the response is essentially determined by three passive components. Although a capacitor with a solid dielectric is not quite as ideal a component as a good resistor, the response is still linear and very stable.

The output signal of the integrator is normally accessible at the "mass position" output. It does not indicate the actual position of the mass but indicates where the mass would go if the feedback were switched off. "Centering" the mass of a feedback seismometer (by adjusting its weight or the spring force) has the effect of discharging the integrator so that its full operating range is available for the seismic signal. The mass-position output is not normally used for seismic recording but is useful as a state-of-health diagnostic, and is used in some calibration procedures.

The relative strength of the integral feedback increases at lower frequencies, while that of the differential feedback decreases. At some frequency, the two contributions are of equal strength but opposite phase ($-\pi/2$ and $\pi/2$, respectively). This is the lower corner frequency of the feedback system, below which the response rolls off with the second derivative of ground velocity. The differential and the integral parts of the feedback cancel at the corner frequency, but the proportional feedback remains and damps the resonance that would otherwise occur. As a result, the feedback system behaves like a conventional electromagnetic seismometer and can be described by the usual three parameters: free period, damping, and generator constant. In fact most of the presently used electronic broadband seismometers follow the simple theoretical response of electromagnetic seismometers more closely than these ever did.

At some high frequency, the loop gain falls below unity. This is the upper corner frequency of the feedback system, which marks the transition between a response flat to velocity and one flat to displacement. A well-defined and nearly ideal behavior of the seismometer as at the lower corner frequency should not be expected there, both because the feedback becomes ineffective and because most suspensions have parasitic resonances slightly above the electrical corner frequency (otherwise they could have been designed for a larger bandwidth). However, the detailed response at the high-frequency corner rarely matters since the upper corner frequency is usually outside the passband of the recorder. Its effect on the transfer function can in most cases be modeled as a small, constant delay (a few milliseconds) over the whole VBB passband.

4.4 The Response of Force-Balance Seismometers

For completeness we give here a slightly simplified formula for the complex frequency response of a broadband force-balance seismometer. The simplification consists in ignoring the mechanical restoring force and damping, which can be absorbed into the electrical parameters, and the impedance of the feedback coil, which can be made small. When X is the Fourier amplitude of the ground displacement and E that of the output voltage, then the response to ground velocity is

$$\frac{E}{j\omega X} = \frac{\omega^2 M/\sigma C}{-j\omega^3 M/\alpha\sigma C - \omega^2 + j\omega/R_1 C + 1/\tau R_2 C} \tag{16}$$

The leading term in the denominator disappears when the responsivity α of the displacement transducer is large. What remains is an ordinary high-pass response with a corner frequency $1/\sqrt{\tau R_2 C}$, damping $\sqrt{\tau R_2/4CR_1^2}$, and generator constant $M/\sigma C$. The response is thus identical to that of a conventional electrodynamic seismometer [Eq. (11)]. The realization of a "very-broadband" type of seismometer, with a free period of 100 sec or more, depends on the ability to build an integrator with a long time-constant τ and low electronic noise (Wielandt and Steim, 1986). At high frequencies, the bandwidth is limited by the finite responsivity α of the displacement transducer. The loop gain falls below unity and the velocity response rolls off with $1/j\omega$ at frequencies above $\alpha\sigma C/2\pi M$; the response is then flat to displacement. The resistance and inductance of the feedback coil, phase delays in the electronic circuit, and parasitic resonances in the suspension make the feedback loop unstable for large α and thus set a limit to the upper corner frequency of the response.

The reader is referred to publications by Usher *et al.* (1978), Wielandt (1983), and Wielandt *et al.* (1982, 1986) for a deeper discussion.

5. Design Criteria for Broadband Seismographs

5.1 The Concept of Very-Broadband Seismometry

Earthquakes and other seismic sources radiate signals in a large range of frequencies and amplitudes. In a logarithmic scale, the seismic frequency band is much broader than that of audible tones. Only a limited range of these signals can be visibly displayed at a time. So any seismograph recording on paper or film has to act as a filter and suppress most of the available information. Quite a number of analog seismographs with different characteristics had to be operated in parallel in order to preserve a reasonable choice of signals. Digital technology now permits the recording of all useful seismic signals on the same medium in a single data stream. Such a system is called a very-broadband (VBB) seismograph. It must meet the following requirements (IRIS, 1985; Wielandt and Steim, 1986):

1. The system must have a sensitivity sufficient to resolve signals at the level of minimum ground noise at all frequencies of interest.
2. Its operating range must be large enough to record the largest earthquakes at regional to teleseismic distances.
3. The largest ground noise, natural or artificial, that is likely to occur in any part of the spectrum must not interfere with the resolution of small signals at other frequencies.

A few explanations are required. A signal is said to be *resolved* when it is present in the seismic record and not masked by instrumental noise. The minimum ground noise is different at each station, but for a uniform system intended for global deployment the instrumental sensitivity must be designed for the global minimum. The present standard is the New Low Noise Model (NLNM) compiled by Peterson (1993). A simple computer program converting noise data into different units and comparing them to the NLNM is available from the author's FTP site (see Section 11). The *operating range* is the maximum signal amplitude that can pass through the system without serious distortion.

The above requirements can be visualized in a doubly-logarithmic diagram like Figure 8, where the levels of ground noise and earthquake signals are expressed in common units, here as average peak values in 1/3 octave or rms values in 1/6 decade (these two measures happen to coincide within a few per cent). In other contexts, noise levels are normally expressed as power spectral densities, which cannot be compared directly with the amplitudes of transient signals. The amplitude ratio between the instrumental noise and the clipping level of a sensor or a recorder is called its *dynamic range*. It is usually expressed in decibels and depends on frequency and on the bandwidth in which it is measured; without this information its specification is meaningless. The reference bandwidth should be chosen with a view to the narrowest bandwidth useful for a waveform plot at each frequency. *Absolute dynamic range* is short for a specification of the dynamic range in absolute signal levels.

5.2 The Sensitivity of Force-Balance Seismometers

The spectral distribution of minimum ground noise is such that noise is difficult to resolve at both the short-period and long-period ends of the seismic band. At short periods, the ground displacements in the NLNM (Fig. 8) are very

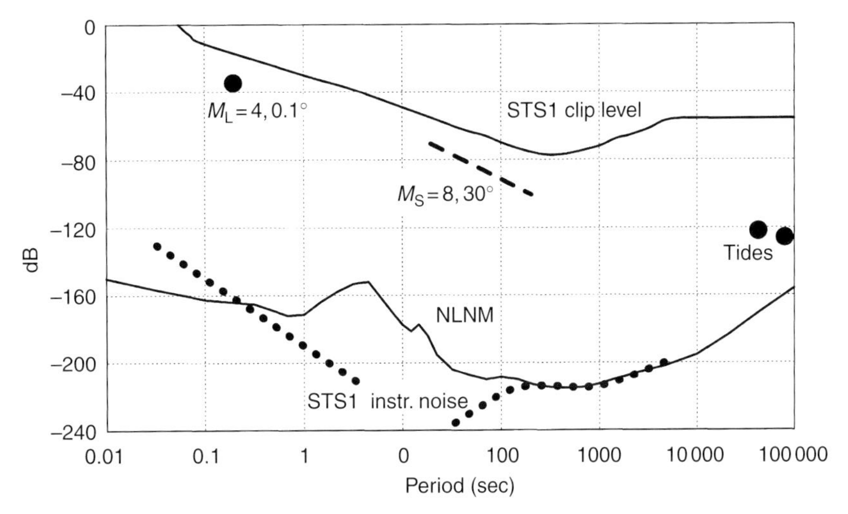

FIGURE 8 A representation of the USGS New Low Noise Model in comparison with the self-noise and operating range of the STS1 very-broadband seismometer. Signal levels are expressed as decibels relative to 1 m sec^{-2} and may be understood as rms values in a bandwidth of 1/6 decade or average peak values in a bandwidth of 1/3 octave. At periods longer than 100 sec, the NLNM is mainly based on STS1 data. Its coincidence with the instrumental noise beyond 200 sec suggests that the STS1 seismometer may not resolve ground noise at the quietest sites (Peterson 1993).

small: 6 pm average peak in a bandwidth of 1/3 octave at 5 Hz, 0.4 pm at 30 Hz. Capacitive displacement transducers (Blumlein bridges) used in VBB seismometers typically resolve between 0.1 and 1 pm at these frequencies. The resolution cannot easily be improved without increasing the undesired effects of power consumption, heat production, electrostatic forces, and viscous (air) damping in the displacement transducer.

At long periods, the seismic noise is relatively large in terms of displacement, but the acceleration is small. Around a period of 300 sec the acceleration associated with the minimum noise is about 15 pm sec^{-1} ($1.5 \times 10^{-12}g$) in 1/3 octave. The relationship $\Delta g = \omega_0^2 \Delta z$ [Eq. (3) with Δg written for the ground acceleration $\ddot{x}$] tells us then that we need a displacement resolution of 0.1 pm when the pendulum has a free period of 0.5 sec, 10 pm at a free period of 5 sec, and 350 pm at 30 sec (the numbers are estimates for the CMG3, STS2, and STS1 seismometers, respectively). The difficulty of building a sensitive VBB seismometer with a mechanical free period substantially shorter than 1 sec is obvious.

A final limit to the sensitivity at long periods is set by the Brownian (thermal) motion of the seismic mass. Air molecules hitting the pendulum not only damp its motion but also exert a force whose spectral noise power density N is given by the same well-known formula that quantifies the voltage noise of a resistor:

$$N = 4kTR \tag{17}$$

where R is the viscous friction experienced by the seismic mass, k is the Boltzmann constant, and T is the absolute temperature. In a typical force-balance seismometer, damping is mainly caused by the air gap of the capacitive displacement transducer. Even with a favorable capacitor geometry and with ventilation holes in the plates, thermal noise is likely to mask the minimum long-period ground noise when the seismic mass is reduced to some 10 g or less. The problem can in principle be solved by evacuating the sensor, but it would be very difficult to maintain a sufficiently high vacuum in a sensor of conventional design.

While failure to meet the basic criteria must spoil the ability to resolve minimum noise, their observation is no guarantee for a sensor to achieve the desired resolution. Additional noise may result from various mechanical, electrical, magnetic, thermal and even chemical effects, and it is difficult to predict which ones will infest a specific design.

5.3 The Operating Range of Force-Balance Seismometers

The electric operating range of a well-designed seismometer is, at least in its passband, limited by clipping at the output, at a signal level slightly below the internal supply voltage. The seismic operating range is the electrical range divided by the amplitude response, and thus in general is frequency-dependent. Clipped waveforms can normally be recognized when the signal has been recorded with adequate bandwidth. However, when clipping occurs in earlier stages of the circuit or in the feedback loop, or when the signal is low-pass filtered in the recorder, clipping may go unnoticed and the record may look quite normal even if it is severely distorted; an example is presented in Wielandt (1983).

The operating range at very low frequencies requires a separate consideration. Due to the presence of an integrator in the feedback loop, the condition that clipping should occur only at the output cannot be maintained at very low frequencies. The operating range is then limited by saturation of the integrator. Since the integrator also generates noise, it determines the dynamic range of the whole system at long periods. This has direct consequences for the ranges of drift and tilt in which force-balance seismometers can be operated. We give an order-of-magnitude estimate: The integrator may have a noise level of 0.1 μV rms (in an appropriate bandwidth) and a saturation level of ± 10 V; the sensor may have been designed to resolve $10^{-12}\,g$ rms. Then the integrator will be saturated by static accelerations of $\pm 10^{-4}g$. A vertical sensor whose suspension has a temperature coefficient of 10^{-5} per Kelvin could be operated with this feedback system in a temperature range of ± 10 K; a horizontal sensor would have to be leveled to within 0.1 mm m^{-1}.

The designer of a force-balance seismometer has a considerable freedom in the choice of the responsivity and thus of the seismic operating range. The dynamic range of sensors and recorders is limited, however, so a decision must be made whether the operating range or the self-noise of a system should be specified in the first place. Recording systems for strong motion usually have a certain level of ground acceleration specified as the operating range. General-purpose seismographs are normally designed to resolve ground noise; their operating range is then made as large, and their gain as small, as this requirement permits.

5.4 Digitizer Noise

An ideal digitizer rounds the samples of the input signal to integer multiples of a quantum q (one count), which can be expressed as an equivalent step of the input voltage. The sequence of rounding errors constitutes the quantization noise. Under certain conditions it is white and has a constant total power of $q^2/12$ in the band from zero to the Nyquist frequency (half the sampling frequency). Its spectral density therefore depends on the sampling rate, and can be reduced by oversampling and subsequent bandwidth reduction. However, digitizers also exhibit ordinary electronic noise both of the white and the $1/f$ variety, which cannot be reduced by oversampling. Filters for bandwidth reduction must have some gain or use floating-point arithmetic in order to avoid additional quantization errors. Nonlinear distortions in modern 24-bit digitizers are normally so small that they can safely be ignored.

5.5 Matching Sensors and Recorders

The clipping level of a seismic recording system—sensor plus recorder—is the minimum of the clipping levels of the components. On the other hand, the supposedly uncorrelated noise levels of the components add up like orthogonal vectors. The overall dynamic range of the system is therefore always smaller than that of each subsystem. (To be comparable, all signal levels must be referred to the same point in the circuit, or expressed as ground motion.) To make best use of the absolute dynamic ranges of the components, the smallest one must be contained in all others. When the dynamic ranges of the sensor and the recorder are of similar magnitude, as is the case with modern broadband sensors and 24-bit digitizers, this condition determines the optimum gain of the sensor within narrow limits.

Similar considerations apply to the seismic input signals. The Earth may in this context be considered as another signal-transmitting system with its own noise and operating range. For a system to resolve ground noise, the sensor must not only have a self-noise below that noise but must also put out the combined ground and sensor noise as an electric signal above the digitizer noise. The necessary gain margin has occasionally been a matter of controversial discussion. A factor of 2 should normally be sufficient; the digitizer noise then increases the noise amplitude by only 12%. Using a larger gain margin would unnecessarily reduce the seismic operating range. A small gain margin does not preclude the extraction of coherent signals from incoherent noise by bandwidth reduction, stacking, or beamforming. When the system resolves incoherent ground noise, it also resolves all coherent signals that can theoretically be extracted from it.

If the digitizer noise were white, the optimum response of the sensor would be inverse to the spectral distribution of the minimum noise. The latter is, however, too complicated to be modeled in a practical sensor. Experience has shown that a response flat to velocity, from short periods to a long-period corner of 100 sec or more, is sufficiently close to the optimum. A more detailed discussion of this topic is found in Wielandt and Steim (1986).

5.6 Scaling Down Seismometers

Seismometers should be as small as possible for easier transportation, installation, and shielding, and because internal air convection (which causes noise) is more easily suppressed in a smaller volume. It is therefore interesting to see what happens if a given design is scaled down by a factor of γ. A simple calculation shows that strict scaling is not possible for vertical sensors that are in equilibrium with terrestrial gravity. The mass, the spring force, and the spring stiffness scale with different powers of γ (3, 2, and 1, respectively). The spring would be too strong in a downscaled version and must be made slightly thinner than to scale; nevertheless, it remains too stiff. The mechanical sensitivity to long-period accelerations is thereby reduced by a factor of γ, and the period becomes shorter by a factor $\sqrt{\gamma}$ as for a simple pendulum. On the other hand, the resolution of the displacement transducer will suffer from the smaller size of the capacitor plates. Air damping in the displacement transducer is also undesirably increased (see Section 5.2 and Jones and Richards, 1973). It is, by the way, unlikely that one would want to reduce the gap between the plates because, if this were practical, one would have done it in the first place. The signal-to-noise ratio at long periods will therefore go down at least with the square of the scaling factor. Undesired forces causing nonlinearity or noise will in general scale with a smaller power of γ than the mass, and thus become relatively larger. Finally, Brownian noise will come up when the mass is too small, as mentioned above. So it appears that miniature seismometers require different designs and are unlikely to reach the sensitivity of presently available VBB seismometers in the near future. There is no doubt, however, that existing designs could be scaled down by a factor of 2 or so, which would make the instruments very handy.

6. Site Selection, Installation and Shielding

6.1 Site Selection

Site selection for a permanent seismic station is always a compromise between two conflicting requirements: infrastructure and low seismic noise. The noise level depends on the geological situation and on the proximity of sources, some of which are usually associated with the infrastrucure. A seismograph installed on solid basement rock can be expected to be fairly insensitive to local disturbances, while one sitting on a thick layer of soft sediments will be noisy even in the absence of identifiable sources. As a rule, the distance from potential sources such as houses and roads should be large compared to the thickness of the sediment layer. Broadband seismographs can be successfully operated in major cities when the geology is favorable; in unfavorable situations, such as in sedimentary basins, only deep mines and boreholes may offer acceptable noise levels.

Seismic noise has many different causes. Short-period noise is at most sites predominantly man-made and somewhat larger in the horizontal components than in the vertical. At intermediate periods (2–20 sec), marine microseisms dominate with similar amplitudes in the horizontal and vertical directions. At long periods, horizontal noise may be larger than vertical noise by a factor up to 300, the factor increasing with period. This is mainly due to tilt, which couples gravity into the horizontal components but not into the vertical. Tilt may be caused by traffic, wind, or the barometric pressure. Large tilt noise is sometimes observed on concrete floors when an unventilated cavity exists underneath; the floor then acts like a membrane. Such noise can be identified by its linear polarization and its correlation with the barometric pressure. The sound of a hammer stroke tells a lot about the solidity of a floor. Even on an apparently solid foundation, the long-period noise often correlates with the barometric pressure (Beauduin *et al.*, 1996). If the situation cannot be remedied otherwise, the barometric pressure should be recorded with the seismic signal and used for a correction. For very-broadband seismographic stations, barometric recording is generally recommended.

Besides ground noise, environmental conditions must be considered. Although seismometers can and must be shielded against these, an aggressive atmosphere may cause corrosion, short-term variations of temperature may induce noise, and seasonal variations of temperature may exceed the drift specifications. As a precaution, cellars and vaults should be checked for signs of occasional flooding.

6.2 Seismometer Installation

We briefly describe the installation of a portable broadband seismometer inside a building, vault, or cave. The first act is to mark the orientation of the sensor on the floor. This is best done with a geodetic gyroscope, but a magnetic compass will do in most cases. The magnetic declination must be taken into account. Since a compass may be deflected inside a building, the direction should be taken outside and transferred to the site of installation. A laser pointer may be useful for this purpose. When the declination is unknown or unpredictable (such as in high latitudes or volcanic areas), the orientation should be determined with a sun compass.

To isolate the seismometer from stray currents, small glass or Perspex plates are cemented to the ground under its feet. Then the seismometer is installed, tested, and wrapped with a thick layer of thermally insulating material. The type of material seems not to matter very much; alternate layers of fibrous material and heat-reflecting blankets are probably most effective. The edges of the blankets should be taped to the floor around the seismometer. Electronic seismometers produce heat and may induce convection in any open space inside the insulation; it is therefore important that the insulation has no gap and fits the seismometer tightly. Another method of insulation is to surround the seismometer with a large box that is then filled with fine styrofoam seeds. For a permanent installation under unfavorable environmental conditions, the seismometer must be enclosed in a hermetic container. A problem with such containers (as with all seismometer housings) is that they cause tilt noise when they are deformed by the barometric pressure. Essentially three precautions are possible: the baseplate is carefully cemented to the floor, or it is made so massive that its deformation is negligible, or a "warp-free" design is used as described by Holcomb and Hutt (1992) for the STS1 seismometers. Some desiccant should be placed inside any hermetic container, even into the vacuum bell of STS1 seismometers.

Guidelines for installing broadband seismic instrumentation are offered at the web site of the Seismological Lab at Berkeley (Uhrhammer and Karavas, 1997). Detailed instructions for the design of seismic vaults are given by Trnkoczy (1998).

6.3 Installation of Strong-Motion Instruments

Although strong-motion instruments are sometimes installed side by side with broadband instruments, their site is normally selected on different criteria. For purposes of seismic engineering, strong-motion records may be desired where one would not normally install a seismic station—for example, in power stations or high-rise buildings. The instruments must be cemented or bolted to the structure; in regions of high epicentral intensity, they should also be protected from falling debris. Other considerations apply when free-field signals are to be recorded. At high frequencies and when the soil is soft, nearby buildings and even the instrument vault itself may modify the signal. The vault should in this case be a small, rigid structure that is firmly coupled to the ground on all sides. Since strong-motion recorders are normally operated in an event-triggered mode and may not record any ground motion for long intervals of time, their state of health must be regularly checked with test signals.

6.4 Magnetic Shielding

Broadband seismometers are to some degree sensitive to magnetic fields because all thermally compensated spring materials are slightly magnetic. This may become noticeable when seismometers are operated in industrial areas or in the vicinity of dc-powered railway lines. When long-period noise is found to follow a regular timetable, magnetic interferences should be suspected. Shields can be manufactured from permalloy metal but are expensive and of limited efficiency. An active compensation is often preferable. It may consist of a three-component fluxgate magnetometer that senses the field near the seismometer, an electronic driver circuit in which the signal is integrated with a short time constant (a few milliseconds), and a three-component set of Helmholtz coils that compensate changes of the magnetic field. The permanent geomagnetic field should not be compensated; the resulting offsets of the fluxgate outputs can be compensated with a permanent magnet or electrically.

7. Deconvolution

7.1 General Deconvolution

It is often necessary to restore the original ground motion from a seismogram. For this purpose, the seismogram must be filtered with the inverse response of the seismograph, a process known as deconvolution. It is rarely carried out in its exact mathematical form because the signal-to-noise ratio deteriorates outside the passband of the seismograph, and the deconvolution must be limited to a passband where the signal-to-noise ratio is still acceptable. Generally speaking, the ground motion is not entirely determined by the seismic record, and its reconstruction is a geophysical inverse problem whose solution must be constrained by a priori information. Similar considerations apply to the case when records from seismographs with different responses must be homogenized. The PREPROC software package (Plešinger *et al.*, 1996) contains different routines for a bandlimited deconvolution and its manual offers a concise introduction to the problem.

A deconvolution can be realized in the frequency or the time domains. The frequency domain is convenient for the construction of approximate inverses with a view to the frequency-dependent signal-to-noise ratio of the data. Causal and acausal solutions are available; waveforms may be better preserved with an acausal inverse, but the resulting precursory signals may give rise to misinterpretation. A division-by-zero problem exists in the low-frequency limit and prevents the construction of an exact inverse. In the time domain, both approximate and quasi-exact inverses can conveniently be realized with recursive (IIR) filters. They are always causal. Short time-series can often be deconvolved with a quasi-exact inverse; no division-by-zero problem is encountered but a quadratic or cubic trend may be generated, which can be removed afterward by polynomial fitting.

7.2 Removing the Noncausal Response of Digital Recorders

The response of digital recorders is normally determined by digital, finite impulse response (FIR) filters that decimate the oversampled data stream. The transfer functions of such filters are transcendental functions, and it is neither correct nor practical to specify them by poles and zeros in frequency domain as if they were rational functions.

Essentially two types of FIR filters are used. Most recorders have zero-phase filters that do not introduce phase shifts; their response can simply be ignored in the overall response of a digital seismograph, except for their limiting the bandwidth. Zero-phase filters are, however, noncausal and can produce spurious precursors of seismic arrivals. This is a problem when first arrivals must be picked, in which case it is preferable to have minimum-phase decimation filters. These, on the other hand, introduce frequency-dependent delays that must be removed from the data in other applications. It is possible to convert a zero-phase filtration into a minimum-phase filtration, and vice versa, by applying a suitable all-pass filter to the data (Scherbaum, 1996). The problem is treated elsewhere in this Handbook.

Problems with acausal zero-phase filters can also be avoided by filtering the data with a causal filter of smaller bandwidth such as a Butterworth low-pass filter, or with the response of a classical electromagnetic seismograph. The influence of the original decimation filter then disappears and only the response of the user-specified filter must be considered.

8. Testing for Linearity and Noise

8.1 Testing for Linearity

Seismographs need not record the ground motion with extreme precision; most methods of interpretation do not depend on minor amplitude errors or waveform distortions. In one respect, however, a broadband seismograph must be precise: it must be a linear system, that is, it must conform to the principle of superposition. The sum of two input signals must produce the sum of the corresponding output signals and nothing else. Then, the presence of any number of input signals in a given frequency band will not produce an output signal in another band. This is a prerequisite for being able to record small long-period signals in the presence of strong short-period signals, or vice versa.

The electrical linearity of a seismometer is usually checked with a two-tone test (Peterson *et al.*, 1980). Two sinusoidal signals of equal amplitude and nearly equal frequency (for example, 1.00 and 1.02 Hz) are applied to the calibration input and their low-frequency (0.02 Hz) intermodulation product is observed at the output. The measurement is facilitated by the fact that the gain of a VBB seismometer at 0.02 Hz is 50 times larger than at 1 Hz, so even a small amount of distortion will be noticeable. Another way of testing linearity is to calibrate the system with the CALEX routine (see Section 9.5) using a sweep or another nearly sinusoidal signal as an input. Nonlinear distortions, if present, are visible in the residual signal provided that the transfer function has been modeled correctly so that the linear distortions (errors in amplitude and phase) are small. Expressing the result in meaningful numbers is another problem. Linearity is not a simple concept in a system with a frequency-dependent response; the test conditions must be precisely defined. Referring the result to the input (acceleration) in place of the output (velocity) may have a large cosmetic effect.

Good electrical linearity unfortunately does not ensure that the system is equally linear for seismic signals. Mechanical forces in the sensor and nonlinear behavior of the force transducer may differ between an electrical and a mechanical experiment. Moreover, a sensor might respond in a nonlinear way to transverse accelerations that cannot be excited electrically. Unfortunately, it is nearly impossible to observe the mechanical nonlinearity of a force-balance seismometer on a shake table. Such experiments invariably end up as tests of the table with the seismometer as the reference. Even a small tilt component in the motion of the table can produce stronger spurious signals than a suspected nonlinearity of the sensor. A seismometer may, however, be considered as sufficiently linear when it produces undisturbed long-period records during local earthquakes (Wielandt and Streckeisen, 1982).

8.2 Measuring Instrumental Noise

Force-balance sensors cannot be tested for noise with the mass locked, so the instrumental noise can only be observed in the presence of seismic noise. This may nevertheless be useful as a first check; a broadband seismograph that resolves tides or free oscillations of the Earth cannot be too bad at long periods. For a quantitative assessment of instrumental noise, however, the two contributions must be separated from each other. The usual procedure is to perform a frequency-domain analysis of the coherency between the output signals of two seismometers, and to assume that coherent noise is seismic and incoherent noise is instrumental. This is, however, questionable because the two instruments may respond coherently to environmental disturbances such as caused by barometric pressure, temperature, the supply voltage, magnetic fields, vibrations, or electromagnetic waves. Nonlinear behavior (intermodulation) may produce coherent but spurious long-period signals. The reference instrument should therefore be one whose (relative) immunity to intermodulation and environmental conditions is known; otherwise two instruments of a different type or at least in a different setup and with different power supplies should be used, hoping that they will not respond to disturbances in a coherent manner.

The analysis for coherency is somewhat tricky in detail. When the transfer functions of both instruments are known, it is in fact theoretically possible to measure the seismic signal and the instrumental noise of each instrument separately as a function of frequency. However, the transfer functions must be known with great precision for this purpose. Alternatively, one may assume that the reference instrument is noise-free; in this case the noise and the transfer function of the other instrument can be determined. Long time-series are required for reliable results: the analysis is made in a set of narrow frequency bands, and the length of the record multiplied with the smallest bandwidth must be a large number. We offer a computer program UNICROSP for the analysis.

8.3 Transient Disturbances

Most new seismometers produce spontaneous transient disturbances, quasi miniature earthquakes caused by stresses in the mechanical components. Although they do not necessarily originate in the spring, their waveform at the output seems to indicate a sudden and permanent (steplike) change in the spring force. Long-period seismic records are sometimes severely degraded by such disturbances. The transients often die out within some months or years; if not, and especially when their frequency increases, corrosion must be suspected. Manufacturers try to mitigate the problem with a low-stress design and by aging the components or the finished seismometer (by extended storage, vibration, or alternate heating and cooling cycles). It is sometimes possible to virtually eliminate transient disturbances by hitting the pier around the seismometer with a hammer, a procedure that is recommended in each new installation.

9. Calibration

9.1 Electrical and Mechanical Calibration

The calibration of a seismograph establishes knowledge of the relationship between its input (the ground motion) and its output (an electric signal), and is a prerequisite for a reconstruction of the ground motion. Since precisely known ground motions are difficult to generate, one makes use of the equivalence between ground accelerations and external forces on the seismic mass [Eq. (2)] and calibrates seismometers with an electromagnetic force generated in a calibration coil. If the factor of proportionality between the current in the coil and the equivalent ground acceleration is known, then the calibration is a purely electrical measurement. Otherwise, the missing parameter—either the transducer constant of the calibration coil, or the responsivity of the sensor itself—must be determined from a mechanical experiment in which the seismometer is subjected to a known mechanical motion or a known tilt. This is called an absolute calibration. Since it is difficult to generate precise mechanical calibration signals over a large bandwidth, one does normally not attempt to determine the complete transfer function in this way.

The present section is mainly concerned with the electrical calibration, although the same methods may also be used for the mechanical calibration on a shake table. Specific procedures for mechanical calibration without a shake table are presented in Section 10.

9.2 General Conditions

Calibration experiments are disturbed by seismic noise and tilt, and should therefore be carried out in a basement room. However, the large operating range of modern seismometers permits calibration with relatively large signal amplitudes, making background noise less of a problem than one might expect. Thermal drift is more serious because it interferes with the long-period response of broadband seismometers. These should be protected from draft and allowed sufficient time to reach thermal equilibrium. Visual and digital recording in parallel is recommended. Recorders must themselves be absolutely calibrated before they can serve to calibrate seismometers. The input impedance of recorders as well as the source impedance of sensors should be measured so that a correction can be applied for the loss of signal in the source impedance.

9.3 Calibration of Geophones

Simple electrodynamic seismometers (geophones) have no calibration coil. The calibration current must then be sent through the signal coil. There it produces an ohmic voltage in addition to the output signal generated by the motion of the mass. The simplest way of circumventing this difficulty is to excite the geophone by interrupting a constant current through

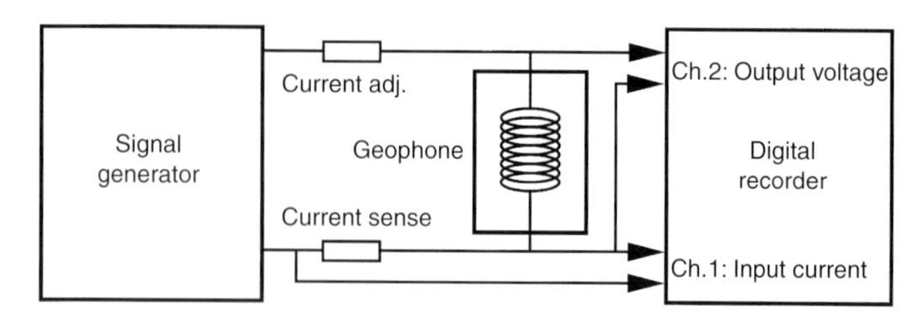

FIGURE 9 Half-bridge circuit for calibrating geophones.

the coil, and to evaluate the resulting transient response either graphically or with any suitable program for system identification (Rodgers *et al.*, 1995). The method will also supply the generator constant when the size of the mass is known and its motion is linear. When other calibration signals are used, the undesired voltage can be compensated in a bridge circuit (Willmore, 1959); the bridge is zeroed with the geophone clamped or inverted. However, when both the calibration current and the output voltage are digitally recorded, it is more convenient to use only a half-bridge (Fig. 9) and to compensate the ohmic voltage numerically. The program CALEX decribed below has provisions to do this automatically.

Eigenfrequency and damping of geophones (and of most other seismometers) can be determined graphically with a set of standard resonance curves on doubly-logarithmic paper. The measured amplitude ratios are plotted on the same type of paper and overlaid with the standard curves (Fig. 10). The desired quantities can be read directly. The method is simple but not very precise.

Geophones whose seismic mass moves along a straight line require no mechanical calibration when the size of the mass is known. The electromagnetic part of the numerical damping is inversely proportional to the total damping resistance [Eq. (13)]; the factor of proportionality is $\sigma^2/2M\omega_0$, so the generator constant σ can be calculated from electrical calibrations with different resistive loads.

9.4 Calibration with Sine Waves

With a sinusoidal input, the output of a linear system is also sinusoidal, and the ratio of the two signal amplitudes is the amplitude response (the modulus of the complex frequency response). An experiment with sine waves therefore permits an immediate check of the response, without any a priori knowledge of its mathematical form and without waveform modeling; this is often the first step in the identification of an unknown system. A computer program would, however, be required for converting discrete values of the response function into a parametric representation; a calibration with arbitrary signals as described later is more straightforward for this purpose.

When only analog equipment is available, the calibration coil or the shake table may be driven with a sinusoidal test signal and the input and output signals recorded with a chart recorder or an X–Y recorder. On the latter, the signals should

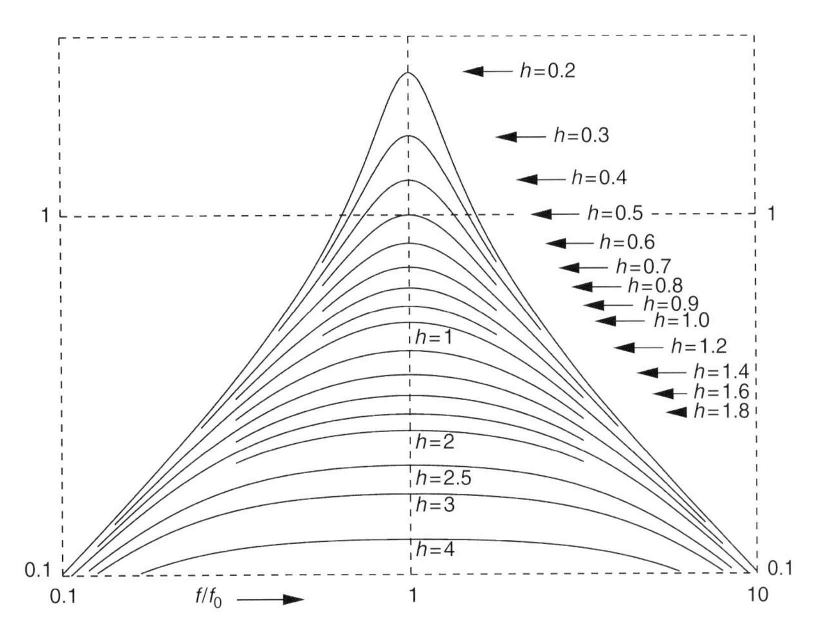

FIGURE 10 Normalized resonance curves.

be plotted as a Lissajous ellipse from which both the amplitude ratio and the phase can be read with good accuracy. The signal period should be measured with a counter or a stop watch because the frequency scale of sine wave generators is often inaccurate.

The accuracy of the graphic evaluation depends on the purity of the sine wave. A better accuracy can of course be obtained with a numerical analysis of digitally recorded data. By fitting sine waves to the signals, amplitudes and phases can be extracted for just one precisely known frequency at a time; distortions of the input signal then do not matter. For best results, the frequency should be fitted as well, the fit should be computed for an integer number of cycles, and offsets should be removed from the data. A FORTRAN program "SINFIT" is offered for this purpose (Section 11).

9.5 Calibration with Arbitrary Signals

The purpose of calibration is in most cases to obtain the parameters of an analytic representation of the transfer function. Assuming that its mathematical form is known, the task is to determine its parameters from an experiment in which both the input and the output signals are recorded. This is a classical inverse problem that can be solved with standard least-squares methods.

The general concept is as follows. A computer algorithm is implemented that represents the seismometer as a filter and permits the computation of its response to an arbitrary input. An inversion algorithm is programmed around the filter algorithm in order to find best-fitting parameters for a given pair of input and output signals. A calibration experiment is then made with a test signal for which the response of the system is sensitive to the unknown parameters but which is otherwise arbitrary. When the system is linear, parameters determined from one test signal will also predict the response to any other signal.

The approximation of a rational transfer function with a discrete filtering algorithm is not trivial (see Section 3.4). For the program CALEX that accompanies this article, we have chosen an impulse-invariant recursive filter after Schuessler (1981). The method formally requires that the seismometer has a negligible response at frequencies outside the Nyquist bandwidth of the recorder, a condition that is severely violated by most digital seismographs; but this problem can be circumvented with an additional sharp low-pass filtration within the program. Figure 11 shows signals from a typical calibration experiment with CALEX. The test signal is a sweep, which permits the residual error to be visualized as a function of frequency even when the whole algorithm works in the time domain. When the transfer function has been correctly parametrized and the inversion has converged, then the residual error consists mainly of noise, drift, and nonlinear distortions. At a signal level of about one-third of the operating range, typical residuals are 0.03% to 0.05% rms for force-balance seismometers and $\geq 1\%$ for passive electrodynamic sensors.

With an appropriate choice of the test signal, other methods such as calibration with sine waves, step functions, random noise, or random telegraph signals can be duplicated and compared to each other. An advantage of the CALEX algorithm is that it makes no use of special properties of the test signal, such as being sinusoidal, periodic, steplike, or random. Therefore, test signals can be short (a few times the free period of the seismometer), and they can be generated with the most primitive means, even by hand. A breakout box or a special cable may, however, be required for feeding the calibration signal into the digital recorder. Alternative routines for

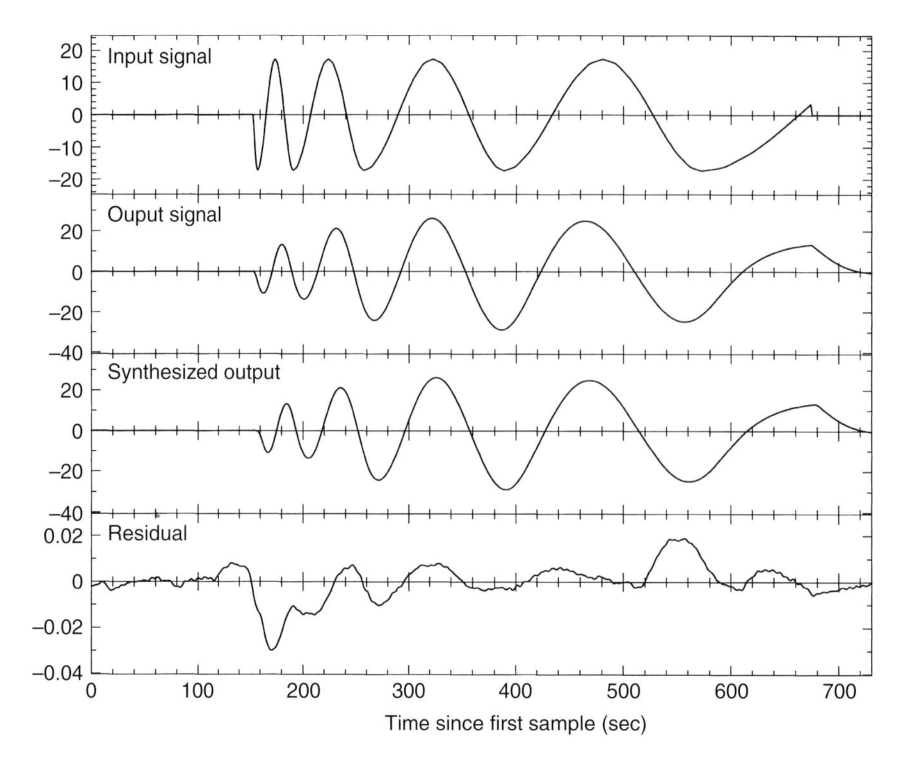

FIGURE 11 Electrical calibration of a STS2 seismometer with CALEX. Traces from top to bottom: input signal (a sweep with a total duration of 10 min); observed output signal; modelled output signal; residual. The rms residual is 0.05% of the rms output.

seismograph calibration are contained in the free PREPROC software package (Plešinger *et al.*, 1996; see Section 11).

9.6 Calibration of Triaxial Seismometers

In a triaxial seismometer such as the STS2 (Fig. 5), transfer functions in a strict sense can only be attributed to the individual *U*, *V*, *W* sensors, not to the *X*, *Y*, *Z* outputs. Formally, the response of a triaxial seismometer to arbitrary ground motions is described by a nearly diagonal 3×3 matrix of transfer functions relating the *X*, *Y*, *Z* output signals to the *X*, *Y*, *Z* ground motions. (This is also true for conventional three-component sets if they are not perfectly aligned; only the composition of the matrix is slightly different.) If the *U*, *V*, *W* sensors are reasonably well matched, the effective transfer functions of the *X*, *Y*, *Z* channels have the traditional form and their parameters are weighted averages of those of the *U*, *V*, *W* sensors. The *X*, *Y*, *Z* outputs can therefore be calibrated as usual. For the simulation of horizontal and vertical ground accelerations via the calibration coils, each sensor must receive an appropriate portion of the calibration current. For the vertical component this is approximately accomplished by connecting the three calibration coils in parallel. For the horizontal components and also for a more precise excitation of the vertical, the calibration current must be split into three individually adjustable and invertible components. These are then adjusted so that the signal appears only at the desired output of the seismometer.

10. Procedures for the Mechanical Calibration

10.1 Calibration on a Shake Table

Using a shake table is the most direct way of obtaining an absolute calibration. In practice, however, precision is usually poor outside a frequency band roughly from 0.5 to 5 Hz. At higher frequencies, a shake table loaded with a broadband seismometer may develop parasitic resonances, and inertial forces may cause undesired motions of the table. At low frequencies, the maximum displacement and thus the signal-to-noise ratio may be insufficient, and the motion may be nonuniform due to friction or roughness in the bearings. Still worse, most shake tables do not produce a purely translational motion but also produce some tilt. This has two undesired side-effects: the angular acceleration may be sensed by the seismometer, and gravity may be coupled into the seismic signal (see Section 2.4). The latter effect can be catastrophic for the horizontal components at long periods since the error increases with the square of the signal period. One might think that a tilt of 10 μrad per mm of linear motion should not matter; however such a tilt will, at a period of 30 sec, induce seismic signals twice as large as those originating from the linear motion. At a period of 1 sec, the same tilt would be negligible. Long-period measurements on a shake table, if possible at all, require extreme care.

Although all calibration methods mentioned in the previous section are applicable on a shake table, the preferred method would be to record both the motion of the table (as measured with a displacement transducer) and the output signal of the seismometer, and to analyze these signals with CALEX. Depending on the definition of active and passive parameters, only the absolute gain (responsivity, generator constant) or any number of additional parameters of the transfer function may be determined.

10.2 Calibration by Stepwise Motion

The movable tables of machine tools like lathes and milling machines, and of mechanical balances, can replace a shake table for the absolute calibration of seismometers. The idea is to place the seismometer on the table, let it come to equilibrium, then move the table manually by a known amount and let it rest again. The total motion can then be calculated from the seismic signal and compared to the actual mechanical displacement. Since the calculation involves triple integrations, offset and drift must be carefully removed from the seismic trace. The main contribution to drift in the apparent horizontal "ground" velocity comes from tilt associated with the motion of the table. With the method subsequently described, it is possible to separate the contributions of displacement and tilt from each other so that the displacement can be reconstructed with good accuracy. This method of calibration is most convenient because it uses only normal workshop equipment; the inherent precision of machine tools and the use of relatively large displacements eliminate the problem of measuring small displacements. A FORTRAN program named DISPCAL is available for the evaluation.

The precision of the method depends on minimizing errors from two sources:

1. The numerical restoration of ground displacement from the seismic signal (a process of inverse filtration) is noncritical for broadband seismometers but requires a precise knowledge of the transfer function of short-period seismometers. Instruments with unstable parameters must be electrically calibrated while installed on the test table. However, when the response is known, the restitution of absolute ground motion is no problem even for a geophone with a free period of 0.1 sec.
2. The effect of tilt can only be removed from the displacement signal when the motion is sudden and short. The tilt is unknown during the motion, and is integrated twice in the calculation of the displacement. So the longer the interval of motion, the more effect the unknown tilt will have on the displacement signal. Practically, the motion may last about one second on a manually operated machine tool, and about a quarter-second on a mechanical balance. It may be repeated at intervals of a few seconds.

Static tilt before and after the motion produces linear trends in the velocity that are easily removed before the integration. The effect of tilt during the motion can, however, only approximately be removed by interpolating the trends before and after the motion. The computational evaluation consists in the following major steps (Fig. 12):

1. The trace is deconvolved with the velocity transfer function of the seismometer.
2. The trace is piecewise detrended so that it is close to zero in the motion-free intervals; interpolated trends are removed from the intervals of motion.
3. The trace is integrated.
4. The displacement steps are measured and compared to the actual motion.

In principle, a single steplike displacement is all that is needed. However, the experiment takes so little time that it is convenient to produce a dozen or more equal steps, average the results, and do some error statistics. On a milling machine or lathe, it is recommended to install a mechanical device that stops the motion after each full turn of the spindle. On a balance, the table is repeatedly moved from stop to stop. Its amplitude may be measured with a micrometer dial but may also be determined from the lever ratio, which in turn is obtained from the ratio of equivalent weights.

From the mutual agreement between a number of different experiments, and from the comparison with a calibration on a precise shake table, the absolute accuracy of the method is estimated to be better than 1%.

10.3 Calibration with Tilt

Accelerometers can be statically calibrated on a tilt table. Starting from a horizontal position, the fraction of gravity coupled into the sensitive axis equals the sine of the tilt angle. (A tilt table is not required for accelerometers with an operating range exceeding $\pm 1g$; these are simply turned over.) Force-balance seismometers normally have a mass-position output that is a slowly responding acceleration output. This output can, with some patience, likewise be calibrated on a tilt table; the small static tilt range of sensitive broadband seismometers may, however, be inconvenient. The transducer constant of the calibration coil is then obtained by sending a direct current through it and comparing its effect with the tilt calibration. Finally, by exciting the coil with a sine wave whose acceleration equivalent is now known, the absolute calibration of the broadband output is obtained. The method is not explained in more detail here because a simpler method exists. In any

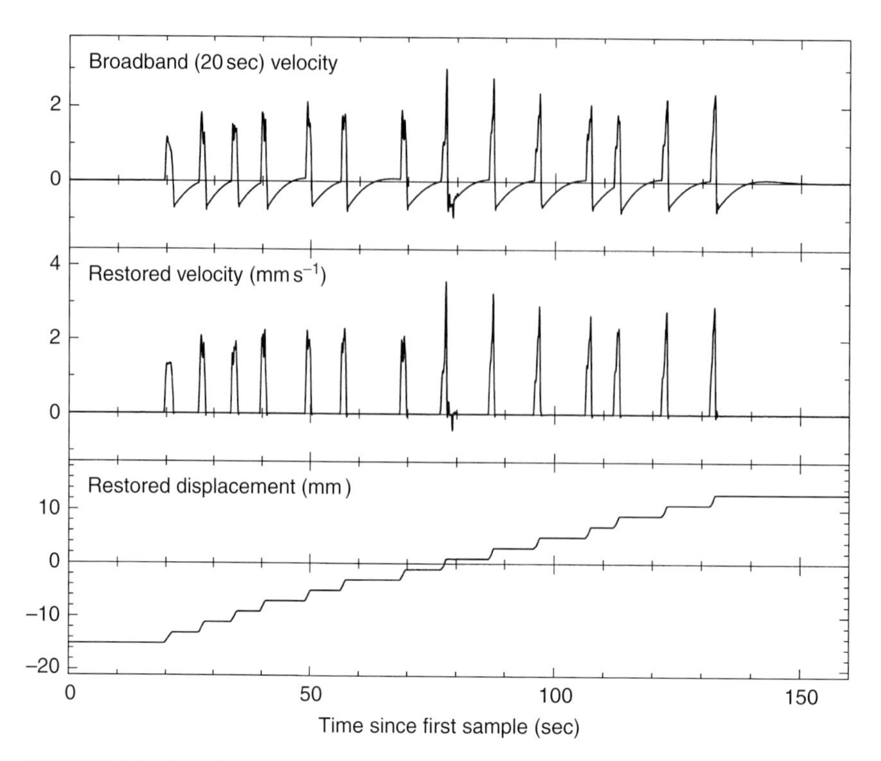

FIGURE 12 Absolute mechanical calibration of an STS1-BB seismometer on the table of a milling machine, evaluated with DISPCAL. Traces from top to bottom: recorded BB output signal; restored and detrended velocity; restored displacement.

case, triaxial seismometers cannot be calibrated in this way because they do not have X, Y, Z mass-position signals.

The method we propose (for horizontal components only; program TILTCAL) is similar to what was described in Section 10.2, but this time we excite the seismometer with a known step of tilt and evaluate the recorded output signal for acceleration rather than displacement. This is simple because we only have to look at the drift rate of the deconvolved velocity trace before and after the step; no baseline interpolation is involved. In order to produce repeatable steps of tilt, it is useful to prepare a small lever by which the tilt table or the seismometer can quickly be tilted forward and back by a known amount. The tilt may be larger than the static operating range of the seismometer; one then has to watch the output signal and reverse the tilt before the seismometer goes to a stop.

11. Free Software

FORTRAN source code of six computer programs mentioned in the text is included on the attached Handbook CD. Both source and executable code can be downloaded from the author's FTP site (see below). These are standalone programs for calibrating and testing seismometers and for standardizing noise data; they do not form a package for general seismic processing such as SAC, SEISMIC UNIX, PITSA, or PREPROC (see below). A README file with explanations, a set of test data, and output files are provided with each program. No graphics are included. The data files have the following format:

- A header line (text only).
- A line in the format (i10, a20, f10.x) that contains the number of samples, their format, and the sampling interval.
- Samples in ASCII.

It is suggested that users modify the input and output routines to conform with their own preferred data format. The programs are as follows:

- CALEX: Determines parameters of the transfer function of a seismometer from the response to an arbitrary input signal (which must be recorded together with the output signal). The transfer function is implemented in the time domain as an impulse-invariant recursive filter. The inversion uses the method of conjugate gradients (moderately efficient but quite failsafe).
- DISPCAL: Determines the generator constant of a horizontal or vertical seismometer from an experiment in which the seismometer is moved stepwise on the table of a machine tool or a mechanical balance.
- NOISECON: Converts noise specifications into all kind of standard and nonstandard units and compares them to the USGS New Low Noise Model (Peterson, 1993).

Interactive program available in BASIC, FORTRAN, or C, and as a DOS-Executable.
- SINFIT: Fits sine waves to a pair of sinusoidal signals and determines their frequency and the relative amplitude and phase.
- TILTCAL: Determines the generator constant of a horizontal seismometer from an experiment in which the seismometer is tilted stepwise.
- UNICROSP: Estimates seismic and instrumental noise separately from the coherency of the output signals of two seismometers.

ftp server: ftp.geophys.uni-stuttgart.de (141.58.73.149)
name: ftp
password: your e-mail address
directory: /pub/ew

Author's address:
Erhard Wielandt
Institute of Geophysics
University of Stuttgart
Richard-Wagner-Str. 44
D-70184 Stuttgart, Germany
e-mail: ew@geophys.uni-stuttgart.de

Free Seismic Software Packages from Other Sources

- SAC: http://www.llnl.gov/sac/SAC_Info_Install/Availability.html
- SEISMIC UNIX: http://www.cwp.mines.edu/cwpcodes/index.html
- PITSA: http://www.uni-potsdam.de/u/Geowissenschaft/Software/haupt_software.html
- PREPROC: ftp://orfeus.knmi.nl/pub/software/mirror/preproc/index.html

If you cannot find these websites, try

- http://www.seismolinks.com/Software/Seismological.htm
- http://orfeus.knmi.nl/other.services/software.links.shtml

References

Agnew, D.C., J. Berger, R. Buland, W. Farrel, and F. Gilbert (1976). International deployment of accelerometers—a network for very long period seismology. *EOS* **57**(4), 180–188.

Agnew, D.C., J. Berger, W.E. Farrell, J.F. Gilbert, G. Masters, and D. Miller (1986). Project IDA: A decade in review. *EOS*, **67**(16), 203–212.

Anderson, J.A. and H.O. Wood (1925). Description and theory of the torsion seismometer. *Bull. Seismol. Soc. Am.* **15**(1), 72.

Beauduin, R., P. Lognonné, J.P. Montagner, S. Cacho, J.F. Karczewski, and M. Morand (1996). The effects of the atmospheric pressure changes on seismic signals. *Bull. Seismol. Soc. Am.* **86**(6), 1760–1769.

Benioff, H. and F. Press (1958). Progress report on long period seismographs. *Geophys. J. R. Astron. Soc.* **1**(1), 208–215.

Berlage, H.P. Jr. (1932). Seismometer. In: "Handbuch der Geophysik," Vol. 4 (B. Gutenberg, Ed.) Chap. 4, pp. 299–526. Gebrueder Borntraeger Verlag, Berlin.

Block, B. and R.D. Moore (1970). Tidal to seismic frequency investigations with a quartz accelerometer of new geometry. *J. Geophys. Res.* **75**(18), 4361–4375.

Bracewell, R.N. (1978). "The Fourier Transformation and its Applications," 2nd edn. McGraw-Hill, New York.

Buttkus, B. (Ed.) (1986). Ten years of the Graefenberg array. In: "Geologisches Jahrbuch Reihe E." Bundesanstalt fuer Geowissenschaften und Rohstoffe, Hannover.

de Quervain, A. and A. Piccard (1924). Beschreibung des 21-Tonnen-Universal-Seismographen System de Quervain–Piccard. *Jahresber. Schweiz. Erdbebendienst*, **1924**.

de Quervain, A. and A. Piccard (1927). Déscription du séismographe universel de 21 tonnes systéme de Quervain–Piccard. *Publ. Bureau Centr. Séismol. Int. Sér. A*, **4**(32).

Dewey, J. and P. Byerly (1969). The early history of seismometry (to 1900). *Bull. Seismol. Soc. Am.* **59**(1), 183–227.

Ewing, J.A. (1884). Measuring earthquakes. *Nature* **30**, 149–152, 174–177.

Galitzin, B. (1914). "Vorlesungen über Seismometrie." B.G. Täubner, Leipzig and Berlin.

Harjes, H.-P. and D. Seidl (1978). Digital recording and analysis of broadband seismic data at the Graefenberg array. *J. Geophys.* **44**, 511–523.

Holcomb, L.G. and C.R. Hutt (1992). An evaluation of installation methods for STS-1 seismometers. *Open File Report 92–302, US Geological Survey*, Albuquerque, NM.

IRIS (1985). "The Design Goals for a New Global Seismographic Network." Incorporated Research Institutions for Seismology, Washington, DC.

Jones, R.V. and J.C.S. Richards (1973). The design and some applications of sensitive capacitance micrometers. *J. Phys. E: Sci. Instrum.* **6**, 589–600.

LaCoste, L.J.B. (1934). A new type long period seismograph. *Physics* **5**, 178–180.

Lehner, F.E. (1959). An ultra long-period seismograph galvanometer. *Bull. Seismol. Soc. Am.* **49**(4), 399–401.

Melton, B.S. (1981a). Earthquake seismograph development: A modern history—part 1. *EOS* **62**(21), 505–510.

Melton, B.S. (1981b). Earthquake seismograph development: A modern history—part 2. *EOS* **62**(25), 545–548.

Melton, B.S. and B.M. Kirkpatrick (1970). The symmetric triaxial seismometer—its design for application in long-period seismometry. *Bull. Seismol. Soc. Am.* **60**(3), 717–739.

Miller, W.F. (1963). The caltech digital seismograph. *J. Geophys. Res.* **68**(3), 841–847.

Oliver, J. and L. Murphy (1971). WWNSS: Seismology's global network of observing stations. *Science* **174**, 254–261.

Oppenheim, A.V. and R.V. Schafer (1975). "Digital Signal Processing." Prentice-Hall International, London.

Oppenheim, A.V. and A.S. Willsky (1983). "Signals and Systems." Prentice-Hall, Englewood Cliffs, NJ.

Peterson, J. (1993). Observations and modelling of background seismic noise. *Open File Report 93–322, US Geological Survey*, Albuquerque, NM.

Peterson, J., H.M. Butler, L.G. Holcomb, and C.R. Hutt (1976). The seismic research observatory. *Bull. Seismol. Soc. Am.* **66**(6), 2049–2068.

Peterson, J., C.R. Hutt, and L.G. Holcomb (1980). Test and calibration of the seismic research observatory. *Open File Report 80–187, US Geological Survey*, Albuquerque, NM.

Plešinger, A. and J. Horalek (1976). The seismic broad-band recording and data processing system FBS/DPS and its seismological applications. *J. Geophys.* **42**, 201–217.

Plešinger, A., M. Zmeškal, and J. Zednik (1996). "Automated Preprocessing of Digital Seismograms: Principles and Software" (E. Bergman, Ed.). Prague and Golden, Prague.

Press, F., M. Ewing, and F. Lehner (1958). A long-period seismograph system. *Trans. Am. Geophys. Union* **39**(1), 106–108.

Riedesel, M.A., R.D. Moore, and J.A. Orcutt (1990). Limits of sensitivity of inertial seismometers with velocity transducers and electronic amplifiers. *Bull. Seismol. Soc. Am.* **80**(6), 1725–1752.

Robinson, E.A. and S. Treitel (1980). "Geophysical Signal Analysis." Prentice-Hall, New York.

Rodgers, P.W. (1968). The response of the horizontal pendulum seismometer to Rayleigh and Love waves, tilt, and free oscillations of the Earth. *Bull. Seismol. Soc. Am.* **58**(5), 1385–1406.

Rodgers, P.W. (1969). A note on the response of the pendulum seismometer to plane wave rotation. *Bull. Seismol. Soc. Am.* **59**(5), 2101–2102.

Rodgers, P.W. (1992). Frequency limits for seismometers as determined from signal-to-noise ratios. Part 1: The electromagnetic seismometer. *Bull. Seismol. Soc. Am.* **82**(2), 1071–1098.

Rodgers, P.W. (1993). Maximizing the signal-to-noise ratio of the electromagnetic seismometer: The optimum coil resistance, amplifier characteristics, and circuit. *Bull. Seismol. Soc. Am.* **83**(2), 561–582.

Rodgers, P.W., A.M. Martin, M.C. Robertson, M.M. Hsu, and D.B. Harris (1995). Signal-coil calibration of electromagnetic seismometers. *Bull. Seismol. Soc. Am.* **85**(3), 845–850.

Savino, J.M., A.J. Murphy, J.M.W. Rynn, *et al.* (1972). Results from the high-gain long-period seismograph experiment, *Geophys. J. R. Astron. Soc.* **31**(1), 179–203.

Scherbaum, F. (1996). "Of Poles and Zeros, Fundamentals of Digital Seismology." Kluwer Academic, Boston.

Schuessler, H.W. (1981). A signal processing approach to simulation. *Frequenz* **35**, 174–184.

Tanimoto, T. (1999). Excitation of normal modes by atmospheric turbulence: source of long-period seismic noise. *Geophys. J. Int.* **136**(2), 395–402.

Trnkoczy, A. (1998). "Guidelines for Civil Engineering Works at Remote Seismic Stations," Application Note 42. Kinemetrics Inc., 222 Vista Av., Pasadena, CA. 91107.

Uhrhammer, R.A. and W. Karavas (1997). "Guidelines for Installing Broadband Seismic Instrumentation," Tech. Rep. Seismographic station, University of California at Berkeley, http://www.seismo.berkeley.edu/seismo/bdsn/instrumentation/guidelines.html.

Usher, M.J., C. Guralp, and R.F. Burch (1978). The design of miniature wideband seismometers. *Geophys. J.* **55**(3), 605–613.

Von Rebeur-Paschwitz, E. (1889). The earthquake of Tokyo, April 18, 1889. *Nature* **40**, 294–295.

Wielandt, E. (1975). Ein astasiertes Vertikalpendel mit tragender Blattfeder. *J. Geophys.* **41**(5), 545–547.

Wielandt, E. (1983). Design principles of electronic inertial seismometers. In: "Earthquakes: Observation, Theory and Interpretation," pp. 354–365. LXXXV Corso, Soc. Italiana di Fisica, Bologna.

Wielandt, E. and T. Forbriger (1999). Near-field seismic displacement and tilt associated with the explosive activity of Stromboli. *Ann. Geofis.* **42**(2/3).

Wielandt, E. and J.M. Steim (1986). A digital very-broad-band seismograph. *Ann. Geophys.* **4B**(3), 227–232.

Wielandt, E. and G. Streckeisen (1982). The leaf-spring seismometer: Design and performance. *Bull. Seismol. Soc. Am.* **72**(6), 2349–2367.

Willmore, P.L. (1959). The application of the Maxwell impedance bridge to the calibration of electromagnetic seismographs. *Bull. Seismol. Soc. Am.* **49**(1), 99–114.

Zürn, W. and R. Widmer (1995). On noise reduction in vertical seismic records below 2 mHz using local barometric pressure. *Geophys. Res. Lett.* **22**(24), 3537–3540.

Editor's Note

The source code of the software described in Section 11 of this chapter is placed on the attached Handbook CD, under the directory of \18Wielandt. A brief summary of the software is given in Chapter 85.18, along with other software packages available on the attached CD. Please see also Chapter 22, Analysis of Digital Earthquake Signals, by Scherbaum.

19

Seismic Noise on Land and on the Sea Floor

Spahr C. Webb
Lamont-Doherty Earth Observatory, Columbia University, New York, USA

1. Introduction

The study of earthquakes or the imaging of the Earth using seismic arrivals requires detecting arrivals above background noise. Differences in noise level between sites within the global seismic network can exceed 60 dB in some frequency bands (see Fig. 1), representing differences in detection threshold for seismic arrivals of four or more magnitudes. The noise level is therefore a very important criterion for choosing sites for seismic stations. The noise level at a site depends both on the quality of the installation and on the natural sources of noise close to and far from the site. This review does not attempt to summarize the many studies of the effect of installation techniques on station quality, but does discuss the impact of borehole installation on signal-to-noise ratios for seismic arrivals both on land and under the sea. Sites poorly coupled to the ground or on poorly consolidated ground will tend to be noisy.

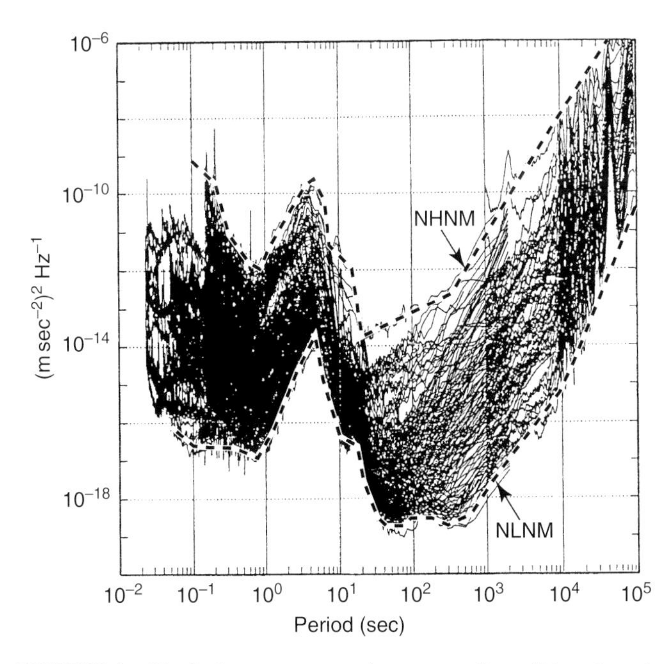

FIGURE 1 Vertical component noise spectra from 75 stations in the global network. The envelope of these spectra defines the new high- and low-noise models (NHNM and NLNM). (Reproduced with permission from Peterson, 1993.)

The perception of the seismic background depends on how one plots the noise spectra. The noise spectrum at every site is very "red" if plotted in displacement, whereas in acceleration the noise spectra have a minimum near 0.04 Hz: "a window for earthquakes" (Savino *et al.*, 1972). Most recent authors have chosen to plot noise spectra in acceleration, minimizing the apparent range. The "microseism peak" near 0.15 Hz (6 sec period) is of overwhelming importance in determining seismic noise levels. The term "microseism" is used variously to describe either any small background disturbance observed on a seismogram, or more usually this noise near 6 sec period. Short period (<1 sec) noise is mostly attributable to man-made ("cultural" sources), whereas long period (>10 sec) noise is almost always associated with natural sources: wind, waves, and water movement.

The frequency band of relevance for most seismology extends from the gravest normal mode of Earth ($\approx$ 0.3 mHz) to an upper limit set by attenuation ranging from 20 Hz for teleseismic events to 100 Hz or more for very local studies. This review discusses seismic noise in the band from the tides (0.02 mHz) to 50 Hz. The energetic microseism peak separates the world of seismology into its traditional short-period and long-period bands. It is usually necessary to filter data into bands either above or below this peak to clearly perceive arrivals above this noise. Until recently, seismometers were constructed as either "long-period" or "short-period" instruments with little response in the microseism band. This review continues the usual practice, and discusses the short-period and long-period bands separately.

Figure 1 shows spectra calculated using records from 75 broadband seismic stations from the global Network (Peterson, 1993). Spectral levels differ by 60 dB or more

INTERNATIONAL HANDBOOK OF EARTHQUAKE AND ENGINEERING SEISMOLOGY, VOLUME 81A ISBN: 0-12-440652-1

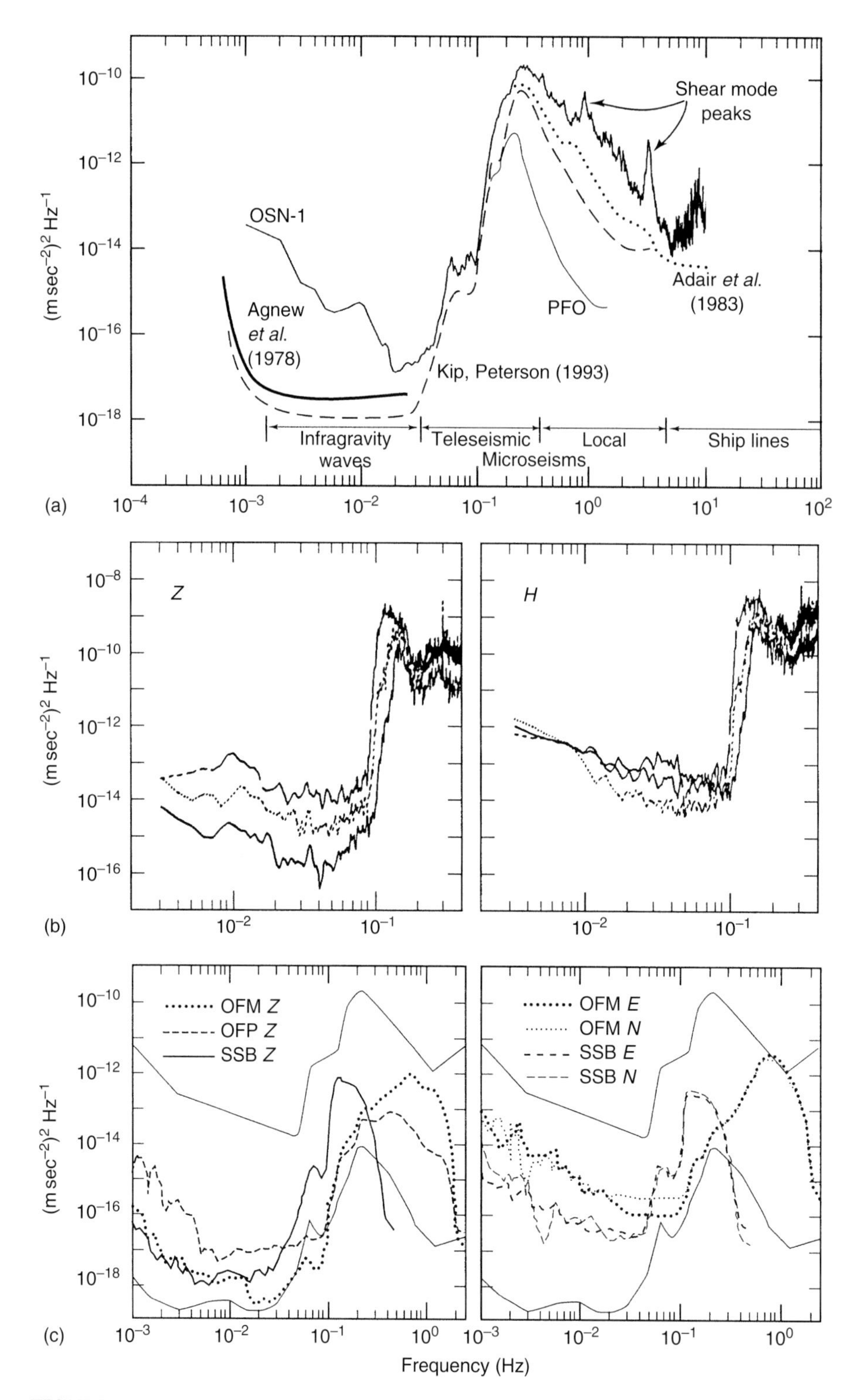

FIGURE 2 (a) Broadband sea floor vertical component noise spectra from Hawaii (Webb, 1998); (b) three-component noise spectrum from offshore of central California (Sutton and Barstow, 1990) and (c) from the central North Atlantic from the seafloor (OFM) and from a sea-floor borehole (OFP) and the station SSB. (Reproduced with permission from Beauduin and Montagner, 1996.)

between the best and worst sites in most bands. The upper and lower bounds of this cloud of noise spectra have been used to derive high- and low-noise model curves, called respectively NHNM and NLNM for "new high-" or "new low-noise models." It should be remembered that nearly all the spectra shown in this report are from preselected data. Data have been winnowed to remove sections with obvious instrument noise problems, atypical noise bursts, and seismic events. The preselection is a recognition that every site will be useless for seismology during some intervals. The spectra shown here are supposed to reflect obtainable noise levels during normal operation.

Although the last decade has seen a resurgence in efforts to put permanent observatories on the sea floor to extend the global arrays to these otherwise inaccessible regions, broadband seismic observations from the sea floor remain rare. Recent results suggest undersea measurements will usually be much noisier than land observations. For this reason, significant improvements in global detection thresholds will only be obtained near the sea-floor stations. The primary motivation for installing stations on the sea floor should be to improve the global ray path coverage for tomographic studies of the Earth and for regional studies of seismicity and structure.

A composite spectrum of short- and long-period vertical component data from the sea floor south of Hawaii shows noise levels considerably higher than at most continental sites at all frequencies (Fig. 2a). The long-period vertical component spectrum from another Pacific sea-floor site is similar, but the horizontal component spectra are two orders of magnitude noisier (Fig. 2b; Sutton and Barstow, 1990). The long-period vertical component spectrum (Fig. 2c) from a site in the North Atlantic is much quieter than these Pacific sites, with noise levels approaching those of good continental sites (Beauduin and Montagner, 1996). This probably reflects a real difference in noise level between these ocean basins, at least in summer. However, the horizontal component noise levels are almost as noisy as the Pacific sites.

2. Long-Period Noise and Long-Period Arrivals (<0.1 Hz)

2.1 Tides and Low-Frequency Ocean Waves (0.01 mHz–0.1 Hz)

The earth tides are below the frequencies normally used in seismology, but provide a framework for a discussion of deformation of the Earth under ocean waves and atmospheric pressure fluctuations. Tides have routinely been observed with gravimeters (long-period seismometers) and tiltmeters for decades. The apparent acceleration due to the tides has three components: (1) the large signal caused by the gravitational attraction of the Sun and Moon, (2) variations in gravity with changes in the distance to the center of the Earth from deformation, and (3) accelerations caused directly by deformation under tidal forces. The tidal signals recorded at seismometers near coastlines or on the sea floor are further affected by ocean tidal loading. Ocean tidal loading also has three components: (1) the gravitational attraction of a changing mass of sea water adjacent to or overhead of a site, (2) variations in apparent gravity due to deformation under the time-varying load of sea water, and (3) accelerations due directly to this deformation (e.g., Lambert, 1970). The direct attraction of the sea water dominates the apparent acceleration signal at the sea floor. The ocean tidal loading signal is rarely in phase with the earth tide, and the observed earth tide at coastal stations will usually be quite different from the theoretical earth tide without ocean loading. Models of ocean tides and Earth structure are sufficiently refined that it is now possible to predict accurately the diurnal and semidiurnal earth and load tides for most sites (Agnew, 1995).

The ocean tides can be thought of as ocean surface gravity waves forced at tidal frequencies. The ocean wave spectrum extends from the periods of tides to periods much shorter than 1 sec. Waves longer than about 25 s (0.04 Hz) are called infragravity waves and the pressure signal from these waves can be observed at any ocean floor site (Filloux, 1980). Wave amplitudes in the infragravity wave band from 0.002 to 0.04 Hz rarely exceed 1 cm in deep water, but the deformation due to infragravity wave loading controls vertical acceleration noise at sea-floor sites in the Pacific in this band (Webb, 1998). Infragravity waves are typically much less energetic in the North Atlantic and may be irrelevant for sea floor seismic observations during at least the summer months in that basin (Beauduin and Montagner, 1996; Webb, 1998).

The problem of deformation of the sea floor under the loading of infragravity waves has been studied extensively (Crawford *et al.*, 1999), and the deformation has been used to determine crustal structure at ridge crests. The pressure signal from ocean waves is evanescent in water depth with an e-folding distance equal to the wavenumber, so that higher-frequency waves are observed at the sea floor in shallower water. Significant pressure fluctuations are observed below about 0.03 Hz in 4 km water depth, and below about 0.045 Hz in 1 km water depth. The amplitude of the deformation increases rapidly with wave frequency so that the infragravity wave deformation signal is much larger in shallow water (Webb and Crawford, 1999). Recent observations of long-period noise off Monterey, California illustrated the twin problems of making long-period measurements on the sea floor near the coastline. Even at 1 km water depth, the infragravity deformation signal dominated the vertical component noise levels below 0.05 Hz, while strong ocean-floor currents overwhelmed the horizontal component observations even though the seismometer was mostly buried below the sea floor (Romanowicz *et al.*, 1998).

Infragravity wave amplitudes become large near coastlines where wave energy becomes trapped as edge waves to the shoreline, but because wavelengths also become short in shallow water, the deformation signal does not penetrate far inland. Agnew and Berger (1978) show that noise levels near 0.01 Hz are high at coastal and island sites due to loading by edge waves. Edge waves are often a problem for sites on small islands and atolls. Borehole installations designed to optimize long-period performance may prove ineffective on atolls (G. Holcomb, personal communication, 1999) partly because of deformation noise from edge waves.

Recent work has shown that low-order normal modes of the Earth are persistently excited by some unknown mechanism. It is possible that edge waves could be the source for this excitation, although recent literature has associated this signal with atmospheric turbulence (Tanimoto, 1999). Infragravity waves coupling to seismic modes could also be the source of the broad spectral peak near 120 sec at some global stations noted by Peterson (1993). The infragravity wave signal otherwise fades quickly inland.

2.2 Atmospheric Pressure Variations, Winds and Currents; Tilt Noise

Ground noise on land at very long period (0.2 to 50 mHz) is usually associated with atmospheric pressure fluctuations. At these very low frequencies, vertical component seismometers react to changes in gravity as the mass of the atmosphere above a site changes with atmospheric pressure. At higher frequencies, the ground deforms under the fluctuating pressures associated with turbulence in the boundary layer (Murphy and Savino, 1975). Seismometers that are not fully sealed may also experience apparent accelerations due to buoyancy effects on the seismometer mass, and pressure fluctuations can drive long-period horizontal component noise by distorting the case of the seismometer or the walls of the vault (Savino *et al.*, 1972). Long-period seismometers are invariably sensitive to temperature changes and most installations include a means to seal the sensor from pressure fluctuations and some kind of insulating box to restrict the rate of change of temperature seen by the seismometer.

Zurn and Widmer (1995) show that vertical component noise levels between 0.2 and 1.7 mHz can be reduced by more than 10 dB by subtracting the scaled, locally recorded pressure signal from the vertical acceleration record, thereby improving the signal-to-noise ratio for normal mode observations. The method reduces noise levels at the station studied to below the Peterson low-noise curve (Fig. 1). Earth tide observations can also be improved using a similar correction (Warburton and Goodkind, 1977). At these long periods, the apparent vertical acceleration is caused by the gravitational attraction of the varying mass of the atmosphere above a site, which is reflected in the changes in barometric pressure.

At shorter periods, turbulence in the atmospheric boundary layer produces pressure fluctuations at the Earth's surface that cause significant deformation to depths a few tens of meters below the surface. This noise source is a primary reason that seismometers at permanent stations are now usually installed beneath the Earth's surface in shallow (100–300 m deep) boreholes. For instruments on the surface, short-wavelength fluctuations produce a large tilt signal that makes the horizontal components particularly noisy.

Turbulent eddies are advected over the ground at velocities comparable to the wind velocity near the top of the atmospheric boundary layer. The characteristic period associated with an eddy is the size (wavelength) of the eddy divided by advection velocity. The e-folding length of decay of the deformation into the ground is about equal to the wavelength of the pressure signal over 2π. A typical wind speed is $10\,\mathrm{m\,s^{-1}}$, for which the e-folding scale is 100 m at 0.016 Hz. Short-wavelength deformation signals are quickly filtered out, leading to large improvements in long-period horizontal component noise levels at shallow depths in boreholes. Noise levels for sensors in boreholes at depths of a few hundred meters become comparable to near-surface sites during calm days, leading Sorrels and Douze (1974) to propose that the remaining noise is due to atmospheric infrasound.

The analogues of wind on the ocean floor are ocean currents. These too set up a turbulent boundary layer with advecting pressure fluctuations (Webb, 1988). The deformation signal decays completely within a few meters depth below the sea floor because much lower velocities are associated with ocean currents than with wind (hence much shorter wavelengths).

Horizontal component seismometers are sensitive to tilt as the acceleration associated with gravity is rotated into each horizontal component. Agnew (1986) provides a review of tiltmeter measurements in the seismic band. It is not possible to discriminate tilt from horizontal acceleration using inertial sensors. For seismometers in a vault, tilt noise is associated with local deformation of the ground due to propagating pressure disturbances or with wind forces on nearby structures (trees, buildings, etc.). The apparent horizontal acceleration A_h is associated with ground tilt $\boldsymbol{\Omega}$ as $\boldsymbol{A}_h = g\boldsymbol{\Omega}$. Noise levels in the band between 0.001 and 0.05 Hz of 10^{-16} $(\mathrm{m\,sec^{-2}})^2\,\mathrm{Hz^{-1}}$ correspond to a root mean square acceleration of about $2 \times 10^{-9}\,\mathrm{m\,sec^{-2}}$ or root mean square tilt of 2×10^{-10}. This tilt corresponds to 1 mm in 4000 km, a very small tilt indeed.

Long-period horizontal component ambient noise levels in surface vaults are usually 10–30 dB above the vertical component levels measured at the same place due to tilt noise. Spectra from shallow (15 m) boreholes are considerably noisier than those from deeper (100 m) boreholes (Peterson *et al.*, 1976). Recent practice has placed three component seismometers in 100–300 m deep boreholes to reduce tilt noise. Murphy and Savino (1975) report that the effect of atmospheric pressure fluctuations on horizontal component noise is reduced by more than 90% at a depth of 150 m. Observations from deep continental boreholes show comparable long-period noise levels on horizontal and vertical components (Fix, 1972; Murphy and Savino, 1975).

It is usually possible on land to bury sensors below the surface to avoid the direct force of wind even for temporary installations. In contrast, ocean-floor instruments are typically dropped on the ocean floor and thus directly exposed to currents. Current-generated noise on the sea floor is usually caused by the direct action of current on the unshielded instrument (Duennebier and Sutton, 1995). Horizontal component noise levels on sea-floor sensors can be as much as 60 dB higher than vertical component noise levels at periods near 100 s due to tilt (Webb, 1998). Tilt noise will appear on vertical component records as well if the vertical component sensor is not very precisely oriented with the true vertical because it will include a horizontal component equal to the sine of the error in orientation. An error of only 0.5° represents a leakage of 1 part in 100 in amplitude between horizontal and vertical components. Maintaining a difference of 60 dB requires orienting the vertical component with the true vertical to better than 0.06°. Most vertical component records from the sea floor show evidence of tilt noise due to currents (including probably the spectra in Fig. 2). A post-recording rotation of the sensor axes using three component data can reduce long-period sea floor vertical component noise levels by 10–20 dB (Crawford and Webb, 2000). The differences between vertical and horizontal noise levels on land are usually much smaller because installations are better and because pressure-induced ground deformations are of longer wavelength and therefore produce smaller tilts. The typical half-degree accuracy in verticality associated with most seismic installations is sufficient to prevent significant leakage of tilt noise into the vertical. Vertical noise generated by atmospheric pressure fluctuations at land sites is mostly due directly to vertical motions of the ground.

Shallow-water sites can be intensely noisy because of strong currents (e.g., Romanowicz *et al.*, 1998). The very short wavelengths associated with pressure fluctuations due to ocean-floor currents suggest that sensors buried a few meters below the ground will be as quiet at long period as sensors at much greater depths (Webb, 1998). Burial of a seismometer at 2 m depth below the sea floor was found to be sufficient to reduce long-period horizontal component noise levels by about 60 dB compared to a surface instrument (Vernon *et al.*, 1998), but incomplete burial of the sensors at a site offshore of Monterey was insufficient to shield a long-period seismometer from strong ocean currents there (Romanowicz *et al.*, 1998). To date, instruments installed in ocean-floor boreholes have been noisier than or comparable to surface sites at long period (Fig. 2c), probably because of convection of sea water in the borehole (Beauduin and Montagner, 1996; Vernon *et al.*, 1998).

Tests on land show that horizontal component noise levels are more than 20 dB higher for seismometers in oil-filled boreholes compared to dry holes. The higher noise levels in the fluid-filled boreholes are attributed to convection currents. The geothermal gradient always provides a temperature gradient that can drive convection. Horizontal component noise levels vary greatly between borehole sites even for dry boreholes. Holcomb *et al.* (1998) found that horizontal component noise levels at noisy holes could be substantially reduced by packing the borehole sensor into the borehole using sand, rather than using the standard locking mechanism. The sand prevents convection in the borehole. They suggest that a test of a good installation is that the long-period horizontal component levels should be comparable to vertical noise levels.

2.3 Single-Frequency Microseism Peak

The single-frequency, or primary, microseism peak occurs as a small peak in the spectrum between 0.05 and 0.1 Hz, just below the main microseism peak in frequency (see Figs. 1–3). This peak is associated directly with ocean waves. Energy is transferred between ocean waves and seismic waves through non-linear interaction of ocean waves and bathymetry (Hasselmann, 1963). Measurements from instruments in coastal regions show spectral peaks that track ocean wave frequencies incident at the coast, although this energy decays rapidly inland (Haubrich and McCamy, 1969). The peak is called the single-frequency peak because it mimics swell frequencies without the frequency doubling observed for the main microseism (or double-frequency) peak. Single-frequency amplitudes are much larger at coastal sites than observed either in the center of continents or on the deep ocean floor. The single-frequency peak seen on Taiwan is 20 dB larger than observed at Pacific sea-floor sites, and also of higher frequency because the lower-frequency swell components are attenuated traveling across the broad shelf around the island (Hedlin and Orcutt, 1989).

In the center of continents or of an ocean basin, the single-frequency peak is related to storm waves on remote coastlines (Cessaro, 1994). An analysis of LASA array from Montana shows energy is primarily carried as fundamental mode Rayleigh waves, although there is often a component of Love wave energy (Lacoss *et al.*, 1969). Most of the energy at LASA originated to the northwest of the array, although other directions were occasionally observed. The spectral amplitude of the single-frequency peak ranges from about 10^{-16} $(\mathrm{m\,sec}^{-2})^2\,\mathrm{Hz}^{-1}$ to 10^{-15} $(\mathrm{m\,sec}^{-2})^2\,\mathrm{Hz}^{-1}$ for the majority of continental sites (Fig. 1). The spectra deviating from this range tend to be coastal or island sites in Figure 1. The relative stability in amplitude for continental sites around the globe suggests persistent, stable sources. Holcomb (1998) reports on a persistent peak in the noise spectrum at 26 sec period that appears on seismometers worldwide. The source of this peak is unknown, but signals are largest during the southern winters and the amplitude of the peak is correlated between sites worldwide, although frequently barely above the background noise level. Smaller peaks of slightly different frequencies may also be present.

2.4 Detection Thresholds for Surface Waves and Long-Period Body Waves

It is evident that noise levels are sufficiently low at most seismic sites to allow the detection of seismic arrivals from distant, large earthquakes. Many of the normal modes of the Earth can be detected at the quieter sites from the largest events. Signal-to-noise ratios are estimated by comparing the root mean square (rms) noise levels with model signal amplitudes in 1/3 octave bands (Fig. 3a). Signal levels are estimated for a source range of 30° in this figure taken from Agnew *et al.* (1986). The 1/3 octave bands represent a useful and achievable bandwidth using digital filters. Signals are detectable in any band where the signal curves significantly exceed noise levels. The solid lines indicate surface wave and body wave amplitudes. The dashed curves extending the surface wave curves represent normal mode amplitudes. Estimates of maximum and minimum earth noise at global seismic stations are shown on the figure.

These curves are very approximate. Surface wave spectra vary greatly with source mechanism and depth. In general, surface waves from shallow events are barely detectable near 0.04 Hz at most sites from events exceeding about $M_W = 4.5$. The detection of lower-frequency surface waves and normal modes requires larger events. Body waves can be detected at frequencies below the microseism peak from events exceeding about $M_W = 4.5$ at most sites. Smaller events can be detected at smaller distances (Fig. 3b; from Heaton *et al.*, 1989).

Spectra from stations on the sea floor suggests similar detection thresholds for surface waves near 25 sec period and for long-period body waves as on land (Webb, 1998). The detection thresholds for pressure measurements for body waves and surface waves in the band from 0.05 to 0.1 Hz are also similar to vertical acceleration detection thresholds. However, horizontal component seismic data from the sea floor are invariably much noisier than vertical component data below 0.03 Hz because of tilt noise. This limits the detection of Love waves and *SH* body waves to periods shorter than about 25 sec. (Shallow burial of the sensor packages should make horizontal component noise levels more comparable with vertical component noise levels and improve detection thresholds enormously.) Vertical component noise levels below 0.03 Hz on the Pacific sea floor are noisier than at most continental sites because of deformation under infragravity waves. This makes detection of long-period (>100 sec) surface waves and normal modes difficult at typical sea-floor stations. Lower infragravity wave amplitudes in the North Atlantic may provide better SNR for these phases, at least in the quiet summer months. Webb and Crawford (1999) show that sea floor pressure data can be used to remove most of the infragravity wave deformation signal from vertical component data in the band from 0.001 to 0.04 Hz, leading to improvements of 10–20 dB in SNR for long-period surface waves recorded at sea-floor stations in the Pacific.

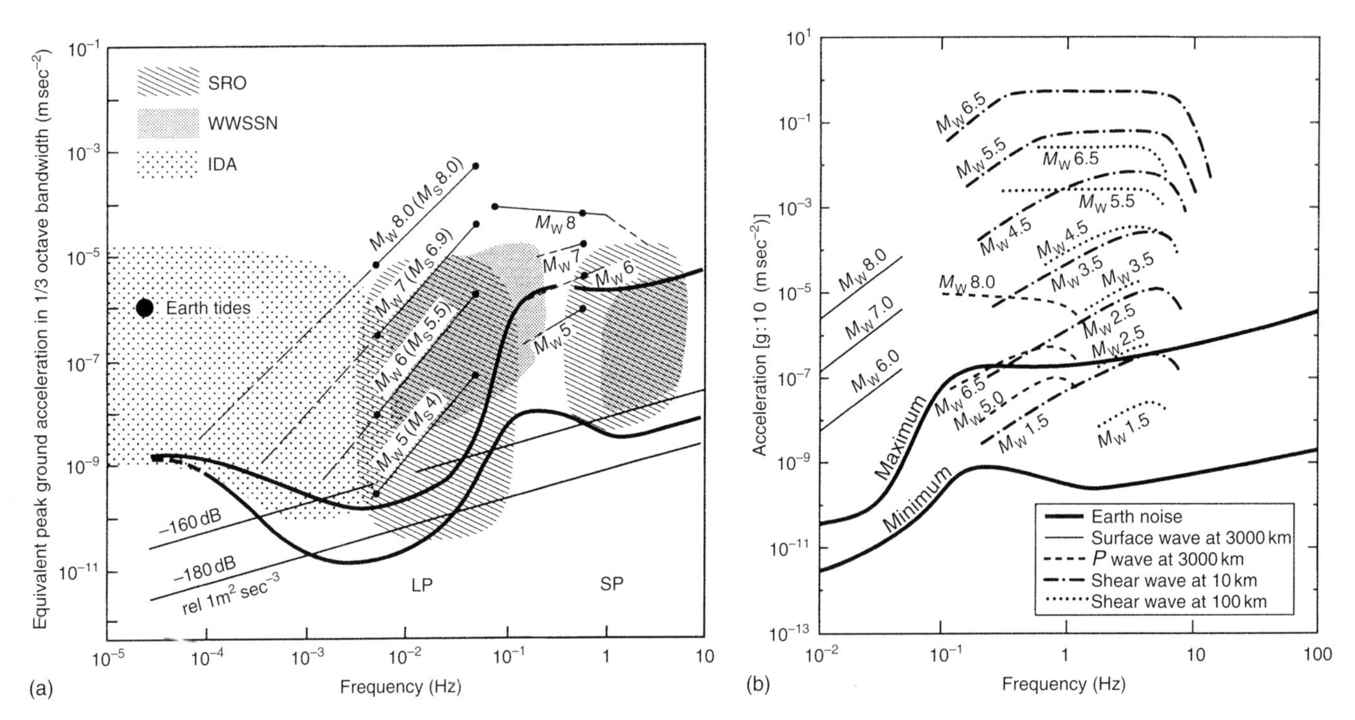

FIGURE 3 (a) Amplitude of seismic arrivals and vertical component noise levels in 1/3 octave bands against low- and high-noise models for Earth noise illustrating typical detection thresholds for surface waves and body waves for arrivals from teleseismic distances. (From Agnew *et al.*, 1986.) The dynamic range of several networks is shown. (b) Short-period noise levels and amplitude of seismic arrival amplitudes at regional and local distances. (Reproduced with permission from Heaton *et al.*, 1989.)

3. The Microseism Peak (0.1–0.5 Hz)

Wind couples energy first into capillary waves on the ocean surface. The ocean wave spectrum evolves toward longer period and higher amplitudes with fetch and with time as nonlinear mechanisms transfer energy toward longer periods. The wind climate and geography of the ocean basins establishes the ocean wave climate. Ocean waves couple into ground motion through several different mechanisms; some are local whereas others couple into elastic waves that travel throughout the Earth. Ocean waves are ultimately the primary source of seismic noise observed on broadband seismometers at continental, island, and sea-floor stations. The microseism peak near 0.15 Hz is generated by the interaction of ocean waves at sea and is prominent in all seismic noise spectra (Figs. 1 and 2).

Ocean waves couple energy into the elastic waves called "microseisms" through a well-understood nonlinear mechanism (Longuet-Higgins, 1950; Hasselmann, 1963). Two ocean waves of similar wavelength and frequency but traveling in nearly opposing directions interact to force an acoustic wave with a wavenumber equal to the difference in wavenumbers and a frequency equal to the sum of frequencies. The ocean waves can couple into an elastic wave of much larger wavelength and higher phase velocity if the propagation directions are nearly opposing because the difference in wavenumbers can be arbitrarily small. The acoustic wave in the ocean couples into elastic waves in the sea bed. There is an approximate frequency doubling as ocean wave energy couples in microseisms at half the wave period. Microseisms propagate primarily as Rayleigh waves in the Earth, but some energy is present as body waves, and heterogeneous Earth structure couples some energy into Love waves.

Microseisms on seismometers worldwide are mostly between 5 and 9 sec period and are the result of energetic ocean waves between 10 and 18 sec period. The limited scale of ocean storms and ocean basins limits the major components of the ocean wave spectrum to periods shorter than about 25 sec, with energetic waves longer than 20 sec fairly rare. In all noise spectra there is a very abrupt change in spectral levels at 0.1 Hz (Fig. 2).

Microseisms in this band propagate long distances as Rayleigh waves, so the noise level at any site depends on the interaction of ocean waves beneath storms and along coastlines all over the world. A typical e-folding scale for attenuation at these periods is thousands of kilometers (amplitudes are also affected by bathymetry and interaction with boundaries such as coastlines). The amplitude of the microseism peak at a particular site depends on its location compared to microseism sources and attenuation along the raypaths from those sources (Webb, 1992). Storms at sea act as prominent microseism sources that can be tracked from land.

4. Short-Period Noise and Short-Period Arrivals (>0.5 Hz)

4.1 Short-Period Band (0.5–10 Hz)—Microseisms

The microseism mechanism dominates sea floor noise up to frequencies of a few hertz. At slightly higher frequencies (5–10 Hz), noise associated with breaking waves may become important (McCreery *et al.*, 1993). The major difference between observations in the microseism band on the sea floor and on land is that there is almost always much more energy at frequencies near 1 Hz on the ocean floor than on land, often by more than 40 dB. (There is little difference in microseism spectral levels between land and the sea floor at longer periods.) This huge difference in spectral levels near 1 Hz has a profound effect on the relative detection thresholds for short-period teleseismic arrivals between the sea floor and on land.

It has been known for a long time that it is advantageous to site seismometer stations away from the coast to avoid the effects of coastal microseisms. A ranking of stations based on noise levels near 0.8 Hz shows a separation between coastal and interior sites (Peterson, 1993). Noise levels at 1 Hz at interior sites range from about 3×10^{-17} to about 5×10^{-14} $(\mathrm{m\,sec^{-2}})^2\,\mathrm{Hz}^{-1}$, while noise levels at coastal sites range from 10^{-13} to 3×10^{-12} $(\mathrm{m\,sec^{-2}})^2\,\mathrm{Hz}^{-1}$ (Fig. 1). Noise levels at the sea floor are yet again higher than at coastal sites most of the time, because of microseisms driven by ocean waves directly overhead ($>10^{-11}$ $(\mathrm{m\,sec^{-2}})^2\,\mathrm{Hz}^{-1}$ Fig. 2). Short-period noise levels on land can also be affected by storm waves on nearby large lakes. Only during exceptional periods when extensive regions of calm winds surround the site can noise levels at sea-floor sites approach noise levels at the typical island site at short period (Webb, 1998).

Noise levels observed at station Kipapa on the island of Oahu are some 30 dB higher than the Peterson low-noise model at 1 Hz, but still 10–15 dB quieter than the sea-floor site south of Hawaii (Fig. 2). Microseisms with frequencies higher than about 0.3 Hz do not propagate efficiently from the oceans onto land, and once on land they attenuate quickly. This noise propagates in the deep ocean primarily as acoustic modes and most of the energy must be reflected or dissipated at the shelf edge before reaching coastal sites. Wilcock *et al.* (1999) found noise levels at 1 Hz diminished by about $0.1\,\mathrm{db\,km^{-1}}$ with distance inland across Iceland, consistent with propagation of the microseism noise as surface wave modes. Stations near the center of the island of Hawaii were 10–15 dB quieter than the Kipapa station close to the coast on Oahu (Li *et al.*, 1994). Noise levels near 1 Hz on Kipapa were sufficiently high to severely limit the detection of short-period phases. In contrast, most of the longer-period energy (0.1–0.3 Hz) is transmitted onto shore, so that island sites tend to only be slightly quieter than sea-floor sites in the main microseism peak. Coastal stations in Greenland are considerably noisier in the summer than in winter because the winter pack ice around Greenland prevents the local generation of microseisms by waves near the coast. In winter, the microseisms at these stations are dominated by distant storm sources (Harben and Hjortenberg, 1993).

4.2 Short-Period Noise (0.5–50 Hz)—Land

Seismic noise levels on land at frequencies above 0.5 Hz depend primarily on wind and on cultural noise, although the movement of water may also contribute. Every site has different characteristics depending on its distance from the ocean and from cultural sources, the wind climate of the site, and depth of burial of the sensor. A seasonal cycle is evident at some sites due to variations in wind, or water flow in nearby rivers. Spring runoff raises noise levels by up to 15 dB between 0.5 and 15 Hz at the NORESS array (Fyen, 1990). Necessary distances for siting stations away from moving water range from 60 km for very large waterfalls and dams to 15 km for smaller rapids (Wilmore, 1979).

High-frequency spectra during quiet intervals are quite similar among the quieter sites. The spectrum tends to be featureless under these conditions. The displacement spectrum follows a power-law dependence in frequency as $1/f^2$ (e.g., Bungum *et al.*, 1985). In acceleration the spectrum rises toward higher frequency as f^2, until limited by attenuation or the coupling of the system to the ground. Wind noise is often the dominant source from 0.5 to about 5 Hz. Wind noise obviously depends on the strength of the wind and the character of the site. It is not unusual to see 20 dB increases in noise level near 1 Hz with changes from calm conditions to winds of 5 m sec^{-1} on instruments in surface vaults. The primary mechanism by which wind couples into high-frequency noise is probably by the direct action of the wind on trees, bushes, and other structures, although at lower frequency the pressure fluctuations associated with wind can directly drive motions of the ground. Wind becomes a detectable source of short-period noise for sites far from cultural sources at wind velocities above about 3 m sec^{-1} at surface sites and at 4 m sec^{-1} for subsurface sensors (Withers *et al.*, 1996).

The term cultural noise is used to describe any seismic noise associated with man or man-made machinery: power plants, factories, trains, highways, and even cattle. Cultural noise is in principle avoidable, although it is often impractical to site stations at sufficient distances away from cities or highways. Some rules of thumb suggest that it may be necessary to site stations as far as 25 km from power plants or rock-crushing machinery, 15 km from railways, 6 km from highways, and a kilometer or more from smaller roads (Wilmore, 1979). Noise travels further across competent rock than through alluvial filled valleys and "safe" distances may be from 1/2 to 2/3 of the values above depending on propagation path (Wilmore, 1979).

Cultural noise varies greatly between sites (e.g., Fig. 1). Noise levels can vary by 30 dB within a day and from day to day, so it is common to present ensembles of spectra by percentile ranking to allow discrimination between infrequent very loud sources and more typical background levels (e.g., Bungum *et al.*, 1985). Noise spectra often show a distinct diurnal or weekly character (e.g., Given, 1990). Noise levels are lowest in the early morning in data from the NORESS array, rising rapidly near 07:00 to levels sustained throughout the working day, to fall gradually through the evening hours (Fyen, 1990). Noise levels are lowest on weekends but increase during holidays because of higher traffic.

Sources such as power plants are narrowband, producing energetic lines in the spectrum at 50 or 60 Hz and harmonics and subharmonics (e.g., 30 Hz, 25 Hz) depending on location. The NORESS array data included lines at 6, 12, and 17 Hz due to reciprocating saws in a nearby sawmill (Fyen, 1990). It may be possible to remove some narrowband noise through application of digital or analog filters. Other types of machinery may also produce narrowband peaks, but the frequencies of the lines may shift with time as motor speed varies. These noise sources are more difficult to remove.

Comparisons of noise spectra from surface vault installations and borehole installations have repeatedly shown significant improvements in SNR for high-frequency phases from borehole installations at depths as shallow as a few meters (e.g., Aster and Shearer, 1991). The signal-to-noise ratio for high-frequency arrivals can be optimized by choosing sites with a lossy blanket of weathered near-surface material (Withers *et al.*, 1996). Large reductions in noise level may be seen in the first 5 m between the weathered rock of the surface and the more competent rock in the borehole. The frequencies of local seismic arrivals are slightly higher in boreholes because attenuation is less, which may also enhance the SNR.

Wind noise couples into open boreholes and covering open boreholes can reduce wind noise considerably; the wind may drive pressure fluctuations that propagate down the borehole. Most comparisons of borehole and vault noise levels show 10–30 dB lower noise levels in the boreholes, with most of the reduction in noise level occurring in the first 100 m (Young *et al.*, 1996). Broadband spectral levels fall with increasing depth into the borehole, but narrow spectral lines due to cultural sources may show less depth dependence (e.g., Galperin *et al.*, 1978) and therefore be of increasing importance with depth. Wind-generated noise couples mostly into surface modes, but cultural noise must couple at least partly into body waves that can propagate to depth (Carter *et al.*, 1991). The difference in noise levels between vault and borehole sensors tends to be largest when data from poorly prepared surface sites are compared with borehole data. Given (1990) found only a factor of 2 improvement in high-frequency SNR for a shallow borehole at a site in the former USSR compared with a well-prepared surface vault.

4.3 Short-Period Noise (0.5–50 Hz)—Sea Floor

The sea floor in the band from 0.5 to 5 Hz is almost invariably noisier than the typical land site because of microseisms generated by small-amplitude, high-frequency ocean waves that can be produced by even quite light winds. Locally generated microseisms maintain sea-floor noise levels some 40 dB above land sites most of the time, but noise levels may fall significantly during long intervals of calm winds.

Noise levels above 1 Hz are directly related to wind and wave conditions over the site, although noise levels "saturate" with little increase in noise levels above a saturation wind velocity that is frequency-dependent (McCreery *et al.*, 1993). Saturation occurs because ocean wave spectra saturate at high frequency, leading to a constant rate of excitation of microseisms, which is balanced by a constant rate of dissipation of the microseisms within a fairly small area (Webb, 1992). The noise spectrum approaches saturation near 1 Hz for a wind velocity of only about 5 m sec^{-1} (10 knots). Records from several sites show relatively constant values of 1 Hz noise except during rare intervals when local winds drop to near calm conditions for at least a day. During these intervals, noise levels can drop by 20 dB or more (Webb, 1998; Wilcock *et al.*, 1999).

One expects to see winds exceeding 5 m sec^{-1} over most sites in the oceans most of the time, so the average sea-floor site is very noisy near 1 Hz most of the time. A notable exception is the Arctic ocean, where the ice cap prevents the excitation of the ocean surface waves that ultimately couple to noise in the microseism peak. The Arctic sea floor has been observed to be very quiet near 1 Hz during the winter (Webb and Schultz, 1992). Parts of the North Atlantic and Indian oceans can also experience long intervals of near calm conditions during the summer months and may provide quiet sites for detection of short-period body waves (Webb, 1998). Most sites in the Pacific seem to be noisy most of the time (McCreery *et al.*, 1993).

It is not obvious that installing seismometers within sea-floor boreholes will necessarily reduce microseism noise levels. Energy is coupled from ocean waves into acoustic waves and then into elastic waves below the sea bed. The wavelength of a 1 Hz acoustic wave in the ocean is 1.5 km, and the wavelengths of the Rayleigh wave modes below the sea bed are even larger. If only these modes were present, noise levels would not be expected to fall until depths of several kilometers were reached, beyond the present range of ocean drilling. However, some improvement in signal to noise is expected at shallow depths because microseisms are scattered by bathymetry and by topography at the sediment–basement interface into short-wavelength (Stoneley) shear modes (Bradley *et al.*, 1997). These slowly propagating shear modes are trapped to the sediment–water interface with displacements that decay away from the sea floor. Correlation lengths measured between elements of a small-aperture array of OBSs were between 100 and 200 m at frequencies above 0.4 Hz (Schreiner and Dorman, 1990), a result that requires significant energy in short-wavelength waves.

Data from seismometers locked at three depths in a borehole near the Bahamas show that the difference between sea-floor and borehole noise levels increases rapidly with frequency and depth, consistent with much of the noise at the sea floor propagating as Stoneley waves (Bradley *et al.*, 1997). There are obvious differences in the structure of borehole and ocean-bottom noise spectra. The deep spectra closely follow a power-law dependence on frequency with a slope of 80 dB/decade. In contrast, surface spectra show significant peaks in the band around a few hertz associated with the Stoneley modes.

The difference between the vertical noise level at 1 Hz on the sea floor and at depth increases from 6 dB at 10 m depth to 10 dB at 100 depth. At higher frequencies, the vertical spectra from the borehole sensors at 10, 70, and 100 m depth differ by less than 3 dB but are 10–15 dB quieter than at the sea floor. Signal levels at depth are also smaller (because of changing impedance) and the resulting improvement in signal-to-noise ratio is near zero at 0.2 Hz, increasing to more than 10 dB at 5 Hz. On the horizontal components, the decrease in noise levels is larger, reaching 15 dB at 70 m depth at 1 Hz and 20 dB at higher frequencies. Most of the improvement occurs in the first 10 m for frequencies above 1.5 Hz, but below 1 Hz the sensor at 10 m depth is only a few dB quieter than at the sea floor. The depth dependence at 0.3 Hz was consistent with propagation as fundamental mode Rayleigh waves with little difference in vertical and horizontal component noise levels between sea-floor and borehole sensors.

High-frequency noise (5–50 Hz) on the sea floor tends to be associated with man-made (cultural) sources. The study of noise in the sea has been much more systematic and extensive than on land as part of the field of ocean acoustics, where noise sources are often military targets. A good recent review of "low-frequency acoustic noise" (<100 Hz) can be found in Richardson *et al.* (1995). The important sources of noise transition from the microseism peak below 5 Hz, through wind-generated noise between 5 and 10 Hz, to primarily cultural (shipping) at higher frequencies. Wind noise becomes important again at much higher frequencies through wave breaking.

The important components of shipping noise include cavitation of propellers and noise from vibration of machinery. Distinct lines associated with generators (50 or 60 Hz) or with other machinery (motors, generators, thrusters and pumps, etc.) make up much of the shipping noise, with noise levels peaking between 10 and 100 Hz. The lowest-frequency components are associated with rotation of propellers. Large ships such as supertankers produce "blade lines" at frequencies of

7 Hz or lower, with many higher harmonics (Richardson *et al.*, 1995). Noise levels vary with intensity of shipping traffic. High-traffic areas can be 20–25 dB noisier than regions remote from shipping. The transition band (5–10 Hz), above the microseisms and below shipping noise, provides relatively quiet conditions. This allows detection of oceanic *Pn* and *Sn* phases (also known as *Po* and *So*) to great distances (Butler *et al.*, 1987) from moderate magnitude oceanic events.

Another low-frequency source for some locations is large whales. Blue and fin whales produce very loud calls at frequencies as low as 16 Hz that are easily seen locally on sea-floor seismometers. In some special locations such as the Gulf of Maine, whale calls can be so frequent as to interfere with active source seismic experiments. Other sources of noise in the seismic band include noise associated with the movement of ice in the polar regions, seismic sources (airguns), and offshore drilling (Richardson *et al.*, 1995).

4.4 Detection Thresholds for Short-Period Body Waves

Approximate detection thresholds for teleseismic (30°) and local short-period *P* waves for sea-floor and continental sites can be estimated from Figures 3 and 4. The amplitudes of short-period arrivals decrease with increasing range because of geometrical spreading, but detection thresholds also depend on the frequency content of the arrivals, which depends on attenuation. Attenuation increases with increasing range but is also path-dependent. The rapid decrease in microseism noise with increasing frequency above the microseism peak means that small changes in frequency content of arrivals may be associated with big differences in detection thresholds. For example, shear waves at teleseismic distances may be too low in frequency to be seen at all above the microseism peak, particularly at sea-floor stations where noise levels are high. Attenuation along the typical raypath for a compressional wave from a distant event to a continental station restricts the frequency content of the typical *P* wave to a few hertz. This is sufficiently high frequency to put the arrival above the significant energy in the microseism peak at a land site, but not at a sea-floor site.

Figure 3 suggests that detection thresholds at most land sites for *P* waves from distant events (>30°) will be roughly magnitude M_W 4.0, with slightly lower values for the quieter sites and much higher detection thresholds for the noisier island sites. This is in rough agreement with more careful estimates of detection thresholds for groups of stations. Recent estimates for the detection threshold for events globally using available seismic arrays range from about M_W 3.5 for heavily instrumented areas such as Europe to greater than M_W 4.5 for remote regions of the world's oceans (e.g., Kvaerna and Ringdal, 1999). Figure 3 suggests that all stations will detect *S* wave arrivals from events as small as about M_W 2.5 at close range (100 km). Efforts to detect small nuclear bomb tests

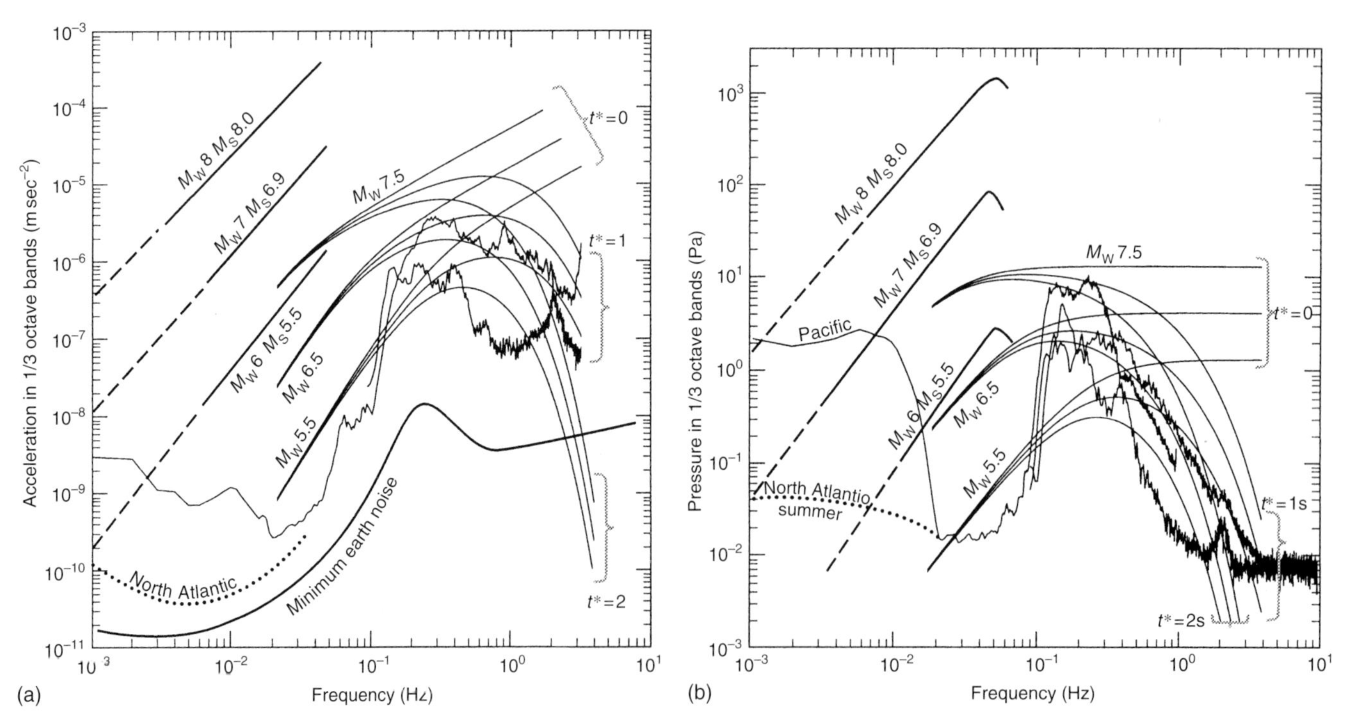

FIGURE 4 Same as Figure 3a for sea-floor sites for (a) vertical acceleration, and (b) pressure with noise levels from an atypically quiet (calm wind) interval and from a more typical noisy (windy) interval. Signal amplitudes are modeled for three different values of attenuation (t^*). High noise levels near 1 Hz make it difficult to detect short-period teleseismic body wave arrivals.

have focused on detecting high-frequency Pn and Sn arrivals at regional distances. Pn is detectable at frequencies up to 50 Hz at 500 km and at distances up to 1500 km at 10 Hz from magnitude M_L 3.0 events.

High noise levels near 1 Hz and high attenuation in the upper mantle (which removes the higher-frequency components of the P wave) conspire to make detection of short-period teleseismic P waves difficult at many sea-floor sites. Only short-period P waves from the very largest events ($M_W > 7$) can be detected (Blackman *et al.*, 1995) and then only with poor SNR. Wilcock *et al.* (1999) have made a systematic study of the wind statistics over the world's ocean-floor ridges to look for regions where long intervals of calm winds are more common. In these regions noise levels will often be 20 dB quieter than the typical site at saturation. The lower noise level is associated with detection thresholds several magnitude units smaller.

Figure 4 shows noise in 1/3 octave bands from a site on the sea floor during a typical noisy period and during one of the rare low-wind days from the MELT experiment. Teleseismic detection thresholds at the sea floor depend on attenuation below a site represented by the value of t^* shown on the model curves for p-wave amplitude for earthquakes of magnitudes 5.5, 6.5, and 7.5. The microseism noise spectrum falls quickly with increasing frequency, so detection thresholds are lower at low-attenuation sites because the P wave has more high-frequency energy. The model curves suggest that only the largest earthquakes ($M_W > 7.5$) will be detected, and then only with very poor SNR at typical noisy sea-floor sites near ridge crests (where attenuation is high). The detection threshold at sites on older oceanic crust, and for arrivals from deep earthquakes for which attenuation is less ($t^* = 1$) are lower ($M_W \approx 6$). Finally at sea-floor sites that see long intervals with calm conditions overhead, the detection threshold may fall as low as $M_W = 5.5$. Detection thresholds at local and regional distances for sites on the sea floor are much lower because the frequency content of the arrivals is above the microseism peak where noise levels are much lower. Local events detected during microearthquake studies on ridge crests range from M_W −1.5 to 2.0.

Lower noise levels within sea-floor boreholes lower detection thresholds for short-period body waves from the M_W 7.5 expected for noisy sea-floor sites (saturated microseism spectrum) to about M_W 6.5 and detection limits for quiet sea-floor noise conditions should be below M_W 5.5. Butler and Duennebier (1989) report on teleseismic earthquakes recorded by a borehole seismometer during a two-month recording period that included 15 nuclear explosions. Only the largest nuclear blast (5.6 m_b) was recorded. The authors note the importance of the high frequencies (6.5 Hz P wave) present in the blast seismogram in the detection of this event. The smallest earthquake observed was 5.4 m_b; all other teleseismic earthquakes seen were greater than 6 m_b. The station was on old oceanic crust with correspondingly low attenuation in the lithosphere so that high-frequency S waves were observed from regional events.

Orcutt and Jordan (1986) report on detection thresholds from a sensor in a borehole in the western Pacific. Detection limits increased from about 5.2 m_b at a distance of 40° to about 5.6 m_b at 80°. This is a site where light winds are expected more than 73% of the time, suggesting that noise levels should be low near 1 Hz. This low detection threshold is consistent with the low-noise model of Figure 4a combined with a 10 dB improvement in signal-to-noise ratio due to the borehole installation, particularly if the observations are associated with deep earthquakes from the subduction zones of the western Pacific, with signals propagating along low-attenuation paths. Regional earthquakes could be detected at a lower threshold (4.6 m_b) because of energy propagating as high-frequency oceanic Pn and Sn waves (also called Po and So). These phases can be detected at great distances (Butler *et al.*, 1987) from moderate magnitude oceanic events because the frequency of the arrival is above the microseism noise.

References Cited in the Text and Bibliography on Seismic Noise

Adair, R.G., J.A. Orcutt, and T.H. Jordan (1984). Analysis of ambient seismic noise recorded by downhole and ocean-bottom seismometers on deep sea drilling project leg 78B. *Initial Report Deep Sea Drill. Proj.* **78**, 767–780.

Agnew, D.C. (1986). Strainmeters and tiltmeters. *Rev. Geophys.* **24**(3), 579–624.

Agnew, D.C. (1995). Ocean load tides at the South Pole: a validation of recent ocean-tide models. *Geophys. Res. Lett.* **22**(22), 3063–3066.

Agnew, D.C. (1997). NLOADF: a program for computing ocean-tide loading. *J. Geophys. Res.* **102**(B3), 5109–5110.

Agnew, D.C. and J. Berger (1978). Vertical seismic noise at very low frequencies. *J. Geophys. Res.* **83**(B11), 5420–5424.

Agnew, D.C., J. Berger, R. Buland, W. Farrell, and F. Gilbert (1976). International deployment of seismometers: a network for very long period seismology. *EOS Trans. Am. Geophys. Union* **57**, 180–188.

Agnew, D.C., J. Berger, W.E. Farrell, J.F. Gilbert, G. Masters, and D. Miller (1986). Project IDA: a decade in review. *EOS Trans. Am. Geophys. Union* **67**(10), 203–212.

Aster, R.C. and P.M. Shearer (1991). High frequency seismograms recorded in the San Jacinto fault zone, Southern California, part 2. Attenuation and site effects. *Bull. Seismol. Soc. Am.* **81**(4), 1081–1100.

Babcock, J.M., B.A. Kirkendall, and J.A. Orcutt (1994). Relationship between ocean bottom noise and the environment. *Bull. Seismol. Soc. Am.* **84**(6), 1991–2007.

Bache, T.C., P.D. Marshall, and J.B. Young (1986). High frequency seismic noise characteristics at the four United Kingdom-type arrays. *Bull. Seismol. Soc. Am.* **76**(3), 601–616.

Beauduin, R. and J.-P. Montagner (1996). Time evolution of broad-band seismic noise during the French pilot experiment OFM/SISOBS. *Geophys. Res. Lett.* **23**(21), 2995–2998.

Blackman, D.J., J.A. Orcutt, and D.W. Forsyth (1995). Teleseismic detection using ocean bottom seismometers at mid-ocean ridges. *Bull. Seismol. Soc. Am.* **85**(6), 1648–1664.

Block, B. and R.D. Moore (1966). Measurements in the Earth mode frequency range by an electrostatic sensing and feedback gravimeter. *J. Geophys. Res.* **71**(18), 4361–4375.

Bradley, C.R. and R.A. Stephen (1996). Modeling of seafloor wave propagation and acoustic scattering in 3-D heterogeneous media. *J. Acoust. Soc. Am.* **100**(1), 225–236.

Bradley, C.R., R.A. Stephen, L.M. Dorman, and J.A. Orcutt (1997). Very low frequency (0.2–10 Hz) seismoacoustic noise below the seafloor. *J. Geophys. Res.* **102**(B6), 11703–11718.

Bungum, H., S. Mykkeltveit, and T. Kvaerna (1985). Seismic noise in Fennoscandia, with emphasis on high frequencies. *Bull. Seismol. Soc. Am.* **75**, 1489–1513.

Butler, K.E. and R.D. Russel (1993). Subtraction of powerline harmonics from geophysical records. *Geophysics* **58**(6), 898–903.

Butler, R. and F.K. Duennebier (1989). Teleseismic observations from OSS IV. *Initial Report Deep Sea Drill. Proj.* **88**, 147–153.

Butler, R., C.S. McCreery, L.N. Frazer, and D.A. Walker (1987). High frequency seismic attenuation of oceanic *P* and *S* waves in the western Pacific. *J. Geophys. Res.* **92**(B2), 1383–1396.

Carter, J.A., F.K. Duennebier, and D.M. Hussong (1984). A comparison between a downhole seismometer and a seismometer on the seafloor. *Bull. Seismol. Soc. Am.* **74**(3), 763–772.

Carter, J.A., N. Barstow, P.W. Pomeroy, E.P. Chael, and P.J. Leahy (1991). High frequency seismic noise as a function of depth. *Bull. Seismol. Soc. Am.* **81**(4), 1101–1114.

Chave, A.D., F.K. Duennebier, R. Butler, *et al.* (1998). H20: the Hawaii-2 observatory. *EOS Trans. Am. Geophys. Union* **79**(45), 65.

Cessaro, R.K. (1994). Sources of primary and secondary microseisms. *Bull. Seismol. Soc. Am.* **84**(1), 142–148.

Crawford, W.C. and S.C. Webb (2000). Removing tilt noise from low frequency (< 0.1 Hz) seafloor vertical seismic data. *Bull. Seismol. Soc. Am.* **90**(4), 952–963.

Crawford, W.C., S.C. Webb, and J.A. Hildebrand (1991). Seafloor compliance observed by long period pressure and displacement measurements. *J. Geophys. Res.* **96**, 16151–16160.

Crawford, W.C., S.C. Webb, and J.A. Hildebrand (1998). Estimating shear velocities in the oceanic crust from compliance measurements by two-dimensional finite difference modeling. *J. Geophys. Res.* **103**, 9895–9916.

Crawford, W.C., S.C. Webb, and J.A. Hildebrand (1999). Constraints on melt in the lower crust and Moho at the East Pacific Rise, 9°48′N, using seafloor compliance. *J. Geophys. Res.* **104**(B2), 2923–2939.

Dougherty, M.E. and R.A. Stephen (1991). Seismic energy partitioning and scattering in laterally heterogeneous ocean crust. *Pure Appl. Geophys.* **128**, 195–229.

Duennebier, F.K. and G.H. Sutton (1995). Fidelity of ocean bottom seismometers. *Mar. Geophys. Res.* **17**(6), 535–555.

Duennebier, F.K. *et al.* (1989). OSS IV: Noise levels, signal-to-noise ratios and sources. *Initial Report Deep Sea Drill. Proj.* **88**, 89–103.

Filloux, J.H. (1980) Pressure fluctuations on the open ocean floor over a broad frequency range: new program and early results. *J. Phys. Ocean. Ogr.* **10**, 1959–1971.

Fix, J.E. (1972). Ambient earth motion in the period range from 0.1 to 2560 s. *Bull. Seismol. Soc. Am.* **62**, 1753–1760.

Fyen, J. (1990). Diurnal and seasonal variations in the microseismic noise observed at the NORESS array. *Phys. Earth Planet. Inter.* **63**, 252–268.

Galperin, E.I., I.L. Nersesov, and R.M. Galperina (1978). "Borehole Seismology and the Study of the Seismic Regime of Large Industrial Centers." D. Reidel, Dordrecht.

Given, H. (1990). Variations in broadband seismic noise at IRIS/IDA stations in the USSR with implications for event detection. *Bull. Seismol. Soc. Am.* **80**(6), 2072–2088.

Gurrola, H., J.B. Minster, H. Given, F. Vernon, J. Berger, and R. Aster (1990). Analysis of high frequency seismic noise in the western United States and Kazakhstan. *Bull. Seismol. Soc. Am.* **80**(4), 951–970.

Harben, P.E. and E. Hjortenberg (1993). Variation in microseism power and direction of approach in northeast Greenland. *Bull. Seismol. Soc. Am.* **83**(6), 1939–1958.

Hasselmann, K.A. (1963). A statistical analysis of the generation of microseisms. *Rev. Geophys.* **1**, 177–209.

Haubrich, R.A. (1970). The origin and characteristics of microseisms at frequencies below 140 cycles per hour. *IUGG Monogr.* **62**, 1753–1760.

Haubrich, R.A. and K. McCamy (1969). Microseisms coastal and pelagic sources. *Rev. Geophys. Space Phys.* **7**, 539–571.

Haubrich, R.A., W.H. Munk, and F.E. Snodgrass (1963). Comparative spectra of microseisms and swell. *Bull. Seismol. Soc. Am.* **53**(1), 27–37.

Hauksson, E., T. Ta-liang, and T.L. Henyey (1987). Results from a 1500 m deep, three level downhole seismometer array: site response, low *Q* values and f_{max}. *Bull. Seismol. Soc. Am.* **77**(6), 1883–1904.

Heaton, T.H., D.L. Anderson, W.J. Arabasz, *et al.* (1989). National Seismic System Science Plan. *USGC Survey Circular* **1031**, 1–42.

Hedlin, M.A. and J.A. Orcutt (1989). A comparative study of island, seafloor and subseafloor ambient noise levels. *Bull. Seismol. Soc. Am.* **79**(1), 172–179.

Herbers, T.H.C., S. Elgar, and R.T. Guza (1995). Generation and propagation of infragravity waves. *J. Geophys. Res.* **100**, 24863–24872.

Holcomb, L.G. (1998). Spectral stucture in the Earth's microseismic background between 20 and 40 seconds. *Bull. Seismol. Soc. Am.* **88**(3), 744–757.

Holcomb, L.G., L. Sandoval, and B. Hutt (1998). Experimental investigations regarding the use of sand as an inhibitor of air convection in deep seismic boreholes. *USGS Open File Report* 98-362.

Hutt, C.R. (1988). Continental downhole installation. In: "Proceedings of a Workshop on Broadband Downhole Seismometers in the Deep Ocean" (G. M. Purdy and A.M. Dziewonski, Eds.). Woods Hole Oceanographic Institution, Woods Hole, MA.

Kvaerna, T. and F. Ringdal (1999). Seismic threshold monitoring for continuous assessment of global detection capability. *Bull. Seismol. Soc. Am.* **89**(4), 949–959.

Lacoss, R.T., E.J. Kelly, and M.N. Toksoz (1969). Estimation of seismic noise structure using arrays. *Geophysics* **34**(1), 21–38.

Lambert, A., (1970). The response of the Earth to loading by the ocean tides around Nova Scotia. *Geophys. J. R. Astron. Soc.* **19**, 449–477.

Lees, J. (1995). Reshaping spectrum estimates by removing periodic noise: application to seismic spectral ratios. *Geophys. Res. Lett.* **22**(4), 513–516.

Li, T.M.C., J.F. Fergunson, E. Herrin, and H.B. Durham (1984). High frequency seismic noise at Lajitas Texas. *Bull. Seismol. Soc. Am.* **74**(5), 2015–2033.

Li, Y., W. Prothero Jr., C. Thurber, and R. Butler (1994). Observations of ambinet noise and signal coherency on the island of Hawaii for teleseismic studies. *Bull. Seismol. Soc. Am.* **84**(4), 1229–1242.

Liu, J.-Y. and H. Schmidt (1993). Effect of rough seabed on the spectral composition of deep ocean infrasonic ambient noise. *J. Acoust. Soc. Am.* **93**(2), 753–769.

Longuet-Higgins, M.S. (1950). A theory for the generation of microseisms. *Philos. Trans. R. Soc. Lond. A* **243**, 1–35.

McDonald, M.A., J.A. Hildebrand, and S.C. Webb (1995). Blue and fin whales observed on a seafloor array in the Northeast Pacific. *J. Acoust. Soc. Am.* **98**(2), 712–721.

McCreery, C.S., F.K. Duennebier, and G.H. Sutton (1993). Correlation of deep ocean noise (0.4–20 Hz) with wind, and the Holu spectrum—a worldwide constant. *J. Acoust. Soc. Am.* **93**(5), 2639–2648.

Merriam, J.B. (1994). The nearly diurnal free wobble resonance in gravity measured at Cantley, Quebec. *Geophys. J. Int.* **116**, 252–266.

Murphy, A.J. and J.R. Savino (1975). A comprehensive study of long period (20–200 s) Earth noise at the high gain worldwide seismograph stations. *Bull. Seismol. Soc. Am.* **65**, 1827–1862.

Orcutt, J.A. and T.H. Jordan (1986). Data from the Ngendei experiment in the southwest Pacific. In: "The Vela Program," (A. Kerr, Ed.), pp. 758–770. Def. Adv. Res. Proj. Agency, Washington, DC.

Oliver, J. (1962). A worldwide storm of microseisms with periods of about 27 s. *Bull. Seismol. Soc. Am.* **52**(3), 507–517.

Peterson, J. (1993). Observations and modeling of seismic background noise. *USGS Open File Report* 93–322.

Peterson, J., H.M. Butler, L.G. Holcomb, and C.R. Hutt (1976). The seismic research observatory. *Bull. Seismol. Soc. Am.* **66**, 2049–2068.

Prentiss, D.D. and J.I. Ewing (1963). The seismic motion of the deep ocean floor. *Bull. Seismol. Soc. Am.* **53**(4), 765–781.

Richardson, W.J, C.R. Greene, C.I. Malme, and D.H. Thomson (1995). "Marine Mammals and Noise." Academic Press, San Diego.

Ringdal, F., S. Mykkeltveit, J. Fyen, and T. Kvaerna (1990). Spectral analysis of seismic signals and noise recorded at the Noress high-frequency element. *Phys. Earth Planet. Inter.* **63**, 243–251.

Romanowicz, B., D. Stakes, J.-P. Montagner, *et al.* (1998). MOISE: a pilot experiment toward long term seafloor geophysical observatories. *Earth Planets Space* **50**, 927–937.

Savino, J., K. McCamy, and G. Hade (1972). Structures in Earth noise beyond twenty seconds—a window for earthquakes. *Bull. Seismol. Soc. Am.* **62**(1), 141–176.

Schreiner, A.E. and L.M. Dorman (1990). Coherence lengths of seafloor noise: effect of ocean bottom structure. *J. Acoust. Soc. Am.* **88**(3), 1503–1514.

Shearer, P.M. and J.A. Orcutt (1987). Surface and near surface effects on seismic waves—theory and borehole seismometer results. *Bull. Seismol. Soc. Am.* **77**(4), 1168–1196.

Sohn, R.A., J.A. Hildebrand, and S.C. Webb (1998). Postrifting seismicity and a model for the 1993 diking event on the CoAxial segment, Juan de Fuca ridge. *J. Geophys. Res.* **103**, 9867–9877.

Sorrels, G.G. (1971). A preliminary investigation into the relationship between long period seismic noise and local fluctuations in the atmospheric pressure field. *Geophys. J. R. Astron. Soc.* **26**, 71–82.

Sorrels, G.G. and E.J. Douze (1974). A preliminary report on infrasonic waves as a source of long period seismic noise. *J. Geophys. Res.* **79**, 4908–4917.

Sorrels, G.G. and T.T. Goforth (1973). Low frequency earth motion generated by slowly propagating partially organized pressure fields. *Bull. Seismol. Soc. Am.* **63**(5), 1583–1601.

Stephen, R.A. *et al.* (1994). The seafloor borehole seismic system (SEABASS) and ULF noise. *Mar. Geophys. Res.* **16**, 243–269.

Sutton, G.H. and N. Barstow (1990). Ocean bottom ultra low frequency (ULF) seismoacoustic ambient noise: 0.002–0.4 Hz. *J. Acoust. Soc. Am.* **87**, 2005–2012.

Suyehiro, K. *et al.* (1995). Ocean downhole seismic project. *J. Phys. Earth* **43**, 599–618.

Tabulevich, V.N. (1992). "Microseismic and Infrasound Waves." Springer Verlag, Berlin.

Tanimoto, T. (1999). Excitation of normal modes by atmospheric turbulence: source of long-period seismic noise. *Geophys. J. Int.* **136**(2), 395–402.

Toomey, D.R., S.C. Solomon, and G.M. Purdy (1988). Microearthquakes beneath median valley of mid-Atlantic ridge near 23° N: tomography and tectonics. *J. Geophys. Res.* **93**, 9093–9112.

Urick, R.J. (1983). "Principles of Underwater Sound," 3rd edn. McGraw-Hill, New York.

Vernon, F.L. *et al.* (1998). Evaluation of teleseismic waveforms and detection thresholds from the OSN pilot experiment. *EQS Trans. Am. Geophys. Union* **79**(45), 650.

Walker, D.A. C.S. McCreery, and G.H. Sutton (1983). Spectral characteristics of high frequency *Pn*, *Sn* phases in the western Pacific. *J. Geophys. Res.* **88**(B5), 4289–4298.

Warburton, R.J. and J.M. Goodkind (1977). The influence of barometric pressure variations on gravity. *Geophys. J. R. Astron. Soc.* **48**(3), 281–292.

Webb, S.C. (1988). Long period acoustic and seismic measurements and ocean floor currents. *IEEE J. Ocean. Eng.* **13**(4), 263–270.

Webb, S.C. (1992). The equilibrium oceanic microseism spectrum. *J. Acoust. Soc. Am.* **94**(4), 2141–2158.

Webb, S.C. (1998). Broad seismology and noise under the ocean. *Rev. Geophys.* **36**, 105–142.

Webb, S.C. and W.C. Crawford (1999). Long period seafloor seismology and deformation under ocean waves. *Bull. Seismol. Soc. Am.* **89**(6), 1535–1542.

Webb, S.C. and A.D. Schultz (1992). Very low frequency ambient noise at the seafloor under the Beaufort sea icecap. *J. Acoust. Soc. Am.* **91**(3), 1429–1439.

Webb, S.C., X. Zhang, and W.C. Crawford (1991). Infragravity waves in the deep ocean. *J. Geophys. Res.* **96**, 2723–2736.

Wilcock, W.S.D., S.C. Webb and I.Th. Bjarnason (1999). The effect of local wind on seismic noise near 1 Hz at the MELT site and on Iceland. *Bull. Seismol. Soc. Am.* **89**(6), 1543–1557.

Wilmore, P.L. (Ed.) (1979). "Manual of Seismological Observatory Practice," World Data Center, Report SE-20. US Department of Commerce, NOAA, Boulder, CO.

Withers, M.M., R.C. Aster, C.J. Young, and E.P. Chael (1996). High frequency analysis of seismic background noise as a function of wind speed and shallow depth. *Bull. Seismol. Soc. Am.* **86**(5), 1507–1515.

Young, C.J., E.P. Chael, M.M. Withers, and R.C. Aster (1996). A comparison of the high frequency ($>$1 Hz) surface and subsurface noise environment at three sites in the United States. *Bull. Seismol. Soc. Am.* **86**(5), 1516–1528.

Zakarauskas, P. (1986). Ambient noise in the sea: a literature review. *Can. Acoust.* **14**(3), 3–17.

Zurn, W. and R. Widmer (1995). On noise reduction in vertical seismic records below 2 mHz using local barometric pressure. *Geophys. Res. Lett.* **22**(24), 3537–3540.

20

US Contribution to Digital Global Seismograph Networks

Charles R. Hutt, Harold F. Bolton, and L. Gary Holcomb
Albuquerque Seismological Laboratory, US Geological Survey, Albuquerque, New Mexico, USA

This chapter summarizes the contribution to digital global seismograph networks by the United States, and provides information about station installation practices. The global effort is coordinated by the Federation of Digital Broad-Band Seismograph Networks (FDSN). Its membership comprises groups responsible for the installation and maintenance of broadband seismographs either within their geographic borders or globally. A brief summary of FDSN is given in Chapter 81, and detailed information of FDSN is given at their Web site *http://www.fdsn.org/*. Any use of trade, firm, or product names is for descriptive purposes only and does not imply endorsement by the US Government.

1. Motivation for Digital Seismograph Systems

1.1 Limitations of Analog Systems

One of the very first seismographs known to exist early in the Christian era, the Chinese *Kan Shin Gi*, could be called digital. During an earthquake, when a suspended ball fell out of a dragon's mouth into a frog's mouth on one side of this instrument, the earthquake was presumed to have occurred somewhere along a line defined by the center point of the instrument and the dragon whose ball had dropped. This amounts to a binary digital indication (1 fallen ball = earthquake, 0 fallen balls = no earthquake). However, most seismographs since that time, until the late 1960s, have been analog recording seismographs. That is, a trace recorded (as a function of time) on a piece of paper or other medium serves as an analog of magnified ground motion.

Analog seismograms gave us a first glimpse of the internal structure of the Earth. In 1911 Mohorovičić viewed the base of the crust. In 1936 Inge Lehmann differentiated the inner and outer core. The travel-time tables established by Jeffreys and Bullen in the 1930s are still in use today. The millions of measurements made from analog recordings compiled by the International Seismological Centre (ISC) continue to be a rich source for research.

There are, however, several limitations of analog recordings:

- **Small dynamic range**. The ratio of the smallest visible trace displacement to the largest capable of being recorded on the medium used is usually no more than a factor of 100 or so (40 dB). For example, the well-known World Wide Standardized Seismograph Network (WWSSN) used 300 mm wide photographic paper. Under the best possible conditions (when the trace was in the center of the paper), and assuming that the smallest signal normally resolvable is 1 mm, then the dynamic range is 300/1, or almost 50 dB. But when the recorded trace is near the beginning or end of the record, the maximum available peak-to-peak amplitude is only 40 mm, so the dynamic range here is only 40/1 (32 dB). Electronic analog recording media such as FM tape suffer from similar small dynamic range capabilities (50–60 dB at best). Note that the range of ground motion from the lowest-noise sites on Earth to the largest accelerations experienced near the epicenter of great earthquakes exceeds 200 dB (a factor of more than 10^{10}). Because high-magnification seismographs are overdriven easily by large events, it is necessary to operate parallel low-gain seismographs whenever possible.
- **Slow to process**. Analog records are normally analyzed by hand. In the early 20th century, there were a limited number of seismograph stations and each station had one or more dedicated professionals who analyzed the seismograms. The WWSSN consisted of about 110 concurrently operating stations and most of these reported phase times, polarities, and amplitudes to the USGS National Earthquake Information Center (NEIC) in the

United States. Even so, NEIC had to employ several analysts to read WWSSN records for those stations not reporting. Manual analysis of analog records is labor-intensive and does not yield as much information as digital recordings.

- **Low information yield**. A majority of analog seismographs recorded narrowband filtered traces, resulting in simplified waveforms. Narrowband seismograms do not yield the complex information and spectra needed for modern source mechanism analyses and underestimate the total energy output of an earthquake. This is especially true for large earthquakes that produce seismic energy outside the preselected bands recorded. The main types of information obtained from analog recordings include phase time, period, amplitude, and polarity of first motion. These parameters allow the calculation of epicenter and magnitude, and give a rough estimate of the source mechanism. Depth phases and core phases also give information about event depth and the gross average velocity of the core and mantle. However, it is quite difficult to use analog recordings to stack many thousands of traces. Data-intensive processing necessary for seismic tomography would be impossible with analog data, as would much other automated processing that is now routine.
- **Difficult to store**. Paper recordings, film, and analog magnetic tapes are examples of analog media that are difficult to store over extended periods of time. Photographic paper and film deteriorate rapidly with age and exposure to light, and must be kept in a dark, cool, dry place. Analog magnetic tape also deteriorates with age. Signals become harder to read and SNRs decrease as apparent background noise levels on the tape increase.

Notwithstanding their limitations, there are millions of analog recordings in existence. It is not too late to convert many of these to digital format for permanent storage. One example is the set of several million film chips of WWSSN seismograms currently in storage at the USGS Albuquerque Seismological Laboratory. The USGS has funded a pilot project to scan at very high resolution several interesting sets of these film chips for evaluation by researchers. If the images are of sufficient quality, and if additional funding and a feasible way of scanning large numbers of the chips can be found, then there will be a large number of potentially quite useful seismograms available in digital form (as scanned images, not digitized traces).

1.2 Advantages of Digital Systems

With the advent of electronic digital computers in the 1950s, many seismologists realized that there would be huge advantages to digitizing their analog records. This would allow automated processing, including the calculation of auto- and cross-correlation functions, Fourier transforms, and a vast number of other things. At the beginning of the digital age, computers were expensive monstrosities relegated to large rooms in research laboratories. Before seismic data had begun to be recorded digitally, many a graduate student ruined his or her eyes on the manual digitization of analog seismograms so that some special digital technique could be applied. Not until the late 1960s had it become feasible to directly digitize seismometer signals at a few selected observatories and thus begin to save the eyes and nerves of graduate students.

There are many advantages of digital seismograph systems:

- **Large dynamic range**. The basic dynamic range of a good seismometer, even without feedback electronics, is usually quite large. For example, the WWSSN short-period seismometer ("Big Benioff") has a clip level of 10^{-3} m peak, and can resolve signals as small as 10^{-9} m, and so has a dynamic range of about 10^6 (120 dB). A good 16-bit digitizer has a dynamic range of 96 dB, which is about 2 orders of magnitude larger than an analog recording system. Modern 24-bit digitizer systems have a dynamic range of about 140 dB or better (a factor of 10^7) (see Fig. 1). (Note that a WWSSN paper record would need to have a width of 10 km to match this dynamic range.) Many modern broadband seismometers utilizing electronic feedback can match or exceed the dynamic range of these digitizers. The combination of high dynamic range digitizers and seismometers results in digital seismograph systems capable of resolving background

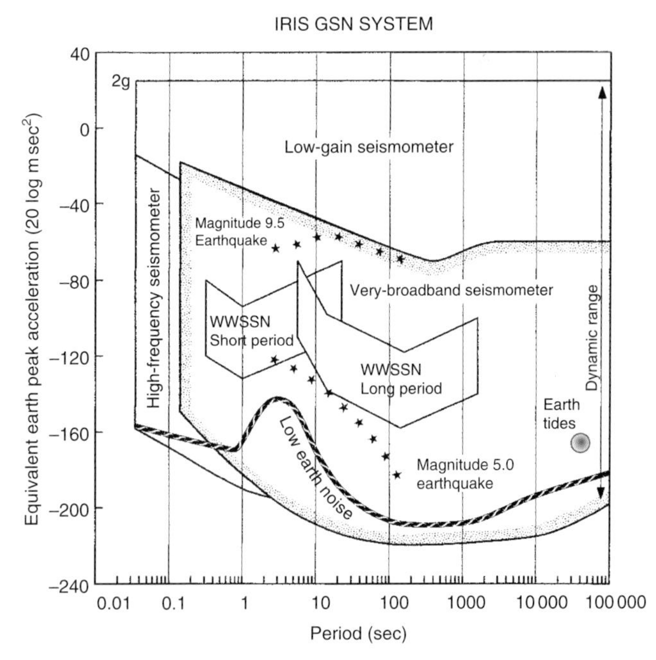

FIGURE 1 Illustration of the relative dynamic range and bandwidth of a well-known analog system (WWSSN short-period and long-period) and a modern 24-bit digital system (the GSN System).

seismic noise at the lowest-noise sites on Earth while remaining on-scale for most of the largest earthquakes, without the requirement to record a separate low-gain channel. This capability has obvious advantages in processing the recorded signals, as two separate channels need not be "patched together" to obtain a continuous time-series.

- **Fast to process**. Automated signal detection and processing by digital computer is possible once the data have been digitized. Such automated signal and phase detections may be used in event associator logic to determine which of many possible events each phase detection belongs to, allowing fully automated epicenter and magnitude determination (to be reviewed later by a human analyst).
- **High information yield**. The use of high dynamic range digitizers and broadband seismometers allow the recording of ground motion without filtering. Older analog seismographs achieved high magnification settings in the short-period and long-period bands by filtering out the main source of noise: the 6–8 sec microseismic band. However, this also filtered out useful information, as energy in the 6–8 sec band is useful in determining the source mechanism of earthquakes. High dynamic range digital seismographs can record across the entire band of interest, including the microseismic band, without filtering because they are capable of resolving ground motion in all frequency bands even in the presence of high microseismic noise. The analyst may then apply digital filters *after* the data have been recorded, without permanent loss of information. In addition, many other types of processing may be applied, because the data are already in digital form. These include spectral analysis, centroid moment tensor analysis, seismic tomography, trace stacking, and background noise analysis at each station.
- **Easy to store**. Once the ground motion data have been recorded in digital format, they may be copied onto any number of media without any loss of fidelity whatsoever. This is not the case with analog recordings, which are degraded every time a copy is made. Thus, as technology continues to bring media with greater capacity and faster access, it is possible retain the full bandwidth of the digital data across each transcription.

2. History of Development of Some US Sponsored Digital Networks

No discussion of modern instrument development can take place without discussing the various types of digital seismographic networks developed over the years. Some of the digital networks developed and installed by the United States are listed below in time order of their development.

TABLE 1 HGLP Stations

Code	Location	Years of Operation
ALQ	Albuquerque, New Mexico, USA	1972–1978
CHG	Chiang Mai, Thailand	1972–1976
CTA	Charters Towers, Australia	1972–1976
EIL	Eilat, Israel	1972–1979
KIP	Kipapa Gulch, Oahu, Hawaii, USA	1972–1978
KON	Kongsberg, Norway	1972–1978
MAT	Matsushiro, Japan	1972–1977
OGD	Ogdensburg, New Jersey, USA	1973–1979
TLO	Toldeo, Spain	1972–1978
ZLP	Zongo Valley, Bolivia	1972–1976

2.1 High-Gain Long-Period (HGLP)—1972 to 1979

This was a network of 10 globally distributed long-period stations developed by the Lamont–Doherty Geological Observatory (now known as the Lamont–Doherty Earth Observatory) for the purpose of earthquake and nuclear monitoring research. Station information is given in Table 1. The Teledyne Geotech long-period seismometers, set to a natural period of 30 sec, were installed inside steel hemispherical sealed tanks for thermal and pressure isolation. These were connected to 100-sec galvanometers in phototube amplifiers (PTAs) that amplified the seismometer signals and shaped the narrowband long-period displacement response with a peak at 40–50 sec period. A 16-bit Astrodata data logger recorded the data at one sample per second on half-inch 7-track computer tape. The HGLP stations were able for the first time on a global scale to resolve the natural low in the Earth's background spectrum near 50 sec period. However, the seismometers in tanks were sometimes installed in relatively shallow tunnels ($\sim$10 m overburden). Also, even though great care was used to secure the tank bottoms rigidly to the tunnel floor or pier, the flat bottoms of these tanks would flex during atmospheric pressure changes. Both the shallow installations and the flexing tank bottoms resulted in noisy horizontal data during periods of moderate to high winds.

2.2 Seismic Research Observatories (SRO)—1974 to 1993

The noisy long-period horizontal data obtained from the HGLP network spurred the development of a broadband borehole seismometer that could be installed at 100 m depth in standard 7-inch oil well casing. Studies had shown that the tilt-generated long-period noise, resulting from wind and passing weather fronts, attenuates rapidly with depth in hard rock. It was also desirable to be able to locate seismic stations at sites other than abandoned mines or expensive hand-dug tunnels. The Teledyne Geotech KS36000 seismometer developed for this program was installed at 13 locations around the world known as the Seismic Research Observatories (see Table 2).

TABLE 2 SRO Stations

Code	Location	Years of Operation
ANMO	Albuquerque, New Mexico, USA	1974–1989
ANTO	Ankara, Turkey	1978–1992
BCAO	Bangui, Central African Republic	1979–1990
BGIO	Bar Giyyora, Israel	1986–1988
BOCO	Bogota, Colombia	1978–1987
CHTO	Chiang Mai, Thailand	1977–1992
GRFO	Grafenberg, Germany	1978–1993
GUMO	Guam, Mariana Islands, USA territory	1975–1991
MAIO	Mashhad, Iran	1975–1978
NWAO	Narrogin, Australia	1976–1991
SHIO	Shillong, India	1978–1985
SNZO	South Karori, New Zealand	1975–1992
TATO	Taipei, Taiwan	1976–1992

The broadband data output had a flat response to acceleration from 0 to 1 Hz, but was amplified and spectrally shaped into two narrow bands before recording: 0.02–0.05 Hz (long period or LP band) and 1–3 Hz (short period or SP band). The LP data were recorded continuously at one sample per second (sps), while only triggered SP events were recorded at 20 sps. The digitizer was a 12-bit gain ranging device with an overall recording range of approximately 132 dB.[1] However, overall system noise levels (including digitizer, seismometer, and filter) were considerably higher than one bit at the highest binary gain step, so the actual recording range was closer to between 110 and 120 dB. Also, there were some noise problems caused by inaccurate binary gain steps, resulting in nonlinear representation of the data. Available dynamic range in the presence of earth noise was further reduced by using a high sensitivity setting. This resulted in the digitizer least significant bit (LSB) at the highest binary gain being about 60 dB below earth noise at a quiet site in the SP band, and about 30 dB below earth noise in the LP band. Available recording range (from quiet site earth noise to clip level) was only about 60 dB in the SP band and about 95 dB in the LP band. The data were recorded on half-inch 9-track computer tapes.

2.3 Modified High-Gain Long-Period Stations (ASRO)—1976 to 1993

A decision was made after the beginning of the SRO program to replace the Astrodata data logger at four of the HGLP stations with SRO-type data loggers, and to change the HGLP amplitude response to more closely match that of the SRO. Also, a short-period vertical Johnson–Matheson seismometer and amplifier/filter were added so that there would be SP data as well as the existing LP data. This made those four stations (plus a new one in Afghanistan) similar to SRO stations except that the seismometers were in seismic vaults rather than in boreholes. These stations (see Table 3) were known as Modified High-Gain Long-Period stations, or Abbreviated SRO stations (hence the ASRO acronym). The digitizer was exactly the same as in SRO stations: a 12-bit digitizer driven by a 11-step binary gain amplifier. The displacement response from the long-period seismometer/amplifier combination peaked at period of about 28 sec. The short-period displacement response peaked at about 3 Hz. The useful dynamic range (from earth noise to clip level) was limited by high-sensitivity settings, as in the SRO systems.

TABLE 3 ASRO Stations

Code	Location	Years of Operation
CTAO	Charters Towers, Australia	1976–1987
KAAO	Kabul, Afghanistan	1977–1982
KONO	Kongsberg, Norway	1978–1991
MAJO	Matsushiro, Japan	1977–1990
ZOBO	Zongo Valley, Bolivia	1976–1993

2.4 Digital World Wide Standardized Seismograph Stations (DWWSSN)—1981 to 1993

A desire for better geographic coverage than provided by the HGLP, SRO, and ASRO networks resulted in the idea of attaching digital acquisition systems to some existing WWSSN stations in appropriate areas. This was accomplished as follows: Amplifiers and filters were attached to the output of the standard WWSSN SP and LP seismometers. These amplified and band-shaped signals were digitized by a 16-bit analog-to-digital converter and recorded on half-inch computer tape under control of a microcomputer system. The dynamic range of the data acquisition system was roughly 90 dB. A digital (manually set) clock module, designed specifically for installation in the WWSSN timing console, provided digital time signals for time tagging the data. In addition to standard narrow SP and LP bands, a band known as "intermediate period" (IP) was recorded. The IP band was basically a low-sensitivity recording of the raw unfiltered output of the Press–Ewing long-period instrument, and was one of the first attempts by the United States to record broadband data. LP data were recorded continuously at 1 sps, whereas the recording of IP (10 sps) and SP (20 sps) data were triggered by a rudimentary event detector. These stations are known as DWWSSN (see Table 4).

[1] The highest-resolution commercially available digitizers available were 12 bits, allowing only 66 to 72 dB recording range. Since a recording range of 120 dB was required for the SRO systems, the digitizer manufacturer attached a 11-step binary gain amplifier (BGA) to the input of the 12-bit digitizer. The gain steps were 2^0 (unity gain) up to 2^{10} (gain of 1024). Small signals resulted in a BGA setting of 2^{10}, thus amplifying these signals to a level large enough for the 12-bit digitizer to resolve. Larger signals resulted in correspondingly smaller BGA gain settings, so that the digitizer would not clip. The largest signal that could be digitized was 10 volts at a BGA setting of 2^0 or a gain of unity.

TABLE 4 DWWSSN Stations

Code	Location	Years of Operation
AFI	Afiamalu, Western Samoa	1981–1991
ALQ	Albuquerque, New Mexico, USA	1981–1982
BDF	Brasilia, Brazil	1982–1993
BER	Bergen, Norway	1981–1984
CMB	Columbia College, California, USA	1987–1992
COL	College Outpost, Alaska, USA	1982–1991
GDH	Godhaven, Greenland	1982–1993
HON	Honolulu Observatory, Hawaii, USA (PTWC)	1983–1990
JAS	Jamestown, California, USA	1980–1984
JAS1	Jamestown, California, USA	1984–1987
KBS	Kings Bay, Spitzbergen	1986–1987
KEV	Kevo, Finland	1981–1993
LEM	Lembang, Indonesia	1982–1988
LON	Longmire, Washington, USA	1980–1994
SCP	State College, Pennsylvania, USA	1981–1992
SLR	Silverton, South Africa	1981–1993
TAU	Tasmania, Australia	1981–1993
TOL	Toledo, Spain	1981–1991

2.5 China Digital Seismograph Network (CDSN)—1986 to 1999

The nine original stations of the CDSN were installed in cooperation with the State Seismological Bureau (SSB) of China beginning in 1986, with the Lhasa, Tibet (LSA) station being added in 1991 (see Table 5). (The SSB has since changed its name to the China Seismological Bureau, or CSB.) The CDSN was the first US-sponsored cooperative network where true broadband seismometers (20-sec Streckeisen STS-1 instrument) were used. Data from the STS-1 seismometers were split into three separate bands that were sampled and recorded as follows:

- VLP, velocity response peaking at 200 sec period, 0.1 sps, continuous
- LP, flat to acceleration 30–500 sec, 1 sps, continous
- BB, flat to velocity 5 Hz to 20 sec, 20 sps, events only

Chinese DJ-1 1-sec seismometers were used as the short-period instrument, and were connected to the digitizer through a short-period amplifier/filter. The SP data were sampled at 40 sps and recorded on an event basis. The digitizer was a gain-ranged type, similar to that used in the SRO and ASRO systems. However, there were only four gain steps: 1, 8, 32, and 128. This gain-ranging amplifier was connected to a 14-bit digitizer, resulting in an overall recording range of 126 dB. A digital clock that was set manually by comparison to WWV time signals provided time tagging. The data were recorded on quarter-inch cartridges.

TABLE 5 CDSN Stations

Code	Location	Years of Operation
BJI	Baijatuan	1986–1994
ENH	Enshi	1986–1997
HIA	Hailar	1986–1994
KMI	Kunming	1986–1996
LSA	Lhasa	1991–1995
LZH	Lanzhou	1986–
MDJ	Mudanjiang	1986–1996
QIZ	Qiongzhong	1992–2000
SSE	Sheshan	1986–1996
WMQ	Urumqi	1986–1995

2.6 International Deployment of Accelerometers (IDA)—1975 to 1995

This section was contributed by Peter Davis, IDA Group, University of California, San Diego

The IDA network was originally established to provide global digital measurements of the Earth's normal modes, at periods from (roughly) 60 to 3600 sec. The network, sponsored by Cecil and Ida Green and the National Science Foundation, was operated by the Institute of Geophysics and Planetary Physics at the University of California, San Diego (see Table 6). The sensor (for vertical motion only) was a Lacoste–Romberg underwater gravity meter, of standard Lacoste zero-length spring design, housed inside a pressure-tight container with temperature control. Electronics (using electrostatic position sensing and feedback) were added to this sensor to make it a force-balance system with a flat response to acceleration. The

TABLE 6 IDA Stations (Gravimeters)

Code	Location	Start Date	End Date
ALE	Alert, Canada	1982,298	1995,189
BDF	Brasilia, Brazil	1977,094	1995,189
BJT	Baijatuan, China	1982,227	1995,189
CAN	Canberra, Australia	1975,122	1978,086
CMO	College, Alaska	1977,214	1995,189
EIC	Easter Island, Chile	1978,236	1980,201
ERM	Erimo, Japan	1980,231	1995,189
ESK	Eskdalemuir, Scotland, UK	1978,258	1995,364
GAR	Garm, USSR	1976,247	1980,343
GUA	Guam, Mariana Islands	1979,314	1995,191
HAL	Halifax, Canada	1976,114	1995,189
KIP	Kipapa, Hawaii	1977,249	1995,189
KMY	Kunming, China	1980,296	1995,189
NNA	Nana, Peru	1975,164	1995,189
PCR	Plaine Des Cafres, Reunion Island (French Territory)	1983,222	1989,219
PFO	Pinon Flat, California, USA	1976,028	1995,189
RAR	Rarotonga, Cook Islands (New Zealand Territory)	1976,271	1995,189
RPN	Easter Island (Rapa Nui), Chile	1987,166	1995,189
SEY	Seychelles	1980,033	1983,010
SJG	San Juan, Puerto Rico (US Territory)	1982,158	1995,189
SPA	South Pole, Antarctica	1978,004	1995,189
SSB	Saint Sauveur en Rue (Saint Sauveur Badole), France	1981,265	1983,102
SUR	Sutherland, South Africa	1975,346	1995,189
TWOA	Adelaide, Australia	1978,099	1995,189

output of this system was passed through two filters: a Tide channel, which was a low-pass filter flat below 16 mHz, and a Mode channel with additional gain and flat response from 0.3 to 16 mHz. Both channels were digitized using a 12-bit fixed-gain digitizer, and recorded on digital cassette tapes. The Tide channel was recorded with a sampling interval of 10 min, and the Mode with an interval of 10 sec. Time tagging was provided by the data-recorder, using an internal digital clock, which was set manually by comparison with WWV time signals when the tapes were changed (roughly every 10 days).

2.7 Global Seismographic Network (GSN)—1988 to the Present

The GSN was the first US effort to collect high-fidelity broadband seismic data at seismological observatories uniformly distributed around the globe. It is sponsored by the Incorporated Research Institutions for Seismology (IRIS), the US National Science Foundation (NSF), and the US Geological Survey (USGS). The US Air Force Office of Scientific Research and the Green Foundation have provided additional funding. The USGS Albuquerque Seismological Laboratory (ASL) and the IDA group at the University of California, San Diego, are the two network operators responsible for installing, maintaining, and collecting data from the stations of the GSN. Although there are several different versions of early prototype seismic data acquisition systems, the relevant features are in general as follows:

- Continuous (rather than event-detected) recording of broadband data at 20 sps.
- Flat response to earth velocity from 0.003 Hz to 5 or 10 Hz.
- The capability to resolve background seismic noise at the lowest-noise sites, without clipping the largest conceivable signals from great earthquakes, over a range of frequencies from earth tides to tens of Hz (see Fig. 1).
- Accurate time tagging without dependence on a human to check or set the clock.
- Real-time telemetry of data to data centers.

These capabilities are achieved through the use of low-noise very broadband (VBB) primary seismometers in conjunction with high-frequency broadband or short-period seismometers and low-gain accelerometers. High-resolution (24-bit) digitizers providing at least 140 dB dynamic range are used. Time tagging, accurate to better than 1 msec, is provided by Global Positioning System (GPS) clocks. Data are recorded on high-capacity cartridge tapes and mailed to data collection centers. Also, as Internet connections and other inexpensive communication links become more available worldwide, an ever-increasing number of GSN stations are sending real-time data to data centers.

3. Seismic Installations in Boreholes

Borehole seismology is a relatively new technology, having developed over the last 30 y or so as compared with the relatively long period over which seismological science as a whole has been evolving. In the early years of seismology, installing a seismometer in a borehole was virtually impossible because of the relatively large physical size of the instruments. As seismological technology matured, the instruments became smaller and it became more practical to consider borehole installations as alternatives to surface vaults or installations in deeper abandoned mine tunnels. There are several reasons for placing seismic instrumentation in boreholes. These include reduced noise levels, temperature stability, and reduced pressure variability.

Experience gained over many years of installing both short- and long-period sensors by many individuals and organizations has shown that sensor systems that are installed at depth are usually quieter than those installed at or near the surface of the Earth. Therefore, abandoned underground mines are frequently used, when available, as sites for low-noise seismological stations. Otherwise, a borehole usually provides a practical solution to the need to install seismic sensors at depth.

A borehole also creates a very stable operating environment for sensitive instruments. Temperature at depth is very stable and the pressure in a cased, sealed borehole is also very constant. Temperature changes and pressure variations with spectral content within the passband of the sensor system are common sources of nonseismic noise in broadband seismic sensors. Systems installed on the surface or in shallow vaults require extensive thermal insulation systems to reduce the influences of temperature to acceptable levels. Similarly, elaborately designed pressure containers are required to eliminate pressure-induced noise in vertical instruments installed near the surface. A sealed borehole of only moderate depth provides excellent temperature stability because of the tremendous thermal mass and inertia of the surrounding Earth. Similarly, most seismic boreholes are lined with steel casing having cylindrical walls that are quite thick. The casing constitutes a quite rigid container and greatly reduces atmospheric pressure variations within the borehole (assuming that both the top and bottom are sealed).

Boreholes are frequently considered to be expensive, but they often represent the only practical alternative if a ready-made abandoned mine is not available. In many cases, a borehole may actually be the cheapest method for achieving an installation at depth.

It is impossible to state exactly how much it will cost to construct either a borehole or a tunnel-type vault because many variable factors are involved. Precise costs will depend on the type of material in which the facility is constructed, raw material costs, local labor costs, and so on. However, we can give some examples of approximate costs that have been encountered in constructing facilities for the IRIS/USGS GSN program over the past 5 to 10 y. In Africa IRIS has excavated three drift-type seismic vaults that extended 25–40 m horizontally into hillsides. The costs of these three projects ranged from approximately \$150 000 to \$250 000. For a typical borehole (100 m deep), costs range from approximately \$25 000 to \$200 000 at large landmass sites. On the other hand, at small isolated Pacific Ocean island sites, borehole costs run in the \$150 000 to \$250 000 range.

3.1 Noise Attenuation with Depth

The primary reason for installing broadband sensors in boreholes is to reduce the long-period tilt noise that plagues horizontal sensors that are installed on the surface. The question commonly asked by seismologists who are contemplating a borehole installation is "How fast does tilt noise decrease with depth?" How deep does the borehole need to be? There is no hard and fast easy answer to this question because the borehole never eliminates all of the long-period tilt noise, regardless of how deep the hole is drilled. In general the noise attenuation rate (dB per unit depth) decreases as the depth increases; therefore, most of the noise reduction occurs in the upper parts of the borehole.

As a simplified illustration of noise attenuation with depth, Figure 2 contains relative power spectral density (PSD) noise level data obtained from the simultaneous deployment of four broadband sensors located near one another at the same site and installed at various depths. The first sensor was installed in a small vault on or near the surface. A second sensor was installed in a very shallow borehole only 4.3 m below the surface. The third sensor was installed in a much deeper borehole 89 m down and the fourth seismometer was 152 m below the surface. The site consisted of about 18 m of unconsolidated (soft/weathered) overburden overlying fractured Precambrian granite bedrock. In Figure 2, noise attenuation data points in dB relative to the noise level in the surface sensor are plotted for 30, 100, and 1000 sec to illustrate how much noise power at depth was reduced as compared to the surface installation. Note the very rapid decay in the noise level over the first few tens of meters followed by the much slower rate of decrease in noise levels at deeper depths. Note that in general, a depth of 100 m is sufficient to achieve most of the practicable reduction of long-period noise.

The data in Figure 2 should be regarded as a singular example of noise attenuation with depth. Apparent surface noise levels at a particular site are frequently highly dependent on the methods used to install the instrumentation. This is particularly true of noise levels at many surface site installations where faulty installation of broadband horizontal sensors causes excessive tilt noise at long periods.

Choosing the optimum depth of a borehole for a particular site involves comparing the cost of a borehole of a given depth against the desired data quality requirements, the anticipated surface noise levels (they are frequently determined by the

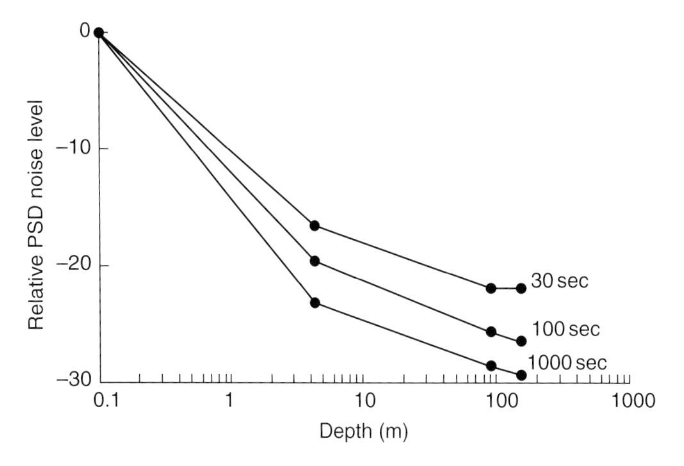

FIGURE 2 Horizontal surface noise attenuation as a function of depth at three selected periods. The depths were 0, 4.3, 89, and 152 m.

anticipated wind speeds and wind persistence at the site), and the depth of the overburden at the site. Unfortunately, studies detailed enough to yield the precise interrelationship between the various factors have never been conducted. Therefore, choosing the depth of a borehole for a particular site usually involves a healthy dose of "sound engineering guesstimate" based on nonquantitative consideration of the various factors involved. Experience over many years has demonstrated that 100 m deep boreholes drilled at sites with a few tens of meters of overburden overlying relatively competent bedrock will provide a sufficiently quiet environment for installing a high-quality borehole instrument. Most broadband IRIS borehole instruments are installed at or near 100 m depth. Sites with excessive overburden and/or softer, lower-quality bedrock are sometimes deeper depending on construction costs and anticipated surface noise levels.

3.2 Site Selection Criteria

There are several criteria for selecting a site for a borehole installation. Ideally, one should select a site at which the seismic background over the band of interest as measured with a sensor system installed on the surface is as low as possible. However, there are other factors, such as better accessibility, availability of power, improved network configuration, the presence of widespread, thick alluvial fill, and/or the presence of cultural activity within the monitored area, that may force the choice of a site with higher background levels.

A good borehole should penetrate well into bedrock (70–100 m); therefore, the site should have bedrock at or near the surface to minimize the need to drill through excessive overburden. If possible, the bedrock should be a relatively hard rock such as granite or quartzite. Relatively soft rocks such as shale, mudstone, or low-grade limestone should be avoided if possible. Harder, more competent rock increases the rate of attenuation of surface noise with depth and it also decreases the chances of collapse problems during drilling operations.

Good bedrock is highly desirable for providing the ultimate benefit from a borehole, but it is not an absolute necessity for achieving benefits from a borehole site. Note that the first data point in Figure 2 (only 4 m down) was obtained in a very shallow borehole that was drilled entirely in loose alluvial fill. Therefore, the lack of shallow bedrock should not preclude the consideration of a borehole installation for a particular site.

Finally, a reliable source of electricity is necessary to power the site. A shelter is needed to house the recording equipment along with some form of communication capability (telephone line, Internet connection, or RF or satellite link). Accessibility for both the drilling equipment and maintenance personnel should also be considered during site selection activities.

Unfortunately, the need to be able to provide adequate security is also becoming a major factor in selecting station sites in many parts of the modern world. There is little point in investing in a good site if it can not be protected from vandalism. Adequate security has many different meanings depending on the particular situation. It may be as simple as a passive protective fence or as elaborate as alarmed fences and entryways with an on-site caretaker depending on the anticipated level of potential damage.

It should be noted that *very* small islands (such as most coral atolls) are not desirable sites for borehole installations because the ground motion generated by ocean wave loading of the beach penetrates rather deeply into the subsurface environment. Likewise, borehole sites should be an adequate distance (at least several kilometers) from any nearby coastlines (Agnew and Berger, 1978).

3.3 Contracting

Seismic boreholes are usually constructed by local private drilling contractors who follow the specifications supplied by the contracting organization. Hiring a local driller helps reduce the mobilization and setup charges that are frequently a significant portion of the cost of a seismic borehole. Specifications should be rigid and specific enough to assure that the final product will be suitable for borehole seismology and yet flexible enough to prevent excessive costs. It is recommended that the contracting agency hire an independent expert with extensive drilling and casing experience whose duties include on-site observation and supervision of all drilling and casing operations. This precaution is advisable to assure that the drilling contractor performs all operations according to the prescribed specifications, because departures from specifications are hard to detect, document, and prove after the project is finished. The contract should be specific about who is responsible for unexpected difficulties that might arise during the drilling and casing operations. The courses of action should be specified if operations are delayed for any reason whatsoever, including weather, availability of materials or crew, and other circumstances beyond the driller's control.

3.4 Suggested Borehole Specifications

The drilling specifications for a seismic borehole should be written in such a manner as to assure that the completed borehole will be suitable for acquiring high-quality earth motion data. Important parameters such as borehole verticality, depth, diameter, and casing type must be clearly specified. For complete drilling and cementing specifications suitable for soliciting bids, go to "ASL Publications" on the USGS Albuquerque Seismological Laboratory's web site at *http://aslwww.cr.usgs.gov/* and look for "Specifications for Seismic Instrumentation Boreholes."

3.5 Instrument Installation Techniques

It is a relatively simple operation to install a borehole sensor, but certain precautions are in order. The sensors are usually fitted with two cables. The first cable is intended to provide sufficient strength to lift the weight of the sensor and also provide any extra pulling force for removing the sensor from the borehole. This cable is usually a steel cable or "wire rope." The second cable contains the electrical connections for power, control of the various mechanical operations within the sensor, and transmitting the seismic signals back up the borehole. Normally, for holes of significant depth, a small, lightweight electrically driven winch and mast assembly is utilized to lower the sensor into the hole and to retrieve it if necessary. Lowering and raising the sensor should be done fairly slowly because the sensor packages sometimes catch on the casing pipe joints as the package moves up or down the borehole. On the way down, this problem is usually temporary, when it results in a short free fall of the sensor and a sudden stop when the load bearing cable becomes taut again. If severe enough, the sudden stop can damage the sensitive instrument. If the sensor catches on a pipe joint on the way up, tension in the load-bearing cable rapidly increases to dangerous levels if the winch is not stopped in time. If the sensor disengages from the pipe joint while the lifting cable is under high tension, the sensor will undergo possibly damaging levels of acceleration. If the sensor does not disengage and if the winch is powerful enough, the lifting cable may break, thereby endangering personnel in the area.

Traditionally, borehole seismometers are rigidly clamped to the inside of the cased borehole with manufacturer-supplied mechanical hardware to assure adequate coupling between the sensor and ground motion. This hardware usually includes a mechanically driven locking mechanism for clamping the sensor to the walls of the borehole. This device sometimes consists of a motor-driven or spring-loaded pawl that is extended on command from the side of the sensor package to contact the borehole wall opposite from the sensor (Geotech GS-21, Guralp CMG-3TB). Sometimes this function is performed by a separate piece of hardware known as a "hole lock" that is clamped into the borehole on which the sensor package is subsequently rested (Geotech KS-36000,

KS-54000, and earlier Guralp sensors). Mechanical clamping mechanisms have been used successfully for many years and have produced satisfactory data from many installations.

Borehole installations that utilize hole locks have almost always produced more long-period noise in the horizontal components than in the vertical component. In some of these installations, the horizontals were orders of magnitude noisier at long periods than the vertical. The source of the excess noise in the horizontal components has been difficult to isolate and eliminate. For many years, workers in this field have suspected that some of this noise was somehow generated by air motion in the vicinity of the sensor package. The conventionally designed (this includes all sensors with garden gate type of suspension such as the STS-1, CMG-3 series, KS-36000, and KS-54000) horizontal components of long-period seismometers are extremely sensitive to tilt. This is due to their inability to separate the influences of pure horizontal acceleration input to the sensor frame (the desired input) from the signal that arises from the tilting of the sensor package (tilt noise). Therefore, fairly elaborate schemes for reducing the potential for air motion around the sensor within the borehole have been devised and utilized with varying degrees of success. Through the process of trial-and-error development, it has become customary to wrap the sensor package (KS-36000s and KS-54000s) with a thin layer of foam insulation in an attempt to somehow modify the flow of heat near the seismometer in the borehole. In addition, it has become common to place long plastic foam borehole plugs immediately above these sensor packages deep in the borehole and near the top of the borehole to block air motion in these sections of the borehole. Additional insulation, which is intended to further reduce air motion within the borehole, is sometimes employed near the top of the sensor package.

Recently, a highly successful method for significantly reducing the long-period noise levels in borehole installed horizontal components has been developed at the Albuquerque Seismological Laboratory. For a complete report on this technique, go to "ASL Publications" on the USGS Albuquerque Seismological Laboratory's web site at *http://aslwww.cr.usgs.gov/* and look for "Experimental Investigations Regarding the Use of Sand as an Inhibitor of Air convection in Deep Seismic Boreholes," USGS Open File Report 98–362.

Determining the orientation of the horizontal components of a seismometer installed in a deep borehole is not a simple matter because one cannot physically get at the instrument once it is installed. One must resort to indirect methods for determining how the instrument is oriented in the hole. For a complete report on this technique, go to "ASL Publications" on the USGS Albuquerque Seismological Laboratory's web site at *http://aslwww.cr.usgs.gov/* and look for "Experiments in Seismometer Azimuth Determination by Comparing the Sensor Signal Outputs with the Signal Output of an Oriented Sensor."

3.6 Typical Borehole Parameters

There are now many broadband borehole installations in use throughout the world. Most of these boreholes are geometrically quite similar because they were designed to accommodate the same type of seismic instruments. All of these boreholes are approximately 16.5 cm in diameter and most of them are drilled to a maximum of a 3.5° departure from true vertical (the borehole at JOHN is rather crooked—5.5° from vertical). They are all cased with standard oilfield grade casing and most of them are watertight.

There is some variation in the depths of these boreholes. As mentioned earlier, the vast majority of these boreholes are approximately 100 m deep. However, some of these boreholes are considerably deeper if they were drilled in areas with thick overburden or poor bedrock. For instance, the borehole sensor at DWPF (Florida) is installed at a 162 m depth because the overburden at DWPF is approximately 46 m thick and the upper layers of bedrock consist of interleaved units of varying grades of soft limestone. The borehole at ANTO, which is drilled in competent rock for most of its depth, is the deepest and oldest IRIS borehole at 195 m. This was the first field borehole that was drilled for the SRO program: As more experience was gained, it became apparent that boreholes that deep were not cost-effective installations. A few of the boreholes are shallower, primarily because severe difficulties were encountered during the drilling operations that necessitated finishing the borehole at a shallower depth than originally desired. For example, the sensor at JOHN is only 39 m down because severe cave-in conditions were encountered during the attempt to drill deeper. Johnson Island is a coral atoll and the surface layers are very poorly consolidated; true bedrock probably lies at very great depths. Drilling in volcanic regions often proves to be very difficult. The borehole at POHA was terminated at 88 m because the drillers experienced severe "loss of circulation" conditions throughout the drilling operation. The surface layers at POHA on the island of Hawaii consisted of badly fractured weathered basalt layers and basalt rubble separate by scoria rubble, ash flows, sand, and other assorted debris produced by an active volcano. Drilling conditions in the volcanic deposits on Macquarie Island proved to be so difficult that it was impossible to complete a borehole.

3.7 Commercial Sources of Borehole Instruments

Currently, there are only two known commercial sources of high-sensitivity, broadband, borehole-installable seismometers. For many years, Teledyne Geotech (renamed Geotech Instruments LLC (*www.geoinstr.com*)) in Dallas, Texas, USA, was the only source of high-sensitivity instruments (KS-36000, KS-54000, GS-21, and 20171) designed specifically for borehole installations. Both the KS-36000 and the KS-54000 are three-component broadband closed-loop force feedback sensors that are designed for deep (up to 300 m) borehole installations. The GS-21 is a conventionally designed short-period, vertical instrument intended for superior high-frequency performance in deep boreholes. The 20171 is a slightly noisier and slightly cheaper version of the GS-21. The KS-36000 is no longer manufactured, but there are many of these instruments still in operation in boreholes around the world. Geotech plans to have a new sensor ready for market sometime during 1999; it will be denoted the KS-2000 and it will be available in both surface and borehole installed versions.

For the past four years or so, Guralp Systems Ltd., Reading, UK (*www.guralp.demon.co.uk*) has been producing a borehole version of the CMG-3T (referred to by some as a CMG-3TB). This instrument is much smaller and lighter than is a KS sensor. The CMG-3TB is also considerably less expensive than a KS sensor. This sensor is a three-component, broadband, closed-loop, force feedback instrument that is very easy to install. Guralp Systems has recently introduced a new borehole sensor that has both velocity and acceleration output and is integrated with its own digitizer. In addition, they are willing to work with the customer to meet specific requirements.

A borehole version of the Streckeisen STS-2 has been under development for several years. Currently, a basic prototype of the instrument exists, but the instrument requires further development of the remote control functions and the final packaging design is yet to be determined. Streckeisen has not announced an availability date for the new instrument.

It is somewhat hazardous to quote sensor prices because they are continuously subject to change by the manufacturer and international currency exchange rates change daily, but the following are some approximate current relative costs for borehole sensors in 1999 US dollars. These prices should be viewed as being approximate; potential buyers should consult the manufacturer for a current quotation.

A Geotech KS-54000 by itself is priced at nearly $65 000 with added cost (about $40 000 for a conventional installation or about $13 000 if installed in sand) if all of the associated installation hardware is purchased. However, this price may be reduced if the instruments are ordered in sufficient quantities (25 or more). A GS-21 is priced at about $8000 and the 20171 weighs in at around $6000 for the instruments themselves. The associated hardware (soft electrical cable, wire rope, winch, etc.) is additional. Estimated delivery time for these instruments is 120 days or more after receipt of order depending on the availability of non-Geotech-manufactured parts. The KS-2000 sensor to be introduced soon will be priced at below $10 000 for the surface system and the borehole version will probably be well below $20 000.

A Guralp Systems CMG-3TB costs about $28 000 if the instrument is to be installed in sand. A hole lock-equipped version is about $39 000. Currently delivery time is about 9 months; however, the manufacturer is trying to decrease delivery time to about 6 months.

3.8 Instrument Noise

It is important to remember that the purpose of installing seismic instrumentation in boreholes is to obtain quiet seismic data. This intent will be foiled if the seismic sensor system itself is too noisy to resolve the lower levels of earth background noise that are expected to be found at the bottom of the borehole. As delivered from the factory, sensor self-noise levels sometimes vary over a wide range and some instruments may be far too noisy to operate successfully in a quiet borehole. It is therefore recommended that the self-noise of all borehole instruments be measured before installation in the field to assure that they will be quiet enough to be able to resolve the background levels of earth motion anticipated at the bottom of the borehole. Self-noise measurements are usually made by installing two or more sensors physically close enough together to assure that the ground motion input to all of the sensors is identical. The data produced by the sensors is then analyzed to determine the level of the incoherent power in each sensor's output. This incoherent power is usually interpreted as the sensor internal noise level. To achieve high-fidelity recording of true ground motion, the seismometer system self-noise level should be well below the anticipated background earth motion levels across the band of interest at the site.

The low-noise models in Figure 3 can serve as guide lines to the instrument noise levels that may be expected from the CMG-3TB and the KS-54000 sensor systems. In this figure, the CMG-3TB low-noise model (CMGLNM) is the longer-dashed line and the KS-54000 low-noise model (KSLNM) is the shorter-dashed line. The solid heavy line in the figure is Peterson's (1993) new low-noise model (NLNM) for the earth background noise at a quiet site. The reader should recognize that there is no single known site in the world whose background power spectral density levels reach NLNM levels across the entire band. Instead, the NLNM is a composite of the lowest earth noise levels obtained from many sites. Similarly, the low-noise models for the instruments should not be regarded as being typical of all instruments, because each seismic sensor has a distinct "personality" of its own. Instead, the low-noise models for the instruments should be regarded as lower limits of instrument noise, just as is the case for the NLNM of earth noise. Individual instruments will in all probability be noisier than the low-noise model for that instrument over at least portions of the spectrum.

The CMGLNM plot in Figure 3 is based on a composite of experimental test data obtained at the Albuquerque Seismological Laboratory over a period of several years. The central portion (from about 0.6 to about 20 sec) of the model was not actually measured because of numerical resolution limits of the current data processing algorithms. This portion of the model is an estimate. As a general rule, many CMG-3TB instrument noise levels approach the CMGLNM at short periods (less than 0.6 sec), while fewer of these instruments achieve the indicated noise levels at long periods (greater than 20 sec).

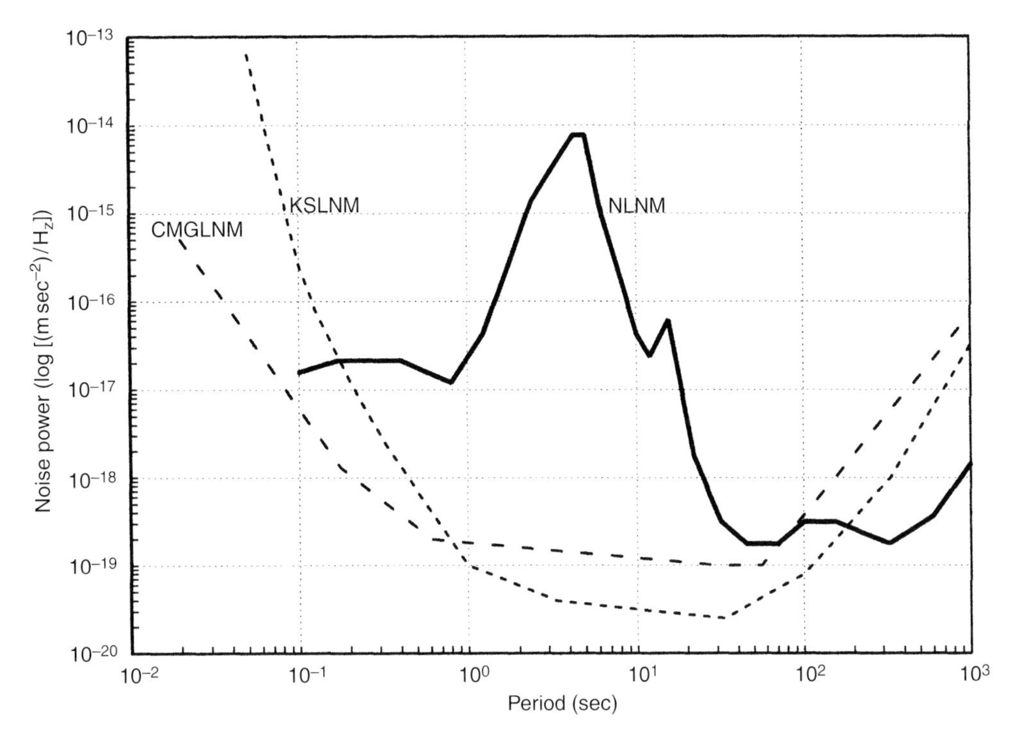

FIGURE 3 Low-noise models for the KS-54000 (KSLNM) and the CMG-3TB (CMGLNM) sensor system self-noise relative to Peterson's (1993) New Low-Noise Model (NLNM) for background earth motion.

The KSLNM plot in Figure 3 is a factory-derived theoretical instrument noise level. As such, it should be regarded as an optimistic estimate of the lower limits of the self-noise in the KS-54000. Most KS-54000 real-world instruments are probably noisier than the levels indicated by the KSLNM curve.

4. Seismic Installations in Tunnels

For many years, abandoned mines have been used as ready-made quiet sites for installing seismic instrumentation. In some cases, even though the site may be somewhat noisy as a result of mining activity during the workday, active mine tunnels have also proved to be successful facilities for seismic stations.

Existing tunnels in solid rock provide very low-cost ready-made facilities that often have convenient existing access and nearly ideal conditions for the installation and operation of high-sensitivity seismic sensors. The bedrock in a mine tunnel is usually already exposed, thereby providing an excellent firm foundation on which to install standard surface instruments. If unventilated, as is usually the case for abandoned mines, a mine tunnel provides an essentially constant-temperature environment that is ideal for seismic sensors. Depending on its thickness, the overburden above the mine tunnels provides varying degrees of isolation of the seismic sensors from the seismic noise that is always present at the surface of the Earth.

Mines are usually geographically located in mineralized zones. Therefore, it is unlikely that an existing mine will be found near the location of many proposed seismic installations. As a result, tunnels are sometimes constructed solely for the purpose of the installation of seismic sensors. Digging tunnels in hard rock without the benefit of a cost-offsetting by-product such as a marketable mineral-bearing ore is a very expensive endeavor.

In many respects, a tunnel installation is quite similar to a surface vault installation. A poured concrete floor or pier is usually constructed on the rough bedrock floor of the tunnel to provide a flat and relatively level surface on which to install the sensors. Despite the improved temperature stability, it is still necessary to provide adequate thermal insulation around the sensors in order to reduce thermally generated noise. A system to reduce the variation in air pressure is also necessary for long-period sensors in underground tunnels. Usually this is accomplished in the same manner as in a surface installation, although sometimes an effort is made to seal off all or parts of the tunnel itself. Sealing a volume enclosed by natural rock walls is difficult because most tunnel walls are riddled with fractures.

In other respects, there are significant differences between surface vault installations and tunnel sites. One of immediate concern is the danger of rock fall in a tunnel installation. Another concern is the danger of the build-up of harmful gasses (bad air) underground if the tunnel is not adequately ventilated.

A common problem in many tunnel installations is the presence of water and high humidity levels in most underground passages. It is very difficult to keep the instrumentation dry, and the wet environment is frequently very unpleasant to work in. The high humidity slowly corrodes the contacts in delicate electrical connectors, which frequently causes poor electrical contact and intermittent operation. The presence of moisture also slowly degrades the effectiveness of thermal insulation materials; therefore, precautions must be taken to prevent moisture accumulation in the thermal isolation system.

Access to power and communications lines is usually more difficult in tunnel installations depending on how far into the tunnel one chooses to position the equipment. Frequently, power and/or communication lines must be installed throughout the entire length of the tunnel. Power can be quite expensive to supply if long distances are involved. Either large-diameter power conductors must be used or a high-voltage line coupled with a step-down transformer must be installed to assure that sufficient voltage is available at the site.

Determining a sensor orientation azimuth line underground is considerably more difficult than in a surface installation. Usually, one must transfer an already known azimuth from outside the tunnel to the installation site within the tunnel using standard land surveying techniques. There are specially designed gyroscopically based systems that can determine the orientation underground, but they are rather expensive.

If a time reference source is required for the installation, it can be difficult to provide this source to a tunnel site. Modern GPS-based timing systems require a relatively short cable between the above-ground antenna and the below-ground receiver. If the tunnel installation is too deep, an in-line radiofrequency amplifier may be required.

5. Future Directions

5.1 Global Seismograph Stations as Geophysical Observatories

There are a number of nonseismic geophysical measurements that are of scientific interest. Some of these, such as temperature, barometric pressure, and wind, may affect the seismic measurements themselves. Others may not have a direct effect on local ground motion but may be of long-term observational interest anyway. Since global seismic observatories typically have a long lifetime at a given site, it makes sense to take advantage of the siting and infrastructure at these observatories and to make long-term observations of any and all geophysical parameters that may be of interest to researchers in other fields. This results in the observatory appealing to a wide range of interests and to concomitant long-term support.

5.1.1 Meteorological Observations

- Temperature measurements are of widespread scientific interest. Temperature can affect seismometer mass positions in shallow installations. Temperature variations may also affect digitizer offsets and noise levels, as well as tape recorders and fixed disks in the seismic data acquisition systems.
- Both absolute pressure measurements and pressure variations are of interest to the meteorological and seismic communities. Absolute pressure measurements give a measure of the total mass of the atmospheric column above the seismic station. These can be used to correct the gravitational attraction of the atmosphere on the mass of the vertical seismometer, resulting in lower vertical seismic noise levels in the long-period band. Variations in atmospheric pressure, especially those caused by wind, are a prime cause of local tilting of the ground surface (attenuating with depth), resulting in increased horizontal seismic noise levels in the long-period band. This is because a typical inertial horizontal seismometer cannot distinguish between tilt and horizontal acceleration. Therefore, recordings of atmospheric pressure variations have potential use in removing some of this tilt noise from the horizontal seismic data.
- Wind velocity (speed + direction) is also of interest to the meteorological and seismic communities. The wind has a strong effect on seismic noise levels, especially in shallow installations. (See also discussion on pressure variations above.)
- Relative humidity is of interest to the meteorological community. When relative humidity is used in conjunction with absolute pressure and GPS information, a measure of precipitable water vapor may be obtained.

5.1.2 Other Geophysical Observations

- Rapid changes in the intensity and/or declination of the Earth's magnetic field are of interest to the geomagnetic community as well as to seismologists. Magnetic storms can affect modern broadband seismometers if they are not magnetically shielded, especially vertical seismometers. (The spring suspending the mass is especially affected.) Example: The borehole seismometer installed at station COLA (College Geophysical Observatory) is installed in a PVC casing rather than the usual steel casing that provides magnetic shielding. At this high latitude (Alaska), magnetic storms are especially intense, and the effects from them can occasionally be seen in the vertical broadband data from this instrument, at periods longer than 30 sec.
- Measurements of the absolute intensity and declination of the magnetic field are much more difficult, because it is necessary to install the magnetometers in a magnetically clean environment (no steel or iron in the area, for

example). Most seismometer installations use a lot of steel or iron: Most borehole casings are steel and most normal buildings have steel or iron in them (nails, reinforcing bars, etc.). However, it may be convenient to colocate magnetic observatories with seismic observatories if care is used in relative placement of instruments and in the materials used (as in the case of College Geophysical Observatory, Alaska).

- Geodetic GPS measurements have become quite important to seismologists as well as to those responsible for accurate location of GPS reference sites, precise determination of GPS orbits, and a host of other considerations. Of special interest to seismologists is the determination of the velocity vectors for the Earth's tectonic plates and the measurement of strain accumulation near major fault systems. In addition, real-time GPS information may be used by meteorologists in conjunction with temperature and relative humidity (see meteorological discussions above).

It may be impossible to specify a complete list of possible long-term observations one could or should make at seismological observatories. However, there are some interesting possibilities, given the semipermanent nature of the established infrastructure at a global seismological observatory:

- Luminous phenomena (LP): possible association with earthquakes
- Electromagnetic phenomena (VLF, RF, etc.): possible earthquake precursory signals
- Strain (both volumetric and linear)
- Tilt (pure rotational measurement)

5.2 Global Communication Systems

Historically, it has not been cost-effective to telemeter seismic data to Data Collection Centers in real time, because expensive dedicated circuits were necessary. The cost of such international circuits can be as much as \$20 000 per month, per station. However, since the early 1990s, the Internet has become almost ubiquitous. It has become possible to use the Internet to telemeter data in near real time for the minimal cost (tens to hundreds of dollars per month) of a "tail circuit" from the nearest Internet Service Provider (ISP) to the seismic station. The downside of using a link like the Internet is intermittent outages, so it is necessary to use a re-transmission protocol to fill in data gaps on the receiving end. Dedicated circuits are still necessary when the data *must* arrive within seconds.

Another relatively inexpensive development in the late 1990s is international VSAT satellite link capability. For an equipment investment of about \$50 000, one can obtain 19.2 kbps full duplex service for about \$200 per month (depending on country tariffs and the satellite used). Further advances may make satellite internet communications as inexpensive as any of the current commercial satellite TV connections.

5.3 Instrumentation Improvements Needed

5.3.1 Bandwidth and Dynamic Range

Since the analog-recording days of the WWSSN, available bandwidth has increased from 2 decades (0.01 to 1 Hz) to 8 decades (10^{-6} to 100 Hz) for a fully capable GSN or equivalent very broadband station. Also, dynamic range has increased from 40 dB to 140 dB (or more, at very low frequencies).

However, earthquake amplitudes span a range in excess of 240 dB (lowest peak earth acceleration at 400 sec period to maximum acceleration very near the epicenter of a large earthquake). Frequency ranges of seismological interest are mostly covered with the above 8-decade range, but one can argue for an even larger frequency range, depending on the application. Perhaps 10^{-8} to 10^{3} Hz is sufficient for practically all foreseeable seismological applications.

With current (2002) technology, it is necessary to use more than one instrument to cover all frequency and amplitude ranges of interest. The situation is much better than it was historically (for example, the prefiltered, oversensitive recording used in the SRO). However, just as was required in using SRO data, one must still patch together waveforms from different instruments in order to derive a complete time history of ground motion from a great earthquake at all distances from epicentral to teleseismic. The ideal way to rectify this situation is to build a single system (seismometer plus acquisition system) that is capable of covering the entire bandwidth and dynamic range requirement *in a single data stream*.

Although this is obviously not an easy task, it should be the ultimate goal. This goal translates into the following nominal system specifications:

- Dynamic range: 250 dB (corresponds to a 42-bit digitizer)
- Noise level: equal to or smaller than -220 dB relative to 1 m sec^{-2} peak-to-peak
- Clip level: at least 39.2 m sec^{-2} (4 g) peak-to-peak
- Frequency range: 10^{-8} to 1000 Hz

5.3.2 Noise Reduction

Installation techniques are always subject to improvement as instrumentation improves (see previous discussion on installation practices). As instrument noise levels decrease, frequency bands increase, and instrument sizes shrink, better installation techniques will become both necessary and more feasible. For example, the development of a high dynamic range "seismometer on a chip" would make it possible to drill very small (perhaps 2 cm diameter) boreholes instead of the 18–25 cm boreholes currently needed. This would make it possible to drill the holes much deeper in order to achieve greater attenuation of surface-generated noise. In addition, verticality requirements, currently typically specified at less than 3–5° because of instrument leveling limitations, should be greatly eased in new seismometers having a very large dynamic range.

Very inexpensive, very small, high dynamic range seismometers should also make possible the easy deployment of massive "arrays" of seismometers. This should, in turn, make possible the use of noise-reducing array signal processing techniques on data from very large numbers of observatory-quality global seismic stations.

5.3.3 New Instruments Needed

- Seismometers capable of 250 dB dynamic range (noise levels less than the USGS New Low Noise Model, clip levels of 2 g peak).
- Digitizers capable of 250 dB dynamic range at 2000 samples per sec (42-bit).
- Pure tilt (rotational) meters with noise levels less than 10^{-12} radians over a 0.0001 to 1 Hz range. Such instruments can be used to remove the noise contamination in horizontal accelerometers (inertial seismometers) caused by tilt noise at or near the Earth's surface, as well as to separate the rotational component of surface waves from the translational component.
- Clocks that do not need an antenna, with absolute accuracy of 1 μsec and drift rate less than 1 μsec per 10 y of operation.
- On-site recording should be on solid-state nonvolatile media with no moving parts, with enough capacity to contain at least one year of data.
- Telemetry of all data from acquisition system should work from any position on Earth (perhaps through LEO satellite systems).
- All of the above should have extremely low power consumption, so that the entire system will operate for at least one year between battery changes or battery charges. Complete seismic station systems consuming higher amounts of power will require connection to reliable commercial power or a very robust backup power system.

Acknowledgments

We thank Rhett Butler for providing the computer file for Figure 1.

Bibliography

Agnew, D.C. and J. Berger (1978). Vertical seismic noise at very low frequencies. *J. Geophys. Res.* **83**, 5420–5424.

Ekstrom, G. and M. Nettles (1997). Calibration of the HGLP Seismograph Network and centroid-moment tensor analysis of significant earthquakes of 1976. *Phys. Earth Planet. Inter.* **101**, 219–243.

Incorporated Research Institutions for Seismology (1984). "Science Plan for A New Global Seismographic Network," December 1984.

Institute of Geophysics, State Seismological Bureau, and the Albuquerque Seismological Laboratory, US Geological Survey (1987). The China Digital Seismograph Network, a Joint Report, October 1987.

Lamont-Doherty Geological Observatory of Columbia University (1971). "High-Gain, Long-Period Seismograph Station Instrumentation," Vols. I, II, 31 March 1971.

Peterson, J., H.M. Butler, L.G. Holcomb, and C.R. Hutt (1976). The Seismic Research Observatory. *Bull. Seismol. Soc. Am.* **66**, 2049–2068.

Peterson, J., C.R. Hutt, and L.G. Holcomb (1980). Test and calibration of the Seismic Research Observatory. *USGS Open File Report 80–187*.

Peterson, J. and C.R. Hutt (1982). Test and calibration of the Digital World-Wide Standardized Seismograph. *USGS Open File Report 82–1087*.

Peterson, J. and E.E. Tilgner (1985). Description and preliminary testing of the CDSN Seismic Sensor Systems. *USGS Open File Report 85–288*.

Peterson J. and C.R. Hutt (1989). IRIS/USGS plans for upgrading the Global Seismograph Network. *USGS Open File Report 89–471*.

Peterson, J. (1993) Observations and modeling of seismic background noise. *USGS Open File Report 93–322*.

Savino, J.M., A.J. Murphy, J.M.W. Rynn, R. Tatham, L.R. Sykes, G.L. Choy, and K. McCamy (1972). Results from the high-gain long-period seismograph experiment. *Geophys. J. R. Astron. Soc.* **31**, 179–203.

Ward, P.L. and G. Hade (1970). Design and deployment of five high-gain, broadband, long-period seismograph stations. Technical Report, Lamont-Doherty Geological Observatory of Columbia University, 28 August 1970.

Editor's Note

Readers are urged to visit the website of the Federation of Digital Broad-Band Seismograph Networks (FDSN), *http://www.fdsn.org/*, for the current information about the global seismograph stations. See also Chapter 18, Seismometry, by Wielandt; Chapter 19, Seismic Noise on Land and on the Sea Floor, by Webb; and Chapter 23, Seismometer Arrays, by Douglas. For a brief history of WWSSN, see Section 3.1.2 in Chapter 17 by Lee.

21

The Structure and Interpretation of Seismograms

Ota Kulhánek
Uppsala University, Uppsala, Sweden

1. Introduction

Interpretation of seismograms is devoted to the art of identification of various types of seismic waves (phases) that appear on seismograms and often generate a rather complex structure. Correct identification of recorded phases with respect to travel paths through the Earth is the doorway to any research using observational seismic data. The appearance of a seismogram reflects the combined effects of the source, the path of propagation, the characteristics of the seismograph, and the ambient noise at the recording site. Understanding the complicated nature of seismograms requires knowledge of seismic source physics, the structure of the Earth, and seismic wave propagation. Above all, it requires long experience in examining seismograms. A veteran seismogram analyst frequently reveals and correctly interprets record features invisible to novice interpreters, recalling Emil Wiechert's "...jede Zacke, jede Zunge zu erklären..." ("every jerk, every wiggle should be explained"). An experienced analyst will usually be able to decipher even complicated records merely by visual inspection. It is, however, desirable to confirm the interpretation by using the proper travel-time tables and synthetic seismograms.

If the studied earthquake exceeds a certain magnitude, its seismic waves can be recorded by seismographs located at places everywhere in the world, on the surface of the Earth, in boreholes or abandoned mines, or on the ocean bottom. Generally speaking, at increasing distance from the epicenter the wave amplitudes diminish due to anelastic attenuation (rocks are not perfectly elastic), geometrical spreading (the area of the wave front increases with increasing propagated distance), and energy loss at interfaces (reflection, refraction, mode conversion, diffraction, scatter). The anelastic attenuation is frequency-dependent (high frequencies are subject to high attenuation) so that high-frequency seismic signals attenuate rapidly and may be recorded only by proper seismic instruments placed at relatively short epicentral distances.

Due to the internal structure of the Earth, at certain distances, e.g., around 20° and 144°, concentration (focusing) of the energy of traveling seismic waves takes place. Seismic signals recorded close to these distances often show an increase of amplitude even with increasing distance from the epicenter.

Source radiation is another factor that influences the amplitude of arriving seismic waves. Tectonic earthquakes, in contrast to underground explosions, cannot be treated as spherically symmetric point sources because the radiated seismic energy transported by certain wave types is beamed in certain directions. It is therefore likely that seismographs deployed at different azimuths will show different amplitude ratios between arriving *P* and *S* waves.

The plot in Figure 1 illustrates some of the basic properties of recorded body and surface waves. There is a sharp *P* onset, followed after approximately $4\frac{1}{2}$ min by a clear *S* onset. About 2 min after the *S* arrival, we observe a gradual increase of the amplitude due to the arriving surface wave of Rayleigh type, *LR* (vertical-component seismogram). At the beginning of the *LR* wave, the wave period is about 40 sec, but it decreases to about 25 sec after three or four swings, clearly demonstrating the dispersive character of the *LR* wave, which in this case manifests a continental propagation path signature. Furthermore, the *LR* wave is the largest signal in the seismogram, indicating that the earthquake occurred at relatively shallow depth.

Since various seismic phases propagate with different velocities along different paths, they arrive at the recording site well separated in time. Theoretically, therefore, there should be no difficulty in recognizing these phases in seismograms. Indeed, in this particular case (Fig. 1, chosen for didactic reasons), the noise level (cf. the portion of the record preceding the *P* onset) is very low when compared with amplitudes of the recorded *P* or *S* waves. Here, the decay of *P* wave amplitudes, so called *P*-coda, is rather rapid so that

INTERNATIONAL HANDBOOK OF EARTHQUAKE AND ENGINEERING SEISMOLOGY, VOLUME 81A ISBN: 0-12-440652-1

not only the P onset but also the S onset can easily be identified. However, it is more common that the analyst has to examine records with high background noise and to identify various wave arrivals masked by noise and/or onsets due to other events (Fig. 2), which often is a difficult task. Note also that, except for the first P onset, all further arrivals are obscured by coda of preceding phases, so that on the seismogram there is virtually no interval of quiescence between individually arriving phases.

To complicate the matter further, seismic waves encountering a discontinuity are reflected and/or refracted and an incident P or S wave gives rise to both P and S waves, so-called mode conversion. Thus seismograms often show a large number of distinct phases, distributed in time, which have traveled along different propagation paths and which have been subjected to different mode conversions P-to-S or S-to-P.

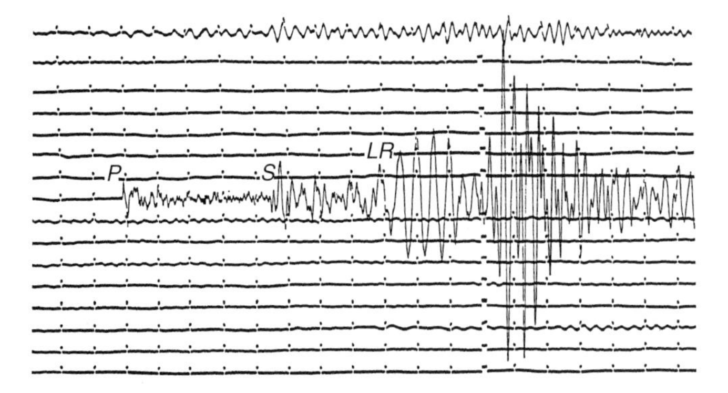

FIGURE 1 Seismogram made at Uppsala, Sweden, by a long-period, vertical-component instrument. This earthquake occurred on Crete, Greece, on 12 June 1969 ($M = 6.2$, $h = 25$ km) at a distance of 26° from Uppsala. The onset of P is distinct. The S wave arrives about $4\frac{1}{2}$ min after P and is also easily recognizable. The trace also shows a clear dispersive Rayleigh wave train, starting 2–3 min after S. The sudden increase of LR-wave amplitude starting approximately 6–7 min after S is due to the arriving Airy phase. Time advances from left to right and there is 1 min between successive time marks (small upward offsets). (Reprinted with modifications from Kulhánek (1990) with permission from Elsevier Science.)

The ground is practically always in motion. Various human activities such as traffic, construction work, industrial activity, etc. generate so-called cultural noise with dominant frequencies usually above 1 Hz. A similar type of noise is also generated by the action of wind, smaller water basins or rivers, and so on. Various interactions between atmospheric effects, oceans, and the solid Earth give rise to microseismic noise, sometimes also called ocean microseisms. The dominant frequencies of microseisms occupy a broad low-frequency range from less than 0.01 Hz to, say, 0.5 Hz, that is, periods from 2 sec to more than 100 sec. The most common microseisms have more or less regular periods of about 6 sec. Cultural noise is recorded with standard instruments at epicentral distances usually not exceeding several tens of kilometers and is, therefore, only of local importance. Microseisms, on the other hand, can travel hundreds of kilometers and, hence, are a continental phenomenon.

2. Seismogram Characteristics

In the sections that follow, we will focus the presentation on seismograms from tectonic earthquakes. For the sake of comparison, several examples from volcanic earthquakes and from unusual seismic sources will also be included.

Let us start with a brief summary of the most important characteristics of seismic waves that deserve our attention and that are invaluable in any seismogram interpretation. First, different waves travel with different velocities. At any epicentral distance, P is recorded first, followed by S, LQ (Love wave) and LR (Rayleigh wave). Second, different waves are polarized in different ways (P linearly, LR elliptically). This provides a means of identifying phase types. Third, various phases are characterized by their amplitude, period, dispersion, etc.

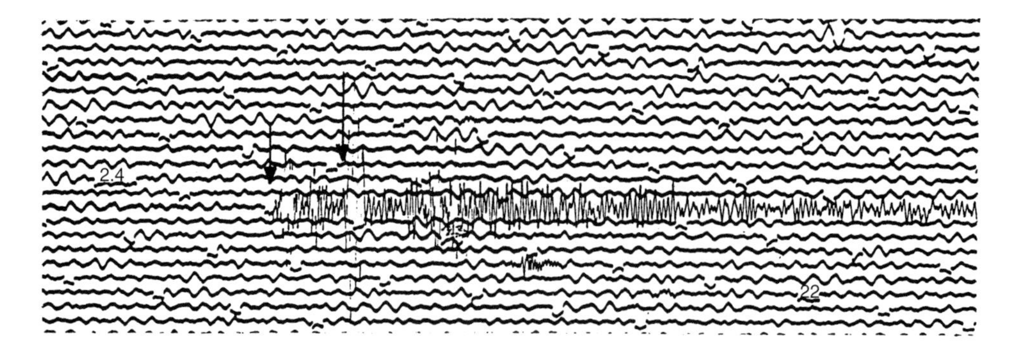

FIGURE 2 Seismogram made at Kiruna, Sweden, by a short-period, vertical-component seismograph. The earthquake occurred in Kuril Islands on 7 May 1996 ($M = 5.7$, h = 52 km) at an epicentral distance of 62°. The P wave arrives at 23 : 30 : 14.5 and is preceded and to some extent masked by a wave train from an explosion, in a local iron ore mine, which arrives about 14 sec earlier. Both arrivals (arrows) can easily be misinterpreted as onsets belonging to the same event.

The propagation of seismic waves through the Earth's interior is governed by exact physical laws similar to the laws of light waves in optics. In general, the wave velocity increases with depth and, consequently, seismic energy follows rays that are curved with the concave side upward providing the shortest time-path through the Earth. For a spherical Earth, it is common practice to express the distance between the focus and the recorded station as the angle (denoted Δ) subtended at the center of the Earth by the arc between the source and receiver ($1^\circ \approx 111$ km).

In the following discussion, we introduce several categories of seismic events. The classification is based on the distance between the event and the recording site (i.e., on the epicentral distance), which in turn governs propagation paths along which seismic waves travel through the Earth's interior. Another categorization used is that according to focal depth. Earthquakes with focal depth between 0 and 70 km (about 80% of all earthquakes) are shallow-focus earthquakes. Intermediate-focus earthquakes have focal depth between 71 and 300 km. Shocks deeper than 300 km are called deep-focus earthquakes. The main reason for these classifications, which do not provide any sharp line of demarcation, is that seismic waves from the different categories may be discriminated from each other due to their different appearances on seismograms.

There are numerous ways of classifying seismogram on the basis of epicentral distances. We will consider events recorded at distances shorter than about 1° as local events. Regional events are events recorded at distances between approximately 1° and 10°. Local and regional distance seismograms are dominated by seismic energy that has propagated through the crust and/or along the Moho discontinuity. Corresponding seismic waves are called crustal waves. Seismic waves propagating through the upper mantle are recorded at distances, roughly speaking, between 10° and 30°. P and S waves recorded at these distances are complex due to interactions with the upper-mantle discontinuities. At recording distances from about 30° to 103°, seismic waves propagate through the lower mantle. Corresponding seismograms are relatively simple due to smooth velocity distribution between 700 km depth and the core–mantle boundary (CMB). For epicentral distances larger than about 103°, seismic waves propagate through the core or are diffracted by the core and seismograms again become complicated. Events recorded at distances larger than about 20° (some workers prefer the distance of 30°) are called teleseismic events or simply teleseisms.

2.1 Crustal Waves; Recording Distances 0–10°

In order to explain the structure of seismic records made at epicentral distances between 0° and 10°, let us first assume a simplified structural model of the crust, depicted in Figure 3. Note that for the distances considered here, we neglect the effects of curvature of the Earth's surface.

Consider waves (rays) leaving the focus F and reaching the recording stations S_1, S_2, and S_3. Since the source radiates both P and S waves, there will be direct longitudinal and transverse waves recorded along the Earth's surface. These waves have ray paths such as FS_2 (see Fig. 3) and are encoded as Pg and Sg or sometimes as $\bar{P}$ and $\bar{S}$. The g indicates the travel path, which for seismic events in the upper crust (most of the crust earthquakes) is entirely confined to the granitic layer. A reflected ray (e.g., ray path FR_1S_1) is also possible from the

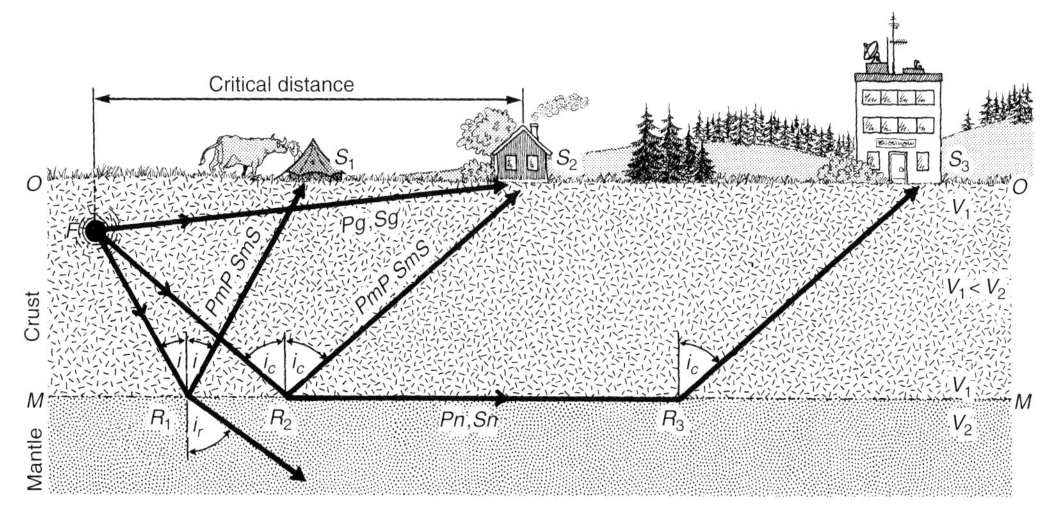

FIGURE 3 Principles of wave propagation from the focus of earthquake F through a simplified one-layer crust model. Symbols O and M designate the Earth's free surface and Moho discontinuity, respectively. S_k is the kth recording seismographic station, i is angle of incidence, i_r is angle of refraction, i_c is critical angle, and V is velocity of propagation for P or S. R_k are the points of reflection at the Moho discontinuity for rays that travel to the kth station. (Reprinted from Kulhánek (1990) with permission from Elsevier Science.)

Moho, and the corresponding reflected P and S waves are labeled as PmP and SmS, respectively.

It follows from Figure 3 that, as the epicentral distance increases, the angle of incidence (i) and the angle of refraction (i_r) also increase. At a certain critical epicentral distance, $i_r = 90°$, the energy of the refracted ray does not penetrate into the mantle but travels along the Moho discontinuity (cf. the ray path $FR_2R_3S_3$). The associated angle of incidence, i, called the critical angle, is denoted i_c. Corresponding P and S waves, called head waves, recorded at S_3 are labeled Pn and Sn, respectively. Waves propagating along discontinuities separating two layers with two different velocities propagate with the higher of the two velocities. Thus, Pn and Sn waves depicted in Figure 3 travel with velocities of the uppermost mantle. As can be seen in the figure, Pg and Sg exist for all epicentral distances from $\Delta = 0$ and outwards; whereas Pn and Sn phases cannot be observed at distances shorter than that corresponding to the location of the station S_2 (Fig. 3), i.e., at distances shorter than the critical distance, which for the continental crust is about 100 km.

It should be stressed that in the above presentation and in Figure 3, a number of simplifications have been made. First, a homogeneous crust is in most cases a rather poor approximation of the true structure. It is common to employ two crustal layers, separated by the Conrad discontinuity, to interpret crustal phases. In special studies, multilayered crustal models are used. Second, the true Moho and Conrad discontinuities are not planar and strictly horizontal boundaries. In reality, they dip and have some degree of undulation. Last, the assumption of Pg or Sg traveling as direct waves (P or S) over large distances, as shown in Figure 3, is again a gross simplification. In practice, direct waves, denoted $\bar{P}$ and $\bar{S}$, are recorded only from local events, i.e., at very short epicentral distances usually not exceeding several tens of kilometers. (See Plate 21.1: Note that all plates are placed on the attached Handbook CD, under the directory of \21Kulhanek.) Pg and Sg waves are then understood as channel or refracted waves traveling along less pronounced boundaries within the granitic layer.

Accepting the two-layer model, we realize that, starting from about 100 km, we record additional phases, namely the refracted P and S traveling along the Conrad discontinuity. An asterisk, P^* and S^*, indicates this phase. An alternative code sometimes used is Pb and Sb. The b refers to basaltic layer.

At very short distances, less than 150 km or so, and for a shallow focus in the upper crust, the first wave arriving at the recording station is $\bar{P}$ or Pg, traveling with a velocity of about 6 km sec^{-1}. For distances larger than critical but less than about 150 km, Pg is followed by P^* and subsequently by Pn. P^* and Pn travel with velocities of about 6.6 and 8.0 km sec^{-1}, respectively, i.e., significantly faster than Pg. Therefore, at distances larger than approximately 150 to 200 km (depending upon the true propagation velocities and the thickness of the granitic and basaltic layers), crustal waves change their order of arrival. For distances larger than about 200 km, the first-arriving phase is Pn, P^* arrives next, and then Pg. Obviously, this is true only for continental travel paths. Seismograms from earthquakes beneath the sea bottom, made at island or coastal stations, will not show Pg or Sg phases since there is no granitic layer in the oceanic crust. Similarly, earthquakes originating in the lower crust, beneath the Conrad discontinuity, do not produce Pg or Sg phases. Hence, first arrivals on records from these earthquakes will be Pn or P^*.

For ease of phase identification, rather than relying on the epicentral distance, which of course may not be available, we can make use of time differences between arrivals of various phases. For example, if the arrival-time difference S–P is less than about 20 sec, the first wave within the P and S group to arrive at the recording site is probably Pg (or $\bar{P}$) and Sg (or $\bar{S}$), respectively. If on the other hand, the difference is more than 25 sec, the first arrival is most likely Pn. Details obviously depend upon the true structure and upon the focal depth. It should be emphasized that only seldom are all the above phases identified on one record. The usual case is that one or several of these waves are too weak or hidden in the background noise to be discernible on the seismogram.

With a certain time delay following the P phases, proportional to the epicentral distance, the crust S waves arrive in the same order as P waves. Thus, for local events (earthquakes, mine explosions, quarry blasts, etc.) the order of S onsets will be Sg, S^*, Sn, while for events from distances larger than about 300 km we observe first Sn followed by S^* and Sg.

P or S waves reflected from Moho are rather scarce and difficult to identify. The best chance of recording PmP or SmS is at short distances where the contamination by Pg and Pn (or Sg and Sn) is not severe. Sometimes, so-called depth phases, arriving between Pn and Pg, are present on seismograms. These leave the focus as P waves, travel upward with a small angle of incidence, are reflected as P at the free surface, and continue further as Pn. Notation for this phase is pPn. Similarly, sPn denotes a depth phase leaving the focus as S, converted through the reflection at the free surface and continuing as Pn. Both pPn and sPn are of great importance in focal depth estimations, but at the same time are rather difficult to identify on actual records.

Short-period S waves multiply reflected between the free surface and Moho, or between other crustal velocity discontinuities, interfere with each other and give rise to a wave group labeled Lg, which follows the Sg arrival. The g again refers to granitic layer. At distances of several hundred kilometers and larger (continental paths), Lg waves, propagate as guided waves through the crust, as they supercritically reflect from the Moho and Earth's surface. Lg waves propagate with a typical velocity of about 3.5 km sec^{-1}, and have large amplitudes, especially on the horizontal components. Lg is usually recorded at epicentral distances of about 5° and larger, although there are reports of Lg propagating over distances of several thousand kilometers (Kulhánek, 1990) and being recorded as prominent phases on the seismograms.

Near-surface regional events (earthquakes, industrial explosions, rockbursts, etc.) also generate short-period surface waves of Rayleigh type, labeled *Rg*. The presence of short-period *Rg* in the seismogram is a reliable indicator of a very shallow event with focal depth of the order of one or a few kilometers. On the other hand, if short-period *Rg* waves are absent (near station, epicentral distance of several hundred kilometers or less), we are concerned with a deeper natural event, i.e., with a crustal earthquake at a depth most likely between about 5 and 25 km, since all types of man-made events as well as triggered mine tremors can be excluded. Short-period *Rg* waves, which travel as guided waves through the crust across continental paths with velocity of 3 km sec^{-1} or slightly higher, are attenuated more strongly than crustal body waves and their range of propagation is therefore limited to distances less than about 600 km (Båth, 1983). However, at short epicentral distances, of say, less than 100 or 200 km, the *Rg* phase from a near-surface event often dominates the recorded wave train (Fig. 4).

Seismic waves from local and regional earthquakes of low or moderate magnitude are of short period and are therefore almost exclusively recorded by short-period seismographs. The seismogram duration depends upon the magnitude, but for events with $M < 5$ generally does not exceed 5 min or so. The number of clear pulses seen on the record, indicating arrivals of the whole group of crust *P* and *S* waves, is often larger than one would expect from models displayed in Figure 3, demonstrating the complex structure of the crust. For continental trajectories, the most prominent phase is usually *Sg*, best recorded by horizontal-component instruments. Generally, *Sg* arrivals start with large amplitudes that successively decrease as the time increases, forming the coda of the event. Coda duration is related to the magnitude. *Rg* phases, best recorded on vertical-component seismograms, often display a clear dispersion. As an example, a record from a regional earthquake is displayed in Figure 5. The first discernible phase is *Pn*, best recorded on the vertical component. It is followed by *Pg*, *Sn*, and *Sg*, which are recognizable on all three channels. Largest amplitudes are exhibited by the *Lg* wave trains on the two horizontal channels. The focal depth of about 15 km prevents the development of short-period *Rg* waves. Note the high-frequency character of all recorded phases.

Strong events (magnitudes about 6 and larger) recorded at local or regional distances ($\Delta < 10°$) will produce seismograms with duration of tens of minutes, especially on long-period records. Associated large amplitudes of ground vibrations often saturate the analog recording system (clipped records) and the resulting seismogram is not of much use except for measuring the arrival time and polarity of the very first recorded phase. In this and many other respects, digital systems with higher dynamic range are superior to analog instruments.

Similar to the low-velocity-layer, LVL, in the upper mantle, there is also a low-velocity channel in the deep ocean. Depending upon the salinity and temperature of water, the sound velocity decreases from the sea surface to a minimum of about 1.5 km sec^{-1} at about 700–1300 m depth and increases again from that depth to the bottom. The depth region of low velocity in the ocean, called SOFAR (sound fixing and ranging), provides extremely favorable conditions for long-distance propagation of a special type of high-frequency seismic wave.

Island and coastal seismographic stations frequently record these waves termed *T* waves (tertiary waves) arriving after *P* and *S* and characterized by propagation within the oceans as ordinary sound waves. Seismic waves emitted by earthquakes near the sea bottom or by submarine volcanic eruptions are refracted through the sea floor and propagate as sound

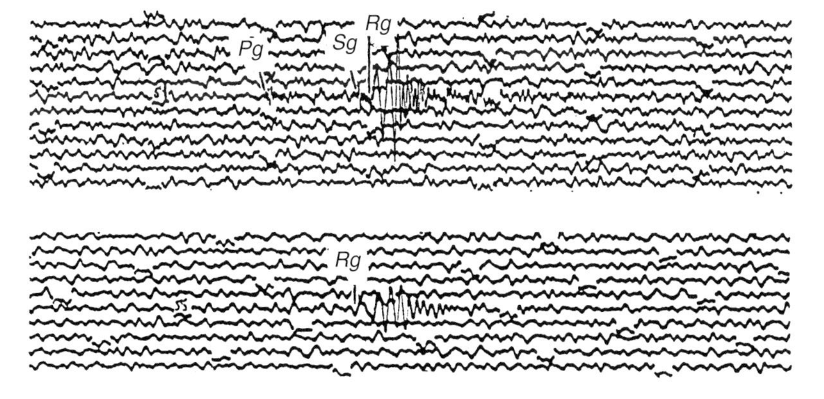

FIGURE 4 Two records with remarkably well developed short-period *Rg* waves with normal dispersion within a period range from 3 to 1 sec. *Rg* dominates the records, emphasizing the shallow focus depth, of probably not more than 1–2 km. Top: rockburst of 19 January 1963, magnitude unknown, recorded at Uppsala, Sweden, $\Delta = 148$ km; bottom: rockburst of 19 January 1963, recorded at Göteborg, $\Delta = 300$ km. (Reprinted with modifications from Kulhánek (1990) with permission from Elsevier Science.)

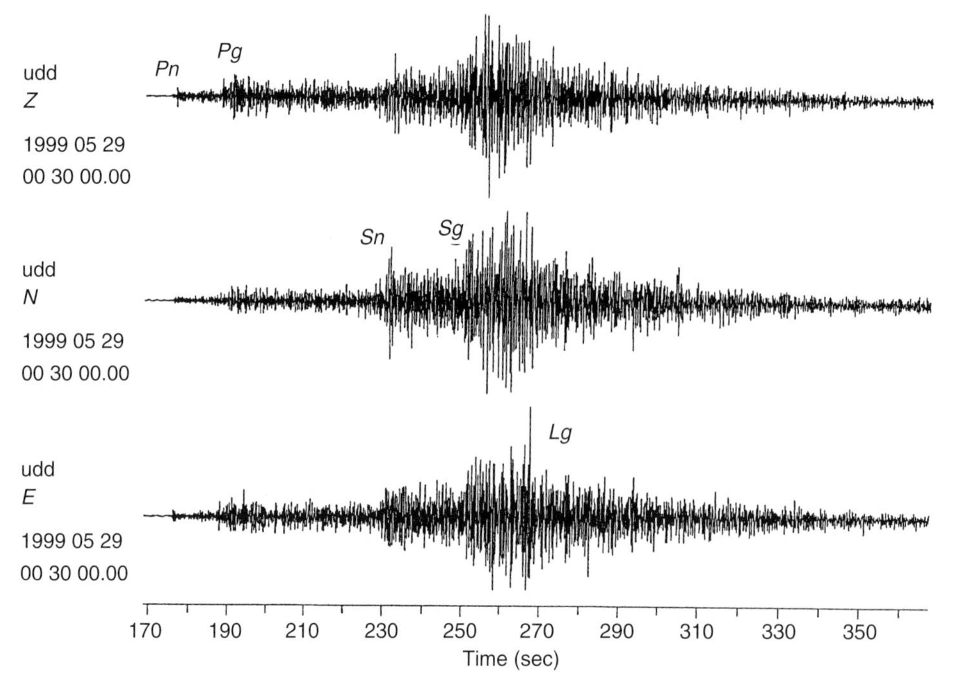

FIGURE 5 Vertical (Z) and horizontal (N, E) component seismograms from a moderate size regional earthquake. The event occurred off coast of southern Norway on 29 May 1999 (magnitude $M_L = 3.9$, $h = 10$ km). The traces are analog displays of broadband digital recording made at Uddeholm, Sweden, at an epicentral distance of 530 km.

(longitudinal) waves through the ocean. The propagation of *T* waves, generally through the SOFAR channel (Bullen and Bolt, 1985) or by multiple reflections between the sea floor and the sea surface (Båth and Shahidi, 1974), is very efficient and observations at distances as large as about 80° have been reported (Kulhánek, 1990). The first observation of *T* waves was made by D. Linehan in 1940.

T waves are best recorded by ocean-bottom seismometers (OBS) and by coastal and island stations (see Plate 21.2). However, instruments deployed further on land sometimes also record clear *T* waves after a water–land conversion of sound waves into *P*, *S* or surface waves propagating over the land portion of the total transmission path. If this is the case, the labeling is *TPg*, *TSg*, and *TRg*, reflecting the fact that the path of propagation over the land is within the crust. Examples of recorded *TSg* phases are given in Figure 6 and in Plate 4. *T* waves are short-period waves, with periods usually less than 1 sec, recorded exclusively by short-period seismographs. On records, they often exhibit rather monochromatic oscillations with a gradual increase and decrease of amplitudes of total duration up to several minutes. When compared, for example, with *P* waves, there is no sharp onset in the *T* wave group, which obviously creates difficulties when reading the *T* arrival times or when identifying phases within the *T* wave group. In general, there is great variety in the appearance of *T* phases due to the dependence upon the bottom topography in the vicinity of generation, oceanic stratification, and water–land conversion and transmission (Båth and Shahidi, 1974).

Observed *T* phases have proved very useful in discriminating between underground nuclear explosions, detonated beneath oceanic islands, and tectonic earthquakes (Adams, 1979). For this type of explosion, the energy is injected directly into the SOFAR channel and recorded *T* phase amplitudes often exceed those of associated *P* wave by a factor of up to 30 (see Plate 4).

As follows from the above description, for epicentral distances less than about 10°, the wave propagation is rather complicated. The seismogram appearance varies from place to place due to regional variations in the crust structure and, consequently, for this distance range it is difficult to list generally valid clues for record interpretation. Nevertheless, some of the following principles may guide the analyst to correctly read seismograms of local and regional earthquakes.

Predominant periods of recorded crust phases such as *Pg*, *P**, *Pn*, *Sg*, *S**, *Sn*, etc. are normally less than one second and hence are best recorded by short-period instruments. *Rg* periods are usually not longer than several seconds.

1. It has often been observed that *Sg* has the largest amplitude (for cases when large short-period *Rg* is missing), best seen on horizontal-component records.
2. For epicentral distances less than about 200 km (depending upon the crust structure and focal depth), the first-arriving phase is *Pg*. For larger distances, *Pn* arrives first.

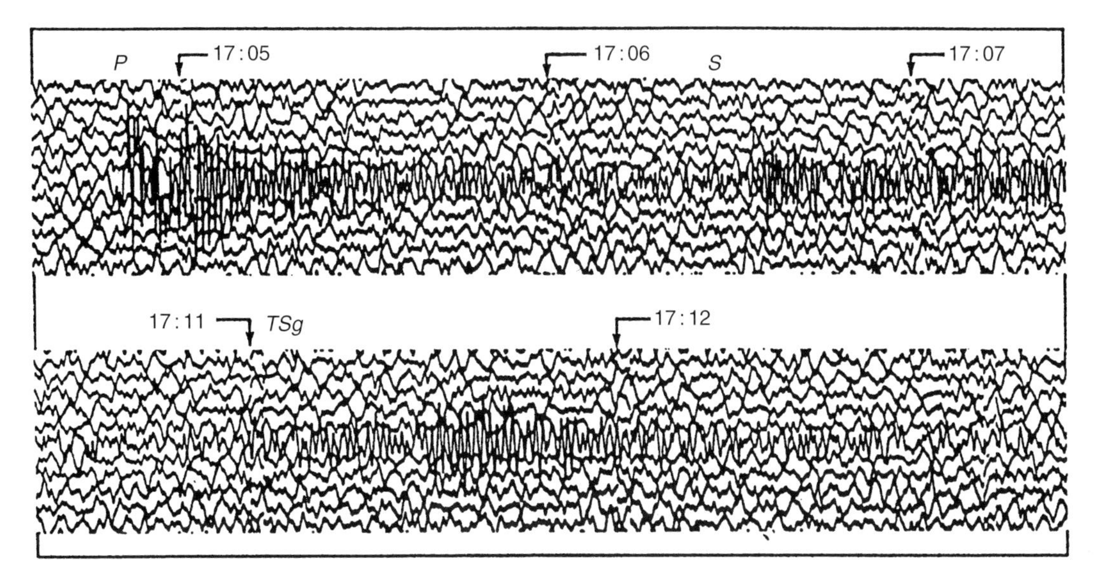

FIGURE 6 Short-period vertical-component record from a shallow ($h = 33$ km) earthquake in Norwegian Sea made at Umeå, northern Sweden. This earthquake occurred on 21 November 1967 ($m = 5.4$) at a distance of 10° from Umeå. The seismogram shows clear *P* and *S* onsets, separated by 107 sec. Approximately 6 min after *P*, an onset labeled *TSg* is identified on the record. It corresponds to a wave propagating through the water as a sound wave and, subsequent to a water–land conversion and refraction, as *Sg* over the land path. In this particular case, the land path is about 1/3 of the total travel length. The *TSg* wave shows oscillations with periods around 1 sec and gradually increasing and decreasing amplitudes. The whole *TSg* wave train lasts for about 2 min. (Reprinted from Kulhánek (1990) with permission from Elsevier Science.)

3. Near-surface events from distances less than about 600 km often generate short-period *Rg* with clear dispersion, best seen on vertical channels.
4. Local and regional earthquakes of low or moderate magnitude are characterized by short total record duration, usually not longer than several minutes.
5. Island and coastal seismographic stations frequently record various kinds of *T* phases.

It is not always possible for analysts to correctly identify crust phases from the records of a single station, although this may be easier if several stations of a network are read together. If there is doubt about correct interpretation, a phase should simply be identified as *P* or *S*.

2.2 Body Waves; Recording Distances 10–103°, Shallow Events

Seismologically speaking, the mantle differs from the overlying crust also in the fact that, in the first approximation, it may be considered as a laterally homogeneous, i.e., as a spherically symmetric body. Seismic wave velocities indeed increase with depth; however, the regional (lateral) irregularities typical for the crust are less distinct in the mantle. Some workers consider the distance range between 20° or 30° and 103° as ideal for recording not only the direct *P* and *S* waves but also the whole family of reflected and converted waves. Travel paths of these waves are dominated by the mantle and corresponding seismograms are relatively simple.

To explain various features of waves traveling through the mantle, let us consider the Earth's cross section, a surface focus event, and travel paths of the more important body waves depicted in Figure 7.

At distances around 10°, *Pn* and *Sn* become difficult to identify in the records, except in some shield areas and other regions with relatively uniform structures. Instead, *P* and *S* phases become visible on seismograms. *P* is usually stronger on the vertical component, while *S* is more clearly seen on horizontal components. *S* often exhibits wave trains with longer periods when compared with corresponding *P*. Large-amplitude *S* waves are often observed at distances of up to about 100°.

Body waves that turn within the mantle and undergo no reflection between the focus and the recording station are labeled with a simple symbol *P* or *S*. Rays corresponding to travel paths of these direct waves (*P* or *S*), also called elementary waves or main waves, are displayed in Figure 7a. They depict paths of least travel time from the focus of the earthquake to the recording site. Waves reflected one or more times from the free surface give rise to singly or multiply reflected *P* or *S*. For example, the direct *P* reflected from the free surface back into the mantle once or twice, is called *PP* or *PPP*, respectively. In the same way, we have also *SS*, *SSS*, etc. Each letter, *P* or *S*, in the phase name defines one leg of the propagation path. Considering also the conversion from *P* to *S*,

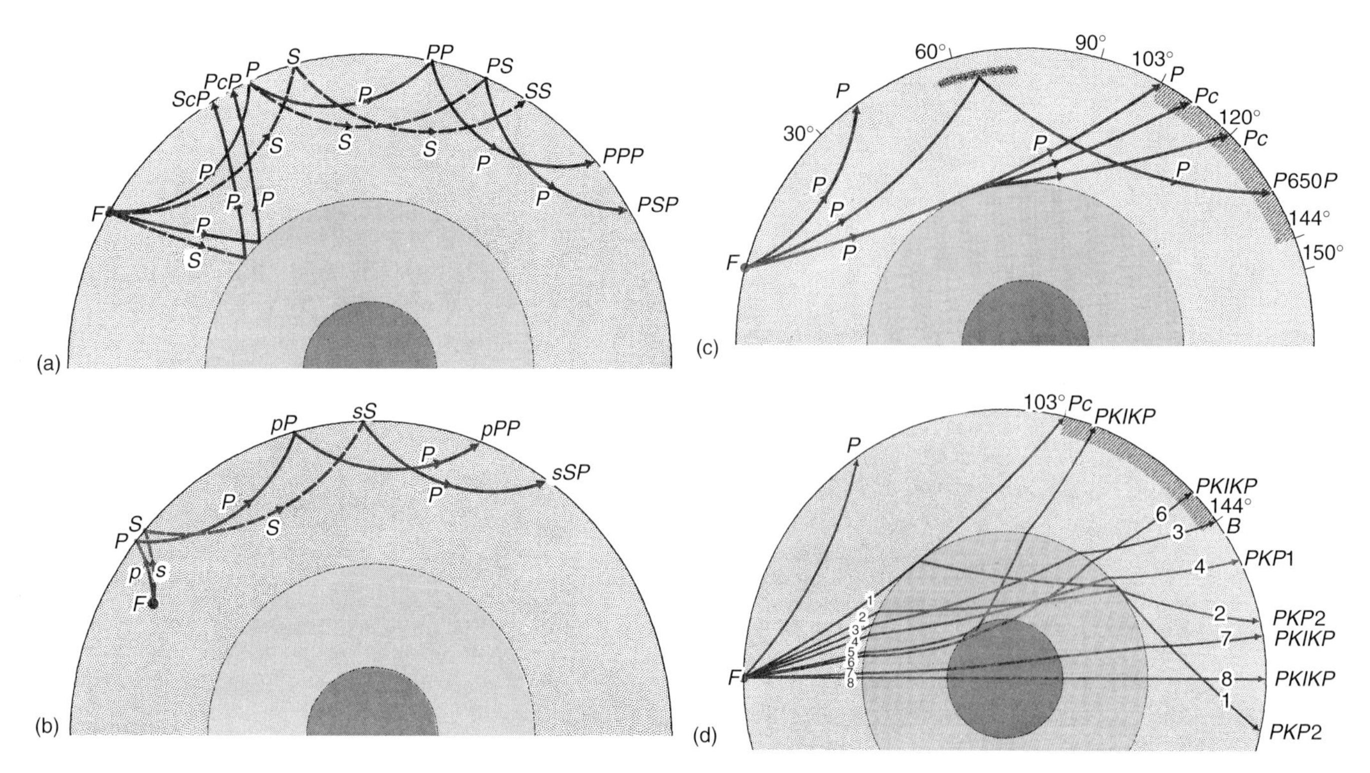

FIGURE 7 Examples of propagation paths of direct and reflected waves in the Earth's interior. Different shadings show the mantle, outer core, and inner core. (a) Examples of waves generated by the surface focus, *F*, of the earthquake, which radiates both *P* and *S* waves. Solid and dashed rays are used to distinguish between *P* and *S* waves, respectively. For notation see the text. (b) Examples of propagation paths of depth phases and their notation. Waves begin at the deep focus, *F*, of the earthquake. (c) Examples of propagation paths of direct *P* waves, *P* waves diffracted around the core–mantle boundary, and *P* waves reflected downward at a discontinuity at 650 km depth. The discontinuity and the shadow zone (103–144°) are shaded. (d) Propagation paths of *P* waves traveling through the Earth's core. The shadow zone is shaded and *B* denotes the caustic point. Rays are numbered in the order of increasing steepness of the initial descent. For details see the text. (Reprinted with modifications from Kulhánek (1990) with permission from Elsevier Science.)

and vice versa, on reflection, we may observe the wave denoted *PS* which travels as *P* from the focus to the reflection point at the free surface and from that point to the recording station as *S*. *PS* and *SP* appear only at distances larger than 40°. For a wave leaving the focus as *P* and twice reflected/converted from the free surface, we have four possible cases, namely *PPP*, *PPS*, *PSP*, and *PSS*. Some of these waves are sketched in Figure 7a. Obviously, we could continue with three and more reflections/conversions and form the corresponding wave symbols. However, from experience we know that it is quite seldom that three and more reflections from the Earth's free surface are clearly visible on actual seismograms. For distances larger than about 40°, the free-surface-reflected phases become very distinct. At distances around 100° and larger, *PP* and *SS* often belong to the largest recorded body waves.

The symbol *c* is used to indicate a single upward reflection, i.e., a reflection back into the mantle from the outer core–mantle boundary. For instance, *ScP* (Fig. 7a) corresponds to an *S* wave that travels down from the focus and is reflected and converted into the *P* type wave at the CMB and finally is recorded at the Earth's surface as *ScP*. A straightforward extension provides *PcP*, *ScS*, and *PcS*. Because these phases emerge steeply, *ScP* is usually stronger on vertical components than *PcS*. Large reflected core phases are usually recorded at shorter epicentral distances, say at 40° or less. At distances around 39°, *ScP* and *PcS* (surface foci) are often contaminated with the arrival of direct *S* and the phase separation is difficult. When the ray path of *PcP* grazes the outer core boundary, the combination of direct *P* and *PcP* is called *P* diffracted. This case is discussed in more detail below. Core-reflected waves together with *PmKP* (see Section 2.3 for notation), recorded from earthquakes at a wide range of distances and focal depths are used to study the properties of the core–mantle boundary.

Waves ascending from the focus to the free surface (in continents) or to the sea bottom (under oceans) and reflected back into the mantle, are commonly called depth phases and are denoted by a lower case prefix: *p* for longitudinal and *s* for transverse waves. We can easily list the four possibilities of reflections near the epicenter, which are *pP*, *sP*, *pS*, and *sS* (Fig. 7b). The first case, for example, denotes a wave that traveled upward from the focus as *P* (short leg) and was reflected back off the free surface again as *P* (long leg). Depth phases, primarily *pP*, are the most important phases routinely used in focal-depth estimations. It is quite obvious that the deeper the focus, the later is the *pP* phase in relation to *P*. Hence, accurately measured arrival-time differences *pP–P* are reliable indicators of the depth of the focus. In the case of a deeper focus, it is sometimes possible to recognize several different reflections from the free surface. Such waves are then labeled *pPP*, *pPS*, *pSP*, and *pSS* in the case of waves with their short leg as *P*. Logically, *sPP*, *sPS*, *sSP*, and *sSS* denote corresponding waves with short leg as *S* (Fig. 7b). Interpretation of depth phases must be done with utmost care since, for example, *pP* from a deep earthquake can easily be erroneously interpreted as *P* when the first arrival (*P*) is weak. Depending on focal orientation and other factors, *sP* may be stronger than *pP* and may be mistaken for it. Depth phases are sometimes stronger than the main *P* wave, and may be the first readable phase.

Events in the upper mantle sometimes generate visible but weak precursors to *pP*. These are Moho underside reflections labeled *pMP*. They are best observed on long-period records from deep-focus earthquakes; they show the same polarity as *pP* (Schenk *et al.*, 1989) but their correct identification is usually difficult. In a similar way, water-surface reflections are denoted *pwP*.

The lower-case symbol *d* (or its value in kilometers) inserted between *PP*, *SS*, etc. has been introduced by B.A. Bolt to indicate seismic waves reflected from "secondary" discontinuities in the upper mantle. For example, symbols *P400P* or *P650P* (Fig. 7c) specify *P* waves reflected at the underside of a discontinuity at a depth of 400 or 650 km, respectively. These phases arrive at the recording station ahead of the expected (calculated) arrival time for the main *PP* phase and are frequently interpreted as reflections from upper-mantle discontinuities (i.e., as *PdP*). However, when the arrival time cannot be explained in terms of known discontinuities as *PdP*, we call these onsets early *PP* or precursors to *PP*.

2.3 Body Waves; Recording Distances 103° and Larger, Shallow Events

It was noticed in the early days of observational seismology that amplitudes of direct *P* waves decay dramatically at distances larger than about 100°. The short-period *P* waves reappear consistently on records, first at distances of about 140° and larger. Correspondingly, the distance range $103° < \Delta < 140°$ is called the shadow zone. Within this zone, there is no penetration of direct high-frequency *P* waves due to the wave diffraction around the Earth's core (Fig. 7c). The last direct *P* wave reaches the Earth's surface at an epicentral distance of about 103° where the shadow zone produced by the Earth's mantle commences. *P* waves traveling beyond this distance creep around (are diffracted by) the core–mantle boundary and lose a large part of their energy there, so that only weak, diffracted *P* phases are observed in this distance range. Similarly to dispersion, diffraction also depends on the wave period (or frequency). The longer-period waves can diffract further into the shadow zone than shorter-period waves. The diffracted *P* waves are labeled *Pdif* or *Pdiff*. Seismological centers like the National Earthquake Information Center (NEIC) in Golden, Colorado, or the International Seismological Centre, ISC, in England use *Pdif*. On seismograms, *Pdif* waves usually show small amplitudes and emergent or gradual onsets, and the energy shifts to the long-period section of the spectrum. Long-period *Pdif* are sometimes observed out to distances of 160° or more. *S* waves are affected at the core–mantle boundary in a similar way; the symbol *Sdif* or *Sdiff* is used for diffracted *S* waves. The shadow zone for *S* waves on the side of the Earth opposite the earthquake extends over all epicentral distances from about 103° to −103° (257°).

Direct *P* and *S* waves and corresponding reflections are easily distinguishable from surface waves. The former usually occupy the period interval from, say, 1 to 10 sec, while the latter show large amplitudes (surface or shallow shocks) and periods in the interval from about 10 to 100 sec. The period of *S* waves increases with distance and, in cases of multiple reflection, may even reach several tens of seconds.

As mentioned above, the limiting ray path, which is tangential to the core in the real Earth, is that corresponding to an epicentral distance of about 103°. *P* and *S* waves recorded at distances greater than 103° graze or strike the surface of the core, the wave pattern becomes rather complicated, and amplitudes decay dramatically. Seismic waves that leave the source with a steeper descent than the grazing waves may either reflect from the core–mantle boundary or refract into the core. The former are reflected back into the mantle as *PcP*, *PcS*, *ScP*, or *ScS*, while the latter, called core phases, are refracted downward and enter the core. The refraction is rather sharp because of the sudden significant drop of *P* velocity beneath the core–mantle boundary.

A *P* wave leg in the outer core is denoted by *K* (from German *Kernwellen* for core waves). It has to be emphasized that the symbol *K* always represents a *P* wave since *S* waves do not enter the fluid outer core. After traveling through the outer core and following another partitioning (reflection or refraction) at the outer core boundary, it emerges at the Earth's surface. Thus, we can form the four symbols for waves that have traveled through the outer core: *PKP*, *PKS*, *SKS*, and *SKP*. For example, the symbol *PKS* corresponds to a wave that starts in the mantle as a *P* wave, is refracted into the outer core as a *P* wave, and is finally, after mode conversion, refracted back in the mantle as *S*. The phase *SKP* is stronger on vertical components than *PKS*. These phases have a caustic near 130° and at this distance are often the only phase recorded on short-period instruments. Some of the seismic rays traversing the core are illustrated in Figure 7d. The notation *PKP* is sometimes abbreviated as P'.

For ray paths that are only slightly steeper than the ray grazing the core surface, corresponding *PKP* waves emerge at the Earth's surface at distances beyond 180° (see Fig. 7d). As the rays (surface-focus event) enter the mantle more steeply, the core refractions become smaller and the rays emerge at the Earth's surface at increasingly shorter epicentral distances, until about 144°. For even steeper take-off angles of the initial ray paths, we see an increase of the distance of emergence up to 165° or so. The phenomenon may be viewed in terms of two *PKP* travel-time branches denoted *PKP1* and *PKP2* for the first and the second arrivals, respectively. Close to 144°, the waves from the two branches coincide and the waves constructively interfere, which gives rise to an energy concentration near that distance. The point of largest energy concentration is called a caustic point or simply caustic. The shape of corresponding travel-time curves is called triplication, indicating that three distinct travel-time branches exist at certain epicentral distance. It has to be stressed that neither *PKP1* nor *PKP2* enter the inner core, i.e., both these wave types have their deepest point of penetration in the outer core.

As we further steepen the initial ray path, we reach the family of rays that enter the inner core (Fig. 7d). These rays progress in a normal way, i.e., the steeper the initial ray path, the greater the distance of emergence from about 110° until at last there is a ray that passes through the Earth's center and reaches the Earth's surface at the antipode of the focus. *P* waves that traverse the inner core are denoted by *I*, giving rise to phases *PKIKP*, *PKIKS*, *SKIKS*, and *SKIKP*, although these are often still simply referred to as *PKP*, *PKS*, etc. Phases with an *S* leg in the inner core would include the letter *J*, such as *PKJKP*, but these have never been unambiguously identified on seismograms. Both the symbols, *I* and *J*, have to be accompanied on both sides by *K*. Rays corresponding to seismic waves reflected at the outside and inside of the inner core are called *PKiKP* and *PKIIKP*, respectively.

If the studied event is weak, *Pdif* is not clearly observed in the entire distance interval $\Delta > 103°$ and the first arrival seen on the record will be that of *PKP*. At epicentral distances 110–120°, *PKIKP* usually provides the first onset discernible on the seismogram.

Near the caustic, i.e., around 144°, the wave train of recorded core phases becomes particularly complicated. It is first at distances beyond the caustic point where observed onsets may be separated into individual *PKP* branches (see Plate 21.18). The energy distribution changes with the increasing distance. *PKP1* is the dominant branch just beyond the caustic, up to about 153°. In records of weaker events (144–153°), *PKP1* is often the first visible onset since *PKIKP*, theoretically preceding *PKP1*, is too weak to be observed. As the distance increases, *PKP1* becomes weaker and vanishes from records at distances of about 160° and larger. For distances beyond, about, 157°, *PKP2* usually dominates the seismogram. Some workers prefer the nomenclature adopted from travel-time charts with branches denoted *AB*, *BC*, and *DF* (Jeffreys and Bullen, 1967). Arrivals associated with these branches are then labeled PKP_{AB}, PKP_{BC}, and PKP_{DF} and correspond to *PKP2*, *PKP1*, and *PKIKP* arrivals, respectively. The *CD* branch (PKP_{CD}) is related to *PKiKP* arrivals corresponding to seismic waves reflected at the outside of the inner core. Waves that are multiply reflected at the underside of the inner core boundary are called *PKmIKP* ($m = 1, 2, \ldots$).

PKIKP in the distance range from about 125° to the caustic is often preceded by early arrivals or precursors that can arrive many seconds ahead of the main phase. These are usually explained by scattering phenomena just above the core–mantle boundary.

In a similar way as for the mantle, we may define new phase names for the family of waves propagating through the core. For example, *PKKP* is a *P* wave that has been reflected from the inside of the core–mantle boundary. *PKKP* is often very pronounced on records made at distances between 60° and 80°. The sharp onset may easily be misinterpreted as a first *P* arrival of another event. *P* waves trapped inside the Earth's liquid core and with multiple *K* legs are called *PmKP*, where $m - 1$ provides the number of reflections from the underside of the CMB. Cases like *P4KP* and *P7KP* have been observed (Bolt, 1982).

PKPPKP, or for short $P'P'$, are *PKP* waves once reflected from the free surface back to a station in the same hemisphere as the focus. Since *PKP* has the caustic at 144°, the best chance to observe $P'P'$ is around distances of $2 \times 144° = 288°$, or 72° if we take the shortest distance from source to station. $P'P'$ is often well recorded, arriving about 30 min after the *P* phase when most of the coda amplitudes of preceding phases have already become faint, and it may in some cases be wrongly interpreted as a new *P* or *PKP*. The value 72° is also equivalent to $3 \times 144° = 432°$ or $(360° + 72°)$ so the phase $P'P'P'$ is also strong at this distance, and may be observed for strong earthquakes about another 20 min after $P'P'$.

In the late 1960s, first observations of forerunners to $P'P'$ were made. These were interpreted as $P'dP'$. Analogous to *PdP*, $P'dP'$ waves are reflected at some discontinuity in the upper mantle. For example, $P'650P'$ travels from the hypocenter to the other side of the Earth, where it is reflected back to the station from a layer 650 km below the surface. $P'650P'$ passes through the core twice and on the seismogram precedes $P'P'$ by about 2 min.

Let us now shift our attention from *P* to *S* waves. Similarly to *PKP*, there are *SKS* waves, i.e., *S* waves traveling from the earthquake source down through the mantle, incident at the outer-core boundary they undergo a mode conversion and as *P* (the *K* leg) transverse the liquid outer core. Following an inverse mode conversion, they again enter the mantle as *S* and emerge at the Earth's surface as *SKS*. Analogous phases to *PmKP* are *SmKS*. First *SKS* waves are observed at distances between 60° and 70° and the range of observations extends out to distances of 180° or so. Depending upon details in the structural model, *SKS* exhibits large amplitudes at a distance near 80° so that the best region for study of *SKS* waves is that between 70° and 90°. However, the phase identification has to be made with utmost care since *SKS* waves recorded in this distance region have similar travel times as direct *S* waves. For Δ greater than about 82°, *SKS* arrives ahead of *S*. For distances shorter than about 95°, *SKS* is usually smaller than *S*; however, at distances beyond 95°, *SKS* amplitudes are often quite large (see Plate 21.14). Mistaking *S* for *SKS* and vice versa will adversely affect the estimation of epicenter location. Since the epicentral distance estimate is frequently governed by the observed arrival-time difference *S–P*, wrong *S* identification on the record will provide a wrong epicentral distance, which in its turn will result in erroneous location. *S* and *SKS* are best recorded on long-period horizontal-component seismograms. However, occasionally these body waves are also observed on short-period records, although the onset of the later of the two phases is usually not obvious due to the contamination by coda of the earlier phase.

2.4 Body Waves from Intermediate- and Deep-Focus Earthquakes

Early in the last century several investigators (Pilgrim, 1913; Turner, 1922) found a number of events with focal depths greater than 100 km. Somewhat later, Japanese scientists (Wadati, 1927; Shida, 1937) presented evidence that Japanese earthquakes occur at practically all depths down to 500 km. They based their conclusions upon observed *S–P* arrival-time differences, intensity distributions, and different appearances of intermediate- or deep-focus earthquakes and of those that take place at shallow depth. Wadati noted very early that the seismograms of intermediate-focus and deep shocks display rather impulsive and large *S* phases, shorter predominant periods, and less well-developed codas.

Later studies confirmed conclusions from Wadati's pioneering work that intermediate- and deep-focus shocks produce simpler seismograms (see, e.g., Plate 21.3) with exceptionally well-recorded impulsive body waves, while surface wave amplitudes are relatively small. Strong depth phases, such as *pP* and *sS*, are also frequently very distinct on records from deep events. However, the duplication of principal phases by surface reflections often complicates the seismogram interpretation. Another important characteristic that accentuates the difference between shallow and deep shocks is the pattern of aftershocks. While large, shallow earthquakes are usually followed by numerous aftershocks, deep events (which may be multiple shocks) rarely show well-developed aftershock series (see, e.g., Wiens *et al.*, 1994).

As an example, Figure 8 displays three-component records made at a teleseismic distance from the Hindu Kush earthquake, which occurred at a depth of 200 km. Note the rather impulsive appearance of *P*, *pP*, *PP*, *PcP*, and *S* and *ScS* (E-W component) which are all easily identified on the seismogram. Practically no surface waves were recorded from this event.

Strong intermediate-focus and deep earthquakes occur in several different seismically active regions. Among these are island arcs such as Tonga-Kermadec Islands, the Marianas, New Hebrides Islands or the Aegean arc; continental margins with deep ocean trenches like Central America and western South America; and mountain chains, e.g., Himalayas (Hindu Kush) or Carpathians. About one-fifth of all reported earthquakes take place at a focal depth exceeding 70 km. Among the deepest known earthquakes are three in the Flores Sea area, on 25 August 1933, 29 June 1934, and 30 June 1943. Their depths are given by Gutenberg and Richter (1938) as 720 km, although other agencies have placed them rather shallower. ISC records contain four recent events in the Fiji-Tonga area with depths greater than 750 km: 15 January 1981 (765 ± 17 km), 21 November 1982 (769 ± 31 km), 25 October 1972

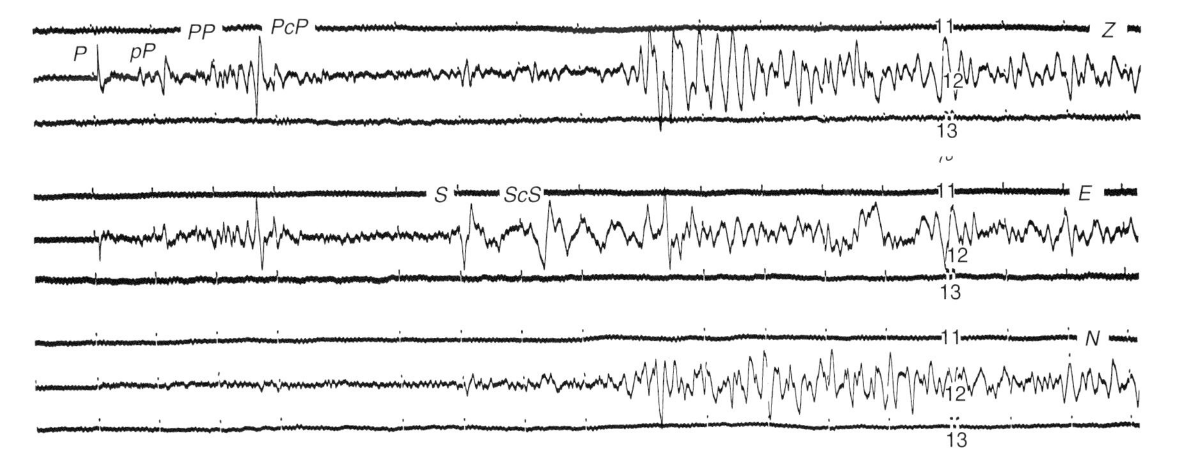

FIGURE 8 Seismograms from an intermediate-focus earthquake in the Hindu Kush region on 4 September 1990, made at the Swedish station Uppsala at an epicentral distance of 41°. The traces exhibit three components of long-period records. Note the rather impulsive character of recorded phases. Virtually no surface waves have been recorded from this event.

(806 ± 84 km), and 7 May 1971 (848 ± 26 km). These events are all small and not widely recorded, and their depths, particularly those of the deepest two, cannot be regarded as well established. Occasionally, deep-focus earthquakes occur in rather unexpected geographical areas. For instance, up to 1954 it was generally accepted that the geographical extent of deep quakes is limited to the Circum-Pacific belt and the areas mentioned above. However, on 29 March 1954, an isolated major earthquake occurred in southern Spain at a depth of 630 km. A smaller earthquake occurred at the same location on 30 January 1973.

Even though several alternative hypotheses to explain deep earthquakes have been launched in the past (e.g., contraction of the Earth due to cooling), today nearly all seismologists agree that deep and intermediate-focus events are associated with the subducting lithosphere, which fits nicely with the idea of plate tectonics. It is of course, still possible that individual shocks may not be related to subduction. Notwithstanding the unifying frame of plate tectonics, it is likely that deep and shallow events are generated by fundamentally different modes of rock failures (see, e.g., Frohlich, 1989; Green, 1994; Kirby *et al.*, 1996).

On the whole, the most distinctive features of a large intermediate-focus or deep earthquakes recorded at teleseismic distances pertinent to seismogram interpretation are the seismogram's simplicity, small amplitudes of or even absence of surface waves, and an impulsive shape of body waves. With some elementary experience, all these characteristics may often be revealed as the eye scans the seismogram. In a more retrospective-type interpretation, the absence of aftershocks will support the classification of the shock as a deep or intermediate-focus event.

2.5 Surface Waves

With exception of very short epicentral distances, surface waves carry by far the largest amount of wave energy radiated by shallow and some intermediate-focus earthquakes. Contrary to other types of surface waves, propagation of *LR* waves is not limited to layered media; they can also be transmitted through a homogeneous half-space (semi-infinite medium). The particle motion of Rayleigh waves follows a retrograde elliptical orbit in the vertical plane containing the direction of propagation. At the Earth's surface, the amplitudes in the vertical and horizontal directions are related roughly as 3:2. Hence, Rayleigh waves are usually best seen on vertical-component seismograms. Amplitudes of *LR* waves decrease rapidly (exponentially) with increasing depth. For example, at a depth equal to one wavelength, the vertical and horizontal amplitudes fall to 0.19 and 0.11 of their free-surface values, respectively. The velocity of Rayleigh waves in a homogeneous medium, c_R, lies between 0.87 and 0.96 times the *S*-wave velocity, v_S, of the half-space. For many rock materials the Poisson ratio is approximately 0.25 which leads to a relation $c_R = 0.92\ v_S$.

Consider now a thin superficial layer superimposed on a homogeneous half-space. Assuming that the *S*-wave velocity in the layer is lower than that in the material below, Love waves designated *LQ*, can be transmitted through the layer without any significant penetration of energy into the lower medium. Love waves may, therefore, also be considered as channel waves, in this particular case transverse waves, trapped in the superficial layer. In contrast to Rayleigh waves, Love waves show no vertical motion, since particles excited by propagating *LQ* waves are polarized in the horizontal plane and perpendicular to the direction of propagation. Consequently, traces of *LQ* waves should be looked for on the horizontal-component seismograms. Love waves propagate faster than Rayleigh waves with velocities limited by *S*-wave velocities in the layer and the half-space. The mean velocities of propagation are 4.43 and 3.97 km sec^{-1} for *LQ* and *LR*, respectively (Bullen and Bolt, 1985).

The analyst has essentially two clues to distinguish *LQ* waves from *LR* waves. The first is the different particle motion. While Rayleigh waves usually show the largest amplitudes on vertical-component records, Love waves are best displayed on horizontal-component seismograms. The second clue is the different propagation velocities. Both *LQ* and *LR* propagate more slowly than *P* or *S*, but since *LQ* propagates faster than *LR*, they are recorded ahead of *LR* waves.

Surface waves traveling through layered media often show appreciable normal dispersion which, as time goes on, continually changes the shape of both *LQ* and *LR*. Due to the dispersion, the original appearance of the wave train becomes disturbed on the seismogram by long-period waves advancing toward the beginning of the wave train as it travels through the medium. Plots of wave velocity as a function of period (frequency) are called dispersion curves. The shape of the curves depends strongly upon the character of the propagation path. In the period range from about 2 to 60 sec, continental travel paths exhibit dispersion curves for *LR* waves with group velocity gradually increasing from 2 to almost 4 km sec^{-1}. For *LQ* waves, the group velocity starts at about 2.5 km sec^{-1} and reaches values close to 4.5 km sec^{-1} at periods of several hundreds of seconds. Dispersion curves associated with oceanic travel paths show a rather abrupt change in the short-period range. At periods around 15 sec, the velocity of *LR* increases sharply from about 1.5 to more than 3 km sec^{-1}. For *LQ* waves, a sudden velocity rise from about 3 to 4 km sec^{-1} is seen at periods around 7 sec (Bullen and Bolt, 1985).

An experienced analyst will distinguish between recorded surface waves that have traveled along pure oceanic or continental paths. Dispersion characteristics of oceanic routes give rise to long wave trains with rather slow and sometimes hardly visible period change over relatively long (5–10 min) record segments. A typical seismogram is displayed in Plate 21.17. In contrast, continental paths generate a characteristic fast period decrease with time, which is often easily recognized by inspecting several minutes of the records (see, e.g.,

Fig. 1 and Plates 21.8 and 21.13). As indicated above, the exact shape of dispersion curves depends upon the structure traversed. This means that empirical dispersion curves for *LQ* or *LR* waves provide the researcher with remarkably effective probes for studying the structure of the traversed medium (so called inverse problem).

Dispersion curves show a rather complicated pattern with several local minima and maxima. Surface waves traveling with these minimal or maximal group velocities are called Airy phases. On seismograms, an Airy phase is characterized by a constant-frequency compact wave train, often with a remarkable amplitude build-up of dispersed surface waves traveling by fundamental-mode propagation (Fig. 9).

At short epicentral distances, it is difficult to identify *LQ* and *LR* waves because they are often contaminated by large-amplitude *S* waves (Plate 21.9). At large distances, the identification is rather simple since *LQ* and *LR* waves dominate the record (shallow events) and are significantly delayed with respect to *S* waves. On seismograms, surface waves may be spread over several hours.

Dispersion curves for *LR* waves (both the oceanic and continental paths) show a local minimum for periods around 200 sec. In the period range from approximately 50 to 200 sec, the group velocity is monotonically decreasing with increasing period. Physically this means that in this period range long-period Rayleigh waves follow the laws of inverse dispersion. Observations of this interesting phenomenon are rather scarce, but one example can be found in Plate 21.10.

Periods of the largest (maximum amplitude) recorded surface waves show a clear positive correlation with epicentral distance. For example, for distances 10°, 50°, and 100°, the expected minimum periods, $T_{\min}$, of the largest continental Rayleigh waves are of the order of 7, 13, and 16 sec, respectively (Willmore, 1979). For oceanic passages, the periods can be somewhat longer. Also, the time occurrence of the beginning of the maximum movement, $T_{R\max}$ (Rayleigh wave), with respect to the first onset of *P* waves, T_P, is obviously distance-dependent. Table 1 gives the time differences $T_{R\max} - T_p$ for various epicentral distances Δ. Hence, the position of the maximum amplitude in the *LR* wave train, with respect to the *P* arrival, and its period offer the interpreter other important information on the approximate epicentral distance of the earthquake.

As mentioned earlier, surface-wave amplitudes are large only close to the Earth's free surface and decrease rather rapidly with increasing depth. Consequently, a shallow-focus earthquake will usually generate large, dominating surface waves, while deep-focus event of the same magnitude will generate abnormally small surface waves. This feature obviously provides the interpreter with a valuable tool for discriminating, at first glance, between shallow events and deep-focus earthquakes. An exception to this rule is provided by seismograms of underground explosions. These events are obviously shallow shocks with focal depths not exceeding several kilometers. Nevertheless, they generate only very weak surface waves even for explosions of body-wave magnitudes around 7 (Plate 21.11).

Comparatively long-period surface waves, so-called mantle waves, have been observed from large distant shocks. These waves can be either Love or Rayleigh wave type with periods from somewhat less than one minute to several minutes. The speed of the *LQ* type mantle waves is often nearly constant at 4.4 km sec^{-1} and the wave has an impulsive shape on the seismogram. The *LR* type mantle waves travel with a velocity between 3.6 and 4.1 km sec^{-1}. Since the wavelength of mantle waves varies from several hundreds to more than a thousand kilometers, these waves are affected by a large part of the Earth's mantle (Plates 21.10 and 21.12). An interesting feature

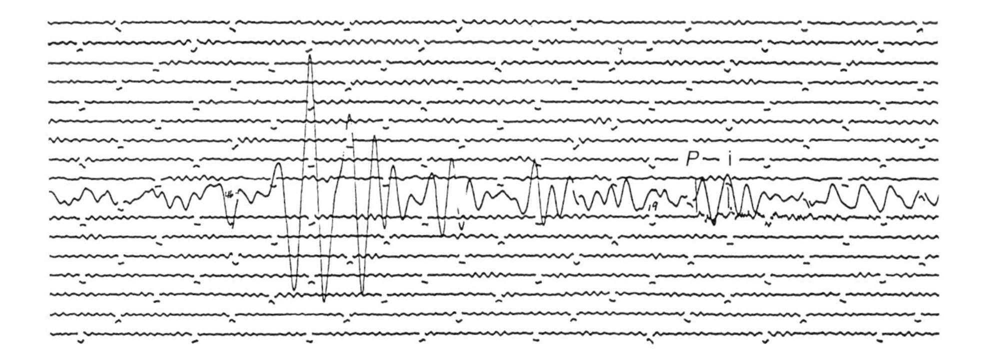

FIGURE 9 Medium-period, vertical-component seismogram from the earthquake of 21 October 1964 ($m = 5.9$, $h = 37$ km) in India–China border region, made at Uppsala, Sweden, at an epicentral distance of 59°. The trace illustrates a record with a distinct Airy phase of the fundamental-mode Rayleigh waves characterized by a practically constant-frequency wave train. Observe the remarkable amplitude build-up and fall-off within three swings. The trace also exhibits the direct *P* and an unidentified phase arriving about 17 sec after *P*. The second phase, when interpreted as *pP*, provides a focal depth of 70 km. (Reprinted with modifications from Kulhánek (1990) with permission from Elsevier Science.)

TABLE 1 Travel-Time Differences, $T_{R\max} - T_P$, and Minimum Periods, $T_{\min}$, for Largest Rayleigh Waves Observed at Various Epicenter Distances Δ

Δ (°)	$T_{R\max} - T_P$ (min)	$T_{\min}$ (sec)
10	3.8	7
15	5.5	8
20	7.4	9
25	9.4	
30	11.5	10
35	13.7	
40	15.9	12
45	18.1	
50	20.3	
55	22.6	
60	24.9	14
65	27.1	
70	29.5	
75	31.8	
80	34.2	16
85	36.7	
90	39.2	
95	41.7	
100	44.2	
105	46.7	
110	49.3	
115	53	
120	55	
125	57	
130	60	
140	64	18
150	70	

From Båth 1947; Willmore, 1979.

of mantle waves is their repeated appearance on records, as they can travel multiple orbits around the Earth. The *LQ* type mantle wave was given the label *G* (after B. Gutenberg) and the *LR* type mantle wave was labeled *R*. The older nomenclature sometimes uses *W* instead of *R* (from German *Wiederkehrwellen*, meaning repeated waves). *G* waves that propagate along the minor and major arc of the great circle path are labeled *G1* and *G2*, respectively. Waves that have traveled additional orbits around the Earth are denoted *G3* and *G4*, and so on. Accordingly, we have *R1*, *R2*, *R3*, *R4*, etc. (cf. Chapter 11 by Romanowicz). On many occasions, observations of up to *G8* and *R8* have been made. As an exceptional case we may mention records of the 1960 Chile earthquake, $M = 8.3$ (surface-wave magnitude). Seismograms made at Uppsala, Sweden, reveal mantle waves *G20* and *R20* that have traveled a total distance equal to that from the Earth to the Moon (Båth, 1979).

2.6 Volcanic Earthquakes and Unusual Seismic Sources

Volcanic earthquakes, discovered by L. Palmieri at the Vesuvius Observatory in 1855, are caused by sudden opening of channels in crustal rocks, rapid changes of motion of magma, excessive accumulation of gas pressure in the crust, roof collapses of subterranean channels emptied of magma, and so on. According to Minakami (1959a,b, 1960) or Tazieff and Sabroux (1983), volcanic earthquakes are classified into three groups: A-type earthquakes with foci between 1 and 10 km deep; B-type earthquakes with foci at depths of 1 km or less; and explosion-type earthquakes taking place at the very surface of the Earth. Another categorization of volcanic earthquakes can be found, for example, in Tokarev (1983). Close to active volcanoes, we also frequently detect so-called volcanic tremor, which is due to long-duration, more or less continuous, volcanic vibration. Whereas volcanic earthquakes are clearly isolated events separated in time from each other, volcanic tremor shows rather spasmodic or harmonic behavior. It is associated with flow of underground magma, oscillations in magma reservoirs, explosions of volcanic gases, and the like. Examples of seismograms from volcanic earthquakes are given in Plate 21.5.

Among other sources of earthquakes one can mention implosion or collapse earthquakes, impact earthquakes due to collisions of meteorites with the Earth's surface (e.g., the Tunguska, Siberia, impact in 1908), low-magnitude icequakes generated by temperature changes in glaciers, earthquakes related to large-scale landslides, etc. These are some of the types of earthquakes that, together with tectonic earthquakes, volcanic earthquakes, and oceanic microseisms belong to the category of natural seismic sources. There is also a variety of man-made seismic sources such as industrial or military explosions and various types of cultural noise (traffic, industry, construction works, sonic booms), which are examples of controlled sources. Other types of man-made seismic sources are induced or triggered events (high dams, mining activity, fluid injections). Examples of records from some unusual seismic sources are exhibited in Plates 21.6 and 21.7.

For earthquakes other than tectonic or volcanic, the source identification is usually a task in itself. A classical example of source identification is the well-known problem of discriminating underground nuclear explosions from earthquakes. Even on a regional scale, reliable identification of conventional explosions is of importance for the correct estimation of seismic properties of an area. The same applies also to other man-made seismic sources. More detailed discussion goes beyond the scope of the chapter and the interested reader is referred to other works for more information.

3. Seismogram Examples with Interpretations

This section comprises a collection of 18 Plates with actual seismograms (see attached Handbook CD). Included are also phase interpretations and short commentaries of specific features reflected in the displayed records. Collected seismograms were made at a number of seismographic stations by a variety of instruments covering different frequency bands.

Exhibited records cover the whole range of epicentral distances, a large range of focal depths and various types of propagation paths. Different seismic sources are presented. More classical records as well as analog displays of digital records are shown. The arrangement of Plates is according to epicentral distance except for records from volcanic earthquakes and exotic sources, which follow the suite of local and regional events. For the ease of the reader, common notations of seismic waves used throughout the text and in the following Plates are listed in Table 2 in the order in

TABLE 2 Nomenclature of Seismic Phases

Symbol	Meaning
Local and regional events	
$\bar{P}$, $\bar{S}$	Direct compressional or shear wave traveling through the upper crust (observed only at very short epicentral distances)
Pg, *Sg*	Compressional or shear wave in the granitic layer of the crust
PmP, *SmS*	Compressional or shear wave reflected at Moho
Pn, *Sn*	Compressional or shear wave traveling along (just beneath) the Moho discontinuity, so called head wave
P^*, S^* (or *Pb*, *Sb*)	Compressional or shear wave traveling along (just beneath) the Conrad discontinuity
pPn	Depth phase that leaves the focus upward as *P*, is reflected as *P* at the free surface and continues further as *Pn*
sPn	Depth phase that leaves the focus upward as *S*, is reflected and converted into *P* at the free surface and continues further as *Pn*
Rg	Short-period crustal surface wave of Rayleigh type
Lg	Guided crustal wave traversing large distances along continental paths
T	Compressional wave propagating through the ocean (tertiary wave). *T* phases are occasionally observed even at large teleseismic distances
TPg (*TSg*, *TRg*)	Wave that travels the ocean and land portion of the total transmission path as *T* and *Pg* (*Sg*, *Rg*), respectively
Teleseismic events	
P, *S*	Direct compressional or shear wave, so called elementary or main wave
PP, *PPP*, *SS*, *SSS*	*P* or *S* wave reflected once or twice at the Earth's surface
SP	*S* wave converted into *P* upon reflection at the Earth's surface
PPS, *PSP*, *PSS*	*P* wave twice reflected/converted at the Earth's surface
PcP, *ScS*	*P* or *S* wave reflected at the core–mantle boundary
PcS, *ScP*	*P* or *S* wave converted respectively into *S* or *P* upon reflection at the core–mantle boundary
pP, *pS*, *pPP*, *pPS*, etc.	Depth phase that leaves the focus upward as *P* (*p* leg), is reflected/converted at the free surface and continues further as *P*, *S*, *PP*, *PS*, etc.
sP, *sS*, *sPP*, *sPS*, etc.	Depth phase that leaves the focus upward as *S* (*s* leg), is reflected/converted at the free surface and continues further as *P*, *S*, *PP*, *PS*, etc.
pMP	*P* wave reflected at the underside of Moho
pwP	*P* wave reflected at the water surface
PdP	*P* wave reflected at the underside of a discontinuity at depth *d* in the upper part of the Earth. *d* is given in kilometers, e.g., *P400P*.
Pc, *Sc* or *Pdif*, *Sdif*	*P* or *S* wave that is diffracted around the core–mantle boundary
PKP (or P')	*P* wave traversing the outer core
PKS	*P* wave converted into *S* on refraction when leaving the core
SKS	*S* wave traversing the outer core as *P* and converted back into *S* when again entering the mantle
SKP	*S* wave converted into *P* on refraction into the outer core
PKP1, *PKP2* or PKP_{BC}, PKP_{AB}	Different branches of *PKP*
PKIKP or P'', PKP_{DF}	*P* wave traversing the outer and inner core
PKiKP	*P* wave reflected at the boundary of the inner core
PKIIKP	*P* wave reflected from the inside of the inner-core boundary
PKKP	*P* wave reflected from the inside of the core–mantle boundary
PmKP ($m=3, 4, \ldots$)	*P* wave reflected $m-1$ times from the inside of the core–mantle boundary
SmKS ($m=3, 4, \ldots$)	*S* wave converted into *P* on refraction at the outer core, reflected $m-1$ times from the inside of the core–mantle boundary and finally converted back into *S* when again entering the mantle
PKPPKP (or $P'P'$)	*PKP* wave reflected from the free surface, passing twice through the core
$P'dP'$	*PKP* reflected at the underside of the discontinuity at depth *d* in the upper part of the Earth. *d* is given in kilometers
LR	Surface wave of Rayleigh type
LQ	Surface wave of Love type
G	Mantle wave of Love type
R	Mantle wave of Rayleigh type
G1, *G2*	*LQ*-type mantle wave that travels the direct and anticenter routes. Waves that have, in addition, traveled once or several times around the Earth are denoted *G3*, *G4*, *G5*, *G6*, etc.
R1, *R2*	*LR*-type mantle wave that travels the direct and anticenter routes. Waves that have, in addition, traveled once or several times around the Earth are denoted *R3*, *R4*, *R5*, *R6*, etc.

which they are cited in the text. If not mentioned otherwise, the interval between time marks on displayed seismograms is 1 min.

Acknowledgments

This chapter was written at the Department of Earth Sciences, Seismology Programme, Uppsala University, Uppsala, Sweden. My thanks are due to my colleague A.J. Anderson who carefully read and commented on various parts of the manuscript. Invaluable indeed are the seismogram examples with commentaries that I have received from V. Barbosa (Plate 21.5), G.Choy (Plates 21.14, 21.15, 21.18), A. Plesinger (Plate 21.12), M. E. Reyners (Plates 21.3, 21.4), D. Seidl (Plate 21.17), R.A. Uhrhammer (Plates 21.6, 21.7, 21.10, 21.16) and M. Yamamoto (Plates 21.1, 21.2). I also wish to thank J. Ritsema and two anonymous reviewers for their constructive comments and suggestions, which improved the text.

References

Adams, R.D. (1979). *T*-phase recordings at Rarotonga from underground nuclear explosions. *Geophys. J. R. Astron. Soc.* **58**, 361–369.

Båth, M. (1947). Travel times of the principal earthquake waves for Uppsala. *Bull. Geol. Inst., Uppsala* **32**, 105–129.

Båth, M. (1979). "Introduction to Seismology." Birkhäuser, Basel.

Båth, M. (1983). Earthquake data analysis: an example from Sweden. *Earth-Sci. Rev.* **19**, 181–303.

Båth, M. and M. Shahidi (1974). *T*-phases from Atlantic earthquakes. *Pure Appl. Geophys.* (*Pageoph*) **92**, 74–114.

Bolt, B.A. (1982). "Inside the Earth." W.H. Freeman, San Francisco.

Bullen, K.E. and B.A. Bolt (1985). "An Introduction to the Theory of Seismology," 4th ed. Cambridge University Press, Cambridge.

Frohlich, C. (1989). Deep earthquakes. *Sci. Am.* **260**, 32–39.

Green, H.W. (1994). Solving the paradox of deep earthquakes. *Sci. Am.* **271**, 50–57.

Gutenberg, B. and C.F. Richter (1938). Depth and geographical distribution of deep-focus earthquakes. *Geol. Soc. Am. Bull.* **49**, 249–288.

Jeffreys, H. and K.E. Bullen (1967). "Seismological Tables." BAAS, London.

Kirby, S.H., S. Stein, E.A. Okal, and D.C. Rubie (1996). Metastable mantle phase transformations and deep earthquakes in subducting oceanic lithospere. *Rev. Geophys.* **34**, 261–306.

Kulhánek, O. (1990). "Anatomy of Seismograms." Elsevier Science, Amsterdam.

Minakami, T. (1959a). The study of eruptions and earthquakes originating from volcanoes: Part 1. *Bull. Volcanol. Soc. Jpn.* **4**, 104–114 (in Japanese).

Minakami, T. (1959b). The study of eruptions and earthquakes originating from volcanoes: Part 2. *Bull. Volcanol. Soc. Jpn.* **4**, 115–130 (in Japanese).

Minakami, T. (1960). The study of eruptions and earthquakes originating from volcanoes: Part 3. *Bull. Volcanol. Soc. Jpn.* **4**, 133–151 (in Japanese).

Pilgrim, L. (1913). Die Berechnung der Laufzeiten eines Erdstosses mit Berücksichtigung der Herdtiefen, gestützt auf neuere Beobachtungen. *Gerlands Beitr. z. Geophys.* **12**, 363–483.

Schenk, T., G. Muller, and W. Brustle (1989). Long-period precursors to *pP* from deep-focus earthquakes: the Moho underside reflection *pMP*. *Geophys. J. Int.* **98**, 317–327.

Shida, T. (1937). Thank-you address at the dedication ceremony of Beppu Geophysical Laboratory. *Chikyu Butsuri Geophys.*, **1**, 1–5 (in Japanese).

Tazieff, H. and J.C. Sabroux (Eds.) (1983). "Forecasting Volcanic Events." Elsevier, Amsterdam.

Tokarev, P.I. (1983). Experience in predicting volcanic eruptions in the USSR. In: "Forecasting Volcanic Events" (H. Tazieff and J.C. Sabroux, Eds.). Elsevier, Amsterdam.

Turner, H.H. (1922). On the arrival of earthquake waves at the antipodes, and on the measurement of the focal depth of an earthquake. *Mon. Not. R. Astron. Soc., Geophys. Suppl.* **1**, 1–13.

Wadati, K. (1927). Existence and study of deep earthquakes. *J. Meteorol. Soc. Jpn. Ser.* 2 **5**, 119–145 (in Japanese).

Wiens, D.A., G.J. McGuire, P.J. Shore, *et al.* (1994). A deep earthquake aftershock series and implications for the rupture mechanism of deep earthquakes. *Nature* **372**, 540–543.

Willmore, P.L. (1979). "Manual of Seismological Observatory Practice," Report SE-20, World Data Center A for Solid Earth Geophysics. NOAA, Boulder, CO.

Editor's Note

All 18 Plates are placed on the attached Handbook CD, under the directory of \21Kulhanek. Willmore (1979) is reproduced in a computer readable file on the attached Handbook CD, under the directory of \81\IASPEI\Training.

22

Analysis of Digital Earthquake Signals

Frank Scherbaum
University of Potsdam, Potsdam, Germany

1. Introduction

The great progress made in the past several decades in seismology has been stimulated mainly by the availability of increasingly plentiful and improved seismic data. Our ability to collect, process, and analyze earthquake data has been aided by rapid advances in electronics and digital computers. Seismograms were mostly recorded using analog techniques up to a few decades ago. Except for very elementary tasks, such records are now considered of rather limited use because of low dynamic range, poor timing accuracy of the recording systems, and the lack of any convenient post-processing.

Modern digital seismograph systems consist of a cascade of analog and digital stages. The analog part of the system involves the seismometer, possible (pre-)amplifiers, and the analog antialias filter. Ideally these components act as high-pass, allpass, and low-pass filters, respectively. The antialias filter is required to assure that sampling is done in accordance with the sampling theorem (Stearns, 1975). The analog to digital (A/D) conversion is nowadays commonly based on the use of oversampling and decimation techniques. The output of the seismometer channels is first sampled at very high sampling frequencies (kHz range) and subsequently decimated (with proper digital antialias filtering) to the desired sampling rate. The reason for this is that the influence of the quantization noise—which directly limits the resolution of the sampling process—becomes less with higher sampling rates. The performance of the A/D conversion is additionally increased if only amplitude differences between a predicted sample value and the actual sample are digitized (differential sampling). For sampling frequencies much higher than the signal frequencies, these differences can become very small, especially if the trace is integrated before sampling. Therefore, low-resolution (down to one-bit) quantizers suffice. This is exploited in sigma–delta modulation, which is commonly used in high-resolution analog-to-digital converters (ADCs) for modern digital recording systems (Proakis and Manolakis, 1992); see Figure 1.

Digital seismograms are merely sequences of numbers that must be processed in the context of analysis. This involves basic processes such as trace editing (spike removal, baseline correction, resampling), integration, differentiation, Hilbert transformation of traces, rotation of components, but also more advanced tasks such as frequency filtering of traces, detection and identification of different wavegroups, determination of exact onset times and true ground motion amplitudes, spectral analysis, polarization analysis, etc. Numerous software tools have become available for these purposes (Scherbaum and Johnson, 1993; Stammler, 1993; Chapter 85 edited by Snoke and Garcia-Fernandez; Chapter 85.5 by Goldstein *et al.*, 2001). In order to apply them correctly in observatory practice, a basic understanding of the fundamentals of system and filter theory is necessary. In other words, digital signal processing and system theory has become an integral part of digital seismometry and seismology. See Appendix 1 for a brief introduction into filter theory and Appendix 2 for a short discussion of the basic elements and characteristics of a digital seismograph system. These appendixes are given on the attached CD Handbook, under the directory \22Scherbaum.

2. Filtering of Digital Seismograms

Extracting relevant information from digital seismograms often requires intensive filtering of the recorded signals. This is especially true for broadband signals. One of the most common tasks is the removal of unwanted noise. This is usually done by restricting the frequency band to that of interest by low-pass and/or high-pass filtering, or by suppressing narrowband noise signals by band-rejection filtering. Numerous well-tested algorithms are widely available for the design of such

INTERNATIONAL HANDBOOK OF EARTHQUAKE AND ENGINEERING SEISMOLOGY, VOLUME 81A ISBN: 0-12-440652-1

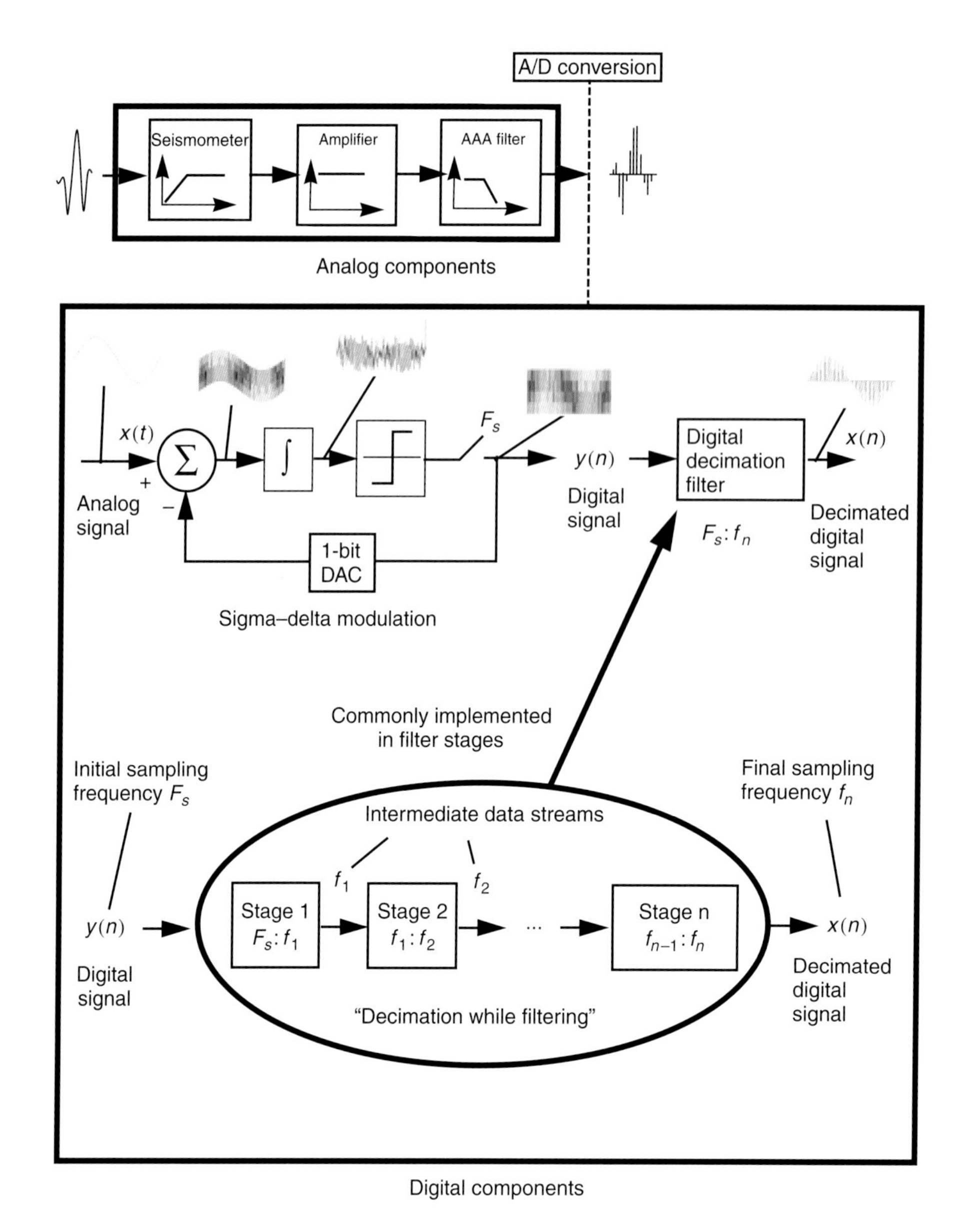

FIGURE 1 Building blocks of modern recording systems based on oversampling and decimation (sigma–delta modulation).

filters (Stearns, 1975; IEEE Committee, 1979; McClellan *et al.*, 1979; Parks and Burrus, 1987; Press *et al.*, 1992; Hamming, 1997).

For the purpose of consistent parameter estimation it is often desirable to convert recorded broadband seismograms into ones from a different seismograph system (simulation filtering). This allows the analyst to determine signal parameters in a manner consistent with other observatories. The simulated systems will most commonly belong to the standard-class instruments described by Willmore (1979). Since there is no single, optimum class of instruments, different instrument types must be considered for different tasks. For example, high-frequency teleseismic body waves are best analyzed on recordings of SP-instruments (class A), long-period body waves and teleseismic surface waves are best analyzed on recordings of LP-instruments (class B), and regional body and surface waves are best analyzed on recordings of intermediate-band (class C) instruments. In addition, the Wood–Anderson instrument is of special historical importance for the determination of local magnitude. The magnification curves of some historical long-period and intermediate-band instruments are shown in Figure 2.

While the design of low-pass or high-pass filters is rather straightforward, the implementation of instrument simulation filters under general conditions is a complex task. To simulate the record of a certain seismograph system, the response of the original recording system, t_1, has to be removed from the

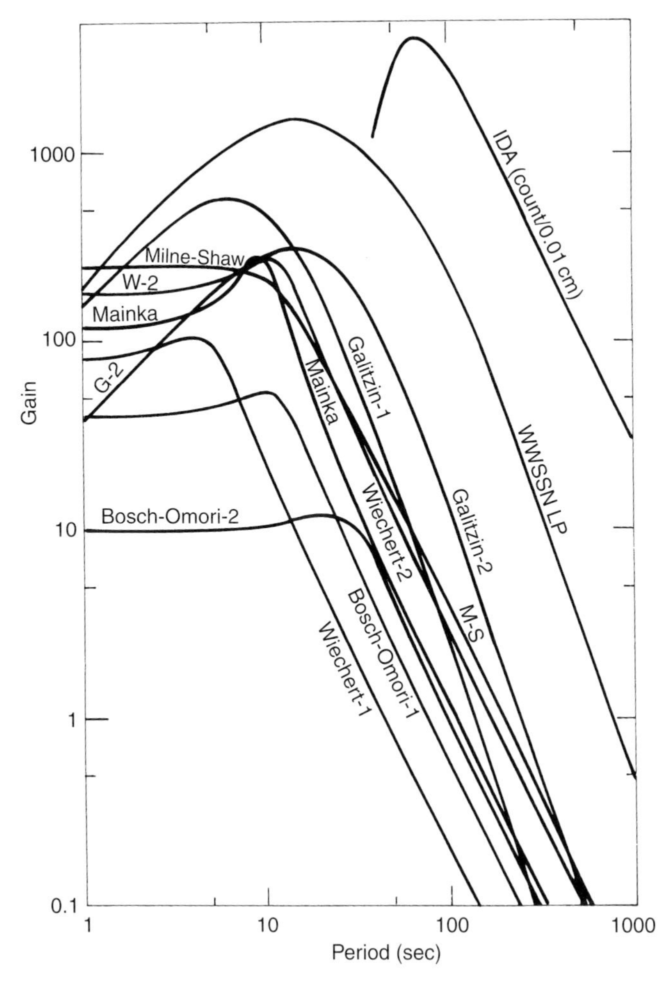

FIGURE 2 Magnification curves of some long period and intermediate band analog seismographs. (From Kanamori, 1988, figure 8.)

record. The resulting signal has to be filtered with the response of the desired instrument:

$$\hat{Y}(z) = \frac{\tilde{T}(z)}{T(z)} \cdot Y(z) = \bar{T}(z) \cdot Y(z) \tag{1}$$

$Y(z)$ and $\hat{Y}(z)$ are the z-transforms of the recorded and the simulated seismograms, respectively. The simulation filter can be thought of as a combination of an inverse filter for the actual recording system—this corresponds to the restitution of the input ground motion signal—and a synthesizing filter for the simulated recording system (Fig. 3).

There is no single best procedure for simulation filtering that performs well under all conceivable conditions (see also "Proceedings of the Seminar on Deconvolution of Seismograms and High-Fidelity Seismometry," *J. Geophys.* **39**, 501–626, 1973). To date, the most advanced software package for automatically performing instrument simulation and deconvolution is the program PREPROC written by Miroslav Zmeškal (Zmeškal, 1996; Zmeškal and Plešinger, 1996). It requires the calibration information be given in GSE format (GSE Waveform Data Format, 1990, 1995) and provides several methods for implementing the filter process. The theoretical background of the techniques used in PREPROC is discussed in Plešinger *et al.* (1996) and a discussion of the stability problems related to simulation filtering in general can be found in Scherbaum (1996).

3. Interpretation of Wave Onsets

Observational seismology is based to a very large degree on the analysis of seismic waveforms, especially on the recognition and interpretation of seismic wave onsets. Onset properties are used in numerous contexts ranging from earthquake location and the determination of focal mechanisms to source modeling and travel-time tomography. Quite recently, the properties of seismic onsets have received additional attention based on the fact that they may allow the distinction of different friction laws discussed in the context of modeling earthquake sources (Iio, 1992, 1995; Ellsworth and Beroza, 1995).

3.1 Onset Properties

An onset can be described by the shape of the discontinuity at the front of a signal. There have been only few quantitative treatments of onset properties (Seidl and Stammler, 1984; Seidl and Hellweg, 1988; Scherbaum, 1996; see Fig. 4). Seidl and Stammler (1984) have used the following definition:

> The discontinuity of the signal front at $t = 0$ is called an onset of order p, if $f^{(p)}(0+)$ is the last nonzero derivative (Seidl and Stammler, 1984).

A causal signal is a signal that is zero for negative times and starts to become nonzero at $t = 0$. In this definition, the onset is defined by two quantities: The onset time at which the signal becomes different from 0 and the onset order, the "rate" at which it starts to increase with time for $t >$ onset time (Fig. 4a). Onset properties are affected in various ways by filters within the recording systems.

3.2 Change of Onset Properties by Causal Filters

One effect is a change in onset order caused by causal low-pass filters. This is best understood from the initial value theorem of the unilateral Laplace transform (Fig. 4b). This theorem states that the shape of the onset is controlled by the values of the Laplace transform times s for s going to infinity. If we consider only the imaginary axis of the s plane, that is, for $s \rightarrow j\omega$, or as it is written in Fig. 4b, $j2\pi f$, we can directly compare time-domain and frequency-domain behavior. In terms of time-domain and frequency-domain behavior, the initial value theorem tells us that if we change the shape of the high-frequency limit of the

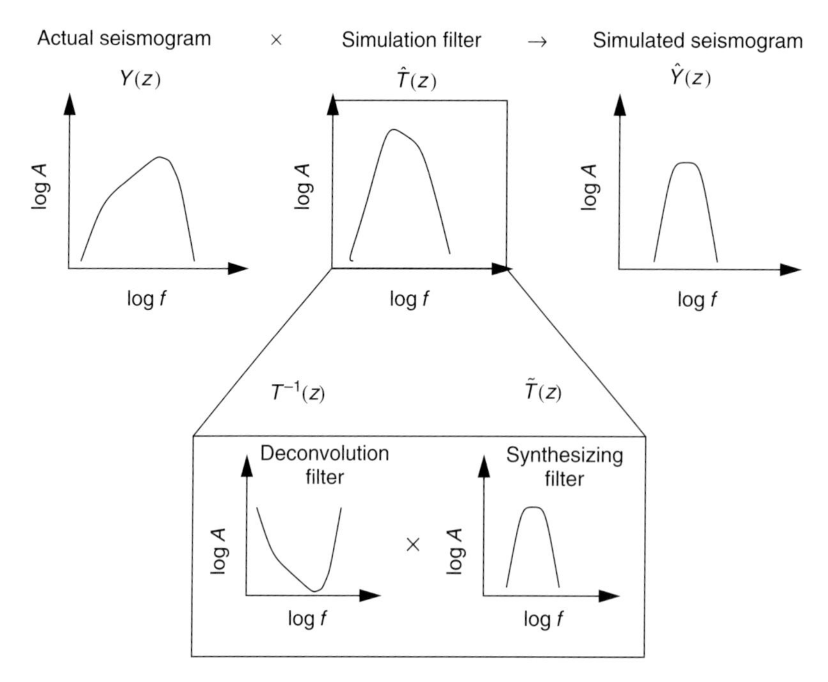

FIGURE 3 Digital simulation of different seismographs systems. The simulation filter can be thought of as a combination of an inverse filter for the actual recording system and a synthesizing filter for the simulated recording system. Displayed are schematic sketches of the amplitude–frequency response functions of the contributing subsystems. (After Scherbaum, 1996.)

spectrum, we automatically change the onset order of a causal signal. Therefore, any causal low-pass filter will increase the onset order, i.e., smoothen the onset. If noise is present, the signal onset may no longer be detectable and the onset seems to be delayed. In other words, changes in the onset order may cause apparent time delays.

In addition to those apparent time delays, however, filters in general will also create real delays to signals. Let us assume a general filter—which could be a seismograph system—with the frequency response function $T(j\omega) = |T(j\omega)| \cdot e^{j\Phi(\omega)}$. For a purely harmonic input signal, $A_i \sin(\omega t)$, the output signal would become $A_i |T(j\omega)| \sin(\omega t + \Phi(\omega))$. Hence, it changes both the amplitude and the phase but not the frequency. The phase $\Phi(\omega)$ gives the amount by which the phase of a harmonic signal with frequency ω is advanced (in radians). This leads directly to the definition of the delay time for a harmonic signal, which is called the phase delay $t_{\mathrm{ph}}(\omega)$ of a filter.

A harmonic signal with frequency ω, which is filtered with a frequency response $|T(j\omega)| \cdot e^{j\Phi(\omega)}$ is delayed by $t_{\mathrm{ph}}(\omega) = -\Phi(\omega)/\omega$ seconds. For signals other than purely harmonic, the situation is more complicated. In the case of narrowband signals, the delay of the signal envelope (or center of gravity) is described by the group delay function $t_{\mathrm{gr}}(\omega)$. For broadband signals, group delay and phase delay have no practical meaning. In this case the parameter of interest is the delay of the onset, the signal front delay t_{fr}. In the case of analog seismograph systems, which can be described by rational transfer functions, the phase response function $\Phi(\omega)$ becomes constant for $\omega \to \infty$. This yields the important results that in this case the signal front delay becomes 0. At first glance this result suggests that in the context of onset time determination we only have to worry about apparent onset time delays caused by changes in the onset order of a seismic wavelet. However, in the context of simulation filtering of digital seismic data this is not true. In that case, the signal front delay depends on the sampling frequency. A more detailed discussion of group phase- and onset time delay can be found in Papoulis (1962).

3.3 Change of Onset Properties by Acausal FIR Filters

The change of onset properties by causal filters is only one part of the story in the context of modern data acquisition systems. Since the good performance of modern data loggers is commonly achieved by oversampling and decimation, these types of data loggers usually contain a cascade of symmetric decimation filters with a linear phase response. Such filters produce a constant time shift, which can be accounted for by changing the time stamp of the output stream. If the time shift is zero or is corrected for, the filter is called a *zero phase* filter. In this case the filter impulse response will begin at negative times and for very impulsive signals will produce acausal precursory oscillations.

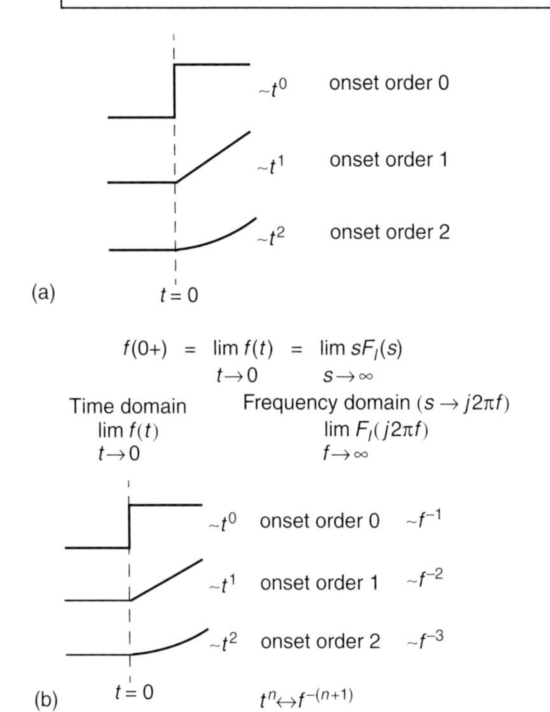

FIGURE 4 Definition of onset order (*a*), and the onset distortions by causal low-pass filters (*b*).

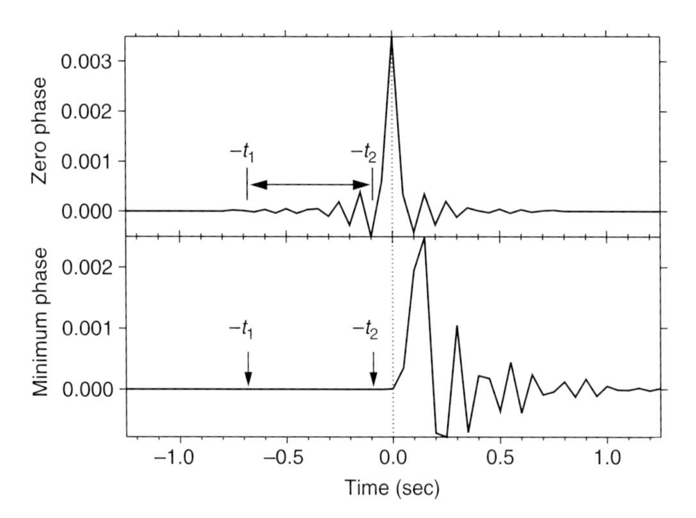

FIGURE 5 Overall impulse response of the 20 Hz data stream of the German Regional Seismic Network (upper panel). The lower panel shows the corresponding minimum phase response.

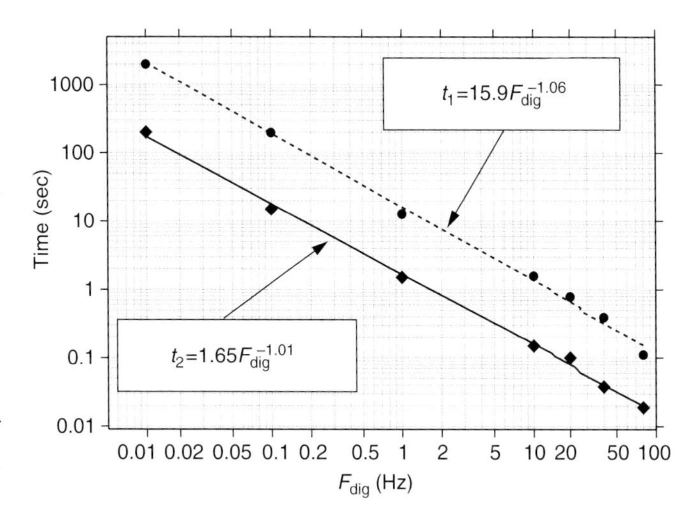

FIGURE 6 Potential onset time biases for different sampling frequencies on a Quanterra data logger. For the definition of t_1 and t_2, see Figure 5. The effects are approximately the same for other data loggers.

As an example, Figure 5 shows in the top panel the overall effective impulse response of the 20 Hz data stream of the German Regional Seismic Network (GRSN). In terms of acausal filter artifacts, this is a worst-case scenario since it describes the distortion of an impulse, the sharpest seismic onset possible. In the lower panel one sees the response of the equivalent minimum phase filter, that is the filter with the same amplitude response but a minimum phase spectrum. In both cases, the input time for the impulse corresponds to time zero.

It can clearly be seen that the zero phase response (in the upper panel) starts out at negative times, which means it is acausal. Here, neither onset time nor onset polarity can be unambiguously determined. The exact onset time corresponds here to the time of the maximum of the impulse response. The onset time in the conventional sense—when the amplitude level exceeds some threshold—will always be determined too early, somewhere between $-t_1$ and $-t_2$. The maximum duration of visible artifact influence t_1 equals roughly half the duration of the impulse response of the last filter stage before decimation. The minimum duration of precursory artifact influence t_2 equals the time from the maximum to the first zero crossing of the overall filter response. In other words, onset times determined on the zero phase filtered trace will be always too early by at least t_2, potentially up to t_1.

In contrast, on the causal filter response, the onset time can be determined exactly, at least if no noise is present. In a noisy situation, the onset is probably determined slightly too late on a causal instrument because of the apparent onset time delay discussed above. This results in systematic differences between onset times determined from classical and modern data loggers (Scherbaum, 1997).

The magnitude of this effect depends strongly on the sampling frequency (Fig. 6). For the Quanterra data logger used in the German Regional Seismic Network, t_1 and t_2 as defined above have been determined empirically for sampling frequencies from 0.01 to 80 Hz. Both scale roughly with the inverse of sampling frequency F_{dig}. This figure allows one to roughly estimate the potential effect of zero phase FIR filters on onset time determinations, but also to judge to what degree a precursory signal might be filter-generated. For example, for a 20 Hz data stream on a GRSN station, FIR filter artifacts are expected to show up for times between 0.08 and 0.7 sec. For the 10-sec data stream, the acausal influence of the filter reaches up to 200 (!) seconds. This plot is intended to illustrate that these times are nonnegligible.

Fortunately, for all purposes involving onset analysis, the FIR filter-generated precursory signals can be removed from the seismograms by post-filtration with specially designed all-pass filters (Scherbaum, 1996; Scherbaum and Bouin, 1997). The discussion of the techniques is beyond the scope of this article, however. More information and correction software is to be found on the accompanying Handbook CD as well as under *http://lbutler.geo.uni-potsdam.de/FIR/fir_daaf.htm.*

In conclusion, we have to worry about three different types of effects of seismic recording systems on onset times (Fig. 7):

1. Apparent time delays that are caused by changes in the onset order, for example by LP filtering. Any filter that changes the high frequency decay of a causal signal changes the onset order at the same time! In this case the amount of noise present in the data determines the apparent time delay.
2. Real-time delays, which are described for purely harmonic signals by the phase delay, for narrowband signals by the group delay, and for broadband signals by the signal front delay. While the signal front delay is zero for analog seismographs, it might be nonzero for digitally simulated seismographs.
3. Time advances produced by zero phase FIR filters. The magnitude of this effect depends strongly on the sampling frequency. It should be noted that zero-phase FIR filters also influence the onset polarity determinations.

Delays		Advances
Apparent delay	**Physical delays**	**Onset advance by acausal FIR filters**
Change of onset order by causal LP filters	Phase delay (harmonic signal)	Depending on recording system and sampling frequency
Depending on SNR	$t_{ph}(\omega)=-\Phi(\omega)/\omega$ (s)	
	Group delay (narrowband signal)	Distorts onset polarity also!
	$t_{gr}(\omega)=-\frac{d}{d\omega}\Phi(\omega)$(s)	
	Signal front delay (BB signals)	
	$t_{fr}=\lim_{\omega\to\infty} t_{ph}(\omega)$(s)	

FIGURE 7 Onset distortions caused by causal and acausal filters.

In addition to onset times, there are other time measurements on seismograms that play an important role in earthquake seismology. The determination of source size as well as of stress release is commonly based on the measurement of characteristic seismogram times, in the form of pulse duration, rise time, or in the spectral domain as spectral corner frequencies. The pulse rise time is commonly defined as the interval between the intersection of the tangent of the steepest rise of a wavelet with the zero-level and peak pulse amplitude (e.g., Gladwin and Stacey, 1974), $t_r = a_{max}/(\text{slope})_{max}$.

For an ideal zero phase low-pass filter that completely suppresses all frequencies above a corner frequency f_c, the minimum rise that can be determined becomes $1/(2f_c)$ (Scherbaum, 1996). Although this is strictly valid only for zero phase low-pass filters, it provide an easy way to get a rough estimate of the minimum rise time observable from a given recording system.

References

Ellsworth, W.L. and G.C. Beroza (1995). Seismic evidence for an earthquake nucleation phase. *Science* **268**, 851–855.

Gladwin, M.T. and F.D. Stacey (1974). Anelastic degradation of acoustic pulses in rock. *Phys. Earth Planet. Inter.* **8**, 332–336.

GSE Wave Form Data Format (1990). Version CRP190/rev.3: Ad hoc group of Scientific Experts, Geneva, Annex D2, D4-1310, D48–D62.

GSE Wave Form Data Format (1995). Version CRP243: Ad hoc group of Scientific Experts, Geneva, OpsAnnex 3.

Hamming, R.W. (1977). "Digital Filters," 2nd Edn. Prentice-Hall, Englewood Cliffs, NJ.

IEEE Digital Signal Processing Committee (1979). "Programs for Digital Signal Processing." IEEE Press, Wiley, New York.

Iio, Y. (1992). Slow initial phase of the *P* wave velocity pulse generated by microearthquakes. *Geophys. Res. Lett.* **19**, 477–480.

Iio, Y. (1995). Observations of the slow initial phase generated by microearthquakes: implications for earthquake nucleation and propagation. *J. Geophys. Res.* **100**, 15333–15349.

Kanamori, H. (1988). Importance of historical seismograms for geophysical research. In "Historical Seismograms and Earthquakes of the World," (W.H.K. Lee, H. Meyers, and K. Shimazaki, Eds.), pp. 16–33. Academic Press, San Diego, CA.

McClellan, J.H., T.W. Parks, and L.R. Rabiner (1979). "FIR Linear Phase Filter Design Program," Programs for Digital Signal Processing. IEEE Press, New York.

Oppenheim, A.V. and R.W. Schafer (1989). "Discrete-Time Signal Processing." Prentice Hall, Englewood Cliffs, NJ.

Oppenheim, A.V. and A.S. Willsky (1983). "Signals and Systems." Prentice-Hall, Englewood Cliffs, NJ.

Papoulis, A. (1962). "The Fourier Integral and Its Application." McGraw-Hill, New York.

Parks, T.W. and C.S. Burrus (1987). "Digital Filter Design." Wiley, New York.

Plešinger, A. (1998). Determination of seismograph system transfer functions by inversion of transient and steady-state calibration responses. *Stud. Geophys. Geod.* **42**, 472–499.

Plešinger, A., M. Zmeškal, and J. Zedník (1996). "Automated Preprocessing of Digital Seismograms—Principles and Software," Version 2.2 (E. Bergman, Ed.). NEIC Golden/GI Prague.

Press, W.H., S.A. Teukolsky, W.T. Vetterling, and B.P. Flannery (1992). "Numerical Recipes in C," 2nd Edn. Cambridge University Press, Cambridge.

Proakis, J.G. and D.G. Manolakis (1992). "Digital Signal Processing Principles, Algorithms, and Applications." Macmillan, New York.

Scherbaum, F. (1996). "Of Poles and Zeros: Fundamentals of Digital Seismology." Kluwer Academic, Dordnecht.

Scherbaum, F. (1997). Zero phase FIR filters in digital seismic acquisition systems: blessing or curse. *EOS* **78**(33), 343–344.

Scherbaum, F. and M.P. Bouin (1997). FIR filter effects and nucleation phases. *Geophys. J. Int.* **130**, 661–668.

Scherbaum, F. and J. Johnson (1992). "Programmable Interactive Toolbox for Seismological Analysis," IASPEI Software Library, Vol. 5. Seismological Society of America, El Cerrito, CA.

Seidl, D. and M. Hellweg (1988). Restoration of broad-band seismograms (Part 11): Signal moment determination. *J. Geophys.* **62**, 158–162.

Seidl, D. and W. Stammler (1984). Restoration of broad-band seismograms (Part 1). *J. Geophys.* **54**, 114–122.

Stammler, K. (1993). SeismicHandler—programmable multichannel data handler for interactive and automatic processing of seismological analyses. *Comput. Geosci.* **19**, 135–140.

Stearns, S.D. (1975). "Digital Signal Analysis." Hayden, London.

Wielandt, E. (1986). Calibration of digital seismographs with arbitrary signals. Internal Report, ETH Zürich.

Willmore, P.L. (Ed.) (1979). Manual of seismological observatory practice. WDCA SE-Report 20, US Department of Commerce, NOAA, Boulder, CO.

Zmeškal, M. and A. Plešinger (1996). "Simulation of Standard Seismograms, Automated Preprocessing of Digital Seismograms—Principles and Software," Version 2.2 (E. Bergman, Ed.). NEIC Golden/GI Prague.

Zmeškal, M. (1996). "PREPROC Library Programmer's Guide," Software Version 2.2. Library Edition, GI, Prague.

Editor's Note

See also Section 11 of Chapter 18, Seismometry, by Wielandt. Appendix 1, Basic Concepts of Filter Theory, and Appendix 2, Effects of the Recording System, are placed on the attached Handbook CD, under the directory \22Scherbaum.

23

Seismometer Arrays—Their Use in Earthquake and Test Ban Seismology

A. Douglas
AWE Blacknest, Brimpton, Berks, UK

1. Introduction

The only way nuclear tests fired underground can be detected at long range is from the seismic signals that such tests generate. It is for this reason that the International Monitoring System (IMS) set up to provide data to enable the Comprehensive Test Ban to be verified has to include a network of seismological stations. Even when the signals are detected, however, it is not possible to recognize underground explosions unambiguously. The purpose of the Monitoring System is to provide data to allow those disturbances that are clearly natural to be identified, and those that have a high probability of being underground tests to be recognized. Should a suspicious disturbance be detected, it is up to the signatories to the Test Ban Treaty to ask the country suspected of carrying out the test to allow an international team to inspect the region around its estimated epicenter (that is, to carry out an On-Site-Inspection) to try to establish whether or not a nuclear test had taken place.

It was recognized when the first talks on banning nuclear tests began in 1958 that there is a need for ways of enhancing seismic signals in noise if a test ban is to be verified down to around a few kilotons (kt). One way proposed for suppressing noise is by the use of arrays of seismometers, and this led to the establishment of the first such arrays in 1960. Since then more than 20 arrays have been put in (and most of these are still in operation), many principally to investigate the contribution that arrays can make to verification whereas others were installed by some nuclear weapon states specifically to detect the signals from foreign tests and so follow at least in part the nuclear weapons test programs of potential adversaries. Experiments have been carried out with a wide variety of array designs. Some of the most significant arrays to operate in the period since 1960 are listed in Table 1 and the designs of some of these are shown in Figure 1.

It is expected that the Monitoring System will allow the detection at long range of underground explosions with yields of 1 kt or more fired anywhere, and those with yields significantly less than this in most continental areas. As well as detecting the signals, the signal-to-noise ratio and bandwidth must be such as to aid discrimination between natural and suspicious seismic sources. The amplitudes of the short-period (SP) *P* waves (the most widely recorded seismic waves from underground tests) from explosions of 1 kt, are around a nanometer at 1 Hz at long range, which is below the seismic

TABLE 1 Examples of Short-Period Seismometer Arrays

Location	Code	Aperture (km)	No. of Seismometers	Operating Period	Remarks
Southeastern Norway	NORESS	3	25	Fall 1984 to date	Small aperture
Wichita Mts., OK, USA	WMSO	5	13	1964–July 1969	Small aperture
Eskdalemuir, Scotland	EKA	10	20	May 1962 to date	Medium aperture
Yellowknife, Canada	YKA	25	19	Nov 1962–May 1963 and Dec 1963 to date	Medium aperture. Gauribidanur (GBA) India and Warramunga (WRA) Australia are arrays of similar design
Billings, MT, USA	LASA	200	525	1965–1978	The large aperture seismometer array; 21 subarrays each of 25 seismometers

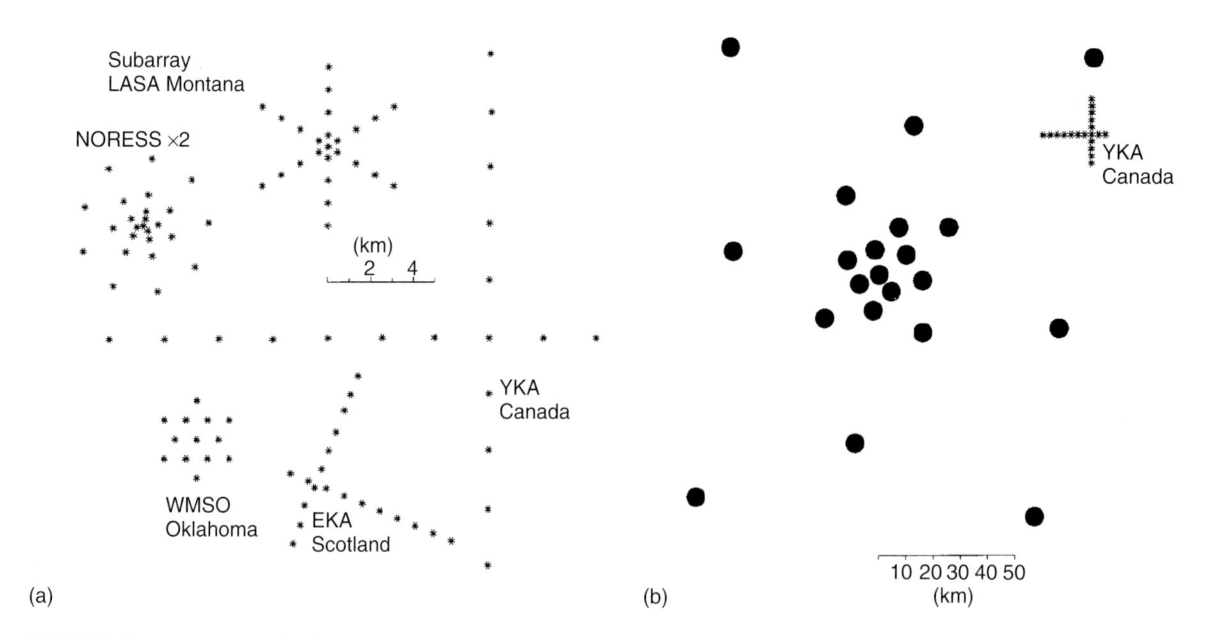

FIGURE 1 Examples of the layout used in short-period seismometer arrays. (a) Seismometer positions in some small- and medium-aperture arrays and a LASA subarray. The NORESS array is shown at twice the scale of the other arrays. (b) The subarrays within the LASA. The layout of YKA is superimposed on the figure to show the relative size of this array and the LASA.

noise level in the SP band over much of the Earth. Thus, it is only by using arrays of seismometers to suppress the noise that the monitoring system can be effective at the 1 kt level.

As well as allowing signals to be enhanced above noise, arrays can be used to determine the apparent velocity (speed and azimuth) of arrivals and hence their slowness, that is effectively the ray parameter, $dT/d\Delta$, where T is travel time and Δ is epicentral distance. If the arrival is a phase for which the slowness is known as a function of distance, the epicentral distance can be estimated and combined with the azimuth estimate to give a rough epicenter (and origin time) if the focal depth is assumed known. Conversely, slowness measurements can be used to test whether an initial identification is correct or can help with the identification of weak or unexpected phases (see, for example, Engdahl *et al.*, 1970; Cleary and Haddon, 1972; Cleary *et al.*, 1975; Barley, 1977). Arrays have also been used to measure $dT/d\Delta$ as a function of Δ for direct P (see, for example, Chinnery and Toksöz, 1967; Corbishley, 1970) and for P travel-time triplications (see, for example, Ram and Mereu, 1977; Simpson and Cleary, 1977).

Because of the large volumes of data generated by arrays, they were among the first seismometer stations to use digital computers to control the operation of the recording systems, and this naturally led to the automatic processing of recordings. The result has been an enormous growth in processing methods, with most of them aimed at detecting signals down to as low a magnitude as possible and providing rough estimates of epicenter and origin time of the signal source (see, for example, Ringdal and Husebye, 1982; Ringdal, 1990). Such automatic processing is necessary for test ban verification where assessments may have to be carried out on most if not all the seismic disturbances detected. This could require investigation of upward of 20 000 disturbances each year.

Although arrays have several uses, in this review I concentrate on the unique feature of recordings from such stations, which is that they allow weak signals to be enhanced above noise. I then go on to show how this aids in the interpretation of seismograms and studies of source mechanisms particularly for test ban seismology—sometimes referred to as forensic seismology—in distinguishing between earthquakes and explosions. I therefore take as the definition of an array a spatially distributed set of seismometers all transmitting their outputs to a central laboratory to be recorded against a common time-base; the principal use of the outputs being to allow signals to be enhanced and noise to be suppressed. Recordings from networks of stations are sometimes combined to simulate an array, but it is only where a group of seismometers has been installed to operate as a unit that such a network can correctly be described as an array.

Although much research has been carried out on methods of distinguishing between earthquakes and explosions using recordings at regional distances ($\Delta < 20°$) there are no discrimination criteria making use of such recordings that are yet accepted as being widely applicable. It turns out that earthquakes and explosions are most easily separated given P recorded at teleseismic distances, that is, distances of more than 30°. Then the body-wave magnitude m_b, can be estimated and, given an estimate of M_S from surface waves recorded at any distance, the $m_b : M_S$ criterion (see, for example, Marshall and Basham, 1972) can be used to identify the source: for a given m_b, explosions have a much lower M_S than most earthquakes. However, there are earthquakes—sometimes called anomalous earthquakes—that generate weak surface waves and on the $m_b : M_S$ criterion look explosion-like. It is in identifying these anomalous earthquakes, that array recordings have proved most useful (see, for example, Douglas *et al.*, 1974a,b). It is for this reason that in what follows most of the discussion is on SP arrays designed to record P at teleseismic distances.

2. Enhancing Seismic Signals in Noise

The use of arrays to enhance seismic signals makes use of the differences in the spatial properties of signals and noise. Ideally the signal at an array would be identical (apart from differences in arrival time) at all seismometers. This is never strictly true, but by restricting the maximum distance between the seismometers of an array to around the largest wavelength of the principal signals of interest, sites can be found where the signal variation is so small that it does not significantly reduce the effectiveness of the array for improving signal-to-noise ratios.

Unfortunately, at some sites the variation in signal shape across the array has turned out to be so great that conventional processing results in loss of signal amplitude. Nevertheless, such recordings can be used for the detection of seismic signals (Ringdal *et al.*, 1975). Processing is carried out on the envelope of the seismograms rather than the seismograms themselves; one way of forming the envelope being to rectify and smooth the seismogram. Such processing has little value for source studies.

To describe how arrays can be used to separate signals from noise, it is usual to represent signals and noise in frequency–wavenumber space. In this representation, array processing can be thought of as wavenumber filtering. A major consideration in designing arrays and processing the data is how they act as wavenumber filters, that is, the filter response as a function of vector wavenumber (see, for example, Birtill and Whiteway, 1965).

White noise that is spatially uncorrelated can be thought of as being uniformly distributed in frequency–wavenumber space. Studies of array recordings show that, provided the separation between seismometers is a kilometer or more, the noise can be assumed to be uncorrelated and random at frequencies above about 1 Hz. At lower frequencies the noise becomes increasingly organized; that is, the noise power is no longer uniformly spread in wavenumber but is concentrated around particular values. The noise is then principally wave trains propagating with well-defined velocities. When a single noise source predominates, as it may do with oceanic microseisms, the noise becomes in effect a coherent signal (see, for example, Douglas and Young, 1981).

Organized noise usually has two components: low-speed surface waves generated at local and regional distances; and body waves with high apparent surface speeds generated at long range (referred to as mantle P waves by Backus, 1966). At sites close to a coast where the noise levels are high, the noise is predominantly the low-speed component and it is only at mid-continental sites, where this component is of low amplitude or absent, that the high-speed noise is seen.

2.1 Delay-and-Sum Processing

The most commonly used method of enhancing signals relative to noise is usually referred to as delay (or shift) and sum: delays are applied to the output of each seismometer (each channel of information) to bring the signals to a common time line, that is, into phase, and the signals are summed with weights n^{-1}, where n is the number of seismometers in the array. The summed signal is often called the array beam.

The noise on the delay-and-sum output is the mean of n numbers (noise amplitudes). If σ^2 is the variance of the noise at the array, then the variance of the mean is σ^2/n, assuming the noise is random and uncorrelated between channels. So the standard deviation of the noise on the sum is $\sigma/n^{1/2}$. If the signal is effectively identical on all channels then it will be almost unaffected by delay-and-sum processing so, as the noise amplitude is reduced by a factor $\sqrt{n}$, the signal-to-noise ratio should show the so-called "root n improvement." Thus, for example, for signals recorded at YKA (Table 1) root n ($n = 19$) improvement is possible at frequencies above 1 Hz, as shown for the P signal from the Marianas Island earthquake of 16 July 1969 (Figs. 2a and 2b).

The root n improvement shown by Figs. 2a and 2b is obtained only if the SP recordings are filtered to cut out frequencies below about 1 Hz. For the unfiltered SP recordings (Figs. 2c and 2d) the noise reduction is only about a factor of 2 rather than the hoped for $\sqrt{19}$. This shows that the low-frequency noise is correlated between channels and thus that the noise is organized at these frequencies. The failure to obtain root n improvement over the whole SP band for delay-and-sum processing is a problem with all arrays and as a consequence it has become common practice to apply a high-pass filter similar to that applied to obtain Figs. 2a and 2b. This, however, fails to exploit fully the wavenumber filtering property of arrays, the very property for which arrays are installed. For, if the wavenumber of the organized noise is sufficiently different from that of the signal, optimum wavenumber filters can be derived to suppress such noise. The application of such filters seems never to have been investigated for YKA recordings, but examples of optimum filtering of organized noise at another array, EKA, are given in the next section.

The advantage of wavenumber filtering is that it should reduce the need for frequency filtering, which as well as attenuating the noise inevitably distorts the signal. Note, for

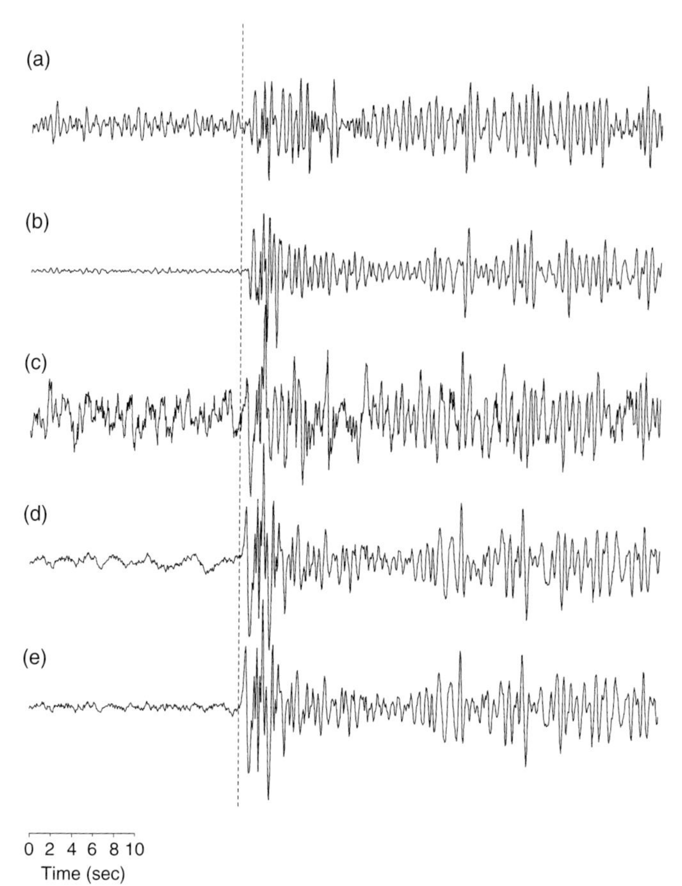

FIGURE 2 Delay-and-sum processing of the P waves from the Marianas Islands earthquake of 16 July 1969 (m_b 5.2) recorded at the Yellowknife array. (a) Single channel after application of a 1–4 Hz (causal) bandpass filter; (b) sum after application of a 1–4 Hz (causal) bandpass filter; (c) single channel unfiltered; (d) sum unfiltered; (e) sum (d) filtered with an optimum frequency filter to attenuate noise at around 4 sec period.

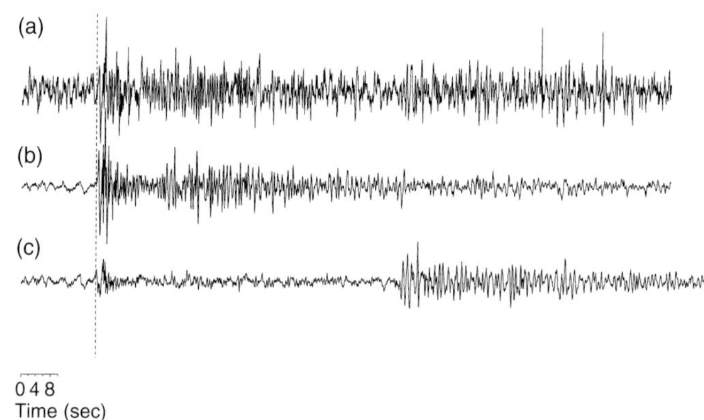

FIGURE 3 Delay-and-sum processing to separate two signals with differing phase velocities using recordings from the Yellowknife array. (a) Single channel showing P from the Marianas Island earthquake of 16 July 1969 (m_b 5.2); (b) sum phased for Marianas Island earthquake (speed 22.1 km sec^{-1}, azimuth 285°); (c) sum phased for speed of 11.8 km sec^{-1} and an azimuth of 183°. Seismogram (c) shows the P signal from the Nevada Test Site explosion ILDRIM (m_b 4.6).

example, as shown in Figure 2, that first motion is reduced by such filtering (cf. Figs. 2b and 2d), making polarity and signal onset difficult to measure reliably. Thus, such high-pass filtering should be avoided if possible. If frequency filtering has to be applied then an optimum filter should be used. Figure 2e shows the result of applying an optimum filter of the type proposed by Douglas (1997) to attenuate the low-frequency noise. Although any frequency filter will inevitably remove signal, as well as noise, the use of such an optimum filter clearly minimizes distortion.

Figure 3 illustrates the ability of an array to separate mixed signals. Figures 3a and 3b are longer sections of the P signal from the Marianas Island earthquake than those shown in Figure 2. The single-channel (Fig. 3a) shows no evidence of other arrivals in the signal coda. However, forming the beam to enhance signals from the Nevada Test Site (NTS) shows (Fig. 3c) a second signal—it is in fact P from the NTS explosion ILDRIM—and the Marianas signal is suppressed.

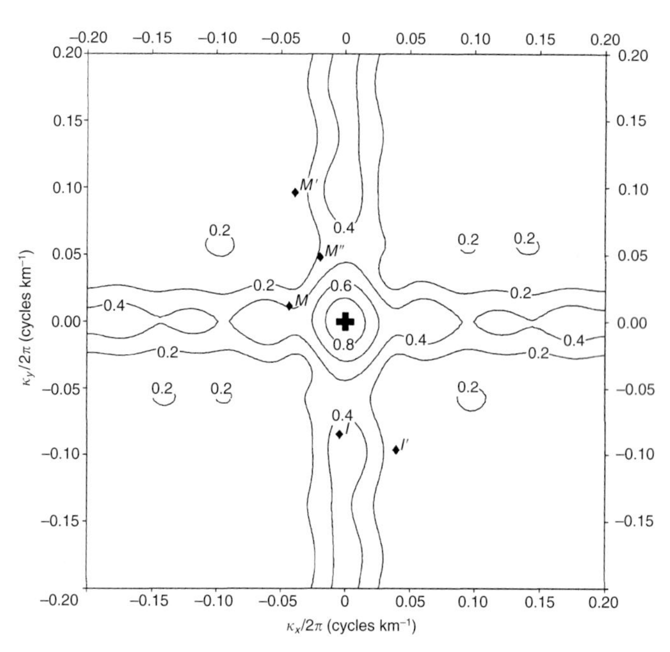

FIGURE 4 The wavenumber response of the Yellowknife array. M is the wavenumber of the signal from the Marianas Islands earthquake at 1 Hz and I that from the NTS explosion ILDRIM. On the delay-and-sum response for the Marianas Islands earthquake, ILDRIM at 1 Hz plots at I'. On the delay-and-sum response for ILDRIM, the Marianas Island earthquake at 1 Hz plots at M', and for 0.5 Hz at M''. κ_x and κ_y are the x and y components, respectively, of angular wavenumber in radians km^{-1}.

The wavenumber response for YKA for delay-and-sum processing is shown in Figure 4. Superimposed on the response is the wavenumber at 1 Hz of the Marianas Islands earthquake (M) and the explosion ILDRIM (I). This shows that if the beam were formed for zero wavenumber (i.e., the sum without time shifts), the 1 Hz amplitude of the earthquake signal would be attenuated by a factor of 2.5 and that of the explosion by about 2. The effect of forming the beam for the earthquake is to move its signal to zero wavenumber so that it is passed unattenuated. The explosion signal at 1 Hz now falls at I' and is predicted to be attenuated by about a factor of 10, and from Figure 3 this could be so. With the array phased for the explosion signal, this signal now lies at zero wavenumber and the earthquake signal at 1 Hz falls at M'. Now it is the earthquake signal that should be attenuated by a factor of $\sim$10. As can be seen from Figure 3c the actual attenuation is only about a factor of $\sim$5. This is because the earthquake signal is not monochromatic with a frequency of 1 Hz: there is clearly a 0.5 Hz component in the signal and for that Figure 4 shows (M'') that the attenuation is only expected to be 5.

The variation in signal shape in the recordings from the individual seismometers of an array is due to scattering at lateral heterogeneities in the crust and upper mantle beneath the array and by the topography at the site. The principal effect of topography is the generation of Rayleigh waves. For, just as a P wave with a curved wave front incident on a plane free surface gives rise to Rayleigh waves—without them the boundary conditions cannot be satisfied—so a plane P wave incident on a nonplanar surface gives rise to surface waves.

One of the best examples of Rayleigh waves generated by a topographic feature is shown by recordings at EKA from earthquakes with epicenters to the north of the array (Key, 1967, 1968). These recordings show arrivals with amplitudes approaching that of initial P and at first sight could be interpreted as a free-surface reflection pP or sP (Fig. 5a). However, summing the array attenuates this arrival (Fig. 5b), showing that its velocity differs from that of P. Investigation shows that the second arrival has a speed of 2.5 km sec^{-1} and an azimuth of 315° (Fig. 5c). The low speed implies that the arrival is a Rayleigh wave, and this is confirmed by the particle orbit, which is a retrograde ellipse. From the speed and azimuth of the surface wave and its arrival time relative to P, it is possible to determine where it was generated: A steep-sided valley, Moffat Water, about 13 km from the array. Rayleigh waves such as that generated by Moffat Water appear large relative to direct P because, although the energy in the waves is small, it is concentrated in a thin zone close to the surface.

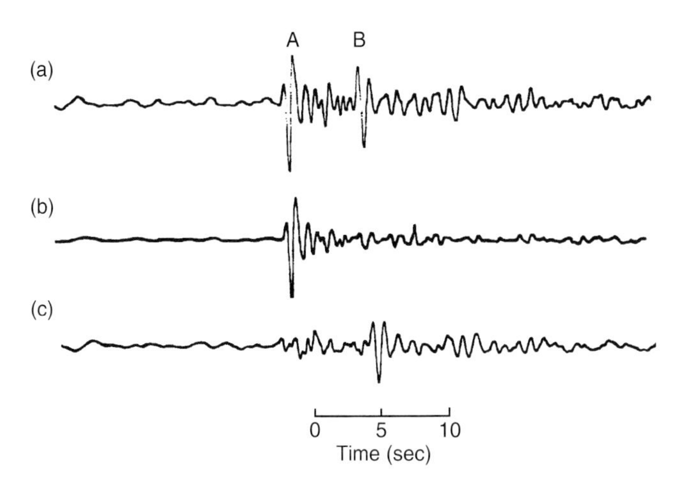

FIGURE 5 Use of the Eskdalemuir array to identify noise generated by P incident at an irregular free surface. (a) Seismogram recorded by one seismometer of the array showing the P waves (A) from an earthquake in Kamchatka. Note the prominent second arrival (B). (b) Summed seismogram after delay-and-sum processing to enhance signals with the velocity of arrival A (19 km sec^{-1} from azimuth 12°). The arrival B is suppressed, suggesting that it has a different velocity from A. (c) Summed seismogram after delay-and-sum processing to enhance signals with speed 2.5 km sec^{-1} on azimuth 315°. Arrival B is now enhanced and A is suppressed. The low-speed of arrival B indicates that it is probably a surface (Rayleigh) wave.

2.2 Optimum Wavenumber Filtering

To introduce the principles of optimum wavenumber filtering, consider the use of an array to enhance P signals at zero wavenumber, in noise that has two components, an organized component that is isotropic and has a speed of 3 km sec^{-1} and a frequency of 0.167 Hz—to represent oceanic microseisms with a period of 6 sec—and a random uncorrelated component at all other frequencies. To ensure that the signal is coherent across the array it is assumed that the maximum separation between seismometers has to be roughly equal to the smallest signal wavelength of interest, say 8 km—the wavelength of a 1 Hz signal with apparent speed of 8 km sec^{-1}.

Now, a plane wave with horizontal wavelength λ will be canceled out by summing the outputs of two seismometers separated by a distance $\lambda/2$ as measured in the direction of propagation. So, as the wavelength of the organized component of the noise is 18 km, a separation of 9 km will enhance signals at zero wavenumber and suppress the oceanic microseisms propagating along the line of the seismometers. Four seismometers at the corners of a square can be used to attenuate microseisms from any azimuth. For such an array, however, the random noise will be reduced only by a factor of $\sqrt{4}$ and can only be attenuated further by increasing the number of seismometers to $n > 4$. If these are to be contained within an overall dimension of $\sim$8 km then the spacing between some pairs of seismometers will be significantly less than $\lambda/2$, so the noise cancellation achieved by spacing the outer seismometers at around half the wavelength of the 6 sec microseisms is degraded by the additional seismometers.

One way out of this problem is to apply a bandstop filter to the delay-and-sum to remove the oceanic microseisms (and any signals at these frequencies). This gives an estimate of the signal at all frequencies except 0.167 Hz. The noise at these frequencies, being uncorrelated, should be attenuated by $\sqrt{n}$. By bandpass filtering the outputs of the outer four seismometers to extract 0.167 Hz and forming the delay-and-sum, an estimate of the signal at this frequency can be obtained and added to the delay-and-sum for the other frequencies. In frequency–wavenumber space this procedure modifies the response of the n seismometer array to place a null at the wavenumber of the microseisms. This is optimum wavenumber filtering for a signal at zero wavenumber for this particular noise model and array design.

The optimum wavenumber filtering described above is a special case of the general process of multichannel filtering. In multichannel filtering, each channel is passed through a filter and the filter outputs are summed. In practice, optimum wavenumber filters are estimated by minimizing the mean-square difference between the actual and the desired outputs. The theory for the estimation of such optimum filters is given in a few seminal papers published in the 1960s (see, for example, Burg, 1964; Backus *et al.*, 1964; Backus, 1966; Capon *et al.*, 1967, 1968). To estimate such filters requires the autocorrelation of the noise at each seismometer and the cross-correlations between all possible pairs of seismometers. These correlations can be derived either from a sample of observed noise ahead of the signal or from a noise model such as the simple model used above.

An example of the use of optimum wavenumber filtering of SP recordings from EKA (Table 1 and Fig. 1) is shown in Figure 6a–c. The filters are minimum-power filters, designed to minimize the power on the beam yet pass the signal unchanged, and are derived using a noise model. The details of the model are given by Douglas (1998) and it is similar to that used above, having a random component over the whole range of frequencies covered by the recordings, combined below 1 Hz with an organized component that is isotropic with a slowness of 0.34 ± 0.05 sec km^{-1} (2.56–3.44 km sec^{-1}). Figure 7 shows signal-to-noise gain as a function of frequency between 0 and 2 Hz. Note that at frequencies below 1 Hz, delay-and-sum processing gives signal-to-noise improvements significantly less than $\sqrt{n}$. Figure 8 shows the wavenumber response for EKA for delay-and-sum and that at 0.2 Hz for the optimum multichannel filters. As expected, with optimum filtering the response is more highly attenuating at the wavenumber of the organized noise than it is for delay-and-sum processing.

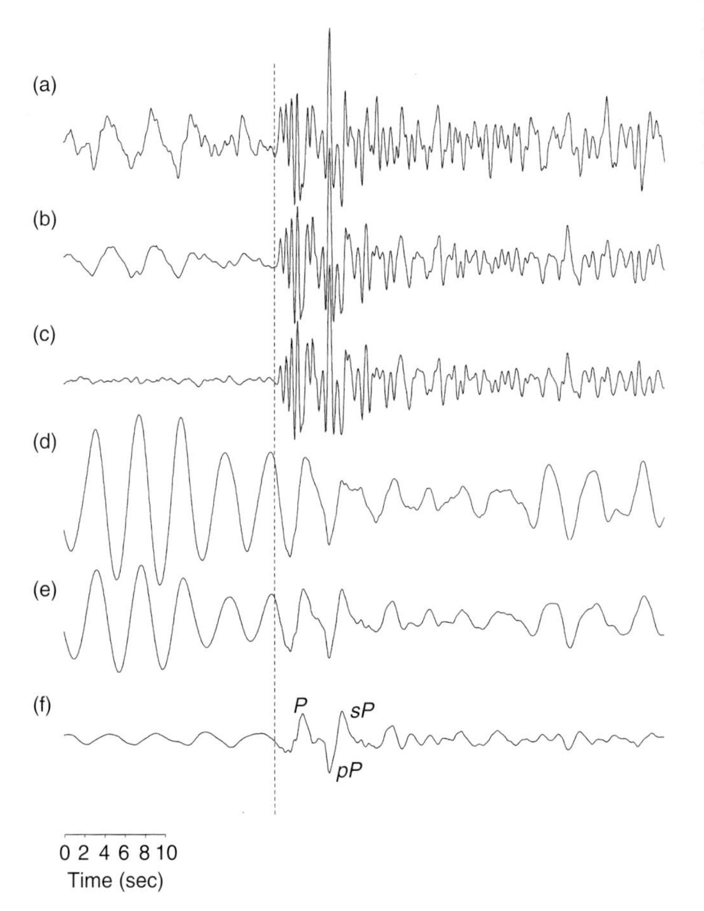

FIGURE 6 The application of various processing methods for signal-to-noise improvement to the short-period P recordings at the Eskdalemuir array from the earthquake of 21 July 1995 in Gansu Province China (m_b 5.6). (a) Single channel short-period; (b) short-period delay-and-sum; (c) short-period minimum-power output; (d) broadband single channel; (e) broadband delay and sum; (f) broadband minimum power output.

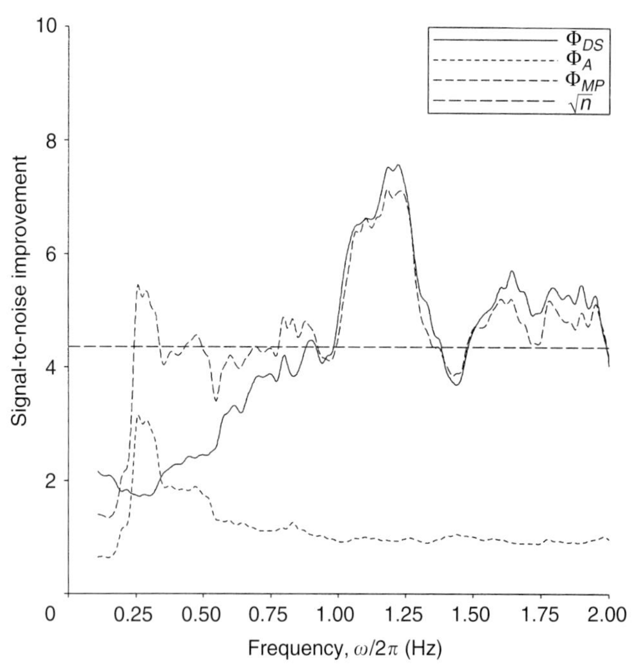

FIGURE 7 Noise reductions as a function of frequency $\omega/2\pi$, obtained for the noise preceding P from the 21 July 1995 Gansu, China earthquake. Φ_{DS} is $\{\sigma^2_{AV}(\omega)/\sigma^2_{DS}(\omega)\}^{1/2}$, where $\sigma^2_{DS}(\omega)$ is the noise power on the delay-and-sum output, $\sigma^2_{AV}(\omega)$ is $\sum \sigma^2_i(\omega)/n$ and $\sigma^2_i(\omega)$ is the noise power on channel i. Φ_{MP} is $\{\sigma^2_{AV}(\omega)/\sigma^2_{MP}(\omega)\}^{1/2}$, where $\sigma^2_{MP}(\omega)$ is the noise power on the minimum-power output. Φ_A is Φ_{MP}/Φ_{DS}. ω is angular frequency in radians sec^{-1}.

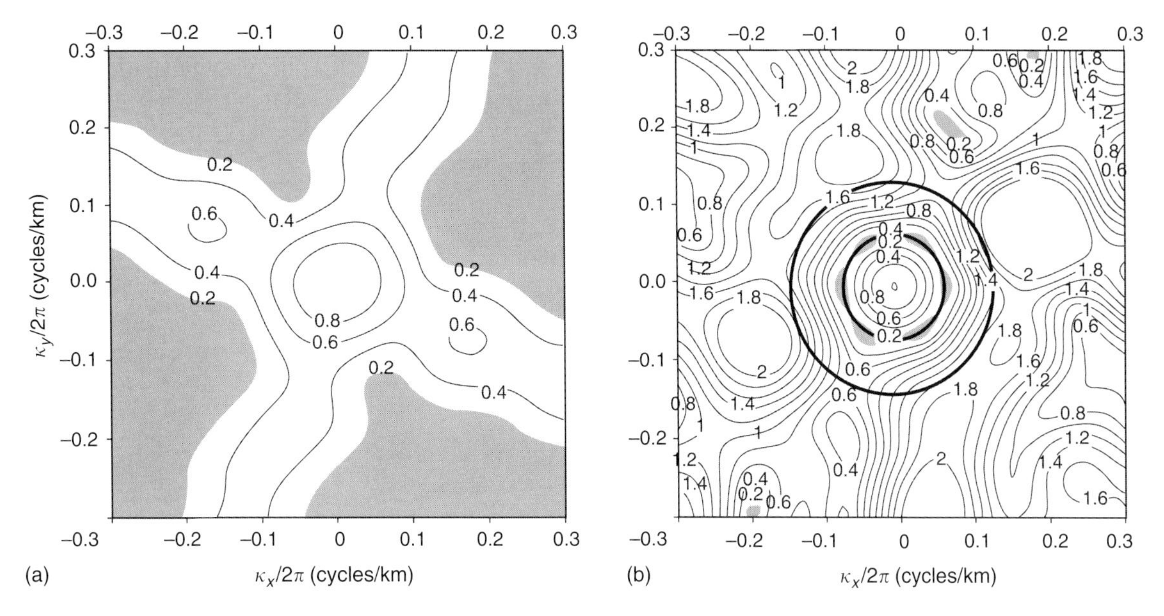

FIGURE 8 Wavenumber responses for the Eskdalemuir array for (a) delay-and-sum processing, (b) minimum power processing (0.2 Hz). The minimum power filters are those used for processing the 21 July 1995 earthquake in Gansu, China. The shaded area shows where the response is less than 0.2. The inner circle marks the wavenumber equivalent to a slowness of 0.34 sec km^{-1} (a speed of $\sim$2.94 km sec^{-1}). The outer circle is the wavenumber equivalent to a slowness of 0.68 sec km^{-1} (a speed of $\sim$1.47 km sec^{-1}). κ_x and κ_y are the x and y components, respectively, of angular wavenumber in radians km^{-1}.

3. History of Arrays

The arrays, installed in 1960–62, had maximum dimensions of 3–5 km and up to 16 SP seismometers. These arrays were designed to record SP P signals from sources at regional distances. The aperture of such arrays is less than the wavelengths of the signals of interest ($\geq$8 km at 1 Hz), so initially the processing scheme was simply to sum the signals without time shifts. Such processing enhances signals with high apparent surface speeds from any azimuth but suppresses low-speed and random noise.

In 1960 the United Kingdom began experimenting with arrays. It seems that from the beginning the United Kingdom planned to apply corrections for the time of propagation of the signals across the array. It was assumed that the signal would be coherent over at least a wavelength, so this defined the aperture of the array. The number of seismometers that could be used was limited by the largest number of channels (24) that could conveniently be recorded on the analog magnetic tapes then in use. With 24 channels, 20 channels of data, time, and error correction could be recorded. As initially it was assumed that arrays would be used for recording at regional distances where the wavelength of a 1 Hz signal is $\sim$8 km, the aperture of the first UK arrays was 8–9 km. The first UK-designed array was set up in cooperation with the United States at Pole Mountain, Wyoming (PMW) in 1961, only 10° from the Nevada Test Site (NTS).

Regional arrays are effective, but the PMW recordings revealed that the signals from NTS explosions at such epicentral distances are complex—they differ little from those from earthquakes even though the explosion source should be simpler than that of an earthquake.

In 1962, PMW recorded P signals from the USSR test site in Eastern Kazakhstan and the French test site in southern Algeria. These signals were simple, being principally a cycle or two of large amplitude followed by a low-amplitude coda. Further, it was realized that the amplitude of P decays only slowly between distances of 30° and 90° so that, unless a station can be much closer than 30° to a test site, wherever a station is between 30° and 90° makes little difference to the detection threshold. The United Kingdom then decided to move to recording at teleseismic distances. In the 30°–90° distance range, apparent horizontal wavelengths range up to 25 km. In consequence, the United Kingdom in cooperation with Canada established in 1963 an array at Yellowknife, Canada (YKA) with an aperture of 25 km. The larger aperture gives YKA a greater resolution for teleseismic P than regional arrays. Two similar arrays that are termed medium-aperture arrays were established as cooperative projects between the United Kingdom and the host countries, at Gauribidanur, India (GBA) and Warramunga, Australia (WRA). Several medium-aperture arrays were also installed by the USSR; an example is the array at Kurchatov (Kim *et al.*, 1996).

When the P signals from these arrays were examined, it was noted that those from earthquakes are often complex; that is, the coda to direct P dies away only slowly and sometimes has amplitude as large as, or larger than, P up to 30 sec after onset. As it appeared that the P signals from explosions are invariably simple, the so-called complexity criterion could be used to identify earthquakes and leave a residue of possible explosions. In one study it appeared that with only one array it was possible to identify up to 90% of the sample of seismic disturbances as definite earthquakes.

One point emphasized in these early studies is that array recordings are required so that the effects of near-receiver scattering can be eliminated and an explosion signal not misclassified as from an earthquake because scattered arrivals make the seismogram appear complex. However, in 1963 the USSR began testing at Novaya Zemlya and it turned out that the signals recorded at YKA from this test site are more complex than is usual for an explosion (Thirlaway, 1966a). The complexity criterion then fell into disuse. However, the differences in form of the P seismograms from earthquakes and explosions are still the basis of some of the most important criteria for recognizing anomalous earthquakes.

The 25 km aperture arrays proved to be remarkably effective for enhancing SP signals and the United States, taking note of this, began to investigate the value of large-aperture arrays. In 1965 the United States opened the LASA in Montana, an array with an aperture of $\sim$200 km (see Fig. 1 and Table 1). The data handling for such a large array was an enormous task; two minicomputers were needed simply to receive the data and write it to tape.

Unfortunately, the SP signals turned out to be highly variable across the array, as did noise amplitudes, so $\sqrt{n}$ improvement ($\sim$23) was never achieved. The signal-to-noise ratio for the beam from the best subarray was often higher than that for the beam for the whole array (Evernden, 1977). There are similar problems with the only other large array ever built, NORSAR in southern Norway.

From 1977 to 1980 negotiations took place between the United Kingdom, United States, and USSR on a Comprehensive Test Ban Treaty. The major component of the proposed verification system at that time was to be stations in each of the three countries. The data from all the stations was to be available to each country. Thus verification that say, the United States was adhering to the treaty would be done mainly using data recorded in the United States. Recordings were thus to be made at regional distances. At these distances the high frequencies in the P signal are less attenuated than at long range and the maximum signal-to-noise ratio is at 2 Hz or above. Ensuring that signals of a few hertz are coherent across the array requires a much smaller array than for recordings at teleseismic distances. Experiments were initiated in Norway on arrays suitable for detecting low-magnitude disturbances at regional distances. The result of this study was the installation of NORESS with an aperture of 3 km (Table 1, Fig. 1). NORESS has seismometers arranged in concentric rings spaced at log-periodic intervals of radius. This design was chosen to make the array resolution independent of azimuth at which the signals cross the array and maximize the ability of the array to pick out the desired signal and suppress interfering signals (Mykkeltveit *et al.*, 1990). Further arrays of similar design have been installed. Such arrays are to be part of the IMS.

Several arrays of long-period (LP) seismometers have been installed. For example, the LASA had LP seismometers at the center of each subarray. Processing of LP recordings is not as well developed as that for SP recordings, but LP array records have been used to study, for example, the origin of scattered arrivals in the coda of surface waves from earthquakes (Capon, 1970). A few arrays have also been installed to record frequencies over a passband that includes much of the conventional SP and LP recording bands. For example, the four-element BNA array in southern England (Douglas and Young, 1981) has a response that is flat for ground velocity in the band 0.1–5 Hz and the 13-element Grafenberg array (GRF) has a passband from 0.01 to 20 Hz. When the BNA array was installed in the 1970s, the band 0.1–5 Hz was considered to be broadband. However, with the improvements in instrumentation, many recording systems have now been installed that, like the system at GRF, encompasses the whole of the LP and SP passbands, and the term broadband has been extended to cover such systems. Here broadband (or wideband) is used generally for any system that covers at least the conventional SP band and extends down to frequencies of around 0.1 Hz.

4. Arrays for Source Studies

SP arrays were first installed to aid the observation of the first motion of P. This was because if an unambiguous downward first motion was observed, the source would be identified as an earthquake. On this criterion, any disturbances for which only positive first motions are seen might be earthquakes or explosions. Inspection of the delay-and-sum P seismograms recorded at long range then showed the often striking differences in the complexity of earthquake and explosion seismograms (Thirlaway, 1963, 1966a). First motion and complexity differ because of the differences in source mechanisms of earthquakes and explosions and although complexity and first motion as such have fallen out of favor, methods for distinguishing between the two types of source that make use of the form and polarity of arrivals are among the most robust discrimination methods available. Thus, procedures have been developed for testing whether a set of P seismograms are consistent with the generally accepted double-couple model of the earthquake source. Initially the assumption is made that the source is an earthquake at significant depth. The amplitude of P and any possible surface (pP and sP) reflections (which can be in arbitrary units) is measured, together with the polarity of the phases where these can be observed unambiguously. A search is then made for orientations of the double-couple sources that are consistent with the amplitude ratios, *pP*/*P*, *sP*/*P*, and *sP*/*pP*. If no orientations are found to be compatible with the observed ratios, the source must be assumed to be shallow if all the seismograms are simple, and (in the absence of a clear negative first motion to *P*) a possible explosion. If there are significant arrivals following *P* that are inconsistent with a double-couple source, then it is possible that the seismograms are from a series of explosions. For this kind of study it is again essential to have array recordings to ensure that first arrivals and coherent arrivals in the coda are enhanced relative to scattered arrivals and, more importantly, that scattered arrivals are not mistaken for surface reflections. Some of the advantages of using arrays to observe the onset of *P* are demonstrated by Douglas *et al.* (1997). A detailed discussion on the testing for consistency of amplitude ratios—the relative amplitude method—is given in Pearce and Rogers (1989). Examples of the application of the method to the identification of anomalous earthquakes are given in Pooley *et al.* (1983), Stewart (1987), and Marshall *et al.* (1989) and to the identification of multiple explosions in Clark and Pearce (1988).

The relative amplitude method is most effective if the array stations used are widely spread in azimuth and distance to ensure that the focal sphere is well sampled. Much effort has gone into finding sites for arrays where the noise is below average, and these sites are usually in the middle of continents. This means that the arrays are not well distributed, particularly for sources near the edges of continents and in the southern hemisphere (Pearce, 1996). The effect of choosing low-noise sites has been to make the range of detection thresholds larger. For forensic seismology, any detection may be better than none; but for seismology in general, arrays would be better employed to bring the threshold for noisy stations down toward that of the best stations. At noisy stations where the noise is organized and of low speed, wavenumber filtering can be particularly effective as shown by the example of EKA (Figs. 6 and 7).

The greatest obstacle to source studies is the oceanic microseism peak at around 0.125–0.167 Hz (8–6 sec period). For it is only for large-magnitude sources that signal is above noise over this band. For earthquake source studies, then, it would be better if arrays had been designed to suppress noise in the microseism band and thus to flatten the spectrum of the noise. Then it would be possible to exploit better the full sensitivity of modern seismometers, the outputs from which are currently filtered into the SP and LP bands where the noise has its lowest amplitude and the signal-to-noise ratio is a maximum. Perhaps surprisingly, however, modern recording systems allow wideband signals to be derived from SP recordings by filtering to enhance the low frequencies. At low-noise sites (e.g., YKA) such wideband signals often allow pulse shapes and polarities to be more clearly seen than on the original SP seismograms even when the wideband signals have lower signal-to-noise ratio than on the original recordings. Such widening of the frequency band can provide information on the fracture processes at the earthquake source as well as aiding in the identification of anomalous earthquakes (see, for example, Douglas *et al.*, 1974a). For noisy sites such as EKA, broadband conversions are usually swamped by noise from the oceanic microseisms, but with wavenumber filtering, as shown by Douglas (1998), these can be reduced significantly. An example of the application of the method of Douglas (1998) to the processing of EKA broadband seismograms derived from the SP recordings is shown in Figs. 6d to 6f. Note that on the broadband seismogram the *P*, *pP*, and *sP* pulses can more clearly be seen than on the SP, so making it possible to estimate a reliable depth.

5. Discussion

The aim of arrays for recording body waves for source studies should be to allow signals to be observed at teleseismic distances from frequencies of a few hertz down as far into the oceanic microseism band as possible. Arrays with apertures of 10 km or more can be used to suppress low-speed noise ($\sim$3 km sec^{-1}), but to suppress mantle *P* wave noise of around 6 sec period requires arrays with apertures of $\sim$50 km on sites where the signal is coherent. This raises the question whether it is possible to find a site where 1 Hz is coherent over such large distances. The experience with the LASA and NORSAR suggests it is not. The recordings from these two arrays show that signals are not coherent over more than a few kilometers. However, for the medium-aperture arrays (YKA, GBA, and WRA) the signal is sufficiently coherent over $\sim$20 km for array processing to be effective at frequencies above 1 Hz. Which sites are the more typical: the highly variable sites of the LASA and NORSAR or those of YKA, GBA, and WRA? There is evidence that there are areas where 1 Hz signals are coherent over large distances (see, for example, Husebye and Jansson, 1966) and this is supported by the similarity of explosion *P* signals recorded at widely separated stations. So arrays significantly larger than 20 km could be effective in extracting signals from noise. The realization that explosion signals are often simple and similar at long range was one of the first results to come from array studies, and this led Thirlaway (1966b) to coin the term "source window" for the distance range 30°–90° where observations at good sites give the clearest picture of the source. However, because of the experience with the LASA and NORSAR, the availability of good sites needs to be investigated. Perhaps mobile arrays could be adapted for this kind of experiment. What is required is an array about twice the size of YKA.

Large arrays of the type discussed above—assuming suitable sites can be found—would be ideal for source studies. However, currently the trend is toward the installation of small arrays to meet the requirements for Test Ban verification in some areas (e.g., former test sites) down to magnitudes well below $m_b 4$. Such low magnitudes can be detected only at regional/local distances. Thus, although the number of arrays will increase with the growth of the IMS, the arrays will be small-aperture and so of little value for extraction of wideband signals.

6. Conclusions

The most effective use of arrays for forensic seismology and seismic source studies in general, at least down to $m_b 4$, is:

1. To widen the frequency band over which signal can be observed above noise.
2. To bring the noise level for recordings at noisy sites down toward that of quiet sites thus increasing the range of distance and azimuth covered by stations recording a given seismic disturbance.
3. To provide from recordings in the distance range 30°–90° estimates of *P* signals free from interfering arrivals and thus signals that are best suited for studies of source mechanisms (including distinguishing between earthquakes and possible explosions) and structure in the source region.

The demands of Test Ban verification, however, for regional arrays means that there is no immediate prospect of increasing the number of teleseismic arrays. This is unfortunate: With a few more such arrays, including one or two of large aperture (assuming suitable sites can be found) seismologists would learn much about seismic source processes. Such information, as well as increasing understanding of the earthquake source, might also strengthen further the ability of forensic seismologists to verify the Comprehensive Test Ban—which was the hope when the first arrays were installed all those years ago.

References

Backus, M.M. (1966). Teleseismic signal extraction. *Proc. R. Soc. Lond. A* **290**, 343–367.

Backus, M.M., J.P. Burg, D. Baldwin, and E. Bryan (1964). Wideband extraction of mantle *P* waves from ambient noise. *Geophysics* **29**, 672–692.

Barley, B.J. (1977). The origin of complexity in some *P* seismograms from deep earthquakes. *Geophys. J. R. Astron. Soc.* **49**, 773–777.

Birtill, J.W. and F.E. Whiteway (1965). The application of phased arrays to the analysis of seismic body waves. *Phil. Trans. R. Soc. Lond. A* **258**, 421–493.

Burg, J.P. (1964). Three-dimensional filtering with an array of seismometers. *Geophysics* **29**, 693–713.

Capon, J. (1970). Analysis of Rayleigh–wave multipath propagation at LASA. *Bull. Seismol. Soc. Am.* **60**, 1701–1731.

Capon, J., R.J. Greenfield, and R.J. Kolker (1967). Multidimensional maximum-likelihood processing of a large aperture seismic array. *Proc. IEEE* **55**, 192–211.

Capon, J., R.J. Greenfield, R.J. Kolker, and R.T. Lacoss (1968). Short-period signal processing results for the Large Aperture Seismic Array. *Geophysics* **33**, 452–472.

Chinnery, M.A. and M.N. Toksöz (1967). *P*-wave velocities in the mantle below 700 km. *Bull. Seismol. Soc. Am.* **57**, 199–226.

Clark, R.A. and R.G. Pearce (1988). Identification of multiple underground explosions using the relative amplitude method. *Bull. Seismol. Soc. Am.* **78**, 885–897.

Cleary, J.R. and R.A.W. Haddon (1972). Seismic wave scattering near the core–mantle boundary: a new interpretation of precursors to *PKIKP*. *Nature* **240**, 549–551.

Cleary, J.R., D.W. King, and R.A.W. Haddon (1975). *P*-wave scattering in the earth's crust and upper mantle. *Geophys. J. R. Astron. Soc.* **43**, 861–872.

Corbishley, D.J. (1970). Multiple array measurements of the *P*-wave travel-time derivative. *Geophys. J. R. Astron. Soc.* **19**, 1–14.

Douglas, A. (1997). Bandpass filtering to reduce noise on seismograms: is there a better way? *Bull. Seismol. Soc. Am.* **87**, 770–777.

Douglas, A. (1998). Making the most of the recordings from short-period seismometer arrays. *Bull. Seismol. Soc. Am.* **88**, 1155–1170.

Douglas, A. and J.B. Young (1981). "The Estimation of Seismic Body Wave Signals in the Presence of Oceanic Microseisms," AWRE Report No. O 14/81. HMSO, London.

Douglas, A., J.A. Hudson, P.D. Marshall, and J.B. Young (1974a). Earthquakes that look like explosions. *Geophys. J. R. Astron. Soc.* **36**, 227–233.

Douglas, A., P.D. Marshall, J.B. Young, and J.A. Hudson (1974b). Seismic source in East Kazakhstan. *Nature* **248**, 743–745.

Douglas, A., D. Bowers, and J.B. Young (1997). On the onset of *P* seismograms. *Geophys. J. Int.* **129**, 681–690.

Engdahl, E.R., E.A. Flinn, and C.F. Romney (1970). Seismic waves reflected from the earth's inner core. *Nature* **228**, 852–853.

Evernden, J.F. (1977). Spectral characteristics of the *P* codas of Eurasian earthquakes and explosions. *Bull. Seismol. Soc. Am.* **67**, 1153–1171.

Husebye, E.S. and B. Jansson (1966). Application of array data processing techniques to the Swedish seismograph stations. *Pageoph* **63**, 82–104.

Key, F.A. (1967). Signal-generated noise recorded at the Eskdalemuir Seismometer Array Station. *Bull. Seismol. Soc. Am.* **57**, 27–37.

Key, F.A. (1968). Some observations and analyses of signal generated noise. *Geophys. J. R. Astron. Soc.* **15**, 377–392.

Kim, W.-Y., V.V. Kazakov, A.G. Vanchugov, and D.W. Simpson (1996). Broadband and array observations at low noise sites in Kazakhstan: opportunities for seismic monitoring of a Comprehensive Test Ban Treaty. In: "Monitoring a Comprehensive Test Ban Treaty" (E.S. Husebye and A.M. Dainty, Eds.), pp. 467–482. Kluwer Academic, Dordrecht.

Marshall, P.D. and P.W. Basham (1972). Discrimination between earthquakes and underground explosions employing an improved M_S scale. *Geophys. J. R. Astron. Soc.* **28**, 431–458.

Marshall, P.D., R.C. Stewart, and R.C. Lilwall (1989). The seismic disturbance on 1986 August 1 near Novaya Zemlya: a source of concern? *Geophys. J.* **98**, 565–573.

Mykkeltveit, S., F. Ringdal, T. Kværna, and R.W. Alewine (1990). Application of regional arrays in seismic verification research. *Bull. Seismol. Soc. Am.* **80**, 1777–1800.

Pearce, R.G. (1996). Seismic source discrimination at teleseismic distances—can we do better? In: "Monitoring a Comprehensive Test Ban Treaty" (E.S. Husebye and A.M. Dainty, Eds.), pp. 805–832. Kluwer Academic, Dordrecht.

Pearce, R.G. and R.M. Rogers (1989). Determination of earthquake moment tensors from teleseismic relative amplitude observations. *J. Geophys. Res.* **94**, 775–786.

Pooley, C.I., A. Douglas, and R.G. Pearce (1983). The seismic disturbance of 1976 March 20, East Kazakhstan: earthquake or explosion? *Geophys. J. R. Astron. Soc.* **74**, 621–631.

Ram, A. and R.F. Mereu (1977). Lateral variations in upper-mantle structure around India as obtained from Gauribidanur seismic array data. *Geophys. J. R. Astron. Soc.* **49**, 87–113.

Ringdal, F. (1990). Introduction to the special issue on regional seismic arrays and nuclear test ban verification. *Bull. Seismol. Soc. Am.* **80**, 1775–1776.

Ringdal, F. and E.S. Husebye (1982). Application of arrays in the detection, location, and identification of seismic events. *Bull. Seismol. Soc. Am.* **72**, S201–S224.

Ringdal, F., E.S. Husebye, and A. Dahle (1975). *P*-wave envelope representation in event detection using array data. In: "Exploitation of Seismograph Networks" (K.G. Beauchamp, Ed.), pp. 353–372. Noordhoff International Publishing, Leiden.

Simpson, D.W. and J.R. Cleary (1977). *P*-signal complexity and upper mantle structure. *Geophys. J. R. Astron. Soc.* **49**, 747–756.

Stewart, R.C. (1987). Comment—A seismic event on August 20, 1983: double explosion or a single earthquake. *Phys. Earth Planet. Inter.* **46**, 381–383.

Thirlaway, H.I.S. (1963). Earthquake or explosion? *New Scientist* **18**, 311–315.

Thirlaway, H.I.S. (1966a). Interpreting array records: explosion and earthquake *P* wavetrains which have traversed the deep mantle. *Proc. Roy. Soc. Lond. A* **290**, 385–395.

Thirlaway, H.I.S. (1966b). Seismology and fundamental geology. *Discovery* **27**, 43–48.

Editor's Note

See also Chapter 24, Seismological Methods of Monitoring Compliance with the Comprehensive Nuclear Test Ban Treaty, by Richards, and Chapter 84, Worldwide Nuclear Explosions, by Yang *et al.*

24

Seismological Methods of Monitoring Compliance with the Comprehensive Nuclear Test Ban Treaty

Paul G. Richards
Columbia University, New York, USA

1. Background

In September 1996, after almost 40 y of negotiations marked by numerous setbacks, the text of the Comprehensive Nuclear Test Ban Treaty (CTBT) was at last agreed and opened for signature at the United Nations. It was signed promptly by all five of the then declared nuclear weapons states—the United States, the Russian Federation, China, the United Kingdom and France—and at the signing ceremony President Clinton called this treaty "the longest sought, hardest fought, prize in nuclear arms control."

The basic obligations of the CTBT are stated in Article I of the treaty text as follows:

1. Each State Party undertakes not to carry out any nuclear weapon test explosion or any other nuclear explosion, and to prohibit and prevent any such nuclear explosion at any place under its jurisdiction or control.
2. Each State Party undertakes, furthermore, to refrain from causing, encouraging, or in any way participating in the carrying out of any nuclear weapon test explosion or any other nuclear explosion.

The CTBT is intended to prevent non-nuclear-weapons states from proliferating nuclear weapons, and restrains nuclear-weapons states from building weapons more sophisticated than those they now deploy. Although the term "nuclear explosion" is undefined, the CTBT is intended to be a zero-yield treaty. As such, it presents the difficulty that any practical monitoring system based upon detection and identification of nuclear explosion signals can be expected to fail at some low level of yield. The question whether the treaty is adequately verifiable or not is therefore a political–military judgment in which the merits of banning *all* nuclear testing—which can be confidently monitored above some yield level—have to be weighed against the possibility that small unidentified tests might occur. The verification system must be designed so that tests at yields large enough to have military significance will be identified with high confidence. In this context, we know that crude fission devices of the types used as bombs on Hiroshima and Nagasaki in 1945, with yields around 15–20 kilotons (kt), would be easily detected and identified today. Good verification capability can act as a deterrent to potential treaty violators to the extent that it becomes clear that violations would be detected and identified.

This arms control prize is not yet in hand, however, because although 165 nations had signed the CTBT as of January 2002, the treaty can enter into force only after being signed and ratified by 44 listed countries that operate nuclear power reactors. As of January 2002, 31 of the 44 listed countries had ratified and 41 had signed (the three exceptions being India, Pakistan, and North Korea). India and Pakistan declared their nuclear capability by a series of nuclear tests in May 1998 (see Barker *et al.*, 1998). In October 1999, the US Senate voted not to give its advice and consent to ratification. President Clinton stated strong support of the CTBT and expectation of its eventual ratification by the US—which would require a new and favorable Senate vote—but his successor President George W. Bush has stated opposition to the CTBT, albeit with support for a continued moratorium on US testing.

This record indicates that many political issues surrounding the CTBT remain contentious. Plans are nevertheless going ahead to establish the verification regime, in which seismology plays a key role. Global monitoring for nuclear

INTERNATIONAL HANDBOOK OF EARTHQUAKE AND ENGINEERING SEISMOLOGY, VOLUME 81A ISBN: 0-12-440652-1

explosions is still needed, whether or not the CTBT has entered into force.

This chapter gives general background on test ban monitoring, describes relevant specialized aspects of seismology, and reviews the procedures being developed by the International Monitoring System (IMS) established by the treaty.

All of the most important technical issues in monitoring a CTBT became apparent between 1958, when the so-called "Conference of Experts" was convened in Geneva to see whether agreement could be reached on a verification system, and 1963, when the Limited Test Ban Treaty (LTBT, which banned testing in the atmosphere, the oceans, and outer space) was quickly negotiated, signed, and entered into force. In 1963 there was a perception, influential in some forums, that seismological methods for monitoring underground nuclear explosions were inadequate—a view that helped to prevent the conclusion of a CTBT in this early period (Richards and Zavales, 1996). The text of the CTBT approved by the United Nations in 1996 is about fifty times longer than the text of the LTBT of 1963, in large part because of the extensive provisions (in the 1996 CTBT) for verification. The formal treatment of verification issues in the CTBT will continue to be developed and documented in extensive detail over the next few years. Much of this work will be specified in six different Operational Manuals, mentioned in the CTBT and its Protocol, that will spell out the technical and operational requirements for the following:

- Seismological monitoring and the international exchange of seismological data
- Radionuclide monitoring and the international exchange of radionuclide data
- Hydroacoustic monitoring and the international exchange of hydroacoustic data
- Infrasound monitoring and the international exchange of infrasound data
- The International Data Centre
- On-site inspections.

The IMS seismographic stations send their data to an associated International Data Centre (IDC) for analysis and also, on request, for circulation to National Data Centers. The US National Data Center is the US Air Force Technical Applications Center.

The IMS can be expected to provide data adequate to monitor for underground explosions down to about magnitude 3.5. For explosions executed in the usual way in shield regions without making special efforts at concealment, this corresponds to a yield of around 0.1 kt. However, it is clear from the negotiating record that some countries desired a better monitoring capability. For example, the Geneva working paper CD/NTB/WP.53 of 18 May 1994 stated the US position that "the international monitoring system should be able to . . . facilitate detection and identification of nuclear explosions down to a few kilotons yield or less, even when evasively conducted, and attribution of those explosions on a timely basis." A few kilotons, evasively tested, could correspond in some cases to seismic signals of about magnitude 3. Assessments as to whether a given monitoring system is adequate for treaty verification, or not, will therefore in practice be driven by perceptions of the plausibility of efforts at treaty evasion.

Up to 1963, when the LTBT was negotiated rather than a CTBT, very few underground nuclear tests had been carried out. Seismographic data then consisted of paper analog recordings from a few hundred stations around the world—data derived almost entirely from earthquakes and chemical explosions—and communications then relied upon standard postal services. But much practical experience in monitoring was acquired following the LTBT, because nuclear explosions were carried out underground at the rate of about one a week for almost thirty years. Yang *et al.* (Chapter 84) list parameters of more than 2000 nuclear explosions conducted in different environments (atmosphere, space, underwater, underground) during the period 1945–1998. A small number (fewer than five) of these explosions were designed to evaluate certain evasion scenarios. At the beginning of a new millennium, seismology is improved in ways hardly imagined in the early 1960s. Thus, seismology today is based upon the operation of tens of thousands of stations, roughly half of them acquiring data that are recorded digitally so that the signals can easily be subjected to sophisticated processing. Such data in many countries are gathered electronically into nationally and internationally organized data centers, for purposes of scientific research and for the study of earthquake hazard and hazard mitigation, as well as for nuclear explosion monitoring. A growing trend since the mid-1980s has been the use of stations equipped with sensors that respond to a broad band of frequencies, with recording systems of high dynamic range so that both weak and strong signals are faithfully documented. Digital seismograms from such systems are now sent to centers from which data segments may be freely accessed by a wide range of users.

The seismic waves of principal interest in CTBT monitoring are the regional phases *Pg*, *Pn*, *Sn*, *Lg*, and *Rg*; the teleseismic phases *P*, *pP*, *pS*, *S*, *sS*, and *sP* (and sometimes *PKP* and *pPKP*); and Love and Rayleigh surface waves. The basic reason for the effectiveness of seismology as a monitoring technology is simply that the different types of seismic sources—earthquakes, explosions (both chemical and nuclear), volcanic eruptions, mine collapses—generate these seismic waves in distinctively different ways. It has therefore been possible to develop objective procedures to identify the nature of the seismic source. The fact that there are so many different types of seismic waves, each with its distinctive characteristics, is on the one hand a burden in explaining the details of monitoring to the many people interested in the subject for policy reasons but not having an interest in seismology *per se*, and on the other hand the essential key to the effectiveness of seismology as a monitoring tool. Having so many different types of seismic wave, each of which can vary strongly over the distance ranges for which it is observable, means that a great deal of information about the seismic source is potentially available.

An understanding of the terms "regional" and "teleseismic" is essential for an appreciation of the practical problems of CTBT monitoring—and of solutions to these problems. The underlying reason for such a grouping of seismic waves is a property of the Earth associated with a zone of lower velocities and/or lower velocity gradient in the upper mantle, beneath which is a region of higher velocity and/or higher gradients at greater depths. The net effect is that the simplest and most important seismic wave—the first-arriving compressional wave (P) traveling downward from the source into the mantle and then back upward—is defocused and thus has low amplitudes at distances within the range about 1000–2000 km from a seismic source. Figure 1 shows schematically the way in which the amplitude of seismic P waves at first decreases and then increases as waves propagate to greater distances. Waves beyond about 2000 km are teleseismic, and P waves throughout most of this range are simple, having very little change in amplitude with distance out to more than 9000 km.

The compressional waves known as Pg and Pn, traveling within the crust or just beneath it, are regional in the sense that they do not propagate to teleseismic distances. But the word "regional" here carries the additional implication that such waves are dependent on local properties of the Earth's crust and uppermost mantle, which can vary quite strongly from one region to another. Regional waves can have large amplitudes (and can thus be easily detected if a seismometer is operated at a regional distance from a source of interest), but they are complex and thus harder to interpret than teleseismic waves. Shear-wave energy can propagate even more efficiently than compressional energy to regional distances. In particular, the largest wave on a seismogram at distances up to 2000 km from a shallow source is usually the Lg wave, which consists of shear waves traveling so slowly that they are trapped by total internal reflections within the low-speed crustal wave guide—similarly to the physics by which light can travel to great distances within a low-speed optical fiber. An extensive dataset of seismic signals from a given region must first be acquired and understood before the regional waves from a new event can be interpreted with confidence. (Sometimes, however, the data from a second, more easily accessed region, can be used to aid in the interpretation, if the two regions are known to be geologically similar.)

The key methods of event identification, all discovered too late to have any impact in the early period of CTBT negotiations (1958–1963), entail comparison of the amplitudes of different seismic waves. For example, slow-traveling surface waves (Rayleigh, Love), with wavelengths of several tens of kilometers, are generated much more efficiently by shallow earthquakes than by explosions. This result was obtained in academic research in the 1960s (e.g., Brune *et al.*, 1963; Liebermann *et al.*, 1996), became thoroughly documented over the subsequent years (e.g., Lambert and Alexander, 1971), and proved during the era of underground nuclear testing to be a reliable method for discrimination using teleseismic signals for sources down to magnitudes in the range

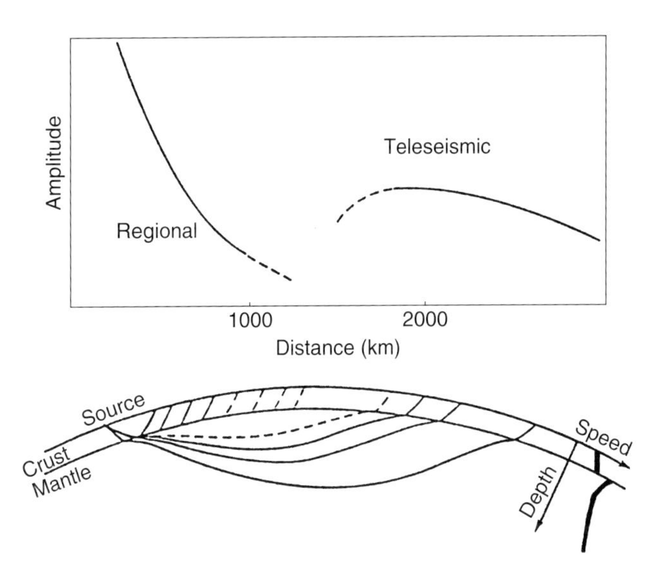

FIGURE 1 This is a generic illustration of the main differences between regional and teleseismic waves. Seismic wave amplitudes at regional distances are strong near the source, for waves such as Pg (propagating within the crust) and Pn (propagating mostly just below the crust/mantle interface), but they decrease and can disappear at about 1000 km. Amplitudes increase at around 1700–1800 km, due to the arrival at these distances of the teleseismic P wave, which has traveled deeply into the mantle and is less affected by shallow structure. The teleseismic P wave varies very little in amplitude out to much greater distances. The reason for the different behaviors of regional and teleseismic waves is the variation in seismic wave speed with depth in the Earth (heavy line, bottom right). The P wave speed in average Earth models typically increases from about 5.8 km sec^{-1} at the top of the crust to about 6.5 km sec^{-1} at the bottom, jumping up to a little more than 8 km sec^{-1} at the top of the mantle, beneath which there is often a layer of lower velocities and then at greater depths an increase in velocity with depth. The layer of lower velocities in the upper mantle causes P wave amplitudes to be low in a range of distances around 1200–1300 km. The lower section of the figure shows ray paths in the crust and upper mantle. For decades following the LTBT of 1963, when nuclear testing was carried out underground and in-country monitoring was not permitted, monitoring was conducted by National Technical Means (NTM) using teleseismic signals. With the CTBT and the permitted use of in-country stations, attention returned to the study of regional waves since they provide the strongest signals. Regional waves may be the *only* detectable seismic signals, from small-magnitude sources. (Adapted from Romney, 1960.)

4–4.5 (Sykes *et al.*, 1983). Over the last 15 years there has been growing recognition that regional *P* waves (such as *Pg* and *Pn*) are often generated much more efficiently by explosions than by earthquakes, in comparison to the excitation of regional *S* waves (such as *Sn* and *Lg*). If signals can be recorded with high enough quality, then the comparison between regional *P* and *Lg* often enables discrimination between earthquakes and explosions to be made at low magnitudes—even lower than magnitude 3 (Hartse *et al.*, 1997; Kim *et al.*, 1997).

2. General Aspects of CTBT Monitoring

This section presents two general points, (1) and (2), and then in (3) describes treaty language on verification and the practical steps such work entails.

(1) An underlying difficulty in treaty monitoring is making the trade-off between simple methods of analysis that are robust, objective, unchanging, and easily explained to people who lack technical training; and those more sophisticated methods which, to a highly trained audience, are demonstrably more effective but that require a greater degree of expert judgment and which will surely change from year to year as methodologies improve. The very existence of so much practical and specialized experience (coming from the decades of monitoring nuclear weapons tests), and of so many seismological resources generating potentially useful data, fundamentally enhances CTBT monitoring capability.

(2) The CTBT will, in practice, be the subject of three types of monitoring, namely:

- That carried out directly by the IDC using data contributed by the IMS
- That carried out by National Technical Means (NTM), i.e., the procedures established unilaterally by various countries using any available information, and all having rights (per CTBT Article IV, ¶34 & ¶37) to use objective information from NTM as a basis for requesting an on-site inspection
- That carried out by numerous private or national organizations, each acquiring and/or analyzing data of some relevance to CTBT monitoring.

The third of these types of monitoring, associated with what the CTBT (Article IV, ¶27 & ¶28) calls supplementary data, is more nebulous than the other two types, but can nevertheless be important, for example, in confidence-building measures and as a source of much basic information and many skills needed by the IMS and IDC. Vast regions of North America, Europe, the Western Pacific, and parts of Central Asia, Northern and Southern Africa, and the Middle East are now being monitored closely for earthquake activity down to magnitude 3 or lower by organizations whose data and methods of analysis are freely available. It is from this type of earthquake monitoring, and from field programs to study the structure of the crust and uppermost mantle, that the broad seismological community has acquired its knowledge of regional wave propagation; and familiarity too with the differences between seismic signals from earthquakes and from blasting associated with the mining, quarrying, and construction industries. Seismology is still a growing science, likely to be driven indefinitely by the need to understand and mitigate earthquake hazards. Individual earthquakes can still kill many thousands of people, and they are one of the few catastrophic phenomena capable of matching the trillion dollar levels of damage that a single nuclear weapon can inflict. It can hardly be emphasized too much that explosion monitoring for the CTBT will be greatly improved in the long term by cross-fertilization between the earthquake monitoring and explosion monitoring communities.

(3) The main part of the CTBT dealing with verification is Article IV, of which ¶1, in its entirety, reads as follows:

> In order to verify compliance with this Treaty, a verification regime shall be established consisting of the following elements:
>
> (a) An International Monitoring System;
> (b) Consultation and clarification;
> (c) On-site inspections; and
> (d) Confidence-building measures.
>
> At entry into force of this Treaty, the verification regime shall be capable of meeting the verification requirements of this Treaty.

Then follow 67 more paragraphs with details on verification. The treaty Protocol further spells out details of the IMS and IDC functions and procedures for on-site inspections, and in an Annex lists the location of 321 IMS stations that are to acquire seismic, radionuclide, hydroacoustic, and infrasound data in support of the treaty.

The work of monitoring for underground nuclear explosions using seismological methods entails the following separate steps:

- Detecting seismic signal.
- Associating into a single group the various seismic signals that are generated from a common seismic source.
- Estimating the location of that source, and the uncertainty in the location.
- Identifying the seismic source (whether earthquake or explosion), together with estimating the source size. The role of the IMS, vis-à-vis identification, is that of "Assisting individual States Parties ... with expert technical advice ... in order to help the State Party

concerned to identify the source of specific events" (CTBT Protocol, Part I, ¶20(c).) The responsibility of identification is left to each State Party, but the IDC assists by carrying out a process called event screening, described further, below.
- Seismology also plays a role in on-site inspections and in confidence-building measures.

It follows that much work of the IMS and IDC focuses on the development of routine procedures for data acquisition and data analysis of all detected events. But historically it is known that monitoring for nuclear explosions results in the need, in practice, to pay special attention to what are called "problem events," which by definition are events that are detected but that cannot be unambiguously identified routinely as not being nuclear explosions. Such problem events can become targeted for special efforts in the acquisition of additional data, and for the execution of additional data analysis. Care will be needed in spelling out appropriate procedures for addressing problem events, since, by their very nature, singling out a detected event as a problem (requiring more than merely routine analysis) entails a preliminary effort to identify the event—which is not an IMS function. Instead, the IMS can carry out event screening, entailing the use of standard procedures to measure agreed-upon parameters that are helpful for characterizing whether an event is a nuclear explosion or a chemical explosion or an earthquake. Different countries will likely reach different conclusions on how to treat a problem event, given the territory on which the event appears to be located, and given the possible predisposition of one or more State Parties to believe that serious efforts might somewhere, someday, be taken to conceal a small nuclear explosion. For example, much has been written on the possibility (or impossibility) of masking a nuclear explosion with a large chemical explosion, and/or reducing its seismic signals by carrying out the nuclear explosion secretly in a large underground cavity that has been sealed to prevent the escape of radionuclides. We may note Article II, ¶51 of the CTBT, which states (with reference to the Technical Secretariat) that "The Director-General may, as appropriate, after consultation with the Executive Council, establish temporary working groups of scientific experts to provide recommendations on specific issues." However, if problem events arise with great frequency, an advisory group convened to deal with them could be overwhelmed. Only time will tell how much of the work of attempting to identify problem events will be done by the IMS and the IDC and how much will be done by States Parties, which can receive IMS data at their National Data Centres (NDCs) as well as having access to their own National Technical Means.

With the above items (1)–(3) as background, let us now turn to brief reviews of the specific steps that must be taken in CTBT monitoring by seismological methods.

3. Technical Issues Arising in Routine Monitoring

3.1 Detection of Seismic Signals

The CTBT Protocol lists 50 sites around the world at which either a three-component seismographic station or an array of seismometers is to be operated, sending uninterrupted data to the International Data Centre (IDC). It is generally understood that this is a digital datastream, available at the IDC in near real time. These 50 stations constitute the primary network, deployed at sites indicated in Figure 2. Much experience is now available from a network of similar size, operated since 1 January 1995 initially as part of the Group of Scientific Experts Technical Test #3 (GSETT-3, carried out under the auspices of the Geneva-based Conference on Disarmament, which negotiated the text of the CTBT), and later under the Comprehensive Test Ban Treaty Organization, headquartered in Vienna. The GSETT-3 IDC, based near Washington, DC, USA became the Prototype IDC (PIDC) shortly after the CTBT was opened for signature in 1996, and served for five years as the testbed for many procedures that were taken up by the Vienna-based IDC in February 2000.

For seismic sources whose signals are detected by the primary network, and for which the location estimate and other attributes of the source are likely to be better quantified if additional signals are acquired, the CTBT Protocol lists an additional 120 sites around the world at which either a three-component seismographic station or an array is to be operated. These stations, constituting the auxiliary network with locations shown in Figure 3, are to operate continuously; but their data is to be sent to the IDC only for time segments that are requested by a message from the IDC. Much experience with such an auxiliary network, of somewhat smaller size, has now been acquired during GSETT-3 and later by the PIDC.

More than half of the 50 primary stations, and some of the 120 auxiliary stations, are listed in the CTBT as operating an array. A seismographic array consists of a number of separate sensors, typically between about 10 and 30 with short-period response, spaced over a few square kilometers or in some cases over tens of square kilometers, and operated locally with a central recording system (see Chapter 23, Douglas). Array data can be processed by introducing a system of signal delays at each sensor and then signal summation, to improve the strength of coherent signals and to reduce the incoherent noise. The resulting improvement in signal-to-noise ratio enables a single array to detect teleseismic signals approximately down to m_b 4–4.5 (Jost *et al.*, 1996), and significantly lower in the case of some well-sited arrays. From the detection of slight differences in the arrival time of a particular seismic wave at the different sensors, it is possible to infer both the azimuth and the horizontal slowness of the arriving signal. IMS array stations also have at least one three-component broadband sensor.

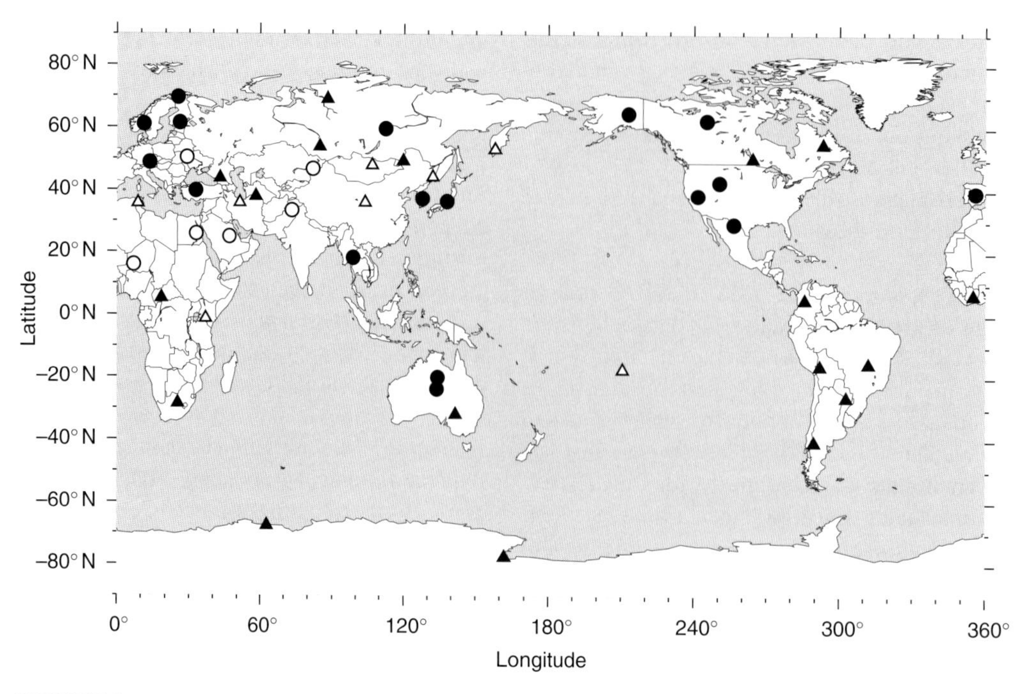

FIGURE 2 Map showing the location of IMS primary seismographic stations (circles and triangles). All these stations are to include at least one three-component broadband instrument. Filled symbols are for stations contributing data in near real time as of July 2000; open symbols are for proposed stations. Triangles indicate stations that operate only a three-component sensor; circles, indicate stations that also operate arrays of short-period vertical sensors. Forty-nine primary stations are shown; the location of one more station is to be decided.

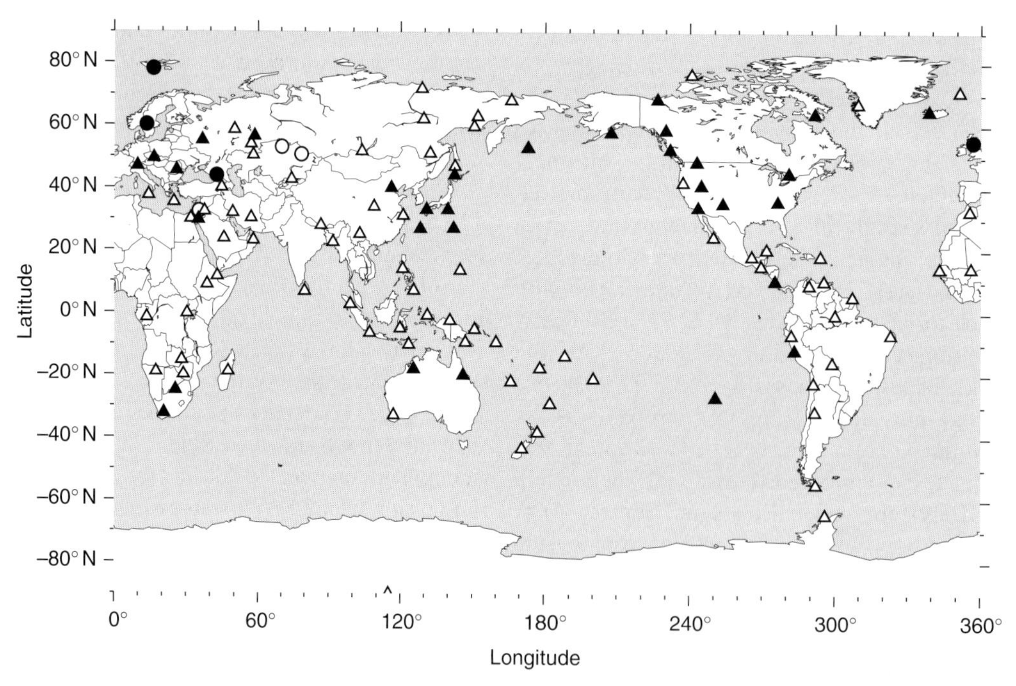

FIGURE 3 Map showing the location of 119 IMS auxiliary seismographic stations, which contribute their data on request to the International Data Centre; the location of one more station is to be decided. Symbols, are as in Figure 2.

What, then, is the expected detection threshold of the CTBT primary seismographic network? (Note that the auxiliary network does not contribute to the IMS detection capability.) For detection of enough signals to provide some type of location estimate, GSETT-3 was planned (CD/1254) to have a threshold detection capability in the magnitude range below 3 for large parts of Eurasia and North America. (It was above magnitude 3.4 in some continental areas of the southern hemisphere, and above magnitude 3.8 in parts of the southern oceans.) The design capability of the CTBT primary network has not been specified. In practice, however, the CTBT 50-station primary network appears capable of providing detection at three or more stations for at least 90% of events with $m_b \geq 3.5$, for almost all of Eurasia, North America, and North Africa. In certain areas, which include the former test sites in Nevada for the United States and in Novaya Zemlya for the Russian Federation, the IMS seismographic network is essentially complete, and detection capability is typically below magnitude 2.5. The IMS station nearest to the Chinese former test site at Lop Nor is an array at Makanchi, Kazakhstan, which began operations in mid-2000 and which can be expected to provide detections down to low magnitudes for Western China.

From January 1995 to February 2000 when operations transferred to Vienna, the stations operating under the auspices of the CTBTO grew from about one-half up to two-thirds of the intended 50-station IMS primary network, and reached about one-half of the intended 120-station IMS auxiliary network. During this five-year period, the PIDC made available (at *http://www.pidc.org*) a number of reports of global seismicity derived from the primary and auxiliary networks. The most important report is the daily Reviewed Event Bulletin (REB), typically published three to five days in arrears, listing the events that have occurred that day, including location estimates, magnitudes, a map of the reporting stations for each event, and arrival times of detected phases at these stations. About 50 events have been reported each day on average, with some wide variations, for a total of about 100 000 events included in the REB over its first five years. The production of reviewed bulletins has to be done for all four monitoring technologies used by the IMS (seismology, hydroacoustics, infrasound, radionuclide monitoring), but for the foreseeable future the most highly developed and most important of these technologies is seismology, and almost all of the events reported each day by the IDC are earthquakes and mining blasts.

The transfer of IDC operations to Vienna in February 2000 became associated with a significant change in practice, namely, that the daily Reviewed Event Bulletin and the seismographic and other data upon which it is based, ceased to be made openly available to the research community, as had been the case during operations under GSETT-3 and by the PIDC. The IDC makes the REB and all IMS data available to National Data Centres, but data can then be subject to Article II, ¶7: "Each State Party shall treat as confidential and afford special handling to information and data that it receives in confidence from the Organization in connection with the implementation of this Treaty." Many voices have been raised in support of a return to open availability of IMS data and IDC data products such as the REB, for example, on grounds that the CTBT itself states in Article IV, ¶10, that "the provisions of this Treaty shall not be interpreted as restricting the international exchange of data for scientific purposes." The quality of IDC operations and products will be greatly enhanced in the long term by having a diverse community of data users. Such users find problems with station calibration, occasional mistakes in the REB and other data products, and continually provide suggestions for improvement, in this way promoting a higher level of monitoring operations at the IDC. Many States Party, including the United States, have indicated their position that all IMS data should be made openly available without delay.

In subsections that follow, further detail is given on the work of seismic signal association, on making seismic location estimates, and on seismic event screening, as carried out by the International Data Centre.

3.2 Association of Signals

Each year, somewhat more than 7000 earthquakes occur with magnitude $m_b \geq 4$; and about 60 000 earthquakes occur with $m_b \geq 3$. While chemical explosions with $m_b \geq 4$ are rare (a few per year, if any), there are probably on the order of a few hundred worldwide at the $m_b \geq 3$ level, and many thousands per year at smaller magnitudes that are detectable at stations close enough. The problem then arises of sorting through the tens of thousands of signals each day that will be detected at a large data center from analysis of array data and three-component station data; and collecting together all the signals that are associated with the same seismic source. This has been a major computational problem of the GSETT-3 IDC, since seismic sources with $m_b > 4$ typically generate more than one detection at many IMS stations (for example, detection of *P*, *pP*, *PKP*, *S*, *SS*, Love, and Rayleigh waves), and the detections from different events can overlap in time at each station. Effective methods of sorting through the interspersed detections have been found, but in practice 40% or more of the detections recorded on a given day remain unassociated. In the work of assembling sets of detections common to the same event, array stations in principle have the advantage, over three-component stations, of permitting determination of the direction of the source from a station at which there is a detection. An array can even indicate two or more different source directions from which signals arrive at the array at about the same time. However, in practice, automated determinations of the azimuth of signal arrival can often be inaccurate. While the azimuth estimate may be useful for association purposes, it is typically not used in event location.

3.3 Location of Seismic Sources

The CTBT Protocol, Part II, ¶3, states that

> The area of an on-site inspection shall be continuous and its size shall not exceed 1000 square kilometers. There shall be no linear dimension greater than 50 kilometers in any direction.

This condition presents a challenge for those at the IDC, who may have to estimate the location of an event that could become the basis of an on-site inspection request. The challenge is especially difficult for small events, when the estimate may have to be based upon regional seismic waves alone.

Note that accurate location estimates are needed not only for events destined to be considered for on-site inspection. They are needed for essentially *all* the events for which a set of detections can be associated, since in practice an interpretation of the location (including the event depth) is commonly used for screening at the IDC, and perhaps for rapid identification by NDCs. It can be crucial to know whether an event is beneath the ocean, or possibly on land; whether it is more than 10 km deep, or possibly near the surface; whether it is in one country or another. One of the first statements of failure to meet these technical challenges was given by North (1996), reporting on GSETT-3 after about 20 months of REB publication:

> Many of the events listed in the (REB) are poorly located. The 90% confidence ellipses exceed 1000 sq. km for 70%, and 10000 sq. km for 30%, of all REB events. Furthermore, comparison by various countries of the locations produced by their own denser national networks with those in the REBs show that the REB 90% confidence ellipses contain the national network location less than half of the time... One feature ... that is particularly troublesome is that many events will be recorded only at teleseismic distances, and that few will be recorded only to regional (< 2000 km) distances. Thus an appropriate means of implementing path-dependent teleseismic travel times, and of transitioning from these to regional travel time curves, will need to be developed.

After more than five years of REB publication, the same problems were still to be found: purported 90% error ellipses were often much larger than 1000 km^2, they did not include true locations 90% of the time, and depth uncertainties did not adequately represent errors in depth.

It is important to understand why REB location estimates using conventional methods have been so poor, since the error in measuring the arrival time of seismic waves is usually less than one second (and is usually less than 0.1 sec when signal-to-noise ratios are good); and the speed of seismic waves is less than 10 km sec^{-1} in the Earth's outer layers where the events of interest occur and where the measurements are made. It would therefore appear that seismic sources can routinely be located to within a few kilometers, and with an areal uncertainty much less than 100 km^2. In principle this conclusion is valid; nevertheless, it does not apply using conventional methods of event location, because these are not based upon a sufficiently good model of the Earth's velocity structure. It is the model errors, not the pick errors, that currently dominate the resulting location errors, at least for events larger than about m_b 4. (For smaller events, the pick errors can be comparable to model errors.) The overall goal of improving locations for events larger than m_b 4 based on seismic arrival times can be achieved only by reducing the effect of model errors.

At depths greater than about 200 km, the Earth's velocity structure is known quite accurately (i.e., to well within 1% at most depths, the biggest exceptions being in regions of subducting lithospheric plates and at certain places just above the core–mantle boundary). The main difficulty is at shallower depths, i.e., within the crust and uppermost mantle, where the actual speed of seismic waves may differ in unknown ways, sometimes by as much as 10%, from the velocity that we often assume (such as that given by *iasp*91 or some other standard Earth model). The actual Earth also has nonhorizontal interfaces, which are not allowed for in simple Earth models, and which can affect the azimuth of an arriving signal. In summary, the measured arrival times and directions of teleseismic waves are influenced in unpredictable ways by unknown Earth structure. But the arrival times and directions of regional waves, which depend much more strongly on shallow structure, can be even more variable than teleseismic waves. For example, if a seismic event is actually 1000 km from an IMS station and generates a regional wave that is detected at the station, and if the arrival is interpreted with a speed that is wrong by 5%, then the event will be estimated from that observation alone as having originated at a distance that is incorrect by about 50 km.

When estimating event locations and their uncertainty, there are essentially only three ways to get around the practical problem of ignorance of Earth structure (i.e., model error): (a) by using numerous stations at different azimuths around the source and thus averaging out the effects of the difference between the Earth's actual velocity structure and that of the model (which traditionally has been taken as a 1D distribution in which velocity depends on depth alone); (b) by building up information about the Earth's velocity structure and thus finding a more sophisticated and significantly more accurate 3D model with which to interpret arrival times; and (c) by "calibrating" the station (or array), so that in effect the source of interest is located with reference to another event, whose location is known accurately, and which preferably is not far from the source of interest. In this latter approach, the data for the unknown event is the difference in arrival times for the two events, as recorded at each station. From such data, we can often estimate accurately the difference in location between the known and unknown locations, and hence estimate the unknown location. Equivalently, for enough calibration events, and using appropriate methods of smoothing, one can use empirically determined travel times for each seismic wave from each potential location to each station. The monitoring community now refers to seismic events whose location is known to within, say, 5 km, as "GT5" events (GT standing for "ground truth," the accurate location information coming, for example, from a local seismographic network, or from field observations of ground rupture in the case of a shallow earthquake, or from special efforts to record the location and origin time of mine blasts).

The USGS and the International Seismological Centre (ISC) rely upon (a) for routine processing, for then the effect of using an incorrect Earth model is reduced. The research community uses one or more of (a), (b), or (c) but only in studying special sets of events, restricted to very limited regions. The IDC cannot use (a) except for large events since the REB is based on IMS stations alone, and for events below m_b 4 only a small subset of IMS stations provide detections. The IDC is therefore beginning to use methods (b) and (c). But even with great efforts based on hundreds of special studies, we will not know the Earth's 3D shallow structure adequately on a global scale for decades, ruling out reliance on (b) alone in the short term, though this method can be used successfully for some regions. This leaves (c), based upon ground truth events in the region for which accurate location capability is required, as the most important short term method. The IDC is likely to be the first organization to publish location estimates, on a global scale, using travel-time curves tailored individually for each station in the network from which the bulletin is prepared. The method will succeed to the extent that GT events of sufficient quality and quantity can be found to calibrate IMS stations. Ideally, one would like to have numerous GT0 events for calibration purposes all over the world. In practice, GT5 events or in some regions GT10 events may be the best that are available.

To conclude this section with a practical example of a day with a typical number of seismic events, Figure 4 shows the location of 58 events reported in the REB for 11 May 1998. One of these events was the nuclear test carried out by India, and announced as consisting of three near-simultaneous nuclear explosions. This test of course attracted headlines, though from the monitoring point of view the initial work of association and event location was done by the PIDC in just the same way as for 57 other events that day. In 1998, event location was based upon interpretation of arrival times using the *iasp*91 model of the Earth. For the Indian event, 62 stations provided detection of 86 different arrivals, and this event was well-located because of the numerous stations that contributed data—see (a) above.

3.4 Identification of Seismic Sources

Once the detections from a seismic event have been associated and an accurate location estimate has been obtained, the next step in monitoring is that of event identification. A review of the best technical methods for carrying out this step is given, for example, by a report of the Office of Technology Assessment of the US Congress (1988). But the actual identification of an event as possibly being a nuclear explosion and hence a treaty violation, perhaps warranting a request for an on-site inspection, entails judgments that bring in political as well as technical assessments. Therefore, as noted above, the IDC's role in event identification is limited to providing assistance to states party to the treaty, rather than actually making an identification. Further detail is given in Annex 2 to the CTBT Protocol, which indicates that the IDC may apply "standard event screening criteria" based on "standard event characterization parameters." To screen out an event, means that the event appears *not* to have features associated with a nuclear explosion. The underlying idea is that if the IDC can screen out most events, then states party to the treaty can focus their attention on the remaining events. For example, an event may have its depth estimated with high confidence as 50 km. Such an event would be screened out, by a criterion that "events confidently estimated as deeper than 10 km are not of concern."

For events detected by the IMS seismographic stations, Annex 2 lists the following parameters that may be used, among others, for event screening:

- Location of the event
- Depth of the event
- Ratio of the magnitude of surface waves to body waves
- Signal frequency content
- Spectral ratios of phases
- Spectral scalloping
- First motion of the *P* wave
- Focal mechanism
- Relative excitation of seismic phases
- Comparative measures to other events and groups of events
- Regional discriminants where applicable.

There is an extensive literature on how each of these characteristics can help to achieve event identification, but it will take a number of years to develop detailed objective criteria in the context of IMS event screening. For the present chapter, a few examples must suffice to demonstrate how methods of identification may become the basis of standard screening by the IDC. Thus, one of the most successful discriminants is based on comparison of the magnitude m_b, measured from the amplitude of short-period P waves, and the magnitude M_S, measured from the amplitude of long-wavelength surface waves. As mentioned above in Section 1, this discriminant was discovered in the 1960s. It exploits the fact that shallow earthquakes are far more efficient than nuclear explosions in exciting surface waves, for events with comparable P waves. As a preliminary result from studies at the

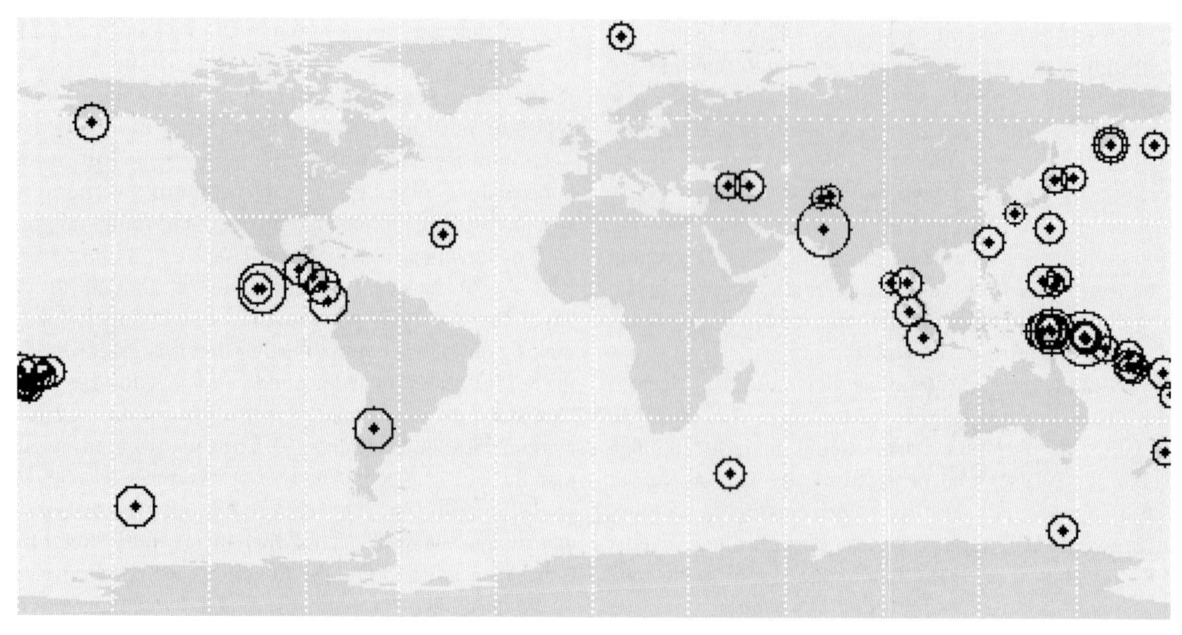

Date	Time	Lat	Lon	Nph	Depth	Mag	Region
1998/05/11	00:16:12.1	13.84S	170.43E	6	604.8	mb 3.1	VANUATU ISLANDS REGION
1998/05/11	00:25:48.8	39.84N	42.33E	11	55.4	mb 3.6	TURKEY
1998/05/11	00:58:11.0	31.98N	130.73E	9	76.4	mb 3.5	KYUSHU, JAPAN
1998/05/11	01:13:36.8	46.61S	42.63E	8		mb 4.0	PRINCE EDWARD ISLANDS REGION
1998/05/11	01:17:47.0	11.26N	97.31E	4		mb 3.8	ANDAMAN ISLANDS, INDIA
1998/05/11	01:34:33.8	8.40N	103.08W	20		mb 4.7	OFF COAST OF MEXICO
1998/05/11	01:37:42.0	8.64N	104.91W	7		mb 3.8	OFF COAST OF MEXICO
1998/05/11	01:54:42.4	23.32N	122.78E	5		mb 3.8	TAIWAN REGION
1998/05/11	02:24:27.0	42.56N	148.46E	16	53.1	mb 3.5	OFF COAST OF HOKKAIDO, JAPAN
1998/05/11	02:48:56.5	33.35S	68.39W	7		mb 4.4	MENDOZA PROVINCE, ARGENTINA
1998/05/11	04:00:20.0	17.40S	174.53W	6		mb 4.1	TONGA ISLANDS
1998/05/11	04:01:37.6	3.07S	142.31E	16		mb 4.7	NEAR N COAST OF NEW GUINEA, PNG.
1998/05/11	04:06:37.7	27.58N	141.12E	11	48.7	mb 3.8	BONIN ISLANDS REGION
1998/05/11	04:41:30.5	2.75S	141.22E	3		mb 3.5	NEAR N COAST OF NEW GUINEA, PNG.
1998/05/11	05:07:24.4	40.28N	48.91E	8		mb 4.0	EASTERN CAUCASUS
1998/05/11	06:02:54.7	2.89S	142.30E	5		mb 3.8	NEAR N COAST OF NEW GUINEA, PNG.
1998/05/11	06:06:08.6	13.46S	166.30E	7	45.6	mb 3.9	VANUATU ISLANDS
1998/05/11	06:18:54.1	16.29S	178.68W	6		mb 3.9	FIJI ISLANDS REGION
1998/05/11	07:13:20.6	18.37S	171.65W	3		mb 3.5	TONGA ISLANDS REGION
1998/05/11	07:33:09.2	22.28S	175.79W	3		mb 3.5	TONGA ISLANDS REGION
1998/05/11	07:36:18.6	39.20S	177.34E	3		mb 3.6	OFF E. COAST OF N. ISLAND, N.Z.
1998/05/11	08:47:31.6	13.77N	91.94W	7		mb 3.8	NEAR COAST OF GUATEMALA
1998/05/11	09:07:50.7	12.08N	87.76W	4		mb 3.8	NEAR COAST OF NICARAGUA
1998/05/11	09:31:16.1	8.34S	159.01E	3		mb 3.6	SOLOMON ISLANDS
1998/05/11	09:39:10.4	36.29N	70.82E	9	223.8	mb 3.4	HINDU KUSH REGION, AFGHANISTAN
1998/05/11	09:48:58.7	63.15S	146.09E	5		mb 3.8	SOUTH OF AUSTRALIA
1998/05/11	10:13:44.2	27.07N	71.76E	72		mb 5.0	INDIA-PAKISTAN BORDER REG.
1998/05/11	10:26:08.4	84.86N	8.72E	10		mb 3.6	NORTH OF SVALBARD
1998/05/11	10:50:46.9	37.07N	73.53E	6	338.5	mb 3.3	TAJIKISTAN
1998/05/11	11:10:23.9	2.19N	98.37E	4		mb 3.8	NORTHERN SUMATERA, INDONESIA
1998/05/11	11:34:39.3	11.57N	139.07E	6		mb 3.7	WESTERN CAROLINE ISLANDS
1998/05/11	11:42:36.9	10.43N	143.19E	3		mb 3.4	SOUTH OF MARIANA ISLANDS
1998/05/11	11:42:41.6	15.67S	176.92E	6		mb 3.7	FIJI ISLANDS REGION
1998/05/11	12:09:26.0	52.79N	160.11E	9		mb 3.7	OFF EAST COAST OF KAMCHATKA
1998/05/11	12:09:45.4	21.06S	177.46W	3		mb 3.7	FIJI ISLANDS REGION
1998/05/11	12:34:48.9	53.13N	160.09E	17		mb 4.1	NEAR EAST COAST OF KAMCHATKA
1998/05/11	12:40:38.4	5.08S	152.50E	39	10.7	mb 4.9	NEW BRITAIN REGION, P.N.G.
1998/05/11	12:49:06.5	4.89S	152.46E	15	50.6	mb 3.8	NEW BRITAIN REGION, P.N.G.
1998/05/11	13:22:14.5	2.89S	139.55E	10		mb 4.2	NEAR NORTH COAST OF IRIAN JAYA
1998/05/11	13:22:27.1	57.09S	142.87W	7		mb 4.4	PACIFIC-ANTARCTIC RIDGE
1998/05/11	13:59:55.6	17.07S	169.73W	4		mb 4.0	TONGA ISLANDS REGION
1998/05/11	14:24:19.1	24.69N	46.36W	6		mb 3.7	NORTHERN MID-ATLANTIC RIDGE
1998/05/11	14:29:48.5	21.95S	177.48W	6	485.9	mb 3.1	FIJI ISLANDS REGION
1998/05/11	14:30:18.6	2.97S	142.41E	16		mb 4.2	NEAR N COAST OF NEW GUINEA, PNG.
1998/05/11	15:32:46.3	8.89N	83.53W	24	21.7	mb 4.1	COSTA RICA
1998/05/11	16:22:03.9	4.41S	151.93E	3		mb 3.5	NEW BRITAIN REGION, P.N.G.
1998/05/11	16:42:20.4	10.67S	165.93E	6	100.2	mb 3.8	SANTA CRUZ ISLANDS
1998/05/11	17:29:59.8	21.65S	179.74E	4		mb 3.6	SOUTH OF FIJI ISLANDS
1998/05/11	18:46:32.0	42.31N	143.14E	31	55.1	mb 3.6	HOKKAIDO, JAPAN REGION
1998/05/11	19:31:06.5	58.12N	156.56W	22	36.6	mb 4.1	ALASKA PENINSULA
1998/05/11	20:03:07.1	11.15N	92.72E	7	167.8	mb 3.4	ANDAMAN ISLANDS, INDIA
1998/05/11	20:25:46.2	12.29N	144.09E	7	71.1	mb 3.7	SOUTH OF MARIANA ISLANDS
1998/05/11	20:28:54.5	5.36S	153.15E	11	43.7	mb 3.8	NEW IRELAND REGION, P.N.G.
1998/05/11	20:48:17.1	20.83S	175.93W	4	700.0	mb 2.7	TONGA ISLANDS
1998/05/11	21:14:39.4	53.28N	173.72E	5		mb 3.7	NEAR ISLANDS, ALEUTIAN ISLANDS
1998/05/11	22:08:42.5	4.72N	82.32W	18		mb 4.5	SOUTH OF PANAMA
1998/05/11	23:28:31.9	5.21S	102.76E	7		mb 4.2	SOUTHERN SUMATERA, INDONESIA
1998/05/11	23:40:13.3	14.21S	167.68E	6		mb 3.9	VANUATU ISLANDS

FIGURE 4 The Reviewed Event Bulletin for 1998 May 11 contained 58 seismic events, of which the 27th was an Indian nuclear test. Shown here is a map and list of these event locations. Larger circles on the map indicate larger magnitudes. (Obtained from *http://www.pidc.org.*)

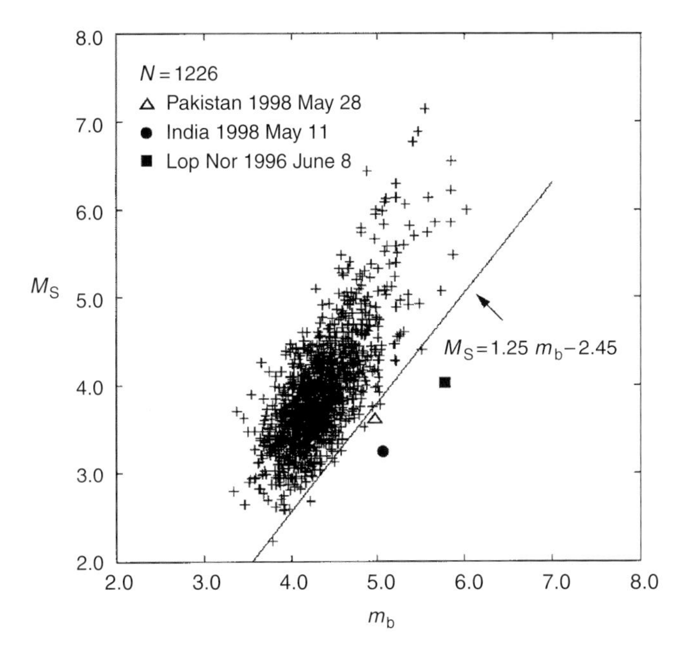

FIGURE 5 The $M_S:m_b$ discriminant, in practical application, depends upon the particular details of how seismic magnitudes are defined. Here are shown magnitudes for 1226 shallow events in the REB, presumed to be earthquakes, in which m_b has been assigned after making corrections for station magnitude bias with respect to network-average m_b values. Three nuclear explosions are included, and a screen that removes events for which $M_S \geq 1.25m_b - 2.45$ is very effective in removing a large fraction of the shallow earthquakes and none of the nuclear explosions. (Adapted from work of David Jepsen and Jack Murphy.)

PIDC, Figure 5 shows that a large percentage of shallow events (but no nuclear explosions, in this example) may be successfully screened out if they have $M_S \geq 1.25m_b - 2.45$.

Another screen, conceptually even simpler, is one based upon depth estimates, recognizing that depths greater than 10 km are in practice inaccessible to human activity, including nuclear testing. Events can thus be screened out if (a) their depth estimate is greater than $10\,\text{km} + 2\sigma_D$, where σ_D is the variance of the depth estimate; and (b) certain criteria are met, to ensure that elementary mistakes have been avoided in making the depth estimate. In work at the PIDC, a conservative approach has been taken to ensure that events that appear to be deep, in terms of interpreting their *P* and *S* arrivals, have additional depth indicators (such as surface reflections *pP* and *sP*, and significant changes in the time interval between *P* and *pP* with increasing distance). The preliminary result is that about 20% of events with magnitude ≥ 3.5 are screened out by a combination of (a) and (b) (D. Jepsen, paper presented at a DTRA-sponsored workshop, Kansas City, June 2000). It is to be hoped that this percentage will rise significantly if more precise estimates of depth can be made with confidence. Note too that discriminants often work well in combination, so that, by making better depth estimates, one can expect to reduce the number of deep events (say, in the range ≥ 50 km) that have to be subjected to the $M_S:m_b$ screen. Such a reduction would help the latter discriminant perform better (i.e., increase the separation between earthquake and explosion populations), since deep earthquakes are less efficient in exciting surface waves and tend to look more explosion-like in an $M_S:m_b$ diagram.

Depth estimates, and measurements of m_b and M_S, can be based upon teleseismic waves. In practice it is important also to develop discriminants, and their specific application as screens, for events too small to be well characterized teleseismically. One example is the use of high-frequency spectral ratios of *P* and *S* regional waves (such as *Pn*/*Lg* ratios). Hartse *et al.* (1997) and Kim *et al.* (1997) have shown in practice that such ratios are abnormally high for small underground explosions as compared to shallow small earthquakes, and this discriminant can be applied successfully down to magnitude 3 and possibly even smaller. The preliminary result is that more than 60% of events with magnitude ≥ 3.5 are screened out on the basis of their *P*/*S* spectral ratios (D. Jepsen, paper presented at a DTRA-sponsored workshop, Kansas City, June 2000). In practice, objective screening based upon such a discriminant may be developed over a period of time and routinely applied at the IDC in ways that may have to be fine-tuned slightly differently for different regions.

Below magnitude 4, mine blasting can result in seismic events detected and located by the IDC that appear similar to small nuclear explosions using criteria such as depth estimates, weak surface waves, and high *P*/*S* spectral ratios. Hedlin *et al.* (1989), Kim *et al.* (1994), and others have found characteristic spectral interference patterns for such events, based presumably on the fact that commercial blasts routinely consist of numerous small explosions that are fired in a sequence of delays. The resulting seismic source is therefore somewhat spread out in space and time. Another potential discriminant of mine blasting is the infrasound signal that can be generated by such sources (Hagerty *et al.*, 2002), and which would not be expected from a small underground nuclear explosion unless it vented significantly—which, in turn, would likely result in a characteristic radionuclide signal. New discriminants for particular types of small seismic events continue to be developed, both in field projects with specially deployed monitoring equipment, and in subsequent adaptation for candidate screening methods to be applied at the IDC to IMS data.

In addition to widely applied screens based upon m_b and M_S and upon depth, more specialized screening desired by a particular State Party may be carried out by the IDC as described in the Protocol, Part I, ¶21:

> The International Data Centre shall, if requested by a State Party, apply to any of its standard products, on a regular and automatic basis, national event screening criteria established by that State Party, and provide the

results of such analysis to that State Party. This service shall be undertaken at no cost to the requesting State Party.

4. Evasion Scenarios

If a nuclear explosion is conducted deep underground (depth around 1 km) at the center of a large spherical cavity in hard rock, with radius greater than or equal to about 25 m times the cube root of the yield in kilotons, then the seismic signals can be reduced by a so-called decoupling factor that may reach up to about 70, i.e., by almost two magnitude units. It is easier to build cavities of the same volume that are elongated rather than spherical, and apparently such aspherical cavities can also achieve high decoupling factors, but they can also increase the concentration of stress on the cavity and make it much more likely that radionuclides will be released into the atmosphere and detected by the radionuclide monitoring network of the IMS. An overall evaluation of the cavity decoupling scenario therefore raises a number of different technical issues:

- Does a country considering an evasive test have access to a suitably remote and controllable region with appropriate geology for cavity construction?
- Can the site be chosen to avoid seismic detection and identification (recognizing that seismic events are routinely reported down to about magnitude 3 by earthquake monitoring agencies for many areas in industrialized countries, and that many countries routinely monitor neighboring territory)?
- Can cavities of suitable size and shape and depth and strength be constructed clandestinely in the chosen region, with disposal of the material that has to be removed from below ground?
- Can nuclear explosions of suitable yield be carried out secretly in sufficient number to support the development of a deployable weapon? (This question covers numerous technical issues, including the ability to ensure the yield will not be larger than planned; and to keep all yields of a test series large enough to learn from, yet small enough to escape identification.)
- Can radionuclides be fully contained from a decoupled explosion?

Each of these questions has been the subject of extensive technical analyses.

My own opinion is that it is technically possible to address a few of these issues in a clandestine program in isolation from the others, but that mastery of all these issues combined could not be achieved except possibly by a full-blown multibillion dollar effort by a nation that already had practical experience with underground nuclear testing. Such a nation would have little to learn technically, much to lose politically, and the treaty evasion program would still be at great risk of discovery since multibillion dollar efforts intrinsically attract attention in the modern world.

Another evasion scenario is the use of mining operations and large chemical explosions to mask or disguise an underground nuclear explosion. There is a limit on the yield of any nuclear explosion that could be hidden in this way. Chemical explosions are almost always delay-fired at shallow depths, so they are inefficient in generating seismic signals, relative to nuclear explosions that are tamped (Khalturin *et al.*, 1998). It therefore appears that only mines with the largest blasting operations are conceivable candidates for hiding militarily significant efforts at CTBT evasion.

There are two types of technical effort that can be carried out to help monitor mining regions for compliance with a CTBT. The first of these is installation of nearby seismographic stations that record digitally at high sample rates. The purpose would be to provide high-quality data that can distinguish between single-fired explosions and the multiple-fired explosions typical of mining operations. The second is provision of technical advice to mine operators so that they execute their blasting activities using modern methods of delay-firing—which have economic advantages, as well as enabling the blasting of rock in ways that do not make the large ground vibrations (and strong seismic signals) typical of old-fashioned methods of blasting.

Addressing the reality of how the IMS and States Party will handle large blasting operations—or the occasional single-fired chemical explosion (whose seismic signals may indeed look just the same as those of a small nuclear explosion), the CTBT Article IV, ¶68, states that

> In order to . . . [c]ontribute to the timely resolution of any compliance concerns arising from possible misinterpretation of verification data relating to chemical explosions . . . each State Party undertakes to cooperate with the Organization and with other State Parties in implementing relevant measures . . . .

These measures are spelled out in the Protocol, Part III. They provide, on a voluntary basis, information on single-fired chemical explosions using 300 tons or more of TNT-equivalent blasting material; and information on mine blasting (such as mine locations) for all other chemical explosions greater than 300 tons TNT-equivalent.

To the extent that mine blasting remains a serious concern, problems can be addressed with nearby installation of infrasound and radionuclide monitoring equipment, and with site visits. These activities would be voluntary under the CTBT, but they could also be made the subject of bilateral agreements.

For a suspected nuclear test, the CTBT contains extensive provisions for on-site inspection, at the request of a State Party and with at least 30 affirmative votes of the 51-member Executive Council of the CTBT Organization.

Some remarkable assessments of evasion scenarios were presented in the US Senate debate of October 1999, leading up to the negative vote on CTBT ratification. Thus, the Senate majority leader stated in executive session that

> A 70-kiloton test can be made to look like a 1-kiloton test, which the CTBT monitoring system will not be able to detect.

He was referring to the decoupling scenario. Treaty opponents gave speeches which included exactly the same sentence:

> While the exact thresholds are classified, it is commonly understood that the United States cannot detect nuclear explosions below a few kilotons of yield.

It was also stated that

> Advances in mining technologies have enabled nations to smother nuclear tests, allowing them to conduct tests with little chance of being detected.

Let us address each of these three statements in turn.

Seventy kilotons would be an enormous explosion, about twice the energy of the combined Hiroshima and Nagasaki bombs, and would indeed be of military significance. The cavity that conceptually might reduce signals from a test of this size to something comparable to 1 kt would have to be deep underground (on the order of 1 km deep), with volume equal to that of a sphere 200 m in diameter. Such a spherical cavity could not be built in practice, not even in the open, let alone in secret. Just supposing it could be built, almost all the 70 kt energy released by the test would go into pumping up gas in the cavity to about 150 times normal atmospheric pressure. These gases would be radioactive, and the high pressure means they would leak through cracks that inevitably would exist in the 125 000 m^2 or more of cavity wall even if the cavity did not collapse. Distant radionuclide sensors of the IMS would detect the test if only 0.1% of the radioactivity escaped. The small fraction of energy released as seismic signals, if comparable to those from a typical (well-coupled) 1-kt test, would travel far beyond the rock surrounding this hypothetical (impossible-to-construct) cavity, and would be large enough for detection at numerous seismometers all over the world, including those of the CTBT monitoring system.

As for the United States not being able to "detect nuclear explosions below a few kilotons of yield," this might have been an accurate representation of the state of affairs in about 1958 when CTBT negotiations began. But it is now more than 40 years later. Unclassified arrays of seismic instruments continuously and routinely monitor vast areas of the globe for earthquakes and explosions down to low seismic magnitudes, reaching down to about magnitude 2 at Russia's former nuclear test site on Novaya Zemlya. This capability translates into the ability to detect down to about 0.01 kt of well-coupled yield.

As for mining technologies, the overall trend of modern methods of mine blasting has resulted in general in smaller seismic magnitudes than were prevalent a few decades ago. There are only a limited number of mining regions where signals from a small (~0.1 kt) nuclear test could be masked by a simultaneous mine blast without taking the additional steps of putting the nuclear test in a cavity—which would be associated with the complications listed above.

To summarize this section: If concern over evasion scenarios is sufficient that problems are targeted for resolution, then the goal of "monitoring ... nuclear explosions down to a few kilotons yield or less, even when evasively conducted" (the US position quoted in CD/NTB/WP.53 of 18 May 1994) is achievable.

5. Concluding Remarks

The work of monitoring the CTBT by seismological methods will be demanding, because of the technical difficulty and the scale of organizational effort that is necessary, and because of the need to interact with political and bureaucratic decision making. Evaluations of CTBT monitoring capability may be carried out as part of the domestic ratification process in each of the different signatory states. Such evaluation during the US Senate debate of 8, 12–13 October, 1999 relied upon a series of unsupported assertions. Evaluations are carried out from time to time, together with budgetary review, by government agencies, not all of which are eager to put resources into a major initiative in nuclear arms control. A review of nuclear explosion monitoring has not been carried out in the US political arena since the 1980s (US Congress, 1988).

It may be noted that there is an ongoing need to monitor all environments for nuclear explosions, regardless of whether the CTBT ever enters into force. The IMS and IDC therefore have an important role in global and national security, independent of the treaty's status.

The work of treaty monitoring is enhanced by the existence of major seismological resources other than the IMS primary and auxiliary networks and the IDC. This point is acknowledged in the CTBT and its Protocol, which make several references to the need to take national monitoring networks into account. In this regard, it is relevant to note that national seismographic networks are not operated by military organizations, but rather by civilian agencies such as the China Seismological Bureau; and, in the United States, the US Geological Survey. National networks are increasing in number and quality. They are associated with the production of detailed bulletins of regional seismicity that typically are intended to meet objectives in the study of earthquake hazard. Such bulletins, and easy access to the underlying digital seismogram data, can be very helpful in CTBT monitoring, for the study of problem events and for improving the IDC's routine seismicity bulletins. In years to come, it is likely that the background of earthquakes will provide a means of calibration and new methods of data acquisition and data analysis, providing useful information that will improve the work of the IMS, and assisting in the attainment of national and international monitoring goals for a CTBT.

Acknowledgments

The work underlying this chapter has been supported by several US federal agencies including the Department of Defense and the Department of Energy. I thank Alan Douglas, Scott Phillips, and Lynn Sykes for constructive criticisms of a first draft, and John Armbruster for help with Figures. This is Lamont-Doherty Earth Observatory Contribution No. 6284.

References

Barker, B. and 18 coauthors (1998). Seismology: monitoring nuclear tests. *Science* **281**, 1967–1968.

Brune, J., A. Espinosa, and J. Oliver (1963). Relative excitation of surface waves by earthquakes and underground explosions in the California–Nevada region. *J. Geophys. Res.* **68**, 3501–3513.

Hagerty, M.T., W.Y. Kim, and P. Martysevich (2002). Infrasound detection of large mining blasts in Kazakhstan. *Pure Appl. Geophys.* in press.

Hartse, H.E., S.R. Taylor, W.S. Phillips, and G.E. Randall (1997). Preliminary study of seismic discrimination in central Asia with emphasis on western China. *Bull. Seismol. Soc. Am.* **87**, 551–568.

Hedlin, M.A., J.B. Minster, and J.A. Orcutt (1989). The time–frequency characteristics of quarry blasts and calibration explosions recorded in Kazakhstan, USSR. *Geophys. J. Int.* **99**, 109–112.

Jost, M.L., J. Schweitzer, and H.-P. Harjes (1996). Monitoring nuclear test sites with GERESS. *Bull. Seismol. Soc. Am.* **86**, 172–190.

Khalturin, V.I., T.G. Rautian, and P.G. Richards (1998). The seismic signal strength of chemical explosions. *Bull. Seismol. Soc. Am.* **88**, 1511–1524.

Kim, W.-Y., D.W. Simpson, and P.G. Richards (1994). High-frequency spectra of regional phases from earthquakes and chemical explosions. *Bull. Seismol. Soc. Am.* **84**, 1365–1386.

Kim, W.-Y., V. Aharonian, A.L. Lerner-Lam, and P.G. Richards (1997). Discrimination of earthquakes and explosions in Southern Russia using regional high-frequency three-component data from the IRIS/JSP Caucasus network. *Bull. Seismol. Soc. Am.* **87**, 569–588.

Lambert, D.G. and S.S. Alexander (1971). "Relationship of Body and Surface Wave Magnitudes for Small Earthquakes and Explosions," SDL Report 245. Teledyne Geotech, Alexandria, Virginia.

Liebermann, R.C., C.Y. King, J.N. Brune, and P.W. Pomeroy (1966). Excitation of surface waves by the underground nuclear explosion LONGSHOT. *J. Geophys. Res.* **71**, 4333–4339.

North, R. (1996). Calibration of travel-time information for improved IDC Reviewed Event Bulletin (REB) locations. Paper presented at the 18th Annual Seismic Research Symposium on Monitoring a CTBT, Annapolis, MD.

Richards, P.G. and J. Zavales (1996). Seismological methods for monitoring a CTBT: the technical issues arising in early negotiations. In: "Monitoring a Comprehensive Test Ban Treaty" (E.S. Husebye and A.M. Dainty, Eds.), pp. 53–81. Kluwer, Dordrecht.

Romney, C. (1960). Detection of underground explosions. In: "Project VELA, Proceedings of a Symposium," pp. 39–75.

Sykes, L.R., J.F. Evernden, and I. Cifuentes (1983). Seismic methods for verifying nuclear test bans. In: "Physics, Technology and the Nuclear Arms Race" (D.W. Hafemeister and D. Schroeer, Eds.), AIP Conference Proceedings, 104. AIP, New York.

US Congress (1988). "Seismic Verification of Nuclear Testing Treaties," OTA-ISC-361. Office of Technology Assessment, Washington, DC, US Government Printing Office.

25

Volcano Seismology and Monitoring for Eruptions

Stephen R. McNutt
Alaska Volcano Observatory, University of Alaska, Fairbanks, Alaska, USA

1. Introduction

Volcanoes are the sources of a great variety of seismic signals that behave differently than those from events on earthquake faults. Nearly every recorded volcanic eruption has been preceded by an increase in earthquake activity beneath or near the volcano, and accompanied and followed by varying levels of seismicity. For this reason, seismology has become one of the most useful tools for eruption forecasting and monitoring. At the present time, approximately 200 of the world's volcanoes are seismically monitored, although the number and quality of stations at each varies considerably. This represents about one out of three of the 538 volcanoes that have erupted in historic times. Over the last several decades, 55 to 70 individual volcanoes have erupted each year (Simkin, 1993). Because erupting volcanoes draw attention, over half of these are seismically monitored.

This chapter reviews some of the developments in volcano seismology over the last 25 years that have led to an improved understanding of volcanoes and the volcanic processes that cause earthquakes and other seismicity. The chapter also describes patterns and relationships in volcano seismology that form the physical basis of contemporary monitoring and forecasting.

1.1 Some Famous Early Eruptions

Throughout history the relation between earthquakes and volcanoes has been close, although there has been confusion about causal relations between the two. This is because of the parallel belts of volcanoes and earthquakes found at subduction zones, where 95% of the Earth's subaereal volcanoes are located. It is common in the historic record, and recently as well, to find an increase in reports of regional earthquakes around the times of eruptions. In spite of this, the reports of the eruption of Vesuvius in AD 79 tell of numerous local earthquakes preceding the eruptive activity. People living near Mt. Nuovo, Italy in 1538 and Mt. Usu, Japan in 1663 fled from the areas before the eruptions because of many felt earthquakes. Early observatories at Vesuvius starting in 1856, Usu in 1910, and Hawaii in 1912 recorded different types of earthquakes and began systematic study of eruption precursors. Three weeks of felt earthquakes heralded the formation of the new volcano Paricutin in Mexico in 1943. These and many other examples illustrate that volcanoes are active seismic features.

1.2 Key Developments in Volcano Seismology

Many of the key developments in volcano seismology have been driven by the different types of seismic events that volcanoes produce. For example, explosion earthquakes and volcanic tremor have required very different theories for their explanation than tectonic earthquakes. Recently, with the advent of broadband seismology, new types of events have been identified with small source sizes that produce very long-period seismic energy. Another key development during the last 25 years has been the proliferation of modern computers, which has led to dramatic improvements in speed and precision in locating earthquakes, and in determining properties of seismic waves, often in near real time. For example, many digital data acquisition systems record at 100 samples per second, so the arrival times can be measured to the nearest 0.01 sec as opposed to 0.1 sec for paper records and 0.05 sec for film recorders. This precision, together with close station spacing, has the effect of improving location accuracy, and the best-located volcanic earthquakes have formal errors as low as 50–100 m horizontally and 100–300 m vertically.

1.3 Volcano Seismology Versus Earthquake Seismology

Many methods or techniques for data analyses are first developed for tectonic earthquakes and then later applied to

INTERNATIONAL HANDBOOK OF EARTHQUAKE AND ENGINEERING SEISMOLOGY, VOLUME 81A ISBN: 0-12-440652-1

volcanoes. This occurs because felt or damaging earthquakes are more common and widespread than eruptions, and consequently there are more seismologists than volcanologists and more research money is spent on earthquakes than on volcanoes. Once techniques are developed by seismologists in general, they are easily applied to volcanoes because volcanoes are small, their locations are known, and seismicity is often concentrated beneath the volcanoes. Analyses of individual seismic events at volcanoes are complicated by the active and passive involvement of fluids, e.g., magma, water, or gases. These produce analytic and computational complexity.

2. Instruments and Networks

Most seismometers used to monitor volcanoes are located on the flanks at distances within 15 km. A typical network is shown in Figure 1. About a third of the monitored volcanoes have at least one station within 1 km of the active vent. Modern networks generally consist of six or more stations. Station spacing of a few kilometers enables better detection of small earthquakes and generally small location errors (<0.5 km). Data are telemetered to a common site and recorded on drum recorders as well as on digital computers.

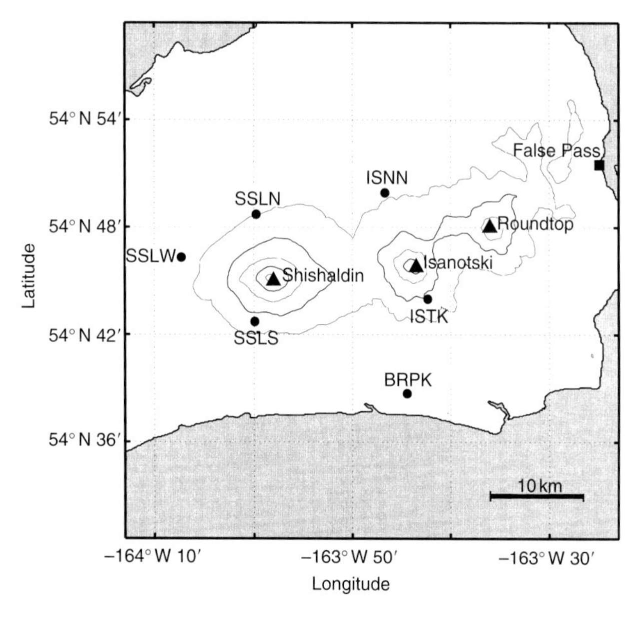

FIGURE 1 Map of local seismic stations near Shishaldin Volcano, Alaska. Circles are seismic stations, triangles are volcanoes, and the square is a town. All seismic stations are short-period single-component (vertical) except SSLS, which has three components. Although designed to monitor Shishaldin, the network also provides data for Isanotski and Roundtop volcanoes. Most well-monitored volcanoes have similar networks of 6–8 stations.

2.1 Short-Period and Broadband Seismometers and Telemetry

Most seismometers at volcanoes are short-period (1 sec) vertical instruments that record useful data between about 0.7 and 30 Hz. Modern three-component, digitally telemetered, high dynamic range, broadband seismometers record over a much wider band of frequencies (0.02–50 Hz). Such broadband seismometers also have a much higher dynamic range, up to 145 dB when using digital telemetry, as opposed to about 40 dB for analog telemetry. Temporary deployments of these broadband instruments have become common recently, and permanent deployments of 1 to 10 instruments at well-monitored volcanoes are becoming more widespread (e.g., Kawakatsu *et al.*, 1992; Falsaperla *et al.*, 1997; Ohminato *et al.*, 1998). Microphones are often added to seismic stations to add an additional independent data channel. They are used to record air waves from explosions (Garces and Hansen, 1998) and the radiated sound field from some volcanic tremors (Garces *et al.*, 1998). They also measure thunder from volcanic lightning and atmospheric pressure fluctuations from wind and storms.

A seismometer in the vicinity of a volcano will typically record seismic signals from four sources: (1) the volcano, (2) local earthquakes (<15 km), (3) regional earthquakes (>15 km), and (4) teleseisms (generally >600 km). Part of the job of the volcano seismologist is to discriminate these. It is helpful to record data from several stations at regional distances of 50–200 km to aid in the identification process. Two examples illustrate why these are helpful. At Nevado del Ruiz volcano, Colombia, in November 1985 and later, many earthquakes from the Bucaramonga nest, a persistent deep source of earthquakes beneath Colombia, appeared on seismograms from stations near the volcano. The Bucaramonga event seismograms resembled many of the volcanic events, and their occurrence created confusion (R. White, personal communication, 1986). At Spurr volcano, Alaska, in 1992, the seismic signal for two eruptions began gradually but the third began sharply and its onset looked like that of a regional earthquake. However, the pattern of arrival times and relative amplitudes at local versus regional stations quickly revealed that the event was an eruption.

The distance from an earthquake to a seismic station and the magnification of the station are the main factors that determine the magnitude detection threshold, the smallest signals that can be recorded. Typical magnifications at volcanoes are 5000 to 50 000, and the corresponding magnitudes of detection are about $M=0$ to $M=1$ for the 1 to 15 km distances involved. It is desirable to record the smallest earthquakes possible, but because volcanoes often have high noise levels, this is usually achieved by putting stations close to the vent rather than operating more distant stations at higher magnifications. A number of volcanoes are monitored indirectly because a nearby (>15 km) seismic station is part of a larger regional seismic network.

2.2 Volcano Monitoring in Near Real Time

To retain small volcanic events and emergent signals like tremor, several observatories continuously record digital data. These can be digitally filtered to remove noise and hence improve the signal-to-noise ratio. Automation of tasks has improved the ability to analyze many types of data in near real time, which leads to better methods of both forecasting and characterizing eruptions in progress. For example, the Real-time Seismic Amplitude Measurement (RSAM) system (Endo and Murray, 1991) and near real time spectrograms (Benoit *et al.*, 1998) measure time-domain amplitudes and frequency-domain amplitudes automatically. These systems produce plots of time series at intervals of 10 min, 1 h, 1 day, 3 days, etc., which are available for inspection on World Wide Web sites administered by the responsible agencies. Calibration information for each seismic station is stored electronically and used to help determine earthquake magnitudes and the reduced displacement (see Section 6.3) of continuous signals such as volcanic tremor. These normalized measures allow comparison between different volcanoes.

2.3 Array Processing

Some volcanic signals have gradual onsets or very long durations (hours to days or more). For these events, traditional location methods, which depend on measuring arrival times for *P* and *S* waves, cannot be used. New techniques use data from dense networks or "arrays" (e.g., Kvaerna and Ringdal, 1992). Here, the stations are very close, a few tens to hundreds of meters apart (e.g., Fig. 2), so it is possible to see the same wave coherently on all the stations (the wave need not be a first or prominent arrival). Because the physics of waves is well known, it is possible to determine what type of wave is present, which direction it is coming from, and how fast it is moving. All these provide information about the sources. Thus, dense arrays can be used like antennae to discern features, such as slowness, of specific seismic waves even when the signals are weak or noise is present (e.g., Kvaerna and Doornbos, 1986). Array methods look at different features of volcanic seismograms and thus extend the information available for analyses (e.g., Goldstein and Chouet, 1994; Oikawa *et al.*, 1994; Metaxian *et al.*, 1997; Chouet *et al.*, 1998).

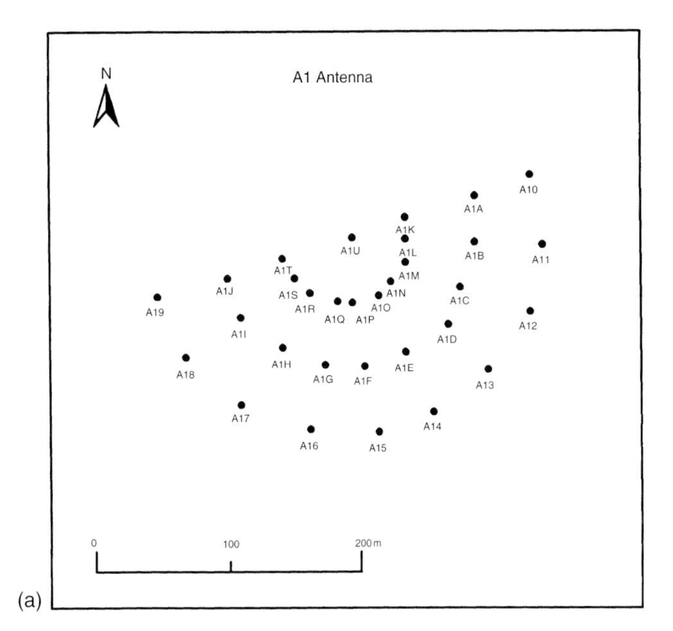

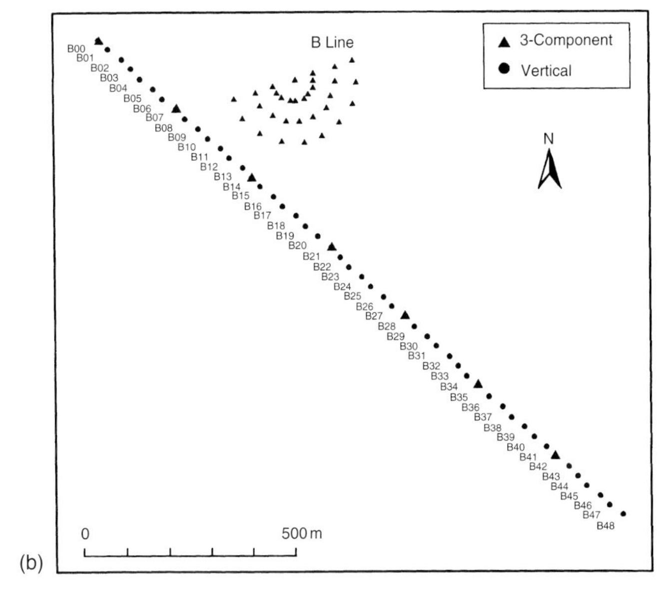

FIGURE 2 Maps of two seismic arrays deployed at Kilauea Volcano, Hawaii, in 1996. (a) The A1 antenna. All seismometers are short-period three-component sensors. (b) The B line. This linear array had vertical sensors (circles) and three-component sensors (triangles).

2.4 Typical Observatories

Volcano observatories vary widely in location and sophistication. The simplest observatory might be a small observation hut, with a single seismograph recorded on paper on a revolving drum. Many larger observatories are located near volcanoes at distances of 10–100 km, and seismic data from local and regional stations are telemetered to them by cable, radio, and telephone. Some observatories are co-located with universities or other research organizations. These may be hundreds of kilometers from the volcanoes of interest, yet the vast array of modern communications make such facilities common and economic. They are, further, safe from large eruptions that may affect closer facilities. There are over 73 members of the World Organization of Volcano Observatories (WOVO, 1997) covering more than 119 volcanoes, most with some kind of seismic monitoring capability (see Appendix on the attached Handbook CD, under directory \25McNutt).

3. Terminology and Types of Events

Active volcanoes are the sources of a great variety of seismic signals. Traditionally, volcanic earthquakes have been classified based on seismogram appearance into four different types: high-frequency or A-type, low-frequency or B-type, explosion quakes, and volcanic tremor (Minakami, 1974). This classification scheme works very well at a large number of volcanoes, but see below for additional discussion. Other local signals, such as those caused by glaciers or landslides, are also recorded at volcanoes. Although Minakami's original classification included restricted depth ranges for various events, these have often been relaxed because of improvements in location accuracy and better understanding of source and propagation effects. Examples of volcanic events from several volcanoes are shown in Figures 3, 4, and 5. In the following sections, the events are described and source processes are briefly discussed.

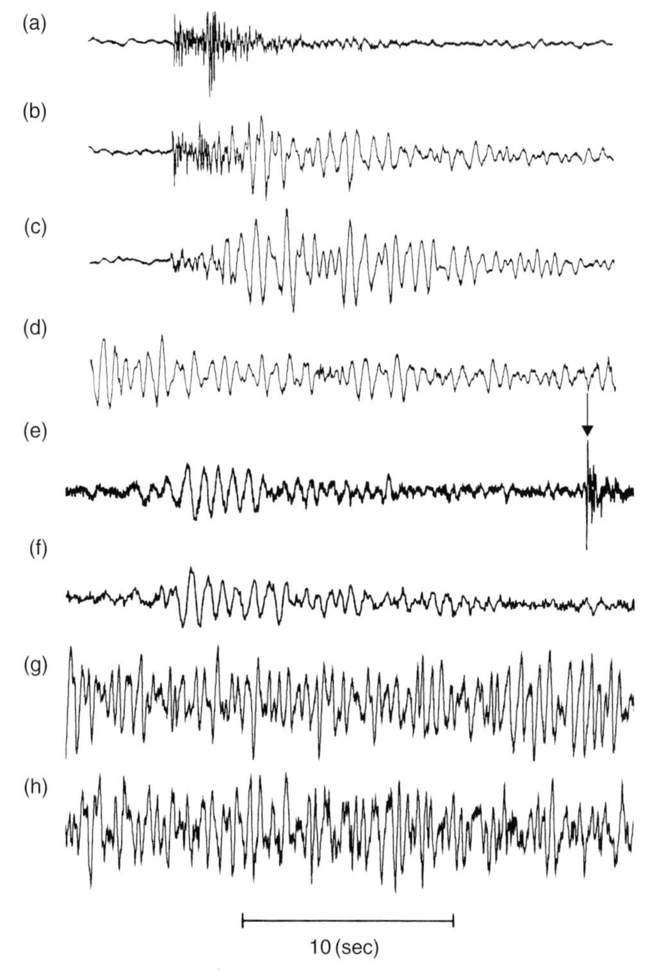

FIGURE 3 Typical waveforms of volcanic earthquakes. (a) High-frequency or volcano-tectonic earthquake, 6.8 km depth, Redoubt Volcano, Alaska, station RED, 8 km from the vent. (b) Hybrid or mixed-frequency event, − 0.6 km depth (0.6 km above sea level), Redoubt Volcano, station RED. (c) Low-frequency or long-period event, −0.4 km depth, Redoubt Volcano, station RED. (d) Volcanic tremor, Redoubt Volcano, station RED. (e) Explosion quake, Pavlof Volcano, Alaska, station PVV, 8.5 km from the vent (note prominent air-wave arrival marked by arrow). (f) Low-frequency or B-type event, Pavlof Volcano, station PVV. (g) Volcanic tremor before eruption, Mt. Spurr, Alaska, station CRP, 4.8 km from the vent. (h) Volcanic tremor during eruption, Mt. Spurr, station CRP. Amplitudes are normalized.

3.1 High-Frequency Events

Most high-frequency (HF) events are thought to be caused by shear failure or slip on faults, and differ from their tectonic counterparts only in their patterns of occurrence. At volcanoes, earthquakes typically occur in swarms rather than mainshock–aftershock sequences. A swarm may be defined as a group of many earthquakes of about the same size and location with no dominant shock. In practical terms, the difference in magnitude between the largest and second largest event of a swarm is 0.5 magnitude units or less, as opposed to 1.0 unit or more for mainshock–aftershock sequences. HF events have clear P and S waves, and dominant frequencies are 5–15 Hz. Higher frequencies may be generated at the source, but are not recorded because of instrumental limitations or high local attenuation. A typical small HF event is shown in Figure 3a.

Some events with the typical frequencies of HF events show clear evidence of non-double-couple focal mechanisms (Julian *et al.*, 1997) or simultaneous shear and tensile faulting (Shimizu *et al.*, 1987), and hence imply fluid involvement. Thus frequency content alone does not uniquely indicate the process. Some studies of HF events have focused chiefly on attributes of swarms (e.g., Okada, 1983; Benoit and McNutt, 1996) as opposed to individual events.

3.2 Low-Frequency Events

Low-frequency (LF) or long-period (LP) events are at the heart of volcano seismology, are an area of active research, and are likely related to fluid flow. Most low-frequency events are thought to be caused by fluid pressurization processes such as bubble formation and collapse, but also by shear failure, tensile failure, or nonlinear flow processes that occur at very shallow depths for which attenuation and path effects play an important role. These events often have emergent P waves, lack S waves, and have dominant frequencies between 1 and 5 Hz, with 2–3 Hz most common. Examples of low-frequency events are shown in Figures 3c, 3f, 4a, and 4b.

Prior to eruption, magma must move from depth to the Earth's surface through conduits, dikes, sills, reservoirs, or combinations of these. Thus, models of LF volcanic earthquakes select a conduit geometry, such as a rectangular crack, a cylindrical pipe, or a fluid sphere, then attempt to reproduce seismograms by appropriate choice of length, width, and thickness, and the velocity and other properties of the material within it, which can be either magma or water (e.g., Chouet,

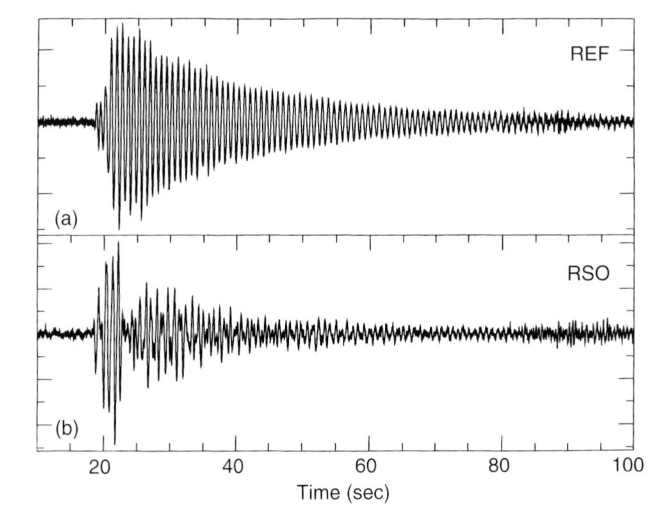

FIGURE 4 Tornillo, also called long-coda event or long-period event. (a) Redoubt Volcano, station REF, 3 km E of the vent. (b) Same event as (a) as recorded at station RSO, 3 km S of the vent (1 April 1992). Amplitudes are normalized.

1992, 1996; Iguchi, 1994; Fujita *et al.*, 1995; Nakano *et al.*, 1998). Seismicity similar to that at volcanoes is found at geysers and geothermal areas (Kedar *et al.*, 1996). A source of mechanical energy is needed; this can be a small earthquake adjacent to a conduit, a flow transient or pressure fluctuation within a conduit, gas bubbles expanding or contracting, a shock wave from choked flow, or other causes. Some workers consider the source to be the mechanical energy alone, while others treat the source as the ensemble of the mechanical energy and the resonant response of the fluid in conduits or dikes (Garces and McNutt, 1997). Low-frequency events are thought to be closely related to volcanic tremor.

Some earthquakes share attributes of both high and low-frequency events. These are called hybrid events (e.g., Power *et al.*, 1994). For example, the event shown in Figure 3b displays the high-frequency onset of an HF event, but the coda is similar to those of LF events. It is thought that such events represent a mixture of processes, such as an earthquake occurring adjacent to a fluid-filled cavity and setting it into oscillation (Chouet, 1992). Others suggest that hybrids are shallower than LF events and thus preserve most of their high-frequency energy, which is attenuated for deeper events (Neuberg *et al.*, 1998).

3.3 Explosion Earthquakes

Explosion quakes accompany explosive eruptions, and many are characterized by the presence of an air-shock phase on the seismograms. There is a partitioning of energy at the source: Part of the energy travels through the ground as seismic waves and part travels through the air as acoustic or air waves (Garces and Hansen, 1998). The air wave then couples back into the ground and is detected by the seismometer. The air wave also shows up clearly on microphones or barographs (Ishihara, 1985). An example of an explosion earthquake is shown in Figure 3e. Source parameters of explosion earthquakes have been studied by Nishimura (1995), who found that the peak amplitude of the modeled single force scaled with the source duration.

3.4 Volcanic Tremor

Volcanic tremor is a continuous signal of duration minutes to days or longer. The dominant frequencies of tremor are 1–5 Hz (2–3 Hz is most common), similar to LF events, and many investigators have concluded that tremor is a series of LF events occurring at intervals of a few seconds. Harmonic tremor and spasmodic tremor are two special cases of more general volcanic tremor. Harmonic tremor is a low-frequency, often single-frequency sine wave with smoothly varying amplitude; or sometimes it consists of a fundamental frequency with many overtones. Spasmodic tremor is a higher-frequency, pulsating, irregular signal. Intermittent or "banded" tremor occurs in regular, periodic bursts separated by periods of quiescence of uniform duration. The resulting pattern looks like stripes or bands on seismograms recorded on rotating drum recorders (Fig. 5). Examples of volcanic tremor are shown in Figures 3d, 3g, 3h, and 5.

Models for tremor are analogous to those for LF events (Section 3.2), and the two are often treated together. While most workers attempt to model the source (e.g., Julian, 1994; Nishimura *et al.*, 1995; Chouet, 1996; Ida, 1996; Neuberg *et al.*, 1998), other workers have elucidated propagation effects of the wave fields (Gordeev, 1992; Ereditato and Luongo, 1994; Kedar *et al.*, 1996; Metaxian *et al.*, 1997) and the relationship to eruptions (Gresta *et al.*, 1991; McNutt, 1994).

The number of earthquakes versus magnitude obeys a power law, known in seismology as the Gutenberg–Richter frequency–magnitude relation. There are more small earthquakes than large ones, and the relative numbers of events form a power-law distribution. Volcanic tremor, however, obeys an exponential law, based on analyses of data from eight volcanoes (Kilauea, Spurr, Pavlof, Redoubt, Karkar, Ulawun, Fuego, and Arenal) and one geothermal area (Old Faithful; Fig. 6) (Benoit, 1998). For tremor, which is a continuous signal, we cannot count the number of events, so we compute the duration at different amplitudes. Synthetic experiments by Benoit (1998) have shown that this is equivalent to counting events. The evidence to date favors a fixed source size at each volcano affected by variable pressure fluctuations; however, the full implications of the exponential law for the source of volcanic tremor are not yet well established.

3.5 Very Long-Period Events

Over the past decade, new broadband seismometers have been deployed at volcanoes. These have the ability to detect

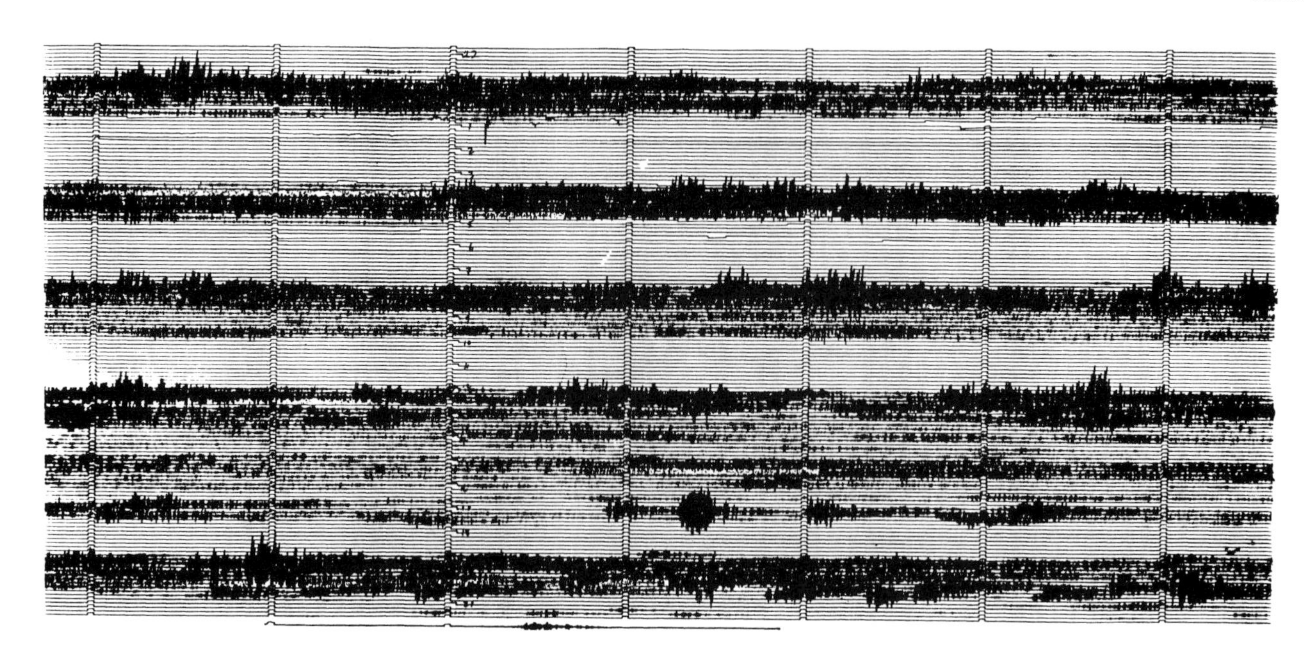

FIGURE 5 Banded tremor from Manam Volcano, Papua New Guinea, station TBL, 4 km from the vent (30 March 1992). Bands correspond to strong phases of subcontinuous Strombolian activity. Tick marks are at 1-min intervals, recording speed is six lines (revolutions) per hour.

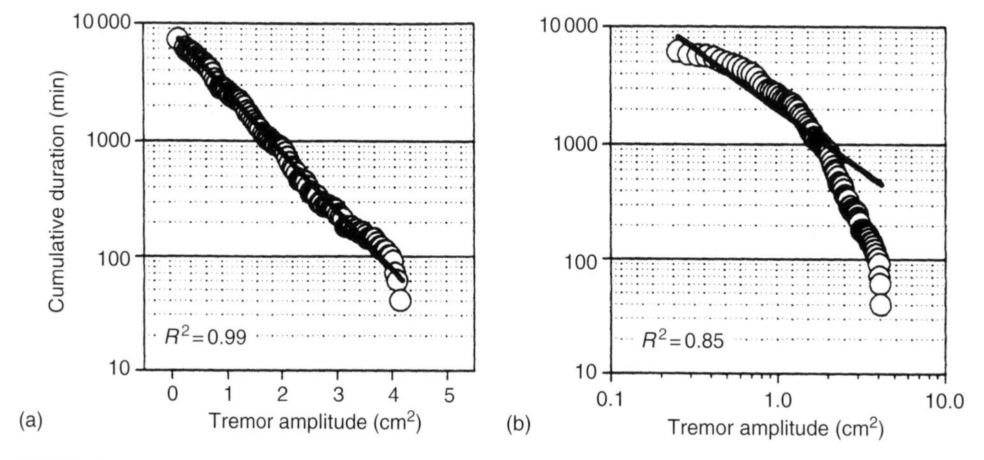

FIGURE 6 Volcanic tremor data from Mount Spurr, Alaska, comparing an exponential (a) and a power law (b) model. Note that the vertical axis is logarithmic in both cases, but the horizontal axis is linear for the exponential model and logarithmic for the power law model. The circles show the duration of tremor observed at various amplitudes; strong tremor occurs less often than weaker tremor. The lines are weighted least-squares mathematical fits of the data to the two models shown. The data fit better using the exponential model, as shown by the higher value of the correlation coefficient R^2. (Modified from Benoit, 1998.)

a wider frequency band, particularly at the low end (down to $f = 0.016$ Hz, or a period of 60 sec). Not surprisingly, a new class of events called very long-period (VLP) events have been observed at some volcanoes. Observations to date show events with periods of 3–20 sec originating from shallow depths of 1.5 km or less under active volcanoes including Kilauea (Ohminato *et al.*, 1998), Stromboli (Neuberg *et al.*, 1994; Chouet *et al.*, 1998), Aso (Kawakatsu *et al.*, 1994; Kaneshina *et al.*, 1996), Sakurajima (Kawakatsu *et al.*, 1992; Uhira and Takeo, 1994), Mount Erebus (Rowe *et al.*, 1998), Satsuma-Iwojima (Ohminato and Ereditato, 1997), and pyroclastic flows at Unzen (Uhira *et al.*, 1994). The VLP events at these volcanoes have been associated with either eruptions or vigorous fumarolic activity, and the events have fairly small amplitudes. An example of a VLP event is shown in Figure 7.

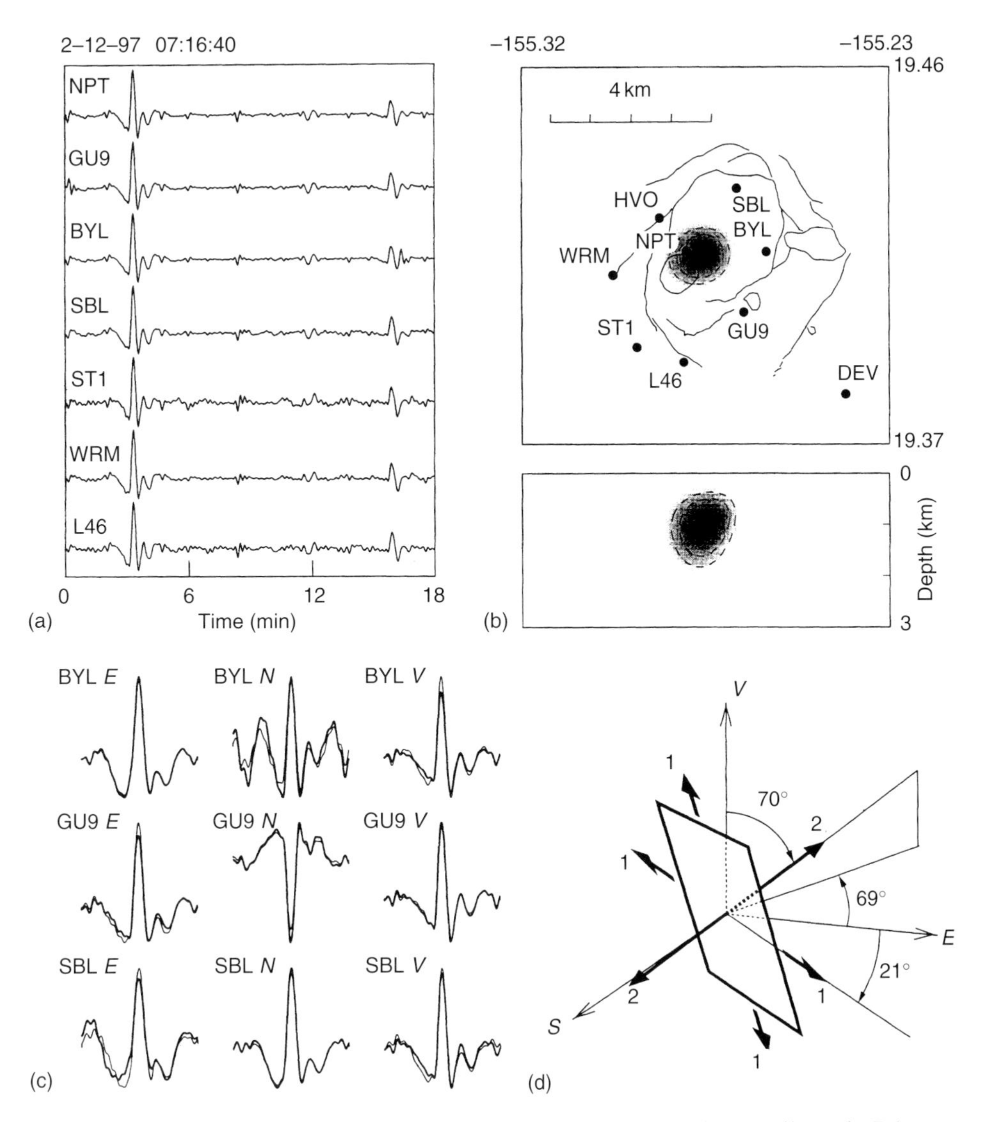

FIGURE 7 Data from a broadband three-component network operating at Kilauea in February 1997. (a) Record section of the vertical component of velocity in band 10–100 s at seven stations (amplitudes normalized in order to emphasize waveform similarities among the stations). (b) Broadband seismic network. Circles are seismic stations, and unbroken lines represent pit craters, faults, and the ring fracture system. Broken lines around shaded region represent 90%, 93%, and 96% semblance. (c) Comparison between observed waveforms (bold lines) and synthetics (thin lines) obtained from inversion of these waveforms (examples depict fits obtained at three stations). (d) Model of source mechanism derived from inversion of network data. This is a rectangular near-vertical dyke, which undergoes a volume change of 700 m^3 associated with the dominant pulse shown in part (a). (Courtesy B. Chouet.)

Normally it takes a large structure, such as a long fault, to produce waves with these long periods. However, many of the VLP events at volcanoes appear to come from small source zones despite the long wavelengths. Some current models suggest that these events may be produced as magma moves through a flapper valve (such as a restriction in a dike or sill) in distinct pulses (e.g., Ohminato *et al.*, 1998). Additionally, theoretical calculations show that it is possible for a crack or tube filled with fluid to produce long-wavelength (and long-period) waves if the acoustic velocity in the fluid is slow compared to that of the rigid walls. The interaction between the fluid and walls produces low phase velocities along the interface (e.g., Ferrazzini and Aki, 1987; Uhira and Takeo, 1994). Such waves are called crack waves or tube waves.

During its climactic eruption on 15 June 1991, Mount Pinatubo, Philippines, produced strong seismic waves with periods of 228–270 sec (Kanamori and Mori, 1992). These are thought to be caused by a type of oscillation in the atmosphere in which thermal energy from the vent travels up and then reflects off the top of the ash column (in this case in the stratosphere). These waves also coupled into the ground and appeared on stations thousands of kilometers away for over 2 h.

Broadband sensors have been deployed at a number of volcanoes that produced no VLP signals, even though several of these were erupting. They include Mount Spurr, Akutan, and Augustine in Alaska (McNutt *et al.*, 1997), Montserrat (Neuberg *et al.*, 1998), and Arenal (Benoit and McNutt, 1997; Hegarty *et al.*, 1997) volcanoes. Further, the scientific community has yet to record any precursors to truly large eruptions on broadband instruments. In spite of this, the VLP signals made visible by the new broadband instruments are an exciting part of volcano seismology, and it is anticipated that many new advances will be made in this area over the next few years.

3.6 Surficial Events

Seismometers located on volcanoes record a variety of local signals caused by shallow processes. These include nonvolcanic processes such as glacial events, shore ice movement, and landslides, as well as volcanic processes such as outburst floods and lahars, pyroclastic flows, and rockfalls from crumbling lava domes. Some of the seismograms are similar, so they must be recognized and properly treated by analysts.

3.6.1 Glacial Events

Volcanoes at high latitude and tall volcanoes host glaciers, which produce seismic events along the ice–rock boundary and within the ice when cracks form. Because the seismic wave velocity in ice is lower than that in rock, the seismic waves may become partially trapped in the ice and set up oscillations of the ice itself. Seismograms from such events resemble those for LF volcanic events. Glacial events are more common in the summer months when the glaciers are moving faster.

3.6.2 Shore Ice

Ice forms from sea water when the air temperature falls below $-12\,^{\circ}\mathrm{C}$. Large blocks of such ice gather on the shores of Augustine Island, Alaska, and move as the tides ebb and flood. When the blocks collide or break, they produce seismic events because the blocks are coupled to the ground. The events are surficial, so the seismic waves are mostly low-frequency surface waves traveling in thin sediment layers. The individual ice events resemble LF volcanic events, and the events occur in swarms of a few hours' duration because of the tides. Both of these features are similar to seismicity preceding eruptions, so one needs to measure air temperature in such situations!

3.6.3 Landslides

Avalanches and landslides of various sizes occur on volcanoes. Those associated with partial melting of ice and snow mainly occur in the spring and summer. The main avalanche or landslide generates a seismic signal of several minutes' duration, depending on the size and runout. The amplitude varies with the amount of material, with larger avalanches generating stronger seismic signals. Some landslides on volcanoes (and elsewhere) are preceded by small discrete seismic events whose rate of occurrence increases up to the time of the main event. This is similar to a common pattern of earthquakes before eruptions so, again, additional information is needed to distinguish between volcanic and nonvolcanic causes of the seismograms.

3.6.4 Rockfalls

Rockfalls are a special type of landslide in which one or a few pieces of bedrock become detached and free fall. They are especially common at volcanic domes and at volcanoes with very steep slopes or cliffs. In the summer of 1995 workers at Mount Spurr, Alaska, observed rocks more than 1 m in diameter falling several hundred meters and impacting on the crater floor. The corresponding seismograms showed strong irregular local signals lasting more than one minute.

3.6.5 Pyroclastic Flows

Pyroclastic flows are products of explosive eruptions and consist of particles of hot molten rock and gases which move rapidly over the ground surface. Large-scale collapses of volcanic domes can also produce pyroclastic flows. These latter events were characterized by complex seismograms that have been modeled as a sequence of forces (Uhira *et al.*, 1994). First, rock is removed from the dome (the dome moves up slightly in response); second, the falling rock collides with the slope below (an impact); and third, the rock breaks apart and the fragments continue to move down the slope as an irregular flow. These seismograms last several minutes and may contain high frequencies, and the larger flows are associated with larger-amplitude signals.

Pyroclastic flows have also been associated with another seismic feature. At Pinatubo volcano, Philippines, pyroclastic flows traveled down a valley between two seismic stations, one of which transmitted its signal to the other. As the ash clouds rose into the air the telemetry path was interrupted for several tens of minutes, after which the signal returned. Thus, the temporary absence of the telemetry could be used to infer that pyroclastic flows were occurring.

3.6.6 Outburst Floods and Lahars

The surface melting of ice and snow, or heavy rains, may cause floods and lahars (volcanic mudflows). Similarly, a volcanic eruption under a glacier will melt the ice, and the flooding

TABLE 1 Selected Volcano Seismology Terminology

This Paper	Minakami[a]	Latter[b]	AVO[c]	Other Names	Example
High-frequency, HF	A-type	Tectonic, volcano-tectonic	Volcano-tectonic, VT	Short-period earthquake	Fig. 3a
Low-frequency, LF	B-type	Volcanic	Long-period, LP		Fig. 3c,f
				Long-coda event, tornillo[d]	Fig. 4a,b
Mixed-frequency	–	Medium-frequency	Hybrid	Medium-frequency	Fig. 3b
Explosion quake	Explosion quake	Volcanic explosion	Explosion	–	Fig. 3e
Volcanic tremor	Volcanic tremor	Volcanic tremor	Volcanic tremor	Harmonic tremor	Fig. 3d,g,h
				Spasmodic tremor	Fig. 5

[a]Minakami (1974).
[b]Latter (1979); only a portion of Latter's full classification scheme is shown.
[c]Alaska Volcano Observatory (1993); also Power *et al.* (1994).
[d]Tornillo is the Spanish word for "screw." The codas of these events resemble a wood screw in profile.

that subsequently occurs is called a jokullhaup. All these floods or lahars cause long-duration seismic signals that resemble volcanic tremor, except that the signal is stronger near the flow channel, as opposed to tremor, which is stronger nearest the volcanic vent. The flood water travels downstream, so the source is moving. Also, the recorded frequencies may be high if the seismometer is near the flow channel, even though this may be far from the vent.

Different groups of investigators and various observatories have used a number of different *local* terminologies. This is similar to the local names of sandwiches, which can be "hoagie," "grinder," or "submarine sandwich" in different parts of the United States. However, no consensus has yet emerged about an appropriate *global* terminology. The lack of a global terminology makes it difficult to evaluate critically the observations from different volcanoes, and creates confusion in the literature. Table 1 compares several common local terminologies. Further, it is often difficult to distinguish between different types of events at a single volcano because the events may grade into one another. In such cases quantitative criteria are applied, such as *P*/*S* amplitude ratios, coda durations, specific frequencies, etc. Recently, Falsaperla *et al.* (1996) applied multilayered neural networks to attempt an automatic classification of seismic events at Stromboli volcano.

Different mechanisms acting at the seismic source are responsible for many of the features of the different events, and path and site effects, which are extreme at many volcanoes, may greatly modify the signals. Three principal seismic sources are (Ferrucci, 1995): shear failure, also called "double-couple" mechanisms; tensile failure, or "non-double-couple" mechanisms; and passive and active fluid involvement in producing low-frequency events and volcanic tremor. Nonlinear flow has also been modeled as a source for volcanic tremor (Julian, 1994). Differences were seen between volcanic and tectonic earthquakes by comparing magnitude using different waves (Okada, 1983), and seismic moment versus fault length (Gorelchik *et al.*, 1990). Such comparisons reveal that many volcanic earthquakes are enriched in low frequencies and occur on smaller faults than tectonic earthquakes of the same magnitude.

4. General Features and Processes Associated with Volcanic Earthquakes and Tremor

The goals of volcano seismology include monitoring the present status of a volcano, forecasting eruptions, estimating the size of eruptions in progress, locating magma chambers, and understanding the physical processes that are occurring. The following topics elucidate some of the features that are common to earthquakes at volcanoes.

4.1 Seismicity Rates, Locations, and Processes

Volcanoes nearly always have a high background level of seismicity, caused by heat; movement of groundwater, volatiles, or magma; glaciers (if present), landslides, and rockfalls; and reactions to stresses from regional tectonics, tides, or other forcing functions. Thus, a baseline of measurements of a few years is necessary to characterize the background seismicity. The background may consist of several different types of events, which are related to different processes. Typical background rates of seismicity are a few to a few tens of events per day, depending on the locations and magnifications of the seismic stations. In contrast, the rates of seismicity before and during eruptions are several tens to several hundreds or more events per day, and include larger-magnitude events. Examples of rates and magnitudes of seismicity prior to selected eruptions are shown in Table 2.

Many successful eruption forecasts have been made because of an increase in seismicity above previously recorded background levels. This is based on the idea that the level of seismicity reflects the level of volcanic activity, so the likelihood of eruption increases when earthquake swarms occur. The main physical processes monitored, however, are those associated with magma intrusion. Thus, there will always be a false alarm rate because some intrusions remain at depth whereas others reach the surface and erupt.

In general, large eruptions are relatively easier to detect than smaller ones. The volume of magma is large, and

TABLE 2 Seismicity Rates for Selected Earthquake Swarms

Location and Date	Detected[a]	$M>1$	$M>2$	$M>3$	$M>4$	M_{max}	LP[b]/Tremor	Eruption
Long Valley								
May 80					≈15/day	6(4)	No	No
May 82				7/3 d		4.1	No	No
Jul. 82				4/10 d		3.4	No	No
Nov. 82	≈100/d		≈9/d			2.8	No	No
Jan. 83	≈800/d			25/h	≈10/day	5.3(2)	No	No
Jun. 89	≈40/d	≈25/d	2/d			3	No	No
Mar. 90	>300/d					3	No	No
Campi Flegri								
Oct. 83	>300/d					4	No	No
Mar. 84	≈500/d		5/d			4	No	No
Rabaul								
Apr. 84	≈1700/d					4.8	No	No
Mt. St. Helens								
Mar. 80	≈600/d		>70/d	≈50/d	1–4/d	4	Yes	Yes
Matsushiro								
Nov. 65	≈2000/d			>200/d		5.0	No	No
Off-Ito								
July 89	≈400/h			40/d		5.5	Yes	Yes
Redoubt								
Dec. 89	≈150/h					<2	Yes	Yes
Fuego								
Jan. 77	2000/d	70/d	≈5/d			2.8	No	No
Augustine								
Feb. 86	>5000/d					2.5	No	Yes
Usu								
Aug. 77	200/h			5/h		3.8	Yes	Yes
Pavlof								
Apr. 86	800/d	400/d				2.1	Yes	Yes
Galapagos								
Jun. 68				90/d	33/d	5.2	?	No
Kilauea								
Jan. 83	>1100/d	92/d	18/d	1/d		3.3	Yes	Yes
Mt. Hood								
Jul. 80	20/h		14/h			2.8	No	No
Medicine Lake								
Sep. 88	80/h			2/h		4.2	No	No

[a]Minimum magnitude not specified but generally $M<1$; d = day; h = hour.
[b]LP, long-period earthquake.

earthquakes tend to be numerous as well as distributed over a large volume. Other precursors such as deformation and steaming are usually observed in conjunction with earthquake swarms. Recent examples include Mount St. Helens, Washington, 1980 (Endo *et al.*, 1981; Malone *et al.*, 1981); Pinatubo, Philippines, 1991 (Newhall and Punongbayan, 1996; Pinatubo Volcano Observatory Team, 1991); and Spurr, Alaska, 1992 (Keith, 1995). Small eruptions and many phreatic (water-driven) eruptions, in contrast, involve much smaller amounts of magma, and have subtle precursors or sometimes no precursors. They are, hence, generally harder to forecast. Recent examples include Galeras, Colombia, January–June 1993 (Munoz *et al.*, 1993); Arenal, Costa Rica, August 1993 (BGVN, 1993); and White Island, New Zealand, February 1992 (BGVN, 1992).

Occasionally "unexpected" eruptions occur; often these take place during the later phases in a sequence of eruptions. Even though these may be large, they may have subtle or undetectable seismic precursors. This is a consequence of the vent remaining "open" or mechanically weak following an initial stage of activity, preventing buildup of stresses that would cause earthquakes. A recent example is the second eruption of Mount Spurr in August 1992, which occurred without immediate detectable precursors (Power *et al.*, 1995). The eruption lasted 3.5 h, produced $52 \times 10^6\,m^3$ of tephra, and ejected its ash cloud into the stratosphere.

Generally, volcanic earthquakes occur beneath the point of eruption. This helps reduce the monitoring problem to that of estimating the time of eruption, since the place is considered known. However, several cases have occurred in which the locations of earthquake swarms did not coincide with the eruptive vent, because different mechanisms were acting (Nakamura, 1984). During the Matsushiro, Japan, earthquake swarm of 1965–1967, which began beneath Mt. Minakami, an extinct volcano, epicenters were aligned NE–SW, but a fissure oriented NW–SE appeared at the ground surface. Water flowed from this fissure but no magmatic eruption occurred. During the Miyake-jima, Japan, basaltic eruption of 1962, the epicenters were located NW of the summit crater but the eruption occurred from a flank fissure on the NE side. More recently, the initial earthquake swarms prior to the dacitic eruption of Pinatubo occurred about 5 km NW of the volcano. Twelve days before the climactic eruption, these earthquakes declined and vigorous shallower swarms began beneath the eventual eruption site (Wolfe, 1992; Harlow *et al.*, 1996). For volcanic systems with large dimensions, earthquakes may be widely separated: for Katmai, Alaska, 1912 there was a 10 km separation between the collapse structure (Mt. Katmai) and the Novarupta vent (Eichelberger *et al.*, 1991); at Rabaul, Papua New Guinea, 1994, two vents, Tavurvur and Vulcan, erupted simultaneously 10 km apart (BGVN, 1994), while precursory earthquakes were distributed in an annulus surrounding the caldera. These cases demonstrate that earthquakes occur where stresses are concentrated, not necessarily where the magma is located.

It is expected that earthquake locations will migrate from depth toward the Earth's surface as magma rises from depths prior to eruption. There are, however, few well-documented cases of such a trend. This overall pattern was first shown for Kilauea volcano, Hawaii, in 1959 (Eaton and Murata, 1960); more recent examples include Augustine, Alaska, 1986 (Power, 1988); Unzen, Japan, 1991 (Ohta *et al.*, 1992); Pinatubo, 1991 (Harlow *et al.*, 1996), and additional 1960–1983 eruptions of Kilauea (Klein *et al.*, 1987). One spectacular case of lateral migration of hypocenters was observed at Krafla in September 1977 (Brandsdottir and Einarsson, 1979); when the earthquakes passed a geothermal borehole, fresh pumice erupted from the hole!

Differences between seismic activity at basaltic versus andesitic and dacitic volcanoes may be subtle. Extremely short-duration swarms of small events are more common at basaltic volcanoes. For example, the January 1991 eruption of Hekla volcano, Iceland, occurred after an earthquake swarm lasting only 0.5 h, with the largest event about $M = 3$ (Gudmundsson *et al.*, 1992; Linde *et al.*, 1993). Most of the handful of known eruptions that occurred without detectable precursors took place at basaltic volcanoes (e.g., Hekla, 1981, Gronvold *et al.*, 1983; Mt. Etna, Italy, flank, March, 1985, McClelland *et al.*, 1989, p.71). Energetic earthquake swarms after eruptions are common at basaltic centers. Spatially, basaltic centers more often produce flank eruptions, whereas andesitic and dacitic volcanoes tend to erupt repeatedly from the same central vent.

4.2 *b* Values at Volcanoes

In general, there are many more small earthquakes than large ones. This is a consequence of the fact that the size distribution of earthquakes obeys a power law, which in seismology is of the form $\log_{10} N = a - bM$ (where N is cumulative number of events, M is magnitude, and a and b are constants) and is known as the Gutenberg–Richter relation. The slope, or b value, determines the relative number of large versus small events. While for tectonic earthquakes the b value has typical values of 0.8–0.9 worldwide, more extreme values have been observed at volcanoes. High-frequency events have b values ranging generally between 0.6 and 1.3, but low-frequency events have b values of up to 3.0 (Fig. 8). This means, for low-frequency events, that small earthquakes dominate, and also implies that the source size is restricted (Wyss, 1973). Laboratory experiments have shown that b value increases with heat (Warren and Latham, 1972) and pore pressure (Wyss, 1973), decreases with applied stress (Scholz, 1968), and

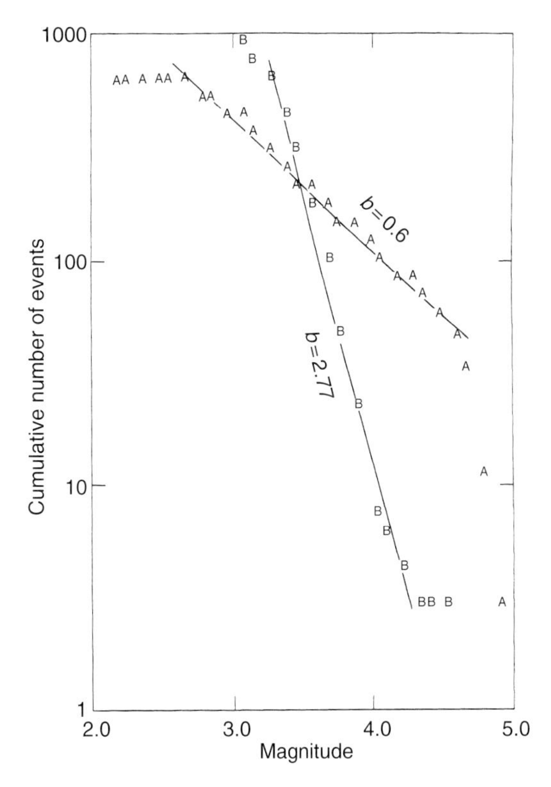

FIGURE 8 Frequency–magnitude relations for high-frequency (A) and low-frequency (B) earthquakes at Mount St. Helens during the March to May 1980 swarm. Regression lines fit to selected portions of the data produce the b values indicated. (From Endo *et al.*, 1981.)

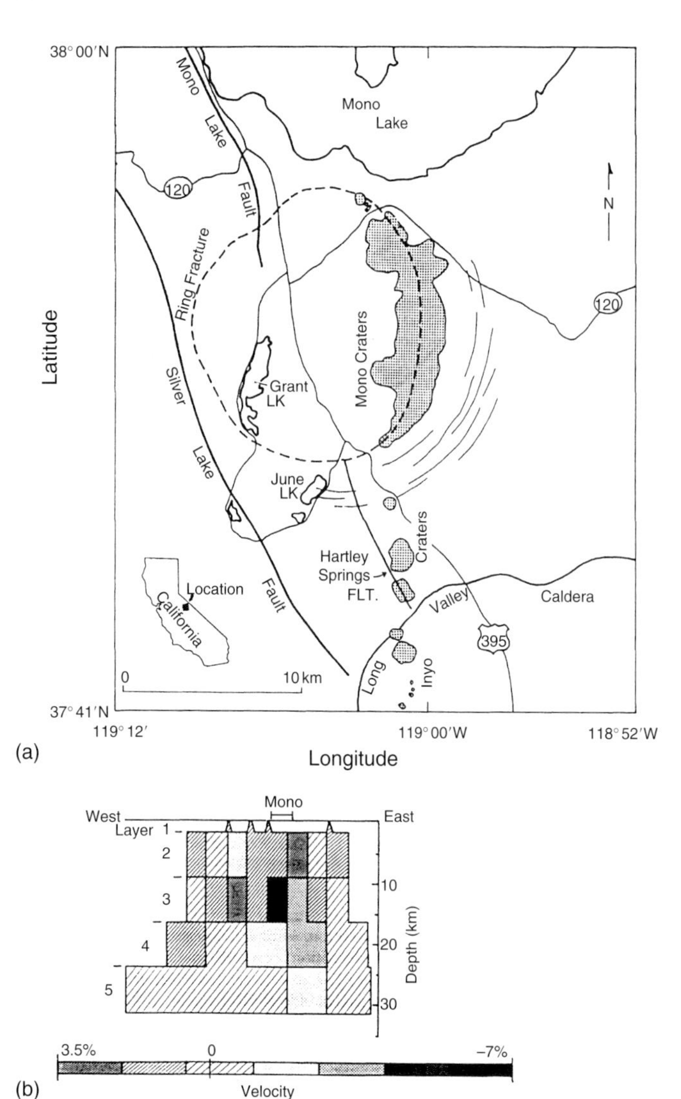

FIGURE 9 Tomography of Mono Craters, California, using *P* waves from distant earthquakes. (a) Map of tectonic and geographic features of the area of Mono Craters, just northeast of Long Valley caldera. Shaded areas are Holocene volcanic flows and domes. (b) An E–W cross section showing tomographic results. The velocities vary from 3.5% higher to 7% lower than the reference values. The strongest anomaly, a 7% lowering of the velocity (black rectangle) occurs 9–16 km beneath the Mono Craters. (From Iyer, 1992.)

increases when material heterogeneity increases (Mogi, 1962). Determination of the *b* value for volcanic earthquakes is thus an important diagnostic tool for identifying changes in seismicity that may indicate changing processes.

4.3 Some Seismic Features of Magma Chambers

Magma is thought to be stored in chambers or reservoirs underground, and seismology is useful to determine the parameters of such chambers. Magma chambers at stratovolcanoes are generally equant in shape, and on the order of 1–20 km^3 in volume. Those at calderas are larger. Their depths are mapped at about 5–20 km or deeper based on several techniques. First, seismic tomography, similar to medical tomography, requires seismic waves to travel through the target region. Thus, a good distribution of earthquakes and seismic stations is needed; both local and distant earthquakes have been used as sources for such studies. The technique exploits the fact that *P* waves speed up in competent rock and slow down in unconsolidated materials or magma. An example of a tomographic image of a magma chamber under Mono Craters, California, is shown as Figure 9; see Chapter 26 by Benz *et al.* for additional examples. Second, *S* wave screening exploits the fact that *S* waves cannot pass through liquids. Thus, the target region will be illuminated by the presence of *P* waves but lack of *S* waves for those waves that pass through magma. Third, post-eruption seismicity is often concentrated at depths of 5–20 km. Here, the removal of magma upward during eruption causes stress changes at depth that induce earthquakes. The idea is that the magma chamber walls collapse inward, and earthquakes are concentrated where these processes are most pronounced.

A new technique that requires well-distributed seismicity (as opposed to concentrated at one spot) looks for systematic spatial variation in the *b* value, discussed earlier. A 2D or 3D grid is made, then the *b* value is determined for the 100 earthquakes nearest to each grid point (Wiemer *et al.*, 1998). The results are displayed in maps or cross sections (see Color Plate 6). It has been shown in the laboratory that heat, high pore pressure, low applied stresses, or heterogeneous materials all produce high *b* values. These are also the conditions that would be expected in the vicinity of magma bodies. Although this technique cannot uniquely identify which process is dominant, strong anomalies have been found at depths of 3–4 km and 7–9 km beneath six volcanoes studied. Examples of *b* value anomalies thought to represent magma chambers beneath Mount St. Helens and Mount Spurr are shown in the Plate.

5. Case Histories

We consider the earthquake activity at six volcanoes to illustrate the variety of seismic activity. The six include small and large eruptions, a noneruption, and systems dominated by basalt, andesite, and dacite magmas. The volcanoes chosen had high-quality seismic data. Parameters of the four cases are summarized in Table 3. The bibliography gives at least one recent reference for each.

5.1 Mount St. Helens, May 1980: A Large Dacite Eruption

Mount St. Helens, Washington, USA, produced a large explosive dacite eruption including a lateral blast on 18 May 1980, following 59 days of increased seismicity (Endo *et al.*, 1981;

TABLE 3 Parameters of Volcanic Activity: Seismic Case Histories

Volcano Name	Date (d/m/y)	VEI[a]	Swarm Duration (days)	Time Since Previous Eruption (y)	Precursor Event Type[b]	Precursor Maximum Magnitude	Strongest Tremor (D_R)[c]
Mount St. Helens	18/5/80	5	59	123	H, L, T	5.0	260
Spurr	27/6/92	3	311	39	H, T	1.8	30
	19/8/92 & 19/9/92						
Arenal	7/68 to present	4	3	~450	H, T	4.5	Not known
	4/94	1–2	NA[d]	NA	L, T	NA	20
Long Valley caldera	10/78 to present	NA	Variable	~600	H, L	6.2	NA
Kilauea	3/1/83 to present	1–2	8	0.3	H, L, T	4.2	~20
Pavlof	16/9/96 to 29/12/96	3	5	8	L, T	~0.5	20

[a]Volcanic Explosivity Index.
[b]H = high-frequency earthquakes; L = low-frequency earthquakes; T = volcanic tremor.
[c]D_R = reduced displacement, which is root-mean-square (rms) amplitude multiplied by distance. Units are cm^2.
[d]NA = not applicable.

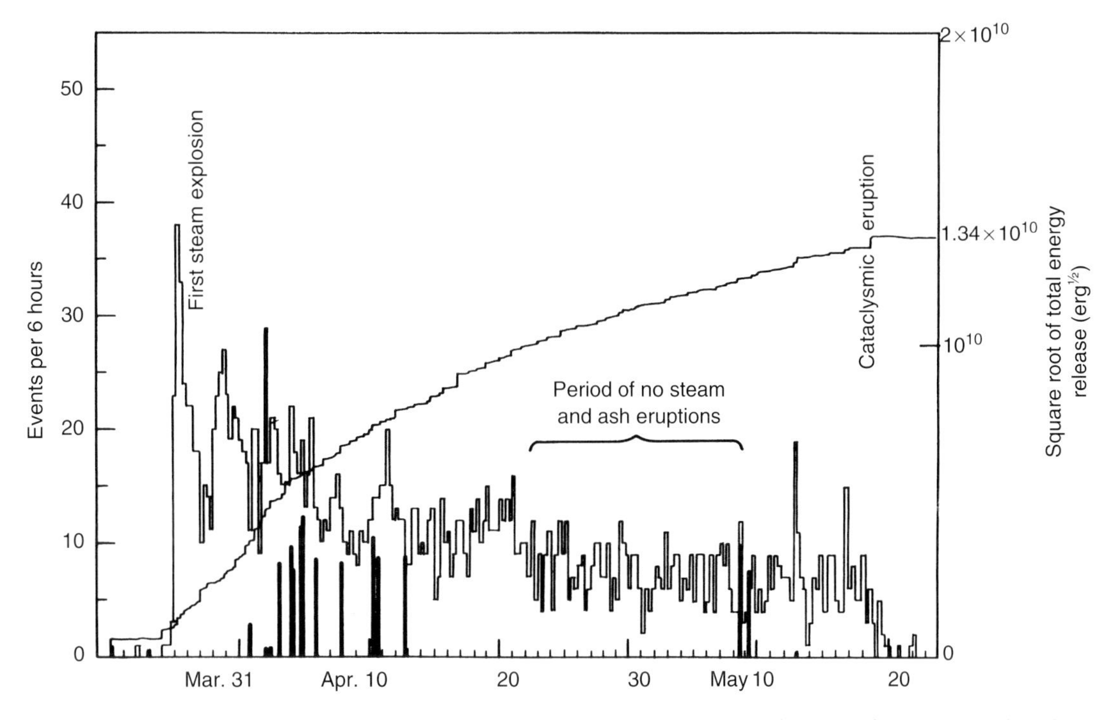

FIGURE 10 Counts of Mount St. Helens earthquakes larger than $M = 2.5$ per 6 h, volcanic tremor (vertical lines, length proportional to tremor duration), and square root of cumulative energy release (a way to convert from magnitude to energy) from 20 March to 24 May 1980. (After Endo *et al.*, 1981, Malone *et al.*, 1981.)

Malone *et al.*, 1981). On 20 March 1980 an $M = 4$ earthquake occurred, followed by many smaller events. Activity increased, reaching eight $M > 4$ events in a single hour on 25 March. Phreatic eruptions began on 27 March. The earthquake sequence eventually included over 10 000 earthquakes, including HF and LF events and volcanic tremor (Fig. 10). Most events were located at depths of 3–7 km beneath the north flank of the volcano, the site of the eventual eruption (Fig. 11). There were no immediate precursors in the hours before the climactic eruption. An $M = 5.1$ earthquake coincident with a landslide initiated the eruption by causing a sector collapse that uncapped the volcano (Kanamori *et al.*, 1984). This is another example of a co-eruptive process, in this case a landslide, that produces seismic signals. Strong volcanic tremor accompanied the main eruption (Table 3). Deep (5–20 km) earthquakes began shortly after the eruption and lasted for several days (Weaver *et al.*, 1981). Later smaller eruptions were generally preceded by swarms of small earthquakes lasting several hours to several days, and were accompanied by volcanic tremor (Malone *et al.*, 1981; Swanson *et al.*, 1983).

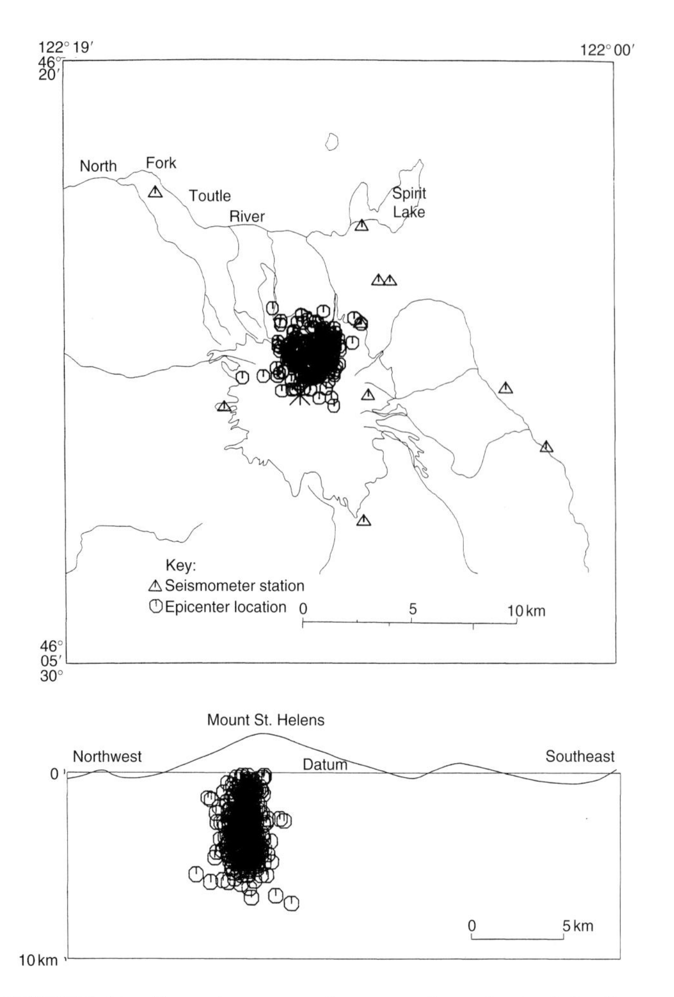

FIGURE 11 Epicenter map and cross section of seismicity from 27 March to 18 May 1980, at Mount St. Helens volcano. Contour on map is 1219 m (4000 ft); datum for cross section is average elevation of seismic stations, approximately 914 m (3000 ft). (Modified from Endo *et al.*, 1981.)

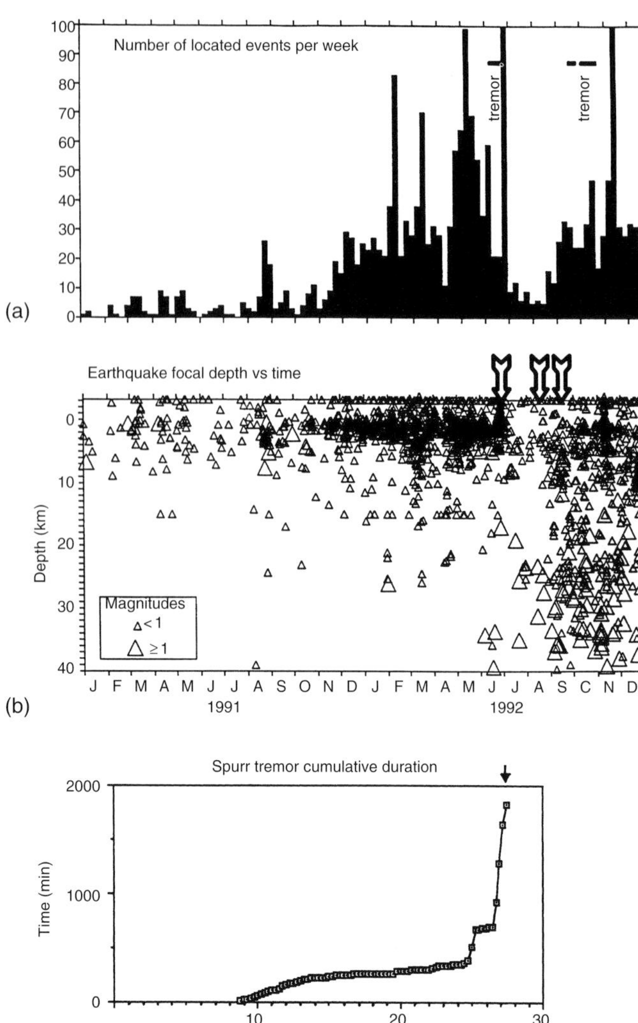

FIGURE 12 Seismicity in 1991 and 1992 associated with the eruptions of Mt. Spurr. (a) Number of events per week. Episodes of volcanic tremor are shown as horizontal bars. (b) Earthquake depth versus time. The three eruptions are shown as arrows. (c) Volcanic tremor cumulative duration versus time for June 1992. Data are computed for 6 h intervals. The eruption occurred on 27 June (arrow). (From Power *et al.*, 1995, McNutt *et al.*, 1995.)

5.2 Mount Spurr, June–September 1992: A Typical Andesite Eruption

The Crater Peak vent of Mount Spurr, Alaska, erupted on 27 June 1992 following a 10-month swarm of HF events (Fig. 12; Alaska Volcano Observatory, 1993). The largest event of this swarm was $M = 1.8$. Bursts of volcanic tremor began three weeks before the first eruption, and continuous tremor began 19 h before (Fig. 12). The final precursor to the first eruption was a 4 h swarm of shallow HF events directly beneath the active vent (Power *et al.*, 1995). The Vulcanian eruption lasted 4 h and produced $44 \times 10^6\ m^3$ of andesite tephra (Keith, 1995). The second, larger eruption on 18 August occurred essentially without precursors. The third eruption, on 17 September, occurred following several weeks of increased deeper (15–40 km) seismicity and 3 h of small shallow events and weak tremor. This last eruption was followed by a vigorous swarm of earthquakes between 5 and 10 km depth (Fig. 12). A shallow swarm of several hundred events occurred on 9 and 10 November without eruption.

5.3 Arenal Volcano, July 1968 and Continuing: Strombolian Activity

Arenal Volcano, Costa Rica, erupted explosively in July 1968 (Minakami *et al.*, 1969) after several days of felt earthquakes (no local seismometers were operating at the time). Four new vents were formed along a fissure. At first activity was strongest at the lower craters, but afterwards activity shifted to a vent near the summit and near steady-state eruption of andesite has occurred ever since (Melson, 1989; BGVN, 1999). The eruption of a block andesite lava flow is punctuated several times per hour by small explosive eruptions of several types. A typical sequence in spring 1994 was an explosion, followed by a "whoosh" that sounded like a jet plane, followed by "chugs" that sounded like a steam locomotive (Benoit and McNutt, 1997). The seismic expression of these is an explosion quake, followed by irregular tremor, then regular tremor with variable amplitude (see Color Plate 6), which gradually fades to background levels (similar data and frequencies are found in acoustic data; Garces *et al.*, 1998). The tremor consists of several integer harmonics or overtones, and these change frequency systematically over the course of several minutes (Color Plate 6). This suggests that the upper few hundred meters of magma in the conduit is changing its pressure or bubble concentration, hence its velocity and frequency (Benoit and McNutt, 1997). Every few years, activity escalates and larger explosions occur, producing pyroclastic flows.

5.4 Long Valley, May 1980 and Continuing: A Restless Caldera

The Long Valley caldera region of eastern California has been a prolific producer of earthquakes from 1980 to 2000. The sequence began on 25 May 1980 when three $M = 6$ earthquakes occurred, along with thousands of smaller events, followed two days later by a fourth $M = 6$ event (Fig. 13; Hill *et al.*, 1985). Several of the events had significant non-double-couple components, suggesting possible fluid injection (Julian and Sipkin, 1985), rupture complexity (Wallace, 1985), or effects from near-source structural inhomogeneities (Wallace, 1985). The relative contribution of these effects remains controversial. Extensive ground deformation had also occurred, with uplift of 0.5 m near the center of the caldera. Later, it was found that measurable deformation actually preceded increases in earthquake activity by about two months, suggesting that earthquakes were a symptom of the deformation (or whatever caused it, such as underground movement of magma) rather than the cause (Langbein *et al.*, 1993). Repeated swarms of earthquakes occurred in the

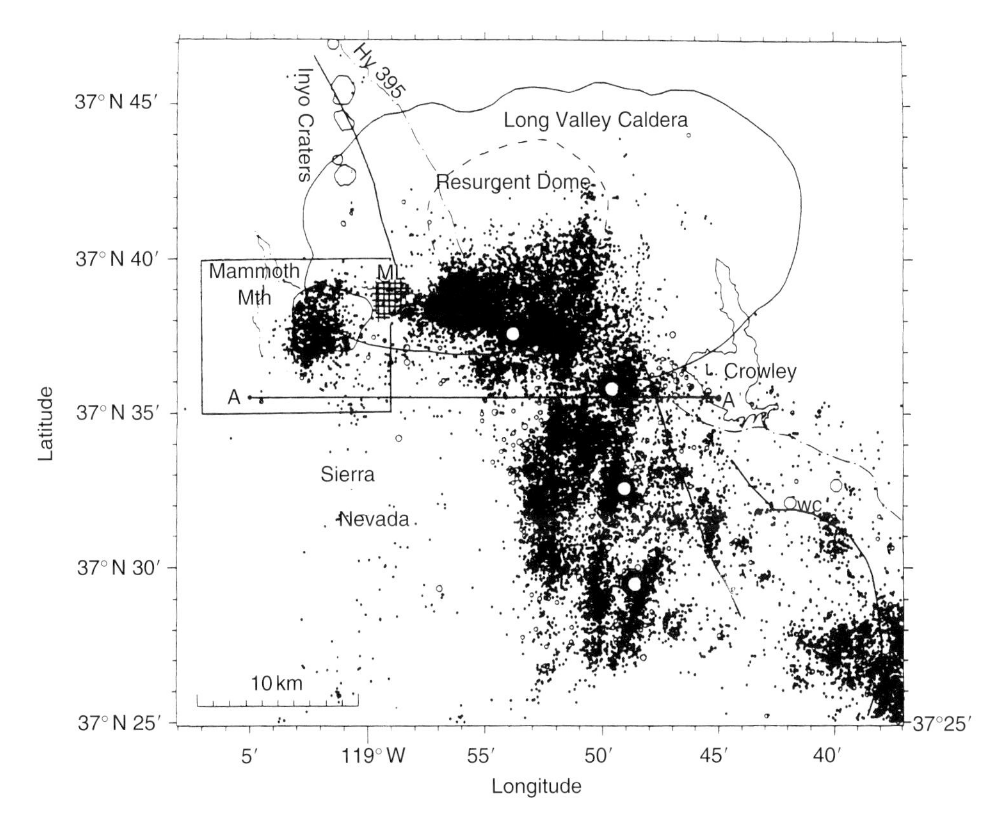

FIGURE 13 Map of Long Valley caldera and Sierra Nevada mountains to the south showing epicenters of $M > 1.2$ earthquakes occurring from January 1978 to June 1995. Each symbol represents one earthquake, and symbol size is proportional to magnitude. The large white circles are the four $M \sim 6$ earthquakes that occurred in May 1980. ML is the town of Mammoth Lakes; WC is the epicenter of the $M = 5.4$ Wheeler Crest earthquake of October 1978; Hy 395 is Highway 395; L. Crowley is Lake Crowley. The caldera outline is shown as a solid line and the uplifted resurgent dome by a dashed line. (From Hill, 1996.)

southern part of the caldera, and the strongest of these, in January 1983, included two $M = 5.3$ events. Altogether 27 earthquakes with $M > 5$ have occurred in and near the caldera over the last 20 years, a very high rate. No eruption has occurred; however, large quantities of CO_2 gas have been measured near Mammoth Mountain and have caused tree kills (Hill, 1996). It is also thought that several of the earthquake swarms accompanied intrusions of magma to shallow depths of about 5 km. Long Valley has become an archetype of a "restless caldera."

5.5 Kilauea, January 1983: A Fissure Eruption

Kilauea Volcano, Hawaii, is a shield volcano with a summit caldera and rift zones to the southwest and east. Earthquakes are numerous and are associated with all three of these features, as well as with a deeper nearly horizontal fault on the south flank of the volcano (Fig. 14). New magma typically rises from depth near the summit, causing earthquake swarms and inflation, then either erupts near the summit or moves laterally into the east rift zone, accompanied by summit deflation. A fissure eruption began on the east rift zone on 3 January 1983, near Napau Crater, extending 8 km downrift toward Puu Kamoamoa (Wolfe, 1988). Prior to this, earthquake swarms had occurred on 26 September and 9 December 1982, near the summit, followed by an intense swarm on the upper east rift from 30 December 1982 to 6 January 1983 (Fig. 15). Most of the Earthquakes were shallow, between depths of 2 and 4 km, and the largest was $M = 4.2$ (Koyanagi *et al.*, 1988). Volcanic tremor first occurred weakly, then was strong at frequencies of 5–10 Hz at the time of fissure opening. Later, tremor was predominantly 2.5 Hz with bursts of stronger activity paralleling the behavior of intermittent high lava fountains. After the initial fissure eruptions, the volcanic activity localized south of Puu Kamoamoa (Fig. 14) and formed a new cone called Puu Oo. In July 1986, activity shifted to a lava lake 3 km downrift. The eruption is continuing as of this writing, and is Kilauea's longest historical eruption. Because of its high activity, Kilauea has been the site of numerous detailed investigations including the deployment of the arrays shown in Figures 2 and 7b.

5.6 Pavlof, September–December, 1996: A Long-Duration Strombolian Eruption Punctuated by Explosive Phases

Pavlof is the most active volcano in North America and has had over 40 eruptions since 1760 (Miller *et al.*, 1998). It is monitored by a network of six seismic stations within 11 km, and three others from 33 to 40 km. An eruption report was received on the morning of 16 September 1996, from residents of Cold Bay, 37 miles to the SW. This was a surprise to scientists because no significant precursors had been seen on seismic records, which were partially obscured by a storm in the area. Later analyses showed that the small-scale seismicity had actually increased on 11 September, which was only revealed after digitally filtering the data (Fig. 16) to remove storm noise. The eruption lasted continuously from 16 September to 4 December, declined, then had two stronger bursts on 10–12 December and 26–29 December (Fig. 17) (McNutt, 1999). The challenge during this eruption was not to forecast the onset, which was missed, but to quickly determine escalations of activity that represented more explosive phases of the eruption. These occurred five times (Fig. 17). The data were analyzed automatically in near real time (a few minutes) to determine reduced displacements (defined in Section 6.3) and these were used to establish an alarm threshold of 6 cm^2. Further, spectrograms were computed automatically in near real time and were used to help identify explosions (with wideband air waves) and escalations in tremor, a strong but narrowband 1–3 Hz signal (Benoit *et al.*, 1998).

6. Forecasting Eruptions and Characterizing Eruptions in Progress

6.1 Generic Volcanic Earthquake Swarm Model

Common observations suggest a general model of earthquake activity at volcanoes (McNutt, 1996). The model, which is termed a generic volcanic earthquake swarm model, is shown as a time-series in Fig. 18. Seismicity rates, types of events, and dominant processes are indicated in the commonly observed temporal sequence. Variations in the involvement of groundwater, as well as variations of parameters such as magma composition, temperature, rate of flow, and volatiles content, can explain many of the observed differences from this model of seismic activity at different volcanoes. A nice example of the sequence is seen accompanying the 1989 Ito-Oki eruption (Ukawa, 1993; Ukawa and Tsukahara, 1996).

In the generic model, HF earthquakes reflect shear fracture of the country rock in the vicinity of volcanoes in response to increasing magmatic pressure. Since magma movement at depth may occur over large distances and long periods of time, the durations of HF swarms are long. At shallow depths of 1–3 km, several significant changes occur. First, gases begin to exsolve from the magma, changing its mechanical behavior, and hence its ability to transmit or reflect seismic waves. For example, exsolution of volatiles will increase the viscosity and lower the acoustic velocity. A seismic event generated by resonance of a magma-filled cavity would thus be richer in lower frequencies. Second, the magma may encounter groundwater, modifying the shallow hydrothermal system and removing some of the available magmatic heat

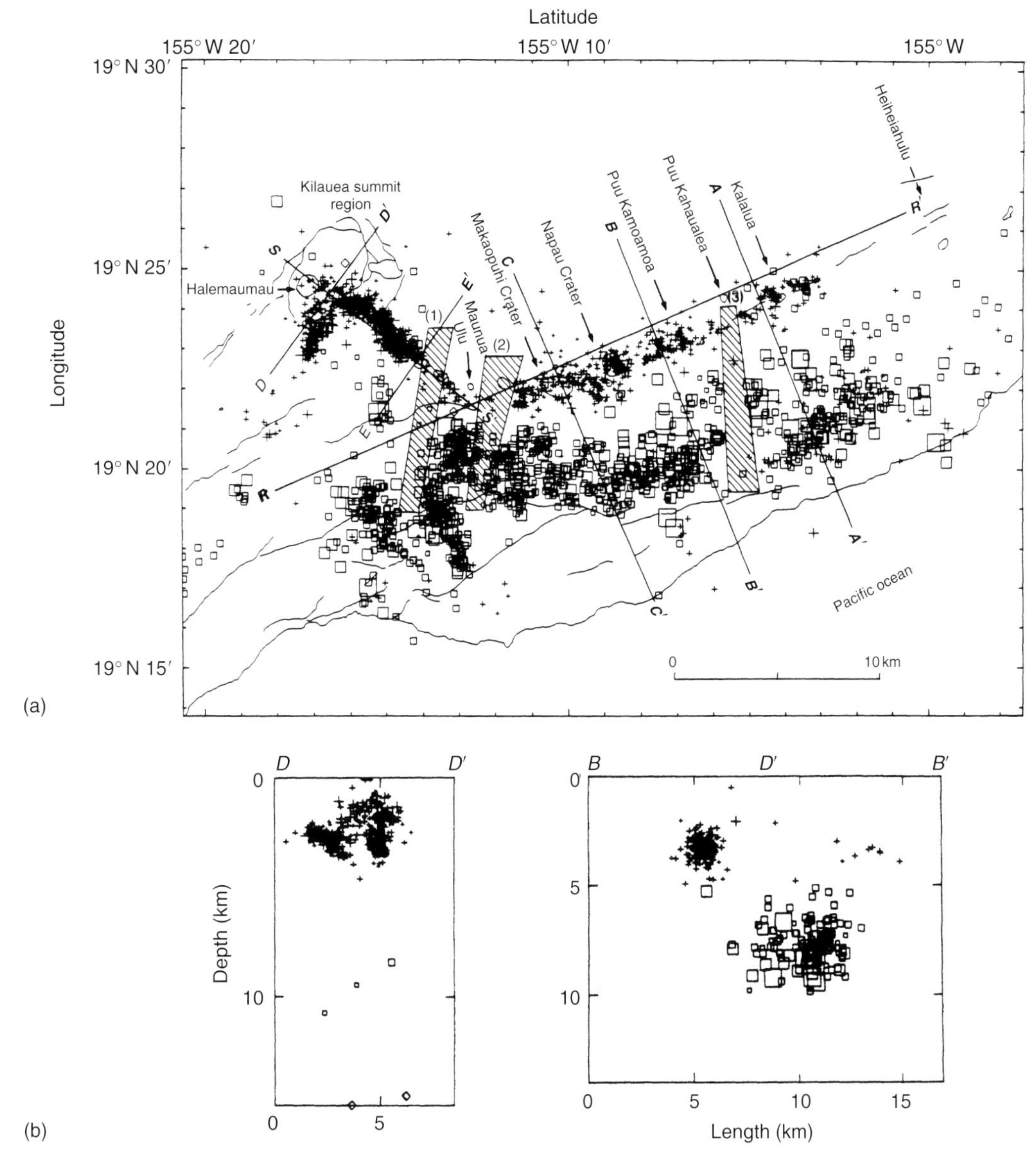

FIGURE 14 (a) Map of Kilauea Volcano and the east rift system showing earthquakes less than 15 km deep for the period September 1982 to January 1983. Shallow earthquakes (<5 km) are shown as crosses, and deeper earthquakes (>5 km) as squares; symbol size is proportional to magnitude. Note the concentrated shallow seismicity near Kilauea's summit and the more diffuse seismicity along the east rift zone. The deeper events occur mostly on a nearly horizontal fault. The 3 January eruption began near Puu Kamoamoa. Zones free of earthquakes are shown with cross-hatching. Lines labeled A–A′, B–B′, etc. show cross sections. (b) Two cross sections (see map) showing the depth distribution of earthquakes near the summit and near the eruption site. (Modified from Koyanagi *et al.*, 1988.)

by conduction and advection. Most LF events and tremor originate at shallow depths, and several models for their origin involve fluids—water, magma, exsolved volatiles, or all three. Tremor caused by boiling of water is generally weaker than tremor involving magma movement. Third, open cracks are found above 3 km depth in the Earth, permitting venting of excess pressure. This may explain the relative quiescence after the peak rate of a swarm as well as the inverse relation between high-frequency events and tremor.

The upper 1–3 km are small distances compared with depths to magma chambers (5–20 km) or crustal thicknesses (about 35 km). Thus, if vertical magma ascent rates are constant, then durations of LF event swarms and tremor sequences will be shorter than those of HF events simply because the

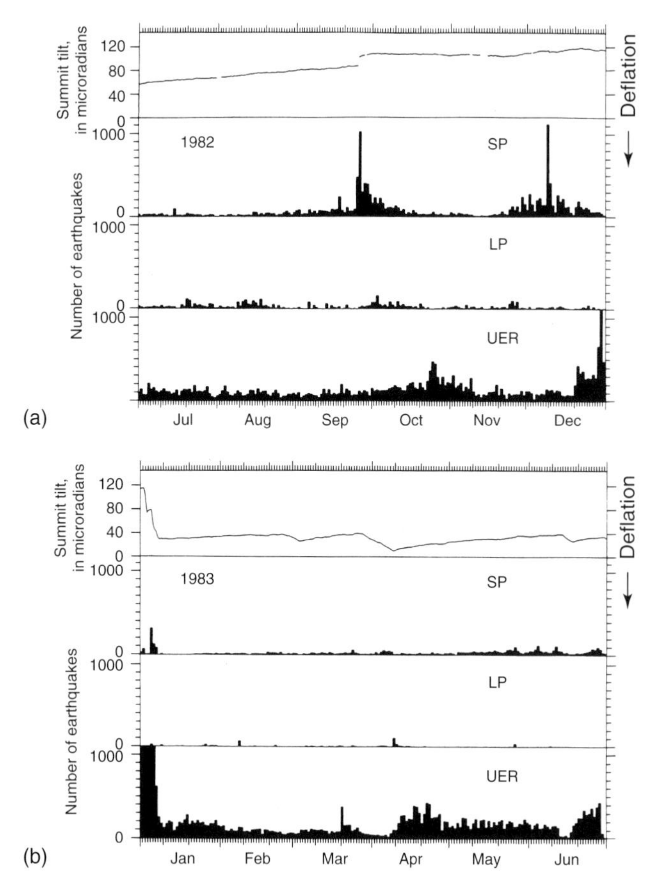

FIGURE 15 Summit tilt and daily number of earthquakes at Kilauea summit and upper east rift (UER) for the periods (a) July to December 1982 and (b) January to June 1983. Summit earthquakes are divided into short-period (SP) and long-period (LP) based on their waveforms. Note the summit swarms in September and December, and the strong UER swarm just before the eruption of 3 January 1983. Summit deflation typically accompanies the migration of magma into the east rift system. (From Koyanagi *et al.*, 1988.)

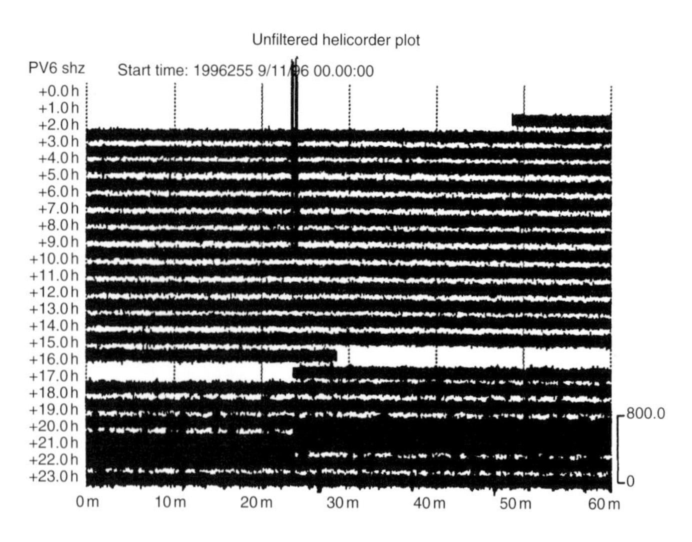

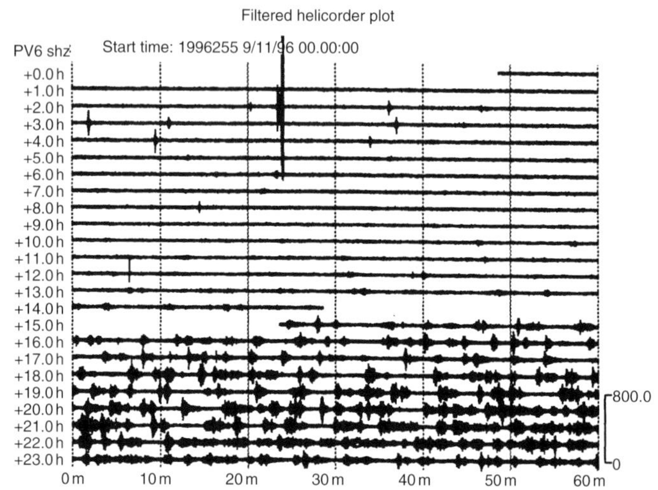

FIGURE 16 Unfiltered and filtered digital seismograms from station PV6 at Pavlof Volcano, 11 September 1996. Each record is for one day, one line per hour. The unfiltered seismogram shows only noise with variable amplitude. The lower seismogram was filtered to remove noise below 0.8 Hz and above 5 Hz. It shows the rapid onset of a swarm of small earthquakes that occurred a few days before the eruption.

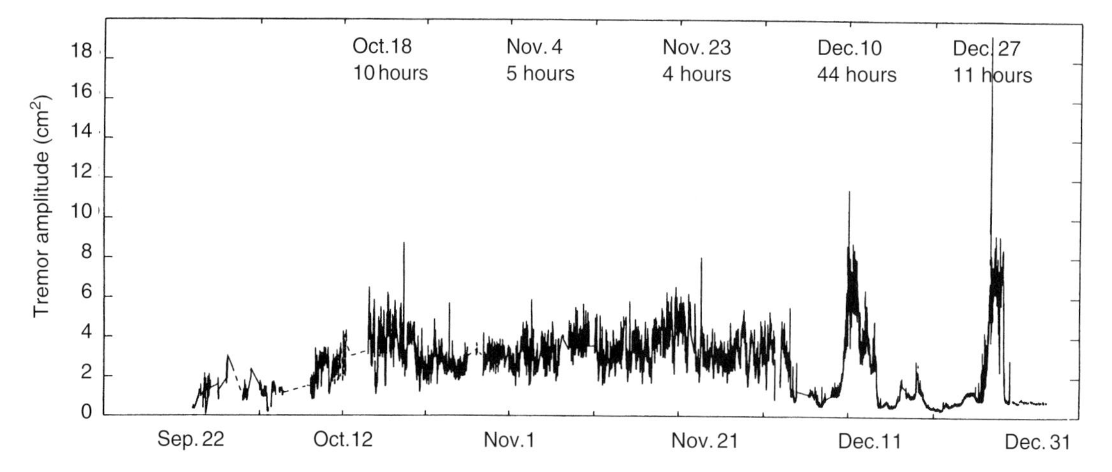

FIGURE 17 Plot showing the reduced displacement of volcanic tremor during the 3½ months of the 1996 Pavlof eruptions. Stronger tremor accompanied more explosive phases of the eruption on 18 October, 4 November, 23 November, 10 December, and 27 December 1996.

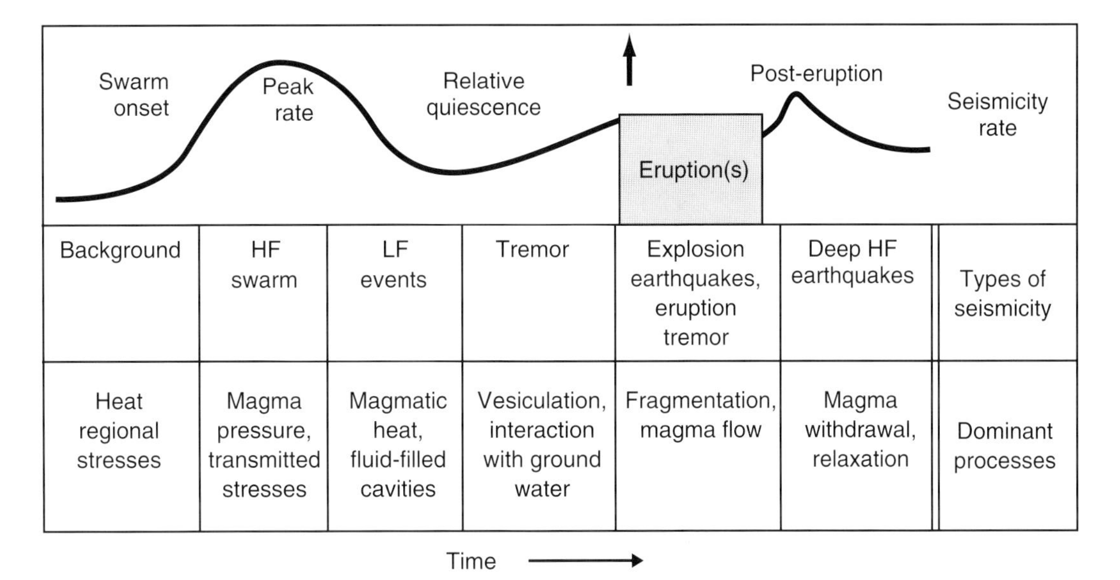

FIGURE 18 Schematic diagram of the time history of a generic volcanic earthquake swarm model. Seismicity rates are shown along with the main types of events observed at each stage. Some ideas about dominant processes are presented at the bottom of the figure.

distances are smaller. Observations from a database of volcanic earthquake swarms (see Section 6.2) support this: LF swarms average 5.5 days and HF swarms 9.3 days. Deep earthquake swarms following eruptions probably reflect stress changes associated with partial evacuation of magma chambers deeper than 5 km.

Relative seismic quiescence following an initial seismicity peak occurred in 25% of a worldwide sample of 192 swarms (Newhall and Endo, 1987; J. Benoit, personal communication, 1994). While 25% quiescence may seem too low to be included in a generic model, it is included because of its importance in terms of eruption forecasting and identifying physical processes. Factors that may cause quiescence include strain hardening; high temperatures near the brittle–ductile transition; increased water content (e.g., groundwater) lowering effective stress; or a reduction in strain rate. It is common for low-frequency events and tremor to occur after seismicity rate peaks, during times of relative quiescence. This may be explained by gradual shoaling of the source, particularly magma interacting with groundwater or exsolution of volatiles.

Each volcano is unique in its seismic behavior, but the observations suggest that variations in a few parameters may account for the main differences. Minor variations in each of several parameters may greatly affect one observed parameter, such as swarm durations. The durations of earthquake swarms are characterized by a log-normal distribution with a mean of 5 days (Fig. 19). Precursory swarms are longer (mean = 8 days) and noneruptive swarms are shorter (mean = 3.5 days). No definitive correlations exist between the duration of a swarm and the size or type of eruptive activity. This is not surprising given the ranges of sizes, rates, compositions, and types of structures observed.

6.2 Swarm Database as a Probabilistic Tool

Recently, parameters of 700 volcanic earthquake swarms have been compiled by Benoit and McNutt (1996) into an electronic database called the Global Volcanic Earthquake Swarm Database (GVESD). The database has one table of basic information on each volcano, such as type of lava, height, location, etc.; one table on eruptions, including volume, time of occurrence, type, etc.; and a table of swarm parameters such as duration, number of earthquakes, depths, magnitudes, etc. Using electronic search commands, it is possible to find all known cases of earthquake swarms that fit a particular profile. For example, we can search for the cases that last 3 weeks, have 60 events per day, and a largest event of $M = 3.3$. The "answer" is a list of similar swarms that can then be analyzed to determine whether eruptions occurred, how long they lasted, and so on. Table 4 gives examples of three query answers that were obtained during the 1996 seismic crisis at Akutan volcano, Alaska, during which earthquakes up to $M = 5.0$ occurred and more than 3000 events were felt These answers are probabilistic in that they present a likelihood of something happening based on past occurrences. Such information is complementary to simultaneous efforts to determine and to constrain physical models of a volcano. The web site for the GVESD can be found at *http://www.avo.alaska.edu/dbases/swarmcat/GVESD.HTML*, and the database is also to be found on the attached Handbook CD, under directory \25McNutt.

TABLE 4 GVESD Query Parameters

(1) Swarms with $M > 500$ felt events: *Returned 4 entries*

Volcano	Date (d/m/y)	No. Felt	No. Detected	Activity
Unzen	15/5/84	519	10 544	No eruption
Usu	1/7/80	1 427	7832	During eruption
Usu	1/7/81	718	3694	During eruption
Tacana	7/5/86	1 000	4000	Before eruption

(2) Swarms with the largest shock $M \geq 5$ and swarms that preceded or accompanied eruptions: *Returned 10 entries*

Volcano	Date (d/m/y)	Duration (days)	M_{max}
Tacana	15/12/85	72	5.0
Izu-Tobu	30/6/89	11	5.5
Sakura-Jima	10/1/14	3	5.2
Soputan	15/3/85	65	5.6
Usu	22/7/10	4	5.1
O-Shima	21/11/86	9	6.0
On-Take	15/8/76	1020	5.3
Miyake-Jima	13/10/83	3	6.2
Mt. St. Helens	20/3/80	59	5.0
Gorely	5/8/85	10	6.0

(3) Swarms with the largest shock $M \geq 5$ and swarms that had no eruption within 3 months: *Returned 9 entries*

Volcano	Date (d/m/y)	Duration (days)	M_{max}
Rabaul	1/7/82	–	5.1
Rabaul	15/9/83	630	5.0
Rabaul	22/4/84	–	5.1
Unzen	15/5/84	270	5.7
On-Take	14/9/84	90	6.8
Norikura	7/3/86	–	5.1
Long Valley	4/10/78	570	5.7
Long Valley	7/1/83	67	5.2
Don Joao de Castro	21/11/88	1	5.8

6.3 Reduced Displacement Versus Volcanic Explosivity Index

Forecasting eruptions has received dominant focus as a goal of volcano seismology. However, characterization of eruptions in progress, including estimates of tephra volume, height of ash column, and end time of eruption, are other important parameters for physical understanding and for hazard mitigation, particularly regarding aviation. Progress has been made recently in estimating the Volcanic Explosivity Index (VEI of Newhall and Self, 1982) based on tremor reduced displacement (D_R) measured during eruptions (Fig. 20; McNutt, 1994). The computed relation is

$$\log_{10}(D_R) = 0.46(\text{VEI}) + 0.08 \tag{1}$$

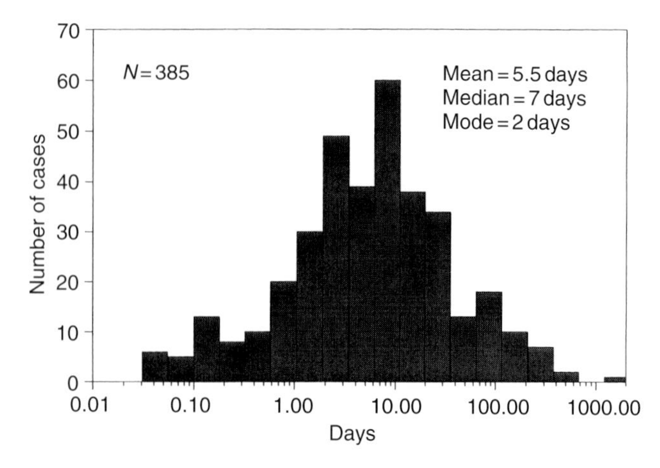

FIGURE 19 Histogram of volcanic earthquake swarm durations. The horizontal axis is logarithmic, and the numbers shown represent the middle value of each bin. The data form a log-normal distribution with mean and median 5.5 and 7 days, respectively. (Modified from Benoit and McNutt, 1996.)

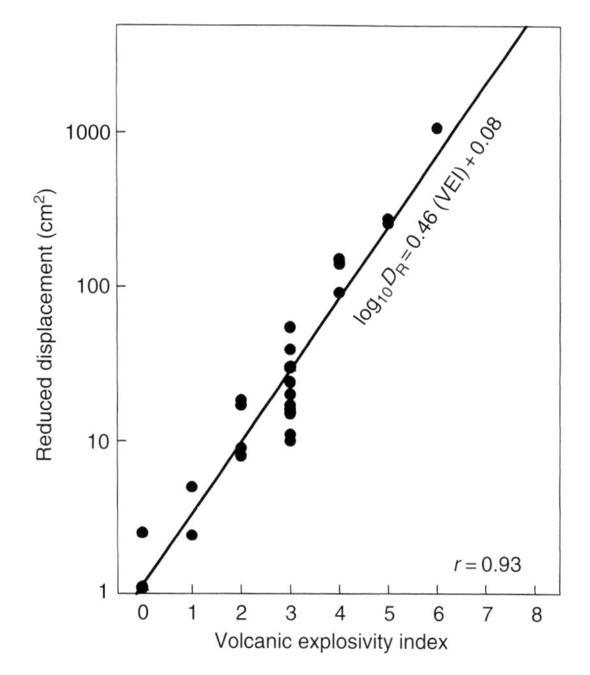

FIGURE 20 Volcanic tremor reduced displacement versus the Volcanic Explosivity Index (VEI) of Newhall and Self (1982). The plotted line is a regression fit to the data.

Reduced displacement is a method of normalizing volcanic tremor to a common scale. For body waves (e.g., *P* waves and *S* waves) the formula of Aki and Koyanagi (1981) is

$$D_R = \frac{A \cdot r}{2\sqrt{2}M} \tag{2}$$

For surface waves (e.g., Rayleigh waves, Love waves, and PL

waves) the formula of Fehler (1983) is

$$D_R = \frac{A\sqrt{\lambda r}}{2\sqrt{2}M} \qquad (3)$$

where A = amplitude in cm peak-to-peak; r = distance from source to seismic station in cm; M = seismograph magnification at the tremor frequency; and λ = tremor wavelength in cm. The factor $2\sqrt{2}$ is the root-mean-square (rms) amplitude correction.

Such information is helpful for estimating plume heights and tephra volumes. The ends of eruptions can be estimated as coinciding with the end of strong tremor associated with eruptions. In many cases tremor declines more or less exponentially on a timescale of minutes to tens of minutes and returns to background levels. In some cases, eruptions of ash end but weak tremor continues for days in association with hydrothermal activity.

The ends of eruption sequences, which may last months to years, are more difficult to determine than the ends of individual eruption pulses which last a few hours to days. Generally volcanic tremor ceases altogether and the number of earthquakes per unit time decreases regularly. Volcanic earthquake swarms decline with a power-law decay similar to aftershocks, except that the timescale is often longer.

7. Conclusions

This chapter has discussed many general features of earthquake activity at volcanoes, and it has been shown that the activity varies widely. This is true because the volcanoes and their eruptions vary, and also because the stresses differ, and volcanoes have different geometries as well as combinations of rock, magma, water, and gases within them. Although seismic data are plentiful, the complexities of the volcanic environment often make interpretations difficult. Different processes can generate similar seismograms, and similar processes can generate different seismograms, although detailed diagnostics can often distinguish between them. This keeps volcano seismology challenging, but also suggests that seismology alone can be a somewhat limited tool.

In spite of this, volcano seismology has helped to answer some fundamental questions: (1) Where is the magma? Tomography and b value anomalies have identified large (km scale) magmatic or near-magmatic structures, while low-frequency events and volcanic tremor are associated with small-scale movement of magma or water. (2) What is "normal" earthquake activity at volcanoes? The case studies and general features sections showed that activity varies widely, but with a number of systematic trends. Successful separation of source, site, and path effects of seismograms helps to elucidate the causes of specific activity. (3) What will happen next? Understanding and modeling of physical processes and study of many case histories are the two main tools used to evaluate this question. It is also possible, and indeed likely, that new high-quality data may lead to fresh insights about the new as well as older examples of earthquake activity at volcanoes.

References

Aki, K. and R.Y. Koyanagi (1981). Deep volcanic tremor and magma ascent mechanism under Kilauea, Hawaii. *J. Geophys. Res.* **86**, 7095–7110.

Alaska Volcano Observatory (1993). Mt. Spurr's 1992 eruptions. *EOS Trans. Am. Geophys. Union* **74**, 217, 221–222.

Benoit, J.P. (1998). Pattern and process in volcano seismology. PhD dissertation, University of Alaska Fairbanks, Fairbanks.

Benoit, J.P. and S.R. McNutt (1996a). Global volcanic earthquake swarm database 1979–1989. *US Geol. Surv. Open-File Report 96–69.*

Benoit, J.P. and S.R. McNutt (1996b). Global volcanic earthquake swarm database and preliminary analysis of volcanic earthquake swarm duration. *Ann. Geofis.* **XXXIX**, 221–229.

Benoit, J.P. and S.R. McNutt (1997). New constraints on source processes of volcanic tremor at Arenal Volcano, Costa Rica, using broadband seismic data. *Geophys. Res. Lett.* **24**, 449–452.

Benoit, J.P., G. Thompson, K. Lindquist, R. Hansen, and S.R. McNutt (1998). Near-real-time WWW-based monitoring of Alaskan volcanoes: the Iceweb system. *EOS Trans. Am. Geophys. Union* **79**(supplement), F957.

BGVN (*Bull. Global Volc. Network*) (1992). **17**, no. 3 (White Island).

BGVN (*Bull. Global Volc. Network*) (1993). **18**, no. 8 (Arenal).

BGVN (*Bull. Global Volc. Network*) (1994). **19**, no. 8 (Rabaul).

BGVN (*Bull. Global Volc. Network*) (1999). **24** (Arenal).

Brandsdottir, E. and P. Einarsson (1979). Seismic activity associated with the September 1977 deflation of the Krafla central volcano in northeast Iceland. *J. Volcanol. Geotherm. Res.* **6**, 197–212.

Chouet, B.A. (1992). A seismic model for the source of long-period events and harmonic tremor. In: "Volcanic Seismology" (P. Gasparini, R. Scarpa, and K. Aki, Eds.), pp. 133–156. Springer-Verlag, Berlin.

Chouet, B.A. (1996). Long-period volcano seismicity: its source and use in eruption forecasting. *Nature* **380**, 309–316.

Chouet, B., P. Dawson, G. De Luca, M. Martini, G. Milana, G. Saccorotti and R. Scarpa (1998). "Array Analyses of Seismic Wavefields Radiated by Eruptive Activity at Stromboli Volcano, Italy." CNR, Gruppo Nazionale per la Vulcanologia.

Eaton, J.P. and K.J. Murata (1960). How volcanoes grow. *Science* **132**, 925–938.

Eichelberger, J.C., W. Hildreth, and J.J. Papike (1991). The Katmai scientific drilling project, surface phase: investigation of an exceptional igneous system. *Geophys. Res. Lett.* **18**, 1513–1516.

Endo, E.T. and T. Murray (1991). Real-time Seismic Amplitude Measurement (RSAM): a volcano monitoring and prediction tool. *Bull. Volcanol.* **53**, 533–545.

Endo, E.T., S.D. Malone, L.L. Noson, and C.S. Weaver (1981). Locations, magnitudes, and statistics of the March 20–May 18 earthquake sequence. *US Geol. Surv. Prof. Paper 1250*, 93–107.

Ereditato, D. and G. Luongo (1994). Volcanic tremor wave field during quiescent and eruptive activity at Mt. Etna (Sicily). *J. Volcanol. Geotherm. Res.* **61**, 239–251.

Falsaperla, S., S. Graziani, G. Nunnari, and S. Spampinato (1996). Automatic classification of volcanic earthquakes by using multi-layered neural networks. *Nat. Hazards* **13**, 205–228.

Falsaperla, S., G. D'Amico, S. Di Prima, A. Pellegrino, L. Scuderi, and O. Torrisi (1997). "The New Permanent Broadband Seismic Network on Stromboli," CNR, Inst. Nazionale di Vulcanologia, Rpt. Tecnico 5/96.

Fehler, M. (1983). Observations of volcanic tremor at Mount St. Helens volcano. *J. Geophys. Res.* **88**, 3476–3484.

Ferrazzini, V. and K. Aki (1987). Slow waves trapped in a fluid-filled infinite crack: implication for volcanic tremor. *Geophys. Res. Lett.* **92**, 9215–9223.

Ferrucci, F. (1995). Seismic monitoring at active volcanoes. In: "Monitoring Active Volcanoes: Strategies, Procedure, and Techniques" (W.J. McGuire, C.R.J. Kilburn, and J.B. Murray, Eds.). University College London Press.

Fujita, E., Y. Ida, and J. Oikawa (1995). Eigen oscillation of a fluid sphere and source mechanism of harmonic volcanic tremor. *J. Volcanol. Geotherm. Res.* **69**, 365–378.

Garces, M.A. and R.A. Hansen (1998). Waveform analysis of seismoacoustic signals radiated during the fall 1996 eruption of Pavlof volcano, Alaska. *Geophys. Res. Lett.* **25**, 1051–1054.

Garces, M.A., M.T. Hegarty, and S.Y. Schwartz (1998). Magma acoustics and time-varying melt properties at Arenal Volcano, Costa Rica. *Geophys. Res. Lett.* **25**, 2293–2296.

Garces, M.A. and S.R. McNutt (1997). Theory of the airborne sound field generated in a resonant magma conduit. *J. Volcanol. Geotherm. Res.* **78**, 155–178.

Goldstein, P. and B. Chouet (1994). Array measurements and modeling of sources of shallow volcanic tremor at Kilauea Volcano, Hawaii. *J. Geophys. Res.* **99**, 2637–2652.

Gordeev, E.I. (1992). Modelling of volcanic tremor wave fields. *J. Volcanol. Geotherm. Res.* **51**, 145–160.

Gorelchik, V.I., V.M. Zobin, and P.I. Tokarev (1990). Volcanic earthquakes of Kamchatka: classification, nature of source and spatio-temporal distribution. *Tectonophysics* **180**, 255–271.

Gresta, S., A. Montalto, and G. Patane (1991). Volcanic tremor at Mount Etna (January 1984–March 1985): its relationship to the eruptive activity and modeling of the summit feeding system. *Bull. Volcanol.* **53**, 309–320.

Gronvold, K., G. Larsen, P. Einarsson, S. Thorarinsson and K. Saemundsson (1983). The Hekla eruption 1980–1981. *Bull. Volcanol.* **46**, 349–363.

Gudmundsson, A., N. Oskarsson, K. Gronvold, *et al.* (1992). The 1991 eruption of Hekla, Iceland. *Bull. Volcanol.* **54**, 238–246.

Harlow, D.H., J.A. Power, E.P. Laguera, G. Ambubuyog, R.A. White, and R.P. Hoblitt (1996). Precursory seismicity and forecasting of the June 15, 1991 eruption of Mount Pinatubo, Philippines. In: "Fire and Mud: Eruptions and Lahars of Mount Pinatubo, Philippines" (C.G. Newhall and R.S. Punongbayan, Eds.), pp. 285–305. University of Washington Press, Seattle.

Hegarty, M., S.Y. Schwartz, M. Protti, M. Garces, and T. Dixon (1997). Observations at Costa Rican volcano offer clues to causes of eruptions. *EOS Trans. Am. Geophys. Union* **78**, 565, 570–571.

Hill, D.P. (1996). Earthquakes and carbon dioxide beneath Mammoth Mountain, California. *Seismol. Res. Lett.* **67**, 8–15.

Hill, D.P., R.A. Bailey and A.S. Ryall (1985). Active tectonic and magmatic processes beneath Long Valley caldera, eastern California: an overview. *J. Geophys. Res.* **90**, 11111–11120.

Ida, Y. (1996). Cyclic fluid effusion accompanied by pressure change: implication for volcanic eruptions and tremor. *Geophys. Res. Lett.* **23**, 1457–1460.

Iguchi, M. (1994). A vertical expansion source model for the mechanisms of earthquakes originated in the magma conduit of an andesitic volcano: Sakurajima, Japan. *Bull. Volcanol. Soc. Jpn* **39**, 49–67.

Ishihara, K. (1985). Dynamical analysis of volcanic explosion. *J. Geodyn.* **3**, 327–349.

Iyer, H. (1992). Seismological detection and delineation of magma chambers: present status with emphasis on the western USA. In: "Volcanic Seismology" (P. Gasparini, R. Scarpa and K. Aki, Eds.), pp. 299–338. Springer-Verlag, Berlin.

Julian, B.R. (1994). Volcanic tremor: nonlinear excitation by fluid flow. *J. Geophys. Res.* **99**, 11859–11877.

Julian, B.R. and S.A. Sipkin (1985). Earthquake processes in the Long Valley caldera area, California. *J. Geophys. Res.* **90**, 11155–11169.

Julian, B.R., A.D. Miller, and G.R. Foulger (1997). Non-double-couple earthquake mechanisms at the Hengill-Grensdalur volcanic complex, southwest Iceland. *Geophys. Res. Lett.* **24**, 743–746.

Kanamori, H., J.W. Given, and T. Lay (1984). Analysis of seismic body waves excited by the Mount St. Helens eruption of May 18, 1980. *J. Geophys. Res.* **89**, 1856–1866.

Kanamori, H. and J. Mori (1992). Harmonic excitation of mantle Rayleigh waves by the 1991 eruption of Mount Pinatubo, Philippines. *Geophys. Res. Lett.* **19**, 721–724.

Kaneshima, S., H. Kawakatsu, H. Matsubayashi, *et al.* (1996). Mechanism of phreatic eruptions at Aso Volcano inferred from near-field broadband seismic observations. *Science* **273**, 642–645.

Kawakatsu, H., T. Ohminato, and H. Ito (1994). 10s-Period volcanic tremors observed over a wide area in southwestern Japan. *Geophys. Res. Lett.* **21**, 1963–1966.

Kawakatsu, H., T. Ohminato, H. Ito, *et al.* (1992). Broadband seismic observation at the Sakurajima volcano, Japan. *Geophys. Res. Lett.* **19**, 1959–1962.

Kedar, S., B. Sturtevant, and H. Kanamori (1996). The origin of harmonic tremor at Old Faithful geyser. *Nature* **379**, 708–711.

Keith, T.E.C. (Ed.) (1995). The 1992 Eruptions of Crater Peak Vent, Mount Spurr Volcano, Alaska. *US Geol. Survey Bull. 2139*.

Klein, F.W., R.Y. Koyanagi, J.S. Nakata, and W.R. Tanigawa (1987). The seismicity of Kilauea's magma system. *US Geol. Surv. Prof. Paper 1350*, 1019–1185.

Koyanagi, R., W.R. Tanigawa, and J.S. Nakata (1988). Seismicity associated with the eruption. In: "The Puu Oo Eruption of Kilauea Volcano, Hawaii: Episodes 1–20, 1983–84" (E. Wolfe, Ed.), pp. 183–235. *US Geol. Surv. Prof. Paper 1463*.

Kvaerna, T. and D.J. Doornbos (1986). An integrated approach to slowness analysis with arrays and three-component stations, Semiann. Tech. Summary, 1 October 1985–31 March 1986. *NORSAR Sci. Rept. No. 2-85/86*, Kjeller Norway.

Kvaerna, T. and F. Ringdal (1992). Integrated array and three-component processing using a seismic microarray. *Bull. Seismol. Soc. Am.* **82**, 870–882.

Langbein, J., D.P. Hill, T.N. Parker, and S.K. Wilkinson (1993). An episode of reinflation of the Long Valley caldera, eastern California: 1989–1991. *J. Geophys. Res.* **98**, 15851–15870.

Latter, J.M. (1979). "Volcanological Observations at Tongariro National Park. 2. Types and Classification of Volcanic Earthquakes." Department of Scientific and Industrial Research, Wellington, New Zealand.

Linde, A.T., K. Agustsson, I.S. Sacks, and R. Stefansson (1993). Mechanism of the 1991 eruption of Hekla from continuous borehole strain monitoring. *Nature* **365**, 737–740.

Malone S.D., E.T. Endo, C.S. Weaver, and J.W. Ramey (1981). Seismic monitoring for eruption prediction. *US Geol. Surv. Prof. Paper 1250*, 803–813.

McClelland, L., T. Simkin, M. Summers, E. Nielsen, and T.C. Stein (1989). "Global Volcanism 1975–1985." Smithsonian Institution, Washington, DC.

McNutt, S.R. (1994). Volcanic tremor amplitude correlated with volcano explosivity and its potential use in determining ash hazards to aviation. *US Geol. Surv. Bull. 2047*, 377–385.

McNutt, S.R. (1996). Seismic monitoring and eruption forecasting of volcanoes: a review of the state-of-the-art and case histories. In: "Monitoring and Mitigation of Volcano Hazards" (R. Scarpa and R. Tilling, Eds.), pp. 99–146. Springer-Verlag, Berlin.

McNutt, S.R. (1999). Volcanic seismicity. In: "Encyclopedia of Volcanoes" (H. Sigurdsson, B.F. Houghton, S.R. McNutt, H. Rymer and J. Stix, Eds.), pp. 1015–1033. Academic Press, San Diego.

McNutt, S.R., G. Tytgat, and J.A. Power (1995). Preliminary analyses of volcanic tremor associated with the 1992 eruptions of Crater Peak, Mt. Spurr, Alaska. In: "The 1992 Eruptions of Crater Peak Vent, Mount Spurr Volcano, Alaska" (T.E.C. Keith, Ed.), pp. 161–178. *US Geol. Surv. Bull. 2139*.

McNutt, S.R., J. Benoit, D. Christensen, *et al.* (1997). Broadband seismology at the Alaska Volcano Observatory, 1993–1997. *EOS Trans. Am. Geophys. Union* **78**(supplement) F429.

Melson, W.G. (1989). Las erupciones del Volcan Arenal 1 al 13 de April de 1989. *Bull. de Vulcan. (Heredia, Costa Rica)* **20**, 15–22.

Metaxian, J.-P., P. Lesage, and J. Dorel (1997). Permanent tremor of Masaya Volcano, Nicaragua: wave field analysis and source location. *J. Geophys. Res.* **102**, 22529–22545.

Miller, T.P., R.G. McGimsey, D.H. Richter, *et al.* (1998). Catalog of the historically active volcanoes of Alaska. *US Geol. Surv. Open-File Report 98-582*.

Minakami, T. (1974). Seismology of volcanoes in Japan. In: "Developments in Solid Earth Geophysics" (L. Civetta *et al.*, Eds), pp. 1–27. Physical Volcanology 6. Elsevier, Amsterdam.

Minakami, T., S. Utibori, and S. Hiraga (1969). The 1968 eruption of Volcano Arenal, Costa Rica. *Bull. Earthq. Res. Inst.* **47**, 783–802.

Mogi, K. (1962). Magnitude-frequency relation for elastic shocks accompanying fractures of various materials and some related problems in earthquakes. *Bull. Earthq. Res. Inst.* **40**, 831–853.

Munoz, F.A., M.L. Calvache, G.P. Cortes, *et al.* (1993). Galeras Volcano: International workshop and eruption. *EOS Trans. Am. Geophys. Union* **74**, 281, 286–287.

Nakamura, K. (1984). Three different mechanisms for elongate distribution of shallow earthquake epicenters. *US Geol. Surv. Open-File Report 84-939*, 482-492.

Nakano, M., H. Kumagai, M. Kumazawa, K. Yamaoka, and B.A. Chouet (1998). The excitation and characteristic frequency of the long-period volcanic event: an approach based on an inhomogeneous autoregressive model of a linear dynamic system. *J. Geophys. Res.* **103**, 10031–10046.

Neuberg, J., R. Luckett, M. Ripepe, and T. Braun (1994). Highlights from a seismic broadband array on Stromboli volcano. *Geophys. Res. Lett.* **21**, 749–752.

Neuberg, J., B. Baptie, R. Luckett, and R. Stewart (1998). Results from the broadband seismic network on Montserrat. *Geophys. Res. Lett.* **25**, 3661–3664.

Newhall, C.G. and E.T. Endo (1987). Sudden seismic calm before eruptions: illusory or real? Abstracts Volume, Hawaii Symposium on How Volcanoes Work, Hilo, Hawaii, p. 190.

Newhall, C.G. and R.S. Punongbayan (Eds.) (1996). "Fire and Mud: Eruptions and Lahars of Mount Pinatubo, Philippines." University of Washington Press, Seattle.

Newhall, C.G. and S. Self (1982). The Volcanic Explosivity Index (VEI): an estimate of explosive magnitude for historical volcanism. *J. Geophys. Res.* **87**, 1231–1238.

Nishimura, T. (1995). Source parameters of the volcanic eruption earthquakes at Mount Tokachi, Hokkaido, Japan, and a magma ascending model. *J. Geophys. Res.* **100**, 12465–12474.

Nishimura, T., H. Hamaguchi, and S. Ueki (1995). Source mechanisms of volcanic tremor and low-frequency earthquakes associated with the 1988–89 eruptive activity of Mt. Tokachi, Hokkaido, Japan. *Geophys. J. Intl.* **121**, 444–458.

Ohminato, T., B.A. Chouet, P. Dawson, and S. Kedar (1998). Waveform inversion of very long period impulsive signals associated with magmatic injection beneath Kilauea Volcano, Hawaii. *J. Geophys. Res.* **103**, 23839–23862.

Ohminato, T. and D. Ereditato (1997). Broadband seismic observations at Satsuma-Iwojima volcano, Japan. *Geophys. Res. Lett.* **24**, 2845–2848.

Ohta K., N. Matsuwo, and T. Yanagi (1992). The 1990–1992 eruption of Unzen volcano. In: "Unzen Volcano, the 1990–1992 Eruption" (T. Yanagi, H. Okada, and K. Ohta, Eds.), pp. 34–37. The Nishinippon & Kyushu University Press, Kyushu.

Oikawa, J., Y. Ida, and K. Yamaoka (1994). Source spectrum and source time function of volcanic tremor determined with a dense seismic network near the summit crater of Izu-Oshima volcano, Japan. *J. Geophys. Res.* **99**, 9523–9532.

Okada, Hm. (1983). Comparative study of earthquake swarms associated with major volcanic activities. In: "Arc Volcanism: Physics and Tectonics" (D. Shimozuru and I. Yokoyama, Eds.), pp. 43–61. Terra Scientific, Tokyo.

Pinatubo Volcano Observatory Team (1991). Lessons from a Major Eruption: Mt. Pinatubo, Philippines. *EOS Trans. Am. Geophys. Union* **72**, 545, 552–553, 555.

Power, J.A. (1988). Seismicity associated with the 1986 eruption of Augustine volcano, Alaska. MS thesis, University of Alaska Fairbanks.

Power, J.A., J.C. Lahr, R.A. Page, *et al.* (1994). Seismic evolution of the 1989–90 eruption sequence of Redoubt Volcano, Alaska. *J. Volcanol. Geotherm. Res.* **62**, 69–94.

Power, J.A., A.D. Jolly, R.A. Page, and S.R. McNutt (1995). Seismicity and forecasting of the 1992 eruptions of Crater Peak vent, Mount Spurr Volcano, Alaska: An overview. In: "The 1992

Eruptions of Crater Peak Vent, Mount Spurr Volcano, Alaska" (T.E.C. Keith, Ed.), pp. 149–159. *US Geol. Surv. Bull. 2139*.

Rowe, C.A., R.C. Aster, P.R. Kyle, J.W. Schlue, and R.R. Dibble (1998). Broadband recording of strombolian explosions and associated very-long period seismic signals on Mount Erebus volcano, Ross Island, Antarctica. *Geophys. Res. Lett.* **25**, 2297–2300.

Scholz, C.H. (1968) The frequency-magnitude relation for elastic shocks and its relation to earthquakes. *Bull. Seismol. Soc. Am.* **58**, 399–415.

Shimizu, H., S. Ueki, and J. Koyama (1987). A tensile-shear crack model for the mechanism of volcanic earthquakes. *Tectonophysics* **144**, 287–300.

Simkin, T. (1993). Terrestrial volcanism in space and time. *Annu. Rev. Earth Planet. Sci.* **21**, 427–452.

Swanson, D.A., T.J. Casadevall, D. Dzurisin, S.D. Malone, C.G. Newhall, and C.S. Weaver (1983). Predicting eruptions at Mount St. Helens, June 1980 through December 1982. *Science* **221**, 1369–1376.

Uhira, K. and M. Takeo (1994). The source of explosive eruptions of Sakurajima Volcano, Japan. *J. Geophys. Res.* **99**, 17775–17789.

Uhira, K., H. Yamasato, and M. Takeo (1994). Source mechanism of seismic waves excited by pyroclastic flows observed at Unzen Volcano, Japan. *J. Geophys. Res.* **99**, 17757–17773.

Ukawa, M. (1993). Excitation mechanism of large amplitude volcanic tremor associated with the 1989 Ito-oki submarine eruption, central Japan. *J. Volcanol. Geotherm. Res.* **55**, 33–50.

Ukawa, M. and H. Tsukahara (1996). Earthquake swarms and dike intrusions off the east coast of Izu Peninsula, central Japan. *Tectonophysics* **253**, 285–303.

Wallace, T.C. (1985). A reexamination of the moment tensor solutions of the 1980 Mammoth Lakes earthquakes. *J. Geophys. Res.* **90**, 11171–11176.

Warren, N.W. and G.V. Latham (1970). An experimental study of thermally induced microfracturing and its relation to volcanic seismicity. *J. Geophys. Res.* **75**, 4455–4464.

Weaver, C.S., W.C. Grant, S.D. Malone, and E.T. Endo (1981). Post-May 18 seismicity: volcanic and tectonic implications. *US Geol. Surv. Prof. Paper 1250*, 109–121.

Wiemer, S. and S.R. McNutt (1997). Variations in the frequency-magnitude distribution with depth in two volcanic areas: Mount St. Helens, Washington, and Mt. Spurr, Alaska. *Geophys. Res. Lett.* **24**, 189–192.

Wiemer, S., S.R. McNutt, and M. Wyss (1998). Temporal and three-dimensional spatial analyses of the frequency-magnitude distribution near Long Valley Caldera, California. *Geophys. J. Intl.* **134**, 409–421.

Wolfe, E.W. (Ed.) (1988). The Puu Oo eruption of Kilauea volcano, Hawaii: Episodes 1 through 20, January 3, 1983, through June 8, 1984. *US Geol. Surv. Prof. Paper 1463*.

Wolfe, E. (1992). The 1991 eruptions of Mount Pinatubo, Philippines. *Earthq. Volcanoes* **23**, 5–37.

WOVO (World Organization of Volcano Observatories) (1997). "Directory of Volcano Observatories 1996–1997." WOVO/UNESCO, Paris.

Wyss, M. (1973). Towards a physical understanding of the earthquake frequency distribution. *Geophys. J. R. Astrom. Soc.* **31**, 341–359.

26

Three-Dimensional Crustal *P*-Wave Imaging of Mauna Loa and Kilauea Volcanoes, Hawaii

H.M. Benz
US Geological Survey, Denver, Colorado, USA
P. Okubo
US Geological Survey, HVO, Hawaii, USA
A. Villaseñor
University of Utrecht, Utrecht, Netherlands

1. Introduction

One of the principal goals and accomplishments of the seismographic monitoring of volcanoes is the tracking of earthquake hypocenters and the identification and interpretation of seismicity patterns that herald impending volcanic unrest (e.g., Koyanagi *et al.*, 1988; Harlow *et al.*, 1996). Traditional volcano monitoring networks are built to locate precisely microearthquakes that are often associated with magma transport. As with microearthquake monitoring networks in tectonically active regions, the ability to register and measure precisely the arrival times of the seismic first arrivals is a basic requirement.

In addition to the precise determination of earthquake hypocenters, it has long been a practice to use earthquake arrival times as constraints for seismic wave speeds. Aki and Lee (1976) first used seismic first arrival times to infer a 3D distribution of seismic *P*-wave speeds in central California. They used *P* wave arrival times from local microearthquakes in Central California, recorded on the northern California regional seismographic network, to image a low-velocity anomaly between the San Andreas and subparallel Calaveras faults. Since their study, many investigators have applied seismic tomographic inverse techniques to image targets in different regions and at different scales. Reviews of seismic tomographic imaging techniques and results are found in Nolet (1987) and Iyer and Hirahara (1993).

As we strive toward developing physically reasonable models of volcanic processes, it becomes increasingly important to resolve and account for details of internal volcanic structure. Arrival-time tomography is becoming a commonly used technique for this purpose. It has been particularly successful in detailed imaging of large, complex volcanic systems, like Yellowstone (Benz and Smith, 1984), Hekla (Toomey and Foulger, 1989), Hawaii (Okubo *et al.*, 1997), and Mt. Etna (Villasenor *et al.*, 1998).

Chouet *et al.* (1994), Lahr *et al.* (1994), and Power *et al.* (1994) describe aspects of the seismicity and sequence of activity culminating in the series of eruptions of Mt. Redoubt Volcano, Alaska, between December, 1989 and April, 1990. Spatial and temporal distributions of long-period (LP) and volcano-tectonic (VT) earthquakes were incorporated into inferences regarding the path along which magma traveled and the physical processes taking place during the eruptive sequence. See Chapter 25 by McNutt for a complete description of LP and VT earthquakes. A 3D tomographic study of Redoubt Volcano using seismic *P* and *S* wave first-arrival times (Benz *et al.* 1996) allowed features observed in the seismicity distributions to be associated with anomalies in V_P, V_S, and, in particular, V_P/V_S. With their tomographic imaging results and those of the earlier studies, Benz *et al.* (1996) were able to propose a detailed dynamic model of the Mt. Redoubt eruptions.

This chapter aims to provide insights into some of the key issues involving the inversion of first arriving *P*-wave data for 3D velocity structure and earthquake locations as applied to volcanic systems. Emphasis is placed on the need for thoughtful parametrization of the velocity model and selection of data, comprehensive error and resolution analysis, and comparison of the results with independent observations and constraints. The reader is referred to Pavlis and Booker (1980), Lees (1992), and Benz *et al.* (1996) for discussions on detailed aspects of the problem, including the inverse formulation, error analysis, and applications to active volcanoes. A general overview of volcanic systems and seismological observations relevant to volcanic systems is provided by McNutt in Chapter 25.

We image the complex upper crustal structure beneath Mauna Loa and Kilauea and the south flank of Hawaii, using abundant well-located earthquakes recorded on the regional seismographic network operated by the Hawaiian Volcano Observatory (HVO) to invert first-arriving *P* phases to determine the 3D *P*-wave velocity structure and hypocenter locations. In this study, we use a robust and efficient method for calculating arrival times in structures with large local velocity variations (Podvin and Lecomte, 1991) and an efficient numerical inverse solver that incorporates smoothing constraints that enables us to optimally parameterize the subsurface structure.

2. Geological and Geophysical Framework for Hawaii

The proximity of active volcanism and faulting and large deformation rates makes the island of Hawaii an ideal place for studying both the dynamics of volcanism and earthquake generation. Located at the southeast end of the nearly 6000 km long Hawaiian-Emperor volcanic chain, Mauna Loa and Kilauea Volcanoes are formed by processes related to Hawaiian hot-spot tectonics (Clague and Dalrymple, 1987). Mantle-derived melts produced by the Hawaiian hot-spot migrate upward in the crust, where they erupt either in the summit calderas or along the east or southwest rifts forming these large, tholeittic shield volcanoes. Buildup of the volcanic shield through episodic volcanism produces crustal deformation that is typically accommodated in gravitationally unstable normal-fault zones along the flank of the volcano. These fault zones periodically fail catastrophically, producing large submarine landslides (Moore, 1964; Moore *et al.*, 1989).

More than 30 years of detailed geophysical investigations have been conducted on the island of Hawaii in order to understand the origins and dynamics of oceanic volcanism. Starting with the first one-dimensional velocity model for the island of Hawaii (Eaton, 1962) seismic investigations have progressed through two-dimensional studies (e.g., Hill and Zucca, 1987) to 3D images of the crust and mantle (Ellsworth and Koyanagi, 1977) or detailed crustal imaging of the Kilauea volcanic system (Thurber, 1984) and the south flank of Hawaii (Rowan and Clayton, 1993).

Much of this previous work has focused on understanding the relationship between subsurface velocity structure and volcanism. However, volcanism and island tectonics, primarily observed in the Kaoiki and Hilina fault systems, are intimately linked as indicated by the proximity of active faulting, volcanism, and deformation. Swanson *et al.* (1976) demonstrate from geologic and geodetic observations that the entire south flank of Kilauea is moving upward and southward due to episodic injection of magma into the rift system. Along the south flank of Kilauea, the Hilina fault system forms as a gravitational slump feature that is reacting to movement within and around, and deformation of, the volcanic system.

Mauna Loa and Kilauea are the two active volcanoes on the island of Hawaii, with Mauna Loa last erupting in 1984 while Kilauea has been in continuous eruption since 1983. Mauna Loa rises to an elevation of more than 4000 m, while Kilauea is situated along the southeast flank of Mauna Loa at an elevation of 1240 m. Both Mauna Loa and Kilauea are characterized by similar structural features; summit calderas, rift zones that extend away from the summits, and long linear zones of steeply dipping normal faults along the flanks of the volcano.

While noted for its prolific volcanism, the island of Hawaii experiences large earthquakes, as evident from the instrumentally recorded 1951 *M* 6.9 Kona earthquake, the 1975 *M* 7.2 Kalapana earthquake, the 1983 *M* 6.6 Kaoiki earthquake, and the 1989 *M* 6.1 Kalapana earthquake, all of which have been modeled and interpreted in terms of predominant seismic moment release along a subhorizontal faulting surface at or near the basal decollement (Ando, 1979; Eissler and Kanamori, 1987; Beisser *et al.*, 1994). How crustal deformation is accommodated between the basal decollement and the Kaoiki and Hilina fault zones is a topic of continuing debate (Swanson *et al.*, 1976; Lipman *et al.*, 1985). Lipman *et al.* (1985) argue that normal faults along the flanks of Mauna Loa and Kilauea extend through the crust to the basal decollement, while Swanson *et al.* (1976) argue that they are listric features that penetrate only a few kilometers of the crust.

We extend these studies by providing additional constraints on the 3D velocity structure, volcanism, and faulting. The 3D velocity results from this study are critical for constraining the location and size of volcanic sources that produce crustal deformation, for understanding the kinematics of magma migration and emplacement, and for providing insights into the interaction between the volcanoes and the development of the Kaoiki and Hilina fault zones. With more complete descriptions of previous geophysical and geological studies of the island of Hawaii available, for example, in Decker *et al.* (1987) and Tilling and Dvorak (1993), we focus our discussions more specifically on structural interpretations.

3. Seismicity and Data

A short-period seismic network that now consists of 52 stations has been continuously monitoring seismicity on the island of Hawaii since the early 1960s (Klein *et al.*, 1987). The majority of stations are located in areas of active volcanism or along the south flank of the island (Fig. 1). Most earthquakes occur beneath Kilauea, along the southwest and east rift zones, and within the Kaoiki and Hilina fault zones. Well-located earthquakes used in this study were selected from the seismicity catalog in the period 1986 to 1992 (Fig. 1). The distribution

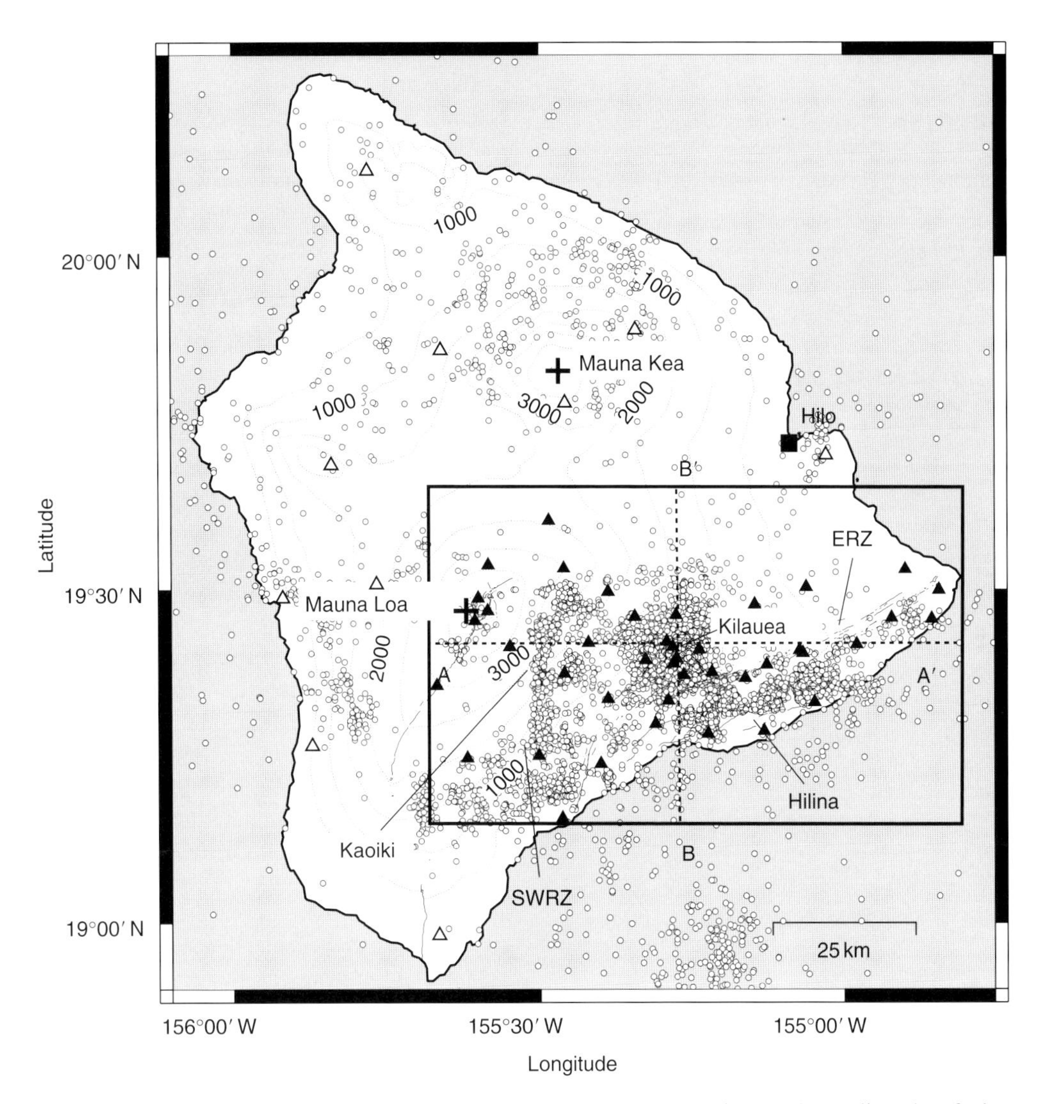

FIGURE 1 Map of the island of Hawaii showing the study region (outlined by heavy line), key faults, vents, and calderas. Seismic stations used in the tomography are shown as filled triangles, while other permanent seismic stations are shown as open triangles. Also shown are the epicenters of the 5500 *M* 2.0 or larger earthquakes (open circles) located by the Hawaiian Volcano Observatory seismic network between 1986 and 1992. Elevation contour interval is 500 m.

of seismicity for this period is similar to the areal distribution of earthquakes located since the beginning of modern network operations in the early 1960s. We selected only earthquakes with a coda duration magnitude of 2.0 or larger, which are generally well-recorded over a large portion of the seismic network and have smaller timing errors due to the impulsive nature of the first *P* arrival. From a catalog of more than 5500 earthquakes *M* 2.0 or larger, we selected 4754 well-located volcano-tectonic earthquakes recorded by 42 stations that lie within the study area shown in Figure 1. To ensure that earthquake swarms in the vicinity of the volcanoes did not dominate the selection of events, we sorted earthquakes so that no two earthquakes were within 0.5 km of each other. Shown in Figure 2 are east–west (A–A′) and north–south (B–B′) cross sections of selected earthquakes in the study area.

Figure 2 shows that seismicity occurs to depths of about 40 km beneath Kilauea, but seismicity is not well distributed below about 20 km. Seismicity beneath Kilauea (distance = 42 km, A–A′; distance = 30 km, B–B′) is well distributed over about a 10 km wide zone to depths of 20 km, while seismicity in the Kaoiki (distance = 20 km, A–A′) and Hilina (distance = 18 km, B–B′) occurs over a depth range of about 8–12 km. The areal and depth distributions of earthquakes are sufficient to adequately image the subsurface velocity structure to depths of 12–15 km beneath the south flank of Kilauea.

For our study area, the seismic stations are roughly 6 km apart with a station spacing of about 3 km within the summit region of Kilauea (Fig. 1). Earthquakes are well distributed to depths of about 10 km beneath the southwest and east rift zone

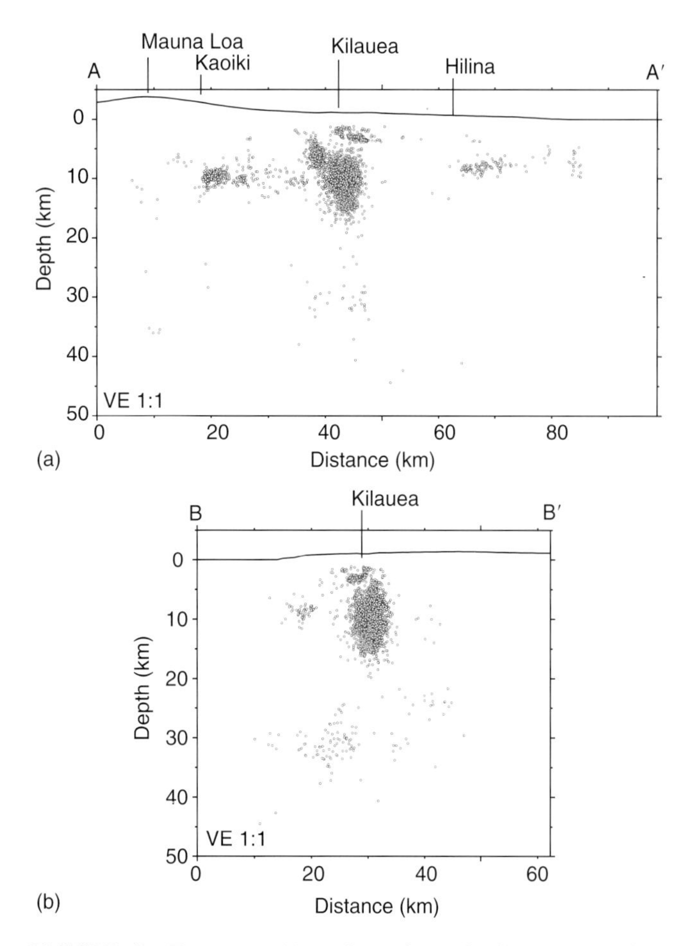

FIGURE 2 East–west (a) and north–south (b) cross sections showing the depth distribution of volcano-tectonic earthquakes used in this study. Only earthquakes that lie within ±5 km of the cross-sections are plotted. See Figure 1 for location of cross sections.

of Kilauea and within the Kaoiki fault zone, and south of the Hilina fault zone (Fig. 2). Deeper seismicity to depths of 20 km within the study area is observed directly beneath Kilauea caldera (Fig. 2). Several steps were taken in our selection of the earthquakes to minimize model errors produced by errors in the observed data. We ensured that the earthquakes were well located by selecting events with a ratio of the earthquake focal depth to the closest station greater than 1.5, an azimuthal gap of less than 180°, and an initial arrival time RMS of less than 0.5 sec.

Using earthquakes with picking weights of 2 or less and recorded by 10 or more stations minimizes errors in the velocity model reconstruction due to poor earthquake locations and picking errors. The quality of the first arrival-time pick is determined using an analyst-defined integer weight between 0 and 4, with 0 being the highest-quality pick. The poorer-quality observations with a weight of 1 or 2 represent a small percentage of the total data used in our study. The distribution of arrival time observations by weight is 78%, 14%, and 8% for weights of 0, 1, and 2, respectively.

Topography in the study region ranges from near sea level to more than 4000 m above sea level, with station elevation ranging from near sea level to 4104 m. In our analyses, the top of the model is 5000 m above sea level and we adopt the convention that negative depths are above sea level and positive depths are below sea level.

4. Formulation and Implementation of the Inverse Problem

Vidale (1988) and Podvin and Lecomte (1991) developed finite-difference techniques that accurately and efficiently compute travel times in media with large, local velocity variations, a necessary requirement for determining the detailed 3D velocity structure and hypocenter locations in complex regions like Hawaii. Benz *et al.* (1996) incorporated the finite-difference formulation of Podvin and Lecomte (1991) into a procedure that simultaneously inverts for 3D velocity structure and hypocenter locations. The reader is referred to Benz *et al.* (1996) for details on the formulation and implementation of the forward modeling and inversion algorithms used in this study.

Simultaneous inversion for velocity structure and hypocenter locations is computationally expensive, while decoupled inversion for the slowness perturbations and hypocenter locations is easily and efficiently implemented. For our application, we solved both the coupled and decoupled velocity structure–hypocenter location problem and found no discernible differences in the final 3D velocity models and hypocenter locations on the scales that we will interpret. Each application will dictate whether such an approach is justified. In either case, the system of equations used to compute the perturbations to the slowness model are solved using an efficient LSQR inverse solver (Paige and Saunders, 1982). Smoothness constraints are added to the system of equations to control the degree of model roughness. Tests are done to determine the appropriate weighting for the smoothness constraints that produce relatively smooth velocity models without severely affecting the minimization of the data misfit.

The hypocenter partial derivatives needed to determine the perturbations to the hypocenter location are computed using a standard approach (Benz and Smith, 1984), while the earthquake hypocenters are determined using an SVD algorithm and damped least squares. Both the slowness model and hypocenters are updated and the procedure is repeated until there is no appreciable reduction of the RMS of the travel time residuals with additional iterations.

Inversion for the velocity structure and hypocenters requires a starting velocity model and initial earthquake locations. In our application to the island of Hawaii, the velocity model is parameterized as constant-velocity cells that are $2 \times 2 \times 1$ km in the horizontal and vertical directions, respectively. Travel times are calculated on a finer 0.5 km square grid of cells, interpolated from the inversion grid, in order to maintain a travel time precision better than ± 0.05 sec. The total number of inversion cells in the model is $46 \times 28 \times 26$ in the east–west, north–south, and vertical directions, respectively. Rays are traced from source to receiver through the travel time grid by back projection along the isochrons using a procedure developed by Hole (1992). Station elevations are correctly modeled by locating the station at its proper location in the model.

4.1 Model Resolution and Error Estimation

The geometry of sources and receivers, quality of observations, and parametrization of the model all contribute to errors in the reconstruction of the subsurface velocity structure. Errors due to model parameterization include inappropriateness of the starting velocity model, coarseness of model grid spacing, accuracy of the raytracing, and corrections for topographic effects like station elevation. Numerous tests are done to find an appropriate combination of sources and receivers and model parameters that produce a robust final 3D velocity structure. It is important to find an optimal inversion grid spacing and starting velocity model. Tests are also necessary to determine how changes in the model parameters translate into model uncertainty. The LSQR solver used here does not allow for the direct computation of formal resolution and model uncertainty. Consequently, resolution analysis is done empirically by recovering a checkerboard input velocity model using synthetic arrival time data that simulates the source and receiver geometry and estimated data error. We provide examples of various tests that were done to insure a robust 3D velocity model of the south flank of the island of Hawaii.

Model resolution is strongly affected by the quality of ray sampling throughout the velocity model. The degree of ray sampling is controlled primarily by the geometry of the sources and receivers and grid spacing of the velocity model. Identifying what portions of the model are well-sampled and how well structures can be resolved throughout the model space is achieved by inversion of synthetic arrival time data based on a checkerboard velocity model. A checkerboard velocity model used here is defined as alternating high- and low-velocity bodies that are 5 km on a side with velocities varying by $\pm 10\%$ relative to a 6.0 km sec^{-1} reference velocity model. Velocities are varied smoothly between the high- and low-velocity zones so that the peak amplitudes of the anomalies are 3.0 km wide. Synthetic arrival times are computed for the checkerboard velocity using the observed source and receiver geometry. The checkerboard is then reconstructed by inverting the synthetic *P*-wave arrival times starting with a homogeneous velocity model of 6.0 km sec^{-1}.

Shown in Color Plate 7 are comparisons of the reference checkerboard model and velocity reconstructions at selected depths and along an east–west cross section. The plate shows that for the 7–8 km depth about 90% or more of the absolute amplitude of the high- and low-velocity anomalies is recovered. For depths shallower than 2 km and deeper than about 12 km, results are best beneath Kilauea caldera, the Kaoiki fault zone, and portions of the southwest and east rift zones. The cross section in the plate shows that the checkerboard model is reproduced to depths of 15 km beneath the south flank of the island of Hawaii. The cross section also shows the effects of smearing along the edges of the model and at deeper depths where there are fewer rays that generally travel in one direction. Smearing causes anomalies to be elongated along the direction of the ray. Results show that for depths of 2–12 km, where most of the seismicity is concentrated, features with wavelengths of about 3–5 km can be imaged for much of the model space. Recovery of the model degrades for shallower (< 2 km) and deeper depths (> 12 km) due to inadequate ray sampling. The exception being beneath Kilauea caldera where sources extend to depths of 20 km.

Another form of synthetic testing is reconstruction of individual synthetic anomalies located at critical positions within the model space. Reconstruction of individual anomalies illustrates the degree of resolution at specific depths and the effects of model smoothing and ray smearing of relatively small model features. In the example shown in Figure 3, a small $2.5 \times 2.5 \times 2.5$ km high-velocity body (10% relative to a reference velocity of 6.0 km sec^{-1}) is centered at a depth of 4.5 km beneath Kilauea caldera. Synthetic arrival times are calculated for this velocity model and the observed source and receiver geometry and then inverted using a homogeneous starting velocity model of 6.0 km sec^{-1} and an inversion and raytracing grid spacing of 0.5 km. The square anomaly shown in Figure 3 was placed at the approximate location of an unusually high-velocity body ($\sim$8.0 km sec^{-1}) with comparable dimensions imaged by Thurber (1984) and Rowan and Clayton (1993).

Figure 3 shows that the lateral and vertical dimensions of the anomaly, which is half the dimension of the anomalies in the checkerboard model, are well recovered in the inversion results. Approximately 70% of the anomaly's amplitude is recovered in the synthetic test. The absolute amplitude of the recovered anomaly will be strongly affected by the degree of model smoothing applied in the inversion and resolvability of the anomaly. For the case shown here, which applies to the observed data, smaller scale features at upper-crustal depths beneath Kilauea are well-resolved and the smoothing used is optimal for recovering most of the anomaly's amplitude without appreciably blurring the anomaly with the background velocities. Recovery of the anomaly's size and shape indicates negligible ray smearing at these depths. These results, combined with the checkerboard test shown in Color Plate 7, demonstrate that small-scale features on the order of a few kilometers in dimension are recoverable in the upper

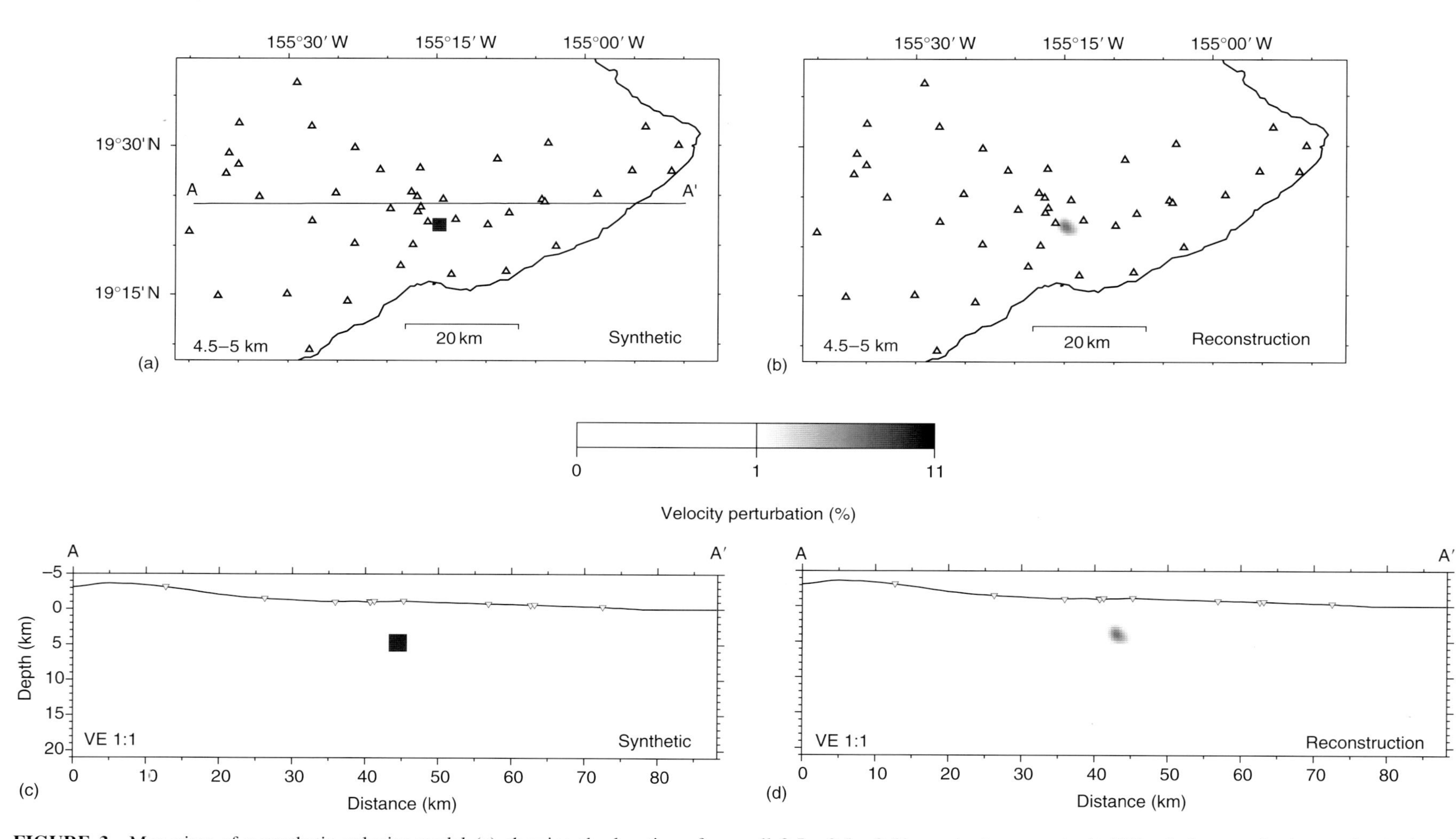

FIGURE 3 Map view of a synthetic velocity model (a) showing the location of a small $2.5 \times 2.5 \times 2.5$ km velocity anomaly (+10% relative to a background velocity of 6.0 km sec^{-1}) and the synthetic reconstruction of the anomaly (b). East–west cross section A–A′ of the synthetic velocity model (c) and the synthetic reconstruction of the anomaly (d). See the text for a description of the generation of the checkerboard velocity model and calculation of the synthetic arrival times.

12–15 km of the crust with the existing seismic network and phase data.

The initial one-dimensional velocity model used in the inversion of the observed data is similar to that used by the HVO in its routine earthquake location procedures (Fig. 4). HVO's velocity model was determined by one-dimensional inversion of arrival time data for earthquakes and stations in the vicinity of Kilauea (Klein, 1981). For depths shallower than about 2 km, the one-dimensional velocity model used in this study is faster than the standard location model, but it has about the same gradient and velocity between 5 and 17 km depth. Our higher upper-crustal velocity model probably reflects higher average velocities in the shallow crust for the study region as compared to shallow upper-crustal velocities in the vicinity of Kilauea. Shown in Figure 4 are different one-dimensional velocity models that we tested for use in this study. These models ranged from the velocity model of Klein (1981) with very slow, shallow velocities and a high-velocity gradient in the upper crust (thin, solid line; Fig. 4) to modified versions of the HVO model. The modified HVO models (dashed lines; Fig. 4) differ primarily by having different upper-crustal (< 5 km depth) velocities and gradients, but general follow the HVO model for depths greater than 5 km. While our velocity model (thick, solid line; Fig. 4) had the smallest initial arrival time rms, results based on the all of the models produced comparable final arrival time rms and similar velocity variations within the study area.

Previous tomographic studies of Kilauea (Thurber, 1984; Rowan and Clayton, 1993; Okubo *et al.*, 1997) have imaged an extremely high-velocity anomaly (~7.8–8.0 km sec^{-1}) at a depth of 5 km beneath Kilauea caldera. Our 3D velocity models based on each of the starting one-dimensional velocity models shown in Figure 4 all show the same feature. In addition, the absolute velocity of the anomaly varies by less than 0.2 km sec^{-1} between models. Within the upper 12 km of the crust, larger anomalies that will be discussed later in this chapter are observed in all inversions based of the starting one-dimensional velocity models shown in Figure 4. These results, combined with the synthetic modeling (Color Plate 7 and Fig. 3) and other tests results (not shown here) using different smoothing parameters and grid spacing, indicate that model uncertainties are ±0.2 km sec^{-1} for depths between about 2 and 12 km and increase to ±0.3 km sec^{-1} for deeper depths. Uncertainties in the velocities at shallower depths (< 2.0 km sec^{-1}) are difficult to quantify due to the poor sampling and large variations in near-receiver velocity structure.

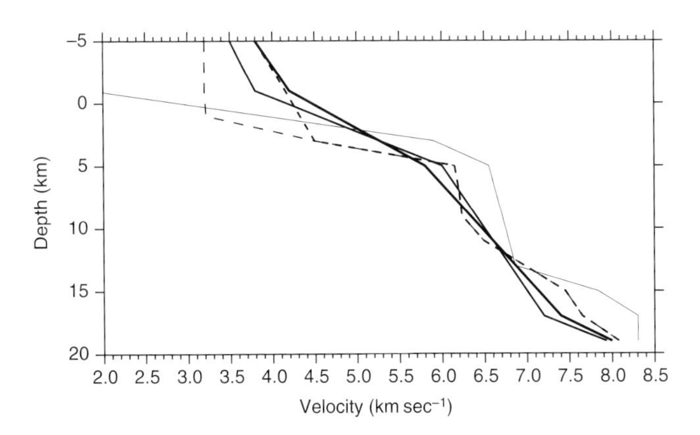

FIGURE 4 The initial one-dimensional *P*-wave velocity model used in the 3D velocity inversion (bold, solid line) and other one-dimensional *P*-wave velocity models tested for use in the inversion.

5. Three-Dimensional *P*-Wave Velocity Model for Hawaii

After 12 iterations, tomographic inversion for the 3D velocity structure resulted in a 77% reduction in the arrival time RMS from a starting residual of 0.39 sec. Shown in Color Plate 8 are map views of the 3D *P*-wave velocity structure at selected depths between 1 and 15 km. Layers shown in the plate are 1 km thick, and cells not sampled by any ray are shown as white. For much of the study area, the resulting 3D velocity structure is consistent with the known geology and previous crustal imaging studies (Hill and Zucca, 1987; Thurber, 1984; Rowan and Clayton, 1993). Development of the Hawaiian shield is primarily from basaltic magmatism; consequently, compositional differences in the rock types are small. Differences in the constitutive properties of the rock and observed *P*-wave velocities are primarily due to hydrothermal alteration, variations in temperature and changes in the porosity or density of microcracks. We will follow the example of Hill and Zucca (1987) and use their table of rock velocities versus rock type in describing the subsurface velocity structure.

The most prominent upper crustal velocity anomaly observed in the inversion results is a northeast-trending high-velocity body that ranges from 5.8 km sec^{-1} at 3 km depth to 6.8 km sec^{-1} at 7 km depth beneath the south flank of Mauna Loa (Color Plate 8). Adjacent to and southeast of this high-velocity region, the Kaoiki fault zone is observed as an approximately 14 km wide zone of low velocities that range from about 3.8 km sec^{-1} at 0 km depth to 5.6 km sec^{-1} at a depth of 5 km. High upper-crustal velocities along the southeast flank of Mauna Loa are consistent with intruded pyroxene gabbros or sheeted-dikes, while relative low velocities beneath the Kaoiki fault zone probably represent highly faulted shallow crust. Prior to the inception of Kilauea, Lipman (1980) argues, the Kaoiki fault zone formed as a gravitational slump feature on the southeast flank of Mauna Loa as the shield grew. High rates of microseismicity within the Kaoiki fault zone and the *M* 6.6 1983 Kaoiki earthquake suggest a continuing development of the fault system.

The velocity structure beneath Kilauea summit is best described as highly complex with velocities ranging from 6.0 km sec^{-1} at 3 km depth to 7.2 km sec^{-1} at a depth of 11 km. Within this region, velocities reach as high as 7.7 km sec^{-1} at a depth of 5 km beneath the south rim of the caldera. Thurber (1984) observed velocities near 8.0 km sec^{-1} at similar depths in his local tomography study of Kilauea. Such high velocities at shallow depths are likely ultramafic cumulates at the core of the conduit system. From about 12 km depth to the bottom of our model at 20 km, a roughly cylindrical relative low-velocity anomaly is observed beneath Kilauea that broadens with depth. This is illustrated at 15 km depth (Color Plate 8) where a velocity of 6.8 km sec^{-1} beneath Kilauea is about 6% slow relative to average layer velocities of 7.2 km sec^{-1}. These slower upper-mantle velocities likely reflect the influence of local, higher temperatures associated with the hot-spot and not compositional or porosity differences that are more prominent at shallow depths. Beneath Kilauea, the Moho is estimated at about 15 km depth and rocks at this depth are likely ultramafic cumulates (Hill and Zucca, 1987). A local zone of high temperatures due to the presence of a mantle hot-spot will produce locally slower upper-mantle/lower-crustal velocities.

Both Thurber (1984) and Rowan and Clayton (1993) imaged a small, shallow magma reservoir beneath the Kilauea summit caldera at depths between 0 and 2 km. Our results reveal velocities of 4.6–4.8 km sec^{-1} at similar depths beneath the summit, which are only slightly slower than velocities of 5.0–5.2 km sec^{-1} outside of the caldera region. Our results are not conclusive in defining a shallow low-velocity magma reservoir. It is likely that the smoothing constraints used in the inversion attenuate small low-velocity zones that are on the order of one or two inversion cell dimensions. Test results showed that we were able to reconstruct shallow low-velocity anomalies in the vicinity of Kilauea to within ± 0.2 km sec^{-1} with wavelengths of about 3–5 km. The shallow magma reservoir beneath Kilauea is smaller than we are able to resolve with the present distribution of permanent stations. In 1995, approximately 100 temporary seismic stations were deployed around Kilauea summit (McNutt *et al.*, 1997) in order to image the shallow structure beneath Kilauea caldera. High-resolution seismic tomography results from that experiment (Dawson *et al.*, 1999) reveal a small, shallow magma reservoir beneath Kilauea caldera.

The southwest rift zone is underlain by high-velocity material between 1 and 9 km depth, where velocities range from 5.4 to 7.0 km sec^{-1}. High velocities are uniformly distributed along the length of the rift zone in the depth range 1 to 5 km, but lie primarily between Puu Koae and Mauna Iki at depths of 7 9 km. The zone of high velocities broadens laterally away from the rift zone with increasing depth from a few kilometers, near the surface, to approximately 12 km wide at depths greater than 5 km. A widening rift zone with depth is similar to that observed by Hill and Zucca (1987), who modeled the rift zones using refraction travel time data as a wedge of high-velocity material, which they interpreted as pyroxene or olivine gabbro sheeted-dike complexes. Beneath the east rift zone, high velocities of 5.6 to 6.8 km sec^{-1} are observed in the depth range from 1 to 8 km, respectively. For much of the east rift zone, high velocities are only observed at depths shallower than about 4 or 5 km. For the depth range 5 to 7 km, high velocities are observed only beneath Makaopuhi. Unlike the southwest rift zone, relative low velocities of 5.8 km sec^{-1} are observed beneath the east rift zone in the depth range 8–11 km between Kalalua and Heiheiahulu.

High velocities between 6.8 and 7.0 km sec^{-1} are observed in the depth range 4–6 km between the southern edge of Kilauea caldera and the Koae fault zone. The Koae fault zone is described by Swanson *et al.* (1976) as an east–west trending, tearaway zone connecting the rift zones that separates the south flank from the volcanic edifice. Eruptions rarely occur within the Koae fault zone, but the 3D velocity results do suggest that dike emplacement occurs in the fault zone as seen by high velocities of about 6.5 km sec^{-1} at upper-crustal depths of 4–6 km. This is consistent with observed magma injection into the Koae fault zone following the 1973 east rift zone eruption, where spreading of the rift zone produced dilation in the Koae fault zone and subsequent dike-intrusion (Koyanagi *et al.*, 1973). An upper-crustal dike complex beneath and north of the Koae fault zone is consistent with the deformation and modeling results of Borgia (1994) and Clague and Denlinger (1994), who argue that a shallow ductile zone is required beneath the Koae fault zone to explain in part south flank deformation observations.

Complex, high velocities observed in the upper 8–9 km of the crust beneath the rift zones and Kilauea and a simpler, roughly cylindrical-shaped anomaly below Kilauea at depths greater than about 12 km, are consistent with the mechanical/structural model of Ryan (1988). Ryan (1988) describes the magma system as a conduit-like feature that raises from the mantle to shallow depths beneath Kilauea. Above about 10 km depth, the magma system spreads out laterally, becoming a zone of sheeted-dike complexes beneath the rift zones with the main conduit remaining beneath Kilauea. Seismicity reveals the conduit to remain nearly vertical beneath Kilauea to depths of 20 km, where it then broadens and dips gently southward in the upper mantle at depths of 20–60 km (Klein *et al.*, 1987).

The largest low-velocity feature observed in the tomographic study is beneath and south of the Hilina fault zone where velocities range from about 4.5 km sec^{-1}, near to the surface, to 5.8 km sec^{-1} at 9 km depth. Interpretation of deformation, bathymetry, and seismicity observations by Denlinger and Okubo (1995) argue that the Hilina fault system must be a through-going feature connected to the basal decollement in order to explain the scales of deformation of the flank and subaerial tectonics. These low velocities are likely due to increased abundance of microcracks from active

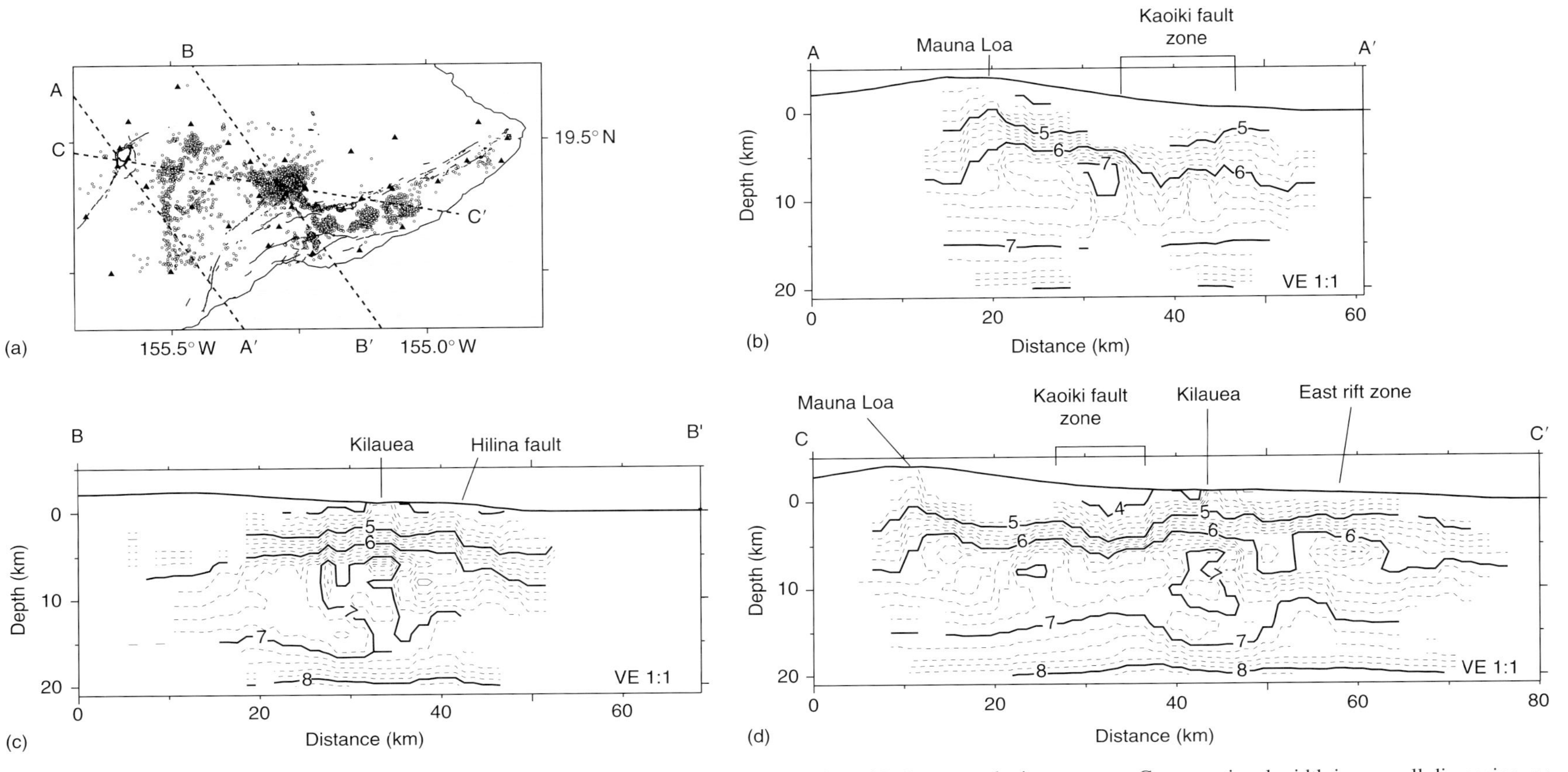

FIGURE 5 (a) Map showing the location of selected cross sections (b–d). Cross sections of the 3D *P*-wave velocity structure. Cross-sectional width is one cell dimension, or 2.0 km. Velocity contours are in km sec^{-1}.

faulting and slumping of the entire volume from the surface to the basal decollement. Laboratory studies (Christensen, 1979) have shown that increased microcracking strongly reduces *P*-wave velocities in the upper crust. Below approximately 11 km depth, velocities beneath the Hilina fault zone do not appear to be anomalous. The south flank velocity variations suggest that the Hilina fault zone penetrates the entire volcanic pile down to, but not below the basal decollement, thought to be at 9–10 km depth.

The complex velocity structure beneath the south flank of the island of Hawaii is well illustrated by selected cross sections through our 3D *P*-wave velocity model (Fig. 5). Cross-section A–A′ shows the velocity structure along a line that cuts through Mauna Loa and the Kaoiki fault zone. High velocities of 7.0 km sec^{-1} are observed beneath the south flank of Mauna Loa in cross section A–A′, while lower upper crustal velocities are observed to depths of 6 km beneath the Kaoiki fault zone. The complexity in the velocity structure beneath Kilauea is clearly seen in cross section B–B′ (distance = 33 km). Velocities as high as 7.7 km sec^{-1} are observed at depths between 5 and 6 km, with most of the structure beneath Kilauea defined by velocities higher than about 6.8 km sec^{-1} between depths of 5 and 12 km. Southward of the Hilina fault zone, velocities are offset downward to depths of about 9 km, showing the depth extent of faulting and direction of fault motion. Cross section C–C′ shows the high velocities of 7.0 km sec^{-1} beneath the flank of Mauna Loa (distance = 24 km), complex high-velocity structure beneath Kilauea (distance = 44 km), and high velocities of 6.6 km sec^{-1} beneath the east rift zone (distance = 58 km).

6. Relocated Seismicity

The average change in the hypocentral location of the 4796 earthquakes relocated in the 3D velocity model is 0.5 km horizontally and −1.3 km vertically. Changes in origin-time average +0.1 sec. Few significant differences are observed between the initial and relocated seismicity for depths less than 5 km (Fig. 6). The relocated seismicity at these depths does not provide any clearer definition of shallow structures. Map views of the initial and relocated seismicity at depths between 5 and 13 km (Fig. 6) show that seismicity between Mauna Loa and Kilauea concentrates in tighter clusters and shows a narrower distribution laterally. Relocated seismicity south of the east rift zone separate into more discrete zones of seismicity, presumably reflecting zones of stress concentration near centers of magma emplacement. For depths greater than 13 km, the initial and relocated seismicity is generally diffuse beneath Kilauea caldera.

Comparison of initial and relocated seismicity (Fig. 7) along a NW–SE cross section centered on Kilauea shows the relocated seismicity beneath Kilauea to be more broadly distributed from about 5 to 20 km depth. This zone of seismicity is also systematically shallower. Generally, the relocated seismicity at shallow depths (−1 to 5 km) halos a relative low-velocity region, which is the preferred location of a shallow magma chamber (Thurber, 1984). Rapid changes in the space–time seismicity patterns in Hawaii (Klein *et al.*, 1987) and the relatively minor spatial changes seen in the relocated seismicity (Figs. 6 and 7) make it difficult for us to use our results to better understand the seismo-volcanic framework of Hawaii. Detailed analysis of the space–time variations of the entire relocated seismicity catalog is necessary for providing a better understanding of Hawaii seismicity and its relationship to active volcanism, which is beyond the scope of this chapter. Importantly, relocating the seismicity in this study is critical for minimizing systematic errors in the observed data due to earthquake mislocations.

7. Discussion

A strong correlation is observed between the 3D *P*-wave velocity structure and the observed magnetic intensity (Hildenbrand *et al.*, 1993) and gravity (Kinoshita *et al.*, 1963) of the island of Hawaii. The southwest and east rift zones, which are underlain at shallow depths by high velocities in the tomography results, are observed as zones of higher magnetic intensity and higher gravity. Hildenbrand *et al.* (1993) interpret high magnetic intensity anomalies in the rift zones as due to shallow dike complexes with magnetization highest at the core of the rift and decreasing laterally away from the rift due to hydrothermal alteration of the intruded material. High velocities in the rift zones suggest that the rocks are relatively dense, which produces a larger gravitational attraction and corresponding anomaly. The Hilina and Kaoiki fault zones are identified as zones of low velocity, low magnetic intensity, and low gravity. Hildenbrand *et al.* (1993) interpret relative low magnetic intensities in the Kaoiki and Hilina fault zones as due to alteration products from groundwater and hydrothermal activity. Lower velocities underlying the Kaoiki and Hilina fault zones are interpreted as highly fractured and perhaps chemically and thermally altered zones of lower density, consistent with Hildenbrand *et al.* (1993).

Constraints on the geometry of zones of magma intrusion and the depth extent of active faulting are important for understanding deformation of the south flank of Hawaii and the potential and size of catastrophic failure of the south flank. An interpretation of the Hilina fault zone based on geomorphic and geodetic observations (Swanson *et al.*, 1976) suggests that the fault zone bottoms into the hyaloclastites and subaerial lava flows at depths of 2–3 km. Lipman *et al.* (1985) and Denlinger and Okubo (1995) argue that the fault zone cuts the volcanic pile through to the basal decollement. Relative low velocities to depths of approximately 9 km south of the surface expression of the Hilina support the more recent interpretations.

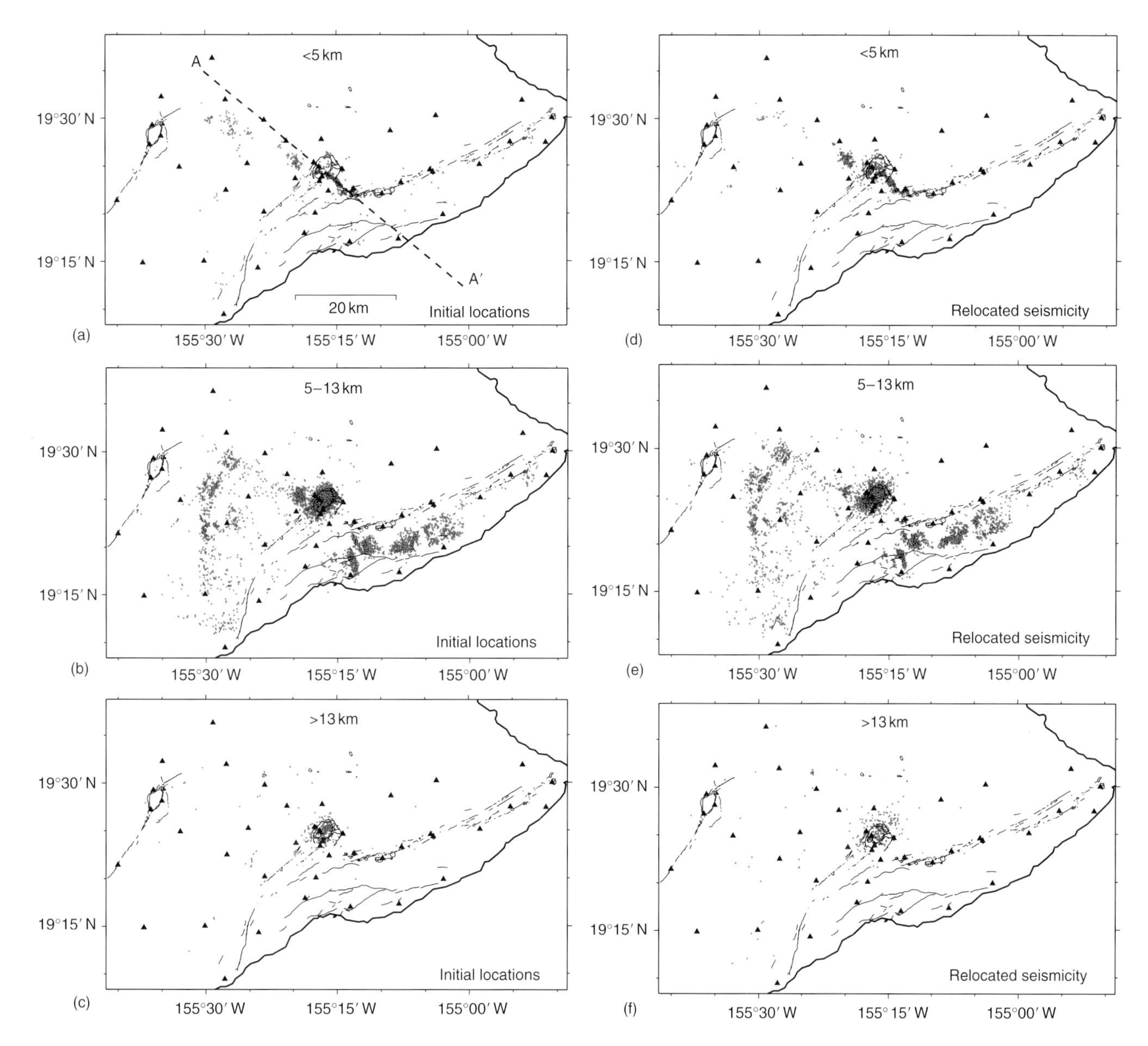

FIGURE 6 Map views of the initial earthquake locations (a–c) and earthquake locations (d–f) based on relocation in the 3D *P*-wave velocity structure (earthquakes are plotted as circles). The seismic stations (triangles) used to monitor seismicity and major faults, fissures, vents, and calderas are shown for reference.

The 1983 *M* 6.3 earthquake beneath the Kaoiki fault zone and a general increase in earthquake swarms along the south flank of Mauna Loa demonstrate that the Kaoiki fault zone is still active. While the seismicity is generally diffuse between Mauna Loa and Kilauea, shallow seismicity does not correlate with the northeast trend of the Kaoiki fault system. This is likely a consequence of the superposition of the crustal strain due to influence of Mauna Loa and its associated flank deformation and that produced by Kilauea.

The 3D velocity model shows the velocity structure beneath Kilauea to be defined as a complex, near-vertical high-velocity conduit approximately 8 km wide between depths of 4 and 12 km. From about 14 km depth, near the Moho, to the base of our model at 20 km, the conduit remains near-vertical but broadens into a relative low-velocity zone with a width of approximately 14 km. At lower crustal and upper mantle depths, a relative low-velocity conduit shows the thermal effects of the hot-spot, where the conduit is defined as a zone of weakness intruded by hot mantle and lower crust partial melt. Using seismicity to depths of 40 km, tomographic results not presented here show the conduit to be a persistent feature that broadens further, dipping gently southward in the upper mantle.

Results from this tomographic study show that high velocities underlying the south flank of Mauna Loa are interpreted

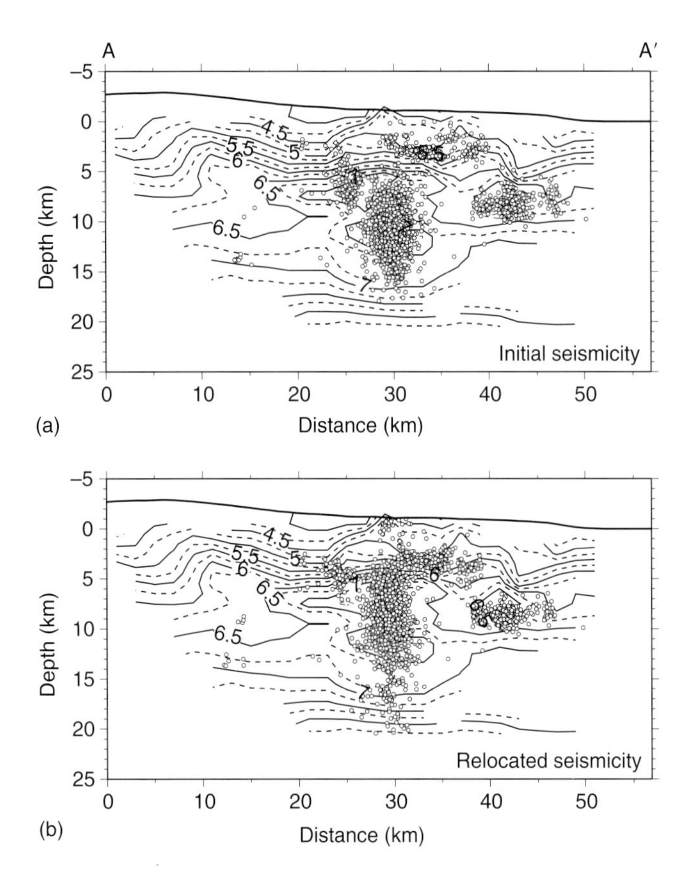

FIGURE 7 Northwest–southeast cross section of the initial earthquake locations (a) and earthquake locations (b) based on relocation in the 3D *P*-wave velocity structure (earthquakes are plotted as circles). See Figure 6 for location of the cross section. Cross-sectional width is one cell dimension, or 2.0 km. Velocity contours are in km sec^{-1}.

as a zone of dense sheet-dikes that have fed previous Mauna Loa eruptions. The Kaoiki fault zone, lying between Kilauea and Mauna Loa, is a zone of normal faults formed by southward migration of the south flank of Mauna Loa prior to the inception of Kilauea volcanism. Much of the flank motion is now being taken up on the Hilina fault system, which lies on the south flank of Kilauea (Swanson *et al.*, 1976). The interaction between Kilauea volcanism and motion on the Hilina fault system is a present-day analogue of the growth of Mauna Loa and the Kaoiki fault system (Lipman, 1980). Beneath Kilauea, the magma conduit is described as a near-vertical upper-mantle and crustal feature interpreted as a sheeted-dike complex that extends as far southward as the Koae fault zone. The Hilina fault system extends to the basal decollement near a depth of 9–10 km.

8. Conclusions

Three-dimensional tomographic imaging of Mauna Loa and Kilauea Volcanoes shows the presence of high-velocity ultramafic rock beneath the volcanoes and the southwest and east rift zones. These high velocities correlate with zones of high magnetic intensity and high gravity and are interpreted as ultramafic cumulates associated with the melt observed at the surface. Episodic injection of new material from beneath produces deformation and inflation of the Kilauea summit and rift zones that destabilizes the upper crust, producing the normal faults observed in the Hilina fault systems. Tomographic imaging reveals that the Kaoiki and Hilina fault zones are observed as relative low-velocity zones that extend to depths of 9–12 km, suggesting that these fault zones penetrate to the basal decollement. This has important implications with regard to long-term stability of the south flank of the island and constraints on the initiation of sudden and mass debris flows off the south flank of the island of Hawaii.

Acknowledgments

This research was supported by the US Geological Survey National Earthquake Hazards Reduction Program and the Volcano Hazards Reduction Programs. Roger Denlinger, Dal Stanley, Fred Klein, Steve McNutt, Wim Spakman, and an anonymous reviewer provided thoughtful comments that improved this manuscript.

References

Aki, K. and W.H.K. Lee (1976). Determination of 3D velocity anomalies under a seismic array using first *P* arrival times from local earthquakes; 1, A homogeneous initial model. *J. Geophys. Res.* **81**, 4381–4399.

Ando, M. (1979). The Hawaii earthquake of November 29, 1975: low dip angle faulting due to forceful injection of magma. *J. Geophys. Res.* **84**, 7616–7626.

Beisser, M., D. Gillard, and W. Wyss (1994). Inversion for source parameters from sparse data sets: test of the method and application to the 1951 ($M = 6.9$) Kona, Hawaii, earthquake. *J. Geophys. Res.* **99**, 19661–19678.

Benz, H.M. and R.B. Smith (1984). Simultaneous inversion for lateral velocity variations and hypocenters in the Yellowstone region using earthquake and refraction data. *J. Geophys. Res.* **89**, 1208–1220.

Benz, H.M., B.A. Chouet, P.B. Dawson, J.C. Lahr, R.A. Page, and J.A. Hole (1996). Three-dimensional *P* and *S* wave velocity structure of Redoubt Volcano Alaska. *J. Geophys. Res.* **101**, 8111–8128.

Borgia, A. (1994). Dynamic basis of volcanic spreading. *J. Geophys. Res.* **99**, 17791–17804.

Chouet, B.A., R.A. Page, C.D. Stephens, J.C. Lahr, and J.A. Power (1994). Precursory swarms of long-period events at Redoubt Volcano (1989–1990), Alaska: their origin and use as a forecasting tool. *J. Volcanol. Geotherm. Res.* **62**, 95–135.

Christensen, N.I. (1979). Compressional wave velocities in rocks at high temperatures and pressures, critical thermal gradients, and crustal low-velocity zones. *J. Geophys. Res.* **84**, 6849–6857.

Clague, D.A. and G.B. Dalrymple (1987). The Hawaiian-Emperor Volcanic Chain: Part I. In: "Volcanism in Hawaii" (R.W. Decker, T.L. Wright, and P.H. Stauffer, Eds.) pp. 5–54. *US Geol. Surv. Prof. Paper 1350*.

Clague, D.A. and R.P. Denlinger (1994). The role of olivine cumulates in destabilizing the flanks of Hawaiian volcanoes. *Bull. Volcanol.* **56**, 425–434.

Dawson, P.B., B.A. Chouet, P.B. Okubo, A. Villasenor, and H.M. Benz (1987). Three-dimensional velocity structure of the Kilauea caldera, Hawaii. *Geophys. Res. Lett.* **26**, 2805–2808.

Decker, R.W., T.L. Wright, and P.H. Stauffer (Eds.) (1987). Volcanism in Hawaii. *US Geol. Surv. Prof. Paper 1350*.

Denlinger, R.P. and P. Okubo (1995). Structure of the mobile south flank of Kilauea Volcano, Hawaii. *J. Geophys. Res.* **100**, 24499–24507.

Eaton, J.P. (1987). Crustal structure and volcanism in Hawaii. *Am. Geophys. Soc. Mon.* **6**, 13–29.

Eissler, H.K. and H. Kanamori (1987). A single-force model for the 1975 Kalapana, Hawaii earthquake. *J. Geophys. Res.* **92**, 4827–4836.

Ellsworth, W.L. and R.Y. Koyanagi (1977). Three-dimensional crust and mantle structure of Kilauea Volcano, Hawaii. *J. Geophys. Res.* **82**, 5379–5394.

Harlow, D.H., J.A. Power, E.P. Laguerta, G. Ambubuyog, R.A. White, and R.P. Hoblitt (1996). Precursory seismicity and forecasting of the June 15, 1991, eruption of Mount Pinatubo. In: "Fire and Mud; Eruptions and Lahars of Mount Pinatubo, Philippines" (C.G. Newhall and R.S. Punongbayon, Eds.), pp. 285–305. Philippine Institute of Volcanology and Seismology, Quezon City, Philippines.

Hildenbrand, T.G., J.G. Rosenbaum, and J.P. Kauahikaua (1993). Aeromagnetic study of the island of Hawaii. *J. Geophys. Res.* **98**, 4099–4119.

Hill, D.P. and J.J. Zucca (1987). Geophysical constraints on the structure of Kilauea and Mauna Loa volcanoes and some implications for seismomagmatic processes. In: "Volcanism in Hawaii," (R.W. Decker, T.L. Wright, and P. H. Stauffer, Eds.), pp. 903–917. *US Geol. Surv. Prof. Paper 1350*.

Hole, J.A. (1992). Nonlinear high-resolution 3D seismic traveltime tomography. *J. Geophy. Res.* **97**, 6553–6562.

Iyer, H.M. and K. Hirahara (Eds.) (1993). "Seismic Tomography; Theory and Practice." Chapman and Hall, London.

Kinoshita, W.K., H.L. Krivoy, D.R. Mabey, and R.R. McDonald (1963). Gravity survey of the Island of Hawaii. *US Geol. Surv. Prof. Paper 475-C*, C114-C116.

Klein, F.W. (1981). A linear gradient crustal model for south Hawaii. *Bull. Seismol. Soc. Am.* **71**, 1503–1510.

Klein, F.W., R.Y. Koyanagi, J.S. Nakata, and W.R. Tanigawa (1987). The seismicity of Kilauea's magma system. In: "Volcanism in Hawaii," (R.W. Decker, T.L. Wright, and P.H. Stauffer, Eds.), pp. 1019–1185. *US Geol. Surv. Prof. Paper 1350*.

Koyanagi, R.Y., J.D. Unger, and E.T. Endo (1973). Seismic evidence for magma intrusion in the eastern Koae fault system, Kilauea Volcano, Hawaii. *Am. Geophys. Union Trans.* **54**, 1216.

Koyanagi, R.Y., W.R. Tanigawa, and J.S. Nakata (1988). Seismicity associated with the eruption. In: "The Pu'u O'o eruption of Kilauea Volcano, Hawaii, episodes 1 through 20, January 3, 1983, through June 8, 1984" (E. Wolfe, Ed.), pp. 183–235. *US Geol. Surv. Prof. Paper 1463*.

Lahr, J.C., B.A. Chouet, C.D. Stephens, J.A. Power, and R.A. Page (1994). Earthquake classification, location, and error analysis in a volcanic environment: implications for the magmatic system of the 1989–1990 eruptions at Redoubt Volcano, Alaska. *J. Volcanol. Geotherm. Res.* **62**, 137–151.

Lees, J.M. (1992). The magma system of Mount St. Helens, nonlinear high-resolution *P*-wave tomography. *J. Volcanol. Geotherm. Res.* **53**, 103–116.

Lipman, P.W. (1980). The southwest rift zone of Mauna Loa: implications for structural evolution of Hawaiian volcanoes. *Am. J. Sci.* **280A**, 752–776.

Lipman, P.W., J.P. Lockwood, R.T. Okamura, D.A. Swanson, and K.M. Yamashita (1985). Ground deformation associated with the 1975 magnitude-7.2 earthquake and resulting changes in activity of Kilauea Volcano, Hawaii. *US Geol. Surv. Prof. Paper 1276*.

McNutt, S.R., Y. Ida, B.A. Chouet, *et al.* (1997). Kilauea volcano provides hot seismic data for joint Japanese–U.S. Experiment. *Eos Trans. Am. Geophys. Union* **78**, 105.

Moore, J.G. (1964). Giant submarine landslides on the Hawaiian Ridge. *US Geol. Surv. Prof. Paper 501-D*.

Moore, J.G., D.A. Clague, R.T. Holcomb, P.W. Lipman, W.R. Normark, and M.T. Torresan (1989). Prodigious submarine landslides on the Hawaiian Ridge. *J. Geophys. Res.* **94**, 17465–17484.

Nolet, G. (Ed.) (1987). "Seismic Tomography; with Applications in Global Seismology and Exploration Geophysics." D. Reidel, Dordrecht.

Okubo, P.G., H.M. Benz, and B.A. Chouet (1997). Imaging the crustal magma source beneath Mauna Loa and Kilauea volcanoes, Hawaii. *Geology* **25**, 867–870.

Paige, C.C. and M.A. Saunders (1982). LSQR: An algorithm for sparse linear equations and sparse least squares. *Trans. Math. Software* **8**, 43–71.

Pavlis, G.L. and J.R. Booker (1980). The mixed discrete-continuos inverse problem: application to the simultaneous determination of earthquake hypocenters and velocity structure. *J. Geophy. Res.* **85**, 4801–4810.

Podvin, P. and L.I. Lecomte (1991). Finite difference computation of traveltimes in very contrasted velocity models: a massively parallel approach and its associated tools. *Geophys. J. Int.* **105**, 271–284.

Power, J.A., J.C. Lahr, R.A. Page, *et al.* (1994). Seismic evolution of the 1989–1990 eruption sequence of Redoubt Volcano, Alaska. *J. Volcanol. Geotherm. Res.* **62**, 69–94.

Rowan, L.R. and R.W. Clayton (1993). The 3D structure of Kilauea Volcano, Hawaii, from traveltime tomography. *J. Geophys. Res.* **98**, 4355–4375.

Ryan, M. (1988). The mechanics and 3D internal structure of active magmatic systems, Kilauea Volcano, Hawaii. *J. Geophys. Res.* **93**, 4213–4248.

Swanson, D.A., W.A. Duffield, and R.S. Fiske (1976). Displacement of the south flank of Kilauea Volcano: the result of forceful intrusion of magma into the rift zones. *US Geol. Surv. Prof. Paper 963*.

Thurber, C.H. (1984). Seismic detection of the summit magma complex of Kilauea Volcano, Hawaii. *Science* **223**, 165–167.

Tilling, R.I. and J.J. Dvorak (1993). Anatomy of a basaltic volcano. *Nature* **363**, 125–133.

Toomey, D.R. and G.R. Foulger (1989). Tomographic inversion of local earthquake data from the Hengill-Grensdalur central volcano complex, Iceland. *J. Geophys. Res.* **94**, 17497–17510.

Vidale, J.E. (1988). Finite-difference calculation of traveltimes. *Bull. Seismol. Soc. Am.* **78**, 2062–2076.

Villaseñor, A., H.M. Benz, L. Fillippi, *et al.* (1998). The three-dimensional *P*-wave velocity structure of Mt. Etna, Italy: implications for volcano and landslide hazards. *Geophys. Res. Lett.* **25**, 1975–1978.

27

Marine Seismology

Kiyoshi Suyehiro
Japan Marine Science and Technology Center, Yokosuka, Japan
Kimihiro Mochizuki
Earthquake Research Institute, The University of Tokyo, Japan

1. Introduction

Seismology is the study of earthquakes and associated phenomena, most notably the physical properties of the Earth's interior. Marine seismology is simply seismology in which the observations are made at sea. Because oceans cover most of the plate boundaries, earthquakes occur mostly beneath the sea floor. This provides a strong incentive to observe and study earthquakes at sea near the sources. Modeling seismic velocity structure provides a basic scheme for understanding the ongoing active processes as well as the history. Global-scale modeling of the Earth's deep interior is presently limited by the lack of broadband seismic data from the ocean floors.

In this chapter we will discuss how these aspects of marine seismology have been tackled by marine seismologists. Our emphasis will be on marine seismological studies at convergent margins. Chapter 55 by Minshull is an important complement to this chapter, since it deals with seismic structures of oceanic crust in general.

Making observations at sea is no easy task, especially for seismic data, which may require an observational window of a large dynamic range of more than 24-bit (144 dB) and wide frequency band from 10^{-4} to 10 Hz, or higher for artificial sources, and a long time span. Seismic sensors of the same type as used on land are usually encapsulated in pressure housings to be emplaced on sea floors often at as deep as 7000 m water depth. Data are recovered by retrieving the data recorder at sea or by telemetering to land via sea-floor cable. Power is supplied by batteries unless cable is available. These factors restrict the capacity and quality of the data. The instrumental challenge of marine seismology has been to overcome these difficulties.

2. Brief History

A brief history of observational aspects of marine seismology is given in this section. We focus here on sea-floor seismic observations that allow us to record both *P* and *S* waves. We will see later that seismic reflection techniques, which grew from oil prospecting efforts, are now indispensable in studying seismic velocity structures.

One of the earliest observations on the sea floor was made in 1937–1940 by the group at the Lehigh University of Pennsylvania, USA (Ewing and Vine, 1938) [Note: Maurice Ewing joined the Columbia University in 1944 and established the Lamont Geological Observatory in 1949.] Since then, groups in the United States, USSR (now Russia), Japan, and others have made efforts in developing ocean-bottom seismographs (OBSs). At first, sea-floor noise measurements were the main objective in order to check the feasibility of sea-floor observations. However, it was difficult to construct a good enough instrument to record the noise level.

A major success of OBS came in the 1970s when a very long profile was shot in the western Pacific to study the deep structure of the oceanic lithosphere (Asada and Shimamura, 1979). This method of employing several OBSs to cover a 1000 km scale distance proved successful in the early days of trying to verify the plate tectonics theory, given that signals from dynamite shots could be recorded at far distances. These results strongly suggested the existence of a large-scale lateral heterogeneity and anisotropy, related to the origin of the lithosphere and subsequent spreading history. However, the seismic characteristics of the oceanic lithosphere are still far from well understood.

OBS technology advanced through the 1970s at several institutions, including more European countries. Each effort employed its own strategy as there were many objectives and choices to be made: geophone or hydrophone, digital or analog, continuous or programmed recording, tethered or pop-up (timed or on-command) for recovery, cassette or open reel, FM or direct recording, and so forth.

Important marine seismological results in the 1960s through the 1970s were mainly not obtained from OBSs, which were essentially in the instrumental development stage. Key results, such as crustal structures around the globe, or uppermost mantle

INTERNATIONAL HANDBOOK OF EARTHQUAKE AND ENGINEERING SEISMOLOGY, VOLUME 81A ISBN: 0-12-440652-1

Pn velocity anisotropy were obtained from refraction seismology by two-ship experiments or sono-buoy systems shooting explosives or airguns (see, e.g., Suyehiro, 1988, for a summary).

In the 1980s, pop-up OBS technology matured in terms of getting the data back to become the main force in crustal structure studies. Tens of OBSs could be deployed for each experiment (e.g., Ewing and Meyer, 1982; Iwasaki *et al.*, 1990).

At the same time, further efforts continued to expand the fidelity of the recordings. One such direction was to utilize deep boreholes drilled by the Deep Sea Drilling Project (DSDP) or its successor the Ocean Drilling Program (ODP) (Adair *et al.*, 1984). The sensors used in these experiments were of short-period type. Their sub-sea-floor data showed improvement in S/N ratio over sea-floor records by up to 30 dB, with better improvements for horizontal components.

The recent direction of marine seismology in the studies of crustal scale has been to utilize as much seismic information as possible, acquiring both reflection and refraction data; that is, to utilize a wide range of incidence angles of seismic waves. For example, in the Japan Sea, the sea bottom is flat in the basin but has a very rugged basement structure. Without the definition of this rugged basement by the reflection method, the perturbation caused by this would propagate to deeper depths if modeled using only the refraction data. In the following section we will see the present status of marine seismology.

3. Seismological Structures

In this section we overview how marine seismology aids our understanding of the tectonics and dynamics of the Earth using convergent margin environments as case studies. By seismological structure, we refer to both seismic activities and seismic velocity structures. How these two results from seismology can be physically connected is an important and unresolved question.

For larger-scale problems we need to obtain the physical properties of the interior of the Earth in terms of temperature, density (pressure), and viscosity (fluid/melt percentage) distribution, since the dynamic equations require these parameters for modeling the Earth. However, we will concentrate on crustal-scale phenomena in this section, since they have been the only attainable targets until very recently.

How the oceanic crust is formed and evolves through time and eventually returns to the mantle is clearly one of the most important questions for which marine seismology can provide critical information. These problems are dealt with in Chapter 55 by Minshull and therefore, are not discussed here. Other important problems for marine seismology include understanding the dynamics of devastating earthquakes at subduction zones, and understanding the structures of oceanic island arcs, plateaus, or backarc basins.

Before the 1970s, marine seismology was really unable to resolve unequivocally the action of plate tectonics. During the last decade or so, improvement of the sensors and the use of many sensors have allowed us to image 2D structures with unprecedented accuracy. Before then, 2D models had been constructed from 1D models from seismic profiles shot in the direction of least lateral structural change. One of our remaining goals is 3D imaging of 3D structures over widely different scales, such as for seamounts or for the deep roots of hot-spots.

Seismic reflection studies and seismic refraction studies have long been made as if they were targeted at two independent structures. Reflection studies often produced results much higher in spatial resolution but limited in depth penetration, while the opposite was often the case for refraction studies. Currently the two kinds of data complement each other to yield a more comprehensive view of the seismic structure. MCS (multichannel seismic) data provide detailed laterally heterogeneous sedimentary structure, which helps correct the refraction data to better constrain the deeper structure. Refraction velocities help to better image the MCS seismic sections.

Just as land seismology must also consider geological or geochemical data in order to infer tectonics and history of any target, marine seismology must also try to incorporate such knowledge. We are now in a much better position than before since, in addition to sea-floor sampling tools that have been in existence for a long time, there are submersibles, swath bathymetry devices, deep coring tools, and so on.

3.1 Plate Subduction Zones

As oceanic plates subside underneath the overriding plates, earthquakes occur and volcanoes erupt. Their modes of occurrence differ in different zones. Plate subduction is closely linked to island arc growth and backarc basin opening. Detailing the characteristics of this variety to understand how and why these processes take place is a major target of marine seismology.

Large earthquakes that occur at plate subduction zones are mostly interplate thrust events, the two largest in the 20th century being the Chilean earthquake in 1960 (M_W 9.5) and the Alaskan earthquake in 1964 (M_W 9.2). The oceanic plates thrust against continents or island arcs at speeds of the order of centimeters per year (10 km My^{-1}), and accumulate displacements to provide meters of earthquake slip over a hundred-year time scale. Thus these interplate earthquakes are explained by the plate tectonics theory. However, only a limited width of the plate boundary experiences such slip and, furthermore, even within that zone less than 100% of the slip is taken up by earthquakes. Such inferences are available from teleseismic observations alone. Where marine seismology can play a major role is in detailing the subduction of the oceanic lithosphere; what goes in and how that changes with depth. What happens to the sedimentary layer (Layer 1) that is only a few hundred meters thick? Does Layer 2 (basalt but porous) experience an increase in seismic velocity with depth? These considerations may then lead to answering questions such as the following: Exactly where and how do the earthquakes occur over the trench area? What are the parameters that control the dynamic rupturing of large interplate earthquakes? Such pieces of information are not only unattainable from land observations but are also challenges to marine seismology.

On the other hand, the structure of the overriding plate is also important to the understanding of the consequences of plate interaction. Oceanic island arcs are believed to have formed from oceanic crust. Would this crust eventually disappear into mantle as normal oceanic crusts would? Once the subduction process starts, it has been postulated that two major environments exist: accretion and erosion (von Huene and Scholl, 1991). How do these processes contribute to the arc growth? These questions are again best answered by marine seismology.

In the western Pacific, there are many marginal seas. Are they different from normal ocean, as we often do not observe obvious magnetic lineation patterns or oceanic ridges? Such regionalized structures can only be studied by local-scale experiments (<100 km horizontal scale for crustal thickness structure). Although the scale is local, its significance can be global.

Figure 1 shows the Pacific Rim and various convergent margins where marine seismological investigations have been conducted.

3.1.1 Japan Trench

The Japan Trench area is one of the best areas to show the present status of the marine seismological understanding of an active *seismogenic zone* (Fig. 1b). Here the subduction of the Pacific plate takes place at a rate of about 8–9 cm y^{-1} with little obliquity. Beneath the east coast of northeast Japan, the top of the Wadati–Benioff zone is reliably determined by a land network at ~50 km depth dipping at ~25°. Historically, great ($M>8$) thrust faulting earthquakes between 38° and 40° N seem to occur irregularly. However, *M*7-class thrust faulting events are frequent (six events since 1968 to 1994 in the same zone). Large thrust type events are known to occur to about 100 km inward from the trench axis, where the intraplate double seismic zone takes over with overlapping with the thrust zone.

Marine seismological studies in the Japan Trench area have been conducted for many years since the 1960s. Many models have been constructed from two-ship and sono-buoy experiments (Ludwig *et al.*, 1966), from MCS experiments (Nasu *et al.*, 1980), and from OBS experiments (Suyehiro and Nishizawa, 1994). These controlled source experiments produced crustal models down to about 20 km from refraction profiles and down to 12 km from reflection profiles. From these studies, it became evident that the accretionary prism is little developed, requiring most of the sediments to be subducted. Interpretation of OBS controlled-source data suggested that subduction of oceanic crust occurs at a dip angle of less than 5° to about 70 km landward of the trench axis (Suyehiro and Nishizawa, 1994).

Figure 2 compiles more recent results and therefore has more detail to larger depths (Hino *et al.*, 2000; Tsuru *et al.*, 2000). The observational features are: (1) the oceanic plate subducts at a shallow dip angle but abruptly increases its dip at about 20 km depth (bending front in Fig. 2); (2) a wide gap exists between the updip end of the seismogenic zone and the trench axis; (3) the deeper end of the seismogenic zone extends into the zone with contact to upper mantle of the continent side.

In 1994, the Sanriku-Haruka-Oki earthquake (M_W 7.7) occurred, after which many OBSs were deployed to observe aftershocks (Hino *et al.*, 2000). Aftershock mechanisms show inconsistent patterns, with interplate thrusts suggesting that many events occur outside the plate boundary. Also, this observation revealed that the aftershock activity was very low where the main shock released most of its seismic moment (bending front in Fig. 2). From an independent seismic structure experiment carried out in 1996 (Takahashi *et al.*, 2000), this zone seems to be where the contact with the continent side wedge mantle begins.

These inferences from marine seismology have important bearings on understanding the seismogenic zone in terms of what controls the shallow and deep ends as well as moment release distribution or asperities.

3.1.2 Nankai Trough

The Philippine Sea plate subduction beneath the Tokai crust has caused *M*8-class interplate earthquakes to recur every ~120 y with about 4 m slip (~3 cm y^{-1}) (Fig. 1b). The last sequence of events were 1944 Tonankai and 1946 Nankaido earthquakes. The eastern limit of the rupture zone of the 1944 Tonankai Earthquake (*M* 8) is somewhat ambiguous, but many studies indicate about 137°–137.5° E, which leaves a possible rupture zone extending into the Suruga Bay along the Nankai Trough, which experienced large interplate earthquakes up until 1854. The seismic efficiency is nearly 100%, in contrast to the much lower Japan Trench area. There seems to be structural control over the rupture areas along the trough axis because all the estimated ruptures seem to start and end at certain areas. We will see in this section what structural control can be obtained from marine seismological results. Conclusive evidence is yet to come, but we are certainly progressing in resolving details that may be related to earthquake generation.

The most conspicuous bathymetric anomaly is the Zenisu Ridge, which is about to subduct beneath the Tokai area. It has been suggested that a so-called proto-Zenisu Ridge is already subducted preceding this ridge (Le Pichon *et al.*, 1996). If this is so, questions arise about what role such bathymetric

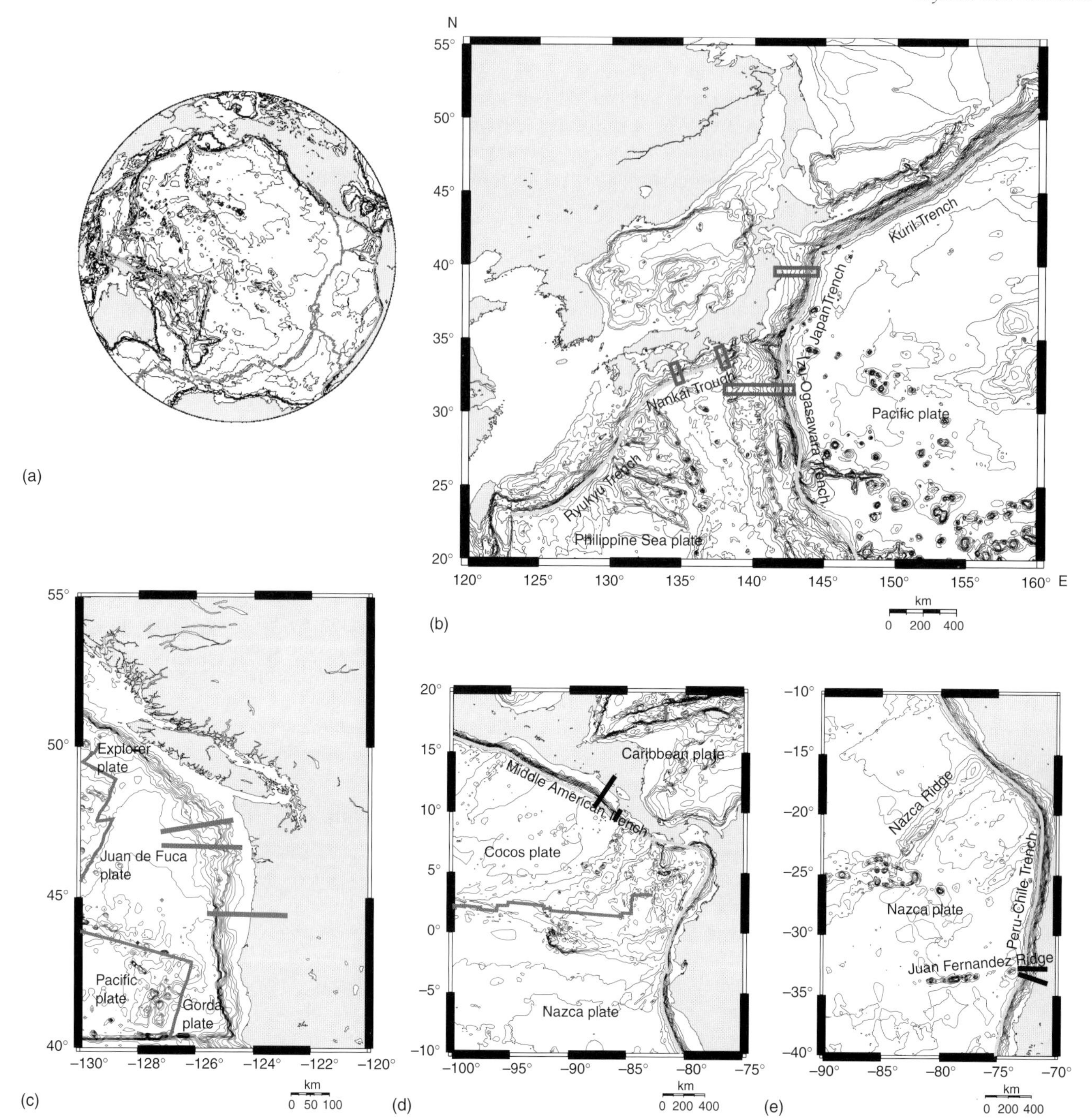

FIGURE 1 Marine seismological investigations are being made at many tectonically different locations where active plate tectonic processes are taking place. (a) Active margins of the circum-Pacific rim. (b) Northwestern Pacific rim. In this sector, the Pacific plate is old (>100 Ma) and subducting at a relatively fast rate (about $10\,\mathrm{cm\,y^{-1}}$) in contrast to the young and slow Philippine Sea plate. (c) Northeastern Pacific rim. Small and young plates are subducting beneath N. America. (d) Eastern Pacific rim. (e) Southeastern Pacific rim. Seismological models at locations indicated by rectangles or thick lines are shown in the following figures.

anomalies play in the subduction tectonics and earthquake cycle.

3.1.2.1 Eastern Nankai District

OBS seismic surveys in the Nankai Trough began in 1992 (Nakanishi *et al.*, 1998). Figure 3 represents a cross section off Tokai district from 25 OBSs where the Zenisu Ridge is approaching the trough (Nakanishi *et al.*, 1998). The seismic velocity structures on both flanks suggest the Ridge to be oceanic crust. The igneous crust section is about 8 km thick. The subduction of the Philippine Sea plate is imaged to about 34° N, where the top of the subducting igneous crust reaches

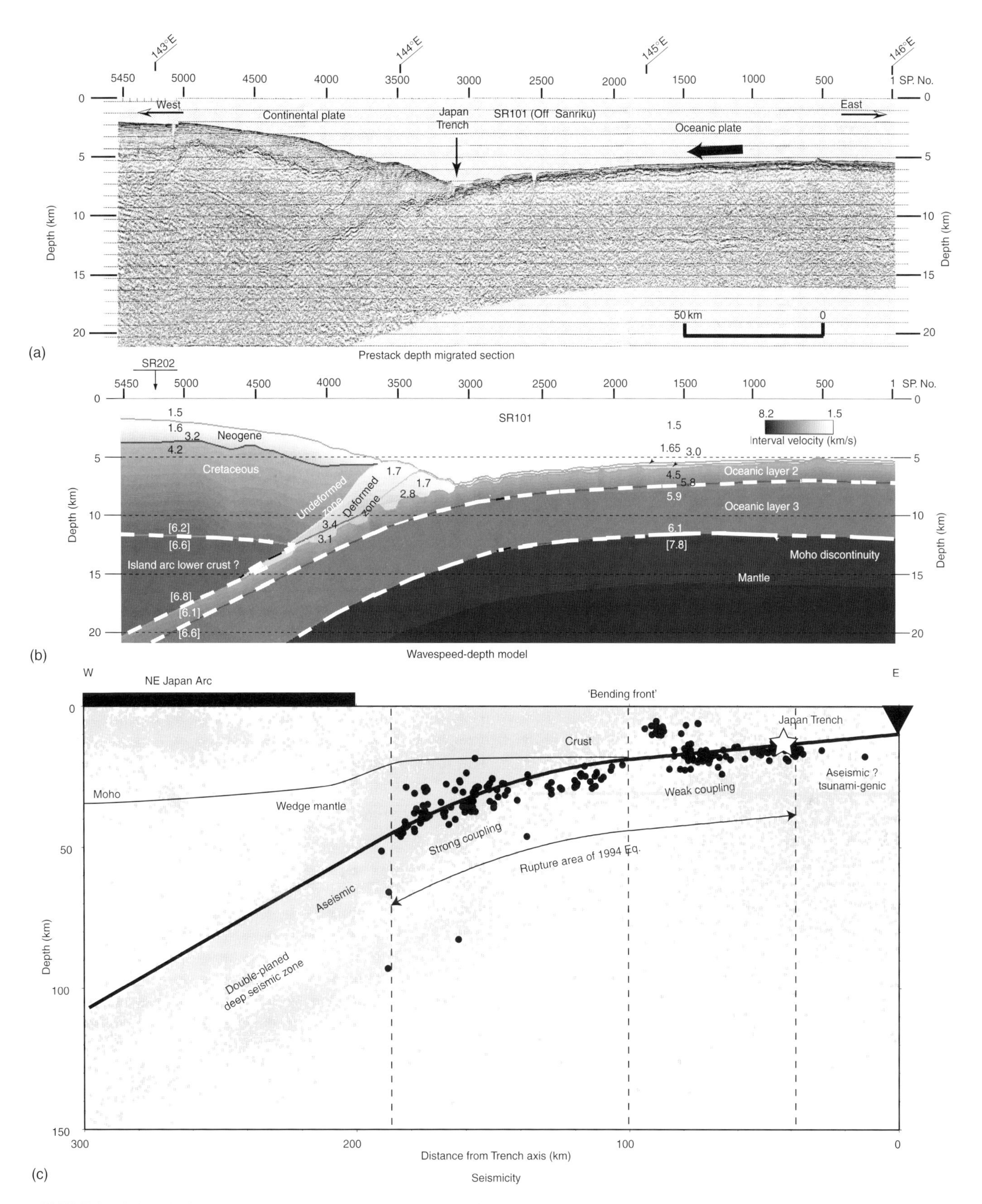

FIGURE 2 Seismological models across the northern Japan Trench (see Fig. 1b). (a) MCS model. Development and destruction of horst and graben structure at the top of Pacific Ocean crust before and after subduction can be observed. (From Tsuru *et al.*, 2000.) (b) Seismic velocities from OBS surveys are shown together with MCS interpretations. (From Tsuru *et al.*, 2000.) (c) Large interplate earthquake (M_W 7.7) in 1994 started from the position of the star and ruptured westward along plate boundary (thick curve) as indicated by arrow. Solid dots are aftershocks determined by a temporary OBS network. Seismicity by land network (gray dots) is biased beneath the marine area. (From Hino *et al.*, 2000.) (a) & (b) copyright by the American Geophysical Union.

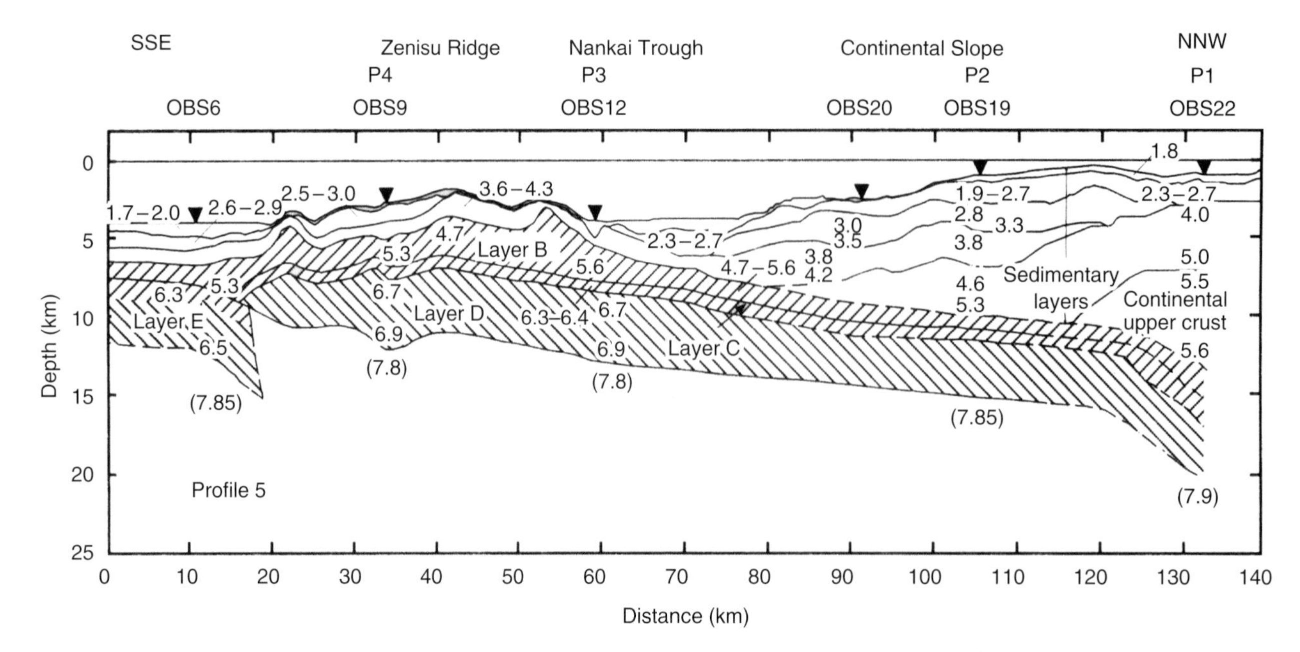

FIGURE 3 Crustal structure across the eastern Nankai Trough where a bathymetric high (Zenisu Ridge) is approaching the trough (see Fig. 1b). (From Nakanishi *et al.*, 1998. Copyright by the American Geophysical Union)

about 12 km depth. The dip angle is only about 4°. Strong seismic coupling occurs below this depth. A dip angle of about 10° can be inferred from seismicity north of 34° N.

Beneath the south flank of the Zenisu Ridge, the Moho seems to become discontinuously shallower landward with an offset of 5 km in depth. This discontinuity coincides with the deeper extension of a major fault zone identified in reflection seismic record near the sea floor (Le Pichon *et al.*, 1996). Such a major fault suggests that the plate motion is accommodated in a much wider area than previously thought. Again, this sort of feature can only be identified by marine seismological techniques.

3.1.2.2 Western Nankai District

In contrast to the eastern part where the bathymetry is more complex, the western part is well known for its textbook-like subduction geometry as defined by MCS seismic records (e.g., Moore *et al.*, 1990). Recently a number of MCS/OBS transects were made in this well-defined accretionary prism area (Mochizuki *et al.*, 1998; Park *et al.*, 1999; Kodaira *et al.*, 2000a,b). Here, the plate dip angle is about 7° (see Color Plate 9(a)). Low seismic velocity prism material exists at least to 10 km depth.

Color Plate 9(a) shows two OBS structure models with seismicity from the land network. The 1946 Nankaido Earthquake fault models ("Earthquake dislocation area" in the plate) extend well into where the plate is in contact with the low seismic velocity accretionary prism. Another interesting feature is the imaging of a subducting seamount by MCS/OBS profiling (Kodaira *et al.*, 2000b). It is estimated to be 17 km in diameter, and seems to have acted as barrier to decelerate the seismic rupture of the 1946 event.

Large earthquakes in the Nankai Trough historically occurred in pairs with little or no time interval between the two events. Mochizuki *et al.* (1998) found structural heterogeneities both above and below the subducting plate boundary near the main shock epicenter of the 1946 event. This zone can be considered as being at the border of the two main rupture areas of Nankai Trough large earthquakes. Whether this feature is the asperity directly causing the pairing characteristics of the large events is not yet clear. However, this offers a structural constraint for any physical modeling of earthquake ruptures.

3.1.3 Other Pacific Rim Convergent Margins

We have seen two contrasting convergent margins above. The northeast Japan Trench is where fast and old plate is subducting, whereas the Nankai Trough is where slow and relatively young plate is subducting. In the following we will overview several of the other convergent margins with different subduction characteristics where modern marine seismological surveys have taken place. From these results, we are clearly gaining valuable information that can be acquired only by marine seismological methods. At this stage the variability among different margin settings or within each setting is still difficult to explain.

3.1.3.1 Cascadia

Cascadia subduction zone is where young (4–8 Ma) Juan de Fuca plate subducts at a slow convergence rate (ENE-ward at 4 cm y^{-1}) with high regional sedimentation rate beneath the North American continent (Fig. 1c). Large earthquakes are rare, but the potential for significant activity seems to exist. Onshore–offshore seismic experiment revealed the subducted oceanic crust subducting at a small angle into well beneath the continental margin. Forearc crustal structure changes significantly along strike (Trehu *et al.*, 1994; Flueh *et al.*, 1998a) (Fig. 4a–c).

Flueh *et al.* (1998a) reported that the dip angle of the subducting plate is only a few degrees at about 47° N. The accretionary wedge is well developed and an increase of seismic velocity to $>4\,\mathrm{km\,sec^{-1}}$ seems to be due to dehydration, compaction, and diagenesis. Velocities $>5\,\mathrm{km\,sec^{-1}}$ may indicate igneous rocks that may sustain strain energy for seismic release.

Trehu *et al.* (1994) compiled an E–W cross section from previous results to show oceanic crust persisting beneath the coast with a larger dip angle than above (Fig. 4c). The Siletz terrane is exposed basement rock of the Cascadia forearc, which consists of accreted oceanic crust and seamounts about 50 Ma. Seismic velocities are much higher than observed by Flueh *et al.* (1998a). The terrane is bordered by lower seismic velocity subduction complex at about 100 km distance (75 km from deformation front). North American mantle starts from more than 40 km depth.

3.1.3.2 Nicaragua and Costa Rica Margins

The Cocos oceanic plate created at the East Pacific Rise is subducting at the Middle American Trench (Fig. 1d). The plate subduction rate is currently about $9\,\mathrm{cm\,y^{-1}}$ nearly perpendicular to the trench axis. The plate is relatively young at about 25–30 Ma.

The convergent margin off Guatemala, Nicaragua, and Costa Rica has been investigated by several MCS and OBS studies including a 3D seismic survey (Shipley *et al.*, 1992). Two hypotheses regarding the making of the margin involve whether the margin is composed of Neogene accretion materials or older (Mesozoic) ophiolitic rocks. Such a distinction is not simple from seismic data alone. Drilled core samples can indicate the origin from a particular spot and the seismic data can be used to extrapolate the ground evidence.

A 335 km long seismic profiling was made by shooting large airguns to Ocean Bottom Hydrophones (OBHs) and temporary land stations off Nicaragua (Walther *et al.*, 2000) (Fig. 4d). The OBHs were spaced at 10–25 km distance. The oceanic crust is 5.5 km thick. Seismic phases interpreted to be from the subducting Moho could be followed to a depth of 26 km, suggesting a clear seismic velocity change across the interface. From land station data, the plate dip angle is inferred to increase strongly at about 25 km depth beneath the continental shelf. Comparison with local seismicity shallower than 25 km depth indicates that many of the events occur above the subducting plate and do not show any clear correlation. However, this may be due to poor location of hypocenters. There is a seismicity gap between 25 and 40 km depth. Below 40 km depth, the earthquakes are located on or mostly below the plate boundary. The gap seems to start as the mantle wedge comes into contact with the subducting plate. On the upper plate, a basement wedge that is 15 km thick beneath the continental shelf and thins to the trench changing seismic velocities from 5.2 to $3.5\,\mathrm{km\,sec^{-1}}$ is interpreted to be of ophiolitic rocks, that is, similar to drilled ophiolites.

Christeson *et al.* (1999) obtained a cross section at the Costa Rica margin transecting the Nicoya Peninsula (Fig. 1d, 4e). The plate dip angle increases less rapidly compared to off Nicaragua (Walther *et al.*, 2000). In comparison with other margins the seismic velocity in the margin wedge is higher, but is lower than in the adjacent Nicoya complex (Fig. 4e). Their interpretation is that the mantle wedge is of the same material as the Nicoya complex but has slow seismic velocity because of fracturing, alteration, or accretion processes. Interplate seismicity seems to start at about 14 km depth.

3.1.3.3 Chile Margin

Large airguns (3×32 liters) were shot in two profiles each of 200 km length at about 100 m spacing to OBHs and land stations to study the subduction structure in the Chile Trench region (Flueh *et al.*, 1998b) (Fig. 1e). Here the tectonic environment changes along the trench axis at about 33° S. To the north, the trench axis is sediment-starved and the deeper slab flattens with no Quaternary volcanism up to 28° S. The southern region is more characteristic of a normal margin with active volcanism and steep subduction plate dip angle. From these refraction profiles and with MCS profiles they could obtain laterally varying structure models to define the seaward extent of continental crust and the volume of recent accretionary prisms. The two profiles about 70 km apart in the same subduction regime showed marked differences without clear correlation to surface geology (Figs. 4f,g).

3.2 Izu-Ogasawara (Bonin) Arc

Oceanic island arcs, such as the Izu-Ogasawara-Mariana, Tonga-Kermadec, or Kurile-Aleutian arcs, are believed to develop from normal oceanic crust at plate subduction boundaries. The scale of an oceanic arc is typically a few hundred kilometers across, with only a small portion rising above sea level. Practically, only marine seismology can provide knowledge of its seismic structure. Eventually, these arcs may accrete to continents and become a component of the continental crust. Thus the crustal structure, as it bears on the process of crustal accretion and its composition, has global significance. We choose the Izu-Ogasawara arc crust as an example of marine seismological investigation of an oceanic island arc. We will see how the seismological results have been used to address the above problems.

The Izu-Ogasawara arc system located off the southern coast of Japan, extending more than 1000 km from north to south and 300–450 km wide, is where active rifting and

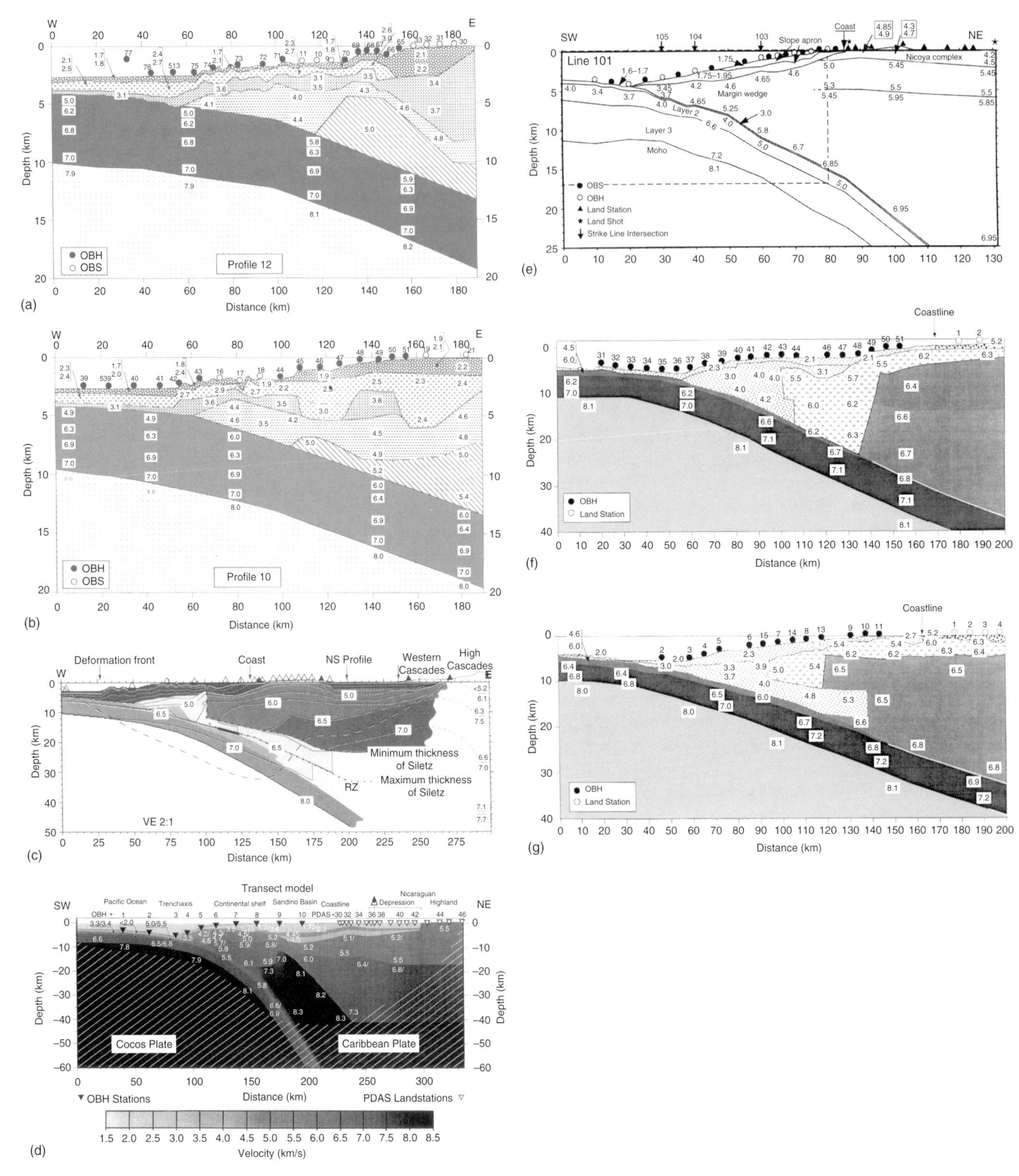

FIGURE 4 Crustal structure models across the northeastern Pacific rim (see Fig. 1c, d, and e). (a) Cascadia forearc at about 47.5° N. (b) Cascadia forearc at about 46.7° N. (From Flueh *et al.*, 1998a.) (c) Cascadia forearc at about 45° N. (From Trehu *et al.*, 1994.) Structure above the dashed line is modeled by onshore–offshore experiments. Open triangles are seismometers and filled triangles are shot points on land. RZ is reflective zone. (d) Nicaragua margin model. (From Walther *et al.*, 2000.) Here, the Cocos plate is subducting northwestward at about 9 cm y^{-1}. (e) Costa Rica margin model. (From Christeson *et al.*, 1999.) (f,g) Chile margin models at about 32.75° S and 33.5° S, respectively. (From Flueh *et al.*, 1998b.) (a), (b), (f), (g) copyright Elsevier Science; (c) reprinted with permission from Trehu *et al.*, 1994, copyright American Association for the Advancement of Science; (d) copy with permission from Blackwell Science Ltd; (e) copyright by the American Geophysical Union.

subduction are occurring (Fig. 5). The arc formation was probably initiated about 48 Ma. Rifting during the Oligocene formed the forearc and back-arc basins, which resulted in the spreading of the Shikoku Basin (to 15 Ma) and the separation of Kyushu-Palau Ridge. The forearc contains an along-arc chain of serpentinite seamounts inferred to result from mantle diapirs.

In 1992, an extensive seismic survey was conducted across the northern Izu-Ogasawara (Bonin) Island Arc system at 32° 15′ N between 138° and 143° E (Suyehiro *et al.*, 1996; Takahashi *et al.*, 1998) (Fig. 5). The marine seismological survey used controlled sources and local earthquakes to model the crust, the upper mantle, and the subducting slab. Seismic reflection profiles using an airgun array (17–60 liters) received by a 24-channel hydrophone consisted of the main transect and two-ship expanding spread profiles (ESP) along strike of the arc. Along-strike profiles were selected to sample the forearc, rift zone, and backarc. A total of 46 OBSs were laid out 15–20 km apart to receive airgun and dynamite shots along these profiles to sample deep structure using wide-angle reflections and refractions. Two OBS seismic arrays were effectively configured to determine local shallow and deep seismicity across the arc, particularly beneath the forearc and rift zone.

From these data, many interesting features were discovered (Suyehiro *et al.*, 1996). The crust has quite different characteristics from an average ocean crust (e.g., White *et al.*, 1992). We will describe and discuss here two key features: (i) the middle crust confined beneath the arc has a velocity of $\sim$6 km sec^{-1} extending as far east as the volcanic front and is thickest ($\sim$20 km) beneath the presently active rift zone; (ii) the Moho becomes obscure east of the VF, and extremely low velocity is observed beneath the forearc serpentinite diapir [see Color Plate 9(b)].

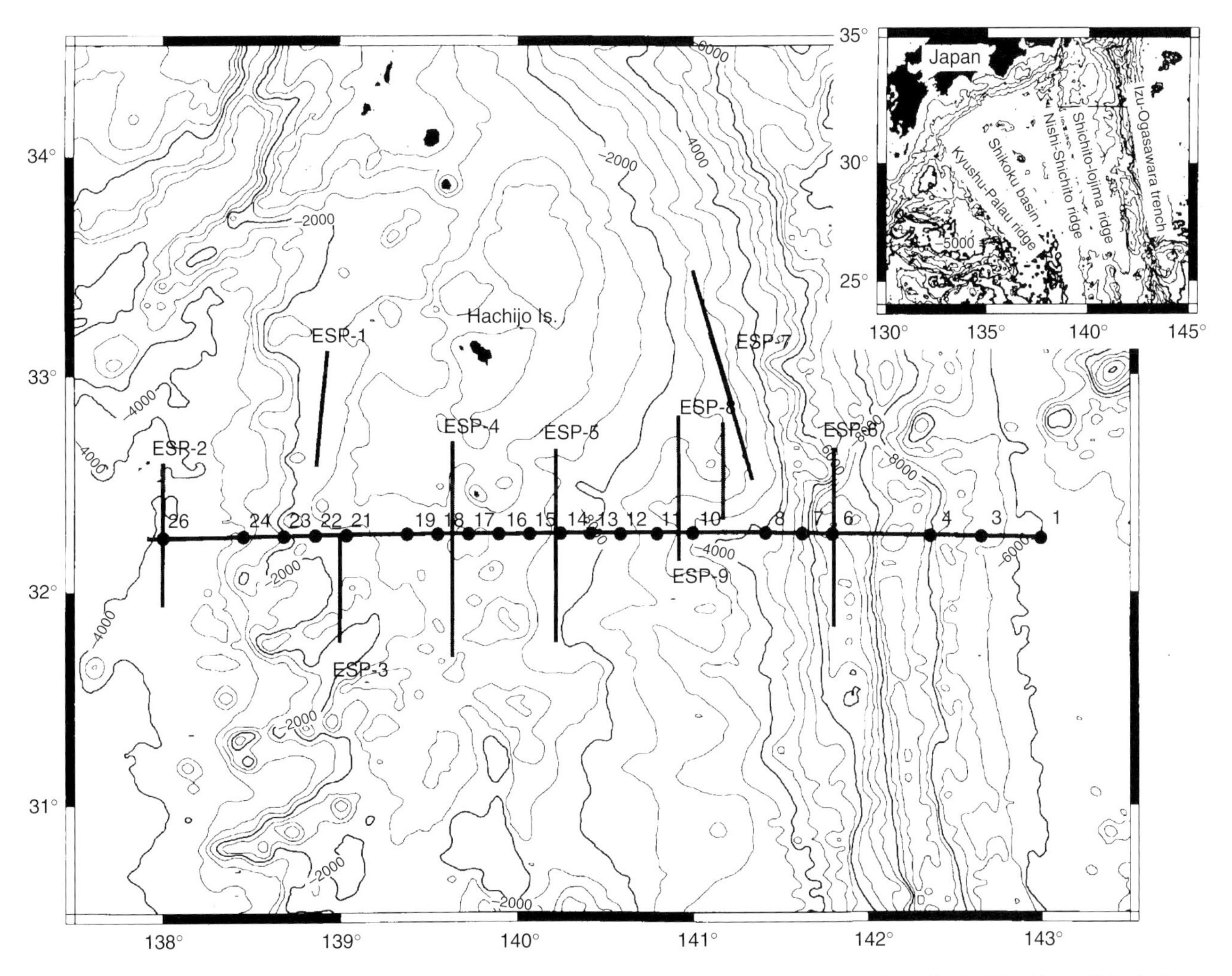

FIGURE 5 An extensive marine seismic survey layout in 1992 is shown. E–W profile transects from the Pacific side to Shikoku Basin side. Dots are OBS positions. Profiles marked ESP are two-ship wide-angle seismic profiles also recorded by OBS. Inset shows the studied Izu-Ogasawara arc.

Once a definitive structure model is established, one can compare with results from other experiments and look for similarities and dissimilarities considering the experimental design that mainly defines the uniqueness and variance of the model.

A ~6 km sec^{-1} middle crust has been found at least north of 31° N all the way to the Izu peninsula from previous results. It is not so clearly observed south of 30° N to the Mariana arc (Ludwig *et al.*, 1973; LaTraille and Hussong, 1980). Considering the technology available when these experiments were carried out, their results should not be regarded as negative evidence. More experiments with higher resolution are required to detail the structure.

The candidate rock to be the main component of this layer was related to granitoid plutons, more specifically tonalite, from exposed rocks at Izu peninsula and dredged tonalite samples from Izu-Ogasawara forearc. We can then estimate that granitoids may be produced in an oceanic island arc setting on the order of 500 $km^3 km^{-1}$ along arc strike in volume and ~10–25 $km^3 km^{-1} My^{-1}$ in growth rate depending on the formation time length of 20–47 My (Suyehiro *et al.*, 1996).

A lower crust with anomalously high *P* velocity (>7 km sec^{-1}) seems to coexist with the 6 km sec^{-1} layer. This characteristic seismic velocity resembles high-velocity lower crustal layers found at passive margins in the north Atlantic (e.g., Mjelde *et al.*, 1993).

It is generally considered that the plate interaction in the Izu-Ogasawara arc is weak, as evidenced by the lack of major earthquakes or accretionary prisms. A low seismic velocity zone in the overriding crust hosting serpentinite seamounts suggests water release from the subducting crust. As the plate subducts further, the mantle wedge has seismic velocity less than 7.5 km sec^{-1}, which may be explained as serpentinization of the mantle peridotite due to further release of water from the subducting plate. At this depth west of 141.5° E, earthquake mechanisms are no longer of down-dip thrust type. This suggests that the seismogenic zone turns to stable sliding at about only 20 km depth, which may be due to the weakened overriding mantle.

The OBS microseismicity revealed that the Pacific Plate subducts at a dip angle of about 10° to as far west as past the forearc basin, where the dip angle increases significantly. Independent seismic evidence for shallow dipping plate comes from dynamite shot waves reflected off the top of the plate boundary. Although other subduction zones show an aseismic zone immediately landward of the trench axis, such a feature is not observed here.

This kind of experiment employing multiple seismic methods and more than one ship is not easy to undertake, but has proven to be powerful in imaging a large area and giving a comprehensive view of the region, in this case an oceanic island arc. However, there remain ambiguities as to how the crustal materials are linked with various igneous activities that have operated in the region. A more detailed image is needed, such as by incorporating seismic scattering, anisotropy, or the attenuation characteristics of the crust.

3.3 Backarc Basins

The backarc basins or marginal seas are mostly developed in the western Pacific area. The Philippine Sea comprises a number of such basins forming the largest entity in this category. Others include the Japan Sea, South China Sea, Okhotsk Sea, Bering Sea, etc. They seem to have formed as a consequence of plate subduction, although there is no direct evidence to support this. One of the first questions to ask is what sort of crust is formed. Is there any significant difference from the crusts of major oceans? Do they evolve differently from normal oceans?

The geological history inferred from magnetic anomalies and drilling indicate rather complex evolution; that is, there does not seem to be a simple ridge system to create these basins. This is why using surface waves observed at long distances is not adequate to address the problems associated with the basin formation and development, although teleseismic observations are useful in capturing the average structure at lithosphere scale. We will see how marine seismological observations have brought the structural complexities to light, taking as an example the case of the Japan Sea.

3.3.1 Japan Sea—Mode of Opening, Present Stress Field

The Japan Sea was created as a consequence of the Japan arc separating from the Asian continent during a period about 20–15 Ma. However, observed magnetic anomalies do not show any clear evidence of sea-floor spreading.

Since 1985, a number of modern seismic surveys have been carried out in various parts of the Sea. The first of the series, in 1985, proved the importance of simultaneous observation of reflection and refraction waves. Beneath the flat bathymetry, the basement structure was rugged. From this and the following surveys, the complexities of the Japan Sea crusts were largely revealed.

Much of the basin areas is found to possess crust more than 10 km thick beneath sediments, except in northeastern Japan Basin where the crust is indistinguishable from that of a normal ocean. *P* wave velocities of about 6.6–7.1 km sec^{-1} characterize the lower 70–80% (~8 km thickness) of the Japan Sea crust. This suggests that a layer characteristic of granitic component of continental origin is thin in the basin areas, if there is any.

How the opening started in the Japan Sea is still uncertain. At passive margins, a number of locations are found to possess a high seismic velocity lower crust, interpreted to be due to magma intrusion during the rifting stage (e.g., White *et al.*, 1987). Its characteristic seismic velocity is ~7.2 km sec^{-1}, with a small gradient. A similar feature is also found in the lowermost part of the Japan Sea crust.

A detailed seismic survey was conducted in the northern Yamato Basin, where an ocean broadband downhole seismometer was installed together with a temporary array of nine OBSs. The upper crust was found to possess azimuthal anisotropy of about 4–7% in *P*-wave velocity, likely due to opening of cracks induced by the present state of stress, which is E–W compression (Hirata *et al.*, 1992). This shows another possibility for studying stress condition where earthquakes are rare.

4. Marine Seismic Instrumentation

In this section we overview the characteristics of marine seismometry, focusing on observing seismic signals beneath the sea water mass. The progress of marine seismology is rapid. We will emphasize the unique challenges marine seismologists are facing in order to obtain high-quality data from the oceans.

4.1 Basics

4.1.1 Ocean-Bottom Seismographs

An ocean-bottom seismograph (OBS) typically consists of the following components; the sensor (e.g., three-component plus hydrophone) with a leveling mechanism, controlling electronics to amplify/filter/digitize as necessary to feed to recorder, a real-time clock, and a power supply. These components are housed in pressure vessels to withstand the pressure of average ocean basin depths (5–6 km) to trench depths (4–10 km). A typical OBS is deployed from a vessel without tethers to sink to the bottom by its own weight. After its planned observation period, it can pop up to the sea surface by releasing the anchor weight on an acoustic command from the surface.

The areas of interest can range from ocean ridges with little sediment and rugged topography to trenches with escarpments. Sea-floor bathymetry must be known before deployment so that steep slopes can be avoided. OBS deployments have been made from vessels small fishing boats up to large vessels, and from helicopters.

4.1.2 Location and Clock Accuracy

The position of deployment can easily be determined within 100 m accuracy by the GPS system, which was not fully available until the early 1990s. The actual OBS location can be further improved by acoustic ranging (transponder communication or controlled-source recordings).

It is now quite easy to obtain 1 μs GPS accuracy anywhere on the globe. The crystal oscillator (typically 10^{-7}–10^{-8} accuracy) inside the OBS package is calibrated against this signal before and after deployments. For long-term deployments, an acoustic link can be designed to calibrate the clock during observation. Direct water wave signals from many controlled shots can also be used for calibration.

TABLE 1 Sensor Type and Power Requirement

Sensor Type	Power Requirement	Remarks
Moving-coil velocity sensor	Nil	Frequency response inadequate for teleseismic studies
Feedback broadband sensor CMG-1 (GSL)	12 V at 32 mA (0.38 W)	FDSN standard
PMD-2023 (PMD)	12 V at 22 mA (0.26 W)	Broadband, robust
KS-54000 (Teledyne)	24 V at 0.1 A (2.4 W)	FDSN standard

4.1.3 Frequency and Dynamic Range

Electromagnetic velocity sensors with 1–4.5 Hz natural frequencies are often used because of their ruggedness, absence of power requirement, and small size. One 17-inch (43 cm) diameter glass sphere (two hemispheres to house instruments) can hold all the components if 4.5 Hz is selected. Adopting a lower-frequency type makes the system larger but allows better registration of low-frequency signals from teleseismic events. Broadband sensors are now beginning to be adapted for marine seismology (Table 1).

The dynamic range in effect is determined by the recording system. Analog recording technology required different gain channels to cover more than 40 dB. Digital technology now allows $>$16-bit ($>$96 dB) range and large-capacity storage ($>$a few Gbytes) to be adopted by an inexpensive OBS.

4.1.4 Power Supply and Data Storage

All the OBSs use batteries of various types such as alkaline or lithium dry cells. A low-power OBS (1 W) can run continuously for more than a month on C-size dry cells packed in a 17-inch pressure housing. For investigations of structures using active sources, a few weeks' observational window would suffice, since shooting ships can cover about 200 km per day. Natural earthquake observation usually needs a certain length of observation period depending on the target events according to the Gutenberg–Richter relation. Newer OBSs with larger pressure vessels are able to store digital data and maintain clock accuracy for several months to a year.

4.2 OBSs for Mobile Arrays

4.2.1 Seismic Profiling with OBS Arrays

By the 1980s, ocean-bottom seismographs were proven to be more effective and reliable for marine deep seismic sounding than the use of surface measurements using hydrophones.

Typically, an OBS record section will show oceanic Moho wide-angle reflection arrivals with a single airgun source with about 10 liter capacity shot at about 100 kg cm^{-2}. With 1 ton of dynamite, mantle waves can be observed over 1000 km distance.

Tens of OBSs can be deployed by the leap-frog method achieving, say, 300 km of profile length with 10–20 km spacings (e.g., Iwasaki *et al.*, 1990). It is not rare that up to 100 instruments are deployed in one operation. This allows two-dimensional OBS surveys with 1–2 km spacing with profiles of a few hundred kilometers.

We cannot overlook the importance of the reflection seismology method, which acquires acoustic data by hydrophone streamers (multiple and a few kilometers long) towed behind a shooting vessel. It is often important to combine both the OBS and seismic reflection datasets to resolve the structure in detail to large depths.

4.2.2 Natural Earthquake Studies Using OBS Arrays

OBS arrays are the best tool for analyzing submerged plate boundaries. OBS arrays can be deployed immediately after a large event to define the main fault plane, as well as its unevenness and complexities. Continuous and real-time seismic monitoring systems are being realized and are increasing, particularly near the coast of Japan.

Operating OBS regional arrays for more than several months to record distant natural earthquakes can reveal important features of the upper mantle. Features such as the deep roots of bathymetric highs (ridges, large igneous provinces, swells, or hot-spots) or how the subducted slabs affect the mantle are now becoming feasible targets.

4.3 Progress toward Permanent Broadband Observatories

The need for long-term ocean-bottom seismic observatories has been recognized since the early days of marine seismology. It is a formidable task, however, to sustain such stations, that is, to supply power and be able to store and retrieve the data. For an FDSN standard station, a broadband sensor is required. Then, together with a three-channel 24-bit digitizer (1.3 W), a real-time clock (5 V at 12 mA), and a data recorder (20 samples/sec for three channels of 24-bit data amount to about 6 Gbytes/year), about 4 W power supply would be a minimum requirement. Also, previous knowledge of noise characteristics is important for a long-term commitment (Webb, 1998).

It is generally conceived that installation in a borehole in intact rocks is superior to sea-floor deployment. Permanent stations of FDSN standard may require borehole installations at about 20 locations around the globe where no land/island exists within 2000 km distance. The efforts in this direction are briefly summarized in Table 2. At Hole 794D, the noise level was generally lower than at island or sea-floor stations, and was as low as about 10^4 (nm sec^{-1})^2 Hz^{-1} at about 0.2 Hz, where normally a noise peak is found. Signals from local events and airguns exhibit less reverberation as compared with OBS records. Surface wave dispersion from an event ($M_b = 5.4$) at 39° distance was clearly observed. The French experiment brought about the first comparative results on broadband seismic signal spectra of borehole (OFP) and buried (OFM) sensors. While the noise levels were generally low, less noise was found below 0.1 Hz with OFM than with OFP. The temperature change in the borehole was larger than on the sea floor. A more comprehensive test was conducted in

TABLE 2 Specifications of Borehole Sensors

	Broadband Sensor	Frequency	Dynamic Range	Resolution	Power	Power Supply	Installation	Operation	Reference
Hole 794D	Guralp Slimline CMG-3	DC-30 Hz	112 dB	10^{-7} m s^{-2}	Approx. 20 W (incl. cable loss)	24 V 38 AH × 20	Japan Sea; drill ship	1989 Sep.	Suyehiro *et al.* (1992, 1995)
Hole 396B	Guralp CMG-3			10^{-9} m s^{-2}			Mid-Atlantic Ridge; manned submersible	1992 May	Montagner *et al.* (1994)
Hole 843B	Teledyne Geotech KS54000	3 m–4 Hz	24-bit	2400 V m^{-1} s^{-1}	sensor-2.4 W	NA	Wireline reentry	1998 Feb.–Jun.	Collins *et al.* (2001)
Holes 1150/1151	Guralp CMG-1	3 m–50 Hz	24-bit	1500 V m^{-1} s^{-1}	<5 W	18 W sea water battery	Drillship	1999 Jul.–Aug.	Araki (2000)
Hole 1172	Guralp CMG-1	3 m–50 Hz	24-bit	1500 V m^{-1} s^{-1}	<5 W	15 W sea water battery	Drillship	2000 Aug.	

1998 off Oahu in Hole 834D. Three-month long broadband records were obtained together with a station in Oahu, on the sea floor and buried in sediment. A significant improvement in noise level was confirmed for the buried and borehole systems compared to the sea-floor system.

In the coming years, it is expected that the number of observatories will increase to fill the global gaps as well as for monitoring active geological processes together with other sensors. The installation of sensor and recording systems can be carried out using one of several existing general- and special-purpose manned submersibles or ROV systems. Modularity, common components, and standardized formats to reduce costs and provide ease of maintenance are desirable.

5. Data Analyses

Seismological data analyses are basically the same as those applied to data obtained on land. For example, when locating earthquakes, one must take into account OBS depths in a similar manner to when handling land station heights. We note in the following some characteristics pertinent to marine controlled-source seismic data.

5.1 Reflection Seismic Surveys

The basic idea of reflection seismic surveys is to make a sectional picture of sub-sea-floor structure by plotting observed seismic waves reflected from interfaces, where seismic impedance (the product of density and seismic velocity) changes. In marine environment, seismic waves generated by artificial seismic sources are received by single or multiple arrays of hydrophone receivers. An airgun array is used to shape the signal waveform into an impulse-like signal. Both source and receiver arrays are towed behind a vessel (Fig. 6a). In general, the wider the aperture of the source–receiver configuration, the deeper the structural target can be. The shorter the channel separation, the higher is the spatial resolution. A basic flow of data analysis consists of the following:

- Data sorting, in which data are grouped by common spatial reflected points (common mid points; CMPs).
- Deconvolution, in which observed waveforms are processed to improve the temporal resolution.
- Seismic velocity analysis, in which seismic velocity structure is determined.
- Normal moveout (NMO) correction, in which traces are stretched according to the velocity structure so that CMP-gather traces having different angles of reflection are mapped onto the vertical axis through the CMP to represent zero-offset reflection (any returned signals are mapped to this single vertical trace as if acoustic impedance changes existed vertically below).
- Stacking, in which NMO-corrected traces are added in order to increase the signal-to-noise ratio (Fig. 6a).
- Migration, in which dipping reflectors and diffracted energy are mapped to their true locations. (This process is intended to move the reflection energy back to its original location (Fig. 6b).)

Detailed explanations of each process can be found in general texts (e.g., Yilmaz, 1987; Bancroft, 1998).

5.2 Refraction Seismic Surveys

In most cases, refraction surveys are carried out using ocean-bottom seismometers with OBSs are deployed on the sea floor along profiles to observe seismic waves generated on the sea surface. This is the main difference from land seismic observations. One must take care of the water depth layer in analyses of observed seismic waves. Formulas for land experiments are often written for a series of seismic signals from a single shot point to many receivers. In the marine case, the data are often assembled in a series of many shots to each OBS. Therefore, when amplitudes are considered, one cannot simply interchange source and receiver from codes for land experiments to correctly calculate amplitude variations. Also, the acoustic basement beneath soft sediments often acts as a good phase conversion interface, more so than the sea floor.

6. Concluding Remarks

In this chapter we have mainly seen the present status of marine seismology from observational examples in the Pacific. Clearly, there is still much room for improvement, particularly in the observational aspects. While most earth scientists will agree that marine seismological investigations should provide fundamental knowledge for understanding the Earth's dynamics, the technical challenge of achieving good-quality observations has been formidable. Starting from scratch, we have come a long way in the last half a century or so.

Controlled-source seismology has often been performed with many shots and a few receivers, which is the opposite of land experiments. However, nowadays, 100-OBS experiments are feasible and are being done together with deep-penetration seismic reflection experiments. Ships allow very flexible and optimal designs of seismic experiments. Steps are being taken to illuminate 3D structures, such as subducted seamounts, magma chambers, or pathways of fluids (e.g., Bangs *et al.*, 1999). High-resolution deep-penetration modeling of laterally heterogeneous structures at plate boundaries will continue to be exciting, with more and more focused targets.

Attempts to establish long-term seismic observatories have been less successful, but recent results are encouraging and with international cooperation it is expected that the global seismic network will finally truly cover the whole globe in

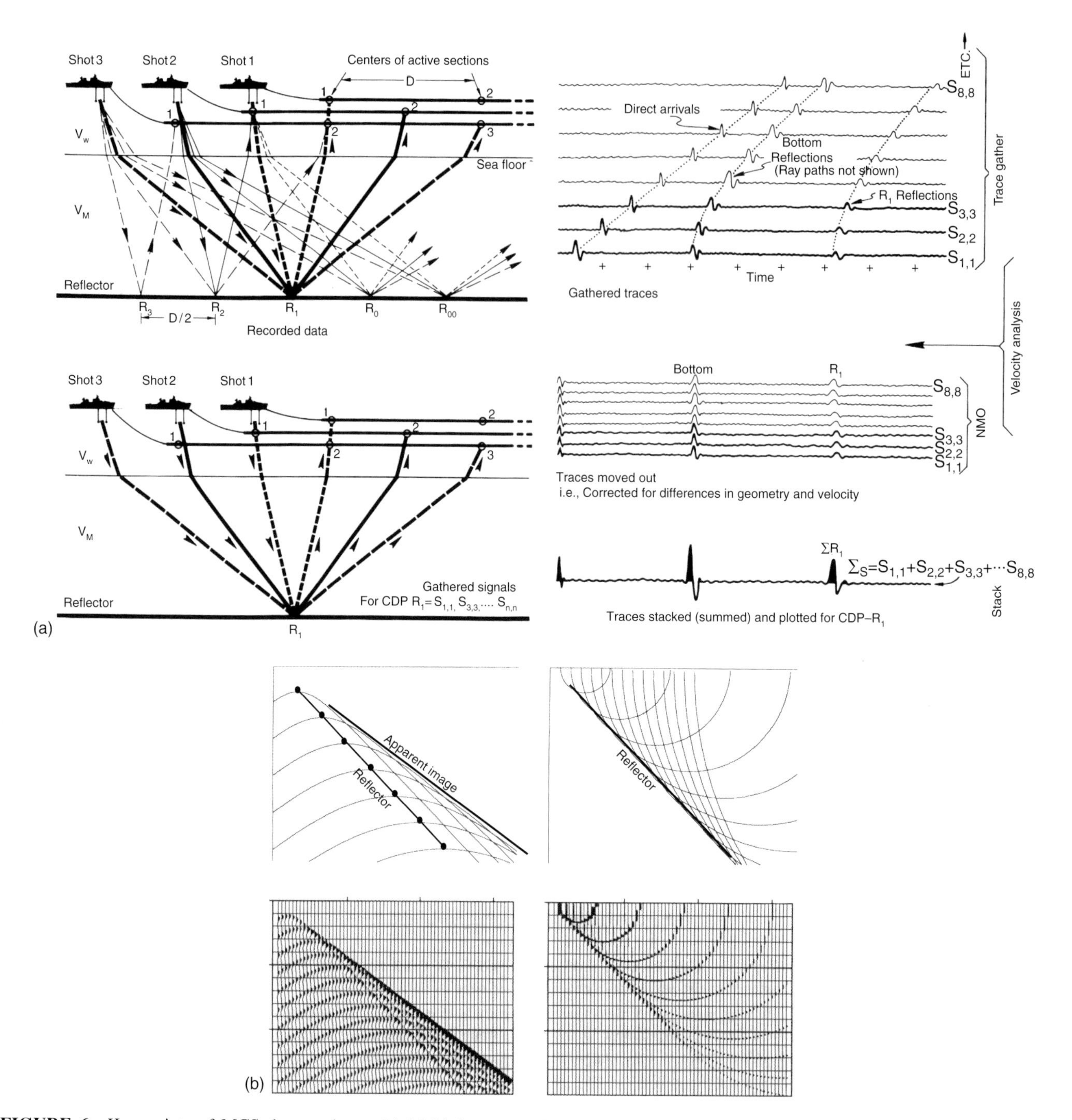

FIGURE 6 Key points of MCS data analyses. (a) Multiple traces with reflections from a common mid point are gathered and stacked to enhance signal-to-noise ratio. A correct velocity estimate is necessary for constructive superposition. (From Talwani *et al.*, 1977; copyright by the American Geophysical Union) (b) Migration. (From Claerbout, 1985.) Left panels: Signals along hyperbolas are diffraction from diffractors indicated by dots at apex. They constructively make an apparent image. The true reflector (connecting dots) is shorter, steeper, and updip relative to the apparent image. They coincide only when reflector is horizontal. Right panels: If a reflection signal is recorded, it is translated vertically beneath along a semicircle. The apparent image is the connection of the semicircle bottom points. The true reflector is the constructive line that appears from superposition of these semicircles.

the early 21st century. At the same time, long-term mobile-array observation will become available to allow year-scale collection of natural earthquake records.

References

Adair, R.G., J.A. Orcutt, and T.H. Jordan (1984). Analysis of ambient seismic noise recorded by downhole and ocean-bottom seismometers on deep sea drilling project Leg 78B. *Init. Rept. Deep Sea Drilling Proj.* **78**, 767–780.

Araki, E. (2000). Geophysical nature of broadband seismic signals in deep oceans. Dr. Sci. dissertation, University of Tokyo.

Asada, T. and H. Shimamura (1979). Long-range refraction experiments in deep ocean. *Tectonophysics* **56**, 67–82.

Bancroft, J. (1998). "A Practical Understanding of Pre- and Poststack Migrations," Vols. 1 and 2. Society of Exploration Geophysicists.

Bangs, N., T. Shipley, J. Moore, and G. Moore (1999). Fluid accumulation and channeling along the northern Barbados Ridge decollement thrust. *J. Geophys. Res.* **104**, 20399–20414.

Christeson, G.L., K.D. McIntosh, T.H. Shipley, E.R. Flueh, and H. Goedde (1999). Structure of the Costa Rica convergent margin, offshore Nicoya Peninsula. *J. Geophys. Res.* **104**, 25443–25468.

Claerbout, J.F. (1985). "Imaging the Earth's Interior." Blackwell Scientific, Oxford.

Collins, J.A., F.L. Vernon, J.A. Orcutt, *et al.* (2001). Broadband seismology in the oceans: lessons from the Ocean Seismic Network Pilot Experiment. *Geophys. Res. Lett.* **27**, **28**, 49–52.

Ewing, J. and R.P. Meyer (1982). Rivera Ocean Seismic Experiment (ROSE) overview. *J. Geophys. Res.* **87**, 8345–8358.

Ewing, M. and A.C. Vine (1938). Deep-sea measurements without wires or cables. *EOS Trans. Am. Geophys. Union* **19**, 248–251.

Flueh, E.R., M.A. Fisher, J. Bialas, *et al.* (1998a). New seismic images of the Cascadia subduction zone from cruise SO108-ORWELL. *Tectonophysics* **293**, 69–84.

Flueh, E.R., N. Vidal, C.R. Ranero, *et al.* (1998b). Seismic investigation of the continental margin off-and onshore Valparaiso, Chile. *Tectonophysics* **288**, 251–263.

Hino, R., S. Ito, H. Shiobara, *et al.* (2000). Aftershock distribution of the 1994 Sanriku-oki earthquake (M_w 7.7) revealed by ocean bottom seismographic observation. *J. Geophys. Res.* **105**, 21697–21710.

Hirata, N., H. Nambu, M. Shinohara, and K. Suyehiro (1992). Seismic evidence of anisotropy in the Yamato Basin. *Proc. ODP Sci. Results* **127–128**(Pt. 2), 1107–1122.

Iwasaki, T., N. Hirata, T. Kanazawa, *et al.* (1990). Crustal and upper mantle structure in the Ryukyu Island Arc deduced from deep seismic sounding. *Geophys. J. Int.* **102**, 631–651.

Kodaira, S., N. Takahashi, J.-O. Park, K. Mochizuki, M. Shinohara, and S. Kimura (2000a). The western Nankai Trough seismogenic zone: results from wide-angle ocean-bottom seismographic survey. *J. Geophys. Res.* **105**, 5887.

Kodaira, S., N. Takahashi, A. Nakanishi, S. Miura, and Y. Kaneda (2000b). Subducted seamount imaged in the rupture zone of the 1946 Nankaido Earthquake. *Science* **289**, 104–106.

LaTraille, S.L. and D.M. Hussong (1980). Crustal structure across the Mariana island arc. In: "The Tectonic and Geologic Evolution of Southeast Asian Seas and Islands" (D.E. Hayes, Ed.), pp. 209–221. *AGU Monograph* 23.

Le Pichon, X., S. Lallemant, H. Tokuyama, F. Thou, P. Houchon, and P. Henry (1996). Structure and evolution of the backstop in the Nankai Trough area: implications for the soon-to-come Tokai earthquake. *Island Arc* **5**, 440–454.

Ludwig, W.J., J.I. Ewing, M. Ewing, *et al.* (1966). Sediments and structure of the Japan Trench. *J. Geophys. Res.* **71**, 2121–2137.

Ludwig, W.J., S. Murauchi, N. Den, *et al.* (1973). Structure of east China Sea–west Philippine Sea margin off southern Kyushu, Japan. *J. Geophys. Res.* **78**, 2526–2536.

Mjelde, R., M.A. Sellevoll, H. Shimamura, T. Iwasaki, and T. Kanazawa (1993). Crustal structure beneath Lofoten, N. Norway, from vertical incidence and wide-angle seismic data. *Geophys. J. Int.* **114**, 116–126.

Mochizuki, K., G. Fujie, T. Sato, *et al.* (1998). Heterogeneous crustal structure across a seismic block boundary along the Nankai Trough. *Geophys. Res. Lett.* **25**, 2301–2304.

Montagner, J.-P., *et al.* (1994). The French pilot experiment OFM-SISOBS: first scientific results on noise level and event detection. *Phys. Earth Planet. Inter.* **84**, 321–336.

Moore, G., T.H. Shipley, P.L. Stoffa, *et al.* (1990). Structure of the Nankai Trough accretionary zone from multichannel seismic reflection data. *J. Geophys. Res.* **95**, 8753–8765.

Nakanishi, A., H. Shiobara, R. Hino, S. Kodaira, T. Kanazawa, and H. Shimamura (1998). Detailed subduction structure across the eastern Nankai Trough obtained from ocean bottom seismographic profiles. *J. Geophys. Res.* **103**, 27151–27168.

Nasu, N., R. von Huene, Y. Ishikawa, M. Langseth, T. Bruns, and E. Honza (1980). Interpretation of multichannel seismic reflection data, Legs 56 and 57, Japan Trench transect. *Init. Rep. Deep Sea Drilling Proj.* **56/57**, 489–503.

Park, J.-O., T. Tsuru, Kaneda, *et al.* (1999). A subducting seamount beneath the Nankai accretionary prism off Shikoku, southwestern Japan. *Geophys. Res. Lett.* **26**, 931–934.

Shipley, T.H., K.D. McIntosh, E.A. Silver, and P. Stoffa (1992). Three-dimensional seismic imaging of the Costa Rica accretionary prism: structural diversity in a small volume of the lower slope. *J. Geophys. Res.* **91**, 2019–2028.

Suyehiro, K. (1988). Controlled source seismology. In: "The Ocean Basins and Margins" (A.E.M. Nairn, F.G. Stehli, and S. Uyeda, Eds.) Vol. 7B, pp. 559–594. Plenum, New York.

Suyehiro, K. and A. Nishizawa (1994). Crustal structure and seismicity beneath the forearc off northeastern Japan. *J. Geophys. Res.* **99**, 22331–22347.

Suyehiro, K., T. Kanazawa, N. Hirata, M. Shinohara, and H. Kinoshita (1992). Broadband downhole seismometer experiment at Site 794: a technical paper. *Proc. ODP. Sci. Results*, **127–128**(Pt. 2), 1061–1073.

Suyehiro, K., T. Kanazawa, N. Hirata, and M. Shinohara (1995). Ocean downhole seismic project. *J. Phys. Earth* **43**, 599–618.

Suyehiro, K., N. Takahashi, Y. Ariie, *et al.* (1996). Continental crust, crustal underplating, and low-Q upper mantle beneath an ocean island arc. *Science* **272**, 390–392.

Takahashi, N., K. Suyehiro, and M. Shinohara (1998). Implications from the seismic crustal structure of the northern Izu-Ogasawara arc. *Island Arc* **7**, 383–394.

Takahashi, N., S. Kodaira, T. Tsuru, *et al.* (2000). Detailed plate boundary structure off northeast Japan coast. *Geophys. Res. Lett.* **27**, 1977–1980.

Talwani, M., C.C. Windisch, P.L. Stoffa, P. Buhl, and R.E. Houtz (1977). Multichannel seismic study in the Venezuelan Basin and the Curacao Ridge. In: "Island Arcs Deep Sea Trenches and Back-Arc Basins" (M. Talwani and W.C. Pitman III, Ed.), pp. 83–98, American Geophysical Union, Washington, DC.

Trehu, A.M., I. Asudeh, T.M. Brocher, *et al.* (1994). Crustal architecture of the Cascadia forearc. *Science* **266**, 237–243.

Tsuru, T., J.-O. Park, N. Takahashi, *et al.* (2000). Tectonic features of the Japan Trench convergent margin off Sanriku, northeastern Japan, revealed by multichannel seismic reflection data. *J. Geophys. Res.* **105**, 16403–16413.

von Huene, R. and D.W. Scholl (1991). Observations at convergent margins concerning sediment subduction, subduction erosion, and the growth of continental crust. *Rev. Geophys.* **29**, 279–316.

Walther, C.H.E., E.R. Flueh, C.R. Ranero, R. von Huene, and W. Strauch (2000). Crustal structure across the Pacific margin of Nicaragua: evidence for ophiolitic basement and a shallow mantle sliver. *Geophys. J. Int.* **141**, 759–777.

Webb, S.C. (1998). Broadband seismology and noise under the ocean. *Rev. Geophys.* **36**, 105–142.

White, R.S., G.D. Spence, S.R. Fowler, D.P. McKenzie, G.K. Westbrook, and A.N. Bowen (1987). Magmatism at rifted continental margins. *Nature* **330**, 439–444.

White, R.S., D. McKenzie, and R.K. O'Nions (1992). Oceanic crustal thickness from seismic measurements and rare earth element inversions. *J. Geophys. Res.* **97**, 19683–19715.

Yilmaz, O. (1987). "Seismic Data Processing." Society of Exploration Geophysicists.

28

Tsunamis

Kenji Satake
Active Fault Research Center, GSJ/AIST, Tsukuba, Japan

1. Introduction

A tsunami is an oceanic gravity wave generated by submarine earthquakes or other geological processes such as volcanic eruptions or landslides. Most tsunamis are caused by shallow large earthquakes and hence are distributed along the subduction zones (Fig. 1). *Tsunami* is a Japanese word meaning harbor wave. Tsunamis are usually small and barely noticed in deep oceans, but they become large and cause damage when they approach coasts or harbors. They are sometimes called seismic sea waves or, erroneously, tidal waves.

Tsunamis have many aspects that are studied by researchers in various fields. Their generation is related to geological processes. Propagation and observation are the domain of oceanographers, and basic hydrodynamics is needed to understand their characteristics. The coastal behaviors, such as run-up on beaches or resonance in bays, are mostly studied by coastal engineers. City and land-use planners need to consider tsunami risks, and public officials are responsible for tsunami warning and evacuation. In this chapter, the wide range of tsunami phenomena will be reviewed from a seismological viewpoint.

1.1 Examples

1.1.1 The 1993 Okushiri Tsunami

Japan has a long history of tsunami damage. The most recent example is that caused by the Southwest Hokkaido earthquake

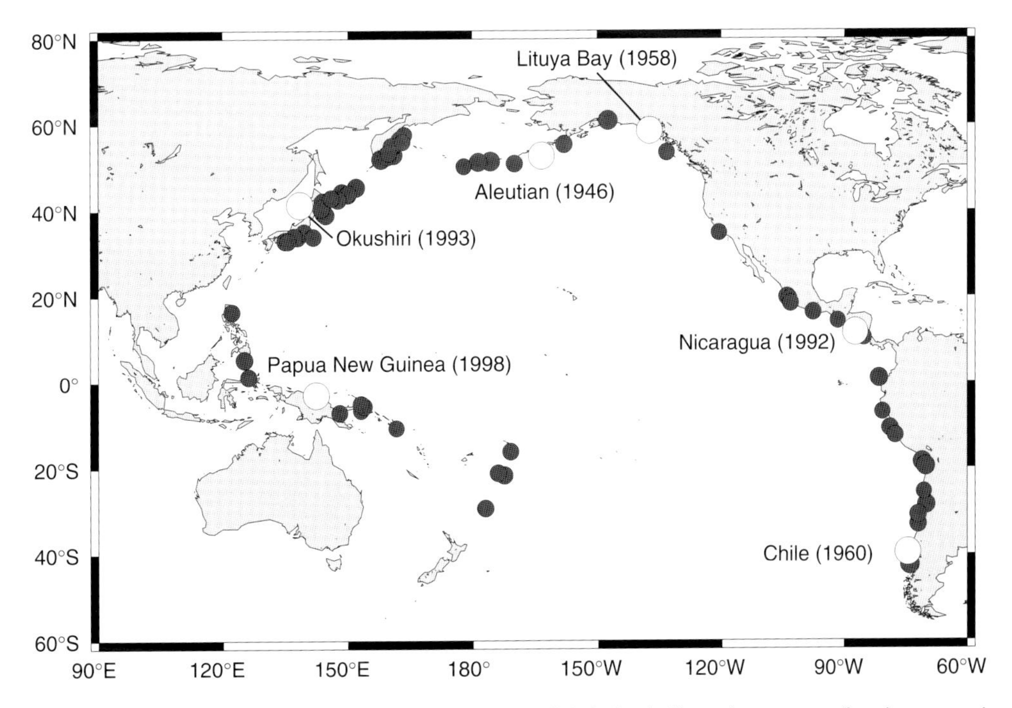

FIGURE 1 Distribution of recent Pacific tsunamis. Solid circles indicate the sources of major tsunamis for 1837–1974 (Abe, 1979). Named open circles indicate the tsunamis described in the text.

INTERNATIONAL HANDBOOK OF EARTHQUAKE AND ENGINEERING SEISMOLOGY, VOLUME 81A ISBN: 0-12-440652-1

on 12 July 1993 ($M_W = 7.8$). Okushiri Island, a small island in the Japan Sea, is located within the source area, hence the strong ground shaking was followed, within a few minutes, by tsunami. The island suffered from fire and floods, causing more than 200 casualties. This was probably the best-studied tsunami in terms of field surveys or modeling (Hokkaido Tsunami Survey Group, 1993; Takahashi *et al.*, 1995; Tanioka *et al.*, 1995). The tsunami height was mostly 5–10 m, while the maximum tsunami run-up was measured as more than 30 m.

1.1.2 The 1960 Chilean Tsunami

The largest ($M_W = 9.5$) earthquake in this century occurred off southern Chile on 22 May 1960. The tsunami caused damage on the Chilean coast, including more than 1000 casualties, then propagated across the Pacific Ocean. About 15 h later, the tsunami reached the Hawaiian islands and took about 60 lives. It continued across the Pacific, reaching Japan in about 23 h, and caused more than 100 casualties. The large tsunami in Japan was partly due to a resonance effect: some bays have a shape whose characteristic period coincided with the predominant period (about 1 h) of the trans-Pacific tsunami.

1.1.3 The 1946 Aleutian Tsunami

The Aleutian earthquake of 1 April 1946 was very unusual; it generated huge tsunamis in the Aleutians and Hawaii, even though M_S was only 7.4. This is a typical example of a "tsunami earthquake" that generates much larger tsunamis than expected from seismic waves. The Scotch Cap lighthouse on Unimak Island, about 100 km from the epicenter, apparently had no damage from ground shaking but was totally demolished by the tsunami. The tsunami traveled the Pacific Ocean southward and reached the Hawaiian Islands in the morning of April Fool's Day. Without any warning, the tsunami took 159 lives in the Hawaiian Islands. A Pacific tsunami warning system was introduced after this tsunami.

1.1.4 The 1958 Lituya Bay Wave

The record of wave height may be the one in Lituya Bay, Alaska. A large strike-slip earthquake ($M_W = 7.9$) occurred on 10 July 1958, and triggered a huge rockslide (total volume of about $3 \times 10^7\ m^3$) in this 10 km long bay. The rockslide generated water waves that surged up the opposite slope to 520 m altitude, but the large waves were mostly confined in the bay. Hence this example is not considered as a typical tsunami.

1.1.5 The 1998 Papua New Guinea Tsunami

The most recent tsunami hazard was from an earthquake ($M_W = 7.1$) along the northern coast of New Guinea Island on 17 July 1998. The tsunami heights were as large as 15 m around Sissano Lagoon, and the total number of casualties was reported to be more than 2000. The large tsunami (> 5 m) was limited to a small region (~40 km) along the coast, suggesting a local effect of tsunami amplification (Kawata *et al.*, 1999). Post-tsunami marine surveys (Tappin *et al.*, 1999) indicate a focusing of tsunamis due to bathymetry and an involvement of submarine landslides.

2. Size of Tsunamis

2.1 Tsunami Magnitude Scales

2.1.1 Imamura–Iida Scale *m*

The traditional tsunami magnitude scale is the so-called Imamura–Iida scale, m. Although the original definition was descriptive, the value is approximately equal to

$$m = \log_2 h \tag{1}$$

where h is the maximum run-up height in meters (Iida *et al.*, 1967). This scale is similar to an Earthquake intensity scale, and is especially convenient for old tsunamis from which no instrumental records exist.

Hatori (1979) extended the Imamura–Iida m scale to include far-field tsunami data. He also considered the effect of distance R and constructed an h–$\sqrt{R}$ diagram in log–log scale. The diagram is calibrated in such a way that m becomes 3 when $h = 0.5$ m and $R = 1000$ km. The value of m increases one unit when the energy becomes five times (or the amplitude is $\sqrt{5}$ times) larger. This can be expressed as

$$m = 3 + \frac{\log\left[(h/0.5)(R/1000)^{1/2}\right]}{\log\sqrt{5}} \tag{2}$$

where h is in meters and R is in km. This indicates that the decay of tsunami height with distances is $R^{-1/2}$, which is theoretically predicted for nondispersive waves traveling a long distance (Comer, 1980).

2.1.2 Tsunami Intensity

Soloviev (1970) pointed out that the Imamura–Iida m scale is more like an earthquake intensity scale than a magnitude. He also distinguished the maximum tsunami height h and the mean tsunami height $\bar{h}$. He then defined tsunami intensity i as

$$i = \log_2(\sqrt{2}\,\bar{h}) \tag{3}$$

Comparison of Eqs. (1) and (3) suggests that the mean tsunami height is given as $1/\sqrt{2}$ times the maximum height. The maximum intensity i on the coast nearest to the source is used to quantify the tsunami source.

2.1.3 Tsunami Magnitude M_t

Another magnitude scale M_t, called tsunami magnitude, is defined and assigned for many earthquakes by Abe. The definition of M_t for a trans-Pacific tsunami is (Abe, 1979)

$$M_t = \log H + C + 9.1 \tag{4}$$

and for a regional ($100\,\text{km} < \Delta < 3500\,\text{km}$) tsunami is (Abe, 1981):

$$M_t = \log H + \log \Delta + 5.8 \tag{5}$$

where H is maximum amplitude on tide gauges in meters, C is a distance factor depending on a combination of the source and the observation points, and Δ is the actual distance in km. The above formulas were calibrated with the moment magnitude scale, M_W of earthquakes. Figure 2 shows comparison of tsunami magnitude M_t and surface wave magnitude M_S for large earthquakes in the last century. Smaller M_S values for events with $M_t > 8$ may be due to the saturation of M_S at around 8.0 (Geller, 1976). Equation (5) indicates that for the same M_t, the tsunami amplitude H decays with $1/\Delta$, which is different from the assumption used for the Imamura–Iida scale m, and may be valid only for regional distances.

2.2 Tsunami Catalogs

The location, size, run-up heights, and damage of past tsunamis have been compiled by several researchers and published as catalogs. For the Pacific tsunamis, Iida *et al.* (1967) contains numerous tsunami height data as far back as AD 173. Russian catalogs (Soloviev and Go, 1974a,b; Soloviev *et al.*, 1992)

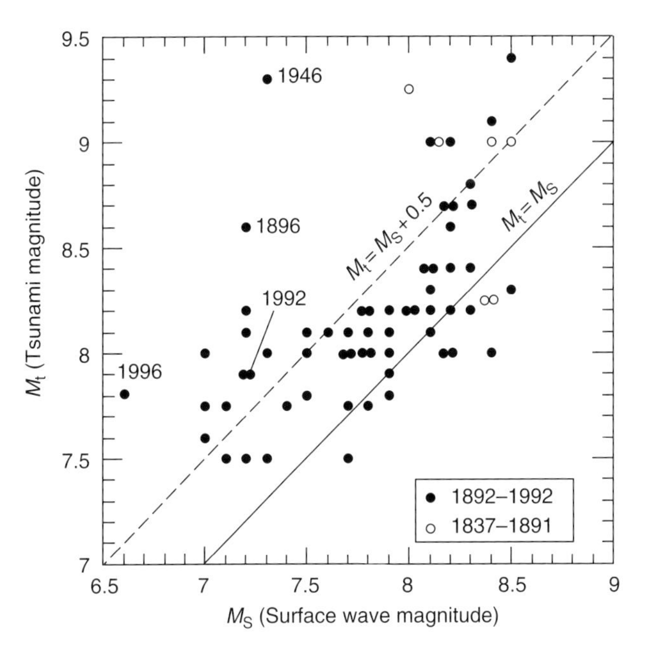

FIGURE 2 Relation between M_t (tsunami magnitude) and M_S (surface wave magnitude). A similar plot by Abe (1979) is updated by replacing with more recent estimates of M_S and by adding data for tsunamis since 1979. "Tsunami earthquakes" (above the dashed line) have M_t larger than M_S by more than 0.5 units.

include Pacific tsunami data (excepting for Russia) with some tide gauge records, tabulated tsunami heights, and descriptions for each tsunami. Regional catalogs including tide gauge records were also published for Alaska (Lander, 1996), the west coast of the United States (Lander *et al.*, 1993), and Japan (Watanabe, 1998).

Some of the numerical data can be accessed at the following web sites:

- *http://www.ngdc.noaa.gov/seg/hazard/tsu.shtml*
- *http://tsun.sscc.ru/htdbpac/*

3. Tsunami Observations

3.1 Instrumental Measurements

3.1.1 Tide Gauges in Ports and Harbors

Instrumental measurements of tsunamis have been done in bays or harbors using tide gauges (Fig. 3). In the United States, the National Ocean Service (NOS) has been operating tide gauge stations since the 1850s.

The most traditional and popular tide gauge is a mechanical type with stilling well or tide well. Vertical motion of a float in the well is transmitted by wire to the recorder. For longer-period motions, water level in the well can be assumed to be the same as that of the outer sea. Although the system is robust, the installation (and construction) of tide wells is expensive and periodic maintenance is essential. A cheaper tide gauge is a pneumatic, or gas-purged, pressure gauge, often called a bubbler gauge.

Because the original purpose of tide gauges is to measure ocean tides, whose typical period is 12 h, the sampling interval

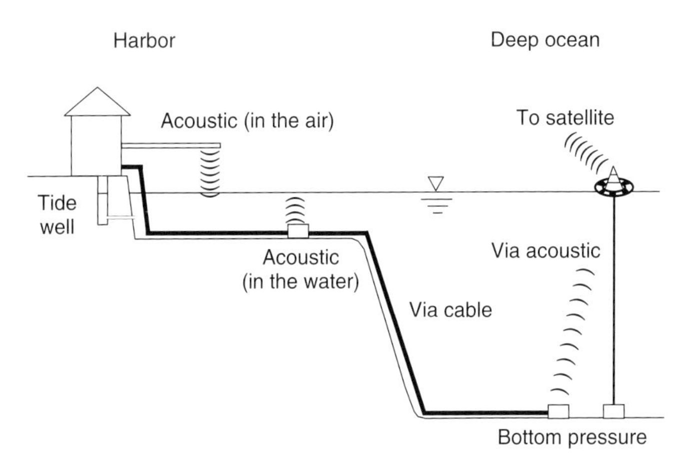

FIGURE 3 Schematic illustration of instrumental tsunami observations. Traditional measurements use tide gauges in harbors. Acoustic sensors, either set on the ocean bottom or in the air, are used near shore. In the deep ocean, bottom pressure gauges are used, and the signals are transmitted either via cable or acoustic modem and satellite.

and system response of tide gauges are often inadequate for tsunamis, whose typical period is 5–30 min (Satake *et al.*, 1988).

3.1.2 Acoustic Gauges and Open Coast Measurements

In recent years, tsunami observations on coastal water, outside harbors or bays, have been made to avoid the topographic effects and for the purpose of early detection. Acoustic gauges can be set on the water bottom or above the water surface. The NOS's Next Generation Water Level Measurement System uses air acoustic sensors. In Japan, Port and Airport Research Institute operates about 50 wave gauges outside ports and harbors (called NOWPHAS), using bottom acoustic sensors.

3.1.3 Pressure Gauges and Deep Ocean Measurements

In the deep ocean (near the tsunami source), tsunami waveforms are expected to be simpler and free from coastal topographic effects, although their amplitudes are smaller. Deep ocean measurements of tsunamis have been made in Japan and the United States using bottom pressure gauges for the purpose of early detection and warnings. The Japanese systems send signals through cables, while the US system uses acoustic modems and satellites. The latter data can be accessed at *http://www.pmel.noaa.gov/tsunami/*.

The quartz crystal pressure transducer is sensitive to changes corresponding to less than 1 mm of sea level change in the tsunami frequency band (Eble and Gonzalez, 1991). However, the pressure transducer is also very sensitive to temperature, hence the temperature correction is critical. Since the instrument is set on the ocean bottom, it also records ground motion and works as an ocean-bottom seismograph; seismic waves sometimes become noise for tsunami records (Okada, 1995).

3.2 Field Survey to Measure Run-up Heights

To supplement instrumental records, field surveys are carried out after large tsunamis. In the surveys, tsunami heights are estimated from various types of evidence, such as traces on walls, damage to constructions, scratches or dead leaves on trees, and seaweed or other floating materials transported by tsunamis (e.g., Hokkaido Tsunami Survey Group, 1993; Kawata *et al.*, 1999). Interviews with local people provide information on tsunami heights, though these are often subjective and less accurate. Interviewers also request the times of earthquake and tsunami, seismic intensity, and other descriptions of tsunamis.

3.3 Study of Historical Documents

Tsunamis and their damage recorded in historical documents are also useful in the study of old earthquakes and tsunamis for which no instrumental data are available. Some countries, such as Japan or Italy, have long histories of written records that have provided accurate dates for past earthquakes and tsunamis (e.g., Tinti *et al.*, 1997). The long-term seismicity or recurrence interval of earthquakes and tsunamis can be studied from these records. The historical tsunami records could be also used for trans-Pacific tsunamis; the date of the most recent great earthquake in the Cascadia subduction zone was found to be 26 January 1700 from Japanese historical documents (Satake *et al.*, 1996).

3.4 Geological Methods

Even older tsunamis can be studied from their deposits by geological means. Geological studies of tsunami deposits have become popular in the last decade or so, and many examples, not only from earthquake-generated tsunamis but also from submarine landslides or impact-generated tsunamis, are reported. In Japan, Minoura and Nakaya (1991) examined the deposits from the 1983 Japan Sea tsunamis by drilling boreholes in lagoons. They further studied paleotsunamis through sedimentological and geochemical analyses, and estimated the dates and run-up heights of prehistoric tsunamis. In Europe, tsunami deposits from the Storegga slides in the North Sea have been found and carefully studied in Scotland, Norway, and Iceland (Dawson, 1994). In the United States, tsunami deposits from prehistoric great earthquakes along the Cascadia subduction zone have been studied (Atwater *et al.*, 1995; Clague, 1997). Geological evidence of late Pleistocene tsunamis has also been studied in Australia (e.g., Bryant *et al.*, 1996).

3.5 Definitions of Tsunami Heights and Sea Levels

3.5.1 Amplitude and Wave Heights

The *amplitude* is measured from zero to peak, either positive or negative. The *wave height*, also called range or double amplitude, is measured from trough to peak. Hence the *tsunami height* is measured from trough to peak on tide gauges, while the *tsunami amplitude* is measured from estimated tidal level at the time (Fig. 4).

3.5.2 Inundation and Run-up Heights

If no tide gauge record is available, the tsunami heights are measured by field surveys. The tsunami height on land measured from sea level at the time of tsunami arrival is called *inundation height*. The horizontal distance measured from the coast (at the time of tsunami arrival) is called *inundation distance*. Tsunami *run-up height* usually refers to the inundation height at the maximum inundation distance. The run-up height is not necessarily the same as the inundation height near the shore.

3.5.3 Sea Levels and Shoreline Problem

Several reference levels are used for measurements of water height or depth. *Mean sea level* is the average value of sea

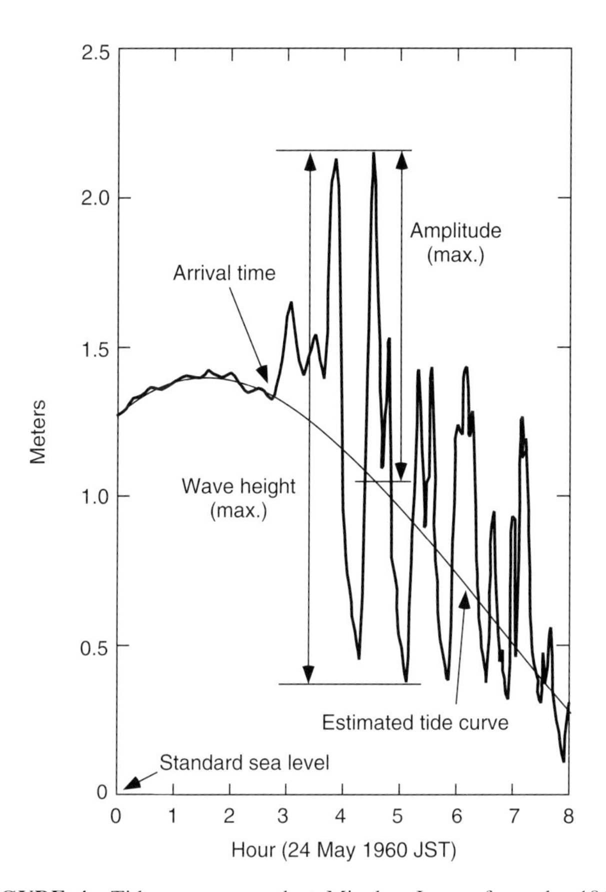

FIGURE 4 Tide gauge record at Miyako, Japan, from the 1960 Chilean earthquake tsunami. The earthquake origin time was about 04:00 on 23 May on Japan Standard Time (JST), hence the travel times is nearly 23 h. Definitions of tsunami amplitude and wave height are also shown.

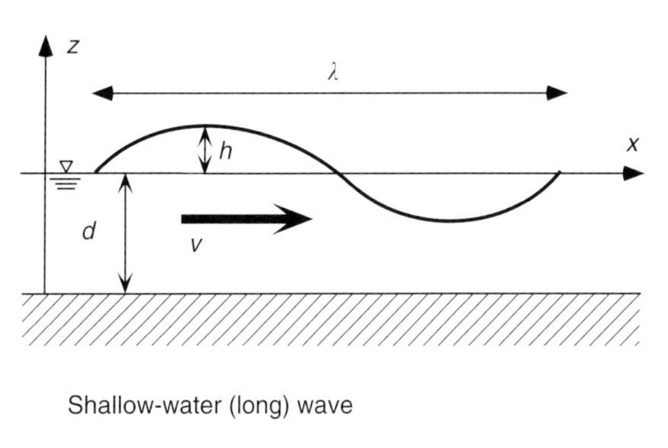

Shallow-water (long) wave

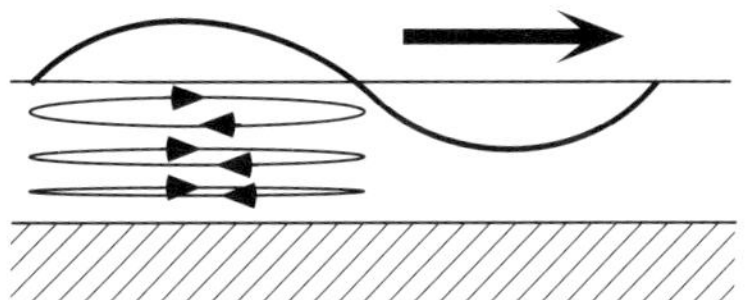

Deep-water (short) wave

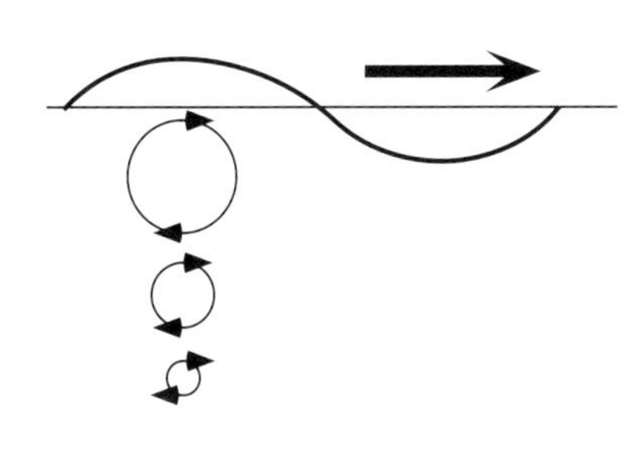

FIGURE 5 Coordinate system and important parameters for hydrodynamic computations of tsunamis (top). The lower figures schematically compare the shallow-water (long) waves and deep-water (short) waves. Particle motions are also shown.

levels observed over a period of years, and is used as a reference level for geodetic surveys. On the other hand, hydrographical measurements, mostly water depth, are referenced to an extreme low water *Chart Datum*. It is *Mean Lower Low Water* in the United States, while it is *Lowest Low Water* in Japan. Hence there is a gap between the geodetic and hydrographic zero lines, known as the "shoreline problem."

4. Hydrodynamics of Tsunamis

4.1 Shallow-Water (Long) Wave Theory

In this section, we take a two-dimensional Cartesian coordinate system with z axis vertical upward (the origin is on the undisturbed water level) and consider propagation of a wave (Fig. 5). When gravity is the restoring force, it is called a gravity wave. Euler's equation of motion can be written as

$$\frac{D\mathbf{V}}{Dt} = -\mathbf{g} - \frac{1}{\rho}\nabla p \tag{6}$$

where $\mathbf{V}$ is the velocity vector (whose x and z components are u and w, respectively), $\mathbf{g}$ is the gravitational acceleration, ρ and p are water density and pressure, respectively, and D/Dt indicates the total derivative

$$\frac{D\mathbf{V}}{Dt} = \frac{\partial \mathbf{V}}{\partial t} + \mathbf{V} \cdot \nabla \mathbf{V}$$

When the horizontal scale of motion, or the wavelength λ, is much larger than the water depth ($\lambda \gg d$), the vertical acceleration of water is negligible compared to gravity. This means that the horizontal motion of water mass is almost uniform from bottom to surface. Such a wave is called a shallow-water wave or long wave. For tsunamis, typical ocean depth is about 5 km and a large earthquake has a source size of several tens to hundreds of kilometers, hence the long-wave approximation is appropriate for most cases.

The horizontal component of expression (6) becomes, by replacing the horizontal pressure gradient with the slope of water surface,

$$\frac{Du}{Dt} = -g\frac{\partial h}{\partial x} \tag{7}$$

For tsunamis, the nonlinear advective term is usually small and can be ignored,

$$\frac{Du}{Dt} = \frac{\partial u}{\partial t} + u\frac{\partial u}{\partial x} \approx \frac{\partial u}{\partial t}$$

Thus the equation of motion can be written as

$$\frac{\partial u}{\partial t} = -g\frac{\partial h}{\partial x} \tag{8}$$

When the amplitude is small compared to the water depth ($h \ll d$), the conservation of mass, or the equation of continuity, can be written as

$$\frac{\partial h}{\partial t} = -\frac{\partial}{\partial x}(du) \tag{9}$$

Such a wave is called a small-amplitude, linear long wave. The assumptions are valid for most tsunami propagation paths except for near-shore regions.

From Eqs. (8) and (9), by assuming that water depth d is constant, we obtain the wave equation

$$\frac{\partial^2 h}{\partial t^2} = c^2\frac{\partial^2 h}{\partial x^2} \quad \text{where} \quad c = \sqrt{gd} \tag{10}$$

in which the velocity is determined by the water depth only.

4.2 General Gravity Waves

We consider a general gravity wave travelling in the x direction, $h = a\cos(kx - \omega t)$, where a is amplitude, k is wavenumber, and ω is angular frequency. The phase velocity is given as

$$c = \frac{\omega}{k} = \left(\frac{g}{k}\tanh kd\right)^{1/2} = \left(\frac{g\lambda}{2\pi}\tanh\frac{2\pi d}{\lambda}\right)^{1/2} \tag{11}$$

and the horizontal and vertical particle velocities are given as follows

$$u = a\omega\frac{\cosh(k(z+d))}{\sinh kd}\cos(kx - \omega t) \tag{12}$$

$$w = a\omega\frac{\sinh(k(z+d))}{\sinh kd}\sin(kx - \omega t) \tag{13}$$

4.2.1 Shallow-Water (Long) Wave Approximation

When the water depth is much smaller than the wavelength ($d \ll \lambda$), we can approximate $\tanh(2\pi d/\lambda) \approx 2\pi d/\lambda$, $\sinh(2\pi d/\lambda) \approx 2\pi d/\lambda$, and $\cosh k(z+d) \approx \cosh kd \approx 1$. Then Eq. (11) becomes $c = \sqrt{gd}$, the long-wave approximation discussed in the previous section. The vertical velocity [Eq. (13)] becomes

$$w = a\omega\left(1 + \frac{z}{d}\right)\sin(kx - \omega t) \tag{14}$$

which shows that $w = 0$ at the bottom, and w increases linearly toward the surface. The horizontal velocity [Eq. (12)] becomes

$$u = \frac{a\omega}{kd}\cos(kx - \omega t) = \frac{h\omega}{kd} \tag{15}$$

and the velocity ratio of vertical to horizontal components becomes $|w|/|u| = 2\pi d/\lambda$, which is very small. In other words, the particle motion is almost horizontal (Fig. 5).

4.2.2 Deep-Water (Short) Wave Approximation

When the water depth is much larger than the wavelength ($d \gg \lambda$), we can approximate

$$\tanh\frac{2\pi d}{\lambda} \approx 1$$

$$\cosh k(z+d) = \tfrac{1}{2}(e^{k(z+d)} + e^{-k(z+d)}) \approx \tfrac{1}{2}e^{k(z+d)}$$

and

$$\sinh kd = \tfrac{1}{2}(e^{kd} - e^{-kd}) \approx \tfrac{1}{2}e^{kd}$$

Then Eqs. (11) through (13) become

$$c = \frac{\omega}{k} = \left(\frac{g}{k}\right)^{1/2} = \left(\frac{g\lambda}{2\pi}\right)^{1/2} \tag{16}$$

$$u = a\omega e^{kz}\cos(kx - \omega t) \tag{17}$$

$$w = a\omega e^{kz}\sin(kx - \omega t) \tag{18}$$

These show that the particle motion is circular, and the amplitude decays exponentially with depth (Fig. 5). This is a surface wave. The phase velocity [Eq. (16)] shows a normal dispersion; the phase velocity is larger for longer wavelengths.

5. Tsunami Propagation

5.1 Ray Theoretical Approach

If the tsunami wavelength is much smaller than the scale of velocity heterogeneity, i.e., the depth change, then we can apply the geometrical ray theory of optics. The wavefronts can be drawn from raytracing results by connecting the locations of rays at the constant travel times. Such a diagram is called a *refraction diagram*. Traditionally, refraction diagrams have been drawn manually. Refraction diagrams can be prepared for major tsunami sources and used for tsunami warning; as soon as

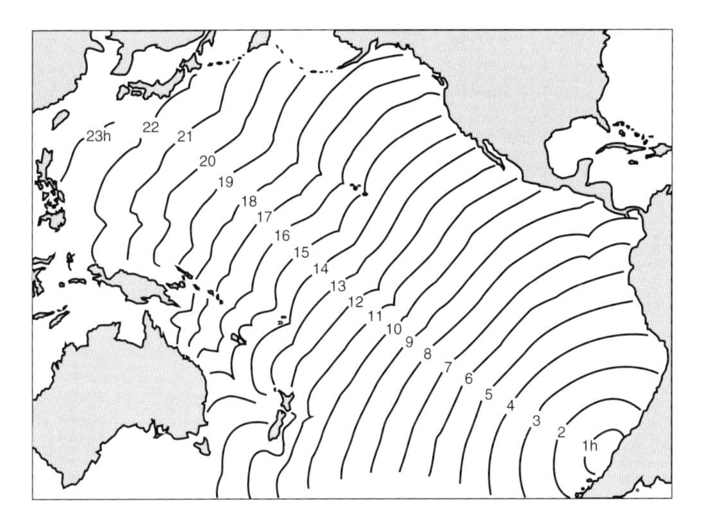

FIGURE 6 Refraction diagram for the 1960 Chilean tsunami. Wavefronts at each hour after the earthquake are shown by curves.

the epicenter is known, the arrival time of tsunamis can be estimated from refraction diagrams.

Figure 6 shows the refraction diagram from the 1960 Chilean earthquake with wavefronts at each hour. It is shown that the tsunami was expected to arrive at Japan in about 23 h and the energy is focused because of refraction.

The refraction diagrams can also be drawn backward from coasts. Such a diagram is called an *inverse refraction diagram* and is used to estimate the tsunami source area; the wavefronts or rays can be traced backward for corresponding travel times. The traced wavefronts from each observed station bound the tsunami source area (see Section 6.4).

5.2 Green's Law

From the conservation of energy along the rays, we can obtain $b_0 d_0^{1/2} h_0^2 = b_1 d_1^{1/2} h_1^2$ where d is the water depth, b is the distance between the rays, h is the tsunami amplitude, and the subscripts 0 and 1 indicate two different locations. If the tsunami amplitude at location 0 (e.g., at the source) is known, the tsunami amplitude after the propagation can be estimated as

$$h_1 = \left(\frac{b_0}{b_1}\right)^{1/2} \left(\frac{d_0}{d_1}\right)^{1/4} h_0 \tag{19}$$

This is known as Green's law. The ratio b_1/b_0 represents the spreading of rays, which can be obtained graphically from refraction diagrams. Note that the amplitude at the coast must be that of the direct wave not of the reflected waves.

5.3 General Gravity Waves

Here we will briefly examine general gravity waves deviating from the small-amplitude, linear shallow-water (long) wave approximations. We have already seen that the phase velocity of a gravity wave is given by Eq. (11). If we expand the hyperbolic tangent function into its Taylor series and include the second term, Eq. (11) becomes

$$c = \sqrt{gd}\left[1 - \tfrac{1}{3}(kd)^2\right]^{1/2} \approx \sqrt{gd}\left[1 - \tfrac{1}{6}(kd)^2\right] \tag{20}$$

The corresponding equation of motion is

$$\frac{\partial u}{\partial t} = -g\,\frac{\partial h}{\partial x} + \tfrac{1}{3}d^2\,\frac{\partial^3 u}{\partial x^2\,\partial t} \tag{21}$$

which is known as the linear Boussinesq equation.

If we relax the small-amplitude ($h \ll d$) assumption, the equations of motion and continuity are given as

$$\begin{aligned} \frac{\partial u}{\partial t} + u\,\frac{\partial u}{\partial x} &= -g\,\frac{\partial h}{\partial x} \\ \frac{\partial h}{\partial t} &= -\frac{\partial}{\partial x}[u(h+d)] \end{aligned} \tag{22}$$

These equations are for the finite-amplitude shallow-water waves. For the linear case (ignoring the advection term), the phase velocity is given by $c = \sqrt{g(d+h)}$. This means that the larger the amplitude, the faster the wave speed. As a consequence, peaks of a wave catch up with troughs in front of them, and the forward-facing portion of the wave continues to steepen. It will then eventually break. Such a phenomenon is sometimes called *amplitude dispersion*.

5.4 Shallow-Water Wave Equations

Including the bottom friction and the Coriolis force, the equation of motion for shallow-water (long) waves can be written for a three-dimensional case as follows:

$$\begin{aligned} \frac{\partial U}{\partial t} + U\,\frac{\partial U}{\partial x} + V\,\frac{\partial U}{\partial y} &= -fV - g\,\frac{\partial h}{\partial x} - C_f\,\frac{U\sqrt{U^2+V^2}}{d+h} \\ \frac{\partial V}{\partial t} + U\,\frac{\partial V}{\partial x} + V\,\frac{\partial V}{\partial y} &= fU - g\,\frac{\partial h}{\partial y} - C_f\,\frac{V\sqrt{U^2+V^2}}{d+h} \end{aligned} \tag{23}$$

and the equation of continuity is

$$\frac{\partial h}{\partial t} + \frac{\partial}{\partial x}[U(h+d)] + \frac{\partial}{\partial y}[V(h+d)] = 0 \tag{24}$$

where the coordinate system is x = East and y = South, f is the Coriolis parameter, C_f is a nondimensional frictional coefficient, and U and V are the average velocities in the x and y directions, respectively. The velocities are averaged from the bottom to the surface,

$$\begin{aligned} U &= \frac{1}{(h+d)}\int_{-d}^{h} u\,dz = \frac{Q_x}{(h+d)} \\ V &= \frac{1}{(h+d)}\int_{-d}^{h} v\,dz = \frac{Q_y}{(h+d)} \end{aligned} \tag{25}$$

where Q_x and Q_y are flow rates in the x and y directions. A typical value of a nondimensional frictional coefficient C_f is 10^{-3} for coastal water and 10^{-2} for run-up on land (see, e.g., Satake, 1995).

5.5 Numerical Computations

Let us consider a simple one-dimensional propagation (on constant depth) of small-amplitude linear long waves. Denoting the average velocity (x component only) as U, the equations of motion and continuity are given as follows:

$$\frac{\partial U}{\partial t} = -g\frac{\partial h}{\partial x} \tag{26}$$

$$\frac{\partial h}{\partial t} = -d\frac{\partial U}{\partial x} \tag{27}$$

Using the staggered (leap-frog) grid system for U and h, and denoting time $t = l\,\Delta t$ as superscripts and space $x = m\,\Delta x$ as subscripts where l and m are integers, these become

$$\frac{\left[U_m^{l+\frac{1}{2}} - U_m^{l-\frac{1}{2}}\right]}{\Delta t} = -g\frac{\left[h_{m+\frac{1}{2}}^{l} - h_{m-\frac{1}{2}}^{l}\right]}{\Delta x} \tag{28}$$

$$\frac{\left[h_{m+\frac{1}{2}}^{l+1} - h_{m+\frac{1}{2}}^{l}\right]}{\Delta t} = -d\frac{\left[U_{m+1}^{l+\frac{1}{2}} - U_m^{l+\frac{1}{2}}\right]}{\Delta x} \tag{29}$$

It can be shown that the error will grow with time unless the following stability condition is met,

$$\sqrt{gd}\left(\frac{\Delta t}{\Delta x}\right) \le 1 \qquad \text{or} \qquad \Delta t \le \frac{\Delta x}{\sqrt{gd}}$$

which is known as the Courant–Friedrichs–Lewy (CFL) condition. The physical meaning of the stability condition is that the time step Δt must be equal to or smaller than the time required for the disturbance to travel the spatial grid size Δx. If Δt is larger, the values other than $(l, m-1)$, $(l-1, m)$, and $(l, m+1)$ will influence the value at $(l+1, m)$. For the two-dimensional case (Fig. 7), the CFL condition becomes

$$\Delta t \le \frac{\Delta x}{\sqrt{2gd}} \tag{30}$$

There are two types of boundaries involved in the tsunami computation. The first is land–ocean boundary. The simplest assumption is a total reflection of energy on the coast. The other kind of boundary is the open boundary to outside the computational region. A typical assumption is the radiation condition, which means that waves go out of the computational region keeping their slopes.

The computer program for one-dimensional finite-difference computation is very simple. Since there is no overlap between the two (height and velocity) grids, in the program the spatial grid points m and $m+\frac{1}{2}$ can be treated by the

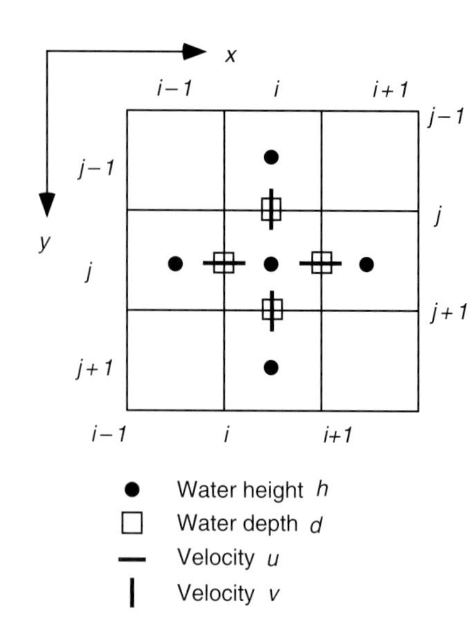

FIGURE 7 Typical staggered grid system used for two-dimensional finite-difference computation of tsunamis.

same argument (e.g., i) and the time $l-\frac{1}{2}$ and l can be treated as the same time (e.g., it). Writing U and h using the two-dimensional array as $U(i, it)$ and $h(i, it)$, Eqs. (28) and (29) become

$$U(i,2) = U(i,1) - g\frac{\Delta t}{\Delta x}[h(i,1) - h(i-1,1)]$$

$$h(i,2) = h(i,1) - d\frac{\Delta t}{\Delta x}[U(i+1,2) - U(i,2)]$$

The algorithm thus becomes as follows:

```
set up parameters
read bathymetry data
read initial condition h(i,1)
while(it<it_end)
  compute U(i,2) from U(i,1)and h(i,1)
  set U(i,2) =0 for land boundary
  compute h(i,2) from U(i,2)and h(i,1)
  adjust h(i,2) for open boundary
  exchange variables
    h(i,2)->h(i,1)
    U(i,2)->U(i,1)
  it+=1
end
```

6. Tsunami Generation by Earthquakes

6.1 Small Effect of the Earth–Ocean Coupling

The earthquake source is located in the elastic Earth, while a tsunami travels in the fluid ocean. Hence we need to first consider coupling between the elastic Earth and the ocean. In the Earth–ocean coupled system, an earthquake source would generate "tsunami waves" both in the elastic Earth and in the ocean, and the propagation of a tsunami in the ocean layer might influence the elastic Earth. On the other hand, the traditional decoupled approach first calculates the elastic deformation on the ocean bottom generated by an earthquake, then uses this as an initial condition on the rigid ocean bottom; the tsunami wave in the ocean would have no influence on the elastic Earth. As we will see, because of the low compressibility of water, both approaches give almost identical solutions. In other words, the coupling between the elastic Earth and the ocean is extremely small, which explains why very large earthquakes are necessary to generate observable tsunamis.

Theoretical treatment of tsunamis is similar to that of Rayleigh waves or spheroidal oscillations. Yamashita and Sato (1974) formulated tsunami generation from a fault model and examined the effects of fault parameters on the tsunami amplitudes. Ward (1980) and Okal (1988) extended the free oscillation approach to solve the generation and propagation of tsunamis in the spherical coordinate system, while Comer (1984) extended the Rayleigh wave approach to solve them in the Cartesian coordinate system. The phase velocity dispersion curve is shown in Figure 8. Except at very low frequencies, the phase velocity is very similar to that of the gravity wave (11), indicating that the coupling effect is negligibly small.

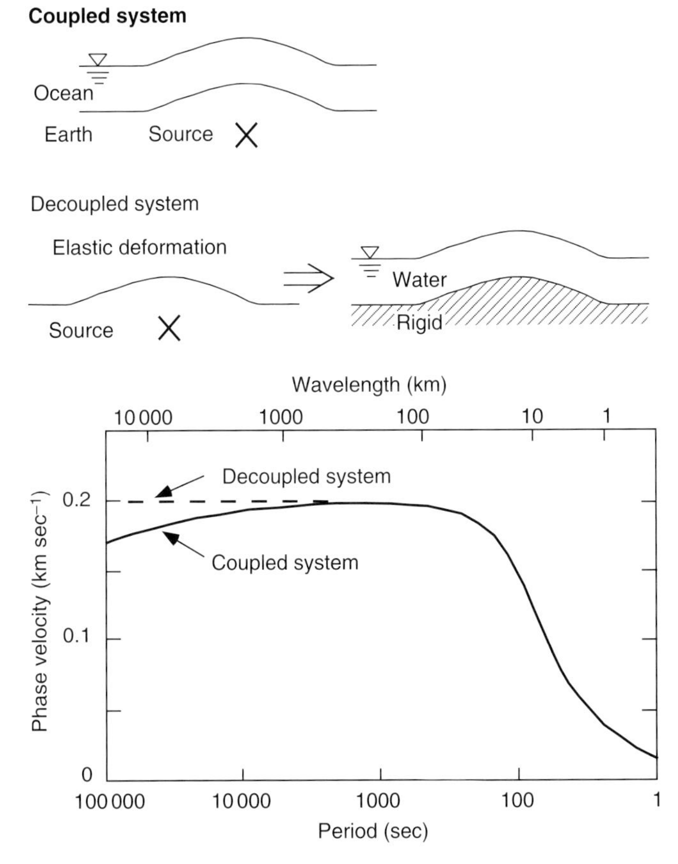

FIGURE 8 Top: Schematic comparison of the coupled and decoupled systems. Bottom: The dispersion curves for the tsunami phase velocity computed for a 4000 m deep ocean. (Reproduced with permission from Comer, 1984.)

6.2 Ocean Bottom and Water Surface Displacements

Analytical solutions for linear gravity waves generated from several types of source have been obtained by various researchers. Kajiura (1963) generalized the solutions by introducing the Green's functions. He showed that the initial conditions given at the ocean bottom and at the water surface differ by a factor of $1/\cosh kd$. As seen before, $\cosh kd = \cosh(2\pi d/\lambda) \approx 1$ if $\lambda \gg 2\pi d$. Therefore, if the wavelength of ocean bottom deformation is much larger than the water depth, it can be assumed that the water surface displacement is the same as the bottom displacement.

Kajiura (1970) examined the energy exchange between the solid Earth and ocean water for tsunamis generated by bottom deformation with finite rise time. He showed that the efficiency of tsunami generation, defined as a ratio of dynamic tsunami energy to static energy, becomes nearly unity if the duration is short compared to the time required for a tsunami to travel through the source area. For an earthquake with $M \sim 8$ at a water depth of 5000 m, the source size is about 100 km and the tsunami travel time over that distance is about 8 min. If the duration is much shorter, e.g., 1 min or less, the efficiency is more than 0.9. This means that the bottom deformation can be considered as instantaneous. If the duration is too short, shorter than $2\sqrt{d/g}$, a part of the energy flux goes into acoustic waves. For the above case, this limit is 45 sec.

6.3 Fault Parameters and Ocean Bottom Deformation

The ocean bottom deformation due to faulting can be calculated using the elastic theory of dislocation. The displacement in an infinite homogeneous medium due to a dislocation across a fault is given by Volterra's theorem (see, e.g., Steketee, 1958) and the explicit formulas are given in Okada (1985), for example. Usually, only the vertical component of the ocean bottom is considered for tsunami generation. When the tsunami source is on a steep ocean slope and the horizontal displacement is large, the vertical displacement of water due to the horizontal displacement of the slope must also be considered (Tanioka and Satake, 1996a).

The fault parameters needed to compute surface deformation are the fault location, geometry (strike, dip, and rake), size (length L and width W), and average slip $\bar{u}$ (or the strike-slip and dip-slip components). The seismic moment M_0 is given as

$$M_0 = \mu \bar{u} S = \mu \bar{u} L W \tag{31}$$

where S is the fault area and μ is rigidity. The moment magnitude is defined as

$$M_W = \frac{\log M_0 - 9.1}{1.5} \tag{32}$$

where M_0 is given in N m. The effects of fault parameters on tsunami generation were examined by Yamashita and Sato (1974) and more recently by Geist (1998).

6.4 Estimation of Earthquake Source Parameters from Tsunamis

Tsunami data can be used to study earthquake source processes in a way similar to that in which seismic waves are used. Abe (1973) analyzed the tsunami arrival times and first-motion data from the 1968 Tokachi-oki earthquake (Fig. 9). By drawing inverse refraction diagrams from each station, he estimated the tsunami source area, which agrees well with the aftershock area. From the first motion of tsunami waves, the initial water surface disturbance was estimated as uplift at the southeastern edge and subsidence at the northwestern edge. This pattern is very similar to the crustal deformation due to the faulting, which was independently estimated from seismological analysis.

Tsunami waveforms recorded on tide gauges can be inverted to estimate the slip distribution on the fault (Satake, 1989). In this method (Fig. 10), the fault plane is first divided into several subfaults and the deformation on the ocean bottom is computed for each subfault with a unit amount of slip. Using this as an initial condition, tsunami waveforms are numerically computed on actual bathymetry. The observed tsunami waveforms are expressed as a superposition of the computed waveforms as follows,

$$A_{ij}(t) \cdot x_j = b_i(t) \tag{33}$$

where A_{ij} is the computed waveform, or Green's function, at the ith station from the jth subfault; x_j is the amount of

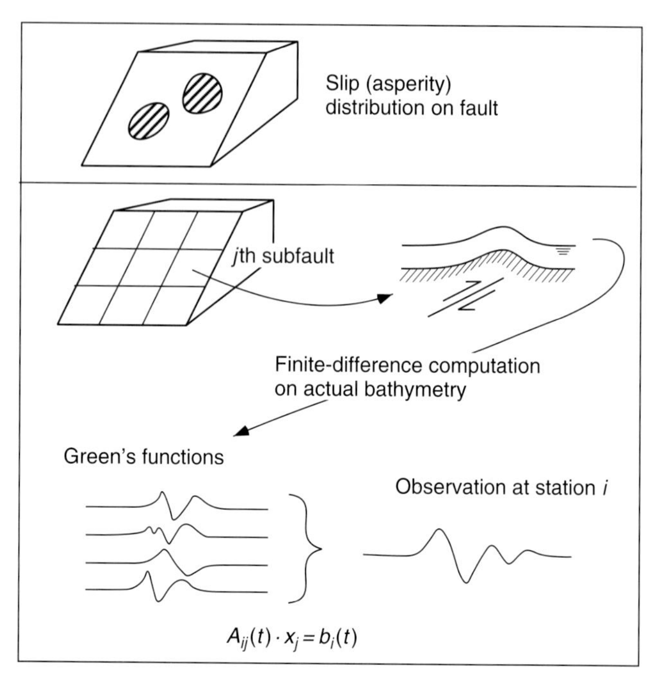

FIGURE 10 Schematic illustration of tsunami waveform inversion.

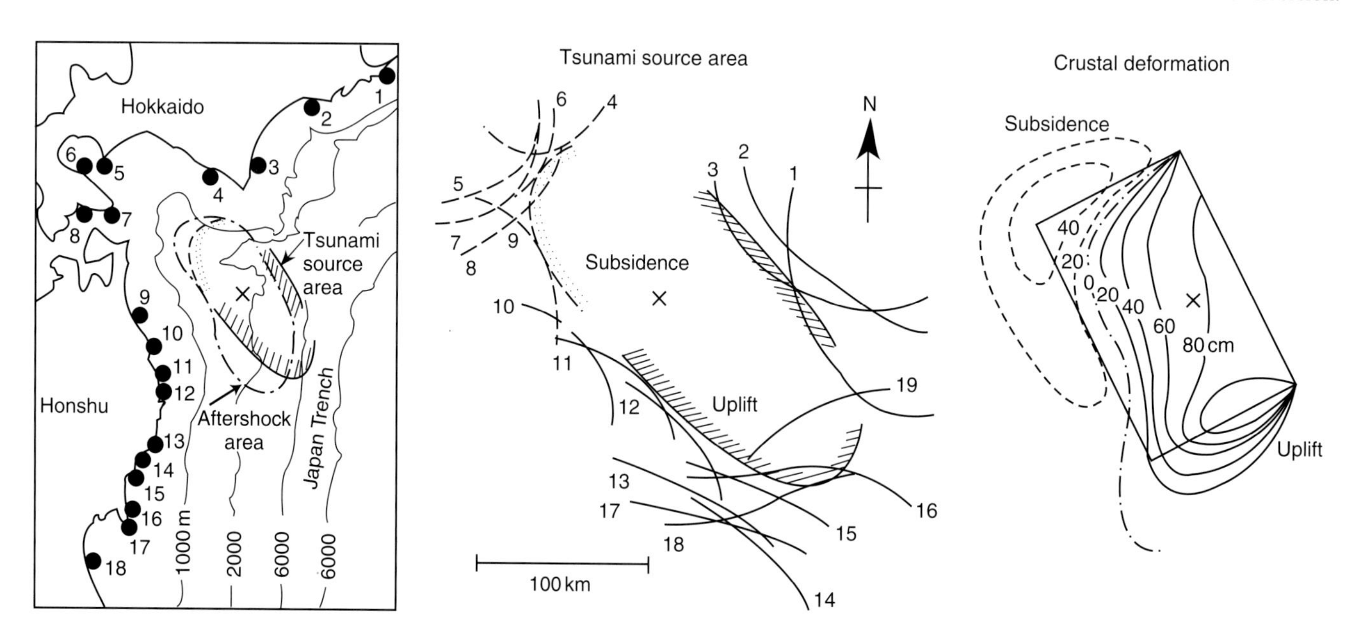

FIGURE 9 The tsunami source area and crustal deformation pattern of the 1968 Tokachi-oki earthquake. The left panel shows the distribution of tide gauge stations (with numbers) where the tsunami waveforms were recorded. The center panel shows the travel-time arcs for each tide gauge station. The attached numbers correspond to the station numbers in the left panel. Solid and dashed arcs represent positive and negative first arrivals. The right panel shows the ocean bottom deformation calculated from a fault model. (Reproduced with permission from Abe, 1973; copyright Elsevier Science Publishers.)

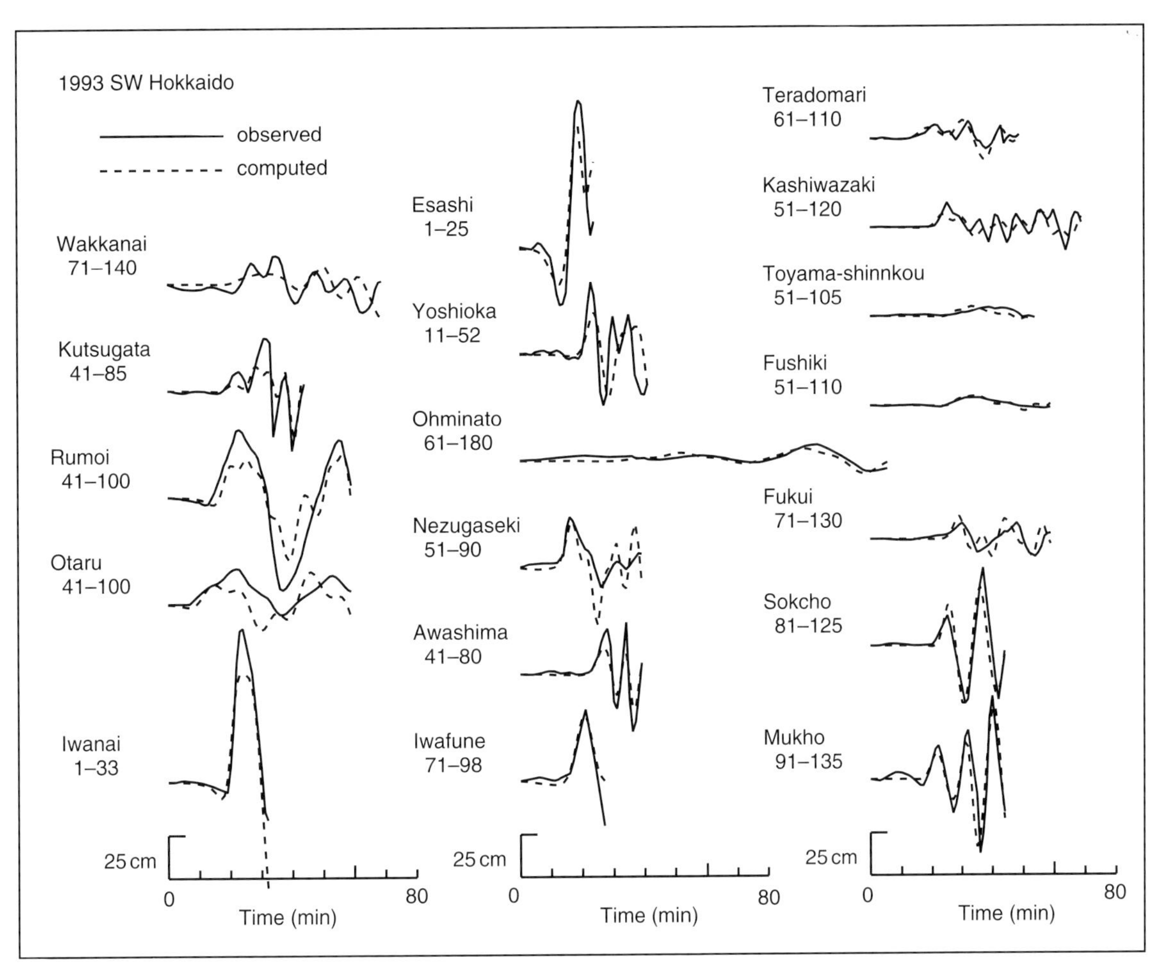

FIGURE 11 Comparison of the observed (solid) and computed (dashed) tsunami waveforms for the 1993 Southwest Hokkaido earthquake. The computation was made from a heterogeneous fault model obtained by the inversion of tsunami and geodetic data. The numbers below the tide gauge station names indicate the time range in minutes after the earthquake. (From Tanioka *et al.*, 1995; copyright American Geophysical Union.)

slip on the jth subfault; and b_i is the observed tsunami waveform at the ith station. The slip x_j on each subfault can be estimated by a least-squares inversion of the above set of equations. Figure 11 shows the comparison of observed and computed tsunami waveforms for the 1993 Southwest Hokkaido earthquake. Computed waveforms are very similar to the observed ones.

The tsunami waveform inversion has been applied to many large and great earthquakes in the Alaska-Aleutians (e.g., Johnson, 1998), northern Japan (e.g., Satake, 1989; Tanioka *et al.*, 1995), and the Nankai trough (Satake, 1993), among others. These results show that the slip distributions are not uniform but are concentrated in small regions that correspond to asperities.

6.5 Tsunami Earthquakes

A "tsunami earthquake" is defined as an earthquake that excites much larger tsunamis than expected from its seismic waves (Kanamori, 1972). The discrepancy can be quantified in terms of the surface wave magnitude (M_S) and tsunami magnitude (M_t) scales; when M_t is larger than M_S by more than 0.5 units, it is called a tsunami earthquake (Abe, 1989; see Fig. 2).

Several proposals have been made for the generating mechanisms. Kanamori (1972) indicated that the large discrepancy between seismic and tsunami waves can be explained by a slow and long rupture process. Fukao (1979) explained the large tsunami excitation by slips in the accretionary wedge. A high-angle subsidiary fault in the wedge may cause a significant tsunami, but contribution to the seismic moment may be small because of low rigidity. Pelayo and Wiens (1992) showed that the tsunami heights were not anomalously large if compared to the seismic moments; the large discrepancy between the tsunami and M_S is due to a saturation of M_S at around 7.3 for those earthquakes that occurred in the accretionary wedge. Okal (1988), using the normal-mode theory, showed that an earthquake source in a shallow sedimentary layer can excite much larger tsunamis than in solid rock for certain geometries.

Recent tsunami waveform modeling shows that the fault parameters and locations of tsunami earthquakes share a common feature: narrow and shallow faulting near the trench axis (Satake and Tanioka, 1999; Fig. 12). The 1992 Nicaragua earthquake ($M_S = 7.2$, $M_t = 7.9$) was the first tsunami earthquake recorded on modern broadband seismic instruments, and seismological studies showed that the duration was very long for its size, about 100 sec (Kanamori and Kikuchi, 1993). Satake (1994) compared the numerically computed tsunami waveforms with tide gauge records and showed that a narrow (~40 km) and shallow (extending only to the upper 10 km of the ocean bottom) fault was responsible for the tsunami generation. A similar feature was found for other tsunami earthquakes: the 1946 Aleutian earthquake ($M_S = 7.3$, $M_t = 9.3$; Johnson and Satake, 1997) and the 1896 Sanriku earthquake ($M_S = 7.2$, $M_t = 8.6$; Tanioka and Satake, 1996b). The large tsunami excitation is attributed to a slow rupture, presumably within the shallow subducted sediments. Tanioka *et al.* (1997) showed that the surface roughness of the ocean bottom near the trench is well correlated with the large earthquake occurrence. The 1896 tsunami earthquake occurred in a region where the ocean bottom topography is rough, characterized by well-developed horst and graben structure.

7. Tsunami Warning Systems

Tsunamis travel very fast as ocean waves, about 800 km h^{-1}, or 0.2 km sec^{-1}, for a water depth of 5000 m, but they are still much slower than seismic waves. This velocity difference between the seismic and tsunami waves makes it possible to issue a tsunami warning after seismic wave detection but before the arrival of the tsunami.

Current tsunami warning systems can be grouped into a Pacific-wide system and regional (or local) systems. The Pacific Tsunami Warning Center located in Hawaii monitors seismic and tsunami waves, issues tsunami warnings, and communicates with foreign agencies for the Pacific-wide tsunamis. It takes at least several hours for a tsunami to travel across the Pacific Ocean, so that a tsunami warning can be issued after the tsunami is actually detected somewhere and it is confirmed that it is dangerous.

The French Polynesia Tsunami Warning Center developed and adopted TREMORS (Tsunami Risk Evaluation through seismic MOment in a Real-time System) in 1987 (Reymond *et al.*, 1991). TREMORS features automatic detection and location of earthquakes using a single three-component long-period seismic station, and estimation of seismic moment through variable-period mantle magnitude, M_m (Okal and Talandier, 1989). The advantage of using long-period seismic waves for tsunami warning is the ability to estimate the overall

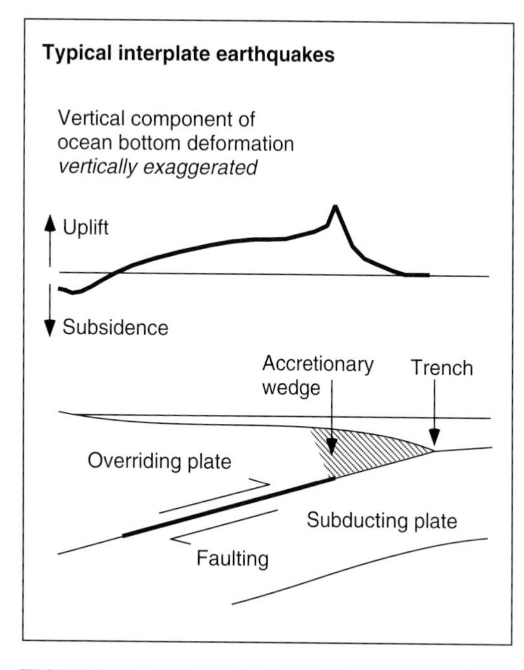

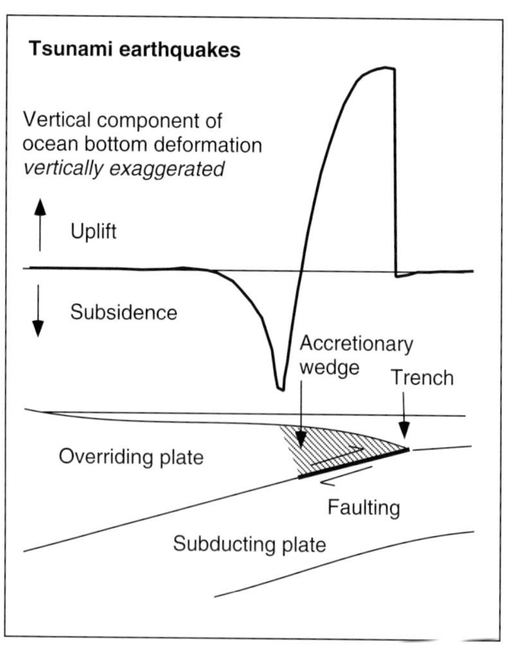

FIGURE 12 Schematic view of the source regions (bottom) and vertical deformation on ocean bottom (top) for typical interplate (left) and tsunami (right) earthquakes. The source region of an interplate earthquake typically extends from a depth of 10–40 km. The ocean bottom above the source region is uplifted and becomes the tsunami source. The source region of a tsunami earthquake is at shallower extension near the trench axis.

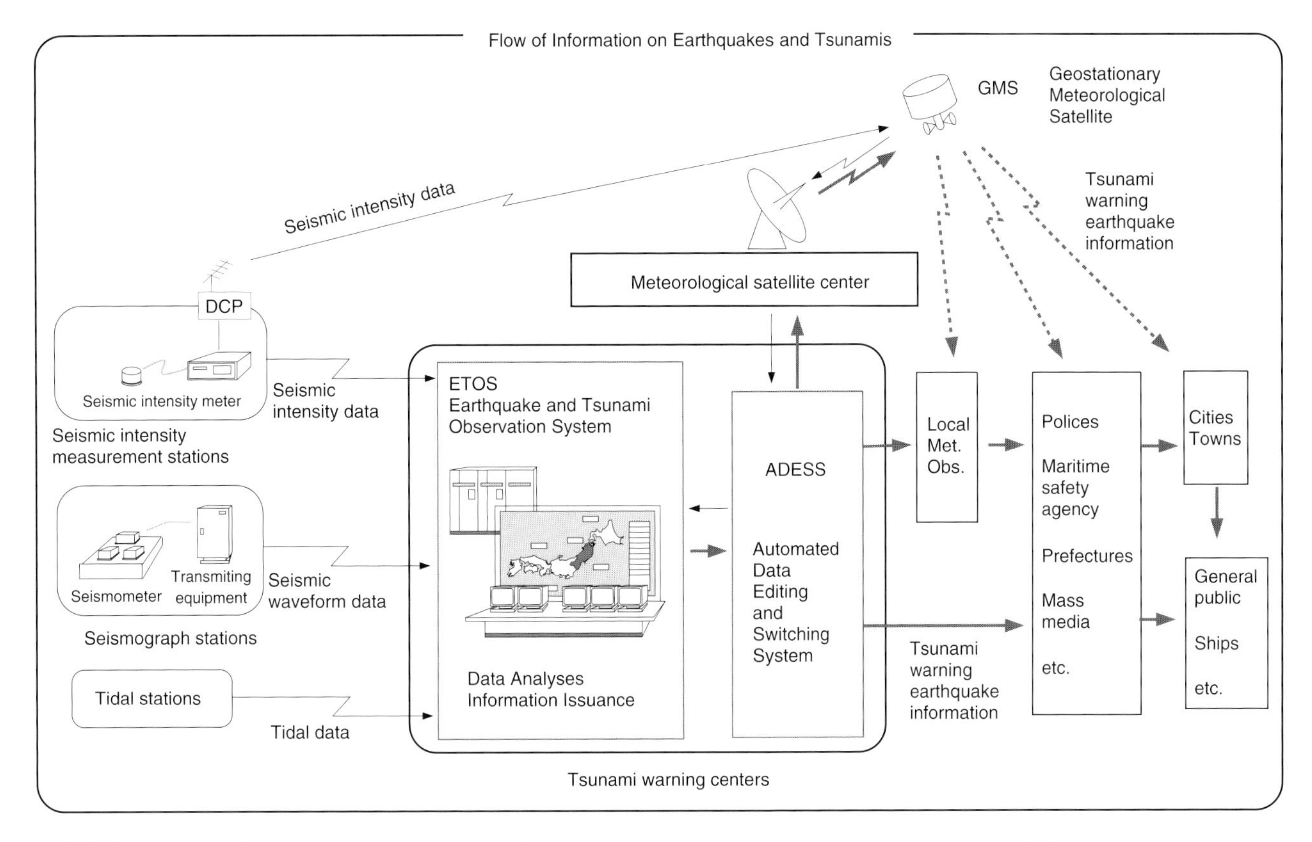

FIGURE 13 Flow diagram of the Tsunami Warning Center system employed at the Japan Meteorological Agency.

size of large or great earthquakes, and the possibility of detecting "tsunami earthquakes" (e.g., Newman and Okal, 1998), but the disadvantage is that longer time is needed before the seismic waves are recorded. TREMORS is a hybrid system; it updates the moment estimate as more seismic data are available.

In the case of regional and local tsunami warnings, a tsunami warning must be issued immediately after the occurrence of an earthquake, because time before the tsunami arrival is shorter—on the order of minutes. Japan Meteorological Agency (JMA) operates a nationwide seismic observation system and has six Regional Tsunami Warning Centers where seismologists are on duty 24 h a day (Fig. 13). When a large earthquake occurs, one of four messages (Major Tsunami, Tsunami, Tsunami Attention, or No Tsunami) is issued for each coastal region, based on the magnitude, location, and depth of the earthquake, and the distance to each coastal region. JMA quickly issues earthquake information (region and intensity, without location or magnitude) within 2 min of an earthquake, followed by a tsunami warning within a minute. The recent warnings include tsunami arrival times and heights estimated from pre-made numerical computations. Several thousands of numerical computations have been used to construct the database (Tatehata, 1997).

References

Abe, K. (1973). Tsunami and mechanism of great earthquakes. *Phys. Earth Planet. Inter.* **7**, 143–153.

Abe, K. (1979). Size of great earthquakes of 1873–1974 inferred from tsunami data. *J. Geophys. Res.* **84**, 1561–1568.

Abe, K. (1981). Physical size of tsunamigenic earthquakes of the northwestern Pacific. *Phys. Earth Planet. Inter.* **27**, 194–205.

Abe, K. (1989). Quantification of tsunamigenic earthquakes by the M_t scale. *Tectonophysics* **166**, 27–34.

Atwater, B.F., A.R. Nelson, and J.J. Clague, *et al.* (1995). Summary of coastal geologic evidence for past great earthquakes at the Cascadia subduction zone. *Earthq. Spectra* **11**, 1–18.

Bryant, E.A., R.W. Young, and D.M. Price (1996). Tsunami as a major control on coastal evolution, Southeast Australia. *J. Coastal Res.* **12**, 831–840.

Clague, J.J. (1997). Evidence for large earthquakes at the Cascadia subduction zone. *Rev. Geophys.* **35**, 439–460.

Comer, R.P. (1980). Tsunami height and earthquake magnitude: theoretical basis of an empirical relation. *Geophys. Res. Lett.* **7**, 445–448.

Comer, R.P. (1984). The tsunami mode of a flat earth and its excitation by earthquake sources. *Geophys. J. Roy. Astron. Soc.* **77**, 1–27.

Dawson, A. (1994). Geomorphological effects of tsunami runup and backwash. *Geomorphology* **10**, 83–94.

Eble, M.C. and F.I. Gonzalez (1991). Deep-ocean bottom pressure measurements in the northeast Pacific. *J. Atmos. Ocean. Technol.* **8**, 221–233.

Fukao, Y. (1979). Tsunami earthquakes and subduction processes near deep-sea trenches. *J. Geophys. Res.* **84**, 2303–2314.

Geist, E.L. (1998). Local tsunamis and earthquake source parameters. *Adv. Geophys.* **39**, 117–209.

Geller, R.J. (1976). Scaling relations for earthquake source parameters and magnitudes. *Bull. Seismol. Soc. Am.* **66**, 1501–1523.

Hatori, T. (1979). Relation between tsunami magnitude and wave energy. *Bull. Earthq. Res. Inst. Univ. Tokyo* **54**, 531–541 (in Japanese with English abstract).

Hokkaido Tsunami Survey Group (1993). Tsunami devastates Japanese coastal region. *EOS Trans. Am. Geophys. Union* **74**, 417–432.

Iida K., D.C. Cox, and G. Pararas-Carayannis (1967). "Preliminary Catalog of Tsunamis Occurring in the Pacific Ocean," Data Report No.5, HIG-67–10. University of Hawaii, Honolulu.

Johnson, J.M. (1998). Heterogeneous coupling along the Alaska-Aleutians as inferred from tsunami, seismic, and geodetic inversions. *Adv. Geophys.* **39**, 1–116.

Johnson, J.M. and K. Satake (1997). Estimation of seismic moment and slip distribution of the April 1, 1946, Aleutian tsunami earthquake. *J. Geophys. Res.* **102**, 11765–11774.

Kajiura, K. (1963). The leading wave of a tsunami. *Bull. Earthq. Res. Inst. Univ. Tokyo* **41**, 535–571.

Kajiura, K. (1970). Tsunami source, energy and the directivity of wave radiation. *Bull. Earthq. Res. Inst. Univ. Tokyo* **48**, 835–869.

Kanamori, H. (1972). Mechanism of tsunami earthquakes. *Phys. Earth Planet. Inter.* **6**, 246–259.

Kanamori, H. and M. Kikuchi (1993). The 1992 Nicaragua earthquake: a slow tsunami earthquake associated with subducted sediments. *Nature* **361**, 714–716.

Kawata, Y., B.C. Benson, J.C. Borrero, *et al.* (1999). Tsunami in Papua New Guinea was intense as first thought. *EOS Trans. Am. Geophys. Union* **80**, 101, 104–105.

Lander, J.F. (1996). "Tsunamis Affecting Alaska: 1737–1996," NGDC Key to Geophysical Research Documentation No. 29. NOAA, Boulder, CO.

Lander, J.F., P.A. Lockridge, and M.J. Kozuch (1993). "Tsunamis Affecting the West Coast of the United States 1806–1992," NGDC Key to Geophysical Research Documentation No. 31. NOAA., Boulder, CO.

Minoura, K. and S. Nakaya (1991). Traces of tsunami preserved in inter-tidal lacustrine and marsh deposits: some examples from northeast Japan. *J. Geol.* **99**, 265–287.

Newman, A.V. and E.A. Okal (1998). Teleseismic estimates of radiated seismic energy: the E/M_0 discriminant for tsunami earthquakes. *J. Geophys. Res.* **103**, 26885–26898.

Okada, Y. (1985). Surface deformation due to shear and tensile faults in a half-space. *Bull. Seismol. Soc. Am.* **75**, 1135–1154.

Okada, M. (1995). Tsunami observation by ocean bottom pressure gauge. In: "Tsunami: Progress in Prediction, Disaster Prevention and Warning" (Y. Tsuchiya and N. Shuto, Eds.), pp. 287–303. Kluwer Academic, Dordrecht.

Okal, E.A. (1988). Seismic parameters controlling far-field tsunami amplitudes: a review. *Nat. Hazards* **1**, 67–96.

Okal, E.A. and J. Talandier (1989). M_m: A variable-period mantle magnitude. *J. Geophys. Res.* **94**, 4169–4193.

Pelayo, A.M. and D.A. Wiens (1992). Tsunami earthquakes: slow thrust-faulting events in the accretionary wedge. *J. Geophys. Res.* **97**, 15321–15337.

Reymond, D., O. Hyvernaud, and J. Talandier (1991). Automatic detection, location and quantification of earthquakes: application to tsunami warning. *Pure Appl. Geophys.* **135**, 361–382.

Satake, K. (1989). Inversion of tsunami waveforms for the estimation of heterogeneous fault motion of large submarine earthquakes: 1968 Tokachi-oki and 1983 Japan Sea earthquakes. *J. Geophys. Res.* **94**, 5627–5636.

Satake, K. (1993). Depth distribution of coseismic slip along the Nankai trough, Japan, from joint inversion of geodetic and tsunami data. *J. Geophys. Res.* **98**, 4553–4565.

Satake, K. (1994). Mechanism of the 1992 Nicaragua tsunami earthquake. *Geophys. Res. Lett.* **21**, 2519–2522.

Satake, K. (1995). Linear and nonlinear computations of the 1992 Nicaragua earthquake tsunami. *Pure Appl. Geophys.* **144**, 455–470.

Satake, K. and Y. Tanioka (1999). Sources of tsunami and tsunamigenic earthquakes in subduction zones. *Pure Appl. Geophys.* **154**, 467–483.

Satake, K., M. Okada, and K. Abe (1988). Tide gauge response to tsunamis: measurements at 40 tide gauge stations in Japan. *J. Marine Res.* **46**, 557–571.

Satake, K., K. Shimazaki, Y. Tsuji, and K. Ueda (1996). Time and size of a giant earthquake in Cascadia inferred from Japanese tsunami records of January 1700. *Nature* **379**, 246–249.

Soloviev, S.L. (1970). Recurrence of tsunamis in the Pacific. In: "Tsunamis in the Pacific Ocean" (W.M. Adams, Ed.), pp. 149–164. East–West Center Press, Honolulu.

Soloviev, S.L. and Ch.N. Go (1974a). "Catalogue of Tsunamis on the Western shore of the Pacific Ocean (173–1968)." Nauka, Moscow. (Can. Transl. Fish. Aquat. Sci. 5077.)

Soloviev, S.L. and Ch.N. Go (1974b). "Catalogue of Tsunamis on the Eastern shore of the Pacific Ocean (1513–1968)." Nauka, Moscow. (Can. Transl. Fish. Aquat. Sci. 5078.)

Soloviev, S.L., Ch.N. Go, and Kh.S. Kim (1992). "Catalog of Tsunamis in the Pacific 1969–1982." Academy of Science, Moscow.

Steketee, J.A. (1958). On Volterra's dislocations in a semi-infinite elastic medium. *Can. J. Phys.* **36**, 192–205.

Takahashi, To., Ta. Takahashi, N. Shuto, F. Imamura, and M. Ortiz (1995). Source models of the 1993 Hokkaido Nansei-oki earthquake tsunami. *Pure Appl. Geophys.* **144**, 747–767.

Tanioka, Y. and K. Satake (1996a). Tsunami generation by horizontal displacement of ocean bottom. *Geophys. Res. Lett.* **23**, 861–864.

Tanioka, Y. and K. Satake (1996b). Fault parameters of the 1896 Sanriku tsunami earthquake estimated from tsunami numerical modeling. *Geophys. Res. Lett.* **23**, 1549–1552.

Tanioka, Y., K. Satake, and L. Ruff (1995). Total analysis of the 1993 Hokkaido Nansei-oki earthquake using seismic wave, tsunami and geodetic data. *Geophys. Res. Lett.* **22**, 9–12.

Tanioka, Y., L. Ruff, and K. Satake (1997). What controls the lateral variation of large earthquake occurrence along the Japan trench? *Island Arc* **6**, 261–266.

Tappin, D.R., T. Matsumoto, P. Watts, *et al.* (1999). Sediment slump likely caused 1998 Papua New Guinea tsunami. *EOS Trans. Am. Geophys. Union* **80**, 329, 334, 340.

Tatehata, H. (1997). The new tsunami warning system of the Japan Meteorological Agency. In: "Perspectives on Tsunami Hazard Reduction" (G. Hebenstreit, Ed.), pp. 175–188. Kluwer Academic, Dordrecht.

Tinti, S., A. Piatanesi, A. Maramai (1997). Numerical simulations of the 1627 Gargano tsunami (southern Italy) to locate the earthquake source. In: "Perspectives on Tsunami Hazard Reduction" (G. Hebenstreit, Ed.), pp. 115–131, Kluwer Academic, Dordrecht.

Ward, S. (1980). Relationships of tsunami generation and an earthquake source. *J. Phys. Earth* **28**, 441–474.

Watanabe, H. (1998). "Catalogue of Hazardous Tsunamis in Japan" (second version; in Japanese). University of Tokyo Press.

Yamashita, T. and R. Sato (1974). Generation of tsunami by a fault model. *J. Phys. Earth* **22**, 415–440.

Earthquake Geology and Mechanics

29

Geology of the Crustal Earthquake Source

Richard H. Sibson
University of Otago, Dunedin, New Zealand

1. Introduction

Earthquakes occur in rock, and rocks of many different kinds provide important information on earthquake processes. The specific focus here is on shallow crustal earthquakes responsible for the vast bulk of earthquake damage; subduction-related events are not discussed to any extent. Primary earthquake effects—incremental fault slip, shaking (strong ground motion), and abrupt changes in the tectonic stress state (including shear stress, τ, mean stress, $\overline{\sigma}$, and fluid-pressure, P_f)—affect many fundamental geological processes (Fig. 1). Earthquake effects are thus widely recorded in rocks, though deciphering the seismological significance of this geological information is in its infancy. However, paleoseismic studies—interpreting the Holocene sedimentary and geomorphic record in terms of seismic increments of fault slip—have made enormous advances in recent years (McCalpin, 1996). These aspects of *earthquake geology* of special relevance to seismic hazard assessment are treated separately in Chapters 30 by Grant and 31 by Jackson.

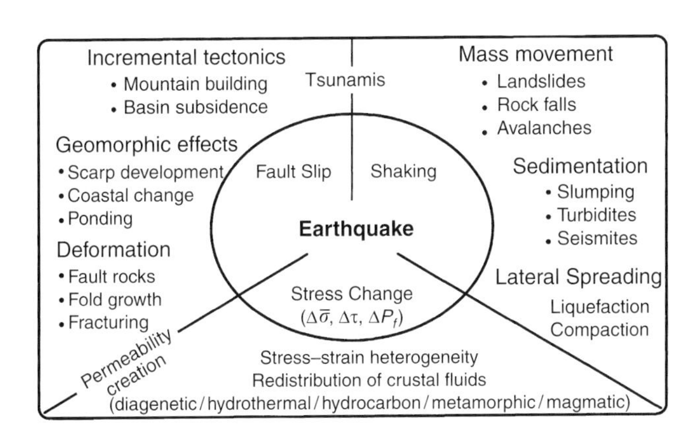

FIGURE 1 Earthquake geology—the involvement of primary earthquake effects (fault slip, shaking, and stress change) in geologic processes.

This chapter focuses specifically on the geological structure and rheology of crustal fault zones in relation to the mechanics of shallow earthquakes.

2. Faulting in Seismogenic Crust

Crustal earthquake activity is concentrated at plate boundaries in association with large, fast-moving fault systems but also occurs, sporadically, far from plate margins, especially in continental intraplate regions which may possess a variety of structures inherited from former tectonic episodes.

2.1 Fault Activity and Slip Modes

The degree of activity on a fault can be classified in terms of the time-averaged slip-rate (v_{ta}) across the structure established from the Holocene (<10 ka) or Quaternary (<1.8 Ma) geologic record (Matsuda, 1975) (Table 1). Plate boundary fault systems tend to be dominated by an isolated Class AA fault and/or a mesh of Class A structures within a hierarchy of still-slower moving faults. As a general rule, intraplate earthquakes occur on far slower moving faults (Class A or less).

Fault creep, though widespread within the San Andreas strike-slip fault system of California where it is accompanied by high levels of microseismic activity (Burford and Harsh, 1980; Hill *et al.*, 1990), is comparatively rare on a global scale (aside from postseismic afterslip) with the exception of some

TABLE 1 Fault Slip-Rate and Earthquake Recurrence

Class	v_{ta} (mm y^{-1})	R.I. (years) ($\bar{u}=1$ m)	($\bar{u}=10$ m)
AA	>10	<100	<1000
A	1–10	100–1000	1000–10 000
B	0.1–1.0	1000–10 000	10 000–100 000
C	0.01–0.1	10 000–100 000	100 000–1 000 000

INTERNATIONAL HANDBOOK OF EARTHQUAKE AND ENGINEERING SEISMOLOGY, VOLUME 81A ISBN: 0-12-440652-1

subduction boundaries. At least onshore, seismic rupturing appears to be the dominant mode of fault slip in the upper crust. Recurrence intervals between successive earthquakes with a characteristic slip value, $\bar{u}$, are then simply anticipated through the recurrence equation:

$$\text{R.I.} = \bar{u}/v_{\text{ta}} \tag{1}$$

suggested by Wallace (1970), though there is no a priori reason why v_{ta} must remain constant over long time periods. Expected recurrence intervals between successive large earthquakes ($1 < \bar{u} < 10\,\text{m}$) on faults with varying activity levels are given in Table 1.

2.2 Seismogenic Crust

Away from areas of active subduction where thermal and fluid-pressure regimes are severely perturbed, the vast bulk of earthquake activity is restricted to the Earth's crust and, in continental regions, mostly to within the upper half of the crust (Meissner and Strehlau, 1982; Sibson, 1982; Chen and Molnar, 1983).

2.2.1 Continental Seismogenic Zone

In areas of moderate to high heat flow (60–$100\,\text{mW}\,\text{m}^{-2}$) within deforming continental crust, microearthquake activity is largely confined to the top 10–20 km of the crust (e.g., Hill *et al.*, 1990). Activity shallows to depths of ~5 km in areas of intense geothermal activity and, though more diffuse, deepens to depths of ~25 km in colder cratonic crust as in eastern North America. Temperatures at the continental seismic–aseismic transition are inferred to vary from ~300°C to ~450°C for, respectively, quartz-dominant and feldspar-dominant rheologies (Sibson, 1984; Scholz, 1988). A notable feature is that larger earthquakes ($M > 6$) tend to nucleate toward the bottom of this microseismically defined seismogenic zone (Fig. 2); it seems likely that larger ruptures occasionally penetrate downwards some distance below the background seismogenic zone (Strehlau, 1986). Minor seismic activity also occurs within the upper mantle below an aseismic lower crust in some areas of continental convergence (Chen and Molnar, 1983).

Similar correlations between regional heat flow and the thickness of the seismogenic zone have been noted in southwest Japan (Ito, 1999), with a temperature of 350 ± 50°C inferred for the base of the seismogenic zone. Moreover, drilling into the contact aureole to a Late Quaternary granitic pluton ($T > 500$°C) within the Kakkonda geothermal field, northeast Japan, has allowed direct correlation of the local microseismicity cut-out at ~2 km depth with the 350°C isotherm (Muraoka *et al.*, 1998). Also in northeast Japan, high resolution studies have revealed a particularly sharp transition between "standard" upper crustal seismicity in the deforming arc and that associated with the subduction interface and downgoing subducted slab (Hasegawa *et al.*, 1994).

2.2.2 Oceanic Seismogenic Zone

In oceanic areas away from spreading ridges it is apparent that both microearthquake activity (e.g., Shen *et al.*, 1997) and large events (Anderson and Zhang, 1991) can nucleate below the crust and propagate through the oceanic lithosphere. The depth of intraplate earthquakes in the oceans appears to be governed by olivine rheology and extends down to temperatures as high as ~750°C, but along major transform faults the depth of rupturing is significantly shallower, corresponding to temperatures in the range of 400–600°C (Engeln *et al.*, 1986).

2.2.3 Varying Seismic Style

Moderate-to-large tectonic earthquakes generally occur as mainshock–aftershock sequences, sometimes with premonitory foreshocks. Studies of moment release and surface

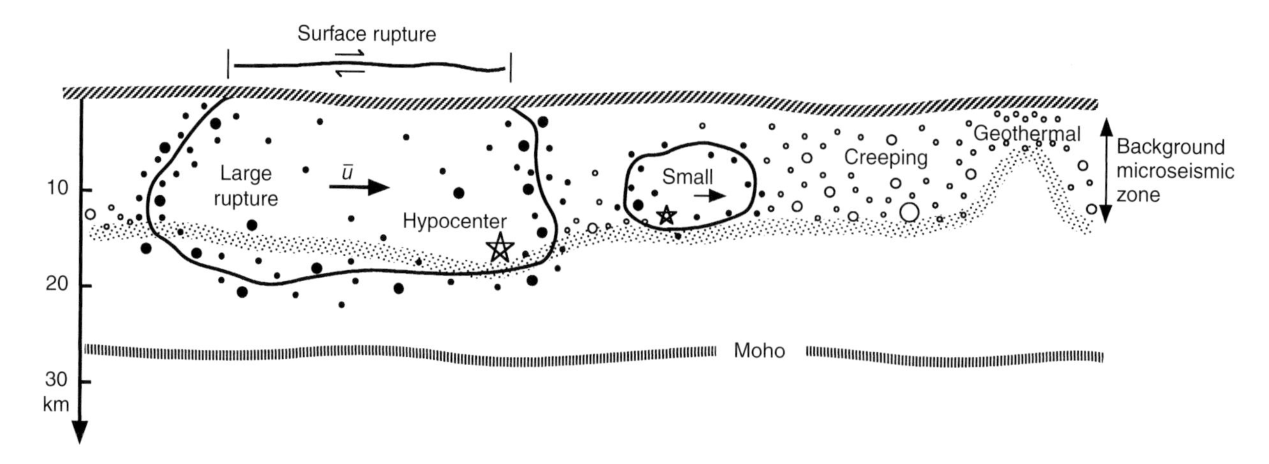

FIGURE 2 Strike-parallel longitudinal section illustrating varying seismic style along a major strike–slip fault zone (modeled on the San Andreas fault) and the relationship of small and large ruptures (perimeters defined by aftershock concentrations-filled circles) to the continental seismogenic zone defined by background microseismicity (open circles).

fault-breaks, coupled with geodetic analyses and precision aftershock locations, suggest that the bulk of slip during rupture is usually confined to a thin tabular zone of indeterminate thickness, though aftershock activity may be distributed through substantial volumes of rock, especially near rupture irregularities. Scholz (1982, 1997) argues that the typical 10–20 km thickness of the seismogenic zone imposes a fundamental length scale on crustal rupturing (Fig. 2). "Small" ruptures confined within the seismogenic layer follow different scaling laws from "large" earthquakes ($M > 6$) that rupture the full depth of the seismogenic zone and then extend along strike to rupture lengths that may be many times the seismogenic thickness (Pacheco *et al.*, 1992).

Earthquake swarms are a variant of seismic activity where numerous small earthquakes occur without a distinct principal shock, the activity waxing and waning through time and sometimes migrating. A key distinction from standard tectonic earthquakes is that swarms tend to be distributed throughout a substantial volume of rock. They are frequently associated with areas of recent volcanic or geothermal activity within extensional and transtensional tectonic regimes; in the latter, swarms are often localized within dilational stepovers. To account for swarm characteristics, Hill (1977) proposed a mechanical model involving migration of magmatic fluids through a "honeycomb" mesh of interlinked minor shear and extension fractures (Fig. 3). Such interlinked fault-fracture meshes can be recognized in rocks on a variety of scales,

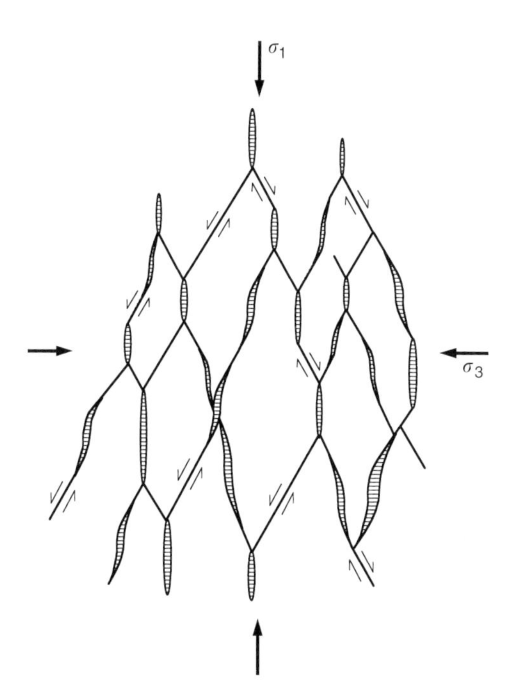

FIGURE 3 Mesh model for swarm seismicity involving interlinked minor faults and extension fractures infilled with dikes, hydrothermal veins, etc. (cross-hatched) developed in relation to principal compressive stresses, $\sigma_1 > \sigma_2 > \sigma_3$ (after Hill, 1977).

functioning as migratory conduits for hydrothermal and hydrocarbon, as well as magmatic fluids (Sibson, 1996).

Major fault zones such as the San Andreas fault system also exhibit significant along-strike variations in seismic style (Wallace, 1970; Hill *et al.*, 1990). Microseismic activity is extremely low along the two segments of the San Andreas Fault which ruptured historically in great $M \sim 8$ earthquakes in 1857 and 1906. In contrast, along segments of the San Andreas and Calaveras faults which are actively creeping, the fault zone is defined by clustered microearthquakes as a subvertical tabular structure extending through most of the seismogenic zone (Fig. 2). Indications from paleoseismology are that these variations in current seismic style represent long-term characteristics of San Andreas fault behavior.

3. Internal Structure and Rheology of Crustal Fault Zones

Surface and mine exposures of ancient fault systems exhumed from different depths provide information on the internal structure, slip processes, and rheology of fault zones at different levels in the crust.

3.1 General Model of a Transcrustal Fault Zone

Major transcrustal fault zones appear as predominantly tabular features in both brittle and ductile regions of the crust with some evidence of flaring with increasing depth in the middle to lower crust (Hanmer, 1988). Their internal structure is often highly heterogeneous with slip and/or strain localized on principal displacement zones (PDZ) which may be planar over considerable distances or anastomose in a mesh-like structure (Wallace and Morris, 1986). In the brittle upper crust, PDZ are commonly localized within or, commonly, at one or other margin to a fault zone of varying thickness ($< \sim 1$ km for large-displacement strike-slip faults) comprising variably fractured and cataclastically deformed material. In the ductile middle to lower crust, high-strain shear zones anastomose around lozenges of comparatively low-strain material. Syntectonic hydrothermal veining may be variably developed at all crustal levels.

3.1.1 Fault Rock Distribution

Fault rocks (reduced in grain size from their protoliths by a range of mechanisms) exhibit a wide variety of microstructural characteristics determined by their parent rock, their original deformation environment, and their exhumation path (Snoke *et al.*, 1998). Their distribution in and around ancient fault zones has been used to build conceptual models of transcrustal fault zones, mostly for quartzo-feldspathic continental crust (Sibson, 1977, 1983; Hanmer, 1988; Passchier and Trouw, 1996). For monolithologic crust, a general depth progression in the predominant fault rock type:

gouge/breccia → cataclasite → mylonite → mylonitic gneiss (often clay-rich)

has been established but it should be noted that geobarometric controls are generally poor. The effect of crustal heterogeneity flanking a transcrustal fault zone is to smear-out this dominant fault-rock progression, forming a mélange of competent and incompetent material (Fig. 4).

3.1.1.1 Frictional Regime (FR)

Incohesive gouge and breccia, and cohesive rocks of the cataclasite series (with increasing grain refinement, *protocataclasite → cataclasite → ultracataclasite*) have deformed largely by brittle cataclastic processes accompanied by varying amounts of hydrothermal alteration. Textures range from essentially random-fabric to foliated with alignment of elongate clasts defining shape fabrics. Associated minor structures include Riedel shear arrays (Tchalenko, 1970; Rutter *et al.*, 1986), and local hydrothermal veining hosted by shear and extension fractures. Discrete slip surfaces may carry frictional striae and other wear-induced features or hydrothermal growth fibers (slickenfibers) indicating slip direction and shear-sense (Doblas, 1998). The predominant fault rocks associated with PDZ are ultracataclasites (Chester *et al.*, 1993). Associated sparse development of pseudotachylyte friction-melt provides relic evidence of transient high power dissipation accompanying seismic slip (see Section 3.2.1) but in general, partly because of structural overprinting, distinguishing seismic from aseismic deformation features is problematic. Evidence for diffusive mass transfer (pressure solution) is sometimes present in finer grained cataclastic detritus with development of phyllonitic microstructure (Bos *et al.*, 2000) but the predominance of cataclastic processes suggests that shearing is governed by pressure-dependent frictional interaction at grain boundaries and along sliding surfaces.

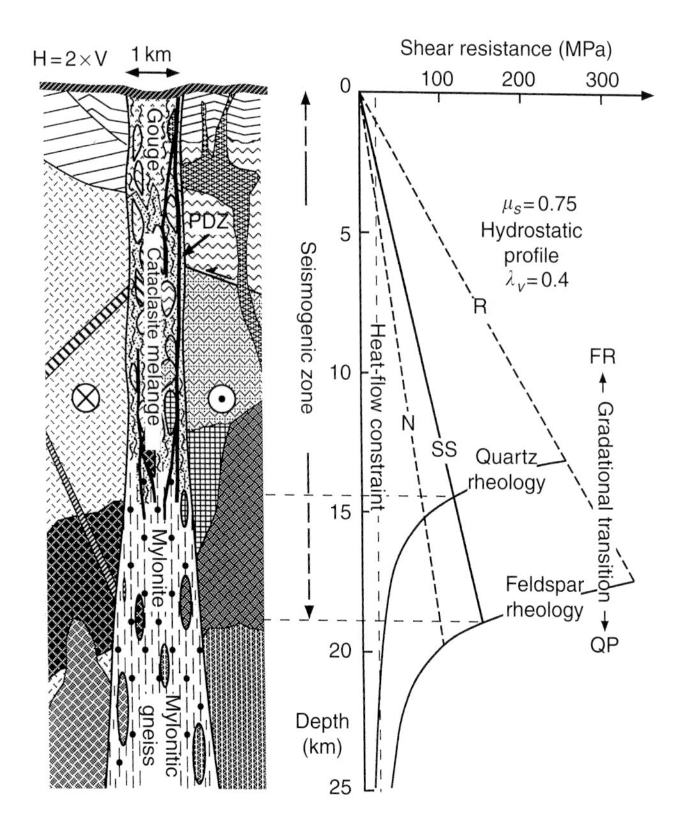

FIGURE 4 General rheological model and schematic strength profile for a transcrustal strike–slip fault zone under a geothermal gradient of $\sim 25°C\,km^{-1}$. Frictional fault strength calculated assuming optimal orientation and hydrostatic fluid pressures (comparative strengths of optimal normal and reverse faults also shown). Average frictional strength imposed by heat flow constraint on the San Andreas fault shown for comparison.

3.1.1.2 Quasi-plastic Regime (QP)

A progressive transition in shearing processes occurs within continental fault zones near the onset of greenschist facies metamorphic conditions ($T > \sim 300°C$), marked by the appearance of fault rocks of the mylonites series (*protomylonite → mylonite → ultramylonite*). Development of penetrative L–S mylonitic fabrics (penetrative foliation with stretch lineation) involves temperature-sensitive crystal plastic flow by dislocation creep which sets in for quartz at $T > 300$–$350°C$ and for feldspar at $T > 450°C$ (Simpson, 1985; Carter and Tsenn, 1987). Greenschist mylonitization therefore generally involves a mixture of deformation mechanisms with plastic flow of quartz accompanied by feldspar cataclasis. Crystal plastic flow may also be aided or supplanted by diffusional mechanisms as grain size is reduced by dynamic recrystallization and/or cataclasis (White *et al.*, 1980; Chester, 1995). As originally defined (Sibson, 1977), the quasi-plastic (QP) regime embraces all these flow mechanisms, with well-ordered mylonitic fabrics largely the product of aseismic shearing.

Textures of mylonitic gneisses (also predominantly L–S tectonites) developed under higher greenschist to amphibolite facies metamorphic conditions in the lower crust ($T > 500°C$) record more continuous shearing deformation within ductile shear zones that commonly range from hundreds of meters to tens of kilometers in thickness (Hanmer, 1988; Corsini *et al.*, 1996).

3.1.1.3 Complexity of the FR/QP Transition

Inevitably, the change from frictional faulting to quasiplastic shearing is gradational across a substantial depth interval, with compositional heterogeneity contributing to the smearing-out of the transition. Fault-rock textures from this transitional region record a mixture of discontinuous and continuous shearing (for example, mylonitic fabrics interlaced with deformed pseudotachylytes, cataclasites, or hydrothermal fault-veins, White, 1996; Takagi *et al.*, 2000). Moreover, once grain size has been reduced by cataclasis or dynamic recrystallization within the temperature range 200–400°C, hydrothermal alteration leads to reaction weakening and the development of phyllonites (hydrated phyllosilicate-rich mylonites) deforming in part by diffusional flow mechanisms (Chester, 1995). Indeed, on the basis of kaolinite–halite analog shearing experiments, Bos *et al.* (2000) suggest that a broad transitional regime may be characterized by mixed frictional–viscous behavior with strength reduced by up to a factor of 3. To complicate the situation further, hydrothermal vein systems provide evidence for substantial fluid-pressure cycling at this structural level in the crust (see Section 5.4.4; Robert *et al.*, 1995; Parry, 1998).

3.1.1.4 Hydrothermal Veining

Syntectonic hydrothermal veins (quartz, calcite, zeolites, etc.) are variably developed at different levels within crustal fault zones (see also Section 5.1). Hydrothermal material may infill breccias and irregular fractures, but may also occupy originally planar fractures in systematic vein-sets. These include fault-veins lying along shearing surfaces and extension veins occupying purely extensional fractures. Both commonly have crystal growth textures recording incremental development. Macroscopic extension veins, usually interpreted as the product of natural hydraulic fracturing, have special significance in that they require fluid-pressure to have exceeded the least compressive stress (i.e., $P_f > \sigma_3$) with differential stress less than four times the tensile strength of the rock ($(\sigma_1 - \sigma_3) < 4\,T$) at the time of their formation (Secor, 1965). Tensile strengths for most sedimentary rocks are less than 10 MPa, and for crystalline rocks rarely exceed 20 MPa (Lockner, 1995). Thus, apart from indicating fluid overpressures in other-than near-surface conditions (see Section 5.3), the presence of extension veining puts an upper limit on differential stress in fault zones (Etheridge, 1983).

Hydrothermal cementation assisted by solution transfer within fine cataclastic detritus may also contribute to "fault-healing" through interseismic periods (Angevine *et al.*, 1982; and Chapter 32 by Lockner and Beeler).

3.1.2 Shear Strength Profiles

Given the field evidence on the changing rheology of a trans-crustal fault zone with depth, a simple procedure may be used to construct bounding profiles of shear resistance versus depth in the crust (Sibson, 1982, 1984; Smith and Bruhn, 1984). The procedure depends critically on input assumptions and, for a variety of reasons, the resulting profiles should generally be regarded as providing an upper limit to the shear strength of natural fault zones.

Shear resistance in the upper crustal frictional (FR) regime is estimated from Amontons Law:

$$\tau_{fr} = \mu_s \sigma_n{}' = \mu_s (\sigma_n - P_f) \qquad (2)$$

where μ_s, the static coefficient of rock friction is generally assumed to lie within Byerlee's (1978) experimentally determined range ($0.6 < \mu_s < 0.85$), σ_n is the normal stress on the fault and P_f is the fluid pressure. Frictional shear resistance at a depth, z, can then be expressed as a function of the effective vertical stress:

$$\tau_{fr} = c\,\mu_s \sigma_v{}' = c\,\mu_s (\sigma_v - P_f) = c\,\mu_s \rho g z (1 - \lambda_v) \qquad (3)$$

where c is a geometrical factor depending on fault attitude and the prevailing stress field, ρ is the average rock density, g is gravitational acceleration, and the pore-fluid factor, $\lambda_v = P_f/\sigma_v$. For the three basic stress regimes with $\sigma_v = \sigma_1$, σ_2, or σ_3 (Anderson, 1951), and with a representative "Byerlee" friction coefficient, $\mu_s = 0.75$, $c = 0.40$, 0.64, and 1.60 for normal, strike-slip, and thrust faults, respectively, when they are optimally oriented for frictional reactivation (containing σ_2 and lying at $\sim 27°$ to σ_1) (Sibson, 1974). Profiles are usually constructed on the assumption of hydrostatic fluid pressures ($\lambda_v \sim 0.4$) throughout the frictional regime.

Support for this procedure comes from borehole measurements made to depths as great as 9 km in crystalline cratonic crust which reveal fluid-pressures close to hydrostatic, and stress levels generally consistent with critical stressing of optimally oriented faults with "Byerlee" friction coefficients (Townend and Zoback, 2000).

Flow shear resistance in the QP regime is generally assumed to be governed by a temperature-dependent power law appropriate to dislocation creep, of general form:

$$\tau_{fl} = [A\,\dot{\gamma}\,\exp(Q/\boldsymbol{R}T)]^{(1/n)} \qquad (4)$$

where $\dot{\gamma}$ is the shear strain-rate, Q is the activation energy, $\boldsymbol{R}$ is the gas constant, T is absolute temperature, and A and n are material constants derived from laboratory experiments (Carter and Tsenn, 1987). Estimates of flow strength at a particular depth, therefore, require specification of the geotherm and the strain-rate in the QP regime. For the commonly observed widths of ductile shear zones in the mid-crust, localized aseismic strain-rates in the range $10^{-13} < \dot{\gamma} < 10^{-10}\,\text{sec}^{-1}$ seem appropriate.

Application of this procedure leads to acute peaks in shear strength and associated strain energy concentration at the FR/QP transition (Fig. 4). However, the limitations of the input assumptions should be kept in mind. For instance, many fault zones are oriented at other than optimal orientations for frictional reactivation (e.g., Fig. 6), leading to higher frictional strength. But frictional strength may be significantly lowered by a high phyllosilicate content in the fault zone (Bos *et al.*, 2000) or by fluid pressure levels elevated above hydrostatic (Streit, 1997). In the QP regime, extrapolation of flow laws derived from fast strain-rate, low-strain laboratory experiments to much slower, large-strain natural deformation is questionable. Complexities arise from the polymineralic character of fault-rock assemblages (Kawamoto and Shimamoto, 1998). Nor is any account usually taken of diffusional mechanisms which may bring about significant strength reduction, especially in the 200–400°C temperature range embracing the FR/QP transition (Chester, 1995; Bos *et al.*, 2000).

3.1.3 Relationship to Seismogenic Zone

Commonly inferred temperatures at the base of the seismogenic zone in continental crust ($T = 350 \pm 50°C$) accord with the recognition of a progressive transition from predominantly frictional (FR) to predominantly quasi-plastic (QP) deformation processes in fault zones (corresponding to the cataclasite–mylonite transition) at the onset of greenschist facies metamorphic conditions. Fault-zone rheology is strongly affected by the quartz/feldspar ratio: the temperature of the FR/QP transition can change from $T \sim 350°C$ for quartz-rich material to $T \sim 450°C$ where feldspar is the dominant load-bearing constituent within the rock-mass. Thus, crustal heterogeneity tends to smear out this complex transition zone involving a mixture of discontinuous and continuous shearing processes. Coupled with variations in regional heat flow, these factors account for much of the observed variation in the depth of the seismogenic zone (Sibson, 1984; Smith and Bruhn, 1984). Complexity at the base of the seismogenic zone, however, may be further compounded by fluid-pressure cycling in the vicinity of the FR/QP transition (Sibson, 1992, 1994).

The base of the seismogenic zone is also sometimes referred to as the brittle-plastic transition (e.g., Scholz, 1988, 1990), a not entirely appropriate term as it takes no account of the potential role played by diffusional flow mechanisms in fine-grained fault rocks. For example, Rutter and Brodie (1987) suggest that shallowing of seismic activity along transform faults in the oceanic lithosphere, in comparison with intraplate regions (Engeln *et al.*, 1986), may be governed by the dehydration of serpentinized ultramafics to a very fine-grained olivine, promoting weak aseismic shearing through grain boundary diffusion.

Another interpretation takes account of constitutive laws recently established for sliding friction in rocks (Dieterich, 1978; Blanpied *et al.*, 1995; Scholz, 1998). Frictional stick–slip instability believed responsible for shallow crustal earthquakes requires velocity-weakening behavior. Increasing temperature leads to velocity-strengthening and corresponds to a change from abrasive to adhesive wear, equated with the cataclasite–mylonite transition (Scholz, 1988, 1998). Marone and Scholz (1988) also suggest that the aseismic character of the upper 2–4 km of major fault zones such as the San Andreas results from velocity-strengthening behavior of unconsolidated granular material at low confining pressure. It is tempting to equate this zone to the anticipated transition from incohesive gouge-breccia (often clay-rich) in the top 3–4 km of mature fault zones to cohesive, well-compacted cataclasites (Fig. 4).

3.2 Seismic Slip Processes

For typical seismic slip-rates of $\sim 1\ \mathrm{m\ sec^{-1}}$, power dissipated over a fault surface during rupture is of the order of 10–100 MW m^{-2} for frictional shear resistances, $\tau_{fr} = 10$ and 100 MPa, respectively. Such high power dissipation can lead to intense disordering, or, if shearing is restricted to a narrow zone, to large increases in temperature. Given the short time periods involved (rise times of 1–10 sec), heating is essentially adiabatic for slip zones thicker than about 10 mm and temperature increase is approximated by:

$$\Delta T = \frac{\tau_{fr}\, u}{c_p\, \rho\, s} \qquad (5)$$

where u is the coseismic slip, s is the thickness of the seismogenic shear zone, ρ is rock density, and the specific heat, $c_p \sim 1000\ \mathrm{J\,kg^{-1}\,°C}$ for most rocks. Anticipated temperature increases are plotted against shear zone thickness, s, for coseismic displacements of 0.1, 1, 5, and 10 m against constant shear resistances of 10 and 100 MPa (Fig. 5a).

For displacement accompanying moderate to large earthquakes (1–5 m) temperature increases of around 1000°C would be expected for $\tau_{fr} = 100$ MPa, provided slip is confined to a zone less than 0.1–0.2 m in thickness, or less than 10–20 mm if $\tau_{fr} = 10$ MPa. Temperature rises of $\sim 100°C$ can be expected for slip zones that are an order of magnitude thicker. Geologically based estimates of the thickness of the seismogenic slip zone, along with crude estimates of applicable depths, are given in Figure 5b. Most suggest that seismic slip on PDZ is commonly localized to within less than 0.1 m, and often to less than 10 mm.

Given the likelihood of significant temperature increase during a seismic slip increment, shear resistance is unlikely to stay constant during rupture. Two temperature-dependent "feedback" mechanisms for lowering shear resistance during slip have been proposed: friction-melting (McKenzie and Brune, 1972), and transient thermal pressurization of fault zone fluids (Sibson, 1973). Both of these mechanisms could lead to a high degree of stress relief during rupture and high seismic efficiency, provided fluids can be contained within the seismic slip zone. However, a variety of dynamic slip-weakening mechanisms that are not temperature-dependent are also under consideration (see Chapter 35 by Brune and Thatcher).

3.2.1 Friction Melting

Taking account of the evidence for slip localization, friction-melting requiring $\Delta T > 1000°C$ might be expected to be widespread in seismically active crustal fault zones (McKenzie and Brune, 1972). Fault-generated pseudotachylyte friction-melts have been reported principally from fault zones developed in crystalline rocks, developed usually on individual faults with comparatively low total displacement (Sibson,

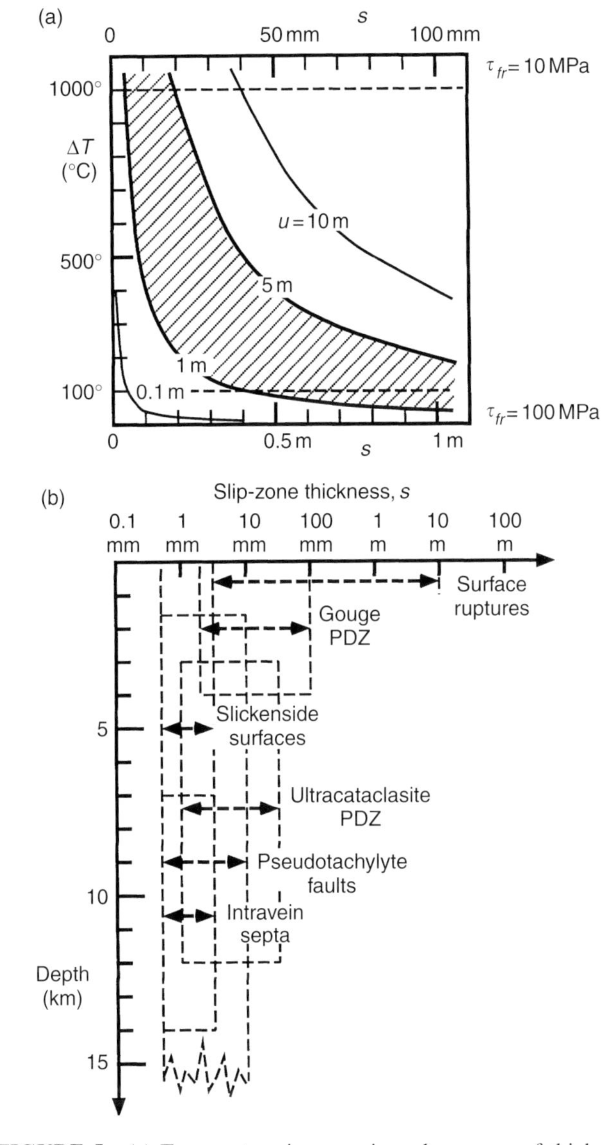

FIGURE 5 (a) Temperature increase in a shear zone of thickness, s, for $\tau_{fr} = 10$ and 100 MPa and various slip values, u, assuming uniform adiabatic heating with specific heat $c_p = 1000\,\mathrm{J\,kg^{-1}\,°C}$. (b) Geological constraints on the thickness of the seismic slip zone (PDZ, principal displacement zone).

1975; Magloughlin and Spray, 1992). They are most commonly associated with cataclasites but are sometimes hosted by mylonites where they appear to have been generated along transient brittle discontinuities within otherwise ductilely deforming shear zones (White, 1996; Takagi *et al.*, 2000). However, pseudotachylytes are comparatively rare in exhumed mature fault zones, raising an important issue as to whether their scarcity is apparent (resulting from cataclastic reworking and hydration), or whether seismic friction-melting in fault zones is a genuinely rare phenomenon inhibited by other mechanisms.

3.2.2 Thermal Pressurization of Fluids

Sudden power dissipation in fluid-saturated fault zones may boost ambient intergranular fluid-pressures by thermal pressurization (possibly aided by dehydration of clay-gouge), reducing kinetic shear resistance in accordance with Eq. (2) and inhibiting the larger temperature increases needed for melting (Sibson, 1973). Under constant volume conditions, $\Delta T \sim 100°\mathrm{C}$ is sufficient to raise initial hydrostatic to near-lithostatic fluid pressure levels. Additional boosting of fluid-pressure may result from dilatant crack closure accompanying shear stress release. Efficacy of all these mechanisms in reducing shear resistance depends critically on permeability adjacent to the slip surface (Lachenbruch, 1980; Mase and Smith, 1987) (see Section 5.2). However, thermal pressurization of fault zone fluids provides a physical explanation for the general scarcity of pseudotachylyte and its restriction to small displacement faults in dry crystalline rocks.

4. Structural Geometry of Crustal Fault Systems

The complex 3D finite structure of fault systems has been investigated by structural mapping, geophysical imaging, and oilfield drilling for over 150 years but only in the past three decades have higher resolution earthquake locations made it possible to correlate mainshock ruptures and aftershock activity with details of fault structural geometry. Earthquake ruptures are often viewed as increments of brittle fault growth because the incremental change in shear strain associated with an individual rupture of length, L_i, and average displacement, d_i, is typically:

$$10^{-4} > d_i/L_i > 10^{-5} \tag{6}$$

(Wells and Coppersmith, 1994). However, the ratio of finite displacement to total fault length on bounded faults generally lies in the range:

$$10^{-1} > d_f/L_f > 10^{-3} \tag{7}$$

(Schlische *et al.*, 1996), implying that faults, over longer time periods, are comparatively "ductile" structures with significant wallrock strains.

Moreover, the complexity of the finite deformation state is generally hard to explain solely in terms of the superposition of successive coseismic slip increments. Fault surfaces are frequently nonplanar; associated wallrock deformation may include large-strain fold structures with penetrative fabric development, as well as subsidiary faulting and fracturing distributed over broad regions. A substantial aseismic component of deformation has clearly contributed to this finite complexity.

4.1 "Anderson–Byerlee" Frictional Fault Mechanics

Anderson (1951) evolved a dynamic classification of faults by assuming: (i) that trajectories of the principal compressive stresses ($\sigma_1 > \sigma_2 > \sigma_3$) in the crust are constrained to be either vertical or horizontal because of the Earth's free surface; and (ii) that faults form according to the Coulomb failure criterion as planar surfaces containing σ_2 and lying at an angle, $\theta_i = 0.5 \tan^{-1}(1/\mu_i)$, to σ_1 where μ_i is the coefficient of internal friction. For most rocks, $0.5 < \mu_i < 1.0$, so that $\theta_i = 27 \pm 5°$ (Jaeger and Cook, 1979). Thrust faults ($\sigma_v = \sigma_3$) and normal faults ($\sigma_v = \sigma_1$) should therefore have initial dips of $27 \pm 5°$ and $63 \pm 5°$, respectively, while strike-slip (wrench) faults ($\sigma_v = \sigma_2$) should be vertical. Note that, strictly, Anderson's classification relates only to the initiation of brittle faults in intact crust.

Some notion of the extent to which larger displacement faults exhibit "Andersonian" behavior may be gained from dip distributions obtained from moderate-to-large ruptures ($M > 5.5$) on reverse and normal faults that are close-to-pure dip slip (Fig. 6) to which two-dimensional (2D) reactivation analysis is applicable (Collettini and Sibson, 2001). Also shown is the ratio of differential stress to effective vertical stress required for reactivation of cohesionless reverse and normal faults at different orientations; note that optimal reactivation is more acutely defined for reverse faults. For the typical range of rock friction coefficients, $0.6 < \mu_s < 0.85$ (Byerlee, 1978), reshear occurs at the lowest stress ratio when a fault is oriented at the optimal reactivation angle, $30° > \theta_r{}^* > 25°$, to σ_1, with frictional lock-up (for $\sigma_3{}' = (\sigma_3 - P_f) > 0$) occurring when $60° > \theta_r = 2\theta_r{}^* > 50°$ (Fig. 7).

On the assumption of horizontal and vertical stress trajectories, it is apparent that the dip ranges for both reverse and normal fault ruptures are consistent with $\mu_s \sim 0.6$, with lock-up occurring at $\sim 60°$ and some evidence for optimal orientation of reverse faults at $\theta_r = 25$–$35°$. Field and seismological observations confirm that strike-slip faults do tend to be subvertical, whereas the majority of reverse faults are thrusts with $\delta < 45°$, and the majority of normal faults have $\delta > 45°$ (as in the rupture dip ranges), all in generally accordance with "Anderson–Byerlee" expectations. Widespread oblique slip suggests, however, that many earthquakes are occurring on inherited faults reactivated in the present stress regime (Zoback, 1992).

Notable exceptions do, however, exist: for example, very low-angle normal faults that were apparently active to large displacements at very low dips ($\delta < 15°$), often associated with metamorphic core complexes (Lister and Davis, 1989; Wernicke, 1995), and high-angle reverse faults ($\delta > 45°$), some clearly developed by compressional reactivation of former normal faults during positive tectonic inversion (Hayward and Graham, 1989). Several

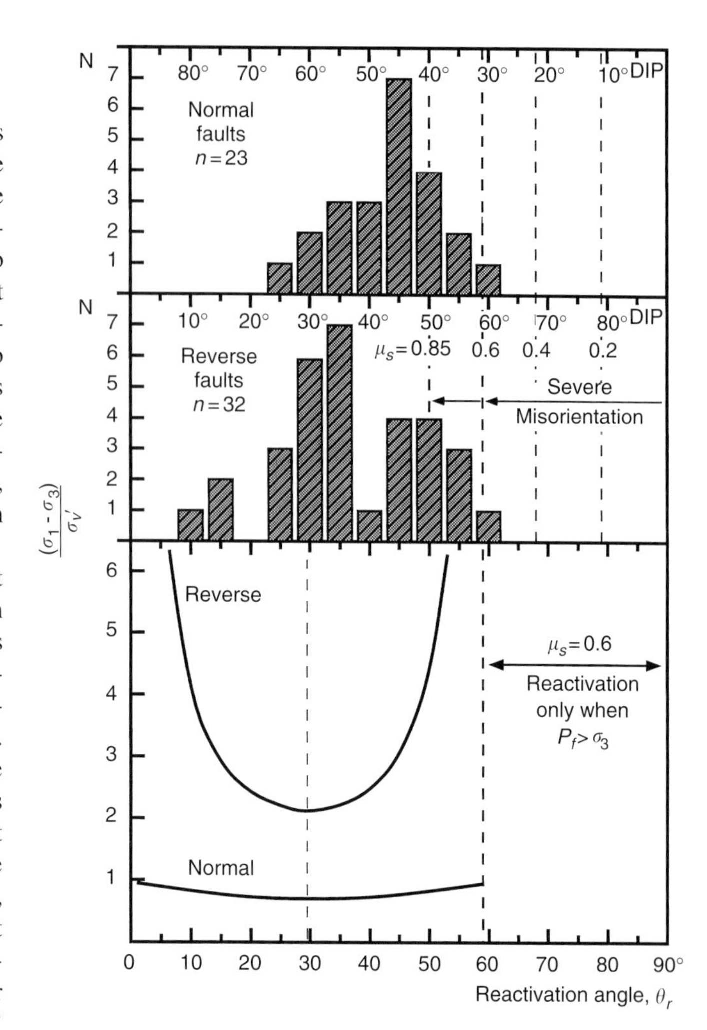

FIGURE 6 Dip distributions for reverse and normal dip–slip $M > 5.5$ ruptures with slip vector lying within $\pm 30°$ of dip direction (data updated from Sibson and Xie (1998), Jackson and White (1989), and Collettini and Sibson (2001)). Dashed lines denote the expected angle of frictional lock-up for different friction coefficients, μ_s, on the assumption of vertical and horizontal stress trajectories. Reactivation parameter $(\sigma_1 - \sigma_3)/\sigma_v{}'$ (calculated for $\mu_s = 0.6$) provides a measure of the relative ease of reactivation of differently oriented faults.

large strike-slip systems, including the San Andreas fault, remain active though apparently oriented at very high angles to regional σ_1 trajectories (Mount and Suppe, 1992).

These structures raise important questions as to whether Byerlee frictional coefficients are universally applicable, or how often and under what conditions stress trajectories depart from the subhorizontal and subvertical within the seismogenic crust (Westaway, 1998). An important related issue is whether seismic increments of deformation that contribute to a complex finite deformation remain close to "Andersonian" or are themselves inherently more complex.

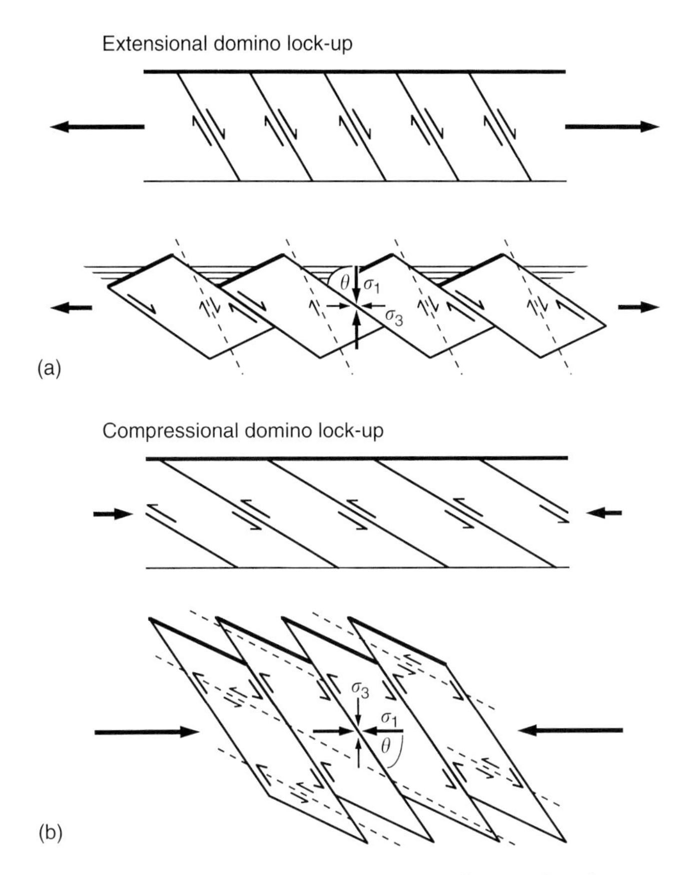

FIGURE 7 Sets of initially "Andersonian" normal and reverse faults undergoing "domino" rotation to frictional lock-up at $\theta \sim 55$–60° to σ_1. Light dashed lines indicate orientations of secondary fault sets expected to develop after lock-up.

4.2 Fault Populations and Scaling Relationships

Crustal deformation is distributed on different-sized faults with larger faults accommodating greater finite displacements than small. Despite earlier suggestions of power-law relationships (Walsh and Watterson, 1988; Marrett and Allmendinger, 1991), a common linear scaling relationship between finite fault displacement, d_f, and fault length, L_f, appears to hold over eight orders of magnitude irrespective of faulting mode, though the bulk of the data comes from normal fault populations (Schlische *et al.*, 1996). On average, fault length increases linearly with displacement according to:

$$d_f \sim 0.01\, L_f \tag{8}$$

a relationship consistent with post-yield fracture mechanics (Cowie and Scholz, 1992).

Size-frequency analysis generally reveals a power-law distribution for faults in continental crust, reflecting a self-similar fractal distribution over at least some scale ranges. An initial fault distribution nucleating on random heterogeneities "self-organizes" by fault growth, interaction, and coalescence to a fractal pattern of clustered faulting where the bulk of the deformation is accommodated by the larger structures which "shield" adjacent areas from fault growth (Sornette *et al.*, 1990; Cowie, 1998). Note, however, that fault populations in deforming oceanic crust near midocean ridges tend to follow an exponential distribution, with a length scale perhaps imposed by the restricted width of the active rift zone or by the thinness of the brittle layer (Cowie, 1998).

4.3 Fault Segmentation and Linkage

Major fault zones at both brittle and ductile levels of the crust are commonly segmented with individual fault segments, either planar or curved, linked in 3D through a variety of subsidiary structures (Davison, 1994). Segmentation may arise from earlier growth and coalescence of previously independent faults; evidence exists that the degree of segmentation on strike-slip faults decreases with finite displacement, affecting seismic style and maximum rupture size (Wesnousky, 1988).

Linkage between adjacent fault segments may be "hard", involving discrete linking faults or "soft," involving distributed deformation (Walsh and Watterson, 1991). For example, en échelon normal faults may be linked by "hard" transfer faults sharing a common slip-vector, or by a "soft" relay ramp involving distributed warping between the segments (Fig. 8). An important distinction may also be made between conservative and nonconservative fault linkages depending on whether or not slip transfer involves local volume change (King, 1986). Transfer faults, sharing a common slip vector with adjacent thrust or normal fault segments, are an example of conservative linkage.

Nonconservative linkages include discontinuous jogs (or stepovers) developed perpendicular to slip vectors between en échelon fault segments (Fig. 8). Jogs may be classified as dilational ($\Delta\bar{\sigma} < 0$) or contractional ($\Delta\bar{\sigma} > 0$), depending on the change in mean stress within the jog accompanying slip transfer, and may develop on all types of fault with cross-fault dimensions ranging from millimeters to kilometers (Segall and Pollard, 1980).

4.4 Crustal Fault Systems

Strictly in a kinematic sense, the dominant modes of faulting at the three fundamental classes of plate boundary (convergent, divergent, and transform) correspond to Anderson's three basic types of fault (thrust, normal, and strike-slip). In reality, however, each type of boundary is generally associated with a range of crustal faulting modes that are concurrently active. The particular aim here is to emphasize those aspects of the structural geometry of commonly encountered crustal fault systems that are relevant to the interpretation of shallow earthquakes.

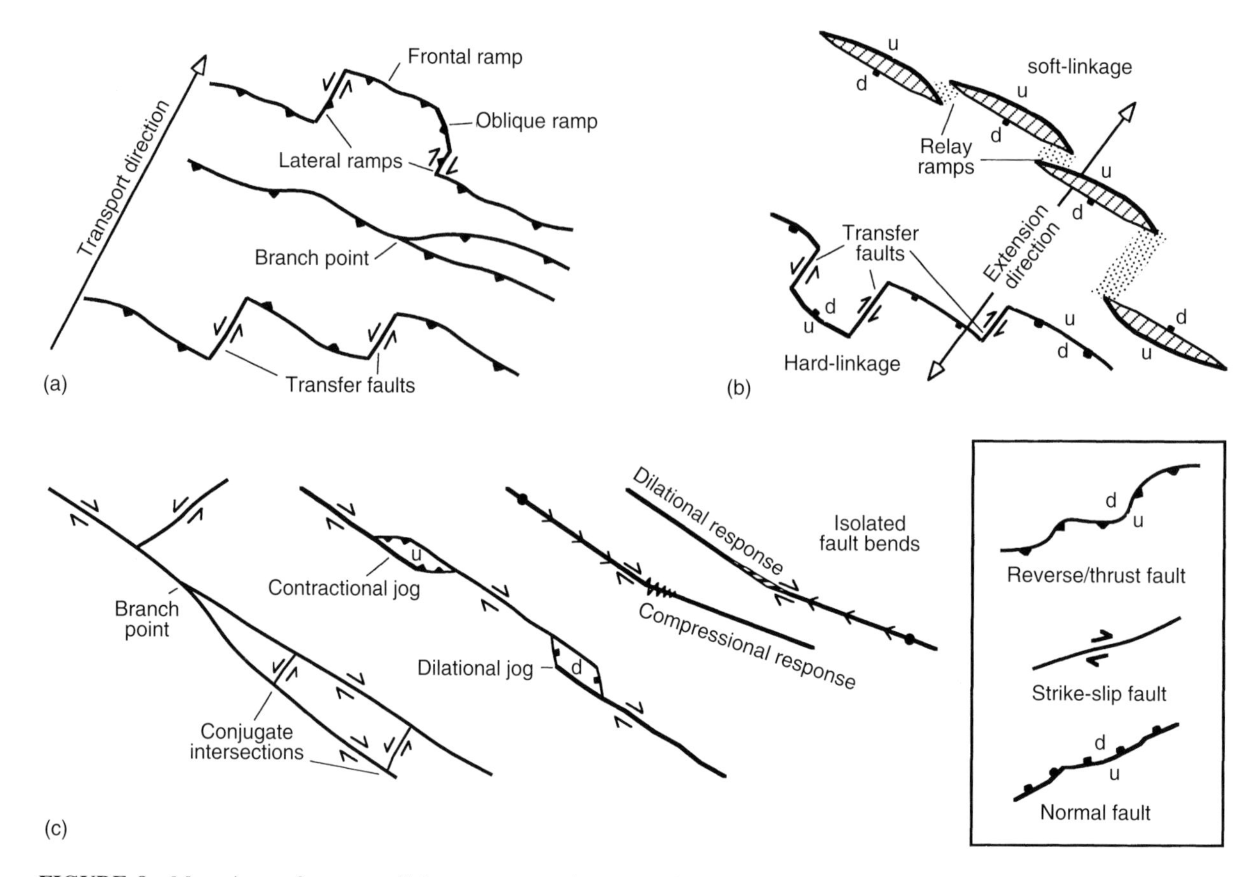

FIGURE 8 Map views of common linkage structures in contractional (a), extensional (b), and strike–slip (c) fault systems. Transfer faults are conservative only when hard-linked and perfectly aligned with the regional transport direction. Non-conservative links such as dilational and contractional jogs may also occur along dip–slip faults. Also shown are possible rupture directivity effects associated with isolated fault bends.

4.4.1 Contractional Reverse Fault Systems

The seismotectonics of reverse fault systems have been reviewed by Yeats *et al.* (1997). In island and mountain arcs, interplate motion is primarily accommodated on the "mega-thrust" at the subduction interface with subsidiary thrusting in the fore-arc basin on the subduction hanging wall and on opposite-vergent thrusts in the back-arc. Thrust faulting likewise dominates the foreland and hinterland flanks of collision orogens. Local areas of thrusting may also develop in association with contractional irregularities on strike-slip faults. Steeper reverse faulting characterizes formerly rifted crust undergoing shortening.

3D complexity revealed within major fold-thrust belts, both across and along strike, by intensive oil–gas exploration has given rise to an equally involved structural terminology (Boyer and Elliott, 1982; McClay, 1992). Deformation may be largely restricted to the sedimentary cover sequence detached from the underlying crystalline basement ("thin-skinned" thrusting) or may actively involve the basement underlying the cover sequence ("thick-skinned" thrusting); the latter commonly involves compressional reactivation of basement-penetrating normal faults inherited from previous crustal extension. Structural complexity arises principally from thrust development at low angles to anisotropy and subhorizontal competence layering within the rock-mass. Characteristic length scales are imposed within thrust assemblages by competent layers which range in thickness from individual beds to the elastic seismogenic lid of the crust.

These high-competence layers serve as σ_1 stress guides at various stages of the evolution of a thrust complex (Fig. 9), deflecting evolving thrusts into flat–ramp–flat assemblages and promoting buckling instabilities. Subsidiary deformation is mostly concentrated in the hanging walls of these "staircase trajectory" thrusts with the development of fault-bend and fault-propagation folds, backthrusts, wedging, and pop-up structures (Suppe, 1985; McClay, 1992). Along-strike complexity arises from transfer (tear) faults and lateral footwall ramps developed subparallel to the overall slip vector, and from oblique ramps.

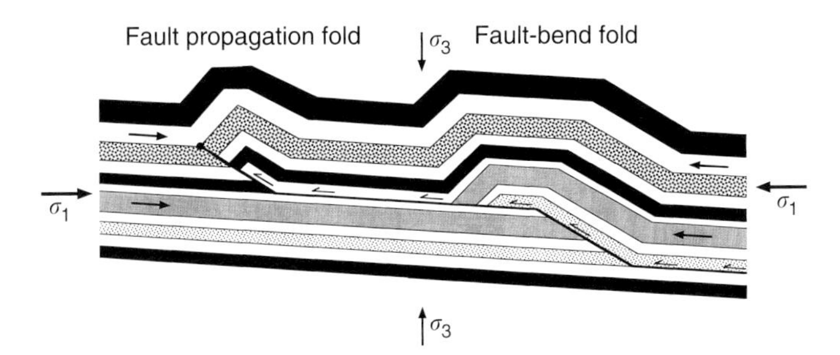

FIGURE 9 Fault bend and fault propagation folds associated with a "staircase" thrust system in layered crust (after McClay, 1992). Thrust ramps form in competent layers which serve as stress guides deflecting principal stress trajectories.

Cross-strike complexity of thrust systems does much to explain the observed dip-distribution of reverse fault ruptures (Fig. 6). The spread of defined reverse rupture dips above optimal orientation ($\delta \sim 30^\circ$) and the notable subsidiary peak at $\delta = 50 \pm 5^\circ$ can be accounted for by progressive "domino" steepening of stacked imbricate thrusts (Fig. 7) with an additional contribution from inherited normal faults reactivated in compression (Sibson and Xie, 1998). Paucity of positively discriminated thrust ruptures with very low dips ($\delta < 15^\circ$) [though widespread in the frontal Himalayas (Molnar and Lyon-Caen, 1989; Powers *et al.*, 1998)] suggests that it is generally the ramps in flat–ramp–flat assemblages that fail in moderate-to-large earthquakes.

4.4.2 Extensional Normal Fault Systems

Reviews of the seismotectonics of extensional fault systems are provided by Yeats *et al.* (1997), Roberts and Yielding (1994), and Jackson and White (1989). At midocean ridges and and in some continental rifts, parallel sets of predominantly inward-facing normal faults tend to be symmetrically disposed about the rift axis. In other regions of distributed continental extension such as the Basin and Range province of western North America, sets of normal faults with consistent facing direction may be associated with series of tilted half-grabens (Fig. 7). In island arc systems, arc-parallel normal faulting may develop in association with magmatic activity in back-arc basins and, in the vicinity of the outer rise, through bending of the subducting slab. Normal faulting may also develop orthogonal to thrusting in collision orogens to allow lateral distension of the uplifted welt.

Although there is considerable evidence from reflection profiling and drilling for the existence of listric normal faults at comparatively shallow depths (<5 km) in sedimentary basins, major seismically active normal faults appear approximately planar in cross-strike profile. However, they are corrugated at various scales along strike, often presenting a somewhat scalloped appearance (Stewart and Hancock, 1991). Extensional stress fields with σ_3 horizontal promote the development of parallel-striking vertical extension fractures. In magmatically active terrain, normal faults are associated with parallel-striking dikes and fissures (Gudmundsson, 1995). Elsewhere, especially in the near-surface, associated fractures may be infilled with hydrothermal veins forming a fault-fracture mesh (Sibson, 1996) (Fig. 3).

In continental regions, individual segments of normal faults typically extend for 10–30 km along strike, distances that are 1–2 times the usual thickness of the seismogenic crust in areas of rifting (Machette *et al.*, 1991), but longer segments are known, for instance in the East African Rift system where the seismogenic zone is notably deeper (Jackson and Blenkinsop, 1997). "Large" normal fault ruptures occupying the full depth of the seismogenic zone generally involve one or more segment lengths within these systems. Individual normal fault segments may be linked to parallel-facing and opposite-facing segments through a variety of linkage structures which include "soft" relay ramps and diffuse accommodation zones, and "hard" linkage in the form of clear-cut transfer faults (Morley *et al.*, 1990; Walsh and Watterson, 1991; McClay and Khalil, 1998).

Progressive crustal extension may be accommodated by the "dominoing" of upper crustal blocks separated by parallel-facing normal faults that extend through the seismogenic layer (Fig. 7). Block rotation may continue until fault dips approach lock-up at $\delta \sim 30^\circ$ and helps to account for the observed distribution of active normal fault dips (Fig. 6). Very low-angle normal faults (detachments) have been usually described in association with metamorphic core complexes (Wernicke, 1995), but more recently in the midocean ridge environment (Blackman *et al.*, 1998). The circumstances under which such structures form and remain active, and their seismogenic role, remain unclear.

4.4.3 Strike–Slip Fault Systems

Comprehensive reviews of strike-slip fault geometry, principally in continental and island arc environments, are provided by Yeats *et al.* (1997), Woodcock and Schubert (1994), and Sylvester (1988). Geomorphic characteristics of oceanic transforms are summarized by Fox and Gallo (1986). Continental strike-slip faulting occurs in a range of tectonic environments and patterns: (1) as often braided systems of subparallel strike-slip faults defining continental transforms; (2) as trench-parallel strike-slip faults contributing to strain partitioning along island and mountain arcs; (3) as indent-related strike-slip faults in collisional orogens; (4) as local transfer structures in predominantly dip-slip fault systems; and (5) as distributed strike-slip faults, developed in either parallel or conjugate sets, accommodating crustal deformation over broad areas, often with significant rotations of crustal blocks about vertical axes (e.g., Luyendyk, 1991).

Seismological studies confirm that the vast bulk of pure strike-slip motion is accommodated on subvertical tabular fault zones. Geological mapping has revealed patterns of ductile strike-slip shear zones ranging up to tens of kilometers in thickness in exhumed middle and lower continental crust, suggesting that in at least some circumstances major strike-slip faults continue downward through the lower crust as localized aseismic shear zones which maintain the anastomosing complexity observed in the upper brittle crust (e.g., Hanmer, 1988; Corsini *et al.*, 1996). Patterns of finite strain associated with strike-slip faults tend to be further complicated by components of transpression and/or transtension which may develop at various stages of the fault's history (Dewey *et al.*, 1998).

In map view, major strike-slip faults are irregular and segmented on a range of scales with isolated fault bends and en échelon stepovers which may be continuous (releasing and restraining bends) or discontinuous (dilational and contractional jogs). Unlike the systematic stepping of Riedel shear arrays which characterize strike-slip rupture traces on cross-strike scales of 1–10 m or so (Tchalenko, 1970), no consistent sense of stepping is usually apparent for these major irregularities which have cross-strike scales of 1–10 km. Such major structural irregularities which, from aftershock studies appear to extend through much of the seismogenic zone (Reasenberg and Ellsworth, 1982), may perturb or arrest rupture propagation and contribute to characteristic earthquake behavior (Harris and Day, 1993; Sibson, 1985). Wesnousky (1988) showed that the frequency of major structural irregularities on strike-slip faults decreases with finite displacement.

Major irregularities induce subsidiary deformation. Thrusting develops in the vicinity of restraining bends and within contractional jogs, and normal faulting occurs near releasing bends and within dilational jogs. Such irregularities serve as local amplifiers of vertical motion within the strike-slip system but many appear to be comparatively short-lived (Brown *et al.*, 1991).

5. Fluid Activity in Fault Zones

The role of aqueous fluids in weakening fault zones and in earthquake rupture processes is contentious but potentially of enormous importance (Hickman *et al.*, 1995). Allowing for cementation or some inherent cohesive strength, C, within the fault zone, frictional fault strength may generally be represented by:

$$\tau_{fr} = C + \mu_s \sigma_n{}' = C + \mu_s(\sigma_n - P_f) \qquad (9)$$

so that fluid overpressures may induce physical weakening by reduction of effective normal stress (Hubbert and Rubey, 1959), as demonstrated by the triggering of seismic activity through deep fluid injection (Nicholson and Wesson, 1990; and Chapter 40 by McGarr *et al.*).

Fault instability may therefore be induced, not just by rising shear stress, τ, as is often assumed, but by increased fluid pressure or decreased normal stress, or by some combination of all three factors. Chemical weakening of faults by aqueous fluids may also occur through processes such as stress corrosion cracking, diffusive mass transfer, reaction softening, and enhanced crystal plasticity in the ductile regime through hydrolytic weakening of silicates (Hickman *et al.*, 1995; and Chapter 32 by Lockner and Beeler).

5.1 Fluid Manifestations

Direct evidence of fluid involvement with earthquakes comes from observations of substantial postseismic discharge in the vicinity of active faults (Muir-Wood and King, 1993), the greatest effects occurring in the vicinity of normal fault ruptures. Other manifestations of fluid/gas migration in and around active fault zones are reviewed by King and Igarishi in Chapter 39. Early suggestions of a link between global CO_2 discharge and seismic belts (Irwin and Barnes, 1980) have been borne out by recent geochemical studies that support the existence of a large CO_2 flux through the San Andreas fault system (averaging >1 tonne per meter strike-length per year) (Kennedy *et al.*, 1997).

Studies of exhumed fault zones provide abundant evidence for their role as fluid conduits at all crustal levels. Hydrothermal alteration assemblages and syntectonic vein systems provide information on fluid compositions at high structural levels (Parry, 1998), whereas metasomatic effects suggest that integrated fluxes of 10^3–$10^5\,m^3\,m^{-2}$ and water/rock ratios $<10^3$ are common in ductile shear zones developed in the middle to deep crust (McCaig, 1997). Compositions of fault fluids are highly variable (Parry, 1998); inferred sources include meteoric waters, diagenesis and metamorphism at depth, local magmatism, and perhaps mantle-derived fluids rich in CO_2.

5.1.1 Fault-hosted Mineralization

Hydrothermal mineral deposits result from focused flow of large fluid volumes and are commonly hosted in and around ancient fault zones. Solubility criteria, incremental vein textures, and the need for rapid transport between different P–T environments to promote instability and localized precipitation suggest that mineralization results from multiple episodes of rapid flow. Although formation of major fault-hosted deposits clearly involves special circumstances, they provide important evidence for fluid redistribution tied to episodes of incremental fault slip and permeability renewal. Two principal habitats are recognized: the shallow *epizonal* environment usually associated with extensional–transtensional faulting near the top of the seismogenic zone, and the *mesozonal* environment developed toward the base of the seismogenic zone in compressional–transpressional regimes.

Abrupt reductions in fluid pressure induced by seismic slip increments appear integral to hydrothermal precipitation in both environments, though the mechanisms differ in detail (see below and Fig. 11).

5.1.1.1 Epizonal Mineralization

Most epizonal deposits (e.g., Au-Ag and Pb-Zn-Ag) occupy high-permeability meshes of steep extension fractures and subsidiary minor faults developed at shallow (<1–2 km) depths in dilational irregularities associated with normal and strike-slip fault systems, which localize ascending hydrothermal plumes and also allow mixing of fluids from different sources (Henley, 1985; Sillitoe, 1993). An association with felsic magmatism is usually apparent for the Au-Ag deposits. Dilational structures typically comprise a mesh of extension veins and subsidiary shears with wallrock breccias derived from multiple episodes of brecciation and hydrothermal cementation.

5.1.1.2 Mesozonal Mineralization

Mesozonal Au-quartz veins, by contrast, are mostly hosted in the subgreenschist to greenschist grade semiductile root zones (10 ± 5 km depth) of fault systems developed in compressonal–transpressional tectonic regimes without clear-cut magmatic associations. The vein systems may extend over 1–2 km vertically, often comprising fault-veins developed on steep reverse or reverse-oblique faults intermeshed in mutual crosscutting relations with arrays of flat-lying extension veins (Sibson *et al.*, 1988; Cox, 1995). These incrementally developed vein systems, inferred to develop through extreme fault-valve action (see below), hold special significance for the shallow earthquake source. They demonstrate that, in at least some instances, near-lithostatic fluid-pressures are localized within fault zones developed in crystalline rock assemblages around the base of the seismogenic zone, with large-amplitude fluid-pressure cycling coupled to the seismic stress cycle (see Fig. 11) (Robert *et al.*, 1995; Parry, 1998).

5.2 Structural Permeability in Fault Zones

Fault permeability is highly variable, depending strongly on the host rock assemblage as well as the level of fault activity. Inactive faults in sedimentary basins variously serve as sealing barriers to fluid migration or as cross-stratal conduits for fluid flow (Jones *et al.*, 1998). Brittle faults developed in initially high-porosity sedimentary rock, through comminution and porosity collapse, tend to be relatively impermeable in comparison with the wallrock (Antonellini and Aydin, 1994), whereas faults formed in initially low-porosity rocks enhance permeability through cataclastic brecciation and subsidiary fracturing. Studies of exhumed fault zones in crystalline rock suggest that a low-permeability core of extremely fine-grained ultracataclasite is often flanked by a comparatively high-permeability "damage zone" of distributed fracturing (Evans *et al.*, 1997).

Evidence from geothermal fields, fault mineralization, and experimental work suggests, however, that fault permeability in active hydrothermal environments can be self-destructive over time periods that are short compared with earthquake recurrence intervals (Morrow *et al.*, 2001; Hickman *et al.*, 1995). Only those fractures that are favourably oriented for shear reactivation remain hydraulically conductive (Barton *et al.*, 1995). Permeability within seismically active fault zones is therefore dynamic, with faulting, fracturing, microcracking and brecciation competing with processes of permeability destruction which include hydrothermal alteration accompanying gouge formation, microcrack healing, hydrothermal cementation of fractures, and solution-precipitation (Parry, 1998). Fault permeability is likely to be highest immediately postrupture (Brown and Bruhn, 1996), progressively diminishing through the interseismic period.

5.3 Fluid Overpressures

Hydrostatic fluid pressures prevail in fluid-saturated rocks where interconnected pore space and fracture systems are freely linked through to the Earth's surface as has been widely demonstrated for crystalline intraplate cratonic crust (Townend and Zoback, 2000). However, fluid pressures in deforming crust are commonly elevated well above hydrostatic at depths of more than a few kilometers (Fyfe *et al.*, 1978). Some earthquakes clearly nucleate in overpressured crust, for example, in fore-arc regions of active thrust faulting (e.g., the 1999 M 7.6 Chi-Chi, Taiwan earthquake), but there is debate on the extent to which major fault zones are routinely overpressured above hydrostatic and whether such overpressuring is restricted to the immediate vicinity of the fault zones (Fig. 10).

Lowered frictional strength from overpressuring [Eq. (2)] could contribute to the apparent weakness of the San Andreas fault system (e.g., Streit, 1997), but there are important questions as to how such overpressures are maintained and the extent of the net fluid flux through the fault zone. Rice (1992) suggested a model where overpressuring is sustained by a continuous upwards fluid flux through the fault zone (large net flux), whereas Byerlee (1993) supposes fault zones to be subdivided into overpressured compartments separated by seals, in each of which the pressure gradient remains hydrostatic, with local fluid redistribution occurring at the time of earthquakes. Sleep and Blanpied (1994) put forward a model in which fluid overpressure is "regenerated" in each seismic cycle through compaction of fault gouge (low net flux).

Evidence for fluid overpressures and large-amplitude cycling of fluid pressure in some exhumed fault zones has been derived from associated syntectonic vein systems developed in both sedimentary and crystalline host rocks and their contained fluid inclusions (Robert *et al.*, 1995; Parry, 1998). Moreover, intense veining in strongly mineralized fault zones suggests focused flow of very large fluid volumes (e.g., Cox, 1995).

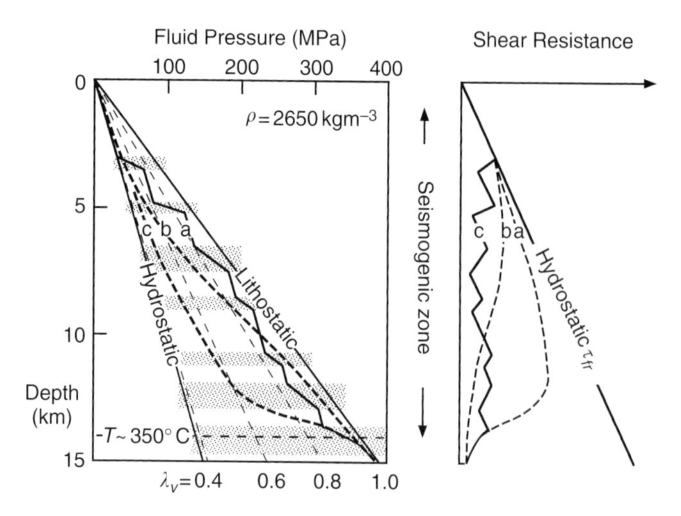

FIGURE 10 Hypothetical fluid-pressure profiles and associated frictional shear resistance (normalized to hydrostatic τ_{fr}) through the FR regime of an active fault zone. Dashed lines (a, b) represent two of a range of smooth progressions from hydrostatic to near-lithostatic fluid pressures towards the base of the seismogenic zone; (c) represents a profile through a compartmentalized fault zone with hydrostatic gradients between each sealing horizon (stippled).

5.3.1 Coseismic Pressure Changes

As previously discussed (Section 3.2.2), frictional dissipation during seismic slip may boost intergranular fluid-pressures significantly by thermal pressurization, thus reducing kinetic shear resistance. Fluid pressure may also be locally reduced by rapid slip transfer across dilational jogs, the transient imbalance in fluid-pressures sometimes inducing hydraulic implosion of wallrock into the dilatant site (Sibson, 1985).

5.4 Mechanisms for Fluid Redistribution

Several dynamic mechanisms have been invoked to account for fluid redistribution linked to the earthquake stress cycle but their relative importance at different crustal levels and in different tectonic regimes is uncertain. More than one mechanism may operate at any time and place.

5.4.1 Dilatancy Pumping

Early suggestions of cyclical, regional-scale microcrack dilatancy operating at high (kilobar) levels of differential stress in the crust adjacent to active fault zones and capable of large-scale fluid redistribution (Scholz *et al.*, 1973) have not been substantiated, though the search for other forms of stress-dependent dilatancy continues (Crampin, 1994). Various other forms of cyclical dilatancy have been proposed, some restricted to the fault zones themselves (Sibson, 1994; and Chapter 32 by Lockner and Beeler). However, until the stress levels driving faulting and appropriate constitutive dilatancy laws are resolved, it is not possible to evaluate their relative contributions to fluid redistribution in the crust.

5.4.2 Mean Stress Cycling

Tectonic shear stress on faults cannot in general be changed without also changing fault-normal stress (altering its frictional strength) and the level of mean stress ($\bar{\sigma} = [\sigma_1 + \sigma_2 + \sigma_3]/3$) (Sibson, 1991). For dip-slip faults, coupled changes in mean stress ($\Delta\bar{\sigma}$) are of the same order as the shear stress drop ($\Delta\tau$), but the coupling is diametrically opposite for reverse and normal faults. Reverse faults are *load-strengthening* with mean stress increasing during loading only to decrease at failure, whereas normal faults are *load-weakening* with mean stress increased postfailure. Closure of subvertical cracks in the near-surface from increased horizontal stress postfailure may account for the large surface effusions sometimes observed following rupture of normal faults (Muir-Wood and King, 1993).

5.4.3 Redistribution around Rupture Irregularities

Rupturing leads to abrupt postfailure changes in mean stress localized around structural irregularities (Segall and Pollard, 1980), with fluids redistributed from areas of raised to areas of lowered mean stress (Nur and Booker, 1972). Aftershock activity associated with fluid influx tends to concentrate in regions of reduced mean stress at fault tips and in dilational jogs and bends which also serve to localize geothermal systems in extensional–transtensional fault systems, forming important sites for mineralization. Dilational bends and jogs (Fig. 8) act essentially as "suction pumps." Rapid slip transfer during rupture propagation leads to abrupt localized reductions in fluid pressure below ambient (hydrostatic?) levels (Fig. 11), triggering brecciation by hydraulic implosion accompanied by episodes of boiling and mineral precipitation at shallow depths in geothermal systems (Sibson, 1985). Induced suctions contribute to rupture perturbation or arrest, promoting time-dependent slip transfer across the jog through the aftershock phase as fluid pressures restore to background levels (Peltzer *et al.*, 1996).

5.4.4 Fault-Valve Action

Valving action occurs wherever ruptures breach impermeable barriers bounding overpressured zones which may be restricted to the fault zones themselves or extend into the surrounding crust. Postseismic fluid discharge upwards along the transient permeability of the rupture zone causes local reversion toward a hydrostatic fluid pressure gradient before hydrothermal self-sealing occurs, and fluid overpressure rebuilds at depth (Fig. 12). Timing of successive failure episodes is then controlled by the cycling of tectonic shear stress, fluid pressure, and frictional fault strength through the interseismic period (Sibson, 1992).

The association of mesozonal gold–quartz vein systems with steep reverse faults that, demonstrably, were severely misoriented for frictional reactivation at the time of mineralization

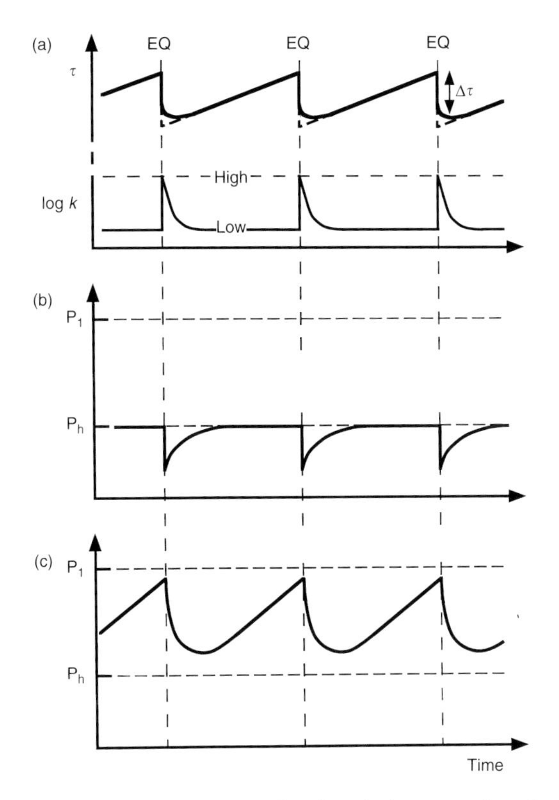

FIGURE 11 Fluid-pressure cycling in relation to earthquake (EQ) rupturing. (a) Schematic relating fault permeability, k, to the earthquake shear stress cycle; (b) "suction-pump" effect at dilational fault jog; (c) fault-valve action in overpressured crust (P_h and P_l are hydrostatic and lithostatic fluid-pressure levels, respectively).

has led to the interpretation that such systems developed through extreme fault-valve action (Sibson *et al.*, 1988; Robert *et al.*, 1995). The compressional stress field served to contain deep-sourced, lithostatically overpressured H_2O–CO_2 fluids until fault rupture, followed by focused discharge along the faults. Valving action may also promote oil and gas migration in overpressured sedimentary basins. Widespread development of syntectonic fault veins, coupled with the growing evidence for fluid-pressure cycling in crustal fault zones (e.g., Parry, 1998), suggest that minor valving action involving small fluid volumes is widespread, leading to significant strength variations that affect rupture nucleation and recurrence.

6. Discussion

The study of exhumed fault zones for the information they can yield on earthquake source processes, both from their structural geometry and from fault-rock assemblages, is in its infancy. Preliminary fault zone models established from such investigations, coupled with the results of experimental rock deformation studies, have already provided insights into factors governing the depth and style of seismic activity in deforming crust. One limitation of most existing models is their monolithologic character (largely based on the deformation of granitoid material at different crust levels); future polylithologic fault zone models should take greater account of compositional heterogeneity over a range of scales. Continued

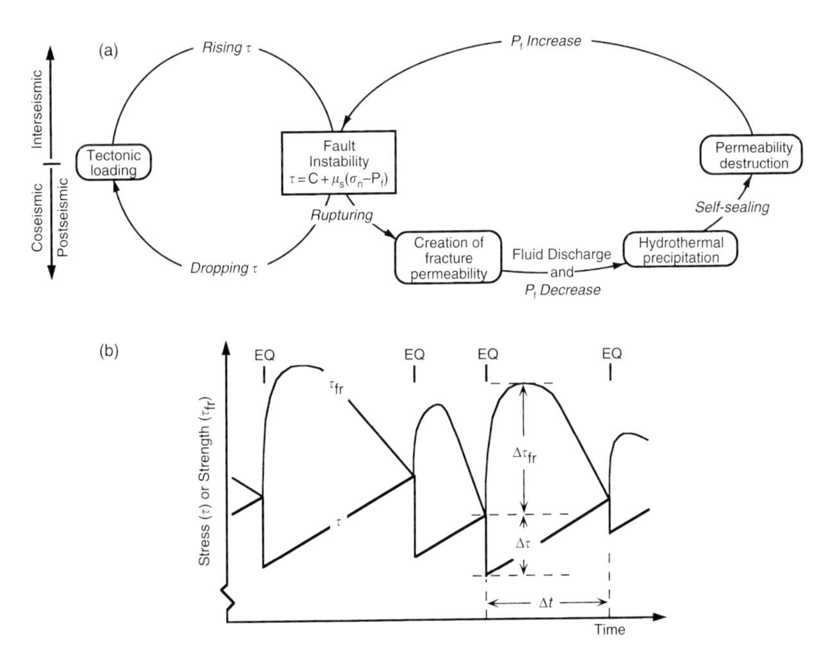

FIGURE 12 (a) Fault failure governed by the coupled cycling of shear stress and fluid pressure through fault-valve action in overpressured crust. (b) Time-dependent fluctuations in shear stress, τ, and frictional strength, τ_{fr}, accompanying successive earthquake ruptures (EQ) in the fault-valve cycle (after Sibson, 1992).

efforts are needed to reconcile shear strength profiles inferred from fault zone models with the apparent weakness of major transcrustal fault zones. However, there are indications that the answer lies in the chemical and physical weakening effects of hydrothermal fluids, both from overpressuring which may be localized within the fault zone itself, and through reaction weakening leading to the development of phyllosilicate-rich material.

Despite problems arising from structural overprinting, developments in microstructural and microthermometric fluid-inclusion techniques will help to discriminate deformation and fluid processes occurring at different stages of the earthquake stress cycle, and should also lead to better geobarometric constraints on fault-rock assemblages. Isolating the deformation characterizing an individual rupture at different crustal levels will vastly improve understanding of dynamic slip processes and the shear stress levels driving faulting.

New seismological techniques, especially high-resolution location of microearthquakes delineating details of fault zone internal structure, coupled with investigatory drilling to sample active fault zone material and make direct measurements of physical conditions around the earthquake source, will help to test and elaborate more advanced fault zone models.

Acknowledgments

I would like to thank Dave Hill, Tom Hanks, Bill Ellsworth, Art McGarr, Steve Hickman, and many others at Menlo Park for their efforts in my seismological education over the past 20 years. Bob Yeats and Toshi Shimamoto provided helpful reviews of the manuscript.

References

Anderson, E.M. (1951). "The Dynamics of Faulting and Dyke Formation with Application to Britain," 2nd edn. Oliver and Boyd, Edinburgh.

Anderson, H.A. and J. Zhang (1991). Long-period seismic radiation from the May 23, 1989, Macquarie Ridge earthquake: evidence for coseismic slip in the mantle. *J. Geophys. Res.* **96**, 19853–19863.

Angevine, C.L., D.L. Turcotte, and M.D. Furnish (1982). Pressure solution lithification as a mechanism for the stick–slip behavior of faults. *Tectonics* **1**, 151–160.

Antonellini, M. and A. Aydin (1994). Effect of faulting on fluid flow in porous sandstones: petrophysical properties. *Am. Assoc. Petrol. Geol. Bull.* **78**, 355–377.

Barton, C., M.D. Zoback, and D. Moos (1995). Fluid flow along potentially active faults in crystalline rock. *Geology* **23**, 683–686.

Blackman, D.K., J.R. Cann, B. Janssen, and D.K. Smith (1998). Origin of extensional core complexes: evidence from the Mid-Atlantic Ridge at Atlantis Fracture Zone. *J. Geophys. Res.* **103**, 21315–21333.

Blanpied, M.L., D.A. Lockner, and J.D. Byerlee (1995). Frictional slip of granite at hydrothermal conditions. *J. Geophys. Res.* **100**, 13045–13064.

Bos, B., C.J. Peach, and C.J. Spiers (2000). Frictional-viscous flow of simulated fault gouge caused by the combined effects of phyllosilicates and pressure solution. *Tectonophysics* **327**, 173–194.

Boyer, S.E. and D. Elliott (1982). Thrust systems. *Am. Assoc. Petrol. Geol. Bull.* **66**, 1196–1230.

Brown, N., M. Fuller, and R. Sibson (1991). Paleomagnetism of the Ocotillo Badlands, southern California, and implications for slip transfer through an antidilational fault jog. *Earth Planet. Sci. Lett.* **102**, 277–288.

Brown, S.R. and R.L. Bruhn (1996). Formation of voids and veins during faulting. *J. Struct. Geol.* **18**, 657–671.

Burford, R.O. and P.W. Harsh (1980). Slip on the San Andreas fault in central California from alinement array surveys. *Bull. Seismol. Soc. Am.* **70**, 1233–1261.

Byerlee, J.D. (1978). Friction of rocks. *Pure and Appl. Geophys.* **116**, 615–626.

Byerlee, J.D. (1993). A model for episodic flow of high-pressure water in fault zones before earthquakes. *Geology* **21**, 303–306.

Carter, N.L. and M.C. Tsenn (1987). Flow properties of continental lithosphere. *Tectonophysics* **136**, 27–63.

Chen, W.-P. and P. Molnar (1983). Focal depths of intracontinental and intraplate earthquakes and their implications for the thermal and mechanical properties of the lithosphere. *J. Geophys. Res.* **88**, 4183–4124.

Chester, F.M. (1995). A rheologic model for wet crust applied to strike–slip faults. *J. Geophys. Res.* **100**, 13033–13044.

Chester, F.M., J.P. Evans, and R.L. Biegel (1993). Internal structure and weakening mechanisms of the San Andreas fault. *J. Geophys. Res.* **98**, 771–786.

Corsini, M., A. Vauchez, and R. Caby (1996). Ductile duplexing at a bend of a continental-scale strike–slip shear zone: example from NE Brazil. *J. Struct. Geol.* **18**, 385–394.

Collettini, C. and R.H. Sibson (2001). Normal faults, normal friction. *Geology* **29**, 927–930.

Cowie, P.A. (1998). Normal fault growth in 3-dimensions in continental and oceanic crust. In: "Faulting and Magmatism at Mid-Ocean Ridges" (R. Buck, P.T. Delaney, J.A. Karson, and Y. Lagabrielle, Eds.), *AGU Monograph* **106**, 325–348.

Cowie, P.A. and C.H. Scholz (1992). Physical explanation for displacement-length relationship for faults using a post-yield fracture mechanics model. *J. Struct. Geol.* **14**, 1133–1148.

Cox, S.F. (1995). Faulting processes at high fluid pressures: an example of fault-valve behaviour from the Wattle Gully Fault, Victoria, Australia. *J. Geophys. Res.* **100**, 12841–12860.

Crampin, S. (1994). The fracture criticality of rock. *Geophys. J. Int.* **118**, 428–438.

Davison, I. (1994). Linked fault systems; extensional, strike–slip, contractional. In: "Continental Deformation" (P.L. Hancock, Ed.), pp. 121–142. Pergamon Press, Oxford.

Dewey, J.F., R.E. Holdsworth, and R.A. Strachan (1998). Transpression and transtension. In: "Continental Transpressional and Transtensional Tectonics" (R.E. Holdsworth, R.A. Strachan, and J.F. Dewey, Eds.), *Geol. Soc. Lond. Spec. Publ.* **135**, 1–14.

Dieterich, J. (1978). Time-dependent friction and the mechanics of stick–slip. *Pure Appl. Geophys.* **116**, 709–806.

Doblas, M. (1998). Slickenside kinematic indicators. *Tectonophysics* **295**, 187–197.

Engeln, J.F., D.A. Wiens, and S. Stein (1986). Mechanisms and depths of Atlantic transform earthquakes. *J. Geophys. Res.* **91**, 548–577.

Etheridge, M.A. (1983). Differential stress magnitudes during regional deformation and metamorphism: upper bound imposed by tensile fracturing. *Geology* **11**, 231–234.

Evans, J.P., C.B. Forster, and J.V. Goddard (1997). Permeability of fault-related rocks, and implications for hydraulic structure of fault zones. *J. Struct. Geol.* **19**, 1393–1404.

Fox, P.J. and D.G. Gallo (1986). The geology of north Atlantic plate boundaries and their aseismic extensions. In: "The Geology of North America; Vol. M, The Western North Atlantic Region" (P.R. Vogt and B.E. Tucholke, Eds.), pp. 157–172. Geological Society of America, Boulder, Colorado.

Fyfe, W.S., N.J. Price, and A.B. Thompson (1978). "Fluids in the Earth's Crust." Elsevier, Amsterdam.

Gudmundsson, A. (1995). Infrastructure and mechanics of volcanic systems in Iceland. *J. Volc. Geotherm. Res.* **64**, 1–22.

Hanmer, S. (1988). Great Slave Lake Shear Zone, Canadian Shield: reconstructed vertical profile of a crustal-scale fault zone. *Tectonophysics* **149**, 245–264.

Harris, R.A. and S.M. Day (1993). Dynamics of fault interaction: strike–slip faults. *J. Geophys. Res.* **98**, 4461–4472.

Hasegawa, A., S. Horiuchi, and N. Umino (1994). Seismic structure of the northeastern Japan convergent margin: a synthesis. *J. Geophys. Res.* **99**, 22295–22311.

Hayward, A.B. and R.H. Graham (1989). Some geometrical characteristics of inversion. In: "Inversion Tectonics" (M.A. Cooper and G.D. Williams, Eds.), *Geol. Soc. Lond. Spec. Publ.* **44**, 17–39.

Henley, R.W. (1985). The geothermal framework for epithermal deposits. *Rev. Econ. Geol.* **2**, 1–24.

Hickman, S., R. Sibson, and R. Bruhn (1995). Introduction to special section: "Mechanical Involvement of Fluids in Faulting." *J. Geophys. Res.* **100**, 12831–12840.

Hill, D.P. (1977). A model for earthquake swarms. *J. Geophys. Res.* **82**, 1347–1352.

Hill, D.P., J.P. Eaton, and L.M. Jones (1990). Seismicity, 1980–86. In: "The San Andreas Fault System, California" (R.E. Wallace, Ed.), *US Geol. Surv. Prof. Pap.*, **1515**, 115–151.

Hubbert, M.K. and W.W. Rubey (1959). Role of fluid pressure in mechanics of overthrust faulting. *Geol. Soc. Am. Bull.* **70**, 115–205.

Irwin, W.P. and I. Barnes (1980). Tectonic relations of carbon dioxide discharge and earthquakes. *J. Geophys. Res.* **85**, 3115–3121.

Ito, K. (1999). Seismogenic layer, reflective lower crust, surface heat flow and large inland earthquakes. *Tectonophysics* **306**, 423–433.

Jackson, J.A. and T. Blenkinsop (1997). The Bilila-Mtakataka fault in Malawi: an active 100-km long normal fault segment in thick seismogenic crust. *Tectonics* **16**, 137–150.

Jackson, J.A. and N.J. White (1989). Normal faulting in the upper continental crust: observations from regions of active extension. *J. Struct. Geol.* **11**, 15–36.

Jaeger, J.C. and N.G.W. Cook (1979). "Fundamentals of Rock Mechanics," 3rd edn. Methuen, London.

Jones, G., O.J. Fisher, and R.J. Knipe, (Eds.) (1998). Faulting, fault-sealing, and fluid flow in hydrocarbon reservoirs. *Geol. Soc. Lond. Spec. Publ.* **147**, 319pp.

Kawamoto, E. and T. Shimamoto (1998). The strength profile for bimineralic shear zones: an insight from high-temperature shearing experiments on calcite–halite mixtures. *Tectonophysics* **295**, 1–14.

Kennedy, B.M., Y.K. Kharaka, W.C. Evans, A. Ellwood, D.J. DePaolo, J. Thordsen, G. Ambats, and R.H. Mariner (1997). Mantle fluids in the San Andreas fault system, California. *Science* **278**, 1278–1281.

King, G.C.P. (1986). Speculations on the geometry of the initiation and termination processes of earthquake rupture and its relation to morphology and geological structure. *Pure Appl. Geophys* **124**, 567–585.

Lachenbruch, A.H. (1980). Frictional heating, fluid pressure, and the resistance to fault motion. *J. Geophys. Res.* **85**, 6097–6112.

Lister, G.S. and G.A. Davis (1989). The origin of metamorphic core complexes and detachment faults formed during Tertiary continental extension in the northern Colorado River region, USA. *J. Struct. Geol.* **11**, 65–93.

Lockner, D.A. (1995). Rock failure. In: "Rock Physics and Phase Relations: a Handbook of Physical Constants." *AGU Reference Shelf* **3**, 127–147.

Luyendyk, B.P. (1991). A model for Neogene crustal rotations, transtension and transpression in southern California. *Geol. Soc. Am. Bull.* **103**, 1528–1536.

McCaig, A.M. (1997). The geochemistry of volatile fluid flow in shear zones. In: "Deformation-Enhanced Fluid Transport in the Earth's Crust and Mantle" (M.B. Holness, Ed.), pp. 227–266. Chapman and Hall, London.

McCalpin, J.P. (1996). *Paleoseismology*. Academic Press, London.

McClay, K.R. (1992). Glossary of thrust tectonics terms. In: "Thrust Tectonics" (K.R. McClay, Ed.), pp. 419–433. Chapman and Hall, London.

McClay, K.R. and S. Khalil (1998). Extensional hard linkages, Gulf of Suez. *Geology* **26**, 563–566.

McKenzie, D. and J.N. Brune (1972). Melting on fault planes during large earthquakes. *Geophys. J. R. Astron. Soc.* **29**, 65–78.

Machette, M.N., S.F. Personius, A.R. Nelson, D.P. Schwartz, and W.R. Lund (1991). The Wasatch fault zone, Utah—segmentation and history of earthquakes. *J. Struct. Geol.* **13**, 137–149.

Magloughlin, J.F. and J.G. Spray (1992). Frictional melting processes and products in geological materials: introduction and discussion. *Tectonophysics* **204**, 197–206.

Marone, C. and C.H. Scholz (1988). The depth of seismic faulting and the upper transition from stable to unstable slip regimes. *Geophys. Res. Lett.* **15**, 621–624.

Marrett, R. and R.W. Allmendinger (1991). Estimates of strain due to brittle faulting: sampling of fault populations. *J. Struct. Geol.* **13**, 735–738.

Mase, C.W. and L. Smith (1987). Effects of frictional heating on the thermal, hydrological, and mechanical response of a fault. *J. Geophys. Res.* **92**, 6249–6272.

Matsuda, T. (1975). Magnitude and recurrence intervals of earthquakes from a fault. *Zisin. J. Seism. Soc. Jpn* **28**, 293–283.

Meissner, R. and J. Strehlau (1982). Limits of stresses in continental crust and their relationship to the depth-frequency distribution of shallow earthquakes. *Tectonics*, **1**, 73–89.

Molnar, P. and H. Lyon-Caen (1989). Fault plane solutions of earthquakes and active tectonics of the Tibetan Plateau and its margins. *Geophys. J. Int.* **99**, 123–153.

Morley, C.K., R.A. Nelson, T.L. Patton, and S.G. Munn (1990). Transfer zones in the East African Rift System and their relevance to hydrocarbon exploration in rifts. *Bull. Am. Assoc. Petrol. Geol.* **74**, 1234–1253.

Morrow, C.A., D.E. Moore, and D.A. Lockner (2001). Permeability reduction in granite under hydrothermal conditions. *J. Geophys. Res.* **106**, 30551–30560.

Mount, V.S. and J. Suppe (1992). Present-day stress orientations adjacent to active strike–slip faults. *J. Geophys. Res.* **97**, 11995–12013.

Muir-Wood, R. and G.C.P. King (1993). Hydrological signatures associated with earthquake strain. *J. Geophys. Res.* **98**, 22035–22068.

Muraoka, H., T. Uchida, M. Sasada, M. Yagi, K. Akaku, M. Sasaki, K. Yasukawa, S.-I. Miyazaki, N. Doi, S. Saito, K. Sato, and S. Tanaka (1998). Deep geothermal resources survey program: igneous, metamorphic, and hydrothermal processes in a well encountering 500°C at 3729 m depth, Kakkonda, Japan. *Geothermics* **27**, 507–534.

Nicholson, C. and R.L. Wesson (1990). Earthquake hazard associated with deep well injection. *US Geol. Surv. Bull.* **1951**, 77 pp.

Nur, A. and J.R. Booker (1972). Aftershocks caused by pore-fluid flow? *Science* **175**, 885–887.

Pacheco, J.F., C.H. Scholz, and L.R. Sykes (1992). Changes in frequency–size relationship from small to large earthquakes. *Nature* **355**, 71–73.

Parry, W.T. (1998). Fault-fluid compositions from fluid inclusion observations and solubilities of fracture-sealing minerals. *Tectonophysics* **290**, 1–26.

Passchier, C.W. and R.A.J. Trouw (1996). "Microtectonics." Springer-Verlag, Berlin, 289pp.

Peltzer, G., P. Rosen, F. Rogez, and K. Hudnut (1996). Postseismic rebound in fault step-overs caused by pore fluid flow. *Science* **273**, 1202–1204.

Powers, P.M., R.J. Lillie, and R.S. Yeats (1998). Structure and shortening of the Kangra and Dehra Dun reentrants, Sub-Himalaya, India. *Geol. Soc. Am. Bull.* **110**, 1010–1027.

Reasenberg, P. and W.L. Ellsworth (1982). Aftershocks of the Coyote Lake, California, earthquake of August 6, 1979: a detailed study. *J. Geophys. Res.* **87**, 10637–10655.

Rice, J.R. (1992). Fault stress states, pore pressure distributions, and the weakness of the San Andreas fault. In: "Fault Mechanics and Transport Properties of Rocks" (B. Evans and T.-F. Wong, Eds.), pp. 475–503. Academic Press, San Diego.

Robert, F., A.-M. Boullier, and K. Firdaous (1995). Gold-quartz veins in metamorphic terranes and their bearing on the role of fluids in faulting. *J. Geophys. Res.* **100**, 12861–12879.

Roberts, A.M. and G. Yielding (1994). Continental extensional tectonics. In: "Continental Deformation" (P.L. Hancock, Ed.), pp. 223–250. Pergamon Press, Oxford.

Rutter, E.H. and K.H. Brodie (1987). On the mechanical properties of oceanic transform faults. *Ann. Tecton.* **1**, 87–96.

Rutter, E.H., R.H. Maddock, S.H. Hall, and S.H. White (1986). Comparative microstructure of natural and experimentally produced clay-bearing fault gouges. *Pure Appl. Geophys.* **124**, 3–30.

Schlische, R.W., S.S. Young, R.V. Ackerman, and A. Gupta (1996). Geometry and scaling relationships of a population of very small rift-related normal faults. *Geology* **24**, 683–686.

Scholz, C.H. (1982). Scaling laws for large earthquakes: consequences for physical models. *Bull. Seismol. Soc. Am.* **72**, 1–14.

Scholz, C.H. (1988). The brittle–plastic transition and the depth of seismic faulting. *Geol. Rundschau* **77**, 319–328.

Scholz, C.H. (1990). "The Mechanics of Earthquakes and Faulting." Cambridge University Press, Cambridge.

Scholz, C.H. (1997). Size distributions for large and small earthquakes. *Bull. Seismol. Soc. Am.* **87**, 1074–1077.

Scholz, C.H. (1998). Earthquakes and friction laws. *Nature* **391**, 37–42.

Scholz, C.H., L.R. Sykes, and Y.P. Aggarwal (1973). Earthquake prediction: a physical basis. *Science* **181**, 803–810.

Secor, D.T. (1965). Role of fluid pressure in jointing. *Am. J. Sci.* **263**, 633–646.

Segall, P. and D.D. Pollard (1980). Mechanics of discontinuous faults. *J. Geophys. Res.* **85**, 4337–4350.

Shen, Y., D.W. Forsyth, J. Conder, and L.M. Dorman (1997). Investigation of microearthquake activity following an intraplate teleseismic swarm on the west flank of the southern East Pacific Rise. *J. Geophys. Res.* **102**, 459–475.

Sibson, R.H. (1973). Interactions between temperature and pore fluid pressure during earthquake faulting—a mechanism for partial or total stress relief. *Nature Phys. Sci.* **243**, 66–68.

Sibson, R.H. (1974). Frictional constraints on thrust, wrench, and normal faults. *Nature* **249**, 542–544.

Sibson, R.H. (1975). Generation of pseudotachylyte by ancient seismic faulting. *Geophys. J. R. Astron. Soc.* **43**, 775–794.

Sibson, R.H. (1977). Fault rocks and fault mechanisms. *J. Geol. Soc. Lond.* **133**, 191–213.

Sibson, R.H. (1982). Fault zone models, heat flow, and the depth distribution of earthquakes in the continental crust of the United States. *Bull. Seismol. Soc. Am.* **72**, 151–163.

Sibson, R.H. (1983). Continental fault structure and the shallow earthquake source. *J. Geol. Soc. Lond.* **140**, 741–767.

Sibson, R.H. (1984). Roughness at the base of the seismogenic zone: contributing factors. *J. Geophys. Res.* **89**, 5791–5799.

Sibson, R.H. (1985). Stopping of earthquake ruptures at dilational fault jogs. *Nature* **316**, 248–251.

Sibson, R.H. (1991). Loading of faults to failure. *Bull. Seismol. Soc. Am.* **81**, 2493–2497.

Sibson, R.H. (1992). Implications of fault-valve behaviour for rupture nucleation and recurrence. *Tectonophysics* **211**, 283–293.

Sibson, R.H. (1994). Crustal stress, faulting, and fluid flow. In: "Geofluids: Origin, Migration and Evolution of Fluids in Sedimentary Basins" (J. Parnell, Ed.), *Geol. Soc. Lond. Spec. Publ.* **78**, 69–84.

Sibson, R.H. (1996). Structural permeability of fluid-driven fault-fracture meshes. *J. Struct. Geol.* **18**, 1031–1042.

Sibson, R.H. and Xie, G. (1998). Dip range for intracontinental reverse fault ruptures: truth not stranger than friction? *Bull. Seismol. Soc. Am.* **88**, 1014–1022.

Sibson, R.H., F. Robert, and K.H. Poulsen (1988). High-angle reverse faults, fluid pressure cycling, and mesothermal gold-quartz deposits. *Geology* **16**, 551–555.

Sillitoe, R.H. (1993). Epithermal models: genetic types, geometrical controls and shallow features. In: "Mineral Deposit Modeling" (R.V. Kirkham, W.D. Sinclair, R.I. Thorpe, and J.M. Duke, Eds.), *Geol. Assoc. Can. Spec. Pap.* **40**, 403–417.

Simpson, C. (1985). Deformation of granitic rocks across the brittle-ductile transition. *J. Struct. Geol.* **7**, 503–512.

Sleep, N.H. and M.L. Blanpied (1994). Ductile creep and compaction: a mechanism for transiently increasing fluid pressure in mostly sealed fault zones. *Pure Appl. Geophys.* **143**, 9–40.

Smith, R.B. and R.L. Bruhn (1984). Intraplate extensional tectonics of the eastern Basin-Range: inferences on structural style from seismic reflection data, regional tectonics, and thermo-mechanical models of brittle–ductile deformation. *J. Geophys. Res.* **89**, 5733–5762.

Snoke, A.W., J. Tullis, and V.R. Todd (Eds.) (1998). "Fault-related Rocks: a Photographic Atlas." Princeton University Press, New Jersey.

Sornette, D., P. Davy, and A. Sornette (1990). Structuration of the lithosphere as a self-organized critical phenomemon. *J. Geophys. Res.* **95**, 17353–17361.

Stewart, I.S. and P.L. Hancock (1991). Scales of structural heterogeneity within neotectonic normal fault zones in the Aegean region. *J. Struct. Geol.* **13**, 191–204.

Strehlau, J. (1986). A discussion of the depth extent of rupture in large continental earthquakes. In: "Earthquake Source Mechanics" (S. Das, J. Boatwright, and C.H. Scholz, Eds.), *AGU Monograph* **37**, 131–146.

Streit, J.E. (1997). Low frictional strength of upper crustal faults: a model. *J. Geophys. Res.* **102**, 24619–24626.

Suppe, J. (1985). "Principles of Structural Geology" Prentice-Hall, Englewood Cliffs, New Jersey.

Sylvester, A.G. (1988). Strike–slip faults. *Geol. Soc. Am. Bull.* **100**, 1666–1703.

Takagi, H., K. Goto, and N. Shigematsu (2000). Ultramylonite bands derived from cataclasite and pseudotachylyte in granites, northeast Japan. *J. Struct. Geol.* **22**, 1325–1339.

Tchalenko, J.S. (1970). Similarities between shear zones of different magnitudes. *Geol. Soc. Am. Bull.* **81**, 1625–1640.

Townend, J. and M.D. Zoback (2000). How faulting keeps the crust strong. *Geology* **28**, 399–402.

Wallace, R.E. (1970). Earthquake recurrence intervals on the San Andreas fault. *Geol. Soc. Am. Bull.* **81**, 2875–2890.

Wallace, R.E. and H.T. Morris (1986). Characteristics of faults and shear zones in deep mines. *Pure Appl. Geophys.* **124**, 107–126.

Walsh, J.J. and J. Watterson (1988). Analysis of the relationship between displacements and dimensions of faults. *J. Struct. Geol.* **10**, 239–247.

Walsh, J.J. and J. Watterson (1991). Geometric and kinematic coherence and scale effects on normal fault systems. In: "The Geometry of Normal Faults" (A.M. Roberts, G. Yielding, and B. Freeman, Eds.), *Geol. Soc. Lond. Spec. Publ.* **56**, 193–203.

Wells, D.L. and K.J. Coppersmith (1994). New empirical relationships among magnitude, rupture length, rupture width, rupture area, and surface displacement. *Bull. Seismol. Soc. Am.* **84**, 974–1002.

Wernicke, B. (1995). Low-angle normal faults and seismicity: a review. *J. Geophys. Res.* **100**, 20159–20174.

Wesnousky, S.G. (1988). Seismological and structural evolution of strike–slip faults. *Nature* **335**, 340–342.

Westaway, R. (1998). Dependence of active normal fault dips on lower crustal flow regimes. *J. Geol. Soc. Lond.* **155**, 233–254.

White, J.C. (1996). Transient discontinuities revisited: pseudotachylyte, plastic instability and the influence of low pore fluid pressure on deformation processes in the mid-crust. *J. Struct. Geol.* **18**, 1471–1486.

White, S.H., S.E. Burrows, J. Carreras, N.D. Shaw, and F.J. Humphreys (1980). On mylonites in ductile shear zones. *J. Struct. Geol.* **2**, 175–187.

Woodcock, N.H. and C. Schubert (1994). Continental strike–slip tectonics. In: "Continental Deformation" (P.L. Hancock, Ed.), pp. 251–263. Pergamon Press, Oxford.

Yeats, R.S., K. Sieh, and C.R. Allen (1997). "The Geology of Earthquakes." Oxford University Press, Oxford.

Zoback, M.L. (1992). Stress field constraints on intraplate seismicity in eastern North America. *J. Geophys. Res.* **97**, 11761–11782.

30

Paleoseismology

Lisa B. Grant
University of California, Irvine, USA

1. Introduction

Paleoseismology is the study of earthquakes decades, centuries or millennia after their occurrence (Yeats and Prentice, 1996). Paleoseismic studies typically focus on prehistoric or pre-instrumental earthquakes (Sieh 1981; Wallace, 1981; McCalpin and Nelson, 1996) to supplement the historic record of seismicity. The time between surface ruptures of active faults is on the order of decades for the fastest-moving faults to thousands or tens of thousands of years for more numerous but less-active faults (see Table 1 in Chapter 29 by Sibson). Modern scientific observation and analysis of earthquakes have occurred for a small fraction of this time, so there are few observations on the spatial and temporal characteristics of fault ruptures over multiple seismic cycles (Sieh, 1996). Paleoseismic investigations provide data on earthquake occurrence and the seismogenic behavior of individual faults that complement modern observations of earthquakes, with important implications for scientific models and practical applications for seismic hazard assessment (Reiter, 1995).

Moderate to large earthquakes may generate surface rupture or induce permanent changes in the landscape and local environment. Under suitable conditions, evidence of earthquakes is preserved in the geologic record. Paleoearthquakes are recognized by detailed observations and analyses of geologic or environmental conditions within fault zones or in tectonically or seismically active regions. Paleoseismic investigations yield information about the recency of fault movement, dates of previous earthquakes, recurrence times, average slip rate, and earthquake effects over time intervals ranging from decades to thousands of years. Paleoseismic data are applied toward estimating the magnitudes of past earthquakes, forecasting the magnitudes of future earthquakes (WGCEP, 1988; Reiter 1991), and identifying areas susceptible to fault rupture.

Paleoseismology is a subspecialty of seismology in that it focuses on earthquakes. However, the primary methods of paleoseismic data collection and analysis are drawn from geology and are distinctly different from analyses of seismic waves conducted by geophysicists. Therefore, paleoseismology is sometimes considered to be a subspecialty of active tectonics or earthquake geology. Paleoseismology began to emerge as a distinct discipline in the late 1960s to mid-1980s. Summaries, compendia or review articles focusing on paleoseismology have been generated by Wallace (1981, 1986), Sieh (1981), Crone and Omdahl (1987), Vittori *et al.* (1991), Pantosti and Yeats (1993), Prentice *et al.* (1994), Serva and Slemmons (1995), Yeats and Prentice (1996), and Pavlides *et al.* (1999). Yeats *et al.* (1997) provide a comprehensive text on the geology of earthquakes with global coverage of active tectonics and paleoseismic studies. For a comprehensive text and reference on paleoseismology, the reader is referred to McCalpin (1996).

The purpose of this chapter is to summarize methods, contributions, and issues in paleoseismology. Topics include paleoseismic investigation methods, models, and applications. Issues in research and application are discussed in each section. Examples and references are chosen to highlight major concepts and findings rather than attempting to provide a comprehensive or geographically inclusive summary. Most examples are from western North America, and heavy emphasis is placed on studies of the San Andreas fault system. Discussions of faulting and seismic hazard assessment are treated more fully by Sibson (Chapter 29) and Somerville and Moriwaki (Chapter 65).

2. Investigation Goals and Methods

With few exceptions, earthquakes are generated by the movement of faults (Bolt, 1999). Therefore, paleoseismic research is directly or indirectly a study of faults and their surface expression. Paleoseismic investigation methods employ techniques from several geologic subdisciplines, including stratigraphic analysis, Quaternary geology, soil science, geomorphology, engineering geology, geochronology, and structural geology. Major objectives of paleoseismic studies include identification of recently active or seismogenic faults,

INTERNATIONAL HANDBOOK OF EARTHQUAKE AND ENGINEERING SEISMOLOGY, VOLUME 81A ISBN: 0-12-440652-1

measurement of displacement (coseismic and long-term average), and establishing catalogs of paleoearthquakes. This section is organized by common goals of paleoseismic studies, rather than by disciplinary techniques, because most investigations employ a variety of techniques.

2.1 Fault Identification

Because earthquakes are generated by faults, the first step in any paleoseismic investigation is identification of a fault for study. An active fault is a fault that may have displacement within a future period of concern to humans (Wallace, 1981). Identification of active faults requires recognizing previous displacement and constraining the age of displacement. Therefore, an active fault is usually identified by associating it with tectonically deformed Quaternary-age materials or surfaces (e.g., Ziony and Yerkes, 1985).

2.1.1 Recognition of Tectonic Deformation

Paleoseimic investigation methods are only applicable to faults that cause recognizable tectonic deformation or environmental changes (e.g., submergence, erosion, anomalous deposition, or mortality) at or near the Earth's surface. Faults may be classified by average slip rate (see Table 1 in Chapter 29 by Sibson), type of displacement (strike slip, dip slip normal, dip slip reverse, thrust, or oblique slip), and depth (surface or blind). The depth, rate, and type of fault slip determine the amount and nature of fault-induced deformation at the surface. The faster the slip rate, the more active the fault and the more likely the fault can be recognized and characterized by paleoseismologists (Slemmons and dePolo, 1986). Faults that reach the surface are easier to recognize than blind faults, but they may rupture to the surface infrequently if they have low slip rates (see Chapter 29 by Sibson).

Paleoseismic data provide an incomplete record of earthquakes and fault rupture. Tectonic deformation due to fault slip may be difficult to recognize if rupture does not reach the surface, especially in intraplate regions (e.g., Lettis *et al.*, 1997). Earthquakes on blind faults may induce subtle surface deformations over broad areas (Stein and Yeats, 1989) or distributed evidence of shaking in disrupted sediments (Reiter, 1995). The existence of an active blind fault can be detected by geomorphic analysis (e.g., Bullard and Lettis, 1993; Keller and Pinter, 1996), by mapping Quaternary sediments and identifying areas of uplift (e.g., Grant *et al.*, 1999) or by recognizing effects of shaking that are not associated with known surficial faults (Reiter, 1995).

Effects of tectonic deformation can be enhanced or obscured by anthropogenic or natural processes such as weathering and erosion. If tectonic changes cannot be readily distinguished from changes induced by other mechanisms, then the results of paleoseismic studies may be biased toward higher or lower estimates of tectonic displacement. Such problems have been noted in a variety of tectonic and human environments (e.g., Ricci Lucchi, 1995; Obermeier, 1996; Nur, 2000).

2.1.2 Application of Geochronology

Geochronology is dating of earth materials, surfaces, and processes. Geochronology is essential for paleoseismology because it constrains dates of paleoearthquakes and average rates of fault displacement. The most useful geochronologic methods for paleoseismic investigations yield high-resolution ages for common, late Quaternary materials such as soils or buried flora and fauna (Lettis and Kelson, 2000). Table 1 summarizes geochronologic methods for dating Quaternary materials in fault zones. A detailed discussion of this complex topic is beyond the scope of this chapter. A comprehensive

TABLE 1 Classification of Quaternary Geochronologic Methods (Adapted from Noller *et al.*, 2000a)

Sidereal	Isotopic	Radiogenic	Chemical and Biologic	Geomorphic	Correlation
Dendrochronology	Radiocarbon	Fission track	Amino acid racemization	Soil-profile development	Stratigraphy
	Uranium series	Thermoluminescence	Rock-varnish cation ratio	Rock-varnish development	Paleomagnetism
Varve chronology	^{210}Pb	Optically stimulated luminescence	Obsidian and tephra hydration	Scarp morphology and landform modification	Tephrochronology
	U-Pb, Th-Pb				Paleontology
Historical records		Electron-spin resonance	Soil chemistry	Rate of deformation	Tectites and microtectites
	K-Ar and ^{39}Ar–^{40}Ar		^{10}Be accumulation in soils	Rate of deposition	Climate correlation
Sclerochronology and growth rings	Cosmogenic isotopes	Infrared stimulated luminescence	Lichenometry	Rock and mineral weathering	Astronomical correlation
					Stable isotopes
				Geomorphic position	Archeology

compilation and summary of Quaternary geochronologic methods and applications is provided by Noller *et al.* (2000b). The most commonly used methods are also described by McCalpin (1996) and Yeats *et al.* (1997).

Radiocarbon dating is the most widely used method for dating Holocene and latest Pleistocene earthquakes. The half-life of radioactive ^{14}C (5730 y) limits the application of radiocarbon dating to organic matter formed from carbon fixed within the last 50 000–60 000 y (Trumbore, 2000). The amount of ^{14}C in atmospheric CO_2 has varied in the past, particularly in the last few centuries due to anthropogenic emissions. To compensate for this variation, radiocarbon ages are calibrated to correspond to calendar ages (absolute ages). Calibrations based on tree rings and glacial varves extend back to the early Holocene. Calibration curves are not linear. Plateaus in the calibration curves limit the precision of radiocarbon dating (Trumbore, 2000). This problem is acute for the last few centuries. For example, Yeats and Prentice (1996) note that the two largest historic ruptures of the San Andreas fault in California, which occurred in 1857 and 1906, are indistinguishable using radiocarbon dating.

All methods listed in Table 1 have limitations and uncertainties. Uncertainty in geochronologic methods is a major source of uncertainty in paleoseismic data (Lettis and Kelson, 2000). In addition to uncertainty in calibration and accuracy of analysis, errors may be introduced in the selection, collection, and interpretation of field samples. Therefore, most paleoseismic studies employ multiple methods of dating to reduce uncertainty and allow cross-checking of results. Methods such as radiocarbon dating that yield accurate, high-precision ages of common or widely distributed materials are preferred. Recent improvements in geochronology methods and in the statistical treatment of dates have reduced uncertainties in previously published ages (e.g., Sieh *et al.*, 1989; Biasi and Weldon, 1994).

2.2 Chronologies of Earthquakes

Large earthquakes create features that may be preserved in the stratigraphic record and recognized by geological, geomorphic, environmental, or archeological analysis. Major effects of earthquakes include surface rupture, ground deformation, and ground failures due to shaking. Chronologies of earthquakes are developed by identifying and dating evidence of multiple paleoearthquakes along a specific fault, or by documenting evidence of shaking or tectonic deformation in a seismically active region. The longest chronology of paleoearthquakes spans 50 000 y, with a mean recurrence time of ~1600 y (Marco *et al.*, 1996).

For most Quaternary faults, the dates of prehistoric earthquakes are not known. Well-studied faults such as the San Andreas fault in California, and the North Anatolian fault in Turkey, have chronologies of only a dozen or so paleoearthquakes. Nonseismic processes can create features that appear very similar to those generated by earthquakes (Ricci Lucchi, 1995; Obermeier 1996). The most commonly reported and least ambiguous paleoseismic data are derived from observations of surface ruptures.

2.2.1 Surface Rupture Investigations

Evidence of surface rupture can be observed directly in natural exposures or, more commonly, in trench excavations (see Figs. 1A and 1B on the Handbook CD-ROM, under the directory \30Grant). Paleoseismologists look for trench sites with distinct evidence of multiple ruptures and abundant material for dating. Trenches are generally excavated with a backhoe or similar excavation equipment. Shallow trenches in unconsolidated sediments may be dug by hand. Deep trenches must be stabilized with shoring, benching or sloping. Stabilization methods are chosen to maximize wall exposure. The maximum practical depth for typical trench investigations is 5 m.

Trench exposures are cleaned by scraping and brushing to highlight stratigraphy and faults. Geologists map the exposures with the aid of a string grid or survey control. The maps are called trench logs. Trench logs are supplemented with descriptive notes of observations and objective records such as photographs. Samples of datable material are collected after their locations have been documented on the trench log. A good trench log documents observations of stratigraphy and structure as objectively as possible. However, it is impossible to make an entirely objective log because construction of a log is partly an act of interpretation.

2.2.1.1 Recognition and Dating of Ruptures

The goal of logging is to recognize and record the number of surface ruptures and their stratigraphic position relative to datable material so that each paleoearthquake can be analyzed and dated as accurately as possible. Evidence of surface rupture is commonly called a paleoseismic event. The stratigraphic unit that was the ground surface at the time of the earthquake is referred to as the event horizon. Figure 1 shows typical stratigraphic and structural relationships that are used for recognizing and dating previous earthquakes (Lettis and Kelson, 2000). Figure 2 shows an example of a trench log from the San Andreas fault, with several paleoseismic events and event horizons marked with letters. The most useful criteria for identifying ruptures include multiple fault terminations at a single stratigraphic horizon, tilted and folded strata overlain by less-deformed strata, and colluvial wedges draped over fault scarps (Yeats and Prentice, 1996). Fault strands may die out and terminate below the surface, or be poorly expressed at the event horizon (Bonilla and Lienkaemper, 1990). Therefore, multiple exposures of a fault should be examined whenever possible to identify the event horizon with maximum confidence.

In some areas a fault cannot be examined with excavations because of a high water table, the presence of human

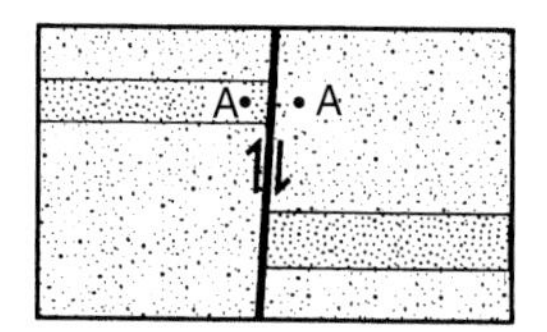

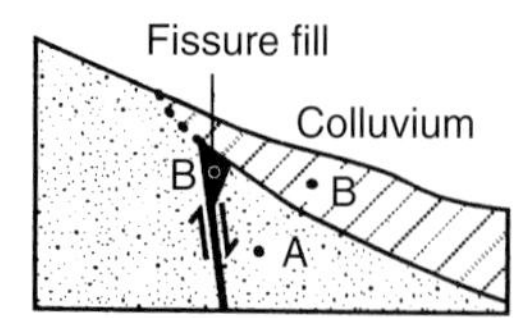

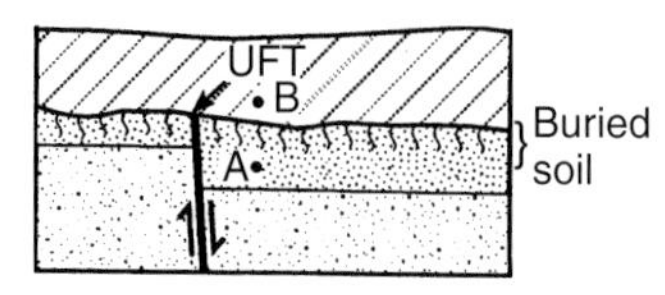

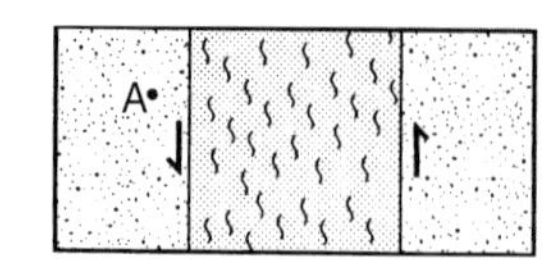

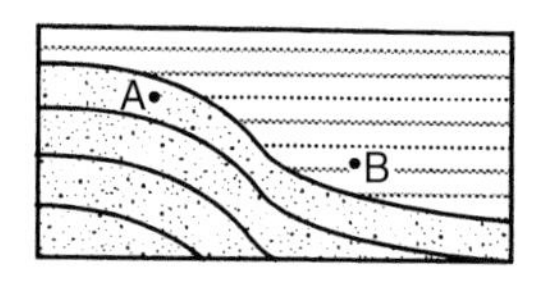

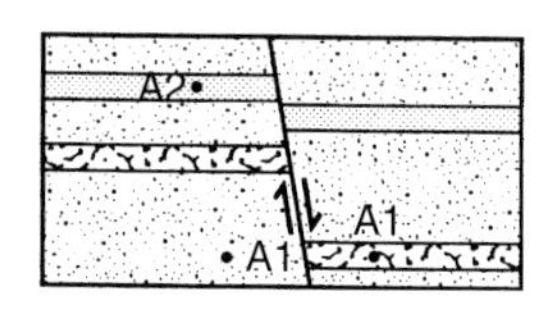

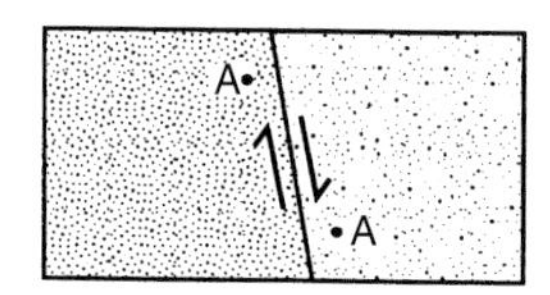

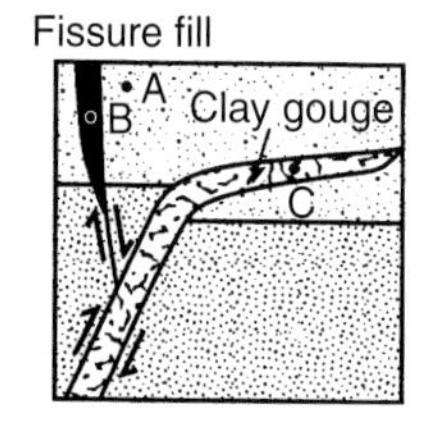

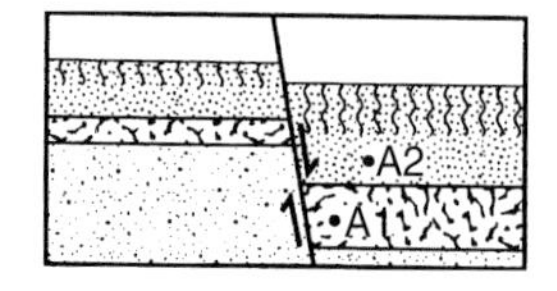

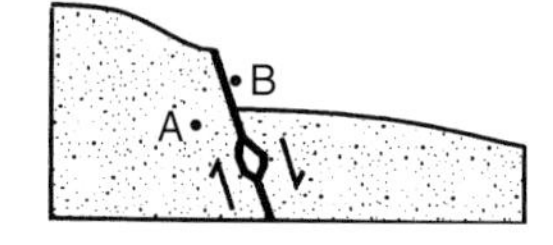

FIGURE 1 Diagrams illustrating stratigraphic and structural criteria used to identify the occurrence and timing of paleoearthquakes. Stratigraphy, faults, and soils are shown schematically. Dated samples at locations marked "A" predate the earthquake. Dated samples at locations "B" postdate the earthquake. Dated samples at "C" are not helpful for deciphering the chronology of past earthquakes because the sample may predate or postdate the earthquake. Where evidence of multiple earthquakes is present, locations A1 predate the earlier event and locations A2 postdate the earlier event and predate the later event. (From Lettis and Kelson, 2000.)

structures, or burial by younger sediments. Subsurface investigation methods such as cone penetrometer testing (CPT) and large diameter borings may be applied for these conditions (Grant *et al.*, 1997; Dolan *et al.*, 2000). These methods must be used in combination with other subsurface investigation methods such as seismic imaging, ground penetrating radar, and standard borings. There are greater

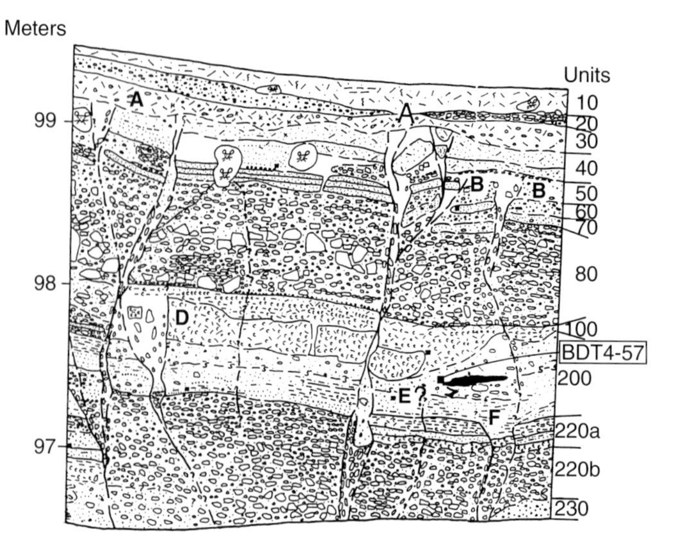

FIGURE 2 Example of a portion of a trench log from the San Andreas fault at Bidart Ranch in the Carrizo Plain, California. Letters indicate the approximate location of evidence for several paleoearthquakes (A, B, D, E?, F). Query indicates uncertainty. (From Grant and Sieh, 1994.)

uncertainties in the identification and dating of paleoearthquakes from subsurface investigations than from surface exposures.

The chronology of surface ruptures revealed by paleoseismic methods is a subset of the actual earthquake history. The paleoseismic record is spatially and temporally incomplete because the conditions required for preservation of earthquakes are not present at all times on all active faults. The minimum earthquake magnitude associated with surface rupture is about M_L 5 (Bonilla, 1988), so $M_L < 5$ earthquakes are rarely recognized in trench exposures. Surface faulting is commonly associated with $M > 6$ earthquakes (Wells and Coppersmith, 1994). Therefore, most paleoseismic events recognized in trenches represent $M \geq 6$ earthquakes. Nor have all faults been studied by paleoseismologists. Early paleoseismic studies focused on examining evidence of larger earthquakes ($M > 7$) because the data were relatively easy to interpret. Therefore, the published paleoseismic record of earthquakes contains less data on the occurrence of smaller ($M < 7$) earthquakes (McCalpin, 1996).

2.2.2 Regional Coseismic Deformation

Earthquakes that generate regional deformation may be recognized and catalogued. Vertical deformation can induce changes in local rates of deposition and erosion that provide evidence of a paleoearthquake, particularly in fluvial (Schumm *et al.*, 2000) and coastal environments (Carver and McCalpin, 1996). Indicators of coseismic uplift in coastal areas include shorelines, platforms, or flights of terraces above mean sea level. Formation of elevated terraces have been reported following historic earthquakes in coastal regions (e.g., Plafker and Rubin, 1978; Matsuda *et al.*, 1978; Pirazzoli, 1991). Sequences of similar terraces and shorelines at higher elevations provide evidence of multiple episodes of tectonic uplift (Fig. 3; and Figs. 4, 5 on the Handbook CD-ROM). In shallow tropical waters, the growth of corals responds to uplift or subsidence relative to sea level and may preserve a record of vertical coseismic motion (e.g., Zachariasen *et al.*, 2000).

Tectonic subsidence induces sedimentation that can record and preserve evidence of an earthquake (Carver and McCalpin, 1996). Buried flora or fauna may provide material for dating the earthquake. For example, in the Cascadia region of northwestern USA, tsunami deposits, drowned forests, and grasslands along the coastline provided material for high precision dating of subduction zone earthquakes (e.g., Atwater *et al.*, 1995). The date of the most recent earthquake (AD 1700) was resolved from radiocarbon dating of submerged deposits, analysis of tree rings (dendroseismology), and historic accounts of tsunami recorded in Japan (Nelson *et al.*, 1995; Satake *et al.*, 1996).

Measurement and interpretation of vertical deformation in coastal regions must be based on knowledge of sea level at the time of uplift. Sea level was 100–150 m below modern levels during the last glacial maximum 15 000–20 000 y BP (Lajoie, 1986). After rising rapidly in the early Holocene, sea level stabilized in the mid-Holocene. However, sea level has fluctuated slightly (± 2 m) during the late Holocene (Lajoie, 1986) and is currently at different levels in different parts of the world (Pirazzoli, 1991). Fluctuations and regional variations have been sufficiently large to complicate interpretations of coseismic uplift or subsidence in areas with low vertical deformation rates.

FIGURE 3 Photo of shorelines and wave erosion platforms along rocky coastline of San Joaquin Hills, California. The person is standing on the active erosion platform. An elevated platform intersects the sea cliff at the paleoshoreline. An older paleoshoreline forms a notch in the sea cliff near the top of the photo. The paleoshorelines were formed by tectonic uplift of the San Joaquin Hills, southern California, USA (Grant and Ballenger, 1999).

2.2.3 Ground Shaking and Secondary Effects

Surface rupture and regional deformation are considered primary effects of earthquakes because they can be directly associated with tectonic movement of specific faults. Secondary effects of earthquakes such as liquefaction, ground failure, slope failure, tsunami, and seismic seiche are induced by shaking. A history of seismic shaking is an indirect history of regional fault activity and is useful for seismic hazard assessment. If secondary effects disturb the environment, and if they are preserved, then these effects can be catalogued to develop a chronology of paleoearthquakes. The dates of shaking events may be constrained by geochronology methods.

Evidence of liquefaction, inundation by tsunami, and seismically induced ground failures may be observed in natural or artificial exposures of young sediments, or in sediment cores. Liquefaction is a loss of bearing strength that occurs when saturated cohesionless sediments are subjected to strong shaking or cyclical loading. Fluidized sediment may mobilize during liquefaction to form distinctive sedimentary or geomorphic structures such as sand blows, sand dikes and craters (see Fig. 6 on the Handbook CD-ROM; Obermeier, 1996). Ground failures such as lateral spreading, slumping, lurching, and cracking may occur in soft sediment. In submarine or lacustrine environments, deposition of turbidites (Obermeier, 1996) or mixed-layer clays (Marco *et al.*, 1996; Doig, 1998) can provide evidence of a paleoearthquake. Nonseismic events can create structures that are virtually indistinguishable from seismically deformed sediments, or seismites (Ricci Lucchi, 1995). Therefore, paleoseismologists must correlate candidate seismites over regions and rule out nontectonic origins before concluding that an earthquake occurred. For example, the widespread presence of features attributed to liquefaction and ground failure provide strong evidence of the AD 1700 Cascadia earthquake (Adams, 1990; Obermeier and Dickinson, 2000).

Earthquake ground motions often trigger slope failures. Mountainous regions with steep slopes or unstable slope materials are particularly susceptible to seismically induced slope failures. Slope failures have been responsible for many fatalities in historic earthquakes (Bolt, 1999). In seismically active regions, slope failures over large areas may be linked to paleoearthquakes. Several methods have been developed to date paleoearthquakes by measuring the age of seismically induced landslide deposits. For example, lichenometry has been applied to date seismically induced rockfalls along the Alpine fault zone, New Zealand (Bull, 1996). The criteria for determining seismic origin, and methods for dating are summarized by Jibson (1996).

2.3 Measurement of Slip

Measurement of slip, or surface displacement, across a fault yields information about the magnitude of paleoearthquakes and the average rate of deformation. The slip rate of a fault provides an upper bound for the rate of seismic moment release (Youngs and Coppersmith, 1985; WGCEP, 1995). Slip from paleoearthquakes can be used in combination with slip rate to estimate the recurrence intervals between earthquakes and to constrain the magnitude of past ruptures.

2.3.1 Slip Rates

The slip rate of a fault, V, is obtained by measuring the displacement, D, across a fault zone during time interval, T, using Eq. (1):

$$V = D/T \tag{1}$$

To measure a slip rate, a geologist must find a feature that crosses the fault, is well defined, and is datable. Such features are called *piercing lines*. The points where they intersect a fault zone are called *piercing points*. Fault slip is measured from the displacement of piercing points and deformation of piercing lines. (Fig. 7 on the Handbook CD-ROM shows a map of an excavated piercing line across the San Andreas fault.) Geological piercing lines include streams, gullies, and linear sedimentary features. Anthropogenic piercing lines include walls, row crops, rice paddy boundaries, fences, and roads. The ideal piercing line crosses the fault at a right angle, is perfectly linear, has the same trend on either side of the fault zone, and is of precisely known age. Such conditions rarely exist. Therefore, most slip rates have considerable measurement uncertainty. For example, a compilation of fault slip rates in California (Peterson and Wesnousky, 1994) shows significant uncertainty in slip rate measurements.

Uncertainty in slip rates are caused by uncertainties in measurements of displacement and in estimates of the time interval over which the displacement occurred. Aperture of measurement affects slip rates because deformation may occur beyond the main fault zone (McGill and Rubin, 1999; Rockwell *et al.*, 2002) and have only subtle effects on piercing lines (Grant and Sieh, 1993; Grant and Donnellan, 1994; Sieh, 1996). Therefore, geologically measured slip rates may provide only minimum values if measured over short apertures. For blind faults, the displacement on the fault plane must be inferred from deformation far outside the fault zone, and the resulting slip rates have even larger uncertainties.

The length and variation of time intervals between surface ruptures may affect measurements of slip rate or slip per event. For faults that do not creep, surface displacements are generated episodically by earthquakes. If the slip rate of a fault is low and the average time between surface ruptures is long—the case for most faults—measurements of slip rate may be significantly greater or less than the average rate over multiple rupture cycles. It is generally assumed that slip rate is constant over the interval of measurement, but for most faults there are insufficient data to test this assumption. Geomorphic analysis of normal fault scarps in the Basin and Range province, USA,

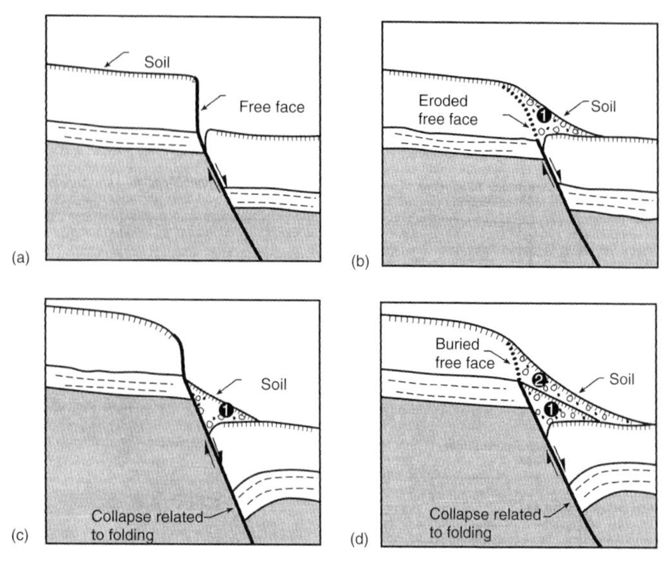

FIGURE 4 Idealized sequence of two normal faulting events and formation of a colluvial wedge, as exposed in trench excavations. (From Lettis and Kelson, 2000.) (a) First surface faulting event; (b) first colluvial wedge (1) and soil formation; (c) second surface faulting event; (d) second colluvial wedge (2) and soil formation.

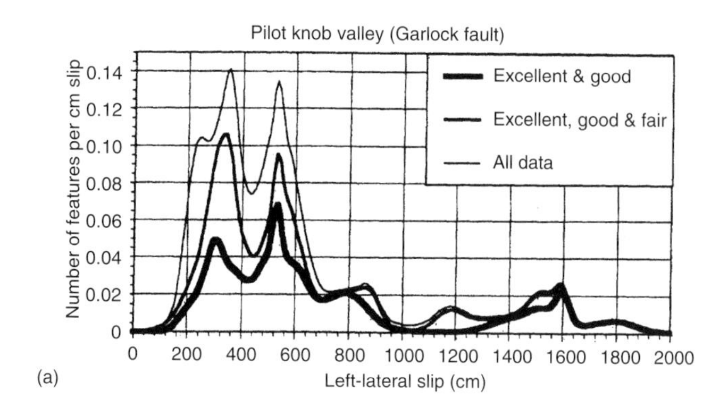

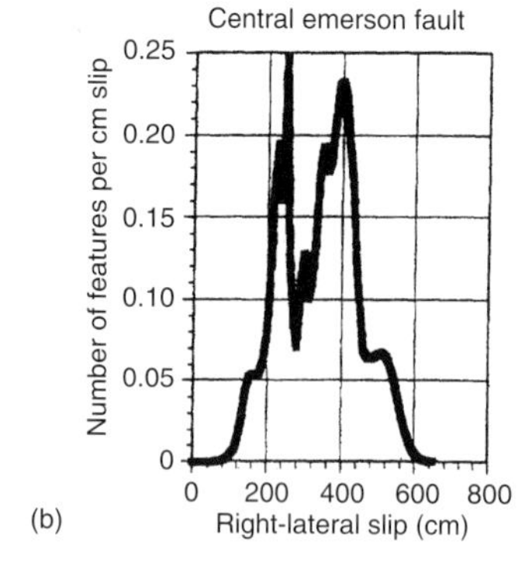

FIGURE 5 Summed Gaussian probability density functions for measured surficial offsets along the Garlock fault (a) (from McGill and Sieh, 1991), and from the 1992 *M*7.3 Landers earthquake on the Emerson fault, California (b). Each peak in the Garlock distribution was originally interpreted to be the result of displacement from a separate surface rupture. Measurements from the Landers earthquake show that surficial offset from a single earthquake can have a multimodal frequency distribution. (From McGill and Rubin, 1999.)

reveals significant temporal variation in slip rates for several faults (Slemmons, 1995).

2.3.2 Slip per Event

The amount of slip per event can be used to estimate the magnitude of previous earthquakes and the time between earthquakes (if the average slip rate is known). The distribution of slip may also reveal the location of segment boundaries, and areas of characteristic slip, if any are present (Ward, 1997).

To measure slip per paleoearthquake, it is necessary to identify a piercing line that has been displaced by a known number of earthquakes. Ideally, this is done by measuring the displacement of a piercing line of known age in an area where the dates of individual surface ruptures are well constrained. For normal and reverse faults, the vertical component of displacement can be measured directly by excavating a trench across the scarp (Fig. 4) because material deposited by collapse or decay of a vertical scarp typically buries and preserves evidence of the rupture. Changes in slope or height of a scarp may indicate multiple ruptures. Variation in the amount of lateral displacement of piercing lines along a strike-slip fault can indicate multiple surface ruptures. In coastal areas, the vertical component of displacement can be measured from elevated marine shorelines (see Fig. 5 on the Handbook CD-ROM) (Matsuda *et al.*, 1978; Plafker and Rubin, 1978). For strike-slip faults, the best measurements are obtained from three-dimensional excavation using multiple trenches (Weldon *et al.*, 1996). An alternate, but less-reliable, method is to measure displacement of surficial features such as laterally offset channels. The total number of earthquakes is then inferred from peaks in the frequency distribution of measurements (Fig. 5), or by the assumption of constant (or "characteristic") slip per event (see Section 3.2). Uncertainty in the resulting slip per event measurements is unconstrained (see Fig. 10 on the Handbook CD-ROM).

2.4 Magnitude of Paleoearthquakes

In general, the effects of an earthquake are proportional to its size (Slemmons and dePolo, 1986). Therefore, the amount of slip, tectonic deformation, or secondary effects preserved in the paleoseismic record helps to constrain the size of a paleoseismic event. The magnitude of paleoearthquakes can be estimated from measurements of slip per event or from rupture length inferred by correlating dates of paleoseismic events at different sites.

Empirical regression relationships between historic earthquake magnitude and rupture length, average slip, or maximum slip (Wells and Coppersmith, 1994) can be applied to estimate the magnitude of paleoearthquakes (Table 2). Hemphill-Haley and Weldon (1999) developed empirical relationships specifically for estimating magnitude from paleoseismic measurements of slip. Their method provides uncertainties based on the number of measurements and percentage of rupture length sampled.

For faults with chronologies of earthquakes from multiple paleoseismic investigations, surface rupture lengths can be

TABLE 2 Hierarchical Classification of Paleoseismic Features (Modified from McCalpin, 1996)

	On Fault Coseismic	On Fault Postseismic	Off Fault Coseismic	Off Fault Postseismic
Primary				
Geomorphic expression	• Fault scarps • Fissures • Folds • Moletracks • Pressure ridges	• Colluvial aprons • Afterslip contribution	• Tilted surfaces • Uplifted shorelines • Drowned shorelines	• Tectonic alluvial terraces • Afterslip contribution
Stratigraphic expression	• Faulted strata • Folded strata	• Colluvial wedge • Fissure fill • Unconformity	• Tsunami deposits	
Abundance of similar nonseismic features	Few	Few	Some	Common
Secondary				
Geomorphic expression	• Sand blows • Landslides • Disturbed trees	• Retrogressive landslides	• Sand blows • Landslides • Fissures • Subsidence from compaction	• Retrogressive landslides
Stratigraphic expression	• Sand dikes	• Rapidly deposited lake or estuarine sediments	• Sand dikes • Filled craters • Soft-sediment deformation • Turbidites	• Rapidly deposited lake or estuarine sediments
Abundance of similar nonseismic features	Some	Very common	Some	Very common

estimated by assuming that ruptures with overlapping time windows occurred at the same time (see Section 3.4). However, there are large uncertainties in correlating ruptures between paleoseismic investigation sites, so the inferred rupture length and resulting magnitudes are poorly constrained. Segmentation models (see Section 3.1) have also been applied to estimate paleoearthquake rupture length and magnitude.

3. Models and Uncertainty

Paleoseismic data provide information about earthquakes over scales of time and magnitude that are useful for seismic hazard assessment and essential for understanding the long-term rupture patterns of faults. Therefore, paleoseismic data have been influential in the development and testing of models that describe fault behavior over multiple rupture cycles, and in forecasting earthquakes. This section describes models that depend critically on observations of paleoearthquakes, and discusses uncertainty in paleoseismic data.

3.1 Segmentation

Segmentation models assume that faults are divided into discrete, identifiable sections that behave distinctively over multiple rupture cycles (e.g., Schwartz and Coppersmith, 1986; Slemmons, 1995). Segment boundaries are thought to control the termination and initiation of fault ruptures and therefore limit the magnitude and rupture pattern of an earthquake. Fault segments are defined based on structural discontinuities, changes in strike, and rheology (Allen, 1968) as well as by paleoseismic data (Chapter 29 by Sibson; Schwartz and Coppersmith, 1984; Sieh and Jahns, 1984; WGCEP, 1988, 1990; Schwartz and Sibson, 1989).

Segmentation models are attractive because they simplify fault behavior. For example, in seismic hazard assessment it is desirable to estimate the size of the largest earthquake that can occur on a fault (Yeats *et al.*, 1997). The maximum magnitude earthquake can be estimated from empirical relationships (Table 3) and the estimated length of maximum rupture (Wells and Coppersmith, 1994). This would be either the maximum length of the fault, or of the segment. Similarly, the size of a paleoearthquake can be estimated by assuming that the entire segment ruptured.

Testing of segment models requires defining segments and then observing multiple ruptures. This would take decades to centuries for the fastest slipping faults, and thousands of years for most faults. Therefore, few segment models have been tested against rupture patterns of historic earthquakes (Sieh, 1996). Results are mixed. For example, two large earthquakes on strike–slip faults in the western US (*M*7 1989 Loma Prieta and *M*7.3 1992 Landers) propagated through previously identified segment or fault boundaries (WGCEP, 1990; Sieh *et al.*, 1993). However, several historic ruptures of normal faults in the Great Basin have terminated at or near fault discontinuities (Zhang *et al.*, 1999). More observations are needed to test segmentation hypotheses and their utility for seismic hazard assessment.

TABLE 3 Selected Empirical Relationships between Moment Magnitude (*M*), Average Displacement (AD), Maximum Displacement (MD), and Surface Rupture Length (SRL)[a]

Equation[b]	Slip Type[c]	a (sa)[d]	b (sa)[d]	Standard deviation	Correlation coefficient
$M = a + b \times \log(AD)$	SS	7.04(0.05)	0.89(0.09)	0.28	0.89
	R[e]	6.64(0.16)	0.13(0.36)	0.50	0.10
	N	6.78(0.12)	0.65(0.25)	0.33	0.64
	All	6.93(0.05)	0.82(0.10)	0.39	0.75
$M = a + b \times \log(MD)$	SS	6.81(0.05)	0.78(0.06)	0.29	0.90
	R[e]	6.52(0.11)	0.44(0.26)	0.52	0.36
	N	6.61(0.09)	0.71(0.15)	0.34	0.80
	All	6.69(0.04)	0.74(0.07)	0.40	0.78
$M = a + b \times \log(SRL)$	SS	5.16(0.13)	1.12(0.08)	0.28	0.91
	R[e]	5.00(0.22)	1.22(0.16)	0.28	0.88
	N	4.86(0.34)	1.32(0.26)	0.34	0.81
	All	5.08(0.10)	1.16(0.07)	0.28	0.89

[a]From Wells and Coppersmith (1994).
[b]Displacement in meters. Surface rupture length in kilometers.
[c]SS, strike-slip; R, reverse; N, normal; All, all fault types.
[d]Coefficients and standard errors.
[e]Relationship is not significant at the 95% confidence level.

3.2 Characteristic Earthquakes

The characteristic earthquake model (Schwartz and Coppersmith, 1984) is probably the most influential model of fault rupture and earthquake recurrence developed from paleoseismic data. The basic tenet of the model is that most surface slip on a fault occurs in *characteristic earthquakes*. Characteristic earthquakes are the result of *characteristic slip*. Figure 6c illustrates the concept of characteristic slip; at a specific location along a fault, the displacement (slip) is the same in successive characteristic earthquakes. This implies that characteristic earthquakes have similar rupture patterns and that a fault can be divided into segments that behave characteristically. Each segment would have a distinctive or "characteristic" rupture pattern and magnitude. Characteristic slip requires variable slip rate along a fault to account for different amounts of displacement.

If characteristic slip occurs on a fault, then most seismic moment is released by repetition of characteristic earthquakes of approximately the same magnitude. Characteristic slip causes a kink in the frequency magnitude relationship, known as characteristic recurrence (Fig. 7), due to the dominance of relatively large magnitude characteristic earthquakes (Schwartz and Coppersmith, 1984). The characteristic recurrence curve appears to fit the frequency–magnitude distribution for some faults (Wesnousky, 1994) but does not fit global seismicity rates as well as other models (Kagan, 1993).

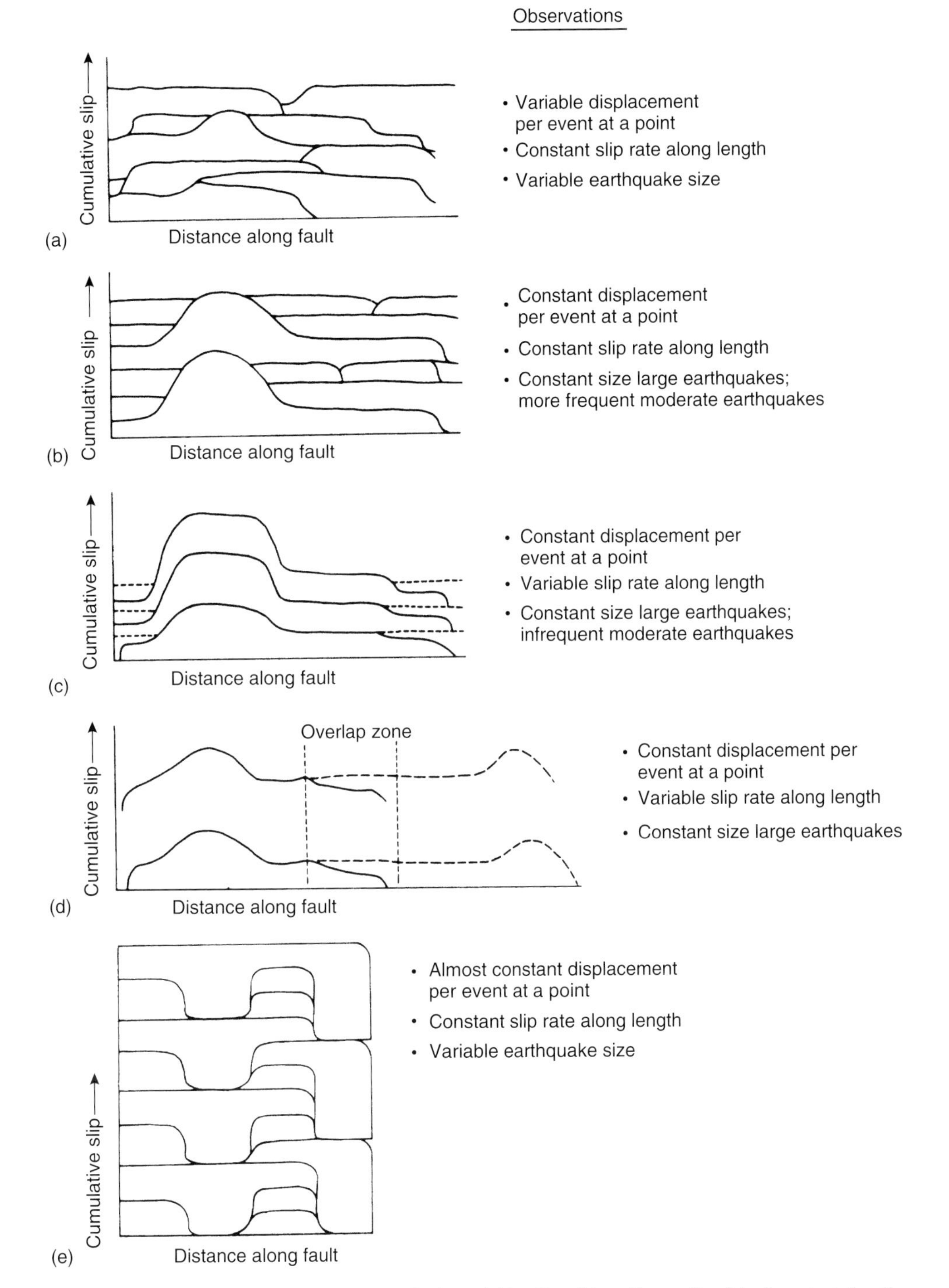

FIGURE 6 Schematic representations of (a) variable slip, (b) uniform slip, (c) characteristic slip, (d) overlap, and (e) coupled slip models. (From Schwartz and Coppersmith, 1984; Berryman and Beanland, 1991. Reprinted from McCalpin, 1996.)

The characteristic earthquake model is convenient for seismic hazard assessment because it assumes that faults can be divided into identifiable segments that rupture with characteristic slip and recurrence. Many studies of active faults have focused on identifying characteristic rupture segments of faults and their properties (e.g., Wesnousky, 1986; Peterson and Wesnousky, 1994; WGCEP 1988, 1990, 1995). For example, earthquake forecasts for the San Andreas fault

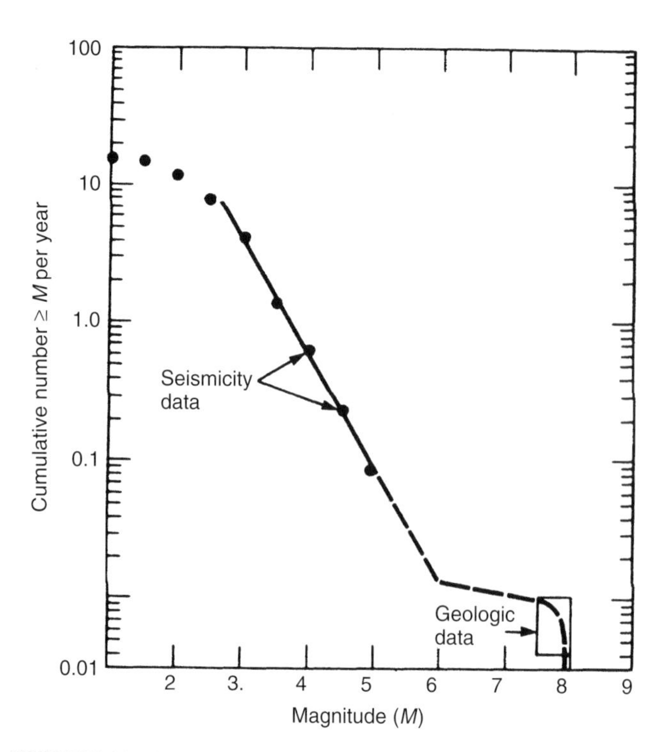

FIGURE 7 Characteristic recurrence model for cumulative frequency–magnitude distribution of seismicity on a specific fault. (From Schwartz and Coppersmith, 1984.)

have been based on a combination of segmentation, time-predictable, and characteristic earthquake models. The Parkfield segment of the San Andreas fault was predicted to break before 1993 with repetition of characteristic earthquake (Bakun and Lindh, 1985; WGCEP, 1988). To date, the expected Parkfield earthquake has not occurred, so the pattern of rupture cannot be tested against the models. The Working Group on California Earthquake Probabilities (WGCEP, 1988, 1990, 1995) divided the San Andreas fault into characteristic segments and issued probabilistic rupture forecasts. In 1989, the *M*7 Loma Prieta earthquake occurred on portions of two segments defined prior to the earthquake (WGCEP, 1990) rather than on a single segment. Hecker and Schwartz (1994) analyzed paleoseismic data for many faults and concluded that most faults exhibit characteristic slip. However, recent paleoseismic studies of the San Andreas fault reveal complexity in spatial and temporal rupture patterns that was not evident when the characteristic earthquake model was first proposed (Grant, 1996). Debate about the applicability of the characteristic earthquake model for forecasting earthquakes is likely to continue (Yeats *et al.*, 1997; Yeats, 2001).

3.3 Time Predictable and Recurrence Models

Recurrence time, T_r, is the time interval between successive ruptures of the same fault. T_r is useful for describing the seismic parameters of an active fault (see Chapter 65 by Somerville and Moriwaki). Paleoseismic investigations are the main source of data for measuring or estimating T_r. Individual recurrence times or average recurrence intervals (the average time interval between ruptures) can be measured directly by dating successive surface ruptures. For well-studied faults such as the San Andreas, average recurrence intervals and their variance are available for several locations along the fault (e.g., WGCEP, 1995, 1999). For most faults, there is insufficient information to describe variability in recurrence times.

If dates of paleoearthquakes are not available, recurrence time may be estimated for a fault using the relationship:

$$T_r = D/V \tag{2}$$

where V is the slip rate and D is average displacement (slip). Assuming the slip rate is constant over the period of observation and there is no creep, then the recurrence time is a linear function of displacement. If displacement occurs in constant-size (characteristic) earthquakes, then the recurrence time can be "predicted". Thus Eq. (2) is referred to as the time-predictable model, after Shimazaki and Nakata (1980) who developed it from observations of uplifted coastal terraces in Japan. If the characteristic earthquake model and time-predictable model are combined, the result is a periodic recurrence model.

Analysis of historic earthquakes and paleoseismic data from several regions shows that occurrence of large earthquakes is irregular (Goes, 1996). Surface ruptures of the San Andreas fault have been proposed to occur irregularly (Jacoby *et al.*, 1988; Sieh *et al.*, 1989) or in clusters (Grant and Sieh, 1994). Temporal and/or spatial clustering of surface ruptures has also been proposed to describe surface ruptures in the Basin and Range (Wallace, 1981; Slemmons, 1995; McCalpin, 1996) and eastern California shear zone, USA (Rockwell *et al.*, 2000). Historic ruptures of the North Anatolian fault zone in Turkey and patterns of late Holocene seismicity in Iran suggest that seismicity may be triggered by stress shadows from previous ruptures to form temporal clusters or sequences of damaging earthquakes (Stein *et al.*, 1997; Berberian and Yeats, 1999). For most faults, the number of documented paleoearthquakes and the precision of their dates are inadequate to test hypotheses of clustered or triggered earthquakes. Larger paleoseismic data sets are needed to conduct statistically significant tests of recurrence models.

3.4 Rupture Patterns in Space and Time

Patterns of fault rupture in space and time yield insights about the physics of faulting (Ward, 1997) and provide templates for estimating the location and size of future earthquakes. Several models of fault behavior (shown in Fig. 6) are based on multiple paleoseismic measurements of slip at different sites along

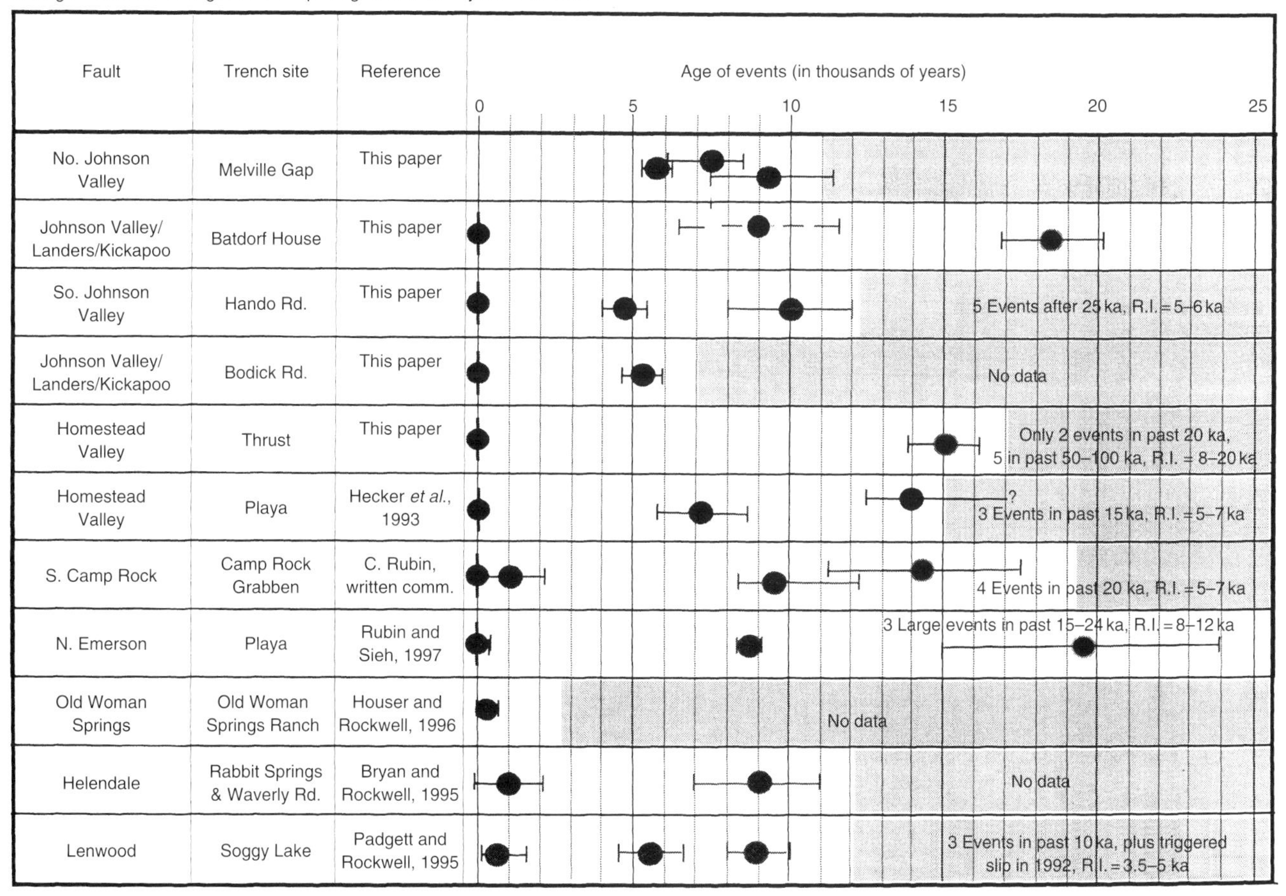

FIGURE 8 Compilation of rupture history for several faults in the Mojave desert, California, USA. The sequence suggests regional temporal clustering of paleoearthquakes. (From Rockwell *et al.*, 2000.)

a fault. The variable slip model (Fig. 6a) assumes a constant slip rate along the fault with variable displacement per event at a location, and variable earthquake size. The uniform slip model (Fig. 6b) is similar to the characteristic slip model (Fig. 6c) because both assume constant-size displacement at a location. The slip rate is constant along the fault in the uniform slip model. Therefore, a range of earthquake sizes is required to account for sections of fault with large or small slip per event. The overlap model (Fig. 6d) assumes constant displacement per event at a point (characteristic slip), variable slip rate along the fault, and repetition of constant size large earthquakes. Finally, the coupled model (Fig. 6e) is a uniform slip model coupled with a segmentation model.

Spatial and temporal patterns of fault rupture can be analyzed by plotting the dates of paleoearthquakes on a space–time diagram. Figures 8 and 9 show examples from a group of faults that ruptured in the 1992 Landers earthquake, and the southern San Andreas fault. Uncertainty in dates ranges from decades to more than a thousand years. Paleoearthquakes at adjacent sites with overlapping dates could have been the same rupture, or separate ruptures that are indistinguishable within the uncertainty. Additional dates of paleoearthquakes would provide better constraints on the locations and magnitudes of past ruptures.

3.5 Uncertainty in Paleoseismic Data

The quality and reliability of paleoseismic data are affected by the investigation method, the characteristics of the study site, and the perspective of the investigator. Many paleoseismic studies are destructive to the study site, so the data cannot be reproduced or compared with a control. Therefore, each paleoseismic publication becomes a repository of data that will be interpreted by scientists who will not be able to evaluate it first-hand. Collection of paleoseismic data requires recognition and interpretation of complex patterns in the geologic record and landscape. Therefore, some interpretation is automatically convolved with the data, and no deconvolution algorithm can be applied after the data are published. Paleoseismologists have developed several methods to reduce

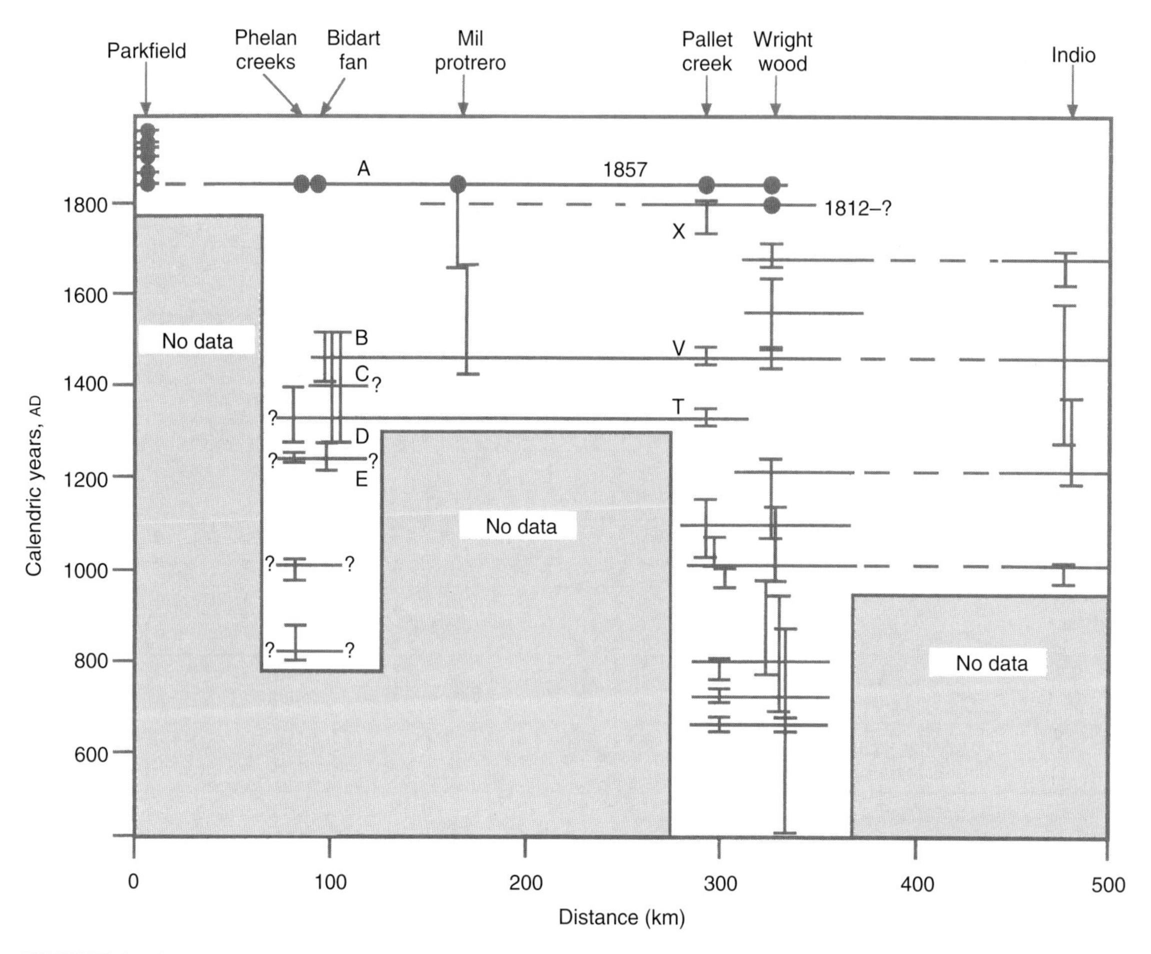

FIGURE 9 Space–time rupture correlation diagram for paleoseismic sites along the southern San Andreas fault, California, USA. Vertical lines show age limits for paleoearthquakes at each site. Horizontal lines represent ruptures, based on proposed correlations between sites. (From Arrowsmith *et al.*, 1997. Modified from Grant and Sieh, 1994 and other sources.)

subjectivity and standardize uncertainty in their data. Fault exposures are generally documented with photographs as well as logs. Photos are objective records, devoid of interpretation. "Trench parties" are held to allow reviewers to examine primary field data prior to publication. When possible, care is taken to fill excavations so that they can be reopened at a later date. Uncertainty in data collection and field interpretations are often reported qualitatively. For example, Sieh (1978) and McGill and Sieh (1991) reported the apparent "quality" of their measurements of offset streams. Paleoseismologists may report high quality measurements with "high confidence". Such assessments are qualitative, and should not be confused with statistical confidence intervals.

4. Applications

Paleoseismic data are used for seismic source characterization in hazard assessment and mitigation (see Chapter 65 by Somerville and Moriwaki). Probabilistic seismic hazard assessment incorporates information on fault location and geometry, slip rates, recurrence intervals, dates of previous ruptures, and maximum earthquakes. Maximum earthquakes are estimated from segmentation models, slip per event, and lengths of previous ruptures. Deterministic hazard assessments use paleoseismic data to develop earthquake scenarios including rupture location, magnitude and surface displacement. Earthquake scenarios are used to estimate expected damage for emergency response planning. Surface rupture hazard can be mitigated by identifying fault zones that are most likely to rupture and restricting land uses within such zones. Paleoseismic investigations provide data on the recency and frequency of fault rupture, as well as the type and width of surface rupture. This information can be used to classify fault zones according to rupture potential and expected displacement in future earthquakes. Critical structures can be sited to avoid rupture hazard zones, or designed to accommodate expected displacement.

5. Future Directions

Paleoseismology is a young science that focuses on understanding the long-term patterns of large earthquakes. Paleoseismic data can be applied toward understanding the behavior of faults, estimating the potential for large earthquakes to occur in populated areas, and for mitigating seismic hazard. As populations increase in seismically active areas, many paleoseismologists are moving toward predictive or applied work in seismic hazard assessment. Establishment of digital databases and standard formats for paleoseismic data allows better integration with more quantitative fields of seismology and earthquake engineering. Investigations of active faults are being extended to offshore areas (e.g., Goldfinger and Nelson, 2000) and paleoseismic data are being compiled for the World Map of Active Faults under the International Lithosphere Program. As paleoseismology matures, it is likely to play an increasingly important role in seismology and society.

Acknowledgments

R. Yeats, M. Bonilla, and R. Sibson provided helpful reviews of the manuscript. This is publication no. 1 in the Department of Environmental Analysis and Design, University of California, Irvine, contribution series.

References

Adams, J. (1990). Paleoseismicity of the Cascadia subduction zone: evidence from turbidites off the Oregon–Washington margin. *Tectonics* **9**, 569–583.

Allen, C.R. (1968). The tectonic environments of seismically active and inactive areas along the San Andreas fault system. In: "Proceedings of Conference on Geologic Problems of the San Andreas Fault System," (W.R. Dickinson and A. Grantz, Eds.) pp. 70–82, *Stanford Univ. Pub. Geol. Sci.* **11**, Stanford.

Arrowsmith, J.R., K. McNally, and J. Davis (1997). Potential for earthquake rupture and *M*7 earthquakes along the Parkfield, Cholame and Carrizo segments of the San Andreas fault. *Seismol. Res. Lett.* **68**, 6, 902–916.

Atwater, B.F., A.R. Nelson, J.J. Clague, G.A. Carver, P.T. Bobrowsky, J. Bourgeois, M.E. Darienzo, W.C. Grant, E. Hemphill-Haley, H.M. Kelsey, G.C. Jacoby, S.P. Nishenko, S.P. Palmer, C.D. Peterson, M.A. Reinhart, and D.K. Yamaguchi (1995). Summary of coastal geologic evidence for past great earthquakes at the Cascadia subduction zone, *Earthquake Spectra* **11**, 1–18.

Bakun, W.H. and A.G. Lindh (1985). The Parkfield, California earthquake prediction experiment. *Science* **229**, 619–624.

Berberian, M. and R.S. Yeats (1999). Patterns of historical earthquake rupture in the Iranian plateau. *Bull. Seismol. Soc. Am.* **89**, 1, 120–139.

Berryman, K.R. and S. Beanland (1991). Variation in fault behavior in different tectonic provinces of New Zealand. *J. Struct. Geol.* **13**, 177–189.

Biasi, G. and R.J. Weldon II (1994). Quantitative refinement of calibrated ^{14}C distributions. *Quat. Res.* **41**, 1–18.

Bolt, B.A. (1999). "Earthquakes," 4th Edn., W.H. Freeman, San Francisco.

Bonilla, M.G. (1988). Minimum earthquake magnitude associated with coseismic surface faulting. *Bull. Assoc. Engr. Geol.* **XXV**, 17–29.

Bonilla, M.G. and J.J. Lienkaemper (1990). Visibility of fault strands in exploratory trenches and timing of rupture events. *Geology* **18**, 153–156.

Bull, W.B. (1996). Prehistorical earthquakes on the Alpine fault, New Zealand. *J. Geophys. Res.* **101**, 6037–6050.

Bullard, T.F. and W.R. Lettis (1993). Quaternary fold deformation associated with blind thrusting, Los Angeles Basin, California. *J. Geophys. Res.* **98**, 8349–8369.

Carver, G.A. and J.P. McCalpin (1996). Paleoseismology of compressional tectonic environments. In: "Paleoseismology" (J.P. McCalpin, Ed.), pp. 183–270. Academic Press, San Diego.

Crone, A.J. and E.M. Omdahl (Eds.) (1987). Directions in Paleoseismology. Proceedings of Conference XXXIX, *U.S. Geol. Surv. Open File Rep.* 87-673, 1–456.

Doig, R. (1998). 3000-yr paleoseismological record from the region of the 1988 Saguenay, Quebec, earthquake. *Bull. Seismol. Soc. Am.* **88**, 1198–1203.

Dolan, J.F., D. Stevens, and T.K. Rockwell (2000). Paleoseismologic evidence for an early to mid-Holocene age of the most recent surface rupture on the Hollywood fault, Los Angeles, California. *Bull. Seismol. Soc. Am.* **90**, 334–344.

Goes, S.D.B. (1996). Irregular recurrence of large earthquakes: an analysis of historic and paleoseismic catalogs. *J. Geophys. Res.* **101**, 5739–5749.

Goldfinger, C. and C.H. Nelson (2000). Holocene seismicity of the northern San Andreas fault based on the turbidite record. In: "3rd Conference on Tectonic Problems of the San Andreas fault system: Program and Abstracts," Stanford, CA.

Grant, L.B. (1996). Uncharacteristic earthquakes on the San Andreas fault. *Science* **272**, 826–827.

Grant, L.B. and L.J. Ballenger (1999). Paleoseismic evidence of a historic coastal earthquake and uplift of the San Joaquin Hills, southern California. *EOS, Trans. Am. Geophys. Union* **80**, F736.

Grant, L.B. and A. Donnellan (1994). 1855 and 1991 surveys of the San Andreas fault: implications for fault mechanics. *Bull. Seismol. Soc. Am.* **84**, 241–246.

Grant, L.B. and K. Sieh (1993). Stratigraphic evidence for 7 meters of dextral slip on the San Andreas fault during the great 1857 earthquake in the Carrizo Plain. *Bull. Seismol. Soc. Am.* **83**, 619–635.

Grant, L.B. and K. Sieh (1994). Paleoseismic evidence of clustered earthquakes on the San Andreas fault in the Carrizo Plain, California. *J. Geophys. Res.* **99**, B4, 6819–6841.

Grant, L.B., J.T. Waggoner, C. von Stein, and T.K. Rockwell (1997). Paleoseismicity of the North Branch of the Newport–Inglewood fault zone in Huntington Beach, California, from Cone Penetrometer Test Data. *Bull. Seismol. Soc. Am.* **87**, 277–293.

Grant, L.B., K.J. Mueller, E.M. Gath, H. Cheng, R.L. Edwards, R. Munro, and G.L. Kennedy (1999). Late Quaternary uplift and earthquake potential of the San Joaquin Hills, southern Los Angeles basin, California. *Geology* **27**, 1031–1034.

Hecker, S. and D.P. Schwartz (1994). The characteristic earthquake revisited: geological evidence of the size and location of successive earthquakes on large faults. *US Geol. Surv. Open File Rep.* 94-568, 79–80.

Hemphill-Haley, M.A. and R.J. Weldon III (1999). Estimating prehistoric earthquake magnitude from point measurements of surface rupture. *Bull. Seismol. Soc. Am.* **89**, 1264–1279.

Jacoby, G., P. Sheppard, and K. Sieh (1988). Irregular recurrence of large earthquakes along the San Andreas fault—evidence from trees. *Science* **241**, 196–199.

Jibson, R.W. (1996). Using landslides for paleoseismic analysis. In: "Paleoseismology" (J.P. McCalpin, Ed.), pp. 397–438. Academic Press, San Diego.

Kagan, Y. (1993). Statistics of characteristic earthquakes. *Bull. Seismol. Soc. Am.* **83**, 7–24.

Keller, E.A. and N. Pinter (1996). "Active Tectonics: Earthquakes, Uplift, and Landscape." Prentice Hall, Upper Saddle River, New Jersey.

Lajoie, K.R. (1986). Coastal tectonics. In: "Active Tectonics: Studies in Geophysics" (Wallace, R.E., Chairman), pp. 95–124. National Academic Press, Washington, DC.

Lettis, W.R. and K.I. Kelson (2000). Applying geochronology in paleoseismology. In: "Quaternary Geochronology: Methods and Applications" (Noller, J.S., Sowers, J.M., and Lettis, W.R., Eds.), AGU Ref. Shelf, ser. **4**, pp. 479–495. American Geophysical Union, Washington, DC.

Lettis, W.R., D.L. Wells, and J.N. Baldwin (1997). Empirical observations regarding reverse earthquakes, blind thrust faults, and Quaternary deformation; are blind thrust faults truly blind? *Bull. Seismol. Soc. Am.* **87**, 1171–1198.

Marco, S., M. Stein, A. Agnon, and H. Ron (1996). Long term earthquake clustering: A 50,000 year paleoseismic record in the Dead Sea graben. *J. Geophys. Res.* **101**, 6179–6191.

Matsuda, T., Y. Ota, M. Ando, and N. Yonekura (1978). Fault mechanism and recurrence time of major earthquakes in southern Kanto district, Japan, as deduced from coastal terrace data. *Geol. Soc. Am. Bull.* **89**, 1610–1618.

McCalpin, J.P. (Ed.) (1996). "Paleoseismology." Academic Press, San Diego.

McCalpin, J.P. and A.R. Nelson (1996). Introduction to paleoseismology. In: "Paleoseismology" (J.P. McCalpin, Ed.), pp. 1–32. Academic Press, San Diego.

McGill, S.F. and K. Sieh (1991). Surficial offsets on the central and eastern Garlock Fault associated with prehistoric earthquakes. *J. Geophys. Res.* **96**, 21597–21621.

McGill, S.F. and C.M. Rubin (1999). Surficial slip distribution on the central Emerson fault during the 28 June 1992, Landers earthquake, California. *J. Geophys. Res.* **104**, B3, 4811–4833.

Nelson, A.R., B.F. Atwater, P.T. Bobrowsky, L.A. Bradley, J.J. Clague, G.A. Carver, M.E. Darienzo, W.C. Grant, H.W. Krueger, R. Sparks, T.W. Stafford, and M. Stuiver (1995). Radiocarbon evidence for extensive plate-boundary rupture about 300 years ago at the Cascadia subduction zone. *Nature* **378**, 371–374.

Noller, J.S., J.M. Sowers, S.M. Colman, and K.L. Pierce (2000a). Introduction to Quaternary Geochronology. In: "Quaternary Geochronology: Methods and Applications" (Noller, J.S., Sowers, J.M., and Lettis, W.R., Eds.), AGU Ref. Shelf, ser. **4**, pp. 1–10, American Geophysical Union, Washington, DC.

Noller, J.S., J.M. Sowers, and W.R. Lettis (2000b). "Quaternary Geochronology: Methods and Applications." AGU Ref. Shelf, ser. **4**, Am. Geophys. Union, Washington, DC.

Nur, A. (2000). Earthquakes, Armageddon, and the Dead Sea scrolls. *Seismol. Res. Lett.* **71**, 261.

Obermeier, S.F. (1996). Using liquefaction-induced features for paleoseismic analysis. In: "Paleoseismology" (J.P. McCalpin, Ed.), pp. 331–396, Academic Press, San Diego.

Obermeier, S.F. and S.E. Dickinson (2000). Liquefaction evidence for the strength of ground motions resulting from late Holocene Cascadia subduction earthquakes, with emphasis on the event of 1700 AD *Bull. Seismol. Soc. Am.* **90**, 876–896.

Pantosti, D. and R.S. Yeats (1993). Paleoseismology of great earthquakes of the late Holocene. *Ann. Geofis.* **36** (3–4), 237–257.

Pavlides, S.B., D. Pantosti, and P. Zhang (1999). Earthquakes, paleoseismology and active tectonics. *Tectonophysics, Special Issue* **308**, 1–2, vii–x.

Petersen, M.D. and S.G. Wesnousky (1994). Review: fault slip rates and earthquake histories for active faults in southern California. *Bull. Seismol. Soc. Am.* **84**, 1608–1649.

Pirazzoli, P.A. (1991). "World Atlas of Holocene Sea-Level Changes." Elsevier Oceanography Series, **58**, Amsterdam.

Plafker, G. and M. Rubin (1978). Uplift history and earthquake recurrence as deduced from marine terraces on Middleton Island, Alaska. *US Geol. Surv. Open File Rep.* 78-943, 687–722.

Prentice, C.S., D.P. Schwartz, and R.S. Yeats (conveners) (1994). Proceedings of the Workshop on Paleoseismology, 18–22 September 1994, Marshall, California. *US Geol. Surv. Open File Rep.* 94-568, 1–210.

Reiter, L. (1991). "Earthquake Hazard Analysis: Issues and Insights." Columbia University Press, New York.

Reiter, L. (1995). Paleoseismology—a user's perspective. In: "Perspectives in Paleoseismology" (L. Serva and D.B. Slemmons, Eds.), *Assoc. Eng. Geol. Spec. Publ.* **6**, 3–6.

Ricci Lucchi, F. (1995). Sedimentological indicators of paleoseismicity. In: "Perspectives in Paleoseismology" (L. Serva and D.B. Slemmons, Eds.), *Assoc. Eng. Geol. Spec. Publ.* **6**, 7–17.

Rockwell, T.K., S. Lindvall, M. Herzberg, D. Murbach, T. Dawson, and G. Berger (2000). Paleoseismology of the Johnson Valley, Kickapoo and Homestead Valley faults: clustering of earthquakes in the eastern California shear zone. *Bull. Seismol. Soc. Am.* **90**, 1200–1236.

Rockwell, T.K., S. Lindvall, T. Dawson, R. Langridge, and W. Lettis (2002). Lateral offsets on surveyed cultural features resulting from the 1999 Izmit and Duzce earthquakes, Turkey. *Bull. Seismol. Soc. Am.* in press.

Satake, K., K. Shimazaki, Y. Tsuji, and Y. Ueda (1996). Time and site of a giant earthquake in Cascadia inferred from Japanese tsunami records of January 1700. *Nature* **379**, 246–249.

Schumm, S.A., J.F. Dumont, and J.M. Holbrook (2000). "Active Tectonics and Alluvial Rivers." Cambridge University Press, Cambridge.

Schwartz, D.P. and K.J. Coppersmith (1984). Fault behavior and characteristic earthquakes: examples from the Wasatch and San Andreas fault zones. *J. Geophys. Res.* **89**, 5681–5698.

Schwartz, D.P. and K.J. Coppersmith (1986). Seismic hazards: new trends in analysis using geologic data. In: "Active Tectonics: Studies in Geophysics" (R.E. Wallace, Chairman), pp. 215–230. National Academic Press, Washington, DC.

Schwartz, D.P. and R.H. Sibson (Eds.) (1989). "Fault Segmentation and Controls of Rupture Initiation and Termination." *US Geol. Surv. Open File Rep.* 89-315, 1–447.

Serva, L. and D.B. Slemmons (Eds.) (1995). "Perspectives in Paleoseismology." *Assoc. Eng. Geol. Spec. Publ.* **6**.

Shimazaki, K. and T. Nakata (1980). Time-predictable recurrence model for large earthquakes. *Geophys. Res. Lett.* **7**, 279–282.

Sieh, K.E. (1978). Slip along the San Andreas fault associated with the great 1857 earthquake. *Bull. Seismol. Soc. Am.* **68**, 1421–1448.

Sieh, K.E. (1981). A review of geological evidence for recurrence times for large earthquakes. In: "Earthquake Prediction: An International Review" (D.W. Simpson and P.G. Richards, Eds.), Maurice Ewing ser. **4**, 209–216. American Geophysical Union, Washington, DC.

Sieh, K.E. and R.H. Jahns (1984). Holocene activity of the San Andreas fault at Wallace Creek, California. *Geol. Soc. Am. Bull.* **95**, 883–896.

Sieh, K.E., M. Stuiver, and D. Brillinger (1989). A more precise chronology of earthquakes produced by the San Andreas fault in Southern California. *J. Geophys. Res.* **94**, 603–623.

Sieh, K.E., *et al.* (20 authors) (1993). Near-field investigations of the Landers earthquake sequence, April to July 1992. *Science* **260**, 171–176.

Sieh, K. (1996). The repetition of large-earthquake ruptures. *Proc. Natl. Acad. Sci.* USA **93**, 3764–3771.

Slemmons, D.B. (1995). Complications in making paleoseismic evaluations in the Basin and Range province, western United States. In: "Perspectives in Paleoseismology" (L. Serva and D.B. Slemmons, Eds.), *Assoc. Eng. Geol. Spec. Publ.* **6**, 19–34.

Slemmons, D.B. and C.M. dePolo (1986). Evaluation of active faulting and related hazards. In: "Active Tectonics: Studies in Geophysics" (R.E. Wallace, Chairman), pp. 45–62. National Academic Press, Washington, DC.

Stein R.S. and R.S. Yeats (1989). Hidden earthquakes, *Sci. Am.* **260(6)**, 48–57.

Stein, R.S., A.A. Barka, and J.H. Dieterich (1997). Progressive failure on the North Anatolian fault since 1939 by earthquake stress triggering. *Geophys. J. Int.* **128**, 594–604.

Trumbore, S.E. (2000). Radiocarbon geochronology. In: "Quaternary Geochronology: Methods and Applications" (J.S. Noller, J.M. Sowers, and W.R. Lettis, Eds.), AGU Ref. Shelf, ser. **4**, pp. 41–60, American Geophysical Union, Washington, DC.

Vittori, E., S.S. Labini, and L. Serva (1991). Palaeoseismology; review of the state-of-the-art. *Tectonophysics* **193**, 9–32.

Wallace, R.E. (1981). Active faults, paleoseismology, and earthquake hazards in the western United States. In: "Earthquake Prediction: An International Review" (D.W. Simpson and P.G. Richards, Eds.), Maurice Ewing ser. **4**, 209–216. American Geophysical Union, Washington, DC.

Wallace, R.E. (Chairman) (1986). "Active Tectonics: Studies in Geophysics." National Academic Press, Washington, DC.

Ward, S.N. (1997). Dogtails versus rainbows: synthetic earthquake rupture models as an aid in interpreting geological data. *Bull. Seismol. Soc. Am.* **87**, 1422–1441.

Weldon, R.J. III, J.P. McCalpin, and T.K. Rockwell (1996). Paleoseismology of strike-slip tectonic environments. In: "Paleoseismology" (J.P. McCalpin, Ed.), pp. 271–329. Academic Press, San Diego.

Wells, D.L. and K.J. Coppersmith (1994). New empirical relationships among magnitude, rupture length, rupture area, and surface displacement, *Bull. Seismol. Soc. Am.* **84**, 974–1002.

Wesnousky, S.G. (1986). Earthquakes, Quaternary faults, and seismic hazard in California. *J. Geophys. Res.* **91**, 12587–12631.

Wesnousky, S.G. (1994). Gutenberg–Richter or characteristic earthquake distribution: which one is it? *Bull. Seismol. Soc. Am.* **84**, 1940–1959.

Working Group on California Earthquake Probabilities (WGCEP) (1988). Probabilities of large earthquakes occurring in California on the San Andreas fault. *US Geol. Surv. Open File Rep.* 88-398, 1–62.

Working Group on California Earthquake Probabilities (WGCEP) (1990). Probabilities of large earthquakes in the San Francisco Bay region, California. *US Geol. Surv. Circ.* **1053**, 1–51.

Working Group on California Earthquake Probabilities (WGCEP) (1995). Seismic hazards in southern California—probable earthquakes, 1994–2024. *Bull. Seismol. Soc. Am.* **85**, 379–525.

Working Group on California Earthquake Probabilities (WGCEP) (1999). Earthquake probabilities in the San Francisco Bay Region: 2000 to 2030—a summary of findings. *US Geol. Surv. Open File Rep.* 99-517.

Yeats, R.S. (2001). "California Earthquakes: A Survivor's Guide," Oregon State University Press, Corvallis, 406 pp.

Yeats, R.S. and C.S. Prentice (1996). Introduction to special section: paleoseismology. *J. Geophys. Res.* **101**, B3, 5847–5853.

Yeats, R.S., K.E. Sieh, and C.R. Allen (1997). "The Geology of Earthquakes," Oxford University Press, New York, 568 pp.

Youngs, R.R. and K.J. Coppersmith (1985). Implications of fault slip rates and earthquake recurrence models to probabilistic seismic hazard estimates. *Bull. Seismol. Soc. Am.* **75**, 939–964.

Zachariasen, J., K. Sieh, F.W. Taylor, and W.S. Hantoro (2000). Modern vertical deformation above the Sumatra subduction zone: paleogeodetic insights from coral microatolls. *Bull. Seismol. Soc. Am.* **90**, 897–913.

Zhang, P.Z., F.Y. Mao, and D.B. Slemmons (1999). Rupture terminations and size of segment boundaries from historical earthquake ruptures in the Basin and Range Province. *Tectonophysics* **308**, 37–52.

Ziony, J.I. and R.F. Yerkes (1985). Evaluating earthquake and surface faulting potential. In: "Earthquake Hazards in the Los Angeles Region" (J.I. Ziony, Ed.), *US Geol. Surv. Prof. Paper 1360*, pp. 43–91.

Editor's Note

Due to the space limitation, the complete set of figures for this Chapter is placed on the Handbook CD-ROM, under directory \30Grant.

31

Using Earthquakes for Continental Tectonic Geology

James A. Jackson
University of Cambridge, Cambridge, United Kingdom

1. Introduction

Seismology contributes to tectonic and structural geology mostly through earthquake source studies or, more specifically, through being able to estimate the location, geometry and size of slip on faults. Earthquake seismology made a fundamental contribution to the discovery and confirmation of plate tectonics through studies that used fault plane solutions to demonstrate plate rigidity (McKenzie and Parker, 1967) and the nature of plate boundaries (Sykes, 1967; Isacks *et al.*, 1968). It is notable how quickly these studies appeared after the installation of the World Wide Standardized Seismograph Network (WWSSN) in the early 1960s made reliable determinations of focal mechanisms possible. In the 1970s earthquake seismology was responsible for important advances in understanding large-scale continental tectonics (e.g., McKenzie, 1972; Molnar and Tapponnier, 1975) but had little influence on more detailed structural geology on the continents until about the 1980s, for two reasons, which are linked. Firstly, the continents deform in ways that are more complicated than the oceans, with earthquakes and faults distributed over wide areas and not confined to the narrow plate boundaries that typify the oceans. There are indeed patterns in the active faulting in such continental regions, but they were not easy to recognize with the relatively few earthquake focal mechanisms available soon after the installation of the WWSSN. Secondly, earthquake source seismology could not contribute much to debates in structural geology through the 1970s, because it was too inaccurate. In particular, first motion fault plane solutions were rarely well-constrained in orientation, and routinely reported focal depths based on teleseismic arrival time data were notoriously unreliable, unable to tell whether an earthquake was in the crust or the mantle, let alone within the crystalline basement or sedimentary cover.

Two advances helped to increase the impact of seismology on structural geology by the late 1970s. The first was the installation of many dense local seismograph networks, often temporary and designed to monitor aftershocks, which greatly improved knowledge of focal depth distributions outside the few areas where permanent dense networks already existed. The second was the development of synthetic seismogram techniques that allowed much more accurate determination of earthquake source parameters from teleseismic waveforms themselves, rather than from just the arrival times and first motion polarities (see Appendix on the Handbook). For earthquakes larger than $M_W \sim 5.5$ it became possible, under favorable conditions, to estimate fault geometry (strike, dip, rake) to within 5–15°, centroid depth to within $\sim\pm3$ km, and seismic moment to within 10–20%.

As more abundant and better quality earthquake data became available on the continents the natural advantages of seismology could then be exploited

- Earthquake source seismology could provide information about the fault geometry underground which, when combined with surface observations, could give an overall three-dimensional image of the fault.
- Faults could be studied as they moved, in their original orientation, unaffected by uplift, rotation or tilting.
- Large areas could be investigated quickly to look for patterns of fault movement, in the sure knowledge that faults which are seismically active now are all part of the same deformational episode.

With these abilities earthquake seismology had something new to offer. Hitherto, structural and tectonic geology had necessarily been based on field or seismic reflection studies of faults that are now inactive. Many of these old structures are no longer in their original orientation, and demonstrating that old faults over a wide region had been active simultaneously was a major effort, if it was possible at all. Moreover, comparative studies between inactive regions were difficult, limited by the area any field geologist could reasonably address.

INTERNATIONAL HANDBOOK OF EARTHQUAKE AND ENGINEERING SEISMOLOGY, VOLUME 81A ISBN: 0-12-440652-1

Earthquake source studies have certainly not supplanted or made redundant the older field- and seismic reflection-based methods of structural investigation, each of which have their own obvious advantages. However, by offering insights that were simply unobtainable by other means, earthquake seismology has profoundly influenced thoughts and concepts in tectonic geology over the last 20 years.

This review summarizes some of the major contributions of earthquake seismology to debates in structural geology and continental tectonics, some of which are still controversial. To keep a perspective, we should realize that earthquake seismology provides a tool that is most effective when used with other tools. Balanced studies of active tectonic processes combine seismological information with data from any other sources that are available, especially surface observations of geology, geomorphology, and geodesy (e.g., Yeats *et al.*, 1997). However, in order to highlight the contribution of earthquake seismology, I have deliberately selected examples illustrating processes that would have been extremely difficult or impossible to demonstrate from nonseismological observations alone or from the study of inactive geological structures. I invite the reader to keep in mind the question: "Could we have discovered that without the earthquakes?" Most of the examples quoted come from the Alpine–Himalayan orogenic belt, partly because I am most familiar with that region but also because it is so large, so active, and provides such a rich store to plunder. This is not a serious bias: the processes illustrated here are general and can often be recognized in the earthquake record of a number of different places.

2. Scale and the Seismogenic Layer

2.1 Focal Depth Distributions and Strength

Once well-determined focal depths were more abundant, it became clear that the confinement of continental seismicity to the upper half of the crust, which had been known in California for some time, was a common global phenomenon (e.g., Chen and Molnar, 1983). In most areas the lower continental crust is virtually aseismic. This pattern is usually interpreted in terms of a temperature-related rheology (e.g., Scholz, 1988), with the implication that the upper crust has greater long- and short-term strength than the lower crust. Until recently, it was thought that occasional rare earthquakes in the uppermost continental mantle beneath the Moho may indicate another strength contrast between the crust and the mantle related to composition (Chen and Molnar, 1983; Chen, 1988), giving rise to a popular view of continental strength profiles in which a weak lower crust is sandwiched between relatively strong layers in the upper crust and mantle. Although it has been influential for 20 years, this view has been challenged by Maggi *et al.* (2000a,b). With the help of better-determined Moho depths from teleseismic receiver functions, they find no evidence for significant seismicity in the continental mantle, though in places earthquakes occur throughout the continental crust down to Moho depths. They also found that variations in the seismogenic thickness (T_s) correlate with variations in the effective elastic thickness (T_e), both of them having similar values, though T_e is usually the smaller of the two (see also McKenzie and Fairhead, 1997). These observations suggest that, at least in some places, the lower crust is stronger than the mantle beneath the Moho, not weaker. If correct, this new view has important implications. In particular, the strength of the continental lithosphere is likely to be contained within the seismogenic layer, with variations in the thickness of this strong layer determining the heights of the mountain ranges it can support. To explain the revised earthquake depth distribution on the continents some other effect is needed in addition to temperature: Maggi *et al.* (2000a,b) suggest that the aseismic nature of the continental mantle and the lower crustal seismicity beneath some shields are probably related to their water contents.

Though the rheological implications of the earthquake focal depth distribution are profound, the more fundamental impact on structural geological thinking is probably the implication of scale. The thickness of the seismogenic upper crust is typically 10–20 km and imposes a scale on all structural observations. Large faults are those with dimensions comparable to that thickness and may rupture the entire layer in earthquakes. Small faults are those that can be accommodated entirely within the seismogenic layer. This framework gives an objective criterion for distinguishing "big" faults and earthquakes from "small" ones. Seismologists have been aware of this distinction in the scaling properties of earthquakes for some time (e.g., Shimazaki, 1981; Scholz, 1982). The geological consequences of a fundamental break in scale at the thickness of the seismogenic layer fall into two main areas.

2.2 Organization of Strain and Faulting

Most maps showing fault plane solutions of moderate-to-large ($M_w > 5.5$) shallow earthquakes on the continents are remarkable for the regular patterns they display (e.g., Figs. 1 and 2 and Color Plate 10), with focal mechanisms uniform or smoothly varying in orientation over very wide regions, with few anomalies. The organization of large active faults into subparallel or regular patterns over areas that are large compared with the lithosphere thickness is an important characteristic of distributed deformation on the continents. It is the reason why the influential early studies of active continental tectonics in the Alpine–Himalayan belt by McKenzie (1972) and Molnar and Tapponnier (1975) were not misled by the relatively sparse focal mechanism data available to them at the time. It is also the organization of the strain fields at long wavelengths that suggests the origin of the large-scale driving forces on the edges of or within the deforming

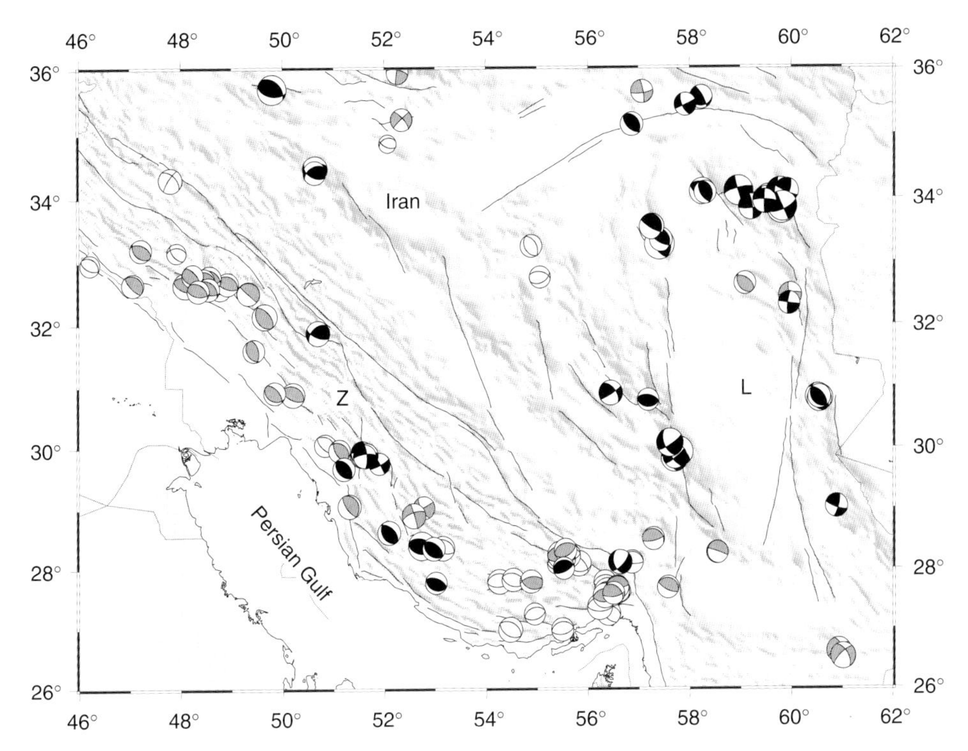

FIGURE 1 Fault plane solutions in the Iranian plateau. Black focal spheres are those constrained by body wave modeling. Dark gray spheres are Harvard CMT solutions for additional earthquakes with $M_w \geq 5.3$ and with more than 70% double-couple component (see Appendix on the attached Handbook CD). Light-gray focal spheres are other earthquakes whose mechanisms are based on long-period first-motion polarities. Main active fault trends are from Berberian and Yeats (1999). Important patterns include: (1) reverse faulting parallel to the topographic trend of the Zagros mountains (Z); (2) NW–SE right-lateral faulting along the NE edge of the northern Zagros (near 34°N 48°E), with slip vectors orthogonal to the nearby thrusts; (3) and N–S right-lateral faulting both east and west of the Dasht-e-Lut (L), changing to E–W left-lateral faulting further north; (4) NW–SE trending reverse faults throughout eastern Iran.

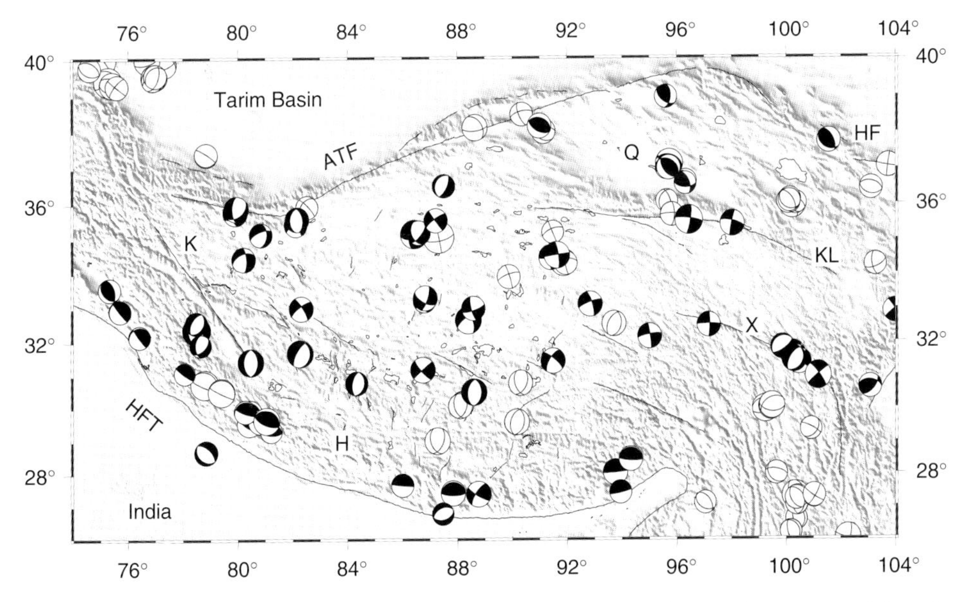

FIGURE 2 Fault plane solutions and active faults in and around Tibet, adapted from Molnar and Lyon-Caen (1989). Black focal spheres are those constrained by body wave modeling. Gray focal spheres are Harvard CMT solutions for additional earthquakes with $M_w \geq 5.3$ and with more than 70% double-couple component (see Appendix on the attached Handbook CD). Important patterns here include: (1) low-angle thrust faults along the Himalayan thrust front (HTF) with divergent slip vectors following the arc of the Himalaya (H); (2) N–S normal faulting, particularly in southern Tibet south of a band of right-lateral faulting running east from the Karakoram Fault (K); (3) a subparallel system of left-lateral faults in eastern Tibet including the Haiyuan (HF), Kun Lun (KL) and Xian Shuihe (X) faults; (4) NW–SE reverse-fault solutions near the Qaidam basin (Q). The Altyn Tagh fault is marked ATF.

regions (e.g., England and McKenzie, 1982; Vilotte *et al.*, 1982).

Such organization is not always seen in small faults and small earthquakes. The focal mechanisms of many small earthquakes are often similar to those of the nearby large ones, but anomalies are quite common, particularly in aftershock sequences, where normal faulting aftershocks can follow reverse faulting mainshocks (e.g., Whitcomb *et al.*, 1973) and vice versa (e.g., Lyon-Caen *et al.*, 1988). Some of these anomalies are no doubt related to local strain inhomogeneities around the main faults, caused by irregularities in shape, slip distribution or fault terminations, and are also seen in minor coseismic ruptures at the surface. Examples include normal faulting in hanging wall anticlines above thrusts (Yielding *et al.*, 1981; Philip *et al.*, 1992) and minor normal faulting perpendicular to major normal faults where the large ones are arranged in en echelon steps or relays (e.g., Jackson *et al.*, 1982). Such instances produce little confusion when the configuration of the larger faults is already known, but small faults on their own may not be a reliable guide to the regional strain pattern. An awareness of the size of the fault compared with the thickness of the seismogenic layer is essential when assessing its tectonic significance.

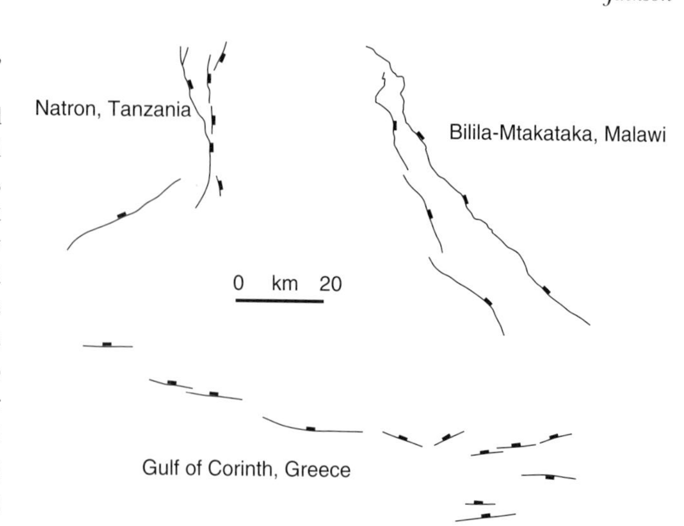

FIGURE 3 A comparison of normal fault segmentation in parts of Tanzania (Foster *et al.*, 1997) and Malawi (Jackson and Blenkinsop, 1997) where earthquakes occur to depths of 25–30 km, and central Greece, where earthquakes are limited to the upper 10–15 km of the crust. In Greece, maximum segment lengths are ~ 25 km, whereas in Tanzania and Malawi they reach lengths of 60–100 km. These examples have been chosen because the author is familiar with all of them in the field, and has used the same criteria for identifying segment terminations in each case.

2.3 Segmentation and Basin Width

The thickness of the seismogenic layer also appears to influence the maximum size of some structures that form within it. This is seen best in the segmentation of normal faults and width of graben in extensional provinces.

The normal fault systems that bound continental rifts are commonly segmented, stepping en echelon or changing the polarity of half graben along strike (e.g., Rosendahl *et al.*, 1986; Crone and Haller, 1991; Roberts and Jackson, 1991). The maximum segment length is a characteristic of any region and appears to correlate with the seismogenic thickness (T_s), being typically ~ 20–25 km long where $T_s \approx 15$ km (Jackson and White, 1989; Wallace, 1989) but increasing to 60–100 km in parts of East Africa (Fig. 3) where, unusually, T_s reaches 30–35 km (Jackson and Blenkinsop, 1997; Foster *et al.*, 1997). Fault-bounded topography within the elastic upper crust is supported by stresses that depend on the wavelength and height of the topographic contrasts as well as on the layer thickness. Thus the seismogenic thickness can be expected to control the maximum width of coherently tilted blocks or half graben if the shear stresses acting on the faults are not to exceed the typical range of 1–10 MPa stress drops seen in earthquakes (Jackson and White, 1989; Foster and Nimmo, 1996). Observations support such a correlation, with maximum half-graben widths of typically 10–25 km where $T_s \approx 15$ km but much greater widths of up to 60 km in those parts of East Africa where $T_s \approx 30$–35 km (Jackson and Blenkinsop, 1993). Thus the thickness of the seismogenic layer seems to control both the maximum width of extensional graben and the continuity of the faults that bound them (Jackson and White, 1989; Hayward and Ebinger, 1996). This influence arises because the thickness of the seismogenic layer controls the scale of the tilting, limiting the displacement on the bounding faults, and thus limiting their length through a displacement–length scaling relationship for faults (Scholz and Contreras, 1998).

It is possible that similar relationships exist for reverse and strike–slip faults. Segmentation can be seen in both types, both in co-seismic surface ruptures and in the body waveforms of earthquakes (e.g., Yielding 1985; Haessler *et al.*, 1992; Sieh *et al.*, 1993; Wald and Heaton, 1994; Berberian *et al.*, 1999). However, the topography associated with strike–slip and particularly with blind reverse faults is often less clear than for normal faults, making objective identification of segment boundaries more difficult in the absence of earthquakes.

3. Fault Dips

3.1 Normal Faults

Ever since the discovery that many continental sedimentary basins are formed by extension (McKenzie, 1978a), there has been intense interest in the geometry and kinematics of normal faulting on the continents. A great variety of normal faulting styles have been seen, or inferred, from geological outcrops or seismic reflection surveys, including steep planar faults, strongly listric faults that become virtually flat within the upper crust, and very low-angle faults dipping at less than $\sim 20^\circ$ (e.g., Wernicke and Burchfiel, 1982). Observations of large normal faults that move in earthquakes show a very simple pattern of dips that are apparently restricted to a range of about 30–65° (Fig. 4a), whether determined by first motion polarities or by body wave modeling (Jackson, 1987; Jackson and White, 1989; Thatcher and Hill, 1991). For coseismic ruptures that break the Earth's surface, the agreement between the centroid dip and the slip vector measured at outcrop makes significant curvature between the surface and the base of the seismogenic layer unlikely (e.g., Jackson and White, 1989), an inference that can be confirmed in some cases by more sophisticated body wave modeling (Braunmiller and Nábělek, 1996) or by coseismic elevation changes (Stein and Barrientos, 1985). It seems probable that the great majority of large, seismically active normal faults on the continents are roughly planar throughout the seismogenic layer, dipping typically at $45 \pm 15^\circ$. The spread in observed dips may be related to the rotation of these faults about a horizontal axis as they move (Jackson and White, 1989), and is entirely compatible with simple views of fault mechanics involving Byerlee friction, vertical and horizontal principal stress orientations, and pore fluid pressures up to, but not exceeding, the least principal stress (e.g., Sibson, 1994). Genuine listric (concave-up) normal faults are common within thick sedimentary sequences, often flattening onto weak horizons of overpressured shale or salt, but they appear not to generate earthquakes when they move. Generalizations of this sort have proved useful in hydrocarbon exploration, as they give some indication of what to expect, depending on whether the large faults of interest are in crystalline basement or in thick sediment sequences containing decoupling horizons.

By contrast, normal faulting earthquakes with dips less than 20°, corresponding to slip on very low-angle "detachment" surfaces of the sort inferred to have been active in the Miocene metamorphic core complexes of the western North America (Wernicke, 1981), are rare (e.g., Abers, 1991). Reconciling this observation with the proposed origin of these currently shallow-dipping surfaces remains controversial (e.g., Jackson, 1987; Wernicke, 1995), with suggestions including the rotation of the surfaces during movement by extreme flexure or lower-crustal flow (e.g., Buck, 1988; Wernicke and Axen, 1988; McKenzie *et al.*, 2000). The point to make here is simply that this subject would not be controversial at all were it not for the observations from earthquakes.

3.2 Thrusts and Reverse Faults

Earthquakes on active thrust and reverse faults on the continents show a much wider range of dips than do active normal faults. Perhaps the most important point here is that centroid

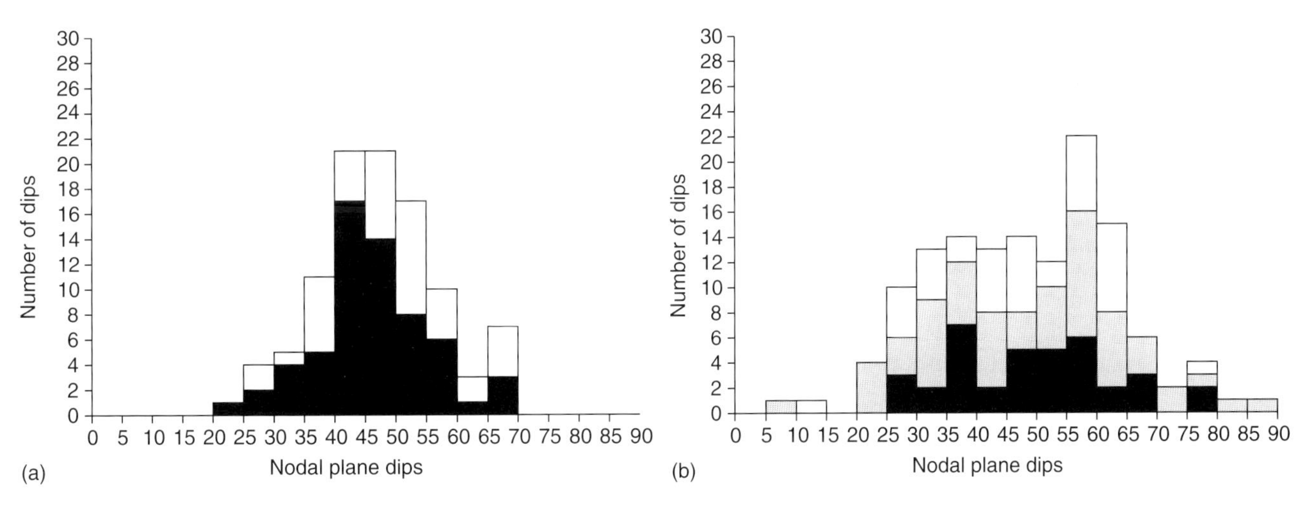

FIGURE 4 (a) Histogram of nodal plane dips for normal faulting earthquakes in the Aegean extension province between 34–44°N 20–32°E. Dips from mechanisms constrained by body wave modeling are in black. In white are dips from Harvard CMT solutions for additional earthquakes with $M_w \geq 5.3$ and with more than 70% double-couple component (see Appendix on the attached Handbook CD). The histogram contains only mechanisms with rakes in the range of -60° to -120° (corresponding to slip vectors within $\pm 30^\circ$ of pure dip slip), and in each case both nodal planes have been included. The concentration between 30° and 65° is a global feature, as is the rarity or absence of very-low-angle dips of less than $\sim 20^\circ$ (Jackson and White, 1989). (b) Histogram of nodal plane dips for reverse or thrust-faulting earthquakes within the Iranian plateau (26–36°N, 46–62°E) (region of Fig. 1). Selection of earthquakes is on the same criteria as in (a), but with the rake range $+60^\circ$ to $+120^\circ$. Dips from body-wave mechanisms are in black, additional earthquakes with Harvard CMT mechanisms are in gray, and other earthquakes with first motion mechanisms are in white. The most common dips are in the range 25–70°, similar to those of normal faults in (a). In this case, however, the lack of low-angle dips is not a global feature, as can be seen in Figure 2 and Color Plate 10.

depths are typically in the range 10–20 km, which clearly implies faulting in "basement" rocks in most regions. There is much interest in how such earthquakes relate to classical concepts of 'thin-skinned' fold-and-thrust belts, in which sediments are decoupled and thrust over an apparently undeformed basement, as described, for example, in the foreland of the Rocky Mountains (Bally *et al.*, 1966). In these thin-skinned systems the thrusts typically follow incompetent horizons at shallow dips ("flats"), cutting up through more competent ones at a steeper angle on reverse faults or "ramps."

Low angle (10–20° dip) thrusts are seen to move in earthquakes on the continents, but are relatively rare (see below). In Asia they occur particularly along the southern margin of the Himalaya (e.g., Molnar and Lyon-Caen, 1989) and on both sides of the Greater Caucasus (Fig. 2 and Color Plate 10). In the case of the Himalaya, such earthquakes occur at depths of 10–15 km, but with epicenters 50–100 km north of the surface outcrop of thrusts in the deforming foreland basin. Particularly rapid uplift occurs in the epicentral regions of these earthquakes (Jackson and Bilham, 1994), and one interpretation is that they represent slip on a basement fault (or ramp) dipping 10–20° that becomes flatter (<10° dip) at shallower depth and is currently aseismic (e.g., Ni and Barazangi, 1984; Molnar and Lyon-Caen, 1988). There is some debate about the precise nature of the very-low-angle surface or decoupling horizon at shallow depths, the degree to which it is locked or creeps aseismically, and how it moves in occasional great earthquakes (e.g., Seeber and Armbruster, 1981; Molnar and Lyon-Caen, 1988; Yeats *et al.*, 1997). But the elements of the "ramp-and-flat" geometry of classic fold-and-thrust belts are at least recognizable, and the basement beneath the shallow decoupling horizon may be relatively undeformed.

More problematic are the earthquakes on relatively steep (30–60° dip) reverse faults, which are actually more common on the continents than earthquakes on gently dipping (10–20°) thrusts (e.g., Jackson and Fitch, 1981; Molnar and Chen, 1982; Triep *et al.*, 1995; Sibson and Xie, 1998). Steep reverse-faulting earthquakes with centroids at depths of 5–20 km occur in many parts of central Asia, Iran (Fig. 4b), southern California, the eastern margin of the Andes, New Guinea and in many intraplate settings. In several places, such as the Zagros of Iran (Jackson and Fitch, 1981; Berberian, 1995), Papua New Guinea (Abers and McCaffrey, 1988) and the Peruvian Andes (Suarez *et al.*, 1983), these seismogenic reverse faults rarely cut the surface, and there is evidence that the steep basement faults are decoupled from their overlying deforming sediments by a weak horizon, such as salt or shale. In these examples, although the deformation in sediments at shallow depths is probably not connected to the basement by through-going faults, it is misleading to think of the system as "thin-skinned" in the same sense as it was in the Rockies or Appalachians, as the basement beneath the sediments is clearly deforming as well. In other places, the steep reverse faults reach, or nearly reach, the surface in earthquakes, often forming anticlines in their hanging walls. In several instances a combination of surface observations, seismology, and sometimes geodesy can be used to demonstrate that these faults remain steep and approximately planar throughout the seismogenic upper crust (e.g., Whitcomb *et al.*, 1973; Yielding *et al.*, 1981; Haessler *et al.*, 1992; Wald *et al.*, 1996).

An important issue with all these steep reverse faults is what happens below the base of the seismogenic zone. It is possible that the steep faults merge with a low-angle, aseismic decoupling horizon within the plastic lower crust (e.g., Namson and Davis, 1988; Shaw and Suppe, 1994). In this view, the ramp-and-flat configuration is retained, with the entire seismogenic upper crust being above the "flat." However, the existence of such low-angle aseismic horizons at depth is conjectural, and it is possible that the reverse faults continue into the lower crust as ductile shear zones, retaining their relatively steep dip. This is a significant issue, as the assumed geometry at depth influences palinspastic reconstructions and the estimates of shortening rates obtained from near-surface folds and faults (Yeats, 1993; Yeats *et al.*, 1997).

Other issues are concerned with why such steep reverse faults should be so common, and how they move at dips that are severely misoriented for frictional reactivation (Sibson and Xie, 1998). Where the reverse faults occur in sedimentary basins that were originally formed by extension, they may represent the reactivation of pre-existing normal faults in the basement (e.g., Jackson, 1980), which are expected to have similar dips (Fig. 4). Where sediment thickness variations across the faults can be seen on reflection profiles, as in the Late Cretaceous inversion province of the North Sea (e.g., Badley *et al.*, 1989), this reactivation mechanism can be verified, but it remains conjectural in areas that are still active.

Thus in the case of fold-and-thrust belts, seismology has focused attention on the fate of the basement, which had been rather neglected by earlier geological investigations that understandably concentrated on the commercially important sedimentary cover. Perhaps the most striking generalization is that any particular mountain belt shows a consistent tendency towards either high-angle reverse (e.g., Zagros, Tien Shan) or low-angle thrust (Himalaya, Greater Caucasus) dips in earthquakes. It is relatively rare to find both together in the same epicentral region. Possible examples of seismic activation of both a "ramp" and a "flat" in subevents of the same earthquake have been suggested (e.g., Nábělek, 1985; Stevens *et al.*, 1998), based on the interpretation of complex body waveforms.

4. The Importance of Earthquake Slip Vectors

Precise slip vector directions on faults are very difficult to determine on old, inactive structures, yet are relatively easy to obtain from earthquakes. Several insights have arisen from the patterns of earthquake slip vector directions on faults over wide-deforming regions. The important conclusion to come out of the examples below is that, as far as big faults are concerned, it is the organization of the slip vectors in a deforming region that is the essential characteristic of the deformation, and the stresses simply adjust to produce slip in the right direction. Scholz (2000) reaches a similar conclusion for the San Andreas fault system in California.

4.1 Partitioning of Thrust and Strike-Slip Motion

The spatial separation, or "partitioning," of strike-slip and thrust components of motion in obliquely converging belts onto parallel fault systems with orthogonal slip vectors was first noticed using earthquakes in island arcs (Fitch, 1972), but can also be seen on the continents. On a relatively local scale, where the strike-slip and thrust faults are only a few kilometers apart, such separation can be seen in fault plane solutions in California (Mount and Suppe, 1987; Ekström *et al.*, 1992), Irian Jaya (Abers and McCaffrey, 1988), Pakistan (Nakata *et al.*, 1990) and Iran (Berberian *et al.*, 1992, and Color Plate 10). It may also occur on a much larger scale, for example, in eastern Turkey and the Caucasus (Color Plate 10), where a deforming zone with a NW–SE strike accommodates roughly north–south convergence between Arabia and Eurasia (Jackson, 1992). Earthquakes in the southern part of this zone, in the Turkey–Iran border region, occur mostly on right-lateral strike-slip faults trending parallel to the zone, with slip vector directions west of the overall convergent motion (Color Plate 10). In the eastern half of the Greater Caucasus, the earthquakes occur mainly on thrusts, with slip vectors directed east of the overall motion, and roughly perpendicular to those on the strike-slip faults. In the intervening region of the Lesser Caucasus the deformation is more complicated (Jackson, 1992), but the spatial separation of the slip vectors and earthquake mechanisms over this broad region of $\sim 600 \times 600\,\text{km}^2$ is quite clear. GPS measurements confirm this separation (Color Plate 10), with velocities of sites in Armenia directed NE (reflecting the thrusting in the Greater Caucasus), and velocities only achieving the NNW direction relative to Eurasia at sites in southeast Turkey, within the strike-slip belt.

In kinematic terms, strain partitioning presents no difficulties, as it is simply one way of accommodating oblique shortening or extension. However, the dynamic reasons for it remain controversial (e.g., Molnar, 1992).

4.2 Rotations about a Vertical Axis

In the late 1970s it became apparent from paleomagnetic studies that large systematic rotations of crustal blocks about vertical axes were a common feature of distributed deformation on the continents (e.g., Beck, 1976; Luyendyk *et al.*, 1980). The way in which such rotations are achieved by faulting, and the relation between the fault geometry and the overall deforming velocity field are of great interest, because they potentially contain information about whether the deformation of the upper crust in such regions is controlled by the strength of the rigid fault-bounded blocks within it or by distributed flow in the aseismic lithosphere beneath (e.g., McKenzie and Jackson, 1983; England and Jackson, 1989). These issues are still controversial (e.g., Jackson, 1994; Molnar and Gipson, 1994; Bourne *et al.*, 1998) but, at least in some regions, we can now see how the faulting achieves the rotations, largely through observations of slip vectors.

For example, the earthquakes in Color Plate 10 show how NE–SW right-lateral strike-slip faulting in the northern Aegean Sea changes into a system of E–W normal faults in central Greece. The strike-slip faulting represents a continuation of the right-lateral shear on the North Anatolian fault system of Turkey, allowing the SW motion of the southern Aegean relative to Eurasia. The strike-slip motion in the northern Aegean must connect with the thrusting in the Hellenic Trench to the SW, or central Greece would be shortening, not extending (e.g., McKenzie, 1972). Thus the overall deformation in central Greece must involve a distributed NE-SW right-lateral shear, which can be seen in the velocity fields obtained from the seismicity (Jackson *et al.*, 1992) and, more accurately, from GPS measurements [(Color Plate 10); Clarke *et al.*, 1998; McClusky *et al.*, 2000]. In a conventional transform fault setting, the slip vectors on the strike–slip and normal faults would be in the same direction, but in Greece they are very different (Color Plate 10), with those on the active normal faults having azimuths in the range S to SSE rather than SW (Taymaz *et al.*, 1991). How can slip vectors to the south achieve an overall motion to the SW? The answer must be that the fault-bounded blocks rotate clockwise as they move, an inference that is supported by paleomagnetic measurements on Mio-Pliocene rocks in the region (Kissel and Laj, 1988). Various simplistic models attempt to illustrate the relation between the faulting and the velocity field (e.g., McKenzie and Jackson, 1983, 1986; Taymaz *et al.*, 1991), all of them suffering from limitations. The point to emphasize here is that it is the slip vectors on the normal faults, apparently in the wrong direction to achieve the overall motion, which indicate that rotations about vertical axes must occur.

Similar arguments have been made elsewhere. In eastern Iran and eastern Tibet the presence of E–W left-lateral faults in regions of probable N–S right-lateral shear strongly suggests that the fault-bounded blocks rotate clockwise (Jackson and McKenzie, 1984; England and Molnar, 1990). In the Western Transverse Ranges of California E–W thrust faults have slip vectors directed N to NE, in the wrong direction to accommodate the NW–SE right-lateral shear that dominates the regional velocity field (Jackson and Molnar, 1990; Molnar and Gipson, 1994). Once again, clockwise rotations, which in this case are well documented by paleomagnetic declinations (Hornafius *et al.*, 1986), can reconcile the faulting with the velocity field.

4.3 Arc-Parallel Extension

The thrust fronts of both the Himalaya (Fig. 2) and the Hellenic arc of the Aegean are strongly curved. In both places slip vectors on low-angle thrusts are directed radially outward around the arc. Such motion requires an along-strike extension on normal faults on the concave side of the thrust front if the underthrusting block is rigid (Molnar and Lyon-Caen, 1989; Hatzfeld *et al.*, 1993). The along-strike extension of the Himalayan thrust front may also be related to strike-slip faulting in southern Tibet, if the entire concave side is analogous to the forearc region in an obliquely converging subduction zone (McCaffrey and Nábělek, 1998). In each case it is the slip vectors that reveal the relationship between the shortening and extension, which would otherwise be obscure. As McCaffrey and Nábělek (1998) point out, it is important to understand such geometric relationships and their consequences before speculating on the dynamic causes of the extension; in their interpretation the normal faulting in southern Tibet need not be related to its altitude, although the association of the style of strain with the topography is nonetheless striking.

5. Fault Scaling, Populations, and Growth

An important aspect of seismology is the ability to study earthquakes, and hence active faults, over a large range of scales. Two important generalizations stand out. First, the population distribution, or relative abundance of active fault sizes, represented by the Gutenberg–Richter frequency–magnitude (or frequency–moment) relation, is well known, and varies little regionally. Second, the ratio of coseismic displacement to length on a fault, which is a measure of the coseismic strain drop, also varies very little (typically between 10^{-4} and 10^{-5}) over a large range of earthquake sizes (e.g., Scholz, 1990). These two observations make it easy to sum the total seismic moment or strain released in a region, as the deficit caused by ignoring earthquakes smaller than a certain size can easily be calculated (e.g., Molnar, 1979; Scholz and Cowie, 1990).

An equivalent geological problem may be, for example, to calculate the total extension along a regional seismic reflection line across a basin, knowing that faults below the level of resolution cannot be detected (e.g., Marrett and Allmendinger, 1992). (This problem triggered an economically important and vigorous debate in the North Sea in the 1980s: see Ziegler 1983.) As in the case of earthquakes, the answer requires knowledge of both the population (relative size) distribution of the faults and a scaling law relating total fault displacement to fault size. But such observations are much harder to obtain from geology because of the difficulty of sampling over size ranges of more than one or two orders of magnitude in the field (e.g., Dawers *et al.*, 1993). As a result, issues related to the scaling of cumulative slip on geological faults remain less certain and more controversial than those related to incremental slip in earthquakes (e.g., Marrett and Allmendinger, 1992; Cowie and Scholz, 1992a).

Nonetheless, this interest in the scaling of faults spawned a fruitful new area of research aimed at understanding how faults grow with time. Various models of fault growth have been produced (Watterson, 1986; Walsh and Watterson, 1988; Cowie and Scholz, 1992b,c) all of which are based on, or constrained by, an attempt to reconcile observations of slip in earthquakes with those of cumulative displacement on faults. As a result, structural geologists are now concerned not just with the pattern of active faults seen in the field, but with how it evolved to its present configuration (e.g., Wu and Bruhn, 1994; Jackson and Leeder, 1994; Jackson *et al.*, 1996; Keller *et al.*, 1998, 1999). It is hard to see how this subject could have developed without a knowledge of earthquake scaling laws.

6. Large-scale Continental Tectonics

Finally, seismology has contributed to the discovery of a number of important large-scale processes in continental tectonics. One of the earliest was the idea that continental material could be expelled sideways from a collision zone on strike-slip faults, thereby accommodating shortening while avoiding crustal thickening. This process was first noticed in Turkey (McKenzie, 1972), where the existence of single right-lateral (North Anatolian) and left-lateral (East Anatolian) fault systems bounding a rigid block (central Turkey) made it quite unambiguous and where it has subsequently been confirmed by GPS measurements (McClusky *et al.*, 2000). It also occurs to some extent in central Asia (e.g., Molnar and Tapponnier, 1975), but is more controversial there because the strike-slip faults occur in subparallel sets within areas that are deforming, allowing the alternative possibility that rotation about vertical axes is important (e.g., England and Molnar, 1990). Resolution of this doubt in Asia will eventually come from better knowledge of the overall velocity field, as in Greece (Color Plate 10). Such velocity fields are best obtained by GPS measurements, but it is worth noting that the first estimates of the deforming velocity fields in Asia, Greece, and Iran were obtained not by geodetic measurements, but from the spatial distribution of seismic strain rates obtained from earthquake catalogues (Holt *et al.*, 1991; Holt and Haines, 1993; Jackson *et al.*, 1992, 1995).

Another important process discovered using earthquakes is the extensional origin of many sedimentary basins on the continents (McKenzie, 1978a). This idea, which now dominates the approach to the geological and hydrocarbon exploration of most passive continental margins and many intracontinental basins, came from noticing the association between active normal faulting, crustal thickness variations and heat flow in the Aegean Sea region (McKenzie, 1978b).

A final example, notable because of the obvious difficulty of proving this from the older geological record, is the localization of active faulting and high seismic strain rates near topographic gradients or, more precisely, near gradients in crustal thickness. This is noticeable in most deforming continental regions (e.g., Figs. 1 and 2), particularly in Iran and central Asia (e.g., Jackson and McKenzie, 1984; England and Houseman, 1986), and has important consequences for the dynamics as it suggests that buoyancy forces resulting from crustal thickness contrasts are an important driving force within the continents (e.g., England and McKenzie, 1982; England and Molnar, 1997).

7. The Future: Is Source Seismology Dead?

Earthquake seismology has contributed much to our understanding of structural processes, but now that other techniques have evolved that can also image slip on active faults on land, such as GPS (e.g., Hudnut *et al.*, 1996) and radar interferometry (e.g., Massonnet *et al.*, 1993), it is worth asking whether it has any more to offer, other than coverage of underwater areas. I believe it has, because seismology will always provide a faster and more easily accessible image of regional fault patterns than geodesy or radar, and because there is still a need to look at such patterns. Deforming velocity fields obtained with GPS are certainly more accurate and have better resolution than those obtained from seismicity, but on their own they offer an incomplete picture of what is happening. A complete description of the kinematics requires knowledge, not just of the overall velocity field, but also of how it is achieved by faulting (e.g., McKenzie and Jackson, 1983). Since the faulting can do this in a variety of ways, there is no alternative but to look, and seismology provides the best way of doing so. For example, the distributed shear that passes through central Greece (Color Plate 10) could be achieved by NE–SW strike-slip faults but is in fact achieved by normal faults that rotate clockwise. Knowing this leads to other questions, such as why the faulting has the orientation it does, what controls its sense and rate of rotation, and how the fault system evolves with time. Questions of this sort have dynamical implications, in particular concerning the relation between discontinuous deformation in the seismogenic upper crust and the (probably) more distributed flow in the aseismic lithosphere beneath (e.g., Jackson, 1994; Molnar and Gipson, 1994; Bourne *et al.*, 1998).

It is the ability of seismology to look at the pattern of active faulting over large regions that will remain its strength. The impressive organization of active faulting at a regional scale was noted at the start of this review. In addition, evidence is now mounting that when the fault patterns change they do so in an organized and rapid way. Once again, Greece provides an example (Color Plate 10). The northwestern Aegean and eastern seaboard of central Greece has a strong NW–SE structural fabric inherited from early Tertiary deformation. This region contains many large normal faults of this orientation that were active in the Quaternary and dominate the trend of the coastline. Yet, almost without exception, all the normal faulting earthquakes with well-constrained mechanisms show E–W striking nodal planes, cutting across this NW–SE trend [(Color Plate 10), Hatzfeld *et al.*, 1999]. This evidence, combined with geomorphological and structural evidence in the field, strongly suggests that activity moved from the NW–SE faults to the E–W faults within the Quaternary (e.g., Caputo and Pavlides, 1993; Leeder and Jackson, 1993; Hatzfeld *et al.*, 1999). What is remarkable is how quickly, and how completely, this change has occurred.

8. Conclusions

For many people, evidence from earthquake seismology has now become an unremarkable part of the structural geologist's armoury. The extent to which this has happened is best gauged by books such as *The Geology of Earthquakes* by Yeats *et al.*, (1997), and by the acceptability of questions that would not have been asked 20 years ago, such as: (1) is this structure large or small compared to the thickness of the seismogenic crust? (i.e., is it a big or a small fault?); (2) is this structure in the same orientation as when it was active, or has it been affected by rotations (about vertical or horizontal axes) or uplift?; (3) is there a modern analog of the structure or style of deformation being inferred?

This absorption of seismological insights into structural and tectonic geology is relatively new. Throughout the 1950s and early 1960s there were still arguments about whether all earthquakes occurred on faults, whether mountain building was still occurring today, and whether earthquakes and mountain building were connected at all. Most of this conceptual uncertainty was swept away by the advent of plate tectonics and by the demonstration by Burridge and Knopoff (1964) that slip on a fault generated a double-couple radiation field. But for a long time, perhaps uniquely among the traditional geological disciplines, structural geology was hampered by not being able to exploit the uniformitarian maxim that "the present is the key to the past," because the present had nothing to offer. With the development of new seismological techniques and the accumulation of better earthquake data in the 1970s the situation finally changed and we could, for the first time, say "here's how faults are moving now: how does that help us interpret the geological record?" Not surprisingly, many processes are much easier to see while they are active than after they have ceased. But there is no reason to believe that the processes we see active today were not also active in the past.

Acknowledgments

I would like to use this opportunity to apologize to all those readers whose work might have been quoted here, but isn't. In any review of this sort the selection of examples is somewhat arbitrary and, by concentrating on geographic areas with which I am most familiar and which I am most confident of discussing, I am aware there may be an appearance of bias. I can only assure those whose research in other regions might reasonably have been quoted also, that their omission arises from my timidity or ignorance, and not from any undervaluing of their work. I am grateful to Dan McKenzie for providing some historical perspective and to Philip England and Bob Yeats for reviews, though I alone am responsible for any misconceptions or errors. Cambridge Earth Sciences contribution 6012.

References

Abers, G. (1991). Possible seismogenic shallow-dipping normal faults in the Woodlark-D'Entrecasteaux extensional province, Papua New Guinea. *Geology* **19**, 1205–1208.

Abers, G. and R. McCaffrey (1988). Active deformation in the New Guinea fold-and-thrust belt: seismological evidence for strike-slip faulting and basement-involved thrusting. *J. Geophys. Res.* **93**, 13332–13354.

Badley, M.E., J.D. Price, and L.C. Backshall (1989). Inversion, reactivated faults and related structures: seismic examples from the North Sea. In: "Inversion Tectonics," (M.A. Cooper and G.D. Williams, Eds.), *Geol. Soc. Spec. Publ. London* **44**, 201–219.

Bally, A.W., P.L. Gordy, and G.A. Stewart (1966). Structure, seismic data, and orogenic evolution of the southern Canadian Rocky Mountains, *Can. J. Earth Sci.* **3**, 713–723.

Beck, M.E. (1976). Discordant paleomagnetic pole positions as evidence for regional shear in the western Cordillera of North America. *Am. J. Sci.* **276**, 694–712.

Berberian, M. (1995). Master "blind" thrust faults hidden under the Zagros folds: active basement tectonics and surface morphotectonics. *Tectonophysics* **241**, 193–224.

Berberian, M. and R.S. Yeats (1999). Patterns of historical earthquake rupture in the Iranian plateau. *Bull. Seismol. Soc. Am.* **89**, 120–139.

Berberian, M., M. Qorashi, J.A. Jackson, K. Priestley, and T. Wallace (1992). The Rudbar–Tarom earthquake of 20 June 1990 in NW Persia: preliminary field and seismological observations, and its tectonic significance. *Bull. Seismol. Soc. Am.* **82**, 1726–1755.

Berberian, M., J.A. Jackson, M. Qorashi, M.M. Khatib, K. Priestley, M. Talebian, and M. Ghafuri-Ashtiani (1999). The 1997 May 10 Zirkuh (Qa'enat) earthquake (M_w 7.2): faulting along the Sistan suture zone of eastern Iran. *Geophys. J. Int.* **136**, 671–694.

Bourne, S., P. England, and B. Parsons (1998). The motion of crustal blocks driven by flow of the lower lithosphere: implications for slip rates of faults in the South Island of New Zealand and Southern California. *Nature* **391**, 655–659.

Braunmiller, J. and J. Nábělek (1996). Geometry of continental normal faults: seismological constraints. *J. Geophys. Res.* **101**, 3045–3052.

Buck, W.R. (1988). Flexural rotation of normal faults. *Tectonics* **7**, 959–973.

Burridge, R. and L. Knopoff (1964). Body force equivalence for seismic dislocations. *Bull. Seismol. Soc. Am.* **57**, 341–371.

Caputo, R. and S. Pavlides (1993). Late Cainozoic geodynamic evolution of Thessaly and surroundings (central-northern Greece). *Tectonophysics* **223**, 339–362.

Chen, W.-P. (1988). A brief update on the focal depths of intracontinental earthquakes and their correlations with heat flow and tectonic age. *Seismol. Res. Lett.* **59**, 263–272.

Chen, W.-P. and H. Kao (1996). Seismotectonics of Asia: some recent, progress. In: "The Tectonic Evolution of Asia," (An Yin and T.M. Harrison, Eds.), pp. 37–62, Cambridge University Press, Cambridge.

Chen, W.-P. and P. Molnar (1983). Focal depths of intra-continental and intraplate earthquakes and their implications for the thermal and mechanical properties of the lithosphere. *J. Geophys. Res.* **88**, 4183–4214.

Clarke, P., *et al.* (14 authors) (1998). Crustal strain in central Greece from repeated GPS measurements in the interval 1989–1997. *Geophys. J. Int.* **135**, 195–214.

Cowie, P.A. and C.H. Scholz (1992a). Displacement–length scaling relationship for faults: data synthesis and discussion. *J. Struct. Geol.* **14**, 1149–1156.

Cowie, P.A. and C.H. Scholz (1992b). Growth of faults by accumulation of seismic slip. *J. Geophys. Res.* **97**, 11085–11095.

Cowie, P.A. and C.H. Scholz (1992c). Physical explanation for the displacement–length relationship of faults using a post-yield fracture mechanics model. *J. Struct. Goel.* **14**, 1133–1148.

Crone, A.J. and K.M. Haller (1991). Segmentation and the coseismic behaviour of Basin and Range normal faults: examples from east-central Idaho and southwestern Montana, USA. *J. Struct. Geol.* **13**, 151–164.

Dawers, N.H., M.H. Anders, and C.H. Scholz (1993). Growth of normal faults: displacement–length scaling. *Geology* **21**, 1107–1110.

Dziewonski, A.M., T.-A. Chou, and J.H. Woodhouse (1981). Determination of earthquake source parameters from waveform data for studies of global and regional seismicity. *J. Geophys. Res.* **86**, 2825–2852.

Ekström, G. (1989). A very broad band inversion method for the recovery of earthquake source parameters. *Tectonophysics* **166**, 73–100.

Ekström G., R.S. Stein, J.P. Eaton, and D. Eberhart-Phillips (1992). Seismicity, and geometry of a 110-km-long blind thrust fault; 1. The 1985 Kettleman Hills, California, earthquake. *J. Geophys. Res.* **97**, 4843–4864.

England, P.C. and G.A. Houseman (1986). Finite strain calculations of continental deformation: 2. Comparison with the India–Eurasia collision zone. *J. Geophys. Res.* **91**, 3664–3676.

England, P. and J. Jackson (1989). Active deformation of the continents. *Annu. Rev. Earth Planet. Sci.* **17**, 197–226.

England, P.C. and D. McKenzie (1982). A thin viscous sheet model for continental deformation. *Geophys. J. R. Astron. Soc.* **70**, 295–321.

England, P.C. and P. Molnar (1990). Right-lateral shear and rotation as the explanation for strike-slip faulting in eastern Tibet. *Nature* **344**, 140–142.

England, P. and P. Molnar (1997). Active deformation of Asia: from kinematics to dynamics. *Science* **278**, 647–650.

Fitch, T.J. (1972). Plate convergence, transcurrent faults and internal deformation adjacent to southeast Asia and the western Pacific. *J. Geophys. Res.* **77**, 4432–4460.

Foster, A. and F. Nimmo (1996). Comparisons between the rift systems of East Africa, Earth, and Beta Regio, Venus. *Earth Planet. Sci. Lett.* **143**, 183–195.

Foster, A., C. Ebinger, E. Mbede, and D. Rex (1997). Tectonic development of the northern Tanzania sector of the East African Rift system. *J. Geol. soc. Lond.* **154**, 689–700.

Haessler, H., A. Deschamps, H. Dufumier, H. Fuenzalida, and A. Cisternas (1992). The rupture process of the Armenian earthquake from broad-band teleseismic records, *Geophys. J. Int.* **109**, 151–161.

Hatzfeld, D., M. Besnard, K. Makropoulos, and P. Hatzidimitriou (1993). Microearthquake seismicity and fault plane solutions in the the southern Aegean and its geodynamic implications. *Geophys. J. Int.* **115**, 799–818.

Hatzfeld, D., M. Ziazia, D. Kementzetzidou, P. Hatzidimitriou, D. Panagiotopoulos, K. Makropoulos, P. Papadimitriou, and A. Deschamps (1999). Microseismicity and focal mechanisms at the western termination of the North Anatolian Fault and their implications for continental tectonics. *Geophys. J. Int.* **137**, 891–908.

Hayward, N.J. and C.J. Ebinger (1996). Variations in along-axis segmentation of the Afar rift system. *Tectonics* **15**, 244–257.

Holt, W.E. and A.J. Haines (1993). Velocity field in deforming Asia from inversion of earthquake-released strains. *Tectonics* **12**, 1–20.

Holt, W.E., J.F. Ni, T.C. Wallace, and A.J. Haines (1991). The active tectonics of the Eastern Himalayan syntaxis and surrounding regions, *J. Geophys. Res.* **96**, 14595–14632.

Hornafius, J.S., B.P. Luyendyk, R.R. Terres, and M.J. Kammerling (1986). Timing and extent of Neogene tectonic rotation in the western Transverse Ranges, California. *Geol. Soc. Am. Bull.* **97**, 1476–1487.

Hudnut, K., Z. Shen, M. Murray, S. McClusky, R. King, *et al.* (1996). Co-seismic displacements of the 1994 Northridge, California earthquake. *Bull. Seismol. Soc. Am.* **86**, S19–S36.

Isacks, B., J. Oliver, and L. Sykes (1968). Seismology and the new global tectonics. *J. Geophys. Res.* **73**, 5855–5899.

Jackson, J.A. (1980). Reactivation of basement faults and crustal shortening in orogenic belts. *Nature* **283**, 343–346.

Jackson, J. (1987). Active normal faulting and crustal extension. In: "Continental Extensional Tectonics," *Geol. Soc. Spec. Publ. London* **28**, 3–17.

Jackson, J. (1992). Partitioning of strike–slip and convergent motion between Eurasia and Arabia in eastern Turkey and the Caucasus. *J. Geophys. Res.* **97**, 12471–12479.

Jackson, J.A. (1994). Active tectonics of the Aegean region. *Annu. Rev. Earth Planet. Sci.* **22**, 239–271.

Jackson, J.A. and T. Blenkinsop (1993). The Malawi earthquake of March 10, 1989: deep faulting within the East African rift system. *Tectonics* **12**, 1131–1139.

Jackson, J.A. and T. Blenkinsop (1997). The Bilila–Mtakataka fault in Malawi: an active 100-km long, normal fault segment in thick seismogenic crust. *Tectonics* **16**, 137–150.

Jackson, J.A. and T. Fitch (1981). Basement faulting and the focal depths of the larger earthquakes in the Zagros mountains (Iran). *Geophys. J. R. Astron. Soc.* **64**, 561–586.

Jackson, J.A. and M.R. Leeder (1994). Drainage systems and the development of normal faults: an example from Pleasant Valley, Nevada. *J. Struct. Geol.* **16**, 1041–1059.

Jackson, J.A. and D.P. McKenzie (1984). Active tectonics of the Alpine-Himalayan belt between western Turkey and Pakistan. *Geophys. J. R. Astron. Soc.* **77**, 185–264.

Jackson, J.A. and P. Molnar (1990). Active faulting and block rotations in the western Transverse ranges, California. *J. Geophys. Res.* **95**, 22073–22087.

Jackson, J.A. and N.J. White (1989). Normal faulting in the upper continental crust: observations from regions of active extension. *J. Struct. Geol.* **11**, 15–36.

Jackson, J.A., J. Gagnepain, G. Houseman, G. King, P. Papadimitriou, C. Soufleris, and J. Virieux (1982). Seismicity, normal faulting, and the geomorphological development of the Gulf of Corinth (Greece): the Corinth earthquakes of February and March 1981. *Earth Planet. Sci. Lett.* **57**, 377–397.

Jackson, J.A., A. Haines, and W. Holt (1992). The horizontal velocity field in the deforming Aegean Sea region determined from the moment tensors of earthquakes. *J. Geophys. Res.* **97**, 17657–17684.

Jackson, J.A., A.J. Haines, and W.E. Holt (1995). The Accommodation of Arabia–Eurasia Plate Convergence in Iran. *J. Geophys. Res.* **100**, 15205–15219.

Jackson, J.A., R. Norris, and J. Youngson (1996). The structural development of active fault and fold systems in central Otago, New Zealand: evidence revealed by drainage patterns. *J. Struct. Geol.* **18**, 217–234.

Jackson, M. and R. Bilham (1984). Constraints on Himalayan deformation inferred from vertical velocity fields in Nepal and Tibet. *J. Geophys. Res.* **99**, 13987–13912.

Keller, E.A., R.L. Zepeda, T.K. Rockwell, T.L. Ku, and W.S. Dinklage (1998). Active tectonics at Wheeler Ridge, southern San Joaquin Valley, California. *Geol. Soc. Am. Bull.* **110**, 298–310.

Keller, E.A., L. Gurrola, and T.E. Tierney (1999). Geomorphic criteria to determine direction of lateral propagation of reverse faulting and folding. *Geology* **27**, 515–518.

Kissel, C. and C. Laj (1988). The Tertiary geodynamic evolution of the Aegean arc: a paleomagnetic reconstruction. *Tectonophysics* **146**, 183–201.

Leeder, M.R. and J.A. Jackson (1993). The interaction between normal faulting and drainage in active extensional basins, with examples from the western United States and central Greece. *Basin Res.* **5**, 79–102.

Luyendyk, B.P., M.J. Kammerling, and R.R. Terres (1980). Geometric model for Neogene crustal rotations in southern California. *Geol. Soc. Am. Bull.* **91**, 211–217.

Lyon-Caen, H., R. Armijo, J. Drakopoulos, J. Baskoutass, N. Delibassis, R. Gaulon, V. Kouskouna, J. Latoussakis, K. Makropoulos, P. Papadimitriou, D. Papanastassiou, and G. Pedotti (1988). The 1986 Kalamata (south Peloponessus) earthquake: detailed study of a normal fault and tectonic implications. *J. Geophys. Res.* **93**, 14967–15000.

Maggi, A., J.A. Jackson, D. McKenzie, and K. Priestley (2000a). Earthquake focal depths, effective elastic thickness, and the strength of the continental lithosphere. *Geology* **28**, 495–498.

Maggi, A., J.A. Jackson, K. Priestley, and C. Baker (2000b). A re-assessment of focal depth distributions in southern Iran, the Tien Shan and northern India: do earthquakes really occur in the continental mantle? *Geophys. J. Int.* **143**, 629–661.

Marrett, R. and R.W. Allmendinger (1992). Amount of extension on "small" faults: an example from the Viking graben. *Geology* **20**, 47–50.

Massonnet, D., M. Rossi, C. Carmona, F. Adragna, G. Peltzer, K. Feigl, and T. Rabaute (1993). The displacement field of the Landers earthquake mapped by radar interferometry. *Nature* **364**, 138–142.

McCaffrey, R. and J. Nábělek (1987). Earthquakes, gravity and the origin of the Bali basin: an example of a nascent continental fold-and-thrust belt, *J. Geophys. Res.* **92**, 441–460.

McCaffrey, R. and J. Nábělek (1998). Role of oblique convergence in the active deformation of the Himalayas and southern Tibet plateau. *Geology* **26**, 691–694.

McCaffrey, R., P. Zwick, and G. Abers (1991). SYN4 Program, *IASPEI Software Library* **3**, 81–166.

McClusky, S., *et al.* (28 authors) (2000). Global Positioning System constraints on plate kinematics and dynamics in the eastern Mediterranean and Caucasus. *J. Geophys. Res.* **105**, 5695–5719.

McKenzie, D. (1972). Active tectonics of the Mediterranean region. *Geophys. J. R. Astron. Soc.* **30**, 109–185.

McKenzie, D. (1978a). Some remarks on the development of sedimentary basins. *Earth Planet. Sci. Lett.* **40**, 25–32.

McKenzie, D. (1978b). Active tectonics of the Alpine–Himalayan belt: the Aegean Sea and surrounding regions. *Geophys. J. R. Astron. Soc.* **55**, 217–254.

McKenzie, D. and D. Fairhead (1997). Estimates of the effective elastic thickness of the continental lithosphere from Bouguer and free air gravity anomalies. *J. Geophys. Res.* **102**, 27523–27552.

McKenzie, D.P. and J.A. Jackson (1983). The relationship between strain rates, crustal thickening, paleomagnetism, finite strain and fault movements within a deforming zone. *Earth Planet. Sci. Lett.* **65**, 182–202.

McKenzie, D. and J. Jackson (1986). A block model of distributed deformation by faulting. *J. Geol. soc. Lond.* **143**, 249–253.

McKenzie, D.P. and D.L. Parker (1967). The North Pacific: an example of tectonics on a sphere. *Nature* **216**, 1276–1280.

McKenzie, D., F. Nimmo, J. Jackson, P. Gans, and E. Miller (2000). Characteristics and consequences of flow in the crust. *J. Geophys. Res.* **105**, 11029–11046.

Molnar, P. (1979). Earthquake recurrence intervals and plate tectonics. *Bull. Seismol. Soc. Am.* **69**, 115–133.

Molnar, P. (1992). Brace-Goetze strength profiles, the partitioning of strike-slip and thrust faulting at zones of oblique convergence, and the stress-heat flow paradox of the San Andreas Fault. In: "Fault Mechanics and Transport Properties of Rocks" (B. Evans and T.-F. Wong, Eds.), pp. 435–459, Academic Press, New York.

Molnar, P. and W.-P. Chen (1982). Seismicity and mountain building. In: "Mountain Building Processes" (K.J. Hsu, Ed.), pp. 41–57, Academic Press, New York.

Molnar, P. and J.M. Gipson (1994). Very long baseline interferometry and active rotations of crustal blocks in the Western Transverse Ranges, California. *Geol. Soc. Am. Bull.* **106**, 594–606.

Molnar, P. and H. Lyon-Caen (1988). Some simple aspects of the support, structure and evolution of mountain belts. *Geol. Soc. Am. Special Pap.* **218**, 179–207.

Molnar, P. and H. Lyon-Caen (1989). Fault plane solutions of earthquakes and active tectonics of the Tibetan Plateau and its margin. *Geophys. J. Int.* **99**, 123–153.

Molnar, P. and P. Tapponnier (1975). Cenozoic tectonics of Asia: effects of a continental collision. *Science* **189**, 419–426.

Mount, V.S. and J. Suppe (1987). State of stress near the San Andreas Fault: implications for wrench tectonics, *Geology* **15**, 1143–1146.

Nábělek, J. (1985). Geometry and mechanism of faulting of the El Asnam, Algeria, earthquake from inversion of teleseismic body waves and comparison with field observations. *J. Geophys. Res.* **90**, 12713–12728.

Nakata, T., K. Otsuki, and S.H. Khan (1990). Active faults, stress field, and plate motion along the Indo-Eurasian plate boundary. *Tectonophysics* **181**, 83–95.

Namson, J. and T. Davis (1988). Structural transect of the western Transverse Ranges, California: implications for lithospheric kinematics and seismic risk evaluation. *Geology* **16**, 675–679.

Ni, J. and M. Barazangi (1984). Seismotectonics of the Himalayan collision zone: geometry of the underthrusting Indian plate beneath the Himalaya. *J. Geophys. Res.* **89**, 1147–1163.

Philip, H., E. Rogozhin, A. Cisternas, J.C. Bousquet, B. Borisov, and A. Karakhanian (1992). The Armenian earthquake of 1988 December 7: faulting and folding, neotectonics and paleoseismicity. *Geophys. J. Int.* **110**, 141–158.

Roberts, S. and J.A. Jackson (1991). Active normal faulting in central Greece: an overview. In: "The Geometry of Normal Faults" (A.M. Roberts, G. Yielding, and B. Freeman, Eds.), *Geol. Soc. Spec. Publ. Lond.* **56**, 125–142.

Rosendahl, B.R., D.J. Reynolds, P.M. Lorber, C.F. Burgess, J. McGill, D. Scott, J.J. Lambiase, and S.J. Derksen (1986). Structural expressions of rifting: lessons from Lake Tanganyika, Africa. In: "Sedimentation in the African Rifts" (L.E. Frostick, R.W. Renaut, I. Reid, and J.J. Tiercelin, Eds.), *Geol. Soc. Spec. Publ. Lond.* **25**, 29–43.

Scholz, C.H. (1982). Scaling laws for large earthquakes: consequences for physical models. *Bull. Seismol. Soc. Am.* **72**, 1–14.

Scholz, C.H. (1988). The brittle-plastic transition and the depth of seismic faulting. *Geol. Rundschau* **77**, 319–328.

Scholz, C.H. (1990). "The Mechanics of Earthquakes and Faulting" Cambridge University Press, Cambridge, 439pp.

Scholz, C.H. (2000). Evidence for a strong San Andreas fault. *Geology* **28**, 163–166.

Scholz, C.H. and J.C. Contreras (1998). Mechanics of continental rift architecture. *Geology* **26**, 967–970.

Scholz, C.H. and P. Cowie (1990). Determination of total strain from faulting using slip measurements. *Nature* **346**, 837–839.

Seeber, L. and J.G. Armbruster (1981). Great detachment earthquakes along the Himalayan arc and long term forecasting. In: "Earthquake Prediction: an International Review" (D.W. Simpson and P.G. Richards, Eds.), American Geophysical Union Maurice Ewing Series, **4**, 259–277.

Shaw, J.H. and J. Suppe (1994). Active faulting and growth folding in the eastern Santa Barbara channel, California. *Geol. Soc. Am. Bull.* **106**, 607–626.

Shimazaki, K. (1986). Small and large earthquakes: the effects of the thickness of the seismogenic layer and the free surface. In: "Earthquake Source Mechanics" (S. Das, J. Boatwright and C.H. Scholz, Eds.), pp. 209–216, *American Geophysical Union.*

Sibson, R.H. (1994). An assessment of field evidence for "Byerlee" friction. In: "Faulting, Friction and Earthquake Mechanics, Part 1" (C.J. Marone and M.L. Blandpied, Eds.), pp. 645–662, Birkhäuser Verlag, Basel, Switzerland (also in *Pure Appl. Geophys.* **142**, 645–662).

Sibson, R.H. and G. Xie (1998). Dip range for intracontinental reverse fault ruptures: truth not stranger than friction? *Bull. Seismol. Soc. Am.* **88**, 1014–1022.

Sieh, K., *et al.* (20 authors) (1993). Near-field investigation of the Landers earthquake sequence, April to July, 1992. *Science* **260**, 171–176.

Sipkin, S.A. (1986). Estimation of earthquake source parameters by the inversion of waveform data: global seismicity, 1981–1983. *Bull. Seismol. Soc. Am.* **76**, 1515–1541.

Stein, R. and S. Barrientos (1985). Planar high-angle faulting in the Basin and Range: geodetic analysis of the 1980 Borah Peak, Idaho, earthquake. *J. Geophys. Res.* **90**, 11355–11366.

Stevens, C., R. McCaffrey, E.A. Silver, Z. Sombo, P. English, and J. van der Kevie (1998). Mid-crustal detachment and ramp faulting in the Markham Valley, Papua New Guinea. *Geology* **26**, 847–850.

Suarez, G., P. Molnar, and C. Burchfiel (1983). Seismicity, fault plane solutions, depth of faulting, and active tectonics of the Andes of Peru, Ecuador and southern Columbia. *J. Geophys. Res.* **88**, 10403–10428.

Sykes, L. (1967). Mechanism of earthquakes and nature of faulting on the mid-oceanic ridges. *J. Geophys. Res.* **72**, 2131–2153.

Taymaz, T., J. Jackson, and D. McKenzie (1991). Active tectonics of the north and central Aegean Sea, *Geophys. J. Int.* **106**, 433–490.

Thatcher, W. and D.P. Hill (1991). Fault orientations in extensional and conjugate strike-slip environments and their implications. *Geology* **19**, 1116–1120.

Triep, E.G., G.A. Abers, A.L. Lerner-Lam, V. Mishatkin, N. Zakharchenko, and O. Starovoit (1995). Active thrust front of the Greater Caucasus: the April 29, 1991, Racha earthquake sequence and its tectonic implications. *J. Geophys. Res.* **100**, 4011–4033.

Vilotte, J.P., R. Madariaga, and M. Dagnieres (1982). Numerical modelling of intraplate deformation: simple mechanical models of continental collision. *J. Geophys. Res.* **87**, 10709–10728.

Wald, D. and T. Heaton (1994). Spatial and temporal distribution of slip for the 1992 Landers, California, earthquake. *Bull. Seismol. Soc. Am.* **84**, 668–691.

Wald, D.J., T.H. Heaton, and K.W. Hudnut (1996). The slip history of the 1994 Northridge, California, earthquake determined from strong-motion, teleseismic, GPS and levelling data. *Bull. Seismol. Soc. Am.* **86**, S49–S70.

Wallace, R.E. (1989). Fault plane segmentation in brittle crust and anisotropy in loading system. In: "Fault Segmentation and Controls of Rupture Initiation and Termination" (D.P. Schwartz and R.H. Sibson, Eds.), *US Geol. Surv. Open-file report* **89-315**, 400–408.

Walsh, J. and J. Watterson (1988). Analysis of the relationship between displacements and dimensions of faults. *J. Struct. Geol.* **10**, 239–247.

Watterson, J. (1986). Fault dimensions, displacements and growth. *Pure Appl. Geophys.* **124**, 365–373.

Wernicke, B. (1981). Low-angle normal faults in the Basin and Range province: nappe tectonics in an extending orogen. *Nature* **291**, 645–648.

Wernicke, B. (1995). Low-angle normal faults and seismicity: a review. *J. Geophys. Res.* **100**, 20159–20174.

Wernicke, B. and G. Axen (1988). On the role of isostasy in the evolution of normal fault systems. *Geology* **16**, 848–851.

Wernicke, B. and C. Burchfiel (1982). Modes of extensional tectonics. *J. Struct. Geol.* **4**, 105–115.

Whitcomb, J., C. Allen, J. Garmany, and J. Hileman (1973). San Fernando earthquake series 1971, focal mechanism and tectonics. *Rev. Geophys. Space Phys.* **11**, 693–730.

Wright, T.J., B.E. Parsons, J.A. Jackson, M. Haynes, E.J. Fielding, P.C. England, and P.J. Clarke (1999). Source parameters of the 1 October 1995 Dinar (Turkey) earthquake from SAR interferometry and seismic bodywave modelling. *Earth Planet. Sci. Lett.* **172**, 23–37.

Wu, D. and R.L. Bruhn (1994). Geometry and kinematics of active normal faults, South Oquirrh mountains, Utah: implication for fault growth. *J. Struct. Geol.* **16**, 1061–1075.

Yeats, R.S. (1993). Tectonics: converging more slowly. *Nature* **366**, 299–301.

Yeats, R.S., K. Sieh, and C.R. Allen (1997). "The Geology of Earthquakes," Cambridge University Press, Cambridge, 568 pp.

Yielding, G. (1985). Control of rupture by fault geometry during the 1980 El Asnam (Algeria) earthquake. *Geophys. J. R. Astron. Soc.* **81**, 641–670.

Yielding, G., J.A. Jackson, G.C. King, H. Sinhval, C. Vita-Finzi, and R.M. Wood (1981). Relations between surface deformation, fault geometry, seismicity and rupture characteristics during the El Asnam (Algeria) earthquake of 10 October 1980. *Earth Planet. Sci. Lett.* **56**, 287–304.

Ziegler, P. (1983). Crustal thinning and subsidence in the North Sea. *Nature* **304**, 561.

Editor's Note

Figures 5 and 6 are in color and are shown on Color Plate No. 10. An Appendix, Source parameters from waveform inversion, is placed on the attached Handbook CD-ROM under directory \31Jackson.

32

Rock Failure and Earthquakes

David A. Lockner and Nicholas M. Beeler
US Geological Survey, Menlo Park, USA

1. Introduction

This chapter summarizes experimental observations and related theoretical developments of faulted and intact rock properties related to earthquake nucleation, failure and dynamic slip. We will be concerned primarily with earthquakes occurring in the brittle crust. Intermediate and deep-focus earthquakes have unique mechanical considerations that are discussed in Section 7. We focus on repeatable laboratory observations and their direct implications for fault strength and stability. Important results that may be new, incomplete or controversial have also been included. To distinguish the well-established findings from others, the latter appear in "new and provisional results" sections.

Natural faults in the Earth's crust are zones of localized shear that are inherently weaker than the surrounding intact rock (Chapter 29 by Sibson). Slip can occur slowly as aseismic creep or rapidly as earthquakes with long interevent times during which little or no slip occurs. Crustal faults have complex and irregular geometry, typically fractal in nature, that includes irregular, interlocked surfaces, offset segments, bends, and junctions. Since compressive stresses increase with depth due to the overburden, the natural tendency for irregular fault surfaces to move apart during sliding is suppressed. Instead, fault slip at depth involves grinding and crushing of grains and must involve continual fracture of asperities or interlocked regions. Breaking of asperities may in fact control the position and timing of earthquake nucleation. For this reason, we will examine the brittle fracture of intact rock as well as frictional properties of preexisting faults.

Experimental and theoretical rock mechanics studies have greatly influenced our understanding of earthquakes and related crustal processes. Many key theoretical advances have either originated or been tested and refined through observations of brittle deformation under controlled laboratory conditions. Although the association of earthquakes with crustal faults was established by the beginning of the 20th century (Lyell (1877) actually suggested a connection before this), important empirical relations for strength of rock and granular material were developed much earlier. The Coulomb failure criterion for initially intact rocks was proposed in the late 18th century and continues to be of great practical use

$$|\tau| = c + \mu_i \sigma_n \qquad (1)$$

where τ and σ_n are shear and normal stress resolved on the eventual fracture plane, c is cohesion, and μ_i is coefficient of internal friction. Here we define fracture as a loss of cohesion across a material plane, and failure as a loss of strength, thus fracture is a subset of failure. For the failure of intact material we conform to the standard definition that the coefficient of internal friction is the slope of the failure surface: $\mu_i = \partial\tau/\partial\sigma_n$ (Jaeger and Cook, 1971), the failure surface or envelope being the solution to the failure criterion, i.e., for Eq. (1) the failure surface is linear and μ_i is a constant. Griffith (1920) demonstrated the importance of flaws, which act as stress concentrators, in controlling the strength of brittle materials, materials whose ability to resist deformation decreases with increasing permanent deformation. Using this approach, fracture mechanics (e.g., Broek, 1982; Atkinson, 1987) has been highly successful in relating defect structures to fracture strength and related properties. Similar ideas have been a fundamental part of frictional contact theory where it is recognized that the real area of contact supporting stresses across a fault is a small fraction of the total fault area (Jaeger and Cook, 1971; Dieterich, 1978b; Dieterich and Kilgore, 1994).

Many of the theoretical advances in rock fracture and friction have their origins in metallurgy. Although the macroscopic phenomenology can be similar, the micromechanical deformation mechanisms are often distinct for metals and framework silicates. Brittle fracture in silicates commonly involves opening of cracks and an accompanying volume increase. Because this dilatational strain must do work against the ambient confining stress, Coulomb materials, unlike metals, are characterized by a pressure sensitivity of yield strength as expressed in Eq. (1).

The basic concept of friction or frictional resistance of two surfaces in contact, the ratio of the shear resistance to the surface normal stress, is referred to as Amontons' law and was developed about eighty years before the Coulomb criterion (Rabinowicz, 1965; Jaeger, 1969). In this case, shear resistance (f_s) is related to the normal force (f_n) acting on the surface by $|f_s| = \mu_f f_n$, where μ_f is coefficient of friction. Dividing by the area of the surface gives

$$|\tau| = \mu_f \sigma_n \tag{2a}$$

For many materials, spanning a wide range of mean stress, μ_f is only weakly dependent on normal stress. Friction will depend both on the material filling the interface (fault gouge) and on the surface geometry (surface roughness, coherence or degree of interlocking of surfaces). For example, joint systems, which are formed by tensile opening of fractures with little net shearing, can result in clean, well-mated surfaces that have higher shear strength than surfaces with uncorrelated roughness. For coherent surfaces, shear strength may not vanish at zero normal stress and a more accurate empirical description of shear strength becomes

$$|\tau| = S_0 + \mu_f \sigma_n \tag{2b}$$

where S_0 is an 'inherent' shear strength similar to cohesion in Eq. (1) (Jaeger and Cook, 1971). The quantity most easily measured and most often used in studies of rock friction is the simple ratio of shear to normal stress. We will refer to this ratio as friction or frictional resistance, denoted by μ, to distinguish it from the coefficient of friction μ_f, so that Eq. (2b) gives

$$\mu = \frac{|\tau|}{\sigma_n} = \frac{S_0}{\sigma_n} + \mu_f \tag{3}$$

High temperature, high pressure or the addition of reactive pore fluids may allow for activation of ductile deformation mechanisms (mechanisms whereby permanent deformation can be accommodated without loss of strength) such as dislocation glide or pressure solution, and can allow deformation that lacks the pressure sensitivity characteristic of friction and fracture. Thus, both μ_i and μ_f are expected to decrease with increasing depth in the crust (Mogi, 1966), and in fact laboratory measurements indicate that μ_i, μ_f and μ decrease with increasing mean stress $(\sigma_1 + \sigma_3)/2$ or normal stress (Fig. 1b). Normal stress increases with depth and provides a convenient independent variable for identifying trends in shear strength and other fault-related properties. However, when considering failure of intact rock, there is no preexisting fault plane and normal stress becomes a less useful concept. In this case, mean stress provides a good alternative and will be used throughout this chapter.

Although it is tempting, based on observations such as shown in Figure 1a, to treat friction as a material property, this assumption may not be correct. Some effects resulting from surface geometry can be accounted for by adjusting S_0.

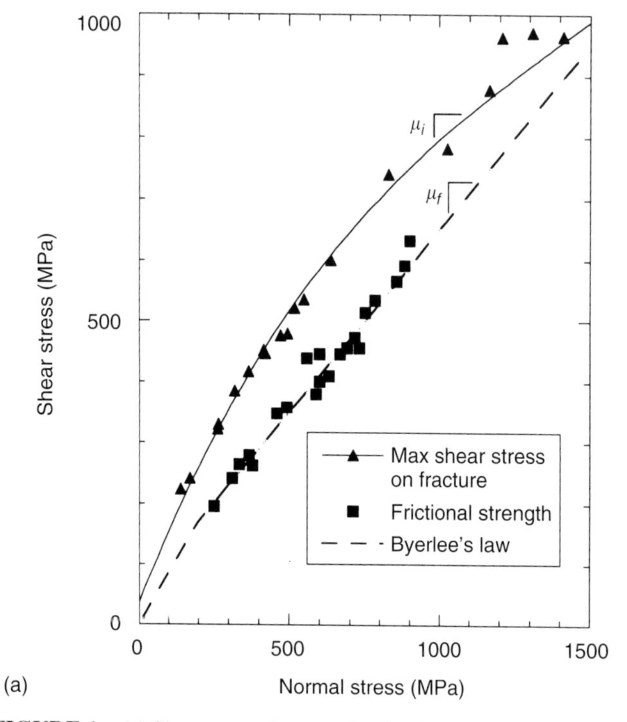

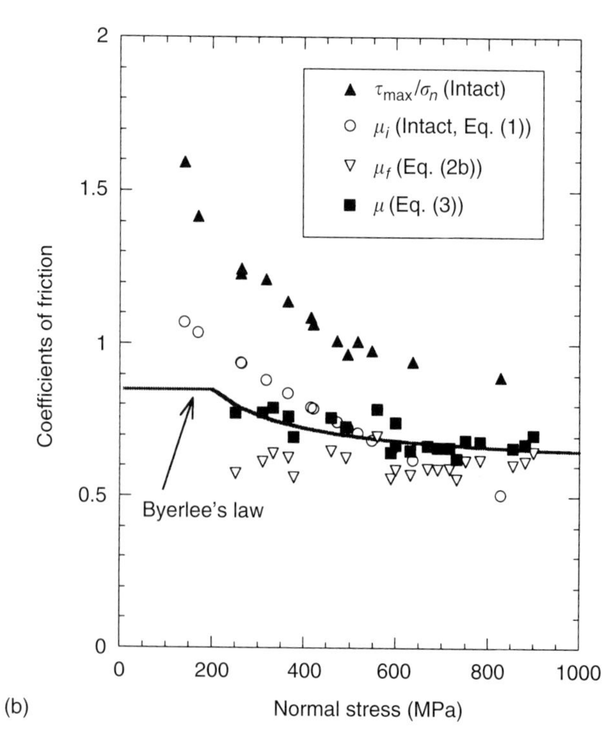

FIGURE 1 (a) Shear-normal stress plot for Westerly granite showing peak intact strength and frictional strength. (b) Alternative measures of strength for intact and faulted Westerly granite.

However, other geometric properties related to the deformation of granular material between rigid fault blocks also affect shear strength. In an idealized sense, a fault zone can be described mechanically as a thin, weak layer sheared between rigid blocks. The boundary conditions for this geometry require that normal strains in the gouge layer in the plane of the fault must be zero and average flow lines must eventually become parallel to the fault walls, imposing constraints on the stress field that develops within the gouge layer (Spencer, 1982; de Josselin de Jong, 1988; Rice, 1992; Savage *et al.*, 1996; Savage and Lockner, 1997). The strains resulting from average simple shear deformation imposed on a gouge layer by the fault walls are inconsistent with strains resulting from flow of a Coulomb material (Byerlee and Savage, 1992). As a result, complex structures such as Riedel shear bands develop in the fault gouge (Gu and Wong, 1994; Savage *et al.*, 1996) which reflect a heterogeneous stress field that develops within fault zones in response to deformation (Mandl *et al.*, 1977; Morgan and Boettcher, 1999).

Laboratory measurements show that numerous parameters in addition to normal stress or confining pressure affect strength of both fault surfaces and intact rock. These include mineralogy, porosity, cementation, packing geometry of gouge, surface roughness, coherence of surfaces, angularity and size of gouge particles, temperature, pore fluid pressure and composition, deformation rate, deformation and stress history, fabric, foliation, or other anisotropic properties. Sample size is reported to affect intact strength, but appears to have little or no effect on frictional strength. Some of these effects are discussed individually in subsequent sections. Pore fluid pressure, in particular, has a first order effect on rock and fault strength in cases where there is well-connected porosity. This effect is discussed in more detail in Sections 2.3.2 and 5. For now, we note that "effective" normal stress should be substituted into Eqs. (1), (2), and (3) where an effective normal stress component is the normal stress reduced by pore pressure: i.e., $\sigma_{n,\mathrm{eff}} = \sigma_n - p$ (Terzaghi, 1929; Hubbert and Rubey, 1959).

2. Brittle Fracture

For theoretical development or numerical modeling, it is common to assume that earthquakes occur in preexisting fault zones. For a number of reasons, however, we will include analysis of the rupture process in both intact and prefractured rock (cf. Section 4.2). Although the Earth's crust is generally permeated with preexisting fractures and joints, it is not known what fraction of earthquakes occur on preexisting faults, what percentage of any single earthquake rupture surface represents new faulting (especially for earthquakes that jump fault segments), or whether rupture nucleation might occur at jogs or locked segments and involve breaking of rock which is at or near the intact rock strength. Also, there are many similarities between the mathematical analysis of intact rock deformation and deformation on faults. The micromechanical mechanisms for fracture and sliding on faults both involve grain breakage, grain rotation, crack propagation, and in many cases plastic deformation mechanisms. Fault zone materials will be altered by passage of pore fluids, but this process also occurs in intact rock. Finally, in the long periods between earthquakes, faults can be expected to heal and restrengthen by an amount which, in the limit, would return to intact rock strength. For these reasons, a proper review of rock friction should also include discussion of rock fracture.

A number of comprehensive studies and reviews of rock failure, faulting and rheology currently exist (Handin, 1966; Jaeger and Cook, 1971; Bock, 1978; Lama and Vutukuri, 1978; Paterson, 1978; Lambe and Whitman, 1979; Kirby and McCormick, 1984; Kirby and Kronenberg, 1987; Mandl, 1988; Scholz, 1990; Evans and Dresen, 1991; Lockner, 1995). Readers are encouraged to refer to these studies for additional information. Many of the effects discussed in this section have direct bearing on how rocks deform along preexisting faults and provide a useful framework for much of the discussion of earthquake processes.

2.1 Basic Relations

The relations between fault orientation angle β, fault normal angle ψ $(=(\pi/2)-\beta)$, and stress components, including maximum compressive (σ_1) and minimum compressive (σ_3) principal stresses (Fig. 2), follow directly from expressing the state of stress on planes which contain the intermediate principal stress axis as a function of the extreme principal stresses

$$\tau = \tfrac{1}{2}(\sigma_1 - \sigma_3)\sin 2\beta \tag{4a}$$

$$\sigma_n = \tfrac{1}{2}[(\sigma_1 + \sigma_3) - (\sigma_1 - \sigma_3)\cos 2\beta] \tag{4b}$$

Here and throughout we take compressive stresses and contractive strains as positive. As indicated by Eq. (4), the orientation of the failure plane is assumed to contain and be independent of the intermediate principal stress.

Laboratory failure tests conducted at different conditions can be used to construct the failure envelope, defined as the locus of points in shear versus normal stress space which define failure (the Mohr plane) (Fig. 2). The simplest laboratory test geometries are uniaxial compression ($\sigma_1 = \sigma_c$, $\sigma_2 = \sigma_3 = 0$) leading to failure stress state "U" in Fig. 2, and uniaxial tension ($\sigma_1 = \sigma_2 = 0$, $\sigma_3 = \sigma_t$) shown as stress state "T". Tensile failure occurs on a fracture oriented perpendicular to the maximum tensile stress direction ($\beta = 90°$), whereas compressive failure involves shearing on a fault inclined at smaller β (typically in the range $20° < \beta < 45°$). More complicated testing conditions are appropriate for deformation tests relevant to earthquake failure, which require higher normal stress and independent variation of one or more

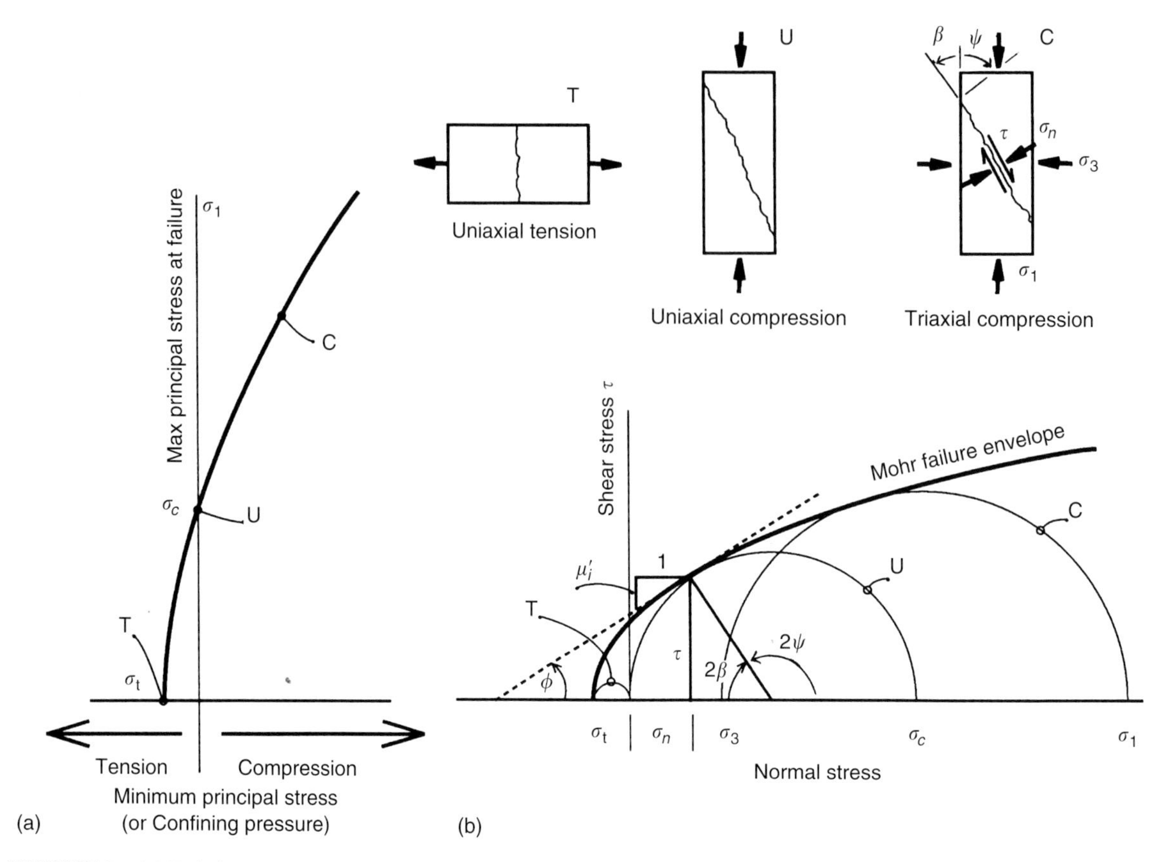

FIGURE 2 (a) Relation between principal stresses at failure. (b) Mohr failure envelope showing relation between stresses and failure parameters. $\beta\ (=(\pi/2)-\psi)$ is the angle between failure surface and direction of maximum principal stress σ_1.

stress components. The two most common test geometries are shear box or direct shear and triaxial. Direct shear machines allow for independent variation of shear and normal stress on the sliding surface. Triaxial machines are the most common method of testing samples at confining pressures as high as 1 GPa, controlled pore fluid pressure, and temperatures relevant to the crust and upper mantle. A restriction on standard triaxial tests is that a fluid is used to apply confining pressure P_{conf} so that $\sigma_2=\sigma_3=P_{\text{conf}}$. Rotary shear machines, where confining pressure, normal stress, shear stress (torque), and pore pressure can be varied independently, are also particularly useful in friction studies since significantly larger displacements can be achieved in this geometry than in any linear displacement apparatus. However, the direct shear and unconfined rotary shear geometries lead to difficulties in controlling pore pressure and temperature.

The steep slope for the failure envelope in Figure 2a shows the strong pressure sensitivity of failure strength for dilatant materials. A Mohr representation in which shear stress is plotted versus normal stress, is shown in Figure 2b (Jaeger and Cook, 1971). For application to the Earth, it is often assumed and observed that failure occurs when Eq. (1) is first satisfied on the most favorably oriented plane. This assumption leads to additional constraints on the stress state and fault orientation,

$$\phi_i = \tan^{-1}\mu_i = \frac{\pi}{2} - 2\beta \tag{5a}$$

$$\tan 2\beta = \frac{1}{\mu_i} \tag{5b}$$

$$\sigma_1 = \sigma_c + [(\mu_i^2+1)^{1/2} + \mu_i]^2\sigma_3 \tag{5c}$$

Equations (5) are Mohr–Navier–Coulomb relations.

2.2 Conceptual Fracture Model

As mean stress increases with depth in the shallow to mid crust due to the increase in overburden, both frictional and intact rock strength can be expected to increase. If special conditions result in near-lithostatic fluid pressure, that is, pore pressure equivalent to overburden (i.e., magma intrusion or trapped pore fluid), then the minimum effective principal stress can be low or negative and the rock can fracture in tension. In the case of earthquakes, however, focal mechanisms indicate shear rather than tensile failure. Rocks are generally about ten times stronger in compression than in tension (Lockner, 1995), reflecting the difficulty of propagating a large-scale shear fracture (generally involving dilatancy) through a compressively loaded brittle material. It is well known that the failure process in rock involves microcrack growth. Open pores, contrasts in elastic properties of adjacent minerals, and weak grain boundaries can all act as stress concentrators in compression (Peng and Johnson, 1972; Tapponnier and Brace, 1976) and shearing along surfaces oblique to the maximum compressive principal stress σ_1 is likely to play an important role in the development of large local stresses (Peng and Johnson, 1972; Horii and Nemat-Nasser, 1986; Sammis and Ashby, 1986; Ashby and Sammis, 1990; Lockner *et al.*, 1992b). The local stresses induced near the crack tips contain a component of tension which leads to local tensile failure before the fracture toughness for failure in shear is achieved (Lawn and Wilshaw, 1975). As a result, tensile "wing" cracks grow aligned subparallel to the σ_1 direction. Unlike the case of remotely applied tensile stress, two important stabilizing processes take place during the loading of cracks in compression. First, as wing cracks extend, stress intensity decreases, and additional deviatoric stress must be applied to cause further crack growth (Costin, 1983, 1987; Horii and Nemat-Nasser, 1986). Second, diagonal flaws, which are favorably oriented to act as nucleation points for shear failure, propagate out-of-plane (parallel to σ_1) and cannot maintain the necessary shear geometry (Lawn and Wilshaw, 1975). Eventually, crack densities become sufficiently large for significant crack interaction to occur, leading to the nucleation of a proto-fault: a shear fracture connecting en-echelon arrays of the dilatant microcracks (Peng and Johnson, 1972; Horii and Nemat-Nasser, 1986; Sammis and Ashby, 1986; Costin, 1987; Kemeny and Cook, 1987; Ashby and Sammis, 1990; Du and Aydin, 1991; Lockner *et al.*, 1992b; Reches and Lockner, 1994). Finally, quasi-static fault growth experiments (Lockner *et al.*, 1991, 1992a) have demonstrated that following nucleation, the shear fault propagates inplane due to its own stress field (Fig. 3). Although the fracture propagates as mixed shear modes II and III, the fracture tip is surrounded by a process zone, much like the earthquake rupture zone described by Rice (1980). In this high-stress region, energy is dissipated by the growth of a halo of microcracks (primarily tensile (Moore and Lockner, 1995)) as the fracture advances through the rock. Similar results have been observed for damage zones surrounding natural faults (Vermilye and Scholz, 1998). The acoustic emission experiments provide an important constraint on nucleation. In past studies it was often argued that the eventual fault plane began forming, as a region of high microcrack damage, well before peak strength. Although this interpretation may hold for samples containing large preexisting flaws, it appears that in homogeneous, isotropic rocks, the location of the eventual fault plane is not correlated in any obvious way to precursory, localized damage. Similar results have been reported in field studies (Linde and Johnston, 1989; Johnston and Linde, 2002) where nucleation zones for moderate-sized earthquakes were inferred to be less than 0.1% of the coseismic rupture surface area.

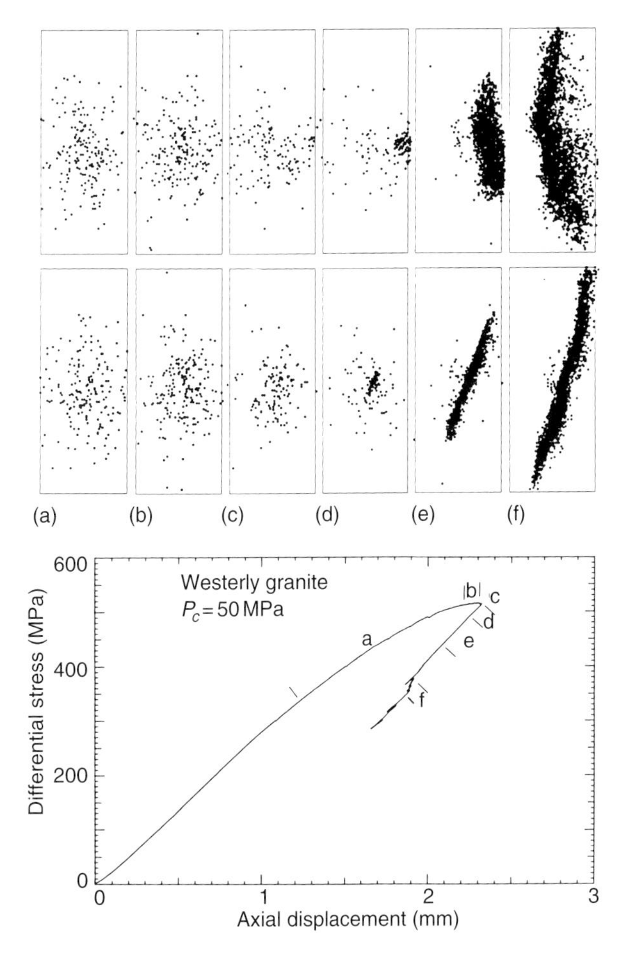

FIGURE 3 Acoustic emission (AE) hypocentral locations during fault formation of initially intact Westerly granite. Time progresses from left to right. Middle: Figures view sample along strike of eventual fault plane which appears as diagonal feature in (e) and (f). Top: Same AE events viewed perpendicular to strike. Bottom: Accompanying stress–displacement curve indicates segments of the experiment from which acoustic emission plots are made. Fault nucleation occurs in (d).

2.3 Factors Affecting Fracture Strength of Rock

2.3.1 Confining Pressure Effect

Confining pressure affects brittle fracture strength by suppressing the growth of dilatant microcracks. Microcracks tend to grow parallel to σ_1 when a sample is loaded in compression by locally overcoming the ambient compressive stress field near crack tips and developing a localized region of tensile stress. This process requires some specialized mechanism such as slip along grain boundaries or bending of a rigid grain that is adjacent to an open pore or more compliant grain (Peng and Johnson, 1972; Tapponnier and Brace, 1976; Sammis and Ashby, 1986; Reches and Lockner, 1994). Such mechanisms are generally enhanced by deviatoric stress and suppressed by mean stress. In addition, slip along grain boundaries will not occur until frictional contact strength, which increases with mean stress, is overcome. The overall strengthening effect can be represented in a shear stress–normal stress plot (Mohr–Coulomb failure envelope) such as Figure 1a for Westerly granite. In general, the failure envelope is concave towards the normal stress axis so that a linear failure criterion such as Eq. (1) may not be a satisfactory approximation. This curvature becomes most important when experimental data span a broad range of normal stress. Some minerals, such as calcite, undergo a brittle–ductile transition within this pressure range and exhibit severe curvature of the failure envelope. Furthermore, the difference between intact failure strength and frictional sliding strength vanishes at high pressure (Byerlee, 1967; Ismail and Murrell, 1990). A good empirical fit to the intact rock strength in Figure 1 (Lockner, 1998) is provided by

$$\sigma_1 = \sigma_3 + (46\,660 + 5128.7\sigma_3)^{1/2} - 8.3\,\text{MPa} \qquad (6)$$

where stresses are in MPa.

The majority of strength measurements have been conducted under uniaxial or "triaxial" stress conditions in which $\sigma_2 = \sigma_3$. A limited number of true triaxial measurements have been performed to explore the effect of intermediate principal stress on failure mode. While the most commonly used failure criteria (e.g., Mohr–Coulomb) assume that failure is independent of intermediate stress, experimental evidence demonstrates that this assumption is not strictly true (Mogi, 1967, 1970; Amadei *et al.*, 1987).

2.3.2 Pore Fluid Pressure and Effective Pressure Law

Pore fluids can affect fracture strength through a direct pressure effect as well as through chemical interactions with the rock matrix. Mechanically, pore pressure acts to reduce the normal stress tensor components throughout the rock mass according to the effective pressure law (Hubbert and Rubey, 1959)

$$\sigma_{n,\text{eff}} = \sigma_n - \alpha p \qquad (7)$$

where α is a constant related to pore geometry. Although variations in α have been measured for the effect of pore pressure on transport properties (Walsh, 1981), $\alpha = 1$ is appropriate for fracture strength of rock with well-connected porosity. If fluid pressures in the crust were low, the increase in mean stress and temperature with depth, due to overburden pressure, should result in large shear strength and the gradual change from brittle to ductile deformation in the mid to lower crust (as suggested by the convergence of frictional and intact strength in Fig. 1). The normal hydraulic head for well-connected porous material will offset the overburden pressure by approximately one-third. However, for isolated porosity at depth, normal consolidation and dewatering processes can compress pore fluids to well above hydrostatic pressures and under proper circumstances may lead to repeated episodes in which the rock is hydraulically fractured. There is currently considerable interest in fluid overpressure phenomena and their relation to oil reservoir partitions (Hunt, 1990; Powley, 1990) as well as to larger-scale crustal processes such as earthquake cycles and the reduction of strength of mature faults (Byerlee, 1990, 1992, 1993; Nur and Walder, 1990; Rice, 1992; Miller *et al.*, 1996; Sibson, 2002) (see Section 3.2).

Since water reacts with silicates to disrupt the Si–O bond, pore fluids can have a profound chemical effect on the physical properties of crustal silicate rocks. CO_2 concentration plays a similar role in the diagenesis and metamorphism of carbonates. Crack healing, cementation, and densification all act to strengthen rock with time (Hickman and Evans, 1992). By contrast, chemically assisted crack growth is the primary mechanism for stress corrosion and static fatigue (subcritical crack growth) at temperatures in the 0–200°C range (Rutter and Mainprice, 1978; Kranz, 1980; Kranz *et al.*, 1982; Lockner, 1993b). Subcritical crack growth is crack extension at crack tip stress below the critical stress intensity K_{IC} needed to split lattice bonds ($K_I < K_{IC}$) (Atkinson and Meredith, 1987; Costin, 1987). Subcritical crack growth rate v is generally related to K_I through a power law

$$v = A_1 K_I^n \qquad (8)$$

or exponential form

$$v = A_2 \exp(b_2 K_I) \qquad (9)$$

where A, b, and n are empirical constants. An alternative view relates crack growth rate to energy release rate G_I ($\propto K_I^2$) (Lockner, 1993b):

$$v = A_3 \exp(b_3 G_I) \qquad (10)$$

Experimental data are insufficient to identify which form fits best. However, there are theoretical arguments, based on reaction rate theory, to prefer Eq. (9) or (10) (Costin, 1983; Lockner, 1993b). Nominal values for n at room temperature and in wet rock (Atkinson, 1987, Table 11.6) are: 15–40 for quartz and quartz rocks; 10–30 for calcite rocks; 30–70 for granitic rocks; and 25–50 for gabbro and basalt. A typical value for b_2 for room temperature granite is 0.04 $(\text{MPa m}^{-1/2})^{-1}$. The corrosive effect of the pore fluid can be altered significantly by varying the activity of dissolved species as well as altering the electrochemical potential (Dunning *et al.*, 1980; Ishido and Mizutani, 1980; Atkinson and Meredith, 1987; Evans and Dresen, 1991; Hickman and Evans, 1991).

2.3.3 Strain Rate

In the brittle deformation field, rocks typically exhibit a pseudo-viscous effect prior to failure which is reflected in a strength increase with increasing strain rate. This effect is best known in the mining industry as static fatigue in which a pillar or other load-bearing structure will fail after some period of time under constant load. Subcritical tensile crack growth has been identified as the principal mechanism responsible for static fatigue in rock (Kranz, 1980; Kranz *et al.*, 1982; Lockner, 1993b) and has been studied in recent years (Atkinson and Meredith, 1987). The strain-rate dependence of rock strength has been measured in the laboratory at rates above $10^{-10}\,\text{sec}^{-1}$ which is generally much faster than geologic rates even in tectonically active regions. However, the lower limit does represent a useful limit for many engineering applications. Experimental results covering a broad range of strain rates are shown in Figure 4. Average strain rates have been obtained from static fatigue tests on Barre granite (Kranz, 1980) by dividing time-to-failure data by total inelastic strain. Various studies (Costin, 1983; Lockner, 1993b) have been successful in relating this macroscopic relation between stress and strain rate to subcritical microcrack growth rate and its sensitivity to stress intensity at crack tips. They determined an apparent activation volume of $0.43 \pm 0.04\,\text{kJ MPa mol}^{-1}$ ($4.3 \times 10^{-4}\,\text{m}^3\,\text{mol}^{-1}$) for crack growth in granite creep experiments. This approach provides a link between energetics of microcrack extension and bulk pseudoviscous response of rocks in creep (Lockner, 1998). It may also prove useful as a way to incorporate time-dependent effects in damage mechanics applications (Costin, 1987). This rate sensitivity is often expressed as a power law dependence of the form

$$\sigma_{\max} = a\dot{\varepsilon}^n \tag{11}$$

Typical rate sensitivities in this regime are $0.02 \leq n \leq 0.04$.

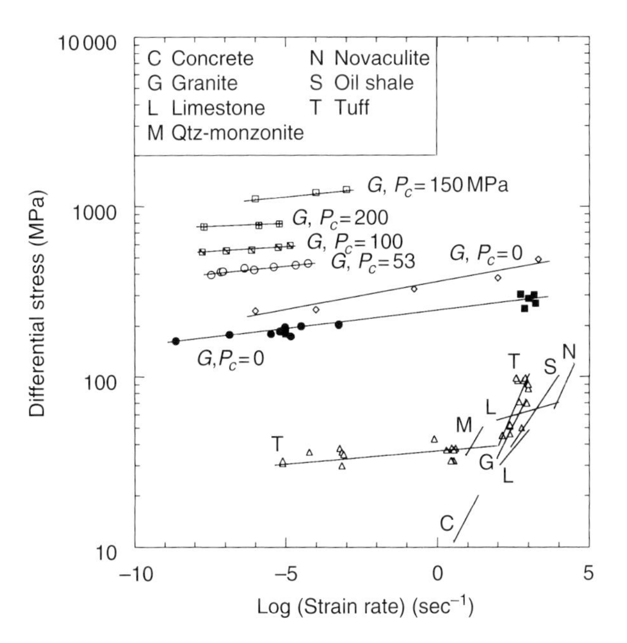

FIGURE 4 Effect of strain rate on brittle fracture strength. Trend lines in lower-right quadrant are from dynamic tensile fracture experiments (Goldsmith *et al.*, 1976.)

By using shock and other impulsive loading techniques, strain rates in excess of $10^4\,\text{sec}^{-1}$ have been achieved for samples failing in tension (Grady and Lipkin, 1980). Although some question remains as to the dependence of the measurement on machine effects and boundary conditions, numerous experiments show a transitional strain rate of approximately $10\,\text{sec}^{-1}$ above which significantly larger rate dependence is observed. High-rate tensile experiments, taken from Grady and Lipkin (1980), are summarized in the lower-right quadrant of Figure 4 as trend lines without accompanying data values. An upper limit of $n \leq 1/3$ is indicated by both theory and experiment (Brace and Jones, 1971; Green *et al.*, 1972; Lipkin and Jones, 1979; Grady and Lipkin, 1980; Olsson, 1991) for rate sensitivity under these conditions.

2.3.4 Temperature

Mechanisms of crystal plasticity involving dislocation motion, diffusion of point defects, and twinning are thermally activated and will dominate brittle cracking as temperature increases (Kirby and Kronenberg, 1987; Evans and Dresen, 1991). Some minerals, such as halite and calcite, will deform ductilely at room temperature if sufficient confining pressure is present to suppress brittle crack growth (Fredrich *et al.*, 1989). However, dry quartz appears to deform brittlely at room temperature even at confining pressures in excess of 1 GPa (Tullis and Yund, 1977). As previously mentioned, water has a significant effect on pressure-assisted grain-boundary deformation mechanisms such as pressure solution (Blanpied *et al.*, 1991; Hickman and Evans, 1991). These fluid-assisted mechanisms will often dominate at intermediate temperatures and, over geologic time scales, may play an important role in determining rock strength at room temperature (Rutter and Mainprice, 1978). Thus, even in the brittle regime, increasing temperature tends to reduce fracture strength (Bragg and Andersland, 1981; Sayles and Carbee, 1981; Wong, 1982a). The effect of temperature on fracture strength of Westerly granite is shown in Figure 5.

2.3.5 Sample Size and Scaling

Laboratory studies in which pressure, temperature and other environmental conditions are controlled have provided strength data relevant to natural faulting and seismicity. However, the scaling of these data to field applications remains problematic (Pinto da Cunha, 1990). For example, the largest laboratory fault models (approximately 1-m^2 surface) are many orders of magnitude less than the fault area involved in a moderate earthquake. A general result of fracture mechanics analysis is that stress intensity at the tip of a flaw scales as the square root of the flaw size. Consequently, larger samples, which can contain larger flaws, are more likely to be weak. This argument assumes, of course, that small samples with visible flaws (and certainly flaws that completely bisect the sample) will be rejected and not included in the strength statistics. However, the degree of weakening has not been well determined and should

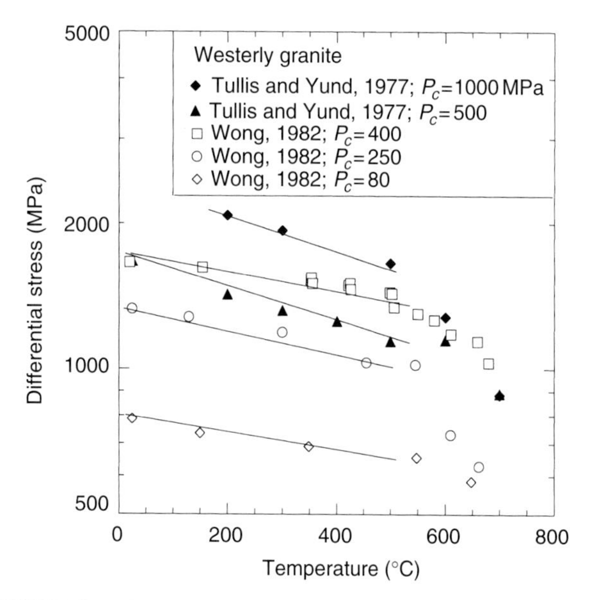

FIGURE 5 Effect of temperature on brittle fracture strength of nominally dry Westerly granite.

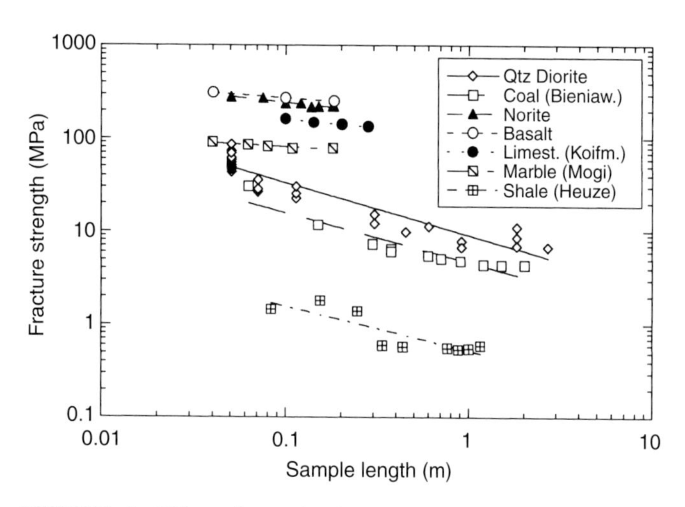

FIGURE 6 Effect of sample size on fracture strength. Observed weakening is up to 70% per decade increase in sample length. All data are for uniaxial experiments.

vary from rock to rock depending on the flaw-size distribution. Scaling procedures have been proposed (Allégre *et al.*, 1982; Madden, 1983) that address this problem. In addition, both laboratory and field studies have been conducted (Mogi, 1962; Koifman, 1963; Bieniawski, 1968; Hodgson and Cook, 1970; Pratt *et al.*, 1972; Singh and Huck, 1972; Herget and Unrug, 1976; Heuze, 1980; Dey and Halleck, 1981) that show a general weakening with increased sample size (Fig. 6). This effect can be large: as much as a 70% strength loss per decade sample size for the weathered diorite studied by Pratt *et al.* (1972). Available data are insufficient to allow a clear understanding of this effect. It is intriguing that the weaker rock types show the greatest size effect. Pratt *et al.* (1972) noted a decrease in size sensitivity for samples greater than 1-m length for both diorite and coal and suggested that this may represent an upper limit to the weakening region. Due to the small amount of existing data, it is not yet clear to what extent these tendencies can be generalized. If flaw size is the correct explanation for the weakening effect, then sample strength will depend as much on crack-size distribution as on mineralogy. Furthermore, the studies cited here are all unconfined. As discussed in Section 3, we may expect large rock masses, especially in tectonically active regions, to support stresses that are comparable to their frictional strength rather than laboratory-measured intact strength (Townend and Zoback, 2000).

3. Friction and Crustal Faults

Numerous theoretical arguments have been proposed to explain frictional properties of rocks; many of which have met with reasonable success in matching experimental observations. Yet, natural faults are complex systems, and many complicated and competing processes probably occur during deformation of fault zones. At present, fault rheologic models should be considered empirical. As more sophisticated and precise experimental observations are made, the corresponding mathematical descriptions have invariably become more complicated. There are many earthquake-related applications that can take advantage of linearized approximations of frictional properties where the non-linear aspects of time-dependent failure or fluid–rock interactions are of less importance.

3.1 Byerlee's Law and Incipient Failure

It has already been noted that Eq. (3) provides a lowest order approximation of rock friction, especially in the lower temperature brittle deformation field. Byerlee (1978) compiled laboratory friction data for rocks and minerals and found that many silicates and carbonates had very similar frictional strengths. He suggested that the characteristic concave-downward trend of shear strength versus normal stress could be adequately approximated by a piecewise linear function often referred to as Byerlee's law:

$$\begin{aligned} \tau &= 0.85\sigma_n & \sigma_n &< 200\,\text{MPa} \\ \tau &= 50\,\text{MPa} + 0.6\sigma_n & 200 &< \sigma_n < 1700\,\text{MPa} \end{aligned} \tag{12}$$

As discussed in the next section, Eq. (12) has proved to be a popular approximation for shallow to mid crustal strength where details of mineralogy or stress state may not be known. It should be noted that many hydrated minerals and especially sheet silicates with good basal cleavage or which accommodate interlayer water have frictional strength significantly lower than that described by Eq. (12) (see Section 3.2). Frictional resistance μ as determined by Eq. (12) is plotted in Figures 1a and b. This function, or values of frictional shear strength given by Eq. (12) will be used as a convenient reference in this paper.

Let us assume for the moment that the strength of the brittle crust can be adequately represented by granodiorite. Although this is an obvious simplification of the range of lithologies that may be found in the continental crust, it should provide a reasonable approximation of both intact and frictional strength for many low-porosity, unaltered crystalline rocks. The intact strength curve shown in Figure 1a, which is derived from granitic samples chosen to be free of macroscopic flaws, provides a reasonable approximation to the upper strength limit of the crust for rock volumes free of faults or joints. For fractured rock masses, containing faults or joints of all orientations, the most favorably oriented faults will fail when the stress state intersects the frictional strength curve (Eq. 12). In this case, the frictional strength would provide a practical upper limit to crustal strength in the brittle field. The limit of differential stress $\sigma_d = \sigma_1 - \sigma_3$ in the crust predicted by Eqs. (2b) and (5b) can be written

$$\sigma_d = \frac{2(S_0 + \mu_f \sigma_{n,\text{eff}})}{\cos(\tan^{-1} \mu_f)} = \frac{2(S_0 + \mu_f \sigma_{n,\text{eff}})}{\cos \phi_f} \tag{13}$$

where the friction angle $\phi_f = \tan^{-1} \mu_f$ and $\tan 2\beta = 1/\mu_f$. Equation (13) specifies the differential failure stress for frictional failure, and by substituting c for S_0 and μ_i for μ_f, Eq. (13) can predict differential failure stress for rock fracture. Although intact rock strength is an appropriate stress limit at the surface of the Earth and in some near-surface applications such as mining, in the case of earthquake faulting, Eq. (13) appears to serve as an upper limit for stress in the upper crust, especially in intraplate regions. This frictional strength limit is confirmed by *in situ* measurements made at depths up to 9 km (Hickman, 1991; Townend and Zoback, 2000) (also see Chapter 40 by McGarr *et al.* and Chapter 34 by Zoback and Zoback). These field observations indicate that σ_d generally does not exceed laboratory measurements of the differential strength of faulted, dry, anhydrous silicates (i.e., Byerlee's law).

Using the same assumptions used to derive Eq. (13), the variation of stress with depth in the brittle part of the crust can be estimated from Eqs. (2b) or (12) provided the magnitude of one of the principal stresses is known. The magnitude of one of the principal stresses can be deduced in normal and thrust faulting environments by using Anderson's assumption that near the Earth's surface one of the principal stresses is vertical due to the lack of shear stress on Earth's free surface (Anderson, 1951), and the vertical stress is given with depth by the overburden. This assumption leads to expressions for normal stress on critically oriented faults in reverse fault environments where $\sigma_v = \sigma_3$:

$$\sigma_n = \frac{\sigma_3 - (S_0 - \mu_f p)\left(\mu_f - \dfrac{1}{\cos \phi_f}\right)}{1 + \mu_f^2 - \dfrac{\mu_f}{\cos \phi_f}} \tag{14a}$$

and for principally oriented normal faults where $\sigma_v = \sigma_1$:

$$\sigma_n = \frac{\sigma_1 - (S_0 - \mu_f p)\left(\mu_f + \dfrac{1}{\cos \phi_f}\right)}{1 + \mu_f^2 + \dfrac{\mu_f}{\cos \phi_f}} \tag{14b}$$

(Fig. 7) (as follows from Sibson (1991)). (Here we have used the definition $\phi_f = \tan^{-1} \mu_f$.) Equations (14a) and (14b) as depicted in Figure 7 indicate the Anderson–Byerlee predictions for depth varying stress in the shallow crust.

3.2 Strength of Crustal Faults

Laboratory measurements of frictional strength indicate that many common silicates should have sliding strengths proportional to effective normal stress as represented by Byerlee's law ($0.6 \leq \mu \leq 0.85$). In fact, field observations of crustal stress,

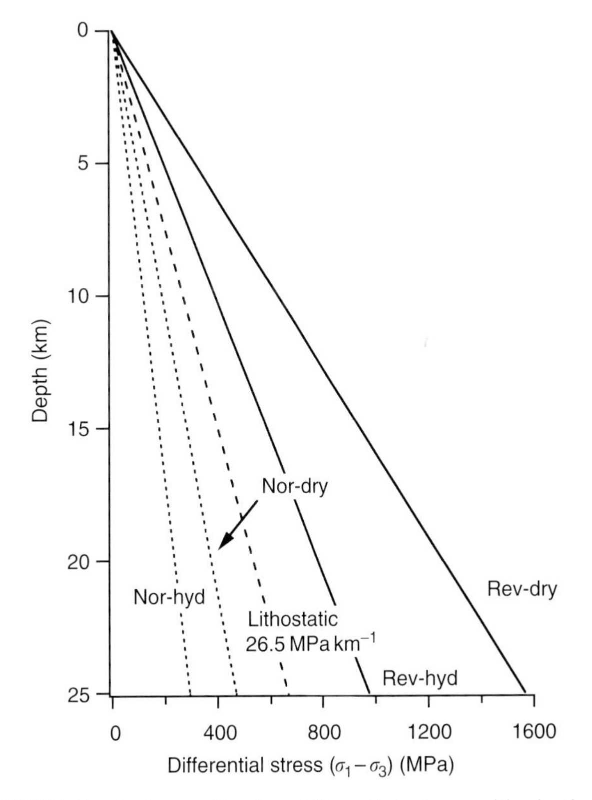

FIGURE 7 Anderson–Byerlee differential stress with depth in reverse (Rev) and normal (Nor) faulting environments, with hydrostatic (Hyd) fluid pressure of 10 MPa km^{-1} and no fluid pressure (Dry).

obtained primarily from shallow borehole stress measurements (see, for example, Zoback *et al.*, 1987; Townend and Zoback, 2000, and other chapters in this volume), show many crustal faults in a stress state consistent with a slightly wider range of friction of $\mu \sim 0.5$–1.0. The occurrence of faults with strengths somewhat below Byerlee's law can be explained by the presence of weak mineral phases (especially at shallower depths). Fault zones are generally wet and contain ground-up and altered gouge with significant hydrated mineral phases. In particular, many clays, micas and other hydrous phases which are common products of chemical reactions between water and anhydrous silicates have lower frictional strength than Byerlee's law (Raleigh and Paterson, 1965; Byerlee, 1978; Wang *et al.*, 1980; Moore *et al.*, 1983, 1986, 1996, 1997; Kronenberg *et al.*, 1990; Morrow *et al.*, 1992, 2000) (Fig. 8). Conversely, faults that are stronger than Byerlee's law may be either partially healed by precipitation reactions, or due to complex fault geometry, require breaking of intact rock to accommodate slip.

Although the strength of many faults seems to conform to Eq. (12), a long-standing debate concerns the strength of major plate-bounding faults and the San Andreas fault (SAF) in particular (see, for example, Hickman, 1991; Scholz, 2000). In what has come to be known as the heat flow paradox, field measurements have failed to detect any significant heat flow anomaly associated with the SAF (Brune *et al.*, 1969; Lachenbruch and Sass, 1980). Whether an active fault dissipates energy either through creep or dynamically during earthquakes, only a small portion of the consumed energy is expended in creating new fractures and surface energy (Wong, 1982b). Similarly, at least for strong faults, seismic energy radiated by earthquakes is also a relatively small fraction of the total energy expended (Lockner and Okubo, 1983; McGarr, 1998). The majority of energy associated with crustal fault deformation should be dissipated as shear heating and, for a fault such as the San Andreas that has been active for millions of years, should produce a measurable heat flow anomaly at the Earth's surface (Brune *et al.*, 1969; Lachenbruch and Sass, 1980). The lack of such an anomaly has been interpreted to imply an average shear strength of the top 14 km of the SAF of less than 20 MPa (or $\mu < 0.2$). Recent stress orientation measurements made at depths of 1–4 km adjacent to the San Andreas (Mount and Suppe, 1987; Zoback *et al.*, 1987; Shamir and Zoback, 1989) indicate that the greatest principal stress is at an angle approaching 90° to the fault which requires the fault to be weak relative to surrounding subsidiary faults that are oriented more favorably relative to the ambient stress field and yet remain locked while the San Andreas accommodates slip. Stress orientations determined along the SAF imply either an average coefficient of friction less than 0.1 or superhydrostatic fluid pressure (Lachenbruch and McGarr, 1990). Recently Scholz (2000) has proposed a model for a strong SAF in which plate-driving forces are applied as horizontal tractions at the base of the crust. This model also predicts rotation of principal horizontal stresses near the SAF and suggests that the inference that large plate-bounding faults such as the San Andreas are anomalously weak, remains open to debate.

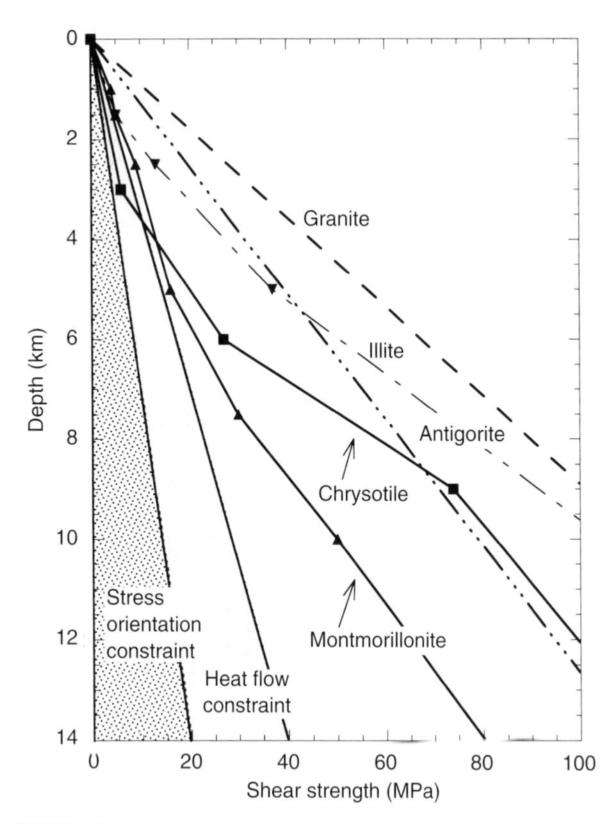

FIGURE 8 Inferred shear strength versus depth on SAF for possible fault zone constituent minerals. Heat flow and stress orientation constraints are indicated by linear strength increase with depth.

If the SAF is in fact weak relative to its surroundings, then a number of possible explanations have been proposed for the apparent discrepancy between field and laboratory strength observations: (1) dynamic rupture effects reduce normal stress across the fault during large earthquakes so that dynamic sliding strength is low; (2) high fluid pressure reduces effective normal stress and therefore shear strength; and (3) intrinsically weak minerals are responsible for low shear strength on mature faults. In terms of field observations of low heat flow, it may be possible that ground-water circulation acts to broaden and suppress a heat flow signature resulting from heating at depth. However, the borehole stress orientations are not subject to this effect and must be explained in some other manner (i.e., Scholz, 2000).

The possibility of low effective normal stress acting across the fault during large earthquakes has significant appeal and continues to be a topic of active research (see Section 6.2). Early models suggested that intense frictional heating would quickly vaporize pore water trapped in the fault zone and lead to high fluid pressure and low effective normal stress (Lachenbruch, 1980). More recent analysis has focused on the possibility that large earthquakes propagate in a pulse-like mode in which dynamic normal stresses unclamp the fault and allow slip at low shear stress (Heaton, 1990; Brune *et al.*, 1993; Ben-Zion and Andrews, 1998). Although these explanations are appealing, they only apply to dynamic slip and do not explain low heat flow along the creeping portion of the SAF in central California which appears to accommodate slip without large earthquakes.

Elevated fluid pressure is known to reduce fault strength (Hubbert and Rubey, 1959) and has been demonstrated to induce seismicity through the reduction of effective normal stress during fluid injection (Raleigh *et al.*, 1976) and reservoir impoundment (Talwani, 1976) (also see Chapter 37 of this volume). Furthermore, fluid overpressured zones, trapped fluids, and pore fluid compartmentalization are commonly observed in boreholes drilled for petroleum exploration and development (Hunt, 1990; Powley, 1990). If fluid migration is localized along faults and mineral precipitation and veining occurs, permeability will decrease and could lead to sealing and isolated fluid compartments. With time, faults can become barriers to flow and time-dependent compaction or dehydration reactions can lead to pore pressure buildup. The steady increase of fluid pressure can eventually lead to hydrofracture of the compartment seal and the escape of fluid. This cycle of slow fluid pressure buildup and then sudden pressure release has in fact been incorporated into a variety of episodic fault models (Sibson, 1981, 1992; Sleep and Blanpied, 1994; Fenoglio *et al.*, 1995; Lockner and Byerlee, 1995; Miller *et al.*, 1996).

The crust surrounding the SAF appears to be in compression (relative motion of Pacific and North American plates suggests a small component of convergence along the SAF) and is stressed critically for thrust failure in some areas such as the central Coast Ranges. Under these conditions, in order to avoid hydrofracture, the pore pressure must be less than or equal to a critical value

$$p_{cr} = \sigma_{H\max} - (\sigma_{H\max} - \sigma_{H\min}) \cos\psi \sin(\psi - \phi_f)/\sin\phi_f \quad (15)$$

where ϕ_f is the friction angle and ψ [$= ((\pi/2) - \beta)$ in Eqs. (4) and (5)] is the angle between the fault normal and the maximum horizontal stress $\sigma_{H\max}$ (Rice, 1992). Ignoring cohesion, for friction coefficients $\mu = \mu_f = 0.6$–0.9, Eq. (15) requires that p will exceed $\mu_{H\min}$ and lead to hydrofracture when ψ is less than 30–40°. Thus, provided the stress state is homogenous across the fault zone, the observed high angle for $\sigma_{H\max}$ along the fault is an argument against high pore pressure as an explanation of the weakness of the SAF.

This problem of excessive pressure buildup resulting in hydrofracture before fault slip occurs can be avoided for a mature fault zone which is wide enough to sustain a pore pressure gradient. Two mechanisms have been proposed for producing pore pressure which increases from the fault zone boundaries to the fault core axis. Byerlee (1990, 1993) suggested that for fine-grained fault gouge (especially containing hydrophilic clays or other minerals) a threshold pore pressure gradient may exist below which pore fluid will not flow. In this case, time-dependent compaction of the fault gouge material would lead to the establishment of a stable pore pressure profile in which pressure increases toward the fault zone axis by an amount equal to the threshold pressure gradient of the gouge material. For small faults that have narrow gouge zones, this pressure gradient is inconsequential. However, for mature fault zones such as the San Andreas where the fault zone may be hundreds of meters wide, a stable pore pressure profile could exist in which there is no pore fluid flow between, for example, hydrostatic pore pressure on the boundaries of the fault zone and near lithostatic pore pressure along the fault zone axis. In this case, the fault would have essentially no shear strength. In this model, rising fluid pressure within the fault zone leads to loss of strength and yielding. The resulting deformation changes the magnitude and orientation of principal stresses within the fault so that the pore pressure remains less than the minimum principal stress everywhere.

Rice (1992) proposed an alternative model in which a deep-rooted fault such as the San Andreas would have higher permeability than the surrounding crustal rock and would act as a conduit for the upward migration of mantle-derived fluids. Based on laboratory observations that permeability is highly sensitive to effective mean stress (Brace *et al.*, 1968; Pratt *et al.*, 1977; Brace, 1978; Morrow *et al.*, 1994), Rice's model predicts that pore fluid pressure at the base of the fault zone increases until a steady-state condition is achieved in which fluid flows up the axis of the fault zone and leaks out into the surrounding country rock. In this case, as in Byerlee's model, a pore pressure gradient is established between the fault zone axis and the country rock, so that pore pressure in the interior of the fault zone can be near-lithostatic and at the same time be less than the minimum principal stress throughout the fault zone. These models must be verified by direct observation over many years since drilling into the fault zone is likely to disturb the local pore pressure and the re-establishment of a steady-state condition may be quite slow.

We now examine the fourth possibility that the San Andreas is weak strictly as the result of the fault gouge mineral strength. Although few minerals have frictional strengths significantly greater than $\mu = 0.85$, there are numerous examples of potential fault gouge materials that are weaker than Byerlee's law. For example, biotite (Kronenberg *et al.*, 1990) and graphite (Morrow *et al.*, 2000) have an easy cleavage and weak interlayer bonding which results in a low shear strength. In this regard, Byerlee's law can be thought of as a limiting shear strength for deformation of a granular aggregate or Coulomb material between rigid driving blocks. When weak minerals are sheared in similar geometries, frictional strength can be ranked according to mineral bond strength (Morrow *et al.*, 2000). Laboratory determinations of friction for a range of rock types and minerals that are found in fault zones are summarized in Figure 9. Most mineral strengths in Figure 9 with friction below 0.7 have been tested wet since the presence of pore water can contribute significantly to low strength of some minerals. Good examples are smectites (montmorillonite clay) and chrysotile (serpentine) which have a strong affinity for interlayer or adsorbed water and exhibit remarkably low shear strength when wet (Moore *et al.*, 1983, 1996, 1997; Reinen *et al.*, 1991, 1994; Morrow *et al.*, 1992). In this case, increased temperature or mean stress tends to drive off adsorbed water and leads to increased shear strength (Fig. 9) (Moore *et al.*, 1997).

Montmorillonite is an expandable clay that, because of its low shear strength and general abundance, has been suggested as a contributor to low strength faults (Wu *et al.*, 1975; Wang, 1984; Morrow *et al.*, 1992). At room temperature wet montmorillonite has a coefficient of friction of about 0.2. As we mentioned, this low shear strength is the result of weakly bonded interlayer water attracted to the charged surfaces of the clay platelets. Under similar conditions, dry montmorillonite has $\mu \sim 0.4$ (Moore *et al.*, 1983; Morrow *et al.*, 2000). Consequently, at shallow depths (low pressure and temperature) montmorillonite comes close to satisfying the heat flow and stress orientation constraints for the SAF. However, shear strength tests on montmorillonite at conditions representing increasing burial depth show two effects. First, the higher temperatures and pressures associated with the deeper fault zone tend to drive off the interlayer water and result in increased frictional strength. Secondly, montmorillonite becomes chemically unstable with increasing temperature and first dewaters and then reverts to illite by approximately 400°C (Wu *et al.*, 1975; Wang, 1984). Illite has a significantly higher shear strength than montmorillonite and cannot satisfy the weak SAF constraints. Above approximately 100°C mixed-layer montmorillonite/illite clay may be expected in the fault zone (Freed and Peacor, 1989; Meunier and Velde, 1989). As a result, the deeper part of the fault is expected to rapidly increase in strength (Fig. 8).

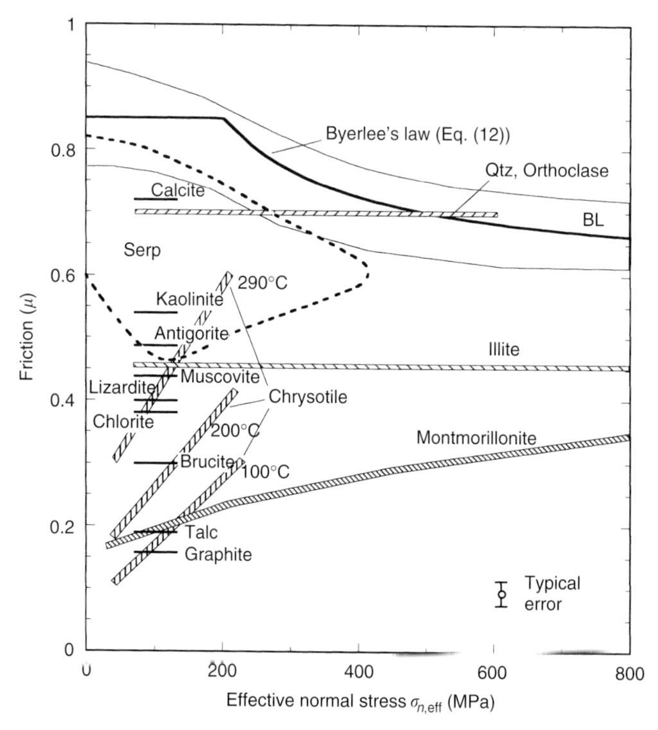

FIGURE 9 Frictional resistance ($\tau/\sigma_{n,\text{eff}}$) plotted for crushed layers of common rocks and minerals sheared between granite or sandstone driving blocks.

Another potential candidate for limiting shear strength of the San Andreas fault is the chrysotile form of serpentine. Due to its strong affinity for adsorbed water, chrysotile is one of the weakest likely fault zone minerals (room temperature friction is $\mu \sim 0.2$ and decreases to a minimum strength of $\mu \sim 0.1$ at 100°C (Moore *et al.*, 1996, 1997). Furthermore, chrysotile is associated especially with the creeping section of the SAF in central California (Coleman and Keith, 1971; Irwin and Barnes, 1975). However, as in the case of montmorillonite, chrysotile also increases strength with temperature and pressure as shown in Figures 8 and 9. Although it is possible that other less-common minerals exist that may retain a low coefficient of friction to relatively high temperature (i.e., brucite, Morrow *et al.*, 2000), the general trend of increasing strength of common fault gouge constituents with increasing pressure and temperature makes it unlikely that a weak San Andreas fault zone can be explained solely in terms of mineralogy.

In summary, a self-consistent workable model for a weak San Andreas fault can be based on elevated fluid pressure. A combination of elevated fluid pressure and inherently weak fault-zone mineralogy should also satisfy the heat flow constraints and has the advantage that less severe pore fluid overpressures are required. Dynamic weakening can also contribute to reduced frictional heating and needs to be understood more fully. It is possible that all three mechanisms will be found to play significant roles in contributing to the apparent low strength of the San Andreas fault.

3.3 Brittle–Ductile Models

Laboratory observations suggest earthquakes should be limited to conditions where the differential stress does not exceed the ductile flow strength of rocks, because under most circumstances ductile deformation, deformation which allows large strain without loss of strength, precludes dynamic stress drop. Ductile yield strength, equal to the differential stress σ_d during yielding, rapidly decreases with temperature as follows from the creep flow law

$$\dot{\varepsilon} = A\,\sigma_d^n \exp\left(\frac{-E}{RT}\right) \tag{16}$$

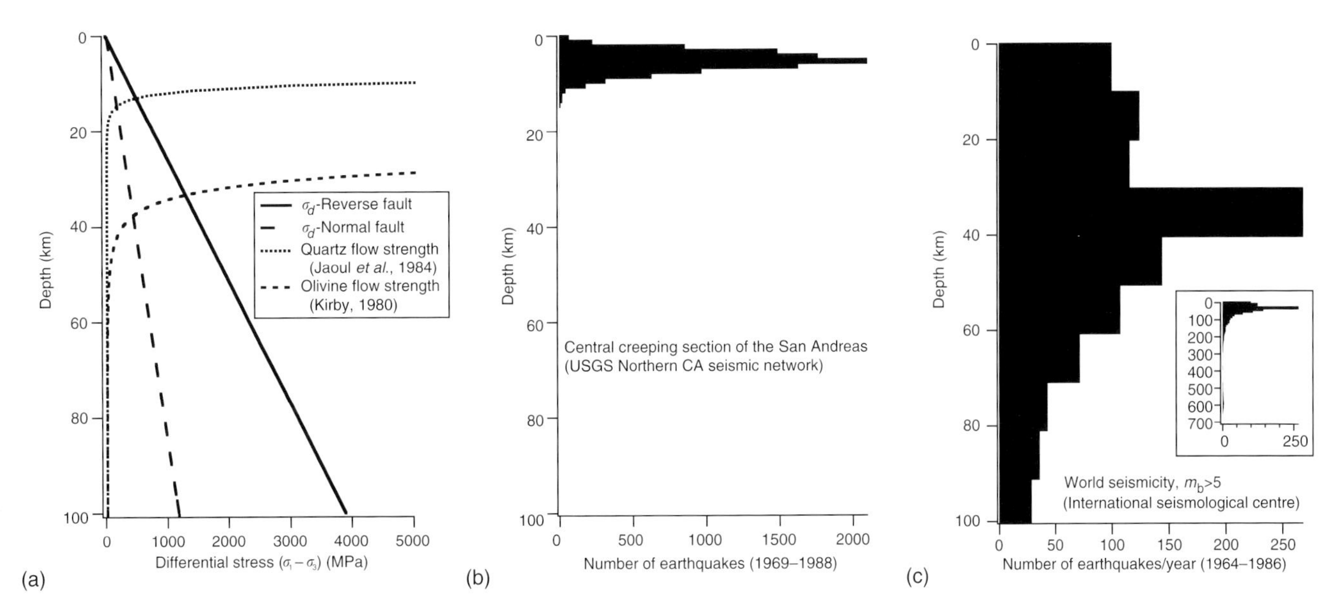

FIGURE 10 The brittle ductile transition and the depth extent of seismicity. (a) Brittle and ductile strength. (b) Seismicity from the central creeping section of the San Andreas fault. (c) World seismicity for events with body wave magnitude $m_b > 5$.

where $\boldsymbol{R}$ is the gas constant, E is the activation energy, n is a dimensionless constant, and A is a constant ($\text{MPa}^{-n}\,\text{sec}^{-1}$). Seismicity occurs at conditions where brittle strength [Eq. (14)] is lower than Eq. (16) (Fig. 10a). Diagrams, originated by Goetze and Evans (1979), such as Figure 10, where quartz strength is assumed representative of the continental crust and olivine of oceanic crust, can explain the general depth dependent features of natural seismicity (Fig. 10b and c). However uncertainties in the form of the flow law (Brace and Kohlstedt, 1980; Kirby, 1980), uncertainty of the Earth's temperature gradient and the strain rate dependence of Eq. (16) do not allow the strength or other material properties to be well constrained by natural seismicity.

An additional consideration in determining the depth extent of seismicity in the Earth is to allow for the possibility of aseismic fault slip within the brittle field. Studies of frictional sliding indicate that faults can slide stably under some circumstances. Thus it is possible that the maximum depth extent of seismicity is not determined by the transition to predominately ductile deformation but instead by a transition from unstable to stable frictional sliding (Scholz, 1990; Blanpied *et al.*, 1991, 1995; Chester, 1994, 1995).

4. Frictional Instability and Earthquakes

If earthquakes correspond to slip on preexisting fault surfaces, as opposed to failure of intact rock, then the conditions leading to unstable fault slip are relevant to studies of earthquake nucleation. Unstable sliding is only possible if fault strength drops following the onset of slip so that stored elastic energy from the surrounding rock can be used to drive the instability. The simplest description of fault rheology [Eq. (2b)] represents shear resistance strictly in terms of (effective) normal stress and precludes the possibility of weakening due to time-, temperature-, slip-, or slip rate-dependent processes. Under proper conditions, Eq. (2b) can allow for instability driven by increasing pore fluid pressure or possibly a rotation of the stress field loading a fault, but there are clear examples of fault instability in laboratory tests that do not involve these mechanisms. For frictional slip, instability depends only indirectly on the absolute level of strength. Rather, instability depends on inequality between fault strength and the stress available to shear the fault, which is supplied by the loading system. In basic terms, elastic energy is stored in the surrounding rock as tectonic processes load the brittle crust. When the strength of a fault is reached, sliding will commence and stored elastic energy is converted into frictional heat or surface energy as the fault rock is crushed. If an increment of slip on the fault surface reduces the fault strength faster than the driving stress is reduced, an unstable condition develops in which fault slip can accelerate into an earthquake and some of the stored elastic energy is radiated as seismic waves (Byerlee, 1970; Dieterich, 1979). In the laboratory, the test sample, piston, and load frame store elastic energy just as the brittle crust stores energy around a fault. Steady loading followed by dynamic fault slip is referred to as the "stick-slip" cycle and has been proposed as the laboratory counterpart of the earthquake cycle (Fig. 11) (Brace and Byerlee, 1966). Indeed laboratory stick–slip events are similar to earthquakes and if considered in terms of the dimension of rupture, earthquakes and other rock seismic faulting events form a continuum over at least 10 orders of magnitude. Great earthquakes occur on faults with linear

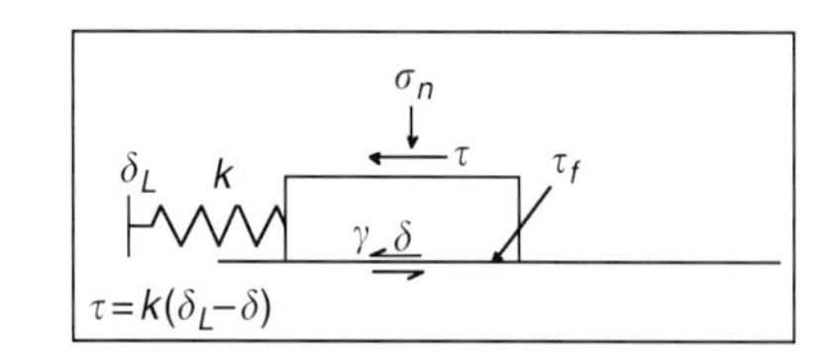

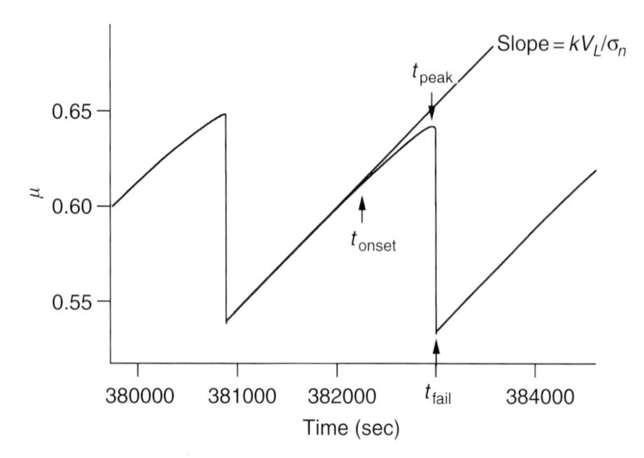

FIGURE 11 Stick–slip. Frictional strength vs. time during sliding on initially bare surfaces of Westerly granite at room temperature, 50 MPa confining pressure and 0.1 μm sec^{-1} loading velocity. The onset of detectable slip (t_{onset}) and the peak stress (t_{peak}) differ from the failure time (t_{fail}). Unstable slip shown by vertical line segments. The inset shows the spring and slider block model appropriate for laboratory experiments.

dimensions of hundreds of kilometers, whereas microseismicity and mining rock bursts represent fault instabilities spanning meters to tens of meters. Typical laboratory faults have lengths of 0.01–0.1 m and for specialized test machines have 2 m fault length (Okubo and Dieterich, 1984). At an even smaller scale, microcracking and associated ultrasonic acoustic emissions occur over a range from micrometers to millimeters and have been used as a laboratory analogue for many earthquake phenomena (Lockner, 1993a).

The elastic energy which drives instability can lead to a net drop in stress as shown from direct high speed stress measurements recorded during model earthquakes (Okubo and Dieterich, 1981, 1984, 1986; Lockner *et al.*, 1982; Lockner and Okubo, 1983; Ohnaka and Kuwahara, 1990). The simplest idealization of this stress drop is static and dynamic friction (Jaeger and Cook, 1971; Rice and Tse, 1986; Scholz, 1990), that is, a fault at rest has static friction represented by Eq. (2b) or (12) but once it begins to slip, strength drops immediately to a characteristic dynamic friction level. This concept proves to be unrealistic when applied to rupture nucleation due to the assumption of an abrupt change in fault strength (Ruina, 1983); strength drop for real materials is not instantaneous, and requires an apparent time or displacement weakening. An improvement over the static and dynamic friction model which well represents many aspects of the laboratory observations (Okubo and Dieterich, 1984, 1986; Ohnaka *et al.*, 1986; Ohnaka and Kuwahara, 1990) is the incorporation of strain weakening, also known as displacement or slip weakening. This slip-weakening behavior has been described theoretically (Andrews, 1976; Rice, 1980; Rudnicki, 1980) and has also been confirmed experimentally for brittle failure of intact laboratory samples (Rice, 1980; Wong, 1982b; Lockner *et al.*, 1991) (see Section 4.2). In this case, stress decreases from static to dynamic friction over a characteristic displacement d_c or a characteristic strain $\varepsilon_c = d_c/w$, where w is the thickness of the deforming layer. Simple slip weakening functions include linear and exponential decay, and because of their mathematical simplicity, are useful in dynamic rupture propagation models.

If the elastic stiffness of the loading system is expressed as $k = d\tau/d\delta$, where δ is fault slip, then a requirement for instability at constant normal stress can be expressed in terms of a critical stiffness (Dieterich, 1979)

$$k_c = \frac{d\mu}{d\delta}\sigma_n \approx \frac{\Delta\mu\sigma_n}{d_c} \tag{17}$$

where $\Delta\mu$ is the drop in coefficient of frictional resistance. A slip instability can then occur for $k < k_c$. This relation has been verified experimentally for samples containing well-characterized displacement weakening (Lockner and Byerlee, 1990). In laboratory experiments, stiffness is the combined stiffness of the intact portion of the sample and the testing apparatus, whereas in the Earth the stiffness is related to the elastic modulus of the rock and the geometry of the portion of the fault that is slipping (e.g., Byerlee, 1970; Walsh, 1971; Dieterich, 1978a, 1979).

4.1 Characteristics of Earthquake Nucleation

In the context of earthquake nucleation, failure criteria such as Amontons' law establish a yield stress but are unable to explain premonitory phenomena such as creep or rupture nucleation. Although some moment inversions of radiated seismic energy suggest that earthquakes have distinct nucleation and propagation phases (e.g., Ellsworth and Beroza, 1995, 1998), only the final, rapid slip portion of the nucleation phase can be studied with seismic wave analysis. Attempts to measure natural fault slip during the early stages of fault nucleation have been unsuccessful (Johnston *et al.*, 1987; Linde and Johnston, 1989; Chapter 36 by Johnston and Linde). Given the high sensitivity of the strain meters in these studies, analysis of limited data suggests premonitory creep for moderate-sized earthquakes is restricted to nucleation patches that are tens of meters or less in linear dimension. Although field measurements of deformation have been unsuccessful in detecting the early strain associated with rupture nucleation, laboratory studies consistently demonstrate a characteristic accelerating slip phase that develops into the dynamic instability. Figure 11 shows the primary characteristics of slowly nucleating laboratory failure. (1) Fault slip δ accrues as the fault is loaded at a constant rate, as evidenced by the deviation from linear increase in stress with time [shear stress on the fault $\tau = k(V_L t - \delta)$]. This behavior, if appropriate for natural earthquake failure allows for the possibility of detectable precursors such as strain or strain-related phenomena. For example, precursory strain should be associated with dilatancy (Nur, 1972; Scholz *et al.*, 1973) and a number of precursory phenomena measured in the laboratory [i.e., changes in acoustic velocity and attenuation (Lockner *et al.*, 1977, 1992a; Yanagidani *et al.*, 1985; Masuda *et al.*, 1990; Satoh *et al.*, 1990), Gutenberg–Richter b-value (Weeks *et al.*, 1978; Meredith *et al.*, 1990), electrical resistivity (Lockner and Byerlee, 1986), and permeability (Zoback and Byerlee, 1975; Morrow *et al.*, 1981)]. (2) Peak stress occurs before failure and post-peak accelerating slip is accompanied by slip weakening (e.g., Okubo and Dieterich, 1984; Ohnaka *et al.*, 1986; Ohnaka and Kuwahara, 1990; Lockner and Beeler, 1999) indicating no unique threshold failure stress. (3) Failure is delayed: there is lag time between the onset of detectable slip (t_{onset}) or the time of peak stress (t_{peak}) and the time of failure (t_{fail}). The reader is reminded that all of these features have direct counterparts in intact rock failure (Fig. 3 and, for example, Lockner *et al.*, 1992a; Lockner, 1998).

4.2 Slip-weakening Model and Unstable Rupture

Any mechanism that results in a loss in shear resistance that is more rapid than the unloading of the surrounding region has the potential for producing slip instability. One obvious example is the fracture of intact rock or, in the case of an irregular or segmented fault, the rupture of an asperity or competent step-over. Laboratory measurements (Fig. 1) indicate that strength reduction resulting from brittle fracture can be 20–50% of peak strength whereas strength loss associated with stick-slip of nominally flat, ground rock surfaces (see Section 4.3) is approximately 5–10% (Okubo and Dieterich, 1986; Ohnaka and Yamashita, 1989). Laboratory measurements have shown that dilatancy-induced pore pressure loss in a hydraulically isolated fault can lead to a 20% increase in effective normal stress (Lockner and Byerlee, 1994). Collapse of this open pore structure with continued deformation (Blanpied *et al.*, 1992) could repressurize pore fluid and provide a mechanism for triggering instability. The idealized flat, parallel fault surfaces used in laboratory tests to determine fault constitutive properties as described in the following section by Eqs. (18) are designed specifically to avoid complicating factors related to geometric fault complexity (and fluid pressure effects). In reality, natural faults are commonly irregular (fractal) over a broad range of scale (Brown and Scholz, 1985b; Scholz and Aviles, 1986; Power *et al.*, 1988; Moore and Byerlee, 1989, 1991). Some aspects of the geometric complexity of faults have been explored (Andrews, 1989) but no framework exists for systematically analyzing this issue.

Ohnaka and co-workers developed an empirical model for unstable rupture that is described solely in terms of fault slip. The model is intended to describe the nucleation and growth of unstable rupture as measured on a biaxial press with granite sample dimension of $280 \times 280 \times 50$ mm (Ohnaka *et al.*, 1986). This apparatus, similar to but smaller than the 2-meter biaxial press (Okubo and Dieterich, 1981), has a fault surface that is large relative to the nucleation patch size for unstable rupture. The idealized slip instability is based on a slip-weakening model as discussed by Rice (1980) and Rudnicki (1980). The model includes a breakdown zone associated with an advancing crack tip in which shear stress τ near the crack tip is a function of shear displacement δ (Fig. 12a). High-speed records of the passage of a rupture front in a 2-meter fault are shown for comparison in Figure 12b (after Okubo and Dieterich, 1986). Note the characteristic stress rise as the rupture front approaches, followed by a rapid weakening to the dynamic sliding strength over displacement d_c. When shear resistance at a point on the fault is plotted as a function of fault

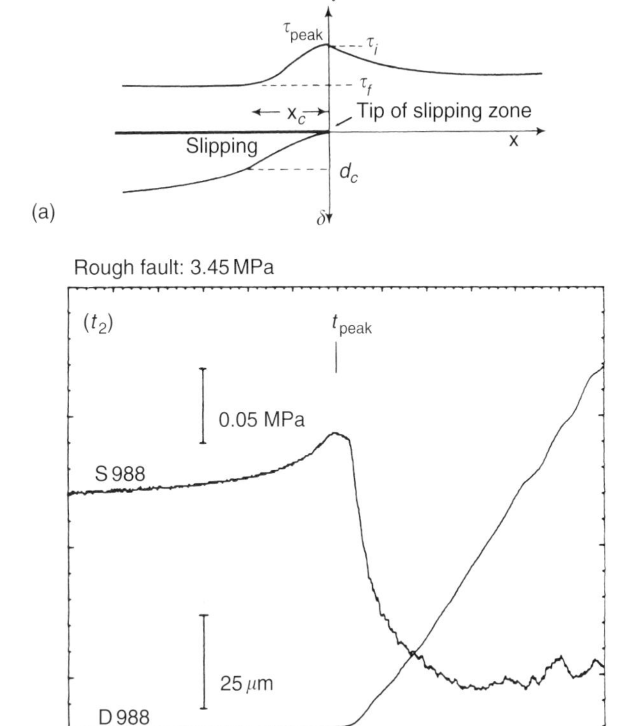

FIGURE 12 (a) Idealized process or breakdown zone associated with an advancing shear rupture. (b) An example of high-speed shear stress and fault displacement recordings during the passage of a shear rupture in granite on a 2-meter biaxial press.

slip (Rice, 1980) (Fig. 13), energy release rate G_{II} associated with the rupture propagation is represented by the area under the stress curve in excess of the frictional sliding resistance E_f. Similar results have been reported on the 0.4-meter biaxial press (Ohnaka *et al.*, 1986; Ohnaka and Yamashita, 1989; Ohnaka and Kuwahara, 1990; Yamashita and Ohnaka, 1991) (Fig. 13b). Scaling relationships have been proposed (Ohnaka and Shen, 1999) for critical slip distance, nucleation patch size and energy release rate assuming a dependence of nucleation on a characteristic surface irregularity size (Ohnaka and Yamashita, 1989; Ohnaka and Kuwahara, 1990; Yamashita and Ohnaka, 1991). The issue of fault heterogeneity arises in many contexts in earthquake processes including nucleation and stopping phases of earthquakes, abundance of high-frequency radiated energy, and earthquake statistics (Rice and Ben-Zion, 1996).

Fracture data for an initially intact sandstone sample are shown in Figure 13c where differential stress is plotted versus inelastic (or permanent) axial strain. This plot might represent, for example, the rupture of an asperity or interlocked region on a fault. Fracture nucleation coincides with the onset of rapid slip weakening. This portion of the breakdown curve is similar to the breakdown curve for the preexisting fault (Fig. 13b). The stress drop can be significantly greater for fracture of crystalline rock such as granite that has significant cohesion. The most notable difference between Figures 13b and c is the large amount of work done in loading the intact rock to peak stress. This work is expended in growing microcracks and in dense crystalline rock is accompanied by a volume increase and changes in a host of related material properties such as wave speed, electrical resistivity and permeability. Since these changes occur quasi-statically in response to tectonic loading, this is the effect most commonly appealed to in searching for premonitory phenomena.

This section ends with a comparison of time–position plots of fracture growth on a preexisting fault and in an intact rock. Figure 14a shows the growth of an instability in a precut granite sample with a 0.4-meter fault (Yamashita and Ohnaka, 1991). Solid circles denote the time the shear stress reaches the peak value, indicating the onset of instability. Open circles denote the time when the shear stress has dropped to the dynamic frictional stress level, indicating the passage of the breakdown zone. Note the finite width of the breakdown zone once the fracture is growing at near shear wave speed. Figure 14b is a space–time plot of acoustic emission rate recorded during shear fracture growth in an initially intact 0.2-meter long granite sample (Lockner and Byerlee, 1992). In this example, the fracture propagated quasi-statically as indicated by the slow advance rate of acoustic emission activity across the sample. The nucleation phase had a width of 2–3 cm in both the intact and faulted samples. Following nucleation, the narrow band of intense acoustic emission activity indicated the position of the advancing process zone.

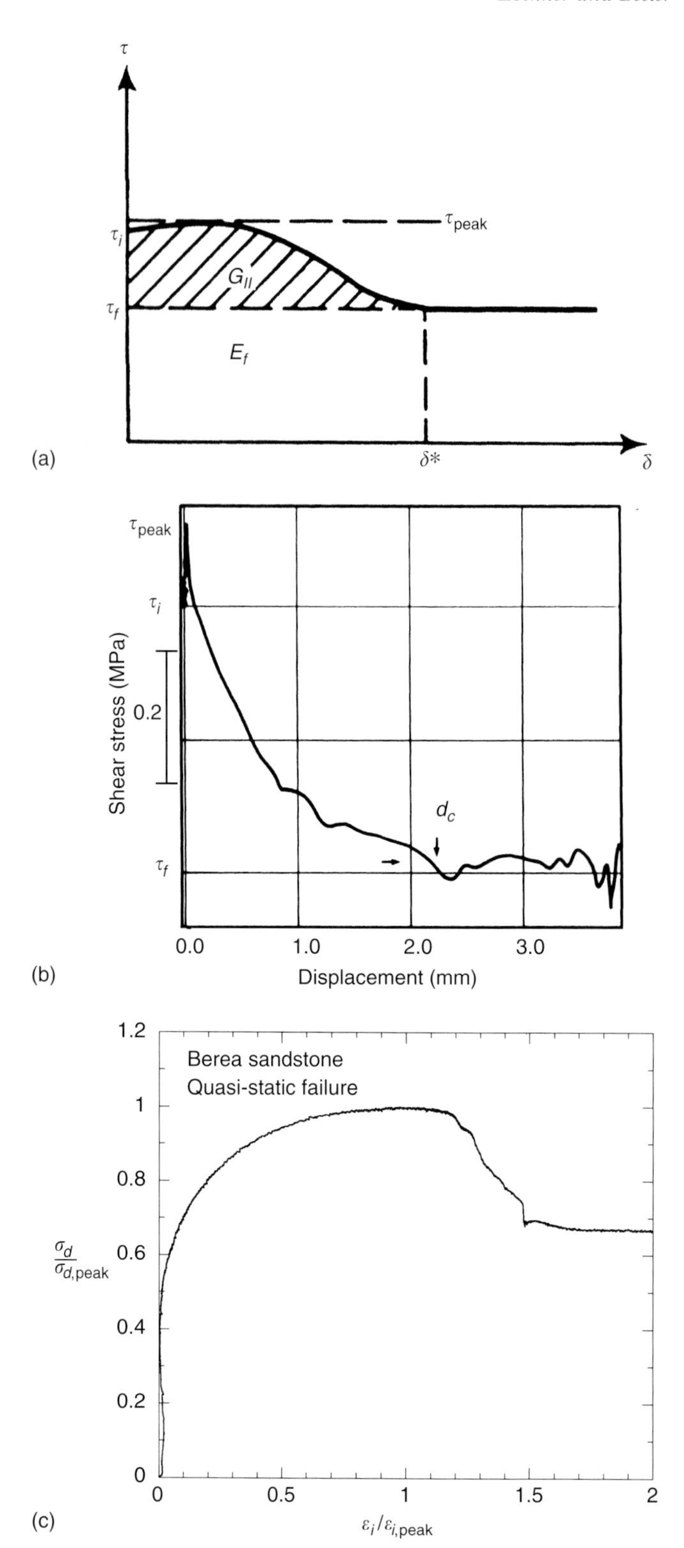

FIGURE 13 (a) Idealized stress–displacement relation for rupture propagation shown in Figure 12a. Energy release rate per unit area of fault surface is represented by G. (b) High-speed data recording showing stress-displacement relation during passage of an unstable rupture. (c) Normalized stress–inelastic axial strain plot for failure and fracture propagation in initially intact Berea sandstone.

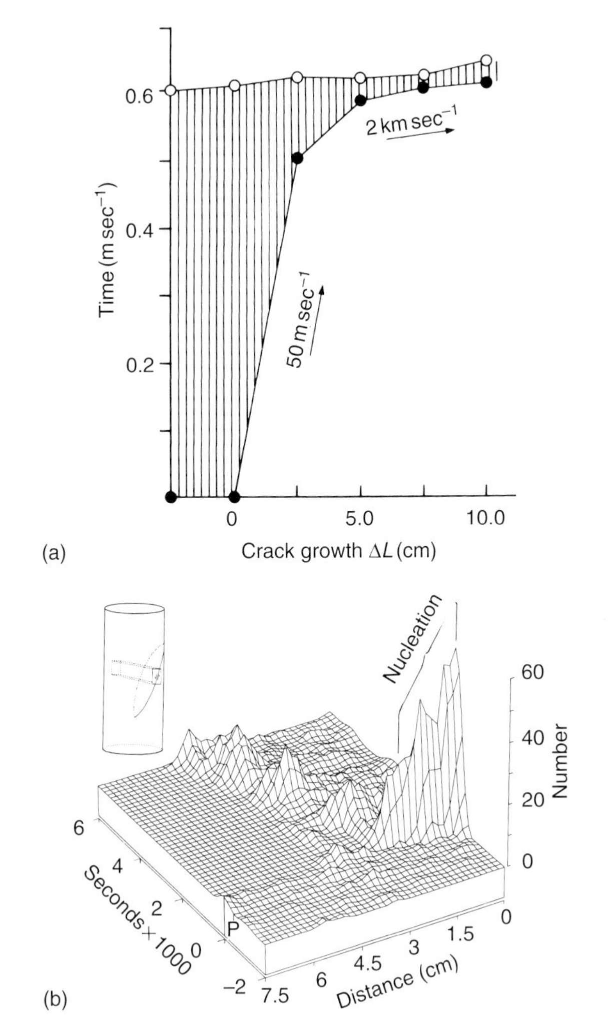

FIGURE 14 (a) Nucleation and initial propagation phase of unstable rupture event on a preexisting fault in granite. (b) Plot of acoustic emission rate in region across center of initially intact granite sample.

4.3 Rate and State Friction Equations

The displacement-dependent rupture model described, for example, in Ohnaka and Shen (1999) describes fault conditions following the onset of rapid slip. Although it can account for failure characteristics such as post-peak loss of strength (Fig. 11), it does not address the characteristic prefailure deformation of slow, time-dependent creep (referred to as phase I in Ohnaka and Kuwahara, 1990). The dependent variables in frictional sliding, especially at slip rates of 10^{-9}–10^{-4} m sec^{-1}, are more easily revealed in rate-stepping tests in a stiff testing machine ($k > k_c$) than in stick-slip or rupture nucleation experiments. Experimental measurements indicate the characteristic response of rock friction to changes in sliding velocity is somewhat complicated, consisting of two apparently additive log velocity-dependent effects: a positive, instantaneous dependence (direct effect) and a negative dependence which evolves approximately exponentially with displacement

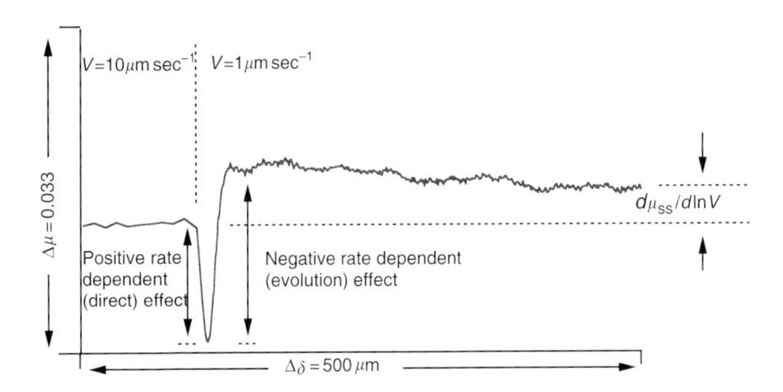

FIGURE 15 Typical slip-rate dependence of fault strength for granite at 25 MPa normal stress and room temperature in response to an imposed change in sliding velocity. The loading velocity is decreased by an order of magnitude at displacement corresponding to the vertical dashed line. The net rate dependence $d\mu_{ss}/d\ln V$ is the sum of two opposing effects.

(evolution effect) (Dieterich, 1978b, 1979) (Fig. 15). These effects are generally small and of similar magnitude (typically a few percent of the total friction for a decade change in velocity (Blanpied *et al.*, 1998a). Since the effects are opposite in sign, they tend to cancel and the net dependence on slip velocity is even smaller and can be either positive or negative (Dieterich, 1978b, 1979; Tullis and Weeks, 1986).

The detailed response of fault frictional strength, at room temperature and modest normal stress (typically 5–200 MPa), to changes in sliding velocity, whether during rate-stepping tests or during stick-slip sliding, can be well described by what are usually referred to as rate and state variable constitutive equations, developed by Ruina (1983) as a generalization of equations of Dieterich (1978a, 1979),

$$\mu = \mu_0 + \boldsymbol{a} \ln\frac{V}{V_0} + \boldsymbol{b} \ln\frac{\theta V_0}{d_c} \tag{18a}$$

$$\frac{d\theta}{dt} = 1 - \frac{V\theta}{d_c} \tag{18b}$$

(Rice and Ruina, 1983; Ruina, 1983). Here μ_0 is the frictional resistance to slip at an arbitrary reference sliding velocity V_0. μ_0 can be thought of as the first order frictional strength, such as described by Eq. (2) or (12). The other two terms in Eq. (18a) are the rate and state dependencies which are second order. The $\boldsymbol{a}\ln(V/V_0)$ term describes the instantaneous rate dependence of strength (direct effect in Fig. 15) and the last term in Eq. (18a) represents the state dependence (evolution effect in Fig. 15). As written here, θ is the "state variable," having the dimension of time. The form of θ prescribed in Eq. (18b) is one of a number of empirical equations that have been used in this context. [See Linker and Dieterich (1992) and Ruina (1983) for detailed discussions of the different forms of state and their properties.] The steady-state value of state, as deduced from Eq. (18b) is $\theta_{ss} = d_c/V$. When substituted in Eq. (18a) this defines the steady-state rate dependence of fault strength $d\mu_{ss}/d\ln V = b - a$ which determines the sliding stability.

In the case where the dependence of steady-state frictional strength on the logarithm of sliding velocity is negative ($d\mu_{ss}/d\ln V < 0$), any perturbation that causes the fault to accelerate will lead to a loss of strength and will encourage further slip acceleration. In this case, slip instability is possible. Conversely, for neutral or positive rate dependence ($d\mu_{ss}/d\ln V \geq 0$) sliding will tend to be stable (Rice and Ruina, 1983). Linearized perturbation analysis of Eq. (18) (Rice and Ruina, 1983; Ruina, 1983) results in an expression for critical stiffness analogous to Eq. (17)

$$k_c = \frac{d\mu_{ss}}{d\ln V}\frac{\sigma_n}{d_c} \tag{19}$$

where μ_{ss} is the steady-state frictional resistance and d_c is the characteristic displacement of the evolution effect. Accordingly, instability is favored by more negative values of $d\mu_{ss}/d\ln V$ and by low values of k (more compliant loading systems) (Rice and Ruina, 1983; Gu *et al.*, 1984).

Laboratory measurements of rate dependence have been conducted using two-fault configurations: initially bare rock surfaces which consist of ground, flat, parallel rock which has been roughened with a coarse (typically 60–80 grit) abrasive, and simulated gouge layers which consist of powdered rock flour (typically $<90\,\mu$m diameter) sheared between rigid blocks. The rate dependence of quartzo-feldspathic rocks, on which the constitutive relations Eq. (18) were based, can depend on whether or not the fault zone contains a thick gouge layer. Thick quartzo-feldspathic gouge layers have positive rate dependence unless shear displacement becomes localized within the layer (Dieterich, 1981; Marone and Kilgore, 1993) (also see review by Marone, 1998). Conversely initially bare rock surfaces of quartzo-feldspathic rock (Dieterich, 1978b; Tullis and Weeks, 1986) are rate weakening. In our review of rock friction measurements relevant to earthquakes we focus primarily on measurements that relate directly to frictional instability. We have therefore omitted studies on the frictional properties of gouge layers which show variability in the sign of the rate dependence (see summary by Beeler *et al.*, 1996), we focus predominately on the results from rate-weakening experiments on initially bare rock surfaces. Slip between such surfaces remains localized even at large displacement (Beeler *et al.*, 1996) and such faults show little tendency for fault zone width growth after an initial period of rapid wear (Biegel *et al.*, 1989) because they are flat at long wavelengths [see Power *et al.* (1988) and as discussed below].

In contrast to experimental faults, a continually expanding fault zone thickness is expected for natural faults (Robertson, 1983). For example, the San Andreas fault zone, with total offset of hundreds of kilometers, has a thickness of as much as a kilometer in places (Chapter 29 by Sibson). Increased thickness of natural faults may result from geometrical and dynamic effects during shear. Natural rock surfaces such as fracture and joint surfaces have a fractal roughness (Brown and Scholz, 1985a). As a result, increasing displacement of an initially matched shear fracture (proto-fault) leads to an increasing volume of mismatched surface irregularity and therefore increasing wear material (gouge) with displacement (Power *et al.*, 1988). During earthquake slip, large dynamic stresses that develop in front of an advancing shear rupture during propagation may be higher on off-fault planes than on the fault plane itself (Rice, 1980). These high stresses may induce branching or off-fault deformation which may lead to nonplanar ruptures, fault surface roughening and gradual growth of the damage zone with successive earthquakes. The experimental observations thus require that slip during earthquake nucleation is localized so that $d\mu_{ss}/d\ln V < 0$, even within very wide fault zones. At least in experiments to date on gouge-filled faults, fault slip at large displacement does not localize to the degree that the rate dependence is negative (Beeler *et al.*, 1996). It is not known how such localization arises in mature natural fault zones.

4.4 Origin of Delayed Failure in Friction and Fracture

The two opposing rate-dependent effects seen in rate stepping tests (Fig. 15) can be qualitatively related to the onset of failure in stick-slip experiments (Fig. 11) in a straightforward manner. Obviously, frictional failure requires that the rate weakening term exceeds the rate strengthening term for the stress to drop as slip accelerates. However, negative rate dependence alone leads to rapid failure following a threshold failure stress and therefore does not allow for delayed failure or precursory slip. It is the long duration of the acceleration to failure (Fig. 16) which most clearly distinguishes observed failure from failure predicted by simple static and dynamic strength. Prior to the peak strength, which occurs when the sliding velocity equals the loading velocity $V = V_L$, the sliding velocity increases (Fig. 16)

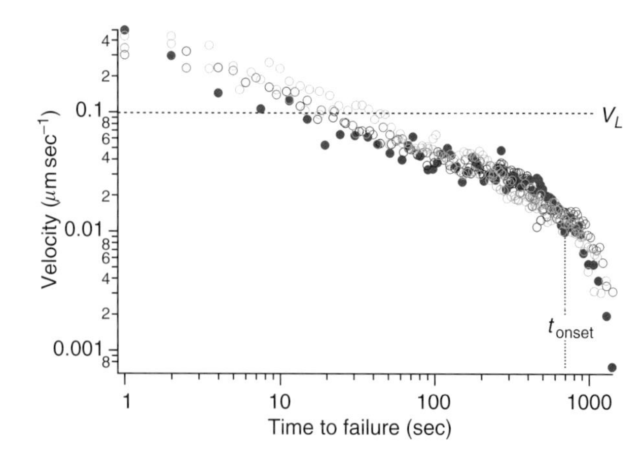

FIGURE 16 Acceleration to failure during slip of granite at room temperature and 50 MPa confining pressure.

and yet the stress on the fault also increases (Fig. 11) indicating that fault strength depends positively on sliding velocity. Thus this precursory creep during delayed failure is directly attributed to the direct effect (Fig. 15). A similar positive dependence of strength on deformation rate is seen during intact failure (e.g., Lockner, 1998). Delayed failure in response to stress perturbations, which is characteristic of both frictional failure and fracture, is consistent with natural observations of Omori aftershock sequences (Scholz, 1968b; Lockner, 1993a; Dieterich, 1994) (see Section 4.5) and other triggered seismic phenomenon (Gomberg *et al.*, 1998) in which failure is delayed following a stress increase.

Brittle fracture at high normal stress is involved in both abrasion during frictional sliding of bare fault surfaces and grain fracture during shearing of granular gouge layers. Therefore, all kinds of experimental shear deformation in the brittle field should share characteristics associated with fracture. Fracture theory as well as experimental evidence suggest log rate dependent resistance during nucleation of a single fracture. The stress–inelastic strain relation for subcritical fracture (i.e., slow, stable fracture growth at stress intensity less than the critical value as discussed in Section 2.3.2), results from equating strain rate (equivalently velocity V) resulting from crack propagation to the rate of chemical reaction at the crack tip (Lockner, 1993b, 1998) where reaction rate is

$$v = v_0 \exp\left(\frac{-E + \Omega\sigma - \Omega_m \gamma/\chi}{RT}\right) \tag{20}$$

(e.g., Charles and Hillig, 1962). E is the activation energy, Ω is the activation volume, Ω_m is molar volume, γ is the surface energy, χ is the radius of curvature of the tip, v_0 is a constant, R is the gas constant, T is temperature and σ is the stress at the crack tip. The stress term is akin to the pressure dependence normally associated with an activation volume, such that the reaction rate is faster if the stress is elevated. As has been suggested by numerous authors (see review by Costin (1987)), this stress sensitivity of crack propagation rate likely underlies the established creep-like behavior of rock at low temperature (Scholz, 1968a) and the log rate dependence of the failure stress of intact rock (e.g., Lockner, 1998). Based on consideration of Eq. (20), and due to the involvement of fracture in frictional sliding, the subcritical component of fracture is a likely cause of the instantaneous positive rate dependence of rock friction.

4.5 Implications of Delayed Failure

Laboratory observations of delayed failure predict that stress changes affect the timing of earthquakes, and many natural observations of delayed failure are consistent with these predictions. For example, assuming that aftershocks result from a stress change superimposed on a constant tectonic loading rate, Omori aftershock sequences indicate that the sensitivity of an aftershock source fault to stress depends on its temporal proximity to failure. That is, because aftershock rates immediately following a main shock static stress change are higher than they are at later times, faults closer to failure are advanced toward failure by a different amount than those that are far from failure. Such a time dependence of aftershock rates follows directly from Eqs. (18) (Dieterich, 1994). Assuming, based on the results of studies of static stress change (Stein and Lisowski, 1983; Harris and Simpson, 1992; Reasenberg and Simpson, 1992), that aftershock failures are triggered by a stress increase and that ongoing tectonic loading provides a constant rate of stress increase, then seismicity rate r as a function of time t following an increase in shear stress $\Delta\tau$ is

$$r(t) = \frac{r_0}{1 - [1 - \exp(-\Delta\tau/a)]\exp(-\dot{\tau}t/a)} \tag{21}$$

where r_0 is the seismicity rate prior to the stress change, $\dot{\tau}$ is the stressing rate, and a is the constitutive parameter which scales the size of the direct effect in Eqs. (18) (Dieterich, 1994). Equation (21) predicts an Omori decay consistent with field observations (Fig. 17); a similar prediction was made by Scholz (1968b) who considered aftershocks as small fracture events, the natural analogue to acoustic emissions, instead of unstable slip of pre-existing faults.

Delay between the application of a stress change and failure, such as seen in laboratory experiments and predicted by constitutive Eqs. (18), are also consistent with the lack of tidal triggering of nonvolcanic earthquakes. Because daily variations in stress due to tidal forces (0.001–0.004 MPa) greatly exceed the daily accumulation of tectonic stress, earthquake occurrence would be strongly correlated with the tides if failure coincided with a stress threshold [e.g., Eq. (2)].

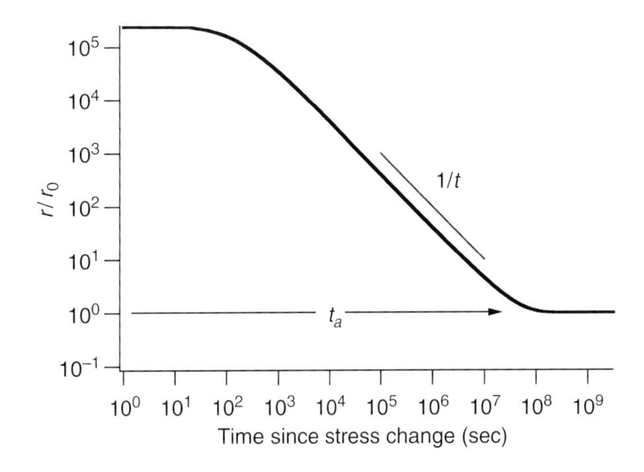

FIGURE 17 Time-dependent effects. Seismicity rate following a stress step from Eq. (21). The constitutive equation in this calculation is a simplification of Eq. (18). Seismicity rate follows an Omori decay and returns to the background rate over a characteristic time $t_a = a\sigma_n/\dot{\tau}$ (Dieterich, 1994).

However, analysis of natural occurrence suggests weak (Vidale *et al.*, 1998) or no (Heaton, 1982; Hartzell and Heaton, 1989) correlation between earthquakes and Earth tides. Numerical simulations of combined tectonic and tidal loading of fault populations obeying Eqs. (18) suggest that failure is insensitive to stress change with the amplitude and frequency of the tidal stress (Dieterich, 1987). Similar conclusions are reached from stick-slip experiments where constant loading was modulated by a small amplitude periodic signal. Correlation between the periodic loading and the time of failure occurred when the periodic amplitude was 0.05–0.1 MPa (Lockner and Beeler, 1999). Simple extrapolation of this result to the Earth suggests that approximately 1% of earthquakes on the San Andreas fault should be correlated with the tides; a result that is consistent with absent or weak correlations in field data.

4.6 Limits of Frictional Instability Predicted by Laboratory Observations

The rate and state variable constitutive Eqs. (18), which well describe many laboratory observations, can qualitatively explain a wide range of averaged seismic observations, e.g., realistic earthquake cycles (Tse and Rice, 1986), Omori aftershock sequences (Dieterich, 1994), the lack of tidal triggering of earthquakes (Dieterich, 1987; Lockner and Beeler, 1999), variations in stress drop with recurrence and loading velocity (Gu and Wong, 1991), foreshocks and earthquake clustering (Das and Scholz, 1981; Dieterich and Kilgore, 1996), earthquake afterslip (Marone *et al.*, 1991), precursory slip (Dieterich, 1994), and remotely triggered earthquakes (Gomberg *et al.*, 1997, 1998). Most important for frictional instability to be a viable explanation for natural seismicity is that transitions from stable (aseismic) and unstable (seismic) frictional sliding seen in experiments correspond to the depth range of observed seismicity. For example, seismicity on the San Andreas fault system is most active between about 4 and 15–20 km (Fig. 10). Two recent complementary experimental studies of quartz gouge and granite gouge attempt to directly predict the depth extent of seismicity by determining the rate dependence of frictional strength at temperatures spanning the entire seismogenic zone (Blanpied *et al.*, 1991, 1995; Chester and Higgs, 1992; Chester, 1994, 1995).

At the top of the seismic zone, corresponding to low temperature, the rate dependence of granite frictional strength is positive, in agreement with room temperature experiments (Byerlee and Summers, 1976; Solberg and Byerlee, 1984; Morrow and Byerlee, 1989; Marone *et al.*, 1990). The rate dependence of granite gouge undergoes a transition to rate weakening between 25 and 50°C (Blanpied *et al.*, 1991, 1995) (Fig. 18). However when converted to depth this prediction corresponds closely to the field observations of a seismic/aseismic transition at around 4 km only if the loading rates are orders of magnitude faster than the plate motion rate on the

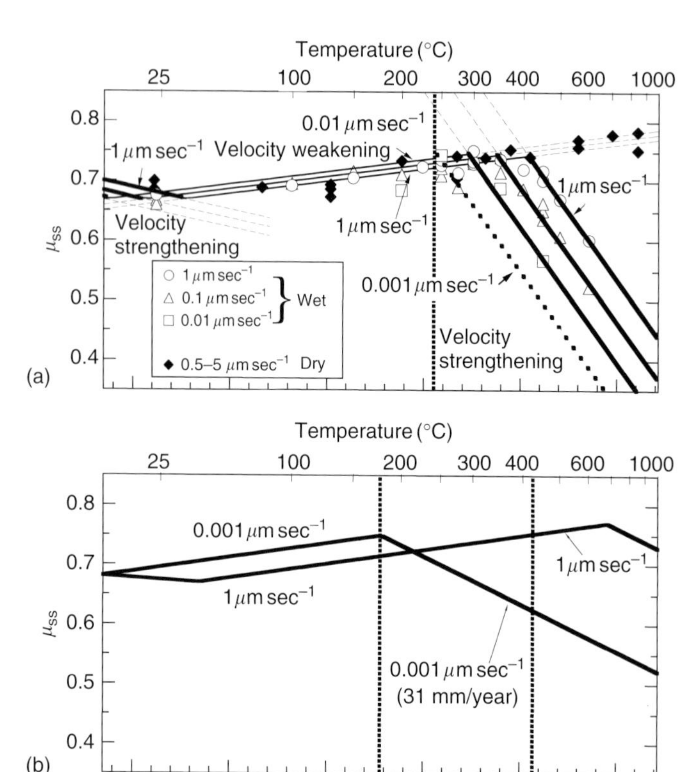

FIGURE 18 Wet frictional strength and rate dependence over a wide range of temperature. (a) Westerly granite (Blanpied *et al.*, 1995). (b) Similar plot showing the fit to the data of Chester (1994) for quartz.

San Andreas. A similar conclusion is reached from application of the quartz data. Quartz does not show rate strengthening at the lowest strain rate and the lowest temperatures (Chester and Higgs, 1992), although thicker gouge layers are rate strengthening at room temperature and slightly elevated temperature (Marone *et al.*, 1990; Chester, 1994). Another explanation of the aseismic to seismic transition, that the depth extent of a thick gouge-filled fault zone controls the lack of shallow seismicity (Marone and Scholz, 1988; Marone *et al.*, 1991), is consistent with field evidence. However, the Marone study is based on data at sliding velocities much higher than plate motion rates and does not consider the temperature dependence observed in the Chester and Blanpied studies.

At higher temperatures, a lower seismic/aseismic transition is also predicted from the laboratory data. However, the depth of the transition is not fully consistent with observed seismicity. Blanpied *et al.* (1991, 1995), Chester and Higgs (1992), and Chester (1994, 1995) studied friction in the presence of water at temperatures up to 600°C. In the Blanpied study, steady-state rate dependence of granite was determined at 100 MPa water pressure and 400 MPa confining pressure. The Chester study involved stress relaxation tests at 100 MPa water pressure and 250 MPa confining pressure. Both studies show a broad range of temperatures where the rate dependence is measured or inferred to be negative, permitting frictional instability from 25–350°C for sliding velocities of 0.1–0.01 μm sec^{-1} (e.g., Blanpied *et al.*, 1991) (Fig. 18). At higher temperatures, ductile processes such as solution transport creep (pressure solution) begin to affect the strength of load-bearing contacts in the experimental fault zones (Chester and Higgs, 1992). This high temperature regime shows strong velocity strengthening and stable sliding. Although these experiments imply a seismic zone that corresponds roughly to natural observations, the sliding rates, corresponding to loading rates on natural faults, used in the steady-state sliding experiments are 10–100 times faster than typical plate motion on the San Andreas fault. If the results of these studies are extrapolated in strain rate to an appropriate natural condition, the base of the seismogenic zone would occur at about 180°C for quartz (Fig. 18b) and about 240°C for granite (Fig. 18a). This is in agreement with the temperature at the base of the seismogenic zone for the Landers rupture, but much less than the 350–400°C inferred for the San Andreas based on heat flow measurements (Williams, 1996). Thus, while existing experimental studies qualitatively predict many aspects of natural seismicity, the field observations are not matched closely. Better knowledge of fault zone mineralogy and the rate dependence of more typical fault zone gouges (e.g., clays and other alteration products) at these deeper hypocentral depths would be useful in providing more accurate representation of fault properties, particularly in instances such as the San Andreas fault where fault strength may deviate significantly from that of quartzo-feldspathic (Byerlee's law) materials. The effects of fluids and the role of fluid pressure at elevated temperatures, not taken into account in the friction studies we have discussed so far, may also be important in frictional failure in the Earth.

5. Pore Fluid Effects

Pore fluid can affect fault failure strength in two fundamental ways: (1) pore fluid pressure can reduce the effective normal stress via the effective pressure law Eq. (7) and therefore reduce the failure shear stress in the brittle portion of the crust [Eq. (2) or (12)] as discussed above in the context of fracture, and (2) pore fluid at elevated temperatures can lead to dissolution and precipitation or other chemical reactions which may alter fault strength (see summary by Hickman *et al.*, 1995). Interest in pore fluids has arisen because dry friction cannot explain all of the field observations of faulting. For a classic example, the lateral dimensions of thrust sheets are large enough relative to sheet thickness that the force necessary to overcome the frictional resistance is sufficient to break the hanging wall block unless high pore fluid pressure reduces the effective stress on the sliding surface (Hubbert and Rubey, 1959). Recently, models involving elevated fluid pressures have been proposed to explain the apparent weakness of the San Andreas (Byerlee, 1990; Byrne and Fisher, 1990; Rice, 1992; Sleep and Blanpied, 1992; Lockner and Byerlee, 1995) (see Section 3.2). Presently there is also interest in the role of pore fluid pressure in Coulomb failure models [based on Eq. (2) or (12)] of triggered seismicity [see reviews by Harris (1998), Stein (1999) and Chapter 73 by Harris]. Although few laboratory studies have directly addressed the role of pore fluid in failure, some useful inferences can be drawn from studies of elastic coupling between stress and pore pressure (summarized in Section 5.1) and from experimental studies carried out on faults under conditions where no excess internal fluid pressure occurs (summarized in Section 5.2). Most evidence for the influence of chemical effects on faulting comes from field observations of healed fractures, pressure solution and veining, and alteration products (Power and Tullis, 1989; Parry and Bruhn, 1990; Parry *et al.*, 1991; Chester *et al.*, 1993; Bruhn *et al.*, 1994). The limited relevant laboratory observations summarized in Section 5.3 do not yet make a strong case for chemical effects in earthquake mechanics.

5.1 Role of Poroelastic Effects in Models of Coulomb Failure (New and Provisional Results)

Studies of triggered seismicity have used the stress changes produced by slip of a large earthquake to explain the spatial distribution of subsequent earthquakes in the surrounding region. Such studies commonly assume that earthquakes in the surroundings obey the Coulomb failure criterion [Eq. (2)] and that changes in proximity to failure of a particular fault plane can be determined if the stress changes are known. The change in proximity to failure is the change in Eq. (2) due to earthquake-induced stress

$$\Delta\sigma_c = \Delta\tau - \mu_f(\Delta\sigma_n - \Delta p) \qquad (22a)$$

Here, S_0 is assumed constant, and the effective stress law has been invoked. Positive values of $\Delta\sigma_c$ indicate that the fault plane in question is closer to failure, whereas for negative values the fault is farther from failure. The stress changes used to evaluate Eq. (22) are commonly calculated from dislocation models of the source earthquake slip, assuming the surroundings are linearly elastic. Although laboratory studies provide limits on μ_f (e.g., Byerlee, 1978), Δp must be estimated using some model of pore fluid response to stress change. In most studies of stress-triggered seismicity, pore pressure change is not considered explicitly and μ_f in Eq. (22a) is replaced by a free parameter,

$$\Delta\sigma_c = \Delta\tau - \mu'\Delta\sigma_n \qquad (22b)$$

The spatial distribution of earthquakes is consistent with positive values of Eq. (22b) if $\mu' = 0.4$ (Stein, 1999).

Current experimental and field observations cannot provide constraints on the choice of appropriate pore pressure response models to use in studies of triggered Coulomb failure.

Experiments suggest that in many cases change in pore pressure in rock and soils is determined by the change in mean stress (Skempton, 1954; Roeloffs, 1996). For a homogeneous isotropic poroelastic medium (Rice and Cleary, 1976; Roeloffs, 1996), the relation between stress change and pore pressure change is

$$\Delta p = B\Delta\sigma_{kk}/3 = B\Delta\sigma_m \tag{23}$$

where σ_{kk} indicates summation over the diagonal elements of the stress tensor and $\sigma_m = \sigma_{kk}/3$ is the mean stress. The parameter B is Skempton's coefficient, and is limited to the range $0 \leq B < 1$. Relatively sparse experimental determinations of B for rocks indicate a range from 0.5 to 0.9 for granites, sandstones, and marbles (Rice and Cleary, 1976; Roeloffs and Rudnicki, 1985; Roeloffs, 1996).

However, Eq. (23) is based on studies of intact rock samples, not faulted rock, and there are no analogous experimental measurements of the poroelastic response of fault zones. Because they may have accommodated large shear strains and are often associated with hydrothermal alteration, it is unlikely that the poroelastic properties of mature fault zones are identical to the surrounding country rock as assumed in Eq. (23). Cocco and Rice (2002) suggest an inhomogeneous isotropic poroelastic response model,

$$\Delta p = B' \frac{K'_u}{M'_u} \left[\frac{G'}{G} \frac{M_u}{K_u} \Delta\sigma_m + \frac{G - G'}{G} \Delta\sigma_n \right] \tag{24}$$

Here the primed and unprimed constants are associated with the fault-zone and country rock, respectively; $M = \lambda + 2G$, $K = \lambda + 2G/3$, and G and λ are the Lame elastic constants. According to Eq. (24), the poroelastic response of fault zones depends on changes in both mean stress and normal stress. In the case where $G = G'$, Eq. (24) reduces to Eq. (23); however if $G \gg G'$, as might be expected for mature fault zones, $\Delta p \propto \Delta\sigma_n$ (Cocco and Rice, 2002) [cf. Eq. (22b)].

Similarly, in the case where the fault zone is anisotropic, stress components other than mean stress may be important in determining the pore pressure response to stress change. For example, simulated laboratory faults, mismatched tensile fractures and natural joint surfaces have large fault-normal compliance (Brown and Scholz, 1985b) because normal stress is supported by a few highly stressed asperities. Although the orientations of fluid-filled fractures along active fault zones at depth are generally unknown, if natural faults and associated fault-parallel fractures have an asperity structure similar to mismatched joints and are hydraulically isolated from the adjacent country rock, then

$$\Delta p \propto \Delta\sigma_n \tag{25}$$

would be the expected pore fluid response (Cocco and Rice, 2002). In this case Eq. (22b) would be an appropriate Coulomb failure model.

While Eqs. (23), (24), and (25) are plausible models of poroelastic response, other measurements indicate that elastic stress–pore-fluid-pressure interactions are more complicated. For example, volumetric changes can be produced by changes in deviatoric stress alone, both in unfaulted (Wang, 1997; Lockner and Stanchits, 2002) and faulted rocks (Woodcock and Roeloffs, 1996). Furthermore, inelastic effects, discussed in the next section, may alter the pore fluid pressure response of seismic fault zones near failure (Segall and Rice, 1995, and references therein).

5.2 The Role of Dilatancy and Inelastic Pore Fluid Effects in Frictional Failure (New and Provisional Results)

Drained measurements of pore volume or fault thickness suggest that fault zone pore pressure will vary with velocity, time, and displacement during frictional sliding and frictional failure. This follows from measurements originally made of gouge volume or fault normal displacement during rate step tests on gouge layers by Morrow and Byerlee (1989), Marone *et al.* (1990), and Lockner and Byerlee (1994) and subsequently also for initially bare surfaces (Wang and Scholz, 1994; Beeler and Tullis, 1997) (Fig. 19a). Following a velocity change there can be both immediate and steady-state changes in dilation or compaction rate within the fault gouge layer. The steady-state changes in dilation or compaction due to displacement rate are generally small compared to changes in friction (Beeler *et al.*, 1996), but immediately after the velocity change can be large (Marone *et al.*, 1990; Sammis and Steacy, 1994). Much like strength evolution, porosity of the fault gouge evolves to steady state over a characteristic displacement. Beeler and Tullis (1997) have argued that rate-dependent changes in porosity are directly related to the state variable in Eqs. (18), at least for initially bare surfaces, based on similar time and slip dependencies between fault thickness and state. Compaction occurs even when the fault sliding rate is negligible, suggesting that fault zone thickness is time dependent (Fig. 19b). The time-dependent compaction under nominally static shear strain follows the same log linear form as fault strength under the same conditions (Dieterich, 1972) (Fig. 19b inset). After a shear zone compacts statically, subsequent shear induces dilatancy (e.g., Fig. 19c). The fault normal displacement δ_n resulting from opposing effects of linear shear-induced dilatancy and log time-dependent compaction can be represented by Eq. (18b) and

$$\delta_n = \delta_{n0} + \varepsilon \ln \frac{V_0 \theta}{d_c} \tag{26}$$

where compaction is reckoned positive, δ_{n0} is the steady-state normal displacement when $V = V_0$, and ε is the negative of the steady-state velocity-dependence of fault thickness $\varepsilon = -d(\delta_n)_{ss}/d \ln V_{ss}$ (Marone and Kilgore, 1993). In Eq. (18b), state changes exponentially with slip and linearly with time and when

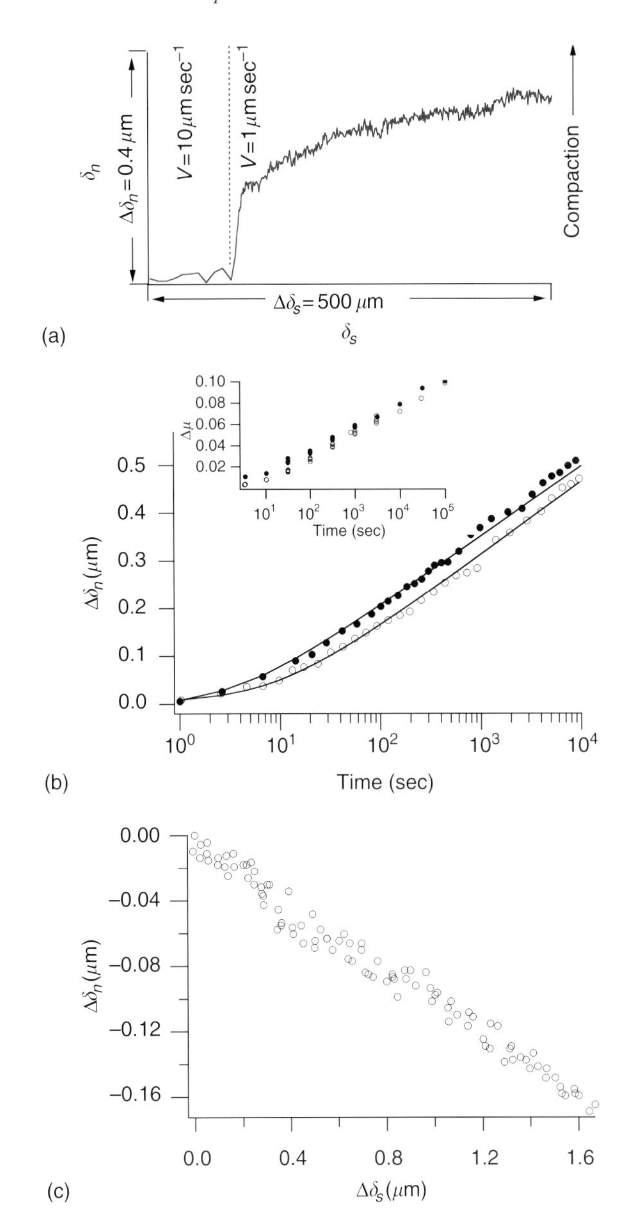

FIGURE 19 Dilatancy during frictional sliding. (a) Changes in fault thickness (proportional to changes in porosity) with sliding velocity. (b) Time dependence of fault thickness. (c) Fault thickness as a function of slip following a hold test for bare surface granite and quartzite (unpublished data courtesy of T. Tullis).

used in Eq. (26) yields the linear slip and log time-dependent thickness changes required by the data (Fig. 19b and c). Equations (26) and (18b) lead to a rate-dependent response of fault thickness to changes in sliding velocity consistent with the observations (Fig. 19a). In experiments on initially bare surfaces, the characteristic displacement of thickness evolution following changes in sliding velocity is approximately equivalent to the characteristic displacement of strength evolution. This association between fault thickness or porosity with state is much less pronounced for thick, gouge layers (Beeler and Tullis, 1997), where the characteristic displacement to recover strength differs from that to recover the porosity.

It should be noted that fault gouge porosity is sensitive to normal or mean stress and that laboratory tests have been carried out at either constant normal stress or constant confining pressure. Loading histories of natural faults may be considerably different from either constant normal stress or confining pressure (although the latter case of increasing normal stress with increasing σ_1 may be closer to the tectonic loading history experienced by thrust or strike-slip faults). Normal faulting may be preceded by a decrease in minimum compressive stress and normal stress. The importance of these naturally occurring stress paths and the effect that they may have on pore pressure and shear strength have not yet been properly explored.

Important implications of experimental observations of rate-dependent porosity change for earthquake nucleation have been considered by Segall and Rice (1995). They extended the original stability analysis of rate and state equations (Rice and Ruina, 1983; Ruina, 1983) to consider the role of pore pressure and conducted numerical simulations of earthquake cycles to determine the effects of pore pressure on strength and fault stability. In this analysis they ignore the small changes in steady-state displacement rate of dilation and assume a steady-state drained porosity

$$\eta_{ss} = \eta_0 + \varepsilon \ln \frac{V}{V_0} \tag{27}$$

where ε is a constant, and a dependence of porosity on state such that porosity evolves to steady-state over the state characteristic displacement d_c, consistent with Eq. (26) and with the laboratory observations. For an undrained porosity, the stiffness for unstable slip must be less than the critical value

$$k_c = \frac{\sigma_{n,\mathrm{eff}}}{d_c}\frac{d\mu_{ss}}{d\ln V} - \frac{\varepsilon\mu_{ss}}{Bd_c} \tag{28}$$

where B is a poroelastic constant (Segall and Rice, 1995). Thus, dilatant faults are found to be more stable in comparison to nondilatant faults; the critical stiffness predicted by Eq. (28) is less in comparison to Eq. (19) due to the additional dilatant term.

Numerical simulations of earthquake cycles using laboratory values of fault strength and porosity conducted by Segall and Rice (1995) do not lead simultaneously to high pore pressure and instability. Thus, to the degree that the Segall and Rice model reflects the dominant poromechanical interactions of fault zones, it implies that high pore pressure is an unlikely cause of the weakness of seismogenic faults such as the San Andreas. However, this conclusion is based on analysis of dilatancy observed during stable sliding of drained faults at room temperature and high porosity (e.g., 11–26%, Marone *et al.*, 1990), conditions which may not be representative of earthquake failure at seismogenic depths. In particular, Eq. (27)

requires that fault thickness (porosity) increases monotonically during failure and that the maximum fault zone porosity is determined by the peak velocity during unstable slip. These predictions are contradicted both by some soil mechanics observations, and by measurements of fault thickness during stress drop. When overcompacted soils are sheared at constant loading rate, porosity may not increase monotonically. Instead, the steady-state thickness can be reached by compaction (e.g., Mandl *et al.*, 1977). In experiments on thick gouge layers where shear stress is artificially reduced at the start of a static hold, stress drop is accompanied by compaction (Karner and Marone, 1998; Nakatani, 1998), presumably consistent with compaction during earthquake stress drop. Compaction accompanying stress drop has also been reported during stick-slip sliding (Beeler and Tullis, unpublished data).

The observations of compaction accompanying shear stress reduction have untested implications for the evolution of fault fluid pressure throughout the seismic cycle. Stress drop will lead to a reduction in effective normal stress under undrained conditions possibly providing a simple "dynamic weakening" of fault strength. This effect would occur at constant normal stress such as in strike-slip environments and may be significant in normal faulting environments where shear stress drop is accompanied by normal stress increase. On the other hand, following compaction, much larger dilatancy occurs during loading to failure (Nakatani, 1998) which should tend to stabilize sliding through dilatancy hardening. Constitutive equations describing the experimental observations have been proposed by Nakatani (1998). However, the implications that this rheological model has for fault stability have not yet been addressed.

5.3 Solution Mass Transport Phenomenon (New and Provisional Results)

Although hydrothermal studies of frictional slip at temperatures greater than 180–240°C and displacement rates corresponding to typical plate motion rates, indicate stable, pressure-solution-like deformation within a fault zone (Blanpied *et al.*, 1995; Chester, 1995), other solution mass transport phenomena such as overgrowths on load bearing contacts and fault zone sealing and compaction could operate at these conditions, and lead to rapid and large changes in fault zone material properties. Laboratory measurements of fault strength (Blanpied *et al.*, 1992) and fault strengthening (Fredrich and Evans, 1992; Karner *et al.*, 1997), and inferences of fault strength (S. Hickman, unpublished data) at hydrothermal conditions (>300°C) where solution transport phenomenon can be rapid, lead to two models where fault failure stress differs from the predictions of conventional rate and state friction. In the following sections we describe these two proposed mechanisms: (1) a solution mass transport compaction and fault sealing model and (2) a solution mass transport lithification model, and the experimental evidence that supports them.

5.3.1 Pressure Solution and Fault Sealing

If minerals are sufficiently soluble in water, then dissolution, transport, and precipitation can occur in response to gradients in chemical potential due to spatial variation of temperature, stress, or surface energy. Numerous studies have reported rapid reduction of permeability of rocks and joints (Moore *et al.*, 1994), and granular aggregates (Tenthory *et al.*, 1998) at such conditions. Additional studies have demonstrated rapid sealing due to flow along a temperature gradient (Summers *et al.*, 1978; Moore *et al.*, 1984; Vaughan *et al.*, 1985). In faulting experiments at 600°C, Blanpied *et al.* (1992) inferred permeability reduction so rapid that in the presence of shear induced compaction, effective pressure increased, leading to a 50% reduction in shear strength. The observations of reduced strength (Fig. 20) suggest a potential earthquake mechanism for weak faults where, within a fault zone hydraulically sealed with respect to its surroundings, shearing leads to compaction and increased fluid pressure. Instability is initiated when the fault failure strength is reached by the combined action of tectonic loading and effective stress decrease (Blanpied *et al.*, 1992), i.e., this failure mechanism differs from standard rate and state models in that porosity and $\sigma_{n,\text{eff}}$ are functions of time and displacement.

A qualitative instability model based by the observations by Blanpied and co-workers (Blanpied *et al.*, 1992; Sleep and Blanpied, 1992, 1994) is not unlike the Segall and Rice (1995) model, and accordingly, its potential for instability will be reduced in comparison to dry friction-based instability models through dilatancy and reduction in the critical stiffness. If dilatancy accompanies increases in sliding velocity (Morrow and Byerlee, 1989; Marone *et al.*, 1990; Lockner and Byerlee, 1994) during failure, the accompanying effective stress increase stabilizes the fault (dilatancy hardening) (Rice, 1975; Rudnicki and Chen, 1988; Lockner and Byerlee, 1994). Decreases in effective stress brought on by compaction during

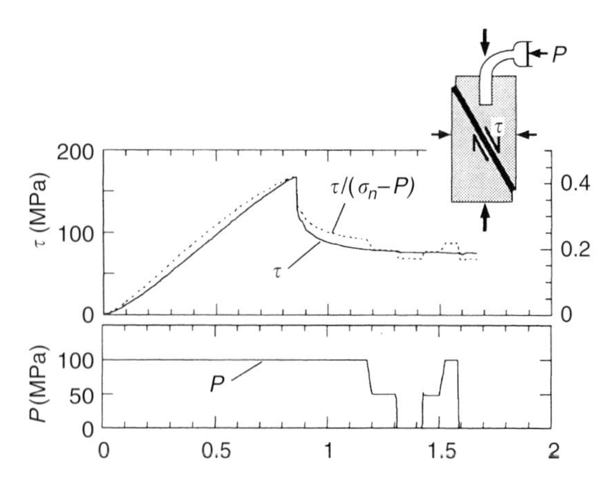

FIGURE 20 Measured shear stress (upper graph solid, left axis scale), $\tau/(\sigma_n - p)$ (upper graph dashed, right axis scale) and pore pressure (lower graph) versus axial displacement for Granite gouge at 600°C and 400 MPa confining pressure.

loading to failure reduce the critical stiffness, stabilizing the fault as in Eq. (17) or Eq. (19). Compaction-induced instability models are untested, as they require creep compaction behavior that has not yet been observed experimentally at the appropriate temperatures (200–400°C). Theoretical studies of creep compaction instability are ambiguous. Sleep and Blanpied (1994) report such instabilities using faults with uniform rate and state-like properties, as do Lockner and Byerlee (1995) using a Burridge–Knopoff-type model with inhomogeneous properties. Dilatancy hardening results in a delay but not an inhibition of dynamic failure for the Lockner and Byerlee model. On the other hand, strict interpretation of the existing experimental compaction data from rate-stepping tests, albeit at much lower temperatures (discussed in Section 5.2) does not lead to both weakness and instability (Segall and Rice, 1995).

5.3.2 Lithification

It has been suggested that at hydrothermal conditions, rates of fault healing may be greatly enhanced due to lithification processes that are not present in room temperature friction experiments (Kanamori and Allen, 1986; Scholz *et al.*, 1986). Increases in static stress drop with recurrence for large and small repeating earthquakes may exceed the rate predicted by low-temperature friction (Kanamori and Allen, 1986; Scholz *et al.*, 1986; Nadeau and McEvilly, 1999; Beeler *et al.*, 2000), consistent with lithification. Accordingly, a fault-strengthening model based on this hypothesis would predict fault-strengthening rates that exceed those predicted by rate and state implementations of room temperature friction. Such rapid fault strengthening has been suggested to occur in experiments at elevated temperatures and hydrostatic pressure on quartz gouge at rates which exceed those observed at low temperature (Fredrich and Evans, 1992; Karner *et al.*, 1997). However, to achieve the apparent enhanced rate of strengthening, fault gouge must be loaded to failure at a lower temperature (230–235°C) than the healing temperature (500–800°C). Experiments where the healing and loading temperature are the same [636°C (Karner *et al.*, 1997)], as would be the case for faults in the Earth, show no time-dependent strengthening. The lack of time dependence indicates that these studies of fault strengthening at very high temperature may have limited relevance to natural earthquake failure. These studies are designed to measure healing due to local redistribution of material, i.e., dissolution from pore walls or fine or angular grains followed by deposition at grain contacts. More rapid or more significant cementation may occur due to mass transport of silica or calcite over large distances within a fracture network. Certainly quartz vein filling is a well-known example of this strength recovery process (Chapter 29 by Sibson).

Lithification processes other than pressure solution which may lead to fault strengthening can be induced at hydrothermal conditions where a reduction in surface energy, not a reduction in contact stress, drives solution mass transport processes (Hickman and Evans, 1992). Recent experiments on single contacts in quartz at $T = 350–530°C$, $P_c = P_{H_2O} = 150$ MPa and non-zero contact normal stress indicate extremely rapid contact overgrowth (S. Hickman, unpublished data). Because of the controlled geometry and *in situ* monitoring, the mass flux as a function of time is known and the results can be extrapolated to other contact geometries. When extrapolated to a contact size typical for laboratory friction experiments, the contact area increase in these experiments greatly exceeds those of frictional strengthening (Fig. 21). The area increase is so great as to imply that fault zones subject to low effective stress may become entirely lithified during the interseismic period. In this case, the fault failure strength would rapidly approach the strength of an intact rock. However, the low contact stress in the experiments prevents easy extrapolation of these results to seismogenic conditions.

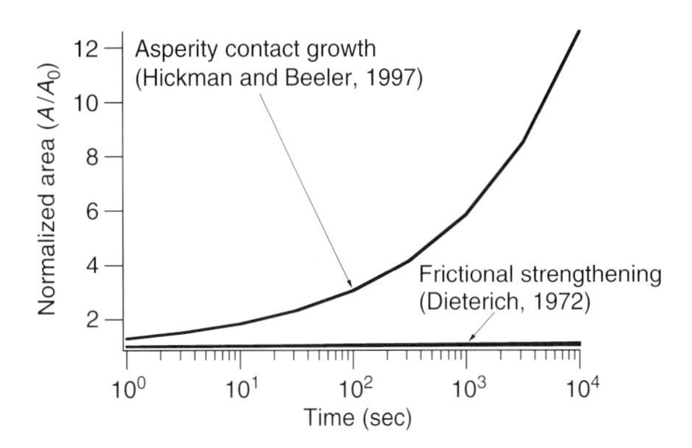

FIGURE 21 Fault strengthening inferred from contact area increase with time at hydrothermal conditions.

6. Dynamic Fault Strength

Although laboratory observations of frictional strength as represented by Eq. (19) have been applied to model the entire seismic cycle (e.g., Rice and Tse, 1986; Boatwright and Cocco, 1996), this assumes shear heating is not significant during dynamic slip. Fault slip speeds during earthquake rupture are on the order of meters per second and for large earthquakes will last many seconds at hundreds of MPa normal stress (appropriate for seismogenic depths). Under these conditions significant shear heating should occur, resulting in dynamic weakening due to shear melting or pore fluid pressurization due to thermal expansion (Chapter 29 by Sibson). Consequently, conventional laboratory measurements of strength are unlikely to apply. Effects unrelated to shear heating but not considered in Eqs. (18) are also possible. Dynamic reduction of fault strength by normal stress reduction has been predicted on theoretical grounds (Weertman, 1980) and observed in numerical models (Ben-Zion and Andrews, 1998).

Many important unresolved issues in seismology are related to or determined by the strength during dynamic slip. If dynamic weakening occurs, dynamic stress drops can be much larger than static drops (Brune, 1970, 1976), and rupture is more likely to propagate as a self-healing slip pulse (Heaton, 1990; Cochard and Madariaga, 1994; Perin *et al.*, 1995; Beeler and Tullis, 1996) rather than as a conventional crack-like rupture (Zheng and Rice, 1998; Chapter 12 by Madariaga and Olsen). If the dynamic stress drop exceeds the static stress drop, larger ratios between the radiated energy and the seismic moment are expected for larger earthquakes than for small events (Abercrombie, 1995).

Stress heterogeneity in the fault zone also depends on dynamic fault strength. Geometrical effects (roughness of the fault surface) and nonuniform material properties lead to stress heterogeneity, but heterogeneity may also result during dynamic rupture propagation and arrest (e.g., Cochard and Madariaga, 1994). Such stress roughening during dynamic propagation is common in numerical models, particularly discrete representations of rupture such as Burridge–Knopoff models, and is associated with simulated earthquake populations which obey the power law Gutenberg–Richter relation between occurrence and event size. Aspects of the association of rough spatial stress distribution and Gutenberg–Richter event populations are summarized by Rice (1993) where it is argued that discrete models are not satisfactory representations of a continuum, and continuum models of dynamic rupture propagation using laboratory values for fault strength generally do not result in a rough residual stress and Gutenberg–Richter populations. The latter is true in particular if Eq. (18) are used (Rice, 1993; Rice and Ben-Zion, 1996). However, more extreme rate dependence of fault strength, such as would result from dynamic weakening (e.g., Zheng and Rice, 1998), does lead to stress roughening (N. Lapusta and J.R. Rice, unpublished data). Thus many open questions and unresolved debates might be resolved by appropriate laboratory determinations of dynamic fault strength.

6.1 Frictional Strength at High Sliding Rates

Measures of fault strength during dynamic rupture events (Okubo and Dieterich, 1986) which have average sliding velocities on the order 0.1 m $\sec^{-1}$ are consistent with frictional strength, i.e., the predictions of Eqs. (18) during stick-slip. However, the implied temperature change in those experiments as estimated from

$$\Delta T = \frac{\tau \Delta\delta}{w\rho c} \tag{29}$$

where w is the fault zone width, ρ is density and c is the specific heat (Lachenbruch, 1980) is not sufficient to induce weakening due to melting (there is no pore fluid) (Lockner and Okubo, 1983). This is primarily due to the small total slip $\Delta\delta$ (80–100 μm) and low normal stress in the experiments. Measurements of fault strength at constant but lower sliding velocity (up to 3.2 mm $\sec^{-1}$), dry, and at much higher cumulative displacement (500 mm), lead to strengths that are also generally consistent with Eq. (18) (Blanpied *et al.*, 1998b) (see also Yoshioka, 1985; Mair and Marone, 1999). However, in the experiments of Blanpied *et al.* (1998b), sliding at the highest velocity and the highest normal stress (25 MPa induced anomalous weakening with displacement. This weakening did not occur at the lower normal stress (10 MPa).

6.2 Dynamic Weakening (New and Provisional Results)

Recently, dynamic weakening has been measured experimentally. The results, summarized in this section, fall into the two classes mentioned at the outset of this section: dynamic weakening due to reduction of normal stress and weakening due to shear heating. However, to date, none of the experimental measurements of weakening are definitive and some of the results summarized in this section are unconfirmed. In some cases the results or their relevance to natural faults are uncertain.

Dynamic weakening due to normal stress reduction was first observed in this context during stick-slip sliding between surfaces of foam rubber (Brune *et al.*, 1993; Anooshehpoor and Brune, 1994). The observations consist of particle trajectories as the rupture passed and indicate opening displacement during the passage of a dislocation-like slip pulse (Fig. 22a). Because the frictional properties of foam rubber are quite unlike rocks [e.g., $1.0 < \mu < 3.7$ (Brune and Anooshehpoor, 1997)], the experiments which are conducted under very low normal stress ($0.9–7 \times 10^{-6}$ MPa) are difficult to scale to the Earth. Some indication of the extreme differences between stick slip of foam rubber and rock is that for foam rubber, stick slip only occurs at the lowest normal stress; increasing normal stress leads to stable sliding (Anooshehpoor and Brune, 1994), in contrast to the observation for friction of rocks and other brittle materials where increased normal stress promotes instability [e.g., Eq. (17) or (19) (Walsh, 1971; Goodman and Sundaram, 1978; Dieterich, 1979)]. These criticisms do not negate the applicability of the foam rubber observations to seismic failure, but they do emphasize fundamental differences between the behavior of foam rubber and rocks at laboratory conditions. The association of true opening displacements and stick slip for foam rubber suggest that this mechanism of instability is encouraged by low normal stress.

Dynamically reduced normal stress has been measured or inferred for materials other than foam rubber (Tostoi, 1967; Bodin *et al.*, 1998; Brown, 1998), suggesting that the foam rubber may be a valid laboratory rock analogue. Significantly, observations relevant to dynamic reduction of normal stress have been made for rock by Brown (1998). Heat generation during stick-slip between Westerly granite surfaces occurs at a lower rate than during stable sliding (Fig. 22b). Brown also found the mean stress during stick slip to be lower than during stable sliding. Assuming that the mean frictional strength is the same for both stick slip and stable sliding, the observations

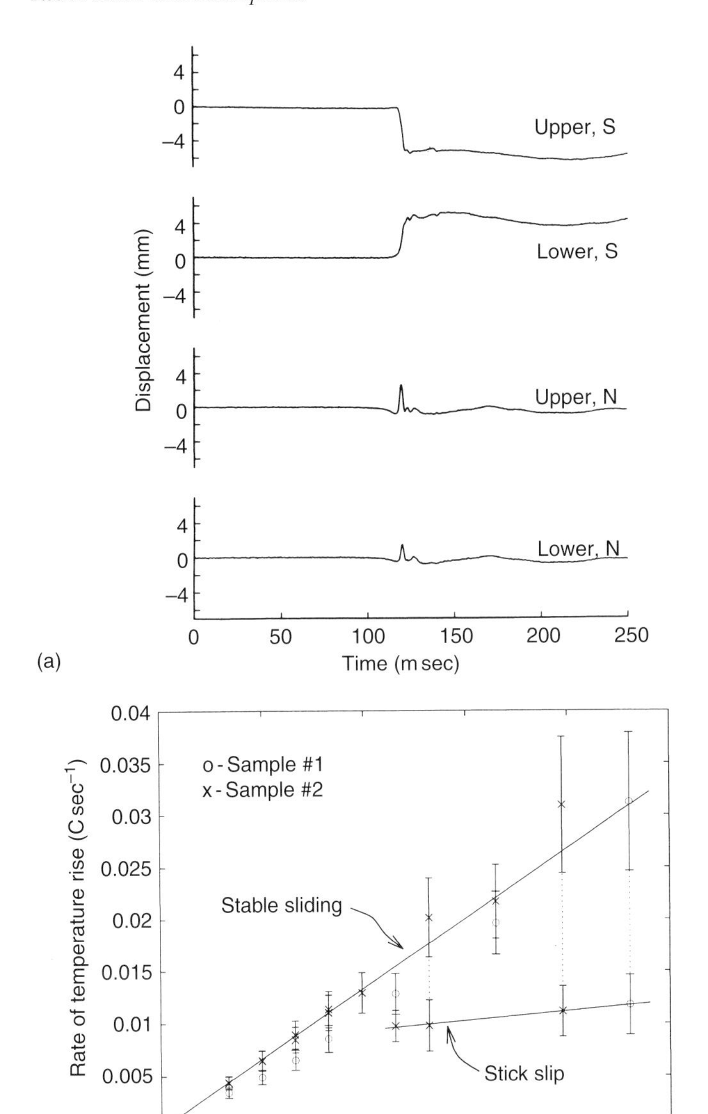

FIGURE 22 Dynamic weakening: normal stress effects. (a) Shear (S) and normal (N) displacements in foam rubber during a stick–slip event. (b) Heat generation during stick–slip versus stable sliding on unconfined, initially bare Westerly granite.

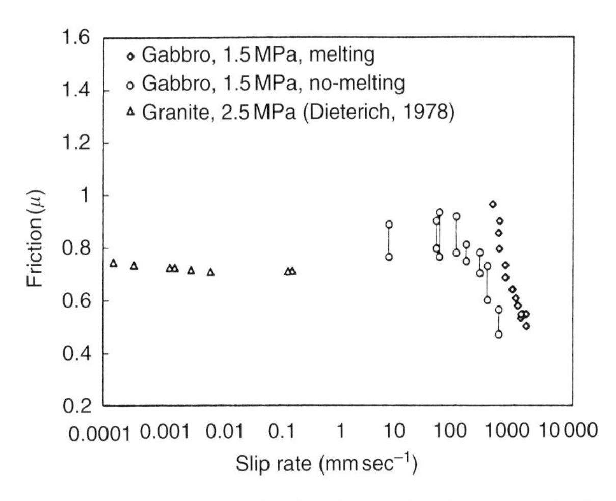

FIGURE 23 Dynamic weakening due to shearing: strength plotted as a function of sliding velocity for unconfined gabbro at 1.5 MPa normal.

can be interpreted to be due to dynamic reduction in normal stress during stress drop.

Dynamic weakening has been reported due to shear heating in unconfined experiments at modest normal stress on gabbro (Tsutsumi and Shimamoto, 1997). Fault strength expressed as frictional resistance drops to the range $\mu = 0.4$–0.5 (Fig. 23) [$\sigma_n = 1.5$ MPa (Tsutsumi and Shimamoto, 1997)] at sliding velocities of 0.43–1.3 m sec^{-1} during melting and at sliding velocities of 100–550 mm sec^{-1} when melting is not observed. The effect in the shear melting experiments is not as dramatic as one might expect, perhaps because of high melt viscosity (Tsutsumi and Shimamoto, 1997), but more likely because the melt is unconfined and is extruded by centrifugal force at these high sliding velocities (experiments are done in a rotary shear geometry). Unconfined experiments conducted by Spray (1987, 1988) on metadolerite and amphibolite at $\sigma_n = 50$ MPa and mean sliding velocity of 0.24 m sec^{-1} show no evidence of weakening though accuracy of the stress determination is poor.

Preliminary results on confined and unconfined surfaces of quartzite and granite at slip rates of only 3.2 mm sec^{-1}, normal stresses of 28–112 MPa, and slip displacements of 0.2–10 m, show that friction can dramatically decrease with slip. In some cases the friction drops below 0.3, or roughly 30% of the value typically found in quartz friction experiments (D. Goldsby and T. Tullis, unpublished data). Calculations of both average surface temperature and local temperatures at contacting asperities (i.e., so-called "flash temperatures") predict maximum temperatures of 100–600°C. In most cases, the dramatic frictional weakening occurs at estimated temperatures below 300°C. Hence, the weakening is difficult to reconcile with melting, either of the entire fault surface or at asperities. It appears, however, that both temperature and displacement are important variables affecting the shear resistance. High temperature triaxial quartzite experiments (restricted to a few millimeter displacements), at temperatures that span the range of frictionally generated temperatures in the Goldsby and Tullis experiments, yield typical values of the friction coefficient of 0.6–0.8 (Stesky *et al.*, 1974). Large displacement experiments (sheared to several meters) at velocities greater than 100 μm sec^{-1}, which generate much less heat than the 3.2 mm sec^{-1} experiments over similar displacements, also result in typical frictional strength. A possible explanation is that the low friction coefficient results from shear heating of profoundly altered, likely amorphous, material along the fault.

7. Deep Earthquakes

Seismicity can extend well below the conventional brittle ductile transition (Fig. 10). This is particularly true in subduction zones where earthquakes occur to depths around 600 km (see Fig. 10c inset). Following Frohlich (1989) we refer to intermediate focus events as those occurring in the depth range 70–300 km and deep focus events as those occurring at depths greater than 300 km. If conventional fracture and frictional instability were the only mechanisms of earthquake failure then seismicity would not exist below depths where the resolved shear stress on a fault is equivalent to 2000 MPa (e.g., Raleigh, 1967), the maximum flow stress above which all rocks and minerals deform by stable creep. For example, in a thrust faulting environment, assuming that σ_3 is fixed at overburden, pore pressure is hydrostatic and $\mu = 0.7$ (ignoring cohesion); failure of optimally oriented faults at this shear stress would occur at about 75 km depth. Obviously, the depth distribution of natural seismicity which indicates significant seismicity at 600 km cannot be interpreted in this simple manner. A number of mechanisms have been proposed to explain deep earthquakes. In this summary we focus on those that are supported by laboratory experiments. Not considered, for example, are models such as those based on a thermal instability (Griggs and Baker, 1969; Hobbs and Ord, 1988) which currently lack experimental support.

Laboratory observations indicate that earthquakes could accompany phase changes in the otherwise ductile portion of the crust. For example, brittle fracture occurs following reduction of the failure stress due to increased pore fluid pressure from a dehydration reaction. This dehydration embrittlement mechanism can explain intermediate events if hydrated minerals are present. Failure can also be triggered by heat from a phase transformation as well as differential stress around inclusions resulting from a phase transformation. Transformation-induced faulting provides a likely mechanism for deep focus events. Radiated stress and inelastic shear displacements result from other phase transformations such as melting (Griggs and Handin, 1960) and amorphization (Meade and Jeanloz, 1991).

7.1 Intermediate Focus Events

The explanation of intermediate focus earthquakes follows directly from the effective pressure law $\sigma_{n,\text{eff}} = \sigma_n - p$, Eq. (7), which was previously discussed in the context of Eq. (1a). Failure strength in excess of cohesion can be reduced to arbitrarily low levels if the pore pressure is high. At intermediate focal depths, where the high overburden stress and high temperature lead to time-dependent creep and recrystallization, porosity and pore connectivity are significantly reduced (Lockner and Evans, 1995). Under these conditions, water, provided by dehydration reactions, can exist at near lithostatic

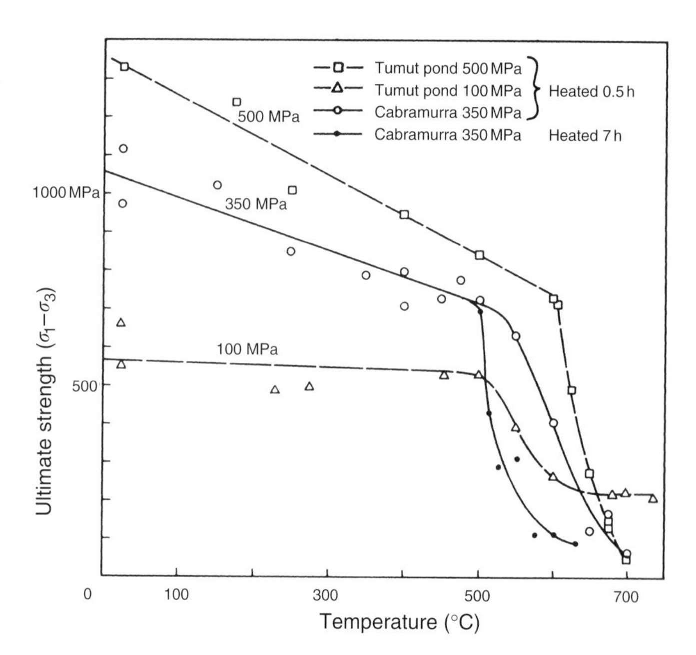

FIGURE 24 Dehydration weakening. Strength of serpentinite deformed at a strain rate of $7 \times 10^{-4}\,\text{sec}^{-1}$, and confining pressures of 100, 350, and 500 MPa as a function of temperature.

pressures and shear fracturing can occur at very low shear stress (Hubbert and Rubey, 1959). Experimental confirmation of this theory was made by Raleigh and Paterson (1965) who observed brittle failure of serpentinite at its dehydration temperature, but ductile behavior at lower temperatures (Fig. 24). Acoustic emissions were detected during dehydration at pressures up to 9 GPa (Meade and Jeanloz, 1991) confirming that this mechanism is capable of the rapid unstable deformation associated with earthquakes.

There are four requirements for dehydration embrittlement to be a viable model for intermediate events:

1. The brittle failure strength must be smaller than the ductile shear strength because the ductile shear strength would otherwise limit the stress available to drive failure.
2. The sliding frictional strength must be smaller than the failure strength so that a coseismic stress drop is possible. These first two requirements are met for serpentinite experiments (Raleigh and Paterson, 1965).
3. Faulting must occur under undrained conditions, since a loss of pore pressure would increase the shear resistance and stabilize the incipient rupture. At intermediate depths where porosity and permeability are extremely low, undrained conditions are expected to be the norm. In this case the stress drop during dehydration faulting is well described by a loss of cohesion.
4. Dilatancy must be suppressed or else the pore pressure will be reduced due to dilatancy hardening (Rice, 1975; Rudnicki and Chen, 1988).

Recent experimental observations of failure of serpentinite even at low confining pressure indicate that deformation is accommodated on shear microcracks that are aligned preferentially with the cleavage direction, as opposed to macroscopic shear fractures initiated from dilatant microcracks (Escartin *et al.*, 1997). Thus little or no dilatancy can be expected during deformation of serpentine at high pressure. Application of the Hubbert and Rubey mechanism to explain intermediate focus events is limited to regions of the crust where dehydration is on-going and also by the constraint that the pore pressure not exceed the least principal stress. However, the conditions necessary for instability have not been rigorously determined for this mechanism.

7.2 Deep Focus Earthquakes

Failure of the crust in deep focus earthquakes occurs well below the depth where the frictional failure strength exceeds the flow strength of rock, and below temperature and pressure conditions of dehydration reactions. Deep focus events therefore require a mechanism that is different from shallow seismic failure in fundamental ways. Among the exotic mechanisms called upon to explain deep focus events is transformation induced faulting (Vaisnys and Pilbeam, 1976), where failure is associated with a phase transformation such as the pressure-driven olivine–spinel transformation. This mechanism is well documented in experiments, notably in water ice I to ice II where faulting is accompanied by release of radiated energy (Kirby *et al.*, 1991), a key requirement for any explanation of deep earthquakes. Transformational faulting also occurs in Mg_2GeO_4 which has the same structure as Si olivine (Fig. 25) (Burnley *et al.*, 1991) and, at temperatures, pressures, and mineralogy consistent with subduction zones, in $(Mg,Fe)_2SiO_4$ (Green *et al.*, 1990). The key observations which distinguish this mechanism from conventional brittle failure are the presence of the transformed phase in the fault zone and in transformed lenses in fault parallel lineations, and the lack of precursory acoustic emissions which are associated with dilatancy during nucleation in conventional frictional failure. That the lenses of transformed material tend to be oriented with their long axis perpendicular to the stress field indicates that the transformation itself is controlled by the deviatoric stress. However, the exact mechanism and mechanics of transformational faulting are not well understood. Suggested mechanisms include the coalescence of transformed material into a shear zone (Kirby, 1987) or transformed material initiating a through-going shear fracture (Ringwood, 1972; McGarr, 1977). Similarly it is not entirely clear whether the weakness of the shear zone which leads to a stress drop in experiments is due to the fault being made of fine-grained transformed material with reduced shear (flow) strength or because of a loss of cohesion during the formation or propagation of the shear zone. Advances have been made in theory to accompany and explain the laboratory observations. Lui (1997) has determined

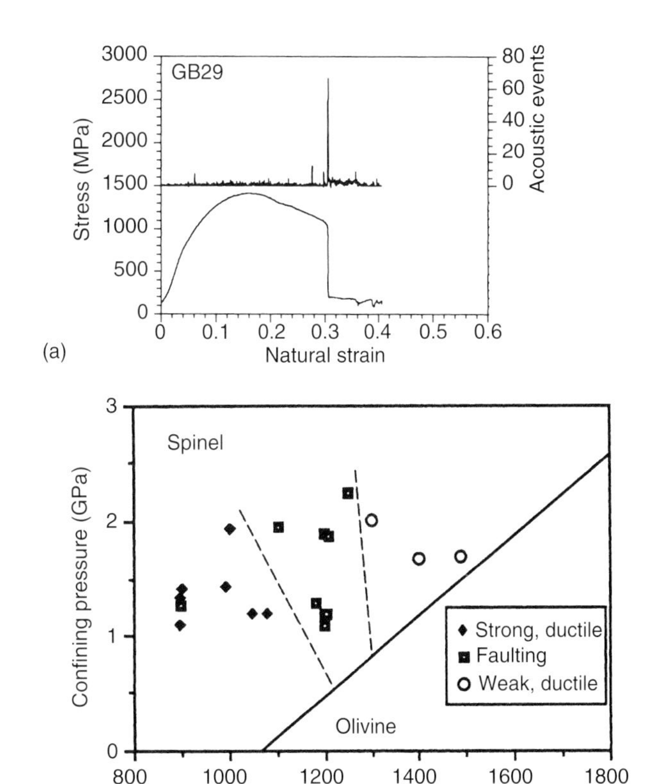

FIGURE 25 Transformation induced faulting in germanate Mg_2GeO_4. (a) Stress versus natural strain. (b) Summary of deformation experiments on olivine–spinel phase transformation at strain rates of 2×10^{-4} and $2 \times 10^{-5}\,sec^{-1}$.

conditions necessary for formation of a transformed shear zone and the associated stress drop based on thermodynamic considerations and the kinetics of the phase transformation. However, the conditions necessary for rapid (seismic) failure have not been determined. Other phase transformations may lead to shear instabilities. Meade and Jeanloz (1991) report acoustic emissions and shear deformation associated with a crystal to glass transformation in serpentinite between 6 and 25 GPa. Other hydrated minerals may also amorphize in this manner.

Acknowledgments

We greatly appreciate the thoughtful reviews provided by J. Byerlee, C. Sammis, C. Scholz, R. Sibson, and W. Stuart. This work was supported in part by the National Earthquake Hazards Reduction Program.

References

Abercrombie, R.E. (1995). *J. Geophys. Res.* **100**, 24015–24036.
Allégre, C.J., *et al.* (1982). *Nature* **297**, 47–49.

Amadei, B., *et al.* (1987). In: "Large Rock Caverns," pp. 1135–1146. Pergamon Press, New York.
Anderson, E.M. (1951). "The Dynamics of Faulting and Dyke Formation with Application to Britain," Oliver and Boyd.
Andrews, D.J. (1976). *J. Geophys. Res.* **81**, 5679–5687.
Andrews, D.J. (1989). *J. Geophys. Res.* **94**, 9389–9397.
Anooshehpoor, A. and J.N. Brune (1994). *Pure Appl. Geophys.* **142**, 735–747.
Ashby, M.F. and C.G. Sammis (1990). *Pure Appl. Geophys.* **133**, 489–521.
Atkinson, B.K. (1987). "Fracture Mechanics of Rock," Academic Press, New York.
Atkinson, B.K. and P.G. Meredith (1987). In: "Fracture Mechanics of Rock," pp. 477–525, Academic Press, New York.
Beeler, N.M. and T.E. Tullis (1996). *Bull. Seismol. Soc. Am.* **86**, 1130–1148.
Beeler, N.M. and T.E. Tullis (1997). *J. Geophys. Res.* **102**, 22595–22609.
Beeler, N.M., *et al.* (1996). *J. Geophys. Res.* **101**, 8697–8715.
Beeler, N.M., *et al.* (2001). *J. Geophys. Res.* **106**, 30701–30713.
Ben-Zion, Y. and D.J. Andrews (1998). *Bull. Seismol. Soc. Am.* **88**, 1085–1094.
Biegel, R.L., *et al.* (1989). *J. Struct. Geol.* **11**, 827–846.
Bieniawski, Z.T. (1968). *Int. J. Rock Mech. Min. Sci.* **5**, 325–335.
Blanpied, M.L., *et al.* (1991). *Geophys. Res. Lett.* **18**(4), 609–612.
Blanpied, M.L., *et al.* (1992). *Nature* **358**, 574–576.
Blanpied, M.L., *et al.* (1995). *J. Geophys. Res.* **100**(B7), 13045–13064.
Blanpied, M.L., *et al.* (1998a). *J. Geophys. Res.* **103**, 9691–9712.
Blanpied, M.L., *et al.* (1998b). *J. Geophys. Res.* **103**, 489–511.
Boatwright, J. and M. Cocco (1996). *J. Geophys. Res.* **101**, 13895–13900.
Bock, H. (1978). "An Introduction to Rock Mechanics," University of North Queensland.
Bodin, P., *et al.* (1998). *J. Geophys. Res.* **103**, 29931–29944.
Brace, W.F. (1978). *Pure Appl. Geophys.* **116**, 627–633.
Brace, W.F. and J.D. Byerlee (1966). *Science* **153**, 990–992.
Brace, W.F. and A.H. Jones (1971). *J. Geophys. Res.* **76**, 4913–4921.
Brace, W.F. and D.L. Kohlstedt (1980). *J. Geophys. Res.* **85**, 6248–6252.
Brace, W.F., *et al.* (1968). *J. Geophys. Res.* **73**, 2225–2236.
Bragg, R.A. and O.B. Andersland (1981). *Eng. Geol.* **18**, 35–46.
Broek, D. (1982). "Elementary Engineering Fracture Mechanics," Martinus Nijhoff, Dordrecht.
Brown, S.R. (1998). *J. Geophys. Res.* **103**, 7413–7420.
Brown, S.R. and C.H. Scholz (1985a). *J. Geophys. Res.* **90**, 12575–12582.
Brown, S.R. and C.H. Scholz (1985b). *J. Geophys. Res.* **90**, 5531–5545.
Bruhn, R.L., *et al.* (1994). *Pure Appl. Geophys.* **142**, 609–645.
Brune, J.N. (1970). *J. Geophys. Res.* **75**, 4997–5009.
Brune, J.N. (1976). In: "Seismic Risk and Engineering Decisions," pp. 140–177, Elsevier, Amsterdam.
Brune, J.N. and A. Anooshehpoor (1997). *Geophys. Res. Lett.* **24**, 2071–2074.
Brune, J.N., *et al.* (1969). *J. Geophys. Res.* **74**, 3821–3827.
Brune, J.N., *et al.* (1993). *Tectonophysics* **218**, 59–67.
Burnley, P.C., *et al.* (1991). *J. Geophys. Res.* **96**, 425–443.
Byerlee, J.D. (1967). *J. Geophys. Res.* **72**, 3639–3648.
Byerlee, J.D. (1970). *Tectonophysics* **9**, 475–486.
Byerlee, J.D. (1978). *Pure Appl. Geophys.* **116**, 615–626.
Byerlee, J.D. (1990). *Geophys. Res. Lett.* **17**, 2109–2112.
Byerlee, J.D. (1992). *Tectonophysics* **211**, 295–303.
Byerlee, J.D. (1993). *Geology* **21**, 303–306.
Byerlee, J.D. and J.C. Savage (1992). *Geophys. Res. Lett.* **19**, 2341–2344.
Byerlee, J.D. and R. Summers (1976). *Int. J. Rock Mech. Min. Sci.* **13**, 35–36.
Byrne, T. and D. Fisher (1990). *J. Geophys. Res.* **95**, 9081–9097.
Charles, R.J. and W.B. Hillig (1962). In: "Symposium sur la resistance due verre et les moyens de l'ameliorer," pp. 511–527. Un. Sci. Continentale du Verr, Belgium.
Chester, F.M. (1994). *J. Geophys. Res.* **99**, 7247–7262.
Chester, F.M. (1995). *J. Geophys. Res.* **100**, 13033–13044.
Chester, F.M. and N.G. Higgs (1992). *J. Geophys. Res.* **97**, 1859–1870.
Chester, F.M., *et al.* (1993). *J. Geophys. Res.* **98**, 771–786.
Cocco, M. and J.R. Rice (2002). *J. Geophys. Res.* **107**, in press.
Cochard, A. and R. Madariaga (1994). *Pure Appl. Geophys.* **142**, 419–445.
Coleman, R.G. and T.E. Keith (1971). *J. Petrol.* **12**, 311–328.
Costin, L.S. (1983). *J. Geophys. Res.* **88**, 9485–9492.
Costin, L.S. (1987). In: "Fracture Mechanics of Rock," pp. 167–215. Academic Press, New York.
Das, S. and C.H. Scholz (1981). *J. Geophys. Res.* **86**, 6039–6051.
de Josselin de Jong, G. (1988). *Geotechnique* **38**, 533–555.
Dey, T. and P. Halleck (1981). *Geophys. Res. Lett.* **8**(7), 691–694.
Dieterich, J.H. (1972). *J. Geophys. Res.* **77**, 3690–3697.
Dieterich, J.H. (1978a). In: "Summaries of Technical Reports," pp. 328–329. US Geological Surveys.
Dieterich, J.H. (1978b). *Pure Appl. Geophys.* **116**, 790–806.
Dieterich, J.H. (1979). *J. Geophys. Res.* **84**, 2161–2168.
Dieterich, J.H. (1981). In: "Mechanical Behavior of Crustal Rocks," pp. 103–120. American Geophysical Union.
Dieterich, J.H. (1987). *Tectonophysics* **144**, 127–139.
Dieterich, J.H. (1994). *J. Geophys. Res.* **99**, 2601–2618.
Dieterich, J.H. and B. Kilgore (1996). *Proc. Natl. Acad. Sci. USA*, 3787–3794.
Dieterich, J.H. and B.D. Kilgore (1994). *Pure Appl. Geophys.* **143**, 283–302.
Du, Y. and A. Aydin (1991). *Int. J. Numer. Anal. Meth. Geomech.* **15**, 205–218.
Dunning, J.D., *et al.* (1980). *J. Geophys. Res.* **85**(B10), 5344–5354.
Ellsworth, W.L. and G.C. Beroza (1995). *Science* **268**, 851–855.
Ellsworth, W.L. and G.C. Beroza (1998). *Geophys. Res. Lett.* **25**, 401–404.
Escartin, *et al.* (1997). *J. Geophys. Res.* **102**, 2897–2913.
Evans, B. and G. Dresen (1991). *Rev. Geophys.* IUGG Report (Suppl.), 823–843.
Fenoglio, M.A., *et al.* (1995). *J. Geophys. Res.* **100**(B7), 12951–12958.
Fredrich, J., *et al.* (1989). *J. Geophys. Res.* **94**, 4129–4145.
Fredrich, J.T. and B. Evans (1992). In: "33rd US Rock Mechanics Symposium," pp. 121–130, Balkema.
Freed, R.L. and D.R. Peacor (1989). *Clay Minerals* **24**, 171–180.
Frohlich, C. (1989). *Annu. Rev. Earth Planet. Sci.* **17**, 227–254.

Goetze, C. and B. Evans (1979). *Geophys. J. R. Astron. Soc.* **59**, 463–478.

Goldsmith, W., *et al.* (1976). *Int. J. Rock Mech. Min. Sci.* **13**, 303–309.

Gomberg, J., *et al.* (1997). *Bull. Seism. Soc. Am.* **87**, 294–309.

Gomberg, J., *et al.* (1998). *J. Geophys. Res.* **103**, 24411–24426.

Goodman, R.E. and P.N. Sundaram (1978). *Pure Appl. Geophys.* **116**, 873–887.

Grady, D.E. and J. Lipkin (1980). *Geophys. Res. Lett.* **7**, 255–258.

Green, H.W., *et al.* (1990). *Nature* **348**, 720–722.

Green, S.J., *et al.* (1972). *J. Geophys. Res.* **77**.

Griffith, A.A. (1920). *Phil. Trans. R. Soc. London Ser. A.* **221**, 163–198.

Griggs, D.T. and D.W. Baker (1969). In: "Properties of Matter Under Unusual Conditions," pp. 23–42, Interscience Publishers, New York.

Griggs, D.T. and J.W. Handin (1960). In: "Rock Deformation, Geol. Soc. Am. Memoir 79," pp. 347–373, Geol. Soc. Am., Boulder, CO.

Gu, J., *et al.* (1984). *J. Mech. Phys. Solids* **32**, 167–196.

Gu, Y. and T.-f. Wong (1991). *J. Geophys. Res.* **96**, 21677–21691.

Gu, Y. and T.-f. Wong (1994). *Pure Appl. Geophys.* **143**, 388–423.

Handin, J. (1966). In: "Handbook of Physical Constants," pp. 223–289, Geological Society of America.

Harris, R.A. (1998). *J. Geophys. Res.* **103**, 347–358.

Harris, R.A. (2002). In: "International Handbook of Earthquake and Engineering Seismology," Chapter 73, Academic Press, London.

Harris, R.A. and R.W. Simpson (1992). *Nature* **360**, 251–254.

Hartzell, S. and T. Heaton (1989). *Bull. Seism. Soc. Am.* **79**, 1282–1286.

Heaton, T.H. (1982). *Bull. Seism. Soc. Am.* **72**, 2181–2200.

Heaton, T.H. (1990). *Phys. Earth Planet. Int.* **64**, 1–20.

Herget, G. and K. Unrug (1976). *Int. J. Rock Mech. Min. Sci.* **13**, 299–302.

Heuze, F.E. (1980). *Rock Mech.* **12**, 167–192.

Hickman, S.H. (1991). *Rev. Geophys.* IUGG Report (Suppl.), 759–775.

Hickman, S.H. and N.M. Beeler (1997). *EOS Trans. Am. Geophys. Un.* **78**, F731.

Hickman, S.H. and B. Evans (1991). *J. Geol. Soc.* **148**, 549–560.

Hickman, S.H. and B. Evans (1992). In: "Fault Mechanics and Transport Properties of Rocks," pp. 253–280. Academic Press, New York.

Hickman, S.H., *et al.* (1995). *J. Geophys. Res.* **100**, 12831–12840.

Hobbs, B.E. and A. Ord (1988). *J. Geophys. Res.* **93**, 10521–10540.

Hodgson, K. and N.G.W. Cook (1970). *Proc. 2nd Congress Int. Soc. Rock Mech.*, pp. 31–34.

Horii, H. and S. Nemat-Nasser (1986). *Phil. Trans. R. Soc. London, ser. A.* **319**, 337–374.

Hubbert, M.K. and W.W. Rubey (1959). *Bull. Geol. Soc. Am.* **70**, 115–166.

Hunt, J.M. (1990). *Am. Assoc. Petrol. Geol. Bull.* **74**, 1–12.

Irwin, W.P. and I. Barnes (1975). *Geology* **3**, 713–716.

Ishido, T. and H. Mizutani (1980). *Tectonophysics* **67**, 13–23.

Ismail, I.A.H. and S.A.F. Murrell (1990). *Tectonophysics* **175**, 237–248.

Jaeger, J.C. (1969). "Elasticity, Fracture and Flow," Methuen, London.

Jaeger, J.C. and N.G.W. Cook (1971). "Fundamentals of Rock Mechanics," Chapman and Hall, London.

Jaoul, O., *et al.* (1984). *J. Geophys. Res.* **89**, 4298–4312.

Johnston, M.J.S. and A.T. Linde (2002). In: "International Handbook of Earthquake and Engineering Seismology," Chapter 36. Academic Press, London.

Johnston, M.J.S., *et al.* (1987). *Tectonophysics* **144**, 189–206.

Kanamori, H. and C.R. Allen (1986). In: "Earthquake Source Mechanics," pp. 227–236. American Geophysical Union.

Karner, S.L. and C. Marone (1998). *Geophys. Res. Lett.* **25**, 4561–4564.

Karner, S.L., *et al.* (1997). *Tectonophysics* **277**, 41–55.

Kemeny, J.M. and N.G.W. Cook (1987). In: "Constitutive Laws for Engineering Materials," pp. 878–887. Elsevier, Amsterdam.

Kirby, S.H. (1980). *J. Geophys. Res.* **85**, 6353–6363.

Kirby, S.H. (1987). *J. Geophys. Res.* **92**, 13789–13800.

Kirby, S.H., *et al.* (1991). *Science* **252**, 216–225.

Kirby, S.H. and A.K. Kronenberg (1987). *Rev. Geophys.* **25**, 1219–1244.

Kirby, S.H. and J.W. McCormick (1984). In: "CRC Handbook of Physical Properties of Rocks," pp. 139–280. CRC Press, Boca Raton.

Koifman, M.I. (1963). In: "Mechanical Properties of Rocks," pp. 109–117. Akademiya Nauk SSSR (English translation).

Kranz, R.L. (1980). *J. Geophys. Res.* **85**, 1854–1866.

Kranz, R.L., *et al.* (1982). *Geophys. Res. Lett.* **9**, 1–4.

Kronenberg, A.K., *et al.* (1990). *J. Geophys. Res.* **95**, 19257–19278.

Lachenbruch, A.H. (1980). *J. Geophys. Res.* **85**, 6097–6112.

Lachenbruch, A.H. and A.F. McGarr (1990). In: "The San Andreas Fault System," pp. 261–277. US Geological Survey.

Lachenbruch, A.H. and J.H. Sass (1980). *J. Geophys. Res.* **85**, 6185–6222.

Lama, R.D. and V.S. Vutukuri (1978). "Handbook on Mechanical Properties of Rocks," Trans. Tech. Pub.

Lambe, T.W. and R.V. Whitman (1979). "Soil Mechanics," Wiley, New York.

Lawn, B.R. and T.R. Wilshaw (1975). "Fracture of Brittle Solids," Cambridge University Press, Cambridge.

Linde, A.T. and M.J.S. Johnston (1989). *J. Geophys. Res.* **94**, 9633–9643.

Linker, M.F. and J.H. Dieterich (1992). *J. Geophys. Res.* **97**, 4923–4940.

Lipkin, J. and A.K. Jones (1979). Proc. 20th US Symposium on Rock Mechanics, pp. 601–606.

Lockner, D.A. (1993a). *Int. J. Rock Mech. Min. Sci.* **30**, 883–899.

Lockner, D.A. (1993b). *J. Geophys. Res.* **98**, 475–487.

Lockner, D.A. (1995). In: "AGU Handbook of Physical Constants," pp. 127–147. American Geophysical Union.

Lockner, D.A. (1998). *J. Geophys. Res.* **103**, 5107–5123.

Lockner, D.A. and N.M. Beeler (1999). *J. Geophys. Res.* **104**, 20133–20151.

Lockner, D.A. and J.D. Byerlee (1986). *Pure Appl. Geophys.* **124**, 659–676.

Lockner, D.A. and J.D. Byerlee (1990). *Pure Appl. Geophys.* **133**, 398–410.

Lockner, D.A. and J.D. Byerlee (1992). *J. Appl. Mech. Rev.* **45**(3), S165–S173.

Lockner, D.A. and J.D. Byerlee (1994). *Geophys. Res. Lett.* **21**, 2353–2356.

Lockner, D.A. and J.D. Byerlee (1995). *Pure Appl. Geophys.* **145**, 717–745.
Lockner, D. and B. Evans (1995). *J. Geophys. Res.* **100**, 13081–13092.
Lockner, D.A. and P.G. Okubo (1983). *J. Geophys. Res.* **88**, 4313–4320.
Lockner, D.A. and S. Stanchits (2002). *J. Geophys. Res.* in press.
Lockner, D.A., *et al.* (1977). *J. Geophys. Res.* **82**, 5374–5378.
Lockner, D.A., *et al.* (1982). *Geophys. Res. Lett.* **9**, 801–804.
Lockner, D.A., *et al.* (1991). *Nature* **350**, 39–42.
Lockner, D.A., *et al.* (1992a). In: "Fault Mechanics and Transport Properties of Rocks," pp. 3–31. Academic Press, New York.
Lockner, D.A., *et al.* (1992b). 33rd US Rock Mechanics Symposium, pp. 807–816. Balkema.
Lui, M. (1997). *J. Geophys. Res.* **102**, 5295–5312.
Lyell, C. (1877). "Principles of Geology," Appleton.
Madariaga, R.I. and K.B. Olsen (2002). In: "International Handbook of Earthquake and Engineering Seismology," Chapter 12. Academic Press, London.
Madden, T.R. (1983). *J. Geophys. Res.* **88**, 585–592.
Mair, K. and C. Marone (1999). *J. Geophys. Res.* **104**, 28899–28919.
Mandl, G. (1988). "Mechanics of Tectonic Faulting," Elsevier, Amsterdam.
Mandl, G., *et al.* (1977). *Rock Mech.* **9**, 95–144.
Marone, C.J. (1998). *Annu. Rev. Earth Planet. Sci.* **26**, 643–696.
Marone, C. and B. Kilgore (1993). *Nature* **362**, 618–621.
Marone, C. and C.H. Scholz (1988). *Geophys. Res. Lett.* **15**, 621–624.
Marone, C., *et al.* (1990). *J. Geophys. Res.* **95**(B5), 7007–7025.
Marone, C., *et al.* (1991). *J. Geophys. Res.* **96**, 8441–8452.
Masuda, K., *et al.* (1990). *J. Geophys. Res.* **95**, 21583–21592.
McGarr, A. (1977). *J. Geophys. Res.* **82**, 256–264.
McGarr, A. (1998). *J. Geophys. Res.* **104**, 3003–3011.
McGarr, A., *et al.* (2002). In: "International Handbook of Earthquake and Engineering Seismology," Chapter 40. Academic Press, New York.
Meade, C. and R. Jeanloz (1991). *Science* **252**, 68–72.
Meredith, P.G., *et al.* (1990). *Tectonophysics* **175**, 249–268.
Meunier, A. and B. Velde (1989). *Am. Mineral.* **74**, 1106–1112.
Miller, S.A., *et al.* (1996). *Geophys. Res. Lett.* **23**, 197–200.
Mogi, K. (1962). *Bull. Earthq. Res. Inst.* **40**, 175–185.
Mogi, K. (1966). *Bull. Earthq. Res. Inst.* **44**, 215–232.
Mogi, K. (1967). *J. Geophys. Res.* **72**, 5117–5131.
Mogi, K. (1970). In: "Rock Mechanics in Japan," pp. 53–55. Japan Society for Civil Engineering.
Moore, D.E. and J.D. Byerlee (1989). *US Geol. Surv. Open File Report 89-347.*
Moore, D.E. and J.D. Byerlee (1991). *Geol. Soc. Am. Bull.* **103**, 762–774.
Moore, D.E. and D.A. Lockner (1995). *J. Struct. Geol.* **17**, 95–114.
Moore, D.E., *et al.* (1983). Proc. 24th US Symposium on Rock Mechanics, pp. 489–500.
Moore, D.E., *et al.* (1984). *US Geol. Surv. Open File Report 84-273.*
Moore, D.E., *et al.* (1986). *US Geol. Surv. Open File Report 86-578.*
Moore, D.E., *et al.* (1994). *Science* **265**, 1558–1561.
Moore, D.E., *et al.* (1996). *Geology* **24**, 1041–1045.
Moore, D.E., *et al.* (1997). *J. Geophys. Res.* **102**, 14787–14801.
Morgan, J.K. and M.S. Boettcher (1999). *J. Geophys. Res.* **104**, 2703–2719.
Morrow, C. and J. Byerlee (1991). *Geophys. Res. Lett.* **18**, 211–214.
Morrow, C., *et al.* (1992). In: "Fault Mechanics and Transport Properties of Rocks," pp. 69–88. Academic Press, New York.
Morrow, C., *et al.* (1994). *J. Geophys. Res.* **99**, 7263–7274.
Morrow, C.A. and J.D. Byerlee (1989). *J. Struct. Geol.* **11**, 815–825.
Morrow, C.A., *et al.* (1981). *Geophys. Res. Lett.* **8**, 325–329.
Morrow, C.A., *et al.* (2000). *Geophys. Res. Lett.* **27**, 815–818.
Mount, V.S. and J. Suppe (1987). *Geology* **15**, 1143–1146.
Nadeau, R.M. and T.V. McEvilly (1999). *Science* **285**, 718–721.
Nakatani, M. (1998). *J. Geophys. Res.* **103**, 27239–27256.
Nur, A. (1972). *Bull. Seism. Soc. Am.* **62**, 1217–1222.
Nur, A. and J. Walder (1990). In: "The Role of Fluids in Crustal Processes," pp. 113–127. National Academy Press.
Ohnaka, M. and Y. Kuwahara (1990). *Tectonophysics* **175**, 197–220.
Ohnaka, M. and L.-f. Shen (1999). *J. Geophys. Res.* **104**, 817–844.
Ohnaka, M. and T. Yamashita (1989). *J. Geophys. Res.* **94**, 4089–4104.
Ohnaka, M., *et al.* (1986). In: "Earthquake Source Mechanics," pp. 13–24. American Geophysical Union.
Okubo, P.G. and J.H. Dieterich (1981). *Geophys. Res. Lett.* **8**, 887–890.
Okubo, P.G. and J.H. Dieterich (1984). *J. Geophys. Res.* **89**, 5815–5827.
Okubo, P.G. and J.H. Dieterich (1986). In: "Earthquake Source Mechanics," pp. 25–35. American Geophysical Union.
Olsson, W.A. (1991). *Int. J. Rock Mech. Miner. Sci.* **28**, 115–118.
Parry, W.T. and R.L. Bruhn (1990). *Tectonophysics* **179**, 335–344.
Parry, W.T., *et al.* (1991). *J. Geophys. Res.* **96**, 19733–19748.
Paterson, M. (1978). "Experimental Rock Deformation," Springer-Verlag, Berlin.
Peng, S. and A.M. Johnson (1972). *Int. J. Rock Mech. Miner. Sci.* **9**, 37–86.
Perin, G., *et al.* (1995). *J. Mech. Phys. Solids* **43**, 1461–1495.
Pinto da Cunha, A. (1990). "Scale Effects in Rock Masses," Balkema.
Power, W.L. and T.E. Tullis (1989). *J. Struct. Geol.* **11**, 879–893.
Power, W.L., *et al.* (1988). *J. Geophys. Res.* **93**, 15268–15278.
Powley, D.E. (1990). *Earth Sci. Rev.* **29**, 215–226.
Pratt, H.R., *et al.* (1972). *Int. J. Rock Mech. Miner. Sci.* **9**, 513–529.
Pratt, H.R., *et al.* (1977). *Int. J. Rock Mech. Miner. Sci.* **14**, 35–45.
Rabinowicz, E. (1965). "Friction and Wear of Materials," Wiley, New York.
Raleigh, C.B. (1967). *Geophys. J. R. Astron. Soc.* **14**, 113–118.
Raleigh, C.B. and M.S. Paterson (1965). *J. Geophys. Res.* **70**, 3965–3985.
Raleigh, C.B., *et al.* (1976). *Science* **191**, 1230–1237.
Reasenberg, P.A. and R.W. Simpson (1992). *Science* **255**, 1687–1690.
Reches, Z. and D.A. Lockner (1994). *J. Geophys. Res.* **99**, 18159–18173.
Reinen, L.A., *et al.* (1991). *Geophys. Res. Lett.* **18**, 1921–1924.
Reinen, L.A., *et al.* (1994). *Pure Appl. Geophys.* **143**, 317–358.
Rice, J.R. (1975). *J. Geophys. Res.* **80**, 1531–1536.
Rice, J.R. (1980). In: "Physics of the Earth's Interior," pp. 555–649. North Holland, Amsterdam.
Rice, J.R. (1992). In: "Fault Mechanics and Transport Properties of Rocks," pp. 475–503. Academic Press, New York.
Rice, J.R. (1993). *J. Geophys. Res.* **98**, 9885–9907.
Rice, J.R. and Y. Ben-Zion (1996). *Proc. Natl. Acad. Sci. USA* **93**, 3811–3818.

Rice, J.R. and M.P. Cleary (1976). *Rev. Geophys. Space Phys.* **14**, 227–241.
Rice, J.R. and A.L. Ruina (1983). *J. Appl. Mech.* **50**, 343–349.
Rice, J.R. and S.T. Tse (1986). *J. Geophys. Res.* **91**, 521–530.
Ringwood, A.E. (1972). *Earth and Planet. Sci. Lett.* **14**, 233–241.
Robertson, E.C. (1983). *Trans. A.I.M.E.* **35**, 1426–1432.
Roeloffs, E. (1996). *Adv. Geophys.* **37**, 135–195.
Roeloffs, E. and J.W. Rudnicki (1985). *Pure Appl. Geophys.* **122**, 560–582.
Rudnicki, J.W. (1980). *Annu. Rev. Earth Planet. Sci.* **8**, 489–525.
Rudnicki, J.W. and C.-H. Chen (1988). *J. Geophys. Res.* **93**, 4745–4757.
Ruina, A.L. (1983). *J. Geophys. Res.* **88**, 10359–10370.
Rutter, E.H. and D.H. Mainprice (1978). *Pure Appl. Geophys.* **116**, 634–654.
Sammis, C.G. and M.F. Ashby (1986). *Acta Metall.* **34**, 511–526.
Sammis, C.G. and S.J. Steacy (1994). *Pure Appl. Geophys.* **142**, 777–794.
Satoh, T., *et al.* (1990). *Tohoku Geophys. J.* **33**, 241–250.
Savage, J.C. and D.A. Lockner (1997). *J. Geophys. Res.* **102**, 12287–12294.
Savage, J.C., *et al.* (1996). *J. Geophys. Res.* **101**, 22215–22224.
Sayles, F.H. and D.L. Carbee (1981). *Eng. Geol.* **18**, 55–66.
Scholz, C.H. (1968a). *J. Geophys. Res.* **73**, 3295–3302.
Scholz, C.H. (1968b). *Bull. Seism. Soc. Am.* **58**, 1117–1130.
Scholz, C.H. (1990). "The Mechanics of Earthquakes and Faulting," Cambridge University Press, Cambridge.
Scholz, C.H. (2000). *Geology* **28**, 163–166.
Scholz, C.H. and C.A. Aviles (1986). In: "Earthquake Source Mechanics," pp. 147–155. American Geophysical Union.
Scholz, C.H., *et al.* (1973). *Science* **181**, 803–810.
Scholz, C.H., *et al.* (1986). *Bull. Seism. Soc. Am.* **76**, 65–70.
Segall, P. and J.R. Rice (1995). *J. Geophys. Res.* **100**, 22155–22173.
Shamir, G. and M.D. Zoback (1989). In: "Rock at Great Depth," pp. 1041–1048. Balkema.
Sibson, R.H. (1981). In: "Earthquake Prediction: An International Review," pp. 593–604. American Geophysical Union.
Sibson, R.H. (1991). *Bull. Seisn. Soc. Am.* **81**, 2493–2497.
Sibson, R.H. (1992). *Tectonophysics* **211**, 283–293.
Sibson, R.H. (2002). In: "International Handbook of Earthquake and Engineering Seismology," Chapter 29. Academic Press, London.
Singh, M.M. and P.J. Huck (1972). Proc. 14th Symp. on Rock Mech., pp. 35–60, *American Society of Civil Engineering*.
Skempton, A.W. (1954). *Geotechnique* **4**, 143–147.
Sleep, N.H. and M.L. Blanpied (1992). *Nature* **359**, 687–692.
Sleep, N.H. and M.L. Blanpied (1994). *Pure Appl. Geophys.* **143**, 9–40.
Solberg, P. and J.D. Byerlee (1984). *J. Geophys. Res.* **89**, 4203–4205.
Spencer, A.J.M. (1982). In: "Mechanics," pp. 607–652. Pergamon, New York.
Spray, J.G. (1987). *J. Struct. Geol.* **9**, 49–60.
Spray, J.G. (1988). *Contrib. Mineral Petrol.* **99**, 464–475.
Stein, R.S. (1999). *Nature* **402**, 605–609.
Stein, R.S. and M. Lisowski (1983). *J. Geophys. Res.* **88**, 6477–6490.
Stesky, R.M., *et al.* (1974). *Tectonophysics* **23**, 177–203.
Summers, R., *et al.* (1978). *J. Geophys. Res.* **83**, 339–344.
Talwani, P. (1976). *Eng. Geol.* **10**, 239–253.
Tapponnier, P. and W.F. Brace (1976). *Int. J. Rock Mech. Miner. Sci.* **13**, 103–112.
Tenthory, E., *et al.* (1998). *J. Geophys. Res.* **103**, 23951–23967.
Terzaghi, K.V. (1929). *Bull. Am. Inst. Min. Eng. Tech. Publ.* **215**, 31.
Tingle, T.N., *et al.* (1993). *J. Struct. Geol.* **15**, 1249–1256.
Tostoi, D.M. (1967). *Wear* **10**.
Townend, J. and M.D. Zoback (2000). *Geology* **28**, 399–402.
Tse, S.T. and J.R. Rice (1986). *J. Geophys. Res.* **91**, 9452–9472.
Tsutsumi, A. and T. Shimamoto (1997). *Geophys. Res. Lett.* **24**, 699–702.
Tullis, J. and R.A. Yund (1977). *J. Geophys. Res.* **82**, 5705–5718.
Tullis, T.E. and J.D. Weeks (1986) *Pure Appl. Geophys.* **124**, 384–414.
Vaisnys, J.R. and C.C. Pilbeam (1976). *J. Geophys Res.* **81**, 985–988.
Vaughan, P.J., *et al.* (1985). *US Geol. Surv. Open File Report 85-262.*
Vermilye, J.M. and C.H. Scholz (1998). *J. Geophys. Res.* **103**, 12223–12237.
Vidale, J.E., *et al.* (1998). *J. Geophys. Res.* **103**, 24567–24572.
Walsh, J.B. (1971). *J. Geophys. Res.* **76**, 8597–8598.
Walsh, J.B. (1981). *Int. J. Rock Mech. Miner. Sci.* **18**, 429–435.
Wang, C.-Y. (1984). *J. Geophys. Res.* **89**, 5858–5866.
Wang, C.-Y., *et al.* (1980). *J. Geophys. Res.* **85**, 1462–1468.
Wang, H.F. (1997). *J. Geophys. Res.* **102**, 17943–17950.
Wang, W. and C.H. Scholz (1994). *Pure Appl. Geophys.* **145**, 303–315.
Weeks, J.D., *et al.* (1978). *Bull. Seism. Soc. Am.* **68**, 333–341.
Weertman, J. (1980). *J. Geophys. Res.* **85**, 1455–1461.
Williams, C.F. (1996). *Geophys. Res. Lett.* **23**, 2029–2032.
Wong, T.-f. (1982a). Mech. Mater. **1**, 3–17.
Wong, T.-f. (1982b). *J. Geophys. Res.* **87**, 990–1000.
Woodcock, D. and E. Roeloffs (1996). *Oreg. Geol.* **58**, 27–33.
Wu, F.T., *et al.* (1975). *Pure Appl. Geophys.* **113**, 87–95.
Yamashita, T. and M. Ohnaka (1991). *J. Geophys. Res.* **96**, 8351–8367.
Yanagidani, T., *et al.* (1985). *J. Geophys. Res.* **90**, 6840–6858.
Yoshioka, N. (1985). *J. Phys. Earth* **33**, 295–322.
Zheng, G. and J.R. Rice (1998). *Bull. Seism. Soc. Am.* **88**, 1466–1483.
Zoback, M.D. and J.D. Byerlee (1975). *J. Geophys. Res.* **80**, 752–755.
Zoback, M.D. and M.L. Zoback (2002). In: "International Handbook of Earthquake and Engineering Seismology," Chapter 34. Academic Press, London.
Zoback, M.D., *et al.* (1987). *Science* **238**, 1105–1111.

Editor's Note

Due to space limitations, three useful items are placed on the Handbook CD-ROM under directory of \32Lockner: (1) symbols used in this chapter are given in LocknerTable1.pdf; (2) references with full citation are given in LocknerFullReferences.pdf; and (3) detailed figure captions are given in LocknerFullFigCaptions.pdf.

33

State of Stress Within the Earth

Larry J. Ruff
University of Michigan, Ann Arbor, USA

1. Introduction

Stress makes geologic processes happen, and geologic processes make stress. Plate tectonics, earthquakes, volcanic eruptions, glacial rebound, landslides, tidal deformation, phase changes, fluid flow, rock folding, and crystallization are examples of processes that generate, modify, and consume stress within the Earth. Since stress is the continuum equivalent of force, it is expected that any mechanical process in the solid or fluid parts of the Earth must involve stress. Gravitational forces acting on and within the Earth are intimately connected to stress state, and earthquakes occur when the shear stress exceeds the failure level for the fault. Some classic results on the stress state of the Earth are valid today, yet stress remains as a topic of intense study across many disciplines in the Earth Sciences.

The central role of stress and the long history of stress studies make it impossible to reference all relevant works. Instead, I will reference several other review papers that contain extensive lists of references. Also, I cannot possibly give comprehensive discussions of all the problems and considerations for all methods and results. Instead, I will resort to simplified summaries and broad sweeping statements as much as possible, though always supported by references. In many instances, I will reference the special issue on a topic, rather than individual contributions. Many subjects and investigators have a long literature history, so I reference just one recent paper that gives the earlier citations. More emphasis will be given to some recent trends in stress studies.

This review of stress first discusses the basic definition of stress and a simplified view of rock strength. The stress state can be divided into two parts: pressure and shear stress. There are no significant controversies today surrounding the state of pressure within the Earth, so that section is relatively short. Then we turn to shear stresses within the Earth. Although shear stress is small compared to pressure in most of the Earth, shear stress is the subject of most interest today. Any discussion or study of shear stress includes constraints from several different methods. This review orders the methods from the most reliable to the more controversial methods and topics, such as earthquake stress drops, plate boundary heat flow, and fault pore pressure. If earthquake hazards are the primary concern, then the most important aspect of stress state is the difference between shear stress and failure stress levels. Although this topic has a long history in seismology, it has attracted several recent studies from both a theoretical view and observations of induced seismicity and earthquake triggering.

The conclusions of this review echo those of previous reviews: support of mountains requires shear stresses of at least 75 MPa somewhere in the lithosphere; stresses associated with plate tectonics are about 10 MPa at plate boundaries, and reach higher values in the strong portions of plate interiors; and although stresses vary wildly in both direction and magnitude in the upper crust at short length scales, there are coherent large-scale trends of stress direction that can be connected to global-scale tectonics. On the other hand, there are significant questions related to low stresses on plate boundaries and major faults. Pore fluid pressure is now a popular mechanism to explain weak faults, but there are still some unresolved contentious issues. Also, the related notions of a uniformly critically stressed crust and earthquake triggering forces us to take a broader view of earthquake hazards.

1.1 Aspects of the Stress Tensor

Stress is a well-defined physical quantity throughout the interior of any fluid or solid material. Basic rules of force balance on a small material element lead to the construction of the stress tensor, which can be viewed as three mutually perpendicular force dipoles acting on the faces of the material element (Fung, 1965; Aki and Richards, 1980). These three dipoles are commonly referred to as the principal stresses, and the units of stress are force per unit area ($1\ \mathrm{Pa} = 1\ \mathrm{N\,m^{-2}}$). The stress tensor accounts for the orientation and strength of the force dipoles, and can vary with time for a given material element or location. Tides cause global variations at a time scale of a day. Earthquakes cause variations at shorter time scales throughout the

INTERNATIONAL HANDBOOK OF EARTHQUAKE AND ENGINEERING SEISMOLOGY, VOLUME 81A ISBN: 0-12-440652-1

Earth due to seismic waves, and at longer time scales near the fault due to the static stress changes. Other geologic processes, such as ice or sediment loading, cause variations at a variety of time scales. Each element inside the Earth is subjected to all these influences.

If all scientists always referred to stress tensor components in the same global coordinate system, then we could avoid most of the confusion and terminology difficulties with stress. But it is more useful to describe stress with respect to local geometry; for example, the trend of a local fault. Furthermore, many studies focus on shear stress size, regardless of orientation. Thus, some terminology inconsistencies are bound to appear. Here, I provide just the notation used in this review.

The most important division of the stress state is into the pressure and shear stresses; this corresponds to splitting the stress tensor into the isotropic and deviatoric parts. The isotropic part is well defined as the "pressure," but unfortunately there is some confusion due to the sign convention. Most formal treatments of stress define a tensional force dipole as positive (e.g., Fung, 1965), which leads to the compressive pressure inside the Earth as being negative pressure. However, there is a long tradition in the Earth Sciences that refers to compressive pressure as positive pressure. Since this overview does not need to follow rigorous derivation of equations, I adopt the common geologic usage and refer to compressional pressure as positive. As shown in Figure 1a, pressure is formally defined as one-third the trace of the stress tensor; or in other words, pressure is the average value of the principal stresses. The "sign" of shear stresses is more difficult than for pressure. Hence, it is common practice in the Earth Sciences to give a description for the shear stress geometry in terms of complete geographic faulting notation: e.g., "shear stress of 10^5 Pa in the sense of right-lateral shear along a plane of N45°W strike and 45° dip to the NE." Figure 1a illustrates how a stress tensor for a single force dipole is decomposed into the isotropic part and deviatoric part. Sometimes the deviatoric part is further decomposed into two "pure shear double couples."

A common description of the stress state is to quote the "normal" and "shear" stresses with respect to a particular plane slicing through the material. In this context, "normal" stress refers to the force component perpendicular to the plane, whereas "shear" stress refers to the force component acting parallel to the plane. For many applications, the plane is a preexisting fault plane. This view is most useful when testing the stress state against some frictional failure criterion. In the matrix representation of stress, "normal" and "shear" stresses are the diagonal and off-diagonal components of the stress tensor after it has been rotated to a new coordinate system that coincides with the plane orientation. Mohr's circles provide a clever geometric visualization of this view of the stress state. The Mohr's circle in Figure 1b also helps to illustrate some of the confusion over shear stress. Since many scientists and engineers use Mohr's circles for stress analysis, it is natural to

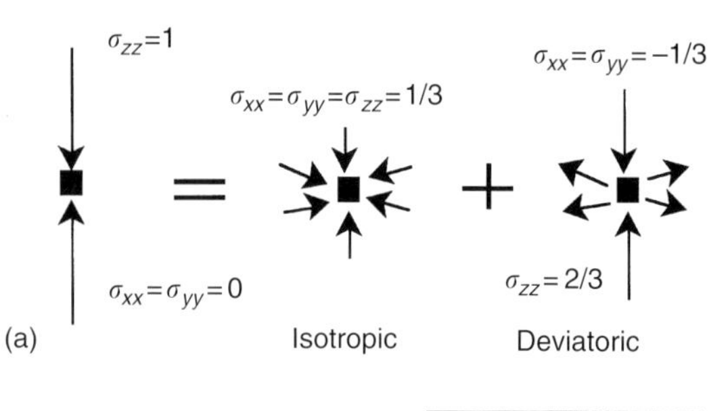

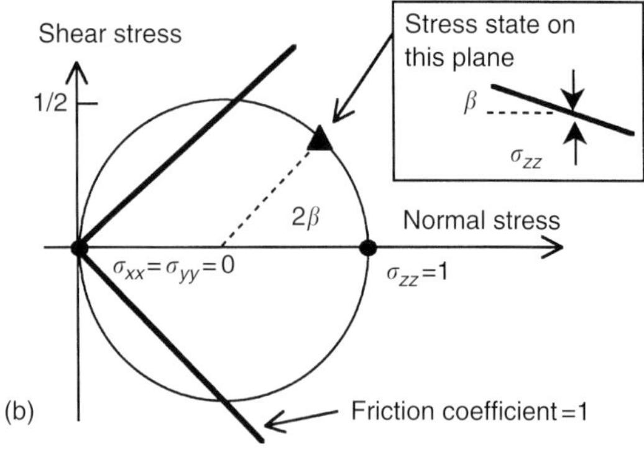

FIGURE 1 Stress tensor for a single vertical force dipole (σ_{zz}) represented by isotropic and deviatoric components (a) or by Mohr's circle (b).

think that shear stress should be defined with respect to a particular plane. As a plane is rotated from an initial orientation that is normal to the maximum principal stress, then shear stress along the plane increases from its initial value of zero.

In the matrix representation, shear stress would be the off-diagonal components of a stress tensor that is rotated from the initial diagonalized form to some arbitrary coordinate system. In the context of the matrix representation, the "natural" coordinate system orientation for any stress tensor is the one that diagonalizes the stress tensor to find the three force dipoles. In this "natural" coordinate orientation, shear stresses are represented by differences in the size of the three principal stresses. Thus, a good invariant, i.e., independent of any particular planar orientations, measure for shear stress is the difference between the greatest and least sizes of the principal stresses; this quantity is commonly called the "stress difference," and I will use this term (it is also referred to as the "differential stress" in some places). But the Mohr's circle in Figure 1b shows us that maximum value of the shear stress resolved onto a single plane is half the stress difference. For the example in Figure 1 where two normal stresses are the same, we can achieve this maximum shear stress value not just on one plane, but on all planes inclined at 45° to the vertical axis. Again, I shall adopt the common geologic usage that refers to shear stress as resolved on a plane of a particular orientation. When I use "shear stress" with no further qualification, it will be the maximum shear stress obtained from a Mohr's circle plot; that is, half the value of the "stress difference." Of course, shear stress values quoted in many of the references might follow other conventions.

How large is a "small material element?" If the stress tensor is homogenous across a large volume, then the entire volume can be considered a "small element." For example, planetary scale calculations might assume that large blocks of the Earth's lithosphere have a uniform stress, then the entire block can be thought of as a single material element with a single stress state. On the other hand, one rock from the middle of this lithospheric block might display significant variations in both orientation and size of stress (see Zoback, 1992, for introduction to special issue). Thus, we must realize that the stress tensor at any one site is the superposition of contributions from different processes that range from global-scale tectonics to microscopic cracks.

1.2 Rock and Fault Strength

Many of the constraints on stress state in the Earth are based on the physical properties of rocks and faults. Mantle and crustal rocks are capable of sustaining shear stresses either from a short-term elastic response, or a fluid flow response for longer times. Laboratory experiments on rocks have established the overall broad constraints on the maximum strength of mantle and crustal rocks (see chapters in Jaeger and Cook, 1979; Scholz, 1990). Here tensional and compressional strength are first briefly discussed followed by shear strength.

Rocks and other materials are weakest for tensional stresses. Tensional failure may be important for some phenomena close to the Earth's surface, and might be part of the microscopic mechanisms of shear failure at depth. Rocks become stronger as isotropic compression increases, unless the pressure becomes so large that they undergo a solid-to-solid "phase change" that increases density. These high-pressure phase changes may be regarded as volumetric failure, though in most cases phase changes occur slowly. Phase changes are known to occur in many crustal rocks as they subside or emerge. Various candidate mantle rock compositions undergo phase changes at the depths of the observed upper-mantle seismic discontinuities (see Chapter 51 by Lay). If rocks are advected through these phase-change boundaries, either in subduction zones or elsewhere, then phase changes are an active process of the present-day earth. Even if phase changes are slow, they might cause earthquakes due to the induced shear stresses. In addition, a current popular explanation for deep Wadati–Benioff zone earthquakes makes use of kinetically delayed phase changes in the subducting slab (Kirby *et al.*, 1996; Green and Houston, 1995).

The shear strength of rocks is directly related to earthquakes, mantle flow, and support of topography and bathymetry. In the laboratory, rock samples can be selected that have no through-going fractures. These experiments give us the upper bound on bulk rock strength (see Chapter 32 by Lockner and Beeler). To simplify the results, divide rock behavior into two regimes that depend largely on homologous temperature: cold rocks are brittle, hot rocks are ductile. In the brittle regime, increasing shear stress deforms the rock elastically until brittle failure begins and eventually the rock breaks. Failure stress is a small fraction of the shear modulus; a typical shear stress value at failure for uniaxial experiment on crustal rocks is 100–150 MPa (Jaeger and Cook, 1979). Failure stress increases with pressure, but in the Earth, the temperature increase eventually places the rock in the ductile regime. In the ductile regime of high temperatures and slow strain rates, the rock acts like a fluid with high viscosity. A ductile rock still behaves elastically to transient perturbations of stress. For example, deep mantle rocks that are flowing as part of mantle convection also transmit seismic waves, and if some sharp stress transient were to occur such as stress loading due to a nearby earthquake, they might show brittle failure. This complex behavior of rocks makes it difficult to assign a single shear stress to rocks as their ultimate strength, but the rock strength envelopes (Goetze and Evans, 1979; Brace and Kohlstedt, 1980) usually show peak shear strength values of about 100–150 MPa at the brittle–ductile transition depth of 15 km or so in continental crust. Other studies (e.g., Molnar, 1992) argue that the peak ductile shear strength of 100 MPa is reached just below the Moho.

Other laboratory experiments are performed on fractured rocks to test the frictional sliding properties on various contact surfaces. In these experiments, a normal stress is applied to the crack surface, then the shear stress is increased until slippage occurs. Over a wide variety of conditions, it is found that the ratio of shear to normal stress is constant; this ratio is the coefficient of friction. For common crustal rocks, "Byerlee's Law" states that the coefficient of friction is about 0.6–1.0 (Byerlee, 1978). If the lithosphere has faults of all orientations, then these results give the upper bound for shear strength of the lithosphere, where the normal stress is approximately the same as the pressure. Byerlee's Law is added to the Mohr's circle plot in Figure 1b. This diagram then shows the range of fault plane orientations that are stable in the sense that shear stress is less than the value required to overcome static friction. One major complication for frictional sliding is that normal stress may be reduced by fluid pore pressure to the "effective normal stress." Pore pressure reduces all principal stresses by the same amount, essentially shifting the Mohr's circle in Figure 1b to the left. Since Byerlee's Law remains the same, this shift reduces the shear stress level required for fault slip. This issue is returned to later in the discussion of weak plate boundaries.

If the pore pressure is greater than the least principal stress, then the left-ward shift of the Mohr's circle places this least principal stress into the tensional stress regime which causes "hydrofractures." There are many technological applications of hydrofracturing in boreholes and it also provides many of the deep direct stress measurements (see Chapter 34 by

Zoback and Zoback). One consequence of hydrofracturing is that fluids flow into the open tensional cracks and pore pressure consequently decreases, unless the crack stops growing and there is a large reservoir of high-pressure fluid. In the discussions of water and weak plate boundaries, we will see that avoidance of hydrofracturing is one of the key constraints on models.

2. Pressure within the Earth

Isotropic stress state within the Earth is based on calculations of the gravitational loading. It can be measured by direct techniques, but estimates based on rock loading are more reliable. The vertical force dipole, σ_{zz}, at a depth of H is the integral of the product of gravitational acceleration (g) and density (ρ) from the surface down to depth H. The Earth's density structure is nearly spheroidally symmetric. The largest lateral variations are near the Earth's surface, and there are small perturbations throughout the mantle. Although the gravitational acceleration inside the Earth depends on the density model, gravitational acceleration through the mantle must be nearly constant for any reasonable earth model that matches the basic constraints of radius, mass, moment of inertia for the Earth, and the core–mantle boundary radius. This convenient coincidence for our planet makes it particularly easy to calculate buoyancy effects anywhere in the crust and mantle. Thus, there is little error in the value of "g", and the subsequent variations in σ_{zz} through the mantle are even smaller. Pressure estimates at the core–mantle boundary have changed from Bullen's Model A (Bullen, 1947) estimate of 137. GPa to the PREM value of 135.7 GPa (Dziewonski and Anderson, 1981). Pressure in the core smoothly increases to a value of 363.8 GPa at the Earth's center. Unless someone discovers major flaws in the constraints or methodology of earth model construction, deep Earth pressure estimates will change by less than 1%. Thus, current interest in pressure is focused on the fluid Earth and the uppermost crust.

2.1 Fluid Pressure Variations in the Earth

The ocean and atmosphere apply pressure to the Earth's surface. Since the lateral density variations and dynamic pressure effects of the atmosphere and oceans are small (i.e., less than 10%), fluid pressure is accurately predicted from either elevation or bathymetry. Atmospheric pressure at sea level is quite close to 10^5 Pa or 1 bar (standard pressure is 1.013 bar). This same pressure can be achieved by the gravitational load of a 10 m tall column of water, or a 4 m rock column. Thus, for most purposes, we can ignore the atmosphere and refer to the Earth's surface as a stress-free boundary. There are a few phenomena where atmospheric pressure must be included, for example, seismic noise generation by weather, and Chandler wobble excitation. Although the dynamic pressure perturbations caused by storms is small, Diakonov *et al.* (1990) review some evidence of induced microseismicity.

At a typical ocean basin depth of about 5 km, the water column pressure is about 50 MPa. This pressure is equivalent to a rock column height of 2 km. Some geologic phase relations change over this pressure interval, thus water overburden pressure is significant for some shallow processes. Dynamic variations in water pressure are due to tides, waves, and currents. These dynamic perturbations are small and can be ignored for most solid-earth applications. Some exceptions are the "microseism" noise which is due to ocean wave pressure fluctuations (see review by Haubrich and McCamy, 1969; and special issue, Hjortenberg and Nikolaev, 1990), and coastal gravity and geodetic measurements that are sensitive to tilting caused by tidal loading of continental shelves. It is even possible that ocean tide loading triggers volcanic eruptions (see Chapter 25 by McNutt) and earthquakes (Tsuruoka *et al.*, 1995).

One other fluid region of the Earth is the outer core. Pressure is accurately predicted by global earth models. Lateral pressure variations in the outer core must be very small, but similarly to the ocean and atmosphere, large fluid motions can occur with tiny pressure variations.

2.2 Pressure Variation in the Solid Earth

The largest fractional variations in σ_{zz} within the Earth are in the uppermost crust. Small rock bodies of unusual density can cause small anomalies, but the strongest variations will be caused by open fractures, or even caves. A cave's free surface requires that the normal stress is zero (or to be more precise, normal stress is 1 bar). Since the rock above the cave must still be supported, σ_{zz} is larger on either side of the cave, whereas the tangential stresses at the top and bottom of the cave can be tensional. Any open crack causes a strong perturbation to the stress field, which decreases away from the crack. At distances that are large compared to the crack size, the stress field is the same as that for no crack. There is a limit to how deep a cave can occur in the Earth due to rock weakness. If a cave generates tensional stresses at top and bottom, the rock will fail with just a modest amount of rock overburden. At deeper depths, the shear failure strength of rocks will be exceeded and the hole will fill with rock fragments. At lower crustal and mantle depths, plastic deformation can eventually close all open cracks. Many boreholes are drilled to depths of a few kilometers, but they are filled with a dense fluid to provide pressure support for the surrounding rock. For deep mines, however, the rock walls are at surface pressure, and these large values of stress difference can cause shear failure (Jaeger and Cook, 1979). In conclusion, any open cracks in the upper crust will have localized large stress perturbations, though this effect can be mitigated with high pressure fluids that fill the cracks.

The maximum stress difference may reach 100 MPa or more in the upper crust, but higher temperatures in the lower crust cause the shear strength of rocks to decrease. Even if rocks could withstand shear stresses of 100 MPa all the way to the core–mantle boundary, this shear stress level fades to a minor perturbation compared to the pressure, which continues to increase to over 100 GPa. Thus, at the bottom of the mantle, the maximum shear stress is less than a factor of one thousand smaller than the isotropic part of the stress tensor. The two horizontal principal stresses are approximately equal to σ_{zz} throughout the mantle and so the pressure is also equal to σ_{zz}.

Subducting slabs may perturb the pressure state of the mantle. Rock is subducted down to a depth of at least 670 km, and it is likely that this rock undergoes phase changes. It is possible for the interior of active phase-change inclusions to be at a different pressure from the surrounding rock, but the shear stress induced in the surrounding rock relaxes to reload the inclusion back to the ambient pressure. Deep earthquakes were once thought to be part of this "relaxation" process (for reviews and recent ideas about deep earthquakes, see Frohlich, 1989; Green and Houston, 1995; Kirby *et al.*, 1996).

In conclusion, the stress tensor is nearly isotropic throughout most of the Earth. Furthermore, there are reliable quantitative estimates of this pressure. However, it is the shear stress, those small deviations away from the isotropic stress state, that is the main interest of geologists and geophysicists. Part of this interest stems from the fact that shear stress is "unknown," but it is also due to the fact that shear stress is fundamental to many interesting geologic processes such as plate tectonics, mountain building, earthquakes, volcanic eruptions, and landslides.

3. Introduction to Shear Stress within the Earth

In general, we want to know both the orientation and size of shear stresses. As summarized in the World Stress Map Project (introduction by Zoback, 1992), there are many more reliable indicators of orientation than of size. In the present review, the brief discussion of orientation is in the "Direct Measurements" section; the related review (Chapter 34 by Zoback and Zoback) contains more information on shear stress directions. The sections below emphasize shear stress size.

Many different methods place constraints on shear stress. This review presents the methods in order from the most reliable to the most controversial. Gravitational loading places the best constraints on minimum shear stress size. This method also provides robust estimates of the minimum shear stresses associated with plate tectonics. Direct measurements give us the somewhat surprising conclusion that shear stresses may be near their maximum allowed value nearly everywhere in plate interiors. We end with the interconnected topics of earthquake stress drops, heat flow, plate boundary stresses, and pore pressure. There is a long-standing debate in Earth Sciences between proponents of "low" (10 MPa) and "high" (>50 MPa) shear stresses. Earlier review papers (e.g., Kanamori, 1980; Hickman, 1991) have offered a compromise of high stresses beneath mountains and low stresses acting along plate boundaries. It seems that this compromise view can accommodate most of the quantitative results acquired in the last two decades. Indeed, much of the debate has now shifted to exactly how plate boundaries are able to operate at low shear stress levels.

4. Gravitational Loading

Gravity produces most of the large stresses within the Earth, and these stresses can be reliably calculated. The best constraints on minimum shear stress size are based on gravitational loading. Lateral variations in density, topography, and bathymetry are responsible for most of the shear stress within the Earth. One of the first and still most powerful statements about stress state is the classic Jeffreys argument about support of mountains. This argument leads to all modern research in mantle density anomalies, geoid, and plate tectonic stresses.

4.1 Classic Jeffreys Argument for Shear Stress

The essence of the Jeffreys argument is that pressure at the bottom of a load sitting on the Earth causes shear stresses somewhere that are about one-sixth to one-quarter the size of the load pressure (Jeffreys, 1970; see McNutt, 1980, for discussion of the Jeffreys argument). Figure 2a shows the trivial case where the stress difference in the rock sample at the base of the cliff is $(\rho g H)$, where g is gravitational acceleration, ρ is rock density, and H is height of rock column. For this rock sample, the shear stress will be $(\rho g H/2)$. With a plateau height of 5 km, this rock sample would have a shear stress of about 63 MPa (or 630 bars). Shear stress in the underlying elastic half-space is 20 MPa along the circular arc in Figure 2a.

A more realistic profile for a mountain chain is the triangular cross-section shown in Figure 2b, where the σ_{zz} loading at the center is $(\rho g H)$. Jeffreys showed that if you apply this load to an elastic half-space, the peak shear stress is $(0.512)(\rho g H)/2$ at a depth that is about one-quarter of the mountain width. Hence, peak mountain height determines the

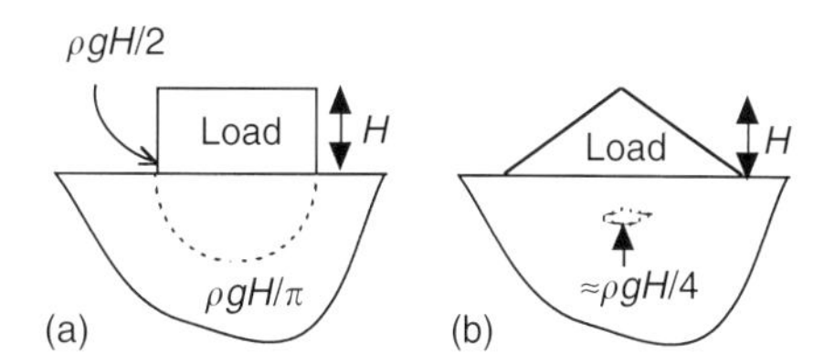

FIGURE 2 Illustration of stresses induced by gravitational loads. In (a), a plateau with height H is placed on the Earth. A triangle-shaped mountain is placed on the elastic half-space in (b).

peak shear stress, regardless of the slope. The analysis becomes more complex if one allows for relaxation and modification of the elastic stress field. Clearly, use of fluid rheology below a certain depth concentrates and increases the shear stresses in the upper layer. Jeffreys discusses various methods to redistribute stress, but notes that reduction of stress in one place necessarily increases it elsewhere. The reduction in peak shear stress from the elastic solution is quite small. One extreme, yet simple, illustration of the efficacy of the Jeffreys argument is the Stokes problem of a high density sphere sinking through a viscous fluid (Turcotte and Schubert, 1982). The gravitational load for a vertical column through the sphere's center is ($\Delta\rho g 2R$), where $\Delta\rho$ is the density anomaly of the sphere with radius R. The peak shear stress in the fluid flowing around the sphere is (1/6)($\Delta\rho g 2R$) (i.e., one-sixth of the maximum load pressure) and is independent of the viscosity and flow velocity. Hence, although the rheology governs the deformation and details of stress distribution, it can not reduce the peak shear stress from the Jeffreys value. To quote from Jeffrey's (1970): "As a general guide it is safe to say that a variation of surface load with range (ρgH) implies a stress difference of about (1/3)(ρgH) to (1/2)(ρgH) somewhere."

To look at mountains from a broader perspective, large mountain chains are isostatically compensated at fairly shallow depth (100 km or less), and this serves to increase shear stress. Jeffreys went on to show that there must be shear stresses of at least 75 MPa beneath the Himalayas, and about half that level beneath other mountains belts such as the Alps, Andes, and the Rockies.

Any departure from spheroidal symmetry is associated with shear stresses. Lateral density anomalies could be calculated theoretically, such as the density variations from a mantle convection calculation, or they might be the results of gravity and geoid anomaly inversion. Variations in the Earth's surface and other boundaries are efficient "generators" of gravitational loads due to density contrasts across boundaries. There is more uncertainty in the stress estimates based on buried loads as compared to loads that we can "see," such as mountains, trenches, and ridges.

An active area of research has been to combine geoid anomalies with other constraints on internal heterogeneity such as seismic tomography (Bai *et al.*, 1992), or the history of subducted slabs (Lithgow-Bertelloni and Richards, 1998) to localize the density anomalies. Once this is done, the shear stresses required to support density anomalies and boundary variations can be calculated. Of course, Jeffreys (1970) already estimated the minimum shear stress level in the mantle required to support early estimates of the long-wavelength geoid anomalies; his stress estimate was about 10 MPa. Modern studies can increase that minimum value if they can better localize the gravitational loads responsible for the geoid anomalies.

One important advance has been the determination of lithosphere thickness that can support elastic shear stresses over geologic time. The advent of digital data sets for topography, gravity, and numerical methods has allowed more accurate description. Oceanic lithosphere provides the best place to perform these studies (see review by McNutt, 1986). Most recent investigations have converged on a consensus view of elastic thicknesses [but see McKenzie and Fairhead (1997) for a contrary view, especially for the elastic thickness of continents]. For the future, geodetic strain rate constraints combined with gravity data promise to provide better rheological models and hence more detailed lithosphere shear stress estimates.

4.2 Stresses that Drive the Plates

Plate tectonics is the most important Earth Sciences paradigm shift in the 20th century. Although plate tectonics is a kinematic description, there has always been a keen interest in the stresses that drive plate tectonics. Most people now agree that mantle thermal convection is the underlying mechanism for plate tectonics, and that the oceanic lithosphere acts like the upper thermal boundary layer on top of this system. In addition, chemical differentiation and phase changes add buoyancy forces, though these are more difficult to estimate. A complete comprehensive simulation of mantle convection in the "real" earth with all buoyancy sources and proper rheology would produce the stress field that drives the plates. This comprehensive simulation has not yet been achieved. But we do not need to know the details of mantle convection to calculate the minimum level of associated shear stresses. Thermal convection produces regions of positive and negative buoyancy that "drive" the system. Hence, if we can see the buoyancy source, a Jeffreys-type calculation gives the minimum shear stress.

Now that we know where to look, we can "see" part of the plate tectonic load by observing the difference between young and old oceanic lithosphere. This is shown in Figure 3, where the two columns show the ridge crest and old oceanic lithosphere. It is easy to calculate the extra load at the ridge by replacing water with hot rock. It gives a loading stress of about 60 MPa. If we then simply apply the Jeffreys rule that minimum peak shear stress should be about (1/6)ρgH, then we

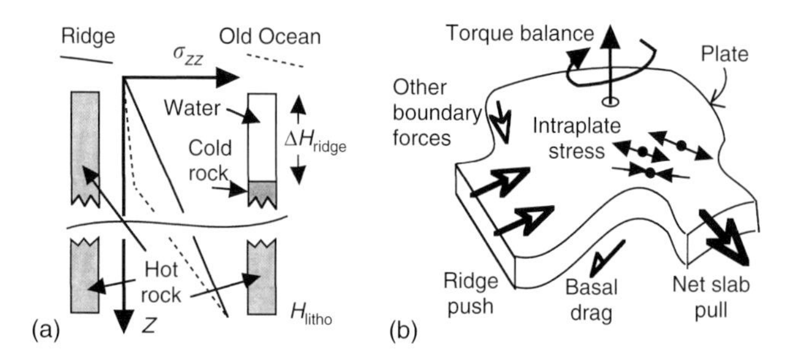

FIGURE 3 Ridge push and plate force balance. A lower bound on peak shear stress associated with plate tectonics is based on the gravitational load of the ridge relative to old oceanic lithosphere (a). The requirement of zero net torque for a tectonic plate connects ridge push to other plate boundary forces (b).

obtain 10 MPa for the minimum shear stress associated with plate tectonics. With more sophisticated arguments, McKenzie and Jarvis (1980) estimated that plate boundary shear stress falls in the range of 10–50 MPa. Some early models of subduction that included the observational constraints of topography and gravity found that boundary shear stress must be less than 70 MPa (Sleep, 1975), with a typical value of 30 MPa. More recent estimates based on much more advanced models find that mantle and plate boundary stresses are "low" (for references, see Bai *et al.*, 1992; Richardson, 1992; Bird, 1998), less than 50 MPa.

4.2.1 Ridge Push and Slab Pull

One of the more popular uses of the ridge loading stress is to turn it into the horizontally directed ridge force. The basic procedure is shown in Figure 3a (see details and references in Artyushkov, 1973). The pressure at a depth of 100 km or so is the same below a ridge and old lithosphere (see Stein and Stein, 1992, for recent model of oceanic lithosphere). Assume a simple stress state of pure isotropic stress within each column so that horizontal stress equals σ_{zz}. Then the horizontal stress at a ridge is larger, but the pressure difference decays to zero at the lithosphere base as required by isostatic equilibrium. If a membrane were placed between these two columns, the ridge column would be pushing on it with the pressure difference as a function of depth. The integrated effect of this push is basically given by the product of the maximum pressure difference and the thickness of the lithosphere, divided by two. In detail, there is little extra push from the height difference between the ridge crest and ocean basin. This integral is then approximately ($g\,\Delta\rho\,\Delta H_{\text{ridge}} H_{\text{litho}}/2$), which equals $3 \times 10^{12}\,\text{N m}^{-1}$ for H_{litho} of 100 km. Note that the units are ridge force per unit ridge length along strike. The number used for horizontal ridge push by several investigators (e.g., Hager and O'Connell, 1980; Richardson, 1992) is the same as above. More sophisticated global calculations, such as those in Richardson (1992), distribute this ridge push across the ocean basin, rather than just concentrate it at the ridge.

It is difficult to make reliable estimates for the gravitational load at subduction zones. Although we can easily "see" ridge push, much of the subduction zone load is buried and geoid interpretation is tricky. Of course, deep-sea trenches give some clue as to the buried mass excess in the subducted slab (Davies, 1983; Zhong and Gurnis, 1995) but we need more clues. The basic thermal contribution to slab buoyancy seems well established (e.g., Turcotte and Schubert, 1982), but the unknown plate boundary stresses affect thermal models (Peacock, 1996). Theoretical calculations of slab buoyancy keep changing as our notions about thermal structure, composition, phase changes, and – most recently – delayed phase changes in subducted slabs evolve (Ito and Sato, 1992). Nonetheless, slab pull forces for typical lithosphere structure are in the range of 4–$10 \times 10^{12}\,\text{N m}^{-1}$ (Spence, 1987; Whittaker *et al.*, 1992). Recent estimates of slab buoyancy with kinetically delayed phase changes emphasize the "parachute" effect that limits subduction velocities (Marton *et al.*, 1999).

4.2.2 Global Force Balance and Plate Boundary Stresses

We now turn to more-specific global models of plate boundary and intraplate stresses. Since the influential paper of Forsythe and Uyeda (1975) on the global force balance on major plates, there have been many versions of global force balance calculations with the observed geometry of the Earth's plates and plate boundary types. Once we assign forces to a certain type of plate boundary, the interconnected global plate system provides an opportunity to calculate the orientation and size of forces on other plate boundaries or areas (Fig. 3b). Forsythe and Uyeda (1975) concluded that "slab pull" force dominates "ridge push" by more than a factor of ten, but also that "slab pull" must be largely balanced by local resistive forces in the surrounding mantle and on the plate boundary. Forsythe and Uyeda (1975) also found that transform fault and subduction zone boundary forces were equal to or less than the ridge push force; hence plate boundary stress averaged over lithosphere thickness is about 30 MPa for transform boundaries and 15 MPa for subduction boundaries. Other global-scale calculations found a closer balance between ridge push and slab pull (e.g., Hager and O'Connell, 1980). Most force-balance models (e.g., Richardson *et al.*, 1979; Hager and O'Connell 1980; Richardson, 1992; Bird, 1998) find that plate boundary resistive stress averaged over the entire lithosphere thickness is approximately 10 MPa.

4.2.3 Global Tectonic Models with Intraplate Stress

The final step in these plate tectonic calculations is to use the buoyancy sources and boundary forces, assume some rheologic structure for the lithosphere, and then calculate plate interior stresses (depicted in Fig. 3b).

Model shear stresses within subducting slabs can be quite variable; typical values easily reach 100 MPa (Vassiliou and Hager, 1989; Wortel and Vlaar, 1988; Whittaker *et al.*, 1992). Outer-rise stresses were a focal point for the low versus high stress debate, but further modeling showed that the non-uniqueness in plate rheology imparts large uncertainties in the model stresses (see Hanks and Raleigh, 1980; Kanamori, 1980). Now the emphasis is on the contributions of bending and time-dependent stresses through the Earthquake cycle to the outer-rise stress state (see Dmowska *et al.*, 1988; Christensen and Ruff, 1988; Liu and McNally, 1993). Also, stress regime of the double seismic zone at intermediate depths continues to present a perplexing problem (Kawakatsu, 1986; Seno and Yamanaka, 1996).

The last topic is the most complex and is an active area of research today: estimation of stress orientation and size in plate interiors for realistic simulations of plate geometry, boundary forces, and rheology (e.g., Richardson *et al.*, 1979; Cloetingh and Wortel, 1986; Govers *et al.*, 1992; Richardson, 1992; Bai *et al.*, 1992; Bird, 1998). Even if the boundary stresses are low, the intraplate stresses can be high due to rheological and geometric effects. It seems that there is some disagreement over the quantitative details, but intraplate shear stress levels are predicted to reach values of 50 MPa in many of these studies. A major emphasis of these current studies is to compare the predicted stress orientations with focal mechanisms, stress provinces, or to geologic history. To conclude, both gravitational loading and boundary force modeling predict high shear stress in some parts of the oceanic lithosphere, but nearly all of these studies have low plate boundary stress values.

5. Direct Stress Measurements

There is no remote sensing instrument that can directly measure stress in the Earth. We must dig a hole to gain access to a rock face, take a "picture," remove more rock to create additional free surface, then take another "picture" or restore the deformation with a known force. Chapter 34 by Zoback and Zoback reviews all the various direct and indirect methods, thus only a few summary statements are made here.

Some methods, such as overcoring and hydrofracture, do give orientation and size information. Many of the other methods, such as volcanic vent alignments, are better characterized as "indicators." There are many excellent review papers on direct and indirect stress measurements (e.g., McGarr and Gay, 1978; McNutt, 1987; Hickman, 1991; Chapter 34 by Zoback and Zoback). As summarized in Zoback (1992) only 4% of the data in the World Stress Map Project database give reliable absolute values of stress. As concluded in earlier reviews (see above), direct measurements indicate that upper crust shear stresses in many locales are at the maximum level predicted from Byerlee's Law.

Most stress measurements in the database only yield a stress "direction": the orientation of one or more principal axes of the stress tensor. Earthquake focal mechanisms provide the best rock volume integrators and the most numerous global indicators of stress direction. The issue of earthquake stress drop is treated below in a separate section. Many investigators have devised clever techniques to use focal mechanisms from regional seismicity to extract more reliable stress directions (see special issue, Zoback, 1992). One key conclusion from the measurements is that the upper crust has tremendous small-scale variations in stress state due to material heterogeneity, open cracks, and short-wavelength stress loading. For many scientific and engineering applications, these shallow rapid variations in stress are the primary interest. But the small length-scale of these stress concentrations does not contribute to large-scale crustal stress averages. Only consistent trends across large distances can be related to large-scale tectonic processes. As reviewed by Zoback (1992) and Chapter 34 by Zoback and Zoback, there are large-scale consistent trends in the stress measurements; these areas are called stress provinces. Furthermore, there are several generalizations that can be made about these regimes: plate interiors are dominated by strike-slip and thrust faulting regimes; normal faulting regimes occur in selected continental and oceanic areas with high topography; most stress provinces coincide with physiographic provinces. Zoback (1992) discusses the alignment of stress orientation in some provinces with absolute plate motions, though it seems that there may be other tectonic explanations for these orientations (Richardson, 1992). The World Stress Map Project provides valuable observations that constrain any detailed model of lithosphere stresses.

5.1 Geologic Indicators of Stress

Earthquakes and other stress measurements provide a present-day map of stresses. Several stress indicators used in the World Stress Map Project are geologic observations. Here, I emphasize the future potential of paleo-stress indicators. There is a long tradition in structural geology of paleo-strain indicators, but finding reliable estimates of stress in exhumed rocks is a much more difficult task. Nonetheless, there has been some progress in a variety of techniques (see related papers in special issues, Engelder and Geiser, 1980; Magloughlin and Spray, 1992; Schmid *et al.*, 1999). Given the current controversy surrounding stress levels on major faults and plate boundaries, studies of frictional melting on exhumed faults could be a particularly productive area for future research.

5.2 Stress from the KTB Project

There have been several special projects and deep boreholes that produced stress estimates that are suggestive of high shear stress in the crust. Some of the most recent and compelling direct stress measurements come from the KTB borehole project in Germany. This borehole reached a depth of 9 km in a plate interior. There are several remarkable features about the stress and related measurements at this site (Brudy *et al.*, 1997). First of all, the shear stress at a depth of about 8 km is about 100 MPa. Also, pore fluid pressure is "hydrostatic" down to this depth. Furthermore, the high shear stress level is close to the maximum shear stress allowed by Byerlee's Law with a friction coefficient of 0.6 and a normal stress that is the lithostatic load minus the pore fluid pressure. There was no tectonic reason to expect high shear stresses in such a setting. The notion that this region is just below the failure stress is further supported by the fluid injection experiment where an incremental pressure increase of roughly 1 MPa induced small earthquakes. Zoback and Harjes (1997) show that focal mechanisms of the induced seismicity are consistent with the measured stress tensor.

To summarize, shear stress orientations in various stress provinces are determined by some aspects of global-scale tectonics. As for stress size, there is evidence from the KTB project and other direct measurements that the upper crust in plate interiors have "high" shear stresses at the frictional failure level. The next section focuses on this aspect.

6. Triggered Earthquakes and Failure Threshold

Earthquakes occur when the shear stress exceeds the failure stress threshold. There is an incubation time between threshold exceedance and earthquake dynamic rupture that ranges from less than a second to years, as documented below [also see special volume edited by Atkinson (1987), Scholz (1990) for discussion, and Ohnaka (1992) for laboratory observations]. This incubation time delay is even more uncertain if triggered creep events or other aseismic processes are responsible for critical loading, and if the failure threshold itself evolves with time (e.g., Sibson, 1990). Empirical studies of earthquake triggering are hampered by the clusters of event occurrence times that are expected from a random Poisson process. Despite all these complexities, there are some well-accepted cases of earthquake triggering at a variety of spatial–temporal time scales. I briefly review these phenomena, starting with the most common example (aftershocks) and ending with the fascinating examples of human-induced seismicity which imply that the Earth's crust must be close to the earthquake failure threshold everywhere.

Aftershocks are the most common and obvious example of triggered seismicity. Aftershocks offer valuable insight to the failure process and there are many empirical and modeling studies (e.g., Yamashita and Knopoff, 1987; Yamanaka and Shimazaki, 1990; Kisslinger, 1993). Although most aftershocks occur on the fault area ruptured by the large earthquake, there is great interest in large earthquakes triggered in adjacent portions of the fault. Scholz (1990) discusses several case histories including the sequence of large earthquakes along the North Anatolian Fault in Turkey (Stein *et al.*, 1997). Many other examples are found in subduction zones, where large-earthquake doublets and multiplets are common in some locales such as the Solomon Islands (Xu and Schwartz, 1993) and Nankai Trough (Ando, 1975). The time delay between these doublets ranges from less than one day to a few years (see Nomanbhoy and Ruff, 1996, for global survey). Since the adjacent plate boundary segments will eventually have a large earthquake, it is fairly easy to understand how stress transfer causes temporal clustering along plate boundaries or major faults. One confusing consequence of large-earthquake doublets and multiplets is that the aftershock areas of adjacent ruptures may overlap (e.g., Tajima *et al.*, 1990), and this hinders the determination of large earthquake recurrence times (see comments by Nishenko and Sykes, 1993; Jackson and Kagan, 1993).

There is more current interest in aftershocks off the main fault. These "aftershocks" tell us that large volumes of crust surrounding the main fault are also close to their failure threshold. Some great underthrusting earthquakes have triggered intraplate events in the outer-rise (see summary in Christensen and Ruff, 1988). One might expect that great earthquakes would also trigger seismicity in the Wadati–Benioff zone, but any effect is rather subtle (Astiz *et al.*, 1988).

Off-fault aftershocks from strike-slip earthquakes have attracted much interest in the last decade. Two well-documented cases are the great (M_w 8) 1989 Macquarie Ridge event (Das, 1993) and the smaller but extensively studied 1992 Landers, California earthquakes (Anon., 1994). Modeling the static stress changes associated with finite fault models of the Landers events has now expanded to re-examination of other older examples of off-fault triggering (Stein *et al.*, 1997; also see special issue, Harris, 1998). One hypothesis that emerges from these studies is that small shear stress increases trigger seismicity in large crustal volumes.

Remote triggering is even more fascinating where the distances are so large that seismic waves must be the agent of triggering. Remote triggering has been accepted in some parts of the international community for a long time (see references in Diakonov *et al.*, 1990), but this phenomenon is now widely accepted after the spectacular triggering of seismicity throughout western North America after the Landers earthquake (Hill *et al.*, 1993). The physics of how seismic wave dynamic stresses might trigger seismicity is still under investigation, but it is worth noting that most of the triggered seismicity occurred in geothermal regions. Perhaps some fluid–solid coupling process is essential for this effect (see papers in Anon., 1994). Also note that most of the triggered events would not have been detected in most parts of the world due to sparse seismograph network coverage.

The last type of triggering is seismicity induced by fluid injection or reservoir impoundment (see reviews by Simpson, 1986; Chapter 40 by McGarr *et al.*). Hydrofracture experiments produce abundant nanoseismicity, but this activity is not surprising since cracks can generate large stresses. On the other hand, reservoir-induced seismicity sends us a clear message that the Earth's crust is close to shear failure in most continental settings. The gravitational load from most reservoirs changes shear stresses by 1 MPa or less. Seismicity may be induced directly by this surface load, or by the changes in fluid pore pressure (Simpson, 1986). Regardless of the details, the fact that 1 MPa water surface loads trigger seismicity compels us to conclude that stress in most of the Earth's crust is close to the failure threshold.

Many different aspects and consequences of earthquake triggering are addressed in the Harris (1998) special issue. The well-known lack of correlation between tidal stresses and earthquake occurrence (Vidale *et al.*, 1998) is puzzling and admits several different interpretations. Theoretical and observational studies on the spatial-temporal evolution of failure thresholds and incubation time should receive more research effort in the future. There is currently great interest in holistic interpretations of the inference that shear stress is close to failure stress in most of the Earth's crust. One perspective is that the Earth's lithosphere presents an example of a system self-maintained at the critical state. In this context, the critical state is shear stress at the failure threshold stress (Bak and Tang, 1989; Sornette *et al.*, 1990; Rundle *et al.*, 1995; Grasso and Sornette, 1998). If we enlarge our definition of failure to include some aseismic processes, then perhaps most of the Earth is "at failure." It is appropriate that one of the more successful quantitative applications has been the critical taper explanation for accretionary prisms in subduction zones – which can be viewed as a big sand pile (see review by Dahlen, 1990). Recall that the detailed tectonic models described in Section 4 do not predict high shear stresses everywhere. Those regions of low stress might still be close to failure due to various processes lowering the failure threshold. Plate boundaries are places of low shear stress and low failure thresholds. The KTB borehole intraplate site is an example of high failure threshold but also high shear stress. In the future, perhaps we should study places of low shear stress and no seismicity to test the notion that the Earth's lithosphere is critical everywhere.

7. Earthquake Stress Drops and Plate Boundary Stresses

Earthquakes are the primary stress failure indicators in the Earth. When shear stress across a locked fault reaches the failure stress, the fault slips to reduce shear stress on the fault plane, though stress might increase in adjacent regions. This overall average reduction in shear stress on the ruptured fault is the static stress drop. Static stress drop provides a lower bound on lithospheric shear stress. Below, both static and dynamic stress drop results for earthquakes are reviewed. The key results are that recent studies reaffirm the "rule" that static stress drops are about 1–10 MPa for large plate boundary earthquakes, and average dynamic stress drops from two techniques – dynamic stress drop and radiated wave energy – are no larger than static stress drops. In detail, there are some systematic trends within the scatter of stress drops, but that should not change the primary conclusion that stress drops of large earthquakes are "low."

Shear stress at a locked plate boundary can slowly increase until it reaches the failure stress. The failure stress level can be represented by the static coefficient of friction. Slip on the fault can occur either as creep or as earthquakes; the behavior is controlled by the dynamic coefficient of friction during slip. If the dynamic coefficient of friction is the same value or larger, than the static coefficient of friction, then creep is the resultant behavior. If the dynamic coefficient of friction drops below the static value fast enough, then we have an earthquake. The friction constitutive laws are covered in Chapter 32 by Lockner and Beeler. The dynamic stress drop is controlled by this complicated frictional behavior, and it is likely to vary in space and time as the rupture front sweeps across the fault. Dynamic stress drop has the potential to reveal more earthquake physics, but current global surveys of dynamic stress drop assume simple models with uniform dynamic stress drop.

7.1 Static Stress Drop

Static stress drop is the simplest measure of the overall reduction in shear stress due to slip on the fault zone. It is the difference between the average shear stress on the fault zone before and after the earthquake (Fig. 4). Since the stress drop of real earthquakes varies across the fault area, the overall static stress drop is a slip-weighted average of the spatially variable stress drop. Seismologists typically use simple constant stress drop models to estimate earthquake stress drops. Regardless of the details of fault geometry and slip distribution, the basic formula for stress drop is:

$$\Delta\sigma_{st} = c\,\mu D/X \tag{1}$$

where D is the average slip over the faulted area (A), X is the characteristic length of the fault area, μ is the elastic shear modulus, and c is a geometric constant that is close to one (if X is properly chosen). Since seismic moment (M_o) for most large earthquakes can be reliably determined from seismic waves, the Eq. (1) can be rewritten as:

$$\Delta\sigma_{st} = c\,\mu DA/(XA) = c\,M_o/(XA) \tag{2}$$

This formula shows that three quantities are needed to calculate stress drop: a measurement of the seismic moment, some estimate of the fault area (A), and then some appropriate choice for the characteristic fault dimension. Whereas the choice of X presents an interpretation problem, it is the estimation of A that presents practical difficulties and introduces large errors into stress drop estimates. The best estimates of A combine geodetic, aftershocks, and seismic wave data, but most estimates of A are based solely on drawing a rectangle or ellipse around aftershocks. This technique is trustworthy for great earthquakes with fault zones more than 100 km across, but its reliability decreases as fault dimensions decrease toward the epicenter location accuracy and as the number of aftershocks decreases. Since the fault area of magnitude 7 earthquakes is expected to be just a little over 10 km across, this aftershock technique is limited just to great earthquakes if we desire a uniform global study.

It is possible to extract the fault length from source-time functions. If a large earthquake ruptures a long fault zone in a unilateral fashion, the directivity seen in the source time

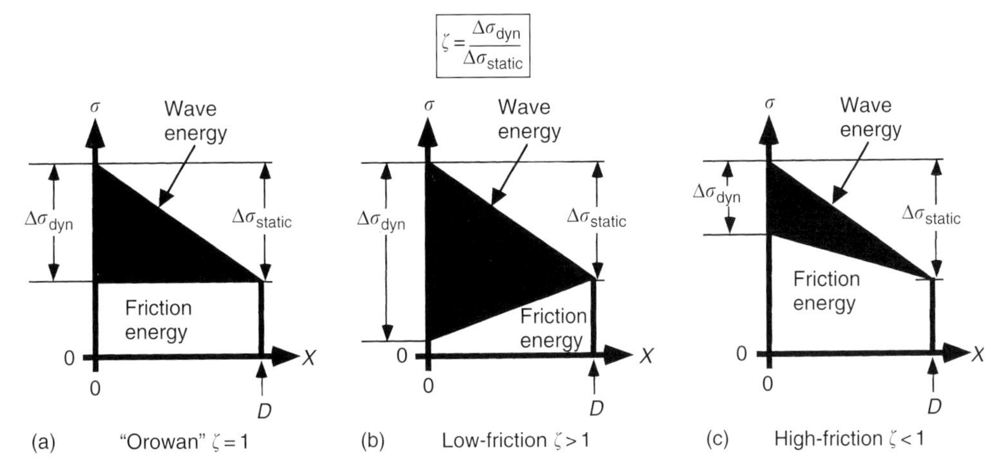

FIGURE 4 Illustration of how the ratio of dynamic to static stress drop partitions earthquake strain energy into wave and friction energy. Three possible behaviors of dynamic fault friction are shown. Each graph shows the fault-averaged history of fault friction stress (σ) as a function of fault slip (X), where final earthquake displacement is D.

functions from azimuthally distributed stations can be quantitatively exploited to determine the longer fault dimension. Other means are usually used to estimate the shorter fault dimension. This rupture process method is time-consuming and difficult, and produces reliable results for just a few earthquakes. Thus, in most studies, just the overall source time function duration is used to estimate fault length. This simple method can be applied to the largest number of events, but the rupture velocity and fault aspect ratio must be assumed.

Previous reviews have combined all the various estimates to compare stress drops of all earthquakes, small to large and deep to shallow (Aki, 1972; Kanamori and Anderson, 1975; Abe, 1982; Hanks, 1977; Purcaru and Berckhemer, 1982; Abercrombie, 1995). These studies have given us one of the key tenets of seismology that stress drop is "approximately constant" for earthquakes of all types and sizes. In detail, "approximately constant" means that stress drop estimates typically fall in the range of 0.1–10 MPa (1–100 bars), with an occasional report of much higher values for smaller events.

There has been an interpretation problem for earthquake stress drop. For small earthquakes, a circular fault shape is thought to be a good model, hence the static stress drop is:

$$\Delta\sigma_{\mathrm{st}} = (7/16)\,M_{\mathrm{o}}/r^3 \tag{3}$$

where r is the fault radius. Indeed, systematic studies of small earthquakes show that the ratio of moment to radius cubed is nearly constant. In detail, most studies of small events use source time function duration as a proxy for radius, thus the empirical result is that duration (d) increases as moment to the one-third power, i.e., $d \propto M_{\mathrm{o}}^{1/3}$.

For large earthquakes, the down-dip fault width is limited by the depth extent of the seismogenic zone. Hence, above a certain size, earthquakes all have the same fault width, but have different along-strike fault lengths. For an elongated fault shape, the characteristic length should be fault width, hence the stress drop formula is:

$$\Delta\sigma_{\mathrm{st}} = c\,M_{\mathrm{o}}/(W^2L) \tag{4}$$

where W and L are fault width and length. For empirical studies, the above formula predicts that for constant stress drop: $L \propto M_{\mathrm{o}}^{1.0}$. In the following subsections, we briefly review empirical results for earthquakes in different tectonic settings.

7.1.1 Shallow Intraplate Earthquakes

Shimazaki (1986) showed observations from intraplate events in Japan that show a change in fault length versus moment scaling at an earthquake size of about $M_{\mathrm{w}} = 6.7$. For the larger earthquakes, Shimazaki (1986) found that the ratio of (M_{o}/L^2) is constant (in detail, $L \propto M_{\mathrm{o}}^{0.44}$). If stress drop is constant for these large events, then Shimazaki's (1986) results imply that L is the characteristic length for stress drop, rather than W (see Scholz, 1990, for extended discussion of this point). Alternatively, if the above expectation in Section 7.1 for $L \propto M_{\mathrm{o}}^{1.0}$ scaling is correct, then stress drop increases with fault length for these events. Regardless of these scaling difficulties, Scholz (1990) reviews evidence that shallow intraplate events have stress drops of about 10 MPa, i.e., at the higher end of the overall observed range.

7.1.2 Strike-Slip Earthquakes

The change in fault aspect ratio from equant to elongated is particularly severe for strike-slip faults. The width of the seismogenic zone is about 15 km, but fault length can be hundreds of kilometers for the largest strike-slip events. For the case of long strike-slip faults that reach the surface, the geometric constant in the above formula is $(2/\pi)$, hence:

$$\Delta\sigma_{st} = (2/\pi)\, M_o/(W^2 L) \tag{5}$$

If this formula is correct, empirical studies of large strike-slip earthquakes should find the $L \propto M_o^{1.0}$ scaling. Romanowicz (1992) shows that this is true for strike-slip earthquakes with vertical fault planes and seismic moments larger than 1×10^{20} Nm (M_w 7.3). If we use the bounds on the $L \propto M_o^{1.0}$ scaling for large events in Romanowicz (1992), then stress drop ranges from 3 to 7.5 MPa. This reaffirms the "rule" that stress drop is constant and falls in the range of 1–10 MPa. One notable exception to this rule is the 1989 Macquarie Ridge earthquake. Although initial stress drop estimates were 30 MPa and higher, later reevaluation by Das (1993) has reduced the overall stress drop to about 12 MPa.

7.1.3 Subduction Underthrusting Earthquakes

The shallow dip angle of the plate interface in subduction zones allows the down-dip seismogenic width to be 150 km or so. Therefore, it is only the greatest underthrusting earthquakes with M_w more than 8 that have elongated rupture areas. For all other large underthrusting events, a good choice for the characteristic length is to take the geometric average of width and length (Aki, 1972). With this choice, $X = A^{1/2}$, and the stress drop formula is:

$$\Delta\sigma_{st} = c\, M_o/A^{3/2} \tag{6}$$

In this case, the empirical scaling would be: $L \propto M_o^{2/3}$. Aki (1972) and Kanamori and Anderson (1975) show that large subduction earthquakes are consistent with a constant stress drop of about 1–10 MPa using the above formula. The greatest subduction earthquakes have ratios of fault length to width greater than 4, which means that stress drop characteristic length should be fault width and the empirical length scaling should be $L \propto M_o^{1}$. However, the best-fit observed empirical scaling between length and moment is $L \propto M_o^{0.4}$ for event size of $2\text{–}2000 \times 10^{20}$ N m. A curious coincidence is that the scaling exponents for large intraplate (see Shimazaki, 1986) and great underthrusting events are 0.44 and 0.4, and this coincidence becomes even more curious since Tanioka and Ruff (1997) found that for recent global seismicity, the scaling exponent for source time function duration versus moment is about 0.4, i.e., $d \propto M_o^{0.4}$. These results imply that the assumed stress drop formula for large events misjudges the width and length scaling, or that stress drops do vary with earthquake size, or both. However, even if the extreme $L \propto M_o^{1}$ scaling is adopted to calculate stress drops, the quantitative results are that stress drop estimates from large to great underthrusting earthquakes are in the range of 0.2–9 MPa.

7.1.4 Deep Slab Earthquakes

Few aftershocks are recorded for deep Wadati–Benioff zone earthquakes (Frohlich, 1989), thus most static stress drop estimates are based on source–time functions. Empirical studies suggest that duration versus moment scaling for deep events is similar to that for shallow events, after correction for the larger shear modulus (e.g., Tanioka and Ruff, 1997; Bos *et al.*, 1998). Three large deep earthquakes have occurred since 1994 to spark new interest in deep earthquake rupture. Aftershock and rupture process studies have defined the rupture area for these events. As reviewed in Goes *et al.* (1997), the resultant static stress drops for these events are: 1994 Fiji, 7 MPa; 1996 Flores, 16 MPa, and 1994 Bolivia, 71 MPa (other estimates for Bolivia are even higher, see special issue, Wallace, 1995). Although all three events have source time function durations that follow the expected scaling rule, the Flores and Bolivia events have smaller than expected rupture areas (see references in Goes *et al.*, 1997), hence higher stress drops. These large deep events have prompted many papers and reevaluation of deep earthquake processes (e.g., Kanamori *et al.*, 1998). For our purposes here, it must be acknowledged that stress drop for large deep earthquakes can be "high," i.e., more than 50 MPa. Most stress drop compilations have a few small events with "high" stress drops, but those values can be ignored after consideration of error estimates. However, the extensive data sets and studies of the 1994 Bolivia event (Wallace, 1995) provide a reliable "high" stress drop.

7.1.5 Summary of Static Stress Drops

In conclusion, there are some problems in large earthquake scaling rules, but the observations still allow the following simple summary statement (with a notable exception): all large earthquakes have a static stress drop of 0.1–10 MPa (except for some large deep-slab earthquakes with static stress drops closer to 100 MPa). These static stress drop values represent a clear lower bound on shear stresses, yet they are close to estimates of total shear stress on plate boundaries from tectonic modeling studies.

7.2 Dynamic Stress Drop and Radiated Wave Energy

Dynamic stress drop measures the difference between initial stress and fault friction stress while the fault is slipping (Fig. 4). If static stress drop is just a partial stress drop and if dynamic friction stress is less than the final stress, then dynamic stress drop is larger than static stress drop. One of the suggestions to escape the "San Andreas heat flow paradox" (discussed in a later section, also see Chapter 35 by Brune and Thatcher) is that dynamic stress drop is much greater than static stress drop. Two methods can be used to estimate the ratio of dynamic to static stress drop: (1) "direct" estimate of dynamic stress drop; (2) measure the ratio of total radiated seismic wave energy to the "Orowan" radiated wave energy (see Fig. 4). It is difficult to reliably determine dynamic stress drop directly, and even more difficult to measure radiated wave energy.

Kikuchi and Fukao (1988) calculated radiated seismic wave energy for several great earthquakes and found that the energy was less than expected. "Effective seismic efficiency" (Kikuchi, 1992) is the wave-to-Orowan energy ratio described above. For the dynamic friction behavior depicted in Figure 4, ζ is the "effective seismic efficiency" [unfortunately, there is some notation confusion with seismic efficiency, see Kikuchi (1992), Kanamori (1994), and McGarr (1999)]. Kikuchi shows that "effective seismic efficiency" is less than 1 for the events in the Kikuchi and Fukao (1988) study, hence dynamic stress drop is less than static stress drop. Choy and Boatwright (1995) used a different method than Kikuchi and Fukao, but they found low values for "apparent stress drop," which implies low dynamic stress drop. Tanioka and Ruff (1997) found that source time function systematics for global seismicity are roughly consistent with the results in Kikuchi and Fukao (1988) and Choy and Boatwright (1995). On the other hand, Kanamori *et al.* (1993) used a empirical–theoretical hybrid technique to study California earthquakes, and found that seismic efficiency is closer to 1. Winslow and Ruff (1999) used a different hybrid technique for large deep Wadati–Benioff earthquakes, and found that "effective seismic efficiency" is less than or equal to 1.

A recent study by Ruff (1999) produced "direct" estimates of dynamic stress drop for global seismicity. Ruff (1999) found that dynamic stress drops were scattered from 0.1 to >10 MPa. Thus, the overall scatter in dynamic stress drops encompasses the scatter in static stress drops discussed above. However, the question remains if there is some systematic shift between stress drops for the same events. Figure 5 addresses this question by plotting both static and dynamic stress drop estimates for the same large underthrusting earthquakes. For most events, the dynamic stress drop estimate is

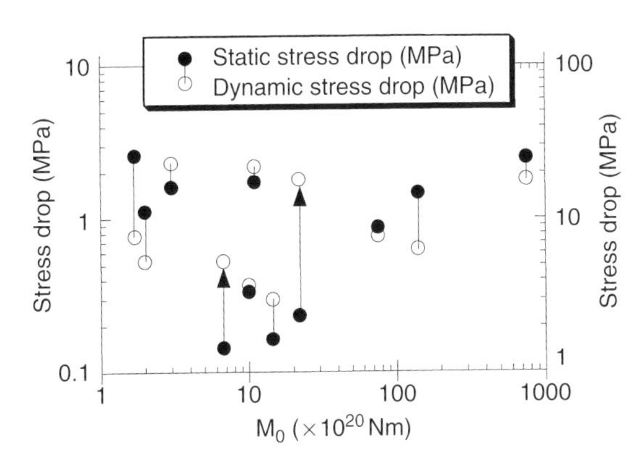

FIGURE 5 Comparison of static and dynamic stress drop estimates for many of the largest underthrusting plate boundary earthquakes that have occurred since 1963. Each earthquake is represented by a connected pair of filled and open dots, representing static and dynamic stress drop, respectively.

equal to or smaller than the static stress drop for the same earthquake (note the two exceptions in Fig. 5). Overall, these results favor the "hi-friction" rupture scenario in Figure 4c.

To summarize all these results that can be translated into the ratio of dynamic stress drop to static stress, observational studies of large earthquakes find that this ratio is less than 1. Since dynamic stress drops are the same or smaller than static stress drops, absolute shear stress levels that are significantly higher than earthquake stress drops must produce significant frictional heating during earthquake slip.

8. Heat Flow Constraints on Plate Boundary Stress

Earthquake stress drops give us a lower bound on shear stress acting along plate boundaries. To find the total shear stress, we also need to measure the average frictional stress level acting on the fault while it slips. In the simplest model of a creeping plate boundary, the average frictional stress level multiplied by the slip velocity gives the heat production per unit area of the fault zone, and this increases the temperature of the fault zone so that sufficient heat energy is conducted away to balance the heat production. Soon after the advent of plate tectonics, it was realized that strike-slip plate boundaries were ideal places to look for the heat flow anomaly from frictional heating. Given the long-term slip rate along a fault and the hypothetical depth distribution of friction shear stress, it is straightforward to predict the shape and strength of the surface heat flow anomaly. Observations show no heat flow anomaly of the correct shape across either the creeping or seismically active portions of the San Andreas Fault. The upper bound on frictional shear stress is about 10 MPa (see Brune *et al.*, 1969; Lachenbruch and Sass, 1992; Sass *et al.*, 1997). This celebrated San Andreas heat flow result is still one of the strongest arguments for low shear stress on plate boundaries.

Dip-slip faulting presents a more difficult heat flow experiment than strike–slip faults (Molnar and England, 1990). Thermal models are more complex for subduction plate boundaries as they must include heat advection and time-dependent thermal structure in the down-going slab. Early thermal models simply assumed high stress levels since one of the objectives was to melt subducted material. Today, most models for island arc volcanism presume that volatiles are fluxed from the slab into the mantle to cause melting (Peacock, 1996), thus there is no a priori reason to assume high shear stresses. If we know the physical properties and thermal structure of the subducting slab, then frictional heating may be determined by matching forearc heat flow observations. In the most comprehensive global effort to date, Tichelaar and Ruff (1993) concluded that the average frictional shear stress level in the world's strongly coupled subduction zones could range from 20 to 35 MPa. This shear stress estimate seems to be significantly above the stress drop of large underthrusting events, and it is also above the shear stress estimates for the San Andreas fault. On the other hand, detailed thermal studies for particular subduction zones may find different values (e.g., Hyndman *et al.*, 1995; Wang *et al.*, 1995). For the future, we need to find new and better constraints on subduction zone thermal structure.

To summarize, plate boundary heat flow studies apparently constrain shear stresses to be less than 10 MPa along the San Andreas fault and less than 35 MPa in subduction zones. These values are less than shear stresses expected for frictional failure from lithostatic pressure and Byerlee's Law. Thus, there has been considerable debate on how to reconcile this discrepancy. If we also consider the modeling results on plate tectonics, geoid anomalies, and boundary forces, it seems that we must accept "low" plate boundary stresses. The next section reviews the efforts to explain why plate boundaries are weak.

9. Why are Plate Boundaries Weak?

Detailed studies of the San Andreas system show that it is a weak fault in two respects: shear stress on the fault is low, and the orientation of the maximum horizontal principal stress is nearly orthogonal to the fault (see Chapter 32 by Lockner and Beeler and papers in Zoback and Lachenbruch, 1992). Thus, the San Andreas fault must be much weaker than nearby faults with an orientation more favorable for failure. The current debate on stress state is largely centered on the question of how can plate boundaries be so weak? Most current answers to this question use water in various roles. Although most of this section is focused on water, some of the alternative interpretations of the low San Andreas heat flux are reviewed first.

9.1 Explanations without High Pore Pressure

One explanation that combines high shear stresses and low heat flow is that dynamic stress drop is a factor of ten higher than static stress drop. Although there may be several theoretical models that can produce these high dynamic stress drops (see references in Brune *et al.*, 1993), observations will ultimately decide the fate of this mechanism. As reviewed above (Section 7), global studies show that dynamic stress drop is *not* larger than static stress drop. Of course, one could still argue for special circumstances along the San Andreas fault. Hence we must wait and measure the dynamic stress drop of future great earthquakes along the San Andreas fault. In the meantime, studies of the 1989 Loma Prieta event suggest low shear stress (e.g., Zoback and Beroza, 1993).

Large dynamic stress drop is not relevant for the creeping sections of the San Andreas. The most direct explanation for both the creeping and seismogenic portions is to just accept that the fault has a very low friction coefficient. The lithostatic vertical normal stress at 15 km depth, i.e., the bottom edge of most strike-slip faults, is about 400 MPa. Then the friction coefficient must be 0.05 to give a failure shear stress of 20 MPa. Even if hydrostatic pressure is subtracted from the above normal stress, a friction coefficient of 0.08 is required to keep shear stress at 20 MPa. In subduction zones where the seismogenic zone extends to 40 km depth, the friction coefficient must be less than 0.10 even for the higher shear stress estimate of Tichelaar and Ruff (1993). For comparison, lubricants such as MoS_2 and Teflon have a friction coefficient of 0.06. Laboratory experiments indicate that candidate geologic materials—even low-friction choices—are unlikely to have such low friction coefficients at crustal conditions (see Chapter 32 by Lockner and Beeler).

One idea to achieve these low friction coefficients is to use "roller bearings" between the fault surfaces. Although there must be low-friction lubricants around the "rollers," it seems that there is both experimental (Brune and Anooshehpoor, 1997) and numerical (Mora and Place, 1998) support that fault "rollers" can reduce the large-scale friction coefficient of the fault. [See Scholz (1996) for arguments against this mechanism as the solution for the San Andreas heat flow paradox.]

The final alternative explanation is that plate boundary shear stress is high and "friction" energy is high, but that this "friction" energy does not appear as high heat flow above the fault. At one extreme, this argument contends that something is wrong with the interpretation of heat flow in the San Andreas region. This series of arguments and counterarguments has a long history, and the following reviews and special issues address this issue: Hanks and Raleigh (1980); Hickman (1991); Zoback and Lachenbruch (1992); Molnar (1992); Scholz (1996). This complicated issue can be easily circumvented in two ways: (1) global-scale tectonic and gravitational modeling prefers low shear stresses on plate boundaries; (2) current ideas that use water in plate boundaries seem to work with low shear stresses. Thus, it seems reasonable to accept the constraint that shear stress is low on the San Andreas fault and subduction zone plate boundaries.

9.2 Water and Weak Plate Boundaries

The admittance of water into plate boundaries opens an area of complex mechanical and chemical interactions that affect stress and earthquake occurrence (see papers in special volumes: introduction by Evans and Wong, 1992; Hickman *et al.*, 1995). Connections between water and faulting appear throughout the geologic literature, but the on-going low/high stress debate has focused more attention on water in the last decade. The compositional complexities of pore fluids are not discussed here; "water" includes all components of pore fluids. The chemical and mechanical role of water is an active research topic, so conclusions are subject to change. A brief overview of current models that may explain weak plate boundaries is given here, together with some of the caveats and complications.

The simplest role of water is that pore fluid pressure reduces the normal stress acting across-fault planes. Pore pressure does not change the shear stress. It is supposed that typical friction coefficient values of 0.5–1.0 can be applied to the ratio of shear stress to effective normal stress (effective normal stress is the normal stress minus the pore pressure). As fluid pressure increases to near the value of the least compressive principal stress, slip can occur with shear stress values less than 10 MPa. Effective normal stress is a simplistic treatment of the solid–water mechanical interaction (see discussion below), but it should work for shallow crustal depths. In most cases, we presume that the relative sizes of the principal stresses follow the Anderson (1942) model for the faulting regime. In a thrust setting, σ_{zz} is the least compressive principal stress, and as fluid pressure on a shallow-dipping fault approaches σ_{zz}, the overlying rock nearly "floats" on the water layer and the thrust fault can move at low shear stress. Hubbert and Rubey (1959) used this device to explain some aspects of shallow thrust faults (see Fig. 6b).

Today, high fluid pressure in subduction zones has become a well-accepted explanation for low shear stress. Although accretionary prisms are quite complex, a wealth of both observational and modeling studies allow the conclusion that de-watering of subducted sediments can sustain high fluid pressures down to a depth of several kilometers, perhaps even ten kilometers (see reviews and extensive references in Von Huene and Scholl, 1991; Moore and Vrolijk, 1992). At the deeper depths of the seismogenic zone, there are few observational and modeling constraints on the existence and nature of interface fluids (Moore and Vrolijk, 1992). Most researchers are willing to accept the notion of high fluid pressures down to a depth of 40 km, though the primary justification is simply that shear stresses are low in the seismogenic portion of subduction zones (see Wang and He, 1999, and references therein). In the future, subduction zones must be the subject of more studies on fluids.

High fluid pressure may explain low shear stress in continental normal and strike-slip faulting environments, but there are some difficulties. First, the fluid pressure must shift from hydrostatic toward lithostatic pressure in the first few kilometers of the crust (see Fig. 6). Since fluid flow should reduce this higher pressure, this topic immediately involves permeability structure and crustal fluid flow. Sibson (1990, and earlier references) discusses how lithostatic fluid pressure may occur in general, and for the San Andreas region in particular. Next, the San Andreas system shows an additional problem in that the maximum horizontal compressive stress is nearly orthogonal to the fault trace. Sibson (1990) shows that in these "severely misaligned" fault situations, high fluid pressure causes hydro-fractures before it fulfills its role of shear stress reduction on the main fault plane. Indeed, it is now a well-accepted conclusion that fluid pressure cannot be at a sufficiently high pressure throughout the crust all the time. To retain water as the explanation for low shear stress, we must keep high fluid pressure in the fault zone without leakage into the surrounding rock. Two related mechanisms that accomplish this feat have been suggested. Byerlee (1990) uses high pressure water trapped in fault zone gouge.

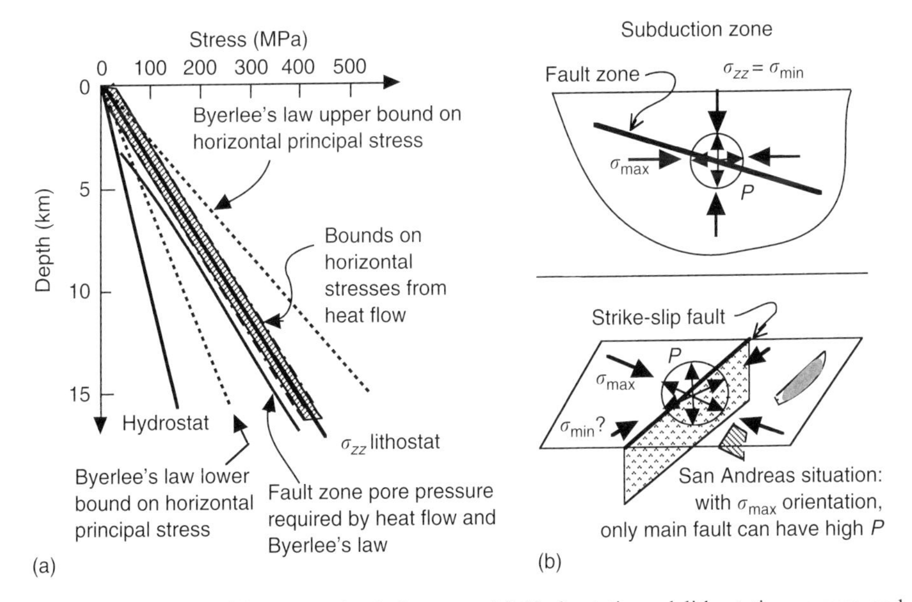

FIGURE 6 High fluid pressure in fault zones. (a) Hydrostatic and lithostatic pressure and various bounds on horizontal principal stresses as a function of depth. For strike–slip faults with unfavorable stress orientation, such as the San Andreas, the high fluid pressure must be contained within the fault zone (b), and this requires certain combinations of permeability structure, fluid flow and sources.

Rice (1992) has developed the most complete model that addresses all aspects of stress regime, anisotropic permeability, fluid flow, and deep crustal source. A key aspect of the Rice (1992) model is that the stress tensor for the narrow fault zone region is different from the stress tensor for the surrounding crust (the Byerlee, 1990, model also makes use of variable stress). The fault zone stress tensor has a different geometry and size of the principal stresses such that the higher pore pressure in the fault zone does not cause hydro-fracture in the surrounding rock. It appears that these two models are viable explanations for low shear stress, though hydrological implications have not been fully tested. The question then becomes: Is fault zone high fluid pressure the correct explanation? Some people have objected to these models because of the requirement that the Earth find these "special" stress/fluid pressures states at all plate boundaries. But in his model, Rice (1992) outlined how fault systems might evolve toward this special stress state. Sibson (1990) and Nur and Walder (1992) and others (see papers in Hickman *et al.*, 1995) have advocated the intrinsic time dependence of the coupled earthquake/fluid system. Although this view makes formal solutions more difficult, it provides opportunities for the system to evolve toward a special stress state. Indeed, it seems that future work in the stress state of plate boundaries must further develop fault evolution scenarios to show that a particular mechanism is not only possible, but that it is a robust system state for "real" fault systems.

9.3 Caveats and Complications with Water in Faults

We now turn to a quick appraisal of the complexities in a time-dependent coupled earthquake–water system (Sibson, 1973; Fyfe *et al.*, 1978; Nur and Walder, 1992). The mechanics of the coupled solid–fluid continuum are complex (e.g., Rice and Cleary, 1976), and allowance for time dependence in stress and material properties such as porosity and permeability make for even greater complexity (Yamashita, 1997). There is a large industrial effort in these calculations for problems such as sedimentary basin development, reservoir management, and other engineering applications. The fully coupled earthquake–fluid problem is also a very complex problem. Here, brief mention is made of one complication that may reduce the utility of fluid pressure in reducing the effective normal stress.

In the usual derivation of Amonton's or Byerlee's Law (see Scholz, 1990), the rock-to-rock contacts across an interface bear both the normal and shear stresses. If the pores are filled with a fluid, then the pores help to support some of the normal stress, but all the shear stress is carried by the rock contacts. As discussed in Jaeger and Cook (1979), the common usage of effective normal stress assumes that the actual area of rock contacts is a small fraction of the total fault area; this is certainly a good assumption at low confining pressure. If a water-filled contact surface is subjected to higher pressure and temperature, pore volume shrinks as rock contact area grows. It is difficult to calculate exactly how the pore surface area shrinks; one must construct various surface roughness models, then use empirical or theoretical models to compress the system (Gavrilenko and Gueguen, 1993). Since the pressure at the seismogenic down-dip edge in subduction zones can be more than 1000 MPa, pore collapse could be a serious problem. In the limit of a small pore area along the fault surface, pore fluid pressure cannot reduce the value of the overall normal stress. In his model for strike-slip faults, Rice (1992) avoids this problem by quoting observational evidence that an extensive fluid region exists in the lower crust, with the immediate consequence that water in these pores must be very close to lithostatic pressure. In subduction zones, it is presumed that any water in the seismogenic zone was carried down in the oceanic crust, thus there is less freedom to make a dense network of pores with the fluid at lithostatic pressure.

9.4 Final Thought on Water in Faults

A final philosophical objection to water in fault zones is that admittance of water makes the problem too complex and obscure. We first seek the most elegant, i.e., simple, solution to scientific problems. For the problem of reconciling low shear stress with Byerlee's Law, I suppose that the elegant solution would be to discover that all plate boundaries are lined with Teflon. However, there are many Earth Science questions that have complex, inelegant, and "messy" answers. "What controls the climate?" provides a good example of how the answer to an important global-scale question involves the interaction of many obscure processes and minor constituents. State of stress on plate boundaries may be an analogous problem, and we have to accept that minor constituents such as water may play a dominant role in the interlinked processes that control shear stress on faults.

10. Conclusions

The support of gravitational loads on and within the Earth provides constraints on minimum shear stress levels. The classic Jeffreys argument shows that isostatically compensated mountains require shear stresses of 75 MPa somewhere in the lithosphere (see Fig. 7).

Buoyancy sources associated with plate tectonics require shear stresses of at least 10 MPa. Global force balance calculations find acceptable solutions with plate boundary stresses of 10 MPa, but some plate interiors may have shear stresses close to 100 MPa. Many models of subducted slab mechanics also produce internal stresses of 100 MPa. Detailed regional modeling efforts of subduction zones with tectonic and buoyancy forces find self-consistent models with low stress levels on the plate interface.

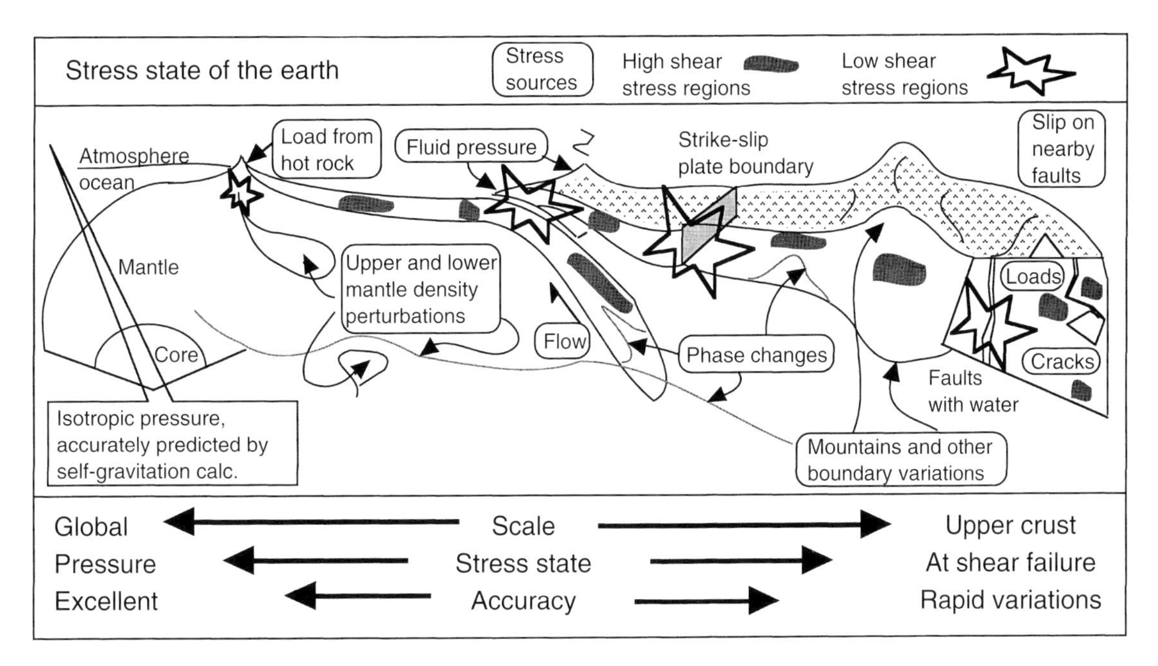

FIGURE 7 Cartoon summary of stress state of the Earth. The length-scale varies from a planetary view at the left to uppermost crust details at the right.

The World Stress Map Project has defined stress provinces where large-volume determinations of the orientation of shear stress are consistent. These directions are mostly consistent with global-scale tectonic models of either lithospheric basal drag or plate boundary forces. Direct stress measurements find shear stresses in the upper crust at the failure value predicted by Byerlee's Law. Many stress indicators show that the maximum horizontal principal stress is nearly orthogonal to the San Andreas fault.

Large earthquake stress drops are in the range of 0.1–10 MPa; intraplate events are biased toward the top of this range, whereas large subduction zone earthquakes are biased toward the lower part of the range. Recent large deep earthquakes have stress drops as high as 70 MPa; this high value is compatible with some models of subduction that predict high intraplate stresses.

The absence of a heat flow anomaly across the San Andreas fault argues for low shear stress, 10 MPa or less, unless there is some other process at work that absorbs the energy. Global-scale subduction zone thermal models produce overall shear stress levels of about 35 MPa, though there is much more uncertainty in this constraint as specific studies (e.g., Cascadia, Wang *et al.*, 1995) propose even smaller stress levels.

The stress state that emerges from these constraints is that plate interiors may have high levels of shear stress, but that plate boundaries and other major active faults have low shear stress. Faults that slip the most are the weakest. It seems that water in fault zones may play a key role by reducing the effective normal stress so that the fault will slip with low shear stresses. There are many other effects of water in fault zones, and thus more exploration is required. There are still many geologic and geophysical questions as to exactly how the Earth system evolves to make this happen.

Many observations indicate that the upper lithosphere of the Earth is close to failure stress. Some of the more intriguing evidence comes from reservoir-induced seismicity, where small stress perturbations cause earthquakes away from plate boundaries. Note that water pore pressure may play a role for induced seismicity. This empirical evidence meshes nicely with some theoretical views that the entire Earth is at the failure threshold stress. This holistic view provides a simple summary of the stress state of the Earth: Stress is everywhere close to the failure stress, whatever that is!

References

Abe, K. (1982). *J. Phys. Earth* **30**, 321–330.
Abercrombie, R. (1995). *J. Geophys. Res.* **100**, 24 015–24 036.
Aki, K. (1972). *Tectonophysics* **13**, 423–446.
Aki, K. and P.G. Richards (1980). "Quantitative Seismology: Theory and Methods," Freeman.
Anderson, E.M. (1942). "The Dynamics of Faulting," Oliver and Boyd, Edinburgh.
Ando, M. (1975). *Tectonophysics* **27**, 63–87.
Anonymous (1994). *Bull. Seismol. Soc. Am.* **84**, 497–498.
Artyushkov, E.V. (1973). *J. Geophys. Res.* **76**, 7675–7708.
Atkinson, B.K. (Ed.) (1987). "Fracture Mechanics of Rock," Academic Press, New York.
Astiz, L., *et al.* (1988). *Phys. Earth Planet. Inter.* **53**, 80–166.
Bai, W., *et al.* (1992). *J. Geophys. Res.* **97**, 11729–11738.
Bak, P. and C. Tang (1989). *J. Geophys. Res.* **94**, 15635–15637.
Bird, P. (1998). *J. Geophys. Res.* **103**, 10115–10130.

Bos, A., *et al.* (1998). *J. Geophys. Res.* **103**, 21059–21065.
Brace, W.F. and D. Kohlstedt (1980). *J. Geophys. Res.* **85**, 6248–6252.
Brudy, M., *et al.* (1997). *J. Geophys. Res.* **102**, 18453–18475.
Brune, J., *et al.* (1969). *J. Geophys. Res.* **74**, 3821–3827.
Brune, J., *et al.* (1993). *Tectonophysics* **218**, 59–67.
Brune, J.N. and A. Anooshehpoor (1997). *Geophys. Res. Lett.* **24**, 2071–2074.
Bullen, K.E. (1947). "An Introduction to the Theory of Seismology," Cambridge University Press.
Byerlee, J.D. (1978). *Pageoph* **116**, 615–626.
Byerlee, J.D. (1990). *Geophys. Res. Lett.* **17**, 2109–2112.
Choy, G.L. and J.L. Boatwright (1995). *J. Geophys. Res.* **100**, 18205–18228.
Christensen, D. and L. Ruff (1988). *J. Geophys. Res.* **93**, 13421–13444.
Cloetingh, S. and M. Wortel (1986). *Tectonophysics* **132**, 49–67.
Dahlen, F.A. (1990). *Annu. Rev. Earth Planet. Sci.* **18**, 55–99.
Das, S. (1993). *Geophys. J. Int.* **115**, 778–798.
Davies, G.F. (1983). *Tectonophysics* **99**, 85–98.
Diakonov, B.P., *et al.* (1990). *Phys. Earth Planet. Inter.* **63**, 151–162.
Dziewonski, A.M. and D.L. Anderson (1981). *Phys. Earth Planet. Inter.* **25**, 297–356.
Engelder, T. and P. Geiser (1980). *J. Geophys. Res.* **85**, 6319–6341.
England, P. and P. Molnar (1991). *Phil. Trans. R. Soc. Lond. A* **337**, 151–164.
Evans, B. and T. Wong (Eds.) (1992). "Fault Mechanics and Transport Properties of Rocks," Academic Press.
Forsythe, D.W. and S. Uyeda (1975). *Geophys. J. R. Astron. Soc.* **43**, 163–200.
Frohlich, C. (1989). *Annu. Rev. Earth Planet. Sci.* **17**, 227–254.
Fung, Y.C. (1965). "Foundations of Solid Mechanics," Prentice-Hall.
Fyfe, W., *et al.* (1978). "Fluids in the Earth's Crust," Elsevier.
Gavrilenko, P. and Y. Gueguen (1993). *Tectonophysics* **217**, 91–110.
Goes, S., *et al.* (1997). *Geophys. Res. Lett.* **24**, 1295–1298.
Goetze, C. and B. Evans (1979). *Geophys. J. R. Astron. Soc.* **59**, 463–478.
Govers, R., *et al.* (1992). *J. Geophys. Res.* **97**, 11749–11760.
Grasso, J. and D. Sornette (1998). *J. Geophys. Res.* **103**, 29965–29987.
Green, H. and H. Houston (1995). *Annu. Rev. Earth Planet. Sci.* **25**, 169–215.
Hager, B.H. and R.J. O'Connell (1980). In: "Physics of Earth's Interior", pp. 464–492, North Holland.
Hanks, T.C. (1977). *Pageoph* **115**, 441–458.
Hanks, T.C. and C.B. Raleigh (1980). *J. Geophys. Res.* **85**, 6083–6085.
Harris, R.A. (1998). *J. Geophys. Res.* **103**, 24347–24358.
Haubrich, R.A. and K. McCamy (1969). *Rev. Geophys.* **7**, 539–572.
Hickman, S. (1991). *Rev. Geophys. Suppl.* 759–775.
Hickman, S., *et al.* (1995). *J. Geophys. Res.* **100**, 12831–12840.
Hill, D.H., *et al.* (1993). *Science* **260**, 1617–1623.
Hjortenberg, E. and A.N. Nikolaev (1990). *Phys. Earth Planet. Inter.* **63**, 144, 1990.
Hubbert, M. and W. Rubey (1959). *Geol. Soc. Am. Bull.* **70**, 115–205.
Hyndman, R., *et al.* (1995). *J. Geophys. Res.* **100**, 15373–15392.
Ito, E. and H. Sato (1992). In: "High Pressure Research: Applications to Earth and Planetary," pp. 257–262, American Geophysical Union.
Jackson, D.D. and Y.Y. Kagan (1993). *J. Geophys. Res.* **98**, 9417–9420.
Jaeger, J. and N. Cook (1979). "Fundamentals of Rock Mechanics," Chapman and Hall.
Jeffreys, H. (1970). "The Earth," Cambridge University Press.
Kanamori, H. (1980). In: "Physics of Earth's Interior," pp. 531–554, North Holland.
Kanamori, H. (1994). *Annu. Rev. Earth Planet. Sci.* **22**, 207–237.
Kanamori, H. and D.L. Anderson (1975). *Bull. Seismol. Soc. Am.* **65**, 1073–1095.
Kanamori, H., *et al.* (1993). *Bull. Seismol. Soc. Am.* **83**, 330–346.
Kanamori, H., *et al.* (1998). *Science* **279**, 839–842.
Kawakatsu, H. (1986). *J. Geophys. Res.* **91**, 4811–4825.
Kikuchi, M. (1992). *Tectonophysics* **211**, 107–113.
Kikuchi, M. and Y. Fukao (1988). *Bull. Seismol. Soc. Am.* **78**, 1707–1724.
Kirby, S., *et al.* (1996). *Rev. Geophys.* **34**, 261–306.
Kisslinger, C. (1993). *J. Geophys. Res.* **98**, 1913–1921.
Lachenbruch, A.H. and J.H. Sass (1992). *J. Geophys. Res.* **97**, 4995–5015.
Lithgow-Bertelloni, C. and M.A. Richards (1998). *Rev. Geophys.* **36**, 27–78.
Liu, X. and K. McNally (1993). *Pageoph* **140**, 211–256.
Magloughlin, J.F. and J.G. Spray (1992). *Tectonophysics* **204**, 197–337.
Marton, F., *et al.* (1999). *Geophys. Res. Lett.* **26**, 119–122.
McGarr, A. (1999). *J. Geophys. Res.* **104**, 3003–3011.
McGarr, A. and N.C. Gay (1978). *Annu. Rev. Earth Planet. Sci.* **6**, 405–436.
McKenzie, D. and D. Fairhead (1997). *J. Geophys. Res.* **102**, 27523–27552.
McKenzie, D. and G. Jarvis (1980). *J. Geophys. Res.* **85**, 6093–6096.
McNutt, M. (1980). *J. Geophys. Res.* **85**, 6377–6396.
McNutt, M. (1987). *Rev. Geophys.* **25**, 1245–1253.
Molnar, P. (1992). In: "Fault Mechanics and Transport Properties of Rocks," pp. 435–459, Academic Press.
Molnar, P. and P. England (1990). *J. Geophys. Res.* **95**, 4833–4856.
Moore, J.C. and P. Vrolijk (1992). *Rev. Geophys.* **30**, 113–135.
Mora, P. and D. Place (1998). *J. Geophys. Res.* **103**, 21067–21089.
Nishenko, S.P. and L.R. Sykes (1993). *J. Geophys. Res.* **98**, 9909–9916.
Nomanbhoy, N. and L. Ruff (1996). *J. Geophys. Res.* **101**, 5707–5724.
Nur, A. and J. Walder (1992). In: "Fault Mechanics and Transport Properties of Rocks," pp. 461–474, Academic Press.
Ohnaka, M. (1992). *Tectonophysics* **211**, 149–178.
Orowan, E. (1960). *Geol. Soc. Am. Bull.* **79**, 323–345.
Peacock, S.M. (1996). In: "Subduction: Top to Bottom," AGU Geophysics Monograph 96, pp. 119–133, American Geophysics Union.
Purcaru, G. and H. Berckhemer (1982). *Tectonophysics* **84**, 57–128.
Rice, J.R. (1992). In: "Fault Mechanics and Transport Properties of Rocks," pp. 475–504, Academic Press.
Rice, J.R. and M.P. Cleary (1976). *Rev. Geophys.* **14**, 227–241.
Richardson, R.M. (1992). *J. Geophys. Res.* **97**, 11739–11748.
Richardson, R.M., *et al.* (1979). *Rev. Geophys.* **17**, 981–1019.
Romanowicz, B. (1992). *Geophys. Res. Lett.* **19**, 481–484.
Ruff, L.J. (1999). *Pageoph* **154**, 409–431.

Rundle, J.B., *et al.* (1995). *Phys. Rev. Lett.* **75**, 1658–1661.
Sass, J., *et al.* (1997). *J. Geophys. Res.* **102**, 27575–27586.
Schmid, S., *et al.* (1999). *Tectonophysics* **303**, 1–319.
Scholz, C.H. (1990). "The Mechanics of Earthquakes and Faulting," Cambridge University Press.
Scholz, C.H. (1996). *Nature* **381**, 556–557.
Seno, T. and Y. Yamanaka (1996). In: "Subduction: Top to Bottom," AGU Geophysics Monograph 96, pp. 347–355, American Geophysics Union.
Shimazaki, K. (1986). In: "Earthquake Source Mechanics," AGU Monograph 37, pp. 209–216, American Geophysics Union.
Sibson, R.H. (1973). *Nature* **243**, 66–68.
Sibson, R.H. (1990). *Bull. Seismol. Soc. Am.* **80**, 1580–1604.
Simpson, D.W. (1986). *Annu. Rev. Earth Planet. Sci.* **14**, 21–42.
Sleep, N.H. (1975). *Geophys. J. R. Astron. Soc.* **42**, 827–857.
Sornette, D., *et al.* (1990). *Geophys. Res.* **95**, 17353–17361.
Spence, W. (1987). *Rev. Geophys.* **25**, 55–69.
Stein, C.A. and S. Stein (1992). *Nature* **359**, 123–129.
Stein, R.S., *et al.* (1997). *Geophys. J. Int.* **128**, 594–604.
Tajima, F. and H. Kanamori (1985). *Phys. Earth Planet. Int.* **40**, 77–134.
Tajima, F., *et al.* (1990). *Phys. Earth Planet. Int.* **61**, 269–290.
Tanioka, Y. and L. Ruff (1997). *Seismol. Res. Lett.* **68**, 386–400.
Tichelaar, B.W. and L.J. Ruff (1993). *J. Geophys. Res.* **98**, 2017–2037.
Tsuruoka, H., *et al.* (1995). *Geophys. J. Int.* **122**, 183–194.
Turcotte, D. and G. Schubert (1982). "Geodynamics," Wiley.
Vidale, J., *et al.* (1998). *J. Geophys. Res.* **103**, 24567–24572.
Von Huene, R. and D.W. Scholl (1991). *Rev. Geophys.* **29**, 279–306.
Wallace, T.C. (1995). *Geophys. Res. Lett.* **22**, 2231–2232.
Wang, K., *et al.* (1995). *J. Geophys. Res.* **100**, 12907–12918.
Wang, K. and J. He (1999). *J. Geophys. Res.* **104**, 15191–15205.
Whittaker, A., *et al.* (1992). *J. Geophys. Res.* **97**, 11933–11944.
Winslow, N.W. and L.J. Ruff (1999). *Phys. Earth Planet. Inter.* **115**, 181–190.
Wortel, M.J. and N.J. Vlaar (1988). *Pageoph* **128**, 625–659.
Xu, Z. and S.Y. Schwartz (1993). *Pageoph* **140**, 365–390.
Yamanaka, Y. and K. Shimazaki (1990). *J. Phys. Earth* **38**, 305–324.
Yamashita, T. (1997). *J. Geophys. Res.* **102**, 17797–17806.
Yamashita, T. and L. Knopoff (1987). *Geophys. J. R. Astron. Soc.* **91**, 13–26.
Zhong, S. and M. Gurnis (1994). *J. Geophys. Res.* **99**, 15683–15695.
Zoback, M.D. and G.C. Beroza (1993). *Geology* **21**, 181–185.
Zoback, M.D. and A.H. Lachenbruch (1992). *J. Geophys. Res.* **97**, 4991–4994.
Zoback, M.D. and H.-P. Harjes (1997). *J. Geophys. Res.* **102**, 18477–18491.
Zoback, M.L. (1992). *J. Geophys. Res.* **97**, 11703–11728.

Editor's Note

Due to space limitations, two useful items are placed on the Handbook CD-ROM under the directory \33Ruff. (1) References with full citation are given in RuffFullReferences.pdf and (2) Detailed figure captions are given in RuffFullFigCaptions.pdf. For an introduction to plate tectonics, please see Chapter 6, Continental drift, sea-floor spreading, and plate/plume tectonics, by Uyeda, and Chapter 7 Earthquake mechanisms and plate tectonics, by Stein and Klosko. Readers interested in heat flow may consult Chapter 81.4.

34

State of Stress in the Earth's Lithosphere

Mark D. Zoback
Stanford University, Stanford, USA
Mary Lou Zoback
US Geological Survey, Menlo Park, USA

The state of stress in the lithosphere is the result of the forces acting upon and within it. Knowledge of the magnitude and distribution of these forces can be combined with mechanical, thermal and rheological constraints to examine a broad range of lithospheric deformational processes. For example, such knowledge contributes to a better understanding of the processes that both drive and inhibit lithospheric plate motions as well as the forces responsible for the occurrence of crustal earthquakes – both along plate boundaries and in intraplate regions.

Although the topic of this chapter is the state of stress in the Earth's lithosphere, the comments below come primarily from the perspective of the state of stress in the brittle upper crust. As defined by the depth of shallow earthquakes, the brittle crust extends to $\sim$15–20 km depth at most continental locations around the world. This perspective is adopted here because nearly all the data available on lithospheric stress come from the upper crust of continents. Furthermore, in the sections that follow, we argue that, to first order, the state of stress in the brittle crust results from relatively large-scale lithospheric processes so that knowledge of crustal stress can be used to constrain the forces involved in these processes.

1. Basic Definitions

Stress is a tensor, which describes the density of forces acting on all surfaces passing through a point. In terms of continuum mechanics, the stresses acting on a homogeneous, isotropic body at depth are describable as a second rank tensor, with nine components (Fig. 1a).

$$\bar{S} = \begin{vmatrix} s_{11} & s_{12} & s_{13} \\ s_{21} & s_{22} & s_{23} \\ s_{31} & s_{32} & s_{33} \end{vmatrix} \qquad (1)$$

The subscripts of the individual stress components refer to the direction that a given force is acting and the face of the unit cube upon which the stress component acts. Thus, in simplest terms, any given stress component represents a force acting in a specific direction on a unit area of given orientation. As illustrated in Figure 1a, a stress tensor can be defined in terms of any arbitrary reference frame. Because of equilibrium conditions

$$s_{12} = s_{21} \qquad s_{13} = s_{31} \qquad s_{23} = s_{32} \qquad (2)$$

so that the order of the subscripts is unimportant. In general, to fully describe the state of stress at depth, one must estimate six stress magnitudes or three stress magnitudes and the three angles that define the orientation of the stress coordinate system with respect to a reference coordinate system (such as geographic coordinates, for example).

We utilize the convention that compressive stress is positive because *in situ* stresses at depths greater than a few tens of meters in the Earth are *always* compressive. Tensile stresses do not exist at depth in the Earth for two fundamental reasons. First, because the tensile strength of rock is generally quite low, significant tensile stress cannot be supported in the Earth. Second, because there is always a fluid phase saturating the pore space in rock at depth (except at depths shallower than the water table), the pore pressure resulting from this fluid phase would cause the rock to hydraulically fracture should the least compressive stress reach a value even as low as the pore pressure.

Once a stress tensor is known, it is possible to evaluate stresses in any coordinate system via tensor transformation. To accomplish this transformation, we need to specify the direction cosines (a_{ij}, as illustrated in Fig. 1b) that describe the rotation of the coordinate axes between the old and new coordinate systems. Mathematically, the equation which

INTERNATIONAL HANDBOOK OF EARTHQUAKE AND ENGINEERING SEISMOLOGY, VOLUME 81A ISBN: 0-12-440652-1

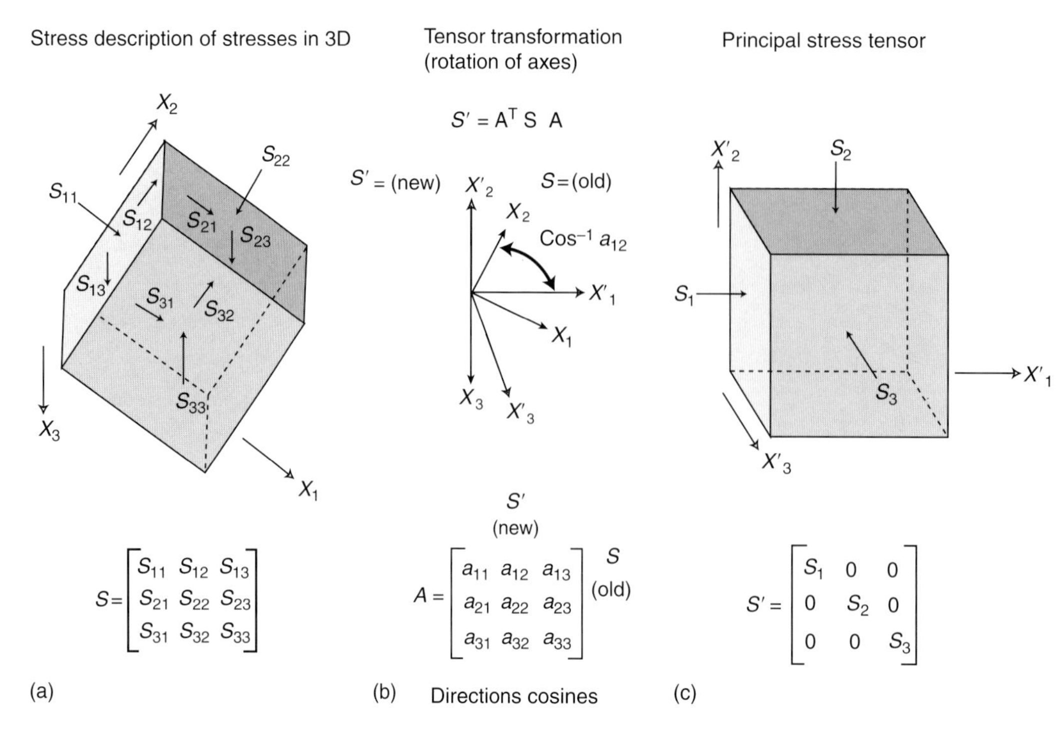

FIGURE 1 Definition of stress tensor in an arbitrary cartesian coordinate system, rotation of stress coordinate systems through tensor transformation and principal stresses as defined in a coordinate system in which shear stresses vanish.

accomplishes this is

$$\overline{S'} = \overline{A}^T \overline{S}\,\overline{A} \tag{3a}$$

where

$$\overline{A} = \begin{vmatrix} a_{11} & a_{12} & a_{13} \\ a_{21} & a_{22} & a_{23} \\ a_{31} & a_{32} & a_{33} \end{vmatrix} \tag{3b}$$

The ability to transform coordinate systems is of interest here because we can choose to generally describe the state of stress in terms of the principal coordinate system. The principal coordinate system is the one in which shear stresses vanish and only three principal stresses, $S_1 \geq S_2 \geq S_3$ fully describe the stress field (as illustrated in Fig. 1c). Thus, we have diagonalized the stress tensor such that the principal stresses correspond to the eigenvalues of the stress tensor and the principal stress directions correspond to its eigenvectors.

$$\overline{S'} = \begin{vmatrix} S_1 & 0 & 0 \\ 0 & S_2 & 0 \\ 0 & 0 & S_3 \end{vmatrix} \tag{4}$$

The reason this concept is so important is that as the Earth's surface is in contact with a fluid (either air or water) and cannot support shear tractions, it is a principal stress plane. Thus, one principal stress is generally expected to be normal to the Earth's surface with the other two principal stresses acting in an approximately horizontal plane. Although it is clear that this must be true very close to the Earth's surface, compilation of earthquake focal mechanism data and other stress indicators (described below) suggest that it is also generally true to the depth of the brittle–ductile transition in the upper crust (Zoback and Zoback, 1980, 1989; Zoback, 1992; Brudy *et al.*, 1997). Assuming this is the case, we must define only four parameters to describe the state of stress at depth; one stress orientation (usually taken to be the azimuth of the maximum horizontal compression, $S_{H\max}$) and three principal stress magnitudes: S_V, the vertical stress, corresponding the weight of the overburden; $S_{H\max}$, the maximum principal horizontal stress; and $S_{h\min}$, the minimum principal horizontal stress. This obviously helps make stress determination in the crust a tractable problem.

In applying these concepts to the Earth's crust, it is helpful to consider the magnitudes of the greatest, intermediate, and minimum principal stress at depth (S_1, S_2, and S_3) in terms of S_V, $S_{H\max}$ and $S_{h\min}$ in the manner originally proposed by E.M. Anderson (Anderson, 1951). This is illustrated in Figure 2. There are a number of simple, but fundamental points about these seemingly straightforward relations.

First, the two horizontal principal stresses in the Earth, $S_{H\max}$ and $S_{h\min}$, can be described relative to the vertical principal stress, S_V, whose magnitude corresponds to the overburden. Mathematically, this is equivalent to integration

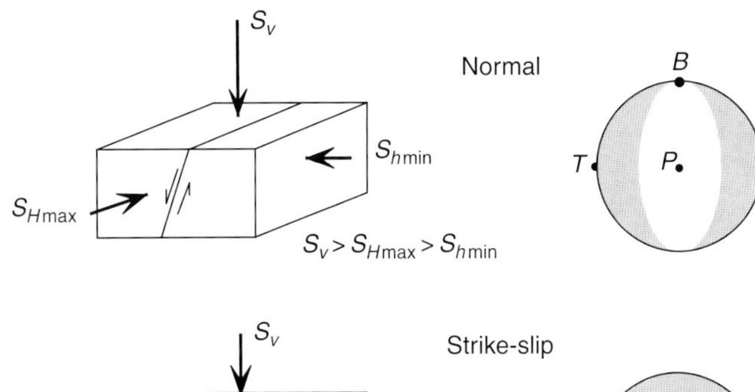

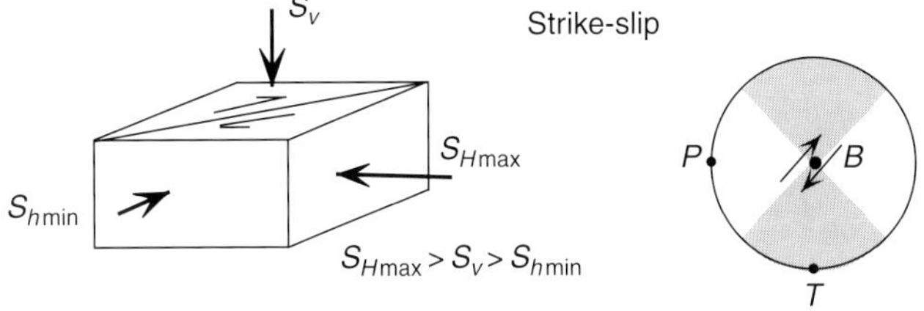

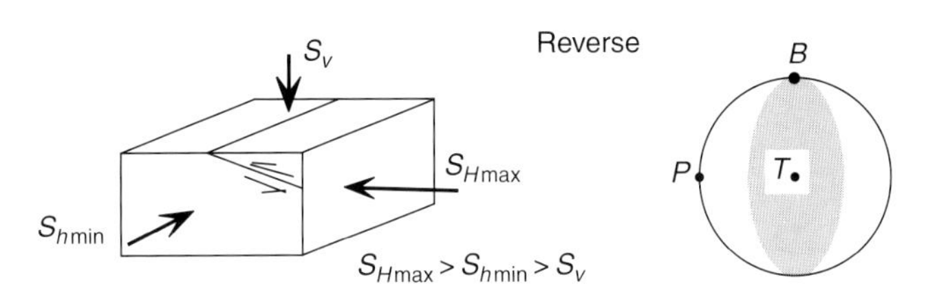

FIGURE 2 E.M. Anderson's classification scheme for relative stress magnitudes in normal, strike–slip and reverse faulting regions. Corresponding focal plane mechanisms are shown to the right.

of density from the surface to the depth of interest, z. In other words,

$$S_v = \int_0^z \rho(z) g\, dz \approx \bar{\rho} g z \tag{5}$$

where $\rho(z)$ is the density as a function of depth, g is gravitational acceleration and $\bar{\rho}$ is mean overburden density. It is, of course, necessary to add atmospheric pressure and the pressure resulting from the weight of water at the Earth surface, as appropriate.

Second, the horizontal principal stresses are almost never equal and may be less than or greater than the vertical stress. In fact, the relative magnitudes of the principal stresses can be simply related to the faulting style currently active in a region. As illustrated in Figure 2 (and following Anderson, 1951), characterizing a region by normal, strike–slip or reverse faulting is equivalent to defining the horizontal principal stress magnitudes with respect to the vertical stress. When the vertical stress dominates in extensional deformational regions ($S_1 = S_v$), gravity drives normal faulting. Conversely, when both horizontal stresses exceed the vertical stress ($S_3 = S_v$) compressional deformation (shortening) is accommodated through reverse faulting. Strike–slip faulting represents an intermediate stress state ($S_2 = S_v$), where the maximum horizontal stress is greater than the vertical stress and the minimum horizontal stress is less ($S_{Hmax} \geq S_V \geq S_{hmin}$).

The concept of effective stress is used to incorporate the influence of pore pressure at depth where a component of effective stress σ_{ij} is related to the total stress S_{ij} via

$$\sigma_{ij} = (S_{ij} - \delta_{ij} P_p) \tag{6}$$

where δ_{ij} is the Kronecker delta and P_p is the pore pressure.

Laboratory studies of the frictional strength of faulted rock carried out over the past several decades indicates that the Coulomb criterion describes the frictional strength of faults. That is, fault slippage will occur when

$$\tau = S_o + \mu \sigma_n \tag{7}$$

where τ is the shear stress acting on the fault, S_o is the fault cohesion, μ is the coefficient of friction on the fault and σ_n is the effective normal stress acting on the fault plane. The maximum shear stress is given by $\frac{1}{2}(S_1 - S_3)$.

Using the concept of effective stress at depth, we can extend Anderson's faulting theory to predict stress magnitudes at depth through utilization of simplified 2D Mohr–Coulomb failure theory. 2D faulting theory assumes that failure is only a function of the difference between the last and greatest principal effective stresses σ_1 and σ_3 as given by Jaeger and Cook (1971)

$$\sigma_1/\sigma_3 = (S_1 - P_p)/(S_3 - P_p) = ((\mu^2 + 1)^{1/2} + \mu)^2 \tag{8}$$

Thus, a third point about crustal stress that can be derived from Andersonian faulting theory is that the magnitudes of the three principal stresses at any depth are limited by the strength of the Earth's crust at depth. In the case of normal faulting, stress magnitudes are controlled by σ_v, and σ_{hmin} which correspond to σ_1 and σ_3, respectively, and σ_{Hmax} corresponds to σ_2 which is intermediate in value between σ_v, and σ_{hmin} and does not influence faulting. Coulomb failure theory indicates that frictional sliding occurs when the ratio shear stress to effective normal stress on preexisting fault planes is equal to the coefficient of friction. As the coefficient of friction is relatively well defined for most rocks and ranges between ~ 0.6 and 1.0 (Byerlee, 1978), Equation (8) demonstrates that frictional sliding will occur when $\sigma_1/\sigma_3 \sim 3.1$–5.8. For the case of hydrostatic pore pressure, and commonly observed friction coefficients of 0.6 (e.g., Townend and Zoback, 2000), in extensional areas $S_{hmin} \sim 0.6\ S_v$, in reverse faulting areas $S_{Hmax} \sim 2.3\ S_v$ and in strike–slip faulting areas [when $S_v \sim 1/2\ (S_{Hmax} + S_{hmin})$], $S_{Hmax} \sim 2.2\ S_{hmin}$. As discussed below, these simple relationships have been confirmed by *in situ* stress measurements to depths of almost 8 km at a number of sites in intraplate areas.

2. Indicators of Contemporary Stress

Information on the state of stress in the lithosphere comes from a variety of sources: earthquake focal plane mechanisms; young geologic data on fault slip and volcanic alignments; *in situ* stress measurements; and stress-induced wellbore breakouts and drilling-induced tensile fractures. A stress measurement quality criterion for different types of stress indicators was developed by Zoback and Zoback (1989, 1991). This quality criterion was subsequently utilized in the International Lithosphere Program's World Stress Map Project, a large collaborative effort of data compilation and analyses by over 40 scientists from 30 different countries (Zoback, 1992). A special issue of the *Journal of Geophysical Research* (vol. 97, pp. 11703–12014, 1992) summarized the overall results of this project as well as presented the individual contributions of many of these investigators in various regions of the world. Today, the World Stress Map (WSM) database has more than 9100 entries and is maintained at the Heidelberg Academy of Sciences and Humanities (Mueller *et al.*, 1997).

Zoback and Zoback (1991) discussed the rationale for the WSM quality criterion used in the World Stress Map project in detail. The success of the WSM project demonstrates that with careful attention to data quality, coherent stress patterns over large regions of the Earth can be mapped with reliability and interpreted with respect to large-scale lithospheric processes.

2.1 Earthquake Focal Mechanisms

Although earthquake focal plane mechanisms are the most ubiquitous indicator of stress in the lithosphere, determination of principal stress orientations and relative magnitudes from these mechanisms must be done with appreciable caution. The pattern of seismic radiation from the focus of an earthquake permits construction of earthquake focal mechanisms (right column of Fig. 2). Perhaps the most simple and straightforward information about *in situ* stress that is obtainable from focal mechanisms and *in situ* stress is that the type of earthquake (i.e., normal, strike–slip, or reverse faulting) defines the relative magnitudes of $S_{H\max}$, $S_{h\min}$ and S_v. In addition, the orientation of the fault plane and auxiliary plane (which bound the compressional and extensional quadrants of the focal plane mechanism) define the orientation of the P (compressional), B (intermediate), and T (extensional) axes. These axes are sometimes incorrectly assumed to be the same as the orientation of S_1, S_2, and S_3.

For cases in which laboratory-measured coefficients of fault friction of ~ 0.6–1.0 are applicable to the crust, there is a nontrivial error of $\sim$15–20° if one uses the P, B, and T axes as approximations of average principal stress orientations, especially if the orientation of the fault plane upon which the earthquake occurred is known (Raleigh *et al.*, 1972). If friction is negligible on the faults in question (but higher in surrounding rocks), there can be considerable difference between the P, B, and T axes and principal stress directions (McKenzie, 1969). An earthquake focal plane mechanism always has the P and T axes at 45° to the fault plane and the B axes in the plane of the fault. With a frictionless fault the seismic radiation pattern is controlled by the orientation of the fault plane and not the *in situ* stress field. One result of this is that just knowing the orientation of the P-axis of earthquakes along weak, plate-bounding strike–slip faults (like the San Andreas), does not allow principal stress orientations to be defined from the focal plane mechanisms of the strike–slip earthquakes occurring on the fault (Zoback *et al.*, 1987). For this reason, it is common practice to omit plate-boundary earthquakes from regional stress compilations (see below).

Principal stress directions can be determined directly from a group of earthquake focal mechanisms (or set of fault striae measurements) through use of inversion techniques that are based on the slip kinematics and the assumption that fault slip will always occur in the direction of maximum resolved shear stress on a fault plane (cf, Gephart and Forsyth, 1984; Michael, 1984; Angelier, 1990). Such inversions yield four parameters, the orientation of the three principal stress and the relative magnitude of the intermediate principal stress with respect to the maximum and minimum principal stress.

The analysis of seismic waves radiating from an earthquake can also be used to estimate the magnitude of stress released in an earthquake (stress drop), although not absolute stress levels (Brune, 1970). In general, stress drops of crustal earthquakes are on the order of 1–10 MPa (Hanks, 1977). Equation (8) can be used to show that such stress drops are only a small fraction of the shear stresses that actually causes fault slip if pore pressures are approximately hydrostatic at depth and Coulomb faulting theory (with laboratory-derived coefficients of friction) is applicable to faults *in situ*. This is discussed in more detail below.

2.2 Geologic Stress Indicators

There are two general types of *relatively young* geologic data that can be used for *in situ* stress determinations: (1) the orientation of igneous dikes or cinder cone alignments, both of which form in a plane normal to the least principal stress (Nakamura, 1977) and (2) fault slip data, particularly the inversion of sets of striae (i.e., slickensides) on faults as described above. Of course, the term *relatively young* is quite subjective but essentially means that the features in question are characteristic of the tectonic processes currently active in the region of question. In most cases, the WSM database utilizes data which are Quaternary in age, but in all areas represent the youngest episode of deformation in an area.

2.3 *In situ* Stress Measurements

Numerous techniques have been developed for measuring stress at depth. Amadei and Stephansson (1997) and Engelder (1993) discuss many of these stress measurement methods, most of which are used in mining and civil engineering. Because we are principally interested here in regional tectonic stresses (and their implications) and because there are a variety of nontectonic processes that affect *in situ* stresses near the Earth's surface (Engelder and Sbar, 1984), we do not utilize near-surface stress measurements in the WSM or regional tectonic stress compilations (these measurements are given the lowest quality in the criteria used by WSM as they are not believed to be reliably indicative of the regional stress). In general, we believe that only *in situ* stress measurements made at depths greater than 100 m are indicative of the tectonic stress field at midcrustal depths. This means that techniques utilized in wells and boreholes, which access the crust at appreciable depth, are especially useful for stress measurements.

When a well or borehole is drilled, the stresses that were previously supported by the exhumed material are transferred to the region surrounding the hole. The resultant stress concentration is well understood from elastic theory. Because this stress concentration amplifies the stress difference between far-field principal stresses by a factor of four, there are several other ways in which the stress concentration around boreholes can be exploited to help measure *in situ* stresses. The hydraulic fracturing technique (e.g., Haimson and Fairhurst, 1970; Zoback and Haimson, 1982) takes advantage of this stress concentration and, under ideal circumstances, enables stress magnitude and orientation measurements to be made to about 3 km depth (Baumgärtner *et al.*, 1990).

The most common methods of determining stress orientation from observations in wells and boreholes are stress-induced wellbore breakouts. Breakouts are related to a natural compressive failure process that occurs when the maximum hoop stress around the hole is large enough to exceed the strength of the rock. This causes the rock around a portion of the wellbore to fail in compression (Bell and Gough, 1979; Zoback *et al.*, 1985). For the simple case of a vertical well drilled when S_v is a principal stress, this leads to the occurrence of stress-induced borehole breakouts that form at the azimuth of the minimum horizontal compressive stress. Breakouts are an important source of crustal stress information because they are ubiquitous in oil and gas wells drilled around the world and because they also permit stress orientations to be determined over a great range of depth in an individual well. Detailed studies have shown that these orientations are quite uniform with depth, and independent of lithology and age (e.g., Castillo and Zoback, 1994).

Another form of naturally occurring wellbore failure is drilling-induced tensile fractures. These fractures form in the wall of the borehole at the azimuth of the maximum horizontal compressive stress when the circumferential stress acting around the well locally goes into tension, they are not seen in core from the same depth (Moos and Zoback, 1990; Brudy and Zoback, 1993, 1999; Brudy *et al.*, 1997; Lund and Zoback, 1999; Wiprut and Zoback, 2000).

3. Distribution of Crustal Stresses

Color Plate 11 (top) shows maximum horizontal stress orientations for North America taken from the WSM database. The legend identifies the different types of stress indicators; because of the density of data, only the highest quality data are plotted (A and B quality, shown by lines of different length). The tectonic regime (i.e., normal faulting, strike–slip faulting or reverse faulting) where known is given by the symbol color. The data principally come from wellbore breakouts, earthquake focal mechanisms, *in situ* stress measurements greater than 100 m depth, and young (<2 My old) geologic indicators. These data, originally presented and described by Zoback and Zoback (1989, 1991), demonstrate that large regions of the North American continent (most of the region east of the Rocky Mountains) are characterized by relatively uniform horizontal stress orientations. Furthermore, where different types of stress orientation data are available (e.g., the eastern US), the correlation between the different types of stress indicators is quite good.

Two straightforward observations about crustal stress can be made by comparison of these different types of stress indicators. First, no major changes in the orientation of the crustal stress field occur between the upper 2–5 km, where essentially all of the wellbore breakout and stress measurement data come from, and 5–20 km where the majority of crustal earthquakes occur. Second, the criteria used to define reliable stress indicators appear to be approximately correct. In other words, data badly contaminated by nontectonic (near surface) sources of stress appear to have been effectively eliminated from the compilations.

4. First-Order Global Stress Patterns

Color Plate 12 shows global maximum horizontal compressive stress orientations based on the 1997 World Stress Map data base. As with Color Plate 11 (top), only data qualities A and B are shown and the symbols utilized in Color Plate 12 are the same as those in Color Plate 11 (top). Although global coverage is quite variable, the relative uniformity of stress orientation and relative magnitudes in different parts of the world is striking and permits mapping of regionally coherent stress fields. Color Plate 11 (bottom) presents a generalized version of the Global Stress Map that is quite similar to that presented by Zoback (1992) and shows, using large arrows, mean stress directions and stress regime based on averages of clusters of data shown in Color Plate 12. Tectonic stress regimes are indicated in Color Plate 11 (bottom) by color and arrow type. Blue inward-pointing arrows indicate $S_{H\max}$ orientations in areas of compressional (strike-slip and thrust) stress regimes. Red outward-pointing arrows give $S_{h\min}$ orientations (extensions direction) in areas of normal faulting regimes. Regions dominated by strike-slip tectonics are distinguished with green thick inward-pointing and orthogonal, thin outward-pointing arrows. Overall, arrow sizes on Color Plate 11 (bottom) represent a subjective assessment of "quality" related to the degree of uniformity of stress orientation and also to the number and density of data.

A number of first-order patterns can be observed in Color Plates 11 (bottom) and 12:

1. In many regions a uniform stress field exists throughout the upper brittle crust as indicated by consistent orientations from the different techniques that sample very different rock volumes and depth ranges.
2. Intraplate regions are dominated by compression (thrust and strike-slip stress regimes) in which the maximum principal stress is horizontal.
3. Active extensional tectonism (normal faulting stress regimes) in which the maximum principal stress is vertical generally occurs in topographically high areas in both the continents and the oceans.
4. Regional consistency of both stress orientations and relative magnitudes permits the definition of broadscale regional stress provinces, many of which coincide with physiographic provinces, particularly in tectonically active regions. These provinces may have lateral dimensions on the order of 10^3–10^4 km, many times the typical lithosphere thickness of 100–300 km. These broad regions of the Earth's crust subjected to uniform stress orientation or a uniform pattern of stress orientation (such as the radial pattern of stress orientations in China) are referred to as "first-order" stress provinces (Zoback, 1992).

5. Sources of Crustal Stress

As alluded to above, stresses in the lithosphere are of both tectonic and nontectonic, or local, origin. The regional uniformity of the stress fields observed in Color Plates 11 (top) and 12 argue for tectonic origins of stress at depth for most intraplate regions around the world. For many years, numerous workers suggested that residual stresses from past tectonic events may play an important role in defining the tectonic stress field (see discussion in Engelder, 1993). We have found no evidence for significant residual stresses at depth. If such stresses exist, they seem to be only important in the upper few meters or tens of meters of the crust where tectonic stresses are very small.

Similarly, no evidence has been found that indicates that horizontal principal stresses result simply from the weight of the overlying rock. This oversimplified theory is based on the "bilateral constraint," the supposition that as a unit cube is stressed due to imposition of a vertical stress it cannot expand horizontally (because a neighboring unit cube would be attempting to expand in the opposite direction). If the bilateral constraint was applicable to the crust, the two horizontal principal stresses at depth would be equal and would always be less than the vertical stress (see Engelder, 1993). As demonstrated above, the dominance of compressional intraplate stress fields indicates that one, or both, horizontal stresses exceed the vertical stress. The broad regions of well-defined $S_{h\max}$ orientations are clear evidence of horizontal stress anisotropy. Thus, the predicted bilateral stress state is generally not found in the crust. In fact, the assumptions leading to the prediction of such a stress state are unjustified as the analysis assumes that an elastic crust exists in the absence of gravity (or any other forces) before gravity is instantaneously "switched" on.

In the sections below, the primary sources of tectonic stress are briefly discussed. Although it is theoretically possible to derive the significance of any particular source of stress, because the observed tectonic stress state is the result of superposition of a variety of forces acting on and within the lithosphere it is difficult to define the relative importance of any one stress source. This can only be resolved by utilizing careful modeling and well-constrained observations.

5.1 Plate Driving Stresses

Sources of the broadscale regions of uniform crustal stress that immediately come to mind are the same forces that drive (and resist) plate motions (e.g., Forsyth and Uyeda, 1975). A ridge push compressional force is associated with the excess elevation of the midocean ridges whereas the slab pull force results from the negative buoyancy of down-going slabs. Both of these sources contribute to plate motion and tend to act in the direction of plate motion. If there is flow in the upper asthenosphere, a positive drag force could be exerted on the lithosphere that would tend to drive plate motion, whereas if cold thick lithospheric roots (such as beneath cratons) may be subject to a resistive drag force that would act to inhibit plate motion. In either case the drag force would result in stresses being transferred up into the lithosphere from its base. There are also collisional resistance forces resulting either from the frictional resistance of a plate to subduction or from the collision of two continental plates. As oceanic plates subduct into the viscous lower mantle additional slab resistive forces add to the collision resistance forces acting at shallow depth. Another force resisting plate motion is that due to transform faults, although, as discussed below, the amount of transform resistance may be negligible. Finally, it has been proposed that a suction force may act on the overriding lithosphere in a subduction zone. This force may tend to "suck" the overriding lithosphere toward the trench and result in back-arc spreading.

Although it is possible to specify the various stresses associated with plate movement, their relative and absolute importance in plate movement are not understood. Many researchers believe that either the ridge push or slab pull force is most important in causing plate motion, but it is not clear that these forces are easily separable or that plate motion can be ascribed to a single dominating force. This is discussed at some length by Richardson (1992) and has been addressed by a detailed series of finite element models of the stresses in the North American plate (Richardson and Reding, 1991).

5.2 Topography and Buoyancy Forces

Numerous workers have demonstrated that topography and its compensation at depth can generate sizable stresses capable of influencing the tectonic stress state and style (Frank, 1972; Artyushkov, 1973; Fleitout and Froidevaux, 1982; Sonder, 1990). Density anomalies within or just beneath the lithosphere constitute major sources of stress. The integral of anomalous density times depth [density moment of Fleitout and Froidevaux (1982)] characterizes the ability of density anomalies to influence the stress field and to induce deformation. In general, crustal thickening or lithospheric thinning (negative density anomalies) produces extensional stresses, whereas crustal thinning or lithospheric thickening (positive density anomalies) produces compressional stresses. In more complex cases the resultant state of stress in a region depends on the density moment integrated over the entire lithosphere. In a collisional orogeny, for example, where both the crust and mantle lid are thickened, the presence of the cold lithospheric root can overcome the extensional forces related to crustal thickening and maintain compression (Fleitout and Froidevaux, 1982).

5.3 Lithospheric Flexure

Loads on or within an elastic lithosphere cause deflection and induce flexural stresses which can be quite large (several hundred MPa) and can perturb the regional stress field with wavelengths as much as 1000 km (depending on the lateral extent of the load) (e.g., McNutt and Menard, 1982). Some potential sources of flexural stress influencing the regional stress field include: sediment loading, particularly along continental margins; glacial rebound; seamount loading; and the upwarping of oceanic lithosphere oceanward of the trench, the "outer arc bulge" (Hanks, 1971; Chapple and Forsyth, 1979). Sediment loads as thick as 10 km represent a potential significant stress on continental lithosphere (e.g., Turcotte *et al.*, 1977; Cloetingh *et al.*, 1982). Zoback (1992) suggested that a roughly 40° counterclockwise rotation of horizontal stresses on the continental shelf offshore of eastern Canada was due to superposition of a margin-normal extensional stress derived from sediment load flexure.

6. The Critically Stressed Crust

Three independent lines of evidence indicate that intraplate continental crust is generally in a state of incipient, but slow, frictional faulting: (1) the widespread occurrence of seismicity induced by either reservoir impoundment [cf., fluid injection (cf, Healy *et al.*, 1968; Raleigh *et al.*, 1972; Pine *et al.*, 1983; Zoback and Harjes, 1997)]; (2) earthquakes triggered by small stress changes associated with other earthquakes (cf. Stein *et al.*, 1992, 1994, 1997); and (3) *in situ* stress measurements in deep wells and boreholes (cf. McGarr and Gay, 1978; Zoback and Healy, 1984, 1992; Brudy *et al.*, 1997; Lund and Zoback, 1999). The *in situ* stress measurements further demonstrate that the stress magnitudes derived from Coulomb failure theory utilizing laboratory-derived frictional coefficients of 0.6–1.0 predicts stresses that are consistent with measured stress magnitudes (Townend and Zoback, 2000). This is well-illustrated in Figure 3 by the stress data collected in the KTB borehole to ~8 km depth. Measured stresses are quite high and consistent with the frictional faulting theory [Eq. (7)]

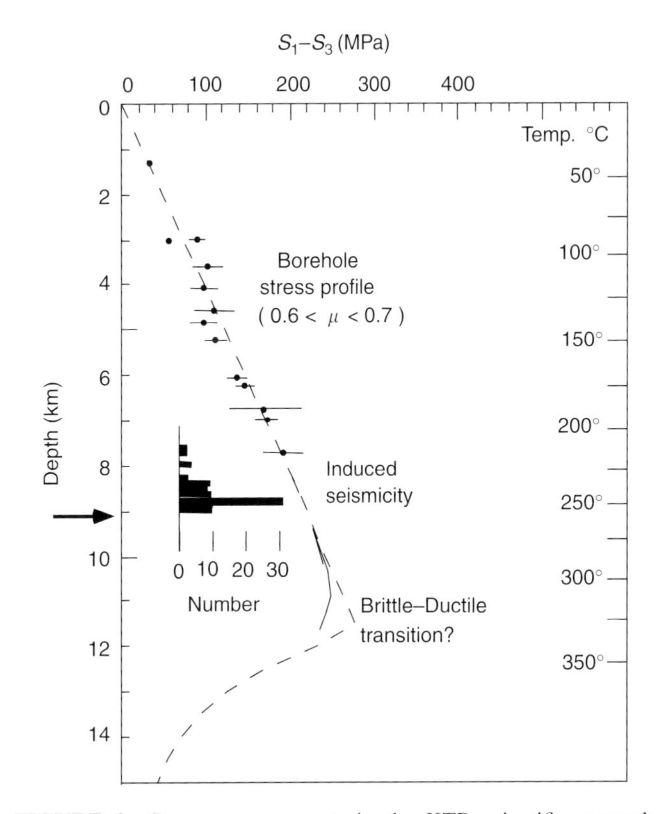

FIGURE 3 Stress measurements in the KTB scientific research well indicate a "strong" crust, in a state of failure equilibrium as predicted by Coulomb theory and laboratory-derived coefficients of friction of 0.6–0.7 (after Zoback and Harjes, 1997).

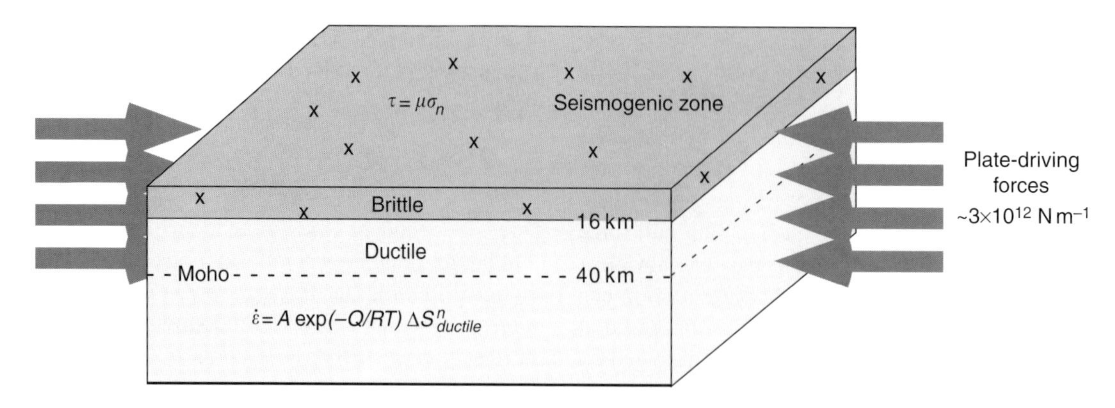

FIGURE 4 Schematic diagram illustrating how the forces acting on the lithosphere keep the brittle crust in frictional equilibrium through creep in the lower crust and upper mantle (after Zoback and Townend, 2001).

with a frictional coefficient of ~0.7 (Brudy *et al.*, 1997). A further demonstration of this "frictional failure" stress state was the fact that a series of earthquakes could be triggered at ~9 km depth in rock surrounding the KTB borehole by extremely low perturbations of the ambient, approximately hydrostatic pore pressure (Zoback and Harjes, 1997).

That the state of stress in the crust is generally in a state of incipient frictional failure might seem surprising, especially for relatively stable intraplate areas. However, a reason for this can be easily visualized in terms of a simple diagram as shown in Figure 4. The lithosphere as a whole (shown simply in Fig. 4 as three distinct layers: the brittle upper crust, the ductile lower crust and the ductile uppermost mantle) must support plate-driving forces. Figure 4 indicates a power-law creep law (e.g., Brace and Kohlstedt, 1980) typically used to characterize the ductile deformation of the lower crust and upper mantle. Because the applied force to the lithosphere will result in steady-state creep in the lower crust and upper mantle, as long as the "three-layer" lithosphere is coupled, stress will build up in the upper brittle layer due to the creep deformation in the layers below. Stress in the upper crust builds over time, eventually to the point of failure. The fact that intraplate earthquakes are relatively infrequent simply means that the ductile strain rate is low in the lower crust and upper mantle. Zoback and Townend (2001) discuss the fact that at the relatively low strain rates characterizing intraplate regions, sufficient plate-driving force is available to maintain a "strong" brittle crust in a state of frictional failure equilibrium.

Although there is appreciable evidence to suggest that the Coulomb criterion and laboratory-derived coefficients of friction are applicable to plate interiors, major plate boundary faults such as the San Andreas fault (and other plate-bounding faults) appear to slip at very low levels of shear stress. Appreciable heat flow data collected in the vicinity of the San Andreas show no evidence of frictionally generated heat (cf, Lachenbruch and Sass, 1980, 1992). This appears to limit average shear stresses acting on the fault to depths of ~15 km to about 20 MPa, approximately a factor of 5 below the stress levels predicted by the Coulomb criterion assuming that hydrostatic pore pressure and laboratory-derived friction coefficients are applicable to the fault at depth. Zoback *et al.* (1987), Mount and Suppe (1987, 1992) and Townend and Zoback (2000) present evidence that the direction of maximum horizontal stress in the crust adjacent to the San Andreas is at an extremely high angle to fault plane. Like the heat flow data, the stress orientation data imply low resolved shear stresses on the fault at depth. Thus, it appears that the frictional strength of plate boundary faults is distinctly lower than intraplate faults which thus enables them to accommodate hundreds of kilometers of relative fault offset and correspondingly high strains.

7. Summary

Large portions of intraplate regions (lateral dimensions of 10^3–10^4 km) are characterized by relatively uniform stress fields suggesting that the state of stress in the brittle upper crust is dominated by large-scale tectonic processes. In general, principal stresses in the crust are in vertical and horizontal planes. In intraplate areas, *in situ* stress measurements and inferences based on topography and flexure all suggest that shear stresses in the upper lithosphere are fairly large and seem controlled by the frictional strength of the faulted crust. Unlike the interiors of plates, plate boundary faults (like the San Andreas fault and subduction zones) appear to be quite weak and slip at low levels of shear stress. At sufficient depth below the brittle upper lithosphere, stresses are likely controlled by the ductile flow properties of the constituent rocks and minerals. Plate driving forces are sufficient in magnitude to maintain the intraplate lithosphere in a state of frictional failure. This frictional failure is manifest as slow, steady-state creep deformation in the lower crust and upper mantle and brittle deformation in the upper crust.

Acknowledgment

We thank Peter Bird for his comments on an early draft of this chapter.

References

Amadei, B. and O. Stephansson (1997). "Rock Stress and its Measurement," 490 pp., Chapman & Hall, London.

Anderson, E.M. (1951). The Dynamics of Faulting and Dyke Formation with Applications to Britain, Oliver and Boyd, Edinburgh.

Angelier, J. (1990). Inversion of field data in fault tectonics to obtain the regional stress – III. A new rapid direct inversion method by analytical means. *Geophys. J. Int.* **103**, 363–376.

Artyushkov, E.V. (1973). Stresses in the lithosphere caused by crustal thickness inhomogeneities. *J. Geophys. Res.* **78**, 7675–7708.

Baumgärtner, J., F. Rummel, and M.D. Zoback (1990). Hydraulic fracturing *in situ* stress measurements to 3 km depth in the KTB pilot hole VB. A summary of a preliminary data evaluation. *KTB Report 90-6a*, pp. 353–400.

Bell, J.S. and D.I. Gough (1979). Northeast–southwest compressive stress in Alberta: evidence from oil wells. *Earth Planet. Sci. Lett.* **45**, 475–482.

Brace, W.F. and D.L. Kohlstedt (1980). Limits on lithospheric stress imposed by laboratory experiments. *J. Geophys. Res.* **85**, 6248–6252.

Brudy, M. and M.D. Zoback (1993). Compressive and tensile failure of boreholes arbitrarily inclined to principal stress axes: Applications to the KTB boreholes, Germany. *Int. J. Rock Mech. Min. Sci.* **30**, 1035–1038.

Brudy, M. and M.D. Zoback (1999). Drilling-induced tensile wall-fractures: implications for determination of *in situ* stress orientation and magnitude. *Int. J. Rock Mech. Min. Sci.* **36**, 191–215.

Brudy, M., M.D. Zoback, K. Fuchs, F. Rummel, and J. Baumgärtner (1997). Estimation of the complete stress tensor to 8 km depth in the KTB scientific drill holes: Implications for crustal strength *J. Geophys. Res.* **102**, 18453–18475.

Brune, J.N. (1970). Tectonic stress and the spectra of seismic shear waves from earthquakes. *J. Geophys. Res.* **75**, 4997–5009.

Byertee, J.D. (1978). Friction of rocks. *Pure Appl. Geophys.* **116**, 615–626.

Castillo, D.A. and M.D. Zoback (1994). Systematic variations in the stress state in the southern San Joaquin valley: Inferences based on wellbore data and contemporary seismicity. *Am. Assoc. Pet. Geol. Bull.* **78**, 1257–1275.

Chapple, W.M. and D. Forsyth (1979). Earthquakes and bending of plates at trenches. *J. Geophys. Res.* **84**, 6729–6749.

Cloetingh, S.A.P.L., M.J.R. Wortel, and N.J. Vlaar (1982). State of stress at passive margins and initiation of subduction zones. *Nature* **297**, 39–142.

Engelder, T. (1993). "Stress Regimes in the Lithosphere," 457 pp., Princeton, Princeton, New Jersey.

Engelder, T. and M.L. Sbar (1984). Near-surface *in situ* stress: introduction. *J. Geophys. Res.* **89**, 9321–9322.

Fleitout, L. and C. Froidevaux (1982). Tectonics and topography for a lithosphere containing density heterogenieties, *Tectonics* **2**, 315–324.

Forsyth, D. and S. Uyeda (1975). On the relative importance of the driving forces of plate motion, *Geophys. J. R. Astron. Soc.* **43**, 163–200.

Frank, F.C. (1972). Plate tectonics, the analogy with glacier flow and isostasy. In: "Flow and Fracture of Rocks," *Geophysics Monograph Series*, vol. 16, pp. 285–292, American Geophysical Union, Washington, DC.

Gephart, J.W. and D.W. Forsyth (1984). An improved method for determining the regional stress tensor using earthquake focal mechanism data: application to the San Fernando earthquake sequence, *J. Geophy. Res.* **89**, 9305–9320.

Haimson, B. and C. Fairhurst (1970). *In situ* stress determination at great depth by means of hydraulic fracturing. In: "11th Symposium on Rock Mechanics," (W. Somerton, Ed.) pp. 559–584, Society of Mining Engineers of AIME.

Hanks, T.H. (1971). The Kuril trench–Hokkaido rise system: large shallow earthquakes and simple models of deformation, *Geophys. J.R. Astron. Soc.* **23**, 173–189.

Hanks, T.H. (1977). Earthquake stress drops, ambient tectonic stresses and stresses that drive plate motions. *Pageoph* **115**, 441–458.

Healy, J.H., W.W. Rubey, D.T. Griggs, and C.B. Raleigh (1968). The Denver earthquakes. *Science* **161**, 1301–1310.

Jaeger, J.C. and N.G.W. Cook (1971). "Fundamentals of Rock Mechanics," 515 pp. Chapman and Hall, London.

Lachenbruch, A.H. and J.H. Sass (1980). Heat flow and energetics of the San Andreas fault zone, *J. Geophys. Res.* **85**, 6185–6223.

Lachenbruch, A.H. and J.H. Sass (1992). Heat flow from Cajon Pass, fault strength, and tectonic implications. *J. Geophys. Res.* **97**, 4995–5016.

Lund, B. and M.D. Zoback (1999). Orientation and magnitude of *in situ* stress to 6.5 km depth in the Baltic Shield. *Int. J. Rock Mech. Min. Sci.* **36**, 169–190.

McGarr, A. and N.C. Gay (1978). State of stress in the earth's crust. *Annu. Rev. Earth Planet. Sci.* **6**, 405–436.

McKenzie, D.P. (1969). The relationship between fault plane solutions for earthquakes and the directions of the principal stresses. *Seismol. Soc. Am. Bull.* **59**, 591–601.

McNutt, M.K. and H.W. Menard (1982). Constraints on yield strength in the oceanic lithosphere derived from observations of flexure. *Geophys. J. R. Astron. Soc.* **71**, 363–394.

Michael, A.J. (1984). Determination of stress from slip data: Faults and folds. *J. Geophys. Res.* **89** (B13), 11517–11526.

Moos, D. and M.D. Zoback (1990). Utilization of observations of wellbore failure to constrain the orientation and magnitude of crustal stresses: application to continental, Deep Sea Drilling Project and Ocean Drilling Project boreholes. *J. Geophys. Res.* **95**, 9305–9325.

Mount, V.S. and J. Suppe (1987). State of stress near the San Andreas fault: Implications for wrench tectonics. *Geology* **15**, 1143–1146.

Mount, V.S. and J. Suppe (1992). Present-day stress orientations adjacent to active strike-slip faults: California and Sumatra. *J. Geophys. Res.* **97**, 11995–12013.

Mueller, B., V. Wehrle, and K. Fuchs (1997). "The 1997 Release of the World Stress Map." *http://www-wsm.physik.uni-karlsruhe.de/pub/Rel97/wsm97.html*

Nakamura, K. (1977). Volcanoes as possible indicators of tectonic stress orientations–principle and proposal. *J. Volcanol. Geotherm. Res.* **2**, 1–16.

Pine, R.J., P. Ledingham, and C.M. Merrifield (1983). *In situ* stress measurements in the Carmenellis granite–II Hydrofracture tests at Rosemanowes Quarry to depths of 2000 m: *Int. J. Rock Mech. Min. Sci. Geomech. Abstr.* **20**, 63–72.

Raleigh, C.B., J.H. Healy, and J.D. Bredehoeft (1972). Faulting and crustal stress at Rangely, Colorado. In: "Flow and Fracture of Rocks," (J.C. Heard, *et al.*, Eds.) pp. 275–284. *Geophys. Monogr. Ser., American Geophysical* Union, Washington, DC.

Richardson, R. (1992). Ridge forces, absolute plate motions and the intraplate stress field. *J. Geophys. Res.* **97** (B8), 11739–11748.

Richardson, R.M. and L.M. Reding (1991). North American plate dynamics. *J. Geophys. Res.* **96**, 12201–12223.

Sonder, L.J. (1990). Effects of density contrasts on the orientation of stresses in the lithosphere: relation to principal stress directions in the Transverse ranges, California. *Tectonics* **9**, 761–771.

Stein, R.S., G.C. King, and J. Lin (1992). Change in failure stress on the southern San Andreas fault system caused by the 1992 magnitude 7.4 Landers earthquake. *Science* **258**, 1328–1332.

Stein, R.S., G.C.P. King, and J. Lin (1994). Stress triggering of the 1994 M = 6.7 Northridge, California earthquake. *Science* **265**, 1432–1435.

Stein, R.S., A.A. Barka, and J.H. Dieterich (1997). Progressive failure of the North Anatolian fault since 1939 by earthquake stress triggering. *Geophys. J. Int.* **128**, 594–604.

Townend, J. and M.D. Zoback (2000). How faulting keeps the crust strong, *Geology* **28**, 399–402.

Turcotte, D.L., J.L. Ahern, and J.M. Bird (1977). The state of stress at continental margins. *Tectonophysics* **42**, 1–28.

Wiprut, D. and M.D. Zoback (2000). Fault reactivation and fluid flow along a previously dormant normal fault in the northern North Sea. *Geology* **28**, 595–598.

Zoback, M.D. and B.C. Haimson (1982). Status of the hydraulic fracturing method for *in situ* stress measurements. In: "23rd Symposium on Rock Mechanics," Society of Mining Engineers (R.E. Goodman and F.E. Heuze, Eds.), pp. 143–156, New York.

Zoback, M.D. and H.P. Harjes (1997). Injection induced earthquakes and crustal stress at 9 km depth at the KTB deep drilling site, Germany. *J. Geophys. Res.* **102**, 18477–18491.

Zoback, M.D. and J.H. Healy (1984). Frictrion faulting and "*in situ*" stresses. *Ann. Geophys.* **2**, 689–698.

Zoback, M.D. and J.H. Healy (1992). *In situ* stress measurements to 3.5 km depth in the Cajon Pass Scientific Research Borehole: implications for the mechanics of crustal faulting. *J. Geophys. Res.* **97**, 5039–5057.

Zoback, M.D. and J. Townend (2001). Implications of hydrostatic pore pressures and high crustal strength for deformation of intraplate lithosphere. *Tectonophysics* **336**, 19–30.

Zoback, M.D. and M.L. Zoback (1991). Tectonic stress field of North America and relative plate motions. In: "The Geology of North America, Neotectonics of North America," (D.B.a.o. Stemmons), pp. 339–366, Geological Society of America, Boulder.

Zoback, M.D., D. Moos, and L. Mastin (1985). Wellbore breakouts and *in situ* stress. *J. Geophys. Res.* **85**, 6113–6156.

Zoback, M.D., M.L. Zoback, V.S. Mount, J. Suppe, J.P. Eaton, J.H. Healy, D. Oppenheimer, P. Reasenberg, L. Jones, C.B. Raleigh, I.G. Wong, O. Scotti, and C. Wentworth (1987). New evidence on the state of stress of the San Andreas fault system. *Science* **238**, 1105–1111.

Zoback, M.D., R. Apel, J. Baumgärtner, M. Brudy, R. Emmermann, B. Engeser, K. Fuchs, W. Kessel, H. Rischmüller, F. Rummel, and L. Vernik (1993). Upper crustal strength inferred from stress measurements to 6 km depth in the KTB borehole. *Nature* **365**, 633–635.

Zoback, M.L. (1992). First and second order patterns of tectonic stress: The World Stress Map Project. *J. Geophys. Res.* **97**, 11703–11728.

Zoback, M.L. and M.D. Zoback (1980). State of stress in the conterminous United States. *J. Geophys. Res.* **85**, 6113–6156.

Zoback, M.L. and M.D. Zoback (1989). Tectonic stress field of the conterminous United States. *Geol. Soc. Am. Mem.* **172**, 523–539.

Zoback, M.D. and J. Townend (2001). Implications of hydrostatic pore pressures and high crustal strength for the deformation of intraplate lithosphere. *Tectonophysics* **336**, 19–30.

Zoback, M.L., *et al.* (28 authors) (1989). Global patterns of intraplate stresses; a status report on the world stress map project of the International Lithosphere Program. *Nature* **341**, 291–298.

Editor's Note

Please see Chapter 7, Earthquake mechanisms and plate tectonics, by Stein and Klosko, and Chapter 33, State of stress within the Earth, by Ruff.

35

Strength and Energetics of Active Fault Zones

James N. Brune
University of Nevada, Reno, USA
Wayne Thatcher
US Geological Survey, Menlo Park, USA

1. Introduction

The strength of active fault zones, i.e., the shear stress level required to cause fault slip, is fundamental to understanding the physics of earthquakes and to assessing earthquake hazard. Although many researchers have concluded that fault zones are weak (shear stresses 10 MPa or less averaged between 0 and ~20 km depth), others maintain that faults are strong (~100 MPa average of an approximate linear increase with depth). Thus, despite 30 y of dedicated research, relevant data remain inconclusive and fault strength remains uncertain by an order of magnitude. In part, this is because the main source of energy release in earthquakes is at depths greater than 5 km, inaccessible to direct instrumental observation. Very large earthquakes rupture to the Earth's surface where direct observation of the shallow rupture process is possible. However, the rupture characteristics at shallow depth may differ from those at seismogenic depths. To date no great ($M \geq 7.75$) earthquakes have occurred within a network of modern strong-motion instruments, but the large, well-recorded 1999 earthquakes in Turkey and Taiwan, both $M = 7.6$, show that this data gap is rapidly being filled. Furthermore, shear stress changes at the earthquake source (and the resulting seismic waves) are nearly linear perturbations of the absolute stress field. Thus, an unknown absolute background stress does not greatly affect the basic characteristics of the observed low frequency seismic waves and the observed geodetic deformation. Other, less-direct data must then be used to infer the physical state and ambient stress levels on active faults.

This chapter discusses available evidence and current ideas about fault zone strength and energetics. In our review we begin by outlining the general physical conditions prevailing in the Earth around active faults and summarize the generally agreed upon observational features of faulting and earthquake occurrence (Section 2). In doing so we make as few assumptions as possible, because making uncertain assumptions may lead to logical inconsistencies and apparent paradoxes. For example, the absence of a measurable, frictionally generated heat flow anomaly near active faults, the so-called stress-heat flow paradox, rests upon assumptions that must somehow be incorrect. We wish to avoid such inconsistencies and to begin we introduce the observations and briefly state what they imply about fault zones. Subsequently, in Sections 3–8, we place these observations in an interpretative context and show how they have been used to infer various measures of fault shear stress. Section 9 summarizes our assessment of the average shear stress state near faults, argues for the general importance of stress heterogeneity in faulting processes, and discusses its implications.

2. General Physical Problem and Observational Constraints

Active faults are stressed by forces applied in the adjacent lithosphere, forces applied at or near plate boundaries due to the motions of the plates as well as those caused by lateral density contrasts within the lithosphere. Parts of some active faults, particularly those at major plate boundaries, slip aseismically, keeping fault zone shear stresses at about the same levels. Others slip primarily in earthquakes, with the fault shear stresses increasing between events and decreasing abruptly when sudden seismic slip occurs. During earthquakes, energy released at depth in the Earth propagates as elastic waves that cause ground shaking when they reach the Earth's surface. We have no direct access to the depths at which the major energy is released. Thus the physical processes must be inferred indirectly from evidence obtained at the surface (proposals for deep drilling into active crustal fault zones are pending, and

INTERNATIONAL HANDBOOK OF EARTHQUAKE AND ENGINEERING SEISMOLOGY, VOLUME 81A ISBN: 0-12-440652-1

would provide the first direct evidence of the physical state at seismogenic depths). What we know about active faults and earthquakes thus comes from measurements made at or near the Earth's surface, and relies on seismology, structural geology, geodesy, and other geophysical data to infer the processes that are thus far inaccessible to direct observation. Studies of these data have established the following general features of stress, faulting, and earthquake occurrence. Detailed discussion follows in the section indicated.

1. Lateral density gradients in the lithosphere and variations in surface and sea-floor topography generate differential stresses of ~10–100 MPa available to drive slip on faults (Section 3).
2. Stresses measured *in situ*, in mines to depths of ~3 km and in boreholes to as deep as 9 km, are consistent with an approximately linear increase in shearing stress with depth of ~5–15 MPa km^{-1}. However, these measurements, particularly those below a few km depths, come from tectonically inactive plate interiors far from large active faults (Section 5).
3. Studies of inactive fault zones exhumed by erosion from seismogenic depths show them to be generally planar features, often with zones of crushed and comminuted rock (fault gouge, microbreccia, and cataclasite) up to 1 km in width. Faulting involves frictional slip to depths corresponding to temperatures of ~350°C; at greater temperatures, deformation is primarily ductile, though still confined to relatively narrow shear zones (see Chapter 29 by Sibson). The fault zones show evidence of having been fluid saturated throughout their depth range (Section 5).
4. Heat flow measurements near active faults show no evidence for the heat generation expected if there were significant frictional resistance to fault slip. Furthermore, the orientations of principal compressive stresses near some major active faults are nearly normal to their strike. These two observations have been used to suggest faults are weaker (support less shearing stress) than the surrounding blocks, with an upper bound on average fault zone shear stress of ~20 MPa or less (Section 6).
5. Data from seismology, geodesy, and geological mapping of earthquake faulting are all consistent with average earthquake stress drops of ~0.1–10 MPa. These same data show, however, that slippage on the fault surface is very spatially heterogeneous, indicating local stress drops up to an order of magnitude greater than the average values (Section 7).
6. Measurements of strong ground motions made near earthquakes as large as $M\,7$ reveal accelerations as large as $1g$ and velocities up to 2 m sec^{-1}, indicating dynamic stress changes of about 30 MPa on the fault (Section 8).

In what follows we amplify on these observations and explain the inferences derived from them. In each case we point out the assumptions implicit in the inference, our own assessment of their reliability, and the implications of the results for fault zone stress and energetics.

3. External Stresses Available to Drive Faulting

The energy that drives active faulting ultimately comes from the stresses imposed on the lithosphere by the forces that drive and resist motions of the major plates and by stresses due to lateral density gradients in the lithosphere. Although at least eight different plate forces are potentially important, only some of these are believed to be decisive in determining the force balance (Forsyth and Uyeda, 1975) and hence the intraplate stress field (Zoback *et al.*, 1989). Here we shall describe only those considered in recent work. The forces are illustrated in Figure 1, which should be referred to in the subsequent discussion of each force.

The best understood force is the "ridge push," F_{rp}, the mean excess pressure (differential stress) ΔP exerted on the lithosphere due to the elevation of the mid-ocean ridges above the surrounding sea floor. It is given by $\Delta P = 0.5g\ (\rho_a - \rho_w)e$, where g is the acceleration of gravity, ρ_a is the asthenospheric density, ρ_w is the density of sea water and e is the elevation of the ridge above the sea floor. All of these parameters are known rather well, and so the resulting differential stress estimate of 30 MPa averaged over an oceanic lithospheric thickness of ~70 km is probably accurate. Several recent studies suggest that the ridge push force is the most important determinant of intraplate stress in the North American plate (e.g., Richardson and Reding, 1991; Zoback, 1992). If so, the field has a particularly simple, predictable form and its magnitude is

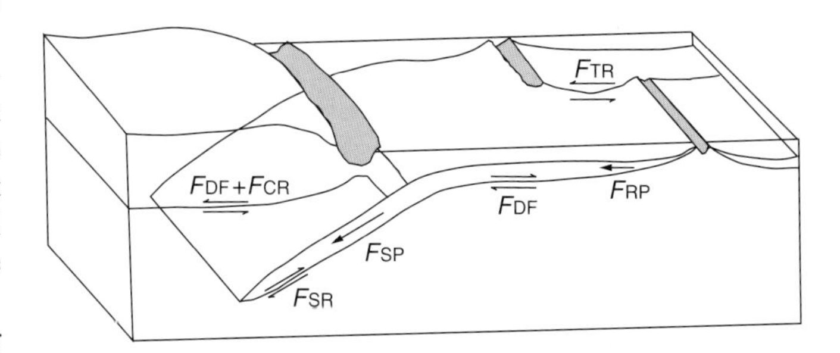

FIGURE 1 Plate driving and resisting forces (modified from Forsyth and Uyeda, 1975). F_{TR} = transform fault resistance; F_{DF} = plate drag force; F_{CR} = plate drag under continents; F_{RP} = ridge push; F_{SP} = slab pull; F_{SR} = slab resistance.

bounded rather well. Since the load-bearing thickness of the continental lithosphere may well be considerably less than 70 km in active regions, differential driving stresses from ridge push could be as much as ~100 MPa in these areas.

The "slab pull" force, F_{sp}, is caused by the excess density of the cold subducting oceanic slab as it sinks into the hotter, more buoyant asthenosphere. Its magnitude can be estimated from the thermal structure of the descending plate (McKenzie, 1969) and it is likely to be the largest of the driving forces. The corresponding force per unit area acting on the lithosphere is about 180 MPa, but this is an upper bound estimate of the differential driving stresses imposed on the lithosphere by subduction. Slab descent into the mantle is almost certainly resisted by poorly known forces of comparable magnitude, F_{sr} in the asthenosphere, and F_{cr} on the intraplate interface of the subduction thrust. As a result, the net drive or resistance from subduction is not well known, although it is likely that this net force importantly influences differential stresses in some plates.

Transform fault resistance, F_{tf}, opposes the strike-slip motion of the plates on both oceanic transforms and continental transcurrent faults like the San Andreas system. Although these frictional and ductile resisting shear stresses could be significant, their magnitudes are poorly known. We discuss them further below in connection with lithospheric rheology. Quasi-static frictional shear resistance on the seismogenic upper crustal faults could average 60 MPa across strike-slip faults and as much as 400 MPa across the deeper portions of subduction thrusts (F_{cr} of Fig. 1). However, as discussed below, heat flow data in both transform and subduction settings place much lower thresholds on the magnitudes of these resisting shear stresses.

Finally, shear stresses imposed on the base of the lithosphere, either as driving or resisting forces, F_{df}, are potentially important but poorly known. F_{df} is a driving force if imposed by general convective flow of the mantle, or resistive if caused by the drag of the plates over a passive asthenosphere. Driving or resisting shear stresses of only a few bars integrated over the large basal area of the plates would have important influences on the intraplate stress field.

Whatever the magnitudes of these largely unknown forces their net effect is to impose a long-wavelength stress field on the interiors of the plates. It is this field, often modified by local perturbations, that supplies the "tectonic" shear stresses that drive active faulting.

Buoyancy forces caused by lateral density contrasts in crust or mantle are the most important local perturbations. Just as the elevated topography at midocean ridges leads to large intraplate stresses, a similar process occurs when elevated topography is caused by thickened continental crust. By considering the force balance for two columns of continental crust, one with thickness y_{cco} and the other thickened so that it has an additional elevation h, and assuming the two columns are in isostatic balance, Turcotte (1982) showed the net horizontal force F_r is

$$F_r = \rho_c gh \left[y_{cco} + \frac{\rho_m h}{\rho_m - \rho_c} \right]$$

where ρ_c and ρ_m are the densities of crust and mantle respectively. If $y_{cco} = 35$ km, $\rho_m = 3300\,\mathrm{kg\,m^{-3}}$, $\rho_c = 2750\,\mathrm{kg\,m^{-3}}$ and we assume F_r is supported over an elastic crust 50 km thick, then we can obtain the resulting compressive stress as a function of elevated topographic height h. If $h > 3$ km then differential stresses exceed 100 MPa. In the absence of any other applied forces we expect these topographic effects will lead to extensional stresses within the elevated region and compressive stresses in the adjacent thinner crust. The same general principles that we have discussed for the crust apply to lateral density gradients in the mantle (see Fleitout and Froidevaux, 1982 for a general formulation), and both lead to stresses that can drive faulting.

3.1 Assessment

Our assessment is that intraplate differential stresses due to plate motion forces and lateral density gradients within the lithosphere can be estimated approximately and lie in the range ~10–100 MPa averaged over the entire lithosphere. They thus provide rough bounds on the magnitude of the long wavelength, far field shear stress that causes faulting.

4. Internal Fault Zone Stress and Energy Balance for Fault Slip

The tectonic stresses discussed above, supported by the lithosphere over long time intervals, provide the boundary stresses which ultimately supply energy for earthquakes and fault slip. Earthquake occurrence modulates the local stress field in an intermittent fashion while steady-state aseismic slip keeps shear stress at about the same level at all times. The fault zone (including gouge) represents an internal boundary on which we seek to infer the stresses. Relatively rapid earthquake fault slip will decrease stress on the fault and its surroundings, radiating seismic waves. In the process, work is done against frictional stresses that resist fault motions. Elastic stress accumulation subsequently restores the stress slowly, over hundreds or thousands of years, to an ambient prefailure level that represents the actual "strength" of the fault. The earthquake stress drop and the absolute ambient pre-earthquake stress, along with the frictional resisting forces active during sliding, determine the partitioning of stored elastic energy released during faulting (E) into seismic waves (E_s) and work done against resisting stresses (E_r). In what follows we outline how this partitioning provides a framework for considering constraints on the magnitudes of stresses acting before, during and immediately after earthquake fault slip. This will in turn provide a bounding framework for evaluating various models and inferences of stress state.

To begin we discuss a relatively small rupture surface over which it may be assumed that the stress is uniform. We assume a confined planar fault surface and uniform stresses and stress changes over the fault. We define u as the total slip averaged over the fault area A. τ_i and τ_f are the initial and final shear stresses on the fault, and the average resisting shearing stress during slip is τ_r. Following Brune (1976) and Lachenbruch and Sass (1980), Figure 2 shows three possibilities for the relative magnitudes of the final stress and average resisting stress ($\tau_r > \tau_f$, overshoot; $\tau_r = \tau_f$, null; and $\tau_r < \tau_f$, locking). The total elastic energy release is the area under the straight line joining τ_i and τ_f in Figure 2,

$$E = \tfrac{1}{2}(\tau_i + \tau_f)uA \tag{1}$$

Then the work done against resisting stresses is the area under the line $\tau = \tau_r$ in Figure 2,

$$E_r = \tau_r\, uA \tag{2}$$

The seismically radiated energy is the difference between Eq. (1) and Eq. (2), the area between the elastic energy release line and the resisting stress line τ_r,

$$E_s = E - E_r = \left[\tfrac{1}{2}(\tau_i + \tau_f) - \tau_r\right] uA \tag{3}$$

Figure 2 graphically shows how the energy partitioning depends on the relative magnitudes of the initial, final, and resisting stresses on the fault. The detailed dynamics of earthquake faulting certainly depend on dynamic shear resistance, as shown by the dashed line in Figure 2, but average values remain useful for intuitive understanding.

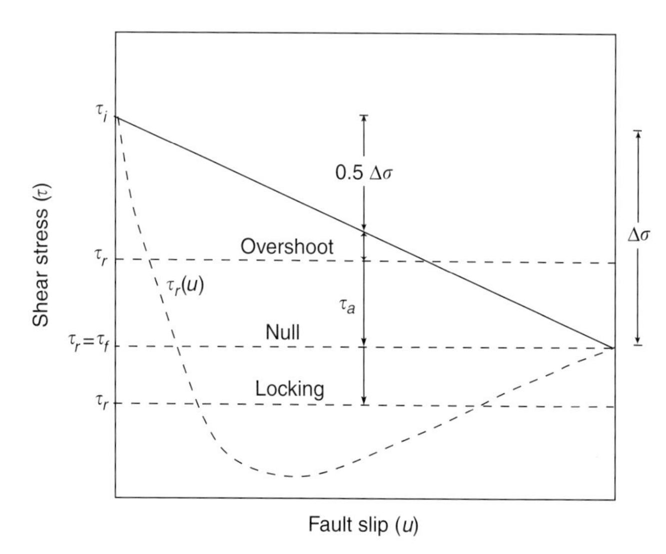

FIGURE 2 Average stress levels during fault slip. Initial stress = τ_i, final stress = τ_f, resisting stress = τ_r, apparent stress = τ_a, stress drop = $\Delta\sigma$. Dashed line shows case of variable dynamic friction during slip. Three possibilities for the relative magnitudes of the final stress and average resisting stress are shown ($\tau_r > \tau_f$, overshoot; $\tau_r = \tau_f$, null; and ($\tau_r < \tau_f$, locking).

We may relate several spatially averaged measures of stress by using Eqs. (1–3) and defining some commonly used stress parameters as follows

$$\text{Average loading stress:} \quad \tau_e = \tfrac{1}{2}(\tau_i + \tau_f) \tag{4}$$

$$\text{Apparent stress:} \quad \tau_a = \tfrac{1}{2}(\tau_i + \tau_f) - \tau_r \tag{5}$$

$$\text{Stress drop:} \quad \Delta\sigma = \tau_i - \tau_f \tag{6}$$

Then since $E = E_s + E_r$ we have

$$\tau_e = \tau_a + \tau_r \tag{7}$$

Using Eqs. (5) and (6) we finally obtain

$$\tau_i = \tau_a + \tfrac{1}{2}\Delta\sigma + \tau_r \tag{8}$$

Equation (8) is useful because it relates stress measures that we can estimate with varying degrees of precision and uncertainty from observational data, permitting intercomparisons and checks on consistency. In what follows we will critically assess the state of knowledge of each stress measure in order to constrain the true strength and energy balance in active fault zones. Our general objective is to understand what physical factors control stresses, to evaluate observations that constrain their magnitudes, and to assess whether these average stresses are relatively high (~100 MPa or greater) or rather low (~20 MPa or less). As we discussed previously, rough bounds can be placed on the initial stress τ_i based on intraplate stress estimates. The apparent stress τ_a can be obtained for individual earthquakes by seismically measuring the moment and radiated energy release. Stress drops ($\Delta\sigma$) for individual earthquakes can be estimated from geodetic measurements, as well as seismically. Stresses τ_r that resist fault motions generate heat, and thermal measurements near active faults constrain the long-term (millions of years) average resisting stress across crustal faults.

5. Quasi-static Lithospheric Rheology

The rheology of the lithosphere, the constitutive laws that determine quasi-static deformation for given applied stresses, may play a decisive role in governing the magnitude and distribution of ambient stress. It is the shear stress component of the ambient field resolved onto the fault that is identified with the initial stress τ_i in the quasi-static model for rupture discussed above. The upper 10–30 km of the crust, where ruptures typically nucleate and propagate, deforms elastically or through brittle failure. Below these depths, aseismic fault slip and distributed bulk ductile deformation occur in the deeper portions of a lithospheric column that may be as much as ~100 km thick (Brace and Kohlstedt, 1980).

The vertical stratification and lateral heterogeneity of lithospheric rheology determine the coupling between the far-field tectonic stress, discussed above, and the local deformation. This variability in rheology affects the ways in which stresses are transferred into the "earthquake machine" in the seismogenic upper crust. Local changes in bulk lithospheric rheology, for example near major faults, could profoundly influence both the stress field and the patterns of active faulting.

The rheological behavior of fault zones, either during earthquake slip or quasi-static slip, is not yet well understood, and in what follows we use extrapolations from laboratory rock mechanics experiments to derive idealized static strength versus depth profiles for the lithosphere. It is not yet certain how far they can be applied to conditions existing in and near active faults. Indeed, as we shall see, observations from the San Andreas fault and elsewhere violate the simplest expectations based on these derived static strength profiles, and so additional complexities are needed to rationalize the field measurements within the rock mechanics framework. In addition, the stress field during faulting in the seismogenic upper crust may be dominated by dynamical effects that are not closely linked to the quasi-static, prefailure stress levels. Thus, although significant uncertainties also exist in understanding the rheology of the deeper, ductile lithosphere, the quasi-static parameters based upon laboratory experiments may more nearly apply there.

Whatever the actual stress state in the Earth, the conventional rock mechanics framework developed here is a useful standard against which to evaluate observations and models of fault zone strength. In what follows we separately consider the rheology of the seismogenic crust and the region that lies beneath it.

5.1 Brittle Upper Crust

Let us define strength as the maximum shear stress that can be supported by rocks. Within the conventional framework, in the upper 10–30 km of the Earth this strength is determined by the stresses required to cause frictional sliding on preexisting faults or fractures. For intact rock masses the brittle fracture strength is generally higher but is more difficult to estimate because it varies widely with rock type, temperature, and strain rate and it is not considered further here.

The general form of the failure condition for frictional slip is given by

$$\tau = \mu(\sigma_n - P) + \tau_o \tag{9}$$

where τ and σ_n are the shear and normal stresses across a planar fault surface, μ is the coefficient of static friction on the surface, P is the ambient fluid pressure, and τ_o is a cohesion term which we subsequently drop for simplicity. Note that high fluid pressure P could considerably reduce the tectonic shear stress τ required to cause frictional failure, a point we will return to in Section 9.

Laboratory measurements of frictional sliding on rock surfaces under the temperature and pressure conditions of the crust provide potentially strong constraints on frictional slip. They lead to the important generalization that the frictional resistance to sliding is (with a few exceptions) independent of rock type and depends only upon confining pressure (Byerlee, 1978). All of the data except those for a few clay-rich minerals can be fit well assuming

$$\mu = 0.75 \tag{10}$$

5.1.1 Triaxial Stress State

If we consider a triaxial stress state with effective principal compressive stresses $(\sigma_1 - P) > (\sigma_2 - P) > (\sigma_3 - P)$ and the fault plane lying at angle θ to the σ_1 axis (Fig. 3), we may rewrite Eq. (9) in terms of the principal effective stresses (see Jaeger and Cook, 1976, p.14) $(\sigma_1' - \sigma_3')\sin 2\theta = \mu[(\sigma_1' + \sigma_3') - (\sigma_1' - \sigma_3')\cos 2\theta]$, where we have replaced all $(\sigma_i - P)$ by σ_i'. After a little algebra we obtain

$$\sigma_1'/\sigma_3' = (1 + \mu \cot \theta)/(1 - \mu \tan \theta) \tag{11}$$

This ratio has a minimum value for fault planes perpendicular to the $\sigma_1 - \sigma_3$ plane intersecting the σ_2 axis and oriented at angles of θ_0 to σ_1. This angle is given by

$$\theta_0 = 45^\circ - \tfrac{1}{2}\tan^{-1}\mu = \tfrac{1}{2}\tan^{-1}(1/\mu) \tag{12}$$

Faults at this optimal orientation thus slip at the minimum value of σ_1'/σ_3' and it is often assumed that such faults exist in the crust when calculating its frictional strength. However, it is not uncommon for pre-existing faults not optimally oriented in this way to be reactivated in the current (different) tectonic stress field, in which case the stress ratio must be greater. Frictional lock-up occurs for faults oriented more than $2\theta_0$ from the σ_1-axis (Fig. 3c), which defines the asymptotic limits for θ (Sibson, 1985). If all available orientations for pre-existing faults lie close to the $2\theta_0$ limit, new faults of optimum orientation may be formed at lower stresses by fracture of previously unfaulted rock.

In principle, failure could occur for $\theta > 2\theta_0$ but this would require negative values of the stress ratio σ_1'/σ_3'. This implies the effective least principal stress $\sigma_3' < 0$, i.e., $P > \sigma_3$, and the rock mass surrounding the fault may show evidence of tensional failure by hydraulic fracturing (see Chapter 29 by Sibson). This condition may apply without pervasive hydrofracture if the region of high pore pressure is confined near the fault zone, permitting slip at low effective stresses on very misoriented faults in a stress field characterized by locally

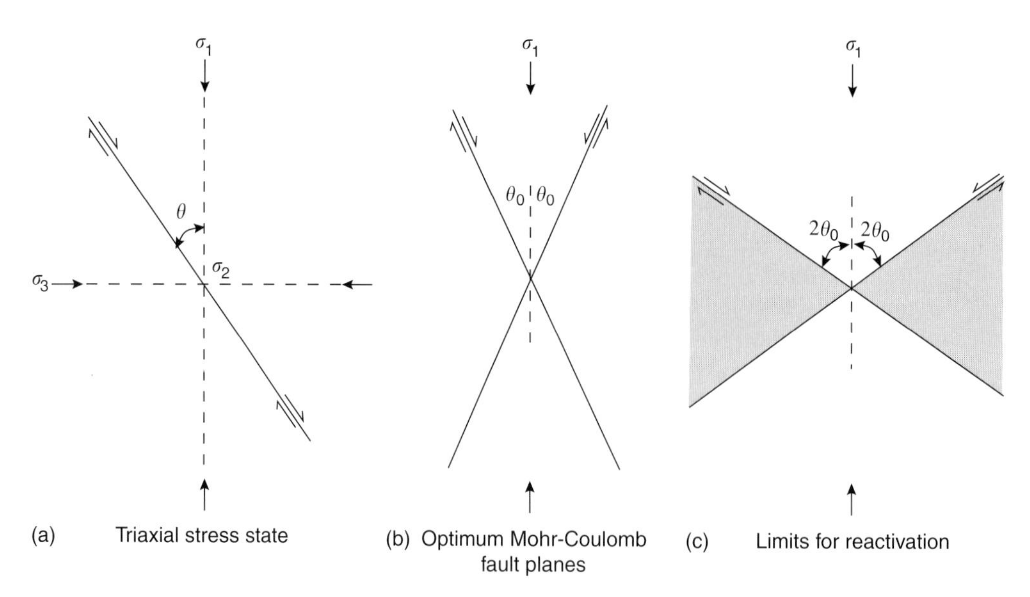

FIGURE 3 2D Mohr–Coulomb faulting theory. (a) Possible fault planes under triaxial stress state ($\sigma_1 > \sigma_2 > \sigma_3$). (b) Optimum faulure planes as given by Eq. (12) in the text. (c) Frictional limits on fault plane orientation. Arrows show orientations of maximum (σ_1) and minimum (σ_3) principal stresses. The intermediate principal stress axis (σ_2) lies on intersection of other two axes and is perpendicular to them. Conjugate planes with opposite senses of slip are shown in each case.

elevated principal stress magnitudes (Rice, 1992). It is also likely that if a fluid phase were present, hydrofracture would keep σ_3' from becoming negative, thereby limiting the ambient stress and promoting lockup.

If we assume the frictional faults are optimally oriented according to Eq. (12) and assume a value for the friction coefficient μ of 0.75 we may use Eq. (11) to relate the magnitudes of the maximum and minimum principal effective stresses. At frictional equilibrium we obtain

$$\sigma_1' = 4.0\,\sigma_3' \tag{13}$$

5.1.2 Andersonian Faulting Types and Strength Versus Depth

In the Earth one of the principal stresses is usually assumed to be vertical. This condition must apply at the surface, where shear stresses vanish, and experimental data from mines and boreholes indicate it is a good generalization to at least 3 km depth (McGarr and Gay, 1978). Then, in the 2D faulting theory of Anderson (1951) the style of faulting depends upon which of the principal effective stresses is vertical. If the maximum effective principal stress σ_1' is vertical, the minimum σ_3' is horizontal and the failure planes are normal faults. When the minimum principal stress σ_3' is vertical, thrust faulting results. If the intermediate principal stress σ_2' is vertical, σ_1' and σ_3' are horizontal and faulting is strike slip.

Given this geometry of the principal stresses and assuming a value for the coefficient of friction, we can determine the orientations of thrust, normal and strike-slip faults that will slip at the lowest differential stress ($\sigma_1'-\sigma_3'$). Assuming $\mu=0.75$, Eq. (12) can be used to obtain $\theta_0=27°$. Thus, optimally oriented thrusts will dip at 27°, normal faults at 63°, and strike-slip faults will be oriented 27° to the direction of maximum compression.

The vertical effective principal stress σ_v' can be estimated independently, permitting the other principal effective stress to be calculated from Eq. (13). σ_v' is simply given by the weight of the overburden reduced by the ambient pore pressure,

$$\sigma_v' = \rho g z - P \tag{14}$$

where ρ is the density of the overburden and z is depth. In a fluid-saturated crust it is often assumed that pores or cracks are interconnected and a part of the rock column is supported by the fluid pressure $P=\rho_w g z$, where $\rho_w = 1000\,\mathrm{kg\,m^{-3}}$ is the fluid density. For $\rho=2700\,\mathrm{kg\,m^{-3}}$, the gradient of σ_v' is $26.5\,\mathrm{MPa\,km^{-1}}$ under dry conditions and $16.7\,\mathrm{MPa\,km^{-1}}$ if pore pressure is hydrostatic.

For each of the main faulting types we can thus use Eqs. (13) and (14) to compute the gradient in resisting shear stress $\tau=0.5\ (\sigma_1'-\sigma_3')\sin 2\theta_0$ on optimally oriented faults under hydrostatic and dry conditions.

Normal:

$$\sigma_1' = \sigma_v' \quad \text{Hydrostatic, dry gradients} = 5.0,\ 8.0\,\mathrm{MPa\,km^{-1}} \tag{15a}$$

Strike slip:

$$\sigma_2' = \sigma_v' \quad \text{Hydrostatic, dry gradients} = 8.0, 12.7\,\text{MPa km}^{-1}$$
$$\text{(strike-slip case assumes } \sigma_v' = 0.5(\sigma_1' + \sigma_3')) \tag{15b}$$

Thrust:

$$\sigma_3' = \sigma_v' \quad \text{Hydrostatic, dry gradients} = 20.0, 31.8\,\text{MPa km}^{-1} \tag{15c}$$

For normal and strike-slip faults extending from the surface to 15 km depth, the average resisting stresses under hydrostatic conditions are thus 38 and 60 MPa. For a thrust fault extending to 30 km, a common earthquake nucleation depth in subduction zones, the corresponding average value is 300 MPa.

Recalling our previous discussion on plate driving and resisting forces, we see that if resisting stresses of this magnitude actually occur on plate boundary faults they will play an important role in determining the force balance of the plates and the intraplate "tectonic" stress field of the lithosphere. However, such high resisting stresses would lead to significant frictional heat generation. Section 6 shows how heat flow measurements sharply constrain the amounts of possible frictionally generated heat along both the San Andreas fault in California and on the Cascadia subduction megathrust offshore of Washington, Oregon, and British Columbia.

5.2 Lower Crust and Lithospheric Mantle Rheology

Above temperatures ranging from 300° to 450°C both laboratory results and field observations show that the rheological behavior of rocks alters drastically. Pressure-sensitive fracture strength and frictional fault slip yield to bulk ductile flow controlled by dislocation creep, a very temperature- and strain rate-sensitive process that is also very dependent on rock type. However, strength in the ductile field is largely independent of both lithostatic pressure and pore pressure. For this deformation mechanism, the strain rate ε and differential stress $(\sigma_1 - \sigma_3)$ are related by an expression of the form

$$\varepsilon = A\,(\sigma_1 - \sigma_3)^n \exp[-Q/RT] \tag{16}$$

or equivalently

$$(\sigma_1 - \sigma_3) = (\varepsilon/A)^{1/n} \exp[Q/nRT] \tag{16a}$$

where A, n and Q are constants that depend on rock type, R is the universal gas constant ($8.316\,\text{J mol}^{-1}\,\text{K}^{-1}$), and T is absolute temperature. For many rock types $n{\sim}3$, implying $(\sigma_1{-}\sigma_3)$ increases by a factor of 2 for each eightfold increase in strain rate. Since temperature normally increases with depth, Eq. (16a) shows that for each rock type $(\sigma_1{-}\sigma_3)$ is greatest at the shallowest depths for which the flow law applies and thereafter decreases exponentially with increasing depth.

The exponential dependence of differential stress on rock type and temperature demonstrates the great influence of these factors on ductile strength. For example, laboratory experiments on ductile flow of dry quartzite, representing quartz-rich crustal rocks, yield $Q = 190\,\text{KJ mol}^{-1}$, whereas similar results for olivine, an upper mantle constituent, give $Q = 520\,\text{KJ mol}^{-1}$. Thus at the same temperatures and strain rates the ductile strength of these two rock types will differ by many orders of magnitude and there may be a significant strength contrast between the lower crust and uppermost mantle in many regions. Similarly, a change of just 100° in temperature of the lower crust or upper mantle will change ductile strength by a factor of 10.

5.3 Composite Quasi-static Strength Profile

If we consider lithospheric rheology to be controlled only by the frictional and ductile processes discussed above, the strength envelope for the crust is then determined by the lowest shear stress at a given depth that satisfies Eq. (9) or Eq. (16a). The depth at which both are satisfied is called the brittle/ductile transition. This image of lithospheric strength is certainly oversimplified, because there is a significant depth range over which a transitional, semibrittle behavior is likely to be important (Kohlstedt *et al.*, 1995) and frictional (i.e., pressure-sensitive) stable sliding may well occur below the seismogenic zone. Recent work suggests that the sharp peak in strength predicted at the brittle–ductile transition (see Fig. 4) may be blunted and modestly decreased by semibrittle processes (e.g., Kohlstedt *et al.*, 1995, Fig. 7, 9). Although these complexities may be important in the earthquake nucleation process, they are likely to be confined to a small fraction of the crustal column and the simple partition into frictional and ductile domains remains useful and intuitively instructive (see also Chapter 29 by Sibson).

Figure 4 shows strength versus depth, plotted as $(\sigma_h{-}\sigma_v)$ for thrust and normal faulting. Dry and hydrostatic pore pressure gradients are shown for each frictional sliding case. Pore pressure is parameterized by λ the ratio of the pore pressure P to the total vertical stress σ_v. Ductile strength curves are shown for both dry quartzite and olivine, a strain rate of $10^{-15}\,\text{sec}^{-1}$ and a geothermal gradient of $15°\,\text{km}^{-1}$ (appropriate for old oceanic lithosphere or stable continental interiors) are assumed.

What general conclusions can be drawn from composite strength profiles like Figure 4? Provided our assumptions that (1) upper crustal rocks are fractured and frictional resistance to slip on optimally oriented faults limits ambient stresses there, and (2) temperature-sensitive creep properties of rocks rich in quartz and olivine determine limiting stresses at greater depths, then lithospheric stresses will lie within the bounds shown in Figure 4. A region of high strength is then expected

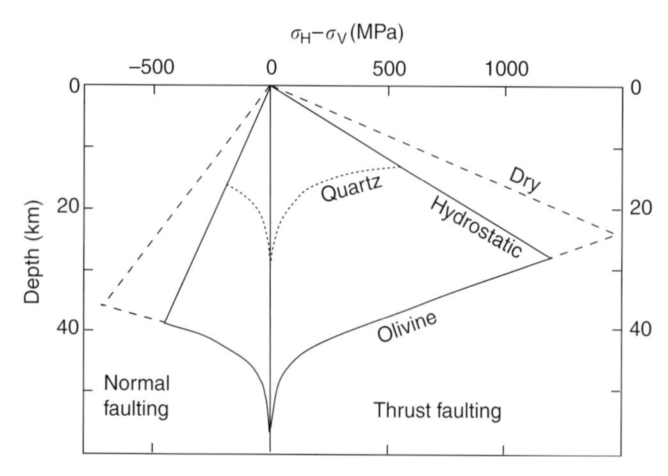

FIGURE 4 Composite strength profile for the continental cratonic lithosphere. Difference between maximum and minimum horizontal and vertical stress versus depth. If $\lambda = $ ratio of pore pressure to vertical (overburden) stress, dry corresponds to $\lambda = 0$ and hydrostatic to $\lambda = 0.37$. Both quartz and olivine rheologies are used for ductile lithosphere and assumed thermal gradient is $15°\text{C km}^{-1}$. A strain rate of $10^{-15}\,\text{s}^{-1}$ is assumed.

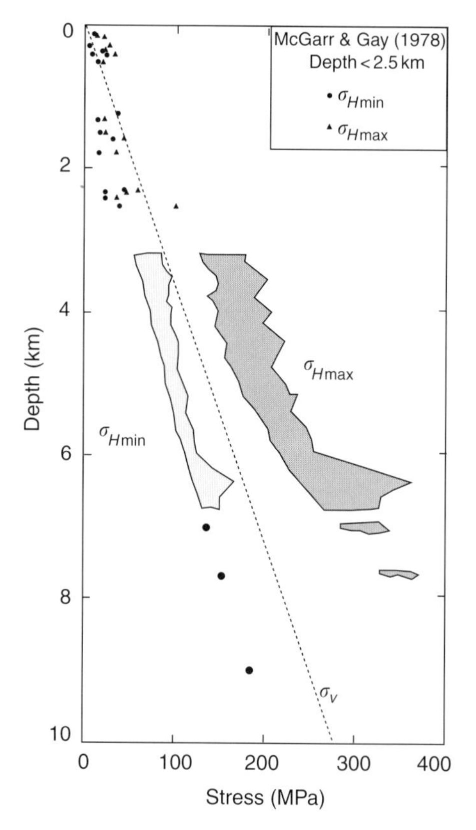

FIGURE 5 Observations of maximum and minimum horizontal *in situ* stress versus depth. Determinations are from mines in southern Africa at depths shallower than 2.5 km (McGarr and Gay, 1978) and from below 3 km in the KTB scientific borehole in southern Germany (Brudy *et al.*, 1997), both in stable continental interiors. The range of possible values from KTB are shown shaded.

in the midcrust, near the transition between the frictional and ductile fields, and possibly also in the uppermost mantle, where a compositional change from quartzo-feldspathic rocks to more basic, olivine-rich compositions is expected. Depending on many factors this region of high strength lies in the depth range 15–40 km, with much lower ambient values above and below this depth interval. Depending on composition, temperature, and to a lesser extent strain rate, it is also possible that the ductile lower crust has a lower strength than either the overlying crust or the underlying upper mantle. For example, the strength curve for quartz (QTZ) in Figure 4 shows a sharp decrease in the 15–20 km depth range that is caused by the normal increase in temperature with depth and the exponential dependence of differential stress on temperature in Eq. (16a). Therefore, in some regions ductile flow may be concentrated in this low strength layer. However, it should be noted that quartz content generally diminishes in the lower continental crust.

5.4 Observational Constraints on Quasi-static Strength Profile

Direct measurements of stress in active fault zones at seismogenic depths have not yet been made. For constraints we must presently rely largely on measurements in plate interiors and on indirect estimates based on the effects of fault zone stress. *In situ* horizontal stresses measured in deep level mines to depths of 3 km (McGarr and Gay, 1978) and from the KTB deep scientific borehole drilled to 8 km depth in Germany (Brudy *et al.*, 1997) are plotted versus depth in Figure 5. The results from the KTB hole in particular support the expectations based on the frictional strength profiles, and the gradient of horizontal differential stress of $14\text{–}19\,\text{MPa km}^{-1}$ (not shown) lies between the strike slip and thrust gradients for hydrostatic pore pressure.

The results of Figure 5 support the "rock mechanics" strength model for frictional slip in the regions sampled, but several indirect observations suggest major faults are considerably weaker. We discuss heat flow constraints in detail in the next section but first consider an independent measure of relative fault strength.

This indicator of fault strength, first suggested by Mount and Suppe (1987) and Zoback *et al.* (1987), is the orientation of maximum horizontal compressive stress direction inferred from earthquake fault plane solutions and borehole elongations in the blocks adjacent to major fault zones. As shown in Figure 3a and Eq. (12), the angle, θ, between fault strike and the σ_1 axis should be about 27° for an expected fault friction coefficient $\mu \sim 0.75$. However, near both major strike-slip faults and subduction zones $\theta \sim 50°\text{–}90°$, suggesting that the shear strength of these faults is much less than that of the adjacent crust (if the fault were a free surface, supporting no shear stress, two principal stresses would lie in its plane and the third would be perpendicular to the fault). Detailed profiles of θ versus distance from the San Andreas fault recently obtained by Hardebeck and Hauksson (1999) from fault plane solutions show a minimum of $\theta \sim 45°–70°$ at the fault and generally support previous observations from strike slip faults and subduction zones. However, the results of Hardebeck and Hauksson also reveal considerable along-strike and strike-normal variability in southern California.

In a related observation, Thatcher and Hill (1991, 1995) have shown that M 5.8–7.3 normal faulting earthquakes have dips that are strongly clustered near 45° (see Chapter 29 by Sibson, Fig. 6). Anderson's faulting theory and Eq. (12) suggest values should be centered near 60°, and the lower value suggests low friction coefficient μ or high pore pressure (or both) in these normal fault zones.

5.4.1 Assessment

The results in Figure 5 indicate that for inactive plate interiors the rock mechanics strength profile applies, and the brittle upper crust supports substantial shear stress. However, horizontal compressive stress orientations show that major fault zones support considerably lower stress, and the clustering of normal fault dips near 45° suggests these faults may also be weak.

6. Heat Flow and Frictional Stresses Resisting Fault Motions at Major Plate Boundaries

As we discussed in Section 4, stresses that resist fault motions generate heat, and measurements of surface heat flux near active faults can thus be used to infer bounds on the magnitudes of these resisting stresses. Using Eq. (2) the rate of work Q done against resisting stress τ_r for average fault slip rate v is $\tau_r v$ per unit fault area, or

$$Q = \tau_r v \tag{17}$$

No significant heat sinks are known (see Lachenbruch and Sass, 1980, p. 6186 for one justification), so the thermal effects of dissipative heating could be substantial. For example if $\tau_r = 100$ MPa and $v = 30\ \mathrm{mm\,y^{-1}}$, then $Q = 96\ \mathrm{m\,Wm^{-2}}$. We will discuss below how such a thermal flux at depth is conducted to the Earth's surface, but clearly sources of this magnitude would contribute significantly to observed heat flow and might violate observational constraints. For example, typical surface heat flow values above subduction thrusts average only about $40\ \mathrm{m\,Wm^{-2}}$ (Hyndman and Wang, 1993), and in a 50–100 km wide region near the San Andreas fault are only $30–50\ \mathrm{m\,Wm^{-2}}$ above background levels (Lachenbruch and Sass, 1980).

6.1 Resisting Stresses on the San Andreas Fault

The vertical strike-slip geometry of the San Andreas fault makes it particularly straightforward to detect any surface heat flux due to dissipative heating at depth on the fault. About 100 heat flux measurements have been made by the US Geological Survey in the vicinity of the fault, and they provide a uniquely detailed picture of the thermal regime near a major active fault. The data possess considerable scatter. Nonetheless, they show a zone of high heat flux averaging $\sim 80\ \mathrm{m\,Wm^{-2}}$, called the Coast Range Anomaly (CRA) by Lachenbruch and Sass (1980). It is spread over a region 50–100 km wide spanning the San Andreas fault zone, and heat flux is $40–50\ \mathrm{m\,Wm^{-2}}$ above background values to the east.

An important feature of the data is that it applies to all of the San Andreas fault system from its inception at the Mendocino triple junction on the north to the "big bend" of the fault 700 km to the south. It thus includes both the seismogenic, predominantly locked segments of the northern San Andreas fault system, which has several active strands, and the central ~160-km long creeping segment of the San Andreas, where nearly all of the motion currently occurs as aseismic slip on a single fault. In neither region is there any obvious local anomaly of the kind expected from dissipative heating on the upper crustal portions of the fault.

To understand the form of this expected thermal anomaly, consider the simple 2D model of dissipative heat generation on the fault shown in Figure 6a. For resisting stress τ_r and slip velocity v acting over depth range d, the rate of heat generation q per unit length of fault is given by

$$q = \tau_r v d \tag{18}$$

Provided the depth of this source, a, is large compared to its width d, following Brune *et al.* (1969) we may write the surface heat flux $Q_0(x, t)$ due to conductive heat transfer as a function of distance x from the fault and time since initiation of slip,

$$Q_o(x, t) = q/\pi[a/(a^2 + x^2)]\exp[-(a^2 + x^2)/4\kappa t] \tag{19}$$

where κ is the thermal diffusivity ($10^{-6}\ \mathrm{m^2 s^{-1}}$).

Assume for the moment that the resisting stress increases roughly linearly with depth, as suggested by Figures 4 and 5, and is greatest near earthquake nucleation depths, about 10 km on the San Andreas fault. The time-dependent exponential term in Eq. (19) will then be negligible after a few million years of slip. The steady-state surface heat flux will have a maximum value $q/\pi a$ at the fault, decrease to half this value at $x = a$, and thereafter decrease rapidly with increasing distance. Therefore if we take $v = 30\ \mathrm{mm\,y^{-1}}$ and $\tau_r = 80$ MPa acting over a depth range of 5 km centered at depth $a = 10$ km, we easily find that the maximum surface heat flow anomaly (i.e., excess over background) is $40\ \mathrm{m\,Wm^{-2}}$ and it decreases

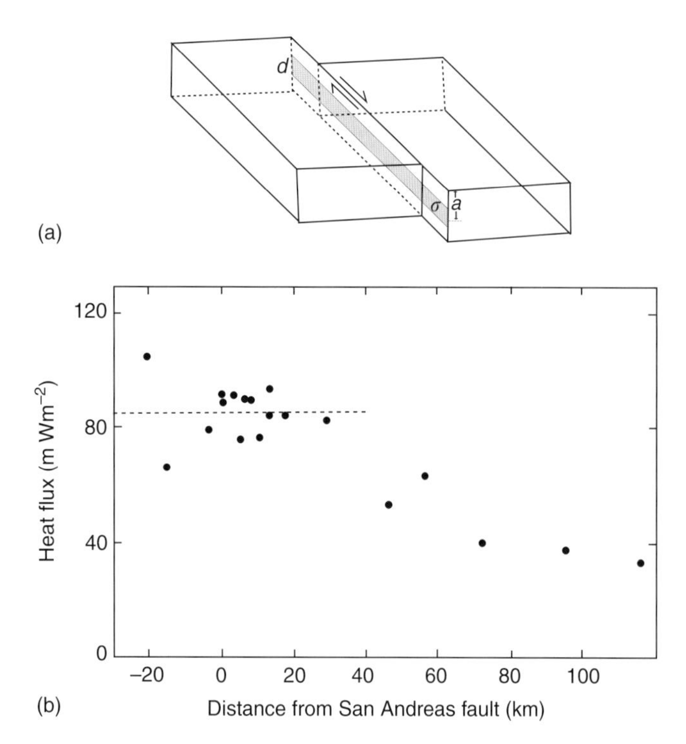

FIGURE 6 (a) Model geometry for heat generation on a vertical strike-slip fault. Resisting stress σ acts on a fault strip of width d centered at depth a below the ground surface. (b) Heat flux versus distance from central San Andreas fault (data from Region 3 of Lachenbruch and Sass, 1980). Mean heat flux within about 40 km of the fault, 86 m Wm^{-2}, is shown by dashed line. This "Coast Range Anomaly" (CRA) lies 40–50 m Wm^{-2} above background values ~100 km east of the San Andreas fault.

rapidly (with scale length a) away from the fault. No such local heat flow anomaly is observed in the data (Fig. 6b). Indeed, individual heat flow profiles across the fault, like that in Figure 6b, fail to show any anomaly larger than the noise in the data, 5–10 m Wm^{-2}, so an upper bound on frictional resistance of about 10 MPa is indicated.

Assuming uniform or linearly increasing distributions of resisting stress with depth (Brune *et al.*, 1969; Lachenbruch and Sass, 1980) makes the analysis more precise. However, unless the dissipative heating is all concentrated below ~10 km depth, the resolvable features of the computed surface heat flow profile do not change and the main conclusion is unaltered. Average resisting stress on the seismogenic portion of the San Andreas fault is thus no greater than about 10 MPa, about a factor of five less than suggested by the extrapolation of laboratory-based estimates of frictional fault strength (Fig. 4).

If frictionally generated heat were concentrated at depths below 15 km, the anomaly at the surface would be considerably broadened, and less obvious. A possibility (Thatcher and England, 1998) is that the CRA (coast range anomally), the regionally high heat flux in Figure 6b, is caused by distributed shear heating due to ductile flow beneath the depths where frictional processes dominate. As the strength profiles of Figure 4 suggest, and Thatcher and England (1998) show, ductile shear heating is concentrated near the ductile strength maximum at the brittle–ductile transition. The resulting heat source thus behaves like a line source concentrated at this depth and its effects can be approximated well by Eq. (19). The surface heat flux distribution predicted by this model is relatively broad and decreases away from the fault with a scale length equal to the depth of the brittle–ductile transition. The ductile shear zone model can generally account for the main features of the CRA (see Thatcher and England, 1998, Figs. 14, 15, 16).

Another proposed explanation of the CRA is that it results from advective transfer of frictionally generated heat by fluid flow away from the fault (Hanks, 1977; Williams and Narasimhan, 1989; Scholz, 2000). In this hypothesis, resisting stresses on the fault could be as high as given in Eq. (15b), with the CRA being smeared out by convective flow of groundwater away from the fault.

6.2 Resisting Stresses on Subduction Thrusts

At subduction zones the geometry of faulting and several potential sources and sinks of heat influence the surface heat flux and affect our ability to distinguish effects of dissipative heating on the fault. These features have been considered in detail with sophisticated numerical models of heat conduction, but the essential features can be intuitively understood and resisting stresses bounded by a simple analytic model of the thermal regime presented by Molnar and England (1990).

Recall from frictional faulting theory that resisting stresses on thrusts could be *very* high (the shear stress gradient for hydrostatic pore pressure is ~20 MPa km^{-1}, implying 600 MPa at 30 km depth!). If shear stress on subduction thrusts was this high it would strongly resist plate motions. Furthermore, the thermal effects of such dissipative heating would be enormous. For example, if $v = 40$ mm y^{-1} and resisting stress was even just half the value suggested by Andersonian faulting theory, then $Q \sim 400$ m Wm^{-2} of fault surface! We will discuss below how such a thermal flux at depth would be conducted to the Earth's surface, but clearly sources of this magnitude would significantly violate observed heat flow constraints. (Note that the heating required to generate arc magmas occurs 100–200 km farther landward and at depths of ~80 km in the slab.)

The geometry of the model is shown in Figure 7a. A plate with dip δ is subducting at velocity v. Its motion is resisted by shear stress τ_r, which may be a function of the depth to the fault plane Z_f. Three sources of heat are important in determining the temperature above the slab and conduction to the Earth's surface: (1) resistive shear heating, as given by Eq. (17), acts on the subduction thrust; (2) the heat Q_o from the descending slab; and (3) heat generated by radiogenic decay in

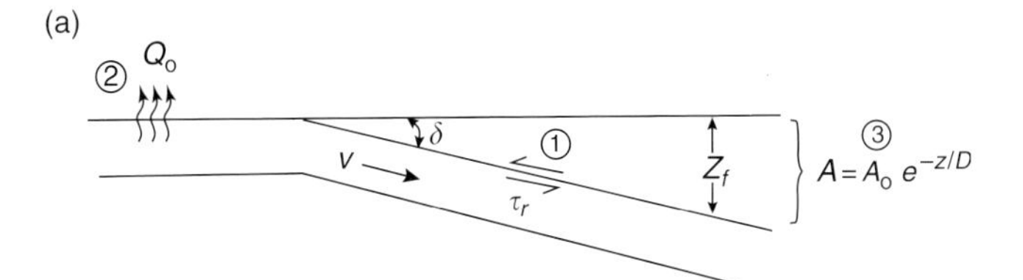

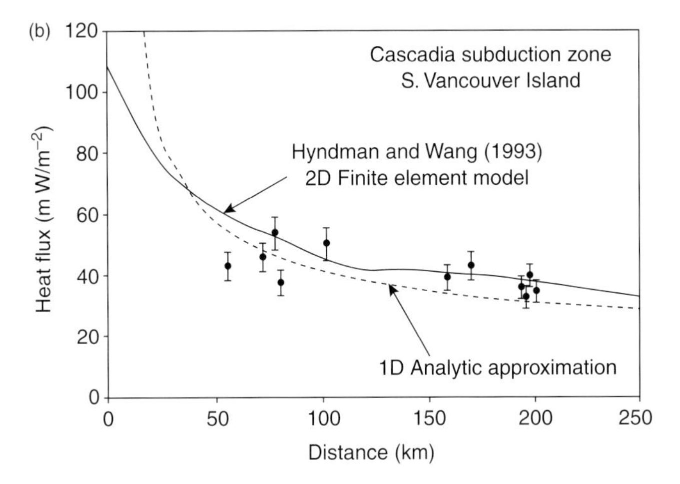

FIGURE 7 (a) Geometry and heat sources of subduction thrust fault model. 1, Shear heating on interplate thrust; 2, heat from subducting oceanic slab; 3, heat from upper plate radiogenic decay. See text for definition of symbols. (b) Cascadia heat flux versus distance from trench axis compared with 1D model discussed in text and with 2D heat finite element model of Hyndman and Wang (1993).

the overlying plate. The advective downward transport of heat by the descending slab must be accounted for in deriving the effect of these three heat sources on the surface heat flux. However, Molnar and England have shown that for gentle slab dips ($\delta < 30°$) it is possible to correct for this advection, ignore the effect of lateral heat conduction in the overlying plate, and obtain 1D analytic expressions for the temperature field. The heat flux at any point on the surface a distance Z_f above the slab is then given by

$$k\frac{dT}{dz} = \frac{\tau_r \nu}{S} + \frac{Q_o}{S} + A_o D[1 - e^{-Z_f/D}] \tag{20}$$

where the advective correction term, S, is

$$S = 1 + \sqrt{\frac{Z_f \nu \sin\delta}{\kappa}} \tag{21}$$

and κ is the thermal diffusivity. The last term in Eq. (20) represents the contribution to surface heat flux of radioactive heat production A_o that decreases exponentially with depth on a scale length D.

Using the Cascadia subduction zone as an example we can show that any contribution to surface heat flux from resistive shear heating must be quite small. The following parameters are taken as representative for Cascadia (Hyndman and Wang, 1993):

Subduction velocity, $\nu = 45\,\text{mm y}^{-1}$
Fault dip, $\delta = 15°$
Heat flux, Oceanic flab, $Q_o = 120\,\text{m Wm}^{-2}$
Crustal radioactive concentration, $A_o = 0.6\,\mu\,\text{Wm}^{-3}$
Decay depth, Crustal radioactivity, D = 10 km
Thermal diffusivity, $\kappa = 10^{-6}\,\text{m}^2\,\text{s}^{-1}$

Hyndman and Wang (1993) show that the surface heat flux at a distance $Z_f = 15$ km above the subducting Juan de Fuca slab is about $50\,\text{m Wm}^{-2}$. For the parameters listed above the advective correction term $S = 3.3$ and the surface flux contribution from the slab is $36\,\text{m Wm}^{-2}$ and from crustal radioactivity is $6\,\text{m Wm}^{-2}$. Thus the last two terms on the right-hand side of Eq. (20) account for most of the observed heat flux, leaving no more than about $8\,\text{m Wm}^{-2}$ to be allocated to shear strain heating. Clearly resisting stresses (either quasi-static or dynamic) on the Cascadia megathrust must be quite small. Figure 7b shows a sample comparison between model and data from southern Vancouver Island. For illustrative purposes we have taken a modest resisting stress gradient $\tau_r = 0.05\,\rho\, g\, Z_f$ MPa, but this contributes no more than $10\,\text{m Wm}^{-2}$ to the surface heat flux. This figure shows that resisting stresses on the Cascadia megathrust are quite low and, as is the case for the San Andreas fault, any thermal contribution is not distinguishable above the noise levels of the heat flow data.

Similar results have been obtained in other subduction zones. Hyndman *et al.* (1995) use the extensive suite of heat-flow data from the Nankai Trough subduction zone of southwest Japan to demonstrate resisting stresses there are as low as in Cascadia. A synoptic review by Hyndman and Wang (1993) of heat-flow data from other circum-Pacific subduction zones suggests the same conclusion (see also Tichelaar and Ruff, 1993). In particular, in Northern Honshu, landward of a segment of the Japan Trench that is subducting predominantly aseismically (Pacheco *et al.*, 1993), no anomalous heat generation attributable to frictional resistance was observed.

6.3 Assessment

We suggest that the heat flux measurements from the San Andreas transform and the Cascadia subduction zone provide strong evidence that stresses resisting slip on major faults are relatively low, on average ~20 MPa or less on subduction thrusts and ~10 MPa or less on the San Andreas. Since this estimate is an average over the seismogenic fault plane, values at any point could be considerably higher (or lower) than this mean. In particular, these resisting stresses could locally be higher on the deeper portions of the fault, at the depths where most large and great earthquakes nucleate.

Despite suggestions that advective transport of frictionally generated heat away from active faults is a quantitatively important process, the preponderance of evidence supports a conductive model of heat transfer near major faults. On the San Andreas fault the integrated heat energy represented by the CRA is about what would be generated by frictional heating on a fault stressed according to Eq. (15b). With a viable mechanism for distributing this locally generated heat across the California Coast Ranges, the CRA might be explained. Williams and Narasimhan (1989) suggest fluid flow induced by the hydraulic gradients due to high topography along parts of the San Andreas, but this mechanism would seem to apply only to restricted segments of the fault. Lachenbruch and Sass (1980, pp. 6196–6198) give qualitative arguments against vigorous hydrothermal circulation near the San Andreas, noting that the integrated output of thermal springs along the fault is orders of magnitude too small to be due to frictional heating from a fault obeying Eq. (15b). However, they do concede that heat transport at low flow rates and moderate temperatures could be significant but very difficult to detect. There is no regionally high heat flow above the upper ~40 km of subduction thrusts (see Fig. 7b). Thus, regardless of heat transfer mechanism, it seems difficult to argue for the frictional heat generation implied by Eq. (15c).

In several regions, the heat flow constraint on frictional resistance applies to major faults that currently slip aseismically. Although it is not possible to be certain that the current aseismic behavior is typical of long-term patterns, the similar behavior of seismogenic and creeping faults suggests the mechanism responsible for weakening these faults may operate for both quasi-static and dynamic slip.

The role of heat generation by ductile shearing is uncertain. Although this mechanism can account for the heat-flow patterns observed in California, the slab window model of Dickinson and Snyder (1979) is at least as successful. In this model, the Juan de Fuca slab north of the Mendocino triple junction is subducting at a shallow angle and the process of triple junction migration thus exposes hot upper mantle at the base of the crust adjacent to the newly created San Andreas fault. Lachenbruch and Sass (1980) showed that this model explains the heat flow data and also the age progression of young volcanics southeast of the Mendocino triple junction. In addition, it matches the observation that the CRA is not symmetrically centered on the San Andreas, as required by models (like ductile shearing) that localize heat sources on a single fault or its downward continuations. However, the extent and history of exposure of the slab window is uncertain, and this heating mechanism may not be generally applicable throughout California.

Ductile shearing has not yet been modeled in subduction settings. However, reference to the 1D model (Yuen *et al.*, 1978) and relevant heat flux data (Fig. 8) suggest the existence of ductile shear heating below ~40 km depth would be difficult to prove (or disprove) from surface measurements.

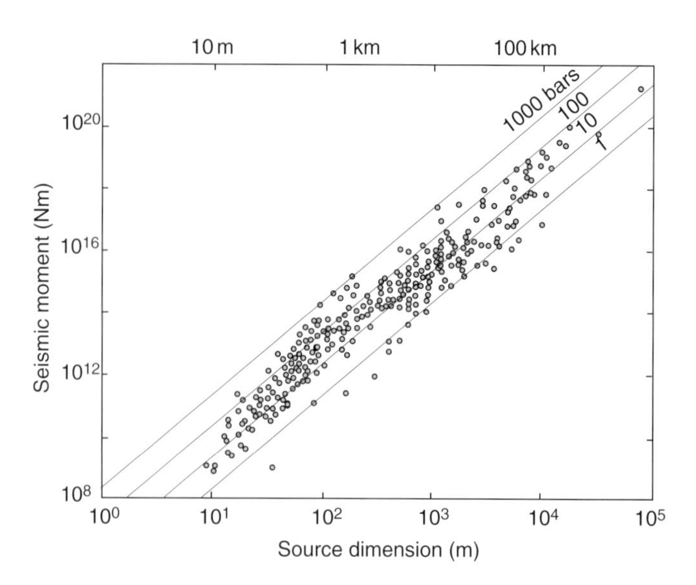

FIGURE 8 Stress drop versus seismic moment obtained from seismic data (modified from Kanamori and Heaton, 2000).

7. Static Stress Drop in Earthquakes

As mentioned in Section 3, observations of the average static stress change due to earthquake faulting supply one of the important stress measures needed to understand fault energetics. Geodetic and seismologic methods can be used to estimate this parameter and its variations with position on the earthquake rupture. The average shear stress drop is given by

$$\Delta\sigma = Gu/L \tag{22}$$

where G is an elastic rigidity modulus, u is fault slip and L is a fault dimension. For small and moderate magnitude earthquakes, $\Delta\sigma$ is measured using the amplitude spectra of seismic body waves (Brune, 1970). For larger events, $\Delta\sigma$ is obtained using independent estimates of earthquake rupture dimensions from aftershock zone size and seismic moment obtained from long period seismograms. Geodetic measurements can be used to measure the strain change (essentially u/L) near the fault rupture, which is typically about 10^{-4}. Multiplied by an elastic modulus typical for the Earth's upper crust (~30 GPa), Eq. (22) yields $\Delta\sigma$ ~3 MPa.

Figure 8 shows average stress drop $\Delta\sigma$ over a wide range of earthquake moment M_o and equivalent magnitude, M_W. This compilation shows that $\Delta\sigma$ ranges from about 0.1 to 10 MPa and is essentially independent of magnitude (the rather lower values at smaller magnitudes may indicate a bias related to attenuation of high frequency seismic waves; see Hanks, 1982). Figure 8 also shows that some values as high as 100 MPa are occasionally observed.

The stress drop values given in Figure 8 are averaged over the entire earthquake rupture. However, both seismological and geodetic methods can be used to infer the spatial

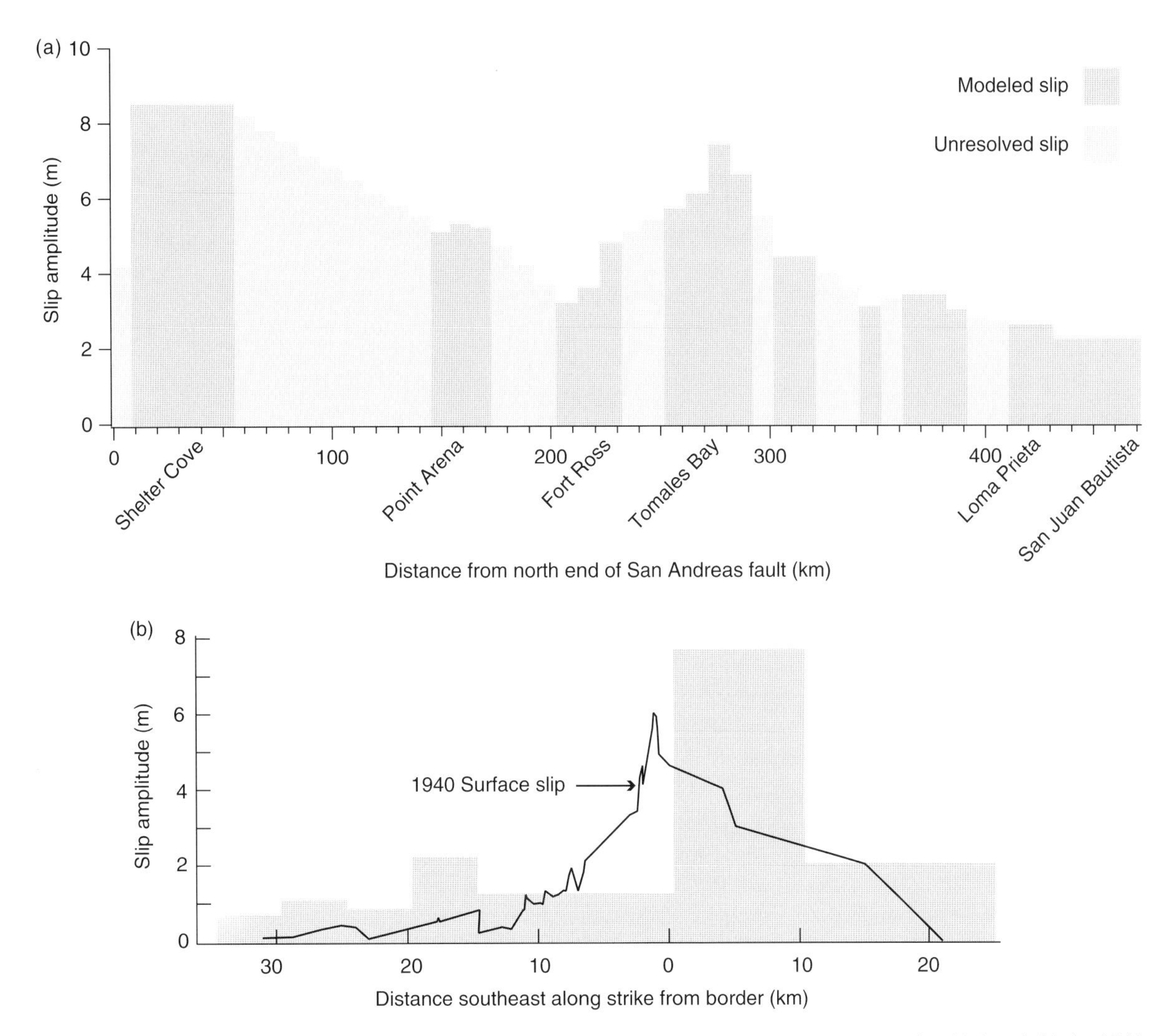

FIGURE 9 Slip distributions in (a) the 1906 $M = 7.8$ San Francisco earthquake (Thatcher *et al.*, 1997) and (b) the 1940 $M = 7.1$ Imperial Valley, California earthquake (King and Thatcher, 1998).

distribution of seismic slip and hence map static stress drop variations on the earthquake fault plane. Two examples, for the great $M = 7.8$ 1906 San Andreas earthquake and the 1940 $M = 7.1$ Imperial Valley earthquake, both strike-slip events with extensive surface faulting and measurements of surface fault offset, are plotted in Figure 9. Inevitable smoothing and nonuniqueness of the fine scale features of the derived slip distributions preclude estimation of the largest static stress changes on the fault plane. However, these results, as well as seismological determinations of slip distribution (e.g., Archuleta, 1984; Cohee and Beroza, 1994; Wald and Heaton, 1994) indicate considerable heterogeneity in stress drop on earthquake rupture planes.

7.1 Assessment

The results summarized in Figure 8 reliably bound average stress drop, which has a mean value of ~3 MPa and ranges from 0.1 to 10 MPa. However, the along-strike and depth variations of slip indicated by slip mappings like Figure 9 show that stress release is quite heterogeneous, with stress drop varying by at least a factor of 5 on the fault.

8. Dynamics of Earthquake Faulting

The occurrence of earthquakes is the result of sudden release of elastic stress stored in blocks adjacent to the fault, and while slip is in progress inertial forces are very large, probably considerably larger than the quasi-static surface forces (stresses) which initially led to rupture nucleation and fault failure. Thus we anticipate that understanding the dynamics of earthquake slip may be crucial to understanding fault energetics, and inferences drawn from quasi-static analyses and observations may be a poor guide to dynamic behavior.

Ideally, we would like to have a firm theoretical and observational accounting of the stress-slip time history at every point on the earthquake fault rupture. Our current knowledge base falls far short of this objective. Instead of comprehensive theoretical understanding we have idealized kinematic and dynamical models and experiments with analogue materials. Rather than spatially complete, broadband seismic records near the largest earthquakes that occur on major faults we have local, usually bandlimited recordings of small and moderate events and distant seismograms from great earthquakes. Here we summarize these data and the inferences based on them and assess what they reveal about dynamic rupture processes.

8.1 Near-field Seismologic Observations

Near-field seismic records (those obtained within a few fault dimensions) from earthquakes as large as $M \sim 7$ permit estimates of ground velocities and accelerations which can be related to local estimates of slip duration and stress changes occurring during slip on the fault. Velocities of $2\,\mathrm{m\,sec}^{-1}$ and accelerations of $1g$ have been observed. If we define dynamic stress drop as the difference between initial stress and average stress during fault slip, such velocities and accelerations imply dynamic stress drops of ~20 MPa. It should be emphasized that these observations are relatively few, may be affected by focusing of energy in the direction of rupture propagation ("directivity"), source–receiver path effects, and local recording site conditions. The peak values obtained may well apply to only restricted parts of the fault rupture and may be biased by sampling relatively more thrust faulting earthquakes.

However, the occurrence of the 1999 $M_W = 7.6$ Chi-Chi (Taiwan) earthquake supplies a large new data set that will be very important in refining our views of earthquake dynamics. This event produced surface rupture over a 125 km-long fault with up to 8 m of reverse dip-slip motion. It was recorded by over 500 free-field strong-motion accelerographs, many of them located in the near field. Preliminary analysis of these records (Mori and Ma, 1999) shows ground velocities of $1\text{–}3\,\mathrm{m\,sec}^{-1}$ near the fault, with clearly larger motions on the upper or hanging wall block of the fault (Ni *et al.*, 1999). These observations reinforce earlier field evidence from thrust faults showing motions larger than $1g$ and $1\,\mathrm{m\,sec}^{-1}$ on the shallow part of the hanging wall, with associated evidence of relatively low accelerations and velocities on the adjacent foot-wall (Allen *et al.*, 1998).

8.2 Estimates of Radiated Energy and Apparent Stress

Although observations of near-field waveforms are rare, more observations exist of total seismically radiated energy (E_s). Using measures of E_s along with seismic determinations of moment (M_o), the apparent stress τ_a is given by

$$\tau_a = G E_s / M_o \tag{23}$$

In the past, accurate estimation of E_s and τ_a has been hampered by the limited frequency bandwidth of seismographs and the unknown effects of high frequency attenuation and scattering. During the past 15 years the first problem has been solved by the deployment of broadband digital seismographic networks both locally and worldwide. The second problem, unknown attenuation effects, may remain and appear as a high frequency saturation of corner frequency and marked dependence of stress drop on seismic moment, M_o (Hanks, 1982). However, this effect is expected to become less important for E_s estimates as earthquake size increases. This follows because the major contribution to radiated energy occurs at frequencies near the spectral corner frequency (Hanks and Wyss, 1972), which is inversely proportional to source dimension and so generally decreases with increasing magnitude. Above $M \sim 7$ or $M_o \sim 10^{19}$ Nm this corner frequency is 0.1 Hz or less and uncertainties due to wave attenuation may be less important.

This contention is supported by the compilation of τ_a versus M_o shown in Color Plate 13. The corner frequency saturation effect at small M_o (Hanks, 1982) would lead to underestimation of the true E_s (and hence an underestimate of τ_a) that becomes smaller with increasing magnitude or moment. Such a bias would cause an apparent increase of τ_a with increasing M_o. This effect is clearly seen at the smaller moments for several of the data sets shown in Color Plate 13, but disappears above $M_o \sim 10^{18}$ Nm. The absence of any dependence of τ_a on M_o is very clear in the teleseismic data of Choy and Boatwright (1995), which show apparent stresses spanning their full range over moments varying from 10^{17} to $10^{20.5}$ Nm.

There is one observational bias that deserves mention. Some existing data suggest teleseismic estimates of radiated energy are significantly less than those based on limited near-field data, especially for thrust faults (e.g., Singh and Ordaz, 1994). More recently, analysis of the uniquely complete data for the 1999 Chi-Chi (Taiwan) earthquake by Ni *et al.* (1999) has confirmed this. They showed that teleseismic estimates E_s and τ_a determined by the method of Choy and Boatwright (1995) yielded an apparent stress value of 0.5 MPa while the near-field value was about 5.0 MPa. Because the near-field data from the 1999 earthquake records are so singular it is not possible to assess how general this bias may be. That the event occurred on a shallowly dipping thrust fault may be significant and is discussed further below. However, we do note that the apparent stress value for the 1999 earthquake lies within the range of values shown for large earthquakes in Color Plate 13. For these events τ_a is small, averaging ~0.5 MPa and ranging from 0.03 to 6.7 MPa.

8.3 Theoretical, Numerical and Analogue Dynamical Models

Dynamical simulations of earthquake rupture are considerably more complex and computationally demanding than kinematic modeling, and much yet remains to be learnt. Elastic dislocation theory is most commonly applied to modeling of earthquake waveforms. However, this approach is essentially kinematic and requires specification either of idealized point sources and rise times or adjustable rupture velocities and distributions of slip-time functions on the rupture surface (e.g., Archuleta, 1984; Cohee and Beroza, 1994; Wald and Heaton, 1994). Fully dynamical models are only now beginning to be applied to earthquake rupture (e.g., Boatwright and Cocco, 1996) and most commonly utilize slip-weakening rules developed for fault slip in rock mechanics laboratories (Dieterich, 1979; Ruina, 1983). Fully 3D models that treat both vertical and inclined faults and account for fault plane heterogeneity are just now being developed and applied to match observed strong motion data.

One simple theoretical result supported by laboratory analogue modeling deserves special note. Both theory and lab results suggest that dynamic fault slip is accompanied by fault opening at the rupture front. Comninou and Dundurs (1977) derived a steady-state rupture model with a fault-opening mode. Freund (1978) criticized this model, showing that the assumed singularities at the front and back of the rupture were physically unrealistic. However, more recently Adams (1999) has argued that the fault-opening mode can occur for different types of assumed singularities. In a related model, Andrews and Ben-Zion (1997) and Ben-Zion and Andrews (1998) have shown such fault opening is a general feature of rupture at the contact of two elastic half-spaces with differing elastic moduli.

Fault opening is commonly observed in laboratory physical analogue experiments. Anooshehpoor and Brune (1994) observed a steady-state propagating rupture with fault opening in a foam rubber physical model with a rough fault between both dissimilar and identical media. Anooshehpoor and Brune (1994) showed that this mode was associated with a strong reduction in the amount of frictional heat generation on the fault. The opening mode between identical media was evidently associated with asperity interactions that fed energy into the opening mode. Fault opening or nearly complete reduction in fault normal stress has also been observed in plastic models of slip between identical media (Brown, 1998; Bodin *et al.*, 1998; Bouissou *et al.*, 1998; Uenishi *et al.*, 1999).

Numerical models have also shown the existence of a fault-opening mode under some conditions. Mora and Place (1994) showed the existence of such a mode in a lattice numerical model with a rough interface. This mode disappears when the roughness is decreased to zero (Mora and Place, 1998; Shi *et al.*, 1998). Mora and Place (1999) have shown that introducing particles on the fault which are allowed to rotate dynamically might also reduce fault friction, in part by causing local fault opening.

For shallow thrusts, fault separation may be greatly enhanced. Brune (1996) observed strong fault opening in a foam rubber model of thrust faulting. The opening increased as the rupture approached the surface, resulting in a spectacular flip of the hanging wall tip of the fault at the surface as it detached from the foot-wall. The ground motions on the foot-wall were much less than those on the hanging wall ($\sim$1/5), resulting in much less energy radiating downward (corresponding to teleseismic radiation in the Earth) compared to the energy trapped in the hanging wall (corresponding to energy typically recorded on near-field accelerograms). Shi *et al.* (1998) reproduced many of the features observed in the foam rubber model using a dynamic lattice numerical model. This asymmetry of ground motion between hanging-wall and foot-wall blocks is supported by seismic observations described above (Allen *et al.*, 1998) and could explain the difference between local and teleseismic energy estimates for the 1999 Chi-Chi (Taiwan) earthquake (Ni *et al.*, 1999).

8.4 Assessment

The dynamics of faulting are the least well understood aspect of fault mechanics. However, seismological observations are suggestive that *on average* the apparent stress, which is a measure of the dynamic stress during slip, is relatively small, averaging $\sim$3 MPa, about the same magnitude as the average static stress drop. This estimate is very sensitive to the accuracy of seismic estimates of radiated energy, which are uncertain by at least a factor of 3 (H. Kanamori, personal communication, 1998).

Collectively, the studies cited above suggest that one of the possible aspects of rupture dynamics, dynamic fault opening or normal stress reduction, may play an important role in earthquake rupture and explain some of the more puzzling aspects of faulting.

9. Summary and Discussion

Although uncertainties and caveats leave room for doubts, the picture that emerges from our review is that plate interiors are strong, supporting shear stresses consistent with the frictional strength of rocks determined in the laboratory (see Townend and Zoback, 2000), whereas major fault zones along plate boundaries are much weaker, for reasons that are not yet well understood.

We may usefully summarize the average stress state near faults by recalling the analysis from Section 4 of stress and energetics at a single point on an idealized fault surface with uniform properties. We showed that the initial, pre-slip shear stress τ_i is

$$\tau_i = \tau_a + \tfrac{1}{2}\Delta\sigma + \tau_r \qquad (8)$$

where τ_a is apparent stress, σ is stress drop and τ_r is stress resisting fault slip. The heat flow constraint (Section 6) gives $\tau_r < 20$ MPa; seismic and geodetic estimates (Section 7) yield $\Delta\sigma = 0.1$–10 MPa. These estimates argue that average values of τ_r and $(\tau_i - \tau_f)$ are small but [see Eq. (8)] leave open the possibility that initial stress τ_i could be large if τ_a were large. This might occur, if for example, the inertia of the accelerating fault blocks causes the resisting stress τ_r to overshoot the final stress τ_f, as sketched in Figure 2. However, such a process would lead to radiation of anomalous seismic energy (E_s) and be reflected in high values of apparent stress τ_a ($\tau_a = E_s/M_o$). Estimates of radiated energy from seismograms (Section 8) give $\tau_a = 1$–10 MPa. These values then yield, using Eq. (8), an upper bound for initial stress τ_i of 20–40 MPa. This is consistent with an earlier upper bound estimate of 20 MPa obtained by Lachenbruch and Sass (1980) for the San Andreas fault.

This upper bound on initial stress averaged over the earthquake rupture plane has important implications for the strength of faults. At subduction thrusts this upper bound is much lower than the fault strength inferred from frictional slip experiments in the laboratory [Eq. (15c)]. It is marginally less than the same strength estimate for strike-slip faulting [Eq. (15b)]. At subduction zones, τ_i could only be as large as suggested by Eq. (15b) (~300 MPa average in the seismogenic upper crust) if τ_a were two orders of magnitude larger than observed. Therefore, the main weakening mechanism relative to the laboratory estimates, at least in the deeper parts of subduction zones, must be quasi-static. Constraints are poorer on the shallower parts of thrust faults and on strike-slip faults, leaving open the possibility that dynamic weakening could be more important there.

9.1 Weakening Mechanisms

The reasons for fault weakness has been debated for over 40 years, dating from the work of Hubbert and Rubey (1959) on the mechanics of slip on low-angle overthrust faults. Though many mechanisms have since been suggested, no single one has yet been universally accepted. Here we mention only some of the candidates we judge to be strong possibilities. Interested readers are referred to additional mechanisms mentioned briefly below and listed in the references section at the end of this chapter.

As shown by Eq. (9), very high fluid pore pressure in fault zones can, formally at least, overcome the effects of overburden pressure and permit slip at arbitrarily low shear stresses. This mechanism was originally suggested by Hubbert and Rubey (1959) to explain how large thrust sheets could be transported tens of kilometers when the weight of the overlying rock imposed frictional resisting stresses of several hundred megapascals or more on the fault. Through increasing elaboration over 40 years, the Hubbert–Rubey mechanism remains one of the leading candidates for reducing fault strength. As applied to a strike-slip setting, Lachenbruch and Sass (1980) pointed out a possible limitation of the Hubbert–Rubey model, that near-lithostatic fluid pressure would produce pervasive hydrofracture normal to the least principal stress in the blocks adjacent to the fault. Rice (1992) pointed out that this difficulty could be overcome if very high fluid pressures were confined to the fault zone only, producing a fault that slips under very low shear stresses bounded by much stronger surrounding blocks where pore pressures were nearly hydrostatic. He suggested upward fluid flow into the fault zone from the ductile roots of faults, where pore pressure may be nearly lithostatic. However, Chery *et al.* (2000) have recently pointed out that unless the highly pressured fault zone is very narrow (~10–100 m wide), such a model implies large, super-lithostatic vertical stresses in the fault zone, leading to plastic extrusion of fault zone rocks toward the Earth's surface. Furthermore, it seems questionable that very high pore pressures could be permanently confined within the fault zone when periodic large earthquakes create fracturing that extends from seismogenic depths to the surface.

A related class of weakening mechanism relies on transient pore pressure changes accompanying earthquake rupture. Sibson (1973) and Lachenbruch (1980) suggested fluid hydraulically confined within the fault zone during earthquake slip could be sufficiently frictionally heated to create high pore pressure via thermal expansion and a phase transition from water to steam. Sibson (1990) later proposed that fluid pressures in the bottom half of the seismic zone could transiently cycle between lithostatic and hydrostatic as a result of fracture permeability created during slip and resealing of these fractures by post-slip precipitation ("fault valve" behavior), and he cited field evidence from exhumed high angle reverse faults to support this model.

Dynamic weakening could result if frictional heat generation were sufficient to actually melt fault zone rocks during slip (McKenzie and Brune, 1972; Kanamori *et al.*, 1998; Kanamori and Heaton, 2000). This inference is supported by restricted occurrences of unusual fault zone rocks ("pseudotachylyte") interpreted as products of melting during slip (Sibson, 1975). For temperature rises to be sufficient to produce melting, the zone of earthquake slip must be very narrow (a few centimeters or less), which may occur in some fault zones. However, the relative rarity of pseudotachylytes generally argues for lower resisting stresses and temperature rises during fault slip.

Another class of weakening models includes dynamic inertial effects. As mentioned in Section 8, theoretical models and analogue experiments show fault opening during dynamical slip, and if this occurs at seismogenic depths in the Earth it would be a weakening mechanism of major importance.

Additional weakening mechanisms include: (1) low intrinsic friction coefficient clays and related rocks (Wu, 1978; Wang *et al.*, 1979); however, note more recent laboratory experiments showing near-normal μ at upper crustal confining pressure (Morrow *et al.*, 1992); (2) rotation of rounded fragments of fault gouge (Brune and Anooshehpoor, 1997; Mora and Place, 1998); (3) dynamic compaction generating high pore fluid pressure (Sleep and Blanpied, 1992); (4) acoustic fluidization of fault gouge during slip (Melosh, 1996).

9.2 Inhomogeneous Fault Strength and Its Implications

Thus far we have discussed models of fault strength that are either explicitly homogeneous or rely largely on observations like heat flux and radiated energy that are integrated averages of stress effects. However, much of what is known and understood about faults indicates they are inhomogeneous in their physical properties and their effects over a wide range of scales. Surface maps show faults are often discontinuous and change in orientation along strike, and seismic and structural evidence often shows these variations persist at depth. Mappings of the distributions of both coseismic slip and moment release from large and great earthquakes show heterogeneity is the rule. Geodetic mappings of earthquake slip show slip magnitude varies by factors of 4–8 on the coseismic rupture (Fig. 9). Using Eq. (22) and a fault zone width of 10 km, these observations indicate stress drop varying from 3 to 25 MPa along fault strike. Analysis of broadband seismograms from great earthquakes (e.g., Kikuchi and Fukao, 1987; Beck and Ruff, 1989) shows moment release rate is very nonuniform. Much of the energy radiated to teleseismic distances originates from restricted portions of the earthquake fault (see compilation of 21 such determinations in Thatcher, 1990, Fig. 4), indicating τ_a is also heterogeneous. Similar results have been reported from analysis of regional strong motion recordings of $M{\sim}7$ earthquakes (Trifunac and Brune, 1971; Cohee and Beroza, 1994; Wald and Heaton, 1994). These observations thus indicate that if the average fault stress is low, the observed heterogeneity in $\Delta\sigma$ and τ_a require that the ambient stress distribution be strongly heterogeneous as well.

9.3 Conceptual Model

These observations suggest a simplistic fault zone strength model shown schematically in Figure 10. Its essential features include patches of strong fault located at the deepest parts of the brittle crust, where large earthquakes nucleate and where Eq. (15) suggests the highest shear stresses occur. These strong

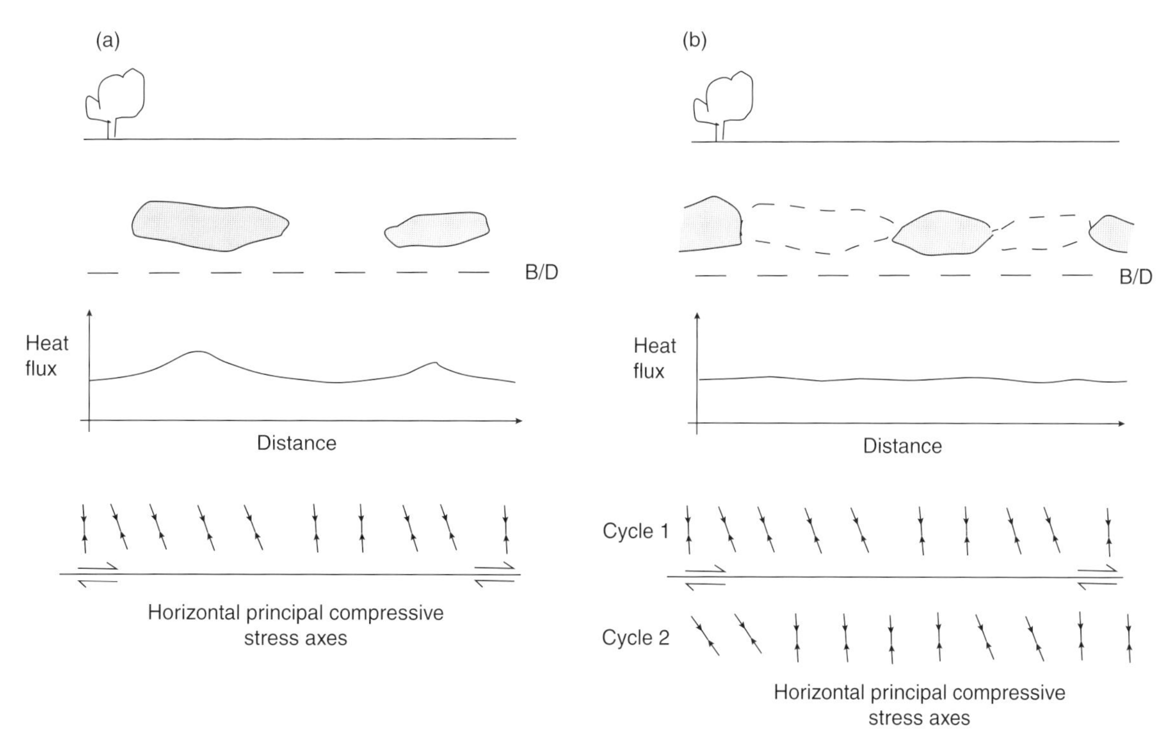

FIGURE 10 Schematic of inhomogeneous strength model: (a) strong and weak regions are permanent; (b) strength varies from cycle to cycle or over long time intervals. Top, strike-parallel longitudinal section view of vertical strike-slip fault, with strong patches shaded. In (b), the strong regions from the previous cycle are shown outlined by dashed lines. Middle, steady-state surface heat flux plotted versus distance along fault strike. Bottom, map view of horizontal principal compressive stress axes along fault strike. In (b), orientations during the preceding cycle (Cycle 1) are shown "north" of fault, and from Cycle 2 to the "south." B/D, brittle–ductile transition.

patches separate much weaker regions at shallower depths and along fault strike at depth.

The model shown in Figure 10 has some similarities with the "asperity" model of faulting inspired by ideas on the distribution of seismic and aseismic slip on plate boundary faults and by seismological mappings of moment release in large earthquakes (e.g., Wesson *et al.*, 1973; Kanamori, 1981; Lay *et al.*, 1982). In this view, the asperities are the locked patches of a fault which slip only during earthquakes, with much of the seismically radiated energy originating from these localized regions.

In the model of Figure 10, the strong patches could have a range of behaviors. They could slip locally with large dynamic stress drops during earthquakes, in which case large amounts of seismic energy will be radiated from these locations. Alternatively, stress drops here may be modest, in which case locally large amounts of energy will be dissipated by frictional heating. The strong patches could be permanent features of the fault zone or they could shift in location from cycle to cycle or over longer time intervals.

Earthquake stress release on high strength patches near the base of the seismogenic zone could be very important in dynamically driving slip at shallower depth. The dynamic model described by Brune (1999), suggested by analogue modeling in foam rubber (Brune, 1996), emphasizes the effect of large driving stresses available at depth due to the higher quasi-static strength there. In this model, near surface slip is driven by dynamic waves from the deeper portions of the fault, and large surface slip results even though ambient stresses there are very low. Similarly, in a lattice dynamics model of thrust faulting (Shi *et al.*, 1998), large displacements at shallow depths occur under very low prestress, again being driven by large dynamic stress drop at depth.

As discussed above, the reason why average fault zone strength is low remains uncertain. However, as suggested by Sibson (1990) and Rice (1992), fault zone stresses on the stronger patches could be moderated by the high (~lithostatic) fluid pressures originating from below the brittle–ductile transition. Results from metamorphic petrology (Fyfe *et al.*, 1978) and study of geothermal systems (e.g., Fournier, 1999) show that ductile rocks are fluid saturated. Furthermore, fluid pressures are likely to be nearly lithostatic, because any cracks or fluid passageways to the brittle crust would be sealed by flow. However, earthquake rupture or steady-state aseismic slip could open contact between the hydrostatic and lithostatic regimes, decreasing resisting stress transiently during seismic slip, as in Sibson's fault valve model or permanently, during steady-state creep. Dynamic weakening during earthquake slip could complement this process or independently supply the fault-averaged low strength required by our inhomogeneous model. This weakening mechanism would not, however, apply for aseismically slipping faults unless the aseismic behavior were a transient feature of the fault zone.

9.4 Role of Average Constraints on τ_r and τ_a

Although successful strength models must simultaneously satisfy the heat flux and radiated energy constraints, these constraints apply only for values of τ_r and τ_a averaged over the fault plane. In the case of heat flux, the constraint is also an average over time and applies over conductive timescales (approximately 100 ka or longer). Thus the requirement that $\tau_r < 10$ MPa on the San Andreas would still permit values consistent with Eq. (15b) over restricted portions of the upper crustal fault surface. If the most dissipative zone were not a permanent feature of the fault, the upper bound value of τ_r might more nearly represent the time-averaged strength of the entire fault zone. The heat flux constraint would appear to be much more stringent on the deeper portions of subduction thrusts, where Eq. (15c) would suggest resisting stresses of ~600 MPa under hydrostatic conditions at 30 km whereas observations permit no more than about 40 MPa at this depth.

The radiated energy constraint applies in the same averaged way as heat flux, and τ_a could locally be much higher than its mean value of 0.5 MPa. Indeed, τ_a values as high as 22 MPa are shown in Color Plate 13. In this regard it may be noteworthy that McGarr (1999) finds the upper bound values of τ_a obtained from seismology are approximately consistent with extrapolations based on laboratory rock mechanics and data from mining-induced seismicity (i.e., 15). This might be expected if some earthquakes ruptured only the high strength portions of an inhomogeneous fault.

9.5 Observation Implications

Observable effects of fault stress heterogeneity will differ depending on whether the strength distribution is permanent or varies from one earthquake cycle to the next or over longer timescales (Fig. 10). Permanent regions of high resisting stress (Fig. 10a) could produce significant frictional heating, but their effect on surface heat flux would be more subtle and difficult to detect than dissipation that is uniform along strike. This follows because laterally confined heat sources generate lower peak steady-state surface heat fluxes that decay more rapidly with distance than do 2D sources. In addition, for a fluid-saturated crust with enhanced fault zone permeability, along-strike advection of heat could spread out the effect of dissipative heating from a local source. If the strength distribution varied temporally (Fig. 10b), dissipative heating would migrate with time, and surface heat flux would become quasi-uniform on thermal conduction timescales.

If high and low strength regions were permanent fault features (Fig. 10a), temporal changes in principal stress orientations would be small or undetectable. If on the other hand, the zones changed from one earthquake cycle to the next (perhaps as slip minima from one earthquake were followed by high slip in a subsequent event), then stress orientations could change from cycle to cycle (Fig. 10b) and also change within each cycle as the fault strengthened by accumulation of elastic stresses.

Mount and Suppe (1987) and Zoback *et al.* (1987) originally proposed that strength differences exist between the central creeping segment of the San Andreas fault (where compressive stresses are nearly normal to the fault), and the currently locked portion of the fault system further north (where these orientations are more oblique). In principle, determinations of principal stress orientations can also be used to map out stress heterogeneities on a more local scale. Hardebeck and Hauksson (1999) used a large suite of earthquake focal mechanisms to infer local variations in compressive stress orientations both along strike and perpendicular to the San Andreas system in southern California. Although it is tempting to attribute these results as evidence for fault strength heterogeneity, facile interpretations are frustrated by the considerable complexity of the derived orientations, which show rapid spatial changes and seldom decay to stable values far from the San Andreas. While perhaps containing evidence for fault stress heterogeneity, we suspect some of the variability is also due to effects other than simply the ambient stresses on the faults themselves. These include: (1) unavoidable limitations of the inversion method (e.g., inadvertently including more than one stress regime in a single stress field determination); (2) biasing of inversions by using focal mechanisms from faults subparallel to the San Andreas; (3) local stress perturbations (Turcotte, 1982) due to significant crustal thickness and lithospheric density variations across southern California; and (4) lateral rheology variations that modify fault-generated stresses (Chery *et al.*, 2000).

9.6 Observational Tests

We do not see simple tests of either the conceptual model of fault zone stress shown in Figure 10 or those previously proposed by others. Like previous reviewers (e.g., Hickman, 1991), we support direct observation of the constitutive properties and physical state of active faults through drilling and sampling. Plans are proceeding to carry out such an experiment to a depth of ~4 km on the San Andreas fault near Parkfield, California (Hickman *et al.*, 1994) and drilling is now expected to begin during 2003.

Because seismically radiated energy is such a diagnostic measure of stress changes during faulting, more precise estimates, particularly those obtained in the near field of large and great earthquakes, would be especially valuable. In particular, such measurements would be one means of testing our suggestion that apparent stresses obtained from radiated energy estimates are low because τ_a is averaged over a rupture surface where quasi-static or dynamic strength is heterogeneous. More generally, high quality near-field recordings of strong ground motion like those now becoming available for the 1999 $M_w = 7.6$ Chi-Chi (Taiwan) earthquake will constrain the dynamics of fault slip through mappings of the stress-slip history over the entire rupture plane.

Acknowledgments

Stimulating discussion was provided by A.G. McGarr and S.H. Hickman, both of whom provided insightful reviews of the manuscript. Thorough reviews were also provided by Y. Fialko, A.H. Lachenbruch, and R.H. Sibson. Apparent stress estimates for teleseismic data, many of which are unpublished, were kindly supplied by G.L. Choy.

References

Adams, G.G. (1999). *ASME J. Tribol.* **121**, 455–461.
Allen, C.R., *et al.* (1998). *Seismol. Res. Lett.* **69**, 524–531.
Anderson, E.M. (1951). "The Dynamics of Faulting," Oliver and Boyd, Edinburgh.
Andrews, D.J. and Y. Ben-Zion (1997). *J. Geophys. Res.* **102**, 553–571.
Anooshehpoor, A. and J.N. Brune (1994). *Pageoph* **142**, 735–747.
Archuleta, R.J. (1984). *J. Geophys. Res.* **89**, B6, 4559–4585.
Beck, S.E. and L.J. Ruff (1989). *Phys. Earth Planet Inter.*
Ben-Zion, Y. and J.D. Andrews (1998). *Bull. Seismol. Soc. Am.* **88**, 1085–1094.
Boatwright, J. and M. Cocco (1996) *J. Geophys. Res.* **101**, 13 895–13 909.
Bodin, P., *et al.* (1998). *J. Geophys. Res.* **103**, 29 931–29 944.
Bouissou, S., *et al.* (1998). *Tectonophysics* **295**, 341–350.
Brace, W.F. and D.L. Kohlstedt (1980) *J. Geophys. Res.* **85**, 6248–6252.
Brown, S.R. (1998). *J. Geophys. Res.* **103**, 7413–7420.
Brudy, M., *et al.* (1997). *J. Geophys. Res.* **102**, 18 453–18 475.
Brune, J.N. (1970). *J. Geophys. Res.* **75**, 4997–5009.
Brune, J.N. (1976). In: "Seismic Risk and Engineering Decisions," pp. 141–171, Elsevier, Amsterdam.
Brune, J.N. (1996). *Proc. Ind. Acad. Sci.* **105**, 197–206.
Brune, J.N. (1999). *Abstract, 1999 GSA Cordilleran Section Centennial*, 2–4 June.
Brune, J.N. and A. Anooshehpoor (1997). *Geophys. Res. Lett.* **24**, 2071–2074.
Brune, J.N., *et al.* (1969). *J. Geophys. Res.* **74**, 3821–3827.
Byerlee, J.D. (1978). *Pageoph* **116**, 615–626.
Chery, J., *et al.* (2000). *J. Geophys. Res.* **106**, 22 051–22 061.
Choy, G.L. and J. Boatwright (1995). *J. Geophys. Res.* **100**, 18 205–18 228.
Cohee, B.P. and G.C. Beroza (1994). *Bull. Seismol. Soc. Am.* **84**, 692–712.
Comninou, M. and J. Dundurs (1977). *J. Appl. Mech.* **44**, 222–226.
Dickinson, W.R. and W.S. Snyder (1979). *J. Geol.* **87**, 609–627.
Dieterich, J.H. (1979). *J. Geophys. Res.* **84**, 2161–2168.
Fleitout, L. and C. Froidevaux (1982). *Tectonics* **1**, 21–56.
Forsyth, D. and S. Uyeda (1975). *Geophys. J. R. Astron. Soc.* **43**, 163–200.
Fournier, R.O. (1999). *Econ. Geol.* submitted.

Freund, L.B. (1978). *J. Appl. Mech.* **45**, 226–227.
Fyfe, W., *et al.* (1978). "Fluids in the Earth's Crust," Elsevier.
Hanks, T.C. (1977). *Pageoph* **115**, 441–458.
Hanks, T.C. (1982). *Bull. Seismol. Soc. Am.* **72**, 1867–1879.
Hanks, T.C. and M. Wyss (1972). *Bull. Seismol. Soc. Am.* **62**, 561–589.
Hardebeck, J.L. and E. Hauksson (1999). *Science* **285**, 233–236.
Hickman, S.H. (1991). *Rev. Geophys. Suppl.* 759–775.
Hickman, S., *et al.* (1994). *EOS, Am. Geophys. Un. Trans.* **75**, 137–142.
Hubbert, M.K. and W.W. Rubey (1959). *Geol. Soc. Am. Bull.* **70**, 115–166.
Hyndman, R.D. and K. Wang (1993). *J. Geophys. Res.* **98** B2, 2039–2060.
Hyndman, R.D., *et al.* (1995). *J. Geophys. Res.* **100**, 15 373–15 392.
Jaeger, J.C. and N.G.W. Cook (1976). "Fundamentals of Rock Mechanics," Wiley.
Kanamori, H. (1981). In: "Earthquake Prediction, An International Review," Maurice Ewing Ser., vol. 4, pp. 1–19, American Geophysical Union.
Kanamori, H. and T.H. Heaton (2000). *AGU Monograph Series* **120**, 147–163.
Kanamori, H., *et al.* (1998). *Science* **279**, 839–842.
Kikuchi, M. and Y. Fukao (1987). *Tectonophysics* **144**, 231–247.
King, N.E. and W. Thatcher (1998). *J. Geophys. Res.* **103**, 18 069–18 078.
Kohlstedt, D.L., *et al.* (1995). *J. Geophys. Res.* **100**, 17 587–17 602.
Lachenbruch, A.H. (1980). *J. Geophys. Res.* **85**, 6097–6112.
Lachenbruch, A.H. and J.H. Sass (1980). *J. Geophys. Res.* **85**, 6185–6223.
Lay, T., *et al.* (1982). *Earthq. Predict. Res.* **1**, 3–71.
McGarr, A. (1999). *J. Geophys. Res.* **104**, 3003–3011.
McGarr, A. and N.C. Gay (1978). *Annu. Rev. Earth Planet,. Sci.* **6**, 405–436.
McKenzie, D.P. (1969). *Geophys. J. R. Astron. Soc.* **18**, 1–32.
McKenzie, D.P. and J.N. Brune (1972). *Geophys. J. R. Astron. Soc.* **29**, 65–78.
Melosh, J. (1996). *Nature* **379**, 601–606.
Molnar, P. and P.C. England (1990). *J. Geophys. Res.* **95**, 4833–4856.
Mora, P. and D. Place (1994). *Pageoph* **143**, 61–87.
Mora, P. and D. Place (1998). *Geophys, Res. Lett.* **103**, 21 067–21 089.
Mora, P. and D. Place (1999). *Geophys. Res. Lett.* **26**, 123–126.
Mori, J. and K.-F. Ma (1999). *EOS, Suppl.* **13**.
Morrow, C., *et al.* (1992). In: "Fault Mechanic sand Transport Properties in Rocks," pp. 69–88, Academic Press.
Mount, V.S. and J. Suppe (1987). *Geology* **15**, 1143–1146.
Ni., S.-D., *et al.* (1999). *EOS, Suppl.* **14**.
Pacheco, J., *et al.* (1993). *J. Geophys. Res.* **98**, 14 133–14 159.
Rice, J.R. (1992). In: "Fault Mechanic sand Transport Properties in Rocks," pp. 475–503, Academic Press.
Richardson, R.M. and L.M. Reding (1991). *J. Geophys. Res.* **96**, 12 201–12 223.
Ruina, A.L. (1983). *J. Geophys. Res.* **88**, 359–370.
Scholz, C.H. (2000). *Geology* **28**, 163–166.
Shi, B., *et al.* (1998). *Bull. Seismol. Soc. Am.* **88**, 1484–1494.
Sibson, R.H. (1973). *Nature* **243**, 66–68.
Sibson, R.H. (1975). *Geophys. J. R. Astron. Soc.* **43**, 775–794.
Sibson, R.H. (1985). *J. Struct. Geol.* **7**, 751–754.
Sibson, R.H. (1990). *Bull. Seismol. Soc. Am.* **80**, 1580–1604.
Singh, S.K. and M. Ordaz (1994). *Bull. Seismol. Soc. Am.* **84**, 1533–1550.
Sleep, N.H. and M. Blanpied (1992). *Nature* **359**, 687–692.
Thatcher, W. (1990). *J. Geophys. Res.* **95**, 2609–2623.
Thatcher, W. and P.C. England (1998). *J. Geophys. Res.* **103**, 891–905.
Thatcher, W. and D.P. Hill (1991). *Geology* **19**, 1116–1120.
Thatcher, W. and D.P. Hill (1995). *J. Geophys. Res.* **100**, 561–570.
Thatcher, W., *et al.* (1997). *J. Geophys. Res.* **102**, 5353–5367.
Tichelaar, B.W. and L.J. Ruff (1993). *J. Geophys. Res.* **98**, 2017–2037.
Townend, J. and M.D. Zoback (2000). *Geology* **28**, 399–402.
Trifunac, M. and J.N. Brune (1970). *Bull. Seismol. Soc. Am.* **60**, 137–160.
Turcotte, D.L. (1982). In: "Mountain Building Processes," pp. 141–146, Academic Press.
Uenishi, K., *et al.* (1999). *Bull. Seismol. Soc. Am.* **89**(5), 1296–1312.
Wald, D.J. and T.H. Heaton (1994). *Bull. Seismol. Soc. Am.* **84**, 668–691.
Wang, C.-Y. and N.-H. Mao (1979). *Geophys. Res. Lett.* **6**, 825–828.
Wesson, R.L., *et al.* (1973). *Stanford Univ. Publ. Geol. Sci.* **13**, 303–321.
Williams, C.F. and T.N. Narasimhan (1989). *Earth Planet. Sci. Lett.* **92**, 131–143.
Wu, F.T. (1978).*Pageoph* **116**, 655–689.
Yuen, D.A., *et al.* (1978). *Geophys. J. R. Astron. Soc.* **54**, 93–119.
Zoback, M.L. (1992). *J. Geophys. Res.* **97**, 11 761–11 781.
Zoback, M.L., *et al.* (1989). *Nature* **341**, 291–298.
Zoback, M.D., *et al.* (1987). *Science* **238**, 1105–1111.

Editor's Note

References with full citation are placed in BruneFullReferences.pdf on the Handbook CD under directory \35Brune. Please see also Chapter 32, Rock failure and earthquakes, by Lockner and Beeler; Chapter 33, State of stress within the Earth, by Ruff; and Chapter 34, State of stress in the Earth's lithosphere, by Zoback and Zoback.

36

Implications of Crustal Strain During Conventional, Slow, and Silent Earthquakes

M.J.S. Johnston
US Geological Survey, Menlo Park, USA
A.T. Linde
Carnegie Institution of Washington, Washington, USA

1. Introduction

Uniform block-slip motion consistent with simple shear on locked fault segments is the primary feature apparent in geodetic measurements of strain accumulation along plate boundaries (Savage, 1983). However, almost every aspect of fault failure is nonlinear in character. This premise derives from theoretical models (Kostrov, 1966; Richards, 1976; Andrews, 1976; Freund, 1979; Rice and Rudnicki, 1979; Rice, 1983, 1992; Stuart, 1979, Stuart and Mavko, 1979; Das and Scholz, 1981; Rundle *et al.*, 1984; Tse and Rice, 1986; Lorenzetti and Tullis, 1989; Segall and Rice, 1995; Shaw, 1997; Miller *et al.*, 1999) and laboratory-generated frictional failure of crustal materials (Dieterich, 1979, 1981; Mogi, 1981; Mogi *et al.*, 1982) which predict accelerating deformation will occur before dynamic slip instabilities, better known as earthquakes.

To investigate this physics of fault failure, it has increasingly become obvious that new techniques, with capability to resolve strain at better than 1 part per billion (ppb), are needed. Geodetic and GPS strain measurements with resolutions with approximately 0.4 parts per million (ppm) resolution on 5-km baselines may help with some broadscale problems such as fault interaction, viscoelastic response (Thatcher, 1983; Pollitz *et al.*, 1998), coseismic and postseismic response of large earthquakes (Langbein *et al.*, 1983; Bock *et al.*, 1997; Burgmann *et al.*, 1997), but issues regarding the type of slip transients, the relative frequency of slip transients, fault creep, earthquake nucleation, pore pressure changes, slow earthquakes and silent earthquakes (where moment release is usually below that typically released for a $M = 5$ earthquake), are unlikely to be resolved with these techniques. Some strain transients, such as those occurring postseismically have been known for a long time (e.g., Langbein *et al.*, 1983; Thatcher, 1983). Many others, including those occurring aseismically, are not well documented or understood even though they may play important roles in the fault-failure process.

From previous strain studies (Johnston *et al.*, 1987; Wyatt, 1988), and discussed further here, it is now generally clear that entire rupture zones do not exhibit nonlinear behavior as implied by "preparation zone" terminology (Sadovsky *et al.*, 1972) and some theoretical models of the rupture process (Mjachkin *et al.*, 1975; Tse and Rice, 1986). Nonlinear (accelerating) deformation is not apparent in arrays of sensitive strain and tilt instruments installed throughout regions that have subsequently ruptured. Instead, rupture over a large fault length appears to be triggered by failure of smaller regions with higher strength, and this rupture draws on the elastic strain energy in the region. The expected signals should have occurred during the period immediately preceding rupture initiation. Also, the total slip moment (preseismic, coseismic, and postseismic) could be determined when each phase of the rupture process is identified.

If failure initiates in small regions of high strength and expands to any arbitrary size until stopped by another mechanical or geometrically strong barrier, this failure still has an easily identifiable form (nonlinear exponential-like strain increase). Unfortunately, detection of any preseismic slip and, more importantly, prediction of the final rupture size, will be difficult to achieve (Brune, 1979) because the total moment release during this time is only a small fraction of that released coseismically.

The timescale of failure is also not clear although the absence of obvious tidal triggering (Heaton, 1982) implies the final stage of failure occurs at periods comparable to, or

INTERNATIONAL HANDBOOK OF EARTHQUAKE AND ENGINEERING SEISMOLOGY, VOLUME 81A ISBN: 0-12-440652-1

shorter than, the major earth tidal periods (12.42, 23.93, 25.82 h, etc.). Numerous observations (Rikitake, 1976; Mogi, 1985) indicate cases when this appears to be true. The shortest timescale for nonlinear strain can be estimated from subcritical crack growth in brittle materials (Atkinson, 1979). However, it is expected that the presence of fluids should stabilize and slow down the process of cascading crack fusion in rock at the temperatures and pressures expected in the upper 10 km of the Earth's crust (Rice and Rudnicki, 1979), as shown in Figure 1. Strain during the final stages before failure accelerates dramatically in comparison with earlier periods, with maximum strain occurring, as expected, at the point of rupture.

Laboratory experiments also indicate that these last stages of failure of crustal rock (sometimes termed tertiary creep) are the end result of a process of cascading crack coalescence which results in strain time histories similar to those shown in Figure 1 (see, for example Mogi, 1985). Large-scale laboratory models with preexisting artificial faults, either with or without gouge material, exhibit similar behavior (Dieterich, 1979; Mogi, 1981, 1985; Mogi *et al.*, 1982).

If the timescale of failure initiation is between one second and one day, continuous measurements of the state of strain and tilt in the Earth's crust near moderate to large earthquakes should quickly provide detailed information on time history, failure mechanics, and size (moment) of the prerupture failure. This assumes, of course, that these measurements are recorded at a sensitivity sufficient to detect failure nucleation in the last few hours to seconds before rupture occurs.

Observations of crustal strain at high sensitivity near large earthquakes during the final stages before rupture are unfortunately rare, and knowledge of the timescale and mechanics of failure is therefore limited. This has resulted largely from the infrequency of earthquakes and the poor areal coverage with adequate instrumentation. With the installation of borehole strain and tilt arrays along the San Andreas fault system and in other seismically active places during the last ten years, data are now accumulating. Along the San Andreas fault, many events have occurred for which we have near-field recordings at high sensitivity although only two for which the magnitude was greater than 7. In addition, we can call upon events recorded with similar instruments in Japan (e.g., Sacks *et al.*, 1978, 1979) and Iceland (Linde *et al.*, 1993).

In this paper we use these data to: (1) document the implications of crustal strain observed during fault creep events; (2) compare the expected form of slip waves on active faults with observations of crustal strain recorded near these faults; (3) compare the coseismic strain generated by earthquakes with calculations from simple elastic dislocation models of the events derived from seismic observations to determine the ratio of total moment release to seismic moment release; (4) discuss some observations of "slow" and "silent" earthquakes, and finally, look at the state of strain during the last few hours to the last few seconds before earthquakes and the implications this has for the issue of earthquake nucleation, earthquake prediction, and current theory. In particular, we use these strain observations to estimate the maximum preseismic slip moment release as a fraction of the observed seismic moment release and to place upper limits on nucleation source size.

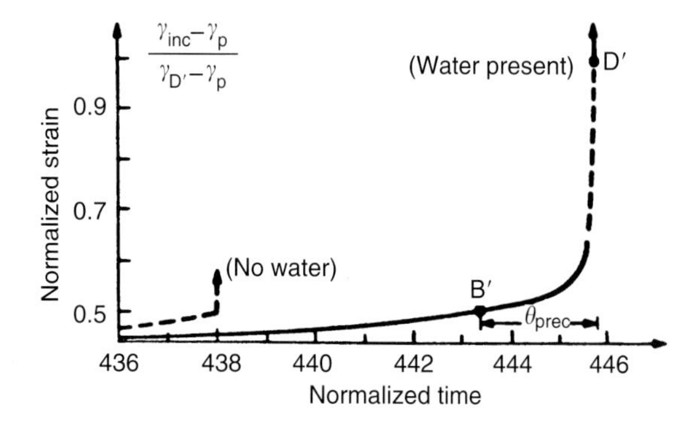

FIGURE 1 Normalized strain time history arising from fault instability models with and without the presence of water (from Rice and Rudnicki, 1979). Θ_{prec} refers to the precursor time defined by Rice and Rudnicki (1979). B′ is the onset of instability, D′ is the onset of dynamic instability and the dashed portions of the curves indicate where the numerical calculations were truncated.

2. Experimental Design and Measurement Precision

2.1 Basic Measurement Limitations

Strain measurements have historically been made along active faults using long baseline geodetic techniques. Sampling was approximately annual and the sensitivity was about 1 ppm. Unfortunately, these techniques lack the sensitivity required to detect short wavelength (5 km) slip waves propagating along faults, coseismic strains for earthquakes with $M \leq 5$, details of aseismic slip, slow earthquakes, and silent earthquakes. New arrays of GPS instrumentation will improve observations of secular motions because of their increased spatial coverage but the sensitivity in the critical band from months to milliseconds is still several orders of magnitude worse than that attainable with borehole strainmeters. Figure 2 shows the comparative resolution of the different measurement techniques as a function of period. Near-surface strainmeters always had more short-term sensitivity than geodetic observations but did not have the long-term stability or spatial averaging to determine long-term strain rates. Records were also compromised at times by effects of rainfall (Wolfe *et al.*, 1981; Yamauchi, 1987). Lower-noise strain data were sometimes obtained by placing strainmeters in deep tunnels and mines. However, a real breakthrough resulted when strainmeters were installed in deep boreholes far from surface noise sources.

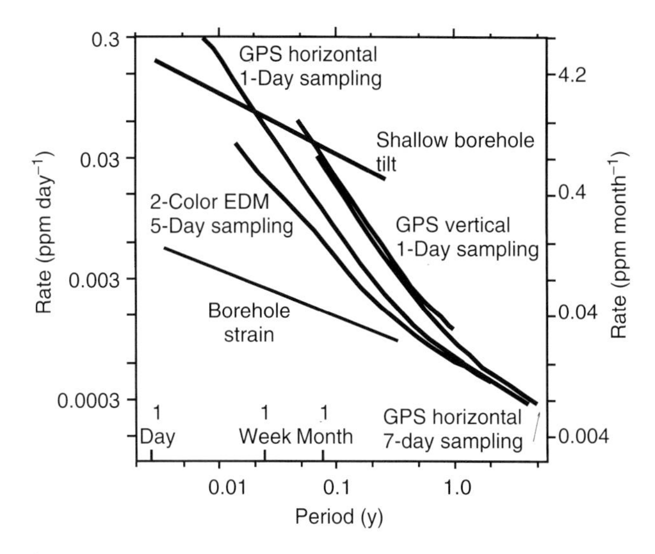

FIGURE 2 Comparative resolutions of crustal strain rates for borehole strainmeters, differential GPS, EDM and tilt sensors as a function of period (after Langbein and Johnson, 1996). (Note: GPS and EDM strain assume an 8 km baseline.)

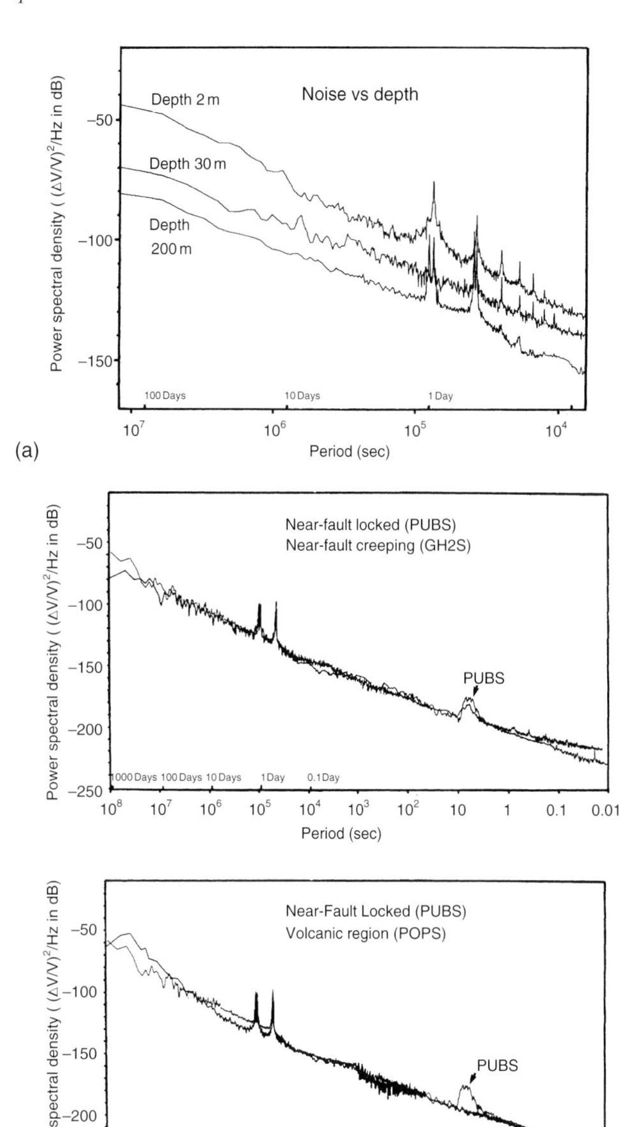

FIGURE 3 (a) Strain noise power as a function of depth beneath the Earth's surface. (b) Comparative strain noise power over ten decades of frequency for different sites in California. PUBA is within 500 m of the San Andreas fault in southern California, GH1S is within 1 km of a creeping segment of the San Andreas fault near Parkfield, and POPA is near an active volcanic caldera in eastern California.

Figure 3a shows strain noise spectra for different depths of burial. At a depth of burial of 200 m noise spectra from strainmeters in a variety of geophysical environments are identical within the errors. Figure 3b shows that the same spectra are obtained whether the strainmeter is in a volcanic region (POPS), within a few hundred meters of a currently quiescent but active region of the San Andreas fault (PUBS), or near a creeping fault (GH2S). Similar spectra are obtained at other sites either near to, or many tens of kilometers from the San Andreas fault.

2.2 Experimental Techniques

The location of the various borehole strainmeters along the San Andreas fault in California is shown in Figure 4. The borehole strainmeters shown as black dots measure dilatational strain (Sacks *et al.*, 1971), whereas those shown as black squares measure 3-component or tensor strain (Gladwin, 1984; Gladwin *et al.*, 1987). The sensitivities of the dilatometer and tensor strainmeters are less than 10^{-11} and less than 10^{-9}, respectively. Each of the strain sensors is installed at depths between 100 and 200 m and each detects strains at levels at least 20 dB below those obtained on near-surface strainmeters (Johnston and Borcherdt, 1984). The dynamic range exceeds 140 dB (Sacks *et al.*, 1971; Sacks, 1979; Gladwin, 1984) over the period band of 0.1 sec to several months. During installation, the strain sensors are cemented in the borehole with expansive grout. The borehole is then filled to the surface with cement to avoid long-term strains from hole-relaxation effects and re-equilibration of the aquifer system.

These broadband instruments record data over ten decades of frequency. Strain from atmospheric pressure loading of the Earth's surface and from the solid earth tides (including effects from ocean loading) are the most obvious signals continuously evident in the data. Superimposed on these are straingrams from local and teleseismic earthquakes. Atmospheric pressure, earth tide and ocean loading effects can be readily predicted and removed from the data (Rabbel and Zschau, 1985; Tamura *et al.*, 1991; Agnew, 1997). Although other corrections, such as for the distortion of strain fields due to nearby extreme topography are sometimes necessary (Harrison, 1976), strain

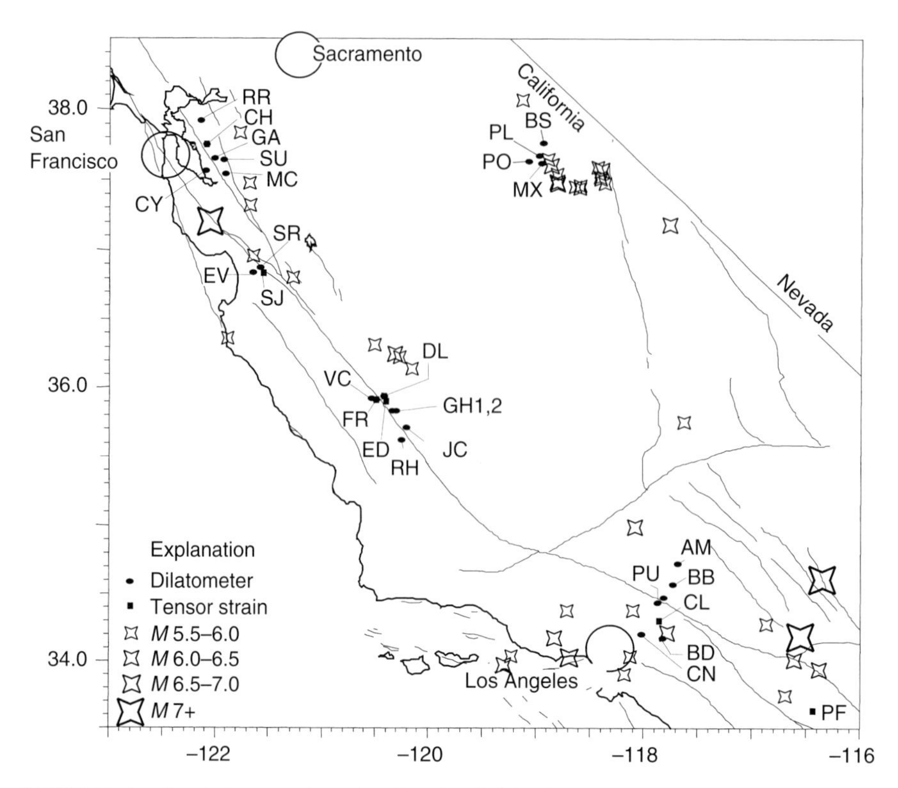

FIGURE 4 Borehole crustal strain sites in California and moderate earthquakes with magnitudes greater than 5 (shown as stars) that were recorded at these sites.

can generally be modeled with dislocations in elastic or viscoelastic half-space (Matsu'ura and Tanimoto, 1980; Okada, 1985, 1992; McTigue and Segall, 1988).

Data from these strain instruments are transmitted in 16-bit digital form through the GOES weather satellite to Menlo Park, California for processing (Silverman *et al.*, 1989). The strain sensors together with the sites are calibrated using ocean-load corrected earth tides (Linde and Johnston, 1989; Linde *et al.*, 1992; Hart *et al.*, 1996). This calibration is generally repeatable to within 1%.

3. Strain Fields and Geophysical Implications

3.1 Fault Creep Events

Surface observations of episodic displacement of active faults are usually termed fault-creep events. These creep events on the San Andreas might be argued to be a special form of slow earthquakes. However, previous work (Johnston *et al.*, 1976, 1977; McHugh and Johnston, 1977, 1978, 1979; Goulty and Gilman, 1978; Evans *et al.*, 1981; Gladwin *et al.*, 1994) indicates that the event-like nature of fault-creep events results primarily from complex failure of the near-surface fault zone materials at depths typically shallower than several hundred meters rather than fault failure at seismogenic depths. Earthquake shaking of the near-surface soils can also trigger creep events (Simpson *et al.*, 1988).

A network of strainmeters [dilational strainmeter (DS) and tensor strainmeter (TS)] installed at Parkfield (Fig. 5a) provides examples of the primary features of crustal strain during fault creep. One such example is the event on 6 October 1997, recorded on XVA1 and XPK1. The rupture length is limited since this event was not recorded at XTA1 just south of XPK1 nor at XMD1 to the northwest of XVA1. Its minimum dimension could be as small as 3 km (the distance between XVA1 and XPK1). If we search the strain records from the strainmeters, FRDS, DLDS, VCDS and EDTS, surrounding this section of the fault, we find a detectable strain transient above the noise only on the closest strainmeter FRDS (Fig. 5b). This signal indicates a compressive signal of about 15 nanostrain. A small negative change of about a nanostrain may have occurred on DLDS. An upper estimate of the depth of this creep event can be determined if we model the event using Okada's (1985, 1992) dislocation formulation using a length of 4 km, a displacement of 2.5 mm (observed at XVA1), and determining the likely depth that would generate a strain transient of 15 nanostrain at FRDS and less than a few nanostrain at the other sites. The lower plot on Figure 5b shows calculated strain amplitude as a function of the width of the creep event. It is apparent that the width cannot exceed a few hundred meters. Similar depths are obtained for numerous other creep events. These data support the notion

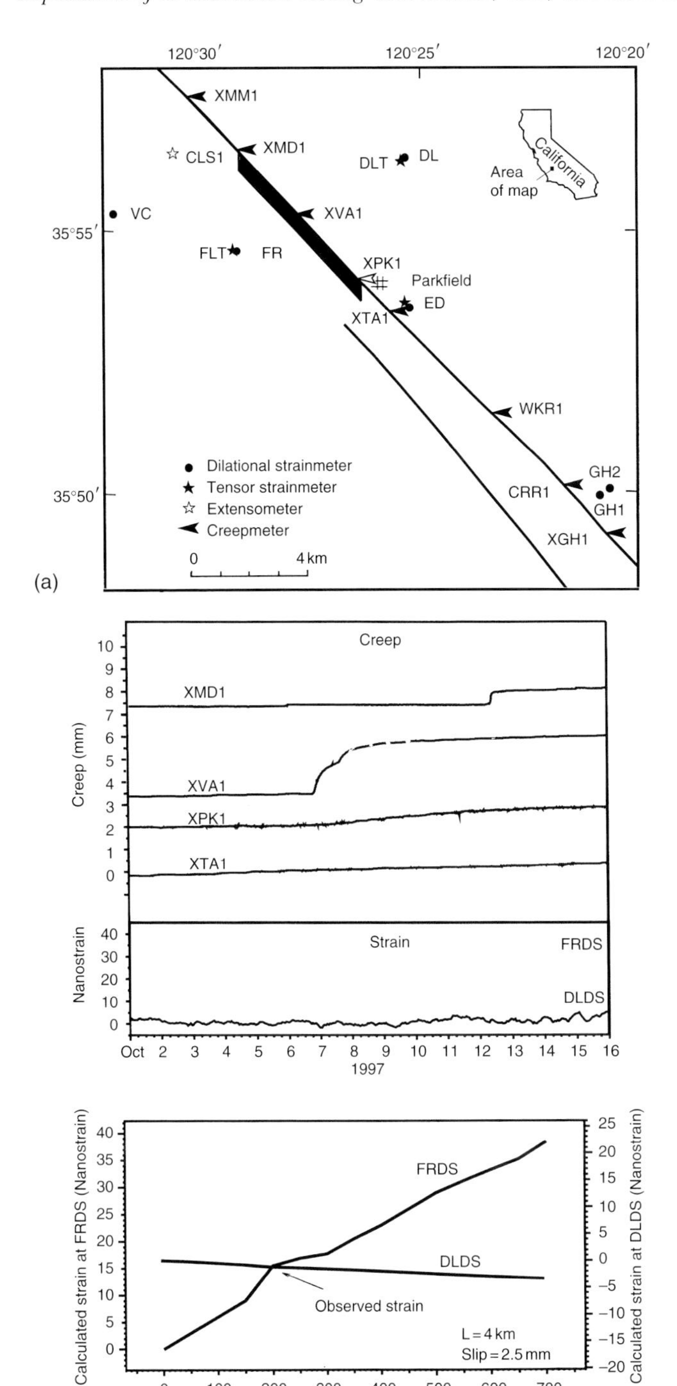

FIGURE 5 (a) Location of the dilatational strainmeters (FR, DL, JC, VC, GH1 and GH2), tensor strainmeters (FLT, DLT and EA), and creepmeters (XMM1, XMD1, XVA1, XTA1, CRR1, XPK1, XGH1, and WKR1) at Parkfield, California. The fault patch shows the location of the creep event discussed in Figure 5b. (b) Strain and creep time histories during the creep events recorded at XVA1 and XPK1 on 6–7 Oct. 1997. Creep records are shown in the top of the upper plot, strain records from FR and DL are shown in the bottom of the upper plot. The lower plot shows the calculated strain as a function of fault slip width at the two strain sites FR and DL.

that creep events result from shallow slip waves in the upper few hundred meters of the fault.

Similar conclusions were evident in data from arrays of creepmeters, strainmeters, and tiltmeters separated by only a few hundred meters in other parts of the creeping section of the San Andreas fault (McHugh and Johnston, 1979; Evans *et al.*, 1981; Goulty and Gilman, 1978). These data also indicated complexity in propagating inhomogeneous slip waves on scales of a few hundred meters. Furthermore, indications of deep slip episodes in near-fault strain data (Johnston *et al.*, 1977; Langbein, 1981; Linde *et al.*, 1996) have not been found to correspond to the times of surface creep events although the long-term fault displacement rates are mostly reflected in long-term creep rate data (Schulz *et al.*, 1982). This view of the event-like character of surface creep events is consistent with the suggestions (Marone *et al.*, 1991) that near-surface fault materials exhibit velocity strengthening but cannot ultimately sustain the rapid slip associated with earthquake source motions. In contrast, materials at seismogenic depths primarily exhibit velocity weakening as expected from theoretical work on rate and state variable friction (Ruina, 1983; Dieterich, 1994). Episodic near-surface creep events are thus not considered to be slow earthquakes.

3.2 Fault Slip Waves

The suggestion that regional strain waves may propagate along active faults has been around since the 1970s (Kasahara, 1979; Savage, 1971; Ito, 1982; Kasahara *et al.*, 1983) although little support has yet been found. Slip waves over the fault face with amplitudes of a few millimeters generate coherent strains exceeding 10^{-7} within a few kilometers of the fault. Signals of opposite form occur on opposite sides of the fault. Figure 6b shows the predicted signals on several instruments at Parkfield for a slip wave with an amplitude of 10 mm between 5 km and 10 km that propagates past several strainmeters installed within a few kilometers of the San Andreas fault, shown in Figure 6a. If these slip waves occurred with propagation velocities from kilometers per month to kilometer per minute they would be readily observable on the strainmeter arrays. Thus far they have not been seen. These slip waves could occur but not be detected on the current strainmeter arrays if the velocities are more than ten times slower (i.e., kilometers per year). Neither would these slip waves be detected in the semicontinuous two-color EDM or continuous GPS measurements in the Parkfield area unless the amplitudes were 10–100 times larger.

Nevertheless, this idea has recently resurfaced in 1995 (Press and Allen, 1995) and, again in 1998, with the suggestion (Pollitz *et al.*, 1998) that slip waves should also occur from time-dependent viscoelastic response from distant subduction earthquakes. The expected amplitudes are 10^{-8} with a timescale of several years.

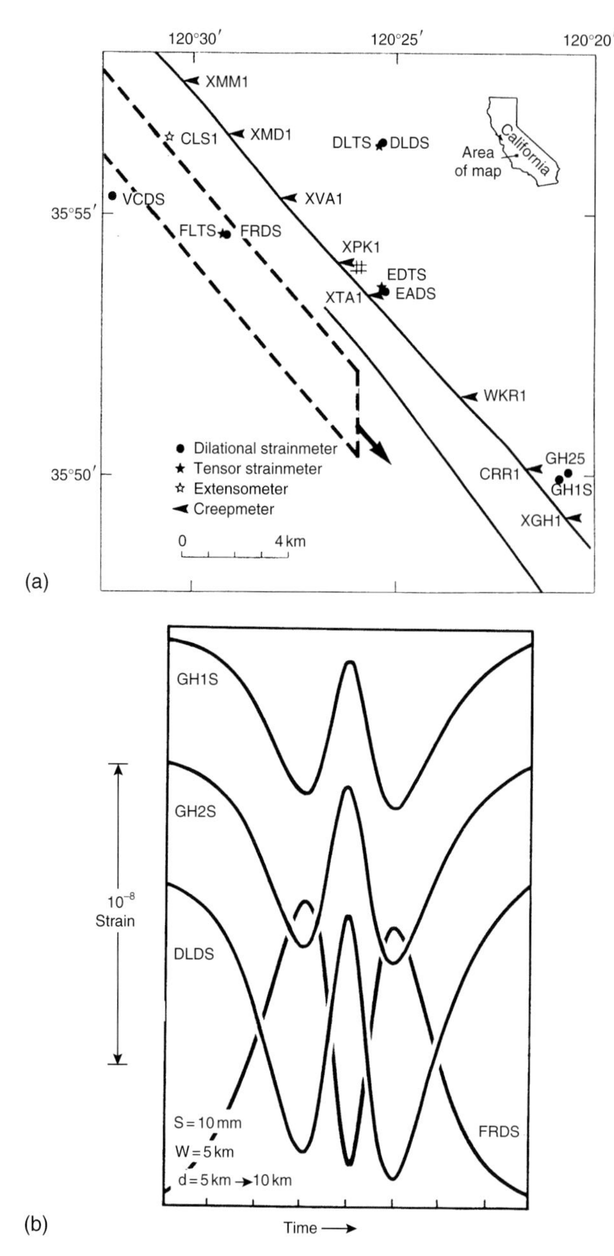

FIGURE 6 (a) Map of the Parkfield region illustrating a hypothetical slip wave propagating between 5 km and 10 km beneath the surface array of strain and creep monitoring sites. (b) Predicted strain time histories for a 5 km wide slip wave propagating between 5 km and 10 km depth past several borehole strainmeters in Parkfield. The slip amplitude is 10 mm. For clarity only four strain–time records are shown.

TABLE 1 List of earthquakes and Observed and Calculated Coseismic Strain Offsets. Details for some earthquakes were Obtained from Eaton (1985 and unpublished), Johnston *et al.* (1986), L. Jones (personal communication), Lester (personal communication), Shimazaki and Somerville (1978), and others from the Northern California Data Center. The code name following the observed strain offsets in column 4 refers to the instrument on which the offset was observed. The locations of these instruments are shown in Figure 4

Earthquake	Magnitude	Date	Obs. Strain ($\mu\varepsilon$)	Calc. Strain ($\mu\varepsilon$)
Homestead Valley	5.2	3/3/79	1.0 (PF)	1.3
Morgan Hill	6.2	4/24/84	−72.0 (EV)?	−140.0
Morgan Hill	6.2	4/24/84	242.0 (SJI)	233.0
Morgan Hill	6.2	4/24/84	−34.0 (SJ2)	−30.0
Morgan Hill	6.2	4/24/84	192.0 (SJ3)	111.0
San Juan Bautista	3.2	5/26/84	1.9 (SR)	1.66
Kettleman Hills	5.5	8/4/84	104.0 (GHI)	108
Kettleman Hills	5.5	8/4/84	166.0 (GH2)	110
Kettleman Hills	5.5	8/4/84	342.0 (EA)	246
Round Valley	5.8	11/23/84	68.0 (PO)	45
Quiensabe	5.3	1/26/86	51.0 (SR)	39
Quiensabe	5.3	1/26/86	50.0 (SJ)	44
Mt. Lewis	5.8	3/21/86	1.5 (SR)	1.4
North Palm Springs	5.8	7/8/86	−16.6 (PU)	−10.1
North Palm Springs	5.8	7/8/86	1.0 (BB)	1.5
North Palm Springs	5.8	7/8/86	15.1 (AM)	12.3
Oceanside	5.8	7/3/86	0.6 (PU)	0.7
Chalfant	6.4	7/21/86	138.7 (PO)	122
Whittier	5.9	10/1/87	27.5 (PU)	25.2
Whittier	5.9	10/1/87	−7.1 (BB)	−6.8
Whittier	5.3	10/4/87	2.3 (PU)	2.0
Supersitition Hills	6.6	11/23/87	13.8 (PF)	13.7
Pasadena	5.0	12/3/88	−11.6 (PU)	−8.6
Sierramadre	6.0	6/28/91	79.1 (PU)	76.7
Landers	7.3	6/28/92	470.0 (PU)	450
Landers	7.3	6/28/92	534.0 (PF)	552
Parkeld	4.8	10/20/92	9.4 (VC)	9.5
Parkeld	4.8	10/20/92	−7.9 (FR)	−6.6
Parkeld	4.8	10/20/92	6.8 (DL)	8.6
Parkeld	4.8	10/20/92	13.2 (EA)	12.1
Northridge	6.7	1/17/94	21.0 (PU)	20
Northridge	6.7	1/17/94	−5.0 (PF)	−3.2
Parkeld	5.0	12/20/94	120.4 (DL)	114.1
Parkeld	5.0	12/20/94	−21.0 (FR)	−21
Parkeld	5.0	12/20/94	30.0 (EA)	41.8

4. Earthquakes

The coseismic observations can be separated into two parts, the dynamic straingram and the static strain field offset generated by the earthquake. Theory of the strain seismogram can be found in Benioff (1935) and Richards (1976) and will not be pursued further here. The observation of strain offsets from earthquakes has a long history. Early hopes that these offsets could be observed teleseismically (Press, 1965) have not been realized. The primary reason seems to be that tilt and strain observations recorded under conditions of high acceleration in the cracked and fractured near-surface material in tunnels, caves and shallow boreholes, are not reliable because the materials and coupling are not stable (McHugh and Johnston,

1977). Installation of instruments in expansive cement in deep boreholes appears to have solved this problem (Sacks *et al.*, 1971, 1979; McGarr *et al.*, 1982). Data now show that coseismic strain steps can be recorded reliably.

Many examples have been recorded with the network of deep borehole instruments along the San Andreas fault. The best examples are listed in Table 1. For all the events shown here, clear strain offsets have been recorded. An example of the observed strain recorded on the nearest borehole strainmeter (PU) for the $M_W = 6.0$, 1 Oct. 1987, Whittier earthquake is shown in Figure 7a. This, and other similar offsets can be compared with the strain expected at each strainmeter location using the simplest model of each corresponding earthquake. These elastic dislocation models (Press, 1965; Okada, 1985, 1992) assume the moment of the earthquake is generated by uniform slip over a rupture plane that is indicated by the aftershock zone. An example of such a simple dislocation model for the 1987 Whittier earthquake is shown in Figure 7b. In this case the seismically determined moment is 1.4×10^{18} N m, the aftershock zone measures 8 km by 13 km at a depth of 15 km, with a dip of 40° down to the northeast and a strike of N62W. For these dimensions and assuming a shear modulus of 3×10^4 MPa, the slip expected for the earthquake is about 57 cm.

Similar models are routinely generated for the many tens of earthquakes with $M \geq 5.0$ for which we have strain offsets recorded. The calculated and observed offsets for the largest are listed in Table 1 and are plotted in Figure 8. The observed and calculated offsets are in good general agreement except for one example (marked with ?) recorded on the instrument EVSS. Because this instrument was installed just above a large fracture zone, nonlinear effects caused by the accelerations from the seismic waves are more likely to be significant. Thus the total moment release for these earthquakes on the San Andreas fault appears on average to be about 20% larger than that released seismically. Important implications of this general correspondence between observed and calculated strain steps are: (1) that observed strain steps recorded immediately after large earthquakes can be used to make a rapid, albeit preliminary, estimate of the moment of these earthquakes, and (2) crustal response at periods of minutes to hours is still largely elastic.

Postseismic responses observed in these data do not occur unless the earthquake magnitude is greater than $M = 7$, although some effects are sometimes apparent for earthquakes in the $M = 6$–7 range. Most of these effects are due to afterslip,

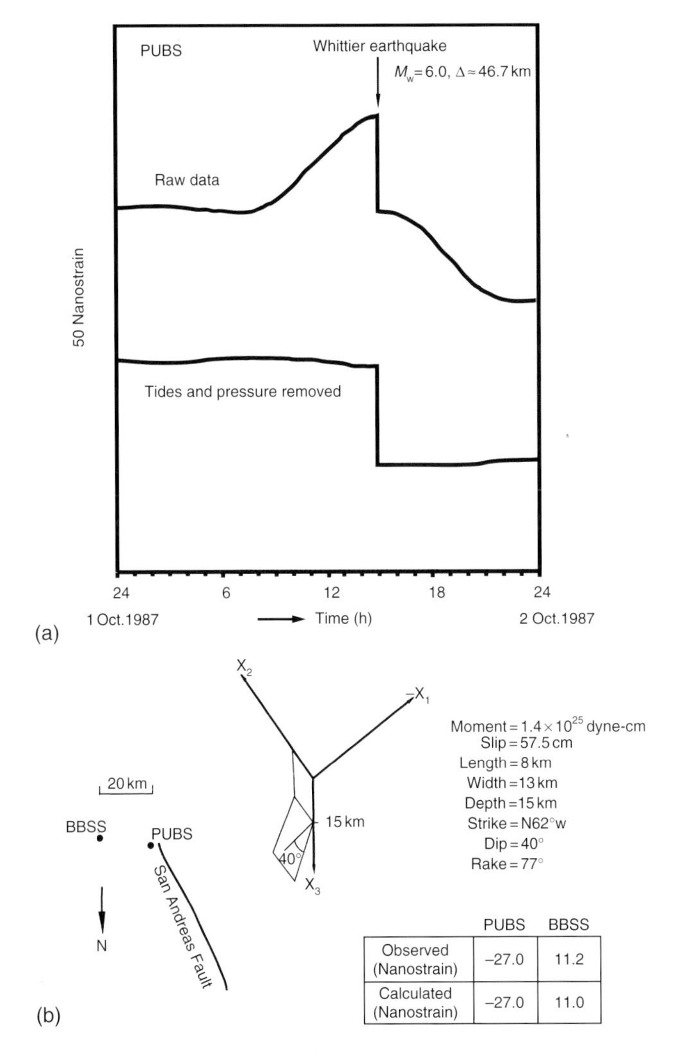

	PUBS	BBSS
Observed (Nanostrain)	−27.0	11.2
Calculated (Nanostrain)	−27.0	11.0

FIGURE 7 (a) Observed strain on 1 Oct. 1987 at PU showing the strain offset generated by the $M_w = 6$ Whittier Narrows earthquake. The variation in the upper plot is due to earth tides. These are predicted and removed in the lower plot. (b) Model of the Whittier Narrows earthquake showing comparison between the observed and predicted strain offsets.

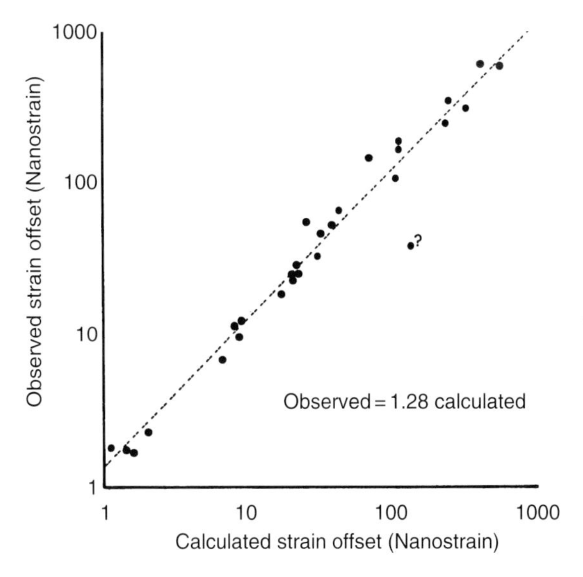

FIGURE 8 Observed coseismic strain offsets as a function of calculated offsets from simple dislocation models of the earthquakes listed in Table 1. See text for comments on point labeled ?.

but other contributions may result from triggered slip on other faults, as suggested by Sacks *et al.* (1979, 1982), to explain postseismic changes following the 1978 Izu, Japan, earthquake. Other short-term and longer-term effects may result from aquifier system response to the incoming seismic waves as observed by Wyatt *et al.* (1994) following the 1992 $M_w = 7.3$ Landers earthquake. In this case the time constant will be determined by the diffusion time constant. Transient strains and tilts related to fluid-diffusion effects triggered by the passage of seismic waves are very evident in measurements taken near the water table (McHugh and Johnston, 1977). The clearest postseismic strain signals have been observed for the larger earthquakes [e.g., Parkfield, 1966 (Smith and Wyss, 1968), Landers, 1992 (Johnston *et al.*, 1994), Loma Prieta, 1989 (Johnston *et al.*, 1990), and Imperial Valley, 1979 (Langbein, *et al.*, 1983)]. These data suggest power law slip in the upper few kilometers with some short-term aquifer re-adjustment effects.

4.1 Slow and Silent Earthquakes

Slow earthquakes refer to fault slip episodes at seismogenic depths that occur with some accompanying seismic radiation in the normal seismic frequency bands. Although long expected to occur (see for example, Kasahara, 1980), little direct evidence for their existence has been found in the few broadband deformation data sets obtained along active faults with enough sensitivity to detect these events. Several different types of slow earthquakes are expected. Slow earthquakes, that precede, and evolve, into "normal" earthquakes have been inferred for the great 1960 Chilean earthquake (Kanamori and Cipar, 1974; Cifuentes and Silver, 1989; Linde and Silver, 1989) and the 1992 $M_w = 7.6$ Nicaragua earthquake (Kanamori and Kikuchi, 1993). Slow earthquakes that follow "normal" earthquakes have been reported following the 1978 $M = 7$ Izu-Oshima earthquake (Sacks *et al.*, 1982), and silent earthquakes with no immediate associated seismicity have been suggested as the cause of strain transients recorded several days before the $M = 7.3$ Japan Sea earthquake (Linde *et al.*, 1988) and with the 1992 event at Sanriku-Oki, Japan (Kawasaki *et al.*, 1995). Large slow earthquakes with related but minor aftershock-like seismicity have occurred on three separate occasions on the San Andreas fault (Linde *et al.*, 1992; Johnston, 1997; Johnston *et al.*, 1998).

Indirect evidence for slow earthquakes has also been proposed on the basis of enhanced tsunami generation (Satake and Kanamori, 1977), anomalous normal modes (Kanamori and Anderson, 1975; Beroza and Jordan, 1990), and anomalous *P* arrivals (Kanamori and Kikuchi, 1993). Overall, when compared to the occurrence rate and moment release of conventional earthquakes, it would be fair to say from the limited near-fault strain data that slow earthquakes are a relatively rare phenomenon on active faults and contribute only in a minor way to total moment release. Nevertheless, they may play an important fundamental role in stress transfer and earthquake triggering (Simpson and Reasenberg, 1994; Harris, 1998).

The most important multiple slow earthquake on the San Andreas fault occurred with about a five-day duration in early December 1992, to the north of San Juan Bautista, California (Linde *et al.*, 1996). Other events were observed in 1996 and 1998. The 1992 sequence was recorded on three near-fault borehole strainmeters and two surface creepmeters along a 10 km section of the fault. The event was accompanied by minor seismicity that was obviously related to the strain data and culminated in surface rupture recorded on a surface creepmeter at the southern end of the strain array. Aseismic moment release for this event totaled about $5\text{–}6 \times 10^{16}$ N m, equivalent to about a $M = 5.1$ earthquake, whereas seismic moment release was more than 100 times less.

Borehole strainmeters at the sites shown in Figure 9a detected large slow coherent strain changes from 10 Dec. to 15 Dec. 1992. The overall event stands out clearly in the

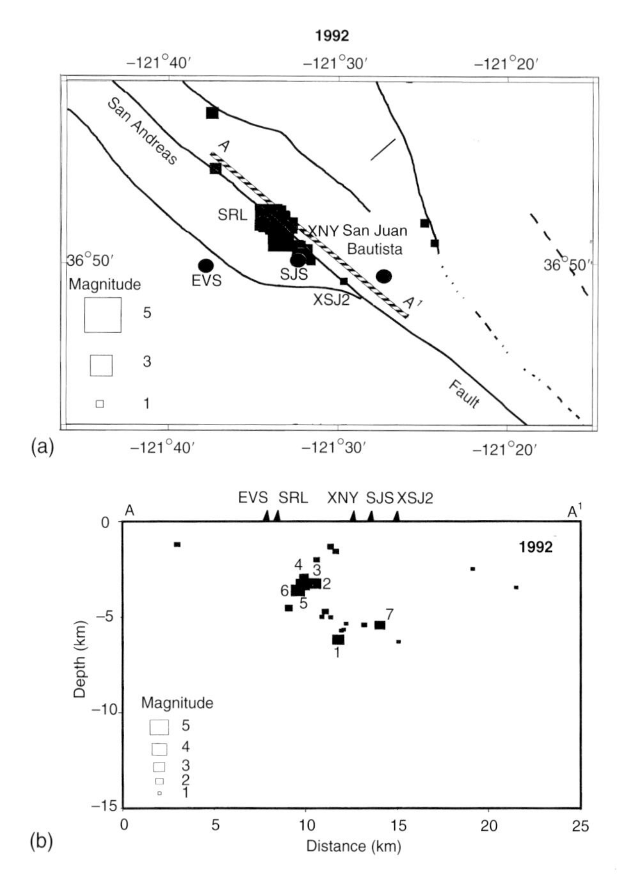

FIGURE 9 (a) Location of borehole dilatational strainmeters EVS, SRL, borehole three-component strainmeters SJS, and creepmeters XSJ2 and XNY with respect to the seismicity (squares) on 11–14 Dec. 1992, associated with a slow earthquake near the town of San Juan Bautista, California (see text). (b) Fault cross-section from A to A′ in (a) showing the locations of earthquakes following the slow earthquake in Dec. 1992, in relation to the position of the strain and creep instruments along the fault.

long-term data as the largest event of this type we have seen on these instruments. A swarm of minor earthquakes (aftershocks?) occurred in this region at the same time. These earthquakes were relocated with a 3D velocity model and their locations are shown on Figure 9a with a symbol size that relates to magnitude and in cross section in Figure 9b. The cross section (indicated by AA′ in Fig. 9b) shows the location of the earthquakes as a function of depth and position relative to the surface instruments. The seismicity outlines a crescent or "donut" shape on the section of fault beneath and between the strainmeters SRL and SJS from a few kilometers to about 8 km deep.

Figure 10a shows the data recorded on the three strainmeters as a function of time. The event began relatively rapidly on 11 Dec. 1992 followed by a slow exponential decay over the next eight days. The long-term decay was interrupted by a series of slow events, labeled A to E, from 11 Dec. to 15 Dec. Some indication of surface creep is evident on XNY following events A, C and E but no surface displacement is evident at XSJ2 until 14 Dec.

A $M=3.1$ earthquake occurred about two hours before the start of event A and two earthquakes of $M=3.3$ and 3.2 occurred within the first and third sample interval on both instruments (earthquakes 1, 2, and 3 in the center of Fig. 10a). Two $M=3.7$ earthquakes (earthquakes 5 and 6) occurred on the next day during event C on both instruments. Smaller earthquakes preceded and followed this event. Coseismic strain changes for all earthquakes are negligible. Because of the instrument sampling intervals, we cannot say with certainty whether these earthquakes preceded or followed the strain events but obviously they are causally related.

Although there are too few observation points to make a robust determination of the source model through formal inversion techniques (Bevington, 1969), we use forward quasi-static modeling to estimate the physical parameters of the simplest physical model that is consistent with the data. Source models are restricted to right-lateral slip segments on the San Andreas fault located in the region of microseismicity and observed surface creep. We search for the simplest quasi-static models that provide consistency with strain amplitudes and duration and surface creep since the strain changes are slow. The best model is shown in Figure 10b and comparison between model predictions (fine lines) and observations (thick lines) are shown on Figure 10a. The model details are generally similar to those given in Linde *et al.* (1996) but it includes three episodes of 3 mm of deeper slip in the region of seismicity, shown in Figure 9b, and slightly different values, shown in Figure 10a, to those used by Linde *et al.* (1996) in the upper section to get a better fit. The event E, coincident with surface creep on XSJ2, is due to surface slip on a patch indicated by W in Figure 10b that has previously failed repeatedly (Gladwin *et al.*, 1994). On this patch, slip of 5 mm with rupture velocity of 0.2 m sec^{-1} provides good agreement with data on SJS.

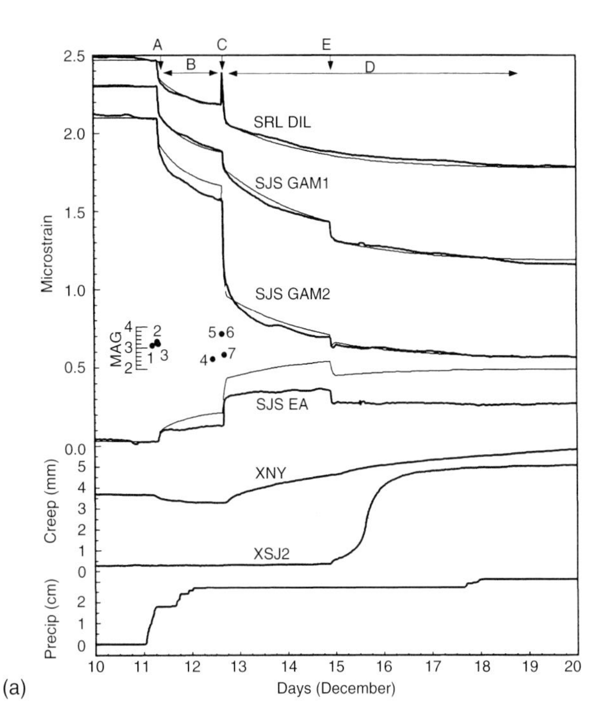

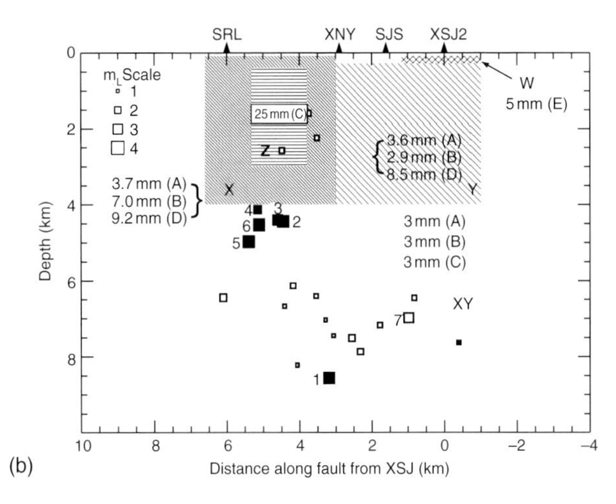

FIGURE 10 (a) Detrended strain data from dilatometers SRL and EVS and tensor strainmeter SJS for the period 10–20 Dec. 1992. Effects from earth tides and atmospheric pressure loading have been predicted and removed from the strain records. The tensor strain data have been converted into fault parallel shear γ_1, NS or EW shear γ_2 and areal strain. Shown also are fault creep data from creepmeter XSJ2 and earthquakes (numbered 1 through 7). Included are comparisons between observed (thin) and calculated (thick) strain-time histories resulting from the simplest model (see text) used to fit the overall strain record and all the slip episodes – subevents A, B, C, D and E discussed in the text. (b) Quasi-static model used to calculate the amplitudes and timescales of the sequence of slow earthquakes beneath San Juan Bautista in Dec. 1992. (See text for description of details.)

Event C has different strain character on SRL than on SJS consistent with changes in the near-surface strain field as the source geometry changes. We have examined all possible model areas between the two instruments and find the only viable solution results from 2.5 cm of slip on a patch Z. The rupture propagates slowly upward with an exponentially decreasing rupture velocity of 0.35 m sec^{-1} resulting in the sign reversal at SRL and the unidirectional changes at SJS.

For the remaining events, strain amplitude ratios cannot be satisfied by uniform slip over a single source area. We require a combination of slip on segments X, Y, and XY shown in Figure 10b. We limit the top of Y to be 0.3 km from the surface for consistency with the lack of creep until 14 Dec. The top of X is 0.1 km deep. We have reasonable control of the extent though the ends could be varied by about 1 km without significantly degrading the fit to the data. Uniform slip that decays with time on X, Y, and XY satisfies the data. This model has slip on X, Y, and XY of 3.7 mm, 3.6 mm, and 3 mm for A, 7.0 mm, 2.9 mm, and 3 mm for B, and 9.2 mm, 8.5 mm, and 3 mm for D (as shown on Fig. 10b) with time constants of 40 min for A, 15 h for B and 43 h for D.

The sequence of events starts with initial slow slip over all XY (event A), followed by slower slip over all XY (event B). Triggered slow upward rupture propagation over Z generates event C, and is followed by continued slow slip over XY (event D) and surface failure recorded on the creepmeter XSJ2 (event E).

The total moment release is 4.5×10^{16} N m. This is approximately equivalent to a $M = 5$ earthquake. In contrast, the seismic moment release was 7×10^{14} N m (a factor of 76 smaller than the total slow earthquake moment).

The microseismicity that was associated with this sequence occurred in a manner that was clearly related to the near-fault strain. Fitting of a simple exponential relaxation function of the form

$$f(t) = C(1 - \exp^{-at}) \qquad (1)$$

to both the strain data and the cumulative earthquake number indicate similar time constants, $(1/a)$, of about 2.5 days. The occurrence of these slow events, and later events in 1996 and 1998 (Johnston, 1999), has reduced strain energy and has thus reduced (at least temporarily) the likelihood of a large damaging earthquake in this area, in contrast to suggestions by Behr *et al.* (1990).

4.2 Strain Redistribution

Strain redistribution has been clearly observed following each of the largest earthquakes in the San Andreas fault system. The best examples have been obtained for the 1992 Landers earthquake (Wyatt *et al.*, 1994; Johnston *et al.*, 1994), the 1989 Loma Prieta earthquake (Johnston *et al.*, 1990), and the 1987 Whittier earthquake (Linde and Johnston, 1989). Further details of strain redistribution following the Loma Prieta earthquake can be found in Burgmann *et al.* (1997) and Mueller and Johnston (2000).

5. Earthquake Nucleation

As pointed out earlier, laboratory observations and modeling efforts suggest that fault failure occurs when displacements exceed those for peak shear stress. Increased deformational "weakening" occurs before the dynamic slip instability that results in an earthquake. In rock mechanics terminology this is called "tertiary creep" (Jaeger and Cook, 1976). In terms of laboratory-determined rate or state-dependent friction (Ruina, 1983; Dieterich, 1994), it is called "slip weakening." That nonlinear strain precedes rupture has provided hope that detection of these strain changes will lead to a method for earthquake prediction. At issue is the scale on which this occurs. Since continuous strain measurements near moderate to large earthquakes thus far have failed to detect any indications of exponentially increasing strain (Johnston *et al.*, 1990, 1994; Wyatt, 1994; Abercrombie *et al.*, 1995), the scale must be small compared to the eventual rupture size. Here we discuss just how small this might be.

5.1 Field Observations

We choose just one typical example of short-term and intermediate-term strain obtained before, and immediately above, a moderate magnitude earthquake to illustrate this scale. The event we choose was one of three $M = 5$ earthquakes that occurred beneath the Parkfield strainmeter array shown in Figure 11a. A cross section of these events is shown in Figure 11b. The first was a $M = 4.7$ earthquake that occurred on 31 Oct. 1992. The second was a $M = 4.8$ earthquake that occurred on 20 Nov. 93 and the third was a $M = 5$ earthquake on 24 Dec. 1994.

Figure 12a shows the strain data from FRS and DLS for the $M = 5$ event on 20 Dec. 94 for the five days before and after this earthquake. The vertical scale is in 10 nanostrain units in the upper plot that shows the strain offsets (+31 nanostrain at FR and −94 nanostrain at DL) generated by the earthquake. The scale is expanded in the lower plot by a factor of 50 (0.5 nanostrain units).

Figure 12b (upper) shows the strain seismograms for this earthquake recorded on these two instruments. The expanded timescale in the lower plot shows the data for these events in the last seconds before rupture. Here the vertical scale is 1000 nanostrain. Figure 12c shows the strain vertical scale expanded by 100 in the upper plot (units of 10 nanostrain) and by 10 000 in the lower plot (units of 100 picostrain). The occurrence time of the earthquake is shown with an arrow. Strain changes, if they occurred before the event, appear to be less than 0.01 nanostrain. Similar observations were obtained for the other two $M = 5$ earthquakes.

It is apparent for these events, and for the 17 or so moderate to large earthquakes for which we have near-field strain data, that none of the strain records shows any indication of accelerating strain prior to rupture at our ability to resolve

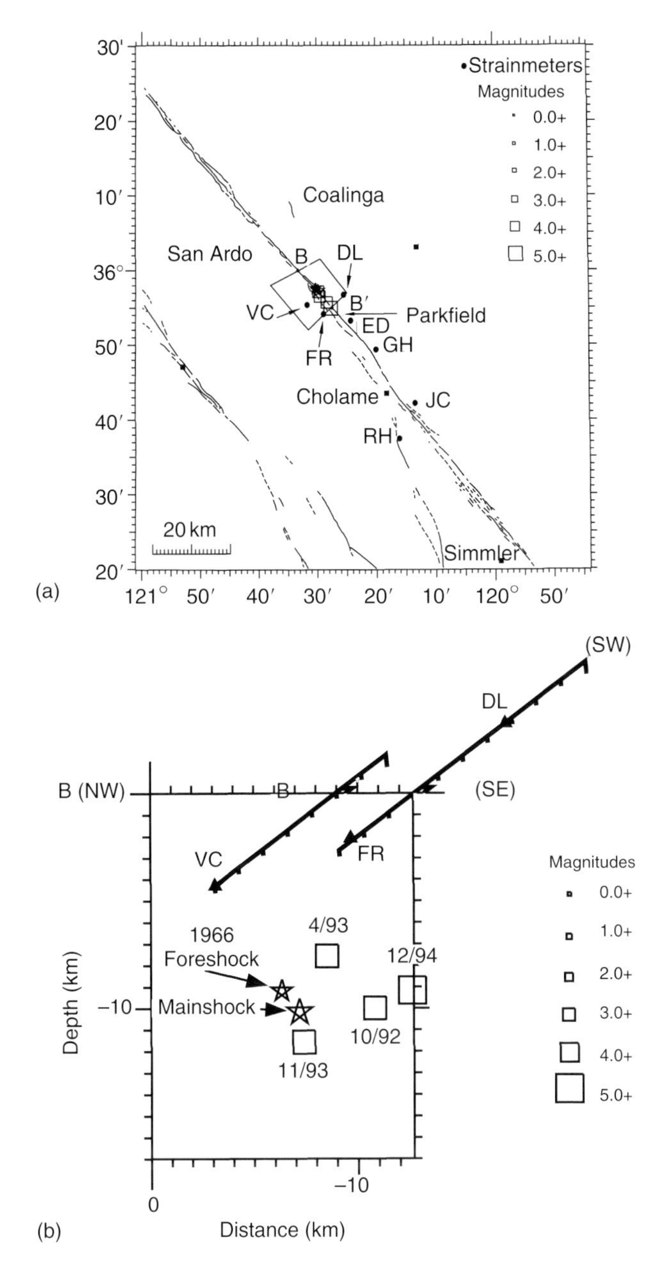

FIGURE 11 (a) Map of Parkfield showing the location of three near-magnitude 5 earthquakes beneath the dilatational strainmeters VC, FR, and DL on 20 Oct. 1992, 14 Nov. 1993, and 20 Dec. 1994. The locations of the various events are shown in cross section in (b). (b) Seismicity cross section showing the location of three near-magnitude 5 earthquakes beneath the dilatational strainmeters VC, FR, and DL on 20 Oct. 1992, 14 Nov. 1993, and 20 Dec. 1994. Also shown is the hypocenter of the 1966 Parkfield mainshock (Star).

strain (0.01 nanostrain over seconds and 1.0 nanostrain over days). This does not mean that accelerating strain is not occurring, only that it is below our measurement resolution. Another way of saying this is that the maximum possible preseismic slip moment in the eventual hypocenter produces strains at the instrument that are less than, or equal to our measurement resolution.

The maximum possible precursive slip moment expressed as a percentage of coseismic moment is plotted against seismic moment for these earthquakes in Figure 13. In general, moment release before these earthquakes is less than a few percent and perhaps less than 0.1% if we take the later data for which the signal-to-noise is best (Johnston *et al.*, 1990, 1994; Wyatt, 1994). What this most likely means is that rocks do indeed fail as expected but that the initial slip is small and localized compared to the final or total slip. The ratio of total slip to nucleation slip is thus large and rupture initiation is inhomogeneous in agreement with a variety of theoretical and seismological observations (discussed below).

Since our ability to resolve strain changes improves at shorter periods, we can place better constraints on the largest precursive slip moment that might occur as a function of decreasing period or increasing resolution. At periods of seconds we can resolve 0.01 nanostrain, over hours it is 0.1 nanostrain and over weeks it is 1.0 nanostrain. Simple elastic dislocation calculations (Okada, 1985, 1992) allow us to calculate strains at these different levels for precursive slip moments at various depths. For example, a slip moment of 10^{12} N m at a depth of 5 km will produce surface strains of 0.01 nanostrain. If the depth is 8 km, the slip moment would be 3×10^{12} N m. Precursive slip moment as a function of resolution (or period) and depth is shown in Figure 14a.

How big might the nucleation zone be? Let us assume displacement, U, is proportional to rupture length, L. Observation and theory (constant stress drop assumption) support this assumption for earthquakes with M_L less than 6 (e.g., see Scholz, 1997). Then,

$$U = KL\Delta\sigma \tag{2}$$

where $K \approx 10^{-7}$ m MPa^{-1} km^{-1}, $L =$ rupture dimension, and $\Delta\sigma =$ stress drop ($\approx$1 MPa). Therefore, the moment M is given by:

$$M \approx \mu 10^{-4} L^3 \tag{3}$$

where μ is the rigidity. It is thus possible to obtain crude limits on the nucleation size since we have limits on the maximum precursory slip moment that can be released over periods of weeks to seconds before seismic rupture. Figure 14b shows the equivalent limit on the source zone size obtained from Eq. (3) as a function of resolution (or period) and depth. For the $M = 5$ earthquakes at Parkfield, where we are within 7 km of the source, the maximum size of the nucleation patch in the last few seconds before seismic radiation occurs, is about 69 m. At longer periods, where the resolution is less, the constraint on size becomes poorer. On time periods of several weeks where our capability to resolve strain changes is not more than a few nanostrain, we cannot resolve nucleation regions unless their dimensions are greater than several hundred meters. However, it is important to note that, in the last hours to seconds before rupture, we would expect the largest nonlinear strain

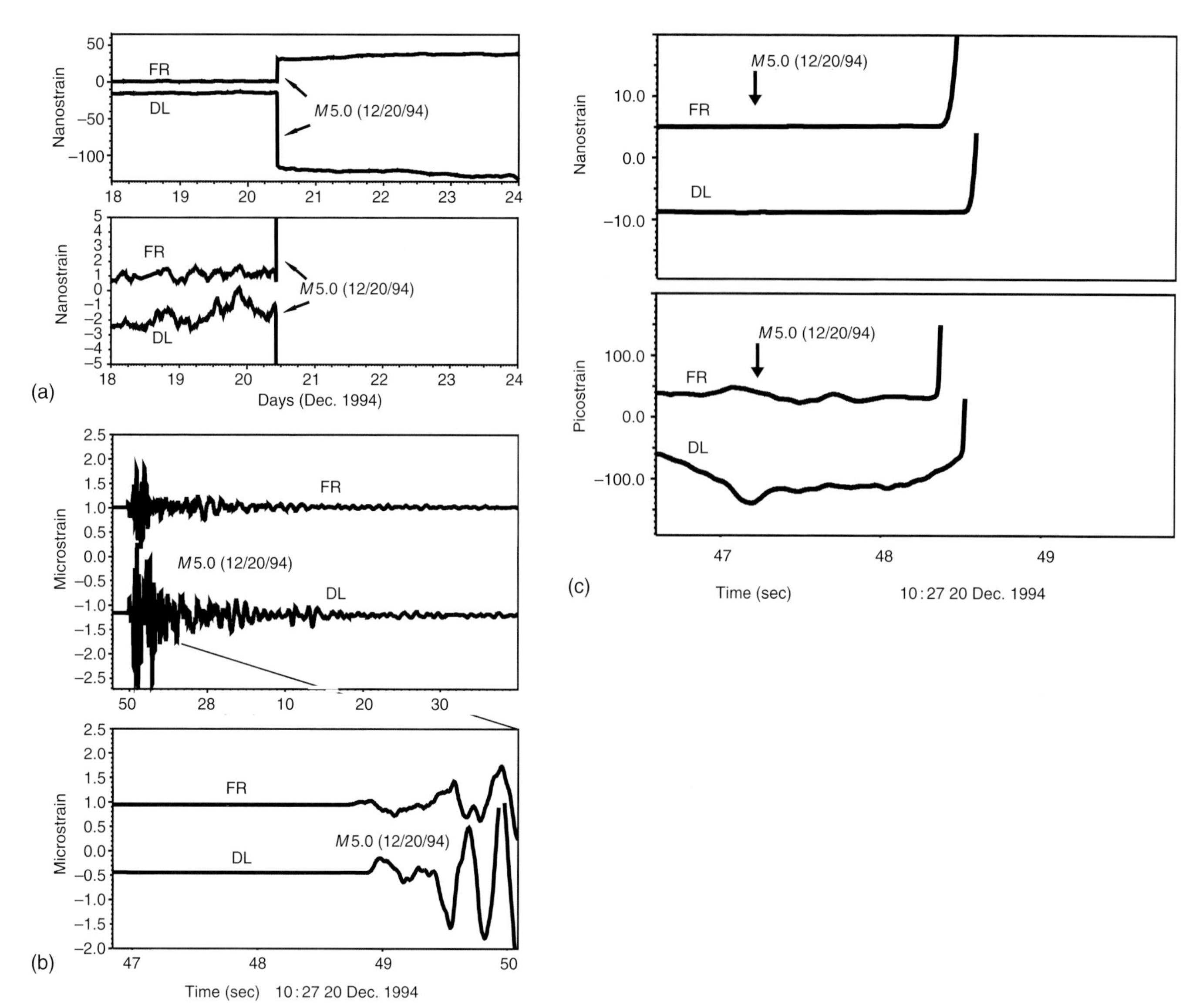

FIGURE 12 (a) Strain time histories from strainmeters FR and DL on opposite sides of the San Andreas fault from 15 to 26 Dec. 1994. Full-scale on the upper plot is 200 nanostrain that on the lower plot is 10 nanostrain. (b) Expanded time-history at 10 : 27 UT on 20 Dec. 1994. The upper plot shows the strain seismogram of the $M = 5.0$ earthquake starting at about 10 : 27 : 49 UT. The lower plot shows an even more expanded section during the 2 sec before the first arrival. The strain units on the vertical scale are 1000 nanostrain. (c) Expanded vertical scale for the time-history at 10 : 27 UT on 20 Dec. 1994 shown in (b) (lower). The upper plot shows the vertical scale expanded by 100 and the lower plot shows the scale expanded by 10 000. The actual occurrence time is shown with a vertical arrow. Similar data were obtained for the other two $M = 5$ earthquakes on these instruments.

(as indicated in Fig. 1). It is during this time period that strain resolution is the best. The fact that no clear indication of accelerating strain is apparent in any of the strain records suggests nucleation patch sizes of perhaps a few tens of meters. Repeating microearthquakes on patches with this dimension have been observed by Nadeau and McEvilly (1997).

These data similarly constrain changes in pore pressure in the fault zone that may be related to the failure process. If we assume no hydraulic connectivity between a region of high pore pressure on the fault zone (such as proposed by Byerlee, 1990, 1993) and ignore for the moment how pore pressure changes in these regions might occur, it is obvious from simple calculations of strain generated by changes in pore pressure for regions with dimensions of a few hundred meters and depths of about 5 km that rapid changes in pore pressure of 0.1 MPa should be readily detected, though not necessarily recognized as such. If hydraulic communication with the surroundings does exist, smaller pore pressure changes might be detected as poroelastic strain as fluids diffuse out to greater distances, in effect increasing the source size. Small changes

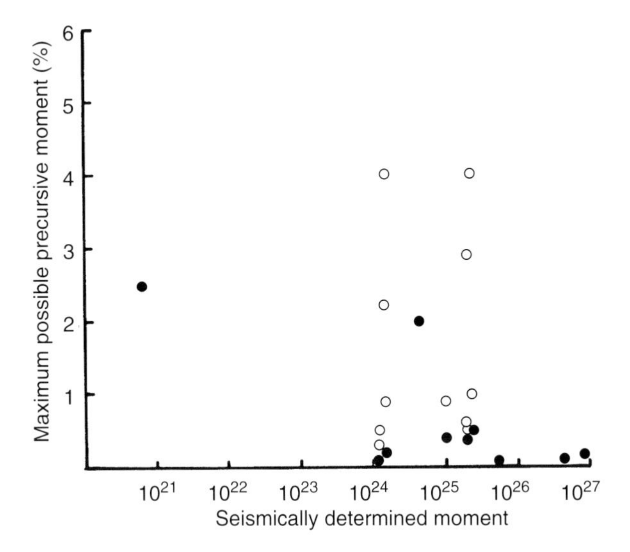

FIGURE 13 Plot of maximum possible prerupture moment as a percentage of the seismically determined moment for the events studied listed in Table 1. Closed circles show data obtained within one rupture length of subsequent earthquakes. Open circles show data obtained at greater distances.

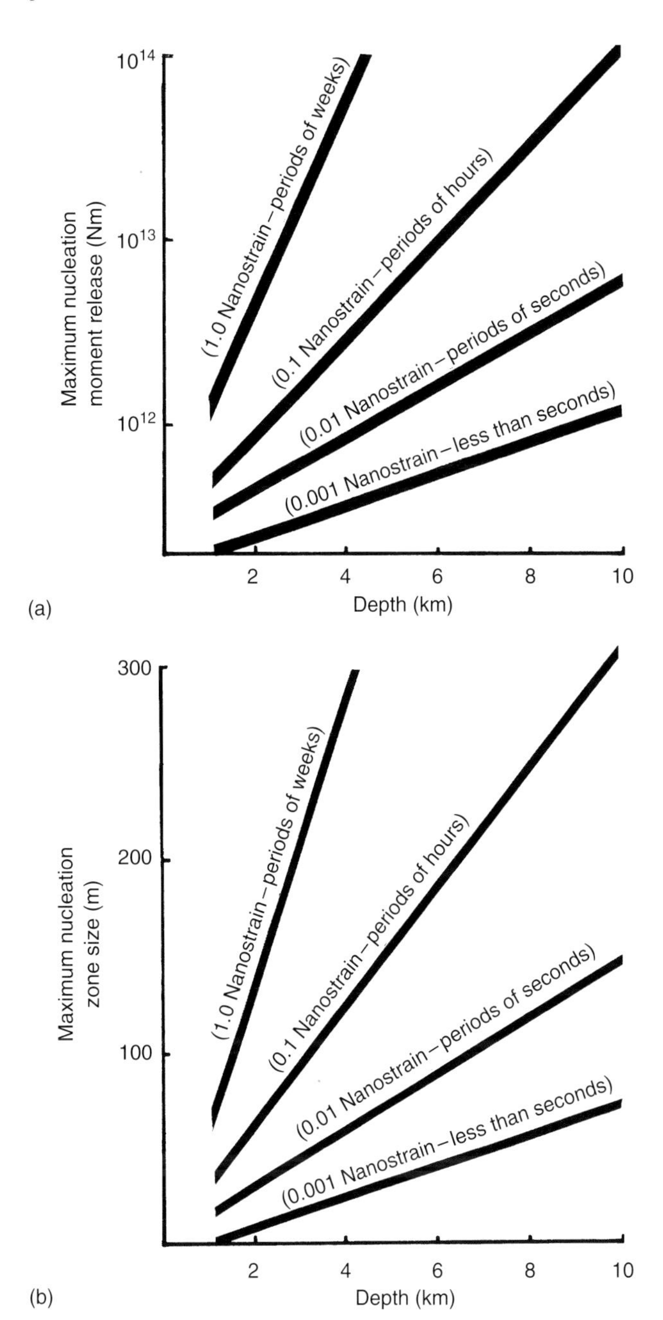

FIGURE 14 (a) Maximum nucleation moment release allowable as a function of nucleation source depth for strainmeter resolution in nanostrain at different periods. Over periods of seconds the resolution is 0.01 nanostrain, over periods of hours it is 0.1 nanostrain and over periods of weeks it is 1 nanostrain. (b) Maximum nucleation zone size allowable as a function of nucleation source depth for strainmeter resolution in nanostrain at different periods. Over periods of seconds the resolution is 0.01 nanostrain, over periods of hours it is 0.1 nanostrain and over periods of weeks it is 1 nanostrain.

in the local geometry of regions with high pore pressure such as, for example, a 100 m by 100 m zone at 5 km depth dropping pressure by doubling its size probably would not be detectable.

It is also apparent in these data that large-scale changes in elastic properties within the fault zone are not occurring near these instruments. If they were, we would expect changes in tidal shear strain and volumetric strain response. Observations of earthtide response before and after the largest earthquakes (Landers, Loma Prieta, etc.) indicate that if measurable changes occur, they are less than 1% (Linde *et al.*, 1992).

5.2 Comparison with Laboratory Observations

It seems unlikely that both the wealth of laboratory data describing anelastic deformation of unfractured crustal rocks before failure and theoretical calculations of anelastic deformation during crack initiation and crack behavior are not relevant to initiation of fault failure. However, it is clear from existing surface and borehole strain data that slip instability over the entire subsequent fault rupture is not possible unless the time constant for rupture initiation is unrealistically short or prerupture deformation is exceedingly small. Laboratory measurements of crustal rocks under high confining pressure indicate anelastic strains before failure are comparable to the elastic strain release upon failure (Jaeger and Cook, 1976). On the other hand, frictional slip on preexisting slip surfaces in large-scale laboratory friction experiments (Dieterich, 1981) indicates nucleation slip does initiate within a small nucleation region. Dieterich (1992) shows numerical calculations that indicate the size of these regions should be 1–10 m for both smooth or rough initial stress distributions, consistent with the upper bounds allowed by the strain data. It would appear from both data that the scale on which rupture initiates is several orders of magnitude smaller than that of the subsequent earthquake. If so, the fault model relevant to this process must be inhomogeneous in character to accommodate these observations that crustal failure does not occur simultaneously across the entire rupture zone, but is apparently triggered by failure of small localized zones.

Comparisons can also be made to the duration of nucleation derived by Dieterich (1994), who determined an expression for the time to failure during nucleation as a simple function of sliding velocity. For reasonable values of stiffness, velocity, normal stress gradient, etc., the duration of nucleation is on the order of 1–10 sec. This is in general agreement with the indications from the strain data. Furthermore, the nucleation moment from these laboratory measurements (Dieterich, 1992) is also only a small fraction of the slip moment as seen in the strain data.

5.3 Comparison with Current Theory

That an inhomogeneous faulting model may best describe the initiation of failure is not at all surprising. Wyss and Brune (1967) and many researchers since then have made similar suggestions. Based on careful analysis of seismic waveforms and on synthetic waveform generation, Kanamori (1981) and Lay *et al.* (1982) have shown that an inhomogeneous rupturing model is required to explain the complex waveforms generated by many large earthquakes. These data have been used to suggest that regions of greater strength (barriers or asperities) can be broken during fault failure and energy release from these regions contributes to the seismic radiation.

Inhomogeneous fault models have been proposed by McGarr (1981) to explain peak ground motion in an earthquake, by Dieterich (1986) to explain simulated faulting in laboratory samples, and, in a more general way, by Rundle *et al.* (1984) to explain seismicity gaps, fault roughness indicated by complex seismic waveforms, and apparent triggering effects. The data reported here are consistent with the suggestion outlined above that rupture initiation most likely occurs at smaller regions of higher strength which, when broken, allow runaway catastrophic failure. An inhomogeneous failure model, in which various areas of the fault plane have different stress-slip constitutive laws, appears necessary to explain both the processes that lead to failure and those that occur following failure. This type of model was proposed by Rundle *et al.* (1984). Other inhomogeneous failure models are also possible.

6. Conclusions

Observations of crustal strain at high sensitivity near active faults allow us to identify the main features of strain transients that occur. On timescales of years, the primary features are uniform strain accumulation with occasional strain steps generated by large earthquakes, followed sometimes by post-seismic strain redistribution (Savage, 1983; Burgmann *et al.*, 1997). Conventional geodetic, EDM instruments and GPS instruments are best used to obtain these data. At shorter timescales (months to milliseconds), higher strain resolution is possible either by spatial averaging with a laser interferometer over about 500 m to suppress surface noise (Wyatt, 1988), or by installing instruments at depth in boreholes (Sacks *et al.*, 1971; Gladwin, 1984).

Strain transients associated with fault creep events indicate that these events result primarily from near-surface soil failure that is driven by deeper smoother fault slip. These events can sometimes be triggered by high accelerations during the passage of seismic waves from large nearby earthquakes (Simpson *et al.*, 1988) and sometimes by rainfall (Schulz *et al.*, 1982). In either case the event-like character is not of much interest in fault mechanics. However, the net creep-time history does appear to reflect the net slip on the fault and long-term changes in fault creep rate are important indicators of change in fault slip rate at depth.

There is little evidence of slip waves propagating slowly along active faults, at least on timescales of months to minutes. Either the amplitudes of these slip waves, if they exist, is very small ($\leq$1 cm), or the period is so large that they cannot be resolved from the longer-term noise. Even at long periods (month to years), evidence of slip waves is not apparent in 2-color EDM data at Parkfield and San Juan Bautista where periodic surface displacements on opposite sides of the San Andreas fault of a millimeter or so should be readily detectable. The resolution of the 2-color EDM is about 0.1 mm (Langbein, 1981).

Coseismic strain offsets recorded on borehole strainmeters for earthquakes with magnitudes from 3 to 7 are generally in good agreement with offsets expected from simple dislocation models of these earthquakes. This suggests that the use of dislocation models for calculating coseismic stress changes (e.g., Reasenberg and Simpson, 1992) is generally valid, or at least far removed from the rupture ends. The variable and general amplified coseismic offset response to earthquakes previously observed in the cracked and fractured material near the Earth's surface, and in caves or tunnels, appears to be avoided by installing instruments in grout in deep boreholes. An important implication of these observations is that strain redistribution from an earthquake is transmitted largely elastically through the complex geology and fault geometry in these regions. A second important implication is that preliminary moment determinations can be made within a few tens of minutes of the occurrence of moderate magnitude earthquakes from observation of the strain offsets generated by these earthquakes.

Unambiguous strain recordings on several borehole strainmeters now exist of several slow earthquakes on the

San Andreas fault near San Juan Bautista and other locations. The clearest event occurred on a 6 km long by 8 km wide segment of the fault just to the north of San Juan Bautista during a 4-day period starting on 11 Dec. 1992. This event is complex with a number of episodes. Evidence exists for both slip acceleration and variable slip geometry for these episodes. This event and later events in 1996 and 1998 are unique in the 15-year record of strain at these sites. The 1992 event generated shear-strains exceeding 1 microstrain on the tensor strainmeter MSJS near the town of San Juan Bautista and dilatational strain exceeding 0.7 microstrain at the dilatational strainmeter SRLS 5 km to the northwest along the fault. A third dilatometer installed some 7 km from the fault recorded strains of about 0.1 microstrain. Quasi-static modeling of this event indicates that it probably ruptured a 50 km^2 segment of the fault in less than 10 min and that episodes of exponential-like slip, either triggered or accelerated by the larger aftershocks, continued with time. Nearly four days after the event started at depth, it reached the surface and generated a 3-day episode of surface fault displacement that was recorded on a surface creepmeter XSJ2 east of the town of San Juan Bautista. The entire sequence generated event moment release that was equivalent to a $M = 5$ earthquake.

Actual recordings of slow earthquakes have been rarely reported in the geophysical literature. The question naturally arises as to whether this is because they are truly rare, or whether the situation has resulted from the paucity of high-sensitivity strain instruments installed near active faults and the relatively short history of high-quality recordings. In any case, as a consequence of these slow earthquakes near San Juan Bautista, the immediate likelihood of moderate magnitude earthquakes in this region has been reduced, in contrast to suggestions by Behr *et al.* (1990).

Popular views of the Earthquake nucleation process and fault kinematics have included suggestions of nonlinear deformation prior to rupture in regional-scale "preparation zones" of earthquakes (Rice and Rudnicki, 1979), strain redistribution by "crustal block interaction" (Jackson and Molnar, 1990), propagating aseismic slip waves (Kasahara, 1979), and variation in the material properties of near-fault materials with time and location (Dobrovolsky *et al.*, 1979).

In contrast, attempts to detect preseismic strain change near several recent moderate earthquakes during the final hours to seconds approaching rupture near many moderate earthquakes in California and Japan have failed. This places important constraints on possible moments and dimensions of nucleation events. First, since short-term nonlinear precursive strains greater than a nanostrain have not been observed in the eventual epicentral region prior to rupture, moment release during nucleation is thus $\approx$300 times smaller than that released statically and $\approx$10 000 times smaller than that released dynamically. For the best cases investigated, the greatest possible preseismic moment is less than 0.1% of the final seismic or geodetic moment. This implies tertiary creep prior to rupture is minimal over the entire eventual rupture zone, as expected for brittle failure.

Secondly, the dimensions of the rupture nucleation region for the largest earthquakes for which we have high-resolution near-field preseismic strain data, are less than several tens of meters. This is consistent with nucleation dimensions suggested by clusters of repeating microseismicity (Nadeau and McEvilly, 1997) and with indications from earthtidal response that fault zone material properties are invariant on timescales of days to months (Linde *et al.*, 1992).

The apparent small size of the rupture initiation moments compared to the total earthquake moments suggests that there is no scaling of the nucleation process with earthquake magnitude. This is inconsistent with suggestions by Ellsworth and Beroza (1995) and Iio (1995) that nucleation scales with earthquake magnitude. In fact, one of the earthquakes used by Ellsworth and Beroza (1995) to demonstrate nucleation is the one shown in Figure 12. The strain data require no processing, have a flat response from 50 Hz to DC and show no indication of nucleation prior to the first seismic wave arrival. The basic failure process thus apparently involves rupture nucleation and runaway. High pore pressure fluids may be associated with this process but if pore pressure changes drive the nucleation process, these changes are less than about 0.1 MPa at seismogenic depths.

An inhomogeneous faulting model for which different areas of the fault plane have either different constitutive properties or different stress-slip constitutive laws most easily explain these data. This type of model was proposed by Rundle *et al.* (1984) and is indicated by observations of complex seismic waveform generation during rupture (Kanamori, 1981; Lay *et al.*, 1982). For these inhomogeneous models, failure initiation occurs in relatively localized zones and expands until barriers of sufficient strength to stop the rupture, are encountered. The processes that lead to failure as well as those that follow failure are inhomogeneous in character. Energy driving the system appears to come from uniformly accumulated elastic strain energy (Prescott *et al.*, 1979). Detection of rupture nucleation apparently will require these highly sensitive and stable instruments be installed even closer than at present to the hypocenters of large earthquakes.

A variety of postseismic responses have been observed. This mostly reflects continuing slip following large earthquakes, with contributions in some cases from the equilibration of the fluid regime. During this postseismic period, slip varies as log (time) consistent with either relaxation of a near-surface region after stress-induced velocity strengthening (Scholz, 1990), or a fluid-diffusion process. Afterslip is common in regions where aseismic fault slip occurs, such as on the San Andreas fault following the 1966 Parkfield earthquake (Smith and Wyss, 1968), and on the Imperial fault following the 1979 Imperial Valley earthquake (Langbein *et al.*, 1983).

Acknowledgments

Numerous colleagues have provided stimulating discussions. We thank Bob Mueller and Doug Myren for help with installation and maintenance of the strain instruments and John Langbein for contributions to Figure 2. Ruth Harris and Nick Beeler provided thoughtful reviews.

References

Abercrombie, R.E., *et al.* (1995). *Bull. Seismol. Soc. Am.* **85**, 1873–1879.
Agnew, D.C. (1997). *J. Geophys. Res.* **102**, 5109–5110.
Andrews, D.J. (1976). *J. Geophys. Res.* **81**, 5679–5687.
Atkinson, B.K. (1979). *Pure Appl. Geophys.* **117**, 1011–1024.
Behr, J., *et al.* (1990). *Trans. Am. Geophys. Un.* **71**, pp. 1645.
Benioff, H. (1935). *Bull. Seismol. Soc. Am.* **25**, 283–309.
Beroza, G. and T. Jordan (1990). *Seismol. Res. Lett.* **61**, pp. 27.
Bock, Y. (1997). *Trans. Am. Geophys. Un.* **78**, F165.
Brune, J.N. (1979). *J. Geophys. Res.* **84**, 2195–2198.
Bevington, P.R. (1969). "Data Reduction and Error Analysis for the Physical Sciences", McGraw Hill.
Burgmann, R., *et al.* (1997). *J. Geophys. Res.* **102**, 4933–4955.
Byerlee, J. (1990). *Geophys Res. Lett.* **17**, 2109–2112.
Byerlee, J. (1993). *Geology* **21**, 303–306.
Cifuentes, I.L. and P.G. Silver (1989). *J. Geophys. Res.* **94**, 643–663.
Das, S. and C.H. Scholz (1981). *J. Geophys. Res.* **86**, 6039–6051.
Dieterich, J.H. (1979). *J. Geophys. Res.* **84**, 2161–2168.
Dieterich, J.H. (1981). *AGU Geophys. Monogr. Ser.* **24**, 103–120.
Dieterich, J.H. (1986). *AGU Geophys. Monogr.* **37**, 37–47.
Dieterich, J.H. (1992). *Tectonophysics.* **211**, 115–134.
Dieterich, J.H. (1994). *J. Geophys. Res.* **99**, 2601–2618.
Dobrovolsky, I.P., *et al.* (1979). *Pageoph* **117**, 1025–1044.
Eaton, J.P. (1985). *US Geol. Surv. Open File Report 85–44*, 44–60.
Ellsworth, W.L. and G.C. Beroza (1995). *Science* **268**, 851–855.
Evans, K.F., *et al.* (1981). *J. Geophys. Res.* **86**, 3721–3735.
Freund, L.B. (1979). *J. Geophys. Res.* **84**, 2199–2209.
Gladwin, M.T. (1984). *Rev. Sci. Instrum.* **55**, 2011–2016.
Gladwin, M.T., *et al.* (1987). *J. Geophys. Res.* **92**, 7981–7988.
Gladwin, M.T., *et al.* (1991). *Geophys. Res. Lett.* **18**, 1377–1380.
Gladwin, M.T., *et al.* (1994). *J. Geophys. Res.* **99**, 4559–4565.
Goulty, N.R. and R. Gilman (1978). *J. Geophys. Res.* **83**, 5415–5419.
Harris, R.A. (1998). *J. Geophys. Res.* **103**, 24347–24358.
Harrison J.C. (1976). *J. Geophys. Res.* **81**, 319–328.
Hart, R.H.G., *et al.* (1996). *J. Geophys. Res.* **101**, 25553–25571.
Heaton, T.H. (1982). *Bull. Seismol. Soc. Am.* **72**, 2181–2200.
Iio, Y. (1995). *J. Geophys. Res.* **100**, 15333–15349.
Ito, T.H. (1982). *Bull. Seismol. Soc. Am.* **20**, 2181–2200.
Jackson, J. and P. Molnar (1990). *J. Geophys. Res.* **95**, 22073–22087.
Jaeger, J.C. and N.G.W. Cook (1976). "Fundamentals of Rock Mechanics," Chapman and Hall.
Johnston, M.J.S. (1997). *Trans. Am. Geophys. Un.* **79**, F156.
Johnston, M.J.S. (1999). *Trans. Am. Geophys. Un.* **80**, F691.
Johnston, M.J.S. and R.D. Borcherdt (1984). *Trans. Am. Geophys. Un.* **65**, 1015.
Johnston, M.J.S., *et al.* (1976). *Nature* **260**, 691–693.
Johnston, M.J.S., *et al.* (1977). *J. Geophys. Res.* **82**, 5683–5691.
Johnston, M.J.S., *et al.* (1986). *J. Geophys. Res.* **91**, 11497–11502.
Johnston, M.J.S., *et al.* (1987). *Tectonophysics* **144**, 189–206.
Johnston, M.J.S., *et al.* (1990). *Geophys. Res. Lett.* **17**, 1777–1780.
Johnston, M.J.S., *et al.* (1994). *Bull. Seismol. Soc. Am.* **84**, 799–805.
Johnston, M.J.S., *et al.* (1998). *Trans. Am. Geophys. Un.* **79**, F600.
Kanamori, H. (1981). *Maurice Ewing Series IV*, pp. 1–19 *American Geophysics Union.*
Kanamori, H. and D.L. Anderson (1975). *J. Geophys. Res.* **80**, 1075–1078.
Kanamori, H. and J. Cipar (1974). *Phys. Earth Planet. Inter.* **9**, 128–136.
Kanamori, H. and M. Kikuchi (1993). *Nature* **361**, 714–716.
Kasahara, K. (1979). *Tectonophysics* **52**, 329–341.
Kasahara, K. (1980). "Earthquake Mechanics," Cambridge University Press.
Kasahara, M., *et al.* (1983). *Tectonophysics* **97**, 327–336.
Kawasaki, I., *et al.* (1995). *J. Phys. Earth* **43**, 105–116.
Kostrov, B.V. (1966). *J. Appl. Math. Mech.* **30**, 1241–1248.
Langbein, J.O. (1981). *J. Geophys. Res.* **62**, 4941–4948.
Langbein, J.O. and H. Johnson (1996). *J. Geophys. Res.* **102**, 591–604.
Langbein, J., *et al.* (1983). *Bull. Seismol. Soc. Am.* **73**, 1203–1224.
Lay, T., *et al.* (1982). *Earthqu. Predict. Res.* **1**, 3–71.
Linde, A.T. and M.J.S. Johnston (1989). *J. Geophys. Res.* **94**, 9633–9643.
Linde, A.T. and P.G. Silver (1989). *Geophys. Res. Lett.* **16**, 1305–1308.
Linde, A.T., *et al.* (1988). *Nature* **334**, 513–515.
Linde, A.T., *et al.* (1992). *Geophys. Res. Lett.* **19**, 317–320.
Linde, A.T., *et al.* (1993). *Nature* **365**, 737–740.
Linde, A.T., *et al.* (1996). *Nature* **383**, 65–68.
Lorenzetti, E. and T.E. Tullis (1989). *J. Geophys. Res.* **94**, 12343–12361.
Matsu'ura, M. and T. Tanimoto (1980). *J. Phys. Earth* **28**, 103–118.
Marone, C.J., *et al.* (1991). *J. Geophys. Res.* **96**, 8441–8452.
McGarr, A. (1981). *J. Geophys. Res.* **86**, 3901–3912.
McGarr, A., *et al.* (1982). *Geophys. J. R. Astron. Soc.* **70**, 717–740.
McHugh, S. and M.J.S. Johnston (1977). *J. Geophys. Res.* **82**, 5692–5697.
McHugh, S. and M.J.S. Johnston (1978). *Bull. Seismol. Soc. Am.* **68**, 155–168.
McHugh, S. and M.J.S. Johnston (1979). *Proc. Eur. Geophys. Soc.*, 181–201.
McTigue, D.F. and P. Segall (1988). *Geophys. Res. Lett.* **15**, 601–604.
Miller, S.A., *et al.* (1999). *J. Geophys. Res.* **104**, 10621–10638.
Mjachkin, V.I., *et al.* (1975). *Pageoph* **113**, 169–181.
Mogi, K. (1981). *J. Soc. Mater. Sci. Jpn.* **30**, 105–118.
Mogi, K. (1985). "Earthquake Prediction." Academic Press.
Mogi, K., *et al.* (1982). *Abstr. Seismol. Soc. Jpn.* No. 1, 128.
Mueller, R.J. and M.J.S. Johnston (2000). *US Geol. Surv. Open File Report 00–146.*
Nadeau, R.M. and T.V. McEvilly (1997). *Bull. Seismol. Soc. Am.* **87**, 1463 1472.
Okada, Y. (1985). *Bull. Seismol. Soc. Am.* **75**, 1135–1154.
Okada, Y. (1992). *Bull. Seismol. Soc. Am.* **82**, 1018–1040.
Pollitz, F., *et al.* (1998). *Science* **103**, 1245–1249.

Prescott, W.H., *et al.* (1979). *J. Geophys. Res.* **84**, 5423–5435.
Press, F. (1965). *J. Geophys. Res.* **70**, 2395–2412.
Press, F. and C. Allen (1995). *J. Geophys. Res.* **100**, 6421–6430.
Rabbel, W. and J. Zschau (1985). *J. Geophys.* **56**, 1–99.
Reasenberg, P.A. and R.W. Simpson (1992). *Science.* **255**, 1687–1690.
Rice, J.R. (1983). *Pure Appl. Geophys.* **121**, 443–475.
Rice, J.R. and J.W. Rudnicki (1979). *J. Geophys. Res.* **84**, 2177–2193.
Rice, J.R. (1992). In: "Earthquake Mechanics and Transport Properties of Rocks", pp. 475–503, Academic Press.
Richards, P.G. (1976). *Bull. Seismol. Soc. Am.* **65**, 93–112.
Rikitake, T. (1976). "Earthquake Prediction," Elsevier.
Ruina, A. (1983). *J. Geophys. Res.* **88**, 10359–10370.
Rundle, J.B., *et al.* (1984). *J. Geophys. Res.* **89**, 10219–10231.
Sacks, I.S., *et al.* (1971). *Papers Meteorol. Geophys.* **22**, 195–207.
Sacks, I.S., *et al.* (1978). *Nature* **275**, 599–602.
Sacks, I.S., *et al.* (1979). *Maurice Ewing Series IV*, pp. 617–628, American Geophysics Union.
Sacks, I.S., *et al.* (1982). *Tectonophysics* **81**, 311–318.
Sadovsky, M.A., *et al.* (1972). *Tectonophysics* **14**, 295–307.
Satake, K. and H. Kanamori (1977). *J. Geophys. Res.* **82**, 5692–5697.
Savage, J.C. (1971). *J. Geophys. Res.* **76**, 1954–1966.
Savage, J.C. (1983). *Annu. Rev. Earth. Planet. Sci.* **11**, 11–43.
Scholz, C.H. (1990). "Mechanics of Earthquakes and Faulting," Cambridge University Press, New York.
Scholz, C.H. (1997). *Bull. Seismol. Soc. Am.* **87**, 1074–1077.
Schulz, S.S., *et al.* (1982). *J. Geophys. Res.* **87**, 6977–6982.
Segall, P. and J.R. Rice (1995). *J. Geophys. Res.* **100**, 22155–22171.
Shaw, B.E. (1997). *J. Geophys. Res.* **102**, 27367–27377.
Silverman, S., *et al.* (1989). *Bull. Seismol. Soc. Am.* **79**, 189–198.
Simpson, R.W. and P. Reasenberg (1994). *US Geol. Surv. Prof. Pap.* **1550-F**, 55–89.
Simpson, R.W., *et al.* (1988). *Pageoph* **126**, 665–685.
Shimazaki, K. and P. Somerville (1978). *Bull. Earthq. Res. Inst.* **53**, 613–628.
Smith, S.W. and M. Wyss (1968). *Bull. Seismol. Soc. Am.* **68**, 1955–1974.
Stuart, W.D. (1979). *J. Geophys. Res.* **84**, 1063–1070.
Stuart, W.D. and G. Mavko (1979). *J. Geophys. Res.* **84**, 2153–2160.
Tamura, T., *et al.* (1991). *Geophys. J. Int.* **104**, 507–516.
Thatcher, W. (1983). *J. Geophys. Res.* **88**, 5893–5902.
Tse, S.T. and J.R. Rice (1986). *J. Geophys. Res.* **91**, 9452–9472.
Wolfe, J.E., *et al.* (1981). *Bull. Seismol. Soc. Am.* **71**, 1625–1635.
Wyatt, F.K. (1988). *J. Geophys. Res.* **93**, 7923–7942.
Wyatt F.K., *et al.* (1994). *Bull. Seismol. Soc. Am.* **84**, 768–779.
Wyss, M. and J. Brune (1967). *Bull. Seismol. Soc. Am.* **57**, 1017–1023.
Yamauchi, T. (1987). *J. Phys. Earth* **35**, 19–36.

Editor's Note

Due to space limitations, references with full citation are given in the file "Johnston1FullReferences.pdf" on the Handbook CD-ROM, under the directory \36Johnston1.

Please see also Chapter 32, Rock failure and earthquake, by Lockner and Beeler; Chapter 33, State of stress within the Earth, by Ruff; Chapter 34, State of stress in the Earth's lithosphere, by Zoback and Zoback; and Chapter 35, Strength and energetics of active fault zones, by Brune and Thatcher.

37

Estimating Earthquake Source Parameters from Geodetic Measurements

Kurt L. Feigl
Centre National de la Recherche Scientifique, Toulouse, France

1. Brief Summary of Applicable Techniques

Figure 1 sketches the geometry of the various geodetic techniques for coseismic surveying. I will emphasize the new INSAR technique because the older techniques have been described well elsewhere and the newer SLR, VLBI, and DORIS techniques apply only to a few earthquakes.

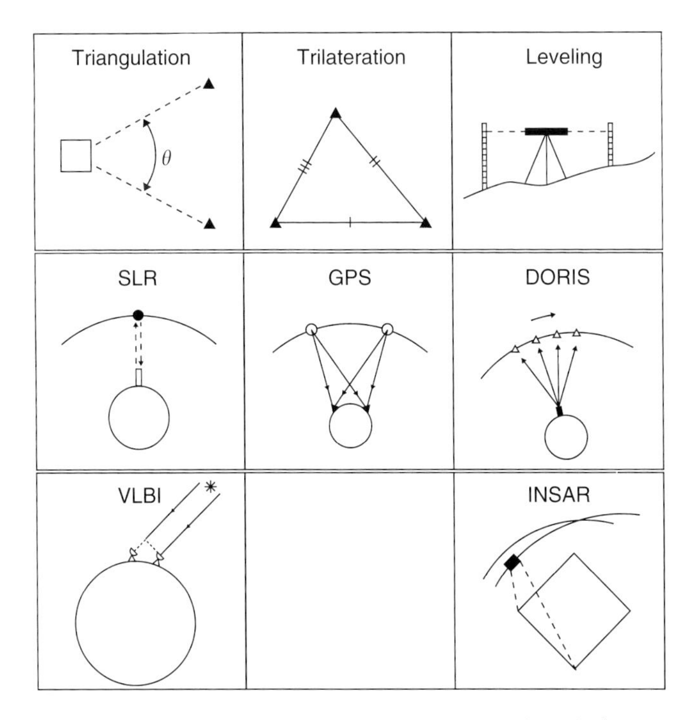

FIGURE 1 Sketch of geometry for various geodetic techniques.

1.1 Ground-based Vertical Techniques

Spirit leveling measures the difference in elevation between pairs of points. Errors accumulate with the square root of the length L of the "line" (almost always a road) between the points. Typical first-order leveling measurements have a standard deviation of the order of $L^{1/2}$ mm with L in km, i.e., ~10 mm over 100 km (Bomford, 1980). Bilham (1991) reviews geophysical applications of this technique.

Leveling measures heights with respect to the geoid (or "mean sea level"). Such "orthometric" heights are not to be confused with the "ellipsoidal" heights provided by some GPS instruments. The difference between the two is typically of the order of 10 m, and represents the undulation of the geoid. In the excitement immediately following an earthquake, scientists often consider remeasuring an old leveling line with GPS. Although this approach will not measure coseismic displacements, it can detect a different quantity — the geoid (Milbert and Dewhurst, 1992; Calais *et al.*, 1993).

Leveling is the ground-based technique of choice for measuring vertical coseismic displacements larger than about a centimeter. It seems to work best for normal-faulting events with magnitude of 6 or larger where surface rupture creates a spectacular offset, for example, the 1983 Borah Peak earthquake (Stein and Barrientos, 1985; Ward and Barrientos, 1986). Leveling data can also provide useful information in thrusting events without surface rupture, such as the 1989 Loma Prieta earthquake (see references in Table 1).

Sea level as recorded by tide gauges can also reveal vertical coseismic motion, as in the great 1964 Alaska earthquake (Holdahl and Sauber, 1994) (see Bilham (1991) for a review). If the coseismic motion is upwards on a coastline, it will kill mollusks by removing them from the water. Their new position above sea level records the vertical displacement with

INTERNATIONAL HANDBOOK OF EARTHQUAKE AND ENGINEERING SEISMOLOGY, VOLUME 81A ISBN: 0-12-440652-1

TABLE 1 Earthquake Parameters Estimated from Space-based Geodesy and Seismology

Year	Month	Day	Name	M_w or M_s	Seismic Moment (10^{18} Nm)	Geodetic Moment (10^{18} Nm)	Data[a]	Reference
1906	4	18	San Francisco, California	7.7	470.00	750.00	TR	Thatcher *et al.* (1997)
1906	4	18	San Francisco, California	7.7	470.00	555.00	TR	Matthews and Segall (1993)
1923	9	1	Kanto, Japan	8.2	573.30	1603.27	LL	Kanamori (1973)
1934	6	8	Parkfield, California	6.0		4.40	TR	Segall and Du (1993)
1944	12	7	Tonanki, Japan	8.1	1500.00	2000.00	TR, tsunami	Satake (1993)
1946	12	20	Nankaido, Japan	8.1	1500.00	3900.00	TR, tsunami	Satake (1993)
1954	9	9	El Asnam, Algeria	6.7		9.80	LL	Bezzeghoud *et al.* (1995)
1959	8	18	Hebgen Lake, Montana	7.3	103.00	120.00		Barrientos *et al.* (1987)
1960	5	22	Chile	9.5	200 000.00	94 000.00	LL, TG, TR	Barrientos and Ward (1990)
1964	3	28	Alaska	9.2	75 000.00	50 000.00	LL, TR, TG	Holdahl and Sauber (1994)
1966	6	28	Parkfield, California	6.0	1.40	4.40	TR, TL	Segall and Du (1993)
1971	2	9	San Fernando, California	6.4			LL, TR, TG	Meade and Miller (1973); Morrison (1973)
1976	7	28	Tangshan, China	7.8	120.00	98.00	LL	Huang and Yeh (1997)
1976	7	28	Luanxian, China	7.8		19.00	LL	Huang and Yeh (1997)
1976	11	15	Ningho, China	7.8		8.90	LL	Huang and Yeh (1997)
1978	6	20	Thessalanoki, Greece	6.4	2.70	4.00	LL	Stiros and Drakos (2000)
1978	11		Asal, Djibouti	5.3	0.17	0.60	TR, LL	Lépine *et al.* (1979); Ruegg *et al.* (1979); Stein *et al.* (1991)
1979	3	15	Homestead Valley, California	5.9	0.36	0.44	TR	Savage *et al.* (1993)
1979	10	15	Imperial Valley, California	6.5				Crook *et al.* (1982); Harsh (1982)
1980	10	10	El Asnam, Algeria	7.3	50.00	62.00	TR, LL	Ruegg *et al.* (1982)
1983	5	2	Coalinga, California	6.5			LL	Hartzell and Heaton (1983); Stein and King (1984); Eberhardt-Phillips (1989)
1983	10	28	Borah Peak, Idaho	7.3	18.50	25.00	LL	Stein and Barrientos (1985)
1985	8	4	Kettleman Hills, California	6.1	1.60	1.25	LL	Ekstrom *et al.* (1992)
1986	7	8	North Palm Springs, California	6.0	0.97	0.69	TL	Savage *et al.* (1993)
1987	11	17	Gulf of Alaska	6.9		66.00	VLBI	Sauber *et al.* (1993)
1987	11	24	Superstition Hills, California	6.2	9.00	9.40	TL	Larsen *et al.* (1992)
1988	3	6	Gulf of Alaska	7.6	1115.00	1220.00	VLBI	Sauber *et al.* (1993)
1989	6	26	Kiluaea South Flank, Hawaii	6.1	5.20	8.00	LL	Arnadottir *et al.* (1991)
1989	6	26	Kalapana, Hawaii	6.1	5.20	8.00	LL	Arnadottir *et al.* (1991)
1989	6	26	Kalapana, Hawaii	6.1	5.20	10.00	GPS	Dvorak (1994)
1989	10	1	Whittier Narrows, California	5.9	0.10	0.10	LL	Lin and Stein (1989)
1989	10	18	Loma Prieta, California	7.1	23.00	29.00	LL	Marshall *et al.* (1991)
1989	10	18	Loma Prieta, California	7.1	23.00	30.00	TL&GPS	Lisowski *et al.* (1990)
1989	10	18	Loma Prieta, California	7.1	23.00	34.00	TL, GPS, LL	Arnadottir and Segall (1994)
1989	10	18	Loma Prieta, California	7.1	23.00	27.00	TL, GPS, LL	Arnadottir and Segall (1994)
1989	10	18	Loma Prieta, California	7.1	23.00	29.00	TL, GPS, LL	Arnadottir and Segall (1994)
1989	10	18	Loma Prieta, California	7.1	23.00	29.00	GPS	Williams *et al.* (1993)
1991	4	22	Valle de la Estrella, Costa Rica	7.7			GPS	Lundgren *et al.* (1993)
1992	4	23	Joshua Tree, California	7.1	2.00	1.84	TL	Savage *et al.* (1993)
1992	4	23	Joshua Tree, California	6.1	2.15	1.70	TL, GPS	Bennett *et al.* (1994)
1992	4	25	Cape Mendocino, California	7.1	44.50	31.00	LL, TL, GPS	Murray *et al.* (1996)
1992	6	28	Big Bear, California	6.3	5.20	4.20	GPS	Murray *et al.* (1993)
1992	6	28	Big Bear, California	6.3	5.20	3.70	GPS	Johnson *et al.* (1994)
1992	6	28	Landers, California	7.3	80.00	79.00	GPS	Murray *et al.* (1993)
1992	6	28	Landers, California	7.3	80.00	80.00	CGPS	Bock *et al.* (1993)
1992	6	28	Landers, California	7.3	80.00	103.00	GPS	Hudnut *et al.* (1994)
1992	6	28	Landers, California	7.3	80.00	77.00	Joint	Wald and Heaton (1994)
1992	6	28	Landers, California	7.3	80.00	99.50	GPS	Johnson *et al.* (1994)
1992	6	28	Landers, California	7.3	80.00	90.00	GPS	Freymueller *et al.* (1994)
1992	12	2	Fawnskin, California	5.1	0.04		INSAR	Feigl *et al.* (1995)
1993	5	17	Eureka Valley, California	6.1	1.20	1.70	INSAR	Massonnet and Feigl (1995b)
1993	5	17	Eureka Valley, California	6.1	1.20		INSAR	Peltzer and Rosen (1995)
1993	8	8	Guam	7.8			GPS	Beavan *et al.* (1994)
1994			Kamchatka				DORIS	A. Cazenave, personal commun. (1999);
1994			Sanriku-Haruka-Oki, Japan	7.5				Miyazaki *et al.* (1996)

(continued)

TABLE 1 *(continued)*

Year	Month	Day	Name	M_w or M_s	Seismic Moment (10^{18} Nm)	Geodetic Moment (10^{18} Nm)	Data[a]	Reference
1994	1	17	Northridge, California	6.7	11.00	13.00	Joint	Wald *et al.* (1996)
1994	1	17	Northridge, California	6.7	11.00	10.07	INSAR	Massonnet *et al.* (1996a)
1994	1	17	Northridge, California	6.7	11.00	15.80	TERRAscope	Thio and Kanamori (1996)
1994	1	17	Northridge, California	6.7	11.00	13.40	GPS	Shen *et al.* (1996) model A
1994	1	17	Northridge, California	6.7	11.00	16.30	GPS	Hudnut *et al.* (1995)
1994	1	17	Northridge, California	6.7	11.00		INSAR	Murakami *et al.* (1996)
1994	2	15	Liwa	6.8			GPS	Duquesnoy *et al.* (1996)
1994	6	18	Arthur's Pass, New Zealand	6.7	13.00	16.00	GPS	Arnadottir *et al.* (1995)
1994	10	4	Hokkaido-Toho-Oki, Japan	8.1		2000.00	GPS	Tsuji *et al.* (1995)
1995	1	17	Hyogo-ken Nanbu (Kobe), Japan	7.2			GPS	Tabei *et al.* (1996)
1995	1	17	Hyogo-ken Nanbu (Kobe), Japan	7.2			INSAR	Ozawa *et al.* (1997)
1995	5	13	Grevena, Greece	6.6	7.60	16.30	GPS	Clarke *et al.* (1996); Clarke *et al.* (1998)
1995	5	13	Grevena, Greece	6.6	7.60	6.40	INSAR	Meyer *et al.* (1996); Meyer *et al.* (1998)
1995	6	15	Corinth, Greece	6.2	4.00	3.90	Joint	Bernard *et al.* (1997)
1995	7	30	Antofogasta, Chile	8.1	1700.00	1500.00	GPS, SL	Ruegg *et al.* (1996)
1995	7	30	Antofogasta, Chile	8.1	1700.00	1420.00	GPS, SW	Ihmlé and Ruegg (1997)
1995	7	30	Antofogasta, Chile	8.1	1700.00	1780.00	GPS	Klotz *et al.* (1999)
1995	10	1	Dinar, Turkey	6.1	3.10	3.18	INSAR	Wright *et al.* (1999)
1995	10	9	Jalisco, Mexico	8.0			INSAR	Vincent (1998)
1995	10	9	Jalisco, Mexico	8.0			GPS	Melbourne *et al.* (1997)
1996	2	26	St. Paul de Fenouillet, France	5.0		0.04	INSAR	Rigo and Massonnet (1999)
1999	8	24	Izmit, Turkey	7.4	195.00	170.00	GPS	Reilinger *et al.* (2000)
1999	10	16	Hector Mines, California	7.1			INSAR	Sandwell *et al.* (2000)

[a]LL = Leveling, TR = Triangulation, TL = Trilateration, TG = Tide guage, SL = Sea level from mollusks, and SW = Surface waves.
Updated at http://bowie.mit.edu/~kurt/coseismicinsar.htm

a precision of the order of a decimeter, as in two Chilean earthquakes (Barrientos and Ward, 1990; Ruegg *et al.*, 1996). This "natural tide gauge" is the only way other than mapping surface rupture to measure coseismic displacements without planning an observation before the earthquake.

1.2 Triangulation and Trilateration Surveying

Triangulation measures the angle between two benchmarks as seen from a third with a precision of 4 μrad at best (Bomford, 1980). Now classic, this technique nonetheless established the relative positions of the vast majority of benchmarks in most national geodetic networks. For many earthquakes, these measurements are the only ones acquired before the earthquake, for example the great 1906 San Francisco earthquake (Thatcher, 1974; Matthews and Segall, 1993; Thatcher *et al.*, 1997) and other studies in California (Savage and Burford, 1970).

Trilateration measures the distance between two benchmarks with a precision of the order of a centimeter (Bomford, 1980). For earthquake studies, its use seems to be limited mostly to California (Prescott *et al.*, 1979), although an early study in the Afar revealed over 2 m of displacement (Ruegg *et al.*, 1979).

Both triangulation and trilateration require a clear line of sight between the benchmarks, limiting their use to distances typically less than 30 or 50 km at most. As a result, many benchmarks were installed on hilltops and mountain tops with difficult access, the most extreme examples being the peaks in the High Karakoram (Chen *et al.*, 1984) and even the summit of Mt. Everest (Bilham, 1998). This offers the advantage of conserving the benchmark from destruction for long periods of time. Interestingly, many of these older benchmarks on summits continue to serve because they also provide an unobstructed line of "sight" to orbiting satellites. Also, these rocky summits provide more stable monuments than do boggy lowland soils.

1.3 VLBI

Very Long Baseline Interferometry (VLBI) measures the position of radio antennas with respect to radio sources in quasars (Smith and Turcotte, 1993). Although capable of submillimeter precision in relative position vectors (Herring, 1992), this technique requires large (~10 m) antennas. As such, it has only measured coseismic displacements for a few earthquakes: Loma Prieta, California (Clark *et al.*, 1990), and the 1987–1988 Gulf of Alaska earthquakes (Sauber *et al.*, 1993).

Nonetheless, VLBI supports earthquake studies by contributing important geometric information to the definition of geodetic reference systems such as the International Terrestrial Reference Frame (ITRF) (Sillard *et al.*, 1998).

1.4 SLR

Satellite laser ranging (SLR) measures the round-trip distance between an instrument on the ground and a reflective, massive, spherical satellite in low (500–1200 km altitude) orbit. The measurement uncertainty is typically 7 cm in distance, which implies subcentimeter uncertainties in all three vector components of relative position between two benchmarks (Tapley *et al.*, 1993). Since SLR instruments are many times heavier than for GPS, they are usually deployed at astronomic observatories, with the exception of a few mobile instruments deployed in California under the auspices of NASA's Crustal Dynamics Program (CDP) (Smith and Turcotte, 1993) and in the Mediterranean region under the WEGENER program (Smith *et al.*, 1994; Noomen *et al.*, 1996). In both of these networks, most of the SLR measurements useful for tectonic studies occurred in the late 1980s and have been largely supplanted by GPS in the 1990s. As a result, I could not find a published example of a coseismic displacement recorded by SLR.

1.5 GPS

The Global Positioning System (GPS) can provide subcentimeter estimates of relative position using an instrument available for less than "10 kg, 10 W, and 10 $K." Since the most precise solutions involve postprocessing data from multiple instruments, it typically requires several days between acquisition and estimate. The constellation of satellites came into use gradually beginning in 1985; it became fully operational in 1992. Data from this early period are typically more difficult to analyze and may yield less precise results than more recent surveys. For reviews of geophysical applications, see Dixon (1991), Hager *et al.* (1991), Hudnut (1995), Larson (1995), and Segall and Davis (1997). For earthquake studies, GPS networks tend to operate in one of two end-member modes: continuous operation of permanently installed, widely spaced antennas (CGPS), or intermittent occupation of densely spaced benchmarks in "campaign" mode. The former offers good temporal resolution (1 measurement/30 seconds = 33 mHz) but poor spatial resolution (>100 km between stations), whereas the latter offers poor temporal resolution (1 measurement/year = 32 nHz) and good spatial resolution (~10 km between stations). This trade-off between temporal and spatial resolution creates a difficult decision in the face of limited resources. Although a compromise "hybrid" strategy could rotate expensive receivers on a roughly monthly basis through several fixed monuments, this approach has yet to be deployed, apparently because it requires more manpower than do permanent installations (Fig. 2).

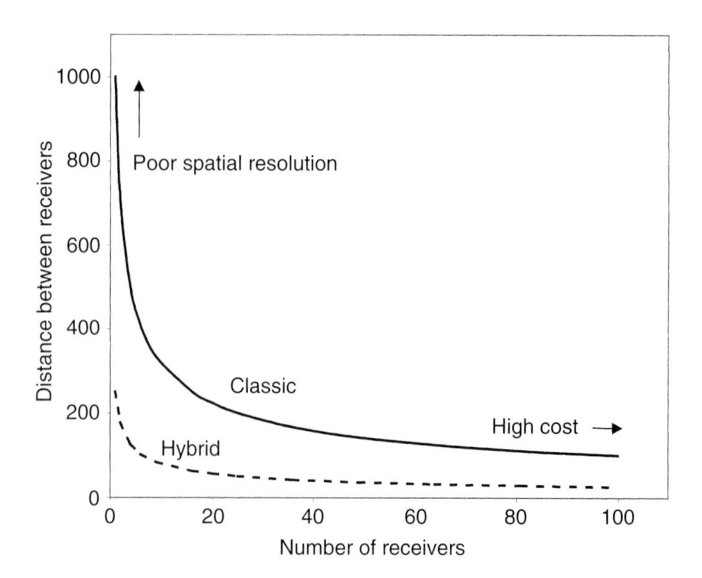

FIGURE 2 Tradeoff between poor spatial resolution and high cost in a GPS network designed to cover an area $a^2 = 1000$ km by 1000 km with n instruments spaced d km apart implies that $d = a/n$. Permanently deploying one receiver at each benchmark in the network in the "classic" approach (solid line) costs more than rotating each receiver through (say) four sites in the "hybrid" approach (dashed line).

1.6 DORIS

Détermination d'Orbite et Radiopositionnement Intégré par Satellite (DORIS) is a Doppler satellite navigation system developed by the French Space Agency (Lefebvre *et al.*, 1996). Designed for tracking satellites in orbit, this system currently flies on three satellites (SPOT2, SPOT4 and Topex/Poseidon). It resembles GPS with three important differences. First, the transmitter is on the ground, not on the satellite. Second, the current design of the space-borne receiver cannot track more than three instruments on the ground within a radius of 1000 km, although this restriction was relaxed with a new design launched aboard Jason in 2000. Third, the current DORIS tracking network covers the globe quite well, with at least one station on each of the 11 lithospheric plates.

These differences imply that DORIS is better suited to measuring plate motions at the global scale than is GPS, which still suffers from a lack of long-term stations in the southern hemisphere. For this application, multiyear time series of DORIS data can determine absolute velocities with uncertainties of 1–2 mm y^{-1} in horizontal components (Crétaux *et al.*, 1998).

On the other hand, DORIS is less well suited to local studies of earthquakes and faulting at scales shorter than 1000 km than is GPS, although the station at Sakhalin did capture the 1994 earthquake in Kamchatcka (A. Cazenave, personal communication, 1999).

1.7 SAR Interferometry

This geodetic technique calculates the interference pattern caused by the phase difference between two images acquired by a space-borne synthetic aperture radar (SAR) at two distinct times. The resulting interferogram is a contour map of the change in distance between the ground and the radar instrument. Each fringe represents a range change of half the wavelength. Thus, the contour interval is 28 mm for C-band radars such as ERS and RADARSAT and roughly four times larger, 125 mm for the L-band JERS satellite. These maps provide an unsurpassed spatial sampling density ($\sim$100 pixels km^{-2}), a competitive precision ($\sim$1 cm) and a useful observation cadence (1 pass/month), as described in a review article by Massonnet and Feigl (1998), which is paraphrased here.

To capture an earthquake, INSAR requires three data sets: a SAR image before the earthquake, one after, and topographic information. The SAR images themselves are rich data sets well documented in the remote sensing literature (Curlander and McDonough, 1991; Henderson and Lewis, 1998).

The topographic information is necessary to model and remove the interferometric fringes caused by topographic relief as "seen in stereo" from slightly different points of view. To handle the topographic contribution, we can choose between the "two-pass" approach (e.g., Massonnet and Feigl, 1998) and the "three-pass" or "double-difference" approach (e.g., Zebker *et al.*, 1994). For earthquake studies, there is usually a trade-off between the two-pass approach, which requires a digital elevation model (DEM), and the three-pass approach, which requires a third SAR acquisition. Further discussion of relative merits of the two- and three-pass approaches are beyond the scope of this chapter.

To interpret an interferogram, one must understand how different effects contribute to the fringe pattern. Many instructive examples appear in review papers by Massonnet and Feigl (1998), Madsen and Zebker (1998) and Burgmann *et al.* (2000). The mathematical details appear in another review (Bamler and Hartl, 1998). Hanssen (2001) has written a thorough textbook. For earthquake studies, the most important effects involve topographic relief, orbital trajectories, and tropospheric refraction, usually in combination.

If the topographic information (a DEM for two-pass, or the "topo pair" in three-pass INSAR) is in error, the interferogram will contain artifactual fringes. They appear in the same location in every interferogram produced using that topographic model. To quantify this effect, Massonnet and Rabaute (1993) define the altitude of ambiguity h_a as the shift in altitude needed to produce one topographic fringe. Indeed, this parameter is inversely proportional to the perpendicular component of the ("baseline") vector separating the two orbital trajectories, conventionally written $B_\perp$, pronounced "B-perp," and given in meters (Zebker and Goldstein, 1986). The number of "topographic" fringes is proportional to $B_\perp$ and inversely proportional to h_a. Thus we seek pairs of orbital trajectories with a small separation, that is, with *small* (absolute) values of $B_\perp$ and *large* (absolute) values of h_a for earthquake studies. It turns out that for the ERS satellites, an acceptably good orbital pair has both $B_\perp$ and h_a approximately equal to 100 m.

A topographic error of ε meters in the DEM will produce a phase error of ε/h_a fringes in the resulting interferogram. Errors in typical DEMs range from 10 to 30 m (Wolf and Wingham, 1992), implying that choosing a pair of images with $|h_a|$ between 20 and 60 m will yield an interferometric measurement with an error better than $\varepsilon/h_a = \pm 1/2$ cycle, or $\pm$14 mm for ERS. Small values of $|h_a|$ can mask even large signals with artifactual topographic fringes. In an extreme (and rare) case, Massonnet and Feigl (1995a) uncovered a topographic error of $\varepsilon \sim 250$ m, roughly 8 times larger than the published precision for the DEM. This artifact resembles the fringe pattern produced by a small earthquake. Avoiding such confusion requires looking at several interferograms with different values of h_a. For an earthquake, the number of coseismic fringes does not depend on h_a.

Atmospheric effects can also complicate the interpretation of an interferogram. Indeed, variations in the refractive index of the troposphere are the current limiting source of error in the INSAR technique (Goldstein, 1995; Massonnet and Feigl, 1995a; Rosen *et al.*, 1996; Tarayre and Massonnet, 1996; Zebker *et al.*, 1997; Hanssen, 2000). Potentially, one could confuse a topographic signature with a displacement, if propagation effects create fringes which "hug" the topography like contour lines, but which measure the change in tropospheric delay. This effect was first observed as several concentric fringes in a 1-day interferogram on Mt. Etna (Massonnet and Feigl, 1998; Beauducel *et al.*, 2000). One can recognize this subtle effect using pairwise logic (Massonnet and Feigl, 1995a) or using a DEM and local meteorological observations (Delacourt *et al.*, 1998; Williams *et al.*, 1998). However, separating the tropospheric noise from the deformation signal can be challenging, particularly when the signal is small, e.g., the magnitude 5.2 earthquake near St. Paul de Fenouillet, France (Rigo and Massonnet, 1999).

1.8 Correlation of Two Remote-sensing Images

It is also possible to detect (large) coseismic displacements by correlating two optical images. The "lag" vectors estimated between the corresponding subpixel cells of a prequake and a postquake image yields the horizontal components of the coseismic displacement vector with meter-level precision and hectometer resolution (Crippen, 1992; Crippen and Blom, 1992; Van Puymbroeck *et al.*, 2000). The same technique also applies to SAR images. By correlating two Single Look

Complex (SLC) SAR amplitude ("backscatter") images acquired at different times, Michel *et al.* (1999) measured ground displacements for the Landers earthquake. Their result is "a two-dimensional displacement field with independent measurements every about 128 m in azimuth and 250 m in range. The accuracy depends on the characteristics of the images. For the Landers test case discussed in the study, the 1-σ uncertainty is 0.8 m in range and 0.4 m in azimuth. [They] show that this measurement provides a map of major surface fault ruptures accurate to better than 1 km and an information on coseismic deformation comparable to the 92 GPS measurements available. Although less accurate, this technique is more robust than SAR interferometry and provides a complementary information since interferograms are only sensitive to the displacement in range." (Michel *et al.*, 1999.)

2. Estimating Earthquake Parameters by Inversion of Geodetic Data

2.1 The Standard Elastic Half-space Model

To explain the observed coseismic deformation, a simple model of a dislocation in an elastic half-space provides a good approximation. Indeed, it has become the conventional model used in most of the case studies considered here. Okada (1985) derives the expressions for the coseismic (permanent) displacement u at the Earth's surface caused by a fault at depth in closed analytic form. Accordingly, the displacement field $u_i(x_1, x_2, x_3)$ due to a dislocation $\Delta u_j(\xi_1, \xi_2, \xi_3)$ across a surface Σ in an isotropic medium is

$$u_i = \frac{1}{F}\int_{\Sigma}\int \Delta u_j \left[\lambda\delta_{jk}\frac{\partial u_i^n}{\partial \xi_n} + \mu\left(\frac{\partial u_i^j}{\partial \xi_k} + \frac{\partial u_i^k}{\partial \xi_j}\right)\right] v_k \, d\Sigma \qquad (1)$$

where δ_{jk} is the Kronecker delta, λ and μ are Lamé's coefficients, v_k is the direction cosine of the normal to the surface element $d\Sigma$, and the summation convention applies. The term u_i^j is the ith component of the displacement at (x_1, x_2, x_3) due to the jth direction point force of magnitude F at (ξ_1, ξ_2, ξ_3). For the complete set of equations see Okada (1985), who also corrects previous derivations. A public-domain computer program performs these calculations (Feigl and Dupré, 1999). Called RNGCHN, this program is included on the Handbook CD-ROM under the directory \37Feigl. Okada (1992) performs this calculation at any point in the half-space.

2.2 Fault Parameters

Here I follow Okada's (1985) notation, as in Feigl and Dupré (1999). To describe a single fault element (also called a "subfault" or "patch") as a dislocation requires ten parameters. The fault patch has length L and width W. The slip on the fault plane is a vector $\mathbf{U}$ with three components, U_1, U_2, and U_3. The position coordinates of the fault patch are E, N, and d, taken positive east, north, and down. The azimuth α gives the strike of the fault, in degrees clockwise from north. Finally, an observer facing along strike should see the fault dip at δ degrees to his right.

The Okada parameters differ slightly from the parameters favored by seismologists. In particular, the origin of Okada's fault patch does not coincide with the centroid at the geometric center of the fault rectangle (Feigl and Dupré, 1999). For a double-couple source, the tensile component vanishes ($U_3 = 0$) and the slip vector $\mathbf{U}$ lies in the fault plane. Seismologists define the rake angle r such that $\tan r = U_2/U_1$ (Aki and Richards, 1980). Inversely, $r = \mathrm{ATAN2}(U_2, U_1)$ where ATAN2 is the usual FORTRAN intrinsic function for arctangent (U_2/U_1) on the range $[-180°, +180°]$. A thrust-faulting mechanism, for example, has $U_2 > 0$ and $r > 0$. A normal faulting mechanism, on the other hand, has $U_2 < 0$ and $r < 0$. Similarly, left-lateral slip implies $U_1 > 0$ and $|r| \leq 90°$, whereas right-lateral slip implies $U_1 \geq 0$ and $|r| \geq 90°$.

2.3 Underlying Assumptions

The standard Okada model assumes that the Earth's surface is flat, corresponding to the bounding plane of the elastic half-space. The Lamé coefficients λ and μ specify the elastic medium. For simplicity, most studies assume that $\lambda = \mu$, so that these parameters drop out of the expressions for surface displacement. Such a medium, called a Poisson solid, has a Poisson's ratio of 1/4, a reasonable approximation to the values of 0.23–0.28 estimated from P- and S-wave velocities in the upper crust (Perrier and Ruegg, 1973; Dziewonski and Anderson, 1981). The so-called "geometric moment" or "potency" simply equals ULW. To obtain the seismic moment, multiply by the shear modulus μ so that $M_0 = \mu ULW$. Typical values (assumed) for μ in the Earth's crust range from 30 to 36 GPa, but values as low as 10 GPa (Dal Moro and Zadro, 1999) and as high as 50 GPa (Barrientos and Ward, 1990) have been used. The simplest assumption takes this value to be constant throughout the half space, although some authors propose increases with depth (Dolan *et al.*, 1995; Cattin *et al.*, 1999). Some authors call μ the "rigidity," whereas others use μ to denote a dimensionless coefficient of friction. To convert moment into the various magnitude scales, use the conventional formulas (Hanks and Kanamori, 1979; Abe, 1995). Empirical relations exist for establishing the size (L, W, and U) of the earthquake from the seismological magnitude or moment (Scholz, 1990; Dolan *et al.*, 1995).

2.4 Particularities of Geodetic Data

Like seismograms, geodetic measurements decompose the displacement vector into components. Although VLBI and GPS both record three components of the coseismic displacement vector (postquake minus prequake position) of a benchmark, INSAR records only the component along the line of sight between the satellite and ground point. Image correlation provides the same line-of-sight (range) component as well as a second component parallel to the satellite trajectory (azimuth). The line of sight between the point on the ground and the radar satellite in the sky defines two angles, the radar incidence (from vertical) and the azimuth of the satellite ground track (from North). For the ERS satellites in California, for example, these quantities are approximately 23° and 13°, respectively. These quantities determine the unit vector $\hat{\mathbf{s}}$ which points from ground to satellite. Then the change in range $\Delta\rho$ or the distance measured along the line of sight between the satellite and ground point is

$$-\rho = \mathbf{u} \cdot \hat{\mathbf{s}} \tag{2}$$

Note that the sign convention is such that an upward movement will produce a positive value of $\mathbf{u}\cdot\hat{\mathbf{s}}$, a decrease in range, and a negative value of $\Delta\rho$. The ray specified by $-\hat{\mathbf{s}}$ is sometimes called the "look vector." INSAR can provide a second component of the coseismic displacement vector if the satellite acquires two images in both the "ascending" (south-to-north) and "descending" (north-to-south) orbital passes. In this case, the east, north, and upward components of the unit vector $\hat{\mathbf{s}}$ are $[x, y, z]$ and $[x, -y, z]$, respectively. To use the radar interferograms as data in an inverse problem requires an unambiguous measurement of the range change, which implies "unwrapping" the interferogram. See Ghiglia (1998) for a review of the techniques involved. For the Landers and Fawnskin earthquakes described below, we simply count and digitize the fringe pattern. Although tedious, this technique avoids errors because the human eye is very good at following colored fringes, even where they are noisy. It also recognizes areas where the fringes become too noisy to count. For the Eureka earthquake, a straightforward algorithm (Tarayre, 1994) performed well because the fringes were clear and simple.

Even unwrapped, radar range changes are still only relative measurements. To make them absolute, we must identify the fringe corresponding to zero deformation. We can do this by trial and error (Feigl *et al.*, 1995), or explicit estimation (Wright *et al.*, 1999), choosing the additive constant which produces the smallest misfit to the observed interferogram. Usually, the null fringe intersects the fault plane.

2.5 Modeling by Trial and Error

The standard Okada model defines the relation between the earthquake source parameters and the geodetic measurements of surface displacement. The goal is to find the values of parameters which best fit the data. This inverse problem seeks to minimize the difference between the modeled displacement field and the one sampled by geodesy. The simplest procedure is trial and error, usually called "forward modeling." We use our best guess for the value of each parameter to calculate a synthetic displacement field. With some clues about the location, geometry, and magnitude of the earthquake, it is not difficult to find a simulation which looks like the observed displacement field. By repeatedly tuning the parameters, we can usually fit the data better than our first guess. This procedure provided the first approximation to the coseismic deformation in most of the studies listed in Table 1.

2.6 Estimating the Focal Mechanism

If we choose to estimate all ten parameters for a single fault patch, the problem is nonlinear because the surface displacement depends strongly on the fault geometry. The approach uses "numerical optimization procedures to determine the best-fitting dislocation surface or surfaces. The methods can generally be divided into two categories: those methods, such as nonlinear least squares or quasi-Newton methods, that make use of the first or second derivatives and Monte Carlo methods that do not require these derivatives," as described by Segall and Davis (1997).

For the derivative-based methods, the RNGCHN program includes analytic expressions for the first derivatives. These allowed us to use an iterative linearized least squares procedure for the 1992 Fawnskin (Feigl *et al.*, 1995) and 1993 Eureka Valley earthquakes (Massonnet and Feigl, 1995b).

The Monte Carlo techniques have the advantage of avoiding local minima and furnishing realistic estimates of uncertainties, as shown for the Loma Prieta (Arnadottir and Segall, 1994) and Cape Mendocino (Murray *et al.*, 1996) earthquakes.

Mixing the two categories of optimization methods in a hybrid, Monte-Carlo, downhill simplex scheme also works (Clarke *et al.*, 1996, 1998; Wright *et al.*, 1999).

2.7 Surface Rupture by Earthquake Faulting

By definition, a mapped fault is a discontinuity separating two blocks of the Earth's crust. If the fault is active, the relative motion (slip) between the two blocks offsets the interferometric fringe pattern. Thus, surface rupture appears as a discontinuity in an interferogram, except where the slip vector is orthogonal to the radar look vector. Offsets as small as a centimeter tear the fringe patterns at Landers (Massonnet *et al.*, 1994; Price and Sandwell, 1998), Hector Mine (Sandwell *et al.*, 2000), and in the South Iceland Seismic Zone (Feigl *et al.*, 2000).

2.8 Fault Slip

Once we know the geometry of the fault, we can estimate the distribution of the slip vector $\mathbf{u}$. This inverse problem is linear. The components of the surface displacement $\mathbf{U}$ are proportional to the components of the slip vector $\mathbf{u}$. As such, it is simple to divide the modeled fault plane into many discrete patches. By varying only the amount of slip on each patch, but not its geometry, we can estimate the distribution of slip on the fault plane. Numerous published examples of this procedure are listed in Table 1. Although most authors approach this problem using discrete fault patches, Bennett *et al.* (1994) use continuous functions. Comparing different solutions to this inverse problem is hindered by the lack of a single standard format for computer files. Worse still, very few authors publish the centroid of their estimate.

2.9 Moment

Geodetic observations of coseismic displacement are difficult to obtain because they require a measurement before the earthquake. Furthermore, such measurements are only possible at the Earth's surface. As a consequence, geodetic data sets tend to be sparse. To extract the most information from these data sets, we want to limit the number of free parameters in the optimization. Taken to the extreme, this approach suggests estimating only a single parameter – the moment – from the available data, as Johnson *et al.* (1994) do for the Landers earthquake.

2.10 Data Covariance Matrix

In solving these inverse problems, we expect to find a more reliable solution and a better estimate of the uncertainties if we account for the full covariance matrix, including the off-diagonal terms of the data. The Loma Prieta earthquake provides a case in point. There, Marshall *et al.* (1991) and Lisowski *et al.* (1990) applied standard least squares methods to the leveling and trilateration data, respectively, neglecting the off-diagonal elements of the data covariance matrix. As a result, they had to resort to a two-patch model to fit the geodetic data. This model came under fire because the second patch did not pass through the locus of aftershock hypocenters. By including the full covariance matrix, however, Arnadottir *et al.* (1992) were able to find an acceptable single-patch model which also fitted the aftershock distribution. The same authors later improved on their first model by allowing the slip to vary in a bootstrap Monte Carlo approach (Arnadottir and Segall, 1994).

Complete knowledge of the data covariance matrix is also necessary in joint inversions to weight the different types of data (Barrientos and Ward, 1990; Holdahl and Sauber, 1994).

3. Case Studies

3.1 Loma Prieta, California, 1989

In addition to revealing the importance of using the covariance matrix when inverting geodetic data, the Loma Prieta earthquake also focused attention on the issue of postseismic deformation. To fit the geodetic observations, Savage *et al.* (1994) proposed a model of "fault collapse" such that the dislocation included a (negative) tensile component akin to reducing the volume of a rectangular prism or dike. Arnadottir *et al.* (1992) later concluded that the "geodetic data do not place useful constraints on the amount of dilatancy for this event," a conclusion reaffirmed later (Bürgmann *et al.*, 1997).

3.2 Landers, California, 1992

Color Plate 14 shows INSAR results for the 1992 Landers earthquake. The slip distribution estimated from the radar data [shown in Plate 21 of Massonnet and Feigl (1998)] agrees qualitatively with those estimated from GPS survey measurements of coseismic displacements (Murray *et al.*, 1993; Freymueller *et al.*, 1994; Hudnut *et al.*, 1994), strong motion accelerations recorded in the near field (Cohee and Beroza, 1994; Cotton and Campillo, 1994, 1995), seismograms in the far field (Wald and Heaton, 1994), a joint inversion of all three data types (Wald and Heaton, 1994), and a combination of INSAR and strong motion data (Hernandez *et al.*, 1997, 1999). All these inversions find relatively little slip (2–3 m) below the epicenter where rupture began, but a maximum of 8–12 m of slip located at 5–10 km depth in the Homestead and Emerson fault segments between 30 and 40 km north of the epicenter. The depth and magnitude of the slip maximum seems to depend on the prior information in the various inversions. All the estimates agree on the seismic moment, in accord with the centroid moment tensor and the bounds estimated from the geodetic data (Johnson *et al.*, 1994).

The estimates of slip distribution contributed to calculations of coseismic stress changes which load the crust and thus trigger subsequent earthquakes (Harris and Simpson, 1992; Jaumé and Sykes, 1992; Stein *et al.*, 1992). By using a fine estimate of slip distribution estimated from several data sources (Wald and Heaton, 1994), Stein *et al.* (1994) predict aftershock locations better than with their original calculation (Stein *et al.*, 1992) which used only a coarse estimate of slip distribution based on GPS measurements alone (Murray *et al.*, 1993).

Both GPS and INSAR also measured postseismic deformation following this earthquake (Massonnet *et al.*, 1994; Shen *et al.*, 1994; Wyatt *et al.*, 1994; Massonnet and Feigl, 1995a; Massonnet *et al.*, 1996b; Peltzer *et al.*, 1996, 1998; Bock *et al.*, 1997; Savage and Svarc, 1997; Pollitz *et al.*, 2000), but these observations and the models needed to explain them exceed the limits of this chapter.

3.3 Eureka Valley, California, 1993

The Eureka Valley earthquake occurred on 17 May 1993 in a remote area of the Mojave Desert at the edge of the Basin and Range province. The normal-faulting $M_w = 6.1$ mainshock and subsequent aftershocks deepened the graben in an oval-shaped coseismic deformation field. The ERS-1 radar images are the only available geodetic measurements. This earthquake is an interesting case study because two different approaches led to different interpretations. Massonnet and Feigl (1995b) calculate this interferogram by stacking two 2-pass interferograms in a combination of three radar images. Peltzer and Rosen (1995) analyze the same three images with the 3-pass technique. Both studies find approximately 10 cm of range increase.

To explain the observed fringe pattern, Massonnet and Feigl use an iterative least-squares procedure and the standard elastic dislocation model to estimate the earthquake focal mechanism. The best-fitting focal mechanism is a normal fault dipping $54° \pm 2°$ to the west and striking S7°W $\pm 2°$. The 16×7 km rectangular fault patch centered at 9 km depth does not cut the surface. The estimated geodetic moment magnitude of 6.1 agrees with the seismological estimates from wave-form inversion. The residual interferogram shows less than one 14 mm cycle in the difference between the observed and modeled fringes.

The location of the centroid estimated from the radar data is less than 6 km horizontally and 2 km vertically from the hypocenter estimated from *P*-wave travel times. The modeled fault patch, however, strikes more westerly than the mapped Quaternary fault or the fault plane estimated from first motions. Indeed, Peltzer and Rosen find that a fault plane striking N7°E, dipping 50° west, but cutting the surface, provides a good fit to their radar interferogram, based on forward modeling. The fault patch estimated by Massonnet and Feigl resembles the locus of aftershocks in dip, length, width and horizontal location, but not depth.

The fault models also disagree about the depth of the slip, in particular, whether or not it breaks the surface. Peltzer and Rosen favor a variable-slip (multi-patch) fault model in which the uppermost fault patches cut the surface, while Massonnet and Feigl's optimized 1-patch model stops some 6 km short of the surface. Furthermore, Peltzer and Rosen observe an offset of approximately 3 cm in their interferogram, whereas Massonnet and Feigl see no such discontinuity longer than 1 km in their interferogram. Furthermore, Peltzer and Rosen observed a fault scarp with 1–3 cm of vertical displacement which they could follow in the field for a few tens of meters. Any surface rupture would have to be small, both in magnitude and spatial extent, to avoid cutting the fringes observed in Massonnet and Feigl's interferogram.

A small, shallow aftershock can explain all the observations and resolve the controversy. Some 3 cm of slip on a fault 1 km^2 in area represents an earthquake of magnitude 4. An earthquake of approximately this magnitude ($M_L = 3.5$) occurred at 0.02 km depth in this area. This location is also less than 1 km from the offset observed by Peltzer and Rosen, well within the uncertainties of the seismological estimates. Such an earthquake could have produced the short scarp observed by Peltzer and Rosen in the field. It could also produce concentric fringes 1 or 2 km in diameter in the interferogram. Two such fringes (28 mm of range) are barely discernible near its epicenter in Massonnet and Feigl's interferogram.

3.4 Grevena, Greece, 1995

A $M_s = 6.6$ normal-faulting event in northern Greece illustrates the complexity of estimating source parameters from diverse data sets with different data types. The earthquake occurred near Kozani–Grevena on 13 May 1995, a decade after a triangulation survey, almost two years after the pre-quake ERS image, and 3–7 days before the field observations of centimeter-scale surface breaks. These observations form the basis of two separate analyses of the coseismic deformation which differ markedly both in their approaches and in their conclusions. Clarke *et al.* (1996) used GPS after the earthquake to measure 91 concrete pillars that had been surveyed prior to the earthquake by triangulation with formal uncertainties of 15 mm in horizontal relative position. Using a "hybrid simplex-Monte Carlo method which requires no a priori constraints," they estimate the focal parameters for a single fault patch. Working separately, Meyer *et al.* (1996) used INSAR to calculate several coseismic interferograms. The best one spans almost two years and is only partially coherent. Using the INSAR results in conjunction with their map of surface rupture, tectonic maps of fault geometry, Meyer *et al.* (1996) determine a model with 22 fault patches.

The two models disagree substantially, leading to a critical exchange of comment (Meyer *et al.*, 1998) and reply (Clarke *et al.*, 1998). The GPS-derived model predicts coseismic INSAR range changes that disagree with the observed interferogram. Similarly, the INSAR-based model predicts coseismic displacements that disagree with the displacements observed by the GPS-triangulation comparison by more than several times their measurement uncertainty. There are several partial explanations for this discrepancy.

Firstly, the two studies invert data which sample the coseismic displacement field in different places. The INSAR coherence breaks down in several crucial areas: both where Meyer *et al.* could measure surface rupture and around many of the pillars where Clarke *et al.* measured large displacements. And the spatial distribution of the data makes a difference in the inversion procedure, as discussed below.

Secondly, the inversion procedure makes a difference, as described in the comment and reply. In particular, constraining the fault plane to pass through the mapped surface break can shift the fault plane significantly, particularly in the presence of irregularly sampled data.

This exercise of independent analyses of independent data sets raises several issues. First, that geodetic and seismological estimates of moment can differ significantly. Second, we still lack good ground-truth evaluations of surface breaks as measured by INSAR, which records them as discontinuities in the fringe pattern. Thirdly, the geometric relationship between geodetically estimated fault planes, the mainshock location, and the distribution of aftershocks is not clear.

4. Synthesis and Conclusions

4.1 Depth Estimates

Several studies suggest that geodetic estimates tend to locate the coseismic slip at a shallower depth than seismological estimates of the mainshock hypocenter or centroid. There are several possible explanations for this discrepancy which became glaringly apparent for the Northridge earthquake (Hudnut *et al.*, 1995).

4.1.1 Rheological Inhomogeneity

For computational simplicity, most geodetic inversions assume an elastic half-space with a constant rheology throughout. Local heterogeneities in crustal rheology clearly violate this assumption. The half-space approximation is not even consistent with the simple layered models routinely used for locating earthquake hypocenters. In particular, if the assumed value of the shear modulus μ is too high, then the geodetic estimate will underestimate the depth, yielding a location which is too shallow (Cattin *et al.*, 1999). This shortcoming may also explain the observation that most of the aftershocks are near, but not on, the mainshock fault plane for Northridge (Hudnut *et al.*, 1995), Cape Mendocino (Murray *et al.*, 1996), and Antofogasta, Chile (Ruegg *et al.*, 1996). Thus Segall and Davis (1997) suggest that "it now seems probable that the effects of inhomogeneity, and perhaps anisotropy in the Earth's crust can no longer be neglected (e.g., Du *et al.*, 1994)." A different solution to this problem allows layering in a spherical earth (Pollitz, 1996; Cummins *et al.*, 1998). The differences with respect to the half-space solution can be of the order of 10–20% at distances of 100–400 km from the source (Cummins *et al.*, 1998).

4.1.2 Nonplanar Fault Geometry

A normal fault which shallows with depth, as Meyer *et al.* (1996, 1998) argue for Grevena, is difficult to approximate with the simple model of a single, planar fault patch. Since the latter is the only feasible geometric parametrization in the nonlinear focal mechanism inverse problem, it may not pass through the mainshock hypocenter unless constrained to do so (Clarke *et al.*, 1996, 1998).

4.1.3 Irregular Distribution of Data

Geodetic networks with benchmarks on rock outcrops do not form regular grids. Even INSAR, which in principle samples the deformation on a regular grid, can break down in certain areas, creating "blots" of missing data. For the Dinar earthquake, for example, all the usable INSAR data fall on the hanging-wall block (Wright *et al.*, 1999). Such an asymmetric distribution of data may tend to "pull" the modeled fault plane towards the data points.

4.2 Distribution of Slip and Aftershocks

The Landers studies show that aftershocks tend to occur near the parts of the fault plane where the amount of slip is small, e.g., Plate 21 of Massonnet and Feigl (1998) and Figure 13 of Cohee and Beroza (1994). In other words, aftershocks correlate spatially with the absence of slip. A similar, but weaker, correlation has also been observed for Loma Prieta (Arnadottir and Segall, 1994). It seems that slip may relieve stress on the fault patch where it occurs, but increase stress in the neighboring patches.

4.3 Geodetic Versus Seismological Estimates of Moment

Figure 3 shows that geodetic estimates of seismic moment tend to exceed seismological estimates by as much as 60%. Again, there are several possible explanations.

4.3.1 Measurement Interval

Geodetic measurements of relative position before and after the earthquake span a much longer period of time than does a seismogram and thus include more deformation, both interseismic and postseismic. Thus geodetic measurements sometimes include the moment released by aftershocks and/or afterslip, whereas the seismological estimate pertains only to the mainshock (Kanamori, 1973; Wyatt, 1988). Whereas historical geodetic measurements may impose an interval of several decades, satellite techniques can reduce the interval to a single 35-day orbital cycle in the case of the ERS satellites or to a single day or less in the case of continuous GPS. In this sense, geodesy, as ultralong period seismology, should follow the rule of thumb that moment estimates tend to increase with the measurement period.

4.3.2 Incorrect Shear Modulus

To convert seismic potency (in m^3) to moment (in N m) requires accurate knowledge of the shear modulus μ. Yet this parameter is rarely measured. Usually, it is assumed to take conventional values between 30 and 36 GPa, a range large enough to explain a 20% discrepancy. Of course, for a rigorous comparison, we should use the same value for the shear modulus μ in the (geodetic) dislocation model and in the (seismological) velocity model. In practice, however, this will require generalizing the dislocation theory to admit layering.

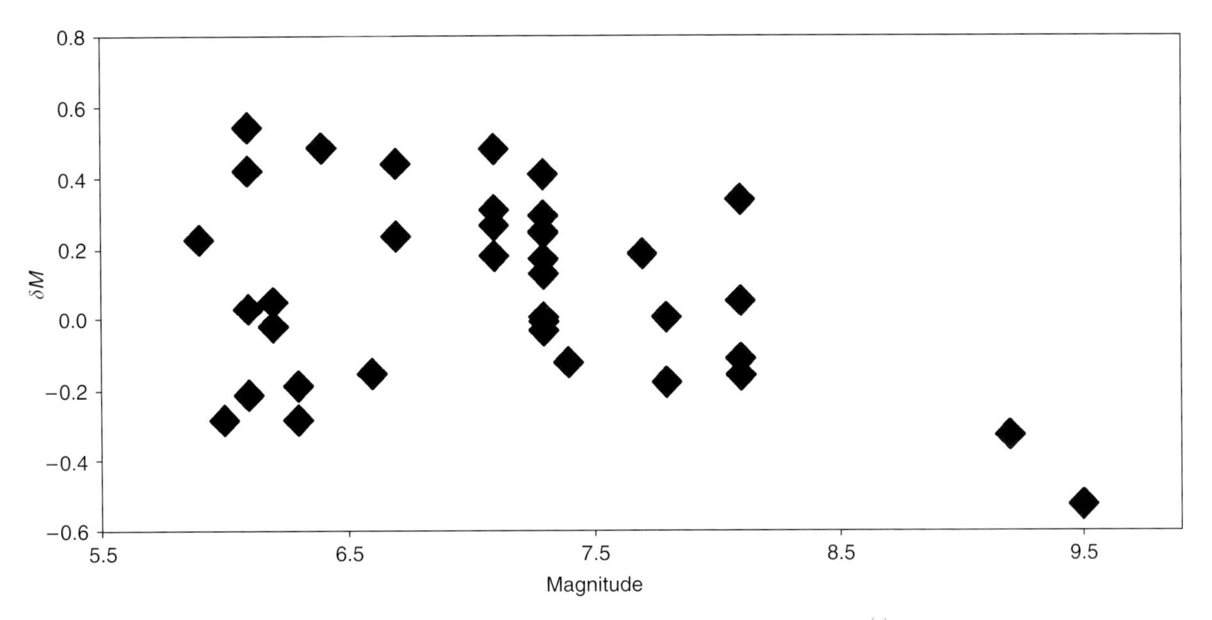

FIGURE 3 Comparison of seismic moment estimated from geodetic measurements $M_0^{(g)}$ and seismic moment estimated from centroid moment tensor solutions $M_0^{(s)}$. The vertical axis shows the ratio $\delta M = (M_0^{(g)} - M_0^{(s)})/M_0^{(s)}$ using the values in Table 1. The geodetic moment is larger than the seismic moment for most events.

4.3.3 Aseismic Deformation

Geodetic estimates of moment will include deformation caused by any phenomenon including creep, fluid injection, or even "silent" earthquakes. For example, the geodetic estimate derived from the 2 m of displacement measured for the 1978 Asal rifting event in the Afar region is 6×10^{17} N m, over three times larger than the 1.7×10^{17} N m estimated for the sum of the two largest $M = 5$ earthquakes (Lépine *et al.*, 1979; Ruegg *et al.*, 1979; Stein *et al.*, 1991).

4.4 Conjugate Faults

For earthquakes of intermediate magnitude, choosing between two possible focal planes is both challenging and interesting. For a point source observed from the far field, neither seismological analysis of first arrivals nor geodetic measurements of surface displacements can tell the difference because the two cases produce rigorously identical results. For example, a fault plane parallel to the subduction zone and a plane perpendicular to it both fit the large coseismic displacements observed by continuous GPS after the 1994 Hokkaido-Toho-Oki earthquake in Japan (Tsuji *et al.*, 1995).

For a dipping, finite fault with dimensions larger than the distance between geodetic observations, this ambiguity should fade as asymmetry begins to appear. In particular, the displacement vectors above the downdip edge of the fault are smaller than above the upper edge. In some cases, the RMS misfit is less than 1 mm better for one fault plane than for its conjugate (Stein *et al.*, 1991; Feigl *et al.*, 1995; Massonnet and Feigl, 1995b). With good sampling, however, it is possible to identify the rupture plane (Hudnut *et al.*, 1995).

4.5 Utility for Assessing Seismic Risk

Since seismologic data yield good estimates of the slip distribution, why estimate it from geodetic data? In remote areas, strong-motion seismological instruments may not exist, whereas a satellite radar interferogram can provide true remote sensing. Second, geodetic surveys record deformation over a much longer period (several months), than the seismological record, revealing any slip which occurred before or after the mainshock rupture. Third, geodetic surveys can capture aseismic slip. Finally, accurate descriptions of the total slip distribution are useful for calculations of coseismic stress changes.

4.6 Future Prospects

To contribute more useful information to the understanding of earthquake source parameters, geodetic analyses should attempt to address the following issues.

4.6.1 Measurement Uncertainty for INSAR

We do not yet understand know the error budget for INSAR measurements. For sophisticated inversion schemes, we should account for the structure of the data covariance matrix. This issue becomes particularly important for joint inversions of different data types, such as INSAR with GPS, or strong-motion seismograms. I emphasize that correlations between data also influence Monte Carlo-like algorithms, which often incorrectly assume independent, random errors.

4.6.2 Routine Application of INSAR

To date, several factors make INSAR measurements of coseismic deformation a hit-or-miss, opportunistic affair. For example, the lack of a good digital elevation model can inhibit the application of INSAR to earthquakes outside the US. This difficulty should be resolved by the Shuttle Radar Topography Mission flown in early 2000. Its data should lead to public distribution of a 90 m DEM by 2004. Similarly, the lack of closely spaced orbital trajectories can force compromises. This difficulty will be partially alleviated by ENVISAT, which will have better orbital control than ERS, JERS, or RADARSAT. Third, capturing earthquakes with INSAR is a major challenge because no-one knows where they will occur. This implies that each INSAR-capable satellite must acquire a catalog of prequake images over all the land areas likely to produce a measurable earthquake. I estimate this area to be of the order of 70 million km^2.

4.6.3 Joint Inversions

Geodetic data seem to help constrain seismological solutions at relatively long temporal periods (days to years) and intermediate spatial scales (within several fault dimensions of the rupture), as demonstrated for Landers by Wald and Heaton (1994). They inverted GPS measurements of coseismic displacements, strong motion seismograms and teleseismic wave forms, both jointly and separately. Including INSAR measurements in this type of inversion is likely to furnish interesting results, as suggested for Landers (Hernandez *et al.*, 1999) and Dinar (Wright *et al.*, 1999).

4.6.4 Related Phenomena

INSAR and CGPS open two new windows in the spatio-temporal spectrum of seismological metrology: INSAR at distance scales between ~1 and ~10 km; CGPS at timescales of days to years. Prior to the introduction of these two techniques, measurements at these scales were prohibitively expensive or prone to drift. Now that both techniques have entered the realm of operational, routine observations, we should expect to see interesting observations of other seismological phenomena, such as slow earthquakes, interseismic strain accumulation, and perhaps even an earthquake precursor.

Acknowledgments

I thank Alexis Rigo, Jean-Claude Ruegg and Didier Massonnet for helpful discussions. Partially financed by l'Institut National des Sciences de l'Univers and GDR INSAR.

References

Abe, K. (1995). In: "Global Earth Physics: a Handbook of Physical Constants," vol. 1, pp. 206–213, American Geophysical Union.
Aki, K. and P.G. Richards (1980). "Quantitative Seismology," Freeman.
Arnadottir, T. and P. Segall (1994). *J. Geophys. Res.* **99**, 21835–21856.
Arnadottir, T., *et al.* (1991). *Geophys. Res. Lett.* **18**, 2217–2220.
Arnadottir, T., *et al.* (1992). *Bull. Seismol. Soc. Am.* **82**, 2248–2255.
Arnadottir, T., *et al.* (1995). *NZ J. Geol. Geophys.* **38**, 553–558.
Bamler, R. and P. Hartl (1998). *Inverse Probl.* **14**, R1–R54.
Barrientos, S.E. and S.N. Ward (1990). *Geophys. J. Int.* **103**, 589–598.
Barrientos, S.E., *et al.* (1987). *Bull. Seismol. Soc. Am.* **77**, 784–808.
Beauducel, F., *et al.* (2000). *J. Geophys. Res.* **104**, 16391–16402.
Beavan, J., *et al.* (1994). *EOS supplement, June 21*, pp. 59.
Bennett, R.A., *et al.* (1994). *J. Geophys. Res.* **100**, 6443–6461.
Bernard, P., *et al.* (1997). *J. Seismol.* **1**, 131–150.
Bezzeghoud, M., *et al.* (1995). *Tectonophysics* **249**, 249–266.
Bilham, R. (1991). *Rev. Geophys.* **29**, 1–30.
Bilham, R. (1998). *Appalachia* **47**, **187**, 79–107.
Bock, Y., *et al.* (1993). *Nature* **361**, 337–340.
Bock, Y., *et al.* (1997). *J. Geophys. Res.* **102**, 18013–18033.
Bomford, G. (1980). "Geodesy," 4th Edn. Oxford University Press.
Bürgmann, R., *et al.* (1997). *J. Geophys. Res.* **102**, 4933–4955.
Bürgmann, R., *et al.* (2000). *Annu. Rev. Earth Planet. Sci.* **28**, 169–209.
Calais, E., *et al.* (1993). *C. R. Acad. Sci. Paris* **317 II**, 1493–1500.
Cattin, R., *et al.* (1999). *Geophys. J. Int.* **137**, 149–158.
Chen, J., *et al.* (1984). In: "The International Karakoram Project," Cambridge University Press.
Clark, T.A., *et al.* (1990). *Geophys. Res. Lett.* **17**, 1215–1218.
Clarke, P.J., *et al.* (1996). *Geophys. Res. Lett.* **24**, 707–710.
Clarke, P.J., *et al.* (1998). *Geophys. Res. Lett.* **25**, 131–134.
Cohee, B.P. and G.C. Beroza (1994). *Bull. Seismol. Soc. Am.* **84**, 692–712.
Cotton, F. and M. Campillo (1994). *J. Geophys. Res.* **100**, 3961–3975.
Cotton, F. and M. Campillo (1995). *Geophys. Res. Lett.* **22**, 1921–1924.
Crétaux, J.-F., *et al.* (1998). *J. Geophys. Res.* **103**, 30167–30182.
Crippen, R.E. (1992). *Episodes* **15**, 56–61.
Crippen, R.E. and R.G. Blom (1992). *Trans. Am. Geophys. Un.* **73**, 364.
Crook, C.N., *et al.* (1982). *US Geol. Surv. Prof. Paper 1254*, 183–191.
Cummins, P.R., *et al.* (1998). *Geophys. Res. Lett.* **25**, 3219–3222.
Curlander, J.C. and R.N. McDonough (1991). "Synthetic Aperture Radar: Systems and Signal Processing," Wiley.
Dal Moro, G. and M. Zadro (1999). *Earth Planet. Sci. Lett.* **170**, 119–129.
Delacourt, C., *et al.* (1998). *Geophys. Res. Lett.* **25**, 2849–2852.
Dixon, T.H. (1991). *Rev. Geophys.* **29**, 249–276.
DMA (1987). *DMA TR 8350.2-B*, Defense Mapping Agency.
Dolan, J.F., *et al.* (1995). *Science* **267**, 199–205.
Du, Y.J., *et al.* (1994). *J. Geophys. Res.* **99**, 3767–3779.
Duquesnoy, T., *et al.* (1996). *Geophys. Res. Lett.* **23**, 3055–3058.
Dvorak, J. (1994). *J. Geophys. Res.* **99**, 9533–9542.

Dziewonski, A.M. and D.L. Anderson (1981). *Phys. Earth Planet. Inter.* **25**, 297–356.
Eberhardt-Phillips, D. (1989). *J. Geophys. Res.* **94**, 15565–15586.
Ekstrom, G., *et al.* (1992). *J. Geophys. Res.* **97**, 4843–4864.
Feigl, K.L. and E. Dupré (1999). *Comput. Geosci.* **25**, 695–704.
Feigl, K.L., *et al.* (1995). *Geophys. Res. Lett.* **22**, 1037–1048.
Feigl, K.L., *et al.* (2000). *J. Geophys. Res.* **105**, 25655–25670.
Fialko, Y., *et al.* (2001). *Geophys. Res. Lett.* **28**, 3063–3066.
Freymueller, J., *et al.* (1994). *Bull. Seismol. Soc. Am.* **84**, 646–659.
Ghiglia, D.C. (1998). "Two-dimensional Phase Unwrapping: Theory, Algorithms, and Software," Wiley.
Goldstein, R. (1995). *Geophys. Res. Lett.* **22**, 2517–2520.
Hager, B.H., *et al.* (1991). *Annu. Rev. Earth Planet. Sci.* **19**, 351–382.
Hanks, T.C. and H. Kanamori (1979). *J. Geophys. Res.* **84**, 2348–2350.
Hanssen, R.F. (2001). "Radar Interferometry: Data Interpretation and Analysis". Dordrecht, Kluwer Academic Publishers.
Harris, R.A. and R.W. Simpson (1992). *Nature* **360**, 251–254.
Harsh, P.W. (1982). *US Geol. Surv. Prof. Paper 1254*, 193–203.
Hartzell, S.H. and T.H. Heaton (1983). *Calif. Div. Mines Geol. Spec. Publ.* **66**, 241–246.
Henderson, F.M. and A.J. Lewis (Eds.) (1998). "Principles and Applications of Imaging Radar," Wiley.
Hernandez, B., *et al.* (1997). *Geophys. Res. Lett.* **24**, 1579–1582.
Hernandez, B., *et al.* (1999). *J. Geophys. Res.* **104**, 13083–13099.
Herring, T.A. (1992). *J. Geophys. Res.* **97**, 1981–1990.
Holdahl, S.R. and J. Sauber (1994). *Pageoph* **142**, 55–82.
Huang, B.S. and Y.T. Yeh (1997). *Bull. Seismol. Soc. Am.* **87**, 1046–1057.
Hudnut, K.W. (1995). *Rev. of Geophys. Suppl.* 249–255.
Hudnut, K.W., *et al.* (1994). *Bull. Seismol. Soc. Am.* **84**, 625–645.
Hudnut, K.W., *et al.* (1995). *Bull. Seismol. Soc. Am.* **86**, S49–S70.
Ihmlé, P. and J.-C. Ruegg (1997). *Geophys. J. Int.* **131**, 146–158.
Jaumé, S.C. and L.R. Sykes (1992). *Science* **258**, 1325–1328.
Johnson, H.O., *et al.* (1994). *Bull. Seismol. Soc. Am.* **84**, 660–667.
Kanamori, H. (1973). *Annu. Rev. Earth Planet. Sci.* **1**, 212–239.
Klotz, J., *et al.* (1999). *Pageoph* **154**, 709–730.
Larsen, S., *et al.* (1992). *J. Geophys. Res.* **97**, 4885–4902.
Larson, K.M. (1995). *Rev. Geophys. Suppl.* 371–377.
Lefebvre, M., *et al.* (1996). *EOS* **77**, 25–29.
Lépine, J.C., *et al.* (1979). *Bull. Soc. Geol. Fr.* **22**, 817–822.
Lin, J. and R.S. Stein (1989). *J. Geophys. Res.* **94**, 9614–9632.
Lisowski, M., *et al.* (1990). *Geophys. Res. Lett.* **17**, 1437–1440.
Lundgren, P.R., *et al.* (1993). *Geophys. Res. Lett.* **20**, 407–410.
Madsen, S.N. and H.A. Zebker (1998). *Manual of Remote Sensing* **2**, 359–380.
Marshall, G.A., *et al.* (1991). *Bull. Seismol. Soc. Am.* **81**, 1660–1693.
Massonnet, D. and K.L. Feigl (1995a). *Geophys. Res. Lett.* **22**, 1537–1540.
Massonnet, D. and K.L. Feigl (1995b). *Geophys. Res. Lett.* **22**, 1541–1544.
Massonnet, D. and K.L. Feigl (1998). *Rev. Geophys.* **36**, 441–500.
Massonnet, D. and T. Rabaute (1993). *IEEE Trans. Geosci. Remote Sens.* **31**, 455–464.
Massonnet, D., *et al.* (1994). *Nature* **369**, 227–230.
Massonnet, D., *et al.* (1996a). *Geophys. Res. Lett.* **23**, 969–972.
Massonnet, D., *et al.* (1996b). *Nature* **382**, 612–616.
Matthews, M.V. and P. Segall (1993). *J. Geophys. Res.* **98**, 12153–12163.
Meade, B.K. and R.W. Miller (1973). In: "San Fernando, California, Earthquake of February 9, 1971," Vol. III, pp. 243–293. NOAA.
Melbourne, T., *et al.* (1997). *Geophys. Res. Lett.* **24**, 715–718.
Meyer, B., *et al.* (1996). *Geophys. Res. Lett.* **23**, 2677–2680.
Meyer, B., *et al.* (1998). *Geophys. Res. Lett.* **25**, 129–130.
Michel, R., *et al.* (1999). *Geophys. Res. Lett.* **26**, 875–878.
Milbert, D.G. and W.T. Dewhurst (1992). *J. Geophys. Res.* **97**, 545–557.
Miyazaki, S., *et al.* (1996). *Bull. Geogr. Surv. Inst. (Japan)* **42**, 27–41.
Morrison, N.L. (1973). In: "San Fernando, California, Earthquake of February 9, 1971," Vol. III, pp. 295–324. NOAA.
Murakami, M., *et al.* (1996). *J. Geophys. Res.* **101**, 8605–8614.
Murray, M.H., *et al.* (1996). *J. Geophys. Res.* **101**, 17707–17725.
Murray, M.H., *et al.* (1993). *Geophys. Res. Lett.* **20**, 623–626.
Noomen, R., *et al.* (1996). *J. Geodynam.* **21**, 73–96.
Okada, Y. (1985). *Bull. Seismol. Soc. Am.* **75**, 1135–1154.
Okada, Y. (1992). *Bull. Seismol. Soc. Am.* **82**, 1018–1040.
Ozawa, S., *et al.* (1997). *Geophys. Res. Lett.* **24**, 2327–2330.
Peltzer, G. and P. Rosen (1995). *Science* **268**, 1333–1336.
Peltzer, G., *et al.* (1996). *Science* **273**, 1202–1204.
Peltzer, G., *et al.* (1998). *J. Geophys. Res.* **103**, 30131–30146.
Perrier, G. and J.C. Ruegg (1973). *Ann. Geophys.* **29**, 435–502.
Pollitz, F., *et al.* (2000). *J. Geophys. Res.* **105**, 8035–8054.
Pollitz, F.F. (1996). *Geophys. J. Int.* **125**, 1–14.
Prescott, W.H., *et al.* (1979). *J. Geophys. Res.* **84**, 5423–5435.
Price, E.J. and D.T. Sandwell (1998). *J. Geophys. Res.* **103**, 27001–27016.
Pritchard, M.E., *et al.* (2002). *Geophysical Journal International*, in press.
Reilinger, R.E., *et al.* (2000). *Science* **289**, 1519–1524.
Rigo, A. and D. Massonnet (1999). *Geophys. Res. Lett.* **26**, 3217–3220.
Rosen, P.A., *et al.* (1996). *J. Geophys. Res.* **101**, 23109–23125.
Ruegg, J.C., *et al.* (1979). *Geophys. Res. Lett.* **6**, 817–820.
Ruegg, J.C., *et al.* (1982). *Bull. Seismol. Soc. Am.* **72**, 2227–2244.
Ruegg, J.C., *et al.* (1996). *Geophys. Res. Lett.* **23**, 917–920.
Sandwell, D.T., *et al.* (2000). *Geophys. Res. Lett.* **27**, 3101–3104.
Satake, K. (1993). *J. Geophys. Res.* **98**, 4553–4565.
Sauber, J.M., *et al.* (1993). In: "Contributions of Space Geodesy to Geodynamics: Crustal Dynamics," Vol. 23, pp. 233–248, American Geophysics Union.
Savage, J.C. and R.O. Burford (1970). *Bull. Seismol. Soc. Am.* **60**, 1877–1896.
Savage, J.C. and J.L. Svarc (1997). *J. Geophys. Res.* **102**, 7565–7577.
Savage, J.C., *et al.* (1993). *J. Geophys. Res.* **98**, 19951–19958.
Savage, J.C., *et al.* (1994). *J. Geophys. Res.* **99**, 13757–13765.
Scholz, C.H. (1990). "Earthquakes and Fault Mechanics," Cambridge University Press.
Segall, P. and J.L. Davis (1997). *Annu. Rev. Earth. Planet. Sci.* **25**, 301–336.
Segall, P. and Y. Du (1993). *J. Geophys. Res.* **98**, 4527–4538.
Shen, Z., *et al.* (1994). *Bull. Seismol. Soc. Am.* **84**, 780–791.
Shen, Z.K., *et al.* (1996). *Bull. Seismol. Soc. Am.* **86**, S37–S48.
Sillard, P., *et al.* (1998). *Geophysical Res. Lett.* **25**, 3223–3226.
Smith, D.E. and D.L. Turcotte (Eds.) (1993). "Contributions of Space Geodesy to Geodynamics: Crustal Dynamics," American Geophysics Union.
Smith, D.E., *et al.* (1994). *Geophysical Res. Lett.* **21**, 1979–1982.

Stein, R.S. and S.E. Barrientos (1985). *J. Geophys. Res.* **90**, 11355–11366.
Stein, R.S. and G.C.P. King (1984). *Science* **224**, 869–872.
Stein, R.S., *et al.* (1991). *J. Geophys. Res.* **96**, 21789–21806.
Stein, R.S., *et al.* (1992). *Science* **258**, 1328–1332.
Stein, R.S., *et al.* (1994). *Science* **265**, 1432–1435.
Stiros, S.C. and A. Drakos (2000). *Geophys. J. Int.* **143**, 679–688.
Tabei, T., *et al.* (1996). *J. Phys. Earth* **44**, 281–286.
Tapley, B.D., *et al.* (1993). In: "Contributions of Space Geodesy to Geodynamics: Earth Dynamics," Vol. 24, pp. 147–173, American Geophysics Union.
Tarayre, H. (1994). PhD thesis, U. P. Sabatier, Toulouse.
Tarayre, H. and D. Massonnet (1996). *Geophys. Res. Lett.* **23**, 989–992.
Thatcher, W. (1974). *Science* **184**, 1283–1285.
Thatcher, W., *et al.* (1997). *J. Geophys. Res.* **102**, 5353–5367.
Thio, H.K. and H. Kanamori (1996). *Bull. Seismol. Soc. Am.* **86**, S84–S92.
Trouvé, E. (1996). PhD thesis, Ecole Nat. Sup. Telecommunications, Paris.
Trouvé, E., *et al.* (1998). *IEEE Trans. Geosci. Remote Sen.* **36**, 1963–1972.
Tsuji, H., *et al.* (1995). *Geophys. Res. Lett.* **22**, 1669–1672.
Van Puymbroeck, N., *et al.* (2000). Measuring earthquakes from optical satellite images. *Applied Optics* **39**, 3486–3494.
Vincent, F. (1998). DEA thesis, U. Nice, Sophia-Antipolis, France.
Wald, D.J. and T.H. Heaton (1994). *Bull. Seismol. Soc. Am.* **84**, 668–691.
Wald, D.J., *et al.* (1996). *Bull. Seismol. Soc. Am.* **86**, S49–S70.
Ward, S. and S. Barrientos (1986). *J. Geophys. Res.* **91**, 4909–4919.
Williams, C.R., *et al.* (1993). *J. Geophys. Res.* **98**, 4567–4578.
Williams, S., *et al.* (1998). *J. Geophys. Res.* **103**, 27051–27068.
Wolf, M. and D. Wingham (1992). *Geophys. Res. Lett.* **19**, 2325–2328.
Wright, T.J., *et al.* (1999). *Earth Planet. Sci. Lett.* **172**, 23–27.
Wyatt, F.K. (1988). *J. Geophys. Res.* **93**, 7923–7942.
Wyatt, F.K., *et al.* (1994). *Bull. Seismol. Soc. Am.* **84**, 768–779.
Zebker, H. and R. Goldstein (1986). *J. Geophys. Res.* **91**, 4993–5001.
Zebker, H.A., *et al.* (1994). *J. Geophys. Res.* **99**, 19617–19634.
Zebker, H.A., *et al.* (1997). *J. Geophys. Res.* **102**, 7547–7563.
Zumberge, J.F., *et al.* (1996). International GPS Service 1995 Annual report, IGS Central Bureau, Pasadena.

Editor's Note

Due to space limitations, references with full citation are given in the Handbook CD-ROM under directory \37Feigl as FeiglFullReferences.pdf. Please see also Chapter 35, Strength and energetics of active fault zones, by Brune and Thatcher; and Chapter 36, Implications of crustal strain during conventional, slow and silent earthquakes, by Johnston and Linde.

38

Electromagnetic Fields Generated by Earthquakes

M.J.S. Johnston
US Geological Survey, Menlo Park, USA

1. Introduction

Independent knowledge of the physical processes that occur with seismic events can be obtained from observations of electric and magnetic fields generated by these complex processes. During the past few decades, we have seen a remarkable increase in the quality and quantity of electromagnetic (EM) data recorded before and during earthquakes and volcanic eruptions. This paper describes the most significant recent data and the implications these data have for different generating mechanisms. We note that, despite several decades of relatively high quality monitoring, clear demonstration of the existence of precursory EM signals has not been achieved, although causal relations between coseismic magnetic field changes and earthquake stress drops are no longer in question. This paper extends discussions of tectonomagnetism and tectonoelectricity, over the various parts of the electromagnetic spectrum from radio frequencies (RF) to submicrohertz frequencies, that are covered in Johnston (1989, 1997), Park *et al.* (1993), Park (1996) special journal issues (Johnston and Parrot, 1989, 1998; Parrot and Johnston, 1993), and books (Hayakawa and Fujinawa, 1994).

2. History

Suggestions that electromagnetic field disturbances are a consequence of the earthquake failure process have been made throughout recorded history. Some of the earliest work (Milne, 1890, 1894) refers to magnetic fields observed during the Great Lisbon earthquake in 1799. However, these early observations, and others during the nineteenth century (Mascart, 1887; Milne, 1894), were recognized as spurious by Reid (1914) who showed that traces recorded by magnetographs located close to earthquake epicenters were produced by inertial effects, not magnetic disturbances. This invalidated all earlier reports based on records of magnetic variometers, which are simply suspended magnets. Later data up through the mid-20th century recorded on fluxgate magnetometers are similarly suspect because these instruments are sensitive to displacements and rotations common in the epicentral regions during the propagation of seismic waves. It was not until the mid-1960s that meaningful results were obtained following the advent of absolute magnetometers, use of noise reduction techniques, and hardening of magnetometer measurement systems against the effects of earthquake accelerations and displacements. This is apparent in a plot of signal amplitude as a function of time, shown in Figure 1. Post-1960s data are generally considered trustworthy provided care has been taken to ensure sensors are insensitive to seismic shaking and are in regions of low-magnetic-field gradient.

Together with better stability against shaking, an impressive improvement in sensors, sensor reliability, data collection

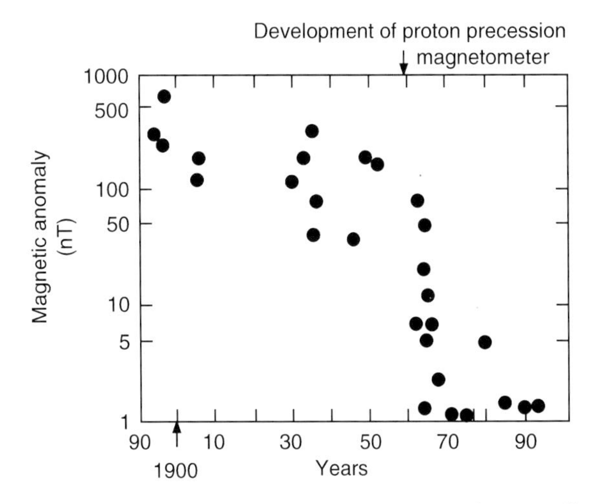

FIGURE 1 Reported amplitudes of tectonomagnetic anomalies as a function of time. The arrow indicates the time at which absolute proton precession magnetometers and noise reduction techniques were introduced (updated from Rikitake, 1968).

techniques, analysis, and international cooperation has also been apparent during the past few years. Largely as a consequence, reliable observations of magnetic, electric and electromagnetic field variations, related to seismic events and tectonic stress/strain loading, have been obtained on faults in Japan, China, Russia, California, and several other locations.

3. Statement of the Problem

This chapter reviews recent results of magnetic, electric, and electromagnetic disturbances apparently associated with earthquakes and discusses the physical mechanisms likely to have produced them. Although some observations are larger than expected, the best field observations are generally in agreement with calculations. Some observations are suggested as precursors yet have no corresponding co-event signals and some have co-event signals yet no precursory signals. More field observations are clearly needed and more careful work needs to be done to demonstrate convincingly causality, or lack thereof, between EM signals and earthquakes. In particular, further improvement is needed with the use of multiple detectors, use of multiple reference sites, application of noise identification/reduction procedures. Most important is the demonstration of consistency and correlation with other geophysical data that independently reflect the state of stress, strain, material properties, fluid content, and approach to failure of the Earth's crust in seismically active regions.

4. Summary of Physical Mechanisms Involved

The loading and rupture of water-saturated crustal rocks during earthquakes, together with fluid/gas movement, stress redistribution, and change in material properties, has long been expected to generate associated magnetic and electric field perturbations. The detection of related perturbations prior to fault rupture has thus been proposed frequently as a simple and inexpensive method to monitor the state of crustal stress and perhaps to provide tools for predicting crustal failure (Wilson, 1922; Kalashnikov, 1954; Stacey, 1964; Stacey *et al.*, 1965; Yamazaki, 1965; Brace and Orange, 1968a,b; Nagata, 1969; Barsukov, 1972; Rikitake, 1968, 1976; Honkura *et al.*, 1976; Fitterman, 1979, 1981; Ishido and Mizutani, 1981; Varotsos and Alexopoulos, 1987; Dobrovolsky *et al.*, 1989; Sasai, 1980, 1991a,b; Park, 1991; Fujinawa *et al.*, 1992; Fenoglio *et al.*, 1995; Utada, 1993). The primary mechanisms for generation of electric and magnetic fields with crustal deformation and earthquake-related fault failure include piezomagnetism, stress/conductivity, electrokinetic effects, charge generation processes, charge dispersion, magnetohydrodynamic effects, and thermal remagnetization and demagnetization effects. Discussion of the different mechanisms will be roughly in order of the degree of attention that has been accorded them.

4.1 Piezomagnetism

The magnetic properties of rocks have been shown under laboratory conditions to depend on the state of applied stress (Wilson, 1922; Kalashnikov and Kapitsa, 1952; Kapitsa, 1955; Ohnaka and Kinoshita, 1968; Kean *et al.*, 1976; Revol *et al.*, 1977; Martin, 1980; Pike *et al.*, 1981). Theoretical models have been developed in terms of single domain and pseudo-single domain rotation (Stacey, 1962; Nagata, 1969; Stacey and Johnston, 1972) and multidomain wall translation (Kern, 1961; Kean *et al.*, 1976; Revol *et al.*, 1977). The fractional change in magnetization per unit volume as a function of stress, can be expressed in the form;

$$\Delta \mathbf{I} \approx K\sigma \cdot \mathbf{I} \tag{1}$$

where $\Delta \mathbf{I}$ is the change in magnetization in a body with net magnetization $\mathbf{I}$ due to a deviatoric stress $\sigma \cdot K$, the stress sensitivity, typically has values of about $3 \times 10^{-3}\,\mathrm{MPa}^{-1}$. The stress sensitivity of induced and remanent magnetization from theoretical and experimental studies has been combined with stress estimates from dislocation models of fault rupture and elastic pressure loading in active volcanoes to calculate magnetic field changes expected to accompany earthquakes and volcanoes (Stacey, 1964; Stacey *et al.*, 1965; Shamsi and Stacey, 1969; Johnston, 1978; Davis *et al.*, 1979; Sasai, 1980, 1983, 1991a,b; Davis *et al.*, 1984; Johnston *et al.*, 1994; Banks *et al.*, 1991). The surface fields ($\Delta \mathbf{B}_P$) at a point, P, can be calculated in two ways: (1) by either integrating the change in magnetization $\Delta \mathbf{I}_Q$ in a unit volume, dv, at a point Q where the stress is σ_{ij}, and r is the distance between P and Q, according to,

$$\Delta \mathbf{B}_P = -\frac{\mu}{4\pi} \nabla \int_V \Delta \mathbf{I}_Q \cdot \frac{\mathbf{r}}{r^3}\, dv \tag{2}$$

as originally done by Stacey (1964), or (2) by using a simpler method pioneered by Sasai (1980, 1994) in which analytic expressions of the surface piezomagnetic potential, W, produced by a known stress distribution in a magnetoelastic half-space are obtained by transforming the stress matrix and integrating over the magnetized region. In this latter case, the surface field can be found from:

$$\Delta \mathbf{B}_P = -\nabla W \tag{3}$$

These models show that magnetic anomalies of a few nanoteslas (nT) should be expected to accompany earthquakes for rock magnetizations and stress sensitivities of 1 ampere/meter ($\mathrm{A\,m^{-1}}$) and $10^{-3}\,\mathrm{MPa}^{-1}$, respectively. As shown below, these signals are readily observed with the correct sign and amplitude.

4.2 Stress/Resistivity and Strain/Resistivity effects

In a like manner, the stress dependence of electrical resistivity of rocks has been demonstrated in the laboratory. Resistivity in low porosity crystalline rock increases with compression as a result of crack closure at about 0.2%/bar (Brace *et al.*, 1965) and decreases with shear due to crack opening at about 0.1%/bar (Yamazaki, 1965; Brace and Orange, 1968a,b; Brace, 1975). More porous rocks have even lower stress sensitivity. The situation is further complicated by the fact that nonlinear strain can also produce resistivity changes (Lockner and Byerlee, 1986). A stress/resistivity relation equivalent to Eq. (1) has the scalar form

$$\frac{\Delta\rho}{\rho} \approx K_r \sigma \tag{4}$$

for homogeneous material, where ρ is resistivity, K_r is a constant, and σ is the stress. Unfortunately, the earth is not homogeneous and many factors including rock type, crack distribution, degree of saturation, porosity, strain level, etc., can localize or attenuate current flow. Nevertheless, this equation provides a starting point for calculating resistivity changes near active faults. Measurements of resistivity change are being made with both active experiments (where low frequency currents are injected into the ground and potential differences, V, are measured on receiver dipoles), or passive telluric and magnetotelluric (MT) experiments where changes in resistivity are inferred from changes in telluric or MT transfer functions. These transfer functions are given by:

$$Z(\omega) = \frac{E(\omega)}{H(\omega)} \tag{5}$$

where ω is angular frequency, $E(\omega)$ and $H(\omega)$ are observed electric and magnetic fields. For active experiments (Park *et al.*, 1993),

$$\frac{\delta\rho}{\rho} = G\frac{\delta V}{V} \tag{6}$$

where δV is the change in potential difference and G is a constant. For MT experiments,

$$\delta\rho = \frac{\delta|Z(\omega)|^2}{\omega\mu} \tag{7}$$

Based on the field observations of stress changes accompanying earthquakes (≈ 1 MPa), resistivity changes of at least 1% might be expected to accompany crustal failure. Field experiments for detection of resistivity changes thus need to have a measurement precision of better than 0.1% (Fitterman and Madden, 1977; Park, 1991; Park *et al.*, 1993). This may be difficult with MT measurements unless remote magnetic field reference measurements are used (Gamble *et al.*, 1979) although measurement precision for telluric electric fields can be made at the 0.1% level (Madden *et al.*, 1993).

4.3 Electrokinetic Effects

The role of active fluid flow in the Earth's crust as a result of fault failure can generate electric and magnetic fields (Mizutani and Ishido, 1976; Fitterman, 1978, 1979; Ishido and Mizutani, 1981; Dobrovolsky *et al.*, 1989; Fenoglio *et al.*, 1995). Electrokinetic electric and magnetic fields result from fluid flow through the crust in the presence of an electric double layer at the solid–liquid interfaces. This double layer consists of ions anchored to the solid phase, with equivalent ionic charge of opposite sign distributed in the liquid phase near the interface. Fluid flow in this system transports the ions in the fluid in the direction of flow, and electric currents result. Conservation of mass arguments (Fenoglio *et al.*, 1995) supported by surface strain observations (Johnston *et al.*, 1987) limit this process in extent and time because large-scale fluid flow cannot continue for very long before generating easily detectable surface deformation.

The current density **j** and fluid flow **v** are found from coupled equations (Nourbehecht, 1963; Fitterman, 1979) given by

$$\mathbf{j} = -s\nabla E - \frac{\xi\zeta\nabla P}{\eta} \tag{8}$$

$$\mathbf{v} = \frac{\phi\xi\zeta\nabla E}{\eta} - \frac{\kappa\nabla P}{\eta} \tag{9}$$

where E is streaming potential, s is the electrical conductivity of the fluid, ξ is the dielectric constant of water, η is fluid viscosity, ζ is the zeta potential, ϕ is the porosity, κ is the permeability, and P is pore pressure.

The current density in Eq. (8) has two components. The second term represents electric current resulting from mechanical energy being applied to the system and is sometimes called the "impressed" current (Williamson and Kaufman, 1981). This term describes current generated by fluid flow in fractures. The first term of Eq. (8) represents "back" currents resulting from the electric field generated by fluid flow. The distribution of electrical conductivity determines the net far-field magnetic and electric fields resulting from these effects. In an extreme case, if the fluid is extremely conducting and the surrounding region is not, current flow in the fluid cancels the potential generated by fluid flow (Ahmad, 1964). At the other extreme, if the fluid is poorly conducting, "back" currents, usually termed "volume currents" (Williamson and Kaufman, 1981) flow in the surrounding region. If the region were homogeneous, magnetic fields would be generated by impressed currents only since the volume currents generate no net field (Fitterman, 1979; Fenoglio *et al.*, 1995). The situation for finite flow in limited fault fractures more closely approximates the second case where the surface magnetic field is approximately given by:

$$\mathbf{B} = \frac{\mu_0}{4\pi} \int_A \frac{\mathbf{j}_i \times \mathbf{r}}{r^2} \, dA \tag{10}$$

where μ_0 is the magnetic permeability in free space. Note that, the physics describing the electric and magnetic fields generated in the human body as blood is pumped through in arteries provides a very good analog to those generated in fault zones (Williamson and Kaufman, 1981). This occurs because the electrical conductivities of bone ($0.001\,\mathrm{S\,m^{-1}}$), muscle ($0.1\,\mathrm{S\,m^{-1}}$) and blood ($1\,\mathrm{S\,m^{-1}}$) and blood velocities are similar to those of rock, fault gouge, fault zone fluids, and the likely fluid velocities determined by Darcy Law fluid diffusion in fault zones. Considerable work has been done in understanding the physics of electric and magnetic field generation in the human body and this can be applied directly to crustal faulting situations. Reasonable fault models, in which fluid flows into a 200 m long rupturing fracture at a depth of 17 km, indicate that transient surface electric fields of several tens of millivolts per kilometer and transient magnetic fields of a few nT can be generated (Fenoglio *et al.*, 1995).

4.4 Charge Generation Processes

Numerous charge generation mechanisms have been suggested as potential current sources for electric and magnetic fields before and during earthquakes. These mechanisms include piezoelectric effects (Finkelstein *et al.*, 1973; Baird and Kennan, 1985), triboelectricity effects produced by rock shearing (Lowell and Rose-Innes, 1980; Gokhberg *et al.*, 1982; Brady, 1992), fluid disruption/vaporization (Blanchard, 1964; Matteson, 1971; Chalmers, 1976), and solid state mechanisms (Dologlou-Revelioti and Varotsos, 1986; Freund *et al.*, 1992). Each of these mechanisms has a solid physical basis supported by laboratory experiments on either dry rocks in insulating environments or single crystals of dry quartz. Each is capable of producing substantial charge under the right conditions. However, at least two fundamental problems need to be studied in the application of charge-generation processes to EM field generation in the Earth's crust. The first concerns the amplitude of each charge generation effect in wet rocks at temperatures and pressures expected in the Earth's crust and the second concerns charge maintenance time and propagation in the conducting crust.

Regarding the first problem, experiments clearly need to be done for each mechanism to quantify the effects expected in wet rocks at temperatures of at least 100°C and at confining pressures of 100 MPa expected at earthquake hypocenters. Experiments on dry rocks at atmospheric pressure are not very relevant to this issue. Piezoelectric effects in dry quartz bearing rocks are less than 0.1% of those observed for single crystals of quartz due to self canceling effects (Tuck *et al.*, 1977; Sasaoka *et al.*, 1998), and effects in wet rocks will likely be smaller still and transient at best. EM generation by fracturing dry rocks (Warwick *et al.*, 1982; Brady, 1992) needs to be extended to wet rocks under confining pressure. Experiments on hole transport of O_- in dry rocks (Freund *et al.*, 1992) and stress charging of dry nonpiezoelectric rocks (Dologlou-Revelioti and Varotsos, 1986) need also to be repeated with wet rocks under confining pressure so that these effects can be quantified. Brady (1992) observed no EM emission during fracture of conductive rocks since the conductor could not maintain charge separation.

The second fundamental problem concerns the discharge time for these processes and just how far EM signals generated by them might propagate. The charge relaxation time τ for electrostatic processes is given by the product of permittivity (ε) and resistivity (ρ). ε is $0.5\text{–}1.0 \times 10^{-10}\,\mathrm{F\,m^{-1}}$ for crustal rocks. If $\rho \approx 10^3$ ohm.m (typical upper value for near fault crustal rock) then,

$$\tau \approx 10^{-6}\ \mathrm{sec} \tag{11}$$

Although polarization effects (Lockner and Byerlee, 1985) will generate somewhat longer timescales (perhaps as much as a second), EM signal generation by charge generation processes must necessarily still be very rapid unless mechanisms can be found for isolating and maintaining large charge densities in a conducting earth. Furthermore, dispersion precludes EM fields propagating very far in a conducting earth (Honkura and Kuwata, 1993).

Attenuation of the magnetic field, B, of a plane electromagnetic wave generated at depth by charge generation/cancellation processes as a function of penetration distance through a conductive medium is given by:

$$B = B_0 e^{-\gamma z} \tag{12}$$

where B_0 is the initial field strength, z is the penetration distance into the medium, and γ is the complex propagation coefficient given by:

$$\gamma = \sqrt{\omega^2 \mu \varepsilon + j\omega\mu s} \tag{13}$$

where ω is the angular frequency of the radiation, μ is the magnetic permeability of the earth, ε is the permittivity, and s is the conductivity of the medium. If s is $0.1\,\mathrm{S\,m^{-1}}$, the frequency is 0.01 Hz, the "skin depth" is 10 km, so fields generated at this depth could be observable at the earth's surface.

If these fields are generated by rock cracking and fracturing, acoustic (seismic) signals should also be generated (see Lockner *et al.*, 1991). Seismic wave attenuation with distance z has the form

$$A(z) = A_0 \exp^{-(\omega z/2cQ)} \tag{14}$$

where ω is the angular frequency, c is the phase velocity, and Q is the quality factor. Taking observed values of 3 km $\sec^{-1}$ and 30 for c and Q, it can easily be shown that seismic waves in the frequency band 1–0.01 Hz are not attenuated significantly in the epicentral area. At higher frequencies, both seismic and EM signals are heavily attenuated. For example, at 10 Hz the EM "skin depth" is 493 m in material with conductivity of 0.1 S m^{-1} and the seismic equivalent "penetration depth" is 2864 m. At 100 Hz the comparative depths are 156 m, 286 m, and at 1 KHz 29 m, 49 m, respectively.

Thus, both high frequency seismic and EM waves are heavily attenuated in the earth's crust. EM sources at 10 Hz should have an acoustic component that is more easily detected over a greater area. In fact, for all EM sources at seismogenic depths capable of propagating to the earth's surface (i.e., with frequencies less than 0.1 Hz), acoustic/seismic consequences of these sources propagate more effectively to the surface and might be used to verify their existence.

4.5 Magnetohydrodynamic (MHD) Effects

The induced magnetic field $\mathbf{B}_i$ generated by the motion v of a fluid with conductivity s in a magnetic field $\mathbf{B}_0$, is governed by the equation:

$$\frac{\partial \mathbf{B}}{\partial t} = \nabla \times \mathbf{v} \times \mathbf{B} + \frac{\nabla^2 \mathbf{B}}{\mu_0 s} + \frac{\nabla s \times \nabla \times \mathbf{B}}{\mu_0 s^2} \qquad (15)$$

where μ_0 is the permeability in a vacuum (Shercliff, 1965). For low magnetic fields and low electrical conductivities in the Earth's crust where the fluid motion is not affected by the induced fields, the induced field is given approximately by the product of the magnetic Reynolds number R_m and the imposed field $\mathbf{B}_0$, i.e.,

$$\mathbf{B}_i \approx R_m \times \mathbf{B}_0 \approx \mu s v\, d\mathbf{B}_0 \qquad (16)$$

where d is the length scale of the flow. Critical parameters here are the likely flow velocities and fluid electrical conductivities in the crust. Flow velocity is determined by rock permeability and fluid pressure gradients according to Darcy's Law. Permeability of fractured rock is not less than $10^{-12}\,m^2$ (Brace, 1980) and pore-pressure gradients cannot exceed the lithospheric gradient. It is difficult to achieve widespread flow velocities of even a few millimeters $\sec^{-1}$ with this mechanism. Furthermore, fluid conductivities are unlikely to exceed that of sea water (≈ 1 S m^{-1}). Using these numbers, fluid flow in fractured fault zones at seismogenic depths (≈ 5 km) with a length scale of 1 km could generate transient fields of about 0.01 nT. This is far too small to be observed at the Earth's surface. As a check on these calculations, we note that fields of a few nT are observed with waves in the ocean where the conductivity is 1 S m^{-1} and wave velocities exceed 100 cm $\sec^{-1}$ (Fraser, 1966).

4.6 Thermal Remagnetization and Demagnetization

Crustal rocks lose their magnetization when temperatures exceed the Curie Point (≈ 580°C for magnetite) and become remagnetized again as the temperature drops below this value. Stacey and Banerjee (1974) describe this process in detail. In crustal rocks at seismogenic depths near active faults, this process is unlikely to contribute to rapid changes in local magnetic fields since the thermal diffusivity of rock is typically about $10^{-6}\,m^2\,\sec^{-1}$ and migration of the Curie Point isotherm by conduction cannot be as much as a meter in a year (Stacey, 1992). At shallow depths in volcanic regions, particularly in recently emplaced extrusions and intrusions, thermal cracking with gas and fluid movement can transport heat rapidly and large local anomalies can be quickly generated (Rikitake and Yokoyama, 1955; Hurst and Christoffel, 1973; Emeleus, 1977; Zlotnicki and Le Mouel, 1988; Hamano *et al.*, 1990; Dzurisin *et al.*, 1990; Zlotnicki and Le Mouel, 1990; Tanaka, 1993, 1995). These anomalies can be modeled as a magnetized slab in a half-space. Good examples of magnetic modeling of anomalies generated by cooling of extrusions can be found in Dzurisin *et al.* (1990) for Mt. St. Helens and in Tanaka (1995) for Mt. Unzen in Japan. Some seasonal variations may result from annual temperature diffusion into magnetic rocks in the upper few meters of the Earth's crust (Utada *et al.*, 2000).

5. Experimental Design and Measurement Precision

5.1 Basic Measurement Limitations

The precision of local magnetic and electric field measurements on active faults varies as a function of frequency, spatial scale, instrument type, and site location. Most measurement systems on the Earth's surface are limited more by noise generated by ionosphere, magnetosphere, and by cultural noise than by instrumental noise. Thus, systems for quantifying these noise sources are of crucial importance if changes in electromagnetic fields are to be uniquely identified. For spatial scales of a few kilometers to an few tens of kilometers, comparable to moderate magnitude earthquake sources, geomagnetic and electric noise power decreases with frequency as $1/f^2$, similar to the "red" spectrum behavior of most geophysical parameters (Johnston *et al.*, 1984). Figure 2 shows an example of a noise power spectrum for two typical sites 10 km apart on the San Andreas fault in regions of low local magnetic gradient and far from sources of cultural noise (after Ware *et al.*, 1985).

Against this background noise, transient magnetic fields can be measured to several nanotesla over months, to 1 nT over days, to 0.1 nT over minutes, and 0.01 nT over seconds. Long-term changes and field offsets can be determined if their amplitudes exceed about 1 nT. Comparable electric field noise

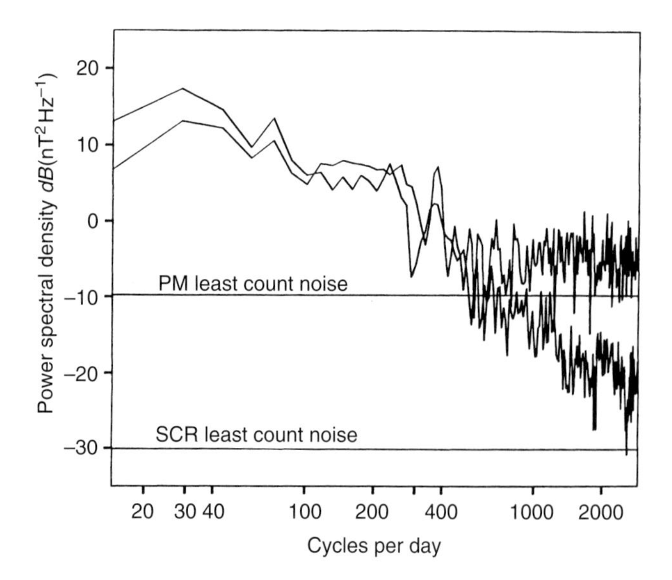

FIGURE 2 Power spectral density of simultaneous differences between sites 10 km apart on the San Andreas fault near San Juan Bautista. The self-calibrating rubidium (SCR) data are consecutive 10 sec measurements with a 0.014 nT least count. The 0.125 least-count proton magnetometer (PM) data are 1.5 sec averages taken every 15 sec (after Ware *et al.*, 1985).

limits are 10 mv km^{-1} over months, several mv km^{-1} over days, 1 mv km^{-1} over minutes and 0.1 mv km^{-1} over seconds (derived from Park, 1991). EM noise increases approximately linearly with site separation (Johnston *et al.*, 1984). Cultural noise further complicates measurement capability because of its inherent unpredictability. This largely precludes measurements in urban areas.

5.2 Experimental Techniques

Thus, unambiguous observations of EM signals originating within the Earth's crust require well-planned experiments to discriminate against these ionospheric, magnetospheric, and cultural noise sources. At lower frequencies (microhertz to hertz) for both electric and magnetic field measurements, the most common technique involves the use of reference sites with synchronized data sampling in arrays using site spacing comparable to the expected source sizes of a few kilometers (Rikitake, 1966; Johnston *et al.*, 1984; Park and Fitterman, 1990; Park 1991; Varotsos and Lazaridou, 1991). This relatively simple system allows as much as a 30 dB reduction in noise (Park, 1991; Johnston *et al.*, 1984). Techniques for further noise reduction such as adaptive filtering (Davis *et al.*, 1981; Davis and Johnston, 1983), use of multiple variable-length sensors in the same and nearby locations (Varotsos and Alexopoulos, 1987; Mori *et al.*, 1993), provide about a factor of three further improvement.

Although these same techniques can be applied to electromagnetic field measurements at higher frequencies (100 Hz to MHz) much less is known about the scale and temporal variation of noise and signal sources at these frequencies. Furthermore, as discussed below, basis physics likely precludes simple generation of high-frequency electromagnetic signals at seismogenic depths (5–10 km) on active faults in the Earth's crust where the electrical conductivity is more than 0.1 S m^{-1}.

6. Recent Results

Although both electric and magnetic fields are expected to accompany dynamic physical processes in the Earth's crust, simultaneous measurements of both fields are not routinely made. I will therefore discuss separately, electric fields, magnetic fields, and electromagnetic fields during and preceding earthquakes. Magnetic and electric fields generated by earthquakes are termed "seismomagnetic (SM)" and "seismoelectric (SE)" effects. Those preceding earthquakes, or occurring at other times, are termed "tectonomagnetic (TM)" and "tectonoelectric (TE)" effects.

If reliable magnetic and electric field observations (i.e., those unaffected by seismic shaking) are indeed source related, clear offsets should occur at the time of large local earthquakes because the primary energy release occurs at this time. These offsets should scale with the earthquake moment (size) and source geometry. In fact, co-event observations provide a determination of stress sensitivity since the stress redistribution and the source geometry of earthquakes are well-determined (Aki and Richards, 1980). With this calibration, tectonomagnetic and tectonoelectric effects can be quantified and spurious effects identified. Observations without consistent and physically sensible coseismic effects are generally considered suspect.

The following examples are restricted to the strongest data: data recorded independently on more than one instrument, data that are independently supported by other stable geophysical measurement systems, and data for which noise levels have been quantified. Reported measurements made with single instruments or time histories of measurements showing data only for a short period before earthquakes with some "precursive" feature but no coseismic signals, are generally suspect and are not included.

6.1 Seismomagnetic Effects

The primary features of seismomagnetic effects have become clear over the past twenty years with many continuous high-resolution magnetic field measurements in the epicentral regions of moderate-to-large earthquakes. Figure 3 shows a summary of observed SM effects as a function of earthquake moment normalized by epicentral distance. The following features are immediately apparent on this plot:

1. SM effects are observed above the measurement resolution in the near field of earthquakes only when the earthquake magnitude is $M=6$ or greater.

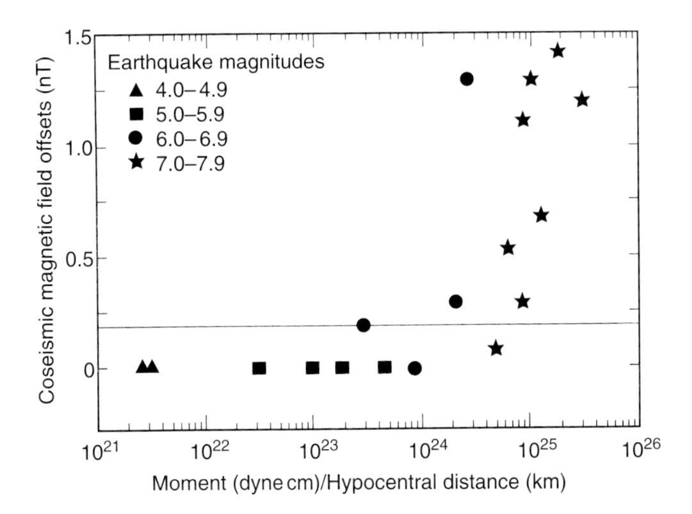

FIGURE 3 Coseismic magnetic field offsets as a function of seismic moment scaled by hypocentral distance. The region below the horizontal line shows the 2-sigma measurement resolution (from Mueller and Johnston, 1998).

2. The amplitudes of SM effects within one rupture length of these earthquakes are not more than 1 nT or so.

Coseismic strain measurements at these same locations for some of these earthquakes are a few microstrain (Johnston *et al.*, 1987). Thus, a good rule of thumb for estimating the expected SM effects, where strain observations are available, is to use a scaling factor of 1 nT per microstrain. A similar scaling factor was obtained from correlation between geodetic strain and local geomagnetic fields during aseismic deformation, uplift, and gravity changes in Southern California in the 1980s (Johnston, 1986).

6.1.1 *M* = 7.3 Landers Earthquake (28 June 1992)

An important set of observations of SM effects were made during the 28 July 1992 Landers earthquake (Johnston *et al.*, 1994). This earthquake had a moment of 1.1×10^{27} dyne cm and a magnitude of 7.3. Two total field proton magnetometers were in operation at distances of 17.1 and 24.2 km from the earthquake and have recorded synchronously sampled local magnetic fields every 10 min since early 1979 using satellite digital telemetry (Mueller *et al.*, 1981). The locations are shown in Figure 4.

The local magnetic field at the magnetometer closest to the earthquake decreased by 1.2 nT while that at the second decreased by 0.7 nT. These values are consistent with a simple SM model of the earthquake in which the fault geometry and slip used are derived from geodetic and seismic inversions of the earthquake data (Johnston *et al.*, 1994). Figure 5a shows the differences between data obtained at the two sites OCHM and LSBM for the period 1 day before and after the earthquake. Note that there is no indication of diffusion-like character in the magnetic field offsets that might indicate these effects were generated by fluid flow, nor are there any indications of enhanced low-frequency magnetic noise preceding the earthquake or indications of changing magnetic fields outside the noise in the hours to days before the earthquake. Figure 5b shows the longer-term data for the previous 7 y. The SM effect from the 1986 *M* 6 North Palm Springs earthquake which occurred beneath these same two instruments is clearly evident (Johnston and Mueller, 1987). Similar coseismic results were found for the 1989 *M* 7.1 Loma Prieta earthquake (Mueller and Johnston, 1990).

6.2 Seismoelectric Effects

Seismoelectric observations that show expected scaling with both earthquake moment release and inverse distance cubed are difficult to make because of the sensitivity of electrode contact potential to earthquake shaking. Earlier work by Yamazaki (1974) showed clear correspondence between local coseismic strain steps and coseismic resistivity steps obtained using a Wenner array with a measurement precision of 0.01%. Strain measurements obtained with deep borehole strainmeters (Johnston *et al.*, 1987) or with geodetic techniques (e.g., Lisowski *et al.*, 1990) do reflect the strain/stress expected from seismically determined models of earthquakes. SE effects should be expected to do likewise. Unfortunately, the effects of shaking on contact potential and on self-potential, as a result of changes in fluid content and fluid chemistry, make these data unreliable reflectors of earthquake stress changes (Ozima *et al.*, 1989). Clearly, colocated strain measurements (at tidal sensitivity) and electric field measurements are needed to demonstrate the sensitivity of the electric field measurements to changes in stress/strain in the Earth's crust. This, and observations on multiple sensors with stable electrodes (Petiau and Dupis, 1980; Perrier *et al.*, 1997) are necessary to have real confidence in SE measurements as earthquake monitors and perhaps as precursor detectors.

Measurements of electrical resistivity to better than 1% have been made since 1988 in a well-designed experiment installed near Parkfield, California (Park, 1997). Although the expected *M* = 6 earthquake in this region has not occurred, several earthquakes with *M* = 5 have occurred since 1990. None of these earthquakes generated any observable changes in resistivity above the measurement resolution. Observations for three of these dipoles during the times of these earthquakes, and other larger but more distant earthquakes, are shown in Figure 6.

Indirect observations of possible SE signals might be obtained using the magnetotelluric (MT) technique to monitor apparent resistivity in seismically active regions. Even with the best-designed systems using remote referencing systems to reduce noise and obtain stable impedance tensors (Gamble *et al.*, 1979), it is difficult to reduce errors below 5% for good soundings and 10–40% for poor soundings (Ernst *et al.*, 1993).

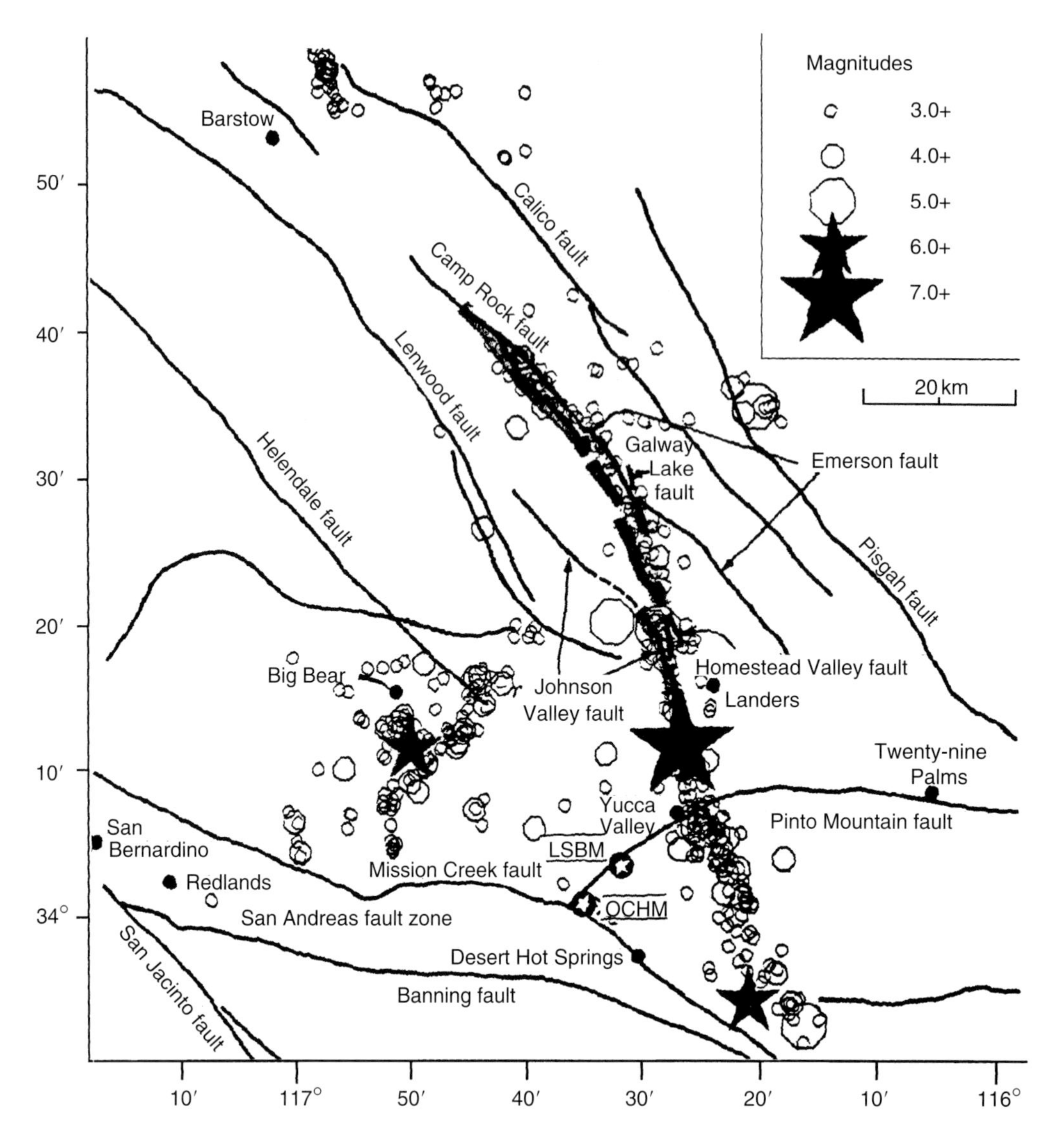

FIGURE 4 Locations of magnetometers, designated LSBM and OCHM (stars inside circles) relative to the epicenter (largest star) of the 28 June 1992 $M = 7.3$ Landers earthquake (from Johnston *et al.*, 1994).

Resistivity changes associated with earthquakes are expected (and are observed) to be only a few percent at best. Thus, it is unlikely that this technique will be used generally for detection of resistivity changes. The pioneering work of Honkura *et al.* (1976) still largely defines the limits of observability for MT observations, though some interesting ways of using MT to detect EM emissions with earthquakes have been explored by Rozluski and Yukutake (1993).

6.3 Tectonomagnetic and Possible Precursory Effects

Few indications of convincing longer-term tectonomagnetic events (i.e., durations greater than minutes to weeks) are apparent in multiple near-field magnetometer records obtained along active faults as a result of strain redistribution prior to moderate/large earthquakes. Only one such observation was made in 20 years of data along the San Andreas fault (Mueller and Johnston, 1998). A typical record of long-term magnetic field near active faults is shown in Figure 5b. Signals of more than a few nanoteslas are rare and are of great interest when they occur. If signal amplitudes near active faults are only about a nanotesla, it is unlikely that any reported precursive signals at great epicentral distances are truly earthquake related. Long-term relatively uniform changes, apparently related to crustal loading have been previously reported (Johnston, 1989; Oshiman *et al.*, 1983).

The situation may be different at higher frequencies (i.e., ≥ 0.01 Hz). Recent efforts have been concentrated at these higher ultralow frequency (ULF) frequencies as a result, primarily, of fortuitous observation of elevated ULF noise power near the epicenter of the $M = 7.1$ Loma Prieta earthquake of

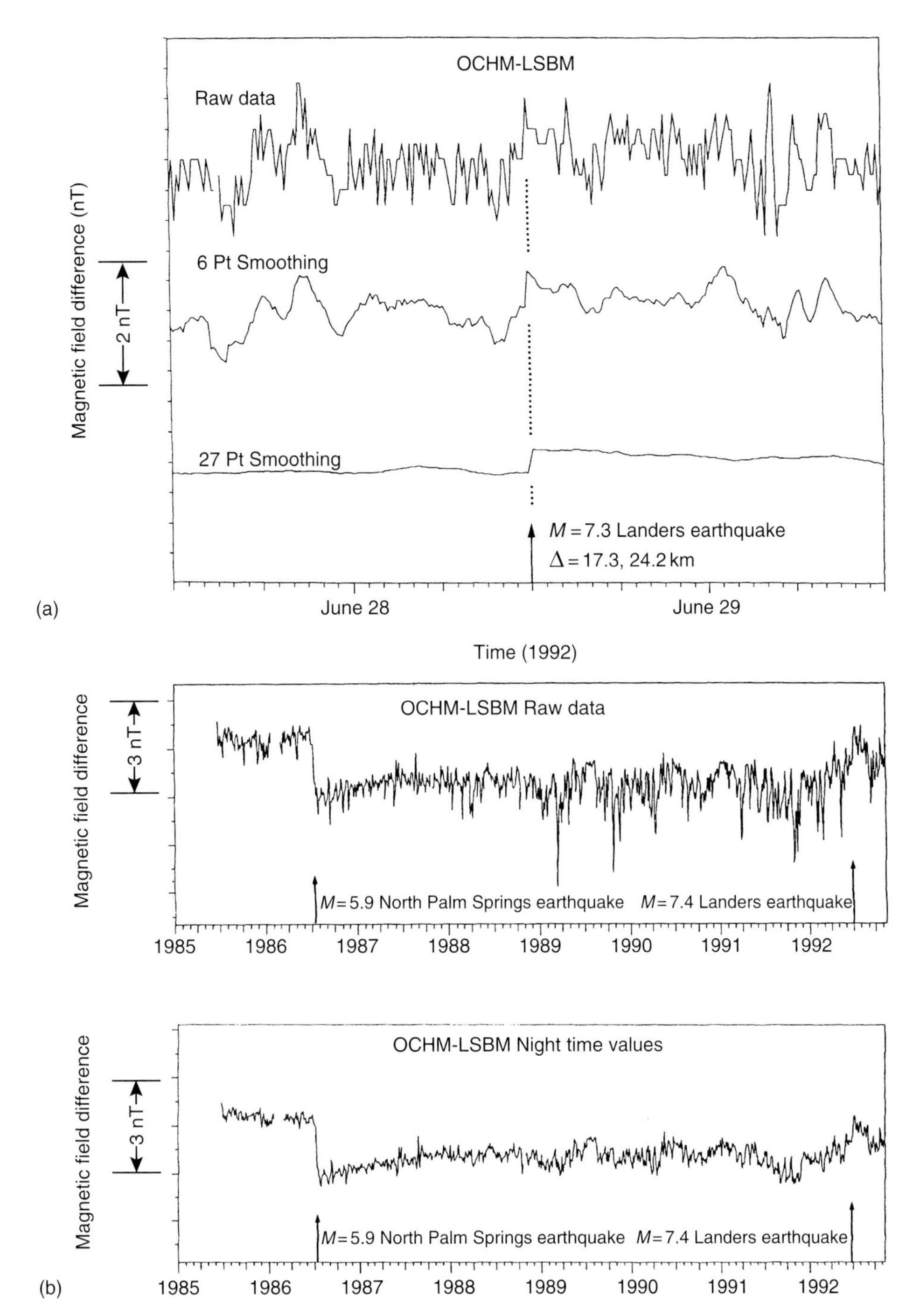

FIGURE 5 (a) Magnetic field differences between OCHM and LSBM (for location see Fig. 4) on the day before and after the Landers earthquake. (b) Similar magnetic field differences from 1985 through 1992 showing the occurrence times of the July 1986 $M = 6$ North Palm Springs earthquake and the June 1992 Landers earthquake (from Johnston *et al.*, 1994).

18 October 1989 (Fraser-Smith *et al.*, 1990). The magnetometer was located only 7 km from the epicenter (Fig. 7) and recorded increased ULF noise reaching 1.5 nT in amplitude during the two weeks before, and a few hours before the earthquake (Fig. 8). A second system recording ULF/VLF (very low frequency) data 52 km from the earthquake showed no corresponding changes. Similar records were not obtained during the $M = 6.7$ Northridge earthquake at a distance of 81 km from the epicenter (Fraser-Smith *et al.*, 1994), for the $M = 7.3$ Landers earthquake, or on magnetic and electric field instruments at the epicenter of the 17 August 1999 $M = 7.4$ Izhmet earthquake in Turkey (Honkura *et al.*, 2000).

Table 1 Significant Parkfield and California Earthquakes

Event	Date	Latitude (°N)	Longitude (°N)	M_b	Depth (km)	Distance (km)
A	Oct. 20, 1992	35.93	120.47	4.5	10.0	6.1
B	Apr. 4, 1993	35.94	120.49	4.3	7.6	8.5
C	Nov. 14, 1993	35.95	120.50	4.0	11.6	9.5
D	Dec. 20, 1994	35.92	120.47	5.0	8.6	4.7
E	Oct. 18, 1989	37.04	121.88	7.0	----	180.0
F	Jun. 28, 1992	34.20	116.43	7.2	----	410.0
G	Jan. 17, 1994	34.19	118.57	6.6	----	250.0
H	May 25, 1989	35.87	120.41	3.6	9.6	3.3

E, Loma Prieta; F, Landers; G Northridge. Distance is measured from array center.

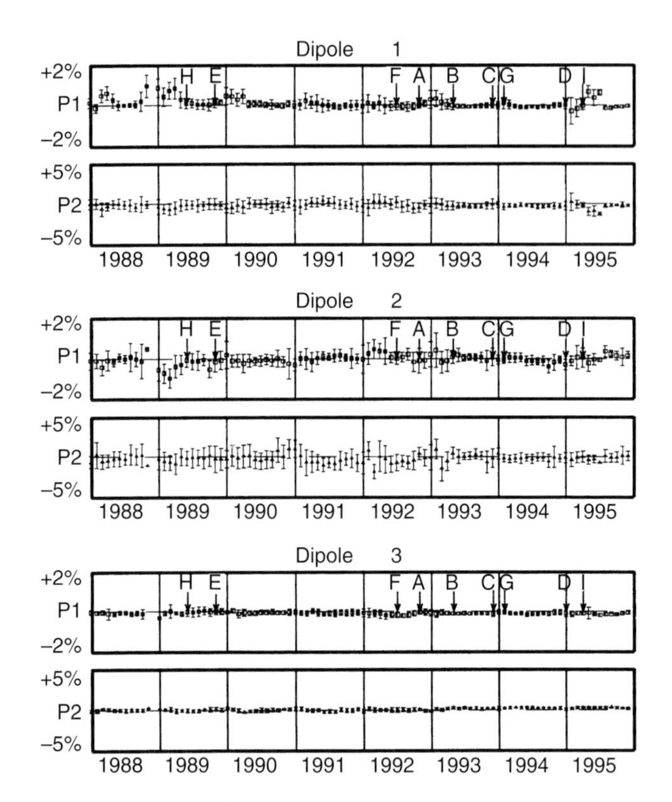

FIGURE 6 Plots of telluric fields at Parkfield, Ca., from 1988 to 1996. Earthquakes A–H (listed in the Table) have occurrence times shown with arrows (from Park, 1997).

Although the Loma Prieta record was recorded on only one magnetometer and therefore violates the multiple instrument requirement discussed above, the instrument was surrounded by independent seismic and strain instrumentation and the data were uncontaminated by cultural noise. For these reasons this example is included here.

For the Loma Prieta observations, Fenoglio *et al.* (1993) have shown that there is no correlation between the ULF anomalies and either the magnitudes or the rates of aftershocks. The absence of signals for these smaller magnitude earthquakes, or for larger earthquakes at greater distances, indicates a localized source for these oscillatory signals. Draganov *et al.* (1991) have suggested a magnetohydrodynamic origin, whereas Fenoglio *et al.* (1995) have suggested an electrokinetic source generated by fluid flow following rupture of high-pressure fluid-filled inclusions in the fault zone.

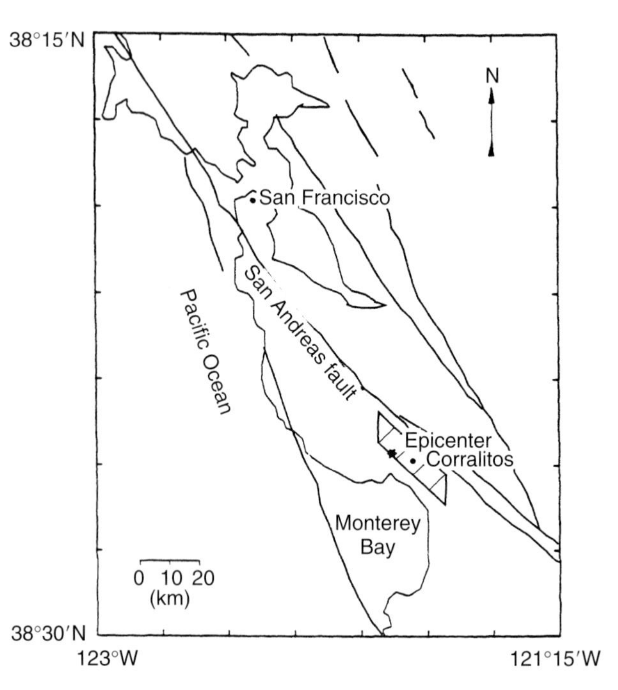

FIGURE 7 Location of ULF receiver at Corralitos, 7 km from the epicenter of the 18 Oct. 1989 $M = 7.1$ Loma Prieta earthquake. A second receiver was located at Stanford University (from Fraser-Smith *et al.*, 1990).

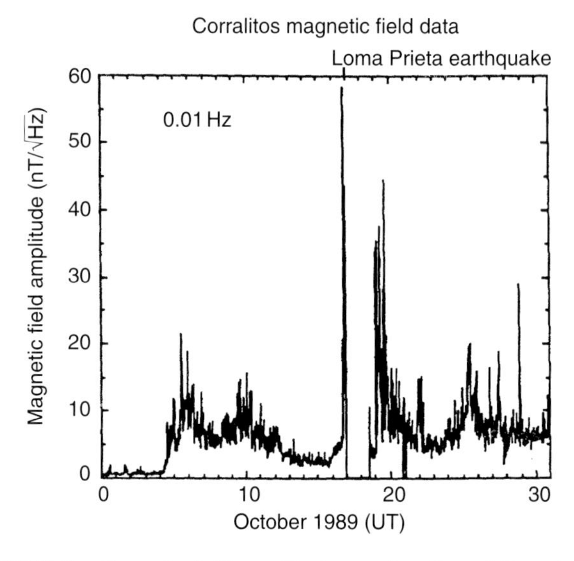

FIGURE 8 Magnetic field amplitude as a function of time during the 17 days before and 14 days after the Loma Prieta earthquake at 00:04 on 17 Oct. 1989 (from Fraser-Smith *et al.*, 1990).

6.4 Tectonoelectric and Possible Precursory Effects

The clearest tectonoelectric changes resulting from changes in crustal resistivity were reported in the Tangshan region of China prior to the 1976 Haicheng earthquake (Qian, 1981).

Other interesting, though unexplained changes occurred in the Palmdale region of the San Andreas fault (Madden *et al.*, 1993) following the infamous "Palmdale Bulge". These changes occurred in the same region where correlated tectonomagnetic, gravity, strain, and elevations had been reported earlier (Johnston, 1986).

Increased, though controversial, interest in tectonoelectric (TE) phenomena related to earthquakes has occurred during the past ten years, primarily as a result of suggestions in Greece and Japan that short-term geoelectric field transients (SES) of particular form and character precede earthquakes with magnitudes greater than 5 at distances up to several hundreds of kilometers (Varotsos *et al.*, 1993a,b; Nagao *et al.*, 1996). These transients are recorded on multiple dipoles with different lengths (10–200 m for short arrays and 1–3 km for longer ones) with signal amplitudes of 20 mv km^{-1} and durations of several minutes. The observation of consistent electric field amplitudes independent of dipole length indicates a local spatially uniform source field. However, there are no corresponding magnetic field transients and no apparent coseismic effects. The SES have been empirically associated with subsequent distant earthquakes in "sensitive" areas (Varotsos *et al.*, 1991, 1993a,b, 1996). An example of an SES recorded on multiple orthogonal dipoles together with parallel recordings of magnetic field rate at Ioannina in northwest Greece on 18 April 1995, is shown in Figure 9. This SES was suggested to have preceded a $M = 6.6$ earthquake on 13 May 1995, some 83 km to the northeast. Two other large earthquakes, the $M = 6.6$ on 4 May at Chalkidiki and the 15 June $M = 6.5$ Eratini earthquake, were also suggested to have been predicted by SES on distant stations several weeks before. Similar experiments have been run in Japan, France, and Italy with various levels of claimed success. Nagao *et al.* (1996) suggest that anomalous SES may have been recorded prior to the $M = 7.8$ Hokkaido earthquake in June 1993.

Careful study of the SES recordings indicates that the SES signals do appear to have been generated in the Earth's crust at the observation sites (Uyeshima *et al.*, 1998). However, the SES signals have the form expected from rectification/saturation effects of local radio transmissions from high-power transmitters on nearby military bases (Pham *et al.*, 1998). Also, no clear physical explanation exists describing how the

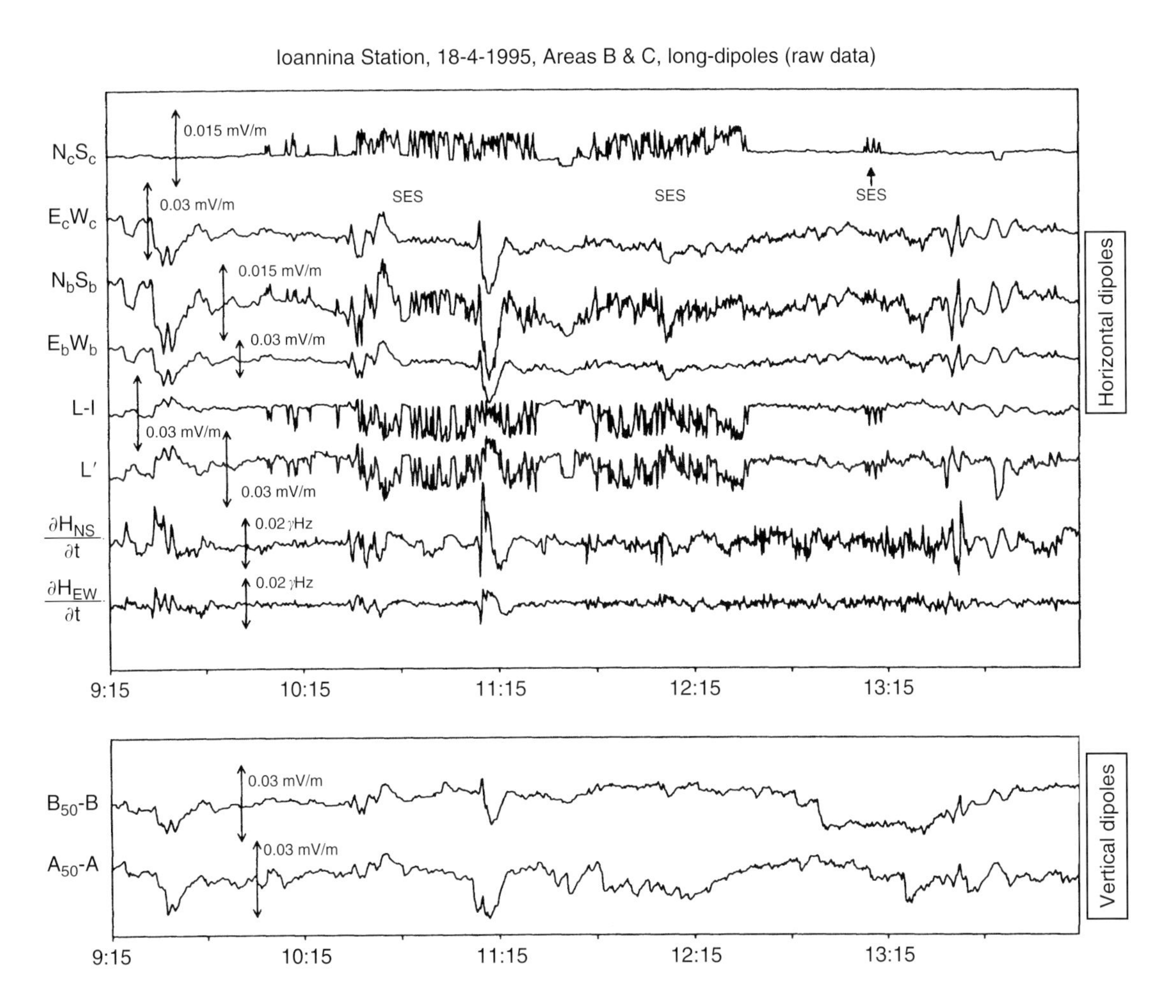

FIGURE 9 Observed SES recorded on multiple dipoles on 19 April 1995 at the Ioannina Station. Simultaneous measurements of magnetic field gradient are shown in the lower two plots (from Varotsos *et al.*, 1996).

SES signals can relate to earthquakes occurring sometimes hundreds of kilometers away (Bernard, 1992) whereas sites closer to the earthquake do not record SESs and do not have coseismic effects corresponding to the primary earthquake energy release. Without a clear causal relation, demonstration of statistical significance is controversial (Mulargia and Gasperini, 1992; Hamada, 1993; Shnirman *et al.*, 1993; Aceves *et al.*, 1996; Varotsos *et al.*, 1996; Debate on VAN, 1996; Lighthill, 1996). Better physical understanding is certainly needed. This could be obtained by careful monitoring of the nearby radio transmissions and study of the response of these transmissions on the measuring systems and on the electrical conductivity structure around sites where SESs are recorded. Independent information on the other local sources can be obtained from measurements of high-precision crustal strain, fluid levels in wells, and local pore pressure. The absence of correlative signals in these parameters would further support Pham *et al.*'s (1998) suggestions that these SESs are spuriously generated by local radio transmissions.

6.5 Electromagnetic Effects

Another enigma concerns the generation of high frequency (≥ 1 kHz) electromagnetic emissions prior to moderate earthquakes. Such emissions are reported to have been detected at great distances from these earthquakes (Gokhberg *et al.*, 1982; Oike and Ogawa, 1986; Yoshino *et al.*, 1985; Yoshino, 1991; Parrot *et al.*, 1993; Fujinawa and Takahashi, 1994; Hayakawa and Fujinawa, 1994; Shalimov and Gokhberg, 1998; Kawate *et al.*, 1998; Ondoh, 1998) and are also reported to have been detected in satellite data (Molchanov *et al.*, 1993; Parrot, 1994), although the statistical significance of these observations is under dispute (Henderson *et al.*, 1993; Molchanov *et al.*, 1993; Parrot, 1994). This area of research has received attention following reports of changes in the level of 81-kHz electromagnetic radiation around the time of a $M = 6.1$ earthquake at a depth of 81 km beneath the receiver (Gokhberg *et al.*, 1982). Radio emissions at 18 MHz were recorded at widely separated receivers in the northern hemisphere for about 15 min before the 16 May 1960 great Chilean $M_w = 9.5$ earthquake (Warwick *et al.*, 1982).

Although generation of high frequency electromagnetic radiation can be easily demonstrated in controlled laboratory experiments involving rock fracture in dry rocks (Warwick *et al.*, 1982; Brady and Rowell, 1986; Brady, 1992), the physical mechanisms for the generation and the method of propagation of very high frequency (VHF) electromagnetic waves through many tens to hundreds of kilometers of conducting crust (and through the ocean) are not at all clear. Furthermore, the absence of significant local crustal deformation in the epicentral regions of earthquakes during the months to minutes before rupture would seem to strongly constrain the scale of precursive failure. These strain data show that precursive moment release is less than 1% of that occurring coseismically (Johnston *et al.*, 1987). As discussed above, simple "skin depth" attenuation arguments preclude sources generating fields, in the kilohertz to megahertz range tens to hundreds of kilometers deep in a conducting crust (1–0.001 S m^{-1}), that would be detectable at the Earth's surface. Appeal to secondary sources at the Earth's surface (Yoshino and Sato, 1993) may avoid this difficulty but the implied large surface strain and displacement fields are not observed.

High frequency disturbances are, of course, generated in the ionosphere as a result of coupled infrasonic waves generated by earthquakes and are readily detected with routine ionospheric monitoring techniques and Global Position System (GPS) measurements (Calais and Minster, 1998). Essentially, the displacement of the Earth's surface by an earthquake acts like a huge piston, generating propagating waves in the atmosphere/ionosphere waveguide (Davies and Archambeau, 1998; see review of gravity waves in the atmosphere by Francis, 1975). Thus, traveling waves in the ionosphere (traveling ionospheric disturbances or TIDs) are a consequence of earthquakes (and volcanic eruptions). EM data at VHF frequencies recorded on ground receivers or by satellite require correction for TID disturbances (and disturbances from other sources) before these data can be identified as direct electromagnetic precursors to earthquakes, or consequences of earthquakes.

7. Conclusions

In summary, improved measurements of magnetic, electric, and electromagnetic fields being made in the epicentral regions of earthquakes around the world, have led to the recognition that observed field perturbations are generated by a variety of EM source processes. It is also clear there are many problems that still need to be resolved. In particular, a clearer understanding of physical processes involved will result from systematic work on:

1. Inclusion of constraints on the various physical mechanisms and models of various processes that are imposed by data from other disciplines such as seismology, geodesy, etc.
2. Demonstration of self-consistency in observations;
3. Determination and inclusion of realistic signal-to-noise estimates.
4. Identification of local noise sources.
5. Checking the implications of these data in other geophysical data obtained in the area. Unusual records can no longer be claimed as precursors just because they precede, or correspond approximately in time, to some local or distant earthquakes. They must be consistent with high-precision seismic and deformation data simultaneously obtained in the near field of each of the earthquakes.

6. Use of reference stations to quantify and remove common-mode noise generated in the ionosphere/magnetosphere.
7. Isolation of the most likely location of signal sources in the Earth's crust consistent with all available data. This appears to be particularly neglected in recent associations of ULF/VLF data with earthquakes.

On the positive side, better understanding has been obtained from well-designed experiments with multiple independent instrument types recording in the same frequency band. This has allowed quantification of at least some of the mechanisms generating electromagnetic fields with tectonic and volcanic activity. It now seems likely that:

1. Stress-generated magnetic effects (piezomagnetic effects) are readily observed as a result of crustal stress drops for earthquakes with $M \geq 6$ such as during the $M = 7.1$ Loma Prieta earthquake on the San Andreas fault in 1989, the $M = 7.3$ Landers earthquake in 1992 and the $M = 6$ North Palm Springs earthquake in 1986.
2. Precursive EM fields are rare in both low frequency and high frequency electromagnetic field data above the measurement resolution of 0.01 nT at 100 Hz to a few nT at hours to days for local magnetic fields and 0.01 mv km^{-1} at 100 Hz to 1 mv km^{-1} at hours to days for electric fields.
3. Electrokinetic effects probably occurred during the $M = 7.1$ Loma Prieta earthquake in 1989.

Finally, it is obvious that more complete observational arrays are needed in critical locations to provide a sound database for EM studies. Earthquakes occur infrequently and great skill is needed to "catch" these events with adequate arrays so we can better understand the physical processes involved.

Acknowledgments

I would like to thank Steve Park, Tony Fraser-Smith, Bob Mueller and other colleagues for stimulating discussions. Larry Byer and Yoshimori Honkura provided useful reviews.

References

Aceves, R.L., *et al.* (1996). *Geophys. Res. Lett.* **23**, 1425–1428.
Aki, K. and P.G. Richards (1980). "Quantitative Seismology," Freeman Press.
Ahmad, M. (1964). *Geophys. Prospect.* **12**, 49–64.
Baird, G.A. and P.S. Kennan (1985). *Tectonophysics* **111**, 147–154.
Banks, P.O., *et al.* (1991). *J. Geophys. Res.* **96**, 21575–21582.
Barsukov, O.M. (1972). *Tectonophysics* **14**, 273–277.
Bernard, P. (1992). *J. Geophys. Res.* **97**, 17531–17546.
Blanchard, D.C. (1964). *Nature* **201**, 1164–1166.
Brace, W.F. (1975). *Pure Appl. Geophys.* **113**, 207–217.
Brace, W.F. (1980). *Int. J. Rock Mech.* **17**, 876–893.
Brace, W.F. and A.S. Orange (1968a). *J. Geophys. Res.* **73**, 1443–1445.
Brace, W.F. and A.S. Orange (1968b). *J. Geophys. Res.* **73**, 5407–5420.
Brace, W.F., *et al.* (1965). *J. Geophys. Res.* **70**, 5669–5678.
Brady, B.T. (1992). In: "Report 92-15, IGPP," University of California, Riverside.
Brady, B.T. and G.A. Rowell (1986). *Nature*. **321**, 488–490.
Calais, E. and J.B. Minster (1998). *Phys. Earth. Planet. Inter.* **105**, 167–181.
Chalmers, A. (1976). "Atmospheric Electricity," Pergamon.
Davies, J.B. and C.B. Archambeau (1998). *Phys. Earth. Planet. Inter.* **105**, 183–199.
Davis, P.M. and M.J.S. Johnston (1983). *J. Geophys. Res.* **88**, 9452–9460.
Davis, P.M., *et al.* (1979). *Phys. Earth Planet. Inter.* **19**, 331–336.
Davis, P.M., *et al.* (1981). *J. Geophys. Res.* **86**, 1731–1737.
Davis, P.M., *et al.* (1984). *Geophys. Res. Lett.* **11**, 225–228.
Debate on VAN (1996). *Geophys. Res. Lett.* **23**, 1291–1452.
Dobrovolsky, I.P., *et al.* (1989). *Phys. Earth Planet. Inter.* **57**, 144–156.
Dologlou-Revelioti, E. and P. Varotsos (1986). *Geophysics.* **59**, 177–182.
Draganov, A.B., *et al.* (1991). *Geophys. Res. Lett.* **18**, 1127–1130.
Dzurisin, D., *et al.* (1990). *Bull. Seismol. Soc. Am.* **95**, 2763–2780.
Emeleus, T.G. (1977). *J. Volc. Geotherm. Res.* **2**, 343–359.
Ernst, T., *et al.* (1993). *Tectonophysics* **224**, 141–148.
Fenoglio, M.A., *et al.* (1993). *Bull. Seismol. Soc. Am.* **83**, 347–357.
Fenoglio, M.A., *et al.* (1995). *J. Geophys. Res.* **100**, 12951–12958.
Finkelstein, D., *et al.* (1973). *J. Geophys. Res.* **78**, 992–993.
Fitterman, D.V. (1978). *J. Geophys. Res.* **83**, 5923–5928.
Fitterman, D.V. (1979). *J. Geophys. Res.* **84**, 6031–6040.
Fitterman, D.V. (1981). *J. Geophys. Res.* **86**, 9585–9588.
Fitterman, D.V. and T.R. Madden (1977). *J. Geophys. Res.* **82**, 5401–5408.
Francis, S.H. (1975). *J. Atmos. Terr. Phys.* **37**, 1011–1054.
Fraser, D.C. (1966). *Geophys. J. R. Astron. Soc.* **11**, 507–517.
Fraser-Smith, A.C., *et al.* (1990). *Geophys. Res. Lett.* **17**, 1465–1468.
Fraser-Smith, A.C., *et al.* (1994). *Geophys. Res. Lett.* **21**, 2195–2198.
Freund, F., *et al.* (1992). In: "Report 92-15, IGPP," University of California, Riverside.
Fujinawa, Y., *et al.* (1992). *Geophys. Res. Lett.* **19**, 9–12.
Fujinawa, Y. and K. Takahashi (1994). In: "Electromagnetic Phenomena Related to Earthquake Prediction," Terra Science Publishers.
Gamble, T.D., *et al.* (1979). *Geophysics* **44**, 53–58.
Gokhberg, M.B., *et al.* (1982). *J. Geophys. Res.* **87**, 7824–7828.
Hamada, K. (1992). *Tectonophysics* **224**, 203–210.
Hamano, Y., *et al.* (1989). *J. Geomagn. Geoelectr.* **41**, 203–220.
Hayakawa, M. and F. Fujinawa (Eds.) (1994). "Electromagnetic Phenomena Related to Earthquake Prediction," Terra Science Publishers.
Henderson, T.R., *et al.* (1993). *J. Geophys. Res.* **98**, 9503–9509.
Honkura, Y. and Y. Kuwata (1993). *Tectonophysics* **224**, 257–264.
Honkura, Y., *et al.* (1976). *Tectonophysics* **34**, 219–230.
Honkura, Y., *et al.* (2000). *Trans. Am. Geophys. Un.* **81**, F891.
Hurst, A.W. and D.A. Christoffel (1973). *NZ J. Geol. Geophys.* **16**, 965–972.

Ishido, T. and M. Mizutani (1981). *J. Geophys. Res.* **86**, 1763–1775.
Johnston, M.J.S. (1978). *J. Geomagn. Geoelectr.* **30**, 511–522.
Johnston, M.J.S. (1986). *J. Geomagn. Geoelectr.* **38**, 933–947.
Johnston, M.J.S. (1989). *Phys. Earth Planet. Inter.* **57**, 47–63.
Johnson, M.J.S. (1997). *Surv. Geophys.* **18**, 441–475.
Johnston, M.J.S. and R.J. Mueller (1987). *Science* **237**, 1201–1203.
Johnston, M.J.S. and M. Parrot (Eds.) (1989). *Phys. Earth Planet. Inter.* **57**, 1–177.
Johnston, M.J.S. and M. Parrot (Eds.) (1998). *Phys. Earth Planet. Inter.* **105**, 109–207.
Johnston, M.J.S., *et al.* (1984). *J. Geomagn. Geoelectr.* **36**, 83–95.
Johnston, M.J.S., *et al.* (1987). *Tectonophysics* **144**, 189–206.
Johnston, M.J.S., *et al.* (1994). *Bull. Seismol. Soc. Am.* **84**, 792–798.
Kalashnikov, A.C. (1954). *Trans. Geofiz. Inst. Akad. Nauk. SSSR, Sb. Statei* **25**, 162–180.
Kalashnikov, A.C. and S.P. Kapitsa (1952). *Proc. Acad. Sci. USSR* **86**, 521–523.
Kapitsa, S.P. (1955). *Izv. Akad. Nauk. SSSR, Ser. Geofiz.* **6**, 489–504.
Kawate, R., *et al.* (1998). *Phys. Earth. Planet. Inter.* **105**, 229–238.
Kean, W.F., *et al.* (1976). *J. Geophys. Res.* **85**, 861–872.
Kern, J.W. (1961). *J. Geophys. Res.* **66**, 3801–3805.
Lighthill, J. (Ed.) (1996). "A Critical Review of VAN," World Science Press.
Lisowski, M., *et al.* (1990). *Geophys. Res. Lett.* **17**, 1437–1440.
Lockner, D.A. and J.D. Byerlee (1985). *Geophys. Res. Lett.* **12**, 211–214.
Lockner, D.A. and J.D. Byerlee (1986). *Pure Appl. Geophys.* **124**, 659–676.
Lockner, D.A., *et al.* (1991). *Nature* **350**, 39–42.
Lowell, F. and A.C. Rose-Innes (1980). *Adv. Phys.* **29**, 947–1023.
Madden, T.R., *et al.* (1993). *J. Geophys. Res.* **98**, 795–808.
Mascart, E. (1887). *Comptes Rendus* **115**, 607–634.
Matteson, M.J. (1971). *J. Colloid. Interface Sci.* **37**, 879–890.
Martin III, R.J. (1980). *J. Geomagn. Geoelectr.* **32**, 741–755.
Milne, J. (1890). *Trans. Seismol. Soc. Jpn.* **15**, 135.
Milne, J. (1894). *Seismol. J. Jpn.* **3**, 23.
Mizutani, H. and T. Ishido (1976). *J. Geomagn. Geoelectr.* **28**, 179–188.
Molchanov, O.A., *et al.* (1993). *Ann. Geophys.* **11**, 431–440.
Mori, T., *et al.* (1993). *Phys. Earth. Planet. Inter.* **77**, 1–12.
Mueller, R.J. and M.J.S. Johnston (1990). *Geophys. Res. Lett.* **17**, 1231–1234.
Mueller, R.J. and M.J.S. Johnston (1998). *Phys. Earth. Planet. Inter.* **105**, 131–144.
Mueller, R.J., *et al.* (1981). *US Geol. Surv. Open-file Report* 81-1346.
Mulargia, F. and P. Gasperini (1992). *Geophys. J. Int.* **111**, 32–44.
Nagao, T., *et al.* (1996). In: "A Critical Review of VAN," pp. 292–300, World Science Press.
Nagata, T. (1969). *Tectonophysics* **21**, 427–445.
Nourbehecht, B. (1963). PhD. thesis, Massachussetts Institute of Technology, Cambridge.
Ohnaka, M. and H. Kinoshita (1968). *J. Geomagn. Geoelectr.* **20**, 93–99.
Oike, K. and T. Ogawa (1986). *J. Geomagn. Geoelectr.* **38**, 1031–1041.
Ondoh, T. (1998). *Phys. Earth Planet. Inter.* **105**, 261–270.
Oshiman, N., *et al.* (1983). *Earthq. Predict. Res.* **2**, 209–219.
Ozima, M., *et al.* (1989). *J. Geomagn. Geoelectr.* **41**, 945–962.
Park, S.K. (1991). *J. Geophys. Res.* **96**, 14211–14237.
Park, S.K. (1996). *Surv. Geophys.* **17**, 493–516.
Park, S.K. (1997). *J. Geophys. Res.* **102**, 24545–24559.
Park, S.K. and D.V. Fitterman (1990). *J. Geophys. Res.* **95**, 15557–15571.
Park, S.K., *et al.* (1993). *Rev. Geophys.* **31**, 117–132.
Parrot, M. (1994). *J. Geophys. Res.* **99**, 23339–23349.
Parrot, M. and M.J.S. Johnston (Eds.) (1993). *Phys. Earth Planet. Inter.* **77**, 1–137.
Parrot, M., *et al.* (1993). *Phys. Earth Planet. Inter.* **77**, 65–83.
Perrier, F.E., *et al.* (1997). *J. Geomagn. Geoelectr.* **49**, 1677–1696.
Petiau, G. and A. Dupis (1980). *Geophys. Prospect.* **28**, 792–804.
Pham, V.N., *et al.* (1998). *Geophys. Res. Lett.* **25**, 2229–2232.
Pike, S.J., *et al.* (1981). *J. Geophys. Res.* **33**, 449–466.
Qian, J. (1981). "Proc. Int. Symposium on Earthquake Prediction," Chinese Seismology Press.
Reid, H.F. (1914). *Terr. Mag.* **19**, 57–189.
Revol, J., *et al.* (1977). *Earth Planet. Sci. Lett.* **37**, 296–306.
Rikitake, T. (1966). *Bull. Earthq. Res. Inst.* **44**, 1041–1070.
Rikitake, T. (1968). *Tectonophysics* **6**, 59–68.
Rikitake, T. (1976). "Earthquake Prediction," Elsevier.
Rikitake, T. and Y. Yokoyama (1955). *J. Geophys. Res.* **60**, 165–172.
Rozluski, C.P. and T. Yukutake (1993). *Acta Geophys. Pol.* **41**, 17–26.
Sasai, Y. (1980). *Bull. Earthq. Res. Inst.* **55**, 387–447.
Sasai, Y. (1983). *Bull. Earthq. Res. Inst.* **58**, 763–785.
Sasai, Y. (1991a). *J. Geomagn. Geoelectr.* **43**, 21–64.
Sasai, Y. (1991b). *Bull. Earthq. Res. Inst.* **66**, 585–722.
Sasai, Y. (1994). *J. Geomagn. Geoelectr.* **42**, 329–340.
Sasaoka, H., *et al.* (1998). *Geophys. Res. Lett.* **25**, 2225–2228.
Shalimov, S. and M. Gokhberg (1998). *Phys. Earth. Planet. Inter.* **105**, 211–218.
Shamsi, S. and F.D. Stacey (1969). *Bull. Seismol. Soc. Am.* **59**, 1435–1448.
Shercliff, J.A. (1965). "A Textbook of Magnetohydrodynamics," Pergamon Press.
Shnirman, M., *et al.* (1993). *Tectonophysics* **224**, 211–221.
Stacey, F.D. (1962). *Philos. Mag.* **7**, 551–556.
Stacey, F.D. (1964). *Pure Appl. Geophys.* **58**, 5–22.
Stacey, F.D. (1992). "Physics of the Earth," 3rd edn. Brookfield Press.
Stacey, F.D. and S.K. Banerjee (1974). "The Physical Principles of Rock Magnetism," Elsevier.
Stacey, F.D. and M.J.S. Johnston (1972). *Pure Appl. Geophys.* **97**, 146–155.
Stacey, F.D., *et al.* (1965). *Pure Appl. Geophys.* **62**, 96–104.
Tanaka, Y. (1993). *J. Volc. Geotherm. Res.* **56**, 319–338.
Tanaka, T. (1995). *J. Geomagn. Geoelectr.* **47**, 325–336.
Tuck, G.T., *et al.* (1977). *Tectonophysics* **39**, 7–11.
Utada, H. (1993). *Tectonophysics* **224**, 149–152.
Utada, H., *et al.* (2000). *Earth, Planets Space* **52**, 91–103.
Uyeshima, M., *et al.* (1998). *Phys. Earth. Planet Inter.* **105**, 153–166.
Varotsos, P. and K. Alexopoulos (1987). *Tectonophysics* **136**, 335–339.
Varotsos, P. and M. Lazaridou (1991). *Tectonophysics* **188**, 321–347.
Varotsos, P., *et al.* (1993a). *Tectonophysics* **224**, 1–38.
Varotsos, P., *et al.* (1993b). *Tectonophysics* **224**, 269–288.
Varotsos, P., *et al.* (1996). *Acta. Geophys. Pol.* **44**, 301–327.
Ware, R.H., *et al.* (1985). *J. Geomagn. Geoelectr.* **37**, 1051–1061.

Warwick, J.W., *et al.* (1982). *J. Geophys. Res.* **87**, 2851–2859.
Williamson, S.J. and L. Kaufman (1981). *J. Magn. Mater.* **22**, 129–201.
Wilson, E. (1922). *Proc. R. Soc. A.* **101**, 445–452.
Yamazaki, Y. (1965). *Bull. Earthq. Res. Inst.* **44**, 783–802.
Yamazaki, Y. (1974). *Tectonophysics* **22**, 159–171.
Yoshino, T., *et al.* (1985). *Ann. Geophys.* **3**, 727–730.
Yoshino, T. (1991). *J. Sci. Explor.* **5**, 121–144.
Yoshino, T. and H. Sato (1993). *JISHIN* **16**, 8–24.
Zlotnicki, J. and J.L. Le Mouel (1988). *J. Geophys. Res.* **93**, 9157–9171.
Zlotnicki, J. and J.L. Le Mouel (1990). *Nature* **343**, 633–636.

Editor's Note

Due to space limitations, references with full citation are given in the file "Johnston2FullReferences.pdf" on the Handbook CD-ROM, under the directory \38Johnston2. Please see also Chapter 32, Rock failure and earthquake, by Lockner and Beeler; and Chapter 72, Earthquake prediction: an overview, by Kanamori.

39

Earthquake-related Hydrologic and Geochemical Changes

Chi-Yu King
Earthquake Prediction Research, Inc., Los Altos, USA
George Igarashi
The University of Tokyo, Tokyo, Japan

1. Introduction

The "solid" earth contains abundant fluids that have been continuously released and partly recycled through geological time. The outgassing process is generally believed to be responsible for the origin of the atmosphere and hydrosphere. The rate of outgassing evidently has not been uniform in either time or space over the Earth's surface but depends on the chemical composition, the physical condition, the geological structure, and the movement of materials in the ground. A considerable amount of study on various terrestrial fluids has been carried out in recent years by scientists working in a wide variety of fields of investigation, including the origin and evolution of the atmosphere and hydrosphere, the Earth's interior, plate tectonics, the mechanism and prediction of earthquakes and volcanic eruptions, the exploration of buried oil, gas, and mineral resources, the disposal of nuclear waste, and the environmental impact of terrestrial gas emissions such as indoor radon pollution (e.g., Ozima and Podosek, 1983; Parnell, 1994, 1998; Metcalfe and Rochelle, 1999). Several collective publications of multidisciplinary studies on subsurface fluid flow in and near fault zones have been published (e.g., Hickman *et al.*, 1995; Jones *et al.*, 1998; McCaffrey *et al.*, 1999; Haneberg *et al.*, 1999).

In this chapter we provide a brief review of studies on hydrologic and geochemical changes observed before, at the time of, and after earthquakes (other than seismic oscillations) during the past several decades. Such studies were usually conducted for the purpose of better understanding the mechanics of earthquakes and the possibility of earthquake prediction, as many geophysical studies. The study of hydrologic and geochemical changes may provide clues for understanding some other earthquake-related phenomena, such as electromagnetic changes (see Chapter 38 by Johnston) and abnormal animal behavior. Earthquake prediction is probably the most challenging problem in earth science today; its feasibility has long been under debate (e.g., Geller, 1997). Crustal fluids may play an active role in generating earthquakes and nonseismic fault movement, but this kind of study is not reviewed here (see Chapter 40 by McGarr *et al.*).

2. Earlier Observations

Hydrologic and geochemical changes, such as anomalous water-level/color changes and occurrence of strange odors, at about the time of earthquakes have been observed since ancient times in many seismic regions. A systematic search for earthquake precursors for the purpose of earthquake prediction with scientific instruments did not begin, however, until the 1960s. Earlier results of such studies, conducted mainly in China, Japan, the former Soviet Union, and the United States, were reviewed by King (1986), Thomas (1988), Ma *et al.* (1990), Igarashi and Wakita (1995), and Roeloffs (1988, 1996), among others. In this section we provide only a brief summary of the general features of the observations obtained before mid-1980s.

Hydrologic and geochemical parameters investigated include water level/pressure, temperature, and electric conductivity at wells, flow rate at springs, concentration of various ions and dissolved gases, and components of soil gas in shallow holes. Earthquake-related changes were reportedly observed beginning hours to years before many destructive earthquakes mainly at certain "sensitive" monitoring sites, among many insensitive sites. The sensitive sites were usually found to be located along active faults, especially at intersections of faults, and may be at unexpectedly large epicentral distances (many source dimensions, or hundreds of kilometers for large events). The long-distance changes were difficult to understand on the basis of dislocation models of earthquake

INTERNATIONAL HANDBOOK OF EARTHQUAKE AND ENGINEERING SEISMOLOGY, VOLUME 81A ISBN: 0-12-440652-1

sources in a homogeneous and elastic medium. Because of this difficulty and because of the existence of many insensitive sites as well as the poor quality of some of the observations, which failed to take proper account of background noises, many researchers have dismissed the precursor claims based on such data. Nevertheless, as pointed out by King (1986) and Ma *et al.* (1990), the difficulty may be partly the result of unrealistic modeling of an earthquake (which occurs in an inhomogeneous crust) by a dislocation in a homogeneous and elastic half-space. If the crustal heterogeneity and a large-scale, episodically increasing tectonic stress field are taken into consideration, the above-mentioned features can be reasonably understood, as discussed in the next section.

3. Mechanisms for Earthquake-related Hydrologic and Geochemical Changes

The Earth's crust contains numerous pores and fractures filled with water, gas, and other fluids that have different chemical compositions at different places. When the crust is deformed in the tectonic process of earthquake generation, the fluids of different chemical composition, especially those in the fault and other weak zones, may be forced to migrate and cause the associated hydrological and geochemical changes.

Several mechanisms were proposed to explain the hydrologic and geochemical changes (at the sensitive sites): increased upward flow of deep-seated fluids to the monitored aquifers or shallow holes; squeezing of gas-rich pore fluids out of the rock matrix into the aquifers; mixing of water from other aquifers through tectonically created fissures in the intervening barriers; increased rock/water interactions; and increased gas emanation from newly created crack surfaces in the rock to the pore fluids (King, 1986; Thomas, 1988).

Some of the explanations were tested by laboratory studies of crushed rock samples and by field studies of underground explosions and groundwater pumping (e.g., Brace and Orange, 1968; Giardini *et al.*, 1976; Honda *et al.*, 1982; Kita *et al.*, 1982; King, 1986; King and Luo, 1990). Most of the laboratory studies involve the prefailure phenomena of volume increase of the specimen and stable sliding along planes of weakness. When a rock specimen is subjected to increasing uniaxial or triaxial compression, it commonly begins to show nonelastic volume increase (dilatancy) at a stress level somewhat below the failure stress due to increasing number and size of microcracks/fissures created in the specimen. The same phenomena were presumed to be occurring only at/near earthquake sources and responsible for the generation of various geophysical and geochemical precursors. However, diligent searches of earthquake-related dilatancy has so far failed to produce credible evidence for its occurrence in nature. Also this hypothesis cannot explain why some earthquake-related changes were observed in certain locations at large epicentral distances but not at other locations closer to the Earthquake. As noted above, to account for such observations, one needs to invoke inhomogeneity of the crust as well as a broadscale and gradually and episodically increasing seismotectonic-deformation field. [Such a field is suggested by the frequent observation of episodic increase of earthquake activity over broad areas (King, 1986; Ma *et al.*, 1990).] Under such a condition, strain may concentrate along some faults and weak zones, where voids and fluids abound, and local dilatancy may occur at the forthcoming-earthquake source as well as at certain sensitive sites where the preexisting stress is near the critical level for the occurrence of fissures and fluid flow.

In a related group of studies, many seismogenic faults, including some buried ones, were found to show spatial anomalies, such as higher concentration of various terrestrial gases (radon, helium, hydrogen, mercury vapor, carbon dioxide, isotopic ratios, etc.) in groundwater and soil air (e.g., King, 1986; Ma, 1990; King *et al.*, 1996). This kind of observation suggests that seismogenic fault zones may be paths of higher fault-parallel permeability such that underground gases from depth may escape to the surface.

4. Origin and Migration of Fluids in the Crust

To help readers better understand the literature on earthquake-related hydrologic and geochemical changes, we review briefly here the current knowledge of origin and migration of subsurface fluids based on studies of a number of so-called tracers (e.g., Fyfe *et al.*, 1978; Ozima and Podosek, 1983; Faure, 1986; Jamtveit and Yardley, 1997).

4.1 Hydrogen and Oxygen Isotopes

A most useful set of tracers to examine the origin and migration of water and other fluids in the crust is the isotopic ratios of hydrogen and oxygen. Hydrogen has a stable isotope called deuterium, D or (^{2}H), whose abundance in natural water is about 0.16% of total hydrogen. Oxygen consists of three stable isotopes, ^{16}O, ^{17}O, and ^{18}O, with corresponding abundances in natural water of about 99.76%, 0.074%, and 0.20%, respectively. Since changes in these isotopic ratios are usually very small, researchers commonly express the deviations of isotopic ratios in their water samples relative to some standard samples in units of "permil" as follows:

$$dR = [(R_A/R_S) - 1] \times 1000 \tag{1}$$

where R_A and R_S are isotopic ratios of the sample and the standard, respectively. For hydrogen isotopes, $R - \mathrm{D/H}$, and for oxygen isotopes, $R = {}^{18}\mathrm{O}/{}^{16}\mathrm{O}$. (^{17}O, being less abundant, is usually not measured because of the analytical difficulty.)

In both cases, sea water (Standard Mean Ocean Water, SMOW) is used as a standard.

Natural waters are generally considered as a mixture of three components: oceanic, meteoric, and magmatic. Sea water evaporates from the ocean surface to the atmosphere, is transported in the atmosphere as water vapor by the wind, is condensed to liquid (rain) or solid (snow), and then falls on the Earth's surface as meteoric water. Part of the meteoric water flows to rivers and lakes, and eventually back to the sea. Other parts of the meteoric water penetrate into the soil, some to greater depths, and circulate through rocks to form deep aquifers of mineral or geothermal waters.

Hydrogen and oxygen isotopic ratios in meteoric waters vary considerably due to the kinetic isotope effect during evaporation and condensation. Craig (1961) analyzed these ratios in water samples gathered from rivers, lakes, rain, and snow all over the world; he found that the ratios follow a linear relationship, which is called the meteoric-water line, as follows

$$dD = 8\, d^{18}\mathrm{O} + 10 \quad (2)$$

The slope of this line can be explained by the ratio of the isotopic fractionation factor of hydrogen to oxygen during condensation of water at a temperature of 25°C, and it has a weak dependence on temperature. The constant term is related to the difference in diffusion velocity of hydrogen and oxygen isotopes and therefore depends on the speed of evaporation at the sea surface (Dansgaard, 1964). Thus, both the slope and the constant term in the above empirical relation can have regional differences, as well as seasonal variations at a given place.

A large part of the magmatic water is believed to be recycled, from subducted oceanic crust as described below. Some sea water is trapped in the ocean-floor sediments as pore water or fixed in oceanic crust to form hydrous minerals. Upon subduction of the sea floor toward the Earth's mantle, the pore waters in the sediments are squeezed out by consolidation at relatively shallow depths, whereas the hydrous minerals in the altered oceanic crust are decomposed as the plate penetrates to greater depths where the temperature and pressure are sufficiently high. The liberated water then tends to migrate upward through the upper mantle and crust to the Earth's surface, because of its lower density than the surrounding rocks. The magmatic water is the water located in deep crust and mantle in equilibrium both chemically and isotopically with surrounding rocks and/or silicate melt.

Some magmatic water may be derived from "juvenile" water, which has never been to the Earth's surface before. By using the hydrogen and oxygen isotopic ratios, it is difficult to distinguish the juvenile water from the recycled water in isotopic equilibrium with the surrounding rocks.

The hydrogen and oxygen isotopic ratios of magmatic waters are usually in the following ranges: $d\mathrm{D} = -80$ to -40 per mil; $d^{18}\mathrm{O} = +5$ to $+10$ per mil, which are distinctly different from those of sea and meteoric waters. Thus hydrogen and oxygen isotopic ratios can be used to tell the origin, or the mixing ratios of the three origins, of groundwater.

Tectonic activities including the earthquake-generating processes are intimately related with large-scale migration of groundwater. Careful examinations of hydrogen and oxygen isotopic ratios of deep groundwater samples can greatly improve our understanding of such migration and its relation to earthquakes.

4.2 Dissolved Noble Gases

Another set of tracers useful for studying the origin and migration of waters in the crust is the dissolved noble gases, such as helium. The origin and geochemical cycle of helium are fairly simple, since helium is chemically inert and nonbiogenic. Helium has two isotopes, ^{3}He and ^{4}He. Most ^{3}He released from the solid earth is primordial, i.e., believed to have been incorporated when the Earth was formed 4.6 billion years ago. In contrast, most ^{4}He is radiogenic, continuously produced by decay of uranium (^{238}U, which has a half-life of 4.5×10^9 y) and thorium (^{232}Th, half-life 1.4×10^{10} y) in the crust and mantle throughout the Earth's history. The terrestrial samples have generally very small ^{3}He/^{4}He ratios, less than 5×10^{-5} (Ozima and Podosek, 1983); their helium contents are mostly radiogenic.

Helium in the Earth's atmosphere is continuously lost to the outer space because of its light atomic weight. Thus the abundance of helium in the atmosphere remains low (5.2 ppm). As a result, in studies of terrestrial helium, the problem of sample contamination by atmospheric gases is generally unimportant.

The ^{3}He/^{4}He ratios observed in groundwaters vary from 10^{-8} to 10^{-5}. These values are usually considered to result from mixing of three helium components: atmospheric, radiogenic, and magmatic with ^{3}He/^{4}He ratios of 1.4×10^{-6}, on the order of 10^{-8}, and 10^{-5}, respectively. The continental crust is enriched in uranium and thorium, with concentrations several orders of magnitude higher than the oceanic crust and the mantle. Thus the ^{3}He/^{4}He ratios in the continental crust are generally much lower than in the atmosphere. On the other hand, the ^{3}He/^{4}He ratios in the oceanic crust and in volcanic regions, which contain magmatic fluids, are usually much higher than the atmospheric value. Since helium diffuses very slowly in rocks by itself, helium migration in the Earth is essentially controlled by the migration of the fluids in which the helium is dissolved (Torgersen and Clarke, 1987). Thus, the dissolved helium can be a useful tracer for studying the origin and large-scale migration of crustal fluids, including groundwater.

A unique component of ^{3}He is that produced by the decay of tritium, ^{3}H. Tritium has a half-life of 12.26 y and was produced mainly by nuclear explosions beginning in the 1950s. Since then, the meteoric water has contained trace amounts of tritium. Once the meteoric water penetrates into the crust, tritium decays monotonically to ^{3}He, and enhances the dissolved ^{3}He/^{4}He ratio in proportion to its residence time. Thus the tritium-generated helium accumulation in groundwater can be used as a chronometer, giving groundwater age since precipitation (Torgersen *et al.*, 1977):

$$T = \ln(^3\mathrm{He}_{\mathrm{excess}}/t + 1)/l \qquad (3)$$

where T is the age, $^3\mathrm{He}_{\mathrm{excess}}$ is the amount of excess ^{3}He due to the decay of tritium, t is the tritium content in the water, and l is the decay constant of tritium (0.05654 y^{-1}).

Another noble gas commonly used as a geofluid tracer is argon. In contrast to helium, the argon isotopes have large atomic weights and thus can accumulate in the atmosphere (to about 0.9%). Argon has three stable isotopes, ^{36}Ar, ^{38}Ar, and ^{40}Ar, with relative compositions in the atmosphere of 0.34%, 0.06%, and 99.6%, respectively. The most abundant Ar isotope, ^{40}Ar, is the decay product of ^{40}K (with a half-life of 1.25×10^9 y). The other isotopes, ^{36}Ar and ^{38}Ar, are primordial and are thought to have largely degassed in Earth's early catastrophic events.

Presently the ^{40}Ar/^{36}Ar ratio is 295.5 in the atmosphere, and generally much higher in the Earth's interior due to accumulation of ^{40}Ar produced by ^{40}K decay. Like ^{4}He, ^{40}Ar in the Earth's interior is mainly transported by subsurface fluid migration. Thus the ^{40}Ar/^{36}Ar ratio in geofluids can be used as a tracer for their (large-scale) migration (Torgersen, 1989).

The heaviest noble gas, radon, has been used as an important geochemical tracer for fluid migration on a small scale in both time and space, and for studying degassing phenomena at/near the Earth's surface. Radon has no stable isotopes, but has three radioactive isotopes: ^{222}Rn (half-life, 3.825 days), ^{220}Rn (half-life, 54.5 sec), ^{219}Rn (half-life, 3.92 sec). Among these isotopes, ^{222}Rn has the longest half-life and is of the most geochemical interest. It is a product of the ^{238}U decay series. When decaying under equilibrium condition, it emits three alpha rays, two beta rays, and numerous gamma rays. ^{220}Rn, also called thoron, is a decay product of thorium, ^{232}Th. Minute amounts of radon can be measured easily and accurately by counting the corresponding radioactivities on a continuous basis (e.g., Noguchi and Wakita, 1977). This and other advantages have made radon one of the most important parameters studied for earthquake prediction. Because of its half-life of 3.825 days, ^{222}Rn is considered an ideal parameter to study in the search of short-term (days) premonitory signals.

5. Recent Observations

Observations of earthquake-related hydrologic and geochemical changes were made more carefully during the past decade, by using better instruments (i.e., continuously recording, higher sensitivity and reliability) and by giving proper consideration to background noise, such as changes caused by changing barometric pressure and solid-earth tide. The results, however, still show largely the same characteristic features observed earlier: (1) sensitivity of monitoring sites, even only meters apart, can be greatly different; (2) at near-field sites, some recorded earthquake-related changes are consistent with poroelastic dislocation models of earthquakes; (3) however, co- and post-seismic changes can be recorded for many moderate and large earthquakes at some sensitive sites too large (up to about 1000 km for magnitude 8) to be explained by the poroelastic models; (4) recorded premonitory changes are relatively few and less certain; (5) sensitive sites are usually located on or near active faults and characterized by some near-critical hydrologic or geochemical condition (e.g., permeability that can be greatly changed by a slight shaking or stress change); (6) the earthquake-related changes recorded for different earthquakes at a sensitive site are usually similar, regardless of the earthquake's location and focal mechanism; and (7) the sensitivity may change with time as the crustal stress changes (e.g., Silver and Vallette-Silver, 1992; Roeloffs, 1996, 1998; Wakita, 1996; King *et al.*, 1999, 2000).

A notable example of the long-distance triggering effect of large earthquakes is the observation of numerous earthquakes triggered by the magnitude 7.3 Landers, California, earthquake in 1992. The triggered earthquakes were located in geothermal and volcanic areas across the western United States at distances of up to 1250 km (Hill *et al.*, 1993). No significant earthquakes were triggered, however, in the nearby San Andreas fault segments outside the aftershock area. Such a long-distance triggering phenomenon cannot be explained by the small static stress changes calculated from elastic models of the earthquake, but can be attributed to seismic-wave-triggered fluid movement in and near some critically loaded faults in the triggered areas. Indeed, various hydrologic phenomena, including gas bubbling, increased spring discharge, and groundwater-level changes, were observed after the earthquake in many of such areas (Roeloffs, 1996).

Another recent example of earthquake-related changes in crustal-fluid movement is provided by the study of eruption-interval changes at a nearly periodic geyser located not far from the San Andreas fault north of San Francisco, California (Silver and Vallette-Silver, 1992). Coseismic and precursory changes were detected for three earthquakes at epicentral distances of 100–200 km, including the magnitude 6.9 Loma Prieta earthquake in 1989.

In Parkfield, California, where an earthquake-prediction experiment has been in progress for two decades, steplike coseismic water-level drops were recorded at four wells at the time of the magnitude 5.8 Kettleman Hills earthquake about 35 km away (Roeloffs and Quilty, 1995). Detailed study of the data also show small water-level changes beginning three days before the earthquake in two of the wells. Also at a shallow well, coseismic water-level rises were observed at the time of three local and five distant earthquakes (as far away as 730 km for a magnitude 7 event) (Roeloffs, 1998).

In Japan, the University of Tokyo has maintained a network of about a dozen continuously monitored hydrologic and geochemical stations along the nearby Pacific coast since 1978 (Wakita, 1996). One of the stations on the Izu peninsula recorded a clear radon-concentration change before the nearby magnitude 7.0 Izu-Oshima-Kinkai earthquake in 1978. Similar preseismic changes in groundwater level, temperature, and strain were observed at three other sites, all at epicentral distances of 25–50 km beginning a month before the event. Also, over the years, only two of the stations (one for radon and one for water level) have been found to be sensitive to distant earthquakes (such as a magnitude 8.1 event 1200 km away). The sensitivity was found to vary with time, probably as the regional stress varies.

Co- and preseismic changes in groundwater level/discharge, geochemistry, radon concentration in groundwater and atmosphere, and strain were observed at several locations up to 220 km away from the magnitude 7.2 Kobe earthquake in 1995, although the earthquake occurred outside of the Japanese intensive-study area of Tokai. Most of the preseismic changes were recorded within the ultimate aftershock zone beginning 3 months before the event (Igarashi *et al.*, 1995; King *et al.*, 1995; Tsunogai and Wakita, 1995; Yasuoka and Shinogi, 1995; Sugisaki *et al.*, 1996). In a questionnaire survey conducted shortly after the earthquake, Koizumi *et al.* (1996) found postearthquake water-level/discharge changes at many sites (among many more sites without such changes) at epicentral distances up to about 300 km. The distribution pattern of these changes cannot be explained satisfactorily by volumetric strain changes derived from dislocation models of the earthquake.

King *et al.* (1999) studied water-level data continuously recorded over a period of ten years at a closely clustered set of 16 wells within 400 m of a fault in Tono, central Japan (Fig. 1). They found that earthquake-related co- and postseismic changes at wells on different sides of the fault show very different features (Fig. 2). The difference is attributable to a seismically induced permeability increase of the normally impermeable fault-gouge zone, which has sustained a considerable hydraulic gradient across the fault. One of the wells showed coseismic changes for more than 20 local and distant earthquakes, including a magnitude 8 event about 1200 km away (Fig. 3). All the coseismic changes and subsequent recoveries have similar shapes, irrespective of the location and focal mechanism of the earthquakes. This well is located on the higher-groundwater-pressure (north) side of the fault. The high sensitivity is attributable to this well's tapping a high-permeability aquifer connected to one of the high-permeability fracture zones that bracket the impermeable fault-gouge zone. Because of the large hydraulic gradient across the gouge zone, a relatively small seismic shaking (or tectonic stress increase) may introduce some quickly healable fissures in the gouge

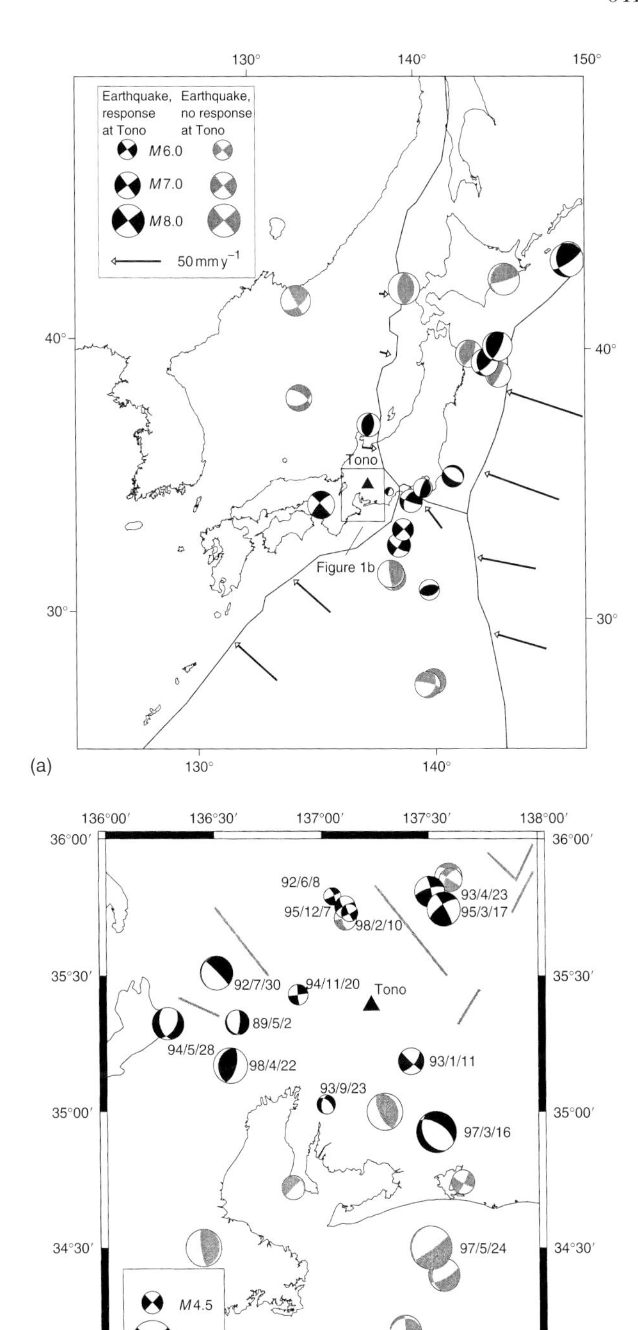

FIGURE 1 (a) Location of Tono study area in central Japan and large earthquakes discussed in this chapter. Bold lines indicate boundaries between the converging Pacific Plate on the east and the Eurasian Plate on the west, with the Philippine Sea Plate (south) and North American Plate (north) in between. Earthquakes for which water-level changes were observed at Tono are indicated by solid-segmented beach balls that show earthquake mechanisms; other earthquakes are indicated by beach balls with shaded segments. (b) Same as above for moderate local earthquakes. The location of Tono is indicated by the triangle, and active faults by shaded lines (after King *et al.*, 1999).

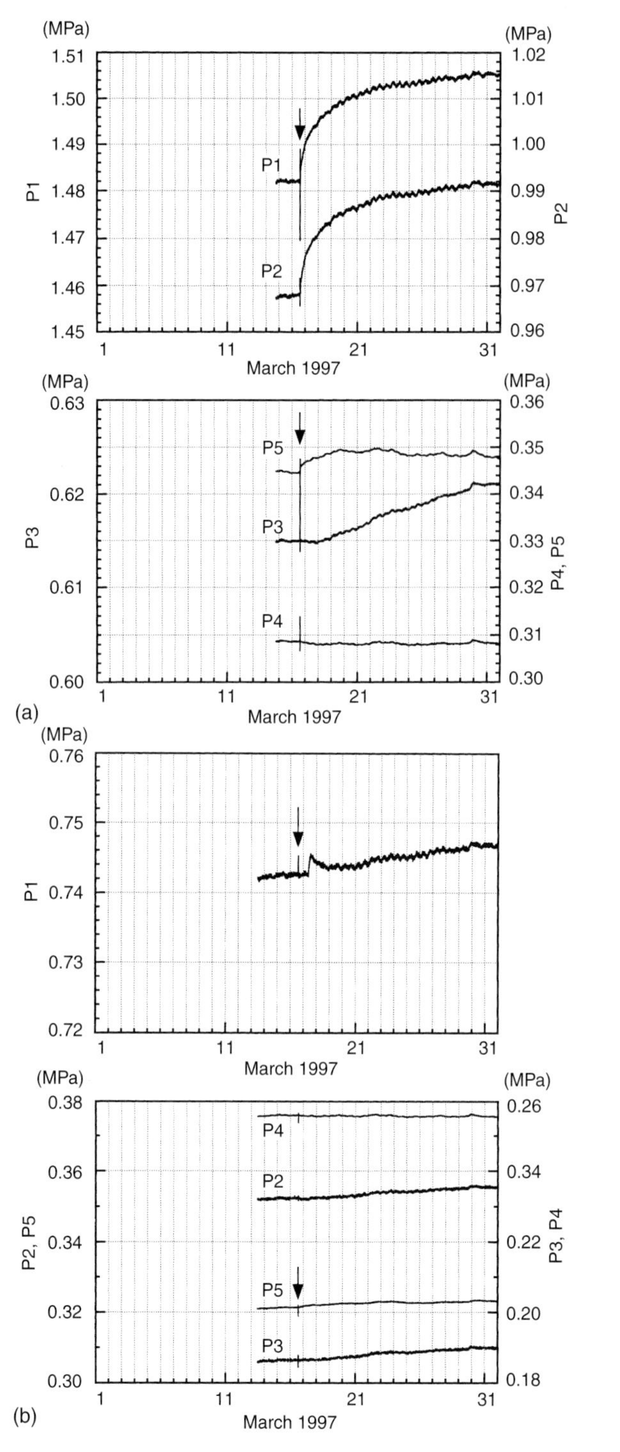

FIGURE 2 (a) Coseismic water-pressure changes recorded during March 1997 at five different depths of TH-8 well (with help of packers) located on the higher-groundwater-pressure (north) side of the Tsukiyoshi fault (see Fig. 1c of King *et al.*, 1999). The pressure data P1 through P5 are numbered sequentially upward and plotted on different vertical scales. The arrow indicates the occurrence of the largest local earthquake since 1983, magnitude 5.8 about 50 km away. (b) The same for comparable depths of TH-7 well located on the lower-pressure (south) side of the fault (after King *et al.*, 1999).

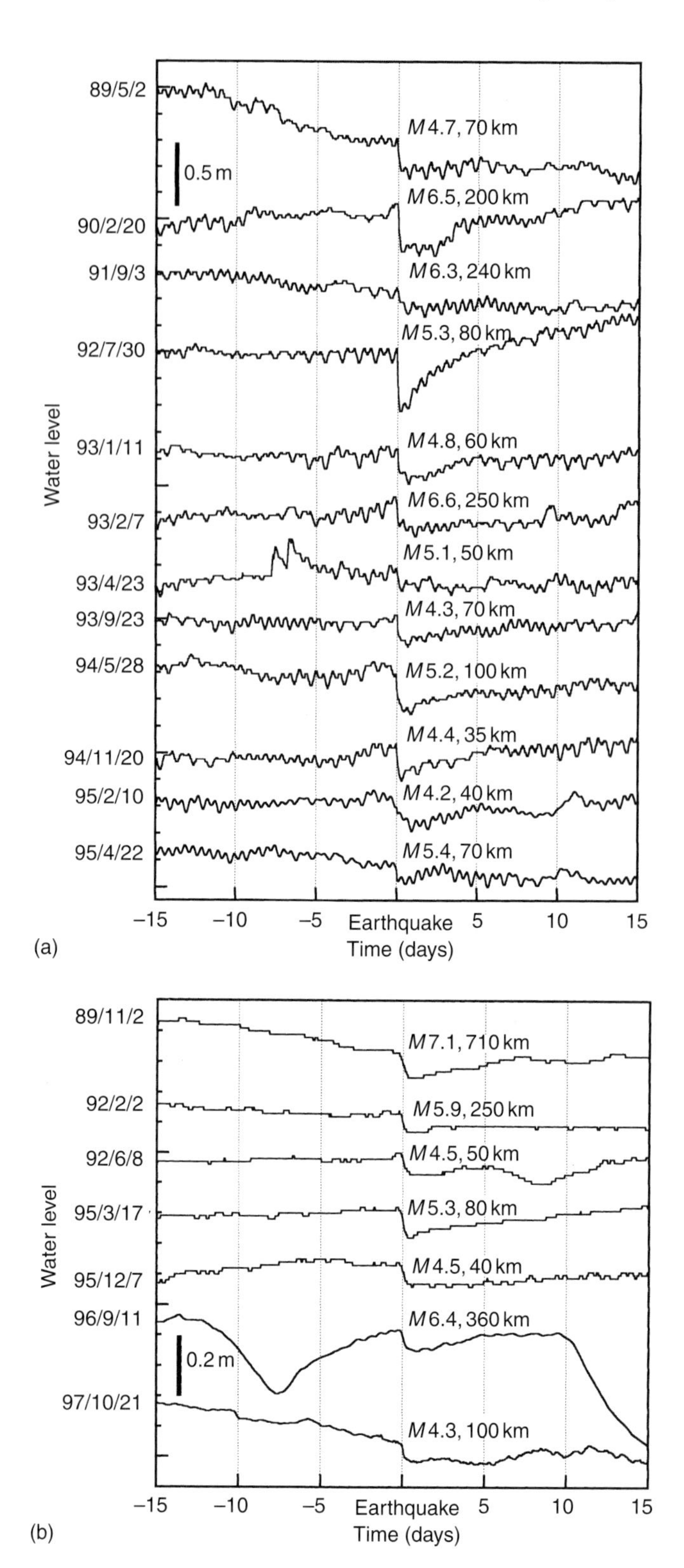

FIGURE 3 Coseismic water-level drops (in the middle of the curves) and subsequent recoveries recorded at the sensitive SN-3 well during periods of 30 days. The corresponding earthquake locations are shown in Figure 1 (solid symbols). The changes in (b) are smaller than those in (a); they become apparent only after the barometric-pressure and tidal effects are corrected (after King *et al.*, 1999).

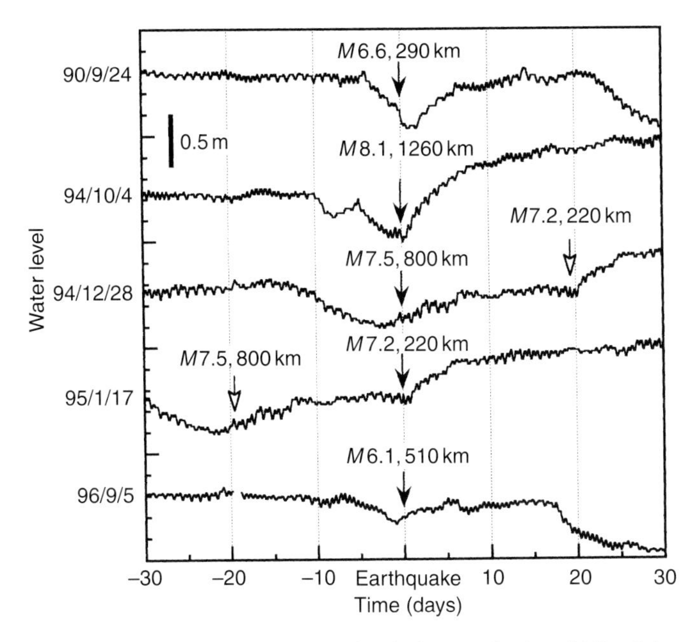

FIGURE 4 Preearthquake water-level changes (in the middle of the 60-day window) recorded at the sensitive SN-3 well for five large distant earthquakes (shown in Fig. 1). The dates of the earthquakes are indicated on the left side of the figure (after King *et al.*, 1999).

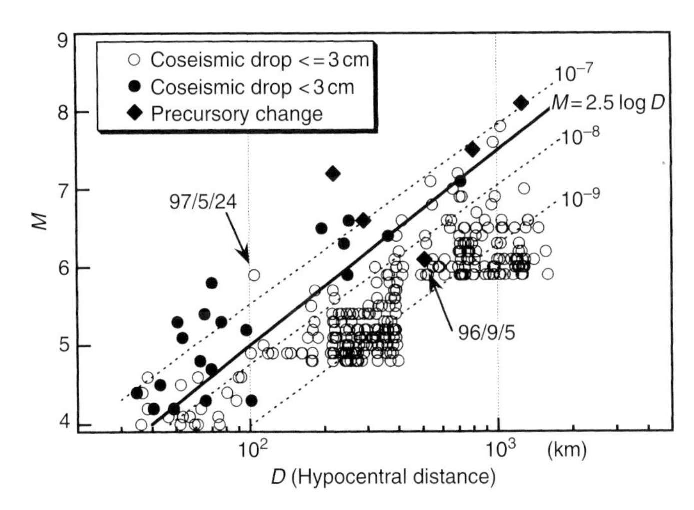

FIGURE 5 Magnitude of earthquakes as a function of hypocentral distance from Tono. The solid line indicates an estimated threshold separating earthquakes associated with a resolvable coseismic water-level change at the SN-3 well (solid circles) from those without (open circles). The diamonds indicate earthquakes associated with pre-earthquake changes. For comparison, the dashed lines indicate volumetric strain levels expected from a dislocation model of earthquakes. Note that, following the magnitude 5.8 earthquake on 16 March, 1997, the earthquake on 24 May 1997 did not cause any coseismic change, as it should according to the empirical relation. The apparently decreased sensitivity of SN-3 at this time is attributed to stress relaxation caused by the earlier event (after King *et al.*, 1999).

zone, resulting in a significant temporary permeability increase causing the observed kind of changes. Water level at this well was also found to have shown preearthquake drops for several earthquakes (Fig. 4). Some of the preearthquake drops were later found to be due to human activities (e.g., drilling of holes across the fault, resulting in a cross-fault water flow), but one of them (see Fig. 7 of King *et al.*, 2000), beginning about six months before a magnitude 5.8 earthquake 50 km away, the largest local event since 1983, may be truly premonitory in nature (King *et al.*, 2000). The sensitivity of the well showed a temporary decrease during a one-year period after the earthquake, presumably due to temporary relaxation of stress to sub-critical level by the earthquake (Fig. 5).

Many gas-geochemical surveys were conducted across active faults worldwide. Most of the results show fault-related spatial anomalies (e.g., King *et al.*, 1996). Based on measurements of radon, CO_2, and O_2, King *et al.*, attributed such anomalies along a fault in California to an upward flow of soil air in the high-permeability fracture zones of the fault.

6. Discussion and Conclusions

Hydrologic and geochemical parameters, like geophysical parameters, have been studied extensively during the past several decades in search of possible premonitory changes useful for earthquake prediction. Nevertheless, relatively few significant precursors have been recorded and the results are still inconclusive. The enormous complexity of the challenge has not been met by matching amount of funding and manpower. Not knowing where the next destructive earthquakes are going to strike, it has been extremely difficult to deploy a sufficient number of reliable instruments at appropriate locations long enough to record sufficient background data and to catch a significant number of possibly true precursors. (So far few significant earthquakes have occurred in areas of intensive studies.) Maintaining monitoring efforts over long periods of time is difficult because of some inevitable problems, such as funding uncertainty, personnel change, and instrument failure. Other difficulties arise from the inaccessibility of fluids at seismogenic depths, inhomogeneity of the crust (and the resultant problem of sensitive versus insensitive sites), and the lack of realistic models that take the imhomogeneity of the crust into account.

In spite of the above-mentioned difficulties, recent studies have shown some credible premonitory changes as well as many co- and postseismic changes. These changes were recorded at a relatively few sensitive sites among many insensitive ones. The sensitive sites can be located hundreds of kilometers away from the corresponding earthquakes, and they are commonly characterized by certain near-critical stress conditions where the local permeability can be greatly changed by a small stress increment or seismic shaking. The sensitivity of a sensitive site may change with time, depending on how close the preexisting stress is to the critical level for fissure production.

Because of the lack of sufficient data to meet the enormous challenge, it seems too early to conclude whether earthquakes are predictable or not in general terms. To overcome the above-mentioned difficulties, a prediction effort should: adopt a multistage, multisite, and multidisciplinary approach; use sensitive telemetered instruments of good long-term stability and fine time resolution; deploy them at sensitive locations long enough to recognize normal background noises and to catch enough target earthquakes; develop sufficiently realistic geophysical and geochemical models for the heterogeneous crust; and use appropriate statistical methods for objective data analysis.

References

Brace, W.F. and A.S. Orange (1968). Electrical resistivity changes in saturated rocks during fracture and frictional sliding. *J. Geophys. Res.* **73**, 1433–1445.

Craig, H. (1961). Isotopic variations in meteoric waters. *Science* **133**, 1702–1703.

Dansgaard, W. (1964). Stable isotopes in precipitation. *Tellus* **16**, 436–468.

Faure, G. (1986). "Principles of Isotope Geology," 2nd edn., Wiley, New York.

Fyfe, W.S., N.J. Price, and A.B. Thompson (1978). "Fluids in the Earth's Crust," Elsevier, Amsterdam.

Geller, R.J. (1997). Earthquake prediction: a critical review. *Geophys. J. Int.* **131**, 425–450.

Giardini, A.A., G.V. Subbarayudu, and C.E. Melton (1976). The emission of occluded gas from rocks as a function of stress: its possible use as a tool for predicting earthquakes. *Geophys. Res. Lett.* **3**, 355–358.

Haneberg, W.C., P.S. Mozley, J.C. Moore, and L.B. Goodwin (Eds.) (1999). Faults and subsurface fluid flow in the shallow crust. *Am. Geophys. U., Geophys. Monogr 113.*

Hickman, S., R. Sibson, and R. Bruhn (1995). Introduction to special section: mechanical involvement of fluids in faulting. *J. Geophys. Res.* **100**, 12831–12840.

Hill, D.P., *et al.* (1993). Seismicity remotely triggered by the magnitude 7.3 Landers, California, earthquake. *Science* **260**, 1617–1623.

Honda, M., K. Kurita, Y. Hamano, and M. Ozima (1982). Experimental studies of H_2 and Ar degassing during rock fracturing. *Earth Planet. Sci. Lett.* **59**, 429–436.

Igarashi, G., T. Saeki, N. Takahata, K. Sumikawa, S. Tasaka, Y. Sasaki, and Y. Sano (1995). Groundwater radon anomaly before the Kobe earthquake in Japan. *Science* **269**, 60–61.

Igarashi, G. and H. Wakita (1995). Geochemical and hydrological observations for earthquake prediction in Japan. *J. Phys. Earth* **43**, 585–598.

Jamtveit, B. and B. Yardley (Eds.) (1997). "Fluid Flow and Transport in Rocks," Chapman and Hall, London.

Jones, G., Q.J. Fisher, and R.J. Knipe (Eds.) (1998). Faulting, Fault Sealing, and Fluid Flow in Hydrocarbon Reservoirs. *Geol. Soc. Special Publ. 147*, London.

King, C.-Y. (1986). Gas geochemistry applied to earthquake prediction: An overview. *J. Geophys. Res.* **91**, 12269–12281.

King, C.-Y. and G. Luo (1990). Variations of electrical resistance and H_2 and Rn emissions of concrete blocks under increasing uniaxial compression. *Pure Appl. Geophys.* **134**, 45–56.

King, C.-Y., N. Koizumi, and Y. Kitagawa (1995). Hydrogeochemical anomalies and the 1995 Kobe earthquake. *Science* **269**, 38–39.

King, C.-Y., B.-S. King, W.C. Evans, and W. Zhang (1996). Spatial radon anomalies on active faults in California. *Appl. Geochem.* **11**, 497–510.

King, C.-Y., S. Azuma, G. Igarashi, M. Ohno, H. Saito, and H. Wakita (1999). Earthquake-related water-level changes of 16 closely clustered wells in Tono, central Japan. *J. Geophys. Res.* **104**, 13073–13082.

King, C.-Y., S. Azuma, M. Ohno, Y. Asai, P. He, Y. Kitagawa, G. Igarashi, and H. Wakita (2000). In search of earthquake precursors in the water-level data of 16 closely clustered wells at Tono, Japan. *Geophys. J. Int.* **143**, 469–477.

Kita, I., S. Matsuo, and H. Wakita (1982). H_2 generation by reaction between H_2O and crushed rocks: an experimental study in H_2 degassing from the active fault zone. *J. Geophys. Res.* **87**, 10789–10795.

Koizumi, N., Y. Kano, Y. Kitagawa, T. Sato, M. Takahashi, S. Nishimura, and R. Nishida (1996). Groundwater anomalies associated with the 1995 Hyogo-ken Nanbu earthquake. *J. Phys. Earth* **44**, 373–380.

Ma, Z., Z. Fu, Y. Zhang, C. Wang, G. Zhang, and D. Liu (1990). "Earthquake Prediction: Nine Major Earthquakes in China (1966–76)," Seismological Press, Beijing.

McCaffrey, K., L. Lonergan, and J. Wilkinson (Eds.) (1999). "Fractures, Fluid Flow, and Mineralization," *Geol. Soc. Special Publ. 155*, London.

Metcalfe, R. and C.A. Rochelle (Eds.) (1999). "Chemical Containment of Waste in the Geosphere," *Geol. Soc. Special Publ. 157*, London.

Noguchi, N. and H. Wakita (1977). A method for continuous measurement of radon in groundwater for earthquake prediction. *J. Geophys. Res.* **82**, 1353–1357.

Ozima, M. and F.A. Podosek (1983). "Noble Gas Geochemistry," Cambridge University Press, Cambridge.

Parnell, J. (Ed.) (1994). Geofluids: Origin, migration, and evolution of fluids in sedimentary basins. *Geol. Soc. Special Publ. 78.*

Parnell, J. (Ed.) (1998). Dating and Duration of Fluid Flow and Fluid-Rock Interaction. *Geol. Soc. Special Publ. 144*, London.

Roeloffs, E.A. (1988). Hydrologic precursors to earthquakes. *Pure Appl. Geophys* **126**, 177–209.

Roeloffs, E.A. (1996). Poroelastic techniques in the study of earthquake-related hydrologic phenomena. *Adv. Geophys.* **37**, 135–195.

Roeloffs, E.A. (1998). Persistent water level changes in a well near Parkfield, California, due to local and distant earthquakes. *J. Geophy. Res.* **103**, 869–889.

Roeloffs, E.A. and E.G. Quilty (1995). Water level and strain changes preceding and following the August 4, 1985 Kettleman Hills, California, earthquake. *Pure Appl. Geophys.* **122**, 560–582.

Silver, P.G. and N.J. Vallette-Silver (1992). Detection of hydrothermal precursors to large northern California earthquakes. *Science* **257**, 1363–1368.

Sugisaki, R., T. Ito, K. Nagamine, and I. Kawabe (1996). Gas geochemical changes at mineral springs associated with the 1995 southern Hyogo earthquake ($M = 7.2$), Japan. *Earth Planet. Sci. Lett.* **139**, 239–249.

Thomas, D. (1988). Geochemical precursors to seismic activity. *Pure Appl. Geophys.* **126**, 241–266.

Torgersen, T. (1989). Terrestrial helium degassing fluxes and the atmospheric helium budget: implications with respect to the degassing processes of continental crust. *Chem. Geol. (Isotope Geosci.)* **79**, 1–14.

Torgersen, T. and W.B. Clarke (1987). Helium accumulation in groundwater, III. Limits on helium transfer across the mantle–crust boundary beneath Australia and the magnitude of mantle degassing. *Earth Planet. Sci. Lett.* **84**, 345–355.

Torgersen, T., Z. Top, B. Clarke, J. Jenkins, and W.S. Broecker (1977). A new method for physical limnology – tritium-helium-3 ages – results for Lake Erie, Huron, and Ontario. *Limnol. Oceanogr.* **22**, 181–193.

Tsunogai, U. and H. Wakita (1995). Precursory chemical changes in groundwater: Kobe earthquake, Japan. *Science* **269**, 61–63.

Wakita, H. (1996). Geochemical challenge to earthquake prediction. *Proc. Natl. Acad. Sci. USA* **93**, 3781–3786.

Yasuoka, Y. and M. Shinogi (1995). Variation of atmospheric radon concentration before and after the southern Hyogo Prefectural earthquake. Atmospheric Radon and Environmental Radiation III, October, pp. 204–206 (in Japanese).

40

Case Histories of Induced and Triggered Seismicity

A. McGarr
US Geological Survey, Menlo Park, USA
David Simpson
The IRIS Consortium
L. Seeber
Lamont-Doherty Earth Observatory

1. Introduction

The study of anthropogenic seismicity began when earthquakes were first felt in Johannesburg in 1894 (McDonald, 1982); by 1908 these events had been attributed to the Witwatersrand gold production, which had commenced in 1886 (Cook *et al.*, 1965). Mine seismicity began to be recognized in Europe in about the same era. The first seismological observatory for monitoring these phenomena was established in Bochum in the Ruhr coal basin, Germany, in 1908 and the first seismic network was installed in the Upper Silesia coal basin, Poland, in the late 1920s (Gibowicz and Kijko, 1994).

Since then, many other types of induced and triggered earthquakes have been either recognized or, at least hypothesized. Seismicity associated with petroleum production became apparent in the early 1920s, with reservoir impoundment in the late 1930s, with high-pressure liquid injection at depth in the mid-1960s, and with natural gas production in the late 1960s. Currently debated is the possible connection between major petroleum or natural gas production and large earthquakes at midcrustal depths.

As used here, the adjective "induced" describes seismicity resulting from an activity that causes a stress change that is comparable in magnitude to the ambient shear stress acting on a fault to cause slip, whereas "triggered" is used if the stress change is only a small fraction of the ambient level (e.g., Bossu, 1996; McGarr and Simpson, 1997). By "stimulated" we refer generally to seismicity either triggered or induced by human activities. As will be seen, most of the case histories reviewed here entail triggered rather than induced seismicity.

Our assumption is that a wide variety of examples of stimulated seismicity can provide independent perspectives regarding the essential problem of the causes of earthquakes (Simpson, 1986). Accordingly, we review case histories exemplary of a broad spectrum of causative mechanisms and ranging from obvious to speculative in the correlation between earthquakes and specific human activities. Hopefully, by surveying a representative cross section of such earthquakes we can gain general insights. To this end, it is important to establish the similarities and differences in the mechanisms responsible for the various types of stimulated earthquakes. Thus, progress in understanding one type of artificial earthquake can be of use in understanding other types and perhaps interactions between natural earthquakes as well (see Harris, 1998; Chapter 73 by Harris).

The mechanisms that have been invoked to account for anthropogenic seismicity include changes in the state of stress, pore pressure changes, volume changes, and applied forces or loads. These mechanisms, of course, are not all independent. Often, for a particular case, the analysis of several mechanisms can provide different perspectives regarding the resulting seismicity. Ideally, if we understand these mechanisms well enough and we also know the background crustal state (tectonic setting), then we can forecast both where triggered seismicity will occur and the maximum likely earthquake magnitude.

Figure 1a,b introduces the types of seismicity reviewed here starting with mining-induced earthquakes, for which cause and effect are established, and ending with large-magnitude seismicity at midcrustal depth whose association with shallow-level hydrocarbon production is only speculative. In general, triggering is a plausible explanation for earthquakes if the corresponding perturbation can be shown to have shifted a fault toward failure at a time that can account for the onset of seismicity. This task is easier where natural seismicity is low and a chance correlation in time and space between a possible trigger and seismicity is unlikely.

INTERNATIONAL HANDBOOK OF EARTHQUAKE AND ENGINEERING SEISMOLOGY, VOLUME 81A ISBN: 0-12-440652-1

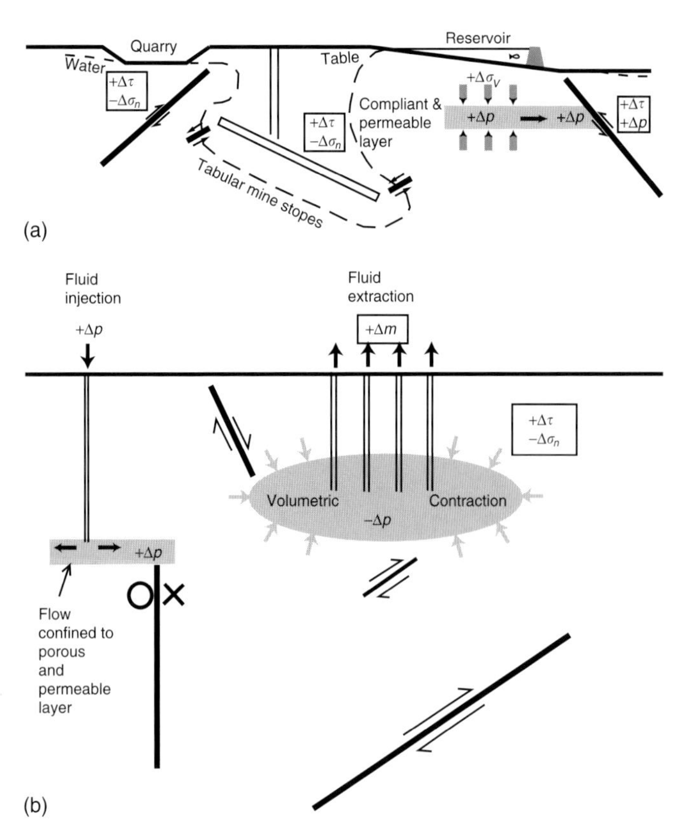

FIGURE 1 (a) Schematic view of seismicity stimulated by quarry, deep mining, and reservoir impoundment. For each case, the important mechanisms for causing seismicity are indicated. Quarries and deep-level mining cause the water table to be depressed because of pumping operations to prevent flooding. (b) Schematic view of seismicity stimulated by liquid injection at depth, which can raise the pore pressure on faults in the environs, and earthquakes caused by the exploitation of major oil and gas fields. Oil or gas production reduces the pore pressure within the reservoir, causing contraction and induced stress changes in the environs. These stress changes can result in a variety of focal mechanisms as indicated. On a larger scale, production can change the crustal loading substantially resulting in major earthquakes at midcrustal depths. A thrust faulting earthquake tends to thicken the crust so as to offset the vertical force imbalance due to production.

2. General Considerations

2.1 Ambient State of Stress and Pore Pressure

In situ stress measurements made in a variety of tectonic settings, both active and inactive, indicate, almost invariably, that the ambient state of stress in the continental crust is quite close to the depth-dependent strength of the crust estimated from laboratory experiments (e.g., Zoback and Healy, 1984; Brudy *et al.*, 1997). Moreover, these same investigations reveal that the ambient pore pressure is nearly always close to hydrostatic. Laboratory estimates of crustal strength (e.g., Brace and Kohlstedt, 1980) are based on stick-slip friction experiments as extrapolated to conditions anticipated at depth in the crust. That is, if the upper seismogenic crust is pervasively faulted then frictional sliding across these faults [Byerlee's (1978) law] will limit the strength of the crust.

The strength of a fault, or the shear stress τ required for failure, can be expressed as

$$\tau = \tau_0 + \mu(\sigma_n - p) \tag{1}$$

where τ_0 is the cohesion, μ is the coefficient of friction, σ_n is the normal stress across the fault and p is the pore pressure within the fault zone. Laboratory measurements of μ are generally in the range 0.6–1.0 (Byerlee, 1978; Dieterich, 1979). Thus, for a given state of stress, the strength of a fault would depend on its orientation, pore pressure, and cohesive strength. In estimating crustal strength it is often assumed that (1) faults exist in the crust that are optimally oriented for failure, (2) the water table is at the surface (hydrostatic pore pressure), and (3) the cohesive strength can be neglected.

2.2 Stress and Pore Pressure Changes Required to Cause Earthquakes

Triggered or induced seismicity occurs when the mechanical state of the seismogenic crust is sufficiently perturbed to cause a fault to fail. As indicated by Eqn. (1), failure can occur either because the stress τ loading the fault increases or the strength of the fault is reduced due to a decrease in the normal stress σ_n or an increase in the pore pressure p.

Figure 2 shows a ternary diagram in which these three components (σ_n, τ and p) form a field in which can be placed different types of induced and triggered seismicity, depending on the dominant cause. In some cases, the isolation of a single mechanism is simple (e.g., increases in pore pressure cause injection-related seismicity). In others (e.g., reservoir triggering) the relationship is more complex. More than one parameter may be involved (shear stress and pore pressure) or parameters may be coupled (e.g., stress and pore pressure). While care should be taken not to oversimplify what is often a complex process, this figure can help as a reference in the following discussion to identify the dominant factors influencing failure.

For a number of the case histories to be discussed here, the stress changes required to trigger seismicity, as well as the corresponding seismic stress drops (of the order of 1 MPa), are small fractions of the shear stress acting to cause fault slip (e.g., McGarr and Simpson, 1997); indeed, numerous studies, some of which are reviewed here, lead to the conclusion that stress changes as small as 0.01 MPa [Eq. (1)] may trigger earthquakes. This general observation is consistent with the idea that the crustal state of deviatoric stress tends to be nearly as high as the crustal strength (e.g., Zoback and Harjes, 1997; Grasso and Sornette, 1998). The exception to this

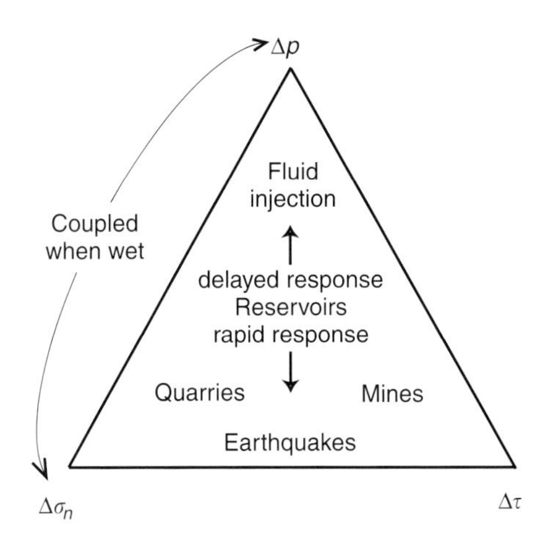

FIGURE 2 A simplified classification of three of the mechanisms controlling stimulated seismicity [see Eq. (1)]. Surface quarries, deep mines and regional earthquakes cause increases in seismicity primarily through modifications to the elastic stress field. Increased pore pressure is the dominant factor for liquid injection at depth. Reservoir loading can entail changes in all three parameters to stimulate earthquakes. Oil and gas field compaction primarily results in changes in the state of stress within the seismogenic rockmass surrounding the reservoirs. (Adapted from Figure 1 of McGarr and Simpson, 1997.)

generalization is mining-induced seismicity (Fig. 1) for which the stress changes causing the earthquakes are of the same order as the ambient crustal stresses loading the mine workings, as will be described.

2.3 How Much Seismic Deformation?

Given that seismicity occurs in response to specific changes in stress or pore pressure [Eq. (1)], then the next question entails the amount of seismic deformation that results. This is an important question because both the maximum magnitude and the seismic hazard are well correlated with the total seismic deformation.

Perhaps the most straightforward factor influencing the potential for seismic deformation is the size of the region over which an activity (e.g., gas production) takes place, as will be illustrated later. Volume changes associated with mining, liquid injection at depth and petroleum exploitation (McGarr, 1976; Pennington *et al.*, 1986) provide another means of estimating the potential seismic deformation. Similarly, large-scale oil and gas exploitation can remove mass and thereby change the crustal loading resulting in an isostatic imbalance; it is straightforward to calculate the seismic deformation needed to restore isostatic balance to the region (McGarr, 1991).

It is important to note that we can only estimate the potential for seismic deformation resulting from a particular human activity, which is not the same as predicting the ensuing seismic deformation. There are, needless to say, many activities that have the potential for producing sizable earthquakes (e. g., large impounded reservoirs) that, in fact, result in little or no detected seismicity. This issue is returned to later.

2.4 Triggered Seismicity in Stable Continental Regions and in Tectonically Active Regions

Even though deformation rates in stable continental regions within the plates are so low that they cannot yet be measured with the current $\approx 10^{-10}$ geodetic resolution (Argus and Gordon, 1996), earthquakes nonetheless occur there which suggests that deformation rates are finite and perhaps coherent over large regions of the continents (Sbar and Sykes, 1973). Faults that can be associated with this seismicity tend to be small and to have little accumulated displacement (Adams *et al.*, 1991; Crone *et al.*, 1992; Machette *et al.*, 1993; Seeber *et al.*, 1996). Geologically, they do not stand out from many other faults scattered over these stable regions. Seismicity and geology, therefore, suggest that potential sources of earthquakes are ubiquitous in these areas in spite of their tectonic stability (Seeber *et al.*, 1996). This is consistent with many sets of *in situ* stress data (e.g., Brudy *et al.*, 1997) indicating states of stress close to the failure state for both stable and tectonically active regions.

Although anthropogenic seismicity in different tectonic environments has not been rigorously compared, it seems higher in stable continental regions because of the combination of low strain rate, which yields a low level of natural seismicity, and a near-failure state of stress, which is the key factor in the occurrence of triggered earthquakes. Assuming that all other factors are equivalent, including the average proximity of the stress to failure, similar absolute levels of triggered seismicity are expected in tectonically stable regions and at plate boundaries. Because of the vast difference in rates of natural seismicity, however, artificially triggered seismicity appears to be far more significant in stable regions than in plate boundary settings.

The tendency of many natural earthquakes in stable tectonic settings to nucleate in the upper few kilometers of the crust (e.g., Ungava, 1990; Adams *et al.*, 1991; Mekering, 1968; Vogfjord and Langston, 1987; Tennant Creek, 1988; Choy and Bowman, 1990; Marryat Creek, 1986; Fredrich *et al.*, 1988) is another factor that may contribute to the relative importance of artificially triggered earthquakes in these areas. Presumably the seismogenic crust in these tectonically stable regions extends to shallow levels and thus may be relatively close to engineering activities that might cause seismicity, such as quarry operations. In contrast, most significant earthquakes in active areas tend to nucleate below the upper few kilometers of the crust which are thought to be too fractured and weak to accommodate significant elastic energy (e.g., Scholz, 1990). Thus the near-surface perturbations from engineering activities may coincide more closely with the seismogenic crust in stable than in active tectonic settings.

2.5 Monitoring Induced and Triggered Earthquakes and their Causes

In contrast to poorly understood natural seismotectonic processes, mechanical changes caused by engineering operations in the seismogenic depth range of the crust can be assessed in some detail in terms of their potential for induced and triggered seismicity. The predictability of triggered seismicity offers the opportunity for a systematic approach to designing seismic experiments that can provide unique insights regarding the earthquakes, their causes and their effects, particularly in areas where natural seismicity is low. Monitoring should start before the perturbation so that an onset of unprecedented seismicity, if it occurs, can be recognized and compared to the evolution of any forcing function. Furthermore, the monitoring needs to continue after the end of the engineering operation because the strongest perturbation may stem from a delayed rise in fluid pressure. Additionally, these observations need to include not only high-resolution earthquake ground motion data, but also geodetic, pore pressure, and water level data, as well as pertinent industrial and engineering parameters.

In many of the early cases of earthquakes triggered near dams, some ambiguity existed because of the lack of detailed monitoring before impounding of the reservoir (e.g., Lake Mead, Roeloffs, 1988). As the possibility of triggered seismicity has become more widely appreciated and acknowledged, it is becoming more common to have seismic stations installed and operating during the early development stages so that a clear picture of background seismicity is available.

2.5.1 Earthquake Monitoring

The instrumentation and techniques for observations of induced and triggered earthquakes are similar to those for any local, shallow seismicity (Chapter 17 by Lee). Dense networks of seismometers are required to determine both the temporal evolution of seismicity, including the time-dependent spatial patterns of earthquake locations. Because the onset of triggered seismicity may involve subtle changes in microearthquakes, it is important to have a monitoring system with adequate sensitivity and high location accuracy, so that low-magnitude events can be well recorded. To avoid the signal degradation associated with near-surface site effects, borehole recording using three-component seismometer packages having wide dynamic range and broad bandwidth is highly recommended (e.g., Malin *et al.*, 1988; Abercrombie, 1995).

With regard to data analysis, precise hypocentral locations and magnitude assignments are essential, but there is much more that can be gleaned from the seismograms. In particular, seismic source parameters including source dimension, seismic moment, stress drop, and radiated energy (e.g., Spottiswoode and McGarr, 1975) can provide useful insights for relating industrial operations to the resulting seismicity. For instance, seismic moment tensors have revealed key information regarding energy budgets of earthquakes and their relationship to mining operations (McGarr, 1992).

Triggered seismicity tends to be close to major engineered structures (dams, wells, mines) and may include damaging earthquakes. Seismic instrumentation deployed on these structures can provide valuable detailed knowledge of their response to strong ground motion.

2.5.2 Monitoring the Causes of Induced and Triggered Earthquakes

If the relationship between the stimulated seismicity and its cause is to be established and understood, it is essential that there be detailed monitoring of the causative agent itself (e.g., water level in reservoirs; pressure and volume of fluids injected or extracted). These data need to be available as time series with sufficient resolution to be compared to the time and size of earthquakes. Measurements of crustal deformation generally have played key roles in relating various human activities to corresponding induced or triggered earthquakes. Such measurements include trilateration, leveling, tilt, and borehole strain. Airborne or space synthetic aperture radar (SAR) interferometry offers new opportunities for detailed monitoring of strain.

2.5.2.1 Reservoirs

To first order, the most obvious parameter to measure at a reservoir is the depth of water at the dam (usually the deepest part of the reservoir). Because of its importance in operation of power plants or irrigation, this parameter is measured precisely and frequently by the dam operator. Daily measurements with a precision of centimeters are usually available. Water volume in the reservoir is also estimated, usually through knowledge of the basin geometry. Often overlooked in studies of reservoir-triggered seismicity is the growth of the reservoir over time and the influence of the reservoir on the regional water table. Less frequently, deep piezometer wells are drilled as part of the assessment of the regional groundwater regime and these may continue to be monitored during exploitation of the reservoir.

2.5.2.2 Liquid or Gas Injection and Extraction

The pressure and volume of fluid in or out of the rock are the key parameters. They are also of prime concern to the facility operator and are carefully monitored. These data are often considered proprietary, however, and so are difficult to obtain because they may have direct economic and liability implications.

2.5.2.3 Mines and Quarries

In cases of seismicity associated with mining or quarrying, the mass of extracted material and the geometry of the excavations are important. In deep mines the opportunity exists to compare seismicity directly with *in situ* stress measurements. Some large mines now include extensive seismic systems as a safety measure for monitoring rockbursts. In addition to being an important part of the mine operations, the data from these systems can provide information of direct relevance to the nature of induced seismicity and earthquake sources in general. Most mines and quarries require continuous pumping of water to prevent flooding. After they are abandoned, pumping ceases, allowing the water table and pore pressure to rebound to original levels. But, the stress is permanently altered by the mining, leading occasionally to triggered seismicity; we review in the next section a case history of this. It is useful, therefore, to continue monitoring seismicity, water level, and pore pressure after the facility is shut down.

2.6 Can Triggered Earthquakes Be of Use?

Earthquakes can be helpful in several ways. For instance, harmless small earthquakes can give information about large potentially damaging ones. In situations where earthquakes are triggered by known hydromechanical changes, analysis of these events can improve our general understanding of processes leading to earthquakes. As an example, the nucleation process, whereby friction on a fault changes from static to dynamic, is recognized as a key to understanding earthquake physics and to make progress in earthquake prediction (e.g., Dieterich and Kilgore, 1996). In the natural laboratory provided at sites of triggered earthquakes, the nucleation process that occurs between the time of the triggering perturbation and the occurrence of the earthquake can be studied in some detail and the dependence of the nucleation time on the characteristics of the earthquake and of the stress perturbation can be explored (e.g., Harris and Simpson, 1998).

Earthquakes can provide information about the changes in hydromechanics that triggered them. Engineering projects that alter conditions in the crust may consider earthquakes as a threat to the integrity of the operation and as a potential liability, but they may also use them for remote monitoring of physical parameters in their field of operation. Applications of modern seismology to engineering problems may include monitoring stress parameters including pore pressure effects (e.g., Reasenberg and Simpson, 1992), illuminating the structure and kinematics of active faults (e.g., Seeber and Armbruster, 1995) and resolving bulk properties such as anisotropy and anelasticity (Hough *et al.*, 1999). Earthquakes may also illuminate flow paths as shear failure tends to increase permeability on joint or fault surfaces (Brown and Bruhn, 1996). Thus seismology is becoming a tool for monitoring hydrocarbon fields, hydrothermal energy recovery systems, and waste-fluid disposal in deep wells. Finally, we note that monitoring of microearthquakes in mines provides information that can influence management decisions regarding production operations so as to enhance underground safety by taking seismic hazard more fully into account.

2.7 Liability for Induced and Triggered Earthquakes

From a scientific perspective, the liability aspect of induced or triggered earthquakes largely entails disadvantages to anyone trying to understand these earthquakes (in contrast to someone seeking to assign blame for them). In most cases, sincere efforts to work closely with professional staff at an engineering operation that could cause or is experiencing triggered or induced seismicity will result in productive cooperation. In some cases, however, especially in areas of high risk, concerns over litigation can stifle cooperation. For example, the potential for litigation may cause industry to be reluctant to release data that, although scientifically valuable, might be used in a court of law to establish liability. Similarly, industry may hesitate to even record data (e.g., seismic) that could be used for purposes of litigation. Moreover, collegial relations between scientists may be compromised by legal proceedings and the exchange of scientific information curtailed. Although self protection by private enterprise is understandable, society in general and the scientific community in particular suffer the disadvantage of being denied potentially beneficial information. Such situations can be mitigated by government regulatory agencies which can apply both "carrot" and "stick." For instance, industrial activities that tend to cause seismicity might be required to collect and release pertinent data. The "carrot" may entail a broader distribution of responsibility for the seismic risk involved in engineering operations which are vital to society. Scientists can also offer a "carrot" to industrial management in the form of improved understanding of the interaction between natural processes and engineering operations that would improve the basis for production decisions from both the economic and the liability standpoint. For the reader interested in a comprehensive discussion of legal issues associated with anthropogenic earthquakes, we recommend the article by Cypser and Davis (1994).

3. Case Histories of Stimulated Seismicity

The following examples have been chosen from many different regions to illustrate various mechanisms that relate human activities to stimulated seismicity. These case histories begin with mining-induced earthquakes for which the relationship between mining and earthquakes is unequivocal, and end with major earthquakes at midcrustal depths beneath oil and gas fields where a connection between the production and the seismicity is controversial.

3.1 Mining Seismicity

We review here five examples of mining seismicity, selected so as to illustrate the effects of factors such as depth, scale, state of stress, pore pressure, and mining technique. A vast literature has accumulated in the last four decades about mining-induced earthquakes (e.g., see bibliography in Gibowicz and Kijko, 1994). This type of seismicity is arguably better understood than other types and, thus, is of particular interest regarding insights into the causes and mechanisms of natural earthquakes. That is, earthquakes caused by mining appear to be largely the same phenomena as natural crustal earthquakes but, as reviewed here, can be analyzed more confidently in terms of causative forces, energy budgets, and hypocentral circumstances.

3.1.1 Deep-level Gold Mining in South Africa

The Witwatersrand gold fields, where the study of induced and triggered earthquakes began early in the 20th century (e.g., Cook *et al.*, 1965), currently involve gold production at depths between 2000 and 4000 m in an extensional, but inactive, tectonic setting. In these mines, gold production is achieved by excavating subhorizontal tabular stopes, with initial widths a little greater than 1 m and lateral extents generally in the range of 100 m to several km. The high level of seismicity, typically within several hundred meters of the active mine faces, is associated with the substantial stress changes in the brittle strata abutting the stopes. The source of energy for this process is the collapse of the tabular voids due to the considerable overburden stress (Cook, 1963; Cook *et al.*, 1965, McGarr, 1976, 1993); magnitudes occasionally exceed 5 (Fernandez and van der Heever, 1984) and levels of ground velocity, especially in the hypocentral environs, can be in excess of several meters per second and highly damaging (Wagner, 1984; McGarr, 1993; Ortlepp, 1997).

The rate of seismic deformation, as measured by estimating seismic moments, has been related quantitatively to the rate of ore production and the attendant stope collapse (McGarr, 1976). Maximum magnitudes or moments, however, are influenced by the mining techniques. Irregular mine geometries give rise to the largest earthquakes (e.g., McGarr and Wiebols, 1977; Gay *et al.*, 1984), a point which is returned to later.

The stress changes that cause the seismicity often involve increases in the shear stress as well as decreases in the normal stress acting on the fault plane and can be estimated using various methods (e.g., Cook, 1963; Cook *et al.*, 1965; McGarr *et al.*, 1975; McGarr, 1993). These changes are substantial, of the order of 25 MPa, which is about an order of magnitude greater than typical stress drops of 0.5–5 MPa for these events (Spottiswoode and McGarr, 1975; Spottiswoode, 1984). Accordingly, these earthquakes are induced, not triggered.

Similarly, volumes of stope collapse associated with mining-induced earthquakes can be used to calculate the energy released by these events for comparison with the energy radiated seismically. It turns out that at most a few percent of the released energy appears in the seismic radiation, which implies that nearly all of the available energy is consumed in overcoming fault friction (e.g., McGarr, 1976, 1994; Spottiswoode, 1980).

The largest mining-induced earthquake recorded in the South African gold fields occurred in 1977 in the Klerksdorp mining district and was assigned a magnitude of 5.2 (Fernandez and van der Heever, 1984; Gay *et al.*, 1984). At that time this district included four large mines covering an area approximately 27 km in extent (Gay *et al.*, 1984), with gold-bearing reefs at depths somewhat in excess of 2000 m; in this district, mine layouts tend to be somewhat irregular because preexisting normal faults offset the gold-bearing horizons often by several hundred meters.

3.1.2 Underground Research Laboratory (URL), Canada

The URL, situated in southeastern Manitoba within the Lac du Bonnet precambrian granite, was established to test the feasibility of nuclear waste storage. This facility, involving a shaft more than 400 m deep and horizontal tunnels extending out at several different levels, has been the site of intense geotechnical and seismological monitoring (Talebi and Young, 1992; Martin and Young, 1993). This example provides some interesting contrasts to that of the deep mines in South Africa with regard to depth (400 m vs. several kilometers), tectonic setting (compressional instead of extensional), and scale (several meters vs. several hundred meters or greater). The seismicity is tightly coupled to the excavation of both shaft (Gibowicz *et al.*, 1991) and tunnels (Martin and Young, 1993). Small events, with magnitudes in the range −4 to −2, are located near the advancing shaft bottom and tunnel faces (Feignier and Young, 1993). In the vicinity of the seismicity caused by shaft development at depths near 400 m (Gibowicz *et al.*, 1991), the ambient principal stresses [(S1, S2, S3) = (55 MPa, 48 MPa, 14 MPa)] are oriented approximately northwesterly, southwesterly, and vertically (Martin and Young, 1993). The seismogenic zones are within several meters of the shaft sidewall where S1 has been amplified by a factor of about 2, from 55 to 110 MPa (McGarr, 1994). Thus, the shaft, with a diameter of 4.6 m, interacts with the horizontal principal stresses to produce earthquakes with magnitudes as high as −1.8 (Gibowicz *et al.*, 1991).

3.1.3 Exceptionally Large Events Caused by "Room and Pillar" Mining

An association between irregular mine geometries and large magnitude events was mentioned above. The 1989 Volkershausen, Germany, rockburst is a spectacular illustration of this effect. This event, with a magnitude of 5.4 (Knoll, 1990; Bennett *et al.*, 1994) occurred in the Ernst Thaelmann Potash Mine and entailed the collapse of about 3200 pillars in the depth range 850–900 m.

Similarly, the 1995 Solvay Trona Mine, Wyoming, event, with a magnitude of 5.1, involved the collapse of pillars at a depth of 490 m over an area of about 1 km by 2 km (Pechmann *et al.*, 1995); the lateral extent of the mine is somewhat greater than 4 km. In both the 1995 Solvay and 1989 Volkershausen earthquakes surface subsidence and waveform modeling indicated a substantial implosive component to the seismic source (Bennett *et al.*, 1994).

These two examples illustrate the importance of mining method (room and pillar in both cases) in determining the maximum event. The unmined pillars are intended to inhibit stope collapse and the attendant seismicity. If one pillar fails, however, many others can fail in a cascade and the result is an exceptionally large earthquake. In contrast, if pillars are not left, stope collapse and seismicity, localized near the advancing face, occur steadily. Thus, the rate of earthquakes is higher but maximum magnitudes are smaller (McGarr and Wiebols, 1977).

3.1.4 Seismicity Caused by Quarrying or Surface Mining

Pomeroy *et al.* (1976) associated an earthquake sequence in 1974, including a mainshock of magnitude 3.3, with open-pit quarrying operations near Wappingers Falls, New York. The mainshock and aftershocks show very shallow hypocenters (0.5–1.5 km deep) with thrusting mechanisms, directly beneath the quarry. The quarry, which has a long dimension of about 1 km, has been excavated since the early 1900s to a depth of 50 m, providing a total stress change, albeit distributed over many decades, of 1.5 MPa near the surface. Studies of stress direction and magnitude (Sbar and Sykes, 1973) have shown southern New York to be a region of high horizontal compression. In such an environment, the removal of surface material decreases normal stress and increases shear stress on an underlying thrust fault so as to bring it closer to failure.

Pumping operations to prevent quarry flooding can have two opposing effects regarding triggered seismicity. First, these operations lower the water table, reduce the pore pressure and, thus, strengthen the fault and suppress earthquakes. Second, the removal of water reduces the effective overburden load on any subjacent fault thereby weakening the fault and encouraging seismicity. These two effects are sufficiently complex that it is difficult to predict, without substantial additional information, which will dominate. In the case of Wappingers Falls, it appears that dewatering was of little consequence whereas in the following example, fault strengthening due to dewatering predominated until the pumping ceased.

The 1992–1997 sequence of earthquakes centered in the Cacoosing Valley near Reading, PA, originated on a well-identified shallow fault located directly below a quarry and appears to have been triggered by it (Seeber *et al.*, 1998). The damaging January 1994 $M_w = 4.3$ mainshock ruptured an oblique-reverse fault in the upper 2 km of the crust. The excavation of an 800 m-wide quarry centered directly above this fault presumably brought the fault closer to failure due to the removal of substantial overburden. This unloading effect, however, had started several decades earlier and had ceased in December 1992, 5 months before the onset of seismicity. Evidently, the unloading effect of the quarry operations was offset by a near simultaneous reduction in pore pressure in the seismogenic zone as the quarry was dewatered to allow production [Eq. (1)]. After the quarry operations ceased, however, the original water table was rapidly restored. Pore pressure on the fault 0.5–1 km below the quarry probably rose almost as fast because permeability of the subjacent Cambrian–Ordovician carbonates is high. The strengthening effect of the lowered pore pressure disappeared leaving the effect of the unloading unmitigated and bringing the fault to failure. Seismicity started in May 1992, and included a damaging $M = 4.3$ event.

Other examples of earthquakes triggered by surface unloading are from the Belchatow region in Poland, where events up to magnitude 4.6 have been related to open-pit coal mining (Gibowicz *et al.*, 1981). The excavation in this case was 1 km by 2 km in area and about 100 m deep.

Finally, at the Lompoc diatomite mine, California, surface mining caused at least four events, with magnitudes up to 2.5, between 1981 and 1995 (Yerkes *et al.*, 1983; Sylvester and Heinemann, 1996); the long dimension of this mining operation is about 1000 m. These Lompoc events differ in several respects from their counterparts in stable continental region settings. First, coastal California is an active tectonic setting. Second, the events at Lompoc produced very distinctive fault scarps on the quarry floor (e.g., Sylvester and Heinemann, 1996), whereas surface faulting was not found at Cacoosing or Wappingers Falls.

3.1.5 Pumping and Flooding in Mines and Quarries

Wetmiller *et al.* (1993) outlined the seismicity history of the Falconbridge, Canada, Mine that was closed in 1984 and then allowed to flood in late 1990. The strong correlation reported between the renewed seismicity and the flooding constitutes quite a credible argument that increasing crustal pore pressure tends to stimulate earthquakes, presumably by reducing the effective normal stress [Eq. (1)].

The Cacoosing Valley sequence, just described, is a shallower example of this. The quarry operations were the essential cause of the seismicity but this only occurred after the quarry was abandoned and the attendant flooding restored the pore pressure in the seismogenic zone to its ambient hydrostatic state.

Finally, pumping to prevent flooding in the deep mines of the South African gold fields has strengthened the surrounding crust by drastically reducing the pore pressure from its ambient hydrostatic state (e.g., McGarr *et al.*, 1975). In the ambient state, the minimum horizontal principal stress is typically about half the vertical overburden stress, the water table is near the surface and pore pressure is close to hydrostatic (McGarr and Gay, 1978; Gay *et al.*, 1984). Based on laboratory evidence, this state of stress and pore pressure is close to failure (Brace and Kohlstedt, 1980). The Witwatersrand tremors, described earlier, occur in spite of the strengthening effect of the large decrease in pore pressure because the change in state of stress from the extensive tabular stopes is sufficient to counteract this effect locally.

3.2 Seismicity Caused by Liquid Injection

3.2.1 Rocky Mountain Arsenal Well

One of the earliest and most spectacular examples of seismicity related to fluid injection occurred near Denver, Colorado in the 1960s (Evans, 1966; Healy *et al.*, 1968). Hazardous wastes were being injected under high pressures at a depth of 3.7 km at the Rocky Mountain Arsenal. Soon after injection started, earthquakes began to be felt in the Denver area, a region that previously had experienced little or no earthquake activity. The seismicity was initially concentrated near the bottom of the injection well, but eventually spread along a linear zone for about 8.7 km. The fault plane solutions for the largest events indicated normal faulting. Of particular interest, however, is that the largest earthquake, of magnitude 4.85 (Herrmann *et al.*, 1981) occurred more than a year after injection had ceased. The injection pressures were of the order of 10 MPa above the initial formation pressure of 27 MPa. Hsieh and Bredehoft (1981) showed that increases in fluid pressure of only 3.2 MPa were sufficient to stimulate activity on favorably oriented faults.

3.2.2 Ashtabula, Ohio

Liquid waste was injected into the 1.8 km deep basal Paleozoic formation of the Appalachian Plateau near Ashtabula Ohio. A magnitude 3.6 mainshock occurred in 1987 a year after the onset of injection and more than 30 km from any other known earthquake. Accurate aftershock hypocenters showed this event to have ruptured, left-laterally, a vertical west-striking fault just below the Precambrian–Paleozoic unconformity (Nicholson and Wesson, 1990; Seeber and Armbruster, 1993). The sequence started where the fault was closest to the well, i.e., 0.7 km from the injection point, and migrated westward about 2 km. The 35 accurately determined hypocenters all cluster along the fault; none are detected at the injection point. This example shows that preexisting structure can play a key role in the spatial distribution of triggered seismicity.

3.2.3 KTB, Germany

By far the deepest example of seismicity caused by liquid injection occurred when 200 m^3 of brine were injected into the 70 m open-hole section at the bottom of the 9.1 km main borehole of the KTB (German Deep Drilling Program) (Zoback and Harjes, 1997). This experiment resulted in several hundred microearthquakes, close to, but somewhat above, the point of injection. The largest event, which occurred 18 h after the start of injection, had a magnitude of 1.4, whereas, for all the remaining events, the magnitudes were less than zero. These KTB events evidently occurred in response to pore pressure changes of 1 MPa or less (Zoback and Harjes, 1997), which is about 0.01 of the ambient pore pressure. The state of shear stress near the bottom of the borehole was estimated, from an extensive set of measurements, to be about 100 MPa (Brudy *et al.*, 1997), which is close to the predicted failure stress at that depth. The results of this experiment in the tectonically stable region of eastern Bavaria is consistent with other measurements in similar settings showing stress levels close to failure. It also suggests that the high deviatoric stress is not limited to near the surface where most of the measurements are taken, but that it extends through the seismogenic part of the crust.

3.2.4 Control of Seismicity

Following the Denver observations, Raleigh *et al.* (1976) performed a partially controlled experiment in Rangely Colorado where water injection for secondary recovery in an oil field was producing low level seismicity ($m < 3.1$). Injection was in a number of wells with depths of up to 2 km. The formation pressure was of the order of 17 MPa. During the experiment, earthquakes could be turned off and on by varying the pore pressure about a critical value of 26 MPa.

Water injected for solution salt mining in Dale, New York (Fletcher and Sykes, 1977) provides another example of a partially controlled experiment, in this case with lower pressures at shallower depths. Earthquakes of magnitude −1 to 1.4 formed a cluster about 650 m across near the bottom of a 426 m injection well. The earthquake activity ceased abruptly when the top hole pressure dropped below 5 MPa.

3.2.5 Summary of Liquid Injection Effects

Earthquakes associated with liquid injection show the following characteristics. First, the seismicity tends to be triggered along preexisting faults that are hydraulically connected with injection points. The earthquake activity is usually concentrated on the portion of the fault with the least hydraulic resistance from the point of injection. Second, there is also evidence of time dependence in injection-related seismicity. Initially, seismicity tends to be concentrated near the injection point and to respond rapidly to changes in injection pressure or rate. As injection proceeds, the zone of influence increases, the upper limit of earthquake magnitudes increases and the response to input pressures becomes more sluggish and subtle. This response tends to lag changes in injection parameters. Activity close to the injection point usually stops immediately after the injection ceases, whereas farther from the injection well earthquakes may continue for some time afterwards.

An excellent review of many more case histories of seismicity related to liquid injection can be found in Nicholson and Wesson (1990).

3.3 Seismicity at Large Impounded Reservoirs

Over the past six decades, numerous cases of seismicity associated with reservoir impoundment have been reported. We briefly describe three examples of reservoir-related seismicity, chosen to highlight different aspects of this complex phenomenon. Koyna produced the largest earthquake associated with reservoir impoundment and presents a clear example of activity lasting for decades. At Aswan, the seismicity was surprisingly deep. At Nurek, detailed observations show how very minor changes in water level can be sufficient to modify the seismicity.

3.3.1 Koyna

The Koyna, India earthquake of 10 December 1967 is the largest earthquake associated with reservoir loading (Gupta and Rastogi, 1976). The earthquake of magnitude 6.5 occurred at the western edge of the Deccan traps, in a region of low natural seismicity. The mainshock was at a depth of less than 5 km, within 10 km of the dam and involved predominantly strike-slip faulting. The water level in the dam at the time of the mainshock was approximately 75 m and the impounded lake was about 52 km long (Gupta *et al.*, 1969). Seismicity continues at Koyna today and a number of significant events since 1967 have been associated with seasonal changes in the reservoir water level (Gupta, 1985). A magnitude 5.2 event in 1973 occurred when the water level in the reservoir had exceeded its previous maximum by only one meter. The seismicity at Koyna – a "main event" of magnitude 6.5 preceded by at least one year of seismicity and followed by persistent activity for more than 30 years – does not fit the usual mainshock–aftershock pattern for tectonic earthquakes. Clearly, the Koyna reservoir has not simply acted to trigger the release of tectonic stress in a normal earthquake sequence. Instead, the reservoir and its annual cycles in water level, continue to interact with the tectonic stress field in a complex manner and is continuing to trigger earthquakes.

3.3.2 Aswan

Lake Nasser, impounded by the 110 m high Aswan dam on the Nile River, Egypt, is one of the largest artificial lakes in the world. A magnitude 5.3 earthquake in November 1981 occurred 60 km upstream from the dam, directly beneath a large embayment that extends west of the former Nile channel (Kebeasy *et al.*, 1987; Simpson *et al.*, 1990). The earthquake was widely felt in upper Egypt, where there is no evidence of prior seismicity of this magnitude in spite of the long historical record. Aswan reservoir is relatively shallow by world standards, with water depths of greater than a few tens of meters being confined to the narrow channel of the Nile River. The water level in the embayment where the earthquake occurred is less than 10 m, but the groundwater level in the epicentral area prior to impoundment is known from well data to be more than 100 m below ground level. Simpson *et al.* (1990) suggest that the flooding of the porous Nubian sandstone beneath the embayment created an effective load whose impact greatly exceeded that of the reservoir itself.

The Aswan earthquake and the long sequence of earthquakes that followed are unique among cases of reservoir-triggered earthquakes in terms of the depths to which they extended. The main shock was well recorded by regional stations and the aftershocks were intensely studied by a local telemetered seismic network (Simpson *et al.*, 1990). The mainshock was located at a depth of 18 km. The earthquakes located with the local network showed two distinct clusters of seismicity: one extending from near the surface to 10 km; the other centered about the mainshock depth and extending from 15 to 25 km. The mainshock and many aftershocks were spatially correlated and show the seismicity to be related to the Kalabsha fault, an ancient right-lateral, strike-slip structure which extends west from the Nile channel and controls the topography in the epicentral area.

3.3.3 Nurek

At 315 m, the Nurek dam, on the Vakhsh River, Tadjikistan is the highest dam in the world and is situated in the Tadjik depression, a region of moderately high seismicity. The reservoir filled in stages, with the first major increase in water level to 100 m occurring in 1972. Two earthquakes of magnitude 4.6 and 4.3 occurred beneath the 50 km-long reservoir, immediately after the water level abruptly stopped rising after reaching the 100 m level in November 1972 (Simpson and Negmatullaev, 1981). Long-term monitoring of the regional seismicity for 20 years before construction of the dam clearly shows the activity following impoundment to be anomalous. Detailed monitoring of the postimpoundment seismicity with a local network revealed a close association of the rate of seismicity with minor changes in both the depth of water in the reservoir and changes in the rate of filling (Simpson and Negmatullaev, 1981). Similar effects observed at Lake Mead (Carder, 1970), on the Arizona–Nevada border, were analyzed by Roeloffs (1988).

The second stage of filling the Nurek reservoir, to over 200 m in 1976, was carried out much more smoothly than the first stage of filling in 1972. A strong burst of low magnitude activity accompanied a rapid, but minor fluctuation in water level, when a tunnel was tested part way through the filling cycle. The maximum level of 200 m was reached gradually, however, and no significant earthquakes or increases in activity accompanied the final filling stage. The seismicity at Nurek is especially sensitive to a sudden drop in water level or a rapid decrease in the filling rate. Simpson and Negmatullaev (1981) showed that water-level changes of a few meters (few $\times$ 0.01 MPa) and changes in the rate of filling of as small as 0.5 m day^{-1} can influence the rate of seismicity.

3.3.4 Triggering Mechanisms

The mechanism of earthquake triggering by large reservoirs is more complex than most other types of triggered seismicity. This is partly a consequence of scale – the mass of water impounded in a large reservoir is typically orders of magnitude larger than in an injection project, for example. The area affected is also much larger and can cover extensive changes in geology and fault structure. In addition, reservoir-triggered seismicity can involve a complex interaction between all three of the factors indicated in Figure 2. The weight of the reservoir can influence both shear and normal stress. Pore pressure increases instantly from the compaction of pore space due to the reservoir load and then from the raised water column with a delay due to diffusion. Because of temporal differences in the response to load and pore pressure, Simpson *et al.* (1988) showed that reservoir-triggered seismicity can be divided into two types: a rapid response related to instantaneous elastic response and a delayed response related to fluid diffusion.

Reservoir loading differs from injection- or mining-related seismicity in that the triggering agent (water in the reservoir) is remote from, and in some cases isolated from, the fault zones on which the triggered activity occurs. The pressure change at the bottom of a 100 m deep reservoir is 1 MPa. In a simple elastic medium, only a small fraction of this stress is propagated to hypocentral depths (Gough, 1978; Bell and Nur, 1978).

If the triggering mechanism is an increase in pore pressure from diffusion of water into permeable zones intersecting the reservoir, the influence at depth is also attenuated, and delayed as well. The amount of attenuation depends on rock properties and geometry of the reservoir, but simple elastic models show that stress at depths of a few kilometers can only be about 10% of the surface value (Bell and Nur, 1978). Under these conditions, a 100 m deep reservoir would thus increase the stress at hypocentral depths by, at most, 0.1 MPa. This is considerably lower than the stress changes we have shown to be acting in cases of mining seismicity, but is of the same order as the stress changes in simple elastic modeling for earthquake–earthquake triggering (e.g., Harris, 1998). For such small changes in stress to be sufficient to initiate failure, the pre-existing state of stress needs to be much less than one stress drop away from failure. This condition is statistically possible on a subset of the potentially seismogenic faults. The relatively common occurrence of seismicity triggered by small stress changes would require many such faults, even in areas where natural seismicity is low (Seeber *et al.*, 1996). Alternatively, the stress change may be locally amplified on the affected faults. Simpson and Narasimhan (1990) have proposed that structural or lithological inhomogeneities can provide one mechanism to produce the necessary stress amplification.

3.4 Hydrocarbon Reservoir Compaction

Seismicity associated with fluid extraction was first recognized in the oil field at Goose Creek, Texas, in 1925 (Yerkes and Castle, 1976). In the meantime, numerous other examples of this phenomenon have been recognized, with the oil field at Wilmington, California producing the most spectacular effects of compaction-caused subsidence, horizontal deformation and attendant earthquakes (Kovach, 1974; Yerkes and Castle, 1976). Yerkes and Castle (1976) demonstrated that differential compaction at depth, due to liquid extraction from the producing formations, could lead to shear failure, including earthquakes.

As reviewed by Segall (1989), the phenomenology of earthquakes associated with hydrocarbon production includes:

1. Seismicity in the immediate environs of a reservoir commences when the pore pressure reduction reaches a level of the order of 10 MPa.
2. The earthquakes tend to occur either immediately above or below the reservoirs and are most prevalent in strata that are especially strong with a tendency toward brittle, rather than ductile, failure.
3. For earthquakes above and below the reservoir, the most common focal mechanisms indicate thrust or reverse faulting; normal faulting mechanisms have been observed for events located in the peripheral region.
4. Earthquakes located in the weaker strata tend to be slow, i.e., the source duration is quite extended compared to that of a typical crustal earthquake of similar magnitude (e.g., Kovach, 1974).

Noteworthy among many examples of seismicity triggered by hydrocarbon extraction are the earthquakes located in the environs of the Rocky Mountain House gas field in Alberta, Canada (Wetmiller, 1986). The substantial reverse slip accumulated on a surface fault at the Buena Vista Hills oil field, in the San Joaquin Valley, California, was aseismic (Kock, 1933; Nason *et al.*, 1968).

Grasso and Wittlinger (1990) described earthquakes triggered by natural gas production in the Pau basin, near Lacq in southwestern France. The gas field, about 10 km in extent with production from depths near 4 km, triggered an exceptionally high level of seismicity for this type of operation, with earthquake magnitudes ranging up to 4.2. These are, so far, the largest events unequivocally associated with reservoir compaction.

Pennington *et al.* (1986) described earthquakes triggered by fluid depressurization in oil and gas fields of south Texas. Earthquakes of $M = 3.9$ in July 1983 (largest) and $M = 3.4$ in March 1984 were located in the Imogene oil field near the town of Pleasanton and in the Fashing gas field adjacent to the town of the same name, respectively. The Imogene field has a long dimension of about 13.5 km whereas that of the Fashing field is about 7 km (Pennington *et al.*, 1986, Figure 1 of that report).

To understand these effects more quantitatively, Segall (1985, 1989) modeled the hydrocarbon reservoir as a poroelastic medium (Biot, 1941) embedded in an impermeable elastic half-space for purposes of calculating the deformation and stress changes due to fluid extraction (see also Geertsma, 1973) assuming a plane strain geometry. Essentially, hydrocarbon production removes fluid (oil, water, or natural gas) from the reservoir, which responds by contracting. This contraction induces both deformation and stress changes in the strata surrounding the reservoir. The calculated deformation can be compared to geodetic measurements of surface subsidence (e.g., Segall, 1985) and horizontal crustal deformation (Yerkes and Castle, 1976). Similarly, calculated stress changes can be compared to focal mechanisms of nearby earthquakes (e.g., Segall, 1989).

Solutions to the linear poroelastic equations for axisymmetric reservoirs, developed by Segall (1992), were applied to the analysis of subsidence and earthquakes associated with the Lacq gas field by Segall *et al.* (1994). Good agreement between computed vertical displacement and geodetically observed subsidence confirmed that the linear model was realistic. Interestingly, the stress changes, computed from the same model, indicate that the earthquakes at Lacq (Grasso and Wittlinger, 1990) are occurring in response to changes of about 0.2 MPa, or less.

Generally, the deformational response of the crust, at least at the surface, to fluid extraction from oil or gas reservoirs seems to be reasonably well understood judging from the success of linear poroelastic models for replicating geodetically measured subsidence (e.g., Segall, 1985; Segall *et al.*, 1994). Such models have, so far, proved to be of more limited use in understanding earthquakes stimulated by oil or gas extraction, partly because the seismic response to hydrocarbon production is highly variable from field to field (e.g., Doser *et al.*, 1992). Evidently, the mechanical characteristics of the strata above and below the producing formation play substantial roles in the seismic response (Volant, 1993). In fact, no seismicity is observed in the vast majority of producing gas and oil fields. Even in cases for which a rich seismic sequence is associated with hydrocarbon production, as in the Lacq field (e.g., Grasso and Feignier, 1990), this seismicity accounts for only a small fraction of the crustal deformation measured geodetically.

3.4.1 Large Midcrustal Earthquakes Beneath Oil and Gas Fields

3.4.1.1 California Earthquakes

Although the 1983 Coalinga earthquake occurred beneath a major producing oil field, the likelihood of a connection between oil production and the earthquake sequence was considered by most seismologists to be quite small, primarily because of the eight vertical kilometers separating the hypocenter from the producing formations (Segall, 1985) and because of the relatively high level of natural seismicity in the area. But the ensuing occurrence of large earthquakes directly beneath two other oil fields from which exceptional amounts of liquid had been extracted, i.e., the 1985 $M=6.1$ Kettleman North Dome and the 1987 $M=6$ Whittier Narrows earthquakes, generated renewed interest in a possible link between oil production and large, midcrustal earthquakes. McGarr (1991) noted that the dimensions of the oil fields, 13 km for Coalinga, 23 km for Kettleman North Dome, and 6 km for the Montebello field above the Whittier Narrows earthquake, are similar to the dimensions of the respective aftershock sequences.

Of particular interest, though, is that the ratios of net liquid production to total seismic moment are nearly the same for all three events. This last observation suggests the following mechanism relating the earthquakes to oil production. Because oil production results in a net removal of mass from an oil field (extracted oil + water − injected liquid), the isostatic response of the crust to this mass removal is to thicken so as to restore the vertical force equilibrium. This crustal thickening can be accommodated by thrust faulting, in a region of compressional tectonics, and so the total seismic deformation can be related to the corresponding liquid production (McGarr, 1991). As presented by McGarr (1991, see Table 1 of that report), the agreement between the seismic deformation and expectations based on liquid production is quite good. This proposed mechanism remains speculative, nonetheless, for a number of reasons including several discussed in the next example.

3.4.1.2 Gazli

The region surrounding the desert town of Gazli, Uzbekistan, was relatively aseismic until April 1976, when a $M=7$ earthquake occurred. This was followed by two more $M=7$ events, one in May 1976 and another eight years later in March 1984, together with a rich sequence of small events spanning an arcuate region, elongated in the east–west direction (Simpson and Leith, 1985).

Simpson and Leith (1985) noted the possibility that this impressive earthquake sequence was related to the enormous natural gas field immediately to the south. They remarked that (1) this area had no history of large seismic events before the gas extraction operations, suggesting that this sequence of $M=7$ events is anomalous and (2) the persistent high level of seismicity, including three $M=7$ events does not form a typical foreshock–mainshock–aftershock pattern. Of particular interest, the arcuate region of seismicity is approximately the same size as the gas field, whose long dimension is about 49 km in the east–west direction (Bossu, 1996).

After numerous recent investigations near Gazli (e.g., Grasso, 1992, 1993; Amorese *et al.*, 1995; Bossu *et al.*, 1996; Bossu and Grasso, 1996; and summarized most recently by Bossu, 1996) more information with which to decide whether the gas field and the earthquakes are related is now available than when Simpson and Leith wrote their report. With depths of 20, 13, and 15 km, in order of their occurrence (Bossu *et al.*, 1996), and with focal mechanisms that largely involve thrust faulting on fault planes that do not extend to the surface (Bossu, 1996), the three Gazli $M = 7$ events are similar to their counterparts beneath the three California oil fields, of the preceding example. If these large-magnitude earthquakes, located at midcrustal depths, are connected to gas production, limited to depths only slightly greater than 1 km, the responsible mechanism is likely to involve the entire upper brittle crust. Might crustal thickening in response to an isostatic imbalance due to production, proposed for the three California earthquakes, also apply to the events at Gazli?

From some of the gas field production information presented by Bossu (1996), one could argue either way. On the one hand, Grasso (1992) noted that either the mass of gas produced from the reservoirs at Gazli or the mass of water infiltrating the reservoirs as gas is produced suffices to account for an appreciable fraction of the seismic deformation (McGarr, 1991). On the other hand, Bossu (1996) pointed out that these two contributions to the crustal load, extracted gas and infiltrated water, tend to cancel one another; in fact, the mass of the infiltrated water apparently predominates. In this way Bossu (1996) concluded that the earthquake sequence and gas production are probably unrelated to the restoration of isostatic equilibrium because gas production added, rather than subtracted, mass to the local crustal load. According to Bossu (1996) about 300 million metric tons of gas was

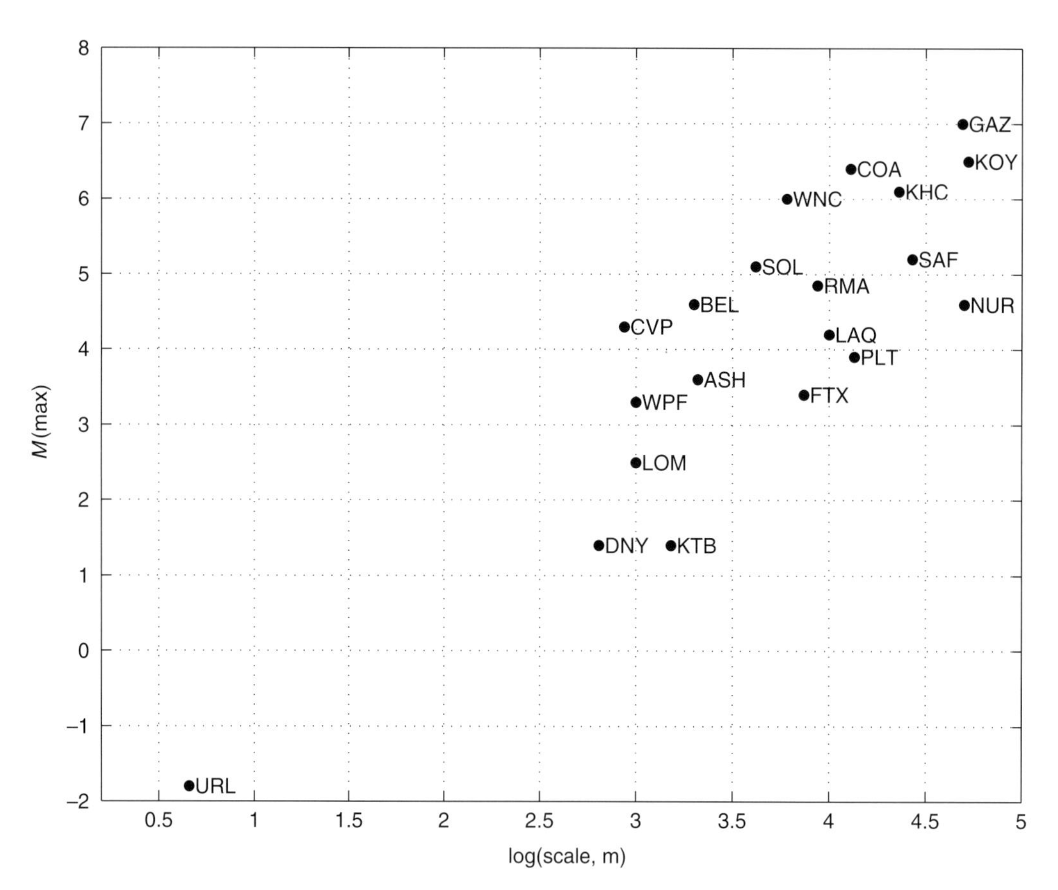

FIGURE 3 Maximum magnitude as a function of scale for 20 case histories. The scale is the maximum dimension of the causative activity as explained in the text. The letter identifications for the case histories, in the same order as they appear in the text, are: underground mining: SAF, deep gold mines in South Africa; URL, Underground Research Laboratory; SOL, Solvay Trona; quarry and surface mining: WPF, Wappingers Falls; CVP, Cacoosing Valley; BEL, Belchatow; LOM, Lompoc Diatomite; Liquid injection: RMA, Rocky Mountain Arsenal; ASH, Ashtabula; DNY, Dale NY; KTB, KTB, Germany; reservoir impoundment: KOY, Koyna; NUR, Nurek; Oil/gas production: LAQ, Lacq; PLT, Pleasanton TX; FTX, Fashing TX; COA, Coalinga; KHC, Kettleman North Dome; WNC, Whittier Narrows; GAZ, Gazli.

extracted whereas, about 1200 million metric tons of water infiltrated the gas field.

4. Effects of Scale

An important question regarding induced or triggered seismicity involves the maximum magnitude earthquake that might result from a particular human activity. For instance, if a dam that impounds a large reservoir is built, what is the maximum credible earthquake that might result?

As described in previous sections, there are several measures of any particular activity that can be related to the total seismic deformation, which, in turn, can be used to estimate the maximum credible earthquake for that activity. Examples include volumes of stope collapse in deep mines (e.g., McGarr and Wiebols, 1977) and the crustal load changes associated with large-scale oil and gas production mentioned in the previous section.

Here, however, we use the maximum dimension of an activity as a simple metric that can be compared to the corresponding maximum magnitude. As described for the case histories reviewed in previous sections, it was straightforward to estimate appropriate maximum dimensions for 20 of them, and Figure 3 shows that the long dimension of an activity correlates fairly well with the corresponding maximum magnitude.

Interestingly, the upper-bound envelope to the data of Figure 3 has an average slope close to 2, suggesting that for an activity of a given size there is an upper limit to the magnitude of the corresponding maximum credible earthquake. For instance, if an activity were 10 km in extent the maximum credible earthquake would presumably have a magnitude near 6. It is important to note, however, that no seismicity is recorded in many instances where human activities affect the stress over large areas, such as below many impounded reservoirs.

5. Conclusions

The case histories reviewed here help to support the following conclusions.

1. The seismogenic continental crust is in a state of stress close to failure, whether the tectonic setting is active or inactive. Because of this, perturbations of the state of stress toward failure [Eq. (1)], even as small as 0.01 MPa, may trigger seismicity. One implication of this is that, for a given activity, triggered earthquakes are as likely in stable as in active tectonic settings, but within stable regions such events are more obvious because the background seismicity is comparatively low.
2. Among the case histories of induced and triggered seismicity reviewed here, perhaps the only examples of induced seismicity (requiring stress perturbations of the order of the ambient deviatoric stress) are from underground mining. This is because pumping operations to prevent flooding strengthen the rock mass by reducing the pore pressure. Extensive mine excavations, especially at depths of several kilometers, amplify the ambient stresses substantially so as to bring the strengthened rock mass back to the point of failure in localized regions. All of the other case histories seem to involve triggered earthquakes inasmuch as the stress perturbations promoting fault slip are tiny compared to the ambient stress level and, in most cases, are much smaller than a typical earthquake stress drop.
3. The size of an activity that perturbs the crustal state of stress appears to be a good predictor of the maximum credible earthquake for that operation. This is evidently the case for all types of induced and triggered earthquake sequences reviewed here including those due to mining, quarries, liquid injection, large reservoir impoundment, and oil and gas exploitation. It is important to note, however, that many large-scale activities result in little or no recorded seismicity.

Acknowledgments

We thank Trudy Closson and Anne Miller for editorial assistance and Joe Fletcher and Anne Miller for graphics support. R.A. Harris, J.B. Fletcher, and S. Gibowicz provided insightful reviews of this manuscript. Encouragement by H. Kanamori and W.H.K. Lee is gratefully acknowledged. This work was partially supported by the National Science Foundation under Cooperative Agreement EAR-9529992 as well as Southern California Earthquake Center under USC P0569934.

References

Abercrombie, R.E. (1995). *J. Geophys. Res.* **100**, 24015–24036.
Adams, J., *et al.* (1991). *Nature* **352**, 617–619.
Amorese, D., *et al.* (1995). *Bull. Seismol. Soc. Am.* **85**, 552–559.
Argus, D.F. and R.G. Gordon (1996). *J. Geophys. Res.* **101**, 13555–13572.
Bell, M.L. and A. Nur (1978). *J. Geophys. Res.* **83**, 4469–4483.
Bennett, T.J., *et al.* (1994). *Maxwell Laboratories Report No. SSS-FR-93-14382*.
Biot, M.A. (1941). *J. Appl. Phys.* **12**, 155–164.
Bossu, R. (1996). PhD Thesis, University Joseph Fourier, Genoble.
Bossu, R. and J.R. Grasso (1996). *J. Geophys. Res.* **101**, 17645–17659.
Bossu, R., *et al.* (1996). *Bull. Seismol. Soc. Am.* **86**, 959–971.
Brace, W.F. and D.L. Kohlstedt (1980). Limits on lithospheric stress imposed by laboratory experiments. *J. Geophys. Res.* **85**, 6248–6252.

Brown, S.R. and R.L. Bruhn (1996). *J. Struct. Geol.* **18**, 657–671.
Brudy, M., *et al.* (1997). *J. Geophys. Res.* **102**, 18453–18475.
Byerlee, J.D. (1978). *Pure Appl. Geophys.* **116**, 615–626.
Carder, D.S. (1970). In: "Engineering Geology Case Histories," No. 8, pp. 51–61, Geological Society of America.
Choy, G.L. and J.R. Bowman (1990). *J. Geophys. Res.* **95**, 6867–6882.
Cook, N.G.W. (1963). In: "Proceedings of the Fifth Rock Mechanics Symposium," pp. 493–516, Pergamon.
Cook, N.G.W., *et al.* (1965). *J. South Afr. Inst. Mining Met.* **66**, 435–528.
Crone, A.J., *et al.* (1992). *US Geol. Surv. Bull. 2032-A.*
Cypser, D.A. and S.D. Davis (1994). *J. Environ. Law Litigation* **9**, 551–589.
Dieterich, J.H. (1979). *J. Geophys. Res.* **84**, 2161–2168.
Dieterich, J.H. and B. Kilgore (1996). *Proc. Natl. Acad. Sci USA* **93**, 3787–3794.
Doser, D.I., *et al.* (1992). *Pure Appl. Geophys.* **139**, 481–506.
Evans, D.M. (1966). *Geotimes* **10**, 11–17.
Feignier, B. and R.P. Young (1993). In: "Rockbursts and Seismicity in Mines," pp. 181–186, Balkema.
Fernandez, L.M. and P.K. van der Heever. (1984). In: "Rockbursts and Seismicity in Mines," pp. 193–198. South African Institute of Mining and Metallurgy.
Fletcher, J.B. and L.R. Sykes (1977). *J. Geophys. Res.* **82**, 3767–3780.
Fredrich, J., *et al.* (1988). *Geophys. J.* **95**, 1–13.
Gay, N.C., *et al.* (1984). In: "Rockbursts and Seismicity in Mines," pp. 107–120. South African Institute of Mining and Metallurgy.
Geertsma, J. (1973). *J. Petrol. Technol.* **25**, 734–744.
Gibowicz, S.J. and A. Kijko (1994). "An Introduction to Mining Seismology," Academic Press.
Gibowicz, S.J., *et al.* (1981). *Eng. Geol.* **17**, 257–271.
Gibowicz, S.J., *et al.* (1991). *Bull. Seismol. Soc. Am.* **81**, 1157–1182.
Gough, D.I. (1978). *Can. J. Earth Sci.* **6**, 1067–1151.
Grasso, J.R. (1992). *Pure Appl. Geophys.* **139**, 507–534.
Grasso, J.R. (1993). In: "Rockbursts and Seismicity in Mines," pp. 187–194, Balkema.
Grasso, J.R. and B. Feignier (1990). *Pure Appl. Geophys.* **134**, 427–450.
Grasso, J.R. and D. Sornette (1998). *J. Geophys. Res.* **103**, 29965–29987.
Grasso, J.R. and G. Wittlinger (1990). *Bull. Seismol. Soc. Am.* **80**, 450–473.
Gupta, H.K. (1985). *Tectonophysics* **118**, 257–279.
Gupta, H.K. and B.K. Rastogi (1976). "Dams and Earthquakes," Elsevier.
Gupta, H., *et al.* (1969). *Bull. Seismol. Soc. Am.* **59**, 1149–1162.
Harris, R.A. (1998). *J. Geophys. Res.* **103**, 24347–24358.
Harris, R.A. and R.W. Simpson (1998). *J. Geophys. Res.* **103**, 24439–24451.
Healy, J.H., *et al.* (1968). *Science* **161**, 1301–1310.
Herrmann, R.B., *et al.* (1981). *Bull. Seismol. Soc. Am.* **71**, 731–745.
Hough, S.E., *et al.* (1999). *Bull. Seismol. Soc. Am.* **89**, 1606–1619.
Hsieh, P.A. and J.D. Bredehoft (1981). *J. Geophys. Res.* **86**, 903–920.
Kebeasy, R.M., *et al.* (1987). *J. Geodynam.* **7**, 173–193.
Knoll, P. (1990). *Gerlands Beitr. Geophys.* **99**, 239–245.
Kock, T.W. (1933). *Am. Assoc. Petrol. Geol. Bull.* **17**, 694–712.
Kovach, R.L. (1974). *Bull. Seismol. Soc. Am.* **64**, 699–711.
Machette, M.N., *et al.* (1993). *US Geol. Surv. Bull. 2032-B.*
Malin, P.E., *et al.* (1988). *Bull. Seismol. Soc. Am.* **78**, 401–420.
Martin, C.D. and R.P. Young (1993). In: "Rockbursts and Seismicity in Mines," pp. 367–371, Balkema.
McDonald, A.J. (1982). MSc Thesis, University of Witwatersrand, Johannesburg.
McGarr, A. (1976). *J. Geophys. Res.* **81**, 1487–1494.
McGarr, A. (1991). *Bull. Seismol. Soc. Am.* **81**, 948–970.
McGarr, A. (1992). *Pure Appl. Geophys.* **139**, 781–800.
McGarr, A. (1993). In: "Rockbursts and Seismicity in Mines," pp. 3–12, Balkema.
McGarr, A. (1994). *Pure Appl. Geophys.* **142**, 467–489.
McGarr, A. and N.C. Gay (1978). *Annu. Rev. Earth Planet. Sci.* **6**, 405–436.
McGarr, A. and D. Simpson (1997). In: "Rockbursts and Seismicity in Mines," pp. 385–396, Balkema.
McGarr, A. and G.A. Wiebols (1977). *Int. J. Rock Mech.* **14**, 139–145.
McGarr, A., *et al.* (1975). *Bull. Seismol. Soc. Am.* **65**, 981–993.
Nason, R.D., *et al.* (1968). In: "43rd Annual Meeting Guidebook, AAPG, SEG, SEPM, Pacific Sections," pp. 100–101, American Association of Petroleum Geologists.
Nicholson, C. and R.L. Wesson (1990). *US Geol. Surv. Bull. 1951.*
Ortlepp, W.D. (1997). "Rock Fracture and Rockbursts," South African Institute of Mining and Metallurgy.
Pechmann, J.C., *et al.* (1995). *Seismol. Res. Lett.* **66**, 25–34.
Pennington, W.D., *et al.* (1986). *Bull. Seismol. Soc. Am.* **76**, 939–948.
Pomeroy, P.W., *et al.* (1976). *Bull. Seismol. Soc. Am.* **66**, 685–700.
Raleigh, C.B., *et al.* (1976). *Science* **191**, 1230–1237.
Reasenberg, P.A. and R.W. Simpson (1992). *Science* **255**, 1687–1690.
Roeloffs, E.A. (1988). *J. Geophys. Res.* **93**, 2107–2124.
Sbar, M.L. and L.R. Sykes (1973). *Geol. Soc. Am. Bull.* **84**, 1861–1882.
Scholz, C.H. (1990). "The Mechanics of Earthquakes and Faulting." Cambridge University Press.
Seeber, L. and J.G. Armbruster (1993). *Géogr. Phys. Quaternaire* **47**, 363–378.
Seeber, L. and J.G. Armbruster (1995). *J. Geophys. Res.* **100**, 8285–8310.
Seeber, L., *et al.* (1996). *J. Geophys. Res.* **101**, 8543–8560.
Seeber, L., *et al.* (1998). *J. Geophys. Res.* **103**, 24505–24521.
Segall, P. (1985). *J. Geophys. Res.* **90**, 6801–6816.
Segall, P. (1989). *Geology* **17**, 942–946.
Segall, P. (1992). *Pure Appl. Geophys.* **139**, 535–560.
Segall, P., *et al.* (1994). *J. Geophys. Res.* **99**, 15423–15438.
Simpson, D.W. (1986). *Annu. Rev. Earth Planet. Sci.* **14**, 21–42.
Simpson, D.W. and W. Leith (1985). *Bull. Seismol. Soc. Am.* **75**, 1465–1468.
Simpson, D.W. and T.N. Narasimhan (1990). *Gerlands Beitr. Geophys.* **99**, 205–220
Simpson, D.W. and S.Kh. Negmatullaev (1981). *Bull. Seismol. Soc. Am.* **71**, 1561–1586.
Simpson, D.W., *et al.* (1988). *Bull. Seismol. Soc. Am.* **78**, 2025–2040.
Simpson, D.W., *et al.* (1990). *Gerlands Beitr. Geophys.* **99**, 191–204.
Spottiswoode, S.M. (1980). PhD Thesis, University of Witwatersrand, Johannesburg.

Spottiswoode, S.M. (1984). In: "Rockbursts and Seismicity in Mines," pp. 29–37, South African Institute of Mining and Metallurgy.
Spottiswoode, S.M. and A. McGarr (1975). *Bull. Seismol. Soc. Am.* **65**, 93–112.
Sylvester, A.G. and J. Heinemann (1996). *Seismol. Res. Let.* **67**, 11–18.
Talebi, S. and R.P. Young (1992). *Int. J. Rock Mech.* **29**, 25–34.
Vogfjord, K.S. and C.A. Langston (1987). *Bull. Seismol. Soc. Am.* **77**, 1558–1578.
Volant, P. (1993). PhD Thesis, University Joseph Fourier, Grenoble.
Wagner, H. (1984). In: "Rockbursts and Seismicity in Mines," pp. 209–218, South African Institute of Mining and Metallurgy.
Wetmiller, R.J. (1986). *Can. J. Earth Sci.* **23**, 172–181.
Wetmiller, R.J., *et al.* (1993). In: "Rockbursts and Seismicity in Mines," pp. 445–448, Balkema.
Yerkes, R.F. and R.O. Castle (1976). *Eng. Geol.* **10**, 151–167.
Yerkes, R.F., *et al.* (1983). *Geology* **11**, 287–291.
Zoback, M.D. and J.H. Healy (1984). *Ann. Geophys.* **2**, 689–698.
Zoback, M.D. and H.-P. Harjes (1997). *J. Geophys. Res.* **102**, 18477–18491.

Editor's Note

Due to space limitations, references with full citation are given in the file McGarrFullReferences.pdf on the Handbook CD, under the directory \40McGarr. Please see also Chapter 32, Rock failure and earthquake, by Lockner and Beeler; Chapter 33, State of stress within the Earth, by Ruff; and Chapter 34, State of stress in the Earth's lithosphere, by Zoback and Zoback.

V

Seismicity of the Earth

41

Global Seismicity: 1900–1999

E.R. Engdahl and A. Villaseñor
University of Colorado, Boulder, USA

1. Introduction

The goal of this chapter is to produce a comprehensive and self-consistent catalog of global seismicity spanning the 20th century and with uniformly computer-determined hypocenters whenever possible. Seismicity data spanning long periods of time are essential for a thorough understanding of earthquake phenomena. Seismic activity is nonuniform over time, and the rate of seismic moment release exhibits large temporal variations. Since the average recurrence interval for great earthquakes along any particular plate boundary is on the order of several decades, with longer intervals in regions away from plate boundaries, at best only one seismic cycle has been recorded by the modern global networks, and only the last three decades of that cycle have been intensively studied. Creation of a digital catalog for the century will support and advance comprehensive seismic hazard analyses, as well as studies of global and regional seismotectonics, the seismic cycle, the rupture zones of large earthquakes, the spatial–temporal pattern of seismic moment release along seismic zones or faults, and the repeat time of large damaging earthquakes.

We approach the problem of assembling this centennial catalog by first combining existing global catalogs of earthquake locations and magnitudes into a single catalog. Second, we assign a single magnitude to each event depending on availability. For shallow earthquakes we preferably use the moment magnitude M_w or the surface-wave magnitude M_S. For earthquakes deeper than 60 km we use the moment magnitude M_w, or the body-wave magnitude m_B (broadband) or m_b (short period). Third, we use these assigned magnitudes to determine the magnitude completeness thresholds of the database and to assign magnitude cut-off values as a function of time.

Finally, using a modern Earth model (ak135) and location algorithm (Engdahl *et al.*, 1998), we relocate all events which are within the magnitude cut-off thresholds of the catalog and for which there are digital phase arrival-time data.

2. Seismicity Catalogs

Excellent primary and secondary sources of catalog information about earthquake hypocenters and magnitudes can be found in the archival systems of the International Seismological Centre (ISC) and the US Geological Survey's National Earthquake Information Center (NEIC). These systems are multi-catalog earthquake databases of source parameters, available in computer-readable format, that are either directly determined by the ISC and NEIC or derived from published papers and institutional contributions.

The most valuable catalog included in these databases for the historical period (before 1964; Lee *et al.*, 1988) is the one derived from Gutenberg and Richter's book "Seismicity of the Earth" (Gutenberg and Richter, 1954), which provides hypocenters and magnitudes for most of the larger earthquakes occurring between 1904 and 1952. Before 1904 the main source of hypocenters and magnitudes is Abe and Noguchi (1983a,b). Other catalogs of large earthquakes, such as Abe (1981, 1984), Abe and Noguchi (1983a,b), Båth and Duda (1979), Geller and Kanamori (1977), Pacheco and Sykes (1992), Rothé (1969), and Utsu (1979, 1982a,b), are used mainly as sources of magnitude information. All these catalogs (listed in Appendix 3) are combined into a single catalog by using a hierarchical scheme to integrate the hypocenters for matched events between catalogs and to carry along the associated magnitude estimates. In the end, all hypocenter and magnitude entries in these catalogs are accounted for.

3. Seismic Phase Bulletins

Bulletins differ from catalogs in that they contain reported arrival times for *P*, *S* and additional seismic phases for each earthquake, which are used to determine a hypocenter for that event when enough consistent phase data are available. A historical account of the international organizations involved in the collection of phase arrival data and global teleseismic

INTERNATIONAL HANDBOOK OF EARTHQUAKE AND ENGINEERING SEISMOLOGY, VOLUME 81A ISBN: 0-12-440652-1

earthquake location can be found in Chapter 2 by Adams. The primary sources of historical phase data are the bulletins of the International Seismological Summary (ISS) and their predecessors. Unfortunately, bulletins for the historical period are mainly preserved in printed form and are not in a computer-ready, digital format. Recently, a project to relocate all instrumentally recorded earthquakes during the period 1900–1963 was initiated (Villaseñor *et al.*, 1997). In this project the printed bulletins are being converted into digital form by scanning the bulletin pages and applying an optical character recognition procedure. We also have obtained a data tape of hand-entered observations from the ISC that contains almost all of the arrival time data for earthquakes reported by the ISS between 1918 and 1942.

From 1953 to 1963 the ISS only processed events with magnitudes greater than or equal to 6. For this period the bulletins of the Bureau Central International de Séismologie (BCIS) fill in the gaps, but the BCIS phase data have not yet been converted in to a digital format and hence were not used in this study.

For the modern period (1964–1999) instrumental phase data for moderate-to-large earthquakes worldwide are already available in digital form from both the ISC and the NEIC. Data from all these sources are integrated and converted into the ISC 96-byte data format adopted for the centennial phase arrival-time database. (Note: please see CENT.DAT on the Handbook CD, under the directory \41Engdahl.)

4. Earth Models and Earthquake Location

Although earthquakes have been instrumentally recorded for more than 100 years, source parameters (locations, magnitudes, and focal mechanisms) for events that occurred before the full implementation of the World-Wide Standardized Seismograph Network (WWSSN) are in general poorly known. In most cases, this is the result of inherent limitations in the station distribution, timing, or low magnification of the instruments. However, many locations for pre-1964 earthquakes are poorly constrained because electronic computers and adequate Earth models were not available at the time the earthquakes occurred and the locations were produced.

In the first part of the century travel-time tables for seismic phases, empirically derived from the historical data, were rudimentary at best. The earliest of these, the Zöppritz–Turner tables (used by the ISS from 1913 to 1929) were inaccurate and incomplete, and valid only for shallow-focus earthquakes. Deep earthquakes had to be accounted for with *ad hoc* corrections. This situation greatly improved with the introduction of the Jeffreys–Bullen (J-B) tables (Jeffreys and Bullen, 1940) which provided a complete, remarkably accurate representation of *P*, *S*, and other later-arriving phases. These tables are still used by the ISC (the successor agency of the ISS) and the NEIC for routine earthquake location.

Earthquake location procedures during the historical period suffered from the lack of electronic computers to implement them. For example, new earthquakes reported to the ISS were commonly assigned epicenters coincident with those determined for previously reported events in the same region. A new epicenter was only adopted if the phase data appeared significantly incompatible with any preexisting locations. The existence of deep earthquakes was discovered by H.H. Turner in the late 1920s during the production of the ISS bulletins (this discovery was also done, almost simultaneously, by K. Wadati). However, depths for approximately 75% of the earthquakes listed in the ISS bulletins were simply reported as shallow (surface focus). A so-called "abnormal" focal depth was calculated only if the phase data were clearly incompatible with a shallow-focus solution. Thus, constraints on the focal depths of many ISS hypocenters were limited.

In an effort to remedy this situation, Villaseñor *et al.* (1997) are extending the current global catalog of computer-determined hypocenters by producing a comprehensive catalog of all globally detected earthquakes during the historical period with locations determined in a uniform fashion. Relocation of historical earthquakes is accomplished by using a new algorithm for teleseismic location (Engdahl *et al.*, 1998). This algorithm, hereafter referred to as EHB, uses travel-time tables derived from an improved Earth model and incorporates *P*, *S*, and other later-arriving phases in the location procedure. The Earth model used is ak135 (Kennett *et al.*, 1995), a derivation of the iasp91 model (Kennett and Engdahl, 1991). Because most seismic stations are in continental areas, the ak135 model was developed with an average continental crust and upper mantle. The most significant differences between the travel times predicted by these models and the older J-B tables are for upper mantle and core phases. The ak135 model more accurately predicts the observed travel times of later-arriving phases, and is in better agreement with *S*-wave data than the J-B tables.

Corrections for lateral variations from the average crust and upper mantle velocities and ellipticity have also been incorporated. The location procedure uses arrival times for the first arriving *P* and *S* phases, core phases (*PKP*) and depth phases (*pP*, *pwP*, and *sP*). By far the most significant improvements provided by the EHB algorithm are in depth determination, through the inclusion of the teleseismic depth phases (with free-surface and sea-bottom bounce points) *pP*, *pwP*, and *sP*. The ISS bulletins reported depth phases, primarily *pP*, but they were not fully used by the ISS to calculate the hypocenter. These phases, reidentified using a new statistical procedure (Engdahl *et al.*, 1998), are now used to minimize mislocation errors introduced by lateral heterogeneity and to provide powerful constraints on focal depth.

The EHB method has already been successfully applied to earthquakes reported by the ISC and NEIC during the modern period, providing a uniform database of well-constrained, significantly improved hypocenters. The application of this method to historical earthquakes will also result in a comprehensive and homogeneous digital earthquake catalog for the entire century. However, it is important to point out that the EHB procedure cannot entirely account for the effects of the Earth's lateral heterogeneity on teleseismic earthquake location. Most deeper-than-normal earthquakes occur in subduction zones where aspherical variations in seismic wave velocities are large (i.e., on the order of 5–10%). Such lateral variations in seismic velocity, the uneven spatial distribution of seismological stations, and the specific choice of seismic data used to determine the earthquake hypocenter can easily combine to produce bias in teleseismic earthquake locations of up to several tens of kilometers (Engdahl *et al.*, 1998). The most accurate earthquake locations are best determined using a regional velocity model with phase arrival times from a dense local network, which may differ significantly (especially in focal depth) from the corresponding teleseismic locations. Similarly, for historical earthquakes (especially before the 1930s), locations obtained from macroseismic or geological data (e.g., surface rupture) can be more accurate than teleseismic locations computed using arrival-time data alone.

5. Seismic Station Information

Creation of a comprehensive digital earthquake catalog also requires a comprehensive global seismic station catalog. The ISC and NEIC maintain listings of seismograph station codes and coordinates for stations distributed worldwide (Presgrave *et al.*, 1985). These listings are an invaluable resource to investigators using cataloged earthquake data, representing the only nearly complete listing of station locations, codes, and dates of operation. We have been assisting NEIC in maintaining and updating this listing, particularly for stations operating prior to the installation of the WWSSN, by examining a variety of alternative sources of historical station information. This effort has resulted in establishing consistent station codes for stations having no prior code assigned, defining date ranges over which station codes and coordinates are available, and compiling lists of alternate spellings for listed stations. (Note: please see CENT.STN on the Handbook CD, under the directory \41Engdahl.)

6. Earthquake Magnitudes and Catalog Completeness

Construction of the earthquake hypocenter catalog must include magnitude information. The ISS bulletins do not list earthquake magnitudes, and other organizations did not start reporting magnitudes until the early 1950s. The main source for magnitudes of historical earthquakes is Gutenberg and Richter (1954) which reports magnitudes for approximately 13% of the earthquakes listed by the ISS and occurring before 1952. Gutenberg and Richter never published the details of the method used to compute magnitudes in "Seismicity of the Earth" (hereafter referred to as M_{GR}) although their relationship with other magnitude scales has been intensively investigated (Geller and Kanamori, 1977; Abe, 1981). For the historical period other commonly reported magnitudes are M_S (surface-wave magnitude as defined by Gutenberg, 1945) and m_B (body-wave magnitude for periods of 5–10 sec as defined in Gutenberg and Richter, 1956). (See Chapter 44 by Utsu for a comparison of magnitude scales.)

For recent earthquakes the main sources of magnitude information are the bulletins of the ISC and the NEIC Preliminary Determination of Epicenters (PDE). The most commonly reported magnitudes in this period for teleseismically recorded earthquakes are M_S (determined using the Prague formula; Vaněk *et al.*, 1962) and m_b (body-wave magnitude for periods around 1 sec as defined by Gutenberg and Richter, 1956). However, from 1964 to 1976 M_S was not systematically reported by the ISC, and to compensate for this we included in the catalog, reliable single station magnitude estimates, such as those reported by Pasadena and Berkeley (California), and Palisades (New York).

When available, moment magnitude (M_w) estimates based on the scalar seismic moment M_0 are preferred (Kanamori, 1977; Hanks and Kanamori, 1979). These values of M_0 are frequently determined with modern methods that use body and/or surface wave waveforms (e.g., Dziewonski *et al.*, 1980; Sipkin, 1982). Reliable estimates of M_w are generally available from the Harvard CMT catalog for most events with M_S larger than about 5.4 during the period 1976–1999. Pacheco and Sykes (1992) report scalar seismic moments for large, shallow events during 1900–1980, and Huang *et al.* (1994, 1997) for large, deep events during 1962–1976.

Combining magnitudes from different types and sources into a single catalog requires some understanding of the relationships between the different magnitude scales. There is an overwhelmingly large set of literature devoted to this subject (for a review see Båth, 1981) but our goal is to produce a set of simple rules for obtaining a single magnitude for each event that results in an earthquake catalog which is representative of the rate of earthquake occurrence for the century at or above specified magnitude thresholds in time. The rules we have used are the following: if M_0 is available the preferred magnitude is M_w; for earthquakes with focal depth $h \leq 60$ km if M_0 is not available then M_S, m_B, or m_b is selected, in this order of preference; for events with $h > 60$ km if M_0 is not available then m_B or m_b is selected, in this order of preference.

Because of the different magnitude scales used, the magnitudes in the resulting catalog are very heterogeneous. In order to evaluate the completeness of the catalog and to establish a cut-off magnitude and select the earthquakes above that cut-off, all magnitudes must be corrected and reduced to a common reference magnitude scale. It is worth noting that

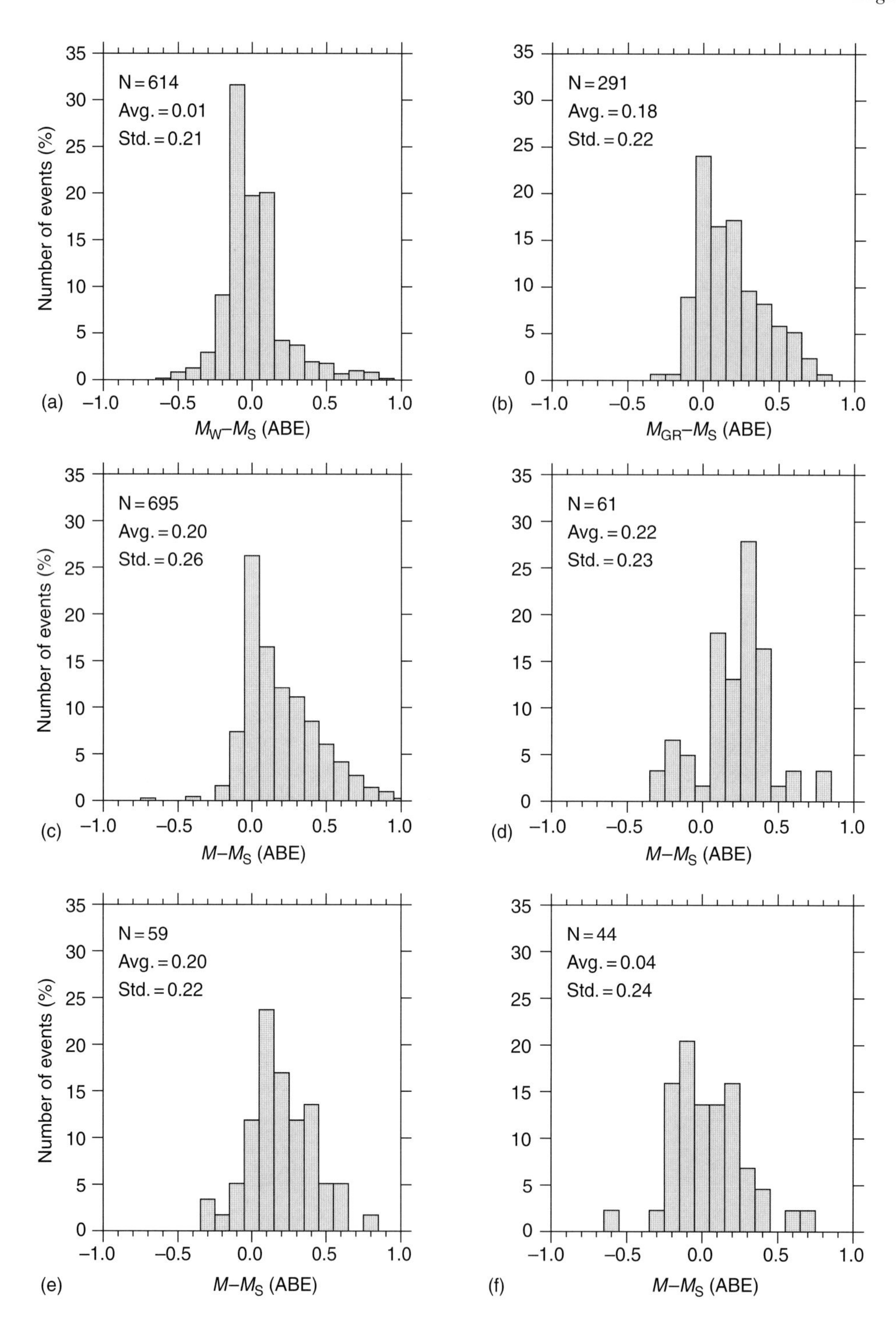

FIGURE 1 Comparison between magnitudes reported by different catalogs relative to surface wave magnitudes (M_S) reported in Abe's catalog (Abe, 1981, 1984; Abe and Noguchi, 1983a,b). The bin width for all histograms is 0.1 magnitude units, and the number of events in each bin is shown as a percentage of the total number of events. N, total number of events; Avg, average residual; Std, standard deviation of the residuals. Catalogs compared: (a) Pacheco and Sykes (1992); (b) Gutenberg and Richter (1954); (c) Båth and Duda (1979); (d) Rothé (1969); (e) Pasadena single-station magnitude for events before 1960; (f) Pasadena magnitudes after 1959.

these corrections are used exclusively to calculate the catalog completeness (as a whole and as a function of time) and to determine which earthquakes are included in the catalog. The magnitudes listed in the catalog are the uncorrected values from the original sources. We have chosen M_S as the reference magnitude scale for the catalog, and investigated its relationship with other magnitude scales.

Moment magnitude M_w agrees very well with M_S for shallow earthquakes larger than a magnitude of ~ 6.5, but for smaller earthquakes M_w is larger than M_S. Ekström and Dziewonski (1988) introduced an empirical moment magnitude relationship which, for each value of M_0 (or M_w), provides a magnitude that approximates the global average of M_S for that scalar moment. We use this empirical magnitude for the purpose of event selection and to obtain frequency–magnitude relationships. For deeper earthquakes the relationship of reported magnitudes to M_S is problematic and, lacking a well-determined empirical relationship, we assume it to be equivalent to M_w, m_B, and m_b (in order of preference).

Figure 1a shows the distribution of magnitude differences between M_w from Pacheco and Sykes (1992) and M_S from Abe (1981) for events common to both catalogs. These events have magnitudes greater than 7.0 and the agreement between both magnitudes is very good (the residuals are small and normally distributed with an average value close to zero). Figure 1b shows the comparison between M_{GR} (Gutenberg and Richter, 1954) and M_S (Abe, 1981). Values of M_{GR} are systematically larger than M_S by 0.2 magnitude units on average, and in this case the magnitude residuals are not normally distributed. Abe (1981) analyzed in detail the nature of these differences and found that the distribution of residuals changed with focal depth. Figures 1c and 1d show comparisons between M_S (Abe, 1981) and magnitudes reported by Båth and Duda (1979) and Rothé (1969), respectively. Figures 1e and 1f show differences between M_S (Abe, 1981) and magnitudes reported by Pasadena before and after 1960, respectively. The distribution of magnitude residuals for Pasadena before 1960 is similar to the distribution for M_{GR}, with predominantly positive residuals. However, after 1960 the agreement with M_S (Abe, 1981) is better, and the residual distribution has zero mean. From this analysis we conclude that magnitudes reported by Gutenberg and Richter (1954), Båth and Duda (1979), Rothé (1969), and Pasadena before 1960 must be decreased by 0.2 magnitude units in order to be reduced to the common reference magnitude of the catalog. A similar magnitude bias has been previously found by Pérez and Scholz (1984). The correction applied here is an obvious oversimplification, but a detailed analysis of the causes of the magnitude differences is beyond the scope of this chapter.

For the historical period (1900–1963) a frequency–magnitude plot (Fig. 2a) shows a gentle roll-off in the number of earthquakes per year for $M_S \leq 7.0$ (both for incremental and cumulative number of events). The irregular shape of the curve for the incremental number of events is caused by magnitudes reported with precisions of 0.25 and 0.5 magnitude units. Most missed earthquakes in the $M_S = 6.5$–7.0 range probably occur during the 1900–1930 period. Hence, to include as many events as possible we have chosen a magnitude of 6.5 as our magnitude cut-off for the historical period.

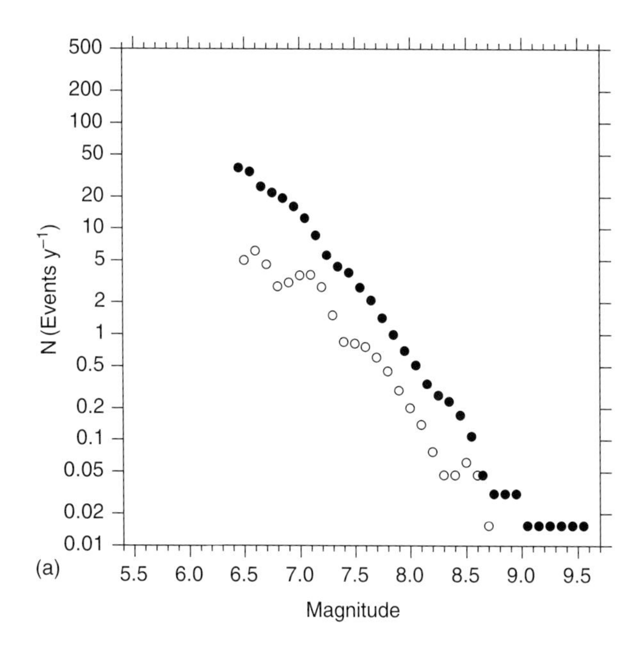

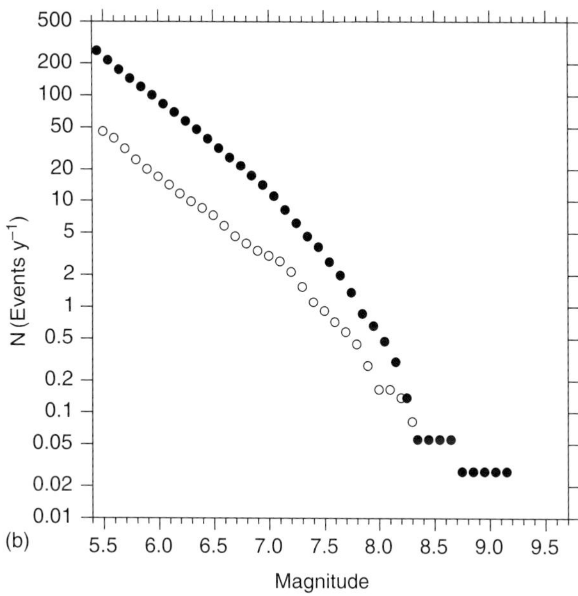

FIGURE 2 Frequency–magnitude (Gutenberg and Richter) relations for the centennial catalog. Open circles represent single frequencies (incremental number of earthquakes with magnitudes in $M \pm \delta M/2$) and filled circles represent cumulative frequencies (total number of earthquakes with magnitudes $\geq M$). The width of the magnitude interval δM is 0.1 magnitude units. The single and cumulative frequencies are normalized to events per year, and the magnitudes have been adjusted to M_S (see text): (a) historical seismicity (1900–1963), and (b) recent seismicity (1964–1999).

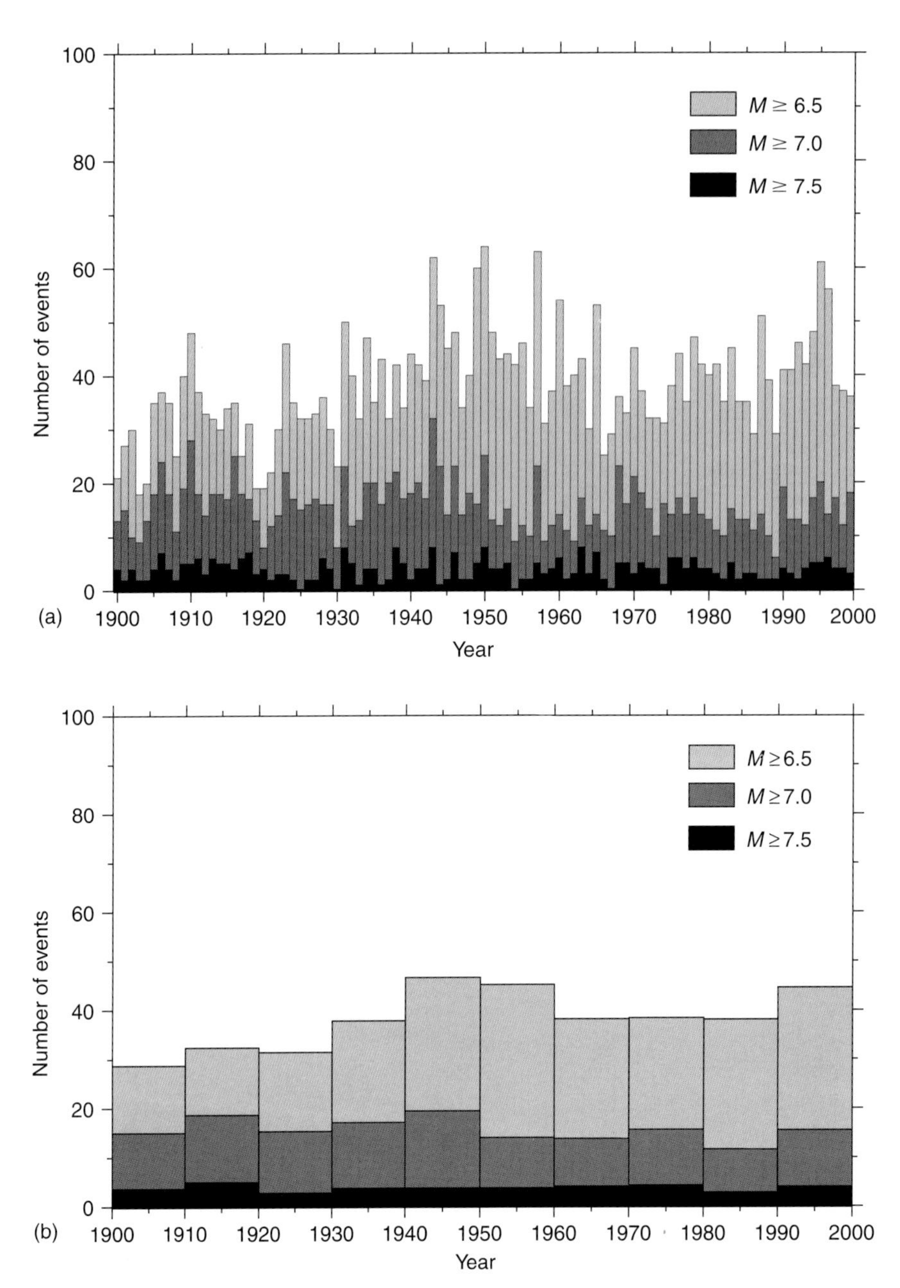

FIGURE 3 Number of events in the centennial catalog as a function of time for the three magnitude levels specified in the legend: (a) number of events per year; (b) number of events in 10 y intervals. The total number of events in each interval is divided by the interval width to allow direct comparison between the two histograms.

For the modern period (1964–1999) Engdahl *et al.* (1998) have shown that the global catalog is nearly complete to a magnitude of 5.2, where the magnitude is defined as (in descending order of preference) either M_w, M_S, or m_b, depending on availability. A frequency–magnitude plot for the modern period (Fig. 2b) shows that the number of earthquakes per year is consistent with a magnitude threshold of at least M_S 5.5. Hence, with the possible exception of the 1960s when less-reliable magnitudes were available and taking into account normal yearly fluctuations in the rate of occurrence (Fig. 3a), the centennial catalog seems to be complete to at least M_S 6.5 during the period 1930–1963 and M_S 5.5 during the period 1964–1999, and complete to M_S 7.0 for the entire century.

If our magnitude assumptions are valid, then the rate of earthquake occurrence at any given magnitude threshold should be constant provided that the global rates of seismicity are assumed constant on a timescale of decades (Pacheco and Sykes, 1992) and the catalog is complete at that threshold.

TABLE 1 List of Earthquakes (Magnitude >= 7) for 1900–1999.
Please See Explanations at the End of the Table

Year	M	d	h:min	sec	Lat.	Long.	Dep.	Mag.	sc	icat	mdo	Region:Earthquake Name
1900	1	5	19:00	0.0	-3.00	102.00		7.0	Ms	ABE	AN2	
1900	1	11	9:07	0.0	-5.00	148.00		7.0	Ms	ABE	AN2	
1900	1	20	6:33	0.0	20.00	-105.00		7.3	Mw	ABE	P&S	
1900	1	31	19:22	0.0	48.00	146.00	450	7.5	Mj	UTSU	UTSU	
1900	5	11	17:23	0.0	38.70	141.10	5	7.0	Mj	UTSU	UTSU	
1900	6	21	20:52	0.0	10.00	-85.50		7.2	Mw	P&S	P&S	
1900	7	29	6:59	0.0	-10.00	165.00		7.6	Mw	ABE	P&S	
1900	9	17	21:45	0.0	-5.00	148.00		7.1	Mw	ABE	P&S	
1900	10	9	12:28	0.0	57.09	-153.48		7.7	Mw	P&S	P&S	
1900	10	29	9:11	0.0	11.00	-66.00		7.7	Mw	ABE	P&S	Venezuela: Caracas
1900	11	9	16:10	0.0	13.00	-90.00		7.0	Ms	ABE	AN2	
1900	11	24	7:57	0.0	43.50	148.00	35	7.0	Ms	UTSU	AN2	
1900	12	25	5:09	0.0	43.00	146.00	35	7.1	Mw	UTSU	P&S	
1901	1	7	0:29	0.0	-2.00	-82.00		7.2	Mw	ABE	P&S	
1901	1	18	4:39	0.0	60.00	-135.00		7.1	Mw	ABE	P&S	
1901	4	5	23:30	0.0	44.50	149.00	35	7.3	Mw	UTSU	P&S	
1901	5	25	0:32	0.0	-10.00	160.00		7.2	Mw	ABE	P&S	
1901	6	15	9:34	0.0	39.00	143.00	35	7.0	Mj	UTSU	UTSU	
1901	6	24	7:02	0.0	28.00	130.00	35	7.2	Mw	UTSU	P&S	
1901	8	9	9:23	0.0	40.50	142.50	35	7.2	Mw	UTSU	P&S	
1901	8	9	13:01	0.0	-22.00	170.00		7.9	Mw	ABE	P&S	
1901	8	9	18:33	0.0	40.60	142.30	35	7.5	Mw	UTSU	P&S	
1901	10	8	2:14	0.0	13.00	-87.00		7.1	Mw	ABE	P&S	
1901	11	25	1:51	0.0	3.00	127.00		7.0	Ms	BJI	BJI	
1901	12	9	2:17	0.0	26.00	-110.00		7.1	Mw	ABE	P&S	
1901	12	14	22:57	0.0	14.00	122.00		7.0	Ms	ABE	AN2	
1901	12	31	9:02	0.0	52.00	-177.00		7.1	Mw	ABE	P&S	
1902	1	1	5:20	0.0	55.00	-165.00		7.0	Ms	ABE	AN2	
1902	1	12	22:18	0.0	3.00	122.00		7.0	Ms	BJI	BJI	
1902	1	24	23:27	0.0	-8.00	150.00		7.2	Mw	ABE	P&S	
1902	4	19	2:23	0.0	14.00	-91.00		7.5	Mw	ABE	P&S	Guatemala
1902	5	2	11:31	0.0	39.00	144.00	35	7.0	Mj	UTSU	UTSU	
1902	8	22	3:00	0.0	40.00	77.00		7.7	Mw	ABE	P&S	China/Kyrgyzstan
1902	9	22	1:46	0.0	18.00	146.00		7.5	Mw	ABE	P&S	
1902	9	23	20:18	0.0	16.00	-93.00		7.8	Mw	ABE	P&S	
1902	10	6	9:15	0.0	36.50	70.50	200	7.2	Ms	BJI	BJI	
1902	12	12	23:10	0.0	29.00	-114.00		7.1	Mw	ABE	P&S	
1903	1	4	5:07	0.0	-20.00	-175.00	400	8.0	UK	B&D	B&D	
1903	1	14	1:47	0.0	15.00	-98.00		7.4	Mw	ABE	P&S	
1903	1	17	16:05	0.0	50.00	-170.00		7.0	Ms	ABE	AN2	
1903	2	1	9:34	0.0	48.00	98.00		7.1	Mw	ABE	P&S	
1903	2	27	0:43	0.0	-8.00	106.00		7.3	Mw	ABE	P&S	
1903	4	29	3:59	0.0	-20.00	-175.00		7.1	Mw	ABE	P&S	
1903	5	13	6:34	0.0	-17.00	168.00		7.0	Ms	ABE	AN2	
1903	8	11	4:32	54.0	36.36	22.97	80	8.3	UK	UK	B&D	
1903	12	28	2:56	0.0	7.00	127.00		7.1	Mw	ABE	P&S	
1904	1	20	14:52	6.0	7.00	-79.00		7.2	Mw	G&R	P&S	
1904	3	19	6:28	0.0	-29.00	-71.00		7.0	Ms	ABE	AN2	
1904	4	4	10:26	0.0	41.75	23.25		7.1	Mw	G&R	P&S	
1904	5	1	15:24	0.0	2.00	130.00		7.0	Ms	ABE	AN2	
1904	6	7	8:17	54.0	39.00	135.00	350	7.4	mB	UTSU	ABE1	
1904	6	25	14:45	36.0	52.00	159.00		7.5	Mw	G&R	P&S	
1904	6	25	21:00	30.0	52.00	159.00		7.6	Mw	G&R	P&S	Kamchatka
1904	6	27	0:09	0.0	52.00	159.00		7.2	Mw	G&R	P&S	
1904	8	24	20:59	54.0	30.00	131.00	35	7.1	Mw	UTSU	P&S	
1904	8	27	21:56	6.0	64.00	-151.00		7.2	Mw	G&R	P&S	
1904	10	3	3:05	0.0	12.00	58.00		7.0	Ms	G&R	AN2	
1904	10	28	13:51	0.0	8.00	113.00		7.0	Ms	BJI	BJI	
1904	12	20	5:44	18.0	8.50	-83.00		7.2	Mw	G&R	P&S	
1905	1	22	2:43	54.0	1.00	123.00	90	7.8	mB	G&R	ABE1	
1905	2	14	8:46	36.0	53.00	-178.00		7.2	Mw	G&R	P&S	
1905	3	18	0:58	0.0	-27.50	-173.00	60	7.5	UK	B&D	B&D	
1905	3	18	23:56	0.0	-10.00	168.00		7.2	Mw	ABE	P&S	
1905	3	22	3:38	0.0	50.00	180.00		7.0	Ms	ABE	AN2	
1905	4	4	0:50	0.0	33.00	76.00		7.8	Mw	G&R	P&S	India: Kangra
1905	6	2	5:39	42.0	34.10	132.50	55	7.0	mB	UTSU	ABE1	
1905	6	14	11:30	0.0	-20.00	-175.00		7.0	Ms	ABE	AN2	
1905	6	30	17:07	0.0	-20.00	-175.00		7.1	Mw	P&S	P&S	
1905	7	6	16:21	0.0	37.40	141.80	35	7.1	Mw	UTSU	P&S	
1905	7	9	9:40	24.0	49.00	99.00		8.5	Mw	G&R	P&S	Mongolia
1905	7	11	8:38	0.0	49.50	97.30	20	7.0	Ms	BJI	BJI	
1905	7	11	15:37	30.0	22.00	143.00	450	7.2	mB	UTSU	ABE1	
1905	7	23	2:46	12.0	49.00	98.00		8.4	Mw	G&R	P&S	Mongolia
1905	9	1	2:45	36.0	45.00	143.00	250	7.3	mB	UTSU	ABE1	
1905	9	15	6:02	0.0	55.00	165.00		7.3	Mw	ABE	P&S	
1905	10	21	11:01	37.0	42.00	42.00	60	7.5	UK	B&D	B&D	
1905	12	17	5:27	0.0	17.00	-113.00		7.1	Mw	ABE	P&S	
1906	1	21	13:49	35.0	34.00	137.00	350	7.7	mB	UTSU	ABE1	
1906	1	31	15:36	0.0	1.00	-81.50		8.6	Mw	G&R	P&S	Ecuador/Colombia
1906	2	19	1:59	0.0	-10.00	160.00		7.1	Mw	ABE	P&S	
1906	3	2	6:15	15.0	43.00	80.00	60	7.3	UK	B&D	B&D	
1906	4	10	21:18	0.0	20.00	-110.00		7.2	Mw	ABE	P&S	
1906	4	13	19:18	0.0	23.60	120.40	5	7.1	Mj	UTSU	UTSU	
1906	4	18	13:12	0.0	38.00	-123.00		7.9	Mw	G&R	P&S	Calif.: San Francisco
1906	6	1	4:30	0.0	0.00	145.00		7.1	Mw	ABE	P&S	
1906	6	24	11:17	49.0	15.00	92.00	60	7.3	UK	B&D	B&D	
1906	8	17	0:10	42.0	51.00	179.00		7.8	Mw	G&R	P&S	Aleutian Is
1906	8	17	0:40	0.0	-33.00	-72.00		8.5	Mw	G&R	P&S	Chile: Valparaiso
1906	8	26	5:59	0.0	-4.00	149.00		7.0	Ms	ABE	AN2	
1906	8	30	2:38	0.0	-21.00	-70.00		7.1	Mw	ABE	P&S	
1906	9	7	18:52	0.0	34.00	141.00	35	7.0	Ms	UTSU	AN2	
1906	9	14	16:04	18.0	-7.00	149.00		8.0	Mw	G&R	P&S	New Britain
1906	9	28	15:24	54.0	-2.00	-79.00	150	7.5	mB	G&R	ABE1	
1906	10	2	1:50	0.0	-4.00	149.00		7.2	Mw	ABE	P&S	
1906	10	17	9:40	0.0	19.00	121.00	60	7.3	UK	B&D	B&D	
1906	11	19	7:18	18.0	-22.00	109.00		7.2	Mw	G&R	P&S	
1906	12	3	22:59	24.0	15.00	-61.00	100	7.2	mB	G&R	ABE1	
1906	12	19	1:14	0.0	-19.00	-172.00		7.2	Mw	ABE	P&S	
1906	12	22	18:21	0.0	43.50	85.00		7.2	Mw	G&R	P&S	China: Shawan
1906	12	23	17:22	0.0	56.85	-153.90		7.2	Mw	P&S	P&S	

(continued)

TABLE 1 (*continued*)

Year	M	d	h:min	sec	Lat.	Long.	Dep.	Mag.	sc	icat	mdo	Region:Earthquake Name
1906	12	26	5:54	0.0	-20.00	-73.00		7.0	Ms	ABE	AN2	
1907	1	2	11:57	0.0	-21.10	-175.10		7.3	Mw	ABE	P&S	
1907	1	4	5:19	12.0	2.00	94.50	50	7.5	Mw	G&R	P&S	Indonesia
1907	2	3	19:34	57.0	-6.00	148.00	60	7.2	UK	B&D	B&D	
1907	3	29	20:46	30.0	3.00	122.00	500	7.2	mB	G&R	ABE1	
1907	3	31	22:00	36.0	-18.00	-177.00	400	7.2	mB	G&R	ABE1	
1907	4	15	6:08	6.0	17.00	-100.00		7.9	Mw	G&R	P&S	Mexico
1907	4	18	20:59	48.0	14.00	123.00		7.1	Mw	G&R	P&S	
1907	4	18	23:52	24.0	13.50	123.00		7.0	Ms	G&R	AN2	
1907	5	4	6:51	0.0	-7.50	153.70	60	7.7	UK	B&D	B&D	
1907	5	25	14:32	8.0	50.50	148.00	600	7.4	mB	UTSU	ABE1	
1907	6	1	8:45	0.0	0.00	-82.00		7.0	Ms	ABE	AN2	
1907	6	5	3:18	0.0	0.00	-86.00		7.0	Ms	ABE	AN2	
1907	6	25	17:54	36.0	1.00	127.00	200	7.5	mB	G&R	ABE1	
1907	8	17	17:27	54.0	52.00	157.00	120	7.2	mB	G&R	ABE1	
1907	9	2	16:01	30.0	52.00	173.00		7.3	Mw	G&R	P&S	
1907	10	16	14:57	18.0	28.00	-112.50		7.2	Mw	G&R	P&S	
1907	10	21	4:23	36.0	38.00	69.00		7.2	Mw	G&R	P&S	Uzbekistan/Tajikistan
1907	12	30	5:26	0.0	12.10	-86.30		7.2	Mw	ABE	P&S	
1908	1	15	12:56	0.0	37.30	141.80	35	7.3	UK	UTSU	B&D	
1908	2	9	18:13	0.0	26.00	100.00	60	7.3	UK	B&D	B&D	
1908	3	26	23:03	30.0	18.00	-99.00	80	7.7	mB	G&R	ABE1	
1908	3	27	3:45	0.0	17.00	-101.00		7.0	Ms	ABE	AN2	
1908	4	22	23:45	0.0	-38.00	48.00		7.0	Ms	ABE	AN2	
1908	5	15	8:31	36.0	59.00	-141.00		7.0	Ms	G&R	AN2	
1908	8	17	10:32	0.0	-60.00	-40.00		7.2	Mw	ABE	P&S	
1908	10	24	21:16	36.0	36.50	70.50	220	7.0	mB	G&R	ABE1	
1908	12	12	12:08	0.0	-14.00	-78.00	60	8.2	UK	B&D	B&D	
1908	12	12	12:54	54.0	26.50	97.00		7.0	Ms	G&R	AN2	
1908	12	28	4:20	24.0	38.00	15.50		7.0	Ms	G&R	AN2	Italy: Messina
1909	1	23	2:48	18.0	33.00	53.00		7.0	Ms	G&R	AN2	Iran
1909	2	22	9:21	42.0	-18.00	-179.00	550	7.6	mB	G&R	ABE1	
1909	3	13	14:29	0.0	34.50	141.50	35	7.6	mB	UTSU	ABE1	
1909	4	10	19:36	0.0	52.00	175.00		7.0	Ms	ABE	AN2	
1909	4	14	19:53	42.0	25.00	122.50	5	7.1	mB	UTSU	ABE1	
1909	4	25	22:36	0.0	4.00	127.00	100	7.1	mB	G&R	ABE1	
1909	4	27	12:44	0.0	0.00	147.00		7.1	Mw	ABE	P&S	
1909	5	17	8:02	54.0	-20.00	-64.00	250	7.0	mB	G&R	ABE1	
1909	5	30	21:01	18.0	-8.00	131.00	100	7.1	mB	G&R	ABE1	
1909	6	3	18:40	48.0	-2.00	101.00		7.2	Mw	G&R	P&S	
1909	6	8	5:46	30.0	-26.50	-70.50		7.2	Mw	G&R	P&S	
1909	7	7	21:37	50.0	36.50	70.50	230	7.6	mB	G&R	ABE1	
1909	7	30	10:51	54.0	17.00	-100.50		7.6	Mw	G&R	P&S	
1909	8	18	0:39	30.0	-22.00	172.00	100	7.2	mB	G&R	ABE1	
1909	9	8	16:49	48.0	52.50	-169.00	90	7.0	mB	G&R	ABE1	
1909	10	20	23:41	12.0	30.00	68.00		7.0	Ms	G&R	AN2	India/Pakistan
1909	11	10	6:13	30.0	32.30	131.10	150	7.5	mB	UTSU	ABE1	
1909	11	21	7:36	0.0	25.50	122.00	5	7.3	UK	UTSU	B&D	
1909	12	9	15:33	0.0	-8.00	161.00		7.0	Ms	ABE	AN2	
1910	1	1	11:02	0.0	16.50	-84.00	60	7.1	Mw	G&R	P&S	

Year	M	d	h:min	Sec	Lat.	Long.	Dep.	Mag.	sc	icat	mdo	Region:Earthquake Name
1910	1	30	3:45	0.0	-22.00	170.00		7.0	Ms	ABE	AN2	
1910	2	4	14:00	0.0	-17.00	168.00		7.0	Ms	ABE	AN2	
1910	2	12	18:10	6.0	33.00	138.00	350	7.2	mB	UTSU	ABE1	
1910	3	30	16:55	48.0	-21.00	170.00	80	7.1	mB	G&R	ABE1	
1910	3	31	18:13	0.0	-71.00	-6.00		7.1	Ms	ABE	AN2	
1910	4	12	0:22	13.0	25.00	123.00	200	7.6	mB	UTSU	ABE1	
1910	4	20	22:22	0.0	-20.00	-177.00	330	7.0	mB	G&R	ABE1	
1910	5	1	18:30	36.0	-20.00	169.00	80	7.1	mB	G&R	ABE1	
1910	5	31	4:54	0.0	10.00	-105.00		7.0	Ms	ABE	AN2	
1910	6	1	5:55	30.0	-20.00	169.00	80	7.3	mB	G&R	ABE1	
1910	6	1	6:48	18.0	-20.00	169.00	80	7.1	mB	G&R	ABE1	
1910	6	16	6:30	42.0	-19.00	169.50	100	7.9	mB	G&R	ABE1	
1910	6	29	10:45	0.0	-32.00	-176.00		7.2	Mw	G&R	P&S	
1910	8	21	5:38	36.0	-17.00	-179.00	600	7.4	mB	G&R	ABE1	
1910	9	1	0:45	0.0	21.00	122.00		7.0	Ms	ABE	AN2	
1910	9	6	19:59	0.0	-25.00	-70.00		7.1	Mw	ABE	P&S	
1910	9	7	7:11	18.0	-6.00	151.00	80	7.2	mB	G&R	ABE1	
1910	9	9	1:13	18.0	51.50	-176.00	25	7.0	Ms	G&R	AN2	
1910	10	4	23:00	6.0	-22.00	-69.00	120	7.2	mB	G&R	ABE1	
1910	11	9	6:02	0.0	-16.00	166.00	70	7.5	mB	G&R	ABE1	
1910	11	10	12:19	54.0	-14.00	166.50	90	7.1	mB	G&R	ABE1	
1910	11	15	14:21	48.0	-58.00	-22.00	60	7.2	Mw	G&R	P&S	
1910	11	26	4:41	18.0	-14.00	167.00	50	7.2	Mw	G&R	P&S	
1910	12	10	9:26	42.0	-11.00	162.50	50	7.2	Mw	G&R	P&S	
1910	12	13	11:37	24.0	-8.00	31.00		7.6	Mw	G&R	P&S	
1910	12	14	20:46	12.0	-21.00	-178.00	600	7.2	mB	G&R	ABE1	
1910	12	16	14:45	0.0	4.50	126.50		7.6	Mw	G&R	P&S	
1911	1	3	23:25	45.0	43.50	77.50		7.8	Mw	G&R	P&S	Kazak./China:Tien Shan
1911	2	18	18:41	3.0	40.00	73.00		7.2	Mw	G&R	P&S	Tajikistan: Pamir
1911	4	4	15:43	54.0	36.50	25.50	140	7.0	mB	G&R	ABE1	
1911	4	10	18:42	24.0	9.00	-74.00	100	7.0	mB	G&R	ABE1	Iran
1911	5	4	23:36	54.0	51.00	157.00	240	7.4	mB	G&R	ABE1	
1911	6	7	11:02	42.0	17.50	-102.50		7.6	Mw	G&R	P&S	Mexico
1911	6	15	14:26	0.0	28.00	130.00	90	8.1	mB	UTSU	ABE1	
1911	7	4	13:33	26.0	36.00	70.50	190	7.4	mB	G&R	ABE1	
1911	7	5	18:40	6.0	-7.50	117.50	370	7.0	mB	G&R	ABE1	
1911	7	12	4:07	36.0	9.00	126.00		7.5	Mw	G&R	P&S	
1911	8	16	22:41	18.0	7.00	137.00		7.7	Mw	G&R	P&S	
1911	8	21	16:28	55.0	-21.00	-176.00	300	7.2	mB	G&R	ABE1	
1911	9	6	0:54	18.0	46.00	143.00	350	7.0	mB	UTSU	ABE1	
1911	9	15	13:10	0.0	-20.00	-72.00		7.1	Mw	G&R	P&S	
1911	9	17	3:26	0.0	51.00	180.00		7.1	Mw	ABE	P&S	
1911	10	20	17:44	0.0	-12.50	166.00	160	7.0	mB	G&R	ABE1	
1911	11	22	23:05	24.0	-15.00	169.00	200	7.3	mB	G&R	ABE1	
1911	12	16	19:14	18.0	17.00	-100.50	50	7.6	Mw	G&R	P&S	
1912	1	31	20:11	48.0	61.00	-147.50	80	7.0	mB	G&R	ABE1	
1912	5	6	19:00	0.0	64.00	-20.00		7.0	Ms	G&R	AN2	
1912	5	23	2:24	6.0	21.00	97.00		7.7	Mw	G&R	P&S	Burma
1912	7	7	7:57	36.0	64.00	-147.00		7.2	Mw	G&R	P&S	
1912	7	24	11:59	0.0	-5.00	-80.00		7.0	Ms	ABE	AN2	Peru

Year	M	d	h:min	sec	Lat.	Long.	Dep.	Mag.	sc	icat	mdo	Region:Earthquake Name
1912	8	6	21:11	18.0	-14.00	167.00	260	7.3	mB	G&R	ABE1	
1912	8	9	1:29	0.0	40.50	27.00		7.6	Mw	G&R	P&S	Turkey
1912	8	17	19:11	48.0	4.00	127.00		7.2	Mw	G&R	P&S	
1912	9	1	4:10	0.0	-4.50	155.00	430	7.0	mB	G&R	ABE1	
1912	9	29	20:51	30.0	7.00	138.00	50	7.5	Mw	G&R	P&S	
1912	10	26	9:00	36.0	14.00	146.00	130	7.0	mB	G&R	ABE1	
1912	11	7	7:40	24.0	57.50	-155.00	90	7.3	mB	G&R	ABE1	
1912	12	7	22:46	50.0	-29.00	-62.50	620	7.3	mB	G&R	ABE1	
1912	12	9	8:32	24.0	15.50	-93.00		7.1	Mw	ABE	P&S	
1913	1	11	13:16	54.0	1.50	122.00		7.2	Mw	G&R	P&S	
1913	2	20	8:58	48.0	41.00	144.00		7.0	Ms	ISS	AN1	
1913	3	14	8:45	0.0	4.50	126.50		7.9	Mw	G&R	P&S	
1913	3	23	20:47	18.0	24.00	142.00	80	7.1	mB	G&R	ABE1	
1913	4	25	17:56	8.0	9.50	128.80		7.2	Mw	ISS	P&S	
1913	5	18	2:08	53.0	14.50	145.50		7.0	Ms	ISS	AN1	
1913	5	30	11:46	46.0	-5.00	154.00		7.7	Mw	ISS	P&S	
1913	6	22	13:49	52.0	48.00	-178.00		7.2	UK	ISS	B&D	
1913	6	26	4:57	12.0	-20.00	-174.00		7.7	Mw	G&R	P&S	
1913	7	28	5:39	18.0	-17.00	-74.00		7.0	Ms	G&R	AN1	
1913	8	1	17:10	57.0	47.50	155.50		7.7	UK	ISS	B&D	
1913	8	6	22:14	24.0	-17.00	-74.00		7.8	Mw	G&R	P&S	
1913	8	13	4:25	42.0	-5.50	105.00	75	7.3	mB	G&R	ABE1	
1913	10	11	4:06	3.0	-7.00	148.00		7.0	Ms	ISS	AN1	
1913	10	14	8:08	48.0	-19.50	169.00	230	7.6	mB	G&R	ABE1	
1913	11	10	21:12	30.0	-18.00	169.00	80	7.2	mB	G&R	ABE1	
1913	11	15	5:27	6.0	-23.00	171.00	150	7.1	mB	G&R	ABE1	
1913	12	21	15:37	48.0	24.50	102.00		7.2	Mw	ISS	P&S	China: E'shan
1914	1	20	12:00	13.0	52.90	159.60	60	7.2	UK	B&D	B&D	
1914	1	30	3:36	0.0	-35.00	-73.00		7.5	Mw	G&R	P&S	
1914	2	26	4:58	12.0	-18.00	-67.00	130	7.2	mB	G&R	ABE1	
1914	3	14	20:00	0.0	39.50	140.40	5	7.0	Ms	UTSU	AN1	Japan: Senpoku-gun
1914	3	30	0:41	18.0	17.00	-92.00	150	7.2	mB	G&R	ABE1	
1914	4	11	16:30	24.0	-12.00	163.00	50	7.2	Mw	G&R	P&S	
1914	5	26	14:22	42.0	-2.00	137.00		7.9	Mw	G&R	P&S	West New Guinea
1914	6	20	7:20	30.0	-12.00	166.00	50	7.0	Ms	G&R	AN1	
1914	6	25	19:07	18.0	-4.50	102.50		7.6	Mw	G&R	P&S	Indonesia
1914	7	4	23:38	54.0	-5.50	129.00	200	7.1	mB	G&R	ABE1	
1914	8	4	22:41	36.0	43.50	91.50		7.2	Mw	G&R	P&S	
1914	10	3	17:22	12.0	16.00	-61.00	100	7.4	mB	G&R	ABE1	
1914	10	3	22:06	34.0	37.50	32.50		7.1	Mw	ISS	P&S	Turkey
1914	10	11	16:17	6.0	12.00	94.00	80	7.2	mB	G&R	ABE1	
1914	10	23	6:18	34.0	6.00	132.50		7.6	Mw	ISS	P&S	
1914	11	22	8:14	18.0	-39.00	176.00	100	7.0	mB	G&R	ABE1	
1914	11	24	11:53	30.0	22.00	143.00	110	7.9	mB	G&R	ABE1	
1915	1	5	14:33	15.0	-15.00	168.00	200	7.3	mB	G&R	ABE1	
1915	1	5	23:26	42.0	25.10	123.30	150	7.3	mB	UTSU	ABE1	
1915	2	25	20:36	12.0	-20.00	180.00	600	7.1	mB	G&R	ABE1	
1915	2	28	18:59	5.0	23.60	123.50		7.7	UK	ISS	B&D	
1915	3	17	18:45	0.0	42.10	143.60	35	7.2	mB	UTSU	ABE1	
1915	5	1	5:00	0.0	47.50	154.50	35	7.9	Mw	UTSU	P&S	Kurile Is.
1915	6	6	21:29	37.0	-18.50	-68.50	160	7.3	mB	G&R	ABE1	
1915	7	31	1:31	24.0	54.00	162.00		7.5	Mw	G&R	P&S	
1915	8	6	13:12	7.0	43.00	150.50		7.1	Mj	ISS	UTSU	
1915	9	7	1:20	48.0	14.00	-89.00	80	7.4	mB	G&R	ABE1	
1915	10	3	6:52	48.0	40.50	-117.50		7.6	Mw	G&R	P&S	Nevada:Pleasant Valley
1915	10	8	15:36	3.0	32.80	139.10	200	7.0	mB	UTSU	ABE1	
1915	11	1	7:24	0.0	38.30	142.90	35	7.5	Mw	UTSU	P&S	
1915	11	1	9:01	0.0	38.00	143.00	35	7.0	Mj	UTSU	UTSU	
1915	11	18	4:04	0.0	37.70	143.10	35	7.0	Mj	UTSU	UTSU	
1915	11	21	0:13	42.0	32.00	-115.00		7.1	Ms	G&R	AN1	
1915	12	3	2:39	19.0	29.50	91.50		7.0	Ms	ISS	AN1	China
1916	1	1	13:20	36.0	-4.00	154.00		7.7	Mw	G&R	P&S	
1916	1	13	6:18	30.0	-3.00	136.00		7.3	Mw	G&R	P&S	
1916	1	13	8:20	48.0	-3.00	135.50		7.6	Mw	G&R	P&S	
1916	1	24	6:55	7.0	41.00	37.00		7.2	Mw	ISS	P&S	
1916	2	1	7:36	22.0	29.50	131.50		7.4	Mj	ISS	UTSU	
1916	2	6	21:51	19.0	48.50	178.50		7.7	UK	ISS	B&D	
1916	2	27	20:20	48.0	12.00	-90.00		7.3	Mw	G&R	P&S	
1916	4	7	9:26	12.0	-30.00	55.00		7.2	Mw	G&R	P&S	
1916	4	18	4:01	48.0	53.25	-170.00	170	7.4	mB	G&R	ABE1	
1916	4	21	11:31	48.0	32.50	141.80	35	7.0	Ms	UTSU	AN1	
1916	4	24	4:26	42.0	18.50	-68.00	80	7.0	mB	G&R	ABE1	
1916	4	24	8:02	12.0	11.00	-85.00		7.2	Mw	G&R	P&S	
1916	4	26	2:21	30.0	10.00	-85.00		7.1	Ms	G&R	AN1	
1916	6	2	13:59	24.0	17.50	-95.00	150	7.0	mB	G&R	ABE1	
1916	6	21	21:32	30.0	-28.50	-63.00	600	7.4	mB	G&R	ABE1	
1916	7	8	9:34	30.0	-18.00	180.00	600	7.1	mB	G&R	ABE1	
1916	7	27	11:52	42.0	4.00	96.50	100	7.0	mB	G&R	ABE1	
1916	8	3	1:30	2.0	-4.00	144.50		7.1	Mw	ISS	P&S	
1916	8	25	9:44	42.0	-21.00	-68.00	180	7.0	mB	G&R	ABE1	
1916	8	28	6:39	42.0	30.00	81.00		7.2	Mw	G&R	P&S	
1916	8	28	7:27	5.0	23.90	120.50	5	7.2	Mj	UTSU	UTSU	
1916	9	11	6:30	36.0	-9.00	113.00	100	7.1	mB	G&R	ABE1	
1916	9	15	7:01	18.0	34.40	141.20	35	7.1	mB	UTSU	ABE1	
1916	10	3	1:26	13.0	-14.00	-74.50		7.1	Ms	ISS	AN1	
1916	10	31	15:30	33.0	45.40	154.00		7.6	Mw	P&S	P&S	
1917	1	30	2:45	36.0	56.50	163.00		7.7	Mw	G&R	P&S	
1917	2	20	19:29	48.0	19.50	-78.50		7.2	Mw	G&R	P&S	
1917	4	21	0:49	49.0	37.00	70.50	220	7.1	mB	G&R	ABE1	
1917	5	1	18:26	30.0	-29.00	-177.00		8.0	Mw	G&R	P&S	
1917	5	31	8:47	20.0	54.50	-160.00		7.5	Mw	ISS	P&S	
1917	6	26	5:49	42.0	-15.50	-173.00		8.5	Mw	G&R	P&S	Samoa Is.
1917	7	4	0:38	20.0	25.00	123.00		7.2	Mw	ISS	P&S	
1917	7	4	5:36	30.0	25.00	123.00		7.2	UK	ISS	B&D	
1917	7	27	1:01	18.0	19.00	-67.50	50	7.0	Ms	G&R	AN1	
1917	7	29	14:32	15.0	41.00	144.00		7.0	Ms	ISS	AN1	
1917	7	29	21:52	24.0	-3.50	141.00		7.6	Mw	G&R	P&S	
1917	7	30	23:54	5.0	29.00	104.00		7.3	Mw	ISS	P&S	
1917	7	31	3:23	10.0	42.00	131.00	500	7.4	mB	UTSU	ABE1	
1917	8	30	4:07	15.0	-7.50	128.00	100	7.3	mB	G&R	ABE1	

(continued)

TABLE 1 *(continued)*

Year	M	d	h:min	sec	Lat.	Long.	Dep.	Mag.	sc	icat	mdo	Region:Earthquake Name
1917	8	31	11:36	24.0	4.00	-74.00		7.1	Ms	G&R	AN1	
1917	11	4	12:03	30.0	4.80	96.80		7.1	Ms	ISS	AN1	
1917	11	16	3:19	30.0	-29.00	-177.50		7.3	Mw	G&R	P&S	
1917	12	29	22:50	20.0	15.00	-97.00		7.7	UK	ISS	B&D	
1918	1	30	21:18	36.4	45.45	136.74	330	7.4	mB	EHB	ABE1	
1918	2	7	5:20	33.9	6.78	126.69	218	7.2	mB	EHB	ABE1	
1918	2	13	6:07	14.0	23.54	117.24	15	7.2	Mw	EHB	P&S	China: Nan'ao
1918	4	10	2:03	53.8	43.48	130.92	565	7.0	mB	EHB	ABE1	
1918	5	20	14:36	5.3	7.34	-34.80	15	7.1	Mw	EHB	P&S	
1918	5	20	17:55	10.9	-28.90	-71.89	35	7.6	mB	EHB	ABE1	
1918	5	22	6:30	46.0	-17.47	-176.49	35	7.1	mB	EHB	ABE1	
1918	7	3	6:52	8.7	-3.47	143.22	35	7.2	Mw	EHB	P&S	
1918	7	8	10:22	7.8	24.81	90.72	15	7.5	Mw	EHB	P&S	
1918	8	15	12:18	16.5	5.65	123.56	35	8.2	Mw	EHB	P&S	Philippines: Mindanao
1918	9	7	17:15	43.9	46.81	150.25	242	7.6	mB	EHB	ABE1	Kurile Is.
1918	10	11	14:14	40.8	18.47	-67.63	35	7.3	Mw	EHB	P&S	Puerto Rico
1918	10	27	17:06	40.4	-1.12	150.28	35	7.2	Mw	EHB	P&S	
1918	11	8	4:38	10.8	43.82	152.77	64	7.5	mB	EHB	ABE1	
1918	11	18	18:41	44.3	-7.17	129.21	35	7.5	mB	EHB	ABE1	
1918	11	23	22:57	48.6	-8.28	126.57	218	7.1	mB	EHB	ABE1	
1918	12	4	11:47	39.0	-27.27	-73.46	25	7.5	Mw	EHB	P&S	
1919	1	1	1:33	50.1	7.18	126.94	35	7.1	Mw	EHB	P&S	
1919	1	1	3:00	4.3	-19.97	-177.91	203	7.7	mB	EHB	ABE1	
1919	3	2	3:26	51.4	-41.40	-72.22	15	7.2	Mw	EHB	P&S	
1919	3	2	11:45	16.2	-42.28	-76.45	15	7.1	Mw	EHB	P&S	
1919	4	30	7:17	12.5	-19.82	-172.21	35	8.2	Mw	EHB	P&S	Tonga Is.
1919	5	3	0:52	6.7	41.01	145.23	35	7.3	Mw	EHB	P&S	
1919	5	6	19:40	38.1	-5.48	152.63	233	7.6	mB	EHB	ABE1	
1919	6	1	6:51	4.8	27.07	123.32	35	7.1	mB	EHB	ABE1	
1919	8	18	16:55	22.3	-19.76	-178.11	215	7.0	mB	EHB	ABE1	
1919	8	29	5:44	5.5	-2.68	127.49	35	7.0	Ms	EHB	ABE1	
1919	8	31	17:20	35.1	-15.88	167.34	35	7.3	mB	EHB	ABE1	
1919	12	20	19:33	0.0	22.50	122.50	35	7.1	Mj	UTSU	UTSU	
1919	12	20	20:37	40.8	23.54	121.21	35	7.0	Ms	EHB	ABE1	
1920	2	2	11:22	14.1	-6.47	151.68	35	7.6	Mw	EHB	P&S	
1920	2	22	17:35	50.6	47.25	146.24	355	7.1	mB	EHB	ABE1	
1920	3	20	18:31	24.0	-35.70	-110.83	15	7.0	Ms	EHB	ABE1	
1920	6	5	4:21	35.4	23.81	122.08	35	7.9	Mw	EHB	P&S	Taiwan offshore
1920	9	20	14:39	5.9	-19.92	168.53	35	7.8	Mw	EHB	P&S	
1920	10	18	8:11	38.5	45.55	149.40	35	7.1	Ms	EHB	ABE1	
1920	12	10	4:25	41.0	-39.46	-74.99	35	7.2	Mw	EHB	P&S	
1920	12	16	12:05	54.7	36.60	105.32	25	8.3	Mw	EHB	P&S	China: Haiyuan
1921	2	4	8:22	43.7	15.41	-90.78	35	7.4	mB	EHB	ABE1	
1921	2	27	18:23	40.0	-18.65	-173.18	35	7.1	Mw	EHB	P&S	
1921	3	28	7:49	33.2	13.36	-87.36	35	7.2	Mw	EHB	P&S	
1921	4	2	9:36	0.0	23.00	123.00	35	7.2	Mj	UTSU	UTSU	
1921	7	4	14:18	12.5	28.89	129.91	35	7.4	mB	EHB	ABE1	
1921	9	11	4:01	26.0	-13.10	110.18	25	7.3	Mw	EHB	P&S	
1921	9	13	2:36	56.4	-56.26	-28.62	35	7.0	Ms	EHB	ABE1	
1921	10	20	6:03	20.6	-20.06	-68.96	114	7.2	mB	EHB	ABE1	

Year	M	d	h:min	Sec	Lat.	Long.	Dep.	Mag.	sc	icat	mdo	Region:Earthquake Name
1921	11	11	18:36	26.2	7.90	127.26	35	7.3	Mw	EHB	P&S	Philippines
1921	11	15	20:36	33.8	36.12	70.72	152	7.6	mB	EHB	ABE1	
1921	12	8	12:31	0.0	36.00	140.20	35	7.0	Mj	UTSU	UTSU	
1921	12	18	15:29	28.8	-4.04	-71.22	545	7.5	mB	EHB	ABE1	
1922	1	6	14:10	43.8	-20.41	-76.39	25	7.1	Mw	EHB	P&S	
1922	1	9	5:09	33.8	23.22	-45.93	15	7.0	Ms	EHB	ABE1	
1922	1	17	3:50	1.5	-6.48	-71.86	359	7.4	mB	EHB	ABE1	
1922	1	31	13:17	28.7	40.70	-125.55	15	7.2	Mw	EHB	P&S	Calif.: Cape Mendocino
1922	3	4	13:07	44.7	52.92	157.18	241	7.1	mB	EHB	ABE1	
1922	3	28	3:58	1.3	-21.45	-68.13	136	7.1	mB	EHB	ABE1	
1922	9	1	19:16	9.2	24.51	122.04	35	7.5	Mw	EHB	P&S	
1922	9	14	19:31	42.5	24.38	122.64	35	7.1	Mw	EHB	P&S	
1922	10	11	14:50	6.1	-16.12	-72.39	160	7.6	mB	EHB	ABE1	
1922	10	24	21:21	3.4	47.27	152.19	35	7.3	mB	EHB	ABE1	
1922	11	7	23:00	15.5	-28.44	-72.19	25	7.1	Ms	EHB	ABE1	
1922	11	11	4:32	45.2	-28.55	-70.75	35	8.7	Mw	EHB	P&S	Chile
1922	12	6	13:55	41.0	36.44	70.94	240	7.3	mB	EHB	ABE1	
1922	12	31	7:20	11.4	45.74	150.80	35	7.0	Ms	EHB	ABE1	
1923	1	22	9:04	19.5	40.49	-125.32	15	7.1	Mw	EHB	P&S	
1923	2	2	5:07	42.7	54.02	161.52	35	7.1	Mw	EHB	P&S	
1923	2	3	16:01	48.8	53.85	160.76	35	8.5	Mw	EHB	P&S	Kamchatka
1923	2	24	7:34	44.2	55.94	162.62	35	7.2	Mw	EHB	P&S	
1923	3	2	16:48	44.6	7.49	124.93	87	7.1	Mw	EHB	P&S	
1923	3	16	22:01	43.7	6.49	127.06	35	7.0	Ms	EHB	ABE1	
1923	3	24	12:40	19.9	30.55	101.26	25	7.2	Mw	EHB	P&S	China: Luhuo-Dawu
1923	4	13	15:31	0.3	55.42	162.59	35	7.1	Mw	EHB	P&S	
1923	5	4	16:26	46.6	55.76	-156.99	25	7.1	Ms	EHB	ABE1	
1923	6	1	17:24	46.5	35.67	141.77	35	7.1	Mw	EHB	P&S	
1923	6	1	20:14	0.0	36.00	142.00	35	7.1	Mj	UTSU	UTSU	
1923	6	22	6:44	38.5	22.59	98.68	25	7.2	Mw	EHB	P&S	
1923	7	13	11:13	41.2	31.00	130.90	35	7.1	Ms	EHB	ABE1	
1923	9	1	2:58	37.0	35.40	139.08	35	7.9	Mw	EHB	P&S	Japan: Kanto
1923	9	1	3:03	0.0	35.10	139.50	35	7.3	Mj	UTSU	UTSU	
1923	9	1	3:48	0.0	35.40	139.80	35	7.0	Mj	UTSU	UTSU	
1923	9	2	2:46	46.5	34.90	140.20	35	7.6	Mw	EHB	P&S	
1923	9	2	9:27	0.0	34.90	140.50	35	7.1	Mj	UTSU	UTSU	
1923	9	9	22:03	53.2	24.94	90.32	35	7.1	Ms	EHB	ABE1	Bangladesh/India
1923	10	7	3:29	42.1	-1.20	129.49	35	7.2	Mw	EHB	P&S	
1923	11	2	21:08	11.5	-4.65	154.23	145	7.2	mB	EHB	ABE1	
1923	11	5	21:28	2.2	29.65	129.64	35	7.1	Mw	EHB	P&S	
1924	1	14	20:50	0.0	35.50	139.20	5	7.3	Mj	UTSU	UTSU	
1924	1	16	21:37	59.2	-22.03	-177.91	350	7.0	mB	EHB	ABE1	
1924	3	4	10:07	52.5	9.56	-83.84	35	7.0	Ms	EHB	ABE1	
1924	4	14	16:20	41.4	7.02	125.95	35	8.2	Mw	EHB	P&S	Philippines: Mindanao
1924	5	4	16:51	51.4	-21.50	-178.44	562	7.2	mB	EHB	ABE1	
1924	5	6	16:09	31.3	15.94	118.95	35	7.0	Ms	EHB	ABE1	
1924	5	28	9:52	0.5	48.38	145.75	432	7.0	mB	EHB	ABE1	
1924	6	26	1:37	32.5	-56.41	158.49	15	8.3	Mw	EHB	P&S	
1924	6	30	15:44	25.8	44.74	147.42	140	7.2	mB	EHB	ABE1	
1924	7	3	4:40	14.8	36.63	83.90	35	7.1	Mw	EHB	P&S	China: Minfeng

Year	M	d	h:min	sec	Lat.	Long.	Dep.	Mag.	sc	icat	mdo	Region:Earthquake Name
1924	7	11	19:44	45.2	36.79	83.96	35	7.1	Mw	EHB	P&S	
1924	7	24	4:55	22.1	-49.86	159.82	35	7.0	Ms	EHB	ABE1	
1924	8	14	18:02	39.8	36.13	141.89	35	7.0	Ms	EHB	ABE1	
1924	8	30	3:05	2.5	8.65	126.53	35	7.2	Mw	EHB	P&S	
1924	10	13	16:17	46.1	36.52	70.83	179	7.2	mB	EHB	ABE1	
1924	12	27	11:21	53.6	44.16	146.37	35	7.0	mB	EHB	ABE1	
1924	12	28	22:55	4.8	43.40	146.58	35	7.0	Ms	EHB	ABE1	
1925	1	18	12:05	57.2	47.91	153.63	35	7.2	Mw	EHB	P&S	
1925	2	20	1:02	0.0	45.50	150.50	35	7.0	Mj	UTSU	UTSU	
1925	3	16	14:42	18.8	25.69	100.49	25	7.0	Ms	EHB	ABE1	China: Dali
1925	3	22	8:42	0.3	-18.58	168.53	35	7.1	Mw	EHB	P&S	
1925	4	16	19:52	43.6	21.80	121.11	35	7.1	Ms	EHB	ABE1	
1925	5	3	17:21	47.5	1.22	126.58	35	7.1	Ms	EHB	ABE1	
1925	5	3	22:59	6.2	-34.11	57.95	15	7.0	Ms	EHB	ABE1	
1925	5	15	11:57	5.3	-25.50	-70.55	35	7.1	Ms	EHB	ABE1	
1925	6	3	4:33	59.6	1.47	126.19	35	7.0	Ms	EHB	ABE1	
1925	6	9	13:40	22.9	-4.14	143.10	35	7.0	Ms	EHB	ABE1	
1925	8	19	12:07	33.7	55.13	167.68	35	7.0	Ms	EHB	ABE1	
1925	10	13	17:40	37.7	10.60	-42.13	15	7.0	Ms	EHB	ABE1	
1925	11	10	13:50	40.9	-0.63	130.27	35	7.2	Mw	EHB	P&S	
1925	11	13	12:14	53.4	12.73	124.85	35	7.2	Mw	EHB	P&S	
1925	11	16	11:55	1.4	18.37	-106.81	25	7.0	Ms	EHB	ABE1	
1926	1	25	0:36	29.3	-9.99	159.03	35	7.2	Mw	EHB	P&S	
1926	2	8	15:17	57.4	12.01	-88.76	35	7.1	Mw	EHB	P&S	
1926	3	21	14:19	14.3	-60.63	-23.67	35	7.1	Ms	EHB	ABE1	
1926	3	27	10:48	21.1	-8.56	160.55	35	7.1	Mw	EHB	P&S	
1926	4	12	8:32	34.1	-10.71	160.63	35	7.2	Mw	EHB	P&S	
1926	4	28	11:13	39.3	-22.67	-68.16	35	7.1	mB	EHB	ABE1	
1926	6	3	4:47	3.8	-15.05	168.19	77	7.1	mB	EHB	ABE1	
1926	6	26	19:46	41.0	36.60	26.89	102	7.7	mB	EHB	ABE1	Greece
1926	6	29	14:27	3.8	26.99	127.47	142	7.4	mB	EHB	ABE1	
1926	8	25	5:44	47.9	-23.01	171.82	35	7.0	Ms	EHB	ABE1	
1926	8	30	11:38	9.7	36.90	23.00	85	7.1	mB	EHB	ABE1	
1926	9	10	10:34	27.8	-9.15	110.70	35	7.0	mB	EHB	ABE1	
1926	10	3	19:38	1.7	-49.81	161.70	15	7.3	Mw	EHB	P&S	
1926	10	13	19:08	16.9	51.77	-175.36	35	7.0	Ms	EHB	ABE1	
1926	10	26	3:44	43.6	-3.22	139.10	35	7.5	Mw	EHB	P&S	
1926	11	5	7:55	30.1	12.67	-86.74	35	7.1	mB	EHB	ABE1	
1927	1	24	1:05	47.5	-18.39	168.23	35	7.1	Ms	EHB	ABE1	
1927	2	16	1:35	27.7	46.62	153.07	35	7.0	Ms	EHB	ABE1	
1927	3	3	1:05	14.9	-6.62	121.81	35	7.0	Ms	EHB	ABE1	
1927	3	7	9:27	42.1	35.80	134.92	10	7.1	Mw	EHB	P&S	Japan: Tango
1927	4	1	19:05	38.5	-19.80	-177.48	35	7.0	mB	EHB	ABE1	
1927	4	14	6:23	30.3	-32.48	-69.68	35	7.2	mB	EHB	ABE1	
1927	4	19	17:30	11.5	15.25	120.45	139	7.1	mB	EHB	ABE1	
1927	5	22	22:32	48.0	37.39	102.31	25	7.7	Mw	EHB	P&S	China: Gulang
1927	6	3	7:12	18.0	-7.07	130.96	169	7.2	mB	EHB	ABE1	
1927	7	1	8:18	58.5	36.38	22.72	35	7.0	mB	EHB	ABE1	
1927	8	5	21:12	58.8	37.87	142.19	35	7.1	Ms	EHB	ABE1	
1927	8	10	11:36	15.7	-0.53	131.67	35	7.1	Mw	EHB	P&S	

Year	M	d	h:min	Sec	Lat.	Long.	Dep.	Mag.	sc	icat	mdo	Region:Earthquake Name
1927	10	24	15:59	56.2	57.44	-136.37	15	7.1	Ms	EHB	ABE1	
1927	11	4	13:51	4.7	34.92	-121.03	15	7.0	Ms	EHB	ABE1	Calif.: Point Arguello
1927	11	21	23:12	30.9	-45.11	-73.62	15	7.1	Mw	EHB	P&S	
1927	11	26	12:54	0.4	-24.32	-66.96	179	7.0	mB	EHB	ABE1	
1927	12	28	18:20	34.2	55.66	160.04	35	7.5	Mw	EHB	P&S	
1928	3	9	18:05	27.8	-2.95	88.80	15	7.7	Mw	EHB	P&S	
1928	3	13	18:31	35.1	-7.21	153.87	35	7.0	mB	EHB	ABE1	
1928	3	16	5:01	10.2	-22.28	170.48	35	7.5	Mw	EHB	P&S	
1928	3	22	4:17	8.4	16.13	-96.50	35	7.5	Mw	EHB	P&S	
1928	3	29	5:06	8.4	31.68	137.73	434	7.0	mB	EHB	ABE1	
1928	5	14	22:14	51.5	-5.26	-78.56	35	7.2	Mw	EHB	P&S	
1928	5	27	9:50	27.5	39.81	143.00	35	7.1	Ms	EHB	ABE1	
1928	6	15	6:12	42.3	12.73	120.82	35	7.0	Ms	EHB	ABE1	
1928	6	17	3:19	33.0	16.03	-97.04	35	7.7	Mw	EHB	P&S	
1928	6	29	22:49	45.5	-15.66	170.70	35	7.1	Ms	EHB	ABE1	
1928	8	4	18:26	20.0	16.42	-98.27	35	7.2	Mw	EHB	P&S	Indonesia
1928	8	12	8:08	44.6	1.83	127.24	35	7.0	mB	EHB	ABE1	
1928	8	24	21:43	31.1	-14.29	168.50	220	7.0	mB	EHB	ABE1	
1928	10	9	3:01	14.2	16.23	-97.55	35	7.5	Mw	EHB	P&S	
1928	12	1	4:06	15.2	-35.09	-71.68	35	7.7	Mw	EHB	P&S	Chile
1928	12	19	11:37	17.2	6.98	124.86	35	7.3	Mw	EHB	P&S	Philippines
1929	1	13	0:03	3.2	50.45	154.87	35	7.4	mB	EHB	ABE1	
1929	2	1	17:14	26.0	36.46	70.93	208	7.0	mB	EHB	ABE1	
1929	2	22	20:41	49.8	10.48	-41.88	15	7.0	Ms	EHB	ABE1	
1929	3	7	1:34	42.8	50.79	-169.52	25	7.8	Mw	EHB	P&S	
1929	5	1	15:37	37.1	37.96	57.69	25	7.1	Mw	EHB	P&S	Iran
1929	5	26	22:40	2.6	51.24	-130.56	15	7.0	Ms	EHB	ABE1	
1929	6	4	15:15	24.4	6.02	125.64	35	7.1	mB	EHB	ABE1	
1929	6	13	9:24	38.7	7.95	126.81	35	7.1	Mw	EHB	P&S	
1929	6	16	22:47	32.2	-41.83	172.29	35	7.5	Mw	EHB	P&S	NZ: West Nelson
1929	6	27	12:47	14.9	-54.71	-29.55	35	8.0	Mw	EHB	P&S	
1929	7	5	14:19	8.7	51.42	-178.14	35	7.0	Ms	EHB	ABE1	
1929	7	7	21:23	16.5	51.35	-177.91	35	7.2	Mw	EHB	P&S	
1929	10	19	10:12	51.4	-23.04	-68.74	35	7.4	mB	EHB	ABE1	
1929	11	15	18:50	32.2	7.52	142.74	25	7.1	Ms	EHB	ABE1	
1929	11	18	20:32	0.8	44.54	-56.01	15	7.4	Mw	EHB	P&S	Atlantic: Grand Banks
1929	12	17	10:58	44.3	53.78	171.51	35	7.7	Mw	EHB	P&S	
1930	1	5	1:19	48.7	49.96	153.58	108	7.0	mB	EHB	ABE1	
1930	5	5	13:45	59.8	17.67	96.54	35	7.2	Mw	EHB	P&S	Myanmar
1930	5	6	22:34	27.8	38.15	44.69	25	7.1	Mw	EHB	P&S	Iran
1930	6	11	0:49	42.6	-5.87	148.89	35	7.1	Ms	EHB	ABE1	
1930	7	2	21:03	44.6	25.64	90.25	35	7.1	Ms	EHB	ABE1	
1930	10	24	20:15	12.9	18.35	146.57	35	7.0	Ms	EHB	ABE1	
1930	12	3	18:51	49.7	17.97	96.42	35	7.3	Mw	EHB	P&S	
1930	12	21	14:51	12.9	19.99	121.58	35	7.0	mB	EHB	ABE1	
1931	1	15	1:50	47.1	16.05	-96.61	35	7.8	Mw	EHB	P&S	Mexico
1931	1	27	20:09	22.1	25.67	96.75	35	7.6	Mw	EHB	P&S	
1931	1	28	21:24	10.9	10.78	144.80	35	7.1	Ms	EHB	ABE1	
1931	2	2	22:46	51.3	-39.77	176.02	35	7.7	Mw	EHB	P&S	NZ: Hawke's Bay
1931	2	10	6:34	33.0	-5.43	102.46	35	7.1	Mw	EHB	P&S	

(continued)

TABLE 1 *(continued)*

Year	M	d	h:min	sec	Lat.	Long.	Dep.	Mag.	sc	icat	mdo	Region:Earthquake Name
1931	2	13	1:27	22.6	-39.48	176.81	35	7.1	Ms	EHB	ABE1	
1931	2	20	5:33	27.3	44.43	135.64	357	7.4	mB	EHB	ABE1	
1931	3	2	2:18	25.2	-21.83	171.15	35	7.0	mB	EHB	ABE1	
1931	3	9	3:48	56.8	40.48	142.66	35	7.7	Mw	EHB	P&S	
1931	3	18	8:02	23.2	-33.83	-71.55	35	7.1	Ms	EHB	ABE1	
1931	3	28	12:38	48.1	-7.37	129.49	151	7.2	mB	EHB	ABE1	
1931	5	20	2:22	53.8	37.47	-16.07	25	7.1	Ms	EHB	ABE1	
1931	8	7	2:11	43.4	-2.86	142.08	35	7.1	Mw	EHB	P&S	
1931	8	10	21:18	47.7	46.57	89.96	35	7.9	Mw	EHB	P&S	China: Fuyun
1931	8	18	14:21	8.0	47.29	90.12	35	7.2	Mw	EHB	P&S	
1931	8	27	15:27	24.6	29.47	67.17	35	7.1	Mw	EHB	P&S	Pakistan
1931	9	9	20:38	28.4	19.15	145.49	172	7.1	mB	EHB	ABE1	
1931	9	25	5:59	52.2	-5.18	102.51	35	7.3	Mw	EHB	P&S	
1931	10	3	19:13	20.8	-10.93	161.02	35	7.8	Mw	EHB	P&S	
1931	10	3	21:55	14.7	-12.68	163.19	35	7.0	Ms	EHB	ABE1	
1931	10	3	22:47	47.5	-10.91	161.40	35	7.1	Mw	EHB	P&S	Solomon Is.
1931	10	10	0:20	1.5	-9.97	161.19	50	7.7	Mw	EHB	P&S	
1931	11	2	10:03	3.5	32.00	131.95	42	7.5	Mw	EHB	P&S	
1932	1	9	10:21	48.2	-6.08	154.45	388	7.2	mB	EHB	ABE1	
1932	1	29	13:41	16.3	-6.81	155.04	54	7.1	Ms	EHB	ABE1	
1932	5	14	13:11	5.7	0.26	126.17	35	8.1	Mw	EHB	P&S	Molucca Passage
1932	5	26	16:09	39.3	-25.40	179.05	569	7.5	mB	EHB	ABE1	
1932	6	3	10:36	53.6	19.46	-104.15	25	7.9	Mw	EHB	P&S	Mexico
1932	6	18	10:12	16.1	19.45	-103.63	54	7.9	Mw	EHB	P&S	Mexico
1932	8	14	4:39	39.3	25.76	95.65	144	7.0	mB	EHB	ABE1	
1932	9	23	14:22	15.1	44.49	138.72	350	7.0	mB	EHB	ABE1	
1932	11	13	4:46	28.9	43.84	136.88	35	7.1	mB	EHB	ABE1	
1932	12	4	8:11	20.9	2.39	121.02	46	7.1	Ms	EHB	ABE1	
1932	12	21	6:10	9.3	38.51	-118.08	15	7.1	Mw	EHB	P&S	Nevada: Cedar Mountain
1932	12	25	2:04	32.0	39.77	96.69	25	7.6	Mw	EHB	P&S	China: Changma
1933	1	1	8:48	45.9	-14.75	167.50	147	7.0	mB	EHB	ABE1	
1933	1	21	19:21	14.1	-33.51	57.73	15	7.1	Ms	EHB	ABE1	
1933	2	23	8:09	20.8	-20.30	-69.96	35	7.2	Mw	EHB	P&S	
1933	3	2	17:31	0.9	39.22	144.62	35	8.4	Mw	EHB	P&S	Japan: Sanriku-oki
1933	3	17	15:55	30.1	54.86	161.60	35	7.0	Ms	EHB	ABE1	
1933	6	18	21:37	37.7	38.28	142.51	35	7.2	Mw	EHB	P&S	
1933	6	24	21:54	49.5	-5.52	104.43	35	7.3	Mw	EHB	P&S	Indonesia
1933	8	25	7:50	32.5	31.81	103.54	25	7.3	Mw	EHB	P&S	China: Diexi
1933	8	28	22:19	46.7	-59.44	-26.64	35	7.2	Mw	EHB	P&S	
1933	9	6	22:08	31.5	-21.95	-179.24	587	7.0	mB	EHB	ABE1	
1933	10	25	23:28	15.7	-23.69	-66.65	207	7.0	mB	EHB	ABE1	
1933	11	20	23:21	35.3	73.00	-70.12	15	7.1	Mw	EHB	P&S	
1933	12	4	19:33	58.9	46.19	144.59	353	7.1	mB	EHB	ABE1	
1934	1	15	8:43	25.4	26.77	86.76	35	8.0	Mw	EHB	P&S	India: Bihar-Nepal
1934	2	14	3:59	41.8	17.40	119.19	35	7.5	Mw	EHB	P&S	
1934	2	24	6:23	46.8	22.88	144.07	35	7.2	Mw	EHB	P&S	
1934	2	28	14:21	55.7	-5.23	149.48	35	7.1	Mw	EHB	P&S	
1934	3	1	21:45	19.5	-40.92	-72.88	35	7.3	mB	EHB	ABE1	
1934	3	5	11:46	19.4	-40.64	175.74	35	7.3	Mw	EHB	P&S	NZ: Pahiatua
1934	3	24	12:04	34.2	-10.06	160.80	35	7.1	Ms	EHB	ABE1	
1934	4	15	22:15	19.5	7.53	126.86	35	7.1	Mw	EHB	P&S	
1934	5	4	4:36	10.6	61.52	-147.60	35	7.1	mB	EHB	ABE1	
1934	6	13	22:10	23.0	27.43	62.59	35	7.0	mB	EHB	ABE1	
1934	6	29	8:26	14.7	-7.08	123.17	648	7.0	mB	EHB	ABE1	
1934	7	18	1:36	28.2	8.05	-82.48	25	7.6	Mw	EHB	P&S	
1934	7	18	19:40	24.0	-11.91	166.73	35	7.8	Mw	EHB	P&S	Santa Cruz Is.
1934	7	19	1:27	32.9	-0.79	133.35	35	7.0	Ms	EHB	ABE1	
1934	7	21	6:18	23.3	-11.14	165.50	35	7.1	Mw	EHB	P&S	
1934	8	12	23:49	23.5	8.03	126.78	35	7.0	Ms	EHB	ABE1	
1934	10	10	15:42	9.9	-23.75	179.78	530	7.2	mB	EHB	ABE1	
1934	11	30	2:05	18.7	18.68	-105.32	25	7.0	Ms	EHB	ABE1	
1934	12	15	1:57	43.6	31.01	89.08	35	7.1	Ms	EHB	ABE1	
1934	12	31	18:45	56.8	32.69	-115.76	15	7.1	Ms	EHB	ABE1	
1935	1	1	13:20	35.2	-15.72	-174.53	35	7.0	mB	EHB	ABE1	
1935	2	22	17:06	1.3	53.27	175.02	35	7.1	Ms	EHB	ABE1	
1935	4	19	15:23	23.6	31.24	15.30	15	7.1	Ms	EHB	ABE1	
1935	4	20	22:02	2.9	24.36	120.61	35	7.1	Ms	EHB	ABE1	Taiwan:Hsinchu-Taichu.
1935	5	30	21:32	56.8	28.89	66.18	35	8.1	Mw	EHB	P&S	Pakistan: Quetta
1935	6	24	23:23	18.5	-15.55	167.87	142	7.1	mB	EHB	ABE1	
1935	7	29	7:38	51.2	-21.54	-177.92	455	7.1	mB	EHB	ABE1	
1935	8	3	1:10	7.2	4.43	96.33	35	7.0	Ms	EHB	ABE1	
1935	8	17	1:44	37.9	-22.10	171.65	35	7.2	mB	EHB	ABE1	
1935	9	4	1:37	46.3	22.26	121.26	35	7.1	Mw	EHB	P&S	
1935	9	9	6:17	38.8	6.23	141.72	35	7.0	Ms	EHB	ABE1	
1935	9	11	14:04	3.2	43.29	146.63	35	7.2	Mw	EHB	P&S	
1935	9	20	1:46	42.7	-3.92	141.33	35	8.1	Mw	EHB	P&S	
1935	9	20	5:23	10.0	-3.58	142.90	35	7.0	Ms	EHB	ABE1	
1935	10	12	16:45	25.3	40.27	143.32	35	7.1	Mw	EHB	P&S	
1935	10	18	0:11	59.6	40.39	144.16	35	7.1	Mw	EHB	P&S	
1935	12	14	22:05	25.3	14.72	-92.48	35	7.2	Mw	EHB	P&S	
1935	12	15	7:07	55.0	-9.59	161.15	35	7.5	Mw	EHB	P&S	
1935	12	17	19:17	36.0	22.52	125.75	25	7.1	Mw	EHB	P&S	
1935	12	28	2:35	30.0	-0.34	98.15	35	7.8	Mw	EHB	P&S	
1936	1	2	22:34	28.5	-0.28	98.87	35	7.0	mB	EHB	ABE1	
1936	1	20	16:56	18.7	6.03	127.13	35	7.0	mB	EHB	ABE1	
1936	2	15	12:47	2.2	-4.36	133.29	35	7.1	Mw	EHB	P&S	
1936	4	1	2:09	27.7	4.16	126.52	35	7.7	Mw	EHB	P&S	
1936	4	19	5:07	19.6	-7.67	156.47	35	7.2	Mw	EHB	P&S	
1936	5	27	6:19	17.6	28.34	83.28	35	7.0	Ms	EHB	ABE1	
1936	6	10	8:23	21.0	-5.50	147.00	190	7.0	mB	ABE	ABE1	
1936	6	30	15:06	45.0	51.27	160.88	35	7.2	Mw	EHB	P&S	
1936	7	5	18:55	14.3	5.86	127.02	35	7.3	mB	EHB	ABE1	
1936	7	13	11:12	17.5	-24.72	-70.02	35	7.1	Mw	EHB	P&S	
1936	8	22	11:09	19.7	21.95	121.19	35	7.2	Mw	EHB	P&S	
1936	8	23	21:12	13.9	5.30	94.76	35	7.1	Ms	EHB	ABE1	Indonesia
1936	9	19	1:01	49.1	3.65	97.55	35	7.1	Mw	EHB	P&S	
1936	10	5	9:44	25.6	1.63	126.24	35	7.0	Ms	EHB	ABE1	
1936	11	2	20:45	59.1	38.34	142.15	35	7.1	Mw	EHB	P&S	
1936	11	13	12:31	34.1	56.04	163.46	35	7.1	Ms	EHB	ABE1	
1937	1	7	13:20	41.2	35.40	97.67	15	7.6	Mw	EHB	P&S	

Year	M	d	h:min	sec	Lat.	Long.	Dep.	Mag.	sc	icat	mdo	Region:Earthquake Name
1937	1	23	10:55	58.5	-5.29	152.76	35	7.0	Ms	EHB	ABE1	
1937	1	25	6:34	5.1	-10.85	162.79	35	7.1	Ms	EHB	ABE1	
1937	2	21	7:02	44.5	44.24	149.60	56	7.3	Mw	EHB	P&S	
1937	4	16	3:01	38.0	-20.77	-177.14	349	7.5	mB	EHB	ABE1	
1937	6	21	15:13	5.9	-8.44	-79.85	42	7.1	Ms	EHB	ABE1	
1937	7	2	2:37	16.7	-13.86	167.05	35	7.0	mB	EHB	ABE1	
1937	7	19	19:35	28.0	-1.98	-76.40	184	7.1	mB	EHB	ABE1	
1937	7	22	17:09	35.3	64.49	-146.85	35	7.2	Mw	EHB	P&S	
1937	7	26	3:47	9.4	18.52	-95.88	35	7.2	mB	EHB	ABE1	
1937	7	26	19:56	34.5	38.40	142.17	35	7.1	mB	EHB	ABE1	
1937	8	11	0:55	48.3	-9.68	117.53	35	7.2	mB	EHB	ABE1	
1937	8	20	11:59	22.3	14.32	121.55	35	7.3	Mw	EHB	P&S	
1937	9	3	18:48	11.0	52.45	-177.40	35	7.2	mB	EHB	ABE1	
1937	9	8	0:40	14.8	-56.54	-26.91	219	7.1	mB	EHB	ABE1	
1937	9	15	12:27	32.0	-10.50	161.50	80	7.2	mB	EHB	ABE1	
1937	9	27	8:55	16.5	-9.03	110.80	35	7.0	Ms	EHB	ABE1	
1937	11	14	10:58	13.1	36.57	70.67	200	7.1	mB	EHB	ABE1	
1937	12	8	8:32	15.0	22.87	121.01	39	7.0	Ms	EHB	ABE1	
1937	12	23	13:18	6.2	17.43	-98.29	35	7.4	Mw	EHB	P&S	
1938	1	24	10:31	50.1	-60.55	-35.58	15	7.0	Ms	EHB	ABE1	
1938	2	1	19:04	21.6	-5.05	131.62	35	8.4	Mw	EHB	P&S	Banda Sea
1938	5	12	15:39	5.8	-6.27	147.80	35	7.3	Mw	EHB	P&S	
1938	5	19	17:08	36.1	-0.37	119.53	49	7.5	Mw	EHB	P&S	
1938	5	23	7:18	32.5	36.46	141.76	35	7.7	Mw	EHB	P&S	
1938	5	23	8:21	54.1	18.05	119.99	35	7.0	mB	EHB	ABE1	
1938	5	30	14:29	52.2	-19.88	169.16	35	7.0	mB	EHB	ABE1	
1938	6	9	19:15	11.4	-3.14	126.93	25	7.0	Ms	EHB	ABE1	
1938	6	10	9:53	44.6	25.21	125.11	35	7.6	Mw	EHB	P&S	
1938	6	16	2:15	19.7	27.58	129.54	35	7.2	Mw	EHB	P&S	
1938	8	16	4:27	56.7	22.96	93.88	35	7.1	Mw	EHB	P&S	
1938	9	7	4:03	23.7	23.74	121.41	47	7.0	Ms	EHB	ABE1	
1938	10	10	20:48	10.6	2.39	126.55	35	7.2	Mw	EHB	P&S	
1938	10	20	2:19	24.5	-9.27	123.24	35	7.3	mB	EHB	ABE1	
1938	11	5	8:43	22.6	37.01	142.04	35	7.9	Mw	EHB	P&S	
1938	11	5	10:50	16.8	37.11	142.08	35	7.8	Mw	EHB	P&S	
1938	11	6	8:53	55.0	37.29	142.28	35	7.7	Mw	EHB	P&S	
1938	11	6	21:38	46.2	36.87	142.62	35	7.0	Ms	EHB	ABE1	
1938	11	10	20:18	47.0	55.33	-158.37	35	8.0	Mw	EHB	P&S	Alaska
1938	11	13	22:31	32.9	36.73	142.36	35	7.0	Ms	EHB	ABE1	
1938	11	17	3:54	40.5	55.22	-157.55	35	7.2	Mw	EHB	P&S	
1938	12	6	23:00	55.6	22.61	121.40	39	7.0	Ms	EHB	ABE1	
1939	1	25	3:32	0.0	-36.20	-72.20		7.7	Mw	P&S	P&S	Chile: Chillan
1939	1	30	2:18	30.0	-7.08	155.39	35	7.7	Mw	EHB	P&S	
1939	2	3	5:26	26.5	-9.48	159.36	35	7.1	Ms	EHB	ABE1	
1939	3	2	7:00	20.1	-4.38	143.15	35	7.0	mB	EHB	ABE1	
1939	3	21	1:11	13.6	-1.28	89.40	15	7.0	Ms	EHB	ABE1	
1939	4	5	16:42	45.0	-19.71	168.98	35	7.0	mB	EHB	ABE1	
1939	4	18	6:22	43.8	-27.14	-70.71	35	7.3	mB	EHB	ABE1	
1939	4	30	2:55	29.1	-9.30	159.23	35	7.9	Mw	EHB	P&S	Solomon Is.
1939	5	1	5:58	32.6	39.84	139.97	35	7.0	Ms	EHB	ABE1	

Year	M	d	h:min	Sec	Lat.	Long.	Dep.	Mag.	sc	icat	mdo	Region:Earthquake Name
1939	6	8	20:46	51.4	-15.27	-173.93	35	7.0	mB	EHB	ABE1	
1939	8	12	2:07	29.2	-16.06	168.15	172	7.0	mB	EHB	ABE1	
1939	10	10	18:32	0.6	38.67	143.15	35	7.2	Mw	EHB	P&S	
1939	10	17	6:22	11.4	-14.81	167.29	101	7.3	mB	EHB	ABE1	
1939	12	16	10:46	31.2	43.62	147.80	35	7.1	mB	EHB	ABE1	
1939	12	21	20:54	53.3	9.99	-84.55	35	7.2	Mw	EHB	P&S	
1939	12	21	21:00	33.2	-0.21	122.57	35	7.8	mB	EHB	ABE1	
1939	12	26	23:57	22.6	39.77	39.53	35	7.7	Mw	EHB	P&S	Turkey: Erzincan
1940	1	6	14:03	29.7	-21.69	170.78	80	7.2	mB	EHB	ABE1	
1940	1	17	1:14	58.7	17.24	148.22	31	7.3	mB	EHB	ABE1	
1940	2	20	2:18	23.6	-13.78	167.14	191	7.0	mB	EHB	ABE1	
1940	4	16	6:43	9.6	52.82	173.35	35	7.1	Ms	EHB	ABE1	
1940	5	19	4:36	48.9	33.22	-115.70	15	7.1	Ms	EHB	ABE1	Calif.:Imperial Valley
1940	5	24	16:33	58.1	-11.12	-77.63	50	7.5	Mw	EHB	P&S	Peru
1940	5	28	9:40	47.8	-2.32	139.15	35	7.0	Ms	EHB	ABE1	
1940	7	10	5:49	58.6	44.39	130.78	578	7.3	mB	EHB	ABE1	
1940	7	14	5:52	54.4	52.12	178.16	35	7.4	mB	EHB	ABE1	
1940	8	1	15:08	26.4	44.51	139.83	35	7.5	Mw	EHB	P&S	
1940	8	22	3:27	18.0	53.00	-165.50	60	7.0	Ms	G&R	ABE1	
1940	9	12	13:17	10.0	-4.50	153.00	40	7.1	Ms	G&R	ABE1	
1940	9	19	18:19	48.0	-24.00	171.00	80	7.0	mB	G&R	ABE1	
1940	9	22	22:51	56.0	8.00	124.00	680	7.0	mB	ABE	ABE1	
1940	10	4	7:54	50.1	-20.59	-70.72	24	7.1	mB	EHB	ABE1	
1940	10	11	18:41	13.7	-42.04	-73.93	15	7.0	Ms	EHB	ABE1	
1940	11	10	1:39	8.4	45.77	26.66	122	7.3	mB	EHB	ABE1	Romania
1940	12	28	16:37	46.6	18.14	147.53	35	7.3	mB	EHB	ABE1	
1941	1	13	16:27	40.1	-4.72	152.32	35	7.0	Ms	EHB	ABE1	
1941	2	4	14:03	19.8	10.08	124.00	602	7.0	mB	EHB	ABE1	
1941	4	3	15:21	30.7	-22.58	-66.64	150	7.2	mB	EHB	ABE1	
1941	4	7	23:29	18.7	17.26	-78.59	15	7.0	Ms	EHB	ABE1	
1941	4	15	19:09	57.0	18.68	-102.96	35	7.6	Mw	EHB	P&S	Mexico
1941	5	17	2:24	58.5	-10.46	166.14	35	7.2	Mw	EHB	P&S	
1941	6	26	11:52	6.6	12.15	92.48	49	7.7	Mw	EHB	P&S	
1941	8	2	11:41	29.1	-29.77	-177.61	35	7.1	Ms	EHB	ABE1	
1941	9	4	10:21	40.9	-5.07	153.69	35	7.1	mB	EHB	ABE1	
1941	9	18	13:14	9.7	-13.93	-72.29	61	7.0	mB	EHB	ABE1	
1941	11	8	23:37	28.4	0.72	122.86	35	7.2	Mw	EHB	P&S	
1941	11	15	4:19	57.3	-59.80	-26.99	35	7.1	mB	EHB	ABE1	
1941	11	18	10:14	41.7	-60.76	-55.37	35	7.1	Ms	EHB	ABE1	
1941	11	18	16:46	0.0	32.00	132.00		7.7	Mw	P&S	P&S	
1941	11	24	21:46	21.9	-28.28	-177.61	35	7.0	mB	EHB	ABE1	
1941	11	25	18:03	58.7	37.17	-18.96	25	8.1	Mw	EHB	P&S	North Atlantic
1941	12	5	20:47	4.4	8.75	-83.16	35	7.3	Mw	EHB	P&S	
1941	12	6	21:24	44.7	8.14	-84.40	35	7.0	Ms	EHB	ABE1	
1941	12	16	19:19	45.7	23.25	120.39	35	7.1	Mw	EHB	P&S	Taiwan: Chiayi
1941	12	26	14:48	3.1	21.08	99.14	25	7.0	Ms	EHB	ABE1	
1942	1	27	13:29	11.3	-4.32	134.84	18	7.1	Ms	EHB	ABE1	
1942	1	29	9:23	48.9	-19.32	169.04	118	7.0	mB	EHB	ABE1	
1942	3	5	19:48	19.0	44.27	141.63	259	7.0	mB	EHB	ABE1	
1942	4	8	15:40	28.8	12.85	120.55	41	7.3	Mw	EHB	P&S	

(continued)

TABLE 1 *(continued)*

Year	M	d	h:min	sec	Lat.	Long.	Dep.	Mag.	sc	icat	mdo	Region:Earthquake Name
1942	5	14	2:13	28.3	0.01	-79.90	35	7.8	Mw	EHB	P&S	Peru/Ecuador
1942	5	28	1:01	55.0	-0.07	123.78	104	7.4	mB	EHB	ABE1	
1942	6	14	3:09	58.0	14.55	148.01	49	7.0	mB	EHB	ABE1	
1942	6	18	9:31	0.9	9.12	140.10	35	7.0	Ms	EHB	ABE1	
1942	6	24	11:16	30.7	-41.53	175.63	35	7.0	Ms	EHB	ABE1	NZ: Wairarapa
1942	7	8	6:55	43.1	-24.60	-70.19	35	7.0	mB	EHB	ABE1	
1942	8	6	23:37	0.4	13.78	-90.91	35	7.7	Mw	EHB	P&S	
1942	8	24	22:50	31.7	-14.98	-74.92	35	7.7	Mw	EHB	P&S	Peru
1942	10	20	23:21	52.2	7.73	122.60	35	7.2	Mw	EHB	P&S	
1942	11	10	11:41	28.1	-49.73	29.95	15	8.0	Mw	EHB	P&S	
1942	11	26	14:27	31.5	45.40	150.18	76	7.4	mB	EHB	ABE1	
1942	11	28	10:38	46.6	7.28	-35.64	15	7.1	Ms	EHB	ABE1	
1942	12	20	14:03	11.1	40.67	36.45	35	7.2	Mw	EHB	P&S	Turkey: Anatolia
1943	2	22	9:20	45.0	17.75	-101.50		7.4	Mw	G&R	P&S	
1943	2	28	12:54	33.0	36.50	70.50	210	7.1	mB	G&R	ABE1	
1943	3	9	9:48	55.0	-60.00	-27.00		7.1	Mw	G&R	P&S	
1943	3	14	18:37	56.0	-20.00	-69.50	150	7.1	mB	G&R	ABE1	
1943	3	21	20:35	43.0	-5.75	152.25		7.1	Ms	G&R	ABE1	
1943	3	25	18:27	15.0	-60.00	-27.00		7.0	Ms	G&R	ABE1	
1943	4	1	14:18	8.0	-6.50	105.50		7.1	Ms	G&R	ABE1	
1943	4	6	16:07	15.0	-30.75	-72.00		8.2	Mw	G&R	P&S	
1943	5	2	17:18	9.0	6.50	-80.00		7.0	Ms	G&R	ABE1	
1943	5	3	1:59	12.0	12.50	125.50		7.2	Mw	G&R	P&S	
1943	5	25	23:07	36.0	7.50	128.00		7.6	Mw	G&R	P&S	
1943	6	8	20:42	46.0	-1.00	101.00	50	7.2	Mw	G&R	P&S	
1943	6	9	3:06	22.0	-1.00	101.00	50	7.5	Mw	G&R	P&S	
1943	6	13	5:11	49.0	42.75	143.25	60	7.1	Mw	G&R	P&S	
1943	7	23	14:53	9.0	-9.50	110.00	90	7.6	mB	G&R	ABE1	Indonesia
1943	7	29	3:02	16.0	19.25	-67.50		7.6	Mw	G&R	P&S	
1943	9	6	3:41	30.0	-53.00	159.00		7.6	Mw	G&R	P&S	
1943	9	10	8:36	53.0	35.25	134.00		7.0	Mw	G&R	P&S	Japan: Tottori
1943	9	14	2:01	12.0	-22.00	171.00	50	7.2	Mw	G&R	P&S	
1943	9	14	3:47	15.0	-22.00	170.00	50	7.1	Ms	G&R	ABE1	
1943	9	14	7:18	8.0	-30.00	-177.00	60	7.4	Ms	G&R	ABE1	
1943	9	27	22:03	44.0	-30.00	-178.00	90	7.0	mB	G&R	ABE1	
1943	10	23	17:23	16.0	26.00	93.00		7.1	Mw	G&R	P&S	
1943	11	2	18:08	22.0	-57.00	-26.00		7.2	Mw	G&R	P&S	
1943	11	3	14:32	17.0	61.75	-151.00		7.2	Mw	G&R	P&S	
1943	11	6	8:31	37.0	-6.00	134.50		7.6	Mw	G&R	P&S	
1943	11	13	18:43	57.0	-19.00	170.00		7.2	Mw	G&R	P&S	
1943	11	26	21:25	22.0	-2.50	100.00	130	7.1	mB	G&R	ABE1	
1943	11	26	22:20	36.0	41.00	34.00		7.5	Mw	G&R	P&S	Turkey: Anatolia
1943	12	1	6:04	55.0	-4.75	144.00	120	7.3	mB	G&R	ABE1	
1943	12	1	10:34	46.0	-19.50	-69.75	80	7.1	mB	G&R	ABE1	
1943	12	23	19:00	10.0	-5.50	153.50	50	7.2	Mw	G&R	P&S	
1944	1	7	2:49	20.0	-4.50	143.50	120	7.0	mB	G&R	ABE1	
1944	1	15	23:49	30.0	-31.25	-68.75	50	7.1	Mw	G&R	P&S	Argentina: San Juan
1944	2	1	3:22	36.0	41.50	32.50		7.2	Mw	G&R	P&S	Turkey: Anatolia
1944	2	29	3:41	53.0	-14.50	-70.50	200	7.1	mB	G&R	ABE1	
1944	2	29	16:28	7.0	0.50	76.00		7.1	Ms	G&R	ABE1	

Year	M	d	h:min	Sec	Lat.	Long.	Dep.	Mag.	sc	icat	mdo	Region:Earthquake Name
1944	3	9	22:12	58.0	44.00	84.00		7.1	Ms	G&R	ABE1	
1944	3	22	0:43	18.0	-8.50	123.50	220	7.2	mB	G&R	ABE1	
1944	3	31	2:51	43.0	-7.00	130.50	60	7.0	Ms	G&R	ABE1	
1944	4	27	14:38	9.0	-0.50	133.50	50	7.2	Mw	G&R	P&S	
1944	5	19	0:19	19.0	-2.50	152.75	50	7.0	Ms	G&R	ABE1	
1944	5	25	1:06	37.0	-21.50	-179.50	640	7.0	mB	G&R	ABE1	
1944	5	25	12:58	5.0	-2.50	152.75		7.3	Mw	G&R	P&S	
1944	6	21	10:58	20.0	-22.00	169.00	50	7.0	Ms	G&R	ABE1	
1944	6	28	7:58	54.0	15.00	-92.50		7.1	Ms	G&R	ABE1	
1944	7	27	0:04	23.0	54.00	-165.50	70	7.1	mB	G&R	ABE1	
1944	9	23	12:13	20.0	54.00	160.00	40	7.1	Mw	G&R	P&S	
1944	10	5	17:28	27.0	-22.50	172.00	120	7.3	mB	G&R	ABE1	
1944	11	15	20:47	1.0	4.50	127.50		7.0	Ms	G&R	ABE1	
1944	11	16	12:10	58.0	-12.50	167.00		7.2	Mw	G&R	P&S	
1944	11	24	4:49	3.0	-19.00	169.00	170	7.4	mB	G&R	ABE1	
1944	11	29	18:51	21.0	-19.00	169.00	170	7.1	mB	G&R	ABE1	
1944	12	7	4:35	42.0	33.75	136.00		8.1	Mw	G&R	P&S	Japan: Tonanakai
1944	12	10	16:24	58.0	-18.00	168.00	50	7.0	Ms	G&R	ABE1	
1945	2	1	10:35	51.0	-22.00	170.00	60	7.0	mB	G&R	ABE1	
1945	2	1	12:13	40.0	-22.00	170.00	60	7.2	mB	G&R	ABE1	
1945	2	10	4:57	56.0	41.25	142.50	50	7.1	Ms	G&R	ABE1	
1945	3	23	23:14	13.0	-62.00	153.00		7.0	Ms	G&R	ABE1	
1945	4	15	2:35	22.0	57.00	164.00		7.1	Mw	G&R	P&S	
1945	7	15	5:35	13.0	17.50	146.50	120	7.1	mB	G&R	ABE1	
1945	8	29	10:22	40.0	-15.00	168.00	50	7.1	Ms	G&R	ABE1	
1945	9	1	22:44	10.0	-46.50	165.50		7.1	Ms	G&R	ABE1	
1945	9	5	21:48	45.0	-5.00	153.50	50	7.1	Ms	G&R	ABE1	
1945	9	13	11:17	11.0	-33.25	-70.50	100	7.0	mB	G&R	ABE1	
1945	10	9	14:36	33.0	43.50	147.50	80	7.0	mB	G&R	ABE1	
1945	11	27	21:56	50.0	24.50	63.00		8.0	Mw	G&R	P&S	Iran/Pakistan
1945	12	8	1:04	2.0	-6.50	151.00		7.1	Ms	G&R	ABE1	
1945	12	28	17:48	45.0	-6.00	150.00		7.6	Mw	G&R	P&S	
1946	1	5	19:57	20.0	-16.00	167.00	50	7.1	Mw	G&R	P&S	
1946	1	17	9:39	35.0	-7.50	147.50	100	7.2	mB	G&R	ABE1	
1946	4	1	12:28	54.0	52.75	-163.50		8.0	Mw	G&R	P&S	Aleutians
1946	4	11	1:52	20.0	-1.00	-14.50		7.2	Mw	G&R	P&S	
1946	5	3	22:23	43.0	-6.00	154.00		7.2	Mw	G&R	P&S	
1946	5	8	5:20	22.0	0.00	99.50		7.1	Ms	G&R	ABE1	
1946	6	23	17:13	22.0	49.75	-124.50		7.6	Mw	G&R	P&S	
1946	7	9	13:13	50.0	-19.00	169.00	170	7.2	mB	G&R	ABE1	
1946	8	2	19:18	48.0	-26.50	-70.50	50	7.1	Ms	G&R	ABE1	
1946	8	4	17:51	5.0	19.25	-69.00		7.9	Mw	G&R	P&S	Dominica/Puerto Rico
1946	8	8	13:28	28.0	19.50	-69.50		7.5	Mw	G&R	P&S	
1946	8	28	22:28	15.0	-26.00	-63.00	580	7.1	mB	G&R	ABE1	
1946	9	12	15:17	15.0	23.50	96.00		7.3	Mw	G&R	P&S	
1946	9	12	15:20	20.0	23.50	96.00		7.7	Mw	G&R	P&S	
1946	9	23	23:30	0.0	-6.00	145.00	100	7.1	mB	G&R	ABE1	
1946	9	29	3:01	55.0	-4.50	153.50		7.6	Mw	G&R	P&S	
1946	11	1	11:14	24.0	51.50	-174.50	40	7.0	Ms	G&R	ABE1	
1946	11	2	18:28	25.0	41.50	72.50		7.3	Mw	G&R	P&S	

Year	M	d	h:min	sec	Lat.	Long.	Dep.	Mag.	sc	icat	mdo	Region:Earthquake Name
1946	11	4	21:47	47.0	39.75	54.50		7.1	Mw	G&R	P&S	Turkmenistan/Kazandzh.
1946	11	12	17:28	41.0	-20.00	-173.50		7.1	Mw	G&R	P&S	
1946	11	28	15:51	35.0	-18.50	-174.00	290	7.0	mB	G&R	ABE1	
1946	12	20	19:19	5.0	32.50	134.50		8.1	Mw	G&R	P&S	Japan: Nankai
1946	12	21	10:18	49.0	44.00	149.00		7.2	Mw	G&R	P&S	
1947	1	26	10:06	46.0	12.50	-86.25	170	7.0	mB	G&R	ABE1	
1947	1	29	8:17	50.0	-26.00	-63.00	580	7.0	mB	G&R	ABE1	
1947	3	17	8:19	32.0	33.00	99.50		7.5	Mw	G&R	P&S	
1947	3	25	20:32	14.0	-38.75	178.50		7.0	Ms	G&R	ABE1	
1947	4	2	5:39	11.0	-1.50	138.00		7.2	Mw	G&R	P&S	
1947	4	14	7:15	33.0	44.00	148.50		7.1	Ms	G&R	ABE1	
1947	5	6	20:30	32.0	-6.50	148.50		7.3	Mw	G&R	P&S	
1947	5	27	5:58	54.0	-1.50	135.25		7.2	Mw	G&R	P&S	
1947	6	12	9:02	30.0	1.50	126.50	40	7.1	Ms	G&R	ABE1	
1947	7	29	13:43	22.0	28.50	94.00		7.3	Mw	G&R	P&S	
1947	9	26	16:01	57.0	24.75	123.00	110	7.4	mB	G&R	ABE1	
1947	10	16	2:09	47.0	64.50	-147.50		7.1	Mw	G&R	P&S	
1947	11	1	14:58	53.0	-10.50	-75.00		7.7	Mw	G&R	P&S	Peru
1947	11	4	0:09	10.0	44.00	140.50		7.1	Ms	G&R	ABE1	
1948	1	24	17:46	40.0	10.50	122.00		8.1	Mw	G&R	P&S	Philippines
1948	1	27	11:58	28.0	-20.50	-178.00	630	7.0	mB	G&R	ABE1	
1948	1	28	3:47	21.0	1.50	126.50	80	7.1	mB	G&R	ABE1	
1948	3	1	1:12	28.0	-3.00	127.50	50	7.1	Ms	G&R	ABE1	
1948	3	3	9:09	54.0	18.50	119.00		7.1	Mw	G&R	P&S	
1948	4	17	16:11	28.0	33.00	135.75		7.2	Mw	G&R	P&S	
1948	4	21	20:22	2.0	19.25	-69.25	40	7.1	Ms	G&R	ABE1	
1948	5	11	8:55	41.0	-17.50	-70.25	70	7.4	mB	G&R	ABE1	
1948	5	14	22:31	43.0	54.50	-161.00		7.3	Mw	G&R	P&S	
1948	5	25	7:11	21.0	29.50	100.50		7.2	Mw	G&R	P&S	China: Litang
1948	6	28	7:13	30.0	36.50	136.00		7.0	Mw	G&R	P&S	Japan: Fukui
1948	7	20	11:02	17.0	-17.00	-75.00	70	7.1	mB	G&R	ABE1	
1948	9	2	23:34	50.0	10.00	125.50		7.0	Ms	G&R	ABE1	
1948	9	8	15:09	11.0	-21.00	-174.00		8.0	Mw	G&R	P&S	
1948	10	5	20:12	5.0	37.50	58.00		7.2	Mw	G&R	P&S	Turkmenistan:Ashkhabad
1948	11	19	1:04	24.0	10.00	-83.50	80	7.0	mB	G&R	ABE1	
1948	11	26	5:36	37.0	-5.00	145.00	70	7.0	mB	G&R	ABE1	
1949	2	23	16:08	8.0	41.00	83.50		7.3	Mw	G&R	P&S	
1949	3	4	10:19	25.0	36.00	70.50	230	7.4	mB	G&R	ABE1	
1949	3	27	6:34	5.0	3.50	127.50		7.1	Mw	G&R	P&S	
1949	4	20	3:29	7.0	-38.00	-73.50	70	7.1	mB	G&R	ABE1	Chile
1949	4	23	11:15	39.0	-8.00	121.00	80	7.0	mB	G&R	ABE1	
1949	4	25	13:54	59.0	-19.75	-69.00	110	7.2	mB	G&R	ABE1	
1949	4	30	1:23	32.0	6.50	125.00	130	7.3	mB	G&R	ABE1	
1949	7	10	3:53	36.0	39.00	70.50		7.6	Mw	G&R	P&S	Tajikistan: Khait
1949	7	23	10:26	45.0	-18.50	170.00	150	7.2	mB	G&R	ABE1	
1949	8	6	0:35	37.0	-18.50	-174.50	70	7.2	mB	G&R	ABE1	
1949	8	22	4:01	11.0	53.75	-133.25		8.0	Mw	G&R	P&S	Queen Charlotte Is.
1949	10	19	21:00	19.0	-5.50	154.00	60	7.5	Mw	G&R	P&S	
1949	11	22	0:51	49.0	-28.50	-178.50	180	7.1	mB	G&R	ABE1	
1949	12	17	6:53	30.0	-54.00	-71.00		7.8	Mw	G&R	P&S	

Year	M	d	h:min	Sec	Lat.	Long.	Dep.	Mag.	sc	icat	mdo	Region:Earthquake Name
1949	12	17	15:07	55.0	-54.00	-71.00		7.8	Mw	G&R	P&S	
1949	12	29	3:03	54.0	18.00	121.00		7.2	Mw	G&R	P&S	
1950	2	2	23:33	39.0	22.00	100.00		7.1	Mw	G&R	P&S	
1950	2	28	10:20	57.0	46.00	144.00	340	7.5	mB	G&R	ABE1	
1950	5	25	18:35	7.0	13.00	143.50	90	7.1	mB	G&R	ABE1	
1950	7	9	4:40	4.0	-8.00	-70.75	650	7.0	mB	G&R	ABE1	
1950	7	9	4:50	5.0	-8.00	-70.75	650	7.0	mB	G&R	ABE1	
1950	7	9	16:10	24.0	36.70	70.50	223	7.5	Ms	ISS	BJI	
1950	7	29	16:46	6.0	2.20	126.90	96	7.0	mB	ISS	ABE1	
1950	8	14	22:51	24.0	-27.25	-62.50	630	7.2	mB	G&R	ABE1	
1950	8	15	14:09	30.0	28.50	96.50		8.6	Mw	G&R	P&S	India: Assam
1950	8	31	7:05	47.0	5.50	126.00	96	7.0	mB	ISS	ABE1	
1950	9	10	15:16	8.0	-15.50	167.00	100	7.1	mB	G&R	ABE1	
1950	10	5	16:09	31.0	11.00	-85.00		7.8	Mw	G&R	P&S	
1950	10	8	3:23	9.0	-3.75	128.25		7.5	Mw	G&R	P&S	
1950	10	23	16:13	20.0	14.50	-91.50		7.5	Mw	G&R	P&S	
1950	11	2	15:27	56.0	-6.50	129.50	50	7.4	mB	G&R	ABE1	
1950	11	8	2:18	12.0	-10.00	159.50		7.1	Mw	G&R	P&S	
1950	12	2	19:51	49.0	-18.25	167.50	60	7.2	Mw	G&R	P&S	
1950	12	2	19:55	27.0	-18.25	167.50	60	7.2	UK	G&R	G&R	
1950	12	4	16:28	3.0	-5.00	153.50	110	7.3	mB	G&R	ABE1	
1950	12	9	21:38	48.0	-23.50	-67.50	100	7.7	mB	G&R	ABE1	
1950	12	10	2:50	42.0	-14.25	-75.75	80	7.1	mB	G&R	ABE1	
1950	12	10	13:23	4.0	-28.00	-178.50	250	7.2	mB	G&R	ABE1	
1950	12	14	1:52	49.0	-19.25	-175.75	200	7.5	mB	G&R	ABE1	
1950	12	14	14:15	51.0	17.00	-97.50		7.3	Mw	G&R	P&S	
1951	2	13	22:12	57.0	56.00	-156.00		7.2	Mw	G&R	P&S	
1951	2	17	21:07	7.0	-7.00	146.00	180	7.2	mB	G&R	ABE1	
1951	3	10	21:57	29.0	-15.00	167.50	130	7.2	mB	G&R	ABE1	
1951	5	21	8:27	20.0	-6.00	154.50	150	7.0	mB	G&R	ABE1	
1951	6	12	22:40	39.0	36.30	71.00	223	7.5	Ms	ISS	BJI	
1951	10	21	21:34	14.0	23.75	121.50		7.5	Mw	G&R	P&S	Taiwan: Hualien
1951	10	22	3:29	27.0	23.75	121.25		7.2	Mw	G&R	P&S	
1951	10	22	4:28	5.0	23.90	121.70		7.0	Ms	ISS	BJI	
1951	10	22	5:43	1.0	24.00	121.25		7.1	Mw	G&R	P&S	
1951	11	6	16:40	5.0	47.75	154.25		7.2	Mw	G&R	P&S	
1951	11	18	9:35	47.0	30.50	91.00		7.7	Mw	G&R	P&S	Tibet
1951	11	24	18:50	18.0	23.00	122.50		7.3	Mw	G&R	P&S	
1951	12	8	4:14	12.0	-34.00	57.00		7.5	Mw	G&R	P&S	
1952	2	14	3:38	12.0	-7.50	126.50		7.2	Mw	G&R	P&S	
1952	3	4	1:22	43.0	42.50	143.00		8.1	Mw	G&R	P&S	Japan: Tokachi-Oki
1952	3	4	1:40	4.0	42.20	143.90		7.1	Ms	ISS	BJI	
1952	3	9	17:03	47.0	42.50	143.00		7.2	Mw	G&R	P&S	
1952	3	19	10:57	12.0	9.50	127.25		7.7	Mw	G&R	P&S	
1952	7	21	11:52	14.0	35.00	-119.00		7.3	Mw	G&R	P&S	Calif.: Kern County
1952	8	17	16:02	7.0	30.50	91.50		7.7	Mw	G&R	P&S	China: Damxung
1952	9	21	2:30	35.0	-21.75	-65.75	260	7.1	mB	G&R	ABE1	
1952	11	4	16:58	26.0	52.75	159.50		9.0	Mw	G&R	P&S	Russia: Kamchatka
1952	11	6	19:47	21.0	-4.60	144.90		7.1	Mw	ISS	P&S	
1952	12	6	10:41	18.0	-8.00	156.50		7.2	Mw	G&R	P&S	

(continued)

TABLE 1 *(continued)*

Year	M	d	h:min	sec	Lat.	Long.	Dep.	Mag.	sc	icat	mdo	Region:Earthquake Name
1952	12	24	18:39	38.0	-5.50	152.00		7.1	Mw	G&R	P&S	
1953	1	5	7:48	20.0	53.00	171.50		7.2	Mw	ISS	P&S	
1953	1	5	10:06	30.0	49.00	155.50	40	7.1	Mw	ABE	P&S	
1953	2	26	11:42	27.0	-11.20	163.90		7.1	Mw	ISS	P&S	
1953	3	18	19:06	13.0	40.00	27.30		7.2	Mw	ISS	P&S	Turkey: Anatolia
1953	3	19	8:27	52.0	14.00	-61.20	128	7.1	mB	ISS	ABE1	
1953	4	23	16:24	22.0	-4.50	153.30		7.6	Mw	ISS	P&S	
1953	5	6	17:16	50.0	-36.50	-72.60	64	7.5	mB	ISS	ABE1	
1953	7	2	6:56	59.0	-19.00	169.00	223	7.4	mB	ISS	ABE1	
1953	8	12	9:23	52.0	38.30	20.80		7.2	Mw	ISS	P&S	Greece
1953	9	29	1:36	46.0	-36.90	177.10	287	7.0	mB	ISS	ABE1	
1953	11	4	3:49	7.0	-13.00	166.40		7.5	Mw	ISS	P&S	
1953	11	10	23:40	23.0	51.20	157.10	33	7.0	mB	ISS	ABE1	
1953	11	25	17:48	54.0	34.00	141.70	33	7.9	Mw	UTSU	P&S	
1953	12	7	2:05	24.0	-22.10	-68.70	128	7.2	mB	ISS	ABE1	
1953	12	12	17:31	25.0	-3.40	-80.60		7.5	Mw	ISS	P&S	
1954	2	11	0:30	18.0	38.80	101.20	33	7.1	Mw	ISS	P&S	
1954	2	20	18:35	5.0	-7.20	124.70	540	7.0	mB	ISS	ABE1	
1954	2	22	12:03	35.0	-56.50	-26.30	96	7.0	mB	ISS	ABE1	
1954	3	21	23:42	17.0	24.20	95.10	223	7.4	mB	ISS	ABE1	
1954	3	29	6:17	5.0	37.00	-3.50	603	7.0	mB	ISS	ABE1	
1954	3	31	18:25	45.0	12.40	57.90		7.1	Mw	ISS	P&S	
1954	4	29	11:34	36.0	29.20	-112.80		7.1	Mw	ISS	P&S	
1954	9	17	11:03	18.0	-21.30	-176.70	223	7.0	mB	ISS	ABE1	
1954	12	16	11:07	12.0	39.20	-118.00		7.1	Mw	ISS	P&S	Nevada: Fairview Peak
1955	1	5	0:50	18.0	-49.70	162.70		7.1	Mw	ISS	P&S	
1955	2	27	20:43	24.0	-28.30	-175.50		7.8	Mw	ISS	P&S	
1955	3	18	0:06	50.0	54.30	161.00	64	7.2	mB	ISS	ABE1	
1955	3	31	18:17	19.0	8.10	123.20	96	7.3	mB	ISS	ABE1	Philippines
1955	4	14	1:29	0.0	30.00	101.70		7.5	Mw	ISS	P&S	China: Kangding
1955	4	15	3:40	52.0	39.90	74.70		7.1	Mw	ISS	P&S	
1955	5	17	14:49	49.0	6.70	93.70		7.2	Mw	ISS	P&S	
1955	5	30	12:31	43.0	24.20	142.50	572	7.1	mB	ISS	ABE1	
1955	8	6	8:31	24.0	-21.10	-177.50	287	7.0	mB	ISS	ABE1	
1955	10	10	8:57	46.0	-5.10	152.80		7.3	Mw	ISS	P&S	
1955	10	13	9:26	49.0	-10.00	160.70	33	7.1	mB	ISS	ABE1	
1955	11	10	1:44	5.0	-15.60	-173.60	64	7.2	Ms	ISS	ROTHE	
1956	2	1	13:41	52.7	18.85	145.06	380	7.0	mB	EHB	ABE1	
1956	2	18	7:34	24.7	30.17	138.27	480	7.1	mB	EHB	ABE1	
1956	5	23	20:48	29.7	-15.52	-178.59	403	7.2	mB	EHB	ABE1	
1956	6	9	23:13	56.4	35.03	67.48	35	7.6	Mw	EHB	P&S	Afghanistan
1956	7	9	3:11	45.2	36.58	26.04	35	7.8	Mw	EHB	P&S	Greece
1956	7	16	15:07	8.9	22.08	95.82	25	7.1	Mw	EHB	P&S	
1956	7	18	6:19	32.6	-5.05	130.36	106	7.2	mB	EHB	ABE1	
1956	10	11	2:24	37.7	45.90	150.61	101	7.3	mB	EHB	ABE1	
1956	10	24	14:42	17.0	11.62	-86.44	35	7.2	Mw	EHB	P&S	
1956	12	27	0:14	11.5	-23.41	-176.88	196	7.0	mB	EHB	ABE1	
1957	2	23	20:26	17.2	23.87	121.49	50	7.2	Mw	EHB	P&S	
1957	3	9	14:22	33.3	51.59	-175.42	35	8.6	Mw	EHB	P&S	Aleutian Is
1957	3	9	20:39	23.6	52.47	-169.48	31	7.2	Mw	EHB	P&S	
1957	3	11	9:58	51.4	52.54	-168.90	36	7.1	Mw	EHB	P&S	
1957	3	12	11:44	59.8	51.74	-176.66	32	7.1	Mw	EHB	P&S	
1957	3	14	14:47	52.0	51.33	-176.62	35	7.2	Mw	EHB	P&S	
1957	3	16	2:34	18.5	51.52	-178.78	32	7.1	Mw	EHB	P&S	
1957	3	22	14:21	13.8	53.69	-165.69	42	7.1	Mw	EHB	P&S	
1957	3	23	5:12	42.3	-5.49	130.91	127	7.1	mB	EHB	ABE1	
1957	4	14	19:18	5.5	-15.40	-173.14	35	7.5	Mw	EHB	P&S	
1957	4	16	4:04	9.6	-4.51	107.27	602	7.2	mB	EHB	ABE1	
1957	4	25	2:25	41.2	36.35	28.70	35	7.1	Mw	EHB	P&S	Turkey/Greece
1957	5	26	6:33	35.3	40.67	31.04	17	7.2	Mw	EHB	P&S	Turkey
1957	6	22	23:50	30.5	-2.01	136.68	35	7.3	Mw	EHB	P&S	
1957	6	27	0:09	32.5	56.42	116.44	15	7.4	Mw	EHB	P&S	
1957	7	2	0:42	28.5	36.07	52.69	35	7.1	Mw	EHB	P&S	Iran
1957	7	14	6:24	2.0	-27.14	-177.74	201	7.0	mB	EHB	ABE1	
1957	7	28	8:40	10.3	16.89	-99.29	41	7.8	Mw	EHB	P&S	Mexico
1957	9	24	8:21	12.5	5.24	127.12	35	7.7	Mw	EHB	P&S	
1957	9	28	14:20	3.0	-20.69	-178.16	587	7.2	mB	EHB	ABE1	
1957	11	29	22:19	42.9	-20.54	-67.23	159	7.4	mB	EHB	ABE1	
1957	12	4	3:37	51.0	45.18	99.22	25	8.1	Mw	EHB	P&S	Mongolia: Gobi-Altai
1957	12	17	13:50	22.4	-12.47	166.67	124	7.2	mB	EHB	ABE1	
1958	1	15	19:14	30.9	-16.78	-72.49	69	7.0	mB	EHB	ABE1	
1958	1	19	14:07	24.8	0.99	-79.49	19	7.8	Mw	EHB	P&S	
1958	3	11	0:25	53.0	23.81	124.16	35	7.2	mB	EHB	ABE1	
1958	4	7	15:30	44.1	65.89	-156.34	15	7.3	Mw	EHB	P&S	
1958	5	31	19:32	34.2	-15.32	168.54	12	7.1	Mw	EHB	P&S	
1958	7	10	6:15	58.2	58.47	-136.28	15	7.8	Mw	EHB	P&S	Alaska
1958	7	26	17:37	14.4	-13.28	-69.41	615	7.2	mB	EHB	ABE1	
1958	11	6	22:58	9.4	44.31	148.65	35	8.4	Mw	EHB	P&S	Kurile Is.
1958	11	12	20:23	32.9	44.05	148.82	35	7.0	Mw	EHB	P&S	
1959	1	22	5:10	31.0	37.56	142.48	32	7.2	Mw	EHB	P&S	
1959	3	1	16:49	14.5	-0.99	134.17	44	7.0	mB	EHB	ABE1	
1959	4	26	20:40	37.9	24.68	122.77	118	7.5	mB	EHB	ABE1	
1959	5	4	7:15	46.1	53.37	159.66	35	8.0	Mw	EHB	P&S	
1959	5	24	19:17	41.7	17.46	-97.18	70	7.0	mB	EHB	ABE1	
1959	6	14	0:11	57.9	-20.27	-68.83	35	7.5	mB	EHB	ABE1	
1959	7	19	15:06	16.2	-15.09	-70.33	210	7.1	mB	EHB	ABE1	
1959	8	15	8:57	5.7	21.98	120.95	27	7.2	Mw	EHB	P&S	
1959	8	17	21:04	49.8	-7.89	156.34	35	7.1	Mw	EHB	P&S	
1959	8	18	6:37	19.9	44.57	-110.65	15	7.3	Mw	EHB	P&S	
1959	9	14	14:09	46.2	-28.73	-177.07	35	7.8	Mw	EHB	P&S	
1959	11	19	11:08	49.9	-5.91	146.49	128	7.0	mB	EHB	ABE1	
1960	1	13	15:40	22.9	-15.88	-73.06	35	7.5	mB	EHB	ABE1	Peru
1960	3	8	16:33	41.5	-17.01	168.71	244	7.2	mB	EHB	ABE1	
1960	3	20	17:07	33.8	39.85	143.40	35	7.8	Mw	EHB	P&S	
1960	5	21	10:02	54.1	-37.83	-73.38	12	8.2	Mw	EHB	P&S	
1960	5	22	18:56	4.0	-38.15	-72.98	35	7.9	Mw	EHB	P&S	
1960	5	22	19:11	17.5	-38.29	-73.05	35	9.6	Mw	EHB	P&S	Chile: Great Chilean
1960	6	6	5:55	49.1	-45.70	-73.50	15	7.2	Mw	EHB	P&S	
1960	6	20	2:01	15.3	-38.26	-73.32	35	7.0	Mw	EHB	P&S	
1960	6	20	12:59	46.0	-39.21	-73.32	25	7.1	Mw	EHB	P&S	

Year	M	d	h:min	sec	Lat.	Long.	Dep.	Mag.	sc	icat	mdo	Region:Earthquake Name
1960	7	25	11:12	9.1	53.65	158.89	139	7.2	mB	EHB	ABE1	
1960	11	20	22:02	0.1	-6.70	-80.62	35	7.8	Mw	EHB	P&S	
1960	11	24	6:52	47.4	-24.71	-176.38	36	7.1	Mw	EHB	P&S	
1960	12	2	9:10	43.3	-24.67	-70.08	14	7.2	Mw	EHB	P&S	
1960	12	13	7:36	17.7	-52.04	160.98	15	7.2	Mw	EHB	P&S	
1961	1	5	15:54	5.0	-4.23	143.21	141	7.1	mB	EHB	ABE1	
1961	2	26	18:10	51.1	31.59	131.51	37	7.7	Mw	EHB	P&S	
1961	3	7	10:10	39.4	-29.26	-175.26	35	7.1	Mw	EHB	P&S	
1961	3	28	9:36	6.0	0.05	123.63	139	7.3	mB	EHB	ABE1	
1961	7	23	21:51	10.6	-18.42	168.40	35	7.3	Mw	EHB	P&S	
1961	8	11	15:51	34.8	42.88	145.33	35	7.2	Mw	EHB	P&S	
1961	8	19	5:09	51.4	-10.83	-70.84	616	7.2	mB	EHB	ABE1	
1961	8	31	1:48	39.0	-10.70	-70.90	598	7.0	mB	EHB	ABE1	
1961	8	31	1:57	9.6	-10.34	-71.00	601	7.3	mB	EHB	ABE1	
1961	9	1	0:09	38.8	-59.26	-26.58	130	7.2	mB	EHB	ABE1	
1961	9	8	11:26	32.1	-56.25	-27.28	104	7.6	mB	EHB	ABE1	
1962	2	14	6:36	4.4	-38.09	-73.00	41	7.5	mB	EHB	ABE1	
1962	3	17	20:47	33.9	10.81	-43.31	15	7.0	mB	EHB	ABE1	
1962	4	12	0:52	43.0	38.08	142.74	25	7.2	Mw	EHB	P&S	
1962	4	26	7:26	26.3	-17.86	-178.69	551	7.5	UK	EHB	BRK	
1962	5	11	14:11	55.9	17.18	-99.64	35	7.3	Mw	EHB	P&S	
1962	5	21	21:15	40.6	-20.00	-177.38	438	7.5	Mw	EHB	HRV	
1962	7	26	8:14	45.7	7.49	-82.78	24	7.2	Mw	EHB	P&S	
1962	8	3	8:56	18.7	-23.29	-68.01	113	7.2	mB	EHB	ABE1	
1962	12	8	21:27	20.9	-25.82	-63.28	589	7.2	Mw	EHB	HRV	
1963	2	13	8:50	4.8	24.35	122.06	35	7.3	Mw	EHB	P&S	
1963	2	26	20:14	11.2	-7.57	146.22	181	7.3	mB	EHB	ABE1	
1963	3	16	8:44	52.6	46.62	154.89	35	7.2	Mw	EHB	P&S	
1963	3	26	9:48	20.1	-30.11	-177.54	35	7.4	Mw	EHB	P&S	
1963	4	16	1:29	23.1	-1.14	128.01	35	7.2	Mw	EHB	P&S	
1963	5	1	10:03	21.0	-19.06	169.07	144	7.1	mB	EHB	ABE1	
1963	8	15	17:25	8.2	-13.72	-69.32	550	7.7	Mw	EHB	HRV	
1963	8	25	12:18	13.7	-17.70	-178.54	571	7.1	Mw	EHB	HRV	
1963	9	15	0:46	54.3	-10.47	165.76	35	7.5	Mw	EHB	P&S	
1963	9	17	19:20	11.8	-10.28	165.41	30	7.5	Mw	EHB	P&S	
1963	10	12	11:27	0.2	44.43	149.27	48	7.1	Mw	EHB	P&S	
1963	10	13	5:17	55.1	44.76	149.80	26	8.6	Mw	EHB	P&S	Kurile Is.
1963	10	20	0:53	10.8	44.76	150.57	27	7.9	Mw	EHB	P&S	
1963	11	4	1:17	5.8	-6.74	129.68	35	7.8	mB	EHB	ABE1	
1963	11	9	21:15	30.3	-8.98	-71.54	577	7.7	Mw	EHB	HRV	
1963	12	15	19:34	48.3	-4.77	108.06	667	7.1	Mw	EHB	HRV	
1963	12	18	0:30	2.6	-24.78	-176.51	35	7.7	Mw	EHB	P&S	
1964	2	6	13:07	22.6	55.64	-156.07	4	7.1	Mw	EHB	P&S	
1964	3	28	3:36	12.7	61.02	-147.63	6	9.2	Mw	EHB	P&S	Alaska: Great Alaska
1964	4	23	3:32	48.5	-5.42	133.94	6	7.1	Mw	EHB	P&S	
1964	5	26	10:59	12.9	-56.24	-27.63	116	7.5	mB	EHB	ABE1	
1964	6	16	4:01	40.1	38.44	139.23	10	7.6	Mw	EHB	P&S	
1964	6	23	1:26	38.4	43.27	146.15	75	7.2	mB	EHB	ABE1	
1964	7	6	7:22	12.8	18.19	-100.51	92	7.2	mB	EHB	ABE1	
1964	7	8	11:55	44.5	-5.60	129.74	207	7.0	mB	EHB	ABE1	
1964	7	9	16:39	51.2	-15.58	167.75	131	7.4	mB	EHB	ABE1	
1964	8	13	0:31	15.4	-5.49	154.28	386	7.0	Mw	EHB	HRV	
1964	10	18	12:32	25.9	-7.18	123.81	582	7.0	Mw	EHB	HRV	
1964	11	17	8:15	41.4	-5.75	150.73	51	7.1	Mw	EHB	P&S	
1965	1	24	0:11	17.1	-2.45	125.96	29	8.2	Mw	EHB	P&S	Indonesia
1965	2	4	5:01	21.7	51.21	178.50	29	8.7	Mw	EHB	P&S	Aleutian Is
1965	2	4	8:40	38.7	51.40	179.56	8	7.8	Mw	EHB	P&S	
1965	3	14	15:53	7.8	36.40	70.71	210	7.5	mB	EHB	ABE1	
1965	3	28	16:33	16.7	-32.49	-71.21	71	7.4	mB	EHB	ABE1	Chile
1965	3	30	2:27	4.8	50.31	177.93	19	7.7	Mw	EHB	P&S	
1965	5	20	0:40	12.2	-14.64	167.50	10	7.6	Mw	EHB	P&S	
1965	6	11	3:33	46.8	44.61	148.94	46	7.1	Mw	EHB	P&S	
1965	8	11	3:40	57.1	-15.47	166.99	17	7.2	Mw	EHB	P&S	
1965	8	11	22:31	52.3	-15.80	167.26	46	7.6	Mw	EHB	P&S	
1965	8	13	12:40	8.3	-15.81	166.96	21	7.3	Mw	EHB	P&S	
1965	8	23	19:46	1.6	16.18	-95.85	11	7.4	Mw	EHB	P&S	
1965	9	12	22:02	35.5	-6.48	70.75	29	7.0	UK	EHB	BRK	
1966	3	12	16:31	19.6	24.31	122.69	28	7.4	Mw	EHB	P&S	
1966	3	20	1:42	51.1	0.85	29.87	15	7.2	UK	EHB	B&D	Uganda/Zaire
1966	6	15	0:59	42.8	-10.35	160.88	3	7.1	Mw	EHB	P&S	
1966	6	15	1:32	53.7	-10.11	161.03	13	7.3	Mw	EHB	P&S	
1966	6	22	20:29	6.3	-7.26	124.65	527	7.0	Mw	EHB	HRV	
1966	7	4	18:33	38.8	51.82	179.87	15	7.3	Mw	EHB	P&S	
1966	9	8	21:15	52.9	2.35	128.38	81	7.2	mB	EHB	ABE1	
1966	10	17	21:41	57.5	-10.80	-78.68	34	8.2	Mw	EHB	P&S	Peru
1966	12	28	8:18	7.2	-25.50	-70.66	30	7.7	Mw	EHB	P&S	
1966	12	31	18:23	11.1	-11.89	166.44	83	7.3	mB	EHB	ABE1	
1966	12	31	22:15	16.8	-12.33	166.68	24	7.2	Mw	EHB	P&S	
1967	1	5	0:14	39.8	48.20	102.92	14	7.0	Mw	EHB	P&S	
1967	2	9	15:24	47.0	2.89	-74.80	41	7.2	Mw	EHB	P&S	Colombia
1967	2	15	16:11	12.8	-9.12	-71.33	601	7.0	Mw	EHB	HRV	
1967	3	13	16:06	53.9	-40.19	-74.84	24	7.3	UK	EHB	BRK	
1967	7	22	16:56	55.3	40.63	30.74	4	7.4	Mw	EHB	P&S	Turkey
1967	10	9	17:21	49.1	-21.30	-179.06	636	7.3	Mw	EHB	HRV	
1967	10	25	0:59	24.1	24.46	122.22	67	7.0	mB	EHB	ABE1	
1967	12	21	2:25	24.7	-21.86	-69.95	45	7.4	Mw	EHB	P&S	
1967	12	25	1:23	34.2	-5.29	153.76	52	7.2	Mw	EHB	P&S	
1967	12	27	9:17	54.7	-21.21	-68.18	116	7.0	mB	EHB	ABE1	
1968	1	29	10:19	7.1	43.59	146.70	39	7.3	Mw	EHB	P&S	
1968	2	12	5:44	43.6	-5.50	153.40	23	7.4	Mw	EHB	P&S	
1968	2	19	22:45	43.8	39.37	24.94	9	7.2	Mw	EHB	P&S	
1968	2	26	10:50	17.9	22.76	121.41	17	7.2	Mw	EHB	P&S	
1968	4	1	0:42	4.9	32.48	132.19	29	7.5	Mw	EHB	P&S	
1968	4	9	2:29	1.8	33.16	-116.19	15	7.0	Ms	EHB	ABE1	
1968	5	16	0:49	0.4	40.90	143.35	26	8.3	Mw	EHB	P&S	Japan: Tokachi-oki
1968	5	16	10:39	0.4	41.60	142.79	11	7.8	Mw	EHB	P&S	
1968	5	16	23:04	52.6	39.88	143.22	9	7.2	UK	EHB	B&D	
1968	5	20	21:09	45.5	44.80	150.36	28	7.1	Mw	EHB	P&S	
1968	5	23	17:24	20.1	-41.74	172.12	47	7.2	Mw	EHB	P&S	
1968	5	28	13:27	20.4	-2.92	139.41	64	7.2	mB	EHB	ABE1	

(continued)

TABLE 1 *(continued)*

Year	M	d	h:min	sec	Lat.	Long.	Dep.	Mag.	sc	icat	mdo	Region:Earthquake Name
1968	6	12	13:41	51.0	39.50	142.89	31	7.1	Mw	EHB	P&S	
1968	7	25	7:23	9.4	-30.81	-178.14	65	7.1	mB	EHB	ABE1	
1968	8	1	20:19	25.0	16.38	122.08	52	7.7	Mw	EHB	P&S	Philippines
1968	8	2	14:06	46.1	16.49	-97.77	49	7.3	Mw	EHB	P&S	
1968	8	3	4:54	36.0	25.65	128.47	29	7.1	Mw	EHB	P&S	
1968	8	10	2:07	4.0	1.42	126.26	19	7.6	Mw	EHB	P&S	
1968	8	14	22:14	20.2	0.06	119.69	17	7.3	Mw	EHB	P&S	Indonesia
1968	8	18	18:38	31.8	-10.20	159.96	543	7.3	Mw	EHB	HRV	
1968	8	31	10:47	40.0	34.04	58.96	12	7.2	Mw	EHB	P&S	Iran: Dasht-i Biyaz
1968	10	7	19:20	21.8	26.29	140.68	519	7.3	Mw	EHB	HRV	
1968	10	23	21:04	42.9	-3.37	143.32	10	7.1	Mw	EHB	P&S	
1969	1	5	13:26	43.2	-7.99	158.97	63	7.3	mB	EHB	ABE1	
1969	1	19	7:02	9.0	44.89	143.21	238	7.3	mB	EHB	ABE1	
1969	1	30	10:29	42.5	4.76	127.44	74	7.1	mB	EHB	ABE1	
1969	2	10	22:58	5.7	-22.80	178.79	659	7.2	Mw	EHB	HRV	
1969	2	11	22:16	12.5	-6.79	126.68	427	7.2	Mw	EHB	HRV	
1969	2	28	2:40	33.7	35.92	-10.58	21	7.8	Mw	EHB	P&S	
1969	5	14	19:32	55.3	51.28	-179.85	17	7.0	Ms	EHB	USCGS	
1969	7	18	5:24	46.7	38.42	119.45	11	7.2	Mw	EHB	P&S	
1969	8	5	2:13	10.2	1.24	126.19	27	7.0	Ms	EHB	USCGS	
1969	8	11	21:27	37.6	43.48	147.82	46	8.2	Mw	EHB	P&S	Kurile Is.
1969	8	11	23:52	57.0	1.83	126.34	27	7.1	Mw	EHB	P&S	
1969	8	17	20:14	59.7	24.84	-109.69	32	7.2	UK	EHB	B&D	
1969	11	21	2:05	35.2	1.97	94.57	10	7.6	Mw	EHB	P&S	
1969	11	22	23:09	35.3	57.73	163.60	9	7.8	Mw	EHB	P&S	
1969	12	25	21:32	29.4	15.73	-59.65	11	7.2	Mw	EHB	P&S	
1970	1	4	17:00	40.3	24.15	102.46	14	7.2	Mw	EHB	P&S	China: Tonha
1970	1	8	17:12	42.9	-34.88	178.85	208	7.0	mB	EHB	ABE1	
1970	1	10	12:07	8.9	6.79	126.69	60	7.3	Ms	EHB	ABE1	
1970	1	20	7:19	48.2	-25.86	-177.17	45	7.2	mB	EHB	ABE1	
1970	3	28	21:02	25.7	39.17	29.55	24	7.4	UK	EHB	B&D	Turkey
1970	3	30	16:46	45.6	6.76	126.62	63	7.1	mB	EHB	ABE1	
1970	4	7	5:34	6.6	15.77	121.66	30	7.2	Mw	EHB	P&S	
1970	4	12	4:01	44.5	15.09	122.01	16	7.0	Ms	EHB	USCGS	
1970	4	20	10:39	13.8	-18.81	169.37	246	7.0	mB	EHB	ABE1	
1970	4	29	14:01	21.8	14.46	-92.76	50	7.3	Mw	EHB	P&S	
1970	5	27	12:05	9.2	27.19	140.23	403	7.1	Mw	EHB	HRV	
1970	5	31	20:23	32.2	-9.25	-78.84	73	7.5	mB	EHB	ABE3	Peru: Peru
1970	6	11	16:46	33.6	-59.42	159.23	15	7.3	Mw	EHB	P&S	
1970	6	15	11:14	50.2	-54.37	-64.11	4	7.0	Ms	EHB	USCGS	
1970	7	25	22:41	13.7	32.24	131.68	45	7.0	Mw	EHB	P&S	
1970	7	31	17:08	6.1	-1.49	-72.56	645	7.5	mB	EHB	ABE1	
1970	8	11	10:22	23.8	-14.10	166.57	40	7.5	Ms	EHB	ISC	
1970	8	30	17:46	10.6	52.35	151.62	650	7.3	Mw	EHB	HRV	
1970	10	31	17:53	7.1	-4.91	145.47	8	7.3	UK	EHB	B&D	
1970	12	2	15:54	17.2	-11.01	163.49	3	7.0	Ms	EHB	USCGS	
1970	12	10	4:34	39.1	-4.08	-80.66	20	7.1	Mw	EHB	P&S	Peru/Ecuador
1971	1	3	17:35	41.6	-55.64	-2.47	27	7.1	Ms	EHB	NEIS	
1971	1	10	7:17	7.7	-3.23	139.74	55	7.7	Mw	EHB	P&S	
1971	2	4	15:33	33.1	0.55	98.67	60	7.1	Ms	EHB	NEIS	
1971	2	8	21:04	20.5	-63.42	-61.41	15	7.0	Ms	EHB	NEIS	
1971	5	2	6:08	25.8	51.43	-177.25	20	7.1	Ms	EHB	NEIS	
1971	6	17	21:00	41.8	-25.42	-69.05	90	7.2	mB	EHB	ABE1	
1971	7	9	3:03	19.8	-32.56	-71.08	59	7.8	Mw	EHB	P&S	Chile
1971	7	14	6:11	30.6	-5.52	153.90	45	8.0	Mw	EHB	P&S	
1971	7	14	7:41	13.2	-5.60	153.77	56	7.1	Mw	EHB	P&S	
1971	7	19	0:14	45.9	-5.75	153.86	31	7.1	Ms	EHB	NEIS	
1971	7	26	1:23	22.2	-4.89	153.18	37	8.1	Mw	EHB	P&S	
1971	7	27	2:02	46.2	-2.83	-77.36	94	7.3	mB	EHB	ABE1	
1971	8	2	7:24	58.5	41.38	143.46	54	7.2	Mw	EHB	P&S	
1971	8	5	1:58	51.9	-0.93	-22.08	22	7.0	Ms	EHB	NEIS	
1971	9	5	18:35	28.6	46.56	141.19	17	7.3	Mw	EHB	P&S	
1971	11	21	5:57	12.8	-11.88	166.62	115	7.1	mB	EHB	ABE1	
1971	11	24	19:35	31.7	52.81	159.17	116	7.4	mB	EHB	ABE1	
1971	12	15	8:29	55.9	56.02	163.17	22	7.8	Mw	EHB	P&S	
1972	1	23	21:17	55.8	-13.23	166.37	52	7.1	Ms	EHB	NEIS	
1972	1	25	2:06	21.4	22.55	122.32	10	7.5	Mw	EHB	P&S	
1972	1	25	3:41	22.1	23.04	122.12	12	7.0	Ms	EHB	NEIS	
1972	2	14	23:29	53.5	-11.43	166.41	106	7.3	mB	EHB	ABE1	
1972	2	29	9:23	1.6	33.37	140.88	59	7.5	Mw	EHB	P&S	
1972	3	30	5:34	52.2	-25.74	179.63	496	7.3	Mw	EHB	HRV	
1972	4	25	19:30	8.4	13.39	120.31	30	7.2	Mw	EHB	P&S	
1972	4	28	23:32	11.2	-5.14	154.23	409	7.2	Mw	EHB	HRV	
1972	5	22	20:45	58.0	-17.83	-175.01	225	7.2	mB	EHB	ABE1	
1972	6	11	16:41	3.3	3.86	124.23	330	7.8	Mw	EHB	HRV	
1972	7	30	21:45	13.8	56.69	-136.10	8	7.6	Mw	EHB	P&S	
1972	8	17	23:44	8.3	-6.00	152.93	15	7.1	Ms	EHB	NEIS	
1972	11	2	19:55	24.8	-20.03	169.01	39	7.0	Ms	EHB	NEIS	
1972	12	2	0:19	54.2	6.46	126.65	83	7.4	mB	EHB	ABE1	
1972	12	4	10:16	13.9	33.33	140.71	71	7.3	mB	EHB	ABE3	
1973	1	30	21:01	13.5	18.45	-102.96	37	7.6	Mw	EHB	P&S	Mexico
1973	2	6	10:37	8.6	31.36	100.50	6	7.4	Mw	EHB	P&S	China: Luhuo
1973	2	28	6:37	55.3	50.48	156.62	60	7.1	Mw	EHB	P&S	
1973	3	17	8:30	54.9	13.40	122.79	45	7.0	Ms	EHB	NEIS	
1973	6	17	3:55	4.0	43.22	145.74	44	7.8	Mw	EHB	P&S	
1973	6	24	2:43	25.0	43.36	146.42	31	7.3	Mw	EHB	P&S	
1973	8	28	9:50	41.0	18.23	-96.61	81	7.3	mB	EHB	ABE1	Mexico
1973	9	29	0:44	1.3	41.91	130.98	570	7.8	Mw	EHB	HRV	
1973	10	6	15:07	37.1	-60.95	-21.63	15	7.5	Ms	EHB	ISC	
1973	12	28	13:41	47.1	-14.51	166.78	24	7.3	Mw	EHB	P&S	
1974	1	2	10:42	33.4	-22.45	-68.32	121	7.1	mB	EHB	ABE1	
1974	1	10	8:51	17.0	-14.53	166.96	54	7.1	Mw	EHB	P&S	
1974	1	31	23:30	7.2	-7.40	155.95	36	7.3	Mw	EHB	P&S	
1974	2	1	3:12	35.6	-7.21	155.74	35	7.4	Mw	EHB	P&S	
1974	3	23	14:28	34.7	-24.02	179.96	520	7.0	Mw	EHB	HRV	
1974	5	8	23:33	28.9	34.57	138.77	12	7.2	UK	EHB	B&D	
1974	7	2	23:26	23.6	-29.15	-175.82	4	7.4	Mw	EHB	P&S	
1974	7	4	19:30	42.1	45.19	93.94	16	7.5	UK	EHB	B&D	
1974	7	13	1:18	22.9	7.74	-77.63	3	7.2	Mw	EHB	P&S	
1974	7	30	5:12	41.9	36.37	70.73	210	7.1	mB	EHB	ABE1	

Year	M	d	h:min	sec	Lat.	Long.	Dep.	Mag.	sc	icat	mdo	Region:Earthquake Name
1974	8	11	1:13	56.1	39.38	73.80	3	7.1	Mw	EHB	P&S	
1974	8	18	10:44	10.3	-38.43	-73.46	9	7.1	Ms	EHB	NEIS	
1974	10	3	14:21	34.5	-12.25	-77.52	36	8.1	Mw	EHB	P&S	Peru: Lima
1974	10	23	6:14	54.8	-8.45	154.14	35	7.1	Mw	EHB	P&S	
1974	11	9	12:59	51.7	-12.53	-77.63	7	7.1	Mw	EHB	P&S	
1974	11	29	22:05	24.1	30.68	138.36	424	7.1	Mw	EHB	HRV	
1975	2	2	8:43	41.5	53.03	173.59	13	7.1	Mw	EHB	P&S	
1975	2	4	11:36	7.1	40.67	122.65	16	7.0	Mw	EHB	P&S	China: Haicheng
1975	5	10	14:27	43.6	-38.21	-73.00	30	7.7	Mw	EHB	P&S	
1975	5	26	9:11	49.0	35.97	-17.65	4	7.9	Mw	EHB	P&S	Azores
1975	6	10	13:47	19.3	43.13	147.65	31	7.0	Ms	EHB	NEIS	
1975	6	29	10:37	41.8	38.73	130.09	556	7.1	Mw	EHB	HRV	
1975	7	20	14:37	42.5	-6.61	155.10	61	7.5	mB	EHB	ABE3	
1975	7	20	19:54	29.7	-7.08	155.21	45	7.3	Mw	EHB	P&S	
1975	10	1	3:30	1.8	-4.86	102.15	38	7.0	Ms	EHB	NEIS	
1975	10	11	14:35	18.0	-24.91	-175.11	21	7.4	Mw	EHB	P&S	
1975	10	31	8:28	4.1	12.54	126.00	51	7.5	Mw	EHB	P&S	
1975	11	1	1:17	34.7	13.85	144.81	108	7.1	mB	EHB	ABE3	
1975	11	29	14:47	40.9	19.45	-155.03	2	7.5	Mw	EHB	P&S	
1975	12	26	15:56	38.1	-16.24	-172.36	15	7.7	Mw	EHB	P&S	
1976	1	1	1:29	39.7	-28.71	-177.43	49	7.3	Mw	EHB	HRV	
1976	1	14	15:56	34.2	-29.21	-177.63	42	7.8	Mw	EHB	HRV	
1976	1	14	16:47	37.7	-29.11	-177.40	36	7.9	Mw	EHB	HRV	Kermadec Is.
1976	1	21	10:05	25.1	44.78	149.14	38	7.2	Mw	EHB	HRV	
1976	2	4	9:01	46.3	15.30	-89.14	13	7.5	Mw	EHB	HRV	Guatemala: Guatemala
1976	3	24	4:46	5.1	-29.87	-177.70	28	7.1	Mw	EHB	HRV	
1976	5	5	4:52	55.2	-29.84	-177.64	59	7.1	Mw	EHB	HRV	
1976	6	3	16:44	40.1	-5.09	153.58	79	7.2	Mw	EHB	HRV	
1976	6	20	20:53	12.7	3.44	96.25	15	7.0	Mw	EHB	HRV	
1976	6	25	19:18	54.6	-4.52	140.10	3	7.2	Mw	EHB	HRV	Indonesia
1976	7	11	20:41	51.3	7.37	-78.07	17	7.3	Mw	EHB	HRV	
1976	7	27	19:42	55.9	39.60	117.89	17	7.6	Mw	EHB	HRV	China: Tangshan
1976	7	28	10:45	35.9	39.72	118.36	18	7.0	Mw	EHB	HRV	
1976	8	16	16:11	11.9	6.29	124.09	59	8.0	Mw	EHB	HRV	Philippines: Mindanao
1976	8	17	4:19	28.9	7.26	122.96	20	7.1	Mw	EHB	HRV	
1976	11	24	12:22	17.1	39.08	44.03	9	7.0	Mw	EHB	HRV	Turkey/Iran
1976	11	30	0:40	58.1	-20.55	-68.87	74	7.6	Mw	EHB	HRV	
1977	3	4	19:21	55.6	45.78	26.70	89	7.5	Mw	EHB	HRV	Romania
1977	3	18	21:43	54.7	16.75	122.27	43	7.3	Mw	EHB	HRV	
1977	4	2	7:15	21.9	-16.77	-171.94	16	7.3	Mw	EHB	HRV	
1977	4	20	23:42	54.6	-9.90	160.46	39	7.3	Mw	EHB	HRV	
1977	4	20	23:49	12.2	-9.57	160.65	16	7.1	Mw	EHB	HRV	
1977	4	21	4:24	12.3	-10.01	160.81	43	7.4	Mw	EHB	HRV	
1977	6	22	12:08	34.4	-22.91	-175.75	64	8.1	Mw	EHB	HRV	Tonga Is.
1977	7	29	11:15	42.2	-8.04	155.58	2	7.2	Mw	EHB	HRV	
1977	8	19	6:08	54.9	-11.12	118.38	21	8.3	Mw	EHB	HRV	Indonesia: Sumbawa
1977	8	26	19:50	0.1	-59.57	-20.51	12	7.2	Mw	EHB	HRV	
1977	8	27	7:12	26.5	-8.14	125.33	44	7.1	Mw	EHB	HRV	
1977	10	10	11:53	51.5	-25.85	-175.28	8	7.3	Mw	EHB	HRV	
1977	10	17	17:26	41.4	-27.99	173.21	32	7.0	Mw	EHB	HRV	

Year	M	d	h:min	Sec	Lat.	Long.	Dep.	Mag.	sc	icat	mdo	Region:Earthquake Name
1977	11	23	9:26	26.3	-31.08	-67.78	17	7.5	Mw	EHB	HRV	Argentina: Caucete
1978	2	9	21:35	20.9	-30.56	-177.52	89	7.0	Mw	EHB	HRV	
1978	3	7	2:48	48.0	31.94	137.46	442	7.1	Mw	EHB	HRV	
1978	3	23	0:31	1.9	44.29	148.95	28	7.1	Mw	EHB	HRV	
1978	3	23	3:14	21.3	44.36	149.12	7	7.5	Ms	EHB	ISC	
1978	3	23	3:15	20.6	44.98	148.52	23	7.6	Mw	EHB	HRV	
1978	3	24	19:47	49.3	44.23	148.92	12	7.5	Mw	EHB	HRV	
1978	6	12	8:14	29.1	38.22	142.02	53	7.7	Mw	EHB	HRV	Japan: Miyagi-ken-oki
1978	6	14	12:32	36.7	8.29	122.39	32	7.0	Mw	EHB	HRV	
1978	6	17	15:11	30.4	-17.10	-172.17	3	7.0	Mw	EHB	HRV	
1978	7	23	14:42	41.3	22.24	121.42	38	7.3	Mw	EHB	HRV	
1978	8	23	0:38	30.0	10.22	-85.20	25	7.0	Mw	EHB	HRV	
1978	9	16	15:35	53.5	33.24	57.38	3	7.4	Mw	EHB	HRV	Iran
1978	11	4	22:29	25.5	-11.33	162.24	42	7.0	Mw	EHB	HRV	
1978	11	5	22:02	10.4	-11.13	162.21	43	7.0	Mw	EHB	HRV	
1978	11	29	19:52	50.1	16.01	-96.60	24	7.8	Mw	EHB	HRV	Mexico: Oaxaca
1978	12	6	14:02	8.4	44.56	146.59	145	7.8	Mw	EHB	HRV	
1978	12	23	11:23	14.5	23.23	122.02	42	7.0	Mw	EHB	HRV	
1979	2	16	10:08	54.7	-16.54	-72.55	55	7.2	Mw	EHB	HRV	
1979	2	28	21:27	8.7	60.66	-141.66	24	7.5	Mw	EHB	HRV	
1979	3	14	11:07	15.0	17.76	-101.22	24	7.5	Mw	EHB	HRV	Mexico: Guerrero
1979	4	10	1:42	23.8	2.99	126.96	40	7.0	Mw	EHB	HRV	
1979	4	15	6:19	44.7	42.00	19.15	15	7.0	Mw	EHB	HRV	Yugoslavia: Montenegro
1979	5	1	13:03	40.2	-21.26	169.87	95	7.4	Mw	EHB	HRV	
1979	7	24	19:31	18.3	-11.19	107.63	13	7.0	Mw	EHB	HRV	
1979	9	12	5:17	56.2	-1.69	135.97	20	7.5	Mw	EHB	HRV	
1979	10	12	10:25	19.5	-46.69	165.82	7	7.3	Mw	EHB	HRV	
1979	10	17	5:43	3.3	18.52	145.41	590	7.0	Mw	EHB	HRV	
1979	10	23	9:51	8.3	-10.64	161.35	22	7.0	Mw	EHB	HRV	
1979	11	23	23:40	31.1	4.79	-76.19	108	7.2	Mw	EHB	HRV	Colombia
1979	11	27	17:10	34.5	34.06	59.76	7	7.1	Mw	EHB	HRV	
1979	12	12	7:59	4.7	1.60	-79.36	24	8.1	Mw	EHB	HRV	Ecuador/Colombia
1980	1	2	20:58	43.0	5.99	126.15	38	7.0	Mw	EHB	HRV	
1980	2	23	5:51	4.1	43.59	146.68	37	7.1	Mw	EHB	HRV	
1980	3	8	22:12	12.7	-22.64	171.45	41	7.2	Mw	EHB	HRV	
1980	3	24	3:59	53.1	52.95	-167.71	35	7.0	Mw	EHB	HRV	
1980	4	13	18:04	40.6	-23.59	-177.22	149	7.6	Mw	EHB	HRV	
1980	7	8	23:19	24.1	-12.48	166.48	56	7.5	Mw	EHB	HRV	
1980	7	16	19:56	48.7	-4.43	143.58	86	7.3	Mw	EHB	HRV	
1980	7	17	19:42	24.6	-12.48	166.00	32	7.8	Mw	EHB	HRV	Santa Cruz Is.
1980	10	10	12:25	25.5	36.14	1.40	12	7.1	Mw	EHB	HRV	Algeria: El-Asnam
1980	10	24	14:53	35.6	18.18	-98.24	65	7.2	Mw	EHB	HRV	Mexico
1980	10	25	7:00	8.5	-22.02	170.13	25	7.1	Mw	EHB	HRV	
1980	10	25	11:00	7.4	-21.94	170.05	38	7.5	Mw	EHB	HRV	
1980	11	8	10:27	35.2	41.11	-124.30	17	7.3	Mw	EHB	HRV	
1981	1	18	18:17	27.7	38.71	142.79	46	7.0	Mw	EHB	HRV	
1981	1	30	8:52	46.5	51.83	176.18	37	7.0	Mw	EHB	HRV	
1981	5	25	5:25	12.2	-48.72	164.65	11	7.6	Mw	EHB	HRV	
1981	7	6	3:08	25.5	-22.25	171.81	30	7.6	Mw	EHB	HRV	
1981	7	15	7:59	8.2	-17.27	167.69	11	7.1	Mw	EHB	HRV	

(continued)

TABLE 1 *(continued)*

Year	M	d	h:min	sec	Lat.	Long.	Dep.	Mag.	sc	icat	mdo	Region:Earthquake Name
1981	7	28	17:22	24.1	29.99	57.77	14	7.3	Mw	EHB	HRV	Iran
1981	9	1	9:29	31.5	-15.11	-173.02	17	7.5	Mw	EHB	HRV	
1981	10	16	3:25	43.1	-33.20	-73.05	32	7.1	Mw	EHB	HRV	
1981	10	25	3:22	15.8	18.12	-102.00	20	7.2	Mw	EHB	HRV	
1981	11	7	3:29	52.1	-32.23	-71.38	64	7.0	Mw	EHB	HRV	
1981	12	26	17:05	32.0	-29.94	-177.65	21	7.1	Mw	EHB	HRV	
1982	1	3	14:09	53.8	-0.99	-21.87	20	7.1	Mw	EHB	HRV	
1982	1	11	6:10	6.7	13.83	124.34	32	7.1	Mw	EHB	HRV	
1982	5	7	5:38	36.8	-60.70	-20.97	11	7.0	Mw	EHB	HRV	
1982	6	19	6:21	58.4	13.34	-89.31	73	7.3	Mw	EHB	HRV	
1982	6	22	4:18	42.3	-7.36	126.05	460	7.5	Mw	EHB	HRV	
1982	6	30	1:57	34.1	44.72	151.12	20	7.1	Mw	EHB	HRV	
1982	7	7	10:43	5.7	-51.30	160.67	15	7.2	Mw	EHB	HRV	
1982	7	23	14:23	53.9	36.30	141.78	24	7.0	Mw	EHB	HRV	
1982	8	5	20:32	51.9	-12.48	166.19	9	7.0	Mw	EHB	HRV	
1982	12	19	17:43	57.0	-24.19	-175.58	33	7.5	Mw	EHB	HRV	
1983	1	26	16:02	21.2	-30.45	-179.21	230	7.0	Mw	EHB	HRV	
1983	3	18	9:05	51.7	-4.89	153.59	91	7.7	Mw	EHB	HRV	
1983	4	3	2:50	4.9	8.72	-83.13	57	7.5	Mw	EHB	HRV	
1983	4	4	2:51	37.4	5.72	94.69	94	7.0	Mw	EHB	HRV	
1983	4	12	12:07	58.2	-4.85	-78.09	126	7.0	Mw	EHB	HRV	
1983	5	26	2:59	59.9	40.47	139.09	15	7.7	Mw	EHB	HRV	Japan: Nihonkai-chubu
1983	7	11	12:56	29.5	-60.92	-53.12	9	7.0	Mw	EHB	HRV	
1983	8	17	10:55	58.7	55.79	161.32	93	7.0	Mw	EHB	HRV	
1983	10	4	18:52	15.8	-26.54	-70.50	24	7.7	Mw	EHB	HRV	
1983	10	22	4:21	34.0	-60.69	-25.50	5	7.1	Mw	EHB	HRV	
1983	10	28	14:06	9.6	44.08	-113.80	16	7.0	Mw	EHB	HRV	
1983	11	24	5:30	37.3	-7.51	128.11	196	7.4	Mw	EHB	HRV	
1983	11	30	17:46	2.4	-6.84	72.06	10	7.7	Mw	EHB	HRV	
1983	12	2	3:09	3.7	14.06	-91.91	35	7.0	Mw	EHB	HRV	
1983	12	30	23:52	41.4	36.39	70.71	215	7.4	Mw	EHB	HRV	
1984	1	1	9:03	41.2	33.65	136.79	386	7.2	Mw	EHB	HRV	
1984	2	7	21:33	21.6	-9.97	160.57	9	7.6	Mw	EHB	HRV	
1984	3	5	3:33	52.9	8.15	123.75	655	7.3	Mw	EHB	HRV	
1984	3	6	2:17	22.4	29.38	138.88	460	7.4	Mw	EHB	HRV	
1984	3	19	20:28	40.0	40.34	63.33	15	7.0	Mw	EHB	HRV	
1984	3	24	9:44	3.7	44.17	148.16	38	7.2	Mw	EHB	HRV	
1984	5	30	7:49	44.9	-4.84	151.59	171	7.1	Mw	EHB	HRV	
1984	8	6	12:01	53.7	-0.08	122.42	244	7.4	Mw	EHB	HRV	
1984	10	15	10:21	8.6	-15.87	-173.56	127	7.1	Mw	EHB	HRV	
1984	11	1	4:48	52.3	8.15	-38.81	12	7.0	Mw	EHB	HRV	
1984	11	15	2:46	23.5	-22.04	171.02	126	7.1	Mw	EHB	HRV	
1984	11	17	6:49	30.8	0.18	97.96	28	7.1	Mw	EHB	HRV	
1984	11	20	8:15	15.2	5.12	125.11	180	7.5	Mw	EHB	HRV	
1985	3	3	22:47	8.4	-33.14	-71.76	35	8.0	Mw	EHB	HRV	Chile
1985	3	4	0:32	24.0	-33.24	-71.74	40	7.4	Mw	EHB	HRV	
1985	4	9	1:57	1.9	-34.12	-71.51	49	7.1	Mw	EHB	HRV	
1985	5	10	15:35	54.3	-5.60	151.06	47	7.2	Mw	EHB	HRV	
1985	7	3	4:36	54.9	-4.45	152.88	47	7.2	Mw	EHB	HRV	
1985	7	29	7:54	45.8	36.16	70.86	100	7.4	Mw	EHB	HRV	
1985	8	23	12:41	60.0	39.44	75.24	20	7.0	Mw	EHB	HRV	China
1985	9	19	13:17	49.6	18.45	-102.37	20	8.0	Mw	EHB	HRV	Mexico: Michoacan
1985	9	21	1:37	13.5	17.83	-101.62	18	7.6	Mw	EHB	HRV	
1985	11	17	9:40	23.3	-1.67	134.94	10	7.1	Mw	EHB	HRV	
1985	11	28	2:25	42.8	-14.03	166.34	25	7.0	Mw	EHB	HRV	
1985	11	28	3:49	57.7	-13.98	166.28	49	7.0	Mw	EHB	HRV	
1985	12	21	1:13	23.7	-14.01	166.63	44	7.1	Mw	EHB	HRV	
1986	1	15	20:17	31.7	-21.46	170.44	140	7.1	Mw	EHB	HRV	
1986	4	30	7:07	19.0	18.37	-103.00	23	7.0	Mw	EHB	HRV	
1986	5	7	22:47	10.5	51.56	-174.81	20	8.0	Mw	EHB	HRV	Andreanof Is.
1986	5	26	19:06	16.9	-20.22	178.95	541	7.1	Mw	EHB	HRV	
1986	6	16	10:48	27.7	-21.99	-178.87	560	7.1	Mw	EHB	HRV	
1986	6	24	3:11	34.6	-4.44	144.01	123	7.2	Mw	EHB	HRV	
1986	8	14	19:39	14.5	1.80	126.48	30	7.5	Mw	EHB	HRV	
1986	8	30	21:28	37.1	45.52	26.27	136	7.2	Mw	EHB	HRV	
1986	10	20	6:46	11.4	-28.16	-176.29	30	7.7	Mw	EHB	HRV	
1986	10	30	1:28	55.9	-21.67	-176.55	190	7.2	Mw	EHB	HRV	
1986	11	14	21:20	5.4	23.98	121.72	25	7.4	Mw	EHB	HRV	
1987	1	30	22:29	38.5	-60.16	-27.01	8	7.0	Mw	EHB	HRV	
1987	2	8	18:33	58.8	-6.02	147.66	42	7.3	Mw	EHB	HRV	
1987	3	5	9:17	5.1	-24.40	-70.10	46	7.6	Mw	EHB	HRV	
1987	3	5	10:55	14.0	-24.44	-70.57	31	7.0	Mw	EHB	HRV	
1987	3	6	4:10	44.8	0.08	-77.79	18	7.2	Mw	EHB	HRV	Ecuador/Colombia
1987	4	1	1:48	7.1	-22.90	-66.25	230	7.0	Mw	EHB	HRV	
1987	6	17	1:32	57.0	-5.59	130.84	85	7.1	Mw	EHB	HRV	
1987	8	8	15:48	58.0	-19.09	-70.01	71	7.2	Mw	EHB	HRV	
1987	9	3	6:40	12.5	-58.94	158.51	15	7.4	Mw	EHB	HRV	
1987	10	6	4:19	8.1	-17.98	-172.17	22	7.3	Mw	EHB	HRV	
1987	10	12	13:57	6.8	-7.27	154.42	28	7.0	Mw	EHB	HRV	
1987	10	16	20:48	0.8	-6.22	149.12	30	7.4	Mw	EHB	HRV	
1987	11	17	8:46	50.6	58.79	-143.20	3	7.2	Mw	EHB	HRV	
1987	11	30	19:23	17.8	58.83	-142.60	15	7.9	Mw	EHB	HRV	
1988	1	19	7:30	30.7	-24.67	-70.51	15	7.0	Mw	EHB	HRV	
1988	2	5	14:01	4.1	-24.79	-70.48	37	7.2	Mw	EHB	HRV	
1988	2	24	3:52	6.7	13.47	124.55	40	7.3	Mw	EHB	HRV	
1988	3	6	22:35	36.8	57.26	-142.75	6	7.8	Mw	EHB	HRV	
1988	4	12	23:19	56.3	-17.25	-72.25	30	7.1	Mw	EHB	HRV	
1988	7	23	15:17	11.7	-6.53	152.81	31	7.0	Mw	EHB	HRV	
1988	8	6	0:36	26.0	25.09	95.11	90	7.3	Mw	EHB	HRV	
1988	8	10	4:38	28.2	-10.32	160.93	34	7.6	Mw	EHB	HRV	
1988	10	8	4:46	26.4	-18.83	-172.32	41	7.1	Mw	EHB	HRV	
1988	11	6	13:03	22.4	22.87	99.57	23	7.0	Mw	EHB	HRV	China: Lancang
1989	2	10	11:15	26.0	2.28	126.68	46	7.1	Mw	EHB	HRV	
1989	5	5	18:28	40.3	-8.31	-71.40	593	7.1	Mw	EHB	HRV	
1989	5	23	10:54	46.1	-52.51	160.60	2	8.1	Mw	EHB	HRV	Macquarie Is.
1989	9	4	13:14	59.1	55.59	-156.86	6	7.1	Mw	EHB	HRV	
1989	11	1	18:25	36.7	39.92	142.79	29	7.4	Mw	EHB	HRV	
1989	12	15	18:43	47.1	8.37	126.69	24	7.5	Mw	EHB	HRV	
1990	3	3	12:16	29.7	-21.96	175.26	36	7.6	Mw	EHB	HRV	
1990	3	5	16:38	14.2	-18.32	168.16	24	7.0	Mw	EHB	HRV	

Year	M	d	h:min	sec	Lat.	Long.	Dep.	Mag.	sc	icat	mdo	Region:Earthquake Name
1990	3	25	13:16	9.4	9.85	-84.77	33	7.1	Ms	EHB	ISC	
1990	3	25	13:22	57.0	9.94	-84.78	22	7.3	Mw	EHB	HRV	
1990	4	5	21:12	38.8	15.16	147.61	23	7.4	Mw	EHB	HRV	
1990	4	18	13:39	22.3	1.20	122.82	37	7.6	Mw	EHB	HRV	
1990	5	12	4:50	10.1	48.99	141.87	613	7.2	Mw	EHB	HRV	
1990	5	20	2:22	1.7	5.11	32.18	7	7.1	Mw	EHB	HRV	
1990	5	24	20:00	9.6	5.34	31.84	16	7.1	Mw	EHB	HRV	
1990	5	24	20:09	24.2	-7.36	120.35	588	7.1	Mw	EHB	HRV	
1990	5	30	10:40	7.7	45.85	26.65	90	7.0	Mw	EHB	HRV	
1990	6	14	7:40	55.7	11.39	122.04	18	7.1	Mw	EHB	HRV	
1990	6	20	21:00	13.2	37.01	49.21	18	7.4	Mw	EHB	HRV	Iran
1990	7	16	7:26	36.0	15.72	121.18	24	7.7	Mw	EHB	HRV	Philippines: Luzon
1990	7	27	12:38	0.9	-15.33	167.54	127	7.2	Mw	EHB	HRV	
1990	8	12	21:25	23.2	-19.44	169.24	140	7.1	Mw	EHB	HRV	
1990	10	17	14:30	13.9	-11.03	-70.74	599	7.0	Mw	EHB	HRV	
1990	11	6	20:14	31.5	53.49	169.82	26	7.1	Mw	EHB	HRV	
1990	12	30	19:14	21.5	-5.09	151.02	188	7.5	Mw	EHB	HRV	
1991	1	5	14:57	13.2	23.57	95.87	18	7.0	Mw	EHB	HRV	
1991	4	22	21:56	53.9	9.67	-83.07	13	7.6	Mw	EHB	HRV	Costa Rica/Panama
1991	4	29	9:12	48.0	42.43	43.67	7	7.0	Mw	EHB	HRV	Georgia
1991	5	30	13:17	43.2	54.52	-161.71	29	7.0	Mw	EHB	HRV	
1991	6	9	7:45	4.5	-20.24	-176.15	278	7.0	Mw	EHB	HRV	
1991	6	20	5:18	54.2	1.21	122.77	33	7.5	Mw	EHB	HRV	
1991	6	23	21:22	30.0	-26.78	-63.27	562	7.3	Mw	EHB	HRV	
1991	8	17	22:17	14.4	41.75	-125.59	7	7.1	Mw	EHB	HRV	
1991	9	30	0:21	47.5	-20.82	-178.50	567	7.0	Mw	EHB	HRV	
1991	10	14	15:58	16.5	-9.05	158.55	38	7.2	Mw	EHB	HRV	
1991	11	19	22:28	53.3	4.55	-77.36	25	7.2	Mw	EHB	HRV	
1991	12	22	8:43	15.1	45.59	151.04	25	7.6	Mw	EHB	HRV	
1991	12	27	4:06	0.9	-55.97	-25.06	22	7.2	Mw	EHB	HRV	
1992	4	25	18:06	6.4	40.35	-124.07	19	7.2	Mw	EHB	HRV	
1992	5	15	7:05	7.7	-6.06	147.62	66	7.2	Mw	EHB	HRV	
1992	5	17	9:49	20.8	7.33	126.67	36	7.1	Mw	EHB	HRV	
1992	5	17	10:15	36.3	7.24	126.75	64	7.2	Mw	EHB	HRV	
1992	5	27	5:13	40.4	-11.08	165.29	20	7.0	Mw	EHB	HRV	
1992	6	28	11:57	38.4	34.18	-116.53	11	7.3	Mw	EHB	HRV	Calif.: Landers
1992	7	11	10:44	21.9	-22.47	-178.32	389	7.2	Mw	EHB	HRV	
1992	8	19	2:04	36.7	42.11	73.61	13	7.2	Mw	EHB	HRV	Kyrgyzstan
1992	9	2	0:16	2.7	11.73	-87.39	40	7.7	Mw	EHB	HRV	Nicaragua
1992	10	11	19:24	28.1	-19.26	169.02	136	7.4	Mw	EHB	HRV	
1992	10	18	15:11	59.8	7.09	-76.76	3	7.1	Mw	EHB	HRV	
1992	12	12	5:29	28.6	-8.49	121.83	33	7.8	Mw	EHB	HRV	Indonesia
1992	12	20	20:52	48.4	-6.59	130.40	76	7.3	Mw	EHB	HRV	
1993	1	15	11:06	7.7	43.03	144.15	100	7.6	Mw	EHB	HRV	
1993	3	6	3:05	52.0	-10.91	164.27	23	7.1	Mw	EHB	HRV	
1993	5	11	18:26	52.8	7.25	126.62	62	7.0	Mw	EHB	HRV	
1993	5	24	23:51	26.7	-23.13	-66.47	218	7.0	Mw	EHB	HRV	
1993	6	8	13:03	36.7	51.19	157.76	61	7.5	Mw	EHB	HRV	
1993	7	12	13:17	12.7	42.90	139.24	12	7.7	Mw	EHB	HRV	Japan: Hokkaido-nanse.
1993	8	8	8:34	25.9	13.00	144.87	57	7.8	Mw	EHB	HRV	S. Mariana Is.: Quam

Year	M	d	h:min	Sec	Lat.	Long.	Dep.	Mag.	sc	icat	mdo	Region:Earthquake Name
1993	8	9	12:42	48.9	36.33	70.87	211	7.0	Mw	EHB	HRV	
1993	8	10	0:51	54.6	-45.22	167.01	29	7.0	Mw	EHB	HRV	
1993	9	10	19:12	56.3	14.76	-92.64	34	7.2	Mw	EHB	HRV	
1993	11	13	1:18	5.8	51.94	158.63	35	7.0	Mw	EHB	HRV	
1993	12	29	7:48	13.2	-20.24	169.92	17	7.0	Mw	EHB	HRV	
1994	1	21	2:24	31.4	1.04	127.77	19	7.0	Mw	EHB	HRV	
1994	2	12	17:58	25.6	-20.55	169.47	30	7.0	Mw	EHB	HRV	
1994	3	9	23:28	7.9	-17.97	-178.36	564	7.6	Mw	EHB	HRV	
1994	3	14	4:30	17.6	-1.09	-23.55	5	7.0	Mw	EHB	HRV	
1994	6	2	18:17	38.5	-10.41	112.93	34	7.8	Mw	EHB	HRV	Indonesia
1994	6	2	18:17	52.2	-10.44	113.18	35	7.1	Ms	EHB	ISC	
1994	6	9	0:33	17.5	-13.88	-67.53	635	8.2	Mw	EHB	HRV	Northern Bolivia
1994	7	13	2:35	58.3	-16.60	167.54	35	7.2	Mw	EHB	HRV	
1994	7	21	18:36	32.6	42.31	132.87	472	7.3	Mw	EHB	HRV	
1994	9	1	15:15	54.5	40.38	-125.78	10	7.0	Mw	EHB	HRV	
1994	10	4	13:23	0.3	43.83	147.33	33	8.3	Mw	EHB	HRV	Kuril Is.
1994	10	4	15:24	17.2	43.60	147.87	17	7.1	Ms	EHB	ISC	
1994	10	4	16:01	4.5	43.78	148.03	18	7.1	Ms	EHB	ISC	
1994	10	9	7:55	41.4	43.97	147.89	34	7.3	Mw	EHB	HRV	
1994	10	9	8:07	5.9	43.75	148.03	36	7.3	Ms	EHB	ISC	
1994	11	14	19:15	31.4	13.54	121.06	28	7.1	Mw	EHB	HRV	Philippines
1994	12	28	12:19	23.3	40.54	143.44	16	7.8	Mw	EHB	HRV	
1995	1	6	22:37	36.2	40.35	142.22	28	7.0	Mw	EHB	HRV	
1995	2	5	22:51	8.8	-37.82	178.88	41	7.1	Mw	EHB	HRV	
1995	4	7	22:06	59.6	-15.34	-173.43	33	7.4	Mw	EHB	HRV	
1995	4	21	0:34	49.9	12.10	125.45	35	7.2	Mw	EHB	HRV	
1995	5	5	3:53	48.4	12.66	125.22	27	7.1	Mw	EHB	HRV	
1995	5	16	20:12	45.3	-22.94	169.99	19	7.7	Mw	EHB	HRV	
1995	5	27	13:03	55.3	52.60	142.82	18	7.1	Mw	EHB	HRV	Russia: Sakhalin
1995	7	3	19:50	52.1	-29.40	-177.45	41	7.2	Mw	EHB	HRV	
1995	7	30	5:11	24.5	-23.34	-70.26	41	8.0	Mw	EHB	HRV	Chile
1995	8	16	10:27	28.2	-5.78	154.29	16	7.7	Mw	EHB	HRV	
1995	8	16	23:10	30.0	-5.77	154.35	75	7.2	Mw	EHB	HRV	
1995	8	23	7:06	4.8	18.82	145.29	607	7.1	Mw	EHB	HRV	
1995	9	14	14:04	33.9	16.85	-98.61	28	7.4	Mw	EHB	HRV	
1995	10	3	1:51	25.2	-2.79	-77.82	24	7.0	Mw	EHB	HRV	
1995	10	9	15:35	54.1	19.05	-104.21	26	8.0	Mw	EHB	HRV	Mexico: Manzanillo
1995	10	18	10:37	28.6	28.07	130.28	32	7.1	Mw	EHB	HRV	
1995	10	21	2:38	58.5	16.83	-93.47	162	7.2	Mw	EHB	HRV	
1995	11	22	4:15	13.7	28.76	34.81	13	7.2	Mw	EHB	HRV	
1995	12	3	18:01	9.6	44.71	149.26	26	7.9	Mw	EHB	HRV	Kurile Is.
1995	12	25	4:43	26.0	-6.92	129.20	145	7.1	Mw	EHB	HRV	
1996	1	1	8:05	12.5	0.71	119.90	25	7.9	Mw	EHB	HRV	Indonesia
1996	2	7	21:36	47.0	45.37	149.90	33	7.2	Mw	EHB	HRV	
1996	2	17	5:59	32.4	-0.92	136.98	36	8.2	Mw	EHB	HRV	Indonesia: West Irian
1996	2	21	12:51	4.3	-9.71	-79.85	15	7.5	Mw	EHB	HRV	
1996	2	25	3:08	16.7	15.93	-98.11	18	7.1	Mw	EHB	HRV	
1996	4	16	0:30	55.3	-24.28	-176.90	110	7.2	Mw	EHB	HRV	
1996	4	29	14:40	43.0	-6.54	155.10	49	7.2	Mw	EHB	HRV	
1996	6	10	4:03	36.3	51.59	-177.59	29	7.9	Mw	EHB	HRV	Andreanof Is.

(continued)

TABLE 1 *(continued)*

Year	M	d	h:min	sec	Lat.	Long.	Dep.	Mag.	sc	icat	mdo	Region:Earthquake Name
1996	6	10	15:24	53.0	51.43	-176.84	27	7.3	Mw	EHB	HRV	
1996	6	11	18:22	53.1	12.68	125.12	39	7.1	Mw	EHB	HRV	
1996	6	17	11:22	19.8	-7.15	122.51	591	7.9	Mw	EHB	HRV	
1996	7	22	14:19	36.9	1.02	120.41	30	7.0	Mw	EHB	HRV	
1996	8	5	22:38	23.3	-20.80	-178.23	557	7.4	Mw	EHB	HRV	
1996	11	12	16:59	43.4	-14.96	-75.56	17	7.7	Mw	EHB	HRV	
1997	1	11	20:28	27.8	18.18	-102.81	35	7.2	Mw	EHB	HRV	
1997	1	23	2:15	23.2	-22.04	-65.66	269	7.1	Mw	EHB	HRV	
1997	2	27	21:08	3.2	29.94	68.19	24	7.1	Mw	EHB	HRV	Pakistan
1997	4	21	12:02	27.3	-12.56	166.80	28	7.7	Mw	EHB	HRV	
1997	5	10	7:57	31.9	33.83	59.80	12	7.2	Mw	EHB	HRV	Iran
1997	5	25	23:22	34.7	-32.26	179.93	341	7.1	Mw	EHB	HRV	
1997	7	9	19:24	12.9	10.47	-63.53	10	7.0	Mw	EHB	HRV	Venezuela
1997	9	20	16:11	33.1	-28.80	-177.44	30	7.0	Mw	EHB	HRV	
1997	10	14	9:53	18.3	-22.26	-176.63	163	7.7	Mw	EHB	HRV	
1997	10	15	1:03	35.2	-30.89	-71.15	62	7.1	Mw	EHB	HRV	
1997	10	28	6:15	18.7	-4.36	-76.60	112	7.2	Mw	EHB	HRV	
1997	11	8	10:02	53.3	35.11	87.37	24	7.5	Mw	EHB	HRV	
1997	11	15	18:59	26.0	-15.11	167.44	127	7.0	Mw	EHB	HRV	
1997	11	25	12:14	35.0	1.20	122.49	24	7.0	Mw	EHB	HRV	
1997	12	5	11:26	56.8	54.80	162.00	37	7.8	Mw	EHB	HRV	
1997	12	5	11:48	41.7	54.42	162.26	28	7.3	Ms	EHB	ISC	
1997	12	22	2:05	52.8	-5.56	147.88	191	7.2	Mw	EHB	HRV	
1998	1	4	6:12	0.2	-22.25	171.01	100	7.5	Mw	EHB	HRV	
1998	1	30	12:16	10.2	-23.85	-70.15	41	7.1	Mw	EHB	HRV	
1998	3	25	3:12	28.3	-62.90	149.61	20	8.1	Mw	EHB	HRV	Balleny Is.
1998	3	29	19:48	16.9	-17.71	-178.91	538	7.2	Mw	EHB	HRV	
1998	4	1	17:56	21.9	-0.56	99.19	32	7.0	Mw	EHB	HRV	
1998	5	3	23:30	22.1	22.43	125.37	24	7.5	Mw	EHB	HRV	
1998	7	16	11:56	35.6	-11.03	166.27	93	7.0	Mw	EHB	HRV	
1998	7	17	8:49	14.3	-2.97	142.69	10	7.0	Mw	EHB	HRV	Papua New Guinea
1998	8	4	18:59	20.9	-0.60	-80.31	26	7.2	Mw	EHB	HRV	
1998	8	20	6:40	55.6	28.98	139.39	425	7.1	Mw	EHB	HRV	
1998	11	9	5:38	45.3	-6.98	128.93	31	7.0	Mw	EHB	HRV	
1998	11	29	14:10	31.9	-1.94	124.82	21	7.7	Mw	EHB	HRV	
1999	1	19	3:35	31.4	-4.55	153.41	75	7.0	Mw	EHB	HRV	
1999	2	6	21:48	1.1	-12.82	166.78	90	7.3	Mw	EHB	HRV	
1999	3	4	8:52	2.8	5.42	121.87	29	7.1	Mw	EHB	HRV	
1999	4	5	11:08	3.9	-5.60	149.65	138	7.4	Mw	EHB	HRV	
1999	4	8	13:10	35.7	43.62	130.34	574	7.1	Mw	EHB	HRV	
1999	5	10	20:33	1.7	-5.16	150.98	124	7.1	Mw	EHB	HRV	
1999	5	16	0:51	18.8	-4.71	152.61	46	7.1	Mw	EHB	HRV	
1999	6	15	20:42	6.9	18.36	-97.45	67	7.0	Mw	EHB	HRV	
1999	8	17	0:01	40.6	40.75	29.94	17	7.6	Mw	EHB	HRV	Turkey: Kocaeli
1999	9	20	17:47	19.7	23.79	120.95	31	7.7	Mw	EHB	HRV	Taiwan: Chi-Chi
1999	9	30	16:31	14.8	16.05	-96.87	40	7.5	Mw	EHB	HRV	
1999	10	16	9:46	46.6	34.51	-116.43	5	7.2	Mw	EHB	HRV	Calif.: Hector Mine
1999	11	12	16:57	21.2	40.78	31.21	10	7.2	Mw	EHB	HRV	Turkey: Duzce
1999	11	15	5:42	45.4	-1.35	88.87	13	7.0	Mw	EHB	HRV	
1999	11	19	13:56	49.5	-6.36	148.81	46	7.0	Mw	EHB	HRV	
1999	11	26	13:21	15.8	-16.39	168.31	24	7.4	Mw	EHB	HRV	
1999	12	6	23:12	34.5	57.35	-154.52	59	7.0	Mw	EHB	HRV	
1999	12	11	18:03	39.0	15.76	119.76	44	7.3	Mw	EHB	HRV	

Explanations:

(1) Earthquake origin time is given in UTC by: Year, Month (M), Day (d), Hour (h), Minute (min), and Second (sec).

(2) Earthquake hypocenter is given by: Latitude (Lat.) in degrees (positive for northern hemisphere, and negative for southern hemisphere); Longitude (Long.) in degrees (positive for eastern hemisphere, and negative for western hemisphere); and Focal Depth (Dep.) in kilometers.

(3) Preferred Magnitude is given by: Magnitude (Mag.) and Magnitude Scale (sc).

Moment magnitude (Mw): Value is computed from the scalar moment (Mo). Mw is related to the scalar moment by the formula Mw = 2/3 log Mo - 10.7 (Hanks and Kanamori, 1979).

Surface-wave magnitude (Ms): Value is computed by ISC/NEIC for earthquakes at depths generally less than 50 km (NEIC) or 60 km (ISC) based on the maximum ground amplitude of surface waves with periods between 18 and 22 seconds that are recorded at distances between 20 and 160 degrees.

Broad-band body-wave magnitude (mB): Value is computed as defined by Gutenberg and Richter (1956) based on the maximum ground amplitude of seismic body-waves with periods of 5-10 s that are recorded at distances greater than or equal to 5 degrees.

Body-wave magnitude (mb): Value is computed by ISC/NEIC based on the maximum ground amplitude of seismic body-waves with periods of 0.1-3 s that are recorded at distances greater than or equal to 5 degrees.

Japan Meteorological Agency magnitude (Mj): Value is computed using either maximum ground displacement or maximum ground velocity.

Unknown magnitudes (UK): The computational method was unknown and could not be determined from published sources. Examples of unknown magnitudes are the ones calculated by Gutenberg and Richter in "Seismicity of the Earth" (MGR).

(4) Source catalog is given in the "icat" column. EHB = EHB origin time and hypocenter (otherwise from other earthquake catalogs).

(5) Source for magnitude (Mag) is given in the "mdo" column.

(6) For some well-known earthquakes, the "Region:Earthquake Name" column gives the geographical region where the earthquake occurred, and a commonly cited name for the earthquake in the literature. "NZ" = New Zealand, "Calif." = California, and "Is." = Island.

(7) Please see Appendix 3 for further explanations of (4) and (5).

TABLE 2 Frequency–Magnitude Distribution for 1900–1999

Incremental			Cumulative	
$\leq M <$		Events y^{-1}	$M \geq$	Events y^{-1}
5.5	6.0	164[a]	5.5	264[a]
6.0	6.5	62[a]	6.0	100[a]
6.5	7.0	22	6.5	38
7.0	7.5	12	7.0	16
7.5	8.0	3	7.5	4
8.0	—	0.7	8.0	0.7

[a]For magnitudes smaller than 6.5 the number of events is based on the period 1964–1999.

Figure 3b shows a histogram of the number of events per 10-year period in different magnitude intervals for the entire century. We note a slight increase in the number of earthquakes with magnitudes greater than or equal to 6.5 during the 1940–1960 period. Most of the magnitudes for events less than magnitude 7 during this period were those listed by Gutenberg and Richter (1954), suggesting that many of the M_{GR} magnitudes, even after applying the correction previously described, may remain too high when compared to other magnitudes such as those reported by Abe (1981).

In summary, our full global earthquake database is given in CENT.CAT on the Handbook CD, under the directory \41Engdahl, and a subset of earthquakes with magnitude greater than or equal to 7.0 is given in CENT7.CAT in the same directory. For general reference, essential information from CENT7.CAT is listed in Table 1 with names for some well-known earthquakes. According to our analysis, this list is complete for the 20th century. Frequency–magnitude distributions for the century have been estimated from our database and are presented in Table 2. We believe these distributions to be reasonably representative of the rate of earthquake occurrence during the last century.

7. Discussion

The earthquake locations listed in our catalog for the historical period are of variable quality, as not all of the phase data listed in ISS bulletins have been converted into digital format and we were only able to relocate events during the 1918–1942 and 1956–1999 periods. Hence, hypocentral errors, especially in depth, may remain quite large during the missing years (as well as for many relocated events before 1930). Maps of the distribution of earthquake locations in our catalog are shown in Color Plates 15 and 16.

Magnitude issues, such as the hierarchical selection process, are more contentious, especially for earthquakes less than about magnitude 6.5. Since the seismic moment M_0 is a more stable measurement than M_S we chose to convert the Harvard M_0 (when available) to an empirical M_S using the relationship derived by Ekström and Dziewonski (1988). This empirical M_S approximates the global average value of M_S observed for that scalar moment and thereby ensures that the assignment of magnitudes for most of the modern period (1976–1999) remains reasonably consistent with reported magnitudes for earlier periods, which are largely M_S. We also confirmed that magnitude estimates listed by Abe (1981, 1984) and by Pacheco and Sykes (1992) for larger events were consistent with Harvard M_w estimates. However, when we made a similar comparison to magnitudes reported by Gutenberg and Richter (1954) and by Båth and Duda (1979), we found that those estimates were consistently about 0.2 magnitude units higher than the Abe (1981, 1984) values. Moreover, it was impossible to match the seismicity rates of the historical period to those of the modern period without making a reduction in the older magnitudes by about 0.2 units. However, final resolution of this problem is presently beyond the scope of this study so that, for example, the apparent higher seismicity rate during the 1940–1960 period (Fig. 3b) will remain problematic.

Our magnitude adjustment and selection process will undoubtedly raise concerns about earthquakes which have been cited in the literature as reputedly higher magnitude than we are able to confirm. However, many of these earthquakes cannot be found in our catalog simply because the selection process put them below the magnitude cut-off of 6.5 for the historical period or 5.5 for the modern period.

Both the location and magnitude issues raised here mean that correcting and updating our catalog will be an ongoing process. Hence, as new versions of the catalog are produced, we will make every effort to have them generally available to the global seismological community via the World Wide Web, CD-ROMs, and technical and general publications.

8. Summary and Conclusions

The goal of this research was to produce a comprehensive digital hypocenter and phase arrival-time database for most globally detected earthquakes during the 20th century, including a complete station list with codes, locations, and dates of operation. For the historical period (1900–1963) we have chosen a magnitude cut-off value of 6.5 although the resulting catalog is complete only to magnitude 7.0. For the recent period (1964–1999) the magnitude cut-off is 5.5, and the catalog is complete at this magnitude threshold.

For the earthquake research community, this database should provide a reliable starting point for a wide range of studies, including (but not limited to) source parameter studies, delineation of rupture zones of large earthquakes from aftershock distributions, further improvements in Earth models, and detailed studies of the seismicity of active regions. The new database may also be of great utility in providing fundamental information for reliable seismic hazard assessment, especially in developing countries located in active seismic belts whose seismic history is poorly known.

Acknowledgments

We are grateful to the ISC for providing unbound copies of the ISS bulletins and for a magnetic tape of punched card images of these bulletins for the period 1918–1942. We thank Willie Lee, Hiroo Kanamori, Jim Dewey, Ray Willemann, and Javier Pacheco for providing helpful reviews of the manuscript. Figures and maps were created using GMT (Wessel and Smith, 1998). This work was partially supported under NSF grant EAR-9506767.

Appendix 1. Centennial Catalog Format Description

The file CENT.CAT, or CENT7.CAT, may be read by the following FORTRAN `read` statement:

```
read(1,100) icat,asol,isol,yr,mon,day,hr,min,sec,
&glat,glon,dep,greg,ntel,(mag(k),msc(k),mdo(k),k=1,8)
100 format(a6,a1,a5,i2,2i3,1x,2i3,f6.2,1x,2f8.3,f6.1,2i4,
&8(f4.1,1x,a2,1x,a5))
```

Variable definitions are:

icat — source catalog
  EHB = EHB origin time and hypocenter (otherwise from other catalogs)
asol — open azimuth of teleseismic stations (delta > 28°) used
  for the period 1900–1963
    blank = unknown
    A = < 180 deg
    B = < 210 deg and >180 deg
    C = < 240 deg and >210 deg
    D = < 270 deg and >240 deg
    F = > 270 deg
  for the period 1964–1999
    blank = < 180 deg
    Z = ≥ 180 deg
isol — solution type
    HEQ = origin time and hypocenter fixed
    DEQ = depth free
    LEQ = depth fixed by program
    FEQ = depth fixed by Engdahl based on independent information
    XEQ = poor solution
other info
    M = focal mechanism available
yr — year
mon — month
day — day
hr — origin hour (UTC)
min — origin minute
sec — origin second
glat — geographic latitude (degrees: negative value = south)
glon — geographic longitude (degrees: negative value = west)
dep — focal depth (kilometers)
greg — Flinn–Engdahl geographic region number (Flinn *et al.*, 1974)
ntel — number of teleseismic observations (delta > 28°) used in solution

magnitudes (up to 8): first listed is magnitude adopted for the event

mag — magnitude
msc — scale
mdo — source

Appendix 2. Magnitude Scale (msc) Descriptors

Moment magnitude (M_w): Value is computed from the scalar moment (M_0). M_w is related to the scalar moment by the formula $M_w = 2/3 \log M_0 - 10.7$ (Hanks and Kanamori, 1979).

Surface-wave magnitude (M_S): Value is computed by ISC/NEIC for earthquakes at depths generally less than 50 km (NEIC) or 60 km (ISC) based on the maximum ground amplitude of surface waves with periods between 18 and 22 sec that are recorded at distances between 20 and 160 deg.

Energy magnitude (M_E): Value is computed by NEIC from the seismic radiated energy (E_S) obtained from energy spectral density of broadband P waves using the formula $M_E = 2/3 \log E_S - 2.9$ (Choy and Boatwright, 1995).

Broadband body-wave magnitude (m_B): Value is computed as defined by Gutenberg and Richter (1956) based on the maximum ground amplitude of seismic body waves with periods of 5–10 sec that are recorded at distances greater than or equal to 5 deg.

Body-wave magnitude (m_B): Value is computed by ISC/NEIC based on the maximum ground amplitude of seismic body waves with periods of 0.1–3 sec that are recorded at distances greater than or equal to 5 deg.

Japan Meteorological Agency magnitude (M_J): Value is computed using either maximum ground displacement or maximum ground velocity.

Unknown magnitudes (UK): The computational method was unknown and could not be determined from published sources. Examples of unknown magnitudes are the ones calculated by Gutenberg and Richter (1954) (M_{GR}).

Appendix 3. Source Catalog (icat) and Magnitude Source (mdo) Descriptors

ABE: Catalog of large earthquakes, mostly magnitude 6.8 and larger, 1897–1980, from Abe (1981, 1984) and Abe and Noguchi (1983a,b). The magnitude sources AN1, AN2, ABE1 and ABE2 listed in the mdo column of Table 1 are from this catalog.

B&D: Catalog of large earthquakes, 1897–1977, compiled by Båth and Duda (1979).

BRK: Magnitudes reported by the Seismographic Station, University of California, Berkeley, USA.

BJI: Catalog of hypocenters and magnitudes reported by the State Seismological Bureau, Beijing, China.

G&R: Catalog of hypocenters and magnitudes, 1904–1952 (Gutenberg and Richter, 1954).

ISC: Hypocenters and magnitudes from bulletins prepared by the International Seismological Centre, Newbury, UK, 1964–1998.

ISS: Hypocenters listed in the International Seismological Summary, 1918–1963. The same code is used for hypocenters listed in the British Association for the Advancement of Science bulletins (1913–1917), the predecessor of the ISS.

JMA: Catalog of hypocenters and magnitudes, 1926-present, reported by the Japan Meteorological Agency, Tokyo, Japan.

MOS: Catalog of earthquakes occurring in the former USSR, 1950–1961.

P&S: Complete and uniform catalog of worldwide earthquakes with M_S magnitudes of 7.0 and larger at depths less than or equal to 70 km, 1900–1989 (Pacheco and Sykes, 1992).

PAS: Magnitudes reported by the California Institute of Technology, Pasadena, USA.

NEIC, NEIS and C&GS: Catalogs of earthquakes located by the US Geological Survey's NEIC and its predecessors.

ROTHE: Catalog of worldwide earthquakes with magnitudes of 5.5 or larger, 1953–1965 (Rothé, 1969).

UTSU: Catalog of earthquakes in the Japan region, 1885–1925 (Utsu, 1979, 1982a,b).

References

Abe, K. (1981). Magnitudes of large shallow earthquakes from 1904 to 1980. *Phys. Earth Planet. Inter.* **27**, 72–92.

Abe, K. (1984). Complements to "Magnitudes of large shallow earthquakes from 1904 to 1980." *Phys. Earth Planet. Inter.* **34**, 17–23.

Abe, K. and S. Noguchi (1983a). Determination of magnitude for large shallow earthquakes, 1898–1917. *Phys. Earth Planet. Inter.* **32**, 45–59.

Abe, K. and S. Noguchi (1983b). Revision of magnitudes of large shallow earthquakes, 1897–1912. *Phys. Earth Planet. Inter.* **33**, 1–11.

Båth, M. (1981). Earthquake magnitude – recent research and current trends. *Earth-Sci. Rev.* **17**, 315–398.

Båth, M. and S.J. Duda (1979). "Some Aspects of Global Seismicity." Report No. 1–79, Seismological Institute, Uppsala, Sweden.

Choy, G.L. and J.L. Boatwright (1995). Global patterns of radiated seismic energy and apparent stress. *J. Geophys. Res.* **100**, 18205–18228.

Dziewonski, A.M., T.A. Chou, and J.H. Woodhouse (1980). Determination of earthquake source parameters from waveform data for studies of global and regional seismicity. *J. Geophys. Res.* **86**, 2825–2852.

Ekström, G. and A.M. Dziewonski (1988). Evidence of bias in estimations of earthquake size. *Nature* **332**, 173–176.

Engdahl, E.R., R.D. Van der Hilst, and R.P. Buland (1998). Global teleseismic earthquake relocation with improved travel times and procedures for depth determination. *Bull. Seismol. Soc. Am.* **88**, 722–743.

Flinn, E.A., E.R. Engdahl, and A.R. Hill (1974). Seismic and geographical regionalization. *Bull. Seismol. Soc. Am.* **64**, 771–993.

Geller, R.J. and H. Kanamori (1977). Magnitude of great shallow earthquakes from 1904 to 1952. *Bull. Seismol. Soc. Am.* **67**, 587–598.

Gutenberg, B. (1945). Amplitudes of surface waves and magnitudes of shallow earthquakes. *Bull. Seismol. Soc. Am.* **35**, 3–12.

Gutenberg, B. and C.F. Richter (1954). "Seismicity of the Earth and Associated Phenomena." Princeton University Press, Princeton.

Gutenberg, B. and C.F. Richter (1956). Magnitude and energy of earthquakes. *Ann. Geofis.* **9**, 1–15.

Hanks, T.C. and H. Kanamori (1979). A moment magnitude scale. *J. Geophys. Res.* **84**, 2348–2350.

Huang, W-C., G. Ekström, E.A. Okal and M. Salganik (1994). Application of the CMT algorithm to analog recordings of deep earthquakes. *Phys. Earth Planet. Inter.* **83**, 283–297.

Huang, W-C., E.A. Okal, G. Ekström, and M. Salganik (1997). Centroid-moment-tensor solutions for deep earthquakes predating the digital era: The WWSSN dataset (1962–1976). *Phys. Earth Planet. Inter.* **99**, 121–129

Jakobsson, M., N.Z. Cherkis, J. Woodward, R. Macnab, and B. Coakley (2000). New grid of Arctic bathymetry aids scientists and mapmakers. *EOS Trans. AGU* **81**, 89, 93, 96.

Jeffreys, H. and K.E. Bullen (1940). "Seismological Tables." British Association for the Advancement of Science, London.

Kanamori, H. (1977). The energy release in great earthquakes. *J. Geophys. Res.* **82**, 2982–2983.

Kennett, B.L.N. and E.R. Engdahl (1991). Traveltimes for global earthquake location and phase identification. *Geophys. J. Int.* **105**, 429–465.

Kennett, B.L.N., E.R. Engdahl, and R. Buland (1995). Constraints on seismic velocities in the Earth from traveltimes. *Geophys. J. Int.* **122**, 108–124.

Lee, W.H.K., H. Meyers, and K. Shimazaki (Eds.) (1988). "Historical Seismograms and Earthquakes of the World." Academic Press, San Diego.

Pacheco, J.F. and L.R. Sykes (1992). Seismic moment catalog of large shallow earthquakes, 1900 to 1989. *Bull. Seismol. Soc. Am.* **82**, 1306–1349.

Pérez, O.J. and C.H. Scholz (1984). Heterogeneities of the instrumental seismicity catalog (1904–1980) for strong shallow earthquakes. *Bull. Seismol. Soc. Am.* **74**, 669–686.

Presgrave, B.W., R.E. Needham, and J.H. Minsch (1985). Seismograph station codes and coordinates. *US Geol. Surv. Open-File Rept. 85–714*.

Rothé, J.P. (1969). "The Seismicity of the Earth, 1953–1965." UNESCO, Paris.

Sipkin, S.A. (1982). Estimation of earthquake source parameters by the inversion of waveform data: Synthetic seismograms. *Phys. Earth Planet. Inter.* **30**, 242–259.

Smith, W.H.F. and D.T. Sandwell (1997). Global sea floor topography from satellite altimetry and ship depth soundings. *Science* **277**, 1956–1962.

Utsu, T. (1979). Seismicity of Japan from 1885 through 1925, A new catalog of earthquakes of $M > 6$ felt in Japan and smaller earthquakes which caused damage in Japan (in Japanese with English abstract). *Bull. Earthqu. Res. Inst.* **54**, 253–308.

Utsu, T. (1982a). Seismicity of Japan from 1885 through 1925 (Correction and supplement) (in Japanese with English abstract). *Bull. Earthqu. Res. Inst.* **57**, 111–117.

Utsu, T. (1982b). Catalog of Large earthquakes in the region of Japan from 1885 through 1980 (in Japanese with English abstract). *Bull. Earthqu. Res. Inst.* **57**, 401–463.

Vaněk, J., A. Zátopek, V. Kárnik, N.V. Kondorskaya, Y.V. Riznichenko, E.F. Savarensky, S.L. Solov'ev, and N.V. Shevalin (1962). Standardization of magnitude scales. *Bull. Acad. Sci. USSR, Geophys. Ser.* 108–111.

Villaseñor, A., E.A. Bergman, T.M. Boyd, E.R. Engdahl, D.W. Frazier, M.M. Harden, J.L. Orth, R.L. Parkes, and K.M. Shedlock (1997). Toward a comprehensive catalog of global seismicity. *EOS Trans. AGU* **78**, 581–588.

Wessel, P. and W.H.F. Smith (1998). New, improved version of the Generic Mapping Tools released. *EOS Trans. AGU* **79**, 579.

Editor's Note

Computer readable data files for this Chapter are given on the Handbook CD, under the directory \41Engdahl.

42

A List of Deadly Earthquakes in the World: 1500–2000

Tokuji Utsu
The University of Tokyo, Tokyo, Japan

A list was compiled containing the earthquakes in the world from 1500 through 2000 in which 50 or more (or *many*) people were reported killed. The data have been collected from various catalogs, research articles, and reports on global and regional earthquakes. They are listed in the Bibliography Section. Many other materials on individual earthquakes have also been used. Since the materials on historical earthquakes accessible to the complier are limited, and some materials may contain various kinds of inaccuracies, this list is neither complete nor accurate, especially for the old-time events.

All values in the list (see Table 1, Figs. 1–3 by century, and Color Plate No. 17) are taken from some previously published

TABLE 1 A List of Deadly Earthquakes in the World for 1500–2000. See Text for Explanation

Year	Month/day	h:min	Lat.	Long.	*M*	Deaths	Remarks
1500	1/04		24.9	103.1	7.0	10000s	China: Yunnan P. [Yiliang EQ] (I ≥ 9) $M \geq 7.0$
1501	1/19		34.8	110.1	7.0	400	China: Shaanxi P. [Chaoyi EQ] (I = 9) D = 170
1502	1/18	10:00L	37.2	138.2	6.8	many	Japan: Niigata P. (Naoetsu)
1502	10/17	20:00L	35.7	115.3	6.5	150	China: Henan P. [Pucheng EQ] D = 50/100 + /many
1505	7/06		34.7	69.2		great	Afghanistan: Kabul, Bala, Hicar, Pangan (I = 9–10)
1508	5/29		35.2	25.1	7.1	300	Greece: Ierapetra, Megalokastron (Crete) [T]
1509			13.2	43.6		many	Yemen: Mawza (1509–1510)
1509	9/14		41.0	28.8	7.4	13000	Turkey: Tsurlu, Istanbul (I = 10–11) [T]
1510	6/10		48.8	10.5		many	Germany: Nordlingen D = 2000?
1510	9/11	4:00L	34.6	135.6	6.8	many	Japan: Osaka P. [T]
1511	3/26	14:40U	46.2	13.4	6.9	many	Slovenia: Idrija/Italy: Friuli D = 3T/6T/12T
1512			46.4	9.0		many	Switzerland (I = 7)
1512	10/08		25.0	98.5	6.8	many	China: Yunnan P. [Tengchong EQ] (I = 9)
1514	4/16		37.7	21.0	6.5	many	Greece: Zakinthos (I = 10)
1514	9/18		25.1	99.1	5.5	many	China: Yunnan P. [Baoshan EQ] Date = 9/18–28
1515	6/17		26.6	100.7	7.8	1000s	China: Yunnan P. [Heqing EQ] (I = 10) D = 100s?
1515	10/23		25.7	100.2	6.0	100s	China: Yunnan P. [Dali EQ] (I = 8)
1518	11/09		37.2	−1.5		many	Spain: Vera (Almeria) (I = 9–11)
1522	9/21		36.5	−2.3		many	Spain: Almeria, Granada (I = 9–11) Date = 9/22?
1522	10/22		38.0	−25.0		5000	Azores: S. Miguel (I = 10)
1530	9/01	14:30U	10.7	−64.1		many	Venezuela: Cumana, Paria (I = 10) [T]
1531	1/26		39.0	−8.0		30000	Portugal: Lisbon/Spain/Morocco D = many/1T [T]

(continued)

INTERNATIONAL HANDBOOK OF EARTHQUAKE AND ENGINEERING SEISMOLOGY, VOLUME 81A ISBN: 0-12-440652-1

TABLE 1 *(continued)*

Year	Month/day	h:min	Lat.	Long.	*M*	Deaths	Remarks
1535	5/		25.8	116.4	4.5	many	China: Fujian P. [Changting EQ] (I = 6) [L]
1536	3/19	2:00L	28.1	102.1	7.5	many	China: Sichuan P. [Xichang EQ] (I = 10) D = 1000s?
1542	6/13	9:00U	44.0	11.4	5.9	500	Italy: Scarperia, Megello (I = 9) D = 120–150
1542	12/10	15:15U	37.2	15.0	6.0	200	Italy: Stracusano, Lentini, Sortino, Mineo
1543			10.6	−64.1		many	Venezuela: Cumana, Nueva Cadiz [T]
1544	4/22		38.8	22.6	6.8	many	Greece: Lamia (I = 9–10) [F?]
1546	1/14	16:00L	32.0	35.5	7.0	many	Israel/Jordan/Lebanon (I = 10–11) [T]
1546	7/18					many	Korea: Ch'ungch'ong P.
1549	2/16	0:00L	33.7	60.0	6.7	3000	Iran: Qayin
1550			38.1	46.4	7.1	many	Iran: Tabriz (I = 9) Date = 1/20?
1551			25.8	105.2	5.0	many	China: Guizhou P. [Qinglong EQ] (summer)
1551	1/28		38.4	−9.1		2000	Portugal: Lisbon (I = 9)
1556	1/23	0:00L	34.5	109.7	8.3	830000	China: Shaanxi P. [Huaxian EQ] (I = 11) [F]
1556	1/24					many	China: Shanxi P. (aftershock) Also 1/26 & 2/01
1560	3/		24.2	102.7	5.5	many	China: Yunnan P. [Tonghai EQ] (I = 7) D = 10s
1560	4/		24.9	103.2	5.5	many	China: Yunnan P. [Yiliang EQ] (I = 7)
1561	7/25		37.5	106.2	7.3	many	China: Ningxia A.R. [Zhongwei EQ] D = 1000/5000
1561	7/31	18:40U	40.6	15.4	6.0	100	Italy: Buccino D = 500/550 (same as 8/19?)
1561	8/19	14:10U	40.5	15.5	6.5	500	Italy: Vallo di Diano (I = 10) Time = 19:00U?
1562	10/28	10:00U	−38.7	−73.2	8.0	many	Chile: Santiago, Arauco (I = 11) [T]
1564	7/20	18:00U	44.0	7.3	5.9	900	France: La Bollene–Vesubie D = 500/650
1566	7/11		39.0	21.7	6.5	many	Greece: Karpenisi (I = 8)
1568			36.0	107.9	5.5	many	China: Gansu P. [Qingyang EQ] (I = 7)
1568	5/15		34.4	109.0	6.8	many	China: Shaanxi P. [NE Xi'an EQ] D = 200 C = 5000
1568	12/27		20.2	−103.5		many	Mexico: Cocula, Tzacoalco (Jalisco)
1570	2/08	5:00U	−36.8	−73.0	8.3	2000	Chile: Concepcion (I = 11) [T]
1570	11/17	19:10U	44.8	11.6	5.5	70	Italy: Ferrara (I = 8) D = 40/125/2000?
1573	1/10	4:00L	34.4	104.0	6.8	many	China: Gansu P. [Minxian EQ] (I = 9)
1574			34.0	51.4		1200	Iran: Fin, Kashan (autumn)
1575	9/08		−0.2	−78.6	7.5	many	Ecuador
1575	12/16	18:30U	−39.8	−73.2	8.5	1500	Chile: Valdivia, Santiago (I = 10) D = 120 [T]
1577	3/13		25.0	98.6	6.8	170	China: Yunnan P. [Tengchong EQ] (I = 8)
1581			−14.1	−71.5	7.6	many	Peru (I = 10)/*M* 6.4? (I = 6?)
1582	1/22	16:30U	−16.6	−71.6	8.2	many	Peru: Arequipa (I = 10) D = 30/40 [T] 7.5W
1584	3/01	8:30U	46.0	6.0		100	Switzerland: Geneva–Yvoire D = many (incl. 3/03–04)
1584	6/17		40.0	39.0	6.6	many	Turkey: Erzincan (I = 9) D = 1500?
1584	9/10	20:30U	43.9	12.0	5.8	196	Italy: San Piero (Bagno), Bagno di Romagna (I = 9)
1586	1/16		36.6	136.9	7.0	many	Japan: Toyama P.
1586	1/18	1:00L	35.3	136.6	7.9	1000s	Japan: Gifu & Aichi P. [Tensho EQ] [T?]
1586	7/10	0:30L	−12.3	−77.7	8.1	many	Peru: Lima, Callao (I = 10) D = 20 8.1W
1586	12/23		14.5	−91.0		many	Guatemala: Antigua
1587	7/09		−5.3	−80.5		many	Peru: Sechura
1587	8/30	20:30L	−0.2	−78.5		many	Ecuador: Quito, Pomasque (I = 9)
1587	11/		37.1	−8.0		170	Portugal: Loule (I = 10)
1588	8/09		24.0	102.8	7.0	1000	China: Yunnan P. [Tonghai EQ] (I ≥ 9) D = many
1590			−17.2	−72.9	7.7	many	Peru: Torata, Camana/Chile *M* = 8.2
1590	7/07		35.4	103.9	5.5	many	China: Gansu P. [Lintao EQ] (I = 7)
1590	9/05		45.9	16.3		many	Croatia: Zagreb/Hungary: Nagy Kanizsa (I = 10)
1590	9/15		48.3	15.5	6.6	many	Austria. Neulengbach, Vienna (I = 9–10) 3 times
1593	9/		27.7	54.3	6.5	3000	Iran: Lar (late summer)
1596	9/04	19:00L	33.3	131.6	7.0	800	Japan: Oita P. (Beppu Bay) Date = 9/01? [T]
1596	9/05	1:00L	34.7	135.6	7.5	1000s	Japan: Hyogo, Osaka & Kyoto P. [F]
1599	2/03	20:00U	10.0	−65.2		60	Venezuela: Bailadores

(continued)

TABLE 1 *(continued)*

Year	Month/day	h:min	Lat.	Long.	*M*	Deaths	Remarks
1600	9/29	20:00L	23.5	117.2	7.0	many	China: Guangdong P. [Nan'ao EQ] (I = 9)
1601	9/08	1:00U	46.8	8.5		many	Switzerland (I = 9) D = 8
1604	11/24	18:30U	−17.9	−70.9	8.4	many	Peru: Arequipa, Camana/Chile: Arica (I = 10+) [T]
1605	2/03	20:00L	33.5	138.5	7.9	1000s	Japan: Tokai–Nankai region [Keicho EQ] [T]
1605	7/13	22:00L	19.9	110.5	7.5	1000s	China: Hainan P. [Qiongshan–Wenchang EQ] (I = 11)
1606	11/30		23.6	102.8	6.3	1000s	China: Yunnan P. [Jianshui EQ] (I = 9)
1608	4/20	12:00L	36.4	50.5	7.4	many	Iran: Taleqan, Rudbarat-i Alamut (I = 10–11) [T]
1609	7/12	0:00L	39.2	99.0	7.3	840	China: Gansu P. [Jiuquan EQ] (I = 10)
1610	2/03	19:30U	8.3	−71.8		60	Venezuela: Merida, La Grita, Bailadores
1611	9/27	9:00L	37.6	139.8	6.9	3700	Japan: Fukushima P. [Aizu EQ] [R]
1611	12/02	14:00L	39.0	144.0	8.1	5000	Japan: Sanriku, Hokkaido D = 1783/3000 [T]
1612	11/08		34.9	25.1	7.5	many	Greece: Crete (I = 8) [T]
1613			35.3	25.1		many	Greece: Iraklion (Crete) (same as 1612/11/08?)
1613	8/25	5:00U	38.1	14.8	5.4	103	Italy: Naso (Messina) (I = 9) [T]
1615	6/26	12:00L	35.7	139.7	6.5	many	Japan: Tokyo
1618	5/20	6:00L	37.0	111.9	6.0	many	China: Shanxi P. [Jiexiu EQ] D = 5000?
1618	5/26		18.9	72.9	6.9	2000	India: Bombay (storm?) (I = 9?) [T]
1618	8/25	20:00L	46.3	9.5		1200	Italy–Switzerland border: Piuro [L]
1618	12/19		35.0	58.0		800	Iran: Dughabad, Khorasan
1619			52.3	14.6		many	Germany: Near Frankfurt an der Oder
1619	2/14	16:30U	−8.9	−79.3	7.7	350	Peru: Trujillo, Piura, Santa (I = 9) D = 400
1619	5/00	12:00L	35.1	58.9	6.5	800	Iran: Dughabad (I = 9–10)
1619	5/01	12:00L	32.5	130.6	6.2	many	Japan: Kumamoto P. (Yatsushiro) [T]
1619	11/30	1:60L	18.5	121.6	8.0	many	Philippines (Luzon): Cagayan, Isabela (I = 10)
1620			40.9	71.4	5.8	many	Uzbekistan: Akhsy (I = 9)
1622	5/05		37.6	21.0	6.6	many	Greece: Ionian Sea, Zakinthos (I = 9) [T]
1622	10/25		36.5	106.3	7.0	12000	China: Ningxia A.R. [Guyuan EQ] (I = 10) D = 6T/15T
1624	2/10	16:00L	32.3	119.4	6.0	many	China: Jiangsu P. [Yangzhou EQ] (I = 8+)
1624	3/18	19:45U	44.7	11.9	5.4	50	Italy: Argenta (I = 8–9) D = 25–50
1625	7/21		32.8	130.6	5.5	50	Japan: Kumamoto P. (Kumamoto)
1626	6/28	2:00L	39.4	114.2	7.0	5200	China: Shanxi P. [Lingqiu EQ] D = 10000s (I = 9)
1627	7/30	10:50U	41.7	15.4	6.8	5000	Italy: San Severo, Lesina, Apricina (I = 10–12)
1628	10/07		40.6	114.2	6.5	61	China: Hebei P. [Xiyanghebao–Dukoubao EQ] (I = 8)
1629			53.6	12.3		many	Germany: Mecklenburg
1629	3/10	9:00L	35.1	23.7	7.0	many	Greece: Crete, Cythera Date = 2/27? 3/07? 3/09? [T]
1630	7/22		38.3	20.9	6.9	many	Greece: Levkas, Cephalonia, Ithaca (I = 9–11)
1631	2/10	0:00L	11.3	41.7	5.7	50	Ethiopia: Waraba, Aussa (I = 9) D = 5000? [V]
1631	8/14	2:00L	29.3	111.7	6.8	many	China: Hunan P. [Changde–Lixian EQ] (I = 8+)
1633	3/01	5:00L	35.2	139.2	7.0	150	Japan: Kanagawa P. [Odawara EQ] D = 237 [T]
1633	11/05	13:00L	37.6	21.0	6.9	many	Greece: Zakinthos (Zante) (I = 10) [T]
1636	9/30		38.0	20.7	7.2	520	Greece: Cephalonia, Zakinthos (I = 10)
1638	3/27	19:30U	39.1	16.3	6.9	30000	Italy: Calabria (I = 11) D = 1580/9571/19000 [T]
1639	10/08	0:35U	42.6	13.3	5.5	500	Italy: Amatrice (I = 10) Another EQ on 10/15
1639	11/		36.1	136.2	6.0	380	Japan: Fukui P. D = doubtful
1640	2/		−1.7	−78.6		5000	Ecuador
1640	2/27		38.2	46.3	6.1	many	Iran: Tabriz (I = 8)
1640	7/31		42.1	140.7		700	Japan: Hokkaido (Komagatake eruption) [VLT]
1640	11/23		36.3	136.2	6.5	many	Japan: Ishikawa P. (Daishoji)
1641	1/11		11.0	−67.0		200	Venezuela: Caracas D = many Date = 6/11?
1641	2/05	18:00L	37.9	46.1	6.8	12613	Iran: Tabriz, Dehkharqan (I = 9) D = 1200/30000
1641	12/21	12:00L	38.7	−28.1		50	Azores: Valas [T]
1642	6/30	2:00L	35.1	111.1	6.0	many	China: Shanxi P. [NE Yuncheng EQ] (I = 8)
1644	1/16	11:00U	7.5	−72.5	7.0	many	Colombia: Pamplona/Venezuela: San Cristobal
1644	2/15	9:20L	44.0	7.3	6.1	many	France: Val Vesubie (I = 8–9) D = 3/150
1644	3/16		7.6	−73.0		many	Venezuela: San Cristobal/Colombia: Pamplona

(continued)

TABLE 1 *(continued)*

Year	Month/day	h:min	Lat.	Long.	M	Deaths	Remarks
1644	10/18		39.4	140.0	6.5	63	Japan: Akita P. (Honjo) D = 117 [T?]
1645	2/19		−1.7	−78.6	7.5	many	Ecuador
1645	11/30	12:30L	15.6	121.2	7.9	600	Philippines (Luzon): Manila D = 500/3000
1647	5/14	2:30U	−33.4	−70.6	8.5	2000	Chile: Santiago (I = 11) D = 1000 [T]
1648	4/01	0:00L	38.3	43.5	6.7	2000	Turkey: Van, Hayotsdzor (I = 9–10)
1649	7/30	1:00L	35.8	139.5	7.0	many	Japan: Tokyo & Saitama P.
1650	1/04		38.0	46.2		many	Iran: Tabriz
1651	6/07		37.7	29.3	6.6	700	Turkey: Laodikea (I = 9) Date = 5/28?
1652	7/13	10:00L	25.2	100.6	7.0	3000	China: Yunnan P. [Midu EQ] (I = 9+)
1653	2/23		38.3	27.1		15000	Turkey: Izmir (I = 10) D = 2500/3000/8000
1654	7/21		34.3	105.5	8.0	31000	China: Gansu P. [Tianshui] (I = 11) D = 12000/10000s
1654	7/23	0:25U	41.6	13.7	6.1	200	Italy: Sorano-Marsica area (I = 10) D = 600
1654	10/20		40.1	3.6		many	Spain: Minorca Is. (I = 9)
1655	11/13	19:45U	−12.3	−77.6	7.7	many	Peru: Lima, Callao D = 11T? Year = 1654? 56? 57?
1657	3/15	23:30U	−36.8	−73.0	8.0	many	Chile: Concepcion, Chillan, Santiago (I = 11) [T]
1657	4/21		31.3	103.5	6.5	many	China: Sichuan P. [Wenchuan EQ] (I = 8)
1658	2/03	20:00L	39.4	115.7	6.0	many	China: Hebei P. [Laishui EQ] (I = 7–8)
1658	2/14		−8.1	−79.0	7.7	many	Peru: Trujillo (I = 7)
1658	8/24		38.3	20.5	6.8	320	Greece: Cephalonia (I = 10)
1659	4/21		37.1	139.8	6.9	50	Japan: Fukushima & Tochigi P.
1659	11/05	22:15U	38.7	16.3	6.4	2035	Italy: Panaia, Polia (C. Calabria) (I = 10)
1660			40.0	41.3	6.5	1500	Turkey: Erzurum (I = 9)
1660	6/21		43.1	0.2		many	France: Lourdes, Bageres de Bigorre (I = 9)
1661	2/15		23.0	120.2	6.5	many	China: Taiwan [Tainan EQ] (I = 8)
1661	3/22	12:50U	44.0	11.9	6.8	400	Italy: Civitella, Galeata (I = 10) D = many
1662	6/16	12:00L	35.2	136.0	7.4	800	Japan: Kyoto & Shiga P. (W of Lake Biwa)
1662	10/31		31.7	132.0	7.6	many	Japan: Miyazaki P. [T]
1664			38.7	112.7	5.5	many	China: Shanxi P. [Xinxian–Daixian EQ] (I = 8)
1664	5/12	9:15U	−14.1	−75.9	7.8	400	Peru: Ica, Piseo (I = 10–11) D = 15/300 7.5W
1664	8/12		35.8	127.4	6.3	50	Korea: Cholla P. (Chinan) (I = 8)
1665	1/		35.0	25.1	6.7	many	Greece: Crete (I = 10)
1665	6/15		35.7	52.1	6.5	many	Iran: Damavand Date = 6/15–7/13
1666	2/01	19:00L	37.1	138.2	6.8	1500	Japan: Niigata P. [Takada EQ] [C]
1667	4/06	8:00U	42.6	18.1	7.2	5000	Croatia: Dubrovnik (I = 10) [T]
1667	11/18		37.2	57.5	6.9	12000	Iran: Shirvan, Shamkha D = 8000
1667	12/17		41.7	47.3	6.5	93	Russia (Dagestan)/Azerbaijan (I = 8) D = 8T?/80T?
1668	1/04		40.5	48.5	7.0	80000	Azerbaijan: Shemakha Date = 1/14? 1667/2/17?
1668	7/25	20:00L	34.8	118.5	8.5	47615	China: Shandong P. [Juxian–Tancheng EQ] (I ≥ 11)
1668	8/17		40.5	35.0	8.0	8000	Turkey: Anatolia (Bolu to Erzincan) D = 18T [F]
1668	9/13		37.9	32.6		many	Turkey: Konia (Konya) (same as 8/17?)
1669	1/14		40.6	48.6	5.7	7000	Azerbaijan: Shemakha (same as 1668/1/04?)
1670	8/19		31.0	122.5		many	China: Jiangsu P. [T]
1671	1/01		40.6	48.6	5.2	many	Azerbaijan: Shemakha (aftershock) (I = 8–9)
1672	4/14	15:45U	43.9	12.6	5.7	200	Italy: Rimini (I = 8) D = 500/1500 [T]
1673	7/30		36.4	59.3	7.1	5600	Iran: Meshed, Neyshabur (Khorasan) D = 4000
1674	1/01		39.5	20.0	6.5	200	Greece: Corfu
1674	2/12		−3.5	128.2		2342	Indonesia (Moluccas): Amboina D = 2323 [T]
1677	8/		33.4	104.9	5.5	many	China: Gansu P. [Wudu EQ] (I = 7) (Aug. or Sept.)
1677	11/04	20:00L	35.0	141.5	8.0	540	Japan: Fukushima, Ibaragi & Chiba P. [T]
1678		0:00L	34.3	58.6	6.5	many	Iran: Gonabad (winter)
1678	6/18	1:45U	−12.3	−77.8	7.9	many	Peru: Lima, Callao (I = 9) Date = 6/17? [T]
1679	6/04	4:00U	40.2	44.7	6.4	7600	Armenia: Garnii, Yerevan, Dvina (I = 9–10)
1679	9/02	11:00L	40.0	117.0	8.0	45500	China: Hebei P. [Sanhe–Pinggu EQ] (I = 11) [F]
1680	9/09	20:00L	25.0	101.5	6.8	2700	China: Yunnan P. [Chuxiong EQ] (I = 8–9)
1680	10/09	7:30U	36.5	−4.4		70	Spain: [Malaga EQ] (I = 9) D = 60

(continued)

TABLE 1 *(continued)*

Year	Month/day	h:min	Lat.	Long.	*M*	Deaths	Remarks
1683	11/22	14:00L	38.7	112.7	7.0	8220	China: Shanxi P. [Yuanping] (I = 9) D = 1000s
1687	10/20	11:30U	−15.2	−75.9	8.2	5000	Peru: Lima, Callao, Ica (I = 10) D = 200/600 [T] 8.4W
1688	4/11	11:30U	44.4	13.0	5.8	50	Italy: Cotignola, Bagnacavallo (I = 9) D = 36/many
1688	5/26		24.0	68.0	7.6	many	India: Delta of the Indus (I = 10) D = 2000
1688	6/05	15:30U	41.3	14.6	6.6	10000	Italy: Campania (I = 11) D = 3311/8000
1688	6/14		26.5	99.9	6.3	193	China: Yunnan P. [Jianchuan EQ] (I = 8) Also 6/16
1688	7/10	11:45U	38.4	26.9	7.0	17500	Turkey: Izmir (I = 10) D = 2T/15T Date = 8/11? [T]
1692	6/07		17.8	−76.7		3000	Jamaica: [Port Royal EQ] D = 2000 [T]
1693	1/09	21:00U	37.1	15.0	6.2	many	Italy: Sicily (Catania foreshock) (I = 8–9)
1693	1/11	13:30U	37.1	15.0	7.4	54000	Italy: Sicily [Catania EQ] (I = 11) D = 18T/93T [T]
1694	6/19	7:00L	40.2	140.1	7.0	394	Japan: Akita P. [Noshiro EQ] [C]
1694	9/08	11:40U	40.9	15.4	6.8	4820	Italy: Basilicata (I = 10) D = 4057/6500
1695	5/11	5:00L	37.1	57.5	7.0	360	Iran: Esfarayen (I = 9–10) D = 100s
1695	5/18	20:00L	36.0	111.5	7.8	52600	China: Shanxi P. [Linfen EQ] (I = 10) D = 30T/100T
1695	6/11	2:30U	42.6	12.1	6.1	50	Italy: Bagnoregio (I = 9–10) D = 31
1696	4/14		39.1	43.9	6.8	many	Turkey: Chaldiran (I = 10)
1698	6/20	6:00U	−1.2	−78.7	7.7	many	Ecuador: Quito, Ambato, Carguayrazo Date = 7/19?
1701	3/19		40.2	19.8	6.4	300	Albania: Tepeleni (I = 8–9) [F]
1702	3/14	5:00U	41.1	15.0	6.3	150	Italy: Benevento (I = 10) D = 40/500
1703	1/14	18:00U	42.7	13.1	6.7	9761	Italy: Norcia, L'Aquila (I = 11) D = 10T/40T
1703	2/02	11:05U	42.4	13.3	6.7	many	Italy: L'Aquila (I = 10) D = 5T Also 3/18 & 3/27
1703	12/31	2:00L	34.7	139.8	8.1	great	Japan: Southern Kanto [Genroku EQ] D = 10T? [TC]
1704	5/27	13:00L	40.4	140.0	7.0	77	Japan: Akita & Aomori P. (Noshiro)
1704	9/18		38.0	116.5	5.5	many	China: Hebei P. [Cangzhou-Dongguang EQ] (I = 7)
1704	9/28		34.9	106.8	6.0	many	China: Shaanxi P. [Longxian EQ] (I = 8)
1705	1/27		38.7	41.7	6.7	many	Turkey: Bitlis (I = 9–10)
1706	11/03	1:30U	42.1	14.1	6.7	2400	Italy: Maiella (I = 10–11) D = 1000/15000
1707			40.4	43.0		1000s	Turkey: Kars D = many
1707	9/17	16:48U	−13.0	−72.0		160	Peru: Cuzco Date = 9/18?
1707	10/28	14:00L	33.0	136.0	8.6	5000	Japan: Tokai–Nankai region [Hoei EQ] D = 20T? [T]
1709	10/14	8:00L	37.4	105.3	7.5	2000	China: Ningxia A.R. [Zhongwei EQ] (I = 9–10)
1710	10/03	14:00L	35.5	133.7	6.5	many	Japan: Tottori & Okayama P. Also 10/04
1713	2/26	20:00L	25.4	103.2	6.8	1000s	China: Yunnan P. [Xundian EQ] (I = 9) D = 2100
1713	9/04	0:00L	32.0	103.7	7.0	many	China: Sichuan P. [N Maowen EQ] (I = 9)
1714	4/28	22:00L	36.7	137.9	6.3	86	Japan: Nagano P. D = 56
1716	2/03		36.9	2.9		20000	Algeria: Algiers Date = 15/5/,8/05,16/5/,17/8/05
1716	2/06		−17.2	−71.2	8.8	many	Peru: Torata (I = 9) [T]
1717	3/12	6:00L	38.1	46.3	5.9	700	Iran: Tabriz
1718	6/19	4:00L	35.0	105.2	7.5	75000	China: Gansu P. [Tongwei EQ] (I = 10) D = 40T+
1718	8/22	14:00L	35.3	137.9	7.0	60	Japan: Nagano P. [R]
1719	5/25		40.8	29.5	7.0	1000	Turkey: Istanbul, Izmit (I = 9–10)
1720	7/12		40.4	115.2	6.8	many	China: Hobei P. [Shacheng EQ] (I = 9) D = 37/1000
1720	7/15		29.0	77.5		many	India: Delhi
1720	10/31		23.0	120.3	6.8	many	China: Taiwan [Tainan EQ] (I = 9)
1721	1/05		23.0	120.3	6.0	many	China: Taiwan [Tainan EQ] (I = 8)
1721	4/26	7:48L	37.9	46.7	7.4	40000	Iran: Tabriz (I = 11) D = 8T/10T/100T/250T [F]
1721	9/		23.0	120.2		1000s	China: Taiwan [Tainan EQ] (I = 8) (same as 1/05?)
1721	9/26		38.1	46.3	6.0	many	Iran: Tabriz (I = 8–9) (same as 4/26?)
1722	12/27		37.0	−8.0		many	Spain/Portugal: Algarve (I = 10) [T]
1723	2/22	2:00L	38.6	20.7	7.0	many	Greece: Ionian Sea, Levkas (I = 9–10) [T]
1725	1/07	4:15U	−9.2	−79.3	7.5	1500	Peru: Trujillo, Ancash (I = 9) Date = 1/25? [L]
1725	1/08	16:00L	25.1	103.1	6.8	894	China: Yunnan P. [Songming–Yiliang EQ] D = 614
1725	8/01	16:00L	30.0	101.9	7.0	many	China: Sichuan P. [Kangding EQ] (I = 9)
1726	9/01	2:15U	38.1	13.4	5.7	250	Italy: Palermo (I = 9) D = 226/6000 [T]
1727	11/18		38.0	46.2	7.2	77000	Iran: Tabriz (I = 10) (same as 1721/4/26?)

(continued)

TABLE 1 *(continued)*

Year	Month/day	h:min	Lat.	Long.	M	Deaths	Remarks
1730	3/28	8:00U	44.0	10.3	5.1	many	Italy: Massa di Carrara (I = 9) Date = 3/29?
1730	5/12	18:30U	42.8	13.1	6.0	500	Italy: Norcia (I = 9) D = 200 Time = 3:15U?
1730	9/30	10:00L	40.4	116.2	6.5	100s	China: Beijing [Beijing EQ] (I = 8) (C = 457/650+)
1731	3/20	0:30U	41.3	15.8	6.6	500	Italy: Foggia (I = 10) D = 600–1500 Date = 4/17?
1732	3/27		40.9	14.8		2000	Italy: Avellino (same as 11/29?)
1732	11/29	7:40U	41.1	15.1	6.6	1942	Italy: Irpinia (I = 10–11) D = 600/1500 Time = 6:30U?
1733	8/02	16:00L	26.2	103.1	7.8	1200	China: Yunnan P. [Dongchuan EQ] (I = 10) D = 26
1735	12/		35.0	34.0	6.5	200	Cyprus: Famagusta
1736	1/30	2:00L	23.1	120.3	6.3	372	China: Taiwan [Tainan EQ] D = 120/327/426/574
1738			33.3	96.6	6.5	115	China: Qinghai P. [NW Yushu EQ] (I = 8) D = many
1739						many	Peru: Santa Catalina
1739	1/03	19:00L	38.8	106.5	8.0	50000	China: Ningxia A.R. [Pingluo–Yinchuan EQ] [F]
1741	8/29		41.6	139.4		2000	Japan: SW Hokkaido [Oshima Tsunami] [TV]
1742	2/14		37.8	20.6	6.5	120	Greece: Zakinthos (I = 9)
1743	2/20	16:30U	39.9	18.8	7.3	180	Italy (Ionian Sea): Nardo (I = 9) [T]
1746	10/29	3:30U	−12.0	−77.2	8.4	18000	Peru: Lima, Callao (I = 10) D = 4800/4941/7141 [T]
1747			−13.9	−70.4		many	Peru: Carabua P., Ayapata (I = 9) Date = 2/15?
1748	3/23		39.5	−0.4		many	Spain: Valencia D = 10000?
1748	5/25	15:00L	38.2	22.2	6.8	many	Greece: Aeghio (I = 9) [T]
1749	3/25		39.5	−0.4		5000	Spain: Coast of Valencia P. (I = 10)
1750	5/24		43.1	0.0		many	France: Juncalas, Tarbes, Lourdes (I = 9)
1750	6/07		36.3	22.8	7.0	2000	Greece: Kythera, Morea, Cerigo (I = 10)
1750	9/15	14:00L	24.7	102.9	6.3	57	China: Yunnan P. [Chengjiang EQ] (I = 8) D = 37
1750	12/17		45.2	14.2		many	Croatia: Rijeka (Fiume) (I = 9) Date = 9/17? [T]
1751	5/21	2:00L	37.1	138.2	7.2	1541	Japan: Niigata & Nagano P. [Takada EQ]
1751	5/25	5:30U	−36.8	−71.6	8.5	65	Chile: Concepcion, Chillan, Talca, Tutuben [T]
1751	5/25	14:00L	26.5	99.9	6.8	1050	China: Yunnan P. [Jianchuan EQ] (I = 9) D = 900/3000
1752	7/21		35.5	35.5	7.0	20000	Syria: Latakia/Lebanon: Tripoli [T]
1754	9/		30.0	32.0		1000s	Egypt: Grand Cairo [Tanta EQ] D = 40000
1755	1/27		24.7	102.2	6.5	345	China: Yunnan P. [Yimen EQ] (I = 8+) D = 270
1755	2/08		23.8	102.7	6.0	74	China: Yunnan P. [E Shiping EQ] (I = 7–8)
1755	4/26		−0.2	−78.5	7.0	many	Ecuador: Quito, Pichincha (I = 9)
1755	6/07		34.0	51.4	5.9	1200	Iran: Kashan, Tabriz (daytime) D = 40T
1755	11/01	10:16U	36.0	−11.0	8.5	62000	Portugal/Spain/Morocco [Lisbon EQ] D = 55T [T]
1755	11/19		34.1	−5.3		3000	Morocco: Mequinez (Mekenes), Fes (I = 10)
1757	2/22		−0.9	−78.6	7.0	1000	Ecuador: Latacunga, Cotopaxi, Tungurahua (I = 9)
1757	4/15		34.0	−6.5		3000	Morocco: Sale, Cap Cantin
1757	7/09		38.4	−28.0		many	Azores: Angra (Terceira), St. George (I = 11) [T]
1757	7/30					many	Korea: Chungchong P.
1758	1/		36.9	10.2		many	Tunisia: Constantine, Tunis (I = 10)
1759	6/22	22:30L	40.7	23.1	6.5	many	Greece: Thessaloniki (I = 9) [F]
1759	10/30	3:45L	33.1	35.6	6.6	2000	Syria: Baalbec, Damascus/Israel: Safad D = 20T [T]
1759	11/25	19:23L	33.7	35.9	7.4	30000	Lebanon/Syria: Baalbec/Israel: Safad D = 40T [FT]
1760	7/11		20.0	−75.5		many	Cuba: Santiago de Cuba
1761	5/23	16:00L	24.4	102.5	6.3	129	China: Yunnan P. [Yuxi EQ] (I = 8) D = 120
1761	11/03	16:00L	24.4	102.5	5.8	82	China: Yunnan P. [Yuxi EQ] (I = 7+) D = 50
1762	4/02		22.0	92.0	7.5	many	Bangladesh: Arakan/Myanmar (I = 10–11) [T]
1763	6/28	5:28U	47.5	18.1	6.2	63	Slovakia: Komarno/Hungary: Gyor D = 83
1763	12/30	22:00L	24.2	102.8	6.5	1000	China: Yunnan P. [Tonghai EQ] D = 800 + many
1764	7/		14.0	−86.0		many	Honduras: Trujillo (I = 8)
1765	9/02	8:00L	34.8	105.0	6.5	2068	China: Gansu P. [Wushan–Gangu EQ] D = 1189
1766	3/08	18:00L	40.7	140.6	7.3	1335	Japan: Aomori P. (Hirosaki) [Tsugaru EQ] [C]
1767	7/11		38.2	20.3	7.2	253	Greece: Lixouri (Cephalonia) (I = 10)
1767	7/14		39.4	16.3	5.6	many	Italy: Luzzi, Rose, Cosenza (I = 8–9) D = 40
1768	10/19	23:00U	43.9	11.9	5.8	54	Italy: Santa Sofia (I = 9) D = 120

(continued)

TABLE 1 (*continued*)

Year	Month/day	h:min	Lat.	Long.	*M*	Deaths	Remarks
1771	4/24	8:00L	24.0	124.3	7.4	12000	Japan: Ishigaki Is. [Yaeyama Tsunami] [T]
1773	6/03		14.6	−91.2		20000	Guatemala: Santiago D = 5000–8000 families [T]
1773	6/10		14.3	−90.4		100	Guatemala: Antigua (D incl. 7/29?)
1773	7/29		14.6	−91.2		100	Guatemala: Santiago, Antigua D = 120
1776	12/		23.5	120.5	6.0	many	China: Taiwan [Chiai EQ] (I = 8) (Dec. or Jan.)
1777	11/		23.5	120.5	5.5	many	China: Taiwan [Chiai EQ] (Nov. or Dec.)
1778	12/15	7:00L	34.0	51.4	6.2	8000	Iran: Kashan D = 30000
1780			34.0	58.0	6.5	3000	Iran: Khurasan
1780	1/08	19:06L	38.1	46.3	7.4	50000	Iran: Tabriz (I = 11) D = 100T/200T [F]
1780	10/02		18.1	−78.1		300	Jamaica: Savanna del Mar [T]
1780	10/28	5:00L	35.0	25.8	7.7	100	Greece: Hierapetra (Crete) Date = 10/22?
1781			36.5	43.0	6.7	80	Iraq: Mosul (1196H) (I = 9–10)
1781	6/03	6:25U	43.6	12.6	6.2	300	Italy: Cagli (I = 10) Time = 10:00U?
1783	2/05	12:00U	38.4	16.0	6.9	35000	Italy: [Calabria EQ] (I = 11) D = 30T/40T/50T [CT]
1784	3/24		16.5	−99.5		many	Mexico: Guerrero, Acapulco [T]
1784	5/13	12:36U	−16.5	−72.0	8.0	400	Peru: Arequipa, Camana (I = 10) D = 54/300 [T] 8.4W
1784	7/18		39.5	39.7	7.6	5000	Turkey: Erzincan, Erzurum D = ? Date = 7/23? [F]
1786	2/05		39.6	19.9	6.6	126	Greece: Corfu, Argos (I = 9–10) D = 120
1786	6/01	12:00L	29.9	102.0	7.8	446	China: Sichuan P. [Kangding–Luding EQ] D = 120/430
1786	6/10	2:00L	29.4	102.2	7.0	10000s	China: Sichuan P. [Luding EQ] [R] (great flood)
1787	5/13	12:22U	10.8	122.5		many	Philippines (Panay): Iloilo, Antique 5/12 22:U?
1787	7/12	22:45L	11.1	122.3	7.4	many	Philippines (Panay): Iloilo, Antique (I = 10) 7/13?
1788	10/12		14.0	−61.0		900	St. Lucia (Lesser Antilles)
1789	5/29		39.0	40.0	7.0	51000	Turkey: Palu (I = 8–9)
1789	6/07	20:00L	24.2	102.9	7.0	1000	China: Yunnan P. [Tonghai–Huaning EQ] (I = 9+)
1789	9/30	10:45U	43.5	12.2	5.8	500	Italy: Cospaia, Selci, Citta di Castello (I = 10)
1790	10/09	1:15U	35.7	0.6		3000	Algeria: Oran (I = 10)
1792	5/21	20:00L	32.8	130.3	6.4	15000	Japan: Nagasaki P. [Shimabara EQ] [LTV]
1792	8/09	17:00L	23.6	120.6	7.0	614	China: Taiwan [Chiai EQ] (I = 9) D = 617
1792	9/07		30.8	101.2	6.8	205	China: Sichuan P. [Qianning EQ] (I = 8)
1793	2/17	14:00L	38.5	144.5	8.3	50	Japan: Sanriku coast D = 39+/720? [T]
1793	5/15		30.6	101.5	6.0	200	China: Sichuan P. [Qianning EQ] (I = 8)
1794	12/14		10.5	−64.0		many	Venezuela: Cumana
1796	4/26		35.7	36.0	6.6	1500	Syria: Latakia (I = 8) Date = 2/26?
1797	2/04	12:30U	−1.7	−78.6	8.3	40000	Ecuador: Quito, Riobamba D = 6000/6406/16000
1797	12/04		10.5	−64.5		16000	Venezuela: Cumana, Cariaco
1799	7/28	22:05U	43.2	13.2	5.8	104	Italy: Camerino (I = 9) D = 60
1799	8/27	0:00L	23.8	102.4	7.0	2251	China: Yunnan P. [Shiping EQ] (I = 9) D = 2030
1803	2/02	2:00L	25.7	100.5	6.3	200	China: Yunnan P. [Xiangyun–Binchuan EQ] (I = 8)
1803	9/01		22.5	88.4		many	India: Barahal, Badrinath, Calcutta (I = 7–9)
1804	1/08		38.3	21.8		many	Greece: Patrai (Patras) (I = 9) [T]
1804	7/10	22:00L	39.1	140.0	7.0	313	Japan: Yamagata & Akita P. [Kisakata EQ] [T]
1804	8/25	7:00U	36.9	−2.4		162	Spain: Granada, Roquetas (Rochetta), Dalias (I = 10)
1805	6/16	8:15U	5.3	−74.6	6.0	200	Colombia: Honda, Mariquita (I = 8–9)
1805	7/26	21:00U	41.5	14.5	6.6	5573	Italy: Molise (I = 10–11) D = 6000/6573
1806	3/25	17:20L	18.9	−103.8	7.5	many	Mexico: Zapotlan, Colima, Oaxaca, Michoacan
1806	6/11		28.2	91.8	7.5	100	China: Tibet [Cuona EQ] (I = 10)
1810	2/16	21:15U	35.7	25.0	7.8	2000	Greece: Iraklion (Crete) (I = 10)
1810	3/20		28.2	−16.6		many	Canary Is.: Teneriffe Is. (I = 9) Date = 3/25? [V?]
1810	9/25	15:00L	39.9	139.9	6.5	60	Japan: Akita P. (Oga Pen.) [T]
1811	9/27	20:00L	31.7	100.3	6.5	481	China: Sichuan P. [Ganze EQ] (I = 9)
1812	3/08	21:00L	43.7	83.0	7.5	many	China: Xinjiang A.R. [Suiding EQ] (I = 11) D = 58 [F]
1812	3/26	20:07U	10.6	−66.9	6.3	20000	Venezuela: Caracas (I = 9) D = 10T/12T/18T/26T/40T
1812	12/07	15:00L	35.5	139.7	6.3	many	Japan: Kanagawa & Tokyo P.
1814	11/24		23.7	102.5	6.0	100s	China: Yunnan P. [Shiping EQ] (I = 7–8) C = 900+

(*continued*)

TABLE 1 *(continued)*

Year	Month/day	h:min	Lat.	Long.	*M*	Deaths	Remarks
1815	10/13	22:00L	25.0	121.3	6.7	111	China: Taiwan [Tanshui EQ] D=93
1815	10/23	0:00L	34.8	111.2	6.8	13090	China: Shanxi P. [Pinglu EQ] (I=9) D=30T+/37T
1815	11/27	18:00U	−8.0	115.2		10253	Indonesia (Bali) D=1200 [T]
1816	12/08	2:00L	31.4	100.7	7.5	2954	China: Sichuan P. [Luhuo EQ] (I=10) D=2854
1817	8/23	8:00U	38.3	22.1	6.8	many	Greece: Aiyion, Vostitsa, Aeghion, Achaia [T]
1818	2/20	18:15U	37.6	15.1	6.2	72	Italy: Catania area (I=9–10)
1819	2/24		36.1	102.3	5.8	126	China: Qinghai P. [Hualong EQ] (I=7)
1819	3/		35.4	0.1		many	Algeria: Mascara, Oran (I=10)
1819	5/26	17:00U	42.5	11.8		many	Italy: Corneto D=not many? Time=5:00U?
1819	6/16	2:00U	23.3	70.0	8.3	1440	India: [Cutch EQ] D=1543/2T/3T Time=23:30U? [FT]
1819	8/02	14:00L	35.2	136.3	7.3	many	Japan: Shiga P. D=100+
1820			29.6	52.5		many	Iran: Shiraz, Babol (I=9) Date=10/19?
1820	8/03		34.1	113.9	6.0	430	China: Henan P. [Xuchang EQ] (I=8)
1820	12/29		−7.0	119.0	7.5	400	Indonesia (Sulawesi): Makassar, Sumbawa [T]
1821	1/06	17:15U	37.8	21.2	6.5	many	Greece: Zakinthos, Lala (Morca) Date=1/09? [T]
1821	7/10	10:00U	−16.4	−72.3	8.2	162	Peru: Camana, Ocona, Caraveli, Arequipa (I=7)
1822	8/13	19:00U	36.7	36.9	7.4	20000	Syria: Aleppo/Turkey: Antakya D=30T–60T [FT]
1822	9/05		35.0	36.0		22000	Syria: Aleppo, Damascus/Turkey (D incl. 8/13?)
1822	11/20	2:30U	−33.1	−71.6	8.5	many	Chile: Copiapo to Valdivia (I=11) D=72? [T]
1823	9/07		36.3	53.3		many	Iran: Mazandaran
1823	9/29		40.0	141.1	5.9	73	Japan: Iwate P. (Mt. Iwate swarm)
1823	10/		42.7	18.2		many	Croatia: Dubrovnik
1824	6/02		29.7	51.5	6.0	150	Iran: Kazirun, Shahpur (I=8)
1824	6/25	5:30L	29.8	52.4	6.4	many	Iran: Shiraz, Kazrum Date=6/23? 6/26?
1825			36.1	52.5	6.7	many	Iran: Haraz (I=9–10)
1825	1/19	11:00U	38.8	20.7	6.8	58	Greece: Levkas, Amaxiki, Preveza D=34 [T]
1825	3/02	7:00U	36.3	2.5		7000	Algeria: Blida (I=10) D=5000
1827	3/23	20:00L	34.9	111.1	5.5	84	China: Shanxi P. (I=7)
1827	9/24		31.6	74.3		1000	India: Punjab/Pakistan: Lahore Date=9/26?
1827	11/16	22:45U	1.8	−76.4	7.0	250	Colombia: Bogota, Neiva, Popayan (I=10–11) D=many
1828	6/06	22:30U	34.2	74.5		1000	India (Kashmir): Srinagar (I=9–10)
1828	8/09	16:00U	40.7	48.4	5.7	8000	Azerbaijan: Shemakha (I=8 or 10)
1828	12/18	7:00L	37.6	138.9	6.9	1681	Japan: Niigata P. [Sanjo EQ] [C] [T?]
1828	12/29		−2.0	121.0		many	Indonesia (Sulawesi): Macassar, Boelekomba [T]
1829	3/21	18:00L	38.2	−0.9	7.0	839	Spain: Murcia, Torrevieja (I=10) D=389/2000/6000
1829	11/18		33.2	117.9	5.5	many	China: Anhui P. [Wuhe EQ] (I=7)
1829	11/19	2:00L	36.6	118.5	6.0	117	China: Shandong P. [Yidu–Linqu EQ] (I=8) D=100s
1830	3/09	11:22U	43.1	46.7	6.8	500	Russia (Dagestan): (I=8–9) D=40/400/410
1830	3/27	12:00L	35.7	52.3	7.1	530	Iran: Damavand, Shamiranat, Teheran (I=9–10)
1830	5/09		35.7	52.1		500	Iran: Teheran, Demavand, Semnan (same as 3/27?)
1830	6/12	20:00L	36.4	114.2	7.5	1000s	China: Hebei P. [Cixian EQ] (I=10) D=5458/10T+
1830	8/19	16:00L	35.0	135.8	6.5	280	Japan: Kyoto P. [Kyoto EQ] D=146/380/438/800
1831	10/11		32.7	117.0		many	China: Anhui P. [NE Fengtai EQ] (I=8)
1832	1/22		36.5	71.0	7.4	great	Afghanistan/Tajikistan/Pakistan: Hindu Kush
1832	3/08	18:30U	39.1	16.9	6.5	234	Italy: Cutro, Catanzaro (I=10)
1833	8/26	17:00L	28.3	85.5	8.0	many	China: Tibet/Nepal (I=10–11) D=414 (Nepal)
1833	9/06	10:00L	25.2	103.0	8.0	6707	China: Yunnan P. [Haoming EQ] (I=11) D=67T?
1833	12/07	15:00L	38.9	139.3	7.7	100	Japan: Yamagata P [Shonai-oki EQ] [T]
1834			28.6	87.1	6.0	many	China: Tibet A.R. (Tirgri, Rongxar)
1834	1/20	11:45U	1.3	−76.9	7.0	80	Colombia: El Tablon/Ecuador: Pasto, Almaguer
1834	5/23	6:00L	31.0	35.5	6.3	many	Jordan/Israel: Jerusalem, Bethlehem
1835	2/20	15:30U	−36.0	−73.0	8.1	many	Chile: Concepcion, Valparaiso [T]
1835	7/20	14:00L	38.5	142.5	7.0	many	Japan: Miyagi P. [T]
1835	8/23	15:00U	38.5	35.5		300	Turkey: Kaisarije, Kumetri, Walkeri (I=10) D=150
1835	10/12	22:35U	39.3	16.3	5.9	115	Italy: Castiglione, Cosenza (I=10) D=27/100/126

(continued)

TABLE 1 *(continued)*

Year	Month/day	h:min	Lat.	Long.	*M*	Deaths	Remarks
1835	11/01		−3.4	128.1		149	Indonesia (Moluccas): Amboina, Haruku, Saparua
1836	4/25	0:20U	39.6	16.7	6.3	239	Italy: Rossano, Crosia (I = 10) D = 590 [T]
1837	1/01	14:34L	33.0	35.5	7.0	5700	Israel: Safad/Syria/Lebanon D = 2T–3T/5T [T]
1837	11/07	11:30U	−39.8	−73.2	8.0	60	Chile: Valdivia, Concepcion [T] Time = 12:51U?
1838			32.0	34.5		3000	Israel: Jaffa
1838	1/23	18:36U	45.7	26.6	6.9	many	Romania: Vrancea region (I = 8) D = 8/42/2000
1839	1/11	5:50L	14.4	−61.0		390	Martinique/St. Lucia
1839	2/07	8:00L	26.1	99.9	6.0	100	China: Yunnan P. [Eryuan EQ] (I = 8)
1839	2/23	10:00L	26.1	99.9	6.0	153	China: Yunnan P. [Eryuan EQ] (I = 8) D = 129
1839	6/28	1:00L	23.5	120.4	6.5	117	China: Taiwan [Chiai EQ] (I = 8+)
1840			−19.0	178.0		many	Fiji: Kandavu (I = 9)
1840	3/22	0:30L	12.9	123.9	6.8	many	Philippines (Luzon): Sorsogon, Masbate [T] D = 17
1840	7/02	16:00U	39.5	43.9	7.4	1000	Turkey: Balikgolu/Armenia: Nakhichevan D = 2063
1840	7/15		39.5	43.0		many	Turkey: Mt. Ararat
1841	4/23		9.5	39.8		100	Ethiopia: Shoa, Ankober (I = 6)
1841	9/02	12:15U	10.0	−84.0		many	Costa Rica: Cartago/Nicaragua
1842	2/19		34.4	70.5		500	Pakistan: Jalalabad, Peshawar/Afghanistan
1842	5/07	21:00U	19.7	−72.8		4500	Dominica: St. Domingo, Santiago/Haiti [T] D = 200/3T
1842	6/11		43.6	93.0	7.0	many	China: Xinjiang A.R. [Barkol EQ] (I = 9) D = 41/107
1842	12/08		9.7	39.8		many	Ethiopia: Ankober (I = 9)
1843	2/08	14:50U	16.5	−61.0	7.8	5000	Guadaloupe/Antigua/Montserrat (I = 9)
1843	4/18	8:00L	38.6	44.8	5.9	1000	Iran: Khoy
1843	10/18		36.3	27.7	6.5	600	Greece: Rhodes, Khalke (I = 9) [T]
1844	5/		11.2	−84.8		200	Nicaragua: Greytown
1844	5/12		41.0	35.0		200	Turkey: Osmancik, Ankara (I = 9)
1844	5/12	18:00L	33.6	51.4	6.4	1500	Iran: Qohrud, Kashan Date = 5/10? 5/11?
1844	5/13	19:00L	37.5	48.0	6.9	great	Iran: Miyaneh, Garmrud (I = 9)
1845	2/08		2.6	122.6	7.0	56	Indonesia (Sulawesi): Menado, Minahassa [T?]
1845	3/04	12:00L	24.1	120.5	6.3	381	China: Taiwan [Changhua EQ] (I = 8)
1846	2/14		0.5	127.3		many	Indonesia (Moluccas): Ternate Is.
1846	8/14	12:00U	43.5	10.6	5.9	60	Italy: Orciano Pisano (I = 10–11) [T] D = 56/384
1847	5/08	21:00L	36.7	138.2	7.4	8174	Japan: Nagano & Niigata P. [Zenkoji EQ] [FR]
1847	8/07	8:15L	29.7	30.8	5.8	85	Egypt: Faiyum, Cairo, Alexandria D = 112/126
1848	12/03	8:00L	24.1	120.5	6.8	2021	China: Taiwan [Changhua EQ] (I = 9) D = 1000/1200
1849	5/03	10:30U	10.6	−71.8		many	Venezuela: Maracaibo
1850	9/12	22:00L	27.8	102.3	7.5	23860	China: Sichuan P. [Xichang EQ] (I = 10) D = 20652
1851	2/07		10.5	−75.7		many	Colombia: Carthagena
1851	2/28	16:58L	36.4	28.6	7.2	many	Greece: Rhodes, Makri/Turkey: Fetiye (I = 10) [T]
1851	6/		36.8	58.5	6.9	2000	Iran: Sarvelayat, Quchan, Ma'dan (& 1852/2/22?)
1851	8/14	13:20U	41.0	15.7	6.3	1000	Italy: Melfi (I = 10) D = 62/671/700/14T? (1:13U?)
1851	10/12	7:00U	40.7	19.7	6.6	2000	Albania: Vlore, Berat, Elbasan (I = 10–11) [T]
1851	10/17		40.7	20.0		400	Albania: Berat (aftershock) (I = 10)
1852	1/24		29.3	68.8		350	India: Upper Sindh, Kahan, Muree Hills D = 250
1852	2/22		37.1	58.4	5.8	2000	Iran: Quchan (I = 9) (D incl. 1851/6/?)
1852	5/26		37.4	105.1	6.0	325	China: Ningxia A.R. [Zhongwei EQ] (I = 8–9)
1853			−21.0	175.0		many	Tonga
1853	3/11	10:00L	35.3	139.2	6.7	100	Japan: Kanagawa & Shizuoka P. [Odawara EQ]
1853	5/04		29.6	52.5		many	Iran: Shiraz (forshocks 5/04–05)
1853	5/05	12:00L	29.6	52.5	6.2	9000	Iran: Shiraz D = 12T
1853	7/15		12.1	−63.6	6.7	800	Venezuela: Cumana (I = 9) D = 110/600/4000
1854	2/12	17:50U	39.3	16.3	6.1	472	Italy: Cosenza (I = 10) D = 232/468
1854	4/16		13.8	−89.2	6.6	1000	El Salvador: San Salvador
1854	7/09	2:00L	34.8	136.0	7.3	1600	Japan: Mie P. [Iga Ueno EQ]
1854	7/30		39.9	20.2	6.4	many	Greece: Souli, Delvine/Albania (I = 9–10)
1854	12/23	9:00L	34.0	137.8	8.4	2000	Japan: Tokai region [Ansei Tokai EQ] [TC]

(continued)

TABLE 1 *(continued)*

Year	Month/day	h:min	Lat.	Long.	M	Deaths	Remarks
1854	12/24	14:00L	29.1	107.1	5.5	many	China: Sichuan P. [Nanchuan EQ]
1854	12/24	16:00L	33.0	135.0	8.4	1000s	Japan: Nankaido [Ansei Nankai EQ] [TC]
1855	2/28	3:00U	40.2	29.1	7.3	1900	Turkey: Bursa, Tayabas (I = 10) D = 300/700
1855	4/11	19:40U	40.2	29.1	7.1	400	Turkey: Bursa (I = 10) D = 1300
1855	4/29		40.2	29.1	6.7	1300	Turkey: Bursa (same as 4/11? or D incl. 4/11?)
1855	11/11	22:00L	35.7	139.8	6.9	7444	Japan: Tokyo [Ansei Edo EQ] D = 10T
1856	3/02		3.5	125.5		3000	Indonesia: Sangihe (Great Sangir Is.) [VT]
1856	6/10	9:00L	29.7	108.8	5.5	150	China: Hubei P. [Xianfeng EQ] [R] D = 100s/1000
1856	10/12	0:45U	35.5	26.0	8.3	538	Greece: Crete, Kasos, Karapathos (I = 11) D = 20 [T]
1857	12/16	21:15U	40.4	15.9	7.0	10939	Italy: Basilicata (I = 11) D = 13488/19000/24000?
1858	4/09	2:00L	36.4	137.2	7.1	426	Japan: Gifu & Toyama P. [Hietu EQ] [LR]
1859	3/22	13:30U	−0.3	−78.5	6.3	5000	Ecuador: Quito, Pichincha (I = 8) Time = 15:42U?
1859	5/21		40.0	41.5		500	Turkey: Erzurum (same as 6/02?)
1859	6/02	10:30L	40.0	41.0	6.4	2000	Turkey: Erzurum (I = 9–10) D = 15000
1859	6/11	13:00U	40.7	48.5	6.1	100	Azerbaijan: Shemakha (I = 8–9)
1859	8/22	12:32U	42.8	13.1	5.7	101	Italy: Valnerina (I = 9) Time = 17:15U?
1861	2/16		0.5	97.5	8.4	great	Indonesia (Sumatra): Lagundi, Padang, Batu [T]
1861	3/09		0.0	98.0	7.0	750	Indonesia (Sumatra): Padang, Batu D = 200 [T]
1861	3/20	23:00U	−32.9	−68.9	7.0	18000	Argentina: Mendoza D = 7000 Time = 20:36L?
1861	5/07	22:00U	13.7	41.6	5.5	106	Ethiopia: Dubbi [V]
1862	6/07	10:00L	23.3	120.2	6.8	1000s	China: Taiwan [Chiai EQ] (I = 9) D = 500
1862	10/16		38.8	30.5	6.1	800	Turkey: Suhut
1862	11/03	3:00L	38.4	27.7	6.6	280	Turkey: Turgutlu (I = 9)
1863			38.5	30.6		800	Turkey: Suhut (I = 10) (same as 1862/10/16?)
1863	6/03	11:25L	14.6	120.9	6.5	298	Philippines (Luzon): Manila D = 320/400/500 [T]
1863	12/30	18:36U	38.1	48.5	6.1	1000	Iran: Ardabil, Nir (I = 8–9) D = 500
1864	1/03		38.2	48.3		500	Iran: Ardabil/Azerbaijan (D incl. 1863/12/30?)
1864	1/17		30.6	57.0	6.0	many	Iran: Kirman
1864	5/23		−3.0	135.0		250	Indonesia (Irian Jaya): Geelvink Bay [T]
1864	12/20		33.0	46.0		100	Iraq: Zurbatiyah
1865	7/19	1:00U	37.7	15.5	5.3	74	Italy: Etna (I = 10) D = 60/64 Date = 7/18–19
1865	7/23	21:30L	39.4	26.2	6.7	many	Greece: Lesbos (I = 9–10)
1865	11/06		24.9	121.6	6.0	many	China: Taiwan (Taipei, Chilung) [L]
1866	1/02	9:00U	40.4	19.5	6.5	60	Albania: Narta, Vlore, Kanina (I = 9) [T]
1866	4/		31.7	99.8	7.3	many	China: Sichuan P. [Garze EQ] (I = 9) (Apr. or May)
1866	6/20	14:00L	38.5	41.0	6.8	many	Turkey: Kulp (I = 8–9)
1867	1/02	7:13L	36.4	2.7		51	Algeria: Mouzaia, El Affroun (I = 10–11) D = 100
1867	2/04	4:19U	38.2	20.4	7.3	224	Greece: Cephalonia (I = 11) D = 200 Time = 4:15L?
1867	3/07	16:00U	39.3	26.2	6.8	500	Greece/Turkey: Lesbos (I = 10) D = 150
1867	6/10		−7.8	110.5		327	Indonesia (Java): Djokjakarta, Soerakarta [T]
1867	11/13		17.2	−62.8		many	St. Christopher
1867	12/18	12:00L	25.3	121.8	7.0	480	China: Taiwan [N off Chilong EQ] D = 100s [T]
1868	4/02	15:40L	19.0	−155.5	8.0	81	USA: South coast of Hawaii [Kau EQ] (I = 10) [T]
1868	8/13	20:45U	−18.5	−71.0	8.5	25000	Chile: Arica/Peru (I = 11) D = 2T/12T/40T [T]
1868	8/16	6:30U	0.3	−78.2	7.7	40000	Ecuador: Ibarra/Colombia: Guayaquil D = 15T/70T
1870	4/11	14:00L	30.0	99.1	7.3	2298	China: Sichuan P. [Batang EQ] (I = 10) D = 100/many
1870	5/17		17.0	−97.0		102	Mexico: Oaxaca, Ocotlan, Lvsicho, Yantepee
1870	8/01	0:41L	38.5	22.6	6.8	117	Greece: Gulf of Corinthos, Phokida, Arachova
1870	10/04	16:55U	39.2	16.3	6.1	117	Italy: Mangone, Cosenza (I = 10) D = 250
1871	3/02		0.0	128.0		400	Indonesia (Moluccas): Tagulandang Is. [T]
1871	12/23	18:00L	37.4	58.4	7.2	2000	Iran: Quchan (I = 9)/ (M 5.6, I = 7–8)
1872	1/06		37.1	58.4	6.3	4000	Iran: Quchan (I = 8–9) D = 30T (incl. 71/12/23)?
1872	1/28	7:00U	40.6	48.7	6.0	118	Azerbaijan: Shemakha (I = 8–9)
1872	3/14	17:00L	35.2	132.1	7.1	555	Japan: Shimane P. [Hamada EQ] [T]
1872	4/03	7:40L	36.4	36.5	7.2	1800	Turkey: Antakya/Syria: Aleppo D = 500–800/1000

(continued)

TABLE 1 *(continued)*

Year	Month/day	h:min	Lat.	Long.	*M*	Deaths	Remarks
1873	2/01		37.5	26.5		many	Greece: Samos (2/01–03)
1873	3/04		13.7	−88.7	6.4	800	El Salvador: San Vicente, San Salvador
1873	4/11		13.6	−89.0		300	El Salvador: San Salvador, St. Vincente
1873	6/29	3:58U	46.2	12.4	6.3	80	Italy: Bellunese (I = 9–10) D = 42/55
1874	9/03	3:02U	14.5	−90.7		116	Guatemala: Itzapo, Antigua, Guatemala D = 200
1874	10/18		35.0	69.0		many	Afghanistan: Kabul, Jabal al Siraj (I = 9)
1875	2/11	20:30L	21.0	−103.8	7.5	70	Mexico: Jalisco, San Cristobal dela Barranca
1875	3/27	22:48L	38.5	39.5	6.7	many	Turkey: Golcuk Golu (I = 8–9) Date = 3/22?
1875	5/03		38.1	30.0	6.7	2000	Turkey: Civril, Dinar (5/03–05) D = 1300/1000s
1875	5/18	16:15U	7.9	−72.5	7.3	16000	Colombia: Cucuta D = 461/1T/10T/14T Time = 10:10U?
1877	5/10	2:16U	−19.6	−70.2	8.3	many	Chile: Iquique, Tarapaca (I = 11) [T]
1878	4/14		10.2	−67.0	5.9	300	Venezuela: Cua D = 400 Date = 4/13?
1878	10/02		13.5	−88.4		many	El Salvador: Jucuapu
1879	3/13		38.2	45.5		many	Iran: Urmia Lake, Marand
1879	3/22	3:42L	37.8	47.9	6.7	2000	Iran: Bozqush, Garmrud, Ardabil D = 500/3200+
1879	3/22	12:35L	38.2	47.5		992	Iran: Mianeh (same as 3:42L?)
1879	4/02		37.5	57.4	6.7	700	Iran: Bodzhnurd (I = 9)
1879	7/01	4:00L	33.2	104.7	8.0	29480	China: Gansu P. [Wudu–Wenxian EQ] (I = 11) D = 20673+
1879	10/09	19:30U	45.1	37.9	6.0	many	Russia: Nizhnyaya–Kuban (I = 7)
1880	7/04		36.4	47.3	5.6	60	Iran: Garrus, Takht-e-Solaiman
1880	7/22		38.1	27.8	6.1	many	Turkey: Izmir
1880	8/	12:00U	27.1	54.1		120	Iran: Bastak
1881	3/04	12:00U	40.7	13.9		126	Italy: Ischia (I = 8–9) D = 118/121 Date = 3/16?
1881	4/03	11:40L	38.3	26.2	6.5	7866	Greece: Khios (I = 9 or 11) D = 3541/4000/4181
1881	5/30		38.6	43.8	6.3	95	Turkey: Van, Tegut (I = 8–9) Date = 6/07? *M* = 7.3?
1881	6/07		38.7	42.4		95	Turkey: Van, Nemrut (same as 5/30?)
1881	7/					70	China: Yunnan P. (Yongsheng) (July or Aug.)
1881	7/20	21:00L	33.6	104.6	6.5	523	China: Gansu P. [Lixian EQ] D = 480
1882	9/07	7:50U	9.0	−77.0	6.5	many	Panama/Colombia [T] (7.3 −77.8 *M* 8.0)
1883	5/03	12:00U	37.9	47.2	6.2	many	Iran: Tabriz, Khvoy (Khoy) (I = 9–10)
1883	7/28	20:25U	40.8	13.9	5.6	2333	Italy: [Ischia EQ] D = 1990/2000/3100/3300
1883	8/27		−5.8	106.3		10000s	Indonesia: Krakatoa Explosion D = 36T+ [TV]
1883	10/15	13:30U	38.4	26.1	7.3	120	Turkey: Izmir, Ayvalik, Urla, Cesme D = 15T?
1884	5/19	18:00L	26.8	55.9	5.4	218	Iran: Qeshm Is., Basidu D = 200/238
1884	12/25	21:08U	37.0	−4.0	6.8	745	Spain (Andalusia): Alhama, Malaga (I = 10) D = 307
1885	1/14	23:00L	34.5	105.7	6.0	80	China: Gansu P. [S Tianshui EQ] (I = 8)
1885	5/30		33.5	75.0		3000	India: Srinagar, Sopur (I = 8) D = 300/2000
1885	8/02	21:20U	42.7	74.1	6.9	54	Kyrgyzstan: Belovodsk, Kalabalty (I = 9–10)
1886	8/27	21:32U	37.1	21.5	7.5	600	Greece: Filiatra, Triphylie (I = 9) D = 300/326 [T]
1886	8/31	21:51L	32.9	−80.0	7.4	60	USA (S.C.): [Charleston EQ] (I = 9–10)
1887	2/23	5:21U	43.9	8.0	6.4	635	Italy: Imperia/France: Menton D = 800/1000 [T]
1887	6/08	23:35U	43.1	76.8	7.3	1800	Kyrgyzstan: Alma–Ata (Verny) (I = 9–10) [L]
1887	12/16	19:00L	23.7	102.5	7.0	2256	China: Yunnan P. [Shiping EQ] (I = 9+) D = 200/2700+
1888	9/22	10:00U	41.3	43.3	6.1	many	Turkey: Okan, Ardahan (I = 8)
1888	11/02		37.1	104.2	6.3	many	China: Gansu P. [Jingtai EQ] D = 32/50
1890	6/28		36.2	59.4		many	Iran: Tasch, Meshed, Teheran (6/27–28)
1890	7/11	0:55L	36.6	54.6	7.2	100s	Iran: Tash, Gurgan, Shahrud (I = 10) D = 121/171 [T]
1890	12/12	7:50L	−6.4	111.0		many	Indonesia (Java): Djoewana (Japara) D = some
1891	4/03		39.1	42.5	5.5	100	Turkey: Malazgirt, Adilcevaz D = 100s Date = ?
1891	10/28	6:38L	35.6	136.6	8.0	7273	Japan: Gifu & Aich P. [Nobi EQ] [F]
1893	3/02	22:51U	38.0	38.3	7.1	885	Turkey: S of Malatya (I = 10) [F]
1893	3/31		38.3	38.5	7.0	1500	Turkey: Malatya D = 400/469 (same as 3/02?)
1893	6/01		36.6	101.8	5.5	many	China: Qinghai P. [S Xining EQ] (I = 7)
1893	6/21	7:30L	7.7	126.1	7.3	many	Philippines (Mindanao): Agusan R. (I = 10) 7:30U?
1893	8/29	6:00L	30.6	101.5	6.8	353	China: Sichuan P. [Qianning EQ] (I = 9) D = 211

(continued)

TABLE 1 (*continued*)

Year	Month/day	h:min	Lat.	Long.	M	Deaths	Remarks
1893	11/17	19:36U	37.1	58.4	7.1	10000	Iran: Quchan (I = 9–10) D = 150/18T Time = 15:06U?
1893	12/		41.7	82.8	6.8	many	China: Xinjiang A.R. (Kuqa) (Dec. or Jan. 1894)
1893	12/25		41.2	80.3	6.5	many	China: Xinjiang A.R. (Akesu) (I = 8)
1894	2/27	0:00L	29.5	53.3	5.9	many	Iran: Karameh, Shiraz
1894	4/20	14:52U	38.6	23.0	6.7	255	Greece: Malesina, Lokris (D incl. 4/27?)
1894	4/27	17:42U	38.7	23.0	7.2	255	Greece: Lokris, Constantinos (I = 8–10) [FT]
1894	4/29	2:45U	8.5	−71.7	7.1	400	Venezuela: Merida (I = 9) D = 300/319
1894	6/28	18:57U	8.2	126.1		many	Philippines (Mindanao): Agusan River
1894	7/10	12:33U	40.6	28.7	6.7	many	Turkey: Geiwe, Istanbul, Adapazari (I = 9–10) [T]
1894	10/22	17:35L	38.9	139.9	7.0	726	Japan: Yamagata P. [Shonai EQ] [C]
1894	11/16	17:52U	38.3	15.9	6.2	101	Italy: Palmi, S. Cristina, Bagnara [T]
1895	1/17	11:30L	37.1	58.4	6.8	1000	Iran: Quchan (I = 9) D = 180/700/11000
1895	5/14	5:00U	39.4	20.5	6.6	many	Greece: Margariti, Paramithia/Albania $M = 7.5$?
1895	5/15		−3.2	40.0		many	Kenya: Malindi
1895	8/30	18:00L	23.5	116.5	6.0	112	China: Guangdong P. [Jieyang EQ] (I = 8) D = 55/72
1896	1/02		37.8	48.4		300	Iran: Sangabad–Khalkhal (foreshock) (I = 8)
1896	1/04	18:28L	37.7	48.3	6.7	1100	Iran: Sangabad–Khalkhal (I = 9–10) D = 800/1600
1896	4/18		−8.3	126.0		250	Indonesia (Timor): Alor (I = 8)
1896	6/15	19:32L	39.5	144.0	8.2	22000	Japan: [Sanriku-oki EQ] [T] 7.2S 8.2W
1896	8/31	17:06L	39.5	140.7	7.2	209	Japan: Akita P. [Rikuu EQ] [F] 7.3S
1897	1/03		6.0	122.7		100	Philippines (Camiguin, Sulu, Mindanao) (I = 10)
1897	1/10	21:00U	26.9	56.0	6.4	1600	Iran: Qeshm Is. D = 750/10T?
1897	3/19		24.7	121.8	6.0	56	China: Taiwan (Yilan) (I = 8) Date = 3/14?
1897	4/29	14:15U	16.2	−61.7	7.6	many	Montserrat, Guadeloupe, Antigua, St. Kitts
1897	6/12	11:06U	26.0	91.0	8.3	1500	India: [Assam EQ] D = 1425/1600 [F] 8.0S
1897	9/21	5:15U	7.1	122.1	8.2	100	Philippines (Mindanao, Sulu, Jolo) [T] 7.5S
1898	7/02	4:20U	43.6	16.7		many	Croatia: Caparice, Grab, Trilj, Vojnic, Losuta
1899	9/20	2:12U	37.9	28.8	6.9	1117	Turkey: Aydin, Nazilli, Buldan (I = 9) [F]
1899	9/29	17:03U	−3.0	128.5	7.4	3864	Indonesia (Moluccas): Ceram [T] 7.1S
1899	12/31	10:50U	41.6	43.6	6.3	248	Georgia: Akhalkalaki (I = 8–9) D = 300/600/800
1900	7/12	6:25U	40.3	43.1	5.9	140	Turkey: Kars, Karakurt, Kagizman, Digor
1900	10/29	9:11U	11.0	−66.0	7.4	many	Venezuela: Caracas (I = 8) D = 20/25 7.7S
1901	2/15		26.0	100.1	6.0	many	China: Yunnan P. [Dengchuan EQ] (I = 8+)
1902	1/16		17.6	−99.5		300	Mexico: Chilpancingo D = 2/many
1902	2/13	9:39U	40.7	48.5	6.9	86	Azerbaijan: Shemakha (I = 7 or 8–9) D = 1000/2000
1902	2/26		13.0	−89.0		185	El Salvador/Guatemala [T]
1902	4/19	2:24U	14.9	−91.5	7.5	2000	Guatemala: Quezaltenango, S. Marcos D = 1.5T 7.5S
1902	8/21	11:17U	6.3	123.6		many	Philippines (Mindanao): Lanao, Cotobato 19:17U?
1902	8/22	11:00L	39.9	76.2	8.3	5650	China: Xinjiang A.R./Kyrgyzstan D = 200/2000 7.7S
1902	12/16	5:07U	40.8	72.3	6.4	4725	Uzbekistan: Andizhan (I = 9) D = 700/4562
1903	4/28	23:39U	39.1	42.7	7.0	3560	Turkey: Malazgirt (I = 9) D = 1700/2200/6000
1903	5/03		38.7	41.5	4.9	many	Turkey: Musch, Erzurum, Bitlis (I = 6)
1903	5/28	3:57U	40.9	42.8	5.4	1000	Turkey: Ardahan, Varginis, Cardahli (I = 7–8)
1903	9/25	1:20U	35.2	58.2	6.4	200	Iran: Kashwar, Turshiz D = 300/350
1904	8/30	19:42L	31.0	101.1	7.0	565	China: Sichuan P. [Daofu EQ] D = 400 6.8S
1904	11/06	4:25L	23.5	120.3	6.3	145	China: Taiwan [Chiai EQ] (I = 8)
1905	4/04	0:50U	33.0	76.0	8.1	20000	India: [Kangra EQ] D = 10600/18815 7.5S 7.8W
1905	6/01	4:42U	42.1	19.5	6.6	120	Albania: Shkoder (I = 9–10)
1905	9/08	1:43U	38.7	16.1	6.8	557	Italy: [Calabria EQ] (I = 10–11) D = 2000/2500 [T]
1905	11/08	22:06U	40.3	24.4	7.5	many	Greece: Chalkidiki, Athoshalb D = 2000 6.8S
1905	12/04	7:04U	38.1	38.6	6.8	many	Turkey: Malatya, Puturge, Celikhan, Rumkale (I = 9)
1906	1/31	15:36U	1.0	−81.5	8.6	1000	Ecuador/Colombia D = 400/600 [T] 8.2S 8.8W
1906	3/17	6:42L	23.6	120.5	6.8	1258	China: Taiwan [Chiai EQ] [F] 6.8S
1906	3/28	6:58L	24.3	118.6	6.3	many	China: Fujian P.
1906	4/18	5:12L	37.7	−122.5	8.3	700	USA (Calif.): [San Francisco EQ] D = 3T? [FC] 7.8S

(*continued*)

TABLE 1 (*continued*)

Year	Month/day	h:min	Lat.	Long.	*M*	Deaths	Remarks
1906	8/17	0:40U	−33.0	−72.0	8.4	3760	Chile: Valparaiso D = 1500/20T [T] 8.1S
1906	12/23	2:21L	43.5	85.0	7.7	285	China: Xinjiang A.R. [Shawan EQ] (I = 10) [F] 7.2S
1907	1/04	5:19U	2.0	96.3	7.8	400	Indonesia (Sumatra): Gunung Sitoli [T] 7.5S
1907	1/14	21:36U	18.2	−76.7	6.5	1000	Jamaica: Kingston, Port Royal D = 1700 [T]
1907	10/21	4:23U	38.5	67.9	7.4	15000	Uzbekistan/Tajikistan: Karatag D = 12T 7.2S
1907	10/23	20:28U	38.1	16.0	6.0	167	Italy: Ferruzzano (I = 9) D = 300/100s [T]
1908	12/28	4:20U	38.2	15.7	7.1	82000	Italy: [Messina EQ] (I = 11) D = 75T/110T [T] 7.0S
1909	1/23	2:48U	33.4	49.1	7.3	5500	Iran: Silakor D = 6000–8000 [F] 7.0S
1909	1/29		35.3	−5.2		100	Morocco: Tetuan, Romars
1909	4/		15.0	44.0		300	Yemen: Bilad al-Waynan (Apr. or May)
1909	6/03	18:41U	−2.5	101.5	7.5	200	Indonesia (Sumatra): Korintji Djambi [T] 7.3S
1909	6/11	21:06U	43.5	5.4	6.2	55	France: St. Cannat (I = 10) D = 40
1909	10/20	23:41U	30.0	68.0	7.2	231	India: Bagh, Shahpur/Pakistan: Bellpat 7.0S
1910	5/05	0:26U	9.8	−83.9	6.1	500	Costa Rica: San Jose, Cartago (I = 9) D = 272/700
1910	6/07	2:04U	40.9	15.4	5.8	50	Italy: Calitri (I = 9–10)
1910	6/24	13:26U	36.3	3.7	6.6	81	Algeria: Aumale, Masqueray (I = 10–11)
1910	10/29		10.0	−67.0		100	Venezuela: Caracas
1911	1/03	23:25U	42.9	76.9	8.2	450	Kazakhstan/China: Xinjiang A.R. (I = 10+) [F] 7.8S
1911	2/18	18:41U	38.2	72.8	7.4	90	Tajikistan: [Pamir EQ] [LR] 7.3S
1911	4/18	18:14U	31.2	57.0	6.2	700	Iran: Ravar D = 60 [F]
1911	6/07	11:02U	19.7	−103.7	7.9	1300	Mexico: Jalisco, Mexico City D = 45 7.7S
1912	7/24	11:50U	−5.6	−80.4	8.1	many	Peru: Huancabamba, Cajamarca/Ecuador (I = 10) 7.0S
1912	8/09	1:29U	40.8	27.2	7.4	2836	Turkey: Saros–Marmara (I = 10) D = 216/1950 7.6S
1912	11/19	13:18U	19.9	−99.8	7.8	many	Mexico: Acambay, Tixmadeje, Timilpan [F] 6.9B
1913	2/24	2:30U	−4.2	−79.4	6.1	many	Ecuador: Gonzanama
1913	3/08	16:05U	14.3	−90.3	6.4	60	Guatemala: Cuilapa D = 32 Time = 15:14U?
1913	6/14	9:33U	43.1	25.8	6.8	500	Bulgaria: Tirnovo, Orahavitza (I = 10–11)
1913	11/04	21:33U	−14.2	−72.9		150	Peru: Aimares, Casaya, Soraya Date = 11/10? 11/14?
1913	12/21	23:37L	24.1	102.4	7.0	1314	China: Yunnan P. [E'shan EQ] D = 942/1900+ 7.2S
1914	3/15	4:59L	39.5	140.4	7.1	94	Japan: Akita P. [Senpoku-gun EQ] 7.0S
1914	5/08	18:01U	37.7	15.1	5.5	69	Italy: Linera, Passopomo (I = 10) D = 12/120
1914	6/10		30.0	109.0		300	China: Hubei P.
1914	6/25	19:06U	−4.0	102.5	8.1	many	Indonesia (Sumatra): Benkulen D = 11/20 7.6S
1914	10/03	22:06U	37.8	30.3	7.0	4000	Turkey: Burdur (I = 9–10) D = few/300 7.1S
1915	1/13	6:52U	42.0	13.7	7.0	32610	Italy: [Avezzano EQ] (I = 11) D = 30T/35T [F] 6.9S
1915	12/03	10:39L	29.5	91.5	7.0	170	China: Tibet A.R. (I = 9) 7.0S
1917	1/05	0:50L	23.9	120.9	6.2	54	China: Taiwan [Puli EQ] D = 70
1917	1/20		−8.3	115.0	6.5	1300	Indonesia (Bali) D = 15T? [T?] Date = 1/21?
1917	7/31	7:54L	28.0	104.0	6.8	1879	China: Yunnan P. [Daguan EQ] (I = 9) 7.5S
1917	12/26	4:30U	14.6	−90.7	5.6	220	Guatemala: Guatemala City D = 70 (& 5:21U, 6:18U)
1918	1/04	4:32U	14.6	−90.6	6.0	100	Guatemala: Guatemala City
1918	1/25	1:20U	14.6	−90.5	6.2	many	Guatemala: Guatemala City (D = 600 Dec.–Jan.)
1918	2/13	14:07L	24.0	117.0	7.3	100s	China: Guangdong P. [Nan'ao EQ] D = 600–700 7.4S
1918	8/15	12:13U	5.5	123.0	8.5	50	Philippines (Mindanao): D = 102 [T] 8.0S
1918	10/11	14:15U	18.5	−67.5	7.5	116	Puerto Rico: Mayaguez, Aguadilla [T] 7.5S
1919	4/28	6:45U	13.7	−89.2	5.9	72	El Salvador: San Salvador D = 100+
1919	6/29	15:06U	44.0	11.5	6.3	100	Italy: Vicchio (Mugello) (I = 10) D = some
1919	11/18	21:54U	39.6	27.7	6.9	many	Turkey: Soma, Pergamos (I = 9) Also 11/27
1920	1/03	16:24U	19.3	−96.9	6.4	648	Mexico: [Puebla EQ] D = 4000 *M* = 7.8?
1920	2/20	11:44U	41.9	44.0	6.2	130	Georgia: Kartli (I = 8–9) D = 40 [F]
1920	9/07	5:55U	44.3	10.3	6.3	171	Italy: Garfagnana (I = 9–10) D = 20/556
1920	11/26	8:51U	40.3	20.0	6.3	200	Albania: Tepelene, Gjirokaster D = 600 [T]
1920	12/16	20:05L	36.7	104.9	8.5	235502	China: Ningxia [Haiyuan EQ] D = 220T/246T [F] 8.6S
1920	12/17	18:59U	−32.7	−68.4	6.0	400	Argentina: Mendoza (I = 8)
1921	1/31		36.0	106.0	5.5	350	China Ningxia A.R. (Guyuan) (I = 7)

(*continued*)

TABLE 1 *(continued)*

Year	Month/day	h:min	Lat.	Long.	*M*	Deaths	Remarks
1921	4/12	17:36L	35.8	106.2	6.5	many	China: Ningxia A.R. [Guyuan EQ] (I = 8) D = 10000
1921	11/11	18:36U	8.0	127.0	7.5	600	Philippines (Mindanao) 7.5S
1922	3/24	12:22U	44.5	20.3	6.0	many	Yugoslavia: Belgrade (I = 9–10)
1922	8/30	19:40L	36.0	106.2	5.5	300	China: Ningxia A.R. (Gyuan) (I = 7)
1922	11/11	4:32U	−28.5	−70.0	8.3	1000	Chile: Atacama D = 600 [T] 8.3S 8.5W
1923	3/24	20:40L	31.5	101.0	7.3	450	China: Sichuan P. [Luhuo–Dawu EQ] D = 4300 7.3S
1923	5/25	22:21U	35.3	59.2	5.5	2219	Iran: Kaj Derakht, Torbat D = 300/770/5000+
1923	6/14		31.3	100.8	5.8	1300	China: Sichuan P. [Kangding EQ] (I = 7)
1923	9/01	11:58L	35.4	139.2	7.9	142807	Japan: Tokyo, Yokohama [Kanto EQ] [CLT] 8.2S 7.9W
1923	9/09	22:03U	25.5	91.5	7.1	50	Bangladesh/India 7.1S
1923	9/17	7:09U	37.9	57.5	6.4	157	Iran: Bodzhnurd (Budjurd)/Turkmenistan? (I = 8–9)
1923	9/22	20:47U	29.5	56.4	6.7	290	Iran: Lalehzar, Qaleh Asgar D = 200+/260
1923	11/13	15:30L	38.7	111.1	5.5	many	China: Shanxi P. C = 1500
1923	12/14	10:31U	1.0	−77.5	7.0	300	Colombia: Ipiales (I = 8–9) D = 85/180
1924	1/04	12:21U	39.2	71.4	5.1	83	Tajikistan: Ura-tope
1924	5/13	1:52U	40.0	42.0	5.3	50	Turkey: Erzurum (I = 8–9)
1924	7/03	12:40L	36.8	83.8	7.2	255	China: Xinjiang [Minfeng EQ] (I = 10) D = 100 7.2B
1924	9/13	14:34U	40.0	42.1	6.8	60	Turkey: Horasan, Erzurum (I = 8–9) D = 20/30/200
1924	11/12		−7.2	109.5		609	Indonesia (Java): Wonosobo D = 60/many
1924	12/02		−7.3	109.9		115	Indonesia (Java): Wonosobo (aftershock) D = 727
1924	12/19		41.0	39.7		210	Turkey: Trabzon
1925	1/09	17:38U	41.2	42.8	5.8	200	Turkey: Ardahan (I = 8) D = 20
1925	2/08		41.8	42.7	5.5	140	Turkey: Ardahan (I = 7) (same as 1/09?)
1925	3/16	22:42L	25.7	100.4	7.0	5808	China: Yunnan P. [Dali EQ] (I = 9) D = 3600 7.0S
1925	5/23	11:09L	35.6	134.8	6.8	428	Japan: Hyogo P. [Tajima EQ] D = 465 [CF]
1925	12/14		34.6	58.1	5.5	500	Iran: Bajestan
1926	6/26	19:46U	36.5	27.5	8.0	110	Greece: Rhodes (I = 11) 7.7B
1926	6/28	3:23U	−0.5	100.5	6.8	222	Indonesia (Sumatra): Padang Highlands [F]
1926	8/31		38.5	−28.6		50	Azores: Fayal Is. D = 14
1926	10/22	19:59U	40.7	43.7	5.7	370	Armenia: Leninakan (I = 8–9) D = 355 [F]
1927	2/14		43.0	18.0	6.1	50	Croatia/Yugoslavia (I = 7–8)
1927	3/07	18:27L	35.6	134.9	7.3	2925	Japan: Kyoto P. [Tango EQ] [CFT] 7.6S 7.1W
1927	5/23	6:32L	37.7	102.2	8.0	41419	China: Gansu P. [Gulang EQ] D = 80T/200T [F] 7.9S
1927	7/11	13:04U	32.2	35.3	6.0	361	Jordan: Damiya, Nablus D = 192/242/268/500
1927	12/01		−0.5	119.5		50	Indonesia (Sulawesi): Donggala (I = 7) [T]
1928	3/31	0:30U	38.5	28.0	6.5	170	Turkey: Tepekoy, Torbali (I = 9–10) D = 30/40/50
1928	4/14	9:00U	42.2	25.3	6.8	107	Bulgaria: Plovdiv P. (I = 9–10) D = many [F]
1928	4/18	19:22U	42.1	25.0	7.0	103	Bulgaria: Plovdiv (I = 8) D = 107 (incl. Apr. 14?)
1928	8/04		−8.3	121.7		226	Indonesia: Paloeweh Is. (Rokatinda Volc.) [LT]
1928	12/01	4:06U	−35.0	−72.0	8.0	225	Chile: Talca (I = 10) D = 218 [T] 8.0S
1928	12/19	11:37U	7.1	124.1	7.3	93	Philippines (Mindanao): Illana Bay [T] 7.5S
1929	1/17	11:45U	10.5	−64.5	6.9	many	Venezuela: Cumana (I = 9) D = 50/200 [T]
1929	5/01	15:37U	37.8	57.8	7.2	3257	Iran: Kopet–Dagh D = 3800/5803 [F] 7.2S
1929	5/18	6:37U	40.2	37.9	6.4	64	Turkey: Susehri (I = 8–9)
1929	5/30	9:43U	−35.0	−68.0	6.8	52	Argentina: Mendoza (I = 9)
1929	7/15	7:44U	32.1	49.6	6.0	many	Iran: Izeh, Andika
1929	11/18	15:32L	44.0	−56.0	7.2	52	Canada: Newfoundland [Grand Banks EQ] [T]
1930	5/05	13:45U	17.0	96.5	7.2	550	Myanmar: Pegu (I = 9) D = 1900/6000 [CT] 7.4B
1930	5/06	22:34U	38.0	44.7	7.3	2514	Iran: Salmas D = 1360/3000 (I = 9–10) [F] 7.2S
1930	7/16		17.4	95.5		50	Myanmar: Tharrawaddy P. near Yangon Date = 7/18?
1930	7/23	0:08U	41.1	15.4	6.7	1404	Italy: Vulture (Irpinia) (I = 10) D = 1883/3000
1930	8/24	18:51L	30.0	100.0	5.5	100s	China: Sichuan P. [Litang EQ]
1930	9/22	16:26U	38.5	69.5	5.7	175	Tajikistan: Dusanbe (I = 8)
1930	11/26	4:02L	35.0	139.0	7.3	272	Japan: Shizuoka P. [Kita–Izu EQ] [F] 7.2S 7.0W
1931	1/15	1:50U	16.0	−96.7	7.9	68	Mexico: Oaxaca (I = 10) D = 114 7.8S

(continued)

TABLE 1 (*continued*)

Year	Month/day	h:min	Lat.	Long.	*M*	Deaths	Remarks
1931	1/28	5:55U	40.6	20.7	5.9	100	Albania: Corce (I = 9) D = 4/90
1931	2/02	22:46U	−39.5	176.9	7.9	256	New Zealand: [Hawke's Bay EQ] D = 225/285 [T] 7.8S
1931	3/31	16:02U	12.2	−86.3	6.0	1000	Nicaragua: [Managua EQ] D = 2000/2450 [CF]
1931	4/27	16:50U	39.4	46.0	6.4	390	Armenia: Zangezur (I = 8–9) D = 300/2890
1931	8/11	5:18L	47.1	89.8	8.0	10000	China: Xinjiang [Fuyun EQ] D = 300 [F] 7.9S 8.0W
1931	8/27	15:27U	29.5	67.2	7.4	200	Pakistan: Mach (I = 7–8) D = 30 7.2S
1931	10/03	19:13U	−10.5	161.7	7.9	50	Solomon Is. [T] 7.9S
1932	3/07		30.1	101.8	6.0	100s	China: Sichuan P. [Kongding EQ] (I = 8) D = some
1932	6/03	10:36U	19.5	−104.3	8.1	60	Mexico: Guadalajara, Colima (I = 10) [T] 8.2S
1932	6/18	10:12U	19.5	−104.3	7.9	52	Mexico: Guadalajara, Colima D = 30 [T] 7.8S
1932	6/22	12:59U	19.0	−104.5	6.9	75	Mexico: Cayutlan D = some [T]
1932	9/26	19:20U	40.5	23.8	7.0	491	Greece: Hierissos (Chalkidiki) (I = 10) [T?]
1932	9/29	3:57U	41.0	23.2	6.2	318	Greece: Sohos (Chalkidiki) (I = 9–10)
1932	12/25	10:04L	39.7	96.7	7.6	275	China: Gansu P. [Changma EQ] D = 400/70T [F] 7.7S
1933	3/03	2:30L	39.2	144.5	8.1	3064	Japan: Iwate P. [Sanriku-oki EQ] [T] 8.5S 8.4W
1933	3/10	17:54L	33.6	−118.0	6.3	115	USA (Calif.) [Long Beach EQ] (I = 9) D = 100/140
1933	4/23	5:58U	36.8	27.3	6.6	74	Greece: Kos (I = 9–10) [T?]
1933	6/24	21:54U	−5.5	104.7	7.5	76	Indonesia (Sumatra): Liwa 7.5S
1933	8/25	15:50L	31.9	103.4	7.5	6865	China: Sichuan P. [Diexi EQ] (I = 10) [R] 7.5S
1933	9/20		29.5	102.5	5.0	100s	China: Sichuan P. [Fulin EQ] (I = 6)
1934	1/15	8:43U	26.5	86.5	8.3	10700	India/Nepal [Bihar–Nepal EQ] D = 7253 8.3S 8.2W
1934	12/03	2:38U	14.65	−89.0	6.3	50	Hondurus: S. Jorge, La Encarnacion, S. Fernando
1935	3/05	10:26U	35.9	53.1	6.0	60	Iran: Alborz, Talar-rud
1935	4/11	23:14U	36.5	53.6	6.4	500	Iran: Kusut, Sari (I = 8–9) D = 690/700
1935	4/21	6:01L	24.3	120.8	7.1	3276	China: Taiwan [Miaoli EQ] [F] 7.1S
1935	5/01	10:24U	40.6	43.7	6.2	500	Turkey: Digor, Kigi (I = 8–9) D = 540
1935	5/30	21:32U	29.5	66.8	7.5	60000	Pakistan: [Quetta EQ] D = 25T/30T 7.6S
1935	12/18	15:10L	28.7	103.6	6.0	100s	China: Sichuan P. [Mabian EQ] (I = 8)
1936	1/09	9:23U	1.1	−77.6	7.0	250	Colombia
1936	4/01		22.5	109.4	6.8	94	China: Guangxi A.R. [Lingshan EQ] (I = 9) D = 150
1936	4/27	7:59L	28.9	103.6	6.8	many	China: Sichuan P. [Mabian EQ] (I = 9) D = 27/100
1936	5/16	15:05L	28.5	103.6	6.8	550	China: Sichuan P. [Leibo EQ]
1936	8/01	14:24L	34.2	105.7	6.0	144	China: Gansu P. [Tianshui EQ] (I = 8) D = 115/134
1936	8/23	21:12U	6.1	94.7	7.3	91	Indonesia (Sumatra): Banda Aceh, Lhok [T] 7.3S
1936	12/20	2:43U	13.7	−88.9	6.1	400	El Salvador: San Esteban, S. Vicente D = 100–200
1937	8/01	4:35L	35.4	115.1	7.0	3833	China: Shandong P. [Heze EQ] (I = 9) D = many/3350
1938	4/19	10:56U	39.1	34.0	6.8	155	Turkey: Kirsehir (I = 9) D = 149/224/800 [F]
1939	1/25	3:32U	−36.3	−72.3	7.8	28000	Chile: [Chillan EQ] (I = 10) D = 30T 7.8S
1939	9/22	0:37U	39.1	26.8	6.5	60	Turkey: Dikili (I = 8–9) D = 41/150
1939	12/26	23:57U	40.1	38.2	7.8	32700	Turkey: [Erzincan EQ] (I = 11–12) D = 45T [F] 7.8S
1940	4/06	21:43L	23.9	102.3	6.0	181	China: Yunnan P. [Shiping EQ] (I = 8)
1940	5/24	16:33U	−12.5	−77.0	8.0	250	Peru: Lima, Callao (I = 10) D = 179 [T] 7.9S 8.2W
1940	7/30	0:12U	39.5	35.2	6.1	many	Turkey: Yazgat, Akdagmadeni (I = 8)
1940	11/10	1:39U	45.8	26.8	7.3	1000	Romania: Vrancea R., Bucharest D = 350 7.3B
1941	1/11	8:31U	16.6	43.3	5.8	1200	Yemen (additional damage Feb. 4 & 23) (I = 9)
1941	2/16	16:38U	33.5	58.9	6.4	680	Iran: Muhammadabad, Chahak D = 640/730 [F]
1941	4/15	19:09U	18.3	−103.3	7.7	90	Mexico: Michoacan, Colima, Jalisco 7.7S
1941	5/05	23:18L	47.0	127.2	6.0	132	China: Heilongjiang P. [Suihua EQ] (I = 8)
1941	9/10	21:53U	39.5	43.0	6.0	500	Turkey: Ericis, Van (I = 7–8) D = 194/430
1941	10/08	23:24L	31.7	102.3	6.0	139	China: Sichuan P. [Heishui EQ] (I = 8) D = 120
1941	12/17	3:19L	23.3	120.3	7.0	357	China: Taiwan [Chiai EQ] D = 358 7.2S
1942	2/01	1:30L	23.1	100.3	6.8	90	China: Yunnan P. [Simao EQ] (I = 8)
1942	5/14	2:13U	−0.7	−81.5	7.9	200	Peru/Ecuador (I = 9) D = many 7.9S 8.2W
1942	12/20	14:03U	40.7	36.8	7.3	3000	Turkey: Niksar, Erbaa D = 1000/1100 [F] 7.3S
1943	1/30	5:33U	−14.4	−72.9	6.9	75	Peru: Yanaoca, Pampamarca D = 200

(*continued*)

TABLE 1 *(continued)*

Year	Month/day	h:min	Lat.	Long.	M	Deaths	Remarks
1943	6/20	15:32U	40.6	30.5	6.3	285	Turkey: Adapazari, Hendek (I = 8) D = 336
1943	7/23	14:53U	−9.5	110.0	8.1	213	Indonesia (Java): Jogyakarta 7.6B
1943	9/10	17:36L	35.5	134.2	7.2	1083	Japan: Tottori P. [Tottori EQ] [CF] 7.4S 7.0W
1943	11/26	22:20U	41.0	34.0	7.6	4020	Turkey: Ladik D = 2824/2900/5000 [F] 7.6S
1943	12/07	1:19U	41.0	35.6	5.6	550	Turkey: Anatolia (I = 7–8) D = ?
1944	1/15	23:49U	−31.5	−68.6	7.4	8000	Argentina: San Juan D = 2900/5000/5600 [F] 7.2S
1944	2/01	3:23U	41.5	32.5	7.6	4000	Turkey: Bolu D = 2381/2790/5000 [F] 7.4S
1944	10/06	2:35U	39.4	26.5	6.8	50	Turkey: Ayvalik (I = 10) D = 27
1944	12/07	13:35L	33.6	136.2	7.9	1251	Japan: C.Japan [Tonanakai EQ] [T] 8.0S 8.1W
1945	1/13	3:38L	34.7	137.1	6.8	2306	Japan: Aichi P. [Mikawa EQ] [FT] 6.8S 6.6W
1945	3/20	7:59U	37.4	35.8	5.7	300	Turkey: Adana (Aelana?), Ceyhan (I = 8) D = 10
1945	7/29	8:56U	38.5	43.3		300	Turkey: Van, Bolgesi
1945	9/23	23:34L	39.5	119.0	6.3	600	China: Hebei P. [Luanxian EQ] (I = 8) D = 17/70
1945	11/27	21:57U	25.0	63.5	8.0	300	Iran/Pakistan: Makran coast D = 4000 [T] 8.0S
1946	2/12	2:43U	35.7	4.8	5.6	277	Algeria: Hodna plain (I = 8–9) D = 246/264
1946	4/01	2:29L	52.8	−162.5	7.4	173	USA (Aleutians): Unimak Is., Hawaii [T] 7.3S
1946	5/31	3:12U	40.0	41.5	6.0	1300	Turkey: Ustukran (I = 7–8) D = 840
1946	8/04	17:51U	19.2	−69.0	8.1	100	Dominica/Puerto Rico [T] 8.0S
1946	11/04	21:47U	39.6	54.9	7.0	400	Turkmenistan: Kazandzhik (I = 8–9) 7.2S
1946	11/10	17:42U	−8.5	−77.5	7.3	1400	Peru: Ancash D = 800/1500/1613 (I = 11) [F] 7.3S
1946	12/05	6:47L	23.1	120.3	6.8	74	China: Taiwan [Tainan EQ] (I = 9) D = 58
1946	12/21	4:19L	33.0	135.6	8.0	1330	Japan: Nankaido [Nankai EQ] [CT] 8.2S 8.1W
1947	9/23	12:28U	33.4	58.7	6.9	500	Iran: Dustabad D = 412/570
1947	11/01	14:58U	−10.5	−75.0	7.3	233	Peru: Satipo, Andamarca, Acobamba (I = 10) 7.3S
1948	1/24	17:46U	10.5	122.0	8.3	72	Philippines: Sulu Sea, Iloilo, Jaro [T] 8.2S
1948	5/25	15:11L	29.5	100.5	7.3	737	China: Sichuan P. [Litang EQ] (I = 10) D = 800 7.3S
1948	6/27	8:08L	26.4	99.7	6.2	280	China: Yunnan P. [Jianchuan EQ] (I = 8) D = 100
1948	6/28	16:13L	36.2	136.3	7.1	3769	Japan: Fukui P. [Fukui EQ] [C] 7.3S 7.0W
1948	10/05	20:12U	37.7	58.7	7.3	19800	Turkmenistan: [Ashkhabad EQ] D = 10T/110T? 7.3S
1949	4/20	3:29U	−38.0	−73.5	7.3	57	Chile: Angol [T?] 7.1B
1949	7/10	3:53U	39.2	70.8	7.4	12000	Tajikistan: [Khait EQ] D = 1000s/20T 7.5S
1949	8/05	19:08U	−1.5	−78.3	6.8	6000	Ecuador: [Pelileo EQ] (I = 11) D = 5050
1949	8/17	18:43U	39.4	40.9	6.7	320	Turkey: Agakevy, Elmalidere (I = 9) D = 450/480
1950	5/21	18:38U	−14.0	−72.0	6.0	120	Peru: Cuzco D = 80/83/94
1950	7/09	2:35U	7.9	−72.6	7.0	211	Colombia/Venezuela (I = 9) D = 20/106/126
1950	8/03	22:18U	9.8	−69.7	6.8	100	Venezuela: Tocuyo D = 15
1950	8/15	14:09U	28.5	96.5	8.6	4000	India: Assam/China: Tibet A.R. D = 2486 [F] 8.6SW
1951	5/06	23:08U	13.5	−88.4	6.0	400	El Salvador: Jucuapa (I = 8) D = 1100
1951	8/03	0:23U	13.0	−87.5	6.0	many	Nicaragua (M 5.8 foreshock 8/02 20:30U) D = 1000
1951	8/13	18:33U	40.9	32.9	6.5	50	Turkey: Kursunlu (I = 9) [F]
1951	10/22	5:34L	23.7	121.3	7.2	68	China: Taiwan [Hualien EQ] D = 47/123 [F] 7.4S
1951	12/21	16:37L	26.7	100.0	6.2	423	China: Yunnan P. [Jianchuan EQ] (I = 9)
1952	1/03	6:03U	39.9	41.7	6.0	103	Turkey: Pasinler (I = 7–8) D = 94/133
1952	6/03		39.8	39.5		94	Turkey: Hasankale, Erzincan (I = 7)
1952	8/18	0:02L	31.0	91.5	7.5	54	China: Tibet A.R. [Damxung EQ] (I = 10) 7.6S
1952	9/30	20:52L	28.3	102.2	6.8	236	China: Sichuan P. [Mianning EQ] (I = 9)
1952	10/08	22:24L	39.0	112.7	5.5	58	China: Shanxi P. [Guoxian EQ] (I = 8)
1952	11/04	16:58U	52.3	161.0	8.5	many	Russia: [Kamchatka EQ] (I = 11) [T] 8.2S 9.0W
1953	2/12	8:15U	35.4	55.0	6.3	973	Iran: Torud [F]
1953	3/18	19:06U	40.0	27.5	7.4	1103	Turkey: Onon D = 224/265/1070 [F] 7.2S
1953	8/12	9:23U	38.3	20.6	7.2	800	Greece: Cephalonia (I = 10) D = 445/455 7.1S
1954	9/09	1:04U	36.3	1.5	6.7	1409	Algeria: [El Asnam EQ] D = 1243 [F]
1955	3/31	18:17U	8.1	123.0	7.6	465	Philippines (Mindanao): Lanao D = 291/432 7.6S
1955	4/14	9:29L	30.0	101.8	7.5	94	China: Sichuan P. [Kangding EQ] (I = 10) [F] 7.4S
1955	9/23	23:06L	26.6	101.8	6.8	593	China: Yunnan & Sichuan P. [Yuzha EQ] (I = 9)

(continued)

TABLE 1 *(continued)*

Year	Month/day	h:min	Lat.	Long.	*M*	Deaths	Remarks
1956	3/16	19:32U	35.6	35.5	6.0	148	Lebanon: Litani D = 136
1956	6/09	23:13U	35.0	67.5	7.3	350	Afghanistan D = 220/400/2000 [L] 7.5S
1956	7/09	3:11U	36.6	26.0	7.5	53	Greece: Santorini, Amorgos (I = 9) [T] 7.7S
1956	7/21	15:32U	23.0	70.0	6.1	111	India: Anjar (Cutch) D = 117
1956	10/31	14:03U	27.2	54.4	5.9	410	Iran: Bastaq D = 347 *M* = 6.8
1957	4/25	2:25U	36.5	28.6	7.2	67	Turkey: Fethiye/Greece: Rhodes (I = 11) D = 18
1957	5/26	6:33U	40.6	31.2	7.1	500	Turkey: Abant (I = 10) D = 25/52/66 [F] 7.1S
1957	7/02	0:42U	36.1	52.7	7.1	1200	Iran: Mazandaran D = 1100/1500/2000 7.0S
1957	7/28	8:40U	16.4	−99.2	7.5	160	Mexico: Guerrero, Mexico City D = 68 [T] 7.5S
1957	12/04	3:37U	45.2	99.4	8.3	many	Mongolia: [Gobi–Altai EQ] D = 30/1200[F] 8.0S 8.1W
1957	12/13	1:45U	34.5	48.0	7.2	2000	Iran: Farsinaj D = 1130/1200/1392/2500 [F] 6.8S
1958	8/16	19:13U	34.4	47.9	6.7	200	Iran: Firuzabad D = 132/137 [F]
1960	1/13	15:40U	−16.0	−72.0	7.5	63	Peru: Arequipa (I = 10) D = 9/57 [T?] 7.5B
1960	2/29	23:40U	30.5	−9.6	5.7	13100	Morocco: [Agadir EQ] (I = 10) D = 12T/14T/15T
1960	4/24	12:14U	27.7	54.4	6.0	420	Iran: Lar D = 380/400/450/500/1000
1960	5/22	19:11U	−39.5	−74.5	8.5	5700	Chile: [Chilean EQ] D = 1743/2231 [T] 8.5S 9.5W
1961	6/11	5:10U	27.9	54.5	7.2	61	Iran: Lar *M* = 6.4/6.6
1962	9/01	19:20U	35.6	49.9	7.2	12225	Iran: Buyin–Zahra [Qazvin EQ] D = 10T [F] 6.9S
1963	2/21	17:14U	32.6	21.0	5.3	300	Libya: Barce (I = 8) D = 290
1963	7/26	4:17U	42.0	21.4	6.1	1070	Macedonia: [Skopje EQ] (I = 9–10) D = 1200
1963	9/02	1:34U	33.9	74.8	5.3	80	India: Kashmir
1964	1/04	22:45U	−1.9	102.3	6.7	110	Indonesia (Sumatra) (confuded with 4/02?)
1964	1/18	20:04L	23.1	120.6	7.0	106	China: Taiwan [SE of Tainan EQ] (I ≥ 9) D = 107
1964	3/27	17:36L	61.0	−147.8	8.3	125	USA: [Alaska EQ] (I = 9–10) D = 131 [T] 8.4S 9.2W
1964	4/02	1:11U	5.8	95.6	7.0	110	Indonesia (Sumatra) [T]
1965	1/24	0:11U	−2.4	126.0	7.5	71	Indonesia: Ceram Sea, Sanana (Sulu) [T] 7.5S
1965	3/28	16:33U	−32.4	−71.2	7.3	337	Chile: La Gigua D = 174/400 7.2S 7.4B
1965	5/03	10:01U	13.5	−89.3	6.3	125	El Salvador: San Salvador D = 120
1966	2/05	23:12L	26.1	103.1	6.5	371	China: Yunnan P. [Dongchuan EQ] (I = 9) D = 200
1966	3/08	5:29L	37.3	114.9	6.8	many	China: Hebei P. [Longyao EQ] (I = 9+)
1966	3/20	1:42U	0.6	30.2	6.8	200	Uganda/Zaire (I = 7) D = 157
1966	3/22	16:19L	37.5	115.1	7.2	8064	China: Hebei P. [Ningjin EQ]/[Xingtai EQ] 7.1S
1966	6/27	10:41U	29.6	80.9	6.0	80	India–Nepal border D = 150
1966	8/19	12:22U	39.2	41.6	6.8	2517	Turkey: Varto (I = 9) D = 2394/2964/3000+ [F]
1966	10/17	21:41U	−10.7	−78.6	7.6	110	Peru: Lima, Huacho (I = 9) D = 125 [T] 7.8S 8.1W
1967	2/09	15:24U	2.9	−74.8	6.8	98	Colombia: Huila (I = 7–9) D = 61/100+ 7.1S
1967	2/19	22:14U	−9.1	113.0	6.8	54	Indonesia (Java): Dampit, Gondang D = 23
1967	4/11	5:09U	−3.3	119.2	5.8	58	Indonesia (Sulawesi): D = 13/37/71 [T]
1967	7/22	16:56U	40.7	30.7	6.8	173	Turkey: Mudurnu (I = 10) D = 86/89 [F] 7.1S
1967	7/26	18:53U	39.5	40.4	6.0	110	Turkey: Tuncali (I = 7–8) D = 97/173
1967	7/30	0:00U	10.6	−67.3	6.5	240	Venezuela: [Caracas EQ] D = 225/266/300
1967	12/10	22:51U	17.7	73.9	6.5	180	India: [Koyna EQ] Reservoir induced D = 177
1968	1/15	2:01U	37.8	13.0	6.5	231	Italy: Val de Belice D = 216/224/281/296/400
1968	4/29	17:01U	39.2	44.3	5.6	61	Iran/Turkey D = 38
1968	5/16	9:48L	40.7	143.6	7.9	52	Japan: Aomori P. [Tokachi-oki EQ] [T] 8.1S 8.2W
1968	8/01	20:19U	16.5	122.2	7.3	207	Philippines: Luzon (I = 9) D = 216 [T] 7.2S
1968	8/14	22:14U	0.1	119.7	7.4	392	Indonesia (Sulawesi) D = 200 [T] 7.3S
1968	8/31	10:47U	34.0	59.0	7.3	15000	Iran: [Dasht-i Biyaz EQ] D = 11588/12100 [F] 7.1S
1968	9/01	7:27U	34.1	58.3	6.4	900	Iran: Firdaus (Ferdow)
1969	1/03	3:16U	37.2	57.8	5.4	50	Iran: Khurasan Prov. (I = 7–8)
1969	2/23	0:36U	−3.1	118.8	6.9	600	Indonesia (Sulawesi) [T] D = 64
1969	3/28	1:48U	38.5	28.4	6.6	53	Turkey: Alasehir, Sarigl (I = 8) D = 11/41 [F]
1969	10/01	5:05U	−11.8	−75.1	6.2	136	Peru: Huaytapallana, Lampa (I = 11) D = 150 [F]
1970	1/05	1:00L	24.2	102.7	7.8	15621	China: Yunnan P. [Tonhai EQ] (I = 10+) [F] 7.3S
1970	3/28	21:02U	39.2	29.5	7.1	1086	Turkey: Gediz (I = 9) D = 1098 [F]

(continued)

TABLE 1 *(continued)*

Year	Month/day	h:min	Lat.	Long.	M	Deaths	Remarks
1970	5/14	18:12U	43.0	47.1	6.6	many	Russia (Dagestan) (I = 8–9)
1970	5/31	20:23U	−9.4	−78.9	7.8	66794	Peru: [Peru EQ] D = 54T/50–70T [L] 7.5S 7.9W
1970	7/30	0:52U	37.9	55.9	6.7	220	Iran: Karnaveh, Marev D = 176/200
1970	12/03	3:12L	35.9	105.6	5.5	117	China: Ningsia A.R. [Xiji EQ] (I = 7+)
1970	12/10	4:34U	−4.0	−80.7	7.6	81	Peru/Ecuador (I = 10) D = 52 7.4S
1971	2/09	14:01U	34.4	−118.4	6.4	58	USA (Calif.): [San Fernando EQ] (I = 11) D = 65 [F]
1971	5/12	6:25U	37.6	29.7	6.2	57	Turkey: Burdur (I = 8–9)
1971	5/22	16:43U	38.8	40.5	7.0	995	Turkey: Bingol (I = 9) D = 755/865/870/878 [F]
1971	7/09	3:03U	−32.5	−71.2	7.5	83	Chile: Illapel, Los Vilos, La Ligua [T] 7.7S
1972	4/10	2:06U	28.4	52.8	6.8	5010	Iran: [Ghir (Qir) EQ] D = 5374/5400 [F]
1972	9/03	16:48U	36.0	73.4	6.5	100	India: Kashmir
1972	12/23	6:29U	12.3	−86.1	6.2	6000	Nicaragua: [Managua EQ] D = 5T/8T/11T [F]
1973	1/30	21:01U	18.4	−102.9	7.5	56	Mexico: Michoacan–Colima D = 30 [T] 7.3S
1973	2/06	18:37L	31.3	100.7	7.6	2199	China: Sichuan P. [Luhuo EQ] (I = 10) [F] 7.2S
1973	8/28	9:50U	18.2	−96.5	7.3	600	Mexico: Puebla, Veracruz D = 539/1000 7.3B
1974	5/11	3:25L	28.2	104.1	7.1	1541	China: Yunnan P. [Yongshan EQ] D = 1641/20T
1974	10/03	14:21U	−12.4	−77.7	7.6	78	Peru: Callao, Lima (I = 9) D = 63 [T] 7.6S 8.1W
1974	12/28	12:11U	35.0	72.8	6.2	5300	Pakistan: Patan D = 700/900
1975	2/04	19:36L	40.7	122.8	7.3	1328	China: Liaoning P. [Haicheng EQ] (I = 9+) [F] 7.2S
1975	9/06	9:20U	38.5	40.7	6.7	2370	Turkey: Lice (I = 9) D = 2100/2385/3000
1976	2/04	9:01U	15.2	−89.2	7.5	22870	Guatemala: [Guatemala EQ] (I = 9) [FL] 7.5S
1976	5/06	20:00U	46.3	13.3	6.1	965	Italy: [Friuli EQ] (I = 9–10) D = 929/978
1976	5/29	20:23L	24.5	99.0	7.3	98	China: Yunnan P. [Longling EQ] (I = 9) Also 22:00L
1976	6/25	19:18U	−4.6	140.0	7.1	6000	Indonesia (Irian Jaya) D = 422 (5T–9T missing)
1976	7/14	7:13U	−8.1	114.8	6.5	563	Indonesia (Bali)
1976	7/28	3:42L	39.4	118.0	7.8	242800	China: Hebei P. [Tangshan EQ] (I = 11) [F] 7.8S
1976	8/16	16:11U	6.2	124.0	7.9	8000	Philippines (Mindanao) (I = 10) D = 6500 [T] 7.8S
1976	10/29	2:51U	−4.5	139.9	7.2	6000	Indonesia (Irian Jaya) (I = 8) D = 108/133
1976	11/24	12:22U	39.1	44.0	7.3	3900	Turkey: Muradiye/Iran D = 3626/10T [F] 7.1S
1977	3/04	19:21U	45.8	26.8	7.2	1581	Romania: Vrancea, Bucharest D = 1387 7.1B 7.5W
1977	3/21	21:18U	27.6	74.0	7.0	167	Iran: Bandar–Abbas area (I = 8) D = 152 6.7W
1977	4/06	13:36U	32.0	50.7	5.9	366	Iran: Shahr Kord D = 348 6.0W
1977	8/19	6:08U	−11.1	118.5	7.9	189	Indonesia: [Sumbawa EQ] [T] 8.1S 8.3W
1977	11/23	9:26U	−31.0	−67.8	7.4	70	Argentina: San Juan P. [Caucete EQ] 7.2S 7.4W
1977	12/19	23:34U	31.0	56.5	5.8	665	Iran: Gisk D = 551/584 [F] 5.9W
1978	6/20	20:03U	40.6	23.3	6.5	50	Greece: [Thessaloniki EQ] (I = 8–9) D = 45 [F] 6.2W
1978	9/16	15:36U	33.4	57.4	7.4	18220	Iran: Tabas D = 15T/20T + [F] 7.2S 7.3W
1978	12/14	7:05U	32.1	49.7	6.3	76	Iran: Izeh, Masjed-e-Solyeman D = 100 6.1W
1979	1/16	9:50U	33.9	59.5	6.7	200	Iran: Boznabad D = 1000 6.5W
1979	4/15	6:19U	42.1	19.2	6.9	129	Yugoslavia: Montenegro/Albania D = 156 6.9W
1979	11/14	2:21U	34.1	59.9	6.5	280	Iran: Karizan, Khwaf D = 240 6.6W
1979	11/23	23:40U	4.8	−76.2	6.3	52	Colombia: Manizales–Armenia area 7.2W
1979	12/12	7:59U	1.6	−79.4	7.9	600	Colombia: Tumaco, Buenaventura [T] 7.6S 8.1W
1980	1/01	16:42U	38.8	−27.8	6.9	56	Azores: Terceira Is., Sao Jorge Is. D = 60 6.9W
1980	7/29	14:58U	29.6	81.1	6.7	100	Nepal India border D = 150–200 6.5W
1980	10/10	12:25U	36.2	1.4	7.3	3500	Algeria: [El Asnam EQ] D = 2590–5000 [F] 7.1SW
1980	10/24	14:53U	18.2	−98.2	6.8	65	Mexico: Oaxaca, Guerrero, Puebla D = 300+ 7.1W
1980	11/23	18:34U	40.9	15.3	6.7	2483	Italy: [Irpinia EQ] D = 2735/3105/4680 [F] 6.9W
1981	1/19	15:11U	−4.6	139.2	6.7	1300	Indonesia (Irian Jaya) D = 261 [L] 6.6W
1981	1/24	5:13L	30 9	101.1	6.8	126	China: Sichuan P. [Daufu EQ] D = 150 [F] 6.5W
1981	6/11	7:24U	29.9	57.7	6.7	3000	Iran: Kerman P., Golbaft [F] 6.6W
1981	7/28	17:22U	30.0	57.8	7.1	1500	Iran: Kerman P. D = 8000 [F] 7.2W
1981	9/12	7:15U	35.7	73.6	6.1	229	Pakistan: Gilgit D = 212 6.1W
1982	12/13	9:12U	14.7	44.4	6.0	2800	Yemen: Dhamar D = 1900/3000 [F] 6.2W
1982	12/16	0:40U	36.2	69.0	6.6	500	Afghanistan: Baghlan D = 450 6.5W

(continued)

TABLE 1 (*continued*)

Year	Month/day	h:min	Lat.	Long.	*M*	Deaths	Remarks
1983	3/25	11:57U	36.0	52.3	4.9	100	Iran: Damavand, Amol D = 30 5.4W
1983	3/31	13:12U	2.5	−76.7	4.9	241	Colombia: Popayan, Piendamo, Moral D = 350 5.6W
1983	5/26	11:59L	40.4	139.1	7.7	108	Japan: Akita P. [Nihonkai-chubu EQ] [T] 7.7W
1983	10/30	4:12U	40.3	42.2	6.9	1400	Turkey: [Narman–Horasan EQ] D = 1300–2000 6.6W
1983	12/22	4:11U	11.9	−13.5	6.2	643	Guinea: Gaoual, Koumbia D = 350/443 [F] 6.3W
1985	3/03	22:47U	−33.1	−71.9	7.8	177	Chile: San Antonio, Valparaiso [T] 7.9W
1985	8/23	20:41L	39.4	75.2	7.3	67	China: Xinjiang A.R. D = 71/80 6.9W
1985	9/19	13:17U	18.2	−102.5	8.1	9500	Mexico: Mexico City [Michoacan EQ] [T] 8.0W
1985	10/13	15:59U	40.3	69.8	5.9	many	Tajikistan: Kairakkum D = 29/1000s 5.8W
1986	10/10	17:49U	13.8	−89.1	5.4	1500	El Salvador: San Salvador D = 1000 [F] 5.7W
1987	3/06	4:10U	0.2	−77.8	6.9	5000	Ecuador/Colombia: Quito, Tulcan, Riobamba 7.1W
1987	11/26	1:43U	−8.3	124.2	6.5	83	Indonesia: Pantar Is., Kabir D = 37 [T] 6.5W
1988	8/20	23:09U	26.8	86.6	6.6	1450	Nepal/India: Bihar P. D = 721 (N) + 280 (B) 6.8W
1988	9/06	0:42U	−6.1	146.2	4.3	74	Papua New Guinea: Kaiapit [L]
1988	11/06	21:03L	22.8	99.6	7.3	748	China: Yunnan P. [Lancang EQ] D = 730/930 7.0W
1988	12/07	7:41U	41.0	44.2	6.8	25000	Armenia: [Spitak EQ]/E. Turkey [F] 6.7W
1989	1/22	23:02U	38.5	68.7	5.3	274	Tajikistan: Gissar area
1989	8/01	0:18U	−4.5	139.0	5.8	120	Indonesia (Irian Jaya): Kurima D = 90 6.1W
1989	10/18	0:04U	37.0	−121.9	7.1	63	USA (Calif.): SF area [Loma Prieta EQ] 6.9W
1990	4/26	17:37L	36.0	100.3	6.9	126	China: Qinghai P. (double shock) 6.4W
1990	5/30	2:34U	−6.0	−77.2	6.5	135	Peru: Moyobamba, Rioja 6.5W
1990	6/20	21:00U	37.0	49.4	7.7	35000	Iran: Manjil, Rudbar D = 30T–50T [F] 7.4W
1990	7/16	7:26U	15.7	121.2	7.8	2430	Philippines: Cabanatuan [Luzon EQ] [F] 7.7W
1991	1/31	23:03U	36.0	70.4	6.4	703	Afghanistan/Pakistan: Hindu Kush D = 1200 6.8W
1991	4/05	4:19U	−6.0	−77.1	6.8	53	Peru: Rioja–Nueva Cajamarca area 6.9W
1991	4/22	21:56U	9.7	−83.1	7.6	75	Costa Rica: Limon, Pandora/Panama [T] 7.6W
1991	4/29	9:12U	42.5	43.7	7.0	184	Georgia: Dzhava, Chiatura D = 114/270 7.0W
1991	7/23	19:44U	−15.7	−71.6	4.7	92	Peru: Maca–Chivay area D = 12+ Missing = 80 5.2W
1991	10/19	21:23U	30.8	78.8	7.0	2000	India: [Uttarkashi EQ] D = 768 6.8W
1992	3/13	17:18U	39.7	39.6	6.8	652	Turkey: Erzincan D = 479/498+ 6.6W
1992	8/19	2:04U	42.1	73.6	7.4	75	Kyrgyzstan: Toluk, Suusamyr [F] 7.2W
1992	9/02	0:16U	11.7	−87.3	7.2	184	Nicaragua (west coast) [T] 7.6W
1992	10/12	13:09U	29.8	31.1	5.3	552	Egypt: Dahshur (S of Old Cairo) D = 545 5.8W
1992	12/12	5:29U	−8.5	121.9	7.5	1740	Indonesia (Flores): Maumere D = 2080 [T] 7.7W
1993	7/12	22:17L	42.8	139.2	7.8	233	Japan: [Hokkaido-nansei-oki EQ] [T] 7.7W
1993	9/29	22:25U	18.1	76.5	6.2	9748	India: [Latur EQ] D = 7601 6.2W
1993	10/13	2:06U	−5.9	146.0	7.0	60	Papua New Guinea: Upper Markham Valley 6.9W
1994	1/17	12:30U	34.2	−118.5	6.8	60	USA (Calif.): [Northridge EQ] 6.7W
1994	2/15	17:07U	−5.0	104.3	7.0	207	Indonesia (Sumatra): Liwa (Lampung P.) 6.9W
1994	6/02	18:17U	−10.5	112.8	7.2	277	Indonesia: SE coast of Java, Bali [T] 7.7W
1994	6/06	20:47U	2.9	−76.1	6.6	800	Columbia: Cauca, Huila, Tolima 6.8W
1994	8/18	1:13U	35.5	−0.1	5.9	159	Algeria: Mascara Province D = 171 5.8W
1994	11/14	19:15U	13.5	121.1	7.1	78	Philippines (Mindro): Calapan [FT] 7.1W
1995	1/17	5:46L	34.6	135.0	7.2	6432	Japan: Kobe [Hyogoken-nanbu EQ] [CF] 6.8S 6.9W
1995	5/27	13:03U	52.6	142.8	7.5	1989	Russia (Sakhalin): Neftegorsk [F] D = 1841 7.1W
1995	10/01	15:57U	38.1	30.1	6.2	101	Turkey: Dinar 6.4W
1995	10/06	18:09U	−2.1	101.4	6.9	84	Indonesia (Sumatra): Jambi P. 6.8W
1995	10/09	15:35U	19.1	−104.2	7.4	58	Mexico: Cihuatlan, Manzanillo (Colima) [T] 8.0W
1995	10/24	6:46L	26.0	102.2	6.4	81	China: Yunnan P. (Wuding) 6.2W
1996	2/03	19:14L	27.3	100.3	6.5	309	China: Yunnan P. [Lijiang EQ] D = 322 6.6W
1996	2/17	5:59U	−0.9	137.0	8.1	166	Indonesia (Irian Jaya): Biak, Supiori [T] 8.2W
1997	2/04	10:37U	37.7	57.3	6.8	100	Iran/Turkmenistan: Bojnurd–Shirvan area 6.5W
1997	2/27	21:08U	30.0	68.2	7.3	57	Pakistan: Harnai–Sibi area 7.1W
1997	2/28	12:57U	38.1	48.1	6.1	1100	Iran: Ardebil D = 965 6.1W
1997	5/10	7:57U	33.8	59.8	7.3	1572	Iran: Qayen, Birjand D = 1640 [F] 7.2W

(*continued*)

TABLE 1 *(continued)*

Year	Month/day	h:min	Lat.	Long.	*M*	Deaths	Remarks
1997	7/09	19:24U	10.6	−63.5	6.8	81	Venezuela: Cariaco–Cumana area 6.9W
1998	1/10	11:50L	41.1	114.5	5.7	70	China: Hebei P. [Zhangbei EQ] D = 49 5.7W
1998	2/04	14:33U	37.1	70.1	6.1	2323	Afghanistan: Rostaq area 5.9W
1998	5/22	4:48U	−17.7	−65.4	6.6	105	Bolivia: Aiquile–Totora area 6.6W
1998	5/30	6:22U	37.1	70.1	6.9	4000	Afghanistan: Badakhshan and Takhar P. 6.6W
1998	6/27	13:55U	36.9	35.3	6.2	145	Turkey: Adana, Ceyhan 6.3W
1998	7/17	8:49U	−3.0	141.9	7.1	2700	Papua New Guinea: Sissano [T] 7.0W
1999	1/25	18:19U	4.5	−75.7	5.7	1900	Colombia: Armenia, Calarca [Quindio EQ] 6.2W
1999	2/11	14:08U	34.3	69.4	5.8	70	Afghanistan: Lowgar & Vardak P. 6.0W
1999	3/29	3:05L	30.5	79.4	6.4	100	India/China border 6.6W
1999	8/17	0:01U	40.8	29.9	7.8	17118	Turkey: [Kocaeli EQ] + many missing [F] 7.5W
1999	9/07	11:56U	38.1	23.6	5.8	143	Greece: Athens area 6.0W
1999	9/20	17:47U	23.8	121.0	7.7	2413	Taiwan: [Chichi EQ] [F] 7.7W
1999	11/12	16:57U	40.8	31.2	7.5	843	Turkey: Bolu-Duzce-Kaynasli area 7.2W
2000	6/04	16:04U	−4.7	102.1	8.0	103	Indonesia (Sumatra): Bengkulu D = 90 7.9W

materials. No new estimation of epicenters and magnitudes has been made. (For revised times, locations, and magnitudes of events since 1900, see Chapter 41 by Engdahl and Villaseñor.) If different values are found in an item among different sources, the one that seems to be most appropriate has been adopted, although an objective assessment of the reliability of data is difficult in many cases. Numbers of deaths for an earthquake from different literature often differ very widely. In such cases, numbers different from the one shown in the column "Deaths" have been included in the column "Remarks" in the form such as D = 850/2T/1000s (2T = 2000, 1000s = thousands). In some cases, casualties are given using the letter C.

Magnitudes shown in the column "*M*" have been taken from various sources. For example, all magnitudes of the earthquakes in Japan are the JMA (Japan Meteorological Agency) magnitudes or equivalents. All magnitudes of the earthquakes in China are taken from recent Chinese works. For most Italian earthquakes, magnitudes given in Boschi *et al.* (2000) have been used. The surface-wave magnitude M_S (or the body-wave magnitude m_B for intermediate-depth earthquakes) and the moment magnitude M_W are given in the column "Remarks." They are marked with S (or B) and W, respectively. The M_S and m_B values are taken from Abe (1981) and Abe and Noguchi (1983a,b). The M_W values are converted from the seismic moments given in the Harvard University's CMT catalog for earthquakes in and after 1977. The M_W values before 1977 are taken from various papers.

The column "Remarks" includes the country and place names where the damage seems to be most severe. Place names represent cities, towns, villages, or, in some cases, provinces, prefectures, and other names of regions according to the sources adopted. Old names are, as a rule, replaced by the present names, such as Istanbul (formerly Constantinople) and Tokyo (formerly Edo). However, some place names not found in the atlas remain as they were. "P." following a place name means Province or Prefecture. "A.R." means Autonomous Region in China. "EQ" is an abbreviation for Earthquake.

The maximum seismic intensity (or estimated intensity at the epicenter) based on a twelve-grade intensity scale (the MM scale, MSK scale, or similar scales used in some countries) is indicated when available and if space allows, such as (I = 9).

Earthquakes accompanied by a great fire, surface faulting, a large landslide, and a tsunami are indicated by marks [C], [F], [L], and [T], respectively. [R] indicates the formation of a dammed lake (and the flood due to the subsequent collapse of the dam in most cases). [V] means that the earthquake occurred during the eruption of a volcano.

Acknowledgments

The compiler wishes to thank W.H.K. Lee, S. Miyamura, and an anonymous reviewer for helpful advice, and S.R. Walter of the US Geological Survey for preparing the figures and color plate.

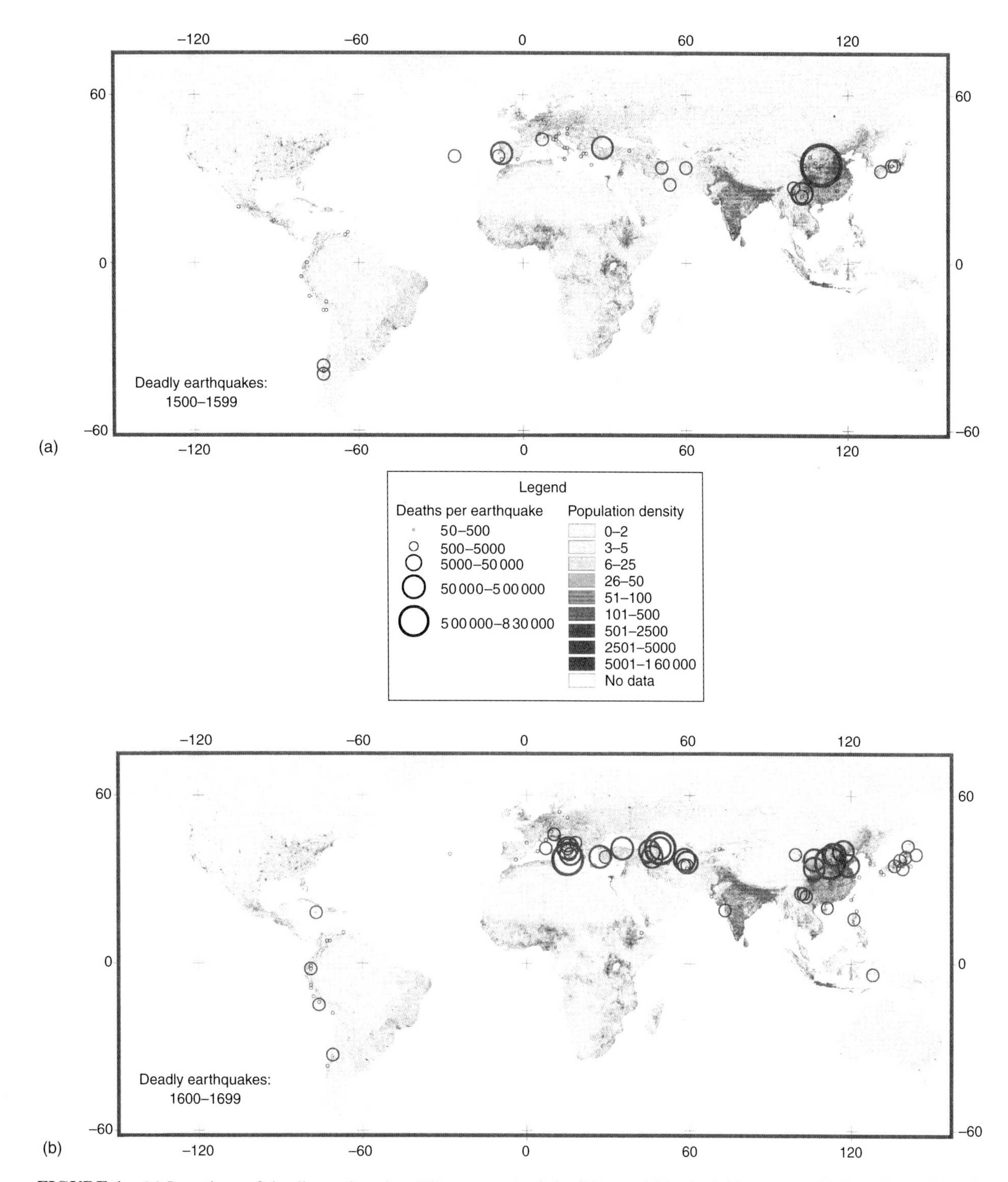

FIGURE 1 (a) Locations of deadly earthquakes (50 or more deaths) of the world in the 16th century. (b) Locations of deadly earthquakes (50 or more deaths) of the world in the 17th century. Population density data was from the US Oak Ridge National Laboratory, 1999.

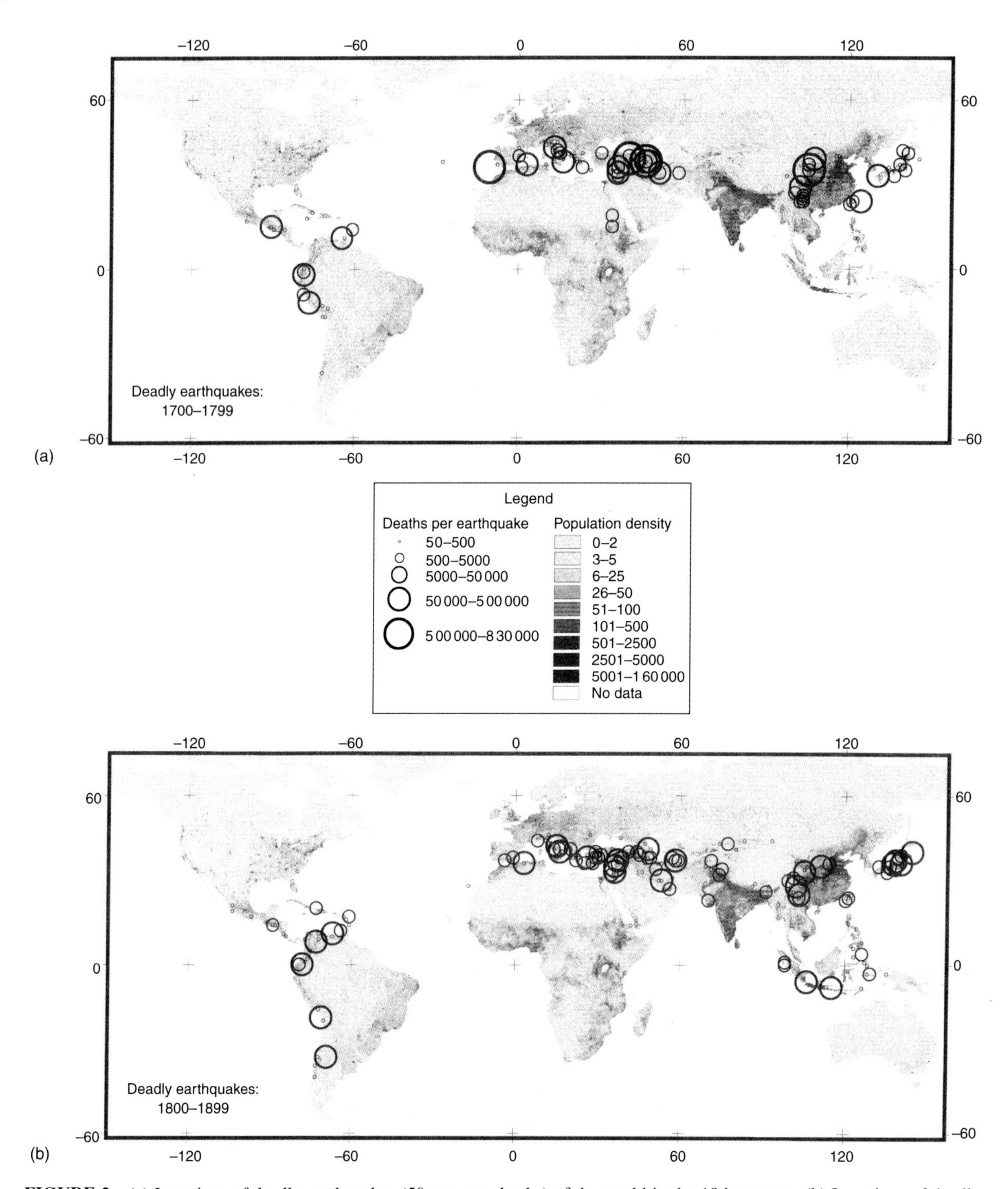

FIGURE 2 (a) Locations of deadly earthquakes (50 or more deaths) of the world in the 18th century. (b) Locations of deadly earthquakes (50 or more deaths) of the world in the 19th century. Population density data was from the US Oak Ridge National Laboratory, 1999.

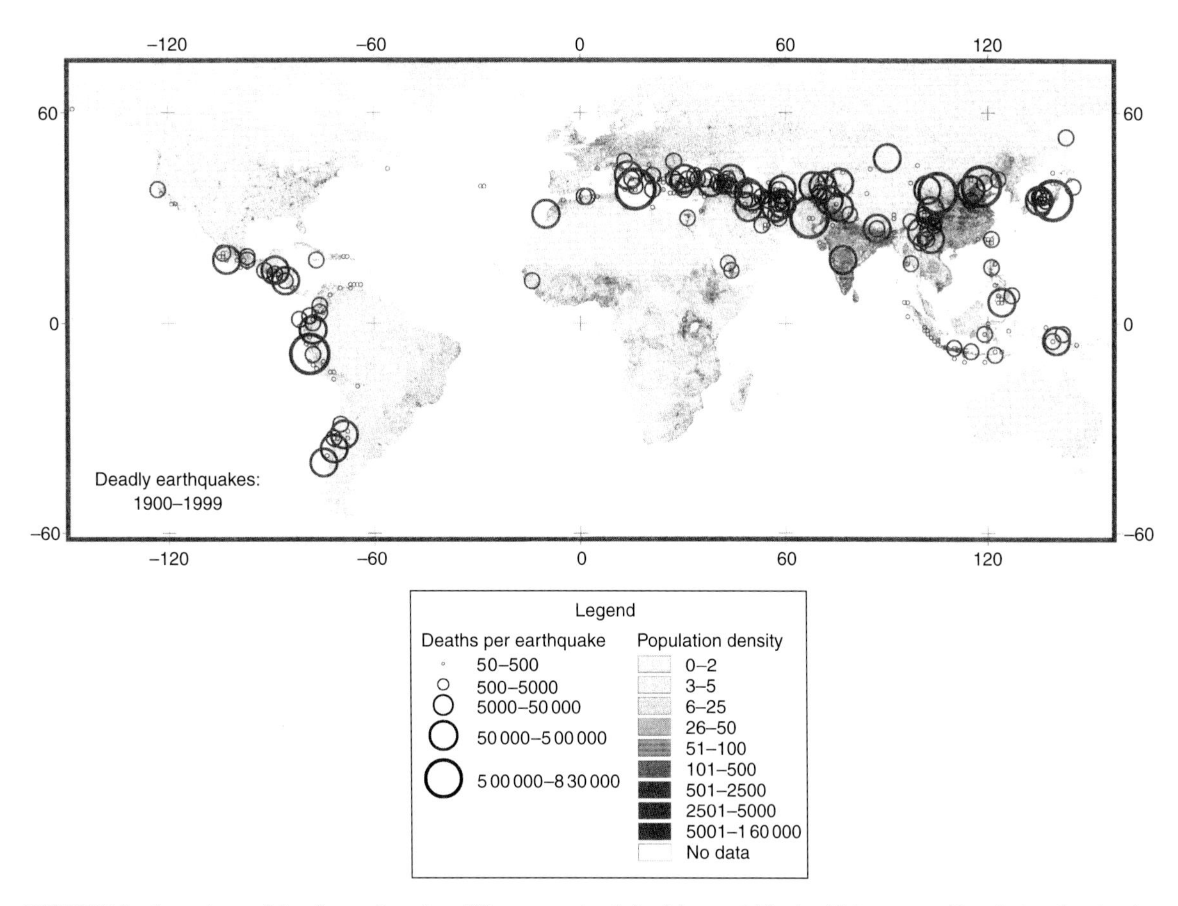

FIGURE 3 Locations of deadly earthquakes (50 or more deaths) of the world in the 20th century. Population density data was from the US Oak Ridge National Laboratory, 1999.

Bibliography Classified According to Regions

Worldwide

Abe, K. (1981). Magnitudes of large shallow earthquakes from 1904 to 1980. *Phys. Earth Planet Inter.* **27**, 72–92.

Abe, K. and S. Noguchi (1983a). Determination of magnitudes of large shallow earthquakes, 1898–1917. *Phys. Earth Planet. Inter.* **32**, 45–59.

Abe, K. and S. Noguchi (1983b). Revision of magnitudes of large shallow earthquakes, 1897–1912. *Phys. Earth Planet. Inter.* **33**, 1–11.

Båth, M. (1979). "Introduction to Seismology." 2nd edn. 428 pp., Birkhauser.

Dunbar, P.K, P.A. Lockridge, and L.S. Whiteside (1992). "Catalog of Significant Earthquakes 2150 B.C. to 1991 A.D. Including Quantitative Casualties and Damage." Report SE-49, 320 pp., World Data Center A for Solid Earth Geophysics.

Ganse, R.A. and J.B. Nelson (1981). "Catalog of Significant Earthquakes 2000 B.C. to 1979 Including Quantitative Casualties and Damage." Report SE-27, 145 pp., World Data Center A for Solid Earth Geophysics.

Ganse, R.A. and J.B. Nelson (1982). Catalog of significant earthquakes 2000 B.C. to 1979, including quantitative casualties and damage. *Bull. Seismol. Soc. Am.* **72**, 873–877. (with microfiches).

Gutenberg, B. and C.F. Richter (1954). "Seismicity of the Earth and Associated Phenomena." 2nd edn. 310 pp., Princeton University Press, Princeton, NJ.

Heck, N.H. (1947). List of seismic sea waves. *Bull. Seismol. Soc. Am.* **37**, 269–286.

Lomnitz, C. (1974). "Global Tectonics and Earthquake Risk." Development in Geotectonics **5**, 231 pp., Elsevier, Amsterdam.

Mallet, R. Third report on the facts of earthquake phenomena – Catalogue of recorded earthquakes from 1606 B.C. to A.D. 1850. *Rep. British Association 1852*, 1–176, *Rep. British Association 1853*, 118–212, *Rep. British Association 1854*, 1–326.

Milne, J. (1911). Catalogue of destructive earthquakes. *Report of 81st Annual Meeting British Association for the Advancement of Science, Appendix No. 1*, 649–740.

Miyamura, S. (1988). Some remarks on historical seismograms and the microfilming project, In: "Historical Seismograms and Earthquakes of the World" (W.H.K. Lee, H. Meyers, and K. Shimazaki, Eds.), pp. 401–419. Academic Press.

National Geophysical Data Center/WDC-A for Solid Earth Geophysics. (1994). "World-wide Tsunamis 2000 B.C.–1992" (Diskette file). Boulder, Colorado.

Richter, C.F. (1958). "Elementary Seismology." 768 pp., Freeman, San Francisco.
Rothé, J.P. (1969). "The Seismicity of the Earth 1953–1965." 335 pp., UNESCO, Paris.
Seismological Society of America. (1911–2000). Seismological notes. *Bull. Seismol. Soc. Am.* **1–84** and *Seismol. Res. Lett.* **66–71** (near the end of almost all issues).
Sieberg, A. (1930). Die Erdbeben. In: "Handbuch der Geophysik," (B. Gutenberg, Ed.). Band IV, Absch. V, 527–686, Gebruder Borntraeger, Berlin.
Sieberg, A. (1932). Die Erdbebengeographie. "Handbuch der Geophysik" (B. Gutenberg, Ed.), Band IV, Absch. VI, 687–1005, Gebruder Borntraeger, Berlin.
Tokyo Astronomical Observatory/National Astronomical Observatory (Ed.) (1925–2000, Annual Publication). Section of Earthquakes: Catalog of Large Earthquakes in the World. In: "Rikanenpyo (Chronological Scientific Tables)." Maruzen, Tokyo (in Japanese).
US Coast and Geodetic Survey/US National Oceanic and Atmospheric Administration/US Geological Survey (1928–1983, annual publication). "United States Earthquakes." US Government Printing Office, Washington, DC.
US Geological Survey, National Earthquake Information Center. "Significant Worldwide Earthquakes (NOAA), 2150 B.C.–1994 A.D." *http://wwwneic.cr.usgs.gov/*
US Geological Survey, National Earthquake Information Center/World Data Center for Seismology. "Significant Earthquakes of the World, 1980–2000." *http://wwwneic.cr.usgs.gov/*

Europe

Ambraseys, N.N. (1962). Data for the investigation of the seismic sea-waves in the eastern Mediterranean. *Bull. Seismol. Soc. Am.* **52**, 895–914.
Ambraseys, N.N. (1985). The seismicity of western Scandinavia, *J. Earthq. Eng. Struc. Dyn.* **13**, 361–400.
Ambraseys, N.N. (1988). Engineering seismology. *J. Earthq. Engin. Struct. Dynam.* **17**, 1–105.
Ambraseys, N.N. (1992). Reappraisal of the seismic activity in Cyprus: 1894–1991. *Boll. Geofis. Teor. Appl.* **34**, 41–80.
Ambraseys, N.N. and J.A. Jackson (1990). Seismicity and associated strain of central Greece between 1890 and 1988, *Geophys. J. Int.* **101**, 663–708.
Ambraseys, N.N., C.P. Melville, and R.D. Adams (1994). "The Seismicity of Egypt and the Red Sea – A Historical Review." 181 pp., Cambridge University Press, Cambridge, UK.
Antonopoulos, J. (1980). Data from investigation on seismic sea-waves events in the eastern Mediterranean from the birth of Christ to 500 A.D. *Ann. Geofis.* **33**, 141–161.
Antonopoulos, J. (1980). Data from investigation on seismic sea-waves events in the eastern Mediterranean from 500 to 1000 A.D. *Ann. Geofis.* **33**, 163–178.
Antonopoulos, J. (1980). Data from investigation on seismic sea-waves events in the eastern Mediterranean from 1000 to 1500 A.D. *Ann. Geofis.* **33**, 179–198.
Antonopoulos, J. (1980). Data from investigation on seismic sea-waves events in the eastern Mediterranean from 1500 to 1900 A.D. *Ann. Geofis.* **33**, 199–230.
Antonopoulos, J. (1980). Data from investigation on seismic sea-waves events in the eastern Mediterranean from 1900 to 1980 A.D. *Ann. Geofis.* **33**, 231–248.
Azzaro, R., M.S. Barbano, A. Moroni, M. Mucciarelli, and M. Stucchi (1999). The seismic history of Catania. *J. Seismol.* **3**, 235–252.
Balan, S., V. Cristescu, and I. Cornea (Eds.) (1983). "Cutremurul de Parint din Romania de la 4 Martie 1977 (The March 4, 1977 earthquake in Rumania)." 516 pp., Bucuresti.
Boschi, E., G. Ferrari P. Gasperini, E. Guidoboni, G. Smriglio, and G. Valensise (1995). "Catalogue of Strong Italian Earthquakes from 461 A.C. to 1980." Instituto Nazionale di Geofisica and Storia Geofis., 973 pp., Ambiente, Bologna.
Boschi, E., E. Guidoboni, G. Ferrari, G. Valensise, and P. Gasperini (1997). "Catalogo dei forti terremoti in Italia dal 461 a.C. al 1990." 644 pp., Instituto Nazionale di Geofisica (with CD-ROM).
Boschi, E., E. Guidoboni, G. Ferrari, D. Mariotti, G. Valensise, and P. Gasperini (Eds.) (2000). Catalogue of strong Italian earthquakes from 461 B.C. to 1977: introductory texts and CD-ROM. *Ann. Geofis.* **43**, 609–868.
Cadiot, B., J. Delaunay, G. Faury, J. Goguel, B. Massinon, D. Mayer-Rosa, J. Vogt, and C. Weber (1979). "Les tremblements de terre en France." 220 pp., Mémoire du Bureau de recherches géologiques et minières, N96-1979.
Carrozzo, M.T., G. de Visintini, F. Giorgetti, and E. Iaccarino (1972). "General Catalogue of Italian Earthquakes." 213 pp., Comitato Nazionale Energia Nucleare.
Comninakis, P.E. and B.C. Papazachos (1982). "A catalog of historical earthquakes in Greece and the surrounding area: 479 B.C.–1900 A.D." 24 pp., University of Thessaloniki Geophysical Laboratory Publication No. 5.
Deresiewicz, H. (1982). Some sixteenth century European earthquakes as depicted in contemporary sources. *Bull. Seismol. Soc. Am.* **72**, 507–523.
Iaccarino, E. and D. Molin (1978). "Atlante macrosismico dell'Italia Nordorientale dall'anno 0 all'aprile 1976." 6 pp. +112 figs, Comitato Nazionale Energia Nucleare.
Iaccarino, E. and D. Molin (1978). "Raccolta di notizie macrosismiche dell'Italia Nordorientale dall'anno 0 all'aprile 1976." 63 pp., Comitato Nazionale Energia Nucleare.
Kárník, V. (1969). "Seismicity of the European Area, Part 1." 364 pp., D. Reidel, Dordrecht.
Kárník, V. (1971). "Seismicity of the European Area, Part 2." 218 pp., D. Reidel, Dordrecht.
Kozak, J. and M.-C. Thompson (1991). "Historical Earthquakes in Europe." 72 pp., Swiss Re.
Maramai, A., and A. Tertulliani (1994). Some events in central Italy: are they all tsunamis? A revision for the Italian tsunami catalog, *Ann. Geofis.* **37**, 997–1008.
Musson, R.M.W. (1994). "A Catalogue of British Earthquakes." 99 pp., British Geological Survey, Tech. Rep. WL94/04.
Papadopoulos, G.A. and B.J. Chalkis (1984). Tsunamis observed in Greece and the surrounding area from antiquity up to the present times. *Marine Geol.* **56**, 309–317.
Papazachos, B.C., Ch. Koutitas, P.M. Hatzidimitriou, B.G. Karacotas, and Ch. Papaioannou (1986). Tsunami hazard in Greece and the surrounding area. *Ann. Geophys.* **4**, 79–90.
Papazachos, B.C., Ch.A. Papaioannou, C.B. Papazachos, and A.S. Savvaidis (1997). "Atlas of Isoseismal Maps for Strong

Shallow Earthquakes in Greece and Surrounding Area (426 BC–1995)." 175 pp., University of Thessaloniki Geophysical Laboratory Publication No. 4.

Papazachos, B.C. and P.E. Comninakis (1982). "A catalog of earthquakes in Greece and the sourrounding area for the period 1901–1980." University of Thessaloniki Geophysical Laboratory Publication No. 5, 146 pp.

Postpischl, D. (Ed.) (1985). "Atlas of Isoseismal Maps of Italian Earthquakes." Progetto Finalizzato Geodinamica, Bologna, 164 pp.

Postpischl, D. (Ed.) (1985). "Catalogo dei Terremoti Italiani dall'Anno 1000 al 1980 (Catalog of Italian Earthquakes from 1000 up to 1980)." 239 pp., Bologna.

Purcaru, G. (1979). The Vrancea, Romania, earthquake of March 4, 1977 – a quite successful prediction. *Phys. Earth Planet. Inter.* **18**, 274–287.

Samardjieva, E., G. Payo, J. Badal, and C. López (1998). Creation of a digital database for XXth century historical earthquakes occurred in the Iberian area. *Pure Appl. Geophys.* **152**, 139–163.

Tinti, S., T. Vittori, and F. Mulargia (1987). On the macroseismic magnitudes of the largest Italian earthquakes. *Tectonophysics* **138**, 159–178.

Tinti, S. and A. Maramai (1996). Catalogue of tsunamis generated in Italy and Cote d'Azur, France: a step toward a unified catalogue of tsunamis in Europe. *Ann. Geofis.* **39**, 1253–1299.

Udias, A.A., L. Arroyo, and J. Mezcua (1976). Seismotectonics of the Azores–Alborn region. *Tectonophysics* **31**, 259–289.

Westaway, R. (1992). Seismic moment summation for historical earthquakes in Italy: tectonic implications. *J. Geophys. Res.* **97**, 15437–15464.

Africa

Ambraseys, N.N. and R.D. Adams (1986). Seismicity of Sudan. *Bull. Seismol. Soc. Am.* **76**, 183–493.

Ambraseys, N.N. and R.D. Adams (1986). Seismicity of West Africa. *Ann. Geophys.* **4**, 679–702.

Ambraseys, N.N. and R.D. Adams (1991). Reappraisal of major African earthquakes, south of 20° N, 1900–1930. *Nat. Hazards* **4**, 389–419.

Ben-Menahem, A., A. Nur, and M. Vered (1976). Tectonics, seismicity and structure of the Afro-Eurasian junction – the breaking of an incoherent plate. *Phys. Earth Planet. Inter.* **12**, 1–50.

Bonouar, D. (1994). Materials for the investigation of the seismicity of Algeria and adjacent regions during the twentieth century. *Ann. Geofis.* **37**, 459–860.

Gouin, P. (1979). "Earthquake History of Ethiopia and the Horn of Africa." 258 pp., IDRC, Ottawa.

Iranga, D.M. (1991). "An earthquake catalogue for Tanzania 1846–1988." Uppsala University Report No. 1–91.

Southwest Asia

Alsinawi, S.A., S.G. Baban, and A.S. Issa (1985). Historical Seismicity of the Arab region. In: "Symposium on Historical Seismograms and Earthquakes of the World, Preliminary Proceedings, Tokyo." pp. 59–83, IASPEI/UNESCO.

Alsinawi, S. and H.A.A. Ghalib (1975). Historical seismicity of Iraq. *Bull. Seismol. Soc. Am.* **65**, 541–547.

Alsinawi, S. and A.S. Issa (1985). On the historical seismicity of Iraq. In: "Symposium on Historical Seismograms and Earthquakes of the World, Preliminary Proceedings, Tokyo." pp. 85–92, IASPEI/UNESCO.

Ambraseys, N.N. (1962). A note on the chronology of Willis' list of earthquakes in Palestine and Syria. *Bull. Seismol. Soc. Am.* **52**, 77–80.

Ambraseys, N.N. (1968). Early earthquakes in north-central Iran. *Bull. Seismol. Soc. Am.* **58**, 485–496.

Ambraseys, N.N. (1988). Engineering seismology. *J. Earthq. Eng. Struct. Dynam.* **17**, 1–105.

Ambraseys, N.N. (1989). Temporal seismic quiescence: SE Turkey. *Geophys. J. Int.* **96**, 311–331.

Ambraseys, N.N. and C.F. Finkel (1987). Seismicity of Turkey and neighbouring regions, 1899–1915. *Ann. Geophys.* **5**, 701–726.

Ambraseys, N.N. and C.F. Finkel (1995). "The Seismicity of Turkey, and Adjacent Areas: A Historical Review, 1500–1800." 240 pp., Eren, Istanbul.

Ambraseys, N.N. and J.A. Jackson (2000). Seismicity of the Sea of Marmara (Turkey) since 1500. *Geophys. J. Int.* **141**, F1–F6.

Ambraseys, N.N. and C.P. Melville (1982). "A History of Persian Earthquakes." 219 pp., Cambridge University Press, Cambridge.

Ambraseys, N.N. and C.P. Melville (1983). Seismicity of Yemen. *Nature* **303**, 321–323.

Ambraseys, N.N. and C.P. Melville (1995). Historical evidence of faulting in eastern Anatolia and northern Syria. *Ann. Geofis.* **38**, 337–343.

Ben-Menahem, A. (1979). Earthquake catalogue for the Middle East (92 B.C.–1980 A.D.). *Boll. Geofis. Teor. Appl.* **21**, 245–310.

Ben-Menahem, A. (1991). Four thousand years of seismicity along the Dead Sea rift. *J. Geophys. Res.* **96**, 20195–20216.

Berberian, M. (1994). "Natural Hazards and the First Earthquake Catalogue of Iran, vol. 1, Historical Hazards in Iran Prior to 1900." 603 pp., UNESCO.

Mortaza Seyed Nabavi (1978). Historical earthquakes in Iran, c.300 B.C.–1900 A.D. *J. Earth Space Phys.* **7**, 70–117.

Poirier, J.P. and M.A. Taher (1980). Historical seismicity in the Near and Middle East, North Africa and Spain from Arabic documents (VIIth–XVIIIth Century). *Bull. Seismol. Soc. Am.* **70**, 2158–2201.

Poirier, J.P., B.A. Romanowicz, and M.A. Taher (1980). Large historical earthquakes and seismic risk in northern Syria. *Nature* **285**, 217–220.

Raid, S. and H. Meyers (1985). "Earthquake Catalog for the Middle East Countries, 1900–1983." Report SE40, World Data Center A for Solid Earth Geophysics, 26 pp.

Ramalingeswara Rao, B. and P. Sitapathi Rao (1984). Historical seismicity of Peninsular India. *Bull. Seismol. Soc. Am.* **74**, 2519–2533.

Tabban, A. (1970). Seismicity of Turkey. *Individual Studies Int. Inst. Seismol. Earthq. Eng.* **6**, 59–74.

Former USSR

Ambraseys, N.N. and R.D. Adams (1989). Long-term seismicity of North Armenia. *EOS* **70**, 145, 152–154.

Berberian, M. (1996). Seismic sources of the transcaucasian historical earthquakes. In: "Historical and Prehistorical Earthquakes in Caucasus." (D. Giardini and S. Balassanian, Eds.), pp. 233–311. Kluwer Academic, Dordrecht.

Giardini, D. and S. Balassanian (Eds.). (1996). "Historical and Prehistorical Earthquakes in Caucasus." 545 pp., Kluwer Academic, Dordrecht.

Kondorskaya, N.V. and N.V. Shebalin (Chief Eds.) (1977). "New Catalog of Strong Earthquakes in the Territory of USSR from Ancient Times to 1975." 536 pp., Nauka, Moscow (in Russian).

Kondorskaya, N.V. and N.V. Shebalin (Chief Eds.) (1982). "New Catalog of Strong Earthquakes in the U.S.S.R. from Ancient Times through 1977." Report SE31, World Data Center A for Solid Earth Geophysics, 608 pp.

Nikonov, A.A. (1994). On the first millennium A.D. large earthquakes in the Greater Caucasus: Revision of original data and catalog. *Izv. Phys. Solid Earth* **30**, 688–693 (English translation.).

Quittmeyer, R.C. and K.H. Jacob (1979). Historical and modern seismicity of Pakistan, Afghanistan, northwestern India, and southeastern Iran. *Bull. Seismol. Soc. Am.* **69**, 773–772.

Shebalin, N.V. and R.E. Tatevossian (Eds.) (1996). Catalogue of large historical earthquakes of the Caucasus. In: "Historical and Prehistorical Earthquakes in Caucasus." (D. Giardini and S. Balassanian, Eds.), pp. 201–232. Kluwer Academic, Dordrecht.

China

Cheng, S.-N. and Y.-T. Yeh (1989). "Catalog of Earthquakes in Taiwan from 1604 through 1988," 255 pp., Institute of Earth Science, Taipei (in Chinese).

Earthquake Disaster Prevention Department of the State Seismological Bureau (Ed.) (1995). "Catalogue of Chinese Historical Strong Earthquakes (23rd century BC–1911 AD)," 514 pp., Seismological Press, Beijing (in Chinese).

Gu, G.-X. (Chief Ed.). (1983). "Catalogue of Chinese Earthquakes (1831 B.C.–1969 A.D.)," 894 pp., Science Press, Beijing (in Chinese). English translation (1989), 872 pp.

Gu, G.-X. (Chief Ed.). (1984). "Catalogue of Chinese Earthquakes (1970–1979)," 334 pp., Science Press, Beijing (in Chinese).

Guo, Z.-J. and L. Chen (Chief Eds.). (1986). "Earthquake Countermeasures," 503 pp., Seismological Press, Beijing (in Chinese).

Lee, P.H. (1988). Historical earthquakes and the seismograms in Taiwan. In: "Historical Seismograms and Earthquakes of the World" (W.H.K. Lee, H. Meyers, and K. Shimazaki, K., Eds.), pp. 241–252. Academic Press, New York.

Lou, B.-T. (Chief Ed.). (1996). "A Comprehensive Compilation of Historic and Recent Earthquake Disaster Status in China," 272 pp., Seismological Press, Beijing (in Chinese).

Seismological Bureau of China. (1986–1999, annual publication). "China Earthquake Yearbook." Seismological Press, Beijing (in Chinese).

Tsai, Y.-B. (1985). A study of disastrous earthquakes in Taiwan, 1683–1895. *Bull. Inst. Earth Sci. Acad. Sinica* **5**, 1–44.

Xie, Y.-S. and M.-B. Tsai (Chief Eds.) (1983–1987). "Compilation of Historical Materials of Chinese Earthquakes." 5 vols., 4912 pp., Science Press, Beijing (in Chinese).

Japan

Earthquake Research Institute, University of Tokyo (Ed.) (1980–1994). "Newly Collected Historical Materials of Earthquakes in Japan." 16 vols, approx. 16000 pp. in total, Earthquake Research Institute (in Japanese).

Japan Meteorological Agency (1983). Table of damaging earthquakes in Japan. In: "Manual for Seismic Observation, Reference Section." Table 5 (in Japanese).

Musha, K. (Ed.) (1941–1943). "Historical Materials of Earthquakes in Japan." vol. 1 (for 416–1693), 606 pp.; vol. 2 (for 1694–1783), 754 pp.; vol. 3 (for 1783–1847), 945 pp., Sinsai-Yobo-Hyogikai (in Japanese).

Musha, K. (Ed.) (1951). "Historical Materials of Earthquakes in Japan." (for 1848–1867 and chronological table), 1107 pp., Mainichi-Shinbunsha (in Japanese).

Tokyo Astronomical Observatory/National Astronomical Observatory (Ed.) (1925–2000, Annual Publication). Catalog of Large Earthquakes in Japan. Section of Earthquakes. in "Rikanenpyo (Chronological Scientific Tables Science)." Maruzen, Tokyo (in Japanese).

Usami, T. (1979). Study of historical earthquakes in Japan. *Bull. Earthq. Res. Inst.* **54**, 299–439.

Usami, T. (1996). "Materials for the Comprehensive List of Destructive Earthquakes in Japan, 416–1995." (Revised and enlarged edition), 495 pp., University of Tokyo Press, Tokyo (in Japanese).

Southeast Asia

Bautista, M.L.P. and K. Oike (2000). Estimation of the magnitudes and epicenters of Philippine historical earthquakes. *Tectonophysics* **317**, 137–169.

Berninghausen, W.H. (1966). Tsunamis and seismic seiches reported from regions adjacent to the Indian Ocean. *Bull. Seismol. Soc. Am.* **56**, 69–74.

Berninghausen, W.H. (1969). Tsunamis and seismic seiches of southeast Asia. *Bull. Seismol. Soc. Am.* **59**, 289–297.

Cox, D.C. (1970). Discussion of "Tsunamis and seismic seiches of southeast Asia" by William H. Berninghausen. *Bull. Seismol. Soc. Am.* **60**, 281–287.

Newcomb, K.R. and W.R. McCann (1987). Seismic history and seismotectonics of the Sunda arc. *J. Geophys. Res.* **92**, 421–439.

Prachuab, S. (1988). Historical earthquakes of Thailand, Burma and Indochina. In: "Historical Seismograms and Earthquakes of the World" (W.H.K. Lee, H. Meyers, and K. Shimazaki, Eds.), pp. 253–226. Academic Press, New York.

Repetti, W.C. (1946). Catalogue of Philippine earthquakes. *Bull. Seismol. Soc. Am.* **36**, 133–322.

Valenzuela, R.G. and L.C Garcia (1988). Studies of Philippine historical earthquakes. In: "Historical Seismograms and Earthquakes of the World" (W.H.K. Lee, H. Meyers, and K. Shimazaki, Eds.), pp. 289–296. Academic Press, New York.

New Zealand

Eiby, G.A. (1968). A descriptive catalogue of New Zealand Earthquakes, Part I, Shocks felt before the end of 1854. *NZ J. Geol. Geophys.* **11**, 16–40.

Eiby, G.A. (1968). An annotated list of New Zealand Earthquakes, 1460–1965. *NZ J. Geol. Geophys.* **11**, 630–647.
Eiby, G.A. (1973). A descriptive catalogue of New Zealand Earthquakes, Part II, Shocks felt from 1846 to 1854. *NZ J. Geol. Geophys.* **16**, 857–907.

North and Central America

Ambraseys, N.N. (1995). Magnitudes of Central American earthquakes 1898–1930. *Geophys. J. Int.* **121**, 545–556.
Ambraseys, N.N. and R.D. Adams (1996). Large magnitude Central American earthquakes, 1898–1994. *Geophys. J. Int.* **127**, 665–692.
Ambraseys, N.N. and R.D. Adams (2001). "The Seismicity of Central America: A Descriptive Catalogue 1898–1995," 309 pp., Imperial College Press, London.
Anderson, J.G., S.K. Singh J.M. Espindola, and J. Yamamoto (1989). Seismic strain release in the Mexican subduction thrust. *Phys. Earth Planet. Inter.* **58**, 307–322.
Berninghausen, W.H. (1964). Tsunamis and seismic seiches reported from the eastern Atlantic south of the Bay of Biscay. *Bull. Seism. Soc. Am.* **54**, 439–442.
Coffman, J.L., C.A. von Hake, and C.W. Stover (Eds.) (1982). "Earthquake History of the United States, Revised edition (Through 1970)." Reprinted 1982 with supplement (1971–80), NOAA and USGS Publication 41–1, 208 + 50 pp.
Leeds, D. (1974). Catalog of Nicaraguan earthquakes. *Bull. Seismol. Soc. Am.* **64**, 1135–1158.
Robson, G.R. (1964). An earthquake catalogue for the Eastern Caribbean. *Bull. Seismol. Soc. Am.* **54**, 785–832.
Singh, S.K., L. Astiz, and J. Havskov (1981). Seismic gaps and recurrence periods of large earthquakes along the Mexican subduction zone: a reexamination. *Bull. Seismol. Soc. Am.* **71**, 827–843.
Singh, S.K. and M. Rodriguez (1984). A catalog of shallow earthquakes of Mexico from 1900 to 1981. *Bull. Seismol. Soc. Am.* **74**, 267–279.
Sutch Osiecki, P. (1981). Estimated intensities and probable tectonic sources of historic (pre-1898) Honduran earthquakes. *Bull. Seismol. Soc. Am.* **71**, 865–881.
White, R.A. and D.H. Harlow (1985). Catalog of significant shallow earthquakes of Central America since 1900. In: "Preliminary Proceedings, Symposium on Historical Seismograms and Earthquakes 1985, Tokyo." pp. 453–459. IASPEI/UNESCO.
Winkler, L. (1979). Catalog of U.S. earthquakes before the year 1850. *Bull. Seismol. Soc. Am.* **69**, 569–602.

South America

Berninghausen, W.H. (1962). Tsunamis reported from the west coast of South America. *Bull. Seismol. Soc. Am.* **52**, 915–921.
CERESIS (Centro Regional de Seismologia para América del Sur) (1985). "Catalogo de Terremotos para America del Sur (Catalog of Earthquakes for South America)." 9 vols, Lima.
CERESIS (Centro Regional de Seismologia para América del Sur) (1985). "Terremotos Destructivos en América del Sur, 1530–1894 (Destructive Earthquakes of South America 1530–1894)." 328 pp., Lima.
Comte, D. and M. Pardo (1991). Reappraisal of great historical earthquakes in the northern Chile and southern Peru seismic gaps, *Nat. Hazards* **4**, 23–44.
Dorbath, L., A. Cisternas, and C. Dorbath (1990). Assessment of the size of large and great historical earthquakes in Peru. *Bull. Seismol. Soc. Am.* **80**, 551–576.

Editor's Note

Table 1 in a computer readable text file is given on the attached Handbook CD, under the directory \42Utsu1. A more extensive text file (for earthquakes with 25 or more deaths) is also provided on the CD.

43

Statistical Features of Seismicity

Tokuji Utsu
The University of Tokyo, Tokyo, Japan

1. Introduction

Statistical studies of earthquake occurrences have frequently been carried out since the early years of seismology. To obtain reliable results from statistical analysis, a sufficient amount of high-quality data is necessary. The data are taken usually from earthquake catalogs, but many of the existing catalogs are inhomogeneous and incomplete. When we use seismicity data, special care must be taken to avoid or correct these defects in the catalogs.

In most statistical studies, earthquakes are represented by point events in a five-dimensional space–time–size continuum. In ordinary earthquake catalogs, the five coordinates are given as longitude and latitude of epicenter, focal depth, origin time, and magnitude. There are many other quantities which characterize an earthquake, fault-plane parameters (or more generally moment tensor components), stress drop, fault rupture length, rupture velocity, etc. Statistical studies involving these quantities are few. This is mainly because the complete data set on these is unavailable especially for small or old earthquakes.

The results of a statistical analysis must be tested for significance. Every method for a significance test is based on certain assumptions. When we use a method, we must remember the underlying assumptions.

It is often the case that we examine two or more data sets and choose the one for which the statistical property in question is most clearly recognized. The effect of such preferential selection of data set must be considered in the significance test. There is no common way of deciding how the significance level must be changed to compensate this effect. Numerical simulation may provide a solution in some cases.

2. Statistical Properties of Earthquake Sequences

2.1 Identification and Classification of Earthquake Sequences

Since the spatial and temporal clustering of earthquakes is the most prominent feature of most earthquake catalogs, we will first consider the properties of earthquake clusters, such as (foreshock–)mainshock–aftershock sequences and earthquake swarms.

There is no universally accepted exact definition of aftershocks, foreshocks, and earthquake swarms. We must give a working definition when we perform a statistical analysis on them. Figure 1a–d shows plots of occurrence rate versus time for typical earthquake sequences. The successive occurrence of mainshock–aftershock sequences of similar size (Fig. 1d) is also called an earthquake swarm, but it is apparently different from an ordinary swarm (Fig. 1c).

In statistical studies of seismicity, we sometimes use declustered catalogs, which contain only independent events (mainshocks and isolated earthquakes; the largest shock in a swarm is considered as the mainshock). Most algorithms for declustering (i.e., identification of earthquake sequences) use either space–time windows to include clustered events or space–time separations to link clustered events. For declustering or cluster identification algorithms, see Davis and Frohlich (1991a), Molchan and Dmitrieva (1992), and references in these papers. Studies using declustered catalogs or lists of earthquake clusters obtained by declustering are found in Utsu (1972a), Prozorov and Dziewonski (1982), Reasenberg (1985), Prozorov (1986), Keilis-Borok and Kossobokov (1990), Frohlich and Davis (1990), Davis and Frohlich (1991a,b), Ogata *et al.* (1995), and many others.

2.2 Aftershocks

2.2.1 Temporal Distribution of Aftershocks

Omori (1894) showed that the frequency of felt aftershocks per day, $n(t)$, following the 1891 Nobi, central Japan, earthquake ($M = 8.0$) decreased regularly with time according to the equation

$$n(t) = K(t + c)^{-1} \quad (1)$$

where K and c are constants and t is the time measured from the mainshock. The cumulative number of aftershocks, $N(t)$,

INTERNATIONAL HANDBOOK OF EARTHQUAKE AND ENGINEERING SEISMOLOGY, VOLUME 81A ISBN: 0-12-440652-1

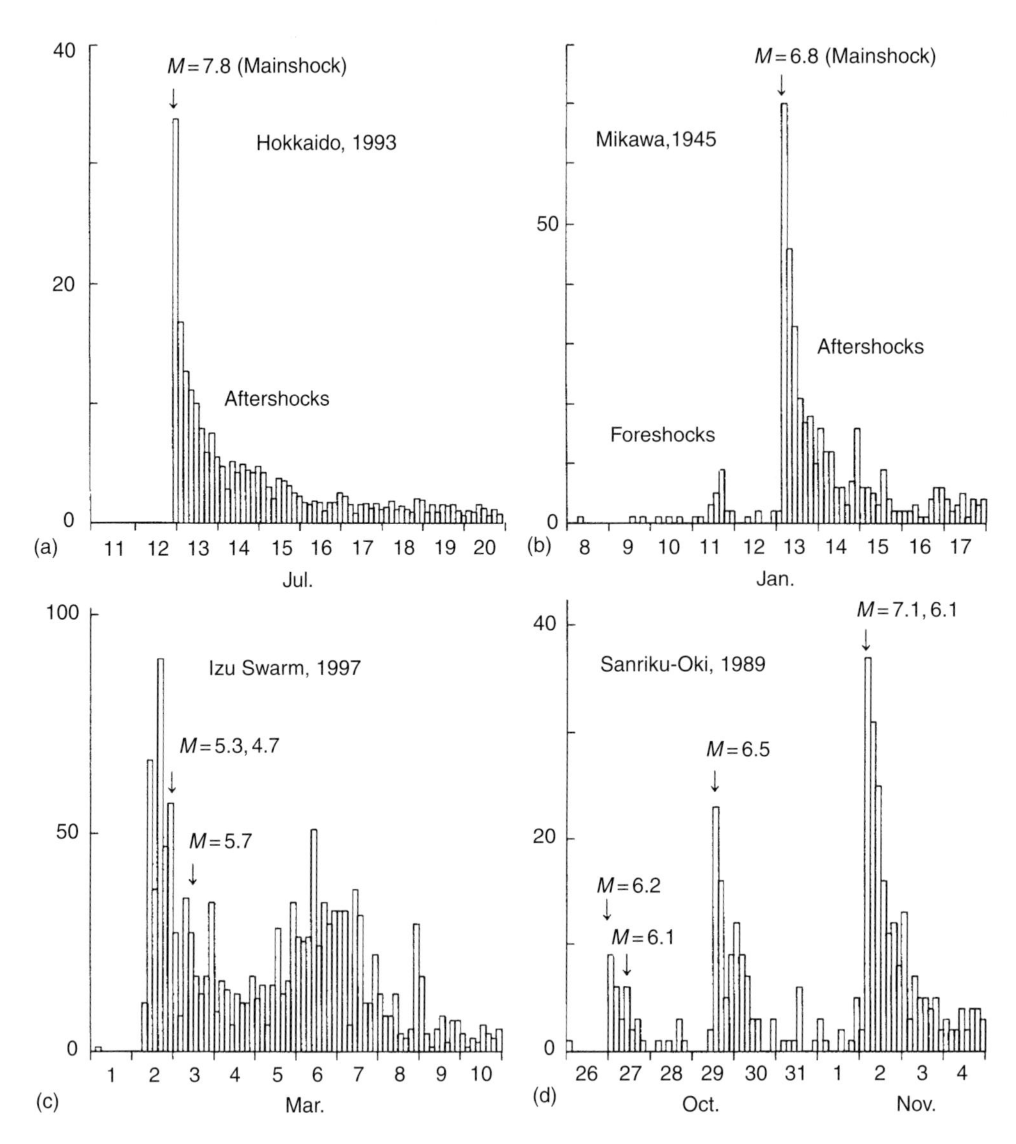

FIGURE 1 Examples of earthquake sequences. Temporal variations in the occurrence frequency per three hours are shown. (a) Mainshock–aftershock sequence; (b) foreshock–mainshock–aftershock sequence; (c) earthquake swarm; (d) successive occurrence of mainshock–aftershock sequences (earthquake swarm of the second kind).

plotted against $\log t$, tends to a straight line with time, because

$$N(t) = \int_0^t n(s)\,ds = K\ln(t/c+1) \tag{2}$$

Utsu (1961) plotted the data of $N(t)$ for 44 aftershock sequences against $\log t$, and showed that the slope of cumulative curve tends to decrease with time for most sequences. He reached the conclusion that Eq. (3) fits the data more closely and called it the modified Omori formula

$$n(t) = K(t+c)^{-p} \tag{3}$$

where p is a constant somewhat larger than 1.0 (mostly 1.0–1.5). $N(t)$ for Eq. (3) is given by

$$N(t) = K\{c^{1-p} - (t+c)^{1-p}\}/(p-1) \qquad (p \neq 1) \tag{4}$$

If the origin times t_i $(i = 1, 2, \ldots)$ are available for all aftershocks of $M \geq M_z$ occurring in a certain period of time, we can compute the maximum likelihood estimates of K, p, and c (Ogata, 1983a). If the data contain the background seismicity of the region, it is recommended to use the equation $n(t) = K(t+c)^{-p} + B$ and estimate the background level B at the same time.

Computer programs for obtaining the maximum likelihood estimates of K, p, c, and B values and their standard errors for a given set of data are available in *IASPEI Software Library* Vol. 6 (Utsu and Ogata, 1997). This package contains the programs for the double sequence (the second sequence starts at $t = t_d$)

$$n(t) = K_1(t + c_1)^{-p_1} + K_2(t + t_d + c_2)^{-p_2} \quad (K_2 = 0 \text{ for } t < t_d) \tag{5}$$

and the triple sequence as well as some other formulas including the Weibull distribution

$$n(t) = K\alpha\beta t^{\beta-1} \exp(-\alpha t^{\beta}) \tag{6}$$

The stretched exponential function used by Kisslinger (1993) is the same as Eq. (6). Gross and Kisslinger (1994) tried more general equations.

The K value in the modified Omori relation depends strongly on the threshold magnitude M_z. The p value seems to be almost independent of M_z (e.g., Utsu, 1962). This is consistent with the stability of mean magnitude $\overline{M}$ (or b value) during an aftershock sequence (e.g., Lomnitz, 1966). The c value often shows strong dependence on M_z. This is mainly due to the deficiency of data. Small aftershocks occurring shortly after the mainshock tend to be missing due to overlapping of seismograms. One recent example supporting the modified Omori relation and the independence of p on M_z is shown in Figure 2 [see also Utsu *et al.* (1995) and Ogata (1999b)]. Nyffenegger and Frohlich (1998) investigated the factors that influence the maximum likelihood estimate of p value. Wiemer and Katsumata (1999) showed spatial variation of p and b values in the aftershock zones of four earthquakes (see also Utsu, 1962). Kisslinger (1996) presented a comprehensive review on aftershock phenomena.

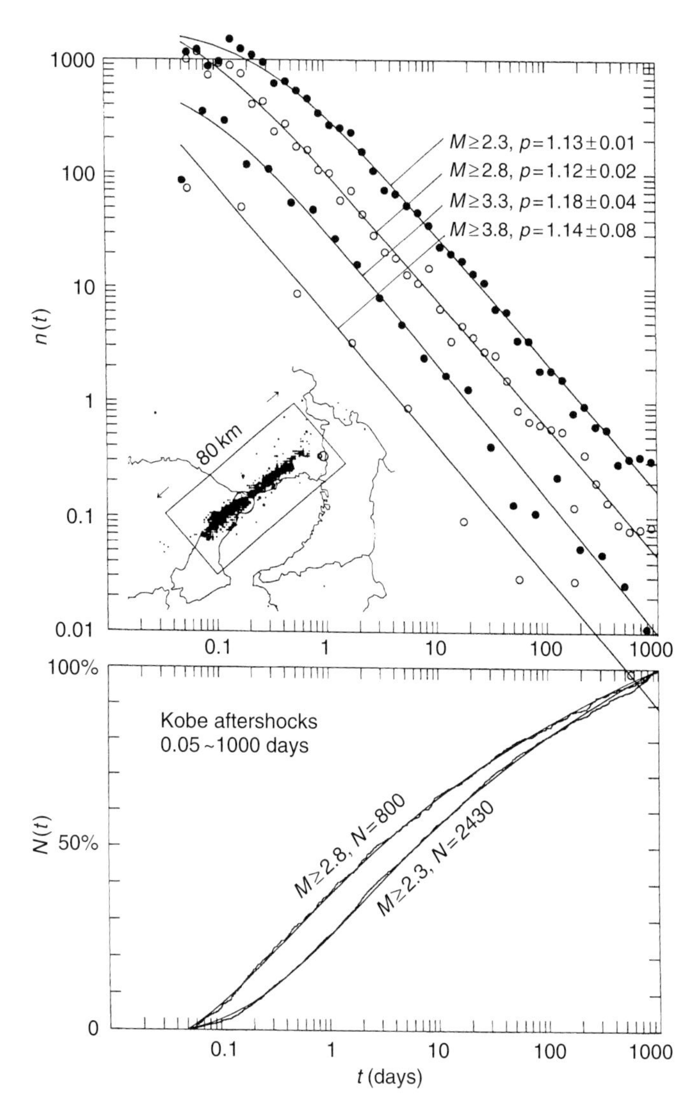

FIGURE 2 Modified Omori curves (top: rate, bottom: cumulative number) fitted to the aftershock sequences of the Hyogoken-Nanbu (Kobe) earthquake of 16 January 1995 (M_w 6.9). Data are taken from the Japan Meteorological Agency catalog. All earthquakes occurring in the rectangular area shown in the inserted map with focal depth less than 40 km are designated as aftershocks. The data and curves corresponding to the four (or two) different threshold magnitudes are shown. Note the independence of p value on the threshold magnitude.

2.2.2 Spatial Distribution of Aftershocks

It had been known before the introduction of earthquake magnitude that the aftershock epicenters scatter more widely for larger mainshocks. Utsu and Seki (1955) introduced the earthquake magnitude in this problem and obtained a relation $\log S = 1.02M_m - 4.01$, where S is the aftershock area in km^2 for a mainshock of magnitude M_m from the data of 40 mainshock–aftershock sequences in Japan ($6 \le M_m \le 8.5$). This equation can be written in a simpler form

$$\log S = M_m - 3.9 \tag{7}$$

If we consider a rectangular aftershock area with length L and width W and simply assume that $W = L/2$, Eq. (8) can be transformed to an equation proposed by Utsu (1961, 1969)

$$\log L = 0.5\,M_m - 1.8 \tag{8}$$

Although an aftershock zone roughly corresponds to the fault ruptured during the mainshock, precise studies indicate that aftershocks are concentrated near the margin of the fault area where the large displacement occurred (e.g., Hartzell and Heaton, 1986; Mendoza and Hartzell, 1988; Takeo, 1988). The relation between surface rupture length and M_m may have different coefficients (e.g., Tocher, 1958; Iida, 1965; Matsuda, 1975; Wells and Coppersmith, 1994), because the surface rupture does not always represent the entire fault rupture responsible for the earthquake.

There are many studies, which relate the size of aftershock zone to the mainshock magnitude. For example, Acharya (1979) obtained similar equations to Eq. (7) for five regions in the Circum-Pacific zone, but the coefficient of M_m scatters from 0.78 to 1.22. Iio (1986) obtained

$$\log L = 0.43 M_m - 1.7 \qquad (9)$$

for smaller earthquakes in Japan ($3.8 \leq M_m \leq 6.2$).

The aftershock zones sometimes show considerable expansion within days to years from the mainshock. For related studies, see Mogi (1968a) and Tajima and Kanamori (1985).

2.2.3 Aftershock Activity

The degree of aftershock activity relative to the mainshock size may be expressed by $\sum_{i=1}^{\infty} E_i/E_m$, where E_m and E_i denote the energy (or moment) of the mainshock and the *i*th largest aftershock, respectively. Since the energy (or moment) of the largest aftershock, E_1, is roughly proportional to the total energy (or moment), the aftershock activity is roughly represented by E_1/E_m, which corresponds to the magnitude difference between the mainshock and the largest aftershock

$$D_1 = M_m - M_1 \qquad (10)$$

D_1 values for many mainshock–aftershock sequences have been investigated for its regional variation and dependence on M_m. Although $D_1 = 1.2$ is known as Båth's law (Richter, 1958), D_1 actually varies very widely from 0 to 3 or more.

The relation between D_1 and M_m has been studied by Utsu (1961) and several other workers. Utsu (1969) expressed $\tilde{D}_1$ (median of D_1 for mainshock magnitude M_m) as $\tilde{D}_1 = 5.0 - 0.5 M_m$ for $8.0 \geq M_m \geq 6.0$ and $\tilde{D}_1 = 2.0$ for $M_m \leq 6.0$ for shallow earthquakes in Japan. Most other workers gave a smaller average D_1 value for M_m less than about 6.5. For example, average D_1 for $M_m = 6.0$ is 0.67, 1.08, 1.14, and 1.05 according to Papazachos *et al.* (1967), Papazachos (1971), Båth (1977), and Kisslinger and Jones (1991), respectively. The difference between Utsu and most other workers may be caused by the difference in data-acquisition principle. Utsu considered the existence of earthquakes whose aftershocks were too small to be observed. Studies of regional variation of D_1 are found in Mogi (1967), Utsu (1969), Tsapanos (1990), Doser (1990), etc.

The aftershock activity has also been discussed by using the number of aftershocks above a certain magnitude level. For example, Singh and Suárez (1988) expressed the average number N of aftershocks of $m_b \geq 5$ accompanying large thrust earthquakes of moment magnitude M_w along the Circum-Pacific subduction zones. The relation is given by $\log N = M_w - 6.34$. Some systematic deviations from the average seem to depend on the degree of interplate coupling and complexities of the source. See Papazachos (1971), Lamoreaux *et al.* (1983), Yamanaka and Shimazaki (1990), and Davis and Frohlich (1991a) for regional trends.

Deep earthquakes are accompanied by few aftershocks (e.g., Wadati, 1931; Frohlich, 1987). Examples of aftershocks of deep earthquakes and depth variation of aftershock activity have been reported by Solov'ev and Solov'eva (1962), Prozorov and Dziewonski (1982), Pavlis and Hamburger (1991), Wiens *et al.* (1997), among others.

2.3 Foreshocks

Foreshocks are generally infrequent as compared with aftershocks. Foreshock activities are highly variable. Many large earthquakes of $M \geq 7.0$ are not preceded by foreshocks, even if the seismograph network is capable of detecting shocks of $M < 3.0$. Therefore the magnitude difference between the mainshock and the largest foreshock must be larger than 4 in these cases.

Temporal distributions of foreshocks are also quite variable. Foreshock sequences may show any pattern shown in Figure 1. Mogi (1985) classified the foreshock sequences into two types, C and D. In type C sequences, the activity increases gradually toward the mainshock. In type D sequences, the mainshock occurs after the foreshock activity dies down. Although no detailed statistics are available, type D seems to form a majority of foreshock sequences.

It is striking that, if we superpose the data of foreshock–mainshock time separations from many foreshock sequences, we obtain the power-law type distribution t^{-p} similar to the aftershock temporal pattern with the time direction reversed (Papazachos, 1974; Jones and Molnar, 1976, 1979; Davis and Frohlich, 1991b; Utsu, 1992; Ogata *et al.*, 1995; Maeda, 1999). Nevertheless, the great majority of individual foreshock sequences do not fit the power-law distribution. Only a limited number of foreshock sequences or precursory seismic activities show a regular increase of activity (power-law or exponential growth) toward the mainshock (e.g., Varnes, 1989; Jaumé and Sykes, 1999; Yamaoka *et al.*, 1999).

The discrimination of foreshock sequences from earthquake swarms is a difficult problem, because no easily recognizable differences in statistical or physical properties are found between them. For further discussion on this issue and other characteristics of foreshocks, including the relation to tectonics and fault types, see Mogi (1963), Jones and Molnar (1979), von Seggern *et al.* (1981), Jones (1984), Ogata *et al.* (1995, 1996), Abercrombie and Mori (1996), and Reasenberg (1999).

2.4 Earthquake Swarms

An earthquake swarm is a cluster of earthquakes in which there is no predominantly large single earthquake. Swarms are roughly divided into two types. The first type is the ordinary swarm in which the activity changes more or less irregularly (Fig. 1c) lasting a few hours to more than one year. Most swarms occurring in volcanic areas show this pattern, reflecting the magmatic activity at depths. The second type is the successive occurrence of mainshock–aftershock sequences of similar size (Fig. 1d), which lasts several days to months.

3. Size Distribution of Earthquakes

3.1 Gutenberg–Richter Relation and its Equivalents

3.1.1 Power-Law Distribution

In general, smaller earthquakes are much more frequent than larger ones. Statistical studies of this property in late 19th century were not so successful because of the lack of adequate scale to measure the earthquake size. For historical review, see Utsu (1999), where the pioneering work of Wadati (1932) has been introduced. He adopted a power-law distribution for earthquake energy E

$$n(E) \propto E^{-w} \quad (11)$$

and estimated the value for exponent w as 1.7–2.1 or smaller by an indirect method. If the magnitude M is related to energy E (erg) by the equation of Gutenberg and Richter (1956)

$$\log E = 1.5M + 11.8 \quad (12)$$

Equation (11) is equivalent to the famous equation by Gutenberg and Richter (1944, 1949)

$$\log n(M) = a - bM \quad (13)$$

or

$$\log N(M) = A - bM, \qquad N(M) = \int_M^{\infty} n(M)dM \quad (14)$$

and $w = 1 + 2b/3$. Numerous studies indicated that Gutenberg and Richter's relation (hereafter called the G-R relation) is approximately valid in most cases and the value of b falls in the range 0.6–1.1. This corresponds to a w value of 1.40–1.73.

Ishimoto and Iida (1939) showed that the maximum amplitude A recorded at a seismograph station has a power-law distribution with exponent $m = 1.74$,

$$n(A) \propto A^{-m} \quad (15)$$

This is equivalent to the G-R relation with $b = m - 1$ (Asada *et al.*, 1951), if magnitude is defined by $M = \log A + f(\Delta)$, where $f(\Delta)$ is the calibrating function for epicentral distance Δ and the b value does not vary regionally. Most m values reported hitherto fall between 1.6 and 2.1.

Since the development of the concept of fractals, many phenomena exhibiting a power-law distribution have been treated as fractals. The G-R relation has often been interpreted in terms of the power-law (fractal) statistics of faults, cracks, and fragments (e.g., Takeuchi and Mizutani, 1968). Aki (1981) related the b value and the fractal dimension D as $D = 2b$. In these interpretations, it is implicitly assumed that each fault (or crack, or fragment) corresponds to one earthquake, i.e., the probability of earthquake occurrence is independent of the fault size. This assumption does not seem so natural.

3.1.2 Estimation of *b* Value

Under the assumption that the magnitude data are random samples from a population obeying the G-R relation, the method of moments (Utsu, 1965) and the method of maximum likelihood (Aki, 1965) both yield the solution

$$b = \log e/(\overline{M} - M_z) \quad (16)$$

where $\overline{M}$ is the mean magnitude of earthquakes of $M \geq M_z$ and $\log e = 0.434294$. M_z is the threshold magnitude above which the data should be complete. The standard error of the maximum likelihood estimate of b is approximately $b/\sqrt{N}$ for large N (the number of earthquakes of $M \geq M_z$). The smallness of error does not necessarily indicate the goodness of fit of the G-R relation to the data.

Since Eq. (16) is obtained for continuous exponential distribution, care must be taken when we use discrete (rounded) magnitude values. If the magnitudes are given at intervals of 0.1 as in most catalogs, and if we use the data with $M \geq 4.0$ for example, we must put $M_z = 3.95$ in Eq. (16), because $M = 4.0$ means $4.05 > M \geq 3.95$. If we use $M_z = 4.0$ in this case, the error in b value may easily reach several percent. There is another kind of error due to rounding (e.g., Tinti and Mulargia, 1987; Vere-Jones, 1989). This error is not large (usually less than 1%) if the magnitudes are given to the nearest 0.1 magnitude unit or less.

3.1.3 Temporal and Spatial Variation of *b* Value

Numerous papers have been published dealing with the spatial and temporal variation (or stability) of b value (e.g., Utsu, 1971; Li *et al.*, 1983; Imoto, 1987; Jin and Aki, 1989; Ogata and Katsura, 1993). For example, temporal changes in b value for background seismicity prior to large earthquakes have occasionally been reported (e.g., Smith, 1986, 1998; Trifu and Radulian, 1991; Imoto, 1991). Smaller b values for foreshocks than those for aftershocks and background seismicity have also been reported (e.g., Suyehiro *et al.*, 1964; Berg, 1968; Wu *et al.*, 1976; Molchan and Dimitrieva, 1990; Molchan *et al.*, 1999). However, we notice that many foreshock sequences have normal b values, and some earthquake swarms with small b values are not followed by future large earthquakes.

Spatial variations of b value have been studied for various regions of the world and for various focal depth ranges (Ogata *et al.*, 1991; Frohlich and Davis, 1993; Okal and Kirby, 1995; Wiemer and Wyss, 1997; Molchan *et al.*, 1997; Wiemer *et al.*, 1998, and many others). However, universally recognizable regularities are relatively few. Some authors have argued against the regional variation of size distribution (e.g., Kagan, 1997, 1999).

To test the significance of the difference in b values between two earthquake groups, a method based on the F-distribution has been used (Utsu, 1966). A method using Akaike Information Criterion (AIC) (see Chapter 82 by Vere-Jones and Ogata) is simpler (Utsu, 1992, 1999). Of course these tests assume the G-R relation. If this assumption is not valid, it is possible that the two groups have nearly equal b values calculated from Eq. (16) but the size distributions are quite different.

3.2 Modified Equations

Although the G-R relation fits the data fairly well in many cases, significant deviation from it has often reported. The $\log n(M)$ (or $\log N(M)$) versus M plots for some data sets show considerable curvature, though the G-R relation predicts a straight line with slope of $-b$.

Various modifications of the G-R relation have been proposed to represent such curved distributions (see a review by Utsu, 1999). The simplest one is the truncated G-R relation, i.e., $\log n(M) = a - bM$ for $M \leq M_{\max}$ and $n(M) = 0$ for $M > M_{\max}$. The maximum likelihood estimate of b in this case is somewhat different from that calculated from Eq. (16) (e.g., Page, 1968). If the truncated G-R relation is valid, the regional variation in b value calculated from Eq. (16) may be caused by the regional difference in $M_{\max}$ rather than the b slope in the range below $M_{\max}$.

The power-law distribution of energy or moment tapered by an exponential function used by Kagan (1991a, 1993, 1997) has the form (Gamma distribution)

$$n(E) = CE^{-w} \exp(-E/E_{\max}) \tag{17}$$

This is equivalent to

$$\log n(M) = a - bM - k10^{1.5M} \tag{18}$$

where $b = 1.5(w-1)$ and $k = 10^{-1.5M_{\max}} \log e$ ($M_{\max}$ is the magnitude corresponding to $E_{\max}$). This is a generalized form of an equation for a branching model (Otsuka, 1972) derived by Saito *et al.* (1973), Vere-Jones (1976), and Maruyama (1978), in which b takes a value of 0.75.

Another modification is the power-law distribution tapered by a logarithmic function

$$n(E) = CE^{-w} \ln(E_{\max}/E), \quad (n(E) = 0 \text{ for } E \geq E_{\max}) \tag{19}$$

which is equivalent to the equation proposed by Utsu (1971)

$$\log n(M) = a - bM + \log(M_{\max} - M) \tag{20}$$

It is possible to obtain the maximum likelihood estimates of the parameters of Eqs. (17)–(20) and some other equations including those proposed by Lomnitz-Adler and Lomnitz (1979), Makjanić (1980), Anderson and Luco (1983), and Seino *et al.* (1989) together with the AIC values for a given set of magnitude data. Computer programs are available in *IASPEI Software Library* Vol. 6 (Utsu and Ogata, 1997). We can compare AIC values for different equations and find the best fitting equation. For model selection criteria using AIC, see Chapter 82 by Vere-Jones and Ogata. Utsu (1999) provides actual examples.

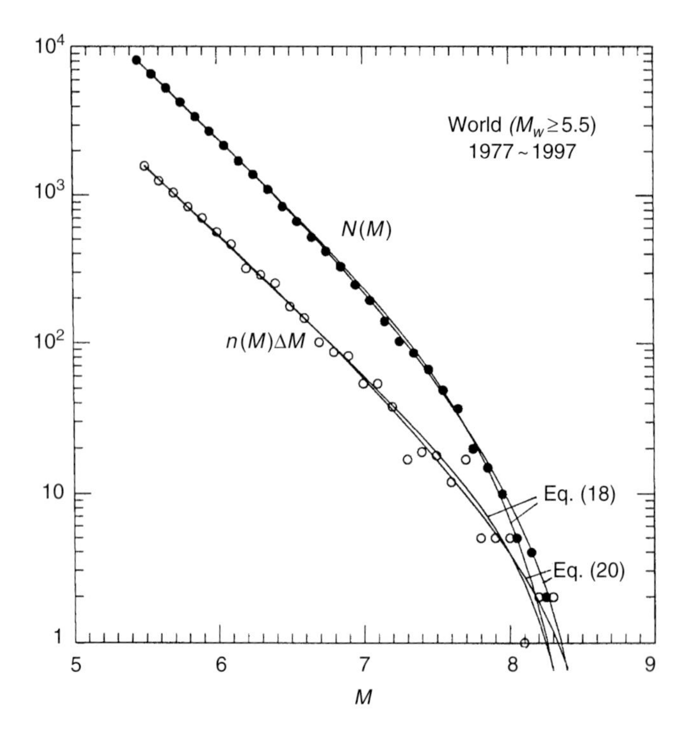

FIGURE 3 Frequency–magnitude distribution for earthquakes in the world for the years 1977–1997. The magnitude used is the moment magnitude (M_w) converted from seismic moments given in the Harvard University CMT catalog. Open and solid circles denote the frequency per 0.1 magnitude unit and cumulative frequency, respectively. Curves for Eqs. (18) and (20) fitted to the data are shown.

Figure 3 shows the distribution of M_w for recent earthquakes in the world. The curves represent Eqs. (18) and (20) fitted to the data. If the complete data in the 1950s and 1960s were available, the distribution would be closer to the G-R relation (straight line), since several great earthquakes of $M_w = 8.7$–9.5 occurred during 1952–1965 (Kanamori, 1977).

4. Temporal Distribution of Earthquakes

4.1 Stationary Poisson Process

The simplest reference model for the distribution of earthquakes in time is the stationary Poisson process, in which all events occur independently and uniformly in time. This process is characterized by only one parameter, the rate of occurrence ν. If N events occur during a sufficiently long time interval T, the rate ν is given by $\nu = N/T$. For the stationary Poisson process, the frequency of events n in an interval of length Δt has a Poisson distribution

$$f(n) = (\nu\Delta t)^n \exp(-\nu\Delta t)/n! \tag{21}$$

The time interval τ between successive events (interevent time) has an exponential distribution

$$w(\tau) = \nu \exp(-\nu\tau) \tag{22}$$

The power spectrum of a point process t_k (occurrence time of the kth event; $k = 1, 2, \ldots, N$) is given by

$$\begin{aligned} S(\omega) &= \left| \sum_{k=1}^{N} \exp(-i\omega t_k) \right|^2 \Big/ N \\ &= \left\{ \left(\sum_{k=1}^{N} \cos \omega t_k \right)^2 + \left(\sum_{k=1}^{N} \sin \omega t_k \right)^2 \right\} \Big/ N \end{aligned} \tag{23}$$

where ω denotes the angular frequency. For a Poisson process, $S(\omega)$ is independent of ω and has an exponential distribution of parameter 1. Therefore the probability that $S(\omega)$ for a randomly chosen ω exceeds a certain value α is $e^{-\alpha}$ (Schuster, 1897).

4.2 Time-dependent Poisson Process

If the rate of occurrence changes gradually with time as expressed by $n(t)$, but can be regarded as a stationary Poisson process for a sufficiently short time span, the process is called a time-dependent Poisson process. For this process, distribution of number of events in a time interval from t_1 to t_2 ($\Delta t = t_2 - t_1$) is given by

$$f(n) = \int_{t_1}^{t_2} \{n(t)\Delta t\}^n \exp\{-n(t)\Delta t\}\, dt/(n!\Delta t) \tag{24}$$

The distribution of interevent times in a time interval from t_1 to t_2 is given by

$$w(\tau) = \int_{t_1}^{t_2} n(t)^2 \exp\{-n(t)\tau\}\, dt \Big/ \! \Big/ \int_{t_1}^{t_2} n(t)\, dt \tag{25}$$

The modified Omori relation [Eq. (1)] is an example of the time-dependent Poisson process. Ouchi (1982) discussed the function

$$n(t) = n_0/[\{1 - (n_0/k)^\alpha\} \exp(-rt) + (n_0/k)^\alpha]^{-\alpha} \tag{26}$$

which is the solution of a logistic equation (Goel *et al.*, 1971)

$$dn(t)/dt = r[1 - \{n(t)/k\}^\alpha]/\alpha \tag{27}$$

Equation (26) can represent various temporal patterns of seismic activity by using different ranges of k, α, r and n_0 [the initial value of $n(t)$]. For $n(t) \gg k$, it is approximately the same as the modified Omori relation with $p = \alpha^{-1}$. For $n(t) \ll k$, $n(t) \propto \exp(rt/\alpha)$ which represents some foreshock activities. For further development along this line, see Ouchi (1993).

Tomoda (1954) reported a power-law distribution of interevent times, $w(\tau) \propto \tau^{-q}$ ($q = 1 \sim 2$), for some volcanic swarms and aftershock sequences. Senshu (1959) interpreted Tomoda's result using Eq. (25) with $n(t)$ of the modified Omori relation. Utsu (1970) showed that Eqs. (24) and (25) with $n(t) = Kt^{-p}$ ($t_1 \to 0$, $t_2 \to \infty$) yield power-law distributions $f(n) \propto K'n^{-r}$ (for $n \geq$ about 4) and $w(\tau) \propto K''\tau^{-q}$, where $r = 1 + p^{-1}$ and $q = 2 - p^{-1}$ ($q + r = 3$).

4.3 Fractal Analyses in Time

Self-similarity or fractal structure in temporal distribution of earthquakes has sometimes been recognized. Various methods have been used for estimating the fractal dimension in time. These include the use of autocovariance function and variance-time curve for time intervals, box counting, periodogram (spectral curve), etc. Relevant papers are Smally *et al.* (1987), Kagan and Knopoff (1987), Ogata and Abe (1991), Kagan and Jackson (1991), Papadopoulos and Dedousis (1992), Godano and Caruso (1995), Lee and Schwarcz (1995), Luongo and Mazzarella (1997), and Wang and Lee (1997), among others.

The fractal dimensions obtained for various catalogs lie mostly between 0.1 and 0.5. Kagan and Jackson (1991) showed long-term clustering (hundreds days or more) with fractal dimension of 0.8–0.9. Ogata and Abe (1991) have also shown long-term dependence.

4.4 Recurrence Models

4.4.1 Renewal Process

Temporal distribution of earthquakes, especially so-called characteristic earthquakes recurring on a specific fault or fault segment, often represented by a renewal process. It is defined as a series of events in which interevent times are independently and identically distributed. In general, the relations between density function of the distribution $w(t)$, hazard rate $\mu(t)$, and survivor function (reliability function) $\phi(t)$ are as follows.

$$w(t) = \mu(t)\phi(t) = -d\phi(t)/dt \tag{28}$$

$$\mu(t) = d \ln \phi(t)/dt = w(t) \Big/ \! \Big/ \int_t^\infty w(t)\, dt \tag{29}$$

$$\phi(t) = \exp\left\{ -\int_t^\infty \mu(t)\, dt \right\} = \int_t^\infty w(t)\, dt \tag{30}$$

The probability that the next earthquake occurs in an interval from t to $t + u$ is given by

$$p(u|t) = 1 - \exp\left\{ -\int_t^{t+u} \mu(t)\, dt \right\} = 1 - \frac{\phi(t+u)}{\phi(t)} \tag{31}$$

TABLE 1 Some interevent distributions for renewal models[a]

Name	Density of time interval $w(t)$	Survivor function $\phi(t)$	Hazard rate $\mu(t)$	Expectancy of time interval E[t]
Poisson	$\alpha\exp(-\alpha t)$	$\exp(-\alpha t)$	α	$1/\alpha$
Weibull	$\alpha\beta t^{\beta-1}\exp(-\alpha t^{\beta})$	$\exp(-\alpha t^{\beta})$	$\alpha\beta t^{\beta-1}$	$\alpha^{-1/\beta}\Gamma(1/\beta+1)$
Lognormal	$1/(\sqrt{2\pi}\sigma t)\exp\{-(\ln t-m)^2/(2\sigma^2)\}$	$1-\Phi\{(\ln t-m)/\sigma\}$	$w(t)/\phi(t)$	$\exp(m+\sigma^2/2)$
Gamma	$c(ct)^{r-1}\exp(-ct)/\Gamma(r)$	$\Gamma(r, ct)/\Gamma(r)$	$c(ct)^{r-1}\exp(-ct)/\Gamma(r, ct)$	r/c
EP (Doubly exponential)	$\exp[(A/b)\{1-\exp(bt)\}+bt]$	$\exp[(A/b)\{1-\exp(bt)\}]$	$A\exp(bt)$	$(1/b)\exp(A/b)\,\mathrm{Ei}(-A/b)$

[a] $\Phi(\cdot)$, $\Gamma(\cdot)$, $\Gamma(\cdot,\cdot)$, and Ei$(\cdot)$ denote the error integral, the gamma function, the incomplete gamma function, and the exponential integral, respectively.

The mean time interval is given by

$$\mathrm{E}[t] = \int_0^{\infty} \phi(t)\,dt \tag{32}$$

The functional forms for five distributions are given in Table 1. The Weibull distribution has been used by Rikitake (1974), Sykes and Nishenko (1984), Utsu (1984), Parvez and Ram (1997), and others. The lognormal distribution has been used by Terada (1918), Nishenko and Bulland (1987), Goes (1996), and others. The Gamma distribution has been used by Udías and Rice (1975), Utsu (1984), Parvez and Ram (1997), and others. In the EP model (Utsu, 1972a,b, 1984), the hazard rate increases exponentially with time as $\mu(t) = A\exp(bt)$. If the stress increases proportionally to time, and the probability of rupture is an exponential function of the stress, this model must be appropriate.

The exponential distribution (Poisson process) is a special case of Weibull distribution ($\beta = 1$), or Gamma distribution ($r = 1$), or the EP model ($b = 0$). The ratio of the standard deviation of τ to its expectancy is given by $I = 1$, $I = [\Gamma(2\beta^{-1}+1)/\Gamma^2(\beta^{-1}+1)-1]^{1/2}$, $I = [\exp(\sigma^2-1)]^{1/2}$, and $I = 1/\sqrt{r}$ for exponential, Weibull, lognormal, and Gamma distributions, respectively. $I \le 0.3$ corresponds to $\beta \ge 3.7$, $\sigma \le 0.3$, and $r \ge 11$. Roughly speaking, events occur fairly regularly (quasi-periodically) if the parameters β, σ, and r take a value in these ranges. Great historical earthquakes of Nankai region, southwest Japan, provide a good example ($\beta = 6$, $\sigma = 0.18$, $r = 30$, Utsu, 1984). Ogata (1999a) discussed the case in which the occurrence times of past events are uncertain, as in the case of paleoearthquakes.

Other distributions found in the seismological literature include normal distribution (must be truncated at $t = 0$) and inverse Gaussian distribution (Kagan and Knopoff, 1987).

4.4.2 Other Recurrence Models

Time-predictable and slip-predictable models (Shimazaki and Nakata, 1980) for large earthquakes are recurrence models involving the earthquake size. The time-predictable model (the time interval between two large earthquakes A and B is proportional to the amount of fault slip in A) seems to provide a better fit in many cases.

Following a paper by Papazachos (1989), his group published many papers providing the empirical equations for the time interval τ to the next earthquake and the expected magnitude M_f of the next earthquake for various regions of the world. These equations express $\log\tau$ and M_f as linear functions of M_z (threshold magnitude adopted), M_p (magnitude of the last earthquake), and $\log\dot{M}_0$ (moment release rate for the region concerned). In early papers, a simple equation in the form of $\log\tau = cM_p + c_0$ was used [see the latest evaluation paper by Papazachos and Papadimitriou (1997) and references listed there].

For the slip-predictable model, see, for example, Kiremidjian and Anagnos (1984).

The stress release model (Vere-Jones and Deng, 1988; Zheng and Vere-Jones, 1991, 1994) is not a recurrence model, but it has some similarity to the above models. It is a point process model with an intensity function expressed as

$$n(t) = A\exp[\beta\{\lambda t - \sigma(t)\}] \tag{33}$$

where $\sigma(t)$ is the stress released by the earthquakes which occurred until time t, and A, β, and λ are model parameters that can be determined by the maximum likelihood method. See also Vere-Jones (1994) for an explanation of this model and more general discussion on statistical models of earthquake occurrence.

4.5 Point Process Models Incorporating the Clustering of Earthquakes

Several point process models in which the clustering of events is taken into account have been used in seismicity studies. These include the trigger model (Vere-Jones and Davies, 1966; Vere-Jones, 1970; Utsu, 1972a,b; Hawkes and Adamopoulos, 1973), a generalized Poisson process (Shlien and Toksöz, 1970; Bottari and Neri, 1983; De Natale and Zoro, 1986) which can be regarded as a simplified version of the trigger model, a kind of branching process (Kagan and Knopoff, 1981), and the ETAS model (Ogata, 1988, 1992, 1999b).

The trigger model successfully represents some non-Poissonian behavior seen in many actual seismicity data (sharp increase of the frequency of short interevent times, increase of the power spectrum toward the zero frequency, etc.) without labeling each event as mainshock, aftershock, or isolated event.

In a self-exiting model used by Hawkes and Adamopoulos (1973), every event produces its offspring events. The occurrence rate (intensity) at time t is expressed by

$$n(t) = \mu + \sum_{t_i<t} g(t - t_i) \tag{34}$$

where μ and $g(t - t_i)$ represent the constant background activity and the activity triggered by an event at time t_i, respectively. They adopted the sum of two exponential functions for $g(t)$.

The ETAS model (epidemic type aftershock sequence model) is a self-exciting model in which the modified Omori relation combined with the G-R relation is used for $g(t)$.

$$g(t - t_i) = K \exp\{\alpha(M_i - M_z)\}(t - t_i + c)^{-p} \tag{35}$$

where M_i and t_i is the magnitude and time of the ith earthquake ($t_1 < t_2 < \cdots < t_N$). N is the number of earthquakes of magnitude M_z and larger. The values of five parameters μ, K, α, p, and c, can be estimated by the maximum likelihood method. Computer programs are included in Utsu and Ogata (1997). The standard errors in the estimated values can be computed using the Hessian matrix (see Chapter 82 by Vere-Jones and Ogata). However, this software does not include the program for the computation of errors because of the limited power of the personal computers in 1993 when this software was prepared.

The ETAS model is useful in objective detection of a period of anomalously increased or decreased activity by comparing the observed rate with the rate computed from the ETAS model fitted to the observed data.

4.6 Periodicity

4.6.1 Power Spectrum for a Point Process

The periodicity of earthquake occurrences has been searched over a century, but only weak periodicity has been found in limited cases. If seismicity has a periodicity of period T_p, the power spectrum calculated from Eq. (23) has a peak at the angular frequency ω_p $(= 2\pi/T_p)$. It is not easy to say what level of the spectral value should be used to establish the existence of periodicity. The Schuster's criterion (Section 4.1) is inadequate when the events show temporal clustering (e.g., Jeffreys, 1938; Vere-Jones and Ozaki, 1982).

Consider the case in which we analyze the events in the time span of length L and compute the power spectrum for m angular frequencies ω_k $(= 2\pi k/L, k = 1, 2, \ldots, m)$. The obtained spectral values are designated by $S_1, S_2, \ldots, S_m$ in order of their strength. Schuster's criterion cannot be applied to S_1 (the highest peak), even if there is no clustering in seismicity. Instead, Fisher's criterion (1929) must be used, which gives the probability $P(g, m)$ that $S_i/\sum_{k=1}^{m} S_k$ exceeds a value g. See Nowroozi (1967) and Shimshoni (1971) for the expression and numerical tables of $P(g, m)$.

4.6.2 Searching Periodicity in Time-varying and Clustered Seismicity

If the earthquakes have a tendency of clustering or the occurrence rate changes with time, the spectral level usually increases beyond the expected level for the Poisson process. For example, the theoretical spectral value for the trigger model increases with decreasing frequency from 1 at very high frequencies to $1 + A + V/(1 + A)$ at zero frequency, where A and V denote the mean and variance of the number of aftershocks following a triggering event (Vere-Jones and Davies, 1966).

Ogata (1983b) examined the periodicity hidden in time-varying seismicity with strong clustering by using a self-exciting model with intensity

$$n(t) = \mu + P_J(t) + C_K(t) + \sum_{t_i<t} g(t - t_i) \tag{36}$$

where $P_J(t)$ is a polynomial of the order J for representing the long-term trend and $C_K(t)$ is a Fourier series of the order K for representing the periodic component of suspected period T_0. For $g(t - t_i)$ the Laguerre polynomial of order L is used.

$$g(t) = \sum_{l=1}^{L} a_l t^{l-1} \exp(-Ct) \tag{37}$$

By using this model, the periodic component can be separated from the trend and cluster components. Maximum likelihood estimates of $J + 2K + L + 2$ parameters and AIC are obtained for various combinations of J, K, and L, then the best combination that gives the smallest AIC can be found. If $J = 0$ in the best combination, there is no trend. If $K = 0$, the periodicity of period T_0 is rejected. Ogata (1983b) confirmed annual periodicity in shallow seismicity in western Japan. Similar studies were made by Matsumura (1986) and Ogata and Katsura (1986) for other regions (see Ogata, 1999b).

4.6.3 Reported Periodicities

Many papers have been published reporting the existence of periodicity in earthquakes of various regions and magnitude ranges. The periods reported are one year, one day, synodic month (29.530589 days), sidereal month (27.32166 days), 11 y (sunspot cycle), and others. A period around 42 min reported by Davison (1938) and later workers may correspond to a normal mode of Earth's free oscillation, but most of the periods reported by him using Schuster's criterion may be attributed to the effect of clusters in the data (see, Jeffreys, 1938).

The annual periodicity is often connected with the seasonal variation of rainfall, snow melting, and subsequent change in groundwater, because the periodicity is recognized predominantly for inland shallow earthquakes (e.g., Oike, 1978; Ogata, 1983b; Costain and Bollinger, 1996). Seasonal variation in amplitudes of microseisms may cause the variation of the number of small earthquakes in certain catalogs. The diurnal periodicity (increased activity in nighttime) reported since old times may be attributed to human activity rather than natural reasons.

Although the periodicity in earthquake occurrences, if it exists, is usually not so strong as to affect earthquake hazard problems, it is interesting to explore the triggering mechanism and some unknown periodic phenomena occurring in the Earth's system. However, since the problem is fairly critical, careful selection and analysis of data are required to draw meaningful conclusions.

5. Spatial Distribution of Earthquakes

5.1 Spatial Patterns of Seismicity

5.1.1 Poisson Process in Space

For independently and uniformly distributed hypocenters in N-dimensional space, the Nth power of the distance r from a hypocenter (or an arbitrarily chosen point) to the nearest neighbor, $s(=r^N)$, has an exponential distribution $f(s)=\lambda\exp(-\lambda s)$, where λ is the mean number of hypocenters falling in an N-dimensional sphere of unit radius. The number n of hypocenters falling in an N-dimensional volume of a fixed size has a Poisson distribution.

The actual hypocenters (or epicenters) are not distributed randomly. Earthquakes are associated with tectonic structures of various scales from long plate boundaries to small local fracture systems. The stationary Poisson process in space seems to be a less plausible model than the stationary Poisson model in time. The nonhomogeneous Poisson process may be a simplest practical model in space.

5.1.2 Representation Using Contour Lines

Spatial variation of the density of hypocenters can be represented by smoothed contour lines. See Ogata and Katsura (1988) for a technique using a Bayesian procedure and B-spline functions. The smoothed spatial distribution of a seismicity parameter can be obtained in a similar way. Spatial variation of b value studied by Ogata *et al.* (1991) is an example. See Vere-Jones (1992) for reviews of these and relevant methods.

5.2 Fractal and Nonfractal Distributions in Space

Tomoda (1952) divided a seismic area of Kanto, Japan, into many narrow zones of equal widths by parallel lines and measured the distance s between the neighboring epicenters in each stripe. The frequency of s has a power-law dependence on s

$$f(s)\propto s^{-q} \qquad (q\cong 1.6) \tag{38}$$

Another power-law distribution was found by Suzuki and Suzuki (1965) for the number n of epicenters of shallow earthquakes in the world and in Japan falling in areas of a fixed size,

$$f(n)\propto n^{-r} \qquad (r\cong 1.4) \tag{39}$$

Equations (38) and (39) suggest a fractal distribution of epicenters. Fractal analyses of spatial distribution pattern of earthquakes are usually made by using either the box counting method (e.g., Sadovskiy *et al.*, 1984; Hirata, 1989; Robertson *et al.*, 1995) or the spatial correlation function (e.g., Kagan and Knopoff, 1980; Henderson *et al.*, 1994; Volant and Grasso, 1994). The slope of spectral curves has also been used (Ogata and Katsura, 1991). The estimates of fractal dimension range from 1.0 to 2.0 for epicenter distribution (2D). Those for hypocenter distribution (3D) scatter widely from 1.0 to nearly 3.0, probably because the vertical distribution of hypocenters is highly variable from region to region.

The spatial correlation function $n(r)$ is the number of hypocenters in unit volume at distance r from a hypocenter. For a stationary Poisson process, $n(r)=\rho$ (constant). For a fractal distribution in N-dimensional space, $n(r)\propto r^{D-N}$, where D is the fractal dimension. The total number of hypocenters within the distance of r from a hypocenter is proportional to r^D.

Some papers discuss the relation between fractal dimension D of hypocenter distribution and the b value of magnitude-frequency distribution for general seismic activity, aftershock sequences, and acoustic emission (AE) events observed in rock fracture experiments [see Hirata (1989), Ogata and Katsura (1991), Henderson *et al.* (1992, 1994), De Rubeis *et al.* (1993), Volant and Grasso (1994), Öncel *et al.* (1996), Guo and Ogata (1997), and Ponomarev *et al.* (1997)].

Other studies of fractal or nonfractal distribution of hypocenters include Tamaki (1961), Francis and Porter (1971), Kagan (1981a,b), Ouchi and Uekawa (1986), Eneva and Pavlis (1988), Frohlich and Davis (1990), Ogata and Abe (1991), Hirata and Imoto (1991), Eneva *et al.* (1992), Hirabayashi *et al.* (1992), and Eneva (1996).

6. Spatiotemporal Seismicity Patterns

6.1 Correlation and Migration of Seismicity

Synchronous or alternate occurrences of large earthquakes or increased seismicity between two adjacent or separate regions have sometimes been reported (e.g., Mogi, 1973; Mantovani and Albarello, 1997). Migration of seismicity is another noticeable phenomenon (e.g., Mogi, 1968b; King and Ma, 1988). However, many of these patterns do not persist permanently and successful predictions based on these patterns are rather rare.

Relatively clear correlation is seen in subduction zones. Tens of papers have been published reporting the correlation between interplate earthquakes and intraplate earthquakes in the overriding plate and between shallow interplate or intraplate earthquakes on the trench side and intermediate or deep earthquakes in the subducting plate (Mogi, 1973; Utsu, 1975; Seno, 1979; Ogata *et al.*, 1982; Mizoue *et al.*, 1983; Ohtake, 1986; Hori and Oike, 1996). Mostly negative results have been reported for correlation between large shallow and intermediate-depth earthquakes in a global survey by Astiz *et al.* (1988).

Since the earthquakes have a tendency to clustering, the correlation coefficient between two regions easily exceeds the level expected for the Poissonian occurrence, even if there is no correlation between the two. To analyze the interaction between earthquakes in two regions, Ogata *et al.* (1982) expressed the seismicity rate of the first region by Eq. (36) with an additional term $\sum_{t_j<t} h(t-t_j)$ for the seismicity excited by earthquakes in the second region. Using a Laguerre polynomial of order M for $h(t)$, they obtained the maximum likelihood estimates of the coefficients of each term for various combinations of J, K, L, and M for a given set of data. If the smallest AIC is attained for a combination in which $M \geq 1$, we can conclude that the seismicity in the first region is affected by the seismicity in the second region. Similar studies were made by De Natale *et al.* (1988) and Mucciarelli *et al.* (1988) (also see Ogata, 1999b).

There are also tens of papers dealing with migration of seismic activity. However, the reported velocities differ very widely from 1 to 5×10^4 km y^{-1}. This scatter is unbelievably large for phenomena caused by a single physical mechanism, though velocities between 2 and 200 km y^{-1} are relatively frequent.

6.2 Regional Seismicity Patterns Preceding Large Earthquakes

Since seismicity varies temporally and spatially in a complex manner, unusual patterns before many large earthquakes can be found rather easily. These patterns include quiescence and activation of various spatial and time scales and more complicated patterns such as doughnuts, migration toward the mainshock hypocenter, increase in clustering, etc. However, the relation between these patterns and the occurrence of large earthquakes is not so definite.

More than 200 papers had been published by the end of 1999, reporting the seismic quiescence preceding some large earthquakes in many parts of the world (e.g., Wyss and Martirosyan, 1998), since Inouye (1965) reported quiescence before three $M = 7$–8 earthquakes in Japan. Only a few cases are known where the quiescence was recognized in advance of the mainshock (e.g., Utsu, 1972b; Ohtake *et al.*, 1977; Wyss and Burford, 1987; Kisslinger, 1988; Matsu'ura *et al.*, 1995).

Due to the existence of remarkable clustering of earthquakes, fluctuation of activity beyond the level expected from the stationary Poisson process easily occurs. To find a period or region of significantly increased or decreased seismicity, we must use a declustered catalog or a model that incorporates the effect of clustering, such as the ETAS model (Ogata, 1992, 1999b).

Precursory activation of seismicity other than the immediate foreshocks has also been studied by many investigators. The precursory seismicity pattern reported by Keilis-Borok and coworkers (e.g., Keilis-Borok and Malinovskaya, 1964; Keilis-Borok and Kossobokov, 1990; Kossobokov *et al.*, 1990; Keilis-Borok, 1996) can be regarded as a kind of activation, though the pattern is complicated to explain in short. Nearly one hundred papers describing the methods for detecting the pattern, case studies in various region of the world, and evaluations of the results have been published. One of the standard algorithms for this approach called M8 which aims the long-term prediction of large earthquakes is included in Kossobokov (1997).

A kind of precursory activation of seismicity near the mainshock epicenter has been observed in Japan (Sekiya, 1976) and New Zealand (Evison, 1977). Evison called this "precursory swarm." See Evison and Rhoades (1997) for a recent re-evaluation of this pattern in New Zealand. Similar patterns have been reported from Central Asia (e.g., Kristy and Simpson, 1980), China (e.g., Zhou *et al.*, 1995), India (e.g., Gupta and Singh, 1989), and some other regions.

6.3 Space-time Point-process Models

There are some point-process models for seismicity involving both space and time. They are inevitably more complicated than the models in time or space domain alone. See the following references for further details: space-time correlation functions by Kagan and Knopoff (1976), Vere-Jones (1978), Chong (1983), and Reasenberg (1985); the branching models by Kagan and Knopoff (1976, 1977, 1978, 1980, 1981) and their extension by Kagan (1991b); the space-time version of a self-exciting model by Musmeci and Vere-Jones (1992) and that of the ETAS model by Rathbun (1993) and Ogata (1998).

7. Concluding Remarks

Most studies of seismicity are more or less statistical, because they deal with earthquake generation in groups. Thousands of papers relevant to statistical features of seismicity have been published. Only a small fraction of them are quoted in this chapter. Some review articles (e.g., Kagan, 1994; Vere-Jones,

1994; Main, 1996; Koyama, 1997; Ogata, 1999b) may supplement this chapter.

The level of seismicity and its statistical properties may be affected by environmental conditions, such as stress, temperature, pore fluid pressure, mechanical and rheological properties of rocks, etc. as well as the large-scale tectonic regimes. Numerous papers have been published discussing the relation of seismicity to these conditions and the physical mechanisms that cause the observed regularities in seismicity. For example, the questions of why the G-R relation and the modified Omori relation are universally observed from shallow earthquakes to very deep earthquakes and from AE events generated in laboratory experiments to large earthquakes, what factors control the b value and p value, etc. have been discussed using various approaches. In spite of these efforts, there remain many problems of describing and interpreting various characteristic features of seismicity.

Acknowledgments

I thank L. Knopoff and D. Vere-Jones for helpful reviews of the manuscript.

References

Abercrombie, R.E. and J. Mori (1996). *Nature* **381**, 303–307.
Acharya, H. (1979). *Bull. Seismol. Soc. Am.* **69**, 2063–2084.
Aki, K. (1965). *Bull. Earthq. Res. Inst. Univ. Tokyo* **43**, 237–239.
Aki, K. (1981). In: "Earthquake Prediction: An International Review," pp. 566–574, American Geophysical Union.
Anderson, J.G. and J.E. Luco (1983). *Bull. Seismol. Soc. Am.* **76**, 273–290.
Asada, T., *et al.* (1951). *Bull. Earthq. Res. Inst. Univ. Tokyo* **29**, 289–293.
Astiz, L., *et al.* (1988). *Phys. Earth Planet. Inter.* **53**, 80–166.
Båth, M. (1977). *Ann. Geofis.* **30**, 299–327.
Berg, E. (1968). *Nature* **219**, 1141–1143.
Bottari, A. and G. Neri (1983). *J. Geophys. Res.* **88**, 1209–1212.
Chong, F.S. (1983). *NZ J. Geol. Geophys.* **26**, 7–24.
Costain, J.K. and G.A. Bollinger (1996). *J. Geodyn.* **22**, 97–117.
Davis, S.D. and C. Frohlich (1991a). *Geophys. J. Int.* **104**, 289–306.
Davis, S.D. and C. Frohlich (1991b). *J. Geophys. Res.* **96**, 6335–6350.
Davison, C. (1938). "Studies on the Periodicity of Earthquakes," Murby.
De Natale, G. and A. Zoro (1986). *Bull. Seismol. Soc. Am.* **76**, 801–814.
De Natale, G., *et al.* (1988). *Geophys. J.* **95**, 285–293.
De Rubeis, V., *et al.* (1993). *Geophys. Res. Lett.* **20**, 1911–1914.
Doser, D.I. (1990). *Bull. Seismol. Soc. Am.* **80**, 110–128.
Eneva, M. (1996). *Geophys. J. Int.* **124**, 773–786.
Eneva, M. and G.L. Pavlis (1988). *J. Geophys. Res.* **93**, 9113–9125.
Eneva, M., *et al.* (1992). *Geophys. J. Int.* **109**, 38–53.
Evison, F.F. (1977). *Phys. Earth Planet. Inter.* **15**, 19–23.
Evison, F.F. and D.A. Rhoades (1997). *NZ J. Geol. Geophys.* **40**, 537–547.
Fisher, R.A. (1929). *Proc. R. Soc. A* **125**, 54–59.
Francis, T.J.G. and I.T. Porter (1971). *Geophys. J. R. Astron. Soc.* **24**, 31–50.
Frohlich, C. (1987). *J. Geophys. Res.* **92**, 13944–13956.
Frohlich, C. and S.D. Davis (1990). *Geophys. J. Int.* **100**, 19–32.
Frohlich, C. and S.D. Davis (1993). *J. Geophys. Res.* **98**, 631–644.
Godano, C. and V. Caruso (1995). *Geophys. J. Int.* **121**, 385–392.
Goel, N.S., *et al.* (1971). *Rev. Mod. Phys.* **43**, 231–276.
Goes, S.D.B. (1996). *J. Geophys. Res.* **101**, 5739–5749.
Gross, S.J. and C. Kisslinger (1994). *Bull. Seismol. Soc. Am.* **84**, 1571–1579.
Guo, Z. and Y. Ogata (1997). *J. Geophys. Res.* **102**, 2857–2873.
Gupta, H.K. and H.N. Singh (1989). *Tectonophysics* **167**, 285–298.
Gutenberg, B. and C.F. Richter (1944). *Bull. Seismol. Soc. Am.* **34**, 185–188.
Gutenberg, B. and C.F. Richter (1949). "Seismicity of the Earth and Associated Phenomena," Princeton University Press, Princeton.
Gutenberg, B. and C.F. Richter (1956). *Ann. Geofis.* **9**, 1–15.
Hartzell, S.H. and T.H. Heaton (1986). *Bull. Seismol. Soc. Am.* **76**, 649–674.
Hawkes, A.G. and L. Adamopoulos (1973). *Bull. Int. Statist. Inst.* **45**(3), 454–461.
Henderson, J., *et al.* (1992). *J. Struct. Geol.* **14**, 905–913.
Henderson, J., *et al.* (1994). *Geophys. J. Int.* **116**, 217–226.
Hirabayashi, T., *et al.* (1992). *Pure Appl. Geophys.* **138**, 591–610.
Hirata, T. (1989). *J. Geophys. Res.* **94**, 7507–7514.
Hirata, T. and M. Imoto (1991). *Geophys. J. Int.* **107**, 155–162.
Hori, T. and K. Oike (1996). *J. Phys. Earth* **44**, 349–356.
Iida, K. (1965). *J. Earth Sci. Nagoya Univ.* **13**, 115–132.
Iio, Y. (1986). *J. Phys. Earth* **34**, 127–169.
Imoto, M. (1987). *NZ J. Geol. Geophys.* **30**, 103–116.
Imoto, M. (1991). *Techtonophysics* **193**, 311–325.
Inouye, W. (1965). *Kenshinjiho (Q. J. Seismol.)* **29**, 139–144.
Ishimoto, M. and K. Iida (1939). *Bull. Earthq. Res. Inst. Univ. Tokyo* **17**, 443–478.
Jaumé, S.C. and L.R. Sykes (1999). *Pure Appl. Geophys.* **155**, 279–305.
Jeffreys, H. (1938). *Gerlands Beitr. Geophys.* **56**, 111–139.
Jin, A. and K. Aki (1989). *J. Geophys. Res.* **94**, 14041–14059.
Jones, L.M. (1984). *Bull. Seismol. Soc. Am.* **74**, 1361–1380.
Jones, L.M. and P. Molnar (1976). *Nature* **262**, 677–679.
Jones, L.M. and P. Molnar (1979). *J. Geophys. Res.* **84**, 3596–3608.
Kagan, Y.Y. (1981a). *Geophys. J. R. Astron. Soc.* **67**, 679–717.
Kagan, Y.Y. (1981b). *Geophys. J. R. Astron. Soc.* **67**, 719–733.
Kagan, Y.Y. (1991a). *Geophys. J. Int.* **106**, 123–134.
Kagan, Y.Y. (1991b). *Geophys. J. Int.* **106**, 135–148.
Kagan, Y.Y. (1993). *Bull. Seismol. Soc. Am.* **83**, 4–24.
Kagan, Y.Y. (1994). *Physica D* **77**, 160–192.
Kagan, Y.Y. (1997). *J. Geophys. Res.* **102**, 2835–2852.
Kagan, Y.Y. (1999). *Pure Appl. Geophys.* **155**, 537–573.
Kagan, Y.Y. and D.D. Jackson (1991). *Geophys. J. Int.* **104**, 117–133.
Kagan, Y.Y. and L. Knopoff (1976). *Phys. Earth Planet. Inter.* **12**, 291–318.
Kagan, Y.Y. and L. Knopoff (1977). *Phys. Earth Planet. Inter.* **14**, 97–108.
Kagan, Y.Y. and L. Knopoff (1978). *Geophys. J. R. Astron. Soc.* **55**, 67–86.

Kagan, Y.Y. and L. Knopoff (1980). *Geophys. J. R. Astron. Soc.* **62**, 303–320.
Kagan, Y.Y. and L. Knopoff (1981). *J. Geophys. Res.* **86**, 2853–2863.
Kagan, Y.Y. and L. Knopoff (1987). *Geophys. J. R. Astron. Soc.* **88**, 723–731.
Kanamori, H. (1977). *J. Geophys. Res.* **82**, 2981–2987.
Keilis-Borok, V.I. (1996). *Proc. Natl. Acad. Sci. USA* **93**, 3748–3755.
Keilis-Borok, V.I. and V.G. Kossobokov (1990). *Phys. Earth Planet. Inter.* **61**, 73–83.
Keilis-Borok, V.I. and L.N. Malinovskaya (1964). *J. Geophys. Res.* **70**, 3019–3024.
King, C. and Z. Ma (1988). *Pure Appl. Geophys.* **127**, 627–639.
Kiremidjian, A.S. and T. Anagnos (1984). *Bull. Seismol. Soc. Am.* **74**, 739–755.
Kisslinger, C. (1988). *Bull. Seismol. Soc. Am.* **78**, 218–229.
Kisslinger, C. (1993). *J. Geophys. Res.* **98**, 1913–1921.
Kisslinger, C. (1996). *Adv. Geophys.* **38**, 1–36.
Kisslinger, C. and L.M. Jones (1991). *J. Geophys. Res.* **96**, 11947–11958.
Kossobokov, V.G. (1997). *IASPEI Software Library* **6**, 167–220.
Kossobokov, V.G., *et al.* (1990). *J. Geophys. Res.* **95**, 19763–19772.
Koyama, J. (1997). "The Complex Faulting Process of Earthquakes," Kluwer Academic Press, Dordrecht.
Kristy, M.J. and D. Simpson (1980). *J. Geophys. Res.* **85**, 4285–4837.
Lamoreaux, R., *et al.* (1983). *Phys. Earth Planet. Inter.* **31**, 193–201.
Lee, H.-K. and H.P. Schwarcz (1995). *Geology* **23**, 377–380.
Li, Q., *et al.* (1983). *Acta Geophys. Sinica* **21**, 101–125.
Lomnitz, C. (1966). *Bull. Seismol. Soc. Am.* **56**, 247–249.
Lomnitz-Adler, J. and C. Lomnitz (1979). *Bull. Seismol. Soc. Am.* **69**, 1209–1214.
Luongo, G. and A. Mazzarella (1997). *Ann. Geofis.* **40**, 1303–1309.
Maeda, K. (1999). *Pure Appl. Geophys.* **155**, 381–394.
Main, I.G. (1996). *Rev. Geophys.* **34**, 433–462.
Makjanić, B. (1980). *Bull. Seismol. Soc. Am.* **70**, 2253–2260.
Mantovani, E. and D. Albarello (1997). *Phys. Earth Planet. Inter.* **101**, 49–60.
Maruyama, T. (1978). *Bull. Earthq. Res. Inst. Univ. Tokyo* **53**, 407–421.
Matsuda, T. (1975). *Zisin (J. Seismol. Soc. Japan) 2nd Ser.* **28**, 269–283.
Matsumura, K. (1986). *Bull. Disas. Prev. Res. Inst. Kyoto Univ.* **36**, 43–98.
Matsu'ura, R.S., *et al.* (1995). *Rep. Coord. Comm. Earthq. Predic.* **54**, 600–601 (in Japanese).
Mendoza, C. and S.H. Hartzell (1988). *Bull. Seismol. Soc. Am.* **78**, 1438–1449.
Mizoue, M., *et al.* (1983). *Bull. Earthq. Res. Inst. Univ. Tokyo* **58**, 25–63.
Mogi, K. (1963). *Bull. Earthq. Res. Inst. Univ. Tokyo* **41**, 615–658.
Mogi, K. (1967). *Bull. Earthq. Res. Inst. Univ. Tokyo* **45**, 711–726.
Mogi, K. (1968a). *Bull. Earthq. Res. Inst. Univ. Tokyo* **46**, 175–203.
Mogi, K. (1968b). *Bull. Earthq. Res. Inst. Univ. Tokyo* **46**, 53–74.
Mogi, K. (1973). *Techtonophysics* **17**, 1–22.
Mogi, K. (1985). "Earthquake Prediction," Academic Press, New York.
Molchan, G.M. and O.E. Dimitrieva (1990). *Phys. Earth Planet. Inter.* **61**, 99–112.
Molchan, G.M. and O.E. Dmitrieva (1992). *Geophys. J. Int.* **109**, 501–516.
Molchan, G.M., *et al.* (1997). *Bull. Seismol. Soc. Am.* **87**, 1220–1229.
Molchan, G.M., *et al.* (1999). *Phys. Earth Planet. Inter.* **111**, 229–240.
Mucciarelli, M., *et al.* (1988). *Tectonophysics* **152**, 153–155.
Musmeci, C. and D. Vere-Jones (1992). *Ann. Inst. Statist. Math.* **44**, 1–11.
Nishenko, S.P. and A. Buland (1987). *Bull. Seismol. Soc. Am.* **77**, 1382–1399.
Nowroozi, A.A. (1967). *Geophys. J. R. Astron. Soc.* **12**, 517–520.
Nyffenegger, P. and C. Frohlich (1998). *Bull. Seismol. Soc. Am.* **88**, 1144–1154.
Ogata, Y. (1983a). *J. Phys. Earth* **31**, 115–124.
Ogata, Y. (1983b). *Bull. Int. Statist. Inst.* **50**(2), 943–961.
Ogata, Y. (1988). *J. Am. Statist. Assoc.* **83**(401), 9–27.
Ogata, Y. (1992). *J. Geophys. Res.* **97**, 19845–19871.
Ogata, Y. (1998). *Ann. Inst. Statist. Math.* **50**, 379–402.
Ogata, Y. (1999a). *J. Geophys. Res.* **104**, 17995–18014.
Ogata, Y. (1999b). *Pure Appl. Geophys.* **155**, 471–507.
Ogata, Y. and K. Abe (1991). *Int. Statist. Rev.* **59**, 139–161.
Ogata, Y., *et al.* (1982). *Ann. Inst. Statist. Math.* **34B**, 373–387.
Ogata, Y. and K. Katsura (1986). *J. Appl. Probab.* **23A**, 291–310.
Ogata, Y. and K. Katsura (1988). *Ann. Inst. Statis. Math.* **40**, 29–39.
Ogata, Y. and K. Katsura (1991). *Biometrica* **78**, 463–474.
Ogata, Y. and K. Katsura (1993). *Geophys. J. Int.* **113**, 727–738.
Ogata, Y., *et al.* (1991). *Geophys. J. Int.* **104**, 135–146.
Ogata, Y., *et al.* (1995). *Geophys. J. Int.* **121**, 233–254.
Ogata, Y., *et al.* (1996). *Geophys. J. Int.* **127**, 17–30.
Ohtake, M. (1986). *Earthq. Predict. Res.* **4**, 165–174.
Ohtake, M., *et al.* (1977). *Pure Appl. Geophys.* **115**, 375–385.
Oike, K. (1978). *Disast. Prev. Res. Inst. Ann.* **20B**-1 35–45 (in Japanese).
Okal, E.A. and S.H. Kirby (1995). *Phys. Earth Planet. Inter.* **92**, 169–187.
Omori, F. (1894). *J. Coll. Sci. Imp. Univ. Tokyo* **7**, 113–200.
Öncel, A.O., *et al.* (1996). *Tectonophysics* **257**, 189–202.
Otsuka, M. (1972). *J. Phys. Earth* **20**, 35–45.
Ouchi, T. (1982). *Geophys. J. R. Astron. Soc.* **70**, 173–189.
Ouchi, T. (1993). *Pure Appl. Geophys.* **140**, 15–28.
Ouchi, T. and T. Uekawa (1986). *Phys. Earth Planet. Inter.* **44**, 211–225.
Page, R. (1968). *Bull. Seismol. Soc. Am.* **58**, 1131–1168.
Papadopulos, G.A. and V. Dedousis (1992). *Pure Appl. Geophys.* **139**, 269–276.
Papazachos, B.C. (1971). *Ann. Geofis.* **24**, 439–456.
Papazachos, B.C. (1974). *Pure Appl. Geophys.* **112**, 627–631.
Papazachos, B.C. (1989). *Bull. Seismol. Soc. Am.* **79**, 77–84.
Papazachos, B.C., *et al.* (1967). *Ann. Geofis.* **20**, 1–93.
Papazachos, C.B. and E.E. Papadimitriou (1997). *Bull. Seismol. Soc. Am.* **87**, 799–808.
Parvez, I.A. and A. Ram (1997). *Pure Appl. Geophys.* **149**, 731–746.
Pavlis, G.L. and M.W. Hamburger (1991). *J. Geophys. Res.* **96**, 18107–18117.
Ponomarev, A.V., *et al.* (1997). *Tectonophysics* **277**, 57–81.
Prozorov, A.G. (1986). *Comp. Seismol.* **19**, 54–58 (English translation).
Prozorov, A.G. and A.M. Dziewonski (1982). *J. Geophys. Res.* **87**, 2829–2839.
Rathbun, S.L. (1993). *Bull. Int. Statist. Inst.* **55**(2), 379–396.
Reasenberg, P.A. (1985). *J. Geophys. Res.* **90**, 5479–5495.

Reasenberg, P.A. (1999). *Pure Appl. Geophys.* **155**, 355–379.
Richter, C.F. (1958). "Elementary Seismology," Freeman.
Rikitake, T. (1974). *Tectonophysics* **23**, 299–312.
Robertson, M.C., *et al.* (1995). *J. Geophys. Res.* **100**, 609–620.
Sadovskiy, M.A., *et al.* (1984). *Izv. Phys. Earth* **20**, 87–96.
Saito, M., *et al.* (1973). *Zisin (J. Seismol. Soc. Japan) 2nd Ser.* **26**, 19–25.
Schuster, A. (1897). *Proc. R. Soc.* **61**, 455–465.
Seino, M., *et al.* (1989). *Zisin (J. Seismol. Soc. Japan) Ser. 2* **42**, 73–80.
Sekiya, H. (1976). *Zisin (J. Seismol. Soc. Japan) Ser. 2* **29**, 299–312.
Seno, T. (1979). *Tectonophysics* **57**, 267–283.
Senshu, T. (1959). *Zisin (J. Seismol. Soc. Japan) 2nd Ser.* **12**, 149–161.
Shimazaki, K. and T. Nakata (1980). *Geophys. Res. Lett.* **7**, 279–282.
Shimshoni, M. (1971). *Geophys. J.* **23**, 373–377.
Shlien, S. and M.N. Toksöz (1970). *Bull. Seismol. Soc. Am.* **60**, 1765–1787.
Singh, S.K. and G. Suárez (1988). *Bull. Seismol. Soc. Am.* **78**, 230–242.
Smally, R.F., *et al.* (1987). *Bull. Seismol. Soc. Am.* **77**, 1368–1381.
Smith, W.D. (1986). *Geophys. J. R. Astron. Soc.* **86**, 815–838.
Smith, W.D. (1998). *Geophys. J. Int.* **135**, 515–522.
Solov'ev, S.L. and O.N. Solov'eva (1962). *Izv. Phys. Earth*, 1053–1060.
Suyehiro, S., *et al.* (1964). *Pap. Met. Geophys.* **15**, 71–88.
Suzuki, Z. and K. Suzuki (1965). *Sci. Rep. Tohoku Univ. Ser. 5* **17**, 9–23.
Sykes, L.R. and S.P. Nishenko (1984). *J. Geophys. Res.* **89**, 5905–5927.
Tajima, F. and H. Kanamori (1985). *Phys. Earth Planet. Inter.* **40**, 77–134.
Takeo, M. (1988). *Bull. Seismol. Soc. Am.* **77**, 1074–1091.
Takeuchi, H. and H. Mizutani (1968). *Kagaku (Science)* **38**, 622–624.
Tamaki, I. (1961). *Mem. Osaka Inst. Tech.* **47**, 45–139.
Terada, T. (1918). *Proc. Tokyo Math.-Phys. Soc. Ser. 2* **9**, 515–522.
Tinti, S. and F. Mulargia (1987). *Bull. Seismol. Soc. Am.* **77**, 2125–2134.
Tocher, D. (1958). *Bull. Seismol. Soc. Am.* **48**, 147–154.
Tomoda, Y. (1952). *Zisin (J. Seismol. Soc. Japan) 2nd Ser.* **5**, 1–6.
Tomoda, Y. (1954). *Zisin (J. Seismol. Soc. Japan) 2nd Ser.* **7**, 155–169.
Trifu, C. and M. Radulian (1991). *J. Geophys. Res.* **96**, 4301–4311.
Tsapanos, T.M. (1990). *Bull. Seismol. Soc. Am.* **80**, 1180–1189.
Udías, A. and J. Rice (1975). *Bull. Seismol. Soc. Am.* **65**, 809–827.
Utsu, T. (1961). *Geophys. Mag.* **30**, 521–605.
Utsu, T. (1962). *Bull. Seismol. Soc. Am.* **52**, 279–297.
Utsu, T. (1965). *Geophys. Bull. Hokkaido Univ.* **13**, 99–103.
Utsu, T. (1966). *J. Phys. Earth* **14**, 37–40.
Utsu, T. (1969). *J. Fac. Sci. Hokkaido Univ. Ser. VII* **2**, 129–195.
Utsu, T. (1970). *J. Fac. Sci. Hokkaido Univ. Ser. VII* **3**, 197–266.
Utsu, T. (1971). *J. Fac. Sci. Hokkaido Univ. Ser. VII* **3**, 379–441.
Utsu, T. (1972a). *J. Fac. Sci. Hokkaido Univ. Ser. VII* **4**, 1 42.
Utsu, T. (1972b). *Rep. Coord. Comm. Earthq. Predict.* **7**, 1–13.
Utsu, T. (1975). *Zisin (J. Seismol. Soc. Japan) 2nd Ser.* **28**, 303–311.
Utsu, T. (1984). *Bull. Earthq. Res. Inst. Univ. Tokyo* **59**, 53–66.
Utsu, T. (1992). In: "Surijishingaku (Mathematical Seismology) (VII)," pp. 139–157, Institute of Statistics and Mathematics.
Utsu, T. (1999). *Pure Appl. Geophys.* **155**, 509–539.
Utsu, T. and Y. Ogata (1997). *IASPEI Software Library* **6**, 13–94.
Utsu, T. and A. Seki (1955). *Zisin (J. Seismol. Soc. Japan) 2nd Ser.* **7**, 233–240.
Utsu, T., *et al.* (1995). *J. Phys. Earth* **43**, 1–33.
Varnes, D.J. (1989). *Pure Appl. Geophys.* **130**, 661–686.
Vere-Jones, D. (1970). *J. R. Statist. Soc.* **B32**, 1–62.
Vere-Jones, D. (1976). *Pure Appl. Geophys.* **114**, 711–725.
Vere-Jones, D. (1978). *Adv. Appl. Probab. Suppl.* **10**, 73–87.
Vere-Jones, D. (1989). In: "Historical Seismicity of Central-Eastern Mediterranean Region," pp. 271–295, ENEA-IAEA, Rome.
Vere-Jones, D. (1992). In: "Statistics in the Environmental and Earth Sciences," pp. 220–246, E. Arnold, London.
Vere-Jones, D. (1994). In: "Proc. First US/Japan Conference on the Frontiers of Statistical Modelling: an Informational Approach," pp. 105–136, Kluwer, Dordrecht.
Vere-Jones, D. and R.B. Davies (1966). *NZ J. Geol. Geophys.* **9**, 251–284.
Vere-Jones, D. and Y.-L. Deng (1988). *Earthq. Res. China* **4**, 8–19.
Vere-Jones, D. and T. Ozaki (1982). *Ann. Inst. Statis. Math.* **34**, 189–207.
Volant, P. and J.-R. Grasso (1994). *J. Geophys. Res.* **99**, 21879–21889.
von Seggern, D., *et al.* (1981). *J. Geophys. Res.* **86**, 9325–9351.
Wadati, K. (1931). *Geophys. Mag.* **4**, 231–283.
Wadati, K. (1932). *Kishoshushi (J. Meteorol. Soc. Japan) 2nd Ser* **10**, 559–568.
Wang, J.-H. and C.-W. Lee (1997). *J. Phys. Earth* **45**, 331–345.
Wells, D.L. and K.J. Coppersmith (1994). *Bull. Seismol. Soc. Am.* **84**, 974–1002.
Wiemer, S. and K. Katsumata (1999). *J. Geophys. Res.* **104**, 13135–13151.
Wiemer, S. and M. Wyss (1997). *J. Geophys. Res.* **102**, 15115–15128.
Wiemer, S., *et al.* (1998). *Geophys. J. Int.* **134**, 409–421.
Wiens, D.A., *et al.* (1997). *Geophys. Res. Lett.* **24**, 2059–2062.
Wu, K.-T., *et al.* (1976). *Acta Geophys. Sin.* **19**, 95–109.
Wyss, M. and R.O. Burford (1987). *Nature* **329**, 323–325.
Wyss, M. and A.H. Martirosyan (1998). *Geophys. J. Int.* **134**, 329–340.
Yamanaka, Y. and K. Shimazaki (1990). *J. Phys. Earth* **38**, 305–324.
Yamaoka, K., *et al.* (1999). *Pure Appl. Geophys.* **155**, 335–353.
Zheng, X. and D. Vere-Jones (1991). *Pure Appl. Geophys.* **135**, 559–576.
Zheng, X. and D. Vere-Jones (1994). *Tectonophysics* **229**, 101–121.
Zhou, S.-Y., *et al.* (1995). *Acta Seismol. Sin.* **8**, 387–397.

Editor's Note

Due to space limitations, References with full citation are given in Utsu2FullReferences.pdf on the attached Handbook CD, under the directory \43Utsu2.

44

Relationships between Magnitude Scales

Tokuji Utsu
The University of Tokyo, Tokyo, Japan

1. Introduction

This chapter is a revision of the author's paper (Utsu, 1982b) having the same title. The reinvestigation using both previously used and newly added data has brought no significant change in the conclusions of the previous paper. Slightly modified curves for the relationships are presented. Since the previous paper was written in Japanese, some of the descriptions are repeated. However, materials that interest only workers of historical earthquakes in Japan are much simplified or omitted. The previous paper deals only with shallow earthquakes. Earthquakes of all depths are included in this chapter.

The magnitude scales used today stem from the one introduced by Richter (1935). This scale was designed for earthquakes in southern California recorded by the network of Wood–Anderson seismometers. This scale has been extended by later researchers in many ways to apply to the data produced in various observational environments. For example, the surface-wave magnitude and the body-wave magnitude were proposed by Gutenberg (1945a) and Gutenberg (1945b,c), respectively. These new scales were constructed, in principle, to provide the equal magnitude values to the same earthquakes or the earthquakes which radiated the equal amount of energy. However, systematic bias actually exists between the newly introduced scales and the reference magnitude scale.

The magnitude determined according to Richter's original definition is usually denoted by M_L, the magnitude calculated from the amplitudes of surface waves using the formula of Gutenberg (1945a) is denoted by M_S, and the magnitude determined from the amplitude/period ratio of P and/or S waves using the table or charts provided by Gutenberg (1945b,c; Gutenberg and Richter, 1956b) is denoted by m_B. These three magnitudes are denoted simply by M in the papers by Gutenberg (1945a,b,c), because these were considered to be equivalent. However, the subsequent studies showed that there are systematic differences between M_L, M_S, and m_B. In a paper by Gutenberg and Richter (1956b), which first used these notations, we find the following equations.

$$m_B = 0.63\,M_S + 2.5 \quad (1)$$

$$M_S = 1.27(M_L - 1) - 0.016\,M_L^2 \quad (2)$$

More than 100 such equations or curves relating the different magnitudes have been published (see Figures in Section 4).

To make matters more complicated, the notations M_L and M_S are sometimes used for magnitudes other than those determined according to the original definition of Richter (1935) and Gutenberg (1945a). The original body-wave magnitude m_B is measured by using medium-period body waves, but the body-wave magnitudes m_b (or M_b) used in PDE (USGS) and ISC catalogs are measured by using the first 5 sec of short-period P waves. This yields a considerable difference between two body-wave magnitudes m_B and m_b.

The problems of relationships between various magnitude scales have been discussed in several papers, some of which contain theoretical consideration on the basis of the seismic source spectra (e.g., Aki, 1967, 1972; Kanamori and Anderson, 1975; Geller, 1976; Noguchi and Abe, 1977; Chen and Chen, 1989). These studies successfully explain some features of magnitude scales, such as the different degrees of saturation of scales for large earthquakes. Here we take only the empirical approach without making the theoretical explanation of the results.

The empirically obtained curves for the relationship between two magnitude scales represent earthquakes with average source characteristics. Significant deviation of a data point from the curve can usually be attributed to anomalous source characteristics of the event such as a nuclear explosion

INTERNATIONAL HANDBOOK OF EARTHQUAKE AND ENGINEERING SEISMOLOGY, VOLUME 81A ISBN: 0-12-440652-1

or a tsunami earthquake, but no further discussion of this problem will be given here.

2. A Brief Survey of Various Magnitude Scales

2.1 Local Magnitude: M_L

This is the magnitude determined by using the equation

$$M_L = \log A + f(\Delta) \tag{3}$$

where A is the maximum trace amplitude in μm (single component) recorded by the standard W-A (Wood–Anderson) seismometer and $f(\Delta)$ is the empirically determined calibrating function of epicentral distance Δ. Richter (1935) gives a table of $f(\Delta)$ for local earthquakes in southern California ($\Delta =$ 30–600 km). $f(\Delta)$ is adjusted to be 0 at $\Delta = 100$ km according to the definition of the scale.

W-A seismometers had been operated for many years in a few regions including southern and central California and New Zealand. Magnitudes reported from these networks were regarded as genuine M_L. It is also possible to determine M_L by using simulated W-A seismograms, which are synthesized from seismograms recorded with seismographs of different response characteristics (e.g., Bakun and Lindh, 1977; Kanamori and Jennings, 1978; Jennings and Kanamori, 1979; Peppin and Bufe, 1980; Uhrhammer and Collins, 1990; Alsaker *et al.*, 1991; Shin, 1993; Savage and Anderson, 1995; Uhrhammer *et al.*, 1996; Margaris and Papazachos, 1999).

However, there are some problems in the determination of M_L. It has been suspected that the actual magnification of W-A seismometers was about 0.7 of the generally used value of 2800 (Boore, 1989; Uhrhammer and Collins, 1990). It has been pointed out that the values of M_L obtained from the data of nearby stations are somewhat smaller than those from data at distant stations. The systematic bias reaches about 0.3–0.4. This means that the attenuation of A with distance is not so strong as indicated by Richter's calibrating function. New calibrating functions have been developed for California by Bakun and Joyner (1984), Hutton and Boore (1987), and Eaton (1992). For further discussions about the calibrating function for M_L, see Haines (1981), Takeo and Abe (1981), Ebel (1982), Hadley *et al.* (1982), Luco (1982), Jennings and Kanamori (1983), Kiratzi and Papazachos (1984), Chávez and Priestley (1985), Bonamassa and Rovelli (1986), Greenhalgh and Singh (1986), Rogers *et al.* (1987), Boore (1989), Shin (1993), and Uski and Tuppurainen (1996).

In many cases (e.g., Haines, 1983; Lee *et al.*, 1990; Hatzidimitriou *et al.*, 1993; Johnston, 1996), the notation M_L is used for magnitudes determined from data of local network of short-period instruments (e.g., accelerograms) by some practical methods without using actual or synthetic W-A seismograms.

2.2 Surface-wave Magnitude by Gutenberg (1945a): M_S

The surface-wave magnitude proposed by Gutenberg (1945a) is calculated from

$$M_S = \log A + 1.656 \log \Delta + 1.818 \tag{4}$$

where A is the amplitude in μm of surface waves of period about 20 sec (combined horizontal amplitudes) and Δ is the epicentral distance in degrees. This equation was designed mainly for the Pasadena station. It may be conceivable that somewhat different equations are more suitable to other stations. Actually, similar equations with different coefficients have been developed for other stations by several investigators.

2.3 Body-wave Magnitude by Gutenberg (1945b,c): m_B

The body-wave magnitude proposed by Gutenberg (1945b,c) is calculated from

$$m_B = \log(A/T) + q(\Delta, h) \tag{5}$$

where A and T are amplitude and period of the maximum waves of P (vertical or combined horizontal components), PP (vertical or combined horizontal components), or S (combined horizontal component), and $q(\Delta, h)$ is the calibrating function given in tables for shallow earthquakes (Gutenberg, 1945b) and by charts of complicated contours for earthquakes of all depths (Gutenberg, 1945c). These charts were revised later (Gutenberg and Richter, 1956b). The calibrating functions for body-wave magnitudes have been the subject of considerable discussion for many years. At least 30 papers have been published dealing with this problem.

2.4 Magnitude Used in Gutenberg and Richter's Book "Seismicity of the Earth" (1949, 1954): M_{GR}

In their book "Seismicity of the Earth," Gutenberg and Richter (1949, 1954) published a catalog of the World's large earthquakes. The catalog in the 1954 edition (hereafter called the G-R catalog) lists about 4000 earthquakes occurring from 1904 through 1952. M values have been assigned to most of them. Before 1917 some earthquakes of $M \geq 7.0$ are missing. Most of the earthquakes with $M \geq 6$ are listed for the period from 1932 through June 1935. Although the M values in this catalog have been used as standards for many years, no clear explanation of M is found in the book. Gutenberg and Richter (1956b) noted that "The final value given for M was a weighted mean between M_B and M_S. This may be taken as defining M without subscript." In this description, M_B represents the value converted from m_B by using the equation

$$M_B - m_B = (M_B - 7)/4 \quad (6)$$

Note that this equation is the same as $M_B = m_B + (m_B - 7)/3$.

However, according to Geller and Kanamori (1977) and Abe (1981a) who examined Gutenberg's worksheets, M in the G-R catalog (hereafter denoted by M_{GR}) is almost equal to M_S for shallow earthquakes ($h < 40$ km). M_{GR} is equal to m_B for earthquakes of $h = 40$–60 km, and M_{GR} is close to $m_B + (m_B - 7)/4$ for $h > 70$ km (Abe, 1984). Lienkaemper (1984) reported that M_{GR} is 0.16 larger than M_S on the average for surface-rupturing earthquakes.

2.5 Magnitude Used in Rothé's Book "The Seismicity of the Earth, 1953–1965" (1967): M_R

The magnitude M listed in the catalog of world's earthquakes of 1953–1965 in Rothé's book (1967) is denoted here M_R. The catalog contains about 5300 earthquakes. Almost all earthquakes of $M_R \geq 6.0$ and some smaller earthquakes are listed. It is not easy to explain briefly how M_R values have been assigned (see the description given in Rothé's book for details). For the earthquakes whose magnitude are found in the catalog of Duda (1965), the same M values have been adopted.

2.6 Surface-wave Magnitude Assigned by USGS and ISC: M_S(PDE), M_S(ISC)

Surface-wave magnitude used in PDE (USGS) and ISC catalogs are based on the formula of Vaněk *et al.* (1962).

$$M_S = \log(A/T)_m + 1.66 \log \Delta + 3.3 \quad (7)$$

where $(A/T)_m$ denotes the combined value of the maximum of amplitude/period ratios of horizontal components of the surface waves (A in μm, T in sec) and Δ is the epicentral distance in degrees. This formula, often called the Prague formula, was recommended by IASPEI at the meeting held in Zürich in 1967 and also called the IASPEI formula. In this recommendation (see e.g., Båth, 1981), the waves are restricted to Rayleigh waves of period 20 ± 3 sec and the range of Δ is limited to 20°–160°, whereas no such restrictions are imposed on the original Prague formula. USGS and ISC have assigned the surface-wave magnitudes of relatively large earthquakes since 1968 and 1978, respectively. They are denoted here M_S(PDE) and M_S(ISC). Until the middle of 1975, the combined horizontal amplitude/period ratios of surface waves (not restricted to Rayleigh waves) of $T = 20 \pm 2$ sec were used in M_S(PDE). Later the vertical component data have been used. In determining M_S(ISC), both horizontal and vertical data in the period range from 10 to 60 sec and the distance range from 5° to 160° are used. Therefore, the original Vaněk *et al.* (1962) formula, the IASPEI recommendation, the conditions adopted by USGS and ISC may yield slightly different M_S values for the same earthquake.

The Vaněk *et al.* (1962) formula [Eq. (7)] is not perfectly consistent with the Gutenberg formula [Eq. (4)]. If $T = 20$ is put in Eq. (7), we obtain $M_S = \log A + 1.66 \log \Delta + 2.0$. This equation gives a larger M_S value than the M_S value calculated from Eq. (4) by about 0.19 for the same data.

More than 20 formulas of the Vaněk *et al.* type with different coefficients have been proposed for specific stations, specific regions, or the whole world.

Panza *et al.* (1989) noted that the Vaněk *et al.* formula yields larger M_S values with increasing epicentral distance. The difference of M_S determined from data at $\Delta = 30°$ and $\Delta = 140°$ is about 0.4. According to Herak and Herak (1993), the average difference of M_S at 20° and 160° reaches 0.5. The coefficient of $\log \Delta$ in Eq. (7), 1.66, seems too large. Herak and Herak (1993) proposed a formula $M_S = \log(A/T)_m + 1.094 \log \Delta + 4.429$ (Δ: 20°–160°). The modified formulas proposed by Nuttli and Kim (1975), Thomas *et al.* (1978), Christoskov *et al.* (1985), and Rezapour and Pearce (1998) also have the coefficients of around 1.1.

2.7 Body-wave Magnitude Assigned by USGS and ISC: m_b(PDE) and m_b(ISC)

Body-wave magnitudes have been published in PDE since 1963 and in ISC since 1964. These are based on the method by Gutenberg (1945c) with the revised chart for $q(\Delta, h)$ of Gutenberg and Richter (1956b). The notations m_b and M_b are used in respective publications. These magnitudes are represented here by m_b(PDE) and m_b(ISC). There is very small but persistent bias between m_b(PDE) and m_b(ISC) (as shown in Section 5 of the full manuscript on the attached Handbook CD, under the directory \44Utsu3).

There are magnitude-dependent systematic differences between m_b and m_B. This is because m_b uses the amplitudes and periods measured on short-period instruments and m_B uses those measured on medium-to-long-period (or broadband) instruments. The periods of P waves used in m_b determinations are around 1 sec, whereas the periods used for m_B are several to about 10 sec. The maximum amplitude of first 5 sec of P waves is used in measuring m_b. In many large earthquakes, the maximum amplitude appears later than 5 sec after the initial arrival.

2.8 Moment Magnitude: M_W

To represent the size of an earthquake as a dislocation phenomenon along a fault, the seismic moment (M_0) is the most adequate measure. The level of the low-frequency end of seismic spectrum is controlled by the seismic moment.

Kanamori (1977) compared the earthquake energy-moment relation

$$E_S = (\Delta\sigma/2\mu)M_0 \quad (8)$$

where $\Delta\sigma$ is the stress drop and μ is the rigidity, with the magnitude-energy relation (Gutenberg and Richter, 1956b)

$$\log E_S = 1.5\,M_S + 11.8 \quad (9)$$

When E_S and M_0 are expressed in erg and dyne cm, respectively, the average value of $\Delta\sigma/2\mu$ in Eq. (8) is approximately equal to 1.0×10^{-4}. If this value is used in Eq. (8), we obtain

$$\log M_0 = 1.5\,M_S + 16.1 \quad (10)$$

It is known that M_S value saturates for great earthquakes (M_0 about 10^{29} dyne cm or more) such as the 1957 Aleutian, 1960 Chile, and 1964 Alaska earthquakes. Equations (9) and (10) do not hold for such great earthquakes. If a new magnitude scale using the notation M_W is defined by

$$\log M_0 = 1.5\,M_W + 16.1 \quad (11)$$

the new magnitude M_W is equivalent to M_S within the limit of saturation and provides a reasonable figure even for great earthquakes (Kanamori, 1977).

Hanks and Kanamori (1979) defined the moment magnitude by

$$\mathbf{M} = \tfrac{2}{3}\log M_0 - 10.7 \quad (12)$$

Equation (12) is the same as $\log M_0 = 1.5\,\mathbf{M} + 16.05$. Both M_W and $\mathbf{M}$ have been used in the name of moment magnitude in the literature, and the notation $\mathbf{M}$ has often been replaced by M_W. If M_W is given to the first decimal place, about 10% of earthquakes have different M_W values by 0.1 depending on whether Eq. (11) or Eq. (12) be used. Moreover, some workers use $\log M_0 = 1.5\,M_W + 16.0$.

There are many empirical equations that relate various magnitudes to M_W (see Figures in Section 4). It may be possible to convert a magnitude to M_W by using these equations. Refined methods have been proposed to obtain a magnitude equivalent to M_W without measuring the seismic moment by a standard procedure (e.g., Mahdyiar *et al.*, 1986). Although, it is quite reasonable to design a magnitude-scale equivalent to the moment magnitude, it may be confusing to use the term "moment magnitude" or the notation M_W for such a converted or simulated magnitude.

2.9 Tsunami Magnitude M_t

Abe (1979a, 1981b) defined the tsunami magnitude M_t for representing the magnitude of a tsunami as well as the magnitude of an earthquake which generates the tsunami. This magnitude is based on the recorded wave heights of a tsunami. The scale is so constructed that M_t agrees with the moment magnitude M_W [see also Abe (1988, 1989, 1994)].

2.10 Magnitude Assigned by the Japan Meteorological Agency M_J

The Japan Meteorological Agency (JMA) (called the Central Meteorological Observatory until 1956) keeps the nationwide seismograph network in Japan and publishes the observational data since 1885. The magnitudes assigned to earthquakes listed in the JMA catalog since 1926 are based on the formulas originally designed to provide equivalent magnitude values to M_{GR}. For details, see "Introductory Notes" of a January issue of Seismological Bulletin of JMA in recent years.

2.11 Kawasumi's Magnitude M_K

The original Kawasumi magnitude (Kawasumi, 1943) was defined as the seismic intensity in the JMA scale at the epicentral distance of 100 km, I_{100}. Kawasumi (1951) devised a formula that converts I_{100} into a value equivalent to M_{GR}. The converted value usually denoted by M_K. The conversion formula is $M_K = 0.5\,I_{100} + 4.85$.

Values of M_K for significant earthquakes in historical times have been published in "Rikanenpyo (Science Chronicle)" since 1952. The M_K values of about 3900 earthquakes in Japan for the years 1885–1943 are included in a JMA catalog published in 1952. However, M_K values in this catalog are unreasonably large, especially for events before 1914. Utsu (1979, 1982b) provided a detailed explanation how and why these M_K values were so strongly biased.

2.12 Magnitude for Earthquakes in Japan in the Years 1885–1925, M_U and M_A

Until 1979 no reliable magnitudes values had been available for most of the earthquakes in Japan during 1885–1925. In the catalog of Utsu (1979, 1982a, 1988), magnitudes determined by the formula used in JMA have been given to all earthquakes of $M \geq 6.0$ and smaller ones accompanied by damage in Japan. Abe (1979b) also published a magnitude catalog of damaging earthquakes in Japan between 1901 and 1925 based on the Vaněk *et al.* (1962) formula. These two magnitudes, denoted by M_U and M_A here, agree fairly closely with almost no systematic differences (Utsu, 1982b).

3. Relationship Between Various Magnitude Scales

3.1 Data Sources

The magnitude values taken from the following sources are considered here.

M_W: The seismic moments given in the CMT catalog of Harvard University have been converted into M_W to the second decimal place by using Eq. (11) for earthquakes of 1977–1997 and $M_W \geq 5.0$. A total of 14416 events.

For 1901–1976, Kanamori's compilation (1977) of M_W of 51 great shallow earthquakes is used.

M_{GR}: Taken from Gutenberg and Richter (1954). A total of 3513 events during 1904–1952. The data are not complete even in the range of $M_{GR} \geq 7.0$, but many events with $M_{GR} < 7.0$ are included.

M_R: Taken from Rothé (1967). A total of 5353 events during 1953–1965. Almost complete for $M_R \geq 6.0$. Many events with $M_R < 6.0$.

M_S: Taken from Abe (1981a, 1984) and Abe and Noguchi (1983a,b). A total of 1047 events during 1901–1980. Mostly $M_S \geq 7.0$.

m_B: Taken from Abe and Kanamori (1979) and Abe (1981a, 1982, 1984). A total of 1290 events during 1901–1980. Mostly $M_B \geq 7.0$.

M_S(ISC), m_b(ISC): Taken from ISC catalog for 1964–1995. A total of 1 13 349 events with m_b(ISC) ≥ 4.5.

M_S(PDE), m_b(PDE): Taken from PDE (USGS) for 1964–1997. A total of 1 09 582 events with m_b(PDE) ≥ 4.5.

M_J: Taken from the catalog of the Japan Meteorological Agency for earthquakes in and near Japan (1926–1997). For earthquakes in 1901–1925, Utsu's catalog (1982a, 1985, 1988) is used. A small portion of the magnitudes given in Utsu's catalog are not determined by the same method as used in JMA. These magnitudes are not used here or replaced by those determined by the JMA method (Utsu, 1982b). A total of 1710 events of $M_J \geq 6.0$ for 1901–1997. A total of 4825 events of $M_J \geq 5.0$ for 1926–1997. A total of 17 178 events of $M_J \geq 4.0$ for 1964–1997.

M_t: Taken from Abe (1979a, 1981b, 1988). A total of 141 events during 1901–1984.

For M_L, the California earthquakes included in the USGS earthquake database (data from CDMG catalog are listed) have been examined for 2396 events of M_L (PAS or BRK) ≥ 4.0 during 1932–1974, but only a small number of shocks are found whose magnitudes were also given in the G-R, or, Rothé, or Abe catalog. Some M_L values for large earthquakes may not be genuine ones determined from the W-A seismograms. For example, the catalog gives $M_L = 7.7$ for the 1952 Kern County earthquake, but this seems to be the M_{GR} value. The simulated W-A seismograms yield $M_L = 7.2$ (Kanamori and Jennings, 1978). No direct comparison between M_L and other magnitudes has been made here. The curve for M_L in Figure 1 has been tentatively drawn from published relations in various papers (see Figures in Section 4), though there are some problems in the definition and determination of M_L as described in Section 2.1.

3.2 Procedure for Constructing the Standard Curves

The main result of the previous paper (Utsu, 1982b) is a graph (called master graph) showing the relationships between M_W and M_x–M_W in a range of M_W between 4 and 9, where M_x represents one of the magnitudes M_S, m_B, m_b, M_L, and M_J. The main purpose of this chapter is to present a new version of the master graph. The new graph is shown in Figure 1. The difference between the old and new curves is small, about 0.1 magnitude unit at the most.

Figures 2–16 are the scattergrams of M_x and M_y, where M_x and M_y are two magnitudes assigned to the same earthquake.

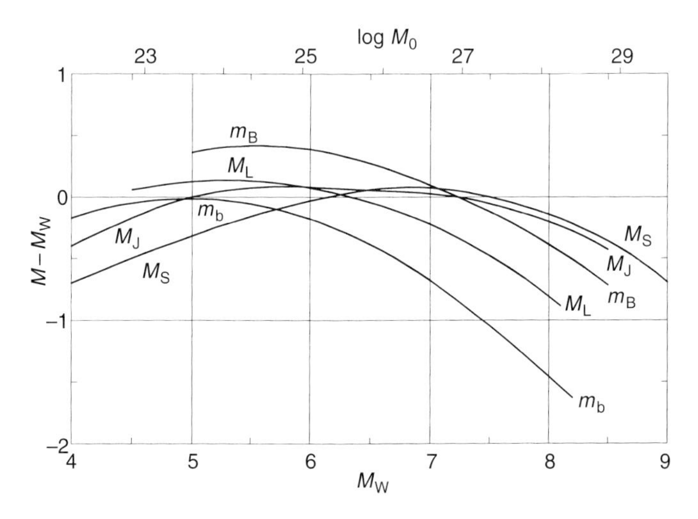

FIGURE 1 Curves for the average magnitude differences. Variations of M_S–M_W, m_B–M_W, m_b–M_W, M_L–M_W, and M_J–M_W with M_W (or log M_0) are shown.

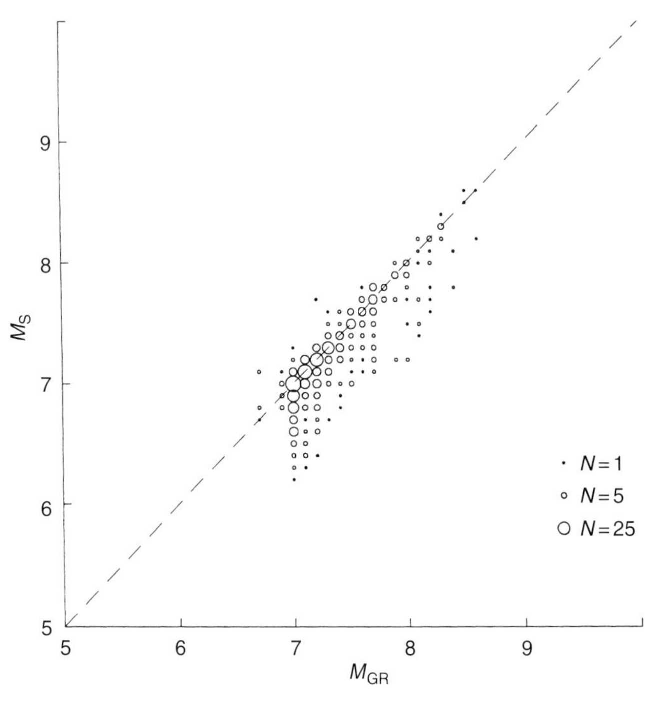

FIGURE 2 Relation between M_{GR} and M_S.

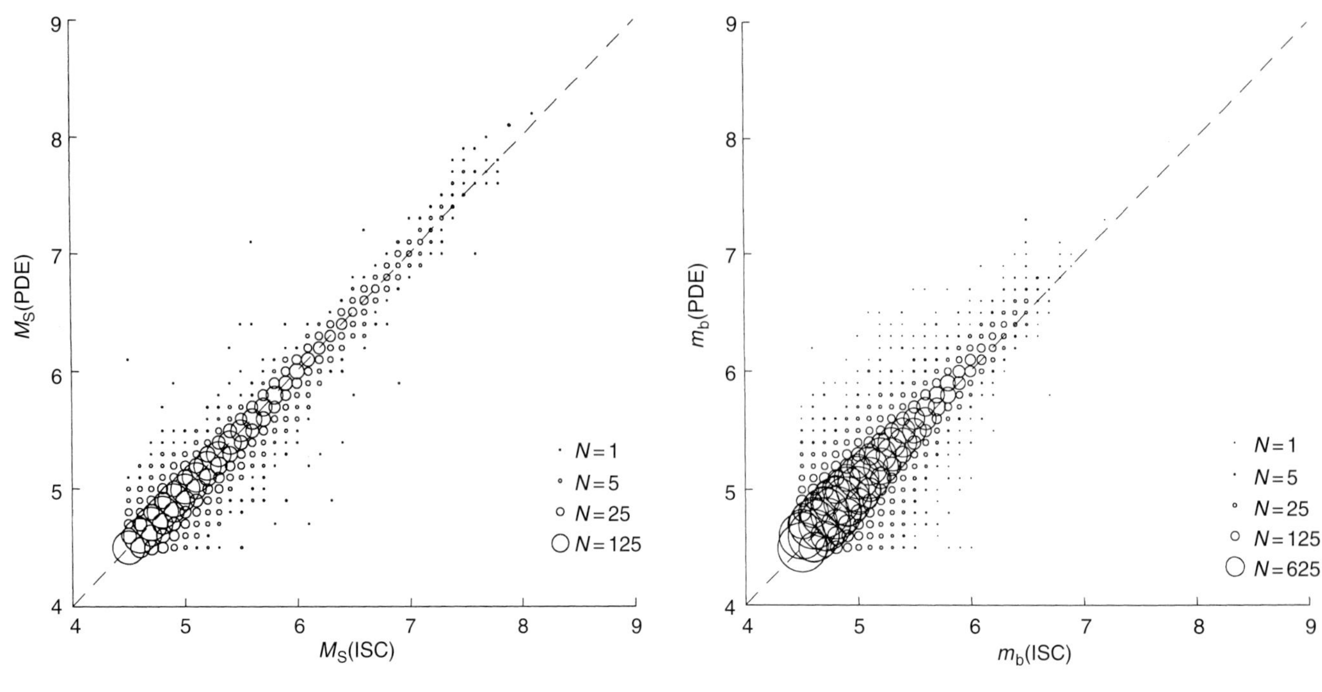

FIGURE 3 Relation between M_S(ISC) and M_S(PDE).

FIGURE 5 Relation between m_b(ISC) and m_b(PDE).

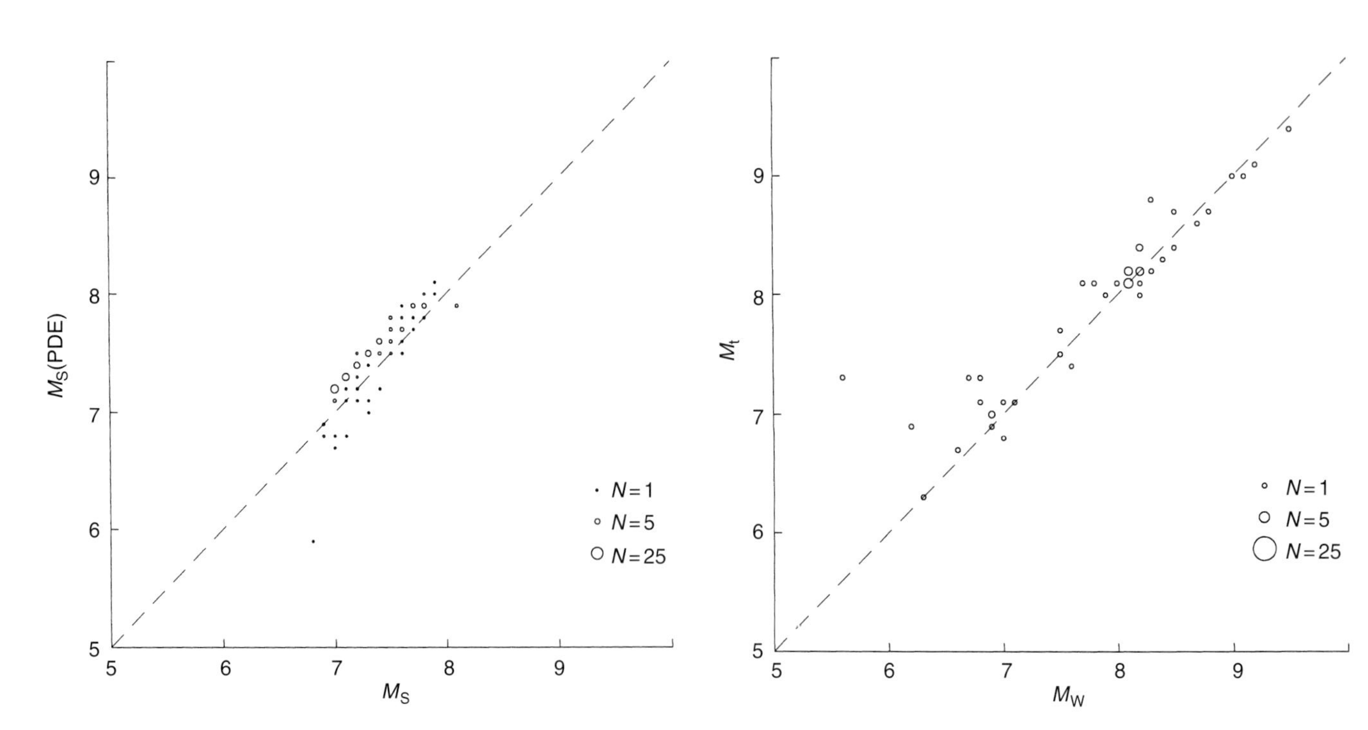

FIGURE 4 Relation between M_S and M_S(PDE).

FIGURE 6 Relation between M_W and M_t.

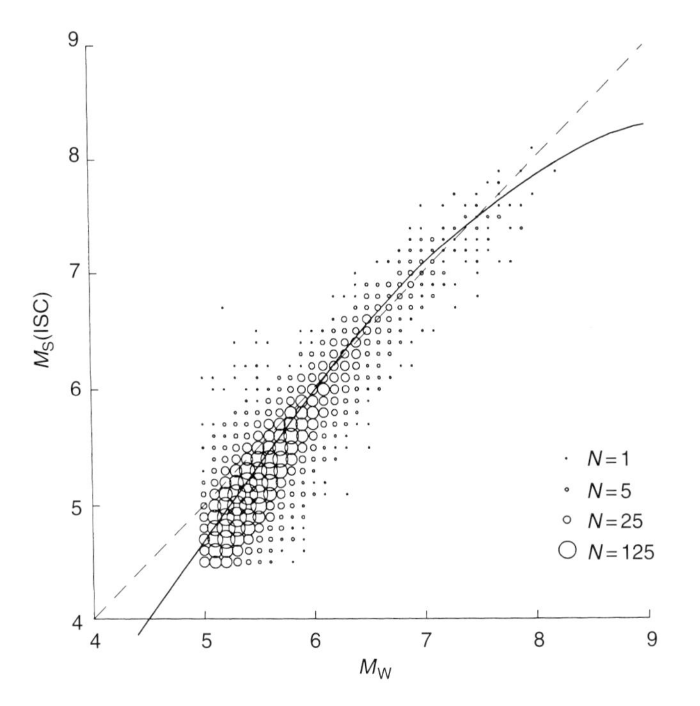

FIGURE 7 Relation between M_W and M_S(ISC).

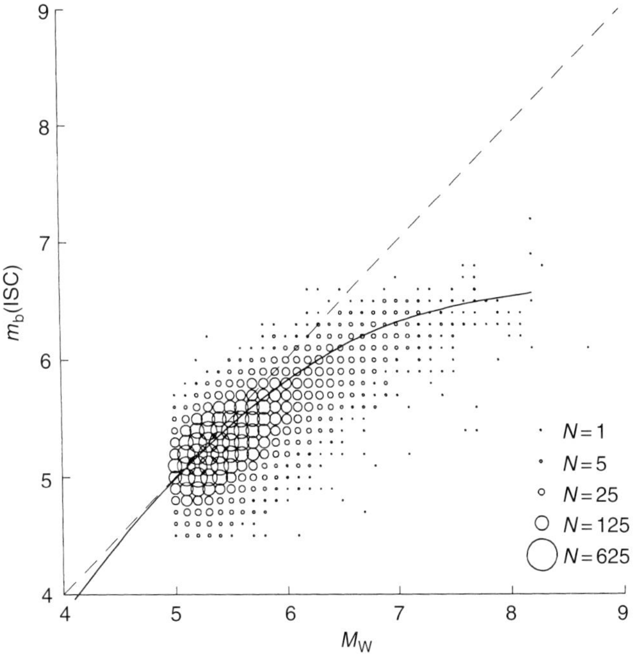

FIGURE 9 Relation between M_W and m_b(ISC).

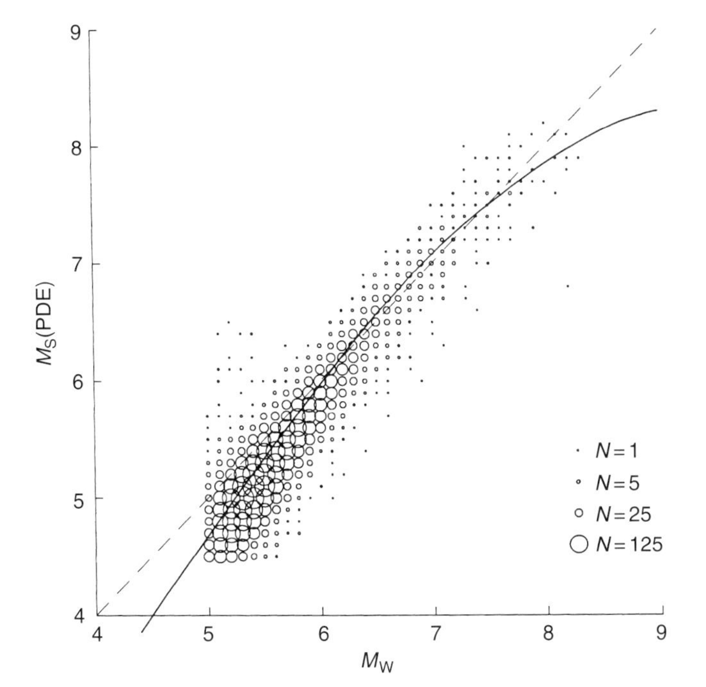

FIGURE 8 Relation between M_W and M_S(PDE).

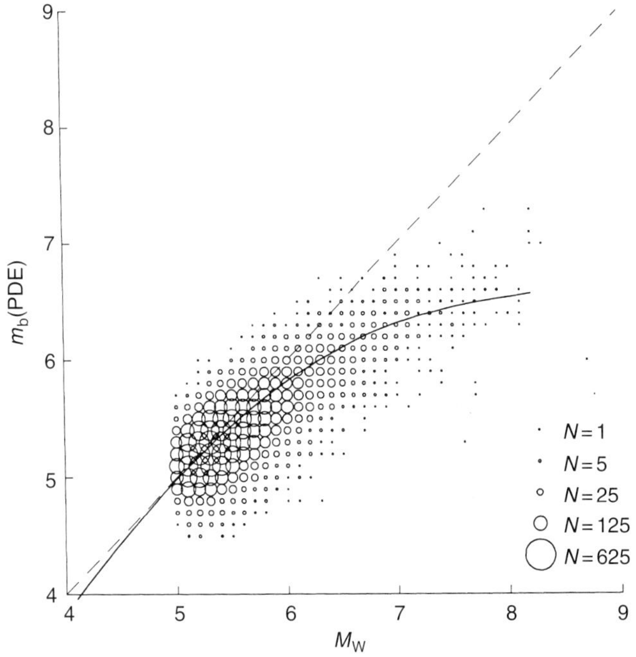

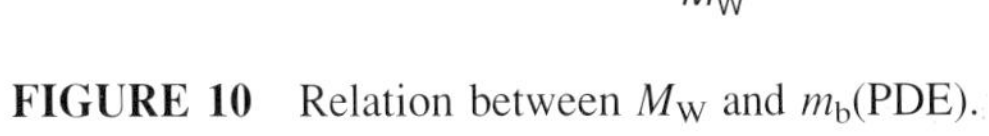

FIGURE 10 Relation between M_W and m_b(PDE).

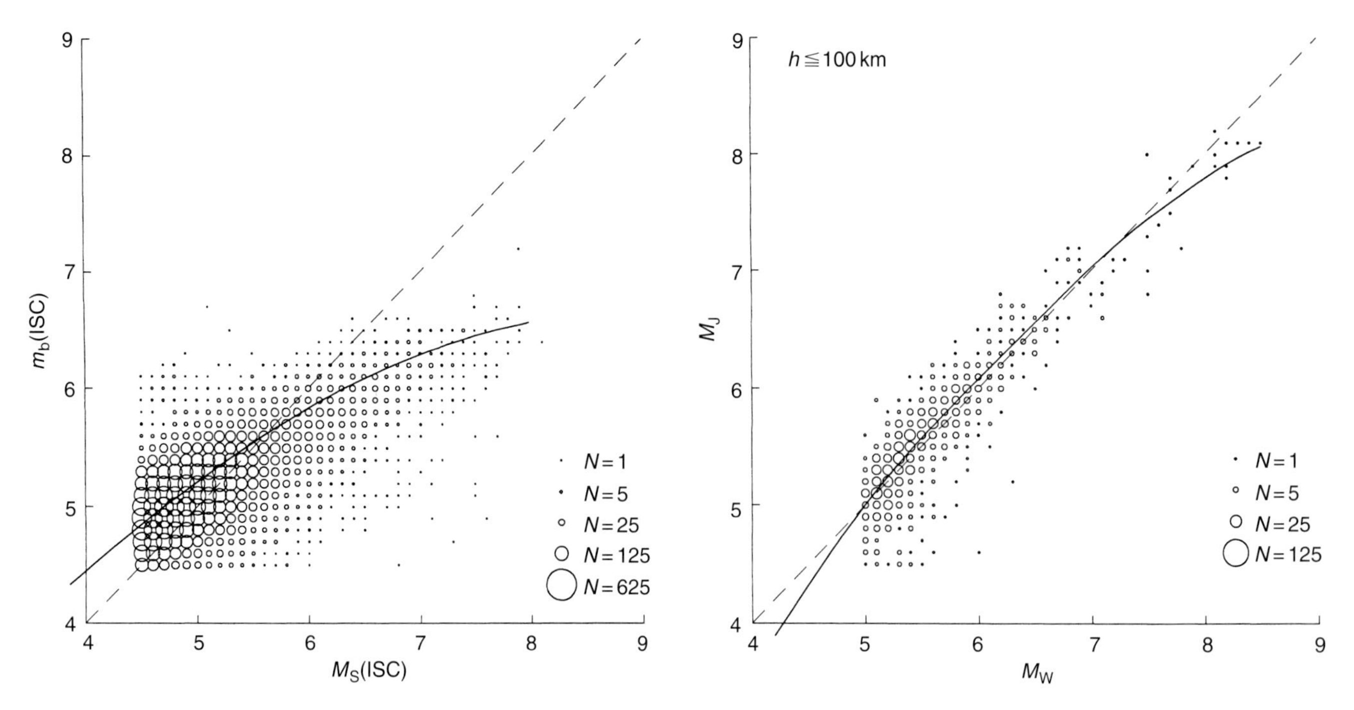

FIGURE 11 Relation between M_S(ISC) and m_b(ISC).

FIGURE 13 Relation between M_W and M_J for focal depth less than 100 km.

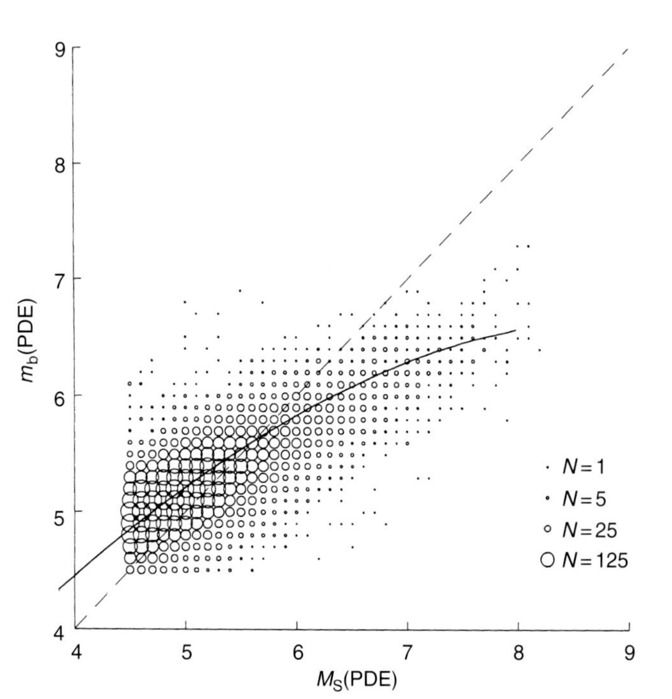

FIGURE 12 Relation between M_S(PDE) and m_b(PDE).

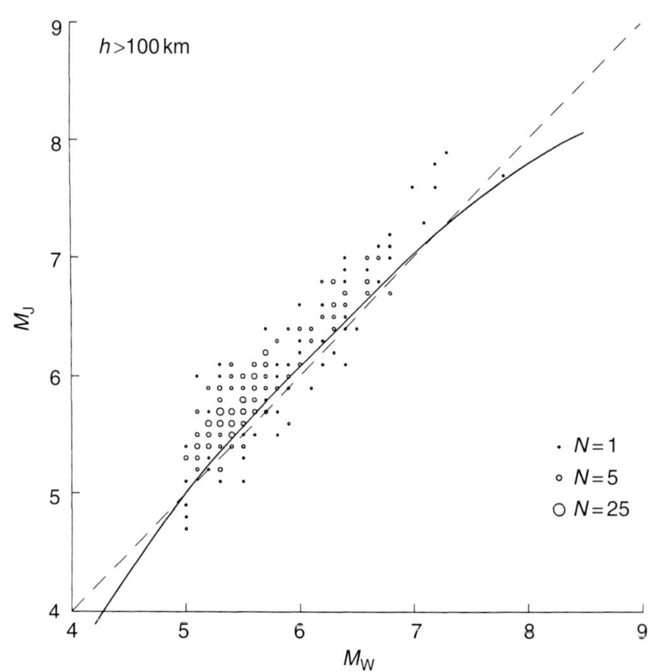

FIGURE 14 Relation between M_W and M_J for focal depths 100 km or larger.

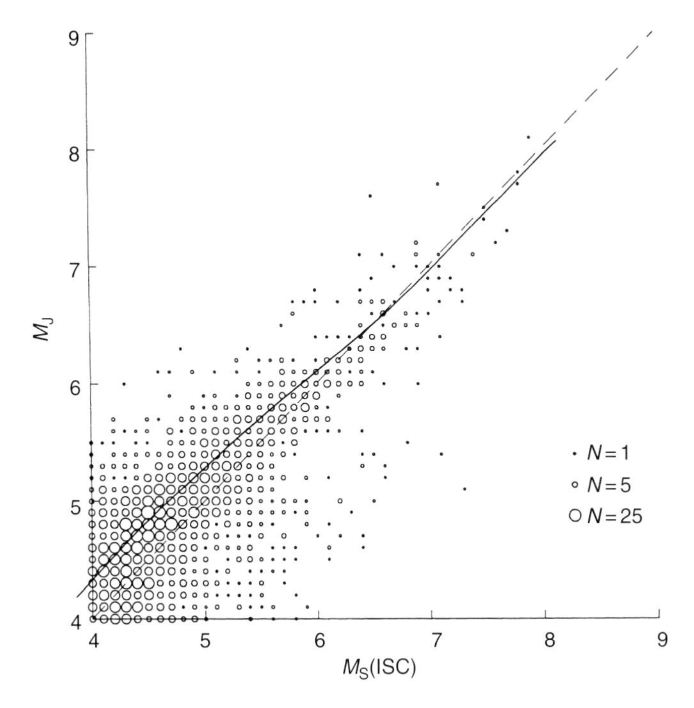

FIGURE 15 Relation between M_S(ISC) and M_J.

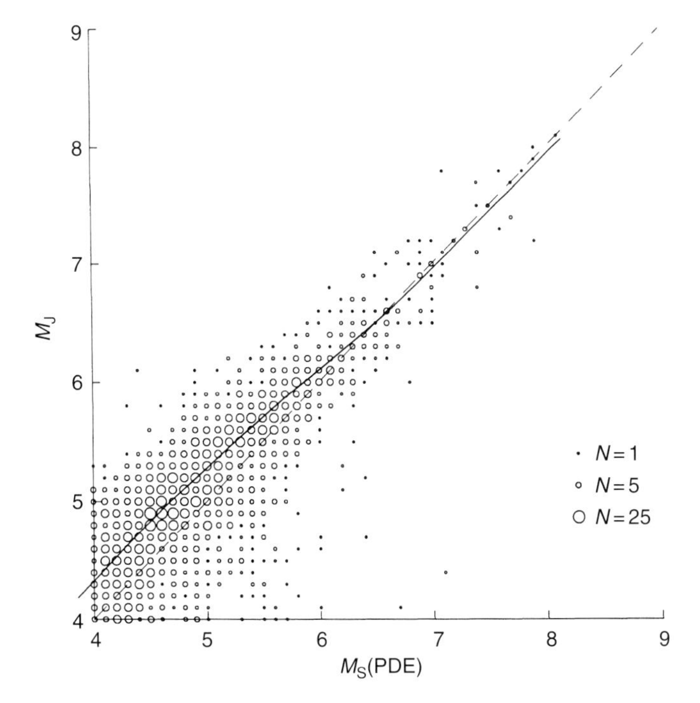

FIGURE 16 Relation between M_S(PDE) and M_J.

The size of a circle is so chosen that its area is proportional to the number, N, of events it represents. The dashed line represents the relation $M_x = M_y$. (Further figures are given on the attached Handbook CD.) The graphs concerning M_K or M_A (Utsu, 1982b) are omitted, because of the reason mentioned in Section 1.

A curve showing the average relation between M_x and M_y computed from the M_x–M_W and M_y–M_W curves in the old master graph was first drawn in the scattergram for M_x and M_y. By watching the goodness of fit of such M_x versus M_y curves, the position and shape of the M_x–M_W and/or M_y–M_W curves in the old master graph have been changed by trial and error to obtain a better fit in the corresponding scattergram. Such a procedure has been repeated for various combinations of magnitudes. The curves shown in Figure 1 are the final ones thus modified from the old ones.

The curves in the scattergrams (Figs. 2–16) (Figs. 2–36 on the attached Handbook CD) are not the ones fitted to individual scattergrams, but they are derived from the curves in the master graph (Fig. 1), which is the result of the adjustment considering every combination of different magnitudes. In some scattergrams, the curve shows a behavior not expected from the plotted data, but this cannot be helped for the sake of overall consistency. For example, the downward curvature in Figures 7 and 8 is not seen from the plotted data. (The main cause of this curvature is due to the data in Figure 20 on the attached Handbook CD.) The curves in Figures 7 and 8, are the same, because they are derived from the M_S–M_W versus M_W curve in Figure 1.

3.3 Relationships Between Two Similar Magnitudes

3.3.1 M_{GR} for Shallow Earthquakes and M_S (1904–1952)

As shown in Figure 2, $M_{GR} = M_S$ for most earthquakes. M_{GR} values larger than M_S values are found for earthquakes occurring before 1910. These are due to overestimation of M_{GR} as remarked by Abe and Noguchi (1983b). It can be said that M_{GR} for shallow earthquakes is equivalent to M_S if these overestimated values are removed or corrected.

3.3.2 M_S(PDE) and M_S(ISC) (1976–1995)

As shown in Figure 3, there is almost no systematic difference between these two magnitudes for earthquakes of M_S(PDE) less than about 7.3 However, M_S(PDE) is larger than M_S(ISC) by about 0.1 on the average for M_S(PDE) larger than 7.4. Since these magnitudes are based on the Vaněk *et al.* (1962) formula, we use the notation M_S(V) to represent both M_S(PDE) and

M_S(ISC). The standard curve labeled M_S in Figure 1 represents M_S(V).

3.3.3 M_S and M_S(V) (1968–1980)

As explained in Section 2.6, it is expected that M_S(V) is systematically larger than M_S by about 0.19. It can been seen from Figure 4 that such a bias actually exists at least between M_S 7 and 8, and probably for M_S less than 7 [see also Abe (1981a)].

3.3.4 m_b(PDE) and m_b(ISC) (1964–1995)

Figure 5 shows that there are small systematic bias between these two magnitudes. This bias can be seen for both shallow and deep earthquakes. m_b(PDE) is about 0.05 higher than m_b(ISC) on the average. This difference persists from 1966 to 1997 (see Fig. 54 in the full manuscript on the attached Handbook CD, under the directory \44Utsu3). In the earliest two years (1964–65) the average difference is about 0.1. Here we use the notation m_b to represent m_b(PDE) or m_b(ISC) + 0.05. For large shallow earthquakes ($m_b > 6.4$) this difference seems to increase, though the data are rather few.

3.3.5 M_W and M_t (1901–1987)

The scale of M_t is so constructed that it gives the value equivalent to M_W (Abe, 1979a, 1989). Direct comparison between M_W and M_t in Figure 6 shows that there is no systematic difference between M_W and M_t except a few events that have special reasons for $M_t > M_W$.

3.4 Relationships between Two Magnitudes of Different Character

Here ten scattergrams are shown for five pairs of magnitudes together with the curves calculated from the master graph (Fig. 1). The remaining 20 scattergrams are included in the full manuscript on the attached Handbook CD, under the directory \44Utsu3.

3.4.1 M_W and M_S(V) (1977–1997)

The relation is shown in Figures 7 and 8. In general, the curve fits the plotted data well, but it yields somewhat large M_S(V) values for M_W less than about 6.

3.4.2 M_W and m_b (1977–1997)

The relation is shown in Figures 9 and 10 for all depths. The curve gives slightly large m_b values for $M_W >$ about 7.

3.4.3 M_S(V) and m_b (1968–1997)

The relation is shown in Figures 11 and 12. The curve fits the data for M_S(PDE) versus m_b(PDE). Slightly large m_b(ISC) values are obtained from the same curve.

3.4.4 M_W and M_J (1901–1997)

The relation is shown for shallow earthquakes (Fig. 13, $h < 100$ km) and deep earthquakes (Fig. 14, $h \geq 100$ km). It is seen that the M_W versus M_J curve fits well to shallow earthquakes, but does not fit for deep earthquakes. The difference $M_J - M_W$ tends to increase with depth (Katsumata, 1996). The M_J curve in Figure 1 should be applied to shallow earthquakes only.

3.4.5 M_S(V) and M_J (1968–1997)

This relation is shown in Figures 15 and 16. The M_S versus M_J curve fits M_S(PDE) data well, but some bias is seen for M_S(ISC) data. Although there are almost no systematic differences between M_S(PDE) and M_S(ISC) for the worldwide data (Fig. 3), some systematic difference of unknown origin exists for earthquakes in the vicinity of Japan as already pointed out by Utsu (1982b). The average difference M_S(PDE) – M_S(ISC) is 0.087 for Japanese earthquakes, whereas it is 0.019 for the worldwide data.

(Note: Additional relationships between two magnitudes of different character are given in the full manuscript on the attached Handbook CD, under the directory \44Utsu3.)

4. Comparison with Relations Proposed by Different Authors

Many papers have been published in which the relationships between two or more magnitude scales have been examined. The empirical relations obtained in these studies are usually represented by a linear equation like Eq. (1) in some limited range of magnitude. More discussions on magnitude-scale relationships have been given by Kárník (1973), Chung and Bernreuter (1981), Kanamori (1983), Nuttli (1983a), Heaton *et al.* (1986), Chen and Chen (1989), and Gusev (1991).

The equations reported in these papers representing the relationships between two magnitude scales, M_x and M_y, are plotted here in the form of M_y–M_x versus M_x diagram. These are shown in Figures 17–22 (or Figs 43–48 in the full manuscript on the attached Handbook CD, under the directory \44Utsu3). The heavy lines in these diagrams represent the M_y–M_x versus M_x curves derived from Figure 1. It can be seen that most of the heavy lines are located near the average position of the lines taken from various papers.

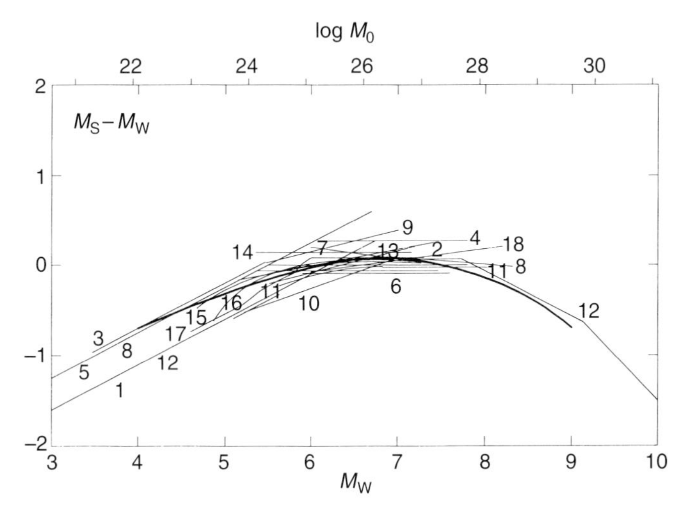

FIGURE 17 Relation between M_W (log M_0) and M_S taken from 18 papers listed below. The thick curve indicates the relation shown in Figure 1. 1: Brune (1968), 2: Udías (1971), 3: Purcaru and Berckhemer (1978), 4: Singh and Havskov (1980), 5: Nuttli (1983b), 6: Dziewonski and Woodhouse (1983), 7: Nowroozi (1985), 8: Nuttli (1985), 9: Kiratzi *et al.* (1985), 10: Bergman (1986), 11: Ekström and Dziewonski (1988), 12: Chen and Chen (1989), 13: Main and Burton (1989, 1990), 14: Papazachos and Kiratzi (1992), 15: Johnston and Halchuk (1993), 16: Johnston (1996), 17: Giardini *et al.* (1997), 18: Pérez (1999).

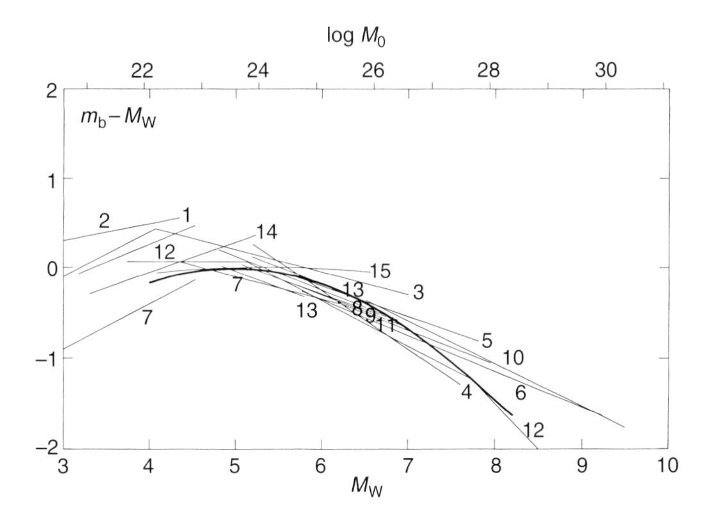

FIGURE 18 Relation between M_W (log M_0) and m_b. 1: Herrmann (1980), 2: Dwyer *et al.* (1983, m_b from Lg), 3: Nuttli (1983b), 4: Dziewonski and Woodhouse (1983), 5: Giardini (1984), 6: Koyama and Shimada (1985), 7: Nuttli (1985), 8: Kiratzi *et al.* (1985), 9: Bergman (1986), 10: Houston and Kanamori (1986), 11: Giardini (1988, for $h > 350$ km), 12: Chen and Chen (1989), 13: Kuge (1992, for $100 < h \leq 300$ km and $h > 300$ km), 14: Patton and Walter (1993, 1994), 15: Johnston (1996).

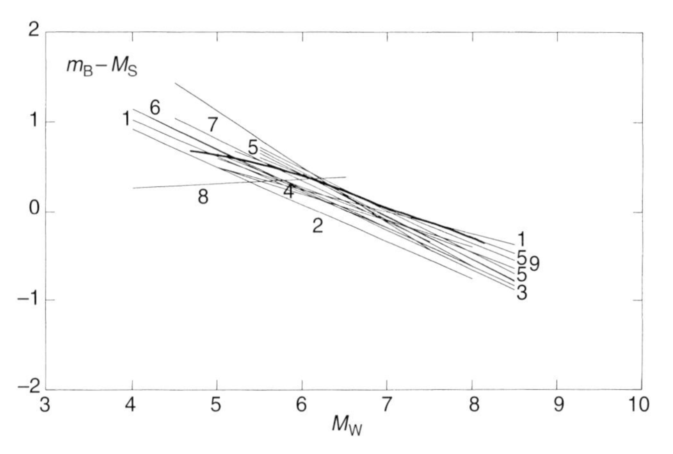

FIGURE 19 Relation between M_S and m_B. 1: Gutenberg and Richter (1956b), 2: Båth (1956), 3: Di Filoppo and Marcelli (1959), 4: Kárník *et al.* (1959), 5: Stelzner (1961), 6: Båth (1966), 7: Kárník (1971), 8: Båth (1977), 9: Thomas *et al.* (1978), 10: Abe and Kanamori (1980).

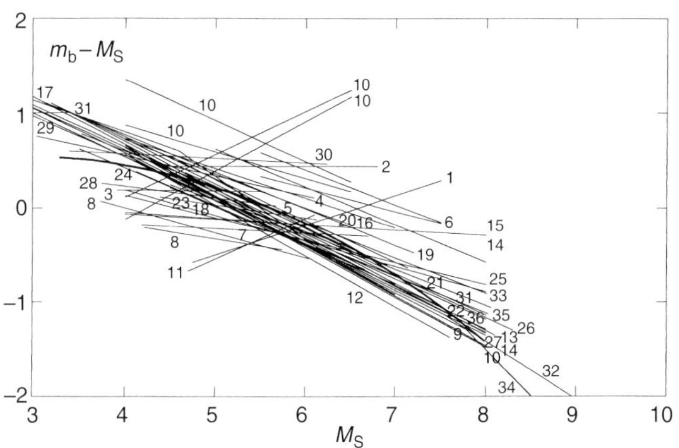

FIGURE 20 Relation between M_S and m_b. 1: Ichikawa and Basham (1963, see Ichikawa, 1966), 2: Romney (1964, see Basham, 1969), 3: Marshall *et al.* (1966, see Gupta *et al.*, 1972), 4: Capon *et al.* (1967, see Basham, 1969), 5: Antonova *et al.* (1968, see Kárník, 1973), 6: Solov'ev and Solove'va (1968), 7: Nagamune *et al.* (1969), 8: Basham (1969), 9: Marshall (1970), 10: Pasechnik *et al.* (1970), 11. Gupta *et al.* (1972), 12: Jordan and Hunter (1972), 13: Nagamune (1972), 14: Gupta and Rastogi (1972), 15: Prozorov and Hudson (1974), 16: Alsan *et al.* (1975, see Makropoulos *et al.*, 1989), 17: Bungum and Tjøstheim (1976), 18: Reichle *et al.* (1976), 19: Street and Turcotte (1977), 20: Thomas *et al.* (1978), 21: Makropoulos (1978, see Makropoulos *et al.*, 1989), 22: Bloom and Erdman (1979), 23: Carter and Berg (1981), 24: Chung and Bernreuter (1981), 25: Makropoulos and Burton (1981), 26: Wyss and Haberman (1982), 27: Prozorov *et al.* (1983), 28: Singh *et al.* (1983), 29: Burton *et al.* (1984), 30: Ambraseys (1985), 31: Nuttli (1985), 32: Wu and Cao (1987), 33: Riad *et al.* (1988), 34: Chen and Chen (1989), 35: Ambraseys (1990), 36: Rezapour and Pearce (1998).

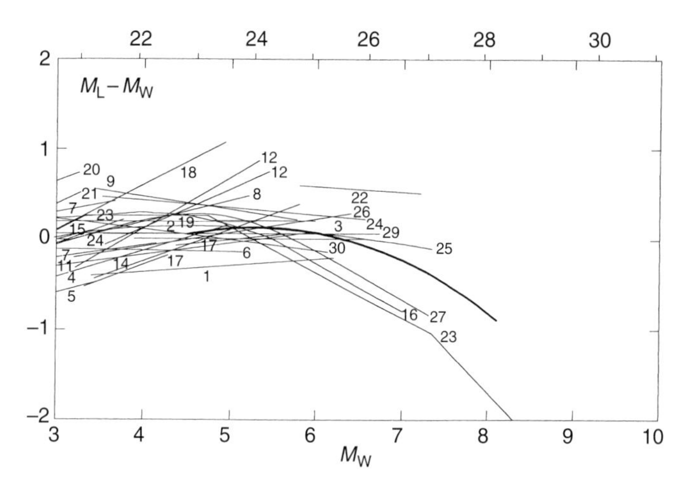

FIGURE 21 Relation between M_W and M_L. 1: Wyss and Brune (1968), 2: Aki (1969), 3: Thatcher and Hanks (1973), 4: Johnson and McEvilly (1974), 5: Spottiswoode and McGarr (1975), 6: Bakun and Bufe (1975), 7: Bakun *et al.* (1976), 8: Bakun and Lindh (1977), 9: Vered (1978), 10: Peppin and Bufe (1980), 11: Chung and Bernreuter (1981), 12: Archuleta *et al.* (1982), 13: Bungum *et al.* (1982), 14: Bolt and Herraiz (1983), 15: Bakun (1984), 16: Hanks and Boore (1984), 17: Fletcher *et al.* (1984), 18: Chávez and Priestley (1985), 19: Console and Rovelli (1985), 20: Zúñiga *et al.* (1988), 21: Sereno *et al.* (1988), 22: Wang *et al.* (1989), 23: Chen and Chen (1989), 24: Vidal and Munguia (1991), 25: Bollinger *et al.* (1993), 26: Shapira and Hofstetter (1993), 27: Fukushima (1996), 28: Nformi *et al.* (1996), 29: Uhrhammer *et al.* (1996), 30: Margaris and Papazachos (1999). Lines given in papers 10, 13, 19, 28 are out of the range in this diagram ($M_W < 3$).

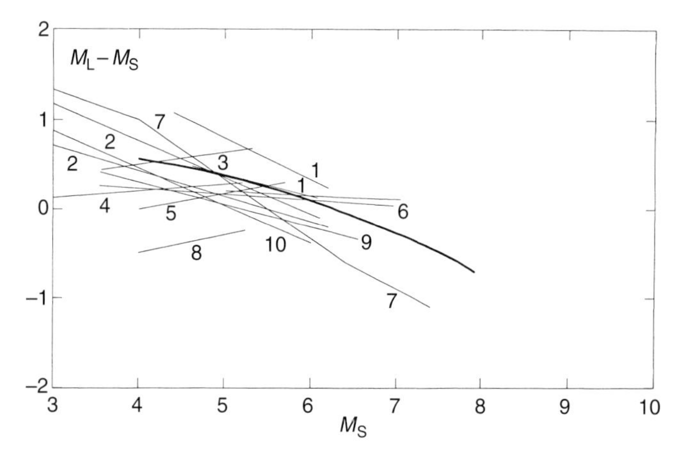

FIGURE 22 Relation between M_S and M_L. 1: Gutenberg and Richter (1956a), 2: Wyss and Brune (1968), 3: Båth (1978, see Båth, 1981), 4: Ambraseys (1985), 5: Bungum *et al.* (1987, see Bungum *et al.*, 1991), 6: Riad *et al.* (1988), 7: Chen and Chen (1989), 8: Alsaker *et al.* (1991), 9: Ambraseys (1990), 10: Papazachos *et al.* (1997).

(Note: Temporal variation of the magnitude relationships is given in the full manuscript on the attached Handbook CD, under the directory \44Utsu3.)

Acknowledgments

I thank David M. Boore and an anonymous reviewer for helpful suggestions.

References

Abe, K. (1979a). *J. Geophys. Res.* **84**, 1561–1568.

Abe, K. (1979b). *Zisin (J. Seismol. Soc. Japan) 2nd Ser.* **32**, 341–353.

Abe, K. (1981a). *Phys. Earth Planet. Inter.* **27**, 72–92.

Abe, K. (1981b). *Phys. Earth Planet. Inter.* **27**, 194–205.

Abe, K. (1982). *J. Phys. Earth* **30**, 321–330.

Abe, K. (1984). *Phys. Earth Planet. Inter.* **34**, 17–23.

Abe, K. (1988). *Bull. Earthq. Res. Inst. Univ. Tokyo* **63**, 289–303.

Abe, K. (1989). *Tectonophysics* **166**, 27–34.

Abe, K. (1994). In: "Global Earth Physics: A Handbook of Physical Constants," pp. 206–213, American Geophysical Union.

Abe, K. and H. Kanamori (1979). *J. Geophys. Res.* **84**, 3589–3595.

Abe, K. and H. Kanamori (1980). *Tectonophysics* **62**, 191–203.

Abe, K. and S. Noguchi (1983a). *Phys. Earth Planet. Inter.* **32**, 45–59.

Abe, K. and S. Noguchi (1983b). *Phys. Earth Planet. Inter.* **33**, 1–11.

Aki, K. (1967). *J. Geophys. Res.* **72**, 1217–1231.

Aki, K. (1969). *J. Geophys. Res.* **74**, 615–631.

Aki, K. (1972). *Geophys. J. R. Astron. Soc.* **31**, 3–25.

Alsaker, A., *et al.* (1991). *Bull. Seismol. Soc. Am.* **81**, 379–398.

Ambraseys, N.N. (1985). *Earthq. Eng. Struct. Dyn.* **13**, 307–320.

Ambraseys, N.N. (1990). *Earthq. Eng. Struct. Dyn.* **19**, 1–20.

Archuleta, R.J., *et al.* (1982). *J. Geophys. Res.* **87**, 4595–4607.

Bakun, W.H. (1984). *Bull. Seismol. Soc. Am.* **74**, 439–458.

Bakun, W.H. and C.G. Bufe (1975). *Bull. Seismol. Soc. Am.* **65**, 439–459.

Bakun, W.H. and W.B. Joyner (1984). *Bull. Seismol. Soc. Am.* **74**, 1827–1843.

Bakun, W.H. and A.G. Lindh (1977). *Bull. Seismol. Soc. Am.* **67**, 615–629.

Bakun, W.H., *et al.* (1976). *Bull. Seismol. Soc. Am.* **66**, 363–384.

Basham, P. (1969). *Geophys. J. R. Astron. Soc.* **17**, 1–13.

Båth, M. (1956). *Trav. Sci. Pub. BCIS* **A19**, 5–93.

Båth, M. (1966). *Phys. Chem. Earth* **7**, 115–165.

Båth, M. (1977). *Ann. Geofis.* **30**, 299–327.

Båth, M. (1981). *Earth-Sci. Rev.* **17**, 315–398.

Bergman, E. A. (1986). *Tectonophysics* **132**, 1–35.

Bloom, E.D. and R.C. Erdman (1979). *Bull. Seismol. Soc. Am.* **69**, 2085–2099.

Bollinger, G.A., *et al.* (1993). *Bull. Seismol. Soc. Am.* **83**, 1064–1080.

Bolt, B.A. and M. Herraiz (1983). *Bull. Seismol. Soc. Am.* **73**, 735–748.

Bonamassa, O. and A. Rovelli (1986). *Bull. Seismol. Soc. Am.* **76**, 579–581.

Boore, D.M. (1989). *Tectonophysics* **166**, 1–14.

Brune, J.N. (1968). *J. Geophys. Res.* **73**, 777–784.

Bungum, H. and D. Tjøstheim (1976). *Geophys. J.* **45**, 371–392.

Bungum, H., *et al.* (1982). *Bull. Seismol. Soc. Am.* **72**, 197–206.
Bungum, H., *et al.* (1991). *J. Geophys. Res.* **96**, 2249–2265.
Burton, P.W., *et al.* (1984). *Geophys. J. R. Astron. Soc.* **78**, 475–506.
Carter, J.A. and E. Berg (1981). *Tectonophysics* **76**, 257–271.
Chávez, D.E. and K.F. Priestley (1985). *Bull. Seismol. Soc. Am.* **75**, 1583–1589.
Chen, P.-S. and H.-T. Chen (1989). *Tectonophysics* **166**, 53–72.
Christoskov, L., *et al.* (1985). *Tectonophysics* **118**, 213–226.
Chung, D.H. and D. Bernreuter (1981). *Rev. Geophys. Space Phys.* **19**, 649–663.
Console, R. and A. Rovelli (1985). *Tectonophysics* **118**, 329–338
Di Filoppo, D. and L. Marcelli (1959). *Trav. Sci. Pub. BCIS* **A20**, 17–30.
Duda, S.J. (1965). *Tectonophysics* **2**, 409–452.
Dwyer, J.J., *et al.* (1983). *Bull. Seismol. Soc. Am.* **73**, 781–796.
Dziewonski, A.M. and J.H. Woodhouse (1983). *J. Geophys. Res.* **88**, 3247–3271.
Eaton, J.P. (1992). *Bull. Seismol. Soc. Am.* **82**, 533–579.
Ebel, J.E. (1982). *Bull. Seismol. Soc. Am.* **72**, 1367–1378.
Ekström, G. and A.M. Dziewonski (1988). *Nature* **332**, 319–323.
Fletcher, J., *et al.* (1984). *Bull. Seismol. Soc. Am.* **74**, 1101–1123.
Fukushima, Y. (1996). *Bull. Seismol. Soc. Am.* **86**, 329–336.
Geller, R.J. (1976). *Bull. Seismol. Soc. Am.* **66**, 1501–1523.
Geller, R.J. and H. Kanamori (1977). *Bull. Seismol. Soc. Am.* **67**, 587–598.
Giardini, D. (1984). *Geophys. J. R. Astron. Soc.* **77**, 883–914.
Giardini, D. (1988). *J. Geophys. Res.* **93**, 2095–2105.
Giardini, D., *et al.* (1997). *J. Seismol.* **1**, 161–180.
Greenhalgh, S.A. and R. Singh (1986). *Bull. Seismol. Soc. Am.* **76**, 757–769.
Gupta, H.K. and B.K. Rastogi (1972). *Geophys. J. R. Astron. Soc.* **28**, 65–89.
Gupta, H.K., *et al.* (1972). *Bull. Seismol. Soc. Am.* **62**, 509–517.
Gusev, A.A. (1991). *Pure Appl. Geophys.* **136**, 515–527.
Gutenberg, B. (1945a). *Bull. Seismol. Soc. Am.* **35**, 3–12.
Gutenberg, B. (1945b). *Bull. Seismol. Soc. Am.* **35**, 57–69.
Gutenberg, B. (1945c). *Bull. Seismol. Soc. Am.* **35**, 117–130.
Gutenberg, B. and C.F. Richter (1949). "Seismicity of the Earth and Associated Phenomena." Princeton University Press, Princeton.
Gutenberg, B. and C.F. Richter (1954). "Seismicity of the Earth and Associated Phenomena." 2nd edn, Princeton University Press, Princeton.
Gutenberg, B. and C.F. Richter (1956a). *Bull. Seismol. Soc. Am.* **46**, 105–145.
Gutenberg, B. and C.F. Richter (1956b). *Ann. Geofis.* **9**, 1–15.
Hadley, D.M., *et al.* (1982). *Bull. Seismol. Soc. Am.* **72**, 959–979.
Haines, A.J. (1981). *Bull. Seismol. Soc. Am.* **75**, 275–294.
Haines, A.J. (1983). *Tectonophysics* **93**, 245–248.
Hanks, T.C. and D.M. Boore (1984). *J. Geophys. Res.* **89**, 6229–6235.
Hanks, T.C. and H. Kanamori (1979). *J. Geophys. Res.* **84**, 2348–2350.
Hatzidimitriou, P., *et al.* (1993). *Tectonophysics* **217**, 243–253.
Heaton, T.H., *et al.* (1986). *Surv. Geophys.* **8**, 25–83.
Herak, M. and D. Herak (1993). *Bull. Seismol. Soc. Am.* **83**, 1881–1892.
Herrmann, R.B. (1980). *Bull. Seismol. Soc. Am.* **70**, 447–468.
Houston, H. and H. Kanamori (1986). *Bull. Seismol. Soc. Am.* **76**, 19–42.
Hutton, L.K. and D.M. Boore (1987). *Bull. Seismol. Soc. Am.* **77**, 2074–2094.
Ichikawa, M. (1966). *Zisin (J. Seismol. Soc. Japan) 2nd Ser.* **19**, 280–282.
Jennings, P.G. and H. Kanamori (1979). *Bull. Seismol. Soc. Am.* **69**, 1267–1288.
Jennings, P.G. and H. Kanamori (1983). *Bull. Seismol. Soc. Am.* **73**, 265–280.
Johnson, L.R. and T.V. McEvilly (1974). *Bull. Seismol. Soc. Am.* **64**, 1855–1886.
Johnston, A.C. (1996). *Geophys. J. Int.* **124**, 381–414.
Johnston, A.C. and S. Halchuk (1993). *Ann. Geofis.* **36** (No.3–4), 133–151.
Jordan, J.N. and R.N. Hunter (1972). *Geophys. J. R. Astron. Soc.* **27**, 23–28.
Kanamori, H. (1977). *J. Geophys. Res.* **82**, 2981–2987.
Kanamori, H. (1983). *Tectonophysics* **93**, 185–200.
Kanamori, H. and D.L. Anderson (1975). *Bull. Seismol. Soc. Am.* **65**, 1073–1095.
Kanamori, H. and P.C. Jennings (1978). *Bull. Seismol. Soc. Am.* **68**, 475–485.
Kárník, V. (1971). Seismicity of the European Area, Part 2. Academia, Praha.
Kárník, V. (1973). Magnitude differences. *Pure Appl. Geophys.* **103**, 362–369.
Kárník, V., *et al.* (1959). *Trav. Sci. Pub. BCIS* **A20**, 67–80.
Katsumata, A. (1996). *Bull. Seismol. Soc. Am.* **86**, 832–842.
Kawasumi, H. (1943). *Zisin* **15**, 6–12.
Kawasumi, H. (1951). *Bull. Earthq. Res. Inst. Univ. Tokyo* **29**, 469–482.
Kiratzi, A.A. and B.C. Papazachos (1984). *Bull. Seismol. Soc. Am.* **74**, 969–985.
Kiratzi, A.A., *et al.* (1985). *Pure Appl. Geophys.* **123**, 27–41.
Koyama, J. and N. Shimada (1985). *Phys. Earth Planet. Inter.* **40**, 301–308.
Kuge, K. (1992). *Bull. Seismol. Soc. Am.* **82**, 819–835.
Lee, V., *et al.* (1990). *Earthq. Engin. Struct. Dyn.* **16**, 1167–1179.
Lienkaemper, J.J. (1984). *Bull. Seismol. Soc. Am.* **74**, 2357–2378.
Luco, J.E. (1982). *Bull. Seismol. Soc. Am.* **72**, 941–958.
Mahdyiar, M., *et al.* (1986). *Bull. Seismol. Soc. Am.* **76**, 1225–1239.
Main, I.G. and P.W. Burton (1989). *Geophys. J.* **98**, 575–586.
Main, I.G. and P.W. Burton (1990). *Tectonophysics* **179**, 273–285.
Makropoulos, K.C. and P.W. Burton (1981). *Geophys. J.* **65**, 741–762.
Makropoulos, K.C., *et al.* (1989). *Geophys. J. Int.* **98**, 391–394.
Margaris, B.N. and C.B. Papazachos (1999). *Bull. Seismol. Soc. Am.* **89**, 442–455.
Marshall, P.D. (1970). *Geophys. J. R. Astron. Soc.* **20**, 397–416.
Nagamune, T. (1972). *Kenshinjiho (Q. J. Seismol.)* **37**, 1–8.
Nagamune, T., *et al.* (1969). *Kenshinjiho (Q. J. Seismol.)* **32**, 103–115.
Nformi, S., *et al.* (1996). *Geophys. J. Int.* **124**, 289–303.
Noguchi, S. and K. Abe (1977). *Zisin (J. Seismol. Soc. Jpn) 2nd Ser.* **30**, 487–507.
Nowroozi, A.A. (1985). *Bull. Seismol. Soc. Am.* **75**, 1327–1338.
Nuttli, O.W. (1983a). *Tectonophysics* **93**, 207–223.
Nuttli, O.W. (1983b). *Bull. Seismol. Soc. Am.* **73**, 519–535.
Nuttli, O.W. (1985). *Tectonophysics*, **118**, 161–174.
Nuttli, O.W. and S.G. Kim (1975). *Bull. Seismol. Soc. Am.* **65**, 693–709.

Panza, G.F., *et al.* (1989). *Tectonophysics* **166**, 35–43.
Papazachos, B.C., *et al.* (1997). *Bull. Seismol. Soc. Am.* **87**, 474–483.
Papazachos, C.B. and A.A. Kiratzi (1992). *Geophys. J. Int.* **111**, 424–432.
Pasechnik, I.P., *et al.* (1970). *Izv. Earth Phys.* 19–24.
Patton, H.J. and W.R. Walter (1993). *Geophys. Res. Lett.* **20**, 277–280.
Patton, H.J. and W.R. Walter (1994). *Geophys. Res. Lett.* **21**, 743.
Peppin, W.A. and C.G. Bufe (1980). *Bull. Seismol. Soc. Am.* **70**. 269–281.
Pérez, O.J. (1999). *Bull. Seismol. Soc. Am.* **89**, 335–341.
Prozorov, A. and J.A. Hudson (1974). *Geophys. J. R. Astron. Soc.* **39**, 551–564.
Prozorov, A., *et al.* (1983). *Geophys. J. R. Astron. Soc.* **73**, 1–16.
Purcaru, G. and H. Berckhemer (1978). *Tectonophysics* **49**, 189–198.
Reichle, M.S., *et al.* (1976). *Bull. Seismol. Soc. Am.* **66**, 1623–1641.
Rezapour, M. and R.G. Pearce (1998). *Bull. Seismol. Soc. Am.* **88**, 43–61.
Riad, S., *et al.* (1988). In: "Historical Seismograms and Earthquakes of the World," pp. 321–334. Academic Press.
Richter, C.F. (1935). *Bull. Seismol. Soc. Am.* **25**, 1–32.
Rogers, A.M., *et al.* (1987). *J. Geophys. Res.* **92**, 3527–3540.
Rothé, J.P. (1967). "The Seismicity of the Earth, 1953–1965." UNESCO.
Savage, M.K. and J.G. Anderson (1995). *Bull. Seismol. Soc. Am.* **85**, 1236–1243.
Sereno, T.J., *et al.* (1988). *J. Geophys. Res.* **93**, 2019–2035.
Shapira, A. and A. Hofstetter (1993). *Tectonophysics* **217**, 217–226.
Shin, T.-C. (1993). *TAO* **4**, 155–170.
Singh, S.K. and J. Havskov (1980). *Bull. Seismol. Soc. Am.* **70**. 379–383.
Singh, S.K., *et al.* (1983). *Bull. Seismol. Soc. Am.* **73**, 1779–1796.
Solov'ev, S.L. and O.N. Solov'eva (1968). *Stud. Geophys. Geod.* **12**, 179–191.
Spottiswoode, S.M. and A. McGarr (1975). *Bull. Seismol. Soc. Am.* **65**, 93–112.
Stelzner, J. (1961). *Gerlands Beitr. Geophys.* **70**, 152–161.
Street, R.L. and F.T. Turcotte (1977). *Bull. Seismol. Soc. Am.* **67**, 599–614.
Takeo, M. and K. Abe (1981). *Zisin (J. Seismol. Soc. Jpn) 2nd Ser.* **34**, 495–504.
Thatcher, W. and T.C. Hanks (1973). *J. Geophys. Res.* **78**, 8547–8576.
Thomas, J.H., *et al.* (1978). *Geophys. J. R. Astron. Soc.* **53**, 191–200.
Udías, A. (1971). *Geophys. J. R. Astron. Soc.* **22**, 353–376.
Uhrhammer, R.A. and E.R. Collins (1990). *Bull. Seismol. Soc. Am.* **80**, 702–717.
Uhrhammer, R.A., *et al.* (1996). *Bull. Seismol. Soc. Am.* **86**, 1314–1330.
Uski, M. and A. Tuppurainen (1996). *Tectonophysics* **261**, 23–37.
Utsu, T. (1979). *Bull. Earthq. Res. Inst. Univ. Tokyo* **54**, 253–308.
Utsu, T. (1982a). *Bull. Earthq. Res. Inst. Univ. Tokyo* **57**, 401–463.
Utsu, T. (1982b). *Bull. Earthq. Res. Inst. Univ. Tokyo* **57**, 465–497.
Utsu, T. (1985). *Bull. Earthq. Res. Inst. Univ. Tokyo* **60**, 639–642.
Utsu, T. (1988). In: "Historical Seismograms and Earthquakes of the World," pp. 150–161, Academic Press.
Vaněk, J., *et al.* (1962). *Izv. Acad. Sci. USSR Geophys. Ser.* 108–111.
Vered, M. (1978). *Israel J. Earth Sci.* **27**, 82–84.
Vidal, A. and L. Munguia (1991). *Bull. Seismol. Soc. Am.* **81**, 2254–2267.
Wang, J.-H., *et al.* (1989). *Tectonophysics* **166**, 15–26.
Wu, J.Y. and X.F. Cao (1987). *Acta Seismol. Sin.* **9**, 225–239.
Wyss, M. and J.N. Brune (1968). *J. Geophys. Res.* **73**, 4681–4694.
Wyss, M. and R.E. Habermann (1982). *Bull. Seismol. Soc. Am.* **72**, 1651–1662.
Zúñiga, F.R., *et al.* (1988). *Bull. Seismol. Soc. Am.* **78**, 370–373.

Editor's Note

Due to space limitations, the full manuscript including additional sections and figures, and references with full citation are given in Utsu3FullManuscript.pdf on the attached Handbook CD, under the directory \44Utsu3.

45

Historical Seismicity and Tectonics: The Case of the Eastern Mediterranean and the Middle East

N.N. Ambraseys
Imperial College, London, United Kingdom
J.A. Jackson
University of Cambridge, Cambridge, United Kingdom
C.P. Melville
University of Cambridge, Cambridge, United Kingdom

1. Introduction

The purpose of this chapter is to describe how historical evidence can be used to address some fundamental questions: when and where have earthquakes happened in the past? Is the instrumental record of this century a guide to past seismicity, and sufficient to anticipate what might be expected in the future? How can accounts of ancient events contribute to our scientific understanding of earthquake activity? The discussion is drawn primarily from our studies of the seismicity and tectonics of the Eastern Mediterranean region and the Middle East (Fig. 1), and is based on our experience of research in this area over the last three decades. In this time, we have observed many problems in the use of historical evidence, but also considerable progress across the field. Our aim is to present a constructive statement of our views on a range of general points, rather than to engage in a detailed critique of individual cases.

As we cannot know what will happen in the future, to estimate likely earthquake hazards we have to find out what happened in the past and extrapolate from there. Previous research has uncovered evidence of destructive earthquakes in areas where only small events have been experienced recently (see for an early statement of this, Ambraseys, 1971). This is not surprising; the timescale of geology is vastly different from that of human history, so some areas will suffer a short period of violent earthquakes only once in a few hundred years. It follows that if we took account only of information about the last century, in which earthquakes have been recorded by instruments (and even then not uniformly throughout the globe), we would have no way of knowing whether an apparently "quiet" area is in fact at risk from a damaging earthquake. The use of the historical record is invaluable, not only in the study of earthquakes but also of the climate and weather, and can guide the engineer to design

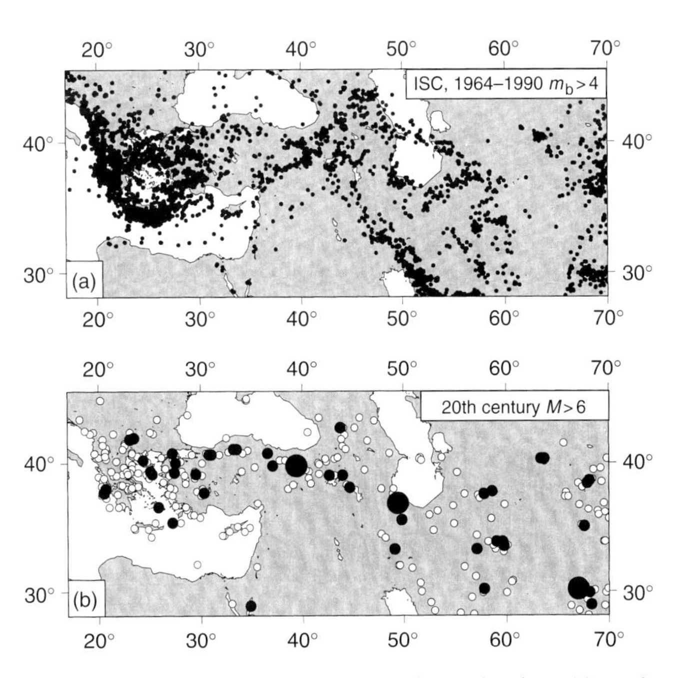

FIGURE 1 The area of our investigations, showing: (a) earthquakes of $m_b > 4$ during the period 1964–1990; (b) the distribution of significant shallow earthquakes this century. Open circles are M_S 6–6.9, solid circles are M_S 7.0 or greater.

INTERNATIONAL HANDBOOK OF EARTHQUAKE AND ENGINEERING SEISMOLOGY, VOLUME 81A ISBN: 0-12-440652-1

structures to resist the forces of nature without being taken by surprise by unanticipated events.

Most of our information comes from local historians and chroniclers. For the better-documented regions where ancient civilizations and developed cultures have flourished, earthquake investigations can go back as far as the last 2500 y or so. Clearly, if such literary sources, which are written in both dead and living languages, are to be useful to modern science, the evidence they provide must be subjected to a rigorous critical analysis, informed by an awareness of the nature of the evidence they provide and of the context in which they were written. If these sources tell us enough about past earthquakes, we can, using modern techniques, estimate the size of these events and their likely effects should they happen again. In the same way, identifying the time intervals between destructive earthquakes in a certain place can help us to establish a pattern for their occurrence and the long-term seismicity in that area.

The use of long-term information, which can best be assessed from an interdisciplinary study, gives a far fuller understanding of earthquake hazard, because it is based on human experience of earthquakes over a much greater segment of the geological timescale, up to 2500 y or more in favorable conditions, than the mere 100 y of instrumental records. Fortunately, the region under consideration here is one of those that lends itself to such a long-term perspective. Similar work has been undertaken for other parts of Asia, such as China and Japan (see Gu Gongxu *et al.*, 1983; Utsu, 1990) and Europe (e.g., Alexandre, 1990; Stucchi, 1993; Albini and Moroni, 1994). Such studies are essential in planning appropriate social and economic development to minimize the risk of unacceptable human and economic losses in areas of high seismicity.

2. Retrieval and Critical Review of Source Data

Historical information can be used to assess earthquake hazard, i.e., the frequency of occurrence of past earthquakes, in terms of their size and location, two of the most important factors in describing seismicity. But the results of historical studies will only be of value to engineers and earth scientists when the data are converted into numbers representing the epicentral location and magnitude of the events concerned, accompanied by an estimate of the reliability of the assessment (for discussions of the appropriate methodology, see, e.g., Ambraseys *et al.*, 1983; Melville, 1984; Guidoboni and Stucchi, 1993, and Chapter 47 by Guidoboni). It is sometimes helpful to divide the period studied on the basis of the sources at our disposal, as these naturally influence the quality of the conclusions that can be reached. The main division comes around the 15th century AD, obviously subject to different subdivisions in different parts of the region.

2.1 Early Period

The sources available for such a study in our area of research are of three main types: literary, epigraphic, and archaeological (cf., Guidoboni *et al.*, 1994; Ambraseys and White, 1997).

2.1.1 Literary Sources

So far as European sources are concerned, in contrast with the period after around 1400, for Classical, Roman, and Byzantine times almost all the sources are well known, relatively limited in number and mostly published. This makes it feasible to reexamine the original sources, rather than rely exclusively on modern catalogers, who may have been working from a variety of different standpoints (see below). The result should be a body of data that is homogeneous and as complete as possible, free of duplications and exaggerations, and therefore suitable for assessing seismicity with an accuracy adequate for scientific and applied purposes.

The main Arabic historical sources too, although relatively numerous in the "classical" age of Islamic civilization, that is from approximately the 8th to the 13th century AD, have generally been identified and published. These are for the most part narrative histories, usually arranged in annals, which report events in a precise chronological framework. The most significant events are frequently recorded, with characteristic details, by several authors. Critical comparisons of the various accounts are normally sufficient to identify and resolve small inconsistencies in dating, which have often found their way into modern earthquake catalogs (see, for one example, Ambraseys and Melville, 1988). Little or no archival material survives from this early period. Many works known to have been composed have not been discovered, but such information as they may have contained about earthquakes has very probably survived in the work of later annalists.

From the point of view of the seismologist or engineer, the principal justification for returning to the primary sources is to improve on previous interpretations, by adopting a consistent and systematic approach to all the pertinent earthquake reports. In so doing, of course, one is not simply looking to verify the information in existing catalogs. Applying a thorough knowledge of the material available for such studies, and of the history of the region investigated (which requires the appropriate specialist skills), has invariably allowed an enormous increase in data, not only for known but also for previously unknown events, in all properly conducted investigations over the last 20 y or so.

Naturally, for the analysis of seismotectonics, the prime purpose is not so much to investigate the historical implications of earthquakes for the social, political, and economic life of past centuries. Thus, historiographic and linguistic problems are relevant only when they have a direct bearing on the understanding of the earthquake(s) being described, for example by revealing any bias or unreliabilty in the author concerned, the quality of his information and the use he made of the sources available to him. Purely historiographical research, interesting though it is, is ancilliary in this context and is not the main end in itself.

2.1.2 Inscriptions

For some of the earliest events, information comes from inscriptions which explicitly mention earthquake destruction or extensive repairs after an earthquake. Epigraphic material may also refer to remission of tribute or taxes following an earthquake. The practice of inscribing such public proclamations on the walls of mosques in Iran, for example, is attested in the late 17th century and doubtless continued beyond that time, though no examples involving earthquakes are known. Since they are almost always contemporary, inscriptions provide valuable and indisputable evidence for the location and, quite often, the effects of earthquakes which, either because of the remoteness of the site, or for other reasons, are not recorded in literary sources. Following the same principle for epigraphy as for literary sources, excursions into linguistic or literary questions are only useful when they contribute to our understanding of the earthquake in question.

2.1.3 Archaeological Evidence

Great caution is needed in using archaeological data to locate and in particular to date earthquakes. Dating earthquake damage is frequently based on, or influenced by, literary sources (as filtered through earthquake catalogs) rather than any precise internal archaeological indicators. Stratigraphy alone, without some hard evidence such as coin finds, is insufficient to establish a chronological association between events that might have been entirely unrelated. The key point here is that the documentary record is grossly incomplete and numerous destructive earthquakes undoubtedly occurred of which no written description survives (see below). Nevertheless, in the past this has developed into a circular process, whereby the conclusions of archaeologists have been taken as factual evidence of earthquake activity and then used in turn by earthquake catalogers to confirm the dates of their entries.

Furthermore, the archaeological evidence for an earthquake is not always clear or unambiguous. Displaced, leaning, damaged, or collapsed walls in an excavation or in extant historical monuments are often explained by archaeologists as being due to an earthquake, invoked as a *deus ex machina* (cf. di Vita, 1995). But they can be due to other, nonseismic, causes such as differential settlements, particularly leaching or weathering of the foundation materials over the ages. This process of deterioration may indeed be assisted by occasional earthquakes, particularly when these structures have been rendered more vulnerable by acts of warfare. Random or coursed rubble masonry walls laid in clay mortar may deform excessively with time, even without the help of earthquakes, and originally badly aligned polygonal and rectangular dry walls can fracture owing to small differential settlements of their lower courses, giving the impression of structural failure as a result of lateral or vertical inertia loading in an earthquake. Deliberate damage and military operations can also leave effects that may be misinterpreted. Furthermore, damage can be the result, perhaps cumulative, of more than one earthquake, even a long time after the abandonment of the site.

Much is often made of evidence from excavations and from the remains of historical monuments, that the ground motion at a given site during an earthquake was principally in the opposite direction to that of the collapse of free-standing walls and colonnades. This evidence is used to determine the attitude of postulated nearby surface faulting. But structures of the Classical, Roman, and early Byzantine period cannot fall easily in any direction other than that of their weaker axis. For colonnades, this is usually at about right angles to the long axis of the cella; for houses, perpendicular to the axis of the course of streets they face, regardless of whether there is a suspicion of seismic origin in their collapse or not. Prostrated colonades have been found in excavations not only near, but also far away from active faults in areas of both high and very low seismicity, such as for instance in Hermopolis Magna in Egypt. This mode of collapse has little to do with "the direction from which the shock came," a notion that has little engineering content. Also, the belief that the direction of fall is parallel with the direction of the near-field ground motion due to strike-slip surface faulting in an earthquake is not necessarily correct.

In recent years, considerable progress has been made in "archaeoseismology" and there is a greater awareness of the need for collaboration between archaeologists, geologists, and workers in other disciplines to evaluate the traces of earthquakes in excavations, both for understanding their effects at the site and for the information they can provide about the nature of the earthquake implicated (see, e.g., Guidoboni and Bianchi, 1995; Guidoboni, 1996; Stiros, 1996).

2.2 Later Period

For more recent periods, and for most areas, information becomes fuller and the sources of historical data more numerous as we approach modern times. Partly, this is a function of the greater survival rate of relevant documents. In the Middle East, the accumulation of material in conditions of relative political stability in great Ottoman imperial centers such as Cairo, Damascus, or Istanbul, greatly extends the opportunities for retrieval of "new" information from old records. Partly, also, it reflects the increased production of written material and, in the European context, the growth in literacy and secular learning associated particularly with the Renaissance. Commensurate with this is a broadening of the range of sources that may preserve accounts or details of earthquake activity.

Chronicles and annals remain the preponderant source of such data in Arabic works, in some areas (such as the Yemen) even well into the 20th century, supplemented only occasionally by biographical, geographical, or topographical works. By contrast, European writings provide an ever-growing volume and range of data. Compilation of chronicles gives way to antiquarian study, travel literature, private diaries, personal letters and official archives, including diplomatic correspondence. By the 18th century in Europe and the 19th century in the Middle East, newspapers (the modern equivalent of the annals of old, in their indiscriminate reportage of ephemera, trivia, and the sensational alongside matters of serious material or moral concern) provide an accurately dated and reasonably full record of the newsworthy event (as differently perceived in different places and at different times). It is symptomatic of cultural changes since the World War I that, as instrumental, electronic or other mechanical reporting of events has grown, and news is increasingly disseminated by radio and television, a parallel decline is visible in both the volume and the quality of documentary and descriptive accounts of earthquakes in the 20th century.

Although such information allows considerable gains to be made, much material is still unpublished, often difficult to access and hard to read. There is no doubt that much remains to be discovered in Middle Eastern archive collections, particularly in Turkey and Egypt. Such work is time-consuming and presents serious hurdles (for a pertinent discussion, see Vogt, 1993). Nevertheless, pursuit of such data in recent decades has spawned a new generation of earthquake catalogs and studies of regional seismicity.

3. Existing Catalogs

Descriptive and parametric earthquakes catalogs are often accepted at face value by earth scientists and engineers and used to assess earthquake hazard. However, their results will necessarily only be as good as the input data. The reliability of these catalogs varies considerably and for the early period it can be very low (see Alexandre, 1990; Vogt 1993, 1996, with reference particularly to Europe, and discussion in Chapter 47 by Guidoboni).

There are many earthquake catalogs that cover the historical (pre-1900) period worldwide, but many of them are not well known outside Europe. Some of the main ones are therefore described briefly in Appendix 1.

There are many more parametric catalogs, which have appeared during the last two decades of electronic information. Though much useful work has been done, many of them are based largely on earlier catalogs and have the relatively modest aim of bringing previous lists up to date. In the process, it is easy for old and new copying errors to be introduced, which slowly accumulate and proliferate in various data-storage systems. If the limitations of these works are not fully appreciated, their use can create more problems than they resolve (cf. Guidoboni and Stucchi, 1993: 203).

The main touchstone for the value of any earthquake catalog is the extent to which it attempts to review primary source material and distil evidence of historical seismicity into a critical parametric form. Some of these modern catalogs are discussed in Appendix 2. There is insufficient space here to evaluate the aims and methods of each one, but earth scientists or engineers wishing to consult them should be aware of the need to judge the principles according to which they have been compiled and ensure that they are suitable for their purposes. It is best not to take any such material purely at face value; even the most rigorous catalogs are putting forward one interpretation of the available historical evidence and others are possible. Ideally, therefore, even the best modern catalogs are a starting point for an evaluation of hazard or seismotectonics in a particular area, appropriate to the specific concerns of the investigation.

As an example of the caution needed, Ambraseys and White (1996) found that by comparing original sources with those in earthquake catalogs used by historians and earth scientists for the whole period BC, of the 153 earthquakes listed, only 60% could be confirmed in the sources and the rest were more or less spurious. These figures do not much differ from those found by Alexandre (1990) for western European earthquakes for the period 394–1259 AD, where the percentage of spurious entries in catalogues varies between 50 and 75%. Similarly, in the case of Great Britain, of the 60 entries before 1620 listed in Davison's (1924) catalog, 25 (42%) can be shown to be either spurious or inaccurate. Of the 84 events in our study region listed for the last 500 y BC, 52 are dubious, leaving on average about six genuine historical earthquakes of all sizes per century. This is in contrast with 188 earthquakes of magnitude equal to or greater than 6.0 that have occurred in the same area during the last 100 y (Ambraseys and White, 1997).

Even without assuming an identical level of seismic activity throughout this long period, it is obvious that the number of earthquakes that can be retrieved from historical sources for the earliest historical past is minute, perhaps 3 out of 100, and demonstrates how many damaging earthquakes are missing from the literary record. This implies that the archeologist should not presume that what is identified as earthquake damage in an excavation is the effect of one of these very few known events, however well attested in the sources, rather than one of the many shocks missing from the record. The attractive and tidy tendency to assign damage to a known earthquake reminds one of Kaplan's parable of the drunkard searching under a street lamp for his house key, which he had dropped some distance away, but he searches there because there is more light.

For later periods, the situation improves, but many lists perpetuate "virus" events which are transmitted from catalog to catalog. A sample of inaccurate or spurious entries in lists for Egypt and the Red Sea region is given by Ambraseys *et al.* (1994: 107–110). Although many corrections lead to only minor improvements in accuracy, it is inevitable that the largest events are the most frequently duplicated, because they are the most widely reported, and the duplication of such large earthquakes creates the most serious distortion of the record of seismotectonic activity. The problem is not the creation of modern authors, but is frequently inherent in the historical sources themselves. Sites may have been damaged or destroyed by separate events which occurred in the same week, month, or year, and for the very early period even in the same century, but not differentiated in contemporary accounts.

One instance of the various levels of problems that can surround a single event is provided by the large earthquake of 18 March 1068 in the Hejaz, which was reported to have caused damage between Jerusalem and Damascus, over 500 km from the epicentral region. In fact, it is probable that two separate events were involved. In addition to being two events reported as one, this large earthquake is duplicated under several dates in contemporary and secondary sources (see Ambraseys and Melville, 1989; Ambraseys *et al.*, 1994: 30–32). It is only recently that the reassessment of such early events is being undertaken in individual case studies or in the process of producing regional catalogs. Thus it is not surprising that many of the existing descriptive and parametric listings of past events attract rather negative criticisms in any critical review of historical seismicity (Vogt, 1996).

We may mention here four of the earliest earthquakes to which modern catalogers invariably give a cosmic dimension, the primary sources for which hardly support such an interpretation. The destruction of Sodom and Gomorrah in the Bible is probably associated with a distant and hazy recollection of ancient earthquakes in the Jordan Rift Valley sometime between 2100 and 1700 BC. The alleged earthquake destroyed Sodom and Gomorrah but not the nearby town of Zoar (Bible, Genesis: xix.24–30). According to a first-century AD source (Strabo: vii.297), the local inhabitants in his time said that "... there were once thirteen towns in the region of which Sodom was the metropolis, but that a circuit of about 60 stadia (11 km) of that town escaped unharmed; and that by reason of earthquakes and eruptions of fire ... the lake burst its bounds ... and, as for the towns, some were swallowed up and others were abandoned by such as were able to ecape ...." Sodom was probably located in the Lisan peninsula, on the east coast of the Dead Sea, not far from Zoar, which would appear from archaeological evidence to be in the vicinity of Bad el-Dra. The location of Gomorrah is not known with certainty. The earthquake should have been local, not necessarily of large magnitude, since, as we are told, destruction did not extend beyond a circuit of 60 stadia, which is less than 2 km radius. However, it seems that its catastrophic effects on these two towns were probably due to the massive liquefaction and slumping of the loose silty clays on which they were built (Harris and Beardow, 1995). It is difficult to see on what grounds this event is classed among the major events in this region (Ben-Menahem, 1991).

The second earthquake is another allegedly great event in Jericho in the Dead Sea Rift sometime between 1500 and 1200 BC. But the collapse of the walls of Jericho is not attributed by the Bible narrative, the only literary source that mentions the event, to a physical agency. Archaeological evidence (Kenyon, 1978) shows that between 1500 and 1200, Jericho had been damaged by one or more earthquakes, but there is no indication as to the date of these events, which may have nothing to do with the narrative in the Bible and which may even have happened since the abandonment of the site. The association of the event mentioned in the Bible with the damage preserved in the ruins of Jericho and in other sites in the region as well as with the damming of the Jordan river at Al Damieh is purely conjectural (see also, Nur and Ron, 1996).

A third catastrophic earthquake is alleged to have taken place circa 1365 BC in the Eastern Mediterranean. The evidence is a tablet from Tel Amarna in Egypt containing a report of the King of Tyre on the situation in Syria sent to Amenhoteph-IV (1353–36) which says that "... Ugarity, the King's town, has been destroyed by fire; half the town has been burnt and the other half is no more ...." The other half of the town was allegedly destroyed by a sea wave, alluded to in other, presumably contemporary, tablets (Dussaud, 1896; Virolleaud, 1935). However, those tablets that refer to the flooding of the land by the sea and also by the rivers, do not necessarily imply that the fire and the flood occurred at the same time, and they do not mention any destruction or association with an earthquake. Archaeological excavations of the Bronze Age site of Ugarit do reflect destruction of the city by earthquakes on several occasions (Schaeffer, 1939), but they provide no indication of seismic activity in this particular instance. It is perhaps significant that other contemporary and near-contemporary tablets that refer to Byblus on the Syrian coast about 160 km to the south of Ugarit, mention no destruction of the city by a sea wave or earthquake (Pomerance, 1970). The available evidence is too tenuous to support the claim put forward by modern writers (Dussaud, 1896; Schaeffer, 1939) that this was a "cosmic" earthquake that marked the end of the Late Bronze Age in this part of the eastern Mediterranean region (Schaeffer, 1948), or that it was the same event as that which destroyed Thera 1000 km away.

Finally, the earthquake of 21 July 365 was a relatively large earthquake, felt throughout the Eastern Mediterranean region, originating somewhere between Crete and the Peloponnese. It is mentioned by many contemporary and later authors, whose accounts, though sometimes confused, permit a reasonable idea of the main effects of the event. In Alexandria, thousands of houses were flooded and people were drowned. It is said that the anniversary of this inundation was commemorated by a yearly festival. The sea wave was reported also from other parts of the region. Along the Peloponnese, the Adriatic, and the Sicilian coasts its effects were more serious than those of the earthquake shaking itself. Indeed, judging by historical and archaeological evidence, ground shaking seems to have caused damage only in southwest Greece, parts of Crete, and in Cyrenaica.

There is no reason to doubt that other regions were affected by this earthquake and its aftershocks, even though it seems improbable they were all destroyed or even damaged. The silence of contemporary and near-contemporary writers on the effects of this earthquake in Greece, Asia Minor, Egypt and in Central North Africa, undermines the case for the global nature of this event. It is relevant to note that not a single town named in historical sources suffered damage or destruction. Of the cities named in near-contemporary texts, Nicaea, Aeropolis and Germe can be shown to have been affected by different earthquakes in 362, 363, and 369 AD, respectively. In Antioch, the shock was felt, apparently without damage, and in Alexandria, as noticed above, it was clearly due to the seismic sea wave; as, indeed, was almost all the reported damage.

The earthquake is used as a point of reference for later writers, who add more sites where the 365 earthquake was allegedly destructive. The evidence is mainly drawn from archaeological observations, but with no clear indication of the simultaneity of the damaging effects in all these places. Inclusion of all these sites would imply an epicentral region extending from Gibraltar to the Dead Sea and from North Italy to northwest Turkey, clearly stretching the size of the event beyond the limits of the possible. Its most damaging effects were confined to a relatively small area, but one implying a magnitude in the upper 7.0s, probably nulceating at some depth, with a fault rupture breaking the surface and causing a substantial sea wave. The shock itself should have been felt across the Eastern Mediterranean, but associated with sporadic damage only to the most vulnerable structures outside this area. Indirect evidence, in need of authentication, suggests that the period 350–550 AD was perhaps one of the most seismically active periods in the region in the last two millennia (Pirazzoli, 1986). This may be so, but the 21 July 365 earthquake was not much different in size and effects from other large shocks in this part of the Eastern Mediterranean, such as those of 1303, 1508, 1609, 1856, and 1926 (see further, e.g., Jacques and Bosquet, 1984; Ambraseys *et al.*, 1994; Guidoboni *et al.*, 1994; Di Vita, 1995).

As in the case of the 365 event, quite often the amalgamation of the effects of two or more different earthquakes stretches the size of the event beyond the limits of the possible. How easily one might report two or more separate earthquakes as one, can be illustrated if we imagine the 20th-century seismicity of, say, Greece being transported back twenty centuries, and recorded in the terse manner typical of that period. The earthquakes of 20 April and 27 April 1894 in Martino and Atalante, for example, occurred so close together in time and space that, if not differentiated clearly in the sources, they could easily have been amalgamated by a later writer into a single major event (Ambraseys and Jackson, 1990). The chances of sources making such a fine distinction in the past, when much information circulated orally and probably without great chronological precision, are slight.

In fact, despite assertions of simultaneity, historical sources rarely give sufficient evidence to establish this beyond doubt. The earthquakes of 1894 in our illustration had surface-wave magnitudes of 6.4 and 6.9, respectively. Their amalgamation, on the other hand, would produce a single major event of macroseismic magnitude well in excess of 7.5. This would seriously mislead the scientist faced with accommodating an earthquake of that magnitude into an area that is tectonically incapable of generating such large events. It is preferable, if in doubt, to report such events as separate earthquakes.

4. Assessment of Intensity Distribution and Isoseoismal Maps

As early as the middle of the 18th century, investigators of earthquakes realized the usefulness of some kind of scale for the measurement of earthquake intensity at a given point. Since then, more than 60 different intensity scales have been devised worldwide, either to express the relative intensity of different earthquakes at a locality, or to trace the variation in intensity of a single earthquake over the affected area in the form of an isoseismal map. The latest scale proposed for Europe is the European Macroseismic Scale 1998 (Grunthal, 1998).

4.1 Intensity

Intensity is a convenient means of conveying (in a single rating of the scale) a measure of the effect of ground motion on artificial structures, and it is a useful parameter. Its assessment, however, in the upper range of the scale is subtle. It is often overlooked that by definition intensity is a rather vague measurement, either of the strength of the earthquake ground motion at a particular point, or the weakness of artificial structures and of the ground itself, or of both. Also, criteria based on secondary effects, such as landslides and faulting, are normally incorporated into the scale. It is impossible to assess how much or how little shaking is required to produce a slide, a type of ground failure that occurs more often without help from an earthquake than with it.

The rating of the upper grades of intensity, which describe damage (say over and above VI or VII), is regionally biased rather than regionally dependent. This is, first, because of regional differences in the geological environment or in the soil profile that may actually exist between different geographical regions or sites. Secondly, there are local, cultural differences in the predominant building stock that is exposed to an earthquake. A third bias lies in the limitations to the application of intensity scales, and the way they are interpreted by the seismologist and engineer, whether experienced or not.

Furthermore, authors of intensity data are seldom explicit about the system they have followed to arrive at their assessments.

Much of the rural and to a lesser extent urban building stock in southern Europe, the Middle East, and Central America is usually relatively old. It has already been weakened by previous earthquakes, by inadequate repairs and by aging, and subsists with an inbuilt vulnerability that is not accounted for in intensity scales. The result is that at intensities VII to VIII the scale "saturates;" that is, all houses are ruined or destroyed and any town or village would thus appear equally, but no more, devastated at so-called higher intensities.

The problem becomes more serious when we consider the assessment of large epicentral intensities, when any scale becomes too subjective and potentially misleading, particularly if it has been designed to describe conditions in a different part of the world, or includes ground effects such a landslides or faulting as diagnostics. A comparison of epicentral intensities assessed independently by different investigators of the same event can show unacceptable differences, up to two grades or more, particularly when intensities are assessed from questionnaires and press reports, without visiting the region, or from the maximum reported intensity and not the mode of the observed estimates. Such a comparison shows how sensitive the method is to the experience of the investigator and the means and time at his or her disposal (see further, Ambraseys and Melville, 1982: 30).

All these problems occur because the description of intensity scale ratings is not precise enough and damage is not consistent enough to do otherwise. In addition, intensities are often assessed by inexperienced observers, "arm-chair" seismologists, and earthquake engineers. It is immaterial, therefore, how precisely one cares to treat the intensity values at one's disposal, since large uncertainties in their original assessment must always exist; in many cases these are so large that their correlation with other measured and more reliably determined earthquake parameters contaminates the final results.

The most recent EMS-1998 intensity scale (Grunthal, 1998) is an attempt to overcome some of these difficulties, but this does not alter the fact that all our existing intensity data come from earlier scales and events. Furthermore, the EMS-1998 scale is too idealized and cumbersome to use in practice. Under the circumstances, one is inclined to sympathize with users of these earlier scales; inasmuch as we are not in a position to be rigorous in our definition of various ratings, then the experienced investigator should have enough leeway to use his or her own judgment without being hemmed in by a scale that is too rigorous or specific.

4.2 Correlation of Intensity with Ground Motions

The practical problem with intensity begins when the engineer, with little hesitation and perhaps with even less understanding of the subtleties involved in its definition and assessment, adopts intensity as a means for the calculation of ground acceleration for design purposes.

In most of the attempts made in the past to correlate intensity with peak ground acceleration, the enormous scatter of data clearly shows, as it should, that there is no physical justification for such a one-to-one relationship (see for instance, Hersberger, 1956; Ambraseys, 1973). The conversion of intensity into peak ground acceleration may be used by the engineer as a last resort, but this must be done at his own risk and with the understanding that such conversion formulas have an enormous scatter and low confidence levels. It is simply not sufficient to derive such correlations without at the same time being explicit about the uncertainties involved.

The prevailing tendency to rely on intensity values to derive empirical relations between intensity and maximum ground acceleration, or between epicentral intensity and magnitude, is rather disappointing. In most scales, sites in the immediate vicinity of a coseismic surface fault-break are assigned epicentral intensities of IX or greater, a value for which empirical formulas predict a magnitude greater than 7.0. In general, the presence or absence of a coseismic surface break is independent of magnitude, though it is rare below about $M=5.5$. It would be preferable not to make assumptions that are not testable nor to advocate on the basis of inadequate physical evidence.

4.3 Isoseismal Maps

Like other empirical measures, intensity, with all the uncertainties in its estimation, when assessed for an earthquake at a large number of points and drawn as an isoseismal map, shows regular distribution patterns and dimensions of the felt area. These indicators have in the past helped seismologists to estimate the size and relative depth of an event.

An isoseismal map provides a measure of the size of the areas over which an earthquake was felt at different intensities; this is an indirect measure of the magnitude or the energy released by the event. The magnitude of an earthquake, therefore, may be assessed from the size of the areas over which a shock was felt, which can then be calibrated against macroseismic information about similar, 20th-century earthquakes for which reliable instrumental magnitudes are available. Such calibration relations may be used to assess the magnitude of historical events for which isoseismal maps can be drawn.

Thus, although intensity at a given point tells the engineer little about ground motion, isoseismal maps, in the construction of which intensity estimates undergo a significant amount of averaging and smoothing of inherent bias, provide a useful tool with which to assess magnitude. Felt areas or average isoseismal radii may be used in calibration relations to estimate magnitude, from which the zero period or spectral ordinates of ground motion can be estimated using modern procedures. In this way ground motions are approached via magnitude and not directly via local Intensity (Howell and Schultz, 1975; Ambraseys, 1985).

It is important, therefore, that isoseismal maps for correlation with magnitude are uniformly drawn and are not regionally biased, which is not always the case. For instance, in the annual US earthquake publications for the period 1929–1969, isoseismal maps were prepared by different staff members of the US Coast and Geodetic Survey from a listing of small towns or suburbs all having the same intensity. The maps were prepared, as was then the custom, by assuming that if in a small town or suburb the intensity varied through a range of values, the highest value was used for the whole area rather than the mode of the observations. In contrast, Russian practice was to use the mode of intensity ratings observed within a particular area.

Regional bias also has an effect on intensity attenuation laws, and is influenced by the personnel responsible for drawing national intensity maps or the agencies involved in their preparation. This becomes apparent with earthquakes affecting more than one country, where individual intensity maps prepared separately for each country frequently do not match across national borders, mismatching by as much as three intensity ratings (Ambraseys and Moinfar, 1988).

5. Assessment of the Size of Early Events

In spite of what has been said earlier about intensities, there is a simple and logical approach to making good use of intensity to assess the size of a shallow earthquake and to discriminate between shallow and intermediate depth events, important considerations in the study of historical events.

The size of a modern or historical earthquake can be assessed reasonably well from the correlation of its "felt radius," that is, the radius of the area within which the shock was felt with a low intensity, say III to IV, with its instrumental surface-wave magnitude. Free-field low intensities are not affected by the regional vulnerability of human-made structures.

Such correlations may be used to assess the surface-wave magnitude of historical events from the size of their felt area. They should be derived from reliable 20th-century macroseismic data and from uniformly assessed surface-wave magnitudes, M_S, using an orthogonal regression technique from which uncertainties in predicted estimates and depth are also determined.

For the preinstrumental period, M_S can be assessed using calibration formulas which can be derived from regional, shallow, 20th-century earthquakes in terms of their radii of isoseismals, r_i, and corresponding intensities, I_i, in the MSK (Medvedev, Sponheuer, Karnik) scale. For our region, the calibration formula used was derived from 123 shallow ($h < 26$ km) earthquakes in the region between 1905 and 1990, and from their corresponding M_S values, which have been recalculated uniformly. The predictive relationship, is

$$M_S = -1.54 + 0.65(I_i) + 0.0029(R_i) + 2.14\log(R_i) + 0.32p \qquad (1)$$

where $R_i = (r_i^2 + 9.7^2)^{0.5}$ and r_i, in kilometers, is the mean isoseismal radius of intensity I_i; p is zero for mean values and one for 84 percentile values.

6. Coseismic Surface Faulting Associated with Historical Events

Evidence for coseismic surface faulting in historical earthquakes in the Eastern Mediterranean and the Middle East is of importance to all modern studies of tectonics and seismicity. Such evidence not only confirms that known tectonic structures are active, but can also identify new ones. Despite shortcomings in the documentary evidence, information about surface faulting can be found in contemporary accounts and this provides a valuable reference point in the paleoseismological record of faults. Such knowledge is particularly important when, for example, the activity of a fault is to be researched by trenching methods, as it allows the completeness of the paleoseismological investigation to be assessed.

Obviously, the most interesting cases of faulting are those that could not have been predicted from the record of 20th-century seismicity alone or, conversely, where surface faulting could be expected but until now is not known to have occurred.

The area we investigated for this reason is within latitude 25 and 45° N and longitude 18 and 70° E (Fig. 2). It comprises the Balkans, Turkey, the Caucasus and the Middle East up to west Pakistan, a region of active tectonics and with a history which is amply, but not uniformly, documented throughout the past two millennia (Ambraseys and Jackson, 1998). Such information is particularly important for studies of active tectonics and for paleoseismology.

We have found 78 cases of faulting pre-1900 and 72 post-1900. Table 1 presents all the cases of pre-1900 coseismic surface faulting known to us at present, which shows that faults which appear to be quiescent today have been active more than once in historical times, as well as a number of hitherto unknown active faults. A few of these cases are also included in the works of Wells and Coppersmith (1994) and Yeates *et al.* (1996). Comparison with the post-1900 record shows that in some cases, a recurrence of historical faulting has not been observed during the instrumental period.

The data are sufficient to allow the derivation of relationships between magnitude and rupture length.

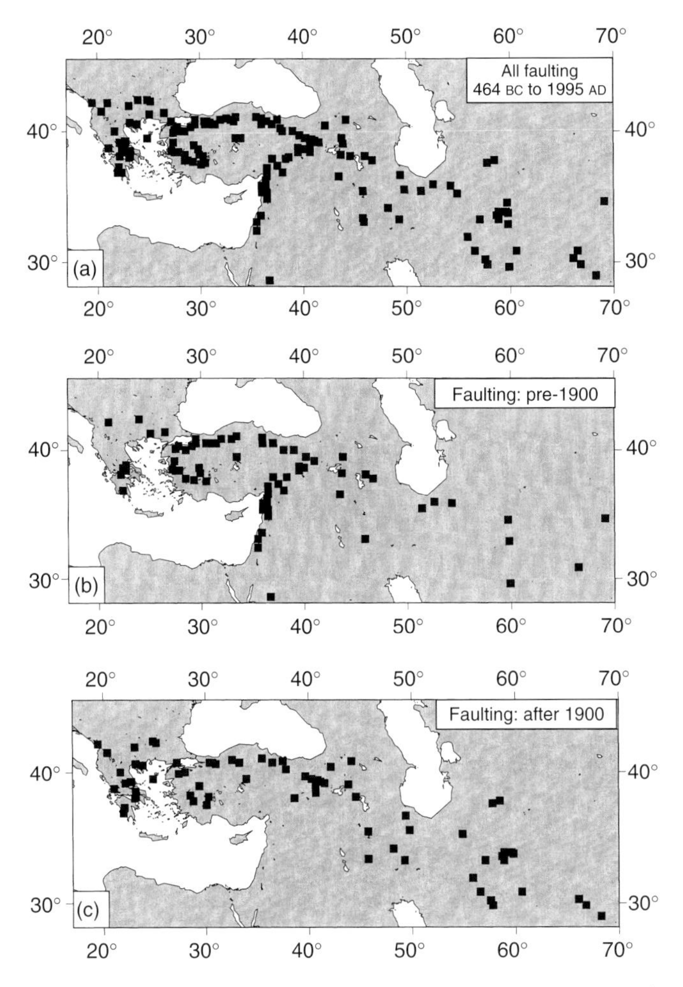

FIGURE 2 (a) The locations of all earthquakes thought to be associated with coseismic surface faulting for the whole period of observations; (b) the locations of pre-20th century earthquakes thought to be associated with coseismic surface faulting; (c) 20th-century earthquakes associated with coseismic surface faulting.

Historical sources record large surface fault ruptures, small ruptures not being spectacular enough to attract attention. Descriptions from which one can deduce faulting are relatively few and hard to verify, particularly when the sources are secondary and the recorded ground deformations are not well described (see, e.g., Ambraseys and Melville, 1995). One of the problems in both early and later descriptions of surface faulting is that one cannot always be certain as to whether ground deformations associated with an earthquake were of tectonic origin or due to landslides, liquefaction, or slumping of the ground. In some cases, ground deformations genuinely of tectonic origin can be identified from the description of ground ruptures which extended continuously or discontinuously along considerable distances, but relative displacements are seldom given for vertical, and never for horizontal, slip. The information which is usually available for this period, therefore, may be classified into three broad categories according to the following criteria.

1. Strong evidence for surface faulting explicitly (A) or implicitly (a) described in the sources. The length of the rupture is rarely given, and only in few cases can it be reckoned from the distances between the localities which it traversed.
2. Surface faulting that is not supported by clear evidence can be inferred from the alignment of a long and narrow epicentral region of a large magnitude earthquake close to or along a known fault. Occasionally the length of a break can be reckoned from the length of the long axis of the epicentral region which contains an assumed rupture. Clearly this would not tell us exactly how far the fault-rupture extended, but would suggest that the shock was probably associated with a surface rupture that can be investigated today in the field. In these cases, historical information will not reveal the exact location and rupture length, but it can help to define the time and the segment of the zone that was probably ruptured.
3. Faulting assumed because of the large size ($M_S > 7.0$) of the associated earthquake and its proximity to a known active fault zone. This category is more tenuous than category 2, but was included to guide further studies. There are many events of $M_S > 7.0$ that might have been associated with faulting, such as those in and around the Marmara Sea area, in Eastern Anatolia and Iran, but these are omitted as their epicentral area is ill defined.

Of these three categories, (1) involves some ruptures which may not previously have been associated with known active or Quaternary faults. Categories (2) and (3) merely date probable breaks of segments of known faults, and help assign size to these events. All these cases indicate recent activity, because the proximity of these earthquakes to known faults was part of the evidence assigning them to these categories.

During the instrumental period, information about both the faulting and the seismological parameters of the associated earthquake improves; there are more detailed field observations and better instrumental data, allowing the uniform reassessment of instrumental M_S magnitudes. However, during the first half of the 20th century, this improvement was very slow and surface faulting continued to be imperfectly reported. Occasionally surface fault ruptures were attributed to landslides and slumping of the ground, and preexisting Quaternary fault scarps were often associated with recent earthquakes.

In this investigation it is important to know the magnitude of the causative earthquake, not only for the development of predictive moment–magnitude relations as a function of the length, slip, and attitude, but also for hazard analysis. For the preinstrumental period, surface-wave magnitudes, M_S, can be assessed from Eq. (1). For the late, preinstrumental period, macroseismic data improve in quality and quantity and this allows the use of Eq. (1) for the assessment of magnitudes.

TABLE 1 List of Earthquakes Associated with Surface Fault Break [See Ambraseys and Jackson (1998) for Details]

No.	Date			Epicentre N	Epicentre E	M_S[a]	Az deg.	Mec[b]	L[c] (km)	H[d] (cm)	V[d] (cm)	Q[e–h]	Location	Ref.[i]
1	−464	–	–	37.0	22.4	M	340	N	20	–	350	Cf	Sparta	GR
2	−426	–	–	38.9	22.7	M	–	–	–	–	–	AAm	Maliac	GGR
3	−280	–	–	35.6	51.4	L	–	–	–	–	–	aA	Sh. Rey	IR
4	17	–	–	38.5	27.8	L	270	N	–	–	–	APf	Gediz R.	TR
5	32	–	–	40.5	31.5	L	080	R	–	–	–	Cf	Gerede	TR
6	37	–	–	36.0	36.0	M	030	L	–	–	–	Cf	Antioch	TR
7	97	–	–	37.3	36.5	M	–	–	–	–	–	C	E. Anatol.	TR
8	110	–	–	39.5	33.5	M	–	–	–	–	–	C	Galatia	TR
9	115	Dec.	13	35.8	36.3	L	010	L	–	–	–	aAf	Oront. R.	SY
10	155	–	–	40.1	27.5	M	100	R	–	–	–	aAf	Manyas	TR
11	181	May	3	40.5	31.0	L	–	–	–	–	–	Bf	Mudurnu	TR
12	236	–	–	40.9	36.0	M	110	R	–	–	–	aPf	Amasya	TR
13	368	Oct.	11	40.5	29.5	L	–	–	–	–	–	Cf	Iznik	TR
14	460	Apr.	7	40.3	27.8	M	060	R	–	–	–	aAf	Manyas	TR
15	499	Sep.	–	40.5	37.0	M	110	R	–	–	–	APf	Niksar	TR
16	518	–	–	42.0	21.0	M	–	–	43	–	–	AP	Macedonia	MC
17	551	–	–	38.5	22.7	M	290	N	–	–	–	APfm	Chaeron	GR
18	554	Aug.	15	40.8	29.5	L	–	–	–	–	–	Bf	Izmit	TR
19	601	Apr.	–	37.0	36.5	–	–	–	–	–	–	aA	E. Anatol.	TR
20	750	–	–	37.0	38.0	–	–	–	–	–	–	aA	Mesopot.	SY
21	856	Dec.	22	36.0	54.3	L	250	T	–	–	–	aAf	Qumis	IR
22	926	Aug.	–	38.5	27.5	M	270	N	–	–	–	aAf	Manisa	TR
23	967	Sep.	–	40.8	32.0	L	080	R	–	–	–	Bf	Gerede	TR
24	995	–	–	38.7	40.0	L	060	L	–	–	–	Cf	Palu	TR
25	1033	Dec.	5	32.5	35.5	L	000	L	–	–	–	Cmf	Jordan	IS
26	1035	May	–	40.8	33.0	M	070	R	–	–	–	aAf	Cerkes	TR
27	1045	Apr.	5	40.0	38.0	L	120	R	–	–	–	aAmf	Erzinc.	TR
28	1050	Aug.	5	41.0	33.5	L	080	L	–	–	–	aAf	Cankiri	TR
29	1068	Mar.	18	28.5	36.7	L	–	–	–	–	–	aA	Hejaz	SA
30	1114	Nov.	29	37.5	37.5	V	040	L	–	–	–	af	Maras	TR
31	1157	Aug.	12	35.0	36.5	V	000	L	–	–	–	af	Hama	SY
32	1170	Jun.	29	35.5	36.5	L	000	L	–	–	–	Cf	Afamiya	SY
33	1202	May	20	33.7	35.9	L	020	L	–	–	–	Bfm	Bekaa	LE
34	1254	Oct.	11	40.0	39.0	L	110	R	150	–	–	APf	Susehri	TR
35	1296	Jul.	17	39.2	27.4	M	050	N	–	–	–	aAf	Soma	TR
36	1336	Oct.	21	34.7	59.7	7.6	155	–	100	–	–	BPf	Khwaf	IR
37	1408	Dec.	29	36.0	36.4	M	010	L	20	–	–	APf	Orontes	SY
38	1419	Mar.	15	40.5	30.5	L	–	–	–	–	–	Cf	Mudurnu	TR
39	1493	Jan.	10	33.0	59.8	7.0	120	T	30	–	–	APf	Birjand	IR
40	1505	Jul.	6	34.8	69.1	7.4	010	–	56	–	300	AP	Kabul	AF
41	1544	Jan.	22	38.0	37.0	M	090	L	–	–	–	BAf	Elbistan	TR
42	1595	Sep.	22	38.5	27.9	M	270	N	–	–	–	aP	Ahmetli	TR
43	1646	Apr.	7	38.3	43.7	L	070	–	–	–	–	aP	Van	TR
44	1651	Jun.	7	37.8	29.3	M	120	N	–	–	–	Bf	Honaz	TR
45	1653	Feb.	22	37.9	28.5	7.1	090	N	70	–	300	APd	Menderes	TR
46	1661	Mar.	15	42.2	24.0	L	–	–	–	–	–	aA	Maritza	BU
47	1666	Sep.	23	36.7	43.5	L	–	–	–	–	–	C	N. Mosul	IQ
48	1668	Aug.	17	40.5	36.0	7.9	090	–	400	–	–	APf	Amasya	TR
49	1721	Apr.	26	37.9	46.7	7.7	125	–	50+	–	–	AP	Tabriz	IR
50	1740	Oct.	5	38.7	22.4	6.6	–	–	20		–	aP	Lamia	GR
51	1752	Jul.	29	41.3	26.5	L	–	–	–	–	–	aA	Evros	TR
52	1759	Nov.	25	33.7	35.9	7.4	020	L	100	–	–	AP	Bekaa	LE
53	1780	Jan.	8	38.2	46.0	7.7	120	N	60+	–	600	AP	Tabriz	IR
54	1784	Jul.	18	39.5	40.2	7.6	110	R	150	–	–	Bfm	Elmali	TR
55	1789	May	28	38.8	39.5	L	–	–	–	–	–	BP	Elazig	TR
56	1796	Apr	26	35.5	36.0	6.6	–	–	–	–	–	aA	Latakia	SY

(*continued*)

TABLE 1 *(continued)*

No.	Date			Epicentre N	Epicentre E	M_S[a]	Az deg.	Mec[b]	L[c] (km)	H[d] (cm)	V[d] (cm)	Q[e–h]	Location	Ref.[i]
57	1822	Aug.	13	36.7	36.5	7.5	020	L	200	–	–	APd	Antakya	TR
58a	1825	–	–	36.1	52.6	6.7	–	–	–	–	–	aPd	Harhaz	IR
58b	1829	May	5	41.2	25.1	7.2	–	–	50	–	–	aPk	Xanthi	GR
59	1837	Jan.	1	33.2	35.5	7.4	000	–	80	–	–	BPkm	Bshara	LE
60	1838	–	–	29.6	59.9	7.0	170	–	70	–	–	APf	Nasratab	IR
61	1840	Jul.	2	39.5	43.8	7.3	140	R	80	–	–	BPf	Kazlgl	TR
62	1855	Feb.	28	40.0	28.5	7.4	270	–	70	–	–	APm	Ulubat	TR
63	1855	Apr.	11	40.3	29.1	6.6	–	–	–	–	–	aA	Gemlik	TR
64	1861	Dec.	26	38.2	22.2	6.6	280	N	13	–	220	APk	Vostiza	GR
65	1864	Dec.	7	33.2	45.9	6.4	–	–	2+	–	50	aPk	Zorbatia	IQ
66	1866	May	12	39.2	41.0	7.2	230	L	45	–	–	APm	Gnek	TR
67	1870	Aug.	1	38.5	22.6	6.7	010	N	6+	–	200	aPm	Fokis	GR
68	1872	Apr.	3	36.4	36.4	7.2	030	–	20	–	–	APk	Amik Gol	TR
69	1874	May	3	38.5	39.5	7.1	250	L	45	–	200	AM	Glcuk 1	TR
70	1875	Mar.	27	38.5	39.5	6.7	250	–	20	–	200	aP	Glcuk 2	TR
71	1875	May	3	38.3	29.9	6.5	040	N	10	–	110	aP	Civril	TR
72	1880	Jul.	29	38.6	27.2	6.5	120	N	10	–	40	aPk	Emiralan	TR
73	1887	Sep.	30	38.7	29.8	6.3	290	N	10	–	50	aPd	Banaz	TR
74	1889	Jan.	17	37.7	30.5	–	–	–	–	–	–	aPk	Isparta	TR
75	1892	Dec.	19	30.9	66.5	6.9	020	L	30	80	30	AMi	Chaman	PK
76	1893	Mar.	2	38.0	38.3	7.1	270	L	–	–	–	Bf	Malatya	TR
77	1894	Apr.	27	38.6	23.2	6.9	300	N	40	–	100	AMdm	Martin	GR
78	1899	Sep.	20	37.9	28.8	6.9	090	N	40	–	100	AMd	Mender	TR

All events are assumed to be crustal.

[a]Magnitude: Magnitudes have been derived from macroseismic information calibrated against 20th century instrumental M_S values. The size of historical events under investigation has been classified under three broad categories: V: very large event $M_S > 8.0$; L: large event $7.0 < M_S < 8.0$; M: medium event $6.0 < M_S < 7.0$.

[b]Fault attitude and mechanism: T = thrust; L = left lateral strike slip; R = right lateral strike slip; N = normal, with a combination of these symbols for oblique motion; * = assumed from regional pattern.

[c]Length of faulting: L = total length of surface rupture, including intermediate unfractured segments in km.

[d]Relative displacements: H = maximum observed lateral offset in cm; V = maximum observed vertical offset in cm; S = indicates small displacements of imperceptible sense of motion; – = no data.

[e]Quality of evidence of faulting Q (first column for Q): A, surface faulting explicity or (a) implicitly, deduced from the sources or field investigations; B, no evidence for faulting in the sources; surface faulting inferred from the elongated shape of the epicentral region; C, faulting assumed because of the large size of the earthquake and its proximity to a known active fault zone.

[f]Location evidence (second column for Q) for quality categories A and a, is subdivided into: G = good, derived from detailed field studies; M = moderate, based on cursory field survey of the facture zone; P = poor, deduced from historical data or, for more recent events from field evidence in need of authentication; A = very poor, exact location of fault-break not known.

[g]Nature of fault zone (third column for Q): d = trace discontinuous or eroded; total length of rupture deduced from few and widely spaced reported observations; U = arcuate trace, graben, or complex fault zone; k = some of the observed or reported ground deformations probably not of tectonic origin; i = only part of the break was accessible or mapped; actual rupture length is probably longer than reported; n = reported ground effects, to the best of our judgment, not of tectonic origin, or associated with a known earthquake; m = multiple shock; observed deformations and rupture length probably enhanced by more than one earthquake.

[h]For quality A, B and C (in any column of Q): f = assumed association of historical event with known Quaternary or recent fault-break.

[i]The name of the location where the event took place is given in the penultimate column, and the last column gives the country: AF, Afghanistan; AL, Albania; BU, Bulgaria; GR, Greece; IQ, Iraq; IR, Iran; IS, Israel; LE, Lebanon; MC, Macedonia; PK, Pakistan; SA, Saudi Arabia; SY, Syria; TR, Turkey; TU, Turkmenistan.

For 62 of the 150 cases of surface faulting retrieved we have both well-determined surface-wave magnitudes from instrumental data, and reasonably reliable rupture lengths from field observations. These events are all in the instrumental period, with 55% of the data coming from strike-slip, 30% from normal and 15% from thrust faults, excluding cases of quality (2) and others for which the rupture length is imperfectly known.

A staightforward orthogonal regression between M_S and $\log(L)$ gives:

$$M_S = 5.13 + 1.14 \log(L) \tag{2}$$

with L in kilometers, with a standard deviation of 0.15 in M_S. Alternatively, regressions of M_S on $\log(L)$, and of $\log(L)$ on

M_S give

$$M_S = 5.27 + 1.04 \log(L) \quad (3)$$

and

$$\log(L) = -4.09 + 0.82 M_S \quad (4)$$

respectively, with almost the same standard deviation of 0.22 in M_S for both cases, whereas a nonlinear fit results in

$$M_S = 5.06 + 1.42 \log(L) - 0.14 [\log(L)]^2 \quad (5)$$

with a slightly larger standard deviation (see Ambraseys and Jackson, 1998).

For 58 of the 62 earthquakes used to derive Eqs. (2)–(5) we have also horizontal (H) and vertical (V) maximum surface displacements but the fit improves little, given by

$$M_S = 5.11 + 0.86 \log(L) + 0.21 \log(R) \quad (6)$$

with a standard deviation of 0.20 in M_S, in which R is the resultant displacement from H and V, in centimeters.

In terms of resultant displacement R alone, M_S may be approximated by

$$M_S = 5.21 + 0.78 \log(R) \quad (7)$$

with a rather large standard deviation of 0.36 in M_S.

We find that the resultant displacement R is about 5.0 ($\pm$4.0) $\times$ $10^{-5}L$, regardless of mechanism. This value for R is no surprise, and is similar to that found in global compilations of more modern datasets (e.g., Scholz, 1982). However, the size of the sample is insufficient and the scatter too large to allow a better estimate of R as a function of mechanism.

However, it is important, particularly for paleoseismological investigations, to have some indication of whether the rupture length and offset estimated from historical sources are likely to be seriously under- or overestimated, given the magnitude of the event or vice versa. This is a principal use of magnitude–length relationships. For an assessment of individual events or particular regions, it may be more informative to make such estimates from a combination of first principles and more closely constrained empirical relationships, along the following lines (Ambraseys and Jackson, 1998).

1. For earthquakes that rupture the entire thickness (d) of the seismogenic upper crust, the down-dip width of the fault is $d/\sin a$, where θ is the fault dip, and the moment is then

$$M_0 = (mcd/\sin\theta)L^2 \quad (8)$$

where m is the rigidity modulus and c is the ratio of average displacement (u) to fault length (L), which is observed to be close to $5 \times 10^{-5}L$ for intracontinental earthquakes (Scholz, 1982).
2. Both observationally and theoretically it is known that for such earthquakes the relationship between moment and magnitude (M, where M_S or M_W) is of the form

$$\log(M_0) = A + BM \quad (9)$$

where A and B are constants, with $B = 1.5$ for events with M not smaller than about 6.0.
3. Combining these expressions gives a relationship between moment and fault length of the form

$$M_0 = \log(mcd/\sin\theta)/B - A/B + 2(\log L)/B \quad (10)$$

For illustration, if we take $m = 3 \times 10^{-3}\,\mathrm{Nm}^{-2}$, $c = 5 \times 10^{-5}$ and $B = 1.5$, then for a seismogenic layer of thickness $d = 15$ km and vertical strike-slip fault ($\theta = 90°$) the relationship is

$$M_W \text{ or } M_S = 4.9 + 1.33 \log(L) \quad (11)$$

with L in kilometers, which is similar to the empirical relationship [Eq. (2)].

7. Discussion

The reappraisal of the seismicity of the Eastern Mediterranean and the Middle East, east of the Adriatic and the Ionian Seas all the way to Pakistan, shows that although the historical record is incomplete, careful reading of the available data can provide valuable insights into the long-term seismicity of the region. The pattern of seismic activity of many areas is seen to have changed little over the past 2500 y, whereas other areas that are at present quiescent can be shown to be capable of generating earthquakes of significant size. The following are some of the lessons we believe we have learned from these studies.

- Archaeological evidence for an earthquake is not always unambiguous and can seldom be used to provide a precise date for the damage caused. Nevertheless, archaeological evidence can provide confirmation of long-term seismicity in a given region and with greater collaboration between disciplines it is likely that many refinements of the existing database will be possible.
- For the earth scientist and earthquake engineer the main objectives of historical research into primary sources are to refine and extend the information contained in secondary studies and catalogs, and to provide an objective measure of the reliability and completeness of the data retrieved.
- It is important to establish unambiguously the simultaneity of damage to different localities in a historical

earthquake. Often one finds cases in which two separate events have been transformed into a large earthquake. This is understandable in view of the tendency of both contemporary and later writers to amalgamate seismic events, whether for lack of sufficiently precise information, from ignorance of the true nature of earthquakes, or from simple convenience. Such an amalgamation of effects will overestimate the size of the damage area, and hence of the size of the event.

- Although some of the problems of dating can be resolved, it is often more difficult to determine a sufficiently accurate location for historical earthquakes. The epicentral area of an historical event is not always certain and judgment has to be exercised to ascertain its location. The primary aim should be to avoid both the amalgamation and duplication of events.
- The size of an historical earthquake can be assessed in terms of its magnitude; such an assessment for historical events can be made only approximately and depends on the reliability of information regarding their effects at large epicentral distances or from the dimensions of their epicentral area. For events in which this information could be estimated, the magnitude of the event should be estimated using a calibration formula derived from 20th-century earthquakes for the region.
- In estimating intensities we find that at large distances an earthquake may cause the collapse of a few important but vulnerable constructions, for which there may be archaeological or historical evidence. This information alone should not always be taken to mean that all the other human-made structures at these sites have been destroyed. The observed effects can be the result of the high vulnerability of long-period structures to sustained ground motions, rather than of the severity of the shock.
- For many historical events, the data are wholly insufficient to permit assessment of intensity in terms of any of the scales currently in use, let alone to reckon the magnitude of an event, except in very general terms. We find that precise local or epicentral intensities assigned by modern catalogers to many historical events, particularly in Greece and the Holy Land, are hypothetical and often inflated.
- Earthquake catalogs are often used by earth scientists and engineers to assess earthquake hazard. A more critical attitude is needed to rely only on those that combine the interpretation of primary sources with estimates of the reliability and completeness of the data provided.
- The location and size of historical earthquakes should not be used for scientific purposes without proper scrutiny of the associated historical material.
- The historical record confirms that some regions that are active today (e.g., the north Anatolian fault zone) were also active 2500 y ago, demonstrating the long-term nature of their seismicity. It also shows that some regions that are at present quiescent (such as the Jordan Rift Valley), are capable of generating relatively large earthquakes. For some of these events this is consistent with their known active tectonic environment.
- Too many modern catalogs of historical seismicity are not sufficiently rigorous to be treated with confidence. This has often been due to the interdisciplinary nature of this field of study, which requires scientists to examine literary texts and historians to glean scientific information from their sources. The result has been the production of a large number of false earthquakes, or of seismic events of a size beyond the limits of the possible, often with a sensationalist tinge. This is of no technical consequence, provided the Earth-scientist and engineer are aware of it.

In view of the uneven value of parametric catalogs that are not based on a reassessment of primary source data, it is important that principles of interpretation should be laid down for the benefit of those working in historical seismicity, since this kind of work will be of value to engineers and other scientists only when the results are reasonably reliable and capable of being converted into figures.

Appendix 1. Historical Earthquake Catalogs

Manetti's work (in late Latin) is the earliest-known compendium of earthquakes and contains an annotated list of earthquakes in the Eastern Mediterranean and elsewhere up to 1456. Manetti does not always cite his sources and quite often the year of an earthquake is recorded only by reference to other events.

Al-Suyut's earthquake catalog (in Arabic) was compiled in the early part of the 16th century and extended by his continuators to the year 1588. It is a reliable source of information for the Muslim world, covering the region from Morocco to Transoxania.

Bonito's large world earthquake catalog (in Italian) is an invaluable compendium of information about earthquakes that ends with 1690. Its 822 pages contain a wealth of information culled from a variety of sources, which Bonito quotes and occasionaly annotates. His work provides an excellent starting point for the identification of earthquakes in Europe and in the New World (Bonito, 1691).

Coronelli's work (in old Italian), although prepared as a global catalog of earthquakes up to 1693, deals mainly with events in the central and eastern Mediterranean. Annotations are kept very brief, making no reference to sources of information and occasionally neglecting to give the full date of an event (Coronelli, 1693).

An anonymous compilation of earthquakes throughout the world was published (in German) in a series of issues of the *Dressdenische Gelehrte Anzeigen* in 1756, and is a useful source of information for earthquakes worldwide during the 16th and 17th centuries up to 1691 (PDGA, 1756).

Hoff's general catalogue of earthquakes (in German) is a valuable work, covering events worldwide for the period up to the end of the 17th century. It is an accurate and methodical study, drawing on a variety of published sources, which are cited (Hoff, 1840).

The compilation of Seyfart's work (in German) on earthquakes was prompted, like many similar works of the mid-18th century, by the large Lisbon earthquake of 1755. It contains interesting entries, mostly extracted from published material in Europe, such as flysheets and newsletters, as well as from the European press (Seyfart, 1756).

Berryat's long chronological list (in French) is an annotated collection of information about earthquakes up to 1760. The author does not cite his sources but they seem to include, among others, earlier catalogs and information from the European press (Berryat, 1761).

Hoff compiled twelve annual earthquake catalogues (in German) for the years 1821–32. He extracted much of the information from press reports, travel diaries and from correspondence. His work is of interest for areas outside Europe (Hoff, 1826–35, 1840–41).

Mallet's catalog occupies nearly 600 pages and contains almost 7000 events worldwide. Although based on several earlier catalogs, especially on those of Hoff and Perrey, his catalog for the period after the 17th century contains a considerable amount of information from relatively early press reports, some of which are useful for investigating the seismicity of the Americas and the Far East (Mallet, 1850–58).

Perrey's annual lists of earthquakes (in French) for the 29 y 1843–71 are invaluable. They occupy 28 papers and the total number of pages in these Mémoires is just over 2500. Perrey collected much of the material by correspondence and also from the international press. His annual lists are a vast storehouse of facts; for the most part he was content to leave discussion of the results to others. There is seldom any attempt to determine the position of the epicenter, none to discover the relation between mainshock and aftershocks or the relation between shocks felt at the same time at different places (Perrey, 1848–75).

Schmidt's catalogs (in German) for the Southern Balkans and Asia Minor is one of the most important sets of data for the region. It depends very little on previous lists or catalogs and from about 1800 onwards, is the result of his own labors. From after about 1858 to the end of 1878, his catalogue contains just under 4000 entries, derived chiefly from correspondence with observers, travelers, and consuls throughout the Eastern Mediterranean and from the Press in Athens, Istanbul, Izmir, and other places in the area (Schmidt, 1867, 1879).

A long memoir (in German) containing lists of earthquakes for the twenty years, 1865–84, was published by Fuchs. These lists include nearly 10 000 entries altogether, containing a substantial amount of information for earthquakes worldwide. In common with some other catalogs, this work must be used with caution, for nowhere does Fuchs cite his sources, and it is accordingly difficult now to appreciate the value of the information that he retrieved (Fuchs, 1886).

Musketoff and Orloff's earthquake catalog (in Russian) for the Russian empire ends in 1888. It is based on previous catalogs but also on contemporary national and local Russian press reports and to a lesser extent on unpublished documents. Events are fully annotated and sources are given in full. This is a very useful source of information (Musketoff and Orloff, 1891).

Appendix 2. Modern Catalogs

Milne's world catalogue of destructive earthquakes up to 1899 is based entirely on previous lists. It is devoid of information from original sources (Milne, 1911).

Montessus de Ballore's world catalog (in French), containing 1 71 434 entries, covers the period up to 1906. Only a small fraction of this enormous volume of information, which covers mainly the second half of the 19th century, has been published, and it remains little known. However, the published information is not of very great value; the unpublished files, kept in the Department des Cartes et Plans, Depot de la Société de Geographie of the Bibliothèque Nationale in Paris, where they occupy 30 m of bookshelf, did not prove, on examination, to be as useful as had been anticipated. Much of the information in these files was extracted from previous catalogs and press reports, with little original material derived from correspondence with observers (Ballore, 1900, 1905, 1924).

Sieberg's annotated world catalog of earthquakes (in German) contains a considerable amount of information, including isoseismal maps for the larger historical earthquakes worldwide up to 1930. However, this catalog gives little indication of his sources of information and contains many errors and duplications in entries. Nevertheless, this highly inaccurate work has for many years been regarded as a standard reference on the subject (Sieberg, 1932a,b).

Stepanian's annotated catalogs (in Russian and Armenian) of earthquakes in Greater Armenia are a useful set of documents. They are based on a considerable number of primary published Armenian sources. These Armenian catalogs of Stepanian are little known; they are accurate and methodical, and contain about 800 events (Stepanian, 1942, 1964). Subsequent work by Guidoboni and Traina (1995) has brought this up to date.

Byus' catalog of earthquakes in the Caucasus and adjacent regions (in Russian) is a systematic compilation of information from previous catalogs, in some cases critically selected, as well as from local Georgian, Armenian, and Russian sources, including local newspapers and reports. This 600-page work contains a wealth of information about events in the Middle East (Byus, 1948).

Rethly's compilation of earthquakes in the Carpathian region and central Europe (in Hungarian) is a serious piece of work. It contains extracts from original sources and is fully referenced. This work is invaluable for the identification of events that affected southeast Europe (Rethly, 1952).

The survey of the seismicity of the Balkan region carried out by UNESCO in the mid-1970s, contributed a summary of the material available at that time for the assessment of regional seismicity. Isoseismal maps for a few events before 1900 and a parametric catalog were published, but must now be used with caution (Shebalin *et al.*, 1974).

The catalogs of earthquakes in the Middle East and along the Dead Sea Rift by Ben-Menahem (1979, 1991) contain information extracted from earlier catalogs of varying quality and from secondary works. These lists, which include a parametric catalog going back to 2050 BC, must be used with caution.

The national earthquake catalog of the former USSR (in Russian) covers a large geographical area for the period before 1977 (Kondorskaya and Shebalin, 1982). It is based chiefly on secondary macroseismic sources but includes a detailed procedure for the systematic quantification of historical events in the region.

The catalog of Poirier and Taher (1980), covers the seismicity of the Middle East, listing nearly 200 events up to 1800. It summarizes information taken from a thorough survey of Arabic source material, presented in Taher's doctoral thesis, Sorbonne, 1979. References are properly identified and cited. Though the catalog contains various errors and duplications, this is a considerable improvement on earlier work. A more extended summary of these primary data, although regrettably without any reference to modern studies of the last two decades, is currently in progress (Taher, 1996).

A useful catalog by Russell (1985) for Palestine in the period to the mid-8th century, presents the texts of the accounts of earthquakes in the region from contemporary sources and attempts to resolve discrepancies in dating. The catalog also provides archaeological evidence of damage that has been adduced to support the dating of some of these events, or to be dated by them.

The works by Ambraseys and Melville (1982) and Ambraseys *et al.* (1994) attempt a thorough re-evaluation of the long-term seismicity of Iran, Saudi Arabia, and the Red Sea, based as far as possible on primary sources. These works present in some detail the methodology proposed to assess historical seismicity by combining instrumental data and macroseismic information.

An authoritative compilation of primary sources for earthquakes in Italy and in the eastern Mediterranean (Guidoboni, 1989; Guidoboni *et al.*, 1994) covers the period from the 8th century BC to 1000 AD. Events are annotated in detail and sources are quoted in full. There is no attempt to determine epicentral areas or to assess the size of the events it describes.

The work by Ambraseys and Finkel (1995) covers Turkey and parts of the Middle East for the period from 1500 to 1800 and is based chiefly on unpublished sources of information.

The catalogs of Papazachos and Papazachou (1989 in Greek, 1997) cover the historical seismicity of Greece and adjacent regions. These are annotated compilations based essentially on previous catalogs.

The work (in Greek) by Spyropoulos (1997) is an exhaustive annotated corpus of extracts from original sources relating to historical earthquakes in Greece.

References

Albini, P. and A. Moroni (Eds.) (1994). "Historical Investigation of European Earthquakes," vol. 2, Milan.

Alexandre, P. (1990). "Les seismes en Europe occidentale de 394 a 1259," Oberv. Royal de Belgique, Ser. Geophysique, Bruxelles.

al-Suyuti "Kashf al-salala 'an wasf al-zalzala," (A. Sa'dani, Ed.), Fez, 1971; French translation by S. Nejjar, Rabat 1974.

Ambraseys, N.N. (1971). Value of historical records of earthquakes. *Nature* **232**, 375–379.

Ambraseys, N. (1973). Dynamics and response of foundation materials in epicentral regions of strong earthquakes, *Proc. 5th World Conf. Earthq. Eng.* vol. 1, Invited papers, pp. cxxvi–cxlviii, Rome.

Ambraseys, N. (1985). Intensity–attenuation and magnitude–intensity relationships for northwest European earthquakes. *J. Earthq. Eng. Struct. Dyn.* **13**, 733–778.

Ambraseys, N. (1997). The little-known earthquakes of 1866 and 1916 in Anatolia, Turkey. *J. Seismol.* **1**, 289–299.

Ambraseys, N. and C. Finkel (1995). "The Seismicity of Turkey and Adjacent Areas; a Historical Review 1500–1800," Eren, Istanbul.

Ambraseys, N. and J. Jackson (1990). Seismicity and associated strain in central Greece between 1890 and 1988. *J. Geophys. Int.* **101**, 663–708.

Ambraseys, N. and J. Jackson (1998). Faulting associated with historical and recent earthquakes in the eastern Mediterranean region. *Geophys. J. Int.* **133**, 390–406.

Ambraseys, N. and C.P. Melville (1982). "A History of Persian Earthquakes." Cambridge University Press, Cambridge.

Ambraseys, N. and C.P. Melville (1988). An analysis of the eastern Mediterranean earthquake of 20 May 1202 "Historical Seismograms and Earthquakes of the World," (W.H.K. Lee, *et al.*, Eds.), pp. 181–200. Academic Press, New York.

Ambraseys, N. and C.P. Melville (1989). Evidence for intraplate earthquakes in Northwest Arabia. *Bull. Seismol. Soc. Am.* **79**, 1279–1281.

Ambraseys, N. and C.P. Melville (1995). Historical evidence of faulting in Eastern Anatolia and Northern Syria. *Ann. Geofis.* **38**, 337–343.

Ambraseys, N. and A. Moinfar (1988). Isoseismal maps across national frontiers. *J. Eur. Earthq. Eng.* **1**, 15–21.

Ambraseys, N. and D. White (1996). A reappraisal of East Mediterranean seismicity before the Christian era. "Seismology in Europe," ESC, Reykjavik.

Ambraseys, N. and D. White (1997). The seismicity of the eastern Mediterranean region 550 BC: a re-appraisal. *J. Earthq. Eng.* **1**, 603–632.

Ambraseys, N., *et al.* (1983). Notes on historical seismicity. *Bull. Seismol. Soc. Am.* **73**, 1917–1920.

Ambraseys, N., C.P. Melville, and R. Adams (1994). "The Seismicity of Saudi Arabia and Adjacent Regions." Cambridge University Press, Cambridge.

Ballore, Montesus de (1900). Sismicite de la peninsule balkanique et de l'Anatolie. *Bull. Com. Geol. Russie* **19**, 31–53.

Ballore, Montesus de (1905). "La Geographie Seismologique." A. Collin, Paris.

Ballore, Montesus de (1924). "La Geologie Seismologique." A. Collin, Paris.

Ben-Menahem, A. (1979). Earthquake catalogue for the Middle East 92 BC–1980 AD. *Boll. Geof. Teor. Appl.* **21**, 245–313.

Ben-Menahem, A. (1991). Four thoudand years of seismicity along the Dead Sea Rift. *J. Geophys. Res.* **96**, B12, 20195–20216.

Berryat, J. (1761). Liste chronologique des eruptions de volcans, des tremblements de terre etc. *Collect. Academique* **6**, 488–676.

Bonito, D.M. (1691). "Terra Tremante." Parrino and Muti, Napoli.

Byus, E.N. (1948–1995). "Seismicheskie usloviya zakavkaziya." 3 vols. Akad. Nauk. Gruz, Tbilisi.

Coronelli, P. (1693). "Epitome Cosmografia." Colonia.

Davison, C. (1924). "A History of British Earthquakes." Cambridge University Press, Cambridge.

Di Vita, A. (1995). Archaeologists and earthquakes: the case of 365 A.D. *Ann. Geofis.* **38**, 971–976.

PDGA. *Dredssdnische Gelehrte Anzeigen*, Dresden 1756.

Dussaud, R. (1896). Voyage en Syrie. *Rev. Archaeol.* ser. 3, 299.

Evangelatou-Notara, F. (1993). Earthquakes in Byzantium, *Parousia*, no. 24, Athens.

Fuchs, C.W. (1886). Statistik der Erdbeben. *Sitzungber. Kais. Akad. Wiss. Math. -Naturwiss. Klass.* **92**, 215–625.

Grunthal, G. (Ed.) (1998). European macroseismic scale 1998, *Cahiers*, no. 15, *Centre Europeen Geodyn. Seismol.* **15**, 99 pp.

Gu, Gongxu, *et al.* (1983). "Catalogue of Chinese Earthquakes," revised Xie Yushou, Beijing.

Guidoboni, E. (1989). Catologo. In: "I Terremoti Prima del Mille in Italia e nell'atea Mediterranea" (M. Guidoboni, Ed.), pp. 574–750. 1st Nazion. Geogys., Rome.

Guidoboni, E. (1996). Archaeology and historical seismology: the need for collaboration in the Mediterranean area. In: "Archaeoseismology" (S. Stiros and R.E. Jones, Eds.), pp. 7–13. Fitch Laboratory Occasional Paper 7, Athens.

Guidoboni, E. and S.S. Bianchi (1995). Collapses and seismic collapses in archaeology: proposal for a thematic atlas. *Ann. Geofis.* **38**, 1013–1017.

Guidoboni, E. and M. Stucchi (1993). The contribution of historical records of earthquakes to the evaluation of seismic hazard. *Ann. Geofis.* **36**, 210–215.

Guidoboni, E. and G. Traina (1995). A new catalogue of earthquakes in the historical Armenian area from antiquity to the 12th century. *Ann. Geofis.* **38**, 85–147.

Guidoboni, E., A. Comastri, and G. Traina (1994). "Catalogue of Ancient Earthquakes in the Mediterranean Area up to the 10th Century." Publ. Istituto Nazion. di Geofisica, Rome.

Harris, G. and A. Beardow (1995). The destruction of Sodom and Gomorrah; a geotechnical perspective. *Q. J. Eng. Geol.* **28**, 349–362.

Hersberger, J. (1956). A comparison of earthquake acceleration with intensity ratings. *Bull. Seismol. Soc. Am.* **46**, 317.

Hoff, K. von (1826–1835). *Annales de Chemie et de Physique*, 1826, **7**, 159–70, 289–304; 1827, **9**, 589–600; 1828, **12**, 555–584; 1829, **15**, 363–383; 1830, **18**, 38–56; 1831, **21**, 202–218; 1832, **25**, 59–91; 1833, **29**, 415–447; 1835, **34**, 85–108.

Hoff, K. von (1840). *Chronik der Erdbeben und Vulkan-ausbrche*, 4 vols, Gotha.

Howell, B. and T. Schultz (1975). Attenuation of modified Mercalli intensity with distance from the epicentre. *Bull. Seismol. Soc. Am.* **65**, 651–665.

Jacques, F. and B. Bousquet (1984). Le raz de marée du 21 juillet 365; du cataclysme locale à la catastrophe cosmique. *Mélanges Ecole de France*, Rome, **96**, 423–461.

Kenyon, K.M. (1978). "The Bible and Recent Archaeology." British Museum, London.

Knudtzon, J.A. (1915). "Die El-Amarna-Tafeln," 2 vols, Leipzig.

Kondorskaya, N. and N. Shebalin (Eds.) (1982). "Nov'ii katalog sil'nikh zemletrasenii na territorii CCCP." Akademia Nauka, Moscow.

Mallet, R. (1852). Report on the facts of earthquake phaenomena, *Reports of the British Association for the Advancement of Science for 1852*, pp. 1–176; 1853, pp. 118–212, 1854, pp. 2–326, London.

Manetti, G. De terraemotu libri tres, Bibl. Apost. Vaticana, cod. Urbin. Lat. 5, 253 fols, tr. C. Scopelliti, ENEA Publ., Rome, 1983.

Melville, C.P. (1984). The use of historical records for seismic assessment. In: "The O.G.S. Silver Anniversary Volume." (A. Brambati and D. Slejko, Eds.), Trieste.

Milne, J. (1911). "A Catalogue of Destructive Earthquakes AD 1 to 1899." British Association for the Advancement of Science, Portsmouth.

Musketoff, I. and A. Orloff (1891). Katalog zemletriasenii Rossiskoi Imperii. *Zapiski Imp. Russ. Geograf. Obches.* **26**.

Nur, A. and H. Ron (1996). And the walls came tumbling down: earthquake history in the Holyland (sic). In: "Arachaeoseismology," (S. Stiros and R.E. Jones, Eds.), pp. 75–85. Fitch Laboratory Occasional Paper 7, Athens.

Papazachos, B. and K. Papazachou (1989). "Earthquakes in Greece" (in Greek), pp. 219–347. Ziti, Thessaloniki.

Papazachos, B. and K. Papazachou (1997). "The Earthquakes of Greece." Ziti, Thessalonioki.

Papazachos, B., Ch. Papaioannou, C. Papazachos, and A. Savvaidis (1997). "Atlas of Isoseismal Maps for Strong Shallow Earthquakes in Greece and Surrounding Area 426 BC–1995." Geophys. Lab. Publ. no. 4, University of Thessaloniki, Thessaloniki.

Perrey, A. (1848). Note sur les tremblements de terre en 1847. *Bull. Acad. Sci. Bruxelles* **15**, 442–454.

Perrey, A. (1850). Memoire sur les tremblements de terre ressentis dans la peninsule turcohellenique et en Syrie. *Mem. Acad. R. Sci. Belg.* **23**(1), 3–73.

Perrey, A. (1851). Tremblements de terre ressentis en 1850. *Bull. Acad. R. Sci. Bruxelles* **18**(1), 291–308.

Perrey, A. (1852a). Tremblements de terre ressentis en 1851. *Bull. Acad. R. Sci. Bruxelles* **19**(1), 353–396.

Perrey, A. (1852b). Supplement la note sur les tremblements de terre ressentis en 1851. *Bull. Acad. R. Sci. Bruxelles* **19**(2), 21–28.

Perrey, A. (1853). Tremblements de terre en 1852. *Bull. Acad. R. Sci. Bruxelles* **20**(2), 39–69.

Perrey, A. (1854). Note sur les tremblements de terre en 1853. *Bull. Acad. R. Sci. Bruxelles* **21**(1), 457–495.

Perrey, A. (1855). Note sur les tremblements de terre en 1854 avec supplements pour les annes anterieures. *Bull. Acad. R. Sci. Bruxelles* **22**(1), 526–572.
Perrey, A. (1856). Note sur les tremblements de terre en 1855, avec suppliments. *Bull. Acad. Sci. Bruxelles* **23**, 23–68.
Perrey, A. (1857). Note sur les tremblements de terre ressentis en 1855, avec supplements pour les annes anterieures. *Bull. Acad. R. Sci. Belgique* **24**, 64–128.
Perrey, A. (1859). Note sur les tremblements de terre en 1856 avec supplements pour les annes anterieures. *Mem. Couron. Acad. R. Bruxelles* **8**(3), 3–79.
Perrey, A. (1860). Note sur les tremblements de terre en 1857. *Mem. Cour. Acad. R. Bruxelles* **10**(4), 3–114.
Perrey, A. (1862a). Note sur les tremblements de terre en 1858. *Mem. Cour. Acad. R. Bruxelles* **12**(4), 3–68.
Perrey, A. (1862b). Note sur les tremblements de terre en 1859. *Mem. Cour. Acad. R. Bruxelles* **13**(3), 3–78.
Perrey, A. (1862c). Note sur les tremblements de terre en 1860. *Mem. Cour. Acad. R. Bruxelles* **14**(3) 3–75.
Perrey, A. (1864a). Note sur les tremblements de terre en 1861. *Mem. Cour. Acad. R. Bruxelles* **16**(5), 2–112.
Perrey, A. (1864b). Note sur les tremblements de terre en 1862. *Mem. Cour. Acad. R. Bruxelles* **16**(6), 3–179.
Perrey, A. (1865). Notes sur les tremblements de terre en 1863. *Mem. Cour. Acad. R. Bruxelles* **17**(5), 1–213.
Perrey, A. (1866). Note sur les tremblements de terre en 1864. *Mem. Cour. Acad. R. Bruxelles* **18**(4), 3–98.
Perrey, A. (1870). Note sur les tremblements de terre en 1866 et 1867. *Mem. Couron. Acad. R. Bruxelles* **21**(5), 3–223.
Perrey, A. (1872). Note sur les tremblements de terre en 1868. *Mem. Cour. Acad. R. Bruxelles* **22**(3), 1–116.
Perrey, A. (1872b). Notes sur les tremblements de terre en 1869. *Mem. Cour. Acad. R. Bruxelles* **22**(4), 1–116.
Perrey, A. (1873). Note sur les tremblements de terre en 1870. *Mem. Cour. Acad. R. Bruxelles* **24**(3), 2–146.
Perrey, A. (1875). Supplments aux notes sur les tremblements de terre ressentis de 1843 a 1868. *Mem. Couron. Acad. R. Bruxelles* **23**(6), 3–70.
Perrey, A. (1875b). Note sur les tremblements de terre en 1871. *Mem. Cour. Acad. R. Bruxelles* **24**(4), 1–143.
Pirazzoli, P.A. (1986). The early Byzantine tectonic paroxsym. *Z. Geomorphol., N. Folg, Suppl.* **62**, 31–49.
Poirier, J.P. and M.A. Taher (1980). Historical seismicity in the Near and Middle East, North Africa, and Spain from Arabic documents (VIIth–XVIIIth century). *Bull. Seismol. Soc. Am.* **70**, 2185–2201.
Pomerance, L. (1970). The final collapse of Santorin (Thera). *Stud. Medit. Archaeol.* **26**, 5–33.
Rethly, A. (1952). "A Karpatmedencék Földrengesei 445–1919." Akad Kiadó, Budapest.
Russell, K.W. (1985). The earthquake chronology of Palestine and northwest Arabia from the 2nd through the mid-8th century. *Bull. Am. Schools Or. Res.* **260**, 37–59.
Schaeffer, C.F. (1939). "The Cuneiform Texts of Ras Shamra – Ugarit." Oxford.
Schaeffer, C.F. (1948). "Stratigraphie Comparee et Chronologie de l'Asie Occidental iii et ii Millenaires." Oxford.
Schmidt, J. (1879). "Studien ueber Erdbeben." Leipzig.
Scholz, C.H. (1982). Scaling laws for large earthquakes: consequences of physical models. *Bull. Seismol. Soc. Am.* **72**. 1–14.
Seyfart, F. (1756). "Allgemeine Geschichte der Erdbeben." Frankurt.
Shebalin, N., V. Karnik, and D. Hadzijevski (1974). "Catalogue of Earthquakes" Part I, 1901–1970; Part II, prior to 1901; Part III, Atlas of isoseismal maps, UNDP/UNESCO Survey Seismicity of the Balkan Region, Skopje.
Sieberg, A. (1932a). Die Erdbeben. In: "Handbuch der Geophysik" (B. Gutenberg, Ed.), vol. 4. Berlin.
Sieberg, A. (1932b). Untersuchugen über Erdbeben und Bruchschollenbau im östlichen Mittelmeersgebeit. *Denskschr. Mediz. -Naturwiss. Gesell. z. Jena* **2**.
Spyropoulos, P.I. (1997). "Chronicle of Greek Earthquakes." Dodoni, Athens.
Stepanian, V.A. (1942). Istoricheskii obzor o zemletriasenikh v Armenii v prilegaiuschikh rayonakh; kratkaia chronologiya. *Zakavkaz. Konfer. po Antiseism. Stroitel'stvu*, pp. 43–72. Akad. Nauk. Izdat. Armen. FAN, Erivan.
Stepanian, V.A. (1964). "Zemletriasenia v Arminskom nagorie i okrestonostiakh." Izdat, Ayastan, Erivan.
Stiros, S. (1996). Identification of earthquakes from archaeological data: methodology, criteria and limitations. In: Archaeoseismology (S. Stiros and R.E. Jones, Eds.), pp. 129–152. Fitch Laboratory Occasional Paper 7, Athens.
Taher, M.A. (1996). Les grandes zones sismiques du monde musulman à travers l'histoire. I. L'Orient musulman. *Ann. Islamologiques* **30**, 79–104.
Utsu, T. (1990). "Table of Disastrous Earthquakes in the World (from ancient time to 1989)." Tokyo.
Virolleaud, Ch. (1935). La revolte de Koser contre Baal. *Syria* **16**, 29.
Vogt, J. (1993). Historical seismology; some notes on sources for seismologists. *Historical Invest. Eur. Earthq.* **1**, 15–24.
Vogt, J. (1996). The weight of pseudo-objectivity. *Ann. Geofis.* **39**, 1005–1011.
Wells, D. and K. Coppersmith (1994). New empirical relationships among magnitude, rupture length, rupture width, rupture area and surface displacement. *Bull. Seismol. Soc. Am.* **84**, 974–1002.
Yeates, R., K. Shieh, and C. Allen (1996). "The Geology of Earthquakes." Oxford University Press, Oxford.

46

Earthquakes and Archaeology

Amos Nur
Stanford University, Stanford, USA

1. Introduction

Although earthquakes have often been associated with unexplained past societal disasters, their impact has been thought to be only secondary for two reasons: inconclusive archaeological interpretation of excavated destruction, and misconceptions about patterns of seismicity. However, new and revised archaeological evidence and a better understanding of the irregularities of the time–space patterns of large earthquakes together suggest that earthquakes have played a role, and at times a key role, in shaping the history of mankind in Central America, Peru, the southwestern USA, China, Japan, and the Eastern Mediterranean.

The Eastern Mediterranean region is one of the most actively excavated areas in the world of archaeology. Many sites in this region are in the form of "Tels" – large, flat-topped, largely artificial mounds, composed of the accumulated layers of collapsed human habitation. Many archaeologists and historians tend to attribute the widespread destructions found in many of these excavations to the actions of man at war and time. However, since this region is also a region of earthquake activity, it is inescapable that earthquakes and archaeology could be intimately related. It is only recently (Stiros and Jones, 1996) that earth scientists and archaeologists have come together to formally address the evidence for earthquakes and their significance in archaeological remains (Nur *et al.*, 1991; Guidoboni, 1994; Nur and Ron, 1996; Nur and Cline, 2000), and resolve the disagreements over the role earthquakes could have played in ancient society.

2. Earthquakes in Human History: Some Examples

Earthquakes have accompanied humans since the beginning of time. However, the time between large earthquakes is usually longer than one generation, often more on the order of ten generations. Thus most people have no personal experience with earthquakes, and they do not factor the possibility of earthquakes into their everyday lives, even if they know intellectually that they live in an earthquake hazard area. It is therefore sometimes hard to appreciate that major earthquake catastrophes are natural facts that have affected people since the dawn of human time. A few examples are described in the following sections.

2.1 Ubeidiya

The oldest human-made structure we know on Earth is found in Ubeidiya, Israel (Bar-Yosef, 1993), a little gully inside a small pull apart basin right on the Dead Sea transform fault south of the Sea of Galilee and near the River Jordan. The "structure" is the simplest possible—a flat floor made of pebbles—that was put there around 1–1.5 Ma by the appropriately named "Pebble Man." Apparently, these humanoids made the floor by pressing small river pebbles into the muddy bank on the Jordan. Thus, instead of working and sleeping in the mud, they created a relatively dry base for their operations.

When this site was first excavated, it was not immediately clear what the surface was, since today it is not flat, but tilted over 60° from the horizontal (Fig. 1). Today it is believed that this large cumulative tilt of the Pebble Man floor is the sum of

FIGURE 1 The 1.5 My-old tilted pebble plateform of the "Pebble Men" at Ubeidiya, Northern Israel. (Photograph: Institute of Archaeology, Hebrew University, Jerusalem.)

INTERNATIONAL HANDBOOK OF EARTHQUAKE AND ENGINEERING SEISMOLOGY, VOLUME 81A ISBN: 0-12-440652-1

small tilts by separate earthquakes on the nearby Dead Sea transform fault over the 1.5 My since the floor was created.

2.2 The Shanidar Cave

The earliest evidence we have for direct interaction between earthquakes and humans is from ca. 60 000 years ago in the Shanidar cave in the Zagros Mountains of north-eastern Iraq. Archaeologist Richard Solecki (1959) dug up four major ceiling collapses and some 20 minor ones. Several of these collapses (Fig. 2), which were almost certainly caused by earthquakes on a nearby seismically active fault, killed Neanderthals living in the cave. We know this because several of their skeletons, crushed by rock falls, were found beneath fallen blocks of stone.

2.3 Nineveh

The oldest written description of an earthquake is found on a tablet from the ancient city of Nineveh, (described by Ambraseys and Melville, 1982, following Thompson, 1937): "On 21 Elul, an earthquake took place. All the back part of the town is down . . . $30\frac{1}{2}$ cubits therefrom being strewn and fallen on the near side of the town. All the temple is down . . . . Let the Chief [architect] come and inspect."

There is a curious story associated with this account. According to the code of Hammurabi, the ancient ruler of Nineveh, a builder was held legally responsible for the integrity of his building. In case of collapse, the punishment was his execution or that of his first-born son. This letter therefore is about the mitigating circumstances that would be

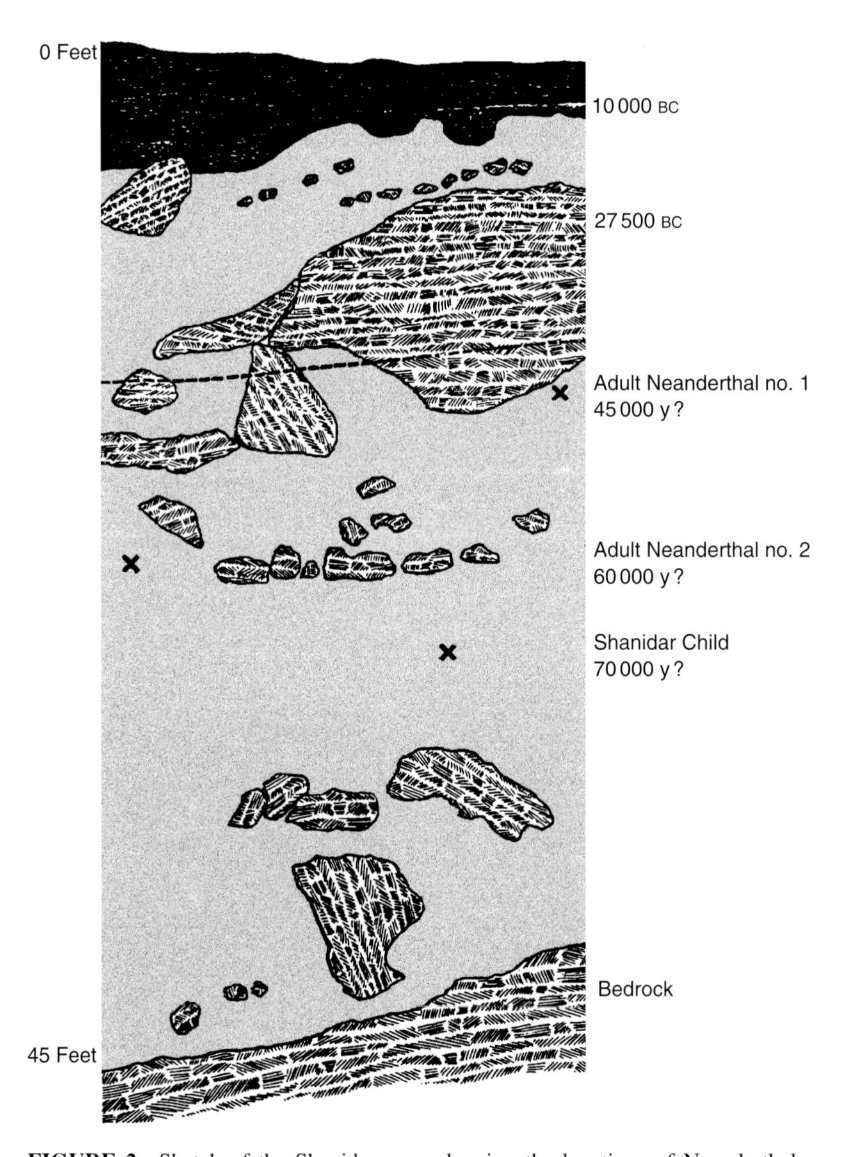

FIGURE 2 Sketch of the Shanidar cave showing the locations of Neanderthals crushed by earthquake induced roof collapses 45 000–70 000 y ago (after Solecki, 1959).

considered in this case, since earthquake-resistant design was probably not considered by Hammurabi.

2.4 Jerusalem circa 760 BC

This is the first clearly written description found in the Bible of an earthquake for which there is also independent archaeological evidence: "The words of Amos, during the reign of Uzziah, King of Judah, two years before the Earthquake." (Amos 1:1). This earthquake devastated Jerusalem, and led to what might be viewed as a written description of strike-slip type fault motion, according to Zechariah (14; 4–5) "... And the mount of Olives shall cleave in the midst thereof toward the east and toward the west, and there shall be a very great valley; and half of the mountain shall remove toward the north, and half of it toward the south. And ye shall flee to the valley of the mountains; ... as ye fled from before the earthquake in the days of Uz-zi'ah king of Judah."

3. Rise and Fall of Cities

3.1 Armageddon circa 1000 BC and 1200 BC

Not only did earthquakes periodically damage cities, they also created the landforms that made some of those cities politically and economically important. The Ancient city of Megiddo in Israel (Armageddon) shown in Figure 3 is a fascinating example of this (Davies, 1986; Nur and Ron, 1997a,b). Megiddo's political and economic importance in the ancient world stemed from its strategic position at the mouth of the Nahal Iron Pass through the Carmel-Gilboa Mountain range, an important route for ancient traffic between ancient Assyria and Egypt. However, the pass and mountain range which controlled traffic in the area at the time were the product of motion along the seismically active Mt. Carmel-Gilboa fault system (a branch of the Dead Sea Fault). The city's proximity to the Mt. Carmel-Gilboa fault system placed it in an area of high earthquake risk. The great importance of Megiddo's strategic location led to some of the greatest ancient battles fought in this region, and was the reason for the maintenance of its fortifications for close to 5 millennia (until ca. 500 BC).

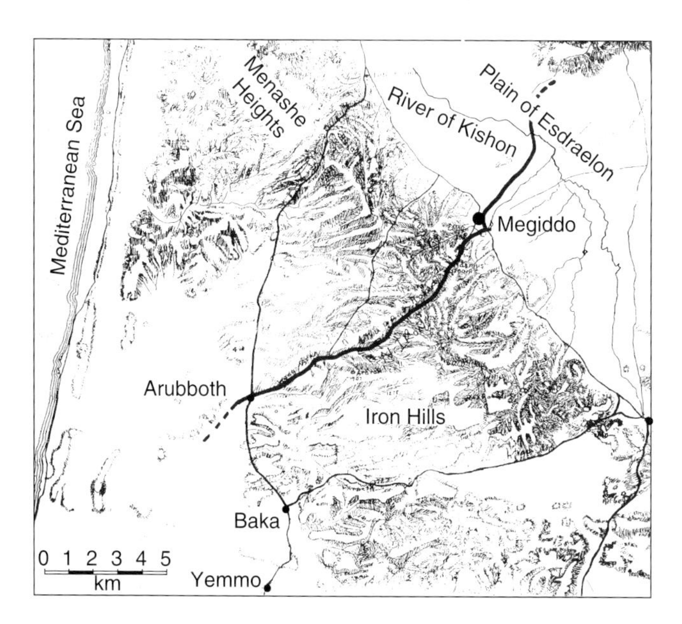

FIGURE 3 Location of Megiddo (Armageddon) (after Nelson, 1913).

Earthquakes were thus responsible for both the rise and the fall of Megiddo as an important walled city. Earthquake activity that shaped the land also explains the over 30 levels of excavated destruction of Megiddo, destruction sometimes assigned, for lack of a better explanation, to unproved battles. For example, King David's often assumed (but not documented) conquest and mindless destruction of Megiddo may have actually been a destructive earthquake in Northern Israel around 1000 BCE (Kempinski, 1993).

4. Earthquake Indicators in Archaeological Sites

There is widespread disagreement about what constitutes archaeological evidence for earthquake destruction especially

FIGURE 4 A row of columns that fell during a strong earthquake in 749 AC, Susita, Northern Israel.

when acts of war, the more exciting explanation, can also be invoked. It is, therefore, helpful to consider some of the most telling features of earthquake damage found in archaeological ruins.

FIGURE 5 The Roman Colosseum: the missing half of the outside wall collapsed in an earthquake in 1349 (Guidoboni, 1994).

4.1 Fallen Columns

An attacker may, by the application of a great deal of manpower or ingenious leverage, topple the columns of a building. However, when an entire row of heavy columns (Fig. 4) is found toppled in the same direction, the sudden ground motion of an earthquake is the most likely explanation. A spectacular example is Susita, an abandoned ruin overlooking the sea of Galilee in Israel. It was a thriving place during Roman and Byzantine times, but began to decline in importance when the country was overtaken by the Arabs in 638 AD. However, the final blow was struck not by the Arab armies, but by the earthquake of 749 AD, which destroyed the temple that stood here, leaving behind the spectacular evidence of a row of a dozen parallel fallen columns, each weighing about 15 tons.

4.2 Collapsed Walls

The uniform toppling of walls is also a sign of earthquakes. When many similarly oriented walls at a site fell in the same direction, particularly when they bury grain, gold, or other valuables in their fall such as at Troy, Megiddo, Jericho, Mycenea, and many other sites, the action of an army is an unlikely cause. An example is the Roman Colosseum where the

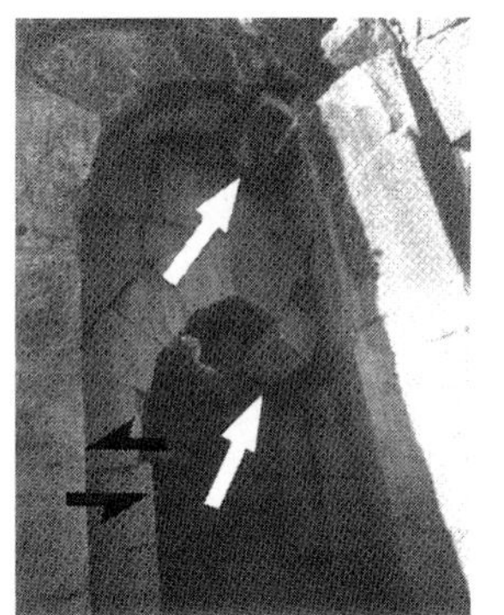

Kalat Namrood, Israel, 1202 AD

Coliseum, Rome, Italy, Ist Millenium AD

Long Beach, California, 1933

Stanford University, California, 1906

FIGURE 6 Comparison of slipped key stones at Kalat Namrood in Northern Israel by the 1202 AD earthquake with slipped key stones in the Roman Coliseum; a high school in Long Beach, California in 1933; Stanford University (1906).

entire southern part of the exterior oval-shaped wall is today missing (Fig. 5). It collapsed during an earthquake in 1349 AD.

4.3 Slipped Keystones

In buildings built with unreinforced arches, earthquakes sometimes cause one or more of the wedge-shaped stones to drop down until the arch is again wedged tight. If the slip is large enough, the entire arch collapses; if not, the dropped keystone remains as clear evidence of the disruption. Four examples, two older ones (Nimrod in Israel ca. 1200 AD, and the Colosseum in Rome, Italy 1349 AD (Guidoboni, 1994)) and two modern ones (Geology corner at Stanford University 1906 and the Long Beach high school entrance, California 1933) are shown in Figure 6. In all four cases, as indicated by the arrows in Figure 6, the bases of the arches slipped sideways due to motion in the earthquake, loosening the key stones in the arches.

5. Crushed Skeletons

People do not intentionally bury their dead in the rubble of collapsed buildings. Likewise, when a building is under attack and soldiers are laboring to topple a wall, few would wait long enough to be crushed beneath the stones. Thus the presence of crushed skeletons beneath ruined buildings is accepted by most archaeologists as some of the best evidence of collapse due to an earthquake. One example, already mentioned, is the Shanidar Cave in Iraq (Solecki, 1959) in which eight Neanderthal skeletons were uncovered. Apparently, some of these people were killed by earthquake-induced rock falls from the cave ceiling. One skull was found crushed on top of a stone, with a pile of rocks on top of it. The nature of the breakage indicates that the victim was probably alive and standing upright when the falling rock mass struck him.

One of the strangest stories of crushed skeletons was uncovered in Arkhanes on the island of Crete, Greece. Yannis Sakellarakis and Efi Sapouna-Sakellaraki (1981) discovered here the ruins of a Minoan temple that had been destroyed around 1700 BC by a large earthquake that also destroyed many other palaces on Crete. In one of the rooms of the temple the skeletons of a priest and his female acolyte were found on the floor before a typical Minoan altar. The priest had his arms thrown up to protect his face. On the altar was the skeleton of a young man reclining on his right side, and probably bound hand and foot. Lying on his bones was a ceremonial knife; indeed, it appears that a destructive earthquake had interrupted a religious rite of human sacrifice in full swing!

A heartbreaking example of earthquake casualties was uncovered by David Soren (1985) and his colleagues in the ancient Kourion, Cyprus. The earthquake destroyed the town at dawn on 21 July 365 AD, when most of the city was still asleep. The excavations revealed entire homes and stables that had collapsed, crushing people and animals inside beneath the rubble. So many skeletons were found, and the site was so undisturbed after the earthquake, that Soren said he and his colleagues "felt like a rescue team arriving 16 centuries too late." The most moving find here was the discovery of a family – a man, a woman, and their one-year-old child – killed under their collapsed home (Fig. 7). It appears that the three had little time to awake and realize what was happening, for their skeletons seem caught in a pose of sleep.

FIGURE 7 A family – man, woman and 1-year-old baby – crushed under their home that collapsed by the 370 AD earthquake in Kourion, Cyprus (Soren, 1985).

6. Regional Earthquake Damage

The effects of earthquakes are not limited to one site but may cause damage to structures over an area on the order of thousands of square kilometers. Although some of the effects of earthquake destruction have been recognized in archaeological excavations for decades, few archaeologists have applied their theories of earthquake destruction outside the boundaries of their own excavations onto a regional scale. However, historic records and archaeological evidence clearly demonstrates the regional extent of damage caused by many past large and even moderate earthquakes. A few examples are described here.

6.1 Holy Land Earthquake in 363 AD

By combining written evidence and dated excavated ruins throughout the region, one can construct a map showing the sites affected by this historically well-documented earthquake (Fig. 8, Guidoboni, 1994). Russell (1980) has suggested that severe damage occurred as far south as the Nabatean city Petra. Recent excavations have revealed the destroyed remains of the oldest church in the world in Aqaba, Jordan—some 120 km further south of Petra (Parker, 1998). The northernmost damaged site is Dan in Israel, 440 km from Aqaba, suggesting a possible magnitude of 7.5–8 for this earthquake.

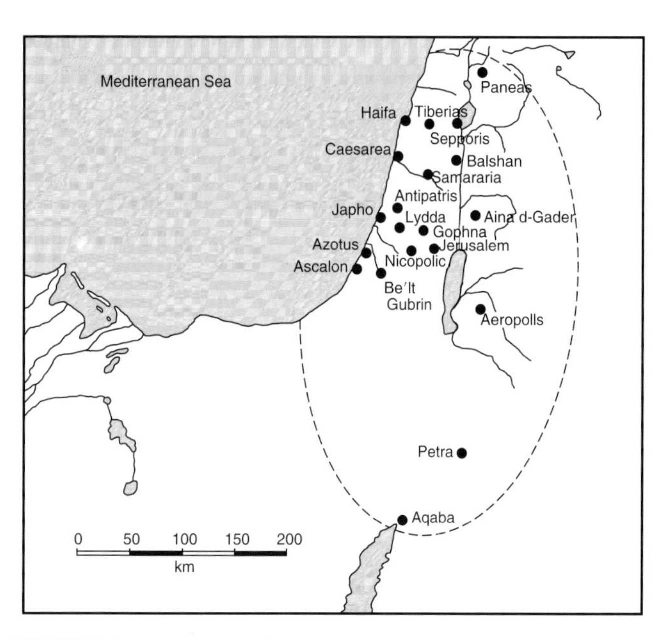

FIGURE 8 The extent of severe damage caused by the strong 363 Holy Land earthquake. The damage zone is 440 km long suggesting a possible magnitude of 7.7–8.0 (after Guidoboni, 1994).

6.2 The North Anatolian Fault System

Two particularly destructive earthquakes of this active fault system in the Eastern Mediterranean are considered here.

The first occurred on 17 August 1668, with a probable magnitude of 7.5–8.2. To the intensity VII region, mapped by Ambraseys and Finkel (1988) from written reports, we have added the location of Hattusas, the ancient capital of the Hittite empire (Fig. 9). Hattusas was of course just a ruin in 1668. However, it was a thriving great power when it was destroyed inexplicably at the end of the Bronze Age ca. 1200 BC (Yon, 1992). This superposition illustrates the possibility that in 1200 BC Hatussas may have experienced an earthquake perhaps like the 1668 event; if so, the city could not have escaped major damage, perhaps even total destruction, and subsquent abandonment (Nur and Cline, 2000).

The second large earthquake struck in Western Turkey on 9 August 1912. To the map of damage caused by this

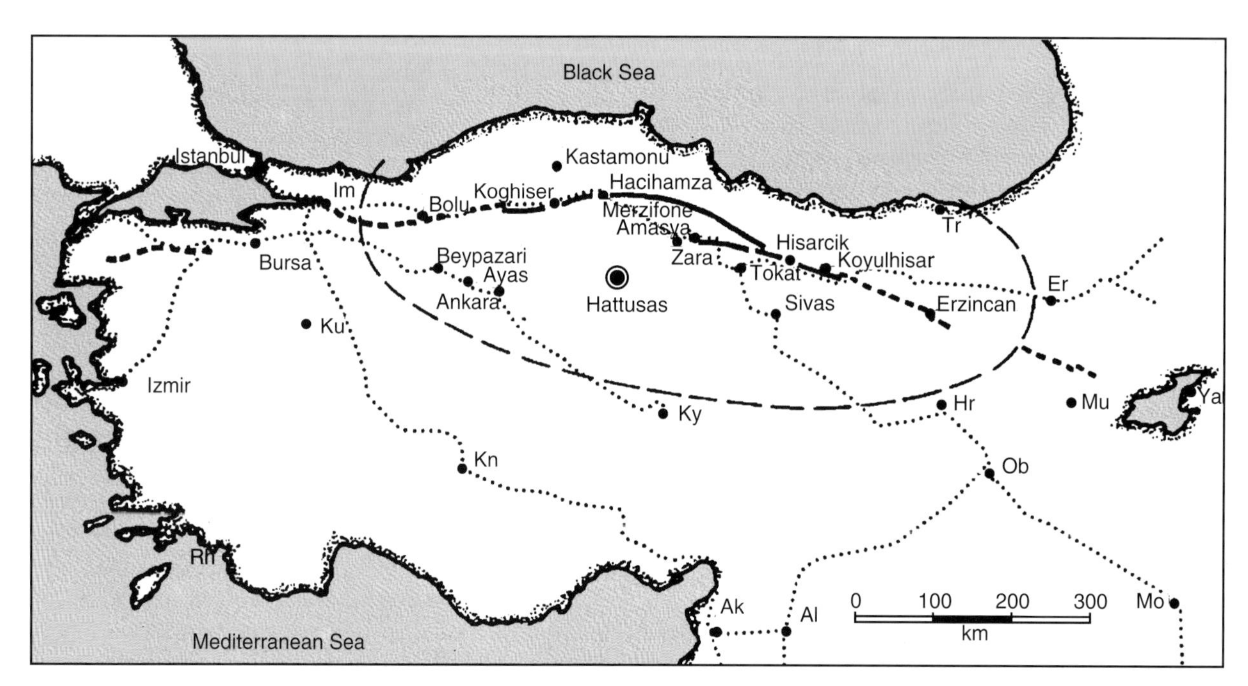

FIGURE 9 Location of Bronze Age Hattusas, the Hittites capitol, well within the intensity VII contour for the August 1668 earthquake on the north Anatolian fault in Turkey (after Ambraseys and Finkel, 1988).

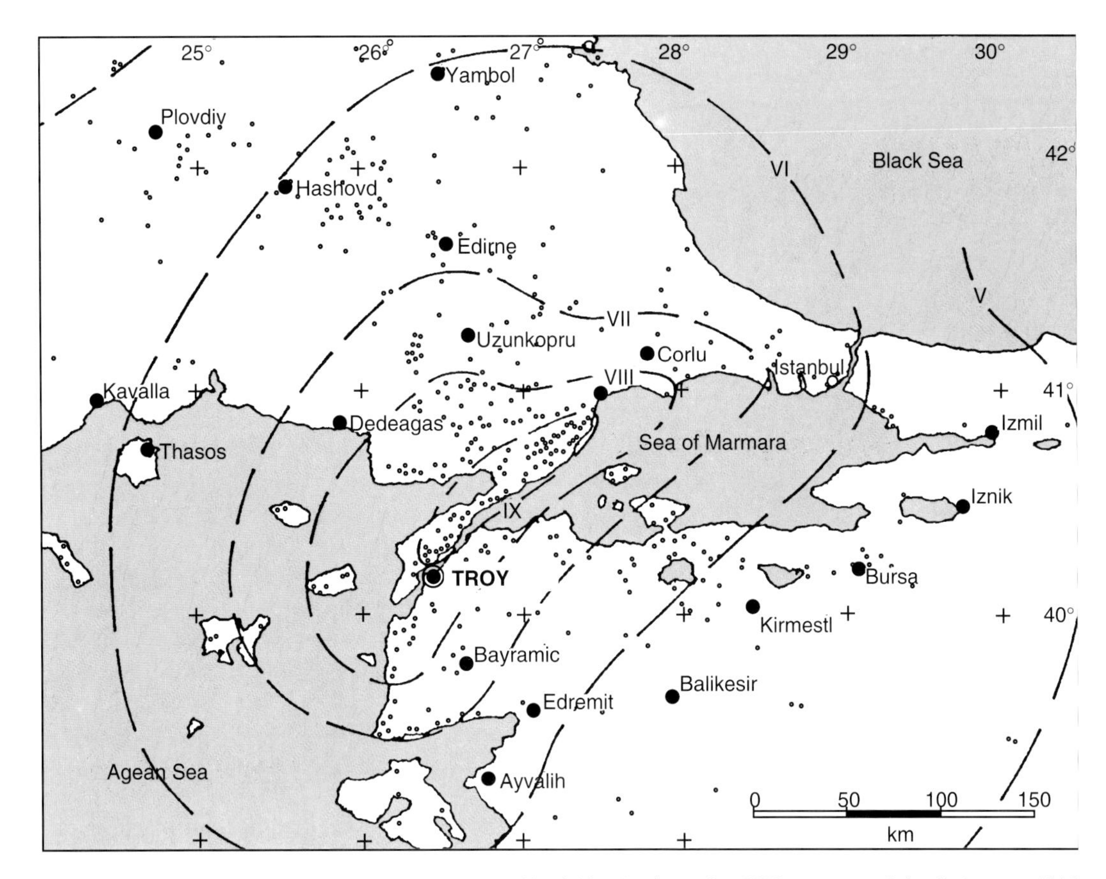

FIGURE 10 Location of Bronze Age Troy, well within the intensity VIII contour of the 9 August 1912 earthquake in north-western Turkey (after Ambraseys and Finkel, 1987).

earthquake, published by Ambraseys and Finkel (1987), we added the location of ancient Troy (Fig. 10) which falls inside the intensity VIII contour. Of course, Troy was only an archaeological ruin in 1912, but if this kind of earthquake ever struck while Troy was a thriving city it would have resulted in great damage, and possibly even total destruction. Archeological evidence does in fact suggest that Troy VI or VII was badly damaged by an earthquake of similar intensity (Blegen *et al.*, 1953, 1958).

As can be seen from these few examples, damage from large earthquakes is regional in extent. It is therefore important to consider the effects of earthquakes in the archaeological record on a regional scale, to attempt correlation between sites, and even to consider the effect of such widespread damage on a civilization's economy and infrastructure.

7. Earthquake "Storms"

One of the most important aspects of the occurrence of large earthquakes in the Eastern Mediterranean and Near East is their apparent tendency to occur in episodic bursts of activity or "storms." The modern history of large earthquakes along the North Anatolian fault illustrates this phenomenon. A series of earthquakes, from 1939 through 1999, together released tectonic strain by unzipping this entire plate boundary over approximately 70 y (Stein *et al.*, 1997). The strain released by the surface slipping of this earthquake sequence (about 4 m on average) must have taken a few hundred years to accumulate. Consequently we should expect such storms to recur only every few hundred years with relatively quiet periods in the interim.

A great "earthquake storm," consisting of a dozen or so earthquakes, appears to have unzipped not just one but several plate boundaries throughout the Eastern Mediterranean over a vast area in the 4th century AD, between 350 and 380 AD (Fig. 11). This 30 y period appears to have been preceded and followed by relatively quiet periods, probably for about 300 y.

It appears therefore that, in general, large earthquakes in the Eastern Mediterranean tend to occur in episodic "storms" that unzip the main plate boundaries during short, 30–100 year, periods separated by 300–500 years of relative quiescence. The implication of this for archaeology is that a great deal of physical damage over a large area and the resulting societal effects are expected to have happened during relatively short periods of time.

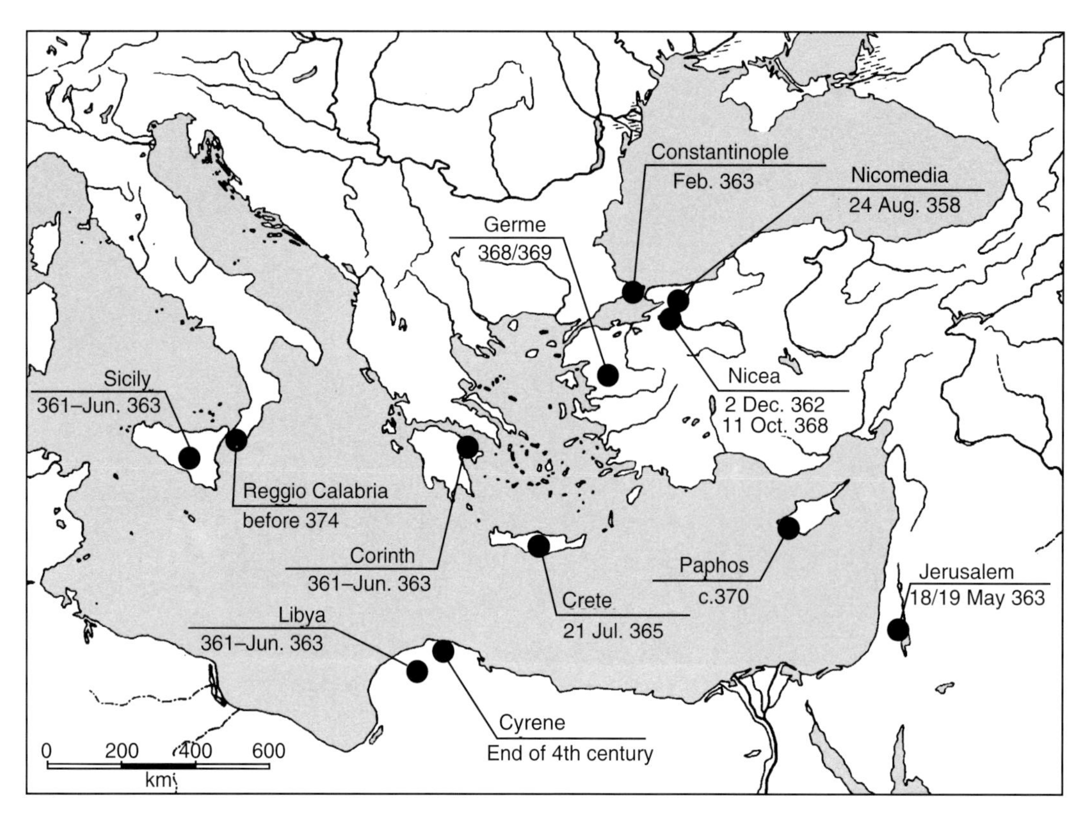

FIGURE 11 The location of sites damaged in the 4th century AD earthquake "storm" in the Eastern Mediterranean (after Guidoboni, 1994).

8. Earthquake Storms, System Collapse, and Human Conflicts

The prevailing opinion among many archaeologists and historians has been that earthquake destruction has not been sufficient to substantially affect the social order. It is true that a healthy city or nation with a robust economy can rebuild relatively quickly after even a major earthquake, possibly achieving complete recovery within a few decades. However, in areas and periods of social tension, unrest, or economic decline, an earthquake can indeed be the trigger for social overturn.

One instance is the Holy Land earthquake of 31 BC. Because of its intensity in Jerusalem, the earthquake caused portions of the city's walls to collapse (Josephus Flavius). Occurring during a state of war between the Arab nation in today's Jordan and the Jewish nation in Jerusalem, this prompted an Arab attack on Jewish-held Jerusalem. In this case, the city held, but had the earthquake damage to the city walls been slightly greater, the outcome could certainly have been different.

In another example, the slave uprising at Sparta in about 469 BC may have been triggered by a large earthquake, which greatly damaged the defenses of the city that sheltered the ruling elite. "... The Spartans began to practice these cruelties... after the great earthquake when the Helots and Messenians together attacked Spartans, causing widespread damage... and putting their city in great danger" (Plutarch, "The Life of Lycurgus").

The grandest historical event (although without written evidence) that may be related to earthquake storms is the catastrophic ending of the Bronze Age (Schaeffer, 1948; Nur, 1998; Nur and Cline, 2000). This system collapse over the entire Eastern Mediterranian region has been well known to historians and archaeologists for over a century, yet the cause remains an enigma to this day.

Figure 12 captures the magnitude of this catastrophe by showing some of the main sites that were physically damaged or destroyed in the Eastern Mediterranean at end of the Bronze Age ca. 1200 BC. Evidently, the collapse of cities and nations occurred over a period of some fifty years between ca. 1225 and 1175 BC.

The prominent French archaeologist Claude Schaeffer proposed in 1948, well before the emergence of the theory of plate tectonics, that one or several earthquakes were responsible for the end of the Bronze Age (Schaeffer, 1948). Schaeffer's idea was widely rejected and even spawned an antiearthquake bias among archaeologists that persists to this day. For example Elizabeth French wrote in 1996: "Archaeologists of my generation, who attended university in the

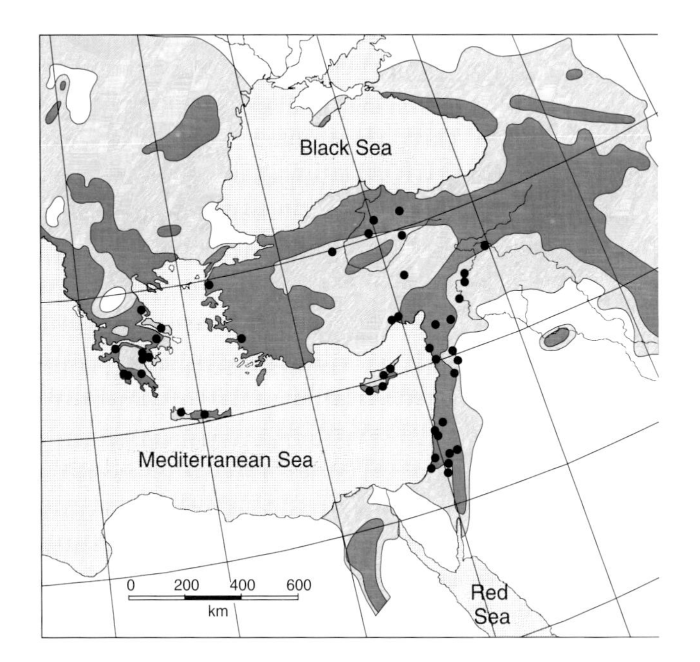

FIGURE 12 The Aegean and Eastern Mediterranean sites destroyed ca. 1225–1175 BC (after Drews, 1993) superimposed on the Aegean and the Eastern Mediterranean regions in which the intensity was greater than VII (after Karnik, 1968).

immediate aftermath of Schaeffer's great work in 1948, were brought up to view earthquakes, like religion, as an explanation of archaeological phenomena to be avoided if at all possible. Thus, it is only recently that an earthquake at Mycenae has begun to be a serious hypothesis."

Figure 12 shows also the contours for Mercalli intensity VII ground motion for large earthquakes in the area (after Karnik, 1968). The result is very suggestive; most of the destroyed sites fall within the high intensity regions, so they have most likely been badly damaged by earthquake at some time in their past.

Although there are no written records going back as far as 1200 BC, there are over a dozen skeletons found crushed beneath collapsed buildings at sites that were destroyed at the end of the Bronze Age (Nur, 1998; Nur and Cline, 2000). As described earlier, this sort of evidence is suggestive of the impact of earthquakes.

As pointed out by Nur (1998) and Nur and Cline (2000), it appears that earthquakes, most likely through an earthquake storm, could have been responsible for ushering in the physical collapse at the end of the Bronze Age. Of course earthquakes alone probably have not destroyed entire societies or powers. More likely they rendered cities and population centers vulnerable, even defenseless to human attack and at times even invited attack, such as in Jerusalem in 31 BC, Sparta in 469 BC, and perhaps in Jericho ca. 1400 BC. In other words, at the end of the Bronze Age, an earthquake storm could have triggered or enabled wars and attacks which then struck the finishing blow to civilization throughout the Eastern Mediterranian.

9. Conclusion and New Questions

There are several reasons why the study of earthquakes in archaeology could be important and should be better incorporated into the interpretation of archaeological sites. First, it provides a much better understanding of the impact of natural disasters on history, when there is no consensus on this (see Tainter, 1988). For example, Ambraseys (1971) adamantly states, "Earthquakes in the past twenty-five centuries have had little, if any, serious influence on historical developments in the Middle and Near East... but they have never caused the ruin of a culturally advanced state, far less the end of a civilization..." Under usual circumstances, Ambraseys is correct. Earthquakes are instantaneous events that cause sudden damage, but societies generally can recover even from the severe damage a large earthquake causes. However, when the effects of a series of earthquakes is compounded with wars or economic difficulties, the end result can be cataclysmic.

While Ambrasey questions the impact of earthquakes on societies, Rapp (1986) questions the evidence itself: "Without direct evidence for seismic destruction, earthquakes should be 'explanations of last resort' in archaeology." However, if an area is known to be susceptible to earthquakes (from geological and geophysical evidence), then earthquake damage is as respectable a hypothesis for destruction, unless historical or archaeological evidence suggests otherwise.

An important geophysical question that may be contributed to by archaeological evidence is how, when, and where historical and prehistoric earthquakes occurred. If earthquake storms are indeed a central feature of the time/space distribution of large earthquakes—their clustering in time is the process that unzips plate boundaries—then our thinking about how earthquakes accommodate plate motions may require important revisions. The storm process implies, for example, that large earthquakes on one fault or fault segment interact with faults many rupture lengths away, much farther than the current theories of plate deformation suggest. For example, can a 70 km long earthquake rupture associated with a magnitude 7 earthquake near Jericho somehow lead to a magnitude 7 earthquake in say Sicily or Western Greece?

The earthquake storm pattern also has an obvious consequence for earthquake prediction research; the probability of a large earthquake during the long "quiet" period may be an order of magnitude lower than during the active "storm" period. This may be applicable, for all we know, not only in the Mediterranean region but also in other areas of complex plate boundaries, e.g., China, Central America, even California, where sufficiently long historic and prehistoric records are not available and therefore we cannot even tell whether activity is governed by storms at all.

The conjunction of earthquake science and archaeology is a field that is ripe for study. Although most of the data to date are anecdotal in nature, improvements in both archaeological methods and geophysical understanding should allow a much better understanding of how earthquakes affected ancient civilizations. Furthermore, both sciences would benefit from collaborative work. Earthquake prediction efforts may be helped by extending the earthquake record back into prehistory, and archaeologists may gain a new understanding as to the types of damage earthquakes or earthquake sequences could have caused at their respective sites.

References

Ambraseys, N.N. (1971). Value of historical records of earthquakes. *Nature* **232**, 375–379.

Ambraseys, M.N. and C.F. Finkel (1987). Seismicity of Turkey and neighbouring regions, 1899–1915. *Ann. Geophys.* **5**, 701–726.

Ambraseys, N. and C. Finkel (1988). The Anatolian Earthquake of 17 August 1668. In: "Historical Seismograms and Earthquakes of the World," (W.H.K. Lee, H. Meyers, and K. Shimazaki, Eds.), pp. 173–180. Academic Press, New York.

Ambraseys, N.N. and C.P. Melville (1982). "A History of Persian Earthquakes." Cambridge University Press, Cambridge.

Bar-Yosef, O. (1993). Ubeidiya. In: "The New Encyclopedia of Archaeological Excavations in the Holy Land." (E. Stern, Ed.), pp. 1487–1488. The Israel Exploration Society, Carta, Jerusalem.

Blegen, C.W., J.L. Caskey, and M. Rawson (1953). "Troy III: The Sixth Settlement." Princeton University Press, Princeton.

Blegen, C.W., C.G. Boulter, J.L. Caskey, and M. Rawson (1958). "Troy IV: Settlements VIIa, VIIb and VIII." Princeton University Press, Princeton.

Davies, G. (1986). "Megiddo." Lutterworth Press, Cambridge.

Drews, R. (1993). "The End of the Bronze Age." Princeton University Press, Princeton.

French, E.B. (1996). Evidence for an earthquake at Mycenae. In: "Archaeoseismology." (S. Stiros and R.E. Jones, Eds.), pp. 51–54. Fitch Laboratory Occasional Paper No. 7, Athens.

Guidoboni, E. (1994). "Catalogue of Ancient Earthquakes in the Mediterranean Area up to the 10th Century." Instituto Nazionale di Geofisica, Rome, 504 pp.

Josephus, Flavius. "Wars of the Jews." Chap. XIX.

Karnik, V. (1968). "Seismicity of the European Area." Academia Press, Prague, Part 2, Maps.

Kempinski, A. (1993). "Megiddo: A city-state and Royal Center in North Israel." Verlag C.H. Beck, Munich, Germany, 208 pp.

Nelson, H.H. (1913). The Battle of Megiddo. PhD thesis. University of Chicago.

Nur, A. (1998). The end of the Bronze age by large earthquakes? In: "Natural Catastrophes during Bronze Age Civilizations" (M. Bailey, T. Palmer, and B.J. Peiser, Eds.), pp. 140–149. British Archaeological Reports, Oxford.

Nur, A. and E. Cline (2000). Poseidon's horses: plate tectonics and earthquake storms in the late Bronze age Aegean and Eastern Mediterranean. *J. Archaeol. Sci.* **27**(1), 43–63.

Nur, A. and H. Ron (1996). And the walls came tumbling down: earthquake history of the Holy Land. In: "Archaeoseismology," (S. Stiros and R.E. Jones, Eds.), pp. 75–85. I.G.M.E, Athens.

Nur, A. and H. Ron (1997a). Armageddon's earthquakes. *Int. Geol. Rev.* **39**, 532–541.

Nur, A. and H. Ron (1997b). "Earthquake! – inspiration for Armageddon." *Biblical Archaeol. Rev.*, July/August, pp. 48–58.

Nur, A., C. MacAskil, and H. Ron (1991). "And the Walls came Tumbling Down: Earthquake History of the Holy Land," Video Documentary, Dept. of Geophysics, Stanford University, 57 min.

Parker, T.S. (1998). An early church, perhaps the oldest in the world, Found in Aqaba. *Near Eastern Archaeol.* **61**(4), 254.

Plutarch, "Life of Lycurgus," (28.11–2) (Ref. in Guidoboni, E. (1994). "Catalogue of Ancient Earthquakes in the Mediterranean Area up to the 10th Century," Instituto Nazionale di Geofisica, Rome, 504 pp.).

Rapp, G. Jr (1986). Assessing Archaeological Evidence for Seismic Catastrophies. *Geoarchaeology* **1**, 365–379.

Russell, K.W. (1980). The earthquake of May 19 A.D. 363. *Bull. Am. School Oriental Res.* **238**, 47–64.

Sakellarakis, Y. and E. Sapouna-Sakellaraki (1981). Drama of death in a Minoan temple. *Nat. Geogr.* **174**, 30–53.

Schaeffer, C.F.A. (1948). "Stratigraphie Comparee et Chronologie de l'Asie Occidentale." Oxford University Press, London.

Solecki, R.S. (1959). Three Adult Neanderthal Skeletons From Shanidar Cave, Northern Iraq. *Annual Report – Smithsonian Institution*, Pub. 4392, pp. 603–635.

Soren, D. (1985). An earthquake on Cyprus. *Archaeology* **38**(2), 52–59.

Stein, R., A. Barka, and J.H. Dietrich (1997). Progressive failure on the North Anatolian fault since 1939 by earthquake stress triggering. *Geophys. J. Int.* **128**, 594–604.

Stiros, S. and R.E. Jones (1996). "Archaeoseismology." Fitch Laboratory Occasional Paper No. 7, Athens, 268 pp.

Tainter, J.A. (1988). "The Collapse of Complex Societies." Cambridge University Press, Cambridge.

Thompson, R.C. (1937). A new record of an Assyrian earthquake. *Iraq* **4**, 186.

Yon, M. (1992). The end of the Kingdom of. In: "The Crisis Years: The 12th Century BC" (W.A. Ward and M. Joukowsky, Eds.), pp. 111–122. Kendall/Hunt Publishing Company. Dubuque, Iowa.

Editor's Note

Please see also Chapter 45, Historical seismicity and tectonics, by Ambraseys *et al.*, and Chapter 47, Historical seismology, by Guidoboni.

47

Historical Seismology: the Long Memory of the Inhabited World

Emanuela Guidoboni
SGA, Storia Geofisica Ambiente, Bologna, Italy

1. What is Historical Seismology?

For a number of years, many seismological, geophysical, and seismological engineering conferences, especially in Europe, have devoted a section specifically to this field of study. It is in fact now accepted that historical data provide valuable information about the "size" and frequency of past earthquakes, and this contributes to the formulation of seismic hazard and risk estimates and the location of active faults, especially where particular geological complications make their identification difficult. Historical seismology (not to be confused, of course, with the history of seismology) is that branch of seismology which uses historical data to identify the effects of past earthquakes and to answer specific seismological questions: when, where, and how large did these earthquakes occur? Nobody teaches this discipline, and (so far, at least) it has no positions in academic institutions or schools; and yet it produces data which require financial investment and continuity of effort. As a discipline, historical seismology can be considered to stand at the point where historical research, from which it takes its research methods and interpretative rules, "intersects" with seismology, from which it takes its problems and aims. It is thus at one and the same time a type of seismology, in that its results relate to that discipline, and a type of history, since its data derive from specifically designed historical research.

The aim of historical seismology, therefore, is to gain an understanding of those past earthquakes, which no longer have a place in human memory. It is true that there were some who studied historical earthquakes even before the days of modern parametric catalogs, but they did so in a quite different way. The change in methodology came with the diffusion of computers, which began to spread into the principal research centers in the 1970s, and proved to be particularly suitable for dealing with all kinds of basic data, including those concerned with earthquakes. The introduction of computerized catalogs brought with it a revolution in the way information about historical earthquakes was dealt with, because it necessitated the conversion even of their less objective aspects into parameters and codes which could be handled in automatic form. Solving this important conceptual problem led to the introduction of far more modern techniques for the analysis of historical seismicity, affecting both the speed with which all the information concerning an area or site of interest could be extracted, and the possibility of dealing with this information not only in a qualitative way, but also using statistical methods.

Data from historical sources have been summarized only in part in traditional parametric earthquake catalogs. The fact is that extremely important assessment factors, which we now try not to lose sight of, are excluded from traditional earthquake lists. The most advanced catalogs are nowadays effectively information systems in which data are presented in a critical way, and grouped according to research levels or within a preestablished logical hierarchy. Among the countless available items of information, some are obviously suitable for calculating traditional parameters; but there are others suitable for "recalculating" them along new and more sophisticated lines, without it being necessary to go back to the basic descriptive data.

Although historical earthquake catalogs vary in quality from one country to another, they nevertheless have in common the fact that they are all based on descriptive data derived from human observation, that is to say, from the long memory of the inhabited world. Nevertheless, those who use these catalogs (whether seismologists, statisticians, engineers, or geologists), usually do not know or have only a hazy idea of the research processes, methods, and problems underlying the data which are available to them. Consequently, they often do not know how these data could be improved, verified, or used to better purpose; and this happens because historical seismology is in essence a multidisciplinary field. It seems quite strange to observe that underlying the relationship between those who produce these data (historians) and those who use them

INTERNATIONAL HANDBOOK OF EARTHQUAKE AND ENGINEERING SEISMOLOGY, VOLUME 81A ISBN: 0-12-440652-1

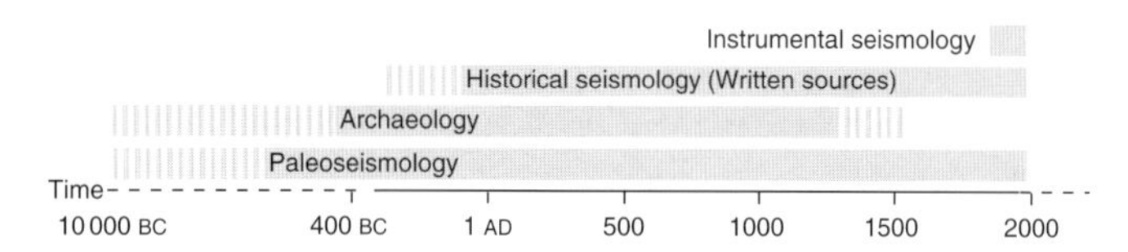

FIGURE 1 Timescale and disciplines which "capture" earthquakes.

(seismologists) there has developed a practice based almost on a "nonformalized epistemology," which to some extent recalls the old antithesis between word and number, or subjective and objective – a sort of "fossil guide" to European positivist culture. Although numerous theories about the complexity of knowledge have on the one hand taken away a certain amount of optimism and self-confidence, they have also, fortunately, made clear the limits of "languages" in the broadest sense of the term (words and numbers, that is to say), and the unending possibility of arriving at new interpretations. These profound changes in our cognitive context are today capable both of encouraging the various disciplines involved in the study of earthquakes to listen to one another more attentively, and of opening up new perspectives (Fig. 1).

2. The Cultural Roots of a Science of Earthquake Observation

The mid-20th century decisions taken at Strasbourg and Utrecht set in motion, perhaps without foreseeing the consequences, an exceptionally influential historical data reassessment process. In many countries in the Far and Near East and the Mediterranean area, there already existed a very ancient tradition of studies and works on earthquake effects and the interpretation of their origin, which provided a sort of "seed bed" for this kind of research.

In the Mediterranean area, for example, earthquake geography already existed as early as Greek and Roman times, and there was a conviction that some areas were more seismic than others. There were numerous naturalists and literary texts describing ancient and contemporary seismic phenomena and their interaction with the inhabited world, thereby demonstrating that there was an interest in the "historical" even in ancient authors writing about earthquakes. But apart from such deliberate use of historical information by writers of the past, there is obviously also "involuntary" written evidence of seismic effects. Data of this kind have managed to survive even from very ancient times, and they are material for historical seismology. The earliest written evidence of an earthquake is the so-called "Letter from Niniveh" (Assyrian Empire), dating to the 13th century BC. It records the restoration of the temple of Emasmas and the tower gate in Niniveh, both of which had been damaged by a violent tremor (Thompson, 1937).

China has an important and very ancient tradition of references to earthquakes, the earliest of which goes back to 1177 BC (Academia Sinica, 1970). Earthquakes, like celestial phenomena, were signs to be interpreted for court horoscopes, and hence their occurrence was recorded in the registers of the astronomical office of the Chinese court. Perhaps for this reason, however, and in spite of their undoubted historical and cultural value (Needham, 1959), the ancient data which have survived are rather slight in quantity and quality and are now difficult to use (Lee *et al.*, 1976; Lee and Brillinger, 1979). But the more recent data, especially from the Ming dynasty (about mid-14th century) onward are useful.

As far as the Mediterranean area is concerned, the earliest information about real earthquakes (that is to say, those for which a location and time are provided, as opposed to those which are in some way myth-related) goes back to about 760–750 BC. It concerns an earthquake, which struck Jerusalem and the valley of Hinnom. Of the rare earthquakes mentioned in the Bible, this is one for which there is other evidence in Hebrew and Greek sources (Guidoboni *et al.*, 1994). In other cases, it is religious symbolism, which predominates. The first earthquake in Persia to be recorded in history, on the other hand, is one, which destroyed the region of Ray in the 4th century BC (Ambraseys and Melville, 1982).

But apart from the rarity and curiosity value of the earliest data, which are often difficult to put to practical use, present-day historical seismology needs to be able to establish detailed seismic effect scenarios, in order to add to the quality and quantity of entries in historical earthquake catalogs. The aim is to bring the image of apparent seismicity (that is to say that part of seismicity which it has been possible to reconstruct by means of historical seismology) as close as possible to actual seismicity. From this point of view, research acquires a special scientific significance when there are substantial quantities of evidence available from sources which were produced on a regular basis and have been well preserved (archive documents and manuscripts). An analysis of the effects of a past earthquake involves various fields of investigation, from the history of administration and institutions (responsible for dealing with the disasters) to that of material culture (the means and techniques of building construction) and that of the history of ideas and culture (the reception and interpretation of the earthquake). It is not always the case that a country's history permits such a systematic and broad approach; but even a few written remains can be illuminating.

The preservation of this particular kind of historical record in modern Europe has a tradition, which began in the Italian Renaissance in imitation of classical texts. Thus, the first known catalog of historical earthquakes was written in 1457 by Giannozzo Manetti (1396–1459). It is preserved in various manuscript codices, and records 70 seismic events from antiquity to the 15th century, together with sources. From the 16th century onwards, an erudite fashion for compiling lists of past earthquakes became widespread, as part of a renewed interest

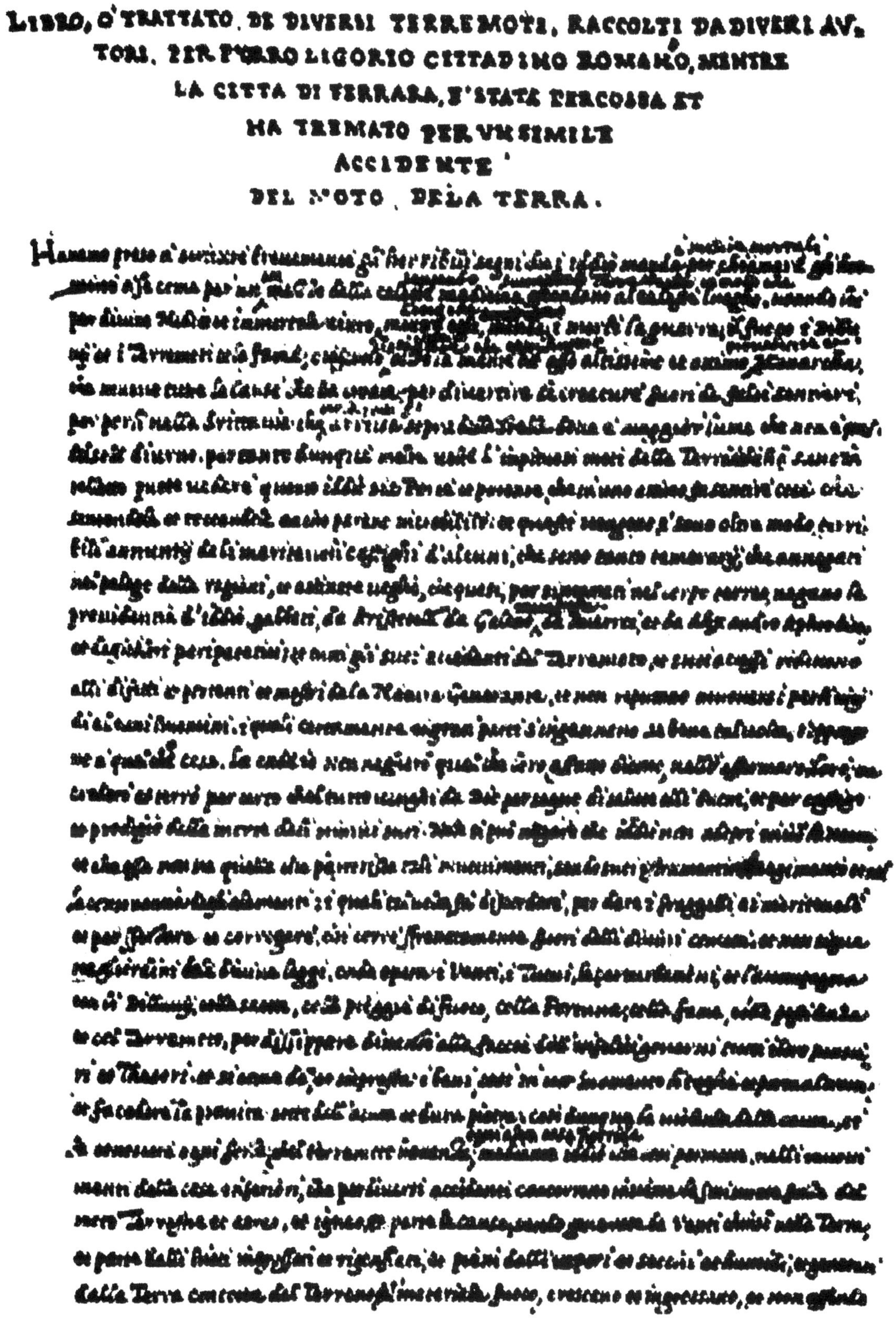

LIBRO, O' TRATTATO. DE DIVERSI TERREMOTI, RACCOLTI DA DIVERSI AV-
TORI. PER PIRRO LIGORIO CITTADINO ROMANO, MENTRE
LA CITTA DI FERRARA, E' STATA PERCOSSA ET
HA TREMATO PER UN SIMILE
ACCIDENTE
DEL MOTO DELA TERRA.

Hanno preso a' scrivere [illegible]

FIGURE 2 Title page of the manuscript treatise (Archivio di Stato di Torino *Trattato di diversi Terremoti*, 1570–1574) by Pirro Ligorio, famous architect of the late Italian Renaissance. This precious catalog contains also a project for a seismic-resistant house (Guidoboni, 1997).

in "signs" and all those phenomena which were generally considered to be "prodigious." This resulted in the production of a number of earthquake lists, though they were obviously not intended for practical use. Among the most important were those of Conradus Lycosthenes 1557 and Pirro Ligorio (manuscript) in 1574 (Fig. 2).

In the more advanced court circles in Italy in the second half of the 16th century, a sort of "theoretical anxiety" was expressed in relation to earthquakes. This undermined previous faith in the Aristotelian theory, and brought into being a new attitude which, in spite of its fragile and contradictory nature, led to the production of extensive and detailed descriptions of seismic sequences. Descriptions of phenomena which were not understood were deliberately passed on "to posterity" in the hope (as is explicitly stated in a manuscript of 1570, Ferrara, Biblioteca Comunale Ariostea, J. de Robertis, cl.I, 294) that someone would find a use for them in the future. There were various earthquake theorists and compilers in the 17th century, but it was above all Bonito's catalog (1691) which became a sort of "ideal" archetype for earthquake catalogs of modern Europe. It is a unique product of literary and historical erudition, with boundless geographical interests (even listing earthquakes in Japan and Latin America) (Fig. 3).

In the early 18th century, central Italy was struck by a very long and violent seismic sequence, consisting of three strong earthquakes. They lasted from January to February 1703, even causing damage in Rome. The first earthquake "diaries" date

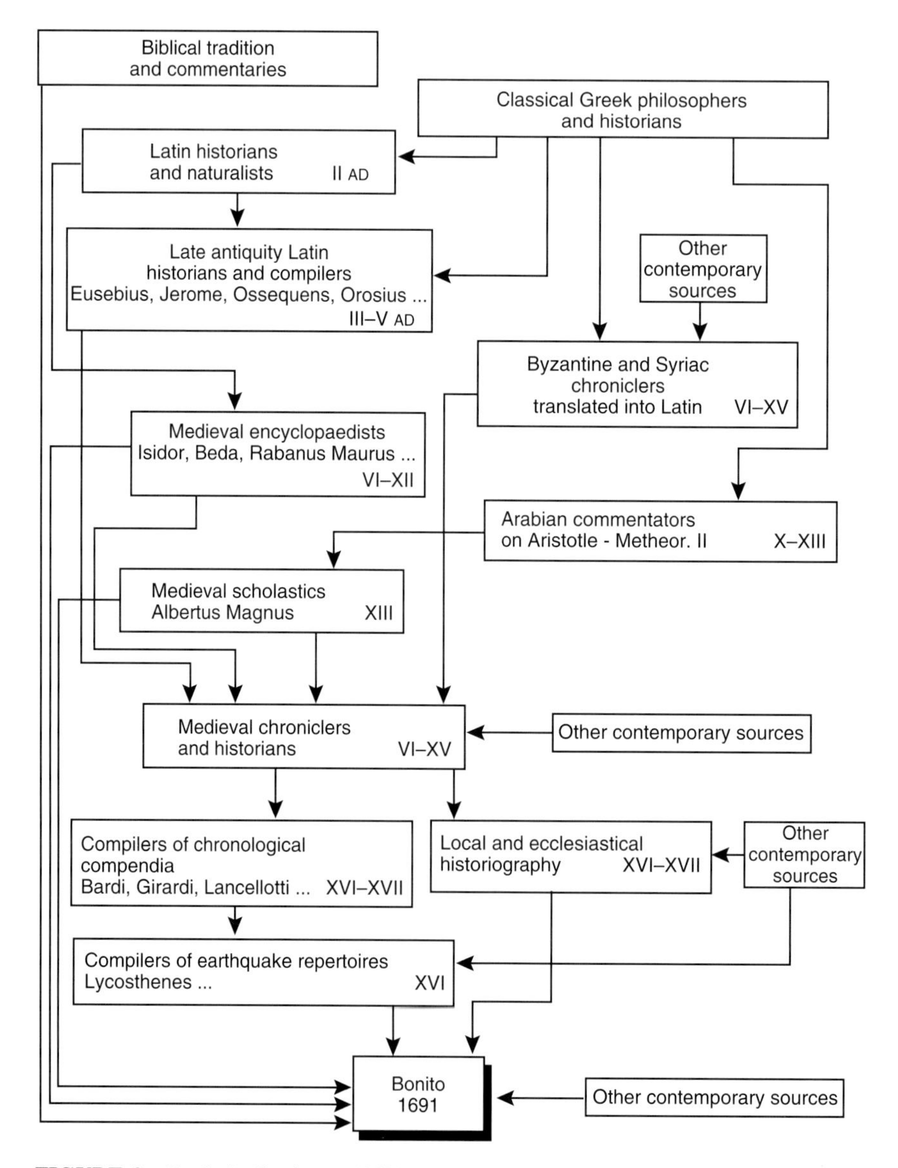

FIGURE 3 Bonito's Catalogue (1691): contribution of different typology of historical sources and various traditions. This is an example of the cultural contexts from which we derive the historical European earthquake catalogs (from Guidoboni and Stucchi, 1993).

from this time. Subsequently, and particularly after the earthquake which destroyed Lisbon in 1755, the attention of physicists and naturalists was particularly drawn to the study and theoretical interpretation of striking seismic events in their individuality, so that the compilation of earthquake catalogs became a matter of secondary concern (Seyfart, 1756; Bertrand, 1757; Moreira de Mendoca, 1758).

This process of observation reached its most intense toward the end of the 18th century (Fig. 4) when the five great earthquakes, which devastated Calabria from February to March 1783, aroused a sort of descriptive frenzy amongst scientists and naturalists. Thousands of pages were written simply to describe that interminable seismic sequence (it lasted for several years) and all its effects. They wanted not so much to interpret, as had previously been the case, but to observe. That extraordinary effort left its mark on European culture and consciousness.

What happened in this field is certainly better known for the 19th century than for earlier periods (see Valone, 1998). To mention only the most famous names, they were David Milne (1805–1890), Robert Mallet (1810–1882), and Karl Ernst Adolf von Hoff (1771–1837). Giuseppe Mercalli (1850–1914), and finally Mario Baratta (1868–1935), who laid the foundations for the first systematic organization of historical earthquake data. Although all these men were primarily specialists, they were also scientists with many scholarly interests, and so they promoted "learning." But multidisciplinary science was still beyond the horizon.

Some aspects of historical seismology research are of principal interest in the seismological and geophysical spheres, but it can also reveal how historical building techniques have been modified with the passage of time in order to mitigate the effects of successive earthquakes.

3. Sources for Historical Seismology

3.1 Types of Sources

Even in the most different historical contexts, geographically and culturally far apart, historical seismology finds its basic data in the same three great record "containers": (1) individuals (chronicles, letters, diaries etc.); (2) institutions (archive sources); and (3) scientific field.

Countries with an ancient written culture and intense seismic activity have almost all developed and consolidated their knowledge of past earthquakes over the centuries, often producing treatises or earthquake catalogs of considerable cultural importance; but they are not always immediately usable today. The old antiquarian and positivist tradition used sources – if it used them at all – only after extracting them from their cultural and territorial context, as though they were fragments of lost records which had been deposited, sometimes by chance, in the tradition of literary or historiographical texts. In the closing decades of the 20th century, however, an effort has been made to go back to the original sources as reliable evidence, to render them usable by means of a process of analysis and treatment, and in doing so to make thoroughly clear the process of evaluation and use of the data involved (Fig. 5).

3.2 Sources Deriving from the Memory and Observations of Individuals

In old and traditional earthquake catalogs, a quite privileged position was occupied by chronicle sources, i.e., sources deriving from the memory of individuals, and from their desire to narrate and describe. The presence of these data sources in

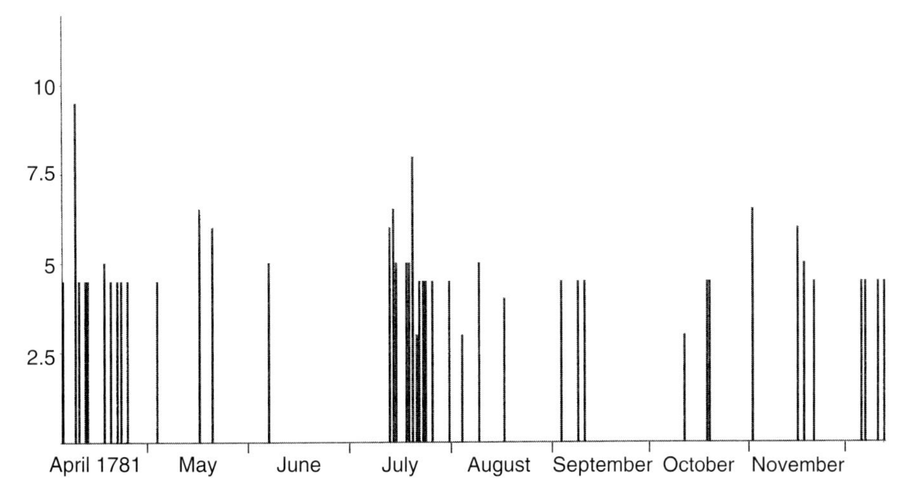

FIGURE 4 This graphic synthesizes available information on the proceeding of the shocks during the earthquake from 4 April to 12 December 1781, which destroyed an area of the Romagna Apennines (Northern Italy). An example of the shocks attested by direct and independent witnesses.

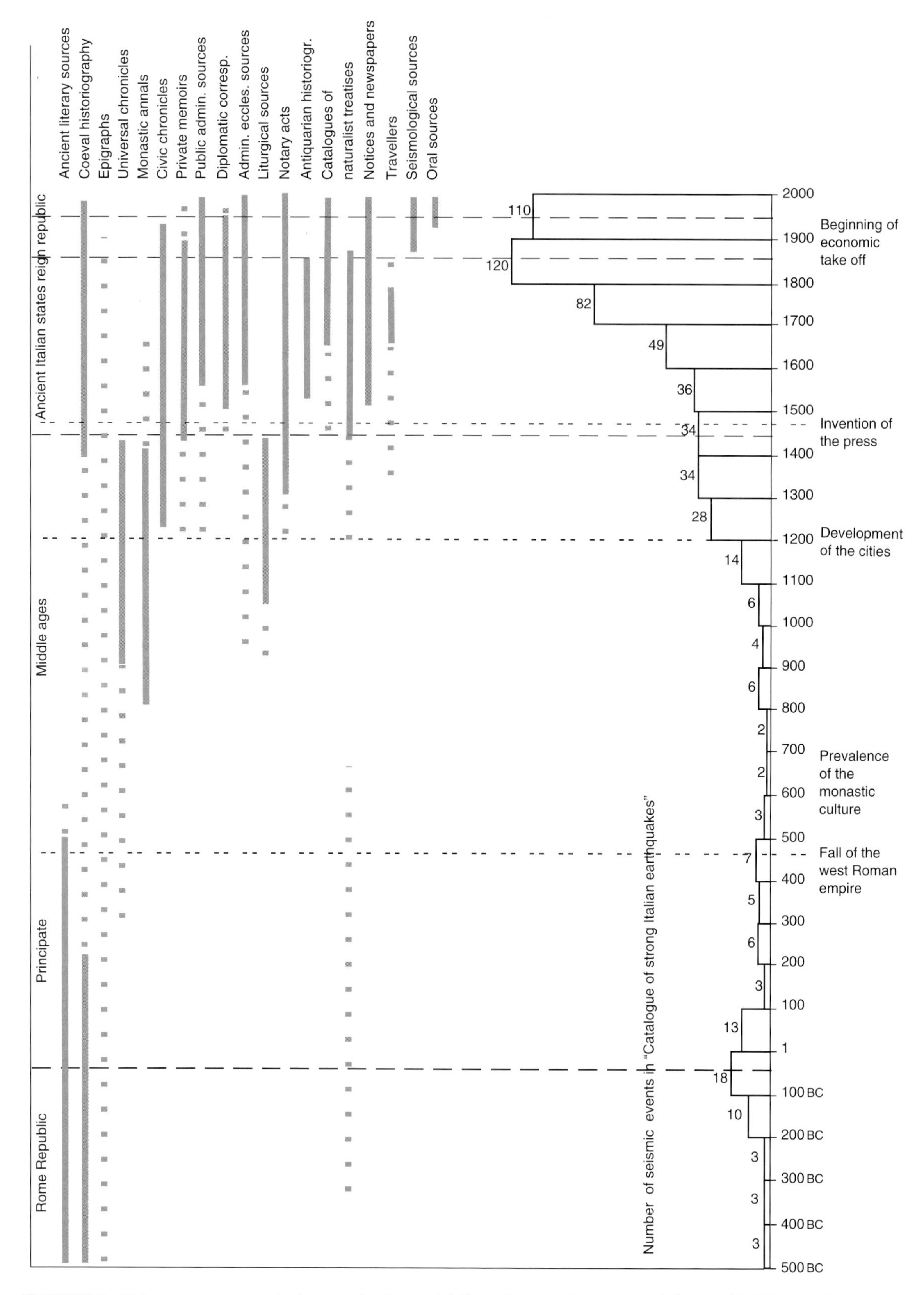

FIGURE 5 Italy represents a particular case for the availability of sources in a temporal frame of 2500 years. The graphic shows the prevailing typologies of sources and their availability in time from the 5th century BC up to the 20th century AD. Since the end of the 16th century various typologies of independent sources guarantee a good level of nationwide, informative coverage. Nevertheless this does not imply the "completeness" of the catalog, which still needs work to be pursued.

old catalogs now necessitates a painstaking analysis of the texts concerned, how they are related and what they derive from, in order to throw light on the possible transmission of errors into the first earthquake lists and from there into earthquake catalogs. Hasty or inexpert critics have often tended to denigrate chronicle sources, thereby mistaking the effect for the cause, so to speak; their own poor or inadequate work being interpreted as a limitation in this kind of source itself, so that the texts concerned were judged to be unimportant or unreliable.

It is certainly true that making use of personal memories requires dedication as well as profound knowledge and a particular critical sensitivity. For such sources are personal universes, immersed in their own time and permeated with their own culture; and so they are very powerful "filters" placed between the actual phenomena and our present hope of understanding them. Only by strictly appropriate means and scrupulous care can we derive from these sources an understanding of many aspects of earthquakes, which would not otherwise be available to us. In the first place, we have to compare the various independent reports of a single event, paying particular attention to those aspects of the individuals' reports which are or are not in agreement with one another, for example, local effects, effects on buildings, and even the perception of the duration of the tremors in a seismic sequence. Secondly, an analysis of these reports makes it possible to correlate data from the individual record with those from public institutions, and hence to pick out complementary information, confirmations, and additions, as though one were dealing with a difficult jigsaw puzzle. This is not a literary exercise, therefore, but rather of a method of work designed to utilize traces of human memory as though man was a very special "seismograph." And the results can justify the effort, because human memory and perceptiveness have in various ages and in quite different parts of the world made a quite extraordinary contribution to our understanding of seismic events, by describing their effects on the environment and on the buildings of the time.

This "network" of data about inhabited areas has today allowed us to learn about large seismic sequences of the past, nondestructive effects in places far from epicentral areas, low-energy local earthquakes, or deep, high energy tremors with very extensive felt areas (Fig. 6).

3.3 Sources Produced by Institutional and Administrative Bodies

Although personal sources are very valuable, historical seismology's most important contribution in recent years has come from research in the field of institutional and administrative sources. Unlike sources based on the individual memory (where there is an intention to leave traces), institutional sources are produced within the framework of administrative, bureaucratic and legal rules and customs, as applying to the government of a particular territory. Precisely because of their "involuntary" nature, these involve fewer subjective influences affecting our attempt to understand the key factors in the effect of a seismic disaster (or at least its damage effects) on the buildings and economic and administrative structures of a society: what can be defined as seismic impact.

Institutional sources are preserved, of course, in archives. These are complex historical memory structures, which partly reflect the structure of organizations, offices, and official committees – in broad terms, that is to say, of the administrative, political, and jurisdictional authorities, which produced them.

Archive sources of interest in the assessment of seismic effects are those concerned with the administrative and financial decisions taken after a destructive earthquake in order to deal with the emergency, control and plan reconstruction, and organize a return to normal habitational conditions. It often happens that the cultural or political importance of the places affected has resulted in earthquakes being mentioned in diplomatic correspondence, as can be seen in the vast number of such diplomatic exchanges preserved in European archives. But of even greater importance for our understanding of the impact of a destructive earthquake within its economic context, are the communications between the various administrative hierarchies, and between central and peripheral authorities, whether political, administrative, or religious. Because we need to understand and make use of documentation of this kind, it is also important to understand the very many extratextual factors concerning both how individual institutions work (a matter of administrative, and political history), and the relationship between institutions and earthquake-affected localities, within the network of relationships which operate after a strong earthquake between central and peripheral authorities, and between authorities and others (whether public, ecclesiastic, or private) who are at the place concerned.

As far as this specific research field is concerned, there are two special stages involved in sound archive work: (1) the translation of seismological questions into historiographical questions; and (2) the translation of historiographical questions into archive searches. Where archives are concerned, one has to be guided by the need not just to find material or subject matter directly connected with earthquakes, but also to identify in an organic way the documentation produced by institutions in relation to their role in the postearthquake phase, i.e., documentation which will be closely connected to administrative, financial, and juridical control over the places affected.

3.4 Sources Produced by the Scientific Institutions of the Past

The third type of source consists of the evidence produced by scientific bodies of the past, which were concerned with earthquakes. In some countries, such as Italy, they came

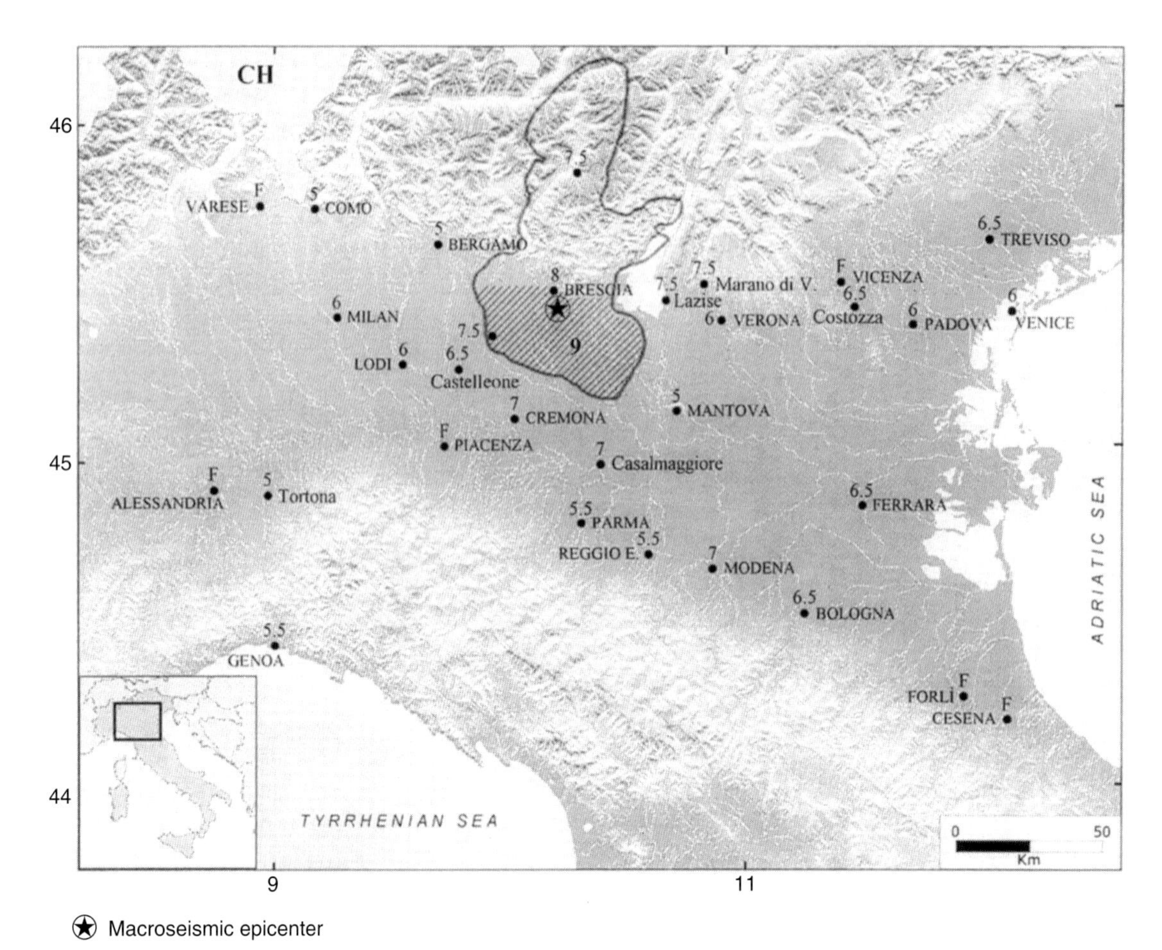

FIGURE 6 Map of effects of the 1222 earthquake in northern Italy. Historical seismology allows us to acquire information on the effects of earthquakes, even the ones from the remote past, and often with a surprising amount of detail. In this case, among the numerous examples now available, the map portrays the effects of the earthquake occurring on 25 December 1222 which hit northern Italy (Io = VIII MCS scale; Me 6.2). This earthquake had for a long time been turned into a legendary event, as a catastrophe of "European" proportions, by historiography and the early seismic catalogs. By the mid-20th century it had entered the parametric Italian catalogs being classified as one of the strongest earthquakes in this area (Io = XI MCS, Me 6.8). But the critical analysis of the accounts of the monks, the notaries and above all of the Pope's chancellory have brought to light less destructive effects, more compatible seismotectonic characteristics of the affected area (from *Catalogue of Strong Italian Earthquakes*: Boschi *et al.*, 2000). The pattern highlights the most affected area, as indicated by the medieval contemporary sources (the area south of the Brescia dioceses).

into being in the mid-18th century through the activities of academies. In the 19th century, centers were set up and organized specifically for this purpose, in the form of observatories or geodynamic and meteorological services run by private individuals who often belonged to religious orders (e.g., Barnabites and Pietists). These bodies produced bulletins, macroseismic postcards (rudimentary but effective questionnaires about local earthquake effects), field surveys, and damage reports, which governments were already beginning to demand. There has so far been no worldwide systematic collection of these texts (there is in fact no data bank where they can be accessed), but together with the first instrumental data, which are material for the history of seismology (Ferrari, 1992, 1997), they constitute a data heritage of great informational value which has not yet been fully utilized.

All these materials, with their different aims and authorship, contain important information about earthquake effects as they occurred, and they now add in a very special and specific way to the information in documents and historical sources of an archival or personal kind belonging to the same period. These sources have been taken into consideration during the infancy of modern seismology, but have often been misunderstood and neglected, even within the seismological sphere (as qualitative data, they have been considered out of date and irrelevant); and to gather them together and utilize them in a systematic way today requires a substantial effort of organization.

4. Earthquake Effects on the Environment

Historical descriptions of the effects of strong earthquakes on the environment are important, but often scattered or neglected. Information of this kind can be found not only in reports belonging to recent centuries but also in ancient texts; and they provide valuable material for seismotectonics and the location of active faults. It is not easy to classify this material, because one has to understand the cognitive framework and means of expression used in producing such descriptions. At times they may appear to be figments of fantasy, but they can be used if decoded and contextualized. Perhaps partly for this reason it is matter of interest that the third and latest edition of the new "Catalogue of Strong Italian Earthquakes" (CSIE) (Boschi *et al.*, 2000) contains more than 2000 codified and georeferenced environmental effects.

Now, although the memory of what happened a few decades ago is already fading, we have access to information about data of this kind as preserved in ancient texts. For in the ancient and late Roman world, the effects of earthquakes on the environment made a great impression on the minds of observers, due to that sense of the sacred and magical with which natural phenomena were imbued, and new attitudes that sometimes led to the idea of "nature" as unchangeable and mysterious. Indeed, seismic effects are often recorded in the sources with such emphasis that the direct cause fades into the background, so that today's reader may sometimes doubt whether the phenomenon described really was caused by an earthquake. In the Roman world, moreover, earthquake effects were seldom even mentioned, especially if they related to smaller settlements.

Evidence about the early Middle Ages was mainly produced in a monastic environment and therefore in very important, though not urban, cultural contexts. One might think that this would lead to a greater number of direct observations of effects on the environment being made. However, the culture that permeated the production of such observations did not favor explicit and direct observation of these phenomena. Take, for example, the description of the effects of the earthquake of 3 January 1117 on the waters of the Po in the heart of the river valley (see the attached Handbook CD, CSIE, in Boschi *et al.*, 2000). A written narrative tells us that the waters came to a standstill and rose up "like a bridge," this being interpreted as a sign from God; but since we now know how such phenomena occur, we may have good reason to examine this information from a different viewpoint.

Language often envelops such phenomena in an aura of legend, telling of mountains that split in two, smoking and roaring. How to reinterpret this type of strongly figurative language within our present cognitive models is not immediately obvious. In general, however, a better understanding of the meaning of often fantastical descriptions only becomes possible when certain effects are repeated in the same areas in later periods. Thus, if the splitting of a mountain can be explained with reference to fractures in the ground and the smoke to dust raised by the rolling of boulders, perhaps the roaring can be identified as the acoustic perception of low frequencies, possibly amplified by particular conformations of the subsoil.

Ever since the beginning of modern times (i.e., since the 17th century), descriptions of environmental effects have become the subject of discussions about the real origin of seismic phenomena; natural philosophers, physicists, and naturalists began to observe the splitting of the Earth, phenomena of liquefaction of the ground (leading to general or differential subsidence of foundations), the emission of flaming gases, poisonous exhalations and even phenomena of ionization of the air ["red sky," see Ferrara, Italy 1570 in Catalogo dei Forti Terremoti in Italia (Boschi *et al.*, 2000)].

The great observational leap toward descriptions which were more accurate and more immediately valuable within current scientific frameworks took place in the 18th century, clearly as the result of huge progress in the sciences and naturalistic thought. Descriptions of the effects of major earthquakes had already in the previous century proved to be aware of the marks left on the environment. An example may be seen in descriptions of the environmental effects of the earthquakes in Calabria (Southern Italy) in 1638, which were also responsible for creating a fairly large area of marshland near Sant'Eufemia (about $180\,km^2$) and a fracture about 11 km, 50 cm wide and 75 cm difference in level of one side compared with the other on the eastern slopes of the Sila mountain. The earthquake at Sannio (Southern Italy) in 1688 is another case in point. Despite the use of fairly general terms like "splitting" and "fissures" in the descriptions, the authors may perhaps be referring to much more complex phenomena, thus providing potentially important information about the seismic source.

4.1 Parameters

Moving from words to numbers is not easy, as can be seen from the whole complex question of parameters in historical earthquake catalogs (a matter, which perhaps deserves more discussion than it has received).

In the scientific world of seismology, macroseismic data are commonly considered to be information of a qualitative kind and therefore somewhat unreliable and generally inferior to instrumental data. This leads many workers in this sphere to reject such information, considering it to be nonscientific. Criteria for the estimation of epicentral and energy parameters for historical earthquakes have in fact rarely been given precise official standing in the past, and qualitative assessment techniques have often been adopted, partly because in many cases the information was not available in a form which lent itself to analytical treatment. Special techniques and algorithms are now used for defining time of origin, epicentral coordinates, epicentral intensity, and equivalent magnitude, as recorded in brief database records. What is probably most important in this process, apart from a clear description of the criteria for assessing the reliability of these estimates, is that decision processes themselves should be open to checking. But let us now consider one by one the principal parameters concerned: date, intensity, epicenter location, and estimated magnitude.

4.2 The Measurement of Time

The time of origin of an earthquake as given in catalogs is often treated as though it was beyond doubt, so that it acquires the status of an identification label. This would be valid in principle if the date and time of an earthquake had indeed been accurately established. In practice, however, this is not the case, since catalogs often do not take into account the date and time conversions necessary to align all historical earthquakes within the same time reference system, conventionally agreed as Greenwich Mean Time (GMT) in earthquake catalogs. In Italy, for example, a series of different dating styles and timescales have been used at different periods and in different places. For this reason, the conversion of historical dates and times to GMT requires careful attention and specific conversion methods, which were unknown to some compilers of previous catalogs. In the most serious cases, this lack of knowledge has led to the duplication even of high-intensity earthquakes. The problem has been approached by few authors (Dominici and Marcelli, 1979; Ferrari and Marmo, 1985), and a solution has not yet been fully absorbed into scientific practice.

Attempts to establish the exact date of occurrence of historical earthquakes from descriptions in the sources have highlighted two types of problem: (1) the difficulty of identifying a universally agreed date; and (2) that of converting it into current dating practice (Fig. 7). Countless pages have been written about the dating problems resulting from ancient calendars, for they have led to mistakes, doublets, and superimpositions in the seismological tradition. Knowledge of ancient dating systems, their equivalents in modern times, the problems raised by various calendar reforms (the Julian, the Gregorian, and that of unified time) is now something, which everyone expects to have. Indeed, there are small computer programs that will carry out a rapid "translation" from one calendar to another. The problem of a correct dating parameter, however, is not just one of correctly establishing the time of origin of the earthquake itself, but also that of dating the many different tremors which make up the overall sequence in large earthquakes. Accurately dating tremors, which preceded the major shock, even though by just a few hours or even minutes, may provide information about the mechanism of its origin. It is also important to be able to distinguish between the effects of separate but chronologically close earthquakes, for they may become superimposed in a single blurred image of a great nonexistent seismic event.

4.3 Intensity: Really So Simple?

Some literature on the historical origins of intensity scales has thrown light on the relationships between the various scales, from Mercalli's first formulation of 1883 to the latest elaboration of the European Macroseismic Scale (Grünthal, 1998). There have been more than 70 macroseismic scales of one kind or another, almost all using from 10 to 12 grades (see Guidoboni and Ferrari, 2000).

As has been pointed out, macroseismic scales first began to be applied to historical earthquakes with the arrival on the scene of the new generation of computerized catalogs early in the second half of the 20th century, thereby creating a series of problems. Work on solving these problems has not always been much exposed to scrutiny, each interpreter assuming complete freedom to apply their own criteria. The more sensitive and expert researchers have often complained of the excessive subjectivity of what was being done, but no "rules" or "protocols" for observance have ever been offered or requested.

The closing years of the 20th century, however, have witnessed an increasing sensitivity in the use of historical data. In countries like Italy, where historical catalogs are very important, being used even for locating active faults

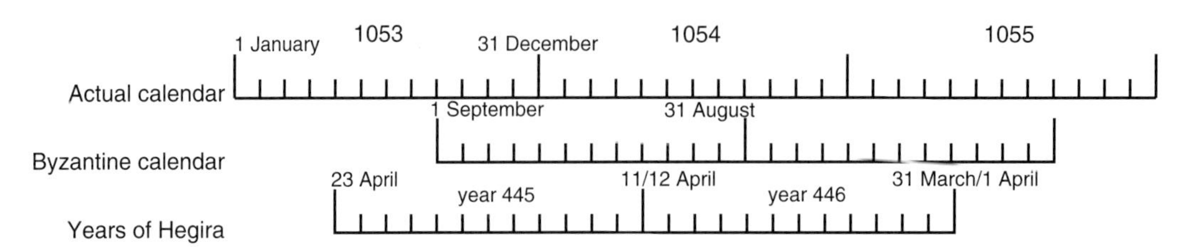

FIGURE 7 Practical example of how the term "year" indicates different chronological time frames depending on the chronological system used by the sources: the correspondence is shown between the contemporary calendar's year 1053–1054 and the Byzantine and Islamic year.

(Gasperini *et al.*, 1998), an attempt has been made to work out a sort of formalization of the criteria to be applied, and to throw light on the selection mechanisms which underlie the famous "subjective criteria" (Monachesi and Moroni, 1993; Molin, 1995; Giuffrè, 1995). There has also been a recent attempt to apply fuzzy sets theory to intensity, with some positive results (see Vannucci *et al.*, 1999, 2000, for the 1920 earthquake in Tuscany); but it is only applicable where there is substantial and rich historical documentation. The problem of assigning intensity values to earthquake effects has many implications, and to discuss them here would take us down too long a road. The factors suggest that a historical catalog is a living thing, which can be analyzed, dismantled, "repaired," and reinterpreted. It provides the base data for calculating seismic hazard and to locate active faults. But if calculations made for purposes of seismic hazard assessment are to be more than statistical exercises, there are a lot of questions to be asked, both from a historical and a seismological point of view.

4.4 Locating Epicenters

Although the definition of an instrumental epicenter is well established in the literature (see Richter, 1958, for example), the same is not true of a macroseismic epicenter. This is mainly due to a typical lack of standardization in macroseismic studies, and to the widely shared opinion that they are in any case unable to provide reliable estimates. Even Richter (1958) stated:

> The practice, in the absence of seismographs, of drawing isoseismals and then locating an "epicenter" at the center of the figure should be discontinued. In the majority of cases the instrumentally located epicenter proves to be at one side of the meizoseismal area.

That the macroseismic epicenter is often considerably distant from the instrumental epicenter should come as no surprise, since the two points clearly have a different physical significance. Indeed, whereas the instrumental epicenter corresponds to the projection onto the surface of the earthquake focus (i.e., the point where the rupture originates), the macroseismic epicenter represents in all likelihood the projection of the Braintree of the rupture zone. It has to be kept in mind, however, that the area of effects is always a matter of interaction between a particular earthquake and a habitational system (and therefore always provides us with additional information about the degree of vulnerability of a particular built-up area).

Establishing the location of a macroseismic epicenter is generally speaking not a problem when data are well distributed across the area concerned. But that is not always the case, especially when one wishes to go a long way back in time, or where the area of origin is at sea or in sparsely inhabited areas (a mountain or a large marshy area are usually a problem). In these cases approximations become rather large. However, the historical background to the area in question will throw light on the status of available information and provide what is necessary to define degrees of uncertainty (Fig. 8).

4.5 Epicentral Intensity and Equivalent Magnitude

The maximum intensity of an earthquake, I_{max}, is often the only parameter that can be related to the energy released by a historical earthquake. In some cases it is indicated as epicentral intensity, I_o, thus presuming that the macroseismic epicenter coincides with the location where maximum effects are observed. This may be acceptable in principle if the intensity has been observed at a single point. When, on the other hand, the location of the epicenter, as described above, is the result of calculating a mean of the coordinates of several points, a different approach must be used, partly because single determinations of intensity are often influenced by local amplification effects and estimation errors which may considerably alter its value.

It is well known, in fact, that the scalar seismic moment (M_0) is a much more significant estimator of the "size" of an earthquake than any definition of magnitude. As Gasperini *et al.* (1999) point out, the main reason for this is that this parameter has a formulation with a specific physical meaning, since it is linked to the dimensions of the source, the amplitude of the coseismic deformation, and the physical characteristics of the material in which the seismogenetic rupture is produced (see Color Plate 18). Thanks to this definition, the value of the seismic moment does not undergo effects of "saturation," as does occur with magnitude, since its seismological estimate is arrived at by making use of the whole spectrum of the elastic energy irradiated, and not just a small portion. In fact, since the seismological determination of the seismic moment requires digital seismograms, routine estimates of this parameter are available only when digital recording of earthquakes became common in the late 1970s.

5. Earthquake Archaeology

5.1 Enlarging the Span Time

In order to extend the temporal space available for observing seismic activity, or else to make up for the lack of written data, the historical seismologist sometimes has recourse to archaeology. Earthquakes are not a novelty in archaeology, since interest in them goes back at least to the second half of the 19th century. For example, de Rossi (1874) and Lanciani (1918) produced the first interesting observations about earthquake effects on the ancient monuments of Rome; Willis (1928) did the same for Palestine, and Sieberg (1932) for ancient Egypt, pointing out seismic effects at ancient temples, etc.

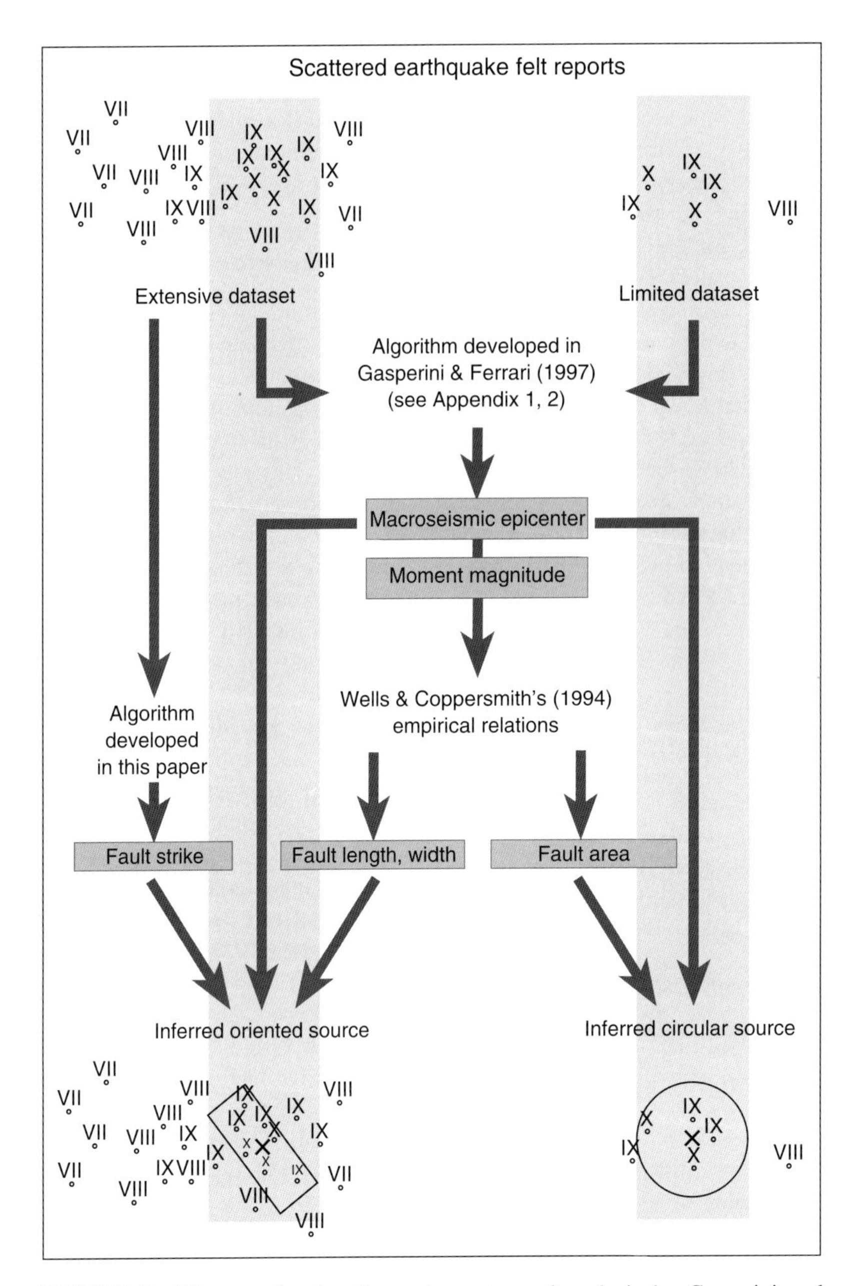

FIGURE 8 Diagram showing the various steps of analysis by Gasperini and Valensise (2000) from macroseismic data to inferred seismic source.

Earthquakes have attracted particular attention in Mediterranean archaeology, partly because of the existence of important and well-preserved ancient traces of their activity, and partly because of the strong seismicity of the area. Until the 1970s, archaeological earthquake studies were somewhat generalized attempts to find or draw attention to traces of seismic activity, or what were thought to be such traces, in ancient monuments or ruins. Only in the closing decades of the 20th century does one find seismologists taking a greater interest in such matters, and a specific orientation in that direction can be said to have begun only with the studies on Israel by Karcz and Kafri (1978, 1981) and Nur and Reches (1979). From the last decade of the 20th century onwards, there is evidence of a broader interest in archaeology. One can see this in the specific nature of some conferences organized either directly by geophysicists or jointly with historical seismologists, and archaeologists (Stiros and Jones, 1996; Boschi *et al.*, 1995; McGuire *et al.*, 2000).

However, even in areas such as the Mediterranean, where earthquakes are powerful and frequent, and the possibility of discovering earthquake effects in an ancient building or a whole archaeological site may not be rare, it is not as frequent as one might expect. There are various factors that limit such possibilities: (1) the history and culture of the sites concerned; (2) the state of preservation of the seismic scenario, or part of it; and (3) the degree to which archaeologists are interested in finding evidence of earthquakes.

Whether or not it is possible to find traces of seismic effects in an archaeological site depends in particular on the economic history of the site and surrounding area at the moment when the earthquake occurred; for if its economy was in a sound or expanding state, it was both possible and obviously desirable to restore and reactivate the damaged habitational and exchange systems, and hence erase any traces of the damage. On the other hand, if the site or settlement network was in a poor economic state, the few available resources would not permit reconstruction *in loco*, and the inhabited area would be completely or partly abandoned. The earthquake/abandonment dynamic is one which affected not only the ancient and medieval worlds, but also marginal areas in much more recent times (see, for example, the case of the 1968 earthquake at Belice in western Sicily, in the "Catalog of Strong Italian Earthquakes" in the attached Handbook CD in Boschi *et al.*, 2000).

5.2 Seismic Indicators in the Archaeological Record

When archaeology is trying to discover traces of earthquakes, one of its central problems is how to define seismic indicators. This is a complex subject with both general and particular aspects, and it has to be assessed within the specific context of an individual excavation. Unfortunately, however, there exists a "seismological common sense" which leads to the selection of elements that may well be the least typically archaeological. Since an earthquake is a phenomenon that always affects an "area" rather than a "point," it is necessary to apply concepts to which seismologists may be unaccustomed, such as the interpretation of landscape and its transformations.

As items of evidence, earthquake indicators typically have the characteristics and ambiguities of all other kinds of archaeological evidence: the nexus between the depositional output revealed by excavations and the action which brought it into being can be hypothesized or inferred, but it cannot be proved in a deterministic way. As a result, there are only a few cases where the literature has adopted a seismic hypothesis to account for a specific situation in the archaeological record. A common characteristic of these cases is the presence of macroindicators, such as collapses, in structures which can be described from a functional and constructional point of view as "prestige buildings," such as villas, baths, and sacred places of the late imperial age. In most of these cases, moreover, the earthquake is assumed to have occurred after abandonment of the site, for it is easier to find indicators of seismic activity where human reaction to seismic damage has not involved restoring the topography and structure of the site. We feel, however, that many pieces of seismic evidence may have been bypassed in this way. For example, data concerning rural settlements and modest buildings regularly escape notice, even though they were characteristic of extra-urban territorial organization in classical Greek and Roman times. The same is true of the various ways in which society reacted to seismic disasters from an economic and recovery point of view.

This complexity requires a few exclusively stratigraphic remarks. The existence of elevated sites introduces the problem of the downslope displacement of archaeological material as a result of erosion and deposition. Secondary deposits of this kind show the need for a careful analysis of possible earthquake effects on slopes. The same may be said of deformation and movement in architectural structures, such as leaning walls and sloping floors, when buildings are situated on steep slopes or areas subject to landslides. A building's resistance to ground movement is obviously conditioned not only by its own structure, but also by the nature of the ground itself. The frequency of structural deformation in monumental buildings (theatres, for example) highlights the problem of lack of solidity in basal strata and the unstable nature of the backfill on which they were built.

5.3 The Dating of Archaeological Indicators from Written Sources: the Birth of "Seismic Monsters"

Among archaeologists who are interested in earthquakes, fairly widespread use is made of a technique that might be called "combinatory," because it makes use of written sources in order to date archaeological evidence. From an epistemological standpoint, this leads to a sort of circular argument in which written sources provide an absolute chronology and support the archaeological sources, which are in turn offered as evidence of local seismic effects caused by a specific earthquake which is mentioned in reliable written sources, but in which there is no mention of damage at the archaeological sites in question. Earthquake damage is thus imposed on the seismic scenario in a rather unscientific and rarely explicit way.

The methodological reason why the combinatory approach cannot be adopted is as follows: written sources and archaeological sources are based on two conceptually different timescales, resulting from the diverse nature of the two types of data concerned. The former may make reference not only to the year, month, and day when the historical event they describe took place, but even to subdivisions of a day, such as hours; the latter make reference to a timescale which has the year as its base, but acquires an identity only within the stratigraphic sequence in which the archaeological record is contained, that is to say only within a span of years. This method of dating archaeological earthquakes by means of written sources has created some seismological "monsters." That is to say, the effects of several earthquakes perhaps some distance apart have been drawn together into a single event. The most famous case is that of the earthquake of 21 July 365 AD to the southeast of the island of Crete, which has been associated with numerous cases of damage caused by different earthquakes occurring in the space of a few decades across the whole Mediterranean area from Cyprus to Sicily and North Africa (Ambraseys, 1965; Di Vita, 1990; Bernabò Brea, 1997). The conceptual problem behind this use of archaeological sources lies, as we have already pointed out, in the habit of dating seismic indicators in an absolute way by anchoring them to a written text.

5.4 Where is Archaeoseismology Going?

Although it offers a host of interesting data and stimulating suggestions, archaeoseismology is not yet a discipline in the strict sense of the term, not only because its interpretations are often not based on a rigorous and transparent method, but also because there does not yet exist a numerically significant community which recognizes itself as seismic archaeology, "takes a stand," and "gives its approval" to methods and results. What we have are simply a few individual scholars. As in the case of historical seismology, nobody teaches it, and it has no recognized academic positions or schools. Nowadays it could be described simply as an important research exercise. So the literature relies for the most part on the goodwill of authors and reviewers who are usually excellent seismologists, geophysicists, and geologists, but often lack relevant experience and archaeological qualifications. But we cannot ignore the fact that research of this kind does happen, and does produce data. Perhaps we can correctly define seismic archaeology as an area where questions about historical seismology are asked, and archaeological sources are used to supply an answer. The usefulness and quality of the results produced by research of this kind, even when it is in essence multidisciplinary, are at the moment proportional to the quality of the dialogue being initiated between archaeologists and those who study earthquakes, and to the correct formulation of the questions to which an answer is being sought.

6. General Conclusions and New Questions

Let us attempt an assessment of the complex research field of historical seismology. It appears to be growing strongly and with increased potential, in spite of its problems (but what research field is free of problems?). As this brief excursus perhaps illustrates, the quality of the research concerned is a function of the clarity and transparent nature of the method of investigation, and of the systematic way in which data are acquired and data sets formed. A substantial refinement in the quality of the historical data produced in the last 20 years has been responsible for increasing the seismological significance of the data themselves, thereby also promoting more refined methods of calculating "traditional" parameters, and also suggesting new applications. More complex and detailed maps of effects have revealed the irregular nature of earthquake propagation as opposed to the regularity that tended to be emphasized by the use of isoseisms. These results have been enhanced by the application of new technologies to the data, and have stimulated a critical attitude that has led to the progressive abandonment of simplificative or simplistic geometries of what was known in the 1970s as the "macroseismic field."

It is clear that Italy has become a testing ground of considerable importance for new methodologies (see the complex introduction to "Catalogue of Strong Italian Earthquakes" Boschi *et al.*, 2000, with several contributions on this topic) both because it has had a large number of strong earthquakes within a context of extraordinary documentary wealth, and because there has been substantial investment in the field of historical seismology.

There are many problems to be grappled within a comprehensive way and one shared by countries which use historical earthquake catalogs. They include the explicit adoption of generally accepted interpretative rules, the use of transparent processes of parameter interpretation, and elaboration, right from the recording of basic data and the drawing up of intensity estimates. The establishment of a research "protocol" acceptable to the experts might also make it possible to enhance the quality of the various current national earthquake catalogs, and lead on to widespread improvements.

Finally, there seems to be a need for stable collaboration between countries in the collection of basic data, even though they are often a great distance apart. The countries of North Africa and Latin America are a good example, for they have many important documentary sources preserved in European cities. And then there are the former Soviet republics, whose many important sources are now in cities which are difficult to reach, or situated in areas of political conflict.

Because of its ability to investigate the cultures of the past and create a sort of *trait d'union* between past and present, historical seismology provides data which are also suitable for scientific popularization purposes and social information. It produces a "deposit" of scientific information which can convey to resident populations the extent to which the characteristics of the environment in which they live are stable, and can be used to good effect in seismic risk reduction programs.

References

Academia Sinica (1970). "Catalog of Chinese Earthquakes," Beijing (in Chinese).

Ambraseys, N.N. (1965). The seismic history of Cyprus. *Rev. l'union Int. Genève* **3**, 25–48.

Ambraseys, N.N. and C.P. Melville (1982). "A History of Persian Earthquakes," Cambridge University Press, Cambridge.

Bernabò Brea, L. (1997). Note sul terremoto del 365 d.C. a Lipari e nella Sicilia nord orientale, In: "La Sicilia dei Terremoti. Lunga Durata e Dinamiche Sociali," (G. Giarrizzo, Ed.), pp. 87–97. Maimone, Catania.

Bertrand, E. (1757). "Mémoires Historiques et Physiques sur les Tremblemens de Terre," La Haye.

Bonito, M. (1691). "Terra tremante, o vero continuatione de' terremoti dalla Creatione del Mondo sino al tempo presente," Napoli (anastatic reprint, Sala Bolognese 1980).

Boschi, E., E. Guidoboni, G. Ferrari, D. Mariotti, and G. Valensise (2000). Catalogue of Strong Italian Earthquakes from 461 BC to 1997. *Ann. Geofis.* **43**, 609–868 (with CD-ROM).

Boschi, E., R. Funiciello, E. Guidoboni, and A. Rovelli (1995). Earthquakes in the past: Multidisciplinary approaches. *Ann. Geofis.* **38**, 5–6.

D'Andria, F. (1994). Report on the 1993 excavation in Hierapolis, "Frontino street" (quoted in Guidoboni *et al.*, 1994, pp. 349–351).

De Rossi, M.S. (1874). La antica basilica di S. Petronilla presso Roma testè discoperta crollata per terremoto. *Bull. Vulcanismo Ital.* **1**, 62–65.

Di Vita, A. (1990). Sismi, urbanistica e cronologia assoluta. Terremoti e urbanistica nelle città di Tripolitania fra il I secolo a.C. ed il IV d.C. In: "L'Afrique dans l'Occident Romain (Ier siècle av. J.-C.– IVe siècle ap. J.-C.). *Collection de Ecole Française de Rome* **134**, 425–494.

Dominici, P. and L. Marcelli (1979). Evoluzione storica delle misure orarie in Italia. *Ann. Geofis.* **32**, 131–212.

Ferrari, G. (1992). "Two Hundred Years of Seismic Instruments in Italy 1731–1940," ING-SGA, Bologna.

Ferrari, G. (Ed.) (1997). Historical seismic instruments and documents: a heritage of great scientific and cultural value. *Cahiers du Centre Européen de Géodynamique et de Séismologie*, **13**, Luxembourg.

Ferrari, G. and C. Marmo (1985). Il "quando" del terremoto. *Quaderni storici* **60**, 691–715.

Gasperini, P. and G. Valensise (2000). From earthquake intensities to earthquake sources: extending the contribution of historical seismology to seismotectonic studies. *Ann. Geofis.* **43** (with CD-ROM).

Gasperini, P., F. Bernardini, G. Valensise, and E. Boschi (1999). Defining seismogenic sources from historical earthquake felt reports. *Bull. Seismol. Soc. Am.* **89**, 94–110.

Giuffrè, A. (1995). Seismic damage in historic town centers and attenuation criteria. *Ann. Geofis.* **38**, 837–843.

Grünthal, G. (Ed.) (1998). European Macroseismic Scale 1998. *Cahiers du Centre Européen de Géodynamique et de Séismologie*, vol. 15. Conseil d'Europe, Luxembourg.

Guidoboni, E. (1997). An early project for an antiseismic house in Italy: Pirro Ligorio's manuscript treatise of 1570–1574. *Eur. Earthq. Eng.* **2**, 13–20.

Guidoboni, E. and G. Ferrari (2000). Seismic scenarios and assessment of intensity: our criteria for the use of the MCS scale. *Ann. Geofis.* **43** (with CD-ROM).

Guidoboni, E. and M. Stucchi (1993). The contribution of historical records of earthquakes to the evaluation of seismic hazard. *Ann. Geofis.* **36**, 201–215.

Guidoboni, E., A. Comastri, and G. Traina (1994). "Catalogue of Ancient Earthquakes in the Mediterranean Area up to the 10th Century," ING-SGA, Bologna.

Karcz, I. and U. Kafri (1978). Evaluation of supposed archaeoseismic damage in Israel. *J. Archaeol. Sci.* **5**, 237–253.

Karcz, I. and U. Kafri (1981). Studies in archaeoseismicity of Israel: Hisham's Palace, Jericho. *Isr. J. Earth Sci.* **30**, 12–23.

Lanciani, R. (1918). Segni di terremoti negli edifizi di Roma antica. *Bull. Commissione Archeol. Comunale di Roma* **45**, 3–28.

Lee, W.H.K. and D.R. Brillinger (1979). On Chinese earthquake history – An attempt to model an incomplete data set by point process analysis. *Pure Appl. Geophys.* **117**, 1229–1257.

Lee, W.H.K., F.T. Wu, and C. Jacobsen (1976). A catalogue of historical earthquakes in China. *Bull. Seismol. Soc. Am.* **66**, 2003–2016.

Ligorio, Pirro, (1574–1577). Libro o Trattato di diversi terremoti, raccolti da diversi Autori per Pyrro Ligorio cittadino romano, mentre la città di Ferrara è stata percossa et ha tremato per un simile accidente del moto della terra. "Archivio di Stato di Torino, Antichità romane," vol. 28.

Lycosthenes, Conradus (1557). "Prodigiorum ac Ostentorum Chronicon," Basel.

Manetti, Giannozzo (1457). "De terraemotu libri tres," Biblioteca Apostolica Vaticana, cod. Urb. Lat. 5.

McGuire, W.J., D.R. Griffiths, P.L. Hancock, and I.S. Stewart (2000). "The Archaeology of Geological Catastrophes," Geological Society special publication no. 71, London.

Molin, D. (1995). Considerations on the assessment of macroseismic intensity. *Ann. Geofis.* **38**, 805–810.

Monachesi, G. and A. Moroni (1993). Problems in assessing macroseismic intensity from historical earthquake records. *Terra Nova* **5**, 463–466.

Moreira de Mendoca, J.J. (1758). *Historia universal dos terremotos*, Lisbon.

Needham, J. (1959). "Science and Civilisation in China. Volume 3: Mathematics and the Science of the Heavens and the Earth," Cambridge University Press, Cambridge.

Nur, A. and Z. Reches (1979). The Dead Sea rift: geophysical, historical and archaeological evidence for strike slip motion. *EOS, Trans. Am. Geophys. Un.* **60**, 322.

Richter, C.F. (1958). "Elementary Seismology," W.H. Freeman, San Francisco.

Seyfart, J.F. (1756). "Algemeine Geschichte der Erdbeben," Frankfurt-Leipzig.

Sieberg, A. (1932). Untersuchungen über Erdbeben und Bruchschollenbau im östlichen Mittelmeergebiet *Denkschriften der medizinsch-naturwissenschaftlichen Gesellschaft zu Jena* **18**, 161–273.

Stiros, S. and R.E. Jones (1996). "*Archaeoseismology*," Fitch Laboratory Occasional Paper 7, Athens.

Thompson, R.C. (1937). A new record of an Assyrian earthquake. *Iraq*, 4.

Vannucci, G., P. Gasperini, G. Ferrari, and E. Guidoboni (1999). Encoding and computer analysis of macroseismic effects. *Phys. Chem. Earth, A* **24**, 505–510.

Vannucci, G., P. Gasperini, and G. Ferrari (2000). Reducing the subjectivity of intensity estimates: the Fuzzy Sets Theory. *Ann. Geofis.* **43** (with CD-ROM).

Willis, B. (1928). Earthquakes in the Holy Land, *Bull. Seismol. Soc. Am.* **18**, 73–103.

Editor's Note

Due to space limitations, the full manuscript including additional text, figures, and references is given in GuidoboniFull.pdf on the attached Handbook CD, under the directory \47Guidoboni. Please see also Chapter 45, Historical seismicity and tectonics, by Ambraseys, Jackson, and Melville; Chapter 46, Earthquakes and archaeology, by Nur; and Chapter 48, Earthquake history: some examples, by Toppozada and Branum (California), Satyabala (India), Usami (Japan), and British Isles (Musson).

48

Earthquake History: Some Examples

48.1 Introduction

W.H.K. Lee
US Geological Survey, Menlo Park, California, USA (retired)

Even before civilization began, earthquakes have caused much suffering to mankind. In order to mitigate earthquake hazards, it is important to learn from past earthquakes. Although seismograms comprise the basic observational data for earthquake studies, instrumental records exist only for slightly over 100 years. We must, therefore, depend on recorded history and related materials to extend our knowledge about the past earthquakes before the instrumental era. Since instrumental coverage of the world was not adequate until the establishment of the WWNSS and the ESSN in the early 1960s (see Chapter 17 by Lee), recorded history and related materials are still critical in many areas of the world up to about 1963. For convenience, we use "historical" to mean the era before 1963.

Because seismograms before 1963 were difficult to access, an effort was made to archive "historical" (i.e., pre-1963) seismograms in the late 1970s and early 1980s. Unfortunately, this effort was not successfully completed due to the lack of continued funding. A symposium on "Historical Seismograms and Earthquakes" was held on 27–28 August 1985, during the General Assembly of the International Association of Seismology and Physics of the Earth's Interior (IASPEI) in Tokyo, Japan, to document what we had accomplished. Sixty papers were presented, of which 51 papers were submitted for publication in a Proceedings volume, "Historical Seismograms and Earthquakes of the World", edited by W.H.K. Lee, H. Meyers, and K. Shimazaki, Academic Press, San Diego, 1988. This volume is placed as a PDF file on the attached Handbook CD, under the directory \48.1Lee2.

In particular, this Proceedings volume contains the following papers, which may be of interest to readers of earthquake history. They are listed below in the order as they appeared in the Proceedings volume, and interested readers can read them on the attached Handbook CD as noted above.

1. "Historical Materials of Chinese Earthquakes and Their Seismological Analysis" by Yushou Xie
2. "The Anatolian Earthquake of 17 August 1668" by N.N. Ambraseys and C.F. Finkel
3. "An Analysis of the Eastern Mediterranean Earthquake of 20 May 1202" by N.N. Ambraseys and C.P. Melville
4. "Preliminary Evaluation of the Large Caracas Earthquake 07 of 29 October 1900, by Means of Historical Seismograms" by G.E. Fiedler
5. "Evaluation of Damage and Source Parameters of the Malaga Earthquake of 9 October 1680" by D. Munoz and A. Udias
6. "Studies of Earthquakes on the Basis of Historical Seismograms in Belgium" by M. De Becker and T. Camelbeeck
7. "Documenting New Zealand Earthquakes" by G.A. Eiby
8. "Historical Earthquakes and the Seismograms in Taiwan" by Pao Hua Lee
9. "Historical Earthquakes of Thailand, Burma, and Indochina" by S. Prachuab
10. "Earthquake History of California" by T.R. Toppozada, C.R. Real, and D.L. Parke
11. "Study of Historical Earthquakes in Japan (2)" by Tatsuo Usami
12. "Studies of Philippine Historical Earthquakes" by R.G. Valenzuela and L.C. Garcia
13. "The History of Earthquakes in the Northern North Sea" by R. Muir Wood, G. Woo, and H. Bungum

This chapter includes four brief subchapters as examples of earthquake history. Their full manuscripts are also given on the attached Handbook CD.

INTERNATIONAL HANDBOOK OF EARTHQUAKE AND ENGINEERING SEISMOLOGY, VOLUME 81A ISBN: 0-12-440652-1

48.2 California Earthquakes of $M \geq 5.5$: Their History and the Areas Damaged

Tousson Toppozada and David Branum
California Division of Mines and Geology, Sacramento, USA

California's documented history started around 1800 with fragmentary writings from the Franciscan missions between San Diego and Sonoma. Two earlier significant earthquakes were also recognized. In 1700 a giant $M \sim 9$ earthquake on the Cascadia subduction zone between Eureka and Vancouver was inferred from studies of Indian legends, submerged soils and trees, and tsunami heights documented in Japan (Satake *et al.*, 1996). In 1769 the Portola expedition felt earthquakes for about a week in the Orange–Los Angeles county area. Their descriptions suggest a $M \sim 6.5$ earthquake and aftershocks in the local area, or a larger earthquake on the San Andreas fault system ~80 km to the east.

Since 1850, 167 earthquakes of $M \geq 6$ have been documented in California and its border regions (bigger border region than in Fig. 1a). At least 91 occurred in the preinstrumental period before 1932, the record being incomplete before about 1910. Seventy-six were located since 1932 with varying seismographic control, indicating an average rate of about 1.1 per year, or 0.9 y between events.

In the vicinity of the border with Mexico, the Imperial and Southern San Jacinto faults region has been active before and after 1932. The San Andreas fault system was very active north west of this vicinity before 1932, but since 1932 only had a few events near Desert Hot Springs, Parkfield, and the San Francisco Bay area (Fig. 1a). The main San Andreas fault south of the 34th parallel has had few earthquakes since 1800.

Five of the largest ($M \sim 7$ or larger) earthquakes since 1800 have occurred on the San Andreas fault (Fig. 1a). The most recent occurred in 1989 near Loma Prieta. The other four events occurred in two pairs of overlapping zones, in Southern California in 1812 (8 and 21 December) and 1857, and in Northern California in 1838 and 1906 (Fig. 1b). The 1812 and 1838 earthquakes generated effects that were generally as strong in their regions as were the effects of the overlapping 1857 and 1906 earthquakes, respectively. Earthquakes that occur only a few decades apart are difficult to differentiate in fault trenches because of the limitations in stratigraphic resolution and radiocarbon dating. Thus, pairs of overlapping major events on the San Andreas fault such as 1812/1857 and 1838/1906 may be more common than paleoseismology would indicate.

The 630 km length of the San Andreas fault between San Francisco and Cajon Pass ruptured in the 1838 and 1857 earthquakes, except for about 75 km between Bitterwater and San Juan Bautista (Figure 1). Probable aftershocks of the 1838 event occurred in 1840 and 1841 near San Juan Bautista. Foreshocks and aftershocks of the 1857 event occurred near Bitterwater until 1885 (Toppozada *et al.*, 2002). Near Parkfield, 40–70 km southeast of Bitterwater, $M = 5.5$ or greater earthquakes have occurred from the 1870s to the 1960s. In the total Bitterwater to Parkfield zone bracketing the northern end of the 1857 rupture, the seismicity and moment release have decreased steadily since 1857, and have tended to migrate southeastward with time.

A major earthquake was thought to have occurred on the Hayward fault in 1836, until Toppozada and Borchardt (1998) determined that it was a smaller event located between Monterey and the San Jose-Santa Clara area, far from the Hayward fault. The $M \sim 7$ Hayward fault earthquake of 1868 was of comparable size to the 1989 Loma Prieta event. The 1868 earthquake was preceded by 13 y in which twelve $M \geq 5.5$ earthquakes occurred within 60 km of the Hayward fault, and was followed by 13 quiet years when only one $M \geq 5.5$ Bay area earthquake occurred, that was not an aftershock. Similarly, five $M \geq 5.5$ Bay area earthquakes occurred in the 10 y leading up to the 1989 Loma Prieta event, which has been followed by 12 quiet years so far. The post-1868 Hayward earthquake quiescence suggests that the present Bay area quiescence following the 1989 Loma Prieta event may not last much longer than 13 years (Toppozada *et al.*, 2002).

INTERNATIONAL HANDBOOK OF EARTHQUAKE AND ENGINEERING SEISMOLOGY, VOLUME 81A ISBN: 0-12-440652-1

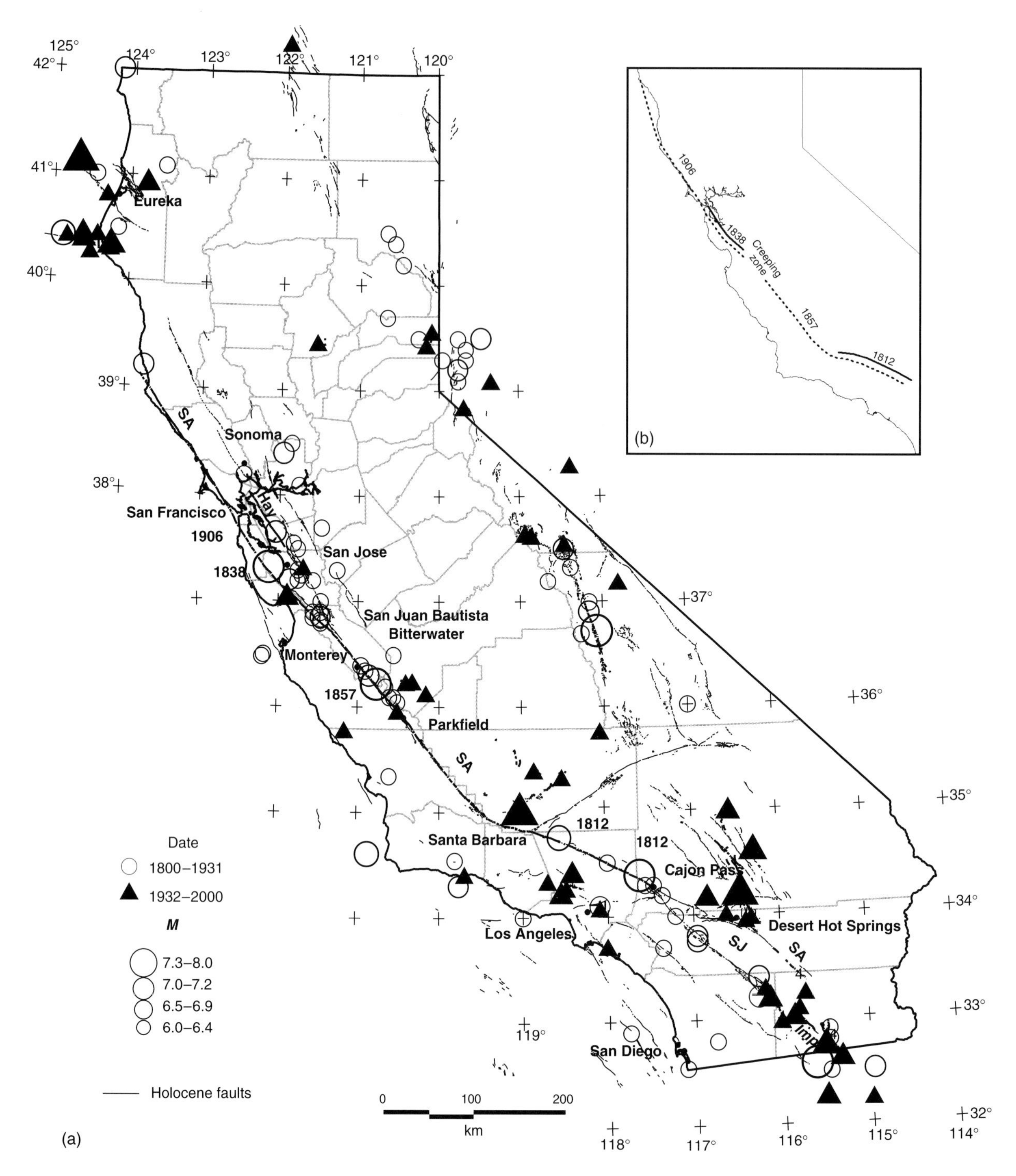

FIGURE 1 (a) $M = 6.0$ and greater California earthquakes, 1800–2000. Triangles indicate the 68-y instrumental record and circles indicate the 132-y preinstrumental record. Faults: SA, San Andreas; Hay, Hayward; SJ, San Jancinto; Imp, Imperial. (b) San Andreas fault ruptures in the overlapping 1812/1857 and 1838/1906 regions.

Toppozada *et al.* (2000) and Toppozada and Branum (2002) provide maps and lists of about 400 $M \geq 5.5$ earthquakes, of which more than half occurred before 1932. The pre-1932 epicenters and magnitudes were estimated predominantly by interpreting and mapping the reported earthquake effects and comparing them with those of instrumented earthquakes (e.g., Toppozada *et al.*, 1981, with additions and revisions). Moment magnitude (M_W) is used when available, otherwise surface-wave magnitude, local magnitude, or magnitude derived from the size of the areas shaken at various intensities

TABLE 1 Significant California Earthquakes ($M \geq 6.5$ and/or Destructive)

Date[a]	Latitude	Longitude	M	Region	Loss of Life and Property
18121208	34.37?	−117.65?	~7.3	Orange County, Los Angeles, Wrightwood	40 deaths at San Juan Capistrano
18121221[b]	34.75?	−118.60?	~7.1	Los Angeles, Ventura, Santa Barbara	1 death
18380600	37.30?	−122.15?	~7.4	San Francisco to San Juan Bautista	Damage from San Francisco to Monterey
18570109	36.20	−120.80	7.9	Great Fort Tejon earthquake	1 death, damage from Monterey Co. to San Bernardino Co.
18651008	37.20	−121.90	6.5	Santa Cruz Mountains	$500 000 in property loss
18681021	37.70	−122.10	7.0	Hayward Fault	30 deaths, $350 000 loss
18720326	36.70	−118.10	7.4	Owens Valley	27 deaths, 56 injuries, $250 000 loss
18731123	42.00?	−124.20?	6.9	Crescent City regoin	Damage in California–Oregon border area
18900209	33.40	−116.30	6.8	San Jacinto fault?	Little damage
18920224	32.55	−115.65	7.3	Laguna Salada, Baja California	Damage San Diego to Imperial Valley
18920419	38.40	−122.00	6.6	Vacaville	1 death, $225 000 loss
18980331	38.20	−122.50	6.4	Mare Island	$350 000 loss
18980415	39.20	−123.80	~6.7	Fort Bragg–Mendocino	Damage from Fort Bragg to Mendocino
18991225	33.80	−117.00	6.7	San Jacinto and Hemet	6 deaths, $50 000 loss
19060418	37.70	−122.50	7.8	Great 1906 earthquake	3000 deaths, $524 million loss (counting fire damage)
19180421	33.75	−117.00	6.8	San Jacinto	1 death, several injuries, $200 000 loss
19230122	40.40	−124.90	7.2	Off Cape Mendocino	Destructive in Humbolt Co., strongly felt to Reno
19250629	34.30	−119.80	6.8	Santa Barbara	13 deaths, $8 million loss
19271104	34.60	−120.90	7.1	40 km west of Lompoc	Damage in Santa Barbara and San Luis Obispo counties
19330311	33.70	−118.00	6.4	Long Beach	115 deaths, $40 million loss
19400519	32.73	−115.50	7.0	Imperial Valley	9 deaths, $6 million loss
19520721	35.00	−119.02	7.3	Kern County earthquake	12 deaths, $60 million loss
19541221	40.93	−123.78	6.6	East of Arcata	1 death, several injuries, $2.1 million loss
19710209	34.41	−118.40	6.6	San Fernando	65 deaths, 2000+ injuries, $505 million loss
19791015	32.61	−115.32	6.5	Imperial Valley	91 injuries, $30 million loss
19801108	41.12	−124.67	7.4	West of Eureka	6 injuries, $2 million loss
19830502	36.23	−120.31	6.4	Coalinga	$31 million loss, 1 death, 47 injuries
19840424	37.31	−121.68	6.2	Morgan Hill	$8 million loss
19871001	34.07	−118.08	6.0	Whittier Narrows	8 deaths, $358 million loss
19871124	33.01	−115.85	6.6	Superstition Hills	$3 million loss
19891018	37.04	−121.88	6.9	Loma Prieta	63 deaths, 3757 injuries, $6 billion loss
19920425	40.33	−124.23	7.2	Cape Mendocino area	356 injuries, $48.3 million loss. Two $M = 6.6$ aftershocks next day
19920628	34.20	−116.44	7.3	Landers	1 death, 402 injuries, $91.1 million (losses include Big Bear earthquake 3 h later)
19940117	34.21	−118.54	6.7	Northridge	57 deaths, 9000+ injuries, ~$40 billion loss
19991016	34.60	−116.27	7.1	Hector Mine	Minimal injuries and damage due to sparse population

Loss from Stover and Coffman (1993).
[a]The date is given as year, month, day (Greenwich).
[b]Proposed epicenter on or near San Andreas fault (Toppozada *et al.*, 2002).

is used. Toppozada and Branum (2002) have developed new analytical relationships between M_W and the areas shaken at or above Modified Mercalli Intensity [MMI] V, VI, and VII. They also list all available estimates of magnitude for earthquakes having more than one estimate. The earthquake information from 1932 onward is, with a few identified exceptions, from the catalogs of Caltech and UC Berkeley, supplemented by USGS information after 1969.

Table 1 lists the significant California earthquakes since 1800, generally of $M \geq 6.5$, and the resulting losses. The greatest losses to lives and property occurred in the great 1906 San Francisco earthquake, and in the $M = 6.4$ to $M = 6.9$ urban earthquakes in 1933 Long Beach, 1971 San Fernando, 1989 Loma Prieta, and 1994 Northridge. Smaller losses resulted from major $M \geq 7$ earthquakes occurring before the population growth in California that followed the 1849 Gold Rush, or in regions that are still sparsely populated. Toppozada *et al.* (2000) and Toppozada and Branum (2002) show maps of the areas shaken at MMI VII or greater, which is the threshold of damage to weak buildings. Such damage has occurred since 1800 at least six times in each of the Eureka, San Francisco Bay, and Los Angeles regions.

References

Caltech/USGS earthquake catalog, [http://www.scecdc.scec.org/catalog-search.html]. "Southern California Earthquake Catalog and Phase Data," Southern California Seismic Network, US Geological Survey, Pasadena; Caltech Seismological Laboratory, California Institute of Technology, Pasadena.

Satake, K., K. Shimazaki, Y. Tsuji, and K. Ueda (1996). Time and size of a giant earthquake in Cascadia inferred from Japanese tsunami records of Janurary 1700. *Nature* **379**, 246–249.

Stover, C. and J. Coffman (1993). Seismicity of the United States. *US Geol. Surv. Prof. Paper 1527.*

Toppozada, T.R. and G. Borchardt (1998). Re-evaluation of the 1836 Hayward and the 1838 San Andreas fault earthquakes. *Bull. Seismol. Soc. Am.* **88**, 140–159.

Toppozada, T.R. and D. Branum (2002). "California Earthquakes of $M \geq 5.5$: their History and the Areas Damaged" (complete report), IASPEI volume.

Toppozada, T.R., C.R. Real, and D.L. Parke (1981). Preparation of isoseismal maps and summaries of reported effects for pre-1900 California earthquakes, *Calif. Div. Mines Geol. Open File Report 81-11 SAC*, 182 pp.

Toppozada, T., D. Branum, M. Petersen, C. Hallstrom, C. Cramer, and M. Reichle (2000). "Epicenters of and Areas Damaged by $M \geq 5$ California Earthquakes, 1800–1999." California Division of Mines and Geology, Map Sheet 49.

Toppozada, T.R., D. Branum, M. Reichle, and C. Hallstrom (2002). San Andreas Fault, California, $M \geq 5.5$ earthquakes 1800–2001. *Bull. Seismol. Soc. Am.* (in press).

UC Berkeley/USGS earthquake catalog, (http://quake.geoberkeley.-edu/ncedc/catalog-search.html). "Northern California Earthquake Catalog and Phase Data," Northern California Seismic Network, US Geological Survey, Menlo Park; Berkeley Seismological Laboratory, University of California, Berkeley.

Editor's Note

Due to space limitations, the full manuscript including additional text, figures, appendix, and references is given in ToppozadaFull.pdf on the attached Handbook CD, under the directory \48.2Toppozada.

48.3 The Historical Earthquakes of India

S.P. Satyabala
National Geophysical Research Institute, Hyderabad, India

1. Introduction

The historical earthquakes of India are important for understanding seismicity in and around India. Documentation of macroseismic effects of earthquakes of India is available for the five great earthquakes (1819, 1897, 1905, 1934, and 1950) and several other major earthquakes. There has been a quiescence of major ($M_s \geq 7.5$) earthquakes in this region during the second half of the 20th century (Satyabala and Gupta, 1996), causing concern for seismic hazard in the region (e.g., Khattri, 1987; Bilham *et al.*, 1995; Yeats and Thakur, 1998). [See Editor's Note.] Great and intermediate-to-small earthquakes form two parallel belts of seismicity along the Himalayan arc. As proposed by Seeber and Armbruster (1981), two major seismogenic faults form concentric arcs along the Himalayan arc: a detachment thrust on which the major/great earthquakes occur and a steeper dipping thrust on which moderate earthquakes occur. Strong evidence that the great earthquakes occurred on a shallow dipping (quasi-horizontal) thrust plane or detachment is the consistent absence of primary faulting for all these earthquakes and the very large area over which the strongest macroseismic effects are manifest.

2. The CD Appendix

The CD appendix includes summaries of the macroseismic effects of some historical earthquakes of India (listed in Table 1 and shown in Color Plate 19). Primary source materials are the Memoirs and Reports published by the Geological Survey of India that report investigations carried out immediately or soon after the earthquake. The material presented for each earthquake includes salient features of the results, summary

TABLE 1 Historical Earthquakes of India [Presented in Detail on the Attached Handbook CD (Text and Graphics)]

	Earthquake		Location			Intensity	Maximum	
S.no.	Date	Region	°E	°N	M_S	Scale	Intensity	Main Reference
1	15 August 1950	Assam	96.76	28.38	8.6	ME	X	Ramachandra Rao (1953)
2	30 May 1935	Quetta	66.4	28.87	7.6	RF	X	R/GSI, vol. LXIX
3	15 January 1934	Bihar–Nepal	87.09	27.55	8.3	ME	X	M/GSI, vol. 73
4	4 December 1930	Pyu	96.5	18	7.6	RF	IX	M/GSI, vol. LXII (Part 1)
5	5 May 1930	Pegu	96.5	17	7.3	RF	IX	R/GSI, vol. LXV
6	2 July 1930	Dhubri	90	25.5	7.1	RF	IX	M/GSI, vol. LXV
7	8 July 1918	Srimangal	91	24.5	7.6	RF	IX	M/GSI, vol. XLVI (Part I),
8	23 May 1912	Burma	97	21	7.7	RF	IX	M/GSI, vol. XLII (Part 1)
9	4 April 1905	Kangra	76	33	8	RF	X	M/GSI, vol. XXXVIII
10	12 June 1897	Shillong	91	26	8	ME	X	M/GSI, vol. XXIX & XXX
11	10 January 1869	Cachar	92.7	26	7.5	MMI	VIII–X	M/GSI, vol. XIX
12	16 June 1819	Cutch	69	24	7.8	MMI	X+	M/GSI, vol. XXVIII and XLVI (Part II)

Note: Brief descriptions are also available on the CD for the 1762 Arakan, 1803 Mathura, 1816 Gangotri, 1833 Nepal, 1839 and 1858 Burma, 1885 Kashmir, 1885 Bengal, 1954 Manipur–Burma, 1956 Burma earthquakes and others.

Abbreviations: M/GSI, R/GSI = Memoirs/Records of the Geological Society of India.

INTERNATIONAL HANDBOOK OF EARTHQUAKE AND ENGINEERING SEISMOLOGY, VOLUME 81A ISBN: 0-12-440652-1

descriptions, and graphic interpretation of the macroseismic effects within each isoseist (movements felt and sounds heard; effects on terrain, e.g., landslides, liquefaction phenomena and faults; and effects on manmade structures, e.g., buildings, roads, and railways). Intensity scales are adopted from the original sources (mostly Rossi-Forel or the Mercalli scale). As discussed in the CD appendix, major limitations in delineating the isoseismals are inaccessibility for investigation and insufficient information across the mountain ranges. An effort has been made to describe/summarize all the macroseismic effects irrespective of whether they are common or rare.

3. Conclusions

The great earthquakes were felt over large areas in India and each individual earthquake produced different macroseismic effects on an unusual scale. The intensity displayed by the 1905 Kangra earthquake contrasts with the other three earthquakes of the Himalaya (1897, 1934, and 1950) that produced cataclysmic effects such as fault-scarps, destruction of railway tracks and iron girder bridges, the bending and snapping of trees, fissures and sand/water vents. The 1905 Kangra earthquake occurred on segments of the sub-Himalaya characterized by dun valleys and by anticlines at the Himalayan front, and is inferred to have occurred on a blind thrust and expressed at the surface as a fold.

The 1934 Bihar–Nepal earthquake, notable for producing the "slump belt" and spectacular sand and water vents, occurred in segments of the central Himalaya characterized by imbricate thrusts. The 1950 Assam earthquake, which caused landslides on an enormous scale, occurred near the Eastern Syntaxis of the Himalaya. The 1905 Kangra, 1934 Bihar–Nepal and 1950 Assam earthquakes occurred on the detachment surface between the Indian shield and Himalayan rocks. The 1897 Shillong earthquake occurred in a region south of the Himalayan arc and produced a bell-shaped meizoseismal area, the strongest effects within which furnished the principal model for the highest grade XII of the Modified Mercalli Intensity Scale (Richter, 1958). The 1819 Kutch earthquake, which produced the "Allah Bund," a ~60 km-long fault scarp with a total vertical displacement of 7–9 m, occurred within the stable continental region of India.

The fact that there is a quiescence of major/great earthquakes in the present modern instrumental era of seismology renders the information available in the Memoirs/Records of the Geological Survey of India invaluable for any possible assessment of earthquake hazard in and around India. These earthquakes best illustrate the possible macroseismic effects and the scale on which these effects could be displayed if earthquakes of such a magnitude should recur.

Acknowledgments

I thank Dr Harsh K. Gupta, Director, National Geophysical Research Institute, India, for his encouragement during this work. I thank Dr Lee and three anonymous reviewers for critical comments that improved this contribution.

References

Bilham, R., P. Bodin, and M. Jackson (1995). Entertaining a great earthquake in Western Nepal: Historic activity and Geodetic test for the development of strain. *J. Nepal Geol. Soc. Kathmandu* **11**, Special Issue, 73–88.

Gansser, A. (1964). "Geology of the Himalayas," Wiley Interscience, London.

Gansser, A. (1983). "Geology of the Himalaya," Birkhauser Verlag, Basel, Boston, Stuttgart.

Khattri, K.N. (1987). Great earthquakes, seismicity gaps and potential for earthquake disaster along the Himalaya plate boundary. *Tectonophysics* **138**, 79–92.

Richter, C.F. (1958). "Elementary Seismology," W.H. Freeman, San Francisco.

Satyabala, S.P. and H.K. Gupta (1996). Is the quiescence of major earthquakes ($M \geq 7.5$) since 1952 in the Himalaya and North-East India regions real? *Bull. Seismol. Soc. Am.* **86**, 1983–1986.

Seeber, L. and J.G. Armbruster (1981). Great detachment earthquakes along the Himalayan Arc and long-term forecasting. In: "Earthquake prediction: An International Review," (D.W. Simpson and P.G. Richards, Eds.), pp. 259–277. Maurice Ewing Series 4, American Geophysical Union, Washington, DC.

Yeats, R.S. and V.C. Thakur (1998). Reassessment of earthquake hazard based on fault-bend fold model of the Himalaya plate-boundary fault. *Curr. Sci.* **74**(3), 230–233.

Editor's Note

Due to space limitations, an Appendix containing 13 chapters of historical earthquakes of India is placed on the attached Handbook CD, under the directory \48.3Satyabala. A disastrous earthquake of magnitude 7.6 occurred on 26 January 2001 near Bhuj (Gujarat State) in western India. About 20,000 persons were reported dead and an economic loss of about US$5 billion was estimated (see *Seism. Res. Lett.* **72**(3), 2001.)

48.4 Historical Earthquakes in Japan

Tatsuo Usami
The University of Tokyo, Tokyo, Japan (retired)

1. Introduction

Japan is one of the most earthquake-prone countries in the world. The earliest earthquake on record is one that hit Yarnato (now Nara Prefecture) on 23 August 416 AD, but no record of the damage has been found. Next oldest is another one that struck Yarnato on 28 May, 599; extant documents indicate that it caused houses to collapse. It would appear to have registered about magnitude 7 on the Richter scale. Since then, there have been 796 destructive earthquakes, or ones causing damage, in the 1397 years up to 1995. In the 128-y period since the Meiji Restoration in 1868, when Japan's modernization began, there have been 453 destructive earthquakes (Usami, 1996); an annual average of 3.5 events.

2. The Study of Historical Earthquakes in Japan

When efforts to modernize the state commenced with the Meiji Restoration in 1868, there was still little information about which areas of the country were frequently afflicted by earthquakes and which were less prone. Observations using seismographs began to be made, but they were too poor both in number and efficiency. Inevitably, the study of earthquakes had to begin by looking into records of past earthquakes and related matters in historical documents. These efforts at material collection are outlined below. Early studies were made by Ichizo Hattori, Edmund Naumann, and John Milne. These studies did indeed provide valuable data for modem seismological studies, which at the time were in their infancy. However, since they all referred to a relatively small range of materials and thus had a limited number of earthquakes to study, their results could not be treated as conclusive. A full-scale study would require a new start and a more comprehensive approach.

2.1 Comprehensive Compilation (Phase I)

In 1892 the Imperial Earthquake Investigation Committee was organized with, as one of its aims, the compilation of data on historical earthquakes. The committee put Minoru Tayama in charge of this task in 1893, and efforts got going under the supervision of Seikei Sekiya. Tayama made an extensive search of historical documents, archives, diaries, and other materials, culminating in publication of the "Catalog of Historical Data on Japanese Earthquakes" (Tayama, 1899). This covered 1896 earthquakes in the period 416–1864, and gave tables in which the date, time, location, and strength of each earthquake was recorded and reference materials were listed. Based on this catalog, Fusakichi Omori compiled a list of 222 major earthquakes, including those that occurred in the Meiji era (1868–1912). In 1904, Tayama published the 1201-page "Historical Data on Japanese Earthquakes" (Tayama, 1904). This was a chronicle of the original documents relating to earthquakes between 416 and 1865, and provided a model for the later "New Collection of Materials for the History of Japanese Earthquakes." Tayama left his commission in 1902 prior to publication of this report.

In 1913, Omori used Tayama's data to examine 95 pre-1876 earthquakes and determine their hypocenter, the area suffering seismic motion, and the occurrence of aftershocks, tsunami, damage etc. (Omori, 1913). Further, in 1919 he prepared a summary table (Omori, 1919) covering the period from the beginning of Japan's recorded history up to 1918.

2.2 Comprehensive Compilation (Phase II)

Efforts at the systematic compilation of earthquake data then came to a halt until 1928,when Kinkichi Musha began work under the supervision of Torahiko Terada at the Earthquake Research Institute. Commissioned by the institute, Musha dedicated himself to the task of compilation and published his results between 1941 and 1951 as expanded and revised

INTERNATIONAL HANDBOOK OF EARTHQUAKE AND ENGINEERING SEISMOLOGY, VOLUME 81A ISBN: 0-12-440652-1

editions of Tayama's "Historical Data on Japanese Earthquakes." His publications consisted of three expanded and revised volumes of "Historical Data on Japanese Earthquakes" (Earthquake Prevention Council, Ministry of Education, 1941–1943), and a volume of "Historical Data on Japanese Earthquakes" (Musha, 1949). These publications total 4000 pages and deal with around 6400 earthquakes.

2.3 Comprehensive Compilation (Phase III)

After Musha, a further period ensued in which no systematic effort at compilation took place. Work only resumed about 20 years ago, when it was decided to focus on the collection of historical materials from which Tayama or Musha did not, in the end, quote. These included materials related to volcanic eruptions and earthquakes in Taiwan and Korea, and to phenomena not clearly identified as related to earthquakes. Generally, they were left out merely because of the huge volume of the documents.

With a budget from the government and assistance from Japan Electric Association, enthusiastic efforts began to recompile these materials. This led to the publication of "New Collection of Materials for the History of Japanese Earthquakes" starting in 1980. A total of 21 volumes of documents had been issued by 1994, filling about 16 000 pages (Usami, 1980–1994). The foundations for studies of historical earthquakes in Japan were thus almost complete. The work of putting all this material into print continues.

3. Examination of Historical Materials

3.1 Interpretation

As mentioned above, records of earthquakes in Japan are in a variety of forms including official historical documents, private historical materials, records, chronicles of local history, letters, and dossiers. All are written by brush on Japanese paper. The first task confronting analysts is to interpret them, which essentially means rendering them into modern script. This work requires a knowledge of archaic terms and writing styles together with certain special training.

3.2 Seismic Intensity Distribution

The second stage of the examination is to sort the various interpreted data chronologically. At this stage records with the same time and locality are considered to belong to the same earthquake. Some earthquakes may have just a single line mention, whereas major ones have descriptions extending over a thousand pages or more.

Archaic documents only include such data as the feelings caused by the ground motion and the damage it caused, as well as related episodes. The seismic intensity at various affected locations is judged mainly from descriptions of the damage, crustal deformation, tsunami, and ground motion.

Intensity is worked out according to the current scale. To do this, it is first necessary to prepare a table of intensities (Usami, 1986) applicable to historical earthquakes, since structures used to be more vulnerable than they are now. On the other hand, the human sensitivity to ground motion remains unchanged. It is also possible to determine seismic intensity comparatively accurately from the damage to ancient structures, such as famous temples and castles, which remain to this day. In the most difficult cases, for example, where nothing that suffered damage remains, as is the case for private houses, the estimation process relies on the experience of those involved.

Once a distribution of seismic intensity is determined by estimating the intensity at various points, the magnitude is estimated and the epicenter determined from the strong motion area (the area within which the intensity is IV, V, and VI on the Japanese scale) using an empirical equation.

This determination of hypocentral characteristics needs to be made by comparison with similar modern earthquakes.

3.3 Trends in Destructive Earthquake Activity

Where an earthquake yields only a few data points, the hypocentral characteristics cannot be determined by the above process, so clearly such analysis is only possible in the case of earthquakes that caused damage. Figure 1 shows the number of destructive earthquakes by decade from the beginning of Japan's recorded history up to 1995. Note that the apparently increasing occurrence of earthquakes results from the ever-improving quality and quantity of historical materials as we approach the present. Although Figure 1 does not show an accurate record of the natural phenomenon itself, it is possible to read the rough trend of seismic activity rising and falling from this type of data.

The number of recorded destructive earthquakes began to rise after around the year 1600. This was because the Edo Shogunate government came into being around that time and produced large numbers of official documents. It also marked the start of a long period of political stability, so the documents have remained intact until the present. A surge in the number of destructive earthquakes is observed starting around 1880, when seismographs began to be used and modern methods of precision recording began. Nowadays, destructive earthquakes occur three to four times a year, but the damage caused is mostly minor. Medium-scale and major damaging earthquakes occur a few times a decade or so.

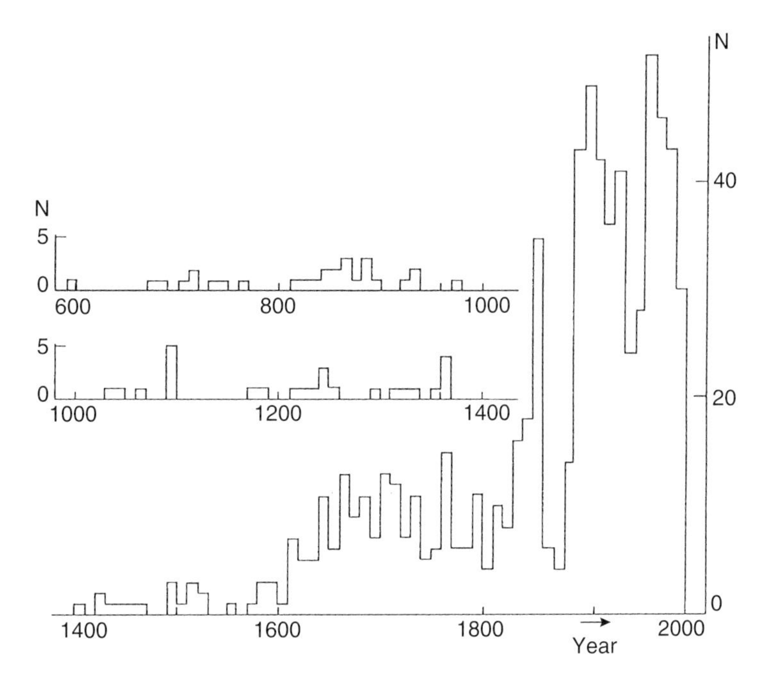

FIGURE 1 Trends in the number of destructive earthquakes by decade in Japan (599–1995).

4. Building a Picture of Historical Earthquakes

In investigating historical earthquakes, it is essential to build up a picture of the event by answering such questions as: Was the hypocenter deep or shallow? Was there a tsunami? Was it an inland earthquake? Was it a major event, or of what magnitude was it? Did a fault appear? This work of developing a picture of the Earthquake progresses in parallel with the effort to determine the hypocentral characteristics mentioned above. In the full manuscript, on the attached Handbook CD, three cases are given to illustrate this process.

5. Social Effects of Earthquakes

Major earthquakes often leave scars of various types on individuals and regions. They wreak havoc over extensive areas of political, social, and economic life, and often have socio-psychological consequences. Although this must have been true of earthquakes that occurred centuries ago, no historical documents have been collected to back up this assertion, nor has any research been done on this question. Just one or two episodes and recorded cases are available for study.

In the full manuscript, two cases of forced migration imposed by earthquakes are discussed. The aim is to shed some light on how seriously earthquakes can affect a society and how difficult it is to prepare for such disasters.

6. Conclusion

The task of summarizing all studies of historical earthquakes in Japan is a difficult one. This chapter simply attempts to provide a taste of such studies by describing some case studies of how an accurate picture of an earthquake can be developed. Two cases of mass migration are also dealt with, highlighting some of the social effects of earthquake. The study of historical earthquakes from the viewpoint of society, as shown in this paper, lags behind other fields, and this is an issue to be addressed in the future.

References

Earthquake Prevention Council, Ministry of Education (1941–1943). "Historical Data on Japanese Earthquakes," expanded and revised, edn., vols. I–III (in Japanese).

Musha, K. (1949). "Historical Data on Japanese Earthquakes," The Mainichi Press (in Japanese).

Omori, F. (1913). "Japanese Great Earthquakes," *Report of the Imperial Earthquake Investigation Committee, No. 68*, Part II, 1–180 (in Japanese).

Omori, F. (1919). "Descriptive Table of Japanese Great Earthquakes," *Report of the Imperial Earthquake Investigation Committee, No. 88*, Part II, 1–71 (in Japanese).

Tayama, M. (1899). "Catalog of Historical Data on Japanese Earthquakes," *Report of the Imperial Earthquake Investigation Committee, No. 26*, 3–112 (in Japanese).

Tayama, M. (1904). "Historical Data on Japanese Earthquakes," *Report of the Imperial Earthquake Investigation Committee, No. 46*, Part I, 1–606, Part II, 1–595 (in Japanese).
Usami, T. (Ed.) (1980–1994). "New Collection of Materials for the History of Japanese Earthquakes," vols. I–V, supplement, supplement II, + 14 separate volumes (21 volumes in total), Earthquake Research Institute, University of Tokyo (in Japanese).
Usami, T. (1986). "Memories of Historical Earthquake Studies," appended table.
Usami, T. (1996). "Materials for Comprehensive List of Destructive Earthquakes in Japan," revised and expanded, edn. University of Tokyo Press, Tokyo.

Editor's Note

Due to space limitations, the full manuscript is given in UsamiFull.pdf on the attached Handbook CD, under the directory \48.4Usami. Additional materials are also included as separate files under the same directory.

48.5 Historical Earthquakes of the British Isles

R.M.W. Musson
British Geological Survey, Edinburgh, United Kingdom

1. Introduction

The British Isles are not a part of the world generally associated in the public mind with earthquakes, and it may come as a surprise that there is so much material for study on the subject of British seismicity. The reasons for this are twofold; in the first case, earthquakes severe enough to cause minor damage do occur from time to time, which means that seismic risk is not completely negligible. Secondly, the intellectual history of Britain has meant that for a long period there have been people interested in recording or studying details of local earthquakes.

This can be seen if one looks at the history of the study of British earthquakes. This can be traced back a very long way; the earliest attempt to compile a chronological list of British earthquakes is that by Fleming (1580). It is no coincidence that 1580 also happens to be the date of a relatively large earthquake in the Dover Straits, which was strongly felt in London, and killed two children. Throughout this subchapter phrases like "relatively large" must be understood as being in the context of British seismicity; the 1580 event had a magnitude of around 5.8 M_L.

London, then as now, was the intellectual center of the kingdom and the location of the bulk of what passed for what would now be termed the media. The earthquake of 1580 stirred up much general alarm and led to the production of numerous pamphlets on the subject of earthquakes rushed out to satisfy the sudden public interest in the subject (Ockenden, 1936). Fleming's purpose in writing was not scientific; although he makes passing reference to theories of earthquake causation (chiefly the classic Aristotelian hypothesis of winds trapped beneath the earth) his main objective in writing is to present earthquakes as warnings from God.

2. Earlier Studies

Truly scientific studies of British earthquakes started in the second half of the 17th century, when the establishment of the Royal Society in 1666 provided a focus for scientific investigation, with the *Philosophical Transactions* as a ready medium for the publication of data and results. Accordingly, the collection and publication of felt reports of the 1666 Oxford earthquake by Boyle and Wallis (Boyle, 1666; Wallis, 1666) appears to be the first recognizable scientific macroseismic study of an earthquake in modern times. A number of these early studies exist, either in print or in manuscript. The English Channel earthquake of 1734 caught the interest of the Duke of Richmond, a member of the Royal Society, who encouraged a local physician to collect reports of the earthquake; although a summary report was published (Richmond, 1738) the original manuscripts still survive (Neilson *et al.*, 1984). These were not particularly large or damaging earthquakes; the low level of seismicity in Britain has always meant that small earthquakes felt at intensities of 4 or 5 EMS (European Macroseismic Scale; Grünthal, 1998) have been considered as notable and newsworthy events. (Note: the EMS used here is a 12° intensity scale similar in equivalence to other 12° scales in general use. The text is available at http://www.gfz-potsdam.de/pb5/pb53/projekt/ems/index.html).

Another boost to interest in British earthquakes was provided by two small shocks (2.6 and 3.1 M_L) on 8 February 1750 and 8 March 1750 with epicenters in central London. This caused a repeat of the pattern of 1580, i.e., a sudden public interest in earthquakes and a rush of publications to meet this interest. Some of these quickly-turned-out works are of little value; the best of the catalogs that resulted is probably that by Zachary Grey, published as "A Chronological and Historical Account of the most Memorable Earthquakes... By a Gentleman of the University of Cambridge" (Grey, 1750). It is often mentioned how scientific interest in earthquakes in Europe received a strong boost from the occurrence of the disastrous Lisbon earthquake in 1755; but in Britain this boost really occurred five years earlier, from these two small London earthquakes, a point noted by Kendrick (1956).

In the 19th century the focus for the study of British earthquakes shifted from the Royal Society to the British

INTERNATIONAL HANDBOOK OF EARTHQUAKE AND ENGINEERING SEISMOLOGY, VOLUME 81A ISBN: 0-12-440652-1

Association for the Advancement of Science (BAAS or simply BA). The occurrence of an earthquake swarm near Comrie, Perthshire, led to the formation of a scientific committee for the study of earthquakes, formed under the auspices of the BA and headed by David Milne (later Milne-Home) and James Forbes. A full account of the work of this committee and its significance in the history of seismology is beyond the scope of this discussion, and the reader is referred to Musson (1993). However, as part of his activities, in addition to collecting numerous felt reports of the larger shocks in the Comrie sequence (especially the largest: 23 October 1839 4.8 M_L), Milne (1842–1844) also compiled a descriptive catalog of previous British earthquakes, the largest such catalog up to this date. This is about the start of the period of the great

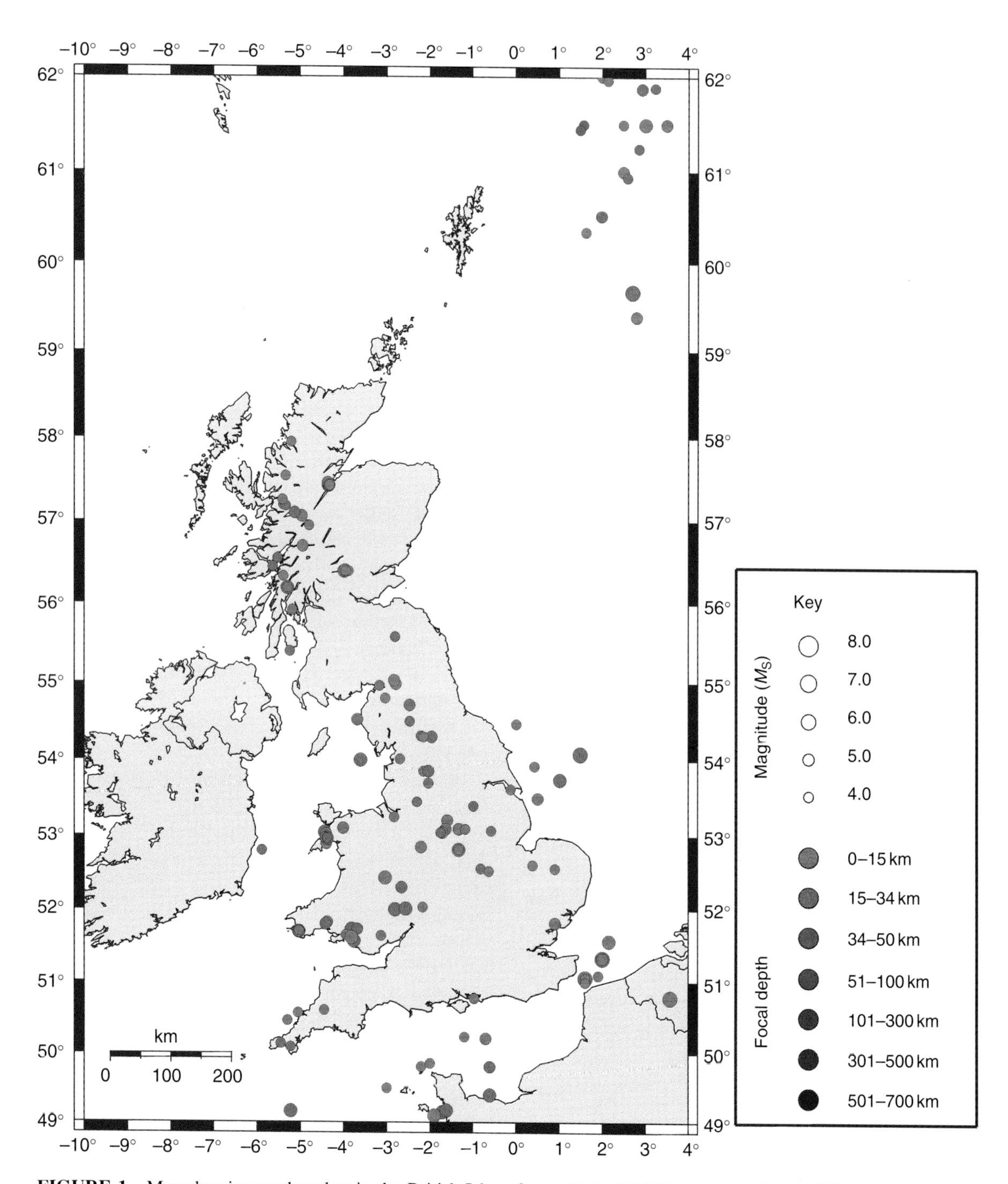

FIGURE 1 Map showing earthquakes in the British Isles of magnitude 4.0 ML or greater from 1382 to 1999.

earthquake catalogs such as von Hoff (1840) and Mallet (1852–1854), which include British earthquakes. Perrey's annual catalogs contain British events and he also published a regional catalog for the British Isles (Perrey, 1849).

The second half of the 19th century saw the rise of the amateur scientific society. These "field clubs" were a recreational focus for amateur naturalists, meteorologists, etc., and usually published their own journals where data could be presented. Local earthquakes, when they occurred, formed a natural subject for study, and some very substantial data sets were published in this way, the best-known being the book-length account of the damaging 22 April 1884 Colchester earthquake (4.6 M_L) by Meldola and White (1885).

3. Modern Studies

The most important amateur contributor to British seismology was Charles Davison, a mathematics teacher, who from 1889 to 1926 single-handedly ran a macroseismic monitoring program in the United Kingdom. In 1896 he started work on a catalog of British earthquakes, which was finally published in 1924. This work (Davison, 1924) contains the first attempt to present a quantitative account of British seismicity, using an intensity scale of Davison's own devising but loosely based on the Rossi–Forel scale. Davison also attempted some critical regard to sources, for example, rejecting a series of spurious earthquakes published first by Thomas Short (1749) as without historical foundation. However, Davison was not a historian, and had a hard time (for example) with the confused dates of medieval earthquakes found in the various chronicles.

Davison's catalog remained the standard source on historical British earthquakes for the next 50 y or so. A first attempt to compile a numerate catalog was started around 1975 by Roy Lilwall of the Institute of Geological Sciences (later the British Geological Survey). Towards the end of the 1970s it was realized that seismic hazard in the United Kingdom needed to be considered in the context of the nuclear power program, and a large amount of effort was devoted to the historical investigation of British earthquakes over the next ten years. The result was a series of reports including notably Principia Mechanica Ltd. (1982), Soil Mechanics Ltd. (1982), Ambraseys and Melville (1983), Burton *et al.* (1984) and Musson (1994). The last of these is an attempt to synthesize all the results of the previous 15 y of research into a single consistent catalog.

4. Discussions

A concept recently introduced into historical seismology is that of the "roots" of an earthquake determination within a parametric catalog. The roots can be classified in terms of quality of supporting data, and a system has been proposed (Stucchi and Camassi, 1997; Stucchi, *et al.*, 1999) grading the quality of any entry in a catalog of historical earthquakes. The best grading situation is that the entry is backed by a study containing intensity data points derived from primary historical sources, and the worst grading situation is that the entry is copied from another parametric earthquake catalog with the original data unknown. A large amount of historical research has been undertaken in the United Kingdom (aided, perhaps, by the fact that the total number of large earthquakes is not very great). Consequently, the situation for the UK catalog is that most earthquake determinations fall into the highest class, and almost all are based on data sets from primary historical data (Musson, 1996; Stucchi *et al.*, 1999). One can now say that our understanding of historical seismicity in Britain is in as good a shape as realistically can be achieved. A plot of earthquakes of magnitude 4.0 M_L or greater is shown in Figure 1. The full manuscript, of which this subchapter is an introduction, is given on the attached Handbook CD under directory \48.5Musson1.

References

Ambraseys, N.N. and C. Melville (1983). "The Seismicity of the British Isles and the North Sea," SERC Marine Technology Centre, London.

Boyle, R. (1666). A confirmation of the former account touching the late earthquake near Oxford, and the concomitants thereof. *Phil. Trans. R. Soc. London* **1**, 179–181.

Burton, P.W., R.M.W. Musson, and G. Neilson (1984). "Studies of Historical British earthquakes", BGS Global Seismology Report No. 237.

Davison, C. (1924). "A History of British earthquakes," Cambridge University Press, Cambridge.

Fleming, A. (1580). "A bright burning beacon, forewarning all wise virgins to trim their lampes against the coming of the bridegroome, conteining a generall doctrine of sundrie signes and wonders, specially earthquakes both particular and generall: A discourse of the end of this world: A commemoration of our late earthquake, the 6 of April, about 6 of the clocke in the evening 1580, and a praier for the appealing of Gods wrath and indignation," Denham, London.

Grey, Z. (1750). "A chronological and historical account of the most memorable earthquakes that have happened in the World from the beginning of the Christian period to the present year 1750, with an appendix, containing a distinct series of those that have been felt in England, and a preface, seriously address'd to all Christians of every denomination," Bentham, Cambridge.

Grünthal, G. (Ed.) (1998). "European Macroseismic Scale 1998," Cahiers du Centre Européen de Géodynamique et de Séismologie No. 15, Luxembourg.

Hoff, K.E.A. von (1840). "Chronik der erdbeben undvulkan-ausbruche," Perthes, Gotha.

Kendrick, T.D. (1956). "The Lisbon Earthquake," Methuen, London.

Mallet, R. (1852–4). Catalogue of recorded earthquakes from 1606 BC to AD 1850. *Br. Assoc. Rep.* **1852**, 1–176, **1853**, 118–212, **1854**, 2–236.

Meldola, R. and W. White (1885). "Report on the East Anglian Earthquake," Essex Field Club Special Memoirs, No. 1.

Milne, D. (1842–4). Notices of earthquake-shocks felt in Great Britain, and especially in Scotland, with inferences suggested by these notices as to the causes of the shocks, *Edin. New Phil. J.* **31**, 92–122; **32**, 106–127; **33**, 372–388; **34**, 85–107; **35**, 137–160; **36**, 72–86, 362–377.

Musson, R.M.W. (1993). Comrie: a historical Scottish earthquake swarm and its place in the history of seismology. *Terra Nova* **5**, 477–480.

Musson, R.M.W. (1994). "A Catalogue of British Earthquakes," BGS Global Seismology Report No WL/94/04.

Musson, R.M.W. (1996). "Roots and References for the UK Earthquake Catalogue," BGS Global Seismology Report No. WL/96/03.

Neilson, G., R.M.W. Musson, and P.W. Burton (1984). "Macroseismic Reports on Historical British Earthquakes VI: The South and Southwest of England," BGS Global Seismology Report No. 231.

Ockenden, R.E. (1936). "Thomas Twyne's discourse on the earthquake of 1580," Pen-in-Hand Publishing, Oxford.

Perrey, A. (1849). Sur les tremblements de terre dans les Iles Britanniques. *Ann. Sci. Phys. Natur. Agric. Ind. Soc. Nat. d'Agr. De Lyon*, 2nd Series **1**, 114–177.

Principia Mechanica Ltd. (1982). "British Earthquakes," PML, Cambridge.

Richmond, Duke of, and Lenox, C. (1738). An account of a shock of an earthquake felt in Sussex on the 25th of October, anno 1734. *Phil. Trans. R. Soc. London* **39**, 361–367.

Short, T. (1749). "A general chronological history of the air, weather, seasons, meteors, etc," Longman & Miller, London.

Soil Mechanics Ltd (1982). "Reassessment of UK Seismicity Data." SML, Bracknell.

Stucchi, M. and R. Camassi (1997). Building up a parametric catalogue in Europe: the historical background. In: "Historical and Prehistorical Earthquakes in the Caucasus." (D. Giardini and S. Balassanian, Eds.), pp. 357–374. Kluwer, Dordrecht.

Stucchi, M., P. Albini, R. Camassi, R.M.W. Musson, and R. Tatevossian (1999). Main results of the project "BEECD": A Basic European Earthquake Catalogue and a Database for the evaluation of long-term seismicity and seismic hazard. *Report to the EC*, Project EV5V-CT94-0497.

Wallis, J. (1666). A relation concerning the late earthquake neer Oxford; together with some observations of the sealed weatherglass, and the barometer, both upon that phaenomenon, and in general. *Phil. Trans. R. Soc. Lond.* **1**, 166–171.

49

Macroseismology

R.M.W. Musson
British Geological Survey, Edinburgh, United Kingdom
I. Cecić
Geophysical Survey of Slovenia, Ljubljana, Slovenia

1. Intensity and Intensity Scales

1.1 The Nature of Intensity

Intensity can be defined as a classification of the strength of shaking at any place during an earthquake, in terms of its observed effects. It can be compared to the Beaufort Scale of wind speed, and is not a physical parameter as such, although it is evidently related to physical parameters of strength of shaking. Also, one can think of it as a sort of shorthand for longer descriptions which may be common from one earthquake to the next. Thus, instead of writing that the earthquake was generally felt indoors at a certain place, and that dishes and cups rattled, one simply writes "intensity 4" (or whatever, according to the scale used) wherever this is the principal description. As a classification, it is essentially an integer quantity when assigned from observed data. Traditionally, Roman numerals have been used to represent intensity values to emphasise this point (it is hard to write "VII½"). Nowadays the use of Roman numerals is largely a matter of taste, and most seismologists find Arabic numerals easier to process by computer.

The word "macroseismic" is used to mean "pertaining to the perceptible effects of earthquakes" in distinction to "microseismic," meaning "relating to instrumental observations of earthquakes." The phrase "macroseismic data" can mean any assemblage of descriptive material on the effects of earthquakes, and need not necessarily include numerical intensities. The word "macroseismology" is applied to the study of earthquake effects.

Intensity scales normally consist of a series of descriptions of typical earthquake effects, each with an accompanying number, and arranged so that increasing numbers reflect increasing severity. This is not always the case; a few scales are arranged so that the lowest numbers express the strongest shaking. Also, other classificatory systems can be applied to macroseismic data, such as simple systems of dividing reports into "no damage," "damage," and "heavy damage".

It must be stressed at the outset that intensity is a classification of effects at a given place, and not a classification of an earthquake as a whole. Any earthquake is likely to produce different intensities at different places, generally decreasing with distance from the epicentre.

1.1.1 The Relation of Intensity to Sensors

Although intensity can be considered purely as a shorthand for a description of effects, it is also intended to be a measure of strength of shaking, and this introduces some important implications. An intensity scale typically consists of a number of a series of descriptions relating to each degree of the scale, and each description will consist of a number of component parts, usually referred to as diagnostics. Thus the description of intensity 5 in the 1956 Modified Mercalli Scale (MM56) reads "Felt outdoors; direction estimated. Sleepers wakened. Liquids disturbed, some spilled. Small unstable objects displaced or upset. Doors swing, close, open. Shutters, pictures move. Pendulum clocks stop, start, change rate" (Richter, 1958). There are therefore eight diagnostics for this degree of the scale, and the investigator will assign intensity 5 to a place if the data he has for that place more closely conform to these diagnostic descriptions than to any other ones in the scale.

Each diagnostic describes the effect of the ground vibration on some form of sensor. In the intensity degree described above, the sensors are: people outdoors, sleepers, liquids in containers, small unstable objects, doors, shutters or pictures, and pendulum clocks. In lower degrees of any scale, the most important sensors are people and domestic objects. In higher degrees, the emphasis moves to buildings as sensors, and higher degrees describe progressively more severe damage.

In the case of low intensity degrees, simple sensors like teacups can be considered to respond to the same degree of

INTERNATIONAL HANDBOOK OF EARTHQUAKE AND ENGINEERING SEISMOLOGY, VOLUME 81A ISBN: 0-12-440652-1

shaking in roughly the same way. Thus, differences from place to place in the number of teacups that rattled or were spilt can be compared directly to differences in the earthquake shaking, and not to differences between the size or type of teacup. Since comparisons in the degree of shaking are what the seismologist wishes to make, items like teacups are good sensors for determining intensity. However, more diverse sensors require more explicit description. If one were to consider the diagnostic "objects fall over," it makes a considerable difference if the "object" is a small, unstable ornament or a large heavy item like a TV set. It is therefore necessary to discriminate between objects that respond to shaking differently, in this case between small unstable objects and large, heavy, stable ones. In this case it is not difficult, and anyone can recognise that it takes more shaking to upset a large heavy object than it takes to knock over a small delicate one.

This becomes crucial when one comes to deal with damage to buildings. For buildings can be of very different strengths, and so when comparing damage from one locality to another it is vital to compare like with like. It was notable, for example, that the Latur earthquake of 1993, which caused many weak rubble masonry houses to collapse, produced very little damage to engineered structures in the same localities. One cannot therefore just use "damage to buildings" as a sensor; it is essential to consider damage to buildings of equivalent strength.

The earliest intensity scales tended to treat all buildings as equivalent, on the assumption that one would always be dealing with the "typical" construction of the country. Whether or not this was justified in the last century, it is not justified today. Later intensity scales used building construction type as an analogue of strength, which for the most part is a good practice. The most recent scale in general use, the European Macroseismic Scale (EMS), further acknowledges that condition and workmanship of a building can alter its strength considerably, and allows for downgrading or upgrading a building with respect to its vulnerability class (the classification of strength used in EMS-98) where the condition of the building requires it (Grünthal, 1998).

The finer the approach to assessing the details of the sensor, the harder it becomes for the seismologist to apply the scale. A key consideration here is robustness. An intensity scale must be a practical instrument, and the construction of the scale needs to steer a course between precision and ease of use. Any categorisation of building type that required the fieldworker to first take a degree in civil engineering before he or she could determine how to apply it to a particular observation, would be very undesirable. The degree of detail present in the EMS-98 is probably the maximum that can reasonably be employed at a practical level.

The hardest area of all concerns the use of effects on nature as sensors. Some outdoor observations are not problematical; for example, there is not so much variation in bushes that the shaking of bushes cannot be used as an intensity diagnostic. But when it comes to effects on ground, the situation is rather different. It is well-known from numerous studies that the proclivity of slopes to failure (as rockfalls or landslides) is hugely dependent on a number of factors, especially those related to the prevailing hydrological conditions. These factors often vary considerably over short distances. As a result, the vulnerability of ground to failure is strongly variable, and has a large effect on the incidence of coseismic observations of slips, slides, fissuring, and so on. Just as it is not helpful to know that "an object fell" if one does not know if the object was a pencil or a typewriter, so also it is not helpful to know that a rockfall occurred if one does not know the strength of the slope beforehand. But whereas it is an easy matter to categorise objects such as pencils as small or large, it is, in practical terms, more or less impossible to usefully classify and recognise different strengths of slopes. As a result, observations of ground processes are of limited use in determining intensity. It has been objected that this restriction makes it practically impossible to derive intensity values for uninhabited areas. The answer to this is that, so far as meaningful values are concerned, it is indeed practically impossible to derive intensity values for uninhabited areas, and this is a fact of life, and it has to be conceded that nature is not arranged expressly for the convenience of seismologists.

This causes a further problem with some older intensity scales. It has been pointed out that in MM56, the description of intensities 11 and 12 depend on ground failure, and as ground failure is not entirely intensity-related, these two intensity values are essentially unusable (Bakun pers. comm., 1999). This is not a problem in EMS-98 where the highest intensity degrees depend on damage to buildings.

It may be the case that intensity values can saturate in some conditions. For instance, in areas where all structures are of very poor construction, such that all are totally destroyed at (say) intensity 9, it is then impossible to recognise intensity 10 in such an area. This is again, unavoidable.

The extent to which patterns of effects, corresponding to increasing degrees of shaking, can be recognised in working practice, determines the fineness of discrimination between different intensity levels that can be made. The general consensus of most major intensity scales on twelve roughly equivalent degrees, to cover the range from not felt to total destruction, is probably more than just conservatism or familiarity. Some authors (Shebalin, 1970; Dengler and Dewey, 1998) have argued for the practicality of discrimination of intermediate levels between the defined levels of a twelve degree scale, but any benefits from this are likely to be outweighed by the dangers of false precision in most cases. Some applications of intensity data do involve treating intensity degrees as if they were real numbers, however (see, for example, Section 3.2 below). It is also sometimes the practice in compiling computer files of intensity data to write an uncertain intensity value, which would normally be written 7–8 or VII–VIII (meaning either 7 or 8) as 7.5.

1.1.2 Assigning Intensity

Although detailed procedures may vary from case to case according to circumstances, the essential stages in assigning intensity can be described as follows.

1. The seismologist gathers all available information about the effects of the earthquake. This is discussed in more detail in Section 2.
2. The seismologist sorts the data by place. The division of places should be large enough for each place to contain a reasonable amount of data, and fine enough so as not to group together observations made in very different local conditions. Typically, data from a single village or small town will be treated together, whereas large towns and cities will be divided into suburbs or districts. One would not want to assign intensity to a single street; nor to a whole county.
3. The seismologist consults the intensity scale to determine which description of those for the various intensity degrees best fits the sum of the data for the particular place under consideration. The key concept here is coherence. It is not helpful to get lost in a pursuit of detail of individual diagnostics; the correct assignment is the one that best expresses the generality of the observations. Note that it is not desirable to try and assign intensities to each individual report and then combine them for a place; the data should be combined for the place first, and then one intensity value should be determined that best expresses the generality of the data.

It is not to be expected that all diagnostics will be satisfied by the data in all cases; for example, some may simply not be present. It is therefore advisable to adopt a flexible approach in seeking the best fit over the range of data available, rather than attempting to set up rigid formulas that depend on one or two key diagnostics only. It is important to look for an element of coherence in the data overall, rather than to rely on any one diagnostic as a yardstick. It is also necessary to be wary of giving too much weight to the occasional extreme observation, which might lead to an overestimation of the intensity at the place in question. For example, overreliance on damage as a diagnostic has in the past resulted in overestimation of intensities in cases where isolated, even anomalous, cases of damage have been used to base assessments on even though the mass of other data suggests a lower value.

Where the data consist of textual descriptions, the effects may be reported in terms far from the wording of the intensity scale. In such cases, it may be useful to consider whether the overall tenor of the description compares with the general character of a degree of the intensity scale.

Information that an effect definitely did not occur is often just as valuable as information that it did occur when determining intensity, and such data should not be neglected. However, to assume automatically that an effect did not occur, just because it was not reported, is dangerous and invalid unless there are specific reasons why such an assumption can be justified. It should also be noted that no single effect is ever certain. This is important when attempting to infer a negative, rather than a positive conclusion. Just as it is possible to get an anomalous instance of damage even when the intensity is actually quite low, it is similarly possible for a vulnerable object to escape undisturbed even though the intensity was quite high.

Where data are simply insufficient to make a clear decision, it is necessary either to resort to the use of a range of values, for example, 7–8 (7 or 8) or >6 (7 or higher), or simply note that the earthquake was felt (usually denoted by the symbol F).

The use of automatic algorithms or expert systems to determine intensity values from a set of data is an interesting topic of current research, although it tends to be tied to a certain questionnaire format. Some examples can be found in Barbano and Salemi (1990) and Gasparini *et al.* (1992).

It is not good practice to try to "correct" intensity values for local soil conditions when assigning values. Soil conditions can be used to explain why certain intensity values are higher or lower than others in the vicinity. But as long as intensity is intended as a measure of what happened, the fact that a strong shaking was influenced in part by local conditions does not detract from the fact that the strong shaking did indeed occur.

In some cases it may be inadvisable to attempt to use certain data for assigning intensities. A particular case in point relates to observations from high buildings. It is well known that people in upper storeys are likely to observe stronger earthquake vibration than those in lower storeys. Various practices, such as reducing the assigned intensity by one degree for every so many floors, have been suggested, but never found general favour. Also, since very tall buildings may behave under earthquake loading in particular ways according to the frequency of the shaking and the design of the building, the increase of severity of shaking with elevation may be irregular. The recommended practice is to discount all reports from observers higher than the fifth floor when assigning intensity; although in practice the actual behaviour of individual buildings will vary considerably, depending especially on the slenderness of the building. In general, the user should be more concerned with effects observed under normal circumstances rather than in exceptional cases.

One special case is the situation where the only reports are from tall buildings, because the shaking was so weak that it was only perceptible on the upper floors of such structures. This sort of datum is typical of the lowest intensities perceptible.

Observations from special structures, such as lighthouses, radio towers, monumental buildings (such as cathedrals) etc., are difficult to use, as the responsiveness of such structures is impossible to categorise. Such data are best restricted to the status of complementing other information on which the intensity assessment depends. If such structures are the only source of information for a place, then it is important to convey the uncertainty inherent in any assignment made from such data. A similar reservation applies to data from observers underground, which are not easily comparable with observations made at the surface.

It is not recommended to derive intensity values from instrumental measures of ground motion. An intensity value is intended as a description of what happened, not an inference from some recorded acceleration. The use of values derived in such a way is deceptive of the reader, who may believe that damage was recorded where in fact it was not. In addition, correlations between such parameters as peak ground acceleration and intensity are notoriously unreliable (Trifunac and Brady, 1975). Accelerations of 1–2*g* have been recorded on many occasions and these are usually not accompanied by especially high damage levels. If it is necessary (e.g., for disaster response purposes) to make predictions of what intensities might have been observed on the basis of instrumental recordings, then it should be made clear that the values are estimates of intensity and not observations.

1.2 The History of Intensity Scales

The use of intensity scales is historically important because no instrumentation is necessary to use them, and useful measurements of an earthquake can be made by an unequipped observer. The earliest recognisable use of intensity was by Egen in 1828, although simple quantifications of damage had been made in the previous century by Schiantarelli in 1783 (Sarconi, 1784), and some earlier Italian examples are said to exist. However, it was only in the last quarter of the 19th century that the use of intensity became widespread; the first scale to be used internationally was the ten-degree Rossi-Forel Scale of 1883. The early history of intensity scales can be found in Davison (1900, 1921, 1933); a later study was made by Medvedev (1961).

The scale of Sieberg (1912, 1923) became the foundation of all modern twelve-degree scales. A later version of it, known as the Mercalli–Cancani–Sieberg Scale, or MCS Scale (Sieberg, 1932), is still in use in Southern Europe. The 1923 version was translated into English by Wood and Neumann (1931) and became the inappropriately named Modified Mercalli Scale (MM Scale). This was completely overhauled in 1956 by Richter (1958) who refrained from adding his name to the new version in case of further confusion with "Richter Scale" magnitudes. Richter's version became instead the "Modified Mercalli Scale of 1956" (MM56) despite the fact that the link to Mercalli was now extremely remote. Local modifications of Richter's MM56 scale have been used in Australia and New Zealand among other places. More recent attempts to modernise the MM scale further, e.g., that of Brazee (1978), have not caught on.

In fact, the 1931 version of the MM scale is still officially in use despite the fact that it is nearly 70 y old and has been effectively obsolete for some time. Dengler and Dewey (1998) note that in fact, routine use of the MM31 scale differs substantially from a literal reading of the 1931 text, so as to incorporate the result of the succeeding six decades of experience of assigning intensities to US earthquakes. This is good from the point of view of the quality of intensity data sets that are produced, but it does mean that procedures are not so transparent as they should be, and the data sets cannot easily be deconstructed or reconstructed by other workers who lack access to the extra rules and conventions being applied.

In 1964 the first version of the MSK Scale was published by Medvedev, Sponheuer and Karnik (Sponheuer and Karnik, 1964). This new scale was based on MCS, MM56 and previous work by Medvedev in Russia (the GEOFIAN scale), and greatly developed the quantitative aspect to make the scale more powerful. This scale became widely used in Europe, and received minor modifications in the mid 1970s and in 1981 (Ad hoc Panel, 1981). In 1988 the European Seismological Commission agreed to initiate a thorough revision of the MSK Scale. The result of this work (undertaken by a large international Working Group under the chairmanship of Gottfried Grünthal, Potsdam) was published in draft form in 1993, with the final version released (after a period of testing and revision) in 1998 (Grünthal, 1998). Although this new scale is more or less compatible with the old MSK Scale, the organisation of it is so different that it was renamed the European Macroseismic Scale (EMS). Since its publication it has been widely adopted inside and also outside Europe. Despite its name, the EMS is easily adaptable for use with the built environment of any place in the world.

As with the MM scale, it was recognised that procedures and conventions were being applied to the use of the MSK scale in best practice, which were often not written down or defined. One of the objectives of the EMS Working Group was to encapsulate these formally into the scale. This was done by incorporating extensive guidelines for use into the text of the scale. In this way, the operation of the scale becomes more transparent, and therefore discrepancies in practice between different users over time, or between region and region, should be reduced.

The one important intensity scale that does *not* have twelve degrees is the seven-degree Japanese Meteorological Agency Scale (JMA Scale). This is based on the work of Omori, and is the scale generally used in Japan and Taiwan. A recent revision (JMA, 1996) has increased the number of degrees in the JMA scale to ten, with intensities 5 and 6 now being subdivided into "lower 5," "upper 5," lower 6," and "upper 6." The historically important Rossi-Forel Scale is no longer much in use, except in the Philippines (Dewey, 2000 pers. comm.).

To some extent the middle years of the 20th century saw a decline in interest in macroseismic investigation, as progress was made in improving the power of instrumental monitoring. When earthquakes could be located accurately by instrumental means, and their magnitude and depth similarly determined, it might well have seemed redundant to pay attention to earthquake intensity. However, since the middle 1970s there has been a strong revival of interest in the subject since macroseismics are essential for the revision of historical seismicity and therefore of great importance in seismic hazard assessments. Macroseismic studies of modern earthquakes are vital for (1) calibrating studies of historical earthquakes; (2) studying local attenuation, and (3) investigations of vulnerability, seismic hazard and seismic risk.

1.3 Intensity and Isoseismal Maps

The presentation of intensity data is usually done in the form of a map. As well as plotting intensity points, it is usually useful to be able to draw contour lines of equal intensity, called isoseismals. An isoseismal can be defined as a line bounding the area within which the intensity is predominantly equal to, or greater than, a given value. Drawing isoseismal lines is therefore a process of contouring the strength of shaking of an earthquake over the area in which it was felt.

No precise instructions can be given for drawing isoseismals as no definitive method has ever been agreed. Some workers adopt a practice of overlaying a grid on the data and taking the modal value in each grid square prior to contouring, others prefer to work directly on the plotted intensity values. Workers have differing preferences for the amount of smoothing, extrapolation, etc., that is to be employed. Thus, at present, the drawing of isoseismals is to some degree subjective.

However, some guidelines can be given. The degree of smoothing employed should reflect the purposes to which the resulting map will be put. If the map is intended for local studies, i.e., to point up areas where seismic hazard may be enhanced owing to local soil conditions, then smoothing will be applied to a lesser extent, and isoseismals will be relatively convoluted. If the map is intended for other purposes (calculation of earthquake parameters, attenuation studies, tectonic studies, etc.) then the curves will normally be smoothed so that only major re-entrants and outliers are shown. The latter case is more common, and can be regarded as the default.

Excessive detail is never justified, given the limits of resolution of the data. As a general rule, re-entrants and outliers should not be drawn unless suggested by a grouping of at least three data points. If isoseismals have to be interpolated or extrapolated across areas of water, or areas without data points, these sections of the lines should be shown as dashed. In cases where, for example, an epicentre is offshore, and only (say) a 120° arc of each isoseismal would fall onshore, it is not correct to project the whole of the remaining 240° of each isoseismal on a map, even as a dotted line. Only the onshore section should be drawn, with each line tailing off with a short dotted section offshore if desired. Plotting isoseismals that are completely offshore and merely projections of an intensity attenuation curve should not be done. For onshore earthquakes with few data, it is not good practice to attempt to draw isoseismals conjectured from one or two points only; at least three mutually supporting data points for one intensity value should exist before one attempts to draw even a partial isoseismal for that value.

The aim of producing isoseismals is a little different from other contouring exercises. Each data point does not necessarily carry the same weight. This is because the purpose of an isoseismal is to delimit areas where the intensity generally exceeded a certain value. Residuals within this area are to be expected, either through soil conditions, sampling effects, or some other cause. These residuals are generally ignored in drawing isoseismals, as it is usually more important to display the general trend of the data under average ground conditions than to point out deviations from this.

As a result, computer contouring programs usually do not give good results with macroseismic data. Whereas the human eye readily spots the residuals to be ignored, computer algorithms cannot do this with the same ease.

It is also recommended that seismologists should preserve and publish tables of intensity data (place, geographical coordinates, intensity) whenever possible. The practice of publishing isoseismal maps without the underlying data either plotted or tabulated, results in the user being unable to determine how much judgement has been exercised in plotting the isoseismal lines.

1.4 Intensity Attenuation

Intensity attenuation, the rate of decay of shaking with distance from the epicentre, can be expressed in two ways. Firstly, there is the drop in intensity with respect to the epicentral intensity. This is shown by the Kövesligethy (1906) formula discussed in Section 3.4; this form of intensity attenuation and depth determination from intensity are closely linked. One can also express intensity attenuation as a function of magnitude and distance. Such formulas usually have the functional form

$$I = a + bM + c\log R + R + dR \tag{1}$$

where R is hypocentral (slant) distance, and a, b, c and d are constants. (The fourth term is sometimes dropped, especially in intraplate areas.) Since most earthquake catalogues include magnitude as a parameter, this form of intensity attenuation is extremely useful in seismic hazard studies. Intensity is sometimes a good parameter to use for expressing seismic hazard, since it relates directly to damage. It yields hazard values which are probably more relevant to planners and insurers than physical ground motion parameters (discussed further in Section 4.1).

Some typical values are:

Interplate (New Zealand):

$$I = 2.18 + 1.41\,M_S - 1.18\ln R - 0.0044\,R \tag{2}$$

Intraplate (SE Australia):

$$I = 4.00 + 1.64\,M_S - 1.70\ln R \tag{3}$$

the sources for these equations being Dowrick (1992) and Gaull *et al.* (1990).

It should be remembered when calculating intensity attenuation, that such equations should represent intensity given normal soil conditions, and therefore high residual data points (resulting from local ground conditions) in a set of macroseismic data should be discounted for this purpose. This is so that site correction factors can be applied when the attenuation equation is being used in a predictive sense, without danger of circularities arising.

2. Macroseismic Surveys

After an earthquake is felt in a region, there are several possibilities as to how to collect the data. The final choice depends on many factors, such as:

1. how strong the earthquake was, and, therefore, how important the data are;
2. whether it was a single shock or a part of a long sequence;
3. if there is already an existing network of earthquake observers in the region, and whether the network is dense enough;
4. the financial situation of the institution that collects the data, etc.

These points will now be considered in more detail.

In many countries there is a practice that macroseismic data are not collected for every single felt earthquake, but a certain threshold is set. It might be that the data are collected only if the magnitude or intensity of the earthquake was larger than some threshold value, or even only in cases that the earthquake has caused some damage. For example, in the United Kingdom surveys are normally done for all earthquakes with onshore epicentres and magnitudes greater than 3 M_L, and for offshore events with a significant felt area. In Italy the magnitude threshold is also set to 3, but the questionnaires are sent even for weaker events if the maximum intensity exceeded 5 MCS (Mercalli–Cancani–Sieberg scale). On the other hand, some smaller countries, as well as countries with low seismicity, such as Hungary, Austria, Slovenia etc., can afford to collect data actively for almost every earthquake felt on their territory. In areas where earthquakes are rare, the data for even a weak earthquake might be much more valuable and useful for the study of seismicity than the data for a stronger event in the region where the average activity is high.

After a strong main shock there is often long and recurrent aftershock activity. In such cases, people feel so many earthquakes that after some time they start to forget what they felt during any particular event. Also, in the aftershock sequence there tend to be several, or even many, events of approximately the same maximum intensity, as well as magnitude and co-ordinates. Getting the macroseismic data for all the aftershocks is impossible; therefore in practice seismologists often have to make decisions and collect the data only for some events.

Several seismological institutions have organised permanent networks of voluntary observers in the field. In the case of an earthquake, the questionnaire forms are distributed to them by mail, or they already have some at home, office etc. For small earthquakes such a method is usually enough. However, for a damaging earthquake even the best network of observers is not sufficient, so additional field work is necessary in order to get as many data as possible from the epicentral area.

Although it might sound trivial, the method that is going to be used in collecting the macroseismic data depends heavily on the financial situation of the institution. Regardless of the scientific reasons, it is unfortunately frequent that in everyday practice data are not collected even for important events due to a momentary lack of funds for stamps, films, daily allowances or other material necessary for field trips.

2.1 Questionnaire Design and Distribution

2.1.1 Questionnaire Design

Macroseismic questionnaires are forms consisting of questions about the earthquake; some forms have built-in already the selection of possible answers, so the observer can choose among them (multiple choice questions); others give the observer the possibility to express his impressions in his own words (free-form questions). The point of having a questionnaire is primarily to have all the data in more or less the same format, which means that all the questions are asked of the whole population of observers in precisely the same way. It makes the collected data comparable within the data set for the same earthquake, as well as between different events (for which the same type of form was used).

There are several types of questionnaire forms being used, depending mostly on the target population and the type of the data that one wishes to collect.

The commonest case is a general form, consisting of a set of questions that follow the chosen macroseismic scale. In many cases the questionnaire includes also some additional questions, that are not strictly included in the scale, but might be useful in having a more complete description of the earthquake effects.

Some institutions have two versions of the same form, with and without the questions on damage, for strong and weak earthquakes, respectively. The point of having the shorter version of the questionnaire is not to burden the observers with too many questions in case of weak events. Since one usually relies on the goodwill of observers to fill in a questionnaire at all, it makes sense not to make the questionnaire unnecessarily long, difficult, or otherwise discouraging.

Generally the questionnaire form will be too complicated for children in primary schools, although they can be a very valuable source of data, especially in rural areas, where children from a relatively large area can be in the same class. Therefore a special form can be arranged for them; instead of the printed form, the questions are read to them one after the other, and all the necessary explanation is given.

After a damaging earthquake, there are a lot of data to be collected in the field; special forms can be constructed in order to make this procedure easier. The use of the EMS scale requires the seismologists to collect detailed data on type of objects and damage. The experience from the recent damaging earthquakes in Europe (1997 Central Italy, 1998 northwest Slovenia) has produced some preliminary forms for describing the damage on single objects, as well as for the whole settlement. An example of the questionnaire form for the assessment of damage on single buildings was proposed by Tertulliani *et al.* (1996/99) and is attached as Appendix 1 on the attached Handbook CD, under directory \49Musson2.

2.1.2 Questionnaire Distribution

There are several ways of distributing questionnaires that are being routinely used by seismologists: (1) by mail, (2) by press, (3) by e-mail, (4) by phone, (5) by making a request through the radio/TV, and (6) by direct distribution.

The most commonly used method for the distribution of questionnaires is surely by mail. In some countries it can be done free of postal charges, whereas others are not so lucky. Generally the method is the following: after an earthquake, the addresses of the observers from some database are printed onto the questionnaires and sent to them; the answers are then sent back using the spare envelope that was included or folding the card according to the instructions. The address database can be one constructed especially for use in macroseismology, or a more general one (phone book etc.) can be used. The ideal procedure is to send questionnaires to a completely random sample of respondents in the affected area, the sample being derived from some source such as electoral rolls. The advantage of this is that accurate statistical conclusions can be drawn about effects on the whole population from the analysis of the sample; the methodology is well known in the social sciences. The drawback is that setting up an appropriate random sample can be time-consuming, and such a task may be difficult to manage in the immediate wake of an earthquake. If one waits too long before distributing questionnaires, the chances are that many people will already have forgotten the details of their experience.

Experience shows that a very good percentage of answers can be achieved, especially with a selected database of respondents, if the database is kept promptly updated. It takes a relatively short time to print and pack the mail, but in cases where the postage has to be paid and in cases of either a large earthquake or a very busy year it might become quite costly. On the other hand, it gives homogeneous and comparable data, and the preparation of the intensity list is fast, because the coordinates of the places are already prepared.

One way to encourage response is to combine the questionnaire survey with some sort of raffle; the sender of one randomly drawn questionnaire wins a small prize, such as a box of chocolates. This gives the respondents an additional motivation for sending in a reply, which might be useful particularly in making sure negative data are adequately represented.

Some countries, such as the United Kingdom, have very good experience with having questionnaires printed in some newspapers (Musson, 1992); in others, such as Italy, the idea seems not to work so well (Riggio and Slejko, 1989). After an earthquake is felt in the region, the questionnaire is printed in one or more local or regional newspapers, and the readers are encouraged to fill in the answers and send them back. The method does not take much time to launch (the questionnaire form is usually already prepared and ready to be used), but it can be quite time-consuming afterwards, especially in the case of large numbers of answers, that need to be sorted and the coordinates for every single place defined. Another disadvantage of this method is that the questionnaire sample is not truly random and may be biased toward those people who felt the earthquake.

Lately there is some experience with collecting macroseismic data using electronic media. Several institutions have their questionnaire forms on their Internet home pages, and they receive some answers by e-mail. It is possible to foresee that this method may become more and more used in future. There is particular interest in this method in the United States, where there is probably a higher proportion of people on-line than in most other countries.

Collecting the data by telephone cannot be treated as a regular questionnaire distribution, but more like polling. Of all the methods mentioned here, it is the most time-consuming and possibly the most expensive as well. However, if a seismologist can not send questionnaires or print one in the newspapers, or go into the field, the telephone is the only available method to collect the data. One should be very careful when asking questions, in order not to suggest the expected answer to the person who is giving the data. In some countries, e.g., USA, this method of collecting the data would not be possible for governmental institutions owing to legal restrictions. Telephone data collection is perhaps best suited for establishing whether a small event at the threshold of perceptibility was felt at all, especially in rural locations, where a few farms and perhaps a hotel are the only places that need be consulted.

After each strong earthquake, there is usually a lot of interest from the media about the details. This gives the seismologist an opportunity to speak (on the radio or TV) directly to the population, explaining briefly what data are of particular interest and asking people to write down their descriptions of the earthquake and send them to the institution that collects the data. An experience from Slovenia showed an unexpectedly good response of listeners to the national radio that broadcast the request several times (for an earthquake of 4.1 M_L and maximum intensity 5 MSK in 1990 approximately 160 letters were received).

Direct distribution of questionnaires after the earthquake can be used if the problem of returning the questionnaires to the institution in charge is solved. Then a pile of questionnaires could be left, for example, in post offices with the sign "Please take one" and clear instructions what to do afterwards. If the questionnaires are free of postal charges, then they can be simply posted back. If not, a box for collecting them or an alternative method should be arranged.

2.2 Field Surveys

Field survey is an obligatory tool that complements the above-mentioned methods in the case of a damaging earthquake (producing intensities of 6 EMS or higher). It can be used also for weaker events, in cases where the expected number and spatial distribution of returned questionnaires would not be enough to make a clear picture of earthquake effects in some part of the felt area.

Field surveys require some necessary equipment, such as good and detailed road maps of the area to be surveyed, questionnaires, notebooks and pencils, copies of the macroseismic scale, camera, and of course sturdy shoes, rainproof clothes, and sunglasses. A portable tape-recorder and a laptop computer might also come handy.

Field survey is usually best done by car, although some other methods of transport can be used in cases where there are several teams working on the same earthquake. The ideal team consists of a seismologist and a driver. In extreme cases, e.g., where the road network is impassable due to bridge damage, a helicopter may be required, preferably in collaboration with civil defence teams, the military, etc.

While interviewing people, one can use several methods; the answers can be filled directly into a questionnaire form, or written in free format in a notebook. The answers can be recorded on tape directly (if the person who is giving the data agrees, which is not often in rural areas) or the seismologist can repeat a summary of what was said into the voice recorder after finishing the conversation. In any case one must be very careful to label all the data carefully, in order to avoid any confusion.

Collecting the data in a small settlement is one as follows: Usually soon after the car is parked people come around and start telling what happened. Their answers, as well as their addresses are written in the notebook or recorded on the tape. If the earthquake caused some damage, this is the fastest way to learn where to find it. If possible, it is very desirable to take photographs of damage and keep records of it. Then a short survey of the settlement should be done, taking notes on the total number of buildings of each category (if EMS is used) and the number of damaged ones, as explained in the scale.

If the locality is still populated after the earthquake, it is usually very useful to visit briefly some points where people come together (shop, post office, parish office, hairdresser, local pub etc.). A visit to a school in rural area can give a lot of data in relatively short time, because in the same class one can find children living in many different villages or isolated farms.

If the intensity in a locality was high and the houses are badly damaged, the inhabitants will usually be living in improvised shelters and not in the village. In that case the entrance to the settlement may be closed and the field team should contact the local authorities in order to obtain the necessary documents to visit the damaged area. They (civil defence, firemen, police etc.) are also a very useful source of data on the damage and usually can point out some interesting cases. Some identification (such as a badge with the name of the member of the team as well as the name of the institution one works for, or at least a letter that certifies that) is very useful to have, as well as a helmet and a battery lamp.

In a case of a large settlement (small or big town) it is impossible to make a complete detailed survey and count every house. One should then contact the local authorities and see which would be the areas of particular interest. If the locality is still populated, the fastest method to collect data is a visit to schools, preferably for older children. The whole address including street and number written on their answers is necessary in order to be able to divide the town into smaller fractions (quarters, wards) when doing more detailed intensity studies. However, where heavily damaged areas have been evacuated and cordoned off, damage data on a house-to-house basis may be ultimately available through co-operation with the civil defence authorities.

It is necessary to stress that a good macroseismic survey for a damaging earthquake can not be done without co-operation with civil engineers. It is especially important to establish such co-operation in cases when the damaged area is large, although it is highly recommended to include a civil engineer in field teams every time there is damage, however slight it might be. Seismologists generally lack training and experience in civil engineering, and identifying the building vulnerability class may be difficult, especially in cases of engineered structures and buildings that have been modified to include some sort of earthquake resistance. Examples from recent earthquakes have shown that without close co-operation with civil engineers, the accurate assessment of damage, as well as intensity, would be difficult, if not impossible (Cecić *et al.*, 1999).

2.3 Review of Some Existing Practices

2.3.1 Italy

In Italy the practice of collecting and evaluating the macroseismic data has been a long and fruitful one. Istituto Nazionale di Geofisica (ING, National Institute for Geophysics) in Rome is at present the largest institution in Italy that collects macroseismic data. However, it is soon to be merged with some other important Italian seismological institutions into the new Istituto Nazionale di Geofisica e Vulcanologia (INGV, National Institute for Geophysics and Vulcanology).

Italy is a country with relatively high seismicity, and every year there are many earthquakes felt on its territory. Therefore at ING, intensity data are not collected for every earthquake, but only for those that exceed 3.0 in magnitude. Even so, the questionnaires are still sent for weaker events if the maximum intensity exceeded 5 MCS. The questionnaires are based on the MCS scale and consist of 79 questions, that cover the effects between 2 and 10 MCS. The questionnaire is shown in Appendix 2 on the attached Handbook CD, under directory \49Musson2 (or visit *http://ing712.ingrm.it/data_www/Macros/questing.html*). Field surveys are made after damaging events.

ING maintains a network of correspondents that report on earthquake effects and fill in the questionnaires (Tertulliani and Maramai, 1992). The network involves some public bodies, such as municipal authorities, the Forest Corps, and the Carabinieri Police Corps, i.e., approximately 13 000 correspondents. This configuration has been operative since 1987.

When evaluating the intensity data an algorithm is used, in order to make the evaluations more objective (Gasparini *et al.*, 1992). The intensity evaluations and the intensity maps are regularly published in quarterly Macroseismic Bulletins. An example of the output for an earthquake is shown in Appendix 3 on the attached Handbook CD, under directory \49Musson2. ING exchanges macroseismic data with neighbouring countries on a regular basis.

2.3.2 Austria

Austria is a country with relatively low seismicity. The history of collecting macroseismic data on the territory of present-day Austria begins at the end of the 19th century, when an "Earthquake Commission" was established in order to compile the earthquake catalogue for the Austro-Hungarian Monarchy, and to establish a network of both seismograph stations and permanent macroseismic observers (Fiegweil, 1989). Since then the macroseismic data have been collected and evaluated on a regular basis. Today the institution in charge is Zentralanstalt für Meteorologie und Geodynamik (ZAMG, Central Institute for Meteorology and Geophysics) in Vienna.

In 1947 an arrangement was made with the Ministry of the Interior providing that, in the case of an earthquake, each police station from the region where it was felt should send a report to ZAMG. Today police stations are still the principal contributors, apart from those people who send their reports spontaneously. In the case of a stronger and larger earthquake, additional questionnaires are sent to schools, municipal offices, post offices etc., and a short announcement with the request for reports is given to Austrian News Agency, the radio, and the TV station.

The reports are interpreted according to the EMS scale. The isoseismal lines are drawn and the focal depth and the macroseismic magnitude of the earthquake are calculated. Finally, the maps and a short description are printed and mailed to everybody who contributed to it (Appendix 4 on the attached Handbook CD, under directory \49Musson2). Austria exchanges macroseismic data with neighbouring countries on a regular basis.

2.4 Macroseismic Studies of Historical Earthquakes

Macroseismology is among the rare branches of seismology that can be applied to the research of events that happened before seismological instruments were invented, and for regions where no such instruments were ever installed. The necessary condition is that the region was populated and that the inhabitants left some record about the earthquakes that happened in their time. Moreover, macroseismic methods are the only means by which one can produce earthquake catalogues for the time before the 20th century, and are very important in studies of hazard (see Section 4.1).

When dealing with preinstrumental earthquakes, the seismologist often encounters several problems that can not be solved without the close co-operation of a trained historian. Institutes in some countries, for example Austria and Italy make it a practice to employ professional historians, who study the historical material (chronicles, newspapers, reports, letters, etc.) and prepare it for seismological interpretation. Many studies of historical earthquakes that were compiled in the earlier part of the 20th century (or in the 19th century) made many errors because of a lack of understanding of the precepts of historical studies on the part of the seismologists who conducted them. This subject is discussed at greater length in the sections on historical earthquakes.

3. Deriving Macroseismic Parameters

There are two basic approaches to deriving most earthquake parameters from macroseismic data. The first is to draw isoseismals and use the enclosed areas, or the average radii. The second is to base the calculations on the intensity data points themselves, without drawing isoseismals. The advantage of the second approach is that any subjectivity in the isoseismal drawing is entirely circumvented. The disadvantage is that the results may be biased by heterogeneity in the distribution of intensity points as a result of variations in population distribution.

3.1 Epicentre and Barycentre

The epicentre of an earthquake as derived from macroseismic data is something, which, in the past, has been approached in different ways (Cecić *et al.*, 1996). The following usage is proposed for the future:

- Macroseismic epicentre: The best estimate made of the position of the epicentre (i.e., the point on the Earth's surface above the focus of the earthquake) without using instrumental data. This may be derived from any or all of the following as circumstances dictate: position of highest intensities; shape of isoseismals; location of reports of foreshocks or aftershocks; calculations based on distribution of intensity points; local geological knowledge; analogical comparisons with other earthquakes, and so on. This is a rather judgmental process with some subjectivity, and does not lend itself to simple guidelines that can be applied uniformly in all cases.
- Barycentre: The point on the Earth's surface from which the macroseismic field appears to radiate. This is often the centre of the highest isoseismal or the weighted centre of the two highest isoseismals. Other methods have also been proposed, such as the point from which application of an attenuation equation to the whole data set produces the lowest residuals (e.g., Peruzza, 1992). Such methods have the advantage of not requiring the drawing of isoseismals at all. However, to determine the barycentre by calculating, say, the geometric mean of the data points equal to the highest and second highest intensity value, would run the risk of the resulting location being pulled toward the area with the highest number of settlements. This is a good illustration of the value of drawing isoseismals. Other synonymous terms proposed are "intensity centre" (Bakun and Wentworth, 1997) and "macrocentre" (Cecić pers. comm., 1994).

The macroseismic epicentre and barycentre are often the same, but need not be. As an example, in the case of the 1989 Loma Prieta earthquake the apparent point of origin of the macroseismic field was, for various reasons, to the north of the actual instrumental epicentre. If one were to attempt to locate a similar event from macroseismic data alone (for example, a historical Californian earthquake) one might be inclined to compensate for this effect by choosing epicentral co-ordinates to the south of the highest isoseismal. This would not affect the location of the barycentre.

Both these concepts have their uses. For any study of the tectonics of an area, the macroseismic epicentre is more useful. For studies of seismic hazard, especially those using a technique like extreme value statistics, the barycentre may give a better indication of the hazard potential of an earthquake.

3.2 Epicentral Intensity (I_o)

Epicentral intensity, usually abbreviated I_o, is a parameter that is commonly used in earthquake catalogues but rarely defined, and it is clear that different usage exists in practice (Cecić *et al.*, 1996). The meaning of the term is clearly the intensity at the epicentre of the earthquake, but since it is likely that there will not be observations exactly at the epicentre itself, some way of deriving this value is necessary. The two main techniques that have been used in the past are:

1. Extrapolation from the nearest observed data to the epicentre without changing the value, or use of the value of the highest isoseismal. Thus if there are a few data points of intensity 9 near the epicentre, the I_o value is also 9. If the epicentre is significantly offshore, I_o cannot be determined.
2. Calculating a fractional intensity at the epicentre from the attenuation over the macroseismic field, using a formula such as that by Kövesligethy (1906) or Blake (1941). In this case, because this is not an observed value (and not a "true" intensity) it may be expressed as a decimal fraction without contravening the rule that intensity values are integers. This value can be determined for earthquakes with sufficient data to draw at least two (preferably three) isoseismals. This is only possible if one is using the concept of the barycentre (see Section 3.1), since the true epicentre may not be central to the macroseismic field. The term "barycentral intensity" might be preferable.

It is recommended that these two methods be discriminated between by the notation used. Thus an integer number (9 or IX) indicates method (1) and a decimal number (9.0 or 9.3) indicates method (2). It is recommended that one should not add arbitrary values to the maximum observed intensity when deriving an I_o value; the arbitrary amount is too subjective. As well as epicentral intensity, a useful parameter is maximum intensity, abbreviated I_{max}. This is simply the highest observed intensity value anywhere in the macroseismic field. For onshore earthquakes, I_o and I_{max} may be equal. For offshore earthquakes it is often not possible to estimate I_o (never if method 1 is used), but I_{max} can be given.

3.3 Magnitude

The use of macroseismic data can give surprisingly robust measures of earthquake magnitude. This is an extremely important part of macroseismic studies, as in this way earthquake catalogues can be extended into historical times with consistent magnitude values. Such extended earthquake catalogues are of great benefit to seismic hazard studies. Early studies attempted to correlate epicentral intensity with magnitude; however, epicentral intensity can be strongly affected by focal depth, so such correlations may perform poorly unless either (1) depths are known and taken into consideration, or (2) one is working in an area where seismogenic depth is narrowly constrained.

The total felt area (A) of an earthquake, or the area enclosed by one of the outer isoseismals (usually 3 or 4), is a much better indicator of magnitude, being not much affected by depth except in the case of truly deep earthquakes. For earthquakes below a threshold magnitude (about 5.5 M_w) magnitude and log felt area scale more or less linearly, and so equations of the form

$$M = a \log A + b \tag{4}$$

can be established regionally by examination of data for earthquakes for which macroseismic data and instrumental magnitude are both available. For larger earthquakes, differences in spectral content may affect the way in which earthquake vibration is perceived, and a different scaling appears to apply. In Frankel (1994) the form

$$M = n \log(A/\pi) + 2m/(2.3\sqrt{\pi})\sqrt{A} + a \tag{5}$$

is used to represent the full magnitude range, where n is the exponent of geometrical spreading and

$$m = (\pi f)/(Q\beta) \tag{6}$$

where f is the predominant frequency of earthquake motion at the limit of the felt area (probably 2–4 Hz), Q is shear wave attenuation and β is shear wave velocity (3.5 km sec^{-1}). Using this functional form and comparing worldwide intraplate earthquakes with interplate earthquakes from one region (California), Frankel found the difference in magnitude for the same felt area to be on average 1.1 units greater for California.

Other forms that have been proposed include

$$M = aI_o + b \ln r + c \tag{7}$$

where r is the radius, rather than the area, of the total macroseismic field, and

$$M = aI_o + \sum b_i \ln r_i + c \tag{8}$$

in which all isoseismals (values for each i) are used as well as the epicentral intensity (see Albarello *et al.*, 1995).

The method of Bakun and Wentworth (1997) is of particular interest in making a joint determination of epicentre and magnitude by drawing contours of goodness-of-fit to the intensity data set of possible epicentres and magnitudes. This method is especially suited in cases where the intensity data set is sparse.

In the above equations, M has been used for generic magnitude; for any particular magnitude equation it is important to specify what magnitude type the derived values are compatible with (M_s, M_L, M_w, etc.). It is also useful to determine the standard error, which will give a measure of the uncertainty attached to estimated magnitude values.

3.4 Depth

The estimation of focal depth from macroseismic data was first developed by Radó Kövesligethy. His first paper on the subject presented the formula

$$I - I_o = 3 \log \sin e - 3\alpha(r/R)(1 - \sin e) \tag{9}$$

where $\sin e = h/r$ and R is the radius of the Earth (Kövesligethy, 1906). Equation (9) was subsequently rewritten and modified slightly by Jánosi (1907), to reach the better-known formula

$$I_o - I_i = 3 \log(r/h) + 3\alpha \log e(r - h) \tag{10}$$

where r is the radius of the isoseismal of intensity I_i, h is depth, and α is a constant representing anelastic attenuation. In this form the equation has been much used (e.g., Burton *et al.*, 1985). The constant value of 3 represents an equivalence value between the degrees of the intensity scale and ground motion amplitudes. Some workers accept it, others prefer to find their own values by fitting to data (Levret *et al.*, 1996). The attenuation parameter α should usually be determined regionally by group optimisation on an appropriate data set rather than for individual earthquakes, since one assumes that this value, a property of the crust, does not alter from earthquake to earthquake.

This technique is also associated with the name of Blake (1941), whose contribution was essentially a reduction and simplification of Eq. (10); Blake's version is still used by some workers today, but Kövesligethy's original equation (in Jánosi's version) is more commonly encountered. Kövesligethy's equation became more widely known, in the form of Eq. (10), through a paper by Sponheuer (1960).

I_o here is properly the barycentral intensity, which has to be solved for in addition to solving for h. This is usually done graphically; one can fit the isoseismal data to all possible values of h and I_o and find a minimum error value consistent with the observed maximum intensity (e.g., Musson, 1996).

A test made by Musson (1993) in comparing macroseismic depths and instrumental depths for modern British earthquakes found that the mean discrepancy between the two depth values was 1.6 km. This represents working with the most favourable sort of data for the method, i.e., from shallow (5–25 km) intraplate events with very small sources and good macroseismic data.

3.5 Source Parameters

It is possible to find on occasional papers that present focal mechanisms for preinstrumental earthquakes. These are commonly the result of a chain of reasoning which can be summarised as: "this earthquake occurred in such-and-such a place, therefore it must have been produced by such-and-such a fault, therefore it must have had a focal mechanism consistent with the type of faulting on this fault." Although the result of this reasoning may sometimes be correct, particularly in cases where the causative fault is obvious from surface rupture, the process cannot be considered as a determination of earthquake parameters so much as educated guesswork.

The first attempt to use a processing of macroseismic data to characterise actual fault parameters can be found in the work of Charles Davison, starting around 1890 in the United Kingdom, which is remarkable since this actually predates the use of instrumental data for determining focal parameters. Davison believed that he could use detailed descriptions of perceived ground motion from human observers to derive analytically the strike and dip of the rupturing fault (Davison, 1891). However, he was unaware of the extent to which perceived direction of motion is distorted by building response (most of his observers being indoors), and his method is overly ambitious and of no practical value.

Modern approaches to the determination of source parameters usually depend on the assumption that the ellipticality of isoseismals is a certain indicator of fault azimuth. The work of Shebalin (1973) seems to be the first attempt to follow this line of approach. Although in some cases it is obviously true that elliptical isoseismals are aligned along the fault axis, there is also the possibility that this simple pattern is distorted by radiation patterns and local effects (especially for smaller events). Subsequent studies along these lines are summarised by Gasperini *et al.* (1999), who points out the absence of adequate testing of the various methods against calibrated modern data.

A recent method, proposed by Gasperini *et al.* (1999), is as follows: (1) locate the epicentre of the earthquake from the highest intensity observations; (2) assess the seismic moment from an algorithm involving the median distance for the observations of each intensity value; (3) infer the source dimensions from the moment; (4) assess the fault azimuth, essentially from the direction of maximum elongation of the highest isoseismals. The authors have tested the method against some recent large Italian earthquakes (5.5 M_s is considered to be the minimum size of earthquake for which the method is reliable) with satisfactory results. It will be an interesting exercise in the future to test this method in other areas.

A study by Parsons *et al.* (2000) applied the method of Bakun and Wentworth (1997) to intensity data for six large earthquakes in the Maramara Sea region of Turkey, and attempted to identify probable causative fault ruptures by comparing contours of epicentre and magnitude goodness-of-fit to the intensity data set with compatible fault segments.

4. Applications of Macroseismology

4.1 Seismic Hazard

Since the late 1970s it has been realised that macroseismic methods are essential for most seismic hazard studies. The corresponding increase of interest in seismic hazard necessitated by expanding nuclear energy programmes throughout the world, has ensured a large resurgence of interest in macroseismology. The principal reason for this is the necessity to expand earthquake catalogues back in time. The instrumental earthquake catalogue available for any region never extends further back in time than the beginning of the 20th century, and in many tectonic regimes this is far shorter than the length of the seismic cycle, and is thus unrepresentative of the long-term hazard. Depending on cultural history, it may be possible to extend the earthquake catalogue back in time for several hundred years more, giving a much more adequate data set for hazard analysis. In Italy, for example, after a long and thorough investigation of historical materials, involving professional historians as well as seismologists, there is now an excellent catalogue providing close on 1000 y of usable data. This was only possible through the application of macroseismic methods.

This project, and others like it in other countries, have also involved increased attention to the methods themselves. On the one hand, there has been much consideration of the processes involved in assigning intensity, in order to make this process as reliable as possible. On the other hand, there has been considerable interest in the subject of deriving earthquake parameters from macroseismic methods.

The focus of much of the work on improving intensity assignments has been the development of the European Macroseismic Scale, intended as a tool suitable for modern macroseismic studies, which would minimise problems of subjectivity and poor practice which have been observed in the past. One of the important advances of the EMS was the provision of methodological guidelines. Before the publication of the EMS anyone starting macroseismic investigations for the first time would find that existing intensity scales consisted of a list of diagnostics, with very little material available in print describing best practice in actually using it. The provision of detailed guidelines (and also pictorial illustrations) should go a long way to reduce inconsistencies in the use of the scale by different investigators.

At the same time, the extensive historical investigations into early earthquake descriptions have provided a large body of experience regarding the ways in which historical evidence can legitimately be treated in the assigning of intensities. The sort of bad practice often encountered in the past, such as the assigning of unqualified high intensity values to single reports of damage to anomalous structures, should be eradicated in the future. As the investigation of historical earthquakes is covered in more detail in a separate chapter of this volume, no more will be said about it here.

The increased interest in macroseismic methods for the determination of earthquake parameters has also pointed the need for modern macroseismic monitoring. Increased and improved macroseismic data sets from modern earthquakes for which good instrumental data are also available, are needed for the purposes of calibration of macroseismic techniques. Intensity can also be used directly in seismic hazard studies as a parameter for expressing the hazard itself, in place of physical measure of ground motion like peak ground acceleration. This is not often done, since most seismic hazard studies are done to establish design values for engineers, and for this purpose physical parameters are usually required (although it was the practice at times in Eastern Europe to stipulate design levels in terms of intensity values). However, there are two reasons why intensity may be used for this purpose.

The first is that there are many parts of the world where there are abundant intensity data available, but no or few strong motion records. In such cases, the attenuation of intensity can be described quite accurately, but the attenuation of physical ground motion parameters is uncertain. By using intensity to express the hazard, a reliable hazard value can be given. However, some way then has to be found to enable engineers to use this in the design process, which might involve some conversion from intensity to other parameters, and such conversions can be contentious.

The second reason is that seismic hazard studies may be conducted for audiences other than engineers, and for whom physical parameters of ground motion are not particularly meaningful. Such groups could include insurers, planners, politicians, and ordinary members of the public. Since intensity can be related directly to earthquake effects, including damage, hazard assessments expressed as intensity values can be readily understood. Peak ground acceleration means little to an insurer unless it can be converted into damage terms, which generally it cannot be.

The introduction of intensity into hazard calculations is straightforward: in place of the usual peak ground acceleration (pga) attenuation function an intensity attenuation function is used. The only other thing that needs to be changed is the way in which scatter in the attenuation is applied; scatter in pga is generally considered to be lognormally distributed, but for intensity it is normally distributed.

It is desirable that intensity attenuation curves should be derived for normal soil conditions, and not be heavily influenced by observations where the intensity value has been increased by soil amplification. This enables a later modification of hazard values to be made due to soil conditions at site.

One objection that is sometimes raised is that intensity distributions can be highly irregular, making attenuation functions unreliable. The situation can be improved by (1) ensuring that the basic intensity assessments are of good quality; (2) not allowing the pattern to be obscured by anomalous values due to soil conditions; and (3) the use of azimuthal equations to deal with isoseismal elongation. The compilation of useful intensity attenuation curves for different parts of the world is one of the most pressing tasks toward the advancement of intensity as a tool in seismic hazard studies. It provides another reason why macroseismic studies of modern earthquakes are important. The fact that some major countries lack coherent macroseismic monitoring programmes is regrettable, and will surely be viewed in the future as having been short-sighted.

4.2 Seismic Risk

The usefulness of earthquake intensity in seismic hazard estimation carries over into seismic risk studies in very similar ways. However, it can be considered to be even more useful, since seismic risk studies are concerned with establishing the probability of damage or loss occurring, and intensity is a direct expression of damage.

In general practice, seismic hazard is defined as the probability that a certain level of earthquake ground motion will occur at a site within a given time interval; seismic risk is the probability that some type of failure will occur due to earthquake, measurable in terms of damage, financial loss, or even casualties. Thus, whereas seismic hazard is purely a product of natural processes, seismic risk is dependent on societal exposure in terms of the built environment or human population. The fragility of structures is expressed in terms of vulnerability, and the relationship between risk and hazard can then be written in terms of a simple equation:

$$\text{Risk} = \text{Hazard} \times \text{Vulnerability} \tag{11}$$

where "vulnerability" is the probability that, given a certain level of ground motion, damage will occur. This equation is sometimes written with an additional term "Exposure," representing the total amount of property liable to be affected.

From the point of view of the seismologist, seismic risk studies present a harder problem than seismic hazard, since data on vulnerability may not be easy to come by. A seismic hazard assessment can be conducted on the basis of knowledge of geophysics and geology; from such data one may construct a seismic hazard map of anywhere in the world. To construct a seismic risk map, one would need to know something about the distribution of buildings, and this information may be more intractable, except at a very generalised level.

However, this task is made somewhat easier by the application of earthquake intensity as a parameter of earthquake shaking, rather than the use of physical parameters such as peak ground acceleration. The use of pga or some similar parameter requires the construction of vulnerability functions relating ground motion to damage. This can be difficult, particularly with pga, which has been shown often not to correlate well with damage. One needs only to consider the very high accelerations recorded in some recent earthquakes, which have not been accompanied by particularly high levels of damage. In contrast, a modern macroseismic scale contains its own implicit vulnerability functions.

An earthquake intensity scale can be used in two different ways. The conventional use (which will be termed here the "forwards" use) is the one that most people will be familiar with: an earthquake occurs; the seismologist goes out and examines the damaged localities. He assigns intensity values to each locality using a macroseismic scale, plots the values on a map, and, from the resulting distribution, earthquake parameters may be derived which augment those derived instrumentally. Further, the seismologist constructs an intensity attenuation equation by analysing observed intensity distributions as a function of magnitude and distance.

The second, or "backwards" use, is as follows. The seismologist projects a design earthquake that may occur some time in the future, with expected location and magnitude. From a knowledge of the regional intensity attenuation, the expected intensity at a given place can be worked out. This expected intensity value then describes the anticipated effects of the earthquake in terms of damage to buildings. Because intensity is defined by observed effects (specifically damage), an intensity attenuation equation is an expression of the expected damage distribution from any earthquake, as a function of magnitude and distance.

The attractive aspect of this from the point of view of seismic risk is that the use of physical parameters of ground motion is simply circumvented. Since the intensity attenuation function is validated in the first place with respect to observed damage, it is automatically applicable in the second case when it comes to estimated damage in the future. Why, then, is this simple and effective approach so underutilised? The most likely reason is a general distrust of intensity as being a subjective and generalised parameter, lacking in scientific precision. However, a generalised approach is sometimes exactly what is needed; a robust instrument is apposite, and an intensity scale operates at the right level for many practical purposes.

The other problem is that many workers are still applying old intensity scales in a modern environment. Most intensity-related work in the USA is still conducted using the Modified Mercalli Scale (MM) of either the Wood and Neumann (1931) or Richter (1958) versions. Both these scales, as formulated, are obsolete. The key objections are that (1) they make no allowances for the probabilistic nature of intensity, where damage is properly represented as a set of distributions; (2) they do not have the capability of dealing with modern engineered structures; and (3) they include miscellaneous diagnostics which have been shown to be inadequate indicators of intensity (Vogt *et al.*, 1994). It is likely that perceptions that intensity is an inadequate tool for modern use are to some extent coloured by a knowledge of the inadequacies of the MM scale.

Most modern data sets prepared using the MM scale are actually much better than the preceding paragraph would imply, since unofficial rules and practices have been built on to the original scale to make allowance for its deficiencies (Dewey *et al.*, 1995). In an institution with long continuity, such as the USGS, conditions are favourable for the preservation and application of wisdom gained from long experience of assigning intensity to local data sets. This does not overcome the problem that such procedures are often not transparent to other users of the data.

Modern intensity scales such as the European Macroseismic Scale are more suitable for application in seismic risk studies. The elimination of unreliable intensity diagnostics based on experience improves the accuracy of intensity assessments. The provision of detailed guidelines reduces undesirable variations in practices from user to user, makes intensity assessment less subjective, and makes the rules used in intensity assessment transparent to the users of the data. A flexible approach to building type and vulnerability allows a wide range of buildings, including modern engineered structures, to be taken into consideration. And importantly, a modern scale needs to be probabilistic in its approach to damage; for any type (strength) of building at a particular level of intensity, damage should be considered as a distribution. It is these factors that make the EMS-98 scale, in particular, suitable for use in seismic risk studies.

Finally, four appendices are placed on the attached Handbook CD, under directory \49Musson2. They are: Appendix 1. Damage inspection form as proposed by Tertulliani *et al.* (1996/99); Appendix 2. Italian macroseismic questionnaire (Internet version); Appendix 3. Excerpt from an Italian macroseismic bulletin; and Appendix 4. Austrian macroseismic questionnaire.

Acknowledgements

The authors would like to thank Andrea Tertulliani and Edmund Fiegweil for their assistance in compiling Section 2 and the Appendices. Tibor Zsiros helped in sorting out the correct history of the Kövesligethy papers. We would also like to thank Jim Dewey for his helpful review comments. The contribution of Roger Musson to this chapter is published with the permission of the Director of the British Geological Survey (NERC).

References

Ad-hoc Panel (1981). Report on the Ad-hoc Panel meeting of experts on up-dating of the MSK-64 intensity scale, 10–14 March 1980. *Gerlands Beitr. Geophys.* **90**, 261–268.

Albarello, D., A. Berardi, C. Margottini, and M. Mucciarelli (1995). Macroseismic estimates of magnitude in Italy. *Pure Appl. Geophys.* **145**, 297–312.

Bakun, W.H. and C.M. Wentworth (1997). Estimating earthquake location and magnitude from seismic intensity data. *Bull. Seismol. Soc. Am.* **87**, 1502–1521.

Barbano, M.S. and G. Salemi (1990). Expert system for the macroseismic intensity degree assessment. In: "Use of PC in Seismology," Proceedings of Workshop WS8 of XXII General Assembly of the ESC, 17–22 Sept. 1990, Barcelona, pp. 3–8.

Blake, A. (1941). On the estimation of focal depth from macroseismic data. *Bull. Seismol. Soc. Am.* **31**, 225–231.

Brazee, R.J. (1978). "Reevaluation of Modified Mercalli intensity scale for earthquakes using distance as determinant," NOAA Tech. Mem EDS NGSDC-4.

Burton, P.W., R. McGonigle, G. Neilson, and R.M.W. Musson (1985). Macroseismic focal depth and intensity attenuation for British earthquakes. In: "Earthquake Engineering in Britain," pp. 91–110. Telford, London.

Cecić, I., R.M.W. Musson, and M. Stucchi (1996). Do seismologists agree upon epicentre determination from macroseismic data? A survey of ESC Working Group "Macroseismology." *Ann. Geofis.* **39**, 1013–1027.

Cecić, I., M. Godec, P. Zupančič, and D. Dolenc (1999). "Macroseismic effects of 12 April 1998 Krn, Slovenia earthquake: An overview." Presented at IUGG 99, Birmingham, UK.

Davison, C. (1891). On the Inverness earthquakes of November 15th to December 14th, 1890, *Q. J. Geol. Soc.* **47**, 618–632.

Davison, C. (1900). Scales of seismic intensity. *Philos. Mag.* 5th Series, **50**, 44–53.

Davison, C. (1921). On scales of seismic intensity and on the construction of isoseismal lines. *Bull. Seismol. Soc. Am.* **11**, 95–129.

Davison, C. (1933). Scales of seismic intensity: Supplementary paper. *Bull. Seismol. Soc. Am.* **23**, 158–166.

Dengler, L.A. and J.W. Dewey (1998). An intensity survey of households affected by the Northridge, California, earthquake of 17 January 1994. *Bull. Seismol. Soc. Am.* **88**, 441–462.

Dewey, J.W., B.G. Reagor, L. Dengler, and K. Moley (1995). Intensity distribution and isoseismal maps for the Northridge, California, earthquake of January 17, 1994. *USGS Open-file Report No 95–92.*

Dowrick, D.J. (1992). Attenuation of Modified Mercalli intensity in New Zealand earthquakes, *Earthq. Eng. Struct. Dyn.* **21**, 181–196.

Egen, P.N.C. (1828). Über das Erdbeben in den Rhein- und Niederlanden von 23 Feb. 1828, *Ann. Phys. Chem.* **13**, 153–163.

Fiegweil, E. (1989). Brief outline of the macroseismic practice in Austria. In: "First AB Workshop on Macroseismic Methods" (Cecić, I., Ed.), 9–11 May 1989, Poljče, Slovenia, SSRS, Ljubljana.

Frankel, A. (1994). Implications of felt area–magnitude relations for earthquake scaling and the average frequency of perceptible ground motion, *Bull. Seismol. Soc. Am.* **84**, 462–465.

Gasparini, C., V. De Rubeis, and A. Tertulliani (1992). A method for the analysis of macroseismic questionnaires. *Nat. Hazards* **5**, 169–177.

Gasperini, P., F. Bernadini, G. Valensise, and E. Boschi (1999). Defining seismogenic sources from historical earthquake felt reports. *Bull. Seismol. Soc. Am.* **89**, 94–110.

Gaull, B.A., M.O. Michael-Leiba, and J.M.W. Rynn (1990). Probabilistic earthquake risk maps of Australia. *Aust. J. Earth Sci.* **37**, 169–187.

Grünthal, G. (Ed.) (1998). "European Macroseismic Scale 1998," Cahiers du Centre Europèen de Gèodynamique et de Seismologie, 15, Conseil de l'Europe, Luxembourg.

Jánosi, I. (1907). Makroszeizmikus rengések feldolgozása a Cancani-féle egyenlet alapján. In: "Az 1906 évi Magyarországi Földrengések" (Réthly, A., Ed.), pp. 77–82. A. M. Kir. Orsz. Met. Föld. Int., Budapest.

JMA (1996). "Explanation Table of JMA Seismic Intensity Scale (February 1996)," Japanese Meteorological Agency leaflet.

Kövesligethy, R. (1906). A makroszeizmikus rengések feldolgozása, *Math. és Természettudományi Értesítő* **24**, 349–368.

Levret, A., M. Cushing, and G. Peyridieu (1996). "Etude des Caractéristiques de Séismes Historique en France: Atlas de 140 Cartes Macroseismiques," IPSN, Fontenay-Aux-Roses.

Medvedev, S.V. (1961). "Engineering Seismology" (in Russian). Academy of Sciences, Inst. of Physics of the Earth, Publ.house for literature on Civil Engineering, Architecture and Building Materials.

Musson, R.M.W. (1992). Routine macroseismic monitoring in the UK. In: "Second AB Workshop on Macroseismic Methods" (Cecić, I., Ed.), pp. 9–13. 15–18 October 1990, Poljče, Slovenia, SSRS, Ljubljana.

Musson, R.M.W. (1993). Macroseismic magnitude and depth for British earthquakes. BGS Global Seismology Report no. WL/93/13.

Musson, R.M.W. (1996). Determination of parameters for historical British earthquakes. *Ann. Geofis.* **39**, 1041–1048.

Parsons, T., S. Toda, R.S. Stein, A. Barka, and J.H. Dieterich (2000). Heightened odds of large earthquakes near Istanbul: An interaction-based probability calculation. *Science* **288**, 661–665.

Peruzza, L. (1992). Procedure of macroseismic epicentre evaluation for seismic hazard purposes. In: "Proc. XXIII General Assembly of the ESC," Prague **2**, 434–437.

Richter, C.F. (1958). "Elementary Seismology," W.H. Freeman, San Francisco.

Riggio, A.M. and D. Slejko (1989). Macroseismic practice at the OGS. In: "First AB Workshop on Macroseismic Methods" (Cecić, I., Ed.), 9–11 May 1989, Poljče, Slovenia, SSRS, Ljubljana.

Sarconi, M. (1784). Istoria dei fenomeni del tremoto avvenuto nelle Calabrie, e nel Valdemone nell'anno 1783. In: "Luce dalla Reale Accademia della Scienze, e delle Belle Lettere di Napoli," Naples.

Shebalin, N.V. (1970). Intensity: on the statistical definition of the term. In: "Proc. X General Assembly of the ESC," 3–11 Sept 1968, Leningrad, Acad. Sci. USSR, Moscow.

Shebalin, N.V. (1973). Macroseismic data as information on source parameters of large earthquakes. *Phys. Earth. Planet. Inter.* **6**, 316–323.

Sieberg, A. (1912). Über die makroseismische Bestimmung der Erdbebenstärke, *Gerlands Beitr. Geophys.* **11**, 227–239.

Sieberg, A. (1923). "Geologische, Physikalische und Angewandte Erdbebenkunde." Verlag von Gustav Fischer, Jena.

Sieberg A. (1932). "Geologie der Erdbeben. Handbuch der Geophysik," Gebr. Bornträger, Berlin, **2**, pt 4, 550–555.

Sponheuer, W. (1960). Methoden zur Herdtiefenbestimmung in der Makroseismik. In: "Freiberger Forschungshefte" C88, Akademie Verlag, Berlin.

Sponheuer, W. and V. Karnik (1964). Neue seismiche skala. In: "Proc. 7th Symposium of the ESC" (Sponheuer, W., Ed.), Jena, 24–30 Sept. 1962, pp. 69–76. Veröff. Inst. F. bodendyn. U. Erdbebenforsch. Jena d. Deutschen Akad. D. Wiss., No 77.

Tertulliani, A. and A. Maramai (1992). Macroseismic practice at ING, Rome. In: "Second AB Workshop on Macroseismic Methods" (Cecić, I., Ed.), pp. 15–26. 15–18 October 1990, Poljče, Slovenia, SSRS, Ljubljana.

Tertulliani, A., I. Cecić, and M. Godec (1996/99). Unification of macroseismic data collection procedures: a pilot project for border earthquakes assessment. *Nat. Hazards* **19**, 221–231.

Trifunac, M.D. and A.G. Brady (1975). On the correlation of seismic intensity scales with the peaks of recorded strong motion. *Bull. Seismol. Soc. Am.* **77**, 490–513.

Vogt, J., R.M.W. Musson, and M. Stucchi (1994). Seismological and hydrological criteria for the new European Macroseismic Scale MSK-92. *Nat. Hazards* **10**, 1–6.

Wood, H.O. and F. Neumann (1931). Modified Mercalli intensity scale of 1931. *Bull. Seismol. Soc. Am.* **21**, 277–283.

Editor's Note

Due to space limitations, four appendices of this Chapter are placed as computer readable files on the attached Handbook CD, under directory \49Musson2. Please see also, Chapter 47, Historical seismology, by Guidoboni; and Chapter 48, Earthquake history, with several subchapters by various authors.

50

USGS Earthquake Moment Tensor Catalog

Stuart A. Sipkin
US Geological Survey, Denver, USA

1. Introduction

Since 1981, the US Geological Survey's National Earthquake Information Center (NEIC) has been computing moment tensor solutions for all moderate-to-large sized earthquakes. From 1981 through the first half of 1982, a moment tensor inversion was attempted for all earthquakes with a magnitude, m_b or M_S, of 6.5 or greater. From the second half of 1982 through 1994, an inversion was attempted for all earthquakes with an m_b magnitude of 5.8 or greater (5.7 for intermediate- and deep-focus earthquakes). Beginning in 1995, an inversion was attempted for all earthquakes with a magnitude, m_b or M_S, of 5.5 or greater. In the earlier parts of the catalog, completeness is compromised because of sparse station coverage for many parts of the Earth. Station coverage has steadily improved, and is mainly responsible for the ability to move the magnitude thresholds lower with time.

The solutions are published on a monthly basis in the *Preliminary Determination of Epicenters*, the *Bulletin of the Seismological Society of America*, and *Seismological Research Letters*. Annual compilations are published in *Physics of the Earth and Planetary Interiors*. These currently include solutions for earthquakes that occurred from 1980 through 2000 (Sipkin, 1986b, 1987; Sipkin and Needham, 1989, 1991, 1992, 1993, 1994a,b; Sipkin and Zirbes, 1996, 1997; Sipkin *et al.*, 1998, 1999, 2000a,b, 2002).

2. Method

The algorithm is described in Sipkin (1982, 1986b). Briefly, the inversion procedure is based on multichannel signal-enhancement theory (Robinson, 1967). In this algorithm the far-field Green's functions are the multichannel input; the observed seismograms are the desired output; and the moment-rate tensor is the convolution filter operating on the input. Because the convolution filter found is not only a signal-enhancement filter but is also a noise-rejection filter, arrivals in the waveform that are not specifically accounted for in the Green's functions (such as those generated by near-receiver or near-source structure, other than *pP* and *sP* phases), and that are not coherent across the suite of seismograms to be inverted, are regarded as noise and do not affect the solution. In this formalism, the system of equations to be solved are in the form of a "block Toeplitz" matrix, so it can be quickly solved using recursive techniques. These techniques are equivalent to using recurrence relations in orthogonal polynomial theory. The details of these recursive techniques can be found in Robinson (1967), Claerbout (1976), or Oppenheim (1978).

This inversion technique is applied to digitally recorded broadband *P*-waveform data from US National Network (USNSN) and Global Seismograph Network (GSN) stations that have good signal-to-noise ratios and that are located at epicentral distances between 30° and 95°. The solution is constrained to be purely deviatoric, but not to be a pure double couple. The source depth is determined by varying the trial depth at 1-km intervals to find the focal depth that minimizes the misfit to the data. Because the *P*-wave forms contain both *pP* and *sP* phases, as well as direct *P*-wave arrivals, this procedure generally is highly sensitive to source depth. The source depth also depends somewhat on the velocity model used in the inversion. The model currently used is ak135 (Kennett *et al.*, 1995). Although this model is quite adequate for computing the effects of mantle structure, its crustal structure may be too simple. This will not affect the mechanism found but may bias the depth estimate. Experiments involving the introduction of more realistic crustal layers yield differences on the order of 2–4 km (Sipkin, 1989), smaller than the intrinsic uncertainty of 5–10 km determined using both inversions of "realistic" synthetic seismograms (Sipkin, 1986a) and comparison with a set of well-located earthquakes (Engdahl *et al.*, 1998). An advantage that using body-wave seismograms has over the use of surface waves is better resolution of source depth for shallow sources. The inversion kernels for two of the moment tensor elements approach zero as the source depth approaches the Earth's surface. This occurs more rapidly at longer periods. Although the use of higher-frequency body waves does not eliminate this problem, it does ameliorate it.

The deviatoric moment tensor can be decomposed into a double couple and a remainder term in an infinite number of ways. We prefer the decomposition suggested by Knopoff and Randall (1970) in which the moment tensor is decomposed into a "best" double couple and a compensated linear-vector dipole (CLVD). This decomposition is preferred to decomposing the moment tensor into a major and minor double couple because it is unique, whereas there are an infinite number of major and minor double-couple combinations. In addition, the "best" double couple and CLVD share the same principal axes; the major and minor double-couples will, in general, have equivalent force systems with differing directions. Using the preferred decomposition, the moment of the "best" double couple, M_0, is $1/2(e_1 - e_3)$ where e_1 is the largest positive eigenvalue, corresponding to the T axis, and e_3 is the largest negative eigenvalue, corresponding to the P axis. The contribution of the CLVD component to the normalized total moment, $\hat{M}_{\mathrm{CLVD}}$, is $2|e_{\min}|/|e_{\max}|$, where $e_{\min}$ is the eigenvalue with the smallest absolute value and $e_{\max}$ is the eigenvalue with the largest absolute value. The percentage double couple is then $100 \times (1 - \hat{M}_{\mathrm{CLVD}})$.

3. Moment Tensor Catalog

The moment tensor solutions from 1981 through 2000, along with the hypocentral parameters determined by the NEIC, are listed in the files MT.LIS and FMECH.LIS included on the attached Handbook CD. These files, as well as updates, can also be downloaded via anonymous FTP at ghtftp.cr.usgs.gov in directory ./pub/momten.

MT.LIS contains the elements of the moment tensors along with detailed event information; FMECH.LIS contains the decompositions into the principal axes and best double-couples; the file FMECH.DOC describes the various fields.

As indicated in the Section 1, the catalog is not uniform over time. Beginning in 1981, the magnitude threshold was 6.5; in the second half of 1982, the threshold was reduced to 5.8 (5.7 for intermediate- and deep-focus earthquakes); and starting in 1995, the threshold was further reduced to 5.5. Since 1981 global coverage with broadband, digitally recording seismic stations has vastly improved. The improved station coverage is mainly responsible for the ability to move the magnitude thresholds lower with time. This is reflected in the annual average number of stations used per event. From 1981 through 1991, the average number of stations per event gradually rose from approximately 8 to approximately 13. By 1995, the average number of stations had rapidly increased to approximately 30. Since 1995 the average number of stations per event has remained at this level.

4. Moment Tensor Inversion Code

An important *caveat* is that the moment tensor source code was not originally intended for general distribution, and so is not very well documented. It is, however, included here (on the attached Handbook CD) for historical documentation purposes.

The inversion software is modular. That is, an independent program handles each step of the process. The programs are bound by a UNIX shell script. This could also be accomplished on different operating systems with a VAX/VMS command file, a DOS batch file, etc. Many of the programs are for preprocessing (filtering, winnowing, and aligning) the data. All of these modules are contained in ***momten.tar.Z***. The code also refers to several library modules—source code for each is contained in compressed tar files. The routines for recursively solving the system of equations can be found in Robinson (1967). ***momcon.tar.Z*** contains routines for converting from one type of source description (double-couple, principal axes, moment tensor, slip vector, etc.) to another (written by R.P. Buland, some routines modified by S.A. Sipkin). ***buplot.tar.Z*** contains the plotting package used (BUPLOT, written by R.P. Buland), but any plotting package of choice will do.

Acknowledgments

The staffs of the USGS National Earthquake Information Center, the USGS Albuquerque Seismological Laboratory, the IDA program at the Scripps Institution of Oceanography, and the Incorporated Research Institutions for Seismology make this work possible by operating and maintaining the US National Seismograph Network and the Global Seismograph Network and its data centers. Madeleine Zirbes wrote much of the preprocessing code and made several valuable suggestions for improving the procedure. Russell Needham, Madeleine Zirbes, and Charles Bufe did most of the routine data processing.

References

Aki, K. and P.G. Richards (1980). "Quantitative Seismology," vol. 1, Freeman, San Francisco.

Claerbout, J.F. (1976). "Fundamentals of Geophysical Data Processing," McGraw-Hill, New York.

Engdahl, E.R., R. van der Hilst, and R. Buland (1998). Global teleseismic earthquake relocation with improved travel times and procedures for depth determination, *Bull. Seismol. Soc. Am.* **88**, 722–743.

Kennett, B.L.N., E.R. Engdahl, and R. Buland (1995). Constraints on seismic velocities in the Earth from traveltimes. *Geophys. J. Int.* **122**, 108–124.

Knopoff, L. and M.J. Randall (1970). The compensated linear-vector dipole: a possible mechanism for deep earthquakes, *J. Geophys. Res.* **75**, 4957–4963.

Oppenheim, A.V. (1978). "Applications of Digital Signal Processing" Prentice-Hall, Englewood Cliffs, NJ.

Robinson, E.A. (1967). "Multichannel Time Series Analysis with Digital Computer Programs." Holden-Day, San Francisco.

Sipkin, S.A. (1982). Estimation of earthquake source parameters by the inversion of waveform data: synthetic waveforms. *Phys. Earth Planet. Inter.* **30**, 242–259.

Sipkin, S.A. (1986a). Interpretation of non-double-couple earthquake source mechanisms derived from moment tensor inversion. *J. Geophys. Res.* **91**, 531–547.

Sipkin, S.A. (1986b). Estimation of earthquake source parameters by the inversion of waveform data: global seismicity, 1981–1983. *Bull. Seismol. Soc. Am.* **76**, 1515–1541.

Sipkin, S.A. (1987). Moment tensor solutions estimated using optimal filter theory for 51 selected earthquakes, 1980–1984. *Phys. Earth Planet. Inter.* **47**, 67–79.

Sipkin, S.A. (1989). Moment-tensor solutions for the 24 November 1987 Superstition Hills California earthquakes. *Bull. Seismol. Soc. Am.* **79**, 493–499.

Sipkin, S.A. and R.E. Needham (1989). Moment-tensor solutions estimated using optimal filter theory: global seismicity, 1984–1987. *Phys. Earth Planet. Inter.* **57**, 233–259.

Sipkin, S.A. and R.E. Needham (1991). Moment-tensor solutions estimated using optimal filter theory: global seismicity, 1988–1989. *Phys. Earth Planet. Inter.* **67**, 221–230.

Sipkin, S.A. and R.E. Needham (1992). Moment-tensor solutions estimated using optimal filter theory: global seismicity, 1990. *Phys. Earth Planet. Inter.* **70**, 16–21.

Sipkin, S.A. and R.E. Needham (1993). Moment-tensor solutions estimated using optimal filter theory: global seismicity, 1991. *Phys. Earth Planet. Inter.* **75**, 199–204.

Sipkin, S.A. and R.E. Needham (1994a). Moment-tensor solutions estimated using optimal filter theory: global seismicity, 1992. *Phys. Earth Planet. Inter.* **82**, 1–7.

Sipkin, S.A. and R.E. Needham (1994b). Moment-tensor solutions estimated using optimal filter theory: global seismicity, 1993. *Phys. Earth Planet. Inter.* **86**, 245–252.

Sipkin, S.A. and M.D. Zirbes (1996). Moment-tensor solutions estimated using optimal filter theory: global seismicity, 1994. *Phys. Earth Planet. Inter.* **93**, 139–146.

Sipkin, S.A. and M.D. Zirbes (1997). Moment-tensor solutions estimated using optimal filter theory: global seismicity, 1995. *Phys. Earth Planet. Inter.* **101**, 291–301.

Sipkin, S.A., M.D. Zirbes, and C.G. Bufe (1998). Moment-tensor solutions estimated using optimal filter theory: global seismicity, 1996. *Phys. Earth Planet. Inter.* **109**, 65–77.

Sipkin, S.A., C.G. Bufe, and M.D. Zirbes (1999). Moment-tensor solutions estimated using optimal filter theory: global seismicity, 1997. *Phys. Earth Planet. Inter.* **114**, 109–117.

Sipkin, S.A., C.G. Bufe, and M.D. Zirbes (2000a). Moment-tensor solutions estimated using optimal filter theory: global seismicity, 1998. *Phys. Earth Planet. Inter.* **118**, 169–179.

Sipkin, S.A., C.G. Bufe, and M.D. Zirbes (2000b). Moment tensor solutions estimated using optimal filter theory: global seismicity, 1999. *Phys. Earth Planet. Inter.* **122**, 147–159.

Sipkin, S.A., C.G. Bufe, and M.D. Zirbes (2002). Moment tensor solutions estimated using optimal filter theory: global seismicity, 2000. *Phys. Earth Planet. Inter.*, **130**, 129–142.

Editor's Note

The moment tensor catalog and software files are given on the attached Handbook CD, under directory \50 Sipkin.

VI

Earth's Structure

51

The Earth's Interior

Thorne Lay
University of California, Santa Cruz, USA

1. Introduction

Earth is composed of minerals and metal alloys under pressure and temperature conditions that allow elastic waves with frequencies ranging from 0.0003 to 30+ Hz to transmit through the planet with relatively little deviation from linear elasticity. In other words, an initial rapid input deformation applied to near-surface rocks, such as that accompanying sudden stress release on a fault or an underground explosion, produces stress imbalances that transmit through Earth in the form of nearly elastic *P* and *S* waves that convey the nature of the source deformations to distant locations in a way that is readily decipherable. This remarkable attribute is akin to the transmission of sound waves through air, for which there is a direct correspondence between atmospheric pressure fluctuations produced at a source, say by specific oscillations of vocal cords, and those detected on our ear drums, allowing the sound to be interpreted (even if there are many echoing reverberations that travel multiple paths to our ears as well as unrelated background noise).

If the source energy is large enough, on the order of a magnitude 5.0 earthquake, elastic wave arrivals can be detected above typical ambient noise levels over the Earth's entire surface, and seismic instruments are now capable of recording the full bandwidth of ground vibrations induced by common sources. The quasi-elasticity of Earth over a wide range of ground motions combines with the predominantly concentrically layered structure of the planet to yield straightforward relationships between observations of seismic waves at various distances from a source and the properties of the deep interior of the planet. For about 100 years seismologists have been systematically accumulating global recordings of ground shaking, locating sources of elastic wave radiation, and extracting characteristics of Earth's interior from the travel times, amplitudes and waveshapes of seismic waves. Seismological models have been interpreted by mineral physicists, geochemists, and geodynamicists for many decades to infer the composition, physical state, and dynamical processes occurring inside the planet. This effort has resulted in a remarkable understanding of our planet's interior, although many important issues have not yet been resolved. This chapter outlines some of our understanding of the Earth's interior derived from seismology.

2. Earth Stratification and Chemical Differentiation

The most fundamental approach to gleaning Earth structure from seismic waves is to measure arrival times of ground vibrations as a function of distance from the source. Seismograms from distant earthquakes are characterized by a sequence of discrete body wave arrivals followed by dispersed surface wave trains. Given networks of seismic instrumentation with relatively uniform ground motion response, it is straightforward to pick consistent arrival times of the discrete body wave phases. Triangulation procedures can be used for natural sources to determine the source location and origin time for an assumed seismic velocity model, and the arrival times can then be plotted as a function of distance from the estimated (or known, in the case of explosions) source location. When this is done on a global scale, as in Figure 1, it is apparent that coherent arrivals associated with distinct travel time branches exist, with behavior that is predominantly dependent on epicentral distance. These arrivals correspond to various portions of the *P*- and *S*-wave fronts sweeping along the surface of the Earth, with the discreteness of the arrivals indicating that simple wave front refraction at depth dominates over 3D scattering effects. This organized transmission of seismic energy through the Earth was recognized very early in the 20th century, and pioneering seismologists began the labor of identifying each travel time branch and deducing the internal structures and source radiation effects that give rise to the complexity of seismograms. The improved structures were used to provide more accurate event locations, allowing the models to be refined in a bootstrapping manner.

INTERNATIONAL HANDBOOK OF EARTHQUAKE AND ENGINEERING SEISMOLOGY, VOLUME 81A ISBN: 0-12-440652-1

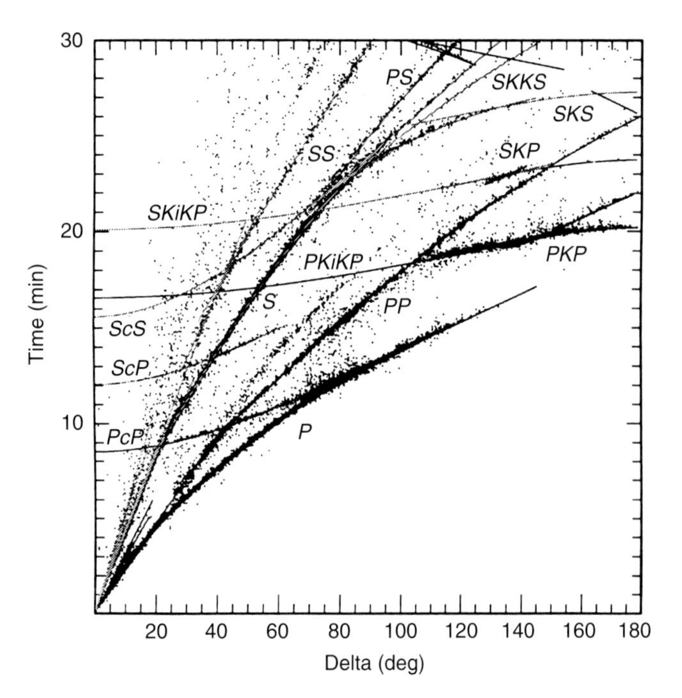

FIGURE 1 Arrival times of P and S seismic phases as a function of epicentral distance in the Earth for shallow earthquake sources, along with predicted travel-time curves for a radially symmetric model of P- and S-velocity variations with depth. P and S are direct phases, PcP, ScP, and ScS reflect from the core–mantle boundary, $PKiKP$ and $SKiKP$ reflect from the inner core–outer core boundary, PP, SS, PS reflect once from the Earth's surface, and PKP, SKP, SKS, $SKKS$ are phases that traverse the Earth's core. (From Kennett and Engdahl, 1991. Reprinted with permission from the Royal Astronomical Society.)

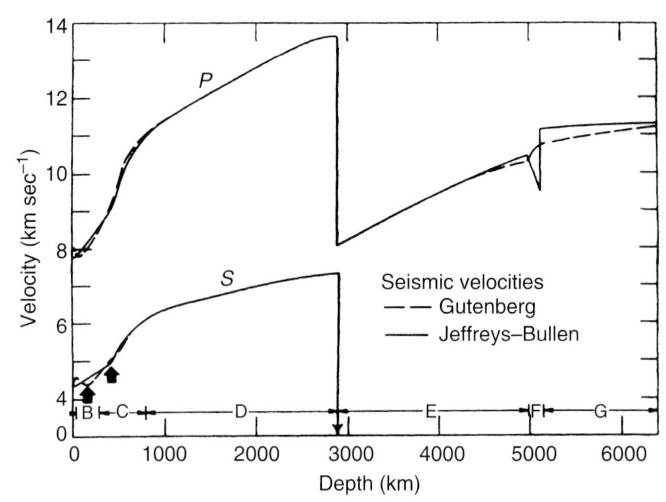

FIGURE 2 P and S velocity as functions of depth in the Earth for the classic Gutenberg and Jeffreys–Bullen earth models. The letters indicate classic subdivisions of the interior based on behavior of the travel time curves and inferred velocity structures. Arrows highlight the low velocity zone in the Gutenberg model and the change in velocity gradient defining the transition zone onset in the Jeffreys–Bullen model (From Anderson, 1963).

Given an understanding of elastic wave behavior based on the theory of elasticity that had been extensively developed in the 19th century (Chapter 8 by Udias; Aki and Richards, 1980; Lay and Wallace, 1995), it was quickly appreciated that the multibranch complexity of Earth's travel-time curve stems largely from the existence of reflections from the surface and from a major internal boundary overlying a region of low seismic velocities in the central region of the planet. The low velocity core of the Earth was detected in 1906 by Oldham. By 1913 the depth to the boundary of the core was quite accurately established by Beno Gutenberg, and with the discovery of the inner core in 1936 by Inge Lehmann, the basic layering of the crust, mantle, and core had been established.

For a source and receiver at the surface and a 1D model of the Earth, with P or S velocities, $c(r)$, that vary only with radius, r, basic linear elasticity yields parametric equations for P or S wave travel times, T, as a function of epicentral angular distance, Δ, given by:

$$T = p\Delta + 2\int_{r_t}^{r_0} \frac{\sqrt{\xi^2 - p^2}}{r} dr \tag{1}$$

$$\Delta = 2p\int_{r_t}^{r_0} \frac{dr}{r\sqrt{\xi^2 - p^2}} \tag{2}$$

where $\xi = r/c(r)$, r_t is the radius to the turning point of the seismic raypath, $p = (r \sin i)/c(r)$ is the seismic ray parameter which is constant for a given raypath, and r_0 is the radius of the sphere. These basic equations, valid for homogenous regions of the Earth with velocity increasing smoothly with depth, can be used to build up piecewise models of the radial P and S velocity structure either by forward calculation or inversion using the Herglotz–Wiechert method (c.f., Lay and Wallace, 1995). All of the necessary information for determining $c(r)$ is provided by the observed travel time curve for waves that turn continuously in a given homogeneous region. The travel time curves must be well enough sampled that stable estimates of the slope of the travel time curve, $dT/d\Delta = p$, can be empirically measured (when seismic arrays began to be deployed in the 1960s this value could be directly estimated, greatly improving the accuracy of inferred velocity models). Internal boundaries separating homogeneous layers can be accounted for by solving boundary value problems that result in Snell's law for raypath kinematics and in reflection and transmission coefficients for amplitudes (c.f., Aki and Richards, 1980; Lay and Wallace, 1995).

From the 1920s to 1940s, interpolated observed travel-time tables such as the Jeffreys–Bullen Tables (Jeffreys and Bullen, 1940) were used in the Herglotz–Wiechert method to produce accurate 1D seismic velocity models for the entire Earth, including the classic Jeffreys and Gutenberg models (Fig. 2).

These particular models served as standard Earth models for decades, and the J-B travel-time tables are still used in earthquake location procedures because they provide accurate predictions of *P* and *S* wave arrival times (including major surface- and core-reflected and converted phases) to within a small fraction of a percent at teleseismic ranges.

Although body wave travel-time analysis remains a primary tool for studying Earth structure, complete waveform analysis has allowed even more information to be gleaned from seismic recordings, and has unraveled the complex interference effects involved in surface waves and normal modes (see Chapter 9 by Chapman, Chapter 10 by Lognonne and E. Clevede and Chapter 11 by Romanowicz). Seismic wave amplitudes and waveforms can now be synthesized for Earth models, allowing comparison of data and synthetics. The strengths of velocity discontinuities, structure of low velocity zones, diffractions, scattering, and complex interference of multiple arrivals can be modeled and constrained by complete waveform modeling.

An example of a contemporary 1D velocity structure for the mantle and core (the thin low-velocity crust is not included) is shown in Figure 3, along with a density structure determined by matching normal mode observations (see Chapter 10 by Lognonne and Clevede), total mass and moment of inertia values, and petrological models of the interior. Seismic velocities increase systematically with depth across the mantle as in the classic models of Jeffreys and Gutenberg, but we are now aware of the existence of abrupt velocity and density increases at the 410 and 660 km seismic discontinuities. The details of the transition zone region from 410 to about 800 km were not resolved prior to the mid-1960s, but it had been recognized as early as 1926 that complexity is present at this depth in contrast to the smoother variations of seismic structure in the lower mantle (below 800 km). Near 2900 km depth the *S* velocity drops to zero, indicating the fluid state of the outer core, while the *P* velocity drops greatly across the core–mantle boundary before increasing smoothly with depth down to the inner core–outer core boundary.

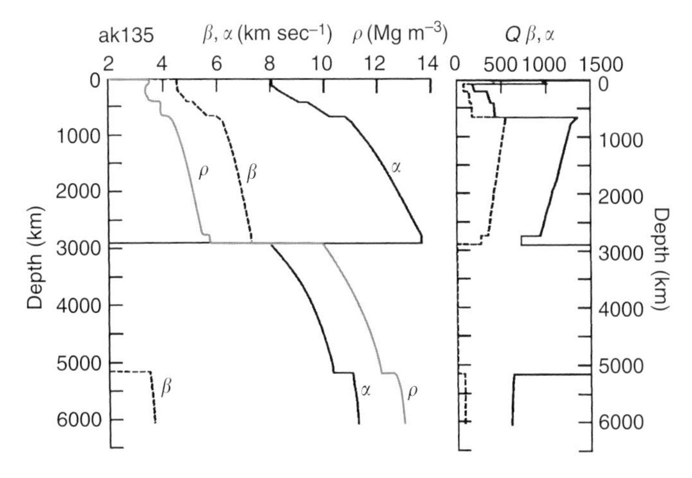

FIGURE 3 The ak135 Earth model indicating *P* velocity, α, and *S* velocity, β, as well as density, ρ and attenuation (Q_α is the solid line, Q_β is the dotted line) (from Kennett and van der Hilst, 1998).

Over vast reaches of Earth the seismic velocities increase very smoothly, i.e., across the lower mantle from 800 to 2700 km depth, and across the outer core and inner core. This is a direct inference from the smooth concave-downward travel-time curves in Figure 1, with lateral variations about the mean velocity at each depth being bounded to only a few percent except in regions near boundary layers (near the Earth's surface, near the core–mantle boundary, and in sinking lithospheric slabs). Thermodynamic calculations indicate that the composition of each major subsection of the Earth may be uniform, with gravitational self-compression fully accounting for the variation in laterally averaged velocity with depth. Solid-state phase transitions in the major Earth mineral olivine provide viable explanations for the discontinuities near 410 and 660 km depth, and cosmochemical arguments and density constraints indicate that the core is predominantly made of iron. The bulk composition of the planet can be approximated by the relative abundance of refractory materials in the Sun or by the bulk composition of undifferentiated chondritic meteorites, with the mantle composition being inferred by allowing for separation of the core alloy and continental crust (for a good review, see O'Neill and Palme, 1998). Thus, the first-order characterization of the Earth is of a chemically differentiated body in which a thin veneer of light materials has segregated into the enduring continental crust, while basalt is extracted by melting from upper mantle material to produce the recycling oceanic crust, with an extensive mantle of crystalline silicates and oxides of nearly uniform overall composition overlying the molten iron-alloy core of which the center has solidified because the geotherm intersects the solidus at the inner core boundary.

If one pauses for a moment to question how much we would know about even these first-order aspects of the interior without seismology, it is clear that predictions could be made about planetary layering for a given cosmochemical model (e.g., the assumption of a chondritic abundance of major rock- and core-forming elements, with the light alloy component of the core being O, predicts a core size of 30.1 wt%, compared to an observed value of 32 wt%, Anderson and Bass, 1986), but we would not necessarily know that there is an inner core, could not preclude a very complex compositionally stratified mantle, and would not have direct means to test and refine the assumed multiphase chemical model (by matching laboratory measurements of appropriate compositions to the observed velocities and densities). As described below, further details of the seismological structure provide the key to understanding the dynamics of the interior as well, which would otherwise remain highly speculative. There is thus no question that seismology plays a major role in our understanding of Earth's interior.

3. Crust

Earth's crust is the most important region of the interior, providing the environment, natural resources, and geological hazards that affect humanity. The complexity of structure and geological history of the continental crust are readily apparent from surface observations, providing important clues in our efforts to understand Earth's interior, however it is essential to know the structure at depth. With very sparse drilling being confined to the upper 10 km of the crust, much of our knowledge of the *in-situ* structure of the oceanic and continental crust has been provided by seismological investigations, as described elsewhere in this volume (see Chapter 54 by Mooney *et al.* and Chapter 55 by Minshull). It is again the properties of elastic waves, particularly their reflection and refraction at interfaces across which there are abrupt changes in material properties (e.g., density, compressibility, rigidity) that enables detailed models of the crustal layering and crust–mantle boundary (a seismically defined compositional boundary called the Mohorovičić or "Moho" boundary in recognition of its discoverer in 1909) to be determined by analysis of dense profiles of ground motion recordings for both natural and human-induced sources. Seismology provides information about the interface geometries, absolute seismic velocities, presence of partial melting, and structural anisotropy of the crust, which can then be interpreted in terms of rock compositions and deformation histories by comparison to laboratory measurements of field samples, accompanied by geological reconstructions.

Gross differences between oceanic and continental crustal properties were first revealed in the 1950s by a combination of refraction and gravity studies and the first analyses of Rayleigh and Love wave dispersion observations in the period range 10–70 sec. Love wave observations in particular provided compelling evidence for an average oceanic crustal thickness of about 6 km, with both Rayleigh and Love waves indicating a typical continental thickness of 35 km or so (e.g. Ewing *et al.*, 1957). Advances in computational capabilities, inverse theory, and data quality have allowed increasing resolution of internal crustal layering by surface wave inversion, culminating in the present day capabilities described in Chapters 11 (by Romanowicz) and 54 (by Mooney *et al.*). Surface waves provide relatively limited resolution models of crustal properties, involving extensive depth and lateral averaging of the actual structure, but the integral constraints from surface waves can be combined with body wave information to give reliable detailed crustal structures.

High resolution of internal crustal properties is attained by using seismic body waves with frequencies of 1–100 Hz, and even higher for very shallow imaging, accompanied by close station spacing (meters to kilometers) to avoid spatial aliasing. The data collection and processing involved in analysis of dense linear and 2D deployments of high frequency seismographs is generally defined as the field of Reflection Seismology, with distinct strategies being needed for analysis of much sparser data sets available for sampling the deeper interior on a global scale (see chapters on crustal imaging). At intermediate levels of resolution, the methods of Refraction Seismology, the study of primary direct, reflected and head-wave arrivals traversing the crust from both natural and human-induced sources, provide constraints on the overall crustal waveguide.

Gross attributes of crustal velocity structure, thickness, and regional variations have been summarized by Christensen

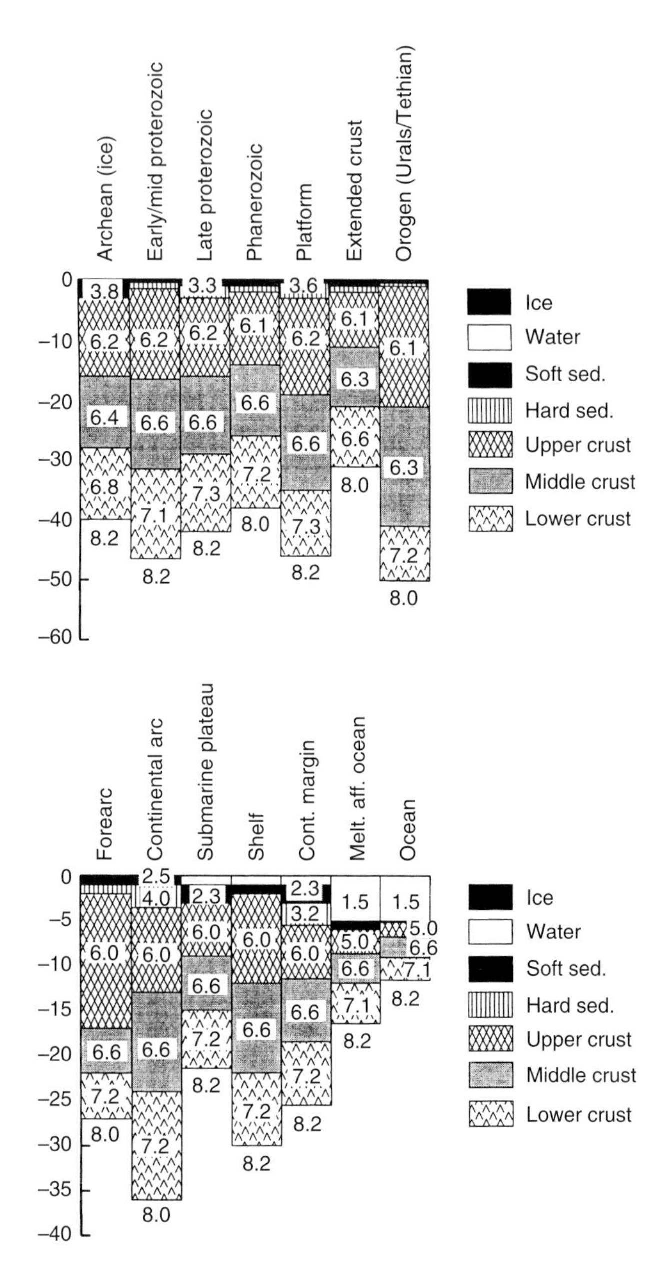

FIGURE 4 Summary of crustal refraction and reflection profiling for diverse crustal regimes indicating thickness and typical *P* velocities of crustal layers (From Mooney *et al.*, 1998).

and Mooney (1995) and Mooney *et al.* (1999). Reflection and refraction studies from hundreds of locations around the world have established that there are characteristic crustal structures in distinct tectonic environments. P velocities and thickness for various crustal types are summarized in Figure 4. Note that water layer thickness is included for oceanic structures. The average P velocity of the crust is $6.45 \pm 0.21\,\mathrm{km\,sec^{-1}}$, the average continental crustal thickness is about 40 km and average oceanic crustal thickness is 12.6 km, including 4.0 km of ocean water (Christensen and Mooney, 1995). A contour map of crustal thickness, including water depth, with $5° \times 5°$ resolution is shown in Figure 5. Seismic models of the crust like these provide a basis for petrological interpretations using laboratory measurements of velocities in plausible crustal materials under appropriate pressures and temperatures. Christensen and Mooney (1995) summarize current inferences about crustal petrology; upper continental crust is matched by diverse lithologies including low-grade metamorphic rocks and silicic gneisses of amphibolite facies grade, middle continental crust is consistent with tonalitic gneiss, granitic gneiss and amphibolite, and lower continental crust is consistent with gabbro and mafic granulite. There appears to be increasing garnet content with depth and mafic garnet granulite comprises the lowermost crust. Localized crustal models also play a key role in unraveling tectonic histories, mountain building and extensional events, shallow volcanic processes, and basin

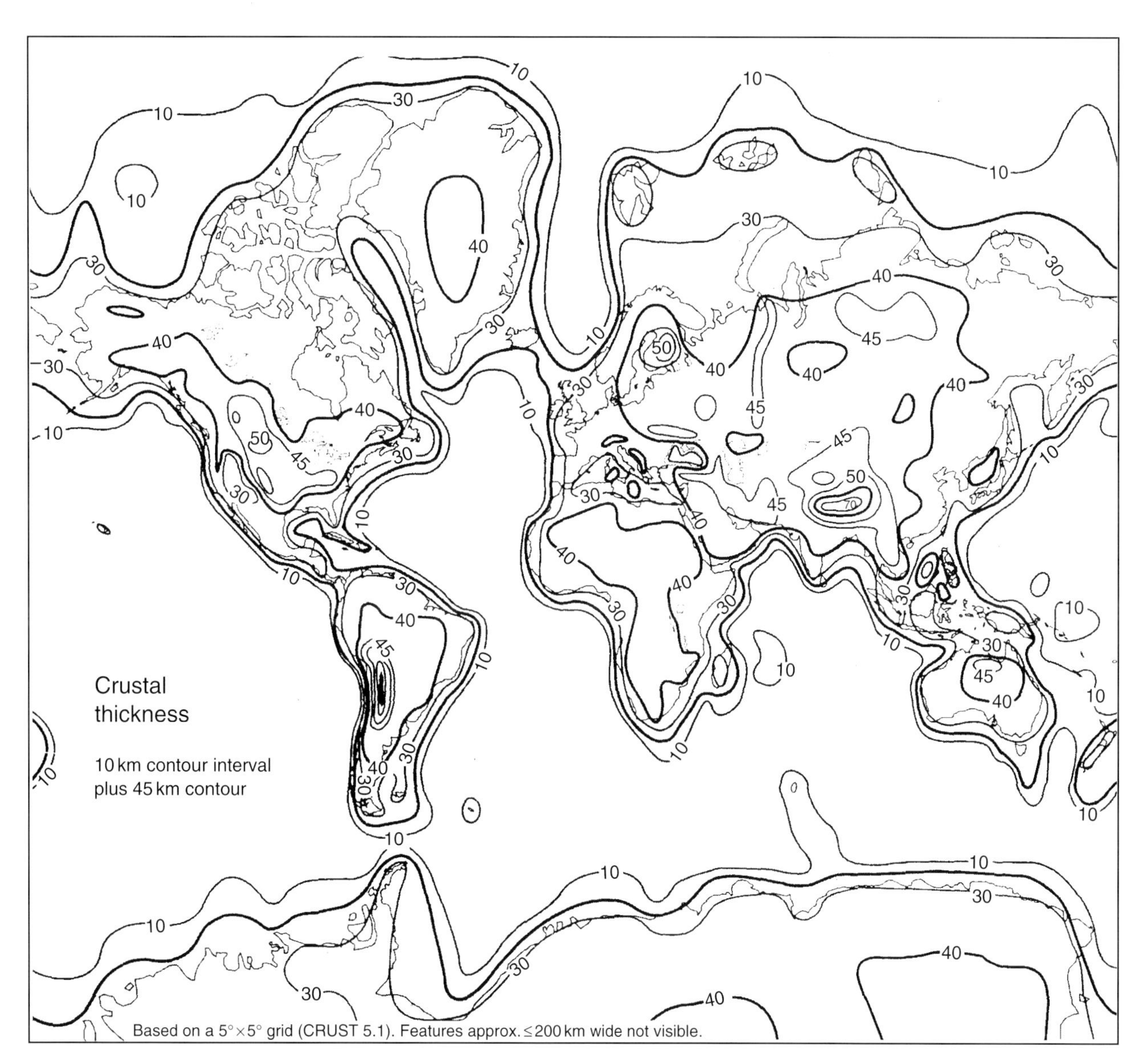

FIGURE 5 Mercator projection of crustal thickness for a 5° model. Extrapolations based on tectonic province and crustal age have been made (from Mooney *et al.*, 1998).

evolution, not to mention the critical role of high resolution models in oil and mineral resource exploration. Many additional details about crustal structure are given in the accompanying chapters on crustal structure. Recognizing that continental and oceanic crust have been extracted from the interior by melting processes, and that the crustal dynamics and history are manifestations of deeper seated processes, the remainder of this chapter will elaborate on the techniques and results of seismological analyses of deeper Earth structure.

4. Upper Mantle and Transition Zone

The existence of complex upper mantle structure is immediately evident from the bimodal distribution of continental and oceanic crust. Preserving highly differentiated rocks dating back more than 3.8 billion years, continental crust clearly has different relationships to its underlying upper mantle than does basaltic oceanic crust, of which very little more than 200 million years old can be found. Given only sparse direct sampling of upper mantle rocks provided by upthrust blocks and xenoliths from deep-seated magmatism, with significant chemical alteration prevalent in both environments, seismology plays a prominent role in constraining our knowledge of upper mantle structure.

The revelations of seafloor spreading in the 1960s provided a context for oceanic upper mantle as an extensive, fairly uniform composition material from which midocean ridge basalts (Ca and Na-rich pyroxene plus an Al-rich phase) are extracted by localized decompression melting in upwellings, and to which both crust and depleted upper mantle material return by subduction of oceanic slabs. Quantification of basalt petrogenesis has played a key role in developing and testing models of upper mantle composition, with the important conclusion that the upper mantle is largely peridotitic (composed of olivine [$(Mg,Fe)_2SiO_4$] and orthopyroxene [$(Mg,Fe)SiO_3$]) rather than eclogitic (mainly clinopyroxene and garnets) in chemistry (e.g., Ringwood, 1975; Green and Falloon, 1998). Ted Ringwood proposed the "pyrolite" model for upper mantle composition as essentially the sum of basalt and peridotite compatible with chondritic abundances. The oceanic crust and the underlying depleted, harzburgitic mantle material are embedded in a conductive thermal boundary layer, the upper half or so of which is relatively cold ($< 650°C$) and stiff, yielding a mechanically competent lithospheric plate (note the distinction between thermal and mechanical boundary layers). As the temperature increases with depth, approaching the melting point of the mantle material, the thermal boundary layer undergoes transitions in mechanical integrity, with viscous lateral shearing, and in thermal transport, with the latter progressively becoming dominated by solid-state advection rather than conduction. Seismological investigations of the suboceanic mantle seek to characterize the layering of the basalt-depleted zone of the lithosphere, to detect any strain-induced fabrics in the oceanic plate or in the strongly sheared region in the lower part of the thermal boundary layer and below it, often called the asthenosphere, and to detect any variations with depth that correspond to effects of increasing temperature and phase transitions.

For subcontinental mantle, the context of plate tectonics provides far fewer well-defined questions, and basic issues such as the thickness of the continental lithosphere, the extent of and nature of compositional depletion of the subcontinental upper mantle, and the thermal history and structure of the subcontinental upper mantle are all open questions, with seismological imaging playing a key role in characterizing the structural elements. As noted below, the distinctions between continental and oceanic upper mantle persist down to depths of 200–350 km, below which any obvious relationships between mantle structure and surface tectonic features rapidly disappear.

4.1 The Seismic Lid

In both continental and oceanic regimes, the uppermost mantle region below the Moho overlies a seismic low velocity zone. The high-velocity layer from the Moho down to the low-velocity zone is called the upper mantle "lid," and this corresponds to the "seismic lithosphere." (Long-term rheological behavior and seismic velocity structure need not correspond directly, for they involve factors such as time of loading, temperature, and stress; however, it is likely that a common underlying influence is the onset of partial melting which affects both seismic velocity and viscosity and defines the bottom of the seismic lid and the rheological transition from the mechanical boundary layer to the ductile asthenosphere.)

The most straightforward aspect of mantle lid velocity structure that can be determined is the velocity just below the Moho discontinuity, as this is manifested in the slope of the *Pn* and *Sn* "headwaves" along the crust–mantle boundary. Usually, a linear fit to the headwave branches is sufficient, with measured *P* velocities of 7.6–8.6 km sec^{-1} and *S* velocities of 4.4–5.0 km sec^{-1} being inferred by many studies from 1950 on (see reviews by Christensen and Mooney, 1995; Christensen and Salisbury, 1979). The average *Pn* velocity beneath continents is 8.09 ± 0.2 km sec^{-1} (Christensen and Mooney, 1995), whereas the average oceanic *Pn* velocity is 8.15 ± 0.31 km sec^{-1} (Christensen and Salisbury, 1979). These velocities are typical of olivine- and pyroxene-rich peridotitic upper mantle rocks from which crustal rocks have been chemically differentiated. Substantial effort has gone into mapping lateral variations in uppermost mantle velocity structure as shown in Figure 6 (e.g., Braile *et al.*, 1989; Hearn *et al.*, 1991; Hearn, 1999), with it being established that the lateral variations in velocity are associated with lateral variations in thermal structure and associated age of tectonic activity. Active rift zones tend to have low *Pn* velocities whereas

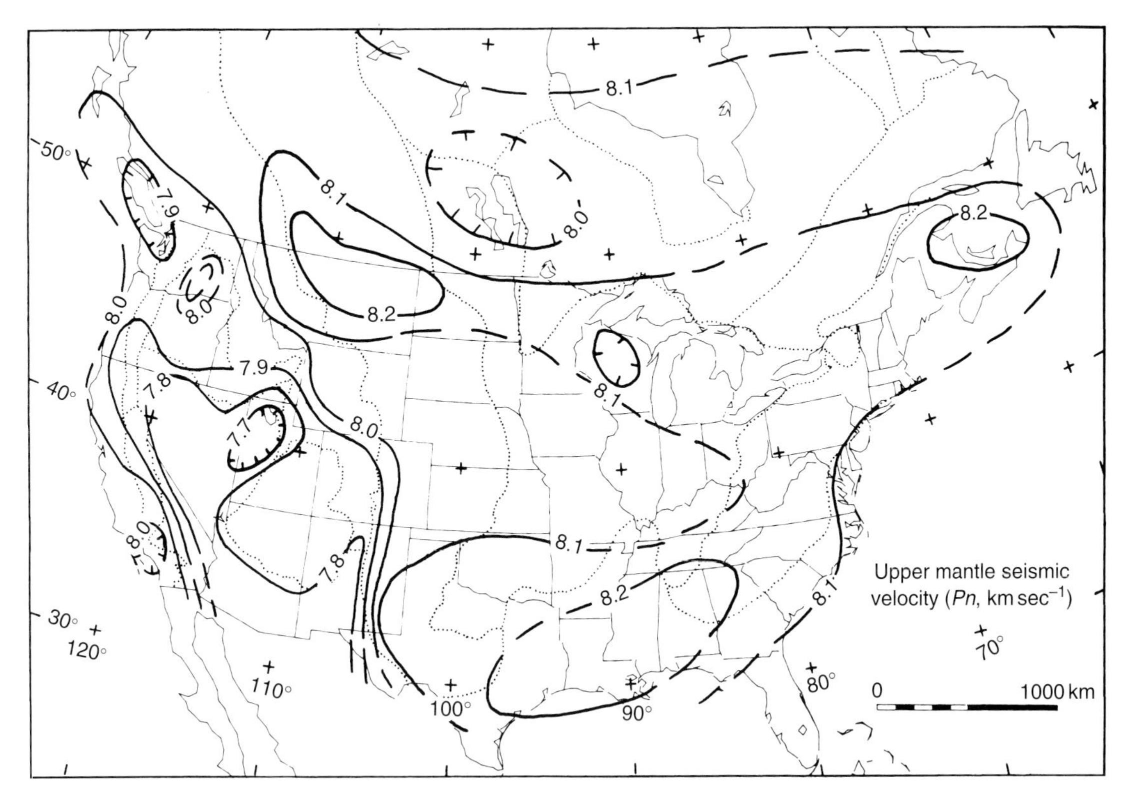

FIGURE 6 Contour map of Pn velocity under the United States based on many localized studies. Note the lower velocities characteristic of the tectonically active western region. (From Braile *et al.* 1989).

stable cratons and platforms tend to have higher *Pn* velocities. A complicating factor is that substantial seismic velocity anisotropy is observed for *Pn* in oceanic (Raitt *et al.*, 1969) and continental areas (Bamford, 1977), presumably as a result of plate-motion or orogenically induced alignments of the upper mantle minerals, particularly olivine. This produces azimuthal dependence of the *Pn* velocity, which must be accounted for when seeking an accurate seismic and petrological model for the uppermost mantle (e.g., Hearn, 1999).

The velocity structure just below the Moho generally involves slowly increasing velocity with depth, which produces a whispering gallery of phases that complicate *Pn* and *Sn* arrivals and the interpretation of the apparent velocities of these phases. Beneath some continental regions, a small abrupt increase in *P* and *S* velocity occurs near 60–90 km (e.g., Hales, 1969; Revenaugh and Jordan, 1991), possibly as a result of a phase transition from spinel to garnet facies in aluminous peridotite. This feature is detected in regional travel-time curves as well as in near vertical reflections. This uppermost mantle region is highly heterogeneous from region to region, largely as a result of chemical and thermal variations in the oceanic and continental lithosphere. In fact, the strongest lateral variations in upper mantle structure are found at mantle lid depths from 50 to 100 km, with greater than 10% variations in *S* velocity and several percent variations in *P* velocity. Continental regions appear to have end-member lid structures associated with stable cratonic and platform regions with relatively high sub-Moho *S* velocities of 4.7–4.8 km sec^{-1}, and *P* velocities of 8.2–8.4 km sec^{-1}, whereas tectonically active areas have *S* velocities of around 4.3 km sec^{-1} and *P* velocities around 7.9 km sec^{-1}. This large variation is believed to be a combined thermal and compositional effect, as cratonic lid is both thicker (150–200 km) and chemically distinct from tectonically active continental lid (which may be only as thick as the crust in active rift environments). There is evidence for significant small-scale heterogeneity in the lid structure, primarily in dense long-range profiles collected in Eurasia, so the average properties given above are likely oversimplified (e.g., Ryberg and Wenzel, 1999). Oceanic regions show systematic increases in lid thickness with plate age, presumably as a result of progressive cooling of the oceanic lithosphere. This has been most systematically characterized by regionalized or tomographic inversions of surface-wave dispersion measurements (see below). Relatively simple thermal models for aging oceanic lithosphere appear to provide a viable basis for characterizing first-order radial and lateral structure of oceanic uppermost mantle lid structure (e.g., Zhang and Lay, 1999).

4.2 The Seismic Low-Velocity Zone

The presence of an upper mantle low-velocity zone is of profound importance for mantle dynamics, as it is commonly associated with a zone of partial melting, strong seismic wave attenuation, and low viscosity. Based on the body wave travel time and amplitude evidence accumulated by Beno Gutenberg for a decrease in velocity below about 60–80 km depth followed by a gradual increase commencing near 100–150 km (Gutenberg and Richter, 1939; Gutenberg, 1948, 1959) it has long been accepted that there is an upper mantle low velocity zone of some type (see the Gutenberg model in Figure 2). Low-velocity zones tend to be difficult to detect and to quantify due to the downward refraction of seismic energy that they cause. The upper mantle seismic low velocity channel is commonly associated with the asthenosphere, a rheologically defined region with strong ductile deformation and possibly partial melting underlying the lithosphere. The onset of the low velocity channel is sometimes characterized by an abrupt velocity decrease [labeled by Revenaugh and Jordan, 1991, as the "G" (for Gutenberg) discontinuity at about 80 km depth]. This relatively sharp feature is typically found in tectonically active regions and under oceanic crust, but in some cases there is a somewhat deeper low velocity zone found under cratons and platforms, which may or may not have an abrupt onset.

Early surface wave dispersion studies of the 1950s indicated that Love and Rayleigh wave-dispersion curves beneath oceanic and continental regions tend to converge for periods longer than about 75 sec and that Rayleigh wave group velocities for "mantle wave" periods of 75–400 sec have an Airy phase minimum near 225 sec that suggests the presence of a decrease in shear velocity with depth in the upper mantle (e.g., Ewing and Press, 1954; Ewing *et al.*, 1957). As inversions for models with multiple layers became viable by the early 1960s a common feature of separate fits of dispersion curves for Love and Rayleigh waves was the presence of a low velocity zone in the upper mantle. Highly precise measurements of great-circle data (multiple passages of the same wavetrain past a given station) began to resolve systematic differences for paths under oceanic, tectonically active continent, and stable continent regions for periods longer than 75 sec (e.g., Anderson and Toksöz, 1963; Toksöz and Ben-Menahem, 1963; Toksöz and Anderson, 1966; Kanamori, 1970), allowing the first attempts at pure-path regionalizations based on percentage path lengths in each tectonic province. Isotropic inversions of the dispersion curves in the range 75–300 sec yielded shear velocity models with low-velocity zones between 80 and 200–350 km depth. An upper mantle lid overlying a low velocity zone with shear velocities of 4.0–4.2 km sec^{-1} with deeper strong positive velocity increases was a common feature of almost all surface-wave inversions of this generation. The regionalized dispersion measurements also suggested that velocity structures under old cratons differ from those under oceans and tectonically active continental regions down to depths exceeding 250 km (e.g., Kanamori, 1970).

However, an important problem was recognized quite early on, as it proved difficult to simultaneously fit precise global and pure-path Love and Rayleigh-wave observations by isotropic models (McEvilly, 1964; Anderson, 1966). This became known at the Love–Rayleigh discrepancy, and it motivated the development of anisotropic models for upper mantle structure (Anderson, 1961; and see Chapter 53 by Cara). The inclusion of anisotropy directly affects the magnitude and nature of the low-velocity zone structure obtained by fitting surface-wave dispersion curves (e.g., Anderson, 1966; Anderson and Dziewonski, 1982). A basic requirement of most laterally and azimuthally averaged dispersion observations is that the *SV* velocity structure of the upper few hundred kilometers of the mantle is a few percent slower than the *SH* velocity structure, and initially this was addressed for regionalized data sets by performing separate isotropic inversions for the *SV*-sensitive Rayleigh waves and the *SH*-sensitive Love waves (e.g., Forsyth, 1975; Schlue and Knopoff, 1977, 1978; Yu and Mitchell, 1979; Mitchell and Yu, 1980). These pseudo-isotropic studies conclusively documented the need for extensive anisotropy of either the lid or the low-velocity zone, but it was subsequently shown that it is important to perform a self-consistent inversion for an anisotropic medium to obtain accurate models (Anderson and Dziewonski, 1982; Regan and Anderson, 1984).

From the early 1980s on, most surface-wave and normal-mode models have allowed for at least a transversely isotropic structure in the upper mantle, with five elastic constants, and this complexity was incorporated into the Preliminary Reference Earth Model (PREM) produced by Dziewonski and Anderson (1981). In PREM, which has served as the background reference model for many subsequent surface-wave tomographic inversions, making them intrinsically transversely isotropic, the horizontal propagating *S* velocity (V_{SH}) is faster than the vertically propagating *S* velocity (V_{SV}), and the horizontally propagating *P* velocity (V_{PH}) is faster than the vertically propagating *P* velocity (V_{PV}) in the upper 220 km of the mantle (Figure 7). In this model, the low-velocity zone is very subdued, and it is not necessary to have very low shear velocities as found in earlier isotropic and pseudo-isotropic inversions.

The parameterization of PREM incorporates a very strong velocity discontinuity near 220 km depth, as has been reported in several continental *P* and *S* wave travel-time studies (e.g., Lehmann, 1959, 1961; Hales *et al.*, 1980; Drummond *et al.*, 1980; Anderson, 1979). This feature, often called the Lehmann discontinuity, was suggested by Lehmann (1959) to correspond to the bottom of the low-velocity zone, a notion that was embraced in the parameterization of PREM. However, this feature is not observed globally, and appears to be primarily

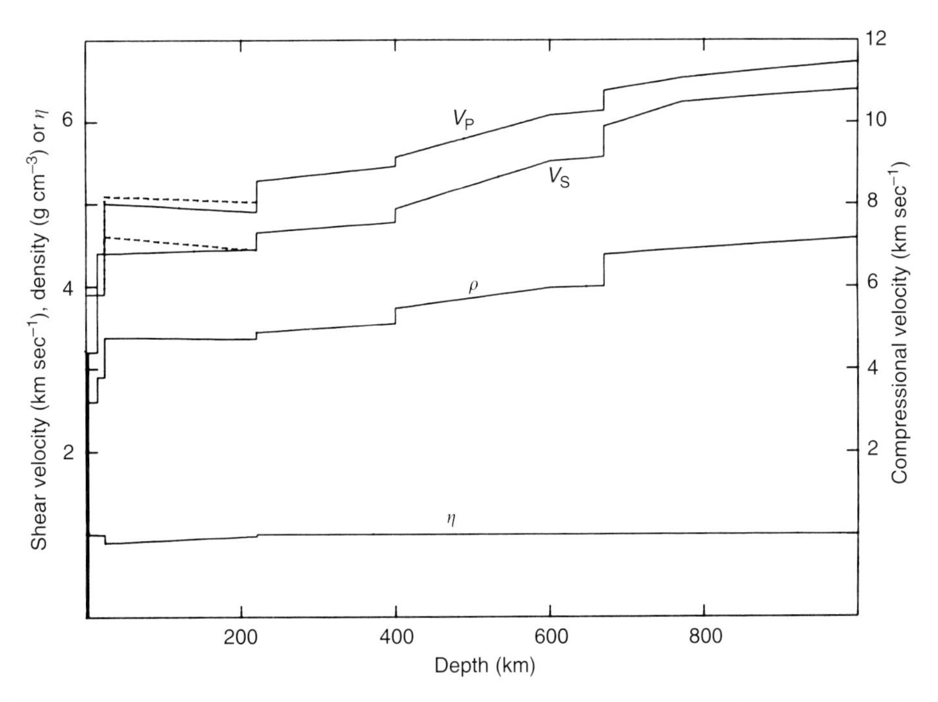

FIGURE 7 Anisotropic upper mantle P velocity (V_p) and S velocity (V_s) for the transversely isotropic, 1 Hz version of the Preliminary Reference Earth Model (PREM), along with density structure (ρ) and the anisotropic parameter η. Dashed lines are the horizontal components of velocity (from Dziewonski and Anderson, 1981).

a subcontinental feature of limited extent, confined to shield and platform regions that do not have strong low-velocity zones or G discontinuities (e.g., Shearer, 1991; Revenaugh and Jordan, 1991). A viable interpretation of the continental 220 km feature is that it corresponds to a critical temperature (about 1200°C) transition from the anisotropic subcontinental mantle to the more isotropic structure below (Revenaugh and Jordan, 1991; Gaherty and Jordan, 1995), possibly embedded within a 300–400 km thick continental root (see below). However, there is evidence for a reflector near 220 km in the vicinity of some subduction zones, which cannot be accounted for by this particular model (e.g., Vidale and Benz, 1992; Zhang and Lay, 1993), so uncertainty in the interpretation of this structure remains.

The modern generation of shear-wave models for oceanic upper-mantle derived from waveform modeling of multimode Rayleigh waves, body waves, or complete body/surface waveforms show relatively pronounced low-velocity zones between 80 and 300 km (Fig. 8), even when transverse isotropy of the lid and low velocity zone is included in the modeling (e.g., Gaherty *et al*., 1996). Both P-wave and S-wave models obtained by body wave travel-time and waveform modeling for upper mantle triplication distances (10–30°) usually include upper mantle low-velocity zones, although the resolution of such structures is limited (e.g., Grand and Helmberger, 1984; Zhao and Helmberger, 1993). Thus, it appears that the seismic low-velocity zone does exist, but is highly variable in its properties depending on tectonic region. Rather than terminating in an abrupt discontinuity as in the PREM model, the low-velocity zone is typically underlain by a relatively steep velocity gradient with depth that persists down to the mantle transition zone. Generally this region from about 200 to 400 km depth has little structure in average Earth models (see Figs. 3 and 7). There is evidence for localized P and S velocity increases near 310 km depth in the vicinity of subduction zones (e.g., Hales *et al*., 1980; Revenaugh and Jordan, 1991; Zhang and Lay, 1993; Zhang, 1994), but this structure is either significantly variable in depth or spatially intermittent, based on analysis of *SS* precursors by Shearer (1991, 1993). Revenaugh and Jordan (1991) label the 310 km feature (actually found in *ScS* reverberation coda at depths ranging from 275 to 345 km) the "X" discontinuity, and, lacking any explanation in terms of standard upper mantle mineralogy, they propose that it is associated with phase reactions in extensively hydrated mantle surrounding zones of extensive subduction.

4.3 The Transition Zone

The region of relatively strong positive velocity gradients below the low-velocity zone is abruptly punctuated by a global velocity increase near 410 km depth, marking the onset of the transition zone which extends from 410 to 800 km. Early body-wave velocity models of Sir Harold Jeffreys in the 1940s

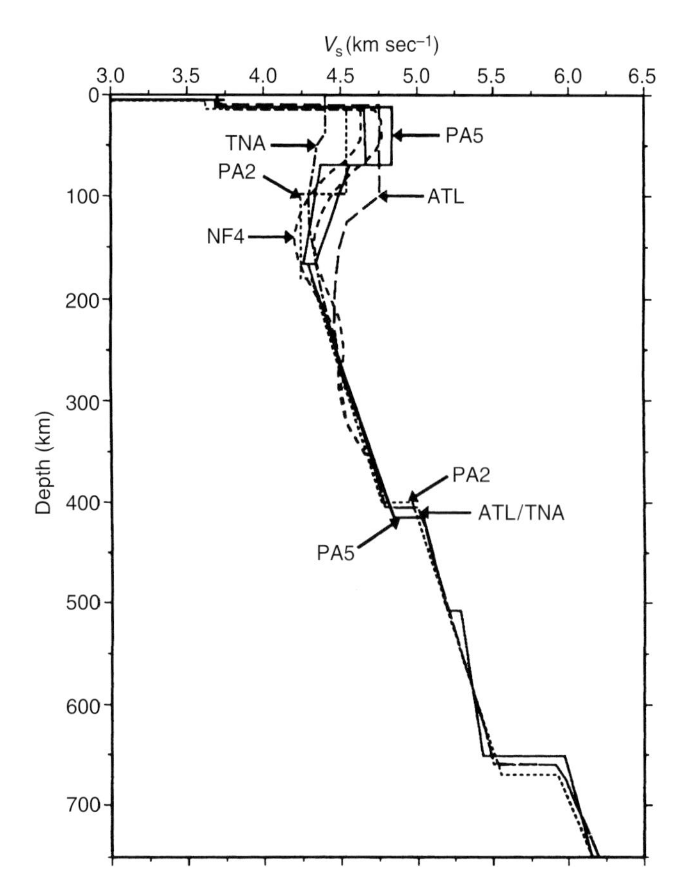

FIGURE 8 Oceanic upper mantle shear velocity structures from waveform modeling studies. Note the presence of a well-developed low velocity zone (from Gaherty *et al.*, 1996).

(see Fig. 2) incorporated a change in velocity gradients above and below 400–500 km depths (based on the so-called 20°-discontinuity first pointed out by Byerly, 1926), but the detailed structure of the transition zone from 410 to 800 km depth only began to be worked out in the 1960s. Modeling of long-period surface-wave dispersion curves first indicated that the upper mantle low-velocity zone is globally underlain by strong positive velocity gradients near 350–450 km and 600–700 km depths (e.g., Anderson and Toksöz, 1963; Toksöz and Ben-Menahem, 1963; Toksöz and Anderson, 1966). This focused attention on body-wave travel-time and apparent-velocity measurements from newly deployed regional seismic arrays for the distance range of 1000–3000 km, and it was soon recognized that two major upper mantle travel-time triplications are produced by abrupt *P* and *S* velocity increases near 410 and 660 km depths (e.g., Niazi and Anderson, 1965; Ibrahim and Nuttli, 1967; Johnson, 1967; Kanamori, 1967). Anderson (1967a) attributed these rapid velocity increases to solid–solid phase changes of $(Mg,Fe)_2SiO_4$ olivine to β-phase (modified-spinel structure) and γ-spinel to more compact oxides, respectively. The precise nature of the post-spinel transformation was not worked out for several years and it is now known to involve a disassociative transformation to the silicate perovskite phase of $MgSiO_3$ plus magnesiowustite (Mg,Fe)O (e.g., Liu, 1976; Ito and Takahashi, 1989; Bina and Helffrich, 1994). The combination of seismological constraints and experimental measurements which has established that these phase transitions in olivine are responsible for the two major upper-mantle discontinuities codified the earlier inferences by Francis Birch and J.B. Thompson that phase transitions were responsible for the zone of steep velocity gradient in Jeffreys' velocity model (Fig. 2) between 400 and 1000 km depth (Birch, 1952). Although there had been suggestions of transition-zone discontinuities dating back to Byerly (1926), global seismic velocity models did not incorporate the 410 and 660 km discontinuities until the surge of observations in the mid-1960s demonstrated their global existence and placed bounds on the depths and velocity increases that are involved.

Throughout the last three decades, numerous seismological procedures have been developed to extract increasing information about the velocity structure in the transition zone. A combination of wide-angle triplication studies and near vertical top–side and bottom–side reflection and conversion studies have yielded extensive information about the contrasts in velocity and density, sharpness (depth extent of the velocity increases), and topography of the 410 and 660 km discontinuities. The process is complicated by the fact that there are strong variations in the lid and low velocity zone structure above the discontinuities, as well as by limitations imposed by the distribution of seismic stations and sources.

Wide-angle triplication studies exploit the complexity of the *P*- and *S*-wave travel-time curves in the so-called "upper mantle" distance range of 12°–30°. This is illustrated in Figure 9, for a *P* velocity model appropriate for the western United States (Walck, 1984), where there is a thin lid and shallow low-velocity zone. As it encounters rapidly increasing velocity with depth the wavefront folds over on itself, with energy that turns above, at and below a discontinuity being refracted to each distance in the associated triplication range. The transition zone discontinuities produce abrupt changes in the slope of the first arrival branch, and the ray parameters for each of the three arrivals of the triplications are different enough that they can be measured, as long as secondary arrivals are recognized amidst the coda generated by the first arrival (Fig. 9). The velocity structure can be determined by matching the timing and amplitudes of the arrivals in the triplication range, as well as by matching the corresponding ray parameter measurements. The origin of the "20°-discontinuity" notion is readily apparent in the overall curvature of the first arrival time curve resulting from energy turning in the much different velocities above and below the transition zone.

Dense sampling of the travel-time curve and recognition of secondary arrivals as triplication branches is required in order to establish that discontinuities are present, and

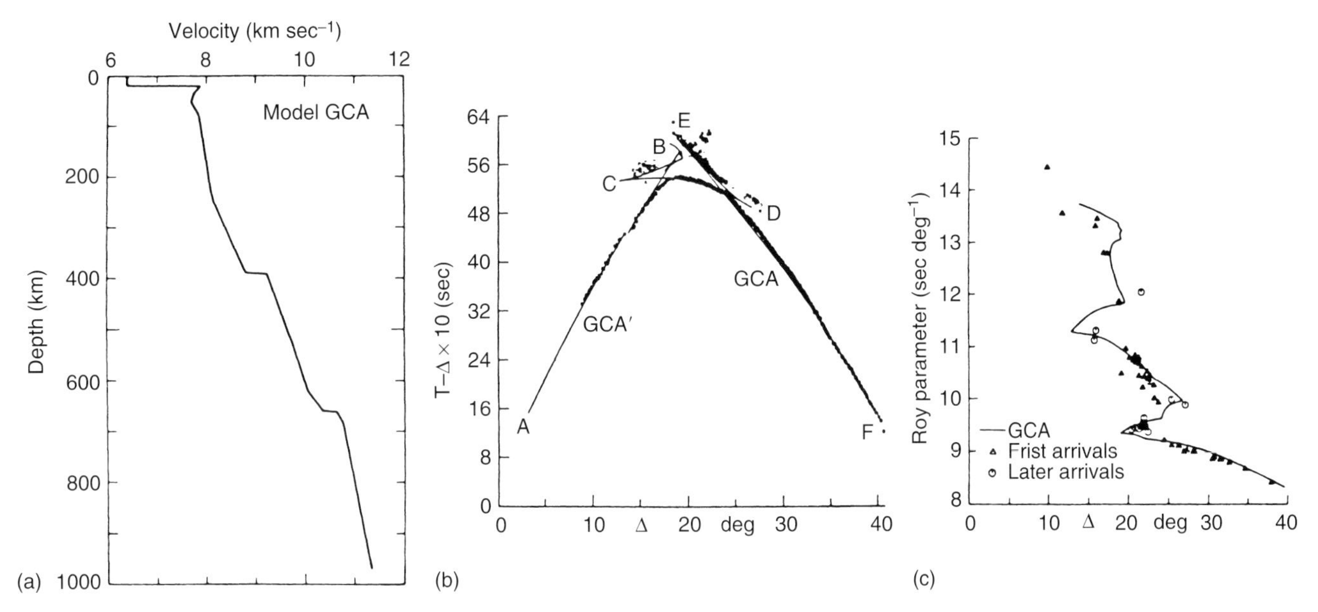

FIGURE 9 (a) A P-wave velocity model found for the tectonically active upper mantle below the Gulf of California. The observed and predicted travel times for this region are shown in (b) and observed and predicted ray parameter values are shown in (c) (modified from Walck, 1984. Reprinted with permission from the Royal Astronomical Society).

typically only limited quantification is possible by classical travel time and ray parameter modeling alone. Waveform modeling methods developed in the late 1960s and throughout the 1970s (e.g., Helmberger, 1968; Helmberger and Wiggins, 1971; Fuchs and Müller, 1971; Chapman, 1976, 1978; Richards, 1973) enabled complete synthesis of the triplication waveform interference and relative amplitudes for 1D, and later 2D structures (e.g., Chapman and Drummond, 1982; Helmberger *et al.*, 1985). With these powerful tools, dozens of wide-angle studies of mantle structure have been carried out, resulting in regionalized velocity structures extending down through the transition zone. Nolet *et al.* (1994) consider the evidence for lateral mantle heterogeneity based on comparison of upper mantle models derived by modeling short-period P, long-period P, and long-period S triplications (Fig. 10). Additional shear-wave models for oceanic areas are shown in Figure 8. The majority of these models have been developed assuming isotropic transition zone structure (the data typically provide too little information to constrain anisotropic structure, but the fact that the data can be well modeled by isotropic models suggests that transition zone anisotropy is probably not very large, if present at all). Note that there are pronounced differences in lid structure, in structure of the low velocity zone, and in structure near 220 km depth, as discussed previously. Some of the variations may result from approximating laterally heterogeneous structure with 1D models. However, all models show 410 km and 660 km discontinuities, and it is accepted that these are globally present as expected for phase transformations of a primary mantle component.

There are apparently significant variations in size and sharpness of the velocity increases. Wide-angle triplication modeling is typically not sensitive to density structure, but the nearly factor of 3 variation in P-velocity jump at 410 km depth seen in these models may be influenced by failure to account for topography on the boundary. The S-velocity jump at 660 km in most models is about 7–8%, and appears to be stronger than that at 410 km (5%), whereas the P velocity contrasts are more similar, averaging 5–6% at 410 km and about 4% at 660 km [Estabrook and Kind (1996) present evidence for a P-velocity jump of only 2% at 660 km]. Very strong evidence for an even stronger jump of as much as 10% in S velocity at 660 km depth, along with reduced density jumps of 4–5%, has recently emerged from the systematic study of underside reflections as a function of angle of incidence (Shearer and Flanagan, 1999) and from amplitudes of top-side conversions (Castle and Creager, 2000). The velocity contrasts at the discontinuities provide a means for assessing the relative abundance of olivine component (for which the contrasts in velocities and densities across phase transformations can be measured in the laboratory) in the transition zone. The observed jumps are smaller than expected for a purely olivine mantle, so it is believed that there is significant presence of orthopyroxene, garnet, and/or clinopyroxene (with the latter two possibly giving an eclogitic component) that "dilute" the percent-wise velocity increases (e.g., Bass and Anderson, 1984; Anderson, 1991). The estimates of large shear-velocity contrast tend to favor an olivine-rich composition of the transition zone. Because there is nonuniqueness in possible mixed compositions that can match the velocities and

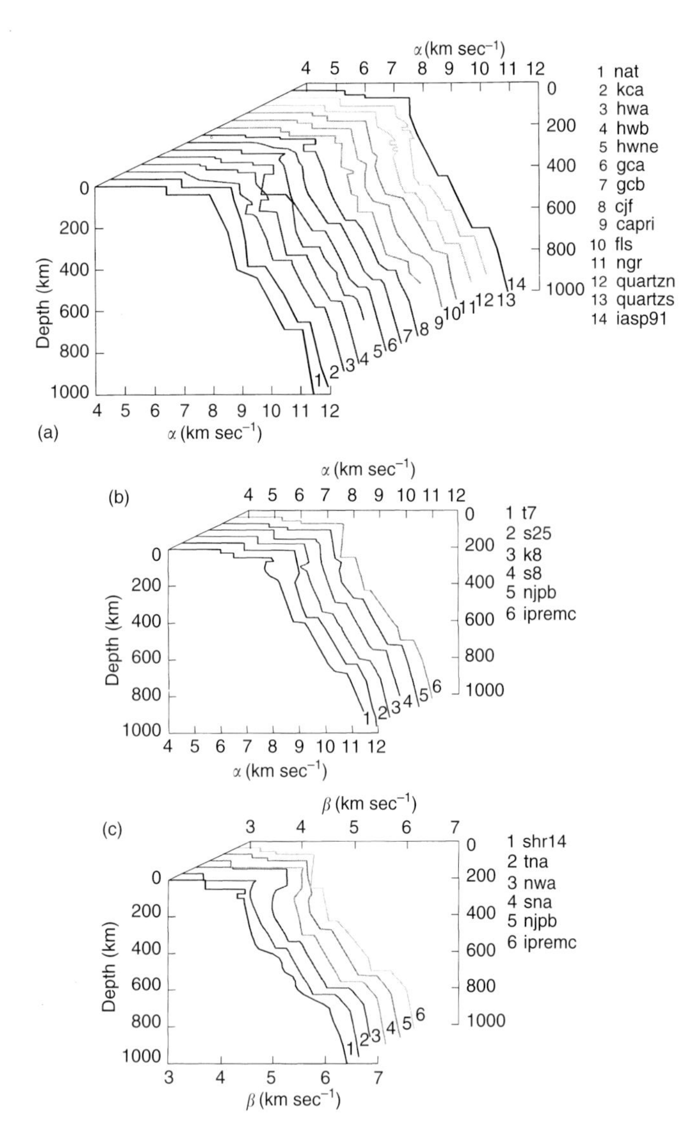

FIGURE 10 (a) Upper mantle P-velocity models derived from short-period P observations. The upper mantle model iasp91 was designed to give a reasonable average. Regional coverage: nat, North Atlantic,; hwb, hwa, hwne, west-central United States; gca, gcb, djf, western North America; capri, fls, ngr, northern Australia; quartzn, quartzs, Asiatic Russia. (b) Upper mantle models for P velocity derived from long-period or broadband observations. Regional coverage: t7, western North America, s25, North American Shield, k8, Eurasia; s8, northeastern United States; njpb, northern Australia, ipremc, Isotropic version of PREM model with continental crust. (c) Upper mantle models for S velocity derived from long-period or broadband observations. Regional coverage: shr14, western United States; tna, tectonic North America; sna, shield North America; njpb, northern Australia, ipremc, isotropic version of PREM model with continental crust. (From Nolet *et al.* 1994.)

densities, details of the bulk chemistry of the transition zone are still unresolved. The possibility that the 660-km discontinuity coincides with a chemical contrast as well as (or instead of) the spinel to perovskite transformation in olivine has been discussed for decades, but it appears possible to fully account for the discontinuity by the phase transition. This still does not preclude the possibility that the bulk composition of the lower mantle differs from that of the transition zone as discussed later.

There is a relatively steep velocity gradient between the two major transition zone velocity discontinuities, with a few models indicating rapid P or S velocity increases near 520 km depth (Fig. 10). Most S-wave models, and some P-wave models indicate a zone of steep velocity gradient extending down from the 660 km discontinuity to about 800 km depth, below which there is a reduced velocity gradient extending smoothly into the lower mantle. This feature is incorporated into recent average Earth models such as PREM (Fig. 7), IASP91 (see iasp91 in Fig. 10) and ak135 (Fig. 3). I take 800 km depth as the lower boundary of the transition zone, for the seismic models generally show only smooth velocity gradients at larger depths and the primary-phase transitions expected for major components of the upper mantle and transition zone should be completed by this depth. However, there is evidence for nonglobal discontinuities at 900 and 1200 km depth as well as deeper transitions in the spectrum of lower mantle lateral heterogeneity (see below), so there may be further localized phase transitions or chemical contrasts at greater depths.

Analyses of converted and reflected waves also constrain properties of the transition zone discontinuities. The first such phases to be exploited were precursors to PKPPKP (P′P′ precursors), involving underside reflections from transition zone discontinuities observed near angular distances of 70°. These near-vertical incidence reflections are particularly sensitive to the sharpness and impedance contrast across the boundary, and were first used to constrain properties of the 410 and 660 km boundaries by Adams (1968, 1971), Engdahl and Flinn (1969) and Whitcomb and Anderson (1970). These phases have subsequently been extensively studied over the years, providing limited sampling of different mantle regions and revealing stronger reflections for 1 Hz signals from the 660 km boundary than the 410 km boundary (e.g., Nakanishi, 1989; Benz and Vidale, 1993). This requires a sharp 660 km impedance contrast, no broader than about 4 km in many areas. In some cases the 410 km feature also gives strong 1 Hz reflections (e.g., Benz and Vidale, 1993; Vidale *et al.*, 1995; Helffrich and Wood, 1996; Neele, 1996), but often it does not, implying that in some regions the 410 km feature may be spread over more than 5 km thickness or is subject to strong lateral variations or small-scale topography. The olivine to β-phase transition may have nonequilibrium effects (e.g., Solomatov and Stevenson, 1994), nonuniform rate of velocity increase with depth (e.g., Bina and Helffrich, 1994; Stixrude, 1997) or complexity due to accompanying gradual transformations of garnet–pyroxene components.

There are many additional top-side and bottom-side reflections and conversions from the mantle discontinuities that can be sought in order to constrain their properties. *P* waves convert to *S* waves (*Pds*) at boundaries under a station, arriving in the coda of direct *P*, and these can be stacked for arrays of sources to quantify the weak phases (e.g., Vinnik, 1977; Stammler *et al.*, 1992; Dueker and Sheehan, 1997; Gurrola and Minster, 1998; Bostock, 1998; Chevrot *et al.*, 1999). *S* waves convert to *P* at boundaries below a station, giving rise to *Sdp* arrivals that precede direct *S* (e.g., Bath and Stefánson, 1966; Jordan and Frazer, 1975, Faber and Müller, 1980; Bock, 1988). *S* waves from deep slab events convert to *P* waves (*SdP*) at boundaries below the sources, arriving in the *P*-wave coda (e.g., Bock and Ha, 1984; Vidale and Benz, 1992; Wicks and Richards, 1993; Niu and Kawakatsu, 1995; Castle and Creager, 1998a), with these phases being particularly useful for determining properties of mantle discontinuities below subducting slabs. Upgoing *P* and *S* phases from deep slab events reflect from the underside of boundaries and are observed at large distances (*pdP*, *sdS*), providing a means to image properties of mantle discontinuities above deep sources (e.g., Vidale and Benz, 1992; Zhang and Lay, 1993; Flanagan and Shearer, 1998b). Near vertical *ScS* reverberations generate a suite of top- and bottom-side reflections from mantle discontinuities which can be stacked to image impedance contrasts and depths (e.g., Revenaugh and Jordan, 1991a,b; Clark *et al.*, 1995). Top- and bottom-side reflections also produce a suite of arrivals between *P* and *PP* and between *S* and *SS* which can be detected and identified by stacking of multiple observations (e.g., Shearer, 1991).

The most important of these boundary interactions phases have been the underside reflectors that are precursors of *PP* (*PdP*) (e.g., Bolt, 1970; King *et al.*, 1975; Shearer, 1991; Neele and Snieder, 1992; Estabrook and Kind, 1996; Shearer and Flanagan, 1999) or *SS* (*SdS*) (e.g., Shearer, 1990, 1993; Petersen *et al.*, 1993). These phases provide much more extensive coverage of mantle discontinuities than most other phases, allowing global maps of topography on the boundaries to be determined (e.g., Shearer and Masters, 1993; Shearer, 1993; Estabrook and Kind, 1996; Flanagan and Shearer, 1998a). Figure 11 shows 10° radius cap-averaged values of depths to the "410-km" and "660-km" boundaries obtained from *SS* precursors, corrected for an upper mantle shear-velocity model. The mean depths of the two main upper mantle discontinuities are 418 km and 660 km. Although there are concerns about biases in these estimates from unresolved small-scale topography (Neele *et al.*, 1997), relatively large-scale coherent regions appear in the topographic maps. Increased depths to the 660 km feature are found in circum-Pacific regions of current and past subduction, supporting the notion that the boundary is associated with the endothermic γ-spinel to perovskite phase change and that subducting slabs deflect and produce broad cool features near this boundary. This is also consistent with studies of localized boundary deflections near slabs (e.g., Richards and Wicks, 1990; Wicks

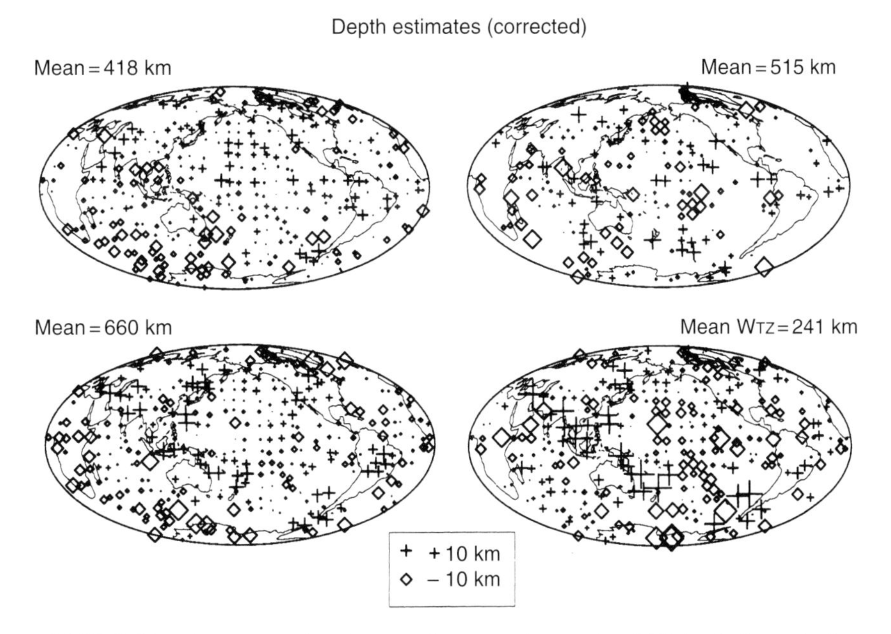

FIGURE 11 Cap-averaged estimates of topography on the "410-km" discontinuity, the "520-km" discontinuity, the "660-km" discontinuity and the transition zone thickness between the "410" and "660" discontinuties, based on underside reflections that arrive ahead of *SS*. The estimates have been corrected for surface topography, crustal thickness, and upper mantle shear wave velocity structure beneath the *SS* bounce points (from Flanagan and Shearer, 1998).

and Richards, 1993; Niu and Kawakatsu, 1995; Vidale and Benz, 1992; Castle and Creager, 1997, 1998b). At long wavelengths the 660 km feature varies in depth by 35–40 km, which is far less than would be expected if the boundary were a compositional contrast. Topography of the 410 km feature is smaller in amplitude (20–25 km) and globally uncorrelated with the 660 km feature [although Revenaugh and Jordan (1991a) present evidence for anti-correlation of the two in some regions]. The olivine to β-phase transition that is associated with the 410 km discontinuity is exothermic, and should be elevated in the vicinity of cold slabs, however, conflicting results have been reported on the near-slab topography of this feature (e.g., Vidale and Benz, 1992; Zhang and Lay, 1993; Collier and Helffrich, 1997; Flanagan and Shearer, 1998b). A complicating factor is that the transformation of olivine may by kinetically inhibited, with untransformed material penetrating well below 410 km (e.g., Sung and Burns, 1976; Rubie and Ross, 1994). The complex thermal structure and kinetic effects near subducting slabs probably result in a rather poor reflector. The fluctuations in distance between the 410 and 660 km features shown in Figure 11 show a pronounced contrast between central Pacific and circum-Pacific regions, but no average difference between continents and oceans. In contrast, Gossler and Kind (1996) have proposed that the discontinuity separation is actually larger under continents, favoring very deep roots of continents.

Long-period *SS* precursors are also extensively observed to originate near a depth of 520 km (e.g., Shearer, 1990, 1996; Flanagan and Shearer, 1998a), and the topographic variations of this reflector can also be imaged (Fig. 11). *ScS* reverberations also consistently show a weak arrival (Revenaugh and Jordan, 1991a). A shear-wave impedance contrast less than half of that found for the 410 km feature is involved, so although the 520 km boundary is probably global in extent, it may produce arrivals below noise levels in some sparsely sampled regions. It is likely that this feature is associated with transformation from β-phase to γ-spinel (e.g., Rigden *et al.*, 1991), which is not expected to produce a very sharp feature in seismic velocities, but should produce a several percent increase in density. As noted previously, several upper mantle *P*- and *S*-wave models obtained by modeling of wide-angle triplications incorporate some increase in velocity gradient near this depth (e.g., Helmberger and Wiggins, 1971; Helmberger and Engen, 1974; Mechie *et al.*, 1993), however many long-period models do not require any velocity structure at this depth, and several studies of short-period *P* waves indicate that no sharp increase in *P* velocity is present (e.g., Jones *et al.*, 1992; Cummings *et al.*, 1992; Benz and Vidale, 1993). This set of observations can be reconciled by the existence of a several percent contrast in density, with little *P*-velocity or *S*-velocity increase, with more than 50% olivine component in the transition zone, which favors an olivine-rich pyrolitic mantle model without much eclogite in the transition zone (Shearer, 1996).

There is also evidence for an impedance contrast near 705–770 km depth, imaged most extensively by *ScS* reverberations (Revenaugh and Jordan, 1991a,b), but with some evidence from wide-angle *P* waves (e.g., Datt and Muirhead, 1976; Muirhead and Hales, 1980) and in $P'P'$ precursors (Sobel, 1978). Although this is not established as a global boundary, it is perhaps closely linked to the bottom of the zone of steep velocity contrast in many *S*-wave models which does appear to have global extent (Fig. 10). The ilmenite to perovskite transformation (Liu, 1977) is a viable candidate for the impedance contrast when it is present (Revenaugh and Jordan, 1991a), which should be in relatively cold (slab-related?) mantle, while gradual transformations of majorite and garnet to perovskite may account for the ubiquitous steep gradient just below the 660 km discontinuity (e.g., Anderson, 1991).

The principal components of upper mantle and transition zone structure are summarized in the schematic in Figure 12. Continents and oceans have significant differences extending from the crust to depths near 350 km (see below), oceans have pronounced shallow low shear velocity layers, there are global discontinuities near 410 km, 520 km and 660 km depth, with temperature variation-induced topography on these boundaries compatible with their interpretation as phase transitions in the olivine component of the mantle. There are intermittent discontinuities associated with continental crust, the vicinity of subduction zones, and the base of the transition zone. All of these features, revealed by seismology, provide a basis for testing and refining models of the composition and state of the upper mantle and transition zone.

4.4 Tomographic Models of the Upper Mantle and Transition Zone

The foregoing discussion has addressed average upper mantle and transition zone structure, with some attention paid to bimodal distinctions between continental and oceanic regions. Given the complexity of Earth's history, with its ongoing dynamical motions, it would be surprising not to find complex lateral variations in structure everywhere in the mantle, which need not be directly coupled to surface geology. The variety of localized 1D models seen in Figure 10 strongly indicates that this is the case. From the early efforts at simple tectonic regionalizations for interpreting great-circle dispersion measurements (e.g., Toksöz and Anderson, 1966; Kanamori, 1970) methods of seismic tomography (see Chapter 52 by Curtis and Snieder) have been developed to image lateral variations relative to a reference Earth model for scales extending from borehole measurements, to crustal features, to lithospheric scales, to global 3D inversions. Seismic tomography, although commonly involving approximations to the

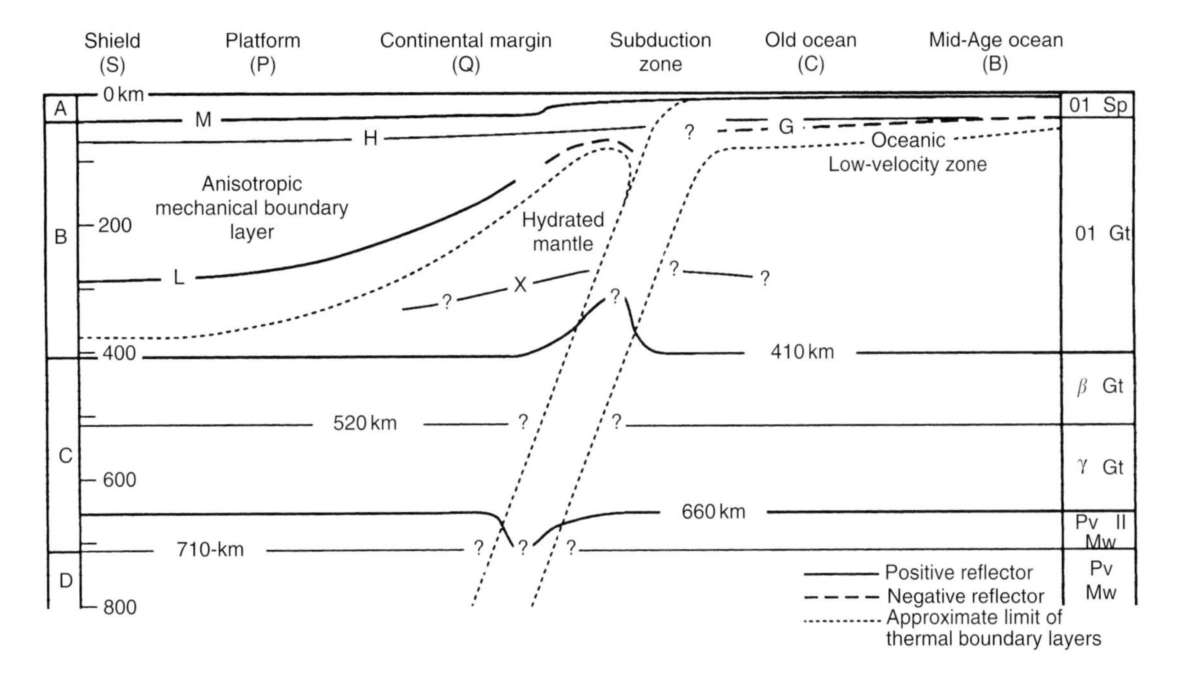

FIGURE 12 Schematic of upper mantle structural variations ranging from continental shield to ocean environments. Major velocity discontinuities are the crust-mantle Moho boundary (M), a continental velocity increase near 80 km depth (H), an oceanic velocity decrease at the base of the oceanic lid (G), a stable continental feature near 200 km (L), an intermittent discontinuity observed near subduction zones (X), and the global structures near 410, 520, and 660 km, as well as a less pronounced feature near 710 km. Cold temperatures near subducting slabs can elevate or depress the phase boundaries at 410 and 660 km, respectively (from Revenaugh and Jordan, 1991b).

propagation effects, is generally applied with few, if any, explicit *a priori* constraints on the structure to be imaged (aside from resolving lengths explicit in the model parameterization), allowing unexpected features to be detected. This enables seismology to characterize the structural elements of complex dynamical systems such as magmatic centers, plumes, subducting slabs, and mantle convective systems. Indeed, the applications are now vast in number, and extensive reviews and books are available providing extensive details about the methodology and results (e.g., Thurber and Aki, 1987; Nolet, 1987; Romanowicz, 1991; Iyer and Hirahara, 1993; Ritzwoller and Lavely, 1995). Without the seismic imaging tool, most of our understanding of dynamical features in the mantle would simply involve conceptual cartoons.

The basic concept of seismic tomography is to use extensive raypath coverage through a given volume of rock to infer heterogeneous properties of the medium such as 3D velocity or attenuation structure. Typically, arrival-time measurements (or amplitudes) for each path are converted into anomalies with respect to predicted times (amplitudes) computed for a background reference model using estimated (or known, in the case of explosions) source origin parameters. The measured anomalies are treated as path integral effects, and are projected onto a spatially parameterized version of the medium, with basis functions in the form of blocks, spherical harmonics, splines, or other general representations. The spatially varying parameters in the medium are inverted for by matching the observations subject to various constraints (smoothness of the medium, designated levels of variance reduction, etc.) with the cross-consistency between paths constructing the image of the heterogeneity (see Chapter 52 by Curtis and Snieder for mathematical details). The resulting images tend to improve in reliability with more uniform raypath sampling of the medium, larger data sets that reduce random errors, improved constraints on source locations, and iterative or nonlinear inversion methods that include the raypath perturbations and fresnel-zone sampling of the medium as the model changes.

Although there had been many earlier studies of velocity heterogeneity in the crust and mantle on various scales, the first (initially reported in 1974) formal application of seismic tomography without *a priori* tectonic regionalizations was to image the lithospheric structure beneath large seismic arrays, where fluctuations in relative arrival times over small distance separations provide constraint on the structure under the array (e.g., Aki and Lee, 1976; Aki *et al.*, 1976, 1977). This led to many inversions for structure under arrays, yielding an understanding that there is small-scale (1–100+ km) velocity heterogeneity almost everywhere in the upper mantle, as reviewed by Aki (1982). In parallel, arrival-time bulletins for large numbers of body waves were also applied to regionalized imaging of mantle downwellings (e.g., Hirahara,

1977; Humphreys *et al.*, 1984) or global mantle P velocity structure (e.g., Sengupta and Toksöz, 1976; Dziewonski *et al.*, 1977; Comer and Clayton, 1983; Dziewonski, 1984). New methods were introduced for solving large matrix problems with as many parameters as necessary for full 3D descriptions of the heterogeneity in the mantle, even when only large-scale structures are allowed for. Rapidly accumulating databases of digital seismic waveforms were also being processed, primarily measuring surface-wave dispersion or free oscillation eigenfrequencies for unprecedented numbers of paths, and these data were also incorporated into inversions for aspherical mantle structure (e.g., Masters *et al.*, 1982; Nakanishi and Anderson, 1982, 1983; Woodhouse and Dziewonski, 1984; Tanimoto and Anderson, 1984).

The first generation of global tomographic models established that surface geology and tectonics are clearly manifested in the velocity heterogeneity of the upper 200 km of the mantle (see Fig. 13). Relatively low velocities underly major upwellings such as midocean ridges and continental rifts (e.g., the Red Sea), systematic increases in velocity are associated with the thickening thermal boundary layer underlying aging ocean crust, low upper mantle velocities are found under continental regions with active tectonic deformation (e.g., the western US), and all large continental cratons have relatively high velocities which may extend as deep as 400 km. The resolution of these early models was only on the order of 5000 km, but it was established that there is strong power in the heterogeneity spectrum at long wavelength, partly as a result of the distribution of continents and ocean, as well as the scale of oceanic plates.

The current generation of tomographic models for the mantle and transition zone have built on these early results, with many advances in data quantity, wave-propagation theory, and types of measurements used in the inversions. P-wave arrival times and fundamental mode surface wave observations have been supplemented by many secondary body-wave arrivals (e.g., *PP*, *PcP*, *PKP*, *PKIKP*, *S*, *SS*, *SSS*, *ScS*, *SKS*) and by higher mode surface waves and split multiplets in the free oscillation spectrum (see Ritzwoller and Lavely, 1995, for extensive references). The numbers of waveforms used for surface-wave inversions have grown to 50 000 and more (e.g., Zhang and Lay, 1996; Trampert and Woodhouse, 1996; Ekström *et al.*, 1997; Laske and Masters, 1996; Boschi and Dziewonski, 1999), and vast body-wave arrival-time data sets have been reprocessed and screened for high quality data (e.g., Engdahl *et al.*, 1998). Global resolution has been improved significantly in the upper mantle, with models having been presented that achieve (or purport to

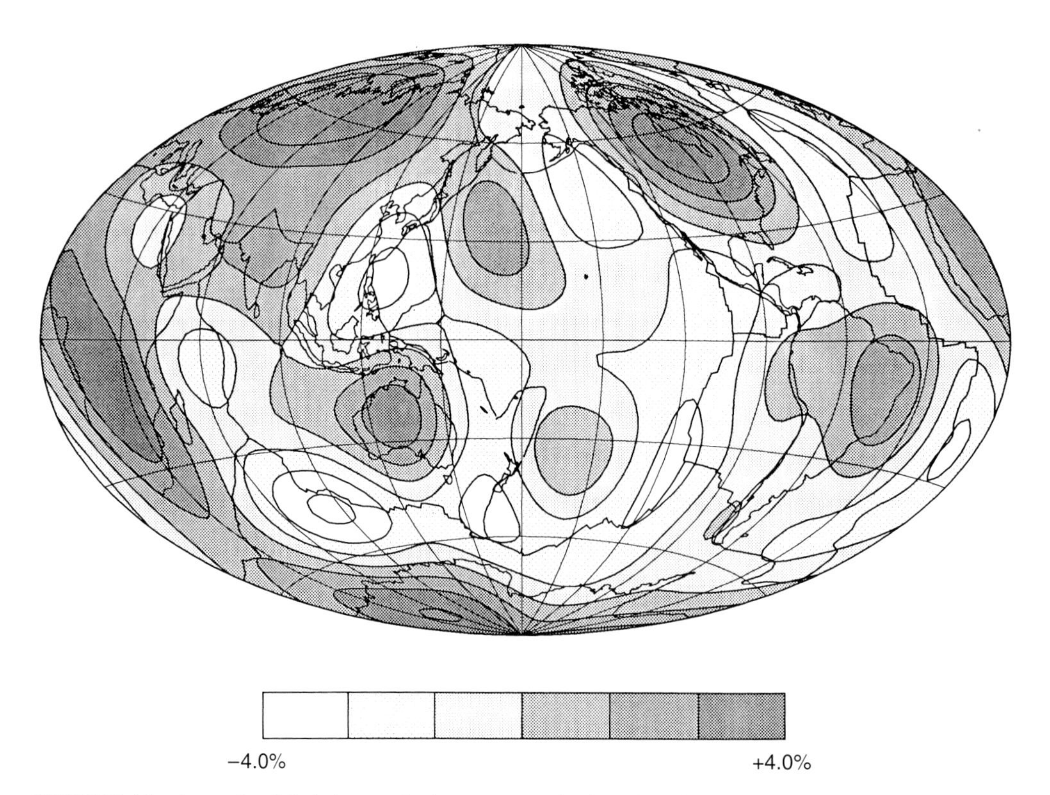

FIGURE 13 An early global shear velocity structure obtained by seismic tomography. This shows the relative perturbations of shear velocity at a depth of 150 km in model M84C. Relatively high velocities underlie old oceanic regions as well as continental regions, while relatively low velocities tend to locate under midocean ridges and in back-arc basins (from Dziewonski, 1989).

achieve) 500–1000 km scale resolution globally (e.g., Inoue *et al.*, 1990; Zhang and Tanimoto, 1991; Vasco *et al.*, 1995; Zhou, 1996; Trampert and Woodhouse, 1996; Ekström *et al*, 1997; Grand *et al.*, 1997; van der Hilst *et al.*, 1997; Boschi and Dziewonski, 1999; Bijwaard, *et al.*, 1999; Ritsema *et al.*, 1999). For the upper mantle there is quite good compatibility between global shear-velocity models expanded in spherical harmonic degrees up to about degree 12–16 (e.g., Masters *et al.*, 1996; Li and Romanowicz, 1996; Su *et al.*, 1994, 1997), and images for a representative recent model are shown in Color Plate 20. This figure shows: improved resolution of the high-velocity structures under continents, which persist to depths greater than 300 km; low velocities at large depths under the fast-spreading mid-Pacific ridge, but not under the slow-spreading mid-Atlantic ridge; a strong decline in velocity fluctuations from the upper mantle to the transition zone, with the latter showing high-velocity features due to slabs in some regions as well as a few localized low-velocity regions under the Pacific. The heterogeneity is still dominated by long-wavelength structure, so the contrast relative to earlier models (see Fig. 13) is not very dramatic.

Higher resolution down to scales of 50–100 km has been achieved in regionalized models that image a limited region involving continental or island arc scale models (e.g., Grand, 1987; Spakman *et al.*, 1989, 1993; Zhou and Clayton, 1990; Spakman, 1991; van der Hilst *et al.*, 1991; van der Hilst *et al.*, 1993, 1995; Wu and Levshin, 1994; Engdahl *et al.*, 1995; Alsina *et al.*, 1996; van der Lee and Nolet, 1997). These regionalized models reveal strong small-scale heterogeneity in structure embedded within the larger provinces imaged by global inversions. There has been steady convergence in the features resolved by high-resolution global inversions and high-resolution regionalized inversions (e.g., Bijwaard *et al.*, 1999). Some of the principal characteristics of upper mantle and transition zone structure revealed by seismic tomography include the deep roots of continents, the structure of subducted slabs and surrounding mantle, and the structure of upwelling plumes.

4.5 Cratonic Roots

The inference that continents have anomalous structure extending down hundreds of kilometers (apparent in Fig. 13 and Color Plate 20), despite their large plate-tectonic motions, suggests the notion that cratonic roots are chemically and thermally stabilized over a thickness of 350 km or more, and this was labeled the tectosphere by Jordan (1975, 1988). Survival of such a deep keel over billions of years of continental drift is surprising, as there is general agreement that the thickness of continental mechanical boundary layer

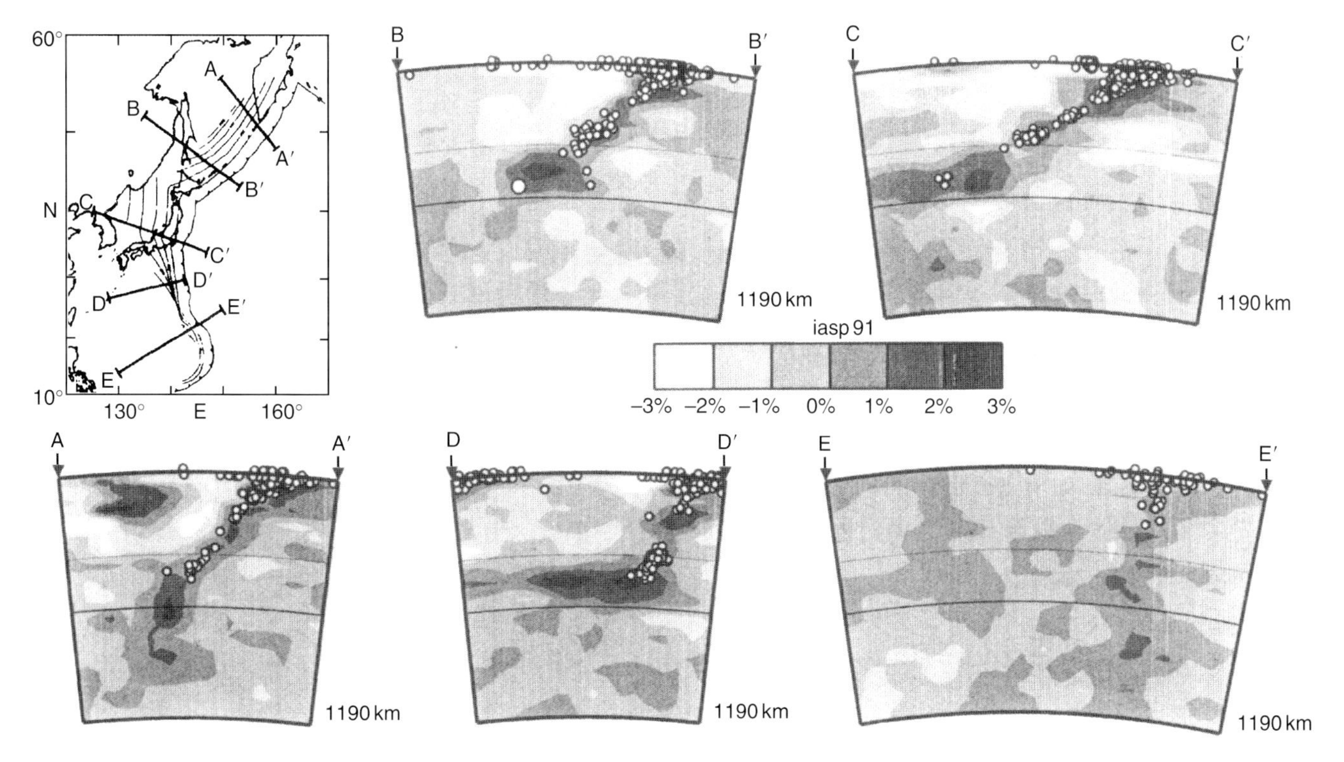

FIGURE 14 Cross sections through three-dimensional tomographic P wave velocity models produced by van der Hilst *et al.* (1991, 1993) for western Pacific subduction zones. The seismic zone in the map is contoured in 100 km intervals. Faster velocity material (relative to the reference iasp91 model) is darker, with shading in 1% velocity variations. Dots are earthquake hypocenters along each of the cross sections. Mantle discontinuities at depths of 410 and 660 km are shown (from Lay, 1994).

(lithosphere) is on the order of 150–200 km thick (e.g., McNutt, 1990). Although the thermal boundary layer may be twice as thick, its deeper portions would normally be sheared and any structure ephemeral, especially for cratonic age continent, unless the deep region is stabilized by unusual chemical buoyancy or by high viscosity, perhaps as a consequence of extensive volatile depletion (e.g., Polet and Anderson, 1995). It is also possible that dynamical processes sustain anomalously cool material beneath continents as a sort of stagnant layer, but this is hard to reconcile with the complex history of motion of the continents.

Statistical analyses support the generalization that cratons have high velocity structure extending to at least 200 km depth (e.g., Wen and Anderson, 1997), but there is controversy over whether the roots extend as deep as the transition zone. Ritsema *et al.* (1998) find that the deep keel of cratons survives during active continental rifting, with the root extending to 300–350 km beneath Eastern Africa, and low-velocity zones beneath the surrounding rifts extend to 300 km or more. This favors the durability of the deep portion of the root even in the presence of active breakup. Upward deflection of the 410-km discontinuity that might be expected for colder than average mantle in the root has not been detected in some careful studies (e.g., Bostock, 1998; Li *et al.*, 1998; Flannagan and Shearer, 1998a; Fouch *et al.*, 2000). Overall, current thinking is that stable archaen and early proterozoic continental crust has an upper mantle root 200–350 km thick; this is a principal feature of the near-surface boundary layer that would not have been recognized without seismic imaging.

4.6 Slabs

Aside from localized regions of partial melting in upwellings, the strongest velocity heterogeneities in the mantle are those associated with subducted oceanic lithosphere. Seismic imaging of descending slabs has involved a great variety of studies (for extensive reviews see Lay, 1994, 1996), many of which have characterized the geometry and magnitude of the velocity heterogeneity of the slab. The velocity anomaly of the slab results from three factors. (1) The relatively low temperature of the slab (with as much as a 1000°C contrast relative to surrounding ambient mantle) intrinsically produces 3–10% high *P*- and *S*-velocity slab signatures. (2) The chemical differentiation and hydration that the oceanic lithosphere has undergone combine with the thermal anomaly to produce distinct phase equilibria in the slab material relative to ambient mantle, which can locally produce 5–6% *P*- and *S*-velocity anomalies, including elevation and depression of transition zone phase boundaries. (3) The subduction process perturbs the local mantle conditions by shear heating, induced flow, and lowering of the melting temperature in the overlying wedge as a result of volatile enrichment caused by hydrous phases extruded from the slab; this can induce partial melting and 5–10% slow *P*- and *S*-velocity anomalies. The purpose of seismic imaging of mantle slabs is thus to study the thermal, chemical, and dynamical structure of subduction.

Seismicity extends as deep as 700 km for rapidly descending slabs (e.g., Wadati, 1935; Isacks *et al.*, 1968; Stark and Frohlich, 1985), and it is believed that all earthquakes below depths of about 100 km must take place in the relatively low temperature environment within subducted material. Pore-fluid assisted brittle failure or frictional sliding are probably responsible for most earthquakes down to depths of 300 km or so, but it is controversial whether there are any free fluids available at greater depths (perhaps as a result of breakdown of hydrous phases), so various other mechanisms have been invoked to account for earthquakes from 400 to 700 km depth. Instabilities associated with transition zone phase transformations are an extensively discussed possibility (e.g., Green and Burnley, 1989; Green *et al.*, 1991; Kirby *et al.*, 1991, 1996; Green and Houston, 1995). Both the distribution of the seismicity and the inferred strain orientations of the Earthquakes have long-been used as direct constraints on the configuration of seismically active deep slab material, the stress orientations in the slab, and the minimum depth of penetration of the downwelling (e.g., Oliver and Isacks, 1967; Isacks *et al.*, 1968; Isacks and Molnar, 1971; Vassiliou, 1984; Apperson and Frohlich, 1987; Burbachk and Frohlich, 1986; Fukao *et al.*, 1987, Chiu *et al.*, 1991).

The cessation of earthquake occurrence at a given depth is not a clear guide as to the fate of the deep slab. Seismicity may terminate due to heating to a critical cut-off temperature for seismicity (e.g., Isacks *et al.*, 1968; Vlaar and Wortel, 1976; Molnar *et al.*, 1979; Wortel, 1982; Brodholt and Stein, 1988; Wortel and Vlaar, 1988), completion of the transition to fine-grained spinel which reduces the strength of the slab (Castle and Creager, 1998a), or, for material that penetrates below 660 km depth, due to the lack of instability for the perovskite phase transformation (e.g., Green and Zhou, 1996). It is known from the history of plate tectonics that far more subducted slab material must be present in the mantle than is illuminated by the current distribution of seismicity, and this material is in varying states of thermal and chemical reassimilation into the mantle (e.g., Richards and Engebretson, 1992; Lithgow-Bertelloni and Richards, 1998). The fate of the vast quantities of aseismic slab material is of central importance to models of chemical, thermal, and dynamical evolution of the planet (e.g., Silver *et al.*, 1988; Jordan *et al.*, 1989; Olson *et al.*, 1990; Lay, 1996). Although it is apparent from both their strain state and geometry that many slabs encounter increasing resistance to descent as they approach 660 km depth, this does not preclude slabs from penetrating into the lower mantle. The endothermic disassociative transformation of spinel-structured (Mg, Fe)$_2$SiO$_4$ into perovskite-structured (Mg,Fe)SiO$_3$ and (Mg, Fe)O near 660 km depth has a negative Clapeyron slope of -2.8 to -4.0 MPaK^{-1} (e.g., Ito and Takahashi, 1989; Ito *et al.*, 1990), which should resist slab penetration, and this is a likely source of resistance to subduction that causes down-dip compression for almost all transition zone earthquakes. The geometry and intrinsic thermal/density anomaly of the slab near 660 km depth will determine whether the resistance from this phase transformation is sufficient to confine the slab to the transition zone or whether the slab components may simply transform to perovskite and penetrate deeper. However, even should the phase transformation be transited, there may also be viscosity increases or chemical contrasts with density increases that further prevent the slab from penetrating deeply into the lower mantle. Imaging of the seismic heterogeneity in the mantle holds the key to determining the fate of subducted slab material.

The anomalous seismic wave transmission properties of slabs were first manifested in anomalous patterns of seismic intensities for deep events across Japan dating back to 1918, but Utsu (1967) and Oliver and Isacks (1967) were the first to clearly articulate the notion that dipping seismogenic zones beneath island arcs involve regions of low seismic attenuation and high seismic wave velocity. This observation prompted a vast number of investigations of relative arrival hyphen;time patterns, relative seismic amplitude and frequency content patterns, and secondary phase observations that constrained gross aspects of slabs and the surrounding mantle near subduction zones (see review by Lay, 1996, and Chapter 42 by Utsu). From these foundations developed the current approaches to imaging slab structures, which involve arrival time tomography, residual sphere modeling using *a priori* slab structures, and analysis of conversions, diffractions, and defocusing effects (Lay, 1996).

Beginning with the work of Hirahara (1977), large-scale 3D models have been developed for all major subduction zones in both regionalized and global tomographic inversions. Images like those in Figure 14 are typical of current inversions that use massive data sets of arrival times from regional and teleseismic observations (e.g., Zhou and Clayton, 1990; Spakman *et al.*, 1993; van der Hilst *et al.*, 1991, 1993; Engdahl *et al.*, 1995; van der Hilst, 1995; Zhao *et al.*, 1995; Widiyantoro and van der Hilst, 1996; and many more references in Lay, 1996). Because the coverage provided by seismic rays is very nonuniform, and because there is structural information lost in the source location process, these models are blurred images of the real structures, and there are many artifacts in the models. Nonetheless, regions of high seismic velocity are found to surround the deep seismicity, and it is accepted that the primary feature being imaged is subducted oceanic slab. In order to better recover the slab velocity structure, *a priori* slab models may be incorporated into the model formulation, leading to improved resolution of the wedge structure (Zhao *et al.*, 1995). Deal *et al.* (1999) and Deal and Nolet (1999) find that many deep artifacts in larger-scale models can also be removed when *a prior is* information about the slab is incorporated into the inversion. This is critical to establishing the depth of slab penetration, for teleseismic raypaths tend to have downward smearing of velocity anomalies. A common tendency is for the velocity anomalies of any possible lower mantle extension of slabs to be significantly reduced relative to the slab anomaly in the transition zone. This actually makes detecting the slab structure much harder.

Global inversions with high-resolution parametrizations in the vicinity of slabs also recover slab images (e.g., Inoue *et al.*, 1990; Fukao *et al.*, 1992; Vasco *et al.*, 1995; Zhou, 1996; van der Hilst *et al.*, 1997; Grand *et al.*, 1997; Bijwaard *et al.*, 1998), and the slab-related features are often consistent with those imaged by regional models. The resulting images do not have a simple end-member behavior for deep slabs. Deep slabs may flatten in the transition zone, as they do underneath the western Mediterranean, the Banda arc, the Solomon Islands, and the Izu-Bonin trench. Beneath Java and the Marianas there appear to be high velocity tabular extensions into the lower mantle, and there may be flattening followed by penetration under the Kuriles and Tonga. In some cases the features imaged in the high resolution tomography appear to connect up to lower mantle features found in lower resolution tomography, as discussed in the next section. The general consensus at this time based on tomographic imaging is that at least some slab material does appear to penetrate deeply into the lower mantle. The complexity of slab structures in the transition zone does appear to reflect the difficulty of penetrating the 660 km discontinuity, and this may be influenced by the slab dip, slab age, extent of trench roll-back, and ambient mantle flow patterns.

Although tomography has proved to be a powerful technique for resolving aspects of slab structure, it appears that increased use of *a priori* constraints in the inversions is required for suppressing artifacts. An alternate strategy for slab imaging that explicitly involves assumption of a slab structure is the residual sphere modeling approach. Residual spheres are simply focal sphere projections of arrival time anomalies at positions corresponding to their raypath azimuths and take-off angles from the source, and such plots have long been used to characterize patterns in seismic data (e.g., Davies and McKenzie, 1969; McKenzie and Julian, 1971; Toksöz *et al.*, 1971; and many other references in Lay, 1996). Observed arrival-time anomalies are strongly affected by the process of locating the event, which removes degree 0 and degree 1 patterns from the observations, and it is important to account for this in the modeling. To further isolate the near-source contributions to the arrival-time patterns, corrections for propagation effects outside the slab (deep mantle and near receiver) must be made, and typically some smoothing is applied to suppress random error. For a specified source location in a slab structure 3D raytracing or numerical methods are used to predict the arrival time at each station, and the model times are processed by an event location filter and any smoothing, with slab model parameters being perturbed to match the data.

Jordan (1977) initiated the complete residual sphere modeling formalism, which was further developed by Creager and Jordan (1984, 1986). These studies demonstrated the sensitivity of the method to both upper mantle and transition zone slab geometry and velocity heterogeneity, as well as to geometry of any steeply dipping slab extension into the lower mantle. Provocative results based on both P and S wave modeling suggested that slab penetration to depths of at least 1000 km with little distortion other than steepening dip occurs in the Kurile, Marianas, and Japan arcs. Additional applications of the method were presented by Fischer *et al.* (1988, 1989), Zhou and Anderson (1989), Zhou *et al.* (1990), Boyd and Creager (1991), Ding and Grand (1992) and Pankow and Lay (1999). The method makes very explicit the limitations of arrival-time data, as event location effects have a huge effect on relative arrival-time anomalies if the data coverage is limited (particularly true if only teleseismic observations are used). Tomographic methods will be strongly biased by this unless the data coverage is such that residual patterns faithfully preserve the slab effects (which may be true when extensive upgoing and downgoing data are included, but not otherwise). Residual sphere modeling also makes clear the importance of deep mantle and receiver corrections, and early applications did not adequately address this issue. In fact, it has been shown that for S waves much of what was initially attributed to near-source effect is eliminated when improved path corrections are applied (e.g., Schwartz *et al.*, 1991; Gaherty *et al.*, 1991; Pankow and Lay, 1999). As global tomographic models improve, this will become less of a problem. Analysis of differential residual spheres for events in the same slab, as first introduced by Toksöz *et al.* (1971) is one approach that has been pursued to suppress distant effects rather completely (e.g., Takei and Suetsugu, 1989; Okano and Suetsugu, 1992; Ding and Grand, 1992; Pankow and Lay, 1999). These studies indicate that in some cases slabs may penetrate to depths of 800 km or more, but significant slab broadening may occur, as well as reduction of velocity heterogeneity to on the order of 2%, much weaker than in early residual sphere studies, and similar to the weak heterogeneity inferred when *a priori* slab structures are introduced into tomographic modeling.

In addition to arrival time imaging, the velocity gradients and internal structure of subducting slabs have been extensively investigated using waveform conversions, diffractions and defocusing effects (see Lay, 1996 for a review). Conversions of P-to-S and S-to-P energy have been used to constrain the velocity contrasts, dip, and sharpness of slab features in many studies since about 1953 (e.g., Katsumata, 1953; Okada, 1971; Nakanishi *et al.*, 1981; Matsuzawa *et al.*, 1990; Helffrich and Stein, 1993; and many other references in Lay, 1996). Such observations are used to explore the detailed structure of the layered slab, with evidence for both high- and low-velocity layers near the top of the slab (possibly involving the eclogitic crust) and the existence of velocity contrasts within the slab that may be associated with double layers of seismicity detected in many regions. Imaging internal slab structure is also of interest for detecting the position of phase boundaries in the slab (key temperature indicators) as well as possible metastable depression of such boundaries (e.g., Solomon and U, 1975; Roecker, 1985; Iidaka and Suetsugu, 1992; Collier and Helffrich, 1997; Koper *et al.*, 1998; Flannagan and Shearer, 1998b). As indicated earlier, there are conflicting results on the nature of major phase-boundary deflections inside the slab, and a thin wedge of metastable olivine has not been resolved. Diffraction and defocusing effects have been used to bound slab velocity gradients in many studies (e.g., Sleep, 1973; Cormier, 1989; Weber, 1990; Sekiguchi, 1992; Silver and Chan, 1986; Vidale, 1987; Gaherty *et al.*, 1991; Vidale *et al.*, 1991), but computational limitations and the difficulties of accounting for wave-propagation effects on amplitudes and waveforms have limited the contribution of such studies.

4.7 Upwellings and Plumes

Seismic methods intrinsically image downwelling high-velocity structures like slabs better than hot, upwelling low-velocity structures for two reasons: slabs often have deep focus earthquake sources in the cold downwelling that provide improved raypath coverage of the surrounding structures, whereas hot upwellings tend to have only shallow activity; and wave propagation through hot, low seismic velocity regions involves significant wave front healing, attenuation, and diffraction effects that obscure the travel-time signature. Nonetheless, regions of expected upwelling beneath major volcanic centers have been the target of many tomographic investigations, for the purpose of ascertaining the size and geometry of the zone of partial melting. Crustal structures beneath volcanic centers are described in Chapter 25 by McNutt and Chapter 26 by Benz *et al.*, and only mantle features are considered here.

Global tomography provides limited resolution of presumed upwelling regions at this point, in part due to the spatially localized nature of most upwellings. The best-imaged regions are the large-scale features associated with the mid-ocean ridge system. Surface wave and long-period body wave tomography indicate that low velocity material underlies almost all of the midocean ridge system on scale-lengths that can be resolved even in models with only 1000 km scale resolution (e.g., Zhang and Tanimoto, 1991; Su and Dziewonski, 1997; Trampert and Woodhouse, 1996; Ekström *et al.*, 1997; Boschi and Dziewonski, 1999; Ritsema *et al.*, 1999). The depth extent of the low-velocity region under ridges is less-well resolved, with some studies indicating concentration of low-velocity material in the upper 100 km (e.g., Zhang and Tanimoto, 1991, 1992), but the more compelling case being that low-velocity material extends down to at least 250 km below fast-spreading ridges such as the Pacific rise (e.g., Su *et al.*, 1992; Su and Dziewonski, 1997). Global *P*-wave tomography models typically have poor sampling of midocean ridge systems, and little vertical resolution of upper mantle structure, in contrast to their resolution of downwellings. The lack of nearby stations causes strong trade-offs between mantle structure and source parameters for near-ridge events. The existence of low velocity material at depths down to 400 km beneath ridges is strongly supported by body-wave analysis of *SS*, *SSS* and *SSSS* phases (e.g., Graves and Helmberger, 1988; Grand *et al.*, 1997). However, transition zone discontinuities beneath the Pacific rise do not appear to have anomalous depths as might be expected if thermal anomalies extend through to the lower mantle (e.g., Shen *et al.*, 1998; Lee and Grand, 1996). The prevailing notion is that most midocean ridge upwellings are relatively passive in nature, with partial melting occurring as a result of pressure reduction as material rises super-adiabatically into the ridge.

Acute heterogeneity and concentrations of low-velocity upwellings are imaged in the upper mantle wedge above subducting slabs, with the best detail being provided by regional-scale tomographic inversions (e.g., Zhao *et al.*, 1992, 1994, 1995, 1997). The low velocity regions are so pronounced that they likely involve significant partial melting, which is thought to be the result of lowering of the melting temperature due to the presence of volatiles released from the subducting slab. Upwellings are also inferred from low-velocity structure found below continental rift zones such as the East African rift. These appear to extend deep into the upper mantle as well, with low velocity *P* and *S* anomalies as deep as 350 km (e.g., Su and Dziewonski, 1997; Ritsema *et al.*, 1998). Deep structures that may correspond to frozen-in plume features have also been imaged by tomography under South America (Van Decar *et al.*, 1995).

There has also been great interest in studying the structure under major hotspots such as Iceland, Yellowstone, and Hawaii. Global surface wave tomography has not resolved the 100–200 scale structures of relevance, but there is some suggestion that many large hotspots are underlain by low-velocity material in the upper 200 km of the mantle (Zhang and Tanimoto, 1991). Tomographic inversion of *P*-wave arrival anomalies suggests that low velocities extend to at least 400 km depths below Iceland as a large cylindrical structure with a radius of about 150 km (Tryggvason *et al.*, 1983; Wolfe *et al.*, 1997). Shear-wave anomalies in this structure are as large as -4% whereas *P*-wave anomalies are -2%. This requires temperature anomalies of on the order of 200–300°C. Shen *et al.* (1998) examine topography of the 410 and 660-km discontinuities beneath Iceland as indicated by *Pds* conversions in receiver functions, and find that the transition zone is about 20 km thinner than average, which they interpret as evidence for a lower mantle origin of the thermal upwelling. Similar deflections of the transition zone discontinuities are observed along the Yellowstone hotspot track (e.g., Dueker, and Sheehan, 1997), although the primary low velocity features imaged by tomography are concentrated in the range 50–200 km beneath the Snake River Plain (Dueker and Humphreys, 1990; Saltzer and Humphreys, 1997). Travel-time anomalies for *P* waves beneath other hotspots have been interpreted as being caused by plumes extending through the transition zone (Nataf and VanDecar, 1993).

4.8 Laterally Varying Anisotropy

The average upper mantle velocity structure can be characterized as transversely isotropic, as described above. A model like PREM provides a first-order fit to global Love and Rayleigh wave dispersion observations, as well as equivalent toroidal and spheroidal eigenfrequencies. However, the very nature of anisotropy, involving large-scale alignments of intrinsically anisotropic minerals such as olivine, systems of oriented cracks, or sheared fabrics or lamellae, is such that lateral variations in anisotropic properties are expected throughout the crust and upper mantle. This is actually found to be the case, with lateral variations in the anisotropic structure of the lithosphere and asthenosphere for both continental and oceanic regions having been imaged by a variety of body-wave and surface-wave methods.

Global and plate-scale tomographic inversions of surface wave dispersion measurements have incorporated anisotropic structure either in the reference model (by using PREM), or explicitly in the parameterization of the model (e.g., Nataf *et al.*, 1984, 1986; Nishimura and Forsyth, 1989; Montagner and Tanimoto, 1990, 1991; Ekström and Dziewonski, 1998). Ekström and Dziewonski (1998) demonstrate convincingly that the contributions to surface-wave travel times from anisotropy variations are significant, and allowance for spatial variations in anisotropy is both justified and necessary. Ekström and Dziewonski (1998) find that the radially varying transverse isotropy of PREM is a good average model for both oceanic and continental regions, even when there are substantial baseline shifts in the average velocities. The main exception is the central Pacific plate where there are strong geographical variations in radial anisotropy, with maximum anisotropy differences being much stronger than in PREM around 150 km deep. It appears that actual anisotropy in the Pacific plate is not simply transverse isotropy, but azimuthal anisotropy (which causes wave speed to vary with azimuth, not just polarization), but mapping of azimuthal anisotropy is still rather poorly constrained (e.g., Nishimura and Forsyth, 1988, 1989; Montager and Tanimoto, 1991; Ekström and Dziewonski, 1988). As constraints on the geometry and variations in anisotropy improve, it will be possible to relate the anisotropic observations to models of shear flow that induce lattice preferred orientations (LPO) or fabrics in the upper mantle lithosphere and asthenosphere (e.g., Tanimoto and Anderson, 1984; Montagner, 1994).

Continental observations of seismic anisotropy have primarily involved body-wave measurements of shear-wave splitting and azimuthal *Pn* travel times, although surface-wave observations in the continents do support the existence of upper mantle anisotropy (e.g., Gaherty *et al.*, 1996). Silver (1996) and Savage (1999) review the observations of upper mantle anisotropy and the basic hypotheses for interpreting anisotropy as a result of either frozen or actively supported fabrics in the rocks. The most extensively analyzed phases have been *SKS* phases, which traverse the core as a *P* phase, and hence have a known initial polarization at the core–mantle boundary on their path to the surface. Vinnik *et al.* (1984, 1989, 1992), Kind *et al.* (1985) and Silver and Chan (1988, 1991) established methods for analyzing the splitting of *SKS* signals to determine the orientation and magnitude of azimuthal anisotropy, typically under the assumption of a horizontal symmetry axis for hexagonal crystals. A large number of analyses of *S* and *SKS* splitting to determine receiver and source-side anisotropic structure have ensued (see Savage, 1999 for many references), demonstrating that the Earth's lithosphere has extensive azimuthal anisotropy, sometimes with large-scale coherence, and sometimes with small-scale regional variations. The rapid variations that are sometimes observed require that the anisotropy be concentrated in the shallow mantle, but the magnitude of splitting (values as large as 2–3 sec have been observed) requires that the anisotropy be as strong as 2.5–3% over the upper 250 km of the mantle. Analysis of splitting for earthquakes at different depths suggest anisotropy of 0.5–2% for the mantle above and below slabs and up to 5% within slabs (e.g., Shih *et al.*, 1991; Kaneshima and Silver, 1995; Fouch and Fischer, 1996; Hiramatsu *et al.*, 1997). Splitting generally does not increase with source depth for events deeper than 400 km (e.g., Kaneshima and Silver, 1995; Fouch and Fischer, 1996; Fischer and Wiens, 1996), however, there is limited evidence for transition zone anisotropy in converted phases (Vinnik and Kind, 1993; Vinnik and Montagner, 1996) and from modeling body-wave and normal-mode observations (Montagner and Kennett, 1996). As yet, there is not a fully satisfying reconciliation of the global model of transverse isotropy provided by PREM and the extensive observations of laterally varying azimuthal anisotropy provided by body-wave studies. It appears that the mantle requires a more general parameterization than transverse isotropy, but the data feeding into global tomographic inversions and reference Earth models are not yet sufficient to constrain the complete anisotropic orientation.

4.9 Attenuation

The foregoing discussion has emphasized elastic properties of the crust and mantle, primarily constrained by elastic wave travel times. However, the Earth is not perfectly elastic, and seismic waves of all types undergo anelastic attenuation as they propagate. This results in amplitude decay at a rate exceeding that caused by geometric spreading, along with slight dependence of seismic velocity on frequency, or dispersion. In general, the mechanisms that cause anelastic losses are thermally activated microscale processes such as dislocation motions and grain boundary interactions (see Anderson, 1967b; Minster, 1980; Minster and Anderson, 1981). One of the primary goals of studying attenuation is to constrain thermal structure inside the planet. The details of microscale processes that cause attenuation are not resolvable with seismic waves, which intrinsically average large volumes, so phenomenological models are used to account for the macroscopic effects of anelasticity (see Lay and Wallace, 1995). The most common parametrization is in the form of the quality factor, Q, defined as the inverse of the fractional loss of energy, E, per cycle of oscillation: $1/Q = -\Delta E/2\pi E$. As Q increases, the attenuation is smaller, and for infinite Q, the elastic solutions are retrieved. Attenuation quality factors can be defined for all types of seismic waves, with the corresponding value depending on the specific path through the Earth, the sense of particle motion involved in the wave, and the frequency of the vibration. Suitably designed experiments have allowed Q values to be estimated for body waves and surface waves for more than four decades (e.g., Sato, 1958; Bath and Lopez, 1962; Anderson, 1963; Press, 1964; Anderson and Archambeau, 1964; Anderson and Kovach, 1964; Ben-Menahem, 1965). Generally, *P*- wave quality factors (Q_α) are higher than *S*-wave quality factors (Q_β) (see Fig. 3), and it is believed that most intrinsic attenuation is related to shear processes associated with lattice defects and grain boundaries (for a Poisson solid with all losses due to shearing mechanisms, $Q_\alpha = 9/4 Q_\beta$) (e.g., Anderson *et al.*, 1965). Observationally, Q for seismic waves in the mantle is not strongly dependent on frequency over the band 0.001–0.2 Hz, with typical Q_β values of 100–500 or so, but at higher frequency the attenuation is lower, and Q increases with frequency.

The existence of attenuation modifies the equations of motion for Earth materials from those for pure linear elasticity. However, for the moderate Q values found in the Earth, good approximations of the solution for the full viscoelastic equations can be obtained by perturbation of the elastic solutions with effective attenuation operators (see Lay and Wallace, 1995). For example, the amplitude spectrum for a propagating *P* or *S* wave is modified by a term like: $A(f) = A_0 e^{(-\pi f t^*)}$, where

$$t^*(f) = \int_s \frac{ds}{v(s)Q(s,f)} \tag{3}$$

with s being a variable along the path, $v(s)$ being the velocity encountered on the path, and $Q(s,f)$ being the spatially and frequency varying attenuation factor. The finite Q encountered by a seismic wave, even if it is approximately constant over a frequency band, results in frequency dependence of seismic wave velocity, with higher frequencies sensing an unrelaxed effective modulus and having higher velocities than lower frequencies, which sense a relaxed effective modulus (e.g., Futterman, 1962; Jeffreys, 1965; Liu *et al.*, 1976; Kanamori and Anderson, 1977; Anderson *et al.*, 1976). The frequency dependence of physical mechanisms that cause attenuation is usually treated using the standard linear solid model, for which each physical mechanism influences a characteristic frequency band defined by a Debye peak, and a propagating wave encounters a distribution of distinct mechanisms that superimpose to produce an effective absorption band over a range of frequencies. The systematic variation of elastic velocities with frequency is observed in the Earth, and must be accounted for in order to reconcile Earth models based on short-period body waves with those based on long-period normal modes (e.g., Hart *et al.*, 1977; Montagner and Kennett, 1996). The PREM model, which has a radially varying constant Q structure, thus has explicitly frequency dependent seismic velocities (Dziewonski and Anderson, 1981).

There have been extensive measurements of attenuation for globally sampling data sets of surface waves and free oscillations (e.g., Kanamori, 1970; Anderson and Hart, 1978; Sailor and Dziewonski, 1978; Nakanishi, 1979; Dziewonski and Anderson, 1981; Masters and Gilbert, 1983; Smith and Masters, 1989; Roult *et al.*, 1990; Romanowicz, 1990; Widmer *et al.*, 1991; Durek *et al.*, 1993; Romanowicz, 1995; Durek and Ekström, 1996). Average Q_β models for the upper mantle are shown in Figure 15. The lithosphere is relatively high Q, but there are Q_β values of 60–70 in the vicinity of the low velocity zone, which suggests a connection between partial melting and strong attenuation. The Q of the lower mantle is much higher (Fig. 3), thus most attenuation is believed to take place along the upper mantle portion of wave paths. The more recent of the global studies have produced

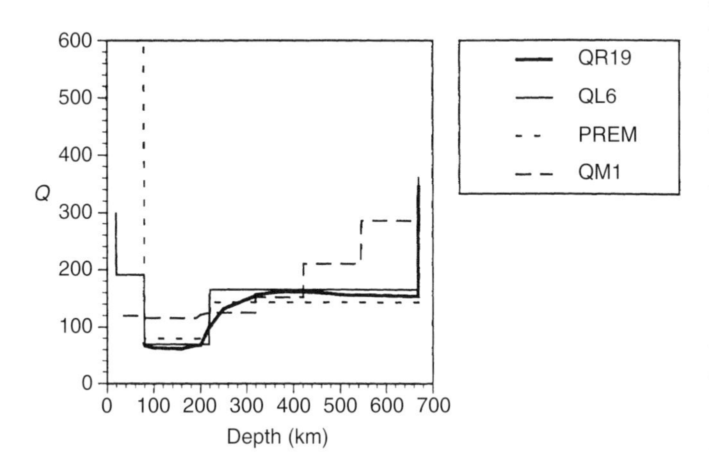

FIGURE 15 Spherically averaged Q models for the upper mantle from various studies of surface wave and normal mode attenuation (from Romanowicz, 1995).

low resolution (up to degree 6) tomographic models of aspherical attenuation that show a negative correlation with models of shear velocity variations as would be expected for temperature effects (e.g. Durek *et al.*, 1993; Romanowicz, 1995). Relatively low Q values are found in the upper mantle under the Pacific and under Eastern Africa. In general, there are higher Q regions underlying most continental areas than underlying oceanic regions (e.g., Dziewonski and Steim, 1983; Romanowicz, 1995). The lateral variations in attenuation are particularly important to account for when inferring thermal anomalies from seismic tomography, because the dispersive effect in low Q regions, which tend to be hot and low seismic velocity, accentuates the velocity anomaly for a given thermal heterogeneity (e.g, Karato, 1993). The tomographic inversions for attenuation are complicated by the need for correcting for focusing and defocusing effects of velocity heterogeneity, and it is likely that joint inversions for velocity structure and attenuation structure will be pursued in the future.

Higher spatial resolution measurements of attenuation for body and surface waves indicate that Q varies laterally by an order of magnitude, especially in the upper mantle. This means that the average radial Q model for PREM, which has an average value of $Q_\beta = 128$ in the upper 400 km of the mantle, is only useful for normal modes and long-path surface waves, which involve extensive lateral averaging. For body waves in the period range 30–1 sec, average values of t^* are 1 ± 0.5 sec for P waves and 4 ± 2 sec for S waves, whereas at shorter periods of 1–0.1 sec the t^* values may decrease to 0.1 or 0.2 sec on specific paths (e.g., Sipkin and Jordan, 1980; Der *et al.*, 1980; Taylor *et al.*, 1986; Chan and Der, 1989; Flanagan and Wiens, 1990; Sheehan and Solomon, 1992; Ding and Grand, 1993; Bhattacharyya *et al.*, 1996; and many more). The exponential form of the t^* operator still means that high frequencies are strongly attenuated on paths through the mantle. The lateral variations in attenuation beneath continental areas have been of particular importance for nuclear explosion monitoring, with estimates of explosion yields trading off directly with estimates of the seismic attenuation of the high frequency P waves from explosions. This has prompted extensive work on the variation and frequency dependence of attenuation for periods near 1 sec (see discussions by Bache, 1985; Burger *et al.*, 1987). Overall, models for attenuation in the mantle remain relatively primitive, but first-order mapping of the structure has been achieved.

5. Lower Mantle

The lower mantle extends from 800 to 2890 km deep, where the Earth's primary internal compositional contrast exists at the core–mantle boundary. For most of this depth range, the structure appears to be relatively uniform, free of major radially symmetric boundaries, and quite plausibly composed of uniform composition of $(Mg_{0.9}\ Fe_{0.1})SiO_3$ perovskite, with

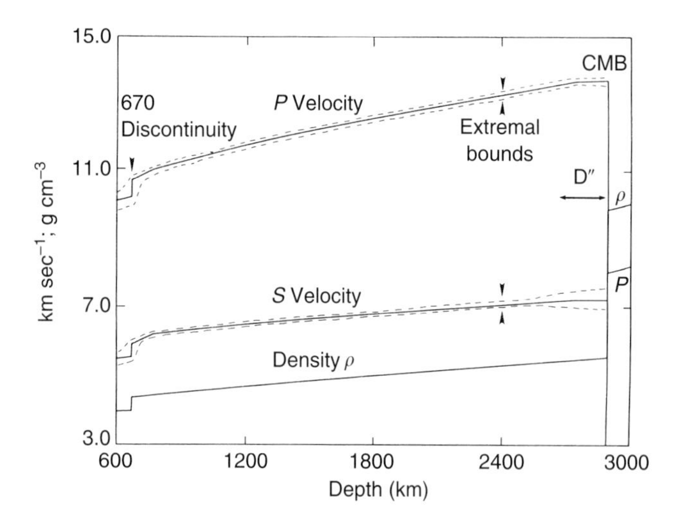

FIGURE 16 Variation of seismic velocities and density through the lower mantle for model PREM, along with extremal bounds that indicate the confidence interval for spherically averaged models based on travel-time data. The D″ region is the lowermost 200 km of the lower mantle overlying the core–mantle boundary (CMB) (from Lay, 1989).

(Mg, Fe)O and minor additional components such as SiO_2-stishovite and calcium perovskite. No major phase changes in these primary components are expected over the pressure range of the lower mantle. Seismic velocity models for the lower mantle (Fig. 16) differ very little from the classical models of Gutenberg and Jeffreys except near the 660 km discontinuity and at the base of the mantle, largely because the travel-time curves for *P* and *S* waves are remarkably free of complexity in the range of 30–100 degrees and therefore provide tight constraints on the structure (Fig. 1). Nonetheless, there have been substantial contributions to our understanding of mantle dynamics and chemical evolution as a result of detailed studies of lower mantle structure, and vigorous research is being pursued to map out small aspherical structures as well as detailed structure of the base of the transition zone and in the lowermost 200 km of the lower mantle, which Bullen (1949) identified as an inhomogeneous zone and labeled the D″ region.

5.1 Radially Symmetric Structure/Discontinuities

The average lower-mantle properties have been determined by both classical arrival time inversion and by normal mode analysis, with the latter refining early estimates of the density structure that had been based on velocity–density systematics, integral constraints on Earth's mass and moment of inertia, and integration of seismic velocity models using the Adams–Williamson equation (Adams and Williamson, 1923). Global observations of body wave travel times, measurements of slopes of the travel-time curves by seismic array analyses, and measurements of normal-mode eigenfrequencies proliferated in the 1960s to 1980s, with many radially symmetric Earth models for the lower mantle being produced (e.g., Chinnery and Toksöz, 1967; Hales *et al.*, 1968; Herrin, 1968; Johnson, 1969; Hales and Roberts, 1970; Randall, 1971; Jordan and Anderson, 1974; Gilbert and Dziewonski, 1975; Dziewonski *et al.*, 1975; Sengupta and Julian, 1978; Uhrhammer, 1978; Dziewonski and Anderson, 1981; Kennett and Engdahl, 1991; Morelli and Dziewonski, 1993). Although the lower mantle variations among these models are less than 1%, there is still great importance in having an accurate reference model both for earthquake location procedures and for use as a background model in tomographic analyses (reference model structures are often optimized in the very process of producing a tomographic model). Thus, efforts to improve the average lower mantle parameters continue, with increasing quantities of data and variety of phase types being incorporated into the analysis (Masters *et al.*, 1999). All of these average lower mantle models have smooth velocity gradients with no significant structures other than reductions of velocity gradients in the D″ region, as in the PREM model (Fig. 16).

The small variations between lower mantle radial velocity models have still received much attention because any departure from homogeneity (as expected for self-compression of uniform composition material) would have major implications for possible chemical layering or phase changes. *P* and *S* velocities throughout the lower mantle above the D″ region are bounded to within about $\pm 0.1\,\mathrm{km\,sec^{-1}}$ in terms of an average model (e.g., Lee and Johnson, 1984). This tight bound (Fig. 16) is consistent with the finding by Burdick and Powell (1980), that small features in ray parameter estimates from seismic arrays tend to vary azimuthally, and are not globally representative, with on average very smooth structure in the lower mantle being preferred as an average model. There have been observations of reflections and converted phases from a velocity or impedance contrast near 900 km depth near subduction zones (e.g., Revenaugh and Jordan, 1991; Kawakatsu and Niu, 1994), but this appears to be a strongly laterally varying structure (Shearer, 1993), and may be associated with steeply dipping mantle heterogeneities (Niu and Kawakatsu, 1997; Kaneshima and Helffrich, 1998; Vinnik *et al.*, 1998; Castle and Creager, 1999). At this time there is no compelling evidence for significant laterally extensive layering of the lower mantle except near the top of the D″ region.

The lower mantle has relatively high *Q* values for seismic waves (see Fig. 3), and mapping any lateral variations is very difficult due to the strong regional variations in the upper mantle. Normal modes and averaged body-wave attenuation measurements place some constraints on the average *Q* values, but it is possible to satisfy most data with extremely simple models (e.g., Dziewonski and Anderson, 1981; Masters and Gilbert, 1983). Although there is some evidence for a low *Q* zone at the base of the mantle, this is not well resolved because of strong trade-offs with velocity gradients in the D″ region.

5.2 D″

The existence of the Earth's magnetic field, caused by ongoing thermochemical convection and an associated geodynamo in the conducting iron alloy core, requires that heat is fluxing from the core into the mantle. While estimates of the total heat budget of the mantle indicate that heating from below comprises only 10–30% of the total mantle heat flux (the balance is from residual internal heat and radioactive decay in the mantle), a thermal boundary layer should exist at the base of the mantle, with a rapid increase in temperature across a conductive boundary layer. Like the Earth's other major thermal boundary layer in the lithosphere, the boundary layer at the base of the mantle is likely to be undergoing strong lateral and vertical flow, as upwellings produced by thermal boundary layer instabilities drain hot material from the boundary layer and downwellings replace it with cooler material. As a hot, low viscosity boundary layer, it is likely that there is much more small-scale structure in the dynamical regime than is found in the relatively stiff lithosphere. It is generally accepted that heterogeneity within the thermal boundary layer is partially responsible for the inhomogeneity detected by Dahm (1934) and confirmed by Bullen (1949) as well as for the reduced seismic velocity gradients in the region found in some of the average Earth models (e.g., Stacey and Loper, 1983; Lay and Helmberger, 1983; Doorbos *et al.*, 1986; Loper and Lay, 1995). However, the juxtaposition of the boundary layer adjacent to the largest density contrast in the Earth (the density jump across the core–mantle boundary is larger than that at the surface of the Earth) heightens the probability that there is also chemical heterogeneity in the D″ region, caused by either density-stratified residue from the Earth's core formation process, ongoing chemical differentiation of the mantle, or even chemical reactions between the core and mantle (e.g., Lay, 1989; Knittle and Jeanloz, 1989; Goarant *et al.*, 1992; Jeanloz, 1993; Manga and Jeanloz, 1996). Because of its importance for unraveling the thermal and chemical processes in the mantle many seismological studies have characterized the structure of the D″ region (see the survey papers in Gurnis *et al.*, 1998, and reviews by Loper and Lay, 1995; Weber *et al.*, 1996; Lay *et al.*, 1998; Garnero, 2000).

The existence of the great velocity reductions across the core–mantle boundary (CMB) causes seismic wave energy to diffract into the geometric shadow zone at distances greater than 100°. Waves diffracted along the core are sensitive to the absolute velocities and the velocity gradients in the D″ region, and have long been studied to constrain average and laterally varying structure (e.g., Alexander and Phinney, 1966; Sacks, 1966; Bolt *et al.*, 1970, Mondt, 1977; Doornbos and Mondt, 1979; Mula and Müller, 1980; Wysession and Okal, 1989; Wysession *et al.*, 1992; Valenzuela and Wysession, 1998, and many more). These studies demonstrate that no single velocity structure sufficiently characterizes D″ everywhere, and that in some cases there are strong negative velocity gradients in D″ while in other places there are near-zero or positive velocity gradients. There are also changes in the relative perturbation of *P* and *S* velocities that are likely due to mineralogical or textural origin (e.g., Wysession *et al.*, 1999). These phases involve extensive lateral averaging of what appears to be a region rich in small-scale structure, but they provide important input into large-scale tomographic models for D″ because of their extensive spatial coverage (e.g., Kuo and Wu, 1997; Kuo *et al.*, 2000; Castle *et al.*, 2000).

Both *P*- and *S*-velocity structure have velocity discontinuities, or zones of rapid velocity increase over depth ranges of no more than 30–50 km, at many locations near the top of the D″ region (the top of D″ is not precisely defined, and many take it to correspond to either where there is a discontinuity or the onset of a change in velocity gradient, which may be from 50 to 350 km above the CMB). The velocity increases are detected by reflections and triplications, which arrive ahead of the core-reflected *PcP* and *ScS* phases (e.g., Wright and Lyons, 1975; Lay and Helmberger, 1983; Wright *et al.*, 1985; Young and Lay, 1987, 1990; Gaherty and Lay, 1992; Weber and Davis, 1990; Houard and Nataf, 1993; Weber, 1993; Kendall and Shearer, 1994; Kendall and Nangini, 1996; Ding and Helmberger, 1997; Thomas and Weber, 1997; Kohler *et al.*, 1997; Reasoner and Revenaugh, 1999; and many others reviewed by Wysession *et al.*, 1998). The reflector varies in depth by several hundred kilometers (e.g., Kendall and Shearer, 1994), and appears to have short wavelength lateral variations of on the order of 100 km that may produce scattering rather than simple reflections (e.g., Weber, 1993; Kruger *et al.*, 1995; Lay *et al.*, 1997; Scherbaum *et al.*, 1997; Yamada and Nakanishi, 1998; Freybourger *et al.*, 1999; Emery *et al.*, 1999). It has been argued that the discontinuities are actually globally extensive, caused by a phase change, with lateral variations in depth and strength (velocity increases vary from 1 to 3%) being the result of lateral temperature variations and interactions with upwelling and downwelling flow (e.g., Nataf and Houard, 1993; Sidorin *et al.*, 1998, 1999). Others have questioned whether there is a first-order discontinuity or simply scattering from strong velocity heterogeneities as imaged in long-wavelength tomography models (e.g., Liu *et al.*, 1998). The latter possibility requires large *ad hoc* increases in the magnitude of the tomographic heterogeneities, and does not appear to be a viable explanation for the broadband reflections that are observed. However, thin high or low velocity lamella models may fit some *P*-wave observations (Weber, 1994; Thomas *et al.*, 1998; Freybourger *et al.*, 1999). Thus, at present, the interpretation of the D″ discontinuity is uncertain, and work continues on characterizing this structure and its dynamical significance. Of particular importance will be determination of whether there is any density increase in D″

that might represent a chemical change or phase change, either of which could strongly affect the dynamics of the boundary layer (e.g., Sleep, 1988; Kellogg, 1997; Montague *et al.*, 1998; Hansen and Yuen, 1988; Tackley, 1999).

The large-scale variations in D″ imaged by seismic tomography have surprising predominant degree 2 and 3 spherical harmonic components (e.g., Su *et al.*, 1994; Li and Romanowicz, 1996; Masters *et al.*, 1996; Dziewonski *et al.*, 1996; Kuo and Wu, 1997; Liu and Dziewonski, 1998; Kuo *et al.*, 2000). These models show consistent high shear velocities rimming the Pacific plate, with low velocities beneath the central Pacific and the southeastern Atlantic and southern Africa. This geometry produces a correlation between areas of slab subduction over the past several hundred million years (e.g., Lithgow-Bertelloni and Richards, 1998) and fast regions of D″, which could result if slabs sink to the base of the mantle and retain enough thermal anomaly to produce high seismic velocities. Similarly, the low-velocity regions of D″ are generally below hotspot regions at the surface, suggesting that D″ upwellings may penetrate all the way to the Earth's surface. This will be discussed further below.

Small-scale variations in D″, with about 1% heterogeneities on scale-lengths of about 10 km are also present. This was first established by interpretation of short-period precursors to *PKP* phases (e.g., Cleary and Haddon, 1972; Haddon and Cleary, 1974; Doornbos, 1976; Bataille and Flatté, 1988; Bataille *et al.*, 1990; Hedlin *et al.*, 1997; Cormier, 1999). It has generally been believed that the levels of heterogeneity increase in D″ relative to the overlying mantle, but there is weak evidence that small-scale structure in D″ is not distinctive (e.g., Hedlin *et al.*, 1997). It is clear that some of the strongest scattering, involving much larger velocity heterogeneities, does arise within D″ (Vidale and Hedlin, 1998; Wen and Helmberger, 1998), and this is likely associated with an intermittent thin layer of partial melt that causes an ultra-low velocity zone (ULVZ) just above the CMB. The bandwidth of the signals used in scattering analyses controls the sensitivity to scatterers of different dimensions, and analysis of broadband data indicates a rich spectrum of scattering scalelengths in D″.

Evidence for an ULVZ at the base of the mantle was first presented by Garnero *et al.* (1993) and Silver and Bina (1993). A layer from a few to tens of kilometers thick with as much as 10% *P* velocity reduction and 30% *S* velocity reduction is found in some regions of the lower mantle (Fig. 17), with the primary evidence (see Garnero *et al.*, 1998 for a review) being delayed $SP_{diff}KS$ phases (e.g., Garnero and Helmberger, 1995, 1998; Helmberger *et al.*, 1998) and the shape of precursors to *PcP* reflections (e.g., Mori and Helmberger, 1995; Revenaugh and Meyer, 1997). The strong velocity reductions tend to be most easily explained by partial melting (Williams and Garnero, 1996), suggesting that some component of the mantle is exceeding its solidus at the hottest temperatures of the thermal boundary layer. There is fairly strong correlation between locations of ULVZ patches and slower than average shear velocities in D″ and the overlying lower mantle, which is suggestive of a relationship between partial melting in D″ and large-scale upwellings (e.g., Williams *et al*,, 1998). There is presently extensive effort to map and interpret the ULVZ feature, as it potentially has significant implications for chemistry and dynamics of D″ (Garnero, 1999).

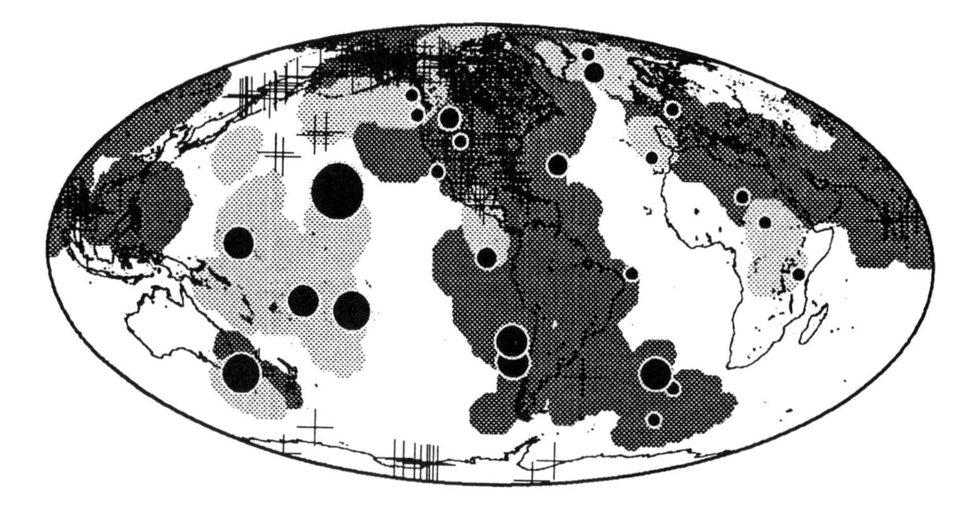

FIGURE 17 Mollweide projection of Earth showing the ULVZ distribution at the base of the mantle. Light shading corresponds to the Fresnel zone regions where a ULVZ has been detected. Dark regions are where no ULVZ has been detected. No shading corresponds to no coverage. Black-filled circles are hot spot locations (where there is ULVZ coverage), scaled to buoyancy flux estimates. Crosses are location of calculated lower mantle density anomalies due to subducted material (From Garnero *et al.*, 1998).

Given the complexity of structure at all scales in D″, it is not surprising that uncertainty remains as to whether there is any topography on the core–mantle boundary itself. Long wavelength topography of the CMB was proposed by Creager and Jordan (1986) and Morelli and Dziewonski (1987) based on studies of bulletin *PcP* and *PKP* arrival times, but it has been demonstrated that allowing for strong heterogeneity in D″ and the limited resolution of the available data make CMB topography models very uncertain (e.g., Doornbos and Hilton, 1989; Rodgers and Wahr, 1993; Pulliam and Stark, 1993; Obayashi and Fukao, 1997; Garcia and Souriau, 2000). As models for the entire mantle improve, this may prove to be a solvable problem, and it is a critical one, for it plays a major role in estimating the extent of mechanical coupling between the core and mantle. For imaging shorter-wavelength topography of the CMB, the primary approach has involved travel-time fluctuations and precursors to underside reflections of internal core reverberations (*PKKP*). These phases provide an upper bound of about 100 m topography on 10 km scale lengths (e.g., Doornbos, 1974, 1980; Chang and Cleary, 1978; Bataille and Flatté, 1988; Earle and Shearer, 1997, 1998).

Although the bulk of the lower mantle does not appear to have large-scale organized anisotropy, the D″ region has been shown to have extensive regions where shear-wave splitting occurs (see review by Lay *et al.*, 1998a, b). Observations of splitting for *ScS* phases have been made for several decades (e.g., Mitchell and Helmberger, 1973; Lay and Helmberger, 1983b), but observations of diffracted waves convincingly demonstrated that anisotropy is present in D″ (e.g., Vinnik *et al.*, 1991, 1995, 1998; Lay and Young, 1991; Kendall and Silver, 1996; Matzel *et al.*, 1996; Garnero and Lay, 1997; Ritsema *et al.*, 1998; Russell *et al.*, 1998). These observations have prompted increased consideration of the anisotropic crystallography of high pressure phases likely to be present in the lower mantle along with what deformation mechanisms are likely to control the formation of fabrics (e.g., Stixrude, 1998; Karato, 1998). At this time, there are substantial uncertainties in the nature of the anisotropy and its cause. In many places it appears that transverse isotropy with a vertical symmetry axis is consistent with the data (producing earlier *SH* arrivals than *SV* arrivals), but there are also clear observations of azimuthal anisotropy, and thus far it has only been possible to characterize the horizontal component of the symmetry axis. Strong shear flows in the boundary layer may induce lattice preferred orientation of the anisotropic lower mantle minerals, but it is not clear why this would not also hold for the overlying lower mantle. Sheared inclusions of chemical heterogeneities and pockets of partial melt may also play a role in generating the seismic anisotropy. As observational and laboratory constraints improve, it is likely that modeling anisotropy in D″ will provide an important constraint on the thermal and dynamical regime in the boundary layer.

5.3 Aspherical Lower Mantle Structure

One of the earliest fundamental contributions of global seismic tomography was the demonstration that large-scale structure exists in the lower mantle and that this unexpected configuration of deep heterogeneity can account for previously unexplained long-wavelength features in the Earth's geoid (e.g., Dziewonski *et al.*, 1977; Clayton and Comer, 1983; Dziewonski, 1984; Hager *et al.*, 1985). This required improved understanding of how mantle heterogeneities induce flow and deflection of boundaries that affect the geoid (Hager, 1984; Richards and Hager, 1984). Although the early low resolution tomographic models for the lower mantle, which have relatively strong spherical harmonic components from degrees 2 to 5, have proved remarkably successful in accounting for the long wavelength geoid (see review by Hager and Richards, 1989), there has been continuing debate about the spectrum of lower mantle heterogeneity. Are the long-wavelength patterns the result of spatial distribution of smaller scale features such as slabs embedded in the lower mantle? If so, then the long wavelength distribution of heterogeneity in the lower mantle is more a consequence of the last few hundred million years of surface tectonics than a fundamental aspect of the lower mantle. Similarly, if the long wavelength patterns in surface hotspots reflect features rising from the core–mantle boundary, then the distribution of D″ boundary layer instabilities may contribute to the present long-wavelength structure of the deep mantle. These are still open questions to a large extent, but there is significant convergence in deep-mantle tomographic shear velocity models of the current generation (e.g., Masters *et al.*, 1996; Li and Romanowicz, 1996; Grand *et al.*, 1997; Liu and Dziewonski, 1998; Ritsema *et al.*, 1999), all of which have substantial long wavelength heterogeneities. The same is true for large-scale *P* velocity models, although the level of agreement between models is not currently as strong (e.g., van der Hilst and Karason, 1999; Bijwaard and Spakman, 1999). The presence of large-scale lower-mantle features enables formulation of simultaneous or iterative inversions for dynamical features such as the geoid and dynamic topography, and this has become a new area of research (e.g., Hager *et al.*, 1985; Hager and Clayton, 1989; Dziewonski *et al.*, 1993; Phipps Morgan and Shearer, 1993; Forté *et al.*, 1993, 1994). The primary additional parameter that is constrained in such geodynamic models is the viscosity structure, and it is generally found in geoid inversions as well as contemporary studies of glacial rebound analyses that the viscosity of the lower mantle is one to two orders of magnitude higher on average than that of the average upper mantle (e.g., Mitrovica and Forté, 1997; Lambeck and Johnston, 1998).

As the resolution of lower mantle structure improves with each new generation of global tomographic model, it has become clear that there are significant intermediate scale features in the lower mantle. This had been deduced for localized regions quite early by array studies or analyses of differential travel times for phase pairs sensitive to lower mantle structure (e.g., Jordan and Lynn, 1974; Lay, 1983), but the geometry and lateral extent of such features was not resolved until tomographic models emerged. Recent high resolution *S*-velocity (e.g., Grand, 1994; Grand *et al.*, 1997) and *P*-velocity (van der Hilst *et al.*, 1997) models resolve a high velocity tabular structure extending vertically beneath North America and South America and a similar elongate body beneath southern Eurasia, both of which extend to at least 1300–1800 km depth (Fig. 18). These are interpreted as relatively cold, sinking slab material that has penetrated into the lower mantle as the Americas moved westward and as the Tethys Sea closed, respectively. The midmantle seismic velocity near 1300 km depth is thus dominated by elongate tabular high-velocity features, and it is likely that these contribute significantly to the strong long-wavelength patterns in spherical harmonic models that have lower resolution (see corresponding features in the models discussed by Dziewonski *et al.*, 1993).

Below about 1800 km depth, tomographic models show less coherence and tabular structures are not clearly imaged (e.g., Grand *et al.*, 1997). Instead, the models become dominated by large horizontally extensive regions of high and low velocity with strong degree 2 and 3 patterns dominating in D″, as noted above. The large-scale low-velocity regions below the Central Pacific and Eastern Atlantic/Southern Africa do appear to extend upward above the D″ region into the lower mantle. These have been identified as "superplumes," given that their scale greatly exceeds that expected for isolated D″ boundary layer instabilities, and Dziewonski *et al.* (1993) call them the "Equatorial Pacific Plume Group" and the "Great African Plume," respectively. Attention has focused on large low velocity features in the lower mantle as possible deep roots of plume upwellings. Ritsema *et al.* (1998) find that velocity anomalies in the structure under southern Africa involves 3% shear velocity anomalies and strong lateral gradients, both of which are more pronounced than in the model of Grand *et al.* (1997). Even stronger anomalies are reported in the D″ region below the southeastern Atlantic, with as much as 10% *S*-velocity reductions in a 300 km thick layer (Wen *et al.*, 2000). Ritsema *et al.* (1999) presented a new *S*-velocity model in which a low velocity region extends from the core–mantle boundary under the southeastern Atlantic Ocean continuously into the upper mantle beneath eastern Africa. They do not find a low shear velocity zone in the lower mantle beneath Iceland, but Bijwaard and Spakman (1999) present a *P* velocity image with low velocity under Iceland all the way to the core–mantle boundary. Goes *et al.* (1999) find a low *P*-velocity structure beneath Europe from 660 to 2000 km depth which they invoke as the source of small plumes in the upper mantle associated with volcanism in

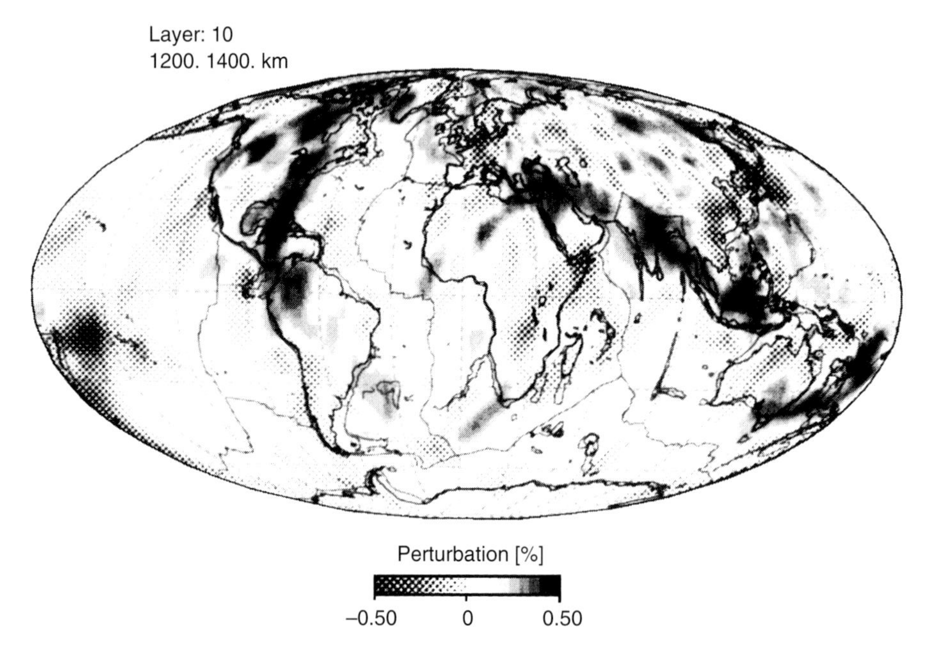

FIGURE 18 Horizontal section through a tomographic model of *P*-wave velocity structure near 1300 km in the lower mantle, obtained by inversion of ISC arrival times after careful processing for depth determination. Solid dark area correspond to high velocities, with two major coherent features that correspond to tabular structures extending vertically below southern Asia and eastern North America. These are inferred to be subducted slab material that has penetrated into the lower mantle (from van der Hilst *et al.*, 1997).

Europe. Smaller-scale plume or slab features may also exist in the lower mantle, that are below the current resolution of seismic tomography. Innovative methods of scattering analysis or array imaging may prove to be the only means by which to constrain such structures (e.g., Ji and Nataf, 1998, Tilmann *et al.*, 1998; Tibuleac and Herrin, 1999).

While the lower mantle appears not to have internal layering, it has been proposed that the downward transition in heterogeneity pattern from a midmantle dominated by slab-like structures to a deep mantle dominated by large-scale high and low velocity features is caused by compositional stratification. In this model (see Kellogg *et al.*, 1999; van der Hilst and Kárason, 1999), the lowermost mantle is compositionally distinct, being composed of undifferentiated, "primordial" mantle material which is the source of isotopic anomalies sampled by major hotspot plumes. Downwelling slabs can depress the chemical boundary by hundreds of kilometers, deflecting it from a depth of around 2000 km. The postulated chemical boundary is not a strong reflector, and does not give rise to coherent features in the radially averaged mantle model. The density increase of the deep layer could be due to enrichment in iron or silica, which have competing effects on the velocity structure. This is a highly speculative model, but can at least reconcile the current observations of the deep mantle seismic structure with geochemical observations. Significant improvement of our understanding of the lower mantle will come with reliable determination of density heterogeneity directly from simultaneous inversion of normal modes and gravity observations. A first step in this direction has been presented by Ishii and Tromp (1999). Although also very preliminary, this study found that high-density material is piled up in regions of uplift beneath the Pacific and Africa, which would require a significant chemical heterogeneity contribution to offset the thermal anomaly of the upwellings. Until the models improve it may even be premature to associate low velocities with upwellings, for chemical heterogeneity may indeed be very important in the deep mantle.

6. Core

The core comprises about 31.5% of the Earth's mass, and the density is such that the only plausible composition is primarily iron. The contrasts in density and viscosity between the lower mantle and the outer core are comparable to those found at the surface of Earth between air and rock and between ocean water and rock, respectively. This is a staggering thought, to step across the core–mantle boundary from ultradense silicates and oxides to relatively superdense iron alloy that has viscosity close to that of water. This profound chemical change was seismically detected early in the 20th century, and many studies have sought to constrain the properties of the molten outer core and the solid inner core. Much of the effort has been motivated by the recognition that a geodynamo resides within the core, producing and sustaining the Earth's magnetic field. The geodynamo involves turbulent flow of the outer core alloy, geometrically constrained by the planet's rotation and the presence of the inner core, with the heterogeneous structure of the base of the mantle producing a variable heat flow boundary condition on the system. A brief summary of principal findings is presented here, with many additional details provided elsewhere (Chapter 56 by song; Jacobs, 1987; Song, 1997; Creager, 2000).

6.1 Radially Averaged Structure

Average velocity and density structures for the core are shown in Figure 19, with smoothly increasing velocities in the outer and inner cores and finite shear velocity only in the inner core. The mean velocities of the outer core are tightly constrained by the well-defined travel-time branches of *PKP* and *SKS* phases (Fig. 1), and once again, the classic models of Jeffreys and Gutenberg (Fig. 2) are very similar to contemporary models for the outer core. The outermost portion of the outer core is best constrained by *SKS* and *SKnS* phases, which have continuous turning points just below the CMB because the *P* velocity of the outer core is higher than the *S* velocity at the base of the mantle (e.g., Hales and

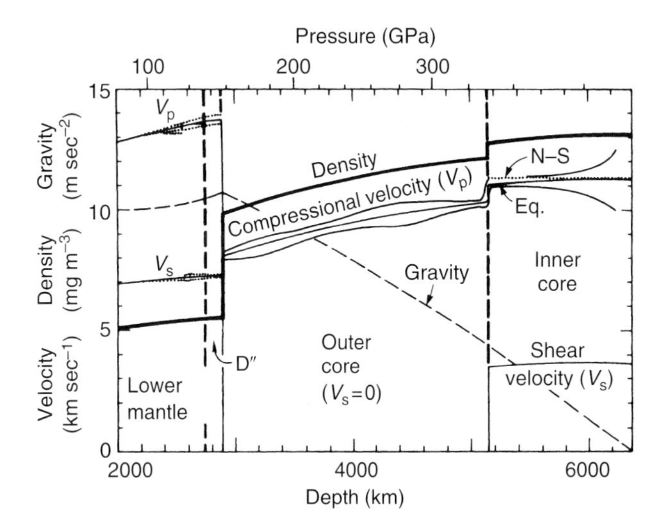

FIGURE 19 Seismologically measured density (bold solid curve), and compressional (*V*p) and shear (*V*s) elastic wave velocities in km sec^{-1} (thin lines) through the core and lowermost mantle shown as functions of depth and corresponding pressure. Extremal bounds on the *V*p profile through the core are included for comparison. The difference between the polar (N-S) and equatorial (Eq.) compressional velocities through the inner core are indicated by dotted lines. Heterogeneity in the D′ region is illustrated by variations in *V*p and *V*s profiles. (Reproduced with permission, from Jeanloz, 1990, © Annual Review Inc.)

Roberts, 1971). Studies of *SKS-S*, *SKKS-SKS* and *SKKKS-SKS* differential times have raised the possibility that a thin region with reduced velocity gradient exists in the outermost 50–100 km of the core (e.g., Hales and Roberts, 1971; Lay and Young, 1990; Souriau and Poupinet, 1991; Garnero *et al.*, 1993; Tanaka and Hamaguchi, 1993, Garnero and Lay, 1998). The issue is still open, but it is clear that if there is any core-side boundary layer it is very subtle and likely to involve only 0.5–1.0% anomalies in a thin layer.

Jeffreys and Bolt interpreted precursors to *PKP* phases in terms of complex inner core–outer core transition models, but these phases are now interpreted as the result of scattered phases from D″, and much simpler models of this transition near 5150 km depth are preferred. Generally, global average Earth models have smooth outer core velocity gradients right down to the inner core boundary, consistent with self-compression of the outer core alloy, however, several studies have suggested the presence of a transition zone with reduced gradients just above the boundary (e.g., Souriau and Poupinet, 1991; Song and Helmberger, 1992;). The contrasts at the boundary are best constrained using waveform modeling of reflections and diffractions, with recent studies preferring simple models (e.g., Cummins and Johnson, 1988; Shearer and Masters, 1990; Song and Helmberger, 1995). The *P* and *S* velocities are almost constant within the inner core on average. The *S* velocity, first constrained by normal mode modeling (e.g., Dziewonski and Gilbert, 1971) is around 3.6 km sec^{-1}. Direct observation of body waves that traverse the inner core as shear waves has been very difficult, but some candidates have been reported (e.g., Julian et al., 1972; Okal and Cansi, 1998; Deuss *et al.*, 2000). The density of the inner core appears to be several percent lower than the density of pure iron, indicating the presence of some lighter elements (Jephcoat and Olson, 1987), whereas the outer core is about 10% less dense than pure iron (Masters and Shearer, 1990), with the light alloying component probably being H, C, S, O, Mg, and/or Si.

The outer core has a very high Q_α value, from 3000 to 10 000+, and short-period seismic energy can travel great distances, including multiple underside reflections from the core–mantle boundary, with little anelastic attenuation (e.g., Engdahl, 1968; Adams, 1972; Cormier and Richards, 1973; Tanaka and Hamaguchi, 1996). However, the inner core has low Q_α, on the order of 200–360, which strongly attenuates short-period phases that penetrate into it (e.g., Doornbos, 1974, 1983; Cormier, 1981; Choy and Cormier, 1983; Bhattacharyya *et al.*, 1993; Song and Helmberger, 1995a). The attenuation of short-period signals may in part be caused by scattering rather than intrinsic attenuation (e.g., Cormier *et al.*, 1998; Vidale and Earle, 2000) because normal modes do not require very low Q values (e.g. Widmer *et al.*, 1991). The inner core indeed displays significant heterogeneity and seismic anisotropy, whereas there is no convincing evidence for either in the outer core.

6.2 Anisotropy and Heterogeneity of the Inner Core

Both body waves and normal modes indicate that the inner core is heterogeneous and anisotropic in its structure. Observations of anomalously split normal modes (e.g., Masters and Gilbert, 1981) and patterns in *PKIKP* arrival times (Poupinet *et al.*, 1983) laid the foundation for the first studies to propose the presence of anisotropy in the inner core (Woodhouse *et al.*, 1986; Morelli *et al.*, 1986). Although alternate models of aspherical structure in the outer core have been considered, there is now consensus that inner core structure can account for the normal mode (e.g., Tromp, 1993, 1995) and *PKIKP* data (e.g. Song, 1997; Creager, 1999). The latter phases travel 5–6 sec faster along the spin axis than across the equatorial plane (e.g., Creager, 1992; Song and Helmberger, 1993, 1995b, 1998; Vinnik *et al.*, 1994; Shearer, 1994; Su and Dziewonski, 1995; McSweeney *et al.*, 1997; Creager, 1999; see the reviews by Song, 1997 and Creager, 2000). The *PKIKP* data have established that there are large-scale lateral heterogeneities in the inner core associated with the anisotropy, with hemispherical patterns that extend all the way to the center of the Earth (e.g., Tanaka and Hamaguchi, 1997; Creager, 1999). The western hemisphere is 2–4% anisotropic on average. The anisotropy may be the result of crystal alignment induced by magnetic fields, convection, crystallization processes, or other effects not yet understood. Some of the structural complexity may also be associated with radial layering of the inner core (e.g., Song and Helmberger, 1998), but mapping out the surprising complexity of the inner core is still underway.

6.3 Rotation Rate of the Inner Core

The existence of heterogeneity and anisotropy within the inner core of the Earth allows a test of the stability of such patterns over time. This is of interest because it is uncertain whether the rotation of the inner core is locked to that of the mantle, and models for the geodynamo predict that there may be some differential rotation due to torques applied to the inner core. Song and Richards (1996) observed temporal variation in *PKIKP-PKP* differential times for observations on a path near the fast direction along the spin axis, with about 0.3 s change over 30 years. Assuming symmetry of the anisotropic fabric in the inner core, they inferred a $1.1^\circ\ y^{-1}$ eastward (faster) relative rotation of the inner core. Su *et al.* (1996) used a lower quality data set with better spatial distribution, and inferred a $3^\circ\ y^{-1}$ relative rotation rate. Subsequent work has extended the high quality data set, along with discarding the inadequate assumption of symmetry of the anisotropic pattern and allowing for small-scale structure in the inner core or near D″ that may enhance or reduce the rate of change of differential time measurements (e.g., Creager, 1997; Souriau *et al.*, 1997; Laske and Masters, 1999; Souriau and

Poupinet, 2000; Poupinet *et al.*, 2000; Song and Li, 2000; Song, 2000; Vidale *et al.*, 2000). The more recent studies, reviewed by Creager (2000), suggest an inner core differential rotation rate of 0.2 to 0.6° y^{-1}, with some evidence favoring no relative rotation, but the evidence for relative rotation at a rate of about 0.2° y^{-1} is quite compelling. More discussion of this topic is found in Chapter 56 by song.

7. Conclusions

This summary of basic attributes of the Earth's interior determined by seismology has only scratched the surface of the detailed knowledge acquired over the past century, and no overview of this length could begin to do full justice to the vast literature and many contributors who have played important roles in building this knowledge. However, this is a good starting point for delving into the topic, and the many references provide multiple pathways to the full body of seismological information. It should be clear that during the past century many first-order questions about the interior have been resolved, and there is good general consensus on basic issues of layering of the planet. However, there are many fundamental questions that remain open, and progress in the future will build upon the foundations described in this text. It is generally accepted that future advances will come from a combination of advances in wave propagation theory, development of new waveform inversion methods (and computing capabilities), increases in high quality seismic data acquired on multiple scales around the world, and by introduction of creative strategies for solving seismological problems. Equally well accepted is the notion that seismologists must increase their interactions and communications with mineral physicists and geodynamicists, building new interdisciplinary approaches to parameterizing and constraining structures and processes in the Earth's interior. The next century will see profound advances in our understanding, of that we can be assured, but all seismologists should study and contemplate the achievements of the many seismologists who contributed to the great first century of exploring the Earth's interior.

Acknowledgments

This research was supported by NSF grants EAR 9418643.

References (abridged set; see Editor's Note below)

Aki, K. and P.G. Richards (1980). "Quantitative Seismology Theory and Methods," Vol. I, W. H. Freeman, San Francisco, 557 pp.

Anderson, D.L. (1963). Recent evidence concerning the structure and composition of the Earth's mantle. *Phys. Chem. Earth* **6**, 1–129.

Burbach, G.V. and C. Frohlich (1986). Intermediate and deep seismicity and lateral structure of subducted lithosphere in the circum-Pacific region. *Rev. Geophys. Space Phys.* **24**, 833–874.

Dahm, C.G. (1934). A study of dilatational wave velocity in the Earth as a function of depth, based on a comparison of the P, P′, and PcP phases. Ph.D. dissertation, St. Louis University, St. Louis, MO.

Dziewonski, A.M. and D.L. Anderson (1981). Preliminary reference Earth model. *Phys. Earth Planet. Inter.* **25**, 297–356.

Ewing, W.M., W.S. Jardetzky, and F. Press (1957). "Elastic Waves in Layered Media," McGraw-Hill Book, Company, New York, 380 pp.

Garnero, E.J. (2000). Heterogeneity of the lowermost mantle. *Annu. Rev. Earth Planet. Sci.* **28**, 509–537.

Gilbert, F. and A.M. Dziewonski (1975). An application of normal mode theory to the retrieval of structural parameters and source mechanisms from seismic spectra, *Philos. Trans. R. Soc. London, Ser. A* **278**, 187–269.

Gurnis, M., M.E. Wysession, K. Knittle, and B.A. Buffett (Eds) (1998). "The Core-Mantle Boundary Region," American Geophysical Union, Washington, DC, 334 pp.

Gutenberg, B. (1959). "Physics of the Earth's Interior," Academic Press, New York.

Herrin, E. (1968). Introduction to "1968 Seismological Tables for P-phases," *Bull. Seismol. Soc. Am.* **58**, 1193–1195.

Isacks, B. and P. Molnar (1971). Distribution of stresses in the descending lithosphere from a global survey of focal mechanism solutions of mantle earthquakes *Rev. Geophys. Space Phys.* **9**, 103–174.

Jacobs, J.A. (1987). "The Earth's Core," Academic Press, San Diego, 413 pp.

Jeffreys, H. and K.E. Bullen (1940). "Seismological Tables," British Association for the Advancement of science London 50 pp.

Kanamori, H. and D.L. Anderson (1977). Importance of physical dispersion in surface-wave and free oscillation problems, Review, *Rev. Geophys. Space Phys.* **15**, 105–112.

Kennett, B.L.N. and E.R. Engdahl (1991). Travel times for global earthquake location and phase identification, *Geophys. J. Int.* **105**, 429–465.

Kirby, S.H., S. Stein, E.A. Okal, and D.C. Rubie (1996). Metastable mantle phase transformations and deep earthquakes in subducting oceanic lithosphere. *Rev. Geophys. Space Phys.* **34**, 261–306.

Lay, T. (1996). "Structure and Fate of Subducting Slabs," Academic Press, San Diego, 185 pp.

Lay, T. and T.C. Wallace (1995). "Modern Global Seismology," Academic Press, San Diego, 521 pp.

Lithow-Bertelloni, C. and M.A. Richards (1998). The dynamics of Cenozoic and Mesozoic plate motions. *Rev. Geophys. Space Phys.* **36**, 27–78.

Loper, D.E. and T. Lay (1995). The core–mantle boundary region. *J. Geophys. Res.* **100**, 6397–6420.

Nolet, G. (Ed.) (1987). "Seismic Tomography with Applications in Global Seismology and Exploration Geophysics," D. Reidel, Dordrecht, 386 pp.

Ringwood, A.E. (1975). "Composition and Petrology of the Earth's Mantle," McGraw-Hill, New York, 618 pp.

Ritzwoller, M.H. and E.M. Lavely (1995). 3D seismic models of the Earth's mantle. *Rev. Geophys. Space Phys.* **33**, 1–66.

Romanowicz, B. (1991). Seismic tomography of the Earth's mantle. *Annu. Rev. Earth Planet. Sci.* **19**, 77–99.

Savage, M.K. (1999). Seismic anisotropy and mantle deformation: What have we learned from shear wave splitting? *Rev. Geophys. Space Phys.* **37**, 65–106.

Silver, P.G. (1996). Seismic anisotropy beneath the continents: Probing the depths of geology. *Annu. Rev. Earth Planet. Sci.* **24**, 385–432.

Silver, P., R.W. Carlson, and P. Olson (1988). Deep slabs, geochemical heterogeneity and the large-scale structure of mantle convection: investigation of an enduring paradox *Annu. Rev. Earth Planet. Sci.* **16**, 477–541.

Song, X. (1997). Anisotropy of the Earth's inner core *Rev. Geophys. Space Phys.* **35**, 297–313.

Thurber, C.H. and K. Aki (1987). 3D seismic imaging. *Annu. Rev. Earth Planet. Sci.* **15**, 115–139.

Editor's Note

A list of complete references cited in this Chapter is given on the attached Handbook CD, under directory \51Lay. See also the next five chapters.

52

Probing the Earth's Interior with Seismic Tomography

Andrew Curtis
Schlumberger Cambridge Research, Cambridge, UK
Roel Snieder
Colorado School of Mines, Golden, USA

1. Introduction

Seismological data recording the oscillations of seismic waves at distinct locations on the Earth's surface can be used to infer variations in elastic properties and density within the Earth. Creating images or models of these variations is called seismic tomography. Figure 1 shows examples of different seismic wave types that are used routinely to estimate Earth structure. Interface waves propagate along interfaces within the Earth. They oscillate with an amplitude that decays with distance from the interface and are effected by (and hence contain information about) structure around that interface. Surface waves are a particular case that are generally classified as either Love waves which oscillate transversely (perpendicular to the path of propagation) or Rayleigh waves which oscillate elliptically in the vertical–longitudinal (parallel to the propagation path) plane. The longer the wavelength of surface waves, the deeper in the Earth they oscillate. The very long wavelength limit of surface waves are normal modes which vibrate the whole circumference of the Earth simultaneously, much like a ringing bell; the resonant frequencies with which they vibrate contain information about the structure of the whole Earth averaged in different ways. Many different body waves are distinguished (at the very least, P-compressional and S-shear waves), but all propagate through the Earth's interior and either refract, diffract or reflect energy due to elastic parameter or density variations. Hence body-wave data contain information about both interface and bulk properties of the Earth.

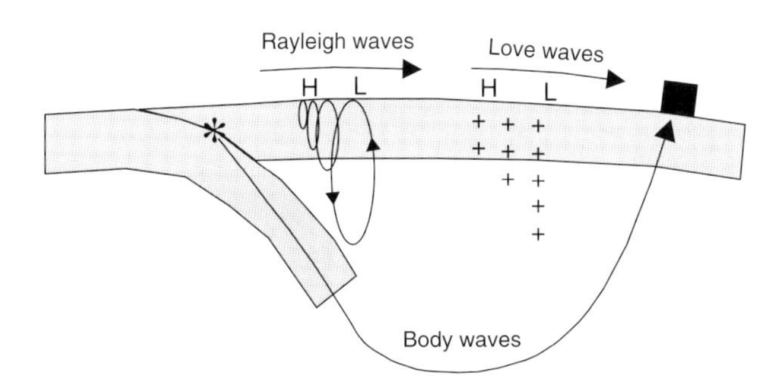

FIGURE 1 Various wave types excited by an earthquake (star) in a subduction zone and recorded at a station (square) that are often used for seismic tomography. Lithospheric plates are shaded in grey. For Rayleigh and Love waves, the general particle oscillation direction beneath the surface is drawn, while the propagation direction is shown above the surface; for the body wave only the path of energy propagation from earthquake to station is shown. "+" denotes particle oscillation perpendicular to the page, "H" denotes high frequency and "L" low frequency components. Note that low frequency surface waves generally travel faster than high frequency surface waves.

This chapter describes in detail the relationships between the character of body or surface wave propagation or of normal mode oscillations, and variations in subsurface Earth structure (the Earth model). It shows how, and under what conditions these relationships can be linearised; this allows linearised inverse theory to be used to estimate the structure from data recording the oscillations of such waves. However, an estimated model of Earth structure may be of little use unless we know the extent to which that estimate reflects the true Earth structure. We discuss how the inverse theory used and how the physics of seismic wave propagation limit the quality of the retrieved Earth model. We will see that constructing model estimates requires excellent data quality and stringent conditions to be fulfilled on the spatial distribution of the data. We show that these conditions can be used to *design* ideal seismological surveys. Finally, we show examples of surface-wave and body-wave tomography and illustrate that

INTERNATIONAL HANDBOOK OF EARTHQUAKE AND ENGINEERING SEISMOLOGY, VOLUME 81A ISBN: 0-12-440652-1

only recently are similar Earth structures "visible" using these different data types.

2. A Brief History of Seismic Tomography

The development of seismic tomography began with attempts to estimate the radial structure of the Earth. The travel times of seismic waves for an Earth model that depends on depth only are related by an Abel transform to the seismic velocity. The inverse Abel transform was used early in this century as a tool to determine the depth-dependent velocity in the Earth from travel-time measurements (Herglotz, 1907; Wiechert, 1910; Schlichter, 1932). Using the WKB-approximation for the structure of surface wave modes, Takahashi (1955, 1957) developed an inversion method to determine the seismic velocity as a function of depth from the phase velocity of Love waves or Rayleigh waves. This method was also based on the inverse Abel transform.

Thus, the field of seismic tomography began with a method where an analytical technique, the inverse Abel transform, was used to infer the seismic velocity. In the Earth sciences, analytical techniques for creating tomographic images are not very useful. The reason for this is that in a realistic tomographic experiment, sources and receivers are distributed inhomogeneously and the reference model used in the tomography is often also inhomogeneous.

This situation is very different from examples in medical tomography where the source–receiver geometry has many degrees of symmetry and where the reference model is usually homogeneous. The resulting symmetry of the wave paths in medical tomography allows analytical methods, such as the Radon transform (Barrett, 1984), to be used to reconstruct the medium (Natterer, 1986). This high degree of symmmetry also makes it possible for the medical equipment to carry out much of the data processing automatically. Because of the lack of symmetry in most seismic tomographic experiments, the reconstruction of the medium requires the solution of very large systems of equations (see Section 4). For this reason, computational challenges have been a central focus of early developments in seismic tomography.

The earliest applications of seismic tomography in its modern form employed array data (Aki and Lee, 1976; Aki *et al.*, 1977). In these studies the local lithospheric structure under the array was inferred from observed anomalies of seismic waves. In contrast to medical imaging, in seismic tomography one has no control over the source locations if earthquakes are used to generate the seismic waves. This has led, for example, Aki and Lee (1976) and Spencer and Gubbins (1980) to include the source location and origin time in the tomographic inversion. Up to the early 1980s, seismic tomography for the 3D Earth structure was based almost exclusively on the arrival times or dispersion properties of seismic waves. The amplitude of seismic waves is influenced by the spatial derivatives of the seismic velocity since these derivatives determine the focusing of seismic waves. This effect was used by Thomson and Gubbins (1982) who developed a method of seismic tomography that treated the amplitude of seismic waves as extra relevant data.

It has been customary to solve the large systems of equations required for seismic tomography iteratively. Such methods are particularly attractive because the linear equations that must be solved are sparse when the seismic velocity is parameterized with a cell model (see Section 4). Initially, row-action methods were used where one row of the matrix at a time is used to update the model vector. Nolet (1985) introduced the more efficient conjugate gradient method to seismic tomography, where search directions in model space are updated in an efficient way. An overview of the numerical methods used to solve the linear equations of seismic tomography is given by van der Sluis and van der Vorst (1987).

With the growth in computing power and in the amount of seismological data available, it soon became possible to carry out seismic tomography on a global scale. Global images of the lower mantle were constructed by Dziewonski (1984) and by Comer and Clayton (1984). A global image for the upper mantle was presented in the same year by Woodhouse and Dziewonski (1984). These studies played an important role in the development of global tomography.

After 1984 the development of seismic tomography was extremely rapid. Refinements and improvements were made to tomographic inversion schemes. These developments include more detailed tomography on a global scale (Tanimoto and Anderson, 1985), the use of algorithms for waveform tomography (e.g., Nolet *et al.*, 1986; Tanimoto, 1987; Snieder, 1988a,b; Nolet, 1991) and the incorporation of ray-bending effects in seismic travel-time tomography (e.g., Bregman *et al.*, 1989; Sambridge, 1990). An essential element in the growth and use of seismic tomography has been the dramatic increase in the size of the tomographic problems that could be solved. Spakman *et al.* (1993) used more than half a million travel times to estimate the subsurface structure under Europe and the Mediterranean with an unprecedented resolution.

This brief overview of the history of seismic tomography cannot do justice to the numerous contributrions to this field that have been made by a large number of researchers. Review articles have appreared that focus on global tomography (Dziewonski and Woodhouse, 1987) as well as regional tomography (Romanowicz, 1991) and on the implications of seismic tomography on the dynamics of the Earth's mantle (Montagner, 1994). In addition, a number of textbooks give a comprehensive overview of the different aspects of seismic tomography (Nolet, 1987; Iyer and Hirahara, 1993; Morelli *et al.*, 1996; Dahlen and Tromp, 1998). The interested reader is encouraged to consult these texts for further information.

3. Linearising the Model–Data Relationships

In general, seismological data depend in a complicated, nonlinear way on the Earth model about which we seek information. The associated nonlinear inverse problem (see Section 4) of constraining the model by the data is usually difficult to solve, and there are no foolproof systematic algorithms that find correct solutions. It is for this reason that linear approximations play a central role in seismic tomography: after linearising the model–data relationship, one can draw on results from linear algebra to constrain the model (Section 4).

In this section three important linearisations are described. In Section 3.1 the Born approximation is introduced; this gives the first-order distortion of the entire wave-field due to model perturbations. The first-order effect of slowness perturbations on the arrival time of waves has its root in Fermat's principle as described in Section 3.2. The linearised imprint of Earth structure perturbations on its normal mode eigenfrequencies is introduced in Section 3.3.

In all cases, linearisations are approximations to the true nonlinear model–data relationships; these approximations are valid when the difference between the true Earth structure and our best initial guess at the structure (our reference model) are relatively small. Whether the true Earth structure is sufficiently close to our current best estimates that correct solutions can be found using linearised methods is still an open question.

3.1 The Born Approximation

The Born approximation is introduced here using the example of the Helmholtz equation that describes acoustic wave propagation:

$$\left(\nabla^2 + \frac{\omega^2}{c^2(\mathbf{r})}\right)p(\mathbf{r}) = F(\mathbf{r}) \tag{1}$$

In this expression $F(\mathbf{r})$ describes the source signal at angular frequency ω that excites the wave-field $p(\mathbf{r})$ which then propagates through a medium with velocity $c(\mathbf{r})$. This equation is not appropriate to describe the propagation of elastic waves in the Earth, but it contains the essential elements to convey the principle of the Born approximation with fewer complications.

Central to the Born approximation is the separation of the model into a reference model and a perturbation. In this example, $1/c^2(\mathbf{r})$ is decomposed into a constant reference velocity c_0 and a perturbation:

$$\frac{1}{c^2(\mathbf{r})} = \frac{1}{c_0^2}[1 + n(\mathbf{r})] \tag{2}$$

Heterogeneity in the model is therefore described by the perturbation $n(\mathbf{r})/c_0^2$. The second ingredient needed for the Born approximation is the response of the unperturbed medium to a point force excitation at any arbitrary location $\mathbf{r}'$. This particular response is called the Green's function; by definition it satisfies the wave Eq. (1) for the reference medium and a point force excitation at $\mathbf{r}'$:

$$\left(\nabla^2 + \frac{\omega^2}{c_0^2}\right)G_0(\mathbf{r},\mathbf{r}') = \delta(\mathbf{r} - \mathbf{r}') \tag{3}$$

This Green's function can be used to find the unperturbed wave $p_0(\mathbf{r})$ that propagates through the reference medium c_0. The unperturbed wave satisfies Eq. (1) with the velocity $c(\mathbf{r})$ replaced by the unperturbed velocity:

$$\left(\nabla^2 + \frac{\omega^2}{c_0^2}\right)p_0(\mathbf{r}) = F(\mathbf{r}) \tag{4}$$

The wave-field $p_0(\mathbf{r})$ at point $\mathbf{r}$ which is excited by the excitation F and which propagates through the reference medium c_0 can then be expressed in terms of this Green's function

$$p_0(\mathbf{r}) = \int G_0(\mathbf{r},\mathbf{r}')F(\mathbf{r}')\,dV' \tag{5}$$

where the integral is over the volume containing all sources of excitation. The life-history of the unperturbed wave can be followed by reading this expression from right to left. At the source location $\mathbf{r}'$ the wave is excited by the excitation $F(\mathbf{r}')$. The unperturbed Green's function $G_0(\mathbf{r},\mathbf{r}')$ describes how this wave propagates from the point of excitation $\mathbf{r}'$ to the point $\mathbf{r}$.

In order to find the waves scattered by perturbations to the reference model we write the total wave field as the sum of the unperturbed wave p_0 and the scattered waves p_S:

$$p(\mathbf{r}) = p_0(\mathbf{r}) + p_S(\mathbf{r}) \tag{6}$$

Inserting this expression and the decomposition (2) in Eq. (1) gives

$$\left(\nabla^2 + \frac{\omega^2}{c_0^2}\right)p_0(\mathbf{r}) + \left(\nabla^2 + \frac{\omega^2}{c_0^2}\right)p_S(\mathbf{r}) + \frac{\omega^2}{c_0^2}n(\mathbf{r})p_0(\mathbf{r}) + \frac{\omega^2}{c_0^2}n(\mathbf{r})p_S(\mathbf{r}) = F(\mathbf{r}) \tag{7}$$

Using Eq. (4) for the unperturbed wave this result can be written as

$$\left(\nabla^2 + \frac{\omega^2}{c_0^2}\right)p_S(\mathbf{r}) = -\frac{\omega^2}{c_0^2}n(\mathbf{r})p_0(\mathbf{r}) - \frac{\omega^2}{c_0^2}n(\mathbf{r})p_S(\mathbf{r}) \tag{8}$$

The last term in the right hand side contains the cross-product of the scattered wave $p_S(\mathbf{r})$ and the perturbation $n(\mathbf{r})$, hence this term is nonlinear in the perturbation of the medium. In the

Born approximation one linearises the relation between the scattered waves and the perturbation of the medium, which means that the last term in Eq. (8) is ignored. Physically this corresponds to ignoring the multiple scattering effects. The resulting expression for the scattered waves is denoted with $p_B(\mathbf{r})$, where B stands for *Born*. In the Born approximation the scattered waves thus satisfy

$$\left(\nabla^2 + \frac{\omega^2}{c_0^2}\right) p_B(\mathbf{r}) = -\frac{\omega^2}{c_0^2} n(\mathbf{r}) p_0(\mathbf{r}) \tag{9}$$

This expression is of the same form as Eq. (4), the only difference is that the right hand side is not given by the excitation $F(\mathbf{r})$ but by a term proportional to $n(\mathbf{r})p_0(\mathbf{r})$. Physically this corresponds to the fact that the perturbation of the medium can be seen as the source of the scattered waves and that this excitation is proporional to the wave field at the location of the scatterers. Equation (9) can also be solved using the unperturbed Green's function to give:

$$p_B(\mathbf{r}) = -\int G_0(\mathbf{r}, \mathbf{r}') \frac{\omega^2 n(\mathbf{r}')}{c_0^2} p_0(\mathbf{r}') dV' \tag{10}$$

Just as for Eq. (5), this expression can be interpreted by reading it from right to left. The wave field $p_0(\mathbf{r}')$ is excited and propagates through the unperturbed medium to the perturbation of the medium at location $\mathbf{r}'$. At this point the wave is scattered with a scattering coefficient $-\omega^2 n(\mathbf{r}')/c_0^2$ that depends linearly on the perturbation. The Green's function then describes how the scattered wave propagates to location $\mathbf{r}$.

It should be emphasised that Eq. (10) is an approximation. In reality the wave field interacts repeatedly with the perturbations of the medium. Physically this corresponds to the double scattered waves, triple scattered waves, etc. The Born approximation retains only the single scattered waves. This is both its strength and its weakness; it leads to a linear relation for the effect of a model perturbation on the recorded data but the approximation has a limited range of validity. A full derivation of the multiple scattering series including the Born approximation is given by Snieder (2001).

The reasoning of this section is not specific to the Helmholtz equation. The only essential elements that are needed are: (1) the wave equation is linear, (2) the model can be divided in a natural way into a reference model and a perturbation and (3) the Green's function of the unperturbed problem can be computed. Exactly the same arguments can be applied to acoustic waves and electromagnetic waves (de Hoop, 1995), quantum mechanics (Mertzbacher, 1970), elastic body waves (Hudson, 1977), elastic surface waves (Snieder, 1986a,b) and the Earth's normal modes (Dahlen and Tromp, 1998). The latter applications are of great importance to seismology.

3.2 Fermat's Theorem

Travel-time tomography uses the arrival times of elastic waves to infer the velocity in the Earth. The basic principle is very simple; the travel time is the integral of the slowness along the ray-path:

$$T = \int_{\mathbf{r}[\mathbf{u}]} u(\mathbf{r}) ds \tag{11}$$

In this expression u is the slowness which is defined as the inverse of the velocity: $u = 1/c$. The travel time T can be measured from recorded data and the aim of travel-time tomography is to infer the slowness u from these data. Equation (11) suggests that this is a simple problem because the travel time and the data are related linearly. There is, however, a catch. Rays are curves that render the travel time measured along the curve stationary with respect to the curve position. This means that the ray along which the slowness in Eq. (11) is integrated also depends on the (unknown) slowness. This makes the relation between the travel time and the slowness in Eq. (11) nonlinear.

Fortunately, the relation between the travel-time perturbation and the slowness perturbation can be linearised in a simple way. Just as in Section 3.1 the model is divided into a reference model and a perturbation

$$u(\mathbf{r}) = u_0(\mathbf{r}) + u_1(\mathbf{r}) \tag{12}$$

and the travel time is decomposed into the travel time T_0 through the reference medium u_0 and a travel-time perturbation δT:

$$T = T_0 + \delta T \tag{13}$$

The traditional way to derive the linearised relation between the travel-time perturbation and the slowness perturbation given in Eq. (14) is to invoke Fermat's theorem. This theorem states that the travel time to first order does not depend on changes in the ray position (e.g., Ben-Menahem and Singh, 1981; Nolet, 1987). The proof of this theorem relies on variational calculus and is unnecessarily complex. An alternative derivation of the linearised relation between δT and u_1 is given in a very short and elegant derivation by Aldridge (1994) who shows that to first order

$$\delta T = \int_{\mathbf{r}_0[u_0]} u_1(\mathbf{r}_0) ds_0 \tag{14}$$

The derivation does not use Fermat's theorem explicity, but does use the fact that the travel time is stationary under ray perturbations. This derivation can also be extended to include higher-order travel-time perturbations (Snieder and Sambridge, 1992; Snieder and Aldridge, 1995). The crux of Eq. (14) is that the integration of the slowness perturbation is now taken over the reference ray $\mathbf{r}_0$ in the reference medium. Since the rays in the reference medium are known (because the reference medium is known), Eq. (14) constitutes a linear relation between the travel-time perturbation and the slowness perturbation.

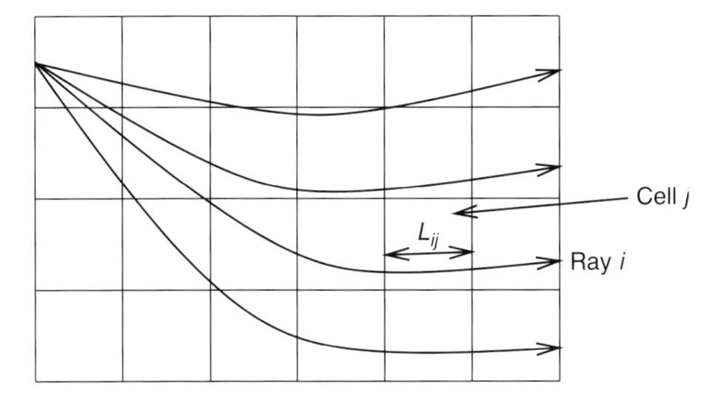

FIGURE 2 Diagram of a tomographic experiment.

The simplest way to implement Eq. (14) in tomographic inversions is to divide the models into cells and to assume that the slowness perturbation is constant in every cell. When the travel-time data are denoted with the subscript i and the cells with the subscript j the discretized form of the integral (14) is given by:

$$\delta T_i = \sum_j L_{ij} u_j \tag{15}$$

where u_j now denotes the slowness perturbation in cell j, and L_{ij} denotes the length of ray number i through cell number j as defined in Figure 2. This means that the inverse problem reduces to the problem of solving a system of linear equations.

By solving Eq. (15) we perform a single linear inversion. The solution comprises a set of slowness perturbations u_j, one in each cell, which can then be added to the slownesses of the reference model. Thus we obtain an updated reference model. Ray paths can be recalculated in the updated medium leading to a new set of linear equations of the form in Eq. (15). Solving these provides a second set of slowness perturbations to be added to the updated reference model, and so on. This iterated, linearized inversion process is often used in ray-path tomography in the hope that the solution accuracy increases with each iteration (e.g., Bijwaard and Spakman, 2000). Note, however, that there is no guarantee that this will be true.

3.3 Rayleigh's Principle

Other data that constrain the internal structure of the Earth are the normal-mode eigenfrequencies of the Earth and the phase or group velocity measurements of surface waves. These data can all be seen as the eigenvalues of a system of differential equations for the motion in the Earth with appropriate boundary conditions. In this section we refer to the normal-mode frequencies or the surface-wave dispersion measurements as λ_i.

In general, the relation between these eigenvalues and the associated velocity model is nonlinear, which complicates the inverse problem. As in Sections 3.1 and 3.2 the Earth model can be decomposed into a reference model m_0 and a perturbation m_1. The reference model has eigenvalues $\lambda_i^{(0)}$ that are perturbed with a perturbation $\delta\lambda_i$. The relation between the perturbations of the eigenvalues and the perturbations of the Earth model can be linearised using Rayleigh's principle (Woodhouse, 1976). This states that, to first order the perturbation of the eigenvalues depends linearly on the perturbation of the Earth model and on the modes in the reference Earth model. Since the modes in the reference Earth model are assumed to be known, the inverse problem can be seen as the solution of a linear system of equations.

For example, for surface waves it can be shown (e.g., Takeuchi and Saito, 1972; Aki and Richards, 1980) that the phase velocity perturbation of surface waves is related to the perturbations in shear velocity $\delta\beta$, compressional velocity $\delta\alpha$ and density $\delta\rho$ by the relation

$$\begin{aligned}\frac{\delta c}{c} = &\int K_\beta(z)\frac{\delta\beta(z)}{\beta(z)}\,dz + \int K_\alpha(z)\frac{\delta\alpha(z)}{\alpha(z)}\,dz \\ &+ \int K_\rho(z)\frac{\delta\rho(z)}{\rho(z)}\,dz\end{aligned} \tag{16}$$

which holds at each location on the Earth's surface. In this expression the kernels $K(z)$ depend only of the reference model and can be assumed to be known (e.g., see Fig. 5). Again, the relation between the data δc and the model perturbations $\delta\beta$, $\delta\alpha$ and $\delta\rho$ is linear. After discretisation this leads to a linear system of equations of the form of Eq. (15) that is relatively easy to solve. Expressions similar to Eq. (16) hold for the Earth's normal modes leading to an equally simple system of equations to solve (Woodhouse, 1980; Woodhouse and Dahlen, 1978).

4. Tomographic Inversion

Performing tomography inevitably involves the use of inverse theory. Since the number of data and model parameters is often very large, linearised physics and the relatively efficient linearised form of inverse theory from the field of linear algebra are usually used. However, to evaluate the quality of any tomographic solution found, it is important that we understand the strengths and limitations of this theory which we illustrate below. Further details can be found in Matsu'ura and Hirata (1982), Tarantola and Valette (1982a,b), Tarantola (1987) or Parker (1994).

4.1 Linearised Inverse Theory

The true Earth structure has more detail than any finite set of observations can resolve. Say we construct an infinite and complete set of basis functions $\{\mathbf{B}_j(\mathbf{x}): j = 1, \ldots, \infty\}$ such that any possible model $\mathcal{M}$ of Earth structure at location $\mathbf{x}$ could be expressed as:

$$\mathcal{M}(\mathbf{x}) = \sum_{j=1}^{P} m_j \mathbf{B}_j(\mathbf{x}) \quad (17)$$

where $P = \infty$. The tomographic problem then consists of estimating coefficients $\mathbf{m}' = [m_1, \ldots, m_P]$. Unfortunately only a finite amount of data is ever measured, so usually we restrict the model to a finite subset of the basis functions (i.e., P finite); these functions span only those features of Earth structure that are important to us and which might be resolvable by the available data. For this purpose, basis functions commonly employed in seismology are mutually exclusive polygonal cells ($B_j(\mathbf{x}) = 1$ if $\mathbf{x} \in$ cell j, 0 otherwise), orthogonal splines, or spherical harmonics for global models.

Suppose seismological data are recorded at a set of seismic receivers, possibly using artificial seismic sources. In the previous section we showed that in some situations such data can be related approximately to an appropriately parameterised Earth model using linearised physical relationships for models close to an initial model estimate $\mathbf{m}_0$. Let the recorded data be denoted by $\mathbf{d}_{\text{rec}}$, and let $\mathbf{F}(\mathbf{m}')$ represent the forward function that predicts the recorded data for a given model $\mathbf{m}'$. The recorded data differ from the data $\mathbf{d}_0 = \mathbf{F}(\mathbf{m}_0)$ calculated for the initial model estimate $\mathbf{m}_0$ because the true Earth structure differs from the initial model, and because the recorded data are contaminated with errors $\mathbf{e}$. In the remainder we refer to the difference between the recorded data and the reference data $\mathbf{d}_0$ for the initial model as "the data" $\mathbf{d}$:

$$\mathbf{d} \equiv \mathbf{d}_{\text{rec}} - \mathbf{d}_0 \quad (18)$$

Let the true Earth model be denoted $\mathbf{m}_\oplus$ [using the infinite-dimensional basis in Eq. (17) with $P = \infty$] and let its projection onto the (usually finite-dimensional) space of basis functions used for $\mathbf{m}_0$ be denoted $\mathbf{m}_t$. The difference between $\mathbf{m}_t$ and the reference model $\mathbf{m}_0$ is referred to as "the model" $\mathbf{m}$:

$$\mathbf{m} \equiv \mathbf{m}_t - \mathbf{m}_0 \quad (19)$$

Using the relation $\mathbf{d}_{\text{rec}} = F(\mathbf{m}_\oplus) + \mathbf{e}$ the relation between the data and model as defined in Eqs. (18) and (19) can be linearised:

$$\begin{aligned} \mathbf{d} = \mathbf{d}_{\text{rec}} - \mathbf{d}_0 &= F(\mathbf{m}_\oplus) - F(\mathbf{m}_0) + \mathbf{e} \\ &\approx \mathbf{A}(\mathbf{m}_t - \mathbf{m}_0) + \mathbf{e} \\ &= \mathbf{A}\mathbf{m} + \mathbf{e} \end{aligned} \quad (20)$$

The matrix $\mathbf{A}$ describes the linearisation of the full physical model–data relationship $\mathbf{F}$ (elements are $A_{ij} = \partial F_i / \partial m_j$) and depends on the data types used, the basis functions and the seismic source and receiver locations (the survey design).

The generalised least-squares solution of Eq. (20) is defined to be the one that minimises the squared misfit function E:

$$E = (\mathbf{d}_{\text{rec}} - \mathbf{F}(\mathbf{m}))^T \mathbf{W}_d (\mathbf{d}_{\text{rec}} - \mathbf{F}(\mathbf{m})) + \lambda \mathbf{m}^T \mathbf{W}_m \mathbf{m} \quad (21)$$

$$\simeq (\mathbf{d} - \mathbf{A}\mathbf{m})^T \mathbf{W}_d (\mathbf{d} - \mathbf{A}\mathbf{m}) + \lambda \mathbf{m}^T \mathbf{W}_m \mathbf{m} \quad (22)$$

E has two components: the first term measures the squared misfit between the observed data (perturbation) $\mathbf{d}$ and the predicted perturbations $\mathbf{A}\mathbf{m}$ that are consistent with model $\mathbf{m}$, where different datum misfits can be weighted with a matrix $\mathbf{W}_d$ to reflect data uncertainty. This is added to λ times the norm of a quadratic combination of model $\mathbf{m}$ weighted by the matrix $\mathbf{W}_m$ (second term). Here $\mathbf{W}_d$ and $\mathbf{W}_m$ are positive definite, and λ is a trade-off parameter: minimising E with λ small gives a solution that preferentially fits the weighted measured data; with λ large the solution preferentially minimises the weighted model norm. The matrix $\mathbf{W}_m$ can be defined such that it causes the model to be either as smooth or as small (damped) as possible for instance. Adding this term is known as regularising the solution (Tikhonov, 1963). A simple piece of algebra (see Snieder and Trampert, 2000) shows that the minimum E occurs at

$$\mathbf{m}_{\text{est}} = (\mathbf{A}^T \mathbf{W}_d \mathbf{A} + \lambda \mathbf{W}_m)^{-1} \mathbf{A}^T \mathbf{W}_d \mathbf{d} \quad (23)$$

When no regularisation is applied ($\lambda = 0$) the matrix inverse taken in Eq. (23) usually does not exist. This reflects the fact that the inverse problem is ill-posed (there are insufficient data to constrain the model). The amount of regularisation that one applies should ensure that the inverse matrix in Eq. (23) is well behaved. As an alternative one can replace the matrix inverse in Eq. (23) by an approximation of the matrix inverse that is devoid of small eigenvalues (see Matsu'ura and Hirata, 1982; Menke, 1989; Parker, 1994). However, stabilising the inverse problem by an explicit regularisation has the advantage that one controls and understands better the effects of regularisation.

4.2 Assessing Quality of the Linearised Solution

The inverse problem is usually treated first as a model estimation problem followed by an assessment of how the estimated model is related to the true Earth model (the appraisal problem, Snieder, 1998). Although the solution [Eq. (23)] solves the problem of estimating Earth structure $\mathbf{m}_{\text{est}}$ from data $\mathbf{d}$, it is necessary to assess how good an approximation to the true Earth structure $\mathbf{m}_\oplus$ that this provides. We now explore the factors that cause the accuracy of $\mathbf{m}_{\text{est}}$ to deteriorate.

First, the data in Eq. (23) are usually contaminated with errors about which our only information is statistical [Eq. (20)]. These errors are mapped onto the estimated model, so model statistics are needed to indicate features in the estimated model that may be caused by errors in the data. The way in which errors in the data propagate into the model depends on the regularization. This follows from Eq. (23): the solution $\mathbf{m}_{\text{est}}$ depends on the type and strength of regularisation applied. The form of matrix $\mathbf{W}_m$ defines the type of regularisation and often there is no objective reason to prefer any particular form. Even once a form for $\mathbf{W}_m$ has been chosen, there is generally no objective way to choose λ unless a physical meaning is attributed to the weight matrices in terms of the probability distributions of noise **e** and models **m** (Tarantola and Valette, 1982a; Matsu'ura and Hirata, 1982; Tarantola, 1987). However, often even this strategy is impossible to apply objectively (Scales and Snieder, 1997), so in practice a degree of subjectivity is inevitable in all tomographic Earth models.

The second source of error comes from the linearised inverse theory itself. Figure 3(a) illustrates the process of finding a solution to a linearised system for one model parameter for $\lambda = 0$. Beginning at the reference model $\mathbf{m}_0$ we calculate $\mathbf{d}_0 = \mathbf{F}(\mathbf{m}_0)$ and hence the required data perturbation **d**, and the linear operator **A** (the gradient) as illustrated previously. Assuming the physical relationship **F** is approximately linear, the gradient is used to extrapolate away from $\mathbf{m}_0$ to estimate the true model $\mathbf{m}_0 + \mathbf{m}$ that minimises the quadratic data misfit E given by Eq. (22). The data uncertainty (here represented by $\pm$ bounds, but more generally a data probability distribution) is mapped into a linearised model uncertainty by projecting it through the (linearised) physics.

Notice, for later on, that for the same data uncertainty, the projected model uncertainty would diminish if $\mathbf{F}(\mathbf{m})$ was steeper (roughly speaking, if **A** was larger in some sense to be defined later). This can be shown explicitly when the data uncertainties and prior model distributions are Gaussian ($\mathbf{C}_m = \mathbf{W}_m^{-1}$ and $\mathbf{C}_d = \mathbf{W}_d^{-1}$ are covariance matrices). Then the model uncertainty after a truly linear Bayesian inversion is Gaussian with covariance

$$\mathbf{C} = (\mathbf{A}^T \mathbf{C}_d^{-1} \mathbf{A} + \mathbf{C}_m^{-1})^{-1} \tag{24}$$

(Tarantola and Valette, 1982a). In the case of one model parameter, **C** is reduced if the (single) gradient in **A** increases.

Figure 3(b) illustrates what can happen when **F** is not truly linear: if the reference model $\mathbf{m}_0$ is "close" to the unknown true solution $\mathbf{m}_t$ then generally the linearised step produces a model $\mathbf{m}_1$ that is closer to $\mathbf{m}_t$. Repeated linearised steps usually converge toward the solution. If the reference model is too far from the true solution (e.g., $\mathbf{m}_{0'}$) or if the problem is far from linear, this procedure may converge to a secondary solution (at a different misfit minimum). The linearised uncertainty estimates in each case give only the projected uncertainties around each solution found (one or other of the shaded areas in Fig. 3). The true uncertainty in both cases is more accurately described by the union of the three shaded projections: this union reflects our true state of uncertainty about the model, representing the complete subset of model space within which the data cannot discriminate. If the misfit E has n distinct minima, then the true uncertainty may comprise the union of n distinct regions of model space. Unfortunately n is seldom known or sought due to the comparitively large computational effort required for such a search. Hence, this source of error may cause not only erroneous model estimates, but invalid model uncertainty estimates.

A third source of error is caused by inherent non-uniqueness in the data recorded. Even if the inverse problem is truly linear and there are no errors in recorded data, there are usually features of the model that cannot be resolved. For example,

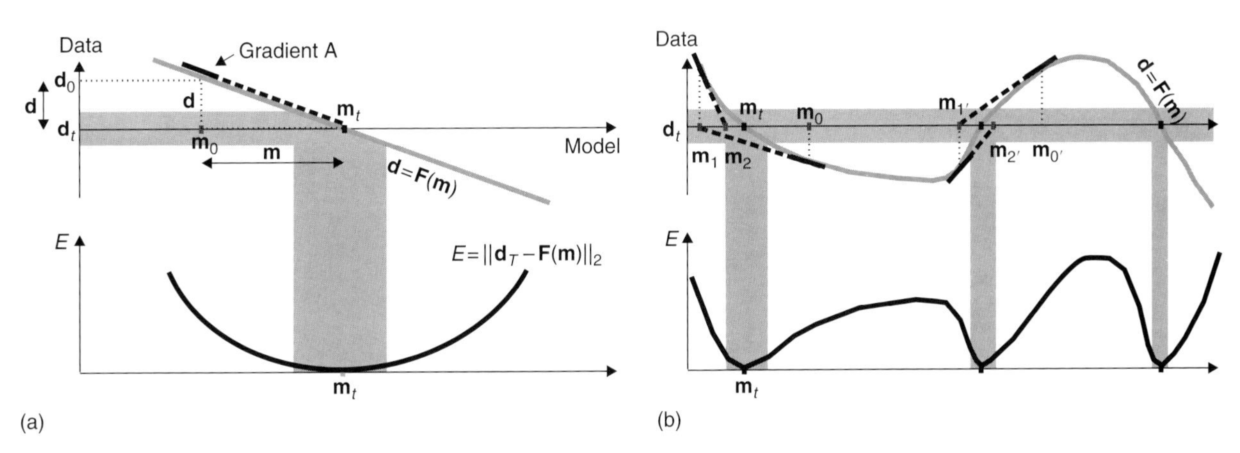

FIGURE 3 One-dimensional example of (a) linear and (b) linearised inversion. In each case, $\mathbf{m}_0$ represents the initial model chosen, and successive models found in the inversion process are shown as $\mathbf{m}_1$ and $\mathbf{m}_2$. $\mathbf{m}_t$ is the true model and the measured datum is $\mathbf{d}_t$ (with no errors in this case). The forward relationship **F** is shown in light grey and its gradient at specific points are represented by bold tangents with dashed linearised projections. The corresponding misfit function E [Eq. (21)] is shown in the lower plot of each pair. Horizontal axes always represent the model space (some labels omitted for clarity). Data uncertainties and their projections into the model space are represented by dark grey shading.

consider trying to constrain the seismic slowness of each cell in Figure 2 from a data set consisting of (noiseless) travel times along only the four ray paths shown; the data set is clearly insufficient to constrain the complete model.

In such cases, regularisation controls the otherwise unconstrained model features in our model estimate. Since the regularisation applied is usually chosen subjectively, these effects must be assessed. This is achieved by calculating the resolution matrix $\mathbf{R}$. This indicates how much information about the true Earth would be contained in estimate $\mathbf{m}_{\text{est}}$ for a given choice of $\mathbf{W}_m$, *if we had perfect data measurements and if the inversion was truly linear*. Hence, it represents a truly optimistic scenario! Its value lies in the fact that it describes the upper limit on our ability to estimate the true model due to the inherent deficiencies in the particular data set used. Substituting Eq. (20) into Eq. (23) one obtains the following relation between the estimated model and the true model:

$$\mathbf{m}_{\text{est}} = \mathbf{R}\,\mathbf{m}_t \tag{25}$$

where

$$\mathbf{R} = (\mathbf{A}^T\mathbf{W}_d\mathbf{A} + \lambda\mathbf{W}_m)^{-1}\mathbf{A}^T\mathbf{W}_d\mathbf{A} \tag{26}$$

If $\mathbf{R} = \mathbf{I}$ the estimated model is equal to the true model; as $\mathbf{R}$ deviates from $\mathbf{I}$ the estimate becomes an average of the true model weighted by resolution matrix $\mathbf{R}$. In nonlinear systems $\mathbf{R}$ is only valid locally around $\mathbf{m}_{\text{est}}$. Equation (26) seems to suggest that regularisation degrades the resolution; in the absence of regularisation ($\lambda = 0$) the resolution operator is given by $\mathbf{R} = (\mathbf{A}^T\mathbf{W}_d\mathbf{A})^{-1}\mathbf{A}^T\mathbf{W}_d\mathbf{A} = \mathbf{I}$. However, in practical problems in the absense of regularisation the inverse $(\mathbf{A}^T\mathbf{W}_d\mathbf{A})^{-1}$ does not exist, so one must use an approximate generalised inverse $(\mathbf{A}^T\mathbf{W}_d\mathbf{A})^{-g}$ instead. Since $(\mathbf{A}^T\mathbf{W}_d\mathbf{A})^{-g}(\mathbf{A}^T\mathbf{W}_d\mathbf{A})$ then differs from the identity matrix the resolution operator in that case also differs from the identity matrix. Usually a range of $\mathbf{W}_m$ is reasonable for any particular inversion so the resolution should be assessed for samples of $\mathbf{W}_m$ within this range (Rezovsky and Ritzwoller, 1999).

One final source of error must be considered, and this is derived from truncating expansion Eq. (17) to a finite set of P basis functions. If the true Earth structure $\mathcal{M}_t(\mathbf{x})$ contains some features that affect the recorded data but that cannot be represented within the truncated basis, these features will cause spurious errors in the finite-dimensional model solution $\mathbf{m}_{est}$ (an example is given later, see Color Plate 22). This process is called spectral leakage, and is caused by an inappropriate choice of the set of basis functions (Snieder *et al.*, 1991). Trampert and Snieder (1996) presented a solution to this problem under certain conditions, but the best way to solve it approximately is to ensure that the P chosen basis functions allow sufficient detail to model those features of $\mathcal{M}$ that have any significant influence on the data (if this is possible).

4.3 Computational Issues

Notice that each linearised iteration i in Figure 3(b) requires the calculation of the synthetic data and its gradient at model $\mathbf{m}_i$, plus a matrix inversion [Eq. (23)] or an approximation to solution (23) if this matrix is too large to be inverted (e.g., Paige and Saunders, 1982; Nolet, 1985; van der Sluis and van der Vorst, 1987). Assessing the true uncertainty in nonlinear problems (e.g., the union of the shaded regions in Fig. 3) requires computationally intensive nonlinear inversion methods such as repeated linearised inversion from many different starting models (Deng, 1997), Monte Carlo (e.g., Hammersley and Handscomb, 1964), or other types of stochastic nonlinear inversion methods (Sen and Stoffa, 1991, 1992; Vinther and Mosegaard, 1996; Douma *et al.*, 1996; Sambridge, 1998; Devilee *et al.*, 1999). All of these typically require the function $\mathbf{F}$ to be evaluated many thousands of times. This is currently impossible for common large inverse systems with complex physics which is the main reason why linearised physics, inversion, and uncertainty estimates are so widely used, despite their shortcomings described above.

5. Optimal Survey Design

We now show how seismic surveys can be designed such that the data collected allow the model to be constrained with minimum uncertainty using linearised inverse methods. We showed above how the effectiveness (quality) of tomography is usually estimated using the model resolution $\mathbf{R}$ and uncertainty $\mathbf{C}$, both of which depend on $\mathbf{W}_m$, $\mathbf{W}_d$, and $\mathbf{A}$ and both of which are valid only locally around any misfit minimum. Once the regularisation $\mathbf{W}_m$ and a value of λ has been fixed, and assuming that the model basis functions have been chosen carefully, the tomographic quality approximately depends only on the number of misfit minima n, on matrices $\mathbf{A}(\check{\mathbf{m}}_i)$ (the gradient of $\mathbf{F}(\check{\mathbf{m}}_i)$) at each misfit minimum $\check{\mathbf{m}}_i$: $i = 1, \ldots, n$), and on data weights $\mathbf{W}_d$. In turn, all of these can only be influenced by varying the survey design.

Let $\mathbf{S} = [\mathbf{r}_1, \ldots, \mathbf{r}_s]$ represent the survey design where the variables $\mathbf{r}_i$ might describe the source or receiver locations, the data bandwidth, data types recorded, the make of equipment used, the number of repeated measurements, etc., all of which might affect n, $\mathbf{A}(\check{\mathbf{m}}_i)$ or $\mathbf{W}_d$. To design surveys optimally requires that we have some measure Φ of survey quality. The design $\mathbf{S}$ can then be varied such that Φ is maximised.

Let us begin by assuming that the number of minima $n = 1$. Therefore, iterated linearised tomographic inversion should be robust (Fig. 3). The easiest way to measure survey quality for linearised inversion would be to define $\Phi[\mathbf{C}, \mathbf{R} - \mathbf{I}]$ to measure some balance between the linearised uncertainty estimate $\mathbf{C}$ and how closely the resolution matrix $\mathbf{R}$ resembles the identity $\mathbf{I}$. These, and related measures are discussed by, for example, Barth and Wunsch (1990), Atkinson and Donev

(1992), Maurer and Boerner (1998), and Curtis (1999a). Increasing the complexity slightly, these linearised techniques can be adapted to measure how information is distributed between particular targets of interest within the model space; by varying the design to alter this distribution, focused surveys can be designed (Curtis, 1999b).

Examples of unfocused and focused crosswell tomographic surveys are shown in Figure 4 and Color Plate 21(a) respectively. Figure 4 shows a tomographic survey and corresponding ray paths through a homogeneous reference model that give a high degree of information about the slowness structure across the whole interwell model space (Curtis, 1999a). Notice that the average density of sources and receivers generally increases down the length of each well, and across the surface the average density is half that within the wells. The survey shown in Color Plate 21(a) uses fewer sources and receivers and hence cannot resolve velocities in all of the model cells. Instead, this plot shows a survey that provides most information about (is focused on) the region of interwell space marked by the crosses (Curtis, 1999b). Notice that the ray paths are mainly focused within the relevant region (but also note that the focusing of ray paths alone is a poor criterion on which to design surveys, see Curtis and Snieder, 1997).

A crucial subtlety exists here: survey design, by definition, is carried out before data have been collected and hence before either a misfit function E or any inversion solution $\mathbf{m}_e$ exist. Hence, instead of evaluating $\mathbf{C}$ and $\mathbf{R}$ at misfit minima, they must be evaluated at our best prior estimate $\mathbf{m}_p$ of the solution. Clearly if the gradients $\mathbf{A}(\mathbf{m}_p)$ and $\mathbf{A}(\mathbf{m}_t)$ differ, the quality estimate will be in error; this is why such methods rely intrinsically on pseudo-linearity of the forward function $\mathbf{F}$.

One method to compensate partially for this problem is offered in the field of Bayesian statistical experimental design (e.g., Atkinson and Donev, 1992; Maurer and Boerner, 1998). Say all of our knowledge about the Earth model prior to inversion can be summarised by a prior probability distribution $\rho(\mathbf{m})$, then our best estimate of survey design quality is given by:

$$\overline{\Phi} = \int_{\mathcal{M}} \Phi(\mathbf{m})\rho(\mathbf{m})d\mathbf{m} \tag{27}$$

where $\Phi(\mathbf{m})$ is any measure of survey design quality evaluated at a specific model $\mathbf{m}$, similar to those measures discussed above for example. $\overline{\Phi}$ is called the Bayesian estimator of survey design quality. Designing experiments to maximise $\overline{\Phi}$ is the best we can do, given our (lack of)

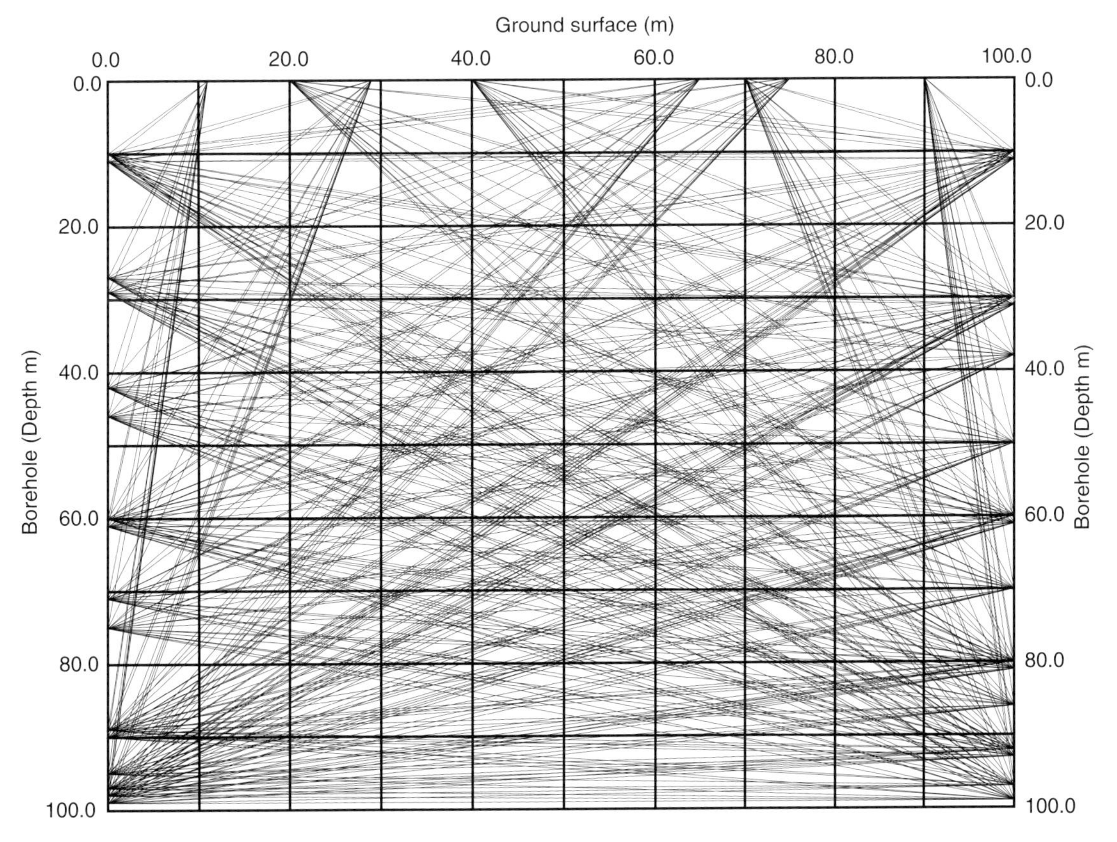

FIGURE 4 Best unfocused crosswell survey geometry found. The left and right sides of the diagram represent vertical boreholes, the top represents the ground surface. Squares represent model cells, ray paths between sources and receivers are shown by straight lines. Twenty sources and twenty receivers were used in this example.

knowledge about the Earth model before we conduct the survey.

The measure in Eq. (27) averages over the different values of Φ which arise due to nonlinearity of $\mathbf{F}$. However, Curtis and Spencer (1999) illustrate that using a standard linearised quality measure for Φ similar to those described above does not circumvent the requirement that there is only a single misfit minimum – this is assumed in Eq. (27) since it is assumed for measure $\Phi(\mathbf{m})$. Hence, this expression only compensates for weak nonlinearity.

As explained earlier, in nonlinear inverse problems with multiple minima ($n > 1$), iterated linearised inversion converges to a single minimum and allows local uncertainty estimation, providing no indication of whether other possible solutions (other minima) exist. The only way to ensure the robustness of such inversions is therefore to change the misfit function such that it has a single minimum, and again this might be achieved by varying the survey design. This approach was demonstrated by Curtis and Spencer (1999) who varied the survey design such that the number of minima n was minimised. However, these methods are computationally very expensive and so far they have only been applied to problems with few model parameters.

If we are prepared to pay the computational cost of using stochastic, nonlinear inversion methods as described earlier then we can relax the requirement that we have a unique misfit minimum since, in principle, multiple minima can be found and explored during the inversion and the true uncertainty can be assessed. In such cases, we might vary the survey design such that a measure of total expected information is maximized (Curtis and Spencer, 1999). However, this generally incurs the highest computational cost of any method proposed in this chapter. For geophysical problems with weak prior information about the model that we wish to estimate, this method is unlikely to be possible other than for low-dimensional problems [$P \sim 10$ in Eq. (17)].

6. Tomographic Applications

Tomography has been applied to many problems at length scales ranging from centimetres to the size of the Earth. There is too little space here to give an overview of all of these, so in the rest of this chapter we will focus on continental and cross-well scale tomography. We will show some examples that illustrate how the theory presented in this chapter has been applied, and how the results may be related to tectonically derived Earth structures.

6.1 Surface Wave Inversion Using the Born Approximation

Although traditionally the Born approximation has been used mostly to analyse reflection data (energy scattered from sharp velocity discontinuities), it can also be used for the analysis of transmission data since it accounts for the complete first-order perturbation of the data by changes in the model. However, it is true that the domain of applicability is larger for reflection data than for transmission data. There is a simple physical reason for this. Suppose the velocity is perturbed with a small perturbation of say 1%. In general the reflected waves will be of the same order of magnitude; they will be around 1% of the incident wave-field and are therefore very weak. Suppose a transmitted wave experiences this velocity perturbation of 1% over 100 wavelengths. This leads to a phase shift of a complete cycle. This is no longer a small perturbation of the wave field so linearised theory will not be applicable.

For surface waves in global seismology the wavelength is fairly large. For example, a Rayleigh wave with a period of 75 sec has a wavelength of about 300 km. This means that these waves propagate over a limited number of wavelengths when they are used for continental-scale inversions. It is for this reason that these waves can be analysed using the Born approximation. A description of this method of analysis, the linearised inversion, and a number of numerical examples is given by Snieder (1988a). This technique has been applied to infer the shear-velocity structure under Europe and the Mediterranean by Snieder (1988b) and under North America by Alsina *et al.* (1996). Shear-velocity perturbations from the latter study in three layers with depths between 25 and 300 km is shown in Color Plate 21(b). Red colours map slow velocities that may indicate high temperature and green colours map high velocities corresponding to low temperature. Note, though, that the identification of velocity anomalies with temperature perturbations should be treated with care. Apart from temperature perturbations the seismic velocity is also affected by variations in composition, the presence of volatiles, such as water, and possibly also pressure. A tectonic interpretation of this model is given by Alsina *et al.* (1996). Images such as this make it possible to see "snaphosts" of plate tectonics in action.

6.2 Body and Surface Wave Comparisons

It was shown earlier that linearizing approximations need to be made both for the estimation and assessment of tomographic models. In general we do not know how nonlinear these tomographic problems are. However, whatever errors are committed by linearisation, we would not expect them to be the same errors in different studies using independent data and different data types. Hence, agreement or disagreement between such studies may be a reasonable indicator of the solution quality.

In continental and global-scale seismology, models have been produced using surface waves (e.g., Trampert and Woodhouse, 1995, 1996; Ekström *et al.*, 1997; Ritzwoller and Levshin, 1998; Curtis *et al.*, 1997; Lee and Nolet, 1997; Laske and Masters, 1998), normal modes (Woodhouse and Dziewonski, 1984), and body waves (Robertson and Woodhouse, 1995; van der Hilst *et al.*, 1997; Bijwaard and Spakman, 2000). However, it was not until recently that these models began to look alike (e.g., Ritzwoller and Lavely, 1995). Before this time, normal mode and most surface-wave models tended to be of very low lateral resolution, whereas body-wave models had very high resolution in some areas of the mantle and around seismically active zones in the lithosphere but very poor resolution elsewhere. Hence, there was no general agreement between the model structures found and it was not clear that the solutions found were really the only such structures that would fit the data.

More recently, surface-wave studies have produced models of continental structure with lateral resolution comparable to that of body-wave models. Color Plate 21(c) shows a comparison of 40 sec period, fundamental mode Rayleigh-wave phase velocity of Curtis *et al.* (1997) and a slice through the body wave P-velocity model of Bijwaard and Spakman (2000) at 53 km depth. To understand the phase-velocity model it is necessary to examine Figure 5 which shows the sensitivity of Rayleigh phase velocities to shear velocity at different depths in the Earth [these are the (normalised) kernels $K_{\beta}(z)$ in Eq. (16)]. Although P velocity and density also effect the phase velocity, shear velocity has a greater influence, hence these kernels show the dominant sensitivity.

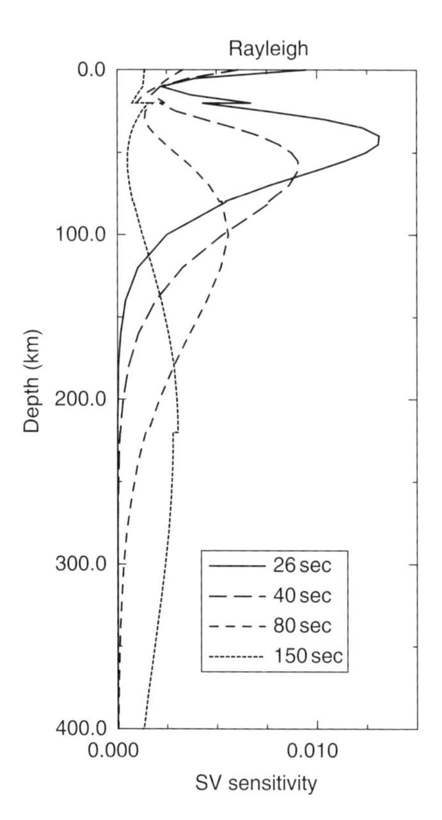

FIGURE 5 Sensitivity of fundamental mode Rayleigh-wave phase velocity to shear velocity variations at depth.

Equation 16 shows that phase velocity anomalies at any point on the Earth's surface are averages of shear velocity anomalies over depth, weighted by these kernels. Hence, the 40 sec period Rayleigh wave is most sensitive to shear velocity anomalies in the depth range 40–100 km with peak sensitivity around 55 km. It makes sense, then, to compare this with the P-velocity model of Bijwaard and Spakman (2000) around 55 km depth: if P and S anomalies are correlated (if they are due to the same Earth structures) then the anomalies should exhibit similar patterns.

Despite the approximations made in the above argument, several features common to both models in Color Plate 21(c) are immediately obvious. The low velocity anomaly running from the Aegean, through the Caucasus and North Iran to the Pamir and Tien Shan north of Tibet is visible in both studies. Low velocities exist beneath central Tibet in both studies, as detected in previous studies using various data types (Romanowicz, 1982; Molnar and Chen, 1984; Curtis and Woodhouse, 1997). Also, the high velocities of the Northwest Indian subcontinent are visible and extend beneath the west-central Himalayan arc in both studies. India is actively subducting beneath the Tibetan plateau but there is still a debate about how far north the subducted plate extends. Both of these studies would suggest that the answer is almost 0 km in the east, increasing to approximately the width of the Karakoram mountain chain in the west, at least around 55 km depth.

7. New Developments and Future Research

Tomographic models using body-wave and surface-wave data now appear to attain comparable resolution within some volumes of the Earth, but their sensitivity to aspects of Earth structure are tremendously different (Figs. 1 and 5). In this sense the two data types are complementary: for example, body wave models tend to have low resolution within much of the Earth's lithosphere but higher resolution in the mantle, whereas fundamental mode surface waves provide better resolution of the lithosphere but very poor resolution of the deeper mantle. Hence, some research effort is currently being spent trying to create tomographic models consistent with both data types simultaneously.

This is just one example of how a new data type can be used to compensate for deficiencies of our current data types. Such strategies would seem to be the most efficient way to add constraints to any current model, since it is often the case that it is impossible to compensate for these deficiencies simply by adding extra data of the same type. (For example, Fig. 5 shows that one could add any number of fundamental mode surface wave observations at periods lower than 40 sec and never be able to constrain the mantle at 400 km depth.) The natural extension to this is to include all possible data types, and recently a Bayesian framework to enable this was presented by Bosch (1999).

Some of the main issues highlighted above were the deficiencies of linearised inversion schemes. Clearly, to remove problems associated with this it is necessary to use nonlinear schemes that explore much more of the misfit surface than just a single misfit minimum. However, for tomographic problems this is still computationally intractible (Curtis and Lomax, 2001). Indeed, it is only recently that iterated linearised inversion has been achieved in global tomography, and even this has only been attempted from a single initial reference model (Bijwaard and Spakman, 2000). Linearised inversion from different reference models and introducing fully nonlinear inversion are areas of future research that will surely follow as computer power increases and more efficient methodologies are found.

Two important areas of tomography that we have not had space to discuss in detail are industrial tomography and seismic anisotropy. Seismic anisotropy is the name used to describe the variation in seismic velocity for waves either travelling or oscillating in different directions. Many studies have been carried out to attempt to constrain anisotropy within the Earth (e.g., Nataf *et al.*, 1986; Montagner and Nataf, 1988; Laske and Masters, 1998), but the data geometries available using earthquake data do not always enable detailed anisotropic structure to be resolved distinctly from the isotropic structure across much of the Earth. Studies in which seismic anisotropy is most likely to be best constrained occur in industrial geophysics where surveys can be designed using the methods presented earlier to provide extremely dense data geometries that are fit specifically for this purpose. Detecting anisotropy is important for hydrocarbon production and in selecting sites for waste disposal since it tends to indicate the orientation of the prevelant fracture field. Such fracture networks produce anisotropic permeability and hence can dramatically change either the optimal network of wells required to drain a reservoir, or our estimates of the directions in which waste-infected groundwater might flow. For these reasons a great deal of industrial research is currently directed toward developing new methods for anisotropy detection.

Color Plate 22 shows an example of the isotropic components of seismic velocity found from cross-well tomography at a site in Europe (see Miller and Chapman, 1991; Miller and Costa, 1992). The upper plot was obtained when only isotropic velocity components were included in model vector **m** in Eq. (17); the lower plot was obtained when **m** included both isotropic and anisotropic velocity components. The Earth structure is clearly fairly constant horizontally between the wells, but the left-hand tomogram shows a superimposed "X" structure. This is characteristic of cases where the data were generated by wave propagation through an anisotropic medium but where the model basis functions [Eq. (17)] cannot represent this anisotropy. The cross appears by spectral leakage of the anisotropic structure into the isotropic model basis components (Section 4). This effect is removed by using a more appropriate choice of basis functions (Color Plate 22, lower plot) that includes the anisotropy of the medium.

It has been impossible to describe all the methods now used for tomographic applications, and there are many more research areas being investigated than the few mentioned here. We have tried to cover those techniques that are common in, or may be useful for, regional or global seismology. For techniques and applications not covered here, we refer the reader to Aki and Richards (1980) or Dahlen and Tromp (1998).

Acknowledgments

We gratefully acknowledge discussions with Wim Spakman and support from Sarah Ryan during the course of this work.

References

Aki, K., A. Christoffersen, and E.S. Husebye (1977). Determination of the three-dimensional structure of the lithosphere. *J. Geophys. Res.* **82**, 277–296.

Aki, K. and W.H.K. Lee (1976). Determination of three-dimensional velocity anomalies under a seismic array using first p arrival times from local earthquakes, part 1. A homogeneous initial model. *J. Geophys. Res.* **81**, 4381–4399.

Aki, K. and P.G. Richards (1980). "Quantitative Seismology, Theory and Methods," Vol. 1. W.H. Freeman, San Francisco.

Aldridge, D.F. (1994). Linearization of the Eikonal equation. *Geophysics* **59**, 1631–1632.

Alsina, D., R.L. Woodward, and R. Snieder (1996). Shear-wave velocity structure in North America from large-scale waveform inversions of surface waves. *J. Geophys. Res.* **101**, 15969–15986.

Atkinson, A.C. and A.N. Donev (1992). "Optimum Experimental Designs." Clarendon Press, Oxford.

Barrett, H.H. (1984). The radon transform and its applications. *Progr. Optics* **XXI**, 217–286.

Barth, N. and C. Wunsch (1990). Oceanographic experiment design by simulated annealing. *J. Phys. Oceanogr.* **20**, 1249–1263.

Ben-Menahem, A. and S.J. Singh (1981). "Seismic Waves and Sources." Springer-Verlag, New York.

Bijwaard, H. and W. Spakman (2000). Nonlinear global p-wave tomography by iterated linearized inversion. *Geophys. J. Int.* **141**, 71–82.

Bosch, M. (1999). Lithologic tomography: from plural geophysical data to lithological estimation. *J. Geophys. Res.* **104**(B1), 749–766.

Bregman, N., R. Bailey, and C. Chapman (1989). Crosshole seismic tomography. *Geophysics* **54**, 200–215.

Comer, R. and R. Clayton (1984). Tomographic reconstruction of lateral velocity heterogeneity in the earth's mantle (abstract). *EOS Trans. Am. Geophys. U.* **65**, 236.

Curtis, A. (1999a). Optimal experiment design: cross-borehole tomographic examples. *Geophys. J. Int.* **136**, 637–650.

Curtis, A. (1999b). Optimal design of focussed experiments and surveys. *Geophys. J. Int.* **139**, 205–215.

Curtis, A. and A. Lomax (2001). Prior information, sampling distributions and the curse of dimensionality. *Geophysics* **66**, 372–378.
Curtis, A. and R. Snieder (1997). Reconditioning inverse problems using the genetic algorithm and revised parameterization. *Geophysics* **62**(5), 1524–1532.
Curtis, A. and C. Spencer (1999). "Survey design strategies for linearized, nonlinear inversion". *Soc. of Expl. Geophys. Expanded Abstracts*, 1775–1778.
Curtis, A. and J.H. Woodhouse (1997). Crust and upper mantle shear velocity structure beneath the Tibetan plateau and surrounding regions from interevent surface wave phase velocity inversion. *J. Geophys. Res.* **102**(B8), 11789–11813.
Curtis, A., B. Dost, J. Trampert, and R. Snieder (1997). Eurasian fundamental mode surface wave phase velocities and their relationship with tectonic structures. *J. Geophys. Res.* **103**(B11), 26919–26947.
Dahlen, F.A. and J. Tromp (1998). "Theoretical Global Seismology." Princeton University Press, Princeton.
de Hoop, A.T. (1995). "Handbook of Radiation and Scattering of Waves: Acoustic Waves in Fluids, Elastic Waves in Solids, Electromagnetic Waves." Academic Press, San Diego.
Deng, H.L. (1997). A complexity analysis of generic optimization problems: Characterizing topography of high-dimensional functions. PhD thesis, Center for Wave Phenomena, Colorado School of Mines, Golden, Colorado.
Devilee, R., A. Curtis, and K. Roy-Chowdhurry (1999). An efficient, probabilistic, neural network approach to solving inverse problems: to inverting surface wave velocities for Eurasian crustal thickness. *J. Geophys. Res.* **104**(B12), 28841–28856.
Douma, H., R. Snieder, and A. Lomax (1996). Ensemble inference in terms of empirical orthogonal functions. *Geophys. J. Int.* **127**, 363–378.
Dziewonski, A.M. (1984). Mapping the lower mantle: determination of lateral heterogeneity in p velocity up to degree and order 6. *J. Geophys. Res.* **89**, 5929–5952.
Dziewonski, A.M. and J.H. Woodhouse (1987). Global images of the earth's interior. *Science* **236**, 37–48.
Ekström, G., J. Tromp, and E.W.F. Larson (1997). Measurements and global models of surface wave propagation. *J. Geophys. Res.* **102**(B4), 8137–8157.
Hammersley, J.M. and D.C. Handscomb (1964). "Monte Carlo Methods." Methuen, London.
Herglotz, G. (1907). Über das Benndorfsche Problem der Fortplanzungsgeschwindigkeit der Erdbebenstralen. *Z. Geophys.* **8**, 145–147.
Hudson, J.A. (1977). Scattered waves in the coda of P. *J. Geophys.* **43**, 359–374.
Iyer, H.M. and K. Hirahara (1993). "Seismic Tomography." Prentice-Hall, London.
Laske, G. and G. Masters (1998). Surface-wave polarization data and global anisotropic structure. *Geophys. J. Int.* **132**, 508–520.
Lee, S. and G. Nolet (1997). Upper mantle s velocity structure of North America. *J. Geophys. Res.* **102**, 22815–22838.
Matsu'ura, M. and N. Hirata (1982). Generalized least-squares solutions to quasi-linear inverse problems with a priori information. *J. Phys. Earth* **30**, 451–468.
Maurer, H. and D.E. Boerner (1998). Optimized and robust experimental design: a non-linear application to em sounding. *Geophys. J. Int.* **132**, 458–468.
Menke, W. (1989). "Geophysical Data Analysis: Discrete Inverse Theory" (Revised edn.), vol. 45 of International Geophysics Series. Academic Press, Harcourt Brace Jovanovich, New York.
Mertzbacher, E. (1970). "Quantum mechanics," 2nd edn. Wiley, New York.
Miller, D. and C.H. Chapman (1991). Incontrovertible evidence of anisotropy in crosswell data. Extended abstracts of the 61st Annual Meeting of the SEG, pp. 858–928.
Miller, D. and C. Costa (1992). Inversion for the devine anisotropic medium. Expanded abstract from the EAGE annual meeting, Paris, pp. 88–89.
Molnar, P. and W.-P. Chen (1984). *s–p* travel time residuals and lateral inhomogeneity in the mantle beneath Tibet and the Himalaya. *J. Geophys. Res.* **89**(B8), 6911–6917.
Montagner, J.-P. (1994). What can seismology tell us about mantle convection? *Rev. Geophys.* **32**, 115–137.
Montagner, J.-P. and H.-C. Nataf (1988). Vectorial tomography—i. Theory. *Geophys. J.* **94**, 295–307.
Morelli, A., E. Boschi, and G. Ekström (1996). "Seismic Modelling of the Earth's Structure." Editrice Compositori, Bologna.
Nataf, H.-C., I. Nakanishi, and D.L. Anderson (1986). Measurement of mantle wave velocities and inversion for lateral heterogeneities and anisotropy 3. inversion. *J. Geophys. Res.* **91**(B7), 7261–7307.
Natterer, F. (1986). "The Mathematics of Computerized Tomography." John Wiley, New York.
Nolet, G. (1985). Solving or resolving inadequate and noisy tomographic systems. *J. Comp. Phys.* **61**, 463–482.
Nolet, G. (1987). Seismic wave propagation and seismic tomography. In: "Seismic Tomography" (G. Nolet, Ed.), pp. 1–23. Reidel, Dordrecht.
Nolet, G. (1990). Partitioned waveform inversion and two-dimensional structure under the network of autonomously recording seismographs. *J. Geophys. Res.* **95**, 8499–8512.
Nolet, G., J. van Trier, and R. Huisman (1986). A formalism for nonlinear inversion of seismic surface waves. *Geophys. Res. Lett.* **13**, 26–29.
Paige, C.C. and M.A. Saunders (1982). LSQR: an algorithm for sparse linear equations and sparse least squares. *ACM Trans. Math. Soft.* **8**, 43–71.
Parker, R.L. (1994). "Geophysical Inverse Theory." Princeton University Press, Princeton.
Rezovsky, J.S. and M.H. Ritzwoller (1999). Regularization uncertainty in density models estimated from normal mode data. *Geophys. Res. Lett.* **26**, 2319–2322.
Ritzwoller, M.H. and E. Lavely (1995). Three-dimensional seismic models of the earth's mantle. *Rev. Geophys.* **33**, 1–66.
Ritzwoller, M.H. and A.L. Levshin (1998). Eurasian surface wave tomography: group velocities. *J. Geophys. Res.* **103**(B3), 4839–4878.
Robertson, G.S. and J.H. Woodhouse (1995). Evidence for proportionality of p and s heterogeneity in the lower mantle. *Geophys. J. Int.* **123**, 85–116.
Romanowicz, B. (1982). Constraints on the structure of the Tibetan plateau from pure path phase velocities of Love and Rayleigh waves. *J. Geophys. Res.* **87**(B8), 6865–6883.
Romanowicz, B. (1991). Seismic tomography of the earth's mantle. *Annu. Rev. Earth Planet. Sci.* **19**, 77–99.
Sambridge, M. (1990). Non-linear arrival time inversion: constraining velocity anomalies by seeking smooth models in 3-D. *Geophys. J. R. Astron. Soc.* **102**, 653–677.

Sambridge, M. (1998). Exploring multidimensional landscapes without a map. *Inverse Problems* **14**(3), 427–440.
Scales, J.A. and R. Snieder (1997). To Bayes or not to Bayes? *Geophysics* **62**(4), 1045–1046.
Schlichter, L.B. (1932). The theory of the interpretation of seismic travel-time curves in horizontal structures. *Physics* **3**, 273–295.
Sen, M.K. and P.L. Stoffa (1991). Nonlinear one-dimensional seismic waveform inversion using simulated annealing. *Geophysics* **56**(10), 1624–1638.
Sen, M.K. and P.L. Stoffa (1992). Rapid sampling of model space using genetic algorithms: examples from seismic waveform inversion. *Geophys. J. Int.* **108**, 281–292.
Snieder, R. (1986a). 3D linearized scattering of surface waves and a formalism for surface wave holography. *Geophys. J. R. Astron. Soc.* **84**, 581–605.
Snieder, R. (1986b). The influence of topography on the propagation and scattering of surface waves. *Phys. Earth Planet. Inter.* **44**, 226–241.
Snieder, R. (1988a). Large-scale waveform inversions of surface waves for lateral heterogeneity, 1, theory and numerical examples. *J. Geophys. Res.* **93**, 12055–12065.
Snieder, R. (1988b). Large-scale waveform inversions of surface waves for lateral heterogeneity, 2, application to surface waves in Europe and the Mediterranean. *J. Geophys. Res.* **93**, 12067–12080.
Snieder, R. (1998). The role of nonlinearity in inverse problems. *Inverse Problems* **14**, 387–404.
Snieder, R. (2001). "A Guided Tour of Mathematical Methods for the Physical Sciences." Cambridge University Press, Cambridge, UK.
Snieder, R. and D.F. Aldridge (1995). Perturbation theory for travel times. *J. Acoust. Soc. Am.* **98**, 1565–1569.
Snieder, R. and M. Sambridge (1992). Ray perturbation theory for travel times and ray paths in 3-D heterogeneous media. *Geophys. J. Int.* **109**, 294–322.
Snieder, R. and J. Trampert (2000). Linear and nonlinear inverse problems. In: "Geomatic methods for the analysis of data in the earth sciences" (Dermanis, A., Grün, A., and F. Sanso, Eds.), Springer, Berlin.
Snieder, R.K., J. Beckers, and F. Neele (1991). The effect of small-scale structure on normal mode frequencies and global inversions. *J. Geophys. Res.* **96**, 501–515.
Spakman, W., S. Van Der Lee, and R. van der Hilst (1993). Travel-time tomography of the European–Mediterranean mantle down to 1400 km. *Phys. Earth Planet. Inter.* **79**, 3–74.
Spencer, C. and D. Gubbins (1980). Travel-time inversion for simultaneous earthquake location and velocity structure determination in laterally varying media. *Geophys. J. R. Astron. Soc.* **63**, 95–116.
Takahashi, T. (1955). Analysis of the dispersion curves of Love waves. *Bull. Earthq. Res. Inst.* **33**, 287–296.
Takahashi, T. (1957). The dispersion of Rayleigh waves in heterogeneous media. *Bull. Earthq. Res. Inst.* **35**, 297–308.
Takeuchi, H. and M. Saito (1972). Seismic surface waves. In: "Seismology: Surface Waves and Earth Oscillations (Methods in Computational Physics, Vol. 11)" (B.A. Bolt, Ed.), Academic Press, New York.
Tanimoto, T. (1987). The three-dimensional shear wave structure in the mantle by overtone waveform inversion—i. Radial seismogram inversion. *Geophys. J. R. Astron. Soc.* **89**, 713–740.
Tanimoto, T. and D.L. Anderson (1990). Lateral heterogeneity and azimuthal anisotropy of the upper mantle: Love and Rayleigh waves 100–250 s. *J. Geophys. Res.* **90**, 1842–1858.
Tarantola, A. (1987). "Inverse Problem Theory." Elsevier Science, Amsterdam.
Tarantola, A. and B. Valette (1982a). Inverse problems = quest for information. *J. Geophys.* **50**, 159–170.
Tarantola, A. and B. Valette (1982b). Generalized nonlinear inverse problems solved using the least squares criterion. *Rev. Geophys.* **20**(2), 219–232.
Thomson, C.J. and D. Gubbins (1982). Three-dimensional lithospheric modelling at norsar: linearity of the method and amplitude variations from the anomalies. *Geophys. J. R. Astron. Soc.* **71**, 1–36.
Tikhonov, A.N. (1963). On the solution of improperly posed problems and the method of regularization. *Dokl. Akad. Nauk SSSR* **151**, 501.
Trampert, J. and R. Snieder (1996). Model estimations based on truncated expansions: possible artifacts in seismic tomography. *Science* **271**, 1257–1260.
Trampert, J. and J.H. Woodhouse (1995). Global phase velocity maps of Love and Rayleigh waves between 40 and 150 seconds. *Geophys. J. Int.* **122**, 675–690.
Trampert, J. and J.H. Woodhouse (1996). High resolution global phase velocity distributions. *Geophys. Res. Lett.* **23**(1), 21–24.
van der Hilst, R.D., S. Widiyantoro, and E.R. Engdahl (1997). Evidence for deep mantle circulation from global tomography. *Nature* **386**, 578–584.
van der Sluis, A. and H.A. van der Vorst (1987). Numerical solution of large, sparse linear algebraic systems arising from tomographic problems. In: "Seismic tomography, with Applications in Global Seismology and Exploration Geophysics" (G. Nolet, Ed.), Reidel, Dordrecht.
Vinther, R. and K. Mosegaard (1996). Seismic inversion through Tabu search. *Geophys. Prospect.* **44**, 555–570.
Weichert, E. (1910). Bestimmung des weges der erdbebenwellen im erdinneren. i. theoretisches. *Phys. Z.* **11**, 294–304.
Woodhouse, J.H. (1976). On Rayleigh's principal. *Geophys. J. R. Astron. Soc.* **46**, 11–22.
Woodhouse, J.H. (1980). The coupling and attenuation of nearly resonant multiplets. *Geophys. J. R. Astron. Soc.* **61**, 261–283.
Woodhouse, J.H. and F.A. Dahlen (1978). The effect of a general aspherical perturbation on the free oscillations of the Earth. *Geophys. J. R. Astron. Soc.* **53**, 335–354.
Woodhouse, J.H. and A.M. Dziewonski (1984). Mapping the upper mantle: three-dimensional modeling of earth structure by inversion of seismic waveforms. *J. Geophys. Res.* **89**, 5953–5986.

Editor's Note

Please see also Chapter 10, Normal modes of the Earth and planets, by P. Lognonne and E. Clevede; Chapter 11, Inversion of surface waves: a review, by Romanowicz; Chapter 16, Probabilistic approach to inverse problems, by K. Mosegaard and A. Tarantola; Chapter 51, The Earth's interior, by T. Lay; and Chapter 53, Seismic anisotropy, by Cara.

53

Seismic Anisotropy

Michel Cara
Institut de Physique du Globe de Strasbourg, France

1. Introduction and Historical Review

Seismologists have been investigating the elastic properties of the Earth's deep interior for more than a century, but observations of seismic anisotropy are quite recent. The first observational evidence can be traced back to the early 1950s when, in connection with the development of seismic exploration techniques, it was found that the velocity of P waves could depend on their incidence angle in layered sediments. It soon became evident, however, that the amount of anisotropy that is required should not substantially affect the exploration techniques in use at that time (Krey and Helbig, 1956). In the subcrustal lithosphere, the first reports of seismic anisotropy were made in the early 1960s with observations of both apparent incompatibilities between the Love- and Rayleigh-wave velocities and azimuthal variations of Pn velocities. These two kinds of observations were independently attributed to the effect of partial melt in the upper mantle (e.g., McEvilly, 1964; Aki, 1968) and to preferred orientation of olivine crystals beneath the oceanic crust (Hess, 1964). For years, these earliest findings were considered more as marginal facts than signs of a situation that could prevail in many parts of the Earth.

Furthermore, many scientists have expressed doubts during the past 30 y on the necessity of invoking anisotropy to explain some of the reported observations. The earliest Love-wave observations were, for example, considered as contaminated by higher modes and of no value with regard to the question of seismic anisotropy (James, 1971). Other authors pointed out the problem of a trade-off between anisotropy and lateral heterogeneities (e.g., Levshin and Ratnikova, 1984).

The number of studies dealing with seismic anisotropy has dramatically increased during the past 10 y. It is out of the scope of the present chapter to present an exhaustive review of the subject. The literature published before 1990 has been reviewed by Babuska and Cara (1991). Since then, mantle anisotropy has been reviewed by Silver (1996), Montagner (1998), and Savage (1999) and a monograph dedicated to various theoretical aspects of seismic anisotropy (Helbig, 1994) is useful in the field of seismic exploration techniques. In the present chapter, we mainly focus our attention on the seismological aspects of the subject and on several results linking the observations of seismic anisotropy to dynamic processes acting in the Earth. Due to the very rapid growth of literature on the subject, it was not possible to list all the recent publications. Only a selected list of references is given, and questions related to mineral physics and rock properties are only marginally addressed.

Let us first recall that the theory of elasticity in homogeneous anisotropic media was already well advanced by the mid-19th century, after the works of Louis Cauchy and George Green. Elasticity of crystals was thoroughly studied later on, and their anisotropic properties have spread light on the 19th century controversy concerning the number of elastic constants to be considered in a general anisotropic solid (e.g., Love, 1944). It is also worthwhile to recall that the development of the theory of elasticity had been intimately linked with research on the propagation of light. The then broadly accepted concept of immaterial "ether" was the ground on which the theory of propagation of the transversely polarized optical waves was developed, and the properties of elastic waves in solids served as a guideline. Finally, let us note that the term "anisotropy" was not adopted in the former English literature. Instead, Kelvin (1904) proposed the word aelotropy: "I see the Germans have adopted the term anisotropy. If we used this in English we should have to say, *An anisotropic solid is not an isotropic solid*; this jangle between the prefix an (privative) and the article an, if nothing else, would prevent us from adopting that method of distinguishing a nonisotropic solid from an isotropic solid."

Elastic anisotropy of homogeneous media is a good start to study the propagation of seismic waves, but more theoretical developments are required to address the question of wave propagation in the Earth. The effects of both inhomogeneities and anisotropy of the medium have to be taken into account. A first step is to consider elastic models with depth-varying properties. At a plane discontinuity between two homogeneous anisotropic media, because of the three possible elastic waves with orthogonal polarizations, reflection and refraction of

INTERNATIONAL HANDBOOK OF EARTHQUAKE AND ENGINEERING SEISMOLOGY, VOLUME 81A ISBN: 0-12-440652-1

a plane wave give birth to three reflected and three refracted waves (e.g., Musgrave, 1960). Transverse isotropy, i.e., symmetry about a fixed axis, is frequently used as an approximation for realistic situations in the Earth. Love's notations for the five elastic parameters of transversely isotropic media (Love, 1944) are broadly used in surface-wave and normal-mode seismology (e.g., Anderson, 1961; Takeuchi and Saito, 1972; Dziewonski and Anderson, 1981). Other terms for transverse isotropy are "radial anisotropy" or "polarization anisotropy" (i.e., differences between SH and SV velocities). For obvious reasons of symmetry, a transversely isotropic medium with invariant properties about the vertical direction is also the only one that is compatible with a spherical symmetry. Travel times in a transversely isotropic spherical model were given by Woodhouse (1981), using the asymptotic behavior of normal modes.

Backus (1962) looked at the important properties of media made of thin isotropic layers with different elastic properties. At long wavelengths, he showed that a stack of parallel plane layers is transversely isotropic with its symmetry axis perpendicular to the plane of the layers. Thomsen (1986) studied the case of weak anisotropy in such transversely isotropic media about a vertical axis of symmetry. He introduced several notations commonly used in the field of exploration geophysics. Within a seismic wavelength, there is thus no way to distinguish between intrinsic anisotropy and anisotropy induced by a thin layering of materials with different rigidities or by small nonspherical isotropic heterogeneities. This is why the term "seismic anisotropy" is well adapted to a description of the material properties at the scale of a seismic wavelength. Note that the seismic properties of materials exhibiting a shaped-induced anisotropy, also called "shape preferred orientation" anisotropy (SPO), play an important role in the crust. In addition to the problem of layered sediments, it soon became evident that distributions of fluid-filled cracks in the crust (Crampin, 1984) could be a major cause of SPO anisotropy. Again, transversely isotropic models, but with horizontal axis of symmetry, became a standard simplified model to investigate crack-induced anisotropy in the crust.

Leaving the field of transversely isotropic structures, let us recall that Stoneley (1963) developed the theory of surface waves in orthorhombic media. Crampin (1970) did the first calculation of surface-wave phase velocities in flat-layered anisotropic media, showing all peculiarities of anisotropic surface waves, including their anomalous polarizations. Backus (1965) used perturbation theory to estimate the azimuthal variation of velocity in refracted *Pn* waves for weakly anisotropic media. In doing so, he opened the way to many theoretical studies of weakly anisotropic media. Smith and Dahlen (1975) and Montagner and Nataf (1986) extended the perturbation theory to surface waves and developed useful tools for interpreting the Love- and Rayleigh-wave velocities in arbitrary anisotropic media. Properties of cracked media, assuming the long wavelengths approximation, have been extensively studied by several authors (e.g., Hudson, 1980; Sayers, 1998).

Although the observational evidence for anisotropic behavior of seismic body waves in fully 3D structures has been rather weak until now, there is a clear need for developing the theory of seismic wave propagation in such structures, in particular for the purpose of seismic exploration techniques (e.g., Rüger, 1997). This raises considerable mathematical difficulties. The earliest theoretical developments of ray theory in 3D anisotropic media were summarized by Cerveny (1972). They made possible the computation of travel times and amplitudes in the high frequency limit. Numerical computations, however, are much more complicated than in isotropic media since an eigenvector–eigenvalue analysis of the Christoffel matrix has to be performed at each step of propagation along a ray. Two basic approaches have been followed to reduce the computer time: perturbation methods (e.g., Farra, 1989; Jech and Psencik, 1989; Chapman and Pratt, 1992) and factorization of the anisotropic media in homogeneous blocks (Cerveny, 1989; Farra, 1990). Together with various high-frequency asymptotic techniques (e.g., Garmany, 1988a,b), these approaches constitute a very active field of research in theoretical seismology. Among several recent studies related to body-wave theory in anisotropic media, Gajewski and Psencik (1990) applied the concept of dynamic ray tracing to compute vertical seismic profile (VSP) synthetics, Babich (1994) made a systematic study of the estimate of amplitudes on wavefronts, Grechka and Obolentseva (1993) looked at the shear-wave behavior near directions of singularity of the Christoffel matrix, and Bakker (1998) looked at the problem of phase shift at caustics. As an alternative to the high-frequency approximation of ray theory, Hung and Forsyth (1998) make use of parallel computing techniques to obtain complete synthetic seismograms in an oceanic inhomogeneous structure.

Finally, before closing this introduction, it should be noted that the way seismic waves attenuate along their propagation path may also depend on their direction of propagation. This seismic "attenuation anisotropy" may occur for example in media with preferred orientations of fluid-filled cracks (Peacock and Hudson, 1990). It may also be due to solid-state processes (e.g., Carcione and Cavallini, 1994).

2. Anomalies in Crustal-wave Polarization and Seismic Anisotropy

One of the most obvious manifestations of seismic anisotropy is shear-wave splitting. Ray theory predicts that a shear wave entering an anisotropic region is split into two shear waves polarized at right angles according to the local symmetry of the medium. In the high frequency limit of ray theory, the two split waves are fully decoupled. They travel independently within the anisotropic region, and the time delay between the split waves varies along the ray, whereas their polarization follows the local symmetry of the anisotropic medium. Observations of split shear waves generated by crustal earthquakes recorded with short-period instruments at small distances from the source have been accumulated during the past 20 y. Interpretation of these observations in terms of alignment of fluid-filled cracks is the most commonly adopted explanation (Crampin, 1978), although there are situations where preferred orientation of minerals may also be put forward (e.g., Kern and Wenk, 1990). When interpreting observations of split shear waves in terms of anisotropic structures, let us note that Coates and Chapman (1990) have shown that great care must be taken if the time delay between the split shear waves is small. They have shown that the high-frequency limit of ray theory may then lead to misleading results when ignoring the coupling between the split shear waves.

In connection with dilatancy theory, Gupta (1973) suggested that observation of temporal changes in *S*-wave splitting might contribute to short-term earthquake prediction. Crampin (1978) developed this idea further. His model called "Extensive Dilatancy Anisotropy" (EDA) closely links the direction of maximum horizontal compressive stress with the direction of polarization of fast shear waves propagating vertically. Considering that crack-induced anisotropy could be a widespread phenomenon, he proposed that the observed temporal changes in *S*-wave splitting might be used as earthquake precursors. Indeed, there are several observations made in tectonically active regions where fast shear-wave polarizations are parallel to the maximum horizontal compressive stress, as in Imperial Valley, California (Zollo and Bernard, 1989), or in the Gulf of Corinth, Greece (Bouin *et al.*, 1996). There are other situations, however, where this is not the case. Paleostress might then better correlate with the fast shear-wave polarization direction (Blenkinsop, 1990; Aster and Shearer, 1992; Diagranes *et al.*, 1996). If the reality of seismic anisotropy in the crystalline crust seems well established, observed time changes between the split shear waves have been much more controversial during the past decade. The interest in such temporal variations of shear-wave splitting is such that many groups attempt to monitor them (e.g., Gao *et al.*, 1998; Crampin *et al.*, 1999).

Shear-wave splitting observations have been attributed to preferred orientation of fluid-filled cracks in geothermal fields (e.g., Evans *et al.*, 1995; Sachpazi and Hirn, 1991) and in a region of extensional tectonics in the Gulf of Corinth (Bouin *et al.*, 1996). In the upper crystalline crust, direct observations of *S*-wave splitting have been reported in the Kola deep continental well from VSP measurements (Diagranes *et al.*, 1996). A shear-wave velocity anisotropy up to 4.5% has been found there together with a direction of propagation of fast *S* waves parallel to the direction of the maximum compressive paleostress. Diagranes *et al.* (1996) favor preferred alignment of hornblende minerals instead of a distribution of cracks to explain their observations. In the deep continental crust, an azimuthal variation of shear-wave velocities has also been reported from two orthogonal refraction profiles in Norway (Mjelde *et al.*, 1995), and shear-wave splitting was observed from teleseismic *P* waves converted into *S* waves at the Moho discontinuity in different tectonic settings [e.g., Herquel *et al.* (1995) in Tibet; Levin and Park (1997) in the Ural Mountains].

In addition, in many circumstances the deep seismic sounding profiles performed in the continental crust have revealed a high reflectivity of fault zones within the crystalline basement that is much larger than would be expected from any plausible velocity contrast between both sides of the fault. Anisotropy of mylonite could explain such a strong reflectivity of some faults by preferred orientation of phyllosilicates, and possibly by laminar structure of alternating high- and low-velocity layers in the fault zone (Jones and Nur, 1984). These authors estimated the thickness of the reflective zone to be 100–150 m in the case of the Wind River and the Pacific Creek thrust.

Finally, let us emphasize that monitoring seismic anisotropy by shear-wave splitting measurements in sedimentary layers is a very challenging prospect for exploration geophysicists (e.g., Cliet *et al.*, 1991). Crampin (1984) proposed that the EDA model could also serve as a guide to explain *S*-wave splitting observations as due to preferred orientation of fluid-filled cracks for reservoir definition.

3. Seismic Anisotropy in the Upper Mantle: Lithosphere, Asthenosphere or Both?

If crack-induced anisotropy is probably an important cause of seismic anisotropy in the crust, lattice-preferred orientation (LPO) of olivine in peridotites is generally considered as the major cause of upper mantle anisotropy (e.g., Nicolas and Christensen, 1987). As stated by Silver (1996), two types of observation have played a major role in detecting upper-mantle anisotropy: observations linked with the polarization of seismic waves, either *S* waves or surface waves, and observations of local variation of the wave velocity with the direction of propagation. Incompatibility between the observed Love- and Rayleigh-wave velocities belongs to the former type of observations. This so-called Love–Rayleigh wave discrepancy played an important role in the earliest studies of upper-mantle anisotropy. Azimuthal variations of *Pn* velocity in the subcrustal oceanic lithosphere belongs to the second type of observations. They were interpreted as due to preferred orientation of olivine crystals in the accretionary process of the oceanic lithosphere (Francis, 1969).

Kind *et al.* (1985) and Silver and Chan (1988) reported observations of transverse components of *SKS* waves on horizontal records, whereas no such transverse motion should be observed in a spherical isotropic Earth. They attributed this anomaly to shear-wave splitting of *SKS* waves in the upper mantle. Since then, thanks to the deployment of broadband seismometers and to the generalization of digital recording (e.g., Vinnik *et al.*, 1989a), the number of such observations has increased dramatically on all continents (for a review see Silver, 1996, and Color Plate 23, and Savage, 1999). In the last 10 y, an impressive amount of observations of shear-wave splitting attributed to upper mantle anisotropy has been reported in the literature. Several attempts have also been made to use other mantle shear waves, either *ScS* (Fukao, 1984; Ansel and Nataf, 1989), direct mantle *S* waves, or *Ps* waves converted at the discontinuities of the upper mantle seismic transition zone (e.g., Vinnik and Montagner, 1996).

The former models used to explain the *SKS* splitting observations in continental regions were rather simple, being made of an olivine-rich homogeneous layer 100–300 km thick. In this simple model, which fits many observations, the polarization of the fast shear wave gives us the direction of preferred orientation of the olivine *a* axes in the horizontal plane, and the time delay between the split shear waves gives us an estimation of the thickness of the layer assuming a reasonable LPO anisotropy in the upper mantle. The two main reasons why a model of lithospheric anisotropy was developed (Silver and Chan, 1988) are: (1) the clear correlation between the observed directions of shear-wave splitting and the trends in surface tectonics; and (2) the fact that there was no evidence for a deeper cause of shear-wave splitting in the lower mantle. Indeed, very weak anisotropy is expected in the deep mantle if one excludes the D$''$ layer in its lowermost part (Meade *et al.*, 1995). It also has been quite obvious that small-scale lateral variations of both the symmetry directions and amplitudes of seismic anisotropy were strongly in favor of a lithospheric origin.

Independently of SKS studies, it has been recognized for a long time that 3D isotropic models of the mantle were unlikely to explain the variations of *P*-wave travel-time residuals observed at a given station for rays with different azimuths and incidence angles (Dziewonski and Anderson, 1983). It was also soon recognized that the amount of seismic velocity variations could be as large as the effect of anisotropy in upper-mantle materials (e.g., Anderson, 1989). The pattern of observed *P*-wave residuals in central Europe has been interpreted in terms of dipping anisotropic bodies in the lithosphere, possibly paleosubducted plates (Babuska *et al.*, 1984). Plomerova *et al.* (1996) made a joint interpretation of *P*-wave residuals and *SKS* splitting observations in Fennoscandia and the western United States. The fact that a single LPO anisotropic model with dipping anisotropic structures may explain both sets of data gives a firm basis for the interpretation of upper mantle anisotropy as frozen in the lithosphere beneath continents.

The question of the precise location of the seismic anisotropy revealed by *SKS* splitting observations, however, gave rise to a controversy that is not yet closed. Indeed, because vertically incident rays lack any depth resolution, only indirect arguments, such as correlations with the local geology or with the directions of present plate motions, have been used in the former debate. A recent re-examination of this question by Kubo and Hiramatsu (1998) shows that *SKS* splitting data correlate with the directions of the present absolute plate motions, which is in favor of an asthenospheric origin, but for plate velocities larger than 1.4 cm y^{-1} only. For slower plates, no correlation is found and the observed *SKS* splitting observations are more likely due to frozen-in anisotropy in the lithosphere. Unfortunately, the use of *P*-wave residual data does not help a lot in this debate because the teleseismic rays follow subvertical paths in the upper mantle and azimuthal variation of *P*-wave velocities have little effect on their travel times. Surface waves are more useful since a rather good depth resolution both in polarization anisotropy and azimuthal anisotropy may be achieved in the upper mantle. Horizontal resolution, however, is much poorer than in *SKS* studies.

Inversion of global normal-mode data also has provided evidence for seismic anisotropy in the upper mantle for a long time. In the broadly used global model PREM (Dziewonski and Anderson, 1981), a few percent polarization anisotropy is required in the upper 220 km. In regional studies, the former interpretation of the observed discrepancies between the Love- and Rayleigh-wave velocities were made in terms of flat horizontal pockets of partial melts (e.g., Aki; 1968; Schlue and Knopoff, 1976). It has been conceded later that they could also be interpreted in terms of LPO anisotropy without invoking partial melt. Both mechanisms can predict the observed discrepancies between the Love- and Rayleigh-wave velocities, but only LPO anisotropy gives us a reasonable explanation for the SKS splitting. In the former studies, only fundamental-mode surface waves were used, lacking in depth resolution, and the question of the depth extent of anisotropy was not easy to address. For example, Schlue and Knopoff (1976) favored an asthenospheric origin for the polarization anisotropy observed in the Pacific, whereas Mitchell and Yu (1980) favored the lithosphere. Forsyth (1975) was the first to use higher-mode surface waves to investigate the upper-mantle structure beneath the Pacific with better depth resolution. He was also the first to resolve the azimuthal variations of Rayleigh-wave velocities and to interpret them in terms of LPO anisotropy. More recently, using higher modes of Love and Rayleigh waves, Cara and Lévêque (1988) found a clear anisotropic layer extending to at least 200 km beneath the Pacific Ocean, extending thus well below the lithosphere. Nishimura and Forsyth (1989), inverting a large set of data related to the Pacific Ocean, found that seismic anisotropy is not confined to a 100 km thick lithosphere but extends well below, within the shear-wave low-velocity zone in the asthenosphere. A model explaining seismic anisotropy of the asthenosphere by finite deformation of upper mantle materials under resistive drag of the sublithospheric mantle has been developed recently by Tommasi *et al.* (1996). It gives a solid basis for the interpretation of the seismic anisotropy observed in the asthenosphere beneath the Pacific.

More powerful tools can now be used to determine the depth extent and lateral variation of seismic anisotropy in the upper mantle. Waveform inversion techniques allow us to interpret the whole shape of a seismogram in terms of depth-varying anisotropic structures. They automatically include the higher-mode content of the surface-wave signal and exploit their great potential for increasing depth resolution. Inverting a set of high quality broadband surface-wave records for many sources and receivers, Lévêque *et al.* (1998) obtained a 3D anisotropic model of the Indian Ocean and its surrounding areas with an unprecedented lateral and vertical resolution. The azimuthal variation of *SV* waves was in particular well resolved on a horizontal scale of 1000 km. The direction of fast *SV* velocities and the directions of the present absolute motion of the plates are well correlated at depth. The correlation breaks down in the lithosphere, above 100 km beneath the Indian Ocean and above 200 km beneath Australia, the only continental region well constrained by the data used in this study. Color Plate 24 shows the map of *SV*-wave velocities and their azimuthal anisotropy at 150 and 200 km. Beneath Australia, a clear rotation of the fast direction of propagation of *SV* waves takes place between the 150 and 200 km depths, with fast *SV* waves in the east–west direction in the lithosphere and fast *SV* waves in a direction parallel to the south–north motion of the Australian plate at 200 km. Using broadband data recorded during the SKIPPY experiment in Australia, Debayle (1999) confirmed this overall rotation pattern beneath central and eastern Australia at about 150 km depth.

In an independent study, Girardin and Farra (1998) have interpreted SKS splitting observations made in Canberra with a two-layer anisotropic model. A similar 90° rotation from a north–south to an east–west direction is found for the polarization of the fast shear waves when passing from the lower asthenospheric layer to the upper lithospheric layer.

All these findings suggest that beneath central and eastern Australia the fast direction of propagation of *SV* waves within the lithosphere could be dominantly oriented east–west, while it could become north–south below. According to the mechanism proposed by Tommasi *et al.* (1996) for the Pacific, a plausible explanation is that the present shearing strain below a 150 km-thick continental lithosphere could cause an LPO anisotropy in an olivine-rich upper mantle. It is interesting to recall that a similar hypothesis had been proposed by Leven *et al.* (1981) to explain the abnormally high *P* velocities observed at about 200–250 km depth beneath northern Australia.

Taking into account all kinds of arguments exchanged during the past 20 y on the location of upper mantle anisotropy, it now seems quite obvious that both frozen-in anisotropy in the lithosphere and dynamic deformation of the root of the lithosphere and the upper part of the asthenosphere are involved. The role of SPO anisotropy due to partial melts of the upper mantle, once a rather popular explanation, does not seem necessary to explain most of the presently available observations, except perhaps beneath active spreading centers. Using data from broadband ocean bottom receivers, Webb and Forsyth (1998), for example, found very low shear-wave velocities down to 100 km depth beneath the East Pacific Rise together with incompatibility between the Love- and Rayleigh-wave velocities. Because *S*-wave splitting observations have also been made on the ocean floor during this experiment (Wolfe and Solomon, 1998), both LPO anisotropy, with olivine *a* axes parallel to the spreading direction, and aligned melt pockets in horizontal directions may be required to fit the data.

At a smaller regional scale, several authors have shown that upper mantle seismic anisotropy is also required in subduction zones. The questions whether the upper lithosphere, the asthenospheric wedge above the subducting slab, or the slab itself is anisotropic has been debated during the past 20 y (e.g., Savage, 1999). Ando and Ishikawa (1980), and Ando *et al.* (1983) have observed local *S*-waves splitting in Japan for intermediate and deep earthquakes. They interpreted this observation as being due to the upper mantle wedge between the surface and the subducting slab, invoking lateral variations over horizontal distances of 50–100 km to explain the changing pattern of their *S*-wave splitting observations. Using *ScS* waves reflected at the core–mantle boundary, Fukao (1984) observed a more homogeneous pattern for the directions of polarization of the fast *S* waves. As these directions are roughly parallel to the plate convergence, he suggested that the cause of the splitting might be preferred orientation of olivine crystals *a* axes within the slab. Since then numerous *S*-wave splitting observations have been made in subduction zones. Many show fast directions of polarization parallel to the trenches (e.g., Silver and Chan, 1991). Fisher and Yang (1994) looked at differential splitting observations between teleseismic *sS* and *S* phases coming from intermediate and deep earthquakes in the Kurils–Kamchatka. Completing this information by local *S*-wave observations, they infer the anisotropy pattern of both the upper mantle wedge and the overriding lithospheric plate. They found a rather complex pattern, invoking both asthenospheric flow in the wedge above the subducted slab and deformation of the upper lithospheric plate in the back-arc area.

Fouch and Fischer (1996) concluded that the variability of *S*-wave splitting observations in the different subduction zones could be explained by preferred orientation of olivine crystals in both the upper lithosphere and the asthenospheric flow above the subducted slab. Seismic anisotropy in the subduction zone could thus be governed by the local conditions of the plate convergence and the tectonic spreading and shearing of the lithosphere in the back-arc area.

4. The Deep Mantle and Core

If the presence of seismic anisotropy in the upper mantle is supported by many seismological investigations, there is presently no evidence of seismic anisotropy in the bulk of the lower mantle (e.g., Meade *et al.*, 1995). Only its upper and lower parts exhibit signs of anisotropy. In their study of subduction zones of the Pacific, Fouch and Fischer (1996) concluded that anisotropy of the lower mantle is very unlikely. Quite convincingly, they compare splitting observations of local *S* waves coming from deep earthquakes in subducted slabs and *SKS* waves coming from the core. They show that deep lower mantle anisotropy can be ruled out, except if anisotropy cancels out along the long ray path between the core–mantle boundary and the upper-mantle transition zone.

At a very large scale, evidence for a weak polarization anisotropy between the upper and lower mantle has been advocated by Montagner and Kennett (1996) from a global inversion of normal mode and body-wave data. According to Karato (1998), who reviewed the different data related to seismic anisotropy in the deep mantle and explored their geodynamical significance on the basis of mineral physics, Montagner and Kennett's (1996) results with a reverse *SH/SV* anisotropy above and below the 660 km discontinuity might be interpreted as due to horizontal shearing in the lower part of the transition region. This would support a partially two-layered convection pattern of the mantle with horizontal shearing between the upper and lower mantle. However, Kiefer and Stixrude (1997), computing values of the elastic constants of Mg_2SiO_4 spinel at high pressure, found a predicted anisotropy for *P* and *S* waves which is not in agreement with the radial anisotropy found in the depth range 410–600 km by Montagner and Kennett (1996). Looking now either at spherical harmonic expansions of global normal mode data (Montagner and Tanimoto, 1991) or at several subduction zones (Fouch and Fischer, 1996), it became evident that if seismic anisotropy is required in the transition zone, large lateral variations do occur there. Anisotropy is for example not required in the transition zone beneath the Izu Bonin but it is required beneath the southern Kurils (Fouch and Fischer, 1996). Interestingly, these lateral variations could be coupled with the mechanical barrier resisting the downward motion of the plates at 600 km depth (Karato, 1998).

Further investigation of the anisotropic properties of the upper mantle transition region, together with their lateral variations, is certainly needed to better understand the dynamics of this important boundary layer of the Earth.

The possibility of radial anisotropy at the bottom of the mantle, within the D″ boundary layer, was mentioned by Doornbos *et al.* (1985). Investigating polarization of diffracted *SV* waves, Vinnik *et al.* (1989b) suggested that azimuthal anisotropy could also take place in D″ layer. The fact that *SH* waves traveling horizontally at the core–mantle boundary are generally faster than *SV* waves has been proposed by several authors (e.g., Maupin, 1994). Pulliam and Sen (1998), however, have found *SV* waves are faster than *SH* beneath the central Pacific, a zone of ultralow seismic velocities (Williams and Garnero, 1996). In fact, D″ appears as a strongly inhomogeneous boundary layer, both for the seismic velocities and for seismic anisotropy. Due to the strong velocity reduction observed in the superswell region beneath the Pacific, a possibility of partial melt is not excluded (Williams and Garnero, 1996). The D″ boundary layer, a region of the Earth which is unevenly sampled by seismic rays, might thus be a complex zone where LPO of both perovskite and magnetowustite could take place together with other SPO anisotropy involving partial melts within a dynamic shearing strain (Karato, 1998; Ritsema *et al.*, 1998).

In the inner core of the Earth, two independent sets of data have led seismologists to consider that it could be strongly anisotropic with hexagonal symmetry and a symmetry axis close to the rotation axis of the Earth. The first set of data is *PKIKP* travel times. Morelli *et al.* (1986) corrected the observed *PKP(DF)* travel times from the effects of the upper-mantle and core–mantle velocity anomalies known at that time. A strong zonal character appeared in the residual inner-core travel times, attaining about 2 sec. Vertically propagating *PKIKP* waves were found to be faster along the rotation axis of the Earth than in the equatorial plane. As no plausible heterogeneity could explain these observations, Morelli *et al.* (1986) proposed that the inner core might be anisotropic with a cylindrical symmetry parallel to the Earth's axis of rotation. A similar observation was made by Cormier and Choy (1986) from the variation in differential travel times between *PKP(DF)* and *PKP(BC)*. In that case no correction for lateral variation of mantle *P* velocities was applied to the data since the two rays have rather close paths in the mantle. The second set of data comes from the analysis of split core modes in the free oscillation spectra of the Earth (Woodhouse *et al.*, 1986). These latter data need 25-km high undulations of the inner-core boundary and large heterogeneities near the upper part of the inner core to be explained by isotropic models (Giardini *et al.*, 1987). A homogeneous spherical model including a hexagonal anisotropy of the inner core with the axis of symmetry coincident with the Earth's rotation axis is an alternative model that fits the splitting observations as well.

The earlier findings on inner-core anisotropy have been confirmed since then by several authors using both differential travel time analysis (e.g., Shearer and Toy, 1991) and normal-mode splitting observations (Tromp, 1993). Furthermore, not only has the inner-core anisotropy been broadly accepted during recent years, but the fact that both its axis of symmetry may be tilted with respect to the Earth's rotation axis (Creager, 1992) and that the *PKP(BC)-PKP(DF)* travel times may have changed during the past 20 y have led Song and Richards (1996) to propose that the inner core could rotate with respect to the Earth's mantle. Souriau *et al.* (1997) critically examined these results. They concluded that even if differential rotation of the inner core could not be ruled out, they doubted that it actually had been detected. Whatever the rotation of the inner core is, the former image of a homogeneous anisotropic inner core with hexagonal symmetry about the Earth's rotation axis appears now as an oversimplified view. Observations of heterogeneities in the inner core (Su and Dziewonski, 1995), if not a tilted axis of symmetry, are now quite broadly accepted. The search for the cause of seismic anisotropy in the inner core is a new challenging problem in solid Earth physics. Weber and Machetel (1992) considered that the inner core can be convective and that dynamically induced anisotropy in high-pressure iron phase is possible. Summarizing our present state of knowledge of the solid-state physics of the inner core, Bergmann (1998) addressed the problem of iron grain size. Within a range between 5 mm and 2000 km, he favored a grain size of the order of 100 m.

Another intriguing manifestation of inner-core anisotropy is the variation of seismic attenuation with the inclination of the *PKIKP* rays with respect to the axis of rotation of the Earth. Souriau and Romanowicz (1996) reported a clear variation of *PKP(DF)/PKP(BC)* amplitude ratios which are about five times smaller for paths close to the rotation axis of the Earth than for nearly equatorial paths. They interpreted this strong amplitude variation as most likely due to anisotropy in seismic attenuation within the inner core.

5. Trade-off between Anisotropy and Heterogeneities

As already stated, evidence for seismic anisotropy may come either from local variation of seismic velocities with respect to the direction of propagation, or from anomalies linked with their polarization. The former type of observation is generally considered as less robust since correction for velocity variations along the ray path is generally done before searching for its local angular variations. In fact, polarization of the waves may also be affected by heterogeneity of the structure. Tilted interfaces, for example, may generate waves propagating in three dimensions, and rays traveling off the source–receiver vertical plane could mimic anomalies in polarization. When using converted *PS* waves at seismic discontinuities, tilted interfaces may also produce transverse shear waves that can be misinterpreted as due to anisotropy of the medium. For example, to avoid such a misinterpretation, Peng and Humphreys (1997), and Savage (1998) simultaneously determined the Moho dip and crustal anisotropy when inverting teleseismic receiver functions.

In surface-wave seismology, many effects can be attributed either to lateral heterogeneities of an isotropic structure or to anisotropic properties of the upper mantle. Levshin and Ratnikova (1984), for example, performed a numerical experiment with a laterally varying isotropic model made of two quarter-spaces in contact. Their results show that the Love- and Rayleigh-wave velocities are not compatible with a single laterally uniform structure. They suggested that it is not necessary to invoke anisotropy to interpret the Love–Rayleigh wave discrepancies that are so frequently observed, but that lateral variation of isotropic structures could also explain these observations. Babuska and Cara (1991) commented on this numerical experiment. They concluded that if such an effect is indeed possible, it requires strong enough lateral variations of structures so that non-linear effects may take place when passing from the local surface-wave velocities to the average velocity along the surface-wave path. Another example where isotropic heterogeneities and lateral variations of seismic anisotropy can equally explain observed anomalies in the seismic wavefield is given by Kobayashi and Nakanishi (1998). They report observations of secondary Love waves in Japan coming from converted Rayleigh waves beneath the Aleutian or Kuril trenches. Referring to a theoretical study of mode coupling effects by Yu and Park (1993), these authors conclude that the Rayleigh- to Love-wave conversion is most probably due to lateral variations of anisotropy in the upper mantle, but that laterally varying isotropic structures could nevertheless also produce the observed effects.

Much care must be taken when performing inversion of seismic data in terms of both 3D inhomogeneities and anisotropy. Ignoring either inhomogeneities along the wave path or anisotropy may lead to erroneous results. Lévêque *et al.* (1998), for example, carefully checked how an isotropic 3D structure could affect the azimuthal anisotropy of *SV*-wave velocities they have obtained from inversion of path-averaged surface-wave velocities. They concluded that there is a slight trade-off between the anisotropic pattern shown in Color Plate 24 and the 3D structure of the upper mantle, but the effect is small and does not alter the major trends in azimuthal variations of *SV* velocities shown in Color Plate 24.

Leaving the field of surface-wave seismology, it is of prime interest to look at the problem of teleseismic *P*-wave travel-time tomography, since most results published to date ignore the possibility that anisotropy may bias the results. To investigate this problem, Sobolev *et al.* (1999) performed several numerical experiments with realistic ray configurations and a plausible LPO anisotropy in the lithosphere and asthenosphere. Direct computations of *P*-wave travel-time residuals are made for several laterally and vertically varying anisotropic structures. The synthetic residuals are then inverted using a 3D isotropic tomographic code. Several anisotropy-induced artifacts are clearly apparent in the results. Even if, in some ray configurations, the apparent pattern of the inverted seismic velocities seems not affected, the amplitude of the velocity variations is. Interpretations of the seismic tomography results in terms of mineral composition or temperature variations may thus be misleading if anisotropy is ignored. When dealing with travel-time inversion, there are other circumstances where remote heterogeneities or anisotropic bodies may affect the results of tomography. As an example, let us mention the study of the structures of the deep mantle and core by Garnero and Lay (1998). These authors pointed out that anisotropy in D″ could significantly affect the isotropic *P*-wave velocity structure of the outer core.

6. What's Next?

The above considerations show that, except in the bulk of the lower mantle and the outer core, seismic anisotropy cannot be ignored any longer in most seismological investigations. From the crust to the inner core of the Earth, lateral and vertical variation of seismic velocities may be closely linked with some level of seismic anisotropy in the structure. In turn, quantifying seismic anisotropy and its lateral and vertical variations in many parts of the Earth is a challenging research goal for seismologists. Many questions regarding the dynamics of the Earth can indeed be addressed by investigating the characteristics of seismic anisotropy. As an illustration, let us mention the question of the nature and thickness of the continental lithospheric roots, a major problem not yet satisfactorily solved. For example, Gaherty *et al.* (1999) advocate a deep continental root beneath western Australia whereas Debayle's (1999) results are in favor of a thinner lithosphere. As another example, Babuska *et al.* (1998) found a global variation of the *SH*/*SV* velocity ratio with regard to the age of continents. If this result is confirmed and its possible trade-off with the different lateral heterogeneities understood, it could convey new important information on the evolution of the continental lithosphere.

At greater depth, the question of anisotropy in the transition zone and its lateral variations is a clue to understanding the coupling between the upper and lower mantle (Karato, 1998). Finally, will we confirm that D″ is the only part of the lower mantle where seismic anisotropy is detectable? Will we be able to better constrain the lateral heterogeneities and seismic anisotropy within the inner core? After a decade where a wealth of new results has been gathered on seismic anisotropy in different parts of the Earth, it is clear that seismologists are faced with a lot of exciting questions to address in the forthcoming years. Thanks to the deployment of broadband digital seismometers with a high-density coverage, we can be confident that much progress will be made on our knowledge of seismic anisotropy and on its connection with the Earth's dynamics.

References

Aki, K. (1968). Seismological evidences for the existence of soft thin layers in the upper mantle, *J. Geophys. Res.* **73**, 585–594.

Anderson, D.L. (1961). Elastic wave propagation in layered anisotropic media. *J. Geophys. Res.* **82**, 277–296.

Anderson, D.L. (1989). "Theory of the Earth," Blackwell Sci. Publ., Oxford.

Ando, M. and Y. Ishikawa (1980). S-wave anisotropy in the upper mantle under a volcanic area in Japan. *Nature* **286**, 43–46.

Ando, M., Y. Ishikawa, and F. Yamazaki (1983). Shear wave polarization in the upper mantle beneath Honshu, Japan. *J. Geophys. Res.* **88**, 5850–5864.

Ansel, V. and H.C. Nataf (1989). Anisotropy beneath 9 stations of the Geoscope broadband network as deduced from shear-wave splitting. *Geophys. Res. Lett.* **16**, 409–412.

Aster, R.C. and P.M. Shearer (1992). Initial shear wave particle motions and stress constraints at the Anza seismic Network. *Geophys. J. Int.* **108**, 740–748.

Babich, V.M. (1994). Ray method of calculating the intensity of wavefronts in the case of a heterogeneous, anisotropic, elastic medium. *Geophys. J. Int.* **118**, 379–383.

Babuska, V. and M. Cara (1991). "Seismic Anisotropy in the Earth," Kluwer Academic, Dordrecht.

Babuska, V., J. Plomerova, and J. Sileny (1984). Large-scale oriented structures in the subcrustal lithosphere of central Europe. *Ann. Geophys.* **2**, 649–662.

Babuska, V., J.P. Montagner, J. Plomerova, and N. Girardin (1998). Age-dependent large-scale fabric of the mantle lithosphere as derived from surface-wave velocity anisotropy. *Pure Appl. Geophys.* **151**, 257–280.

Backus, G.E. (1962). Long-wave elastic anisotropy produced by horizontal layering. *J. Geophys. Res.* **67**, 4427–4440.

Backus, G.E. (1965). Possible forms of seismic anisotropy of the uppermost mantle under oceans. *J. Geophys. Res.* **70**, 3429–3439.

Bakker, P.M. (1998). Phase shift at caustics along rays in anisotropic media. *Geophys. J. Int.* **134**, 515–518.

Bergmann, M.I. (1998). Estimates of the Earth's inner core grain size. *Geophys. Res. Lett.* **25**, 1593–1596.

Blenkinsop, T.G. (1990). Correlation of paleotectonic fracture and microfracture orientations in cores with seismic anisotropy at Cajon Pass drill hole, southern California. *J. Geophys. Res.* **95**, 11143–11150.

Bouin, M.P., J. Téllez, and P. Bernard (1996). Seismic anisotropy around the Gulf of Corinth, Greece, deduced from three-component seismograms of local earthquakes and its relationship with crustal strain. *J. Geophys. Res.* **101**, 5797–5811.

Cara, M. and J.J. Lévêque (1988). Anisotropy of the asthenosphere: the higher mode data of the Pacific revisited. *Geophys. Res. Lett.* **15**, 205–208.

Carcione, J.M. and F. Cavallini (1994). A rheological model for anelastic anisotropic media with applications to seismic wave propagation. *Geophys. J. Int.* **119**, 338–348.

Cerveny, V. (1972). Seismic rays and ray intensities in inhomogeneous anisotropic media. *Geophys. J. R. Astron. Soc.* **29**, 1–13.

Cerveny, V. (1989). Ray tracing in factorized anisotropic media. *Geophys. J. Int.* **99**, 91–100.

Chapman, C.H. and R.G. Pratt (1992). Traveltime tomography in anisotropic media-I. Theory. *Geophys. J. Int.* **109**, 1–19.

Cliet, Ch., L. Brodov, A. Tikhonov, D. Marin, and D. Michon (1991). Anisotropy survey for reservoir definition. *Geophys. J. Int.* **107**, 417–427.

Coates, R.T. and C.H. Chapman (1990). Quasi-shear wave coupling in weakly anisotropic 3-D media. *Geophys. J. Int.* **103**, 301–320.

Cormier, V.F. and G.L. Choy (1986). A search for lateral heterogeneity in the inner core from differential travel time near PKP-D and PKP-C. *Geophys. Res. Lett.* **13**, 1553–1556.

Crampin, S. (1970). The dispersion of surface waves in multilayered anisotropic media. *Geophys. J. R. Astron. Soc.* **21**, 387–402.

Crampin, S. (1978). Seismic wave propagation through a cracked solid: polarization as a possible dilatancy diagnostic. *Geophys. J. R. Astron. Soc.* **53**, 467–496.

Crampin, S. (1984). Anisotropy in exploration geophysics. *First Break* **2**, 19–21.

Crampin, S., T. Volti, and R. Stefanson (1999). A successfully stress-forecast earthquake. *Geophys. J. Int.* **138**, F1–F5.

Creager, K. (1992). Anisotropy in the inner core from differential travel times of the phases PKP and PKIKP. *Nature* **356**, 309–314.

Debayle, E. (1999). SV-wave azimuthal anisotropy in the Australian upper mantle: preliminary results from automated Rayleigh waveform inversion. *Geophys. J. Int.* **137**, 747–754.

Diagranes, P., Y. Kristoffersen, and N. Karajev (1996). An analysis of shear waves observed in VSP data from the superdeep well at Kola, Russia. *Geophys. J. Int.* **126**, 545–554.

Doornbos, D.J., S. Spilopoulos, and F.D. Stacey (1985). Seismological properties of D″ and the structure of the thermal boundary layer. *Phys. Earth Planet. Inter.* **41**, 225–239.

Dziewonski, A.M. and D.L. Anderson (1981). Preliminary reference earth model. *Phys. Earth Planet. Inter.* **25**, 297–356.

Dziewonski, A.M. and D.L. Anderson (1983). Travel times and station correction for P waves at teleseismic distances. *J. Geophys. Res.* **88**, 3295–3314.

Evans, J.R., B.R. Julian, G.R. Foulger, and A. Ross (1995). Shear wave splitting from local earthquakes at the Geysers geothermal fields, California. *Geophys. Res. Lett.* **22**, 501–504.

Farra, V. (1989). Ray perturbation theory for heterogeneous hexagonal anisotropic media. *Geophys. J. Int.* **99**, 723–737.

Farra, V. (1990). Amplitude computation in heterogeneous media by ray perturbation theory: a finite element approach. *Geophys. J. Int.* **103**, 341–354.

Fisher, K. and X. Yang (1994). Anisotropy in Kuril–Kamchatka subduction zone structure. *Geophys. Res. Lett.* **21**, 5–8.

Forsyth, D.W. (1975). The early structural evolution and anisotropy of the oceanic upper mantle. *Geophys. J. R. Astron. Soc.* **43**, 103–162.

Fouch, M.J. and K.M. Fisher (1996). Mantle anisotropy beneath northwest Pacific Subduction zones. *J. Geophys. Res.* **101**, 15987–16002.

Francis, T.J.G. (1969). Generation of seismic anisotropy in the upper mantle along the midoceanic ridges. *Nature* **221**, 162–165.

Fukao, Y. (1984). Evidence from core-reflected shear waves for anisotropy in the Earth's mantle. *Nature* **309**, 695–698.

Gaherty, J.B., M. Kato, and T. Jordan (1999). Seismological structure of the upper mantle: a regional comparison of seismic layering. *Phys. Earth Planet. Inter.* **110**, 21–41.

Gajewski, D. and I. Psencik (1990). Vertical seismic profile synthetics by dynamic ray tracing in laterally varying layered anisotropic structures. *J. Geophys. Res.* **95**, 11301–11315.

Gao, Y., P. Wang, S. Zheng, M. Wang, Y. Chen, and H. Zhou (1998). Temporal changes in shear-wave splitting at an isolated swarm of small earthquakes in 1992 near Dongfang, Hainan Island, southern China. *Geophys. J. Int.* **135**, 102–112.

Garmany, J. (1988a). Seismograms in stratified anisotropic media – I. WKBJ theory. *Geophys. J. R. Astron. Soc.* **92**, 365–377.

Garmany, J. (1988b). Seismograms in stratified anisotropic media – II. Uniformly asymptotic approximations. *Geophys. J. R. Astron. Soc.* **92**, 379–389.

Garnero, E.J. and T. Lay (1998). Effects of D″ anisotropy on seismic velocity models of the outermost core. *Geophys. Res. Lett.* **25**, 2341–2344.

Giardini, D., X.D. Li, and J.H. Woodhouse (1987). Three-dimensional structure of the Earth from splitting in free-oscillation spectra. *Nature* **325**, 405–411.

Girardin, N. and V. Farra (1998). Azimuthal anisotropy in the upper mantle from observations of P to S converted phases: application to southeast Australia. *Geophys. J. Int.* **133**, 615–629.

Grechka, V.Y. and I.R. Obolentseva (1993). Geomctrical structure of shear-wave surfaces near singularity directions in anisotropic media. *Geophys. J. Int.* **115**, 609–616.

Gupta, I.N. (1973). Premonitory variations in S-wave velocity anisotropy before earthquakes in Nevada. *Science* **182**, 1129–1132.

Helbig, K. (1994). Foundation of anisotropy for exploration seismics. In: "Handbook of Geophysical Exploration, Section 1, Seismic Exploration" (K. Helbig and S. Treitel, Eds.), vol. XXII, Pergamon, New York.

Herquel, G., G. Wittlinger, and J. Guilbert (1995). Anisotropy and crustal thickness of northern Tibet. New constraints for tectonic modelling. *Geophys. Res. Lett.* **22**, 1925–1928.

Hess, H. (1964). Seismic anisotropy of the uppermost mantle under oceans. *Nature* **203**, 629–631.

Hudson, J.A. (1980). Overall properties of a cracked solid. *Math. Proc. Camb. Philos. Soc.* **88**, 371–384.

Hung, S.H., and D.W. Forsyth (1998). Modelling anisotropic wave propagation in oceanic inhomogeneous structures using the parallel multidomain pseudo-spectral method. *Geophys. J. Int.* **133**, 726–740.

James, D. (1971). Anomalous Love wave phase velocities. *J. Geophys. Res.* **76**, 2077–2083.

Jech, J. and I. Psencik (1989). First-order perturbation method for anisotropic media. *Geophys. J. Int.* **99**, 369–376.

Jones, T.D. and A. Nur (1984). The nature of seismic reflections from deep crustal fault zones. *J. Geophys. Res.* **89**, 3153–3171.

Karato, S. (1998). Seismic anisotropy in the deep mantle, boundary layers and the geometry of mantle convection. *Pure Appl. Geophys.* **151**, 565–587.

Kelvin, Lord (W. Thomson) (1904). "Baltimore Lectures on Molecular Dynamics and the Wave Theory of Light," Cambridge University Press, London (course delivered in 1884).

Kern, H. and H.R. Wenk (1990). Fabric related anisotropy and shear-wave splitting in rocks from the Santa Rosa mylonite zone, California. *J. Geophys. Res.* **95**, 11213–11223.

Kiefer, B. and L. Stixrude (1997). Calculated elastic constants and anisotropy of Mg_2SiO_4 spinel at high pressure. *Geophys. Res. Lett.* **24**, 2841–2844.

Kind, R., G.L. Kosarev, L.I. Makeyeva, and L.P. Vinnik (1985). Observations of laterally inhomogeneous anisotropy in the continental lithosphere. *Nature* **318**, 358–361.

Kobayashi, R. and I. Nakanishi (1998). Location of Love-to-Rayleigh conversion due to lateral heterogeneity or azimuthal anisotropy in the upper mantle. *Geophys. Res. Lett.* **25**, 1067–1070.

Krey, Th. and K. Helbig (1956). A theorem concerning anisotropy of stratified media and its significance for reflection seismics. *Geophys. Prospect.* **4**, 294–302.

Kubo, A. and Y. Hiramatsu (1998). On presence of seismic anisotropy in the asthenosphere beneath continents and its dependence on plate velocity: significance of reference frame selection. *Pure Appl. Geophys.* **151**, 281–303.

Leven, J.H., I. Jackson, and A.E. Ringwood (1981). Upper mantle seismic anisotropy and lithospheric decoupling. *Nature* **289**, 234–239.

Lévêque, J.J., E. Debayle, and V. Maupin (1998). Anisotropy in the Indian Ocean upper mantle form Rayleigh- and Love-waveform inversion. *Geophys. J. Int.* **133**, 529–540.

Levin, V. and J. Park (1997). Crustal anisotropy in the Ural Mountain foredeep from teleseismic receiver functions. *Geophys. R. Lett.* **24**, 1283–1286.

Levshin, A. and L. Ratnikova (1984). Apparent anisotropy in inhomogeneous media. *Geophys. J. R. Astron. Soc.* **76**, 65–69.

Love, A.E.H. (1944). "A Treatise on the Mathematical Theory of Elasticity," Cambridge University Press, Cambridge, 1st edn. in 1892; (republication in 1944 of the 1927, 4th ed. by Dover, New York).

Maupin, V. (1994). On the possibility of anisotropy in the D″ layer as inferred from the polarization of diffracted S waves. *Phys. Earth Planet. Inter.* **87**, 1–32.

McEvilly, T.V. (1964). Central U.S. crust-upper mantle structure from Love and Rayleigh wave phase velocity inversion. *Bull. Seismol. Soc. Am.* **54**, 1997–2015.

Meade, C., P.G. Silver, and S. Kaneshima (1995). Laboratory and seismological observations of lower mantle isotropy. *Geophys. Res. Lett.* **22**, 1293–1296.

Mitchell, B.J. and G.K. Yu (1980). Surface wave dispersion, regionalized velocity models, and anisotropy of the Pacific crust and upper mantle. *Geophys. J. R. Astron. Soc.* **63**, 497–514.

Mjelde, R., M.A. Sellevoll, H. Shimamura, T. Iwasaki, and T. Kanazawa (1995). S-wave anisotropy off Lofoten, Norway, indicative of fluids in the lower continental crust? *Geophys. J. Int.* **120**, 87–96.

Montagner, J.P. (1998). Where can seismic anisotropy be detected in the Earth's mantle? In boundary layers.... *Pure Appl. Geophys.* **151**, 223–256.

Montagner, J.P. and B.L.N. Kennett (1996). How to reconcile body-wave and normal mode reference earth models. *Geophys. J. Int.* **125**, 229–248.

Montagner, J.P. and H.C. Nataf (1986). A simple method for inverting the azimuthal anisotropy of surface waves. *J. Geophys. Res.* **91**, 511–520.

Montagner, J.P. and T. Tanimoto (1991). Global upper mantle tomography of seismic velocities and anisotropies. *J. Geophys. Res.* **96**, 20337–20351.

Morelli, A., A.M. Dziewonski, and J.H. Woodhouse (1986). Anisotropy of the inner core inferred from PKIKP travel times. *Geophys. Res. Lett.* **13**, 1545–1548.

Musgrave, M.J.P. (1960). Reflexion and refraction of plane elastic waves at a plane boundary between aelotropic media. *Geophys. J. Astron. Soc.* **3**, 406–418.

Nicolas, A. and N.I. Christensen (1987). Formation of anisotropy in upper mantle peridotites – A review. In: "Composition, Structure and Dynamics of the Lithosphere–Asthenosphere System" (K. Fuchs and C. Froidevaux, Eds.), Geodynamics Series, AGU, Vol. XVI, pp. 111–123.

Nishimura, C.E. and D.W. Forsyth (1989). The anisotropic structure of the upper mantle in the Pacific. *Geophys. J. Int.* **96**, 203–229.

Peacock, S. and J.A. Hudson (1990). Seismic properties of rocks with distributions of small cracks. *Geophys. J. Int.* **102**, 471–484.

Peng, X. and E.D. Humphreys (1997). Moho dip and crustal anisotropy in Northwestern nevada from teleseismic receiver functions. *Bull. Seismol. Soc. Am.* **87**, 745–754.

Plomerova, J., J. Sileny, and V. Babuska (1996). Joint interpretation of upper mantle anisotropy based on teleseismic P travel-time delays and 3-D inversion of shear-wave splitting parameters. *Phys. Earth Planet. Inter.* **95**, 293–309.

Pulliam, J. and M.K. Sen (1998). Anisotropy in the core-mantle transition zone may indicate chemical heterogeneity. *Geophys. J. Int.* **135**, 113–128.

Ritsema, J., T. Lay, and E.J. Garnero (1998). Seismic anisotropy in the lowermost mantle. *Geophys. Res. Lett.* **25**, 1229–1232.

Rüger, A. (1997). P-wave reflection coefficients for transversely isotropic models with vertical and horizontal axis of symmetry. *Geophysics* **62**, 713–722.

Sachpazi, M. and A. Hirn (1991). Shear-wave anisotropy across the geothermal field of Milos, Agean Volcanic arc. *Geophys. J. Int.* **114**, 759–777.

Savage, K.S. (1998). Lower crustal anisotropy or dipping boundaries? Effects on receiver functions and a case study in New Zealand. *J. Geophys. Res.* **103**, 15069–15087.

Savage, M.K. (1999). Seismic anisotropy and mantle deformation: what have we learnt from shear wave splitting? *Rev. Geophys.* **37**, 65–106.

Sayers, C.M. (1998). Misalignment of the orientation of fractures and the principal axes for P and S waves in rocks containing multiple non-orthogonal fracture sets. *Geophys. J. Int.* **133**, 459–466.

Schlue, J.W. and L. Knopoff (1976). Shear-wave anisotropy in the mantle of the Pacific Basin. *Geophys. Res. Lett.* **3**, 359–362.

Shearer, P.M. and K.M. Toy (1991). PKP(BC) versus PKP(DF) differential travel times and aspherical structure in the Earth's inner core. *J. Geophys. Res.* **96**, 2233–2247.

Silver, P.G. (1996). Seismic anisotropy beneath the continents: probing the depths of geology. *Annu. Rev. Earth Planet. Sci.* **24**, 385–432.

Silver, P.G. and W.W. Chan (1988). Implications for continental structure and evolution from seismic anisotropy. *Nature* **335**, 34–39.

Silver, P.G. and W.W. Chan (1991). Shear-wave splitting and subcontinental mantle deformation. *J. Geophys. Res.* **96**, 16429–16454.

Smith, L.N. and F.A. Dahlen (1975). Correction to "The azimuthal dependence of Love and Rayleigh wave propagation in a slightly anisotropic medium." *J. Geophys. Res.* **80**, 1923.

Sobolev, S., A. Grésillaud, and M. Cara (1999). How robust is isotropic delay time tomography for anisotropic mantle? *Geophys. Res. Lett.* **26**, 509–512.

Song, X. and P.G. Richards (1996). Seismological evidence for differential rotation of the Earth's inner core. *Nature* **382**, 221–224.

Souriau, A. and B. Romanowicz (1996). Anisotropy in inner core attenuation: a new type of data to constrain the nature of the solid core. *Geophys. Res. Lett.* **23**, 1–4.

Souriau, A., P. Roudil, and B. Moynot (1997). Inner core differential rotation: facts and artefacts. *Geophys. Res. Lett.* **16**, 2103–2106.

Stoneley, R. (1963). The propagation of surface waves in an elastic medium with orthorhombic symmetry. *Geophys. J. R. Astron. Soc.* **8**, 176–186.

Su, W.J. and A.M. Dziewonski (1995). Inner core anisotropy in three dimensions. *J. Geophys. Res.* **100**, 9831–9852.

Takeuchi, H. and M. Saito (1972). Seismic surface waves. In: "Methods in Computational Physics" (B.A. Bolt, Ed.), Vol. XI, pp. 217–295. Academic Press, New York.

Thomsen, L. (1986). Weak elastic anisotropy. *Geophysics* **51**, 1954–1966.

Tommasi, A., A. Vauchez, and R. Russo (1996). Seismic anisotropy in ocean basins: resistive drag of the sublithospheric mantle? *Geophys. Res. Lett.* **23**, 2991–2994.

Tromp, J. (1993). Support for anisotropy of the Earth's inner core. *Nature* **366**, 678–681.

Vinnik, L.P. and J.P. Montagner (1996). Shear-wave Splitting in the Mantle from Ps Phases. *Geophys. Res. Lett.* **23**, 2449–2452.

Vinnik, L.P., V. Farra, and B. Romanowicz (1989a). Azimuthal anisotropy in the Earth from observations of SKS at Geoscope and NARS broadband stations. *Bull. Seismol. Soc. Am.* **79**, 1542–1558.

Vinnik, L.P., V. Farra, and B. Romanowicz (1989b). Observational evidence for diffracted SV in the shadow of the Earth's core. *Geophys. Res. Lett.* **16**, 519–522.

Webb, S.C. and D. Forsyth (1998). Structure of the upper mantle under the EPR from waveform inversion of regional events. *Science* **280**, 1227–1229.

Weber, P. and Ph. Machetel (1992). Convection within the inner-core and thermal implications. *Geophys. Res. Lett.* **19**, 2107–2110.

Williams, Q. and E.J. Garnero (1996). Seismic evidence for partial melt at the base of Earth's mantle. *Science* **273**, 1528–1530.

Wolfe, C.J. and S.C. Solomon (1998). Shear-wave splitting and implications for mantle flow beneath the MELT region of the East Pacific rise. *Science* **280**, 1230–1232.

Woodhouse, J.H. (1981). A note on the calculation of travel rimes in a transversely isotropic Earth model. *Phys. Earth Planet. Inter.* **25**, 357–359.

Woodhouse, J.H., D. Giardini, and X.D. Li (1986). Evidence for inner core anisotropy from free oscillations. *Geophys. Res. Lett.* **13**, 1549–1552.

Yu, Y. and J. Park (1993). Upper mantle anisotropy and coupled-mode long-period surface waves. *Geophys. J. Int.* **114**, 473–489.

Zollo, A. and P. Bernard (1989). S wave polarization inversion of the 15 October 1979, 23:19 Imperial Valley aftershock: evidence for anisotropy and a simple source mechanism. *Geophys. Res. Lett.* **16**, 1047–1050.

Editor's Note

Please see also Chapter 11, Inversion of surface waves: a review, by Romanowicz; Chapter 51, The Earth's interior, by T. Lay; and Chapter 52, Probing the Earth's interior with seismic tomography, by Curtis and Snieder; and the following 3 Chapters.

54

Seismic Velocity Structure of the Continental Lithosphere from Controlled Source Data

Walter D. Mooney
US Geological Survey, Menlo Park, CA, USA
Claus Prodehl
University of Karlsruhe, Karlsruhe, Germany
Nina I. Pavlenkova
RAS Institute of the Physics of the Earth, Moscow, Russia

1. Introduction

The purpose of this chapter is to provide a summary of the seismic velocity structure of the continental lithosphere, i.e., the crust and uppermost mantle. We define the crust as the outer layer of the Earth that is separated from the underlying mantle by the Mohorovičić discontinuity (Moho). We adopted the usual convention of defining the seismic Moho as the level in the Earth where the seismic compressional-wave (*P*-wave) velocity increases rapidly or gradually to a value greater than or equal to 7.6 km sec^{-1} (Steinhart, 1967), defined in the data by the so-called "*Pn*" phase (*P-n*ormal). Here we use the term uppermost mantle to refer to the 50–200+ km thick lithospheric mantle that forms the root of the continents and that is attached to the crust (i.e., moves with the continental plates).

This summary has been preceded by 90 y of intense scientific activity. Mohorovičić (1910) was the first to publish an

TABLE 1 Published summaries of crustal structure by region covered

Year	Authors	Areas covered	J/A/B[a]
1961	Closs and Behnke	World	J
1963	Hill	Oceans	B
1966	James and Steinhart	World	B
1966	McConnell *et al.*	World	J
1969	Hart	World	B
1970	Ludwig *et al.*	Oceans	A
1970	Maxwell	Oceans	B
1971	Heacock	N-America	B
1973	Meissner	World	J
1973	Mueller	World	B
1975	Makris	E-Africa, Iceland	A
1977	Bamford and Prodehl	Europe, N-America	J
1977	Heacock	Europe, N-America	B
1977	Mueller	Europe, N-America	A
1977	Prodehl	Europe, N-America	A
1978	Mueller	World	A
1980	Zverev and Kosminskaya	Europe, Asia	B
1982	Soller *et al.*	World	J
1984	Prodehl	World	A
1986	Meissner	Continents	B
1987	Orcutt	Oceans	J
1989	Mooney and Braile	N-America	A
1989	Pakiser and Mooney	N-America	B
1991	Beloussov *et al.*	Europe, Asia	B
1991	Collins	Australia	A
1992	Blundell *et al.*	Europe	B
1992	Holbrook *et al.*	Continents	A
1992	Mooney and Meissner	Continents	A
1993	Mueller and Kahle	Mediterranean	A
1994	Durrheim and Mooney	Continents	J
1995	Olsen	Continental rifts	B
1995	Christensen and Mooney	World	J
1995	Rudnick and Fountain	World	J
1995	Tanimoto	World	A
1996	Pavlenkova	Europe, Asia	A
1997	Prodehl *et al.*	Near East, Africa	J
1998	Li and Mooney	China	J
1998	Mooney *et al.*	World	J

[a]J, Article published in an international journal; A, Article published as part of a book; B, Book publication.

INTERNATIONAL HANDBOOK OF EARTHQUAKE AND ENGINEERING SEISMOLOGY, VOLUME 81A ISBN: 0-12-440652-1

estimate for crustal thickness (54 km near Zagreb, Croatia), and to describe the seismically defined boundary between the crust and mantle that now bears his name (often shortened to "the Moho"). The fact that the oceanic crust is significantly thinner than continental crust (5 km versus about 40 km) was documented 40 y later (e.g., Hersey *et al.*, 1952). Numerous later studies demonstrated that the continental crust varies in thickness from about 15 km to greater than 70 km beneath the Tibet plateau. Jarchow and Thompson (1989) provide a useful summary of early crustal studies, and Table 1 provides additional references. We emphasize results from active-source seismic refraction profiles that provide detailed *P*-wave velocity information. The shear-wave (*S*-wave) structure of the crust and uppermost mantle, as determined by surface waves, is discussed, for example, by Ekström *et al.* (1997) and Ritzwoller *et al.* (1998).

2. Data and Methods of Analysis

The properties of the Earth's lithosphere are particularly well suited to measurements using seismological data due to the existence of the pronounced variations in seismic wave speeds both laterally and with depth. Beneath continents, the Moho is characterized by a more or less rapid increase of *P*- and *S*-velocities with depth. However, because *S*-waves travel more slowly than *P*-waves and therefore cannot be detected as first arrivals, they are often hidden in the coda of the *P*-waves. Thus, the study of *S*-waves for crustal research has been a limited, but rapidly growing, area of research, and a worldwide mapping of crustal and uppermost mantle *S*-wave results from seismic refraction data is not yet possible.

The chief advantages of controlled-source seismic-refraction data are: (1) The exact time and position of the seismic sources are accurately known; (2) It is possible to plan, in relation to the specific geologic target, the number, spacing, and geometry of sources and receivers. The main disadvantages of these seismic-refraction data are: (1) The seismic sources are limited to the surface of the Earth, whereas natural seismicity extends to a depth of 10–20 km within the crust, and to a maximum depth of 670 km in subduction zones; (2) There is a practical limit to the strength of chemical or other controlled sources, thereby limiting the depth of investigation to the lithosphere.

It is important to note that vertical-incidence seismic reflection profiles, which are not reviewed here, provide the highest resolution of structural detail within the crust. These data provide an image of the entire crust that is similar to the pictures obtained within sedimentary basins by the oil exploration industry. The application of seismic reflection profiles to the study of the continents began in Germany (cf., discussion in Meissner, 1986), and was adopted in the United States of America in 1974 by COCORP (Barazangi and Brown, 1986a,b). Many other countries soon followed suit with their own national programs, such as BIRPS (UK, Matthews, 1986), DEKORP (Germany, Meissner *et al.*, 1991). Crustal reflection profiles show that the crust has many fine-scale impedance contrasts that are related to compositional layering, shear zone and fractures, and other geological features, such as dikes and sills (Brown *et al.*, 1983; Matthews, 1986; Meissner, 1986; Barazangi and Brown, 1986a,b; Mooney and Meissner, 1992; Klemperer and Mooney, 1998a,b). Deep reflections from the subcrustal lithosphere have also been recorded in many regions. Consequently, the advent of deep seismic reflection profiling has led to a major new understanding of the evolution of the continental lithosphere. Optimal information is obtained when coincident seismic refraction and vertical-incidence reflection profiles are collected, as has been the case for the Canadian LITHOPROBE program (Clowes *et al.*, 1999).

Since the early 1950s a large number of seismic profiles have been collected around the world. Advancing technology has led not only to better instrumentation, but also to improved data processing and interpretation techniques. Useful recent reviews of seismic methods for determining crust and upper mantle structure have been provided by Mooney (1989) and Braile *et al.* (1995), whereas Giese *et al.* (1976) summarize older methods of data analysis. Table 1 lists previously published summaries of crustal structure by region covered. The papers and books most comparable to the present work are Soller *et al.* (1982), Prodehl (1984), Meissner (1986), Tanimoto (1995), Pavlenkova (1996), Chuikova *et al.* (1996), and Mooney *et al.* (1998).

Resolution is a major consideration when discussing seismic data. Because of the variety in seismic techniques, however, general statements about resolution are difficult. In seismology, resolution is related both to data quality and physical laws. The quality of a data set is usually considered to be a function of the strength of the signal relative to noise (expressed as signal-to-noise ratio) and the number or density of measurements depending on the number of sources and receivers and their relative spacing. Generally, the more data available, the higher is the resolution. However, physical laws limit the resolution of even near-perfect data sets. The Earth strongly attenuates high frequencies so that signals penetrating deeper will contain relatively low frequencies. Typically, signals traversing the whole crust have peak frequencies of 5–20 Hz resulting in absolute accuracy of depth determination of not better than 2–5% of the depth (e.g., 1–2 km for a 40 km-thick crust).

It is also of interest to use seismic velocities to infer the composition of the lithosphere. This can be accomplished by comparison with laboratory measurements of the seismic velocity of rock specimens believed to have once resided at depth within the lithosphere. This research was pioneered by Francis Birch, who measured numerous igneous rock samples at confining pressures up to 100 MPa and compared these results with early field determinations of seismic velocities (Birch,

1960, 1961). Such measurements have been expanded by others to include metamorphic rocks, the measurement of shear-wave velocities, and the determination of temperature effects (Simmons, 1964; Christensen, 1965, 1966, 1979; Kern, 1978). These studies are reviewed in Holbrook *et al.* (1992), Rudnick and Fountain (1995), and Christensen and Mooney (1995). The interplay of temperature and pressure is worth noting here; a temperature increase of 100°C at constant pressure will decrease seismic velocity by 0.05 km sec^{-1}. However, the influence of increased temperature with depth is compensated for by increased pressure. Thus, seismic velocities will generally increase with depth only if there is a change in rock composition, or an unusually low temperature gradient. An exception is the uppermost (0–5 km) crust, where the closing of cracks in the crystalline crust will lead to an increase in both *P*- and *S*-wave velocities.

3. Main Features of Continental Crustal Structure

Over the past 50 y a vast amount of data has been accumulated regarding the seismic structure of the continental crust. A typical data example is shown in Figure 1. The basic features of continental crustal structure were recognized by the 1960s (Steinhart and Meyer, 1961; James and Steinhart, 1966; Ludwig *et al.*, 1970; Pavlenkova, 1973). The seismic velocity distributions vary widely in different geographic localities, and crustal models generally consist of two, three, or more layers separated by velocity discontinuities or gradients. In relatively stable continental regions the thickness of the crust is between 30 and 50 km. Seismic velocities in the upper crustal layer are usually 5.6–6.3 km sec^{-1}. At a depth of 10–15 km the seismic velocity commonly increases to 6.4–6.7 km sec^{-1}. This velocity increase is sometimes referred to as the Conrad discontinuity, a term that is now in disfavor since the velocity structure of the crust has been found to be sufficiently complex that the use and definition of this term is ambiguous. In many stable continental interiors there is a third, basal crustal layer with a velocity of 6.8–7.2 km sec^{-1}. The seismic velocity below the Moho (*Pn* velocity) is typically about 8 km sec^{-1}. The exact nature of the velocity–depth distribution is not always well defined. The evidence for distinct layers within the continental crust almost exclusively depends on the interpretation of second-arrival phases. In some regions, clear evidence of later arrivals confirms that the velocity increases discontinuously

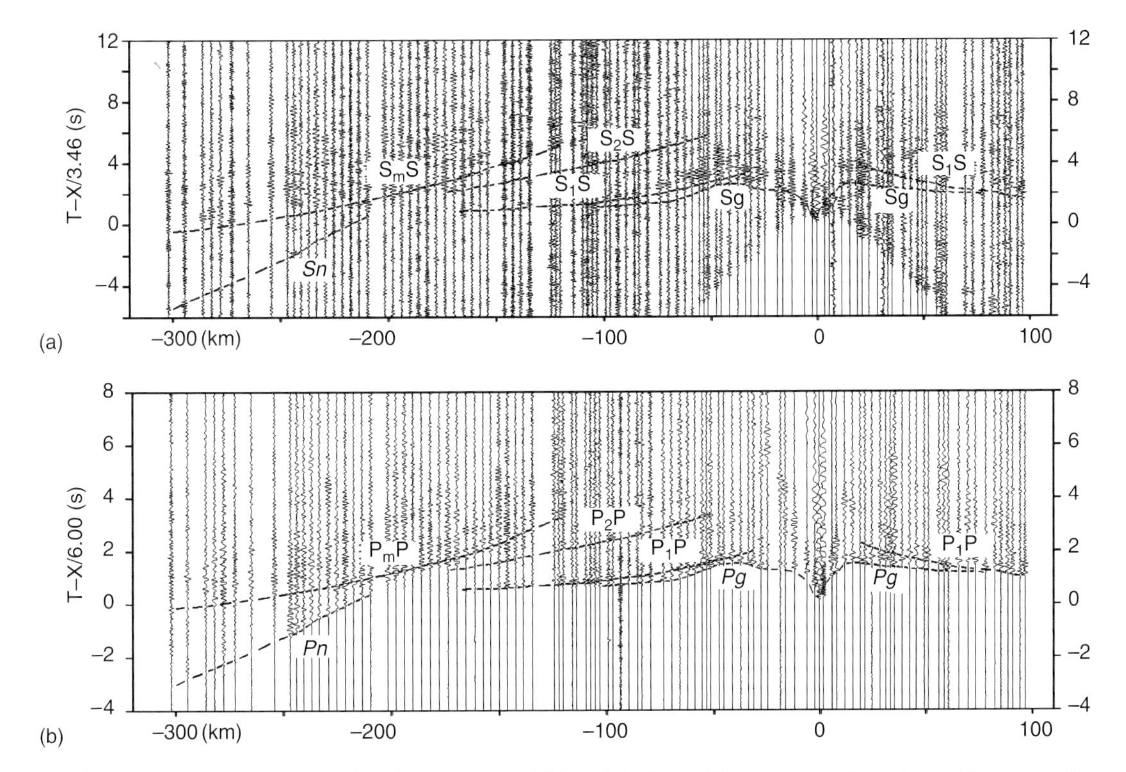

FIGURE 1 Sample record section from western China. (a) Shear (*S*) wave record section with a reduction velocity of 3.46 km sec^{-1}. Shot point corresponds to 0 km; *Sg*, direct arrival; S1S, reflection from boundary between upper and middle crust; S2S, reflection from middle and lower crust; SmS, reflection from Moho; *Sn*, refraction from uppermost mantle. (b) Compressional (*P*) wave record section with a reduction velocity of 6.0 km sec^{-1}; *Pg*, direct arrival; P1P, P2P, PmP, *Pn*, *P*-wave arrivals as defined for *S*-waves above. (Wang, Y.X. *et al.*, submitted 2002).

through intermediate layers within the crust, whereas in other regions velocity may increase gradually with increasing depth producing no distinct intracrustal reflection (Levander and Holliger, 1992).

The compilation of seismic crustal data around the world has defined the characteristic primary crustal types connected with specific tectonic settings (Fig. 2). Each primary crustal type is defined by the average seismic model of the crust of a specific age or tectonic setting (Fig. 3).

Despite the abundance of refraction-seismic experiments having been carried out throughout the world, the coverage is very selective (Fig. 4). For example the continents of the northern hemisphere are reasonably well covered, whereas in the southern hemisphere only selective seismic surveys dealing with particular target areas, such as the Andes of South America and the East African Rift System have been studied. Figure 3 shows a world map indicating the main tectonic units of the continents. The map also indicates the location of the seismic cross sections that are discussed later.

Figure 4 shows that Eurasia is densely covered with seismic profiles, as is the central portion of North America (the United States of America and southern Canada). Only limited data are available for northern Canada, Mexico, Central America, and Greenland. The South American and African continents have a concentration of data points in select locations, but vast regions remain without seismic control. Data are lacking for significant portions of the Middle East and Southeast Asia. Data coverage is good within Australia, New Zealand, and the coastal portions of Antarctica. These data points have been used to map the thickness of the Earth's crust (Fig. 5). Crustal thickness is estimated in regions lacking data by determining the geologic setting (Fig. 3) and assuming a typical crustal structural for that setting (Fig. 2). Short period (35–40 sec) Love-wave phase and group velocity maps (Ekström *et al.*, 1997; Ritzwoller *et al.*, 1998), which are mainly sensitive to crustal thickness and average crustal shear velocity, provide a useful guide to estimating crustal thickness in regions lacking active-source seismic profiles (cf., Mooney *et al.*, 1998).

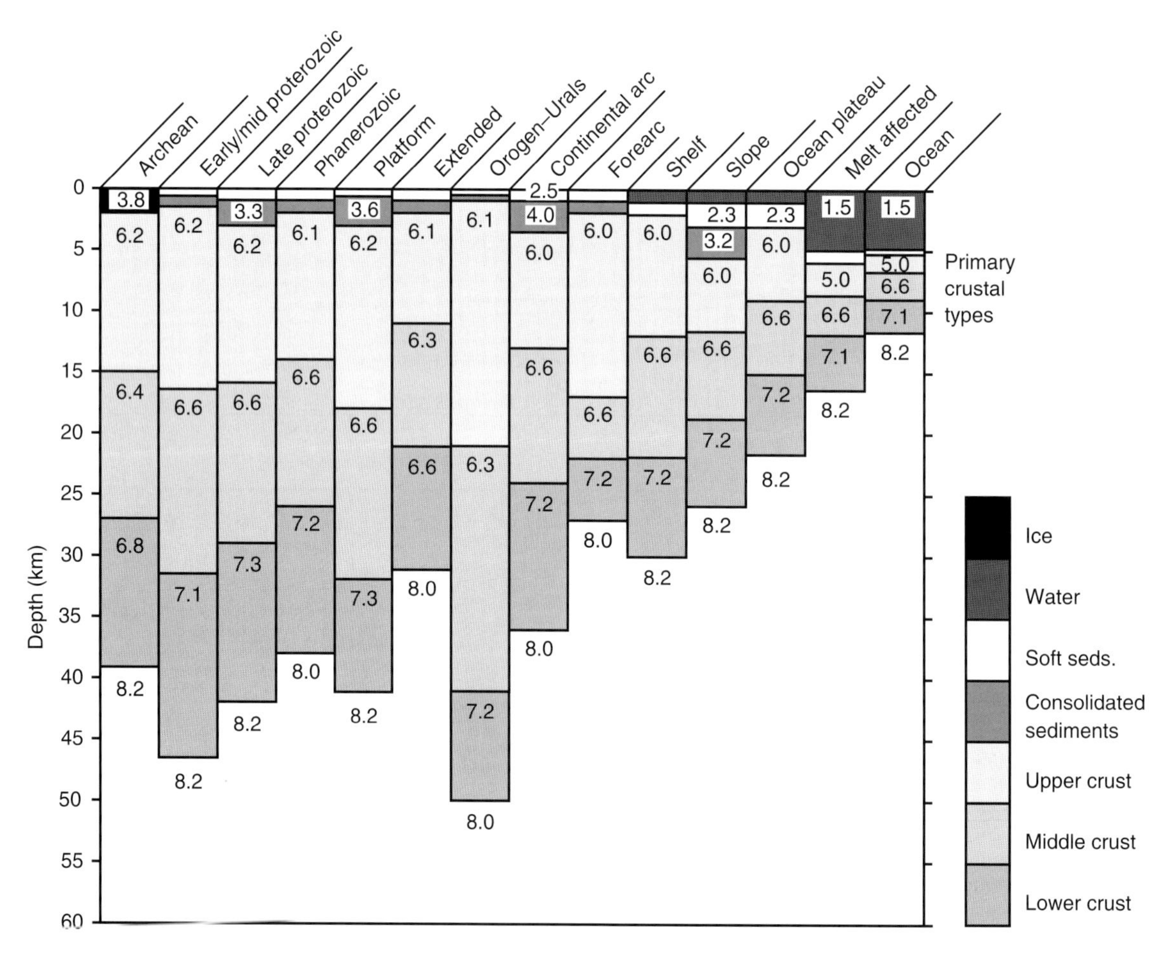

FIGURE 2 Fourteen common continental and oceanic crustal types (Mooney *et al.*, 1998). Typical *P*-wave velocities are indicated for the individual crustal layers and the uppermost mantle. Velocities refer to the top of each layer, and there is commonly a velocity gradient of 0.01–0.02 km sec^{-1} km^{-1} within each layer. The crust thins from an average values of 40 km in continental interiors to 12 km beneath oceans.

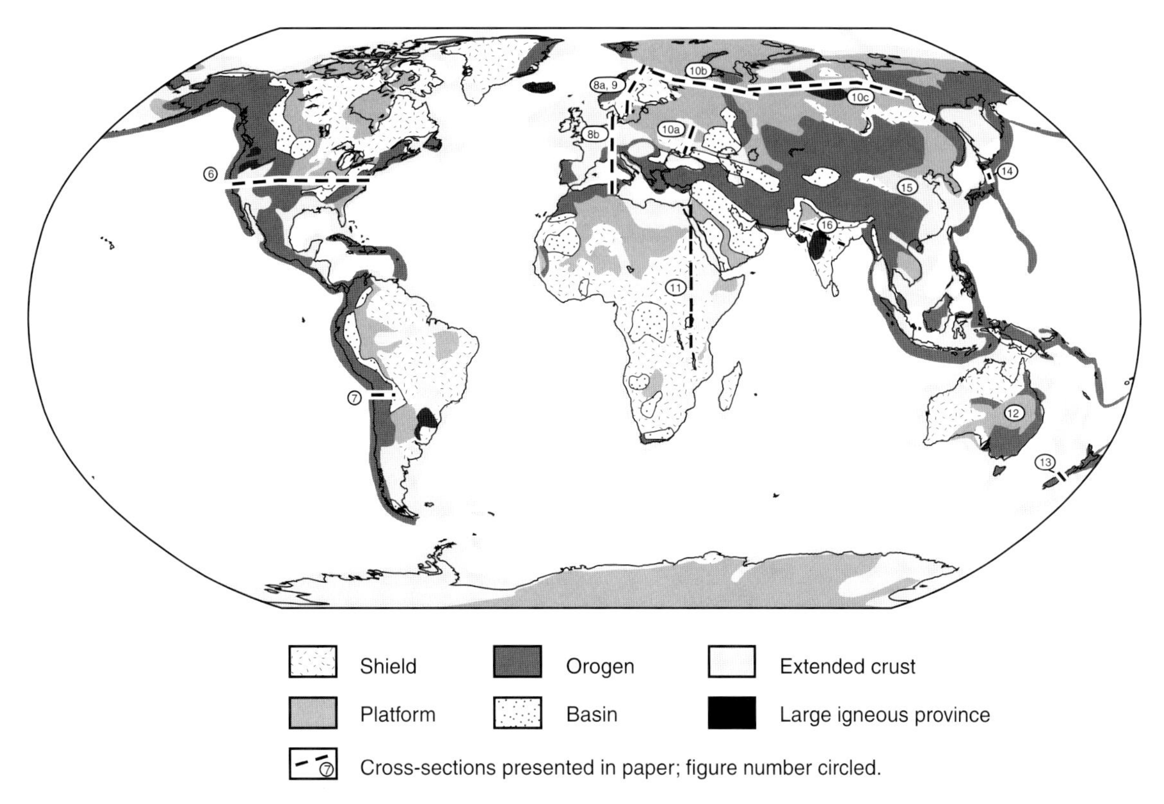

FIGURE 3 World map showing six primary Geological provinces and the Figure numbers of cross sections presented in this chapter. Index of locations shown on the map: N-America (Fig. 6) (38° N); N-Chile (Fig. 7); European Geo-Traverse (EGT; Figs. 8 and 9); Russia (Fig. 10); Afro-Arabian rift system (Fig. 11); Australia (Fig. 12); New Zealand (Fig. 13); Japan: northern Honshu (Fig. 14); China (Fig. 15); India (Fig. 16).

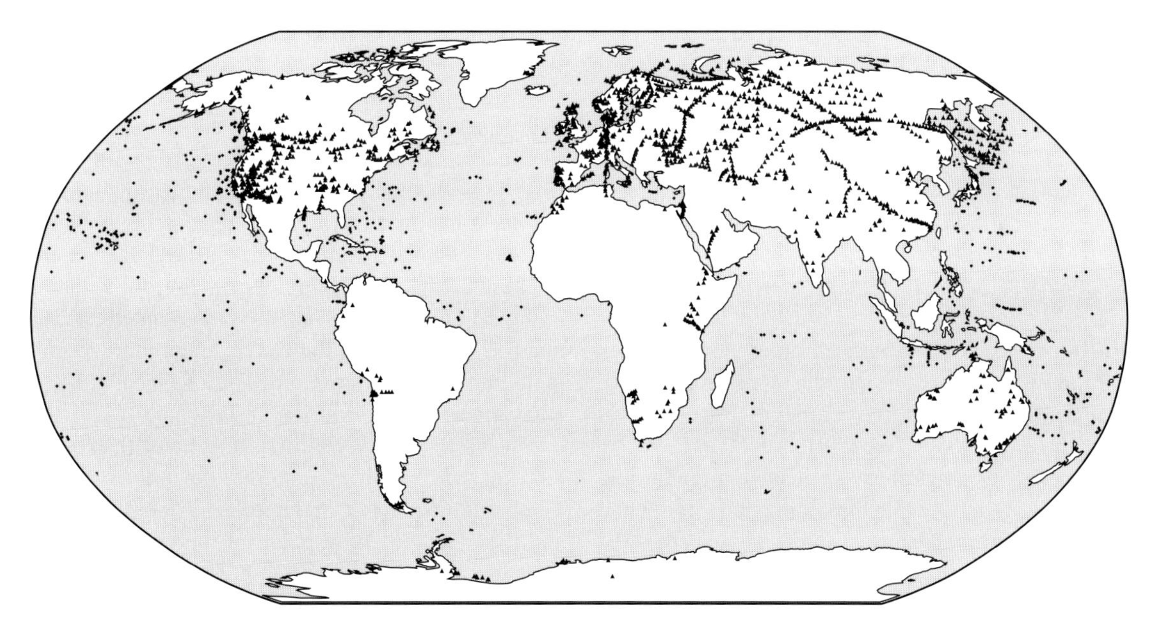

FIGURE 4 World map showing the location of seismic refraction profiles used to compile the maps of crustal thickness (Fig. 17) and mean crustal velocity (Fig. 18) (Mooney *et al.*, 1998). Each triangle corresponds to a point along a seismic refraction profile where a crustal column has been extracted.

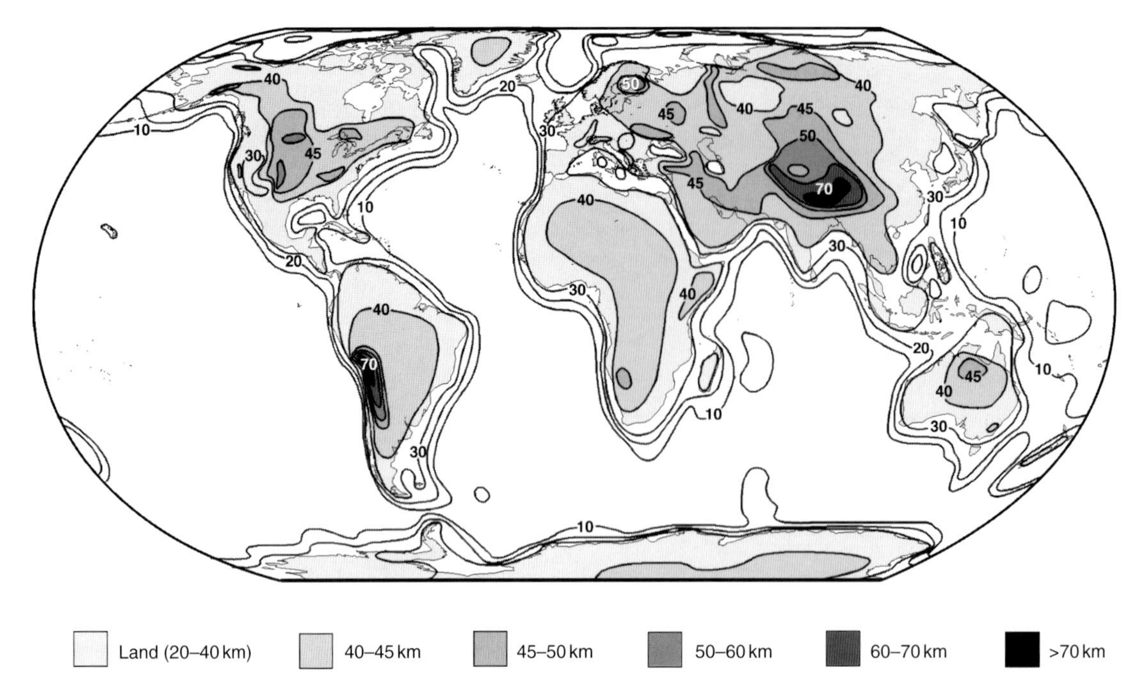

FIGURE 5 Global crustal thickness (Mooney *et al.*, 1998). The average thickness on continents is 40 km, and generally increases in thickness from 30 km near the margins toward the interior. In this figure, crustal thickness under narrow features (e.g., continental rifts, midocean ridges and oceanic fracture zones) is not visible.

The thickest crust (70+ km) occurs beneath the Tibetan Plateau and the South American Andes. The thickness of continental crust with an elevation above sea level is generally between 28 km and 52 km, and the average thickness (excluding continental margins) is 40 km. Aside from the two regions mentioned above, very little of the continental crust is thicker than 50 km. Likewise, continental crust thinner than 30 km is generally limited to rifts and highly extended crust including continental margins. It is important to bear in mind that the Earth's crust is very complex, and there are numerous local exceptions to these generalizations. In the following sections we provide a detailed description of the structure of the crust on a continent-by-continent basis.

3.1 North America

The deep structure of the North American lithosphere has been the subject of intense investigation since the pioneering work of M.A. Tatel and H.E. Tuve at the Carnegie Institution of Washington in the early 1950s (e.g., Tatel and Tuve, 1955). These and other early investigations established the main feature of a thick (~45 km) crust within the continental interior and thin (~30 km) crust at the margins. They also discovered unexpectedly thin (~30 km) crust beneath the Basin and Range Province, where a high average surface elevation of 1–2 km can be found.

Several summaries of the crustal structure of North America have been published over the past 25 y (Pakiser and Steinhart, 1964; Hart, 1969; Heacock, 1971, 1977; Prodehl, 1977; Pakiser and Mooney, 1989; Mooney and Braile, 1989). A great deal of crustal structure data has also been published by Canadian researchers over the past 15 y under the LITHOPROBE Program (Clowes *et al.*, 1999). These Canadian investigations have used coincident seismic reflection and refraction data to obtain higher-resolution data than previously was possible. A recent compilation of North American data contains more than 1400 measurements of crustal structure (Chulick and Mooney, 2002).

The contour map of crustal thickness of North America (Fig. 5) shows the previously mentioned pattern of increasing thickness toward the interior. The average thickness of crust is 36.7 km, with a standard deviation (SD) of 8.4 km. The thickest crust (50–55 km) occurs in four relatively isolated regions. The thinnest crust is associated with continental margins and regions that have undergone recent extension, including the Western Cordillera of Canada, the Rio Grande Rift, and the Basin and Range Province.

The upper mantle velocity, *Pn*, beneath North America is 8.03 km sec^{-1} with a standard deviation of 0.19 km sec^{-1}. Temperature plays a major role in determining *Pn* velocity, and thus the cold lithosphere of the continental interior typically has a *Pn* velocity of 8.1–8.2 km sec^{-1}, whereas velocities of 7.8–8.0 km sec^{-1} are measured beneath the warm crust of the Western Cordillera and the Basin and Range Province.

The average *P*-wave velocity of the crust is related to crustal composition. The global average is 6.45 km sec^{-1}

(SD 0.23), and corresponds to an average crustal composition equivalent to a diorite. The North American continental interior has an average crustal velocity of 6.5–6.7 km sec^{-1}, indicating that its bulk composition is denser and more mafic than the global average. The continental margins, Basin and Range Province, Sierra Nevada (California) and the Canadian Cordillera all have an average crustal velocity of 6.2–6.3 km sec^{-1}. This can be attributed to a missing, or very thin high-velocity (6.9–7.3 km sec^{-1}) lower crustal layer. Elevated temperature may also play a role in reducing the average crustal velocity in these regions.

Many of the features of the lithospheric structure of North America can be seen in a cross section from San Francisco, California, to Washington, DC at 38° N latitude (Fig. 6). This cross section shows that the crust is highly variable in structure. The western USA has a relatively thin crust, averaging 30 km, with a low average crustal velocity (6.2 km sec^{-1}). Surprisingly, a pronounced crustal root does not underlie the topographically high Rocky Mountains. Indeed, the thickest crust (45–50 km) underlies the low-lying Great Plains. The crust thins modestly to the east beneath the Appalachian Foreland, and then thins beneath the Atlantic Coastal Plain. The velocity of the uppermost mantle increases from 7.8–7.9 km sec^{-1} beneath the western USA to more typical continental values of 8.0–8.1 km sec^{-1} beneath the central and eastern USA. A laterally variable *S*-wave low-velocity zone exists at a depth of 60–180 km beneath North America, and the top of this zone is generally taken as the depth of the seismic lithosphere. The thermal lithosphere is commonly defined as the depth to the 1300°C isotherm and can be estimated from heat flow data. In Figure 6 we consider both the seismic and thermal lithosphere and estimate lithospheric thickness based on the summary of seismic results by Iyer and Hitchcock (1989) and the thermal estimates of Pollack and Chapman (1977) and Artemieva and Mooney (2001). The lithosphere thickens from about 60 km beneath the Basin and Range province to about 180 km beneath the Precambrian continental interior. We estimate a lithospheric thickness of 100 km at the Atlantic coastal margin.

3.2 South America

Few seismic refraction measurements have been made in South America, with the exception of the Andean region. East of the Andes the crust consists of a collage of Precambrian shields, platforms and paleorifts (Goodwin, 1991, 1996). Sparse refraction data indicate an average crustal thickness of 40 km, with typical shield velocity structure (Fig. 7). This estimate is consistent with phase velocity measurements of surface waves (Ekström *et al.*, 1997) and determinations of crustal structure from passive seismology (Myers *et al.*, 1998).

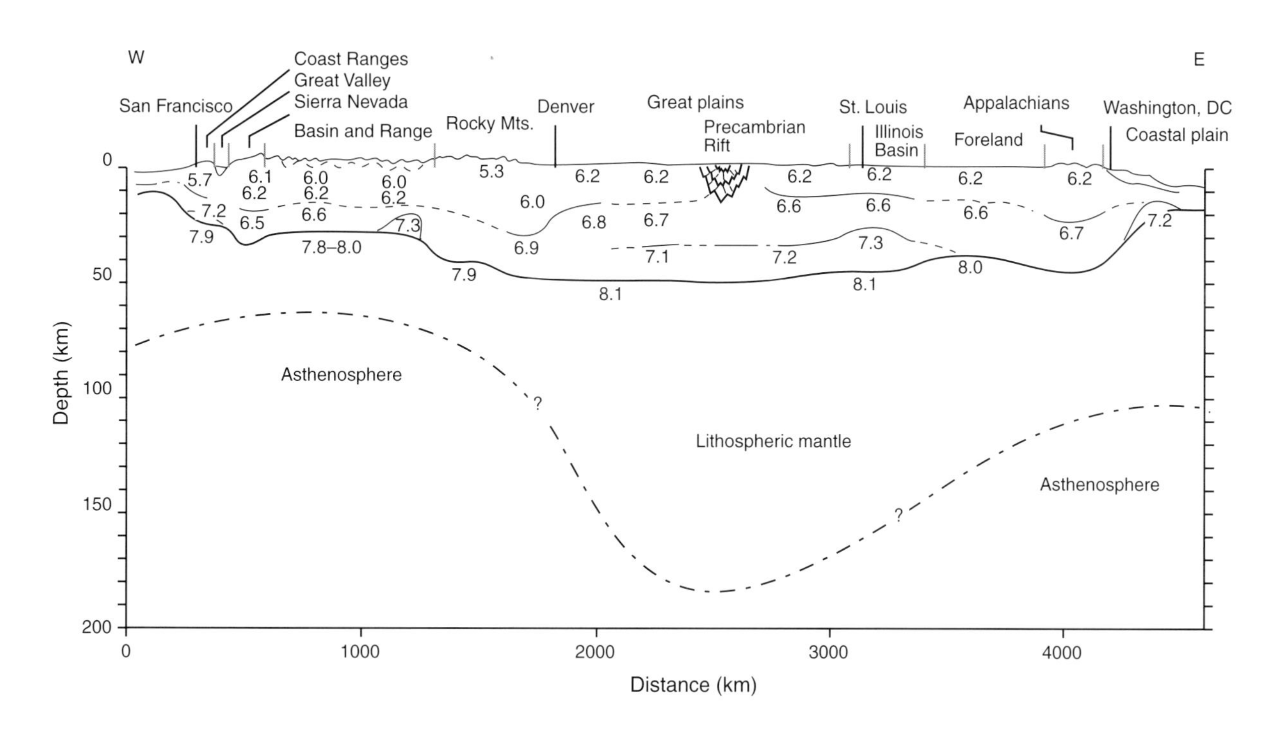

FIGURE 6 Lithospheric section through the conterminous United States. The section traverses the continent from east to west at approximately 38° N. Solid lines indicate velocity discontinuities where well-determined, dashed lines were inferred. The seismic velocity structure is generalized, and fine-scale structure is not included. The base of the lithosphere is from Iyer and Hitchcock (1989), Pollack and Chapman (1977) and Artemieva and Mooney (2001). Vertical exaggeration is 10 : 1.

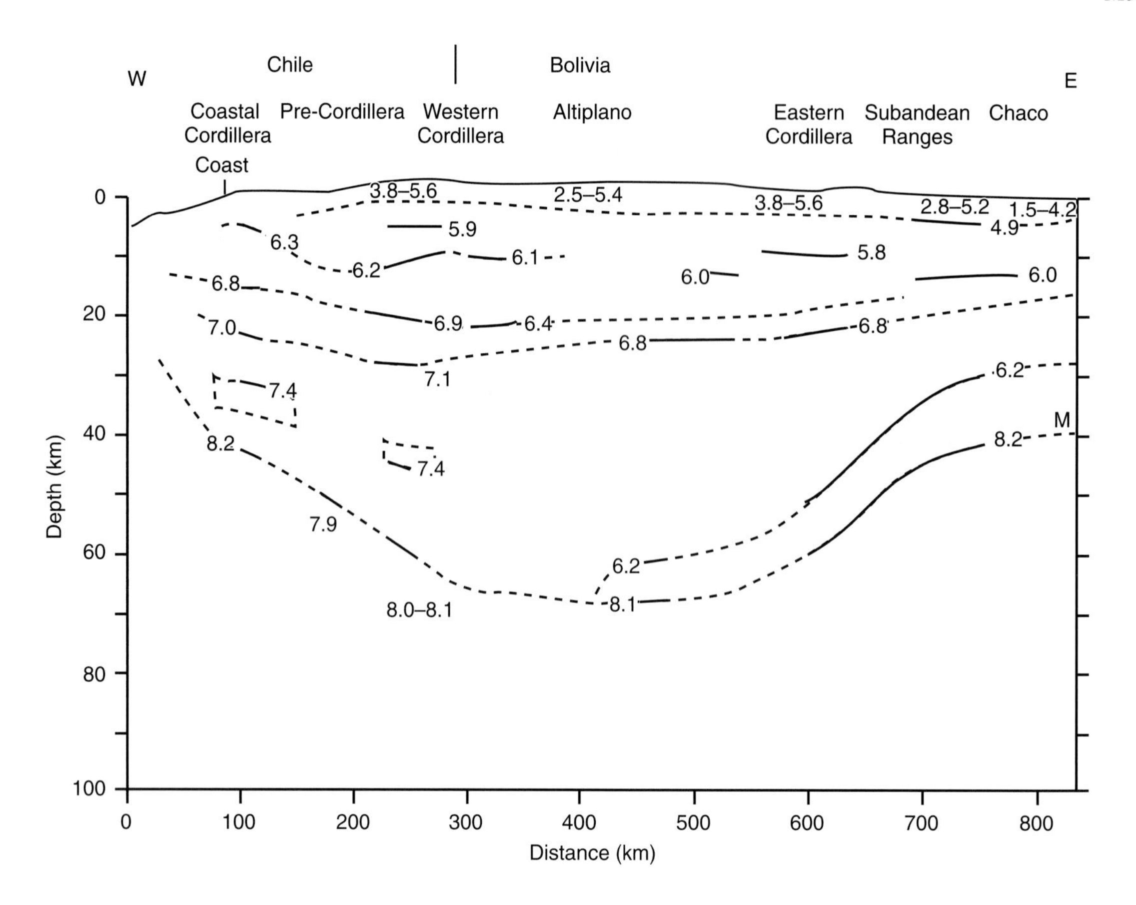

FIGURE 7 East–west lithospheric section through South America at approximately 21° S (see Fig. 4). Vertical exaggeration is 5 : 1. Source: Wigger *et al.* (1994), Giese *et al.* (1999), Patzwahl *et al.* (1999).

The structure of western South America is dominated by mountain building and volcanism associated with the subduction of the Pacific Plate beneath the continent. The structure of the central Andes was first determined from surface wave observations in a classic paper by James (1971) that demonstrated that the high Andes of Peru are underlain by a remarkably thick crust (~60 km). These results have been confirmed by more recent surface-wave phase-velocity measurements (Ekström *et al.*, 1997). In the northern Andes of Colombia, Meyer *et al.* (1976) present a cross section from the Pacific basin into the Western Cordillera. The plate dips at ~15°, and the velocities within the continental crust are higher than average due to the Cenozoic accretion of a basaltic oceanic plateau.

The southern Andes between 19° and 25° S were the subject of numerous seismic investigations during the 1990s by South American, German, and American researchers (e.g., Wigger *et al.*, 1994; Giese *et al.*, 1999; Patzwahl *et al.*, 1999). These studies have contributed a great deal of information regarding the deep structure, and confirm the existence of thick crust south of Peru (Fig. 7). Several detailed seismic offshore and onshore refraction and reflection profiling campaigns (e.g., Wigger *et al.*, 1994; Patzwahl *et al.*, 1999) were undertaken to investigate the subduction of the oceanic Nazca plate under the South American continent between the Peru–Chile trench and the Central Andes in northern Chile between 19° and 25° S. The active seismic campaigns provided details of crustal and upper mantle structure to depths of 70 km and more, and were supported by long-term passive seismological investigations that resolved structure to more than 100 km. The data clearly show the Nazca plate subducting under an increasing angle of 9–25° down to 30–50 km depth near the coast, and they show a portion of the Moho dipping eastward from 30–50 km near the coast to 55–64 km up to about 240 km inland.

Crustal thickness beneath the western Cordillera and the Altiplano is at least 70 km (Fig. 7). However, only weak Moho arrivals are observed by active seismic measurements beneath the Altiplano and the Western Cordillera. Broadband seismologic data are more reliable in these regions and indicate an approximately 60–70 km thick crust. In the Precordillera a pronounced Moho discontinuity is detected at a depth of 50 km. Along the coast, the oceanic Moho boundary can be identified at a depth of 40–45 km. The seismic measurements have revealed a clearly pronounced asymmetric structure and evolution of the Central Andean crust.

In the Andean backarc region, tectonic compression has produced pronounced crustal thickening and, locally, a doubling of the Moho discontinuity (Giese *et al.*, 1999). In the forearc region, dehydration of the lower plate may have caused hydration and metasomatism of the overlying mantle wedge that is associated with a decrease in density and seismic velocity to values typical of the crust. Such processes, if real, may reduce the velocity contrast at the Moho and result in a blurring of the crust–mantle boundary from a geophysical point of view. The paleo-Moho today appears as an intracrustal boundary. Magmatic intrusions and underplating also have the potential to produce new seismic boundaries, including one that appears to be the Moho discontinuity. Giese *et al.* (1999) summarize their interpretation of the Central Andes by suggesting that there exists a variety of different Moho-discontinuities, which they refer to as a "Moho-menagerie". This term may be understood to refer to a highly complex seismic velocity structure in the lower crust and uppermost mantle.

3.3 Europe and Mediterranean

Europe becomes progressively younger in tectonic age from north to south. The northern Baltic Shield consists of Archean and Proterozoic rocks; to the west of the Baltic Shield the Caledonian orogeny has added crust of Paleozoic age. The basement rocks of central Europe are of Paleozoic age and formed during the Variscan orogeny, covering the area between Scotland and Portugal. The terranes that comprise this region have undergone strong orogenic collapse (extension), as evidenced by a relatively thin lithosphere and crust. The Tornquist Zone separates Paleozoic Western Europe from the extensive Precambrian platform areas of Eurasia. The Neogene Alpine orogeny shaped southern Europe, from the Spanish Pyrenees to the Carpathians, and provide a physiographic separation between central Europe and the Mediterranean.

The European Geotraverse (EGT) was a multinational geologic and geophysical lithospheric investigation carried out in the 1980s that reached from the northernmost tip of Scandinavia to Tunisia in Northern Africa. This investigation include very detailed seismic-refraction surveys, the results of which are summarized in Figure 9.

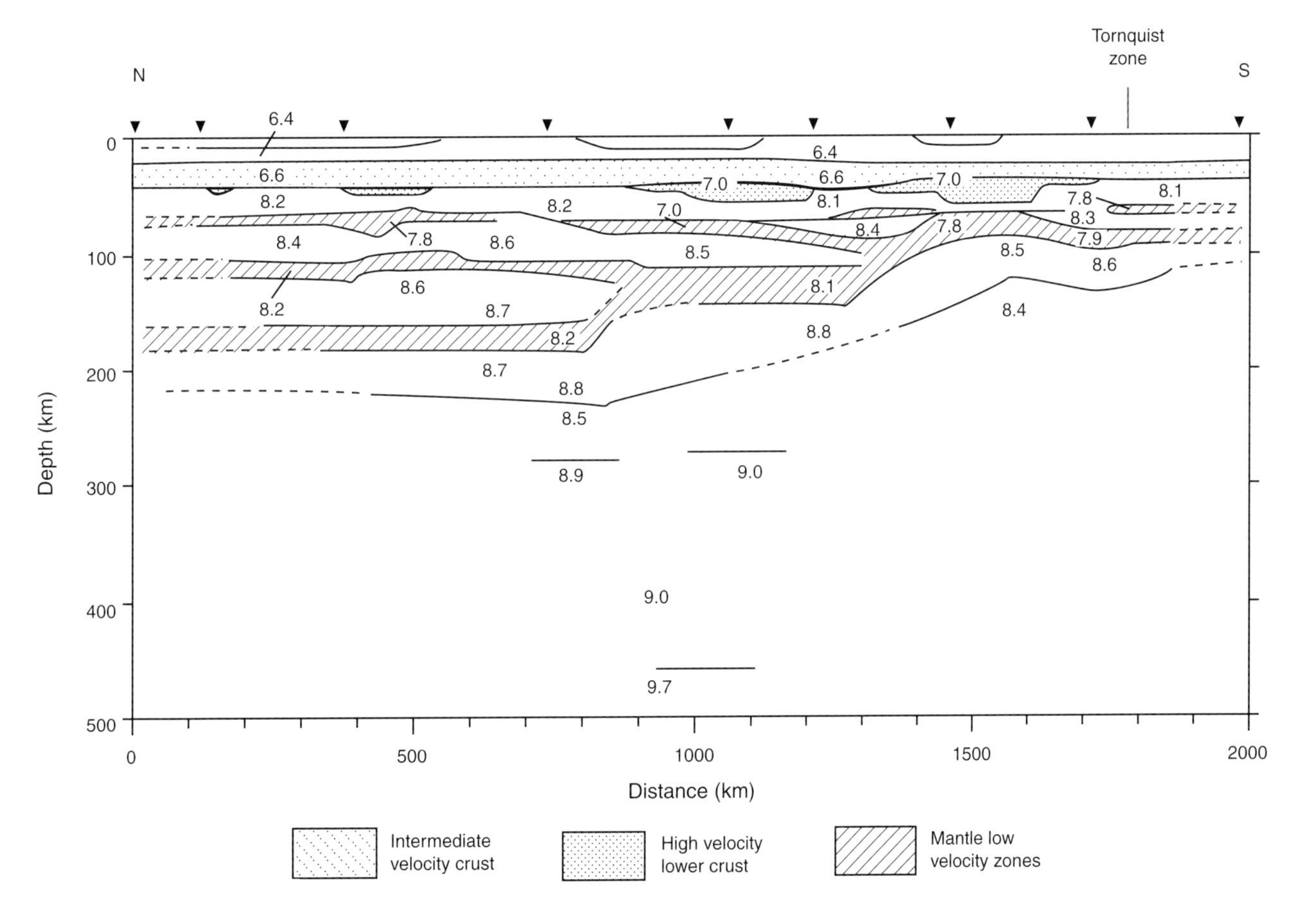

FIGURE 8 (a) Lithospheric cross section through northern Europe (EGT) (after Ansorge *et al.*, 1992, Figs. 3–18). The section traverses Scandinavia from north to south at approximately 20°E. Depth versus distance is 2:1.

The Baltic Shield has almost no sedimentary cover and is characterized by a generally thick crust with rather high *Pg*-velocities near the surface. Crustal thickness is 40–50 km and the lower crust shows rather high velocities (>7 km sec^{-1}). In southern Sweden, several 100 km north of the Tornquist Zone, the crustal thickness decreases to less than 35 km (Fig. 9a), a value that is typical of the Caledonian and Variscan regions of central Europe. Seismic velocities within the crust of the Baltic shield increase gradually from slightly over 6 km sec^{-1} near the surface to 7.0–7.5 km sec^{-1} at 40–50 km depth. It is notable that an intermediate crustal boundary, where the velocity increases by 0.1–0.2 km sec^{-1}, can be followed near 20 km depth throughout the Baltic Shield.

Sub-Moho velocities along the Baltic portion of the EGT are 8.0–8.2 km sec^{-1}, but velocities as high as 8.5 km sec^{-1} are measured on the adjacent Polar Profile (about 50 km further east). Due to very quiet noise conditions and efficient energy propagation (up to 2000 km), Scandinavia is the only place in Western Europe where seismic penetrations to a depth of 400 km by explosion sources has been achieved (Fig. 8). The model derived from upper mantle arrivals beneath Scandinavia shows several distinct velocity inversions between Moho and 200 km depth (Guggisberg and Berthelsen, 1987). Corresponding *S*-wave velocities were derived from surface wave studies that were calculated for all of western Europe and thus cover the whole area of the EGT (Ansorge *et al.*, 1992; Mueller and Panza, 1986)

South of the Tornquist Zone, the crust of the Danish and the North German Basins thins significantly to 26 km. Deducting the voluminous sediments, only 15–20 km of crystalline crust remain. A distinct feature underlying the southern end of the North German Basin is the Elbe line. It separates the lower crust into two regions with 6.9 km sec^{-1} average velocity in the north and 6.4 km sec^{-1} in the south (EUGENO-S Working Group, 1998; Prodehl and Aichroth, 1992).

The internal crustal structure throughout the area of the European Variscides (Fig. 9b) is complex and shows considerable lateral variations. Distinct crustal blocks differing in their internal velocity structure can be assigned to geologically defined terranes of the Variscan orogeny. A subdivision of upper and lower crust by a well-defined boundary (Conrad discontinuity) is often, but not always seen. Towards the Alps the average velocity of the lower crust is as low as 6.2.km sec^{-1}, in contrast to the area north of the Swabian Jura where the velocities above the Moho vary between 6.8 and 7.2 km sec^{-1}. The total crustal thickness under the Variscan part of Germany is fairly constant between 28 and 30 km, except under the Rhine Graben area with 25–26 km and beneath the central part of the Rhenish Massif where an anomalous crustal thickening to 37 km is observed. An important point is that the crust consists of only two layers, in comparison with shield crust that is thicker and has three layers, including a high-velocity (6.9–7.3 km sec^{-1}) lower crustal layer. A joint consideration of seismic velocity structure and petrological information from xenolith studies indicates an intermediate-to-mafic composition for the deeper levels of the crust, at least for the central part of the Variscan crust along the European Geotraverse in Central Europe (Prodehl and Aichroth, 1992). More-recent combined reflection/refraction seismic studies have concentrated on the detailed crustal structure of the Saxothuringium terrane of the Variscan orogen in Saxony, southeast Germany (Enderle *et al.*, 1998; Krawczyk *et al.*, 1999) and on seismic signatures of the Variscan Front in southeast Ireland (Masson *et al.*, 1998, 1999).

The southern segment of the European Geotraverse crosses the Swiss Alps, the Po Plain, the Ligurian Sea and the continental fragments of Corsica and Sardinia, and thus covers units with extremely varying lithospheric structure (Fig. 9b) reflecting the collision of Africa and Europe. In combination with seismic-reflection measurements through Switzerland, the EGT data resulted in a complex and detailed model from the Alps to the Apennines, and demonstrates their asymmetric crustal structure (Valasek *et al.*, 1991, Blundell *et al.*, 1992, Pfiffner *et al.*, 1997). The major element in the upper crust of the central Alps is the relatively low velocity beneath the Aar Massif compared with the well-documented higher velocities in the metamorphic crystalline nappes of the Penninic zone (Pfiffner *et al.*, 1997). From the northern margin of the Alps, the crust steadily thickens southwards to about 60 km underneath the central Alps. The image obtained from combined reflection and refraction data indicates that the European lower crust is a pre-Alpine, possibly Variscan, feature and extends relatively undisturbed from the northern Molasse basin to about 120 km south of the northern margin of the Alps where it is subducted beneath the Adriatic microplate. A fault-like step of the Moho is indicated underneath the Insubric line, which is the tectonic boundary between the central and southern Alps, and as such is possibly the surface expression of the boundary between the European and African crust. Seismic reflection and refraction data indicate that the Adriatic lower crust has been driven northward as a wedge between the European upper and lower crust (Ansorge *et al.*, 1992; Mueller and Kahle, 1993; Valasek *et al.*, 1991; Waldhauser *et al.*, 1998). The complicated arrangement of the various plates in the Southern Alps can best be viewed in enlarged Moho contour maps and 3D block diagrams for this area (Ansorge *et al.*, 1992; Waldhauser *et al.*, 1998).

Egger (1992) provides a detailed discussion of the seismic refraction data of the southern segment of the EGT from northern Apennines into Tunisia. The results support the hypothesis that the opening of the Provencal basin, the rotation of Corsica and Sardinia, and the opening of the Tyrrhenian Sea belong to a system of back-arc spreading, with subduction as a counterpart that is active beneath the Calabrian arc in southern Italy. Complete crustal surveys across the margins and deep basins have complimented the EGT seismic survey to elucidate the nature of the crust at the incipient and ongoing

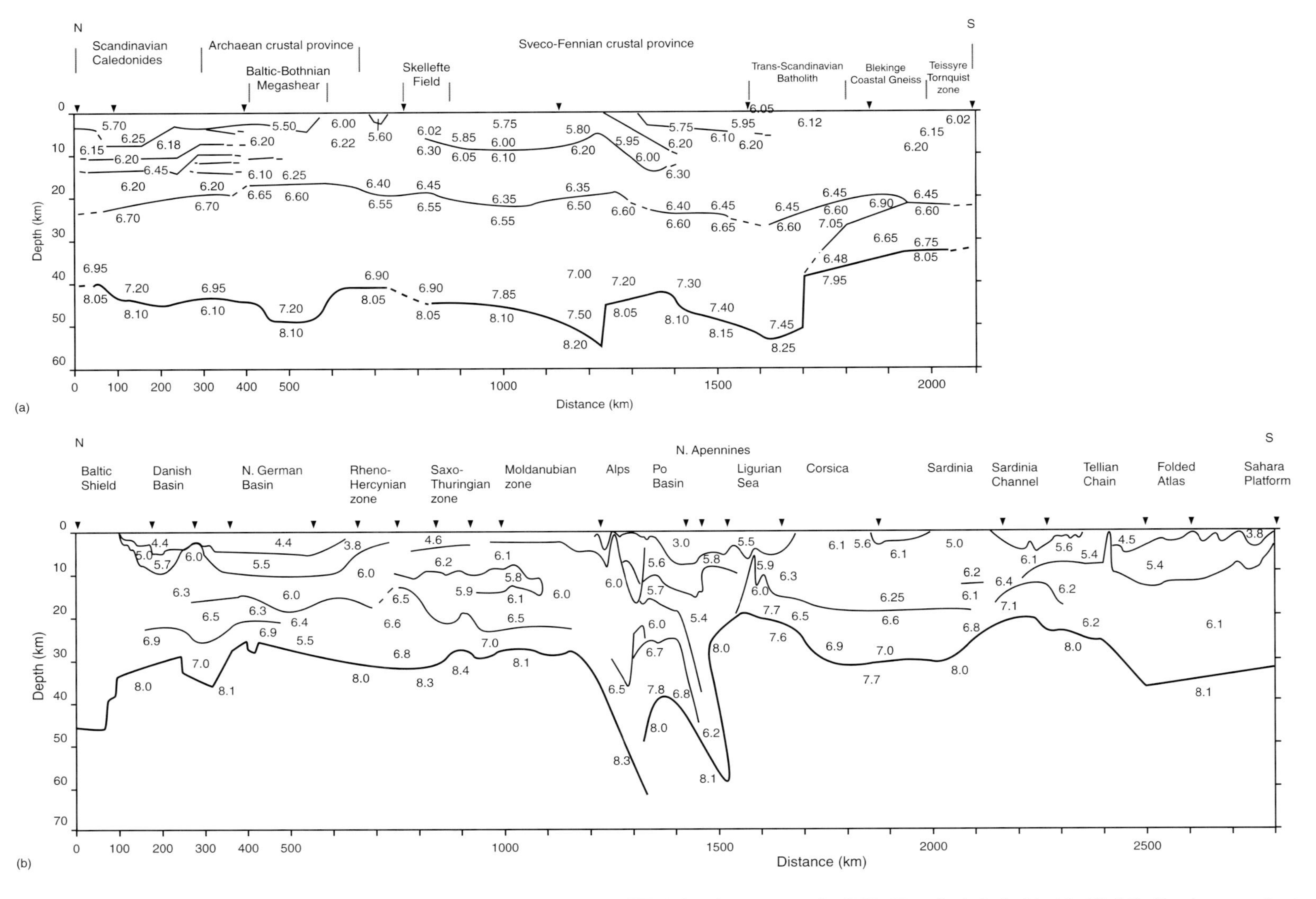

FIGURE 9 Crustal cross section through Europe along the European Geotraverse (EGT) (after Ansorge *et al.*, 1992, Figs. 3–4, 6, 9, 11, 12, 15, 16). Depth versus horizontal distance is 10 : 1. (a) Section through Scandinavia at approximately 20° E, (b) Section through central and southern Europe traversing central Germany, the Alps, the Appenines of northern Italy, Corsica, Sardinia and Tunisia at approximately 10° E.

rifting in the Ligurian Sea and the structure of the Sardinia Channel north of Tunisia. Corsica and Sardinia show a typical, 30-km thick Variscan crust (two layers with average velocities of 6.2 and 6.8 km sec^{-1}), underlain by *Pn*-velocities increasing gradually from 7.6 km sec^{-1} under northern Corsica to 8.0 km sec^{-1} under southern Sardinia and the Sardinia Channel (Fig. 9b). The crust thins to 20 km under the Ligurian Sea and Sardinia Channel.

The EGT reaches African continental crust in Tunisia. The crustal thickness increases steadily from 21 km under the center of the Sardinian Channel to 38 km in the central folded Atlas, from where it decreases southward to 33 km under the southern end of the traverse (Fig. 9b). The velocity of the uppermost mantle is 8.0–8.1 km sec^{-1}. There are thick post-Paleozoic sediments varying in thickness from 8 km in the north to 3 km in the south. A velocity of 6 km sec^{-1} is found everywhere at unusually great depths of 8–14 km. Other striking features are the low average velocity of the entire crust of 6.05 km sec^{-1} and the lack of any clear intracrustal layering (Ansorge *et al.*, 1992; Buness *et al.*, 1992).

Several large-scale seismic investigations have provided further details on the crustal structure of Europe. To the east of the EGT profile, Abramovitz *et al.*, (1999) describe the crustal structure of the North Sea (that portion due north of the Netherlands) and western Ireland. Both regions have thin (30–32 km) crust with three layers. To the east of the EGT profile, the POLONAISE and EUROBRIDGE projects investigated the crust of Precambrian and Variscan Europe, as described by Guterch *et al.* (1999), Grad *et al.* (1999), and the EUROBRIDGE Seismic Working Group (1999). These data document that the crust thickness from 32 km beneath the Paleozoic Variscan Europe to 42 km beneath the Precambrian East European Platform. The Precambrian crust has a very well-defined three-layer structure with seismic velocities in the upper, middle, and lower crust of 6.2, 6.6, and 7.1 km sec^{-1} respectively (Sroda *et al.*, 1999). This is in contrast to the two-layer crust of central Europe, as described above (Fig. 8).

3.4 Northern Eurasia

Seismic refraction profiles cover all of the major geologic structures of northern Eurasia, including Precambrian and younger platforms and shields, deep depressions (e.g., Caspian Sea), orogenic belts (e.g., Urals and Tien Shan Mountains), rifts, and marginal and inner seas. The crustal structure for three long-range profiles is summarized in Figure 10: (a) Black Sea to Dnjeper-Donetz basin, (b) Baltic Shield to West Siberian Plate (passing through the Urals); and (c) West Siberian Plate to Siberian Craton. The first profile was carried out in the 1960s on a closely spaced profile with a 100 m distance between seismometers and chemical explosions at 50–60 km intervals ("continuous profiles"; Pavlenkova, 1973). The long-range profiles from the Baltic Shield across the Urals to the West Siberian Plate and from the West Siberian Plate to the Siberian Craton were recorded with a 10 km interval between seismometers and two types of explosions: chemical and Peaceful Nuclear Explosions (PNEs). The latter profiles provided arrivals from the upper mantle to a depth of 700 km (Egorkin *et al.*, 1987; Beloussov *et al.*, 1991; Mechie *et al.*, 1993). For all of these profiles, the crustal structure can be described by four layers with the following seismic velocities: 2.0–5.5 km sec^{-1} (sediments), 5.8–6.4 km sec^{-1} (upper crust), 6.5–6.7 km sec^{-1} (middle crust) and 6.8–7.4 km sec^{-1} (lower crust).

The first profile (Fig. 10a) crosses the Black Sea basin, Crimean Mountains, Sivash Basin, Ukrainian Shield, and the Dnieper-Donetz Basin (Fig. 3). The Ukrainian Shield has a crustal thickness of more than 40 km and consists of three crustal layers, each with a thickness of 10–15 km. Beneath the basins the Moho is uplifted and the upper crustal layer thins. Such changes in crustal structure become even more dramatic in the Black Sea area where the consolidated crust is only 20 km thick and where the upper and lower crustal layers are absent. The adjacent Crimean Mountains have a deep crustal root, and the crustal average velocity is slightly lower than in the Ukrainian Shield.

The second profile (Fig. 10b) crosses the Baltic Shield, Russian Platform, Timan–Pechora plate, Ural Mountains, and West Siberian Platform. These tectonic units have similar crustal structure, an exception being the Timan–Pechora plate. This plate is characterized by considerably lower average crustal velocity, because a lower-crustal high-velocity layer ($Vp > 7$ km sec^{-1}) is absent. The Urals have a pronounced crustal root, whereas the Timan ridge is compensated by an increased thickness of the upper crust. Below the crust, strong reflections are recorded from a velocity discontinuity (referred to in Russian literature as the "N discontinuity") at depths of 70–90 km. The relief on this boundary appears to be opposite to that of the Moho. The seismic velocity between the Moho and the N discontinuity varies from 8.0 beneath the Pechora and West Siberian plates to 8.4 beneath the Urals. These changes clearly correlate with measured heat flow, with the lower velocities corresponding to higher heat flow areas. An anomalous high-velocity zone was revealed near the boundary between the old Russian plate and the younger Timan–Pechora plate. The dotted line in Figure 10b traces this zone from a high-velocity intrusion in the crust to the north where it is subducted beneath the Pechora Plate.

The third profile (Fig. 10c) crosses the West Siberian plate and the Siberian Platform. Though of differing age, these platforms have similar crustal thickness. Lateral variations in crustal structure are mainly noted beneath deep sedimentary basins. For example, beneath the West Siberian basin and the Tunguss basin the thickness of the middle crustal layer is nearly constant, whereas the thickness of the lower crust increases. The Vilui basin, which is younger than the Tunguss basin, has a clear Moho uplift. The upper mantle velocities hardly change along this profile. The only characteristic feature is the region with lower velocity (8.0 km sec^{-1}) in the

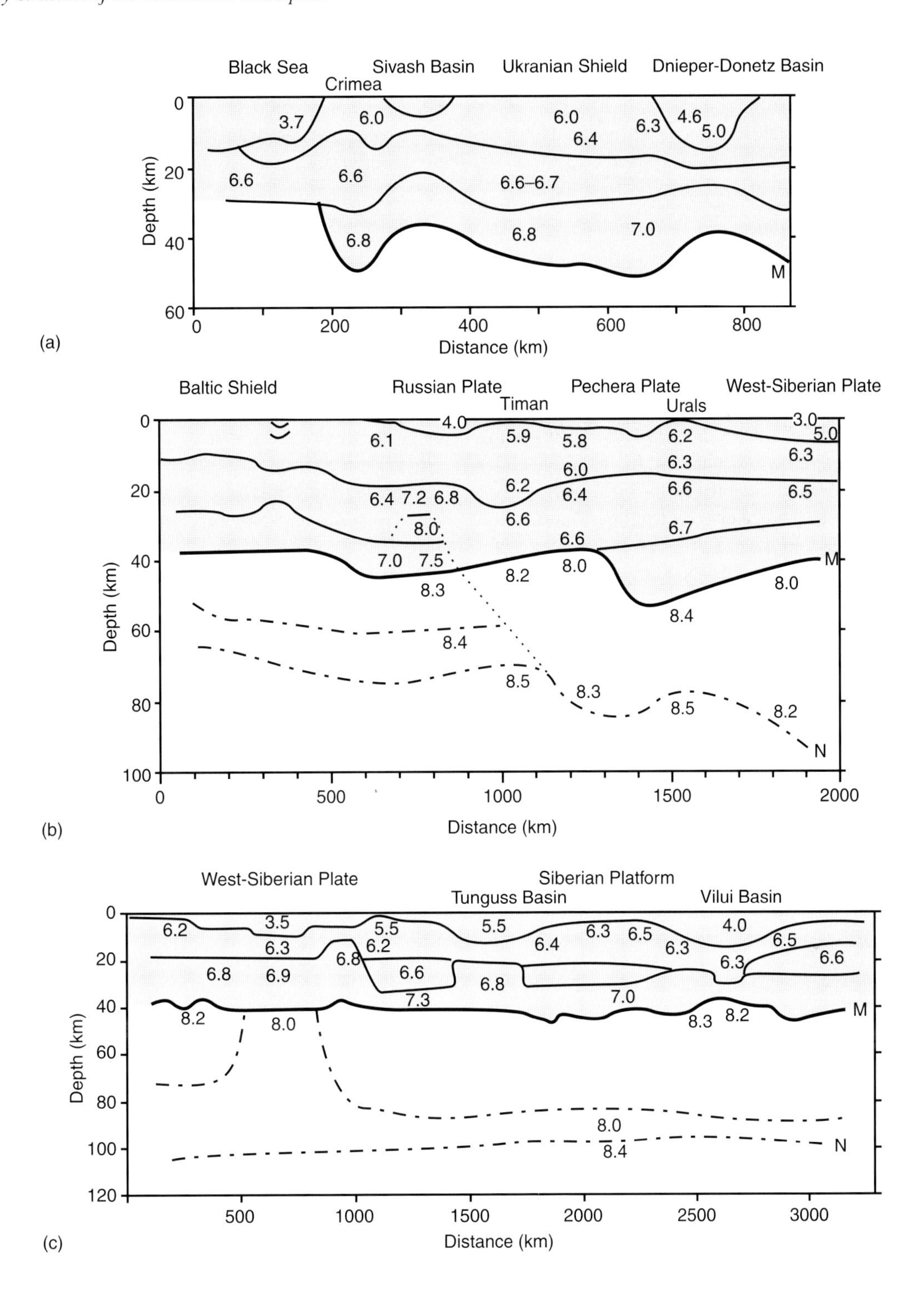

FIGURE 10 Crustal cross sections through Eurasia (Egorkin *et al.*, 1987; Beloussev *et al.*, 1991). (a) Black Sea–Dnjeper-Donetz basin. The section traverses southeastern Russia in a SSW–NNE direction from about 45 to 53° N and 34 to 40° E. Depth versus horizontal distance is 5 : 1. (b) Baltic Shield – Urals – West-Siberian Plate. The section traverses northwestern Russia in a WNW–ESE direction from about 69 to 62° N and 30 to 72° E. A seismic discontinuity ("N") within the uppermost mantle is often reported beneath northern Eurasia (e.g., Pavlenkova, 1996) but has not been well determined elsewhere. Depth versus horizontal distance is 6 : 1. (c) West-Siberian Plate–Siberian Craton. The section traverses northeastern Russia in a W–E direction from about 65 to 62° N and 66 to 132° E. Depth versus horizontal distance is 10 : 1.

central West Siberian plate. This location corresponds to a prominent rift zone that crosses the plate from north to south. Below the crust, an important feature is an upper mantle low-velocity zone (8.0 km sec^{-1}) immediately above the N boundary, which is at a depth of 100 km.

3.5 Near East and Africa

The Near East and Africa are largely underlain by Precambrian cratons, with, in comparison to other continents, a very small fraction of the crust being of younger age. The younger crust occupies the Cape, Mauritanide, and Atlas fold belts, located at the south, northwest, and north margins, respectively. The geologic evolution of the African continent began in the Early Archean Eon, 3.2–3.65 Ga with the formation of the Kaapvaal craton of southern Africa. The entire landmass was assembled and stabilized by 0.6 Ga. The main tectonic events since the end of the Precambrian have been: (1) the Mesozoic and Cenozoic rifting at the west margin during the opening of the south Atlantic Ocean; and (2) rifting in East Africa and the Red Sea in the late Cenozoic.

As is evident in Figure 4, large portions of Africa have not been investigated by seismic refraction profiles. Major crustal seismic research has been carried out in the Afro-Arabian rift system and its immediate surroundings of eastern Africa (Prodehl *et al.*, 1997). The crustal and uppermost-mantle structure has been investigated by seismic refraction surveys in the Jordan–Dead Sea rift, the Red Sea, the Afar depression and the East African rift of Kenya (Fig. 11). With the exception of the Jordan–Dead Sea transform, the entire Afro-Arabian rift system is underlain by anomalous mantle with *Pn*-velocities less than 8 km sec^{-1}, while under the rift flanks the *Pn*-velocity is clearly equal to or above 8.0 km sec^{-1}. Various styles of rifting have been found. Oceanic crust floors the axial trough of the southern Red Sea rift, thinned continental crust underlies the margins of the Red Sea, Afar depression, and northern Kenya rift. In contrast, 30–35 km thick continental crust is found both under the Jordan–Dead Sea rift, an area of strike-slip rifting, and under the central Kenya rift, where updoming driven by a buoyant mantle is apparently the controlling feature. The transition from thinned continental crust to 5–6 km thick oceanic crust in the center of the Red Sea appears to be gradual. In contrast, the transition from the thin crust of the East Africa rift to undisturbed continental crust of 40 ± 5 km thickness is mostly rather abrupt. The seismic data indicate various stages of rifting, and imply the presence of hot uppermost mantle under most parts of the rift system, possibly related to plume activity. Volcanism may disrupt and/or underplate the crust in places, altering particularly the lower crust.

Portions of northern Africa have been investigated with seismic profiles. Measurements in Tunisia were completed as part of the European Geotraverse of 1982–1985, and are summarized in Section 2.3 and Figure 9. In addition, a geological and geophysical transect was completed through the High Atlas Mountains in Morocco. These mountains were formed over a major intracontinental rift system that had extended from what is now the Atlantic margin of Morocco to the Mediterranean coast of Tunisia (~2000 km) during the convergence of the African and European plates.

Crustal thickness variations in the region highlight one of the more remarkable aspects of this orogen. Despite topography that is locally in excess of 4 km, there does not seem to be crust that is thicker than 40 km. Beauchamp *et al.* (1999) report thin crust (18–20 km) beneath the Moroccan shelf and a thickness of 30 km south of the High Atlas Mountains. The crust thins to 24 km toward the Atlantic margin, and attains an average crustal thickness of 35 km to the north of the mountains. The crust is 39 km thick beneath the mountains. There is also evidence for a low-velocity zone at depths of 10–15 km beneath the High and Middle Atlas Mountains. This may be due to a crustal detachment located at the base of the Paleozoic (Beauchamp *et al.*, 1999).

3.6 Australia, New Zealand, and Antarctica

Australia is a stable continent that is largely isolated from the immediate effects of plate tectonics. Within its interior it contains no active rift systems, and only at its offshore northern domain does an active convergent margin exist. The central and western portions of the Australian continent are very ancient. Its western Archean cratons (Pilbara and Yilgarn) have yielded some of the oldest geochronologic ages (>4.0 Ga) determined so far, and the midcontinent is composed of Proterozoic crust. The eastern third of the continent and the Tasmanides, are Paleozoic and Mesozoic in age (Drummond, 1991).

Since the early 1950s, a considerable body of data has accumulated on the crustal structure of the Australian continent. Cleary (1973), Finlayson and Mathur (1984), Drummond and Collins (1986) and Collins (1988, 1991) summarize this work. These active-source results cluster at specific areas (Fig. 12). However, a recent study based on receiver functions has made possible the determination of a comprehensive Moho map (Clitheroe *et al.*, 1999).

The crust is thickest beneath the Proterozoic North Australian Craton (45–50 km), Central Australia (>50 km), and under the Paleozoic Lachlan Fold Belt (40–50 km). It is shallowest under the Archean Pilbara and Yilgarn Cratons of Western Australia (28–37 km), and possibly Tasmania (25–35 km). *Pn*-velocities are generally greater than 8 km sec^{-1}. They are around 8.2 km sec^{-1} and higher in central Australia, slightly lower under the thinner western Australian Pilbara and Yilgarn cratons, and seem to be lowest under eastern Australia along the Great Eastern Divide and the Tasman Geosyncline.

New Zealand lies across the Australian/Pacific plate boundary that transects the South Island. Davey *et al.* (1998)

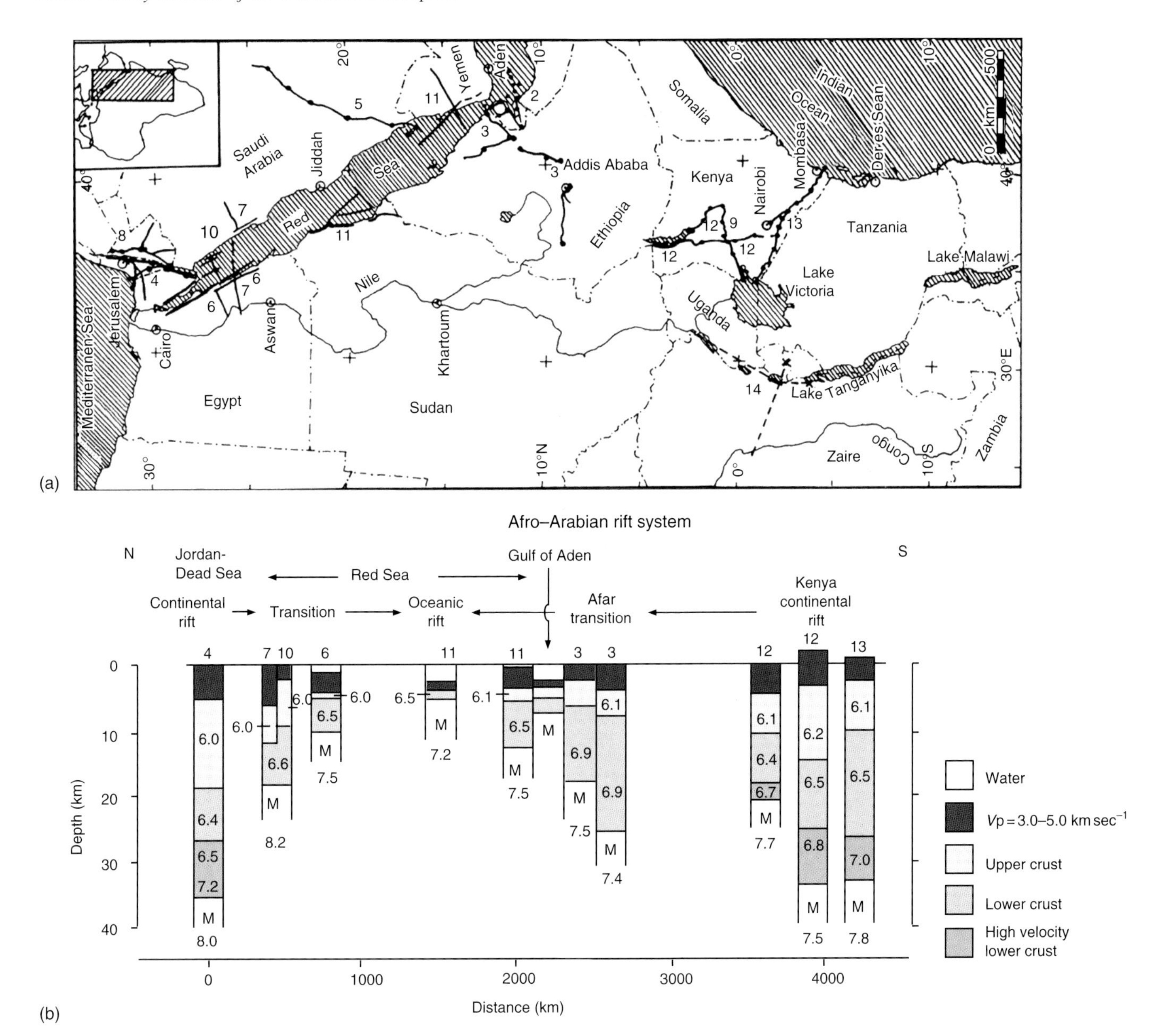

FIGURE 11 (a) Location of long-range seismic lines in the Afro-Arabian rift system (Prodehl *et al.*, 1997). Continuous lines: seismic-refraction surveys, full circles mark shotpoints where locations were published. Dashed lines: approximate lines through epicenters of local earthquakes, crosses: IRSAC (Institut pour la Recherche Scientifique en Afrique Centrale) network at Zaire. (b) Velocity–depth columns through the Afro-Arabian rift system. The evolutionary sequence [for locations see Fig. (a)] illustrating the variation in crustal thickness under the Afro-Arabian rift system from the Jordan–Dead Sea transform system through the Red Sea, Gulf of Aden and Afar triangle to the southern end of the Kenya rift. M, Moho. Depths refer to sea level. Numbers of crustal columns refer to (a) (Prodehl *et al.*, 1997).

and Stern and McBride (1998) present the crustal structure of the South Island based on recently recorded seismic refraction and near-vertical reflection data (Fig. 13). These results indicate that the crust has been formed by the accumulation of low-grade metasedimentary rocks in a convergent margin setting. Modeling of both onshore and offshore data indicate a crust of fairly constant P-wave velocity (5.9–6.2 km sec^{-1}) overlying a 5–10 km-thick lower crust with seismic velocity of about 7 km sec^{-1}. The crustal structure (Fig. 13) is remarkable for the unusual thickness of low-velocity materials that extend along the Alpine fault zone. The total crustal thickness varies from about 30 km at the east coast to about 42 km under the western mountains, the Southern Alps.

Antarctica is divided by the Trans-antarctic Mountains into East Antarctica, a Precambrian craton, and West Antarctica, which has experienced significant Cenozoic tectonic activity, including crustal extension. Relatively few measurements of

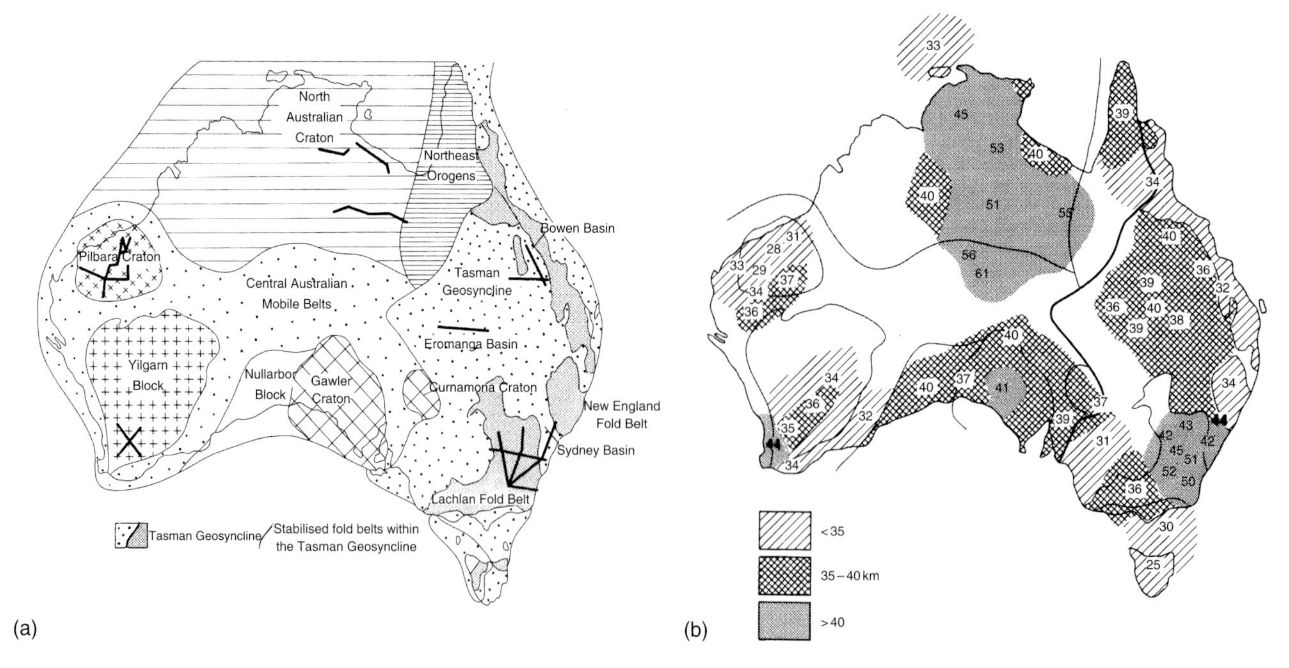

FIGURE 12 Crustal structure of Australia (Collins, 1991): (a) Location of major geological provinces and selected seismic profiles (solid lines); (b) Moho depths in Australia; patterns show regions with similar values, numbers indicate point values. Western Australia is composed of Precambrian crust, including some of the most ancient cratons on Earth (Archean Pilbara and Yilgarn blocks). These Archean cratons have a crustal thickness of 28–37 km, which is less than the global continental average of 40 km.

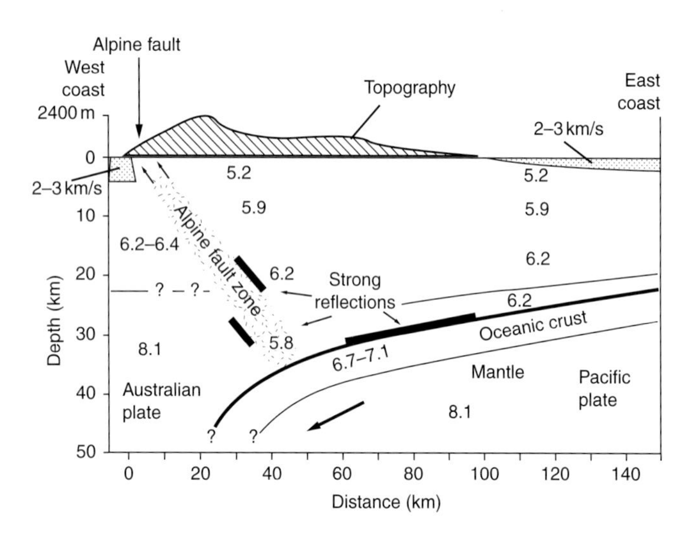

FIGURE 13 Seismic section across the South Island of New Zealand (Stern and McBride, 1998) showing western-thickening crust and east-dipping Alpine fault zone. The crustal structure is unusual in showing low seismic velocities (5.2–6.2 km sec^{-1}) throughout the crust. This indicates that the entire crust, above the hypothesized oceanic crustal layer (6.7–7.1 km sec^{-1}), is composed of low-grade metasedimentary rocks.

crustal structure have been reported for the continental crust of Antarctica. The top of basement in western Antarctica lies at least 2 km below sea level and drops an additional 2 km beneath the deeper parts of the Byrd Subglacial Basin. In the east, the basement is found very close to sea level, and the overlying section is typically no more than a few hundred meters thick. Seismic velocities in the upper crust range from 5.7 to 6.4 km sec^{-1} (Bentley, 1991).

The best information on deeper crustal structure comes from seismic refraction profiles. In the east, two profiles in Maud Land show a depth to Moho of 38–40 km inland, which decreases to about 30 km near the coast. On one of the profiles, a pronounced intracrustal (Conrad?) discontinuity was found at a depth of 18 km between layers with seismic velocities 6.0–6.2 km sec^{-1} above and 6.7–6.9 km sec^{-1} below. The apparent *P*-wave velocity in the upper crust ranges from 6.0 to 6.4 km sec^{-1} and appears to gradually increase with depth (Bentley, 1991).

In the Amery ice shelf, Bentley (1991) describes a layered continental crust with thickness ranging from 30 to 40 km. The upper crustal layer is about 20 km thick. The seismic velocity seen in this section of upper crust is somewhat higher than expected, and may be due to the great age of the crystalline rocks in the basement.

In western Antarctica, a deep sounding in front of the Weddel Sea found the Moho at 33–36 km depth, but it shallows to about 30 km beneath the Crary Trough, then quickly drops to 40 km beneath Coats Land. The lower crustal layer is about 10 km thick. In the east, however, 15 km from the shoreline, the crust–mantle boundary is found to lie 25 km below sea level. From this point, the Moho depth increases to 28 km toward the west (beneath Ross island), and to 35 km eastward (at the shoreline). There is only one additional unpublished seismic profile available for western Antarctica, and this profile shows Moho depth at 24 km below sea level.

In summary, the mean thickness of the crust in East Antarctica is approximately 40 km compared to 30 km in West Antarctica, and the boundary between the two is abrupt. There is a precipitous drop in topography at the Trans-Antarctic Mountains, and geophysical evidence also implies that this change in crustal thickness is sudden.

3.7 Japan and Southeast Asia

The opening of the Sea of Japan in the early Miocene rifted Japan from the Asiatic margin (Finn *et al.*, 1994). Present-day Japan has a composite structure composed of Mesozoic metamorphic and plutonic rocks, and Cenozoic volcanic and plutonic rocks associated with active subduction and magmatism. The seismic structure of Japan has been investigated for many years (e.g., Matsuzawa, 1959; Asano *et al.*, 1979), and abundant seismicity within the crust and upper mantle has permitted detailed modeling of lithospheric structure (Zhao *et al.*, 1990). Iwasaki *et al.* (1994) provide a detailed interpretation of the *P*- and *S*-wave structure of the crust of northern Honshu (Fig. 14). The crust is about 33 km thick, and shows a nearly uniform increase in velocity with depth from 6.0 km sec^{-1} (beneath a 1-km thick layer of sediments and/or volcanic rocks) to 7.0 km sec^{-1} at the base of the crust. The seismic velocity of the uppermost mantle is unusually low, 7.5–7.6 km sec^{-1}, a value that is found only in regions of very high heat flow and/or active volcanism, such as the Afro-Arabian rift system (Fig. 11).

Indonesia has had a complex geologic history, and the physical properties of the crust are largely undetermined. Most likely the area consists of a continental or borderland type of crust of intermediate (30 km) thickness. It has been proposed that cratonization of mobile shelves occurs by sedimentary and calcalkaline accretion, following in the wake of island arc migration by backarc extension (Curray *et al.*, 1977).

The Gulf of Papua has been the focus of study by a number of researchers. Drummond *et al.* (1979) found sediments to be 5 km thick on the Papuan Plateau, but 10 km to the west and northwest along the axes of the Moresby and Aure Troughs. Offshore, sedimentary basins as thick as 11 km have been measured. Beneath these sediments, a layer with *P*-wave velocity of 6.1 km sec^{-1} was inferred over the region, but was thought to be underlain to the south by a layer with a velocity of 6.9 km sec^{-1}. This second, faster, layer is also thought to be present beneath the Eastern Plateau. This lower-crustal layer is noticeably absent from the Aure Trough in the north, though velocity increases are expected with depth.

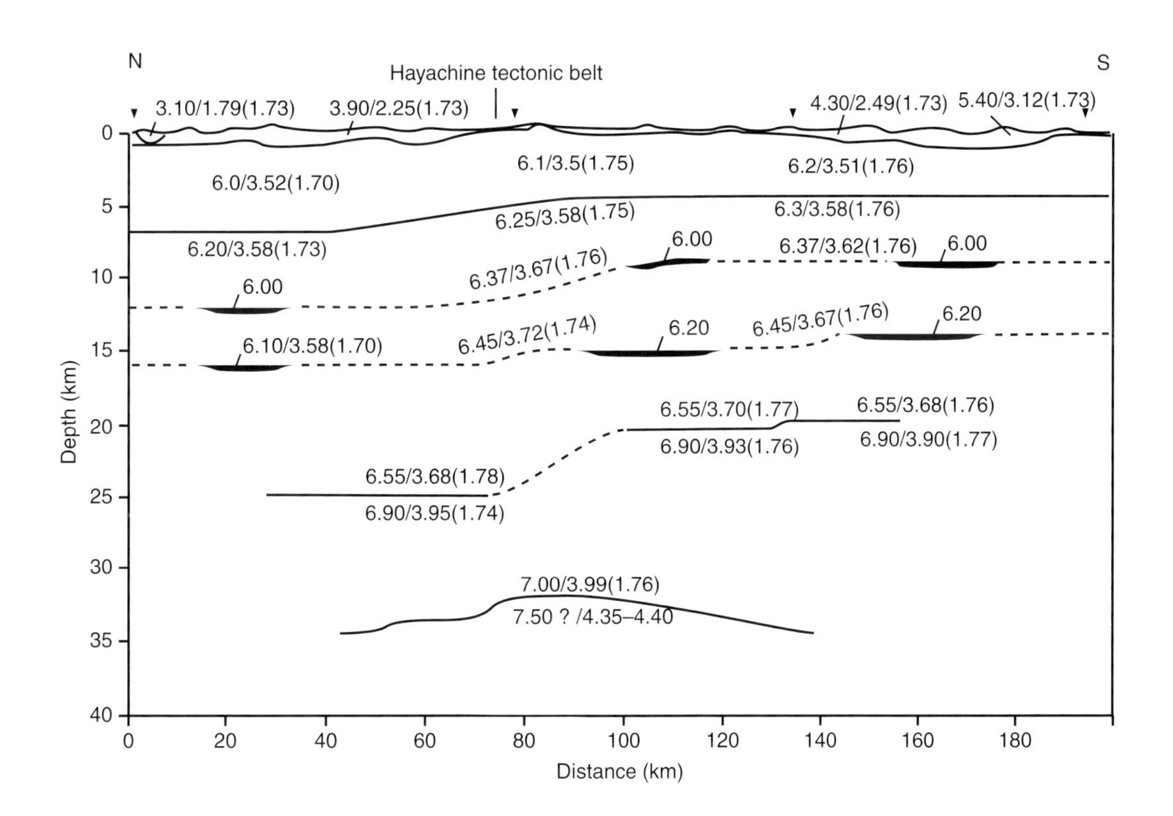

FIGURE 14 Crustal *P*-wave and *S*-wave velocity structure through a typical island arc: the Kitakami region, northern Honshu, Japan. Indicated are *P*-velocity/*S*-velocity, and their ratio. Modified from Iwasaki *et al.* (1994).

The Moho is 27–29 km deep along the southwestern coast of the peninsula and shallows beneath the Moresby Trough to 19 km. This depth is greater than normal for the Moho beneath oceanic crust, but it is not unreasonable when compared to other measurements in the area. Curray *et al.* (1977) found the Moho to be at 18 km depth south of the Island of Bali and 21–25 km south of the Island of Java.

Beneath the Eastern Plateau, the Moho again deepens to 25 km (Drummond *et al.*, 1979). With a lower crustal layer of 9 km in thickness, the crust beneath the Papuan Peninsula is

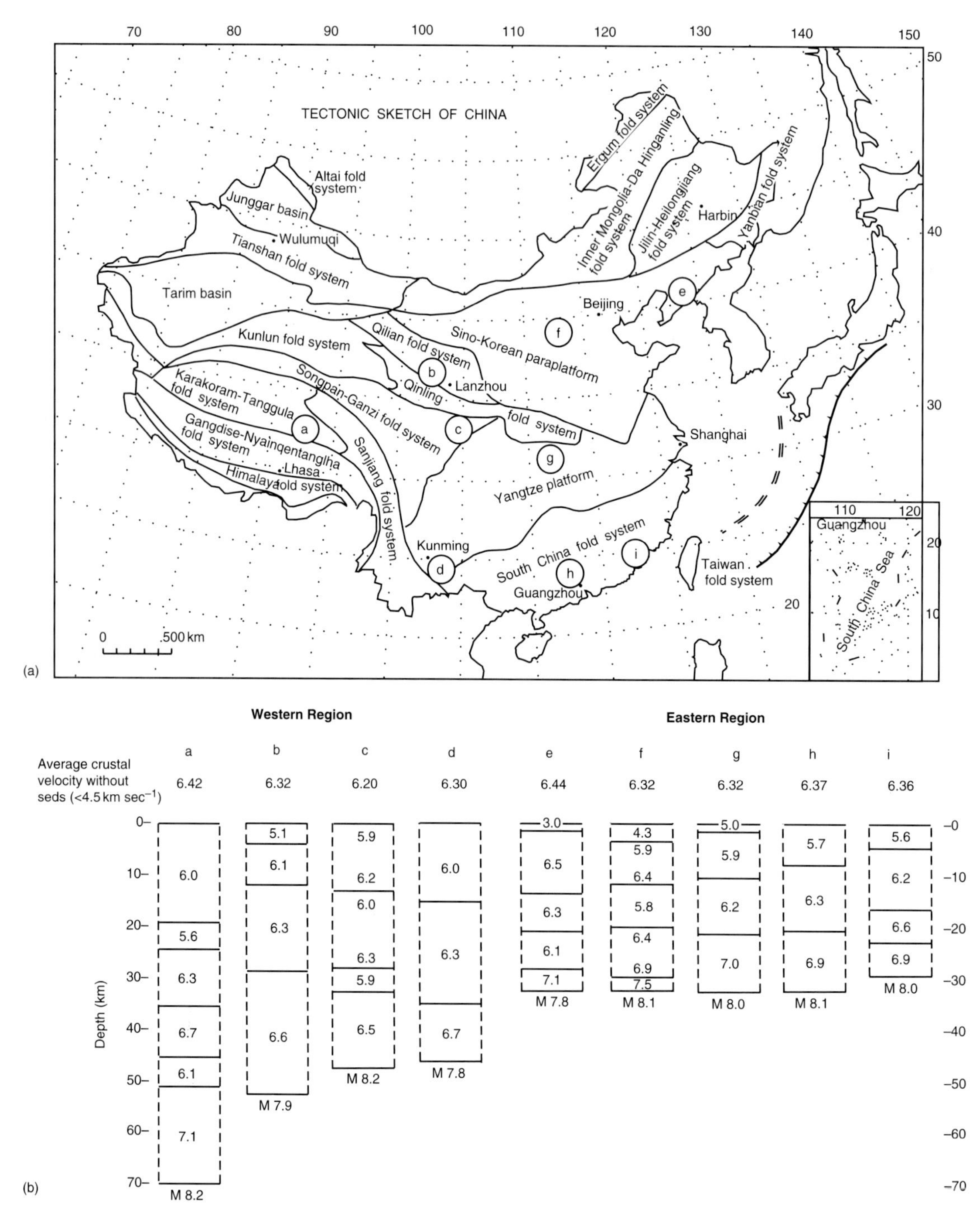

FIGURE 15 Crustal structure of China (Li and Mooney, 1998): (a) Tectonic sketch map of China with location of seismic crustal surveys; (b) Representative seismic velocity–depth functions for nine regions of China. The crust thickens from about 35 km in the east to greater than 70 km beneath the Tibetan Plateau to the west.

therefore continental. Offshore, the crust in the Moresby Trough is about 19 km thick, with sediments comprising about the top 10 km. Therefore, the crust below the sediments has a thickness typical of oceanic crust. The northern extent of the thin crust beneath the Moresby Trough is uncertain. Similar measurements of crustal thickness to those of Drummond *et al.* (1979) were made in Papua by Finlayson *et al.* (1976), who detected a low-velocity zone below the Moho and a return to normal upper-mantle velocities at a depth of about 50 km.

3.8 China and India

China has undergone a long geologic evolution, from the formation of the Archean Sino-Korean platform in the north to the active continent–continent collision in Tibet. More than 36 000 km of active-source seismic refraction profiles (referred to as Deep Seismic Sounding, DSS, profiles, in the Russian, Chinese, and Indian literature) have been collected in China since 1958. The most remarkable aspect of China's crustal structure is the >70 km thick crust of the Tibetan Plateau (Fig. 15). All of western China is underlain by crust that is 45–70 km thick and is separated from 30–45 km thick crust of Eastern China by a seismic belt which trends roughly north–south at 105° E longitude. The crustal structure of western China (Fig. 15b, columns a–d) show thick crust that is characteristic of young orogenic belts worldwide. Three of the four columns differ from the thick crust of shields in that they lack a high-velocity (7.0–7.5 km sec^{-1}) lower crustal layer (cf. Christensen and Mooney, 1995). A high velocity (7.1 km sec^{-1}) lower crustal layer has been reported in western China beneath the southernmost portion of the Tibetan Plateau. Several investigators have identified this layer as the cold lower crust of the subducting Indian plate (cf. Li and Mooney, 1998).

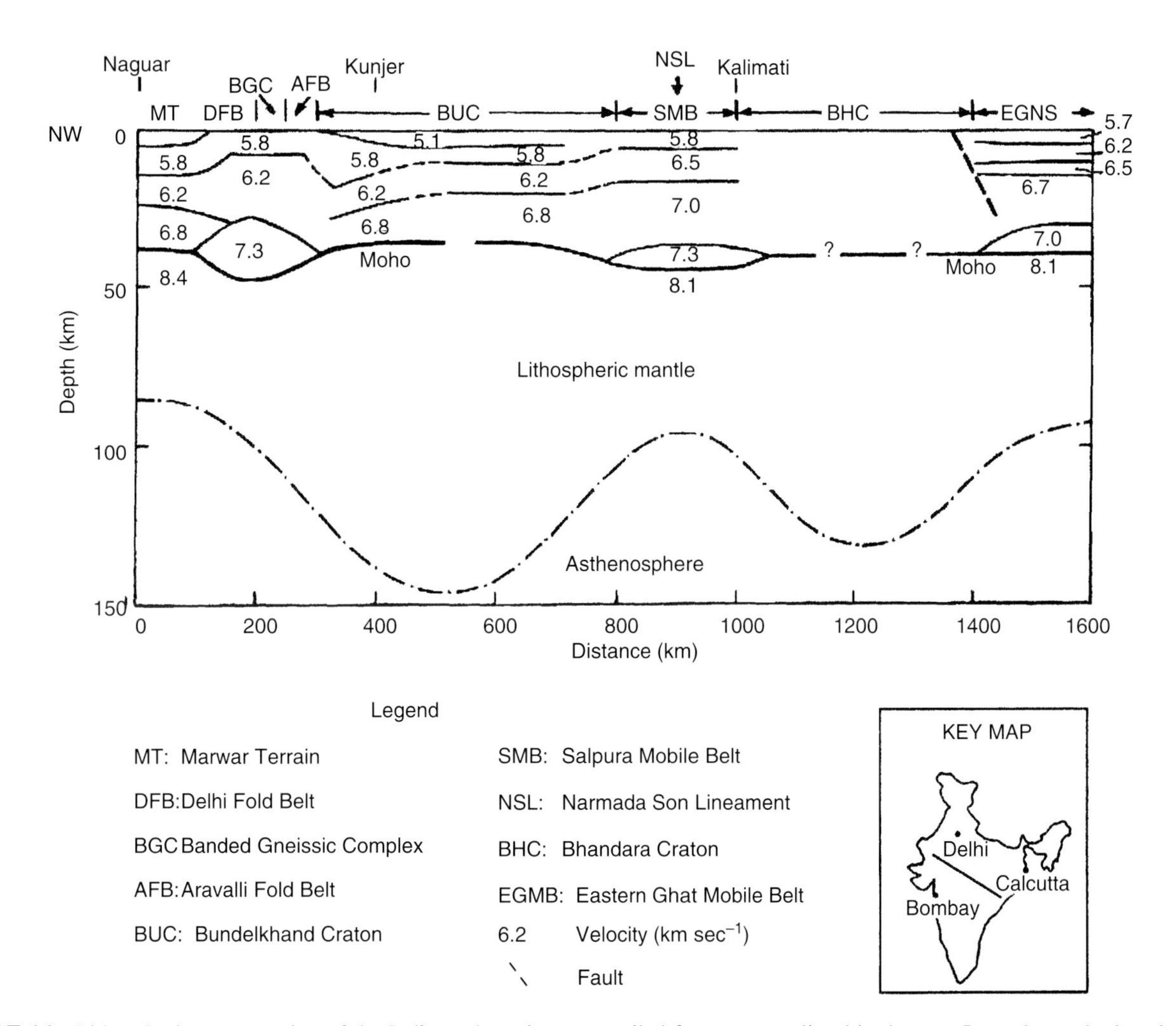

FIGURE 16 Lithospheric cross-section of the Indian subcontinent compiled from sources listed in the text. Inset shows the location of the cross section. The crustal structure is highly variable laterally, but the crustal thickness, where measured, is relatively uniform at about 40 km, equal to the global average for Precambrian cratonic crust. Estimated lithospheric thickness ranges from 80 to 150 km, and is greatest beneath the Bundelkhand (BUC) and the Bhandara (BHC) cratons.

Recently recorded near-vertical and wide-angle reflection data indicate a low-velocity, perhaps fluid-rich or partially molten zone in the middle crust of the Tibetean Plateau (Zhao *et al.*, 1993; Nelson *et al.*, 1996; Makovsky *et al.*, 1996a,b).

The crustal columns for eastern China (Fig. 15b, columns e–i) all indicate relatively thin (29–33 km) crust, close to the continental global average for extended crust (30 km), but significantly thinner than the global average of 39 km (Christensen and Mooney, 1995).

The average crustal velocity of China (6.15–6.45 km sec^{-1}, Fig. 15b) indicates a felsic to intermediate bulk crustal composition. Upper mantle *Pn* velocities (about 8.0 km sec^{-1}) are equal to the continental global average. These results have been interpreted in terms of the most recent thermotectonic events that have modified the crust (Li and Mooney, 1998). In eastern China, Cenozoic extension has created a thin crust with low average crustal velocities while in western China, Mesozoic and Cenozoic arc–continent and continent–continent collisions have led to crustal growth and thickening.

India consists of Archean cratons and associated Archean to Proterozoic high-grade mobile belts, Proterozoic sedimentary basins, horst and graben belts associated with Gondwana sedimentation, Deccan volcanism, and the Himalayas. The investigation of the crustal structure of India by seismic refraction profiling was begun in 1972 as an Indo-Soviet cooperation. Since then, more than 20 long-range refraction, near-vertical reflection and wide-angle reflection profiles have been carried out. Results are reviewed by Kaila (1982), Kaila and Krishna (1992), Mahadevan (1994), Reddy (1999) and Reddy and Rao (2000). Estimates of crustal thickness are in the 35–40 km range, with the exception of the Himalayas that are known to have thicknesses of about 70 km (Mahadevan, 1994). A lithospheric cross section compiled from several seismic profiles illustrates the main properties of the seismic structure of the crust and uppermost mantle of India (Fig. 16).

4. Summary Discussion

4.1 Histograms of Crustal Thickness and Average Crust Velocity

We have divided the crust into six geologic provinces (Fig. 3) and present histograms for two basic properties: crustal thickness (Fig. 17) and mean crustal velocity (Fig. 18; Mooney *et al.*, 1998). These histograms reinforce many of the points made earlier regarding the relationship between the seismic structure of the crust and geologic setting. Shields and platforms have an average crustal thickness of about 40 km, with a standard deviation (SD) of 7 km. Shields have the highest average *P* velocity of any geologic province, 6.49 km sec^{-1} (SD 0.23 km sec^{-1}). Orogenic crust has a wide range of thickness of 20–80 km, and an average thickness of 43 km (SD 10 km). Isostatic forces uplift crust thicker than about 50 km, leading to erosion and crustal thinning. Most thick (>50 km) crust is young orogenic crust that is undergoing active tectonic compression. The average *P* velocity of orogenic crust is 0.13 km sec^{-1} lower than shield crust. This may be due to thickening of low-velocity upper crustal layers during compression, as seen in the European Alps, intrusion by silicic magma, such as are seen in the Andes, or by the delamination of the high-velocity lower crust, as found beneath the southern Sierra Nevada, California.

Extended crust refers to crust that has experienced localized rifting and/or regional extension. Examples include the Basin and Range province of the western USA (Fig. 6) and much of Western Europe. The average thickness of extended crust is about 30 km, and the mean *P*-wave velocity (6.16 km sec^{-1}) is 0.33 km sec^{-1} lower than shield crust. Forearc crust refers to those regions that were formed in front (oceanward) of a volcanic arc, such as many parts of the west coast of North America. Only 26 seismic refraction measurements are presently available for forearcs. These regions have an average crustal thickness of 27 km (SD 8 km) and a very low mean crustal velocity (6.09 km sec^{-1}). These values reflect the abundant sedimentary accumulations in forearc crust. Magmatic (volcanic) arcs, such as Japan (Fig. 14) and the Aleutian Islands, Alaska, have an average crustal thickness of 31 km (SD 8 km) and a mean *P*-wave velocity of 6.14 km sec^{-1} (SD 0.23 km sec^{-1}).

4.2 General Characteristics of Continental Crustal Structure

The continental lithosphere shows several important regularities. (1) Crustal thickness and average velocities increase from continental margins toward the interior; (2) The crust of old, stable shields and platforms is 35–45 km thick, whereas young, nonorogenic crust is significantly thinner (25–35 km); (3) An inverse correlation is clearly evident between the depths to the basement and to the M boundary: the crust is thicker under orogenic belts, and thinner under basins; (4) An average petrological model may be presented for stable continental interiors in terms of three layers. The two upper layers have silicic-to-intermediate composition, but the middle crust is composed of rocks with a higher degree of metamorphism. The third layer is composed of mafic rocks of granulite-grade metamorphism. The composition of the continental crust is discussed in detail by Rudnick and Fountain (1995) and Christensen and Mooney (1995); (5) The upper mantle, including the subcrustal lithosphere and possibly the asthenosphere, is characterized with fine-scale seismic and rheological stratification. Rheologically weak layers appear in the middle and lower crust, and the crust/mantle boundary. In northern Eurasia, a thin low-velocity zone is reported at a depth of 80–100 km based by a clear "N" reflector

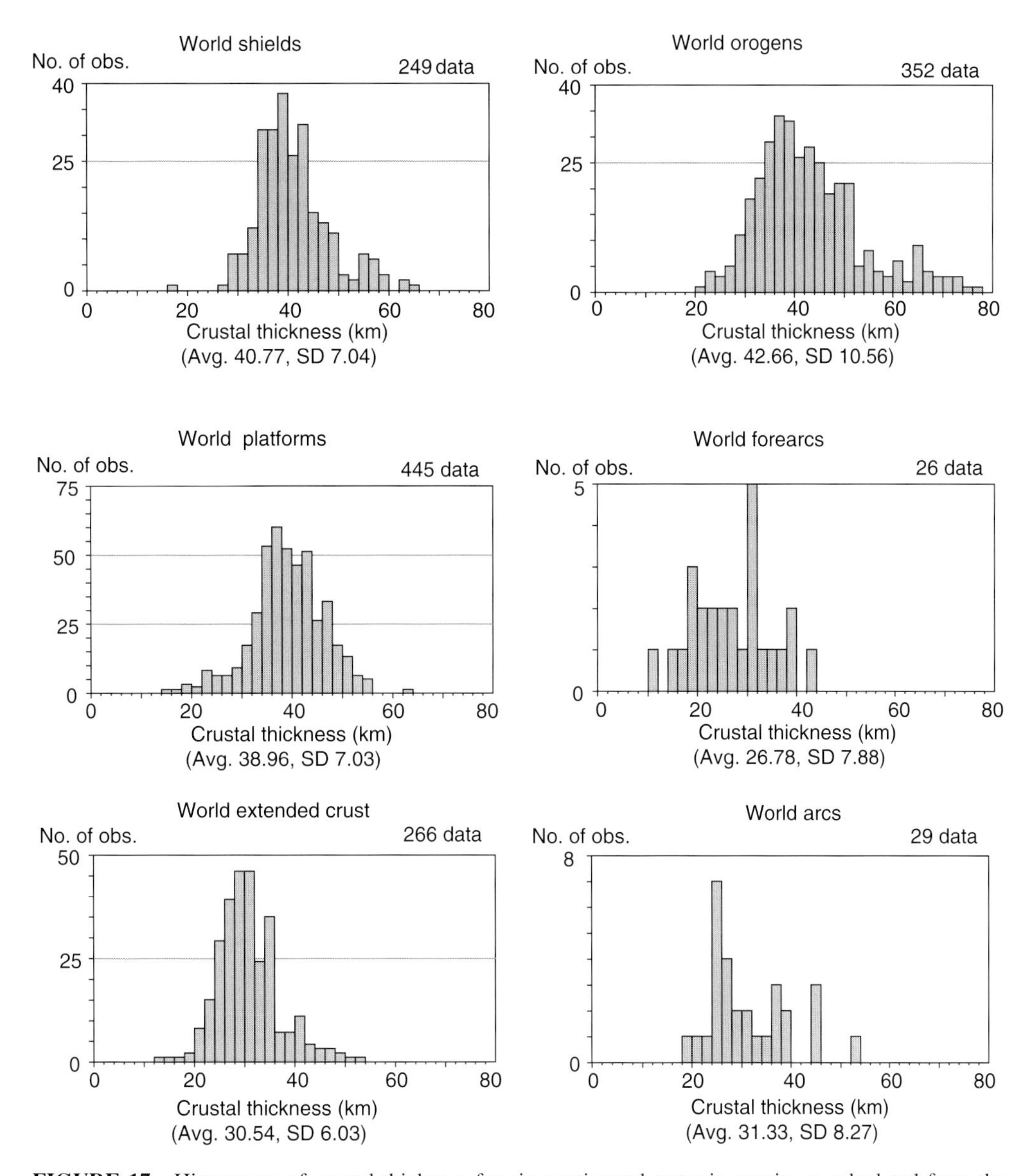

FIGURE 17 Histograms of crustal thickness for six continental tectonic provinces calculated from the individual point measurements (triangles) of Figure 4 (Mooney *et al.*, 1998). Average and SD are indicated. These histograms indicate systematic differences among tectonic provinces, and provide a basis for extrapolating crustal thickness into unmeasured regions.

(Pavlenkova, 1973, 1996; Zverev and Kosminskaja, 1980; Beloussov and Pavlenkova, 1984; Kozlovsky, 1987; Beloussov *et al.*, 1991; Krilov *et al.*, 1991, Puzyrev, 1993); (6) The thickness of the crust usually increases under orogenic belts. The depth of crustal roots varies in different kinds of orogens. The thickest crust is under young mountains, such as those situated on Alpine geosynclines (Caucasus, Pamir). In contrast, orogens developed on ancient platforms have smaller roots (e.g., the northern Tian-Shan and the Altai Mountains). There are also orogens that lack roots, such as the Canadian Cordillera (Clowes *et al.*, 1999), and that are held up by low density, buoyant mantle.

In most cases there is an inverse correlation between sediment thickness and depth to the Moho. For example, crustal thickness decreases under platform depressions, the basins of the inner and marginal seas, and under rifts. There is an inverse correlation between crustal thickness and heat flow. This correlation is observed both in active regions and on stable platforms.

As noted above, stable continental crust is 40 ± 7 km thick with three main crustal layers, each 10–15 km thick. This crustal type covers the interior of Eurasia, North America, Africa, and Australia. Thin (25–35 km) two-layer continental crust is observed on the marginal parts of Eurasia, in Western Europe, in the Pechora Plate of north central Russia, in the Far

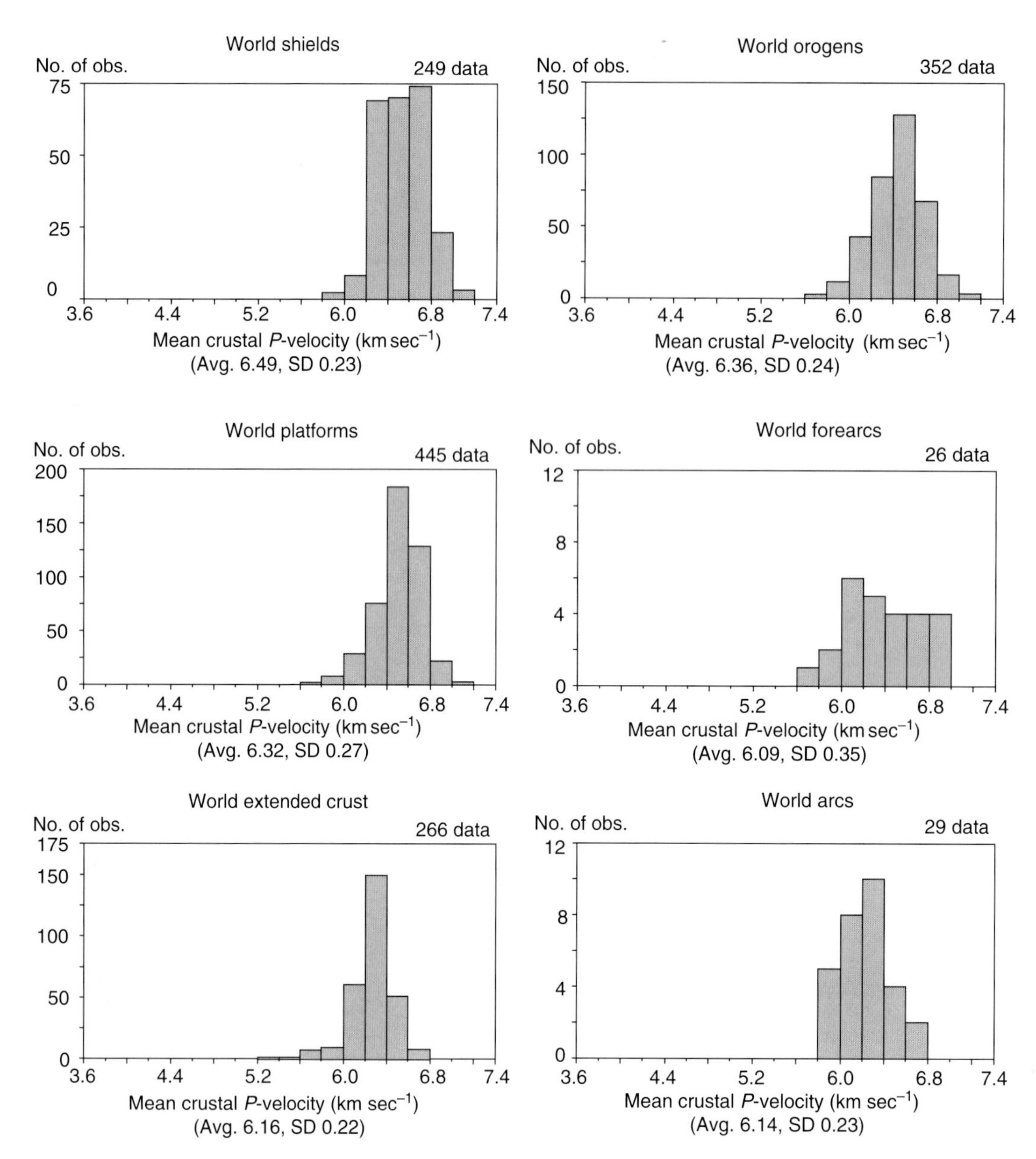

FIGURE 18 Histograms of mean crustal velocity for six continental tectonic provinces calculated from the individual point measurements (triangles) of Figure 4 (Mooney *et al.*, 1998). Average and SD are indicated. These histograms indicate systematic differences among tectonic provinces, and provide a basis for extrapolating mean crustal velocity into unmeasured regions.

East, Eastern China, and the Basin and Range Province, western United States of America. It differs from the stable crust not only in thickness but also by virtue of a lower mean crustal velocity (6.16 km sec^{-1} versus 6.49 km sec^{-1}). An important characteristic feature of this crustal type is the absence of the high-velocity layer (7 km sec^{-1}) in the lower crust.

A consideration of more than 1000 seismic refraction profiles worldwide lends strong support to the definition of the seismic Moho as that depth where seismic velocity first exceeds 7.6 km sec^{-1}. The Moho is nearly always evident as a refraction horizon, and can be clearly distinguished from the lower crust. The uppermost mantle, including the lithosphere and asthenosphere, is sometimes characterized by fine stratification; high velocities (up to 8.4+ km sec^{-1}) alternate with lower ones (7.8–8.0 km sec^{-1}). There is a close correlation between the properties of the crust and upper mantle; stable shield and platform crust exists above cold, high-velocity uppermost mantle, and thin, extended or young orogenic crust exists above low velocity regions. This correlation indicates that mantle processes play an important role in crustal evolution, such as providing a driving force for crustal extension.

This brief review of the seismic velocity structure of the continental lithosphere highlights only the general features of its structure. The focus has been on the structure of the upper portion of the lithosphere, the continental crust. A great deal of new, high-quality data are recorded each year, and these show that the lithosphere to be highly complex and subject to multiple processes, including compression accompanied by brittle faulting and dutile deformation, extension, displacement by strike-slip faulting, and modification by igneous processes. Controlled source seismic data have been important in developing an understanding of the structure, composition, and evolution of the lithosphere, but future progress will depend on investigations that combine geophysical, geologic, and geochemical constraints.

Acknowledgments

It is a pleasure to acknowledge our colleagues at the US Geological Survey, numerous US academic institutions, the Lawrence Livermore and Los Alamos National Laboratories, the Universities in Karlsruhe (Germany), Paris, and Strasbourg, the Institute of the Physics of the Earth, and the Russian Ministry of Natural Resources, and Russian Academy of Sciences. Review comments by R. Meissner, H. Kanamori, P.R. Reddy, I. Artemieva and an anonymous reviewer are appreciated. Special thanks to W.H.K. Lee for his encouragement, and to K. Favret (USGS) for making the figures. S. Detweiler (USGS) assisted with preparation of the final manuscript.

References

Abramovitz, T., *et al.*, (1999). *Tectonophysics* **314**, 69.

Ansorge, J., *et al.* (1992). In: "A Continent Revealed – the European Geotraverse," (D. Blundell *et al.*, Eds.) Cambridge University Press p. 33.

Artemieva, I.M. and W.D. Mooney (2001). *J. Geophys. Res.* **106**, 16387–16414.

Asano, S., *et al.* (1979). *J. Phys. Earth*, **27**, supplement, S1.

Bamford, D. and C. Prodehl (1977). *J. Geol. Soc. London* **134**, 139.

Barazangi, M. and L. Brown (Ed.) (1986a). *Am. Geophys. Un. Geodynam. Series 13*, Washington, DC.

Barazangi, M. and L. Brown (Ed.) (1986b). *Am. Geophys. Un. Geodynam. Series 14*, Washington, DC.

Beauchamp, W., *et al.* (1999). *Tectonics* **18**, 163.

Beloussov, V.V. and N.I. Pavlenkova (1984). *J. Geodynam.* **1**, 167.

Beloussov, V.V. *et al.* (Ed.) (1991). "Deep Structure of the USSR Territory." Nauka, Moscow (in Russian).

Bentley, C.R. (1991). In: "The Geology of Antarctica," (R.J. Tingey, Ed.), pp. 335. Oxford University Press.

Birch, F. (1960). *J. Geophys. Res.* **65**, 1083.

Birch, F. (1961). *J. Geophys. Res.* **66**, 2199.

Blundell, D., *et al.* (Ed.) (1992). "A Continent Revealed – the European Geotraverse," Cambridge University Press.

Braile, L.W., *et al.* (1995). In: "Continental Rifts," (K.H. Olsen, Ed.), pp. 61. Elsevier, Amsterdam.

Brown, L.D., *et al.* (1983). *Geol. Soc. Am. Bull.* **94**, 1173.

Buness, H., *et al.* (1992). *Tectonophysics* **207**, 245.

Christensen, N.I. (1965). *J. Geophys. Res.* **70**, 6147.

Christensen, N.I. (1966). *J. Geophys. Res.* **71**, 3549.

Christensen, N.I. (1979). *J. Geophys. Res.* **84**, 6849.

Christensen, N.I. and W.D. Mooney (1995). *J. Geophys. Res.* **100**, 9761.

Chuikova, N.A., *et al.* (1996). *Izv. Phys. Solid Earth* **34**, 701.

Chulick, G.S. and W.D. Mooney (2002). *Bull. Seism. Soc. Am.*, **92**, in press.

Cleary, J. (1973). *Tectonophysics* **20**, 241.

Clitheroe, G.M., *et al.* (1999). *J. Geophys. Res.* **105**, 13697.

Closs, H. and C. Behnke (1961). *Geol. Rundschau* **51**, 315.

Clowes, R., *et al.* (1999). *Episodes* **22**, 3.

Collins, C.D.N. (1988). *Australian Bureau of Mineral Resources, Geol. Geophys. Report 277.*

Collins, C.D.N. (1991). *Geol. Surv. Aust.* **17**, 67–80.

Curray, J.R., *et al.* (1977). *J. Geophys. Res.* **82**, 2479.

Davey, F.J., *et al.* (1998). *Tectonophysics* **288**, 221.

Drummond, B. (Ed.) (1991). *Geol. Soc. Aust. Special Publ. 17.*

Drummond, B. and C. Collins (1986). *Earth Planet. Sci. Lett.* **79**, 361.

Drummond, B., *et al.* (1979). *J. Aust. Geol. Geophys.* **4**, 341.

Durrheim, R.J. and W.D. Mooney (1994). *J. Geophys. Res.* **99**, 15359.

Egorkin, A.V., *et al.* (1987). *Tectonophysics* **140**, 29.

Egger, A.P. (1992). PhD Thesis, ETH Zurich.

Ekström G., *et al.* (1997). *J. Geophys. Res.* **102**, 8137.

Enderle, U., *et al.* (1998). *Geophys. J. Int.* **133**, 245.

EUGENO-S Working Group (1988). *Tectonophysics* **150**, 253.

EUROBRIDGE Seismic Working Group (1999). *Tectonophysics* **314**, 193.

Finlayson, D. and S.P. Mathur (1984). *Ann. Geophys.* **2**, 711.

Finlayson, D. *et al.* (1976). *Geophys. J. R. Astron. Soc.* **44**, 45.

Finn, C., *et al.* (Ed.) (1994). *J. Geophys. Res.* **99**, 22135.

Giese, P., *et al.* (Ed.) (1976). "Explosion Seismology in Central Europe." Springer, Berlin.

Giese, P., *et al.* (1999). *South Am. J. Geosci.* **12**, 201–220.

Goodwin, A.M. (1991). "Precambrian Geology," Academic Press, San Diego.

Goodwin, A.M. (1996). "Principles of Precambrian Geology," Academic Press, San Diego.

Grad, M., *et al.* (1999). *Tectonophysics* **314**, 145.

Guggisberg, B. and A. Berthelsen (1987). *Terra Cognita* **7**, 631.

Guterch, A., *et al.* (1999). *Tectonophysics* **314**, 101.

Hart, P.J. (Ed.) (1969). *Geophys. Monograph 13*, American Geophysics Union Washington, DC.

Heacock, J.G. (Ed.) (1971). *Geophys. Monograph 14*, American Geophysics Union Washington, DC.

Heacock, J.G. (Ed.) (1977). *Geophys. Monograph 20*, American Geophysics Union Washington, DC.

Hersey, J., *et al.* (1952). *Bull. Seismol. Soc. Am.* **42**, 291.

Hill, M.N. (Ed.) (1963). "The Sea," vol. 3, Interscience Publishers, New York.

Holbrook, W.S., *et al.* (1992). In: "Continental Lower Crust," (D.M. Fountain *et al.*, Eds.), pp. 1. Elsevier, Amsterdam.

Iwasaki, T., *et al.* (1994). *J. Geophys. Res.* **99**, 22187.

Iyer, H.M. and T. Hitchcock (1989). *Geol. Soc. Am. Memoir 172.*

James, D.E. (1971). *J. Geophys. Res.* **76**, 3246.

James, D.E. and J.S. Steinhart (1966). *Am. Geophys. Un. Geophys. Monograph* **10**, 293.

Jarchow, C.M. and G.A. Thompson (1989). *Annu. Rev. Earth Planet. Sci.* **17**, 475.
Kaila, K.L. (1982). *Geophys. Res. Bull.* **20**, 309–328.
Kaila, K.L. and V.G. Krishna (1992). *Curr. Sci.* **62**.
Kern, H. (1978). *Tectonophysics* **44**, 185.
Klemperer, S.L. and W.D. Mooney (Ed.) (1998a), *Tectonophysics* **286**, 1.
Klemperer, S.L. and W.D. Mooney (Ed.) (1998b). *Tectonophysics* **288**, 1.
Kozlovsky,Ye.A. (Ed.) (1987). "The Superdeep Well of the Kola Peninsula," Springer-Verlag, Berlin.
Krawczyk, C.M., *et al.* (1999). *Tectonophysics* **314**, 241–253.
Krilov, S.V. *et al.* (1991). *Geodynamics* **13**, 87.
Levander, A.R. and K. Holliger (1992). *J. Geophys. Res.* **97**, 8797.
Li, S.L. and W.D. Mooney (1998). *Tectonophysics* **288**, 105.
Ludwig, W.J., *et al.* (1970). In: "The Sea," (A.E. Maxwell, Ed.), vol. 4, part I, pp. 53. Wiley-Interscience, New York.
Mahadevan, T.M. (1994). *Geol. Soc. India Memoir 28*, Bangalore.
Makovsky, Y., *et al.* (1996a). *Tectonics* **15**, 997.
Makovsky, Y., *et al.* (1996b). *Science* **274**, 1690.
Makris, J. (1975). In: "Afar Depression of Ethiopia," (A. Pilger and Rösler, Eds.), pp. 379. Schweizerbart, Stuttgart.
Masson, F., *et al.* (1998). *Geophys. J. Int.* **134**, 689.
Masson, F., *et al.* (1999). *Tectonophysics* **302**, 83–98.
Matsuzawa, T. (1959). *Bull. Earthq. Res. Inst. Univ. Tokyo* **37**, 123.
Matthews, D.H. (1986). *Geol. Soc. London Special Publ.* **24**, 11.
Maxwell, A.E. (Ed.) (1970). "The Sea – New Concepts of Sea Floor Evolution," vol 4, Wiley-Interscience, New York.
McConnell, R.K., *et al.* (1966). *Rev. Geophys.* **4**, 41–100.
Mechie, J., *et al.* (1993). *Phys. Earth Planet. Inter.* **79**, 269.
Meissner, R. (1973). *Geophys. Surv.* **1**, 195–216.
Meissner, R. (1986). "The Continental Crust," Academic Press, London.
Meissner, R., and the DEKORP Research Group, 1991. The DEKORP surveys: Major achievements for tectonical and reflective styles. In: Meissner, R., Brown, L., Dürbaum, H.-J., Franke, W., Fuchs, K., and Seifert, F. (Eds.), Continental Lithosphere: Deep Seismic Reflections. *Am. Geophys. Un., Geodyn. Ser.*, **22**, 69–76.
Meyer, R.P., *et al.* (1976). *Am. Geophys. Un. Geophys. Monograph* **19**, 105.
Mohorovicic, A. (1910). *Jahrbuch Meteorol. Obs. Zagreb für 1909* **9**(4), 1.
Mooney, W.D. (1989). *Geol. Soc. Am. Mem.* **172**, 11.
Mooney, W.D. and L.W. Braile (1989). In: "The Geology of North America – An Overview." (A.W. Bally and A.R. Palmer, Eds.), pp. 39. Geological Society of America, Boulder, CO.
Mooney, W.D., *et al.* (1998). *J. Geophys. Res.* **103**, 727.
Mooney, W.D. and R. Meissner (1992). In: "Continental Lower Crust," (D.M. Fountain *et al.*, Eds.), pp. 45. Elsevier, Amsterdam.
Mueller, S. (Ed.) (1973). *Tectonophysics* **20**, 1.
Mueller, S. (1977). *Am. Geophys. Un. Geophys. Monograph* **20**, 289.
Mueller, S. (1978). In: "Tectonics and Geophysics of Continental Rifts," (I. Ramberg and E. Nuemann, Eds.), vol. 2, pp. 11. Reidel, Dordrecht.
Mueller, S. and H.-G. Kahle (1993). *Am. Geophys. Un. Geodynam. Series* **23**, 249.
Mueller, S. and G.F. Panza (1986). *Dev. Geotecton.* **21**, 93.
Myers, S.C., *et al.* (1998). *J. Geophys. Res.* **103**, 21233.
Nelson K.D. *et al.* (1996). *Science* **274**, 1684.
Olsen, K.H. (Ed.) (1995). "Continental Rifts." Elsevier, Amsterdam.
Orcutt, J. (1987). "Scattering and Q within the Anza Array: A Final technical report," University of California, San Diego.
Pakiser, L.C. and W.D. Mooney (Ed.) (1989). "Geophysical Framework of the Continental United States," *Geol. Soc. Am. Memoir 172*.
Pakiser, L.C. and J.S. Steinhart (1964). In: "Research in Geophysics," (J. Odishaw, Ed.), 2, pp. 123. M.I.T. Press, Cambridge, MA.
Patzwahl, R., *et al.* (1999). *J. Geophys. Res.* **104**, 7293.
Pavlenkova, N.I. (1973). "Wave Fields and Crustal Models of Continental Type." Naukova Dumka, Kiev, (in Russian).
Pavlenkova, N.I. (1996). *Adv. Geophys.* **37**, 1.
Pfiffner, O.A., *et al.* (Ed.) (1997). "Deep Structure of the Swiss Alps: Results of NRP 20." Birkhäuser, Basel.
Pollack, H.N. and D.S. Chapman (1977). *Tectonophysics* **38**, 279.
Prodehl, C. (1977). *Am. Geophys. Un. Geophys. Monograph* **20**.
Prodehl, C. (1984). In: "Physical Properties of the Interior of the Earth, the Moon and the Planets," (K. Fuchs and H. Soffel Eds.), pp. 97. Springer, Berlin.
Prodehl, C. and B. Aichroth (1992). *Terra Nova* **4**, 14.
Prodehl, C., *et al.* (1997). *Tectonophysics* **278**, 1.
Puzyrev, N.N. (Ed.) (1993). "Detailed Seismic Studies of the Lithosphere with P- and S-Waves," Nauka, Novosibirsk, (in Russian).
Reddy, P.R. (1999). *Curr. Sci.* **77**, 1606.
Reddy, P.R. and V. Vijaya Rao (2000). *Curr. Sci.* **78**, 899.
Ritzwoller, M.H., *et al.* (1998). *Geophys. J. Int.* **134**, 315.
Rudnick, R. and D.M. Fountain (1995). *Rev. Geophys.* **33**, 267.
Simmons, G. (1964). *J. Geophys. Res.* **69**, 1123.
Soller, D.R., R.D. Ray, and R.D. Brown (1982). *Tectonics* **1**, 125.
Sroda, P. and the POLONAISE Profile P3 Working Group (1999). *Tectonophysics* **314**, 175.
Steinhart, J.S. (1967). In: "International Dictionary of Geophysics," (K. Runcorn, Ed.),vol. 2, pp. 991.
Steinhart, J.S. and R.P. Meyer (1961). *Carnegie Inst. Washington, DC, Publ. 622*.
Stern, T.A. and J.H. McBride (1998). *Tectonophysics* **286**, 63.
Tanimoto, T. (1995). In: "Global Earth Physics: A Handbook of Physical Constants," (T. J. Ahrens, Ed.), pp. 214. American Geophysics Union, Washington, DC.
Tatel, H.E. and M.A. Tuve (1955). *Geol. Soc. Am. Special Paper* **62**, 35.
Valasek, P., *et al.* (1991). *Geophys. J. Int.* **105**, 85.
Waldhauser, F., *et al.* (1998). *Geophys. J. Int.* **135**, 264.
Wang, Y.X., *et al.* (2002). *J. Geophys. Res.* (in Press).
Wigger, P.J., *et al.* (1994). In: "Tectonics of the Southern Central Andes," (K.J. Reutterr, *et al.*, Eds.), pp. 23. Springer, New York.
Zhao, D., *et al.* (1990). *Tectonophysics* **181**, 135.
Zhao, W.J., *et al.* (1993). *Nature* **366**, 557.
Zverev, S.M. and I.P. Kosminskaya (Ed.) (1980). "Seismic Models of the Lithosphere for the Major Geostructures on the Territory of the USSR," Nauka, Moscow (in Russian).

Editor's Note

Due to space limitations, references with full citation are given in MooneyFullReferences.pdf on the attached Handbook CD, under the directory \54Mooney. See also Chapter 55, Seismic structure of the oceanic crust and passive continental margins, by Minshull.

55

Seismic Structure of the Oceanic Crust and Passive Continental Margins

T.A. Minshull
Southampton Oceanography Centre, Southampton, UK

1. Introduction

The oceanic lithosphere covers approximately 60% of the Earth's surface. Well-constrained measurements of the seismic velocity structure of the crust and upper mantle beneath the ocean floor are sparse. However, the crustal structure appears to vary systematically with a rather small number of parameters: the age of the lithosphere; the spreading rate at which the crust was formed; its position with respect to offsets in the ridge axis; and the proximity or otherwise of thermal or chemical perturbations such as mantle plumes at the time of crustal formation or later in its history. Significant portions of this parameter space remain to be explored. However, the available data do allow various empirical correlations to be made, and numerical modelling of the processes governing the formation and evolution of the oceanic crust allow some of these correlations to be tied to the physics of these processes, so they can be used as predictive tools in poorly explored areas. The printed version of this chapter is abstracted from a longer, fully referenced version on the attached Handbook CD, to which the reader is referred for further details.

2. Methodology

Single and multichannel seismic reflection profiling (Fig. 1a) has provided considerable detail on the thickness and internal structure of the <1–2 km sediment layer which typically covers oceanic basement, and limited information on structures within the basement. Some information on seismic velocities can be gleaned from multichannel reflection data, from vertical seismic profiles and from local earthquake seismology. However, the vast majority of our knowledge of the seismic velocity structure of the oceanic crust and uppermost mantle comes from seismic refraction (or "wide-angle seismic") experiments. The earliest experiments in the 1950s and 1960s used two ships, one of which fired explosive shots at intervals of several kilometers while the other held station with a hydrophone

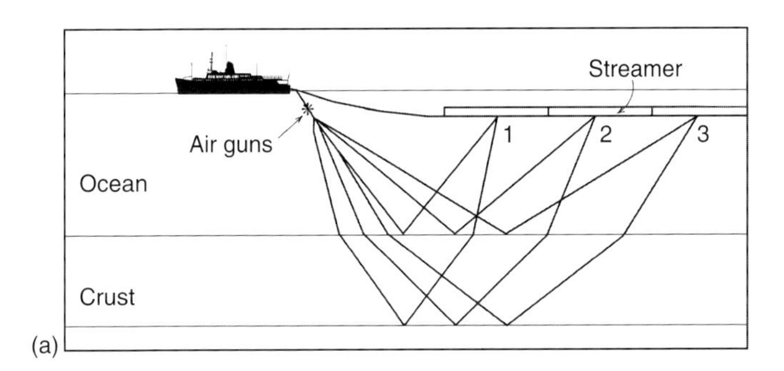

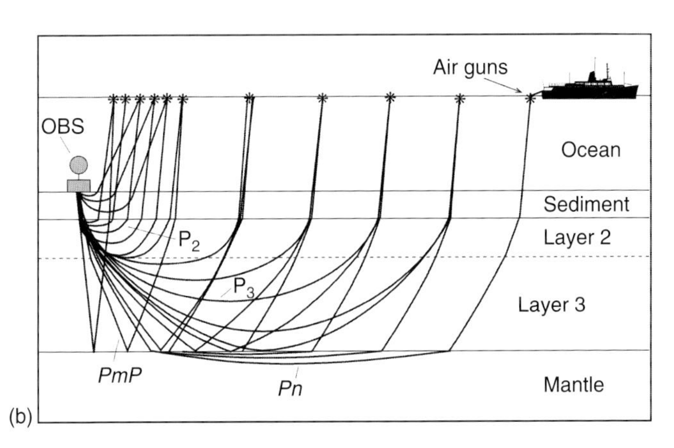

FIGURE 1 Methodology for seismic experiments on oceanic crust. (a) Multichannel seismic reflection. (b) Seismic refraction using ocean bottom seismometers or hydrophones. Ray paths for the main crustal signals P_2 (turning in Layer 2), P_3 (turning in Layer 3) and *PmP* (reflecting at the Moho) are shown.

INTERNATIONAL HANDBOOK OF EARTHQUAKE AND ENGINEERING SEISMOLOGY, VOLUME 81A ISBN: 0-12-440652-1

receiver. The hydrophone signal was recorded directly onto paper rolls, first-arrival travel times were picked and velocities were inferred using simple graphical methods (e.g., Raitt, 1963). This early exploration phase provides the only available data from many areas of the oceans.

The use of internally recording free-floating sonobuoys later allowed seismic refraction experiments to be conducted from a single ship. By the early 1980s these were replaced by ocean bottom hydrophones (OBH) and seismometers (OBS) anchored on the sea floor (Fig. 1b); the sea floor provides a quieter environment and data analysis is much simpler if the receiver is fixed. An alternative approach was the "expanded spread profile" (ESP), which involved a shooting ship with explosives or airguns and a receiving ship with a multichannel hydrophone streamer travelling at the same speed on reciprocal courses to maintain a common midpoint. Along with these technical developments came more sophisticated interpretation techniques using synthetic seismograms and waveform inversion. A parallel development was the use during seismic reflection profiling of disposable sonobuoys which transmit the seismic signal by radio back to the shooting ship. By the late 1980s, the data storage capacity of OBHs and OBSs had become sufficiently large that hundreds of closely spaced airgun shots could be recorded, and the use of explosives in marine refraction experiments declined. Recent experiments have used large numbers of OBHs and OBSs laid out in a grid to allow analysis of 3D structures by sophisticated travel-time inversion techniques.

3. Structure of the Oceanic Crust

3.1 *P*-wave Velocity Structure

The basic *P*-wave velocity structure of the oceanic crust was established in the early days of marine geophysics, before the discovery of sea-floor spreading. It was recognised that oceanic crust was thinner than continental crust, and that the same basic structure was present throughout the ocean basins. The average thickness of normal oceanic crust, away from regions considered for various reasons to be anomalous, was found to be about 6.5 km, and this figure has changed little in more recent compilations, despite the use of a variety of different definitions for what is normal oceanic crust. The igneous crust has traditionally been divided into two layers, a 2–3 km Layer 2 and a 3–5 km Layer 3 (e.g., Raitt, 1963), with the overlying sediments labelled as Layer 1, and an underlying mantle layer.

In early compilations the main crustal layers were assigned constant velocities (5.07 ± 0.63 km and 6.69 ± 0.23 for Layer 2 and Layer 3 respectively in Raitt's compilation), as the data analysis methods available (least-squares slope-intercept solutions for picked first-arrival travel times) did not allow for velocity gradients. However, once synthetic seismogram modelling had come into widespread use (e.g., Spudich and Orcutt, 1980; Fig. 2), it was recognised that the high amplitude signals of Layer 2 corresponded to a high velocity gradient (typically 0.5–1.0 sec^{-1}), whereas the lower amplitude signals of Layer 3 corresponded to considerably lower

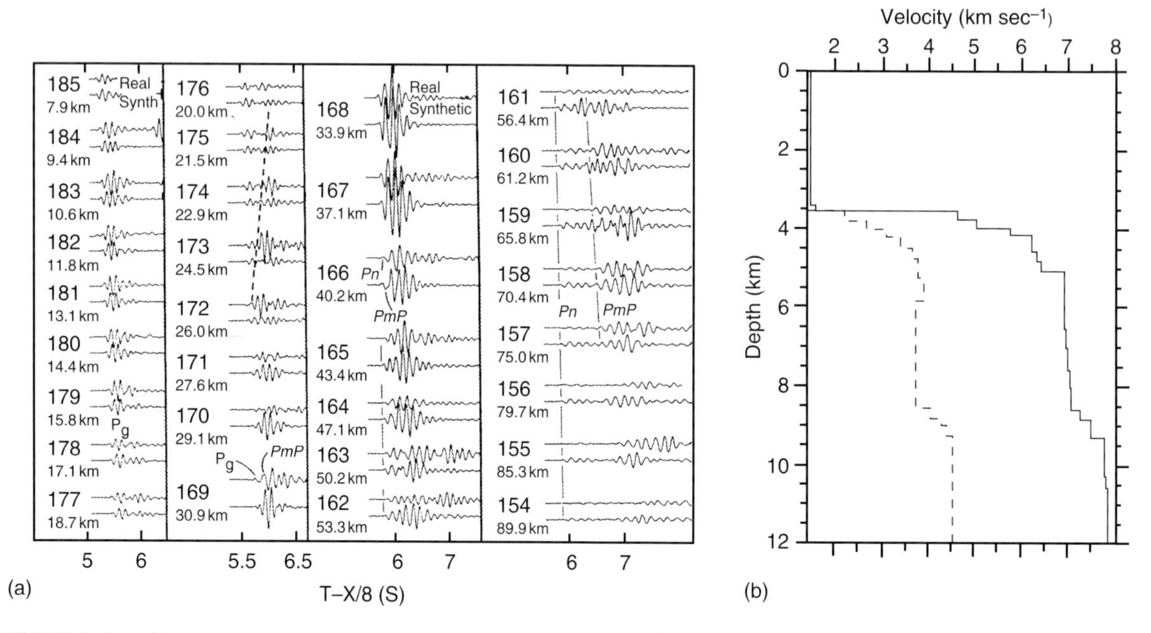

FIGURE 2 (a) Real and synthetic seismograms for an oceanic crustal site near Guadelupe. The source–receiver range is marked beside each pair of seismograms, and a reduction velocity of 8 km sec^{-1} has been applied (reprinted from Spudich and Orcutt, 1980). (b) Seismic *P*-wave (solid) and *S*-wave (dashed) velocity model corresponding to the synthetic seismograms in (a).

gradients (typically 0.1–0.2 sec^{-1}). Reanalysis of early data by modern synthetic seismogram methods has shown that slope-intercept solutions systematically underestimate oceanic crustal thickness, because the deeper, faster parts of layers are observed only as second arrivals (White *et al.*, 1992).

The Layer 2 and Layer 3 phases are easily recognised on modern record sections from oceanic crust (Fig. 3), which typically show a strong Layer 2 arrival, a weak Layer 3 arrival, a strong Moho reflection and weak mantle arrivals. Both Layer 2 and Layer 3 have also been further subdivided. The terms "Layer 2A" and "Layer 2B" are still commonly used to describe the upper oceanic crust. Layer 2A was defined as a thin layer with low velocities at the top of the crust (Houtz and Ewing, 1976), and Layer 2B was a deeper, higher-velocity layer with high gradients. The structure of Layer 2A appears to vary systematically with age (see Section 4), and variations close to midocean ridge axes have formed a recent focus of attention (see Section 6). Subdivisions of Layer 3 appear to vary between locations.

The geological interpretation of Layers 2 and 3 remains controversial, because direct sampling of the igneous stratigraphy by deep drilling has proved elusive. Based on studies of ophiolites, which are interpreted to be sections of oceanic crust which have been emplaced tectonically onto continents, Layer 2 is commonly identified with extrusive basaltic lavas and a sheeted dyke complex, whereas Layer 3 is identified with intrusive gabbroic rocks. However, many recent studies have emphasised the role of porosity over that of lithology in

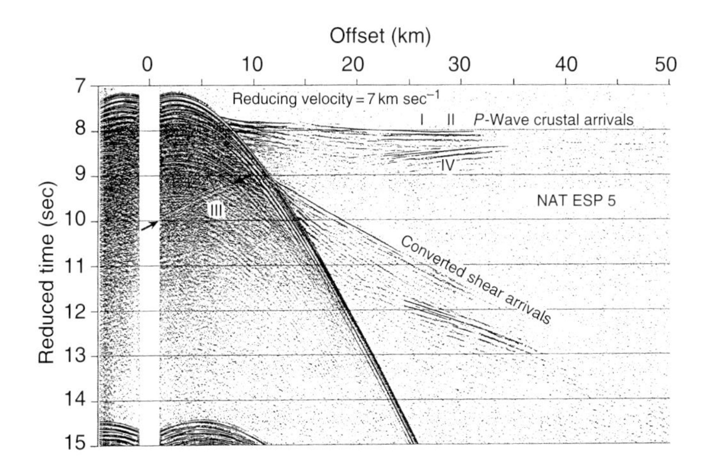

FIGURE 3 Expanded spread profile 5 from the North Atlantic Transect (reprinted from Mithal and Mutter 1989 with permission from Blackwell Science Ltd). This exceptionally clear record section shows features typical of oceanic crustal structure: high amplitude first arrivals from Layer 2 at 6–11 km range, weaker Layer 3 arrivals at 11–32 km range, a strong postcritical reflection from the Moho labelled IV, weak first arrivals from the mantle at 40–50 km range, and an equivalent set of *S*-wave arrivals. The phase labelled III is a precritical Moho reflection which is rarely seen in wide-angle data from oceanic crust.

the oceanic crust. In the Bay of Islands ophiolite, one of the best-preserved ophiolites, there is an abrupt increase in seismic velocity to 6.7–6.8 km sec^{-1} at the boundary between brecciated and nonbrecciated dykes, and no sharp increase and the boundary between dykes and gabbros (Fig. 4). This has led many authors to identify the Layer 2/3 boundary with the limit of dyke brecciation. A combination of seismic refraction and downhole logging at ODP Hole 504B on the Costa Rica Rift also shows that the Layer 2/3 boundary occurs within the sheeted dykes (Detrick *et al.*, 1994), though it seems unlikely that the results of Hole 504B can be extrapolated to oceanic crust everywhere.

The significance of the seismic Moho in the oceans, and the composition of Layer 3, have also remained controversial. The recovery of large amounts of serpentinised peridotite by dredging, ODP drilling and submersible diving on the Mid-Atlantic Ridge and other slow-spreading ridges has led some to the suggestion that Layer 3 consists of isolated gabbroic intrusions in serpentinised peridotite and to question the conventional interpretation that the Moho marks a petrological boundary between mafic rocks above and ultramafic rocks beneath (Fig. 5). However, the good agreement between seismically determined crustal thicknesses and geochemically determined melt thicknesses (White *et al.*, 1992), suggests that serpentinised peridotite is not the dominant component in Layer 3 of normal oceanic crust.

3.2 *S*-wave Velocity Structure

Few data are available on *S*-wave velocities for the oceanic crust and uppermost mantle, and even fewer data are available for the overlying sediment layer. This is because marine seismic sources are commonly within the water column and therefore can only generate *P* waves. *S* waves are only generated if fortuitous mode conversions occur, for example, at the base of the sediment layer. Detectable *S* waves cannot be generated in this way if the sediments are very thick, since then the *S*-wave velocity in the igneous crust is lower than the *P*-wave velocity in the sediments above and therefore any converted *S* waves do not turn within the crust. *S*-wave velocities in oceanic crust are most commonly characterised by Poisson's ratio, which for isotropic rocks is a function of the *P–S*-velocity ratio. Poisson's ratio is controlled by lithology, degree of alteration, porosity and the distribution of pore aspect ratios; a value of 0.25 is commonly assumed in crustal rocks where no *S*-wave data are available. Spudich and Orcutt (1980) compared Poisson's ratios derived from synthetic seismogram modelling of converted shear waves in seismic refraction data with measured values from ophiolites and dredge samples to constrain the composition of the oceanic crust. Values of 0.32–0.38 were found in the upper 200 m, indicating the presence of significant fracture porosity which is not present in dredge samples. Anomalously low values of Poisson's ratio (0.22–0.26) are commonly found at around

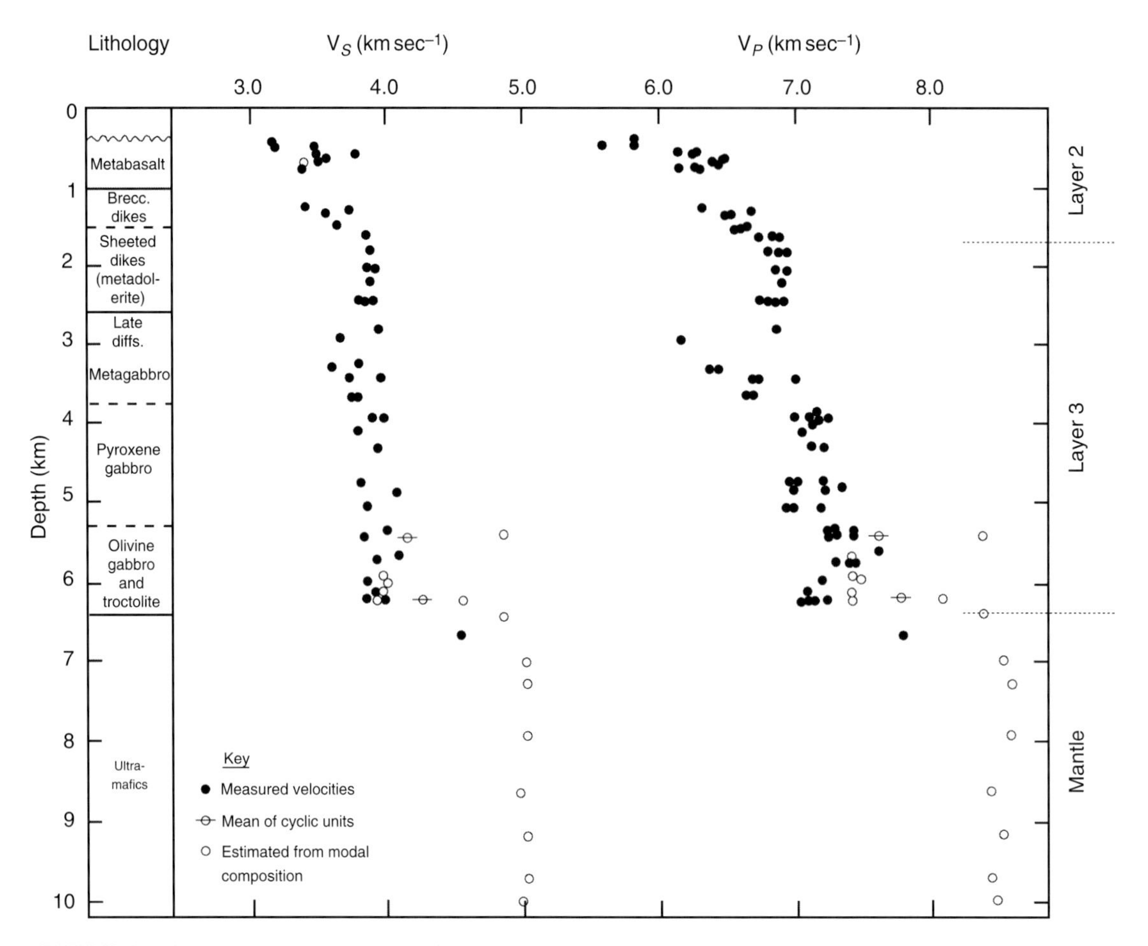

FIGURE 4 Lithology and seismic velocities, based on laboratory measurements, of a section through the Bay of Islands ophiolite (simplified from Salisbury and Christensen, 1978).

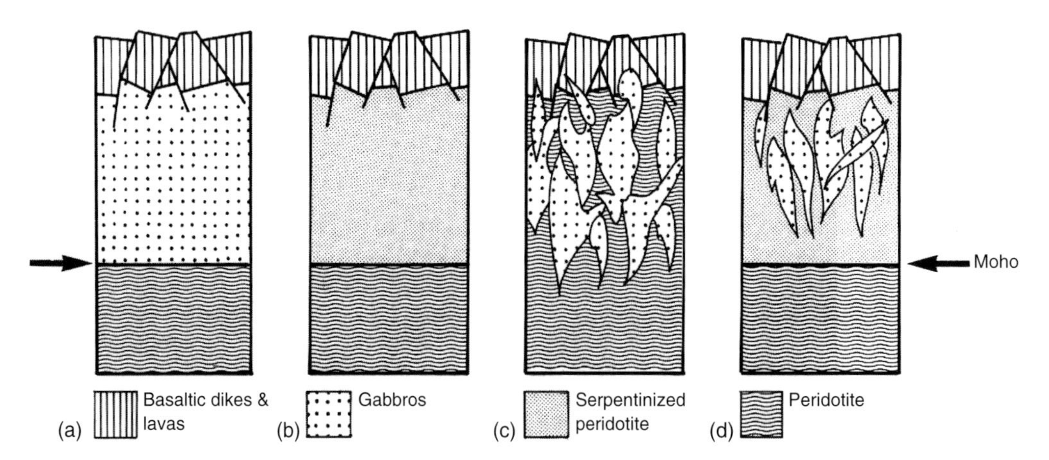

FIGURE 5 Four possible models of oceanic crustal stratigraphy (reprinted from Minshull *et al.*, 1998 with permission from the Geological Society of London). (a) Layer 3 consists entirely of gabbro. (b) Layer 3 consists entirely of partially serpentinised peridotite. (c) Layer 3 consists of gabbroic intrusions within unserpentinised to partially serpentinised peridotite. (d) As (c), but the limit of serpentinisation is deeper than the deepest gabbroic intrusion.

1.0–1.5 km depth in oceanic crust, particularly in young oceanic crust. These values might be explained by the presence of thick cracks resulting from hydraulic fracturing of the rock during dyke injection (e.g., Shearer, 1988), or as an artifact of the normal assumption that the crust is isotropic if differing *P*- and *S*-wave anisotropy is present (Fryer *et al.*, 1989).

3.3 Anisotropy

Constraints on seismic anisotropy in the oceanic crust are sparse. Since *S* waves are generally only present if there are fortuitous mode conversions, and coupling problems commonly make *S*-wave polarisation data unreliable, observations of *S*-wave splitting at oceanic sites are rare. Early wide-angle seismic experiments (e.g., Shearer and Orcutt, 1985) found a few percent anisotropy in the upper crust for both *P* and *S* waves, with a fast direction commonly parallel to the paleo-ridge axis. This anisotropy was attributed to the presence of aligned cracks in the upper crust. Several recent tomographic experiments have also suggested a few percent *P*-wave anisotropy in the upper 2 km of the crust at ridge axes, but often an isotropic model can also explain observed travel times. Another form of anisotropy which has been suggested for the oceanic crust is one with a horizontal plane of symmetry, arising from a fracture system with a preferred horizontal alignment which would be very hard to detect by conventional seismic experiments. The lower oceanic crust is not thought to be significantly anisotropic because the crack porosity is thought to be greatly reduced by the confining pressure.

3.4 Attenuation

Attenuation is typically quantified as an effective *Q* value, which includes both intrinsic and scattering attenuation. Scattering attenuation is likely to be the dominant physical process in low porosity, but heterogeneous, oceanic crustal rocks. Most estimates of *Q* for oceanic crust at seismic frequencies come from reflectivity modelling of high-quality expanding spread profile data, where a 1D assumption is approximately valid. Typical values for *P*-waves found to be consistent with observed data away from ridge axes are 400 for crustal rocks and 1000 for the uppermost mantle (e.g., Spudich and Orcutt, 1980). These values must be considered to have an uncertainty of at least a factor of two. Laboratory measurements at ultrasonic frequencies have given significantly lower values of less than 100 at all depths except the lower part of Layer 2 (Wepfer and Christensen, 1991).

4. Variation of Crustal Structure with Spreading Rate

The gross thickness of the oceanic crust varies remarkably little with spreading rate (Fig. 6) because of the balance

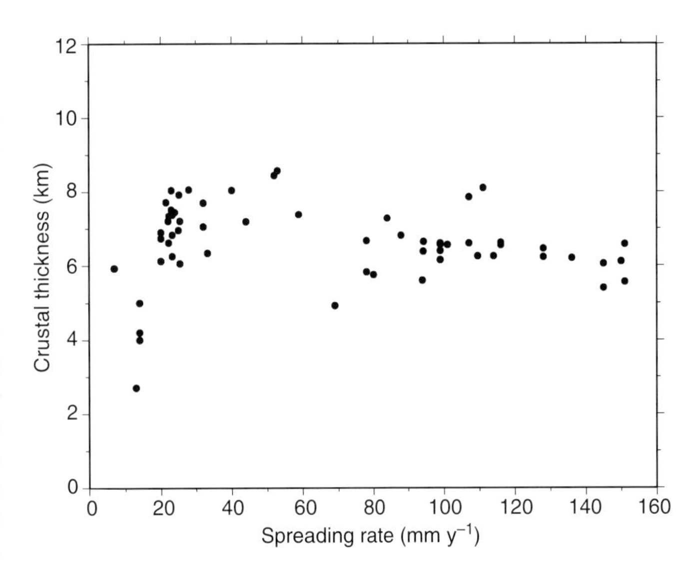

FIGURE 6 Crustal thickness versus spreading rate for oceanic crust away from the influence of mantle plumes and fracture zones, and omitting anomalous crust adjacent to nonvolcanic passive margins and at extinct spreading centres. Only values constrained by synthetic seismogram modelling or modern 2D travel time inversion techniques are plotted (modified from Bown and White, 1994).

between mantle upwelling rate, which controls the rate of melt production at a midocean ridge, and plate separation rate, which controls the surface area of crust to be created. A slight decrease in crustal thickness, when measurements at fracture zones are excluded, at higher spreading rates may be due to the more efficient distribution of melt along segments at fast-spreading ridges. Most models of melt generation at ridges predict a decrease in crustal thickness at very slow spreading rates. Recent results from the Mohns Ridge (spreading rate ~16 mm y^{-1}) and the Southwest Indian Ridge (~12–16 mm y^{-1}) suggest a crustal thickness of ~4 km and therefore support these models. Crustal thicknesses are generally more variable at slow spreading rates, but most of this variability can be attributed to the effects of ridge segmentation (see Section 7). At extinct rifts such as the Labrador Sea spreading centre, where the spreading rate has slowed to zero, the crust is thinner than normal oceanic crust and is underlain by material with velocities intermediate between Layer 3 velocities and upper mantle velocities, interpreted to be partially serpentinised peridotite. Despite indications from seismic reflection profiles of substantial spreading-rate-controlled differences in the internal architecture of oceanic crust, and differences in composition indicated by geological studies of midocean ridge axes, the seismic velocity structure appears to vary little with spreading rate (e.g., White *et al.*, 1992). At very slow spreading rates, where crustal thickness is reduced, the reduction appears to be predominantly at the expense of Layer 3, as would be expected if the thickness of Layer 2 is determined predominantly by the porosity and alteration structure of the upper crust.

5. Variation of Crustal Structure with Age

Except in the immediate vicinity of midocean ridge axes (see Section 6), the seismic structure of oceanic crust also varies rather little with age. The dominant age-dependent variation is in the velocity structure of the uppermost crust. Houtz and Ewing (1976) suggested that Layer 2A velocities increased to values typical of Layer 2B (around 5.2 km sec^{-1}) over 30–40 My and its thickness decreased with age, so that Layer 2A gradually disappeared. This systematic variation was attributed to progressive hydrothermal infilling of the upper crustal porosity. More recent analyses suggest that Layer 2A velocities increase rapidly in the first 10 My of evolution and change little thereafter, never reaching Layer 2B values and that there is no systematic variation in Layer 2A thickness with age (e.g., Grevemeyer *et al.*, 1998).

6. Seismic Structure of the Moho and Uppermost Oceanic Mantle

Reflections from the Moho are often strong and continuous in wide-angle record sections from normal oceanic crust (e.g., Fig. 3), suggesting a transition from crustal velocities of around 6.8–7.2 km sec^{-1} to upper mantle velocities of around 8 km sec^{-1} over a depth interval which is short compared to seismic wavelengths. However, in seismic reflection data the Moho reflection is often discontinuous, and observations from ophiolites suggest that the crust–mantle transition may sometimes occur over an interval of several kilometres in which the ultramafic content increases with depth. Such transition zones may be spatially localized and therefore not resolved by the wide-angle survey geometries commonly used prior to the 1990s. Mantle arrivals are commonly weak in airgun refraction experiments from oceanic crust (e.g., Fig. 3). Where good constraints are available, the *P*-wave velocity of the uppermost mantle in mature oceanic crust varies little from an average of 8.0 km sec^{-1}, though some azimuthal variation due to preferential alignment of olivine crystals has been observed (e.g., Shearer and Orcutt, 1985). The low *Pn* amplitudes suggest the velocity gradient in the uppermost mantle is very low. Lower velocities have been measured close to ridge axes; for example, velocities appear to increase from around 7.5 km sec^{-1} at the axis of the East Pacific Rise to around 8.0–8.2 km sec^{-1} over a period of a few millions years (e.g., Grevemeyer *et al.*, 1998). Upper mantle *S* wave arrivals are even more rare in marine refraction data; however, the sparse data available are consistent with an *S*-wave velocity of around 4.6 km sec^{-1} and a Poisson's ratio of 0.25.

7. Crustal Structure of Midocean Ridges

7.1 Fast-spreading Ridges

Over the last decade, the seismic velocity structure of the fast-spreading East Pacific Rise has been intensely studied. A localised low velocity zone, interpreted as resulting from the presence of a magma chamber containing molten or partially molten rock, was first detected on the northern East Pacific Rise by OBS refraction experiments in the 1970s. Subsequent ESPs and seismic tomography experiments (e.g., Toomey *et al.*, 1990) have indicated that the low-velocity zone extends from 1–2 km below the seabed to the base of the crust, and increases in width from ~2 km wide at its top to 10–12 km near the Moho (Fig. 7a). The lowest velocities (below 5 km sec^{-1}), attributed to the presence of melt, were constrained by these data to lie within a zone less than 1 km thick about 2 km beneath the rise axis, whereas below this region velocities were found to be considerably higher, though still up to 1 km sec^{-1} below normal Layer 3 velocities. OBS experiments have also suggested the presence of a several-kilometer-thick low-velocity zone immediately beneath the crust, also interpreted as due to the presence of partial melt. A limitation in the interpretation of these results is that there are few good laboratory studies of the elastic properties of partially molten basalt, and these studies have contradictory results about how seismic velocities vary close to the solidus. Multichannel seismic reflection techniques have demonstrated the presence of a high-amplitude and continuous reflector along large sections of the East Pacific Rise. Synthetic seismogram modelling and waveform inversion studies of this reflector have indicated that it consists of a thin lens (<100 m) of molten rock with a width varying from around 250 m to 4 km.

Based on multichannel reflection and bottom shot experiments, the structure of the upper few hundred meters of the crust at the East Pacific Rise axis appears to vary remarkably little along strike, with a <100 m thick zone where the velocity is around 2.5 km sec^{-1} and the velocity gradient is very low, underlain by a 100–200 m zone where the velocity increases sharply to around 5.0 km sec^{-1}, underlain by a shallower gradient again. At ridge axes, the term "Layer 2A" is commonly applied to the low-gradient zone near the seabed and the steep gradient zone beneath it. This gradient is sufficiently steep to generate wide-angle reflections in multichannel reflection data, which have allowed the layer thickness to be mapped along large regions of the East Pacific Rise. These studies have shown that Layer 2A is 100–300 m thick on-axis, and doubles in thickness within 1–4 km of the rise axis (e.g., Christesen *et al.*, 1994). Layer 2A is generally interpreted as

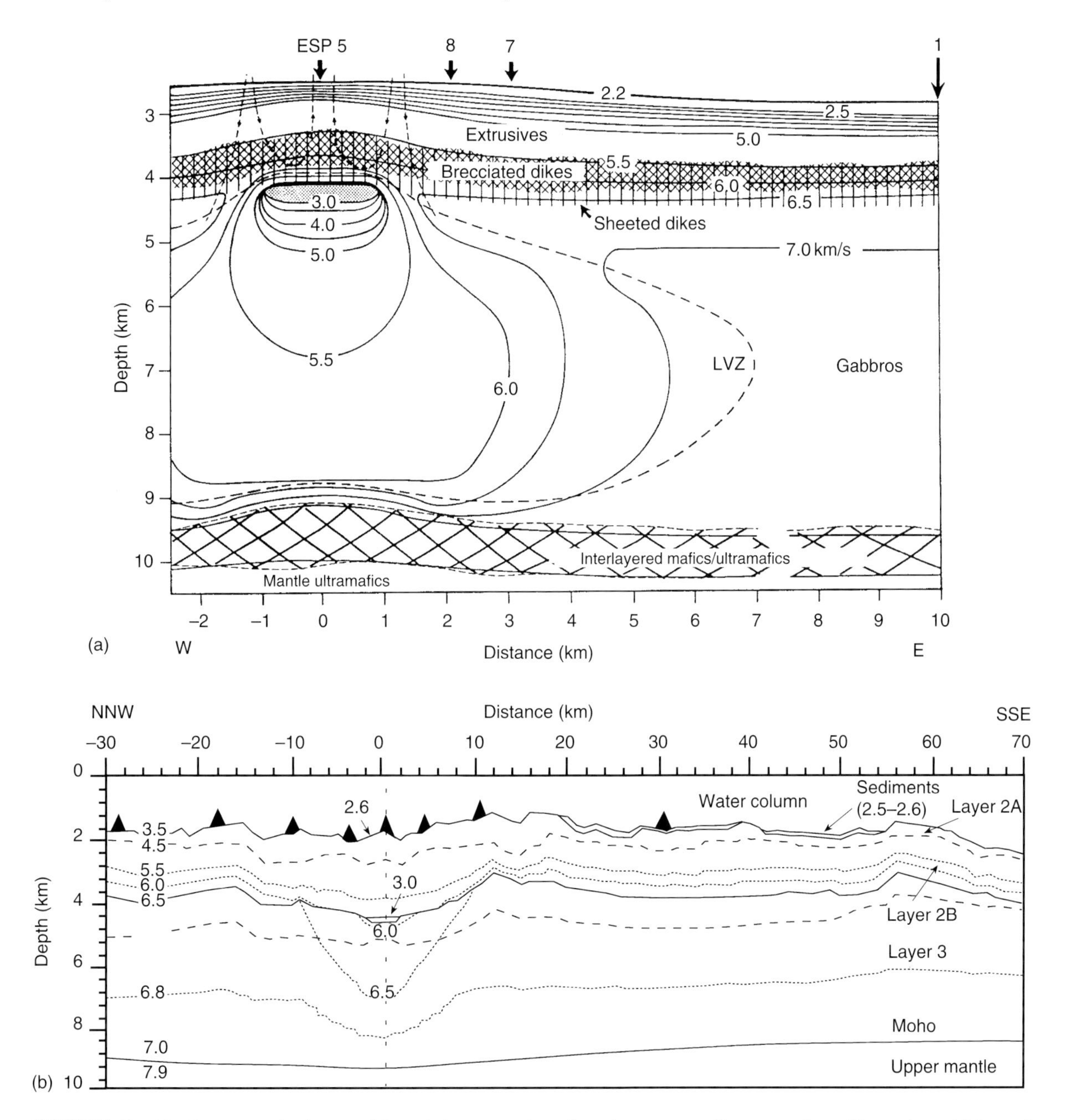

FIGURE 7 (a) *P*-wave velocity model and interpretation, based on expanding spread profile and multichannel reflection data, of the fast-spreading East Pacific Rise axis at 9° N. Arrows mark ESP locations (redrawn from Vera *et al.*, 1990). (b) *P*-wave velocity model and interpretation, based on OBS and seismic reflection data, of the slow-spreading Reykjanes Ridge axis at 57°45′N. Triangles mark OBS locations and dashed lines mark changes in velocity gradient (redrawn from Navin *et al.*, 1998 with permission from Blackwell Science Ltd).

the extrusive layer at fast-spreading ridges, with the steep velocity gradient at its base corresponding to a transition to lower-porosity sheeted dikes, and off-axis thickening resulting from further accumulation of lava flows.

7.2 Intermediate-spreading Ridges

There have been many fewer seismic studies of intermediate-spreading ridges. A seismic reflection survey of the back-arc Valu Fa spreading centre found a widespread, high amplitude reflector at a depth of around 3.2 km with a width of 0.6–4 km (Collier and Sinha, 1992). The reflection coefficient for this event corresponds to a *P*-wave velocity of less than 3.8 km sec^{-1}, and coincident wide-angle seismic data indicate the presence of a low-velocity zone, so the reflector is interpreted as marking the top of a crustal magma chamber which is significantly deeper than the magma chamber on the East Pacific Rise. OBH studies conducted on various parts of the

Juan de Fuca Ridge have not found convincing evidence for a crustal magma chamber, but have found a seismic Layer 2A similar to that found on the East Pacific Rise, with an abrupt doubling of *P*-wave velocity at 0.25–0.60 km below the seabed. Layer 2A is thinnest at the ridge axis, but off-axis the thickness varies irregularly, perhaps due to an episodic form of magmatism, with tectonic rotations occurring between magmatic episodes.

7.3 Slow-spreading Ridges

Early OBS refraction experiments on the Mid-Atlantic Ridge found neither a low-velocity zone corresponding to a magma chamber, nor a strong velocity contrast at Moho depths; rather, the crust–mantle transition was characterised by a gradual increase in velocities. However, modern experiments have found clear Moho reflections corresponding to a more abrupt step in velocity, and a detailed tomographic study focused on the upper crust at 35° N (Barclay *et al.*, 1998) found a zone of anomalously low velocities immediately beneath the seabed with a width of about 2 km and a depth extent of about 1 km, interpreted as due to near-solidus temperatures or possibly the presence of partial melt. To date, the most convincing seismic evidence for a magma chamber at a slow-spreading ridge comes from an experiment at 57°40 N on the Reykjanes Ridge (Fig. 7b). Here, a high-amplitude, reversed polarity reflection was imaged in low-fold reflection data and a low-velocity zone is required to match shadow zones seen in wide-angle seismic data. The dimensions of the low-velocity zone are weakly constrained, but the data are consistent with a thin lens of low-velocity material between 2 and 3 km depth. Seabed velocities at slow-spreading ridge axes appear to be around 2.5 km sec^{-1}, similar to those at faster spreading rates, indicating similar porosities at the ridge axis. However, Layer 2A does not appear to be significantly thinner on-axis than immediately off-axis; accretion of the uppermost crust on slow-spreading ridges is a more irregular process, and Layer 2A may not be equivalent to the extrusive section.

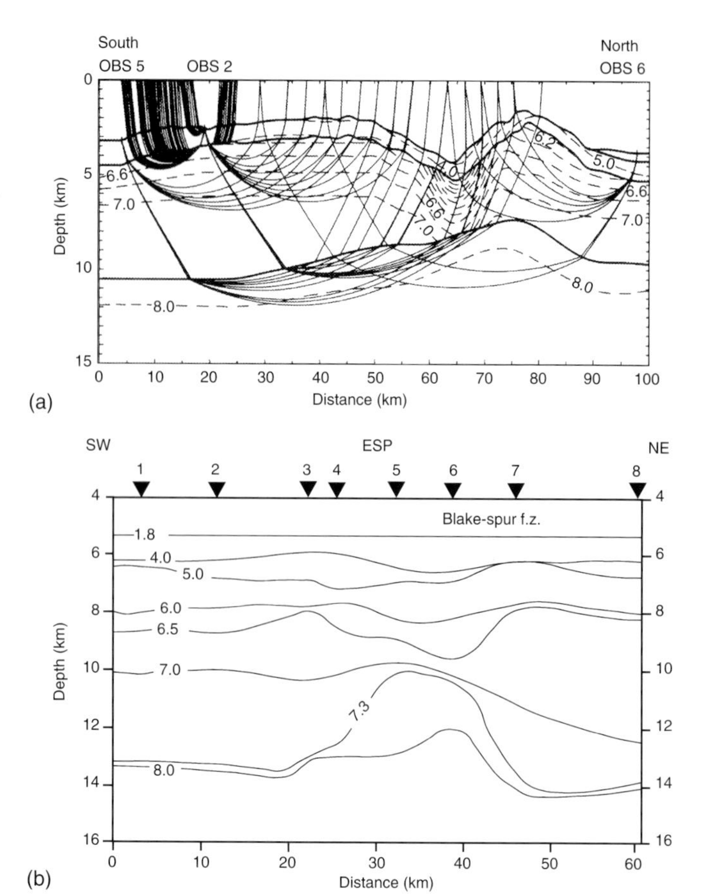

FIGURE 8 Seismic velocity structure of oceanic fracture zones at slow-spreading ridges. (a) A large-offset fracture zone: the Charlie-Gibbs Fracture Zone. OBS locations are marked at the top of the figure and rays traced to shot positions (redrawn from Whitmarsh and Calvert, 1986, with permission from Blackwell Science Ltd). (b) A small-offset fracture zone: the Blake Spur Fracture Zone. Arrowheads mark ESP locations (redrawn from Minshull *et al.*, 1991).

8. Crustal Structure at Oceanic Fracture Zones and Segment Boundaries

8.1 Segmentation of Slow-spreading Ridges

Several major studies of large-offset Atlantic fracture zones showed that they are characterised by anomalously thin crust and anomalously low seismic velocities (Fig. 8a). Fracture zone crustal structures generally fall between two end-members, referred to as "type A" and "type B" by Sinha and Louden (1983); both types of crust have been found at several large-offset Atlantic fracture zones. Type A crust is typically 4–5 km thick and characterised by lower seismic velocities than normal for oceanic crust, particularly at lower crustal depths. Type B crust is less than 3 km thick and has a velocity structure that is fundamentally different from normal oceanic crust. Typically, velocities in the uppermost crust are anomalously low, and the velocity gradient is high throughout the crustal section. The low upper crustal velocities have been attributed to highly fractured basaltic and possibly gabbroic material, though serpentinite has been dredged frequently in fracture zone valleys and must be present in the upper crust of the valley walls and/or floor. Another common feature is the presence of velocities in the range 7.2–7.5 km sec^{-1}, intermediate between normal oceanic crust and upper mantle velocities, interpreted as due to the presence of serpentinised upper mantle at the base of the crust. Seismic experiments have shown large along-axis variations in crustal thickness even where the offset in the ridge axis is relatively small (Fig. 8b). In addition to the dramatic changes in crustal structure which occur toward segment ends, some experiments have also shown a steady increase in crustal thickness to anomalously large values of 8–10 km toward segment centres.

8.2 Segmentation of Fast-spreading Ridges

The only fracture zone on the East Pacific Rise which has been studied in some detail is the Clipperton Fracture Zone (Begnaud *et al.*, 1997), with an age offset of 1.5 My, similar to that of small-offset transforms and nontransform discontinuities in the Atlantic. Here, there is no apparent crustal thinning at the fracture zone. Rather, the crustal thickness is normal in the active transform valley and the crust is slightly thickened at the ridge–transform intersection, where the addition of extrusive material from the adjacent spreading centre results in a local bathymetric high. Both in the active transform and at the ridge–transform intersection, seismic velocities are lower throughout the crust than in the adjacent oceanic crust. The reduced velocities are attributed to enhanced fracturing associated with the strike-slip motion.

9. Hotspots, Ocean Islands, Aseismic Ridges and Oceanic Plateaux

The majority of crustal thickness variations in the oceans which cannot be attributed to the effects of ridge axis segmentation are attributed to the influence of so-called "hotspots." Hotspots take a variety of forms, but are generally thought to be associated with anomalies in mantle temperature and/or chemistry due to anomalous material rising from deep in the mantle.

9.1 Ridge–hotspot Interactions

Hotspots have their greatest effect on crustal structure when they are located at or near a midocean ridge. The clearest and best-studied example is the Iceland hotspot, which thickens the crust sufficiently to lift the spreading centre above sea level. The thickness of the crust in Iceland was thought for many years to be only around 10–15 km, but a series of recent refraction experiments has shown that the crustal thickness varies between 20 and 40 km, and is greatest at the inferred present-day location of the hotspot. The upper crust in Iceland has velocities increasing from about 3.2 km sec^{-1} at the top to about 6.4 km sec^{-1} at the base. The thickness of the upper crust varies between about 2 km and 10 km, and reaches a minimum beneath central volcanoes (Fig. 9). The Icelandic lower crust has a smooth gradient between about 6.4 km sec^{-1} and 7.2–7.35 km sec^{-1} and a thickness varying between 10 and 30 km. Dramatic variations in crustal thickness may result from a history of ridge jumps as well as from variations in

(a)

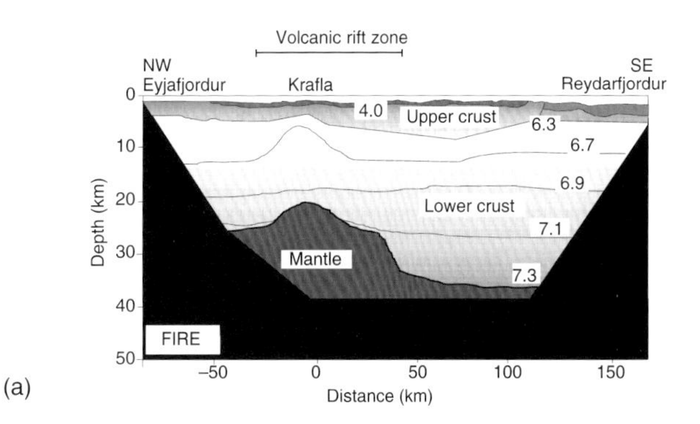

(b)

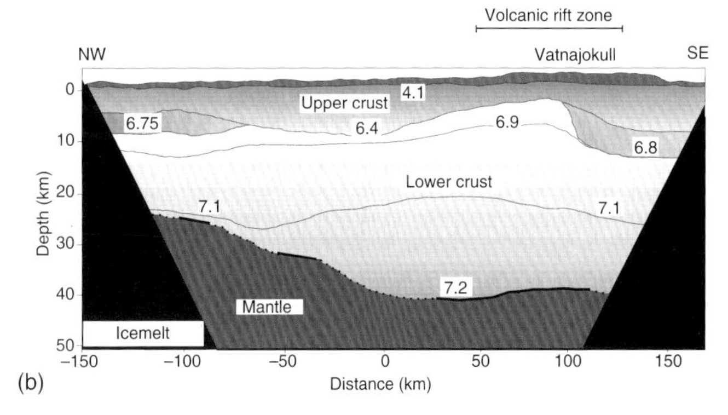

FIGURE 9 Seismic velocity structure of Icelandic crust (redrawn from Darbyshire *et al.*, 1998 with permission from Blackwell Science Ltd). (a) Seismic velocity model across the Northern Volcanic Zone in northeast Iceland; (b) seismic velocity model across the Eastern Volcanic Zone in southeast Iceland.

proximity to the core of the inferred underlying mantle plume. In contrast to the elusive evidence for crustal melt bodies elsewhere on slow-spreading ridge axes, several such melt bodies have been detected seismically in Iceland. For example, a low-velocity zone detected beneath Krafla central volcano in northeast Iceland and inferred to represent a magma chamber has horizontal dimensions of 2–3 km by 8–10 km and a thickness of 0.75–1.8 km, and its top lies at about 3 km depth (Brandsdóttir *et al.*, 1997).

9.2 Aseismic Ridges and Oceanic Plateaux

Aseismic ridges are thought to mark the off-axis trace of ridge-centred hotspots which have stayed close to the spreading centre. Some of these ridges probably have been subaerial for much of their history, and therefore have had their upper parts removed by erosion. The most prominent aseismic ridges, such as the Madagascar Ridge, have crustal thicknesses of up to 25–30 km (Fig. 10a). Other less-prominent features such as the Madeira-Tore Rise in the central Atlantic have crustal thicknesses of 10–15 km (Fig. 10a). The high-gradient upper crust at these ridges is only slightly thicker than oceanic Layer 2, at typically ~4 km, whereas the underlying low-gradient layer is much thicker than normal oceanic Layer 3.

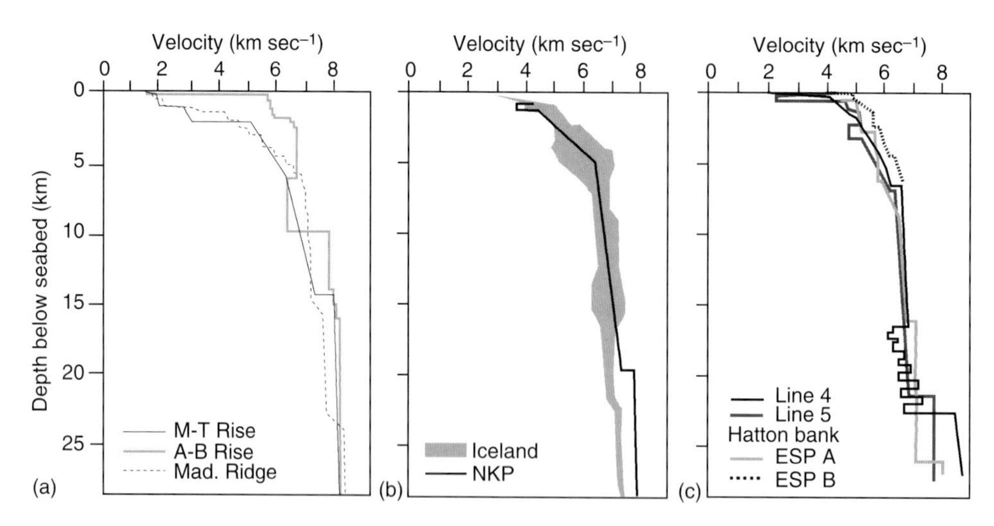

FIGURE 10 Crustal structure of aseismic ridges and oceanic plateaux. (a) Madagascar Ridge (Sinha *et al.*, 1981), Madeira-Tore Rise (Peirce and Barton, 1991), Azores-Biscay Rise (Whitmarsh *et al.*, 1982). (b) Crustal structure of northern Kerguelen plateau compared with Iceland (redrawn from Operto and Charvis, 1996). (c) Crustal structure of southern Kerguelen plateau compared with the Hatton Bank rifted margin (redrawn from Operto and Chavis, 1996).

Anomalously high velocities of 7.2–7.5 km sec^{-1}, interpreted as due to a high magnesium content of the igneous material, have been detected along some aseismic ridges. Oceanic plateaux represent vast outpourings of magma which may also result from near-ridge hotspot activity, but on a larger scale than for aseismic ridges. A type example is the Kerguelen Plateau in the southern Indian Ocean. The southern part of Kerguelen Plateau has recently been interpreted as a stretched continental fragment, and its velocity structure is very similar to that of volcanic rifted margins (Fig. 10c). The northern Kerguelen Plateau is interpreted to be purely oceanic and has a 20–22 km thick crust with a very thick, low-velocity gradient lower crust; the overall structure is very similar to that of southwest Iceland (Fig. 10b). The Kerguelen Isles themselves appear to have a structure which closely resembles that of large oceanic intraplate volcanoes (see Section 9.3).

9.3 Oceanic Intraplate Volcanism

Oceanic intraplate volcanism takes place on a variety of scales, from large islands such as La Reunion to the very many small seamounts which dot the ocean floor. The most well-known example is the Hawaiian chain, where the motion of an underlying mantle plume relative to the overlying Pacific plate is clearly marked by a linear chain of islands and seamounts. Seismic experiments at islands and seamounts have commonly focused on quantifying the deformation of the underlying lithosphere in response to the volcanic load. These experiments have confirmed the suggestion based on gravity studies that there is commonly a depression of both the top of the oceanic crust and the Moho (Fig. 11), and also shown the presence of a high-velocity basal layer attributed to underplated basaltic melt. The significance of underplating remains controversial. The results of an experiment close to the Hawaiian Island of Oahu, were initially interpreted to indicate an underplated body 3–6 km thick (Fig. 11a), but later the same data were reinterpreted to suggest that no underplated body was present. An experiment around the Canary island of Tenerife found no evidence for underplating (Fig. 11b), but a thick underplated body has been inferred beneath the Marquesas Islands (Fig. 11c).

Rather little is known about the internal velocity structure of ocean islands. The basic picture which emerges from seismic experiments is of a central high-velocity core, which probably consists primarily of intrusive rocks, and an outer shell of lower velocity material consisting primarily of lava flows, pyroclastic deposits and the products of mass wasting. Ocean islands often have multiple rift zones cored by high-velocity intrusives, but such detailed structure is rarely resolved by seismic data. Good constraints have been obtained for a few oceanic intraplate volcanoes which are entirely submarine; these data suggest that small volcanic edifices contain a much larger proportion of low-velocity materials than their larger counterparts. On the island of Hawaii, an intensive effort using mainly passive seismic techniques has allowed the detailed structure of the crustal magma body supplying the main active volcanoes, Mauna Loa and Kilauea, to be imaged (e.g., Ryan, 1988).

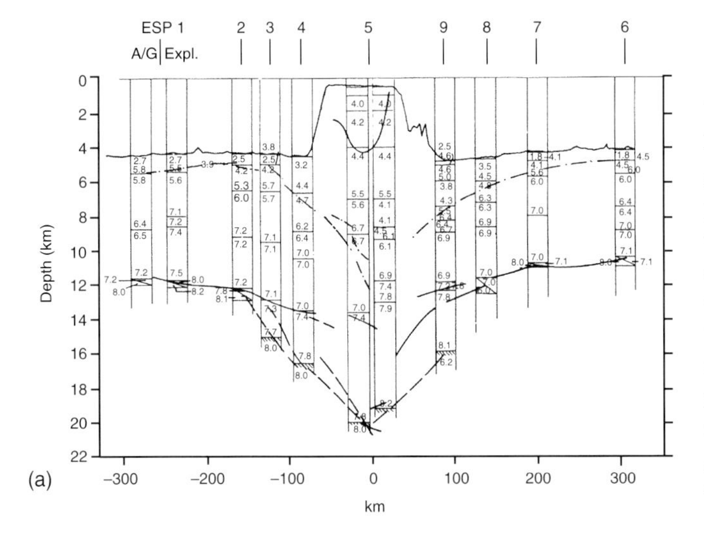

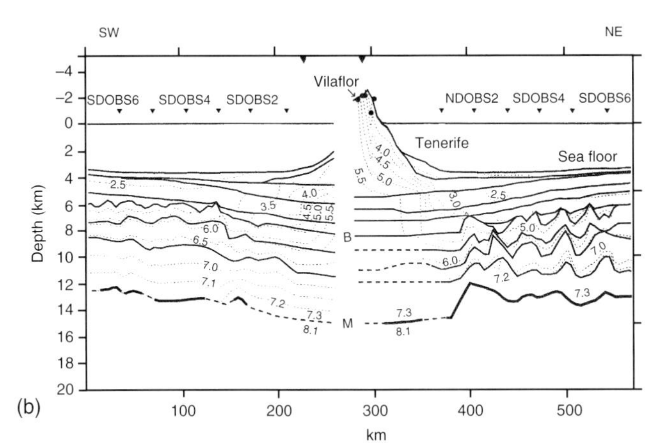

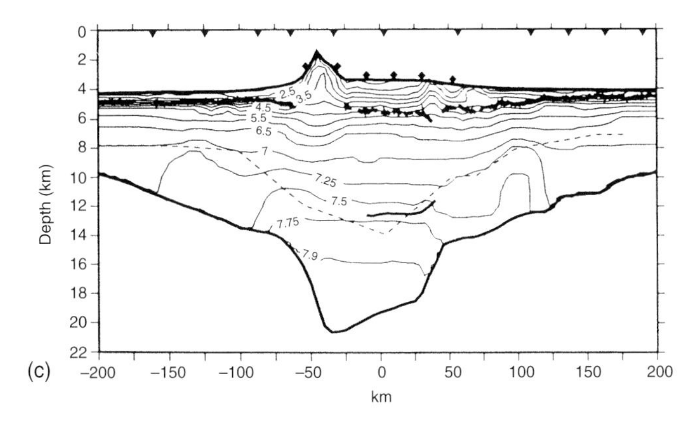

FIGURE 11 Seismic velocity structure of ocean islands. (a) Oahu in the Hawaiian islands, with ESP locations indicated [reprinted with permission from Watts *et al.* (1985). *Nature* **315**, 105–111, Copyright 1985 MacMillan Magazines Limited]. (b) Tenerife in the Canary Islands. Arrowheads mark OBS locations [reprinted from Watts *et al.* (1997). Copyright 1997, with permission from Elsevier Science]. (c) A transect across the Marquesas Islands. Triangles mark sonobuoy locations and diamonds mark OBS locations [reprinted with permission from Caress *et al.* (1995). *Nature* **373**, 600–603, Copyright 1995 MacMillan Magazines Limited].

10. Passive Margins and Ocean–Continent Transitions

Passive continental margins may be divided into rifted margins, where the direction of extension is approximately perpendicular to the resulting margin; and transform margins, where the relative plate motion was along the margin. Rifted margins have been further divided into "volcanic" margins, where there is extensive evidence for syn-rift magmatism, and "non-volcanic" margins, where syn-rift magmatism is weak or absent. The origin of this variation remains controversial: some authors have emphasised the role of the temperature of the underlying mantle, whereas others have emphasised the role of small-scale convective circulation in the mantle in generating volcanic margins.

10.1 Volcanic Rifted Margins

Of the three categories described above, volcanic rifted margins may be the most abundant and are certainly the best studied. Two main seismic characteristics have been identified. Firstly, distinctive arcuate seaward dipping reflectors are observed in seismic reflection profiles, confirmed by ODP drilling to be the result of the impedance contrasts between successive subaerial basalt flows. Secondly, volcanic margins are characterised by the presence of a thick wedge of high-velocity material at the base of the thinned continental crust, interpreted as the result of mafic underplating during continental breakup. In the northeast Atlantic, the thickness of this wedge varies systematically along the margin, with the thickest underplating in the vicinity of the inferred location of the Iceland plume at breakup time and thinner underplates in more distal regions (Fig. 12). This systematic variation provides strong evidence for a mantle plume control on the degree of magmatism, but is not seen in the other well-studied group of nonvolcanic margins, off the east coast of the United States of America. At volcanic margins, the transition to oceanic crust is fairly abrupt, and any transition zone is inferred to be the result of intense magmatic intrusion of the thinned continental crust, since seismic velocities are intermediate between those of the thinned continental crust and those of the adjacent oceanic crust. The first-formed oceanic crust seaward of the volcanic margins is also often anomalously thick.

10.2 Nonvolcanic Rifted Margins

Until recently, nonvolcanic rifted margins were thought to be typified by seismic velocity structures such as that inferred for the Goban Spur margin (Fig. 13a): continental crust thins from a full thickness of around 30 km to a thickness of 5–6 km, and adjacent to this lies a slightly thin oceanic crust with a velocity structure very similar to normal oceanic crust. However, several recent seismic refraction experiments, with

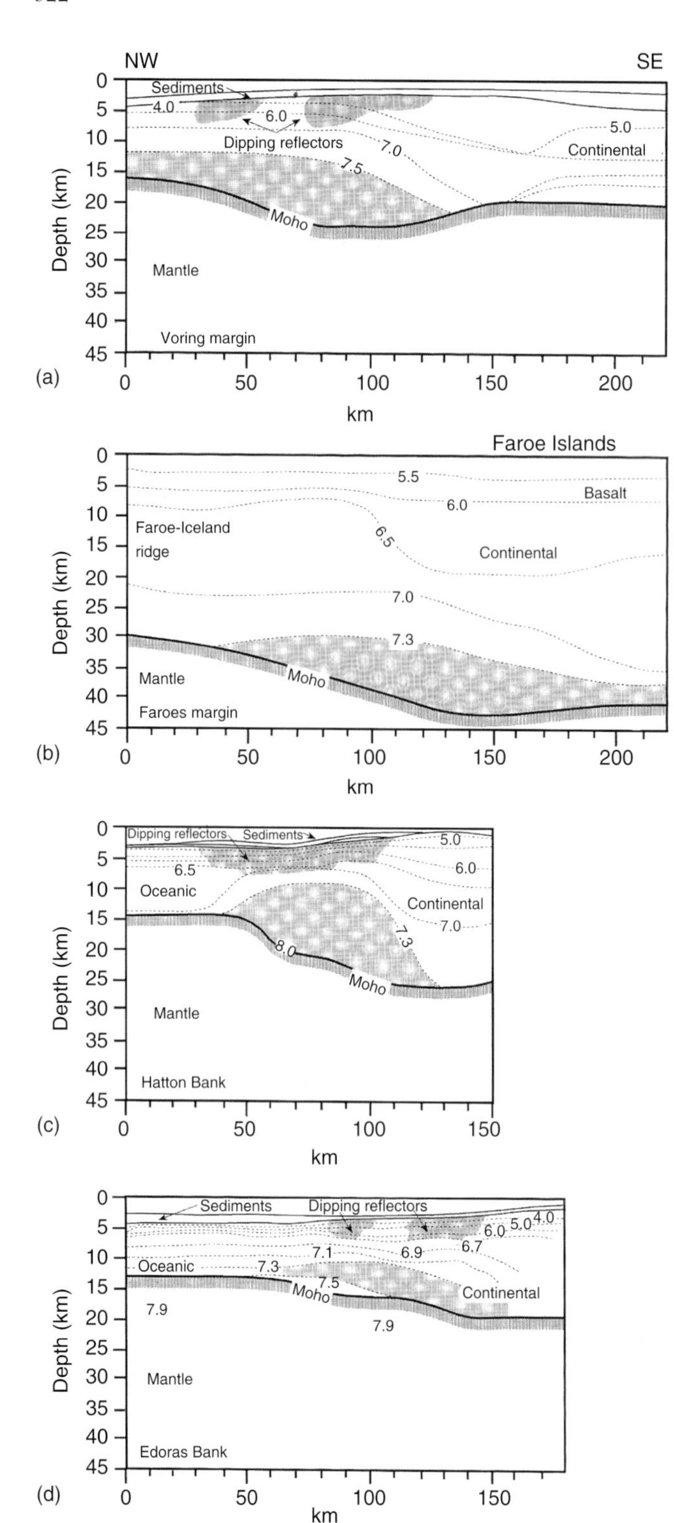

FIGURE 12 Seismic velocity models for volcanic continental margins off northwest Europe. Cross-hatching shows areas where seaward-dipping reflectors, interpreted as lava flows, have been imaged in seismic reflection profiles, and shaded regions are high-velocity lower crustal bodies interpreted as underplated igneous material [reprinted from Richardson *et al.* (1998). Copyright 1998, with permission from Elsevier Science].

a large number of closely spaced receivers to define the velocity structure in much greater detail, have detected a broad "transition zone" between thinned continental crust and unequivocal oceanic crust with an anomalous velocity structure which cannot be interpreted as the result of mafic intrusion of continental crust. This transition zone typically has low velocities in the upper part of the basement and high velocities (in the range 7.2–8 km sec^{-1}) at depths of 3–5 km (Fig. 13b). In this case the high basal velocities are commonly interpreted as due to partial serpentinisation of the uppermost mantle. Seismic reflection profiling at nonvolcanic margins has revealed much about the way the continental lithosphere deforms during the final stages of continental breakup. Tilted basement blocks overlain by wedges of synrift sediments are commonly observed; on some margins the faults bounding these blocks are also imaged, and these faults appear to sole out into a low-angle detachment fault marked by a high-amplitude subhorizontal reflector (e.g., Reston *et al.*, 1996).

10.3 Transform Margins

In contrast to rifted margins, transform margins have received rather little attention from seismic experiments. A detailed study of the southern Exmouth Plateau transform margin off northwest Australia found a thick high-velocity body is present at the base of the crust which resembles the mafic underplating observed at volcanic rifted margins. However, other studies at transform margins have not found underplated bodies, and instead have found a narrow zone of rather thin crust adjacent to the continent (e.g., Edwards *et al.*, 1997). In some cases this thin crust has high velocities at its base, and these velocities are interpreted as due to the presence of partially serpentinised mantle in an analogy to oceanic fracture zones.

11. Concluding Remarks

Modern marine seismic experiments are complex and very expensive operations, increasingly difficult to justify in a world of intense competition for research funds. The reconnaissance work of the 1950s to 1970s gave us our basic knowledge of the thickness and seismic structure of oceanic crust, but most profiles were acquired without an understanding of their plate tectonic setting, methods of analysis were limited, and spatial sampling was not sufficient to resolve subtle lateral variations which are now considered to be important. The more sophisticated experiments of the last two decades have provided us with insights into the fundamental processes which cause variations in the seismic velocity structure of the ocean basins, but by any standard the sampling of oceanic structure remains sparse, and heavily concentrated in the North Atlantic and northeast Pacific. Our view of what is typical may well be changed in the future by more extensive exploration of the

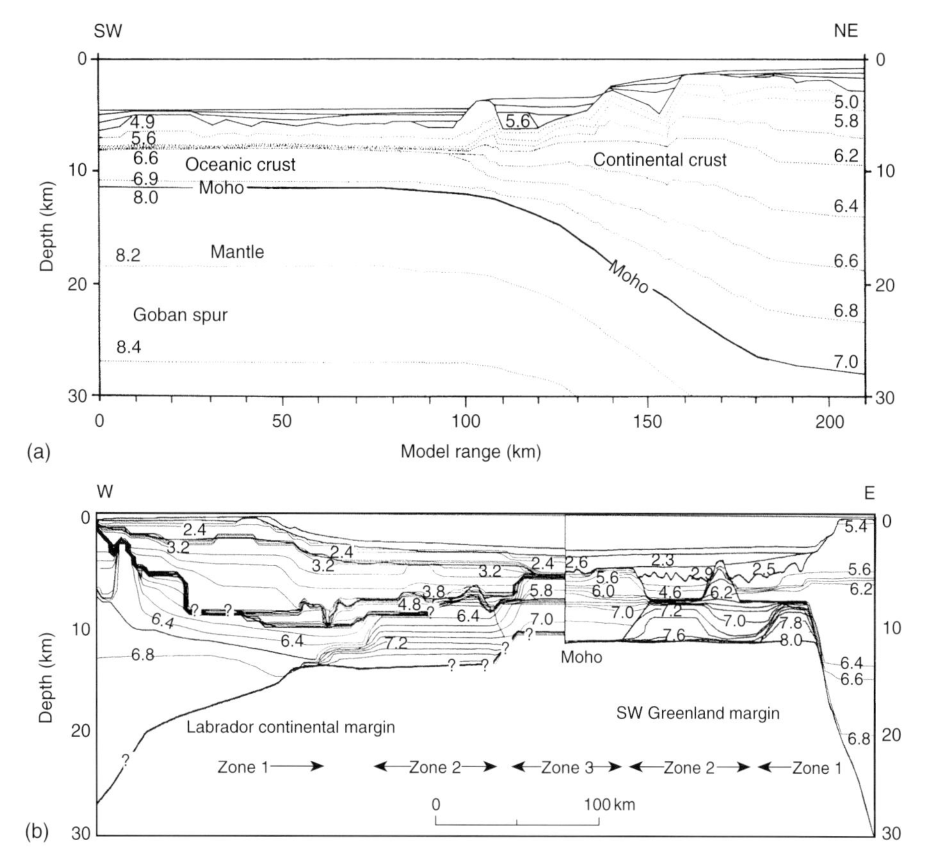

FIGURE 13 Seismic velocity models for nonvolcanic continental margins. (a) The Goban spur margin, with no interpreted ocean–continent transition zone [redrawn from Horsefield *et al.* (1993) with permission from Blackwell Science Ltd]. (b) Conjugate margins of the Labrador Sea (redrawn from Chian *et al.*, 1995). Here, zone 1 is thinned continental crust, zone 2 is the transition zone and zone 3 is oceanic crust.

Arctic Ocean, the South Atlantic, the Indian Ocean basins, the Southern Ocean, the southwest Pacific and the margins of these oceans. Even in the North Atlantic, it is clear that finer-scale sampling of the seismic structure is required to fully understand the complexities revealed by geological sampling of the ridge axis.

References

Barclay, A.H., D.R. Toomey, and S.C. Solomon (1998). Seismic structure and crustal magmatism at the Mid-Atlantic Ridge, 35° N. *J. Geophys. Res.* **103**, 17827–17844.

Begnaud, M.L., J.S. McClain, G.A. Barth, J.A. Orcutt, and A.J. Harding (1997). Preliminary velocity structure from forward modeling of the eastern ridge-transform intersection area of the Clipperton Fracture Zone, East Pacific Rise. *J. Geophys. Res.* **102**, 7803–7820.

Bown, J.W. and R.S. White (1994). Variation with spreading rate of oceanic crustal thickness and geochemistry. *Earth Planet. Sci. Lett.* **121**, 435–449.

Brandsdóttir, B., W. Menke, P. Einarsson, R.S. White, and R.K. Staples (1997). Faeroe–Iceland Ridge experiment 2. Crustal structure of the Krafla central volcano. *J. Geophys. Res.* **102**, 7867–7886.

Caress, D.W., M.K. McNutt, R.S. Detrick, and J.C. Mutter (1995). Seismic imaging of hotspot-related crustal underplating beneath the Marquesas Islands. *Nature* **373**, 600–603.

Chian, D., K.E. Louden, and I. Reid (1995). Crustal structure of the Labrador Sea conjugate margin and implications for the formation of nonvolcanic continental margins. *J. Geophys. Res.* **100**, 24239–24253.

Christeson, G.L., G.M. Purdy, and G.J. Fryer (1994). Seismic constraints on shallow crustal emplacement processes at the fast spreading East Pacific Rise. *J. Geophys. Res.* **99**, 17957–17973.

Collier, J.S. and M.C. Sinha (1992). Seismic mapping of a magma chamber beneath the Valu Fa Ridge, Lau Basin. *J. Geophys. Res.* **97**, 14031–14053.

Darbyshire, F.A., I.Th. Bjarnason, R.S. White, and O.G. Flóvenz (1998). Crustal structure above the Iceland mantle plume imaged by the ICEMELT refraction profile. *Geophys. J. Int.* **135**, 1131–1149.

Detrick, R., J. Collins, R. Stephen, and S. Swift (1994). In situ evidence for the nature of the seismic layer 2/3 boundary in oceanic crust. *Nature* **370**, 288–290.

Edwards, R.A., R.B. Whitmarsh, and R.A. Scrutton (1997). The crustal structure across the transform continental margin off Ghana, eastern equatorial Atlantic. *J. Geophys. Res.* **102**, 747–772.

Fryer, G.J., D.J. Miller, and P.A. Berge (1989). Seismic anisotropy and age-dependent structure of the upper oceanic crust. In: "The Evolution of Mid-Oceanic Ridges" (J.M. Sinton, Ed.), American Geophysical Union, Washington, DC.

Grevemeyer, I., W. Weigel, and C. Jennrich (1998). Structure and ageing of oceanic crust at 14° S on the East Pacific Rise. *Geophys. J. Int.* **135**, 573–584.

Horsefield, S.J., R.B. Whitmarsh, R.S. White, and J.-C. Sibuet (1993). Crustal structure of the Goban Spur rifted continental margin, NE Atlantic. *Geophys. J. Int.* **119**, 1–19.

Houtz, R. and J. Ewing (1976). Upper crustal structure as a function of plate age. *J. Geophys. Res.* **81**, 2490–2498.

Minshull, T.A., R.S. White, J.C. Mutter, P. Buhl, R.S. Detrick, C.A. Williams, and E. Morris (1991). Crustal structure of the Blake Spur Fracture Zone from expanding spread profiles. *J. Geophys. Res.* **96**, 9955–9984.

Minshull, T.A., M.R. Muller, C.J. Robinson, R.S. White, and M.J. Bickle (1998). Is the oceanic Moho a serpentinisation front? In: "Modern Ocean Floor Processes and the Geological Record" (R.A. Mills and K. Harrison, Eds.), Geological Society of London, Special Publications **148**, 71–80.

Mithal, R. and J.C. Mutter (1989). A low velocity zone within the layer 3 region of 118 Myr old oceanic crust in the western North Atlantic. *Geophys. J.* **97**, 275–294.

Navin, D., C. Peirce, and M.C. Sinha (1998). The RAMESSES experiment – II. Evidence for accumulated melt beneath a slow spreading ridge from wide-angle refraction and multichannel reflection seismic profiles. *Geophys. J. Int.* **135**, 746–772.

Operto, S. and P. Charvis (1996). Deep structure of the southern Kerguelen Plateau (southern Indian Ocean) from ocean bottom seismometer wide-angle seismic data. *J. Geophys. Res.* **101**, 25077–25103.

Peirce, C. and P.J. Barton (1991). Crustal structure of the Madeira-Tore Rise, eastern North Atlantic – results of a DOBS wide-angle and normal incidence seismic experiment in the Josephine Seamount region. *Geophys. J. Int.* **106**, 357–378.

Raitt, R.W. (1963). The crustal rocks. In: "The Sea," vol. 3 (M.N. Hill, Ed.), pp. 85–102. Wiley-Interscience, New York.

Reston, T.J., C.M. Krawczyk, and D. Klaeschen (1996). The S reflector west of Galicia: results of prestack depth migration – detachment faulting during continental breakup. *J. Geophys. Res.* **101**, 8075–8091.

Richardson, K.R., J.R. Smallwood, R.S. White, D.B. Snyder, and P.K.H. Maguire (1998). Crustal structure beneath the Faeroe Islands and the Faeroe-Iceland Ridge. *Tectonophysics* **300**, 159–180.

Ryan, M.P. (1988). The mechanics and three-dimensional internal structure of active magmatic systems: Kilauea Volcano, Hawaii. *J. Geophys. Res.* **93**, 4213–4248.

Salisbury, M.H. and N.I. Christensen (1978). The seismic velocity structure of a traverse through the Bay of Islands ophiolite complex, Newfoundland, an exposure of ancient oceanic crust and upper mantle. *J. Geophys. Res.* **83**, 805–817.

Shearer, P.M. (1988). Cracked media, Poisson's ratio, and the structure of the upper oceanic crust. *Geophys. J.* **92**, 357–362.

Shearer, P.M. and J.A. Orcutt (1985). Anisotropy in the oceanic lithosphere – theory and observations from the Ngendei seismic refraction experiment in the south-west Pacific. *Geophys. J. R. Astron. Soc.* **80**, 493–526.

Sinha, M.C. and K.E. Louden (1983). The Oceanographer fracture zone, I. Crustal structure from seismic refraction studies. *Geophys. J. R. Astron. Soc.* **75**, 713–736.

Sinha, M.C., K.E. Louden, and B. Parsons (1981). The crustal structure of the Madagascar Ridge. *Geophys. J. Astron. Soc.* **66**, 351–377.

Spudich, P. and J.A. Orcutt (1980). Petrology and porosity of an oceanic site: results from waveform modeling of seismic refraction data. *J. Geophys. Res.* **85**, 1409–1433.

Toomey, D.R., G.M. Purdy, S.C. Solomon, and W.S.D. Wilcock (1990). The three-dimensional seismic velocity structure of the East Pacific Rise near latitude 9°30 N. *Nature* **347**, 639–645.

Vera, E.E., J.C. Mutter, P. Buhl, J.A. Orcutt, A.J. Harding, M.E. Kappus, R.S. Detrick, and T.M Brocher (1990). The structure of 0- to 0.2-My-old oceanic crust at 9° N on the East Pacific Rise from expanded spread profiles. *J. Geophys. Res.* **95**, 15529–15556.

Watts, A.B., U.S. ten Brink, P. Buhl, and T.M. Brocher (1985). A multichannel seismic study of lithospheric flexure across the Hawaiian-Emporer seamount chain. *Nature* **315**, 105–111.

Watts, A.B., C. Peirce, J. Collier, R. Dalwood, J.P. Canales, and T.J. Henstock (1997). A seismic study of lithospheric flexure in the vicinity of Tenerife, Canary Islands. *Earth Planet. Sci. Lett.* **146**, 431–447.

Wepfer, W.W. and N.I. Christensen (1991). Q structure of the oceanic crust. *Mar. Geophys. Res.* **13**, 227–237.

White, R.S., D. McKenzie, and R.K. O'Nions (1992). Oceanic crustal thickness from seismic measurements and rare earth element inversions. *J. Geophys. Res.* **97**, 19683–19715.

Whitmarsh, R.B. and A.J. Calvert (1986). Crustal structure of Atlantic fracture zones – I. The Charlie–Gibbs fracture zone. *Geophys. J. R. Astron. Soc.* **85**, 107–138.

Whitmarsh, R.B., A. Ginzburg, and R.C. Searle (1982). The structure and origin of the Azores-Biscay Rise, north-east Atlantic Ocean. *Geophys. J. R. Astron. Soc.* **70**, 79–107.

Editor's Note

Due to space limitations, the full manuscript including additional text, figures, and references is given in MinshullFull.pdf on the attached Handbook CD, under the directory \55Minshull. Please see also Chapter 27, Marine seismology, by Suyehiro and Mochizuki, and Chapter 54, Seismic velocity structure of the continental lithosphere from controlled source data, by Mooney, Prodehl, and Pavlenkova.

56

The Earth's Core

Xiaodong Song
University of Illinois, Urbana, IL, USA

1. Introduction

The purpose of this chapter is to present a brief review of our basic understanding of the Earth's core, which was acquired almost entirely during the 20th century. The Earth's core plays an important role in the formation and chemical differentiation of our planet, in global earth dynamics and earth rotation, and in the generation of the magnetic field (see Table 1 for a summary of our most basic knowledge about the core).

The review emphasizes seismological accomplishments. For in-depth discussions on various aspects of core studies, see previous reviews on seismology (Bolt, 1987; Masters, 1991; Tromp, 1995a; Song, 1997; Creager, 2000; Richards, 2000; Romanowicz and Durek, 2000), mineral physics (Jeanloz, 1990; Boehler, 1996; Stixrude and Brown, 1998), core formation (Ringwood, 1984; Jones and Drake, 1986; Jacobs, 1987; Stevenson, 1987; Labrosse *et al.*, 1997), light elements in the core (Jeanloz, 1990; Poirier, 1994a; Allègre *et al.*, 1995), physical properties of the core (Poirier, 1994b; Anderson, 1995; Braginsky and Roberts, 1995), core dynamics (Glatzmaier and Roberts, 1998; Buffett, 2000), and core–mantle boundary (CMB), and core–mantle interactions (Young and Lay, 1987; Lay *et al.*, 1998; and Chapter 51 by T. Lay).

2. Discovery of the Earth's Core

The discovery of the Earth's core was contemporaneous with the birth of modern instrumental seismology in the early part of the 20th century. An elegant account of the discovery and characterization of the Earth's core was presented in a historical memoir by Brush (1980).

In 1896 Wiechert proposed a two-layered model of the Earth with an iron core and a stony shell, inferred from the Earth's rotation and gravity. He argued that a difference in chemical composition in the Earth's interior is required to explain the difference, by a factor of two, between the average density of the Earth and the density of rocks near the surface. The direct evidence for the Earth's core came from seismology in 1906 when Oldham (1906) observed abrupt changes in P and S velocities beyond about 120°. He postulated that a low-velocity region in the interior, the central core, produced a shadow zone. Gutenberg (1913) accurately estimated a depth to the core of 2900 km in 1912 from P-wave travel times. By 1926 Jeffreys had showed that the combination of tidal deformation and the rigidity of the mantle inferred from seismic velocities implies a low rigidity (possibly zero) in the central core. A liquid core explains the absence of observable S waves traversing the core.

TABLE 1 A Summary of the Earth's Core

- The Earth's core was formed very early in Earth's history as heavy molten iron alloy migrated toward the center of the planet. High temperatures (~5000 K) keep the bulk of the core liquid.
- As the Earth cooled and dissipated its internal heat toward the surface through mantle convection, molten iron began to solidify under enormous pressure (over 3 Mbar) to form the solid inner core.
- The fluid outer core is predominantly iron but cannot be purely iron; the inner core is almost pure iron.
- The size of the outer core is more than a half of the Earth in radius. The size of inner core is about one fifth of the Earth in radius and slightly smaller than the moon.
- The Earth's core was discovered by Oldham in 1906 and the inner core was discovered by Lehmann in 1936 using seismic waves.
- Jeffeys proved the outer core is liquid in 1926 and Dziewonski and Gilbert proved the inner core is solid in 1971.
- The gravitational energy release from the slow growth of the inner core as the liquid iron freezes drives the convection in the outer core which generates the Earth's magnetic field.
- The inner core is elasticly anisotropic with the fast direction approximately parallel to the spin axis. The anisotropy comes from preferred alignment of iron crystals in the inner core.
- The inner core is suggested to be rotating faster than the mantle at a fraction of a degree per year.

INTERNATIONAL HANDBOOK OF EARTHQUAKE AND ENGINEERING SEISMOLOGY, VOLUME 81A ISBN: 0-12-440652-1

TABLE 2 Notation Used for Seismic Core Phases

Notation	Description
P	Compressional wave in the mantle or compressional wave in general
S	Shear wave in the mantle or shear wave in general
K	Compressional wave in the core (after "Kern," a German word meaning core)
I	Compressional wave in the inner core
J	Shear wave in the inner core
c	Topside reflection from the core–mantle boundary
i	Topside reflection from the inner core boundary
PcP	Compressional wave reflected from the core–mantle boundary
ScS	Shear wave reflected from the core–mantle boundary
PKP or *P′*	Travelling from the source through the mantle, entering the core, and coming back up through the mantle again to the station at the Earth's surface all as compressional wave
PKP(AB)	A branch (*AB*) of *PKP* that turns in midouter core
PKP(BC)	A branch (*BC*) of *PKP* that turns near the bottom of the outer core
PKP(CD) or *PKiKP*	A branch (*CD*) of *PKP* that is reflected from the inner core boundary
PKP(DF) or PKIKP	A branch (*DF*) of *PKP* that transverses the inner core as compressional wave
PmKP	Short hand for *PKP* multiples (*PKP*, *PKKP*, *PKKKP*, etc.). As an example, *PKKP* has one underside (i.e., internal) reflection at the core–mantle boundary (*K*-to-*K*).
PKJKP	Compressional wave that converts to shear wave from the outer core to the inner core that converts back to compressional wave as it re-enters the outer core
PKIIKP	One underside reflection from *I* to *I* at the inner core boundary
SKS	Shear wave through the mantle, compressional wave through the core, and shear wave back up through the mantle
SmKS	Short hand for *SKS* multiple (*SKS*, *SKKS*, *SKKKS*, etc.).

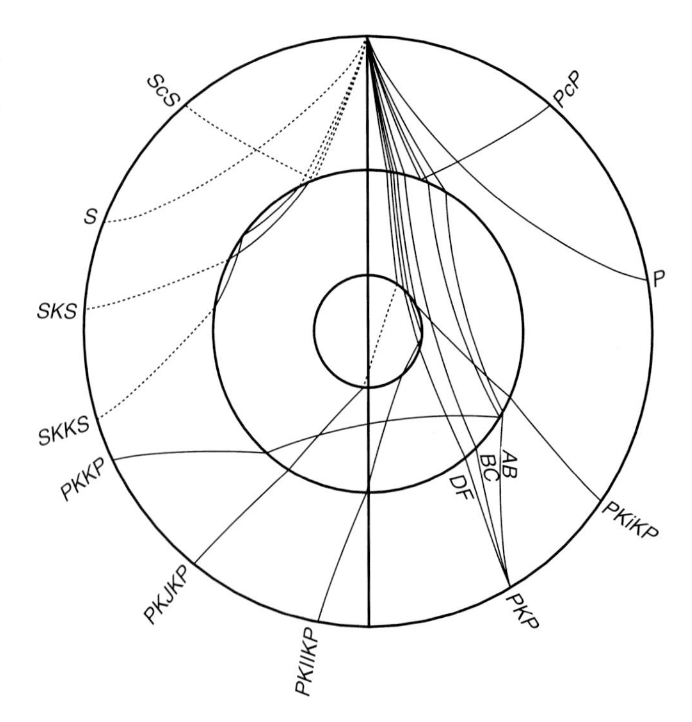

FIGURE 1 Ray paths of selected seismic phases. The three circles represent the surface, the core–mantle boundary, and the inner core boundary, respectively. The solid segments of the ray paths represent compressional waves and the dashed segments represent shear waves. See also Table 2 for notation of the phases.

The inner core was hypothesized by Lehmann in 1936 from the presence of waves in the core shadow zone. The waves (P', to denote the compressional waves that pass through the mantle and the core) could not be explained by the alternative explanation of diffraction. See Table 2 for the notation used for selected seismic core phases and Figure 1 for their ray paths. Assuming a simplified model with constant velocities in the mantle, outer core, and the inner core, she inferred the radius and the velocity of the inner core that explained the observed P' travel times reasonably well.

Birch (1940) conjectured that the inner core might be solid; it might simply be a solidified phase of iron. Bullen (1946) argued that the inner core must be solid from the marked increase of P velocities across the inner-core boundary (ICB). The P velocity $V_P = [(K + 4/3\mu)/\rho]^{\frac{1}{2}}$, where K, μ, and ρ are bulk modulus, shear modulus, and density, respectively. Since $\mu = 0$ in the outer core and ρ would not decrease with depth and K would change only slowly at high pressures, the only plausible way to increase V_P markedly is to give μ in the inner core a finite value. It was not until the beginning of 1970s that solidity of the inner core was demonstrated by the existence of finite rigidity affecting the periods of core-sensitive normal modes (Dziewonski and Gilbert, 1971). The more recent discovery that the inner core is anisotropic (see discussion below) also requires finite rigidity in the inner core.

3. Composition

The distinct difference in composition between meteorites and rocks from the Earth's mantle virtually requires a major repository for iron and nickel, i.e., in the Earth's core. Using sound velocity and density measurements obtained from shock-wave experiments for various elements, Birch (1961, 1964) conclusively showed that iron is the only element compatible with the density and pressure of the core. The idea that the Earth's core consists mainly of iron and nickel alloys has since been well accepted. On the basis of cosmic abundances, the nickel content of the core is a few percent of the iron content. However, the Earth's core cannot be made of pure iron based on the comparison of the density and sound velocity determined from seismology with laboratory measurements and the equation of state for pure iron. The outer core is about 10% less dense than pure iron at the core conditions, requiring the presence of one or more light elements in the outer core (Birch, 1952). The nature and proportions of these light elements have been the subject of an ongoing active debate (e.g., Brett, 1976; Stevenson, 1981; Jacobs, 1987, 1992; Ringwood, 1984; Jeanloz, 1990; Wood, 1993; Poirier, 1994a,b; Allègre *et al.*, 1995). Possible light elements include Si, S, O, H, and C. Two major considerations on the choice of light element are cosmic abundances and metallurgical constraints. The nature of the light elements in the core has important implications for the temperature profile in the core, for the compositional convection in the outer core (which is thought to drive the geodynamo), and for the formation and evolution of the core.

In contrast with the outer core, the elastic properties and density of the inner core determined from seismology are consistent with those of pure solid iron with some nickel (and perhaps with some, but much smaller, amount of light elements) (e.g., Anderson, 1986; Jephcoat and Olson, 1987; Stixrude *et al.*, 1997).

4. Earth's Core and Geodynamo

It was long ago recognized that the Earth is magnetic. For example, the first compass was invented in China two thousand years ago (Needham, 1962). However, the origin of the Earth's magnetic field had not been convincingly explained until the 20th century [see review by Glatzmaier and Roberts (1998)]. The seismological discovery of the fluid core (Oldham, 1906) marked a key turning point in the understanding of the origin of the geomagnetic field. The combination of fluidity and electrical conductivity soon allowed Larmor (1919) to propose that the Earth's magnetic field is generated by fluid motions in the core through self-excited dynamo action. The concept of a geodynamo is well accepted today, notably through pioneering work by Elsasser, Bullard, and others in 1940s and 1950s.

The dynamo problem is notoriously difficult, partly due to the famous Cowling (1934) theorem, which states that an axisymmetric magnetic field cannot be maintained by a symmetric motion. Therefore, any realistic dynamo model has to be 3D, a task far more difficult to handle. Nevertheless, thanks to the rapid advancement of computer technology, numerical simulations of self-consistent 3D models of the geodynamo have been obtained (Glatzmaier and Roberts, 1995a; Kuang and Bloxham, 1997). The magnetic field produced possesses several features resembling the geomagnetic field. Most remarkably, geodynamo simulations containing a finite conducting inner core have been shown to undergo a magnetic reversal (Glatzmaier and Roberts, 1995a,b), similar to those observed in the geomagnetic field over the Earth's history. The Glatzmaier and Roberts (1995a) dynamo also predicted a fast superrotation of the inner core, which was later observed by seismology (see below).

Despite its small size compared to the outer core, the inner core plays a role in the geodynamo. The favored source of energy for driving the geodynamo is the gravitational energy derived from the release of light constituents as the liquid iron freezes to form the solid inner core (Braginsky, 1963; Gubbins, 1977; Loper and Roberts, 1978; Loper, 1978; Lister and Buffett, 1995). The solid inner core has been shown to stabilize the dynamo and establish a constant polarity of the field such as is found on the Earth (Hollerbach and Jones, 1993; Glatzmaier and Roberts, 1995a; Sakuraba and Kono, 1999).

5. Outer Core Structure

Due to its vigorous convection, the liquid outer core is generally considered to be laterally homogeneous in density and seismic velocity (Stevenson, 1987), except perhaps near the boundaries (Wahr and de Vries, 1989). By 1939, Jeffreys (1939a,b) had developed a velocity model, which is still used as a standard reference today [known as the J-B table from a later summary by Jeffreys and Bullen (1940)]. The difference in *P*-wave velocity in the bulk of the outer core between the Jeffreys's model and models proposed later is quite small (within about 1%). The main areas of contention are near the boundaries (the ICB and the CMB) (e.g., see a comparison of *P*-velocity models in Song and Helmberger, 1992).

The nature of the ICB has been one of the main topics since the discovery of the inner core in 1936. The key issues include the sharpness of the boundary, the velocity and density jump at the boundary, the velocity gradient above and below the boundary, and the *Q* structure near the boundary (e.g., Bolt, 1987; Cummins and Johnson, 1988; Song and Helmberger, 1992). The inner core appears sharp since short-period (1 sec) vertical reflections from the boundary *PKiKP* can clearly be observed (Engdahl *et al.*, 1970). These short-period *PKiKP* reflections provide excellent observations for estimating the inner-core radius (Engdahl *et al.*, 1970, 1974) and for estimating the density jump at the ICB (using *PKIKP*/*PcP* ratios) (Bolt and Qamar, 1970; Bolt, 1972; Souriau and Souriau, 1989; Shearer and Masters, 1990). Observations of *PKP* travel times and waveforms seem to suggest a low to zero velocity gradient at the base of the fluid core (Souriau and Poupinet, 1991; Song and Helmberger, 1992, 1995a,b; Kennett *et al.*, 1995). Such a low velocity gradient implies inhomogeneity at the base of the fluid core (Song and Helmberger, 1995a,b). A transition zone structure at the base of the fluid core with distinct discontinuity about 420 km above the ICB (the so-called F layer) was proposed (Bolt, 1962, 1964) to explain short-period precursors to *PKP* at distances near 130°. The idea of a distinct discontinuity in the fluid core is now generally discarded and the *PKP* precursors have been successfully interpreted as scattering from the lowermost mantle region (Cleary and Haddon, 1972; Haddon and Cleary, 1973, 1974) or perhaps from the whole mantle (Hedlin *et al.*, 1997).

The light elements released from the compositional convection driven by the crystallization of the solid inner core may form a low-density layer at the top of the outer core (Braginsky, 1963; Fearn and Loper, 1981; Lister and Buffett, 1998; Braginsky, 1999), which has consequences for the core convection and the geodynamo (Bloxham, 1990). Seismological evidence for such a stably stratified layer has been reported but remains inconclusive (Lay and Young, 1990; Souriau and Poupinet, 1990, 1991; Kohler and Tanimoto, 1992; Tanaka and Hamaguchi, 1993; Garnero *et al.*, 1993; Garnero and Helmberger, 1994; Garnero and Lay, 1998). Because of the sharp decrease in *P* velocity from the mantle to the core, the only seismic phases with ray paths bottoming the outermost 200 km of the core are the suite of *SmKS* (*SKS*, *SKKS*, etc.). Studies of the outermost outer core rely on differential *SKS* times (*SKKS–SKS*, etc.), which reduces biases from upper mantle heterogeneity. However, *SKS* differential times are still subject to strong lateral variations on the mantle side of the CMB, making inferences on core-side structure difficult.

The quality factor Q of the outer core is apparently extremely high since high-frequency *PmKP* phases (up to *P7KP*) are observable (Qamar and Eisenberg, 1974). The Q_α in the outer core is at least several to ten thousand and probably higher from spectral ratios of seismic core phases (Buchbinder, 1972; Sacks, 1972; Qamar and Eisenberg, 1974; Cormier and Richards, 1976; Tanaka and Hamaguchi, 1996).

6. Inner Core Structure

Discovered by Lehmann (1936), the inner core has turned up a number of surprises in recent years: it is anisotropic, it stabilizes the Earth's magnetic field, it is rotating relative to the mantle, it is layered, and it is strongly heterogeneous.

6.1 Inner Core Anisotropy

The first evidence for strong aspherical structure within the core came from anomalous splitting (not explainable simply by Coriolis and ellipticity effects) of the normal modes (Masters and Gilbert, 1981), and anomalous *PKIKP* arrival times from the International Seismological Centre (ISC) Bulletins (Poupinet *et al.*, 1983). Anisotropy of the inner core was first proposed in 1986 to explain observations of ISC arrival time anomalies (Morelli *et al.*, 1986) and anomalous normal mode splitting (Woodhouse *et al.*, 1986). Morelli *et al.* (1986) and Woodhouse *et al.* (1986) hypothesized that the inner core is anisotropic with cylindrical symmetry around the Earth's spin axis.

By the mid-1990s, the presence of significant anisotropy in the inner core was well established with evidence from both travel time studies (Creager, 1992; Song and Helmberger, 1993; Shearer, 1994) and normal mode data (Tromp, 1993, 1995b,c). The anisotropy appears to be dominated by cylindrical anisotropy with the axis of symmetry aligned approximately with the north–south spin axis. Compressional waves traversing the inner core along the north–south direction are about 3% faster than those propagating parallel to the equatorial plane. For a summary of the models of inner-core anisotropy parameters, see Song (1997). The strongest evidence for the inner-core anisotropy comes from differential travel times *PKP*(*BC*) – *PKP*(*DF*) (or simply *BC-DF*) (Creager, 1992; Song and Helmberger, 1993) (an example is given in Fig. 2). Quasi north–south paths show consistent *BC-DF* travel time residuals, relative to the Preliminary Earth Model (PREM) (Dziewonski and Anderson, 1981), as large as 4 sec at a distance of around 150° (where *BC* is observable). Because the *BC* and *DF* phases have very similar ray paths in the mantle and the outer core, using the differential times greatly reduces the biases from event mislocations and mantle heterogeneity. Although small-scale heterogeneity in the lowermost mantle may still influence the *BC-DF* times (Breger *et al.*, 1999), the observed large *BC-DF* anomalies require significant anisotropy in the inner core.

The inner-core anisotropy appears to vary both laterally and with depth. Differential *AB-DF* times suggest that significant anisotropy appears to extend to the center of the Earth (Vinnik *et al.*, 1994; Su and Dziewonski, 1995; Song, 1996; McSweeney *et al.*, 1997). However, the outermost ~150 km of the inner core is nearly isotropic (Shearer, 1994; Song and Helmberger, 1995a,b). There appears to be a sharp transition from the isotropic upper inner core to the anisotropic lower inner core producing a seismic triplication along certain north–south paths (Song and Helmberger, 1998). The inner-core anisotropy appears to vary laterally on a hemispherical scale (Tanaka and Hamaguchi, 1997; Creager, 1999) and on a few hundred kilometers scale (Creager, 1997; Song, 2000a). Vidale and Earle (2000) observed 200 sec of waves scattered from the inner core following the *PKiKP* arrivals, suggesting strong heterogeneity even on a scale of a few kilometers. If one can assume uniform anisotropy with cylindrical symmetry (despite the strong heterogeneity the inner core seems to exhibit), *PKP* travel time data seem to favor a tilt of about 10° from the spin axis (Su and Dziewonski, 1995; Song and Richards, 1996; McSweeney *et al.*, 1997), although the resolution of the tilt is poor at present (Souriau *et al.*, 1997).

Mineralogical studies favor ε (hexagonal close-packed) over γ (face-centered cubic) iron as the stable phase of iron in the inner core (e.g., Brown and McQueen, 1986; Anderson, 1986; Jephcoat and Olson, 1987; Sayers, 1989; Stixrude and Cohen, 1995). The cause of the inner-core anisotropy is believed to be the preferred orientation of ε iron. The

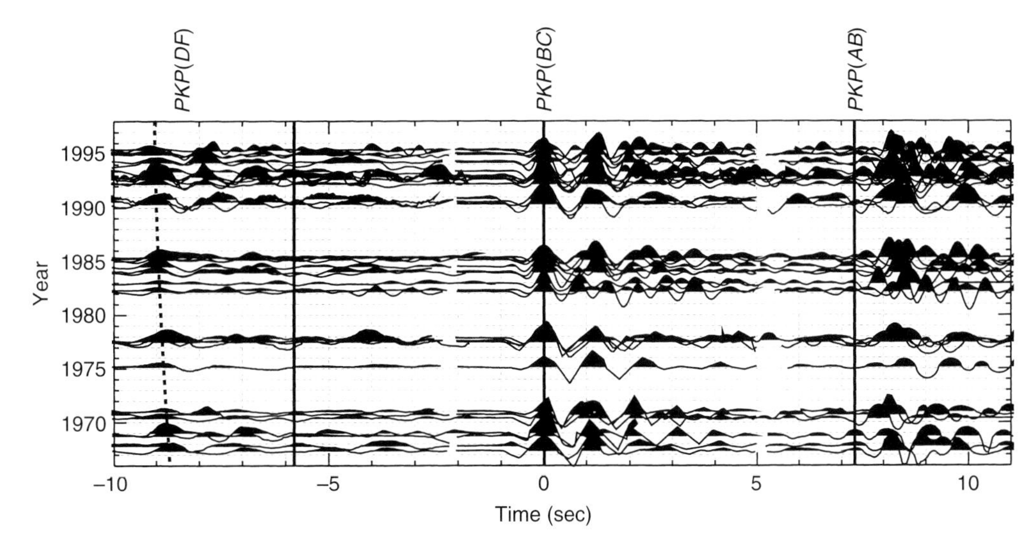

FIGURE 2 A set of short-period (1 sec) *PKP* seismograms from earthquakes in South Sandwich Islands recorded at College, Alaska station over a period of 28 y. Three branches of *PKP* (as labeled) are visible in all the traces: *PKP*(*DF*) that traverses the inner core, *PKP*(*BC*) that turns near the base of the outer core, and *PKP*(*AB*) that turns in the midouter core. The records are plotted with respect to earthquake origin times (from 1967 to 1995) and are aligned with *PKP*(*BC*). The *PKP*(*DF*) and *PKP*(*AB*) waveforms are windowed out and corrected to a reference distance of 151° and depth of 0 km using the Preliminary Reference Earth Model (PREM) (Dziewonski and Anderson, 1981). The *PKP*(*DF*) amplitudes are enlarged by 5 times and the *PKP*(*AB*) arrivals are corrected by $\pi/2$ phase shift. The positive amplitudes of the traces are highlighted for better visual comparisons. The three vertical solid lines mark the predictions of the relative times between the three branches of *PKP* for PREM at the reference distance and depth. The set of seismograms are unusual in the following way. (1) The *PKP*(*DF*) waves arrive anomalously earlier than the PREM predictions (by about 3 sec), which is believed to be the results of the inner-core anisotropy. (2) It is clear that the *PKP*(*DF*) waves arrive progressively earlier from the late 1960s to the 1990s (by about 0.3 sec over the 28 y). The time-dependence of the *PKP*(*DF*) waves from these 38 seismograms was the strongest evidence for the differential rotation of the inner core originally reported by Song and Richards (1996). The dashed line shows the predictions for their best-fitting model of inner core rotation. (3) The *PKP*(*DF*) amplitudes are anomalous small (about 10 times smaller than the *PKP*(*BC*) amplitudes) and the *PKP*(*DF*) waveforms are rather complex, which is observed in many other seismograms along north–south paths (Creager, 1992; Song and Helmberger, 1993). The complexity may be caused by the transitional structure within the inner core (Song and Helmberger, 1998), but no convincing interpretation has been provided.

mechanism, however, for producing such a preferred alignment of iron crystals is uncertain. The proposed models of preferred alignment include two categories: (1) alignment established during the solidification of iron crystals at the surface of the inner core (Karato, 1993; Bergman, 1997); (2) alignment caused by large-scale convective flow in the inner core, under thermal convection (Jeanloz and Wenk, 1988), under differential growth of the inner core (Yoshida *et al.*, 1996), or under Maxwell stress from the magnetic field (Karato, 1999).

6.2 Inner Core Attenuation

The inner core is strongly attenuating: The attenuation is comparable to two other highly attenuating regions of the Earth: the asthenosphere and the shallow crust. The estimated Q_α for the *P* waves in the inner core is around 200–600 from amplitude ratios between *PKP* branches (Buchbinder, 1971; Sacks, 1972; Doornbos, 1974; Anderson and Hart, 1978; Bolt, 1977; Cormier, 1981; Choy and Cormier, 1983; Bhattacharyya *et al.*, 1993; Niazi and Johnson, 1992; Souriau and Roudil, 1995; Song and Helmberger, 1995a,b; Cormier *et al.*, 1998) and from the underside inner-core reflected waves *PKIIKP* (Masse *et al.*, 1974; Doornbos, 1974; Bolt, 1977).

The estimate of Q_μ from inner core shear modes ranges from 3000 (Buland and Gilbert, 1978; Masters and Gilbert, 1981; Suda and Fukao, 1990) to a much lower value of 110 (Widmer *et al.*, 1991). If the inner core was formed by slow solidification of the outer core as the Earth cooled, one may expect a "mushy zone" of dendrites in the inner core (Fearn *et al.*, 1981; Loper and Fearn, 1983) and *Q* to increase with depth. The depth dependence of *Q* in the inner core was inferred (Doornbos, 1974; Choy and Cormier, 1983; Song and

Helmberger, 1995a,b) but remains inconclusive (Cormier, 1981; Niazi and Johnson, 1992; Bhattacharyya *et al.*, 1993).

6.3 Inner Core Shear Waves

Although the solidity of the inner core is well accepted, the direct identification of inner core shear waves, such as *PKJKP* with a shear wave leg *J* in the inner core, remains one of the outstanding problems of seismology. The difficulty of detecting *PKJKP* is caused by the inefficiency of *P*-to-*S* (and *S*-to-*P*) conversion at the inner-core boundary. Detection of *PKJKP* is also hindered by high attenuation at short periods, causing Doornbos (1974) to conclude that the amplitude of *PKJKP* is too small to be observed.

The reported identification by Julian *et al.* (1972) was later thought not to be *PKJKP* because its travel times do not agree with those predicted by normal-mode models. The inner core *S*-wave velocity deduced by Julian *et al.* (1972) is $2.95 \pm 0.1\,\mathrm{km\,sec^{-1}}$, more than 10% slower than the 3.5–3.6 $\mathrm{km\,sec^{-1}}$ reported by Dziewonski and Gilbert (1972) from free oscillations. Recently, Deuss *et al.* (2000) and Okal and Cansi (1998) reported observations of inner core shear waves consistent with an inner core shear velocity of 3.6 $\mathrm{km\,sec^{-1}}$.

7. Inner Core Rotation

The idea that the solid inner core, at the center of a much larger fluid outer core of low viscosity (Gans, 1972; Poirier, 1988), may rotate differently from the daily rotation of the mantle came from studies of the geodynamo (Gubbins, 1981; Glatzmaier and Roberts, 1995a,b). Electromagnetic torque between the magnetic field generated in the liquid outer core and the conducting inner core is expected to drive the inner core to rotate (Gubbins, 1981; Glatzmaier and Roberts, 1995b, 1996; Aurnou *et al.*, 1996; Kuang and Bloxham, 1997).

Song and Richards (1996) first reported evidence for differential rotation of the inner core from time-dependent *BC-DF* differential times along certain earthquake-station pathways. In particular, they found that the *BC-DF* differential times along the pathway from earthquakes in the South Sandwich Islands (SSI) to the station COL at College, Alaska, have increased systematically over 28 y by about 0.3 sec (Fig. 2). They interpreted the temporal change as evidence for a differential inner-core rotation, which shifts the orientation of the inner core's anisotropy. Assuming a homogeneous inner-core anisotropy model with the anisotropy symmetry (fast) axis tilted from the spin axis (Su and Dziewonski, 1995; Song and Richards, 1996; McSweeney *et al.*, 1997), they estimated the rotation rate to be about $1^\circ\,\mathrm{y^{-1}}$.

However, determining the inner-core rotation from the tilt of anisotropy axis may be problematic because the anisotropy symmetry axis may not be reliably determined (Souriau *et al.*, 1997; Dziewonski and Su, 1998). A much faster rotation rate of $3^\circ\,\mathrm{y^{-1}}$ was reported by Su *et al.* (1996) from inversions for locations of the anisotropy axis using International Seismological Centre (ISC) arrival times. But a reassessment by Dziewonski and Su (1998) of the temporal variation of the total pattern of the inner core anomalies suggests that the inner-core rotation is not detectable from the ISC data.

Repicking the differential times at COL, Creager (1997) confirmed the time-dependence observed by Song and Richards (1996). He also observed a steep lateral gradient in the anisotropy of the part of the inner core sampled by SSI–Alaska ray paths. He provided an alternative interpretation that the temporal change at COL is caused by a shift of the local velocity gradient from the inner-core rotation and obtained much smaller differential rotation rate of 0.2–$0.3^\circ\,\mathrm{y^{-1}}$.

More recent body-wave studies generally support inner-core rotation with a rotation rate of a few tenths of a degree per year. The data at COL were extended to cover a period of 45-years and the time-dependence was found to persist to earlier years (Song, 2000a). In addition, three other stations in Alaska also show systematic increases in differential *BC-DF* times from SSI earthquakes during the past two decades. Surprisingly, the inner-core rotation seems detectable from digital records at the dense Alaska Seismic Network from SSI earthquakes in the 1990s alone (spanning only 8 years). Significant systematic changes in differential *BC-DF* times were also observed from paths from nuclear explosions in Novaya Zemlya in the former Soviet Union to station NVL in Antarctica (Ovchinnikov *et al.*, 1998), and from Alaska earthquakes to the South Pole station (Song and Li, 2000). Weak time-dependence was observed by Li and Richards (1998) in the differential *AB-DF* times from nuclear explosions in Novaya Zemlya recorded at Scott Base, Antarctica. Using a new approach with newly identified inner-core scattered waves, Vidale *et al.* (2000) observed small but coherent shifts in the scattered waves recorded at LASA from two nearby nuclear tests at Novaya Zelmya spanning 3 years. The pattern of the scattered waves suggests a slow inner-core rotation rate of $0.15^\circ\,\mathrm{y^{-1}}$.

The claim that the inner core is rotating relative to the mantle faces serious challenges and is under debate. Firstly, some attempts have failed. Souriau (1998a) examined *BC-DF* times from Novaya Zemlya nuclear explosions recorded at DRV, Antarctica over 24 y and did not observed robust temporal change. She concluded that an eastward rotation of $3^\circ\,\mathrm{y^{-1}}$ (as in Su *et al.*, 1996) is not possible, but a rate of $1^\circ\,\mathrm{y^{-1}}$ or lower could not be ruled out. Results from normal-mode studies on the inner-core rotation are uncertain. Sharrock and Woodhouse (1998) appear to favor a westward rotation component; but Laske and Masters (1999) found that inner-core rotation is essentially zero (to within 1σ error of $\pm 0.2^\circ\,\mathrm{y^{-1}}$) over the last 20 y. Secondly, the detection of the inner-core rotation has been questioned on the grounds of potential biases from event mislocations, mantle heterogeneity, and heterogeneous event magnitudes (Souriau *et al.*, 1997; Souriau, 1998a,b,c; Poupinet *et al.*, 2000).

In particular, Poupinet *et al.* (2000) used a novel "doublet" technique to examine 12 pairs of *PKP* seismograms from the SSI–COL path. They concluded that the apparent variations with time along the SSI–COL path observed by Song and Richards (1996) are artifacts due to poor earthquake locations. Since the temporal change along the SSI–COL path remains as one of the most convincing observations, the conclusion by Poupinet *et al.* (2000) is very significant in that it questioned directly the existence of a differential inner-core rotation. These criticisms were addressed by Song (2000b, 2001) and Song and Li (2000). Song (2001) suggested that the analyses of Poupinet *et al.* (2000) were flawed; and a re-examination of the SSI–COL path by Song (2000b) using the technique of Poupinet *et al.* (2000) concluded the opposite. Thirdly, theoretical calculations suggest that the inner core may be "locked" to the mantle by gravitational coupling between the mantle and small topography of the inner core induced by mantle heterogeneity (Buffett, 1996) unless the inner core is "soft" so that it continually adjusts its shape as it rotates (Buffett, 1997) or the electromagnetic torque is stronger (Aurnou and Olson, 2000).

With a unique opportunity to observe motions at depth within the core, a detectable inner-core rotation has direct implications for geodynamo studies (e.g., Aurnou *et al.*, 1996; Glatzmaier and Roberts, 1996; Hollerbach, 1998; Sakuraba and Kono, 1999; Aurnou and Olson, 2000).

8. Looking Ahead

From the discovery of the Earth's core less than a 100 years ago to the recent claim of the inner-core rotation, a fascinating journey through the center of the Earth was witnessed in the last century. The breakthroughs made as recently as in the 1990s on many fronts, including core dynamics, high-pressure experiments, and seismology, testify to an equally exciting time ahead. Seismology is likely to continue to play a key role in this journey. We have yet to answer some of the most fundamental questions concerning the core. Is there a low-density stratified layer at the topmost outer core? What is the nature of the low-velocity gradient region at the lowermost outer core? Can seismology resolve the likely small lateral variations in the fluid outer core? Can seismologists directly detect the fluid motions in the core? Can we improve the estimate of the density increase across the ICB? What is the cause of the anisotropy of the inner core? Why are the amplitudes of the inner core arrivals along north–south paths anomalously small and the waveforms complex (Creager 1992; Song and Helmberger, 1993; Fig. 2)? Can we develop high-resolution 3D images of the inner-core anisotropy and heterogeneity? Can we reliably detect inner core shear waves and thus constrain inner core shear wave structure? What is the nature of the inner core attenuation? Can we confirm the differential inner-core rotation and improve estimates of the rate? Is the inner-core rotation variable on a human timescale? Can we reconcile apparent differences between body-wave studies and normal-mode studies? For example, travel-time data suggest weak anisotropy in the top ~150 km of the inner core whereas strong anisotropy at the top seems to be required to explain the splitting of core-sensitive modes (Durek and Romanowicz, 1999). Would the conclusion of Laske and Masters (1999) change when more high-quality mode data become available in the future? The key to these questions is our ability to isolate influences from mantle and surface structure. So far we have relied on differential travel-time measurements of core phases and/or corrections for mantle structure. With the accumulation of high-quality digital seismic data and the improvement of 3D mantle models, our ability to image the core will undoubtedly improve.

Acknowledgments

I thank Jay Bass for comments on the early version of the manuscript. I thank the critical reviews from three anonymous reviewers and editorial comments from Hiroo Kanamori. The work is supported by NSF grant EAR 00-96025, NASA grant NAG 5-8871, and Natural Science Foundation of China grant 49825503.

References

Allègre, C.J., *et al.* (1995). *Earth Planet. Sci. Lett.* **134**, 515–526.
Anderson, D.L. and R.S. Hart (1978). *J. Geophys. Res.* **83**, 5869–5882.
Anderson, O.L. (1986). *Geophys. J. R. Astron. Soc.* **84**, 561–579.
Anderson, O.L. (1995). *Rev. Geophys.*, suppl., 429–441.
Aurnou, J.M. and P.L. Olson (2000). *Phys. Earth Planet. Inter.* **117**, 111–121.
Aurnou, J.M., *et al.* (1996). *Geophys. Res. Lett.* **23**, 3401–3404.
Bergman, M.I. (1997). *Nature* **389**, 60–63.
Bhattacharyya, J., *et al.* (1993). *Geophys. J. Int.* **114**, 1–11.
Birch, F. (1940). *Am. J. Sci.* **238**, 192–211.
Birch, F. (1952). *J. Geophys. Res.* **57**, 227–286.
Birch, F. (1961). *Geophys. J. R. Astron. Soc.* **4**, 295–311.
Birch, F. (1964). *J. Geophys. Res.* **69**, 4377–4388.
Bloxham, J. (1990). *Geophys. Res. Lett.* **17**, 2081–2084.
Boehler, R. (1996). *Annu. Rev. Earth Planet. Sci.* **24**, 15–40.
Bolt, B.A. (1962). *Nature* **196**, 122–124.
Bolt, B.A. (1964). *Bull. Seismol. Soc. Am.* **54**, 191–208.
Bolt, B.A. (1972). *Phys. Earth Planet. Inter.* **5**, 301–311.
Bolt, B.A. (1977). *Ann. Geofis.* **30**, 507.
Bolt, B.A. (1987). *EOS* **68**, 73–81.
Bolt, B.A. and A. Qamar (1970). *Nature* **228**, 148–150.
Braginsky, S.I. (1963). *Dokl. Akad. Nauk SSSR* **149**, 8–10.
Braginsky, S.I. (1999). *Phys. Earth Planet. Inter.* **111**, 21–34.
Braginsky, S.I. and P.H. Roberts (1995). *Geophys. Astrophys. Fluid Dyn.* **79**, 1–97.

Breger, L., *et al.* (1999). *Geophys. Res. Lett.* **26**, 3169–3172.
Brett, R. (1976). *Rev. Geophys. Space Phys.* **14**, 375–383.
Brown, J.M. and R.G. McQueen (1986). *Geophys. J. R. Astron. Soc.* **91**, 7485–7494.
Brush, S.G. (1980). *Am. J. Phys.* **48**, 705–724.
Buchbinder, G.G. (1971). *Bull. Seismol. Soc. Am.* **71**, 429–456.
Buchbinder, G.G. (1972). *Earth Planet. Sci. Lett.* **14**, 161–168.
Buffett, B.A. (1996). *Geophys. Res. Lett.* **23**, 3803–3806.
Buffett, B.A. (1997). *Nature* **388**, 571–573.
Buffett, B.A. (2000). In: "Earth's Deep Interior" (S. Karato *et al.*, Eds.), AGU Monograph.
Buland, R. and F. Gilbert (1978). *Geophys. J. R. Astron. Soc.* **52**, 457–470.
Bullen, K.E. (1946). *Nature* **157**, 405.
Choy, G.L. and V.F. Cormier (1983). *Geophys. J. R. Astron. Soc.* **72**, 1–21.
Cleary, J.R. and R.A.W. Haddon (1972). *Nature* **240**, 549–551.
Cormier, V.F. (1981). *Phys. Earth Planet. Inter.* **24**, 291–301.
Cormier, V.F. and P.G. Richards (1976). *J. Geophys. Res.* **81**, 3066–3068.
Cormier, V.F., *et al.* (1998). *Geophys. Res. Lett.* **25**, 4019–4022.
Cowling, T.G. (1934). *Mon. Not. R. Astron. Soc.* **94**, 39–48.
Creager, K.C. (1992). *Nature* **356**, 309–314.
Creager, K.C. (1997). *Science* **278**, 1284–1288.
Creager, K.C. (1999). *J. Geophys. Res.* **104**, 23127–23139.
Creager, K.C. (2000). In: "Earth's Deep Interior" (S. Karato *et al.*, Eds.), AGU Monograph.
Cummins, P. and L. Johnson (1988). *J. Geophys. Res.* **93**, 9058–9074.
Deuss, A., J. Woodhouse, H. Paulssen, and J. Trampert (2000). The observation of inner core shear waves. *Geophys. J. Int.* **142**, 67–73.
Doornbos, D.J. (1974). *Geophys. J. R. Astron. Soc.* **38**, 397–415.
Durek, J.J. and B. Romanowicz (1999). *Geophys. J. Int.* **139**, 599–622.
Dziewonski, A.M. and D.L. Anderson (1981). *Phys. Earth Planet. Inter.* **25**, 297–356.
Dziewonski, A.M. and F. Gilbert (1971). *Nature* **234**, 465–466.
Dziewonski, A.M. and F. Gilbert (1972). *Geophys. J. R. Astron. Soc.* **27**, 393–446.
Dziewonski, A.M. and W.-J. Su (1998). *Eos. Trans. AGU* **79**(17), Spring Meet. Suppl., S218.
Engdahl, E.R., *et al.* (1970). *Nature* **228**, 852–853.
Engdahl, E.R., *et al.* (1974). *Geophys. J. R. Astron. Soc.* **39**, 457–463.
Fearn, D.R. and D.E. Loper (1981). *Nature* **289**, 393–394.
Fearn, D.R., *et al.* (1981). *Nature* **292**, 232–233.
Gans, R.F. (1972). *J. Geophys. Res.* **77**, 360–366.
Garnero, E.J. and Helmberger (1995). *Phys. Earth Planet. Inter.* **88**, 117–130.
Garnero, E.J., *et al.* (1993). *Geophys. Res. Lett.* **20**, 2463–2466.
Garnero, E.J. and T. Lay (1998). *Geophys. Res. Lett.* **25**, 2341–2344.
Glatzmaier, G.A. and P.H. Roberts (1995a). *Phys. Earth Planet. Inter.* **91**, 63–75.
Glatzmaier, G.A. and P.H. Roberts (1995b). *Nature* **377**, 203–209.
Glatzmaier, G.A. and P.H. Roberts (1996). *Science* **274**, 1887–1891.
Glatzmaier, G.A. and P.H. Roberts (1998). *Int. J. Eng. Sci.* **36**, 1325–1338.
Gubbins, D. (1981). *J. Geophys.* **43**, 453–464.
Gubbins, D. (1981). *J. Geophys. Res.* **86**, 11695–11699.
Gutenberg, B. (1913). *Phys. Zeits* **14**, 1217–1218.
Haddon, R.A.W. and J.R. Cleary (1973). *Phys. Earth Planet. Inter.* **7**, 495.
Haddon, R.A.W. and J.R. Cleary (1974). *Phys. Earth Planet. Inter.* **8**, 211–234.
Hedlin, M.A.H., *et al.* (1997). *Nature* **387**, 145–150.
Hollerbach, R. (1998). *Geophys. J. Int.* **135**, 564–572.
Hollerbach, R. and C.A. Jones (1993). *Nature* **365**, 541–543.
Jacobs, J.A. (1987). "The Earth's Core" (2nd edn.), Academic Press, San Diego.
Jacobs, J.A. (1992). "Deep Interior of the Earth," Chapman and Hall, London.
Jeanloz, R. (1990). *Annu. Rev. Earth Planet. Sci.* **18**, 357.
Jeanloz, R. and H.R. Wenk (1988). *Geophys. Res. Lett.* **15**, 72–75.
Jeffreys, H. (1939a). *Geophys. Suppl.* **4**, 548–561.
Jeffreys, H. (1939b). *Geophys. Suppl.* **4**, 594–615.
Jeffreys, H. and K.E. Bullen (1940). "Seismological Tables," British Association for the Advancement of Science, London.
Jephcoat, A. and P. Olson (1987). *Nature* **325**, 332–335.
Jones, J.H. and M.J. Drake (1986). *Nature* **322**, 221–228.
Julian, B.R., *et al.* (1972). *Nature* **235**, 317–318.
Karato, S. (1993). *Science* **262**, 1708–1711.
Karato, S. (1999). *Nature* **402**, 871–873.
Kennett, B.L.N., *et al.* (1995). *Geophys. J. Int.* **122**, 108–124.
Kohler, M.D. and T. Tanimoto (1992). *Phys. Earth Planet. Inter.* **72**, 173–184.
Kuang, W.J. and J. Bloxham (1997). *Nature* **389**, 371–374.
Labrosse, S., *et al.* (1997). *Phys. Earth Planet. Inter.* **99**, 1–17.
Larmor, J. (1919). *Elec. Rev.* **85**, 412.
Laske, G. and G. Masters (1999). *Nature* **402**, 66–69.
Lay, T. and C.J. Young (1990). *Geophys. Res. Lett.* **17**, 2001–2004.
Lay, T., *et al.* (1998). *Nature* **392**, 461–468.
Lehmann, I. (1936). *Publ. Bur. Cent. Assoc. Int. Seismol. A* **14**, 87–115.
Li, A.Y. and P.G. Richards (1998). *Eos. Trans. AGU*, suppl. **79**(45), F80.
Lister, J.R. and B.A. Buffett (1995). *Phys. Earth Planet. Inter.* **91**, 17–30.
Lister, J.R. and B.A. Buffett (1998). *Phys. Earth Planet. Inter.* **105**, 5–19.
Loper, D.E. (1978). *Geophys. J. R. Astron. Soc.* **54**, 389–404.
Loper, D.E. and D.R. Fearn (1983). *J. Geophys. Res.* **88**, 1235–1242.
Loper, D.E. and P.H. Roberts (1978). *Geophys. Astrophys. Fluid Dyn.* **9**, 289–321.
Masse, R.P., *et al.* (1974). *Geophys. Res. Lett.* **1**, 39–42.
Masters, G. (1991). *Rev. Geophys.* **29**, 671–679.
Masters, G. and F. Gilbert (1981). *Geophys. Res. Lett.* **8**, 569–571.
McSweeney, T.J., *et al.* (1997). *Phys. Earth Planet. Inter.* **101**, 131–156.
Morelli, A., *et al.* (1986). *Geophys. Res. Lett.* **13**, 1545–1548.
Needham, J. (1962). "Science and Civilisation in China," Vol. 4, Part 1, Cambridge University Press, Cambridge.
Niazi, M. and L.R. Johnson (1992). *Phys. Earth Planet. Inter.* **74**, 55–62.
Okal, E.A. and Y. Cansi (1998). *Earth Planet. Sci. Lett.* **164**, 23–30.
Oldham, R.D. (1906). *Q. J. Geol. Soc.* **62**, 456–475.
Ovchinnikov, V.M., *et al.* (1998). *Dokl. Akad. Nauk.* **362**, 683–686.
Poirier, J.-P. (1988). *Geophys. J.* **92**, 99–105.
Poirier, J.-P. (1994a). *Phys. Earth Planet. Inter.* **85**, 319–337.

Poirier, J.-P. (1994b). *C.R. Acad. Sci.*, Paris **318**, 341–350.
Poupinet, G., *et al.* (1983). *Nature* **305**, 204–206.
Poupinet, G., *et al.* (2000). *Phys. Earth Planet. Inter.* **118**, 77–88.
Qamar, A. and A. Eisenberg (1974). *J. Geophys. Res.* **79**, 758–765.
Richards, P.G. (2000). *Astron. Geophys.* **41**, 20–24.
Ringwood, A.E. (1984). *Proc. R. Soc. Lond.* A **395**, 1–46.
Romanowicz, B. and J. Durek (2000). In: "Earth's Deep Interior" (S. Karato, *et al.*, Eds.), AGU Monograph.
Sacks, I.S. (1972). *EOS Trans. AGU* **53**, 601.
Sakuraba, A. and M. Kono (1999). *Phys. Earth Planet. Inter.* **111**, 105–121.
Sayers, C.M. (1989). *Geophys. Res. Lett.* **16**, 267–270.
Sharrock, D.S. and J.H. Woodhouse (1998). *Earth Planets Space* **50**, 1013–1018.
Shearer, P.M. (1994). *J. Geophys. Res.* **99**, 19647–19659.
Shearer, P.M. and G. Masters (1990). *Geophys. J. Int.* **102**, 491–498.
Song, X.D. (1996). *J. Geophys. Res.* **101**, 16089–16097.
Song, X.D. (1997). *Rev. Geophys.* **35**, 297–313.
Song, X.D. (2000a). *J. Geophys. Res.* **105**, 7931–7943.
Song, X.D. (2000b). *Phys. Earth Plant. Inter.* **122**, 221–228.
Song, X.D. (2001). *Phys. Earth Plant. Inter.* **124**, 269–273.
Song, X.D. and D.V. Helmberger (1992). *J. Geophys. Res.* **97**, 6573–6586.
Song, X.D. and D.V. Helmberger (1993). *Geophys. Res. Lett.* **20**, 2591–2594.
Song, X.D. and D.V. Helmberger (1995a). *J. Geophys. Res.* **100**, 9805–9816.
Song, X.D. and D.V. Helmberger (1995b). *J. Geophys. Res.* **100**, 9817–9830.
Song, X.D. and D.V. Helmberger (1998). *Science* **282**, 924–927.
Song, X.D. and A.Y. Li (2000). *J. Geophys. Res.* **105**, 623–630.
Song, X.D. and P.G. Richards (1996). *Nature* **382**, 221–224.
Souriau, A. (1998a). *Geophys. J. Int.* **134**, F1–5.
Souriau, A. (1998b). *Science* **281**, 55–56.
Souriau, A. (1998c). *Science* **282**, 1227a.
Souriau, A. and G. Poupinet (1990). *Geophys. Res. Lett.* **17**, 2005–2007.
Souriau, A. and G. Poupinet (1991). *Geophys. Res. Lett.* **18**, 2023–2026.
Souriau, A. and P. Roudil (1995). *Geophys. J. Int.* **123**, 572–587.
Souriau, A. and M. Souriau (1989). *Geophys. J. Int.* **98**, 39–54.
Souriau, A., *et al.* (1997). *Geophys. Res. Lett.* **24**, 2103–2106.
Stevenson, D.J. (1981). *Science* **214**, 611–619.
Stevenson, D.J. (1987). *Geophys. J. R. Astron. Soc.* **88**, 311–319.
Stixrude, L. and J.M. Brown (1998). *Rev. Mineral.* **37**, 261–282.
Stixrude, L. and R.E. Cohen (1995). *Science* **267**, 1972–1975.
Stixrude, L., *et al.* (1997). *J. Geophys. Res.* **102**, 24729–24739.
Su, W.J. and A.M. Dziewonski (1995). *J. Geophys. Res.* **100**, 9831–9852.
Su, W.J., *et al.* (1996). *Science* **274**, 1883–1887.
Suda, N. and Y. Fukao (1990). *Geophys. J. Int.* **103**, 403–413.
Tanaka, S. and H. Hamaguchi (1993). *J. Geomag. Geoelectr.* **45**, 1287–1301.
Tanaka, S. and H. Hamaguchi (1996). *J. Phys. Earth* **44**, 745–759.
Tanaka, S. and H. Hamaguchi (1997). *J. Geophys. Res.* **102**, 2925–2938.
Tromp, J. (1993). *Nature* **366**, 678–681.
Tromp, J. (1995a). *Geophys. J. Int.* **121**, 963–968.
Tromp, J. (1995b). *Rev. Geophys.*, suppl., 329–333.
Tromp, J. (1995c). *GSA Today* **5**, no. 7.
Vidale, J.E., *et al.* (2000). *Nature* **405**, 445–448.
Vidale, J.E. and P.S. Earle (2000). *Nature* **404**, 273–275.
Vinnik, L., *et al.* (1994). *Geophys. Res. Lett.* **21**, 1671–1674.
Wahr, J. and D. de Vries (1989). *Geophys. J. Int.* **99**, 511–519.
Widmer, R.W., *et al.* (1991). *Geophys. J. Int.* **104**, 541–553.
Wood, B.J. (1993). *Earth Planet. Sci. Lett.* **117**, 593–607.
Woodhouse, J.H., *et al.* (1986). *Geophys. Res. Lett.* **13**, 1549–1552.
Yoshida, S., *et al.* (1996). *J. Geophys. Res.* **101**, 28085–28103.
Young, C.J. and T. Lay (1987). *Annu. Rev. Earth Planet. Sci.* **15**, 25–46.

Editor's Note

Due to space limitations, references in full citation are given in SongFullReferences.pdf on the attached Handbook CD, under the directory \56Song. Please see also Chapter 51, The Earth's interior, by Lay.

Index for Part A

Locators in **bold** indicate main discussion or chapter pages by an author; those in *italic* indicate table or figure. Locators followed by 'p' refer to a colour plate, e.g. 13p refers to colour plate 13.

F

G

H

I

J

N

Q

R

S

International Geophysics Series

EDITED BY

RENATA DMOWSKA
Division of Engineering and Applied Science
Harvard University
Cambridge, MA 02138

JAMES R. HOLTON
Department of Atmospheric Sciences
University of Washington
Seattle, Washington

H. THOMAS ROSSBY
University of Rhode Island
Kingston, Rhode Island

Volume 1 BENO GUTENBERG. Physics of the Earth's Interior. 1959*

Volume 2 JOSEPH W. CHAMBERLAIN. Physics of the Aurora and Airglow. 1961*

Volume 3 S. K. RUNCORN (ed.) Continental Drift. 1962*

Volume 4 C. E. JUNGE. Air Chemistry and Radioactivity. 1963*

Volume 5 ROBERT G. FLEAGLE AND JOOST A. BUSINGER. An Introduction to Atmospheric Physics. 1963*

Volume 6 L. DUFOUR AND R. DEFAY. Thermodynamics of Clouds. 1963*

Volume 7 H. U. ROLL. Physics of the Marine Atmosphere. 1965*

Volume 8 RICHARD A. CRAIG. The Upper Atmosphere: Meteorology and Physics. 1965*

Volume 9 WILLIS L. WEBB. Structure of the Stratosphere and Mesosphere. 1966*

Volume 10 MICHELE CAPUTO. The Gravity Field of the Earth from Classical and Modern Methods. 1967*

Volume 11 S. MATSUSHITA AND WALLACE H. CAMPBELL (eds.) Physics of Geomagnetic Phenomena. (In two volumes.) 1967*

Volume 12 K. YA KONDRATYEV. Radiation in the Atmosphere. 1969*

Volume 13 E. PALMÈN AND C. W. NEWTON. Atmospheric Circulation Systems: Their Structure and Physical Interpretation. 1969*

Volume 14 HENRY RISHBETH AND OWEN K. GARRIOTT. Introduction to Ionospheric Physics. 1969*

Volume 15 C. S. RAMAGE. Monsoon Meteorology. 1971*

Volume 16 JAMES R. HOLTON. An Introduction to Dynamic Meteorology. 1972*

Volume 17 K. C. YEH AND C. H. LIU. Theory of Ionospheric Waves. 1972*

Volume 18 M. I. BUDYKO. Climate and Life. 1974*

Volume 19 MELVIN E. STERN. Ocean Circulation Physics. 1975*

Volume 20 J. A. JACOBS. The Earth's Core. 1975*

*Out of Print

Volume 21 DAVID H. MILLER. Water at the Surface of the Earth: An Introduction to Ecosystem Hydrodynamics. 1977

Volume 22 JOSEPH W. CHAMBERLAIN. Theory of Planetary Atmospheres: An Introduction to Their Physics and Chemistry. 1978*

Volume 23 JAMES R. HOLTON. An Introduction to Dynamic Meteorology, Second Edition. 1979*

Volume 24 ARNETT S. DENNIS. Weather Modification by Cloud Seeding. 1980*

Volume 25 ROBERT G. FLEAGLE AND JOOST A. BUSINGER. An Introduction to Atmospheric Physics, Second Edition. 1980

Volume 26 KUG-NAN LIOU. An Introduction to Atmospheric Radiation. 1980*

Volume 27 David H. Miller. Energy at the Surface of the Earth: An Introduction to the Energetics of Ecosystems. 1981*

Volume 28 HELMUT G. LANDSBERG. The Urban Climate. 1981

Volume 29 M. I. BUDYKO. The Earth's Climate: Past and Future. 1982*

Volume 30 ADRIAN E. GILL. Atmosphere-Ocean Dynamics. 1982

Volume 31 PAOLO LANZANO. Deformations of an Elastic Earth. 1982*

Volume 32 RONALD T. MERRILL AND MICHAEL W. MCELHINNY. The Earth's Magnetic Field. Its History, Origin, and Planetary Perspective. 1983*

Volume 33 JOHN S. LEWIS AND RONALD G. PRINN. Planets and Their Atmospheres: Origin and Evolution. 1983

Volume 34 ROLF MEISSNER. The Continental Crust: A Geophysical Approach. 1986

Volume 35 M. U. SAGITOV, B. BODKI, V. S. NAZARENKO, AND KH. G. TADZHIDINOV. Lunar Gravimetry. 1986*

Volume 36 JOSEPH W. CHAMBERLAIN AND DONALD M. HUNTEN. Theory of Planetary Atmospheres: An Introduction to Their Physics and Chemistry, Second Edition. 1987

Volume 37 J. A. JACOBS. The Earth's Core, Second Edition. 1987*

Volume 38 J. R. APEL. Principles of Ocean Physics. 1987

Volume 39 MARTIN A. UMAN. The Lightning Discharge. 1987*

Volume 40 DAVID G. ANDREWS, JAMES R. HOLTON AND CONWAY B. LEOVY. Middle Atmosphere Dynamics. 1987

Volume 41 PETER WARNECK. Chemistry of the Natural Atmosphere. 1988

Volume 42 S. PAL ARYA. Introduction to Micrometeorology. 1988

Volume 43 MICHAEL C. KELLEY. The Earth's Ionosphere. 1989*

Volume 44 WILLIAM R. COTTON AND RICHARD A. ANTHES. Storm and Cloud Dynamics. 1989

Volume 45 WILLIAM MENKE. Geophysical Data Analysis: Discrete Inverse Theory, Revised Edition. 1989

Volume 46 S. GEORGE PHILANDER. EL NIÑO, LA NIÑA, and the Southern Oscillation. 1990

Volume 47 ROBERT A. BROWN. Fluid Mechanics of the Atmosphere. 1991

Volume 48 JAMES R. HOLTON. An Introduction to Dynamic Meteorology, Third Edition. 1992

Volume 49 ALEXANDER A. KAUFMAN. Geophysical Field Theory and Method.
Part A: Gravitational, Electric, and Magnetic Fields. 1992
Part B: Electromagnetic Fields I. 1994
Part C: Electromagnetic Fields II. 1994

Volume 50 SAMUEL S. BUTCHER, GORDON H. ORIANS, ROBERT J. CHARLSON, AND GORDON V. WOLFE. Global Biogeochemical Cycles. 1992*

Volume 51 BRIAN EVANS AND TENG-FONG WONG. Fault Mechanics and Transport Properties of Rocks. 1992

Volume 52 ROBERT E. HUFFMAN. Atmospheric Ultraviolet Remote Sensing. 1992

Volume 53 ROBERT A. HOUZE, JR. Cloud Dynamics. 1993

Volume 54 PETER V. HOBBS. Aerosol-Cloud-Climate Interactions. 1993

Volume 55 S. J. GIBOWICZ AND A. KIJKO. An Introduction to Mining Seismology. 1993

*Out of Print

Volume 56 Dennis L. Hartmann. Global Physical Climatology. 1994

Volume 57 Michael P. Ryan. Magmatic Systems. 1994

Volume 58 Thorne Lay and Terry C. Wallace. Modern Global Seismology. 1995

Volume 59 Daniel S. Wilks. Statistical Methods in the Atmospheric Sciences. 1995

Volume 60 Frederik Nebeker. Calculating the Weather. 1995

Volume 61 Murry L. Salby. Fundamentals of Atmospheric Physics. 1996

Volume 62 James P. McCalpin. Paleoseismology. 1996

Volume 63 Ronald T. Merrill, Michael W. McLhinny, and Phillip L. McFadden. The Magnetic Field of the Earth: Paleomagnetism, the Core, and the Deep Mantle. 1996

Volume 64 Neil D. Opdyke and James E. T. Channell. Magnetic Stratigraphy. 1996

Volume 65 Judith A. Curry and Peter J. Webster. Thermodynamics of Atmospheres and Oceans. 1998

Volume 66 Lakshmi H. Kantha and Carol Anne Clayson. Numerical Models of Oceans and Oceanic Processes. 2000

Volume 67 Lakshmi H. Kantha and Carol Anne Clayson. Small Scale Processes in Geophysical Fluid Flows. 2000

Volume 68 Raymond S. Bradley. Paleoclimatology, Second Edition. 1999

Volume 69 Lee-Lueng Fu and Anny Cazanave. Satellite Altimetry and Earth Sciences. 2000

Volume 70 David A. Randall. General Circulation Model Development. 2000

Volume 71 Peter Warneck. Chemistry of the Natural Atmosphere, Second Edition. 2000

Volume 72 Michael C. Jacobson, Robert J. Charlson, Henning Rodhe and Gordon H. Orians. Earth System Science: From Biogeochemical Cycles to Global Change. 2000

Volume 73 Michael W. McElhinny and Phillip L. McFadden. Paleomagnetism: Continents and Oceans. 2000

Volume 74 Andrew E. Dessler. The Physics and Chemistry of Stratospheric Ozone. 2000

Volume 75 Bruce Douglas, Micheal Kearney and Stephen Leatherman. Sea Level Rise: History and Consequences. 2000

Volume 76 Roman Teisseyre and Eugeniusz Majewski. Earthquake Thermodynamics and Phase Transformations in the Earth's Interior. 2000

Volume 77 Gerold Siedler, John Church and John Gould. Ocean Circulation and Climate. 2001

Volume 78 Roger Pielke. Mesoscale. Meteorological Modeling, 2nd Edition. 2001

Volume 79 S. Pal Arya. Introduction to Micrometeorology, 2nd Edition. 2001

Volume 80 Saltzman. Dynamical Paleoclimatology. 2001

*Out of Print